# American College of Veterinary Internal Medicine Forum & Canadian Veterinary Medical Association Convention 2009



**Montreal, Quebec, Canada**
**3 – 6 June 2009**

ISBN: 978-1-61782-710-5

**Printed from e-media with permission by:**

Curran Associates, Inc.
57 Morehouse Lane
Red Hook, NY 12571

**Some format issues inherent in the e-media version may also appear in this print version.**



Printed by Curran Associates, Inc. (2011)

For permission requests, please contact the American College of Veterinary Internal Medicine
at the address below.

American College of Veterinary Internal Medicine
1997 Wadsworth Boulevard Suite A
Lakewood CO 80214-5293

Phone: (800) 245-9081
Fax: (303) 231-0880

acvim@acvim.org

**Additional copies of this publication are available from:**

Curran Associates, Inc.
57 Morehouse Lane
Red Hook, NY 12571 USA
Phone: 845-758-0400
Fax: 845-758-2634
Email: curran@proceedings.com
Web: www.proceedings.com

## WELCOME TO MONTRÉAL AND THE 2009 ACVIM FORUM/CANADIAN VMA CONVENTION

Welcome to Montréal! I am pleased and honored that you have chosen to attend the 2009 ACVIM Forum/Canadian Veterinary Medical Association Convention, the premier veterinary internal medicine continuing education meeting! Much of the credit for this distinction goes to the ACVIM Forum Program Committee and the Canadian Veterinary Medical Association Convention Program Committee, as well as the many specialists who deliver state-of-the-art clinical science and veterinary internal medicine advances to our attendees.

This year's agenda is full, with a core program of approximately 280 scientific lectures, more than 350 research abstracts, and a technician lecture series. All sessions of the core program are open to every attendee. Thursday through Saturday you'll find a variety of special interest group (SIG) sessions (workshops, case discussions, and labs) that address diverse topics in veterinary internal medicine, many of which are available in the evenings. Technicians will be pleased to see a scientific program and workshops designed just for them, but they are also welcome to attend any veterinary sessions.

Make plans to attend the Opening Night Reception and Attendee Party, sponsored by Bayer on Thursday evening. On Friday morning, all attendees are invited to the Keynote address. Relevant to all, this year's topic is "Creating *Buy-In* to Avoid a *Bail-Out* During Challenging Economic Times," presented by Dr. Pat LeBlanc, Director of the Veterinary Teaching Hospital at Michigan State University. After the Keynote address, please stay for the Small Animal or Large Animal Consensus Statement panels.

Research posters are on display throughout the meeting and can be discussed with their authors on Thursday morning during the Welcome Coffee sponsored by Hill's Pet Nutrition, Inc., Friday during the morning break, and Friday evening during the Wine and Cheese Reception sponsored by Nestlé Purina.

There is an abundance of quality session topics on the daily program grid. However, if you are unable attend a session of interest these Proceedings will help to fill in the gaps.

Make sure to leave plenty of time to meet with our many exhibitors in the Exhibit Hall. You will be surprised by the number and the diversity of companies that come to bring you the very latest in veterinary technology, pharmaceutical advances, superior nutritional products, and educational materials.

The ACVIM is indebted to the ACVIM Forum Program Committee (chaired by Dr. Virginia Buechner-Maxwell) and the Canadian Veterinary Medical Association Convention Program Committee (chaired by Dr. Jeanne Lofstedt) for the incredible effort it takes to put together a program of this magnitude. We also want to reiterate our thanks to our many speakers, facilitators, and moderators who are willing to share their knowledge, talent, and time. We are very grateful to our sponsors for their recognition, presence, and support of our efforts to provide an unequaled educational experience for all attendees. The ACVIM Forum continues to grow substantially in both size and quality due to the wide range of our sponsors' support.

The dedicated ACVIM staff members and the Canadian VMA Convention staff members have handled all of the organizational details needed to bring this combined meeting to you, and they are available at the registration desk to help you with any special needs or requests. Please take a few minutes to personally thank them for a job well done.

The members of the ACVIM Board realize that you have a choice in where to spend your continuing education dollars, and we appreciate the fact that you have chosen to join us in Montréal. We hope that you have an outstanding educational experience, and that your attendance at the 2009 ACVIM Forum/Canadian Veterinary Medical Association Convention will facilitate your continued commitment to the highest quality of service in veterinary practice, academia, or industry. Mark you calendars now to join us for the 2010 ACVIM Forum, June 9–12, in Anaheim, California.

Wishing you a stimulating and productive meeting,

Philip Bergman, Diplomate ACVIM (Oncology)
Chair, Board of Regents of the ACVIM

# ACVIM FORUM

As far as you want to go.

# ACVIM

As far as you want to go.

Address correspondence to:
AMERICAN COLLEGE OF VETERINARY INTERNAL MEDICINE (ACVIM)
1997 WADSWORTH BLVD., SUITE A
LAKEWOOD, CO 80214-5293
www.ACVIM.org

**2009 ACVIM FORUM & CANADIAN VETERINARY MEDICAL ASSOCIATION CONVENTION SPONSORS**


The ACVIM appreciates the generous support provided by our sponsoring companies:

**Diamond Sponsor**
Hill's Pet Nutrition

**Platinum Sponsor**
Nestlé Purina

**Gold Sponsors**
Infiniti Medical™
Pfizer Animal Health

**Silver Sponsors**
Bayer HealthCare LLC, Animal Health Division
Royal Canin Veterinary Diet™

**Bronze Sponsors**
IDEXX Laboratories
Intervet/Schering-Plough Animal Health
Lilly
Merial
Novartis Animal Health US, Inc.
Scotiabank
VIN
Virbac

**General Program Support**
Aloka
Eklin Ultrasound Systems
Karl Storz Veterinary Endoscopy
Nutramax Laboratories, Inc.
Nutrience
Sound Technologies
SoundVision
Universal Ultrasound
VetImaging™

**Media Sponsors**
*Canadian Veterinary Journal (CVJ)*
*International Veterinary Information Systems (IVIS)*
*Journal of Veterinary Internal Medicine (JVIM)*
*The Horse*
*Veterinary Learning Systems (VLS)*

# SPEAKER DISCLOSURE INDEX

| Session # | Speaker | Disclosure |
|---|---|---|
| 001 | N. Sydney Moise | Nothing Disclosed |
| 002 | Roberto Santilli | Nothing Disclosed |
| 003 | Bruce Kornreich | Nothing Disclosed |
| 004 | Jonathan Levine | Research Support: American Kennel Club National Institutes of Health Scott and White Health Sciences Center; Grants support research in the areas of spinal cord injury and disk herniation; MR imaging is a secondary focus of this work. Consultant South Texas Veterinary MRI; Practice in San Antonio, TX which provides MRI services for small animals on a referral basis. |
| 005 | Ronaldo da Costa | Nothing Disclosed |
| 006 | Amy Fauber | Work with Animal Orthopaedics, Inc. on the development of titanium spinal implants. |
| 007A | Tim Fan | Nothing Disclosed |
| 007B | Steve Suter | Nothing Disclosed |
| 008A | Carlos Rodriguez | Nothing Disclosed |
| 008B | Kim Selting | Nothing Disclosed |
| 009A | Enrico Spugnini | Nothing Disclosed |
| 009B | Dennis Bailey | Nothing Disclosed |
| 010 | Sanjay Kapil | Nothing Disclosed |
| 011 | Sanjay Kapil | Nothing Disclosed |
| 012 | Katrin Hartmann | Nothing Disclosed |
| 013 | Janice Dye | Nothing Disclosed |
| 014A | Brendan McKiernan | Nothing Disclosed |
| 014B | Elizabeth Rozanski | Nothing Disclosed |
| 015A | Michael Davis | Nothing Disclosed |
| 015B | Carol Reinero | Nothing Disclosed |
| 016A | Robert Washabau | Nothing Disclosed |
| 016B | Stanley Marks | Nothing Disclosed |
| 017 | Diane Shelton | Nothing Disclosed |
| 018A | Stanley Marks | Nothing Disclosed |
| 018B | Keith Richter | Nothing Disclosed |
| 019 | Ted Whittem | Nothing Disclosed |
| 020 | Tomás Martín-Jiménez | Research Support: Morris Animal Foundation, Pfizer Animal Health; Consultant: Elanco, Pharmaceutical Solutions Inc. |
| 021 | Mark Papich | Consultantships: Pfizer Animal Health Bayer Corporation Schering-Plough Animal Health Boehringer-Ingelheim Central Life Sciences (formerly Farnam Company) Idexx Corporation Vetoquinol PetMedLabs (PetMedicus Laboratories, Inc.); Funding Sources: Bayer Corporation Schering-Plough Animal Health Pfizer Inc. Central Life Sciences (formerly Farnam Company) Morris Animal Foundation AAEP Foundation; Royalties: Elsevier (formerly WB Saunders), Blackwell Publishing; Other: Member of the Council of Experts for the United States Pharmacopeia (USP). Member of the FDA Veterinary Medicine Advisory Committee (term expired in 2007). Member of the CLSI (NCCLS), Veterinary Antimicrobial Susceptibility Testing subcommittee (VAST), Editorial Board of Journal of Veterinary Pharmacology and Therapeutics, Editorial Board of Journal of the American Animal Hospital Feline Medicine and Surgery |
| 022 | Nathan Slovis | Nothing Disclosed |
| 022 | Alan Loynachan | Nothing Disclosed |
| 023 | Martin Furr | Nothing Disclosed |
| 024 | Thomas Divers | Nothing Disclosed |
| 025 | Robert Larson | Nothing Disclosed |

| | | |
|---|---|---|
| 026 | Robert Larson | Nothing Disclosed |
| 027 | Peter Constable | Research Support: Boehringer-Ingelheim, Fort Dodge Animal Health; Consultant Bayer Boehringer-Ingelheim, Fort Dodge Animal Health |
| 028 | Maria Prado | Research Support: Center of Excellence, CVM, U of Tennessee USDA - NRI Pfizer Animal Health |
| 029 | Simon Peek | Nothing Disclosed |
| 030 | Sanjay Kapil | Nothing Disclosed |
| 031 | Louise Dunn | Nothing Disclosed |
| 032 | Louise Dunn | Nothing Disclosed |
| 033 | Louise Dunn | Nothing Disclosed |
| 034 | Dennis O'Brien | Nothing Disclosed |
| 035 | Scott Schatzberg | Nothing Disclosed |
| 036 | Joan Coates | Nothing Disclosed |
| 037 | Charles Vite | Nothing Disclosed |
| 038 | Barbara Kitchell | Nothing Disclosed |
| 039A | Pamela Lucas | Nothing Disclosed |
| 039B | Carrie Tupper Hume | Nothing Disclosed |
| 040 | William Kisseberth | Nothing Disclosed |
| 041 | Carolyn Henry | PetScreen Ltd - collaborative research Bion Diagnostics - research support for V-BTA study |
| 042 | Michael Lappin | Nothing Disclosed |
| 043 | Jody Gookin | Nothing Disclosed |
| 044 | Kristy Dowers | Research support from the Winn Feline Foundation |
| 045 | Vanessa Barrs | In-kind research support has been provided by Schering-Plough |
| 046 | Scott Friedman | Nothing Disclosed |
| 047 | Scott Friedman | Nothing Disclosed |
| 048 | Amy Warren | Nothing Disclosed |
| 049A | David Twedt | Nothing Disclosed |
| 049B | Mark Rondeau | Nothing Disclosed |
| 050 | Lauren Trepanier | Nothing Disclosed |
| 051 | Dawn Boothe | Nothing Disclosed |
| 052 | Dawn Boothe | Nothing Disclosed |
| 053 | Mark Papich | Consultants: Pfizer Animal Health Bayer Corporation Schering-Plough Animal Health Boehringer-Ingelheim Central Life Sciences (formerly Farnam Company) Idexx Corporation Vetoquinol PetMedLabs (PetMedicus Laboratories, Inc.); Funding Sources: Bayer Corporation Schering-Plough Animal Health Pfizer Inc. Central Life Sciences (formerly Farnam Company) Morris Animal Foundation AAEP Foundation; Royalties: Elsevier (formerly WB Saunders), Blackwell Publishing; Other: Member of the Council of Experts for the United States Pharmacopeia (USP). Member of the FDA Veterinary Medicine Advisory Committee (term expired in 2007). Member of the CLSI (NCCLS), Veterinary Antimicrobial Susceptibility Testing subcommittee (VAST), Editorial Board of Journal of Veterinary Pharmacology and Therapeutics, Editorial Board of Journal of the American Animal Hospital Feline Medicine and Surgery |
| 054 | Ahmed Galal | Nothing Disclosed |
| 055 | Stephanie Smith | Nothing Disclosed |
| 056 | Kathy Lunn | Study of liposomal clodronate supported by the Morris Animal Foundation; Consultant for Veterinary Information Network |
| 057A | Tamara Wills | Nothing Disclosed |
| 057B | Marjorie Brooks | Nothing Disclosed |
| 058 | Marc Prentki | Nothing Disclosed |
| 059 | Marc Prentki | Nothing Disclosed |
| 060 | André Marette | Nothing Disclosed |

| | | |
|---|---|---|
| 061 | Constantin Polychronakos | Nothing Disclosed |
| 062 | Nathalie Kirschvink | Nothing Disclosed |
| 063A | Michael Davis | Nothing Disclosed |
| 063B | Melissa Mazan | Nothing Disclosed |
| 064 | Melissa Mazan | Nothing Disclosed |
| 065A | Mathilde Leclére | Nothing Disclosed |
| 065B | Virginia Buechner-Maxwell | Nothing Disclosed |
| 066 | Derek Foster | Nothing Disclosed |
| 067 | Jeff Tyler | Large Animal Program Coordinator for the Western Veterinary Conference; National Board of Veterinary Medical Examiners |
| 068 | Michael Riggs | Research Support: National Institutes of Health-NIAID U.S. Department of Agriculture-NRI Boehringer Ingelheim Vetmedica Inc ioGenetics LLC |
| 069 | Michael Riggs | Research Support: National Institutes of Health-NIAID U.S. Department of Agriculture-NRI Boehringer Ingelheim Vetmedica Inc ioGenetics LLC |
| 070 | Louise Dunn | Nothing Disclosed |
| 071 | Louise Dunn | Nothing Disclosed |
| 072 | Louise Dunn | Nothing Disclosed |
| 073 | Louise Dunn | Nothing Disclosed |
| 074A | Stephen Ettinger | Nothing Disclosed |
| 074B | Andrew Beardow | Nothing Disclosed |
| 075A | Heidi Kellihan | Nothing Disclosed |
| 076 | Vakis Papanastassiou | Nothing Disclosed |
| 077 | Vakis Papanastassiou | Nothing Disclosed |
| 079A | Steve Suter | Nothing Disclosed |
| 080 | Mitchell Halperin | Nothing Disclosed |
| 081 | Mitchell Halperin | Nothing Disclosed |
| 082 | Patty Lathan | The Barry and Savannah French-Poodle Memorial Fund provided financial support for ACTH stimulation test research |
| 083 | JP Schoeman | Research support: South African National Research Foundation, Waltham International, Hill's Pet Foods, ECVIM Clinical Studies Fund, Morris Animal Foundation, Cambridge Commonwealth Trust, Wyeth International, CPD Honoraria, Pfizer Pharmaceutical, Intervet, Fort Dodge, Merial, Royal Canin; Consultant: Idexx Laboratories |
| 084 | Jan Suchodolski | Associate Director of the GI Laboratory at Texas A&M University, a nonprofit institution that offers gastrointestinal function assays for dogs and cats; Research Support: The IAMS Company, Nestle Purina, Waltham, Morris Animal Foundation, ACVIM Foundation, and Nutramax Laboratories |
| 085 | Al Jergens | Nothing Disclosed |
| 086 | Clint Lothrop | Nothing Disclosed |
| 087 | Clint Lothrop | Nothing Disclosed |
| 088 | Dominique Peeters | Nothing Disclosed |
| 089 | Dominique Peeters | Nothing Disclosed |
| 090 | Kate Hurley | Nothing Disclosed |
| 090 | Patricia Pesavento | Nothing Disclosed |
| 091 | Jeff Bender | Research Support: Centers for Disease Control and Prevention, Minnesota Department of Health, United States Department of Agriculture, United States Department of Homeland Security, National Institutes of Health; Consultant: Minnesota Department of Health |
| 092 | Elizabeth Davis | Partially funded by: Fort Dodge Animal Health, Inc. |
| 093 | Julia Flaminio | Nothing Disclosed |
| 094 | Martin Nielsen | Nothing Disclosed |
| 095 | Martin Nielsen | Nothing Disclosed |
| 096 | Michael Gomez | Researcher-Medimmune, Synagis Prophylaxis in Preterm Infants Researcher, Advisory Panel-Discovery Labs, Surfaxin (lucinactant) |

| | | |
|---|---|---|
| 097 | Michael Gomez | Researcher-Medimmune, RSV Prophylaxis in Preterm Infants (Synagis) Researcher, Advisor-Discovery Labs, Surfaxin (Lucinactant) to prevent BPD in preterm infants |
| 098 | Sylvain Nichols | Nothing Disclosed |
| 099 | André Desrochers | Nothing Disclosed |
| 100 | Michael Cockram | Research support: Sir James Dunn Animal Welfare Centre, Animal Welfare Foundation of Canada, Deer Commission for Scotland, UK Department for Environment, Food and Rural Affairs, UK, ITI Techmedia, UK; Other: Editorial Board of Animal Director of Atlantic Poultry, Research Institute Member of CFIA, Advisory Group on Humane Destruction of Swine, Chair of CVMA Expert Group to advise CFIA on an animal welfare issue |
| 101 | Michael Cockram | Research support: Sir James Dunn Animal Welfare Centre, Animal Welfare Foundation of Canada, Deer Commission for Scotland, UK Department for Environment, Food and Rural Affairs, UK, ITI Techmedia, UK; Other: Editorial Board of Animal Director of Atlantic Poultry, Research Institute Member of CFIA, Advisory Group on Humane Destruction of Swine, Chair of CVMA Expert Group to advise CFIA on an animal welfare issue |
| 102 | Stanley Nattel | Nothing Disclosed |
| 103 | Stanley Nattel | Nothing Disclosed |
| 104A | Anna Gelzer | Nothing Disclosed |
| 104B | Christophe Amberger | Nothing Disclosed |
| 105A | Eric de Madron | Nothing Disclosed |
| 105B | Nicole Van Israel | Nothing Disclosed |
| 106 | James Buchanan | Nothing Disclosed |
| 107 | Justin Thomason | Nothing Disclosed |
| 108 | Rebecca Stepien | Consultant: Boehringer Ingelheim Vetmedica, Idexx |
| 109 | Andrew Beardow | Nothing Disclosed |
| 109 | Stephen Ettinger | Nothing Disclosed |
| 109 | Kristine Yee | Nothing Disclosed |
| 111A | G. Diane Shelton | Nothing Disclosed |
| 111B | Thomas Flegel | Nothing Disclosed |
| 112A | Annie Chen | Nothing Disclosed |
| 112B | Clare Rusbridge | Nothing Disclosed |
| 113A | Natasha Olby | Nothing Disclosed |
| 113B | Stephane Blot | Nothing Disclosed |
| 116A | Maude Touret | Nothing Disclosed |
| 116B | Chand Khanna | Nothing Disclosed |
| 117A | Anthony Moore | Nothing Disclosed |
| 117B | Angela Frimberger | Nothing Disclosed |
| 118 | Barbara Biller | Nothing Disclosed |
| 119 | William Kisseberth | Research Support: American Kennel Club Canine Health Foundation, Morris Animal Foundation, OSU Targeted Investment in Excellence Grant |
| 120 | Melissa Paoloni | Nothing Disclosed |
| 121 | Jamie Modiano | Nothing Disclosed |
| 122A | Els Acke | Nothing Disclosed |
| 122B | Wayne Carter | Speaker Employed by: Hill's Pet Nutrition |
| 123A | Elizabeth Rozanski | Nothing Disclosed |
| 123B | Anant Radhakrishnan | Nothing Disclosed |
| 124A | Michael Lappin | Nothing Disclosed |
| 124B | Karin Allenspach | Nothing Disclosed |
| 125A | Nicole Luckschander | Nothing Disclosed |
| 125B | Oliver Garden | Nothing Disclosed |
| 126 | Charlotte Reinhard Bjornvad | Governmental research grant (Health and Disease), Research grant from the University of Copenhagen |
| 127 | Audrey Cook | Consultant for Dechra PLC, and Abbott Animal Health. Occasional collaboration with Idexx Laboratories. No |

| | | |
|---|---|---|
| | | financial support has been provided for the material presented in this talk |
| 128 | Lucy Davison | Nothing Disclosed |
| 129 | Cynthia Ward | Research Support: National Institutes of Health, James Parker Foundation, Robert Winn Foundation, Morris Animal Foundation |
| 130 | Joanne Messick | Nothing Disclosed |
| 131 | Anne Barger | Nothing Disclosed |
| 132 | Anne Avery | Nothing Disclosed |
| 133 | Anne Barger | Nothing Disclosed |
| 134 | Kate Hopper | Nothing Disclosed |
| 135 | Kate Hopper | Nothing Disclosed |
| 136 | Kate Hopper | Nothing Disclosed |
| 137 | Kate Hopper | Nothing Disclosed |
| 138 | Gad Baneth | Nothing Disclosed |
| 139 | Leah Cohn | Research related to this topic funded by the ALSAM Foundation and the Winn Feline Foundation |
| 140 | Adam Birkenheuer | Nothing Disclosed |
| 141 | Douglass Macintire | Nothing Disclosed |
| 142 | Jeffrey Mogil | Nothing Disclosed |
| 143 | Jeffrey Mogil | Nothing Disclosed |
| 144 | Chris Sanchez | Research Support: Veterinary Product Laboratories, A division of Farnam Companies, Inc., University of Florida, University of Copenhagen |
| 145A | Kristen Wegner | Nothing Disclosed |
| 145B | Ashley Stokes | Nothing Disclosed |
| 146 | Hal Schott | Research Support: Waltham Centre for Pet Nutrition, Boehringer Ingelheim Animal Health; Consultant: Wyeth Pharmaceuticals |
| 147 | Ramiro Toribio | Nothing Disclosed |
| 148 | Steeve Giguere | Research Support: Florida Thoroughbred Breeder's & Owner's Association; Consultant: Pfizer Animal Health, Merial |
| 149A | Gayle Hallowell | Nothing Disclosed |
| 150A | Dawn Morin | Nothing Disclosed |
| 150B | Gayle Hallowell | Nothing Disclosed |
| 151A | Michel Lévy | Nothing Disclosed |
| 151B | Geoff Smith | Nothing Disclosed |
| 152A | Dawn Morin | Nothing Disclosed |
| 152B | Mireille Meylan | Nothing Disclosed |
| 153A | Amelia Woolums | Nothing Disclosed |
| 153B | Fiona Maunsell | Nothing Disclosed |
| 154 | Robert Van Saun | Research Support: Pennsylvania Department of Agriculture Consultant: Mikrop Cebin Feeds, Czech Republic |
| 155 | Thomas Herdt | Nothing Disclosed |
| 156 | Thomas Herdt | Nothing Disclosed |
| 157 | Robert Van Saun | Research Support: Pennsylvania Department of Agriculture Consultant: Mikrop Cebin Feeds, Czech Republic |
| 158 | Carol Morgan | Nothing Disclosed |
| 159 | Carol Morgan | Nothing Disclosed |
| 160 | Carol Morgan | Nothing Disclosed |
| 161 | Carol Morgan | Nothing Disclosed |
| 162 | Ray Dillon | Nothing Disclosed |
| 163 | Anne French | Nothing Disclosed |
| 164 | Teresa DeFrancesco | Research support: IDEXX, Waltham Foundation, Norvartis, Barry and Savannah French-Poodle Memorial Fund; Consultant:IDEXX, Speakers Bureau, Boehringer Ingelheim Pharmaceuticals |
| 165 | Amara Estrada | Nothing Disclosed |
| 166 | Sonya Gordon | Nothing Disclosed |
| 167 | Etienne Coté | Nothing Disclosed |
| 168 | Rebecca Stepien | Consultant: Boehringer Ingelheim Vetmedica, Idexx |

| | | |
|---|---|---|
| 168 | Virginia Luis Fuentes | Nothing Disclosed |
| 169 | Paul Pion | Nothing Disclosed |
| 169 | Mark Rishniw | Nothing Disclosed |
| 170 | Dennis O'Brien | Nothing Disclosed |
| 171 | Andy Shores | Nothing Disclosed |
| 172 | John Speciale | Nothing Disclosed |
| 173 | Filippo Adamo | Nothing Disclosed |
| 174 | Nicola Mason | Research support: Alliance for Cancer Gene Therapy, Canine Health Foundation, Charitable giving office of Kindy French |
| 175 | J. Paul Woods | Nothing Disclosed |
| 175 | Anthony Mutsaers | Nothing Disclosed |
| 176 | Cheryl London | Consultant: Pfizer Animal Health |
| 177 | Timothy Fan | Nothing Disclosed |
| 178 | Kenny Simpson | Research support: Crohn's Colitis Foundation of America, Morris Animal Foundation, Nestle-Purina, NIH, Waltham Weill-Cornell Roberts, IBD Center Consultant: Nestle-Purina |
| 179 | Bud Tennant | Nothing Disclosed |
| 180 | Stephen Withrow | Nothing Disclosed |
| 181 | Matthew Breen | Partially funded by the National Institutes of Health, the AKC Canine Health Foundation, Morris Animal Foundation |
| 182 | Leah Cohn | None Current Funding Sources include the ALSAM Foundation and the Mark Morris Animal Foundation |
| 183 | Carol Reinero | Research support: Morris Animal Foundation, Winn Feline Foundation, Sepracor Pharmaceuticals, Kansas City Area Life Sciences Institute, Charitable Giving Office of Kindy French Trudell Medical International, The Joan Scarangeilo Foundation to Conquer Lung Cancer, University of Missouri: College of Veterinary Medicine, Research Council, Richard Wallace Research Incentive Grant and Phi Zeta |
| 184 | Deborah Fine | Nothing Disclosed |
| 185 | Ray Dillon | Data partially funded by: Pfizer, Merial, Idexx |
| 186 | Craig Thompson | Nothing Disclosed |
| 187 | Craig Thompson | Nothing Disclosed |
| 188 | Rose Raskin | Nothing Disclosed |
| 189 | Rose Raskin | Nothing Disclosed |
| 190 | Jane Quandt | Nothing Disclosed |
| 191 | Jane Quandt | Nothing Disclosed |
| 192 | Marie Kerl | Nothing Disclosed |
| 193 | Marie Kerl | Nothing Disclosed |
| 194 | Robert Backus | Nothing Disclosed |
| 195 | Kelly Swanson | I will do my best to present an unbiased presentation on the topic. However, I want to disclose that I am a consultant for the pet food industry and current/past funding sources include Nestle Purina Petcare, Natura Manufacturing Inc, and Nature's Variety |
| 196 | Kelly Swanson | I will do my best to present an unbiased presentation on the topic. However, I want to disclose that I am a consultant for the pet food industry and current/past funding sources include Nestle Purina Petcare, Natura Manufacturing Inc, and Nature's Variety. |
| 197 | Robert Backus | Nothing Disclosed |
| 198 | Michelle Barton | Commercial Enterprise: The Glass Horse Project, LLC and the University of Georgia Research Foundation, Inc. Copyright 2007 |
| 199 | Michelle Barton | Commercial Enterprise: The Glass Horse Project, LLC and the University of Georgia Research Foundation, Inc. Copyright 2007 |
| 200 | Bruce McGorum | The literature review upon which this presentation was based was sponsored through an educational grant by Pfizer Animal Health. |
| 201A | Bonnie Barr | Nothing Disclosed |

| | | |
|---|---|---|
| 201B | Paul Morley | Nothing Disclosed |
| 202 | Yvette Nout | Nothing Disclosed |
| 203 | Kathy Seino | Nothing Disclosed |
| 204A | Gisella Hussey | Nothing Disclosed |
| 204B | Gisella Hussey | Nothing Disclosed |
| 205A | Lara Maxwell | Nothing Disclosed |
| 206 | Christopher Cebra | Research Support: Alpaca Research Foundation, Willamette Valley Llama Foundation, Morris Animal Foundation |
| 207 | Christopher Cebra | Research Support: Alpaca Research Foundation, Willamette Valley Llama Foundation, Morris Animal Foundation |
| 208 | Jeff Lakritz | Nothing Disclosed |
| 209 | Pam Walker | Nothing Disclosed |
| 210 | Subramaniam (Sri) Srikumaran | Nothing Disclosed |
| 211 | Subramaniam (Sri) Srikumaran | Nothing Disclosed |
| 212 | Anthony Confer | Research Support: USDA, Noble Foundation, Boehringer-Ingelheim, Pfizer; Consultant: Boehringer-Ingelheim |
| 213 | Anthony Confer | Nothing Disclosed |
| 214 | Charlotte Lacroix | Nothing Disclosed |
| 215 | Charlotte Lacroix | Nothing Disclosed |
| 216 | Charlotte Lacroix | Nothing Disclosed |
| 217 | Charlotte Lacroix | Nothing Disclosed |
| 218 | Mark Kittleson | Nothing Disclosed |
| 219 | Mark Kittleson | Past research support:Intervet to examine the effect of ramipril on feline HCM |
| 220 | Mark Kittleson | Nothing Disclosed |
| 221 | Mark Kittleson | Nothing Disclosed |
| 222 | Nick Jeffery | Nothing Disclosed |
| 223 | Nick Jeffery | Nothing Disclosed |
| 224 | Nick Jeffery | Nothing Disclosed |
| 225 | Jonathan Levine | Research Support: National Institutes of Health, American Kennel Club Scott and White Health Sciences **NIH grant was in direct support of this work; other grants support related spinal cord injury research. Consultant: Capital Area Veterinary Specialists South Texas Veterinary MRI |
| 226 | Angela Frimberger | Nothing Disclosed |
| 226 | Antony Moore | Nothing Disclosed |
| 227 | Angela Frimberger | Nothing Disclosed |
| 227 | Antony Moore | Nothing Disclosed |
| 228 | Angela Frimberger | Nothing Disclosed |
| 228 | Antony Moore | Nothing Disclosed |
| 229 | Angela Frimberger | Nothing Disclosed |
| 229 | Antony Moore | Nothing Disclosed |
| 230 | Jeanne Barsanti | Nothing Disclosed |
| 231 | Annette Litster | Support for some of the work reported in this presentation was provided by: Bayer Animal Health, Leverkusen, Germany; Consultant: Bayer USA |
| 232 | Tony Buffington | Nothing Disclosed |
| 233 | Tony Buffington | Nothing Disclosed |
| 234 | Jacquie Rand | Purina Petcare, Waltham/Royal Canin, Iams, Hill's Pet Nutrition, IDEXX Pharmaceuticals, Pfizer, Abbott Animal Health, Intervet, Support for Research projects or speaking engagements has been received from the above companies |
| 235 | Jacquie Rand | Purina Petcare, Waltham/Royal Canin, Iams, Hill's Pet Nutrition, IDEXX Pharmaceuticals, Pfizer, Abbott Animal Health, Intervet, Support for Research projects or speaking engagements has been received from the above companies |
| 236 | Jacquie Rand | Purina Petcare, Waltham/Royal Canin, Iams, Hill's Pet Nutrition, IDEXX Pharmaceuticals, Pfizer, Abbott Animal Health, Intervet, Support for Research projects or speaking engagements has been received from the above companies |

| | | |
|---|---|---|
| 237 | Jacquie Rand | Purina Petcare, Waltham/Royal Canin, Iams, Hill's Pet Nutrition, IDEXX Pharmaceuticals, Pfizer, Abbott Animal Health, Intervet, Support for Research projects or speaking engagements has been received from the above companies |
| 237 | Kirsten Roomp | Nothing Disclosed |
| 238 | Mike Willard | Prior research support and/or sponsorship from: ACVIM Foundation, Morris Animal Foundation, Purina Pet Foods, Hill's Pet Nutrition, IAMS, Royal Canin |
| 239 | Mike Willard | Prior research support and/or sponsorship from: ACVIM Foundation, Morris Animal Foundation, Purina Pet Foods, Hill's Pet Nutrition, IAMS, Royal Canin |
| 240 | Mike Willard | Prior research support and/or sponsorship from: ACVIM Foundation, Morris Animal Foundation, Purina Pet Foods, Hill's Pet Nutrition, IAMS, Royal Canin |
| 241 | Mike Willard | Prior research support and/or sponsorship from: ACVIM Foundation, Morris Animal Foundation, Purina Pet Foods, Hill's Pet Nutrition, IAMS, Royal Canin |
| 242 | Debra Zoran | Research Support: Purina PetCare, Pfizer Animal Health, GINN funds (local agency); Consultant: Pfizer Animal Health |
| 243 | Thomas Schermerhorn | Nothing Disclosed |
| 244 | Sherry Sanderson | Pfizer Animal Health, Proctor and Gamble Pet Care, Nestle Purina, Veterinary Learning Systems |
| 245 | Margarethe Hoenig | Nothing Disclosed |
| 246 | Anthony Carr | Speaking sponsorship: DVM Solutions, Vetoquinolol, Novartis, S+B MedVet Equipment donations, S+B MedVet Ramsey Medical, VMed Technologies |
| 247 | Joerg Steiner | Director of the Gastrointestinal Laboratory at Texas A&M University Consultant: Idexx Laboratories; Research Support: Morris Animal Foundation, Nutramax Laboratories, P&G Petcare, Royal Canin Speaker for: P&G Pet Care; Hill's Pet Nutrition; Nestle Purina; Idexx Laboratories |
| 248 | Sheila McCullough | Nothing Disclosed |
| 249 | Sheila McCullough | Nothing Disclosed |
| 250A | Larry Cowgill | Nothing Disclosed |
| 250B | George Lees | Nothing Disclosed |
| 251A | David Polzin | Nothing Disclosed |
| 251B | Shelly Vaden | Nothing Disclosed |
| 252A | Jody Lulich | Nothing Disclosed |
| 252B | Joseph Bartges | Nothing Disclosed |
| 253A | Dennis Chew | Nothing Disclosed |
| 253B | Larry Adams | Nothing Disclosed |
| 254 | Robert Mattrey | Nothing Disclosed |
| 255 | Robert Mattrey | Nothing Disclosed |
| 256 | Gabriela Seiler | Nothing Disclosed |
| 257 | Shannon Holmes | Nothing Disclosed |
| 258 | Catherine McGowan | Research funded by: University of Queensland's Internal Grants Scheme |
| 259 | David Horohov | Nothing Disclosed |
| 260A | Catherine McGowan | Nothing Disclosed |
| 260B | Dianne McFarlane | Nothing Disclosed |
| 261A | Nicholas Frank | Nothing Disclosed |
| 262 | Colin Schwarzwald | Nothing Disclosed |
| 263 | Colin Schwarzwald | Nothing Disclosed |
| 264 | Colin Schwarzwald | Nothing Disclosed |
| 265 | Colin Schwarzwald | Nothing Disclosed |
| 266 | Tawni Silver | Nothing Disclosed |
| 267 | Tawni Silver | Nothing Disclosed |
| 268 | Gilles Fecteau | Nothing Disclosed |
| 269 | Melanie Boileau | Nothing Disclosed |

| | | |
|---|---|---|
| 270 | Robert Whitlock | Research Support: USDA-CREES-NRI USDA-Cooperative Agreement Tetracore, Inc; Consultant: ELANCO Animal Health |
| 271 | Marie-Eve Fecteau | Nothing Disclosed |
| 272 | Christian Gerspach | Nothing Disclosed |
| 273 | Elizabeth Doré | Nothing Disclosed |
| 274 | Charlotte Lacroix | Nothing Disclosed |
| 275 | Charlotte Lacroix | Nothing Disclosed |
| 276 | Charlotte Lacroix | Nothing Disclosed |
| 277 | Charlotte Lacroix | Nothing Disclosed |

# 2009 ACVIM Forum & Canadian VMA Convention Proceedings

## TABLE OF CONTENTS

# 2009 ACVIM Forum & Canadian VMA Convention Proceedings

## TABLE OF CONTENTS

## ACVIM BUSINESS MANAGEMENT



## ACVIM CARDIOLOGY

## ACVIM-EQUINE

## ACVIM-FOOD ANIMAL INTERNAL MEDICINE

## ACVIM-NEUROLOGY

## ACVIM-ONCOLOGY

## ACVIM-SMALL ANIMAL INTERNAL MEDICINE

## ACVIM/INFECTIOUS DISEASE

## ASVCP



## AVHTM



## CGS

## CVMA ANIMAL WELFARE



## CVMA BUSINESS MANAGEMENT



## LIVER STUDY GROUP

## SCE



## SVN/U



## VCRS (SA & EQUINE)

## VECCS



## VUS

**BACK TO THE FUTURE: USING METABOLIC PROFILES TO IDENTIFY PERIPARTURIENT DISEASE RISK**

**Robert J. Van Saun, DVM, MS, PhD, DACT, DACVN**
**University Park, PA**

## INTRODUCTION

A "metabolic profile" is defined as a series of specific analytical tests run in combination and used as a herd-based, rather than individual-based, diagnostic aid.[1] Use of a metabolic profile is the result of technologic improvements in analytical instrumentation, which can complete multiple analyses in a short time period. The Compton Metabolic Profile (CMP), first introduced in the early 1970's, has traditionally been used in this approach.[2,3] The original intent of the CMP was to monitor metabolic health of the herd, help diagnose metabolic problems and production diseases, and identify metabolically superior cows. Interpretation issues and a lack of specificity in differentiating normal from problem herds coupled with high inherent costs with little diagnostic returns has limited the application of the CMP test, especially within the United States. Research since the time of CMP development has clarified many metabolic issues of the transition cow and its relationship to periparturient disease.[4-6] In concert with this improved understanding of integrated transition metabolism there has been improvement in technical methods to assess metabolic status. Additionally, the shift to increasing herd size and recognition of significant health, production, and economic consequences of periparturient disease has led to renewed interest in a revised metabolic profile application in monitoring transition cow health and disease risk.[7,8] This presentation will review the application of metabolic profiling and provide new insights to its use in diagnosing herd disease risks.

## HISTORICAL PERSPECTIVES ON METABOLIC PROFILING

The CMP involved collecting 7-to-10 blood samples from 3 predefined groups of dairy animals, i.e., dry, peak lactation and midlactation, and having 13 selected metabolites measured.[2] Seven animals were considered the minimum number sampled to be statistically significant for interpretation. From the test results, means for each metabolite are calculated for each respective group and compared to reference values. Reference values for metabolites are determined statistically from means across all herds and production stage data with the range including ± 2 standard deviations around the mean.[3,9] As one might expect completing 13 biochemical tests on a minimum of 21 individual samples was extremely expensive, even with automated equipment. Smaller scale versions of the CMP scheme have been proposed to address cost concerns and maintain some diagnostic capability.[10,11] The CMP has generally received positive endorsements as a diagnostic aid from studies outside the United States.[12-14] In contrast, US based studies have generally been less than enthusiastic about their potential diagnostic value of metabolic profiles.[15-17] Reasons for this difference in perception are varied and include feeding and management differences between European and US dairies (e.g., pasture-based vs. confinement feeding), study design and study objectives. Studies completed in the US focused on the use of metabolic profiles as determinants to predict the possibility or presence of metabolic disease, assess fertility status, or assess nutritional status.[1] These are different from the intended use of the CMP as defined by the original authors who had suggested the metabolic profile test is not a nutritional test and did not associate results with fertility potential.[9]

This test procedure, besides being expensive, had perceived limitations as a diagnostic procedure. Data reported by Payne and colleagues show herds with no recognizable problems were not well differentiated from problem herds.[3] With a random selection process, selected dry cows could have been at any point relative to expected calving. At the time, all dry cows were considered metabolically similar, contrary to current knowledge. Peak and mid-lactation cows were specifically chosen as they were considered "homeostatically stable" compared to recently fresh cows. However, within the current framework of understanding relationships between periparturient disease and metabolic status, it would be cows currently experiencing perturbed homeostasis that should be selected for analysis. The selected populations of cows preferred for the CMP test may have limited the diagnostic sensitivity of the testing procedure.

A problem with interpretation of the metabolic profile is the desire for simplicity. Unfortunately, determination of normal and abnormal is not as black and white as we would desire. A confounding issue with the CMP was the interpretation of "mean" samples compared with the traditional individual sample. Abnormal metabolite values are defined as those outside of the predefined normal reference range. However, there are discrepancies among studies as to what should be defined as the "normal range". The mean ± 2 standard deviations (95%) had been the standard definition; though some studies have advocated ± 1.3 or 1 standard deviation depending upon the metabolite of concern.[1] Wide reference ranges result in a failure to identify abnormal herds (false negative) whereas more narrow ranges may result in calling normal herds abnormal (false positive). A lack of sensitivity and specificity in differentiating normal from abnormal herds had been a primary detractor to accepting this procedure as a useful diagnostic aid.

## RATIONALE AND DIAGNOSTIC APPROACH: NEW CONCEPTS

Although similar samples and analytical methods are used in assessing disease diagnosis or metabolic profiling, the approach to sample collection and results interpretation are different.[18,19] With disease diagnosis, one selects a small population of representative clinically affected animals for blood analyses. Results are compared to laboratory reference ranges for interpretation. Reference ranges for a given analyte are 95% confidence intervals determined from a population of healthy animals. Disease conditions often induce dramatic changes in analyte variability, much greater than inherent variability due to biologic randomness and other factors.[20,21] Disease diagnosis is based on a recognized pattern to changes in one or more blood analytes. In contrast for metabolic profiling, one collects a greater number of samples from clinically "normal" individuals. A greater number of samples are required to highlight analyte differences due to environment, namely nutrition, as its contribution to analyte variation is of a smaller magnitude compared to biologic randomness. As a consequence, metabolic profiling should not be a "random" selection process; rather samples need to be appropriately selected to minimize variation resultant from controllable factors.

In contrast to methods of the past, metabolic profiling should be considered just one tool in the nutritional diagnostic toolbox. In the broadest sense of the terminology, a metabolic profile should also include evaluation of herd records, animals, facilities and rations and integrate the information into a final diagnostic assessment.[14,19,22] Properly applied metabolic profile testing, as defined by appropriate animal selection and sample collection, can potentially provide direct confirmatory evaluation of metabolic health and disease risk or evaluate nutritive status on a herd basis. Results of a metabolic profile should be used to direct or focus attention of a diagnostic process (screening tool) or to help confirm presence of a herd disease process (diagnostic tool).

The screening tool approach is consistent with traditional metabolic profiling methods where multiple analytes are determined within selected group or groups of cows. Determination of multiple analytes is predicated on the concept that periparturient metabolic disease is a result of the cow's inability to maintain coordinated interrelationships between lipid, glucose and amino acid metabolism. A screening tool approach to metabolic profiling can be used as a broad-based diagnostic evaluation of herd nutritive status, assessment of disease risk factors, or indicator of potential factors responsible for disease conditions. Limitations to the screening tool approach are high testing costs and potential interpretation issues.

The targeted diagnostic approach utilizes well defined diagnostic analytes to determine herd risk for specific "gateway" periparturient diseases. In this approach, specific analyte concentration is determined and compared to specific threshold criteria. Percent of individuals above or below the threshold value are used to interpret herd disease risk. Other specific analyte values might be interpreted as a mean value for the individuals within a defined group. Individual testing, lower testing costs, and ease of interpretation are strengths of this approach. Limitation of this approach is scope of analytes determined.

## INDICATORS OF PERIPARTURIENT DISEASE RISK

Over the past 20 years our understanding of transition cow metabolism and its relationship to pathogenesis of periparturient diseases has expanded greatly. In concert with this improved understanding of integrated transition metabolism there has been an increase in technical methods to assess metabolic status. Together, these advancements have reinvigorated interest in metabolic profiling. Periparturient disease-associated culling and reproductive infertility, coupled with the obvious health and production concerns, are driving interest in predicting potential risk for disease. Most work in this area has focused on specific disease entities (i.e., milk fever, ketosis, left displaced abomasum); however, it is well documented that periparturient diseases are interconnected and not single entities.[23,24] Our preventive medicine focus should be directed to increasing the number of cows that complete the transition period without disease events of any degree. Currently, more than 50% of cows experience one or more periparturient diseases.[25,26] In most clinical situations there is tremendous need to assess risk of any disease versus no disease problems, as opposed to defining specific disease risks.

### Energy Balance

Energy balance is by and far one of the most critical nutritional factors impacting on animal health, lactation, and reproductive performance. Traditionally, we have monitored changes in energy balance via body weight and condition score changes over time. This procedure may not be a sensitive enough tool when dealing with the transition cow. Nonesterified fatty acid (NEFA) and β-hydroxybutyrate (BHB) concentrations have become the mainstay in determination of energy balance. Many research studies have shown good correlations between energy balance and serum NEFA concentrations. Serum NEFA concentration is the result of adipose tissue breakdown of fat in response to negative energy balance. Circulating NEFAs are absorbed and metabolized for energy by the liver and other tissues. Concentration of NEFA directly reflects amount of adipose tissue breakdown taking place. Another parameter useful in assessing aberrant energy status is ketone body concentrations, primarily as BHB. However, BHB can come from dietary sources (poorly fermented silage) and not reflect aberrant metabolism.

Excessively high NEFA concentrations due to negative energy balance results in fatty infiltration of the liver, which is associated with higher incidence of periparturient metabolic diseases. Prior to calving, BHB concentrations generally do not exceed 6-8 mg/dl, unless the animal is in negative energy balance or consuming ketogenic silage. Following calving, BHB concentrations can become greatly elevated. Elevated prepartum NEFA concentration and postpartum BHB concentration are recognized risk factors for ketosis and left displaced abomasum.[8,27-29] The original reference values for NEFAs were based on data from Michigan State University Clinical Nutrition Laboratory suggesting NEFA concentrations above 0.4 (Close-up dry) and 0.6 (Fresh) mEq/L (mmol) are associated with 4 and 5-fold increase of disease risk, respectively. These criteria have been supported by similar findings in other studies, and show increasing risk with higher concentrations.[26,27] Concentrations of BHB $< 26$ mg/dl and $> 14.5$ mg/dl represent animals with subclinical ketosis. Those with concentrations $\geq 26$ mg/dl are defined with clinical ketosis. However, recent studies have shown cows with BHB concentrations above 10 or 14 mg/dl are 3.2 and 4.3 times at greater risk for postpartum disease.[26]

### Protein Evaluation

At present there is no single metabolite that can be measured which directly reflects protein status. As a result, multiple parameters are needed to assess protein status including BUN, creatinine, total protein, albumin and creatine kinase (Ck). Urea nitrogen concentrations are influenced by a wide variety of interrelated parameters including: dietary protein intake and rumen degradability, dietary amino acid composition, protein intake relative to requirement, liver and kidney function, muscle tissue breakdown and dietary carbohydrate amount and rumen degradability. Total protein and albumin reflect availability of protein and their concentration decline in the face of protein deficiency. However, this occurs over a period of time. Albumin has a relatively short half-life and can reflect protein deficiency problems over a period of a month or two. Creatine kinase is released from muscle when it is catabolized or injured.

Albumin was found to be associated with postpartum disease and can be used to predict disease risk in close-up and fresh periods.[26] Fresh cows that could maintain serum albumin concentrations ≥ 3.5 g/dl were less likely to have postpartum disease. Serum albumin concentrations ≤ 3.25 g/dl in close-up dry cows resulted in a three-fold greater risk for postpartum disease. In spite of concerns about variables confounding albumin interpretation, it seems to be a good disease risk indicator, possibly reflecting availability of amino acids from the labile protein pool. These results are similar to others reporting low total protein (<5.0 g/dl) in recently calved cows being associated with increased postpartum disease risk.

**Liver Function**

Liver function can be evaluated using a variety of enzyme activities (i.e., gamma-glutamyltransferase (GGT), aspartate aminotransferase (AST), and sorbitol dehydrogenase (SDH)) and total bilirubin concentrations in the blood. Unfortunately, an elevation in any of these parameters does not suggest anything more than some insult has occurred to the liver and may not be specific for hepatocellular disease. Since these parameters are not specific to liver function, other liver function indices have been advocated. A liver activity index parameter that accounts for changes in liver-specific proteins over the first 28 days following calving has been defined.[30,31] Although a robust diagnostic tool, it use requires multiple samples from the same cow over a period of time. This would preclude its use within a typical metabolic profiling approach. Calculating the NEFA to Cholesterol ratio (molar basis) to assess the liver's ability to export incoming NEFA has been advocated.[32] Calculated NEFA to Cholesterol ratio was predictive for postpartum disease in the close-up dry and fresh cows.[33,34]

**Macromineral Evaluation**

Assessment of Ca concentrations around the time of calving is a useful indicator of how well the Ca regulatory system is working and potential for clinical or subclinical hypocalcemia problems.[8] It is perceived that other than the 2 weeks prior to and following calving, blood Ca is not a very diagnostic value as a result of the intact regulatory system. Therefore macromineral blood concentrations will need to carefully interpret in light of whether or not the homeostatic system is in proper operation. In a recent retrospective study, a number of significant predictive relationships were found between serum mineral concentrations in the four weeks prior to or following calving and periparturient disease.[35,36] Most significant was the association between cows with pre- or postpartum calcium concentrations below 8 mg/dl. Either pre- or postpartum cows with serum total calcium < 8.0 mg/dl were 4 times more likely to have postpartum disease problems. Pre- and postpartum sodium concentrations were also highly associated with postpartum disease risk. These were interesting findings, and suggest macromineral concentrations may be more predictive than previously thought. Associations between trace mineral concentrations and risk of periparturient disease were minimal, though high prepartum iron and low postpartum zinc concentrations tended to be associated with infectious disease problems.[37] Ratios of copper to zinc or iron, potentially reflecting changes indicative of an acute phase inflammatory response, tended to be associated with increased mastitis and metritis risk. Further work is needed to validate the relationships identified in this preliminary study.

**SUMMARY**

Traditional metabolic profiling of the dairy herd resulted in tremendous financial investment, with subsequent unsatisfactory results in many situations. A number of factors are responsible for individual and herd variation in blood metabolite concentrations, confounding interpretation. Additionally the cow has an exquisite system of checks and balances, which maintains normal physiologic function within a wide array of dietary and environment insults. As a result of these physiologic regulatory mechanisms, simple blood concentration analysis has not been highly rewarding in accurately assessing nutritional and fertility status. Improved understanding of transition metabolism and a new approach to metabolic profiling have with other economic factors revitalized interest in using serum metabolite concentrations as a herd-based diagnostic tool. Most importantly, it must be remembered that metabolic profiles are almost useless without being coupled with animal and facility evaluations, body condition scoring and ration evaluation. It is only when the whole picture is evaluated will the uses of metabolic profiles produce useful diagnostic information.

**REFERENCES**

[1]Ingraham RH, Kappel LC. **Vet Clinics NA:Food Anim Pract** 1988;4(2):391-411. [2]Payne JM, et al. **Vet Rec** 1970;87:150-158. [3]Payne JM, et al. **Brit Vet J** 1973;129:370-381. [4]Bell AW. **J Ani Sci** 1995;73:2804-2819. [5]Grummer RR. **J Anim Sci** 1995;73:2820-2833. [6]Drackley JK. **J Dairy Sci** 1999;82:2259-2273. [7]Herdt TH, et al. **Proc Am Assoc Bov Pract** 2001;34:27-34. [8]Oetzel GR. **Vet Clinics NA: Food Anim Pract** 2004;20:651-674. [9]Rowlands GJ. **Wld Rev Nutr Diet** 1980;35:172-235. [10]Blowey RW. **Veterinary Annual** 13 (1973):21-30. [11]Blowey RW. **Vet Rec** 1975;97(17):324-327. [12]Bogin E et al. **J Dairy Res** 1982;49:13-23. [13]Ghergariu S et al. **Br Vet J** 1984;140:600-608. [14]Bertoni G, Trevisi E. **Proc V$^{th}$ Congress Internat Soc Anim Clinical Biochem**, Sept 2-6, pp. 167-177, 1992. [15]Adams RS, et al. **J Dairy Sci** 1978;61:1671-1679. [16]Lee AJ, et al. **J Dairy Sci** 1978;61:1652-1670. [17]Jones GM, et al. **J Dairy Sci** 1982;65:683-688. [18]Van Saun RJ. **Bov Pract** 1997; 31(2): 43-50. [19]Herdt TH, et al. **Proc Am Assoc Bov Pract** 2001;34:27-34. [20]Herdt TH. **Vet Clinics NA: Food Anim Pract** 2000; 16(2): 387-403. [21]Herdt TH, et al. **Vet Clinics NA: Food Anim Pract** 2000; 16(3): 423-444. [22]Van Saun RJ, Wustenberg M. **Bov Pract** 1997; 31(2): 37-42. [23]Curtis CR, et al. **J Dairy Sci** 1985;68:2347-2360. [24]Goff JP. **J Dairy Sci** 2006;89:1292-1301. [25]Bigras-Poulin M, et al. **Prev Vet Med** 1990;10:79-89. [26]Van Saun RJ. **Proc Am Assoc Bov Pract** 2004;37:212-213. [27]LeBlanc SJ, et al. **J Dairy Sci** 2005;88:159-170. [28]Geishauser T, et al. **J Vet Med, A** 1988;45:499-511. [29]Geishauser T, et al. **Comp Cont Ed Pract Vet** 2001;23(8):S65-S70. [30]Bertoni G, et al. **J Anim Sci** 2006;84(Suppl. 2):84. [31]Bertoni G, et al., J Dairy Sci 2008;91:3300-3310. [32]Holtenius P, Hjort M. **Bov Pract** 1990;25:91-. [33]Van Saun RJ. **Proc 23$^{rd}$ World Buiatrics Congress**, pg 87, 2004. [34]Van Saun RJ. **Production Diseases in Farm Animals: 12th International Conference**, pg 41, 2006. [35]Van Saun RJ, et al. **Production Diseases in Farm Animals: 12th International Conference**, pg 221, 2006. [36]Van Saun RJ, et al. **Proc Am Assoc Bov Pract** 2005;38:178-179. [37]Van Saun RJ, et al. **Proc XXIV World Buiatrics Congress**, Nice, France, 2006.

**APPROACHES TO HERD-LEVEL INTERPRETATION OF NUTRITIONAL VARIABLES IN DAIRY HERDS**

**Thomas H. Herdt, DVM, DACVIM, DACVN**
**East Lansing, MI**

Analysis of blood is a fundamental means by which the health of individuals is assessed, but when the patient changes from an individual to a herd how does that affect the interpretation of blood values? On one hand the problem becomes more complex because there will be variation within the herd. On the other hand, if this variation is appropriately managed it can be a useful tool that adds statistical power to our ability to interpret herd-level laboratory data.

On an individual patient basis we analyze a blood sample and expect to compare the results to a reference interval. Clinical pathologists generate reference intervals based on the frequency density distribution of blood concentration variables in a reference population. We use that reference population as a framework for statistical inference. Thus, establishing the reference population is a critical part of interpreting blood values. Biological, as well as statistical, information goes into selecting the reference population so as to minimize the size of the reference interval, i.e. the reference interval may be based on animals that have been fasted, or are of a given age range, or gender, or some other factor that we know or expect would add variability to the frequency distribution. We then compare the value from our individual patient to the reference interval asking the question "is it likely that another source of variation, other than those incorporated into the reference interval, is influencing this value." In veterinary clinical pathology, the additional source of variation in which we are interested is the presence of disease.

If the values from the reference population have a frequency distribution that is approximately Gausian, then the reference interval may be chosen based on a symmetrical range around the mean, usually two standard deviations above or below the mean. This range incorporates 95% of the variability within the reference population. If our patient value lies within this range, we accept that there are probably no additional sources of variability acting on our patient, other than those in place in the reference population. As clinicians we temper this laboratory observation with our other clinical observations, knowing that the single value from our patient is subject variation. However, we have only our judgment on which to rely when modifying the interpretation of a laboratory result that is within the reference range, because based on a single sample we have no information about the variability of our patient value.

If we measure a blood component (or any other type of variable) from a number of animals in a herd we will have additional information not available from our single sample, namely we will have information on the variability of the value within the herd. In general, two types of tests may be done on blood samples: those with dichotomous results (positive or negative) and those with continuous results, such as the measurement of a metabolite or nutrient concentrations. Dichotomous tests typically involve the presence of absence of an infectious agent that may be inducing either clinical or subclinical disease. The methods for statistical appraisal of such data are highly developed and widely applied. Two types of testing are done with these types of assays: testing of the entire herd, in which the variability factors involve primarily the accuracy of the tests, and testing a portion of the herd to estimate the prevalence across the entire herd. In each case herd-level confidence limits can be calculated for the results, but those limits apply to the proportions of animals that are truly infected. The underlying assumption is still that an animal is either infected or it is not; the uncertainty factor is our ability to make a correct classification.

Clinicopathological or nutritional testing for continuous variables, when applied on a herd basis, is sometimes evaluated by similar statistical means as are dichotomous data, but much of the power of continuous data and replicate samples from a single herd are lost in this process. A frequent approach to evaluating such data is to use logistic regression to link continuous data to some dichotomous event, usually the occurrence of a disease. For example, in a large multi-herd study logistic regression indicated that for every incremental increase of 0.1 mEq/L in serum NEFA concentration before calving, the odds ratio (OR), or relative risk, of a cow developing left displacement of the abomasum increase by a factor of about 1.0. (LeBlanc et al, 2005) The application of such information has frequently been to select a "cut point" in the distribution of the continuous variable and to use that single value as a predictor of disease occurrence. In the example with prepartum serum NEFA concentrations, a cut point of 0.5 mEq/L was chosen as an optimal predictor of LDA. By applying this cut point the prepartum serum NEFA concentration in an individual cow predicts the postpartum occurrence of LDA with a sensitivity of 46% and a specificity of 82%. While these data were analyzed appropriately for correlative effects due to herd, all herd information was sacrificed for the development of an individual cow test.

To apply such information at a herd level, the cut point could be applied to samples of cows in order to estimate the proportion of cows in the herd with values above the cut point. A confidence interval for this estimated proportion can be calculated. "Action levels" for herd decisions can then be applied based on the estimated proportion of individuals above the cut point. The "action levels" are arbitrary and based on the tolerance for risk for the predicted disease. Sample size calculations can be done based on the confidence desired for predictions. This approach of dichotomizing continuous data has been used by production medicine clinicians at the University of Wisconsin in the interpretation of serum BHB concentrations in early-lactation dairy cows (Oetzel, 2004). Figure 1 illustrates the sensitivity with which a proportion of cows with serum BHB concentrations above 1400 uM could be predicted from a sample of twelve cows. The approach is clearly a step forward in the herd-level application of clinicopathological data, particularly because it develops a rational basis for sample size calculation. However it assumes that the distribution of values above and below the cut point is unimportant. This assumption sacrifices the power available from sampling multiple animals from a continuous distribution.

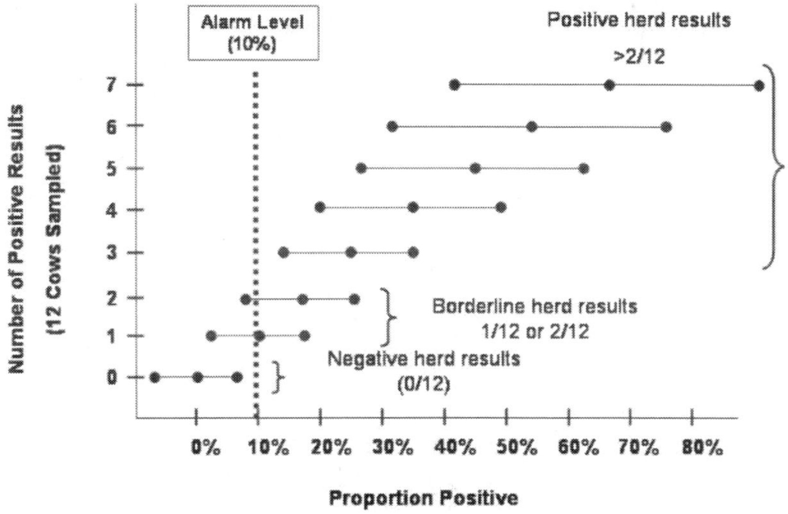


**Figure 1.** Herd-level interpretation of serum beta-hydroxybutric acid (BHB) concentrations. Bars indicate the 75% confidence intervals for estimated proportions of animals with values above the "action level." Redrawn from Oetzel GR, Vet Clin Food Anim 20:651, 2004.

The approach of dichotomizing continuous data means that the distribution of values above and below the cut point is ignored. Consider the hypothetical example in figure 2 of a variable with Gaussian distribution. Using the "cut point" approach, each herd would be evaluated as having the same predicted proportion of cows with values above the cut point, twenty percent. However, the true distribution within the Herd A is more likely to include a higher proportion of cows above the cut point than in Herd B, as shown by the shaded areas in the figure. This illustrates the potential loss of information resulting from dichotomizing the data.

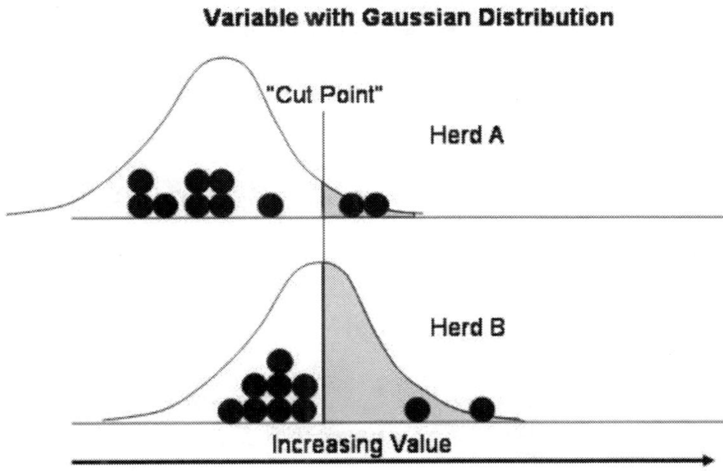


**Figure 2.** Hypothetical distribution of a variable with Gaussian distribution in two herds. The dichotomous estimation would be that each herd had 20% of animals above the cut point, whereas the Gaussian distribution would predict a much higher proportion in Herd B than in Herd A, as illustrated by the shaded areas.

An alternative approach is to evaluate the distribution of values in the "patient" herd among the distribution of values in a population of reference herds by a means similar to that used for evaluating values for individual patients. The challenge is in devising a method that appropriately considers the variability within and among herds in the reference population. An approach might be to compare the mean value in the "patient" herd to the same reference range used in evaluating individual patients. This is clearly inappropriate, however, because if we expect the concentration of the analyte to vary among herds, then we should also expect the variance of a population of individuals to be greater than the variance of herd means within the population. Thus, the reference range used for the evaluation of individuals is expected to be too broad to use as a reference range for herd means. Thus, the appropriate reference range must be based on a population of herds, rather than a population of individuals and the comparison of the "patient" value to the reference range must take into account the variance in each. Figure 3 illustrates the relationship of variation within individual herds to the variation among the population of herds. The confidence with which a herd may be accurately placed within the population of herds depends on the relative amounts of within and among herd variation.

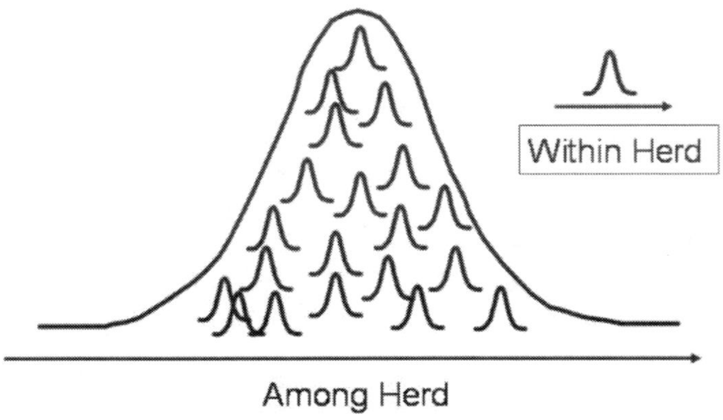


**Figure 3.** This figure illustrates the relationship of among herd to within herd variation.

Random-effects linear models provide a means by which to separate within and among herd effects in a population of herds. A statistical model is a mathematical means of providing potential explanations for why data follow an observed pattern of distribution. A random-effects linear model applied to herd-level clinicopathological data estimates an overall mean for the population and an "effect" of each herd (Littell RC et al., 1996). A herd effect is the estimated difference between the herd mean and the population mean. A confidence interval for the herd effect and a probability that the herd mean is different than the population mean can be calculated. To use this information clinically what is needed is a large population of herds and a biological concept of how much of a deviation from the population mean is meaningful from a clinical standpoint. The biological decision need not necessarily be based on predicted disease incidence, although that may be important. For nutritional data the decision may also be based on predicted effects on animal performance characteristics, which may be affected by smaller deviations in herd means than are disease risks. As data are accumulated the population of reference herds may be refined by the retrospective examination of clinical data. This is not different than the means by which reference intervals for individual patients are generated and offers a means by which data collected from clinical testing can be economically applied for the establishment of reference populations. As such herd-level clinical data continue to accumulate random-effects linear models promise to be a powerful means for interpretation.

**REFERENCES**

[1.] LeBlanc SJ et al., **J Dairy Sci.** 88:159, 2005. [2.] Littell RC, SAS System for Mixed Models, **SAS Institute Inc**. Cary NC, 1996. [3.] Oetzel GR, **Vet Clin Food Anim** 20:651, 2004.

**HERD-LEVEL ASSESSMENT OF MINERAL NUTRITURE IN CATTLE**

**Thomas H. Herdt, DVM, DACVIM, DACVN**
**East Lansing, MI**

The consequences and pattern of depletion of minerals, particularly trace minerals, from the animal body generally can be envisioned based on the stylized graph presented in figure 1 (Suttle N, 2004). It indicates that minerals have a storage pool, a functional pool, and one or more transport pools that I have described as excretory and functional pools to distinguish those transport functions related to the excretion of excess mineral from those related to transport of minerals from storage sites to functional pools.

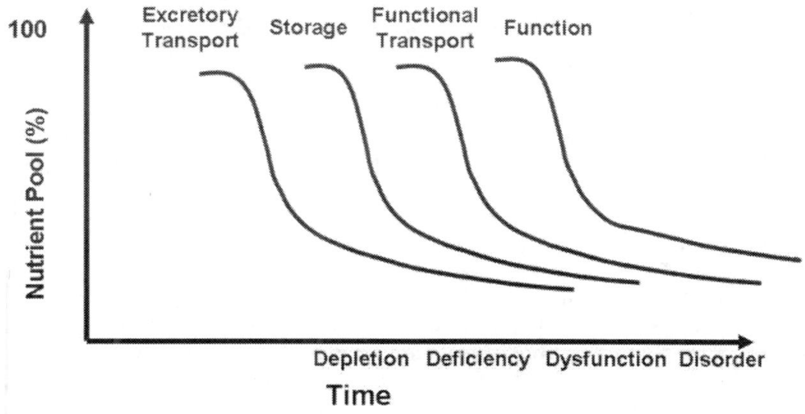


**Figure 1.** Each curve on the graph represents the time frame of depletion of a mineral from various physiologic compartments following the removal of the mineral from the diet. (Modified from the figure of Suttle N, 2004)

Assessment of mineral nutriture is aimed at detecting depletion of mineral from storage or transport pools, before disturbances in function have occurred. The challenge in diagnosing mineral depletion is that the various pools generally are not anatomically distinct and they may not be practically accessible. Furthermore, not all minerals have metabolic patterns that follow exactly this conceptual frame work.

Sampling of blood and liver tissue may provide practical means for assessing storage and transport pools, depending on the specific mineral in question. Liver is the specific site of copper storage and thus forms a storage pool that is easily accessed. The liver is also a site of iron storage and while not the only site of iron storage hepatic iron concentration may at least reflect the size of the overall iron storage pool. Liver also is the site of cobalamin storage, and thus may reflect the cobalt status of ruminnats. However, not all hepatic cobalt is in the form of cobalamin, which is the only biologically active form of cobalt.

Blood and blood serum are convenient samples and can be expected to form part of the functional transport pool for minerals, and for some minerals blood is an important part of the excretory transport pool. Whether or not blood is part of the excretory pool is dependent on the type of homeostatic mechanisms in place to regulate body mineral stores. For several minerals including iron, zinc, and calcium the primary site of homeostatic control is at the site of absorption in the small intestine. As body stores become replete, the efficiency of absorption is substantially reduced such that little additional mineral is absorbed. In these cases, dietary mineral in excess of requirements never reaches the blood stream, and thus blood serum is not involved in excretory transport. In other cases, such as for magnesium, selenium, and molybdenum, intestinal absorption is not highly regulated and absorption continues even in the face of full body stores. In such instances, the mineral must traverse the blood vascular system in route to excretion, usually via the kidney. In these cases, blood serum could be expected to be a much better indicator of early mineral depletion than in those cases in which blood is not involved in excretory transport

Interest in the assessment of mineral nutrition is often at a herd level, which may offer some advantage in the early diagnosis of mineral depletion. Sampling multiple animals reduces the effect of random variation in the evaluation of mineral concentrations and thus increases the sensitivity with which mineral depletion may be detected. Examination of herd-level characteristics in the distribution of variation in mineral concentrations may also aid in selecting and evaluating tests for mineral nutriture.

Examining the distribution of variation within and among herds may provide insight to the relative effects of environment, particularly the nutritional environment or diet, on mineral concentrations in blood or liver. Examination of variance components may be envisioned as a means of determining the "signal to noise ratio" in the evaluation of herd-level sampling for mineral status. If there is substantial individual variation among animals, even within individual herds, that represents noise or unwanted variation in our interpretation of mineral concentrations. If, on the other hand, the variation between herds is substantially greater than the variation within herds, that represents "signal" and implies that herd environment has a substantial effect on the measured mineral concentration. Presumably, although not necessarily, the primary environmental factor affecting mineral status within herds is diet.

One approach to making this determination is to examine the proportion of variability attributable to herd in a population of samples from multiple herds. A brief discussion of within- and among-herd variability can be found elsewhere in these proceedings (T. Herdt: "Approaches to herd-level interpretation of nutritional variables in dairy herds"). The information in table 1 represents data from herds from which three or more samples from separate animals were submitted to the Diagnostic Center for Population and Animal Health (DCPAH) at Michigan State University for mineral analysis. The serum information represents data from 165 herds and the liver information is from 12 herds. All animals included were adults, but little other detailed information about the individual animals was available. The table lists the proportion of variability attributable to herd effects, which are likely to be largely nutritional.

| Mineral | Proportion of variation attributable to herd effects | |
|---|---|---|
| | Serum Concentration | Liver Concentration |
| Copper | 48% | 65% |
| Cobalt | 58% | 64% |
| Iron | 30% | 59% |
| Manganese | 12% | 16% |
| Selenium | 82% | 70% |
| Zinc | 82% | 47% |

**Table 1.** Herd variance components for various mineral concentrations measured in blood serum and liver tissue. Serum data are from 165 herds and liver data from 12 herds. Calculations were made using random-effects linear models (Proc Mixed – SAS System).

This information has several ramifications for herd-level assessment of mineral status in cattle. First it demonstrates that whether serum or liver are used as the sample, the sensitivity for detecting differences among herds varies with the specific mineral. For example the sensitivity with which herd differences in selenium status can be detected is much greater than that for manganese status, given an equal sample size per herd. Furthermore, the "signal to noise ratio" for specific minerals differs between serum and liver, and not always in the same direction. For example herd differences in hepatic copper concentrations will be detected with more sensitivity using liver samples rather than serum samples, and the converse is true with respect to selenium. Moreover, this information has ramifications on the sample sizes, or number of animals that should be tested for assessment of mineral status. In this set of 165 herds, herd effects on serum copper concentration as small as -0.1 ug/mL from the intercept (which approximates the overall mean) could be reliably ($p<0.01$) detected by testing as few as 5 animals. Likewise, herd effects on serum selenium concentration as small as -15 ng/mL from the intercept could be reliably detected ($p<0.01$) by testing as few as 3 animals. These observations illustrate the power gained from a large data base and the use of random-effects linear models for the herd-level assessment of mineral status in cattle.

**REFERENCES**

[1.] Littell RC, SAS System for Mixed Models, **SAS Institute Inc**. Cary NC, 1996. S [2]Suttle NF, **In Practice** 26:553, 2004.

# POOLED SERUM PROFILES & PERIPARTURIENT DISEASE RISK: CLINICAL PATHOLOGY HERESY OR PRACTICAL HERD DIAGNOSTICS?

**Robert J. Van Saun, DVM, MS, PhD, DACT, DACVN**
**University Park, PA**

## INTRODUCTION

Metabolic profiling as originally defined using the Compton Metabolic Profile (CMP) procedure was insufficient in identifying herd problems. More recent research has better characterized metabolic dynamics around the time of calving and their relationship with high prevalence of metabolic and infectious disease problems. This, coupled with a series of factors including improved diagnostic tests, increasing herd size and recognition of transition cow health impact on herd performance, have revived interest in new testing protocols to better assess herd periparturient disease risk. Cost and interpretation have been the limiting factors to widespread adoption of a metabolic profiling procedure. New strategies to reduce costs and yet maintain some useful diagnostic value have evolved either herd-based diagnostic or screening approaches. Use of pooled samples collected from predefined physiologic groups, and testing a wide array of analytes, has been advocated as a practical and economic method of screening the dairy herd for periparturient disease risk. This presentation will describe methods, validation, and interpretation of a pooled sample approach in the application of metabolic profiling as a herd diagnostic tool.

## POOLED SAMPLING APPROACH

Few would argue the strength of individual analysis in metabolic profile analysis. Indeed the gold standard for analytical analysis would be to measure a large percent of the population of interest as individuals. Statistical modeling would suggest at least 8 individuals from a population is representative (mean analysis), though 12 or 13 samples are best for threshold analysis using the targeted diagnostic approach.[1-3] Though statistics may show a minimum number of samples needed to represent a population, the best way to truly characterize a population is to obtain more samples. The inherent problem with this approach is economics. With individual sampling strategies, a decision is often made to either reduce sample numbers, analyte selection or some combination. The goal in metabolic profile testing is to appropriately balance quality of information derived from the testing process with analytical costs.

### Rationale

A pooled-sample approach was an attempt to address both cost concerns and increased scope of testing in assessment of herd nutritional or disease risk status.[4,5] In the individual sampling approach using either cow or herd level interpretation, single variables are being used to assess specific disease risks.[2,3,6] A pooled-sample approach should be considered more of a herd-level screening process and not a specific diagnostic test. The approach to this application is to evaluate various aspects of integrated metabolism as well as changes that occur over the transition period. Testing is broad based and includes analytes reflecting energy balance (nonesterified fatty acids [NEFA], β-hydroxybutyrate [BHB]), protein status (blood urea nitrogen [BUN], albumin, total protein), liver function (various enzymes, cholesterol, and triglycerides) and macromineral homeostasis.[5] This can also be expanded to include microminerals and vitamins, if deemed appropriate for the diagnostic investigation. To measure this range of analytes, one must reduce costs. One approach that has been described is to measure analytes in pooled samples of individuals and evaluate pooled samples over a series of predefined cow groupings.[4,5]

Mean analyte values within physiologic groupings were used for interpretation in the original CMP protocol. These mean values were arithmetically determined from individual samples. Use of individual sampling resulted in the high associated costs of this procedure. Can pooled samples be used to reduce the cost and provide some valid method of herd assessment? There are essentially two issues to be resolved in answering this question. First, can pooled samples accurately reflect the arithmetic mean of the individuals? Second, can one adequately define valid interpretation criteria for pooled sample values? When these questions were posed to clinical pathologists, responses were short and derisive. No response was supportive of the idea, but undaunted we moved forward on the challenge. The first question was the easiest to answer, but a protocol to generate pooled samples was necessary.

### Pooled Profile Procedure

Proper sample selection and pooling technique are keys to success with pooled samples. In selecting samples we need to ensure that true herd or nutritional differences will be borne out and not masked by random or other sources of controllable variation.[7] Samples can be pooled by appropriate physiologic states to allow interpretation of dynamic changes in "population" means over a period of time.[5] For example to address a fresh cow problem, pooled samples can be collected from recently dry cows (>10 days following dry-off up to 30 prior to calving), close-up dry cows (3 to 21 days prior to calving) and fresh cows (3 to 21 days in milk). Other appropriate sample pools can be determined given the specific problem to be addressed. If a herd was experiencing increased prevalence of retained fetal membranes (RFM), you could pool both clinical and non-clinical animals within the fresh cow group for comparison. Obviously you would not know which animals go on to have RFM prior to calving. Also within these groupings, first lactation animals are kept separate from second and greater lactation animals. This physiologic grouping strategy is based on findings of significant differences in mean analyte concentrations across these groups in either healthy or sick cows.[8] Most clinical pathology laboratories do not alter their reference ranges relative to physiologic criteria or stage of lactation.

Sample collection and handling are important if one is to expect useful diagnostic information in return. Blood samples should be taken from either jugular or coccygeal veins with a minimal amount of stress. Blood samples from the mammary veins are not appropriate given the loss of nutrients into the mammary gland. Lower concentrations of phosphorus and potassium have been documented in jugular compared to coccygeal blood samples as a result of salivary gland uptake.[9,10] Time of sampling relative to feeding and feeding management may also influence metabolite concentrations and should be considered in the decision process of when to sample. If herds are being repeatedly sampled as a monitoring tool, samples should be taken at approximately the same time of day to minimize diurnal and prandial variation.

Vacuum tubes are the most common and easiest form of sample collection. A variety of vacuum tubes are available. Vacuum tubes are colored coded for specific diagnostic test procedures based on the specific anticoagulant or additive present in the tube. Most commonly a single serum (red top, no anticoagulant) or plasma (green top, sodium heparin) sample is collected; however, in some cases a whole blood sample may be desired. Extreme efforts should be taken to prevent hemolysis of the sample. All samples should be properly identified with animal and group identification and date of collection. Other pertinent information for interpretation of the metabolic profile would include: animal age, lactation number, milk production level, milk composition, days in milk, pregnancy status and body condition score.

By pooling samples you are obtaining information from a greater number of animals for reduced cost. Rather than the standard 21 samples to calculate 3 group means, you may submit 3 pooled samples, which represent means of 10 to 20 animals each. For data from pooled samples to be relevant, all cows should be equally represented. In preparing pooled samples one must be meticulous in precisely measuring equal amounts of serum from each individual to be included in the pooled sample. Depending upon the total number to be included, typically between 100 and 500 µl (0.1 to 0.5 ml) from each individual are mixed into a new clean vacuum tube (7 to 10 ml capacity). This process is best completed with use of a micropipetter or a TB syringe for precision. Pooled samples should be adequately mixed then directly submitted to the laboratory or frozen and shipped.

**Validation**

In a number of preliminary studies, individual samples were used to determine if pooled sample results for any given analyte were different from the arithmetic mean of the individuals within the pool.[4,11,12] Using a variety of statistical methods, we found that pooled samples do accurately represent individual means from pool sizes ranging from 5 to 20. When one uses sample sizes of 25 to 30 we found more statistical differences between arithmetic mean and pooled values, but numerical values were not greatly different. This statistical difference may have been more of a "statistical artifact" due to the larger "n" and smaller error term. In plotting arithmetic mean versus pooled value for various analytes most of the graphs showed a nearly 1-to-1 relationship. These results are consistent with another study looking at a pooled sampling technique.[13]

Does this mean pooled values precisely represent the mean of individuals? No, there is some variation, though it depends upon the analyte. Fortunately, key metabolic indicators for transition cows showed less than a 5% difference between arithmetic mean and pooled values.[12] In our experience, pooled sample values are always in the same direction relative to the reference population mean, but may vary in magnitude in distance away from the mean. Samples that were not collected and handled under the best conditions (various degrees of hemolysis present) resulted in increasing differences between pooled and mean values. We have also found that pooled samples held at room temperature for a period of time (hours) resulted in analytical problems resulting from sample gelling. Herds that were experiencing more disease problems had more statistical differences between pooled and arithmetic mean values.[12] From the preliminary studies, it would seem feasible that analyte analysis of a pooled sample does reasonably represent the arithmetic mean of the sampled individuals.

**INTERPRETATION OF POOLED SAMPLES**

The real challenge for metabolic profiling is defining appropriate and sensitive reference values. This is even more of a concern in interpreting mean samples, especially pooled values. This has been the stumbling block for adoption of this procedure. One must recognize that this process, however well refined, will never be as sensitive as individual animal analysis. Interpretation criteria for pooled samples are still in the process of development and refinement, but basic concepts will be discussed.

**Basic Concepts**

For individual animals, metabolite values are compared to standard, laboratory-dependent reference ranges. These reference values generally represent a 95% confidence interval. This means that 95% of normal animals should have a given metabolite concentration within this range. This also suggests that 5% of the population will be outside of this reference range and still be normal, emphasizing the need to clinically evaluate the animal. A number of factors, most notably physiologic state and age, have been shown to influence blood metabolite concentrations.[7,10] Most reference ranges do not account for these differences and thus may confound direct interpretation. Having a thorough understanding of the physiologic regulation of a given nutrient is crucial to interpretation.

In contrast to individual animal samples, pooled mean metabolite values cannot be directly compared to reference ranges in the same way. When interpreting pooled samples one needs to remember that measured value represents a population with individuals above and below the mean. As a general rule, means of pooled samples should be near the midpoint of the reference range to be considered normal. For example, if serum total calcium (Ca) concentration for fresh cows is 9 mg/dl and the reference range is 9 to 12 mg/dl, this might be interpreted to suggest a potential problem with subclinical hypocalcemia whereas it would be considered normal in an individual. The measured mean of 9 mg/dl represents a population with approximately 50% of the individual values above and below. This suggests that a number of individuals would have serum Ca concentrations below the normal range. Of course interpretation of metabolic profile results has to be considered in light of presenting problems in the herd. If the herd is experiencing clinical signs consistent with subclinical hypocalcemia, e.g., slow increase in feed intake and milk production, displaced abomasum and ketosis problems, this would be supportive evidence of the metabolic profile results.

Without population variance determinations in pooled samples, you cannot really determine how significant mean differences are. Yet, with many metabolites, like BUN, calcium, magnesium or glucose, you can eliminate the possibility that a single sample was sufficiently low or high to skew the mean. For low BUN values, it is difficult to have values approaching zero whereas for other metabolites, if the sampled cow had an extremely skewed value, it would have been exhibiting clinical signs and would not have been sampled. Metabolites with high variability (wide range of values; liver enzymes) will be of less diagnostic value as compared to low variability metabolites (minerals).

**Methods of Interpretation**

In following the original concept promoted in interpreting CMP samples, pooled samples can be compared to an expected healthy population mean (or median) and the associated population variance.[14,15] Population statistics were determined for various blood analytes in dairy cows having no evidence of periparturient disease at different defined periods relative to calving (far off dry, close-up dry and fresh).[8] The number of standard deviations away from the healthy population mean was determined for a number of pooled samples, using both arithmetic means and pooled values. The percent of abnormal values within a pool was determined using individual samples, compared to standard reference values for respective analytes. A linear relationship, though specific for each analyte, was found between percent abnormal values within a pool and number of standard deviations different from the mean resulting. In general, if a pooled sample value was less than 0.25 (range 0.1-0.5) standard deviations from the healthy population mean, then < 10% of the individuals had abnormal values.[14,15] Analyte-specific criteria can then be developed to estimate number of abnormal values with a pooled sample. This analysis is consistent with previous work on defining criteria for mean values and what many laboratories are using for metabolic profile evaluation criteria. Further work to establish specific pooled sample criteria is ongoing.

A second method to use and interpret pooled samples is by a modification of statistical process control. In larger herds, where sampling can be done more repetitively, one could collect samples on a monthly or bimonthly basis and plot the pooled sample results over time. Standard deviations from a healthy population as described above could be used to initially generate various sigma levels for evaluation. One can then graphically monitor metabolic status using one or more key analytes over time to assess potential risk for periparturient disease.

Results of pooled sample metabolic profiles in 37 dairy herds were retrospectively evaluated relative to effects of time period and health status.[16-19] Pooled values among herds were influenced by time period and health status, suggesting pooled samples could provide diagnostic information and showed differential patterns relative to disease problems within a herd.[16,17] Pooled values within a defined sampling group (e.g., early dry, close-up, fresh) were compared to a healthy population median value and divided by the population standard deviation. Number of standard deviations from the reference population median was influenced by health status with magnitude and time effects being specific to the given disease process.[18,19] These data would suggest pooled samples could potentially provide useful screening information about periparturient health risks in dairy herds.

## SUMMARY

Though some variation may be masked, pooled sampling may be used as an economic approach to a herd metabolic status screening tool. Most important measures of metabolic status (Albumin, BUN, NEFA, BHB) showed minimal differences between pooled and individual samples. Pooled samples are statistically equivalent to the mean of pooled individuals. Observed effect of herd on sample differences may suggest poor sample handling or processing practices. The real challenge to using pooled samples is interpretation. Empirically one can interpret pooled samples by determining how far they deviate from the midpoint of the reference range for healthy individuals. A new approach to pooled samples was described and requires further refinement and validation, but is a reasonable starting point. Use of statistical process control (SPC) may be a reasonable approach to using pooled samples for on-going herd monitoring, especially for larger herds that are more likely to complete repetitive sampling. Pooled samples may provide an economic alternative to traditional metabolic profiling.

## REFERENCES

[1]Payne JM, et al. **Vet Rec** 87:150, 1970. [2]Oetzel GR. **Vet Clinics NA: Food Anim Pract** 20:651-674, 2004. [3]LeBlanc SJ. **Proc 24th World Buiatrics Congress**, Nice, France, pp. 460-471, 2006. [4]Van Saun RJ. **Production Diseases in Farm Animals: 12th International Conf**, pg 47, 2006. [5]Van Saun RJ. **Bov Pract** 1997; 31(2): 43-50. [6]Duffield T. **Proc Am Assoc Bov Pract** 2007;40:213-218. [7]Herdt TH. **Vet Clinics NA: Food Anim Pract** 2000; 16(2): 387-403. [8]Van Saun RJ. **Proc 23rd World Buiatrics Congress**, Quebec, Canada, pg 87, 2004. [9]Maas J. **Modern Vet Prac** 1983;64:963-967. [10]Rowlands GJ. **Wld Rev Nutr Diet** 1980;35:172-235. [11]Tornquist SJ, Van Saun RJ. **Vet Path** 1990;36(5):487, Abstr. [12]Van Saun, R.J. **Proc XXIV World Buiatrics Congress**, Nice, France, Abstract #50/OS07-3, 2006. [13]Lehwenich T. Dissertation, Berlin University, Germany, 1999. [14]Van Saun RJ. **Proc Am Assoc Bov Pract**, 2005;38:180. [15]Van Saun, R.J. **Proc XXIV World Buiatrics Congress**, Nice, France, Abstract #117/PS3-015, 2006. [16]Davidek, J., Van Saun, R.J. **Proc 25th Jubilee World Buiatrics Congress**, Budapest, Hungary, pp 23-24, 2008. [17]Davidek, J., Van Saun, R.J. **Proc 25th Jubilee 25th World Buiatrics Congress**, Budapest, Hungary, pg. 24, 2008. [18]Davidek, J., Van Saun, R.J. **Proc 25th Jubilee 25th World Buiatrics Congress**, Budapest, Hungary, pg. 24, 2008. [19]Van Saun, R.J., Davidek, J. **Proc Am Assoc Bov Pract**, 2008;41:286.

## CONTROVERSY OVER CARBOHYDRATE IN DIETS FOR CATS

**Robert C. Backus, MS, DVM, PhD, DACVN**
**Columbia, MO**

### INTRODUCTION

Extruded dry-type diets have become the most common of diet types fed to cats in the United States as a result of advancements in pet food technology over 50 years ago. The popularity of the diets appears to stem from their being convenient to use and low in cost relative to other diet forms. The unique extrusion technology used to produce extruded dry-type diets requires that ingredients contain sufficient starch to form a gelatinized-dough. After cooking under high pressure in moist heat, the starch-containing dough of extruded diets expands into stable shapes when cut and dried. The starch component in the diets in recent years has been argued by some veterinarians and animal owners to be unhealthy for cats. The basis of the health concern appears somewhat diverse. Some assert that because starch is of plant origin it is unnatural for a carnivore to eat and therefore unhealthy for cats to consume. Some individuals appear more disinclined to grains than the starch that grains contain. In appearing to accommodate a consumer bias, some in the pet food industry are now producing extruded dry-type diets using potato rather than grain as a starch source. Perhaps the most common of concerns about use of starch in diets for cats is the challenge it reputedly presents to the physiology and metabolism of a carnivore.

Comparative reviews of feline anatomy, physiology, biochemistry, and nutrition indicate that cats are carnivores [1]. The behavior of cats indeed is consistent with their being carnivores. Studies of gut and fecal contents of feral cats show that domestic cats choose a diet of prey when food provided by humans is lacking. Small mammals, birds, reptiles, amphibians, and insects are consumed in variable proportions depending on the feral habitat. Given the macronutrient composition of animals that feral cats might consume, it can be surmised that the diet of choice by cats is low in carbohydrate, < 10% of metabolizable energy (ME). By contrast, the carbohydrate content of manufactured dry-type diets is considerably greater than that expected in prey, with a few exceptions. As indicated by recent sales volume statistics, the extruded dry-type cat foods most commonly used in the United States contain approximately 35 % of ME as carbohydrate. In recent years, inclusion of carbohydrate in cat foods beyond that expected in prey has occurred in some wet-type, canned foods.

The objective of the present review is to discuss recent findings in light of past findings that are relevant to health concerns raised about feeding diets to cats that contain carbohydrate at arguably "unnaturally" high levels, such as those present in extruded dry-type diets. The review particularly will focus on the reputed carbohydrate intolerance of cats and suggestions that obesity and diabetes mellitus are consequences of feeding diets high in carbohydrate. Reports of increasing incidence of obesity and diabetes in cats and believed benefits of low carbohydrate diets in the management of obesity and diabetes mellitus appear to have contributed to suggestions of the unhealthfulness of dietary carbohydrate. Whether evinced as unhealthy or not, some pet food manufacturers have responded by reducing the amount of starch in their extruded dry-type diets. Advances in extrusion technology during the last two decades have made this feasible, while at the same time allowing reciprocal increases in the dietary protein, with some of the added protein being meat rather than meat-meals.

### REPUTED CARBOHYDRATE INTOLERENCE

Cats start out their lives being well able to utilize carbohydrate. Glucose serves as the principal fuel during gestational development because lipoproteins cannot pass the syncytial cellular barrier of the placenta. Non-esterified fatty acids do pass the barrier but their energy contribution is believed to be low compared to that of glucose. For a few weeks following parturition, carbohydrate contribution to dietary energy is lower, but it still is substantially present as lactose in milk. Though varying somewhat with the stage of lactation and fat content of the diet of the dam, the contribution of carbohydrate to ME in feline milk is approximately 10 to 20 % [2,3]. Similar proportions of energy in lactose are observed in the milk consumed by some omnivore species. It is during the weaning period that a carbohydrate-intolerance may develop in cats because of declining intestinal lactase activity. However, intolerance to lactose as a matter of course in postnatal maturation is not unique to cats or carnivores. Past weaning, the dietary carbohydrate proportion of ME is increased well in excess of that in milk when a typical extruded dry-type is fed. Nonetheless, the dietary carbohydrate in extruded dry-type diets appears tolerated.

Feeding trials show that dietary starch is effectively digested by kittens and supports normal growth without evidence of digestive upset if the starch is adequately ground and cooked. Comparative observations indicate that starch may be less completely digested in cats than some in omnivores. At the level of the ileum, apparent digestibility of starches in cats is lower than that found in dogs. The meaning this difference is not clear because the same species trend may be found for dietary protein. The species digestibility difference appears to depend on the degree to which starch is processed and the relative abundance of alimentary enzyme specific for starch digestion. The health significance of undigested starch in cats is not clear in amounts that do not result in diarrhea. One study shows that undigested starch is fermented in the large but not in the small intestine of cats, despite a reportedly abundant of small intestinal microflora in cats. Fermentation of undigested residue in cats is not unique to diets containing starch. As indicated from breath hydrogen measurements, intestinal fermentation may occur with feeding of low carbohydrate wet-type diets to cats [4]. The site of such fermentation and responsible substrates are unknown. Curiously, substantial small intestinal fermentation is observed when a high protein diet is given to dogs [5].

When cats are weaned to a low carbohydrate diet, the need for dietary energy is met from catabolism of protein and fat. For wet-type and newer dry-type diets that are low in carbohydrate, energy from carbohydrate is exchanged with energy from protein. In the context of metabolic carbohydrate tolerance by cats, it should be appreciated that a substantive amount of carbohydrate, i. e., glucose, is formed in the catabolism of protein. Nearly all amino acids found in dietary proteins may be in part converted into glucose during their catabolism, which may be quite rapid with a highly digestible protein. Between 50 and 80 grams of glucose are estimated to be formed from catabolism of 100 grams of protein [6]. The amount of glucose arising from dietary protein and the abundance of this glucose relative to the glucose that may be derived from alimentary starch does not seem to have been reported in cats.

Amino acid conversion to glucose presumably occurs principally in the liver of cats with minor kidney and intestinal contributions. As with glucose derived from alimentary carbohydrate, glucose derived from protein in cats is well stored in liver as glycogen, a substance that has been called animal starch. When given a high protein or high carbohydrate diet, the glycogen content of found in the liver of cats is similar to that in the liver of rats given diets with similar macronutrient proportions [7]. The glycogen in liver of cats as in other species appears to buffer changes in blood glucose that might occur with ingestion of starch or protein. Only small absorptive-phase swings in blood glucose are reported after ingestion of starch in amounts found in extruded dry-type diets [8]. Yet, it does appear that cats are less able to dispose of glucose than omnivores when challenged.

Intravenous glucose tolerance tests (IVGTT) show that an intravenous load of glucose requires about one-and-a-half to twice as long to clear in cats as a similar glucose load clears in dogs [9,10]. This species difference appears consistent with research showing an absent or low liver glucokinase activity, poor *in vitro* conversion of glucose to glycogen, and lower *in vitro* uptake of glucose by hepatocytes in cats compared to dogs, rats, and humans [11,12]. However, the *in vivo* capacity of liver to dispose of glucose in cats relative to other species and species differences in the ability to spare glucose between meals should be more closely evaluated. Cats may be as proficient as omnivores in disposing glucose during the adsorptive phase. Liver hexokinase activity in cats is reportedly over twice as great as that in dogs [12]. In the fed state, liver and muscle glycogen are as replete in cats as they are in rats when a high carbohydrate diet is fed [13]. Recent hyperglycemic glucose clamp studies indicate that the rate of glucose disposal after a period of food deprivation in cats is less than that in dogs and humans [14]. After a period of food withholding, the carnivore metabolism of cats may be especially adapted to spare glucose and thereby slowly clear an unphysiological glucose load as in an IVGTT or glucose clamp.

**DIABETES MELLITUS**

The hypothesis that unnaturally high dietary carbohydrate leads to diabetes in cats appears to at least in part owe its origin to a 1994 article titled "The carnivore connection: dietary carbohydrate in the evolution of NIDDM" [15]. Though written to explain diabetes predisposition of certain people groups, the article infers that cats as carnivores are insulin resistant and that diabetes develops in cats as a result of feeding high dietary carbohydrate. Unfortunately, these conditions of cats are not supported by references cited in the article [16,17]. Nonetheless, the human thesis about dietary carbohydrate has been applied to cats. In essence, a high carbohydrate diet is suggested to chronically present an unnaturally high glycemic load against a background of insulin resistance. The result is suggested to be an extraordinary stimulation of pancreatic β cells that progressively causes dysfunction, exhaustion, and loss of the cells. Indeed, experimentally induced hyperglycemia is observed to be detrimental to pancreatic β cells in cats [18]. However, such hyperglycemia has the same effect on omnivore species, and the level of starch present in extruded dry-type feline diets does not appear to produce such hyperglycemia [19].

Exchanging dietary carbohydrate for protein is observed to increase insulin concentration in the peripheral venous plasma of cats following meals [20,21]. The effect on insulin concentration appears to indicate greater β cell stimulation by dietary carbohydrate than protein in cats, but the cause and health implications are not evident from reported studies. With respect to cause, insulin sensitivity, glucose effectiveness, and the peripheral venous glucose response to meals do not vary significantly in cats if dietary carbohydrate is substituted for protein [22,23]. Hence, arterial differences in glucose concentration may underlie a greater insulin response to dietary carbohydrate than protein. With respect to health implications, a greater insulin response to carbohydrate than protein is also observed in humans [24], though humans are not generally considered insulin resistant or intolerant of carbohydrate. It could be argued that β cells in cats are less tolerant of glucose stimulation that those in humans, but research supporting of this are lacking. It is noteworthy that effect of exchanging dietary carbohydrate with protein on insulin kinetics has not been evaluated in cats. The fractional extraction of insulin by liver in rats is reportedly greater on a high protein compared to a high carbohydrate diet [25]. Given that many amino acids are potent secretogogues of insulin in cats, it is conceivable that insulin secretion in response dietary protein is nearly as robust as the response to dietary carbohydrate but the condition is not detected in systemic venous insulin because of insulin removal by liver.

Insufficient insulin secretion and impaired insulin sensitivity are the major abnormalities of feline diabetes. As hypothesized by the "Carnivore Connection" paradigm, these abnormalities are predicted results of long-term feeding of high dietary carbohydrate. Results of two recent population studies are consistent with extruded dry-type diets not being causal of these abnormalities in cats. In one study, an age crossection of IVGTT insulin profiles of intact female colony cats were evaluated [26]. Though all cats were given extruded dry-type diets lifelong, no significant difference between young (< 3 years) and middle-aged cats (4-7 years) of similar body weight were observed in early- and late-phase insulin responses to intravenous glucose. Amplitudes of the IVGTT early- and late-phase insulin responses are considered indicative of pancreatic β cell function and insulin resistance, respectively. The second study evaluates questionnaires completed by owners of diabetic and healthy cats seen by practitioners in the Netherlands [27]. After controlling for body weight, body condition, gender, and age, the study investigators found that the proportion of dry food used by owners was not a risk factor for development of diabetes. This negative finding contrasts with other study findings indicating that physical inactivity and glucocorticoids or progestagen treatment increase diabetes risk.

**OBESITY**

Recent surveys of risk factors for disease in clinically presented cats confirm earlier reported associations of overweight body condition with increased risk for development of diabetes [28,29]. This relationship seems well explained in observations of reduced glucose tolerance, insulin resistance, and signs of pancreatic β cell dysfunction with gain body fat by cats. The means by which obesity imparts these diabetogenic changes is not completely understood. It does appear that the effects of obesity are mediated by altered lipid metabolism and hormonal signaling brought on by expansion of adipose. Apart from suggested direct effects, high dietary carbohydrate is postulated to be diabetogenic by virtue of its promotion of body fat deposition in cats. This suggestion contrasts with what is observed in omnivores, where high dietary fat is more promoting of body fat gain than high dietary carbohydrate. Against the background of insulin resistance of a carnivore, the greater potency of dietary carbohydrate compared to protein and fat in stimulating insulin release amidst abundant adipogenic substrate (e. g., glucose) is suggested to particularly favor fat deposition in cats. Indeed, feeding of extruded dry type diets rather than lower carbohydrate wet-type diets reportedly increases risk for obesity. However, attributes other than dietary carbohydrate content such as feeding practices and energy density may account the diet difference in risk. Findings in recent studies in which fat was substituted for carbohydrate do not appear to support dietary carbohydrate as a cause for obesity in cats. Dietary carbohydrate over a range of 24 to 64 % ME was not found to significantly affect body weight in young mature sexually intact cats given for *ad libitum* consumption purified diets of similar protein-energy ratio [30]. In contrast, high dietary fat at 64 but not ≤ 44 % ME induced overconsumption and gain in body weight the cats. This effect of dietary fat is consistent with past epidemiological survey findings indicating obesity risk increases with increasing fat in dry-type feline diets. Unfortunately, manipulation of dietary ratio of carbohydrate to fat does not appear to prevent the undesired weight gain of neutering (orchiectomy or ovariohysterectomy). However, increasing dietary carbohydrate in exchange for fat may moderate post-neutering weight gain with *ad libitum* feeding [31]. The benefit in this macronutrient exchange is not consistent and when observed the mechanism is not clear.

Exchanging dietary carbohydrate for protein appears to be useful for weight loss treatment and management of non-insulin dependent diabetes in cats. Benefits of the macronutrient exchange for these conditions may be reduced demand of and dependence on insulin secretion. Given these benefits and the apparently lower energy utilization efficiency of protein compared to carbohydrate [23], it might be expected that exchanging dietary carbohydrate for protein will prevent development of weight gain in cats. Unfortunately, a recent study in young female cats fed *ad libitum* indicates that exchanging dietary carbohydrate for protein may not prevent or even mitigate against post-neutering weight gain [32]. Results of this work are consistent with previous observations reporting that energy restriction is still required for achieving weight loss in obese cats when dietary carbohydrate is exchanged for protein.

## REFERENCES

[1]Morris J. **Nutr Res Rev** 2002;15(1):153. [2]Keen C, et. al. **J Nutr** 1982;112(9):1763. [3]Jacobsen KL, et. al. **J Anim Physiol Anim Nutr** 2004;88(1-2):46. [4]Backus R, et. al. **J Nutr** 2002;132:1763S. [5]Banta C, **J Nutr** 1979;109(9):1592. [6]Gannon M, et. al. **J Clin Endocrinol Metab** 2001;86(3):1040. [7]Kettelhut I, et. al. **Am J Physiol** 1980;239(5):R437. [8]de-Oliveira L, **J Anim Sci** 2008;86(9):2237. [9]Hoenig M, et.al. **J Vet Intern Med** 2002;16:529. 10Mattheeuws D, et.al. **Am J Vet Res** 1984;45(1):98. [11]Ballard F, **Comp Biochem Physiol** 1965;14:437. [12]Tanaka A, et.al. **Vet Res Commun** 2005;29:477. [13]Kettelhut I, et.al. **Am J Physiol** 1980;239(5):R437. [14]Slingerland L, et.al. **Horm Metab Res** 2007;39(5):341. [15]Brand-Miller J, Colagiuri S, **Diabetologia** 1994;37:1280. [16]O'Brien T, et.al. **Vet Path** 1985;22:255. [17]MacDonald M, et.al. **Ann Rev Nutr** 1984;4:521. [18]Zini E, et.al. **Diabetologia** 2009;52:336. [19]de-Oliveira L, et.al. **J Anim Sci** 2008;86:2237. [20]Hoenig M, et.al. **Comp Cont Ed Pract Vet** 2001;23(9A):98 (abst). [21]Rand J, et.al. Asia Pac J Clin Nutr 2003;12Suppl:S6. [22]Leray V, et.al. **J Nutr** 2006;136:2028S. [23]Hoenig M, et.al. **Am J Physiol Regul Integr Comp Physiol** 2007;292:227. [24]Nuttall F, et.al. **J Am Coll Nutr** 1985;4(4):437. [25]Demigné C, et.al. **J Nutr** 1985;115(8):1065. [26]Backus R, et.al. **ACVIM Forum proceedings**, 2007 (abst). [27]Slingerland L, et.al. **Vet J** 2009;179(2):247. [28]Prahl A, et.al. **J Feline Med Surg** 2007;9(5):351. [29]Lund E, et.al. **Int J Appl Res Vet Med** 2005;3:88. [30]Backus R, et.al. **Br J Nutr** 2007;98(3):641. [31]Nguyen P, et.al. **Am J Vet Res** 2004;65(12):1708. [32]Vester B, et.al. **Nestle Purina Forum proceedings**, 2007.

# MATERNAL NUTRITION: EFFECTS ON LONG-TERM HEALTH OF THE OFFSPRING

**Kelly S. Swanson, Ph.D.**
**Urbana, IL**

## INTRODUCTION

The important role diet plays on the health of a mother and her offspring has been recognized for decades. Early experiments pertaining to canine and feline reproductive performance and management were performed in the 1930s, '40s and '50s[1-4]. Although the majority of these initial experiments used crude measurements of success (e.g., conception rate; litter size, birth weight, and mortality) and were not focused solely on maternal diet, they provided the foundation for subsequent experiments testing specific nutritional interventions[5-7]. Numerous such studies have been performed over the past few decades, which have been the basis for the current AAFCO recommendations. In addition to feeding the mother a highly-digestible, balanced diet that meets AAFCO standards for reproduction or all life stages, recommendations for feeding increased dietary protein and fat concentrations are common today[8]. Effective strategies for meeting the changing energy needs of gestating and lactating dogs and cats have also been devised[9]. More recently, the role of essential fatty acids on reproductive performance has been described[10]. Finally, the use of dietary supplements (e.g., essential fatty acids, vitamins, minerals, probiotics, etc.) is often promoted, but is controversial due to the risks associated with oversupplementation and/or dietary imbalances[11]. In general, current AAFCO recommendations appear to be adequate for reproduction and avoidance of fetal malformations. Dietary formulations for optimal reproductive performance and offspring health, however, are yet to be determined. Moreover, the majority of published research thus far has focused on the impact of dietary intervention at or before mating, through gestation, and/or through lactation. Based on the available published literature, the long-term effects of maternal nutritional status have been largely untested in dogs and cats.

## EPIGENETIC AND METABOLIC PROGRAMMING

Recent experiments have demonstrated that long-term health may be affected by maternal diet/nutritional status during fetal development and early postnatal growth. The term "epigenetic programming" has been used to describe the specific mechanisms involved with the maternal and environmental factors influencing fetal development, while "metabolic programming" is often used to describe outcomes specific to nutritional intervention. For the past 60 years scientists have extensively studied the genetic code, a system based on the alphabet of only 4 letters, and have revealed its importance on animal phenotype. There is no doubt that the implications of the canine and feline genome sequencing initiatives led by the National Institutes of Health are wide reaching and will play a crucial role in the health care of humans and pets in the future. However, the fact that all mammalian genomes contain approximately 20,000 to 30,000 genes, many of which are highly similar in sequence and function, points to the importance of gene regulation and expression. For example, all tissues in the body contain the same set of genes, yet are very specialized. This phenotypic specialization is due to the epigenetic code, a system that controls the expression of a select set of genes at certain developmental stages. Epigenetics, originally defined as the developmental processes by which genotype gives rise to phenotype[12], is now used to describe the inheritance of information based on gene expression levels rather than on gene sequence[13]. In contrast to genetic inheritance, epigenetic inheritance is the result of heritable changes in gene expression and regulation that are independent of changes in DNA sequence. The primary mechanisms of epigenetic inheritance involve chromatin modification and chromatin remodeling. The epigenetic code comprises several levels of interconnected and interdependent codes: the DNA methylation code, the histone code (methylation, acetylation, phosphorylation), and the co-regulator code[13]. These codes define a process involving the recruitment of a myriad of chromatin-remodeling complexes, enzymes, and co-regulators and effectors, directing appropriate chromatin remodeling[14]. Epigenetic patterns resulting from the combination of these intricate systems lead to the expression or silencing of genes.

Interactions of nutrients with the epigenetic machinery lead to epigenetic changes associated with chromatin remodeling and regulation of gene expression, many of which underlie the developmental programming of pathological consequences in adulthood[12]. The cancer field has been the most widely studied area of epigenetics, dating back to as early as 1983[15]. Nonetheless, research pertaining to metabolic programming continues to gain interest due to its link to chronic diseases such as obesity, diabetes, heart disease, and behavioral disorders. Evidence now exists supporting the notion that both prenatal and postnatal nutrient intake plays a role in metabolic programming. Barker's fetal origins hypothesis[16] was based off the observation that disproportionate size at birth of the newborn, due to an adverse intrauterine environment, was highly correlated with an increased risk of adult-onset ill health outcomes. Epidemiological evidence has substantiated this hypothesis, demonstrating that maternal dietary or placental insufficiency "programs" susceptibility in the fetus to later development of these chronic diseases[17]. Dietary manipulations during the early postnatal period have also been shown to lead to physiological and anatomical changes that have a life-long impact on metabolism. Rodent studies have reported the metabolic programming of several organs (e.g., pancreatic islets, hypothalamus, liver, muscle), which were associated with increased incidence of adult-onset diseases[18-20]. In rats, for example, a life time growth trajectory can be programmed by reducing litter size so that pups in small litters receive more milk than pups in large litters[18]. In that particular experiment, permanent programming was accomplished after only 3 weeks of such dietary intervention. Other experiments have reported that malnutrition, induced by either protein or caloric restriction during gestation and lactation in female rats, was accompanied by adult-onset diseases in those offspring[19,20].

Human evidence of metabolic programming has also been published, with some of the most profound results being obtained from the survivors of the Dutch famine during World War II[21]. Offspring from women restricted of food during gestation were reported to have an increased incidence of adult obesity, glucose intolerance, and hypertension[22,23]. Smaller birth weight was also associated with higher risk of adult onset diabetes in this population[24]. The most remarkable finding from these experiments was that increased risk of disease was detected in the third generation (i.e., grandchildren), demonstrating that maternal diet may not only impact a female's offspring, but those of future generations. Experiments studying disadvantaged populations in the U.S., South Africa, India, and the Caribbean have reported similar findings. Gestational restriction of micronutrients, including calcium[25], iron[26], and zinc[27] have also been investigated in rodent models.

The effects of overnutrition or dietary imbalance during the fetal or neonatal periods have also been an area of interest. In humans, Kaati et al.[28] aimed to study the effects of overeating during the slow growth period, before prepubertal peak in growth velocity, and its influence on descendents' risk of death from cardiovascular disease and diabetes. Results from this study demonstrated that when food was not readily available during the father's or grandfather's slow growth period, offspring tended to be protected from cardiovascular disease and diabetes as a cause of death. However, diabetes mortality was increased 4-fold if the paternal grandfather was exposed to a surfeit of food during his slow growth period[28]. Research using animal models, including large animal models (sheep is commonly studied), has produced similar findings. Taken together, human and animal research suggests that both under- and over-nutrition, in utero or during the early postnatal period, may result in permanent changes to anatomy and/or physiology that impact life-long metabolism. Because many models of over- or undernutrition demonstrate long-term programming effects without having differences in fetal weight at birth[29], the utility of epidemiological studies is limited. Thus, studies designed to identify mechanisms of action are needed.

## MECHANISMS OF METABOLIC PROGRAMMING

Numerous biological mechanisms have been proposed to contribute to metabolic programming, including the impact of several nutrients and hormonal factors. DNA methylation patterns are the best-understood form of epigenetic inheritance. DNA methylation occurs on both strands of palindromic CpG dinucleotides (the "p" denotes the intervening phosphate group in the dinucleotide), and is critical for mammalian development[30]. Specific patterns of CpG methylation are established in early development and are propagated during DNA replication by DNA-methyltransferase-1[31]. DNA methylation may impact transcription directly by influencing the binding of methyl-sensitive DNA-binding proteins or indirectly by influencing regional chromatin conformation. Maternal nutrition is able to influence DNA methylation because one-carbon metabolism, which ultimately provides methyl groups for all methylation reactions, is highly dependent on dietary methyl donors and cofactors[32]. Dietary sources of folate and other donor methyl groups provide the most compelling data for the interaction of nutrients and DNA methylation because they are directly involved in DNA methylation via one-carbon metabolism. Waterland and Jirtle[33] demonstrated the impact of dietary methyl donor supply on DNA methylation in agouti mice by supplementing diets with folate, vitamin B12, choline, and betaine during gestation and lactation. The fur of agouti mice can be yellow, brown, or a calico-like mixture, depending on the level of DNA methylation. Offspring from dams fed the diet containing a rich supply of donor methyl groups had increased CpG methylation, altering the coat color of these animals. Mice receiving supplements had offspring with mostly brown fur, while unsupplemented mothers gave birth to mostly yellow pups. Folate supplementation effectively prevents the majority of neural tube defects in newborn humans and has been suggested in some canine breeds (albeit with controversy). The data reported by Waterland and Jirtle[33] and similar follow-up studies, however, suggest that oversupplementation of folate and/or other donor methyl groups may result in unintended consequences arising from epigenetic gene-regulatory mechanisms during embryonic development.

In addition to folate and other donor methyl groups that have a direct impact on DNA methylation, many other micronutrients and vitamins are known to be indispensable in DNA metabolic pathways[34,35]. Deficient dietary vitamin (e.g., vitamin B12, vitamin C, vitamin D, vitamin E, niacin) and mineral (e.g., magnesium, iron, zinc, selenium) concentrations have been reported to influence DNA damage and/or DNA methylation patterns[36]. Such nutrient imbalances are thought to affect gene expression, cell growth rates, and tissue differentiation, influencing long-term health outcomes including cancer, the aging process, and other complex diseases. Given that most of the vitamin and mineral requirements for dogs and cats are still poorly defined[9], the results from recent epigenetic studies justify future research focused on determining optimal dietary micronutrient concentrations, especially in gestating and lactating females.

Numerous rodent experiments have suggested that glucose load/metabolism contributes to metabolic programming. Glucose has been demonstrated to act as a low-grade mutagen[37,38] and is suspected in the etiology of certain birth defects during the first trimester in the offspring of diabetic mothers[38]. Recent studies have shown that high-carbohydrate (HC) milk formula fed early in life can metabolically program pancreatic islets in rat pups[39]. Consumption of HC formula results in an immediate onset of hyperinsulinemia that persists throughout the period of dietary intervention. This hyperinsulinemic state results in increased hexokinase activity and increased gene expression of preproinsulin and related transcription factors and kinases in pancreatic islets. While the acute effects of feeding a HC diet seems logical, a permanent alteration on long-term metabolism is quite remarkable. In those experiments, the early adaptations were permanently programmed and expressed in adulthood, leading to chronic hyperinsulinemia (throughout life) and adult-onset obesity. Similar to human data, female rats fed the HC formula have been shown to spontaneously transmit the "HC phenotype" to their progeny[40]. Similarly, Kozak et al.[41] reported increased neuropeptide Y release from the hypothalamic paraventricular nucleus in rats fed a HC diet, suggesting that intrauterine and perinatal programming of the central regulatory mechanisms may also be partially responsible for the obesity and comorbidities that develop during adulthood.

There is some evidence suggesting any form of undernutrition or nutrient imbalance activates a relatively small number of common pathways. For instance, hypertension, glucose intolerance, insulin resistance, and obesity have been reported in many experiments using a diverse array of animal models and nutritional insults[29]. Further research is needed to confirm these findings. Despite the growing body of evidence documenting the physiologic outcomes of specific dietary treatments, the field is in its infancy and many responses are likely going unnoticed. Furthermore, identifying specific mechanisms (genes; gene-gene interactions) involved with phenotypic changes is difficult and very complex. It will likely take decades of intense research to develop a clear understanding of these complex processes.

## ADVANCES IN CANINES AND FELINES

Genome sequence data and molecular biological tools needed to study epigenetic mechanisms, including those influenced by maternal diet, are now available and should be applied to canine and feline research. Currently a paucity of information pertaining to canine and feline epigenetics exists. A handful of epigenetics studies in dogs[42,43] and cats[44,45] have been published, but were not focused on evaluating the effects of maternal diet.

Our lab recently performed an experiment to evaluate the effects of in utero and postnatal exposure to high- or moderate-protein diets in cats[46]. Today's domestic cats are often fed diets containing high amounts of digestible carbohydrate. Dietary protein:carbohydrate ratio has been suggested as a contributor of obesity in many species[47] and is often blamed, in part, for the increasing incidence of feline obesity. However, there are few data to support this belief. Feline diets containing protein levels well above the established metabolic requirement have been demonstrated to increase heat production in lean cats[48], maintain insulin sensitivity of fat metabolism in obese cats[48], and maintain lean body mass during weight loss[49]. High-protein diets have primarily been tested in adult cats. Limited research has been performed in kittens and pregnant and lactating cats. In rat and sheep experiments, feeding energy- or protein-deficient diets to the dam has resulted in a greater propensity for insulin resistance in the offspring[50,51]. Rats prenatally exposed to high-protein diets were reported to have decreased activity levels compared to rats exposed to an adequate-protein diet prenatally[52]. Few studies have tested the occurrence of metabolic changes in cats as it pertains to maternal nutrition. Although complete and balanced commercial feline diets contain the recommended protein concentrations for kittens or reproducing females as suggested by AAFCO (30% crude protein, DM basis), they are often lower than what a cat would consume in the wild (rat carcass 55% CP on DM basis)[53].

The objective of our study was to measure the changes in body composition, adipose tissue gene expression, blood metabolite and hormone concentrations, and insulin sensitivity in kittens exposed to high- (~53% CP, DM basis) or moderate-protein (~34% CP, DM basis) nutrition in utero and through the growth period. Our data suggest that feeding a high-protein diet influences blood metabolite and hormone concentrations (e.g., leptin, triglycerides, protein, albumin) without changes in BUN or creatinine. Furthermore, the high-protein diet increased light:dark ratio of physical activity and subcutaneous adipose mRNA abundance of hormone-sensitive lipase, insulin receptor, leptin, and uncoupling protein-2. Because maternal nutritional status has been shown to cause life-long changes in metabolism in other species, further study of these cats throughout adulthood in our colony may shed light on the effects of exposure to a high-protein diet during early development. These results may have implications on feline diets formulated for cats during reproduction and rapid growth. Further research focused on the effects of exposing kittens to a high-protein diet to kittens in utero is therefore warranted.

**REFERENCES:**


[1]Hall V and Pierce G. **Anatom Rec** 1934;60:111. [2]Elvehjem C et al. **J Pediatr** 1944;24:436. [3]Campbell J and Phillips P. **J Nutr** 1952;47:621. [4]Ontko J and Phillips P. **J Nutr** 1958;65:211. [5]Scott P and Thompson S. **Proc Nutr Soc** 1969;28:8A. [6]Romsos D et al. **J Nutr** 1981;111:678. [7]Switzer B et al. **J Nutr** 1986;116:689. [8]Johnson C. **Theriogenol** 2008;70:1412. [9]NRC. Natl Acad Press, 2006. [10]Lepine A and Kelley R. **Proc Pre-Congress Symp World Congress WSAVA** 2004;34. [11]Greco D. **Theriogenol** 2008;70:393. [12]Waddington C. **Nature** 1942;152:563. [13]Junien C. **J Inherit Metab Dis** 2006;29:359. [14]Spotswood H and Turner B. **J Clin Invest** 2002;110:577. [15]Feinberg A and Vogelstein B. **Nature** 1983;301:89. [16]Barker D. **Br Med J** 1995;311:355. [17]Armitage J et al. **J Physiol** 2004;561:355. [18]McCance R. **Lancet** 1962;2:271. [19]Berney D et al. **J Pathol** 1997;183:109. [20]Ozanne S and Hales C. **Proc Nutr Soc** 1999;58:615. [21]Lumey L. **Paediatr Perinat Epidemiol** 1992;6:240. [22]Ravelli G et al. **N Engl J Med** 1976;295:349. [23]Ravelli G et al. **Lancet** 1998;351:173. [24]Rosebloom T et al. **Paediatr Perinat Epidemiol** 2001;15:220. [25]Bergel E and Belizan J. **Brit J Obstetr Gynaecol** 2002;109:540. [26]Gambling L et al. **J Physiol** 2003;552:603. [27]Beach R et al. **Science** 1982;218:469. [28]Kaati G et al. **Eur J Hum Genet** 2002;10:682. [29]Langley-Evans S et al. **Matern Child Nutr** 2005;1:142. [30]Li E et al. **Cell** 1992;69:915. [31]Jones P. **Trends Genet** 1999;15:34. [32]Van den Veyver I. **Annu Rev Nutr** 2002;22:255. [33]Waterland R and Jirtle R. **Mol Cell Biol** 2003;23:5293. [34]Ames B. **Mutat Res** 2001;475:7. [35]Fenech M and Ferguson L. **Mutat Res** 2001;475:1. [36]Friso S and Choi S-W. **J Nutr** 2002;132:2382S. [37]Lee A et al. **Mutat Res** 1990;238:185. [38]Lee A et al. **Diabetes** 1995;44:20. [39]Srinivasan M et al. **Exp Biol Med** 2003;228:15. [40]Vadlamudi S et al. **Am J Physiol** 1995;269:E731. [41]Kozak R et al. **Eur J Neurosci** 2005;21:2887. [42]Cooley D et al. **Cancer Epidemiol Biomarkers Prev** 2002;11:1434. [43]Pelham J et al. **Res Vet Sci** 2003;74:101. [44]Cho S et al. **Cloning Stem Cells** 2007;9:477. [45]Gomez M et al. **Cloning Stem Cells** 2008;10:469. [46]Vester B et al. **Brit J Nutr** 2009(submitted). [47]Simpson S and Raubenheimer D. **Obes Rev** 2005;6:133. [48]Hoenig M et al. **Am J Physiol** 2007;292:R227. [49]Laflamme D and Hannah S. **Int J Appl Res Vet Med** 2005;3:62. [50]Erhuma A et al. **Am J Physiol** 2007;292:E1702. [51]Burt B et al. **Soc Reprod Fertil Suppl** 2007;64:411. [52]Daenzer M et al. **J Nutr** 2002;132:142. [53]Zoran D. **JAVMA** 2002;221:1559.

# NOVEL PROTEINS & DIET FORMATS FOR DOGS & CATS

**Kelly S. Swanson, Ph.D.**
**Urbana, IL**

## INTRODUCTION

Pet diets have changed dramatically since the inception of the modern pet food industry approximately 150 years ago. From James Spratt's biscuits of the 1860s to the dog cakes and milk-bones of the 1920s to the canned meat diets of the 1930s to the extruded diets first produced in the 1950s to the wide selection of diets currently available today, the industry has continued to evolve[1]. Diet quality has vastly improved over the years, yet diet format, ingredient selection, and chemical composition continue to change even today. Over the past few decades, societal changes, the humanization of pets, and most recently the internet has had a significant impact on the industry. A recent study revealed that ethical considerations played an important role in the selection of petfoods for some consumers[2]. In a time when consumer opinion is largely formed from information read on the internet or heard on the television, many times overriding scientific evidence, those working in the petfood industry and/or veterinary profession face unique challenges. For the industry, great opportunities come with the challenges, as many consumers are willing to pay for high-priced foods perceived to be the best for their pet. For veterinarians and nutritionists, just keeping a handle on the wide variety of diet types and marketing strategies used by the industry may be difficult. Differentiating real vs. perceived benefits for clients, however, may be an even greater challenge. Novel diet formats and protein sources are a major topic of discussion in the petfood industry and in veterinary medicine. The rationale for feeding such diets and research supporting or opposing their use will be discussed.

## RATIONALE FOR FEEDING NONTRADITIONAL PETFOODS

Although the commercial petfood industry is still dominated by extruded kibble and canned foods, today's consumers may now choose from a variety of novel diet formats, many of which closely resemble human foods. The feeding of home-prepared meals and/or raw diets also continues to increase despite the controversies surrounding them. A recent survey of Australian and US pet owners revealed that noncommercial foods, including table scraps, raw foods, and home-prepared diets, comprised at least 25% of the diet for 17.3% dogs and 6.3% cats[3]. This survey also revealed that bones and raw foods were provided at least weekly to 23.6% of dogs and 10.5% of cats as a main meal or treat. Although the likelihood of feeding raw foods was greater in Australian pet owners, the trend to feed noncommercial foods is also increasing in the US. Widespread petfood recalls due to melamine, salmonella, and mycotoxin contamination in recent years has tarnished the industry and has been enough motive for many owners to abandon the mainstream commercial diets. Home-prepared or raw diets have numerous proposed benefits, including owner control over dietary ingredients, preventing the destruction of natural enzymes or phytonutrients, and avoiding artificial preservatives[4-6]. While these claims may contain some truth, there are currently no peer-reviewed published reports supporting their use. Nutritional deficiencies[7] and case reports of a variety of illnesses[8-10], however, have been published in respectable journals. Despite the current body of evidence that clearly favors the mainstream diets, it behooves petfood professionals and veterinarians to understand the rationale behind feeding novel diet formats. Because it is clear that an increasing number of pet owners are feeding home-prepared and/or raw diets, performing proper research on such diets is also of importance. Thus, rather than simply dismissing these diets and/or owners as irrational (as I have heard countless times at scientific meetings), it is time to perform and publish the appropriate research in this area so that decisions may be based on facts rather than anecdotes.

The anthropomorphism of pets continues to increase and has influenced the way companies formulate and market pet foods and treats. Recent surveys have revealed pertinent information in that regard. Laflamme et al.[3] summarized the data from 1104 respondents (635 and 469 dog and cat owners, respectively). In that study, a large percentage of dog owners claimed to cuddle with (95%), talk to (89%), walk/run with (79%), use toys to play with (67%), groom (64%), play fetch with (62%), and watch TV with (54%) their pet. Cat owners had similar tendencies, as cuddling (96%), talking to (91%), sleeping with (60%), watching TV with (60%), grooming (58%), and using toys to play with (52%) their cat were common. It was also noted that approximately 22% and 26% of dog and cat owners, respectively, ate meals with their pets. Finally, while veterinarians were listed as a primary source of information pertaining to pet care, a significant number of respondents also listed family and friends, media, internet, and pet paraprofessionals (breeders, trainers, and groomers).

A smaller part of that survey, published separately, focused on the attitudes of dog and cat owners as it pertains to pet foods and feeding management[11]. That article reported important data pertaining directly to the opinions of owners and how they affect the philosophies for feeding their pets. Based on their feeding practices, owners were divided into "noncommercial feeders" (>50% of pet's diet was from foods other than commercial pet foods) and "commercial feeders" (>75% of pet's diet was in form of commercial pet foods). The data revealed that all owners were in agreement that the care and nutrition of their pet was important and that they trust their veterinarian for pet health care advice. However, noncommercial feeders were less likely to trust veterinarians to provide sound nutritional advice. In general, noncommercial feeders reflected greater mistrust of commercial pet foods, food processing, and the pet food industry compared with commercial feeders. More specifically, noncommercial feeders were more likely to feel that processed foods for pets are unhealthy, that cooking destroys nutrients, that genetically modified foods are unsafe, and that organic foods are safer and healthier than other foods. A greater percentage of noncommercial feeders felt that dogs and cats required a meat-based diet because they were carnivores and that commercial foods did not contain enough meat and/or meet all of the nutrition needs of their pet. These owners also thought that information on the pet food label is misleading and that additives have negative side effects on their pets. Finally, noncommercial feeders were more likely to have a positive attitude towards raw and home-prepared diets, feeling that raw meat provides better nutrition than cooked foods, raw bones can safely be fed to pets, and that foods sold for human consumption provide better nutrition than commercial pet foods[11]. Although more research is needed, these initial studies have revealed interesting owner perspectives and the rationale behind nontraditional feeding practices.

**CANINE & FELINE RESEARCH ON NOVEL PROTEINS OR DIET FORMATS**

The purpose of this review is not to defend any particular feeding practice, but to present the existing data pertaining to the topic at hand. Published data on the effects of feeding novel diet formats, including raw, pasteurized, or other lightly-processed diets, is severely lacking. Feeding nontraditional diets is controversial, but both sides of the argument rely heavily on anecdotal evidence or a small number of case reports. In my opinion, both sources of information have their flaws. While veterinarians must make decisions based on fact rather than anecdotes, they cannot rely on case reports alone. For every case report demonstrating the risks associated with raw diets, there is a media report describing the latest petfood recall. Until unbiased, properly designed treatment-based experiments testing such diets have been performed and published in peer-reviewed journals, the argument will continue to be largely based on emotion and anecdotes rather than fact. Over the past 10-15 years, researchers have studied a myriad of common plant and animal protein-based ingredients used in petfoods, but have been primarily focused on extruded diets. The research group at the University of Illinois has determined the chemical composition, in vitro digestibility, palatability, and in vivo ileal and total tract nutrient digestibility of protein sources fed to dogs[12-20]. The researchers at the University of Kentucky have also generated several reports focused on canine protein sources during this time[21-26]. A few other researchers have also published similar reports in dogs[27-29] and cats[30,31]. Thus, the literature contains a vast knowledge base on common plant- and animal-based ingredients in petfoods in regards to their palatability, chemical composition, and effects on nutrient digestibility and fecal characteristics. Very few articles, however, have been published on raw dietary protein sources and/or with a focus on feline nutrition and health. Our laboratory has recently performed several experiments measuring the macronutrient digestibility, nitrogen metabolism, fecal microbial populations, and fecal fermentative compounds of domestic cats and captive exotic felids fed raw diets[32-36]. We are currently performing a similar study in dogs as well[37]. Our results from these studies are briefly described below.

Only a handful of experiments have been performed in captive exotic felids in past years. Thus, our initial experiment[32] focused on evaluating differences in nutrient digestibility and fecal characteristics of bobcats, jaguars, cheetahs, Malayan tigers, and Amur tigers fed a beef-based raw diet (Nebraska Brand® Special Beef Feline). Fecal scores were greater ($P<0.01$; looser stools) in Malayan tigers when compared to all other species, and cheetahs had greater ($P<0.01$) fecal scores than jaguars and bobcats. Apparent total tract fat digestibility was greater ($P<0.01$) in Amur tigers, Malayan tigers, and bobcats (96%) compared to cheetahs and jaguars (94%). Total tract digestible energy was greater ($P<0.05$) in bobcats and Malayan tigers at 93.5 and 92.9%, respectively, compared to cheetahs and jaguars, 91.6%. Fecal indole concentrations were greater ($P<0.05$) in cheetahs and jaguars compared to bobcats and Malayan tigers, while fecal ammonia concentrations were greater ($P<0.05$) in cheetahs compared to all other species. Overall, the beef-based raw diet was highly digestible in all species, but small differences in nutrient digestibility and fecal fermentative end-products were noted. Based on these data, a follow-up experiment[33] was performed using cheetahs, Malayan tigers, jaguars, Amur tigers, and domestic cats. Raw meat diets [beef-based diet (57% protein; 28% fat); horse-based diet (51% protein; 30% fat)] were tested using a crossover design. In that study, food intake did not differ due to diet, but fecal output was greater when cats consumed the horse-based diet. Total tract apparent dry matter digestibility was higher ($P<0.05$) and organic matter and crude protein digestibilities were lower ($P<0.05$) when cats were fed the beef-based diet compared to the horse-based diet. Crude protein digestibility was similar between domestic cats and cheetahs and was greater ($P<0.05$) than that of Amur tigers. Fecal scores were lower and fecal dry matter was greater ($P<0.05$) when cats consumed the horse-based diet compared to the beef-based diet. Domestic cats had lower ($P<0.05$) fecal ammonia concentrations compared to all other species. Fecal ammonia concentrations were lowest ($P<0.05$) when cats were fed the horse-based diet. Fecal total short-chain fatty acid, branched-chain fatty acid, and butyrate concentrations were higher ($P<0.05$) when cats consumed the beef-based diet. Total short chain fatty acids were lower ($P<0.05$) when cats consumed the horse-based diet. Our results suggest that the domestic cat serves as an appropriate model for large exotic felid species, but differences among species exist. Decreased nutrient digestibility by tigers and jaguars should be considered when developing feeding recommendations for these species based on domestic cat data. Dietary fiber source was different between the beef-based (beet pulp) and horse-based (cellulose) diets tested in that experiment. Future experiments strictly controlling for fiber are being planned so that differences due to protein source alone may be identified.

The African wildcat is one of the closest ancestors to the domestic cat and is believed to have similar nutrient requirements, but research is lacking. The objective of our next experiment[34] was to determine the effects of feeding a high-protein extruded kibble diet versus a raw meat diet on nutrient digestibility, nitrogen metabolism, and blood metabolite concentrations in this species. Our results revealed greater ($P<0.05$) total tract apparent crude protein digestibility in cats fed the raw meat diet versus the kibble diet. Nitrogen balance was positive for all cats but was not different between diets. Although fecal butyrate concentrations were not different between diets, the molar ratio of butyrate was higher ($P<0.01$) in cats fed the kibble diet. Blood analysis showed few differences between diets, but alanine aminotransferase activity was higher ($P<0.05$) when cats were fed the raw meat diet. This study indicated that feeding an extruded kibble diet to African wildcats is a suitable replacement to a raw meat diet. A consequent experiment in domestic cats, testing extruded, raw beef-, and cooked beef-based diets was recently performed[35]. Results from that experiment were similar to that in African wildcats. Total tract apparent dry matter, organic matter, and crude protein digestibilities were higher ($P<0.05$) in cats fed the raw or cooked beef-based diets versus cats fed the kibble diet. Fecal scores and dry matter percentage were not different among treatments. Fecal ammonia concentration was lower ($P<0.05$) in cats fed the raw and cooked beef-based diets versus cats fed the kibble diet Fecal total short-chain fatty acid concentrations were not different among treatments, but the proportion of butyrate and propionate were altered. Fecal butyrate was lower ($P<0.05$) in cats fed the raw or cooked beef-based diets versus cats fed the kibble diet, while propionate was higher ($P<0.05$) in cats fed the cooked beef-based diet versus cats fed the kibble diet. Fecal isobutyrate, isovalerate, valerate, and total branched-chain fatty acid concentrations were lower ($P<0.05$) in cats fed the raw or cooked beef-based diets versus cats fed the kibble diet. Nitrogen balance was positive for cats fed all diets, but was lower ($P<0.05$) in cats fed the cooked beef-based diet versus cats fed the kibble diet. Urinary output was the major route of nitrogen excretion for all treatments. Serum chemistry analysis identified few differences among treatments. Triglycerides were higher ($P<0.05$) in cats fed the cooked beef-based diet versus cats fed the kibble diet, while cats fed the raw beef-based diet tended to have higher ($P<0.10$) triglycerides than cats fed the kibble diet. Serum glucose levels tended to be higher ($P<0.10$) in cats fed the cooked beef-based diet versus cats fed the kibble diet. Serum creatinine was higher ($P<0.05$) in cats fed the raw or cooked beef-based diets versus cats fed the kibble diet. Circulating phosphorus was lower ($P<0.05$), and chloride was higher ($P<0.05$) in cats fed the cooked beef-based diets versus cats fed the kibble diet.

The objective of our most recent experiment was to evaluate the differences in total tract apparent macronutrient digestibility of domestic cats fed beef-, bison-, elk-, and horse-based raw diets[36]. Apparent dry matter and organic matter digestibilities were lower ($P<0.05$) for cats fed the beef-based diet (84.1%; 87.4%) as compared to those fed the bison-based (88.1%; 90.4%) and horse-based (87.1%; 89.8%) diets, and lower ($P<0.05$) for those fed the elk-based diet (84.3%; 87.6%) as compared to the bison-based diet. Apparent fat digestibility was lower ($P<0.05$) for cats fed the elk-based (88.7%) diet as compared to those fed the beef-based (94.4%), bison-based (96.8%), and horse-based (96.9%) diets. Apparent fat digestibility and was also lower ($P<0.05$) for cats fed the beef-based diet versus those fed the bison- and horse-based diets. Apparent crude protein digestibility was lower ($P<0.05$) for cats fed the beef-based (96.6%) diet as compared to those fed the elk-based (97.3%) diet. Although dietary macronutrient composition was highly variable and statistical differences were noted, all diets were highly digestible in all cats. We are hopeful that these initial studies will establish a foundation of knowledge on novel diet formats and act as a springboard for consequent experiments. There is still a great need for further research pertaining to the rationale behind owner feeding practices and the nutritional and health effects of feeding nontraditional petfoods.

## REFERENCES

[1]Bicks J. **Pet Business** 2001;59. [2]Wakefield L et al. **JAVMA** 2006;229:70. [3]Laflamme D et al. **JAVMA** 2008;232:687. [4]Pitcairn R and Pitcairn S. **Dr. Pitcairn's complete guide to natural health for dogs and cats** 1995; Rodale Press, Inc. [5]Remillard R et al. **Small Animal Clinical Nutrition** 2000; Walsworth Publishing Co. [6]Berschneider H. **Clin Tech Small Anim Pract** 2002;17:1. [7]Streiff E et al. **J Nutr** 2002;132:1698S. [8]Niza M et al. **J Feline Med Surg** 2003;5:271. [9]Stiver S et al. **J Am Anim Hosp Assoc** 2003;39:538. [10]Polizopoulou Z et al. **J Feline Med Surg** 2005;7:363. [11]Michel K et al. **JAVMA** 2008;233:1699. [12]Zuo Y et al. **J Anim Sci** 1996;74:2441. [13]Murray S et al. **J Anim Sci** 1997;75:2497. [14]Johnson M et al. **J Anim Sci** 1998;76:1112. [15]Bednar G et al. **Arch Anim Nutr** 2000;53:127. [16]Clapper G et al. **J Anim Sci** 2001;79:1523. [17]Dust J et al. **J Anim Sci** 2005;83:2414. [18]Folador J et al. **J Anim Sci** 2006;84:2752. [19]Cramer K et al. **J Anim Sci** 2007;85:3285. [20]Faber T et al. (unpublished data). [21]Yamka R et al. **J Anim Sci** 2003;81:2279. [22]Yamka R et al. **Anim Feed Sci Technol** 2003;109:121. [23]Yamka R et al. **Anim Feed Sci Technol** 2004;116:239. [24]Yamka R et al. **J Anim Sci** 2005;83:393. [25]Yamka R et al. **Anim Feed Sci Technol** 2005;120:79. [26]Yamka R et al. **J Anim Sci** 2007;85:204. [27]Hill R et al. **J Anim Sci** 2000;78:2633. [28]Hill R et al. **J Anim Sci** 2001;79:2162. [29]Zentek J et al. **J Nutr** 2004;134:2158S. [30]Funaba M et al. **Am J Vet Res** 2002;63:1247. [31]Cave N and Marks S. **Am J Vet Res** 2004;65:1427. [32]Vester B et al. **Zoo Biol** 2008;27:126. [33]Vester B et al. (unpublished data). [34]Vester B et al. (unpublished data). [35]Kerr K et al. (unpublished data). [36]Kerr K et al. (unpublished data). [37]Beloshapka et al. (unpublished data).

# COULD DIETARY TAURINE SUPPLEMENTATION IN DOGS BE MASKING A PROBLEM?

**Robert C. Backus, MS, DVM, PhD, DACVN**
**Columbia, MO**

## INTRODUCTION

The need for dietary protein relates to its constituent amino acids. Perhaps the most recognized need for dietary protein is for provision of amino acids used in the synthesis of body proteins, which constantly turnover, and the synthesis of non-protein substances, such as signaling molecules, pigments, antioxidants. Perhaps less well recognized about the need for dietary protein is for the provision of amino acids for support of metabolic pathways and products of amino acid catabolism, such amino nitrogen, methyl carbon, reduced sulfur, and the carbon- and nitrogen-containing skeletal elements of amino acids. The basis of the need for dietary protein is in essence reflective of needs for amino acids in a great variety of functions. Determining dietary adequacy in providing amino acids is then complicated by there being so many different functions that depend on the supply of amino acids. Hence, at times the dietary amino acid content needed to meet one function may not be great enough to meet the need for another function. For this reason, dietary protein and amino acid adequacies are evaluated from outcomes of a few different variables. Nitrogen balance and optimal growth rate and nitrogen retention are the most commonly evaluated variables. In some cases, other additional variables are evaluated because a physiological or metabolic function with a recognized health consequence requires a greater dietary concentration for optimization. One example used in dogs is urinary excretion of orotic acid. When dietary arginine is adequately provided, urinary excretion of orotic acid is minimized. Another variable that has come to light recent years is blood taurine concentration in dogs. Dietary protein provision that supports a minimal concentration of taurine in the blood of dogs is targeted to prevent development of cardiomyopathy. In the information presented below, the basis for evaluation of blood taurine in dogs will be described and the practice of taurine supplementation of diets for normal healthy dogs will be discussed.

## TAURINE DEFICIENCY

Taurine is unique among dietary amino acids in that it is structurally a β-sulfonic amino acid and not bound in protein. Body needs for taurine may be met from dietary sources replete with taurine and by synthesis from sulfur-containing amino acids that are variably abundant in dietary protein, i.e., methionine and cyst(e)ine. Body needs for taurine have been realized from metabolic aberrations, dysfunction, and disease observed in experimental and natural depletions of body taurine caused by inadequate intake of methionine, cyst(e)ine, and or taurine. Body taurine reserves affect many functions, including fetal development, growth, reproduction, neuromodulation, vision, hearing, immunity, and cardiac performance.

Nutritional research that determined the cause of taurine deficiency in cats led to recognition that the synthesis of taurine from dietary methionine and cyst(e)ine is limited in strict carnivores. The limitation is so great that taurine must be included in the diets of cats and other carnivores to prevent deficiency disease. Low activity of a liver enzyme in the taurine biosynthetic pathway (cysteine dioxygenase and cysteine cysteinesulfinate decarboxylase) is among suggested causes for limited production of taurine in cats [1] (Fig. 1). Most herbivores and omnivores, among which dogs are included, appear capable of meeting body taurine through *de novo* synthesis of taurine when dietary methionine and cyst(e)ine are adequately provided. It is for this reason that taurine is not considered to be an essential nutrient in dogs.

Nonetheless, taurine deficiency has been observed in dogs given diets apparently adequate in methionine and cyst(e)ine. At this time, taurine deficiency appears most clearly to affect heart function in dogs, seen clinically as a reversible dilated cardiomyopathy [2,3]. Retinal degeneration as indicated from fundic exams has been described in few dogs [4]. Poor reproduction, small litters, and short-statured puppies have been anecdotally reported [3]. Correlation with low activity, health problems, and need for medications are also described [5]. Some of the taurine deficiency disease may be breed-specific. A dilated cardiomyopathy that appears to respond to taurine supplementation has been described in American Cocker Spaniels with low blood taurine concentration [6]. Low blood taurine concentration has been described in Newfoundlands with cystinuria [7]. In this case, urinary loss of biosynthetic precursor of taurine is surmised to explain the taurine deficiency. A defect in renal tubular reabsorption of cystine from an inherited mutation of an amino acid transporter appears to underlie Newfoundland cystinuria [8].

During the last 10 years, a taurine deficiency of nutritional origin has been described in dogs. One case in research Beagles clearly could be attributed to long-term feeding of a low protein diet [9]. Methionine and cyst(e)ine in the experimental diet were evidently inadequate because blood taurine concentrations among the dogs were low, and two dogs showed signs consistent with dilated cardiomyopathy that was improved by taurine supplementation. In other cases, clinically presented dogs of a variety of breeds are reported to have low blood taurine concentration and varying degrees of myocardial failure [2,3,10,11]. Many of these dogs were of large body size and maintained on commercial diets in which lamb-meal and rice were the principal protein sources. The cause for the taurine deficiency among the dogs was initially unclear. Breed differences in protein digestibility have been reported but the differences did not seem consistent with body size [12]. Chemical analyses of suspect diets revealed more than adequate dietary protein, methionine, and cyst(e)ine, and varying amounts of dietary taurine. However, taurine status could be improved by oral methionine or dietary change [3]. It seemed that bioavailabilities of either dietary protein, methionine, and or cyst(e)in were low.

The cause for the susceptibility of large-breed dogs to taurine deficiency has been investigated in recent years. The approach for this has involved comparisons between Beagles and larger dogs, about 3 times or greater in body weight. This research has produced several important findings. Gastrointestinal and not urinary taurine loss was found to be greater in large compared to small dogs given the same diet [13]. The cause for the size-difference was not clear but as indicated from fecal bile acid analysis it may have been related to differences in efficiency of bile acid reabsorption. Like cats, dogs conjugate their bile acids with taurine. Comparisons of taurine synthesis between large and small dogs revealed an impact of food intake and energy requirement of maintenance [13, 14].

Unlike that which is observed in cats, taurine status in dogs was found to increase with food intake and energy needed to maintain a healthy body weight [15,16]. Together these findings indicated a greater need for taurine biosynthesis and precursor to support the biosynthesis in large compared to small dogs, especially in large dogs with a low maintenance energy requirement. This conclusion meant that a greater risk for taurine deficiency may be expected in large- than small-breed dogs if fed diets that are marginal protein content and or sulfur-containing amino bioavailability.

**DIETARY SUPPLEMENTATION**

As a result of finding causes for taurine deficiency in dogs, taurine was recognized to be a dispensable nutrient for dogs when inadequate quantities of sulfur-containing amino acids are provided [17]. An adequate intake of taurine for dogs was recommended for diets low in protein or known to be low in sulfur-containing amino acids. For other diets, the sulfur-containing amino acid allowance for dogs was substantially increased by the NRC [17,18]. Based on the extent of the increase in the allowance, it would seem that much of the sulfur-containing amino acid requirement in dogs is devoted to taurine synthesis. This condition can be appreciated by comparing recommendations for dogs with those for cats. The recommended allowance for methionine + cystine in diets for dogs (1.65 g/1000 kcal) in the maintenance state are now about twice that of cats (0.85 g/1000 kcal), a species in which taurine needs cannot be met by biosynthesis. Recommended allowances for most other amino acids in dogs are lower than those in cats.

In apparently responding to reports of taurine deficiency in dogs, pet food manufacturers have been supplementing some diets for dogs with taurine. The taurine supplementation is not limited to diets low in protein. This practice unfortunately eliminates application of a very useful means for evaluation of sulfur-containing amino acid bioavailability. As pointed out by Odle et al. over 15 years ago [19], the monitoring of taurine status is potentially useful for determining the adequacy dietary sulfur-containing amino acid content in animals that synthesize taurine, such as dogs. The adequacy of dietary provision of methionine and cystine is not sensitively indicated by changes in plasma concentrations of methionine and cyst(e)ine. This appears to be a beneficial consequence of homeostatic regulation of the amino acids by liver. When in excess, methionine and cysteine compared to other dietary amino acids have a greater potential to be toxic. In contrast, taurine has low toxicity and its concentrations in plasma are observed to vary substantially with dietary sulfur amino acid concentration. End products of methionine and cysteine metabolism other than taurine could be monitored; however, they may not be as sensitive or as desirable to measure as taurine. Glutathione is a good example for comparison because glutathione synthesis utilizes methionine and cyst(e)ine. Glutathione, like taurine, occurs in high concentrations in many cell types, serves as an antioxidant, circulates in plasma, and has many vital functions. Low glutathione concentrations in tissues are believed to compromise cell function, to promote tissue damage, and increase morbidity under various disease conditions [20]. Though glutathione is arguably important to consider in suspected dietary deficiency of sulfur-containing amino acids, taurine is a more sensitive indicator to monitor than glutathione for a least two reasons. First, the liver regulates concentrations of circulating glutathione like that observed for methionine and cyst(e)ine. And, second, the synthesis of taurine appears to have a lower priority than synthesis of glutathione. The Michaelis constant ($K_m$) for the rate controlling enzyme of glutathione synthesis is less than that of taurine synthesis [1].

The relative sensitivity of taurine as an indicator of adequacy of dietary sulfur-containing amino acids is evident from measurements in large dogs given commercial dog foods believed to have low sulfur-containing amino acid bioavailability [15,16]. Plasma concentration of taurine was found to be significantly depressed while plasma glutathione concentration appeared unaffected and whole blood glutathione appeared only mildly depressed.

**MONITORING ADEQUACY OF SULFUR-CONTAINING AMINO ACIDS**

Commercial dog foods that have produced taurine deficiency appear to adequately support protein synthesis. Labeling claims for some diets indicate that the diet passed a feeding protocol in which acceptable changes in body weight, albumin, and hemoglobin concentration were observed. Evidence that protein synthesis is supported by such diets is still consistent with a low level of sufur-containing amino acid intake. Under such a condition, catabolism of methionine and cysteine as with taurine synthesis is restricted while remethylation of homocysteine back to methionine and incorporation of cysteine into protein and glutathione given high priorities [1] (Fig. 1). Hence, the monitoring of taurine status is tool that may be used to specifically assess sulfur-containing amino acid adequacy beyond that needed to achieve nitrogen balance. Such monitoring would seem worthwhile given the potential for producing diets deficient in sufur-containing amino acids.

Methionine is often the first or second limiting amino acid in diets for dogs [17] and cystine is usually present in diets in a limited range of proportions (~ 1:1 to 1:2) relative to methionine [1]. Thus, because typical amino acid analyses of diets do not indicate bioavailability [21], unrecognized, less than optimal amounts of methionine and cystine may occur in dog foods. This has been substantiated by studies of diets and ingredients in ileal-cannulated animals, where variable and low digestibilities of cystine are reported [22,23]. The heat processing of diets and including the rendering process used in producing meat-meals is suggested to especially affect the cystine bioavailability. It should be noted that though heat processing has this effect, a few factors are mitigating. Digestibility effects of heat processing vary considerably with the temperature, time, and nature of the food matrix [e.g., 24]. Also, heat processing can increase digestibility of some proteins that contain sulfur-amino acids, such as protein from soybeans [17].

For non-invasive convenient monitoring of taurine status, taurine concentrations in urine, blood, and plasma are useful. Of these concentrations, taurine to creatine ratio in free-catch urine samples appears to change most acutely with a dietary change [13]. Unfortunately, reference ranges are not reported for urine taurine to creatine ratio, and urine creatinine concentration varies with muscle mass and may scale differently with body size than taurine excretion. With daily urine collections over 5 days, taurine synthesis rate in dogs may be estimated after a single oral administration of stable isotopic taurine [15,16]. Determination of taurine synthesis rate is useful for distinguishing the cause of low taurine status. High loss of taurine could reduce taurine status, but so could low synthesis, as would occur with low intake or bioavailability of sulfur-containing amino acids. Low values thought to be indicative of taurine deficiency have been reported for blood and plasma taurine concentrations [2,5,6]. These values are based on prevention of dilated cardiomyopathy. Greater values than these may be needed to optimize other functions dependant on taurine and other sulfur-containing amino acids. In the maintenance state, dietary amino acids concentrations that optimize variables like the immune response, wound healing, and health in old age have not been determined [17].

## CORRECTING DEFICIENCY

For diets that are not made intentional low in protein, as with some therapeutic diets, it seems inappropriate to add taurine to correct low taurine status induced by a diet. Correction by increasing methionine and cysteine bioavailability seems most appropriate. In making such a correction, monitoring of taurine status as described above may be used assess sulfur-containing amino acid bioavailabilities. Increasing the concentration of dietary protein with low sulfur-containing amino acid bioavailability is also not a desirable means of correction. This manipulation may increase fermentable residue if the protein is not well digested. Of relevance here is a report of many years ago demonstrating distal small intestinal fermentation in dogs given a high protein diet [25]. The health significance of such fermentation is not known. Better options than increasing dietary protein would be to substitution of higher quality protein or supplementation with DL-methionine. It is interesting that the latter measure may especially support taurine synthesis while filling other needs for sulfur-containing amino acids. Research in rats show higher taurine formation and excretion in animals fed excess methionine than in those fed excess protein [26]. With methionine compared to protein supplementation, the switch to formation of taurine more than sulfate is suggested to have the advantage of less perturbation of acid-base balance [1].

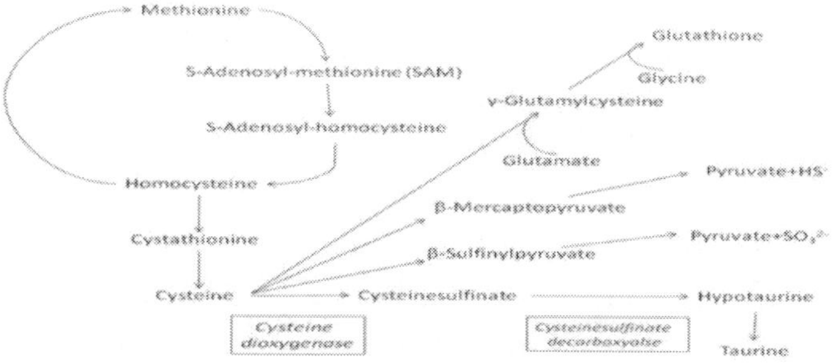


**Figure 1.** Metabolic pathways to taurine synthesis

## REFERENCES


[1]Stipanuk M, et.al. **Annu Rev Nutr** 2004;24:539. [2]Fascetti A, et.al. **J Am Vet Med Assoc** 2003;223(8):1137. [3]Backus R, et.al. **J Am Vet Med Assoc** 2003;223(8):1130. [4]Pion P, et.al. **Vet Clin North Am Small Anim Pract** 1998;28(6):1495. [5]Backus R, et.al. **J Nutr** 2006;136(10):2525. [6]Gavaghan B, et.al. **Aust Vet J** 1997;75(12):862. [7]Sanderson S, et.al. **J Vet Intern Med** 2001;15(2):94. [8]Henthron P, et.al. **Hum Genet** 2000;107(4):295. [9]Sanderson S, et.al. **Am J Vet Res** 2001;62(10):1616. [10]Bélanger M, et.al. **J Am Anim Hosp Assoc** 2005;41(5):284. [11]Dukes-McEwan J, et.al. **BSAVA proceedings**, 2001. [12]Hannah S, et.al. **Vet Clin Nutr** 1995;2:145. [13]Torres C, **PhD Thesis**, Univ. Calif. Davis 2004. [14]Tôrres C, **J Anim Physiol Anim Nutr** 2003;87(9-10):359. [15]Backus R, **J Nutr** 2006;136(10):2525. [16]Ko K, **J Nutr** 2007;137(5):1171. [17]NRC, **Nutrient Requirements of Dogs and Cats** 2006. [18]NRC, **Nutrient Requirements of Dogs** 1985. [19]Odle J, **Adv Exp Med Biol** 1992;315:55. [20]Lyons J, et.al. **PNAS** 2000;97(10),5071. [21]Rutherfurd S, Moughan P, **J AOAC Int** 2008;91(4):907. [22]Johnson M, et.al. **J Anim Sci** 1998;76:1112. [23]Gajda E, **J Anim Sci** 2005;83:160. [24]Piva G, **Ani Feed Sci** Tech 2001;89(1-2);59. [25]Banta C, **J Nutr** 1979;109(9):1592. [26]Bella D, Stipanuk M, **Am J Physiol** 1995;269:E910.

## RATIONAL ANTIMICROBIAL PROPHYLAXIS BEFORE SURGERY.

**Ted Whittem, BVSc, PhD, DACVCP, FACVSc & Pierre Moissonnier, DMV, PhD, DECVS**
**Melbourne, Australia**

### INTRODUCTION

The peri-operative administration of antimicrobial agents to surgical patients in order to minimise the opportunity for microbial infection is often called "surgical antimicrobial prophylaxis" (SAP). Our human medical colleagues are enabled by their prolific research and publication to use SAP in an evidenced-based manner. They have demonstrated that post-surgical infections in people can be minimised by use of particular antibiotic regimens which vary between factors such as populations, surgeons, hospitals and tissues. They have also demonstrated that the regimens need to be monitored and altered appropriately as patterns of infection change with corresponding changes in associated factors.

In veterinary medicine our use of SAP is not similarly evidence-based. The totality of our evidence for SAP is minimal and insufficient. Furthermore, to assume that the human evidence applies to veterinary species is an example of the "fallacy of inappropriate extrapolation".[1]

In the absence of substantial evidence, we are forced to rationalise our habits in SAP, i.e. to base them on fundamental principles.

### IS SURGICAL ANTIMICROBIAL PROPHYLAXIS BENEFICIAL OR NECESSARY?

Generally, surgical site infection (SSI) occurs in approximately 2 to 10% of cases in dogs and cats. This incidence appears to vary in association with several independent factors. Factors attributed by some as associated with SSI rates include; age, morbidity, ASA preoperative score, wound contamination score, duration of anaesthesia, hypothermia, endocrine diseases, pre-anaesthetic surgical site preparation, surgical suite air quality, number of people in the surgical suite, surgeon, surgical skill and antimicrobial drug administration. [2-6] Insufficient data are available to evaluate covariance between these factors. Causal associations between these factors and SSI rates have not been established in veterinary medicine. Evaluation of SAP as a factor has included studies of different types and designs which have yielded some conflicting and equivocal results.

#### Subordinate trials

As bases for so-called rational practice, case-controlled and cohort studies or epidemiological surveys and studies using experimental animals in the target species or laboratory species are acceptable but all are subordinate to well-designed randomised prospective clinical trials. Many such pre-clinical studies are available for evaluation, but the results do not consistently support the worth of SAP. Some examples are discussed.

In a fundamental study carried out on skin-wounds in a rat model, the timing of administration was evaluated for potential effect on the efficacy of SAP. Three groups were used. Rats received; 1) wound contamination with staphylococcus and no antibiotic treatment (control), 2) contamination by killed staphylococcus (placebo) or 3) contamination and a treatment by penicillin at different times after contamination. The diameter of inflammatory reaction was measured 24 hours after treatment. It was shown that only the early injections of penicillin can reduce the size of inflammatory reaction. Dosing with antibiotic more than three hours after contamination did not reduce the inflammatory reaction. In this study the decisive period was defined as between 2 and 6 hours.[7]

In a retrospective analysis of 1,992 cases of dogs and cats at the University of California-Davis, the SSI rates according were associated with the expected degree of wound contamination. The administration of antibiotics did reduce the frequency of wound infection in clean surgical procedures that required 90 minutes or less to complete when performed by senior veterinary students, but not when by conducted by faculty or resident surgeons. It was concluded that local factors of tissue management, including the way the tissues are respected during surgery, can play a major role in the local outcome.[8]

The fluoroquinolone antibiotic marbofloxacin was evaluated in an experimental study in which 18 healthy dogs were divided in three groups and two time periods (3 dogs per period). Tissue cages were implanted for collection of tissue fluids. A bacterial strain of staphylococcus intermedius isolated from naturally occuring canine pyoderma was used. Its MIC for marbofloxacin was 0.5 µg/ml and concentration of bacteria was adjusted to $10^8$ to $3 \times 10^8$ colony forming units. Two mL of bacterial suspension was injected into the cage 15 min after the injection of experimental SAP treatment. Treatments groups were single injection of 4, 2 or 1 mg marbofloxacin/kg IV. Dogs treated with 2 and 4 mg/kg had significantly lower bacterial counts in the harvested tissue fluid. All the clinical criteria were similar in all groups. It was concluded that marbofloxacin might be suitable for SAP when staphylococci were a likely contamination.[9]

Another 1997 study was a prospective survey of postoperative surgical site infections (SSI) in dogs and cats (1574 cases) at the University of Pennsylvania teaching hospital found that 5.5% cases became infected. In this study 'infection' was defined simply as occurrence of purulent discharge up to the time of suture removal. Infection rates varied as expected in accordance with wound contamination classification. Importantly, for animals with clean wounds appropriate SAP did not change infection rates (no SAP= 4.4, SAP=2.2%) whereas off-protocol use of SAP increased the rate to 6.3%.[3]

In a subsequent retrospective re-evaluation of the same cases the authors found similarly that SAP did not decrease the infection rate for clean-contaminated wounds.[5] These two studies ought to cause reflection on whether, in similar cases at that hospital, SAP is cost effective and rational.

A similarly designed prospective survey of SSI was conducted at the University of Bern in Switzerland in dogs and cats (1010 cases). In this study a single occurrence of purulent discharge, abscess or fistulae resulted in recording SSI or alternatively the simultaneous occurrence of >2 signs from the list including redness, swelling, pain, heat, serous discharge and wound dehiscence was recorded as "inflammation/infection". The incidence of SSI and "inflammation/infection" in this study were 3.0 % and 5.8%, respectively. These results are similar to those of other studies. In this study using univariate analysis the SSI rate increased when pre- or post-operative antimicrobials were used but the study did not show peri-operative SAP as beneficial in reducing SSI. This failure was attributed to the study design's lack of statistical Power through the relative paucity of controls. Using a logistic regression model the SAP was included as a significant factor for reduction of both SSI and "inflammation/infection" with odds ratios of 0.15 (p=0.035) and 0.26 (p=0.021), respectively.

The use of SAP during cranial cruciate repair in dogs was examined by retrospective analysis of cases between 2003 and 2005 at the Ontario Veterinary College. Approximately 90% of dogs (n=83) received SAP but the efficacy of such administration could not be ascertained. The principal and important finding of this study was that, in the absence of a Standard Operating Procedure at the hospital, individual habits of drug choice and dose regimen varied widely. Frequently these habits contrasted markedly with 'rational' SAP dose guidelines if 'rational' is based on either the human or veterinary clinical trials available for evaluation.[10]

**Randomised controlled clinical trials**

Ideally evidence-based medicine is founded primarily upon repeatable, prospective randomised controlled clinical trials for each indication. Such evidence is not available for evaluation of SAP. We have identified only a handful of such veterinary clinical trials of SAP conducted over the last 30 years. It must be remembered also that not all clinical trials provide reliable information, often suffering from serious design faults. Evaluation of clinical trial publications has been a subject of discussion at the ACVIM Forum previously.[11] Publication bias, also known as the "fallacy of positive results",[1] also conspires to mislead even the careful reader.

In the 1980s Vasseur and colleagues at the Univerisity of California-Davis pioneered the field with a randomised, prospective, blind, placebo-controlled clinical trial for evaluation of SAP in 122 dogs and 7 cats. The frequency of SSI after clean surgical procedures was compared in dogs and cats given ampicillin with those given a placebo. The 2 groups were similar in terms of other potential covariates. The infection rates in the 2 groups were not significantly different. Unfortunately the study suffered from inadequate statistical Power.[12]

Another prospective study was described in 1991 in which the authors utilised a randomised placebo-controlled study design evaluating SAP in clean and clean-contaminated surgeries of less than 3 hours of duration in 329 dogs and 544 cats. Neither pre nor postoperative antibiotics were administered and all the animals were ASA I or II. The animals received either 0.9% NaCl (placebo) or SAP as cefalexine (30mg/kg). Three subgroups were made using the surgeons' experience as an evaluation criterion (low-students, middle-interns or high-senior surgeon). Variables assessed included inflammation state which was scored as 0 to 3, suppuration scored as 0-3, lymphadenitis and body temperature. Infection was diagnosed when pus was observed, or when hyperthermia occurred simultaneous to a score of >2 in one other parameter superior to 2. The rate of SSI was not significantly different between the placebo group (9.4%) and the SAP group (8.5%)(p= 0.661) despite adequate case numbers. It was demonstrated that the surgeon's experience confounded the possible influence of other factors such as age and anaesthetic class. It was concluded that surgical skill is more influential than SAP in prevention of SSI.[13]

Does SAP reduce the incidence of SSI in dogs undergoing elective orthopaedic procedures? A randomized, controlled, blinded, intention clinical trial was conducted at the University of Illinois teaching hospital in 1999. Dogs were randomly assigned to 1 of 3 groups: treatment with saline solution, treatment with potassium penicillin G, and treatment with cefazolin. Dogs were monitored for 10 to 14 days after surgery for evidence of infection. Criteria for infection were similar to the Daude-Lagrave study described above. Despite a Power analysis requirement for > 250 dogs, after the first 112 dogs were enrolled in the study, it was found that infection rate for control dogs (5/32 dogs) was significantly higher than the rate for dogs treated with antimicrobials (3/80 dogs). Therefore, due to ethics requirements no more dogs were enrolled in the study. A total of 126 dogs completed the study. Use of SAP significantly decreased SSI rate in dogs undergoing elective orthopaedic surgery, compared with infection rate in control dogs. Cefazolin was no more effective than potassium penicillin G in these dogs.[4] It is interesting to note that the use of cefazolin in preference to penicillin G continued at the University of Illinois, despite their own evidence that the more expensive and less safe drug was not more effective. This is an example of fallacious application of 'rational' therapeutics; the failure to use strong evidence to change prescribing habits.

Finally, one other clinical trial of SAP in dogs was identified for review. This study evaluated SAP for 73 dogs receiving diagnostic, de-bulking or amputation surgery followed by cancer chemotherapy with doxorubricin. This well designed study found that SAP with trimethoprim-sulfadiazine administered every 12 hours for 14 days following surgery reduced morbidity and hospitalisation of surgical patients (dogs) receiving doxorubricin for treatment of lymphoma or osteosarcoma. This study illustrates that when the choice of antimicrobial agent and dose regimen are tailored to the clinical scenario in which the SAP is applied the outcome can be beneficial.[14]

Some work has been published on the use of SAP in cattle. Eighty-three calves with umbilical hernia repair were prospectively randomized to one of two groups. The first group received antibiotics for one day, the second group for four days. The SSI rate in group one was significantly higher than in group two, suggesting that ongoing antibiotics were helpful in this cohort. Male calves were more susceptible to infection than female calves.[15]

Following rumenotomy in 29 Angus steers, both SAP with penicillin G and a protracted course of procaine-penicillin G proved effective when compared to untreated controls. Steers receiving antibiotics had significantly greater postoperative feed intake, lower rectal temperatures, and fewer abscesses at the surgical site than those receiving no antibiotics but no difference was found in SSI rates between the short SAP versus the protracted dose regimen.[16]

**The "bundle of sticks" fallacy**

The 'bundle of sticks fallacy' is the belief that a collection of twigs can provide the strength, form and function of a large branch of a sturdy tree.[1] Our small bundle of evidentiary twigs do not provide the trunk of evidence needed to maintain with certainty that SAP is of value in veterinary surgery. The lack of evidence prevents us from making sound decisions.

## CLINICAL RELEVANCE AND GUIDELINES

Despite the lack of overwhelming evidence for efficacy of SAP, prudent rational SAP therapy both persists as common practice and is recommended by 'authorities' (We note here that Skrabanek and McCormick are quite illustrative in their examples of "The Fallacy of Authority'[1]). There is also no evidence that use of antimicrobial drugs in SAP protocols poses any risk for development of antimicrobial drug resistance. Some antimicrobial drugs do cause adverse reactions but knowledge of these allows their management or avoidance. Therefore, it is appropriate in veterinary surgery to apply SAP. The following three major points are worth taking home.

1. Ideally each major surgical practice should evaluate the theoretical rational SAP protocols for major groups of surgery in their populations, at their hospital sites and with their own surgeons. Any major changes to these variables should prompt re-examination of the protocols. The evaluations should be in the form of controlled, randomised and masked clinical trials. For ethical reasons the controls should be positive. Smaller practices should emulate their 'nearest' major hospital.
2. Since the aim of SAP is to reduce bacterial contamination to a point where the host immune mechanisms can prevent establishment of an infection,[2] it is probably not necessary to use wide spectrum antibiotics. It is rational to choose an antibiotic with a good safety profile, that is cheap, that is not fundamental in the treatment of zoonoses or major human infections and that is active against the most likely contaminating organism. It would be unwise and imprudent to choose otherwise.
3. The recommendation, for administration of the chosen antibiotic immediately prior to surgery and to stop administration after completion of surgery, is supported by subordinate studies and in one veterinary clinical trial.

## REFERENCES

[1]Skrabanek P & McCormick J. **Follies & fallacies in medicine**. Tarragon Press, Glasgow, Scotland, Glasgow, 1989. [2]Wilcke JR. Probl **Vet Med** 1990;2:298-311. [3]Brown DC et al. **J Am Vet Med Assoc** 1997;210:1302-1306. [4]Whittem T. et al. **J Amer Vet Med Assoc** 1999;214 [5]Nicholson M, et al. **Vet Surg** 2002;31:577-581. [6]Eugster S, et al. **Vet Surg** 2004;33:542-550. [7]**Burke JF. Surgery** 1961;50:161-168. [8]Vasseur PB, et al. **Vet Surg** 1988;17:60-64. [9]Gruet P, et al. Vet Rec 1997;140:199-202. [10]Weese JS & Halling KB. **J Am Vet Med Assoc** 2006;229:92-95. [11]Whittem T. **Proceedings of the Seventeenth Annual Veterinary Medical Forum, ACVIM**, Chicago, IL., 1999. [12]Vasseur PB, et al. **J Amer Vet Med Assoc** 1985;187:825-857. [13]Daude-Lagrave A, et al.**Vet Comp Orthop** Traumatol 1991;14:146-150. [14]Chretin JD, et al. **J Vet Intern Med** 2007;21:141-148. [15]Klein WR & Firth EC. **Vet Rec** 1988;123:564-566. [16]Haven ML, et al. **J Am Vet Med Assoc** 1992;200:1332-1335.

**ANTIMICROBIAL DRUG-DRUG INTERACTIONS – SYNERGY & ANTAGONISM**

**Tomás Martín-Jiménez, DVM, PhD, DACVCP, DECVPT.**
**Knoxville, TN.**

## INTRODUCTION

When two or more medications are administered concurrently, drug-drug interactions may take place that can affect their pharmacological response. These interactions may or may not result in altered therapeutic efficacy or increased toxicity. One of the main difficulties in understanding the practical relevance of drug-drug interactions in domestic animals is the dearth of in vivo clinical data. Most of the published information comes from in vitro experiments that provide very valuable mechanistic information and can be used to make reasonable predictions about possible or likely interactions in animals. In practice, however, many factors influence the clinical significance of drug-drug interactions; these include size and frequency of dose, duration of treatment, time-course of the interaction, drug therapeutic index, disease status, pharmacokinetic differences between interacting drugs, status of the immune system, inter-animal variability, etc.

Drug-drug interactions among concurrently administered antibiotics or between antibiotics and other agents may result in synergistic or antagonistic antimicrobial effects, combined organ toxicity, or altered drug concentrations that may result in increased toxicity or decreased efficacy. Knowledge of the mechanism or mechanisms by which a particular drug-drug interaction takes place may provide clues as to the expected clinical relevance, time course, and methods used to avoid the interaction or palliate its clinical consequences.

## MECHANISMS OF DRUG-DRUG INTERACTIONS

Mechanisms by which drugs may interact are generally categorized as pharmaceutic, pharmacokinetic, pharmacodynamic, and combined organ toxicity.

### Pharmaceutic interactions

May occur when the chemical stability and conformation of drugs in a mixture are affected by other drugs.

### Pharmacokinetic interactions

These occur when drugs affect the absorption, distribution or elimination profile of concurrently administered drugs.

The gastrointestinal (GI) **absorption** of one agent may be affected by other agent that alters GI pH or motility, chelate other chemicals, or affect active transporters such as P-glycoprotein. While both rate and extent of absorption may be altered, interactions that affect the systemic availability of drugs are more likely to be of clinical importance. The most common type antibiotic drug interaction affecting absorption is the chelation of the antibiotic by concurrently administered metal ions, which may be present in food or antacid medications. Separating the administration times of the antibiotic and the chelating agent may reduce the extent of interaction. A reasonable approach is to leave 2-4 hours of separation between both.

The **distribution** of concurrently administered drugs may be affected by competitive displacement from plasma and tissue binding sites or alteration of active tissue transporters (e.g. blood-brain barrier), with the latter being more likely to result in clinically relevant consequences.

The **elimination** of concurrently administered drugs can be altered through mechanisms affecting metabolism, secretion, or excretion. Induction or inhibition of cytochrome P450 isozymes in the small intestine and liver may result in increased or decreased drug clearance, respectively. For example, chloramphenicol is known to inhibit cytochrome P450 enzymes in the liver of dogs resulting in prolonged barbiturate sedation and potential toxicity[1-6]. On the other hand, phenobarbital is a well known inducer of enzymes involved in Phase I and Phase II drug metabolism, and as such it decreases the plasma concentration of drugs like chloramphenicol[7,8] or doxycycline[9]. One of the most commonly P450 enzyme involved in metabolic drug-drug interactions is CYP3A4, which represents the primary route of metabolism for many of the drugs used in domestic animals. In practice, the clinician has to consider the time course of these processes, for while enzyme inhibition occurs relatively quickly, enzyme induction commonly requires more time. It takes approximately 1-2 weeks for the maximum inductive effect to occur and about the same time for it to cease once drug administration has stopped. Rifampin may be an exception to the rule given that it may elicit enzyme induction just after a few doses. The elimination half-life of the drug is another important factor, since it will take about 3-5 half-lives to reach the new steady state concentration after drug clearance is changed.

The renal elimination of drugs can be affected by concurrently administered agents that modify urinary pH, or that compete for binding to transporter proteins involved in active tubular secretion.

### Pharmacodynamic interactions

Concurrently administered drugs may exhibit additive or synergistic effects by acting on the same or different receptor systems. In general, a synergistic effect can be expected when the mechanisms of action are complementary, for example, when drugs act at different places of the same biochemical pathway (sulfonamides and diaminopyrimidine), or when the presence of one antibiotic increases the bacterial penetration of another (beta-lactams and aminoglycosides). For example, synergism has been reported to occur between flouroquinolones and either aminoglycosides, 3rd generation cephalosporins, extended-spectrum penicillins, carbapenems, vancomycin, or trimethoprim against some bacterial pathogens, including *Pseudomonas aeruginosa* (e.g. ciprofloxacin and amikacin or imipenem), *Staphylococcus aureus* (e.g. levofloxacin and oxacillin) *Enterococcus spp.* (e.g. ciprofloxacin and amikacin)[10] and *Escherichia coli* (e.g. trimethoprim and ciprofloxacin)[11]. On the other hand, antibiotic combinations characterized by overlapping mechanisms of action or suppression of important efficacy elements (e.g. lack of active bacterial growth) may result in antagonism characterized by the clinical efficacy of the drug combination being less than that of the individual agents.

**Combined organ toxicity**

Combination of drugs that elicit deleterious effects on the same organ can greatly increase the probability of toxicity. Moreover, drugs without intrinsic toxicity on one organ may increase the toxicity of a concurrently administered drug (e.g. diuretic action can increase the nephrotoxicity of aminoglycosides).

The following table shows selected examples drug-drug interactions involving antibiotics commonly used in domestic animals.

**Table 1**. Selected antibiotic drug-drug interactions

| Interacting Drugs or Drug Groups | Species* | Mechanism and Effect | Evidence |
|---|---|---|---|
| **Chloramphenicol**/Phenobarbital | Dog, cat | Chloramphenicol inhibits liver CYP2B11[12]. Decreased phenobarbital clearance, increased elimination half-life and increased plasma and tissue concentrations. Prolonged sedation and toxicity potential. | In vitro, In vivo PK and PD[1-6] |
| **Chloramphenicol**/Propofol | Dog | CYP2B11 inhibition by chloramphenicol. Decreased clearance and prolonged recovery time. Breed differences are possible. | In vitro, In vivo PK and PD[13-14]. |
| **Fluoroquinolones**/Theophylline | Dog | Some flouroquinolones inhibit liver CYP1A2. Decreased theophylline clearance and potential for increased toxicity. In vivo effect more evident with enrofloxacin than with marbofloxacin. In vitro effect observed with ofloxacin, orbifloxacin, and ciprofloxacin. | In vivo PK[15-16] In vitro[17]. |
| **Fluoroquinolones**/NSAIDs | Dog, Buffalo, Sheep | Enrofloxacin and flunixin meglumine decrease each other clearance in dogs. Enrofloxacin decreases diclofenac clearance in sheep and increases Vd in buffalo. Unknown interactions between FQ and other NSAIDs. | In vivo PK[18-20]. |
| **Phenobarbital**/Doxycycline | Human | Induction of doxycycline metabolism by phenobarbital. Clearance of doxycycline doubled in patients undergoing long-term phenobarbital therapy. Decreased doxycycline efficacy is likely. | In vivo PK[9]. |

**Table 1 (Cont'd).** Selected antibiotic drug-drug interactions

| Interacting Drugs or Drug Groups | Species* | Mechanism and Effect | Evidence |
|---|---|---|---|
| **Antacids**/Fluoroquinolones, Tetracyclines, Azithromycin. | Dog, cat, horse, other | The systemic availability of many drugs is consistently decreased by their adsorption to concurrently administered antacids. The extent of this decrease is variable but usually significant. | In vivo PK[21-23] |
| **Erythromycin-Chloramphenicol**/Cisapride | Human, Dog | Chloramphenicol, Erythromycin and to lesser extent Clarithromycin inhibit CYP3A4 resulting in increased cisapride concentrations in humans. In one study in dogs erythromycin did not | In vivo PD[24] |

| | | | |
|---|---|---|---|
| | | modify cisapride cardiovascular pharmacodynamics. | |
| Clindamycin/Metronidazole | Human isolates | Synergistic effect has been observed in vitro against *Bacteroides fragilis*. | In vitro[25]. |
| Beta-lactams/Fluoroquinolones | Human isolates. Human. Mice. | Ceftazidime or cefepime plus a flouroquinolone (ciprofloxacin, levofloxacin, gatifloxacin, or moxifloxacin) resulted in additive effect when strains were susceptible to both agents in the combination (less than 10% of synergy) but showed synergy 92% of the cases when strains were resistant to one or both agents. No significant differences were found between the various combinations. Clinical relevance depends on PK. | In vitro[26]. |
| Beta-lactams/Aminoglycosides | Human isolates. Human. Mice. | Classical synergistic interaction described in vitro for many organisms, including strains of *Staphylococcus aureus* that are susceptible to each drug alone. One mechanism is enhanced bacterial penetration of the aminoglycoside by the beta-lactam in organisms such as streptococci. This synergism may carry to gram-negative organisms like Enterobacteriaceae, including *Pseudomonas aeruginosa*. Even though synergistic effect is expected when using this combination, it is often possible that the effect cannot be clinically achievable[27,28]. | In vitro[29], In vivo[30] |

## REFERENCES

[1] Sanders JE, Yeary RA, Fenner WR, Powers JD. Interaction of phenytoin with chloramphenicol or pentobarbital in the dog. **J Am Vet Med Assoc**. 1979;175(2):177-80. [2] Teske RH, Carter GG. Effect of chloramphenicol on pentobarbital-induced anesthesia in dogs. **J Am Vet Med Assoc**. 1971;159(6):777-80. [3] Adams HR, Dixit BN. Prolongation of pentobarbital anesthesia by chloramphenicol in dogs and cats. **J Am Vet Med Assoc**. 1970;156(7):902–5.[4] Houston DM, Cochrane SM, Conlon P. Phenobarbital toxicity in dogs concurrently treated with chloramphenicol. Nonrefereed case report. **Can Vet J.** 1989;30:598.[5] Adams RH, Isaacson EL, Masters BSS. Inhibition of hepatic microsomal enzymes by chloramphenicol**. J Pharmacol Exp Therap**. 1977;203(2):388-96.[6] Campbell CL. Primidone intoxication associated with concurrent use of chloramphenicol. **J Am Vet Med Assoc**. 1983;182(9):992-3.[7] Palmer DL, Despopoulos A, Rael E. Induction of chloramphenicol metabolismo by phenobarbital. **Antimicr Ag Chemother**. 1972;1(2):112-5.[8] Graham RA, Downey A, Mudra D, Krueger L, Carroll K, Chengelis C, Madan A, Parkinson A. In vivo and in vitro induction of cytochrome P450 enzymes in beagle dogs. **Drug Metab Dispos** 2002;30(11):1206–13.[9] Neuvonen PH, Penttila O. Interaction between doxycycline and barbiturates. **Br Med J**. 1974;1:535-6.[10] Eliopoulos GM, Moellering RC, Jr: Antimicrobial combinations, Chapter 9**.** In **Antibiotics in Laboratory Medicine**, 4th ed. 1996; Lorian V: Williams and Wilkins.[11] Huovinen P, Wolfson JS, Hooper DC: Synergism of Trimethoprim and Ciprofloxacin in Vitro against Clinical Bacterial Isolates. **Eur J Clin Microb and Infectious Diseases** 1992, **11**:255-257[12] Trepanier LA. Cytochrome P450 and its role in veterinary drug interactions. **Vet Clin Small Anim**. 2006:975-85.[13] Mandsager R, Clarke C, Shawley R, et al. Effects of chloramphenicol on infusion pharmacokinetics of propofol in Greyhounds. **Am J Vet Res** 1995;56(1):95–9.[14] Hay Kraus BL, Greenblatt DJ, Venkatakrishnan K, Court MH. Evidence for propofol hydroxylation by cytochrome P4502B11 in canine liver microsomes: breed and gender differences. **Xenobiotica**. 2000;30(6):575-88. [15] Hirt RA, Teinfalt M, Dederichs D, et al. The effect of orally administered marbofloxacin on the pharmacokinetics of theophylline. **J Vet Med A Physiol Pathol Clin Med** 2003;50(5):246–50.[16] Intorre L, Mengozzi G, Maccheroni M, et al. Enrofloxacin-theophylline interaction: influence of enrofloxacin on theophylline steady-state pharmacokinetics in the beagle dog. **J Vet Pharmacol Ther** 1995;18(5):352–6.[17] Regmi NL, Abd El-Aty AM, Kuroha M, Nakamura M, Shimoda M. Inhibitory effect of several fluoroquinolones on hepatic microsomal cytochrome P-450 1A activities in dogs. **J Vet Pharmacol Ther**. 2005;28(6):553-7.[18] Ogino T, Mizuno Y, Ogata T, Takahashi Y. Pharmacokinetic interactions of flunixin meglumine and enrofloxacin in dogs. **Am J Vet Res.** 2005;66(7):1209-13. [19] Kumar N, Singh SD, Jayachandran C. Pharmacoknetic study of diclofenac and its interaction with enrofloxacin in buffalo calves. **J Vet Sci**. 2003;42):155-9. [20] Rahal A, Kumar A, Ahmad AH, Malik JK. Pharmacokinetics of diclofenac and its interaction with enrofloxacin in sheep. **Res Vet Sci.** 2008;84(3):452-6.[21] Lehto P, Kivisto KT. Effect of sucralfate on absorption of norfloxacin and ofloxacin. Antimicrob. **Agents Chemother**. 1994;38:248–51.[22] Jaehde U, Sorgel F, Stephan U, Schunack W. Effect of an antacid containing magnesium and aluminum on absorption, metabolism and mechanism of renal elimination of pefloxacin in humans. **Antimic Ag Chemother**. 1994;38(5):1129-33.[23] Sadowski DC. Drug interactions with antacids. Mechanisms and clinical significance. **Drug Saf.** 1994;11(6):395-407.[24] Al-Wabel NA, Strauch SM, Keene BW, Nakayama T, Hamlin RL. Electrocardiographic and hemodynamic effects of cisapride alone and combined with erythromycin in anesthetized dogs. **Cardiovasc Toxicol**. 2002;2(3):195-208.[25] Busch DF, Sutter VL, Finegold SM. Activity of combinations of antimicrobial agents against Bacteroides fragilis. **J Infect Dis**. 1976;133(3):321-8.[26] Fish DN, Choi MK, Jung R. Synergic activity of cephalosporins plus flouroquinolones against Pseudomonas aeruginosa with resistance to one or both drugs. **J Antimicr Ther**. 2002;50:1045-9.[27] Miranda-Novales G, Leaños-Miranda BE, Vilchis-Perez M, Solórzano-Santos F. In vitro activity effects of combinations of cephalothin, dicloxacillin, imipenem, vancomycin and amikacin against methicillin-resistant Staphylococcus spp. strains. Ann **Clin Microb Antimic**. 2006;5:25-9.[28] Paul M, Benuri-Silbiger I, Soares-Weiser K, Leibovici L. Beta-lactam monotherapy versus beta-lactam/aminoglycoside combination therapy for sepsis in immunocompetent patients: systematic review and meta-analysis of randomised trials. **Br Med J.** 2004;328:668-72.[29] Snyder RJ, Wilkowske CJ, Washington JA. Bactericidal activity and combinations of gentamicin with penicillin or clindamycin against Streptococcus mutans. **Antimic Ag Chemother**. 1975;7(3):333-5.[30] Klastersky J, Meunier-Carpentier F, Prevost JM. Significance of antimicrobial synergism for the outcome of gram negative sepsis. **Am J Med Sci**. 1977 Mar-Apr;273(2):157-67.

## ANTIMICROBIAL DRUG THERAPY: THE POPULAR MYTHS & MISCONCEPTIONS

**Mark G. Papich, DVM, MS, Diplomate ACVCP**
**Raleigh, North Carolina**

### INTRODUCTION

In veterinary antimicrobial therapeutics there are several unsubstantiated recommendations, misconceptions, and myths that persist. Data is now available through microbiologic and pharmacologic studies to dispel many of these old thoughts and concepts. For the instances when there is not specific data to prove or disprove the value or harm of a drug treatment, we can make some educated predictions based on our evaluation and interpretation of experimental data, even when it has been derived from human or laboratory animal studies.

This presentation will focus on antibiotic drugs. One of the major controversies is that of administration of antibiotics to food-producing animals and its effect on bacterial resistance. However, this discussion will focus on companion animals, primarily dogs and cats. In addition, time will not allow an exploration of other areas for which there is controversy, and perhaps unproven treatments: alternative and complimentary medical therapies and holistic treatments.

### CAN ANTIBIOTICS BE COMBINED THAT ARE BACTERICIDAL & BACTERIOSTATIC?

Drug combinations can improve therapy by providing a synergistic effect and broadening the spectrum. Occasionally, it is necessary to administer drug combinations because the infecting pathogen is not known and the clinician desires to target several different potential pathogens. When prescribing combinations of antibiotics, a frequent concern is that drugs administered may produce *both* a bactericidal and bacteriostatic effect. Will this produce an interaction that will compromise effective therapy? There are no documented situations in veterinary medicine in which these combinations produce a less effective interaction. This recommendation is primarily taken from a clinical study in children treated with penicillin and tetracycline published over 50 years ago. This study has not been repeated in other clinical situations. The line separating *bacteriostatic* and *bactericidal* antibiotics is becoming more blurred. Drugs now are usually referred to as *time-dependent* or *concentration-dependent*. Macrolides, chloramphenicol, and tetracyclines are more bacteridical than once thought, but are time-dependent in their action. There is no documented problem with combining these drugs with other drugs such as beta-lactam antibiotics or fluoroquinolones in the same patient. For example, there is no known interaction from administering doxycycline to a patient (to treat a suspected tick-borne pathogen) in combination with a fluoroquinolone, penicillin, aminoglycoside, or cephalosporin. Likewise, combining two drugs with a different spectrum (clindamycin + fluoroquinolone) broaden the spectrum of activity with no evidence of interaction.

### ANTIBIOTIC SUSCEPTIBILITY INFORMATION. IS IT ALWAYS RELIABLE?

The susceptibility data derived from a bacterial isolate should help to guide therapy with the appropriate antibiotic. Unfortunately, sometimes this data is not as reliable as we would like. The breakpoints for susceptibility testing in the United States are set by the Clinical Laboratory Standards Institute (CLSI), formerly called NCCLS (CLSI, 2007). The Veterinary Antimicrobial Susceptibility Testing (VAST) subcommittee of CLSI deals specifically with veterinary drugs and pathogens. For companion animals, veterinary-specific MIC breakpoints have only been established for the four licensed fluoroquinolones (enrofloxacin, difloxacin, marbofloxacin, and orbifloxacin), gentamicin, cefpodoxime proxetil, ampicillin (urinary tract infections only), and clindamycin (dogs only). Until veterinary-specific breakpoints are established for other antibiotics used in companion animals, we will continue to rely on the human breakpoints for drugs such as amikacin, amoxicillin-clavulanate, other cephalosporins, chloramphenicol, erythromycin, carbapenems (imipenem), penicillins, sulfonamides, potentiated sulfonamides, and tetracyclines. Similarities in pharmacokinetics and pathogen susceptibilities between humans and animals allow for an acceptable approximation to extrapolate human breakpoints to animal situations for many drugs until veterinary-specific standards are available. However in other situations, human-derived breakpoints are not appropriate for interpretation because the breakpoint is much higher than concentrations that can be achieved clinically. For example, there is evidence that cephalosporin breakpoints should be lower that the criteria currently used (Kahlmeter, 2008). Subsequently, an organism considered "sensitive" by a human standard breakpoint, may not be susceptible under veterinary use conditions. There also is evidence that the current standard for reporting methicillin-resistant staphylococci, particularly *Staphylococcus pseudintermedius* should be updated (Bemis et al, 2009). The current standard (CLSI M31-A3, 2007), lists the breakpoint for oxacillin as $\leq 2$ µg/mL for *Staph. aureus*, and *S. intermedius (pseudintermedius)*, but not coagulase-negative staphylococci (CNS). For other staphylococci and CNS, the breakpoint of $\leq 0.5$ µg/mL is used. This recent study (which has been confirmed by other yet unpublished papers) suggests that we should be using a breakpoint of $\leq 0.5$ µg/mL for all of our non-*S. aureus* isolates, including *S. pseudintermedius*. Although this is a deviation from the current standard, laboratories should consider using this recommendation.

### IS AN ANTIBIOTIC WITH HIGH VOLUME OF DISTRIBUTION / HIGH LIPOPHILICITY SUPERIOR?

The pharmacokinetic term *volume of distribution* is only a measure of the proportion of drug concentration in plasma relative to the dose (Lees et al, 2008). A high volume of distribution implies that the drug has distributed outside of plasma, possibly intracellular. Highly lipophilic drugs have the ability to penetrate membranes and these are the antibiotics most often associated with a high volume of distribution. A high volume of distribution does not necessarily imply that the drug has high tissue penetration.

Tissue concentrations are rarely measured for many antibiotics, but the plasma concentration usually reflects the drug concentration in extracellular tissue fluid.

Many effective drugs have low lipophilic characteristics and a relatively low volume of distribution, yet they are highly effective. The explanation for this observation is that most infections encountered in veterinary medicine are *extracellular*. It is not necessary for an antibiotic to penetrate cells or membranes to be effective (provided that if it is an oral drug it can be absorbed). Studies in our laboratory showed that volume of distribution and lipophilicity did not determine the extent of antimicrobial drug distribution to the tissue fluid (Bidgood & Papich, 2005). Lipophilicity and high volume of distribution are highly over-rated terms to measure the effectiveness of antibiotics.

**WILL HIGHLY PROTEIN BOUND ANTIBIOTICS CAUSE A DRUG INTERACTION FROM DRUG DISPLACEMENT FROM A BINDING SITE?**

Protein-binding displacement reactions are rare and undocumented in veterinary medicine. Certain drugs are known to displace drugs from protein binding sites and increase the fraction of drug unbound. But for most drugs, the amount of protein in the plasma (and subsequently the number of available drug binding sites) greatly exceeds the concentration of drug in the plasma and binding is rarely saturated. Interactions that involve displacement of protein-bound drugs are therefore rare unless there is severe hypoproteinemia or the drug is so highly protein bound that it occupies most of the binding sites.

It is listed on some highly protein bound antibiotic labels that an interaction may be possible when combining the antimicrobial with another highly protein bound drug such as ketoconazole, NSAIDs, doxycycline, or furosemide. But *in vivo* interactions from these drugs are not documented. The activity of a drug in plasma is dependent on the concentration of free (unbound) drug in plasma, not necessarily the free fraction. Although protein binding displacements and interactions may affect the *free fraction*, they rarely affect the *free concentration*. Only drugs that are highly protein bound (approximately greater than 85%), exhibit high clearance, administered IV, and have a low therapeutic index are likely to be involved in protein binding interactions of clinical significance. Few drugs meet these criteria. Frequently cited reviews (Toutain & Bousquet-Melou, 2003; Benet & Hoener, 2002) have illustrated that drug protein binding interactions may not have consequences that were once thought to be important. As concluded in their paper titled *Changes in plasma protein binding have little clinical relevance*, the authors stated, "… although changes in plasma protein binding can have an important influence on individual pharmacokinetic parameters, we have shown that changes in plasma protein binding will usually not influence the clinical exposure of a patient to a drug." (Hoener & Benet, 2002).

**WILL ADMINISTRATION OF ANTIBIOTICS TO SMALL ANIMALS PRODUCE DRUG RESISTANCE?**

The effect of antibiotic administration to small animals on drug resistance requires additional on-going investigation and surveillance. This is a controversial subject with strong opinions on both sides of the argument. It is inaccurate to suggest that administration of antibiotics *causes* bacterial resistance. However, antibiotic administration – if not active enough to eliminate resistant isolates – can select for these resistant strains, which can multiply and flourish. There are examples of some drugs that can induce resistance mechanisms, but these are rare compared to selection of drug resistant strains. Many of the aspects of this argument were covered thoroughly in a recent book (Guardabassi, et al, 2008). Inadequate antibiotic treatment consisting of doses too low, infrequent administrations, or selection of a poorly active drug, is probably the most common reason for emergence of drug resistance. As it was stated in one of the chapters of the book cited above, "It is exposure, and especially exposure to sub-optimal drug concentrations that is the most important single factor in resistance emergence and its subsequent spread." (Lees et al, 2008). Because inadequate treatment may produce resistance, a strategy has been advocated that employs the use of highly active drugs, administered using appropriate regimens to attain pharmacokinetic-pharmacodynamic (PK-PD) targets, for as short of a duration as possible (Amyes et al, 2007). This suggests that in companion animals, veterinarians should not be reluctant to administer highly active drugs, provided appropriate regimens are used for only as long of a duration as necessary. When possible, therapy then can be *de-escalated* to less active, more narrow spectrum drugs, based on culture and susceptibility results and clinical response. This type of two-tiered approach to therapy has been successful in human medicine.

If administered rationally, and at correct dosages, the administration of an antibiotic should eradicate the infection and actually decrease resistance. However, what we have observed over the 60 years (approximately) of antibiotic use is that resistant strains have been selected. Many of these resistant strains are now common and contributing to the difficulty in selecting effective antibiotics for some patients. In small animals the most common bacteria producing resistant infections are *Escherichia coli, Pseudomonas aeruginosa, Staphylococcus* species, and *Enterococcus* species. These resistant strains have most likely been selected because of antibiotic pressure and inadequate exposure. The resistant strains emerge because the competition from more susceptible bacteria is reduced or eliminated following antibiotic administration. However, the extent to which prescribing practices can influence this trend is not straight-forward. The notion that antibiotic treatment causes resistance is too simplistic and does not reflect the complexity of the issue. Although antibiotic administration is certainly associated with resistance, it is not always clear. A recent review stated "... Thus, the emergence of resistance to antibiotics is associated with their use, although the precise correlation can be highly variable." (Hawkey, 2008). For some bacteria, infection control may be more important than restrictions on antibiotic use (Harbarth & Samore, 2008).

Resistance emerges both through horizontal spread – through transfer of genetic elements carrying genes for resistance – as well as mutations arising during treatment through selection of resistant strains. Reports have shown that resistant strains are more likely when the animal has previously been treated with antibiotics. However, consistent evidence is elusive for antibiotic use in companion animals. In a report from Europe, resistance was low to the antibiotics that were most frequently prescribed (Pedersen et al, 2007). Although resistance to enrofloxacin has been associated with prescribing practices (Cooke et al, 2002), resistance to fluoroquinolones in other studies has not been associated with use (Meunier et al, 2002).

One of the resistant bacteria that may be associated with antibiotic use is methicillin-resistant staphylococci. Various species of methicillin-resistant *Staphylococcus (*MRS) have been identified in companion animals – primarily horses and dogs. These have been reported more frequently in recent years (Soulsby 2008). As reviewed by Harbarth & Samore (2008), in people there is a relationship between antibiotic use and MRSA rates. The drugs most often cited for driving MRSA acquisition and transmission is the use of cephalosporins and fluoroquinolones (Harbarth & Samore, 2008; Dancer, 2008), both of which are frequently prescribed in small animals (Guardabassi et al. 2008).

The risk of increased bacterial resistance in pets as a result of veterinary prescribing practices should be taken seriously. In addition to increased difficulty treating resistant infections in pets, there may be a public health consequence. Transfer between animals and people of resistant bacteria, or genetic elements coding for resistance is possible (Guardabassi et al, 2004). The best documented resistant bacteria that may be transferred from pets to people are *Salmonella, Campylobacter,* and methicillin-resistant *Staphylococcus* (Jensen et al, 2008).

## REFERENCES

[1.] Amyes SGB, Walsh FM, Bradley JS. Best in class: a good principle for antibiotic usage to limit resistance development? **Journal of Antimicrobial Chemotherapy** 59: 825-826, 2007. [2.] Bemis DA, Jones RD, Frank LA, & Kania SA. Evaluation of susceptibility test breakpoints used to predict *mecA*-mediated resistance in *Staphkylococcus pseudintermedius* isolated from dogs. **J Vet Diagn Invest** 21: 53-58, 2009. [3.] Benet LZ, Hoener B. Changes in plasma protein binding have little clinical relevance. **Clin Pharm Ther** 71: 115-121, 2002.[4.] Bidgood TL, Papich MG. Plasma and interstitial fluid pharmacokinetics of enrofloxacin, its metabolite ciprofloxacin, and marbofloxacin after oral administration and a constant rate intravenous infusion in dogs. **J Vet Pharmacol Ther**. 2005 28(4):329-41. [5.] Clinical Laboratory Standards Institute. CLSI. *Performance Standards for Antimicrobial Disk and Dilution Susceptibility Tests for Bacteria Isolated From Animals; Approved Standard—Third Edition.* CLSI document M31-A3. Wayne, PA: **Clinical and Laboratory Standards Institute**; 2007.[6.] Cooke CL, Singer RS, Jang SS, & Hirsh DC. Enrofloxacin resistance in *Escherichia coli* isolated from dogs with urinary tract infections. **J American Veterinary Medical Association** 220: 190-192, 2002.[7.] Dancer SJ. The effect of antibiotics on methicillin-resistant *Staphylococcus aureus*. **Journal of Antimicrobial Chemotherapy** 61: 246-253, 2008.Guardabassi L, Houser GA, Frank LA & Papich MG. Guidelines for antimicrobial use in dogs and cats. (Chapter 11). *In* Guardabassi L, Jensen LB & Kruse H (editors). **Guide to Antimicrobial Use in Animals**. Blackwell Publishing, pages 183-206, 2008.[8.] Guardabassi L, Schwarz S, Lloyd DH. Pet animals as reservoirs of antimicrobial-resistant bacteria. **J Antimicrob Chemother** 54: 321-332, 2004.[9.] Harbarth S, & Samore MH. Interventions to control MRSA: high time for time-series analysis? **Journal of Antimicrobial Chemotherapy** 62: 431-433, 2008.10. Hawkey PM. The growing burden of antimicrobial resistance. **Journal of Antimicrobial Chemotherapy** 62 (Suppl 1): i1-i9, 2008.[11.] Jensen LB, Angulo FJ, Mølbak K, Wegener HC. Human health risks associated with antimicrobial use in animals. (Chapter 2). *In* Guardabassi L, Jensen LB & Kruse H (editors). **Guide to Antimicrobial Use in Animals**. Blackwell Publishing, pages, 13-26, 2008[12.] Kahlmeter G. Breakpoints for intravenously used cephalosporins in Enterobacteriaceae – EUCAST and CLSI breakpoints. **Clinical Microbiology and Infection 14** (Suppl. 1): 169-174, 2008.[13.] Lees P, Svendsen O, & Wiuff C. Strategies to minimize the impact of antimicrobial treatment on the selection of resistant bacteria. (Chapter 6). Guardabassi L, Jensen LB & Kruse H (editors). **Guide to Antimicrobial Use in Animals**. Blackwell Publishing, pages, 77-101, 2008.[14.] Meunier D, Acar J-F, Martel J-L, Kroemer S, & Valle M. A severn-year survey of susceptibility to marbofloxacin of pathogenic strains isolated from pets. **International Journal of Antimicrobial Agents** 24: 592-598, 2004.[15.] Pedersen K, Pedersen K, Jensen H, Finster K, Jensen VF, & Heuer OE. Occurrence of antimicrobial resistance from diagnostic samples from dogs. **Journal of Antimicrobial Chemotherapy** 60: 775-781, 2007.[16.] Soulsby L. The 2008 Garrod Lecture: Antimicrobial resistance—animals and the environment. **Journal of Antimicrobial Chemotherapy** 62: 229-233, 2008.Toutain PL, Bousquet-Melou A. Free drug fraction vs free drug concentration: a matter of frequent confusion. **J Vet Pharmacol Therap** 25: 460-463, 2002.

## APPROPRIATE EMPIRICAL ANTIMICROBIAL THERAPY: MAKING DECISIONS WITHOUT A CULTURE

**Lauren Trepanier, DVM, PhD, DACVIM, DACVCP**
**Madison, WI**

**Is there good evidence of bacterial infection?**

The first step to appropriate empirical antimicrobial therapy in veterinary patients is to first critically ask whether there is actually evidence of a bacterial infection. Too often, antimicrobials are prescribed on a "just in case" basis or because an owner resists additional diagnostics. There are several very real drawbacks of "just in case" antimicrobials: they add to the cost of the visit or hospitalization, without contributing to a diagnosis; they may lead to diarrhea, inappetance, or vomiting that obscure the underlying problem; they can cause adverse reactions or drug interactions; and importantly, they encourage the selection of resistant bacteria, both globally, in your practice, and in an individual patient.

*Fever alone is inadequate criteria for prescribing an antimicrobial.* If the probable source of fever cannot be localized during a physical exam, it is impossible to choose an appropriate spectrum of bacterial coverage or decide how long to treat. Physical exam findings that support the use of antimicrobials without additional diagnostics are straightforward, such as a carnasial tooth root abscess, skin pustules or epidermal collarettes, or a traumatic wound with purulent exudate. Cytology should be performed on all exudates in areas endemic for fungal infections, prior to prescribing antibacterials.

Leukocytosis alone is also an inadequate justification for antimicrobials, since leukocytosis can result from stress, inflammation, glucocorticoid administration, or hyperadrenocorticism. If a left shift and toxic change are present with leukocytosis, then a source of infection or significant inflammation should be pursued. Fever with neutropenia, however, is an established indication for empirical antimicrobials, at least in humans. Meta-analyses of studies in humans suggest that the benefit of antibacterials in neutropenic patients, even prior to fever, outweighs the negative effects of selecting for bacterial resistance. [1] A beta lactam and fluroquinolone combination is recommended in humans.

Antimicrobials are inappropriate in most cats with lower urinary tract signs. There is a < 5% incidence of positive urine cultures in cats with lower urinary tract disease overall. [2] Cats at higher risk for bacterial urinary tract infections are those with diabetes mellitus (13% prevalence),[3] or perineal urethrostomies, or in older cats with dilute urine.[2] Antimicrobials are also not indicated in many cases of upper respiratory infections in cats. Although Mycoplasma is commonly isolated from pharyngeal swabs, underlying viral infection is typical, and cats have similar recovery rates whether antimicrobials with and without activity against Mycoplasma are used. [4]

Similarly, acute diarrheas in dogs and cats are usually not caused by pathogenic bacteria; for example, the prevalence of Salmonellosis (2%), Campylobacter (5%), and Clostridium difficile toxin (10%) is low in dogs with acute diarrhea. [5] Empirical antimicrobials, such as amoxicillin or fluoroquinolones, are not indicated. Fiber and a short-term diet change may be more rational. Finally, pancreatitis is usually sterile in dogs and cats. Antimicrobials are not indicated unless peritonitis, pancreatic abscess, or loss of intestinal mucosal integrity develops. In humans, antimicrobials in severe (necrotizing) pancreatitis do not reduce the incidence of secondary pancreatic infection or decrease mortality. [6]

**Helpful diagnostics in lieu of culture**

Pyuria with bacteria in a urine sediment provides a strong indication for antimicrobials, although bacteriuria can be overdiagnosed. If cocci are frequently diagnosed in your in-house urine sediments, be cautious; stain precipitates can mimic cocci. The majority of urinary pathogens in dogs are gram negative rods. [7]

The finding of intracellular bacteria on cytology is a good rationale for antimicrobials. A gram stain, which is immediate and cheap, narrows the spectrum to gram positive, gram negative, or mixed. Gram stains are underutilized in companion animal practice. This simple technique is described at:http://www.life.umd.edu/classroom/bsci424/LabMaterialsMethods/GramStain.htm.

**Deciding about cultures**

Cultures are not, in the practical sense, mandatory for first time empirical treatment of many routine bacterial infections, to include acute contaminated wounds, carnasial tooth abscess, infectious tracheobronchitis, superficial pyoderma, cat bite abscess, or first time bacterial cystitis in an adult dog.

Cultures are mandatory for any second line antimicrobial treatment, to include lack of response to empirical treatment, relapse after treatment discontinuation, or waxing and waning signs. Examples of this include suspected bacterial bronchopneumonia, urinary tract infections, or non-responding pyoderma. Avoid antibiotic roulette in these cases! With recurrent urinary tract infections, serial cultures can be very helpful. Repeated culture of the same organism suggests inadequate clearance (e.g. poor compliance, uroliths, pyelonephritis, or accompanying prostatitis with inadequate drug penetration). Repeated culture of different organism suggests new (usually ascending) infection, due to an anatomic defect, vulvar fold pyoderma, or poor perineal hygiene in debilitated or obese animals.

Cultures and cytology are required for first time treatment of long-standing infections, such as fistulous tracts, possible chronic foreign bodies, or non-healing wounds, as well as for serious or life-threatening infections, such as pyothorax, suspected endocarditis, osteomyelitis, joint sepsis, renal failure with suspected pyelonephritis, or suspected sepsis. Cultures are also recommended for suspected hospital-acquired infections, since nosocomial bacteria may have multi-drug resistance patterns, [8] and hospitalized patients may be more susceptible to sepsis because of IV lines, urinary catheters,[9] invasive procedures, and immunocompromise.

**Practical culture techniques**

For aerobic culture, our microbiologist at UW-Madison recommends BBL CultureSwab Plus (Amies gel formulation without charcoal; available from Fisher Scientific). This system is not ideal for anaerobes in our hands (despite the label). Urine is ideally set up for culture within 15 to 30 minutes of collection, but this is impractical in many settings. Alternatively, a sterile syringe containing urine can be capped and refrigerated immediately, for up to 12 hours prior to culture. While some fastidious bacteria may not survive storage > one hour, this approach is adequate in most situations, and also allows quantitative culture by the laboratory. Fluids for culture should not be placed in heparin or EDTA.

For anaerobic (along with aerobic) cultures, which are recommended for bile, body fluid, and pus, transport media should be used. A.C.T. II agar tubes (Remel) contain a nonnutritive semi-solid medium to which you can add fluid, or, if using swabs, insert one swab each for aerobic and anaerobic culture. Organisms should be stable for 24 hours for both aerobic and anaerobic culture set up. Mycoplasma cultures, which are recommended for tracheal wash samples, [10] can be run from either BBL CultureSwab Plus (without charcoal) or A.C.T. II agar tubes.

If your available laboratory does not provide cultures at an affordable price or with a useful turnaround time, talk to the laboratory manager about your concerns. In-house cultures are also an option, and may not require an incubator. One author recommends inoculation of blood agar and McConkey plates with a sterile calibrated loop or pipette, followed by incubation eleven cm under a 60W bulb for 24 hours. [11] Positive plates are submitted for identification and susceptibility, while negative plates are discarded. This can reduce costs to the client and make cultures an affordable habit in your practice.

**Empirical first line regimens**

There are several principles to follow when using empirical antimicrobials. First, use the narrowest spectrum drug for the suspected organism. For example, choose amoxicillin or penicillin, instead of amoxicillin-clavulanate, for a cat abscess, and doxycycline, rather than a fluoroquinolone, for suspected Bordetella infectious tracheobronchitis. [12] Fluoroquinolones are over-used as first line antimicrobials in both human and veterinary medicine. Common reasons for misuse in humans include the wrong disease indication and the availability of narrower spectrum first line agents. [13]

Second, consider tissue penetration. It should be assumed that urinary tract infections in male dogs involve the prostate, and antimicrobials with good prostatic penetration, such as fluoroquinolones,[14] doxycycline, or potentiated sulfonamides, should be chosen. For bronchitis without pneumonia, drugs that achieve high concentrations in bronchial secretions should be prescribed, to include fluoroquinolones, doxycycline, azithromycin, or potentiated sulfonamides. For endophthalmitis or ocular trauma, azithromycin and fluoroquinolones show excellent ocular penetration. Beta lactams and aminoglycosides, which are relatively polar, have poor penetration into the prostate, eye, testes, or bronchial secretions.

Finally, treat for the shortest effective period possible. Shorter courses of appropriate antibiotics decrease the emergence of resistant organisms in humans, and there is a trend towards the use of shorter courses of antimicrobials, with equivalent efficacy compared to longer regimens, in human patients. [15] For example, acute sinusitis,[16] pneumonia,[17] or uncomplicated urinary tract infections [18] are treated effectively with only a 3 to 7-day course of antibiotics in humans. Bacterial otitis in children can be treated with a single dose of azithromycin (as effective as multiple dosing over 7 days), and for community-acquired pneumonia, treatment regimens are tailored to patient response, and are continued for only 2 to 3 days beyond resolution of fever.[15] In veterinary medicine, the recommended duration of treatment for many infections is entirely empirical, and there has been little work to determine whether the longer courses that are recommended in textbooks are actually necessary. Consider using these shorter regimens, with a follow-up recheck and culture one week after discontinuation. Shorter treatment regimens are less expensive for clients (allowing more resources for diagnostics and follow-up), and are associated with better compliance. Most importantly, shorter courses of antimicrobials lead to significantly less bacterial resistance. [15]

**Common isolates from bacterial infections** (* Cultures strongly recommended at first presentation)

| **Indication** | **Most common organisms** | **Empirical antimicrobial** |
|---|---|---|
| **Dogs** | | |
| Bacterial cystitis | E. coli (51%) [7] | Amoxicillin/clavulanate (female)<br>Fluoroquinolone (male) |
| Endocarditis* | Gram positives (51%; esp. Strep canis)[19]<br>Gram negatives (22%)<br>Bartonella (20%) | Cephalexin plus fluoroquinolone, awaiting culture and serologies |
| Hepatobiliary | 72% negative cultures (bile)<br>E. coli, gram positives, anaerobes [20] | Amoxicillin/clavulanate plus fluorquinolone |
| Joint sepsis* | Staph. sp. [21, 22] | Cephalexin |
| Osteomyelitis* | Staph. and Strep [23] | Cephalexin |
| Pneumonia | Young dogs: Bordetella, other gram negatives [24] | Doxycycline |
| Prostatitis | E. coli | Fluoroquinolone |
| Pyometra | E. coli [25] | Fluoroquinolone |
| Superficial pyoderma | Staph intermedius | Cephalexin |
| **Cats** | | |
| Abscess | Pasteurella, anaerobes [26] | Amoxicillin (95% efficacy)[26] |
| Hepatobiliary | 64% negative cultures (bile)[20]<br>Mixed gram positives, negatives, and anaerobes | Amoxicillin/clavulanate plus fluoroquinolone |

| | | |
|---|---|---|
| Pyelonephritis* | E.coli, Enterococcus [27] | Base on urine sediment |
| ax* | Anaerobes, Pasteurella [28] | Penicillin (awaiting culture) |

**Typically effective antimicrobials for different microorganisms**

| | |
|---|---|
| **Gram positive aerobes**<br><u>Commonly effective</u><br>Penicillin<br>Amoxicillin, ampicillin<br>Clindamycin (not Enterococcus)<br>Cephalexin (not Enterococcus)<br>Chloramphenicol<br><br><u>Typically ineffective</u><br>Metronidazole | **Beta-lactamase producing gram positive aerobes**<br><u>Commonly effective</u><br>Amoxicillin/clavulanate<br>Cephalexin<br>Clindamycin<br>Fluoroquinolones (Staph > Strep)<br>Potentiated sulfonamides<br><br><u>Typically ineffective</u><br>Penicillin<br>Amoxicillin, ampicillin<br>Metronidazole |
| **Gram negative aerobes**<br><u>Commonly effective</u><br>Fluoroquinolones<br>Aminoglycosides<br>Amoxicillin/clavulanate (57% E. coli sensitivity)[29]<br>Cephalexin<br>Chloramphenicol<br>Potentiated sulfonamides<br><br><u>Typically ineffective</u><br>Clindamycin<br>Azithromycin<br>Metronidazole | **Anaerobes**<br><u>Commonly effective</u><br>Metronidazole<br>Amoxicillin/clavulanate<br>Penicillin<br>Clindamycin<br>Azithromycin<br>Chloramphenicol<br>(Penicillin, if not beta lactamase producing)<br><br><u>Typically ineffective</u><br>Fluoroquinolones<br>Aminoglycosides<br>Cephalexin |
| **Spirochetes, rickettsiae, protozoa**<br><u>Borrelia</u><br>Doxycycline<br>Pencillin, ampicillin, amoxicillin<br>Azithromycin<br>Fluroquinolones (but $MIC_{90}$ > 2 μg/ml for cipro)[30]<br><br><u>Leptospira</u><br>Doxycycline<br>Pencillin, ampicillin, amoxicillin<br>Fluoroquinolones<br>Chloramphenicol (higher MICs than for others)<br><br><u>Rickettsiae</u><br>Doxycycline<br>Chloramphenicol (inferior to doxycyline in humans)[31]<br>Fluoroquinolones (RMSF and Anaplasma, not E. canis) [32]<br><br><u>Toxoplasma</u><br>Clindamycin<br>Azithromycin | **Other organisms**<br><u>Mycoplasma,</u><br>Doxycycline<br>Chloramphenicol<br>Clindamycin<br>Fluoroquinolones<br><br><u>Chlamydia</u><br>Doxycycline<br>Amoxicillin/clavulanate<br>Azithromycin (does not eliminate Chlamydia carrier) [33]<br><br><u>Bordetella</u><br>Doxycycline<br>Chloramphenicol<br><br><u>Bartonella</u><br>Doxycycline<br>Azithromycin<br>Fluoroquinolones |

**Initial concept for table courtesy of Dr. Jill Maddison, Royal Veterinary College, University of London**

## REFERENCES

[1.] Paul, M., et al., **Mayo Clin Proc,** 2005. 80(9): 1122-5. [2] Bailiff, N.L., et al., **Vet Clin Pathol**, 2008. 37(3): 317-22. [3.] Bailiff, N.L., et al., **J Vet Intern Med,** 2006.20(4): 850-5. [4] Ruch-Gallie, R.A., et al., **J Feline Med Surg**, 2008. 10(6): 542-50. [5.] Chase, H.P., et al., **Diabetes Care**, 2005. 28(5): 1101-6. [6.] Hart, P.A., et al., **South Med J,** 2008. 101(11): 1126-31. [7] Ball, K.R., et al., **Can Vet J,** 2008. 49(10): 985-90. [8.] Ogeer-Gyles, J., et al., **J Vet Emerg Crit Care**, 2006. 16(1): 1-18. [9.] Ogeer-Gyles, J., et al., **J Am Vet Med Assoc**, 2006. 229(10): 1584-90. [10.] Chandler, J.C., et al., **J Am Anim Hosp Assoc,** 2002. 38(2): 111-9. [11.] Bartges, J.W., **Vet Clin North Am Small Anim Pract,** 2004. 34(4): 923-33, vi. [12.] Speakman, A.J., et al., **Vet Microbiol,** 2000. 71(3-4): 193-200. [13.] Mean, M., et al., **Eur J Clin Microbiol Infect Dis,** 2006. 25(12): 757-63. [14.] Dorfman, M., et al., **Am J Vet Res,** 1995. 56(3): 386-90. [15.] Niederman, M.S., **Int J Antimicrob Agents,** 2005. 26 Suppl 3: S170-5. [16.] Falagas, M.E., et al., ***Br* J Clin Pharmacol**, 2008. [17.] Dimopoulos, G., et al., **Drugs,** 2008. 68(13): 1841-54. [18.] Lutters, M., et al., **Cochrane Database Syst Rev,** 2008(3): CD001535. [19.] Sykes, J.E., et al., **J Am Vet Med Assoc**, 2006. 228(11): 1723-34. [20.] Wagner, K., et al., **J Vet Intern Med,** 2007. 21(3): 417-424. [21.] Clements, D.N., et al., **J Small Anim Pract**, 2005. 46(4): 171-6. [22.] Marchevsky, A.M., et al., **Aust Vet J,** 1999. 77(4): 233-7. [23.] Caywood, D.D., et al., **J Am Vet Med Assoc**, 1978. 172(8): 943-6. [24.] Radhakrishnan, A., et al., **J Am Vet Med Assoc,** 2007. 230(10): 1493-7. [25.] Hagman, R., et al., **Vet Rec,** 2005. 157(7): 193-6. [26.] Roy, J., et al., **Can Vet J,** 2007. 48(6): 607-11. [27.] Litster, A., et al., **Vet Microbiol,** 2008. [28.] Walker, A.L., et al., **J Am Vet Med Assoc,** 2000. 216(3): 359-63. 29. Oluoch, A.O., et al., **J Am Vet Med Assoc,** 2001. 218(3): 381-4. [30.] Kraiczy, P., et al., **Antimicrob Agents Chemother,** 2001. 45(9): 2486-94. [31.] Buckingham, S.C., **Paediatr Drugs**, 2005. 7(3): 163-76. [32.] Branger, S., et al., **Antimicrob Agents Chemother,** 2004. 48(12): 4822-8. [33.] Owen, W.M., et al., **J Feline Med Surg,** 2003. 5(6): 305-11.

## THE PITFALLS OF CULTURE & SUSCEPTIBILTY TESTING

**Dawn Merton Boothe, DVM, PhD, DACVIM, DACVCP**
**Auburn, AL**

### SAMPLE COLLECTION

The pitfalls of C&S testing are many. Culture data is only as good as the sampling methods of collection; the importance of proper culture techniques can not be overemphasized (See information box). For skin wounds, the surface always contains commensals; normal flora causing background noise must be "filtered" out. Swabs are less than ideal cultures for a variety of reasons (Ford 2007). Only 3 out of 100 CFU will actually make it to the culture stage. For anaerobes in particular, air between the fibers inhibits growth. Despite the greater level of difficulty in sample acquisition, tissue is none-the-less the preferred sample. This might be an aspirate of fluids, or macerated tissues. The risk of improper anaerobic collection is greater than that for aerobic collection and the absence of anaerobes may simply reflect improper techniques. An anaerobic infection should be suspect if clinical signs are supportive (foul smell, adjacent to mucosal membranes, gas, etc). Note that facultative anaerobes may be cultured aerobically, and demonstrated to be susceptible to drugs under aerobic conditions only to fail to respond because the infection in the patient is taking place in anaerobic conditions.

### CLUES FROM THE CULTURE GROWTH

Even a properly collected culture may not confirm infection or identify the infecting microbe. The chances of proper identification are greatest if vibrant growth is obtained in an otherwise sterile environment. However, for tissues characterized by a normal flora, culture may not be able to discriminate colonization from infection by normal opportunistic organisms that have become pathogenic. The culture may give some indications. For example, several organisms, such as *Bacillus* sp and *Corynebacteriu* are common contaminants and their presence in wounds may be indicative of contamination and thus, potentially, is a poorly representative sample. The same may be true for a non-hemolytic *Staphylococcus*. The location of culture may also be important. For example, whereas beta hemolytic *Streptococcus sp* collected from a wound may be important, it is a likely contaminant if cultured from the ear. The number of organisms may be helpful in identifying the cause-effect of microbial presence and infection. Isolation of multiple organisms may represent floral colonization rather than a polymicrobial infection. For example, *Pasteurella* as one of several organisms collected from a nasal swab may not be relevant, but if cultured as a pure isolate, is probably indicative of infection. In the presence of polymicrobial infections, those isolates characterized by light growth might be de-emphasized in favor of organisms with significant growth; indeed, controlling the latter might facilitate control of the former. If multiple organisms are cultured, and the culture was improperly collected, cleansing of the site (if possible) and reculture may facilitate correct identification of the pathogen. If the culture was properly collected, then the isolate with the heaviest growth might be initially targeted, with reculture indicated after therapy to assure the correct organism was targeted. In contrast to polymicrobial growth, pure growth is generally indicative of infection and the potential need for therapy. Another key indicator of infection is the intensity of growth. For countable tissues, the number of CFU per ml of tissue should be considered when assessing whether or not the inoculum represents an infection. Thus, cultures yielding successful growth only after incubation in enriched nutrient broth might be indicative of colonization rather than infection, with notable exceptions including blood cultures, cerebral spinal fluid and others. For those tissues, any non-contaminating growth is considered indicative of infection for such tissues. For example, $>10^5$CFU must be cultured from site that is easily contaminated (eg, wounds, clean catch urines) before the growth is considered an inoculum. On the other hand, only two colonies of *Pseucomonas* cultured from a properly collected bronchial alveolar lavage might be significant. Cytological examination of the sight can be revealing if phagocytized organisms are present. The susceptibility pattern may also help confirm infection: complex patterns of resistance are more indicative of infection compared to colonization. However, it should not be assumed that an isolate susceptible to most drugs is not a pathogen.

The veterinary clinical microbiologist can be a powerful ally in determining the significance of isolates yielded from a sample culture. A cytologist provided with a well collected cytological sample (ie, a slide) may be able to provide relevant preliminary results that may guied initial antimicrobial therapy. The microbiologist that is trained in veterinary medicine will be of most benefit in providing guidance regarding the relevance of the cultured microbe. However, the effectiveness of the clinical microbiologist will be markedly curtailed if an insufficient history is provided regarding the patient from whom the sample was collected.

### CLUELESS WITHOUT CONTROL

As an *in vitro* test, eventually the data generated with C&S testing must be applied to *in vivo* conditions. Even with a properly collected and submitted sample, the methods of *in vitro* testing offer many opportunities for mistakes that might lead to incorrect interpretation or inappropriate patient application. The sophisticated nature of C&S procedures mandates standardization and marked attention to quality control. The Committee on Laboratory Standards Institute (CLSI; previously the National Committee for Clinical Laboratory Standards [NCCLS]) validates methods protocols, guidelines and interpretive standards for C&S testing; one of its subcommittees addresses veterinary –specific standards for both animal approved, and when data is available, human-approved antimicrobials. These standards and guidelines are applicable throughout the nation and are promulgated only after careful and exhaustive review of pharmacodynamic (microbial response to drug) and pharmacokinetic (host handling of drug) data regarding each specific drug and microbe. Just as microbial populations are dynamic, standards and guidelines addressing their culture and susceptibility are dynamic. CLSI re-examines intermittently new data as it emerges and adjusts criteria accordingly. In this manner, microbial response to drugs across time, or new pharmacokinetics can be incorporated into the standards. CLSI publishes its findings

(M31-A2, M100-S16) such that clinical microbiological laboratories can access and implement the standards. An important caveat to C&S testing is that no mechanisms exist to assure that clinical micribiological laboratories that serve the veterinary profession adhere to these standards. Some laboratories generate their own standards. Yet, i is only CLSI standards that undergo national peer review assessment and discussion among unbiased experts representing government, industry, academia and practitioner. CLSI has generated guidelines for a variety of C&S testing methods. The pharmacodynamic information varies with the susceptibility procedures, with disk diffusion (Figure 8-4) and broth dilution (Figure 8-5) offering excellent examples of contrasts in advantages. It is the latter that provides the minimum inhibitory concentrations (MIC) necessary for comparison among drugs and desing of dosing regimens. Both methods require rapid growth of organisms; however, broth dilution procedures are particularly dependent on rapid growth and may not be available for some organisms. The data generated from culture and susceptibility testing represents the "pharmacodynamic" portion of PK-PD integration in that in indicates "what is needed" to target the microbe.

Because of the inherent risks of inaccuracy associated with C&S testing, results yielded from procedures that are not based on CLSI standards should be interpreted with caution. Aspects subject to varability include pH, cation content and osmolality of the media, inoculum size, media volume, temperature and duration of incubation, humidity, and for broth dilution, the method of observing growth (Amsterdam 2005). Accordingly, in practice culture methods should be considered less than ideal unless CLSI protocols are followed. Further, "preliminary data", quick "snap" tests or other methods intended to generate rapid results must be interpreted cautiously; the role of the organisms in causing infection and the susceptibility of the organisms (unless identified as a multi-drug resistant microbe) may require full C&S testing.

**FROM BENCHTOP TO BEDSIDE**

Among the pitfalls of C&S testing are the step-wise dilutions and the range of concentrations tested for each drug. Although at lower concentrations, the clinical relevance of differences between wells (ie, 0.25 versus 0.5 mcg/ml) is not significant, at higher concentrations, the difference between 8 and 16 or 128 and 256 μg/ml are substantial and do not lend themselves well to precise dose design. This lack of precision reflects, in part, the nature of testing. It also limits the range of drugs tested. While automated C&S techniques increase accuracy, precision, and ease of C&S testing, as well as reduce the time to results automation also limits the number of drugs and dilutions to be tested. Ideally, concentrations tested by broth dilution procedures should span a range that approximates the range of MICs established in a sample population of isolates of the organisms, thus allowing the clinician to assess how susceptible that isolate is compared to the rest of the population. Ideally, concentrations tested should exceed by at least one dilution, the highest range of concentrations to be expected in biologic fluids such that the MIC of the infecting microbe can be considered in the context of what is achieved in the patient (Amsterdam, 2005). However, the limitations of automated systems result in a very narrow range being tested, and these concentrations are often close to the threshold of susceptibility (ie, the resistant MIC breakpoint) established by CLSI. As such, these methods are more indicative of "resistance" rather than susceptibility. In human medicine, a third testing system has been approved by the FDA that circumvents this problem of narrow testing ranges. The "E test®" (Epsilon-test®) combines the simplistic advantages of disk diffusion with the informative nature of broth dilution. The E- test ® system (www.abbiodisk.com ) is available only in a limited number of veterinary clinical microbiology laboratories, primarily those located at academic institutions. The system is comprised of a strip that contains increasing concentrations of the drug of interest (one drug per strip) which, when placed on an agar plate that has been inoculated and incubated with the infecting organism (again, using CLSI based standards), is released at a logarithmic rate. The result is a tear drop zone of inhibition is yielded, with the wide zone of the tear drop indicating the higher concentrations of the drug and the point of the tear drop the MIC of the drug (Figure 8-6). The size of the test strip limits the number of drugs that can be tested per large plate (3 strips for a large plate, one per small plate), which contributes to the cost of the system increasing the cost of the system. However, because the drugs can be chosen for each patient, the system lends itself well to expanded susceptibility testing in the presence of a multi-drug resistant isolate. Although E-tests® are tedious and expensive, the system offers a distinct advantage to the narrow range of testing of broth dilution procedures. The wider range is allows identification of a very susceptible isolate, and, if the MIC is slightly higher than the CLSI threshold of susceptibility, provide support of using a drug at a higher dose. In contrast, broth dilution (or disk diffusion) methods would simply identify this isolate as "R". E-testing® also addresses a second disadvantage of broth dilution procedures in that the concentrations tested using the E-strip® are more precise, allowing more accurate design of the dosing regimen (Figure 8-6). In general, MIC generated by the E-test ® correlate well with MIC generated from broth dilution procedures

**MISCELLANEOUS CAVEATS**

1. CLSI does not provide interpretive standards for all drugs; as such, these drugs do not appear on C&S testing, and $MIC_{BP}$ are not available. This may be a decision on the part of the manufacturer of the drug to not pursue CLSI validation, the lack of adequate data for CLSI to determine criteria, or failure of the data to correlation with patient response. For such drugs, population pharmacodynamic and pharmacokinetic data might be used to help assess antimicrobial efficacy (see later discussion). 2. For some drug classes, CLSI has established criteria for a model drug which serves to reflect patterns of susceptibility to other members in the same class. In some instances, cross susceptibility and resistance justify this approach (e.g., fluorinated quinolones might represent all veterinary fluorinated quinolones, ampicillin accurately predicts amoxicillin, amoxicillin-clavulanic acid accurately represents ampicillin-sulbactam and sulfamethoxazole / trimethoprim appears to predict other "potentiated" sulfonamides). However, exceptions to the relevance of model drugs to other members in the class occur. Generally, CLSI has indicated the exceptions in its interpretive guidelines and the clinical microbiological laboratory should indicate these exceptions in the C&S report. For example, cephalothin (which is no longer available) represents first generation cephalosporins, yet cefazolin generally is less effective against

*Staphylococcus aureus* and more effective against *E coli* than is cephalothin (cephalexin). The spectrum of third and fourth generational cephalosporins is markedly disparate and thus the class can not be well represented by a model drug. For such drugs, comparison of MIC 90 and Cmax might be a reasonable assessment of potential use. Among the newer aminoglycosides, gentamicin is more effective than tobramycin against Serratia spp., whereas, tobramycin and amikacin are more effective than gentamicin against *Pseudomonas aeruginosa*.3. For selected drugs and organisms combinations, in vitro behavior does not manifest itself in the patient. For example, facultative anaerobes sampled from an anaerobic environment but cultured aerobically may respond better in vitro compared to in vitro, particularly to aminoglycosides. *Enterococcus* sp are often susceptible *in vitro* to aminoglycosides, but respond poorly *in vivo*. Another example is provided by extended spectrum beta-lactamases (ESBL) that destroy selected 3rd and 4th generation cephalosporins. These enzymes require special (expensive) methods of detection that many laboratories have yet to incorporate in testing procedure. 5. For any C&S method, generally only parent drug is included in the interpretive standards, yet an active metabolite may contribute markedly to activity. For some drugs (eg, ceftiofur), interpretive criteria include the metabolite, but for others, activity of the metabolite is not addressed. For example, most animals metabolize enrofloxacin to its de-ethylated metabolite, ciprofloxacin. 6. As organisms are exposed to microbes, MIC increase across time. CLSI should re-evaluate and adjust interpretive criteria to address these changes. However, new criteria depend upon the generation of new data. Current antimicrobial resistance surveillance systems focus on human medicine and thus largely address food animals (eg, NARMS ). Thus, the lack of new data needed by CLSI to promulgate new guidelines may prevent timely reassessment. Clearly, there is a need for a coordinated surveillance system for monitoring antimicrobial resistance of companion animal pathogens. Additionally, the relevance of population pharmacodynamic data provided on labels and through scientific literature will decline with the passage of time and caution might be taken when basing the use of a drug on population data that is more than a decade old. 7. Ideally, as MIC change, drug dosing regimens also should change. Because changes in also impact $C_{max}$, CLSI guidelines in $MIC_{BP}$ should change if the labeled dose of an antimicrobial changes. However, manufacturers may not pursue new interpretive standards or providers of test kits may be slow to incorporate the new CLSI guidelines into their products. Enrofloxacin offers an example. Originally approved at 2.5 mg/kg, the accompanying $MIC_{BP}$ was $\geq 2$ μg/ml. The approved dose is "flexible", ranging as high as 20 mg/kg which will result in a $C_{max}$ of 4 μg/ml. In response to the dose change, Bayer Animal Health pursued new interpretive standards and accordingly, CLSI has determined the appropriate $MIC_{BP}$ to be $\geq 4$ μg/ml. However, laboratories or manufacturers of C&S materials have been slow to incorporate the new standards into their interpretation. 8. Ideally, both pharmacokinetic and pharmacodynamic data upon which CLSI bases $MIC_{BP}$ should be collected from and promulgated for the target species to be treated. However, much of this data simply does not exist for animals. For drugs approved for use in animals, assuming the manufacturer suppled the data, interpretive standards often do exist and are published separately (M016) from those established for human medicine. However, depending on the amount of pharmacokinetic and pharmacodynamic data available for review, CLSI also has established interpretive standards for selected human drugs for animals. However, for other drugs, human interpretive standards are used.

**TABLE 1.** CULTURE SAMPLING TECHNIQUES

1. Commonalities regardless of site
   a. Site preparation: Don sterile gloves. . Thoroughly cleanse wound by removing excessive debris, flushing with saline, blotting with sterile gauze. Change to sterile gloves prior to collection. Do not culture purulent or necrotic debris
   b. Tissue aspiration and body fluids: Clean intact skin with antiseptic (eg, 70% alcohol and 10% povidone-iodine). Allow to air dry (do not fan). Expel air from a 10 ml syringe to which is attached a 22 ga needle. Insert needle into intact skin at the deepest portion of the lesion. Aspirate approximately 0.5 ml fluid. Needle can be moved back and forth at different angles in skin. Remove needle with hemostat. Discard needle and recap syringe with blood-gas cap. c. Swab technique are acceptable only for eyes, ears and uterine cultures. Note that only 3/100 CFU collected in a swab are likely to be successfully cultured. Use a swab in appropriate carrier media. The swab should be moistened with sterile preservative free solution if wound is not moist. Sample should be collected without touching the edge of the wound or skin. Rotate swab over 1 cm area of open wound for 4 seconds. Place swab aseptically in transport sleeve making sure tip contacts liqud transport media (break ampule if present). One swab should be collected for each sample type (ie, aerobic, anaerobic, and a third if cytology is of interest).
2. Skin or wound biopsy (swab strongly discouraged; aspirate acceptable, macerated tissue preferred): Clean intact skin with antiseptic and allow to air dry (do not fan). Collect biopsy aseptically. Place in transport tube containing liquid media. Clinical microbiology Laboratory will macerate.
3. Bone: place in transport system. Moisten with sterile physiologic saline as necessary.
4. Drain tube site: Treat as a contaminated wound. Care must be taken to not culture the biofilm associated with the foreign body. The tube should be removed, the site surgically cleaned and flushed, and the area cultured . Ideally, tissue will be collected at the presumed site of infection.
5. Respiratory tract: Bronchoscopy specimens include bronchoalveolar lavage, bronchial washing, bronchial brushing, and transbronchial biopsy specimens The bronchscope should be passed transorally in nonintubated patients or via the endotracheal tube in intubated patients. Bronchial wash or bronchoalveolar lavage specimens should be obtained before brushing or biopsy to minimize blood in the recovered fluid.
6. Urine: Samples should be collected by cystocentesis only. Catheterized samples generally not preferred. Sample can be submitted in a red-top serum collection tube. Samples should be kept cold, submitted on ice such that samples are received by the laboratory within 24 to 26 hrs of submission.
7. Blood culture, CSF or Joint Fluid: Liquid media is indicated. Volume is critical; anticipate only one CFU is needed for a patient to be bacteremic. Sterile prep the collection site, and for blood cultures, at least 3 collections at 3 different times are indicated. Collection during a fever spike is recommended. For blood (but not CSF or joint fluid), the volume should be 1 part blood to 10 parts broth.

8. Ocular: One or 2 drops of topical anesthetic are generally instilled. Organisms are more readily detected in conjuctival scrapings than from a swab.
9. Gastrointestinal tract: Fecal specimens are submitted primarily for the detection of *Campylobacter, Shigella,* and *Salmonella* species, *Clostridium difficile* and in certain cases to detect *Yersinia, Vibrio,* and *Aeromonas* species and enterohemorrhagic *Escherichia coli.* Care should be taken to make sure sample is not contaminated with urine. Fecal WBC should be ordered on liquid stools to indicate degree of inflammation. Stool specimens should be mixed with transport medium to maintain viability of pathogens which may be present. Sample should be collected digitally wearing a sterile glove or using a sterile fecal loop.S tool specimens should be mixed with transport medium such as Cary Blair to maintain viability of pathogens which may be present and submitted to laboratory within 36 hours of collection. Sample should be collected digitally wearing a glove or using a clean fecal loop.

**REFERENCES**

[1] The information in this proceedings including this table is From Boothe DM. **Small Animal Clinical Pharmacology and Therapeutics**. WB Saunders/Elsevier (In Press)

## ANTIMICROBIAL DOSING REGIMENS:A DYNAMIC CHALLENGE

**Dawn Merton Boothe DVM, PhD, DACVIM, DACVCP**
**Auburn University**

Dosing regimens for antimicrobials exemplify the integration of pharmacokinetics (what the body does to the drug) and pharmacodynamics (what the drug does to the body). For antimicrobial therapy, the "body" is the microbe. Integration is based upon what is needed to achieve the pharmacodynamic response – in this case, the minimum inhibitory concentration of the drug of interest for the infecting microbe – and comparing it with what will be achieved at the chosen dose. For this discussion, we will build a dose around a desired pharmacodynamic index (PDI). If the MIC of the infecting is not known, then the MIC 90 is a reasonable surrogate. The MIC90 is the MIC at or below which 90% of the isolates is an sample population of the organism is inhibited.

**Relationship between MIC, Plasma and tissue Drug Concentrations**

The parameters that are most predictive of antimicrobial efficacy and lack of resistance are the ratio of $C_{max}$ / MIC, the area under the inhibitory curve (AUC/MIC); and the percent time that PDC are above the MIC [T> MIC]. Based on these relationships, two generally categories of drugs have been described. Exceeding the efficacy targets decreases resistance. Dosing regimens can be designed based on population statistics, using MIC 90 (eg, packaged inserts) as a surrogate indicator of what is needed, or using MIC data from a culture report. The design of the dosing regimen depends upon the drug and its relationship between plasma drug concentrations, the MIC of the infecting organisms and whether or not the drug has a substantial post antibiotic effect (PAE). The PAE refers to the continued inhibition of microbial growth after a short exposure of the organisms to the drug. The impact is particularly profound for concentration-dependent drugs, and allows some drugs to be administered at long dosing intervals. The PAE may be absent for some organisms or some patients (e.g., some immunocompromised patients). In general, the PAE of concentration- dependent drugs is increased as the Cmax/MIC increases. Time versus Concentration Dependent Drugs. The relationship between MIC and the magnitude and time course of PDC allows drugs to be categorized as to either concentration-dependent (sometimes referred to as dose dependent) or time-dependent; these definitions are supported by primarily by in vitro but also in vivo studies. Concentration dependent drugs, best represented by the fluoroquinolones and aminoglycosides, are characterized by efficacy best predicted by the magnitude of plasma drug concentration ($C_{max}$ ) compared to the MIC of the infecting organism. For such drugs, the magnitude of the ratio generally should be 10 to 12, but ideally is higher for more difficult infections (eg, Pseudomonas aeruginosa, or infections caused by multiple organisms. The duration that PDC is above the MIC is not as important; in fact, efficacy may be enhanced by a drug-free period (i.e., a long interval between doses). For concentration dependent drugs, a dose that is too low is particularly detrimental. As such, concentration-dependent drugs generally can be administered at longer intervals, ie, once a day. Package inserts can be used to demonstrate the design of a dose for a FQ. Using the marbofloxacn MIC 90 for E. coli (0.06 mcg/ml) or Proteus mirabilus (o.125), the target concentration would be 0.06 X 10 or 0.6 mcg/ml and 1.25 mcg/ml, respectively. A dose of 2.75 mg/kg yields a Cmax of 2.0 mcg/ml, or a ratio of 2.0/0.06 or 33 for E coli and 16 for Proteus. As such, theoretically the low dose would be appropriate for treatment of both. For Staph intermedius, with an MIC 90 of 0.25 mcg/ml, 2.5 mcg/ml is the target. The higher dose of 5.5 mg/kg would be more prudent. For organisms with an MIC of 0.5 mcg/ml, the target of 5 mcg/ml could not be reached at 5.5 mg/kg. However, for the fluorinated quinolones (FQ), efficacy also is predicted in vitro by AUC/MIC: a ratio of < 60 renders the drugs bacteriostatic, whereas > 125 results in (slow) killing but also decreases the risk of resistance and > 250 causing more rapid bacterial killing. Thus, resistance might be less likely to develop for FQ characterized by longer half-lives (or for ENR, by the production of an active metabolite). Twice daily administration of an FQ might be indicated for organisms already characterized by low level resistance (see MPC below); however, the once daily dose should be given twice daily in such situations. The use of a second drug in combination with the FQ might also be considered for isolates whose MIC are sufficiently high that a Cmax/MIC >10 is difficult to achieve. Note that difloxacin might be appropriate at the low dose to treat E coli but not Proteus; for the latter, a plasma drug concentration of 18 mcg/ml would be necessary. This is not achievable at reasonable doses.

**Pharmacokinetic and Pharmacodynamic data for Marbofloxacin and Difloxacin**

| Marbofloxacin | Dog | Dog | Cat |
|---|---|---|---|
| Dose (mg/kg) | 2.5 | 5.5 | 5.5 |
| Cmax | 2 | 4.2 | 4.5 |
| AUC (24) | 31 | 64 | 79 |
| t1/2 | 11 | 11 | 13 |
| | n | MIC 50 | MIC 90 |
| S intermedius | 135 | 0.25 | 0.25 |
| E coli | 61 | 0.03 | 0.06 |
| Proteus | 35 | 0.06 | 0.125 |
| P. multocida | 13 | 0.015 | 0.06 |
| E faecalis | 11 | 2 | 2 |
| P.aeruginosus | 7 | range | 0.25-1 |
| **Difloxacin** | **Dog** | **Dog** | |
| Dose (mg/kg) | 5 | 10 | |
| Cmax | 1.8 | 2.8 | |
| AUC (24) | 14.5 | | |
| t1/2 | 9.3 | 11 | |
| | n | MIC 50 | MIC 90 |
| Staphylocccous sp | 193 | 0.23 | 0.46 |
| E. coli | 28 | <0.05 | 0.11 |
| Proteus | 15 | 0.92 | 1.83 |
| Pasteurella | 8 | ≤0.05 | ≤0.05 |
| P.aeruginosus | 5 | 0.11 | 0.92 |

**Time dependent drugs**

In contrast to concentration dependent drugs, efficacy of time-dependent drugs (eg, β-lactams) is enhanced if PDC remain above the MIC for the majority (50 to 70%) of the dosing interval; efficacy is best predicted by percent time that PDC are above the MIC [T> MIC]. For such drugs, simply achieving the MIC (Cmax/MIC =1) is insufficient because PDC (and certainly tissue concentrations) fall below the MIC immediately. With time-dependent drugs, generally a Cmax/MIC of 4 is a good starting point because it assures 2 half-lives will lapse before T=MIC. Two more half-lives can then be added to the dosing interval before the next dose must be given if T>MIC 50%. However, while this sounds like a long time, for amoxicillin and cephalexin, with a half-life of about 1.5 hr, the dosing interval can only be 6 hrs if Cmax/MIC = 4. For example, the MIC 90 for *Staph intermedius* and **amoxicillin-clavulanic acid** is < 0.5 mcg/ml. Cmax of 5.5 mcg/ml will be achieved at the labeled dose of amoxicillin-clavulanic. The duration of the dosing interval with this dose depends on the number of half-lives that can lapse as drug concentrations decline to the MIC. In one half-life, plasma drug concentrations will be 2.75 mcg/ml; in two half-lives, 1.35 and in three half-lives, 0.65, which is just above the target. The half-life of amoxicillin is at best 1.5 hrs resulting in 4.5 hrs of T>MIC. The dosing interval can be twice this long. Thus, the next dose should be administered at 9 hrs (8 hr). To reach a 12 hr dosing interval, 3 more hours, or two half-lives are needed. T>MIC is needed for one more half-life; thus, the dose needs to be doubled. Thus, to treat an organism with an MIC of 0.5 mcg/ml with amoxicillin-clavulanic acid, a dose of 13.5 mg/kg every 8 hrs, or 27 mg/kg every 12 hrs must be given. The dose would need to be further modified for drug, microbial and host factors. *Staphylococcus intermedius* is characterized by a low MIC; if we repeated the process for *Staph. aureus*, its MIC 90 is 4 mcg/ml. At 13.5 mg/kg, drug concentrations will reach the MIC before one half-life lapses. Even for E. coli, with an MIC90 for amoxicillin-clavulanic acid at 1 mcg/ml, a dose of 26 mg/kg every 8 hrs is the minimum that should be considered. The process can be repeated for **cephalexin**, with a half-life of 1.3 hr. At 25 mg/kg PO, 15 mcg/ml is achieved. The MIC 90 for *Staphylococcus intermedius* is 2 mcg/ml; a dose of 25 mg/kg achieves 15 mcg/ml. The amount of time that can lapse can be calculated as follows: the Cmax/MIC (15/2)= 7.5. This equates to essentially 8, or 3 half-lives (2*2*2). Thus, 4 hours can lapse during T>MIC; a dosing interval of 8 hrs is indicated. (To check: 15 mcg/ml > 7.5 > 3.85 > 1.9 mcg/ml = 3 half-lives). For *Staphylococcus aureus,* the MIC90 is 8 mcg/ml; the Cmax/MIC = 15/8 = essentially 2. One half-life can lapse during T>MIC; the dosing interval can be 2 half-lives, or 3 hours long. Increasing the dose to treat St. aureus at a convenient dosing interval is not practical. For E coli, with an MIC 90 of 16 mcg/ml, not even one half-life can lapse. Cephalexin should not be used to treat E coli. In general, for time dependent drugs, especially if the half-life is short, adding an additional dose is more cost effective than increasing the dose. This is in contrast for time dependent drugs that have a long half-life. For example, once daily dosing may be appropriate for cefpodoxime, depending on the organism. According to the package insert, the MIC 90 for both Staph intermedius and E coli is 0.5 mcg/ml. The Cmax at 10 mg/kg achieves 16 mcg/ml. The number of half-lives that can lapse is 16/0.5 = 32 = 5 half-lives (2X2X2X2X2), or 16 > 8 > 4 > 2 > 1 > 0.5. The half-life of **cefpodoxime** is 4.5 hrs, thus the dosing interval can be 25 hr X 2, or (theoretically) every 2 days. However, the variability in drug concentrations is marked, and prudence suggests that a 24 hour dosing regimen, as is indicated on the label, is appropriate. For Staph aureus, the MIC 90 is 2 mcg/ml. Three half-lives can lapse during T>MIC; 6 half-lives or essentially a day can lapse before the next dose. However, because this facilitates efficacy, but not necessarily avoids resistance, and because these calculations assume all drug in plasma makes it to the site of infection, a 12 hour dosing interval might be more prudent.

**5.5 yr Male Lab**
**Chronic wound**
***Pseudomonas aeurginosa***

| Antibiotics | pseaer (1) |
|---|---|
| Amikacin | 16 S |
| Amoxicillin/CA | >=32 R |
| Ampicillin | >=32 R |
| Carbenicillin | 64 S |
| Cefotaxime | |
| Cefotetan | |
| Cefoxitin | |
| Ceftazidime | >=32 R |
| Ceftiofur (Vet.) | >=8 R |
| Ceftizoxime | |
| Cephalothin | >=32 R |
| Chloramphenicol | 16 I |
| Ciprofloxacin | <=0.5 S |
| Enrofloxacin (Vet. | 1 I |

**6Y M Aust Shep Osteomyelitis with fracture at surgical Site. External fixation of tibia**

| | |
|---|---|
| Amikacin | <u><4</u> |
| AmoxiClav | R |
| Ampicillin | R |
| Cefazolin | R |
| Cefoxitin | R* |
| Ceftiofur | R* |
| Cephalothin | R |
| Chlormaph | S (8) |
| Clindamycin | R |
| Enrofloxacin | R |
| Erythromycin | R |
| Gentamicin | S (4) |
| Marbofloxacin | R |
| Orbifloxacin | R |
| Penicillin | R |
| Rifampin | S(≤1) |
| Tetracycline | R |
| Trimethoprim: | R |

**Methicillin resistant *Staphylococcus aureus***

For **cefovecin,** with a 133 hr half-life (due to protein binding which slowly releases the drug), for each 2X Cmax/MIC, 6 days can lapse (assuming time dependency is valid for periods beyond 24 hrs). However, the Cmax must be based on unbound, not bound drug. The Cmax of unbound drug in dogs approximates 4.0 mcg/ml. The MIC 90 for Staph intermedius is 0.25 mcg/ml; approximately 4 half-lives can lapse during T>MIC; approximately 8 half-lives (40 days) can lapse before the next dose is given. However, if the target organism is Staph aureus, with an MIC of 2 mcg/ml, T>MIC for only one half-life and dosing should occur (if indicated) in one week or less. Constant rate infusion or slow release products might be ideal for time dependent drugs with short-half-lives in the critical patient. Slow release products might be considered for time dependent drugs; however, the dose must be designed to assure that the MIC is achieved for the older slow release products because MIC have changed through the years. Azithromycin is another example of a drug with a very long half-life (72 hr) because of tissue distribution and accumulation. Although the drug can be administered at 2 day intervals. depending on the target MIC, note that the drug may not reach steady-state concentrations for 6 to 14 days. Indeed, care must be taken to remember that the maximum effect of any drug with a long half-life will not be achieved for 3 to 5 half-lives and a loading dose might be indicated for such drugs. Finally, some time dependent drugs have a very long half-life.

**Designing a dosing regimen based on MIC from a culture report**

Based in the *Pseudomonas aeruginosa*, the dose for amikacin should be sufficient to achieve Cmax/MIC = 10 or 10 X 16 mcg/ml = 160 mcg/ml. Table 1 indicates that 22 mg/kg achieved 64 mcg/ml in the blood stream. One could calculated the dose needed to achieve 160 mcg/ml based on either a proportion of : (160 mcg/ml)/(64 mcg/ml) = 2.5 X 22 mg/kg = 55 mg/kg; or one could calclulate the dose based on target (160 mcg/ml) X Vd = 37 mg/kg. Either calculation is likely to result in the same conclusion: the drug may not be safet at this high dose, even with once daily dosing, and the addition of a second drug is indicated. For enrofloxacin, the target is 1 mcg/ml * 10 = 10 mcg/ml. At 20 mg/kg, enrofloxacin achieves 4 mcg/ml and its active metabolite ciprofloxacin, achieves 2.9 mcg/ml for a total bioactivity of approximately 7 mcg/ml. Interestingly, despite an "I" designation, enrofloxacin comes closer to achieving the target Cmax/MIC of 10 (7 is achieved). For enrofloxacin, a second 20 mg/kg dose could be added. However, the combination of enrofloxacin and amikacin would be a wiser choice. For the MRSA, note that chloramphenicol, despite an "S" designation, requires an MIC of 8 mcg/ml. The dosing table indicates that 55 mg/kg PO will achieve a Cmax of 10 mcg/ml. Thus, not even one half-life can lapse before T>MIC is reached. A dose of 80 mg/kg would achieve approximately 16 mcg/ml, which would allow one half-live of T>MIC, or a 2 half-life dosing interval. The reported half-life is variable; we will us an average of 4 hours. Thus, if the dog could tolerate it (unlikely), a dose of 80 mg/kg every 8 hours would be indicated. However, this would only result in bacteristatic concentrations. The combination with rifampin would at least increase the changes of therapeutic success.

The previous demonstrations have been based on the assumption that all drug in plasma makes it to the site of infection. However, a variety of host, drug and microbial factors should cause the dosing regimen to be modified even further.

**Host Factors**

The impact of host response to infection can be profound. Problems contributing to therapeutic failure include immunocompromise (design a dosing regimen that will assure bactericidal concentrations of the chosen drug reach the site of infection), inflammatory response (debride or otherwise appropriate clean/drain accessible infections, select a drug that distributes into tissues well and ideally accumulates in phagocytes and increase the dose appropriately). Interpretation of C&S is based on the assumption that the MIC should be achieved in plasma. Basing MIC interpretation on plasma drug concentrations (PDC) might result in over or under estimation of drug efficacy. For tissues which concentrate the drug (or if the drug can be applied topically), and for drugs which can be concentrated by phagocytes and thus transported to the site of infection, concentrations may markedly exceed PDC, resulting in underestimation of efficacy for several reasons. Much of the data for water soluble drugs (volume of distribution [Vd] generally < 0.3 L/kg) suggests antimicrobial concentrations may be 30% or less of PDC in some tissues, particularly those characterized as sanctuaries, ie, non-fenetrated capillaries. In humans, recommended doses of beta-lactams drugs (water soluble) are increased 5 to 10 fold when treating infections of the central nervous system. Even tissues traditionally considered "well perfused" might be of concern. For example, drugs do not penetrate bronchial secretions well, despite the fact that the lungs are well perfused. Amoxicillin is often used to treat respiratory tract infections. Yet, only 30% of the amoxicillin that is in plasma is distributed to bronchial secretions. Theoretically, one must dose amoxicillin 3X the recommended dose to achieve targeted PDC in bronchial secretions. Most water soluble drugs (beta-lactams and aminoglycosides) reach only 20 to 25% of PDC in bronchial secretions whereas over 50% of lipid soluble drugs reach bronchial secretions. Dosing adjustments also are necessary for those infections that are intracellular or complicated by host response to infection. In the presence of marked inflammation, use of a drug that accumulates in phagocytes (eg FQs, macrolides, lincosamides) is likely to increase distribution of the drug to the site.

**Microbial Factors**

Microbial resistance is addressed in another manuscript in this same proceedings. Materials released from microbes facilitate invasion, impair cellular phagocytosis, and damage host tissues. Most staphylococci associated with canine pyoderma produce "slime," a material that facilitates bacterial adhesion to cells. Soluble mediators released by organisms (hemolysin, epidermolytic toxin, leukocidin) may damage host tissues or alter host response. Staphylococcal organisms contain protein A, which impairs antibody response, activates complement, and causes chemotaxis. Nocardia stimulates the formation of calcium-containing "sulfur granules" that impair drug penetration to the organisms Pseudomonas and other gram-negative organisms produce a glycocalix, or biofilm, that protects the organism.. Biofilms are microcolonies of pathogenic and host microbes embedded in a polysaccharide matrix ("slime" or "glycocalyx") produced by the bacteria; dental plaque is the prototypic example. Normal microflora of the skin or mucous membranes in the biofilm are lost with shedding of the skin surface or by the excretion of mucus; new cells and mucus are rapidly colonized by biofilm forming bacteria. Translocation of the normal microflora to otherwise sterile tissues (which can be facilitated by the presence of foreign bodies) may lead to acute infections (again, associated with biofilm) and accompanying inflammatory response. Persistent, chronic bacterial infections may reflect biofilm producing bacteria; persistant inflammation associated with immune complexs contributes to clinical signs. Unfortunately, bacteria growing in biofilms more easily resist antimicrobial killing and immune defenses of the host. In addition to debridement or other methods of cleansing should facilitate antimicrobial penetration; dose modification (increase) may be indcated to compensate for debris. Attention to PDC is important not only for efficacy, but also in order to reduce the risk of resistance. For drugs in which resistance emerges as a result of point mutations, dosing regimens should be designed to target the MPC. The mutant prevention concentration (MPC) is defined as the highest MIC identified in a population ($\geq 10^7$) infecting the patient (see sister manuscript). The MPC, rather than the MIC, should be the targeted concentration of drug at the site of infection if resistance is to be avoided. Unfortunately, determining the MPC of an isolate cultured from a patient requires culture techniques based on $\geq 10^7$ organisms, which currently is not possible. In one study, the ratio of MPC to MIC for various FQs toward human pathogens ranged from a low of 6 to 10 for E. coli (ATCC 8739) but 23 to 50 and as high as 125 for selected drugs toward Staphylococcus aureus (ATCC 6538).

**Drug Factors**

In addition to drug characteristics previously addressed (eg, concentration versus time dependent, static versus cidal, drug distribution), pharmaceutical manufacturers have been able to manipulate antimicrobial drugs in a variety of ways such that efficacy and thus bacterial killing is enhanced such that resistance might be reduced. For example, efficacy has been decreased by synthesizing smaller molecules that can penetrate smaller porins (e.g., the extended spectrum penicillins ticarcillin and piperacillin); "protecting" the antibiotic (e.g., with clavulanic acid, which "draws" the attention of the β-lactamase away from the penicillin); modifying the compound so that it is more difficult to destroy (e.g., amikacin, which is a larger and more difficult to reach molecule than gentamicin); and developing lipid-soluble compounds that are more able to achieve effective concentrations at the site of infection (e.g., doxycycline compared with other tetracyclines). However, with each innovative approach to reducing resistance, microbes are able to circumvent the drug in a disconcertingly short time. The use of pro-biotics or pre-biotics to minimize emergence of resistance in the gastrointestinal tract is controversial and requires additional scientific evidence.

**Duration of dosing**

Increasingly, in an effort to reduce antimicrobial resistance, the duration of dosing in human medicine is being limited. Durations of 5 days or less are recommended for non-complicated infections. High and/or frequent dosing is intended to result in rapid kill. For slow grow ing organisms, or infections complicated by poor local immunity or prolonged healing, longer durations are indicated. Note that pulse dosing at high, frequent doses intermittently might be preferred to long term dosing at lower doses. The former should be approached such that mutants are killed. In such cases, recurrent infection might be caused by organisms that are not resistant.

| Drug | Vd (l/kg) | Half-Life (hr) | Dose (mg/kg)[1]/ Cmax (μg/ml) |
|---|---|---|---|
| Amikacin | 0.23 (D) 0.18-0.38** 0.14 § | 1 [D]§ 0.8-1.3 [C]§ | 20 (IV)/65 14/ 10 (IM)/14 10(SC, D)/14 § 5/16.4±2; 18.5 [IM, C]§ |
| Amoxicillin | 0.2 (D) | 1-1.5 | 20 (IV)/13 11 (PO)/6 11 (SC, D)/7 10 (SC, C)/7 |
| Amoxicillin with clavulanic acid (Vree 2003)+ | 0.2 | 1-1.5 +1.5 amoxcillin ; 0.71 clavulanic acid | 20 (IV)/13 11 (PO)/6 11 (SC,D)/7 15 (SC, C)/10 +16.7/4.2 (PO, D)/11.4 amoxcillin ; 2.06 clavulanic acid |
| Ampicillin | 0.2-0.4 [D] 0.12§-0.22 [C] | 0.5-1.5 0.2 [D]§ 1.25 [C]§ | 20 (IV)/50 12 (SC)/14 6.6 (SC)/7 30 (PO)/10 10 (PO)/3 |
| Azithromycin (Hunter 1995)+ | 12 (Vdss, D) 23 (Vdss, C) | 29 (D)§ 35 (C)§ | 24 (IV, D)/4.2 (F=0.97) 5 (PO, C)/0.97 |
| Cefazolin[1st] | 0.3-0.7 | 0.75-1.4 | 15 (IV)/45 30 (IV)/90 15 (SC)/25 30 (SC)/50 |
| Cefotaxime[3rd] | 0.48 [D] 0.4 [D] (Vdss) 0.18§ [C] | 0.75-0.8§ [D] 1 [C]§ | 50 (IV,IM, D)*/41 50 (IM, D)/47§ 50 (SC, D)/30§ |
| Cefoxitin[2nd] | 0.32 [D]§ | 0.7-1.3 [D]§ | 60 (IV)/20 30 (IV)/10 30 (SC)/20 10 (SC)/15 |
| Cefpodoxime | 0.15 [D] | 5.6 (PO, D) 4.7 (IV, D) | 10 (PO, D )/16 7 (PO, D) 78.2 |
| Doxycyline (90% eliminated in bile | 0.93 (VDss§)-1.5 (D) | 7-10 (D)§ 4.6 (C)§ | 5 (PO, D)/5 2.5 (PO, D/3) 5 (PO, C)/6 2.5 (PO, C)/3 |
| Enrofloxacin +(Boeckh 200) | 2.6 3.7 -7 (D, Vdss)§ 4 (C, Vdss) | 0.92 (2.5)' 2.02 (5)' 2.4/3.9 (at 5 mg/kg,D) + )+ | 2.5 (PO, D)/1 5 (PO, D)/1.6-2 10 (PO,D)/1.7 (1.2)** 20(PO, D)4.2 (1.9)** 2.5 (PO, C)/1.3 5 (PO, C)/2.5 |
| Erythromycin | 2.7 (Vdss, D)§ | 1-1.5 1.7 (D)§ | 20 (PO)/3.5 |
| Florfenicol | 0.6 (C) | 9.2 (IM, D) < 1 (IV, D) 4 (IV, C) | 20 (IM, D)/1.64) 22 (IM, C)/20 22 (PO, C)/27 |
| Gentamicin | 0.23§-0.34 [D] 0.14-0.2 [C] | 0.87-1.36 [D] 1-1.36 [C] | 4 (IV)*/27 8 (IV)*/44 10 (IV, D)/28 10 (IV, C)/28 |
| Imipenem/cilas tin (Barker 2003) | 0.32 l/kg (D) | 0.83-0.92 (IM) 1.5 (SC) | 30 (IV)/180 10 (IV)/65 5 (IM)/13.2 (D) 5 (SC)/8.8 (D) |
| Marbofloxacin (Lefebvre 1998)+ | 1.2-1.37 [D]+ | 10.8-12.4 [D]§ + 11.5 (at 1 mg/kg, SC, D) | 2 (IV, D)/2.5 (extrapolated)+ 1 (PO, D)/0.83§ 2 (PO, D)/1.38§ 2 (PO, D for 8 [D])/1.4§ |
| Meropenem (Bidgood 2002) | 0.4 | 45 | 20 (IV) / 60 (extrapolated) and 24 in ICF 20 (SC)/26 (plasma) and 11 ICF |
| Metronidazole | 0.95 (D)§ | 4.3 4.5(D)§ | 44 (IV)/60 44(PO,D)/42 |
| Orbafloxacin | | 4.5 -5.2 (C)§ 5.4 – 5.6 (D)§ | 2.5 (PO, C)/2§ 2.5 (PO, D)/1.4-2.3§ |

# ANTIMICROBIAL DRUG OPTIONS FOR DIFFICULT INFECTIONS IN DOGS & CATS

**Mark G. Papich, DVM, MS, Professor of Clinical Pharmacology, Diplomate ACVCP**
**Raleigh, NC**

## INTRODUCTION

Guidelines for treatment of common infections in small animals has been reported, and many regimens established for effective therapy. Drug manufacturers also have produced several important drugs to treat the most common infections encountered in small animals. However, the drugs and approaches to therapy are more limited when the infection is more refractory, resistant, or is associated with another complicating factor. As with many novel treatments, there are few veterinary clinical studies to provide "evidenced-based" support. Many of the approaches discussed in this article are derived from sound pharmacologic principles and/or extrapolated from human medicine or laboratory animal studies.

## LONG-TERM CHRONIC THERAPY OR "PULSE THERAPY"

Some of the approaches to recurrent infections have been to either administer long-term, low dose antibiotics on a chronic regimen, or, to administer regular therapeutic doses intermittently (eg, once per week) as a "pulse therapy" on a long-term basis. This approach has been used to treat recurrent urinary tract and skin infections, especially in immunosuppressed patients. This approach has not been evaluated rigorously. In one preliminary report, the investigators used weekend therapy (two days per week at 15 mg/kg q12h orally) of cephalexin to prevent relapses of recurrent idiopathic superficial or deep pyoderma in dogs. This approach was effective for preventing relapses and did not produce resistance.

In one report, management strategies for persistent urinary tract infections and re-infections were evaluated in 100 dogs (Seguin et al, 2003). In that study, almost an equal number of dogs with persistent urinary tract infections had failure to clear the original infection as did those with new infections. Strategies included conventional antibiotic therapy, log-term low dose antibiotics, or administration of methenamine or urine acidifier. When correction of the underlying disorder was not possible, more dogs responded to low dose, long term antibiotic therapy than the other regimens. Methenamine was somewhat effective, but was only evaluated in a small number of animals. (Urinary antiseptics such as methenamine are discussed in more detail below.)

Other than these limited reports and anecdotal testimonials, these approaches for long-term treatment have not been rigorously evaluated for efficacy, long term safety, or effect on emergence of resistance.

## CONSTANT RATE INFUSIONS FOR ANTIBIOTICS

Antibacterial drugs can be classified by their action as *time-dependent* or *concentration-dependent*. If the drug is concentration-dependent, usually a once-daily regimen can be administered effectively as long as a sufficiently high dose is administered. When treating a pathogen for which emergence of resistance is a concern, a higher dose may be justified, but the interval remains the same (eg, once-a-day). A time-dependent antibiotic, on the other hand, does not produce better bacterial killing with higher concentrations. Activity is improved with longer exposure. Therefore, for these drugs, a long half-life is beneficial. If the drug has a short half-life (eg, a cephalosporin or penicillin drug) a constant rate infusion may improve therapeutic success. It can be shown that for cephalosporin antibiotics frequently administered intravenously, a constant rate infusion (CRI) delivered over 24 hours actually requires less total drug to maintain the drug concentrations above the MIC compared to intermittent doses. For a time-dependent antibiotic such as a β-lactam, maximizing the time above MIC (T>MIC) improves therapy. If the drug has a short half-life, the T>MIC can be maximized with a constant rate infusion (Nicolau, 2003). This effect has been more successful than intermittent therapy in febrile neutropenic patients, and the benefit greater with the more severe neutropenia. Continuous rate infusion can also increase the rate of response.

## THERAPEUTIC DRUG MONITORING

Therapeutic drug monitoring (TDM) is practical for aminoglycosides (primarily gentamicin and amikacin) and vancomycin. Although specialized laboratories can measure fluoroquinolones, chloramphenicol, and antifungal drugs, the method of measure (typically HPLC) is not practical for clinical laboratories. Nevertheless, monitoring of aminoglycosides can be helpful. Because the pharmacokinetic and pharmacodynamic (PK-PD) targets are known for these drugs, one can individualize therapy based on monitoring blood samples. Once-daily dosing regimens are the standard approach for both small animals and equine patients. Doses of aminoglycosides that result in peak plasma concentrations ($C_{max}$) approximately 8-10 times MIC confer optimal bactericidal effect ($C_{max}$/MIC ratio of 8 to 10).

Therapeutic drug monitoring (TDM) in animals receiving aminoglycoside treatment has been used to optimize therapy. If TDM can identify situations in which the dose can be lowered, there are also cost benefits. TDM can be useful because there may be a wide variation in drug disposition among sick patients, especially in critically ill neonates. Therefore, TDM can identify those patients with parameters that do not represent the average.

## STRATEGIES FOR URINARY TRACT INFECTIONS

Urinary tract infection represents a common infection in small animals, and can be recurrent and refractory in immunocompromised patients, including patients with diabetes mellitus and those receiving corticosteroid drugs (Torres et al, 2005; Hess et al, 2000; Kerr et al, 1999; McGuire, et al. 2002). The secondary UTI associated with diabetes and other underlying problems often are often occult (McGuire, et al, 2002). Clinical signs may be minimal or may be attributed to underlying disease or drug therapy, and white blood cells (WBC) and bacteria may be found in very minimal concentration in dilute urine. Treatment of complicated UTI must be based on urine culture and sensitivity, and even with this information it can be very difficult to eradicate infection (Sequin et al, 2003). After initiating antimicrobial therapy, follow-up urine culture should be obtained after approximately one week to be sure of antimicrobial efficacy in vivo. Culture should be repeated approximately one week after antibiotics are discontinued, and again several weeks later if immunocompromise continues.

**Use of Urinary Antiseptics**

For frequent recurrent UTI, once acute infection is resolved urinary antiseptics or daily antimicrobial administration has been used to prevent re-infection. However, the ideal urinary antiseptic has not been identified in veterinary medicine (Sequin et al 2003). Drugs used for this purpose include nitrofurantoin (Macrodantin) and methenamine (methenamine hippurate or methenamine mandelate). Nitrofurantoin has caused adverse gastrointestinal effects in animals. Methenamine is converted to formaldehyde—which is naturally antibacterial—in acidic urine. However, for the conversion to be effective the urine pH must be low, in the range of 5 to 6. Depending on the animal's diet and other medications, urinary pH consistently in this range may not occur in all animals.

Drugs or natural compounds that decrease the virulence of uropathogens, such as fosfomycin tromethamine (Monural), or cranberry juice also have been used in dogs with recurrent, or refractory, urinary tract infections. Cranberry juice (or its extract) contains a proanthocyanindin that inhibits the attachment of bacteria to the urinary mucosa (Howell et al, 2007). Its efficacy has not been reported in animals, but in women it had similar effectiveness as trimethoprim for prevent of recurrent urinary tract infections (McMurdo et al, 2009). Fosfomycin tromethamine can be antibacterial and also decrease virulence of uropathogens. In people it is used as a one-time administration for uncomplicated urinary tract infections in women (Monural 3-gram packet taken once with water). The efficacy has not been evaluated in animals.

Despite various treatment protocols for urinary tract infection, it should also be noted that it may not be necessary, or practical, to eradicate bacteria altogether as long as the animal remains asymptomatic (including absence of calculi or renal pelvic dilation). In such cases, a wait and watch approach may be most useful. In a review, one of the veterinary internists with expertise in urinary tract disease suggested that with refractory, resistant urinary infections the correct approach may be to discontinue antibiotic treatment if the patient is asymptomatic (Lulich 2007). He suggests that in some patients, the bacteria became less resistant, or replaced with bacteria that were more susceptible, over time (several months). This approach has also been followed in some human patients with asymptomatic bacteriuria (Andriole 2002; Harding et al. 2002).

**SELECTING MORE ACTIVE DRUGS**

This approach may seem self-evident if the problem is treating a resistant or refractory infection in animals. However, it is sometimes overlooked to simply select a more active drug. If a drug can be selected that has greater activity – and therefore lower MIC values – it is easier to achieve pharmacokinetic-pharmacodynamic (PK-PD) targets. The most simple approach is to select drugs with the lowest MIC values. Drugs may be technically "sensitive" on the basis of *in vitro* susceptibility tests (CLSI 2007), but can vary within the susceptible range. Strains with high MIC values near the breakpoint may have already acquired some resistance mechanisms (eg, a mutation). If other factors (eg, cost, route of administration, are similar), selecting a drug that has a lower MIC value ("best in class") is a preferred approach.

The principle of "best in class" for selecting effective antibiotics and limiting resistance development has been promoted in human medicine (Amyes et al, 2007). As this author points out, "...using less active drugs first has long been part of microbiology folklore." Instead, a principle of using highly active drugs during the initial treatment has been advocated. Using a more active drug initially will limit emergence of resistance, particularly if the drugs are bactericidal. To select for more active drugs requires difficult decisions. These drugs are almost always more expensive, and are valuable to human medicine. Nevertheless, some of the more active drugs used by veterinarians are briefly listed below:

**Carbapenems**

The carbapenems are β-lactam antibiotics that include imipenem-cilastatin sodium (Primaxin), meropenem (Merrem), and most recently, ertapenem (Invanz). Imipenem is administered with cilastatin to decrease renal tubular metabolism. Cilastatin does not affect the antibacterial activity. Imipenem has become a valuable antibiotic because it has a broad spectrum that includes many bacteria resistant to other drugs (Edwards & Betts, 2000). The high activity of imipenem is attributed to its stability against most of the β-lactamases (including ESBL) and ability to penetrate porin channels that usually exclude other drugs (Livermore 2001). The carbapenems are more rapidly bactericidal than the cephalosporins and less likely to induce release of endotoxin in an animal from gram-negative sepsis. Resistance to carbapenems has been extremely rare in veterinary medicine.

Meropenem, one of the newest of the carbapenem class of drugs has antibacterial activity approximately equal to, or greater than imipenem. Other characteristics are similar to imipenem. Its advantage over imipenem is that it is more soluble and can be administered in less fluid volume and more rapidly. Dosage regimens for dogs have been published (Bidgood & Papich, 2002).

**Aminoglycosides**

Aminoglycosides include gentamicin, amikacin, and tobramycin. They are active against most gram-negative bacilli, including *Pseudomonas aeruginosa*, and against many gram-positive cocci. Because they must be injected (or administered topically) and can be nephrotoxic, there has been a reluctance to rely on this group of drugs in veterinary medicine. However, they have a long history of effectiveness and can be used as monotherapy. Amikacin and tobramycin are more active than gentamicin, and resistance is less likely to these drugs. They have been administered once-daily for systemic infections IV, IM, or SC. Despite the risk of nephrotoxicosis, if used carefully with monitoring of renal clinicopathologic parameters or therapeutic drug monitoring (see above), they have been used for many days, and even weeks in some patients.

**New Fluoroquinolones**

Of the current, more familiar, fluroroquinolones, (eg, enrofloxacin, marbofloxacin, orbifloxacin, difloxacin) there are deficiencies in the spectrum of activity. They have less activity against gram-positive cocci, especially enterococci (*Enterococcus faecalis* and *Enterococcus faecium*), and anaerobic bacteria. The newest generations of fluoroquinolones (referred to by some authors as the 3rd-generation fluoroquinolones) include gatifloxacin and moxifloxacin. The new generation of fluoroquinolones, with substitutions at the C-8 position, (C-8 methoxy for example) have as their advantage a broader spectrum that includes anaerobic bacteria and gram-positive cocci. The difference in spectrum of activity is largely caused by increased activity against the DNA-gyrase of gram-positive bacteria, rather than activity against Topoisomerase IV, which is the target in gram-positive bacteria for the older quinolones (Pestova et al, 2000). Pradofloxacin (Veraflox), a veterinary 3rd generation quinolone has been evaluated in dogs and cats, but the experience thus far is limited to research abstracts and clinical reports (Stephan et al, 2008; Spindel et al, 2008; Hartmann et al, 2008; Litster et al, 2007; Mueller & Stephan, 2007; Ganiere, et al, 2005; Silley et al, 2007). Because it is active against two targets of fluoroquinolones (Topoisomerase IV and DNA gyrase) development of resistant mutants may be less likely (Wetzstein 2005). At the time of this writing (January 2009) pradofloxacin was not available. Therefore, a very similar drug used in humans, moxifloxacin has been used in clinical situations where pradofloxacin would be a rational choice. Moxifloxacin (Avelox) is a human drug of this group and has been used on a limited basis for treatment of infections in dogs and cats caused by bacteria that have been refractory to other drugs.

## REFERENCES

[1.] Amyes SGB, Walsh FM, Bradley JS. Best in class: a good principle for antibiotic usage to limit resistance development? **J Antimicrob Chemother** 59: 825-26, 2007.[2.] Andriole VT. Asymptomatic bacteriuria in patients with diabetes--enemy or innocent visitor? **N Eng J Med**. 2002;347:1617-1618.[3.] Barker CW, Zhang W, Sanchez S, et al. Pharmacokinetics of imipenem in dogs. **American Journal of Veterinary Research** 64: 694-699, 2003.[4.] Bidgood T, Papich MG. Plasma pharmacokinetics and tissue fluid concentrations of meropenem after intravenous and subcutaneous administration in dogs. **American Journal of Veterinary Research** . 2002;63(12):1622-1628.[5.] Carlotti DN, Jasmin P, Gardey L, & Sanquer A. Evalation of cephalexin intermittent therapy (weekend therapy) in the control of recurrent idiopathic pyoderma in dogs: a randomized, double-blinded, placebo-controlled study. **Veterinary Dermatology** 15 (Supplement 1): 8, 2004 [abstract 13].[6.] Clinical Laboratory Standards Institute. CLSI. **Performance Standards for Antimicrobial Disk and Dilution Susceptibility Tests for Bacteria Isolated From Animals; Approved Standard—Third Edition**. CLSI document M31-A3. Wayne, PA: Clinical and Laboratory Standards Institute; 2007.[7] Edwards JR, Betts MJ: Carbapenems: the pinnacle of the Beta-lactam antibiotics or room for improvement? **J Antimicrob Chemother** 2000; 45: 1-4.[8.] Ganiere JP, Medaille C, Mangion C. Antimicrobial drug susceptibility of Staphylococcus intermedius clinical isolates from canine pyoderma. **J Vet Med B Infect Dis Vet Public Health**. 2005 Feb;52(1):25-31. [9.] Harding GK, Zhanel GG, Nicolle LE, et al. Antimicrobial treatment in diabetic women with asymptomatic bacteriuria. **N Eng J Med.** 2002;347:1576-1583.[10] Hartmann A, Krebber R, Daube G, Hartmann K. Pharmacokinetics of pradofloxacin and doxycycline in serum, saliva, and tear fluid of cats after oral administration. **J Vet Pharmacol Ther.** 2008 Apr;31(2):87-94. [11.] Hartmann AD, Helps CR, Lappin MR, Werckenthin C, Hartmann K. Efficacy of pradofloxacin in cats with feline upper respiratory tract disease due to Chlamydophila felis or Mycoplasma infections. **J Vet Intern Med**. 2008 Jan-Feb;22(1):44-52. [12.] Hess RS, Saunders HM, Van Winkle TJ, et al. Concurrent disorders in dogs with diabetes mellitus: 221 cases (1993-1998). **J Am Vet Med Assoc**. 2000;217:1166-1173.[13.] Howell AB. Bioactive compounds in cranberries and their role in prevention of urinary tract infections. **Molec Nutr Food Res**. 2007;51:732-737.[14.] Kerr KG. The prophylaxis of bacterial infections in neutropenic patients. **J Antimicrob Chemother**. 1999;44:587-591.[15.] Litster A, Moss S, Honnery M, Rees B, Edingloh M, Trott D. Clinical efficacy and palatability of pradofloxacin 2. 5% oral suspension for the treatment of bacterial lower urinary tract infections in cats. **J Vet Intern Med**. 2007 Sep-Oct;21(5):990-5.[16.] Livermore DM. Of Pseudomonas, porins, pumps, and carbapenems. **J Antimicrob Chemother** 47: 247-250, 2001.[17.] Lulich J. Solving difficult urinary tract infections. **Compend Contin Educ Pract Vet** 2007;29(3A): 15-19. [18.] McGuire NC, Schulman R, Ridgway MD, et al. Detection of occult urinary tract infections in dogs with diabetes mellitus. **J Am Anim Hosp Assoc**. 2002;38:541-544.[19.] McMurdo MET, Argo I, Phillips G, Daly F & Davey P. Cranberry or trimethoprim for the prevention of recurrent urinary tract infections? A randomized controlled trial in older women. **J Antimicrobial Chemotherapy** 63: 389-395, 2009.[20.] Mueller RS, Stephan B. Pradofloxacin in the treatment of canine deep pyoderma: a multicentred, blinded, randomized parallel trial. **Vet Dermatol**. 2007 18(3):144-51.[21.] Nicolau DP. Optimizing outcomes with antimicrobial therapy through pharmacodynamic profiling. **J Infect Chemother** 9: 292-296, 2003.[22.] Periti P, Mazzei T. New criteria for selecting the proper antimicrobial chemotherapy for severe sepsis and septic shock. **Int J Antimicrob Agents** 1999; 12: 97-105.[23.] Pestova E, Millichap JJ, Noskin GA, et al. Intracellular targets of moxifloxacin: a comparison with other fluoroquinolones. **J Antimicrob Chemother** 45: 583-590, 2000.[24.] Seguin MA, Vaden SL, Altier C, Stone E, and Levine JF. Persistent urinary tract infections and reinfections in 100 dogs (1989-1999). **Journal of Veterinary Internal Medicine** 17: 622-631, 2003.[25.] Silley P, Stephan B, Greife HA, Pridmore A. Comparative activity of pradofloxacin against anaerobic bacteria isolated from dogs and cats. **J Antimicrob Chemother**. 2007 Nov;60(5):999-1003.[26.] Spindel ME, Veir JK, Radecki SV, Lappin MR. Evaluation of pradofloxacin for the treatment of feline rhinitis. **J Feline Med Surg**. 10: 472-479, 2008. [27.] Stephan B, Greife HA, Pridmore A, Silley P. Activity of pradofloxacin against Porphyromonas and Prevotella spp. Implicated in periodontal disease in dogs: susceptibility test data from a European multicenter study. **Antimicrob Agents Chemother**. 2008 Jun;52(6):2149-55. [28.] Torres SM, Diaz SF, Nogueira SA, et al. Frequency of urinary tract infection among dogs with pruritic disorders receiving long-term glucocorticoid treatment. **J Am Vet Med Assoc**. 2005;227:239-243.[29.] Wetzstein HG. Comparative mutant prevention concentrations of pradofloxacin and other veterinary fluoroquinolones indicate differing potentials in preventing selection of resistance. **Antimicrob Agents Chemother**. 2005 Oct;49(10):4166-73.

# COST OF A SUIT: NON-ECONOMIC DAMAGES DO YOU KNOW WHO THE OWNER IS?©

**Charlotte A. Lacroix DVM, JD.**[1]
**Flemington, NJ**

For the past several years, pet owners and animal care industries have been confronted with new legal principles, which challenge how animals have historically been viewed under the law. While it is no surprise that society's growing compassion for animals and the almost universal embracement of the human-animal bond have lead to a greater sensitivity of the needs of animals and changes in how the law protects animals, it by no means justifies a blanket change in the laws that govern the relationship between humans and non-humans.

Attempts by special interest groups and animal law rights advocates to categorically turn the current animal related laws on their head and apply laws that have formerly only applied to humans without a careful and detailed analysis of the pros and cons is imprudent. While most would agree, animals should receive greater protections under current laws and there should be more supervision to ensure animals are treated humanely, the recent proposals to change the status of owners to "guardians" and award non-economic damages akin to those awarded for the loss of a human life are excessive measures which create more problems than solutions. Such proposals appear innocuous on their surface but reflect monumental changes in the law akin to throwing the baby out with the bath water.

Two main questions which are currently challenging the veterinary profession, pet owners and other animal care givers include, 1) whether animal owners should be considered guardians under the law, and 2) whether the human animal bond has monetary value.

## GUARDIANSHIP

*Black's Law Dictionary* defines property right as "an aggregate of rights which are guaranteed and protected by the government . . . the highest right a man can have to anything."[i] The rights that humans have over non-humans have been well established for centuries and have reflected society's long-term relationship with animals. As this relationship has evolved, the public has become more sensitive to the needs of animals and this has been reflected in the increase in legislative activity on animal law issues in both the United States and Europe. Such legislation, has attempted and succeeded in some cases, to impose restrictions on owners' property rights and/or remove animals from being treated as property under law.

While the use of the word "property" offends many, and while it may seem appropriate to use a different term to refer to animals in day-to-day conversation, one must examine whether a change in the terminology as used in <u>legal contexts</u> would be wise in the long run. Some have argued that changing "owners" to "guardians" is just a semantic change with no legal significance. Such a statement is disturbing and wrong, as those familiar with legal arguments know full well that there is no such thing as "semantics" under the law. Lawyers routinely argue before judges on what was or was not meant by simple words and because guardianship laws are well founded and have universally adopted legal principles, one wonders why a change in terminology would not be interpreted in the same way courts have interpreted its meaning for decades.

Furthermore, while everyone is focusing on whether or not the term "guardian" has the same meaning as "owner", the debaters neglect to address whether the "guardian-owner" change in semantics extends to a semantic change between "animal" and "ward." Even if one were to argue that the "guardians" have the same meaning as "owners", no one has stated that "ward" has the same meaning as "animal" under current property laws. For example, the change in terminology that was adopted in Rhode Island in 2001, did not address whether a "ward" was created by the adoption of the term "guardian" in lieu of "owner."[ii] Perhaps this was left out intentionally for animal lawyers to establish the point at a later date, when representing pets in lawsuits against their "guardians" or any other defendant. To say that a parallel semantic change between "animal" and "ward" is implicit and self-evident is unpersuasive. The following analogy, demonstrates the faulty logic. If one it told that the glass if ½ full with water, it does not mean that the other ½ is automatically filled with air, it could be water too.

The laws that govern guardianship are extensive and well established and it is no coincidence that the term was selected as the substitute for "owner" in current animal related laws. Under current legal theory, guardians are obligated to care for their "wards" in a manner that subordinate's their personal interests. One can only speculate on how this relationship would be applied to animals, but the starting point likely would be the current long standing definition, that owners have a high duty to exercise care and diligence *viz a vie* their relationship with their animals. Some may argue that guardianship laws would not apply to the owner-animal scenarios because the change is limited to animal anti-cruelty statutes. Such a narrow view of the issue is akin to saying that the study of cardiovascular disease does not fall within the study of physiology because cardiovascular disease falls within cardiology not physiology.

While Rhode Island is the only state to have adopted such change in terminology, multiple municipalities and cities nationwide have made similar changes, including, Wanaque, NJ; Windsor, Ontario Canada; Berkley, CA; San Francisco, CA; West Hollywood, CA; Boulder, CO; Colorado Springs, CO; Sherwood, AR; Amherst, MA, St. Louis, MO; and Menomonee Falls, WI. While it is not yet known how the changes will be interpreted in the future, a complex series of questions invariably come to mind.

---

[1]Dr. Charlotte Lacroix owns and operates Veterinary Business Advisors, Inc., a specialized firm based in Flemington, New Jersey (908) 782-4426; www.veterinarybusinessadvisors.com. Dr. Lacroix assists veterinarians and their attorneys nationwide with their business related legal matters, including employment contracts, purchase and sale of veterinary practices, specialty practice issues, mediation and start-ups.

**NON-ECONOMIC DAMAGES**

Closely related to the issue of guardianship, is the issue of whether the intentional or negligent cause of an animal's death (or injury) should permit owners to obtain some form of monetary award to compensate them for their loss. Since veterinarians have been leaders in promoting the human-animal bond and since we understand, perhaps better than most, the strong relationships owners can have with their pets, it would seem appropriate to allow for some form of remedy when a pet has suffered from the wrongful acts of another. Before reacting too quickly in an attempt to do the right thing, however, the profession must carefully examine the current model of non-economic damages in the human context to determine if and when such remedies are awarded, and when they are not, why not.

Generally, in the case of human death or injury, courts award both economic and non-economic damages. Economic damages typically compensate victims for medical bills and the aggregate future lost income the injured or deceased persons would have earned had they not been harmed. Non-economic damages, which come in the form of "loss of compassion", "pain and suffering" and "emotional distress damages" can be awarded by courts in cases where the conduct that caused the death or injury was caused by the wrongdoer's intentional act, but also in some situations as a result of a tortfeasor's negligent acts, such as medical malpractice.

Courts are inconsistent when awarding non-economic damages for negligence, because, since negligent acts are based on one's failure to act within the standard of care and typically do not involve intentional misconduct, there is little deterrent effect. Non-economic damages are also only awarded to certain "classes" of plaintiffs. For example, if an automobile driver's negligence kills a person, none of the person's siblings or best friends are permitted to bring an action for non-economic damages. The plaintiffs eligible for such awards varies from state to state and in some states, parents cannot recover for the death of their child.

Since animals are considered personal property under the law, the amount of monetary damages courts have awarded owners has been based on the fair market value of the animal. This is the same long-standing rule that has been applied to the loss of other forms of personal property. This economic remedy historically has been low, unless the animal is a breeding or performance animal. Within the last decade there have been attempts at the legislative and court levels to change the law to allow for greater recovery for the loss suffered by those who have lost their animals to the intentional or negligent acts of others.

There only are a small number of states, including, Tennessee and Illinois that have adopted laws that allow for a non-economic remedy for owners who have had the animals cruelly injured or killed by violent perpetrators. Additionally, there are a few states, including, FL, KY, NJ, NY, PA and OR that allow owners to recover "non-economic" damages for the "intentional infliction of emotional distress" that the owners can prove they suffered from the death and/or injury of their animal as a result of a wrongdoer's reckless, extreme and outrageous misconduct. However, except for Tennessee, the author is unaware of any laws that permit non-economic damages as a remedy for negligent acts and most courts that have addressed the issue of whether "non-economic" damages are appropriate for actions based on "negligence" where there is no cruelty, recklessness or intentional misconduct, have concluded that such a remedy is inappropriate and not in the best interest of the public.

For example, in a September 2004 court decision that was based on a veterinary malpractice case where the veterinarian had been found negligent for having failed to timely diagnose a gastro-intestinal obstruction, the Superior Court of New Jersey concluded that based on "public policy reasons, non-economic damages for the negligent death of a pet should not be recognized as a matter of law."[iii] In arriving at its opinion, the Superior Court's decision referenced another court's findings:

> "We are particularly concerned that were such a claim to go forward, the law would proceed upon a course that had no just stopping point. Humans have an enormous capacity to form bonds with dogs, cats, birds and an infinite number of other beings that are non-human. Were we to recognize a claim for damages for the negligent loss of a dog, we can find little basis for rationally distinguishing other categories of animal companion . . . it is difficult to define with precision the limit of class of individuals who fit into the human companion category . . . [o]ur vast ability to form these bonds adds to the richness of life. However, . . the public policy concerns relating to identifying genuine claims of emotional distress, as well as charging tortfeasors with financial burdens that are fair, compel the conclusion that [limiting the liability to the human companion] suggested by Plaintiff will not definitively meet public policy concerns."[iv]

Whether the conclusion by most courts not to award non-economic damages for the negligent loss or injury of an animal, is the right thing to do, is a question that is currently being debated among our colleagues. However, when reviewing the current laws allowing such awards for loss of human lives, and considering that such awards are not available for the loss of a person's sibling, best friend, or same sex partner, it would be inequitable and troublesome to allow such a remedy for the loss of an animal.

**CONCLUSION**

The issues of whether pets should be treated as "children" under the law and whether "non-economic" damages should be awarded for negligent acts, are challenging veterinarians' ability to conduct their practices "as usual", and the ethical paradigm to which they have grown accustomed. The danger in responding to these issues lies not only in burying our heads in the sand and failing to recognize that social change is upon us, but, more importantly, reacting too quickly and making the wrong decision because of incorrect assumptions about the social change. Fortunately, the AVMA, AAHA and many state associations, including the PVMA, are looking at these issues, in order to help the industry do the right thing.

[i] *Black's Law Dictionary* 6th edition (1990 West Publishing Co.) pg 1216.

[ii] The 2001 Cruelty to Animals Law defines guardian as a person(s) having the same rights and responsibilities of an owner, and both terms shall be used interchangeably. A guardian shall also mean a person who possesses, has title to or an interest in, harbors or has control, custody or possession of an animal and who is responsible for an animal's safety and well-being

[iii] *Frampton v. Allenwood Animal Hospital*, (Docket No. A-2154-03T3)

[iv] *Rideau v. City of Racine*, 243 Wis.2d 486, 627 N.W.2d 795, 798-802 (2001).

**STANDARDS OF CARE: MEDICAL RECORDS©**

**Charlotte A. Lacroix DVM, JD.**[1]
**Flemington, NJ**

In face of these litigious times, veterinarians are seeking methods to limit their liability and transfer the responsibility of medical decision-making to their clients. The one method that has been ignored for far too long is the proper and complete documentation of patients' medical information and client communications. We know this to be true, because the large majority of sanctions imposed by veterinary state boards, are a result of medical record keeping violations. For example, of 51 disciplinary actions taken by the Florida board of Veterinary Medicine between Oct. 2001 and Oct. 2003, 53% involved a failure to maintain proper records.[1]

The medical record is an evidentiary document generated for the main purpose of communicating to others what was done, why it was done and if not done, why it was not done. The information within medical records must explain and substantiate a veterinarian's actions or omissions. By way of example, the Pennsylvania Practice Act under section 31.22 states this clearly,

> "Veterinary medical records serve as a basis for planning animal care and as a means of communication among members of the veterinary practice. The records furnish documentary evidence of the animal's illness, hospital care and treatment and serve as a basis for review, study and evaluation of the care and treatment rendered by the veterinarian . . ."

The shortcomings of most medical records is that they neither contain sufficient information to access the veterinarian's actions or inactions nor is the information presented is such a way that the reader can follow the cognitive reasoning of the veterinarian. Additionally, rarely do veterinarians record clients' refusals to follow their recommendations or clients' non-compliance with important treatments and follow-ups. And even if some of this information is documented, far too frequently, it is rendered meaningless because it is illegible.

**MEDICAL RECORD INFORMATION**

Medical records typically encompass the patient records of individual pets, but also include the business and legal documents of the hospital, such as, estimates, invoices, hospital handouts, financial records, and all logs. The amount and type of information recorded, should be consistent with the level and complexity of the services rendered. The first place to look as to what is required is in your state's practice act, which will stipulate what is required. Another good comprehensive resource is the American Animal Hospital Association (AAHA) guidelines for an accredited hospitals on the AAHA website.

State Practice Acts vary, but as a general rule, they require:

1. Separate record for each patient
2. Accuracy – the vet is responsible for this, even if use dictation
3. Complete – normals, abnormals, AND not examined
4. Legible – can dictate, doesn't have to be your handwriting
5. Unalterable – set lock-up time and don't alter original after
6. Easily accessible/retrievable
7. Retained for 3-5 years

Generally, The most common information veterinarians fail to include in the medical record is communications with clients, staff and colleagues regarding the care of the patient. This is likely because when the communication occurs, the medical record is nowhere to be found and most practices are just not sufficiently well organized to make locating the record any easier than finding a needle in a haystack. Fortunately, advances in computer and intercom technology are helping reduce this administrative nightmare. As to information that should not be recorded, under no circumstances should medical records include derogatory statements about clients, patients or colleagues, as such comments invariably come back and crush the credibility of the author.

---

[1]Dr. Charlotte Lacroix owns and operates Veterinary Business Advisors, Inc., a specialized firm based in Flemington, New Jersey (908) 782-4426; www.veterinarybusinessadvisors.com. Dr. Lacroix assists veterinarians and their attorneys nationwide with their business related legal matters, including employment contracts, purchase and sale of veterinary practices, specialty practice issues, mediation and start-ups.

## PATIENT MEDICAL RECORD INFORMATION

1. Client admission forms – client (name, address, contact info) & patient ID (name, species, age, sex) – emergency contact person & contact info

2. Medical History & Treatment chart
   (1) Date & Time of hospital visit
   (2) Vaccination & Medical History, Chief Complaint
   (3) Physical Exam – abnormal & normal, not examined - weight
   (4) Diagnosis – master problem list w/diagnosis, w/ prognosis if applicable
   (5) Treatment – Medical & surgical, medications (dosages in mg, not ml) – include all options, let clients make treatment decisions and have sign off on them
   (6) Final assessment of patient
   (7) Home care instructions to clients

3. Client consents – include recommendations, requests contrary to recommendations, refusals/acceptance of recommendations

4. Reports/Communications
   (1) Radiographs, lab data, surgical and dental reports
   (2) Client education materials – preventative healthcare list/professional advice rendered
   (3) Client communications/correspondence
   (4) Colleague communications

5. Large Animal Specific
   (1) Conversation Logs, date & time & summary of discussion
   (2) Dates & dosages of all medications administered to animals, including route of administration & concentration of dosages, withdrawal periods
   (3) Instructions left with clients
   (4) Feed additives prescribed with dosages and withdrawal periods
   (5) Documentation that the client understands the requisite withdrawal periods
   (6) Procedures used to prevent transfer of zoonotic disease from one farm to another
   (7) ID of all treated animals – encourage producers to permanently ID
   (8) Recommendations to clients of potential contamination risks – observations & conversations

**CONFIDENTIALITY AND RELEASE OF MEDICAL INFORMATION:**

Protecting personal information of clients is an ethical duty for all veterinarians, and a legal duty in about 22 states. For example, Illinois Practice Act, 225 ILLCS 115/25/17, stipulates: "No veterinarian shall be required to disclose any information concerning the veterinarian's care of any animal except on written authorization or other waiver by the veterinarian's client or an appropriate court order or subpoena….When communicable disease laws, cruelty to animals laws, or laws providing for public health or safety are involved, this privilege is waived."

The 2002 revision of the AVMA Principles of Veterinary Medical Ethics under Principle II (L) states, "Veterinarians and their associates should protect the personal privacy of patients and clients. Veterinarians should not reveal confidences unless required to by law or unless it becomes necessary to protect the health and welfare of other individuals or animals." Even in those states which do not have explicit prohibitions on divulging client confidential information, veterinarians should still exercise caution before revealing client information to a third party. Always attempt to obtain the client consent and in cases of emergency such as when the health of a person or pet is at risk, be sure there is a paper trail, which justifies the disclosure of the confidential information.

To maintain confidentiality of medical records, veterinarians must establish specific procedures for the release of information and adhere to them. Generally, it is preferable to send copies of medical records directly to follow-up veterinarians. The use of a release form simplifies this process.

## REQUEST FOR RELEASE OF MEDICAL RECORDS

To: **[Insert practice name and address of sender]**

I, the owner of or agent for the animal(s) named **[insert pet(s) name(s)]**, hereby request that copies or summaries of the medical records of these animals be released to **[Insert recipient practice name, doctor name, address, telephone and fax]**

Owner's signature: ________________ Date:________________

Payment of $ _______ is provided/enclosed for you to photocopy and mail this information as directed.

**CONCLUSION**

Relative to other professionals, veterinarians as a whole have had few state board complaints and negligence lawsuits filed against them. This has led to an industry wide problem of complacency with respect to medical record keeping practices. Unfortunately, with the rising interest in animal law related issues, veterinary professions will need to adopt practices that ensure their medical records are complete, accurate and easily accessible.

**REFERENCES**

[1.] **JAVMA** Vol 198; #1; 1/1/91; pg 67 [2.] **JAVMA** Vol 197; #1; 7/1/90; pg 48 [3.] **JAVMA** Vol 178; #3; 2/1/81; pg 202 [4.] **JAVMA** Vol 172; #2; 1/15/78; pg 176

---

[i] Florida Dept of Business and Professional Regulation, Board of Veterinary Medicine. Disciplinary Cases. Newsletter (Spring 2004) pp 4-7.

## COST OF A SUIT: WHAT IT TAKES DISCUSSION ON STANDARD OF CARE ©

**Charlotte A. Lacroix DVM, JD.**[i]
**Flemington, NJ 08822**

In the law of professional negligence, the standard of care is the benchmark by which others assess a veterinarian's competence. To be within the standard of care, veterinarians must perform their duties with an average degree of skill, care and diligence exercised by colleagues practicing under the same or similar circumstances. Unfortunately, this is a general rule and not always helpful when one is trying to determine whether or not to do something in a given situation. For example, when is it or isn't it necessary to refer a patient?

In general, compared to other professionals, veterinarians are minimally regulated. Aside from the state board of examiners, DEA and OSHA, few governmental agencies interfere with how we practice medicine. This is a good thing, because veterinarians can still exercise independent judgment. It is a bad thing, however, because it is not clear as to what and what is not the standard of practice. For this reason, authors of this manuscript have compiled a set of veterinary care standards for various exotic species. We believe that it is better such standards are articulated and published "within" and "for" the industry, rather than waiting around and having the lawyers and courts determine the standards one by one, each at the expense of a veterinarian's career.

There are primarily two areas of law that regulate the conduct of veterinarians and help ensure that veterinarians act prudently and reasonably in their dealings with clients and their animals. The first is the civil court system that adjudicates claims made by clients who allege that their veterinarians have acted negligently. The second is the state board of examiners which is an administrative office charged with enforcing a state's veterinary practice act which sets forth laws with which veterinarians must comply to obtain and maintain their veterinary licenses. In performing their daily clinical duties, veterinarians should be cognizant of these two areas of law since they represent the two principle avenues by which clients may have complaints addressed.

Receiving letters from the state board of examiners and or a disgruntled client's attorney can be very distressing, causing veterinarians to respond impulsively and not always in their best interests. This is especially the case with veterinarians who have been practicing for only a few years since they are not likely ever to have been named in a lawsuit or reprimanded by a regulatory agency. It is important for veterinarians to realize that how they initially respond to such allegations can have a significant impact on the outcome. For this reason it behooves us to become knowledgeable about the processes by which state boards and the courts adjudicate such allegations. The following scenario illustrates how these procedures work in real life.

**Martinique's Case**

Mrs. Bridges brings in Martinique, her 2 year-old female iguana, to Dr. Steel, a small animal practitioner who prides himself in being knowledgeable about exotics as he is a member of the Association of Avian Veterinarians. Mrs. Bridges informs Dr. Steel that Martinique has not eaten for 2 weeks and after performing a physical examination, Dr. Steel diagnoses "Egg Binding", a common condition in both birds and reptiles. Without advising Mrs. Bridges of his limited experience with reptiles, nor offering her a referral to a colleague who has more experience, Dr. Steel obtains Mrs. Bridges' consent to perform a spay surgery on Martinique. Dr. Steel has performed many spay surgeries on birds to treat "Egg Binding", and performs the same surgical procedure on Martinique whereby he removes the shell glands without removing the ovaries. Martinique is sent home and recovers well from surgery after 3 weeks of antibiotics and hand feeding.

Next spring, Martinique goes off her food again, but this time, she is lethargic and has a markedly swollen abdomen. Mrs. Bridges is concerned this is a recurrence of the same problem and brings Martinique to a new vet in town, who advertises herself as an exotic veterinarian with expertise in reptiles, birds and pocket pets. Dr. Zoo examines Martinique, performs an x-ray and ultrasound, and diagnoses "Egg Binding." Mrs. Bridges is puzzled and frustrated as to why Martinique has the exact same condition, when the surgery performed by Dr. Steel, a year ago, should have permanently fixed the problem. Dr. Zoo, also is confused, and with Mrs. Bridges' consent obtains a copy of Martinique's medical records and contacts Dr. Steel.

The medical records are vague at best. The only notation relative to the surgery is "spayed, surgery routine, recovery uneventful." Dr. Steel confirms the information in the medical records and informs Dr. Zoo, in a patronizing manner, that his experience is that reptiles are just like birds, and their shell glands can be removed without removing their ovaries. In a defensive tone, he tells her that he has been performing these procedures on reptiles long before she was ever admitted to veterinary school.

Dr. Zoo responds to Dr. Steel by informing him of her residency training in exotics, experiences working at various zoos around the country, and tells him that reptiles are not at all like birds and must have their ovaries removed. She further explains that since the ovaries were not removed during the first surgery, Martinique has re-presented with the same symptoms because she has ovulated eggs, which are now in the coelomic cavity. In fact, she tells Dr. Steel that it is likely that Martinique was just pregnant a year ago and did not have "Egg Binding" condition as he had diagnosed and due to his misdiagnosis, Martinique will need a second surgery.

After speaking with Dr. Steel, Dr. Zoo informs Mrs. Bridges that Martinique will need a second surgery and discusses the procedure and fees involved. Mrs. Bridges, who is angry that she has to pay a second time for the exact same procedure, asks Dr. Zoo to discount her fees. While sympathetic and understanding, Dr. Zoo explains to Mrs. Bridges that a discount is not possible.

---

[i]Dr. Charlotte Lacroix owns and operates Veterinary Business Advisors, Inc., a specialized firm based in Flemington, New Jersey (908) 782-4426; www.veterinarybusinessadvisors.com. Dr. Lacroix assists veterinarians and their attorneys nationwide with their business related legal matters, including employment contracts, purchase and sale of veterinary practices, specialty practice issues, mediation and start-ups.

Frustrated, Mrs. Bridges leaves Dr. Zoo's hospital with Martinique. Later that night Martinique became weak, started mouth breathing and was unresponsive to touch. Mrs. Bridges rushed her pet to the local emergency clinic where the doctor on duty performed an emergency exploratory. Sadly, Martinique died during the recovery.

Three months later, Dr. Steel receives two letters; one from the State Veterinary Board of Examiners and another from Mrs. Bridges' attorney. The state board letter requests Dr. Steel to respond to Mrs. Bridges' assertions that Dr. Steel was negligent in (a) failing to inform her that the he was not experienced in treating reptiles; (b) failing to offer a referral to a veterinarian who was qualified and experienced in treating reptiles; (c) misdiagnosing Martinique's condition; (d) performing a surgical procedure below the standard of care; and (e) failure to maintain appropriate medical records. The correspondence from Mrs. Bridges' attorney includes a copy of a complaint filed with the state court alleging malpractice and a demand for $100,000, for economic and emotional distress damages.

What should Dr. Steel do?

**Responding to Allegations of Professional Malpractice**

How veterinarians address such accusations will in part depend on whether the allegations are in the form of a lawsuit, state board complaint, or both. Regardless of the form in which the allegation is made, the first step veterinarians should take is to carefully read the complaint and determine what is being requested of them and in what time frame. Once this information has been assimilated, they should gather the pertinent medical records and any other documentation relating to the services in question and write down in chronological order their recollection of the events.

In this case, the complaints allege that Dr. Steel performed a procedure for which he had insufficient knowledge, failed to refer the case and had poor medical records. The facts indicate that Dr. Steel examined Martinique, made a diagnosis and performed the spay to treat the "Egg Binding." Dr. Steel should carefully review the medical records to corroborate his recollection of the events. Unfortunately in this case, because the documentation is poor, it will be a scenario of Dr. Steel's word against Mrs. Bridges'. For example, it will be difficult for Dr. Steel to claim that he informed Mrs. Bridges of alternatives, including a referral, as there is no such notation in the records. Since Dr. Steel has a legal obligation to maintain medical records, the fact that he hasn't will imply that he also was careless with his medicine. As he reviews the records, Dr. Steel should write down the events that led to Mrs. Bridges' complaint. Most veterinarians will find this helpful since it will refresh their memories, help them develop a consistent "story" as to what happened, and provide a draft from which to develop a written response.

So as not to compromise his defense in the lawsuit, Dr. Steel should immediately upon receiving the complaint, contact his professional liability insurance carrier and ask for advice. However, if Dr. Steel suspected earlier that Mrs. Bridges was likely to pursue legal action, he should have contacted his insurance carrier at that time. Insurance carriers may differ in how they handle negligence actions, but usually require the defendant to fill out a claims form in which the veterinarian describes the circumstances that led to the claim. A claims representative then reviews the facts, makes a recommendation as to a course of action and may assign an attorney to the case if the complaint cannot be settled quickly. In this case, if Mrs. Bridges is offered a settlement and rejects it, it is likely an attorney would be assigned to defend Dr. Steel since in this case it appears that Dr. Steel's care was substandard in several respects.

In dealing with the letter from the state board, Dr. Steel should be aware that he will most likely be defending his conduct at his own expense, since professional liability insurance carriers generally do not provide coverage for state board actions (exception: AVMA-PLIT now offers a limited policy insuring against state board actions). While Dr. Steel may respond on his own, it is usually advisable for him to obtain legal advice as to how he should respond to the allegation(s) and at the very least have an attorney review his letter. In drafting his response, Dr. Steel should not underestimate the time and effort it will take to address all the issues in the complaint, in an organized and articulate manner. Responses that are, disorganized, incomplete and difficult to follow, often lead to further investigation by the board as opposed to an early dismissal of the charges. Additionally, Dr. Steel may find it helpful to consult with other veterinarians to determine whether they use inform their clients of alternatives and/or "wing it" on exotic patients. This will assist Dr. Steel in determining whether he acted within the standard of care and provide an indication as to his liability.

**Negligence**

The burning issues for Dr. Steel of course, are whether he was negligent in failing to (a) refer Martinique, (b) remove the ovaries and, (c) maintain proper medical records. Our courts and juries decide negligence on a case by case basis in light of the specific facts and circumstances of each situation, but, veterinarians should be aware of a few general principles. First and foremost, it is important to note that a veterinarian can be found negligent even if he or she did not intend to cause harm. Simply put, "I didn't mean to" is no defense to "you should have known better". A simple mistake can lead to liability.

Secondly, veterinarians can be found negligent even if the rest of their colleagues would have acted in the exact same way. Judges can determine that the entire industry is at fault if it is in the public's interest. Judge Leonard Hand, a famous judge once wrote in his opinion "[c]ourts must in the end say what is required; there are precautions so imperative that even their universal disregard will not excuse their omission."[ii] Hence it is a false security to rely on what the rest of your colleagues are doing.

To recover damages from a veterinarian based on negligence, a client must prove four elements by a preponderance of the evidence, meaning it is more likely then not that the veterinarian erred:

**Duty of Care**

Clients must show that their veterinarians "owed" them a duty of care to provide veterinary services of a certain standard.

[ii] The T.J. Hooper [60 F.2d 737 (2d Cir. 1932)]

This element is easy to prove, because courts almost always find that once a veterinarian has agreed to provide veterinary services, the veterinarian also has assumed the legal duty to take reasonable care in providing such services. In our scenario, Dr. Steel clearly owed Mrs. Bridges a duty to take reasonable care in providing veterinary services to Martinique.

**Breach of Standard of Care**

A duty to provide services within the standard of care is breached when veterinarians fail to meet the standard of care as established by the veterinary profession, that is, when they fail to act with the level of skill and learning commonly possessed by members of the profession in good standing. Mrs. Bridges probably will be able to prove breach of duty if her attorney can show that veterinarians routinely remove ovaries when spaying iguanas. Conversely, Dr. Steel will attempt to establish that he did not breach his duty of care, by showing that most general practitioners do not remove the ovaries and the patients do just fine. It is at this stage that expert witnesses are hired to testify as to what is the standard in the case at issue.

**Proximate Cause**

Clients must then prove that the veterinarian's failure to provide services within the standard of care "proximately" or "closely" caused the harm suffered by the clients. If the harm suffered by the client is not a result of the veterinarian's actions or omissions, it would be unfair to hold the veterinarian responsible. In our case, it is clear that Martinique's death was caused by Dr. Steel's initially performing the incorrect surgical procedure. Suppose, however, that Mrs. Bridges is suing Dr. Steel because Martinique died of renal failure a month after spay procedure. It will be a lot harder to prove that Martinique's death resulted from *anything* Dr. Steel did or failed to do during the procedure.

**Damages**

Even after they have proved negligence, clients also must establish that they suffered harm resulting from such negligence. Since animals are considered as property under the law and most state courts do not recognize loss of companionship, this harm is usually in the form of an economic loss. As a result veterinary malpractice awards are usually much lower than in human malpractice cases and clients usually only recover the fair market value of the animal, costs incurred for veterinary care, and loss of income or profits in cases where the use of the animal is lost. However, we are seeing more and more states entertain the possibility of awarding non-economic damages and this is likely to increase the scrutiny with which standards of care are evaluated as well as the number of lawsuits filed against veterinarians.

**Responding to Client Complaints**

Veterinarians often can avoid receiving letters from clients' attorneys and state boards by addressing client complaints long before client dissatisfaction leads to legal recourse. Clients often resort to litigation and or state board action when they believe their veterinarian either acted negligently or failed to respond appropriately to their concerns. When faced with a client complaint, veterinarians should consider the following:

1. Listen to the client.

a. Clients who have complaints are often angry and need the opportunity to "vent".

b. Veterinarians should show their clients that they are taking the matter seriously by listening carefully to what their clients have to say and taking notes of the conversation.

c. Do not interrupt the clients since this will only anger them further and likely interfere with a clear understanding of the facts.

2. Remain calm and objective.

a. Avoid becoming defensive and emotional, since this may inadvertently reinforce the client's belief that the veterinarian acted inappropriately with respect to the care of the client's pet.

b. A client's criticism of a veterinarian's actions, even when fully justified, does not necessarily mean that any negligence occurred. Veterinary medicine is an imperfect science and veterinarians are not omnipotent.

3. Communicate, communicate, communicate.

a. Many lawsuits are filed because veterinarians fail to adequately communicate with their clients. Often the client does not fully understand the diagnosis or proposed treatment and has unrealistic expectations as to the veterinarian's services and the respective outcome.

b. Veterinarians can enhance communication and reduce potential misunderstandings by 1) obtaining informed consents, 2) providing fee estimates, 3) encouraging questions, and 4) providing handouts explaining the contemplated services.

c. Veterinarians should use "plain English" when communicating to clients since medical jargon may not only confuse clients but also intimidate them, making them reluctant to ask important questions.

4. Show sympathy and concern.

a. Clients whose pets have died are often emotionally distraught and under certain circumstances may seek to blame someone, sometimes their veterinarian, for their pet's death. Veterinarians who are compassionate and attempt to comfort their clients are more likely to diffuse their client's perception that the veterinarian should be held accountable for their pets death.

b. Veterinarians should not hesitate to recommend grief counseling for clients who appear to have difficulty coping with the loss of their pet. Several veterinary schools have such hotlines, including, the University of California at Davis, University of Florida and Colorado State University.

5. Coach the staff.

a. Staff members can help diffuse client complaints and should be coached in what to do and say, if anything, when a client complains.

b. The staff should remain professional at all times and avoid "offensive - defensive" discussions with clients who may be less intimidated by staff members and therefore more hostile to the staff as compared to the veterinarian.

6. Do not admit fault or offer a settlement.

a. Veterinarians should avoid making apologetic statements or excuses and should not admit fault, since this would compromise their case in the event a lawsuit was later filed. Veterinarians with only a few years of experience are more likely to feel guilty and accountable for bad outcomes, even though there was no negligence. Remember that "feeling guilty" is NOT the same thing as "being guilty."

b. Veterinarians should not offer to settle a malpractice charge or agree to any settlement offered by the client without first contacting their insurance carrier and attorney since it may be interpreted as an admission of fault, thereby prejudicing their case. Under certain circumstances it may be appropriate to reduce the client's bill in an attempt to amicably and expeditiously resolve a dispute, but without admitting liability.

**Avoiding Client Complaints**

Just as an ounce of prevention is worth a pound of cure, the best practice to avoid being dragged into a lawsuit or state board investigation is to take measures to avoid client complaints. Even if successful, Dr. Steel will spend a lot of time, effort and money, defending himself in court and before the state board. In retrospect, it would have been far less costly and burdensome if Dr. Steel had informed Mrs. Bridges that he had limited experience with reptiles and Martinique should be referred to an exotic veterinarian with expertise in treating reptiles—at a minimum Dr. Steel should have done his research before performing surgery.

Veterinarians will save themselves a lot of grief if they periodically evaluate their practices to identify areas where preventive measures and procedures will help avoid complaints before they start. Additionally, veterinarians should regularly consult with the staff, their colleagues and perhaps even their insurance carrier to ensure that they are aware of the latest preventive measures adopted by other practitioners. Keeping abreast of developments in the legal liability field should be an integral part of any veterinarian's continuing professional education. Because people are people, there is no way to prevent client complaints entirely. But in this area like many others, ignorance is dangerous and a preventive attitude is the best approach.

**CONCLUSION**

Being accused of malpractice can be a disconcerting experience for any veterinarian, but especially for associates who have been in practice for only a few years. These allegations can come in the form of a civil law suit or state board action and require veterinarians' immediate attention so as not to compromise their defense. Preparing a defense against such allegations is facilitated by having knowledge of the law of negligence and an understanding of the adjudicatory process. Nonetheless, the best defense lies in addressing client complaints when they first arise by using honed listening and communication skills, keeping abreast of the standard of care within the industry and adopting preventative measures.

**SPECIALTY EMPLOYMENT RELATIONSHIPS PARTS I & II©**

**Charlotte A. Lacroix DVM, JD.**[i]
**Flemington, NJ**

The following is an overview of the importance of contracts and the common problems that arise when business relationships rely too much on verbal promises or past business practices and too little on clear written agreements. The main focus of the article will be on specialty employment contracts with some discussion of buy-sell agreements. Mistrust and disputes arise from a mis-match in expectations and thoroughly addressing legal issues before getting hitched in an employment relationship or partnership goes a long way in vaccinating against ugly fights and the nasty lawyers that come with them.

**WHAT IS A CONTRACT?**

A contract is a set of bargained for promises between two or more people, where one party promises to do X in exchange for another party's promise to do Y. Courts require that an enforceable promise meet certain conditions. For example, the parties must be of age (no minors), of sound mind, and not under duress; there must be no fraud or mutual mistake over an important aspect of the transaction, and the deal must not be so one-sided as to be "unconscionable."

**Consideration**

To distinguish binding promises from charity or gifts (you can't sue Santa Claus because he didn't give you enough presents last year), the law requires that the party to whom the promise is made give "consideration" for the promise in the form of a benefit to the promissor and/or detriment to the promisee. Thus, Dr. Specialist promises to work 35 hours per week in consideration for an annual salary of $110,000 (i.e., a benefit to Dr. Specialist and detriment to Dr. Owner ). Dr. Owner promises to pay such salary to Dr. Specialist in consideration for Dr. Specialist's labor (benefit to Dr. Owner and detriment to Dr. Specialist). Consideration exists for each promise, which is therefore enforceable.

**Avoid Oral Contracts**

Oral contracts generally are binding only if their performance lasts less than a year, because the law assumes that the parties' recollections of what was agreed to become unreliable over time, increasing the tendency to remember events in a self-serving way. Few disagreements are less productive than the "you promised X," "I don't remember X but you promised Y" litany. Prevent such wasteful bickering by always insisting on a written contract, regardless of its terms.

**Contract formation**

Legal theory provides that a contract is formed once an offer is accepted. Real life usually is a lot messier. An offer can be oral or written (e.g., employer advertisement in a professional journal, on a bulletin board or mailed to the applicant). Typically, the prospective employee will ask for clarification and wish to change the terms of the original offer by making a counter-offer. The employer counters such counter-offer with his own counter-counter-offer. This confusing and frustrating process continues until either the parties reach an agreement or, realizing they can't make a deal, go their separate ways.

Legally, the contract is formed as soon as the offer is accepted. This can be a trap for an impulsive party who accepts an offer, but who later (like Columbo) asks for "just one more thing." After acceptance, it's too late and the other party can sue for damages if the impulsive party doesn't perform his or her obligations under the originally accepted offer. Ideally, an accepting party will clearly indicate his acceptance to the offering party, at best by signing an employment agreement or acknowledging acceptance in writing on the offer. More difficult to prove, but still unambiguous is an oral "I accept" or words to that effect.

**Avoid unclear contract formation situations**

Courts have created the so-called "action in reliance" (promissory estoppel) doctrine to find enforceable contracts even when one of the parties thought no contract existed. Courts have found valid contracts in cases where an:

- Employer knew or should have known that the employee had acted " in reliance upon the offer" such as incurring expenses to move to the job location, searching for lodging thereat, and informing other employers they no longer are job applicants; and

- Employee made the last offer or counter-offer, and such employee knew or should have known that in reliance thereon, the employer ceased advertising for the position, informed candidates that the job was filled, or bought new expensive specialty equipment or hired additional support staff in anticipation of the employee's arrival.

Accordingly, a party considering an offer should not talk or act in a way it knows or should know will lead the other party to believe that such offer was accepted and should make sure that the other party is not taking action "in reliance" on anything it did or said.

**EMPLOYMENT CONTRACT TERMS**

Turnover of veterinary specialist associates is caused principally by the failure of practice owners and employees to properly articulate their respective expectations and negotiate and document the employment relationship. Time and effort invested up front

---

[i]Dr. Charlotte Lacroix owns and operates Veterinary Business Advisors, Inc., a specialized firm based in Flemington, New Jersey (908) 782-4426; www.veterinarybusinessadvisors.com. Dr. Lacroix assists veterinarians and their attorneys nationwide with their business related legal matters, including employment contracts, purchase and sale of veterinary practices, specialty practice issues, mediation and start-ups.

will help avoid mismatched expectations, misunderstandings and separation down the road.

**Can the practice even afford another full-time veterinarian?**

Management consultants estimate that a small animal practice vet needs to produce a minimum of $180,000-$250,000 gross income (excluding OTC product sales) to be worth his salary. This number is far greater for veterinary specialists…..probably at least 3-4 times, depending on the specialty.

Assuming that the offer, counter-offer, counter-counter offer, etc. ballet results in the bliss of acceptance, the employment contract terms contain the nuts and bolts of the "meeting of the minds" of the parties. Following is a list of the main questions addressed in a proper employment agreement:

**How Long?**

Is there a fixed term (period) of employment (six months, one year, two years, or is it "at-will" (i.e., the contract continues until a party decides to terminate it)? Is the term automatically renewed on the expiration date? Is the one year term of the agreement superceded by a 90 day termination provision? In which case, how long really is the term? (ans: 90 days).

**Work Schedule**

How many scheduled hours per week must the employee work, and beyond the schedule, how many additional hours will employees actually spend phoning clients, performing diagnostics, interpreting laboratory work, overseeing patient care, etc. What is the schedule for any required emergency work? Is it equitable? Is it articulated as "minimum" hours?

**Duties**

What are the associate's responsibilities? May employees decline (without penalty) to perform procedures they deem ethically wrong? How much emergency duty is required?

**Compensation: How Much? Is It Enough?**

Serious job applicants must know the relevant "comparables" in their labor market, i.e., what compensation is paid to other starting associate veterinarians in the area were they are seeking employment. Currently, the salaries paid to different specialists are not well documented. The AVMA "2005 Economic Report on Veterinarians and Veterinary Practices" and ACVIM proceedings from the Hill's Practice Health Symposium, titled "Insights on Veterinary Specialty Practice Productivity" are a sampling of the few publications that have salary figures. Regardless of the trends, however, debt ridden veterinary specialists cannot assume that current salaries will permit them to survive (let alone live comfortably). So the first question isn't really "how much are they paying?" but rather "what do I need to pay my debts, buy cold cereal and go to a few movies?"

The only way to answer this question is by doing a budget. Budgets undoubtedly are one of the most boring tasks in the world, but boring beats finding out that you can't make ends meet six months after you've been hired. Technology has reduced the pain of budgeting, so there is no excuse for not doing it. Any financial software program worth its salt will permit veterinary graduates to establish a budget. See attached form to assist in determining one's budget. There also are several websites, which can help you with a budget…..google "budget" and you'll find plenty.

**Tips for Making More Money**

As discussed below, a common way for associate veterinarians to increase their compensation is to join a practice, which pays them a percentage of the collected income they generate. Other ways include working additional shifts, and working at another practice is the employer will permit it. Generally, there are three types of veterinary associate compensation: (1) flat salary; (2) commissions based on a percentage of the income generated by the associate; and (3) a hybrid of flat salary and commissions.

**Flat Salary**

Flat salary (a fixed amount per year), is a common form of associate compensation. A fixed salary provides the veterinary associate with the security of a predictable income. It is also simple to keep track of. Associate veterinarians earning flat salaries, however, cannot increase their compensation, no matter how much income they generate for the practice or how hard they work. Flat salaries are not preferred by associates looking for the opportunity to increase their compensation in exchange for a greater contribution to the practice.

**Straight Commission**

The straight commission system simply replaces the flat salary with a commission. The straight commission scheme link the dollars veterinary associates earn with their contribution to practice revenues. Because practice revenues (and the commissions) will vary month to month, associates may have a more difficult time managing the repayment of their student debt. This scheme also requires the associate to trust that management will equitably assign revenue generating cases among specialists of the same discipline.

**Hybrid Systems**

Under a hybrid compensation system veterinary associates are paid a guaranteed base salary plus an income production bonus equal to the percentage of the collected income they generate in excess of a certain target. The base salary provides security, as well as a predictable income stream with which to service student debt. This is a significant advantage over the straight commission system.

**Production Compensation Pitfalls**

While production compensation usually permits associates to increase their compensation, the system does have its problems and pitfalls. By carefully examining practice operations and asking the right questions, prospective associate veterinarians should be able to either avoid these pitfalls or at least reduce their impact.

- <u>Assigning Cases and Receptionist Gate Keepers</u>.

- Staff Efficiency and Leverage.
- Data Processing and Definition Issues.
- Competition and Distrust Among Veterinarians.
- Exclusions from the Production Calculation

**EMPLOYEE BENEFITS**

Practices usually offer at least some of the employee benefits described below to their employees. The cost of many benefits (such as health, professional, and disability insurance, qualified retirement plans) are tax deductible business expenses to the employer and are not included in the employee's income, resulting in a savings to the employee of 25 to 40%. Not taking advantage of this juicy gift from Uncle Sam is wasteful. On the other hand, employees must realize that the practice may not be able to afford all the benefits they desire.

**Health Insurance**

Does the employer offer health insurance? If not, what does the employer do when *he* gets sick? If so, what kind of medical plan is it (e.g., fee for service, HMO, PPO)? What about pre-existing conditions, vesting, eligibility, deductibles and co-payments?

**Disability Insurance**

Employees at age 25 have a 58% chance of becoming disabled for more than three months (with an average disability duration of three years), so employees need disability insurance to protect their greatest asset: the ability to work. If the employer does not offer disability insurance, employees are well advised to get it on their own (after asking, of course how the employer, protects himself or herself against disability).

**Professional Liability Insurance and License Defense Coverage**

Do employers pay the premiums on the employees' professional liability insurance? Does the policy protect against complaints brought before the state board of examiners?

**Retirement Plans**

Has the employer established a retirement plan for the employees? (Profit sharing plans are the most common type of retirement plan offered by veterinary practices.) When do employees become "vested" or "eligible?" If the employer does not offer a retirement plan, employees will need to save on their own (and that means more than just the annual IRA contribution).

**Vacation**

One work week? Two work weeks? More? How many consecutive days may be taken? How much advance notice must be given? May unused vacation days be carried forward to next year? How are vacation days paid for percentage compensated employees?

**Sick Leave and Disability**

Does the employer offer paid sick leave? Disability leave? After how long can disabled employees be terminated? May unused sick days be carried forward?

**Continuing Education**

How many CE leave days are granted and are they paid? To what extent do employers reimburse CE expenses?

**Association Dues**

Are national, state, local and specialty veterinary association dues reimbursed?

**Veterinary License Fees and DEA Registration**

Are these fees paid by the employer? Should the employee register with the DEA so she is permitted to prescribe and order controlled substances (rather than just administer them under the supervision of a DEA licensed veterinarian)?

**Relocation (moving) expenses**

Many prospective employers provide some form of moving expense. Sometimes a "signing bonus" or short term loan can cover all or part of these costs.

**Vehicle allowance or mileage payments**

Employees using their personal vehicles for practice business should be reimbursed for a pro-rata portion of their insurance, general maintenance, registration and inspection fees, fuel, repairs, depreciation, and lost opportunity costs.

**Performance Evaluation**

Will the employer provide written and/or oral performance evaluations? How often? Will these be used to modify compensation?

**Non-Competition**

Many employers require their employees to sign non-competition clauses (also called restrictive covenants) forbidding terminated employees from competing with the employer. Such clauses must be limited in time (e.g., 2-3 years after termination) and geographic area (e.g., 25-50 air-miles from the practice) to be enforceable. The precise limits on the scope of such clauses vary from state to state. Specialists typically have a larger radius as the trade area for specialty practices is much larger than a generalist's. Many of these clauses, also will prohibit terminated associates from soliciting employees and clients.

**Termination**

Does the contract have a specific term (e.g., "this agreement will expire after one year") or is it employment "at-will", in which case, either party can terminate the relationship at any time, for any reason? Contracts with no term are deemed to be "at-will" in most states. If there is a term, then an employee leaving or an employer firing before the term would constitute a breach unless the contract provides otherwise. Most contracts, which provide for termination before the expiration of the term require that the terminating party give advance notice (e.g., 90 days) to the other party. Such contracts usually also contain a list of situations (e.g., suspension of the associate veterinarian's license) permitting the employer to fire the employee at any time without notice (a.k.a. termination "for cause").

Employees should make every effort to leave their employer on good terms even if they are not requesting a reference. The veterinary industry is quite small, and an employee's reputation can easily suffer through casual conversation among colleagues.

**Option to Buy-In**

Experienced specialists that have their own clientele, may not wish to enter into an employment agreement with a non-compete, without also being provided with an opportunity to buy an interest in the practice after a 1-3 year "try-out" period. These often are complex provisions to negotiate depending on the amount of security the associate wants up-front, and should not be undertaken without consultation with an attorney that has experience with medical practice transactions. Too often associates lock themselves into a non-compete, and agree to an "option" provision that turns out to be a smoke screen.

**LAWYER REVIEW**

Negotiating and drafting an employment contract can be long, painful and complicated. It therefore makes as much sense to seek professional help in this endeavor as it does to take a pet to a qualified veterinarian when it is sick. Lawyers who have experience negotiating transactions in the veterinary industry are probably most qualified, since they understand the customs of the industry and how best to protect the interests of their clients. Lawyers are expensive, of course, just as much as specialists.

**SAMPLE BUDGET**

| **HOUSEHOLD EXPENSES** | | | **BUDGET ($)** | |
|---|---|---|---|---|
| | | | **Monthly** | **Yearly** |
| **Fixed or Committed Expenses** | | | | |
| | | | | |
| 1. | Mortgage or Rent | | | |
| 2. | Real Estate or Property Taxes | | | |
| 3. | Homeowners or Renters Insurance | | | |
| 4. | Savings | | | |
| 5. | Debt Reduction - Personal (Credit Cards, etc.) | | | |
| 6. | Debt Reduction - Student Loans - Monthly Payment<br>Total Student Loan Debt Owed at Graduation | | | |
| 7. | Automobile (Monthly Payment) | | | |
| | **Total Fixed Expenses** | | | |
| | | | | |
| **Variable Expenses** | | | | |
| 8. | Food (staples, milk, meat, veggies, etc.) | | | |
| 9. | Telephone | | | |
| 10. | Electricity, gas, or oil | | | |
| 11. | Water | | | |
| 12. | Cleaning - Laundry | | | |
| 13. | Auto - gas | | | |
| | a. | Car washes, maintenance, and repairs | | |
| | b. | Insurance, license, registration, etc | | |
| 14. | Physician's Medical Bills | | | |
| 15. | Dentist's Bills (Include Dental Prophy) | | | |
| 16. | Optometrists (Eyecare) | | | |
| 17. | Prescriptions, Vitamins, Aspirin, etc. | | | |
| 18. | Clothing - Self or - | | | |
| | a. | Husband, | | |
| | b. | Wife, and | | |
| | c. | Children | | |
| 19. | Personal (haircuts, hairdressers, cosmetics, etc.) | | | |
| 20. | Repairs and Home Improvements (appliances, painting, lawncare, etc.) | | | |

| | | | | |
|---|---|---|---|---|
| | 21. | Home Furnishings | | |
| | 22. | Education (books, newspapers, periodicals, etc.) | | |
| | 23. | Recreation (sitters, meals out, movies, concerts, sports, etc.) | | |
| | 24. | Charitable Contributions (church, schools, veterinary school) | | |
| | 25. | Gifts and Flowers | | |
| | 26. | Animal Food, Care, and Veterinary Bills | | |
| | 27. | Miscellaneous (Bank Service Charges) | | |
| | 28. | Vacation | | |
| | 29. | Moving Expenses and Security Deposit | | |
| | 30. | | | |
| | | **Total Variable Expenses** | | |
| | | | | |
| | **TOTAL FIXED & VARIABLE HOUSEHOLD EXPENSES** | | | |
| **Multiplied by Multiplier from Chart** | | | | |
| | **INCOME NEEDED BEFORE TAXES** | | | |
| | | | | |
| | **Variable (Nontaxable) Benefits Hopefully Provided By Employer** | | | |
| | 31. | Dues & Licenses (Veterinary associations, etc.) | | |
| | 32. | Continuing Education | | |
| | 33. | Health Insurance | | |
| | 34. | Pension | | |
| 35. Relocation reimbursement | | | | |
| 36. Other Misc. Benefits (Health club membership, dental/vision coverage, etc.) | | | | |
| | | **Total Nontaxable Expenses Provided By Employer** | | |
| | | | | |
| **Taxes** | | | | |
| | 35. | **Federal Taxes (IRS)** 22-25% | | |
| | 36. | **FICA (Social Security)** OUCH! 7.65% | | |
| | 37. | **STATE TAXES** UGH! 2-5% (Depends on state) | | |
| | | | | |

**SPECIALTY EMPLOYMENT RELATIONSHIPS: RETAINING YOUR RESIDENTS©**

**Charlotte A. Lacroix DVM, JD.**[1]
**Flemington, NJ**

Finding the right resident for your practice is a lot of time and work. Protect your investment and keep your resident from straying to other practices by including proper retention provisions in your residency agreements.

The basic bargain you make with your resident is simple: you agree to pay all or part of your resident's training and residency living expenses; and your resident agrees to work at your practice for a minimum period—the "retention period"-- after she is board certified.

To implement this bargain, residency provisions typically use a stick and/or carrot approach. The stick requires your resident to pay back the residency expenses you fronted if she fails to timely pass her residency or if she leaves your practice before the agreed upon retention period expires. The carrot, which is optional, pays your resident a bonus at the end of the retention period.

Here's a list of the principal issues your residency provisions should address, bearing in mind that the main variable affecting their structure and content is whether or not the residency will be in-house.

***

**Residency Rules**

If you are sponsoring the residency, both you and the resident should agree to respectively follow the residency rules and guidelines set by the applicable veterinary college ("Residency Rules"). Accordingly, you would be prudent to ensure that compliance with Residency Rules will not unduly burden your practice *before* you commit to sponsoring the residency. The residency provisions should also: (a) require the resident to keep track and inform you of any rule changes; and (b) allow you to terminate the residency without penalty in the unlikely event a rule change makes compliance too burdensome for your practice.

**In-House Residencies**

With in-house residencies, your resident will invariably be your employee during the residency period, so all the usual employee issues relating to compensation and benefits apply. However, you will need to adjust your standard employment agreement to give your resident time to: (a) complete whatever externships are required by the Residency Rules; and (b) study for her boards. (In this regard, you may consider reducing her compensation accordingly.) Do not forget to check that your insurance and benefit plans will cover your resident during this period.

**Off-Site Residencies**

If a veterinary school or other hospital is going to sponsor the residency, you need to consider whether you want your resident to be your employee during this time. If the resident will be attending a veterinary school or other hospital far from your practice, you may not want her to be your employee, since an employer is generally liable for their employees while they are acting within the scope of their employment. Since you have deeper pockets than your (normally) impecunious resident, plaintiffs will be motivated to sue you if your resident gets into trouble.

If this liability worries you, then you will need to loan your resident the amounts she needs, with the understanding that you will employ her as soon as she is permitted to take the boards (and then forgive the loan at the end of the retention period).

Loaning residency expenses to your resident with an employment agreement to follow will be more complicated to structure, negotiate and document, than simply employing your resident from the outset. This will take more time and cost more in legal fees. You will need to balance this increased cost against the risk of incurring liability for your employee's acts and omissions while a resident. Such balancing will require weighing various factors, including the extent to which the veterinary school or other institution is liable for their residents and to which your resident employee will be deemed to be acting within the scope of her employment. Common sense would indicate that the school or institution should be primarily liable for all resident activities and that the risk of your incurring such liability is remote. But all bets are off in our sue-happy society. It may also be possible to cost-effectively insure against any residual liability. As a precaution, your resident should agree to seek your permission before engaging in any remunerative activity while a resident, (e.g., working at a shelter or temping at an emergency clinic), so that you can evaluate the risk thereof.

Finally, depending upon the residency program's schedule and how far away it is from your practice, consider whether you want your resident to work for you during week-ends, holidays and/or residency program breaks.

**Ensuring Diligence**

Whether or not your resident will be completing her residency at your practice, she should agree to diligently pursue her residency to completion, which will include studying and sitting for the boards as soon as she is permitted to do so. Your resident should commit to become board-certified by a certain deadline (which can be extended for a limited time if your resident becomes disabled). If the residency program is held off-site, your resident should agree to provide you with adequate documentation to monitor her progress.

---

[1] Dr. Charlotte Lacroix owns and operates Veterinary Business Advisors, Inc., a specialized firm based in Flemington, New Jersey (908) 782-4426; www.veterinarybusinessadvisors.com. Dr. Lacroix assists veterinarians and their attorneys nationwide with their business related legal matters, including employment contracts, purchase and sale of veterinary practices, specialty practice issues, mediation and start-ups.

**Residency Expense Tracking**

Your residency provisions will also need to specify the residency expenses for which you will be responsible and how they will be documented and paid. Consult with your tax advisor to ensure that this is done in a tax efficient manner.

**Residency Expense Repayment**

Now for the stick. The residency provisions typically will provide that the resident will repay the residency expenses you have advanced, unless she works at the practice through the end of the agreed upon retention period. In essence, your resident is "working off" her "debt" to you. (The "debt" being your advance of residency expenses to her.) Thus, during the retention period, your former resident's compensation should be less than market to reflect this "repayment."

This loan analogy cuts both ways however, because a savvy resident will demand that the repayment obligation be suitably pro-rated, so that if she leaves, say, in the middle of the retention period, she need reimburse only half of the residency expenses.

The provisions should provide for a repayment schedule and an interest rate (or specify that the resident will owe no interest). In attempting to reduce interest as much as possible or eliminate it all together, a savvy resident will argue that the practice is benefiting from the services of a "captive" specialist, who cannot leave without incurring a substantial reimbursement obligation. This benefit is above and beyond what an un-affiliated lender would receive, and in consideration for this benefit your practice should not charge interest. (Be advised, however, that charging below market interest or no interest may subject you to tax liability.)

**Disability and Early Termination**

Proper residency provisions must also cover life's more foreseeable contingencies. The two principal intervening events that should be addressed are disability, and employee termination before the expiration of the retention period.

Resident disability generally will extend the deadline for obtaining board certification and also length of the retention period. If your resident's disability lasts longer than this extension, she normally would be terminated, just like any other employee subject to long-term disability. But what about her repayment obligation? Should she still owe you for the residency expenses you advanced? If you're tough you might say yes. If you're nicer you might want to forgive your disabled resident's repayment obligation in whole or in part. (Note that you might be able to insure against this risk.)

What happens if you terminate your resident before the end of the retention period? If you terminate for the usual "for good cause" reasons, then the resident should still repay you for the residency expenses you advanced. In this regard, the residency provisions should allow you to terminate your resident "for good cause" if she fails to become board-certified by a specified date.

But if you terminate your resident at your discretion, i.e., for any reason other than "for good cause", then the provisions should extinguish your resident's obligation to repay you for the residency expenses.

If your resident leaves before the end of the retention period she will owe you the residency expenses you advanced. That is after all the whole point of having retention provisions in the first place. But heads up: a savvy resident will require the provisions to address what happens if she terminates her residency agreement because of your breach of that agreement.

**Retention Bonus**

If you wish to motivate your resident with a carrot in addition to the stick, the residency provisions can provide that you will pay your resident a specified bonus if she stays through a specified date (which need not coincide with the end of the retention period, and could even be paid periodically in installments). As with your resident's obligation to repay residency expenses, the bonus provisions will need to deal with disability and early termination (either implicitly or explicitly).

**Residencies as CLE**

Residency programs at veterinary schools or other institutions can constitute a valuable educational resource for your practice. Accordingly, do not forget to require your resident to provide copies of all interesting residency documents and give your practice periodic presentations of residency activities.

***

Like many things in life, residency provisions are simple in concept but complex in implementation. You will need to invest some time and effort, and incur some expense in preparing a proper residency agreement. Accordingly, it makes sense to temporarily employ your future resident at your practice *before* committing to any residency obligation---just to make sure that she is worth the investment.

# BUYING INTO A PRACTICE: WHAT SHOULD THE SELLER AND BUYER BE DOING ©

**Charlotte A. Lacroix DVM, JD.**[i]
**Flemington, NJ**

Buying or selling a veterinary practice generally is a long and arduous process. Preparation and a good lawyer are key to smoothing the bumps along the way.

## PLANNING

How well you plan determines whether you will control the process or vice versa. Beyond identifying what is to be sold and the purchase price therefor, here is an overview of the main planning issues:

## WHAT SELLER SHOULD BE DOING

Future retirees should note that it can take a decade to properly prepare a practice for sale.

### Tax

Find out how the sale price will be taxed. Tax planning may require transformation of the practice legal structure and long waiting periods to optimize tax on sale. Bear in mind that the buyer may not accept the structure you wish to use for the transaction (e.g., he may prefer to buy assets rather than an interest in the practice entity.)

### Practice Appraisal

To increase practice profitability and sales price, it may be worthwhile to get an appraisal. Implementing the appraisal's recommendations may take several years. From the buyer's perspective, a purchase without an appraisal may mean the practice does NOT have the cash flow to fund the purchase payments. If the buyer defaults on the payments, seller may have to take back practice, long after retirement.

### Cleaning Up

Consult with your lawyer to see if there are any liabilities that should be taken care of before the sale (e.g., settle a claim). You also need to address any personal guarantees you have given to the business, and third party claims on any assets to be transferred (such as spouse and secured lenders). If the practice entity owns the real estate, you may need to divest it (at what tax cost?) back to yourself or another entity you own because the buyer may be unable to afford to buy both.

### Financing

Many potential buyers can't afford to pay cash and can't get bank loans on acceptable terms. Consider whether you should provide financing by taking buyer's promissory note (secured of course) for a portion of the purchase price.

## WHAT BUYER SHOULD BE DOING

Buyer's planning phase usually is much shorter than seller's. Try to minimize preparation expenses until you are reasonably sure that you have a deal with the seller. If the deal is aborted, you want to be as little "out-of-pocket" as possible.

### Hiring Counsel

When hiring counsel, trust your instinct. If you don't feel comfortable with him or her, get another one, there are plenty around. You find a good lawyer the same way you find a good doctor: by word of mouth. Your lawyer should be at least as smart as you are. Avoid lawyers that not nothing about assisting buyers purchase medical practices.

### Tax

Find out how the sale can be structured to minimize taxes down the road for the new owner(s) and the practice, including taxes when you sell or withdraw from the practice. (Remember every buyer is destined to be a seller.)

### Due Diligence

You don't buy a house without visiting it, nor a car without a test drive. Buyers need to spend time at the practice to "kick the tires." One of the most important goals of due diligence is to make sure buyers are not picking up hidden or poorly identified liabilities. Due diligence allows early detection of potential problems, so that the parties can address them...or separate. Not infrequently, good due diligence leads to a reduction in purchase price.

### Financing

If you can't pay cash, you need to line up financing (or be prepared to ask the seller for same) well before the sale.

### Co-Ownership Issues

If you are just buying a portion of a practice or joining with other partners to buy a practice, there are many co-ownership issues that need to be addressed. Like marriage, co-ownership works a lot better if the principal terms governing the relationship *before* getting hitched. At least as much time and effort should be allocated to properly establishing *and* documenting your relationship with your partners, as to the purchase itself.

---

[i]Dr. Charlotte Lacroix owns and operates Veterinary Business Advisors, Inc., a specialized firm based in Flemington, New Jersey (908) 782-4426; www.veterinarybusinessadvisors.com. Dr. Lacroix assists veterinarians and their attorneys nationwide with their business related legal matters, including employment contracts, purchase and sale of veterinary practices, specialty practice issues, mediation and start-ups.

**Investment Vehicle/Practice Structure**

Unless you are buying into a pre-existing structure, consult with your lawyer to find the investment vehicle or practice structure best suited to fulfill your tax, financing and (co)ownership goals. (In making this choice, do not omit to factor in you succession or exit strategy.) Even if buying into a pre-existing structure, be aware of the consequences. For example, stock purchases are riskier than asset purchases because all of the liabilities of a corporation automatically are transferred with the stock, whether or not such liabilities were disclosed to buyer.

## WHAT SELLER AND BUYER SHOULD BE DOING *TOGETHER*

**Employees**

The practice's employees constitute its most valuable asset by far and it's in the interest of both parties that the transition take place smoothly. For the buyer, smooth transition is critical, since the departure of one or more key employees can cripple the practice. While legally (to avoid trailing liabilities), the buyer will insist that all employees be terminated prior to the sale (and to the extent the buyer wants to retain them, rehired after the sale), it is up to the parties to work together to ensure that the ownership change disrupts the employees as little as possible.

**Contracts**

In asset sales where contracts are to be transferred (e.g. real estate and equipment leases), the consent of the other party to the contract is required. Obtaining such approvals is often easier when both buyer and seller approach such other party together (although as a contractual matter, it is almost always the seller's responsibility to get these consents).

## NEGOTIATING THE DOCUMENTS

**Preliminary Documents**

At some point during the preparatory stage, the parties usually negotiate some form of preliminary document which can range from: (a) a short letter of intent, pursuant to which the parties agree to negotiate in good faith with each other (in some cases exclusively) and confidentiality; to (b) a detailed memorandum of understanding detailing the process for going forward; to (c) an at least partially binding detailed letter of offer which sets forth the main terms of the deal. The parties should make absolutely sure they understand to what extent these documents bind them to go forward, and what liability, if any, a party will incur if he withdraws from the deal.

**Purchase and Sale Agreement**

While the preliminary documents set forth an intent, a process or a framework, the purchase and sale agreement (called stock or asset purchase agreement depending upon the transaction) is the contract in which the buyer binds himself to purchase the practice and pay the price, and the seller binds himself to sell the practice, in each case subject to a limited number of conditions precedent to closing the deal. Some conditions precedent are applicable to both parties obligation to close (e.g. no litigation exist challenging the transaction). Some are applicable to only one party (such as buyer obtaining satisfactory bank financing).

Other than the price and conveyance language, the most important clauses of the purchase agreement are seller's representations and warranties, pursuant to which seller makes a series of declarations about the business (there are no hidden liabilities, there is no outstanding litigation against the practice; all inventory is saleable in the ordinary course, etc.) These provisions are designed to protect the buyer in case the practice isn't what it appears to be, but the devil is in the wording, and both parties should make sure that they are comfortable with the language.

Another key clause is the non-competition provision (a.k.a. restrictive covenant) which prohibits the seller from competing with the practice for a certain period of time (e.g., five years) within a certain radius (e.g., 15 miles). This is an important provision to protect the practice's goodwill.

## TOWARDS CLOSING

Signing the purchase agreement means that the parties commit to consummate the transaction once all of the conditions to closing have been satisfied. Closing the transaction means actually transferring legal title to the shares or assets and paying the purchase price. Between these two dates, the parties and their lawyers do what is necessary to ensure that the closing conditions in the purchase agreement for which they are responsible will be satisfied by the closing date. (For example, buyer may have to obtain bank financing and seller get certain third party consents to the deal.)

In many cases, the signing of the purchase agreement and the actual closing of the transaction take place simultaneously, in which case there are no closing conditions.

***

The preceding was merely a simplified overview. It seems that there are a billion issues and details that seller and buyer must keep track of when selling and buying a practice. That's why competent legal advice is crucial. But if you feel overwhelmed, just ask yourself how you'd behave if the transaction involved a house or a car. Usually the solution will come to you as if by magic.

## PRACTICE ENTITY-WHICH ORGANIZATION IS BEST FOR YOU & WHY IT MATTERS ©

Choosing the correct structure for your veterinary practice is an important decision with consequences reaching far into the future. Selecting your practice structure is definitely *not* a "do it yourself" project. Substantial tax, legal and accounting expertise is required. Veterinarians nevertheless need to stay active in the process to ensure the experts' narrow technical proposals get folded into a coherent plan that reflects *your* needs and goals.

**It's Mostly About Tax**

Tax considerations are the primary drivers in choosing a legal structure for a veterinary practice. The two key aspects are taxation of income/profits and taxation upon the sale or transformation of the practice. Don't paint yourself into a corner by choosing a business structure without establishing a succession or exit strategy. Exit strategies should focus not only on your richly deserved retirement, but also on contingencies such as death or disability). Since the transformation of an existing business structures in anticipation of a sale or the buy-in of a new partner usually triggers adverse tax consequences, it is usually better to choose an initial structure with the necessary flexibility to handle new arrivals, departures and divestitures at minimum fiscal cost.

**Liability Shield**

In some structures such as partnerships, the owners are personally liable on their individual assets for the debts of the business. In others their personal assets generally are not at risk. Business structures, however, do not insulate veterinarians from liability arising from malpractice claims.[1] But the shield works for almost all other claims, which in our litigious society are increasingly frequent. Unless you are an equine or food animal veterinarian, you generally have greater exposure to claims from your client's "slipping and falling" in your hallway, than malpractice.

**Flexibility and Formalities**

Some structures allow more management flexibility and/or are less burdensome to administer than others. Veterinarians generally tend to ignore formalities which is a serious mistake. Courts regularly have looked past the liability shield and held owners personally liable when the owners have failed to observe the formalities separating their personal affairs from those of the practice entity.

**AN OVERVIEW**

The accompanying table compares the more common business structures from a liability, management and formality perspective (in simplified form). Following is a brief and much simplified overview of the tax characteristics of each entity.

**Sole Proprietorships**

Since sole proprietorships are not legally separate from the single owner, there is no separate tax return. The practice's profits are included in owner's total income and are taxed at his ordinary income tax rate. In addition to federal and (if applicable) state income tax, the owner must also pay self-employment tax equivalent to the payroll taxes due as if the owner were an employee of the practice.

Upon the sale of the sole proprietorship practice's assets, the IRS will recapture all depreciation/amortization deductions taken by the owner/seller thereof and tax such amount at the seller's ordinary income tax rates. In the unlikely event that any gain remains on the assets (after adding back any depreciation/amortization to their respective "bases"[2]) they will be taxed at the lower 20% long term capital gains rate (assuming the relevant holding period is met).

The buyer receives a "step-up" (increase) in his basis in the assets proportional to the amount of (purchase price allocated thereto) allowing him to re-depreciate/amortize them. Thus, asset sales usually are a better deal tax-wise for the buyer than for the seller, and all other things being equal, buyers will prefer to purchase assets rather than stock (in a C corp).

**Partnerships**

Partnerships are "pass-through" or "flow-through" entities for tax purposes, meaning that each partner includes in his own taxable income the profits (or losses) of the partnership, which are taxed as ordinary income at the partner's individual rate (much like the owner of a sole proprietorship). Note that each partner's share of partnership income is taxable each year, whether such share was distributed to the partner or retained in the partnership. If the latter, then the partner may not have the cash to pay the tax.

A consequence of the pass-through principle is that the sale of partnership interests are treated for tax purposes similarly to the sale of the underlying assets of the partnership (i.e., the assets are subject to depreciation recapture as in sole proprietorships).

**Corporations**

All corporations must file separate tax returns.

**"S" Corporations**

"S" corporations are corporations that elect to be taxed as a partnership. As "pass-through" entities, profits will be taxed in the hands of the shareholders whether distributed or not. An advantage of S corporations is that shareholders may take a portion of their profits as "S corporation profit," free of payroll or self-employment tax (i.e., subject only to income tax). Profit corresponding to what the veterinarian shareholder would have earned as an employee is subject to payroll taxes in addition to income tax. (Sole proprietorships on the other hand must pay self-employment tax on all profits.) S corps are popular with veterinarians for this reason.

**"C" Corporations**

"Plain vanilla" corporations (called "C" corporations to distinguish them from "S" corps) are not "pass-through" entities and are subject to corporate income tax, usually at the 35% rate for veterinary practices.[3] Distributed profits (dividends) are taxed as ordinary income in the hands of the shareholders. This "double taxation" discourages the distribution of C corporation profits. On the plus side, C corp profits are not taxed until distributed, pension plan contributions are not subject to the S corp limits, and employee-shareholders' health benefits are not taxed. Veterinarians wishing to maximize their benefits will choose a C corp over an S corp.

If the holding period requirement has been met, the sale of C corporation stock is taxed at the favorable 20% long term capital gains rate. The buyer does not receive a step-up in the basis of the underlying assets since he is buying the corporation stock. (The buyer can under certain circumstances elect to treat the transaction as an asset sale for tax purposes (a.k.a. a Section 338 election).)

**Limited Liability Companies**

Limited Liability Companies are very quite tax-wise. Single member LLCs can elect to be taxed either as a C corp or a sole proprietorship. Multi-member LLCs can elect to be taxed either as a C Corp or a partnership. Unfortunately, not every state allows veterinarians for form LLC (ie, California).

**A Word Regarding Real Estate**

If the practice owns its own real estate it's better placed in a separate entity held by the owner(s) or held individually by the practices owner(s). This allows the owners to receive rent (which will be deductible from the practice's income). Moreover, placing the real estate and the practice in the same legal entity frequently leads to problems because the buyer can't afford to buy the real estate in addition to the practice.

***

Choosing the correct business structure for your practice is important. Don't treat it lightly.

**SIMPLIFIED PARTIAL COMPARISON OF DIFFERENT BUSINESS STRUCTURES**
**(*Ex* tax issues)**

| **Structure or Entity Type\Issue** | **Liability** | **Formalities/Flexibility** |
|---|---|---|
| Sole Proprietorship<br>No entity; business co-mingled with personal assets | **No liability shield** | None. Just open your door and you're in practice! |
| Corporations ("C" or "S" Corp) A Professional Corporation ("PC") is identical to a C Corp in all respects except that only members of the same profession (e.g., vets) can own its shares | Shareholder not liable for debts/liabilities of corporation (unless "corporate veil is pierced" because shareholders fail to separate their personal affairs from corporations (e.g. by ignoring formalities) | Must file documents with state secretary of state. Formalities are the most cumbersome of all entities. Less formal flexibility re management/profit sharing issues |
| Limited Liability Company (LLC) (Created to provide more management flexibility than S Corp and "pass through" tax treatment ) | Member not liable for debts/liabilities of LLC (subject to piercing corporate veil doctrine) | Must file documents with state secretary of state; but management, profit sharing can be flexible. |
| General Partnership[4] | Partners liable for debts/liabilities of Partnership; **no liability shield** | Must file documents with state secretary of state, but management; profit sharing can be flexible. |

[1] Salvation lies in adequate malpractice insurance.

[2] The basis of an asset is it's original cost to the owner, as adjusted pursuant to IRS rules.

[3] Because veterinary practices usually are personal service corporations.

[4] Limited partnerships are different from general partnerships. An LLP generally is formed among several limited partners who are normally passive financial investors and one general partner responsible for managing the enterprise. Limited partners normally are not liable for the debts/liabilities of the LLP, whereas the general partner is. Contrary to the motion picture business, real estate or oil and gas exploration, LLPs may not be appropriate for a veterinary practice where all the members are actively engaged in the enterprise.

USE OF HOLTER ANALYSIS IN CLINICAL MEDICINE & CLINICAL RESEARCH

N. Sydney Moïse, DVM, MS, DACVIM
Ithaca, NY

## INTRODUCTION

Holter analysis of canine, feline and equine patients is common in clinical practice. This diagnostic procedure also is used in the determination and characterization of disease in research.

## VALIDATION OF DIAGNOSTIC TESTS

Holter analysis has not been validated for its accuracy in our patients. Furthermore, analysis may be performed by those unfamiliar with the actual process of evaluating the recordings or by those who are not familiar with animal electrocardiography. The vastness of the data also may lead to acceptance of a degree of inaccuracy that may in fact preclude accurate conclusions. We have worked individually with this diagnostic tool, but perhaps we need to inquire whether we should have this diagnostic test validated and our results open for evaluation of accuracy. In human medicine, it is required that Holter systems used in patient care be validated through several banks of test ECG Holter recordings such as the MIT-BIH Arrhythmia Database (http://www.physionet.org).[1-3] In veterinary medicine there is no such requirement. Should we consider that Holter analysis be validated at least for the publication of information whereby Holter analysis was used? We cannot use the validation for human ECG Holters as the evaluation of canine, feline and equine recordings is different. We would need to establish some test recordings to judge accuracy. However, it is not that simple because the completely automated approach to analysis is only one step. Editing of the automated analysis requires human intervention and this ability can vary. Moreover, editing may demand an excessive amount of time. Also, editing tools in different Holter systems vary and this can influence the time and consequently the accuracy. We are investigating means by which selective beat annotation can be used within < 3% error of complete annotation to assist in this evaluation. We have used Holter analysis extensively and will be testing algorithms.[4-12]

## BEAT ANNOTATION & PATTERNING

Correct beat annotation is most challenging with rapid and closely coupled premature complexes. Unfortunately the most difficult Holters to correctly annotate are those that are of the greatest importance for the determination of the frequency and type of the arrhythmias. Knowing the severity of an arrhythmia at baseline followed by a recheck examination after treatment is vital to document an antiarrhythmic effect. True, elimination of an arrhythmia may not insure that sudden death will not occur; however, it is important in many cases by decreasing the risk. In veterinary patients the two most common types of arrhythmias monitored are atrial fibrillation and ventricular arrhythmias. I will emphasize that although there are patients for whom the risks of a baseline Holter recording is too great because antiarrhythmic treatment is deemed immediately necessary, the vast majority of animals will benefit from a baseline recording. If a baseline Holter is not performed, it is almost impossible to know if the drug is working or if it is having a proarrhythmic effect. We use Holter monitoring to determine if adequate rate control of atrial fibrillation has been obtained, if a suppression of ventricular or supraventricular arrhythmias has occurred, or if disease has progressed. Another important point is that Holter only provides information about the arrhythmia that a patient may have at that time. Most of the time, a baseline recording gives an accurate reflection of the danger that a patient develops unfortunate consequences of the abnormal rhythm. However, this is not always true. For example, a few boxers may have very rare premature complexes, but, ventricular tachycardia is documented under certain circumstances of stress. Despite these limitations we have found that Holter analysis has been vital in the decision regarding treatment of a variety of arrhythmias.

Another important component in Holter analysis is looking beyond specific beat annotation. Patterning of normal and abnormal rhythms gives clues to the autonomic influence on the heart rhythm. Such patterns of beats also are altered by medication.

## HOLTER ANALYSIS FORM

Cornell University
College of Veterinary Medicine

**HOLTER ANALYSIS REPORT**

| | | | |
|---|---|---|---|
| Holter system: | | Date of recording: | |
| Owner contacted? | | Holter (initial, R1, R2, ...) | |
| Rdvm contacted? | | Diagnosis: | |
| Edited by : | | Time required for editing: | |
| Final report compiled and written by | | | |

| Patient information: | | | | | | |
|---|---|---|---|---|---|---|
| Case number | Owner's name | Patient's name | Breed | Age | Sex | Weight |
| | | | | | | |

| Owner information: | | | | | |
|---|---|---|---|---|---|
| Street address | City | State | Zip | Phone | Email |
| | | | | | |

| Referring veterinarian information: | | | | | |
|---|---|---|---|---|---|
| Veterinarian's name | | | | | |
| Clinic name | | | | | |
| Street address | City | State | Zip | Phone | Email |
| | | | | | |

| | |
|---|---|
| **Quality of recording:** | A total of 24 hours and 15 minutes were recorded and evaluated. Interactive editing was required in order to have greater than 99% accuracy in beat identification and annotation. The quality of the recording was good with <1% artifact, although channel 3 was mostly artifact. |
| **History (see chart below for medications):** | |
| **Purpose of Holter:** | |
| **Rate assessment:** | |
| | |

| HOLTER (HLTR) DATA – Ventricular ectopy (VE), Supraventricular ectopy (SVE) | | | | | | | | | | | |
|---|---|---|---|---|---|---|---|---|---|---|---|
| Holter | Date of HLTR | Duration of HLTR (Hrs) | Drug | Dosage | HR >120 (Hrs) | HR <50 (Hrs) | HR avg | HR min | HR max | Longest Pause (Sec) | # Pauses (>2s) |
| Baseline | 4/22/08 | 24:37 | X | X | 04:07:43 | | 84 | 37 | 235 | 2.3 | 8 |
| Reck 1 | 8/13/08 | 24:22 | Sotalol | 80mg BID | 01:04:56 | | 73 | 37 | 231 | 2.5 | 40 |
| Reck 2 | 9/4/08 | 24:12 | X | X | 02:22:17 | | 81 | 38 | 231 | 2.4 | 19 |
| Reck 3 | 10/28/08 | 24:15 | Sotalol | 100mg BID | 00:53:08 | 03:06:16 | 71 | 37 | 235 | 2.4 | 25 |

| Holter | Total VE Singles | VE Singles /Hr | Total VE Couplets | VE Couplets /Hr | Total VE Runs | VE Runs /Hr | Total QRS | Total VE | VE % |
|---|---|---|---|---|---|---|---|---|---|
| Baseline | 3215 | 134 | 13 | 0.5 | 0 | 0 | 121,187 | 3241 | 3 |
| Reck 1 | 10,411 | 434 | 63 | 2.6 | 6 | 0.25 | 104,656 | 10,556 | 10 |
| Reck 2 | 14,523 | 605 | 803 | 34 | 39 | 1.6 | 114,014 | 16,247 | 14 |
| Reck 3 | 2347 | 98 | 48 | 2 | 4 | 0.2 | 99,716 | 2457 | <1 |

| Holter | Total SVE Singles | SVE Singles /Hr | Total SVE Couplets | SVE Couplets /Hr | Total SVE Runs | SVE Runs /Hr | Total QRS | Total SVE | SVE % |
|---|---|---|---|---|---|---|---|---|---|
| Baseline | 6 | 0.25 | 2 | 0.08 | 0 | 0 | 121,187 | 10 | <1 |
| Reck 1 | 85 | 3.5 | 28 | 1.2 | 138 | 5.6 | 104,656 | 1401 | 1 |
| Reck 2 | 181 | 7.5 | 147 | 6.1 | 283 | 11.8 | 114,014 | 2042 | 2 |
| Reck 3 | 87 | 3.6 | 9 | 0.4 | 19 | 0.8 | 99,716 | 245 | <1 |

**Rhythm assessment:**

**Conclusions and Treatment:**

| Drug | Current dose | New dose |
|---|---|---|
| | | |
| | | |
| | | |
| | | |
| | | |

## REFERENCES

[1] MIT-BIH Arrhythmia Database at **http://www.physionet.org**[2] Goldberger AL, Amaral L, Glass L, Hausdorf JM, Ivanov PC, Mark RG, Mietus JE, Moody GB, Peng CK, and Stanley HE. **PhysioBank, PhysioToolkit**, and PhysioNet: Components of a New Research Resource for Complex Physiologic Signals, Circulation. 2000:101, e215-e220.[3] AHA-DB. AHA database series 1. **The American Heart Association**, 1997.[4] Moïse NS and Pariaut R: Examen Holter 24 **heures au rythme du Coeur. Cardiologie**. 2002:3-11.[5] Moïse NS: **Diagnosis and management of canine arrhythmias**: In: Canine and Feline Cardiology. 2nd edition. Fox, P. R., Sisson, D. D., Moïse, N. S., editors. Philadelphia, W. B. Saunders Company 1999, pp. 331-385.[6] Moïse NS, De Francesco T: Twenty-four hour ambulatory electrocardiography. In: **Current Veterinary Therapy** XII. Kirk, R. W., editor. Philadelphia, W. B. Saunders Company, 1995, 792-799.[7] Moise NS, Gelzer ARM, Kraus MS. Treatment of ventricular arrhythmias. In **Current Veterinary Therapy** XIV, 2008.[8] Meurs KM, Spier AW, Wright NA, Atkins CE, DeFrancesco TC, Gordon SG, Hamlin RL, Keene BW, Miller MW, Moise NS. Comparison of the effects of four antiarrhythmic treatments for familial ventricular arrhythmias in Boxers. **J Am Vet Med Assoc**. 2002 221(4):522-527.[9] Gelzer ARM, Kraus MS, Moïse NS, Pariaut R, Jesty SA, Rishniw M, Erb H, Hemsley SA. Assessment of antiarrhythmic drug efficacy to control heart rate in dogs with atrial fibrillation using 24 hour ambulatory electrocardiographic (Holter) monitoring. In review. **J Vet Int Med.** December 2008.[10] Moïse NS, Meyers-Wallen V, Flahive WJ,et al: Inherited ventricular arrhythmias and sudden death in German shepherd dogs. **J Am Coll Cardiol** 1994, 24:233-243.[11] Moïse NS, Gilmour RF Jr, Riccio ML, et al: Diagnosis of inherited ventricular tachycardia in German shepherd dogs. **Am J Vet Med Assoc**1997, 210:403-410.[12] Oxford EM, Everitt M, Coombs W, Fox PF, Kraus MS, Gelzer ARM, Saffitz J, Taffet SM, Moïse NS, Delmar M. Molecular composition of the intercalated disc in a spontaneous canine animal model of arrhythmogenic right ventricular dysplasia/cardiomyopathy. **Heart Rhythm**, 2007, 4(9): 1196-1205.

# TREATMENT OF CANINE ARRHYTHMIAS WITH RADIOFREQUENCY CATHETER ABLATION

**Roberto Santilli, DVM, DECVIM-CA (Cardiology)**
**Samarate, Varese, ITALY**

## INTRODUCTION

In 1985 Huang and co- investigators used radiofrequency current as an alternative energy source in the experimental treatment of cardiac arrhythmias in a closed-chest canine model.[1] In this experiment they showed that it was possible to safely and effectively ablate the AV junction with a conventional 2-mm-tipped electrode catheter. Several studies in animal models followed the initial use of radiofrequency catheter ablation (RFCA) assessing the effects on different tissues such as the ventricle, the coronary sinus, the atrium and the tricuspid annulus. Nowadays RFCA is the treatment of choice for different cardiac arrhythmias such as atrioventricular reciprocating tachycardia (AVRT), atrioventricular nodal reciprocating tachycardia (AVNRT), focal atrial tachycardia (FAT), cavo-tricuspid dependent atrial flutter (AFL), atrial fibrillation and ventricular tachycardia[2]. In veterinary clinical cardiology the first report of RFCA is dated 1993[3], since then the use of RFCA has been described in the treatment of orthodromic AVRT, FAT and AFL.[4-9]

Radiofrequency currents present unmodulated sinusoidal waveforms at a high frequency (0.3 to 30 kHz). They can be used in medicine for ablating, coagulating and cauterizing tissue. For cardiac ablation, frequencies lower than 10 kHz are avoided to prevent stimulation of excitable muscular and cardiac tissues. Radiofrequency current is applied at endocardial or epicardial surface and the voltage induces a current to flow between the electrode inside the cardiac cavity to a large, grounded, dispersive electrode on the surface. The absorption of RF energy in tissues is governed by the electrical properties of tissue, and in particular the dielectric permittivity and tissue conductivity.[10] Dielectric permittivity is the degree to which a medium resists the flow of electric charge divided by the degree to which free space resists such charge.The applied electric field causes a mobile ions accumulation within the intra- and extracellular spaces. During RF delivery an insufficient time is allowed to permit charging of the cell membranes with a decrease in the value of dielectric permittivity. At frequencies above 1 kHz, a short-circuit of cellular membranes occurs with an increase of the participation of intracellular fluid in current conduction.[10] The current flow throughout the tissues is impeded by tissue resistances and decreases as the fourth power of distance from the active electrode. Once absorbed the energy is converted in heat that warm tissue. The tissue heating is localized to a very short distance from the electrode-tissue interface. Dissecation and coagulation of tissue close to the electrode promote a decrease of electrical conductivity and a raise of the resistance to the current flow. These factors further impede the effective tissue heating and limit the size of RF-induced lesions. Maximal diameter of RF-induced lesion is achieved during the initial period of power application, subsequently the rate of increase of the lesion diameter diminishes rapidly as the resistance rises at the electrode-tissue interface and the current flow falls.[11] Lesions that occur in the vicinity of the electrode-tissue interface are the result of passive heat transfer from the shallow high temperature region.

The variation of temperature in cardiac tissue depends, in addition to RF energy absorption, on the tissue composition, blood perfusion, thermal conductivity of the tissue, and heat generation due to metabolic process. Temperature at the electrode-tissue interface and the temperature in the tissue can be indirectly assessed during RF application using tip temperature. Tip temperature is a valuable tool that provides important information regarding the adequacy of tissue heating, minimizing the development of coagulum and maximizing lesion size.[12]

The range of tip temperatures used in RFCA is 50°C to 90°C, although temperatures comprise between 60°C and 70°C are the target for many temperature-controlled catheters. These temperatures induce a rise of electrode-tissue interface temperature (70°C-90°C) that induces myocardial damage through a resistive heating. The temperature reached at deeper tissue is significantly lower and at this level the tissue damage is induced by conductive heating. When tissue temperature is lower than 50°C a reversible loss of cellular excitability occurs; irreversible tissue injury can be obtained only with tissue temperature $\geq$50°C. Abnormal automaticity has been observed at temperature > 45°C, whereas conduction block occurs at temperatures comprised between 51.7°C and 54.4>°C.[13] Most electrophysiological laboratories use to perform RFCA a thermocouple-tipped steerable 7F catheter connected to a RF generator. Maximal catheter tip temperature and power output are usually set respectively at 65°C and 75 watts. Each RF application is performed in controlled temperature and maximal ablation time for a particular target is 60 seconds.

Histological studies run in dogs after catheter ablation of the atrioventricular junction showed well-circumscribed areas of coagulation necrosis surrounded by a peripheral zone of hemorrhage and mononuclear cells and neutrophils infiltration 4 to 5 days after catheter ablation,. Two months after RFCA, areas of fibrosis, granulation tissue, fat cell deposition, cartilage formation and chronic inflammatory cell infiltration were observed.[14-15] Mature RF lesions within the ventricular myocardium have a maximal lesion width located intramurally because the endocardial surface is cooled by the intracavitary blood flow.

The most important determinants of lesion formation are: electrode-tissue contact and orientation, power, temperature and time set, electrode radius and length and type of RF generator. The dimension of a lesion created by RFCA primarily depends on the amount of tissue heated above the critical temperature of 50°C that creates an irreversible myocardial damage.[16] Because RF-induced tissue damage is the result of resistive heating, good electrode-tissue contact is mandatory. Appropriate contact can be revealed via a rapid rise in tip temperature and a drop in the tissue impedance (more than 10 Ohm). Resistive heating at electrode-tissue interface causes and exponential raise of tissue temperature to greater than 15°C within 4 seconds using 20 watt of RF power, whereas conductive heating present at deeper level require 1 to 2 minutes to equilibrates.[17] Temperatures between 70°C and 80°C, low power and longer time enable to obtain a good rise in tissue temperature preventing coagulation and increase in resistance. The electrode size can varies from 2 to 12 mm. Increase in size can augment the volume of the lesion through the increase electrode-tissue contact and decrease heat dissipation. This is not completely true for electrodes with a radius larger than 8 mm, because of the variability of the endocardial trabeculations and curvature, tissue perfusion, and intracardiac blood flow, which affect the heat dissipation and contact.

[18]. The use of cooled-tip electrode prevents tissue overheating, tip coagulation and rise of tissue impedance and permits to ablate area with low flow (i.e. coronary sinus).[19]

## CLINICAL APPLICATION & TECNIQUE

Radiofrequency catheter ablation is performed after obtaining a detail endocardial mapping to assess the arrhythmogenic mechanism, the site of origin of the arrhythmias, and to assess the part of the circuit or the automatic focus target of ablation.

In dogs, electrophysiological studies are performed under general anesthesia. The animals are placed in dorsal recumbence. Three multipolar electrode catheters are inserted (EP) in the venous system, with a modified Seldinger technique, using fluoroscopic and intra-cardiac ECG guidance. A decapolar catheter is inserted through the right external jugular vein into the coronary sinus; two quadripolar catheter through the right femoral vein, one to record the His bundle potential and one for programmed electrical stimulation, are placed in the high right atrium and right ventricular apex. For AFL a decapolar electrode catheter is positioned through the right external jugular vein into the coronary sinus while a second decapolar electrode catheter is inserted through the right femoral vein and placed around the tricuspid valve annulus to record with the proximal electrodes pairs the antero-lateral aspect of the crista terminalis potentials, with the middle pairs the middle lateral right atrium (potentials and with the distal pairs the low postero-lateral right atrium potentials. All intra-cardiac electrograms, 12-lead surface ECG are displayed and analyzed on an EP recorder.[3-9] Normal conduction intervals (A-H and HV), effective antegrade and retrograde refractory periods (ERP) and Wenckebach point of the normal AV node are obtained initially.

During EP study, presence and conduction properties of AV bypass tract can be assessed. Antegrade conduction is considered present in case of ventricular pre-excitation (either abnormally short, or negative His to ventricular interval) during normal sinus rhythm or while atrial pacing at progressively shorter cycle lengths. The presence of retrograde conduction through the accessory pathways (AP) is demonstrated by the evidence of eccentric VA activation during ventricular programmed pacing or tachycardia, and retrograde non-decremental conduction by programmed ventricular stimulation.[5,7-8]

Similarly to AV nodal ERP analysis, programmed atrial stimulation delivered at twice the pacing threshold is used to induce tachycardia and to study its location, type and mechanism. Tachycardia induction is considered reproducible if initiated with the same S1-S2 interval. Reentrant tachycardia such as AVRT, AFL or reentrant ventricular tachycardia, can be initiated and terminated as soon as S1-S2 interval encounter the refractory period of one of the two limb of the circuit.[1-6] Focal atrial tachycardia due to triggered activity or enhanced abnormal automatism can be initiated and terminated with rapid atrial pacing, atrial bursts and isoproterenol infusion.[8] Focal atrial tachycardia cannot be predictably and reproducibly initiated and terminated with programmed stimulation.[5,7-8,20] Focal atrial tachycardia can be differentiated from reentrant tachycardia by measuring first VA interval after cessation of overdrive atrial pacing.[5,7-8,20] In the latter case, it is usually fixed to testify the presence of a predetermined circuit. Resetting is the interaction of single extrastimulus with the tachycardia circuit resulting in a less than a compensatory pause (S1-S2 interval + pause = double of tachycardia cycle length). Resetting is used to prove that AP is part of supraventricular tachycardia circuit, and to differentiate atrioventricular nodal tachycardia from orthodromic AVRT with concentric VA conduction. A single ventricular extrastimulus during tachycardia is given while His is refractory, looking for pre-excitation of the atrium with identical activation sequence.[5,7-8,20] The entrainment technique is useful when study anatomical reentry. It is defined as a stable conduction that derives from the interaction of two rhythms.[20] The degree of interaction depends on the distance between the site of stimulation and the circuit, electrophysiological properties of tissues and presence of excitable gap. Pacing the atrium with a cycle length shorter than tachycardia cycle length accelerates the tachycardia cycle with the same activation sequence and different degree of fusion depending from the distance between the site of stimulation and the circuit. If the site of stimulation is within the circuit, no fusion appears (concealed entrainment).[20] In case of apparent entrainment, the first beat post-interruption of pacing could be used to assess the location of the tachycardia circuit. The length of this interval is proportional to this distance.

The sequence of atrial and ventricular activation during tachycardia is crucial to define the type and the location of tachycardia. Atrioventricular reciprocating tachycardia usually has a fixed ventriculo-atrial interval, whereas FAT has VA dissociation and variable VA interval.[5,7-8,20] Very often FAT presents the warm-up phenomenon: when atrial tachycardia starts, the cycle length tends to progressively shorten for several cycles until its ultimate rate is achieved. Retrograde atrial activation during orthodromic AVRT starts at atrial insertion of accessory pathways. It could be defined, either during tachycardia or ventricular pacing as the point of earliest atrial activation of bipolar signal. With the exception of antero-septal bypass tracts, atrial activation is eccentric and not decremental.[5,7-8,20] Concomitant unipolar recordings are used to optimize the accuracy of accessory pathways localization. Sharp and negative unipolar waveform indicates the exact site of impulse spreading.[5,7-8,20]

In human medicine, RFCA of accessory pathways has a global success rate of 97.4%, and significant complication rate of 0.14 %.[21] Success rate of RFCA of atrioventricular nodal reciprocating tachycardia resulted 96.1 % with 1% of iatrogenic atrioventricular block occurrence, whereas RFCA of accessory pathways has a 94 % success with a complication rate of 1.83%. Acute success rate of RFCA of AFL was 86% with a complication rate of 1.34%, whereas the success rate of RFCA of atrial tachycardia was 80% for right atrial focus, 72 % per left atrial focus and 52 % for septal focus. Success rate of RFCA of ventricular tachycardia was higher for those with idiopathic ventricular tachycardia compared to those with ventricular tachycardia due to ischemic heart diseases or cardiomyopathy. The differences in success rate or incidence of complications were not significantly different according to the volume of the center (< 100 ablation/year) or the presence of teaching activity. Initial studies reported an higher incidence of complications of radiofrequency catheter ablation of arrhythmias: global 5.1 %, atrial tachycardia atrial/atrial flutter 5 %, ablation of the AV junction 3.2 %, modification of atrioventricular junction in atrioventricular nodal reciprocating tachycardia 8 %, ablation of accessory pathways 4.4 %, ablation of ventricular tachycardia 7.5 %. Iatrogenic atrioventricular block was the most common complication deriving from RFCA (5.1%), whereas procedure-related death ranged from 0.08 to 0.13 %).[20-23] Recurrence rate reported for accessory pathways RFCA, FAT and AFL was respectively 5.5 %, 8%, and 7%.[23]

In our electrophysiological laboratory, so far we have a global success rate of 86 % with a complication rate of 8 % (1.7% stroke, 3.4% anesthesia-related deaths, and 5 % iatrogenic atrioventricular block that require pacemaker implantation) and a recurrence rate of 9%, only in cases of focal atrial tachycardia. Only three cases requires transeptal puncture with no complications noted. Success rate of RFCA of accessory pathway is 91 % with three 3rd degree atrioventricular block occurring post-ablation of 1 antero-septal and 2 mid-septal accessory pathways in three small breed dogs (body weight < 10 kg) with orthodromic AVRT resistant to medical therapy and provoking tachycardia-induced cardiomyopathy. Acute success rate of RFCA for focal atrial tachycardia is 74 %, since 26 % of focal atrial tachycardia was not-inducible with programmed or incremental atrial stimulation, atrial burst or isoproterenol infusion. So far we had only one minor stroke with spontaneous resolution during RFCA of a pulmonary vein ostium automatic focus. Other procedures including few cases of RFCA of AFL and ventricular tachycardia arising from the right ventricular outflow tract in a dog with segmental arrhythmogenic right ventricular cardiomyopathy showed a high acute success rate with low complication and recurrence rate.

In the veterinary literature, accessory pathways are the most common target of ablation.[3-9] In dogs, most accessory pathways are right-sided, have unidirectional retrograde conduction, and are associated with various arrhythmias, including orthodromic AVRT and atrial fibrillation without evidence of pre-excitation.[7] Anatomical distribution of accessory pathways reported in dogs include most commonly postero-septal and right-free wall locations. For AVRT ablation catheter is placed at the atrial endocardial insertion of accessory pathways around the tricuspid valve annulus or within the coronary sinus. Ablation procedures are guided by intra-cardiac bipolar recording, searching for the shorter VA interval, or bypass tract potentials, during orthodromic AVRT or ventricular pacing, and with unipolar recording with sharp and negative waveform, which permit to assess the pole closest to the bypass tract.[5,7-8,20] For ablation within the coronary sinus, particularly in case of postero-septal AP, a cool-tipped ablation system should be used.

We found that the most common sites of origin of automatic tachycardias in dogs are along the crista terminalis, right atrial appendage, triangle of Koch, pulmonary veins, and the coronary sinus. Mapping these sites of origin is usually performed using both bipolar and unipolar recordings. The goal of mapping is to identify the electrical epicenter of atrial activation characterized by the earliest high-frequency and high-amplitude signal on a bipolar recording timed to the surface P waves (usually precedes the onset of P waves by a minimum of 20 ms, mean 38 ms).[24] Sharp and negative unipolar recording are also useful in this setting.

Radiofrequency catheter ablation of the cavo-tricuspid (CTI) dependent atrial flutter is considered the treatment of choice of CTI-dependent AFL in people.[1] The chronic success rate after the induction of bidirectional block in the CTI is 90-100%.[25] Ablation can be performed during sustained atrial flutter or sinus rhythm. The target of ablation is the cavo-tricuspid isthmus, an area of slow conduction in the low posterior right atrium between the tricuspid valve orifice and the inferior vena cava and the Eustachian ridge. Endpoint of ablation is the induction of bidirectional conduction block at the CTI demonstrated by a collision of waveforms, while pacing the coronary sinus, with a supero-inferior activation of the right free wall post-ablation, by the presence of a slow conduction from low right free wall and coronary sinus during pacing of controlateral site, and by recording double potential at the site of linear ablation at the CTI.[9,20,25] Cavo-tricuspid dependent atrial flutter is a rare arrhythmia in dogs and so far there is only one report of a long-term successful ablation of CTI in a dog with bidirectional AFL.[9]

In human beings RFCA has been used to ablate arrhythmic substrate of different ventricular tachycardia both in presence of heart diseases (coronary artery diseases and cardiomyopathy) and in structurally normal heart. Radiofrequency catheter ablation is performed in patients with low risk of sudden cardiac death with sustained monomorphic ventricular tachycardia resistant to drug therapy or who are drug intolerant.[26] Due to the short life expectancy of dogs with dilated cardiomyopathy or hypertrophic cardiomyopathy and the rarity of coronary artery disease, RFCA of ventricular tachycardia in dogs with heart disease is usually not performed. In our experience, dogs with a particular type of sustained monomorphic ventricular tachycardia associated with localized right ventricular outflow tract aneurisms caused by segmental arrhythmogenic right ventricular cardiomyopathy are good candidates to ablation therapy.[9] In these dogs, ventricular tachycardia arising from the right ventricular outflow tract can be mapped within the localized aneurisms searching for fractionated complex electrograms and diastolic potentials as ablation targets. Pace-mapping can be use in case of rapid ventricular tachycardia with hemodynamic compromise.[27-29]

## REFERENCES

[1] Huang SK, et al. **J Am Coll Cardiol** 1987;9:349. [2] Guidelines Am Coll Cardiol/Am Heart Assoc. **Circulation** 1995; 92:673. [3] Sherlag BJ, et al. **J Am Vet Med Assoc** 1993;203:1147. [4] Wright KN, et al. **J Vet Intern Med** 1999;13:361. [5]Santilli RA, et. al. **J Vet Cardiol** 2006. [6] Foster SF, et al. **Aus Vet J** 2006;84:326. [7] Santilli RA, et al. **J Am Vet Med Assoc** 2007;231:393. [8] Santilli RA, et al. **J Vet Intern Med** 2008:22:915. [9] Santilli RA, et al. **Proceedings 18th ECVIM-CA congress**, 2008;1:222. [10] Wasgal AB, et al. In Lin JC, ed. Plenum press, NY, 1997;1:205. [11] Blouin LT, et al. **PACE** 1989;12:136. [12] Dinerman JL, et al. **J Cardiovasc Electrophysiol** 1996;7:163. [13] Simmers TA, et al. **J Am Coll Cardiol** 1995;18:323. [14] Hoyt R, et al. **J Appl Cardiol** 1986;1:469. [15] Langberg J, et al. **PACE** 1990;13:1242. [16] Nath S, et al. Circulation 1993;88:1826. [17] Avitall B, et al. **J Am Coll Cardiol** 1995;41:705. [18] Haines D, et al. **Circ Res** 1990;8:293. [19] Ruffy R, et al. **J Cardiovascul Electrophysiol** 1995;6:1089. [20]Josephson ME, Clinical Cardiac electrophysiology. Technique and interpretations,(ed 3), Philadelphia PA: Lippincott Williams & Wilkins 2002,p. 322 .[21] Scheinman MM, et al, **Pacing Clin Electrophysiol**. 2000;23:1020. [22] Hindricks G, et al. **European Heart Journal** 1993;14:1644. [23] Kay GN, et al. **J Cardiovasc Electrophysiol** 1993;4:371. [24] Walsh EP, et al. **Circulation** 1992;86:1138. [25] Chen SA, et al **J Am Coll Cardiol** 1996;27:860. [26] Zipes PD, et al. **Europace** 2006;8:746. [27] HottKamp H,et al. **J Cardiovasc Electrophysiol** 2006;17:477-479. [28] Yao Y,et all. **Pacing Clin Electrophysiol** 2007;30:526-533. [29] Nogami A, et al. **J Cardiovasc Electrophysiol** 2008;19:681-688.

# CARDIAC ION CHANNELS IN HEALTH & DISEASE

**Bruce G. Kornreich DVM, PhD, DACVIM (Cardiology)**
**Ithaca, NY**

Ion channels form the basis for excitability in cardiac tissue. These membrane bound proteins are composed of subunits arranged around a central channel pore that allows the passage of current in the form of ions when the channel assumes an open conformation. The stimulus that induces the channel to make the transition from the closed conformation found in the quiescent cell to the open conformation that passes current (i.e. that "gates" the channel) may take the form of a change in transmembrane voltage, the binding of an extracellular agonist, the activity of an intracellular second messenger, or stretch of the cell membrane. This presentation will focus primarily on the structure and function of voltage gated sodium, calcium, and potassium channels as they pertain to cardiac physiology and pathophysiology.

A central theme in voltage gated channel structure is the evolution of a scheme in which multiple subunits coalesce around a water filled central channel pore that traverses the lipid bilayer of the cell membrane. Each individual subunit is an integral membrane protein composed of amino acids linearly linked by protein bonds. This initially linear structure traverses the membrane several times to form a gross transmembrane topography, and the structural complexity of ion channels is compounded by specific protein folding that is vital to their normal function. The portions of the subunits that traverse the membrane (transmembrane, or TM domains) are generally hydrophobic and form helical structures (primarily α helices), while those portions protruding into either the intracellular or extracellular space are hydrophilic and are likely to be in a less ordered conformation than the TM domains (i.e. random coil). The intracellular and extracellular portions of a subunit linking TM domains are often referred to as "loops", while the ends (termini) are termed amino terminal (N terminal) and carboxy terminal (C terminal) in reference to the chemical moieties found at these sites (Figure 1). Generally speaking, α subunits form the portions of a protein that line the channel pore upon assumption of an open conformation, while auxiliary subunits (i.e. β and γ ) may interact with other proteins important for processes including trafficking, anchoring, and/or modulation of channel function. The number of subunits that aggregate to form a functional channel varies among subgroups, but voltage gated channels generally possess four subunits with six TM domains in each subunit. Channels within a given subtype are usually composed of homologous subunits that may be expressed in varying combinations, providing a complex mechanism by which channel function may be modified in response to various stimuli. Voltage gated ion channels are often the target of pharmacologic therapeutic agents (i.e. sodium channel blockers, calcium channel blockers).

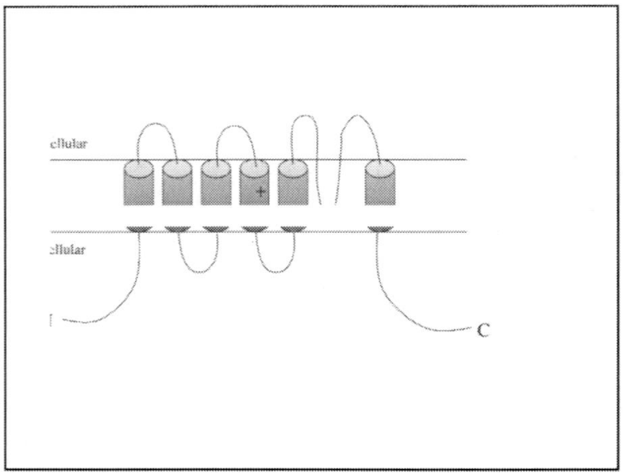

**Figure 1.** Transmembrane topology of a stereotypic voltage gated ion channel subunit. Barrels represent transmembrane domains, lines represent loops and amino (N) and carboxy (C) termini. P refers to the semipermeant loop that forms the channel pore. + + indicates the voltage sensor domain.

The action potential, which describes the change in membrane voltage over time, is the primary electrical signal generated by excitable cardiac cells. This wave of electrochemical activity is initiated and propagated through cardiac tissue via changes in membrane permeability that are mediated by the opening and closing of ion channels. The protein structure of these ion channels largely determines the resistance (R) to flow across them at any given time and, along with the concentration of permeant ions on either side of the membrane (which comprises the transmembrane voltage, V), determines the amount and direction of current (I, charge/unit time) that flows across the channel. The relationship between these parameters is described by Ohm's Law:

$$V=IR$$

The mechanism of voltage sensitivity of a voltage gated ion channel lies in what has been termed the voltage sensor domain of ion channels within this subgroup. Composed of a region dense in (usually positively) charged resides located in one of the TM regions, the voltage sensor is moved by attractive/repulsive forces induced by changes in membrane potential. If of sufficient magnitude, this movement, which is proportional to changes in membrane potential, can induce ion channel proteins to assume a conformation that energetically favors the assumption of a particular state (i.e. open or closed). Evolved over eons, voltage sensitivity provides an eloquent mechanism by which excitable cells can propagate electrical activity in the intact heart.

Depolarization of cardiac tissue results from the influx of positively charged ions (cations) down their electrochemical gradient from the extracellular space to the intracellular space once the voltage gated channel responsible for depolarization in a given anatomic region opens. Generally speaking, voltage gated sodium channels are responsible for the initial depolarization (phase 0) of atrial myocardial cells, ventricular myocardial cells, and of cells within the rapidly conducting His-Purkinje system upon stimulation. Cells within the sinoatrial (SA) and atrioventricular (AV) nodes are dependent upon the flow of calcium through voltage gated calcium channels for phase 0 depolarization. In addition to this gross anatomic variation in distribution of voltage gated channels within the heart, significant anatomic variation in the expression of channel subtypes with respect to right / left, apical / basal, and endocardial / epicardial symmetry provides an exquisite system for increasing the complexity of electrophysiologic properties across the different tissues of the heart. An additional level of complexity in cardiac electrophysiology is imparted by modulation of channel function by posttranslational modification such as phosphorylation by kinases, and many cardiac disease states are characterized by the altered expression and/or function of ion channels, resulting in abnormal electrical (and often resultant mechanical) function. Several examples of this phenomenon will be discussed.

Once a voltage gated ion channel makes the transition from the rest state (i.e. quiescent at resting membrane potential) to the open state (passing current), it remains in this conformation for a period of time (μsec to msec) before making the transition to a third, inactive conformation. Like the rest state, the inactive state does not pass current, but it is structurally distinctive from the rest state conformation of the channel protein. After residing in the inactive state for a period of time (usually msec), the channel returns to the rest state and is then available to be excited (i.e. to gate) once again, provided that it is allowed to return to its physiologic resting potential (Figure 2). It is important to note that the channel, in most cases, must cycle through the inactive state to return to the fully excitable rest state once it has opened, and that attempts to depolarize the cell while the channel responsible for depolarization of that cell (i.e. voltage gated sodium/calcium channels) is not fully recovered from inactivation will result in a failure to depolarize the cell or in a decreased current amplitude across the channel. This altered excitability forms the basis for a number of clinically significant electrophysiologic phenomena, including aberrancy and conduction block. Studies have shown that certain antiarrhythmic agents have a higher affinity for one state of a voltage gated channel than for others. Lidocaine, for example, has a higher affinity for the inactive state of the voltage gated sodium channel, making it a better therapeutic option for rapid ventricular tachycardia (VT), during which the channel spends a greater percentage of time in the inactive conformation. This increase in affinity of an antiarrhythmic for a conformation of an ion channel associated with tachycardia is termed use dependence. Procainamide, conversely, has a higher affinity for the rest state of the voltage gated sodium channel. Since the channel spends a greater percentage of its time in the rest state during slower rhythms, procainamide may be a good therapeutic option for the treatment of relatively slow VT. Increased affinity of an antiarrhythmic agent for a conformation of an ion channel associated with bradycardia (i.e. the rest state) is termed reverse use dependence. The time spent within one state and the rate of transition between states of an ion channel are described by channel kinetics (i.e. activation rate, inactivation time constant), which may have important implications for mechanisms of arrhythmias and antiarrhythmic therapy.

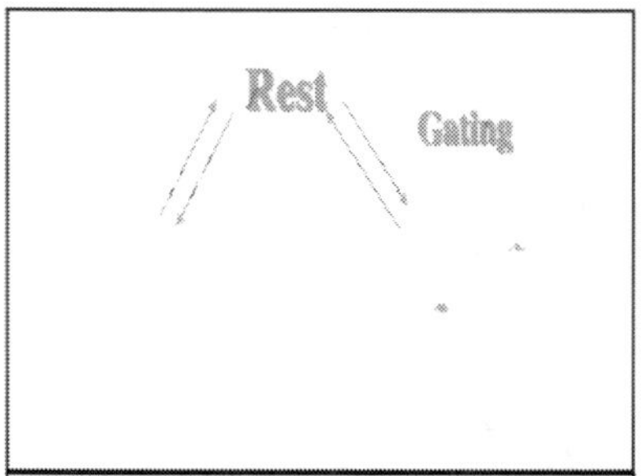


**Figure 2.** Simple Markov model describing the transition of a voltage gated channel among three protein conformations (states).

The plateau of the cardiac action potential is largely a balance between the inward flow of calcium ions through L-type calcium channels and the outward flow of potassium ions through voltage gated potassium channels. The calcium that enters during this portion of the action potential serves as a signal transduction molecule for the release of sarcoplasmic reticulum calcium stores, which the diffuses to the actin-myosin apparatus to initiate contraction. The duration of this portion of the action potential, combined with the rate of removal of calcium from the intracellular space via a number of mechanisms, can have important implications for both systolic and diastolic function. Given the importance of calcium in both the electrophysiologic and mechanical function of the heart, modulators of transmembrane calcium current and of intracellular calcium handling are important tools in the management of a number of clinically important cardiac diseases.

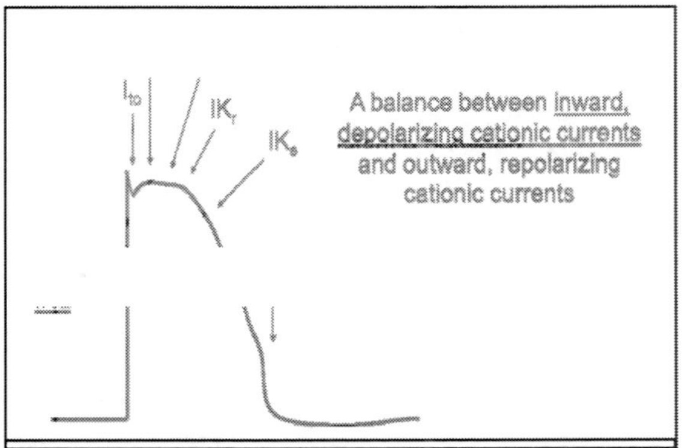


**Figure 3.** Action potential morphology is largely the result of a balance between inward, depolarizing currents (underlined), and outward repolarizing currents, as shown for this prototypical sodium channel.

Cardiac potassium currents have been subdivided into rapid and slow components of a delayed rectifier (IKr and $I_{Ks}$, respectively), an inward rectifier ($I_{K1}$), and rapidly and slowly recovering components of a transient outward current ($I_{tof}$ and $I_{tos}$, respectively) based largely upon activation and inactivation kinetics and voltage dependence. These repolarizing currents have important implications for action potential duration (APD). APD is roughly correlated with QT interval on the surface electrocardiogram, and a number of clinically important naturally occurring and iatrogenic cardiac arrhythmias are characterized by abnormal potassium channel function. Figure 3 presents a stereotypic sodium channel action potential demonstrating the cationic cardiac currents that will be discussed.

With recent advances in molecular biology, protein biochemistry, nuclear magnetic resonance spectroscopy, x-ray crystallography and other techniques, our understanding of the role of ion channels in health and disease has expended considerably. These modalities have shed new light on the importance of ion channel protein structure in their normal function, and on the potentially devastating effects that mutations can have on the structure and function of ion channels at the molecular level and on cardiac function at the organ level. This discussion will review recent advances in our understanding of cationic cardiac ion channels as it pertains to normal cardiac function and to the etiology, diagnosis, and treatment of a number of clinically important diseases.

## REFERENCES


[1.] Clancy, CE, et. Al. **J Clin Invest** 2002;110(8):1078; Nerbonne LM, et. al. **Physiol Rev** 2005;85(4):1205; [2.] Tamargo J et. al. **Cardiovasc Res** 2004;62(1):9; [3] Abbott GW et. al. **J Electrocardiol** 2007;40 (6 Suppl):S42; [4.] Kambouris et. al. **Circulation** 1997;97:640; [5.] Bean BP et. al. **J Gen Physiol** 1983;81:613; [6.] Doyle DA et. al. **Science** 1998;280:69; [7.] Delisle BP et. al. **Circ Res** 2004;94:1418; [8.] Roden DM et. al. **Ann Rev Physiol** 2002;64:431;[9.] Nattel S **Physiol Rev** 2007;87:425; [10] Hulme JT et. al. **J Mol Cell Cardiol** 2004;37(3):625; [11.] Schram G et. al. **Circ Res** 2002;90:939; [12.] Modell SM et. al. **Genet Med** 2006;8(3):145

**DEPENDENT AND INDEPENDENT VARIABLES AFFECTING CANINE N-TERMINAL PRO-B-TYPE NATRIURETIC PEPTIDE LEVELS AND THEIR IMPACT ON THE CLINICAL ACCURACY OF THE ASSAY.**

**SJ Ettinger[1], SD Forney[1], A Beardow[2], G Farace[2], K Yeung[2].**
**1. California Animal Hospital Veterinary Specialty Group, Los Angeles, CA. 2. IDEXX Laboratories Inc., Westbrook, ME.**

The traditional methods available for diagnosing cardiac disease in canines consist of physical examination including auscultation, electrocardiography, thoracic radiography, echocardiography and doppler blood flow/tissue analysis. Recently, cardiac biomarkers have become commercially available and offer an additional objective assessment of underlying cardiac health in the canine patient. Our goal was to clinically validate the accuracy of this assay and determine the dependent and independent variables affecting canine N-terminal pro-B-type natriuretic peptide (NT-proBNP) levels. Potential confounding variables on NT-proBNP concentrations were also assessed. Serum NT-proBNP levels were analyzed from 1130 dogs. The diagnostic accuracy of serum NT-proBNP levels was determined by correlating results with physical, radiographic and echocardiographic indicators of cardiac disease severity.

The effect of age, weight, gender, reproductive status, body condition score (BCS) and indicators of renal function were evaluated using regression analysis. Serum concentrations of NT-proBNP correlated to the underlying disease status (LA/Ao $p< 0.0001$) and were not significantly impacted by age, weight, and BCS and weakly impacted by the gender ($p= 0.04$), and reproductive status ($p= 0.01$) of the patient. There was a slight positive association between increasing concentration of NT-proBNP and elevated creatinine levels ($p= 0.0007$).

Canine NT-proBNP levels are a practical biomarker of cardiac disease and a potentially beneficial supplement to the basic cardiac diagnostic workup. These values should be interpreted in consideration of the patient's renal function. Additional studies are needed to better understand the possible benefits of using this assay to monitor disease progression and response to therapy.

## CANINE N-TERMINAL PRO-B-TYPE NATRIURETIC PEPTIDE AND ECHOCRDIOGRAPHIC MEASUREMENTS IN SMALL BREED DOGS WITH AND WITHOUT CARDIAC DISEASE.

**A Beardow[1] SJ Ettinger[2], SD Forney[2], G Farace[1], K Yeung[1].**
**1. IDEXX Laboratories Inc., Westbrook, ME. 2. California Animal Hospital Veterinary Specialty Group, Los Angeles, CA.**

Cardiac markers such as N-terminal pro-B-type natriuretic peptide (NT-proBNP) have been shown to have utility in diagnosing heart disease in the dog. The goal of this study was to retrospectively examine a cohort of 213 small breed dogs that had been enrolled as part of a larger study evaluating the performance and utility of NT-proBNP in the clinic and to determine if there was a correlation between the echocardiographic parameters and NT-proBNP.

NTproBNP concentration, LA/Ao ratio and LA all show a correlation with the severity of the underlying cardiac disease. Indeed NTproBNP and LA/Ao seem to increase in concert and show a moderate but significant correlation with each other ($p<0.0001$). Other echo parameters such as fractional shortening and Ao, and the vertebral heart score measured from lateral chest radiographs did not show any correlation with disease severity.

There is no difference in NTproBNP and LA/Ao between those individuals where disease only involved the mitral valve and those where disease involved both mitral and tricuspid valves, however, both were significantly elevated over those dogs with no valve disease. ROC analysis shows that 92% specificity is achieved with an NT-proBNP cut-off of 585pmol/L while an LA/Ao ratio of 1.36 gives 91% specificity; sensitivities are low - 58% and 67% respectively. This suggests that NTproBNP and LA/Ao can both be used to 'rule-out' valve disease in these patients.

This study shows that there is a correlation between LA/Ao and NT-proBNP and that these two factors can be correlated with valve disease.

**NT-PROBNP ELEVATION IN DOGS WITH NON-CARDIAC RELATED PULMONARY HYPERTENSION.**

**HB Kellihan, BA MacKie, RL Stepien;**
**Department of Medical Sciences, University of Wisconsin School of Veterinary Medicine, Madison, Wisconsin.**

N-terminal pro-B-type natriuretic peptide (NT-proBNP) concentration has been used to discriminate between cardiac disease and respiratory disease in dogs. However, [NT-proBNP] have not been reported in dogs with non-cardiac related pulmonary hypertension (PH). In people, [NT-proBNP] is elevated in the presence of non-cardiac related PH, and is utilized to stratify disease severity, monitor response to treatment, and serves as a prognostic parameter. The purposes of this study were to examine [NT-proBNP] in dogs with non-cardiac related PH and to evaluate if [NT-proBNP] reflect the degree of non-cardiac related PH.

Fourteen dogs with non-cardiac related PH were examined prospectively (n=11 primary pulmonary disease, n=1 heartworm disease, n=1 reverse patent ductus arteriosus). All dogs had no evidence of echocardiographically apparent left-sided cardiac abnormalities and had a diagnosis of PH based on Doppler assessment of peak tricuspid regurgitation (TR) gradient, which was used to indirectly assess the systolic pulmonary artery pressure (SPAP). Plasma samples for NT-proBNP assay (Canine CardioCare, Veterinary Diagnostic Institute, Irvine, CA, USA) were obtained from dogs at the time of diagnosis.

The median age was 11 years (range: 2-15 years, n=14). Seven were female spayed and 7 male neutered dogs. The median weight was 9.5 kilograms (range: 3.1-38 kilograms). [NT-proBNP] was elevated above the currently accepted reference range (566 pmol/L) in 10/14 dogs with a median concentration of 1780 pmol/L (range: 248->3000 pmol/L) for all dogs. Dogs were grouped by severity of PH as estimated by peak TR gradient (mild: SPAP 31-50mmHg, moderate: SPAP 51-75mmHg, severe: SPAP >75mmHg). [NT-proBNP] in dogs with mild PH was significantly lower (median: 468 pmol/L, range: 248-1026 pmol/L) than [NT-proBNP] in dogs with severe PH (median: 2446 pmol/L, range: 1326->3000 pmol/L) (p=0.0025). A significant correlation was found between SPAP and [NT-proBNP] (r=0.61, p=0.0217). [NT-proBNP] are elevated in dogs with non-cardiac related PH, and concentrations correlate with severity of PH.

**IN-HOSPITAL ELECTROCARDIOGRAPHY OVERESTIMATES 24-HOUR VENTRICULAR RATE IN DOGS WITH ATRIAL FIBRILLATION.**

**AR Gelzer, M Rishniw, MS Kraus.**
**Cornell University, Ithaca, NY.**

Determining response to rate-control therapy in canine atrial fibrillation (AF) relies on accurate assessment of the ventricular rate (VR). Clinicians most commonly determine VR by auscultation or brief in-hospital electrocardiography (ECG). However, 24-hour Holter ($Holter_{24H}$) monitoring more accurately reflects VR during the patient's normal daily routine, and is used in humans with AF to evaluate drug efficacy. No veterinary studies have documented the ability of in-hospital ECG to predict the patient's VR in the home environment. Thus, we sought to determine if brief in-hospital ECG-derived VR could substitute the more cumbersome and expensive Holter-derived VR in managing AF patients.

We acquired 81 $Holter_{24H}$ and a 1-minute ECG (just prior to Holter application) from 34 dogs with AF of multiple etiologies, receiving various rate-control and heart failure therapies. We compared average VR obtained by each method by Limits of Agreement analysis. The ECG-derived VR overestimated $Holter_{24H}$ VR by an average of 20bpm at VR of 90-140bpm, by 30bpm at VR of 140-200bpm, and by 50bpm at VR 200-250bpm; 95% limits of agreement ranged from -40 to +40bpm throughout the VR range.

Our results show that ECG assessment of VR in dogs with AF does not reliably estimate the patient's VR in their home environment; wide limits of agreement prevent simple derivation of $Holter_{24H}$ VR from ECG VR. Thus, optimal rate control in AF requires Holter-based assessment of VR.

**PHENOTYPICAL VARIATION OF HYPERTROPHIC CARDIOMYOPATHY IN DOMESTIC SHORTHAIR CATS, ECHOCARDIOGRAPHIC CLASSIFICATION.**

**D Brizard, C Amberger, S Hartnack, M Doherr, C Lombard.**
**Geneva reference centre and Vetsuisse University, Berne CH.**

Purpose of the study: to use an echocardiographic classification of feline hypertrophic cardiomyopathy in relation to the variation of phenotypical expression of myocardial hypertrophy.

Animals and methods: 111 DSH with clinical HCM based on heart murmur and clinical signs of heart disease, has been classified into 5 pattern of hypertrophy in relation to anatomical variation and necropsy findings.

Inclusion criteria: Domestic shorthair with heart murmur or other signs of heart diseases.

Exclusion criteria: Cats with congenital heart diseases, hyperthyroidism or other metabolic diseases.

Hypothesis: The structural variation of feline hypertrophic cardiomyopathy is the key of understanding and treatment of this common disease.

Results: The 111 studied cats were classified into five basic patterns of hypertrophy. Seven cats had hypertrophy of papillary muscles only (A: 6.3%), 13 cats had hypertrophy of the basal and mid-anterior part of the septum (B: 11.8%), 16 cats had generalized septal hypertrophy (C: 13.6%), 67 cats had hypertrophy involving septum and free wall (D: 60.9%) and 8 cats had hypertrophy confined to the free wall (E: 7.2%).

Differences in LA dimensions between phenotypes were significant (P<0.001. Enlargement was extreme in cats belonging to phenotype E.

*Clinical presentation:* Significant differences were found among cats in the 5 phenotypes of hypertrophy in regard to age (p=0,007), heart rate (P=0,003) and ISACHC class.

Heart rate tended to be higher in phenotypes C, D and E comparing medians among groups. All 111 cats with HCM presented a murmur, 59 percent had a grade 2 and 34 percent had a high-grade murmur. No significant differences were found among phenotypes but high-grade murmurs were frequently found in cats with phenotypes C and D.

*Turbulences in the left ventricle:* Aortic turbulences without other concomitant turbulences occurred more frequently in cats with hypertrophy confined to the septum (19 of 29 cats with type B and C of hypertrophy [65.5 percent] ) than in all the other phenotypes (P<0.001). Cats with hypertrophy involving both septum and free wall had the highest prevalence of aortic turbulences together with mitral regurgitation (24 of 67 cats with type D [36 percent]; P=0.004). Mitral regurgitation alone was over-represented in phenotypes where hypertrophy involves the papillary muscles (6 of 7 cats with type A and 7 of 8 cats with type E; P=0.027 and P=0.049). Mitral regurgitation results in left atrium dilatation in all cats with phenotype E.

**COMBINATION OF PIMOBENDAN, BENAZEPRIL, AND FUROSEMIDE IN THE TREATMENT OF CONGESTIVE HEART FAILURE DUE TO MITRAL REGURGITATION IN DOGS.**

**E De Madron. Alta Vista Animal Hospital, Ottawa, ON, Canada.**

Both pimobendan and benazepril have proven benefits in the treatment of dogs in CHF due to mitral regurgitation (MR). The purposes of this retrospective study were to evaluate cardiac status at onset of CHF and survival times (onset of CHF to death of any cause) using a combination treatment of pimobendan, benazepril and furosemide.

We reviewed the data of 51 dogs with MR and CHF. Echocardiographic (end systolic volume index: ESVI, regurgitant fraction: RF) and radiographic (Vertebral Heart Score: VHS) data at onset of CHF was available in 24 dogs. Kaplan-Meier survival curves and median survival times were calculated for the 28 dogs weighing 5 to 20kgs, and compared to data from the QUEST study (Haggstrom *et al*, 2008).

Mean ( IQR) survival times were 525 (206-719) days, which is almost double the value of 267 (122-523) days of the QUEST pimobendan cohort. The ESVI was normal in 71 % of dogs < 20 kg, and increased in 83% of dogs > 20 kg. The mean RF was 57%. There was no correlation between VHS and RF and VHS and ESVI.

We conclude that combining benazepril, furosemide and pimobendan may improve survival time over furosemide and pimobendan alone. Systolic function appears still normal in most dogs < 20 kg at onset of CHF, and radiographs are not good predictors of systolic function.

**SURVIVAL CHARACTERISTICS OF 50 DOGS WITH CONGESTIVE HEART FAILURE SECONDARY TO DEGENERATIVE MITRAL VALVE DISEASE TREATED WITH STEPWISE POLYPHARMACOTHERAPY.**

**N. Van Israël, Animal CardioPulmonary Consultancy (ACAPULCO), Masta, Belgium.**

Congestive heart failure (CHF) secondary to degenerative mitral valve disease (DMVD) is the most common cardiac reason for death in dogs. Although all the available survival data from comparative trials have added a lot of useful information, they do not reflect current specialty practice. It is now believed by most specialists that the ideal therapy of CHF includes at least furosemide, ACEI and inodilators. There is currently very little evidence to underpin this. Additionally, the impact of specialist follow-up of out-patients with CHF has now been established in multiple human clinical trials. No data on this is available in veterinary medicine.

The aim of this retrospective study was to determine the survival characteristics of a population of dogs with CHF secondary to DMVD, treated with stepwise polypharmacotherapy (ACEI and furosemide and later pimobendan and other cardiac drugs) and rigorously followed by a specialist.

The case records of all of the patients diagnosed with CHF secondary to DMVD at one referral practice, treated with at least triple therapy and followed over time by the same specialist were reviewed. All dogs with other medical conditions that could influence survival were excluded from the analysis. The animals were divided in those that died for cardiac reasons (natural or euthanasia), those that died for non-cardiac reasons and those that were still alive (censored).

Dogs diagnosed between February 2003 and April 2008 were included (n=50). The breeds most commonly represented were Bichon (8), Cavalier King Charles Spaniel (6) and Chihuahua (4). Twenty-six dogs were male (9 neutered) and 24 female (15 neutered). Eleven dogs were less than 10 years of age at the time CHF was confirmed. Eight dogs weighed more than 20 kg. Seventy-eight percent of the dogs also had evidence of degenerative tricuspid valve disease (DTVD), 54% developed pulmonary arterial hypertension (PAH) at some stage, 44% ruptured a chord (CR) and 18% developed atrial fibrillation (AF). The Kaplan-Meier estimate of the median survival time of this study population was 640 days (95% CI 472 – 824 days). Similar estimates for dogs younger than 10 years and older dogs at time of diagnosis of CHF were respectively 371 and 705 days (p-= 0.011). The median survival time for dogs that develop atrial fibrillation was 472 days as compared to 729 days for dogs without (p=0.012). There were no evidence of any effect of gender, body weight, the presence of DTVD, the development of PAH and the occurrence of CR on survival. The above findings were confirmed by a Cox proportional hazard model.

This study suggests that staged polypharmacotherapy, based on correct diagnosis and initiated following the development of radiographic evidence of CHF, tailored throughout follow-up to suit the needs of the individual patient and monitored carefully by experienced veterinarians, improves survival in dogs with CHF secondary to DMVD.

# PATENT DUCTUS ARTERIOSUS FROM A TO Z

**James W. Buchanan, DVM, MMedSci, DACVIM (Cardiology)**
**Philadelphia, PA**

Images for this text can be seen at: http://www.vin.com/library/general/JB110pda.htm

## INTRODUCTION

A patent ductus arteriosus (PDA) is an arterial shunt between the pulmonary artery and the aorta. It is normally open in fetuses but should constrict and close within hours after birth. Incomplete closure (patency) yields a variable size channel whose minimum diameter primarily determines the amount and direction of blood flow through the shunt. PDA is the most common congenital cardiovascular abnormality in dogs and is caused by a structural defect in the ductus wall.

## NORMAL ANATOMY AND FUNCTION

The ductus arteriosus is commonly called the arterial duct, arterial canal, or ductus Botalli in Europe. It extends from the bifurcation of the main pulmonary artery to the ventral aspect of the descending aorta between the left subclavian and intercostal arteries. In fetuses, it carries relatively oxygenated blood from the right heart into the descending aorta thus bypassing the nonfunctioning lungs. The ductus is functionally closed by constriction in the first few hours after birth and later by anatomic closure in 2-4 weeks.

## ANGIOGRAPHY and ECHOCARDIOGRAPHY

PDAs vary considerably in diameter, shape and length. The majority of PDAs are funnel shaped with maximal narrowing at the pulmonary artery orifice. They are categorized as long or short and tubular or tapering based on angiography or echocardiography for the purpose of selecting an occluding device or recommending surgery. These distinctions are arbitrary labels in a continuum of the histologic abnormality that causes PDA. Most of the ductus arteriosus lies within the wall of the aorta in dogs (not in humans) and the aortic wall segment may not be evident externally. Dogs with medium or large PDAs usually have enlarged pulmonary arteries due to left-to-right PDA flow and they have a ventral and bilateral bulge in the aorta. The dilated portion is commonly called a ductus diverticulum or ampulla but it is essentially a ductal aortic aneurysm and remains aneurysmal after surgical or transcatheter PDA closure. Doppler echocardiography reveals turbulent left-to-right flow into the main pulmonary artery as well as left heart enlargement. Dogs with large PDAs, pulmonary hypertension and right-to-left shunting may not have an aneurysm and usually do not have significant pulmonary artery enlargement.

## GROSS PATHOLOGY

External examination of gross specimens does not reveal the true size of the defect since most of the ductus length lies within the wall of the aorta. Internal PDA diameter varies with location but the narrowest segment is almost always at the pulmonary artery orifice where an intimal ridge or short tunnel defines the minimum diameter. Vary rarely, the narrowest segment is found in the middle or towards the aortic end of the ductus. Generally, the external PDA length varies inversely with the size of the aneurysm. When the PDA is very short and the aneurysm is large, the external segment resembles an aorta-pulmonary window although it is not the same defect embryologically.

## HISTOPATHOLOGY

Consistent histopathologic changes are found in dogs with PDA. The ductus muscle is hypoplastic and eccentrically located and the ductus wall adjacent to the aorta has non-contractile, inappropriate aorta-like elastic tissue. (See details in the etiology section)

## DIAGNOSIS

A continuous (machinery) murmur at the left heart base is the cardinal sign of a PDA. Additional studies are performed to assess secondary changes, determine the urgency of PDA closure and to rule out coexisting abnormalities. The intensity, frequency, duration and radiation of the murmur are determined by the diameter and length of the PDA and the degree of pulmonary hypertension. Dogs with a small PDA have a focal, relatively high pitched, continuous murmur heard only at the left heart base. The continuous murmur in dogs with a medium or larger size PDA usually can be heard all over the thorax and is accompanied by a precordial thrill. When the heart is enlarged, a systolic murmur of mitral regurgitation due to annular dilation also may be heard at the left apex. Dogs with a large PDA may have so much flow through the PDA that diastolic blood pressures equalize in the aorta and pulmonary artery and the diastolic murmur is abbreviated or not present. The pathognomic radiographic sign of a medium or large PDA is a leftward bulge of the aorta in a DV or VD radiograph at the one-o'clock position caused by a ductal-aortic aneurysm. Other radiographic signs of a medium or large PDA include left heart enlargement and increased pulmonary vascular markings. Pulmonary edema is seen when congestive heart failure develops. Electrocardiograms usually have high amplitude R waves indicative of left ventricular hypertrophy. Atrial fibrillation is the most common arrhythmia. Echocardiograms typically show an enlarged, hyperdynamic left ventricle. Doppler interrogation shows continuous, turbulent, retrograde flow into the main pulmonary artery from the aorta.

The clinical signs of PDA can be grouped into four types of cases:

**Type 1** (Small PDA)
- Asymptomatic left-to-right shunt
- High frequency continuous murmur only at the left heart base
- Precordial thrill is faint or absent at the left heart base
- Heart rate and pulse quality are normal
- Radiographs and ECG normal even at 1-2 years of age
- Focused, turbulent, continuous retrograde PA flow in color Doppler echo.
- PDA closure is not urgent but is recommended for a normal life span

**Type 2** (Medium size PDA)
- Asymptomatic left-to-right shunt
- Coarse continuous murmur and at the left heart base and also at the left apex
- Precordial thrill easily palpable at the left heart base
- Pulses normal or slightly bounding
- Mild or moderate left heart enlargement before 1 year of age
- Small or medium size ductal-aortic aneurysm
- Borderline increase in pulmonary vascular markings
- Electrocardiogram lead II R waves usually exceed 3 mV
- Color Doppler shows turbulent continuous flow in the dilated main pulmonary artery
- PDA closure is recommended but it can still wait a few weeks

**Type 3a.** (Large PDA before congestive heart failure)
- Usually reduced exercise capacity
- Coarse continuous murmur and thrill over most of the left thorax
- Systolic murmur of mitral regurgitation often heard at the left apex and usually persists a few days after PDA closure
- Pulses bounding due to wide pulse pressure
- Marked left heart enlargement before 6 months of age
- Medium-to-large ductal aortic aneurysm usually present
- Definite increase in pulmonary vascular markings
- ECG lead II R waves may exceed 5 mV
- Color Doppler shows large turbulent flow continuous flow in dilated MPA
- PDA closure recommended without delay

**Type 3b.** (Large PDA plus congestive herart failure)
- All the 3a features plus dyspnea due to pulmonary edema
- Usually poor body condition (cachexia)
- Atrial fibrillation occasionally present
- Pulmonary edema must be resolved as much as possible by cage rest, oxygen, and diuretics before anesthesia and PDA closure
- Animals with atrial fibrillation should be digitalized before anesthesia to slow heart rate.

**Type 4.** (Large PDA with right-to-left or bidirectional shunt)
- Two weeks to 12 years old
- Hind leg weakness or collapse with excercise
- Cyanosis usually limited to caudal part of the body
- Pulses normal or weak
- Polycythemia (packed cell volume as high as 80%)
- Usually no murmur or precordial thrill after one month old
- Split or prominent 2$^{nd}$ heart sound
- Right side apex beat stronger than left
- Right axis deviation in ECG
- Large right heart and main pulmonary artery
- Pulmonary vessels may appear normal or decreased with slight tortuosity
- PDA closure is contraindicated
- Treat polycythemia by periodic phlebotomy or chemotherapy

**EPIDEMIOLOGY**

PDA was the most frequent cardiovascular abnormality in a tabulation of 1000 dogs with congenital heart disease. It can occur in any breed was diagnosed more frequently in small, relatively non-muscular breeds and predominantly in females (3:1)

## ETIOLOGY

This information is based on serial section histology of the ductus and adjacent vessels in 145 fetuses and neonates from a mixed poodle strain of dogs with hereditary PDA compared to 55 unrelated normal fetuses and neonatal dogs and 8 dogs of various breeds with sporadic PDA. Sections were routinely stained for elastic tissue to differentiate muscular and elastic elements. Early observations revealed inconsistencies between segments so a standardized orientation and grouping method was developed to allow different regions of the ductus to be selectively compared. These studies revealed a consistent pattern of abnormality in dogs with hereditary as well as sporadic PDA. The ductus was shorter than normal; its smooth muscle was hypoplastic and eccentric and aorta-like elastic tissue was present in segments that should have been muscular. The changes varied in degree and were classified into 6 histologic phenotypes. The histologic abnormalities recognized in post-natal dogs with PDA were also detectable in fetuses in the family predisposed to PDA.

## SIGNIFICANCE

The size of a PDA determines its significance. Very small PDAs may not affect life span. However, the majority of dogs with PDA develop cardiac signs before middle age and life span would be shortened without treatment. Therefore PDA closure is always recommended except when right-to-left shunting is present. The only other instance when PDA closure is contraindicated is the presence of a co-existing ductus dependent heart abnormality such as pulmonary atresia or cyanotic tetralogy of Fallot.

## TREATMENT

### Medical treatment

PDA closure with a prostaglandin synthetase inhibitor such as indomethacin or aspirin requires responsive ductus muscle and is only effective in children in the first week after birth. In dogs with PDA, the small amount of ductus muscle and its eccentric distribution preclude closure in response to drugs even in the neonatal period. The diagnosis of PDA in dogs usually is not made until they are several weeks or months old and degeneration has already occurred in the small amount of muscle they have.

### Transcatheter occlusion

Transarterial or transvenous deployment of coils and other devices to occlude a PDA has gained favor in many hospitals in recent years. Coils usually work well in tapered PDAs but are less reliable in tubular types. More recently, a Duct Occluder device designed for dogs has been developed. It is considerably more expensive than a coil.

### Surgical treatment

PDA closure by surgery was the main treatment option for 40 years before the advent of transcatheter devices. Currently most PDAs are closed with transcatheter devices but about 10% of dogs with PDA still require thoracic surgery because the dogs are too small to insert transvascular devices or very large and/or their PDA is tubular and coils will not stay in place. Ductus closure is done through the left 4$^{th}$ intercostal space. Triple ligation is preferred with the middle ligature transfixing the PDA lumen. A division method is used when the PDA is very short or has a diameter greater than 10 mm. PDA division is mandatory in order to relieve esophageal constriction when a persistent right aortic arch is present.

## COMPLICATIONS

Complications can occur at any stage in the management PDA. The two most common ones are congestive heart failure and uncontrollable hemorrhage during ductus dissection. Congestive heart failure manifested as pulmonary congestion or edema may occur early in dogs with large PDAs and is sometimes the first sign of heart disease. It is important to treat congestive heart failure medically for a few days before operating because necessary manipulation of the lungs during thoracotomy is not tolerated. Hemorrhage during ductus dissection is often uncontrollable and occurs most frequently during the initial phase in the learning curve of a surgeon. Other complications include arrhythmias, pulmonary hypertension, endarteritis, thrombosis, and aortic or ductal rupture. Patient size is not a complication for surgery. The surgical mortality rates in very small dogs in the first reported series in 1968 was actually less than in larger dogs. Complications also occur with transcatheter occlusion of PDAs. The most frequent one is pulmonary artery embolism of one or more coils. Others include hematoma formation at vascular access sites and systemic coil embolism.

## REFERENCES

[1] Buchanan JW, **In *Canine and Feline Cardiology***. 2$^{nd}$ ed. Fox P, Sisson D, and Moise NS, eds, WB Saunders Co. 1999, p 457. [2] Buchanan JW and Patterson DF, **J Vet Intern Med** 2003;17:167. [3] Buchanan JW, **Birth Defects: Original article series** 1978;14:349. Buchanan JW and Chmielewski S, **Microscopy** 1973;99:353. [4] Knight DH, et..al. **Circulation** 1973;42:127. Buchanan JW, **J Vet Cardiology** 2001;3:7. Pyle RL, **JAVMA** 1971;158:202. Buchanan JW, **J Sm Anim Pract** 1968; 9:409. [5] Holt D, et. al. **Vet Surg** 2000;29:264. Goodrich KR, et. al, **Vet Surg** 2007:36:43.

# NEUROCARDIOGENIC BRADYCARDIA-RELATED SYNCOPE IN DOGS

**Justin D. Thomason, DVM, DACVIM (Internal Medicine)**
**Stillwater, OK**

## INTRODUCTION

Neurocardiogenic bradycardia-related syncope is a common cause of syncope in Doberman pinschers, Boxers, and in elderly small breed dogs with advanced degenerative valvular disease.[1,2] Episodes are triggered by activities or situations that result in a sympathetic surge (i.e. exertion, excitement, or startle). Surges of sympathetic activity trigger intracardiac receptors that respond to loading or contractile conditions by evoking reflex vagal afferent traffic to the brainstem. The final pathway is afferent vagal stimulation of the medullary vasomotor (vasodepressor) center. This center then responds with sympathetic withdrawal and mild or severely accentuated vagal efferent traffic.[3-7]

### Doberman pinschers

Neurocardiogenic bradycardia-related syncope is a marker for dilated cardiomyopathy in Doberman pinschers. Proposed mechanisms for neurocardiogenic bradycardia-related syncope in this breed include: excessive stimulation of the mechanoreceptor C-fibers located in the left ventricle during hypercontractile states, accentuated ventricular mechanoreceptor activity due to increased left ventricular wall stress associated with cardiomyopathy, and/or autonomic dysfunction.[1]

Given that neurocardiogenic bradycardia-related syncope is a marker for cardiomyopathy, ventricular tachycarrhythmias often coexist.[1] The differential diagnosis between neurocardiogenic bradycardia and ventricular tachycardia can be difficult. The minimal database should include: echocardiogram, static ECG, +/- Holter recording. The echocardiogram of Doberman pinschers with neurocardiogenic bradycardia and/or ventricular tachycardia is almost always abnormal.[1] However, early in the evolution of cardiomyopathy, the echocardiogram may be normal or equivocally normal. If the static ECG contains many ventricular premature complexes and ventricular tachycardia, then the cause of the syncope is presumed to be the result of rapid ventricular tachycardia. If ventricular premature complexes are not evident or are infrequent on static ECG, a Holter recording should be performed. If the Holter recording contains no or few ventricular premature complexes, then the syncope is likely due to neurocardiogenic bradycardia. If ventricular tachycardia is evident on the Holter recording, the safest assumption is that the syncope is due to rapid ventricular tachycardia and treatment aimed at suppressing ventricular tachycardia should be initiated. If the syncope and ventricular tachycardia are eliminated, as assessed by repeat Holter recording, then a therapeutic diagnosis has been accomplished. If ventricular tachycardia is suppressed, but syncope persists, then neurocardiogenic bradycardia is likely the cause of the syncope. Neurocardiogenic bradycardia-related syncope does not require specific therapy in Doberman pinschers. Avoiding instigating situations is the best treatment. However, patients should be monitored for progressive dilated cardiomyopathy and arrhythmia.

### Boxers

Neurocardiogenic bradycardia-related syncope in Boxers is likely triggered from the heart; but the exact predisposition is uncertain. A genetic basis is likely given the incidence of this syndrome in this breed and may be related to a subtle autonomic imbalance.

Since both cardiomyopathy and neurocardiogenic bradycardia are common in Boxers, both conditions may coexist in individual patients.[2] Therefore, a minimum database should include an echocardiogram to evaluate for cardiomyopathy, static ECG, +/- Holter recording. Boxers can have many ventricular premature complexes and even ventricular tachycardia, and yet have syncope secondary to neurocardiogenic bradycardia.[2] Similar to Doberman pinschers, if ventricular premature complexes are not evident or are infrequent on static ECG, a Holter recording should be performed. If sotalol is administered to suppress ventricular tachycardia or presumed ventricular tachycardia and the syncopal episodes become more frequent, then neurocardiogenic bradycardia is the probable cause of the syncope, especially if a Holter recording documents suppression of ventricular tachycardia.[1,2] For frequent episodes of bradycardia and syncope, pacemaker implantation and anticholinergic drug therapy are options. However, anticholinergic drugs are ill advised in Boxers with comorbid ventricular premature complexes. Avoiding the instigating sympathetic surge that triggers the episodes sometimes is practical.

### Degenerative Valvular Disease

Syncope in geriatric small breed dogs with advanced degenerative valvular disease is commonly secondary to neurocardiogenic bradycardia. Stretch receptors in the left atrium-pulmonary vein junctions are sensitive to loading conditions and may serve as triggers for neurocardiogenic bradycardia. Most patients experience their first episode when they are volume expanded. The combination of volume dilation of the left heart, with resultant Frank-Starling effect on contractility, and the rapid decompression of the ventricle via the mitral regurgitation appears to predispose these patients to this syndrome. Such a trigger may be similar to the "empty ventricle syndrome" described in people.[3,7]

The diagnosis of neurocardiogenic bradycardia-related syncope is based on exertion, excitement, or startle-induced syncope which is not associated with coughing and a subsequent ECG contains no or minimal pathologic arrhythmia. Holter or event recording would be required for definitive diagnosis. Because this syndrome is associated with volume overload and impending congestive heart failure, cardiac unloading therapy, with at least an ACE inhibitor and furosemide, is indicated. When cardiac unloading and client education do not eliminate the syncopal episodes, an anticholinergic drug can be added to the treatment regimen.

## REFERENCES

[1]Calvert CA, et al. **J Vet Int Med** 1996;10:88-93.[2] Thomason JD, et al. **J Vet Int Med** 2008:22;931-936.[3]Abboud FM. **N Engl J Med** 1993;328:1117-1120.[4]Dickinson CJ. **Lancet** 1993;342:970-972.[5]Rea R, et al. **J Cardiovasc Electrophys** 1993;4:587-595.[6]van Lieshout JJ, et al. **Clin Sci** 1991;81:575-586.[7]Sumiyoshi M, et al. **Am J Cardiol** 1998;82:1117-1118.

# THE FORGOTTEN TARGET ORGAN: THE HEART IN HYPERTENSION

**Rebecca L. Stepien DVM, DACVIM (Cardiology)**
**Madison, WI**

## INTRODUCTION

The response of the heart to systemic hypertension (SH) is well documented in people and is thought to involve both functional changes (e.g. diastolic dysfunction) and anatomic/pathologic changes (e.g. left ventricular hypertrophy [LVH] and myocardial fibrosis)[1,2]. These changes are thought to result from interactions between genetic and hemodynamic factors, since only a proportion of hypertensive people (~20-50%) develop LVH (as assessed by LV mass calculations)[2]. In affected hypertensive patients, congestive heart failure, arrhythmias and abnormalities of myocardial perfusion may occur, increasing the risk of coronary artery disease, stroke and sudden death. Many of the etiologic and pathologic features of systemic hypertension are similar between people and companion animals, but some differences exist. For example, coronary artery disease is not a common cause of morbidity in dogs and cats, and despite increasing recognition of SH in the clinical veterinary population, CHF due to SH does not appear to occur frequently.

Organized clinical studies regarding the cardiac effects of SH in dogs and cats are sparse, but more information regarding the effects of naturally occurring or chronic experimental SH on the heart has been gathered in cats[3-6,11] than in dogs[7-9,14]. Until more species-specific information regarding the cardiac response to naturally occurring SH in dogs and cats becomes available, better understanding of the process as seen in people may facilitate clinical evaluation and therapeutic planning.

## FUNCTIONAL CHANGES

Systolic and diastolic dysfunction has been documented in human hypertensive patients and there is likely significant interaction between these functional abnormalities. In addition, the contributions of co-existing coronary or other heart disease complicates systolic function evaluation. Recent information suggests that early diastolic dysfunction (preceding the presence of LVH) can be detected in hypertensive patients[10,12] and may provide an early target for therapy. Improvement of diastolic dysfunction is important in successful management of cardiac diseases of many types. Although the direct effects of therapy on outcome of diastolic dysfunction in hypertensive patients (human or veterinary) in unknown, assessment of diastolic abnormalities may help better characterize the disease, provide information for risk stratification and, in conjunction with assessment of LVH, provide information helpful in the choice of anti-hypertensive therapies.

Diastolic dysfunction involves abnormalities of LV filling, LV relaxation and LV compliance. When diastolic dysfunction exists, LV pressure increases disproportionately with increases in LV volume, resulting in elevated end-diastolic pressures at rest. Diastolic dysfunction may be clinically silent and occur in the absence of notable LVH, but degree of diastolic dysfunction increases in hypertensive patients in a linear fashion in relation to degree of LVH. In dogs and cats, where clinical pulmonary congestion due to the diastolic dysfunction of hypertensive heart disease is uncommon, diastolic dysfunction may be manifest as fatigue or decreased exercise tolerance due to decreased stroke volume.

The presence of LVH (which can be diagnosed by routine echocardiography) is a risk factor for diastolic dysfunction due to altered myocardial compliance in the thickened ventricular wall. LV systolic function (assessed echocardiographically as ejection fraction or fractional shortening) is typically preserved in hypertensive patients, so if the combination of LVH, normal systolic function and left atrial dilation is noted on routine echocardiographic examination of hypertensive patients, diastolic dysfunction can be suspected. However, diastolic dysfunction may exist prior to development of the architectural changes of LVH, and the most sensitive method of diagnosis of diastolic dysfunction in clinical patients usually involves evaluation with Doppler echocardiographic techniques[10,12]. Left ventricular filling can be assessed using transmitral E wave (early ventricular filling) and A wave (late ventricular filling) velocities, E/A ratios and measurement of transmitral E wave deceleration time (typically prolonged in patients with relaxation abnormalities). Other functional parameters that may be assessed include pulmonary venous inflow patterns and myocardial (tissue) Doppler analysis (systolic and diastolic function)[10-12].

## STRUCTURAL CHANGES

Left ventricular hypertrophy due to chronic SH differs from "athletic" hypertrophy. In both cases, mechanical deformity from stretch or shear forces is the stimulus for increased production of myofibrillar proteins resulting in cardiomyocyte hypertrophy, but pathologic circumstances, this hypertrophy is accompanied by fibrosis and vascular change (medial thickening of intramyocardial coronary arteries). If these accompanying abnormalities are not present, myocardial hypertrophy has little deleterious effect and typically recedes with cessation of the stimulus (e.g. the transition from Tour de France winner to couch potato). The degree to which fibrosis and vascular change occurs in conjunction with LVH is influenced by the interplay of genetics, the sympathetic nervous system (SNS) and the renin-angiotensin-aldosterone (RAAS) systems[1].

Local and circulating hormones directly affect myocardial fibrosis. Collagen synthesis and degradation are ongoing in the heart, and the balance is affected by local and systemic influences. Substances promoting fibrosis (RAAS [especially aldosterone], SNS, inflammatory cytokines, growth factors) offset those substances with collagenolytic effects (natriuretic peptides, bradykinin, nitric oxide and glucocorticoids)[13]. Since the activities of the RAAS and SNS can be manipulated with medications, these findings provide reasons to believe that intervention in the LVH process may be possible. Not all successful antihypertensive agents lead to regression of LVH, suggesting that contributions to LVH by neurohormonal systems are significant and regression of LVH with antihypertensive medications may not be due to decreased blood pressure alone.

The coronary dysfunction documented in hypertensive patients is related to endothelial dysfunction in which there is incomplete response to endothelial dependent vasodilators[2]. In addition, increased resistance is present as the vascular wall thickens as an adaptive response to elevated wall stress. In patients with LVH, the coronary microcirculation does not increase in proportion to LVH and exertional ischemia may occur and contribute to myocardial fibrosis.

## VETERINARY PATIENTS

A variety of structural echocardiographic findings have been documented in cats with SH[4-6] and are inconsistent from patient to patient. In one study of cats with SH, LVH regression occurred in some cats with successful management of the SH with amlodipine[3]. Recent information indicates that similar functional changes (as measured by tissue Doppler analysis) occur in cats with SH as occur in cats with hypertrophic cardiomyopathy[11]. Little documentation of hypertensive cardiac changes in dogs with naturally occurring SH is available[9, 14], although unexplained LVH has been suggested as an indication for BP measurement in dogs[15].

## CURRENT STUDY

A study of myocardial adaptation to naturally occurring SH was recently completed. Many of the hallmarks of structural and functional changes noted in other species were noted. Further studies may address changes in structure and function in response to successful antihypertensive therapy.

## SUMMARY

The cardiac response to SH in dogs and cats is likely to be qualitatively similar to that of other species, but may differ quantitatively. Although the severe complications of hypertensive heart disease in people (e.g. CHF or sudden death) do not appear to occur commonly in hypertensive dogs and cats, further exploration and understanding of the cardiac responses of dogs and cats to naturally-occurring hypertension may allow better understanding of the disease and the ability to make better therapeutic choices.

## REFERENCES

[1]Diamond JA et.al. **Hypertens Res** 2005; 28:191-202, [2]Gradman AH, et. al. **Prog Cardiovasc Dis** 2006;48(5):326-341, [3]Snyder PS, et. al. **J Vet Intern Med** 2001;15(1):52-56, [4]Henik RA, et.al. **J Am Anim Hosp Assoc** 2004;40(5):359-363, [5]Chetboul V, et. al. **J Vet Intern Med** 2003;17(1):89-95, [6]Nelson OL, et. al. **J Vet Intern Med** 2002;16:418-425, [7]Morioka S, et.al. **Am J Cardiol** 1982;49(8):1890-1895, [8]Massabuau P, et. al. **Arch Mal Coeur Vaiss** 1997;90(8):1033-1035, [9]Nicolle AP, et. al. **J Vet Med A Physiol Pathol Clin Med** 2005;52(2):83-87, [10]Aeschbacher BC, et. al. **Am J Hypert** 2001;14:106-113, [11]Sampedrano CC, et. al. **J Vet Intern Med** 2006;20:1106-1115, [12]de Simone G, et. al. **Circulation** 2000;101:152-157, [13]Cuspidi C, et. al. **Nephrol Dial Transplant** 2006;21: 20-23, [14]Cortadellas O, et. al. **J Vet Intern Med** 2006;20:941-947, [15]Rapoport GS, et. al. 1998-2001 (abstract) **11th Congress, European Society of Veterinary Internal Medicine**. Dublin, Ireland, 2001:62.

## CARDIAC BIOMARKERS MEASURED IN 500 CATS WITH HEART DISEASE, HEART FAILURE & OTHER DISEASE

**Stephen Ettinger, Andrew Beardow**
**Los Angeles, CA & IDEXX Laboratories, Westbrook, ME**

The ability to measure cardiac biomarkers in dogs has been established. In the canine, the utility of cardiac biomarkers was convincingly reported by our group in 2008 (this paper has been submitted to a refereed journal for publication). The certainty of this utility in cats remains unresolved. In many smaller cardiac biomarker studies in cats, the markers that were chosen and the results varied, the conditions under which the studies were completed varied and the ability to utilize the results have not been clear. In this report, cats were studied under clinical conditions but the biomarkers themselves were taken with research level consistency. The blood samples were drawn and immediately preserved with enzyme inhibitors; plasma samples were separated after refrigerated centrifugation. Samples were quickly separated, flash frozen and stored frozen, maintaining a constant temperature of -80° C. Samples were shipped to the laboratory for analysis under dry ice conditions, always arriving frozen. Measures were undertaken prior to the study to guarantee sample safety and consistency.

We will report on the normal levels of troponins (cTnI), NT-proANP and NT-proBNP in this study of approximately 500 cats from one hospital, evaluated between 2007 and 2009. We choose to measure biomarkers in a large number of cats in a manner that would permit us to determine the true normal levels for these animals and to do so under circumstances that allowed us to simultaneously evaluate the cats utilizing methods accepted by veterinary cardiologists.

When possible, multiple examinations (separated by weeks to months) were made on some of the cats so that repeatability and or changes in biomarker levels could be evaluated to test the reliability of the data or in cases of advancing medical disease, the effects of advancing disease and or medication on the data.

Perhaps the most difficult portion of this process turned out to be the ability to clearly and consistently evaluate the stage of heart disease, if present. Although multiple reports and recommendations are available in the veterinary literature on the scoring or classification of heart disease in dogs, we did not find any accepted grading system for cats with heart disease. The ISACH veterinary scoring system suggested for the canine did not clearly separate the various levels of disease short of overt heart failure. Our newly developed grading system for the dog was modified for the cat. This grading permitted us to repeatably define each patient's class of disease for the purpose of this trial. The system was subjected to blind evaluation by other board certified veterinary clinicians to validate that our system was robust. We do not know if the scoring system we use is better than others, but for the purpose of this trial it did allow us to separate cats sufficiently to determine if there was a difference in biomarker levels between grading class.

We will show what we believe to be the repeatable and reliable normal levels of cardiac biomarkers as well as the levels of these markers in the various states of disease as determined through clinical testing. We will show the variations of these biomarkers, where they exist, by age, breed, sex, and weight. We also looked at normal cats and those with cardiac and non cardiac disease. We will note the correlations made with other diseases, the relationship to specific cardiac and non cardiac conditions that also became apparent after the blinding period was ended. The repeatability of the studies in normal and cardiac states was determined. Other than the in house traditional cardiac information we had no access to the biomarker results until we were unblinded (March, 2009) when we began to identify individual cases to prepare this information.

We did not begin the project with the intent of defining pre-answered questions that we needed to validate but instead entered the project asking questions. We are still doing so and anticipate that the collection of such a large amount of data (and continuing) will provide us with many answers regarding the relevance of cardiac biomarkers as it relates to the knowledge of cardiac disease in the feline.

### CLASSIFICATION OF FELINE HEART DISEASE

*Caveat: This classification is intended to be universal, easily used by every veterinarian in a similar manner, leaving as little room for variation in the clinical determination of heart disease class as possible.*

**PHASE 0** No heart disease present.

There are no signs or findings of heart disease. This includes clinical signs, physical examination, or ancillary study findings (laboratory, electrocardiographic, radiographic, or ultrasonographic) that indicate or suggest a cardiac problem.

**PHASE 1** No clinical signs. A murmur, arrhythmia, or mild cardiac changes are present. There are no clinical signs of congestive heart failure present. A heart murmur or extra heart sounds may be present but without clinical signs of disease. There may be an arrhythmia. There is no echocardiographic evidence of compensatory structural change.

A: None to minimal cardiac structural changes; i.e. change in echogenicity or valvular abnormalities with no secondary changes in chamber size or function.

B: Pathophysiology is present with no secondary compensatory changes. i.e. LVH with no atrial enlargement

**PHASE 2** No clinical signs but with evidence of cardiac dysfunction and compensation.

A: Pathophysiology is present. Echocardiogram reveals mild left or right atrial enlargement (LA:Ao of >1.5 and <=1.7) with one or more of the following present: abnormal FS, abnormal chamber size (LVH; LV dilation; segmental myocardial changes), abnormal structure or function of one or more of the four cardiac valves.

B: Same as A with moderate or severe left or right atrial enlargement (LA:Ao > 1.7)

**PHASE 3** Congestive heart failure present that is mild or controlled.

Clinical signs, physical exam and cardiac study abnormalities prove the presence of clinical heart failure requiring therapy. Signs are:

A: Well- controlled with drug therapy such as: ACEI, digoxin, pimobendan, furosemide, spironolactone, beta-blockers, calcium channel blockers (CCB)

B: Observed most days suggesting that sub-clinical or no congestive heart failure is present.

**PHASE 4** Congestive heart failure is present and overt.

Heart failure is observed by clinical signs, physical examination and anamnesis; supported by radiography and / or echocardiography. Signs observed can occur in association with other diseases and must be differentiated when making a diagnosis of congestive heart failure.

A: Left sided heart failure alone as indicated by pulmonary edema, dyspnea, tachypnea, or pleural effusion with or without pulmonary edema in the absence of right atrial enlargement.

B: Pleural effusion, pericardial effusion, or ascites with evidence of right atrial enlargement. Concurrent pulmonary edema and left sided heart failure may or may not be present.

APPLICATION OF CARDIAC MRI TO DOGS

**AR Dillon[1] DVM, MS, MBA, DACVIM, J Hathcock[1] DVM, PhD, DACVR, DM Tillson[1] DVM, MS, DACVS, T Denney[1] BS, PhD, S Barney[1] BS, C Killingsworth[2] DVM, PhD, DACVS, LJ Dell'Italia[2] MD, DACC**
**Auburn, AL & Birmingham, AL**

## GENERAL INTRODUCTION TO CARDIAC MRI

The application of a strong magnet field and sequentially measuring how the electrons in tissues reorient has lead to many advances in static tissues such as the brain and axial tissues. The cardiac anatomy is complex, and cardiac structures have different appearances depending on the imaging plane and other options selected as it moves though the cardiac cycle. The most useful imaging planes are those parallel and perpendicular to the cardiac axes. These planes are not based on external landmarks. Obtaining images in these double-oblique planes requires the use of multiple localizing sequences to basically find a perfect perpendicular and parallel plane to the dog LV. These images have no comparison to images via radiographs or echocardiogram and orientation is critical. The use of rigid protocols is ill advised because of the varied orientation of the LV in the chest of dogs . Each dog imaging session must be customized to accommodate individual variations and distortions caused by cardiac disease and varied dog chest symmetry..

The choice of pulse sequences and parameters that affect the appearance of the images obtained with a given sequence greatly depends on the clinical question. The limitations and advantages inherent in particular imaging methods such as spin-echo (SE) or gradient-echo (GE) techniques must be considered in the selection of pulse sequences. Techniques are basically considered "bright blood", "dark blood", or CINE loop techniques. Cardiac morphology and anatomy are appreciated better with SE images, which show higher contrast. Contrast is related to signal intensity differences, which are chiefly determined by the proton density, T1 and T2 relaxation time differences, flow and motion effects, and magnetic susceptibility, as they are related to the pulse sequence applied. In general, tissue characteristics and pathologic changes are generally more pronounced on SE images.

### Blood pool appearance

The appearance of the blood pool is a factor in selecting the protocol. In general, blood flows too quickly through an imaging section for it to be exposed to both the 90° and 180° pulses. Because of this, SE images can fail to depict structures that move out of the imaging plane. Because signal generation at SE imaging depends on refocused spins (protons stimulated by a 90° excitation pulse and refocused by a 180° inversion pulse), the blood pool, by virtue of not being exposed to both pulses, appears as a region of signal void; the result is termed black-blood imaging. Darkness of the blood pool can be further promoted by applying a saturation signal and/or a double inversion recovery (DIR) technique.

Black-blood imaging improves margins of the endocardial borders and helps demarcate intracardiac masses from normal structures. The improved delineation of vessel walls with SE also enables superior anatomic evaluation. If the application involves only one section-selective excitation, after that excitation, blood contributes to the image even after it moves out of the imaging plane. Such contrast is termed bright-blood imaging. Complex branching vessels, such as the pulmonary veins or arteries, are better depicted with this technique. The white-blood technique simulates cine angiographic techniques common to veterinary medicine. A simulated MR angiogram can be created by obtaining the same views with both GE and DIR. Subtracting the DIR image from the GE image leaves a depiction of the blood pool. This technique helps in the distinction of small-caliber or hard-to-visualize vascular structures without the intravenous administration of contrast material.

### Spatial & temporal resolution

Spatial resolution is a concern with any imaging modality. Fortunately, the size of the major cardiac structures is well within the limit for MRI. Spatial resolution generally is determined by the field of view (FOV) and the number of phase-encoding steps. For comparable FOVs and matrix sizes, spatial resolution is equal with many techniques. For typical static SE imaging, one may use a matrix of 256 x 256 and a FOV of 34 cm. To make a movie of the beating heart, one may select a GE technique and reduce the number of phase-encoding steps, for example, to 128. The spatial resolution decreases from 1.33 x 1.33 mm to 1.33 x 2.66 mm accordingly. That resolution is sufficient to depict cardiac wall motion and to calculate the ejection fraction, but higher resolution may be required if detailed evaluation of the cardiac valves is desired.

## ELECTROCARDIOGRAPHIC GATING & ACQUISITION IN DOGS

The sequence of images are acquired and triggered from the R wave of the ECG. Trigger times and averaging of images are controlled but typically for cine-loops, 15 images per cardiac cycle are collected from each slice over a period of multiple heartbeats to determine a statistical average. Flowing blood creates a signal void on SE images. However, the complex pathway of blood flow through the heart results in disordered flow patterns. Focal areas of slow flow create signal, and the blood pool does not appear homogeneous. Depending on the imaging subject, the degree of in homogeneity can degrade image quality and limit interpretation of the findings as would occur in very low cardiac outputs. To counteract this affect, two 180° presaturation pulses are used before the SE acquisition is initiated. This results in images on which the blood is truly black termed double inversion recovery (DIR) images. This technique provides superior delineation of the endocardial margin and vessel walls.

Cine GE sequences are used for functional assessment. Cine imaging involves multiple acquisitions or cardiac phases throughout a cardiac cycle in each imaging section. The result is a collection of images in each section that depict differing degrees of ventricular contraction. The images in each section can be stacked and displayed in a loop to permit visualization of the cardiac chamber and valvular motion. At Auburn University, these images have been combined to reconstruct complete cardiac images of the heart at every part of the cardiac cycle and calculate dimension and strain at each area of the LV in all parts of the cardiac cycle. Thus each area of the heart can be visualized at 15 different parts of the systole-diastole cycle and geometrically calculated strain assigned to each area as a color grid..

Typically the wait time after a triggering R wave is zero to enable to visualize diastole as the start image. Non-ferric leads and connectors are required. The popular option of veterinarians placing identifying "chips" in the back-shoulder area of dogs does distort signal. Most cardiac MRI software evaluates multiple R wave signals to optimize the triggering signal. To negate the effects of chest motion during acquisition, static positive pressure ventilation simulates a "breath hold" used in humans. Each "slice" (none overlapping 8 mm) is the average of multiple acquisitions and this is collected for each slice for all the 15 different phases based on the cardiac cycle after the R wave. Each breath hold is typically less than 1 minute, depending on heart rate. Further, multiple slices across the LV from the base (typically atria) down to the LV apex requires no gaps and depending on the size of the heart, typically 11-15 slices are needed to assure complete LV meaurements. Even by negating the chest wall motion with "breath holds", the twisting motion of the LV from diastole to systole may require, after reconstruction, following a specific wall area from one slice in diastole up to the next higher slice in systole. To reconstruct 3D images, the LV is also "sliced" longitudinally in similar fashion to assess apex and wall thicknesss. The ventricular septum is used as the guide for the perpendicular cross LV slices and longitudinal guide for long slices. Thus, each 8 mm slice is reconstructed as a cine loop of the LV with all other cardiac structures in the field from diastole through systole. All images are cross-referenced for perpendicular planes for localization. This allows for evaluation of such structures as mitral valve annulus in both planes with the ability to make accurate measurements.

### SPECIFIC GEOMETRIC ANALYSIS- AUBURN UNIVERSITY METHODS

Three-dimensional (3D) LV geometric parameters were measured from endocardial and epicardial contours manually traced on cine images acquired near end-diastole and end-systole. The contours were traced to exclude the papillary muscles. Cubic B-spline surfaces were fit to the endocardial and epicardial contours for each time frame. Three-dimensional (3D) surface curvatures were computed from these surfaces using standard formulas at the wall segments (excluding the apex) defined in (Cerqueira et al. 2002). Three-dimensional wall thickness was computed at the same segments by measuring the distance from a point on the epicardial surface to the closest point on the epicardial surface along a line perpendicular to the epicardial surface.

### Filling & Ejection Rate Analysis- Auburn University Methods

Peak ejection and filling rates were computed from a volume-versus-time curve computed by automatically propagating the contours manually traced at end-systolic and end-diastolic to the remaining 18 cardiac phases with a non-rigid registration algorithm (Feng, et al. 2007). The peak ejection rate was defined as the maximum negative time derivative during the systolic interval. The early diastolic peak filling rate was defined as the maximum positive derivative during the early diastole.

### Strain Analysis- Auburn University Methods

Three-dimensional LV strain was measured from tagged images at end-systole, which was defined by visual inspection of the image data as the time frame with maximum contraction. Three-dimensional Lagrangian strain and torsion were computed using an in-house software package (Denney, et al. 2003, Li, et al. 2006). Torsion was defined as the relative rotation angle of the apex relative to the base in degrees. Twist was defined as the torsion angle divided by LV length.

Two-dimensional strain rates and times-to-peak strain were measured using harmonic phase (HARP) analysis (Osman, et al. 1999, Osman, et al. 2000, Liu et al. 2004). HARP analysis measures the local, two-dimensional strain of the myocardium based on the local spatial frequency of the tag lines. When the myocardium contracts, the tag lines become closer to each other, and the tag frequency increases in proportion to that contraction. All strains, strain rates, times-to-peak, torsion indices, and geometric parameters were computed at the wall segments (excluding the apex) defined in Cerqueira et al. 2002.

### Application of Cardiac MRI to Canine Cardiac Studies

Over the past 3 years, over 100 cardiac MRI have been performed on research dogs at Auburn University. Direct hemodynamic parameters were collected for each dog before MR induction, immediately after MR induction and 4- 12 months after induction depending on research protocol. Using a Picker Vista 1.0 T magnet, cardiac magnetic resonance 8 mm thick images (DICOM) were collected during the cardiac cycle in each dog before and after MR induction. Paired T tests allowed each dog to serve as own control for changes in hemodynamics and MRI compared to baseline data. The geometric changes were calculated on 3D reconstructions of the left ventricle. Based on the cine MRI and hemodynamics at baseline and sacrifice, the wall stress (complex geometric multicentric calculations) was calculated at various points during the cardiac cycle. In 20 of the dogs, tagged cardiac MRI were performed to evaluate directional motion of each area of the left ventricle. Some dogs had repeated MRI via another magnet to compare with human data collects in same magnet and software.

### Cardiac MRI very early in the remodeling process suggested dysfunctional remodeling.

Geometrically based on cardiac MRI, significant changes ($p<0.05$) compared to baseline included: 1) increase in ED volume 2) increase in total SV (forward plus regurgitant volumes), 3) decrease in LV wall , 4) increase in wall stress at ED and ES, and 5) increase in LV mass. Mitral annular area and LA were enlarged, and ejection fraction was increased. Changes ($p<0.05$) in LV sphericity, radius/wall ratios, and circumferential curvature also suggested dysfunctional remodeling. Thus although the visual contractility appeared to increase and the total stroke volume increased significantly reflecting the enlarged left ventricle, the volume of blood actually being ejected out the aorta and contributing to functional cardiac output continues to decrease over the initial course of cardiac remodeling. Further, the early subtle geometric changes in the apex of the LV (sphericity and radios/wall) which cannot be typically accessed via even 3-D echocardiographic analysis would indicate that local strain and LV distortion results in progressive and dysfunctional dilation in the earliest phases of mitral regurgitation in the dog. The notation of early strain in the apex of the heart with MR would correlate with the MRI demonstrated change in the radius and sphericity of the LV especially in the relatively thin walled LV apex. MRI strain analysis and non-global collagen loss lead to investigation of this dysfunctional mechanism of LV remodeling. Molecular changes in the mRNA evaluated before and after MR would suggest that the increased strain noted with MRI and confirmed by collagen loss is associated with a downregulation of extracellular matrix and profibrotic growth factor genes.

**REFERENCES**

[1.] Cerqueira MD. Weissman NJ. Dilsizian V. Jacobs AK. Kaul S. Laskey WK. Pennell DJ. Rumberger JA. Ryan T. Verani MS. American Heart Association Writing Group on Myocardial Segmentation and Registration for Cardiac Imaging. Standardized myocardial segmentation and nomenclature for tomographic imaging of the heart: a statement for healthcare professionals from the Cardiac Imaging Committee of the Council on Clinical **Cardiology of the American Heart Association. Circulation**. 105(4):539-42, Jan 29 2002. [2.] Feng, W., Gupta, H., Lloyd, S., Dell'Italia, L., Denney Jr., T.,"Myocardial Contour Propagation in Cine Cardiac MRI," **Proceedings of the 15th Meeting of the International Society for Magnetic Resonance in Medicine**, Abstract # 766, Berin, Germany, May 2007. [3.] Denney Jr., TS, Gerber, BL, and Yan, L,"Unsupervised Reconstruction of Three-Dimensional Left Ventricular Strain from Parallel Tagged Cardiac Images," **Journal of Magnetic Resonance in Medicine**, v49, n4, pp.743-754 April 2003. [4.] J. Li and T.S. Denney Jr., "Left Ventricular Motion Reconstruction From Planar-Tagged Magnetic Resonance Images With A Prolate Spheroidal B-Spline Model," **Physics in Medicine and Biology**, v51, n3, Feb 2006, pp 517-537. [5.] Osman NF, Kerwin WS, McVeigh ER, Prince JL, Cardiac motion tracking using CINE harmonic phase (HARP) magnetic resonance imaging, **Magn Reson Med**, 1999 Dec;42(6):1048-60. [6.] Osman NF, McVeigh ER, Prince JL, Imaging heart motion using harmonic phase MRI, **IEEE Trans Med Imaging**, 2000 Mar;19(3):186-202. [7.] Liu W, Chen J, Ji S, Allen JS, Bayly PV, Wickline SA, Yu X, "Harmonic phase MR tagging for direct quantification of Lagrangian strain in rat hearts after myocardial infarction," **Magn Reson Med.** 2004 Dec;52(6):1282-90. [8.] Zheng J, Chen Y, Pat B, Dell'Italia L, Tillson M, Dillon AR, Powell P, Shi K, Shah N, Denney T, Husain A, Dell'Italia L. Microarrya identifies extensive downregulation of noncollagen extracellular matrix and profibrotic growth factor genes in chronic isolated mitral regurgitation in the dog. **Circulation** (in press 09).

# 3D ECHOCARDIOGRAPHY IN CONGENITAL HEART DISEASE

**Anne T French MVB PhD CertSAM DVC DipECVIM-Ca (Cardiology) MRCVS FHEA**
**Edinburgh, UK**

## BACKGROUND

Three dimensional echocardiography (3DE) was first described by Dekker et al. in 1974[1]. The past 35 years has seen major advances in this field. The earliest machines acquired multiple cross sectional two dimensional (2D) images which were then volume rendered into a 3D data set. Images were obtained using a transducer attached to a mechanical arm. Free hand imaging was then developed and resulting 3D wire frame images acquired[2]. Acquisition of these images required complex cardiac and respiratory gating and time consuming offline reconstruction. Current day probe technology uses a fully sampled matrix array probe with 3000 individual elements which allows real time 3D echocardiographic (RT-3DE) images to be displayed[3]. Reduced acquisition time decreases the need for respiratory and cardiac gating by linking to the cardiac cycle. Increased computer processing power allows rapid processing, storage and analyses of data. RT-3DE data can be acquired in different modalities live restricted volume, thick slices with rotational capabilities, full volume or colour Doppler and then navigated, cropped and analysed using off line analysis packages. Transesophageal (TE) 3DE probes were available 10 years ago, however it is only in the past two years that RT-3DE TE probes have been produced. As technology advances RT-3DE is rapidly moving towards a place in routine clinical cardiac evaluation in human medicine. Whilst some of the original research was undertaken using experimental dogs, sheep and pigs there is little information published on the use of 3DE in veterinary medicine. What follows is based on publications from the human literature [4, 5].

## GENERAL APPLICATION OF 3DE

3DE has been used for the evaluation of both acquired and congenital cardiac disease. It has many applications: 1) the provision of non-invasive realistic views of cardiac valves and congenital abnormalities; 2) the evaluation of chamber volumes without the need for geometric modelling; 3) volumetric quantification of regurgitant lesions, shunts and cardiac output using 3D colour Doppler; 4) pre operative, intra-operative guidance and post operative assessment of surgical or transcatheter interventions; 5) assessment of regional LV wall motion and quantification of systolic asynchrony; 6) the evaluation of cardiac tumours. It has been validated against the gold standard MRI in many studies.

## 3DE & CONGENITAL HEART DISEASE

3DE has been used to evaluate congenital abnormalities of the aortic and pulmonic valves and has been shown to provide accurate assessment of valvular and sub or supra-valvular pathology[6]. 3DE studies of the mitral and tricuspid valves have increased understanding of the geometry of the valves and the relationship between the valve leaflets, the chordae tendinae and the papillary muscles. A major advantage of 3DE is the ability to obtain a perpendicular en-face cut plane of the valve orifice allowing accurate valve area measurements. Transesophageal RT-3DE has been shown to be superior to TE - 2DE in the assessment of mitral valve pathology[7].

Transthoracic 3DE has been shown to equal TE-2DE echocardiography in assessing the morphology of atrial, ventricular, and atrio-ventricular septal defects[8]. Furthermore multiplanar review of 3D data has been shown to be of additional benefit and to provide extra information relevant for surgical management[9]. Transesophageal 2DE has been used to guide transcatheter closure of septal defects however recent case reports suggest that because of better spacial orientation transesophageal RT-3DE may be superior.

Color Doppler RT-3DE has been used to produce an 'echo angiogram' of the aorta, pulmonary artery and other vascular structures in patients with congenital disease. Post processing was used to suppress the black and white image and the resulting color image could be cropped and rotated to allow unique perspectives. Results were validated against MRI, CT angiography or direct visualisation during surgery[10].

The assessment of systolic and diastolic function is important in evaluation of congenital heart disease as in acquired cardiac disease. 3DE has been shown to be superior to 2DE in accuracy and reproducibility of LV mass, volume and EF in congenital cardiac disease and shown it to be comparable to MRI[11]. Recent work would suggest that the same is true for analysis of RV and LA volume[12].

## APPLICATIONS FOR VETERINARY MEDICINE

Congenital heart disease represents approximately 5% of cardiac diseases seen in small animals. 2DE echocardiography provides a reasonable assessment in the majority of cases however angiography is required in complex cases and in guidance of interventional procedures. Cardiac MRI is the gold standard for diagnosis of complex congenital cardiac disease in human medicine. The high cost and limited availability of cardiac MRI equipment limits its use in veterinary practice. MRI and angiography have the disadvantage of requiring general anaesthesia. The recent advance in RT-3DE makes this a more affordable and safer modality to improve diagnosis and evaluation of congenital cardiac disease in veterinary patients.

## REFERENCES

[1]Dekker DL, et al. **Comp Biom Res** 1974; 7: 544. [2]Legget ME, et al. **IEEE Trans Biomed Eng** 1998; 45: 494. [3] Kleijn SA, et al. **Neth Heart Journal** 2009; 17: 18. [4]Houck RC, et al. **Amer J of Roent** 2006; 187: 1092. [5]Lang RM, et al. **J Am Coll Card** 2006; 48: 2053. [6] De Castro S, et al. **Echo** 2006; 23 (7): 553. [7] Grewal J, et al. **J Am Soc Echo** 2009; 22:34. [8] Morgan GJ, et al. **Eur J Echo** 2008; 9 (4): 478. [9] Bharucha T, et al, **Ann Thorac Surg** .2008; 86:875. [10]Hlavacek A, et al. **Echo** 2006; 23: 860. [11] Lu X, et al. **Am Heart J** 2008; 155: 946. [12] Grison et al. J Am Soc Echo. 2007; 20 (8): 921.

# MECHANICAL VENTILATION FOR HEART FAILURE: LET THEM REST OR REST IN PEACE?

**Terri DeFrancesco, DVM, DACVIM (Cardiology), DACVECC**
**Raleigh, NC**

*Let them rest or rest in peace?* Quite frankly, this little question reminds me of two key considerations for mechanical ventilation in dogs and cats with acute, severe medically-refractory decompensated congestive heart failure.

**The first consideration (let them rest) or better said, indication for mechanical ventilation is to reverse acute heart failure-induced respiratory muscle fatigue and to decrease the work of breathing.**

Respiratory muscle fatigue may be diagnosed by a decrease in respiratory rate, associated with hypercapnea and declining level of consciousness. Invasive mechanical ventilation should only be used after acute respiratory failure does not respond to diuretics, vasodilators, oxygen therapy and in humans, non invasive ventilatory support.(see below) General goals of mechanical ventilation in these patients is to support the pulmonary system to maintain an adequate level of alveolar ventilation and to restore normal acid-base balance and oxygenation to the organs and tissues. Additionally, a general goal of mechanical ventilation is to reduce the work of breathing (WOB) until the cause of respiratory failure can be reversed. Normally in a heart failure patient, the ventilatory therapy buys you more time, that is, time to let aggressive medical therapy to take effect or time for the left atrium to stretch if an acute mitral valve chordae tendinae has ruptured. Normally, WOB accounts for 1 - 4 % of the total oxygen consumption at rest. Oxygen consumption by the respiratory muscles increases during respiratory failure of any cause. Oxygen consumption by the respiratory muscle increases to as much as 35% - 40% in some patients. This increased WOB can lead to respiratory failure secondary to respiratory muscle fatigue causing significant tissue hypoxemia and acid base abnormalities. By taking over breathing, one can divert this cardiac output away from the muscles of respiration to more vital organs and tissues.

**The second consideration (rest in peace) reminds me that the reversibility or the prognosis for a meaningful recovering from an episode of severe and medically refractory heart failure is an extremely important one, especially in veterinary medicine.**

One is more likely to offer mechanical ventilation to a medically refractory heart failure patient if it is its first time in heart failure versus a patient with chronic heart failure on high end drug doses. A clinical example of a "good ventilator candidate" would be a dog with first time severe heart failure due to acute chordae tendinae rupture of the mitral valve. These patients often have peracute fulminant pulmonary edema, and given a day or two, may do well with ventilator support. Unfortunately, no single parameter can determine the final outcome of treatment. Thoughtful clinical judgment is essential. First, one needs to consider the nursing, and equipment resources in your hospital as well as your familiarity with ventilatory therapy. Ventilatory therapy has complications (see below) and even a few hours of mechanical ventilation can put a significant strain of available personnel. If your hospital setting is unable to realistically perform long term ventilator therapy, you can potentially transfer the patient to another hospital that is able to perform long term ventilator therapy. In addition to the labor costs, the financial costs to the pet owner are significant for even 1- 2 days on the ventilator. At our institution, the initial cost for set up and the first day on the ventilator is approximately $1,500, with an additional $750-1000 for each additional day. It's my experience that a straight forward heart failure patients are typically relatively "easy" to ventilate and usually only require 1-2 days on the ventilator.

Another consideration for ventilation is to allow further diagnostic workup in a declining patient with suspect heart failure and lack of improvement with conventional heart failure medication. One is always concerned that a medically refractory heart failure patient may actually not have cardiogenic or increased hydrostatic pulmonary edema. Mechanical ventilation in these patients will support of the respiratory system and increase the safety of further diagnostic testing such as an endotracheal wash with cytology and culture and thoracic imaging. We will also offer mechanical ventilation for humane relief of suffering in animals in severe medically refractory respiratory distress whose owner wishes to be present for euthanasia.

In humans, heart failure guidelines recommend that invasive mechanical ventilation with endotracheal intubation should be used primarily to reverse acute heart failure induced respiratory muscle fatigue after non invasive techniques (CPAP or NIPPV) have failed. Invasive mechanical ventilation is not commonly performed even in tertiary care human hospitals. In a study from the 1990's, assessing trends in hospital mortality, length of stay and costs in hospitalized human heart failure patients, mechanical ventilation was reportedly used in approximately 3% - 6 % of these patients. Multiple human studies have evaluated the used of ventilatory support without endotracheal intubation or non-invasive ventilation. **Two techniques are used for non invasive ventilatory support in humans: CPAP or continuous positive airway pressure ventilation; and NIPPV or non invasive positive pressure ventilation.** There is strong consensus on the human side that one of these two techniques should be used before endotracheal intubation and mechanical ventilation. These are safe and effective means of improving gas exchange in many patient with acute respiratory failure of various causes. Application of CPAP or continuous positive airway pressure allows recruitment of collapse alveoli and is associated with an increase in function reserve capacity. The improved pulmonary compliance, reduced transdiaphragmatic pressure swings and decreased diaphragmatic activity can lead to a decrease in the overall work of breathing and therefore a decrease in metabolic demand from the body. NIPPV is a more sophisticated technique that requires a ventilator. The physiological benefits of this mode of ventilation are the same as for CPAP but also includes the inspiratory assist that further reduces the work of breathing and overall metabolic demand. Several studies have shown that utilization of these non invasive techniques dramatically reduces the need for endotracheal intubation and mechanical ventilation. These techniques also improve clinical signs of heart failure and improve oxygenation but interestingly do not appear to improve short term survival. Unfortunately, these techniques may have limited applicability in our veterinary patients as it requires a conscious patient willing to wear a tight-fitting face mask. Even in humans, discomfort with these non invasive ventilatory techniques is reported in 5- 8 % of patients. A small pilot study evaluating continuous non invasive ventilation in cats showed that it is feasible but provided no cardiovascular benefit over traditional invasive mechanical ventilation because the sedation drug requirements were the same.

Positive pressure ventilation is an invasive procedure and as such has the potential for several associated complications. Developing an understanding of potential complications of mechanical ventilation is essential for anyone involved in ventilation management. Briefly, positive pressure ventilation alters the normal physiological pressures in the thorax. **The primary cardiovascular consequence of positive pressure ventilation is decreased cardiac output.** As a result of increased vascular pressures transferred by the positive pressure in the thorax, the pressure gradient for blood flow back to the heart is reduced. Less blood returns to the thorax. The decreased venous return causes a lowering of right ventricular filling (preload) and right heart stroke volume. The degree of depression of cardiac output with positive pressure depends on several factors such as lung and chest wall compliance, airway resistance and the duration and magnitude of the positive pressure. Another complication of positive pressure ventilation is pulmonary damage from overdistention. This most commonly occurs because of regional differences in pulmonary diseases with refractory hypoxemia. When high levels of positive pressure are applied to the more compliant areas of the lungs, the resulting overdistention causes acute lung injury. Low volume ventilation with prevention of collapsing alveoli using PEEP are strategies to protect from lung injury. Complications can also arise with the use of artificial airways. These include infection of the airways (ventilator associated pneumonia), excessive or inadequate heat or humidification . Care must be employed when handling all parts and pieces of the ventilator. Clean exam gloves should always be used. Animals who are sedated and intubated can not maintain normal oral hygiene and inflammation will develop rapidly. Most animals with significant pre-existing gingivitis or periodontal disease will develop marked exudate within 24 hours. In many patients, the endotracheal tube and airway will need to be suctioned at regular intervals. Good nursing care is essential to successful management of a ventilator patient.

**Patient and Ventilator Set Up**

1. The patient should have at minimum a peripheral IV catheter to allow induction medications to allow intubation. Possible induction protocols include: a) morphine 1-2 mg/kg + midazolam 1 mg/kg $\pm$ atracurium 0.1 mg/kg IV. One could used Fentanyl instead of morphine at 5 mcg/kg IV. One should be ready with suction as the pulmonary edema will usually flood the oral pharynx once induction drugs are administered. A sample of the pulmonary edema should be saved for measurement of total protein and comparison to serum total protein. A ratio of edema TP: Serum TP of 0.5 or less is suggestive of cardiogenic edema. Once the patient is intubated, an AMBU bag with 100 % oxygen is attached and hand ventilation is initiated at 20 breaths per minute. The endotracheal tube should be the largest tube that comfortably fits within the trachea to decrease airway resistance. Cautiously inflate the endotracheal high volume low pressure cuff with a minimal leak technique. Attach an ECG if not already being monitored. Transfer the patient to the ventilator.
2. The settings of the ventilator should be individualized for each and every patient and are dynamic. However, reasonable initial settings include: 100 % inspired oxygen, respiratory rate of 20-30 per minute, tidal volume of approximately 6-12 ml/kg or peak airway pressure limit of 20 cm $H_2O$ or less. A small amount of Positive End Expiratory Pressure (PEEP) , usually less then 2- 3 cm $H_2O$, is typically used to maintain open alveoli and enhance oxygenation. For a heart failure patient, I recommend completely taking over ventilation for at least a few hours with a mandatory control ventilation mode. This mode allows the ventilator to do all the WOB. Over time, assisted and supported breaths can be introduced especially when assessing the patient's ability to discontinue mechanical ventilation. Additionally, for a heart failure patient, the minimal amount of airway pressure should be employed to maintain adequate oxygenation to reduce the effects of diminished cardiac output. One can maintain a patient on 100% oxygen safely for approximately 6 hours before one has concerns of oxygen toxicity. Usually with a heart failure patient, inspired oxygen will be easily reduced to 50% over the next few hours. One can leave the patient on 50% inspired oxygen long term.
3. Continuous sedation drugs will soon need to be administered to maintain intubation. We will usually start with fentanyl 5 mcg/kg/hr and midazolam 0.5 mg/kg/hr. For a dog, additional sedation with low dose propofol may need to be used especially to allow a minimally invasive procedure such as the placement of an arterial catheter or oral hygiene protocols. The use of multiple syringe pumps will facilitate the administration of multiple drugs. Eye and tongue lubrication are applied periodically.
4. Non invasive monitoring is started as soon as feasible. An ECG should already be started. Noninvasive blood pressure monitoring is usually performed via an oscillometric device. A pulse oximeter and capnometer should be available. A continuous rectal temperature probe should be employed to maintain thermal homeostasis. Most heart failure patients will be hypothermic initially and require a heating blanket or other warming devices.
5. The decision to use invasive monitoring is tailored to the patient. Most patients should have a second long "bleed back" type catheter placed to allow repeated venous sampling and drug administration. We will usually place either a jugular catheter or a long saphenous catheter. Ideally, an arterial catheter should be placed to allow both invasive blood pressure monitoring and blood gas analysis. However, placement of an arterial catheter in a cat or toy breed dog is difficult. Thus the risk vs. benefit should be weighed. Arterial cut down of the dorsal pedal artery may be performed. If quantification of urine output is important, then a urinary catheter is placed. In the typical heart failure patient, we don't usually place a urinary catheter and just monitor urine output semiquantitatively by assessing the amount of urine produced on the diaper pads.
6. In addition to ventilator therapy, conventional heart failure medications are given. If oral medications are desired, such as pimobendan or enalapril, one can give via an orogastric tube.
7. The decision to discontinue mechanical ventilation is multifactorial. Improvement in pulmonary auscultation (crackles should be resolved), improvement in blood pressure and tissue oxygenation, improvement in pulmonary patterns on thoracic

radiographs, improvement in static compliance and improved ratio of PaO2:FiO2 of $\geq$ 300. Additionally one can test the patient's strength with assisted breaths with a lighter sedation protocol and a Simultaneous Intermittent Mandatory Ventilation (SIMV) mode of ventilation. In a dog or cat, there is a limit to lightening of the sedation as they will begin to chew the endotracheal tube. Just prior to extubation, one should perform oral and pharyngeal hygiene as well suction out the endotracheal tube, stomach and esophagus. One should also reverse the opiate with naxolone and midazolam with flumazenil. One should always be ready to re-intubate if respiratory failure recurs.

## REFERENCES

[1] Execute Summary of the guidelines for diagnosis and treatment of acute heart failure. **European Heart Journal** 2005; 26:384-416. [2] Antolnelli M et al. A comparison of noninvasive positive pressure ventilation vs. conventional mechanical ventilation in patients with acute respiratory failure. **N Engl J Med** 1998;339:429-35. [3] Polanczyk CA et al. Ten-year trends in hospital care for congestive heart failure. **Arch Intern Med** 2000;160:325-332.[4] Pilbeam SP. **Mechanical Ventilation: physiology and clinical applications**. 3rd ed. Mosby: St Louis, MO. 1998.[5] Brown JE et al. Non-invasive ventilation in cats – a pilot study. **J Vet Emerg Crit Care** 2008 ;18 :410-411.[6] Hopper K et al. Indications, management and outcome of long-term positive-pressure ventilation in dogs and cats: 148 cases (1990-2001). **J Am Vet Med Assoc** 2007;230:64-75. [7] Gray A. Non invasive ventilation in heart failure. **New Engl J Med** 2008; 359:142-51.

## STEM CELL THERAPY FOR HEART DISEASE

**Amara Estrada, DVM, DACVIM (Cardiology)**
**Gainesville, FL**

### INTRODUCTION

The heart, with a scarce number of resident stem cells and a limited capacity of cardiomyocytes to re-enter into the cardiac cycle, has an inadequate capacity to repair itself. An exciting and new field in research termed stem cell therapy has recently emerged as a novel means to repair diseased hearts. The term 'stem cell' arose from the concept that these cells have properties analogous to those of the stem of a plant. They have the ability to not only renew themselves for long periods of time through cell division, but can also differentiate into a number of different cell types under the right conditions. Within the past 6 to 7 years more than half-dozen early clinical studies have been published, ranging from case reports to formal trials and deploying a range of differing cell-based therapies for the acute and chronically infarcted/ischemic heart[1-5]. There is less data on the effects of cell transplantation in the setting of nonischemic heart disease such as idiopathic dilated cardiomyopathy. Another field, still very much in its infancy, where stem cells have been employed is biological pacing. Several groups have reported a variety of methods for developing biological pacemakers whose ultimate function would be to supplement/replace electronic pacemakers.

### CELL TYPES

The ability to differentiate into other cell types is termed the 'potential' of a cell. The stem cell with greatest potential to differentiate is the zygote, the earliest cell. This cell type is called 'totipotent' as all cells in the body are derived from this single cell. An embryonic stem cell, which arises from division of a zygote, is termed 'pluripotent' and it is capable of differentiating into any cell type from the three germ layers (endoderm, mesoderm, ectoderm). Adult stem cells are 'multipotent' and present in many tissues. These cells are able to differentiate into different tissue types depending on their origin. Finally, the committed progenitor cell is termed 'unipotent' as it is destined to become one cell type.

One of the questions regarding stem cell therapy is which type of stem cell should be used in which setting. Embryonic stem cells have the advantage of being pluripotent and are also relatively easily grown ex vivo. They are difficult to isolate, however, and also carry with them vigorous ethical objection. Adult stem cells on the other hand, are easily obtainable, ethically not as objectionable, and may be pluripotent or multipotent. Additionally, lack of differentiation potential of these cells may actually be beneficial as their potential for neoplastic differentiation is lower than with embryonic stem cells.

Currently, a variety of adult stem cells are undergoing research and clinical evaluation – all autologous, so that tissue rejection is obviated. Bone marrow derived stem cells were one of the first discovered and researched. The cell types from the bone marrow which have generated the most interest are those found to be precursors of endothelial cells[6]. Following intravenous delivery, these endothelial precursor cells have been found to have the potential to migrate to an infarcted region of the myocardium[7]. Additionally, these cells can be detected in low circulating numbers as well, especially post-infarction in human patients[8-10]. Another stem cell found in the bone marrow is the mesenchymal precursor cell. This cell is present in very low numbers and is able to differentiate into bone, cartilidge, fat or muscle cells depending upon the environmental stimuli. One study has shown the ability for mesenchymal stem cells to differentiate into cardiomyocytes[11]. The first clinically relevant cells to be proposed for use in cardiac disease were skeletal muscle myoblasts[12-14]. These cells are found between the basement membrane and sarcolemma of skeletal muscle fibers. The major advantage of this cell type is the ability to obtain autologous grafts with easily obtainable muscle biopsies, and cell expansion as well as the ability of cells in the muscle lineage to improve cardiac function. A number of clinical trials in people with ischemic cardiac disease have been reported[1,15,16]. One real concern of using skeletal myoblasts is the development of malignant ventricular arrhythmias. The electrical properties of skeletal myoblasts are very different from those of cardiomyocytes and thus transplanted cells may create a region of electrical heterogeneity which allows for the development of re-entrant arrhythmias. It does appear, however, that arrhythmias which develop following myoblast transplantation decrease and disappear over time. Whether this is due to integration of the cells into myocardial tissue or due to diminished viability of the transplanted cells is currently debated[17,18]. Adult stem cells have also been found to reside in many organs, including the heart[19]. The role of these resident stem cells is unknown but much interest exists regarding the potential isolation and expansion of these cells from cardiac muscle biopsies.

### DELIVERY METHODS

Thus far, cells have been delivered to the heart in 3 basic ways.: (1) vascular delivery with either direct intracoronary arterial injection or retrograde coronary sinus delivery, (2) injection of the ventricular wall via percutaneous endocardial injection or (3) injection during a surgical epicardial approach. The advantage of intracoronary infusion is that cells can travel directly into myocardial regions in which nutrient blood and oxygen supply are preserved. However, unperfused regions, such as that with ischemic heart disease, are targeted far less efficiently. Additionally, intracoronary infusion of progenitor cells may obstruct the microcirculation and lead to embolic myocardial damage. Animal studies have demonstrated rising troponin levels, ST segment changes and T wave inversion following intra-coronary infusion of stem cells. Histologic analysis in these research animals confirmed the presence of micro-infarction at these regions[20].

By contrast, direct delivery of progenitor cells to the endocardium by percutaneous catheter based needle injections, direct epicardial injection during open heart surgery and minimally invasive thoracoscopic procedures are not limited by cell uptake from the circulation or by embolic risk. Direct epicardial injection can be difficult in reaching all areas of the heart, the interventricular septum being an important example. Direct injection requires either an open chest or thoracoscopic procedure for delivery of cells. For this reason, percutaneous catheter based techniques for delivery of cells to the endocardium, were developed. This route, as compared to direct epicardial injection, has a number of advantages. It allows the delivery of a large number of cells specifically into the myocardium without the need for a major surgical procedure. Additionally, it allows the delivery of multiple injections over time and carries low procedure related risks. In a small clinical trial in humans with stable angina treated with catheter based endocardial injection of stem cells, there were no procedure related complications such as ventricular or atrial arrhythmias, myocardial ischemia, pericardial effusion or bleeding[21]. Another study in human patients with chronic ischemic heart disease also confirmed the safety and feasibility of this approach[22]. Direct delivery of cells endocardially via catheter based delivery systems have also appeared safe and shown to improve cardiac function in dog models of coronary artery disease[23,24]. The functional benefits of intramyocardially transplanted skeletal myoblasts have been established in various animal models that have involved ischemia as the primary mechanism of injury[14, 25,26].

While delivery method of cells is important, a more critical step in stem cell therapy is cell retention and viability following delivery. Cell retention is defined as the fraction of transplanted cells retained in the myocardium for a short period of time (minutes to hours) while engraftment is the amount of cells viable after a longer time period. Strategies for improving cell engraftment include not only successful delivery, but also successful survival of cells following delivery. In several animal studies, it has been shown that the number of cells retained is increased by either decreasing the total volume injected or distributing the volumes into multiple small aliquots[27-29]. This approach appears to be a safer technique than smaller numbers of injections of larger cell quantities and results in greater cell retention at injection site.

Another strategy for improving cell survival and engraftment is optimization of environmental conditions under which cells are injected. Several approaches including use of angiogenic growth factors, cytokines and gene therapy are currently being investigated. The efficacy of combining cellular cardiomyoplasty with angiogenic growth factors have been shown in previous studies using hepatocyte growth factor[30] and vascular endothelial growth factor[31]. It is speculated that this combined strategy works better than either treatment alone because the growth factors provide a better environment for the transplanted cell by increasing the blood supply and improving cytoskeletal organization. Other studies have proven that cell signaling and cytokines also play an important role in cell homing and engraftment[32-37]. Viral transfection of stem cells with genes that may improve function or capacity to migrate to an affected region is also being investigated[38-40].

## CONCLUSION

While there is great potential for stem cell therapy and heart disease, there is still much to be learned and many hurdles to overcome before this therapy can replace current medical management. Selection of the appropriate cell type, delivery method, maximizing environmental conditions and achieving long term viability of cells are only some of the many issues. Thus far, the various case series and clinical trials have demonstrated that it is a safe treatment which provides improvements in cardiac function but perhaps is no better than current medical therapy. There are still many questions which need to be answered and too many gaps in our knowledge which need to be filled before skeletal myoblasts, bone marrow stem cells or resident cardiac stem cells will find their way into routine clinical practice. Nevertheless, the prospect and promise of this exciting field of intense research and investigation is just gearing up.

## REFERENCES

[1]Menasche P,et al. **Lancet** 2001; 357: 279. [2]Strauer BE, et al. **JACC** 2005;46:1651. [3]Wollert KC, et al. **Lancet**. 2004;16(364):141. [4]Assmus B, et al. **Circulation** 2002;106:3009. [5]Perin EC, et al. **Circulation** 2004;110:II213. [6]Romagnani P, et al. **Circ Res** 2005; 97:314. [7]Kocher AA, et al. **Nat Med** 2001; 7:430. [8]Shintani, et al. **Circulation** 2001; 103:2776. [9]Massa M, et al. **Blood** 2005; 105:199. [10]Wojakowski W, et al. Circulation 2004;110:3213. [11]Makino S, et al. **J Clin Invest** 1999; 103:697. [12]Koh GY, et al. **J Clin Invest** 1993; 92: 1548. [13]Chiu RC, et al. **Ann Thorac Surg** 1995; 60: 12. [14]Taylor DA, et al. **Nat Med**. 1998; 4:929. [15]Menasche, et al. **JACC** 2003; 41:1078. [16]Smits, et al. **JACC** 2003; 42: 2063. [17]Abraham MR, et al. **Circ Res** 2005; 97: 159. [18]Makkar RR, et al. **JACC** 2003; 42: 270. [19]Messina E, et al. **Circ Res** 2004; 95: 911. [20]Vulliet PR, et al. **Lancet** 2004; 363: 783. [21]Tse HF, et al. **Lancet** 2003; 361: 47. [22]Fuchs S, et al. **JACC** 2003; 41: 1721. [23]Linke A, et al. **Proc Nat Acad Sci** 2005; 102:8966. [24]He KL, et al. **J Heart Lung Transpl** 2005; 24: 1940. [25]Murray CE., et al. **J Clin Invest**. 1996; 98: 2512. [26]Ghostine S, et al. **Circulation,** 2002; 106(suppl I): I-131. [27]Grossman PM, et al. **Catheter Cardiovasc Interv** 2002; 55: 392. [28]Teng CJ, et al. **J Thorac Cardiovasc Surg** 2006; 132: 628. [29]Muller-Ehmsen J, et al. **J Mol Cell Cardiol** 2002; 34: 107. [30]Kondoh H, et al. **Ann Thorac Surg** 2007; 84: 134. [31]Suzuki K, et al. **Circulation** 2001; 104 (Suppl 1) 207. [32]Askari AT, et al. **Lancet** 2003; 362: 697. [33]Yamaguchi J, et al. *Circulation* 2003; 107: 1322. [34]Schomig K, et al. **Eur Heart J** 2006; 27: 1032. [35]Kocher AA, et al. **J Mol Cell Cardiol** 2006; 40: 455. [36]Chavakis E. **J Mol Cell Cardiol** 2006; 40: 442. [37]Minatoguchi S, et al. **Circulation** 2004; 109: 2572. [38]Seeger FH et al. **Nat Clin Pract Cardiovasc Med** 2007; 4: S110. [39]Abbot JD, et al. **Circulation** 2004; 110: 3300. [40]Askari AT, et al. **Lancet** 2003; 362: 697.

THE HEART IN CONTEXT: CARDIO-RESPIRATORY INTERACTIONS IN K-9 HEART DISEASE

**Sonya Gordon DVM, DVSc, DACVIM**
**College Station, TX**

## INTRODUCTION

The significance of cardio-respiratory relationships arises from the relatively high incidence of both primary cardiac and primary respiratory diseases in small breed geriatric dogs and in particular the frequency with which they co-exist in the same patient. The clinical presentations for primary respiratory and cardiac diseases are similar (e.g. cough, dyspnea and tachypnea +/- exercise intolerance). The most common causes of respiratory disease (RD) leading to cough, dyspnea and tachypnea in small breed older dogs are small airway disease (a.k.a. chronic bronchitis) and large airway disease (a.k.a collapsing trachea). The most common causes of cough or dyspnea or tachypnea due to chronic degenerative atrio-ventricular valve disease (CVD) include; cardiogenic pulmonary edema (left sided congestive heart failure), left mainstem bronchus compression due to left atrial enlargement (LMSBC), pulmonary hypertension (PH) and less commonly a combination of ascites and pleural effusion (right sided congestive heart failure).

Common diagnostic tests such as thoracic radiographs can be difficult to interpret and are often not able to definitively discriminate between clinical signs due to RD and those related to CVD. Newer tests including the circulating cardiac biomarker NT-proBNP may be of particular value in this scenario. Ultimately diagnosis may be established by evaluation of response to staged trial therapy. Therapeutic goals should be realistic and communicated to the client. The author's therapeutic goal often involves palliation of clinical signs, not cure, particularity in dogs with one or any combination of RD, LMSBC or pulmonary hypertension (PH). Therapy is typically multimodal including a poly-pharmacy approach in combination with appropriate weight loss (RD) and environmental modification when appropriate (RD).[1]

Determination of the cause(s) of cough and respiratory embarrassment in older small breed dogs is frequently difficult even when based on signalment, history, physical examination and thoracic radiographs. Additional non-invasive diagnostic tests may include fluoroscopy (tracheal and mainstem bronchus evaluation), biochemistry panel, complete blood cell count, heartworm antigen test and an echocardiogram. The relative routine availability of Doppler echocardiography and its utility in the diagnosis of PH through estimation of right ventricular systolic pressures has lead to an increasing appreciation for both the incidence and clinical significance of PH in dogs with CVD. Newer Doppler echocardiographic techniques such as tissue Doppler may ultimately contribute to discrimination of etiology of cough and respiratory embarrassment in dogs. More invasive diagnostic tests such as tracheobronchoscopy, bronchoalveolar lavage and fine needle lung biopsy are also employed with some frequency. Evaluation of plasma N-terminal pro-B-type natiuretic peptide (NT-proBNP) concentration has more recently been added to the list of available non-invasive diagnostic tests.

Plasma NT-proBNP represents the only routinely available novel diagnostic modality for the evaluation of cough and respiratory embarrassment in the dog. NT-proBNP is a peptide hormone that is cleaved from the pro-hormone BNP which is produced primarily in response to increases in cardiac (ventricular) filling pressures and ventricular wall stress. Production of NT-proBNP is enhanced by up regulation of the rennin angiotensin aldosterone system, sympathetic nervous system, and myocardial hypoxia. Concentrations of NT-proBNP are elevated in dogs with heart disease (HD) and heart failure (HF) and measurement of NT-proBNP concentration in dogs with clinical signs of respiratory embarrassment has been shown to help differentiate congestive heart failure (CHF) from primary RD with relative high sensitivity and specitivity.[2-5] In future, additional circulating biomarkers such as pro atrial natiuretic peptide may also be demonstrated to be useful for the discrimination of the etiology of cough and respiratory embarrassment in dogs. These novel biomarkers will not replace conventional diagnostic modalities but rather compliment them and ultimately improve the overall accuracy of diagnosis.

### Congestive Heart Failure due to CVD

Goals of both right (ascites +/- pleural effusion) and left sided (cardiogenic pulmonary edema) CHF therapy include: 1) resolution of clinical signs of congestion though appropriate abdominocentesis, diuresis, inhibition of the rennin angiotensin aldosterone system (RAAS), venodilation, +/- dietary sodium restriction, 2) preservation or improvement of peripheral perfusion through tolerable afterload reduction (arterial dilation), increase left ventricular ejection fraction and ideally reduce severity of mitral regurgitation, 3) prolongation of survival through inhibition of the RAAS, reduction in the levels of maladaptive cytokines such as tumor necrosis factor alpha, tolerable inhibition of the sympathetic nervous system (SNS) when appropriate and most importantly resolution or reduction in the severity of clinical signs. Thus the current therapeutic approach involves triple therapy with furosemide. pimobendan and an angiotensin converting enzyme inhibitor (ACEI).[6]

### Left Mainstem Bronchus Compression due to Left Atrial Enlargement

Left mainstem bronchus compression (LMSBC) frequently contributes to clinical signs in dogs with CVD. The most common clinical sign of LMSBC is a persistent often progressive and debilitating cough. This condition may occur concurrently with CHF or may represent the solitary underlying etiology for cough in dogs with CVD. Clinical trials in symptomatic CVD focus on left sided CHF and thus there is no prospective data available to guide therapeutic recommendations for the treatment of symptomatic LMSBC. One of the most important aspects of therapy for cough due to LMSBC involves realistic well communicated goals to the client. In this condition cure is unlikely and thus a more realistic goal is palliation of the cough to a 'tolerable' level for the dog and client.

Recommended therapy is variable between clinicians including cardiologists. When managing LMSBC it is important to give due consideration to potential important co-morbidities such as collapsing trachea and chronic bronchitis. The author typically uses a combination of some or all of the following in a staged approach so response to therapy can be evaluated. An ACEI, furosemide (low dose), pimobendan, arterial dilators (e.g. amlodipine), bronchodilators and cough suppressants.

**Pulmonary Hypertension**

There are numerous etiologies of PH which may be linked by a single unifying hypothesis: PH develops in patients with an underlying genetic predisposition following exposure to specific stimuli(ae) which serve as triggers. The World Health Organization classification of PH (1998) is based on common pathobiological features and was developed to aid in assessment and treatment of human patients with PH and serves as a useful guide for dogs. 1) Pulmonary arterial hypertension; primary (e.g. idiopathic/familial), those associated with other conditions (e.g. persistent pulmonary hypertension of the newborn, toxic response to drugs etc.), 2) Pulmonary venous hypertension; due to chronic left atrial hypertension (e.g. canine CVD), 3) PH associated with chronic hypoxia; chronic obstructive pulmonary disease (COPD), interstitial lung disease, high altitude disease, alveolar hypoventilation disorders (e.g. canine brachycephalic disease & obesity/pickwickian), 4) PH due to chronic thrombotic or embolic diseases (e.g Cushing's, protein loosing nephropathy or enteropathy etc.), 5) PH due to disorders directly affecting the pulmonary vasculature (e.g canine heartworm disease).

Pulmonary venous hypertension (PVH) resulting in pulmonary arterial systolic pressures in excess of 50 mmHg has been reported in nearly 50% of human patients with isolated severe mitral regurgitation a disease similar to canine CVD. Two PH veterinary studies reported chronic left atrial hypertension as the most common underling etiology of PH accounting for 41-43% of PH.[7,8] The true prevalence of PH in dogs with CVD has not been reported however the author and others (personal communication) suspect that mild to moderate PH is present in at least 50% of dogs with advanced preclinical and clinical stages of CVD and ultimately contributes to clinical signs in many patients with CVD. It is important however to consider that dogs with PH due to CVD may have concurrent morbidities that contribute the their PH such as chronic bronchitis, collapsing trachea, obesity, heartworm disease, Cushing's and other causes of a pro-coagulable state (eg. protein loosing nephropathy or enteropathy etc.). Determination of the presence and relative contribution of co morbidities to the presence and or severity of PH may significantly impact therapeutic strategies.

Many patients with PH due to CVD are asymptomatic for PH and may or may not have CHF (cardiogenic pulmonary edema) due to CVD. Signs of right heart failure (ascites +/- pleural effusion) may predominate. Alternatively, dyspnea unrelated to cardiogenic pulmonary edema may be present with or without reports of exercise intolerance or syncope with activity or excitement. Physical examination findings are typically dominated by the murmur associated with CVD.

Diagnosis of PH due to CVD can frequently be established by echocardiographic estimation of systolic, mean and diastolic pulmonary artery pressures using tricuspid and pulmonic insufficiency signals respectively when present. These estimates however, do not take cardiac output into consideration and thus do not truly reflect pulmonary resistance. In the absence of pulmonic or tricuspid regurgitation, determination of the ratio of pulmonic acceleration time to ejection time may help establish a diagnosis. Many dogs with PH due to CVD have no obvious evidence of right ventricular hypertrophy, right atrial enlargement or main pulmonary artery dilation.

Pulmonary vasculature is regulated by a number of factors and vascular mediators including; adrenergic tone, hypoxia, altitude, aging, exercise, prostaglandins, nitric oxide, endothelin and serotonin. The recognition of these mediators and their potential roles in the development and progression of PH has lead to the development of pharmacologic strategies for PH. In CVD development of chronic left atrial hypertension leads to pulmonary venous hypertension which triggers the release of vascular mediators that lead to pulmonary arteriolar vasoconstriction, mitogenesis, vascular dysfunction and potentially insitu thrombogenesis.

Palliation of clinical signs associated with both right and left CHF represents a therapeutic cornerstone. In addition, the following therapeutic categories could be considered alone or in combination when appropriate. Many of these have not been demonstrated to have efficacy in the treatment of PVH specifically but are being evaluated in many of the other categories of human PH. In veterinary medicine pimobendan and sildenafil are frequently used alone or in combination to chronically palliate clinical signs of PH due to CVD and other etiologies. 1) Reduction in left atrial pressure (left atrial preload) through reduction in plasma volume [e.g. diuretics such as furosemide, RAAS inhibition with an ACEI, and or aldosterone antagonist, and dietary sodium restriction when tolerated], pulmonary venous dilation [e.g. pimobendan, ACEI], reduction of mitral regurgitation fraction [tolerable afterload reduction with agents such as pimobendan, ACEI, amlodipine, hydralazine, inotropic augmentation with pimobendan, surgery or device therapy]. 2) Pulmonary arterial dilation through phosphodiesterase V inhibition [e.g. sildenafil, pimobendan?, theophylline?], oxygen supplementation, endothelin antagonist [e.g. bosentan], clopidogrel [via limiting serotonin release], nitric oxide, prostaglandin analogues [e.g. epoprostenol], serotonin antagonists, calcium channel antagonists [e.g.diltiazem]. 3) Antimitogenic (reverse remodeling agents) such as endothelin antagonists [e.g bosentan] and serotonin antagonists. 4) Antithrombotic agents such as clopidogrel and aspirin, epoprostenol, pimobendan? 5) Inotropic support with agents such as pimobendan, epoprostenol endothelin antagonists [e.g. bosentan].

The list of potential therapies for PH reflects the plethora of therapeutic targets in PH. Many of these agents are currently available to veterinarians including pimobendan, sildenafil, clopidogrel, aspirin and theophylline as well as oxygen for acute support and stabilization.

In a small (N=30) unpublished retrospective study of moderate canine PH (median Doppler estimated right ventricular systolic pressure 92 mmHg) from TAMU where the etiology was predominantly CVD (75%) the Kaplan Meier estimated median survival time was 200 days following the initiation of sildenafil (average dose approximately 1-3 mg/kg BID). All dogs were also receiving pimobendan (0.25-0.3 mg/kg) twice of three times per day. Eighty percent of dogs died of causes related to their cardiac disease, the majority electing euthanasia when clinical signs were no longer adequately palliated. Dogs with PH due to CVD may represent a subset of patients that derive additional benefit from pimobendan beyond its beneficial effects in straight forward CHF. Pimobendan may work well in combination with selective phosphodiesterase V inhibitors such as sildenafil. Prospective evaluation of these agents or others listed above either alone or in combination is warranted and feasible given the prevalence of PH due to CVD.

**Collapsing Trachea**

The etiology of tracheal collapse in the dog is unclear. Many theories exist including: genetic, nutritional, neurological and inflammatory. Medical management is palliative and symptomatic. Dogs are commonly managed with a combination (usually not all) of cough suppressants, bronchodilators, corticosteroids, antibiotics and weight loss. There are to date no controlled clinical studies to support any particular medical management strategy. Reparative (ring placement) and salvage (permanent tracheotomy) surgical procedures have been reported and more recently and non-surgical repair options (stents) have been reported. However studies are limited and most of these procedures are reserved for severe cases (Stage IV tracheal collapse) at this time.[1]

**Chronic Bronchitis**

Chronic bronchitis is a insidious incurable disease that predominantly affects middle aged to older small breed dogs. It is characterized clinically by persistent cough (+/- dyspnea, tachypnea, etc.) due to chronic airway inflammation and excessive mucus secretion. Diagnosis is usually presumptive based on history and signalment and the exclusion of other obvious causes for cough (e.g. CHF, bacterial or fungal pneumonia, pulmonary neoplasia, heartworm disease, collapsing trachea etc.). Therapy is guided by the individual patient's severity of clinical signs. Therapeutic strategies typically include one or all of the following at some point in time during the management of this insidious chronic disease. Modification of contributing environmental factors, weight loss (if obese), cough suppressants, control of secondary infections with antibiotics (lipophilic agents ideally selected based on culture and sensitivity testing; author's drug of choice is doxycycline), oxygen supplementation in emergencies, control of inflammation (corticosteroids), +/- bronchodilators to improve airflow characteristics. Appropriate management is critical to avoid and or delay the development of potentially devastating complications such as: COPD, bronchopneumonia, bronchiectasis, and PH leading to cor pulmonale.[1]

**REFERERENCES**

[1]Lesley G. King Ed., **Elsevier Health Sciences** (Elsevier) 2004. [2]Oyama MA et.al.**J Vet Int Med** 2008;22(3). Abstract Number 66, p: 719 [Oyama MA et.al. **J Am Vet Med** Assoc 2009 manuscript submitted], [3]Boswood A et.al. **J Small Animal Pract** 2008;49:26, [4]Oyama MA et.al. **J Am Vet Med Assoc** 2008;232:1496. [5]Fine DM et.al. **J Am Vet Med** Assoc 2008;232:1674, [6]Haggstrom J et.al. **J Vet Intern Med** 2008;22:1124, [7]Johnson L, et.al. **J Vet Intern Med** 1999;13:440-447. [8]Pyle RL et.al. Intern **J Appl Res Vet Med** 2004 99-109.

## SEEKING THE PERFECT BALANCE: CASE STUDIES IN CARDIORENAL PATIENTS

**Etienne Côté, DVM, DACVIM (Cardiology, SAIM)**
**Charlottetown, PEI, Canada**

### INTRODUCTION

The unifying feature of the following cases is that they all involve the dilemma of encountering heart disease and kidney disease simultaneously in the same patient. Most structural cardiac diseases, such as cardiomyopathy and valvular heart disease, increase the likelihood of clinical signs if a patient receives an excess of total body fluid, whereas most chronic renal tubular diseases, when they cause overt clinical signs, present a need to restore total body fluid.[1,2] A patient with both disorders would seem to need more fluid and less fluid simultaneously. Further complicating this dilemma is the difficulty associated with determining exactly when the euvolemic state has been reached, to find a balance that avoids too little (inadequate parenteral fluids, excessive diuretic drugs) or too much (excessive fluids or insufficient diuretics). The following cases identify key principles and apply them to practical clinical situations as examples of managing heart and kidney disorders simultaneously.

### CASE 1.

**Signalment**

13 year-old male castrated Shih-Tzu dog.

**History**

Myxomatous mitral valve disease detected on incidental finding of heart murmur 3 years ago, confirmed at that time with echo (early stage without left atrial enlargement), no treatment. No cardiac follow-up since then. Presents today for cough of 2 days' duration, and dyspnea and restlessness of 24 hours' duration.

**Physical exam**

Bright, alert, responsive, but anxious dog with moderate dyspnea at rest. Systolic murmur difficult to detect over very loud breath sounds. Regular cardiac rhythm @ 160/minute. Pulse fair, synchronous.

**Problem overview and working diagnosis**

Dyspnea and cough with longstanding myxomatous mitral valve disease in older small-breed dog. Suspect congestive heart failure.

**Diagnostic testing**

Thoracic radiographs identify left-sided cardiomegaly (including left atrial enlargement, but without obvious compression of a mainstem bronchus) and severe interstitial to alveolar infiltrates consistent with pulmonary edema. A complete blood count, serum biochemistry profile, and urinalysis drawn immediately at admission are unremarkable, including a urine specific gravity of 1.046.

**Treatment**

Furosemide 3 mg/kg IV once, then again 2h later. Good response, with reduction in respiratory effort and rate (8h after admission), sleep (12h after admission), and some eating (20h after admission). Oxygen supplementation for first 12 hours, then tapered and stopped based on substantially improved respiratory character. Furosemide continued @ 3 mg/kg q 6 hours for 1st 48 hours.

**In-hospital follow-through**

Repeated serum biochemistry profile and urinalysis 48h after admission: BUN = 84 mg/dl (normal: 7-26), creatinine = 2.9 mg/dl (normal: 0.6-1.4); urine specific gravity = 1.012.

**Therapeutic questions and dilemmas**

Does azotemia with isosthenuria indicate renal failure? Should an ACE inhibitor be started at this time? Will continuing treatment with diuretics worsen the azotemia and possibly lead to renal failure? What would be an appropriate interval for the first routine recheck?

The inability to produce concentrated urine –isosthenuria- together with concurrent azotemia is the clinical laboratory hallmark of intrinsic renal failure, with 3 exceptions: diuretic use, diabetes insipidus, and hypoadrenocorticism. These conditions produce prerenal azotemia and concurrent isosthenuria, even when renal function is preserved. Specifically, a mild or moderate BUN elevation is expected in patients receiving diuretics, and a mild creatinine elevation also is possible.[3,4] These mild elevations can be caused by diuretic-induced prerenal azotemia, with isosthenuria caused by the urinary dilutional effect of the diuretic. The degree of diuretic-induced BUN and creatinine elevations is variable, and acutely, as in the case described above, an elevation of 3- or 4-fold above normal for BUN and 2-fold above normal for creatinine, may frequently be observed in the initial 24-48 hours of treatment, when robust dosages of furosemide are often used in order to eliminate pulmonary edema quickly. When the acute heart failure patient shows overt clinical improvement, such as normalizing respiratory effort, general comfort, and return of appetite, such azotemia does not alter the diuretic treatment plan (see below).

Angiotensin-converting enzyme (ACE) inhibitors were initially introduced in small animal practice in the 1990s with substantial concern regarding their renal effects. Time and experience have shown that such concerns were unfounded. Excessive caution may have arisen from a fear that widespread adverse renal effects in dogs could negatively impact the human market for these medications. In hindsight, this class of drug has been used safely in large numbers of clinical veterinary patients. The benefit of ACE inhibitors is measured on the scale of weeks to months or years of treatment. Therefore, in the acute setting, initiation of ACE inhibitors should be withheld until the patient is breathing comfortably and eating adequately. These clinical features are also used as surrogate markers for avoiding the administration of ACE inhibitors to patients who are hypovolemic and/or hypotensive, situations where the vasodilation conferred by ACE inhibitors could be unpredictable (variable perfusion of the gastrointestinal tract during hypovolemia) or excessive if compounded with the hypovolemia possibly caused by acute, lifesaving doses of diuretics. The advantages and drawbacks of different ACE inhibitors (enalapril, benazepril, ramipril, lisinopril) continue to be debated in veterinary cardiology, without a clear consensus supporting one drug or another.

Ongoing diuretic treatment *versus* adverse renal effects. The optimal outpatient approach is to use the lowest amount of diuretic that effectively keeps congestive heart failure at bay (returns the patient to an "asymptomatic" state). What is the lowest effective amount of diuretic? Many factors vary from patient to patient to influence the answer to this question, including type and severity of heart disease, amount of dietary salt intake, and renal adaptation to diuretic use and development of diuretic resistance. Once a patient has successfully responded to acute treatment for congestive heart failure (as above), a typical discharge plan includes furosemide ≈ 2 mg/kg PO q 12h, and an ACE inhibitor. The role of pimobendan is still being defined for patients with valvular heart disease and in the context of the present case, the author reserves it for states that are more advanced than the first episode of congestive heart failure. With a co-operative and proactive client/owner, a patient's diuretic dosage may gradually be reduced to the lowest effective level. Doing so should ***not*** come at the expense of the owner's patience or financial abilities, since an end to either of these resources can lead to euthanasia just as quickly as would terminal heart failure or renal failure. A typical process first involves rechecking the patient to be sure that signs of congestive heart failure have indeed resolved. The first recheck after successful discharge from acute congestive heart failure is typically 7-10 days later. In a patient who is eating well, breathing comfortably, and has resumed normal activities and mentation –an optimal response- a thoracic radiograph is taken to confirm resolution of edema, and the diuretic dosage may be reduced by 25% and the patient monitored at home by the owner for signs of recurrence of congestive heart failure. If such signs occur, the patient is rechecked, including thoracic radiographs to confirm pulmonary edema (versus clinical signs from an impostor disorder such as upper airway abnormalities). If a recheck is not possible, the owner may give an additional single dose of oral diuretic and return to the previous dosage regime (25% higher). When the first diuretic dosage reduction is tolerated well, then a further reduction can be considered. Whether or not to do so should depend on, among other factors: the severity of the cardiac disorder, where advanced stages characterized by such changes as marked atrial enlargement would discourage extensive dosage reductions; and whether or not pre-existing kidney disease was present prior to diuretics, and if present, whether it was reversible (e.g., pyelonephritis) or not (e.g., sterile chronic kidney disease). An important aid in keeping a patient from having recurrent congestive heart failure is to minimize sodium intake once the patient is eating and feeling well. Since the fundamental mechanism of most diuretics is natriuresis, less salt intake leads to less fluid retention for a given dose of diuretic. Such dietary changes are not indicated prior to congestive heart failure because diuretics are not administered for heart diseases prior to congestive heart failure. When a patient has recovered from congestive heart failure, and the appetite is good, low-sodium foods can be substituted gradually (wean onto). Options include commercial prescription diets, home-cooked diets as recommended in publications,[5] or as formulated by veterinary nutritionists on a case-by-case basis.[6] Overall, the concept of starting with a mid-range diuretic dose and reducing to the lowest level is the exact reverse of the approach recommended for human heart failure patients, where the diuretic is started low and gradually increased until no symptoms are present.[7] This difference can be explained by the limitations inherent in veterinary medicine: clinical signs of heart failure may only be apparent when moderate or severe fluid retention is present, and owners may not be objective assessors of their companions' clinical signs.As indicated above, recovery from acute congestive heart failure should be followed by a recheck 7-10 days after discharge, sooner if complications or uncertainty arise.

**CASE 2.**

**Signalment**

8 year-old male castrated Persian cat.

**History**

Polycystic kidney disease (PKD) identified during routine screening at age 1 year. Polyuria, polydipsia, and anorexia at age 7 years, with azotemia and isosthenuria. Diagnosis: uremia due to PKD. Good response to supportive care. Three months prior to current presentation: heart murmur noted on routine physical exam. Echocardiography: "left ventricular thickening, no left atrial enlargement". Presents today for inappetence of 4 days' duration, worsening lethargy and weakness of 24 hours' duration. No known inciting cause beyond documented medical problems.

**Physical exam**

Quiet but responsive, tractable cat with no apparent visual or neurologic deficits. Normal respiratory effort at rest. Thin body condition, body weight = 4 kg. Mucous membranes pale pink, tacky; together with skin tent, suggestive of 7% dehydration. Regular cardiac rhythm, 240 beats/minute. Grade III/VI systolic murmur loudest over the left parasternal region. Pulse weak but synchronous. Moderate to marked renomegaly bilaterally, markedly irregular kidney surfaces.

**Problem overview and working diagnosis**

Suspect recurrent uremia due to PKD; clinically compensated heart disease.

**Diagnostic testing**

Complete blood count, serum biochemistry profile, urinalysis identify nonregenerative anemia (hematocrit = 19%; normal: 24-45%), hypokalemia (2.9 mEq/l; normal: 3.4-6.0 mEq/l), azotemia (BUN = 174 mg/dl [normal: 13-32], creatinine = 6.5 mg/dl [normal: 0.7-1.9]), and isosthenuria (USG = 1.009) with an inactive sediment. Thoracic and abdominal radiographs show bilateral renomegaly consistent with the abdominal palpation, mild diffuse cardiomegaly, and no other abnormal findings. Interpretation: renal failure with associated hypokalemia and anemia.

**Therapeutic dilemma**

How to correct this cat's fluid deficits without triggering congestive heart failure?

**Additional diagnostic testing**
In order to address this dilemma, several additional sources of information could be useful. Specifically, arterial blood pressure, fundic exam, and follow-up abdominal ultrasound and echocardiography are indicated.
**Results of additional tests**
Arterial blood pressure (systolic, Doppler): 174 mm Hg while cat is calm; fundic exam reveals chronic pinpoint retinal hemorrhages; abdominal ultrasound demonstrates extensive polycystic renal changes bilaterally and no other relevant findings; and echocardiography performed after fluid resuscitation (since dehydration can cause ventricular pseudohypertrophy in cats[8]) confirms left ventricular thickening that is asymmetric, with the interventricular septum almost exclusively affected (6.8 mm thickness in diastole, vs. 5.9 mm left ventricular free wall thickness also in diastole). Interpretation: Despite the low prevalence of systemic hypertension in feline PKD,[9] this cat with typical fundic lesions is hypertensive, likely to a degree that is more severe than that indicated by the present measurement as a consequence of hypovolemia. The asymmetry of left ventricular concentric hypertrophy raises the possibility of idiopathic hypertrophic cardiomyopathy (HCM) in addition to systemic hypertension-associated left ventricular thickening. It is to be expected that a generalized disorder such as systemic hypertension should produce uniform, symmetrical left ventricular thickening. While some human hypertension patients demonstrate asymmetrical left ventricular thickening, such a phenomenon has only partially been investigated in cats:[10] hypertensive cats have been found to have asymmetrical left ventricular thickening, but extended follow-up after correction of hypertension remains to be conducted, to document resolution of the asymmetrical hypertrophy. In conclusion, the asymmetry of left ventricular thickening in this cat suggests that even prolonged normotension with arteriodilator drugs may not produce a normal heart, since concurrent HCM may be present. The incidental finding of *E. coli* on the urine culture offers the possibility of reversing some degree of renal azotemia if low-grade pyelonephritis is present. A positive culture result like this one is the basis for routine urine culture in dogs and cats with chronic kidney disease, even when the sediment is inactive, and if clinically silent pyelonephritis is present (a real possibility), it represents an opportunity to improve renal function by eradicating the infection.
**Revised problem overview and additional dilemmas**
Anemia and hypokalemia are likely secondary to renal disease; how should they best be corrected? Borderline systemic hypertension in a hypovolemic cat suggests hypertension is likely with fluid therapy; what is the best approach?
**Treatment**
Hypovolemia requires correction using fluids that reflect what is missing from the body. Replacement-type fluids such as lactated Ringer's solution, Normosol-R, and Plasmalyte-148 are appropriate. The standard calculation is used for calculating fluid deficits: dehydration (expressed as decimal, so 7% is 0.07) X body weight (in kg) = deficit in liters. Here, 0.07 X 4 = 0.28 liters = 280 ml, which is administered along with maintenance fluid requirements over the first 12 hours. Vigorous potassium supplementation (e.g., 40 mEq/l administered at these supramaintenance rates) is also indicated, since correction of hypokalemia may restore muscular strength and improve demeanor; contraindications are oliguria or anuria, which can be assessed in response to the first 12 hours of fluid therapy. During this time, and thereafter during parenteral fluid administration, it is important to monitor not only for the onset of tachypnea or respiratory distress as markers of iatrogenic fluid excess and congestive heart failure, but for the onset of a gallop heart sound. A gallop sound (sometimes misleading termed a gallop rhythm) occurs due to vibrations of the ventricular walls. These vibrations are most likely to occur when the ventricle fills under duress, such as in situations of excessive diastolic filling pressures. Thus, the appearance of a gallop sound during fluid therapy in a cat with pre-existing heart disease indicates the risk of impending congestive heart failure. Of note is the frequent presence of gallop sounds in cats in general; therefore, it is the new onset of a gallop sound that should be suspected of indicating a pre-congestive state, justifying fluid rate reduction and further monitoring. When fluid deficits have been corrected, it is appropriate to change not only the fluid infusion rate but also the fluid type. Ongoing losses during maintenance fluid therapy consist of urine, breath, and other low-sodium, high-potassium fluids. Consequently, ongoing administration of replacement-type fluids during the maintenance phase of fluid support risks fluid overload and ongoing hypokalemia. Instead, maintenance-type fluids are indicated.[1] Examples include Plasmalyte 56% with 5% dextrose, and less optimally, 0.45% NaCl with 2.5% dextrose and 20 mEq/l KCl. Hypokalemia and anemia both may predispose to ventricular arrhythmias. Therefore, vigorous potassium supplementation is indicated, and oxygen carrying capacity may or may not need to be increased. Blood transfusions are rarely administered for dilutional anemia that is expected to self-correct, but persistent anemia of renal origin may require correction using [ideally recombinant feline] erythropoietin, for example. Oral treatment of systemic hypertension is indicated in this cat when appetite and demeanor are restored, given the documented hypertension and end-organ (ocular) damage. ACE inhibitors alone may be effective for this purpose in PKD,[9] but amlodipine generally is more effective. Routine antibiotic treatment of the urinary tract infection is appropriate in hospital (parenteral) or at home (oral), with interdosage interval adjusted to reflect decreased renal function.

**CASE 3.**
**Signalment**
12 year-old female spayed German shepherd dog.
**History**
Myxomatous mitral valve disease X 1 year (incidentally-found heart murmur 1 year prior to this presentation. First episode of congestive heart failure occurred 5 months ago, with 2 subsequent relapses. Current medications: furosemide 4 mg/kg PO q 8h, enalapril 0.6 mg/kg PO q 12h. Presented on emergency due to anorexia of 2 days' duration and acute, severe dyspnea.
**Physical exam**
Anxious, markedly dyspneic patient. Generalized cyanosis.

**Diagnostic testing**

Thoracic radiographs: consistent with severe cardiogenic pulmonary edema. Laboratory tests: BUN= 130 mg/dl (normal: 7-26), creatinine= 5.2 mg/dl (normal 0.6-1.4); Na, K within normal limits.

**Problem overview and working diagnosis**

Acute relapse of severe left-sided congestive heart failure. Concurrently, severe azotemia.

**Diagnostic and therapeutic dilemmas**

More diuretics to treat congestive heart failure will surely worsen azotemia, but less diuretic or none at all will worsen pulmonary edema.

**Treatment**

First, identify which problem is producing the most severe overt clinical signs. With dyspnea and cyanosis, the respiratory system is a greater concern than the urinary system. Treatment needs to be aimed at pulmonary edema; despite the severe azotemia, this dog is more likely to die of pulmonary edema than renal failure based on its presenting signs. The urinary system can be monitored with urine production during hospitalization. Furosemide constant rate IV infusion was initiated to deliver slightly above current doses: 4 mg/kg q 8h equals 12 mg/kg daily; this dosage was increased to 16 mg/kg to address the severe pulmonary edema and divided over 24 hours: 0.67 mg/kg/hr IV CRI. Oxygen therapy was provided as needed, and the ACE inhibitor was stopped until the appetite returned (avoidance of oral medications in severely dyspneic heart failure patients). This regime produced a good response, with improved respiratory effort and mentation in the first 12 hours. Further treatment and prognosis can be guided by identifying precipitating factors. Echocardiography revealed no appreciable change over the last several months, suggesting that a ruptured chorda tendineae or other form of deterioration of the valvular disease were less likely. Further history-taking revealed that the dog had eaten a sodium-rich treat (bowl of mashed potatoes), which represented a marked deviation from the usual sodium-restricted diet. A one-time sodium load carries a more favorable long-term prognosis than spontaneous worsening of the primary valve disorder, since survival of the acute episode can mean a return to the previous status quo barring permanent changes incurred during the crisis. This dog responded well to treatment and was discharged on the 4th day after admission with a higher dosage of diuretics for one week (5 mg/kg PO q 8h), reduced to the 4 mg/kg PO q 8h dosage thereafter. The azotemia persisted but at lower levels, and the dog showed no overt signs of renal failure (despite 2 subsequent recurrences of congestive heart failure) in the following 7-month period.

**REFERENCES**

[1]Bonagura JD et al. In DiBartola SP, ed. **Fluid, Electrolyte, and Acid-Base Disorders in Small Animal Practice**, 3rd ed (St. Louis, MO: Elsevier, 2006), 490. [2]Chew DJ, et al. In DiBartola SP, ed. **Fluid, Electrolyte, and Acid-Base Disorders in Small Animal Practice**, 3rd ed (St. Louis, MO: Elsevier, 2006), 518. [3]Nicolle AP, et al. **J Vet Intern Med** 2007; 21 (5): 943. [4]Boswood A, et al. **J Vet Cardiol** 2006; 8:1. [5]Strombeck DR. **Home-Prepared Dog & Cat Diets** (Ames, IA: Wiley-Blackwell, 1999). [6]www.petdiets.com. Accessed 2/1/09. [7]Mann DL. In Libby P et al, eds. **Braunwald's Heart Disease**, 8th ed (Philadelphia, PA: Saunders Elsevier, 2008), 625. [8]Campbell FE et al. J Vet Intern Med 2007; 21: 1008. [9]Miller RH et al. **Am J Vet Res** 1999; 60: 1516. [10]Nelson OL et al. **J Vet Intern Med** 2002; 16: 418.

# COMPLICATED CARDIOVASCULAR CASES

**Rebecca L. Stepien DVM, DACVIM (Cardiology)[1], Virginia Luis Fuentes MRCVS, DACVIM (Cardiology)[2]**
**[1]Madison, WI, USA; [2]London, UK**

## INTRODUCTION

Interactions between the cardiovascular system and abnormalities of other body systems are well-recognized.[1-5] Contrary to Occam's Razor (paraphrased: "All other things being equal, the simplest solution is the best") and the frequently cited "common things occur commonly", a given patient may, despite longstanding and severe cardiovascular disease, develop other significant unrelated diseases. Alternatively, animals with severe non-cardiovascular disease can be presented with signs consistent with cardiovascular disease. Clinical acumen and focus allows recognition of the origin of new clinical signs in known cardiac patients, and the origin of cardiovascular signs in patients with non-cardiac disease. These complicated situations require clinical vigilance, high clinical suspicion of complicating disease and an open mind.

## PATIENTS WITH KNOWN CARDIOVASCULAR DISEASE AND NEW CLINICAL SIGNS

Most veterinarians and physicians have been taught to look for horses, rather than zebras, when developing differential diagnoses lists for new problems or clinical signs in a patient.[6] For reasons of simplicity and economics, it is tempting and often correct to assume that new signs represent new developments in the significant disease already present, especially when the new signs are "typical" of known complications. While it is important to rule out "expected" complications when a patient with chronic disease is presented, it is equally important to simultaneously develop alternative differentials that are not directly or indirectly related to the known disease.

**Each new problem is a new patient.**

The first step in maintaining an open mind to unsuspected diagnoses is to listen carefully to the owner giving the history and perform a complete physical examination as if it were the first time the patient was presented. Similarly, the caretaker of the pet should be asked for a complete history, and chronic drug therapy/recent changes should be recorded. Although the new problem may be a new disease, it may also be a remote systemic response to the medication or combinations of medications that the animal is receiving. In this regard, it is prudent to be sure that the patient is not receiving supplements or treatments of which the clinician is not aware. This problem is particularly acute if the patient sees several veterinarians over time and records are not carefully scrutinized.

**Each new problem MAY be a new disease.**

This approach is as easy as recalling that non-cardiovascular diseases should be on your differential diagnosis list. For example, a listless cat with a new heart murmur that has a known history of a gallop rhythm and a diagnosis of hypertrophic cardiomyopathy may have a new heart murmur because she is now anemic, rather than it being a development related to the heart disease. A dog receiving furosemide that has developed changes in its urinary habits may have developed diabetes mellitus (with increased urination) or a urinary tract infection (with pollakiuria).

**Diagnostic testing should be based on today's information as well as the known disease.**

Any time that a chronic cardiovascular disease patient is presented, whether on recheck or urgently, it makes sense to check on the chronic response to ongoing therapy (e.g. monitoring potassium and renal function in patients on chronic heart failure therapies). However, additional testing indicated by the new findings should be pursued rather than assuming there is only once disease process at work. For example, if an otherwise stable heart failure patient on a stable medication and nutrition regimen suddenly begins to lose weight, other causes of weight loss, including neoplasia, should be considered. In this case, thoracic radiography may be helpful to identify metastases and should be considered even when the overt respiratory status of the patient is normal and radiographs might otherwise be considered unnecessary from a cardiac point of view.

## PATIENTS WITH CLINICAL SIGNS OF HEART DISEASE WITH NO COMPATIBLE HISTORY

**Remember the relationship between plasma volume and heart failure.**

Any process that increases pulmonary capillary pressures (eg. an increase in left atrial pressures or an expanded plasma volume) will lead to extravasation of fluid into the pulmonary interstitium. With a chronic moderate increase in pulmonary capillary hydrostatic forces, the fluid may be removed by an increase in pulmonary lymphatic clearance. There is insufficient time for recruitment of additional pulmonary lymphatic drainage with an acute increase in pulmonary capillary pressures, such as with sudden persistent tachycardia in hypertrophic cardiomyopathy, or with an acute intravenous fluid load. It is well recognized that over-zealous IV fluid therapy can precipitate signs of congestive heart failure in animals with cardiac disease, but rapid infusion of large volumes of IV fluids can, on occasion, lead to pulmonary edema and small volumes of pleural effusion even in animals with normal hearts. Where vascular integrity is compromised, pulmonary edema will occur at lower pulmonary hydrostatic pressures.

**ANY cause of 'arterial under-filling' will lead to neurohormonally-mediated sodium and water retention**

The process by which heart disease leads to congestive heart failure can occur in non-cardiac disease. Baroreceptor-mediated detection of 'arterial under-filling' triggers a sequence of events that includes an increase in sympathetic tone, activation of the renin-angiotensin-aldosterone system and other sodium- and water-retaining neurohormones, and eventual expansion of plasma volume to increase preload and increase cardiac output, restoring normal systemic arterial filling. This is an advantageous acute response in hypovolemia, but a less advantageous chronic response in cardiac disease. Other causes of arterial under-filling include disease processes that result in an abnormal reduction in systemic vascular resistance (anemia, arterio-venous fistulae, hyperthyroidism). Plasma volume may be expanded in these conditions, leading to cardiomegaly and congestive signs, even in the absence of prior structural heart disease.

**Remember the relationship between plasma volume and cardiac chamber size**

Expansion of plasma volume can lead to dilation of (generally) all 4 cardiac chambers, but hypovolemia can lead to a decrease in size of cardiac chambers.[7] This is a well-recognized sign in hypovolemic shock or hypoadrenocorticism, but the effect of reduced filling pressures on ventricular wall thickness is sometimes forgotten. If ventricular mass is unchanged, then a reduction in ventricular diameter must result in an increase in wall thickness. This is seen as 'pseudohypertrophy' in volume-depleted cats or dogs with pericardial effusion. Simple measurement of left ventricular (LV) wall thickness is insufficient to distinguish between LV hypertrophy versus LV pseudohypertrophy; assessment of history, blood pressure and size of other chambers is necessary. Left atrial size is usually reduced in LV pseudohypertrophy.

## PATIENTS WITH CARDIAC DISEASE THAT IS WELL-HIDDEN

**When the heart seems normal(ish) but the signs look like congestive heart failure**

Abnormalities are frequently evident on physical examination in patients with cardiac disease (eg. murmurs, gallop sounds, arrhythmias). In those cardiac patients where physical examination is unremarkable (eg. some cats with myocardial disease), then echocardiography will frequently reveal the underlying problem. Occasionally the physical and echocardiographic findings are at odds with each other. In this situation, it is worth considering the reliability of different signs of heart failure. The closest to a universally accepted 'gold standard' for left-sided heart failure is measurement of pulmonary capillary wedge pressure. This is usually not practical, so radiographic evidence of pulmonary infiltrates with left atrial enlargement is usually the accepted alternative. Pulmonary infiltrates may be hilar or generalized in dogs, but distribution may be very unpredictable in cats.[8] Conditions that may lead to left-sided heart failure that are not associated with left atrial enlargement include infective endocarditis (murmurs may be low grade if acute), rupture of a chorda tendinea in previously mild myxomatous mitral valve disease (murmur intensity may be reduced with increased filling pressures), heart base tumors obstructing pulmonary venous inflow (no murmur), or cor triatriatum sinister in cats (again, no murmur). All of these conditions will have structural abnormalities evident on echocardiography. Right-sided heart failure is usually associated with jugular distension, and hepatic venous distension on ultrasound. A distended caudal vena cava on radiographs may also be highly suggestive of right-sided failure, and is more reliable than apparent right atrial enlargement on radiographs. Conditions which commonly present with right-sided heart failure that may have confusing signs include cor triatriatum dexter (no murmur, no jugular distension) and constrictive pericardial disease (jugular distension and hepatic venous distension, but no murmur and subtle echocardiographic changes).

**Intermittent arrhythmias**

Clinical signs associated with arrhythmias may include syncope OR congestive heart failure. Bradyarrhythmias generally present with syncope, although with prolonged asystole then manifestations of syncope may overlap with hypoxic seizures. With intermittent tachyarrhythmias, a tachycardia-mediated cardiomyopathy may masquerade as an idiopathic cardiomyopathy. It is important to detect these arrhythmias, as systolic dysfunction due to an intermittent supraventricular tachycardia may be reversible with prevention of the arrhythmia.

## REFERENCES

[1]Atkins CE. **Vet Clin North Am Sm Anim Pract** 1991;21;1035, [2]Peterson ME et. al. **J Am Vet Med Assoc** 1983;183:103, [3]Dillman WH, **Thyroid Today** 1983;6:1, [4] Henik RA, et.al. **J Am Anim Hosp Assoc** 2004;40(5):359-363, [5]Panciera DL. **J Am Vet Med Assoc** 1994;205:996, [6]Groopman J. How Doctors Think, Mariner Books: 2008, [7]Campbell FE, et. al. **J Vet Intern Med** 2007:21:1008, [8]Benigni L, et. al. **J Small Anim Pract** 2009;50:9.

# WHY CURRENT RESEARCH AND PUBLICATION PRACTICES MAY NOT BE SERVING YOUR OR YOUR PATIENTS' BEST INTERESTS

**Paul D. Pion, DVM, DipACVIM (Cardiology) & Mark Rishniw, BVSc, DipACVIM (Internal Medicine and Cardiology)**
**Veterinary Information Network, Inc., Davis, CA**

To quote Lou Reed: "Don't believe half of what you see and none of what you hear." Can we believe what we read? And if not, what can we believe?

The bottom line of our musings (this presentation) are "what we (as a profession) know, how we know it, how we judge it, and how willing we are to admit we don't know as much as we think we do".

The ACVIM letter accepting this talk for the Forum included the following (slightly paraphrased):
"Your lecture has been assigned as a Comprehensive Review session. Comprehensive Reviews provide an extensive, critical, up-to-date overview of a topic that is directly or indirectly applicable to Veterinary Internal Medicine. Speakers should avoid delivering empirical information and interpretation. Lecture content is derived from evidence based reports and similar sources with minimal inclusive of empirical information." On the surface, that reads as sound, professional and reasonable instructions to authors. But is it constructive? The intent is good. It is Sergeant Friday (from the TV show and later movie Dragnet) reminding the witness to keep to "just the facts ma'am."

However, both these reasonable statements, that few would choose to argue, ignore the reality that whether examining the role of eyewitnesses,[1] expert witnesses, or even physical evidence in the justice system, or the role of expert or literature-based evidence in scientific or medical decision making, it is just as dangerous to accept them unquestioningly as it is to ignore them. Whether we are talking about the law, politics, economics or the medical sciences, the literature and all sources of "evidence" are potentially flawed and can be used to maliciously or innocently misdirect decisions. And while we are certain that the Forum committee did not intend to encourage propagation and validation of flawed information and thinking, we fear that is a potential unintended outcome, if these words are taken too literally. Contrary to what some might think, we don't believe there is evil or ill-intent behind these unintended consequences. We believe they are a part of the natural evolution of our profession and, to some extent mathematically provable.

Recently the field of "credibility analysis" has exploded, largely due to several biostatisticians (Altman, Bland, Ioannidis and others). These investigators have begun to look critically at published data, and analyze the statistical probability of obtaining the sorts of results that are being routinely reported in the medical/veterinary literature. Their insights are illuminating. In addition to examining these events, these authors have begun to identify causes or reasons for their findings. This presentation summarizes some of their data.

**Most research findings are false [2]**

This premise can be examined mathematically. With a low enough prior probability of outcome, and enough statistical testing, it is mathematically probable that many chance differences will be found to be statistically significant.

**Many publication practices promote the publication of exceptional results [3]**

Using a "Winner's Curse" model of publication behavior, it is easy to demonstrate that articles that show the biggest findings are the most highly sought after prizes, and therefore are given the most prestigious publication positions. The Winner's Curse model, however, mandates that the winning bidder has "over-valued" the prize. Thus, these articles often represent either irreproducible results, or have outcomes that lie at the far end of the bell curve of possible outcomes for the particular study.

**Accentuate the positive, eliminate the negative**

One could think Bing Crosby was singing about medical publications. Data are accruing that clearly demonstrate a bias to publication of positive findings. This distorts the clinical reality, and allows attribution of greater effects than really exist. Additionally, it allows new or naïve investigators to repeat unsuccessful studies endlessly, because they are unaware of the unpublished negative results.

Why are positive findings published more readily? They are more appealing to the reader – finding that some new intervention produced a desired outcome excites the reader, and offers clinicians hope of improving their patient's lives. Many clinical studies are controlled by pharmaceutical companies.[2, 4, 5] Thus, a vested interest exists in publishing positive findings, and suppressing negative findings. Finally, authors who recognize the publication bias (tendency to publish positive findings) will accentuate their positive findings, and suppress their negative findings within a manuscript – i.e. cherry-picking their findings for the good ones.[6, 7] Or, having found largely negative results, after considerable blood, sweat and toil, they want to have something to account for their effort, especially when the alternative hypothesis at the outset was one in which they believe.[8]

**Discredited information lingers on and on and on…**

Studies have demonstrated that observational information, found to be inaccurate by subsequently well-performed clinical trials, is still propagated in the literature. Similarly, studies that have been retracted for presenting fraudulent information are retracted much more slowly than they are published. This is a direct effect of the "winner's curse" predicament – a true example of "overpaying" for your prize. Malcolm Gladwell, in his book "The Tipping Point" illustrates that much of what influences a population's behavior can be attributed to the effects of select individuals, so-called "mavens" who are trusted by the adopters to be information providers. If these "mavens" perpetuate in disseminating out-dated information, their views will continue to be adopted by the masses.[9]

**Lies, damned lies & statistics!**

Statistical abuse in veterinary and medical literature is rampant. This is driven by a lack of statistical knowledge of the reviewers and authors, who misapply statistical methodology.[7, 10, 11] Additionally, little is understood about statistical significance, and its confusion with clinical significance. For example, virtually every study of prognostic indicators in oncology shows a statistically positive association between that factor and survival.[12] This makes it very difficult to determine which indicator is better than another, and which is truly clinically relevant.

Very few studies highlight measures of confidence of their estimates, despite years of pleading for such estimates to be routinely included. Instead, the almighty p-value determines whether a difference exists or not. And where differences are not found, studies often do not report the *a priori* power calculations to demonstrate what difference might have been anticipated.[13]

**I saw it – it must be true!**

Most observational studies have greater effects than are found in randomized clinical trials. And studies that initially report a finding often present outcomes that are substantially greater than is later found to be the case. In many cases, this is due to an underpowered study, where the anticipated difference must necessarily be larger than "average" in order to be detected.[Error! Bookmark not defined.] For example, the use of gabapentin as an antidepressant or for managing bipolar disorder showed 100% efficacy in open label studies, but demonstrated a benefit in only 16% of double-blinded placebo-controlled studies.[14]

**If it ain't new, I don't wanna hear about it**

Despite the recognition that independent replication of results either strengthens the credibility of the initial observation, or refutes it, investigators and editors are both loathe to publish replicative studies. This often forces the reader to base their decision on very imprecise information, which, in veterinary medicine, is often based on a very limited sample population. Furthermore, funding for replicative studies, especially by pharmaceutical companies, is rarely forthcoming, because one positive result is much more lucrative than one positive and one negative result. Very few clinical veterinary trials have been replicated – indeed the ACE-I trials of the past decade reflect one such example, because the initial observation was negative![15, 16]

**I didn't get a fair hearing!**

The review process, far from being fair, is flawed. New information can be, at least for a time, suppressed by biased reviewers whose own views disagree with those being submitted for publication, even if on technical minutiae.

**My house, in the middle of the street**

Not only are published studies likely to be skewed towards the far end of an anticipated range of results, but the studies are often performed in strictly controlled settings, which reduce the sample variability, and therefore, the sample size necessary to obtain a result. Consequently, the study participants (subjects) are often managed better than the "average" patient, regardless of which arm of a trial they are enrolled in. Data acquisition is more closely monitored, so diagnostic accuracy may not reflect the ability of the average clinician. This makes extrapolation of clinical outcomes to routine practice extremely tenuous – the clinician with a practice in the "middle of the street" cannot ever expect to replicate results observed in the trial.

**A change is gonna come…**

The antidote to what ails us today has been bottled in the field of Evidence Based Medicine (EBM).[17, 18] The tenants of EBM hold promise for formalizing the explicit use of a range of information/data for making clinically relevant decisions. However, given that in the standard EBM pyramid where meta-analysis of multicenter, randomized, double-blinded, placebo controlled clinical trials are at the pinnacle and expert opinion are at the bottom, few can deny that Evidence Based Veterinary Medicine could be largely redefined as EBM in a void of evidence.

Despite this lack of type IA data, a growing list of Evidence Based Veterinary Medicine articles are appearing from unbiased and biased sources. "Evidence Based" in the title of a paper has almost become the new Good Housekeeping seal that makes it "all OK." Nothing could be further from the truth. Evidence Based grading does not negate the old adage of Garbage In, Garbage Out.

Perhaps it is time to put aside all the fancy grading and debating and simply look at each other and ask: "what Do we know and How do we know it?" More than what grade a study fits into, we should be asking "how much do we believe that result" and "how applicable is it to our patients?" And no matter how ugly the outcome of the process, we should feel that the process was beneficial.

**We know a lot!**

On the surface this presentation and premise might make some believe that we are suggesting that our professional knowledge base is something to be ashamed of. Quite the contrary, our profession has done a wonderful job of answering many of the basic questions and moved from the days of James Herriot to the days of truly scientifically based medicine. We have a wonderful base of "Background Information" and we have now moved to the point where we are asking truly important questions about "Foreground Information"—questions that only come with true expertise in a field.

This is something to be proud of. But to get to the answers we seek, we must commit to respecting the data – no matter what the data telling us.

## REFERENCES

[1] Engelhardt L. **Stanford J Legal Studies** 1999;1(1):25 [2] Ioannidis J. **PLoS Medicine** 2005;2(8):e124. [3] Young NS, et al. **PLoS Medicine** 2008;5(10):1418 [4] Landefeld CS, et al. **N Engl J Med** 2009;360(2):103 [5] Pion PD, et al. **ACVIM Proceedings**, 2008 [6] Dwan K, et al. **PLoS one** 2008;3(8):e3081 [7] Pocock SJ, et al. **N Engl J Med** 1987;317(7):426 [8] Brown DC. **Am J Vet Res** 2007;68(7):694 [9] Gladwell M. **Little Brown**, 2000 [10] Plant JD, et al. **Vet Dermatol** 2007;18(4):260 [11] McCance I. **Aust Vet J** 1995;72(9):322 [12] Kyzas PA, et al. **Eur J Cancer** 2007;43(17):2559 [13] Hofmeister EH, et al. **J Small Anim Pract** 2007;48(2):76 [14] Carey TS, et al. **J Psychiatr Pract** 2008;14 Suppl 1:15 [15] Kvart C et al. **J Vet Intern Med** 2002;16(1):80 [16] Atkins CE, et al. **J Am Vet Med Assoc** 2007;231(7):1061 [17] Sackett DL et al. **Churchill Livingstone** 2000 [18] Cockroft P et al. **Wiley Blackwell** 2003

# FELINE HYPERTROPHIC CARDIOMYOPATHY – GETTING INTO THE THICK OF IT

**Mark D. Kittleson, DVM, PhD, DACVIM (Cardiology)**
**Davis, CA**

## DEFINITION

Hypertrophic cardiomyopathy (HCM) is a primary myocardial disease that primarily affects left ventricular (LV) myocardium and is characterized by mild to severe thickening (concentric hypertrophy) of the LV wall (septum and/or free wall) and papillary muscles. The thickening may be global or regional. In feline HCM, the LV papillary muscles are consistently enlarged. The diagnosis is, to some degree, one of exclusion. Several possible secondary causes of LV concentric hypertrophy exist in cats including systemic arterial hypertension, hyperthyroidism and aortic stenosis. When LV hypertrophy is caused by any of these diseases, the cardiac abnormality is not called HCM since it is not cardiomyopathy (primary heart muscle disease). Instead it is called concentric hypertrophy secondary to the primary abnormality (e.g., concentric hypertrophy secondary to hyperthyroidism). However, it should be noted that no other primary or secondary cause of LV hypertrophy can generally increase the LV wall thickness to more than 1.5 times the upper limit of normal thickness. Consequently, when the LV wall is 7 mm thick or more in a cat, this is almost invariably severe HCM. The one possible exception to this rule is lymphoma, a rare cause of increased LV wall thickness produced by infiltration of the myocardium.

## NOMENCLATURE

According to a recent American College of Cardiology/European Society of Cardiology Clinical Expert Consensus document, "HCM is now widely accepted as the preferred term because it describes the overall disease spectrum without introducing misleading inferences that LV outflow tract (LVOT) obstruction is an invariable feature of the disease, such as is the case with hypertrophic obstructive cardiomyopathy (HOCM), muscular subaortic stenosis, or idiopathic hypertrophic subaortic stenosis. Indeed, most patients with HCM do not demonstrate outflow obstruction under resting (basal) conditions, although many may develop dynamic subaortic gradients of varying magnitude with provocative maneuvers or agents."[1] It is suggested by the author that veterinary medicine should follow suit on this recommendation.

## CLINICAL MANIFESTATIONS

### Sex Predilection

A male predilection for HCM is commonly reported. However, in cat breeds where the mode of inheritance is known to be autosomal dominant (i.e., Maine Coon and Ragdoll cats) the incidence of the mutation is obviously the same in males and females. However, in Maine Coon cats it is known that males often develop a more severe form of the disease and do so at an earlier age than females.[2] Consequently, it is likely that the male predominance seen clinically is purely because males develop worse disease and so present to a veterinarian.

### Breed Predilection

The genetic cause of HCM in Maine Coon cats and Ragdoll cats has been determined and will be discussed in detail in a subsequent lecture. Numerous other breeds also have HCM as a prevalent problem, including, but not limited to, Persians, American and British Shorthairs, Siberians, Norwegian Forest Cats, and Turkish Vans. Most HCM is diagnosed in mixed breed cats (domestic shorthair, domestic longhair). The mutations identified so far in cats are inherited as autosomal dominant traits since they alter structural proteins and almost all mutations in humans are inherited in an autosomal dominant fashion. Given the breeding patterns of domestic cats, it is not difficult to see how an autosomal dominant trait could disseminate into the population at large. It may also be that the portions of the feline genome that are responsible for sarcomere development are more prone to mutation than other species.

### Natural History

HCM is a progressive disease in Maine coon cats, as it is in humans.[2] Although most cats with HCM are not or have not been followed over long periods like Maine Coon cats have, the disease is almost certainly progressive in other cats as well. In Maine Coon cats a typical scenario is for a cat to have no evidence of HCM during the first 2-5 years of life and then to develop the maximum stage of its HCM (mild to severe HCM) over a year or more. Cats that only develop mild to moderate HCM usually have no clinical sequelae to their disease (although a small percentage may die suddenly). This means that it is possible that when a veterinarian examines an older cat and identifies, for example, moderate HCM, the disease has been present for much of the cat's life. Cats that develop severe disease may go on to have severe sequelae (i.e., heart failure, sudden death, systemic thromboembolism) within a short period while others may stabilize for a remarkably long time. It would appear, and it would be logical, that cats that are homozygous for a mutation that causes HCM have an accelerated course to their disease and so have echocardiographic evidence of HCM at an earlier age and more frequently develop severe disease, again at an earlier age.

### History & Clinical Signs

Cats with HCM may have no clinical evidence of the disease, many have a heart murmur, a few have a gallop sound (if the disease is severe), some are presented to a veterinarian with subtle signs of heart failure, others have moderate to severe heart failure and so tachypnea and/or dyspnea, and some are presented because of systemic thromboembolic disease. Cats with no clinical signs can have mild to severe left ventricular thickening, however, those with severe thickening usually go on to develop heart failure. Cats with severe disease that appear to have no clinical signs may show subtle signs of heart failure (e.g., tachypnea) that may be detected by an observant owner but more often goes undetected. The respiratory rate is often increased in these cats at rest and they may become more tachypneic or even dyspneic if stressed. Cats with mild to moderate HCM may never develop clinical signs and may live normal lives. In others, the left ventricular wall may thicken further and complications may develop when they are older and develop a complicating disease such as systemic hypertension or hyperthyroidism. Cats with severe HCM and moderate to severe heart failure are usually presented to a veterinarian because of respiratory abnormalities (tachypnea and/or dyspnea) due to pulmonary edema, pleural effusion, or both. Since household cats are generally sedentary, owners usually do not notice that they are having respiratory difficulty until it is advanced. At that time, the onset of the disease commonly appears to be acute to the owner, whereas the disease actually has been present for years and the heart failure gradually worsening for weeks.

Cats with HCM may die suddenly, often with no prior clinical signs referable to heart disease or failure. The cause of the sudden death in these cats is unknown but may be due to a fatal arrhythmia, large thrombus lodging in the left ventricle or proximal aorta, or acutely worsening outflow tract obstruction associated with severe stress. The incidence of sudden death in feline HCM is probably underrepresented in the literature because cats that die suddenly are generally not presented or reported to veterinarians.

### Physical Examination

Cats with severe HCM commonly have auscultatory abnormalities. A systolic murmur, heard best over the mid sternum or left apex beat is common and is usually due to SAM. The murmur is often dynamic, increasing in intensity when the cat becomes excited and heart rate and myocardial contractility increase (especially when contractility of the papillary muscles increases) and decreasing in intensity or disappearing when the cat is calm. It should be noted change in contractility, not heart rate, is the cause of the change in murmur intensity. Heart rate change is just along for the ride and is an index of changes in sympathetic tone.

A gallop sound may be present in cats with severe HCM. It should be noted that not all three heart sound rhythms are due to gallop sounds since systolic clicks also occur with some frequency in cats. A systolic click is more likely to vary in intensity (although often it does not) and should be the assumed cause when there is mild or no apparent cardiac disease.

## PATHOPHYSIOLOGY

HCM almost always causes diastolic dysfunction rather than systolic dysfunction. Diastolic dysfunction is usually either a problem with chamber compliance (1/stiffness) or chamber relaxation. Severe concentric hypertrophy by itself increases chamber stiffness. In addition, blood flow and especially blood flow reserve to severely thickened myocardium is compromised which causes myocardial ischemia, cell death, and replacement fibrosis. This has been documented in cats by showing that cardiac troponin I concentration is increased in cats with severe HCM.[3] Increased concentrations of circulating neurohormones may also stimulate the production of collagen, which produces interstitial fibrosis. Fibrosis increases chamber stiffness (increased $\Delta$pressure/$\Delta$volume) further and is the primary reason for the marked diastolic dysfunction seen in this disease. The stiff ventricular chamber causes a greater increase in pressure for any given increase in volume when the ventricle fills in diastole. The increased diastolic pressure backs up into the left atrium and pulmonary veins and produces congestive heart failure (CHF) (i.e., pulmonary edema and pleural effusion).

When LV hypertrophy is severe, it is common for the LV wall thickness to be twice the normal thickness (increased from 3-5 mm up to 7-10 mm). The end-systolic volume and diameter are almost always reduced, often to zero (end-systolic cavity obliteration). Studies in humans have shown that global myocardial contractility is normal in patients with HCM.[4] Instead the reduction in end-systolic volume is due to a decrease in afterload (systolic wall stress) brought on by the increase in wall thickness (Laplace's law).

Left heart failure is manifested as either pulmonary edema or pleural effusion in cats. The pleural effusion is thought to be produced from the surface of the lungs since the visceral pleural veins drain into pulmonary veins (and so the left heart).[5] Of course, pleural effusion can also be produced by right heart failure since the parietal pleura (pleura lining the chest cavity) drains into the systemic venous system but in the vast majority of cats with heart failure due to HCM there is absolutely no evidence of right heart failure (e.g., an enlarged right atrium, jugular or hepatic vein distension).

### Systolic Anterior Motion (SAM) of the Mitral Valve

SAM of the mitral valve is common in cats with HCM. Cats with HCM and SAM are commonly said to have the obstructive form of HCM or hypertrophic obstructive cardiomyopathy (HOCM). In one survey of 46 cats, SAM was present in 67%.[6] However, as in humans, SAM comes and goes depending on the contractile state of the heart so it is likely that SAM can be provoked in many cats that do not have SAM at rest by just taking them to see a veterinarian and it is also likely that many cats that have SAM in a veterinary clinic do not have it when asleep. As noted previously, it has been recommended that the term HOCM be dropped.

Systolic anterior motion of the mitral valve is the process of the septal (anterior) mitral valve leaflet or the chordal structures inserting on this leaflet being pulled into the LVOT during systole by the enlarged papillary muscles. Here it is caught in the blood flow and pushed toward and often ultimately against the interventricular septum. The initial pulling of the mitral valve leaflet toward the LVOT in systole can clearly be seen on many echocardiograms from cats with HCM to be due to the grossly enlarged papillary muscles encroaching on the LVOT (the region of the LV between the anterior leaflet of the mitral valve and the interventricular septum) and pulling the mitral apparatus structures into the basilar region of the outflow tract. SAM produces a dynamic subaortic stenosis that increases the velocity of blood flow through the subaortic region and usually produces turbulence. Simultaneously, when the septal leaflet is pulled toward the interventricular septum, a gap in the mitral valve is produced creating mitral regurgitation. These abnormalities are by far the most common causes of the heart murmur heard in cats with HCM.

## PATHOLOGY

### Gross Pathology

Cats with severe HCM have severe thickening of the LV myocardium (the interventricular septum and free wall), with the LV wall measuring 7-10 mm thick. The hypertrophy may be symmetric, involving the entire circumference of the LV, but is often asymmetric. In some cats the interventricular septum is significantly thicker than the free wall while in others the free wall is thicker (asymmetric hypertrophy). In those cats with primarily septal hypertrophy the hypertrophy most commonly involves the entire septum but it may be confined to the basilar or apical regions of the septum. Isolated free wall hypertrophy most commonly occurs in the region between the papillary muscles. As in many pathologic specimens, hearts from cats with HCM may undergo contraction (rigor) following death resulting in a wall thickness that is closer to the end-systolic wall thickness in life rather than the end-diastolic thickness (falsely thickened). Consequently, heart weight must be combined with subjective or objective evidence of LV wall thickening to make the diagnosis of HCM post mortem. To weigh a cat heart the pericardium should be removed and the aorta and pulmonary artery transected so that no more than 2-4 cm are left. Most normal cats have a heart that weighs less than 20 gm and so most cats with HCM have a heart that weighs more than 20 grams. Cats with severe HCM almost always have a heart that weighs more than 25 gm, usually weigh over 30 gm, and can be as heavy as 38 gm.

The left atrium is often enlarged in cats with severe HCM, often markedly so. However, even with early, severe disease we have identified normal left atrial size in some Maine Coon cats. Occasionally a thrombus is present, most commonly within the left auricle. Cats with mild to moderate HCM have lesser wall thickening and a more normal sized LV chamber. The left atrium is usually normal in size but may be enlarged. Papillary muscle hypertrophy is often the predominant lesion.

### Histopathology

Histopathologically, there is a wide range of abnormalities from only myocyte hypertrophy to moderate to severe interstitial and replacement fibrosis and dystrophic mineralization (20 to 40% of cases). In humans, myocardial fiber disarray (MFD) that involves at least 5% of the myocardium in the interventricular septum is found in 90% of patients with familial HCM. In cats with HCM, myocardial fiber disarray in the interventricular septum of the same magnitude observed in humans is only identified in 30% to 60% of cases. However, myocardial fiber disarray is a consistent feature of HCM in Maine coon cats.[2] Unfortunately, most veterinary pathologists are not trained to characterize MFD.

## DIAGNOSIS

### Echocardiography

The diagnosis of HCM is almost always made using echocardiography where cats with severe HCM have severe papillary muscle hypertrophy, a markedly thickened LV wall (7-10 mm), and often, but certainly not always, an enlarged left atrium. The hypertrophy can be global, affecting all areas of the LV wall or can be more regional.[6] Because it can be regional, HCM is a diagnosis that should be made by examining several different two-dimensional echocardiographic views and measuring wall thickness in diastole from the thickest region or regions. M-mode echocardiography may miss regional thickening unless it is guided by the two-dimensional view and so should never be used without two-dimensional guidance, if at all. End-systolic cavity obliteration is commonly present in cats with severe HCM.

### Systolic anterior motion (SAM) of the mitral valve

Mitral valve SAM can often be most easily identified using two-dimensional echocardiography, either in real time or in slow motion of a cine loop (see http://www.vmth.ucdavis.edu/cardio /cases/case5/echodopp.htm). Color flow Doppler echocardiography can be used to demonstrate the hemodynamic abnormalities associated with SAM. With this modality one observes two turbulent jets originating from the LVOT - one regurgitating back into the left atrium (mitral regurgitation) and the other projecting into the aorta (dynamic subaortic stenosis). Spectral Doppler is used to determine the pressure gradient across the region of dynamic subaortic stenosis to characterize the severity of the SAM although it can be quite labile, changing with the cat's level of excitement. Care must be taken not to record the high velocity mitral regurgitation jet. Usually, the dynamic subaortic stenosis jet is narrower (shorter period of systole) than the mitral regurgitation jet; often becoming even more narrow in mid - to late systole.

### Diastolic dysfunction

Diastolic dysfunction in cats with severe HCM has been documented using tissue Doppler imaging (TDI) and measures of transmitral flow and relaxation time.[7] Cats with severe HCM routinely have a decrease in early diastolic wall motion of the LV free wall and mitral valve annulus using DTI. In addition, on pulsed wave Doppler echocardiography peak E wave velocity may be reduced, peak A wave velocity increased, isovolumic relaxation time prolonged, and rate of deceleration of early inflow reduced in cats with severe HCM when the heart rate is slow enough that E and A wave velocities can be identified.[8]

## REFERENCES

[1] Maron BJ, et al. **J Am Coll Cardiol** 2003; 42(9):1687. [2] Kittleson MD, et al. **Circulation** 1999; 99(24):3172. [3] Herndon WE, et al. **J Vet Intern Med** 2002; 16(5):558. [4] Saito T, et al. **Jpn Circ J** 1987; 51:511. [5] McLaughlin RF, et al. **Am J Anat** 1961; 108:149. [6] Fox PR, et al. **Circulation** 1995; 92:2645. [7] Gavaghan BJ, et al. **Am J Vet Res** 1999; 60(12):1478. [8] Carlos SC, et al. **J Vet Intern Med** 2006; 20(5):1106.

## TREATMENT OF FELINE HYPERTROPHIC CARDIOMYOPATHY (HCM) – LOST DREAMS

**Mark D. Kittleson DVM, PhD, DACVIM (Cardiology)**
**Davis, CA**

### TREATMENT OF CATS WITH HCM AND NO CLINICAL SIGNS

There currently is no evidence that any drug alters the natural history of hypertrophic cardiomyopathy (HCM) in cats until they are in heart failure. Diltiazem, atenolol, ACE inhibitors and, possibly, spironolactone are commonly administered by veterinarians to cats with mild to severe HCM that are not in heart failure on an empirical basis in the hope that they will slow the progression of the disease. A recent survey of veterinary cardiologists revealed that less than 20% use diltiazem in these cases.[1] However, 80% indicated that they did use some drug to try to slow the progression in those cats even with moderate disease.

Recent studies by the author have shown that ramipril, an ACE inhibitor, and spironolactone have no beneficial effects on HCM prior to the onset of heart failure.[2,3] In addition, spironolactone produced severe skin lesions in some cats. In the first study we administered ramipril, an ACE inhibitor, to 13 Maine Coon and Maine Coon cross cats with HCM but without evidence of heart failure from the author's research colony at UCDavis for one year and placebo to the same number of cats in a blinded fashion.[2] Left ventricular mass was quantified using MRI at baseline and again at the end of the study. Tissue Doppler imaging (TDI) was used to assess diastolic function over the year. Plasma neurohormones and ACE activity were measured. Ramipril adequately reduced (97%) plasma ACE activity, meaning the dose administered was effective. Ramipril did not reduce LV mass nor did it improve diastolic function. There was no change in plasma aldosterone and BNP concentrations. We concluded that ACE inhibition is not beneficial for cats with HCM that are not in heart failure.

In the second study, we examined spironolactone in the same colony in the same types of cats.[3] It was hypothesized that spironolactone would reduce myocardial fibrosis and in so doing improve diastolic function and possibly also reduce LV mass. Two-dimensional echocardiography and TDI were used to assess LV mass and diastolic function respectively in 13 Maine Coon and Maine Coon cross cats administered spironolactone and were compared to 13 cats on placebo. Cats were studied over a 4 month time frame. Plasma aldosterone concentration did increase in the cats on spironolactone, which documented that the dose was appropriate. There was no change in LV mass or improvement in LV diastolic function. In addition, 4 of the 13 cats on spironolactone developed a severe ulcerative dermatitis on their face. We concluded that spironolactone is not an appropriate drug in this population.

**Beta-adrenergic receptor blockers**

Beta blockers are primarily used to reduce systolic anterior motion (SAM) of the mitral valve and heart rate in cats with HCM. In humans, beta blockers are almost routinely used in an attempt to improve symptoms due to HCM although the exact means by which symptoms are improved is poorly understood.[4] However, it is a consensus that beta blockers, when given in high doses to humans, reduce SAM during exertion. To quote a recent consensus statement by the American College of Cardiology and European Society of Cardiology regarding the use of beta blockers in human patients with HCM, "Substantial experience suggests that standard dosages of these drugs can mitigate disabling symptoms and limit the latent outflow gradient provoked during exercise when sympathetic tone is high and heart failure symptoms occur. However, there is little evidence that beta-blocking agents consistently reduce outflow obstruction under resting conditions. Consequently, beta-blockers are a preferred drug treatment strategy for symptomatic patients with outflow gradients present only with exertion."[4]

There is little doubt that the practice of using beta blockers in cats with HCM is a direct extrapolation from human medicine. That may be problematic. First of all, cats generally don't exert themselves. Instead, cats spend 85% of their life asleep and sleep certainly reduces sympathetic activity in cats better than any beta-adrenergic blocking drug. There is little doubt that cats with mild to moderate SAM in a veterinary clinic have no or milder SAM at home and the same can be said for tachycardia. However, it is also apparent that cats can get into very stressful situations (e.g., cat fights, chased by a dog) at times and it is possible that a beta blocker in those brief instances could be beneficial in preventing acute heart failure. Second, the dose used in cats by veterinarians is generally small in comparison to that used in humans.

Beta blockers are effective for reducing the severe SAM observed in clinical situations. At least one study has examined the effects of esmolol, a short-acting $\beta_1$ - adrenergic blocking drug, in cats with HCM and obstruction to left ventricular outflow due to SAM and shown a reduction in the pressure gradient across the outflow tract. In that study the degree of outflow tract obstruction decreased and the heart rate slowed and esmolol was more effective than diltiazem.[5]

Atenolol is a specific $\beta_1$ - adrenergic blocking drug that needs to be administered twice a day, usually at a total dose of 6.25 to 12.5 mg PO q12h. In cats, atenolol has a half-life of 3.5 hours. When administered to cats at a dose of 3 mg/kg, atenolol attenuates the increase in heart rate produced by isoproterenol for 12 but not for 24 hours.[6]

In a recent survey of veterinary cardiologists regarding their preferences for treating cats with HCM, 70% used atenolol to treat cats with severe SAM while 50% used it to treat cats with mild SAM.[1]

### TREATMENT OF CATS IN HEART FAILURE DUE TO HCM

Cats that present in left heart failure primarily have clinical signs referable to pulmonary edema and/or pleural effusion. Consequently, therapy is generally aimed at decreasing left atrial and pulmonary venous pressures in these cats and physically removing pleural effusion. In some cats with severe heart failure, clinical evidence of hypoperfusion (low-output heart failure) may be apparent in addition to the signs of CHF (e.g., cold extremities, total body hypothermia).

Pulmonary edema is primarily treated with diuretics (almost exclusively with furosemide) acutely and chronically and an ACE enzyme inhibitor chronically, although there is some evidence to suggest that ACE inhibition may not be that helpful in prolonging survival in cats with HCM.[7] Diltiazem and beta adrenergic blockers, usually atenolol, have been commonly used as adjunctive agents. Again, there is some evidence to suggest that diltiazem is not helpful in prolonging survival in cats with heart failure due to severe HCM and that atenolol may actually shorten survival time.[3] Pleurocentesis is most effective for treating cats with severe pleural effusion. Furosemide is usually helpful at slowing effusion reaccumulation but repeated pleurocenteses are often needed to keep signs of heart failure in check.

**Acute Therapy**

Cats that present with respiratory distress suspected of having heart failure secondary to HCM need to be placed in an oxygen enriched environment. If possible the cat should be initially evaluated by doing a cursory physical examination, taking care not to stress the patient during this or any other procedure since stress exacerbates dyspnea and arrhythmias and often leads to death. Most, but not all, cats with severe HCM that are in heart failure will have a heart murmur and many will have a gallop sound (gallop rhythm). A butterfly catheter should be used to perform thoracentesis on both sides of the chest to look for pleural effusion as soon as possible. Generally this should be done with the cat in a sternal position so that it does not become stressed during the procedure. Clipping of the hair is not needed. If fluid is identified, it should be removed. If none is identified, a lateral thoracic radiograph to identify pulmonary edema may be taken with the veterinarian present to ensure that the cat is not stressed.

Cats present in respiratory distress due to a number of respiratory and thoracic cavity diseases in addition to heart failure. Differentiation between left heart failure and respiratory failure may be difficult. Almost all cats with left heart failure will have moderate to severe left atrial enlargement. Consequently, if differentiation is difficult, assessment of left atrial size via radiography or echocardiography is one of the most useful diagnostic tests available. This is not always possible radiographically because of the presence of pleural effusion or other complicating factors. In general, it is relatively easy to determine left atrial size echocardiographically from the right side of the chest. Visualizing the left atrium and comparing it to the size of the aorta from a right parasternal short-axis view at the base of the heart is the most accurate means of doing this. A recent report has suggested that measurement of NT-proBNP serum concentration may be a reasonable means of making this differentiation.[8] Unfortunately, no cageside test yet exists for doing this.

**Furosemide**

Furosemide should initially be administered IV or IM to the cat in severe respiratory distress due to pulmonary edema. Cats that can tolerate an IV injection may benefit from the more rapid onset of action (within five minutes of an IV injection vs. 30 minutes for an IM injection). The initial furosemide dose to a cat in distress should generally be in the 2-4 mg/kg range IM or IV. This dose may be repeated within 1 hour to 2 hours. Dosing must be reduced sharply once the resting respiratory rate starts to decrease to avoid severe dehydration.

High-dose parenteral furosemide therapy commonly produces electrolyte disturbances and dehydration in cats. Cats with severe heart failure that require intensive therapy are often precarious. They may be presented dehydrated and electrolyte-depleted because of anorexia. They may remain anorexic and consequently dehydrated and depleted of electrolytes once the edema and/or the effusion are lessened. Judicious intravenous or subcutaneous fluid administration may be required to improve these cats clinically. Overzealous fluid administration will result in the return of CHF. If fluid administration is required, the furosemide administration must be discontinued for that time.

**Nitroglycerin**

Nitroglycerin cream might be beneficial in cats with severe pulmonary edema secondary to feline HCM. However, no studies have examined any effects of this drug in this species and its efficacy is suspect. Nitroglycerin is certainly safe and some benefit may occur with its administration in some cats. Consequently, 1/8 "-1/4" of a 2% cream may be administered to the inside of an ear every 4-6 hours for the first 24 hours as long as furosemide is being administered concomitantly. However, one should never rely on nitroglycerin to produce a beneficial effect and its use is by no means mandatory. Nitroglycerin tolerance develops rapidly in other species and probably does so in the cat. Consequently, prolonged administration is probably of even lesser benefit.

Once drug administration is complete the cat should be left to rest quietly in an oxygen enriched environment. Care should be taken not to distress the cat. A baseline measurement of the respiratory rate and assessment of respiratory character should be taken when the cat is resting. This should be followed at 30 minute intervals and furosemide administration continued until the respiratory rate starts to decrease (a consistent decrease of the respiratory rate from 70 to 90 breaths/minute into the 50 to 60 breaths/minute range is a general guide) and/or the character of the cat's respiratory effort improves. When this occurs, the furosemide dose and dosage frequency should be curtailed sharply.

**Sedation or anesthesia**

In some cats sedation with acepromazine (0.04 - 0.1 mg/kg subQ) may help by producing anxiolysis. Oxymorphone (0.04-0.1 mg/kg q6h IM, IV, or subcutaneously) or butorphanol tartrate (0.08 mg/kg IV or 0.36 mg/kg q4h subcutaneously) may also be used but can produce respiratory depression.

In some cats with fulminant heart failure, anesthesia, intubation, and ventilation are required to control the respiratory failure. Although this method is not preferred for most severely dyspneic cats, it can be life-saving in some.

**Chronic Therapy**

Many aspects of chronic therapy of HCM are controversial. All therapy is palliative and ultimately futile in most cases. Furosemide is the only drug that has a clearly beneficial effect chronically on survival in cats with HCM. An ACE inhibitor is usually given concomitantly. In a recent survey of veterinary cardiologists, all that returned the survey treated cats with HCM that were in heart failure with furosemide and an ACE inhibitor.[1]

**Pleurocentesis**

Many cats with HCM are dyspneic because of pleural effusion that reaccumulates despite appropriate medical therapy. These cats need periodic pleurocentesis.

**Furosemide**

In cats with CHF due to HCM, furosemide administration, once initiated, should almost always be maintained for the rest of the cat's life. In a few cases, furosemide can be discontinued gradually once the cat has been stabilized. This usually only occurs in a cat that has had a precipitating stressful event that does not have underlying severe HCM and does not have severe left atrial enlargement.

The maintenance dose of furosemide in cats usually ranges from 6.25 (½ of a 12.5 mg tablet) once a day to 12.5 mg PO q8h although the dose may be increased further if the cat is not responding to a conventional dose. We have administered higher doses (up to 37.5 mg q12h) than commonly recommended to a few cats with severe heart failure without identifying severe consequences as long as the cats were eating and drinking. Cats on high-dose furosemide therapy are commonly mildly dehydrated and mildly to moderately azotemic. However, they often continue to maintain a reasonable quality of life and so even moderate azotemia can often be ignored. The azotemia in this situation is prerenal in origin.

The furosemide dose needs to be titrated carefully in each patient. The owner should be taught how to count the sleeping respiratory rate at home and instructed to keep a daily written log of the respiratory rate. This is highly beneficial for making decisions regarding dosage adjustment in individual patients. Normal sleeping respiratory rate in a cat is usually in the 15-30 breaths/minute range but some cats will go up to as high as 40 breaths/minute.

**ACE inhibitors**

Although furosemide has been used to treat heart failure secondary to feline HCM for decades, the use of ACE inhibitors in cats with HCM is relatively recent because veterinarians shared the fears of their human medical counterparts that they would worsen SAM. Over the past 10 years it has become obvious to most veterinary cardiologists that ACE inhibitors do not worsen the clinical signs referable to HCM and a study has documented that SAM is not worsened by enalapril administration in cats.[9] Many have believed and one study has suggested that ACE inhibitors improve the quality and quantity of life of cats with HCM. Preliminary evidence from a recent placebo-controlled and blinded clinical trial suggests that enalapril produces little to no benefit when compared to furosemide alone in cats with heart failure due to HCM.[7] However, this study also included cats with unclassified (restrictive) cardiomyopathy and both cats with and without SAM. Subgroup analysis failed to change the conclusions of the study but the subgroups were small. In addition, this study has only been presented in abstract form (2003). Consequently, it is the recommendation of this author to continue to use an ACE inhibitor such as enalapril in cats in heart failure due to HCM at a dose of 1.25 to 2.5 mg PO q24h. Generic enalapril maleate is available in 2.5 mg tablets. **Diltiazem -** In cats with severe HCM that have or have had evidence of CHF, diltiazem or a beta adrenergic blocking agent are often administered. Beta blockers and calcium channel blockers (usually verapamil) provide symptomatic benefit in human patients. Their utility in cats with HCM is controversial although there is little doubt that neither drug produces dramatic benefits. Diltiazem, however, appears to produce no harm. Diltiazem is a calcium channel blocker previously reported to produce beneficial effects in cats with HCM when dosed at 7.5 mg q8h.[10,11] Beneficial effects that have been reported include lessened edema formation and decreased wall thickness in some cats. In the author's experience only the rare cat appears to experience a clinically significant decrease in wall thickness and it is impossible to tell if this is due to drug effect or time. Rarely does it appear clinically that diltiazem controls CHF on its own or helps control pulmonary edema or pleural effusion when added on to furosemide therapy. Diltiazem does improve the early diastolic relaxation abnormalities seen in feline HCM. Whether this helps decrease diastolic intraventricular pressure and so decrease edema formation is unknown. Theoretically it should have little benefit in the resting cat with a slow heart rate. Slower myocardial relaxation during rapid heart rates may not allow the myocardium enough time to relax, resulting in increased diastolic intraventricular pressure. Consequently, diltiazem may help protect a cat that undergoes a stressful event. Incomplete relaxation and decreased compliance, however, are more plausible explanations for increased diastolic pressure due to diastolic dysfunction in feline HCM. In humans, diltiazem does not change left ventricular chamber stiffness and so does not alter passive diastolic function.[12] Diltiazem decreases SAM, which may decrease the amount of mitral regurgitation, but beta blockers generally produce a greater decrease in the amount of SAM.(5) Recent evidence suggests that diltiazem has no effect on survival time in cats with severe HCM and heart failure. Consequently, there currently appears to be no ethical mandate for its use in cats with heart failure due to HCM and it would appear that many veterinary cardiologists have abandoned its use.

In addition to its regular formulation (30-mg tablets in the USA), diltiazem is supplied as slow-release (long-acting) products. Cardizem CD is supplied as 180 mg capsules that contain hundreds of small capsules. The larger capsule can be opened and a number of the smaller capsules divided into groups of four (45 mg each) and placed in smaller gelatin capsules for administration. One capsule is then administered q24h. Dilacor XR capsules can be opened to yield 2, 3, or 4 60-mg tablets. This drug is dosed at 30 mg per cat PO q12h and produces a significant decrease in heart rate and blood pressure in cats with HCM for 12-14 hours.[6]

**Pimobendan**

Pimobendan is contraindicated in cats with HCM. A positive inotropic agent should increase the severity of SAM and worsen end-systolic cavity obliteration. Pimobendan's effects on myocardial hypoxia are potentially detrimental.

## REFERENCES

[1] Rishniw M, et al. **J Vet Intern Med** 2006; 20[6], 1534. [2] Macdonald KA, et al. **J Vet Intern Med** 2006; 20(5):1093. [3] Macdonald KA, et al. **J Vet Intern Med** 2008; 22(2):335. [4] Maron BJ, et al. **J Am Coll Cardiol** 2003; 42(9):1687. [5] Wey AC, et al. **J Vet Intern Med** 2000; 14(3):335. [6] Keene BW. **Proceedings of the Fifteenth Annual Veterinary Medical Forum.** ACVIM, 1997: 155. [7] Fox PR. **J Vet Intern Med** 2003; 17[3]: 398. [8] Wess G, et al. **J Vet Intern Med** 2008; 21[3]: 707. [9] Oyama MA, et al. **J Vet Intern Med** 2003; 17(3):400. [10] Bright JM, et al. **J Vet Intern Med** 1991; 5:272. [11] Bright JM, et al. **Vet Clin North Am Small Anim Pract** 1991; 21:1023. [12] Betocchi S, et al. **Am J Cardiol** 1996; 78(4):451.

# THE GENETICS OF HCM –MUTANT CATS

**Mark D. Kittleson DVM, PhD, DACVIM (Cardiology)**
**Davis, CA**

## SARCOMERIC GENE MUTATIONS IN HUMANS WITH HCM

In humans, the familial nature of HCM was first reported in 1958. It has been demonstrated that over 70% of human cases of HCM are inherited in an autosomal dominant pattern, with most other cases being sporadic (although often still genetic in origin). Since 1989, over 450 mutations in 11 genes that encode for sarcomeric proteins have been identified in human families with HCM.[1-5] They include the β-myosin heavy chain, α-tropomyosin, cardiac troponins I, C, and T, myosin binding protein C, essential and regulatory light chains, titin, actin, and α-myosin heavy chain genes.[1,6-10] The genes with the most mutations described and the ones that most commonly produce disease are the β-myosin heavy chain and cardiac myosin binding protein C genes. It is now known that sarcomeric gene mutations actually cause HCM since several mutations that have been identified in human families with HCM have been placed in transgenic mice and the disease reproduced, thus fulfilling Koch's postulates.[11-13]

Several families with HCM have also been shown to have mutations in the gene that codes for LIM protein.[14] Another set of families has been identified with mutations in the gene coding for AMP-activated kinase that produces glycogen accumulation in the myocardium, mimicking HCM.[15] Interestingly, mutations in β-myosin heavy chain, α-tropomyosin, titin, and troponin T genes also cause dilated and restrictive cardiomyopathies along with left ventricular noncompaction, a relatively new form of cardiomyopathy in humans (although the mutations are at different sites than those that cause HCM) and mutations in the LIM protein gene cause DCM in mice.[16]

The pathophysiology of HCM production by sarcomeric mutations is relatively unknown and controversial. One theory is that the abnormal protein produced by a mutated gene results in dysfunctional sarcomeres that contract poorly and this forces the functional sarcomeres to bear a larger load. The myocardium compensates by replacing the dysfunctional sarcomeres with additional functional as well as dysfunctional sarcomeres.[7,17-19] The net result is that the heart may end up with almost twice as many sarcomeres as it should have. Because sarcomeres make up a large part of the heart muscle, the heart muscle ultimately almost doubles in thickness in severe cases.

HCM severity and penetrance of mutations appears to be determined by the type of mutation, although this is still somewhat controversial. Cardiac myosin binding protein C (MYBPC3) mutations, however, are well known to produce a less aggressive form of HCM with reduced penetrance and age related penetrance in heterozygous individuals. This means that humans with a MYBPC3 mutation may never have echocardiographic evidence of the disease and if they do it's not uncommon for it to first show up when the individual is 50 or 60 years of age.[20]

## FAMILIAL FELINE HCM

The first "family" of cats with an inherited form of HCM was identified in Maine coon cats in 1992 and reported in 1999.[21] The disease is inherited as a simple autosomal dominant trait in this breed and is 100% penetrant in experimental cats that are beyond 5 years of age. However, in Maine coon cats in the real world it appears that the disease is not 100% penetrant, since both parents of an affected cat may have no echocardiographic evidence of the disease. The disease has been reproduced in Maine coon cats by mating affected to unaffected and affected to affected cats as well as by breeding affected Maine coon male cats to domestic shorthair female cats. The course of the disease is accelerated in affected cats produced by mating affected to affected cats where homozygous cats are probably more common. The disease is progressive. In most affected Maine coon cats HCM is not apparent during the first year of life but becomes apparent by 2 years of age in males. Females tend to get the disease later, with many manifesting the disease by 3-4 years of age but some not showing evidence of the disease until 6 or 7 years of age. When both parents have HCM, an affected Maine coon kitten may have echocardiographic evidence of the disease as early as 6 months of age and have severe disease by one year of age, whether male or female. Again these are assumed to be cats that are homozygous for the mutation subsequently identified.

In 2005, the first gene mutation responsible for HCM was identified in Maine Coon cats.[22] This mutation is in exon 3 of MYBPC3. The exact location is codon 31 of the gene where a single point base pair mutation changes alanine to proline in the encoded protein (A31P). This region of the gene is highly conserved across species and the resultant amino acid change results in a change in protein structure and function.

In 2007 a mutation in the same gene (MYBPC3) was identified in Ragdoll cats with HCM (C820T).[23] This mutation is at a completely different location on the gene but it again occurs in a highly conserved region. Ragdoll cats have long been known to have a particularly malignant form of the disease where they often die before reaching one year of age. This now appears to be due to the presence of a larger number of homozygous cats in this highly inbred population.

A family of American shorthair cats, primarily with SAM of the mitral valve, but with other evidence of HCM as well, has also been identified.[24] The disease in this breed also appears to be inherited as an autosomal dominant trait. In addition to these breeds, there is anecdotal evidence of HCM being inherited in numerous other breeds, including Persian, British shorthair, Norwegian forest, Ragdoll, Turkish van, and Scottish fold cats, along with others. HCM is most likely inherited when it is identified in a specific breed. However, HCM is most commonly identified in domestic (mixed-breed) cats. Whether the disease is inherited in these cats, is due to a de novo mutation, or is associated with a different disease process is unknown although suspicion of inheritance has been reported in mixed breed cats.[25-27] Since the disease is inherited as an autosomal dominant trait in some purebred cats, it is not hard to imagine how these mutations could make their way into the mixed breed population and disseminate.

A recent study (abstract) appears to try to cast some doubt on the validity that the A31P mutation in Maine Coon cats causes HCM.[28] This study looked at a group of Maine Coon cats during a screening clinic and tried to correlate the genetic findings with echocardiographic findings. They found some Maine Coon cats with the A31P mutation did not have echocardiographic evidence of HCM and some cats without the A31P mutation that had HCM. The veiled conclusion by the authors was that the A31P mutation might not be causal and that genetic testing for the mutation in this breed may not be warranted. However, the study design was too flawed to make any such conclusion. What was shown instead is that the penetrance of this mutation may be relatively low, at least in young cats, which is not surprising since most MYBPC3mutations in humans have a low and an age-related penetrance. And that there appears to be at least one more cause of HCM in this breed. So the correct interpretation of this study is that it appears (if the genetic analysis was correct) that the A31P mutation is not 100% penetrant (not all cats with the mutation have echocardiographic evidence of HCM) and that there is at least one more cause for HCM in Maine Coon cats that is responsible for HCM. That there is at least one more cause for HCM in Maine coon cats has been apparent in the colony at UCDavis for a number of years.

Even more recently, a study was published that also made preliminary observations on the echocardiographic appearance of hearts from Maine coon cats with the A31P MYBPC3 mutation.[29] In this study, echocardiography was performed on 96 Maine coon cats presented for screening for HCM. Both two-dimensional and tissue Doppler imaging echocardiography were performed. Cats had to have an LV wall thickness greater than 6 mm to make the diagnosis of HCM. Of the 96 cats, 44 of the cats had the A31P mutation (38 were heterozygous and 6 were homozygous). Unfortunately, 45 of the 96 cats were less than 2 years of age and so too young to have evidence of HCM if they were heterozygous. These young cats probably should have been excluded from analysis if they were heterozygous. Of the 38 heterozygous cats, four had clear evidence of at least moderate HCM. However, only 10 of the 34 heterozygous cats that did not have HCM were over 4 years of age. Of the 44 cats that had the mutation, 13 were male and 31were female. This produced an additional bias since female Maine coon cats get HCM at a later age and get less severe disease. Still, this study clearly suggests that the A31P MYBPC3 mutation is not 100% penetrant in Maine coon cats when a cat is heterozygous for the mutation. As expected, homozygous cats appear to be a different story. Of the 6 cats that were homozygous for the A31P MYBPC3 mutation in this same study, 4 had clear echocardiographic evidence of HCM and the other two had abnormal diastolic function as assessed by TDI, an abnormality previously shown to be present in cats that go on to develop HCM.[30] Consequently, it appears that all 6 of these cats had HCM, which once again proves the mutation to be causal. Two cats in this study without the known mutation also had HCM. This once again documents that there is at least one more cause of HCM in Maine coon cats.

Both of the previously referenced studies make only a preliminary attempt to look at penetrance of the A31P mutation in Maine coon cats that are heterozygous for the A31P mutation. What really needs to be done is a longitudinal study looking at the same mutated cats until they are at least 7-8 years of age to tell what the true penetrance is for heterozygous cats rather than looking at a group of cats once at one point in time.

We are currently examining the cellular effects of this mutation and have preliminary evidence to show clear abnormalities produced by this mutation (e.g., myocardium from cats homozygous for the A31P mutation has no myosin binding protein C). Our preliminary findings clearly show that the A31P mutation causes intracellular derangement.

There are currently several labs that test for the A31P MYBPC3 mutation in Maine Coon cats (one in the USA and at least two in Europe). This service has been set up so that breeders can try to rid their breed of this mutation and the HCM caused by this mutation. Many breeders have been reluctant to do test and even more reluctant not to breed mutated cats. To be fair, the method of dealing with this problem is controversial. In the lab at Washington State University, approximately 35% of the DNA samples submitted have had the mutation.[31] This suggests that the mutation is very prevalent in this breed. Concern has been expressed about what would happen if all of these cats were removed from the breeding pool. The concern has to do with decreasing the size of the gene pool and so producing more recessive traits in this breed. In other words, if 35% of the cats could no longer be bred the remaining cats would have to make up the breeding pool and by shrinking that pool the breed may be worse off because of increased inbreeding. The author's counter to that argument is as follows. Breeders commonly only use around 10% of cats in a purebred population for breeding. That means they already exclude 90% of cats from being bred. So if we recommend that they don't breed any cat with a mutation that means we're recommending that they not use 35% of 10% or 3.5% of the entire population. Yes, that does mean that 35% of the cats they would normally breed (the good-looking 10%) would be no longer eligible for breeding. But it also means that they could still use the 65% of the cats they normally don't breed (the less than good-looking 90%) to breed. What would they sacrifice? Maybe their breed wouldn't be quite as visually attractive as before – but it also might be healthier without the mutation. Of course, getting breeders to breed cats they don't think look quite as good as some others is probably an impossible task. So one recommendation has been made that if a cat is heterozygous for the A31P mutation and it is very good looking, it can be bred once. Its kittens then should be tested for the mutation and only those without should be bred. The question then comes, what do you do with the mutant kittens that have been produced in this process? And this recommendation was first made 3 years ago meaning breeders have had plenty of time to do this and should no longer be relying on this method. Consequently, the author's current recommendation is that no cat with the A31P mutation should be bred.

## REFERENCES

[1] Marian AJ. **Curr Opin Cardiol** 2002; 17(3):242. [2] Jarcho JA, et al. **N Engl J Med** 1989; 321:1372. [3] Rosenzweig A, et al. **Mol Biol Med** 1991; 8:159. [4] Tanigawa G, et al. **Cell** 1990; 62:991. [5] Geisterfer-Lowrance AA, et al. **Cell** 1990; 62:999. [6] Thierfelder L, et al. **Cell** 1994; 77:701. [7] Marian AJ, et al. **Circulation** 1995; 92(5):1336. [8] Watkins H, et al. **Nat Genet** 1995; 11:434. [9] Bonne G, et al. **Nat Genet** 1995; 11:438. [10] Poetter K, et al. **Nat Genet** 1996; 13:63. [11] Geisterfer-Lowrance AA, et al. **Science** 1996; 272:731. [12] Vikstrom KL, et al. **Molecular Biology of the Cell** 4, 39a. 1993. [13] Berul CI, et al. **J Interv Card Electrophysiol** 1998; 2(1):7. [14] Geier C, et al. **Circulation** 2003; 107(10):1390. [15] Blair E, et al. **Hum Mol Genet** 2001; V10(N11):1215. [16] Morimoto S. **Cardiovasc Res** 2008; 77(4):659. [17] Lankford EB, et al. **J Clin Invest** 1995; 95:1409. [18] Cuda G, et al. **J Clin Invest** 1993; 91:2861. [19] Sweeney HL, et al. **J Biol Chem** 1994; 269:1603. [20] Michels M, et al. **Neth Heart J** 2007; 15(5):184. [21] Kittleson MD, et al. **Circulation** 1999; 99(24):3172. [22] Meurs KM, et al. **Hum Mol Genet** 2005; 14(23):3587. [23] Meurs KM, et al. **Genomics** 2007; 90(2):261-264. [24] Meurs KM, et al. **J Vet Intern Med** 1997; 11(2):138. [25] Nakagawa K, et al. **J Vet Med Sci** 2002; 64(7):619. [26] Baty CJ, et al. **J Vet Intern Med** 2001; 15(6):595. [27] Fujii Y, et al. **J Vet Med Sci** 2001; 63(5):583. [28] Wess G, et al. **J Vet Intern Med**. 22[3], 717. 2008. [29] Sampedrano CC, et al. **J Vet Intern Med** 2009; 23:91. [30] Macdonald KA, et al. **J Vet Intern Med** 2007; 21(2):232. [31] Fries R, et al. **J Vet Intern Med** 2008; 22(4):893.

## FELINE UNCLASSIFIED CARDIOMYOPATHY – WHAT'S IN A COGNOMEN?

**Mark D. Kittleson, DVM, PhD, DACVIM (Cardiology)**
**Davis, CA**

### UNCLASSIFIED CARDIOMYOPATHY

A relatively common echocardiographic finding in cats with heart disease and heart failure is a normal to mildly abnormal left ventricular chamber size, wall thickness, and wall motion with left atrial enlargement. The left heart is most commonly affected but there may also be concurrent right heart changes. It is commonly believed that since there is no evidence of systolic dysfunction, left-to-right shunting, or obvious valvular regurgitation that these cats must have diastolic dysfunction. Consequently, these cats are often assumed to have restrictive cardiomyopathy (RCM). One study has suggested that this is not always the case.[1] In this study, diastolic function was examined in 9 cats with the aforementioned findings using tissue Doppler imaging (TDI). TDI is a pulsed Doppler technique that measures the velocity of ventricular or annular wall motion and is an accurate means of detecting RCM in humans.[2] Six of the cats in the aforementioned study had clear evidence of diastolic dysfunction (i.e., lower than normal early diastolic LV wall velocity) and so fit the criteria for diastolic dysfunction due to RCM. However, three cats clearly had normal diastolic function on DTI and so did not have RCM. Consequently, not all cats with the echocardiographic findings described above have RCM. Therefore the author feels that the more appropriate term for this disease is unclassified cardiomyopathy unless it can be documented that diastolic dysfunction exists. In this scheme, RCM then is a subclassification of unclassified cardiomyopathy while the other subclass is unknown or idiopathic.

What could cause this two-dimensional echocardiographic pattern of a normal left ventricle and large left atrium other than RCM? Some form of primary atrial myopathy is one possibility and atrial standstill is occasionally identified in these cats (see http://www.vmth.ucdavis.edu/cardio/cases/case18/case18.htm). Misdiagnosis is another possibility. In our clinic we have seen many cats with primary mitral regurgitation, mitral and tricuspid valve dysplasia, and hyperthyroidism placed into this category (called either unclassified or RCM), probably because it was convenient and gave the clinician a false sense of security of making a "definitive" diagnosis. Significant mitral regurgitation is most often missed on color flow Doppler examination when a high frequency ultrasound probe is used and so one should take care to also examine cats in this category with a lower frequency transducer and from several views. Unfortunately, many ultrasound machines do not have a frame rate high enough to adequately detect apparent severe mitral regurgitation in cats.

In some cats, more clear evidence of endocardial scarring can be seen on the echocardiogram, either as a discrete lesion most commonly at the level of the mid-left ventricle or as diffuse scarring of the endocardium. In these cases, the diagnosis of restrictive cardiomyopathy may be made using two-dimensional echocardiography.

The terms intermediate and intergrade cardiomyopathy have been used previously to describe this class of cardiac disease in cats. They are probably inappropriate because they suggest that the characteristic changes seen in these cats share clinical and echocardiographic features of both dilated and HCM, which is not the case.

### LEFT VENTRICULAR NONCOMPACTION

Left ventricular noncompaction (LVNC) is a relatively new type of cardiomyopathy in humans and has not previously been described in cats. It was first described in humans in 1984.[3] However, the vast majority of medical literature on the subject has been written since 2000. The World Health Organization classifies it as an unclassified cardiomyopathy (http://www.orpha.net/data/patho/GB/uk-LVNC.pdf). But in 2006, the American Heart Association decided to classify it as its own form of cardiomyopathy.[4]

LVNC is characterized by the presence of deep trabeculations in ventricular myocardium. The myocardium has been described as "spongy" and the trabeculations as sinusoids.[5] The vast majority of the disease has been recognized in the left ventricle but it also can occur in the right ventricle. It most commonly occurs in the apex of the left ventricle but it can also occur in the mid wall region of the left ventricular free wall.

One theory of how LVNC develops is that this is a situation of arrested development.[5] During embryonic life, prior to the development of the coronary circulation, the ventricular myocardium is initially heavily trabeculated so that oxygen can diffuse the shortest distance to developing myocytes. Once the coronary circulation does develop the trabeculations disappear and the myocardium is said to become "compact". Consequently, it may be that the myocardium fails to go on to become compact for whatever reason resulting in LVNC.

Until recently the cause of LVNC was unknown. Within the last several years, however, several genetic mutations have been identified in humans with the disease.[6] Most of the genes in which mutations have been identified are genes where other mutations have been identified that also cause other forms of cardiomyopathy or skeletal muscle myopathy (muscular dystrophy). These include taffazin, LDB3, alpha dystrobrevin, beta myosin heavy chain, alpha actin and troponin T. Mutations in taffazin also cause dilated cardiomyopathy, endocardial fibroelastosis, hypertrophic cardiomyopathy and Barth syndrome (cardiomyopathy-neutropenia syndrome). Mutations in LDB3 also cause dilated cardiomyopathy. Mutations in dystrobrevin cause a form of muscular dystrophy. And the last three on the list are all sarcomeric protein genes associated with hypertrophic cardiomyopathy.[7] The finding that mutations in these genes also cause LVNC muddies the water quite a bit since it now appears that genes associated with one form of cardiomyopathy can also be associated with another. However, that news isn't totally new. Within the past 5 years, sarcomeric gene mutations have been associated with dilated cardiomyopathy and restrictive cardiomyopathy as well as with hypertrophic cardiomyopathy.[8] So things just keep getting more interesting in the world of cardiomyopathy and genetics.

The diagnosis of LVNC is usually made using echocardiography. On echo there most commonly are deep recesses in the left ventricle in the apex of the ventricle. These recesses communicate with the left ventricular chamber and so color flow Doppler images often show blood flow within them.

LVNC has not been previously described in cats. However, the author has recently identified one cat in his research colony of Maine Coon cats with hypertrophic cardiomyopathy due to a mutation in myosin binding protein C that he believes has the characteristic lesions of this form of cardiomyopathy and several others that may be affected, all with the A31P MYBPC mutation. Echocardiographic images will be presented for discussion.

## ARRHYTHMOGENIC RIGHT VENTRICULAR CARDIOMYOPATHY

ARVC is a rare disease in cats that was first reported in 2000.[9] It apparently can present in any cat over 1 year of age (age range = 1-20 years). In young cats it can be mistakenly diagnosed as tricuspid valve dysplasia since most cats with the severe form of the disease have tricuspid regurgitation. Cats with the disease most commonly present in right heart failure (i.e., ascites or hepatosplenomegaly due to congestion) and have a systolic right parasternal heart murmur. An arrhythmia is usually present. A variety of arrhythmias have been recorded including ventricular tachycardia, atrial fibrillation, supraventricular tachycardia, and premature ventricular complexes (of RV and LV origin). Right bundle-branch block is common and first-degree AV block has been identified in some cats. Although a ventricular tachyarrhythmia is common in a cat with ARVC it is not always present on a resting ECG. On two-dimensional echocardiography the right ventricular and atrial chambers are enlarged, most often severely enlarged. The tricuspid valve is anatomically normal although tricuspid regurgitation is almost always identified if color flow Doppler is utilized and is due to tricuspid valve annular dilation. Although the tricuspid regurgitation is most commonly mild to moderate it can be severe. Areas of thinning of the myocardium, especially at the apex where an aneurysm may form, may be identified with careful examination in some affected cats. The left ventricle is usually normal although the left atrial chamber may be enlarged. Intracardiac thrombi are rare but have been identified in the right ventricle and left atrium. On post mortem examination the right ventricular cavity is enlarged and the right ventricular free wall is thinner than normal. The thinning may be diffuse or segmental. In some cases it is so thin in the aneurysmal regions right ventricular wall that the wall is translucent. As in Boxer dogs and humans, the hallmark feature of the disease is the presence of replacement fibrosis and fatty infiltration of the right ventricular free wall. Most cats also have evidence of inflammation, apoptosis, and dying and atrophied myocytes. Fibrous replacement is also commonly present in the left ventricular free wall and interventricular septum although it is not nearly as extensive and the left ventricular free wall is not thinner than normal. Cats in right heart failure have a poor prognosis with an average survival time of 1 month (range = 2 days to 4 months in one study). Thromboembolism may also occur.

## MYOCARDIAL INFARCTION

Myocardial infarction is a rare cause of left ventricular dysfunction and heart failure in cats. Echocardiographically it is evidenced by a region of the left ventricular wall that is hypokinetic (moves poorly in systole), akinetic (doesn't move at all in systole) or dyskinetic (moves outward in systole). The region of wall is also often thinner than normal. The cause of myocardial infarction in cats is unknown. However, it is unlikely to be due to atherosclerosis (as in humans) and is most likely due to thromboembolic disease. It is not unusual to see myocardial infarction in association with some form of cardiomyopathy, most commonly HCM. Prognosis depends on the extent of the infarction.

## REFERENCES

[1] Gavaghan BJ, et al. **Am J Vet Res** 1999; 60(12):1478. [2] Kushwaha SS, et al. **N Engl J Med** 1997; 336(4):267. [3] Engberding R, et al. **Am J Cardiol** 1984; 53(11):1733. [4] Maron BJ, et al. **Circulation** 2006; 113(14):1807. [5] Chrissoheris MP, et al. **Clin Cardiol** 2007; 30(4):156. [6] Moric-Janiszewska E, et al. **Clin Cardiol** 2008; 31(5):201. [7] Klaassen S, et al. **Circulation** 2008; 117(22):2893. [8] Morimoto S. **Cardiovasc Res** 2008; 77(4):659. [9] Fox PR, et al. **Circulation** 2000; 102(15):1863.

**LEPTOSPIROSIS: FUNDAMENTAL PRINCIPLES OF DISEASE**

**Alan T. Loynachan, DVM, PhD, DACVP and Nathan M. Slovis, DVM, DACVIM, CHT**
**Lexington, KY**

Pathogenic *Leptospira* are ubiquitous bacteria not restricted by host or geographic boundaries. This zoonotic microorganism remains to be an important cause of equine septicemia, renal disease, uveitis, and abortion. Over the past 10 years, great strides have been made to further elucidate the virulence mechanisms and pathophysiology associated with leptospirosis. This manuscript will provide a brief update and review of the microbiology, epidemiology, pathophysiology, clinical manifestations, and diagnosis of *Leptospira interrogans* infection in horses.

## MICROBIOLOGY

The genus *Leptospira* is composed of gram-negative spiral-shaped bacteria of the family *Leptospiraceae*. These aerobic 6-20 X 0.1-0.2 μm microorganisms can be morphologically differentiated from other spirochetes based on the presence of 1 or 2 terminally hooked ends and the frequency and amplitude of its spirals. Special media are typically required to grow and isolate this fastidious organism. Traditional nomenclature is based on serologic classification. This taxonomy scheme places the pathogenic *Leptospira* into the species *interrogans* and the non-pathogenic species into *biflexa*. *Leptospira interrogans* has been subdivided into 30 serogroups composed of over 250 serovars based on variance in the organism's antigenic lipopolysaccharide. Molecular biology has allowed for reclassification of leptospires based on DNA homology. Over 20 species have now been identified based on genotypic classification. Nine pathogenic species of *Leptospira* are currently recognized. Unfortunately, the classical serologic classification scheme does not easily translate into the updated genotypic scheme.

## EPIDEMIOLOGY

Leptospirosis is a zoonotic disease of worldwide distribution; however, serovar prevalence varies by country and region. The organism is maintained in the kidneys of chronically infected wildlife, domestic animals, amphibians, and reptiles. Following urinary excretion, the organism can survive for hours to months in fetal membranes, soil, and water. Occurrence of leptospirosis is higher in temperate to subtropic climates based on the organism's ability to survive in the environment. Warm stagnant water remains to be an important environmental factor for survival of leptospires and explains the increased incidence of disease in times of increased rainfall.

Exposure to pathogenic leptospires is thought to be common based on serologic surveys, but development of disease occurs with less frequency. Infection most commonly occurs through invasion of mucus membranes and abraded skin. Additional documented routes of exposure include ingestion, inhalation, transplacental infection, venereal infection, and through animal bites. Incidence of disease is influenced by route of exposure, infectious dose, host immunity, and the associated serovar. Serovars vary in pathogenicity based on the presence of different virulence factors and the host which they infect.

Infected animals can serve as either maintenance or incidental hosts based on their susceptibility to certain serovars. A maintenance host is defined as a species in which infection is endemic. These hosts are chronically infected, harbor the organism in their kidneys, and intermittently shed the pathogen in urine. Maintenance hosts generally do not exhibit clinical signs and may or may not develop an antibody titer. Incidental hosts are highly susceptible to infection, frequently develop systemic disease, and generally mount a significant antibody titer. Generation of an antibody response will frequently clear the infection from the incidental host. An animal can serve as a maintenance host for one serovar but an incidental host for other serovars.

Horses are susceptible to infection and are considered incidental hosts to a large number of *Leptospira* serovars. Serologic titers to serovars Bratislava, Grippotyphosa, Icterohaemorrhagiae, Pomona, Kennewicki, Canicola, Ballum, Copenhageni, Autumnalis, Kirschneri, and Hardjo have been reported. Additionally, horses are speculated to be a maintenance host for *Leptospira interrogans* serovar Bratislava based on the high seroprevalence of the organism in certain countries. Abortion and recurrent uveitis remain the most common manifestations of equine leptospirosis.

Equine leptospiral infection has been identified as an important and emergent cause of placentitis and abortion. Reports have indicated that *Leptospira* sp. account for approximately 1.5-5.9% of equine abortions. Infected mares do not generally exhibit premonitory signs, but frequently develop high antibody titers. Infected mares may abort the fetus, deliver a stillborn fetus, or may give birth to a weak foal. Abortions most often occur from 6 months of gestation to full term. Immunocompetent fetuses are able to mount an *in utero* humoral response to the pathogen, but this response may not be sufficient to clear the infection. Serovar Kennewicki is most commonly associated with abortion in central Kentucky; however, serovars Grippotyphosa, Pomona, Kennewicki, Hardjo, and Australis have also been implicated.

Recurrent uveitis is major cause of ocular disease and blindness in horses and humans worldwide. Also known as periodic opthalmia and moon blindness this chronic sequela of leptospirosis develops months to years after initial infection, and is named after the characteristic cycles of acute uveal inflammation followed by quiescent periods. Dwyer et al. indicated that seropositive horses were over 13 times more likely to develop uveitis as compared to seronegative horses, and that 56% of horses with uveitis were seropositive for *L. interrogans* serovar Pomona. It is important to note that a majority of seropositive horses do not develop recurrent uveitis. Recurrent uveitis is likely induced through a combination of immune mediated factors and direct interaction with leptospiral organisms. Persistent intraocular infection remains debatable as various reports contradict one another on the direct presence of leptospires or leptospiral DNA in affected eyes. Serovars most commonly associated with recurrent uveitis include Pomona, Muenchen, Bratislava, Lora, and Ricardia.

## PATHOPHYSIOLOGY

*Leptospira* serovars vary in the number of virulence factors they posses and the degree of disease they inflict. The infectious dose 50 has been estimated, in an intradermal hamster model, to be 6 to 35 organisms. Until recently, leptospiral virulence mechanisms have remained elusive due to the unique properties of this organism.

Pathogenic leptospires have recently been determined to express a number of adhesion molecules on their outer membrane that aid in infection of host cells. Leptospiral surface proteins designated lipoprotein 32(LipL32), surface adhesion 21 (Lsa21), and Lig binding proteins (LigA, LigB, and LigC) aid in penetration of host cells and bind to the host's extracellular matrix proteins (collagen, laminin, fibronectin, and fibrinogen). Acute leptospiremia (septic phase) ensues once the organism penetrates the host and enters the vascular system. Naive hosts primarily rely on the classical (CCP) and alternative (ACP) complement pathways to reduce bacteria in the blood. To circumvent this process pathogenic leptospires produce factor H-binding protein (LfhA) which binds to factor H, an ACP regulator protein, and ultimately disables the bactericidal affects of the ACP. The CCP is also inhibited by pathogenic leptospires which bind C4BP through an unknown mediator. The organism is then disseminated through the blood where it adheres to and infects susceptible cells utilizing the previously described binding proteins. Once in these target tissues, pathogenic *Leptospira* produce virulence factors such as sphingomyelinases, proteases, collagenase, and lipopolysaccharide that aid in tissue destruction and evasion of phagocytosis. Up regulation of the humoral immune response (convalescent/immune phase) clears leptospires from the blood. The organism can persist and induce chronic disease in immunoprivileged locations such as the renal proximal convoluted tubule, uterus, and eye.

Systemic vasculitis is responsible for the majority of pathology associated with acute leptospiremia. Endothelial cytotoxicty and hemolysis have been attributed to a sphingomyelinase-like protein (Lk73.5). Acute renal disease manifests secondarily from vasculitis, direct leptospiral disruption of the renal tubules, and collateral damage produced by the host's immune response. Acute microscopic disease is characterized as tubulonephritis, interstitial edema and hemorrhage, tubular epithelial necrosis, and disruption of the basement membrane. This renal pathology clinically presents as azotemia and hypokalemia. Less frequently leptospires may induce hepatic disease characterized as hepatocellular apoptosis and necrosis, hemorrhage, edema, and bile cannulicular disruption. Cannulicular disruption results in reabsorption of bile and clinically manifests as hyperbilirubinemia. Acute disease in humans can also manifest as aseptic meningitis, pulmonary hemorrhage, rhabdomyolysis, and uveitis.

Persistence of leptospires in immunoprivileged locations such as kidney, uterus, and eye predispose the host to chronic disease. Colonization of the proximal convoluted tubules can result in chronic fibrosing interstitial nephritis and induction of a subclinical carrier state. Colonization of the uterus may incite fetoplacental disease and abortion. Gross fetal lesions may include icterus, hemorrhagic foci, an enlarged mottled liver, and allantochorionic edema and necrosis. Microscopic lesions are frequently observed and consist of hepatocellular dissociation and giant cell production, portal triad inflammation, tubulointerstitial nephritis with fibrosis, and placental thrombosis, inflammation, and vasculitis. Ocular invasion and colonization occasionally results in recurrent uveitis. *Leptospria* express membrane proteins Lru A and Lru B upon infection of the eye. These proteins serologically cross-react with normal equine ocular structures and are likely involved in the immune mediated pathogenesis of recurrent uveitis. Other reports indicate the direct involvement of leptospires, leptospiral protein, and/or leptospiral DNA in affected eyes. Microscopically, recurrent uveitis is characterized as cyclic infiltration of CD4 T lymphocytes, plasma cells, and macrophages into the ciliary body and iris. As disease progresses cataracts, synechiae, chorioretinitis, and iridial pigment alteration may develop.

## CLINICAL MANIFESTATIONS

Equine leptospirosis commonly manifests as acute disease, chronic disease, or abortion. The incubation period can range from a few days to over 3 weeks. Vague signs of pyrexia, dullness, anorexia, and weakness typically ensue upon acute infection. Leukocytosis and neutrophilia will also manifest. As the organism replicates and colonizes the kidneys acute renal failure can develop and result in azotemia, dysuria, and hematuria. Typical urinalysis will indicate a low specific gravity (<1.020) without bacteriuria. Hepatic disease, although not common, can also arise from infection and replication in the liver, and if present, hyperbilirubinemia and mildly increased hepatic enzymes may develop. Uveitis or inflammation of other tissues may also develop during acute infection due to vasculitis.

Recurrent uveitis is an important intermittent manifestation of chronic leptospirosis and is a major cause of equine blindness. One or both eyes may be affected. Clinical signs associated with recurrent uveitis include: miosis, blephrospasm, photophobia, aqueous flare, keratitis, iritis, and ciliary injection.

## DIAGNOSTICS

Antemortem diagnosis of leptospirosis remains problematic due to the special growth requirements of this organism and difficulty in interpretation of serologic assays. Leptospires are best identified in blood and urine during the acute leptospiremic phase. Darkfield microscopy (DF), fluorescent antibody tests (FAT), culture (C), and polymerase chain reaction (PCR) assays are capable of identifying leptospiral material in blood and urine. Darkfield microscopy is used to visually identify leptospires in the sampled material, but this test has low sensitivity and low specificity. The FAT has increased sensitivity and specificity, as compared to DF, and is the most widely used test in veterinary diagnostic laboratories. Culture is considered the reference method for identification of leptospires but requires special media and is time consuming as the organism can take several weeks to grow. With the development of molecular techniques more diagnostic labs have begun using PCR for the detection of leptospiral DNA. *Leptospira* sp. PCRs are quick, cost efficient, highly sensitive and specific, and do not require the presence of live leptospires. Molecular leptospiral techniques have been developed to differentiate pathogenic from non-pathogenic *Leptospira*, but cannot differentiate samples based on serovar.

A number of serologic tests have been developed for identification of leptospiral exposure based on the presence of antibody. Interpretation of serologic data remains problematic as these tests detect previous exposure to the organism, and high antibody titers can remain for months to years after exposure. The microscopic agglutination test (MAT) is considered the gold standard, but is difficult to perform as it requires live *Leptospira* for serovar differentiation. A single MAT titer of 1:800 is generally considered positive, but a 4 fold titer increase in paired sera is considered more definitive. Complement fixation (CF) and macroscopic agglutination (MT) tests are more easily performed and are currently used by some diagnostic labs. A commercially available enzyme-linked immunosorbent assay (ELISA) test is not available for use in horses, however, research based ELISA to proteins of interest are frequently used in academia. The clinical diagnosis of recurrent uveitis is frequently complemented by use of DF, C, PCR, and MAT on ocular aqueous and vitreous fluids.

A summary of the most commonly used tests for diagnosis of acute disease, chronic disease, and abortion is list in table 1. In addition to the tests mention above, immunohistochemistry (IHC) and silver stains (S) have been used to identify leptospires in tissue sections. The IHC is specific but lacks sensitivity, and the silver stains lack both sensitivity and specificity. Usually, a combination of multiple tests is used to aid in the diagnosis of leptospiral abortion.

**Table 1.** Tissues to collect and potential diagnostic tests based on disease state.

| Disease | Tissues to collect | Diagnostic tests |
|---|---|---|
| Acute | Blood | DF, FAT, C, PCR |
| | Urine | DF, FAT, C, PCR |
| | CSF* | DF, FAT, C, PCR |
| Chronic | Urine | DF, FAT, C, PCR |
| | Serum | MT, CF, MAT |
| | Ocular fluid | DF, C, PCR, MAT |
| Abortion | Fetal fluids | MAT |
| | Fetal tissues | DF, S, FAT, IHC, C, PCR |
| | Placenta | DF, FAT, IHC, C, PCR |

*If CNS signs are evident. DF, darkfield microscopy; FAT, florescent antibody test; C, culture; PCR, polymerase chain reaction; MT, macroscopic agglutination test; ELISA, enzyme-linked immunosorbent assay; CF, complement fixation; MAT, microscopic agglutination test; S, silver stain, IHC, immunohistochemical stain.

## SUMMARY

Leptospirosis remains to be an important disease associated with equine abortion, uveitis, nephropathy, and systemic infection. Although this pathogen is distributed worldwide the importance of individual serovars varies by country and region. Numerous recent studies have identified new virulence mechanisms induced by this pathogenic spirochete to mask itself from the host and aid in the development of disease. New diagnostic tests are continuously being developed and modified to aid in the antemortem and postmortem diagnosis of leptospirosis. Intervention strategies encompass decreased interaction amongst horses and maintenance hosts and prevention of environmental contamination. The conference lecture will additionally provide updated serovar data in regards to leptospiral abortion and disease in central Kentucky.

## REFERENCES

[1.] Artiushin S, et.al. **Infect Immun** 2004; 72(2):742. [2.] Atzingen MV, et.al. **BMC Micro** 2008; 8:70. [3.] Barbosa AS, et.al. Infect Immun 2008; ahead of print. Brandes K, et.al. **J Vet Med A Physiol Pathol Clin Med** 2007; 54(4):270. [4.] Divers TJ, et.al. **JAVMA** 1992; 201(9):1391. [5.] Djadid ND, et.al. **Diagn Microbiol Infect Dis**. 2009; ahead of print. [6.] Donahue JM, et.al. **J Vet Diagn Invest** 1995; 7:87. [7.] Donahue JM, et.al. **Emerg Infect Dis** 2000; 16(3)443. [8.] Dwyer AE, et.al. **JAVMA** 1995; 207(10):1327. [9.] Gilger BC, et.al. **Am J Vet Res** 2008; 69(10):1329. [10.] Hartskeerl RA, et.al. **J Vet Med B** 2004; 51:110. [11.] Hauk P, et.al. **Infect Immun** 2008; 76(6):2642. [12.] Hines MT. In *Eq Infect Dis* 2006. Ch 34. Hong CB, et.al. **J Vet Diagn Invest** 1993; 5:56. [13.] Levett PN. Clin Microbiol Rev 2001; 14(2):296. [14.] Lin YP, et.al. **J Vet Sci** 2008; 9(2):133. [15.] Mori M, et.al. **Jpn J Med Sci Biol** 1974; 27(6):297. [16.] Pearce JW, et.al. **J Vet Diagn Invest** 2007; 19(6):686. [17.] Plank R, et.al. Microbes Infect 2000; 2:1265. [18.] Verma A, et.al. **Infect Immun** 2005; 73(11):7259. [19.] Verma A, et.al. **Infect Immun** 2006; 74(5):2659. [20.] Yang CW. **Kid Int** 2007; 72:918.

# UPDATE ON EPM DIAGNOSIS

**Martin Furr, DVM, DACVIM, PHD**
**Leesburg, VA**

## INTRODUCTION

The clinical diagnosis of EPM remains a challenging task for veterinarians. Variable and sometimes subtle presentation of clinical signs and the inherent complexities of the diagnostic tests make a clear cut and confirmed diagnosis of EPM in the clinical patient difficult. Given these constraints, careful interpretation of clinical and ancillary diagnostic laboratory testing are key to making a diagnosis. Diagnosis is dependant upon:

1) confirming the presence of clinical signs which are consistent with EPM
2) ruling out other potential causes of the observed clinical signs by appropriate means (spinal tap, radiography, serology, etc)
3) confirming the presence of *S. neurona* specific antibodies by one of a variety of means (i.e. immunodiagnostic testing)

### Clinical examination

A thorough physical and neurological examination remain the keystone of the EPM diagnosis. A full clinical examination to document the presence or absence of other clinical abnormalities, and collection of historical information, are always warranted. A lameness examination is very important, and should be carefully done to eliminate disease of the musculoskeletal system, as it is much more commonly affected than the CNS. For the diagnosis of EPM to be supported conclusive evidence of CNS disease must be present. The classic clinical signs of EPM have been well described and include gait abnormalities, and/or signs of cranial nerve, brainstem or cerebral disease. Asymmetry of clinical signs is a hallmark of the disease, although symmetrical deficits can be seen. In addition, muscle atrophy is common. The finding of ataxia completes the diagnostic triad of the "three A's of EPM"- ataxia, asymmetry and atrophy. While these may not exist in every case, they are compelling findings in the EPM suspect when present.

### Eliminate other potential causes

A differential diagnosis list should be constructed to ensure that additional diagnoses are considered, and appropriate tests conducted to support, or eliminate, their consideration. Numerous diseases of the CNS can present similarly to EPM, including bacterial or verminous meningitis, neoplasia, spontaneous epidural hemorrhage, EHV- 1 myelitis, etc. Survey radiographs, serology, and cerebrospinal fluid (CSF) collection are all important components of a full evaluation. If all such testing is negative, the probability of EPM increases, particularly if immunodiagnostic testing supports the diagnosis.

Cervical radiographs are indicated for the evaluation of any horses which have neuroanatomic localization of the clinical signs to the cervical region. While EPM does not cause radiographic changes, it is important to rule out other causes of spinal ataxia, such as cervical compression, fracture or cervical facet arthritis, which may be present.

Collection and analysis of CSF is a common component of the neurologic evaluation and should not be ignored in evaluating the EPM suspect. The observation that CSF false positive tests can occur has led some to question the value of analyzing CSF in horses being evaluated for EPM. It is important to remember that the CSF is not collected for the sole purpose of doing an EPM test; additional information is obtained from the sample. In the authors experience, several neoplasia and meningitis cases were detected when fluid was collected "for an EPM test" in a neurologic horse. If the sample is not collected and evaluated, the correct diagnosis may be missed. Therefore, while it is true that some false positive tests occur using CSF, this should not preclude the use of this diagnostic test. Horses exposed to *S neurona* will not necessarily and invariably develop a false positive test on CSF. This appears to be associated with a high serum titer, and is often transient. In addition, a negative CSF test has great value in eliminating EPM in most cases. A weak or suspect positive reaction in CSF is not compelling evidence for a diagnosis of EPM, while a stronger test reaction on immunoblot, or a titer exceeding the reference laboratories high-cut-off is strong support for the diagnosis. Cerebrospinal fluid collection and evaluation is also recommended for full evaluation of the horse with CNS disease. The CSF in horses with EPM is usually normal- although in fulminant cases abnormalities may be seen- this appears to be exceedingly rare, however. However, one should remember that much information about the CNS may be gained from evaluation of the CSF, and testing for EPM is not the only information obtained when a CSF sample is analyzed.

### Immunodiagnostic testing

The most important aspect of the diagnosis of EPM is the detection of *S neurona*-specific antibody.[1,2] Although there has been discussion about the relative merits of the various immunodiagnostic tests available, it is important to remember that all currently available tests are assaying the same thing, ie., the presence of *S neurona*-specific antibody in the sample fluid. Hence, even though the testing platforms may differ, they are going to provide very similar information. The western blot (WB) was the first specific immunodiagnostic test for EPM developed. Immunoblot testing of CSF samples demonstrated greater than 90% specificity and sensitivity among approximately 100 horses with neurologic disease that received postmortem examination.[3] Only approximately one half of the cases were histologically confirmed as EPM, however. The non-quantitative nature of the WB test makes some clinicians uncomfortable, and some clinicians prefer tests with definable end-points.

Presence of antibodies in the serum only indicates exposure of the horse to the parasite, but not necessarily active disease. The presence of *S. neurona* specific antibodies in CSF is more suggestive of infection, but is not definitive. It was originally presumed that the presence of *S. neurona* antibodies in CSF could only be present if the organism was present within the CNS. Several years of experience with this test have determined that this is not completely accurate. While *S. neurona* specific antibodies are present in the CSF in active disease, false positive results have been widely reported.[4,5] Blood contamination during collection of the sample is commonly incriminated as causing false positives. More importantly, however, antibodies to non-replicating protein has been found in the CSF of vaccinated horses, directly correlating to plasma concentrations.[6] This shows that in the normal, healthy horse, serum-derived antibody can pass the normal blood-brain barrier, giving rise to false positive tests. This observation suggests that examination of paired serum and CSF WB results may be beneficial. If the serum reaction is strong, the corresponding CSF response should be robust to support a diagnosis of EPM.

The $IgG_{index}$ has been advocated as a method to more carefully evaluate the CSF tests results. The $IgG_{index}$ was found in one study to be higher than normal in horses with EPM[7], however subsequent research has not supported this initial observation.[8] It appears that the $IgG_{index}$ provides only limited diagnostic information, and its routine use as a diagnostic for EPM is not recommended.[9] Additional antigen specific measures of determining intrathecal antibody production have been proposed, and are termed the C-value (Goldman-Witmer coefficient) or the antibody index.[6,10] At present these tests await further validation, but are readily performed using any test methodology that provides an antigen specific titer; in addition, CSF and serum immunoglobulin and albumen concentrations are required. Using these techniques and the snSAG 2 ELISA assay, a sensitivity and specificity of 83% and 90%, respectively was found in several well documented (ie. histologically confirmed) cases.

The standard or conventional WB (cWB) test has been modified at one testing laboratory (Michigan State) by blocking the 30-kD band, a suspected cross reacting protein, with pooled serum with a high anti-*S. cruzi* antibody titer.[11] This test was reported to have a sensitivity and specificity approaching 100%.[11] However, this study was performed on only a small number of EPM positive horses (6). Subsequent evaluation of the mWB by other investigators has found a much lower sensitivity/specificity of 89% and 69%, respectively.[12]

A whole organism indirect fluorescent antibody test (IFAT), has been developed and is currently available from the University of California Diagnostic Laboratory.[13] Diagnostic performance of the IFAT was tested and compared to the cWB and mWB, and was reported to have a (slightly) better diagnostic efficiency than either test.[12] Serum titers of greater than 1:100 and CSF titers of greater than 1:5 were considered positive, and diagnostic of active infection. Some investigators have demonstrated, however, that the test was unable to differentiate between *S. neurona* and *S. fayeri* (a non-pathogenic *Sarcocystis*) infections.[14] Therefore, the superiority of the IFAT as a diagnostic for EPM is not clearly established. An additional benefit of the IFAT, however, is that the laboratory commercially performing this test also runs an IFAT for Neospora hughesi as part of the EPM testing protocol.

Another diagnostic test described for EPM is an enzyme linked immunoassay (ELISA) for antibodies to the snSag-1 protein. This test is currently commercially available from Antech, Inc. This test provides a titer, and values greater than 1:100 in serum are reported to indicate active infection.[15] Use of this test is not without concern, however. The test has not been rigorously evaluated for sensitivity and specificity, and the cut-off values for a positive diagnosis appear to be arbitrary. Further, it has been clearly demonstrated that not all *S. neurona* isolates produce the SAG-1 protein, hence false negative results are possible with this test.

A stall-side *S. neurona* screening test is currently available for the confirmation of antibodies to *S. neurona* (S neurona, Endocrine Technologies) in serum. This test uses the ELISA platform and snSAG1 as the antigen, as well as a second testing "spot" for mixed surface proteins. Presence of antibodies to both spots is considered diagnostic for EPM caused by *S. neurona*. The sensitivity and specificity of this test kit has not been determined. While it may have value as a screening test, its use as a definitive testing method is undetermined at the present.

In conclusion, it seems that the ideal diagnostic test does not yet exist for the clinical diagnosis of EPM. Diagnosis in a clinical patient is best supported by the finding of definitive CNS abnormalities on a neurologic exam, the elimination of other neurologic diseases by radiography and CSF evaluation, and the finding of *S. neurona* specific antibodies in appropriately collected CSF by WB or IFAT.

## REFERENCES

[1]Granstrom D. 1993. **Proc 21[tst] Annual Meet ACVIM**, 587. [2]Miller MM, et al. **J Am Vet Med Assoc** 1999,215:67-71. [3]Granstrom DE. **Proc 21[tst] Annual Meet ACVIM** 1993,587. [4]Bentz BG, et al. **Compend Contin Educ Pract Vet** 1999,21:975. [5]Cook AG, et al. **J Am Vet Med Assoc** 2002;220:208. [6]Furr M. **J Vet Intern Med** 2002;16:588. [7]Andrews FM, et al. **Proc Annu Conv Am Assoc Equine Pract** 1997;43:10-2. [8]Furr M, et al. **Vet Ther** 2001;2:215. [9]Furr M, et al. **J Vet Intern Med** 2002;16:618. [10]Furr M. **J Vet Intern Med** 2007;21:806. [11]Rossano MG, et al. **J Vet Diagn Invest** 2000;12:28. [12]Duarte PC, et al. **J Vet Diagn Invest** 2003;15:8. [13]Duarte PC, et al. **J Parasitol** 2004;90:379. [14]Saville WJ, et al. **J Parasitol** 2004;90:1487. [15]Ellison SP, et al. **J Appl Res Vet Med** 2003,1:318.

**EQUINE LYME DISEASE: SIGNIFICANT OR COINCIDENCE**

**Thomas J. Divers, DVM, DACVIM, DACVECC**
**Ithaca, NY**

Lyme disease is a common tick-borne infectious disease in parts of North America, Europe and Asia . It is caused by a helical shaped gram negative spirochete *Borrelia burgdorferi*. In North America only, *B. burgdorferi* sensu stricto causes Lyme disease.[1] Deer and the white-footed mouse (Peromyscus leucopus) are the most common mammals involved in maintaining the organisms 2 year life cycle in North America. *Ixodes scapularis/I. pacificus*, *Ixodes ricinus* and *Ixodes persulacatus* are the most important vectors in North America, Europe and Asia, respectively.[2] Once tick feeding begins, the *Borrelia* organism begins both up and down regulation of certain genes to allow transmission and survival. OspA may be down-regulated, while other surface proteins(OspE, OspF and Vis-E/C6), that are in low concentration in the tick gut are up-regulated during transmission to enhance immune invasion in mammals.[3,4] The organism may also reside in the host by residing in connective tissue and collagen and having no requirement for iron.[5] A large percentage of adult horses in the more eastern parts of the Northeast and Mid-Atlantic United States are or have been infected with *B. burgdorferi*.[6] Infection may also common in Wisconsin and Minnesota in the USA. Seroprevalence in horses in many other parts of the United States has not been reported, but would be expected to fluctuate in a manner similar to that seen with the human Lyme disease. There have been at least 3 retrospective clinical studies (with control groups) that attempt to correlate seropositivity for *Borrelia* with clinical signs. One study in Germany looked specifically at equine recurrent uveitis (79 horses with ERU and 153 controls) and seropositivity to *Borrelia* and found no correlation between the two.[7] Although, *Borrelia* was recently found in the chamber fluid of a 3 year old horse with recurrent uveitis at Cornell. Two studies looked at correlation between seropositivity to *Borrelia* and signs of musculoskeletal disease, often attributed to *Borrelia* infection. Those signs included: shifting lameness, stiffness, behavioral changes and inability to perform work. In 1 of the studies in the U.S. there was a significant correlation between the two,[8] while a retrospective owner/survey study in Europe did not find an association between *Borrelia* seropositivity and signs commonly attributed to Lyme disease.[9] Neurologic dysfunction and disease attributed to *Borrelia* infection have been reported in a horse and a pony.[10,11] Recently, I have helped in the evaluation of 2 horses with ataxia, neck stiffness and behavioral changes that were attributed to Lyme infection. In one case, marked lymphocytic pleocytosis was found in the CSF and, upon necropsy, a marked rostral cord lymphocytic (mixed size and type) meningitis with thickened dura and nerve root involvement was found. In the other horse, there was a high CSF *Borrelia* antibody (ELISA of 393 units; 412 in the serum). Using the same samples, comparison of serum to CSF antibody for EHV1 (1:32 vs 1:4), EPM (positive serum, negative CSF), IgG (3,200 mg/dl serum vs. <200 CSF) strongly suggested intrathecal production of *Borrelia* antibodies. The horse responded favorably to tetracycline and doxycycline treatments. The neurologic signs, absence of other causes for the CNS signs, strong evidence of intrathecal antibody production, and response to therapy would meet the criteria for diagnosing neuroborreliosis in humans.[12] Our experimental pony studies support a possible preferred migration of the organism through connective, perineural, and perivascular tissue in the skin, fascia, muscle and synovial membranes, with the synovial membrane most commonly affected.[13,14] The organism was found in the meninges in one experimentally infected pony. There was mild pathology in these areas which could cause hyperesthesia, lameness and neurologic signs; the first two being the most commonly reported (by owners and veterinarians) clinical signs. None of the experimentally infected ponies had obvious clinical signs; it may be that clinical signs rarely exist in infected equines or that the clinical signs are mild and/or can only be recognized in certain horses due to unique immune reactions or clinical use of the horse. Unfortunately, response to antibiotic treatment has not helped with the question proposed in this title, since we have found that the most commonly used antibiotics for treating Lyme disease in the equine have anti-inflammatory properties which may cause symptomatic improvement. In experimental ponies the synovium was the most common site of culture for *Borrelia* and tetracycline and doxycycline has its primary matrix metalloproteinase inhibition/anti-inflammatory effect on the synovium as opposed to articular cartilage cells.[15] It has been our limited experience that clinical signs believed to be associated with Lyme disease are most commonly reported in performance horses that are used for eventing purposes. Although not the only possible explanation, one explanation for this observation would be that hyperesthesia and mild lameness/stiffness are more likely to be noticed in this type of horse. We have almost never been referred a race horse with a presumptive diagnosis of Lyme disease. We have observed muscle wasting and markedly swollen joints in one horse with *Borrelia* infection and a dramatic response to treatment, but these signs have not been common in horses diagnosed with Lyme disease. Fever and edema, both responsive to tetracycline therapy, may be found in horses just prior to Lyme seroconversion, but these signs are most likely a result of *Anaplasma phagocytophilia* infection. *A. phagocytophilia* is found concurrently with *B. burgdorferi* in many *Ixodes* ticks resulting in dual infection in many horses. Heavy tick infection and local cellulitis has resulted in low grade fever in some horses.

Lyme disease is diagnosed on the basis of the horse being from an endemic area, compatible clinical signs, ruling out other causes for these signs, and a high enzyme-linked immunosorbent assay (ELISA) titer, other serologic test or positive western blot (WB) test results for anti-*B. burgdorferi* antibodies. I do not recommend treating horses based solely upon a positive test! The WB is most useful in horses with moderate ELISA titers (150 to 300 units) or when there is suspicion the horse may have received a commercial canine Lyme vaccine. Time from infection to seroconversion appears to be 3 to 10 weeks.[13] One potential limitation of serologic tests is that they do not distinguish between active infection and previous exposure. For this reason, tests (e.g., synovial biopsy) that detect the spirochete directly are more conclusive of current infection. It is also possible that many infected horses may be infected for a prolonged duration, even life or are repeatedly reinfected since prior infection may not produce protective antibody. These possibilities are supported by the fact that field horses with high ELISA values rarely have marked decline in ELISA titer when monitored for months or even years. Treatment of naturally-occurring cases rarely results in a decline in antibody as was seen in treated experimentally-infected ponies (experimentally infected ponies had dramatic decline in serum antibody to *Borrelia* after 4 weeks of treatment with antibiotics.[14] Antibodies rose again within 4 weeks after discontinuing antibiotics only in ponies that were not cured of the infection (absence of organism on post mortem exam by both culture and PCR). Some form of molecular mimicry can not be ruled out as a cause of persistent titers in field infections. More recently, an on-site C-6 ELISA SNAP test marketed for detection of *Borrelia*, *Ehrlichia canis*, and *Anaplasma phagocytophilum* infection in dogs was evaluated for detection of *Borrelia* infection in horses; sensitivity of the test is similar to that of a whole cell ELISA.[16] An advantage of the C-6 snap test and Western blot test is that vaccination with OSPA antigen or whole cell vaccines do not result in a positive test result. Most infections do not cause a strong WB band reaction associated with OSPA antigen, but a small number of horses may produce antibody to this antigen following natural infection.

The two most commonly used antibiotics for treating Lyme disease in horses are tetracycline, given intravenously, and doxycycline, given orally. Horses with what are often believed to be more typical signs of Lyme disease (e.g., chronic stiffness, lameness, and hyperesthesia) are most frequently treated with doxycycline (10 mg/kg, PO, every 12 hours). Duration of treatment is often 1 month, but this duration of treatment is mostly empirical. Diarrhea develops in a low percentage of doxycycline treated horses. A clinical response of less stiffness or lameness is often reported following doxycycline treatment, but this could be a nonspecific anti-inflammatory response. Tetracycline (6.6-11.0 mg/kg, intravenously [IV], every 24 hours) may be more efficacious because of higher tissue concentrations of IV tetracycline as opposed to poorly absorbed doxycycline given orally. In the experimental infection studies, Oxytetracycline given intravenously was superior to doxycycline given orally.[14] Oxytetracycline should not be administered in high doses or for prolonged periods if dehydration is present or there is preexisting renal dysfunction! Prevention of Lyme disease in endemic areas would involve preventing tick exposure or prolonged (more than 24 hours) attachment. Decreased tick infection might be helped by clipping tall grasses, clearing shrubs and bushes, and preventing the horse from entering forest and woodlands. Topical tick sprays that do not have adverse reactions could be used when exposure to ticks is expected. Spraying is most commonly used when adult ticks are noticeable in late summer, fall, and the early part of winter, but infection with larval or nymphal stages earlier in the year should also be considered. If ticks are found on the horse, they should be identified to determine whether they are *Ixodes* spp. Ponies have been protected by vaccination with an OspA vaccine,[17] but a vaccine approved for use in horses is not commercially available. Efficacy of administration of the canine vaccines in the horse is unknown.

Many questions regarding Lyme disease in the horse remain unanswered. Our experimental studies and previous cross-sectional studies on seroprevelance compared to clinical signs provide some specific basis for understanding the infection but do not definitively determine the significance of Lyme disease in the horse. This question is unlikely to be answered any time soon, I am sorry to say.

## REFERENCES

[1]Lo Re, V III., et al. **Am Fam Physician** 2004;69:1937. [2]Hovius, J., et al. **Trends Parasitol** 2007;23:434. [3]Pal, U., et al. **J Immunol** 2001;166:7398. [4]Kraiczy, P., et al. **Int J Med Microbiol** 2002;291:141. [5]Nanagara, R., et al. **Hum Pathol** 1996;27:1025. [6]Magnarelli, L.A., et al. **J Am Ve. Med Assoc** 2000;217:1045. [7]Gerhards, H., et al.. **Berl Munch Tierarztl Wochenschr** 1996;109:273. [8]Marion, T.B., et al. **Proc Am Assoc Equine Pract** 1998;44:144. [9]Engvall, E.O., et al.. **Int J Med Microbiol** 2002;291:100-103. [10]Hahn, C.N., et al. **Equine Vet J** 1996;28:84. [11]Burgess, E.C., et al. **J Am Vet Med Assoc** 1986;189:1340. [12]Lanska, D.J. **Neurology** 2008;71(2):150. [13]Chang YF, et al. **Vet Pathol** 2000;37(1):68. [14]Chang YF, et al. **Vet Microbiol** 2005;107(3-4):285. [15]Fortier, L.A. (In preparation 2009). [16]Johnson, A.L., et al. **J Vet Diagn Invest** 2008;20:321. [17]Chang, Y., et al. **Vaccine** 1999;18:540.

## DETERMINATION OF INTRACELLULAR CYTOKINE EXPRESSION IN EQUINE PERIPHERAL BLOOD MONONUCLEAR CELLS

**Elizabeth Davis, DVM, PhD, DACVIM**
**Manhattan, KS**

### INTRODUCTION

Effective immune activation involves stimulation of innate and adaptive immune responses. Innate immunity provides immediate host protection, yet this activity does not provide a lasting response. Adaptive responses provide broad-based immune protection through the combined actions of immunologic memory and antigenic specificity. Antigen presenting cells (APCs) include several cellular elements: dendritic cells, macrophages and B lymphocytes that orchestrate key adaptive functions. APCs process and present antigen peptides in the context of Major Histocompatibility Complex molecules, type I or II. MHC I with peptide typically interact with CD8+ lymphocytes, whereas MHC II molecules with peptide interact with CD4+ (T helper; Th lymphocytes). These cells can be distinguished phenotypically by the expression of surface cell molecules using a panel of monoclonal antibodies and flow cytometry. In addition to cellular phenotype, determined by surface antigen expression, cellular function based on cytokine production is an important factor that determines immune activation status and immunologic response to antigenic challenge. Dual staining of activated lymphocytes with flow cytometry (Figure 1) is an efficient method to determine host response to antigen stimulation[1].

### DETERMINATION OF CELLULAR CYTOKINE SYNTHESIS

Characterization of lymphocyte subsets can be made based on intracellular cytokine synthesis. The T helper type I (Th1) cell profile is characterized by cytokines interleukin (IL) 2, IL-3, IL-12, IL-18 and interferon (IFN) gamma[2]. T helper 2 (Th2) cytokine profiles typically include IL-4, IL-5, IL-10, and IL-13[1]. The importance of knowing the cytokine profile expressed by a cell type relates to the inflammatory cascade induced by their activation. For instance, Th1 profiles are associated with strong cytolytic cell-mediated responses that are highly effective at pathogen clearance, particularly in association with intracellular viral and bacterial infections. In contrast, Th2 responses are noted for their effectiveness at persistence and the induction of strong antibody responses. Although this response is effective under certain clinical situations such as parasitic disease, it is more common for Th2 responses to be detrimental to the host. Allergic conditions are frequently characterized by the persistence of cytokines IL-4, IL-5, IL-10 and IL-13. Not only does this induce a perpetual cycle of chronic inflammation that may be detrimental to the host, but it may also be inhibitory to the induction of a Th1 response. Clearance of severe intracellular disease (E.g. *Rhodococcus equi*) requires a pronounced Th1 immune profile[1,3]. Therefore, it is to the advantage of the host to have a strong Th1 response induced early in the course of intracellular disease, so that pathogen clearance can be managed efficiently and effectively.

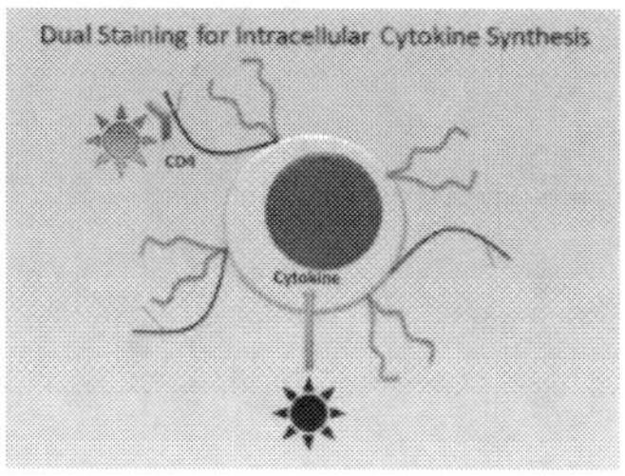


**Figure 1:** Surface staining can be performed to determine the lymphocytes subset. Cellular permeabilization and staining for cytokine can be used to determine intracellular protein synthesis.

Although the standard used to determine vaccine efficacy has classically involved measurement of humoral responses, a more cosmopolitan approach to determining vaccine efficacy includes the determination of cell-mediated induction as an important indicator of the installment of Th1 responses. Specific cell mediated activation that is required includes IFN-gamma synthesis from CD4+ lymphocytes and IFN-gamma production from CD8+ along with pronounced cytolytic activity, which defines the latter cell type as a cytolytic T lymphocytes (CTL).

Based on the knowledge that cell-mediated immunity is a critical factor for intracellular pathogen clearance, we have utilized a previously described assay to determine intracellular cytokine expression profile in equine peripheral blood mononuclear cells (PBMC)[1,4]. Healthy horses were evaluated for cytokine production in CD4+ and CD8+ lymphocytes following administration of a vaccine that contained an inactivated antigen. It is well accepted that humoral immunity is a key host response following such an immunization protocol and was established in this model, in addition we identified a pronounced IFN-gamma response resulting from CD4+ and CD8+ lymphocyte synthesis, consistent with a Th1 response[4]. In a separate investigation, we determined that bovine cell-mediated lymphocyte responses were activated following immunization with a modified live product, yet in addition we determined that gamma-delta T-lymphocytes contributed to a pronounced IL-4 response following vaccination for infectious bovine rhinotracheitis.

When investigations are designed to identify the origin of cytokine production, intracellular cytokine analysis may provide superior results when compared with molecular techniques that measure cytokine production from the entire PBMC population[5,6].

**WHAT IS THE UTILITY OF THIS ANALYSIS?**

Figure 2 provides an example of flow cytometric data for dual staining of PBMC. Molecular techniques, such as RT-PCR, effectively measure cytokine expression from PBMC, whereas a major advantage of intracellular cytokine analysis is the opportunity to determine the cellular subset responsible for cytokine synthesis in an antigen specific manner.

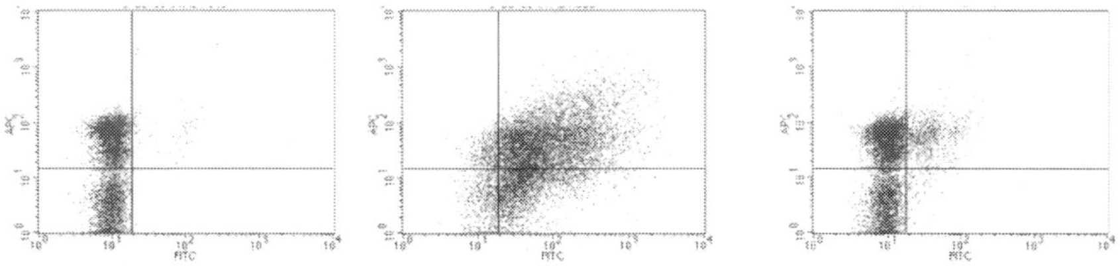

**Figure 2:** The left panel shows a population of CD4+ cells (APC stained) in the upper left quadrant. The middle panel shows a PHA stimulated sample where the majority of CD4+ cells are also expressing IFN-gamma in the upper right quadrant (APC and FITC positive staining). Right panel demonstrates an antigen specific dual staining CD4+ IFN-gamma expressing population of cells in the upper right quadrant.

In addition to determination of cellular responses in a particular group of animals, determination of activation of cellular subset regarding cytokine production can be compared between groups (Figure 3). From this data we determined that younger horses demonstrated a mixed lymphocyte response following vaccination with an inactivated product, whereas adults had a more pronounced Th1 vs. Th2 response following vaccination, represented by the IFN-gamma expression in this cellular subset that exceeded the magnitude of IL-4 expression.

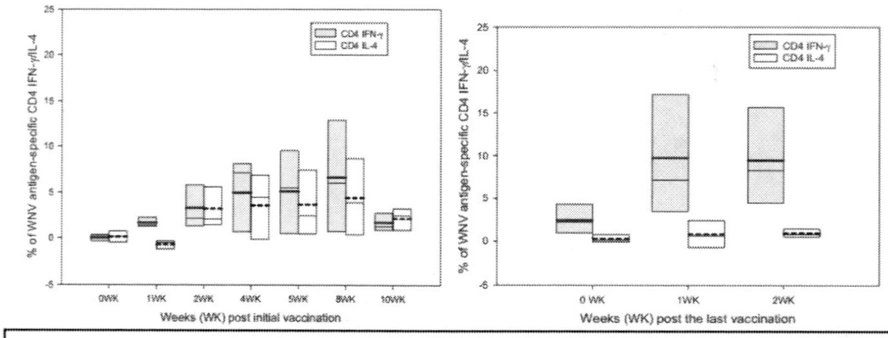


**Figure 3**: Comparison of CD4+ IFN-gamma and IL-4 synthesis in yearling (left panel) and adult horses (right panel). Younger horses showed a mixed T lymphocyte response, while adult horses demonstrated a pronounced antigen specific IFN-gamma response

**CONCLUSION**

Intracellular cytokine analysis is a useful technique to determine the source of cellular cytokine production. This method of analysis can be expanded to use additional staining techniques that include measures of cellular activation, yet protocol design may depend on species specific reagent availability. The described method of dual cell staining can be efficiently used to determine cellular response for specific situations where host immunity is impaired (immunodeficiency) or as a measure of immune function to determine vaccine or immune activation efficacy (adjuvant or immunostimulant administration).

## REFERENCES

[1]Coombs DK et al. **Vet Immunol Immunopathol** 2006; 1-2:109-116 [2]Kolls et al., **Nat Rev Immunol** 2008 11:829-835. [3]Giguere S and Prescott JF. **Infect Immun** 1998; 5:1848-1854. [4]Davis EG et al., **Vet Immunol Immunopathol** 2008; 3-4:293-301. [5]Platt R, et al. **Am J Vet Res** 2006; 12:2050-2058. [6]Platt R, et al. **Vet Immunol Immunopathol** 2008; 1-2:8-15.

**CLINICAL & IMMUNOLOGICAL ASPECTS OF COMMON VARIABLE IMMUNODEFICIENCY IN HORSES**

**M. Julia B. F. Flaminio, DVM, MS, PhD, Diplomate ACVIM**
**Ithaca, NY**

## INTRODUCTION

Primary immunodeficiencies comprise quantitative or qualitative defects in cells or components of the immune system that impair immune cell development, function or structure. Most disorders occur during cell differentiation in the bone marrow due to single or complex mutations/deletions in essential developmental genes[1]. For many types of immunodeficiency, the gene mutation is not known, and the pattern of inheritance may not be clear. In general, manifestation of a primary immunodeficiency occurs in the young, but late-onset manifestations (i.e. in adulthood) are possible. Humoral deficiency is the most commonly diagnosed primary immunodeficiency in mammals, including the horse. In contrast, secondary immunodeficiencies may occur at any time in life due to immunosuppressive treatment; infectious, tumoral, or metabolic/endocrine disorders; immaturity; aging; and malnutrition. In secondary immunodeficiencies, the disturbance may be transient or chronic, and one or several elements of the immune system may be affected at the same time. Primary immunodeficiency leads to susceptibility to infections, importantly recurrent infections, or infections with opportunistic organisms. The type(s) of organism(s) isolated from an immunodeficient patient often suggests the faulty immune mechanism: encapsulated bacterial infections are associated with humoral deficiencies, whereas intracellular pathogens suggest cellular immune disorder. However, infection alone does not support a disorder of the immune system, and non-immunologic causes of infection should always be considered, such as conditions that disrupt pathogen clearance mechanisms (e.g. viral infections, catheterizations, allergies, auto-immunity, lymphoma/lymphosarcoma).

## COMMON VARIABLE IMMUNODEFICIENCY IN HORSES

CVID is a humoral immunodeficiency described in the horse, characterized by late-onset B cell lymphopenia or depletion, hypo- or agammaglobulinemia, inappropriate response to tetanus toxoid vaccination, recurrent fevers and bacterial infections[2]. Pneumonia is the most common complaint, and bacteremia and supurative meningitis are also frequent[3]. Other clinical manifestations include sinusitis, peritonitis, diarrhea, gingivitis, skin abscesses, and hepatitis[4,5]. Septicemia and meningitis in the adult horse are rare clinical presentations in equine referral practices, and warrant immunologic testing for the investigation of underlying immunodeficiency[6]. Occasionally, meningitis may be under-diagnosed if neurological signs are not obvious or misinterpreted as depression from illness. CVID has been diagnosed in horses between 2 and 19 years of life, with no gender, breed or blood line predilection. Over a period of 6 years, 11% of the clinical samples from horses analyzed in our laboratory for immunologic testing revealed abnormalities that were compatible with CVID. Although the incidence of the disease seems higher than expected, it is too early to determine its prevalence in the total horse population.

### The diagnosis of common variable immunodeficiency

Although immunological testing is critical for the diagnosis and monitoring of horses with CVID, the clinical history of recurrent bacterial infection is decisive to support the laboratorial findings. If humoral immunodeficiency is considered based on the clinical history, the first step is to test serum immunoglobulin concentrations for IgG, IgM and IgA (using radial immunodiffusion test). Serum immunoglobulin concentrations correlate with B cell function, and at least 2 immunoglobulin isotypes must be below reference ranges to suspect CVID. In affected horses, serum IgM concentration is often less than 25 mg/dL and does not change through time[7]. Serum IgG concentrations are often below 800 mg/dL, and may oscillate between 200 and 800 mg/dL for months before levels become below detectable ranges. The low serum total IgG concentration reflects the low production of all IgG subtypes ($IgG_1$, $IgG_{4-7}$, $IgG_{3-5}$). Serum IgA concentrations are often within normal range (between 75 to 120 mg/dL) despite low serum IgM and IgG concentrations, but progressively decrease with severity of disease. Whenever possible, rechecking abnormal values in 30 days is advised to account for dynamic changes during disease in an immunocompetent system. *In vivo* immunoglobulin production should be evaluated using pre- and 15 to 20 day post-tetanus toxoid vaccination serum samples for IgG titers against tetanus $C_3$ fragment using an ELISA. Although the magnitude of tetanus toxoid response varies individually in immunocompetent horses, horses with CVID show no increase in the titers after vaccination.

Peripheral blood lymphocyte phenotyping is used to evaluate B cell distribution, which is often below 2% in horses with CVID, whereas CD4+ and CD8+ T cells are normal to slightly increased. Persistent B cell lymphopenia reflects the B cell depletion observed in the lymphoid tissues of affected horses. Occasionally, low CD4+ T cell distibution ($45.9 \pm 14.6\%$ cells), with a low CD4/CD8 ratio ($1.4 \pm 0.4$) accompanies the B cell lymphopenia. When present, CD4+ T cell lymphopenia seems to increase the susceptibility to fungal or intracellluar pathogen, with clinical disease (e.g. *Pneumocystis carinii* or *Rhodococcus equi* pneumonia). Other less specific laboratorial findings include intermittent or persistent lymphopenia (< 1,500 lymphocytes/uL). Subpopulation deficiencies are even more significant when total peripheral blood lymphopenia is present. Neutrophilia and hyperfibrinogenemia reveal infectious/inflammatory episodes; and hypoglobulinemia is caused by hypogammaglobulinemia, despite these infectious/inflamatory processes.

### Clinical management and prognosis

CVID is a progressive disease in which affected horses are apparently normal for many years before clinical manifestation of disease. At the time of clinical signs, the humoral system is already impaired. Therefore, horses with CVID rarely live beyond 6 months from the time of diagnosis. Some horses maintain subnormal production of IgG and can be managed with antibiotic therapy for a few years; nevertheless, underweight condition, recurrent fevers and infections are expected. The majority of the affected horses require hospitalization for appropriate clinical diagnosis and treatment of severe pneumonia and/or meningitis. Periodic intravenous or subcutaneous plasma therapy to maintain serum IgG concentrations around 500 mg/dL is impractical and expensive for the adult horse, and there is no purified formula of equine IgG for continued clinical use. Euthanasia is often elected by the owner due to clinical complications of pneumonia and/or meningitis.

**Serum immunoglobulin concentrations**
(radial immunodiffusion, mg/dL)

A

| | Horses with CVID | Control Horses |
|---|---|---|
| IgM | < 25 | 103 ± 5 |
| IgG | 456 ± 199 | 1,913 ± 754 |
| IgA | 78 ± 44 | 214 ± 100 |

**B Serum pre- and post-vaccination tetanus toxoid titers**
(ELISA, units/ml)

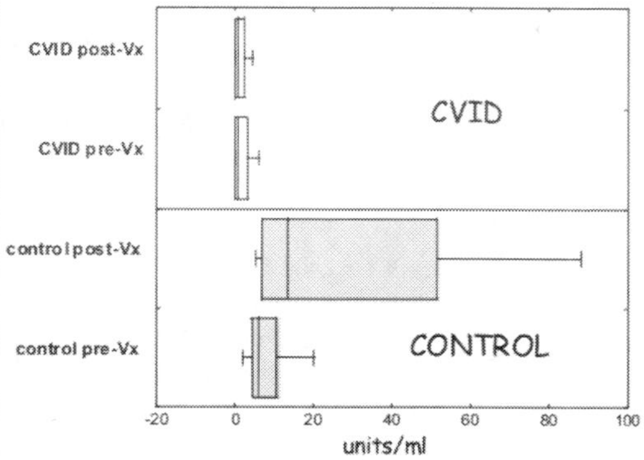


**C Peripheral blood lymphocyte phenotyping**
(flow cytometry, % + cells)

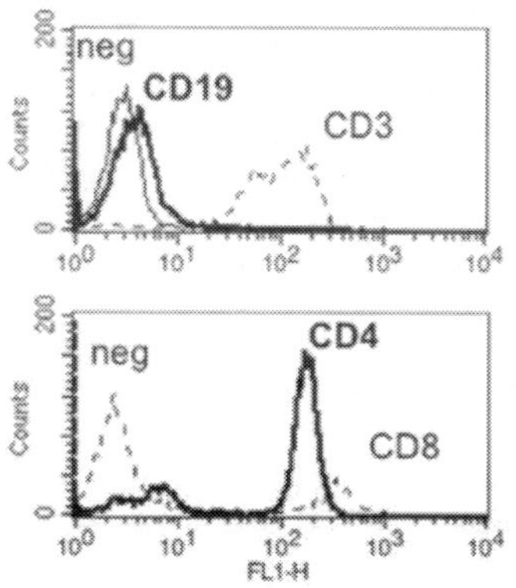


**Immunological testing for the diagnosis of CVID in horses: (A)** Humoral immunodeficiency is confirmed by the low **serum immunoglobulin concentrations** measured by radial immunodiffusion. At least 2 immunoglobulin isotypes should be persistently below the reference range, i.e. in two independent tests performed 2 months apart. In general, horses with CVID have low serum IgM and IgG concentrations, whereas IgA may be low or within normal reference range. **(B)** The inappropriate immunoglobulin production should be confirmed by an in vivo test using tetanus toxoid **vaccination** (Vx). Tetanus toxoid titers are measured in pre- and 15 to 20 day-post-vaccination serum samples using ELISA. Horses with CVID do not change their specific antibody titer after vaccination. **(C)** The cause of inappropriate immunoglobulin production is verified by measuring the distribution of **lymphocyte subpopulations**; the reduced number of circulating B cells (< 2%) suggests paucity of this cell population in lymphoid tissues; often the distribution of CD4+ and CD8+ T cells is normal, but decreased circulating CD4+ T cells can be identified in some patients, which predisposes them to infections with fungus and intracellular pathogens.

Initially, prolonged parenteral antibiotic therapy seems to control infection and improve the clinical condition of non-hospitalized or hospitalized horse patients, but clinical signs tend to reoccur with time or when antibiotic therapy is discontinued. Horses with CVID should be monitored closely for signs of infection, so antibiotic therapy can be initiated immediately. Routine blood work to measure neutrophil and lymphocyte counts, and fibrinogen help to detect subtle infections or monitor response to treatment. Serum IgG and IgA concentrations monitored every 3 months indicate changes that reflect progression of the disease and, therefore, increased susceptibility to infections. In case of inadequate response to antibiotic therapy, infections with intracellular pathogens or fungal organisms should be investigated. Depression or neurologic signs may reflect bacteremia and meningitis; in these cases, blood culture and spinal fluid analysis can be used to confirm the diagnosis.

Minimizing stress to prevent secondary immunodeficiency is advised in these patients. More frequent deworming is necessary to control gastro-intestinal parasites in affected horses. The use of immunosuppressive drugs (e.g. corticosteroids) is contraindicated in horses with CVID in order to preserve cellular immunity. Regular vaccination with inactivated vaccines is warranted; even though a humoral response may not be elicited, many vaccines promote a cellular immune response. Indeed, despite the fact that horses with CVID do not respond with immunoglobulin production to tetanus toxoid vaccination, tetanus has not been diagnosed in these patients. Vaccination of affected horses with modified-live vaccines is not advised, since some vaccines may pose risk of vaccine-organism replication and virulence recovery in immunocompromised individuals.

**Our current research on CVID in horses**

Immunohistochemical and histopathological analyses of lymphoid tissues collected during necropsy from horses with CVID reveals no positive B cells in the bone marrow and spleen, although occasional positive cells may be present in the lymph node. The lymph node structure lacks the typical cortical (B cell)-paracortical (T cell) distribution, and germinal centers. In addition, there is paucity or absence of plasma cells in all tissues. It is possible that peripheral blood lymphopenia results from the abnormal structure and cell interaction in the secondary lymphoid tissues due to the lack of germinal centers.

It is apparent that B cells become progressively depleted or reduced to almost undetectable numbers in affected horses, in both primary and secondary lymphoid tissues, which consequently translates to a progressive depletion of circulating immunoglobulins. Therefore, a failure in the differentiation and/or survival of B lymphocytes is possible, including defects in the expression of B-cell essential molecules, increased rate of spontaneous apoptosis, impaired DNA repair, and the presence of somatic mutations affecting B-cell regulation. It is unclear if this condition involves different mechanisms among patients, and our laboratory has been studying the expression of essential genes for B cell development for comparisons.

B lymphocytes are comprised of 2 populations of cells known as B1 and B2 cells. The B2 cells originate from hematopoietic stem cells in the bone marrow mainly after birth[8]. Many identified regulatory genes control the development and differentiation of B2 cells, which leave the bone marrow expressing surface immunoglobulin M, and migrate to the lymphoid tissues. They interact with CD4+ T cells and undergo immunoglobulin class switching, somatic hypermutation, and differentiation into plasma and memory cells[9]. B1 cells are distinguished from conventional B2 cells by their cell surface markers, anatomical location, self-replenishing activity and production of natural antibodies. B1 cell development occurs during fetal and perinatal lives, and these cells function in the innate immunity[10]. With age, B1 cells decrease in number in blood and spleen, and the conventional B2 cells increase concomitantly. B1 cells can produce antibodies upon multivalent crosslinking and activation threshold of B cell receptor (BCR), a mechanism independent of T cell receptor (TCR) interaction on T helper cells[11]. We have gathered some evidence for the presence of B1 cells in the peritoneal cavity of horses with CVID, and functional response upon polysaccharide vaccination (a T-Independent mechanism). We hypothesize that these cells contribute for the production of IgA despite IgG and IgM hypogammaglobulinemia, and the absence of the classic B2 cells. We are currently developing a model of study of B1 and B2 cells of the horse to, hopefully, identify the faulty mechanism that leads to B2 cell extinction in the horse, and perhaps target the resistant B1 cells for immunologic protection.

## COMMON VARIABLE IMMUNODEFICIENCY IN HUMAN PATIENTS

In humans, more than 150 different types of primary immune disorders have been described, and their prevalence is one in 500 to one in 500,000 in the general population; these numbers vary depending on the diagnostic skills and medical resources available in different countries[12]. The delay and inappropriate diagnosis of primary immunodeficiencies are still limitations in human medicine[13]. CVID in human patients is characterized by recurrent bacterial infections, hypogammaglobulinemia involving at least two isotypes (e.g. IgG and IgA), and impaired humoral response to protein antigen vaccination[14]. B cells may be present but dysfunctional, either due to intrinsic abnormalities or defects in helper T-cell signaling[15]. The incidence of this disorder is greater during the second and third decades of life (estimated at 1 case per 10,000-50,000 population), but the diagnosis can be made during childhood or advanced age. Both sexes are affected, and sporadic inheritance pattern (i.e. single family case) is often observed[16]. Initial diagnosis is usually made by exclusion of other primary causes of immunodeficiency, and a small proportion of cases have been identified with genetic abnormalities. CVID in human patients has been associated with the mutation of five genes: ICOS (inducible co-stimulatory receptor), CD19, MSH5, TNFRSF13B (encoding TACI), or TNFRSF13C (encoding BAFF-R)[17]. In addition, splenomegaly, granulomas, autoimmune disorders and malignancies occur in a subset of patients[18]. A multicenter study has recently suggested categorizing human patients with CVID based on the B cell population distribution: absent or < 1% B cells (which likely involve faulty mechanisms in early B-cell differentiation), severely reduced or < 2% switched memory B cells (e.g. defective germinal center development due to lack of T cell co-stimulation), and > 10% transitional or $CD21^{low}$ B cells (defects in the expansion of B cells)[19]. CVID in horses seems to be similar to a subgroup of human CVID patients who present low numbers of IgM+ B cells, and an increased rate of infections. Clinical management of primary humoral immunodeficiencies in human patients includes periodic administration of purified immunoglobulin, antibiotic and/or antifungal therapy, bone marrow transplantation and gene therapy according to the etiology of disease[20].

## ACKNOWLEDGEMENTS

The author thanks the referring DVMs and their clients for supporting this research with samples and clinical history of their horse patients. This research is supported by NIH 1R03AI079796-01.

## REFERENCES

[1]Wood P., Primary antibody deficiency syndromes. Ann Clin Biochem 2009 [Epub ahead of print]; [2]**Flaminio** et al., Common variable immunodeficiency in a horse. J Am Vet Med Assoc 2002;221:1296; [3]**Pellegrini-Masini** et al., Common variable immunodeficiency in three horses with presumptive bacterial meningitis. J Am Vet Med Assoc 2005;227:114; [4]**Freestone** et al., Acquired immunodeficiency in a seven-year-old horse. J Am Vet Med Assoc 1987; 190:689; [5]**MacLeay** et al., Acquired B lymphocyte deficiency and chronic enterocolitis in a 3-year-old-Quarter Horse. Vet Immun Immunopathol 1997;57:49; [6]**Overturf** , Indications for the immunological evaluation of patients with meningitis. Clin Infec Dis 2003;36:189; [7]**Flaminio** et al., Common Variable Immunodeficiency in Horses is Characterized by B Cell Depletion in Primary and Secondary Lymphoid Tissues. J Clin Immunol 2009;29:107; [8]**Medina** KL et al., Gene regulatory networks orchestrating B cell fate specification, commitment, and differentiation. Curr Top Microbiol Immunol 2005;290:1; [9]**McHeyzer-Williams** LJ, et al., Antigen-specific memory B cell development. Annu Rev Immunol 2005;23:487; [10]**Martin** F, et al., B1 cells: similarities and differences with other B cell subsets. Curr Opin Immunol 2001;13:195; [11]**Fagarasan** et al., T-Independent immune response: new aspects of B cell biology. Science 2000;290:89-92; [12]**Park** et al., Common variable immunodeficiency: a new look at an old disease. Lancet. 2008;372:489; [13]**Brandau** et al., Common variable immunodeficiency: an uncommon disease with high mortality. J Insur Med 200;739:71; [14]**Cunningham-Rundles** et al., Common variable immunodeficiency: clinical and immunological features of 248 patients. Clin Immunol 1999;92:34; [15]**Giovannetti** et al., Unraveling the complexity of T cell abnormalities in common variable immunodeficiency. J Immunol 2007;178:3932-3943; [16]**Cunningham-Rundles** et al., Common variable immune deficiency: reviews, continued puzzles, and a new registry. Immunol Res 2007;38:78-86; [17]**Kopecký** et al., Genetic defects in common variable immunodeficiency. Int J Immunogenet 2007;34:225; [18]**Brandt** et al., Common variable immune deficiency and autoimmunity. Autoimmun Rev 2006;5:465; [19]**Wehr** et al., The EUROclass trial: defining subgroups in common variable immunodeficiency. Blood 2008;111:77-85; [20]**Yong** et al., Common variable immunodeficiency: an update on etiology and management. Immunol Allergy Clin North Am 2008;28:367.

# DIAGNOSIS & MANAGEMENT OF ANTHELMINTIC RESISTANCE IN EQUID PARASITES

**Martin K. Nielsen, DVM, PhD & Ray M. Kaplan, DVM. PhD, DEPVC**
**Copenhagen, Denmark**

## DIAGNOSIS OF ANTHELMINTIC RESISTANCE

Over the past decade there have been numerous reports documenting high levels of anthelmintic resistance throughout the world. Despite this evidence, most veterinarians and horse owners continue to blindly equate treatment with control and do not monitor the effectiveness of anthelmintic treatments. This false sense of security brings with it a substantial risk to horse health. Consequently, routine evaluation of anthelmintic efficacies in equine establishments is increasingly important. Several attempts have been made to develop *in vitro* and molecular assays for detecting resistance, but none of these have proven to be useful or practical at the practice level. Thus, there is presently only one diagnostic method available for usage in practice; the fecal egg count reduction test (FECRT). This method evaluates the reduction of parasite egg output following an anthelmintic treatment.

The FECRT is simple to perform and interpretation of results is usually straight forward. Fecal egg counts (FEC) are performed just before treatment and again 10-14 days after (see figure 1 for a thorough description of the technique). Detailed guidelines for performing scientifically and statistically valid resistance surveillance on horse farms are currently under development by an expert committee and will be published in due course. A brief highlight of the most important features will be given here.

### Anthelmintic Resistance is a Farm Diagnosis

It is important to emphasize that parasites should be viewed as a population being shared by all horses on the premises. This implies that if a reduction in anthelmintic efficacy is detected in a given sample of the horse population, it will invariably be found in other horses on the same farm. However, measurement of FEC have inherent variability, which often will produce a non-uniform response among individual animals. Thus, FEC reduction should always be calculated as an average for the group of treated horses, and it is recommended that 5-10 horses be tested per farm for each drug. One should not make the mistake to interpret the results of each horse separately; the worms on the farm as a whole are resistant or they are not resistant (or results are equivocal). There are reasons other than resistance which can cause a poor efficacy in a single animal, so when egg count reduction proves unsatisfactory in a single horse, it is recommended to test several other horses on the farm, or if other horses are not available for testing, then the same horse should be tested again at a later date. Anthelmintic resistance does not revert to susceptibility, so once detected, it should be reproducible throughout the farm. Nonetheless, because usually only a small sample of horses is tested, normal variability will cause the actual percentage reduction in FEC to vary with each subsequent test.

### Interpretation of the FECRT

For most anthelmintic formulations, FECRTs of 95% or above should be expected. Suggested cut-offs for various anthelmintics are presented in figure 1. Provided at least 5 horses are tested, FECRTs in the 80-90% range provide convincing evidence for resistance, but is important to understand that many horses can still benefit from treatments with such drugs, given that it still removes a large majority of the worm burden. Using drugs with emerging levels of resistance against them, however, will greatly increase the rate of further development of resistance.

Although equine nematodes as a group have now shown resistance to all three major drug classes on the market[1,2], it is important to note that no single anthelmintic formulation has become useless in equine practice. In fact, it appears that the drugs can now be categorized into two groups; a) those that maintain efficacy against cyathostomins, but fail to control ascarids, and b) those that maintain efficacy against ascarids, but fail to control cyathostomins. Thus, these drugs, all of which had broad spectrum nematocidal activity when first introduced to the market, in many cases now need to be viewed as complementary narrow spectrum drugs. For instance, benzimidazole drugs generally fail to control cyathostomins in horse establishments world-over, but they remain a fine choice for treatment of *Parascaris equorum*. However, more recent reports indicate emerging ivermectin resistance in cyathostomin parasites[3,4], and one study suggests pyrantel resistance in *P. equorum*[5], and thus no single drug can claim to be effective against all relevant parasite species.

Before the proper choice of anthelmintic can be made, it is essential to know which types of parasites are present. First of all, it is crucial to know whether *P. equorum* is present or not. Secondly, it is imperative to establish an anthelmintic resistance profile for the farm. The most important age groups to monitor are foals, yearlings and youngsters of 2 and 3 years as these animals can harbor both ascarids and strongyles. A rough way to monitor treatment efficacy is simply to collect representative samples 10-14 days after treatment. If parasite eggs are found in these samples, FECRTs should be done to diagnose resistance properly.

## CONTROL OF ANTHELMINTIC RESISTANCE

The key for managing anthelmintic resistance is parasite *refugia*, which is the proportion of the parasite population which is not exposed to the anthelmintic drug at the time of treatment. This term covers all stages of eggs and larvae residing on pasture. Parasite stages in *refugia* are not under any selection pressure for drug resistance. Thus, they can dilute out resistant genetic alleles which are selected among the parasite stages exposed to the drug.

Parasites are always overdispersed among their hosts. As a rule of thumb, 20% of the horses harbor 80% of the parasites and shed 80% of the total egg output. Often about half of the adult population sheds very low numbers of eggs and this trend has been shown to be a consistent feature among individual horses over time[6]. By treating only horses exceeding a predetermined threshold FEC, total egg shedding can be reduced virtually as efficiently as if the entire herd were treated. In addition, this has the advantage of maintaining a low parasite burden to stimulate immunity, enabling horses to withstand future infections [7,8]. By depriving horses of parasite stimulation of their immune systems, they practically become dependent on frequent anthelmintic treatments, and are at risk if exposed to significant levels of parasites at some point in the future.

Targeting treatments so that only those horses that really need treatment are dewormed is referred to as Selective Therapy. Although Selective Therapy has been recommended for the past 20 years[9,10] it has found very little adoption[11]. Since 1999, Denmark has been the exception to the rule. Here, anthelmintic drugs are available on a prescription-only basis, requiring veterinary involvement before drugs can be prescribed. As a result, Danish equine veterinarians routinely perform surveillance using fecal egg counts, and Selective Therapy is widely used[12]. Other countries of the European Union have recently implemented similar legislation, and more can be expected to follow.

**Drug Rotation**

Rotation among anthelmintic drug classes has long been assumed to be a solid strategy for preventing resistance[13]. However, there is virtually no evidence behind this prevalent assumption. Drug rotation was initially recommended in the 1960s because the drug types available at that time were not broad-spectrum[14] . Today, calendar-based drug rotation is still widely practiced without any type of evaluation of efficacy of the drugs used[11]. It should be emphasized that such rotation-based treatment regimen cannot substitute for surveillance of parasite levels and efficacy of treatments. It carries an inherent risk of creating a false sense of security.

A sound, scientifically based anthelmintic program requires knowledge of the parasite species present and the efficacy levels of the anthelmintic drugs against these. Anthelmintic resistance is an inevitable biological consequence of treatment. It cannot be prevented it, but the rate of development can be greatly reduced by using treatment strategies that reduce selection pressure for resistance. By implementing such strategies, the useful life of the current anthelmintic products can be significantly prolonged.

**REFERENCES**

[1] Kaplan RM **Vet Res** 2002; 33:491. [2] Kaplan RM (2004). **Trends Parasitol** 2004; 20(10):477. [3] Molento MB, et al.**Vet Rec** 2008; 162(12):384. [4] Lyons ET, et al. **Parasitol Res** 2008; 103(1):209. [5] Lyons ET, et al. **Parasitol Res** 2008; 103(2); 287. [6] Nielsen MK, et al. **Vet Parasitol** 2006; 135(3-4):333. [7] Monahan CM, et al. **Vet Parasitol** 1997; 73(3-4):277. [8] Monahan CM et al. **Vet Parasitol** 1998; 74(2-4):229. [9] Duncan JL, Love S **Eq Vet J** 1991; 23:226. [10] Gomez HH, Georgi JR **Eq Vet J** 1991; 23:198. [11] NAHMS. Equine '98, Part I: Baseline reference of 1998 equine health and management. Fort Collins, National Animal Health Monitoring System, USDA:APHIS:VS. [12] Nielsen MK, et al. **Vet Parasitol** 2006; 135(1):47. [13] Kelly J, et al. **Aust Vet J** 1981; 57(4):163. [14] Drudge JH, Lyons, ET **J Am Vet Med Assoc** 1966; 148(4):378

**Figure 1:**

**Fecal egg count reduction test (FECRT)**

The FECRT evaluates the efficacy of an anthelmintic drug based on its ability to reduce the fecal egg output after treatment. Fecal egg counts (FEC) are performed just before and 14 days after treatment, and the egg reduction is calculated according to the formula:

$$((FEC_{pre}-FEC_{post})/FEC_{pre}) \text{ X } 100\%$$

Any egg count technique can be employed for the FECRT, but it is recommended to use a technique with a detection limit of $\leq$25 eggs per gram. Always use the same technique consistently.

The FECRT should be established on the farm level by calculating the FECR for a number of individual horses and subsequently calculating the average FECR for the treated group. It is recommended to include at least 5-10 horses on each farm if possible.

Suggested cut-off values for resistance depend on the drug tested and the number of horses investigated, but for the range of 5-10 horses, the following cut-off values are recommended as general guidelines for strongyle nematodes:

| | |
|---|---|
| Benzimidazoles: | 90% |
| Pyrantel: | 90% |
| Ivermectin: | 95% |
| Moxidectin: | 95% |

If the farm average FECR falls below these values, anthelmintic resistance should be suspected. However, it is important to rule out other causes of decreased efficacy, such as inappropriate dosing, inadequate storage etc.

More detailed guidelines for diagnosis of anthelmintic resistance in horses are currently under development, and will be published in due time.

Currently, there are no available methods for diagnosis of anthelmintic resistance in equine tapeworms.

**WHAT'S WRONG WITH ANIMAL MODELS OF PAIN?**
**Dr. Jeffrey Mogil**
**Montréal, Québec**

1. Mogil, J.S. and Crager, S.E. What should we be measuring in behavioral studies of chronic pain in animals? **Pain**, 112:12-15, 2004.

2. Mogil, J.S. and Chanda, M.L. The case for the inclusion of female subjects in basic science studies of pain. **Pain**, 117:1-5, 2005.

3. Langford, D.J., Crager, S.E., Shehzad, Z., Smith, S.B., Sotocinal, S.G., Levenstadt, J.S., Chanda, M.L., Levitin, D.J., and Mogil, J.S. Social modulation of pain as evidence for empathy in mice. **Science**, 312:1967-1970, 2006.

4. Mogil, J.S., Ritchie, J., Sotocinal, S.G., Smith, S.B., Croteau, S., Levitin, D.J., and Naumova, A.K. Screening for pain phenotypes: analysis of three congenic mouse strains on a battery of nine nociceptive assays. **Pain**, 126:24-34, 2006.

5. LaCroix-Fralish, M.L., Ledoux, J.B. and Mogil, J.S. The *Pain Genes Database*: an interactive web browser of pain-related transgenic knockout studies. **Pain**, 131:3.e1-3.e4, 2007.

6. Callahan, B.L., Gil, A.S.C., Levesque, A., and Mogil, J.S. Modulation of mechanical and thermal nociceptive sensitivity in the laboratory mouse by behavioral state. **J. Pain**, 9:174-184, 2008.

7. Mogil, J.S. Animal models of pain: progress and challenges. **Nature Rev. Neurosci.**, in press.

**THE NATURE & NURTURE OF PAIN**
**Dr. Jeffrey Mogil**
**Montréal, Québec**

[1] Mogil, J.S., Wilson, S.G., Chesler, E.J., Rankin, A.J., Nemmani, K.V.S., Lariviere, W.R., Groce, M.K., Wallace, M.R., Kaplan, L., Staud, R., Ness, T.J., Glover, T.L., Stankova, M., Mayorov, A., Hruby, V.J., Grisel, J.E., and Fillingim, R.B. The melanocortin-1 receptor gene mediates female-specific mechanisms of analgesia in mice and humans. **Proc. Natl. Acad. Sci. USA**, 100:4867-4872, 2003.

[2] Mogil, J.S., Ritchie, J., Smith, S.B., Strasburg, K., Kaplan, L., Wallace, M.R., Romberg, R.R., Bijl, H., Sarton, E.Y., Fillingim, R.B., and Dahan, A. Melanocortin-1 receptor gene variants affect pain and μ-opioid analgesia in mice and humans. **J. Med. Genet.**, 42:583-587, 2005.

[3] Mogil, J.S., Miermeister, F., Seifert, F., Strasburg, K., Zimmermann, K., Reinold, H., Austin, J.-S., Bernardini, N., Chesler, E.J., Hofmann, H.A., Hordo, C., Messlinger, K., Nemmani, K.V.S., Rankin, A.L., Ritchie, J., Siegling, A., Smith, S.B., Sotocinal, S.G., Vater, A., Lehto, S.G., Klussmann, S., Quirion, R., Michaelis, M., Devor, M., and Reeh, P.W. Variable sensitivity to noxious heat is mediated by differential expression of the CGRP gene. **Proc. Natl. Acad. Sci., U.S.A.,** 102:12938-12943, 2005.

[4] Langford, D.J., Crager, S.E., Shehzad, Z., Smith, S.B., Sotocinal, S.G., Levenstadt, J.S., Chanda, M.L., Levitin, D.J., and Mogil, J.S. Social modulation of pain as evidence for empathy in mice. **Science,** 312:1967-1970, 2006.

[5] LaCroix-Fralish, M.L., Ledoux, J.B. and Mogil, J.S. The **Pain Genes Database**: an interactive web browser of pain-related transgenic knockout studies. **Pain,** 131:3.e1-3.e4, 2007.

[6] LaCroix-Fralish, M.L. and Mogil, J.S. Progress in genetic studies of pain and analgesia. **Annu. Rev. Pharmacol.**, 49:97-121, 2009.

**TESTING PAIN MANAGEMENT IN HORSES: DO THE MODELS WORK?**

**L. Chris Sanchez, DVM, PhD, DACVIM**
**Gainesville, FL**

## OVERVIEW

Colic and lameness are two of the most clinically and economically important medical problems facing horses and their owners. Pain is a critical component of each disease process and its alleviation is often critical to a successful outcome. But, a limited number of analgesics are available for use in horses and many of these options are associated with significant untoward effects. Thus, one's ability to objectively assess new pharmacologic options or new combinations of previously available options is critical to the advancement of equine medicine. This talk will address the strengths and weaknesses of currently used equine models of visceral and somatic nociception.

When assessing pain in animals or non-verbal humans, one must remember the differences from pain assessment in human beings. In the latter, pain is typically evaluated by self-assessment, or the patient's perception. In the former, the assessment is always based upon the observer's perception. Thus, when addressing an animal's status, one must remember inherent differences based upon species, age, sex, genetics, environment, source (visceral, somatic, neuropathic), and duration (acute versus chronic) of the stimulus. For research models of pain in animals, Gebhart and Ness put forth the following necessary criteria: the subject is not anesthetized, the experimental stimulus mimics a natural stimulus, the stimulus is minimally invasive and ethically acceptable, the stimulus is controllable, reproducible, quantifiable, and the responses are reliable and quantifiable.[1]

## RESEARCH MODELS OF VISCERAL NOCICEPTION

Most models of visceral pain in horses and other species (including humans) involve the use of some device resulting in distension of a portion of the gastrointestinal tract. As such, information gained from these models can be applied only to clinical conditions that cause pain as a result of visceral distension, and these are typically models of acute pain. Because many naturally occurring conditions involve distension of one form or another, such models have provided clinically meaningful information regarding analgesic medications commonly used in the horse. The first equine models involved balloon distension of the cecum via a surgically implanted cecal cannula. This model was used to demonstrate the visceral analgesic potential of several commonly used medications.[2,2,3] The distension can be applied in such a way to determine a threshold (typically pressure) at which a given response (typically pawing or flank-watching) is observed. Alternatively, the pressure at which this behavior is typically observed is applied for a set period of time (typically 1-5 minutes) and the number of behavioral events observed is tabulated. Threshold models offer the advantage of discontinuing the stimulus when the nociceptive behavior begins.

A method of duodenal distension, involving placement of a mylar balloon into the duodenum via a surgically implanted gastric cannula, has also been used as a model of visceral discomfort in the conscious horse. This model used a barostat for computer-controlled distension of a non-elastic balloon to a set pressure. This system allows measurement of motility between distensions and involves little risk to the subject. Various treatments have been evaluated in clinically normal conscious horses with this model, including xylazine, flunixin, acupuncture, detomidine, fentanyl, lidocaine, butorphanol, and N-butylscopolammonium bromide.[4-9]

Colorectal distension (CRD) is widely used in visceral pain research using human and animal subjects to determine an individual's threshold of visceral discomfort[10,11,11] In humans, CRD can be used to visceral hypersensitivity, such as that associated with the Irritable Bowel Syndrome. CRD offers the advantage over the cecal and duodenal methods of distension in that it does not require surgical manipulation of the subject animals. CRD with a modified barostat has been used to evaluate the effect of intravenous xylazine and intrarectal lidocaine administration on nociceptive threshold and rectal compliance[12] and the effect of systemic detomidine, fentanyl, lidocaine, butorphanol, and N-butylscopolammonium bromide on distension threshold.[5-8] A modified version of this technique has been used to evaluate the effect of electroacupuncture and butorphanol on colorectal distension threshold.[13]

The primary advantage of the cecal and duodenal distension models is that both the stimulus (distension) and associated nociceptive behaviors (pawing, flank watching, etc.) mimic the clinical syndrome of colic. With the CRD, the associated behavior (expulsion of balloon) is not as clear and results may be more closely associated with the "urge to defecate" response in humans and therefore not truly nociceptive in nature. One potential disadvantage of these models is the potential effect of repeated distension on the nociceptive threshold.

## RESEARCH MODELS OF SOMATIC NOCICEPTION

A variety of models are currently in use for the objective evaluation of somatic nociception, mostly associated with acute pain. As for visceral nociceptive testing, a variety of threshold-based models are also available for somatic nociception. Currently, heat (thermal) and mechanical (pressure algometry) stimuli are most commonly used. Thermal threshold testing has previously been used in a variety of species and has been used in the horse to evaluate the effects of lidocaine, fentanyl, detomidine, acepromazine, and butorphanol.[5-8] The main advantages of this method are ease of use and repeatability. Disadvantages include the need for a cord tether to the testing device and the potential for drug effect on skin temperature to affect results, such as occurs with the alpha-2 adrenoreceptor agonists.[6] Mechanical stimuli include the use of pneumatic devices to apply pressure to the coronary band or a pressure algometer to determine a mechanical nociceptive threshold (MNT) at either the pastern or along the thoracolumbar musculature. The pneumatic pressure models have been used to assess the somatic nociceptive effects of xylazine, detomidine, romifidine, butorphanol and levomethadone.[14,15] The pressure algometer has been used to assess the effect of induced back pain, induced synovitis and osteoarthritis, chiropractic therapy, phenylbutazone, massage therapy, sacroiliac dysfunction and distal forelimb irritants in Tennessee Walking Horses.[16-21] The use of both mechanical and thermal stimuli to evaluate analgesic therapy is useful in that both types of stimuli are seen in clinical situations.

Reversible models of lameness, such as a specialized shoe fitted with an adjustable screw to induce solar pressure, have been used to evaluate the analgesic effects of non-steroidal anti-inflammatory agents and topical capsaicin.[22-24] Similar models have also been used to evaluate the physiologic effects of induced lameness on compensatory load adjustments and thoracolumbar kinematics.[25,26] These models offer the advantage of a naturally occurring stimulus and quantifiable nociceptive effects. Other models of somatic nociception include induced synovitis, osteoarthritis, and hemarthrosis.[16,27]

## SCORING SYSTEMS FOR CLINICAL RESEARCH

In order to be useful, a pain scoring system should meet the following criteria: it should include clearly defined assessment criteria, be suitable for all observers, be simple and quick to use, be sensitive, have identified strengths and weaknesses, and be validated. Possible deficiencies include bias, inter- and intra-observer variability. A lack of agreement between observers is one of the flaws of simple scoring systems such as the visual analog scale (VAS), in which pain is scored on a numerical scale, when used in humans. A modified VAS has been suggested for use in the horse, but has not yet been validated.[28]

One of the most useful clinical applications of integrated pain scoring systems is the assessment of interventions (i.e. assess pain score, apply treatment, then re-assess pain score at various intervals). When critically assessing scoring systems, the investigator should control for signalment (age, breed, sex), observer (veterinarian, student, owner/trainer), procedure, and other effects (food withdrawal, anesthesia, management, etc.). Continuous video assessment allows for quantification of either time budgets (locations within the stall, ear position, head position, eating, lying down, etc.) or events (vocalizing, stomping feet, shifting weight, etc.). Time budget and event analyses has been performed on horses following colic surgery, or arthroscopy.[29,30] The numerical rating system used following colic surgery was also used to assess the effect of a butorphanol constant rate infusion on physiologic and outcome variables in that population of horses.[31] All horses received flunixin following surgery. Interestingly, horses receiving butorphanol had improved behavior scores in the 24 hours following surgery, decreased cortisol concentration, lost less weight, and were discharged from the hospital faster than horses receiving a placebo infusion. These results correspond to work in humans in which primary negative outcomes associated with untreated postoperative pain include increased sympathetic activity, a neurohormonal stress response and weight loss.

Objective measures such as vital signs, plasma cortisol concentration, and force plate analyses alleviate the subjective nature of assessment. But, vital signs and cortisol are affected by a variety of factors in addition to pain, including hydration status, perfusion, sepsis and/or endotoxemia, fear, and anxiety.

## SUMMARY

Overall, the research models available for nociceptive testing in the horse allow for the assessment of analgesia in the horse via a variety of nociceptive mechanisms. While all have advantages and limitations, they can be used in a complimentary fashion for objective assessment of currently used and future therapies. Alternatively, clinical pain assessment in the horse currently lacks a consensus approach. Validation of currently available scoring systems and development of newer systems will likely be a focus of future research efforts.

## REFERENCES

[1.] Gebhart GF, Ness TJ . Central mechanisms of visceral pain. **Can.J.Physiol Pharmaco**l. 1991;69: 627-634. [2.] Kohn CW, Muir WW, III . Selected aspects of the clinical pharmacology of visceral analgesics and gut motility modifying drugs in the horse. **J.Vet Intern.Med** 1988;2: 85-91. [3.] Muir WW, Robertson JT . Visceral analgesia: effects of xylazine, butorphanol, meperidine, and pentazocine in horses. **Am J.Vet Res** 1985;46: 2081-2084.[4.] Merritt AM, Burrow JA, Caldwell S, et al. Development and characterization of a highly controllable equine colic model. **Proc Equine Colic Res Symp** 6, 43. 1998. [5.] Sanchez LC, Elfenbein JR, Robertson SA . Effect of acepromazine, butorphanol, or N-butylscopolammonium bromide on visceral and somatic nociception and duodenal motility in conscious horses. **Am J Vet Res** 2008;69: 579-585.[6.] Sanchez LC, Elfenbein J, Mock J, et al. Effect of detomidine on somatic and visceral nociception and duodenal motility in conscious horses. **J Vet Intern.Med.** 19[3], 487 (Abstract 316). 2005. [7.] Sanchez LC, Robertson SA, Maxwell LK, et al. Effect of fentanyl on visceral and somatic nociception in conscious horses. **J Vet Intern.Med** 2007;21: 1067-1075.[8.] Robertson SA, Sanchez LC, Merritt AM, et al. Effect of systemic lidocaine on visceral and somatic nociception in conscious horses. **Equine Vet J** 2005;37: 122-127.[9.] Merritt AM, Xie H, Lester GD, et al. Evaluation of a method to experimentally induce colic in horses and the effects of acupuncture applied at the Guan-yuan-shu (similar to BL-21) acupoint. **Am J.Vet Res** 2002;63: 1006-1011.[10.] Verne GN, Robinson ME, Price DD . Hypersensitivity to visceral and cutaneous pain in the irritable bowel syndrome. **Pain** 2001;93: 7-14.[11.] Al Chaer ED, Kawasaki M, Pasricha PJ . A new model of chronic visceral hypersensitivity in adult rats induced by colon irritation during postnatal development. **Gastroenterology** 2000;119: 1276-1285.[12.] Sanchez LC, Merritt AM . Colorectal distention in the horse: visceral sensitivity, rectal compliance and effect of i.v. xylazine or intrarectal lidocaine. **Equine Vet J** 2005;37: 70-74.[13.] Skarda RT, Muir WW, III . Comparison of electroacupuncture and butorphanol on respiratory and cardiovascular effects and rectal pain threshold after controlled rectal distention in mares. **Am J.Vet Res** 2003;64: 137-144.[14.] Moens Y, Lanz F, Doherr MG, et al. A comparison of the antinociceptive effects of xylazine, detomidine and romifidine on experimental pain in horses. **Veterinary Anaesthesia and Analgesia** 2003;30: 183-190.[15.] Schatzman U, Armbruster S, Stucki F, et al. Analgesic effect of butorphanol and levomethadone in detomidine sedated horses. **J Vet Med.**A Physiol Pathol.Clin.Med. 2001;48: 337-342.[16.] Haussler KK, Hill AE, Frisbie DD, et al. Determination and use of mechanical nociceptive thresholds of the thoracic limb to assess pain associated with induced osteoarthritis of the middle carpal joint in horses. **American Journal of Veterinary Research** 2007;68: 1167-1176.[17.] Sullivan KA, Hill AE, Haussler KK . The effects of chiropractic, massage and phenylbutazone on spinal mechanical nociceptive thresholds in horses without clinical signs. **Equine Veterinary Journal** 2008;40: 14-20.[18.] Haussler KK, Erb HN . Mechanical nociceptive thresholds in the axial skeleton of horses. **Equine Veterinary Journal** 2006;38: 70-75.[19.] Haussler KK, Behre TH, Hill AE . Mechanical nociceptive thresholds within the pastern region of Tennessee Walking Horses. **Equine Veterinary Journal** 2008;40: 455-459.[20.] Haussler KK, Erb HN . Pressure algometry for the detection of induced back pain in horses: a preliminary study. **Equine Veterinary Journal** 2006;38: 76-81.[21.] Varcoe-Cocks K, Sagar KN, Jeffcott LB, et al. Pressure algometry to quantify muscle pain in racehorses with suspected sacroiliac dysfunction. **Equine Veterinary Journal** 2006;38: 558-562.[22.] Foreman JH, Barange A, Lawrence LM, et al. Effects of single-dose intravenous phenylbutazone on experimentally induced, reversible lameness in the horse. **Journal of Veterinary Pharmacology and Therapeutics** 2008;31: 39-44.[23.] Seino KK, Foreman JH, Greene SA, et al. Effects of topical perineural capsaicin in a reversible model of equine foot lameness. **Journal of Veterinary Internal Medicine** 2003;17: 563-566.[24.] Foreman JH . Phenylbutazone and flunixin meglumine used singly or in combination ("stacking") in experimental lameness in horses. **Journal of Veterinary Internal Medicine** 2007;21: 166.[25.] Alvarez CBG, Wennerstrand J, Bobbert MF, et al. The effect of induced forelimb lameness on thoracolumbar kinematics during treadmill locomotion. **Equine Veterinary Journal** 2007;39: 197-201.[26.] Weishaupt MA, Wiestner T, Hogg HP, et al. Compensatory load redistribution of horses with induced weight-bearing forelimb lameness trotting on a treadmill. **Veterinary Journal** 2006;171: 135-146.[27.] Judy CE, Galuppo LD . Evaluation of iatrogenic hemarthrosis of the metacarpophalangeal joint as a method of induction of temporary reversible lameness in horses. **American Journal of Veterinary Research** 2005;66: 1084-1089.[28.] Muir WW (2004), Recognizing and treating pain in horses. In: Reed SM, Bayly WM, and Sellon DC, ed. **Equine Internal Medicine.** Philadelphia: W.B. Saunders; 2004:1529-1541.[29.] Pritchett LC, Ulibarri C, Roberts MC, et al. Identification of potential physiological and behavioral indicators of postoperative pain in horses after exploratory celiotomy for colic. **Applied Animal Behaviour Science** 2003;80: 31-43.[30.] Price J, Catriona S, Welsh EM, et al. Preliminary evaluation of a behaviour-based system for assessment of post-operative pain in horses following arthroscopic surgery. **Veterinary Anaesthesia and Analgesia** 2003;30: 124-137.[31.] Sellon DC, Roberts MC, Blikslager AT, et al. Effects of continuous rate intravenous infusion of butorphanol on physiologic and outcome variables in horses after celiotomy. **J Vet Intern.Med.** 2004;18: 555-563.

# CLINICALLY-APPLICABLE PAIN SCORING IN HORSES.

K Wegner
New Bolton Center, University of Pennsylvania, Kennett Square, PA, USA.

Quantitating the subjective experience of pain is a challenge for both physicians and veterinarians. However, in order to determine the need for, and the effectiveness of analgesic interventions, some objective measure of pain is necessary. Challenges in developing and implementing a pain scoring system include the unique behavioral and physiologic qualities of the species of interest, the nature of the pain, and the ability of one or more scorers to produce consistent, sensitive, specific results. Simple scoring systems such as the visual analog scale, numeric, and simple descriptive scales have been adapted from human to veterinary medicine. Validated acute pain scoring systems incorporating observed and interactive behavioral elements have been published and are in clinical use in small animals. Among these are the Glasgow Composite Pain Scale (CPS), and the composite acute pain score for dogs and for cats developed at Colorado State University.

Equine lameness scoring, such as the Obel laminitis pain scale, and the lameness score adopted by the American Association of Equine Practitioners, has been employed for many years to offer an inter-individual objective measure of pain. Several comprehensive studies on post-operative pain behavior in horses have yielded important behavioral markers of pain unique to both the species and type of pain. Descriptive and composite pain scales utilized in these studies correlated closely with surgical and analgesic interventions. The relative complexity of the behavior descriptions and scoring process may limit the practical application of these pain scales.

Drawing upon the models of small animal and equine composite pain scoring systems, a simplified scoring system has been developed to measure both dynamic, interactive, and static, non-interactive lameness-related pain behaviors in horses. This system employs a numeric score of 1 – 4 for interactive, dynamic behaviors, and a 1 – 10 scale for observed, static behaviors in both acutely and chronically lame horses. Dynamic behavior descriptors are based on the Obel laminitis pain scale, while static behavior descriptors are modified from several sources, including the Glasgow CPS. While not yet experimentally validated, this stall-side scoring system was used to follow the response to therapy in horses suffering both acute and chronic lameness. Serial assessments by multiple scorers were made based on inclusive yet narrow behavior descriptors to reduce subjectivity and allow for identification of trends. The time necessary to complete a score was < 3 minutes in most cases. Scoring may be done as part of a running medical record, or on a separate sheet specifically designated for serial pain scoring.

This simple pain scoring method can be adapted to other pain states by modification of behavior descriptors in the dynamic and static categories. With modest training, consistent scoring between individuals can be expected. Analgesic efficacy can be monitored in both hospital and field conditions prompting quick, effective intervention for many pain states in horses.

**STATUS OF THE TOLL-LIKE RECEPTOR SIGNALING PATHWAY IN HORSES WITH LAMINITIS.**

**AM Stokes, ML Keowen, M McGeachy, K Carlisle, F Garza.**
**Equine Health Studies Program, Veterinary Clinical Sciences, Louisiana State University School of Veterinary Medicine, Baton Rouge, LA.**

The link between primary diseases such as GI disruption (colitis) or metabolic disease (Equine Cushing's Disease) and the development of equine laminitis is lacking and discovery would allow targeted development of therapeutics. The toll-like receptor (TLR) signaling pathway is important in orchestrating innate immunity (i.e. inflammation and apoptosis) in response to many factors including bacterial components (i.e. endotoxin) as well as decreases in insulin receptor stimulation. Equine laminitis pathogenesis is thought to include an acute inflammatory response in the hoof and may be modulated by mediators released as part of the TLR signaling pathway. In the cases of GI disruption, portal blood flow may lead to strong stimulation of the hepatic TLR signaling pathway with the subsequent systemic release of inflammatory mediators (i.e. cyclooxygenase 2 (COX2), IL-1, and TNFalpha) known to play a role in laminitis. The role of metabolic diseases and their potential link with TLR signaling pathway mediators is unknown and deserves investigation.

Our objectives were to design equine primers/probes to evaluate key mediators of the TLR signaling pathway to determine its role in laminitis. Specifically we examined cellular membrane bound pathway activators, (TLR4, cluster of differentiation 14 (CD-14), TLR2), cytoplasm pathway regulator,(phosphatidylinositol-3-kinase regulatory subunit 1 (PIK3R1)), nuclear transcription regulator (inhibitor of NF-κB (INFKB)) and a down-stream mediator (cyclooxygenase 2 (COX2)).

Primers and probes were designed from cloned equine samples, predicted sequences, and multi-species compilations and were validated for horses. Liver tissue (normal horses [n=6], laminitic horses [n=15], and horses with clinical signs of metabolic disease [n=4]) were used in TaqMan RT-PCR.

CD14, TLR2, PIK3R1 and INFKB were significantly ($p<0.05$) upregulated in liver tissue in horses with laminitis and horses with clinical signs of metabolic disease (highly correlated with laminitis development). Upregulation of the TLR signaling pathway may play a key role in laminitis pathogenesis and more information about this pathway would provide possible therapeutic intervention.

# DISORDERS OF SODIUM & POTASSIUM BALANCE IN THE FOAL & ADULT HORSE

**Harold C. Schott II, DVM, PhD, DACVIM**
**East Lansing, MI**

## INTRODUCTION

Abnormal serum concentrations of sodium ($Na^+$) and potassium ($K^+$) can accompany a variety of disease processes in foals and adult horses. These clinicopathologic abnormalities can be attributed to disorders of both water balance and electrolyte balance but for the most part are found in patients with a total body deficit of water and electrolyte stores.

## WATER BALANCE

Appropriate water balance maintains plasma osmolality in a relatively narrow range and is achieved by matching daily water intake with water loss. Total body water (TBW) is about 65% of body weight, or about 325 L in a 500 kg horse. Water is provided from three sources: (1) drinking; (2) water in feed; and (3) metabolic water. Most water is consumed by drinking (about 85%), but feed and metabolic water provide about 5% and 10% of daily water, respectively. Water is lost by three routes: (1) in urine; (2) in feces; and (3) as insensible losses (evaporation) across the skin and respiratory tract. Investigations of water balance in horses have revealed a maintenance water requirement of 60 to 65 mL/kg per day or 27 to 30 L/day for a 500-kg horse.[1,2] Urinary and fecal water losses vary with diet and can range from 20% to 55% and 30% to 55%, respectively, of the total daily water loss.[3] The remaining (insensible) loss accounted for up to 15% to 40% of daily water loss, despite mild ambient conditions and the lack of observed sweating in most studies of water balance.

Water drinking and urine production are the mechanisms by which water balance is finely tuned; however, they are affected by age, environmental conditions, level of exercise, and diet. For example, neonatal foals daily consume milk in excess of 20% of their body weight.[4] This equates to water intake approaching 250 mL/kg per day. Next, equids on all-roughage diets drink more and have greater daily fecal water loss (owing to greater daily fecal volume) than animals fed a large amount of concentrate or complete pelleted diets.[3] Further, most water is drunk within the hour after feeding. There are two main stimuli for thirst: increased plasma osmolality and hypovolemia/hypotension.[5] The former, mediated through osmoreceptors in the hypothalamus that have a rather high threshold for activation (about 295 mOsm/kg in humans), is the most potent stimulus for thirst. Hemodynamic changes are mediated by both low- and high-pressure baroreceptors but are a less potent thirst stimulus. Both osmotic and hemodynamic stimuli can produce their dipsogenic effect, in part, by activating a local renin-angiotensin-aldosterone system within the central nervous system.

Renal water reabsorption is controlled primarily by the action of arginine vasopressin (antidiuretic hormone) on the collecting ducts. As for thirst, an increase in plasma osmolality and hypovolemia/hypotension are the stimuli for vasopressin release. Osmoreceptors for vasopressin release are also located in the hypothalamus, adjacent to the osmoreceptors mediating thirst. The threshold for vasopressin release in humans is significantly lower (270 to 285 mOsm/kg) than that for thirst. Thus, vasopressin release is the initial line of defense against a mild increase in plasma osmolality, whereas thirst and drinking are secondary responses to even greater increases in plasma osmolality.[5]

Although balance of daily water intake and output is critical for maintenance of homeostasis, it warrants mention that equidae tolerate water deprivation quite well and replace water deficits fairly rapidly when water becomes available. For example, after horses were deprived of water for 72 hours (which resulted in body weight loss in excess of 10%), the majority of the weight lost (90% of which was assumed to be water) was recovered within the first hour after being provided access to water.[6,7] An important reason for their ability to tolerate water deprivation appears to be a substantial intestinal reserve of water and electrolytes that can be called upon during periods of dehydration for the maintenance of plasma volume.[8] In contrast to water deprivation, horses that become dehydrated as a consequence of disease (colic or diarrhea) often do not drink. This can be attributed to the fact that these conditions produce near isosmotic fluid losses of both water and electrolytes in the form of gastric reflux or diarrhea. As a result, plasma osmolality does not increase and an osmotic thirst stimulus is not produced.

## ELECTROLYTE BALANCE

Intake and loss of electrolytes must also be appropriately matched to maintain body ion contents within relatively narrow ranges. This balance is most important for the exchangeable ions ($Na^+$, $K^+$, and $Cl^-$), as these have minimal tissue (skeletal) reserves that can be called on during times of need. An exception is the fluid and electrolyte reserve in the lumen of the gastrointestinal tract, which may be able to provide replacement of 10% or more of the body content of these electrolytes.[8] Electrolytes are not evenly distributed throughout TBW: more than 90% of exchangeable $Na^+$ and $Cl^-$ is contained within the extracellular fluid space (ECF, about one-third of TBW) while more than 98% of $K^+$ is contained within the intracellular fluid space (ICF, about two-thirds of TBW).

Electrolytes are absorbed from feed and are also lost by three routes: urine; feces; and sweat (insensible losses). Investigations of electrolyte balance have revealed that most horses that eat predominantly hay or pasture grass ingest excess $K^+$ and $Cl^-$. In contrast, $Na^+$ intake is variable and with some diets may be marginal.[1,2] The maintenance requirement for sodium is 0.4 to 0.8 mEq/kg per day or 200 to 400 mEq per day (6 to 12 g per day for a 500-kg horse); however, exercising horses, which can lose 500 to 1000 mEq of sodium per hour in sweat have greater dietary requirements to replace such losses.

Equine kidneys have a remarkable capacity to conserve $Na^+$ when dietary intake is low such that urine [$Na^+$] can approach 0 mEq/L. In contrast, urine [$Na^+$] can increase dramatically (to >100 mEq/L) with dietary salt supplementation or administration of polyionic fluids. Urine is also is the major route of loss of $K^+$ and $Cl^-$. Although dietary intake of potassium is usually excessive, equine kidneys do not appear to have a great capacity to conserve potassium during periods of food and water deprivation or with anorexia associated with disease.[6,7] Thus, urine [$K^+$] and total excretion can remain substantial in the face of decreased intake.

Consequently, with anorexia horses can develop significant total body potassium depletion due to ongoing urinary losses. Electrolyte balance is not as precisely or rapidly regulated as water balance to match intake with losses. For example, when salt is available *ad libitum*, horses typically consume more than their maintenance needs. The excess is eliminated by increasing urinary $Na^+$ and $Cl^-$ excretion.

**PREVALENCE OF ABNORMAL SODIUM AND POTASSIUM CONCENTRATIONS IN HOSPITALIZED HORSES**

Although abnormal serum [$Na^+$] and [$K^+$] values are well recognized in sick horses, there is little data on the prevalence of these electrolyte alterations. In horses presented to Michigan State University's Veterinary Teaching Hospital during the years 1997, 2000, 2003, and 2006, the prevalence of these electrolyte alterations are summarized below (**Table 1**). Because laboratory reference ranges varied somewhat as equipment changes were made over these years, absolute cutoff values for hypo- and hypernatremia and hypo- and hyperkalemia were selected on the basis of what most clinicians would consider significant. The number of horses in the table refers to the number of patients (both in- and outpatients) for which blood samples were submitted for serum chemistry analyses (generally about 20% of the hospital caseload). Hyponatremia was the most common abnormality detected with diarrhea (33%) and colic (25%) being the most common presenting complaints, as compared to primary renal disease being found in only 7% of cases with hyponatremia. Hyperkalemia, generally considered the most life-threatening electrolyte alteration, was the next most common abnormality detected in horses presented for colic (35%), diarrhea (11%), and various other problems (but only 4% with primary renal disease). Hypokalemia and hypernatremia were the least common abnormalities detected with diarrhea (38%) and colic (26%) being the most common complaints, respectively.

**Table 1.** Prevalence of abnormal electrolyte concentration in horses presented to Michigan State University in the years 1997, 2000, 2003, and 2006 (see text for further details).

| Year | number of horses | Na<130, % (range) | Na>145 | K<2.5 | K>5.0 |
|---|---|---|---|---|---|
| 1997 | 397 | 6.3% (107-129) | 3.5% (146-154) | 2.8% (0.8-2.4) | 5.0% (5.1-9.1) |
| 2000 | 423 | 7.6% (114-129) | 1.7% (146-152) | 1.4% (2.0-2.4) | 3.5% (5.1-7.0) |
| 2003 | 546 | 8.8% (114-129) | 1.1% (146-168) | 1.3% (1.3-2.4) | 1.8% (5.1-7.1) |
| 2006 | 380 | 14.5% (108-129) | 0.8% (146-149) | 2.6% (1.9-2.4) | 2.4% (5.1-7.6) |
| **all years** | **1746** | **9.3% (n=160)** | **1.8% (n=30)** | **2.0% (n=34)** | **3.2% (n=54)** |

**HYPONATREMIA**

Abnormal values for serum [$Na^+$] should be used to assess a patient's water balance. In other words, hyponatremia is indicative of an excess of water in body fluids.[5] Nevertheless, the majority of horses presenting with hyponatremia (usually patients with diarrhea or colic) are clinically dehydrated and have signs supporting a decrease in TBW and effective circulating volume (e.g., prolonged skin tent, poor distensibility of the jugular veins, "tucked-up" abdomen, tachycardia, poor perfusion). Thus, hyponatremia in these patients is usually a consequence of loss of both water and electrolytes (both $Na^+$ from ECF and $K^+$ from ICF), generally in the form of diarrhea or gastric reflux. When measured, however, [$Na^+$] of diarrhea or gastric reflux fluid is generally lower than that of plasma. Thus, for hyponatremia to develop, additional water must be added to body fluids. There are two sources of excess water: (1) drinking plain water as dehydration and thirst initial develop, and (2) enhanced renal water reabsorption due to action of vasopressin on collecting ducts. In addition, because body fluid losses are initially derived from ECF, a subsequent shift of $K^+$-rich/ $Na^+$-poor ICF to partially replace the ECF deficit further dilutes the remaining $Na^+$ in the ECF and further contributes to hyponatremia.

Hyponatremia in dehydrated patients typically resolves with rehydration therapy using isotonic to mildly hypertonic intravenous (IV) fluids, enteral fluids, and hypertonic oral electrolyte pastes. A return of appetite is also important for restoration of water and electrolyte balance. In fact, over the course of several days of fluid therapy, $Na^+$ replacement is often excessive despite ongoing fluid loss and increased urine $Na^+$ excretion. The end result of excessive $Na^+$ replacement is edema formation, especially when the primary disease process results in decreased plasma oncotic pressure and/or increased vascular permeability.

The main clinical concern about hyponatremia is the potential for development of neurological disease as a complication to the primary disease process.[9,10] Acute, severe hyponatremia (48 hours or less and serum [$Na^+$] <115 mEq/L) can lead to cerebral edema, especially when rehydration is accomplished by administration of fluids that are low in $Na^+$ (0.45% NaCl or 5% dextrose). Treatment of neurological deficits that may develop in affected patients involves judicious use of hypertonic saline to increase serum [$Na^+$] by 3-5 mEq/L over 1-2 hours and subsequent use of isotonic fluids with a [$Na^+$] >135 mEq/L. In horses, the risk of this complication is minimal as long as the loss of body fluid has been fairly acute and serum [$Na^+$] is >115 mEq/L. Further, horses are rarely treated initially with fluids that are low in $Na^+$.

In patients with longer-lasting (>48 hours) and more severe hyponatremia ([$Na^+$] <115 mEq/L), the approach to correcting serum [$Na^+$] must be more cautious because these patients may no longer have ongoing fluid losses and may be euhydrated. With chronic hyponatremia, brain cells will lose both electrolytes and organic osmoles to prevent cell swelling.[9,10] If serum [$Na^+$] is corrected too rapidly (>0.5 mEq/L per hour or >10 mEq/L per day), further dehydration of brain cells and disruption of the blood-brain-barrier can lead to a syndrome of cerebral and central pontine myelinosis (osmotic demyelination). Although osmotic demyelination has not been documented in horses, this latter situation would be of greatest risk in foals with renal disease and/or "inappropriate" vasopressin secretion in the face of continued nursing.[11,12]

## HYPERKALEMIA

Hyperkalemia is considered the most life-threatening of this group of electrolyte abnormalities due to effects on cardiac electrical activity.[13,14] Serum [$K^+$] can increase to dangerous values (>8 mEq/L) as a consequence of decreased urine $K^+$ excretion (renal failure or uroperitoneum) or tissue catabolism (e.g., rhabdomyolysis). It also warrants mention that moderate to severe hyperkalemia is a common consequence of strenuous exercise in horses that is attribute to K+ release from active skeletal muscle at a rate exceeding reuptake by Na/K/ATPase pumps.[15] Fortunately, concurrent acidemia and catecholamine release appear to be cardioprotective (antiarrhythmic) during and following exercise.[16] Of interest, primary renal disease was an uncommon cause of hyperkalemia in the cohort of horses reviewed supporting that tissue catabolism in horses with colic and diarrhea may be a more important source of $K^+$ release than previously recognized.

Treatment of moderate ([$K^+$] >6 mEq/L) to severe ([$K^+$] >8.0 mEq/L) hyperkalemia is considered a medical emergency.[14] The initial step is to provide $K^+$-free IV fluids supplemented with calcium gluconate. Increasing serum [$Ca^{++}$] will rapidly increase cardiac muscle threshold potential and decrease the risk of arrhythmia. Next, glucose can be administered to produce an endogenous insulin response leading to stimulation of Na/K/ATPase pumps and translocation of $K^+$ from ECF to ICF. In horses with uroperitoneum, it is be equally important to drain the $K^+$-rich fluid from the abdomen. Once diuresis has been induced and retained urine drained, hyperkalemia typically resolves over a few hours. In the face of anuric to oliguric acute renal failure, hyperkalemia may persist despite these treatments. Although the prognosis for these patients is poor, additional measures to counteract hyperkalemia can include insulin infusion and possibly peritoneal dialysis to assist with $K^+$ removal.

## HYPOKALEMIA

Hypokalemia, with total body depletion of $K^+$ stores, has not been well documented to cause serious clinical signs in horses. However, $K^+$ depletion is usually underestimated in horses that are anorectic or have ongoing fluid losses. Further, because most polyionic IV fluids have minimal $K^+$ (5 mEq/L), potassium depletion is generally exacerbated by the increased urine output (and $K^+$ excretion) produced by IV fluid therapy. In theory, $K^+$ depletion could contribute to muscle weakness and prolong conditions with compromised intestinal motility including ileus and enterocolitis. To counteract $K^+$ depletion, IV fluids should be supplemented with KCl (at least 50 mEq/h to a completely anorectic 500 kg horse) while hypertonic oral KCl pastes can be administered to horses with diarrhea.

## HYPERNATREMIA

Hypernatremia was the least common electrolyte abnormality detected in horses. As with hyponatremia, serum [$Na^+$] should again be considered an indication of water balance in horses with hypernatremia. In fact, hypernatremia is almost always an indicator of pure dehydration or a deficit in TBW that can be supported by concurrent increases in packed cell volume and $Cl^-$ and albumin concentrations. Treatment of hypernatremia consists of correcting problems leading to water deprivation (e.g., frozen or electrified water source) or anorexia (e.g., dysphagia or esophageal disorders) and providing fresh water or hypotonic enteral fluid therapy.

## CONCLUSION

Although problems with $Na^+$ and $K^+$ balance are not always reflected by abnormal serum concentrations of these electrolytes, detection of abnormal serum [$Na^+$] or [$K^+$] warrants further assessment of possible causes. Further, specific modification of fluid therapy plans is required to correct these abnormalities.

## REFERENCES

1. Tasker JB. **Cornell Vet** 1967;57:649. 2. Groenendyk S, et al. **Equine Vet J** 1988;20:189. 3. Cymbaluk NF. **Equine Pract** 1989;11(1):19. 4. Martin RG, et al. **Equine Vet J** 1992;24:295. 5. Robertson GL, et al. In Brenner BM, Rector FC, eds: **The Kidney**, ed 4, Vol 1. Philadelphia, WB Saunders, 1991, p 667. 6. Carlson GP, et al. **Am J Vet Res** 1979;40:982. 7. Rumbaugh GE, et al. **Am J Vet Res** 1982;43:735. 8. Meyer H, et al. **Proc Equine Nutr Physiol Symp** 1989;11:3. 9. Adrogué HJ, et al. **N Engl J Med** 2000;342(21):1581. 10. Vachharajani TJ, et al. **J Intens Care Med** 2003;18(1):3. 11. Wong DV, et al. **J Vet Emerg Crit Care** 2006;17(3):275. 12. Arroyo LG, et al. **Can Vet J** 2008;49:494. 13. Acker CG et al. **Arch Intern Med** 1998;158:917. 14. Weisberg LS, et al. **Crit Care Med** 2008;36(12):1. 15. Schott HC, et al. **Eq Vet J** 2002;Suppl 34:532. 16. Paterson DJ **J Appl Physiol** 1996;80(6):1853.

## DISORDERS OF THE EQUINE CALCIUM & MAGNESIUM METABOLISM

**Ramiro E. Toribio, DVM, MS, PhD, DACVIM**
**Columbus, OH**

### CALCIUM

Calcium has essential physiological functions that include muscle contraction, hormone secretion, enzyme activation, cell division, cell membrane stability, neuromuscular excitability, and blood coagulation. Calcium also participates in pathological processes such as cell necrosis, apoptosis, free radical production, cytokine release, protease activation and vasoconstriction. There is a fine line between the physiological and pathological roles of calcium, thus, maintaining extracellular calcium concentrations within a narrow limit is extremely important. To regulate extracellular calcium concentrations there is homeostatic system that includes 3 body systems: gastrointestinal tract, kidney, bone; 3 hormones: parathyroid hormone (PTH), calcitonin (CT), and 1,25-dihydroxyvitamin D (1,25-D); and a calcium-sensing receptor (CaR) in the chief cells of the parathyroid gland, C-cells of the thyroid gland, and renal tubular cells. Another endocrine factor that has minimal calcium-regulatory functions in the adult but is important in fetal calcium homeostasis is parathyroid hormone-related protein (PTHrP). In the extracellular compartment total calcium consists of calcium bound to protein, complexed to anions (phosphate, lactate, sulfate), and in an active, free, ionized form ($Ca^{2+}$). A decrease in plasma pH increases $Ca^{2+}$ while the opposite occurs with alkalosis. In horses ionized calcium represents ~55% of the total calcium. A decrease in $Ca^{2+}$ concentrations increase PTH secretion while an increase in $Ca^{2+}$ decreases PTH and increases CT secretion. $Ca^{2+}$ has minimal effect on PTHrP secretion. Disorders of calcium homeostasis include hypocalcemia and hypercalcemia.

#### Hypocalcemic Disorders

Conditions associated with hypocalcemia in horses include sepsis/endotoxemia, gastrointestinal disease, acute renal failure, lactation, intense exercise, alkalosis, rhabdomyolysis, retained placenta, cantharidin intoxication, and loop diuretics. The pathogenic mechanisms by which horses develop hypocalcemia include hypoparathyroidism, calcium sequestration in soft tissues, gastrointestinal and renal losses, and alkalosis. Clinical signs of hypocalcemia include muscle fasciculations, tremors, tetany, seizures, hyperexciability, ataxia, ileus, synchronous diaphragmatic flutter, dyspnea, cardiac arrhythmias, and sudden death.

#### Hypocalcemia in critically ill horses

Sepsis/endotoxemia is the most common cause of hypocalcemia in critically ill equine patients. Total and ionized hypocalcemia are frequent in horses with severe gastrointestinal disease. Hypocalcemia during sepsis increases serum PTH concentrations in most horses and foals, but not all. Animals that have low PTH concentrations are less likely to correct their hypocalcemia. In our experience, most horses with enterocolitis and intestinal strangulation have both ionized hypocalcemia and hypomagnesemia. A number of these animals have an inappropriately low PTH response to hypocalcemia. A similar situation occurs in septic foals. We have also shown that endotoxemia induces hypocalcemia and hypomagnesemia, and that interleukin (IL)-1 and IL-6 are the cytokines responsible for this process. In equine parathyroid cells, IL-1 and IL-6 decrease PTH mRNA expression, PTH secretion, and increase CaR mRNA expression. By increasing CaR mRNA, parathyroid chief cells become more sensitive to extracellular calcium, thus resetting the calcium set-point to a lower value, which leads to less PTH secretion and hypocalcemia. We consider hypocalcemia from sepsis to be a cytokine-induced hypoparathyroidism. Another mechanism by which severe illness can result in hypocalcemia is by shifting calcium to the intracellular compartment (sequestration) or by increasing gastrointestinal and renal loses. Tissue sequestration remains to be documented in the horse, and renal loses of calcium as a cause of hypocalcemia is unlikely as most critically ill horses have decreased urinary excretion of calcium. Based on our studies, increased CT concentrations as a cause of hypocalcemia in sick horses is unlikely.

#### Hypocalcemia in Foals

Hypocalcemia is also a frequent finding in critically ill foals. The pathogenic mechanisms for the development of hypocalcemia in neonates are likely to be similar to the adult, including parathyroid gland dysfunction from systemic inflammation. The neonate is also susceptible to hypocalcemia from decreased dietary calcium intake, in particular when sick.

#### Hypoparathyroidism

Hypoparathyroidism is a condition characterized by hypocalcemia, hyperphosphatemia, hypomagnesemia, and decreased serum intact PTH concentrations. **Primary hypoparathyroidism** results from decreased synthesis and secretion of PTH, while **secondary hypoparathyroidism** most commonly results from hypomagnesemia. These horses/foals present with clinical signs of hypocalcemia. The diagnosis in based on the determination of serum concentrations of $Ca^{2+}$, $Mg^{2+}$, intact PTH, and phosphorus. Secondary hypoparathyroidism presents in critically-ill horses and foals (cytokine-induced hypoparathyroidism, $Mg^{2+}$ depletion).

#### Exercise-induced hypocalcemia

Horses under intense physical activity can develop hypocalcemia from different mechanisms, including soft tissue calcium sequestration and metabolic alkalosis. It remains unclear how calcium is sequestered in soft tissues, in particular skeletal muscle. During exercise there is hyperventilation and excessive chloride loss in the sweat, both leading to alkalosis, ionized hypocalcemia and hypomagnesemia. Total calcium and magnesium in these animals are often normal.

#### Treatment of hypocalcemia

Rapid calcium therapy is more critical in horses that develop acute hypocalcemia. The decision to treat these horses should be based on the presence of hypocalcemia and not in the presence of clinical signs. Most horses with ionized hypocalcemia do not show signs or the signs are very subtle to be detected, and the lack of therapy may result in additional complications (ileus, retained placenta, diaphragmatic flutter). Horses with mild hypocalcemia often restore their normocalcemia without calcium administration; however, calcium treatment should be considered. An evolutionary advantage of horses compared to other domestic animals is their ability to eliminate large amounts of calcium in the urine. Standard formulas to calculate electrolyte deficits / replacement do not apply to calcium. The other important point is that most equine practices do not measure $Ca^{2+}$ but rather total calcium. The following formula can be used to calculate the calcium deficit:

$$\frac{(6.5 - Ca^{2+})(0.3)(BW)(10)}{Ca^{2+}\text{ratio}} = Ca^{2+}\text{deficit} \qquad \frac{(6.5 - 4.5)(0.3)(450\text{ kg})(10)}{0.45} = 6000\text{ mg}$$

where the difference between measured $Ca^{2+}$ and normal $Ca^{2+}$ (6.5 mg/dL) is multiplied by the extracellular fluid volume, body weight (kg), and a factor of 10 ($Ca^{2+}$ is expressed in dL, but body weight is in kg = L), and then divided by the $Ca^{2+}$ ratio ($Ca^{2+}$/total calcium). For a 450-kg horse with a serum $Ca^{2+}$ concentration of 4.5 mg/dL and a total calcium of 10 mg/dL, the estimated $Ca^{2+}$ deficit will be 6000 mg. This is a deficit of elemental calcium. Calcium gluconate or borogluconate are the salts of choice for parenteral administration. Calcium gluconate contains 9.3% of elemental calcium; in other words, every 100 ml of calcium gluconate 23% solution contains 2.14 g of elemental calcium or 21.4 mg/ml. The horse in this example would require approximately 300 ml of calcium gluconate 23% solution to replace the $Ca^{2+}$ deficit. $Ca^{2+}$ conversion factors: mg/dL=mmol/L×4; mmol/L=mg/dL×0.25. From a practical standpoint, calcium doses of 2 mg/kg/h are safe in horses. This represents approximately 50 ml of calcium gluconate 23% solution for a 500 kg horse in the first hour of treatment. We have used up to 8 mg/kg/h. In most horses with mild hypocalcemia we add 50 ml of calcium gluconate to each bag 5 L of LRS until calcium concentrations are in the normal range. In few cases 100 ml/5L of fluids might be required. Horses with chronic or recurrent hypocalcemia should receive oral supplementation of calcium salts such as calcium carbonate (limestone) or dicalcium phosphate. Affected animals can be supplemented with 100-300 grams/day of these salts. Another consideration with calcium therapy is concurrent hypomagnesemia. A number of horses with severe gastrointestinal disease also develop hypomagnesemia and the calcium status will not be restored until magnesium therapy is implemented.

**Hypercalcemic Disorders**

Hypercalcemic disorders are divided in two categories: 1) associated with increased PTH secretion, and 2) independent of PTH secretion. This distinction is clinically important in the diagnosis and treatment of hypercalcemia. Diagnostic tests to investigate hypercalcemia include measurement of intact PTH, PTHrP, $Ca^{2+}$, phosphorus, and vitamin D concentrations.

**Primary Hyperparathyroidism (PHPT)**

In PHPT, the parathyroid gland chief cells secrete large amounts of PTH in an autonomous way, and the parathyroid cells are refractory to the negative feedback of $Ca^{2+}$. High PTH concentrations increase renal $Ca^{2+}$ reabsorption, hyperphosphaturia, increased bone resorption, and increased $1,25(OH)_2D_3$ synthesis. The end result is loss of cortical bone and osteodystrophia fibrosa. Hypercalcemia, hypophosphatemia, phosphaturia, and increased intact PTH concentrations are laboratory findings reported in these horses. Additional tests to rule out other conditions associated with hypercalcemia include measurement of PTHrP and vitamin D metabolite concentrations.

**Secondary Hyperparathyroidism**

In secondary hyperparathyroidism, excessive secretion of PTH is the response of the parathyroid gland to hypocalcemia and/or hyperphosphatemia, and/or hypovitaminosis D, from renal dysfunction or from nutritional imbalances.

**Chronic Renal Failure**

Horses with CRF develop hypercalcemia that is associated with calcium retention from decreased GFR rather than from renal secondary hyperparathyroidism as occurs in other species.

**Nutritional Secondary Hyperparathyroidism (NSHPT)**

Horses fed diets low in calcium, high in phosphorus, or rich in oxalates may develop NSHPT (bran disease, big head, osteodystrophia fibrosa). Clinical signs of NSHPT in young horses include enlargement of facial bones, upper respiratory noise, and lameness. In older horses, facial enlargement is not evident. Both hypocalcemia and hyperphosphatemia increase PTH secretion. The urinary excretion of calcium is low while that of phosphorus is high.

**Hypervitaminosis D**

The administration of vitamin D or ingestion of plants containing $1,25(OH)_2D$-like compounds (*Solanum glaucophyllum, Solanum sodomaeum, Cestrum diurnum, Trisetum flavescens)* causes clinical signs of vitamin D intoxication. Hypervitaminosis D increases intestinal absorption of calcium and phosphorus. Hyperphosphatemia is the most consistent laboratory finding in affected horses. Vitamin D inhibits parathyroid cell function and proliferation, and PTH secretion. Serum calcium concentrations are normal or elevated, while PTH concentrations are normal or low. Radiographically, these horses have increased bone density, decreased size of the medullary cavities, and increased calcification of soft tissues.

**Humoral hypercalcemia of malignancy (HHM)**

This term refers to a paraneoplastic condition in which some malignancies secrete parathyroid hormone-related protein (PTHrP) that interacts with the PTH receptor. Horses with HHM have hypercalcemia, hypophosphatemia, increased serum concentrations of PTHrP, decreased serum concentrations of PTH, decreased urinary excretion of calcium and increased urinary excretion of phosphorus. HHM has been reported in horses with gastric, preputial, and vulvar squamous cell carcinomas, adrenocortical carcinoma, lymphosarcoma, multiple myeloma, and ameloblastoma.

**Treatment of hypercalcemia**

Most conditions associated with hypercalcemia in horses have a guarded prognosis. Horses can live for a long time with hypercalcemia and the prognosis often relates to the primary problem. Surgical resection and chemotherapy can be attempted with some tumors. For chronic renal failure diets low in calcium content are recommended. For horses with evidence of nutritional secondary hyperparathyroidism, increased calcium intake together with less phosphate or oxalate intake are recommended. Glucocorticoids have been suggested to treat hypervitaminosis D as they decrease gastrointestinal calcium absorption and renal reabsorption.

**MAGNESIUM**

Magnesium is an important cofactor for enzymatic activation, metabolism of carbohydrates, fats, and proteins, energy metabolism, cell membrane function, nerve function, muscle contraction, and cell proliferation. As $Ca^{2+}$, ionized magnesium ($Mg^{2+}$) is the active form of magnesium. Calcium regulation, including PTH release, PTH receptor activation, and renal vitamin D synthesis requires $Mg^{2+}$. Hypomagnesemia and $Mg^{2+}$ depletion can lead to hypocalcemia. Any cellular process associated with ATP requires $Mg^{2+}$. Thus, maintenance of the cell membrane potential, $K^+$ and phosphate transport, and carbohydrate/lipid metabolism are highly dependent on the magnesium status. Unlike calcium regulation, there is not a precise $Mg^{2+}$ homeostatic system and its concentrations are regulated by gastrointestinal absorption, renal reabsorption, and to a lesser extend by calcium-regulating hormones. Other hormones that affect $Mg^{2+}$ concentrations include insulin, vasopressin, ß-adrenergic agonists, and aldosterone. Insulin signal transduction is $Mg^{2+}$-dependent. $Mg^{2+}$ is an essential regulator of neuroexcitability by blocking signal transmission via inhibition of $Ca^{2+}$-dependent, presynaptic excitation-secretion coupling. There is evidence that in the central nervous system $Mg^{2+}$ modulates N-methyl-D-aspartate (NMDA) receptors, antagonizes the cytotoxic effects of intracellular calcium, and protects against free radical injury. Hypomagnesemia is the most common magnesium disorder in the horse.

**Hypomagnesemia**

$Mg^{2+}$ is a cofactor for the $Na^+/K^+$ ATPase that maintains the cell membrane potential. Magnesium depletion leads to loss of intracellular $K^+$, intracellular $Na^+$ accumulation, lowers the resting membrane potential, causing hyperexcitability, seizures, arrhythmias, muscle weakness, muscle fasciculations, synchronous diaphragmatic flutter, and cardiac arrhythmias. In critically horses hypomagnesemia is associated with hypokalemia, in particular in those with gastrointestinal disease, and foals and ponies receiving parenteral nutrition. Of interest, hypophosphatemia is also common finding in these animals. Conditions associated with hypomagnesemia in horses include sepsis, gastrointestinal intestinal disease, respiratory and metabolic alkalosis, acute renal failure, hyperlipemia, and parenteral nutrition.

**Magnesium and brain injury**

As $Mg^{2+}$ blocks $Ca^{2+}$-dependent excitation-secretion coupling, Mg depletion contributes to tetany by increasing acetylcholine release and delaying its degradation by acetylcholinesterases. Even though $Mg^{2+}$ infusions (CRI) are used in humans with brain and spinal injury (and more recently in horses and foals), the benefit of this treatment remains unclear. The main rationale is to decreases free radical injury, NMDA receptor activity (glutamate), and calcium-mediated cytotoxicity. NMDA receptor activation by glutamate allows intracellular $Ca^{2+}$ influx resulting in neuronal excitability and death, and this process is blocked by $Zn^{2+}$ and $Mg^{2+}$. In addition, $Mg^{2+}$ in smooth muscle prevents $Ca^{2+}$ entry, resulting in vasodilation. Under this premise, $Mg^{2+}$ has been used to treat foals with evidence ischemic encephalopathy and horses with acute brain and spinal cord injury.

**Hypermagnesemia**

Increased serum magnesium concentrations are rare in sick horses. Conditions associated with hypermagnesemia include chronic renal failure, cellular injury and necrosis (myonecrosis/rhabdomyolysis, hemolysis, rhabdomyolysis, carcinomatosis, severe sepsis), and administration of magnesium-containing products (MgO, $MgCl_2$, $MgCO_3$, $MgSO_4$). Signs of hypermagnesemia include sweating, hyperexcitability, muscle tremors, depression, recumbency, flaccid paralysis, tachycardia, and tachypnea.

**Assessment of magnesium status**

The easiest way to assess the Mg status is to determine serum total Mg and $Mg^{2+}$ concentrations. As for $Ca^{2+}$, serum $Mg^{2+}$ is a more reliable marker of magnesium activity. In animals with respiratory or metabolic alkalosis (prolonged exercise, proximal enteritis) serum $Ca^{2+}$ and $Mg^{2+}$ concentrations may be low from increased protein binding. Methods to assess total body Mg status include 24-hour urinary Mg excretion, % Mg retention tests, and red blood cell Mg content. These tests have been validated in horses but are time consuming.

**Treatment of hypomagnesemia**

$MgSO_4$ and $MgCl_2$ are the solutions of choice for intravenous administration. For oral treatment magnesium oxide (MgO), magnesium carbonate ($MgCO_3$) and magnesium sulfate (MgSO4; Epsom salt) can be used. For $MgSO_4$ (9.7% of $Mg^{2+}$), a dose of 100 mg/kg will provide 9.7 mg/kg of $Mg^{2+}$ while for $MgCl_2$ (25.5%) a dose of 100 mg/kg will provide 25.5 mg/kg of $Mg^{2+}$. This is critical as overdosing can be fatal. Excessive oral (nasogastric) supplementation with $MgSO_4$ can be laxative and induce depression. Recommended IV dose rates for $MgSO_4$ in adult horses are 25-150 mg/kg/IV/day diluted in 0.9% NaCl, dextrose or polyionic isotonic solutions. Constant rate infusions (CRI) of $MgSO_4$ at 100-150 mg/kg/day should meet the daily requirements of foals and adult horses. MgSO4 is also used to treat ventricular arrhythmias, in particular quinidine intoxication (torsades de pointes), and should be considered in horses with refractory ileus and synchronous diaphragmatic flutter. Doses of 50-100 mg/kg over 30 min are considered safe. This is equivalent to 25-50 g of $MgSO_4$ for an adult horse. In foals with ischemic encephalopathy, an initial $MgSO_4$ dose of 50 mg/kg in the first hour, followed by a CRI of 25 mg/kg/hr has been proposed. This regime can be continued for several days, adjusting the dose based on serum $Mg^{2+}$ concentrations. Oral supplementation with MgO, $MgCO_3$ or $MgSO_4$ should be considered in animals with chronic hypomagnesemia, malabsorption, renal disease, and exercise-associated hypomagnesemia. Recommended doses are 30-50 mg/kg/day for MgO; 60-80 mg/kg/day for $MgCO_3$; or 80-100 mg/kg/day for $MgSO_4$. These doses are safe for horses, in particular when compared to the cathartic doses of $MgSO_4$ (0.5-1.0 g/kg).

**Treatment of hypermagnesemia**

Administration of 0.9% NaCl, furosemide, parenteral calcium administration, and control of the primary problem.

**REFERENCES**

Toribio RE, et.al. **J Vet Intern Med** 2005;19:223-231; Toribio RE, et.al. **Am J Vet Res** 2007;68:543-554; Toribio RE, et.al. **Am J Vet Res** 2001;62:938-947; Hurcombe SDA, et.al. **J Vet Intern Med** 2009; Garcia-Lopez JM, et.al. **Am J Vet Res** 2001;62:7-12; Wilkins PA, **ACVIM 19th Forum**, 2001, 242-244; Toribio RE, et.al. **J Mol Endocrinol** 2003; 31:609-620; Johansson AM, et.al. **J Vet Intern Med** 2003;17(6):860-867; Stewart AJ, et.al. **Am J Vet Res** 2004;65:422-430.

# RESEARCH & CLINICAL EXPERTISE IN HEMODYNAMIC MONITORING OF CRITICALLY ILL FOALS

**Steeve Giguère, DVM, PhD, DACVIM**
**Gainesville, FL**

## INTRODUCTION

The goal of hemodynamic monitoring in critical care medicine is to ensure adequate oxygen delivery to tissues in an attempt to maintain normal organ function. Early recognition of metabolic or physiologic derangements will allow prompt medical intervention and may contribute to improving survival. Selection of monitoring aids should be based on reliability, availability, cost, and the value of information gained by their use. Once selected, these measurements should be obtained at regular intervals, because trends and temporal changes are more important than individual measurements at a given point in time. This text briefly reviews indications, usefulness, and limitations of common monitoring practices in critically ill foals.

## CARDIAC OUTPUT

Cardiac output (CO) is the best available parameter to assess overall cardiovascular function.[1] Measurement of CO, along with blood hemoglobin concentration and oxygen saturation of hemoglobin, also allows calculation of global tissue oxygen delivery and consumption. Accurate measurement of CO in critically ill foals would lead to a better detection of cardiovascular derangements and aid in optimal titration of fluid therapy as well as titration of inotropic or vasopressor agents. Because measurement of cardiac output is currently considered impractical for routine use in foals, arterial blood pressure is commonly used as an estimate of blood flow.[2] However, arterial pressure is a poor indicator of blood flow in situations when vascular impedance (resistance or compliance) is altered.[3] Systemic vascular resistance is often altered during sepsis, therefore limiting the usefulness of blood pressure as an indicator of blood flow in critically ill neonatal foals. Data from human medicine also indicates that it is impossible to accurately predict cardiac output from heart rate, blood pressure and physical examination findings. Noninvasive methods of CO measurement include thoracic electrical bioimpedance, pulse contour analysis, partial carbon dioxide ($CO_2$) rebreathing, and echocardiography.[4] Several methods of measuring CO have been investigated in foals.

### Thermodilution

The thermodilution method has been regarded by some as the gold standard for CO monitoring and is the most commonly used method in human critically ill patients.[4] Thermodilution has also been used to measure the CO of foals in a research setting.[5] However, the thermodilution method requires placement of a catheter in the pulmonary artery and, therefore, is not practical for routine clinical use in foals. The use of pulmonary artery catheters has been reported to increase mortality and predispose to pulmonary thromboembolism in people. Pulmonary artery catheterization has also been associated with traumatic endocardial lesions in the right heart and pulmonary artery of adult horses.[6] Complications associated with thermodilution have led to growing interest in the development and validation of less invasive means of hemodynamic monitoring.

### Lithium dilution

The lithium dilution method of CO measurement only requires administration of a small bolus of lithium chloride in a peripheral vein and measurement of lithium concentration by a sensor attached to a catheter placed in a peripheral artery. Cardiac output is calculated from the time-concentration curve of lithium in arterial blood. Lithium dilution cardiac output has been shown to compare closely with thermodilution in people, dogs, adult horses and foals.[5] However, the difficulty in maintaining an arterial catheter in non-anesthetized foals makes lithium dilution CO measurement less practical for monitoring over a prolonged period of time.

### Partial $CO_2$ rebreathing (NICO)

Partial $CO_2$ rebreathing is a simple and completely noninvasive approach to measure cardiac output. The technique only requires measurement of end-tidal $CO_2$ and hemoglobin saturation to derive CO using the Fick principle. The technique automatically provides a CO measurement every 3 min. However, the need for stable $CO_2$ elimination prevents the use of this method in awake, spontaneously breathing subjects. In addition, pulmonary pathology decreases the accuracy of the technique compared to values obtained in subjects with normal pulmonary function.[7] Ventilator settings are also important to assure the accuracy of partial $CO_2$ rebreathing. In humans, tidal volumes of 10 ml/kg result in good agreement between partial $CO_2$ rebreathing and thermodilution whereas lower tidal volumes in the order of 6 ml/kg resulted in underestimation of CO. This is probably due to higher baseline end-tidal $CO_2$ that interfere with the rebreathing cycle controlled by the computer software.[8] In a recent study in foals, a tidal volume of 10-15 mL/kg was used to maintain end-tidal $CO_2$ between 40 and 50 mmHg. Despite these optimal conditions, relative bias of the method varied significantly with the level of CO (Figure 1).[9] As a result, the technique should not be used interchangeably with the lithium dilution method. However, partial $CO_2$ rebreathing may be useful to follow trends in CO during general anesthesia in foals with normal respiratory function.

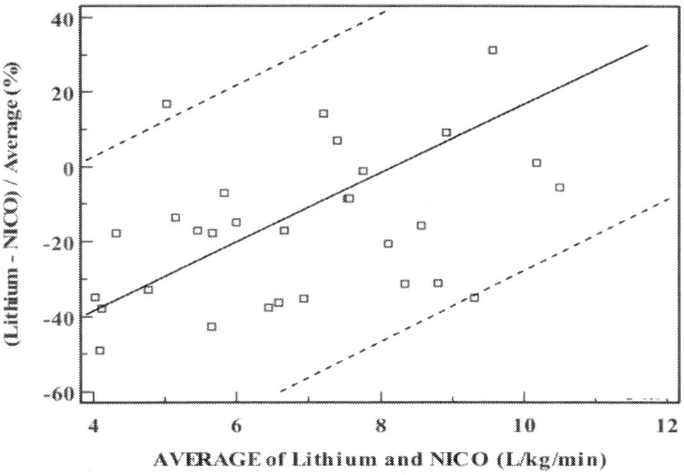


**Figure 1-** Bland-Altman analysis of CO values measured concurrently by lithium dilution and CO2 rebreathing (NICO) in anesthetized neonatal foals. Ten neonatal foals were anesthetized and CO was manipulated by varying the depth of anesthesia and infusion of dobutamine. Concurrent CO measurements were obtained by lithium dilution (reference method), and partial $CO_2$ rebreathing. The solid line represents the least squares mean relative bias whereas the dashed lines represent the upper and lower limits of agreements (1.96 x s.d.). Note that bias varies according to the level of CO. Adapted from Giguère et al., 2005.

**Transthoracic echocardiography**

Transthoracic echocardiography has been used to measure cardiac output in people, small animals and adult horses using various methods. Ultrasonography is widely available in equine neonatal intensive care units and is routinely use to examine the lungs and abdomen of critically ill foals. Most ultrasound units have packages allowing calculation of various cardiovascular parameters including CO. Therefore, transthoracic echocardiography would represent a convenient and noninvasive mean of measuring CO in neonatal foals. Echocardiography can be used to measure CO in a number of ways that can be divided into volumetric or Doppler based methods. The two main limitations of volumetric methods are resolution and geometry. In people, a 10% change in stroke volume corresponds to only a 0.7 mm change in ventricular radius while resolution of two dimensional echocardiography ranges from 0.3 to 1.5 mm. In addition, geometric assumptions rely on smooth circular or elliptical shapes and the ventricle is assumed to contract uniformly. As a result, these methods are less accurate in patients with segmental ischemic left ventricular involvement.[10] Doppler-based methods integrate the blood velocities over time to calculate the distance traveled by a column of blood during a cardiac cycle. By multiplying this distance by the cross-sectional area through which the column moves, a volume can be obtained. In adult horses, transthoracic Doppler echocardiography was found to agree well with thermodilution.[11]

Volumetric echocardiography using the Bullet method provides an accurate and noninvasive estimate of CO in neonatal foals.[9] A major advantage of echocardiography over other methods is that, in the hands of an appropriately trained clinician, it provides a vast amount of functional and morphological information in addition to CO measurement including indices of systolic and diastolic dysfunction, regional wall abnormalities, valve regurgitation, pericardial effusion, and chamber dilatation. Disadvantages of echocardiography are the capital cost associated with the initial purchase of the equipment and the fact that the accuracy of the technique may vary with the skill of the user.

**Other methods**

Other noninvasive methods of CO measurement include thoracic electrical impedance, ultrasound velocity dilution, and pulse contour analysis. Electrical impedance measures basal chest electrical impedance or resistance to flow. The change of impedance across the chest wall is related to the change in blood flow in the thoracic cavity. With some devices pulmonary edema or pleural effusion may interfere with impedance signal. Pulse contour analysis allows measurement and monitoring of stroke volume on a beat to beat basis based on the arterial pulse pressure waveform. The fluctuations of blood pressure around a mean value are caused by the stroke volume forced in arteries during systole. The magnitude of this change in pressure (pulse pressure) is a function of the magnitude of stroke volume. There are currently 2 commercially available methods allowing measurement of CO based on pulse contour analysis. Results of a study comparing these 2 methods in foals along with the results of a study validating the value of ultrasound velocity dilution to assess CO in newborn foals will be presented.

**BLOOD PRESSURE**

Arterial blood pressure monitoring is routine practice in equine neonatal intensive care units, allowing recognition of some cardiovascular derangements and titration of therapy with intravenous fluids, vasopressors, and inotropic agents. Mean arterial pressure (MAP), rather than systolic or diastolic pressure, is the true driving pressure for systemic blood flow and organ perfusion.[12] In addition, the MAP is less likely to change as the pressure waveform moves distally (from the aorta to a peripheral artery), nor is it altered by distortion generated by recording systems.[2] As a result, MAP is superior to systolic or diastolic pressure for arterial pressure monitoring. Arterial blood pressure can be measured by direct and indirect methods. Direct monitoring via cannulation of a peripheral artery is accurate and provides a continuous display of blood pressure results. However, the difficulty in maintaining an arterial catheter in non-anesthetized foals makes direct blood pressure measurement less practical for routine monitoring. Indirect methods of measuring blood pressure include manometric measurement as well as Doppler and oscillometric methods. Oscillometric techniques offer the advantage of providing systolic, diastolic and mean arterial pressures whereas other indirect methods do not provide MAP. Several automated oscillometric blood pressure devices are currently available for use in human or veterinary medicine. Most of these monitors provide a reasonable assessment of MAP in foals.[3,13] Sites of cuff placement that have been recommended for indirect blood pressure measurement in foals include the coccygeal, dorsal metatarsal or median arteries. Adequate sites for cuff placement depend on the type of monitor used.[3] In general, cuff placement over the coccygeal artery results in more accurate results. The metatarsal artery is a reasonable alternative.[3]

## OXYGEN SATURATION

Knowledge of hemoglobin concentrations in conjunction with oxygen tension and saturation allows for the calculation of oxygen content of blood. Pulse oximetry provides an adequate estimate of the arterial saturation in oxygen of hemoglobin in humans. However, the accuracy of measurement of oxygen saturation by pulse oximetry in foals varies according to the equipment used and the site of probe placement. Similarly, many methods of calculating oxygen saturation based on arterial oxygen tension are inaccurate in foals. The results of studies evaluating the accuracy of pulse oximetry in foals will be discussed.

## END TIDAL $CO_2$ MONITORING

Infrared $CO_2$ detectors can be used to measure end-tidal $CO_2$ in intubated or nonintubated patients. Because a decrease in cardiac output will decrease $PetCO_2$, capnometry can be used to monitor trends in cardiac output during volume resuscitation in hypovolemic shock and during cardiopulmonary resuscitation.

## CLINICAL APPLICATIONS

There are many clinical situations when accurate measurement of the cardiovascular parameters would be helpful. Specific clinical examples of how these data are used in patient management will be discussed.

## REFERENCES

[1]Corley KT, et al. **J Vet Intern Med** 2003;17:262. [2]Magdesian KG. **Vet Clin North Am Equine Pract** 2004;20:11. [3]Giguère S, et al. **J Vet Intern Med** 2005;19:571. [4]Tibby SM, et al. **Curr Opin Pediatr** 2002;14:303. [5]Corley KT, et al. **Equine Vet J** 2002;34:598. [6]Schlipf JW, et al. **Proc 5th Int Congress of Vet Anesthesia** 1994;71. [7]Maxwell RA, et al. **J Trauma** 2001;51:849. [8]Tachibana K, et al. **Anesthesiology** 2002;96:96. [9]Giguère S, et al. **J Vet Intern Med** 2005;19:737. [10]Teichholz LE, et al. **Am J Cardiol** 1976;37:7. [11]Blissitt KJ, et al. **Equine Vet J** 1997;29:18. [12]Hollenberg SM, et al. **Crit Care Med** 1999;27:639. [13]Nout YS, et al. **J Vet Emerg Crit Care** 2002;12:75.

**AORTIC VALVE PROLAPSE IN THE HORSE: EPIDEMIOLOGY AND THE EFFECTS OF TRAINING.**

**GD Hallowell and IM Bowen;**
**School of Veterinary Medicine and Science, University of Nottingham, UK.**

Aortic valve prolapse (AVP) in humans may be a precursor to aortic valve disease later in life[1]; mitral valve prolapse, which has been more extensively studied, is a precursor to mitral valve disease in humans[2] and dogs[3]. Up until now, the prevalence of AVP in the horse has only been reported in a small group of racehorses (20%; n=12)[4] and risk factors have not been investigated. The aims of this study were to evaluate prevalence of AVP in different types of horse and assess the effect of training and age on the development of this condition.

Echocardiography was performed on groups of horses (abattoir population (n=120; age 14±6.8 years), foals (n=29; age 10.7±4.3 weeks), older unexercised horses (n=19; age 18.8±4.7 years), army horses undergoing moderate intensity exercise (n=75; age 5.2±0.4 years), flat racehorses (n=25; 2 years) and National Hunt (NH) horses (n=32; age 4.5±0.8 years)) to assess prevalence. Examinations were repeated every 6 months to assess effects of age and training. AVP was only diagnosed if identified echocardiographically on both short and long axis views of the aortic valve. Effects of water deprivation and use of twitches were examined in a sub-group of abattoir horses.

Prevalence of AVP in the subgroups was as follows: foals-0.03%, flat-29%, NH-67%, army-13%, abattoir population-26% and older-5%. Water being withheld for 8 hours due to transport did not affect AVP prevalence. Application of a twitch induced AVP in 50% of horses. Training affected the prevalence of AVP. In the flat racehorses (n=15) AVP developed in 70% of horses that did not have AVP previously (prevalence increased from 29% to 50%) and in the army horses (n=47) AVP developed in 44% of horses (prevalence increased from 13 to 62%). Differences were identified using a McNemar's test (flat racehorses p=0.023 and army horses p<0.0001). A logistic regression model revealed that age and sex had no effect upon the prevalence of AVP; however racehorses were more likely (p=0.0001; RR=9.9) to have AVP compared to other breeds and horses that had never worked (p=0.001; RR=0.03), were in early training (p=0.001; RR=0.1) or moderate training (p=0.021 RR=0.2) were less likely to have AVP than those in full training.

In conclusion, different sub-groups of horse have different prevalence of AVP and this appears to be an effect of type and duration of training on valvular dynamics; AVP may predispose these horses to developing aortic regurgitation later in life due to abnormal blood flow and changes in haemodynamics. Application of a twitch appears to affect valvular function and should be a consideration if performing echocardiograms under these conditions.

1.Huang, H.C., S.H. Lo, and S.S. Lee. *J Formos Med Assoc*, 1999. **98**(10): p. 668-71. 2. Cheng, T.O. and J.B. Barlow. *Angiology*, 1989. **40**(2): p. 77-87. 3.Olsen, L.H., M. Fredholm, and H.D. Pedersen. *J Vet Intern Med*, 1999. **13**(5): p. 448-56. 4.Marr, C.M. and V.B. Reef. *Equine Vet J Suppl*, 1995(19): p. 56-62.

# NORMAL ULTRASONOGRAPHY OF THE HORSE ABDOMEN ILLUSTRATED BY 3D COMPUTERIZED ANIMATIONS

**Michelle H. Barton DVM, PhD, DACVIM**
**Athens, GA**

## INTRODUCTION

The recent introduction of more affordable and portable ultrasound equipment makes it possible to perform transabdominal ultrasonography in the field situation on equine patients. The unique feature of this review does not lie in the information provided, but in the *way* that the information will be provided.* During the presentation, standard ultrasound images of the healthy equine abdomen will be accompanied by three-dimensional computer-generated animations that were designed to orient the viewer to normal anatomy.

## GETTING STARTED

There are several different types of ultrasound transducers, though most imaging of the adult horse's abdomen is done with a 2 to 5 MHz curvilinear array transducer.[1] With the curvilinear transducer, ultrasound waves radiate from a curved footprint at the point of contact on the patient, and generate a curved "pie-shaped" image through the plane of projection into the horse. Ideally, prior to ultrasonography, the patient's hair should be clipped with a number 40 blade and the skin should be cleansed with isopropyl alcohol. Couplant gel should be liberally applied. If clipping the hair is not an option, soaking the hair with isopropyl alcohol will often suffice. Many horses will tolerate transabdominal ultrasonography without sedation. If however, sedation is needed, be mindful that alpha two agonists, such as xylazine and detomidine, will induce a transient state of ileus, and thus intestinal motility may be reduced and the luminal diameter of the small intestine may appear more dilated than in a patient that is not sedated.[2,3]

When performing ultrasonography, there are four key elements that will greatly facilitate the examination: being aware of the orientation of the probe with respect to the patient, using information on the depth of the viewing field, understanding how the image is generated by the ultrasound machine, and having knowledge of normal anatomy with respect to probe placement.

## ORIENTATION OF THE PROBE

Ultrasound transducers have a physical mark on them that provides orientation of the transducer's placement on the patient relative to the projected ultrasound image on the viewing screen.[1] Because much of abdominal ultrasonography is performed with the footprint of the transducer in an intercostal space, most imaging of the abdomen is performed with the transducer oriented in a slightly oblique transverse plane with the probe marker "up." In other words, the transducer marker is at the 11 to 12 o'clock position or toward the "dorsal" aspect of the patient. When the image is displayed on the viewing screen, most ultrasound machines display the position marker on the screen. This screen position marker is intended to orient the ultrasonographer to the displayed image. For example, if the orientation marker on the transducer is 11 to 12 o'clock relative to the patient to obtain a transverse plane image (slicing across the long axis of the body), and if the ultrasound machine normally displays the probe orientation mark in the upper left hand corner of displayed image, then the ultrasonographer knows immediately that the portion of the image on the left hand side of the screen represents the dorsal aspect of the captured image. Knowing the orientation of the transducer marker relative to the patient and the way the ultrasound machine normally displays its images greatly facilitates orientation to the structures that appear in the image on the screen. Be mindful of the fact that you can manually "flip" the image display on most ultrasound machines by pressing a left/right or top/bottom inversion key.

## DEPTH OF VIEW

In any ultrasound examination, it is important to be mindful of the depth of the field of view. The selection of the appropriate frequency for a transducer is the key to producing a high quality image that is most suitable for the depth of display. Recall that higher frequency probes provide sharp images; however, this clear resolution is compromised as the depth of the viewing field increases. If using a fixed single frequency probe, the 3.0 or 3.5 MHz curvilinear probe is the most suitable compromise to accommodate imaging of the majority of the abdominal viscera.[4] You can simply orient yourself to the depth of view by looking at the scale of the displayed image. You can be easily tricked into believing that something is missing from the field of view, only to realize that the depth setting is too shallow to identify the structure of interest. It is also easy to fool yourself into thinking that a lesion is enormous, only to realize that the depth of view is only a few centimeters and therefore magnified.[5]

## HOW THE IMAGE IS GENERATED

As ultrasound waves project through the body, they are reflected by tissue interfaces and sensed as 'echoes'. If adjacent tissues have the same acoustic impedance, no sound is reflected and sound waves penetrate into the deeper tissues. While denser tissues have greater acoustic impedance, it is the *interface* between adjacent tissues, or between tissues within the same organ, that determines how much of the sound wave is reflected back to the transducer.[1] The more sound reflected to the transducer, the "whiter" the image appears on the viewing screen. These tissue interfaces are called "echogenic" or "hyperechoic." In contrast, less dense tissues reflect less sound and are called "anechoic" or "hypoechoic" and appear blacker. Knowing that tissue interface difference is what is primarily responsible for reflecting sound back to the transducer, more sound waves should echo back to the transducer if two adjacent interfaces have markedly different acoustic impedances. The interface between soft tissues and gas is an excellent example of this concept. The soft tissue of the gastrointestinal walls has an acoustic impedance that is several thousand fold greater than that of the free gas inside the adjacent lumen.[1] Consequently, the image at this soft tissue to gas interface appears as a fuzzy hyperechoic border. Since most of the sound waves at this interface are reflected, and the free gas in the lumen has extremely low impedance, the rest of the lumen appears darker as sound is neither penetrating nor reflecting from the lumen. Remember, gas within a large viscus is actually one of the greatest limitations to gastrointestinal ultrasonography: the gas prevents visualization of deeper structures. While imaging a patient, keep in mind that fluid and heavier structures will fall to the dependent side and gas will float to the nondependent side and obstruct deeper views.

With a very high resolution linear array transducer, one may be able to see up to 5 layers to the gastrointestinal wall.[4] Though most typically, when using the standard approach with a 3 to 3.5 MHz transducer, and depending on the surrounding tissue and the contents of the lumen, only three or less layers are normally visible (hyperechoic serosa, hypoechoic muscularis to mucosa, hyperechoic interface with lumen).

## INTEGRATION OF KNOWLEDGE OF NORMAL ABDOMINAL ANATOMY

When scanning the equine abdomen it is helpful to use a systematic approach, scanning from top to bottom, rostral to caudal, left and right sides. Careful attention should be paid to the spatial relationship of the viscera, as this may be important in the distinction between normal and abnormal.[5] The walls of some sections of the gastrointestinal tract appear strikingly similar and without knowledge of where the transducer is placed on the abdomen, they may not be distinguishable.[6] Transabdominal ultrasonography not only provides structural information, but it can provide some idea of function i.e. motility. Recall that heavy sedation can cause transient ileus and mildly dilated small intestine.[3]

### Ultrasonographic Anatomy of the Left Side of the Abdomen

If imaging is started on the left rostral side of the abdomen, the **stomach** should be located deep to the spleen between the ninth to the thirteenth intercostal spaces at approximately the level of the shoulder. In this location, the only part of the stomach that normally can be seen is the wall of the greater curvature, which can be reliably identified as a curved line with proximity to the adjacent spleen and the gastrosplenic vein.[7] If the stomach extends beyond the 14th intercostal space in a horse that has not recently eaten, it would be an indication of gastric distension or displacement by other viscera. The stomach has the thickest wall of the gastrointestinal tract, measuring roughly 7 mm from the serosal to the mucosal/lumen interface. When the stomach is empty, the wall may be up to 1 cm thick. Since only the dorsal portion of the greater curvature can be seen and the lumen generally contains gas in this location, often the contents of the stomach are not visible and the curved wall appears hyperechoic. If gastric fluid is present ventrally, a distinct gas/fluid interface may be apparent in the lumen.

The size and location of the **spleen** is highly variable, though it should be identifiable immediately adjacent to the body wall, from the left ventral eight intercostal space to the paralumbar fossa. It may remain to the left of the midline, or extend slightly beyond the right of the ventral midline. The only measurement of the spleen that can be reliably obtained is its central thickness or depth, which usually is less than 15 cm.[5] In some horses, in the rostral ventral left abdomen, the most rostral aspect of the spleen can be seen either lateral or medial to the liver.[5] Normally the spleen's ultrasonographic architecture is homogenous with vessels rarely visible. The general echogenicity of the spleen should be greater than that of the liver and kidney.

The **left kidney** can be found between the sixteenth to seventeenth intercostal space and the first to third lumbar vertebra, medial or deep to the spleen, between the level of the tuber coxae and the tuber ischii.[8,9] Rarely, the left kidney may directly appose the left body wall.[5] Gas in the small colon or left colon or lung may preclude transabdominal viewing of the left kidney. The left kidney is 15 to 18 cm long, though this measurement is difficult to obtain in its long axis (i.e. dorsal plane that is parallel to the spine) because of interference from the ribs.[9] The height (11 to 15 cm in slightly oblique transverse plane) and thickness or depth (5 to 6 cm) measurements are more reliably obtained. The corticomedullary junction should be distinct, with the cortex approximately 1 cm thick. The renal cortex is more echogenic than the adjacent medulla, except in areas of the medulla where interlobar vessels course centrally to form the renal pyramids, which are most readily visible in the middle regions of the kidney, as compared to the poles.[9] Adjacent to the renal pyramids, centrally located at the rostal and caudad regions of the kidney are the terminal recesses wherein urine is collected and carried to the pelvis. In these areas, the renal pyramids appear as distinct hypochoic "circles," converging on the hyperchoic and indistinctly parallel lines of the terminal recesses. The walls of the renal pelvis are best imaged in the hilus and also appear as parallel to diverging hyperechoic lines that are often accentuated by the presence of fat in the renal pelvis.[9] The renal artery and vein can sometimes be identified medial to the kidney at the hilus in transverse planes. The normal left ureter cannot be imaged.

The **left colon** is located ventromedial to the spleen. The left ventral colon is sacculated and has "sluggish" motility. The wall of the colon should measure less than 4 mm. The left dorsal colon is not sacculated and may be located dorsal, lateral, medial, or even ventral to the left ventral colon. Gas in the left ventral colon may preclude identification of the left dorsal colon when it lies medial or dorsal to the left ventral colon. Gas in the colons typically generates a hyperechoic appearing wall with an indistinct luminal border and intraluminal acoustic shadowing that precludes identification of the contents and the medial walls.

The **small colon** is located in the left paralumbar fossa medial or ventral to the spleen. With its small diameter, sacculations, and packed serpentine loops that suspend from the dorsal mesocolon, often only small sections of the surface of loops are visible ultrasonographically as short sharply curving hyperechoic lines. Like the left colon, the motility is slow and luminal gas typically prevents visualization of the contents and the distal walls.

The **small intestine** is hard to visualize in normal horses unless a peristaltic wave generates transient expansion of the lumen from movement of fluid contents. The medial location of the ileum precludes distinct identification. The jejunum most reliably is found in the left inguinal area, medial to spleen and the left ventral colon. The small intestine has the most visible motility of any part of the gastrointestinal tract, with peristaltic waves producing rhythms contractions. The fluid contents of the lumen enables distinction of the wall thickness (2 to 4 mm) and visualization of the distal wall in either its long or short axis. In the normal horse, luminal diameters rarely exceed 3 cm and should be seen contracting down to obscurity. Fasting and sedation with alpha-two agonists will individually and additively decrease motility of the small intestine.[2,3]

**Ultrasonographic Anatomy of the Right Side of the Abdomen**

Three anatomic structures that provide characteristic proximity that can be identified in the right rostral abdomen at the level of the shoulder are the liver, the descending duodenum, and the right dorsal colon. The liver can be located from the sixth to the fourteenth intercostal spaces between the diaphragm and the right dorsal colon. Air in the lungs dorsally often interferes with hepatic imaging. Only a small portion of the right side of the **liver** can be imaged and estimates of size rely on its expanse across intercostal spaces.[5] It is unusual for the liver to be seen beyond the fifteenth intercostal space or in the same transverse plane as the right kidney, except at the most rostral aspect of the kidney.[4] The ventral edges of the normal liver should be distinctly sharp. Like the spleen, the architecture of the liver is relatively homogenous, though more vessels are visible in the liver and the general echogenicity of the liver is less than the spleen. Portal veins have more connective tissue in their walls and thus have more echogenic walls than the hepatic veins.[5] Often short segments of smaller portal veins appear as hyperechoic parallel lines. In some smaller horses, the portal vein can be seen entering the hilus deep on the medial side of the image. The common bile duct and its branches within the liver are not normally visible. [4,5]

The position of the **duodenum** is fixed by its suspending mesoduodenum. It can reliably be found descending the right middle abdomen at about the level of the shoulder and is located between the liver and the right dorsal colon where it can be imaged transversely in short axis. Like the jejunum, patience must be exercised when looking for the descending duodenum as one must wait for a peristaltic contraction to deliver fluid through the lumen. Otherwise, the duodenum appears flattened. It would be unusual for the entire duodenal diameter to exceed approximately 4 cm in normal horses during peristaltic propulsion of ingesta.[10] The duodenum contracts between 1 to 4 times per minute in fed horses,[10] though contractions will be reduced in anorexic, starved or heavily sedated horses. The duodenum can be followed to the level of the ventral right kidney, wherein it crosses medially into the abdomen and is no longer distinguishable. The wall of the duodenum is less than 4 mm in thickness.

The **right dorsal colon** has no sacculations and consistently appears as a hyperechoic curved line adjacent to the liver. Once the right dorsal colon is located, if one slides the transducer ventrally, often the junction between the right dorsal and right ventral colons is identifiable. The right ventral colon has sacculations. Like the left colons, the right colons' wall thickness should be less than 4 mm, motility is sluggish, and the contents and far walls are normally obscured by the presence of luminal gas.

The **cecum** extends from the right paralumbar fossa to the ventral midline. It is sacculated and motility is usually more apparent than in the colons. Wall thickness is less than 4 mm and gas in the lumen precludes imaging the contents and far wall.

Gas in the cecum, right dorsal colon, or lung sometimes obscures visualization of the **right kidney** which can normally be found in the rostral right paralumbar fossa to the sixteenth intercostal space.[8,9] The height (slightly oblique transverse plane) of the right kidney is 13 to 18 cm; its length (dorsal plane) and thickness are 13 to 15 cm and 5 cm, respectively.[9] Normally it is difficult to image the ureter, though the proximal right ureter can sometimes be seen as a hyperechoic circular structure in the hilus. The right kidney architecture is similar to that described for the left kidney.

The **urinary bladder**, nongravid uterus, and ovaries are best imaged in the adult horse transrectally. When full, the urinary bladder may be found ventrally at the caudal most aspect of the abdomen near the pelvic brim. Owing to the presence of mucous and calcium, urine in the adult horse often appears very echogenic.

**REFERENCES**

[1.] Barton MH. **Equine Compendium**. Submitted 2009. [2.] Freeman S et al. **Equine Vet J** 33:570, 2001. [3.] Mitchell C et al. **Canadian Vet J** 46:134, 2005. [4.] Reef VB in **Equine Diagnostic Ultrasound**, Philadelphia, PA, 1998, WB Saunders. [5.] Rantanen N. **Vet Clin of North Amer** 2:67, 1986. [6.] Hendrickson E et al. **Canadian Vet J** 48:289, 2007. [7.] Cannon J et al. **Annual AAEP Convention** 1995, page 39. [8.] Kiper M et al. **Comp Contin Educ Pract Vet** 12:993, 1990. [9.] Hoffman K et al. **Am J Vet Res** 56:1403, 1995. [10.] Kirkberger R et al. **Vet Rad** 36:50, 1995.

*This presentation will use 3 dimensional computer-generated animations from the Glass Horse, Equine Colic CD; The Glass Horse Project, LLC and the University of Georgia Research Foundation, Inc., Copyright 2007. As one of the contributing content experts for this version of the Glass Horse, Dr. Barton was granted permissions from Glass Horse Project to use the animations for this presentation at ACVIM.

# ULTRASONOGRAPHIC ABNORMALITIES OF EQUINE INTESTINAL DISEASES ILLUSTRATED WITH 3D COMPUTERIZED ANIMATIONS

**Michelle H. Barton DVM, PhD, DACVIM**
**Athens, GA**

## INTRODUCTION

Now that normal ultrasonography of the abdomen has been reviewed, let's put it into practice. There are many indications for performing transabdominal ultrasonography in the horse, including acute or chronic abdominal pain, weight loss, inappetance, and fever of unknown origin, to name a few. You may be using transabdominal ultrasonography because there are specific clinical signs or physical findings that are directing your attention to a particular organ or area (i.e. stomach, liver, or kidney) that you would like more information on, or it may be that the signs are nonspecific, and the entire abdomen is being scanned. Reviewing all abnormalities of the equine abdomen is beyond the scope of this review. However, since "colic" is a frequent clinical complaint of the horse, this review will focus on use of transabdominal ultrasonography as a guide to the diagnosis of equine abdominal pain. It is important to note, the transabdominal ultrasonography only occasionally leads to a definitive cause for the painful equine abdomen. In other words, transabdominal ultrasonography of the equine abdomen is an ancillary diagnostic aide that must always be interpreted in conjunction with the signalment, history, clinical signs, and other diagnostic findings. The unique feature of this review does not lie in the information provided, but in the *way* that the information is provided. During the presentation, the conditions discussed below will be accompanied by three-dimensional computer-generated animations that were designed to orient the viewer to the alterations in ultrasonographic anatomy created by the disease state.*

## THE STOMACH

In general, the ultrasonographic view of the stomach is limited, and thus diagnostically, the information gained is also somewhat limiting. **Gastric distension,** as evidenced by gas or excessive fluid reflux when a nasogastric tube is passed, is a sign of delayed gastric emptying or proximal obstruction or inflammation. Gastric distension can be verified by transabdominal ultrasonography simply by the size or position of the stomach.[1] Normally, the stomach wall does not extend beyond the 13$^{th}$ intercostal space. If the stomach is distended with fluid, in addition to the stomach wall being visualized beyond the 13$^{th}$ rib, a gas/fluid interface may be present. **Gastric impactions** are very rare and may also result in gastric distension, depending on the size of the ingesta bolus. However, because gas is usually also present in the stomach when a gastric impaction is present, the impaction per se may be obstructed by gas. Because **gastric ulceration** in horses commonly occurs along the margo plicatus, especially at the lesser curvature of the stomach, an area difficult to capture ultrasonographically, gastric ulceration is difficult to document by ultrasonography in the mature horse. However, thickening, edema (distinct hypoechoic layer in the wall), or "pitting" of the wall are findings suggestive of gastric ulceration.[2] In general, gastric impactions and ulceration are less reliably detected on transabdominal ultrasonography and 3-meter endoscopy is recommended to secure these diagnoses.

## THE SMALL INTESTINE

There are several differential diagnoses for thickened wall or distended small intestine. Ultrasonographically, thickened or distended small intestine often looks the same, despite the etiology. However, using a combination of clinical signs and some distinguishing ultrasonographic features can be helpful in securing a diagnosis, especially if the affected portion of small intestine cannot be felt by transrectal palpation. The wall of the small intestine can be **thicker** than normal (i.e. > 4 mm) from inflammation, edema, hypertrophy, or neoplasia. Some causes of inflammation and edema are acute in nature, i.e. **"proximal or anterior enteritis,"** and some causes are chronic or insidious in nature, such as the **inflammatory bowel syndromes (i.e. granulomatous enteritis, eosinophilic enteritis)** or idiopathic intestinal hypertrophy. In addition to the clinical signs of proximal enteritis (low grade fever, gastric reflux, inflammatory leukogram, pain followed by depression after gastric decompression), the mucosa wall of the small intestine may appear corrugated or wavy,[2] like the edge of a lasagna noodle. The proximal portions of the small intestine, especially the duodenum) and the stomach will be distended symmetrically with fluid. The motility of these loops may vary from low to apparently no movement, to hypermotile. If hyperechoic gas sounds are seen within the wall of the small intestine, it is indicative of either necrosis or anaerobic infection (clostridium) and is a poor prognostic sign. If the signs of abdominal pain are more insidious and/or weight loss is a component, symmetrically and diffusely or regionally thickened small intestine may be a sign of infiltrative or hypertrophic disease, such as the inflammatory bowel diseases, neoplasia, especially lymphosarcoma, or idiopathic ileal hypertrophy.[3] In these circumstances, the thickened wall often appears uniformly hyperechoic with a narrow lumen, which when viewed across the small axis, appears like a "bagel." Cases of inflammatory bowel disease may also have hypoproteinemia. Definitive diagnosis of inflammatory bowel disease or idiopathic hypertrophy requires histopathologic evaluation.

**Obstructive or strangulating lesions** of the small intestine may also result in edema and inflammation, and thus thickening of the bowel wall. Here the combination of clinical signs (persistent pain, perhaps sanguineous peritoneal fluid) will be helpful signs. However, clearly distinguishing acute severe enteritis from strangulating obstructions can be difficult. With obstructive or strangulating lesions, the thickened small intestinal wall tends to be focal (i.e. at the local of the obstruction), with relatively uniformly thickened walls. The intestine proximal to the obstruction will be distended with either gas for fluid. With complete obstructions, the distended proximal loops will often be seen to have a "hair-pin" appearance, as distended loops of bowel stack upon each other. The motility should be poor to absent with obstructive lesions. Thus, with regionally, uniformly thickened small intestine, with other evidence of obstruction on ultrasonography (hypomotile, U-turn stacks of turgidly round and distended loops of small intestine), an obstruction should be the top consideration. In one study, focally thickened and distended small intestine with no motility was 100%

sensitive and specific for strangulation.[4] Also bear in mind that most obstructive and strangulating lesions of the equine small intestine involve the distal jejunum and ileum, thus if the affected segments are most readily visualized in the lower left flank or adjacent to the spleen and stomach on the left or next to the cecum, these locations are areas in which gastro-splenic, epiploic, and inguinal incarcerations are readily found. Rarely, with intraluminal obstructions (ileal impaction, ascarids in foals) the obstructing contents of the small intestine may be visualized in the lumen.

Finally, **intussusceptions** of the small intestine are most common in foals and young horses and have a characteristic "bulls-eye" or target lesion appearance when imaged across the short axis of the small intestine.[5] The intussuscepted intestine tends to fall toward the dependent portion of the abdomen. Like other obstructive diseases, the proximal small intestine will be distended, with reduced motility and hairpin turns, depending on the degree of obstruction and the patient may present with either chronic intermittent signs (incomplete obstruction) or acute intense signs (complete obstruction).

## THE CECUM

Most diseases affecting the cecum can be sufficiently identified by transrectal palpation. However, ultrasonography can be helpful in validating some of the diseases. With **typhylitis**, or acute inflammatory disease of the cecum, the wall will be thickened and hypermotile. Often gas normally present in the body or base of the cecum precludes visualization of its contents. With typhylitis, the cecum is sometimes distended with fluid that is readily seen swirling inside of the hypermotile cecum. The presence of fluid cecal contents also enables discrete visualization of cecal haustral folds in the lumen that sometimes appear as "finger-like" projections from the cecal wall. Like the stomach, gas contents or gas surrounding ingesta often interferes with the diagnosis of an impaction. The size of the cecum and loss of sacculations may be signals that the cecum is distended. Unlike the stomach, cecal distension or impaction could be readily identified by transrectal palpation. **Cecocecal or cecocolic intussusceptions** are often palpable on transrectal evaluation as indistinct large masses in the right or mid caudal abdomen. Like small intestinal intussusceptions, the characteristic **"target lesion"** of the cecal apex moving into the cecal body or if it continues, the cecal wall in the right ventral colon can be seen. In long standing intussusceptions, the walls are very thick and may appear abscessed or "mass-like."

## THE LARGE INTESTINE

Like the cecum, many disorders of the large colon may be palpable by rectal examination, and thus transabdominal ultrasonography is an ancillary tool. However, in areas in which the colon is not reachable rectally or with some more obscure diseases of the colon, ultrasonography may be the essential tool that secures the diagnosis. For example, sometimes a horse presenting in the early stages of colitis, before diarrhea develops, is quite colicky and little is palpably abnormal on a rectal examination. On ultrasonography, the colon wall will be thickened and edematous in some cases of **colitis**. Motility will be variable. If NSAID toxicity is suspected, the right dorsal colon wall will be particularly affected.[6]

Displacements and impactions of the colon are mostly palpable per rectum. If the **left colon is dorsally displaced** and entrapped in the nephrosplenic ligament, gas within bowel will be seen dorsal to spleen, between spleen and left kidney, which usually obliterates the left kidney from view or lateral to the spleen.[7] With a **right dorsal colon displacement,** ultrasonography is rarely helpful as a sole diagnostic tool. Use of ventral transabdominal ultrasonography was helpful in the diagnosis of 75% of horses with colon volvulus when the colon wall reached at least 9 mm thickness.[8] As with any colon displacement or torsion, the finding of colon wall thickening (edema) warrants a guarded prognosis. In some cases of colon torsion, nonsacculated colon replaces sacculated colon in the left ventral abdomen.[9] With a right dorsal displacement or colon torsion, nonsacculated (left dorsal colon) may be visualized in the right abdomen, or distended vessels in colonic mesentery with reduced blood flow may be seen. Gas within the colon often precludes accurate ultrasonographic diagnosis of impaction. Occasionally, an **impaction** can be suspected by the presence of a hyperechoic border at the mucosal surface that casts an acoustic shadow. Heavy impactions may also flatten sacculations. **Sand impactions** are best seen in the ventral abdomen and often compress the mucosal surface, thus sacculations are lost. Sand and gravel may cast small acoustic shadows. Due to their location in the transverse and small colon, enteroliths can rarely be imaged by transabdominal ultrasonography.

## THE SMALL COLON

The contents of the small colon are often hard to image transabdominally in full sized horses. The same general principles for impactions and colitis discussed for the large colon also apply to the small colon. Transrectal imaging may facilitate visualization of distal sections of the small colon or the urinary bladder can be helpful as a sonographic window to the small colon with a transabdominal technique.

## THE PERITONEAL CAVITY

Peritonitis should be suspected if there is increased volume or echogenicity of peritoneal fluid. In some cases, free floating or adherent fibrin or fibrous tags may be visible. An abdominocentesis would be an integral set to further characterize peritoneal disease.

## INTRA-ABDOMINAL MASSES

Abscesses and neoplasia would be two differential diagnoses for intra-abdominal masses and they can be just about anywhere in the peritoneal cavity, in or attached to bowel or other intra-abdominal viscera (i.e. spleen, kidney, liver). The basic concept is that there are soft tissue densities of variable shapes and acoustic properties that distort normal sonographic architecture. Although a definitive etiology of an intra-abdominal mass is often not discernable ultrasonographically, a "classic" abscess appears loculated with mixed echogenicity and a cobweb-like appearance. A hematoma may appear similar to an abscess. An abscess or hematoma more commonly is isolated to a single location, whereas metastatic neoplasia should be considered if variable masses are present in multiple locations. Although the ultrasonographic appearance of a mass may not be distinguishing to etiology, ultrasonography may be a helpful guide to biopsy or aspiration of an intra-abdominal mass.

## MISCELLANEOUS CAUSES OF ABDOMINAL PAIN: OTHER VISCERA

Occasionally, acute renal or liver disease may be accompanied by enlargement of the organ and therefore pain. Reviewing normal organ size will be helpful in detection of organ enlargement. The caudal edge of the liver is normally distinctly sharp and may appear rounded with hepatomegaly. Obstructing nephroliths or choleliths may rarely be a cause of colic in the horse and appear as variable sized hyperechoic foci with an anechoic acoustic shadow in respective organs. Hydronephrosis, hydroureter and dilated biliary ducts may be present with nephroliths, ureteroliths and choleliths, respectively.[10] Most obstructive choleliths that cause colic in the horse are located in the common bile duct, which cannot be imaged ultrasonographically by the transabdominal approach in the horse.

## REFERENCES

[1.] Lores et al. **AJVR** 68(2): 153, 2007. [2.] Reef in **Equine Diagnostic Imaging,** Saunders, 1998. [3.] Dechant et al Vet Rad Ultrason 49(3): 300, 2008. [4.] Klohnen et al. **JAVMA** 209 (9): 1597, 1996. [5.] Bernard et al. **JAVMA** 194:395, 1989. [6.] Jones et al **JAVMA** 222: 1248, 2003. [7.] Santschi et al. **Vet Surg** 22(4):281, 1993. [8.] Pease et al. **Vet Rad and Ultrason** 45:220, 2004. [9] Abutarbush et al. **JAVMA** 228 (3):409, 2006. [10.] Reef et al. **JAVMA** 196:1836, 1990.

*This presentation will use 3 dimensional computer-generated animations from the Glass Horse, Equine Colic CD; The Glass Horse Project, LLC and the University of Georgia Research Foundation, Inc., Copyright 2007. As one of the contributing content experts for this version of the Glass Horse, Dr. Barton was granted permissions from Glass Horse Project to use the animations for this presentation at ACVIM.

# ANTIMICROBIAL ASSOCIATED DIARRHEA

**Bruce McGorum, PhD, MRCVS, DECEIM & R. Scott Pirie, PhD, MRCVS, DECEIM**
**Edinburgh, Scotland**

## INTRODUCTION

Antimicrobial associated diarrhoea (AAD) is the most commonly recognised adverse effect of antimicrobial treatment in horses, although its incidence is low given the frequency of antimicrobial administration. Symptoms vary from transient self-limiting diarrhoea to rapidly fatal toxic enterocolitis. AAD prolongs the duration of hospitalisation, increases diagnostic and therapeutic costs, and was associated with a lower case survival rate compared with other types of acute diarrhoea in one study (Cohen and Woods 1999). Virtually all antimicrobials have been implicated in AAD, but some pose a greater risk than others. Unfortunately our understanding of AAD is hampered by the following considerations;

### Lack of consistent case definition for AAD

There is no consistent case definition for human or equine AAD. A presumptive diagnosis of AAD is commonly made based on the occurrence of diarrhoea, which is otherwise unexplained, and which is temporally associated with antimicrobial therapy. Furthermore, there is a lack of consistency in the criteria used to indicate a temporal association between the onset of antimicrobial administration and diarrhoea. While a diagnosis of AAD is more likely to be made if diarrhoea occurs shortly after the onset of antimicrobial administration, some human studies include cases which develop diarrhoea up to 8 weeks after receiving antimicrobials (McFarland 2008).

### Factors other than antimicrobials may cause diarrhoea

In the clinical setting, it is rarely possible to establish a direct causal link between antimicrobial administration and the onset of diarrhoea. Many horses which develop diarrhoea while receiving antimicrobials have additional risk factors for developing enterocolitis including intercurrent disease, transportation, hospitalisation, surgery, alterations in food intake and diet, and other medical treatments.

### A variety of pathogens & pathogenic mechanisms may contribute to AAD

AAD is often subdivided into those cases which are associated with intestinal pathogens such as *Clostridium difficile*, *Clostridium perfringens* and salmonella, and those in which a recognised pathogen cannot be demonstrated. Furthermore, the aetiology of equine AAD appears to differ with geographical location, perhaps reflecting regional differences in the prevalence, toxigenicity and/or antimicrobial susceptibility of intestinal pathogens.

### Relative paucity of adequately designed equine studies

In contrast to human AAD, only a few well designed and adequately populated studies have investigated equine AAD. Thus, there is a paucity of objective data to support some clinicians' strongly held beliefs regarding the association of particular antimicrobials with AAD.

## EPIDEMIOLOGY & RISK FACTORS

Compared with the human field, less is known about the overall prevalence of equine AAD, and the proportionate risks associated with the use of individual antimicrobial agents in horses. Wilson et al (1996) reported the prevalence of diarrhoea after administration of trimethoprim-potentiated sulphonamides to be 3% (9/261 horses), which was not significantly different from that in the general hospital population. The prevalences of diarrhoea after oral administration of pivampicillin and trimethoprim/sulphadiazine were 3 and 7%, respectively (Ensink et al 1996). The prevalence of diarrhoea in foals given erythromycin (36%, 26/73), was higher than that for foals receiving other antimicrobials (4%, 3/70) (Stratton-Phelps et al 2000). Conversely, the proportion of mature horses with diarrhoea or colitis which had a temporal association with antimicrobial administration varies considerably among studies, being reported as 22% (Cohen *et al* 1999), 24% (Sutton *et al* 1999), 29% (Weese 2000), 52% (Herholtz *et al* 1999), 79% (Gustafsson 2004) and 94% (McGorum *et al* 1998).

Equine AAD may be associated with the initiation or cessation of antimicrobial therapy, but most commonly occurs in the first few days after the start of treatment, indicating that the disturbance of the intestinal microflora may occur relatively soon after drug administration.

## PATHOGENESIS OF AAD

The most important of the many potential mechanisms which contribute to AAD is antimicrobial-mediated disruption of the normal enteric microflora and ecosystem (Figure 1). The microflora has a key role in microbial digestion, producing a wide array of metabolites which can influence the survival and colonisation of bacteria within this ecosystem. The enteric flora also influences fluid and ion transport across the large intestine. Indeed as microbially derived volatile fatty acids are one of the major solutes absorbed from the equine large intestine, the ionic composition and water content of large intestinal contents is largely dependent on the rate of microbial digestion. Consequently, many of the clinically mild AAD cases may reflect functional disturbances of intestinal carbohydrate, volatile fatty acid and bile acid metabolism, which in turn lead to osmotic diarrhoea. The normal balanced enteric microflora, and in particular the anaerobes, also contribute to '*colonisation resistance*' which maintains the stability of the constituent populations, thereby inhibiting intestinal colonization by opportunistic and pathogenic bacteria. Many cases of antimicrobial-associated severe toxic enterocolitis in horses are likely attributable to intestinal colonisation with *C. difficile*, *C. perfringens* and salmonella. These pathogens release exotoxins which cause mucosal damage and inflammation and alter the absorptive and secretory functions of the large intestine. However, a putative causal organism is identified in only approximately 50% of equine AAD cases (Gustafsson 2002). Additional mechanisms potentially contributing to development of AAD include a direct effect of antimicrobials on toxin production by enteric pathogens, a direct intestinal prokinetic effect, and hypersensitivity and toxicity reactions.

The composition of the intestinal flora, and in particular the presence of potential pathogens such as *C. difficile, C. perfringens* and salmonella, their toxin profile(s) and their antimicrobial susceptibilities, are important determinants of a horse's susceptibility to develop AAD. The effect of an antimicrobial on the number of potential pathogens in the bowel is determined by the balance of its effect on the microflora that confers colonisation resistance and its direct effect on the potential pathogens. However, perhaps counter-intuitively, the likelihood of a particular antimicrobial inducing *C. difficile* or salmonella-associated disease is not predicted by its efficacy against those pathogens. Thus aminopenicillins can precipitate *C. difficile* associated AAD even though most *C. difficile* strains are susceptible to aminopenicillins *in vitro*. This is because the intestine is readily recolonised by *C. difficile* spores or salmonella which have not been eliminated by antimicrobial treatment, or which are ingested after cessation of antimicrobial treatment. As the large intestine is an important reservoir of potential pathogens such as *C. difficile*, many apparently hospital acquired infections are caused by overgrowth of pathogens that were present within the horse's intestine prior to hospitalisation, as a result of loss of colonisation resistance.

**Figure 1:** Schematic flow diagram highlighting the potential pathogenic mechanisms underlying equine AAD.

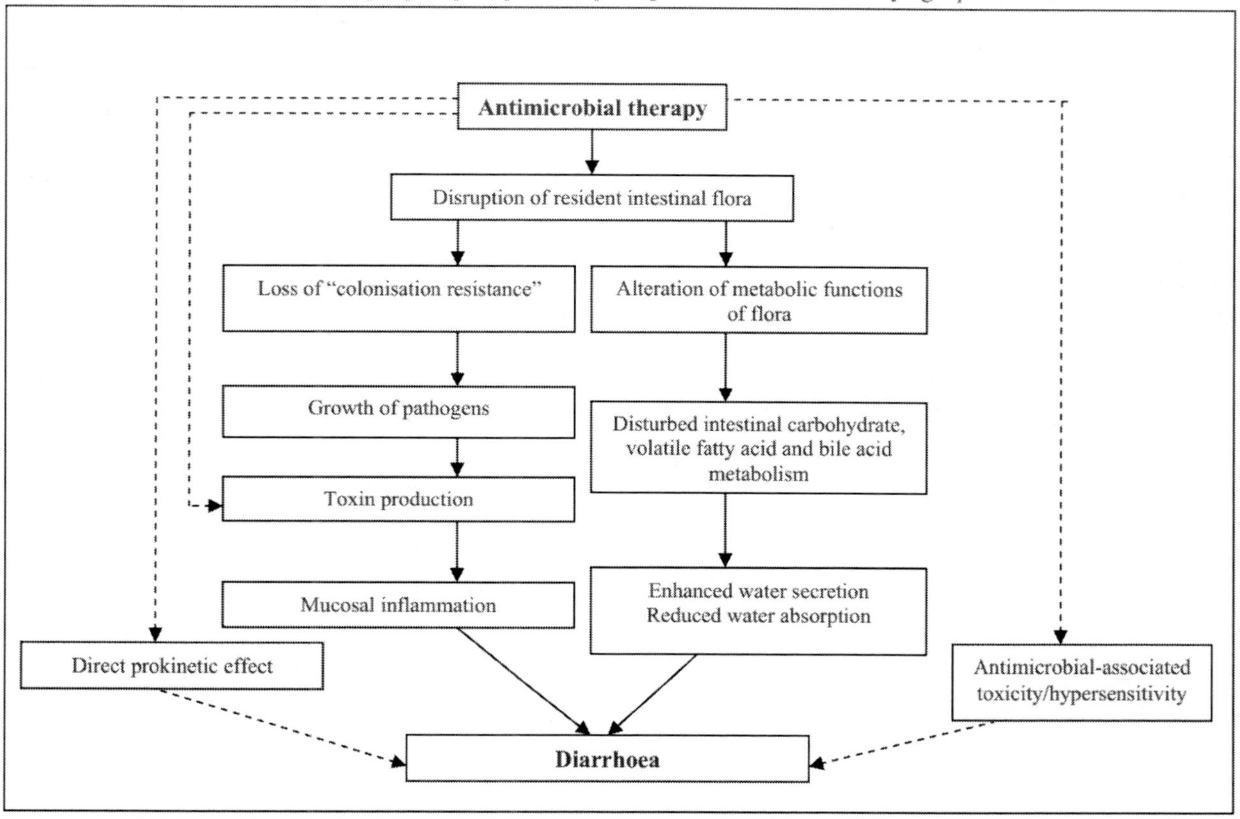


## WHICH ANTIMICROBIALS ARE ASSOCIATED WITH AAD IN THE HORSE?

Almost all oral and parenterally administered antimicrobials have been associated with AAD in both humans and horses, although some antimicrobials clearly pose a higher risk. The relative risks posed by particular antimicrobials however may change over time because of acquisition of antimicrobial resistance by enteric pathogens. Thus while the risk of inducing human *C. difficile*-associated AAD was previously lower for fluoroquinolones than for cephalosporins, as a consequence of emergence of high level fluoroquinolone resistance in some strains of *C. difficile,* fluoroquinolones are now a major risk factor for infection with hypervirulent strains of *C. difficile* (Monaghan *et al* 2008). Similarly, the relative risks of particular antimicrobials for *C. difficile*-associated AAD in horses may change with alterations in the use of antimicrobials.

Anecdotally, there appear to be geographical differences in the susceptibility of the local equine population to develop AAD after administration of a particular antimicrobial. Consequently knowledge of the antimicrobials which are associated with enterocolitis in a particular region is helpful when selecting antimicrobials. This may reflect regional differences in the composition of the enteric flora, the frequency of intestinal carriage of pathogens such as *C. difficile, C. perfringens* and salmonella, and the toxigenicity and antimicrobial spectrum of these pathogens. These factors in turn will be dependent in part on diet and management practices, which vary geographically.

In contrast to the human field, the relative risks of individual antimicrobials for equine AAD have not been evaluated systematically and consequently there is a paucity of objective data. Hollis and Wilkins (2008) considered that the antimicrobials commonly implicated in equine AAD included enrofloxacin, trimethoprim-sulphonamides, oxytetracycline, penicillin, ceftiofur, doxycycline, erythromycin, neomycin, lincomycin, clindamycin, moxifloxacin, florfenicol and tylosin. Magdesian *et al* (1997) described equine *C. difficile*-associated diarrhoea associated with trimethoprim-sulphamethoxazole, ampicillin, metronidazole, penicillin and erythromycin, especially when used in combination. Weese (2000) reported that trimethoprim-sulphonamides, penicillin G and ceftiofur sodium, alone or in combination, were most commonly implicated in a series of 40 cases of AAD. Many authors report that tetracyclines, lincomycin, erythromycin and clindamycin are associated with a high risk of inducing equine AAD. However, when comparing the relative risks of antimicrobials, it is essential to take into consideration the frequency of usage of the antimicrobials so that an apparent increased risk does not simply reflect increased use of that antimicrobial. Thus, for human *C. difficile*-associated AAD, clindamycin and vancomycin pose a greater proportionate risk than other antimicrobials, but cases are more frequently associated with ampicillin and cephalosporins because these drugs are used more frequently. Furthermore, to meaningfully compare the relative risks of individual antimicrobials in relation to equine AAD, patient selection bias must be avoided by comparing the incidences of AAD in populations of horses which do not differ with respect to other risk factors for diarrhoea, such as intercurrent disease, hospitalisation, dietary changes and other medical and surgical treatments. Alternatively, a large scale longitudinal case-controlled study with subsequent multivariate risk factor analyses may facilitate the identification of the relative risks associated with individual antimicrobials and with non-antimicrobial-associated factors. The oral presentation will provide a detailed review of the literature regarding the association between equine AAD and individual antimicrobials.

The potential of an antimicrobial agent to induce AAD is largely dependent on its effect on the enteric anaerobes, which in turn reflects its spectrum of antibacterial activity, and the concentration of active drug within the intestine. Thus, antimicrobials which have poor efficacy against anaerobes, such as trimethoprim/sulphonamides, fluoroquinolones and aminoglycosides, are less likely to disrupt the enteric microflora than lincosamides, macrolides and β-lactams which have efficacy against anaerobes. The intestine appears to be a relatively independent pharmacokinetic compartment, and the concentration of active drug within the intestinal contents is determined by the route of administration, the dose administered, the proportion of drug which is absorbed from the intestine, and excreted in saliva, bile or mucus, and the extent to which the drug is inactivated by the intestinal contents. Antimicrobials which are incompletely absorbed from the gastrointestinal tract, or are excreted in the bile predominantly in active forms, have been implicated in equine AAD. While orally administered antimicrobials would intuitively appear to yield higher intestinal concentrations of antimicrobial, most case studies have shown equal or increased association of AAD with parenterally administered antimicrobials (Weese 2000; Gustafsson 2002). Administration of antimicrobials after withholding food may reduce the risk of AAD, because the bioavailability of orally administered antimicrobials is reduced when they are administered with food, as a result of drug binding to ingesta and reduced gastrointestinal transit time.

## REFERENCES

[1.] Cohen ND et.al. **JAVMA** 1999;214:382. [2.] Ensink JM et.al. **Vet Rec** 1996;138:253. [3.] Gustafsson A **Equine Vet Educ** 2002;14:186. [4]Gustafsson A et.al. **Equine Vet J** 2004;36:180. [5.] Herholz C et.al. **J Clin Micro** 1999;37:358. [6.] Hollis AR et.al. **Equine vet Educ** 2008;20, electronic doi:10.2746/095777308X321107. [7] Magdesian KG et.al. **Rev Med Micro** 1997;8(Suppl 1):S46. [8.] McFarland LV **Future Micro** 2008;3:563. [9.] McGorum BC et.al. **Vet Rec** 1998;142:635. [10.] Monaghan T et.al. **Gut** 2008;57:850. Sutton GA et.al. In: Proceedings of the eighth international conference on equine infectious diseases, Dubai, 1998, Eds: U. Wernery, J.F. Wade, J.A. Mumford and O.-R. Kaaden, R & W Publications, Newmarket, pp191-195. Weese JS Ontario Veterinary College, University of Guelph, Canada. DVSc thesis, 2000. [11.] Wilson DA et.al. **JAVMA** 1996;10:258.

**EPIDEMIOLOGIC FEATURES & FECAL PATHOGENS IN HORSES WITH ANTIMICROBIAL ASSOICATED DIARRHEA: PRELIMINARY RESULTS.**

**BBarr[1], CClark[2], JDonecker[3].**
**1. Rood and Riddle Equine Hospital, Lexington, KY. 2. Peterson and Smith Equine Hospital, Ocala, FL. 3. Pfizer Animal Health, New York, NY.**

Acute diarrhea in the horse has been reported to be associated with the use of antimicrobials and is presumed to be due to secondary disruption of the normal gastrointestinal flora. This change in the flora can lead to proliferation of enteropathogens such as *Clostridium difficile*, Salmonella spp and *Clostridium perfringens*. Although all antimicrobials have the potential to alter gastrointestinal flora and cause diarrhea the ones most often incriminated are the macrolides and tetracyclines. In some cases the geographic incidence of antimicrobial associated diarrhea appears to differ markedly, which may be due to regional variation of the intestinal microflora dependent on different feeding practices, climate, or management.

The aim of this prospective study is to gather information on the types of antimicrobials administered and to identify the antimicrobials most often associated with acute diarrhea. Fecal samples are being collected and analyzed to identify types of pathogens associated with antimicrobial induced diarrhea. Patients included in the study will be any horse weanling age or older admitted for diarrhea, which had been treated with antimicrobials prior to admission for a condition other than diarrhea. Epidemiological data to be collected include signalment, clinical history, findings on initial physical examination, result of clinicopathologic testing, treatment, complications and outcome. Fecal samples are being tested for *Clostiridium difficile* toxin A and B, *Clostridium perfringens* enterotoxin and Salmonella spp. At the time of the writing of this abstract only a few cases have been identified. These horses had a history of treatment with enrofloxacin, gentamycin or doxycycline. All of these cases were tachycardic, dehydrated and had watery diarrhea at admission. Response to treatment was favorable. No enteropathogens were isolated from fecal samples. This is an on-going study with multiple participants from various geographical locations and more results will be presented at the forum.

**SALMONELLA SHEDDING IN HOSPITALIZED HORSES & THE SUBSEQUENT RISK OF DIARRHEA IN STABLE-MATES AFTER DISCHARGE.**

**AK Hartnack, DC Van Metre, PS Morley.**
**Animal Population Health Institute, Colorado State University, Fort Collins, CO.**

*Salmonella enterica* is the agent most commonly identified in association with outbreaks of nosocomial infections at veterinary teaching hospitals. Accordingly, large animal hospitals often use active surveillance for *Salmonella* to aid efforts for reducing the risk of nosocomial infections. Although this type of surveillance is used with increasing frequency, the value of this surveillance has only been evaluated subjectively. Further, advising owners regarding the long term consequences of shedding is difficult as it is not known whether these horses or their stable-mates are more likely to experience GI disease after discharge. The purpose of this study was to evaluate risk factors for *Salmonella* shedding among hospitalized horses and to determine if shedding was associated with differences in owner-reported gastrointestinal disease after horses returned to their home environment.

This study was designed as a retrospective cohort study nested within a cross-sectional study that included horses admitted to the CSU-VTH from 2002-2008. For the cohort evaluation of shedding consequences, *Salmonella*-positive horses had ≥1 positive fecal culture after admission, and unexposed horses had ≥3 negative cultures. A total of 11,937 fecal samples were cultured from 4,690 equine patients during the study period. Overall, 1.9% samples collected from 2.5% of patients were culture-positive. Subjective characterization of the severity of systemic illness was the factor most strongly associated with shedding while hospitalized. Eighty-two *Salmonella*-positive horses were alive at the time of discharge, had complete datasets, and were enrolled, and 3 unexposed horses were randomly selected for enrollment in the cohort study for each *Salmonella*-positive horses (total n=244), matching unexposed subjects to *Salmonella*-positive horses on month and year of admission. A scripted interview was developed to collect data regarding health of these patients, as well as the health of stable-mates. Owners of 222 (68%) horses were successfully contacted and agreed to participate in the study. Significant differences in the occurrence of colic after discharge were not associated with shedding status while hospitalized ($P$>0.30). Similarly, there were no differences in the reported occurrence of diarrhea among stable-mates of horses in the two exposure groups (OR=0.9, 95%CI=0.3-3.2, $P$=0.91).

These data suggest that *Salmonella* shedding by hospitalized horses is not generally associated with an increased risk of GI illness after discharge among patients or their stable-mates. The health risk would likely be increased when *Salmonella* strains were considered to be highly contagious or highly pathogenic, and consequences would be diminished when good biosecurity precautions were employed in the home environment.

# PATHOPHYSIOLOGY & METHODS FOR THERAPEUTIC INTERVENTIONS FOR TRAUMATIC CENTRAL NERVOUS SYSTEM INJURY

**Yvette S. Nout, DVM MS PhD DACVIM DACVECC**
**San Francisco, CA**

## INTRODUCTION

Trauma to the central nervous system (CNS) is the most common cause of neurologic disease in horses. A recent study from Europe reported CNS trauma to account for 22% of neurologic disorders (1), which is consistent with the findings from an older report out of Australia that reported CNS trauma to account for 24% of neurological cases (2). Both studies reported trauma to the spinal cord to be more prevalent than brain injury. Feige *et al* (1) reported a diagnosis of traumatic brain injury (TBI) in 5 out of 22 (23%) horses that were presented for traumatic neurologic disease, whereas 17 of the 22 (77%) were diagnosed with spinal cord injury (SCI). Tyler *et al* (2) reported 47 cases of TBI and cranial nerve disease in 107 (44%) horses examined, whereas 60 horses in this group had SCI (56%). One study examining head fractures in 21 horses demonstrated that 24% had neurologic deficits secondary to their injuries (3). Although mechanisms of CNS cell damage, to a certain degree, are similar after TBI and SCI, there are a few processes with regards to etiology and pathophysiology that are different. Clinical syndromes as a result of traumatic injury to the CNS can vary, but the most common are coma, vestibular disease, tetra- and paraparesis or –plegia, and cauda equina syndrome. Treatment for CNS injury is aimed towards reducing inflammation and swelling, halting secondary injury, and promoting regeneration and recovery of function. Prognosis depends primarily on severity of primary injury and on the extent and neuro-anatomic location of CNS damage.

## REVIEW PATHOPHYSIOLOGY

Although traumatic injury to the head is reported to be common in horses, subsequent TBI occurs only in 25 – 50% of cases (3,4). The remaining horses typically present for bone fractures of the skull (3-5). A recent study involving 34 horses with TBI showed that the most common cause (44%) of injury was rearing and falling backwards resulting in poll injury (6). Another type of head injury that may lead to CNS damage is fracture of the petrous temporal bone associated with temporohyoid osteoarthropathy (7,8). TBI may occur with or without fracture of the calvarium; however, some of the most severe injuries to the brain occur when the injury is contained within the closed calvarium (9). Trauma to the vertebral column is typically caused by incidents such as collision with an immovable object or falling down. Trauma to, or fractures of, the cervical vertebrae are the most common (1,2).

After head trauma, most severe damage generally takes place at the place of impact (coup), and/or opposite to the side of impact (contrecoup). Additionally, the brain is subjected to other forces after trauma, such as rotational and shock wave forces. TBI and SCI are a result of both direct, immediate mechanical disruption of CNS tissue (primary injury) and indirect, delayed (secondary) injury mechanisms. The primary damage is a result of the biomechanical effects of the injury and is characterized by immediate and often irreversible damage to neuronal cell bodies, dendritic arborizations, axons, glial cells, and brain vasculature. In TBI, this initial brain injury may be focal, multifocal, or diffuse (10). SCI secondary to trauma is a dynamic process of which the severity is related to the velocity, degree, and duration of the impact. These secondary injury processes do not necessarily coincide with the clinical picture, as pathologic changes may progress in severity for weeks to months, even in the face of clinical improvement.

For TBI, principles of raised intracranial pressure are described in the Monro-Kellie doctrine which states that, once the fontanelles and sutures of the skull are closed, the brain is enclosed in a non-expandable case of bone, the brain parenchyma is nearly incompressible, the blood volume in the cranial cavity is therefore nearly constant, and a continuous outflow of venous blood from the cranial cavity is required to make room for continuous incoming arterial blood (11). Blood flow to the brain is controlled by changes in diameter of resistance blood vessels and cerebral blood flow is controlled by autoregulation whereby the perfusion pressure is maintained within a range of approximately 50–150 mmHg (12). Increased intracranial pressure leads to a decreased cerebral perfusion pressure and subsequently reduces cerebral blood flow. Reduced cerebral blood flow results in areas of ischemia and subsequent restriction of delivery of substrates such as oxygen and glucose to the brain. Reduced cerebral blood flow has been associated with an unfavorable neurological outcome in humans and has been implicated in increased susceptibility of the brain to secondary injury (13,14). Brain swelling after TBI due to edema and/or hematoma formation within the skull is the most common cause of increased intracranial pressure. Subarachnoid hemorrhage, or hemorrhage into the CSF, occurs commonly in horses (5).

Secondary injury involves both necrotic and programmed cell death and is a complex cascade of molecular, cellular, and biochemical events that can occur for days to months following the initial insult, resulting in delayed tissue damage. Multiple mechanisms are involved in this process: 1) vascular compromise leading to reduced blood flow, loss of autoregulation, loss of microcirculation, vasospasm, thrombosis, and hemorrhage; 2) electrolyte shifts, permeablility changes, loss of cellular membrane integrity, edema, and loss of energy metabolism; 3) biochemical changes including neurotransmitter accumulation, arachidonic acid release, free radical and prostaglandin production, and lipid peroxidation; and 4) apoptosis (15,16). Uncontrolled glutamate release and failure of energy systems in neuronal and supporting tissues lead to elevated intracellular calcium concentrations and subsequent cell death. The consequence of secondary injury is enlargement of the area of cell death. The phase of secondary injury is widely studied since this process progresses from minutes to months after injury and is, thus, considered to be a target for therapeutic interventions. Minimizing secondary injury through protection of neural elements that initially survived the mechanical injury would increase the quantity of spared tissue and could lead to reduced functional impairment.

## THERAPEUTIC CONSIDERATIONS

Based on the pathophysiology of events that occur after CNS trauma it is likely that single drug intervention would not be effective. Treatment to reverse primary injury does not currently exist and it is the assortment of pathophysiological processes that occur during the period of secondary injury that are considered the target for pharmaceutical intervention. In SCI, the period of secondary injury can be divided into three therapeutic windows (17). The first 48 hours after acute spinal cord injury are dominated by the vascular and biochemical changes that occur within the spinal cord. The second period is a result of the effects of inflammatory cells that occurs within hours of injury and peaks around four days after injury. The third period occurs approximately one week after injury and is characterized by axonal regeneration and lesion repair. The goals of treatment are to stop the cascade of cellular events initiated by the traumatic insult, to protect spared neural tissue, and to promote regeneration. Surgical intervention is warranted when there is need for stabilization, fracture repair, or evidence of a compressive lesion, however, is not routine practice. The use of medical treatment to stabilize the patient should always be instituted before surgery is performed.

Both TBI and SCI have a complex multifactorial pathophysiology and likely require a combinational treatment intervention for successful outcome. In TBI initial treatment is aimed at optimizing delivery of oxygen and substrates to brain tissue in order to salvage brain tissue that is undamaged or reversibly damaged. This requires optimizing cerebral blood flow, i.e. optimizing mean arterial blood pressure and hemoglobin concentration, and ensuring intracranial pressure is not elevated. Mean arterial blood pressure should be maintained within normal. Blood transfusion may be indicated in cases of severe hemorrhage. Crystalloid fluids are recommended as fluid therapy of choice, particularly in light of findings of the serum versus albumin fluid evaluation (SAFE) study (18) that determined no difference between outcomes between administering albumin versus normal saline in the intensive care unit, and in which subgroup analysis demonstrated an increased mortality in TBI patients that were treated with albumin. However, isotonic crystalloid fluids administered in typical shock doses of 40-90 mL/kg/hr may produce worsening of cerebral edema and increased intracranial pressure (19). Comparison of isotonic crystalloid fluids with hypertonic saline solutions has shown hypertonic saline to be the fluid of choice for fluid support of head trauma patients in shock (20). Hypertonic saline is associated with significant decreases in intracranial pressure and cerebral water content compared with isotonic fluid treatment. Furthermore, hypertonic saline has positive effects on cerebral blood flow, oxygen consumption, and inflammatory response at a cellular level (21). Hypertonic saline can be administered intravenously to head trauma horses in shock as 5% or 7% NaCl solutions (4-6 mL/kg) over 15 minutes. Isotonic fluids can then be used for maintenance if needed. Contraindications to the use of hypertonic saline include dehydration, ongoing intracerebral hemorrhage, hypernatremia, renal failure, hyperkalemic periodic paralysis, and hypothermia. Systemic side effects include coagulopathies, excessive intravascular volume, and electrolyte abnormalities. Monitoring central venous pressure and maintaining it within normal limits (5-7 $cmH_2O$) is therefore important as well as monitoring serum sodium and potassium concentrations if hypertonic saline is used frequently (20). Also in SCI systemic hypotension is considered detrimental to recovery. Hypotension may exacerbate spinal cord hypoperfusion and ischemia, and maintaining systemic blood pressure has been shown to improve spinal cord perfusion. The current recommendation is to maintain euvolemic normotension, and because of sympathetic outflow disruption after cranial SCI, pressor therapy is commonly indicated in the treatment regimen. Maintaining normal mean arterial blood pressure is also important to consider during stabilization of the acutely injured horse, and is particularly important when horses are placed under general anesthesia for various diagnostic/therapeutic procedures.

To reduce edema hyperosmolar substances such as hypertonic saline and/or mannitol are used with or without furosemide. In addition to this, treatment with antiinflammatory and free-radical scavenging drugs such as NSAID's, DMSO, and corticosteroids have a place in the treatment of CNS trauma. Many agents, that could target different aspects of the secondary injury mechanisms, have been investigated for use in SCI and have been reviewed (22,23). Until now only methylprednisolone sodium succinate (MPSS) was shown to be efficacious in both animal models and humans (24-27), however, based on conflicting reports and small neurologic improvements, a number of reviews of the data have been published expressing concern with regards to the true beneficial effect of MPSS in acute SCI (28,29). Controlling fever and seizure activity is important as well as maintaining appropriate antibiotic coverage and nutritional status. For the subacute to chronic phase, currently investigated treatment strategies are those that promote regeneration and restoration of circuitry. Modifying the lesion environment directly through cellular transplantation techniques, injections of genetically modified virus to produce growth factors, and/or use of scaffolds that contain drugs/substances that can be released to the lesion environment over a prolonged period of time are techniques investigated for this purpose.

**THERAPEUTIC STRATEGIES**

Acute phase: Hypertonic saline attenuates cord swelling and severity of hemorrhage and edema detected by magnetic resonance imaging in the acute phase of unilateral cervical spinal cord injury in rats. This was a prospective, randomized placebo-controlled study in which 12 adult female rats underwent a unilateral 12.5mm SCI at cervical vertebra 5. Animals were administered either 0.9% NaCl (n=6) or 5% NaCl (n=6) at 1.4ml/kg intravenously (IV) every hour starting 30min after SCI. Immediately after SCI, rats were placed in a 4.7T Bruker MRI system and images were obtained continuously for 8hrs using a home-built transmitter/receiver 3cm Helmholtz coil. Rats were sacrificed at 8hrs after SCI. Quantification of cord swelling and volumes of hypointense and hyperintense signal within the lesion were determined from MRI. At 36 minutes after SCI, significant swelling of the spinal cord at the lesion center and extending rostrally and caudally was demonstrated by MRI. Also, at this time point, a hemorrhagic core was identified on T1, PD, and T2 weighted images. Over time this hemorrhagic core reduced in size and in some animals was no longer clearly recognizable as hemorrhage by 8hrs after SCI. A prominent ring of T2 weighted image hyperintensity, characteristic of edema, surrounded the core of hemorrhage. At the level of the lesion center, this rim of edema occupied the entire unilateral injured cord and in all animals extended to the contralateral side. Administration of hypertonic saline resulted in increased serum [Na], attenuation of cord swelling, and decrease of volumes of hemorrhage and edema at the last time points. We were able to use MRI to detect rapid and acute changes in the evolution of tissue pathophysiology, and show beneficial effects of hypertonic saline in acute cervical SCI.

Subacute phase: We used a delayed combined transplant strategy that included transplantation of glial restricted precursor (GRP) cells and administration of cAMP. We asked whether combining cAMP with a GRP cell transplant would improve functional recovery, similar to that observed after combining cAMP with Schwann cell transplants (30). We used 51 adult, male rats. Groups: 1: age-matched control; 2: operated control; 3: GRP control; 4: cAMP control; 5: GRP cAMP. We used 12 rats in a short-term survival part of this study. Groups: 1: uninjured control; 2: operated control; 3: cAMP control. Our moderate contusion injury produced a very consistent lesion with the area of remaining tissue at the lesion center in the OP control group ($12.4 \pm 2.7\%$) in the same range as described previously for this level of injury ($9.9 \pm 4.8\%$). The presence of GRP cells resulted in a significant increase of tissue throughout the lesion region. The GRP cells survived, differentiated, and formed extensive transplants that were well integrated with host tissue in all animals that received a transplant. In animals that received exogenous cAMP, however, the area occupied by GRP cells at the lesion center and the volume of the GRP transplant throughout the lesion region were reduced. The histopathological alterations we found in the lesion region did not appear to affect serotonergic input to the DM and DL nuclei in the lumbosacral spinal cord, however, animals that were given exogenous cAMP showed greater reduction of amount of serotonin immediately distal to the lesion center compared to the control group. As far as the author is aware, this is the only study that has quantified extent of transplant within the lesioned spinal cord. Certainly there are no other reports on the effect of cAMP on survival of transplanted cells within an injured spinal cord. Our study suggests that elevation of cAMP is not necessarily beneficial to survival of GRP cells. In conclusion, our study showed no beneficial effects on functional outcome measurements of transplanting GRP cells and administering cAMP after SCI. Although elevation of cAMP has been reported to have beneficial effects in neural regeneration and may have beneficial effects on lineage differentiation of GRP cells, from our study it appears that cAMP are not always reduced after SCI and that administering cAMP may reduce survival or proliferation of GRP cells following transplantation into the injured spinal cord. Transplant strategies appear promising; however it seems appropriate to use combinatorial strategies in which cell transplants are combined with other therapies such as treatments that induce differentiation of progenitor cells into populations with desired functional characteristics, treatments that affect regeneration, and/or treatments that alter the spinal cord milieu.

**REFERENCES**


[1.] Feige K *et al*. Traumatic injury to the central nervous system in horses: occurrence, diagnosis and outcome. **Equine Veterinary Education** 2000;12(4):220-224.[2.] Tyler CM DR, Begg AP, Hutchins DR, Hodgson DR. A survey of neurological diseases in horses. **Aust Vet J** 1993;70(12):445-449.[3.] Little CB *et al*. A retrospective study of head fractures in 21 horses. **Aust Vet J** 1985;62(3):89-91.[4.] Reed SM. Management of head trauma in horses. **Comp Cont Educ Pract Vet** 1993;15:270-273.[5.] Reed SM. Intracranial Trauma. In: Robinson NE, editor. **Current Therapy in Equine Medicine.** Philadelphia: W.B. Saunders Co.; 1987. p 377-380.[6.] Feary DJ *et al*. Traumatic brain injury in horses: 34 cases (1994-2004). **J Am Vet Med Assoc** 2007;231(2):259-66.[7.] Johnson PJ *et al*. The vestibular system. Part II: Differential diagnosis. **Equine Veterinary Education** 2001;June:185-194.[8.] MacKay RJ. Brain injury after head trauma: pathophysiology, diagnosis, and treatment. **Vet Clin North Am Equine Pract** 2004;20(1):199-216.[9.] Reed SM. Medical and surgical emergencies of the nervous system of horses: diagnosis, treatment, and sequelae. **Vet Clin North Am Equine Pract** 1994;10(3):703-15.[10.] Nout YS. Central Nervous System Trauma. In: Reed SM, Furr MO, editors. **Equine Neurology.** Ames, Iowa, USA: Blackwell Publishing; 2007.[11.] Andrews PJ *et al*. Intracranial pressure. Part one: historical overview and basic concepts. **Intensive Care Med** 2004;30(9):1730-3.[12.] Steiner LA *et al*. Monitoring the injured brain: ICP and CBF. Br **J Anaesth** 2006;97(1):26-38.[13.] Narayan RK *et al*. Improved confidence of outcome prediction in severe head injury. A comparative analysis of the clinical examination, multimodality evoked potentials, CT scanning, and intracranial pressure. **J Neurosurg** 1981;54(6):751-62.[14.] Golding EM. Sequelae following traumatic brain injury. The cerebrovascular perspective. **Brain Res Brain Res Rev** 2002;38(3):377-88.[15.] Beattie MS *et al*. Review of current evidence for apoptosis after spinal cord injury. **J Neurotrauma** 2000;17(10):915-25.[16.] Tator CH *et al*. Review of the secondary injury theory of acute spinal cord trauma with emphasis on vascular mechanisms. **J Neurosurg** 1991;75(1):15-26.[17.] Olby N. Current concepts in the management of acute spinal cord injury. **J Vet Intern Med** 1999;13(5):399-407.[18.] Finfer S *et al*. A comparison of albumin and saline for fluid resuscitation in the intensive care unit. N Engl **J Med** 2004;350(22):2247-56.[19.] Gunnar W *et al*. Head injury and hemorrhagic shock: studies of the blood brain barrier and intracranial pressure after resuscitation with normal saline solution, 3% saline solution, and dextran-40. **Surgery** 1988;103(4):398-407.[20.] White H *et al*. The use of hypertonic saline for treating intracranial hypertension after traumatic brain injury. **Anesth Analg** 2006;102(6):1836-46.[21.] Kempski O. Cerebral edema. **Semin Nephrol** 2001;21(3):303-7.[22.] Kwon BK *et al*. Pathophysiology and pharmacologic treatment of acute spinal cord injury. **Spine J** 2004;4(4):451-64.[23.] Baptiste DC *et al*. Pharmacological approaches to repair the injured spinal cord. **J Neurotrauma** 2006;23(3-4):318-34.[24.] Hall ED *et al*. Effects of a single large dose of methylprednisolone sodium succinate on experimental posttraumatic spinal cord ischemia. Dose-response and time-action analysis. **J Neurosurg** 1984;61(1):124-30.[25.] Behrmann DL *et al*. Modeling of acute spinal cord injury in the rat: neuroprotection and enhanced recovery with methylprednisolone, U-74006F and YM-14673. **Exp Neurol** 1994;126(1):61-75.[26.] Bracken MB *et al*. Administration of methylprednisolone for 24 or 48 hours or tirilazad mesylate for 48 hours in the treatment of acute spinal cord injury. Results of the Third National Acute Spinal Cord Injury Randomized Controlled Trial. **National Acute Spinal Cord Injury Study. Jama** 1997;277(20):1597-604.[27.] Bracken MB *et al*. Methylprednisolone or naloxone treatment after acute spinal cord injury: 1-year follow-up data. Results of the second National Acute Spinal Cord Injury Study. **J Neurosurg** 1992;76(1):23-31.[28.] Coleman WP *et al*. A critical appraisal of the reporting of the National Acute Spinal Cord Injury Studies (II and III) of methylprednisolone in acute spinal cord injury. **J Spinal Disord** 2000;13(3):185-99.[29.] Sayer FT *et al*. Methylprednisolone treatment in acute spinal cord injury: the myth challenged through a structured analysis of published literature. **Spine J** 2006;6(3):335-43.[30.] Pearse DD *et al*. cAMP and Schwann cells promote axonal growth and functional recovery after spinal cord injury. **Nat Med** 2004;10(6):610-6.

# THE PROTECTIVE IMMUNE RESPONSE TO WEST NILE VIRUS IN HORSES

**Kathy K. Seino, DVM, MS, PhD**
**Pullman, WA**

## INTRODUCTION

West Nile Virus (WNV) is a member of the *Flaviviridae* family of nearly 70 positive-sense, single-stranded enveloped RNA viruses. This family of viruses is distinguished with several mosquito-borne pathogens that cause endemic or epidemic disease worldwide including, Yellow Fever Virus, Dengue Virus, Japanese Encephalitis Virus, and Murray Valley Encephalitis Virus. Historically WNV was isolated in 1937 from a febrile woman in the West Nile region of Uganda and was the cause of periodic outbreaks of fever and encephalitis in horses and people in Africa and the Middle East. West Nile emerged in the United States in the fall of 1999, in New York City, with the first fatalities being avian inhabitants at the Bronx Zoo and resulting in the deaths of nine horses, seven people, and hundreds of local birds (corvids) in the initial outbreak. Investigation confirmed the encroachment of a virulent strain of West Nile Virus, closely related to a neurotropic Lineage I strain of WNV isolated from Israel.

WNV relies on an avian reservoir host and transmitting mosquito, with horses and humans being incidental dead-end hosts. A majority of humans and horses infected by WNV remain asymptomatic. In people, 10-20% may develop West Nile Fever (WNF), characterized typically with flu-like symptoms, fever, and malaise, with some developing GI symptoms, viral exanthem ("rash"), neck pain, headaches, or cognitive symptoms. Less than 1% of humans and horses succumb to infection to develop meningitis, encephalitis, or poliomyelitis-like flaccid motor paralysis within 7 to 14 days after infection.[1,2] Clinical presentation in horses may be variable but behavioral changes and muscle fasciculations are hallmark signs. Affected horses may also exhibit cranial nerve deficits (particularly tongue paresis, head tilt, muzzle deviations) and spinal abnormalities (ataxia and paresis) in which 30% progress to complete paralysis in one or more limbs. Horses typically improve in 3 to 7 days with 30% that have recrudescence in signs. In horses with significant improvement, full recovery by six months was reported in 90% with the remainder having residual deficits.[3] In humans, 70% of survivors experience persistent neurologic deficits for months to years after recovery.[1,2]

Since its emergence in 1999 and the subsequent years, West Nile Virus (WNV) has spread throughout the United States, Mexico, and Canada. In the United States, WNV has killed a total number of 965 humans, approximately 6,000 horses, and an estimated hundreds of thousands of birds. This encephalitic virus has caused clinical disease in 25,332 humans (both West Nile Fever and West Nile Encephalitis (WNE)) and 24,985 horses (WNE). In its wake, there was heightened attention surrounding this serious pathogen leading to the development of diagnostic tests, therapeutics, and prophylaxis including three equine WNV (killed, recombinant, and live chimera) vaccines. Though the virus is now considered endemic with cases still reported annually, WN remains a pathogen of concern. An effective antiviral treatment as well as a human vaccine for WN is still currently unavailable, and it is becoming clear that competent host immunity is a factor in severe WN infection. Understanding the protective immune response to WN infection is still an area of intensive scientific and medical focus.

## IMMUNOPATHOGENESIS OF WNV

The immunopathogenesis of WNV has been extensively studied in murine models. Infection occurs in the muscle or other cells near the site of the mosquito bite with salivary antihemostatic molecules that prevent vertebrate hemostasis and enhance the initial replication of the virus in the skin by dampening the innate immune response. Other studies provide evidence of up-regulation of MHC class I molecule expression with WNV infection, preventing natural killer cell activation. Localized virus replication results in the migration and activation of Langerhans cells (MHC I and MHCII+, CD80 and CD86 expressing) to the site and subsequent migration to draining lymph nodes. Viral replication in these tissues leads to a primary viremia, resulting in infection of tissues associated with the vascular system. Experimental murine models demonstrate that WNV has a primary predilection for neural tissues. Intraperitoneal injection of WNV ($10^2$ PFU) into 8 to 12 week-old mice results in dissemination into the CNS by 4 to 6 days after inoculation. Experimental infection of horses results in viremia at days 3 to 5 and clinical signs in 7 to 10 days. WNV inoculation into the CNS results in direct infection of nerve cell bodies. In rodent models, initial replication occurs in the basal ganglia, with subsequent dissemination to the cortex, cerebellum, and hippocampus. The large neurons of the ventral or anterior horns are infected later in the course of the disease.

Two routes of neuroinvasion are proposed for WNV infection. In the first route, WNV gains entry into the CNS across the blood brain barrier (BBB). It is hypothesized that systemic viral infection results in local cytokine responses (such as IFN$\gamma$) that causes increased upregulation of intracellular adhesion molecules (ICAM) expression allowing for increased CNS trafficking of infected lymphocytes. ICAM-1 has been detected several days prior to WNV in the CNS. WNV infection induced expression of cytokines may also cause leakiness of the BBB for viral entry. In particular, tumor necrosis factor alpha (TNF-$\alpha$) mediated by toll-like receptors (TLR), particularly TLR3, stimulated by WNV in murine models increased vascular permeability and allowed infection of peripheral nerves. The second route of entry of WNV proposes direct axonal retrograde transport from infected peripheral neurons or infection of the olfactory neurons and spread to the olfactory bulb. Several studies have been published data that provide evidence for and against both routes.[1,2,3]

## IMMUNE RESPONSE TO WNV

A successful response to primary infection by a viral pathogen, such as WNV, would be expected to rely on the interplay of innate and adaptive immune responses. WNV replication in the periphery has been shown to be limited by the antiviral actions of interferon-alpha/beta (IFN-$\alpha/\beta$). Mutation in the IFN inducible Oas1b gene in WN susceptible mice results in the truncation of the

2'5' oligoadenylate synthetase protein that normally mediates the activation of cellular RNase L which cleaves viral ds RNA.[4] Complement activation of all three pathways (classical, lectin, and alternative pathways) and IFNγ secreting γδ Tcells, have also been shown in mice to be important in the initial control of WNV infection and in the priming of humoral and cellular immune responses.[1,2,4]

Protective immunity to WNV infection relies on induction of adaptive immunity. After primary infection, WNV infected Langerhans cells travel to localized lymph nodes to stimulate activation of B cells and the induction of IgM antibody and WNV specific IgG. Passive transfer of WNV specific IgG was shown to protect mice if administered before or shortly after lethal challenge and, in another study, B cell deficient were protected from lethality by passive transfer of immune sera. These humoral immune responses have been shown to be crucial in controlling peripheral viral dissemination but cell mediated and T helper type 1(Th1) responses are also needed for survival and recovery. WNV cytotoxic T cells (CTL) are observed in mice at 5 to 6 days post infection and adoptive transfer experiments of these CD8 T cells have shown these cells to be important for decreased mortality to eliminate WNV infected cells via cytolytic mechanisms. $CD4^+$ T cells have a critical role in sustaining these WNV-specific effector $CD8^+$ T cell responses and for maturation of antiviral IgG production.[1,2]

The immune response to WNV in the CNS is still completely unknown and this knowledge may be critical to understanding the protective immune response to WNV. WNV specific CD 8 and CD4 cells have been shown to also be important for the clearance of WNV in the CNS in mice.[1] Individuals with a homozygous deletion in the CCR5 gene (CCR5Δ32) have a significant increased risk for fatal WNV encephalitis due to decreased leukocyte trafficking into the CNS from deficient chemokine CCR5 receptor expression.[5] Cell-mediated immunity, however, may also play a role in immunopathology as it has been also shown that lesions in brains of mice with fatal WNV are predominantly composed of $CD8^+$ T cells.[1,2,3] In human patients that developed acute flaccid paralysis from WNV, the gray matter of the cervical and lumbar regions that were severely affected had a predominance of activated microglia and T cells in immunologic reaction with the virus. Infiltration of the CNS with non-specific T cells is speculated to be a potential factor. The immune response of the CNS must be carefully regulated in a balance between clearance of virus infected neurons and damage to bystanding neuronal tissue. How that balance is maintained is still unknown with WNV. Regulation of this balance in the CNS, based on other viral CNS pathogens, is likely to involve the resident CNS microglia and astrocytes.[6] Specifically microglia, essentially act as resident macrophages to express TLR, process and present antigens, produce TNF, nitric oxide, and other mediators and astrocytes express TLR and cytokines in response to virus infection. The response of these cells to WNV is still being studied.[6]

## IMMUNE RESPONSE TO WNV IN HORSES

### Equine WNV Vaccines

Early epidemiological studies predicted that vaccines would be the best preventative strategy in the face of the outbreak in 1999. Since, then four equine WNV vaccines have been licensed by the USDA, with three commercially available. The first licensed vaccine (K-WN; Innovator, Fort Dodge Animal Health, Overland Park, KS), available since 2001, is composed of an adjuvanted formalin-inactivated whole virion. Licensed in 2004, a second vaccine was marketed featuring the expression of the WNV membrane (prM) and envelope (E) genes under control of the canarypox virus vector (CP-WN; Recombitek™, Merial, Duluth, GA) and packaged with an adjuvant. Since 2006 a nonadjuvanted, single-dose, attenuated West Nile Virus, live flavivirus chimera (WN-FV) vaccine (PreveNile®, Intervet, DeSoto, KS) has been also available. This vaccine is based on the technology used to create an attenuated human vaccine for Japanese Encephalitis virus (Chimerivax-JE). The recombinant chimera expresses the envelope (E) and membrane (prM) of WNV in a Yellow Fever vector (YF17D). This last vaccine has been shown experimentally in rodent and primate models to induce both humoral and cell mediated immunity (WNV specific T cell responses). Vaccines have shown to be highly effective in prevention of WNV infection in horses.[7,8,9]

### Immune Responses of Vaccinated Horses

Various studies have demonstrated the humoral and cell mediated immune responses elicited in horses vaccinated with these vaccines. The WNV recombinant canarypox virus vaccine was shown to elicit WNV prM/E insert-specific IFN-g producing cells in vaccinated horses that remained detectable to day 91 and detectable after booster injections. Neutralizing antibody to WNV was not detected by PRNT after the primary vaccination series at high levels (2.4 log10+ 0.16 log 10) with booster injections (3 and 5 months).[10] Antigen specific antibody responses and cell mediated (CD4 and CD8) responses were measured in horses vaccinated with inactivated WNV vaccine at 3 week intervals. Vaccinated horses produced increased levels neutralizing antibody and of several immunoglobulin isotypes (IgM, IgA, IgGa, IgGb,IgGc, and Ig (T)) after completion of the vaccination series. IFNγ and IL-4 expressing $CD4^+$ T lymphocytes detected with intracellular cytokine staining of PBMC were significantly increased 14 days following initial vaccination compared to unvaccinated horses. IFNγ $CD8^+$ T cells were also increased after vaccination and were detected for 110 days. WNV antigen lymphocyte proliferation of PBMC collected from vaccinated horses were elicited in cells up to 90 days after the third vaccination.[11]

### Response of Vaccinated Horses to Virulent WNV Challenge

A severe challenge model that produces clinical West Nile Virus (WNV) disease in horses was used to characterize and compare the clinical efficacy and immune responses in the CNS and periphery of three equine WNV vaccines. Twenty-four healthy, WNV-seronegative horses of varying ages and gender were placed into three randomized, blinded trial groups consisting of eight horses with two horses in each group receiving 1) an inactivated (K-WN) WNV vaccine, 2) a modified-live vaccine (CP-WN), containing the WNV prM and E proteins expressed by a canarypox vector 3) a live-chimera vaccine (WN-FV) containing WNV prM and E proteins expressed in a YF17D vector, or 4) diluent control. Challenge by this model caused grave neurological signs, viremia, and an outcome of 0% survivorship in all six control horses. In contrast, challenge in horses at between 28 days post-vaccination of the chimera vaccine or 56 days post vaccination of the commercial inactivated or modified-live vaccines resulted in 100% survivorship, protection from the onset of WN encephalitis and viremia. Horses vaccinated with the live-chimera vaccine showed significantly fewer clinical signs than horses vaccinated with inactivated vaccine ($p=0.035$) and the control horses ($p\leq 0.01$).[9]

**Humoral Immune Response of Vaccinated Horses**

Similar to other studies, earliest detection of neutralizing titers using the plaque reduction neutralization test (PRNT) was in the K-WN horses at the time of the second immunization. At the time of challenge, K-WN had a geometric mean titer (GMT) of 224, the CP-WN of 26 and the WN-FV of 5. After WNV challenge, the GMT of the K-WN horses increased to 320 and remained the same to the end of the study. Four out of the five CP-WN horses had a slight increase of neutralizing titer and a GMT of 258 two weeks after challenge. Infection with WNV in the WN-FV elicited a rapid ramping of neutralizing antibody titer at 7 days to 122 GMT and peak of 320 at 14 days. WNV antibodies for IgG total, IgGa, IgGb with a capture IgG isotype ELISA were detected in the K-WN (Mean endpoint titer (MET) for IgG total 26K, IgGa 20K, IgGb 11K) and CP-WN vaccinates (MET IgG total 150, IgGa 120, IgGb 260) after vaccination. There were marked increases in antibodies in the CP-WN but most in the WN-FV (MET IgG total 117K, IgGa 91K, IgGb 97K) horses by D14PC. On D7PC, all six control horses developed low levels of total IgG to WNV (MET of 683, range 50 to 1,600). Only one out the six control horses had a detectable WNV IgGa antibody titer of 400 and no WNV IgGb antibody was detected in any of the control horses.[9]

**Cellular Immune Response of Vaccinated Horses**

Lymphocyte subsets, lymphoproliferative responses of equine peripheral blood monocytes (PBMC), *ex vivo* and *in vitro* cytokine mRNA profiles, of the control horses and WN-FV horses were studied to characterize their immune response. Protective recall responses were seen in vaccinated horses with WNV challenge and data was supportive of the hypothesis that vaccinated horses have WNV antigen specific $T_H1$ $CD4^+$ T cell response. WNV proliferative responses were most significantly increased in the WN-FV vaccinates at 21 days after challenge, with an increased IFN-γ/IL-4 ratio and had increased $CD4^+$ T cell populations. A protective response was also characterized by a rapid induction of an antibody response mediated by IL-5. Successful response also appeared to be correlated with low expression levels of TNF-α. In contrast, naïve control horses had significant expression of TNF-α after virulent challenge and had no WNV antigen specific lymphocyte proliferation. Immune response characterization of the K-WN and CP-WN vaccinates is pending.

**Immune Response in the CNS of Vaccinated Horses**

In the severe challenge model of infection, all six control horses had moderate to severe histopathologic lesions in the brain and spinal cord, whereas none of the vaccinated horses in each group had any lesions consistent with WN polioencephalomyelitis. A few of the vaccinates in each group had mild inflammatory lesions in the CNS. WNV RNA was not detected in any CNS tissue collected from the vaccinates, but was amplified in all of the control horses. Viral load was quantified to be $10^2$ PFU from thalamus from control horses. In contrast to the peripheral expression, significant increases in the TNF-α and IFN-γ cytokine expression were measured in WN-FV and K-WN vaccinates in CNS tissues (thalamus, midbrain, and cervical cord) collected at D21PC. Characterization of the immunophenotypes of the cells in the CNS is in work.

## SUMMARY

Murine models provide evidence to support that a protective immune response to WNV infection relies on both humoral and cell mediated immunity. With horses we have a unique opportunity in that a model of clinical WN disease has been developed in the horse and also the availability of three clinically effective WNV vaccines to which the immune responses can be studied. Examination of the immune responses of vaccinated horses to WNV support murine models of WNV infection and appears to rely on both humoral immunity of neutralizing antibody but also reliant on antigen specific cell mediated responses with a bias towards Th1 type. WN disease is associated with the presence of virus in the CNS, whereas no virus nor histologic lesions were found in any of the vaccinated horses. Further characterization of the immune response in the CNS is needed.

## REFERENCES

[1] Klein RS, et.al **Trends Mol Med** 2008;14(7):286. [2] King NJC, et.al. **Immunol Cell Bio** 2007;85:33. [3]. Long MT 2007 Flavivirus Infections. In **Equine Infectious Diseases**. Sellon DC and Long MT, eds,. P 198. WB Saunders, St. Louis. [4]. Wang T, et al. **Current Opin Immunology** 2004;16:519. [5] Diamond MS, et. al. **Trend Microbio** 2006;14(7) 288. [6]. Griffin DE **Nature Reviews Immunol** 2003;3:493. [7]. Gardner IA, et al. **Vet Res** 2007;38(1):109-16. [8]. Epp T, et al. **Equine Vet J** 2007;39:498. [9]. Seino KK, et al. **Clin Vacc Immunol** 2007;14(11):1465. [10]. El Garch H, et. al. **Vet Immunol Immunopath** 2008;123:230. [11]. Davis EG et. al. **Vet Immunol Immunopath** 2008;126:293.

**GENERATION OF TETRAMERS TO STUDY EHV-1 IMMUNE MODULATION OF CELLULAR IMMUNE RESPONSES.** **G.Soboll [1*], R Case[2], S MacRae[2], DP Lunn[1], JH Kydd[3].**
**1. Colorado State University, 2. AHT Newmarket, UK, 3. Nottingham University, UK.**

The importance of EHV-1 infection in the equine population is well documented worldwide.[1] As for many viruses the outcome of EHV-1 infection is significantly influenced by the molecular mechanisms of pathogen-host interactions and elucidating these mechanisms is crucial for disease prevention. Despite years of research, our understanding of why immune responses following EHV-1 infection or vaccination are short lived is limited. Like other α-herpesviruses, EHV-1 has developed multiple strategies to evade the host's immune responses including interference with ADCC and cytotoxic T-lymphocyte (CTL) responses, all of which may explain the short lived immune responses that follow infection.[9] Of particular importance is EHV-1's ability to evade the host immune response by down-regulating MHC-1 antigen presentation,[3,4,7] because it has been convincingly established that protection of horses from EHV-1 infection is critically dependent on generating strong CTL responses.[2,5,6]

It is currently extremely laborious to perform quantitative studies on EHV-1-specific CTLs in horses, making it impossible to measure the impact of viral immune evasion strategies such as MHC-1 down-regulation. We have recently made a critical advance in understanding the antigen-specificity of EHV-1-specific CTL responses that offers the opportunity to overcome this limitation. In a recent study we have identified the immediate early (IE) gene product as a potent inducer of MHC I restricted CTL responses in horses expressing the equine MHC class I haplotype A3 [8]. Using this information, the goal of this project is to exploit these advances to develop tetramer staining technology that will rapidly quantitate antigen-specific equine CD8+ CTL precursors in horses using flow cytometry and without the need for complex and time-consuming conventional CTL assays. This model system, based on the IE-specific CTL response in MHC-1 A3 expressing horses, will provide a platform for the study of interactions between EHV-1 and the equine immune system. In particular, it will allow us to study the effect of different EHV-1 genes on MHC-1 down-regulation and the development of CTL responses.

For generation of tetramers the first step has to be the identification of the minimal peptide epitopes that induce CTL responses. Two strategies have been chosen to identify these fragments of the IE gene of EHV-1 that are recognized by CTLs from A3 positive horses. For strategy A, we generated plasmids expressing IE fragments based on the functional domains of the IE gene and are currently screening them for their ability to induce EHV-1 (IE) specific IFN-γ, perforin and granzyme B responses. For this purpose IE-fragment expressing plasmids are electroporated into PBMCs from A3+ ponies or P815 cells expressing the A3 MHC I allele and used as stimulators with PBMCs from A3+ ponies previously infected with EHV-1. Induction of IFN-γ, perforin and granzyme B responses is measured by real-time PCR. For strategy B, IE-fragment containing plasmids were cloned and are currently used to generate stable transfectants via electroporation of P815 cells expressing the A3 MHC I allele. The stable transfectants will be used as target cells in CTL assays with PBMCs from A3+ ponies previously infected with EHV-1. Upon identification of IE fragments that induce EHV-1 specific CTL responses peptide libraries will be generated and screened as described to identify the minimal peptide epitopes for generation of tetramers to elucidate the role of CTL responses in EHV-1 pathogenesis.

As for other alphaherpesviruses of veterinary importance, EHV-1 evades the host immune response by down-regulating MHC-1 antigen presentation. The importance of this phenomenon depends on the crucial role of CTLs in mediating protection from herpesviral infection. Development of tetramer technology for equine studies, as well as the identification of a number of EHV-1 genes by us and others that interfere with MHC-1 antigen presentation will make it possible to measure the impact of viral immune evasion strategies such as MHC-1 down-regulation and leave us ideally placed to advance our understanding of this critically important area of host-pathogen interaction.

**REFERENCES**

[1.] Allen GP, et al. In: Coetzer JAWT, R.C., ed. Infectious Diseases of Livestock. Newmarket: Oxford University Press, 2004:829. [2.] Allen GP, et al. **Journal of Virology** 1995; 69;606. [3.] Ambagala APN, et al. **Journal of General Virology** 2004;85:349. [4.] Koppers-Lalic D, et al. **Proceedings of the National Academy of Sciences of the United States of America** 2005;102:5144. [5.] Kydd JH, et al. **Veterinary Immunology & Immunopathology.** 2003;96:207. [6.] O'Neill T, et al. **Veterinary Immunology and Immunopathology** 1999;70:43. [7.] Rappocciolo G, et al. **Journal of General Virology.** 2003;84:293. [8.] Soboll G, et al. **Journal of General Virology** 2003;84:2625. [9.] van der Meulen K, et al. **Veterinary Immunology and Immunopathology** 2006;111;31.

# VACCINATION OF PONIES WITH THE IE GENE OF EHV-1.

**G.Soboll [1*], C.C. Breathnach[2], J. Kydd[3] S.B. Hussey[1], R Mealey[4], D.P. Lunn[1].**
**1. Colorado State University, 2. Gluck Equine Center, University of Kentucky, 3. Nottingham University, UK, 4. Washington State University.**

Equine herpesvirus infection-1 (EHV-1) causes enormous suffering to horses and extensive economic losses through frequent outbreaks of a range of clinical signs including respiratory disease[8,9], abortion[2], neonatal foal death[18], myeloencephalopathy[10] and chorioretinopathy[19]. Primary EHV-1 infection typically occurs at the respiratory epithelium, resulting in erosion of the upper respiratory mucosal surface and viral shedding for 4-7 days post-infection. Cell-to-cell spread results in the presence of virus in respiratory tract lymph nodes within 12-24 hours post-infection[15]. A lymphocyte-associated viremia is then established which can persist for up to 21 days, and is directly responsible for the delivery of EHV-1 to the pregnant uterus or the spinal cord resulting in abortion or myeloencephalopathy. It has been shown that there is a correlation between levels of viremia and occurrence of myelitis[1,11]. Control of viremia is therefore critical for the prevention of both abortion and myeloencephalopathy.

While circulating and mucosal antibody each play some role in protection[7], the most important component of a protective immune response to control intracellular EHV-1 is the cytotoxic T lymphocyte (CTL) response[2,3,16]. It has been shown that EHV-1-specific CTL precursor frequencies may be the most accurate measurable correlate of immunity[16]. CTLs recognize foreign peptides derived from viral proteins and expressed on the surface of antigen presenting cells in combination with highly polymorphic MHC I molecules. The equine equivalent of MHC class I is called the Equine Leukocyte Antigen (ELA) and three loci, A, B and C have been detected (Antczak 1992 Equine infectious disease proceedings, Cambridge). For EHV-1, we have previously demonstrated that CTLs from horses that carry the ELA-A3 haplotype could recognize components of the EHV-1 immediate early (IE) protein[20], which is the first viral protein expressed following EHV-1 infection[2]. This was the first identification of an ELA-A restricted EHV-1 CTL target antigen in the horse. In independent experiments, a molecular approach was used to identify CTL target proteins and their genetic restriction elements. These experiments confirmed the original observation that the A3 haplotype is the genetic restriction element in CTLs that are directed against peptides within the EHV-1 IE protein[14]. Importantly, this work also identified that within the A3 haplotype, it is the B2 allele, which presents a peptide(s) derived from the EHV-1 IE gene to CTL.

In a pilot study to demonstrate proof of principle, vaccination of A3/B2 positive ponies with a vaccinia virus construct which expressed the IE gene resulted in the stimulation of high levels of virus specific CTL activity (Paillot et al., 2006). This study was however not designed to test the efficacy of the CTL-mediated protective immunity to challenge infection. Thus in order to determine whether CTL responses directed against specific components of EHV-1 are protective against challenge infection, it is important to employ vaccination experiments designed for this purpose that use horses of the appropriate ELA-A genotype to generate the relevant cell mediated immune responses. This was the objective of this study.

The induction of cell mediated immune responses represents a challenge for any vaccination strategy, therefore we have adapted recombinant modified vaccinia Ankara (MVA) vectored vaccination, to deliver the antigens of interest to the horse[5,6]. In mice, this strategy results in the generation of high levels of specific immunity [17] and the endogenous route of antigen processing leads to high levels of CTL activity[13 4] making it very attractive for application to vaccination of horses against EHV-1. The goal of this study was to show that, we could induce strong cellular immune responses *in vivo* that are associated with protection of ponies from EHV-1 disease, viremia and viral nasal shedding. To achieve this goal we used the IE gene of EHV-1 for vaccination of A3/B2 positive ponies, and then performed a challenge infection, monitoring the humoral and cellular immune responses and clinical and virological signs before and after vaccination and challenge. Our hypothesis was that vaccination of A3/B2 ponies with a rMVA construct which expresses the IE gene will result in the induction of cell mediated immune responses leading to a reduction in the duration and magnitude of the clinical and virological signs, in particular the cell associated viremia induced by challenge infection, compared to responses in ponies of a different genetic background.

The IE gene was cloned into an rMVA vector for vaccination of horses carrying the A3/B2 MHC-I haplotype and horses with a different MHC I haplotype. The experiment included 4 experimental groups: 1. A3/B2 vaccinates (n=10), 2. A3 non-B2 vaccinates (n=5), 3. Non-A3 vaccinates (n=6) and 4. Non-vaccinated controls. Ponies were vaccinated on 3 occasions with the rMVA-IE vaccine via i.m., i.d. and i.v route. Vaccinates and unvaccinated controls were challenge infected 7 weeks after the last vaccination with $5*10^7$ PFU of EHV-1 strain Army 183 via intranasal installation. Interferon gamma (IFN-γ) responses and VN titers were determined throughout the study and clinical signs, nasal shedding, and viremia were determined following challenge infection. Viremia was determined using a quantitative real-time PCR system developed in our laboratory that allows for sensitive analysis of daily samples in a large group of horses.

Vaccination of A3/B2 ponies with rMVA-IE conferred significant clinical protection from EHV-1 challenge and resulted in a significant reduction in viremia and nasal viral shedding when compared to the other experimental groups. IFN-γ mRNA increased significantly following vaccination in the A3/B2 vaccinates. Antibody titers remained low until after challenge infection, indicating that no accidental field acquired or recrudescent EHV-1 infection had occurred.

This study shows that vaccination of horses expressing the ELA-A A3/B2 genotype with an rMVA- IE vaccine confers protection from challenge infection in terms of clinical disease and cell associated viremia and nasal virus shedding as quantified by real time PCR. It shows for the first time, that cellular immune responses to the IE gene alone directly reduces the number of gB copies in PBMC and therefore most probably the cell associated viremia. Control of the cell-associated viremia is hypothesized to be a prerequisite for prevention of abortion and neurological disease. Recent studies have confirmed that the incidence of neurological disease is linked to both the magnitude and the duration of cell associated viremia, as measured by real time PCR[11,12]. In addition they suggested that EHV-

1 strains of higher neuropathogenicity may preferentially target specific lymphocyte subsets and thereby achieve a more robust level of replication and thus a cell associated viremia of greater magnitude, either through more virus particles per cell or more cells infected[1,11]. This finding makes the induction of immune responses that reduce viremia all the more important and to our knowledge, the results of this study are the only data for any form of EHV-1 vaccination in horses that provide evidence for a reduction of viral DNA in PBMC and hence cell associated viremia consequent to challenge infection.

Lastly, while a significant elevation in IFN-γ cDNA levels was achieved when compared to pre-vaccination levels in the A3/B2 vaccinates, the post vaccination levels were not significantly higher then levels in the other vaccination groups. We believe that this may be a sensitivity issue and plan to address this problem by developing EHV-1 specific tetramers and perforin/granzymeB based assays to study cellular immune responses to EHV-1 in genetically characterized horses.

## REFERENCES


1. Allen GP, et al. **Equine Veterinary Journal** 2006;38:252. 2. Allen GP, et al. In: Coetzer JAWT, R.C., ed. **Infectious Diseases of Livestock. Newmarket**: Oxford University Press, 2004:829. 3. Allen GP, et al. **Journal of Virology** 1995;69:606. 4. Allen TM, et al. **Journal of Immunology** 2000;164:4968. 5. Breathnach CC, et al. **Vaccine** 2006;24:1180. 6. Breathnach CC, et al. **Veterinary Immunology and Immunopathology** 2004;98:127. 7. Breathnach CC, et al. **Equine Veterinary Journal** 2001;33:651. 8. Bryans JT, et al. In: Wittmann G, ed. **Herpesvirus diseases of cattle, horses, and pigs**. Boston: Kluwer Academic Publishers, 1989:176. 9. Cullinane AA. In: Robinson NE, ed. **Current Therapy in Equine Medicine**. Philadelphia: W.B. Saunders Company, 1997:443. 10. Donaldson MT, et al. **Journal of the American Veterinary Medical Association** 1998;213:671. 11. Goodman LB, et al. **PLoS Pathogens** 2007;3:e160. 12. Goodman LB, et al. **Vaccine** 2006;24:3636. 13. Hanke T, et al. **Journal of Virology** 1999;73:7524. 14. Kydd JH, et al. **Journal of General Virology** 2006;87:2507. 15. Kydd JH, et al. **Equine Veterinary Journal** 1994;26:470. 16. Kydd JH, et al. **Veterinary Immunology & Immunopathology**. 2003;96:207. 17. Meseda CA, et al. **Journal of Infectious Diseases** 2002;186:1065. 18. Murray MJ, et al. **Journal of Veterinary Internal Medicine** 1998;12:36. 19. Slater JD, et al. **Veterinary Record** 1992;131:237. 20. Soboll G, et al. **Journal of General Virology** 2003;84:2625.

**EFFICACY OF VALACYCLOVIR AGAINST CLINICAL DISEASE FOLLOWING EHV-1 CHALLENGE.**
**LK Maxwell[1], BG Bentz[2], LL Gilliam[1], JW Ritchey[1], RW Eberle[1], TC Holbrook[1], D McFarlane[1], GB Rezabek[1], CG MacAllister[1], CL Goad[1], GP Allen[3].**
**1. Oklahoma State University, Stillwater, OK. 2. Lexington, KY. 3. University of Kentucky, Lexington, KY.**

Recent outbreaks of equine herpes virus type-1 (EHV-1) have been associated with devastating neurological disease, even in well-vaccinated horses. Administration of antiviral drugs, such as valacyclovir, may protect horses from clinical signs of EHV-1 infection. However, the timing of drug administration may affect the efficacy of antiviral therapy. This study tested whether prophylactic administration of valacyclovir versus initiation of therapy at the onset of fever would offer different levels of protection against viral disease. Eighteen aged mares were randomized to treatment: no therapy (control), therapy at onset of fever, or prophylactic therapy. Oral valacyclovir (Valtrex®, GlaxoSmithKline) administration was continued through 7 days (Period 1) or 14 days (Period 2) after challenge with EHV-1. The effects of treatment on clinical score, rectal temperature, nasal shedding of virus, viremia, and development of ataxia were tested by ANOVA. The prophylactic-therapy horses had lower clinical scores from days 2-7 after inoculation as compared to control horses ($P$=0.03). Between days 1-7 post-challenge, valacyclovir administration decreased body temperatures ($P$=0.01), viral shedding ($P$=0.006), and viremia ($P$<0.001) in both drug-therapy groups versus the control horses. Although valacyclovir administration did not reduce the risk of ataxia ($P$=0.12), it did decrease the severity of ataxia ($P$=0.02) as compared to control horses. In summary, valacyclovir therapy significantly decreased viral replication and signs of EHV-1 disease. The protective effect was greatest when therapy was initiated before viral infection. However, therapy that was delayed to as late as two days after inoculation also decreased viral shedding, viremia, and signs of disease.

**UTILITY OF TWO SEROLOGIC TESTS FOR ANTEMORTEM DIAGNOSIS OF EQUINE PROTOZOAL MYELOENCEPHALITIS (*SARCOCYSTIS NEURONA* INFECTION) IN NATURALLY OCCURRING CASES.**

**AL Johnson, RW Sweeney.**
**New Bolton Center, University of Pennsylvania School of Veterinary Medicine, Kennett Square, PA.**

Antemortem diagnosis of equine protozoal myeloencephalitis (EPM) is challenging and always presumptive. Limited information is available regarding the performance of a recently developed commercial test (SAG-1 ELISA). The primary goal of this prospective study was to evaluate the SAG-1 ELISA using naturally occurring cases of EPM; a commercially available indirect fluorescent antibody test (IFAT) was also evaluated in the same patient population. Cases were recruited over 16 months and classified into 4 groups. Confirmed positive cases (n=6) had asymmetric or multifocal neurologic deficits and post-mortem lesions consistent with EPM. Confirmed negative cases (n=13) had variable clinical signs and post-mortem lesions consistent with another neurologic disease (or no lesions). Suspected positive cases (n=8) had asymmetric or multifocal deficits, marked improvement after treatment for EPM, and exclusion of other likely diseases (via radiography, myelography, CSF analysis, and serology). Suspected negative cases (n=16) had no neurologic deficits and a diagnosis of orthopedic disease. Serum and/or CSF were submitted for SAG-1 ELISA and IFAT; not all results were available for all cases. Overall (confirmed + suspected) sensitivity of the SAG-1 ELISA was 0.17 (95% CI 0.03-0.49) and specificity was 0.79 (95% CI 0.59-0.91) using a cutoff value of 1:16. Sensitivity of the IFAT was 1 (95% CI 0.70-1) and specificity was 0.5 (95% CI 0.24-0.76) using serum samples; sensitivity was 1 (95% CI 0.66-1) and specificity was 0.86 (95% CI 0.56-0.97) using CSF samples. Low sensitivity of the SAG-1 ELISA limited its usefulness for antemortem diagnosis of EPM in this patient population.

# UPDATE ON AGED HORSE HEALTH, MANAGEMENT & WELFARE

**Catherine M McGowan BVSc, MACVSc, DECEIM, DEIM,PhD & Thomas W McGowan BSc, DVM, PhDcandidate**
**Liverpool, UK**

## INTRODUCTION

Gerontology is well established in human and small animal medicine, yet only relatively recently has an interest in aged horses been apparent in the veterinary literature. As highlighted in the preface to the first major publication on geriatric medicine, there is a paucity of published material on the subject of equine geriatric medicine.[11] Human gerontology has advanced during the second half of last century due to dramatic changes in life expectancy and population demographics; largely due to a reduction in mortality from infectious disease. It is not known if a similar change in demographics has occurred in equine populations, but there is little doubt that the control of infectious diseases and advances in medicine and surgery may also have altered the demographics of horses. In the past this was limited by the requirement of most horses to have continued athletic or reproductive ability. However, veterinarians are increasingly recognizing that aged horses may have value to their owners over and above these abilities and many horses are kept as companion animals. In support of this is the increasing evidence to suggest horses are presented more frequently in veterinary hospitals. Between 1989-1999 one referral hospital in the USA saw admissions of equine patients over 20 yrs of age increase from 2.2% to 12.5%.[2] A similar trend was also found at another US referral veterinary hospital where the proportion of patients 15 years and over increased linearly four-fold from 1973 (3%) to 2000 (12%).[20]

## DEFINITION OF AGED

Aging or senescence has been defined as a progressive deterioration of the organism after maturity of size, form and function have been reached and which is universal, intrinsic, progressive, and deleterious with time.[7] Overall, aging can be considered to be a generalised impairment of physiological functions, a decreased ability to respond to a wide range of stresses, an increased risk of age-associated disease and an increased likelihood of death.[10] Aging is commonly defined based on physiological aspects of aging. However, aging can also be defined in chronological or demographic terms.

### Physiological Age

Physiological age refers to the functional capacity of an animal, where senescence begins once it has reached peak physiologic function. In the horse, physiological definitions of aging are problematical with the physiological peak often determined by the function. The onset of senescence, or decline in physiological function in the horse may be at around 15 years of age in reproductive terms, yet could be less than 5-10 years of age in racing horses. Much is still unknown about decline in physiological function in horses, yet most would agree that the onset of senescence is around 15 years of age.

### Chronological age

Chronological age refers to an animal's numerical age compared to its expected longevity. Many horses have been reported to live into their 40's or even longer. Chronological age may therefore overestimate an aged horse, as 75% of a horses' natural lifespan of 40 years is 30 years of age, double that of physiological age.

### Demographic Age

Demographic age is determined by survival relative to that of a population in a region. The age at which an animal becomes demographically old is defined as the age at which there is only 25% survivorship at or above that specific age. In a demographic study in the UK, 25% of the equine population were over the age of 15 years.[14] It is interesting to note that this corresponds to physiological decline and supports that 15 years of age is a suitable definition point for an "aged" horse.

### Aged, Old, Very Old & Geriatric

Based on physiological and demographic definitions, for our research we considered a horse as *aged* if it was *15 years of age or greater*. Horses have also been considered *old* if they are between 20 and 29 years of age, and *very old* if they are 30 years or greater.[2] The use of the word geriatric is usually avoided due to the offence it causes many aged horse owners.

## DEMOGRAPHICS

The proportion of aged horses is likely to be approximately 25% based on the population pyramids from UK research.[8,14] This proportion does contradict the research from the USA[21] where only 7.6% of the population sampled were over the age of 20 years. However, it must be pointed out that this NAHMS research may have inadvertently selected for horses with continued athletic or reproductive ability as opposed to companions since it sampled commercial equine 'operations' with a minimum of 5 horses. In comparison, UK demographic research has shown that the median number of horses owned per person was only 2,[14] and our research has shown that the majority of aged horses are kept on properties where the total number of horses are less than 3 horses. Further supporting the importance of small equine 'operations' in contributing to equine demographics was the AVMA pet Survey, targeting US households in a census style survey, where the average number of horses per horse owing household was 3.5 in 2006 and of these, 38% of horse owners considered their horses the be a family member.[1]

**DISEASES OF AGED HORSES**

There has been limited research into the epidemiology of disease of aged horses. However, age has been reported to be a risk factor in the occurrence of a number of diseases, including lameness, Equine Cushing's Syndrome (ECS), chronic laminitis and colic. The table below shows previous studies and the results of the authors that have specifically looked into the prevalence of disease in aged horses. An important factor is that different study methodologies are not directly comparable. Two factors are of note. Firstly, of the 4 surveys of apparently normal aged horses (B, C, D, E) there is a clear under-recognition of clinical signs of disease by owners (B, D) compared to when the horses are examined by veterinarians (C, E). Secondly, cases of aged horses referred to a veterinary hospital (A) differed substantially from disease prevalence in the field. Of note is the overrepresentation of colic in the hospital-based study while dental, ocular, cardiovascular and skin disorders were far more common in veterinary examinations in the field.

**Table 1.** Prevalence of disease in aged horses using owner reporting or veterinary examination.

| Disease prevalence | A[2] | B[3] | C[5] | D[13] | E[13] |
|---|---|---|---|---|---|
| Number of horses | 467 | 165 | 23 | 974 | 339 |
| Age group | >20 | >20 | >20 | ≥15 | ≥15 |
| Signs detected by: | Veterinarian | Owner | Veterinarian | Owner | Veterinarian |
| Type of study | Retrospective referral veterinary hospital survey | Survey via veterinary clinics | Examination of "normal" aged horses | Survey via Equestrian group | Examination of "normal" aged horses |
| Colic | 54% | NR* | NR | 8% | 0% |
| Lameness/ musculoskeletal (not laminitis) | 24% | NR* | NR | 28% | 50% |
| Respiratory disorders | 16% | NR | 39% | 19% | 21% |
| Equine Cushing's syndrome (ECS) | 11% | 8%* | 39% | 2% | 15% |
| Eye disorders | 11% | NR | NR | 22% | 88% |
| Cardiovascular disorders | 7% | NR | 65% | 0.4% | 43% |
| Reproductive disorders | 6% | NR | NR | 1% | NR |
| Skin disorders | 6% | NR | 65% | 21% | 40% |
| Neurological disorders | 4% | NR | NR | 1% | 10% |
| Urinary disorders | 3% | NR | NR | 3.7 | 0% |
| Hemo-lymphatic | 2% | NR | 17% | 0% | 0% |
| Dental disorders | NR | 10%* | 48% | 10% | 95% (55% moderate – uncorrectable) |
| Laminitis | NR | NR | NR | 4.8% | 6% |
| Weight loss | NR | 4% | 35% | 17% | 25% |
| Behavioral problems | NR | NR | NR | 4.5% | NR |
| Tumors | NR | NR* | NR | 4% | 11% |

*Legend; NR = not reported, *significantly greater in older horses compared with 53 younger horses sampled from the general population.*

**Colic**

There is also an increased risk of colic with increasing age and evidence to suggest increasing severity when colic episodes do occur.[9,17] Further, horses 15 years or greater had an increased risk of an episode of colic requiring surgical rather than a medical treatment.[16] There was also a significant decrease in surgical survival with increasing age.[15,16] As well as epidemiological associations with age, specific types of colic have been associated with occurrence in aged horses, especially abdominal neoplasia. In study A above[2], the lesion involved the large colon in 45% and the small intestine in 40%; of those with small intestinal disease, 44% had strangulating lipomas.[2] This supports another study showing that geldings greater than 14 years were at a significantly higher risk for pedunculated lipomas than younger horses.[6] Our research has shown a history of colic was associated with an increased mortality in aged horses using multivariable analysis.

**Dental Disorders**

There has been a paucity of age-specific dental research and no age associations have been found on descriptive UK-based studies. In a retrospective survey of records on 537 horses presented for first opinion dentistry in Australia, just over 20% were aged

16 or older and despite them not being overrepresented, the severity of disease in this group was greater. The percentage of horses that needed only sharp points removed was approximately 50%, which decreased to 25% in horses that were 21 yrs or older.[22] There was a marked increase in periodontal disease as horse age increased (31% in the 21yr or older group). In the authors' study, dental disorders were associated with a number of clinical signs of disease, with multivariable analysis showing an association with diarrhea, difficulty eating and weight loss. Further, both weight loss and a history of difficulty eating were risk factors for mortality in the multivariable model.

### Ocular Disease

Repeated episodes of minor trauma and low grade infections lead to changes in the eye, sometimes considered 'normal aging'. The effects of these episodes are often cumulative and result in ocular pathology with the severity increasing with age. Ocular changes have been shown to be highly prevalent in aged horses with 67 out of 83 (80.7%) in a UK-based study, but only 7 horses were reported by owners to have evidence of visual problems. The two most common findings were vitreal degeneration and senile retinopathy. Of the 7 horses with visual disturbances, all had senile retinopathy.[4]

### Musculoskeletal Disorders

Lameness not associated with laminitis is one of the most frequently observed and reported problems in aged horses yet little research into risk factors or even types of lameness affecting aged horses are available. In a recent review it was suggested that age may be a risk factor for degenerative joint disease in horses as well and that prior use may hasten this process.[18] However, no data were referenced in the latter review and there are currently no epidemiological studies linking osteoarthritis and age in horses. Despite this musculoskeletal disease remains a leading reason for euthanasia of aged horses.

### Cardiovascular Disease

Cardiac murmurs are common in horses of all ages, with the prevalence of auscultable murmurs in racehorses over 80%. However, approximately 70% of these murmurs are associated with turbulent blood flow rather than valvular regurgitation. In aged horses there is a reduced prevalence of physiological murmurs and increased prevalence of valvular regurgitation, especially left sided valvular regurgitation (LSVR), yet the presence of LSVR has not been found to be a risk factor for mortality.[19]

### Skin and Respiratory Disease

Immunology of aging will be covered elsewhere but chronic immunological respiratory and skin diseases have been shown to increase in prevalence with age, notably recurrent airway obstruction (RAO) and *Culicoides* hypersensitivity. The increase in prevalence with age is most likely a combination of increased exposure to the predisposing conditions and, especially in the case of RAO, chronic, irreversible remodeling of the pulmonary parenchyma.

### ECS

ECS has been shown to be a dopaminergic neurodegenerative disease associated with ageing, that shares pathologic similarities to Parkinson's disease in humans.[12] The disease is rarely reported in horses less than 15 years of age and the median age of most studies has been around 20 years. The prevalence of ECS in aged horses $\geq$ 15 years was shown to be 15% based on elevations of both ACTH and alpha MSH[13] and supports the high prevalence of this disorder in aged horses. While ECS was of relatively low prevalence in aged horses compared to the disorders above, horses with both ACTH and alpha MSH values indicative of ECS and horses with clinical signs indicative of chronic laminitis had an increased risk of mortality in the multivariable model in the authors' research.

## ROLE OF VETERINARIANS

Apart from a general under-recognition of disease and disorders present in their horses, our research also revealed that most (61%) horses had not seen a veterinarian in the previous 12 months. Many of the most common diseases are not immediately life threatening and not apparent without a veterinary examination, yet many of these also impact significantly on the risk of mortality, especially dental disease, Cushing's syndrome and weight loss. Colic, when it does occur is associated with an increased risk of mortality in aged horses. Veterinarians need to be aware of common conditions that affect aged horses and owners should be aware of the very high frequency of preventable and treatable disease and be encouraged to have yearly health examinations in their aged horses.

## REFERENCES

[1]AVMA 2007, U.S. Pet Ownership & Demographics Sourcebook, AVMA, Illinois. [2]Brosnahan MM & Paradis MR 2003, **J Am Vet Med Assoc**; 223, 93-98. [3]Brosnahan MM & Paradis MR 2003, **J Am Vet Med Assoc**; 223, 99-103. [4]Chandler KJ et al.2003, **Vet Rec**; 153, 319-322. [5]Chandler KJ & Mellor DJ 2001, In: BEVA, 40th Congress, Birmingham. p 217. [6]Garcia-Seco E et al. 2005, **J Am Vet Med Assoc**; 226, 1529-1537. [7]Green M 1981, **Clin Endocrinol Metab**: 10, 207-228. [8]Hotchkiss JW et al. 2007, **Equine vet J**: 39, 301-308. [9]Kaneene JB et al. 1997, **Prev Vet Med**; 30, 23-36. [10]Kirkwood TB 1996 **Bioessays**; 18, 1009-1016. [11]MacLeay JM 2002, **Vet Clin Nth Am: Equine**; 18, xi-xii. [12]McFarlane D 2007, **Ageing Res Rev**; 6, 54-63. [13]McGowan TW, 2009 PhD thesis, [14]Mellor DJ et al. 1999, **Vet Rec**; 145, 299-304. [15]Proudman CJ et al. 2005, **Equine vet J**; 37, 366-370. [16]Reeves MJ et al. 1989, **Prev Vet Med**; 7, 149-160. [17]Reeves MJ et al. 1996, **Prev Vet Med**; 26, 285-301. [18]Schlueter AE & Orth MW 2004, **Equine & Comp Ex Physiol**; 1, 221-231. [19]Stevens KB et al. 2009 **Vet Rec.;** 164:6-10. [20]Traub-Dargatz JL et al. 2006, In: Equine geriatric medicine and surgery, Ed: J.J. Bertone, WB Saunders. pp 1-4. [21]USDA:APHIS:VS 2005, Fort Collins. [22]Wilson GJ & Liyou OJ 2005, **Aust Equine Vet**; 24, 79-83.

# INFLAMM-AGING AND IMMUNOSENESCENCE IN THE HORSE

**Amanda A. Adams, Ph.D. & David W. Horohov, Ph.D.**
**Lexington, KY**

## IMMUNOSENESCENCE

Advancing age in horses, as with other species, is associated with significant alterations in immune function. This age-related decline in immune function is referred to as immunosenescence and involves changes in multiple components of cellular and humoral immunity with the hallmark change being a decline in T cell function [1; 2]. Decreased proliferation or replicative senescence of T cells with age contributes to reduced responsiveness to vaccination which leads to increased susceptibility to infectious agents and cancers [1; 3; 4; 5]. While the phenomenon of immunosenescence in aged humans and experimental animals is widely recognized, the mechanisms underlying this process remain largely unknown. Aging-associated impairment of several signal transduction pathways has been observed in experimental animals and in humans. These include alterations in initial intracellular signaling events following T cell receptor engagement [6; 7; 8; 9; 10]. In addition, downstream signaling events have also been shown to be altered in aged individuals [11; 12; 13; 14]. These changes culminate in decreased gene expression of cell-surface and cytokine receptors [15; 16; 17]. These impaired signal transduction pathways affect the immunological synapse between aged T cells and antigen presenting cells leading to decreased proliferation of T cells and impaired immune responses [18; 19; 20; 21]. Thus, immunosenescence is now considered a state of immune dysregulation.

Advanced age in the horse is likewise associated with alterations in the immune system characteristic of immunosenescence; including decreased T cell proliferation [22; 23]. While decreased proliferative responses could be the result of apoptosis in aged lymphocytes due to impaired intracellular signaling [24; 25], there was no significant differences in the percent apoptotic positive peripheral blood mononuclear cells from old versus young horses [26]. Instead, stimulated peripheral blood mononuclear cells from older horses remain undivided or in the parental generation when compared to those from younger horses indicating a failure of these cells to progress through the cell cycle [26]. While a large percentage of PBMC from the aged horses remained non-dividing, these cells were capable of cytokine production as a significant percentage produced IFNγ. It is known that as T cells undergo replicative senescence, the *in vitro* production of pro-inflammatory cytokines such as IFNγ increases (Effros, 2007).

## INFLAMM-AGING

Paradoxically, advanced age is also associated with increased production of pro-inflammatory cytokines both *in vivo* and *in vitro,* a process referred to as "Inflamm-aging" [27]. Serum levels of inflammatory cytokines including IL-6 [28; 29] and TNFα [30] are significantly elevated in the elderly, while the anti-inflammatory cytokine IL-10 is decreased [31]. Furthermore, it has been shown *in vitro* that as T cells undergo replicative senescence, the production of pro-inflammatory cytokines increases [1]. *In vitro* studies of human mononuclear cells have shown an aging-related increased IL-1, IL-6, IFNγ and TNFα production [32]. Inflamm-aging cytokine profiles measured *in vitro* likely reflect the dysfunctional activity of senescent cells *in vivo*. This increased production of inflammatory mediators has been associated with increased morbidity and mortality in the elderly [30]. Moreover, chronic inflammation is characteristic of the pathological processes of age-related diseases such as atherosclerosis [33], osteoarthritis [34], Alzheimer's [35], Parkinson's [36] osteoporosis [37] and type 2 diabetes [38]. Though, it remains controversial whether inflammatory mediators have a primary causal relationship or simply contribute and/or aggravate the pathologies or are the consequence of age-related disease [27; 30; 39; 40].

While the inflamm-aging process has been recognized as a part of the aging process in humans and experimental animals, it has only recently been described for the horse [26]. Thus, there is increased production of IFNγ from lymphocytes collected from old horses. Older horses also have significant levels of IL-1, IL-15, IL-18 and TNFα gene expression in peripheral blood, increased levels of TNFα protein in their serum and increased frequencies of IFNγ and TNFα producing cells in their circulation. Together these results indicate that aged horses, like aged humans, exhibit signs of inflamm-aging which could impact upon their overall health status.

## ADIPOSITY & ITS EFFECT ON INFLAMM-AGING

While a number of factors probably contribute to the increased inflammatory response in the elderly, recent attention has focused on the role of increased body fat [41]. Though the relationship between adipose tissue and the immune system in aging remain largely unknown; it has been suggested that elevated body fat in older individuals may play a role in contributing to increased inflammation with age [42]. Obesity is now considered to be a chronic mild inflammatory state [43; 44]. Its incidence in the elderly is rising and will likely continue to increase [45]. Fat mass not only increases with age but is redistributed, resulting in greater abdominal fat or white adipose tissue (WAT) [46]. WAT is composed of adipocytes, fibroblasts, endothelial cells and macrophages [47; 48]. These resident adipocytes and macrophages have been shown to secrete a variety of pro-inflammatory cytokines such as TNFα, IL-6 and IL-1 in humans and rodents [48; 49; 50; 51]. While the exact source of circulating inflammatory cytokines remains unclear; there is no doubt that obese individuals have increased amounts of pro-inflammatory cytokines and reductions in body fat can decrease their levels [52; 53; 54; 55; 56; 57]. Adipocytes from the horse are also capable of producing inflammatory cytokine TNF-α[58]. In addition, obese adult horses have increased circulating levels of inflammatory cytokines [59]. Furthermore, PBMC from old horses have increased inflammatory cytokine production compared to young adults [60]; moreover, fat old horses compared to thin old horses have even greater frequencies of lymphocytes and monocytes producing inflammatory cytokines [60]. Therefore, we propose that decreasing adiposity in old horses would reduce age-associated increases of inflammatory cytokines both *in vitro* and *in vivo*, and likely diminish the adverse health effects of inflamm-aging.

## REFERENCES

[1] R.B. Effros, et al., In vitro senescence of immune cells. **Exp Gerontol** 38 (2003) 1243-9. [2] G. Pawelec, et al., T cells and aging, January 2002 update. **Front Biosci** 7 (2002) d1056-183. [3] R.L. Walford, The immunologic theory of aging, Munksgaard, **København**., 1969. [4] G. Pawelec, et al., Human immunosenescence: is it infectious? **Immunol Rev** 205 (2005) 257-68. [5] V.J. Cristofalo, et al., Cell senescence: an evaluation of replicative senescence in culture as a model for cell aging in situ. **J Gerontol A Biol Sci Med Sci** 58 (2003) B776-9; discussion 779-81. [6] R.L. Whisler, et al., Phosphorylation and coupling of zeta-chains to activated T-cell receptor (TCR)/CD3 complexes from peripheral blood T-cells of elderly humans. **Mech Ageing Dev** 105 (1998) 115-35. [7] T. Fulop, Jr., et al., Age-related impairment of p56lck and ZAP-70 activities in human T lymphocytes activated through the TcR/CD3 complex. **Exp Gerontol** 34 (1999) 197-216. [8] B. Chakravarti, et al., Effect of age on mitogen induced protein tyrosine phosphorylation in human T cell and its subsets: down-regulation of tyrosine phosphorylation of ZAP-70. **Mech Ageing Dev** 104 (1998) 41-58. [9] M.A. Pahlavani, et al., Activation of p21ras/MAPK signal transduction molecules decreases with age in mitogen-stimulated T cells from rats. **Cell Immunol** 185 (1998) 39-48. [10] M. Li, et al., Impaired signal transduction in mitogen activated rat splenic lymphocytes during aging. **Mech Ageing Dev** 113 (2000) 85-99. [11] T. Fulop, Jr., et al., Cellular distribution of protein kinase C isozymes in CD3-mediated stimulation of human T lymphocytes with aging. **FEBS Lett** 375 (1995) 69-74. [12] J.J. Proust, et al., Age-related defect in signal transduction during lectin activation of murine T lymphocytes. **J Immunol** 139 (1987) 1472-8. [13] M. Utsuyama, et al., Impairment of signal transduction in T cells from old mice. **Mech Ageing Dev** 93 (1997) 131-44. [14] R.A. Miller, Calcium signals in T lymphocytes from old mice. **Life Sci** 59 (1996) 469-75. [15] T. Fulop, et al., Cytokine receptor signalling and aging. **Mech Ageing Dev** (2006). [16] R. Mo, et al., T cell chemokine receptor expression in aging. **J Immunol** 170 (2003) 895-904. [17] M. Fann, et al., Gene expression characteristics of CD28null memory phenotype CD8+ T cells and its implication in T-cell aging. **Immunol Rev** 205 (2005) 190-206. [18] A. Tamir, et al., Age-dependent alterations in the assembly of signal transduction complexes at the site of T cell/APC interaction. **J Immunol** 165 (2000) 1243-51. [19] I. Beckman, et al., T cell activation in the elderly: evidence for specific deficiencies in T cell/accessory cell interactions. **Mech Ageing Dev** 51 (1990) 265-76. [20] J. Plowden, et al., Innate immunity in aging: impact on macrophage function. **Aging Cell 3** (2004) 161-7. [21] J. Plowden, et al., Impaired antigen-induced CD8+ T cell clonal expansion in aging is due to defects in antigen presenting cell function. **Cell Immunol** 229 (2004) 86-92. [22] D.W. Horohov, et al., Effect of exercise on the immune response of young and old horses. **Am J Vet Res** 60 (1999) 643-7. [23] D.W. Horohov, et al., The effect of aging on T cell responses in the horse. **Dev Comp Immunol** 26 (2002) 121-8. [24] G. Pawelec, et al., Decreased proliferative capacity and increased susceptibility to activation-induced cell death in late-passage human CD4+ TCR2+ cultured T cell clones. **Exp Gerontol** 31 (1996) 655-68. [25] S. Gupta, et al., Life and death of lymphocytes: a role in immunesenescence. **Immun Ageing 2** (2005) 12. [26] A.A. Adams, et al., Advanced age in horses affects divisional history of T cells and inflammatory cytokine production. **Mech Ageing Dev** 129 (2008) 656-664. [27] C. Franceschi, et al., Inflamm-aging. An evolutionary perspective on immunosenescence. **Ann N Y Acad Sci** 908 (2000) 244-54. [28] T.B. Harris, et al., Associations of elevated interleukin-6 and C-reactive protein levels with mortality in the elderly. **Am J Med** 106 (1999) 506-12. [29] R. Roubenoff, Catabolism of aging: is it an inflammatory process? **Curr Opin Clin Nutr Metab Care 6** (2003) 295-9. [30] H. Bruunsgaard, et al., Age-related inflammatory cytokines and disease. **Immunol Allergy Clin North Am** 23 (2003) 15-39. [31] M. Saurwein-Teissl, et al., An imbalance between pro- and anti-inflammatory cytokines, a characteristic feature of old age. **Cytokine** 12 (2000) 1160-1. [32] U. Fagiolo, et al., Increased cytokine production in mononuclear cells of healthy elderly people. **Eur J Immunol** 23 (1993) 2375-8. [33] R. Ross, Atherosclerosis is an inflammatory disease. **Am Heart J** 138 (1999) S419-20. [34] M.B. Goldring, et al., Osteoarthritis. **J Cell Physiol** 213 (2007) 626-34. [35] H. Akiyama, et al., Inflammation and Alzheimer's disease. **Neurobiol Aging 21** (2000) 383-421. [36] R.J. Dobbs, et al., Association of circulating TNF-alpha and IL-6 with ageing and parkinsonism. **Acta Neurol Scand** 100 (1999) 34-41. [37] R. Pacifici, Aging and cytokine production. **Calcif Tissue Int** 65 (1999) 345-51. [38] G. Paolisso, et al., Advancing age and insulin resistance: new facts about an ancient history. **Eur J Clin Invest** 29 (1999) 758-69. [39] C. Franceschi, et al., Inflammaging and anti-inflammaging: a systemic perspective on aging and longevity emerged from studies in humans. **Mech Ageing Dev** 128 (2007) 92-105. [40] K.S. Krabbe, et al., Inflammatory mediators in the elderly. **Exp Gerontol** 39 (2004) 687-99. [41] B.K. Pedersen, et al., Possible beneficial role of exercise in modulating low-grade inflammation in the elderly. **Scand J Med Sci Sports** 13 (2003) 56-62. [42] M. Pedersen, et al., Circulating levels of TNF-alpha and IL-6-relation to truncal fat mass and muscle mass in healthy elderly individuals and in patients with type-2 diabetes. **Mech Ageing Dev** 124 (2003) 495-502. [43] P. Dandona, et al., Inflammation: the link between insulin resistance, obesity and diabetes. **Trends Immunol** 25 (2004) 4-7. [44] M.W. Rajala, et al., Minireview: The adipocyte--at the crossroads of energy homeostasis, inflammation, and atherosclerosis. **Endocrinology** 144 (2003) 3765-73. [45] D.E. Arterburn, et al., The coming epidemic of obesity in elderly **Americans. J Am Geriatr Soc** 52 (2004) 1907-12. [46] D.T. Villareal, et al., Obesity in older adults: technical review and position statement of the American Society for Nutrition and NAASO, The Obesity Society. **Am J Clin Nutr** 82 (2005) 923-34. [47] A.H. Berg, et al., Adipose tissue, inflammation, and cardiovascular disease. **Circ Res** 96 (2005) 939-49. [48] S.P. Weisberg, et al., Obesity is associated with macrophage accumulation in adipose tissue. **J Clin Invest** 112 (2003) 1796-808. [49] G.S. Hotamisligil, et al., Adipose expression of tumor necrosis factor-alpha: direct role in obesity-linked insulin resistance. **Science** 259 (1993) 87-91. [50] H. Tilg, et al., Adipocytokines: mediators linking adipose tissue, inflammation and immunity. **Nat Rev Immunol** 6 (2006) 772-83. [51] E.E. Kershaw, et al., Adipose tissue as an endocrine organ. **J Clin Endocrinol Metab** 89 (2004) 2548-56. [52] P. Dandona, et al., Tumor necrosis factor-alpha in sera of obese patients: fall with weight loss. **J Clin Endocrinol Metab** 83 (1998) 2907-10. [53] J.P. Bastard, et al., Elevated levels of interleukin 6 are reduced in serum and subcutaneous adipose tissue of obese women after weight loss. **J Clin Endocrinol Metab** 85 (2000) 3338-42. [54] L. Fontana, et al., Aging, adiposity, and calorie restriction. **Jama** 297 (2007) 986-94. [55] A.M. Xydakis, et al., Adiponectin, inflammation, and the expression of the metabolic syndrome in obese individuals: the impact of rapid weight loss through caloric restriction. **J Clin Endocrinol Metab** 89 (2004) 2697-703. [56] J.M. Bruun, et al., Regulation of adiponectin by adipose tissue-derived cytokines: in vivo and in vitro investigations in humans. **Am J Physiol Endocrinol Metab** 285 (2003) E527-33. [57] K. Clement, et al., Weight loss regulates inflammation-related genes in white adipose tissue of obese subjects. **Faseb J** 18 (2004) 1657-69. [58] M.M. Vick, et al., Effects of systemic inflammation on insulin sensitivity in horses and inflammatory cytokine expression in adipose tissue. **Am J Vet Res** 69 (2008) 130-9. [59] M.M. Vick, et al., Relationships among inflammatory cytokines, obesity, and insulin sensitivity in the horse. **J Anim Sci** 85 (2007) 1144-55. [60] A.A. Adams, et al., Effect of body condition, body weight and adiposity on inflammatory cytokine responses in old horses. **Vet Immunol Immunopathol** (2008).

# MORALITY RATES AND RISK FACTORS FOR EUTHANASIA/DEATH IN AGED HORSES.

**TW McGowan[1], GL Pinchbeck[2], CM McGowan[2];**
**[1] The University of Queensland, QLD Australia, [2]Univeristy of Liverpool, UK.**

Mortality and survival studies are important in order to allow us to determine expected death rates, assess risk factors for death, or the effect of interventions on death rates. Yet to date there have been no studies specifically relating to mortality and survival of aged horses. The aim of this study was to determine the mortality rates of aged horses in Queensland, Australia; the causes of mortality and reasons for euthanasia; and to evaluate demographic and health factors associated with probability of survival in horses aged over 15 years.

Owners of horses aged 15 years or greater were sent a questionnaire via an Equestrian newsletter and information on signalment, management and health of horses that they owned obtained. They were contacted again 2-4 years later. The start date for estimating days to event (death) was defined as the date the owner completed the questionnaire. The end date was either the last date the horse was known to be alive or the date the horse died or was euthanased. If the horse/s had died, owners were asked if they could provide the date of death/euthanasia and the cause/reason of death/euthanasia. Analysis was performed using univariable and multivariable Cox proportional hazards analysis.

Follow up contact was made with 171 owners and survival data obtained on 513 horses. Of the 513 horses, 1.0% (5/513) did not have a recorded age, 2.4% (12/508) were aged 10-14 years, 44.9% (228/513) 15-19 years, 31.5% (160/513) 20-24 years, 15.6% (79/513) 25-29 years and 5.7% (29/513) were 30 years or older. The mean time of censor/death was 686 days (median = 670 days; Range = 6 - 1379). There were 91 deaths during 965 available horse making an average mortality rate of 943 deaths/10,000 horse years at risk. Of the 91 deaths, 11.0% of horses were found dead with the remaining 89.0% euthanased. The four most frequently listed reasons for euthanasia were lameness (including arthritis, unspecified and laminitis), weight loss, colic and equine Cushing's syndrome (ECS).

Multivariable analysis, using 4 separate models, revealed age, breed, region, use of horse for 'Riding for the Disabled', historical clinical signs of difficulty eating and colic, body condition score, previously diagnosed lameness, seedy toe and a positive test for ECS were associated with an increased hazard ratio.

This study is the first to investigate mortality rates in aged horses in Australia. The most import risk factors involved are signalment (age, breed and region), use, signs of dental disease, colic, lameness, appropriate body condition score and ECS.

# FECAL EGG COUNTS (FEC) FOLLOWING ANTHELMENTIC ADMINISTRATION IN AGED HORSES AND HORSES WITH EQUINE PITUITARY PARS INTERMEDIA DYSFUNCTION (PPID).

**D McFarlane, GM Hale, EM Johnson, LK Maxwell.**
**Oklahoma State University, Stillwater OK.**

Aged horses and horses with PPID are thought to be more susceptible to endoparasitism than are healthy adult horses, despite a lack of evidence to support this view. The objective of this study was to determine if the fecal egg count (FEC) is increased and the egg reappearance period is shorter in aged horses and horses with PPID as compared to healthy adult horses. Forty-two horses in the Stillwater, OK area were studied. Horses with PPID were identified by clinical signs and confirmed by plasma α-melanocyte stimulating concentration (α-MSH $\geq$45 pmol/L). Healthy horses were selected from the same or nearby farms. A modified Wisconsin sugar centrifugation fecal float was performed before and at two week intervals following ivermectin treatment. Prior to treatment, horses with PPID had a greater FEC compared with healthy horses (PPID: 918.6 ± 380.2 epg, n=13 versus Controls: 301.9 ± 119.3 epg, n=29, P=0.05). Multiple linear regression determined that the presence of PPID and post-anthelmentic treatment interval significantly affected FEC ($P \leq 0.05$), whereas age did not ($P > 0.05$). The median FEC from horses with PPID was greater than that of the closest aged control from the same farm at 0, 8, 10 and 12 weeks post-treatment ($P \leq 0.05$). In conclusion, horses with PPID had a greater FEC and a shorter time to egg reappearance following anthelmentic treatment as compared to healthy horses; age had no effect. Based on this data, increased diligence in parasite monitoring and preventative management may be warranted in horses with PPID.

**EFFECTS OF SEASON AND PASTURE GRASS CONSUMPTION ON BLOOD HORMONE AND METABOLITE CONCENTRATIONS IN HORSES WITH PITUITARY PARS INTERMEDIA DYSFUNCTION.**

**Frank N, Elliott SB, Tóth F, Chameroy K, McClamroch R, Etheridge EA. Department of Large Animal Clinical Sciences, College of Veterinary Medicine, University of Tennessee, Knoxville, TN 37996.**

Horses with pituitary pars intermedia dysfunction (PPID) may be at greater risk for developing pasture-associated laminitis because they respond differently to seasonal or dietary challenges. We therefore hypothesized that horses with PPID would exhibit hormonal and metabolic responses throughout a 12-month period that differed significantly from unaffected animals.

Blood samples were collected every month from August 2007 until July 2008 from 17 horses (9 mares, 8 geldings) ranging in age from 8 to 30 years (14 horses $\geq$ 20 years of age) located at the same farm. At each monthly sampling point, blood was collected between 0800 and 1000 immediately after horses were brought in from pasture. Horses were subsequently returned to pasture and then moved back into individual stalls overnight. Blood was collected again the following morning before hay or grain was fed. Blood hormone and metabolite concentrations were measured and pasture grass samples were collected and analyzed to determine carbohydrate content.

After samples were analyzed, horses were allocated to a PPID group (n = 8) if plasma adrenocorticotropin hormone (ACTH) concentrations exceeded 35 pg/mL on 3 or more occasions between December and June (outside of the late summer-fall period). Plasma ACTH concentrations were persistently elevated (> 35 pg/mL) in 5 of 8 horses in the PPID group. Nine horses were allocated to the control group. Analysis of variance analysis (ANOVA) for repeated measures was performed after logarithmic transformation of insulin data. Pasture, month, and pasture × month effects were significant ($P < 0.05$) for glucose, insulin, triglyceride (TG), and non-esterified fatty acids (NEFAs). The highest insulin concentrations were detected on pasture in September. A significant PPID status × pasture interaction was detected for TG and NEFAs. Results indicated that pasture grazing significantly influenced blood hormone and metabolite responses across a 12-month period, but differences between control and PPID groups were not revealed by ANOVA. However, plotted insulin concentrations were also examined and peaks $\geq$ 30 mU/L were identified for each horse. For samples collected after pasture grazing, peaks were identified in 7 of 8 horses in the PPID group (1-3 peaks/horse; total of 13) compared with 3 of 9 horses in the unaffected group (1-4 peaks/horse; total of 8). Insulin peaks were most common in September (all 10 horses with peaks) and April/May (6 horses). Only 3 horses from the PPID group exhibited insulin peaks when samples were collected after stall confinement. Alterations in blood variables did not correlate with percent carbohydrate composition of pasture grass, which suggests that horses responded to increases in the amount of grass available on pasture.

# PHYSIOLOGY OF CARDIAC MECHANICAL FUNCTION: SYSTOLIC & DIASTOLIC FUNCTION OF THE HEALTHY HEART

**Colin C. Schwarzwald, Dr. med. vet., PhD, DACVIM**
**Zurich, Switzerland**

## INTRODUCTION

The ability of the ventricles to eject blood into the circulation depends on systolic and diastolic ventricular function, atrial function, as well as heart rate and rhythm.[1-3] Structural and functional competency of the cardiac valves and ventricular septa influence ventricular function. Appreciation of the physiology of cardiac mechanical function and cardiovascular hemodynamics is central for the understanding of the pathophysiology of cardiac disease. Lack of familiarity with common cardiac disorders and unawareness regarding their influence on cardiac function and hemodynamics may lead to misdiagnosis, incomplete or incorrect clinical assessment, and possibly wrong decisions about prognosis and treatment. This lecture will provide an overview on some aspects of cardiac mechanical function that are important for the clinician performing a cardiovascular examination.

## VENTRICULAR SYSTOLIC FUNCTION

Normal cardiac activation is one of the prerequisites for normal heart function. The electrical excitation of the cardiac myocytes is transformed into mechanical activity in a process that is commonly referred to as electro-mechanical coupling or excitation-contraction coupling. It depends on the function of the L-type calcium channels (influx of extracellular calcium into the cell), the sarcoendoplasmatic reticulum (calcium-induced calcium release), and the myofilaments (cross-bridge cycling). Ventricular systolic performance, quantified by stroke volume, depends on myocardial contractility, preload, and afterload.[1-3] The traditional approach to regard contractility, loading conditions, and heart rate effects as independent determinants of systolic myocardial performance is outdated, because they are closely interrelated and a change in one variable will subsequently alter the others (Fig. 1).

**Figure 1:**
Determinants of stroke volume, cardiac output, and blood pressure.

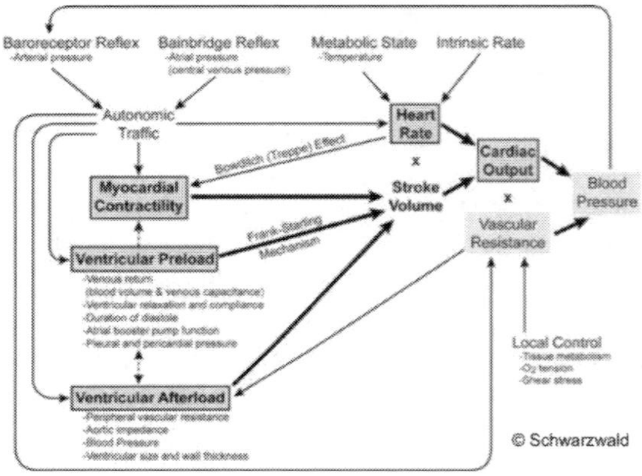


### Contractility (inotropy)

Myocardial contractility or inotropy is the inherent ability of cardiac contractile proteins, actin and myosin, to interact and to develop a contractile force in the presence of calcium. The three basic mechanisms that acutely regulate the contractile state under physiologic conditions are autonomic control by sympathetic or parasympathetic systems, preload, and heart rate. Beta-adrenergic stimulation enhances calcium cycling and sensitizes contractile proteins, thereby increasing the rate of contraction (and relaxation). An increase in preload causes sensitization of the myosin heads to calcium (Frank-Starling mechanism[4]), whereas an increase in heart rate progressively enhances the force of ventricular contraction due to a rate-dependent accumulation of cytosolic calcium (Bowditch or "treppe" effect). In the failing heart, contractility will decrease due to alterations in regulatory signal transduction mechanisms.

### Preload

Ventricular preload, the stretch of the ventricular myofilaments before ejection, depends on venous return and is a positive determinant of ventricular systolic function. The relationship between end-diastolic ventricular volume and systolic ventricular performance (force or pressure development) is known as Frank-Starling mechanism or Starling's law of the heart. The mechanical function of a healthy ventricle is strongly preload-dependent. Hypovolemia and subnormal venous filling pressures are two common causes of reduced cardiac output. In heart failure, preload is increased due to volume retention, in an attempt to recruit the Frank-Starling mechanism to maintain normal stroke volume despite the loss of intrinsic contractility.

### Afterload

Ventricular afterload represents the forces that resist the ejection of blood into the aorta and is closely related to the tension (or stress) in the ventricular wall during systole. According to Laplace's law, wall stress ($\sigma$) in a thin-walled sphere is defined as $\sigma = p \cdot r / 2d$, where p = pressure, r = radius, and d = wall thickness.[1,2] This formula assumes steady conditions and does not account for the pulsatile nature of blood flow and the dynamic aspects of afterload (aortic input impedance). However, it demonstrates that afterload increases with increased vascular resistance or aortic impedance (e.g., due to neurohumoral response to heart failure), ventricular dilation (e.g., due to chronic volume overload or dilated cardiomyopathy), or decrease in wall thickness. Conversely, hypotension (due to vasodilation) or increased wall thickness (due to physiologic hypertrophy, chronic pressure or volume overload, or hypertrophic cardiomyopathy) can decrease wall stress and reduce ventricular afterload. High afterload limits ventricular systolic performance and decreases stroke volume, thereby reducing cardiac output. The failing ventricle is more sensitive to afterload than the normal ventricle.

**Ventricular interdependence**

Due to the serial association of the pulmonic and the systemic circulation and the close anatomic association of the two ventricles within a common pericardial sack, the function of one ventricle directly influences the function of the other ventricle. This interaction between the two ventricles is commonly referred to as ventricular interdependence.[5,6] Direct interaction is caused by forces transmitted through the interventricular septum and by pericardial constraint. Series interaction is equally important and refers to the fact that, over time, cardiac output from the left ventricle (LV) has to equal that of the right ventricle (RV) because ventricular filling depends on venous return and output from the contralateral ventricle. Under physiologic conditions, ventricular interdependence is responsible for the respiratory variations in stroke volume and systemic blood pressure. Its effects become more pronounced and clinically relevant in patients with right ventricular volume or pressure overload, pericardial disease, and congestive heart failure.

## VENTRICULAR DIASTOLIC FUNCTION

Diastolic ventricular function determines ventricular preload and influences coronary perfusion, which primarily occurs during diastole. Factors that affect diastolic function include active ventricular relaxation (affecting isovolumetric relaxation and early filling), ventricular wall thickness and compliance (affecting passive ventricular filling in mid-to-late diastole), atrial rhythm and atrial mechanical function (determining atrial contribution to ventricular filling), heart rate and rhythm (determining filling time), sympathetic activity (improving active relaxation and increasing filling pressures), venoconstriction and venous return (influencing filling and preload), coronary perfusion (assuring oxygen and energy supply needed for the active relaxation process), pericardial disease (restraining cardiac filling), and intrapleural pressure (opposing ventricular filling pressures, thereby reducing transmural pressures responsible for distention of the ventricles). [1-3,7-9] Generally, horses with abnormal diastolic function depend more on heart rate and require higher filling pressures to distend the ventricles and to maintain cardiac output.

**Ventricular relaxation (lusitropy)**

Myocardial relaxation, also referred to as lusitropy, is an active, energy-consuming process linked to re-uptake of calcium from the cytoplasm into the sarcoendoplasmatic reticulum and extrusion of calcium through the cell membrane. Lusitropy is closely linked to inotropy. Sympathetic stimulation exerts a positive lusitropic effect. Conversely, relaxation is impaired with hypoxia, acidosis, and many cardiac diseases, including disorders leading to ventricular hypertrophy, reduced coronary perfusion, and ischemia.

**Ventricular compliance**

Myocardial compliance (the inverse of stiffness) reflects the passive diastolic pressure volume relationship. It is defined as the change in ventricular volume produced by a given change in ventricular filling pressure (dV/dP). Alterations in ventricular compliance primarily affect passive filling during mid to late diastole. Myocardial fibrosis, ventricular hypertrophy, pericardial disease, or severe ventricular dilation reduce compliance.

## ATRIAL FUNCTION

Atrial mechanical function can be divided into 3 distinct phases: (1) During ventricular systole, the atria act as a reservoir to receive blood from venous return; (2) during early- and mid-diastole, the atria serve as a conduit, allowing emptying of the stored blood into the ventricles and continued passage of venous return from the large veins into the ventricles; (3) during late diastole, the atria actively contract and serve as booster-pumps that establish the final ventricular end-diastolic volume and contribute to preload.[10-12] It is determined by the complex interrelation between atrial loading conditions, atrial and ventricular inotropic state, the rate and extent of atrial and ventricular relaxation, atrial and ventricular compliance, heart rate, and intraatrial electrical conduction.

At resting heart rates, the loss of atrial pump function is not necessarily associated with a deterioration of hemodynamics and is usually clinically irrelevant if no other cardiac diseases coexist. During high-intensity exercise or in patients with a failing ventricle and impaired ventricular diastolic function, sudden atrial dysfunction can cause acute decrease of ventricular systolic performance, leading to poor performance or clinical signs of heart failure.

## CONTROL OF VENTRICULAR MECHANICAL FUNCTION

Ventricular mechanical function in the healthy heart is largely preload-dependent (Frank-Starling mechanism[4]). The vasomotor centers in the medulla oblongata, receiving input from the baroreceptors and from a variety of other peripheral receptors, are responsible for central regulation via the autonomic nervous system that innervates the heart via parasympathetic (negative chronotropic, dromotropic, inotropic, and lusitropic) and sympathetic (positive '-tropic' effects) branches.[1-3]

## REFERENCES


[1]Berne RM, Levy MN. **Cardiovascular physiology**, 8th ed. Mosby, St. Louis, 2001. [2]Opie LH. **Heart Physiology**, 4th ed. LWW, Philadelphia, 2004. [3]Schwarzwald CC, et al. **The cardiovascular system**. In: Muir WW, Hubbell JAE. Equine anesthesia, 2nd ed. Saunders St. Louis, 2009:37-100. [4]Fuchs F, Smith SH. **News Physiol Sci** 2001;16:5-10. [5]Belenkie I, et al. **Ann Med** 2001;33:236-241. [6]Santamore WP, **Gray L. Chest** 1995;107:1134-1145. [7]Lester SJ, et al. **J Am Coll Cardio** 2008;51:679-689. [8]Hamlin SK, et al. **Am J Crit Care** 2004;13:453-466. [9]Villars PS, et al. **Am J Crit Care** 2004;13:394-403. [10]Pagel PS, et al. **Anesthesiology** 2003;98:975-994. [11]Stefanadis C, et al. **Eur Heart J** 2001;22:22-36. [12]Hoit BD. **Europ Heart J Suppl** 2000;2:K9-K16.

# ECHOCARDIOGRAPHIC ASSESSMENT OF CARDIAC DIMENSIONS & MECHANICAL FUNCTION: CONVENTIONAL METHODS, CURRENT ISSUES, & CONTROVERSIES

**Colin C. Schwarzwald, Dr. med. vet., PhD, DACVIM**
**Zurich, Switzerland**

## INTRODUCTION

Over the last 30 years, echocardiography has evolved to become a standard diagnostic procedure in equine cardiology. Development of M-mode echocardiography and introduction of 2-dimensional (2D) real-time gray-scale imaging provided the basis for non-invasive evaluation of cardiac structures, chamber dimensions, and chamber function in horses. Doppler echocardiography led to the ability to assess blood flow characteristics in the equine heart. Many publications have addressed the clinical use of echocardiography in horses, including 2D, M-mode, and Doppler echocardiography. However, the assessment of chamber dimensions and mechanical function of the heart is still challenging and is limited by a variety of technical, anatomical, and physiological issues that need to be considered when performing echocardiographic examinations.[1-3] The examinations should be performed in a consistent and systematic manner, using a variety of standardized imaging planes and measurements. The findings should be critically assessed in the light of clinical findings and should not be 'over- or misinterpreted'. Also, despite the availability of quantitative echocardiographic measurements, subjective assessment remains a cornerstone of echocardiography and should not be neglected.

This presentation aims to provide an overview on the echocardiographic assessment of cardiac dimensions and mechanical function in horses, to discuss some of the current issues and controversies in the field of equine echocardiography, and to delineate the limitations of conventional echocardiographic methods.

## ASSESSMENT OF CARDIAC DIMENSIONS

One of the main goals of echocardiography is the detection of cardiac chamber dilation or hypertrophy and the grading of the enlargement, if present, as mild, moderate, or severe. Cardiac dimensions can be quantified using linear, area, and volumetric methods that are based on 2D and M-mode echocardiography, but all of them are complicated by a variety of issues that should be considered.[1]

### Left Ventricle

*Eccentric hypertrophy*, characterized by left-ventricular (LV) dilation and normal wall thickness, is seen with volume overload due to valvular regurgitation, left-to-right shunt, or compensated cardiomyopathy. *Simple dilation* without hypertrophy of the LV can be seen with severe dilated cardiomyopathy or myocarditis. *Concentric hypertrophy*, characterized by thickening of the LV wall at the expense of its lumen, is rare in horses and may be seen with hypertrophic or infiltrative cardiomyopathy (e.g., amyloidosis) or as a consequence of misuse of anabolic steroids. A volume-depleted LV may also appear hypertrophied ('pseudohypertrophy'). Mixed patterns are possible. The assessment of LV dimensions is a crucial part of every echocardiographic examination and provides important information on hemodynamics, LV remodeling, and severity of disease.

LV dimensions are usually measured from a right-parasternal short-axis view or from a right-parasternal four-chamber view. The systolic (s) and diastolic (d) LV internal diameter (LVID), the thickness of the interventricular septum (IVS), and the thickness of the LV free wall (LVFW) are measured at the chordal level using 2D or M-mode recordings. These linear measurements of the LV minor dimension may not well describe true LV size, as asymmetric LV dilation and dimensional changes along the major axis of the ventricle are neglected. Similarly, linear measurement of wall thickness may not be accurate estimates of LV mass, particularly if asymmetric thickening is present or if the M-mode cursor crosses a papillary muscle or fails to sample the thickest part of the wall. Furthermore, standardized placement of the M-mode cursor (i.e., bisecting the LV at the chordal level, parallel to the mitral annulus) may be challenging and is difficult to verify in all three dimensions. Measures that consider the short-axis area and the length of the LV (i.e., various area-length models and the Simpson's method of discs) may be more accurate estimates of internal LV dimensions and LV mass. However, true LV long-axis in systole and diastole may be difficult to measure by 2D echocardiography due to the LV shape and the translational motion of the LV during contraction. Currently, volumetric assessment of the LV cavity and the LV mass is primarily used in research settings, but is not routinely employed in the workup of horses with cardiac disease.

### Left Atrium

Enlargement of the left atrium (LA) develops in association with mitral valve disease, left-to-right shunts, cardiomyopathies, or other causes of systolic or diastolic LV dysfunction. The magnitude of enlargement is often used to assess the severity of hemodynamic consequences of chronic disease and the extent of myocardial remodeling.

In horses, assessment of LA size has traditionally been limited to subjective evaluation and measurement of the LA diameter. However, the two methods traditionally used to measure the LA diameter have significant limitations. M-mode recordings from a right-parasternal short-axis view at the level of the heart base largely depend on the placement of the cursor line and often provide a measure of the LA appendage rather than the LA body. 2D echocardiography in a left-parasternal long-axis view often does not allow imaging of the LA in its entirety due to interference with the ventral lung border; this often results in measurements of the LA diameter that are not parallel to the mitral valve annulus or that are made too close to the annulus and thereby underestimate the true maximum atrial diameter. Also, the timing of LA measurements is often not specified in the equine literature, leading to uncertainties when comparing results from different studies.

With contemporary echocardiographic systems, the LA can usually be displayed in its entirety from a right-parasternal window in long-axis and short-axis, which allows displaying of the entire LA and provides sufficient anatomical landmarks for consistent measurements of the largest LA dimensions in several directions and at any instance of the cardiac cycle. In addition to LA diameter, assessment of LA area should be considered because enlargement may not occur in a uniform fashion and LA geometry may change over time.[4] The LA size is best measured at the end of the ventricular systole, 1 frame prior to mitral valve opening, when the LA chamber is at its greatest dimension.

**Right Atrium & Ventricle**

Causes of right-atrial (RA) and right-ventricular (RV) dilation include severe tricuspid valve disease (regurgitation, rarely atresia), pulmonary hypertension, atrial septal defect, large ventricular septal defect, right heart failure, and biventricular heart failure. Subjective assessment of the right heart chambers in multiple image planes is important, but requires some experience. Quantification of the RA and RV dimensions is difficult, because the geometrical shape of the right heart is complex and the internal dimensions of the RA and the RV cavity largely depend on transducer placement and imaging plane. When measured in M-mode from a right-parasternal short-axis window, the internal diameter of the RV should be smaller than the diameter of the LV.

## ASSESSMENT OF SYSTOLIC VENTRICULAR FUNCTION

The echocardiographic detection of myocardial failure and deterioration of systolic function has great prognostic implications. It is important to realize that the echocardiographic indices of systolic ventricular function generally do not reflect contractility per se, but are (to a variable extent) influenced by preload, afterload, heart rate, and rhythm. Generally, it is advised to use a variety of echocardiographic variables to assess LV systolic function and to interpret them in the context of the clinical findings.[2,3]

**Two-dimensional Ejection Phase Indices**

The 2D ejection phase indices are based on measurements of LV dimensions, and the same limitations as discussed above apply to these indices. Left ventricular ejection fraction (EF% = SV/LVEDV·100; where SV is stroke volume and LVEDV is LV end-diastolic volume) has traditionally been the standard index of LV systolic function. The LV fractional shortening (FS% = [LVIDd-LVIDs]/LVIDd·100), is an approximation of the EF% and is the most commonly used index of LV systolic function in horses. In fact, often times it is the only one used in routine echocardiography, because it can be easily calculated from M-mode recordings. However, reliance on this index as a single measurement of LV function may be problematic. The FS% actually represents the relative shortening of the LV short axis in only one single dimension, disregarding the fact that the LV contracts in all three dimensions. Also, it may lack accuracy when the IVS and the LVFW do not contract synchronously or when the cursor line is not placed optimally. In horses with mitral regurgitation (MR) and normal myocardial function, the FS% can actually be increased due to the increased preload and decreased afterload. Conversely, a normal FS% in the presence of severe MR may indicate (but does not necessarily prove) myocardial failure.

Area-based measurements such as the LV fractional area change (FAC%) are less sensitive to asynchronous wall motion and allow assessment of shortening in two dimensions. Volume-based measurements (ejection fraction, stroke volume, cardiac output) can be calculated using a variety of geometrical models and are generally considered more accurate and least affected by altered chamber geometry. However, there are differences and limitations related to the geometrical model used. Currently, area- and volume-based measurements of LV systolic function are primarily used in research settings.

**Doppler-derived Indices of Stroke Volume & Cardiac Output**

Doppler-based estimates of stroke volume (SV) and cardiac output (CO) are indices of global ventricular function. The SV is calculated by multiplying the velocity-time integral (VTI) of the aortic or pulmonic flow signal by the cross-sectional area of the respective vessel. There are many limitations of this echocardiographic estimate of SV, including the angle-dependency of Doppler-based flow measurements, inaccuracies in the determination of the flow area, and uncertainties as where exactly to measure flow and cross-sectional area (or diameter) of the vessel.[5] In adult horses, adequate alignment of the Doppler beam with blood flow by transthoracic echocardiography (TTE) is difficult and Doppler estimates of SV are neither considered very accurate nor reliable. Transesophageal echocardiography (TEE) allows parallel alignment between Doppler beam and aortic blood flow in adult horses, and CO measurements by TEE in adult, anesthetized horses have been shown to be in better agreement with thermodilution CO than CO measurements by TTE.[5,6] However, TEE requires expensive, custom-made transesophageal ultrasound probes and is not routinely available. In foals, alignment of the ultrasound beam with blood flow can be achieved using apical imaging planes. Although this may improve the accuracy of CO measurements by Doppler echocardiography, in foals volumetric measurements by 2D echocardiography using the bullet method have been shown to provide better agreement with CO by lithium dilution than Doppler estimates of CO.[7]

Generally, measurements of CO by echocardiography may be more accurate for detecting acute changes in CO over time than for obtaining absolute measurements of CO. Assessment of SV and CO by Doppler is not routinely performed in horses with cardiac disease. In fact, myocardial dysfunction generally must be very severe for blood flow indices to be affected, because cardiac output is maintained in the failing heart until the compensatory mechanisms (e.g., neuro-humoral activation, volume overload, Frank-Starling mechanism) are overwhelmed. Therefore, estimates of SV and CO may not be clinically relevant in horses with heart disease.

**Measurement of Systolic Time Intervals**

Systolic time intervals (STI), including the LV pre-ejection period (PEP), the LV ejection time (LVET), and the LV pre-ejection period-to-ejection time ratio (PEP/LVET), can be measured from M-mode images of aortic valve motion and from Doppler tracings of aortic blood flow, respectively. The STIs may serve as alternative indicators of LV function that are independent of ventricular shape and geometry and that may be superior to the calculation of the FS%. However, they are also variably influenced by heart rate and loading conditions. Accurate time intervals may be difficult to obtain due to inability to clearly identify the onset and the end of ejection. Their clinical value has not been well established in horses with cardiovascular disease.

**ASSESSMENT OF DIASTOLIC FUNCTION**

Diastolic ventricular function is more complex than systolic function. Ventricular filling results from the dynamic interplay of active ventricular relaxation, ventricular compliance, filling pressures, pericardial restraint, ventricular interaction, and atrial function. Afterload and contractility will further influence diastolic function. Therefore, the echocardiographic assessment of diastolic ventricular function and filling pressures (which are commonly assessed together) is difficult. In people, diastolic dysfunction is common to virtually all forms of cardiac failure and associated with impaired ventricular relaxation and increased ventricular stiffness.[8,9] In horses, diastolic ventricular dysfunction may certainly play a role in pericardial disease, but the clinical relevance of diastolic dysfunction in other types of cardiac disease is not clear and the prevalence of diastolic heart failure is unknown.

Doppler-derived transmitral flow velocities (E wave, A wave, E/A ratio, E-wave deceleration time and deceleration slope) are commonly used in humans and small animals for assessment of diastolic LV function and filling pressures. The E/A ratio can be used to classify filling patterns into 'normal', 'impaired relaxation' (reduced E/A ratio), and 'restrictive' (increased E/A ratio). However, the usefulness of this index is limited, because impaired relaxation coexisting with increased filling pressures may lead to an apparently normal E/A ratio (so called 'pseudonormalized' pattern).[2] Furthermore, PW Doppler recordings of transmitral flow velocities strongly depend on alignment with blood flow and are sensitive to placement of the sample volume relative to the position of the mitral valve. Transmitral flow velocity profiles can be recorded in horses, but optimal alignment with blood flow from an apical window is not possible in adult horses. Transmitral flow velocity measurements in horses are relatively unreliable and may not be suitable to detect minor changes in diastolic LV function.[4,10,11] Other indices of LV diastolic function, including the isovolumic relaxation time (IVRT) and pulmonary venous flow patterns are even more difficult to obtain consistently and are not used in horses.

**HEMODYNAMIC ASSESSMENT**

The hemodynamic load placed on the heart by cardiac lesions can be estimated by echocardiography, combining information of chamber size, cardiac motion, LV systolic and diastolic function, and intracardiac blood flow. Doppler studies can be used to assess intracardiac pressures and pressure gradients. Normal pressure gradients that generate red blood cells to flow range between 0.25 and 1.5 m/s. Abnormally high velocities can be found in many cardiac conditions, including ventricular septal defect and valvular regurgitation, in which a pressure gradient drives blood across a restrictive orifice. The pressure gradients are either reflections of normal intracardiac pressures or consequences of pathologically increased pressures. Pressure gradients can be estimated using continuous wave Doppler by employing the simplified Bernoulli equation ($dp = 4 \cdot v_{max}^2$, where dp is the pressure gradient in mmHg and $v_{max}$ is the peak velocity in m/s). The main indications to measure pressure gradients in horses are the diagnosis of pulmonary hypertension (by interrogation of regurgitant flow at the tricuspid and pulmonic valves) and the assessment of ventricular septal defects (by interrogation of the shunt flow). When pressure gradients are to be estimated, it is important to optimize alignment with blood flow. Excessive angle of interrogation (i.e., > 20 degrees to blood flow) will lead to an underestimation of flow velocities and pressure gradients. The automated angle correction feature that is available on most imaging systems should not be used, because the true direction of flow is unknown and angle correction may lead to overestimation of true velocity and severity of the lesion.

**ASSESSMENT OF VALVULAR REGURGITATION**

Assessment of valvular regurgitation should be achieved using an integrated qualitative and quantitative approach, combining clinical examination (including auscultation of a typical murmur) and 2D, M-mode and Doppler echocardiography. Measurement of cardiac chamber dimensions provides information on the hemodynamic relevance of chronic valvular regurgitation. Abnormal timing and direction of transvalvular flow as well as flow turbulences can be detected by color flow mapping (CFM). The regurgitant signal in the 'receiving chamber' can be interrogated. Multiple imaging planes must be used to accurately identify the origin and to assess the extent of the regurgitation. The duration of the regurgitation should be timed relative to the ECG using spectral Doppler or color M-mode techniques. It is important to realize that CFM only represents blood flow direction and velocity but not absolute volumetric flow. Quantification of regurgitation by assessing signal strength of the spectral Doppler regurgitant signal or measuring the area of regurgitation within the receiving chamber is therefore problematic. The Doppler-derived regurgitant signal is largely influenced by gain settings, direction of flow, orifice size and shape, driving pressure, flow, and characteristics of the receiving chamber. Therefore, receiving chamber analysis of the regurgitant jet should not be used as a sole measure of severity.

Determination of the regurgitant orifice area by CFM is considered an appropriate approach for the assessment of regurgitant severity, if the true orifice can be measured. The 'vena contracta' is the narrowest part of the CFM regurgitant signal and corresponds to the effective orifice area. The regurgitant flow rate or volume and the regurgitant orifice area can also be calculated by analyzing the proximal isovelocity surface area (PISA). However, alignment of the ultrasound beam with regurgitant flow is often inadequate in horses and frequently neither the vena contracta nor the isovelocity shells in the zone of flow convergence can be identified.

The Doppler technology of current echocardiographic systems is very sensitive to detect valvular regurgitation and care must be taken not to overinterpret the echo findings (particularly in otherwise healthy animals without abnormal clinical findings and in the absence of heart murmurs). A scoring system can be used for assessment of valvular regurgitation, combining information on the duration of the regurgitant signal, high-velocity jet area and flow disturbance, vena contracta width, proximal flow convergence surface area, regurgitant signal duration, and the number of imaging planes in which the high-velocity jet can be observed.[12]

## INFLUENCE OF SEDATION

Some horses do not tolerate the echocardiographic procedure and require sedation to allow conducting the exam safely and with sufficient quality. Many of the echocardiographic changes seen with sedation are small and, in the light of marked variability, the clinical relevance of the effects of sedation may in fact be minor. Nonetheless, sedation should be avoided during cardiovascular examinations, because cardiac function will be altered due to drug effects on contractility, preload, afterload, rate, and rhythm.

A few studies have addressed the influence of sedation on echocardiographic indices of cardiac size and function.[13-15] In healthy horses, alpha-2 agonists (detomidine, 0.01 mg/kg bwt i.v.; romifidine, 0.04 mg/kg bwt i.v.) cause a reduction in heart rate, an increased incidence of 2nd degree AV block, and a reduction in LV systolic performance (increased end-systolic LV dimensions, decreased FS% and FAC%, decreased systolic thickness of LVFW and IVS, increased PEP/ET). Changes in end-diastolic LV dimensions and the LA diameter are generally minor. Acepromazine (0.03[15] and 0.1[14] mg/kg bwt i.v.) appears to have less effect upon standard echocardiographic variables in healthy horses[14,15] It has also been reported that alpha-2 agonists, but not acepromazine, increase the incidence of (supposedly physiological) valvular regurgitation detected by echocardiography in healthy horses without audible murmurs.[14] Acepromazine may therefore be the preferred sedative for quantitative echocardiography.

However, it is important to realize that the results of studies performed in healthy horses may not be directly translated to horses with cardiac disease. The function of the healthy heart mainly depends on preload (Frank-Starling mechanism). Conversely, in the failing heart, preload is increased (volume overload) and afterload becomes more important in determining (limiting) cardiac performance. Therefore, the drug effects may have different consequences in diseased hearts compared to healthy hearts. Furthermore, the valvular regurgitant fraction largely depends on the driving force, i.e. on the pressure difference across the affected valve. Whenever the pressures on either side of the valve change, the regurgitant fraction will change (increase or decrease). Therefore, alterations in filling pressures (influencing preload), systemic arterial pressures (influencing afterload), and contractility will invariably change the regurgitant fraction and therefore the appearance of valvular regurgitations on echocardiography. Generally, alpha-2 agonists tend to worsen regurgitation[16], while acepromazine may actually reduce regurgitation due to a decrease in afterload.

## STRESS ECHOCARDIOGRAPHY

Cardiac disease may be an important cause of poor performance in horses, but the diagnostic options for the detection of cardiac causes of poor performance are limited. Heart murmurs are commonly found in athletic horses, even in the absence of clinical signs. Unless severe valvular regurgitation, significant chamber enlargement, obvious myocardial dysfunction, or marked cardiac dysrhythmia can be diagnosed at rest, it is challenging to prove an etiologic relationship between an echocardiographic finding (i.e., mild or moderate valvular regurgitation) and impaired performance. Furthermore, myocardial disease can exist in the absence of any detectable murmur or rhythm disturbance, and can be subclinical at rest.

Stress echocardiography has been used in people for diagnosis of myocardial ischemia and coronary artery disease, assessment of myocardial viability, prognosis and risk stratification after acute myocardial infarction, and detection of myocardial diseases that are not evident at rest.[17] The primary goal of stress echocardiography is the detection of stress-induced regional wall motion abnormalities. Only in selected cases stress echocardiography is used for evaluation of valvular regurgitation. Stress echocardiography has also gained interest in equine medicine to provide additional information on global and regional myocardial function and valvular competency during exercise. However, coronary artery disease and myocardial infarction are rare in horses and the indications for stress echocardiography and the target diseases to be detected are not as well defined as in humans. Generally, the indications would include the presence of mild- to moderate valvular regurgitation, mild to moderate cardiac chamber enlargement, or suspected myocardial disease. However, the true clinical value of stress echocardiography in horses is largely unknown to date.

Several issues need to be considered that may limit the clinical value of stress echocardiography. Echocardiographic assessment of LV function in horses after exercise is currently limited to measurement of chamber dimensions, calculation of ejection phase indices derived from 2D or M-mode images, manual tracking of endocardial motion, and subjective evaluation of regional wall motion. Quantitative approaches to evaluate regional myocardial function have not been well established in horses. The reliability of cardiac function assessment by stress echocardiography, the influence of heart rate on indices of cardiac function, the optimal heart rates for stress echocardiography, and the ideal stressors are unknown. Certainly, differences related to the type of stress must be considered. Treadmill exercise represents true physical stress and is associated with peripheral vasodilation in the working muscles. Therefore, cardiac activation and hemodynamics are likely clinically relevant. The disadvantages of treadmill exercise are that the echocardiogram has to be performed during the recovery period after exercise. Therefore, the time available for the echocardiographic examination is limited and the heart rates are changing rapidly. Pharmacological stress testing (e.g., using a combination of atropine and dobutamine) may overcome these limitations, as HR can be kept increased over a prolonged period of time.[18] It mimics exercise testing reasonably well in healthy horses. However, hemodynamic alterations during pharmacologic stress testing differ from physiologic alterations during treadmill exercise (e.g., lack muscle activity and peripheral vasodilation, leading to differences in preload and afterload). It is currently unknown if pharmacologic stress testing is useful and safe to replace treadmill exercise testing.

## REFERENCES

[1]Lang RM, et al. **Europ J Echocardiogr** 2006;7:79-108. [2]Otto CM. **Textbook of clinical echocardiography**, 3rd ed. Saunders, Philadelphia, 2004. [3]Boon JA. **Manual of veterinary echocardiography**. LWW, Baltimore, 1998. [4]Schwarzwald CC, et al. **Am J Vet Res** 2007;68:735-747. [5]Blissitt KJ, et al. **Equine Vet J** 1997;29:18-25. [6]Young LE, et al. Br **J Anaesth** 1996;77:773-780. [7]Giguere S, et al. **J Vet Intern Med** 2005;19:737-743. [8]Lester SJ, et al. **J AmColl Cardiol** 2008;51:679-689. [9]Kass DA, et al. **Circ Res** 2004;94:1533-1542. [10]Schwarzwald CC, et al. **J Vet Int Med** 2007;21:820-827. [11]Blissitt KJ, Bonagura JD. **Equine Vet J Suppl** 1995:38-46. [12]Young LE, et al. **J Vet Intern Med** 2008;22:418-426. [13]Patteson MW, et al. Equine Vet J Suppl 1995:33-37. [14]Buhl R, et al. **Vet Anaesth Analg** 2007;34:1-8. [15]Menzies-Gow **NJ. Vet Rec** 2008;163:21-25. [16]Gehlen H, et al. **Schweiz Arch Tierheilk** 2004;146:119-126. [17]Douglas PS, et al. **J Am Coll Cardiol** 2008;51:1127-1147. [18]Sandersen CF, et al. **J Vet Intern Med** 2006;20:575-580.

# ECHOCARDIOGRAPHIC ASSESSMENT OF CARDIAC DIMENSIONS & MECHANICAL FUNCTION: RECENT DEVELOPMENTS & CURRENT INVESTIGATIONS

**Dr. Colin C. Schwarzwald, Dr. med. vet., PhD, DACVIM**
**Zurich, Switzerland**

## INTRODUCTION

Echocardiographic measurement of chamber dimensions and assessment of chamber function are important aspects of cardiac evaluation. Besides subjective visual assessment, a variety of 2-dimensional (2D) and M-mode measurements have traditionally been used to quantify chamber dimensions and function in horses. Normal values have been reported, and applicability and reliability have been established for several echocardiographic indices. However, routine assessment of chamber size has largely been based on 1-dimensional, linear measurements, which may not accurately reflect changes in chamber geometry and actual chamber size. Furthermore, the functional characteristics of the cardiac chambers have been incompletely studied in horses, and most of the indices used in clinical practice serve to evaluate global systolic left-ventricular (LV) function, while diastolic LV function, left-atrial (LA) function, right heart function, regional myocardial function, and myocardial synchrony are rarely considered. This presentation provides an overview on recent developments in equine echocardiography and gives some insights into current investigations studying the use of novel echocardiographic methods for assessment of cardiac size and function in horses.

## RECENT DEVELOPMENTS IN ECHOCARDIOGRAPHY

Our understanding of the physiology of cardiac function is becoming increasingly complex, and discoveries made over the last few decades indicate that cardiac performance depends on a multitude of interrelated factors, including systolic and diastolic function, ventricular synchrony, and regional myocardial function, that can be altered by hemodynamic load, neurohumoral activation, and cardiac injury. Many of the newer concepts of cardiovascular physiology and pathophysiology have been developed using novel cardiovascular imaging methods. In particular, a variety of advanced echocardiographic modalities recently have been introduced in human cardiology, many of which may also improve the future diagnostic capabilities in equine cardiology.

### Area- and Volume-based Measurements

Recent human imaging guidelines emphasize the advantages of volume-based assessment of chamber dimensions and function.[1] Conventional, 1-dimensional (linear) measurements of chamber dimensions are easy to obtain, but may be somewhat insensitive in the detection of chamber dilation and dysfunction, particularly in the presence of non-uniform chamber enlargement, alterations in chamber geometry, and presence of regional wall motion abnormalities.[1] Conversely, volume determinations allow accurate assessment of asymmetric remodeling of the cardiac chambers.[1] The use of volume measurements for quantification of chamber dimensions, chamber function, and ventricular mass should probably not be considered a recent development in echocardiography, because volume estimates based on conventional 2D gray-scale images have been used for quite some time. However, estimation of chamber volumes by 2D echocardiography and subsequent calculation of ejection phase indices is not commonly employed in horses, mostly because the procedure is time consuming and often requires precise identification of at least two perpendicular imaging planes that are difficult or even impossible to obtain in horses. All conventional volumetric indices are calculated based on several geometrical assumptions and approximations. Although the inaccuracies related to the calculations have been considered inevitable and of minor clinical importance, many situations require accurate measurements, particularly when the course of a disease or the response to treatment is to be followed with serial examinations.[1] Volume estimates based on single-plane 2D images can be obtained in horses, but are considered inaccurate and may not provide superior information compared to linear or single-plane area measurements. With the introduction of real-time 3-dimensional (3D) echocardiography, volume-based assessment of chamber dimensions and chamber function will possibly replace the conventional linear or 2-dimensional echocardiographic indices in the near or mid-term future. However, because of the size of a horse's heart and thorax, the currently available echocardiographic systems do not allow real-time, 3D investigation of an entire ventricle in adult horses. Therefore, the time for routine volumetric measurement of cardiac dimensions in horses has not yet arrived. However, in addition to conventional linear measurements one may consider measuring the cross-sectional areas of the cardiac chambers and of the myocardium in long-axis and short-axis views, particularly in cases where the subjective evaluation and the conventional measurements give equivocal results.

### Anatomical M-mode

The conventional M-mode technique is an "old" technique that has the advantage of a very high pulse repetition frequency (200-1000 Hz), leading to a very high temporal resolution that allows studying mobile structures. Its major limitation is that the cursor line can only be rotated around the apex of the imaging sector. Therefore, optimal placement of the cursor may be challenging or may not be possible at all. The development of fully digital echocardiography systems with highly improved spatial and temporal resolution of 2D images allowed the introduction anatomical M-mode (AMM; sometimes called differently by certain manufacturers), allowing free orientation of the cursor line within the imaging plane.[2] AMM can be applied to every myocardial structure and in every spatial direction. In practice, AMM greatly facilitates optimal placement of the cursor line. The AMM is a post-processing technique that allows offline analysis of conventional 2D cineloop recordings. Therefore, the time required for data collection can be significantly shortened and measurements can be performed at a later time. Obviously, this can be of great advantage when dealing with unsedated and, at times, uncooperative horses. In offline mode, AMM even allows tracking of moving structures. For example, it allows following the spatial position of the aortic valve on 2D cineloop recordings in order to better identify the time of aortic valve opening and aortic valve closure, providing the basis for measuring of systolic time intervals. For optimal results, 2D recordings should be performed at the highest possible spatial and temporal resolution (i.e., by minimizing the sector width and imaging depth). The spatial resolution may become a limiting factor when the angle between the cursor line and the line of insonation is too high and the imaging depth is too deep. The temporal resolution of AMM recordings depends on the recording frame rate and becomes most relevant when time intervals are to be measured. The AMM technique is available on most contemporary echocardiography systems and can easily be applied to equine echocardiography. Clinical applications of AMM include the assessment of LV function and LV wall motion and the evaluation of aortic and mitral valve motion.

**Tissue Velocity, Strain, & Strain Rate Imaging**

*Tissue Doppler imaging (TDI)* allows quantifying myocardial wall motion velocity and transmural velocity gradients.[3] In TDI mode, the echocardiograph is set to filter out the low-intensity, high-velocity signals (i.e., blood flow) and to amplify the high-intensity, low-velocity signals (i.e., myocardial motion). The tissue velocity signals can be displayed as spectral tracings (pulsed-wave = PW TDI mode) or as color-coded maps that are superimposed on a 2D gray-scale image (color = cTDI mode). Similar to other Doppler-based methods, velocity measurements by TDI depend on the angle of insonation and can only be performed in a single dimension. *Tissue velocity imaging* is used in humans and small animals for assessment of global systolic and diastolic LV function, detection of regional wall motion abnormalities, diagnosis of ventricular dyssynchrony, and estimation of ventricular filling pressures.[4] Echocardiographic strain and strain rate imaging provides additional indices for assessment of regional and global ventricular function.[3,5] *Strain (ε)* is a measure of deformation of a myocardial segment, expressed as percentage of change from its original dimension. For normal myocardium, systolic ε is an analog of regional ejection fraction. *Strain rate (Sr)* is the temporal derivative of ε and describes the rate of deformation. Systolic Sr reflects regional contractile function. In people, ε and Sr have been valuable for diagnosing subclinical myocardial disease and coronary artery disease, and for differentiating hypertrophy caused by hypertension or cardiomyopathy.[5,6] Strain and strain rate imaging may further be useful to detect subclinical systolic dysfunction in valvular heart disease, and it may allow objective assessment of myocardial performance during stress testing to identify stunning, acute ischemia, and myocardial infarction. Finally, detection of ventricular dyssynchrony may be relevant in patients with heart failure. Strain and SR can be calculated using TDI and 2D speckle tracking (2DST).[3,5-7] The latter is based on conventional 2D recordings. It analyzes the gray scale "speckles" as they move during contraction and relaxation and allows tracking of myocardial motion throughout the entire cardiac cycle. The 2DST method offers a Doppler-independent approach to the measurement of ε and Sr that may overcome some of the limitations of strain measurements by TDI.[3,5-7] Importantly, 2DST is independent of the angle of insonation, allows simultaneous measurement of ε and Sr in 2 dimensions, and provides data on rotational motion, derived from circumferential strain at different levels of the LV.[3,5] Thus, if complex regional wall motion patterns are to be investigated, 2DST is considered a preferred method. Only few reports on TDI in equine echocardiography are available,[8-11] and the use of 2DST has not been reported in this species to date. To our current knowledge, horses rarely suffer from clinically relevant myocardial ischemia, coronary artery disease, and myocardial infarction, and the prevalence of congestive heart failure is certainly low. The clinical relevance of regional wall motion abnormalities and ventricular dyssynchrony is largely unknown in horses. Therefore, it is uncertain whether these new echo modalities will offer any advantage compared to the conventional 2D, M-Mode, and Doppler methods commonly used in horses. However, myocardial diseases may be underappreciated in horses, at least in part because of the limitations of the currently used diagnostic methods. Therefore, there is a need to investigate novel echocardiographic methods such as TDI and 2DST, to better quantify wall motion abnormalities and ventricular dyssynchrony that may indicate occult myocardial disorders.

## RECENT & ONGOING INVESTIGATIONS IN HORSES

**Assessment of Left Atrial Size & Function**

One of our recent studies aimed to show the feasibility, describe the techniques, and determine the reliability of transthoracic echocardiography for noninvasive assessment of LA size and LA mechanical function.[9] Six standing, unsedated, healthy horses underwent repeated echocardiographic examinations performed by 2 independent observers using 2-dimensional (2D) echocardiography, flow Doppler, and tissue Doppler imaging (TDI) techniques. Test reliability was determined for all echocardiographic variables. The results indicated that LA size and LA mechanical function can be reliably assessed in standing, unsedated horses by use of 2D echocardiography, transmitral Doppler flow velocity profiles, and analyses of LA wall motion by TDI. Clinically, the use of 2D variables of LA size and function, including the LA diameter, LA area, and LA fractional area changes, is straightforward and may be most useful. Transmitral flow velocities and TDI variables may provide additional information on LA function, but are more difficult to record and are less reliable compared to 2D indices.

**Detection of Atrial Dysfunction After Treatment of atrial Fibrillation**

The results of the previous study were subsequently applied to investigate the effects of AF on LA mechanical function in horses after conversion of AF to sinus rhythm (SR).[10] Five Standardbreds with AF and 6 healthy Standardbreds were included. Echocardiographic indices of LA function were measured in normal horses (once) and in AF horses (24 h and 72 h after conversion) using 2D, transmitral flow Doppler, and TDI techniques. The 2D and TDI indices of LA mechanical function indicated significant decrease of LA contractile function and LA reservoir function 24 hours after cardioversion. This decrease was no longer statistically significant 72 hours after cardioversion, but changes in echocardiographic variables between 24 and 72 hours varied among horses. The results of this study showed that LA contractile dysfunction can be detected in horses by use of 2D echocardiography, transmitral Doppler flow velocity profiles, and analyses of LA wall motion by TDI. The decrease of LA contractile function after conversion from

AF to SR was consistent with AF-induced atrial 'stunning', although residual treatment effects or influence of underlying primary myopathy could not be ruled out.

**Assessment of Left Ventricular Radial Wall Motion Using Tissue Doppler Imaging**

The goal of another study was to show the feasibility, describe the techniques, and determine the reliability of TDI for characterization of LV radial wall motion in horses.[11] Six healthy, adult horses were included in the study. Repeated echocardiographic examinations were performed by 2 observers in standing, unsedated horses. The LV was imaged in a right-parasternal short-axis view at the chordal level. PW TDI and cTDI examinations were performed on a segment of the LV free wall. Systolic and diastolic radial wall motion velocities and systolic and diastolic time intervals were measured. Test reliability was determined for all variables. Assessment of LV radial wall motion by TDI was feasible in all horses. The results showed that radial wall motion of the LV free wall can be reliably characterized in standing, unsedated horses by use of PW TDI. The use of cTDI is technically more challenging, requires specific analysis software, appears less reliable, and does not seem to offer significant advantages over PW TDI. Calculation of cTDI-based transmural velocity gradients and radial strain was highly variable and thus unreliable.

**Left Ventricular Radial & Circumferential Wall Motion Analysis by 2D Speckle Tracking**

The goal of this study was to show the feasibility, describe the techniques, and determine the reliability of 2DST for characterization of LV radial and circumferential wall motion that could be useful for assessment of regional and global LV systolic and diastolic function in horses.[12] Six healthy, adult horses were included. Repeated echocardiographic examinations were performed by 2 observers in unsedated horses. A right-parasternal short-axis view of the LV at the level of the chordae tendineae was imaged in 2D mode at a frame rate between 59 and 73 FPS. 2DST measurements included circumferential and radial peak-systolic strain ($\varepsilon_C$, $\varepsilon_R$); circumferential and radial strain rate during systole ($Sr_{C-S}$; $Sr_{R-S}$), early diastole ($Sr_{C-E}$; $Sr_{R-E}$), and late diastole ($Sr_{C-A}$; $Sr_{R-A}$); and maximum systolic radial displacement ($D_R$). All measurements were reported by the software as average over each of 6 wall segments. Global strain, global strain rate, and global displacement were then calculated as averages over all segments. The 2DST analyses were feasible in 16 of 18 echocardiographic studies, and global $\varepsilon_R$, $Sr_{R-S}$, and $D_R$ by 2DST were reliable. Circumferential strain and strain rate measurements were less reliable. Diastolic strain rate measurements were invalid due to inaccurate tracking, possibly related to undersampling. The use of longitudinal strain and strain rate by 2DST will have to be investigated in a future study.

**Quantitative Assessment of Stress Echocardiograms**

In an ongoing study we are investigating the use of AMM, TDI and 2DST for quantitative assessment of LV wall motion in horses undergoing stress echocardiography.[13] The goals are to describe the techniques and determine the rate-dependence of echocardiographic variables in horses after high-speed treadmill exercise. Five healthy Warmblood horses in athletic condition underwent treadmill exercise testing. Echocardiography was performed before (baseline) and over 5 min after peak exercise. Right-parasternal short-axis images of the LV at the chordal level were recorded in 2D mode at a frame rate of 54 fps and analyzed offline. Preliminary measurements were performed at baseline and post-exercise at target heart rates of 120, 100, and 80/min. AMM was used to measure LV fractional shortening (FS). Segmental peak $\varepsilon_R$ and peak $Sr_{R-S}$ were measured by 2DST and global $\varepsilon_R$ and $Sr_{R-S}$ were calculated. Time to peak SR ($t_{SR}$) was measured in each segment and the standard deviation of measurements ($SD_{tSR}$) was calculated to assess synchrony. AMM and 2DST analyses were possible for all recordings. The preliminary results show that FS by AMM and global $\varepsilon_R$, $Sr_{R-S}$, and $SD_{tSR}$ by 2DST can be used to characterize LV wall motion and LV synchrony in horses after treadmill exercise.

## CONCLUSIONS & FUTURE DIRECTIONS

Our studies aimed to formulate preliminary recommendations for the echocardiographic assessment of LA and LV function by conventional 2D and Doppler echocardiography as well as novel TDI and 2DST techniques. The results can be used in future studies to investigate the applicability and the clinical value of these echocardiographic methods in horses with cardiac disease, at rest and during stress testing. Reference values for different populations will have to be generated, "normal" and "abnormal" findings will have to be defined, and the diagnostic and prognostic relevance of abnormal findings will have to be determined.

## A WORD OF CAUTION

Some of the results of our studies, particularly those concerning the cTDI and the 2DST methodology, may only pertain to the equipment used in this study (GE Vivid 7 and EchoPac Software), because different algorithms used by other manufacturers may provide slightly different measurements of deformation. Furthermore, despite the encouraging results of recent studies, one needs to be careful not to over-interpret them and not to be too enthusiastic about the potential clinical value of these new imaging methods. The importance of the clinical findings and the conventional echocardiographic indices should not be ignored. It is well possible that some of the novel echocardiographic indices will prove to be of clinical value in future studies. Nonetheless, many or even most of them will probably not withstand the test of time and will not make an entrance into the routine echocardiographic protocols.

## REFERENCES

[1] Lang RM, et al. **Eur J Echocardiogr** 2006;7:79-108. [2] Carerj S, et al. **Echocardiography** 2003;20:357-361. [3] Stoylen A. Website **(http://folk.ntnu.no/stoylen/index.html)** 2008. [4] Yu CM, et al. **J Am Coll Cardiol** 2007;49:1903-1914. [5] Marwick TH. **J Am Coll Cardiol** 2006;47:1313-1327. [6] Abraham TP, et al. **Circulation** 2007;116:2597-2609. [7] Thomas JD, Popovic ZB. **J Am Coll Cardiol** 2006;48:2012-2025. [8] Sepulveda MF, et al. **Equine Vet J** 2005;37:222-226. [9] Schwarzwald CC, et al. Am J Vet Res 2007;68:735-747. [10] Schwarzwald CC, et al. **J Vet Int Med** 2007;21:820-827. [11] Schwarzwald CC, et al. **J Vet Int Med** 2007;21:590 (Abstract). [12] Schwarzwald CC, et al. **J Vet Int Med** 2008;22:759 (Abstract). [13] Schwarzwald CC, et al. 3rd **Congress of the ECEIM**, Barcelona, Spain 2009 (Abstract).

# APPLYING EVIDENCE-BASED VETERINARY MEDICINE TO POPULATION-BASED CLINICAL QUESTIONS

**Robert (Bob) L. Larson, DVM, PhD, ACT, ACVPM-Epi**
**Kansas State University**

In their daily roles, many veterinarians encounter clinical questions that are based on animal populations such as livestock on a herd or region basis, wildlife in their natural environment, zoological populations, companion animal populations in breeding colonies or shelters, and human populations. The role of an evidence-based approach to veterinary involvement with populations of animals is in many ways similar to an evidence-based approach to individual animals, but with a couple of clarifications. The first is that in populations of animals, the distribution of outcomes (a bell-shaped or skewed distribution) is observed closer to its entirety, and the second is that the economic measures of biologic effects and interventions are more removed from constraints associated with close identification with individual animals. Because of these differences, in many ways it is simpler for a population-based veterinarian to perceive the benefits of an evidence-based or clinical epidemiological approach to veterinary practice.

## DAILY PRACTICE WITH THE BELL-SHAPED CURVE

The normal (aka Bell-Shaped or Gaussian) distribution or near-normal distributions are the most commonly observed distributions of measurements of almost everything in daily life - and is the starting point for modeling many natural processes. The term normal distribution refers to the particular way that measured observations tend to pile up around a particular value rather than to be spread evenly across a range of values – or to take a single value. A normally distributed variable can be described by two parameters, the mean (the point at which the curve is its maximum height) and the standard deviation (how wide it is). Normal distributions are found in events that are the aggregation of many smaller, but independent (maybe unobservable) random events. The conditions that tend to produce bell-shaped curved outcomes are:

1. The presence of many factors contribution to the variable
2. The factors are independent in occurrence
3. The factors are independent in their effects
4. The factors make equal contributions to the variance (variability)

The reasons that the normal distribution is so common is that genetic control of living organisms is made up of many binomial (yes/no) outcomes because each gene has two alleles and each allele randomly comes from one parent. The outcome of multiple independent binomial outcomes is a "bell-shaped curve". Another reason that many outcomes are bell shaped is even if some of the factors listed above are not true – i.e. the factors don't make equal contributions to the variance – as the number of individuals measured increases, the distribution in outcomes becomes more "bell shaped". In addition, even for outcomes that are not "bell shaped" – but are "skewed", the average outcomes of many skewed distributions takes on a normal distribution.

One of the foundational tenets of veterinary medicine is to realize that no biological processes (growth, health, reproduction, behavior, etc.) results in a single level of output, but rather an output that has a distribution. Even if every aspect under human influence can be controlled, the output (or measurements) will not all be identical. Many outputs have a normal distribution, meaning that the output is symmetrical around the mean (½ of the output is above the mean and ½ of the output is below the mean). Other outputs may be skewed (greater than ½ of the output is either above or below the mean). In veterinary medicine, measures such as rectal temperature, packed cell volume, average daily gain, and weight have a distribution that is close to normal (symmetrical around the mean). In contrast, percent morbidity, percent mortality, and fecal egg counts have a skewed distribution.

Standard deviation is a mathematical depiction of the amount of variation in the output of interest. If the standard deviation is small, each individual output measurement is similar to the mean of all the output measurements. If the standard deviation is large, any individual measurement may be very different (higher or lower) than the mean of all the measurements. Regardless of the amount of variation in the output, in a normal distribution, roughly 68% of all the output falls within one standard deviation of the mean, 95% fall within two standard deviations of the mean, and 99.7% fall within three standard deviations of the mean. With skewed output, the percentage of the output that fall within a given standard deviation to the mean is similar to a normally distributed output.

### Quincunx Machine

The Quincunx machine was originally invented by Sir Francis Galton to demonstrate normal distribution and the law of error. The machine consists of a vertical board with interleaved rows of pins. Balls are dropped from the top, and bounce randomly left and right as they hit the pins. Every time a ball hits a pin it has a probability of 50 percent to fall to the left of the nail and a probability of 50 percent to fall to the right of the nail. Eventually, they are collected into one-ball-wide bins at the bottom. The height of ball columns in the bins approximates a bell curve. The final position of each ball is determined by many (number of rows in a Quincunx machine) independent, random events of whether to drop to the left or the right of the pin, thus the (approximate) normal distribution. Therefore, this simulation provides an illustration of the central-limit theorem, which states that the distribution of the sum of $n$ random variables approaches the normal distribution when $n$ is large

Galton's interest was not in the mean but in the distribution of deviations from it. He saw the normal distribution as a tool for analyzing populations in terms of their members' variations from a mean -the kind of variations inevitably manifest in biology and all other natural sciences - for example, body temperature, breaths per minute, height, weight, daily gain in weight, etc. from any randomly selected group of animals. He saw (as do I) that one of the most amazing things about life was the wonderful order in the variation.

Veterinarians who daily work with populations see the distribution in biologic outcomes and come to intuitively expect variation in output even as inputs appear to be the same. Thus, they are accustomed to asking about population measures such as mean, median, and variance and tend to realize that asking about individual performance is not a valid biologic assessment. To use the quincunx machine example, a population-based, or clinical epidemiology-based veterinarian understands the absurdity of asking why a particular ball landed in a particular bin. Rather, a population-based veterinarian is only interested in the distribution in output from a defined set of variables and avoids over-interpreting the final outcome of a single animal. This perspective may allow population-based veterinarians to more easily embrace the rigorous and exclusive nature of evidence-based veterinary medicine principles.

**Economic Measures of Biologic Effects**

Veterinarians are given the responsibility to optimize the utilization of scarce resources (money, time, labor, etc.) to enhance the health, welfare, and productivity of animal populations. Economic measures of livestock production provide a fairly good estimate of biologic cost of disease to populations of animals. These biologic costs include death loss (reduced income), clinical disease (treatment costs and reduced growth and reproductive efficiency), and subclinical disease (reduced growth and reproductive efficiency). Labor, facilities, pharmaceutical inputs, vaccines, and diagnostic testing are the costs that can be quantified with economic measures incurred by veterinarians to prevent or treat disease.

Many (but not all) costs of disease occurrence can be captured at the farm level including production loss, direct treatment costs, and labor costs including hourly wages, workman's compensation, and overtime costs. Within this limitation, by evaluating how the biologic cost of a particular disease or condition (as quantified by economic measures) compares to the economic cost of preventing and treating that disease or condition, veterinarians can determine where the scarce resources of labor, time, and money are best spent to alleviate the biologic cost due to disease. A number of tools are available to approximate the cost of a disease, and the biologic characteristics of the disease determine the proper economic analysis. Partial budgets are appropriate for diseases that are sporadic or horizontally transmitted and immunity or other responses (death, sterility, removal from population, etc.) confine the negative effect of the disease to a short period of time. Multi-year enterprise analyses are necessary to estimate the economic cost of diseases that are vertically transmitted, due to an environmental source, or for diseases that have a chronic production-losing component.

The total biologic or economic loss associated with a disease condition is not the estimate on which veterinarians should focus when evaluating health interventions – veterinarians should focus on the avoidable loss. It should be recognized that there are technical and biologic limits to the amount of loss that can be avoided.[1,2] Within the context of avoidable loss, specific interventions can be evaluated for their effectiveness in reducing the biologic loss (as quantified by economic cost) compared to their economic cost. The optimum set of interventions will reduce economic losses while at the same time conserve resources that can be used more effectively on other important problems.

Each available health intervention can be evaluated for its potential to avoid biologic loss compared to its economic cost. The first intervention implemented should be the one that provides the greatest biologic (as measured by economics) advantage per unit of economic cost, assuming the benefit:cost ratio is greater than one. Further health interventions may be implemented if they add to the biologic benefit at a benefit:cost ratio greater than one. Because the most cost-effective interventions are implemented first, one will expect diminishing returns from increasing expenditures on disease prevention. At the point where additional expenditures on health interventions equals the benefit, further interventions should cease and any remaining scarce resources (time, money, labor, etc.) should be targeted to other important problems.

**Using Probabilistic Models**

Probabilistic decision making is the everyday challenge of making decisions in situations of uncertainty. Unfortunately, not even highly trained experts are able to model probabilistic decision making in their minds. Even committees of experts are unable to do probabilistic decision-making. Computers are required to do stochastic simulation when several uncertain inputs and their myriad of interactions are evaluated. Probabilistic models provide the decision maker with the range in outcomes that will occur based on the inputs of the model.

An analysis is considered deterministic when a variable is treated as having a single value even when it is known to vary, or when its true value is not known with certainty. Variability is the natural phenomena of inherent dispersion. This type of dispersion is not reduced by more research or better knowledge. In contrast, uncertainty is used to describe the fact that we have incomplete knowledge. An uncertain variable has a knowable distribution, but we have not done the necessary research to arrive at an accurate estimate. Uncertainty can be reduced with better knowledge. Both variability and uncertainty are ignored in a deterministic model, which promotes a false sense of certainty, and allows one's biases to influence the outcome of the model.

It is difficult for the human mind or spreadsheets to deal with the amount of variability inherent in biological decisions. The term Monte Carlo simulation refers to the process of using computer simulation to appropriately select a value from the distributions of values for each variable in a model based on the probability of that value occurring, and then calculating the net return. The computer program repeats the entire procedure thousands of times and tracks the results. After many simulations, it is possible to graph the distribution of net returns and to examine the descriptive statistics of the output such as the mean, standard deviation, and the probability of a particularly negative or positive return. If the computer plays the game long enough, we will have a very good idea of the distribution of possible results.[3]

The advantage of probabilistic models is that the outcome of the model is not a single value, but an expected range of values with the probability for each level of output. This appropriate uncertainty in the output allows the veterinary decision maker to not only determine the most commonly expected outcome, but also the range of expected outcomes and the likelihood that the outcome will be particularly negative or positive.

**SUMMARY**

Evidence-based practice for both population-focused and individual animal-focused veterinarians is gaining increasing attention. A key foundation for understanding biology is its inherent and predictable variability. Evidence-based practices grounded in clinical epidemiologic principles account for this variability and help practitioners avoid inaccurate interpretations of biologic data. Population-based veterinarians have the advantage of working with population-based data on a daily basis and should have an inherent understanding the "bell-shaped curve" and its implications to practicing high-quality medicine.

In addition, population-based veterinarians are routinely called upon to allocate scarce resources among competing health, welfare, and production goals. The appropriate allocation of economic resources is a contentious issue across veterinary and human medicine and a rigorously-applied evidence-based approach allows for the best outcome for a given population given the resources available.

**REFERENCES**

[1] McInerney J.P., et. Al. *Prev. Vet. Med.*,1992: 13: 137; [2] Ott S.L., et al. *Prev. Vet. Med.*, 1995: 24: 203 [3] *Making Hard Decisions: An Introduction to Decision Analysis* (1991), R.T. Clemen ed. Belmont, CA, Duxbury Press, pp. 167

## APPLYING EVIDENCE-BASED VETERINARY MEDICINE TO INDIVIDUAL-BASED CLINICAL QUESTIONS

**Robert (Bob) L. Larson, DVM, PhD, ACT, ACVPM-Epi**
**Kansas State University**

### INTRODUCTION

Evidence-based medicine (EBM) was introduced to the medical literature in a 1992 article by the Evidence-based Working Group at McMaster University Health Sciences Centre in Canada to label the clinical learning strategy they had been developing for over a decade.[1] The basic concepts of EBM are being applied to the veterinary profession under the term evidence-based veterinary medicine (EBVM) as the formal and explicit incorporation of research evidence into the veterinary practice setting.[2,3,4] Evidence-based veterinary medicine is an effort to increase the occurrence of basing clinical decisions on a dispassionate review of published trials that adequately meet *a priori* standards of experimental design and experimental execution. There are a number of definitions for EBM available, but one I find most complete is from the International Epidemiological Association Dictionary of Epidemiology, "the consistent use of current best evidence derived from published clinical and epidemiological research in management of patients, with attention to the balance of risks and benefits of diagnostic test and alternative treatment regiments, taking account of each patient's unique circumstances, including baseline risk, co-morbid conditions and personal preference."[6,7]

### LEVELS OF EVIDENCE

Evidence used to assist clinical decision making can be very strong (rigorously tested in the target-species under natural conditions in experiments designed to prove a theory to be false), very weak (not tested), or intermediate.[9,10,11] The hierarchy of evidence is based on the strength of evidence of causation, the ability of the study to control bias, and the similarity between the study population and the population currently being considered in a clinical setting. Sources regarded as strong evidence for therapeutic questions include randomized controlled clinical trials in the target species under a similar husbandry environment or systematic reviews of more than one trial that meet these criteria. These are followed by epidemiological studies such as cohort studies and case-control studies, and then controlled experimental challenge trials, and case-series. The weakest evidence for making clinical decisions includes extrapolations based on *in vitro* research, pathophysiologic rationale, and uncontrolled observations. Unfortunately, the number of studies and amount of available literature decreases as one moves from weak to strong evidence.

Veterinarians considering the strength of evidence must use several perspectives to determine the reliability of research for clinical use. The first consideration is the internal validity of the research which is determined by the study method and appropriate use of controls for bias. Research reports with good internal validity provide assurance that the results represent an unbiased estimate of the true direction and magnitude of the treatment effect. The second consideration is the population used in the research and its appropriateness as a model for the population that generated the clinical question. Generally, the target species in similar housing and husbandry environments provides stronger evidence than the target species in significantly different housing and husbandry environments, related species, unrelated species, or in-vitro methods. And thirdly, the clinical relevance of the outcomes of the research should be considered with patient-/herd-oriented outcomes providing more direct evidence than disease-oriented outcome measurements. Using these considerations, the highest rating in all three dimensions would provide the highest level of evidence.

These three considerations can be used to evaluate the strength and quality of evidence in a published study, by asking the following questions: First, are the results of this individual study valid? This can be addressed by determining if the assignment of patients to treatments was randomized and if the caregivers and clinicians were blind to those assignments. It is also important that all animals in a group or treatment were evaluated, and that all groups were treated equally except for the experimental therapy. In addition, it is important that the trial last long enough to recognize both potential positive and negative outcomes. Randomization of study subjects to treatment group is critical for study validity. Studies in which treatment is allocated by any method other than randomization tend to show larger and frequently false-positive treatment effects than do randomized trials.[12,13,14]

If it seems that the study results are valid, the second question one needs to ask is, are the valid results of this individual study important? Or asked another way, how large was the treatment effect? This question may be addressed by a report of the magnitude of treatment effect, such as relative risk reduction, absolute risk reduction, or number needed to treat. To determine if a particular magnitude of effect is important, the effect produced by the therapy in question should be compared to the magnitude of effect for other interventions and durations of therapy. The true magnitude of effect of a therapy can never be known, but a valid trial provides a point estimate around which the true effect is expected to lie. The 95% confidence interval of the point estimate is commonly reported to communicate the precision of the treatment effect estimate, and indicates the range that includes the true treatment effect 95% of the time. Confidence intervals reflect the level of uncertainty in point estimates and indicate the expected range of values within which the true effect is likely to fall. The larger the sample size and the larger the number of outcome events, the greater is one's confidence that the true treatment effect is near the point estimate (narrow 95% confidence interval). To determine the importance of a reported positive treatment effect, one should look at the lower boundary of the confidence interval. If this lower boundary is important to you as a clinician, then you are confident that the treatment in question only has a 2.5% chance of failing to reach this lower boundary and the treatment can be deemed important. In contrast, if the treatment effect indicated by the lower boundary is not clinically important to you, the study cannot definitively indicate an important treatment effect even if the results are statistically significant. Similarly, to determine the risk of a reported neutral or negative treatment effect, one should look at the upper boundary of the confidence interval. If the upper boundary is clinically important, you cannot exclude the possibility that an important treatment effect is possible even though the trial indicated either no difference between treatment and controls or a negative treatment effect.

And finally, if a study appears to be valid and important, are the valid, important results of this study applicable to my client's animal or herd?[15] If the study animals or treatment setting is sufficiently different from the patient/herd that initiated the question, or if the treatment is either not feasible or not consistent with the client's/manager's values and expectations, a valid and important therapy would not be applicable in your current situation. For a reported trial to influence treatment strategies, one should ask whether all clinically important outcomes were considered. And most importantly, were the outcomes of most importance to you, the clinician, and the owner/manager reported? Trials that report substitute endpoints such as blood cell counts, blood chemistry, or body temperature are not as valuable for evaluating treatment options as indications of reduced mortality risk or improved production (weight gain, conception rates, etc.). Even when a report indicates one or more positive outcomes associated with a treatment, one should consider whether other negative outcomes are also associated with that treatment. This is particularly important if the negative outcomes are more important to the owner/manager than are the positive outcomes.

**USING EBVM IN CLINICAL PRACTICE**

Until recently, efficiently finding the current best evidence to a clinical question was a difficult challenge. However, developments have permitted a new reality. The first of those developments has been the creation of searchable databases (such as Medline, CAB, etc.) that are continually updated and available via the web. The other development is the increased education of veterinarians in the areas of experimental design, statistics, and epidemiology in order to be able to appraise evidence for its validity and relevance. In many ways, EBVM is a method to improve information literacy in the veterinary profession and to extend learning beyond formal classroom (and continuing education) settings and into daily practice. The definition of information literacy from the Association of College & Research Libraries (ACRL) is "A set of abilities requiring individuals to recognize when information is needed and to have the ability to locate, evaluate, and use effectively the needed information."[16]

According to the ACRL standards, an information literate individual is able to:

- Determine the extent of information needed
- Access the needed information effectively and efficiently
- Evaluate information and its sources critically
- Incorporate selected information into one's knowledge base
- Use information effectively to accomplish a specific purpose
- Understand the economic, legal, and social issues surrounding the use of information, and access and use information ethically and legally

In human medicine, the creation of systematic reviews and the development of information systems that bring these reviews to a practitioner's desktop computer almost instantaneously have greatly enhanced the ability of medical practitioners to practice evidence-based medicine. A systematic review is a structured review of the literature that includes a clear statement of the purpose of the review, a comprehensive search and retrieval of the relevant research, explicit selection criteria, critical appraisal of the primary studies, and reproducible decisions regarding relevance, selection, and methodological rigor of the primary research.[17,18] In a systematic review, emphasis is placed on results from studies of higher quality in preference to results than from lower quality studies; and for questions with an adequate number of studies, may exclude lower quality experimental designs. This rigorous method is in contrast to a traditional review whereby the completeness of the literature search and evaluation rigor of the primary research is not obvious to the reader. Veterinarians need to demand this level of rigor in systematic reviews in the veterinary literature, as well as be willing to pay for these types of resources for use in day-to-day practice.

**OPPORTUNITIES AND CHALLENGES OF EBVM**

Even though EBVM has logical merit, it has not been substantiated in controlled trials to give clinical results superior to other methods of practice because such evaluations have not taken place. In addition, EBVM focuses on gathering and evaluating the best available evidence; but using the best evidence available will not likely provide an improved clinical outcome if the veterinarian making the clinical decisions does not have good critical thinking skills. In addition, one of the greatest challenges to EBVM is that by knowing the strength of evidence for potential interventions, the veterinary practitioners will often be less certain of his/her diagnostic, therapeutic, and prognostic pronouncements compared to client expectations for greater certainty. This puts the expectations and values of the client at odds with the expectations and values of the EBVM-practitioner. It appears that many clients gravitates toward the clarity of an untested (or under-tested), but strongly stated pronouncement in preference to the skeptical, methodical, testing-rejecting-retesting pace of the scientific method.

While EBM has gained interest and integration into medical teaching and practice, EBVM has lagged behind because of several challenges. Compared to human-patient-centered clinical and field trials, veterinary-patient-centered clinical and field trials are very rare.[3] In addition to the slight investment that veterinary medicine has made in research that creates the higher levels of evidence, veterinary medicine must stretch those investments over a great many species. Therapeutic questions that are investigated in one species may or may not provide accurate evidence in other, even related species. Also, in human medicine, there are two practical factors that would appear to favor the adoption of EBM, at least in the U.S. One is the pressure of cost efficiency by medical insurance and health maintenance organizations which restrict payment to those procedures that have met some standard. The other is the cost of malpractice insurance and lawsuits, which have negative consequences for those using procedures that haven't been validated by sound and scientifically-based (EBM) research. These discrepancies and challenges are not new to veterinary medicine and the advent and development of EBVM has great potential to bring the best available evidence into use in veterinary practices thereby improving the effectiveness of veterinary services to patients and clients.

## REFERENCES

[1.] Evidence-Based Medicine Working Group, *J Am Med Assoc* 1992, 268:2420; [2.] Keene BW, (editorial) *J Vet Intern Med* 2000, 14:118; [3] Doig GS, *Aust Vet J* 2003, 81:412; [4] Roudebush P., et al. *J Am Vet Med Assoc* 2004, 224:1766; [5.] Naldi L,et. al. *Dermatology* 2000, 204:1; [6.] Sackett DL, et al. *Br Med J* 1996, 312:71; [7.] Last JM (editor). *Dictionary of Epidemiology. 4th Edition*. International Epidemiological Association, Oxford: Oxford University Press 2001; [8.] *Merriam-Webster Dictionary* http://www.merriam-webster.com/. [9.] Sackett DL, *Can J Cardiol* 1993, 9:487; [10] Dans AL, et al. *J Am Med Assoc* 1998, 279:545; [11.] Berg AO, In: Geyman JP, Deyo RA, Ramsey SD, editors:. *Evidence-based clinical practice: concepts and approaches*, Boston, 2000, Buterworth-Heinemann. [12.] Sacks HS et al. *Am J Med* 1982, 72:233; [13.] Chalmers TC, et al. *N Eng J Med* 1983; 309:1358; [14.] Colditz GA, et al. *Stat Med* 1989, 8:441; [15.] Sackett DL, et al. Diagnosis and screening. In Sackett DL, Straus SE, Richardson WS, Rosenberg WA, Haynes RB, editors: *Evidence-based medicine: how to practice and teach EBM, 2nd ed*, Philadelphia, 2000, Churchill-Livingstone. [16] American Library Association, Association of College and Research Libraries http://www.ala.org/ala/acrl/acrlstandards/informationliteracycompetency.cfm accessed Feb 1, 2008. [17.] Cook D, et al. *J Clin Epidemiol* 1995, 48:167; [18.] Cook D, et al. *Crit Care Med* 1996, 24:1757.

# TOP TEN EVIDENCED-BASED PAPERS IN FOOD ANIMAL MEDICINE AND SURGERY FOR 2008

**Peter Constable, BVSc(Hons), MS, PhD, DACVIM**
**West Lafayette, IN**

Peer-reviewed food animal medicine and surgery articles published in 2008 were examined. English-language articles published in the top 50 veterinary journals and 40 animal agricultural journals based on impact factor (www.isiwebofknowledge.com), and articles published in other journals based on recommendations from colleagues, were considered. The title and abstract of eligible articles were reviewed and more than 100 articles selected for further consideration. Ten evidenced-based articles that dealt with an important topic were subsequently selected for discussion. Criteria used for selection included diversity in subject matter and geographic location, excellence in study design and statistical analysis, and a high likelihood that the results of the study will impact food animal practice. The majority of the 10 papers were hypothesis-testing studies that contained a clearly stated rationale and hypothesis, and a well described materials and methods section that will permit anyone, anywhere, to reproduce the study. Most of the studies used an appropriate method of randomization and had appropriate statistical power, internal validity, and external validity (generalizability).

The intent of limiting the list to 10 was to focus attention on a small number of evidenced-based papers that are likely to become difference makers. The selected papers increase our understanding of the pathophysiology of disease or change the way we diagnose, treat, and control disease or perform surgery. Many excellent papers were not able to be listed because of the desire for diversity in subject matter and geographic location. The emphasis on diversity meant that the 10 selected papers were published in 9 different journals and originated from 9 countries: Belgium, Canada, Czech Republic, France, Germany, Ireland, Sweden, Switzerland, and the United States. Two of the 10 papers were open-access articles. It is rare for a single case report to be viewed as a difference maker; however, the paper by Richt and Hall provides valuable insight into the potential origin of the Bovine Spongiform Encephalopathy (BSE) pandemic in cattle and the relationship between BSE and variant Creutzfeldt-Jacob disease (CJD) in humans.

The ten papers are listed below in alphabetical order within each category:

## RANDOMIZED CONTROLLED STUDIES

1. Bergeron R, Elsener J. Comparison of postvaccinal milk drop in dairy cattle vaccinated with one of two different commercial vaccines. *Vet Ther* 2008; 9:141-146.

2. Chigerwe M, Tyler JW, Schultz LG et al. Effect of colostrum administration by use of oroesophageal intubation on serum IgG concentrations in Holstein bull calves. *Am J Vet Res* 2008; 69:1158-1163.

3. Klein P. Preventive and therapeutic efficacy of halofuginone-lactate against *Cryptosporidium parvum* in spontaneously infected calves: a centralized, randomized, double-blind, placebo-controlled study. *Vet J* 2008; 177:429-431.

4. Pang WY, Earley B, Gath V, and Crowe MA. Effect of banding or burdizzo castration on plasma testosterone, acute-phase proteins, scrotal circumference, growth, and health of bulls. *Livestock Sci* 2008; 117:79-87.

## OBSERVATIONAL STUDIES IN TARGET SPECIES

1. Arcangioli MA, Duet A, Meyer G, et al. The role of *Mycoplasma bovis* in bovine respiratory disease outbreaks in veal calf feedlots. *Vet J* 2008; 177:89-93.

2. Fournier C, Kuhnert P, Frey J, et al. Bovine *Staphylococcus aureus*: association of virulence genes, genotypes and clinical outcome. *Res Vet Sci* 2008; 85:439-448.

3. Kutzer P, Schulze C, Engelhardt A, et al. *Helcococcus ovis*, an emerging pathogen in bovine valvular endocarditis. *J Clin Microbiol.* 2008; 46:3291-3295.

4. Pringle M, Bergsten C, Fernstrom LL, et al. Isolation and characterization of *Treponema phagedenis*-like spirochetes from digital dermatitis lesions in Swedish dairy cattle. *Acta Vet Scand* 2008; 50:40 doi:10.1186/1751-0147-50-40.

5. Vercruysse J, Deprez P, Everaert D, et al. Breed differences in the pharmacokinetics of ivermectin administered subcutaneously to Holstein and Belgian Blue calves. *Vet Parasitol* 2008; 152:136-140.

## SINGLE CASE REPORT IN TARGET SPECIES

1. Richt J, and Hall SM. BSE case associated with prion protein gene mutation. *PLoS Pathogens* 2008; 4:e1000156.

# SUAM: AN IMPORTANT *STREPTOCOCCUS UBERIS* VIRULENCE FACTOR

**Maria E. Prado, DVM, Ph.D., DACVIM & Stephen P. Oliver, Ph.D.**
**Knoxville, TN**

## INTRODUCTION

Mastitis is the most economically significant disease affecting the dairy cattle industry worldwide. During the past 20 years, the incidence of contagious clinical and subclinical mastitis cases has decreased following the implementation of control practices such as the five-point-plan. However, well-managed dairy herds with low somatic cell counts are now experiencing a major increase in mastitis cases of environmental origin. Thus, currently available protocols would appear to have little effect in the control and prevention of environmental mastitis. *Streptococcus uberis* accounts for the majority of environmental streptococci isolated from mastitis cases and it is the second most common pathogen causing clinical mastitis in the dairy cow. To date, information regarding how *S. uberis* causes mastitis and which methods are effective in the control or prevention of this bacterium is limited. It has been shown, that many *S. uberis* intramammary infections (IMI) that originate during the nonlactating period and near calving result in clinical and subclinical mastitis during early lactation. Control programs for reducing *S. uberis* IMI should focus on periods adjacent to the nonlactating period where opportunities exist to develop strategies to reduce the impact of *S. uberis* infections in the dairy herd. The purpose of this presentation is to bring participants up to date on novel research being conducted in our lab regarding host-pathogen interaction of *S. uberis* which led to the identification of virulence factors with the potential of becoming vaccine candidates.

## *STREPTOCOCCUS UBERIS* RECENT DEVELOPMENTS: IS IT EXCLUSIVELY AN ENVIRONMENTAL PATHOGEN?

*Streptococcus uberis* is an amazingly versatile mastitis pathogen. It can affect lactating cows and dry cows, heifers and multiparous cows, causes clinical or subclinical mastitis, and even causes persistent colonization without an elevation in the somatic cell count[1,2]. It is a major cause of mastitis in pasture-based management systems with seasonal calving, and in housed cattle with year-round calving. It has been described as an environmental pathogen[3-6] and as a potentially contagious pathogen[1-2,7-8,].

Understanding the sources and transmission of *S. uberis* is important for mastitis control. Standard mastitis prevention measures may reduce the prevalence of contagious mastitis, resulting in a shift to predominantly environmental mastitis. Overemphasis on environmental mastitis may subsequently result in a re-emergence of contagious mastitis. Research using molecular typing techniques has greatly contributed to insight into sources and transmission of *S. uberis* mastitis[9]. A variety of strain typing methods have been used around the world. In some countries, such as New Zealand and the UK, the bacterial population within a herd is usually highly heterogeneous, which indicates that the bacteria come from many different sources[3,6]. In New Zealand, climatic conditions, farm management, and heavy cattle traffic areas work together to create high bacterial loads and mastitis incidence when cows are most vulnerable, during transition from the dry period to lactation[10]. Research in the USA has shown that many environmental sources and vectors can harbour *S. uberis*, including bedding, soil, plant matter, water, and insects[11]. Fecal shedding of *S. uberis* by dairy cattle contributes to presence of *S. uberis* in the environment. Certain environments, such as straw, allow for amplification of *S. uberis* to large numbers and straw is often considered a risk factor for *S. uberis* mastitis[12]. However, the high incidence of *S. uberis* in New Zealand's pasture based system, and the peak incidence of *S. uberis* in The Netherlands, which occurs during the summer pasture season rather than the winter housing season, show that *S. uberis* can also thrive in environments without straw[13]. In addition to environmental *S. uberis*, contagious *S. uberis* can occur if conditions permit transmission from animal to animal[2,8]. Transmission can take place via the milking machine, and possibly also via the environment, blurring the distinction between the two modes of spread. Contagious transmission via the environment may occur if an infected cow leaks milk into the bedding of a stall that is subsequently used by another cow. DNA-sequence based typing of an international collection of *S. uberis* isolates suggests that certain *S. uberis* strains may be more host-adapted than others. One of these host adapted strains causes infections that are more chronic than the average infection with environmental *S. uberis*[9]. The longer duration of infection creates a window of opportunity for the bacteria to spread to other animals in the herd. The mechanism behind chronicity is not known. Quite possibly, adherence of the bacteria to udder epithelial cells, invasion of the bacteria into these cells and subsequent intracellular survival, protects *S. uberis* from the immune system of the cow, and from treatment with antibiotics[14-16]. To reduce reliance on antibiotics for mastitis control, development of a vaccine is highly desirable. Immunogens that prevent adherence of *S. uberis* in the udder may play a role in mastitis prevention. While strain typing techniques are not routinely available for use in veterinary practice, insights gained through their use can be applied in dairy herd management. Combined use of epidemiological knowledge, herd level data and culture results will contribute to judicious implementation of mastitis control measures on the farm, and to improvement of animal health and milk quality.

## *STREPTOCOCCUS UBERIS* VIRULENCE FACTORS: FOUNDATION FOR POTENTIAL IMMUNOGENS

For IMI to occur, mechanisms associated with avoidance of phagocytic defenses, rapid growth of bacteria, adherence of bacteria to epithelial cells, and/or bacterial colonization of mammary tissue are probably present. During the last decade, several potential virulence factors of *S. uberis* have been identified. Some of these factors are cell-associated while other factors are extracellular.

### Capsule

Mastitis pathogens must overcome the initial encounter with phagocytic cells, either by avoiding and/or resisting phagocytosis, for infections to occur. Bacterial capsules have been considered an important antiphagocytic factor. Bacterial capsules exert their antiphagocytic functions through mechanisms such as electrostatic repulsion, antigenic mimicry and low immunogenicity, antigen masking, physical barrier between bacteria and phagocyte, binding and depletion of opsonic factors, and suppression of phagocyte cell activity. Matthews et al. reported that about 50% of *S. uberis* isolates were encapsulated[17]. Capsular material of *S. uberis* was composed primarily of hyaluronic acid[18]. Work from our laboratory confirmed that S. *uberis* capsule was an antiphagocytic factor since the phagocytosis index of encapsulated strains by bovine mammary macrophages was 48% compared to 75% for nonencapsulated strains[19]. These findings were further confirmed by inhibition of the antiphagocytic effect of the capsule when opsonized with anticapsular antibodies[19]. In a subsequent study, we showed that resistance to phagocytosis was a common feature among encapsulated strains of *S. uberis*, and that cross-reactivity existed between encapsulated strains[20]. Released hyaluronic acid capsule may impede binding of opsonic antibodies on the phagocytes' membranes, consequently blocking binding and engulfment of opsonized bacteria. Thus, it is possible that extracellular capsular material may be one means of protecting nonencapsulated strains of *S. uberis* by blocking phagocytic and killing activity of bovine mammary macrophages, thus explaining the survival of nonencapsulated strains of *S. uberis* in the mammary gland. It is likely that hyaluronic acid capsular material represents a shield that protects bacteria from phagocytes. Another characteristic that could explain the antiphagocytic attributes of hyaluronic acid capsule is antigenic mimicry. Hyaluronic acid is a ubiquitous substance of mammals. Therefore, hyaluronic acid capsules are not easily detected by the host immunological surveillance system, thus providing bacteria time to grow, to adhere[14] and to invade mammary epithelial cells[17]. Conversely, Field et al. recently showed that *S. uberis* wild type and a mutant that lacked the ability to produce a hyaluronidase capsule were able to withstand killing by bovine neutrophils[21]. This study concluded that *S. uberis* resistance to killing by bovine neutrophils might not be mediated by the capsule.

**Plasminogen Activator**

Rapid growth of mastitis pathogens is considered an important virulence factor, and achieving this in mammary secretions is a formidable task for an auxotrophic bacterium such as *S. uberis*. Activation of plasminogen with subsequent production of plasmin and degradation of casein has been proposed as a mechanism utilized by *S. uberis* to obtain essential nutrients required for optimal growth in milk. Oliver reported that *S. uberis* did not grow well in milk obtained during late lactation, but growth increased markedly in mammary secretions obtained during early- and mid-involution[22]. In vitro growth of *S. uberis* in colostrum and milk obtained during early lactation decreased significantly in comparison to growth observed in mammary secretions obtained during the nonlactating period[22]. *Streptococcus uberis* is highly auxotrophic and depends on acquisition of several amino acids, vitamins, trace elements and a carbohydrate source to successfully grow[23]. Since the lactating mammary gland is deficient in free or peptide associated amino acids, *S. uberis* must posses a mechanism to acquire amino acids for growth. Marshall and Bramley indicated that *S. uberis* was capable of converting plasminogen to plasmin which in turn degraded casein and generated by-products needed for optimal bacterial growth[36]. In a subsequent study, a plasminogen activator (PauA) was isolated from *S. uberis*[23]. The plasminogen activator was an extracellular protein. In addition to facilitating bacterial growth, it is possible that plasmin, through its fibrinolytic activity, could influence attachment and spreading of *S. uberis* in the bovine mammary gland. However, the relationship between activation of plasminogen and virulence is not clear since a low pathogenic strain, designated EF20, which showed caseinolytic activity[24] was susceptible to killing by bovine neutrophils after growth in medium supplemented with casein[25]. Furthermore, subcutaneous immunization of dairy cows with concentrated culture supernatant containing PauA resulted in partial protection against heterologous challenge[26]. In contrast, cattle immunized with supernatant devoid of PauA developed mastitis following experimental challenge in the same study. In addition, PauA appears to be highly conserved among the great majority of strains isolated from clinical mastitis cases; and strains from UK, USA and Denmark share a high level of homology[27],

***Streptococcus uberis* Adhesion Molecule (SUAM)**

Research from our laboratory has focused extensively on development of in vivo and in vitro models to study host-pathogen interactions, and on identification and characterization of virulence factors associated with the pathogenesis of *S. uberis* mastitis and other environmental streptococci[28]. We have shown that *S. uberis* was able to adhere to epithelial cells and that was followed by internalization into the host cell via exploitation of host cell machinery. Our lab demonstrated that *S. uberis* used host elements like extracellular matrix proteins to achieve increased adherence, probably utilizing these as a molecular bridge to attach to host cell membranes. Another of these host cell factors appears to be lactoferrin (LF), a whey protein found in milk. Use of molecular biology tools such as proteomics, genomics and bioinformatics has led to the discovery of a novel protein produced by *S. uberis* referred to as *Streptococcus uberis* Adhesion Molecule or SUAM.

We have conducted numerous studies on SUAM and a brief summary of these studies follows. Collectively, experiments from our laboratory have provided evidence that: (1) SUAM bound to lactoferrin (LF), a whey protein found in milk. (2) binding of LF through SUAM enhanced adherence of *S. uberis* to bovine mammary epithelial cells[29]. Lactoferrin likely functions as a bridging molecule between *S. uberis* and bovine mammary epithelial cells facilitating adherence to host cells[30]. (3) SUAM in the absence of LF influenced adherence to and internalization of *S. uberis* into bovine mammary epithelial cells. (4) SUAM was isolated, purified and partially characterized[31]. (5) Antibodies against SUAM (whole protein) and to a peptide of the N-terminus of SUAM designated pepSUAM cross-reacted with homologous proteins present in other strains of *S. uberis* demonstrating the ubiquity of SUAM across all strains of *S. uberis* evaluated[31]. (6) Antibodies against pepSUAM inhibited adherence to and internalization of *S. uberis* into bovine mammary epithelial cells suggesting that pepSUAM is biologically active[31]. (7) The theoretical DNA sequence of SUAM was determined and confirmed by PCR and restriction digests[32]. Further confirmation of the theoretical SUAM sequence was obtained when the SUAM gene (*sua*) from *S. uberis* UT888 was amplified, cloned and sequenced. Sequence analysis demonstrated that *S. uberis* UT888 SUAM had 99% sequence identity to the theoretical SUAM identified in the Sanger *S. uberis* genomic database by homology to the reverse translated peptide sequence. When the SUAM DNA sequence was compared to GeneBank, no homologies as an entire gene were found demonstrating that SUAM is a unique *S. uberis* protein[32]. (8) The SUAM gene (*sua,* GenBank accession number DQ232760) was found in all strains of *S. uberis* evaluated from geographically diverse areas. (9) *sua* was amplified by PCR from genomic DNA of *S. uberis* and sub-cloned into pBAD-directional TOPO vector. Pilot rSUAM purification trials were performed, and a method for obtaining milligram quantities of affinity purified rSUAM for in vivo use was developed[33-34]. (10) Bovine antibodies to pepSUAM to rSUAM significantly decreased adherence to and internalization of *S. uberis* into mammary epithelial cells[35]. (11) Vaccination of dairy cows with rSUAM during physiological transitions of the mammary gland either from or to a state of active milk synthesis increased antibody titers in serum at times when the udder is highly susceptible to new *S. uberis* IMI. Antibodies to SUAM were also detected in colostrum of cows vaccinated with rSUAM[34]. (12) Serum from cows vaccinated with rSUAM significantly decreased adherence to and internalization of the homologous and a heterologous strain *S. uberis* into mammary epithelial cells[35]. Thus, we have much in vitro data and some data in dairy cows to suggest that SUAM may be a promising antigen that can be used in vaccination strategies for the control of *S. uberis* IMI.

## CONCLUSIONS

Results of studies published thus far provide an excellent foundation towards a more fundamental understanding of the pathogenesis and early interactions between bovine mammary epithelial cells and *S. uberis*. Research directed towards characterization of host-pathogen interactions that take place during the early stages of *S. uberis* IMI, and manipulation of host defenses that could neutralize the early pathogenic mechanisms of infecting bacteria are needed to enhance our chances of developing methods to control this important mastitis pathogen.

## REFERENCES

[1]Oliver SP, et al. **FEMS Microbiol Lett** 1998;160:69. [2]Zadoks RN, et al. **Epidemiol Infect** 2003; 130:335. [3]Douglas VL, et al. **Vet Microbiol** 2000;75:27. [4]Leigh JA, **Vet J** 1999;157:225. [5]McDougall S, et al. **J Dairy Sci** 2004;87:2062. [6]Wieliczko RJ, et al. **J Dairy Sci** 2002;85:2149. [7]Cullor JS, et al. **Proc NMC** 2006; 145-146. [8]Phuektes P, et al. **J Clin Microbiol** 2001;39:1460. [9]Zadoks, RN and Schukken YH. **Vet Clin North Am Food Anim Pract** 2006;22:229[10]Lopez-Benavides M, et al. **Proc 4th IDF International Mastitis Conference** 2005.[11]Zadoks RN, et al. **Vet Microbiol** 2005;109:257.[12]Ward WR, et al. **Vet Rec** 2002;151:199.[13]Barkema HW. **Tijdschr. Diergeneeskd** 1999;124:338.[14]Almeida RA, et al. **Zentralbl Veterinarmed** [B] 1996;43:385.[15]Almeida RA, et al. **FEMS Microbiol Lett** 1999;177:313.[16]LammersA, et al. **Vet Microbiol** 2001;80:255.[17]Matthews KR, et al. **Infect Immun** 1994;62:5641.[18]Almeida RA and SP Oliver. **J Vet Med B** 1993;40:697.[19]Almeida RA and SP Oliver. **J Vet Med B** 1993;40:707.[20]Almeida RA and SP Oliver. **J Vet Med B** 1995;42:331.[21]Field TR, et al. **Infect Immun** 2003;71(1):132.[22]Oliver SP. **J Vet Med B** 1991;38:538.[23]Leigh JA. **FEMS Microbiol Lett** 1994;118:153.[24]Leigh JA, et al. **Res Vet Sci** 1990;49:85. [25]Leigh JA and Field TR. **Vet Res Comm** 1991;15:1.[26]Leigh JA, et al. **Vaccine** 1999;17(7-8):851.[27]Leigh JA. **Adv Exp Med Biol** 2000;480:307.[28]Oliver SP, et al. **J Vet Med B** 1998;45:461. [29]Fang W, et al. **Am J Vet Res** 2000; 61:275.[30]Patel D, et al. **Vet Microbiol** 2009; in press.[31]Almeida RA, et al. **Vet Microbiol** 2006;115:186.[32]Luther DA, et al. **Vet Microbiol** 2008;128:304.[33]Prado ME, et al. Abstract **CRWAD** St Louis MO 2005.[34]Prado ME, et al. Abstract **2nd Annual Comparative and Experimental Medicine Research Symposium** Knoxville, TN 2008.[35]Almeida RA, et al. Abstract **CRWAD** St Louis MO 2007.[36]Marshall VM and Bramley AJ. **J Dairy Res** 1984;51:17.

## JEJUNAL HEMORRHAGIC SYNDROME – SHORT & LONG TERM OUTCOME IN DAIRY CATTLE

**Simon F. Peek BVSc, MRCVS, PhD, DACVIM, DECEIM**
**Madison, WI**

### INTRODUCTION & REVIEW

Following the initial report of jejunal hemorrhage syndrome (JHS) in 1992 by Ruggles et al[1], the disease has subsequently been reported in dairy cows in the US[2], dairy and beef cows in Canada[3,4], dairy cows in Europe[5,6] and dairy cows in the Middle East[7]. It has emerged as an economically important disease, particularly within the higher density dairy areas of the United States. The published literature on JHS has concentrated on retrospective case series from a number of countries[1-6] and investigations into potential risk factors for the disease in dairy cattle in the United States[8,9].

JHS is more common in mature dairy cows than in first calf heifers and bulls according to the published literature[1,2,4], but herd investigations at our institution have involved dairies that also report problems in cattle of all ages and stages of lactation. The largest retrospective published to date reported that JHS showed a predilection for older cattle 100-110 days in milk[2] but we have seen the disease at any stage of lactation, in dry cows and even in mature breeding bulls at stud. Most cattle appear to be in good health immediately prior to the development of JHS. There does not appear to be a consistently increased level of other production related diseases such as subclinical milk fever (hypocalcemia), mastitis or ketosis on dairies that experience JHS but there is undoubtedly much about this condition that we do not yet understand.

The typical cow with JHS becomes depressed, goes acutely off feed, and frequently has cold ears and extremities. Mild, ventral abdominal distension may develop, manure production is scant or absent and pain may be demonstrated by teeth grinding, kicking at the abdomen, or getting up and down frequently. The condition usually advances quite rapidly from initial onset. Affected cattle usually have no fever but heart and respiratory rates are high commensurate with the degree of pain and dehydration. Upon rectal examination, the veterinarian usually finds scant manure that contains fresh blood, blood clots and digested blood. Palpable, distended small intestine is not a common feature per rectum but transabdominal ultrasound examination typically reveals numerous distended loops of small intestine in the right ventral, mid and caudal abdomen. JHS can be difficult to distinguish from other causes of small intestinal obstruction particularly when the animal is presented peracutely, prior to the presence of blood clots in the rectum. Many cows with JHS die within the first 6-36 hours of showing clinical signs. Some progress more slowly over a couple of days and others appear to improve over the first 24 hours only to dwindle and do poorly over the following days. We have seen affected cattle that survive for up to 10 days from the initial episode of pain and abnormal manure without surgical intervention. These individuals may survive longer because the intestine is not completely blocked but death results due to intestines that gradually lose their integrity and leak resulting in peritonitis.

Because JHS cannot be experimentally reproduced in cattle, the definitive cause is as yet unknown, although considerable attention etiologically has been given to potential roles for the bacterium, *Clostridium perfringens* type A, and the mold, *Aspergillus fumigatus*. While both theories about cause have strong supporters, it is difficult to discern what truly initiates the syndrome from what happens as a result of the disease. *Clostridium perfringens* bacteria are normal inhabitants of the intestinal tract of healthy cattle and are commonly found in the environment. *C. perfringens* type A, the type incriminated as playing a role in JHS, can also be found in haylage under some storage conditions. While the organism is ubiquitous and generally considered to be innocuous, it can become problematic with dietary stress, injury, changes in management, parasitism or other unusual circumstances that set up a favorable growth environment that allows the bacteria to proliferate and produce potent toxins. *Clostridium perfringens* type A has been associated with conditions similar to JHS of cattle in poultry (necrotic enteritis of fowl) and in people (pigbel), making further investigation of its role either as a cause or a consequence of JHS important. As a group, these bacteria can proliferate after death and invade tissues of the cow's body other than the intestine, creating problems in interpretation of a positive culture from post mortem specimens where there has been a delay between the time of death and sample collection and/or submission. In our veterinary hospital, we have obtained pure cultures of specific toxin producing *Clostridium perfringens* type A bacteria from fresh, appropriately handled post mortem specimens and from intestinal contents from JHS-affected cattle undergoing surgery to remove the blood clot or the abnormal portion of intestine. While the presence of toxins (alpha and beta-2) can help distinguish the potentially normal bacteria from disease producing strains, it is impossible in an affected cow to determine whether the toxins are a primary or secondary factor. Management practices may also play a role in the condition via their impact on the microflora of the small intestine, creating an environment that could enhance the growth of toxin producing *Clostridium perfringens* to disease producing levels. The mold, *Aspergillus fumigatus*, has also been associated with JHS. This organism, commonly found in soil, can be detected in feeds and forages and has been found in the blood and tissues of JHS-affected cattle. Although a causal relationship has not been established, some researchers believe that a primary insult like rumen acidosis, abnormal GI motility, metabolic disease, stress or an immune suppression, creates an environment for *Aspergillus* to gain entrance into the blood in high numbers. Alternatively, a heavy load of *Aspergillus fumigatus* may gain entrance into the blood through an injured intestinal or abomasal lining.

Management and nutritional factors have also been implicated and recent publications by Godden et al[8] and Berghaus et al[8] have suggested that high milk production and the consumption of high energy diets may be risk factors associated with the condition. Although *Clostridium perfringens* type A appears to be frequently isolated from feces or intestinal contents of affected cattle[2,4,10] and both alpha toxin producing and alpha and beta-2 toxin producing strains are more likely to be obtained from cattle with JHS than abomasal displacements[10], it has not proven possible to reproduce the disease with a beta-2 toxin producing isolate from a field case experimentally[11], further supporting earlier doubts that this organism on its own is not capable of causing the disease.

Because the exact cause and/or contributing factors that lead to JHS are currently uncertain, consistently effective preventive strategies are yet to be determined. The interventions that have seemed to decrease the prevalence of the disease on problem dairies include vaccination, feed additives and nutritional management. There has been considerable interest amongst some herd owners and veterinarians in the development of autogenous vaccines (bacterin-toxoids) to *Clostridium perfringens* type A isolated from affected herds. To be effective, products made under fermentation processes maximized for bacterial growth and toxin production using a genotypically appropriate *Clostridium perfringens* type A are necessary. At this point, testimonials are the best evidence we have for the efficacy of autogenous vaccines. Other herds have used *Clostridium perfringens* type C and D toxoid or 7- and 8-way *Clostridium* bacterin-toxoids. Some herds report a reduction in prevalence by vaccinating 2 to 3 times per year with these products. With the recent availability of a USDA licensed *Clostridium perfringens* type A, toxoid (Novartis Animal Health, Greensboro, NC) there is now a specific product available commercially, designed to protect against diseases associated with the alpha toxin. Again, only testimonial reports currently exist for the beneficial effect of this product in protecting against JHS. Much of the preventative effort has been directed towards nutritional management on problem dairies. Control of subacute rumen acidosis, metabolic disease, diseases like Salmonellosis that affect gastrointestinal motility, and consistency of feed components, mixing, and delivery are logical targets when trouble-shooting a problem herd. Clostridial bacteria as a group tend to multiply rapidly in response to specific opportunities such as carbohydrate overload, an overly acidic environment (either in the abomasum or rumen), high concentration protein or sugar overspill from the rumen into the small intestine or motility disturbances. Insuring a healthy rumen microflora, limiting oral medications and maintaining appropriate fiber length with high quality storage and fermentation of total mixed rations are also important. Inappropriately stored forages can be a source of pathogenic Clostridia and may also pose an added risk for contaminating molds like *Aspergillus fumigatus*. Feed additives have been used in problem herds. Omni-Gen AF™ (Prince-Agri Products Inc, Quincy, Ill) claims to tie up *Aspergillus fumigatus* in moldy feed, while mannan oligosaccharides, fragments of yeast cell walls, are included in hopes of reducing toxin load, binding harmful bacteria like *Salmonella* and *Clostridia* perhaps stimulating some immunity.

Treatment of cattle with JHS is challenging. High mortality rates (between 77%[2] and 100%[3,4,6]) have been documented for JHS in the retrospective case studies published to date. There are anecdotal reports of successful medical treatment that has alleviated the need for surgery but we consider the condition to indicate surgery at our institution. Medical treatment may include oral or ruminal fluids, catharctics, pain-killers or anti-inflammatory drugs, antibiotics and *Clostridium perfringens C and D* antitoxin. Unfortunately, there is no specific antitoxin for *Clostridium perfringens* type A, but there may be limited cross protection from the antitoxin of *Clostridium perfringens type C* and *D*. Surgical relief of the jejunal obstruction may involve aborad massage of occlusive blood clots or their removal at enterotomy. In situations where bowel compromise is deemed significant, a resection and anastomosis may prove necessary. Post-surgical management commonly involves many of the same medical therapies indicated previously. There is an observed risk of reformation of blood clots with subsequent re-obstruction that can be evident clinically immediately intra-operatively or in the hours to days following surgery.

**Retrospective Study**

In 2008 we performed a retrospective study designed to report on the signalment, operative findings, short, medium and long term outcome of cattle undergoing celiotomy for JHS, as well as recurrence rate and an analysis of potential risk factors influencing survival and reoccurrence. The study used case material from the in-patient population in the large animal clinic at the University Of Wisconsin's Veterinary Medical Teaching Hospital. Cattle were determined to have JHS at surgery by virtue of intra-luminal blood clots that occupied the jejunum. Follow-up telephone interviews with the owners of cattle that were discharged from the hospital as short term survivors provided longer term survival data.

Sixty one percent of cattle identified with JHS survived to initial discharge, with 52% and 41% surviving to 6 and 12 months post-admission to the hospital respectively. A relatively high rate of recurrence (33%) was documented in short term survivors that were discharged from the hospital. A significant proportion of affected cattle were of the Brown Swiss breed. Manual massage aborad alone was associated with higher survival rates compared to cattle that underwent either enterotomy or enterectomy. The survival rates that we documented were higher than previously reported. Prompt surgical exploration and resolution of intraluminal clots by manual massage was associated with higher survival rates.

**References**

1. Ruggles AJ, Sweeney RW, Freeman DE, et al.. Intraluminal hemorrhage from small intestinal ulceration in two cows. **Cornell Vet**. 1992;82(2):181-186. 2. Dennison AC Van Metre D.C, Callan RJ, et al. Hemorrhagic bowel syndrome in dairy cattle: 22 cases (1997-2000). **J Am Vet Med Assoc** 2002;221:686-689. 3. Abutarbush SM, Carmalt JL, Wilson DG, et al. Jejunal hemorrhage syndrome in 2 Canadian beef cows. **Can Vet J** 2004;45:48-50. 4. Abutarbush SM and Radostits OM. Jejunal hemorrhage syndrome in dairy and beef cattle: 11 cases (2001-2003). **Can Vet J** 2005;46:711-715. 5. Rademacher von G, Lorenz I, and Haenichen T. Jejunumanschoppung mit koaguliertem Blut infolge blutender Darmulzera bei Kuehen. **Sonderdruck Aus Tieraerztliche Umschau** 2002;57:399-411. 6. Muskens J, Veldhorst GJ, Snoep JJ et al. High mortality in a herd with signs of jejunal hemorrhage syndrome. **Tijdschr Diergeneeskd**. 2007;132:116-119. 7. Breener J, Orgad U, Tiomkin D et al. Adult jejunal hemorrhage syndrome – an emerging dairy cow disease. **Isr J Vet Med** 2002;57:41. 8. Godden S, Frank R, and Ames T. Survey of Minnesota dairy veterinarians on the occurrence and potential risk factors for jejunal hemorrhage syndrome in adult dairy cattle. **Bov Pract** 2001;35:104-116. 9. Berghaus RD, McCluskey BJ and Callam RJ. Risk factors associated with hemorrhagic bowel syndrome in dairy cattle. **J Am Vet Med Assoc** 2005;10:1700-1706. 10. Dennison AC, Van Metre D.C, Morley PS, et al. Comparison of the odds of isolation, genotypes and in vivo production of major toxins by Clostridium perfringens obtained from the gastrointestinal tract of dairy cows with hemorrhagic bowel syndrome or left displaced abomasums. **J Am Vet Med Assoc** 2005;227:132-137. 11. Ewoldt JM and Anderson DE. Determination of the effect of single abomasal or jejunal inoculation of Clostridium perfringens Type A in dairy cows. **Can Vet J** 2005;46:821-824.

# BOVINE RESPIRATORY CORONAVIRUS

**Sanjay Kapil, DVM, MS, PhD, ACVM, Professor, OADDL**
**Stillwater, OK**

## INTRODUCTION

Bovine respiratory disease (BRD), a major cause of morbidity and mortality in feedlot cattle, results in major economic losses to the beef industry, worldwide. Multiple bacterial and viral pathogens are involved in the development of BRD. The most common viral agents of BRD are pestivirus, infectious bovine rhinotrachitis, bovine respiratory syncytial virus, and parainfluenza-3. Bovine respiratory coronavirus (BRCV) may also play a role in BRD. Coronaviruses have three antigenic groups: Group I have no hemagglutinin-esterase protein and contain feline infectious peritonitis and transmissible gasteroenteritis virus; Group II have hemagglutinin-esterase and contain bovine coronavirus; and Group III has avian virus-like infectious bronchitis virus. SARS human coronavirus contains HE protein. Bovine coronavirus has been clinically classified as bovine enteric coronavirus (BECV), bovine respiratory coronavirus (BRCV), and winter dysentery (WD).

### Incidence

Bovine enteric coronavirus (BECV) is responsible for 36% of the mortality associated with cases of bloody diarrhea in calves up to 9 weeks of age. Respiratory disease is responsible for another 26% of the mortality in calves. BRCV is a common cause of upper respiratory tract infections in calves along with intestinal infection and diarrhea. Infection in the respiratory tract is limited to nasal passages, the upper one-third of the trachea, and a low level of focal infection of the lungs. Respiratory disease can be consistently reproduced in young calves (Kapil et al., 1990) with a pneumoenteric bovine coronavirus, Minnesota-1988 (MN-1988). BRD was responsible for 70% of the mortality reported in Ontario, Canada feedlots[1]. Viral respiratory disease accounted for 20% of the mortality in feedlot cattle. BRCV was found in about 2% of the cases based on virus isolation in cell culture. The actual incidence may have been higher if fluorescent antibody testing or RT-PCR were used for detection of BRCV. In a recent study, we found that the incidence of BRCV in feedlot cattle was much higher than in the Canadian study and that the presence of BRCV was second in incidence after bovine herpesvirus[2].

## EPIDEMIOLOGY

BCV has a world wide distribution. Based on BCV ELISA data from Kansas State University, Manhattan, KS, one in three case of calf diarrhea in the age group of 1-9 weeks is due to BCV. There is seasonal variation in the incidence of BCV diarrhea. Young calves shed BCV at high titer and develop diarrhea along with presence of the virus in the respiratory tract. In adult cattle, the virus is shed at low levels and the animals are subclinically affected. The periparturiant shedding of BCV becomes the source for the virus for the next generation of calves. Other farm animals (horses, water buffalo, camel and New World camelids) and wildlife (deer, elk) also have BCV-like viruses that can also infect calves because these viruses are genetically related to BCV. The infection of small ruminants with coronavirus is less common. Stressors, including inclement weather and shipping, are important contributing factors that may exacerbate disease from BCV and BRCV infections.

### Route of transmission

Infection is primarily transmitted orally and to a lesser extent by respiratory route. Most often, infection is passed from the adult cattle/cow to newborn calves. Calf-to-calf transmission of BCV occurs commonly. Coronaviruses are enveloped viruses but are stable in the presence of organic material. Clay soils bind BCV efficiently, concentrating them on the surface of the soil. Dogs may play a role in BCV infection; canine respiratory coronavirus is genetically related to BCV and has been found in kennel cough cases in Europe and the USA. Dogs may be a passenger of the BCV on farms

### Clinical Syndromes Associated with BCV

BCV is associated with calf diarrhea, calf respiratory disease, winter dysentery, and adult cattle respiratory disease. Viruses isolated with all of these syndromes are virtually identical and cannot be distinguished genetically or antigenically. Based on tracheal-bronchial lavage from calves, 12% of the respiratory samples were positive for BRCV. In this study, 25% of calf herds were involved. At OADDL, we have found BRCV by direct fluorescent antibody testing of lungs, trachea, and nasal turbinate samples of feedlot cattle in the late stage of respiratory disease. According to a published report[3], shedding of BRCV correlates statistically to reduction in weight gain in feedlot cattle.

## DIAGNOSIS

Spiral colon is a sample of choice for BECV. Antemortem feces, nasal swabs, trachea (upper-one third), and lungs can be submitted for BRCV. The tests of choice are antigen-capture ELISA that is available at the Kansas State University Diagnostic Laboratory, Manhattan, KS. The cost of testing is $6.50 per sample and turn around time is fast. Another useful diagnostic reagent is the 8F2 monoclonal antibody that has activity against nucleocapsid protein of bovine coronavirus, the most predominant protein of BCV. The 8F2 reacts with the viral antigen in formalin-fixed tissues (Intestines). The 8F2 MoAb also reacts with BCV-like viruses (alpaca coronavirus, equine coronavirus, camel coronavirus, and elk coronavirus). It is important that a diagnostic laboratory with good quality BCV reagents is utilized for this virus.

## VACCINES

### Correlates of Protective Immunity

Antibody present in the colostrum of immune cows provides complete protection against BECV diarrhea. When colostral antibody wanes, calves become susceptible to BCV infections. Colostral antibodies are of IgG1 isotype and have activity against the spike protein of BCV. The mechanism of protection against BRCV and pathogenesis of BRCV is not known.

### Vaccines for BECV and BRCV

Immunizing the pregnant cattle to produce high titer colostrum will protect the newborn calf if there is transfer of passive antibodies within 48 hours after birth. These antibodies will provide solid protection but gut antibodies are short lived because it is an open ended system. Protection of the respiratory tract from BRCV has not been studied, thus, it is difficult to speculate about immunization strategies and routes of immunization for BRCV.

### Antigenic Variation among BECV and BRCV

BRCV and BECV are antigenically related and cross-reactive, thus, vaccines developed for BECV should provide adequate protection against BRCV. However, BECV isolates are antigenically classified among three subgroups using a polyclonal antiserum against the Mebus strain of BCV (Antigenic Group I). Antigenic Group III is the predominant BECV presently circulating in the USA. Calf diarrhea isolates of BECV are not blocked by polyclonal antiserum prepared against antigenic Group I (Mebus isolate of BECV). One commercial BECV vaccine (Guardian, Schering Plough Animal Health, Elkhorn, NE, USA) claims to contain both the antigenic variants of BECV. Other commercial calf diarrhea vaccines contain Group I BECV that is almost extinct in nature[4].

### Pathology

BECV has an incubation period of about 20 hours after oral infection of colostrum-deprived calves. After experimental infection, mucus and blood appear in the feces; loose feces occur for several days. Diarrhea leads to dehydration and calves soon develop sunken eyes. Pneumoenteric BCV (Minnesota 1988 isolate) isolate attacks the respiratory tract.. Calves develop open mouth breathing and respiratory distress 3-4 days oral infection. About 50% of BCV infections in dairy calves are pneumoenteric, 25% are enteric only, and 25% are respiratory only based on fluorescent antibody test on nasal swabs and spiral colon.

### Disinfection

BCV is an enveloped RNA virus and it is extremely susceptible to disinfectants, soap, pH, and lipid solvents. However, BCV can survive for a long time in soil, feces and bedding materials.

### Differential Diagnosis

Other causes of calf diarrhea, such as rotavirus and *E. coli*, must be considered in the differential diagnosis. BCV is responsible for diarrhea in calves up to 9 weeks of age. If blood and mucus are present in the feces, BCV should be considered as a cause especially if there is a history of colostrum transfer failure based on serum antibody/IgG testing.

## CONCLUSIONS

BCV is responsible for a more severe form of diarrhea than rotavirus, which infects only the small intestine. Bovine coronaviruses associated with all clinical syndromes are antigenically and genetically similar. I believe that 3 disease syndromes are actually caused by tropism of the 3 distinct sites (whole intestine, nasal passages, upper-trachea) by a closely related coronavirus. Stress is an important part of respiratory infection. Lack of colostral immunity is a factor in intestinal infections. Inclement weather is an important environmental stressor contributing to outbreaks of winter dysentery in dairy cattle. All these disease syndromes can be experimentally reproduced by related BCV-like viruses. I believe that enteric calf diarrhea vaccines can be used for BRCV infection in calves and feed-lot cattle to reduce the burden of BRD when no other major viral agent is detected in cases of acute death associated with viral pneumonias. Vaccines containing broad antigenic composition (antigenic groups I and III) are recommended to control calf diarrhea and pneumonia.

## REFERENCES:

[1]Gagea et. al. **J. Vet. Diagn. Invest.** 2006; 18:18; [2]Kapil et. al. **Proc. Amer. Soc. Virol.** 2008; 27 th Annual Meet.:P19-1; [3]Martin et. al. **Can. J. Vet. Res.** 1998; 62:257; 4 Kapil et al. **Vet. Microbiol.** 1999;67:221.

**PATHOPHYSIOLOGY OF DIARRHEA IN CALVES**

**Derek M. Foster, DVM, MS, DACVIM**
**Raleigh, NC**

**ENTEROTOXIGENIC *E. COLI***

Diarrhea due to enterotoxigenic *E. coli* (ETEC) occurs almost exclusively in calves less than 4 days of age, and is the most common pathogen isolated from calves of this age. Infection occurs via the fecal-oral route as other bacteria are also colonizing the gastrointestinal tract. Two virulence factors are critical to the pathogenesis of ETEC infections: presence of the K99 fimbria and production of heat stable toxin (STa). K99 mediates attachment to the intestinal epithelium and allows ETEC to proliferate. As K99 is only expressed at an environmental pH of greater than 6.5, the distal ileum is the initial site of colonization,[1] but then the organism spreads proximally. The binding of K99 to the intestinal epithelium is age dependent and gradually decreases over the first two weeks of age.[2]

Once ETEC has colonized the ileum, STa, the predominant mediator of ETEC diarrhea in calves, is secreted in alkaline conditions.[3, 4] Once it is secreted into the lumen, STa binds to the brush border enzyme guanylyl cyclase-c (GCC).[5] Activated GCC increases intracellular cyclic guanylyl monophospate (cGMP), which is the second messenger that activates cGMP-dependent protein kinase II (cGKII). cGKII in turn activates the cystic fibrosis transmembrane conductance regulator (CFTR), the primary channel through which chloride is lost in secretory diarrhea.[6] STa can also upregulate bicarbonate secretion that is independent of GCC and CFTR[7] and decrease sodium absorption by inhibiting the sodium-hydrogen exchanger in the proximal intestine.[8] The net effect of these changes is an increase in luminal fluid and pH, leading to diarrhea, greater expression of K99, and increased production of STa.

***CRYPTOSPORIDIUM PARVUM***

Shedding of *Cryptosporidium parvum* oocysts occurs as early as 3 days of age, peaks at 14 days of age, and can continue throughout life, but clinical disease is rarely seen beyond 3 weeks of age. Once oocysts are ingested, *C. parvum* has a complicated life cycle involving multiple stages which allows it to proliferate exponentially. Multiple proteins appear to mediate attachment of the organism to the intestinal epithelium,[9] which can be inhibited by colostral oleic acid.[10, 11]

*C. parvum* infection leads to a malabsorptive diarrhea due to the intestinal epithelial cell loss and severe villous atrophy that occurs. The cell loss appears to be due to an increase in apoptosis, but which cells are lost and when is tightly regulated. It is currently unclear if this regulation of apoptosis benefits the host to maintain epithelial continuity or the pathogen to maintain its intracellular location.[12] Because differentiated, absorptive villous epithelial cells are lost, the normal secretion of the crypts overwhelms the absorptive capacity of the villi.[13]

Villous atrophy cannot account for all of the fluid loss in C. parvum infections, and prostaglandin-mediated secretion and inhibition of sodium chloride absorption have been documented.[14, 15] $PGE_2$ causes approximately 25% of the secretion by acting on the enterocyte, while $PGI_2$ leads to the majority of the secretion by stimulating the enteric nervous system.[15] Secretion due to prostaglandin production is mediated by an intracellular increase in cyclic adenosine monophosphate and calcium.[16, 17]

**ROTAVIRUS**

Rotavirus diarrhea occurs after oral exposure to the virus with a peak incidence at 6 days of age, and occurring most commonly in calves less than 3 weeks of age. Originally, the pathophysiology of rotavirus diarrhea was believed to be due to malabsorption, similar to *C. parvum* diarrhea. The virus does preferentially infect the villous epithelial cells and causes cell loss and villous atrophy.[18] Yet, clinical signs do not always correspond to histologic damage, and additional research demonstrated the production of an enterotoxin by rotavirus that appears to mediate most of the clinical signs.[19]

Nonstructural glycoprotein 4 (NSP4) is produced during intracellular viral replication and acts on both the infected cell and neighboring cells. In the infected cell, NSP4 leads to the movement of calcium from the endoplasmic reticulum into the cytoplasm,[20] while luminal exposure leads to binding of NSP4 to a transmembrane protein, caveolin-1.[4, 21] This activates phospholipase C (PLC) which increases intracellular inositol 1,4,5-triphosphate and leads to an increase in intracellular calcium.[22] Through the intracellular increase in calcium and direct actions, NSP4 is able to modulate the movement of nutrients, electrolytes, and water across the intestinal epithelium by inhibiting the translocation of disaccharidases to the luminal surface[23] and blocking SGLT1, the primary sodium-glucose cotransporter.[24] Furthermore, NSP4 increases epithelial secretion primarily by activating the enteric nervous system,[25] though prostaglandin secretion also contributes to this response.[26]

**CORONAVIRUS**

The epidemiology of coronavirus diarrhea is similar to rotavirus in that it occurs most commonly in the first 3 weeks of life and peaks around 7 days of age. The clinical signs typically last longer (3-6 days) as it can infect the villous and crypt epithelium and coloncytes. Using the Spike and Hemagglutinin glycoproteins (S and HE gp), coronavirus binds to the luminal surface of the intestinal epithelium, fuses with the cell membrane, and enters the cell.[27, 28] Intracellular viral replication leads to cell death and villous blunting. Yet, diarrhea begins at the time of viral entry into the epithelial cells, two days prior to histologic evidence of cell loss. The mechanism by which this occurs is currently unknown.

## REFERENCES

[1]Francis DH, et al. **Am J Vet Res** 1989;50(6):822. [2]Runnels PL, et al. **Infect Immun** 1980;28(1):298. [3]Mitchell ID, et al. **J Med Micro** 1974;7:395. [4]Parr RD, et al. **J Virol** 2006;80(6):2842. [5]Giannella RA, et al. **Trans Am Clin Climatol Assoc** 2003;114:67. [6]Golin-Bisello F, et al. **Am J Physiol Cell Physiol** 2005;289(3):C708. [7]Sellers ZM, et al. **Am J Physiol Gastrointest Liver Physiol** 2005;288(4):G654. [8]Lucas ML, et al. **J Membr Biol** 2005;206(1):29. [9]Yao L, et al. **Exp Parasitol** 2007;115(4):333. [10]Johnson JK, et al. **J Parasitol** 2004;90(5):980. [11]Schmidt J, et al. **Mol Biochem Parasitol** 2008;162(1):32. [12]Mele R, et al. **Infect Immun** 2004;72(10):6061. [13]Argenzio RA, et al. **Gastroenterology** 1990;98(5 Pt 1):1129. [14]Gookin JL, et al. **J Vet Intern Med** 2002;16(1):12. [15]Argenzio RA, et al. **Gastroenterology** 1993;104(2):440. [16]Argenzio RA, et al. **J Pharmacol Exp Ther** 1996;279(3):1109. [17]Blikslager AT, et al. **Am J Physiol Gastrointest Liver Physiol** 2001;281(3):G645. [18]Ramig RF. **J Virol** 2004;78(19):10213. [19]Ball JM, et al. **Science** 1996;272(5258):101. [20]Berkova Z, et al. **Cell Calcium** 2003;34(1):55. [21]Storey SM, et al. **J Virol** 2007;81(11):5472. [22]Dong Y, et al. **Proc Natl Acad Sci U S A** 1997;94(8):3960. [23]Martin-Latil S, et al. **Cell Microbiol** 2004;6(8):719. [24]Halaihel N, et al. **J Virol** 2000;74(20):9464. [25]Lundgren O, et al. **Science** 2000;287(5452):491. [26]Lorrot M, et al. **Virol J** 2007;4:31. [27]Schultze B, et al. **J Virol** 1991;65(11):6232. [28]Schultze B, et al. **Virology** 1991;180(1):221.

# CRYPTOSPORIDIOSIS IN CALVES, PEOPLE & THE ENVIRONMENT

**Jeff W. Tyler, DVM, MPVM, PhD; Clemence Z. Chako, BVSc; Brenda Beerntsen, PhD**
**From the MU Public Health Program (Chako, Tyler, Beerntsen) & the Departments of Veterinary Medicine & Surgery (Chako, Tyler) & Veterinary Pathobiology (Chako, Beerntsen), College of Veterinary Medicine, University of Missouri, Columbia, MO**

## BACKGROUND & INTRODUCTION

Cryptosporidiosis is a common causes of diarrhea in people and livestock. Cryptosporidiosis is caused by coccidial parasites of the genus *Cryptosporidium*. Transmission occurs through the ingestion of infectious oocysts in contaminated water or food and by direct contact with fecal material from individuals or animals actively shedding oocysts (1). Oocysts shed in feces are fully sporulated and immediately infective (2). This has critical ramifications in the development of control strategies for cryptosporidiosis. In the case of Eimeria and Isospora sp., common coccidial pathogens of livestock, sporulation of oocysts occurs in the environment, rather than within the intestinal lumen of the host (2). Consequently, the oocysts contained in the fecal material shed by an Eimeria or Isospora sp. infected host poses minimal risk unless the fecal material remains in contact with new hosts for the obligatory period of sporulation. As a direct consequence, frequent removal of feces or movement to clean sites will halt transmission of these pathogens, unlike for cryptosporidia. Cryptosporidiosis most commonly affects neonatal calves. However, infections have been reported in sheep, goats, horses, swine, camels, deer, raccoons, dogs, cats and birds (3). Regardless of the species affected, young animals are at greater risk for either infection or disease. The primary clinical sign in calves is a profuse watery diarrhea. Over 90% of dairy farms and 40% of beef farms are endemically infected with cryptosporidiosis (4). Approximately half of dairy calves between the ages of 1 and 3 weeks of age are actively shedding oocysts at any time (5,6,7). Cryptosporidiosis also is one of the most common causes of infectious diarrhea in humans (8). Although clinical manifestations of cryptosporidiosis vary with age and immune status, they include voluminous, self-limiting diarrhea of 1-14 days duration, often accompanied by abdominal cramps, fatigue, vomiting, fever and malaise. The prevalence and incidence of cryptosporidiosis is higher in people infected with HIV (9-17) and in residents of developing countries whereas in developed countries it is sporadic with large scale outbreaks associated with contaminated water and food sources, recreational exposures to contaminated waters, and horizontal transmission in child care settings. In immunodeficient humans, cryptosporidiosis can cause a chronic, life- threatening diarrhea (18) with case fatality rates of approximately 50% (19). No drug or drug combination effectively treats cryptosporidiosis in humans.

## RECENT CHANGES IN OUR UNDERSTANDING OF CRYPTOSPORIDIUM HOST RANGE

There are at least 16 recognized species and more than 40 genotypes of Cryptosporidia (20). Disease in people is frequently caused by either *Cryptosporidium hominis* or *Cryptosporidium parvum (20)*. *Cryptosporidium hominis* was previously thought to be a genotype of *Cryptosporidium parvum* (3,21). Other cryptosporidia that have been identified in humans include *Cryptosporidium meleagridis, Cryptosporidium felis, Cryptosporidium canis, Cryptosporidium muris* and *Cryptosporidium suis* (22,23). Although cryptosporidium was once viewed as a singular pathogenic species with a broad host range, recent advances in molecular epidemiology have demonstrated that Cryptosporidia are a diverse group with distinct and divergent host ranges. Some are specific to people, others are specific to animals whilst some are true zoonotic pathogens which freely move between people and animals. Perhaps the most illustrative example is our changing understanding of *C. parvum. Cryptosporidium parvum* was once thought to be a single species affecting people and a broad range of mammals. Later we recognized the presence of 2 distinct genotypes, one of which was a specific human pathogen and another which affected both people and animal hosts. The former was referred to as *C parvum* genotype I and the latter *C. parvum* genotype II (21). Presently, we recognize that *C.parvum* genotype I isolates were, in fact, *C.hominis*, a specific human pathogen (33), whereas, *Cryptosporidium parvum*, previously referred to as *C.parvum* genotype II, is a pathogen of several mammals including cattle, sheep, goats, deer, raccoons, horses, dogs and people (20,23,23). Finally, there exists a wide range of newly recognized Cryptosporidium species, some of which are specific to animals and others which have the potential for human disease.

The most common Cryptosporidium species responsible for human disease in the US is *C. hominis* a human adapted pathogen (24); however, a substantial number of human isolates and disease are caused by *C parvum. Cryptosporidium meliagridis* is increasingly recognized as an important pathogen in people (18). Other species of Cryptosporidium (*C.felis, C.canis, C. suis* and *C.muris*), are infrequent causes of human disease and are diagnosed primarily in people with immunodeficiencies (18). Our understanding of cryptosporidiosis probably has been biased by the experiences of people with significant occupation contact with livestock. This experience creates the impression that cryptosporidiosis is zoonotic and that livestock are the principal reservoir and source of infectious oocysts. Although this experience is descriptive of occupational risk, it does not reflect the general population's risk. A clearer understanding of cryptosporidium host range is needed to develop preventive strategies for the general public. We can anticipate this understanding will evolve as reports are published which include definitive identification of incriminated Cryptospodium. For example, *Cryptosporidium hominis* (genotype Ia) is more closely related to *C.parvum* (genotype IIa) than it is to other genotypes (genotypes Ib and Ie) of *C. hominis* (3). We anticipate future reclassification whereby genotype Ia is recognized as a distinct species.

## POTENTIAL FOR CRYPTOSPORIDIOSIS ASSOCIATED WITH CONTAMINATED WATER SOURCES

More than 15 large scale cryptosporidiosis outbreaks associated with contaminated municipal water source have occurred since 1985 in US. Immunocompromised persons should never drink water directly from lakes or rivers. Even well and municipal water supplies should not be considered risk free. Surface water sources like lakes, rivers, beaches, swimming pools, recreational water parks, and ornamental water fountains are often contaminated with human or animal waste containing cryptosporidium oocysts as a result of contaminated run-off off or inadequate treatment of waste waters. Wild and feral animals have access to the surface water sources. Surface waters are more likely to be contaminated with Cryptosporidium than are wells. The prevalence of cryptosporidia oocyst contamination of surface waters ranges from 39 to 87% in USA (25-29). Efforts to reduce contamination of surface waters are premised upon the prevention of water source contamination with human or animal effluent. Vegetative barriers are highly effective in reducing the number of Cryptosporidium oocysts in run-off (30). Consequently, the restriction of livestock access to streams and waterways is a rational strategy to prevent contamination of water sources. Animal housing should be strategically located so as to prevent watershed contamination. Dairy calf rearing areas should be considered a high risk effluent source. Water derived from wells is less likely to be contaminated than surface water. Soil serves as a passive filtration system for oocysts. In a small survey in the United Kingdom, only 5.8% of 258 groundwater samples were positive for Cryptosporidium (31). It should be noted that all samples positive for Cryptosporidium were also positive for coliform bacteria, suggesting contamination with animal or human effluent (26). Although wells pose a lower risk of cryptosporidiosis, clustering of houses that use septic tanks may increase the risk for contamination of wells (32).

**FOOD AS POTENTIAL EXPOSURE TO CRYPTOSPORIDIOSIS**

Cryptosporidium oocysts do not replicate or multiply within foodstuffs. The potential risks posed by the foodstuffs relates to either fecal contamination by effluent or the use of contaminated water during foodstuff preparation. Consequently, the focus in the prevention of food-borne cryptosporidiosis is on the prevention of contamination.

**RECREATIONAL EXPOSURES**

Because of its resistance to chlorination, Cryptosporidium has become the leading cause of gastroenteritis outbreaks associated with recreational waters. Cryptosporidiosis accounted for approximately 60% of the outbreaks reported to CDC during 1995-2004(33). Majority of the outbreaks have been attributed to fecal contamination of pools by swimmers. Conventional chlorination is inadequate to control cryptosporidiosis associated with recreational water use (34). Surface recreational waters are potentially contaminated by inadequately treated human effluent, agricultural run-off and free-roaming animals. Run-off from a dairy calf rearing facility poses the greatest risk (35).

**EXPOSURES ASSOCIATED WITH ANIMAL CONTACT**

Contact with animals and animal effluent are important risk factors for zoonotic strains of cryptosporidium which include *C. parvum, C. meliagridis, C.felis, C.canis, C. suis* and *C.muris*. Cryptosporidium parvum and meliagridis are by far the most common zoonotic cryptosporidia (51). Occupational or recreational contact with cattle, sheep, goats, deer, horses, pigs, calves, birds, cats, dogs and rodents creates a potential risk for infection. Individuals experiencing occupational risk include farm employees or veterinarians and their family members. Young animals typically shed oocysts in higher numbers. Calves shed cryptosporidium from approximately day 5 to day 30 with a peak at day 12 (35). It is therefore advised that immunocompromised employees should not work with calves. Petting zoos and farm tours where young people are exposed to farm livestock are a particular risk. Immunologically naïve urban populations are at a greater risk of developing the disease. Teachers who arrange trips for school children to go to farms should liaise with the farm managers and veterinarians with regard to exposure to zoonotic pathogens. Pupils should not come into direct contact with kids, lambs, calves or fecal material. The inclusion of young animals in tour or pet zoo displays should be discouraged and no food should be consumed during the tour and other hygienic precautions such as hand washing should be practiced. Immunocompromised individuals should be advised that young pets may pose a limited risk for transmitting cryptosporidium, but they should not be advised to destroy or give away healthy pets. Persons contemplating acquiring a new pet should avoid acquiring any animal that has diarrhea. Families with immunocompromised members should avoid purchasing a dog or cat aged less than 6 months. Immunocompromised individuals who wish to assume the limited risk for acquiring a puppy or kitten aged less than 6 months should request that their veterinarian examine the animal's stool for *Cryptosporidium* before they have contact with the animal.

**EXPOSURES ASSOCIATED WITH HUMAN CONTACT**

A substantial number of people are at risk of being infected by *Cryptosporidium* originating from human sources. This group includes doctors, nurses and child care workers. A cryptosporidiosis outbreak which occurred in rural Missouri in 2005 was associated with attendance at child care centers. Children attending child care facilities were the source of a communitywide outbreak (36).Among the commercial disinfectants that appear most effective are those that contain hydrogen peroxide, chlorine or ammonia (37).

Immunocompromised childcare and health care workers should work in areas where they are less likely to be exposed to Cryptosporidium and other infectious diseases. They may choose not to work in areas with high numbers of children or adults with clinical diarrhea. A risk management plan should be drawn and agreed upon by the management and employees. Because of potential nosocomial cryptosporidium infections, immunocompromised individuals should not share rooms with patients with cryptosporidiosis.

**DIAGNOSIS OF CRYPTOSPORIDIOSIS AND THE APPLICATION OF RESULTS IN PUBLIC HEALTH**

Historically, the diagnosis of cryptosporidiosis has been premised upon detection of the organism in feces. Variations on this strategy include direct microscopic examination of feces, microscopic examination after sucrose floatation, acid fast staining or use of fluorescent-labeled antibodies and ELISA on fecal samples. All these techniques are capable of detecting oocysts. However, it should be noted that these techniques will not permit differentiation of Cryptosporidium species (38-40). Given that our current understanding of cryptosporidiosis links species of Cryptosporidium with more defined host ranges, more specific assay systems are recommended. The diagnosis of *C. hominis* directly implicates a human source and the diagnosis of *C. parvum* may suggest either a human or animal source. Recognition of *C. hominis*, particularly in clustered large numbers of patients, should encourage physicians and public health officials to consider person to person transmission, contamination of food or water by human effluent or the presence of contaminated point sources such as recreational waters. Recognition of *C. parvum* raises the possibility of animal effluent contaminated food or water and occupational or recreational exposures of animal feces in addition to the previously listed sources of exposure to human reservoirs. Similarly, recognition of *C. meliagridis* suggests exposure to avian effluent. Recognition of less common species of Cryptosporidium not only tends to implicate an alternative, animal source of infection, but raises the distinct possibility that the affected person may be immunosuppressed. The less common Cryptosporidium species (muris, felis, suis, canis) are more common among people infected with HIV (3) The application of diagnostic tools to identify Cryptosporidium species as a mechanism to incriminate the reservoir hosts will require the application of genomic based assay systems.

## LESSONS FROM MILWAUKEE, WISCONSIN

In the spring of 1993 there was a widespread outbreak of cryptosporidiosis in Milwaukee, Wisconsin. An estimated 403,000 people comprising 25% of the population experienced clinical diarrhea. Laboratory confirmed diagnoses of cryptosporidiosis were available in 285 patients. Of these patients, 48 (17%) were immunocompromised and the case fatality rate among these individuals approached 50% (41). The Milwaukee outbreak was initially attributed to contamination of the municipal water supply by animal effluent contaminated run-off. Retrospective studies revealed a different picture. Waste water samples (n=179) were analyzed for the presence of Cryptosporidium (42). Twenty-four samples were positive for *C. hominis*, a human specific pathogen. Five samples were positive for *C. parvum*; however, only 2 of the 5 isolates were genotype IIa, the recognized zoonotic pathogen of calves and people. All human stool isolates in this outbreak were *C. hominis*. Indeed, the water supply probably was contaminated by both human and animal effluent; however, a human specific genotype of Cryptosporidium (*C. hominis* 1b) was the likely cause of this outbreak (42). The vast majority of mass water borne Cryptosporidium outbreaks in the United States have been caused by human-adapted genotypes (24). Although animal effluent creates the potential for human disease (43), public health officials should be encouraged to broaden their concerns to include human effluent contamination and recognize that human effluent poses a greater risk.

## REFERENCES

[1] Nichols G. Epidemiology.In: Fayer R, Xiao L, eds. **Cryptosporidium and Cryptosporidiosis, 2nd ed. Boca Raton: CRC Press**; 2008: 79-101. [2] Ebeid M, et al. **Parasitology Research** 2003; 90(3):232-235. [3] Xiao L, et al. Clin Microbiol Rev 2004; 17: 72-97. [4] Harp JA, et al. **J Dairy Sci** 1998; 81:289-294. [5] Brandoniso O, et al. **J res Health Sci** 2007;7:20-24. [6] Farrington M, et al. **Vet Parasitology** 1995; 60:7-17. [6] McCluskey BJ, et al. **Vet Parasitology** 1995; 60: 185-190. [7] Sturdee AP, et al. **Vet Parasitology** 2003; 116(2):97. [8]Ajjampur SS, et al. **J Clin Microbiol**. 2007; 45: 915–920. [9] Wheeler, et al. **B M J** 1999;318:1046-1050. [10] Bushen OY, et al. **Trans R Soc Trop Med Hyg**. 2007; 101: 378–84. [11] Gatei W, et al. **Am J Trop Med Hyg.** 2006; 75: 78–82. [12] Gatei W, et al. **Infect Genet Evol**. 2007; 7: 197–205. [13] Leav BA, et al. **Infect Immun**. 2002; 70: 3881–90. [14] Peng MM, et al. **J Eukaryot Microbiol**. 2003; 50 (Suppl):557–559. [15] Raccurt CP, et al. **Trop Med Int Health**. 2006;11:929–34. [16] Samie A, et al. **Exp Parasitol**. 2006;114: 314–22. [17] Tumwine JK, et al. **Am J Trop Med Hyg**. 2005; 73: 921–5. [18] Cama V,et al. **J Eukaryot.Microbiol**. 2003; 50 suppl: 531-533. [19] Rose **JB..Annu. Rev.Public Health**.1997; 18: 135-161. [20] Xiao L, et al. **Current Opin Infect Diseases**. 2004; 17: 483-490. [21] Morgan-Ryan UM, et al. **J Eukaryot.Microbiol**.2002; 49: 433-440. [22] Caccio SM. Et al. **Parasitologia** 2005; 47: 185-192. [23] Xiao L, et al. **Clin Microbiol Rev** 2004;17:72-79. [24] Xiao L, et al. In: **Cryptosporidiosis,2nd ed**. Boca Raton: **CRC Press**;2008:119-163. [25] LeChavallier MW, et al. **Appl. Environ. Microbiol** 1991; 57: 2617-2621. [26] LeChavallier MW, et al. **JAWWA**. 1995; 87: 54-68. [27] LeChevallier MW, et al. **Appl.Envirol.Microbiol**. 2003; 69: 971-979. [28] McTigue NE, et al. **AWWA and AWWA** RF. Denver, CO 1998. [29] Rose JB, et al. **Environ.Sci**. Technol. 1991; 25:1393–400. [31] Lisle JT, et al. Aqua - **Journal of Water Supply: Research and Technology**. 1995; 44(3): 103-117. [32] McQuillan D. accessed on **http://wrri.nmsu.edu/conf/conf06/conf.html on 5-23-08**. [33] CDC. MMWR 2007; 56 (29):729-732. [34] Korich DG, et. Appl Environ Micribiol 1990;56:1423-1428. [35] Nydam DV, et al. **Am J Vet Res** 2001; 62: 1612-1615. [36] Turabelidze G, et al. **Arch Pediatr Adolesc Med**.2007; 161:878-883. [37] Fayer R, et al. **General Biology**. In: Fayer R, Xiao L, eds. Cryptosporidium and Cryptosporidiosis, 2nd ed. Boca Raton : **CRC Press**; 2008:1-42. [38] Smith HV, et al. **Trends Parasitol**.2006; 22:160-166. [39] Smith HV, et al. **In Mara D and Horan N eds. Elsevier Science Limited**, London: Oxford; 2003: 619-781. [40] Smith HV, et al **The Royal Society of Chemistry**, Cambridge;2001:1-43. [41] Mac Kenzie WR, et al. N Engl J Med 1994; 331(22): 1529-30. [42] Zhou Ling, et al. **Journal of Clinical Microbiology**. 2003;14(11): 5254-5257. [43] Steiner TS, et al. **Annu Rev Med** 1997; 48: 329-340.

# THE IMMUNOBIOLOGY OF CRYPTOSPORIDIOISIS

**Michael W. Riggs, D.V.M., Ph.D., DACVP**
**Tucson, AZ**

## INTRODUCTION

The increasing recognition of *Cryptosporidium spp.* which infect the intestinal tract as parasites of veterinary and human medical importance has provided a catalyst for advancing our understanding of the immunobiology of cryptosporidiosis. Three valid enteric species are currently recognized in calves: *C. parvum*, *C. bovis* and *C. ryanae*. However, only *C. parvum* is pathogenic and zoonotic, and is the most prevalent species causing diarrheal disease in confined calves.[1] There is currently a lack of approved, consistently effective parasite-specific drugs, vaccines, and immunotherapeutics for use against cryptosporidiosis in calves and other agriculturally important livestock species, although significant advances have been made in understanding the parasite's immunobiology.[2-8] Parasite-specific and innate immune responses are known to control *C. parvum* infection in immunocompetent hosts. Following recovery, such hosts are relatively resistant to reinfection and clinical disease. These basic observations have provided the rationale for continuing investigation of the immunobiology of cryptosporidiosis and development of immunologic control strategies. Identification and characterization of host immune responses to *Cryptosporidium spp.* which are responsible for clearance of primary infection and resistance to reinfection is essential for the rational design of vaccines and immunotherapeutics. In this article, recent advances in knowledge of host immune responses in mammalian cryptosporidiosis are briefly reviewed. Several earlier[9, 10] and more recent[2-6, 8] reviews are available for comprehensive coverage of the immunobiology of cryptosporidiosis.

## HUMORAL IMMUNE RESPONSES IN CRYPTOSPORIDIOSIS

### Role of antibodies in controlling *C. parvum* infection and their use for identification of functional antigens

The relative contribution of antibody (Ab) responses in controlling *C. parvum* infection in the immunocompetent host appears to be minor but remains unclear. Serum and mucosal Ab responses coincide with resolution of primary infection and diarrhea, and may contribute to resistance to reinfection; however, these observations are largely correlative and additional studies may be needed to evaluate Ab responses against known neutralization-sensitive *C. parvum* epitopes. For example, mucosal Abs are produced against an immunodominant neutralization-sensitive zoite protein designated P23 and are secreted into the intestinal lumen during recovery from experimental *C. parvum* infection in calves.[9] Ab responses could represent biologically significant but redundant mechanisms contributing to resolution of infection and acquired resistance, or may only be temporal markers that other protective immune responses have been generated.[9, 11-14]

Abs produced against *C. parvum* in response to natural infection or parenteral immunization have been used to identify biologically relevant antigens (Ags) as vaccine candidates, and to characterize functional epitopes defined by neutralizing Abs. The emphasis has been on apical complex and surface Ags of the infective, extracellular sporozoite and merozoite stages which are known to be involved in motility, attachment, invasion, and/or intracellular development. Further, there is a high degree of antigenic conservation between the sporozoite and merozoite life cycle stages of *C. parvum*, although differences between *Cryptosporidium spp.* have been reported.[15] Multiple *C. parvum* Ags capable of inducing neutralizing Abs have been identified (Table 1). These include multiple high $M_r$ sporozoite and merozoite glycoproteins (gps). The relationship between these high $M_r$ gps is presently unclear although distinct differences have been reported.[9, 15, 19] It should also be noted that CP15/60[23] and CP15[24] are molecularly distinct from each other, as well as from gp15 and gp40 which are derived from a gp60 precursor.[25-30] Studies evaluating these and other Ags as vaccine candidates are the subject of an accompanying review.[16] The diversity and biochemical characteristics of known or putative functional Ags identified to date (Table 1) suggests that *C. parvum* has redundant mechanisms for infection, or uses multiple ligands recognizing multiple host receptors in a multistep infection process.

### Antibody–mediated mechanisms of *C. parvum* neutralization

Neutralizing Abs against *C. parvum* bind functional surface-exposed antigens on the infective sporozoite and merozoite life cycle stages and prevent their attachment and invasion.[6, 9] Additionally, stages bound by Abs while extracellular, but which retain the ability to invade, may undergo Ab-mediated arrest of intracellular development.[9, 17] Agglutination does not appear to be an important mechanism in neutralization. In vitro studies have shown that motility of *C. parvum* zoites involves attachment of surface adhesion molecules to host cells, followed by posterior translocation and release, and forward locomotion, leaving trails of deposited Ag along the path traveled. Similarly, attachment and posterior translocation of the moving junction between zoite and host cell membranes has a role during invasion. *C. parvum* Ags which are deposited during gliding motility have been identified (Table 1). gp15 and gp40 are also shed during invasion in vitro. Neutralization by Abs against each of these Ags is thought to involve interference with motility and/or invasion. Neutralizing Abs against CSL bind zoites and elicit the circumsporozoite precipitate (CSP)-like reaction, characterized by release of CSL bound in membranous complexes with Ab. Because CSL contains a sporozoite ligand involved in attachment to a surface receptor on bovine and human intestinal epithelial cells, zoite infectivity, but not viability, is lost after occurrence of the CSP-like reaction.

**Table 1.** *C. parvum* Ags Recognized by Neutralizing MAbs

| Antigen | Characteristics/Localization | Function(s) | References |
|---|---|---|---|
| CSL | ~1300 kDa gp; dense granules, micronemes, sporozoite and merozoite surface | ligand for intestinal epithelial cell receptor | 6, 18-20 |
| GP900 | >900 kDa gp; micronemes, sporozoite surface | ligand for host cells; motility | 6, 21 |
| CP47 | ~47 kDa protein; sporozoite apical subpellicle | ligand for host cells | 4, 22 |
| CP15/60 | ~15/60 kDa gps; sporozoite surface | motility | 23 |
| CP15 | ~15 kDa gp; sporozoite surface | motility | 24 |
| gp15 | ~15 kDa gp; surface of sporozoites and merozoites | ligand for host cells; motility | 25-30 |
| gp40 | ~40 kDa gp; apical complex, sporozoite/merozoite surface | ligand for host cells | 25-30 |
| TRAP C1 | ~76 kDa protein; sporozoite apical pole, micronemes | motility, attachment | 31, 32 |
| P23 | ~23 kDa gp; surface of sporozoites and merozoites | motility, attachment | 33-35, 57 |
| GP25-200 | ~25-200 kDa gps; apical complex, sporozoite/merozoite surface | motility | 18, 35 |
| Gal/GalNAc lectin | ~30 kDa lectin; apical sporozoite | ligand for host cells | 36, 37 |
| CPS-500 | glycolipid; sporozoite and merozoite surface | motility | 38, 39 |

## CELL –MEDIATED IMMUNE RESPONSES IN CRYPTOSPORIDIOSIS

### Role of cell–mediated response in controlling infection

The critical role of cell-mediated immune (CMI) responses in controlling *C. parvum* infection in immunocompetent hosts has been clearly established.[2, 4, 5, 6, 8, 9] A variety of genetically characterized rodent models have been used to determine the nature and role of CMI responses to *C. parvum* despite the sometimes confounding influence of strain particulars on interpretation of data. Immune responses in such genetically defined rodent models are important to define and understand but may not always be applicable to outbred populations comprising the target species, namely neonatal calves. These studies have demonstrated the central importance of CD4$^+$ T cells, IFN-gamma (T cell and non-T cell derived), and IL-12 in resistance to and recovery from *C. parvum* infection, and suggested that mechanisms of resistance to primary infection and recovery from established infection differ. While the results of some rodent studies are conflicting, peripheral CD8$^+$ T cells are generally considered to have a minor or redundant role, whereas CD8$^+$ intraepithelial lymphocytes (IEL) may be more important. Studies in immunocompetent mice suggest that CD8$^+$ IEL may be more important in the elimination of established infection while CD4$^+$ IEL may be more important in controlling early infection. However, it has recently been shown that early resistance to *C. parvum* in mice is correlated with increased mucosal expression of IFN-gamma from naïve CD8$^+$ T-cell receptor (TCR) alpha beta$^+$ IEL.[40] In immunocompetent humans, studies suggest that mucosal IFN-gamma production may be an important part of the memory response and resistance to reinfection in previously exposed individuals.[5] Results of more recent studies involving lymphocyte and cytokine depletion/deficiency, adoptive transfer, and genetically engineered/knockout rodent models have largely confirmed and extended the above findings. Both TCR alpha beta$^+$ and TCR gamma delta$^+$ T cells may have significant but overlapping roles in controlling *C. parvum* infection in mice. Such studies have also indicated that CD154 (a TNF-like molecule expressed on activated CD4$^+$ cells) and CD40 (the cell surface receptor for CD154) are both required to resolve *C. parvum* infection in mice, confirmed the importance of an intact CD154–CD40 signaling pathway, and demonstrated that CD154-dependent, antigen-independent effector functions mediated by activated CD4$^+$ T cells, in concert with innate responses, can clear infection in the absence of *C. parvum* specific CD4$^+$ T cells. CD154–CD40 interaction may stimulate production of nitric oxide, IL-12, and IFN-gamma, and stimulate T cell responses. Cytokine depletion/deficiency studies have indicated that IFN-gamma is important in controlling the severity of infection, and that either IFN-gamma or IL-4 can accelerate recovery from infection, but that recovery can occur in the absence of either cytokine in adult C57BL/6 mice although mouse strain genetic differences have been observed.[41] Nevertheless, the critical importance of IFN-gamma produced by *C. parvum*-specific CD4$^+$ T cells, antigen-nonspecific CD4$^+$ T cells, and antigen non-specific natural killer cells in controlling *C. parvum* infection has been clearly established.[2, 5, 42]

Studies on the early intestinal mucosal immune response to *C. parvum* in experimentally infected naïve 3 day old calves terminated at day 3 of infection identified increased numbers of CD4$^+$alpha beta$^+$, CD8$^+$ alpha beta$^+$, and gamma delta$^+$ T cells in ileal lamina propria and Peyer's patches. A trend towards increased expression of CD8 and gamma delta TCR in ileal IEL was also observed.[2, 9, 43] Mucosal responses in later stages of infection, during peak clinical disease, was studied in experimentally infected naïve 3 day old calves terminated at days 6 to 9 of infection. These studies identified increased numbers of CD2$^+$, CD3$^+$, CD4$^+$, and CD8$^+$ alpha beta$^+$ T cells in ileal IEL, and increased CD3$^+$, CD4$^+$, and CD8$^+$ T cells in lamina propria. Increased IFN-gamma and IL-12 expression by IEL and lamina propria T cells was also observed, and IFN-gamma expression was correlated with CD4$^+$ and CD8$^+$ IEL.[2, 9] Following complete recovery at days 13 to 19 post infection, CD4$^+$, CD8$^+$ and gamma delta$^+$ T cells, and cytokine expression in ileal mucosa had reverted to the normal phenotype found in uninfected age-matched controls.[2, 9] Peripheral blood CD4$^+$ T lymphocytes from recovered calves were shown to proliferate and produce IFN-gamma when incubated in vitro with recombinant P23[43] as well as whole oocyst antigen extracts.[2, 9]

In recovered immune calves challenged ~30 days following primary infection, no infection was detected, indicating that primary infection conferred protective immunity.[2, 9] A mucosal memory response was evident by 3 days post-challenge, characterized by increased CD8$^+$ alpha beta$^+$ T cells closely associated with ileal epithelial cells, and increased CD4$^+$ alpha beta$^+$ T cells in Peyer's patches. These findings suggest that CD8$^+$ T cells have an important role in protection against reinfection in calves, unlike conclusions from most rodent model studies. No changes in ileal gamma delta$^+$ T cells were observed in immune calves, suggesting that these cells may be more important in controlling primary infection in calves.

**Cell–mediated mechanisms of *C. parvum* neutralization**

IFN-gamma has multiple systemic effector functions in CMI responses to *C. parvum*, and has also been shown to be locally produced in intestinal mucosa during infection and directly induce resistance to *C. parvum* in enterocytes.[9] Increased expression of intestinal mucosal IL-15 has been demonstrated in *C. parvum* infected humans and is thought to be involved in control of acute primary infection via activation of NK cells to lyse *C. parvum* infected epithelial cells.[44] While beyond the scope of this brief review, other antigen independent innate responses, and their possible roles in resistance to *C. parvum* as well as mucosal lesion development, have been reported. These include responses during infection by neutrophils and other granulocytes, dendritic cells, macrophages and other inflammatory cells; nitric oxide production; and expression of chemokines, cytokines, innate adapter proteins, antimicrobial peptides, pro-inflammatory mediators and repair mediators,[9, 14, 45-53] as well as innate responses induced by administration of exogenous immunomodulators.[54, 55] Finally, bovine colostrum is known to contain multiple antimicrobial peptides. Recent studies have shown that long-chain monounsaturated fatty acids such as oleic acid and linolenic acid, generated following digestion of bovine colostrum by calf intestinal fluid, may serve as natural, potent bioactive inhibitors of *C. parvum* sporozoite attachment to host cells.[56]

**REFERENCES**

[1]Fayer R, et.al. **Vet Parasitol** 2008; 156:191. [2]Fayer R, et.al. **Ed. *Cryptosporidium* and Cryptosporidiosis**, 2nd Ed. 2008; CRC Press. [3]Sanderson SJ, et.al. **Proteomics** 2008; 8(7):1398. [4]Borowski H, et.al. **Trends Parasitol** 2008; 24(11):509. [5]Pantenburg B, et.al. **Infec Immun** 2008; 76(1):23. [6]Boulter-Bitzer JI, et.al. **Biotech Advances** 2007; 25(1):13. [7]O'Connor RM, et.al. **Mol Biochem Parasitol** 2007; 152(2):149. [8]Wanyiri J, et.al. **Future Microbiol** 2006; 1(2):201. [9]Riggs MW. **Microbes Infect** 2002; 4:1067. [10]Tzipori S, et.al. **Microbes Infect** 2002; 4:1047. [11]Jakobi V, et.al. **Parasite Immunol** 2008; 30(3):151. [12]Preidis GA, et.al. Amer J Trop Med Hygeine 2007; 77(3):583. [13]Ehigiator HN, et.al. **Parasitol Res** 2007; 101(4):943. [14]Lean IS, et.al. **Infect Immun** 2006, 74(7):4379. [15]Sturbaum GD, et.al. **Mol Biochem Parasitol** 2008; 159(2):138. [16] Riggs MW. **27th Annual ACVIM Forum Proceedings** 2009. [17]Jakobi V, et.al. **Microbes Infect** 2006; 8(8):2186. [18]Riggs MW, et.al. **J Immunol** 1997; 158:1787. [19]Langer RC, et.al. **Infect Immun** 1999; 67:5282. [20]Langer RC, et.al. **Infect Immun** 2001; 69:1661. [21]Barnes DA, et.al. **Mol Biochem Parasitol** 1998; 96:93. [22]Nesterenko MV et.al. **Biochim Biophys Acta** 1999; 1454:165. [23]Jenkins MC, et.al. **Infect Immun** 1993; 61:2377. [24]Jenkins MC, et.al. Mol **Biochem Parasitol** 1995; 71:149. [25]Priest JW, et.al. **J Clin Micro** 1999; 37:1385. [26]O'Connor RM, et.al. **Mol Biochem** Parasitol 2007; 156(1):80. [27]Cevallos AM, et.al. **Infect Immun** 2000; 68:4108. [28]Strong WB, et.al. **Infect Immun** 2000; 68:4117. [29]Cevallos AM, et.al. **Infect Immun** 2000; 68:5167. [30]Priest JW, et.al. **Mol Biochem Parasitol** 2000; 106:261. [31]Spano F, et.al. **Mol Biochem Parasitol** 1998; 92:147. [32]Putignani L, et.al. **Mol Biochem Parasitol** 2008; 157(1):98. [33]Shayan P, et. al. **Parasitol Res** 2008; 103(5):1207. [34]Perryman LE, et.al. **Vaccine** 1999; 17:2142. [35]Schaefer DA, et.al. **Infect Immun** 2000; 68:2608. [36]Ward H, et.al. **Adv Parasitol** 1998; 40:151. [37]Bhat N, et.al. **J Biol Chem** 2007; 282:34877. [38]Riggs MW, et.al. **Infect Immun** 1999; 67:1317. [39]Priest JW, et.al. **J Biol Chem** 2003; 278(52):52212. [40]Leav BA, et.al. **Infec Immun** 2005; 73(12):8425. [41]McDonald SAC, et.al. **J Infec Dis** 2004; 190(5):1019. [42]Singh I, et.al. **Parasitol Res** 2005; 97(4):309. [43]Wyatt C, et.al. **J Parasitol** 2005; 91(5):1239. [44]Dann SM, et.al. **J Infec Dis** 2005; 192(7):1294. [45]Takeuchi D, et.al. **Infect Immun** 2008; 76(8):3657. [46]Lacroix-Lamande S, et.al. **Microbes Infect** 2008; 10(4):390. [47]Auray G, et.al. **Microbes Infect** 2007; 9(5):574. [48]Zadrozny LM, et.al. **Infect Immun** 2006; 74(10):5497. [49]McDonald V, et.al. **Clin Exp Immunol** 2006; 145(3):555. [50]Rogers KA, et.al. **Infect Immun** 2006; 74(1):549. [51]Kirkpatrick BD, et.al. **Clin Infect Dis** 2006; 43(3):289. [52]Gookin JL, et.al. **Amer J Physiol** 2006; 290(1):G164. [53]Carryn S, et.al. **Int J Antimicrob Agents** 2004; 24(2):S117. [54]Barrier M, et.al. **J Infect Dis** 2006; 193(10):1400. [55]Turanek J, et.al. **Parasitol** 2005; 131(5):601. [56]Schmidt J, et.al. **Mol Biochem Parasitol** 2008; 162(1):32. [57]Yao LQ, et.al. **Exp Parasitol** 2006; 115(4):333

# ADVANCES IN THE DEVELOPMENT OF IMMUNOLOGIC CONTROL STRATEGIES FOR CRYPTOSPORIDIOISIS IN CALVES

**Michael W. Riggs, D.V.M., Ph.D., DACVP**
**Tucson, AZ**

## INTRODUCTION

There are currently no vaccines or immunotherapeutics approved for use against cryptosporidiosis in calves or other agriculturally important livestock species. Three enteric *Cryptosporidium* species are presently recognized in calves, but only *C. parvum* is pathogenic and zoonotic, and is also the most prevalent species causing diarrheal disease in confined calves.[1] An increased understanding of host immune responses to *Cryptosporidium parvum* that are responsible for clearance of primary infection and resistance to reinfection, and characterization of the parasite molecules to which they are directed, are essential for discovery of effective immunization strategies against cryptosporidiosis. In this article, progress in the development of vaccines and parasite-specific biotherapeutics for bovine cryptosporidiosis are briefly reviewed. Several earlier[2, 3] and more recent[4-9] reviews are available for comprehensive coverage of the immunobiology of cryptosporidiosis and perspectives on immunologic control strategies.

## PASSIVE IMMUNIZATION

Early studies demonstrated that hyperimmune polyclonal antibodies (Abs) produced against whole *C. parvum* antigens (Ags) have significant anticryptosporidial activity in vitro and in various rodent models of infection.[2, 3] These studies also demonstrated that polyclonal Abs have significant prophylactic efficacy against *C. parvum* infection and clinical disease in neonatal calves, the principal target species for application of passive immunization strategies. Because of the ubiquity of *C. parvum*, and the fact that infected calves may shed up to $10^7$ infectious oocysts per gram of diarrheic feces, infection in confined calves is common and may be endemic in some operations.[4] Neonatal calves may be exposed to *C. parvum* soon after birth and typically become infected during the first week of life, with peak infection levels observed in 1 to 3 week old calves.[10] Neonatal calves have relatively low percentages of $CD4^+$ and $CD8^+$ T cells in the intestinal mucosa, both of which are involved in recovery from primary infection and resistance to reinfection.[2,4] Because of the immaturity of the intestinal mucosal immune system in calves at birth, active immunization strategies in such unprimed newborns is unlikely to induce protective responses prior to exposure. Thus, strategies for immunologic control of *C. parvum* infection in newborn calves have focused on passive immunization via maternal colostrum or administration of exogenous Ab preparations. Protection against many enteropathogens during the first few days of life in calves is mediated locally by ingested colostral Ab, and thereafter by colostrum-derived serum Ab, which is secreted back into the intestinal lumen over the first month of life. Given the complex, asynchronous life cycle of *C. parvum*, it is not presently considered feasible to completely prevent infection in calves by passive immunization. However, studies have indicated that this approach can reduce infection to subclinical levels during the relatively narrow period of susceptibility to clinical disease caused by *C. parvum*. Additionally, reducing infection to subclinical premunition type levels by passive immunization may facilitate induction of active cellular immune responses that could contribute to resolution of primary infection and resistance to reinfection. Collectively, early studies in rodent models and neonatal ruminants provided the rationale for further investigation of passive prophylactic and treatment regimens for cryptosporidiosis. Because of limitations inherent in polyclonal Ab preparations produced against antigenically uncharacterized whole *C. parvum*, subsequent studies have investigated functionally defined Ags for production of polyclonal as well as monoclonal antibodies. Such biologically relevant *C. parvum* Ags have been identified as rational candidate targets for passive immunization based on recognition by neutralizing Abs. Studies on the immunobiology of these parasite molecules, including CSL, GP900, CP15/60, CP15, gp15, gp40, P23, and GP25-200 are the subject of an accompanying review.[11]

### Polyclonal antibodies

The immunodominant P23 antigen previously defined by neutralizing MAbs and recognized by convalescent serum antibodies from *C. parvum* infected mammals[2, 8, 12, 13] has been evaluated in a seminal study as a target for passive immunization of calves.[14] In this study, late gestation Holstein cows were immunized subcutaneously three times with recombinant P23[15] to produce monospecific anti-P23 colostrum (P23 Ab titer >100,000). Newborn calves were fed anti-P23 colostrum from these cows within 2 hours of birth and at 12 and 24 hours of age, and challenged with $10^7$ oocysts at 12 hours of age. None of six calves fed anti-P23 colostrum developed diarrhea. In contrast, all six identically challenged control calves fed non-immune colostrum (P23 Ab titer <100) developed severe diarrhea. Further, calves fed anti-P23 colostrum shed significantly fewer total oocysts than control calves, representing a 99.8% reduction.[14] The absence of diarrhea and marked reduction in oocyst shedding further validated passive immunization as a viable strategy and indicates that immune bovine colostrum prepared against recombinant P23 can protect calves against cryptosporidiosis. Recombinant P23 is therefore a lead vaccine candidate for further research and development, and evaluation in large scale clinical trials. Recombinant P23 has also been shown to elicit antigen-specific responses by spleen and mesenteric lymph node cells from *C. parvum* infected mice,[4] and peripheral blood $CD4^+$ T lymphocytes from recovered calves which proliferated and expressed IFN-gamma in vitro.[16] This finding is of interest because immune bovine colostrum contains cytokines which may have anti-cryptosporidial activity. It is therefore conceivable that cell-mediated immune responses may contribute to passive protection by colostrum, although previous studies have clearly ascribed the predominant anti-cryptosporidial activity in immune colostrum to *C. parvum* specific neutralizing Abs induced by immunization.[4]

The molecularly distinct CP15/60 Ags defined by neutralizing MAbs [17] have been evaluated as a DNA vaccine to prepare monospecific polyclonal Abs for passive immunization studies. Holstein-Fresian cows were immunized by the intramammary route with recombinant plasmid DNA encoding the CP15/60 Ags and produced *C. parvum* specific serum and colostral Abs.[18] The immunoglobulin fraction from anti-CP15/60 colostrum was administered orally to immunosuppressed adult C57BL/6 mice before and after oocyst challenge, and reduced infection levels in the ileum ~50% by day 4 post-challenge compared to identically challenged control mice administered non-immune colostral immunoglobulin.[18]

Recombinant plasmid DNA encoding the CP15 antigen[19] (unrelated to CP15/60[17]) has been evaluated in adult BALB/c mice immunized by the mucosal route.[20] Immunized mice developed *C. parvum* specific intestinal mucosal IgA consistent with a Th2 response. Serum Abs recognizing *C. parvum* sporozoites and a native ~15 kDa antigen were also produced and were predominantly IgG2a and IgG2b, consistent with a Th1 response. Splenocytes from immunized mice proliferated and produced IFN-gamma when stimulated with *C. parvum* Ag in vitro, also consistent with a Th1response. The same recombinant CP15 plasmid was subsequently evaluated in pregnant Alpine goats.[21] Following intranasal immunization, goats produced monospecific serum and colostral Abs recognizing both recombinant CP15 and a native ~15 kDa *C. parvum* antigen. Kids were fed anti-CP15 colostrum from immunized dams and challenged with *C. parvum* oocysts at one day of age to evaluate protective ability. Kids fed immune colostrum had significantly reduced (~8-fold) oocyst shedding, patent periods, and duration of diarrhea compared to identically challenged control kids fed colostrum from vector only-immunized dams.[21]

**Monoclonal antibodies**

Efforts to improve the efficacy of passive immunization against cryptosporidiosis have involved the use of polyvalent formulations of neutralizing MAbs to selectively target defined epitopes on Ags known to have an important role in the infection process. To this end, a panel of ~150 mouse MAbs was generated against immunoaffinity-purified CSL, GP25-200, and P23 from *C. parvum.*[12] MAbs having the highest sporozoite neutralizing activity in vitro and reactive with distinct epitopes were identified and then evaluated, individually and in multiple epitope-specific combinations, for the ability to prevent or reduce infection in oocyst-challenged neonatal mice. Anti-CSL MAb 3E2 had the highest protective activity of all MAbs in the panel, and was able to reduce intestinal infection levels by 62 to 92% when administered at the time of, and after oocyst challenge. A combination of 3E2, anti-GP25-200 MAb 3H2 and anti-P23 MAb 1E10 demonstrated significant additive protection over that of the individual MAbs, and reduced infection levels by 86 to 93%. Infection was completely prevented in up to 40% of mice administered MAb 3E2, alone or in combination with 3H2 and 1E10. These findings suggested that polyvalent neutralizing MAb formulations may improve the efficacy of passive immunoprophylaxis against primary infection. In a subsequent study, the same three MAbs were evaluated for therapeutic efficacy against persistent *C. parvum* infection in IFN-gamma-depleted adult SCID mice.[22] Treatment with MAb 3E2, alone or in combination with 3H2 and 1E10, had a significant therapeutic effect on an established intestinal infection.

Based on the anti-cryptosporidial efficacy of MAbs observed in the above studies, a proof-of-principle pilot study was conducted to evaluate MAb 3E2 for passive immunization against experimental *C. parvum* challenge in newborn Holstein calves.[23] Following oocyst challenge, MAb 3E2-treated and control calves were evaluated for development of cryptosporidiosis using previously described methods and outcome parameters.[14] All control calves developed severe diarrhea and shed large numbers of oocysts. 3E2-treated calves also developed diarrhea, but of significantly less severity, and shed significantly fewer oocysts. These results support the rationale for design and optimization of MAb-based passive immunization strategies for bovine cryptosporidiosis. Passive immunization or immunotherapy against *C. parvum* in calves using mouse hybridoma-derived MAbs faces two major limitations, both of which can likely be overcome. First, production costs for hybridoma derived MAbs using commercial bioreactors (~$300 per gram MAb) would not be economically feasible for veterinary applications. However, a number of recombinant-based expression systems exist for MAb production at a cost and scale which are economically feasible for commercial use in calves. Secondly, rapid gastrointestinal transit occurring with cryptosporidial diarrhea is an impediment to passive immunization or immunotherapy with orally administered Abs, effectively reducing the exposure time of Abs to extracellular infective stages. This variable and the asynchronous nature of the *C. parvum* life cycle would require a multiple dose treatment regimen. Bovinization of recombinant MAbs to the IgG1 isotype would likely allow these obstacles to be overcome. Such bovinized MAbs could be used as a colostral additive for newborn calves, an approach which could be integrated into existing management practices, and which would be expected to allow continuous transmucosal delivery of bovinized MAbs in susceptible calves.

**Monoclonal antibody–anticryptosporidial fusion proteins**

Recombinant fusion proteins that genetically link anti-cryptosporidial peptides and enzymes[24] to high-affinity *C. parvum* specific MAbs[12] for delivery targeted specifically to parasite stages have recently been produced.[25] In effect these represent combinatorial anticryptosporidials which join protective elements from both the innate and acquired immune responses. Studies in the neonatal mouse model indicate that MAb-anticryptosporidial fusions have superior efficacy than either component alone when administered at equimolar amounts.[25] Further, the effective dose of some fusions has been 50-fold or lower than that of the stand alone neutralizing MAb. Lead MAb-anticryptosporidial fusions identified in neonatal mouse studies will next be evaluated in calves for efficacy against *C. parvum* infection and clinical disease.

**Cytokines**

The key role of IL-12 in resistance to and recovery from *C. parvum* infection has been established[2, 4, 8, 9] and this cytokine was recently evaluated in calves for its ability to reduce *C. parvum* infection.[26] Subcutaneous treatment of *C. parvum*-challenged calves with recombinant bovine interleukin-12 stimulated a strong intestinal immune response characterized by an increase in mucosal T lymphocytes and expression of IFN-gamma but did not have any effect on the course of infection.[26]

## ACTIVE IMMUNIZATION

Several classical approaches have been explored for active immunization of calves against *C. parvum*. Efforts to develop a precocious strain of *C. parvum* as a live vaccine for calves were unsuccessful.[4] Other studies investigated the use of orally administered killed (lyophilized) oocyst preparations to immunize newborn calves.[27] Initial results suggested that oocyst shedding and diarrhea could be reduced following subsequent challenge with viable *C. parvum* at 1 week of age.[27] However, when field tested on a large dairy operation with endemic *C. parvum* infection, the killed vaccine failed to provide protection when administered at birth.[27] It was thought that failure was related to exposure of immunized calves to environmental oocysts within hours of birth, typical of natural exposure.[27] More recently, day-old dairy calves inoculated orally with viable *C. parvum* oocysts which had been attenuated by gamma irradiation had reduced oocyst shedding and diarrhea during the primary infection and were resistant to reinfection with non-irradiated virulent oocysts at 21 days of age.[28] These promising results warrant further investigation. However, any attenuated *C. parvum* vaccine is subject to limitations imposed by the current inability to propagate *C. parvum* in quantity in vitro[4] and cold-chain dependency between production and use.

A DNA vaccine encoding *C. parvum* P23[15] has recently been evaluated in adult IL-12 knockout mice and elicited both humoral and cellular parasite specific responses.[29] Subcutaneous immunization induced significant anti-P23 IgGl and lgG2a antibody responses and specific CMI responses based on proliferation of spleen cells from immunized mice when stimulated with recombinant P23. Long-term memory responses were also detected in immunized mice.[29]

It is of relevance to both passive and active immunization strategies that the biochemical composition of *C. parvum* is complex, being comprised of carbohydrates, lipids, proteins, glycoproteins, and other glycoconjugates.[2-4, 8, 9] The glycosylated state of many *C. parvum* molecules and carbohydrate-dependency of multiple epitopes recognized by neutralizing antibodies has presented challenges to development of traditional recombinant vaccines in established prokaryotic or eukaryotic systems. Recent reports of successful expression of appropriately glycosylated *C. parvum* proteins in stably transfected *Toxoplasma gondii* may represent the next milestone accomplishment in development of a recombinant vaccine against cryptosporidiosis.[30-32]

## REFERENCES

[1]Fayer R, et.al. **Vet Parasitol** 2008; 156:191. [2]Riggs MW. **Microbes Infect** 2002; 4:1067. [3]Tzipori S, et.al. **Microbes Infect** 2002; 4:1047. [4]Fayer R, et.al. Ed. ***Cryptosporidium* and Cryptosporidiosis**, 2nd Ed. 2008; CRC Press. [5]Sanderson SJ, et.al. **Proteomics** 2008; 8(7):1398. [6]Borowski H, et.al. **Trends Parasitol** 2008; 24(11):509. [7]Pantenburg B, et.al. **Infect Immun** 2008; 76(1):23. [8]Boulter-Bitzer JI, et.al. **Biotech Advances** 2007; 25(1):13. [9]Wanyiri J, et.al. **Future Microbiol** 2006; 1(2):201. [10]Santin M, et.al. **Vet Parasitol** 2004; 122(2):103. [11]Riggs MW. **27th Annual ACVIM Forum Proceedings** 2009. [12]Schaefer DA, et.al. **Infect Immun** 2000; 68:2608. [13]Shayan P, et. al. **Parasitol Res** 2008; 103(5):1207. [14]Perryman LE, et.al. Vaccine 1999; 17:2142. [15]Perryman LE, et.al. **Mol Biochem Parasitol** 1996; 80:137. [16]Wyatt C, et.al. **J Parasitol** 2005; 91(5):1239. [17]Jenkins MC, et.al. **Infect Immun** 1993; 61:2377. [18]Jenkins MC, et.al. **Vaccine** 1999; 17:2453. [19]Jenkins MC, et.al. **Mol Biochem Parasitol** 1995; 71:149. [20]Sagodira S, et.al. **Parasite Immunol** 1999; 21:507. [21]Sagodira S, et.al. **Vaccine** 1999; 17:2346. [22]Riggs MW, et.al. **Antimicrob Agents Chemother** 2002; 46:275. [23] Riggs MW, Schaefer DA, Dryden VC, unpublished data. [24]Carryn S, et.al. **Int J Antimicrob Agents** 2004; 24(2):S117. [25]Imboden M, Riggs MW, Schaefer DA, Homan EJ, Bremel RD, unpublished data. [26]Pasquali P, et.al. **Vet Parasitol** 2006; 135:259. [27]Harp JA, et.al. **J Dairy Sci** 1998; 81:289. [28]Jenkins M, et.al. **J Parasitol** 2004; 90(5):1178. [29]Ehigiator HN, et.al. **Parasitol Res** 2007; 101(4):943. [30]O'Connor RM, et.al. **Infect Immun** 2003; 71(10):6027. [31]Shirafuji H, et.al. **J Parasitol** 2005; 91(2):476. [32]O'Connor RM, et.al. **Mol Biochem Parasitol** 2007; 152(2):149.

**PHYSIOLOGY & TREATMENT OF CONGENITAL CONDITIONS IN THE NEWBORN INTENSIVE CARE UNIT OF THE EASTERN OKLAHOMA PERINATAL CENTER**

**Michael R. Gomez, M.D.**
**Tulsa, OK**

Neonatology has advanced tremendously during my twenty years in the field. The early years of Neonatology resulted in treatment methodologies that dramatically improved survival among term and preterm infants. Medically disastrous conditions such as congenital group B streptococcus infections, severe respiratory distress syndrome (RDS), meconium aspiration syndrome with respiratory failure, and neonatal/perinatal hypoxic-ischemic encephalopathy have become more rare through prevention and prophylaxis employed using evidence based strategies, and are much more easily treatable despite an ever increasing number of preterm births.

The problem of an increasing frequency of preterm births is not well understood and is multifactorial. In fact, we have become so good at what we do as Neonatologists, that Obstetricians no longer fear the complications of preterm births. In a way, we are victims of our own success. Nonetheless, not being perfect gives us a target to shoot for. The two most effective antenatal treatment strategies employed by Obstetricians that have helped Neonatologists is the use of antenatal steroids (betamethasone, dexamethasone) to reduce respiratory complications of extreme preterm birth (25-34 weeks gestation compared to 37-40 weeks at term) by reducing the incidence and severity of RDS and other concomitant complications such as intraventricular hemorrhage, necrotizing enterocolitis and hypotension. First described in 1972[1], it took almost 25 years for a collaborative effort of Obstetricians and Neonatologists to agree that this simple measure, when done in a two-dose regimen within a week of a preterm delivery, could reduce mortality by 67% and significantly reduce morbidities as well. Another simple antenatal strategy to reduce complications was the use of intrapartum antibiotics prophylaxis against group B streptococcus (GBS or Strep algiacitae). GBS emerged in the 1960's as the leading infectious cause of infant mortality, replacing Staph aureus, when births moved from a week-long hospitalization to a drive-by procedure. GBS infection can vary from mild to severe, but most infants suffer rapid cardiovascular and respiratory collapse when septic. 20-35% of pregnant females have rectovaginal colonization with GBS and expose their infants to it during a vaginal delivery or with rupture of amniotic membranes. GBS has a surface antigen structure that does not produce immunological memory nor does it incite much of an immune response, so maternal serum antibody levels are typically very low. Attempts to create a vaccine to date have been unsuccessful because of this characteristic. Studies done by giving infants a prophylactic dose of penicillin G intramuscularly showed that is was partially effective at reducing the incidence of neonatal sepsis. By extension, a strategy to given a dose of penicillin G or other similarly effective antibiotic to a mother 4 hours prior to delivery while in labor was developed, and to date, has reduced the incidence of GBS sepsis in neonates by about 70% while inducing a minimal number of resistant organisms or infections according to the Active Bacterial Core (ABC) surveillance database.[2] IAP also provides some prophylaxis against Listeria monocytogenes, which is rare, but still the third most common congenital infection. Gram-negative organisms are unaffected, thought this IAP strategy has yet to result in a severe problem with congenital gram-negative infection except in a small subset of very preterm infants.

RDS is a condition of a deficiency of surfactant typically associated with preterm birth or infant's of diabetic mothers with delayed surfactant production, but can also result from a consumption of surfactant in excess of production as seen in sepsis syndromes and meconium aspiration syndrome. Prior to the development of a commercially viable surfactant by Fujiwara (Survanta®) and Burroughs Wellcome (Exosurf®), RDS was treated by tracheal intubation and the application of intermittent positive airway pressure using crude, air driven ventilator devices. As surfactants were being commercially developed, refinements in mechanical ventilator devices and ventilation strategies were refined so that they provide synergy in the treatment of RDS. Many infants more than 27 weeks gestational age, now routinely never receive IPPV any longer, but rather are treated with continuous positive airway pressure (CPAP) alone or in combination with surfactant. The incidence of complications such as airleak syndromes or chronic complications such bronchopulmonary dysplasia have fallen from over 50% to less than 15% for many populations of preterm infants with continued improvement. Surfactant products such as Surfaxin® (Lucinactant by Discovery Labs) are being used not only to treat or prevent RDS, but have an active component that may reduce the incidence of BPD[3] instead of using powerful systemic agents such as supraphysiologic doses of dexamethasone[4] with significant side-effects and long term complications on a rapidly growing infant.

Meconium, or fetal stool, is passed in about 10-15% of term infant deliveries. In most cases, it is a benign event coincident with birth and is not a definite marker of fetal stress. In a small number of cases, meconium is passed in utero before or at delivery. It can be acutely or chronically aspirated in the fetal or neonatal lung and result in severe, life-threatening cardiopulmonary failure. It seemed rational that one could avoid meconium aspiration syndrome (MAS), as it became to be known, by aspirating the trachea with a suction tube before baby's first breath, but ultimately, it was found to be of no benefit unless the meconium was associated with a baby who was not breathing and could potentially obstruct the trachea[5]. Not even suctioning on the perineum is routinely done. The earlier recognition of stress in utero with fetal monitoring, more induced pregnancies at 39 weeks or less and the use amnioinfusions has vastly reduced the frequency of severity of MAS, even with the rate of meconium passage remaining constant or increasing. Each delivery in most U.S. hospitals is attended by a team trained and competent in the use of neonatal resuscitation methods taught in a regimented way and is felt to contribute to the decreasing frequency of such complications.[6] The philosophy of doing less invasive therapies both at delivery and afterwards, with gentle ventilation, avoiding high positive pressure ventilation, allowing permissive hypercapnia (pCO2 50-65 mm Hg with lower blood pH), and avoiding vigorous airway suctioning has produced better outcomes that ever before, especially when combined with treatments such as inhaled nitric oxide for pulmonary hypertension[7] and extracorporeal membrane oxygenation.

Hypoxic ischemic encephalopathy occurs in infants whenever in utero blood flow, often just prior to delivery, is below the threshold to provide adequate oxygen and perfusion to tissues with nutrients needed for aerobic metabolism. Despite the protective dive reflex, enough tissue hypoxia occurs in some cases that fetal end organs began to suffer beginning with the non-essential areas, but gradually progressing to the more vital organs of the kidneys, liver, heart and brain. There are few good prenatal markers for an increased risk of fetal hypoxemia and even fewer reliable signs. What has been interesting to follow is that as awareness of the condition through education about HIE, the use of relatively simple prevention techniques such as proper fetal monitoring techniques, more operative deliveries, and earlier deliveries (before 41 weeks post menstrual age), as well as providing a regimented program of evidence based resuscitation education with a dedicated team to provide infant resuscitation immediately available in every delivery room, the incidence of HIE has decreased dramatically. Unfortunately, even under the best circumstances, HIE cannot be prevented. For those infants who develop encephalopathy, they are truly at risk for permanent brain injury and cerebral palsy. Whole body and selective head cooling techniques have shown some potential for reducing the severity of moderate to severe neonatal encephalopathy and without increasing the incidence of neonatal death or the rate of poor outcomes among survivors. In other words, it does not pull some babies back from the brink of survival only to be permanently and irreversibly impaired.[8]

The transfer of medical therapeutic applications has traditionally been from animal in the form of a treatment model, to humans. In Neonatology, though much of the early work is done in an animal model, there is the ability to translate results back. As human technology is more widely utilized, its utility in veterinary medicine becomes more obvious and the high costs often associated with its development have been amortized making its use in animals more feasible.

## REFERENCES

1. Liggins GC, Howie RN. **Pediatrics** 1972;50:515-25. 2. Committee on Infectious Diseases and Committee on Fetus and Newborn. **Pediatrics** 1997,99:489-496. 3. Moya FR,et al. **Pediatrics** 2005, 115, No. 4: 1018-1029. 4. Cummings JJ, et al. **New England J Med** 1989, 320:1505-1510. 5. Keenan WJ, **Pediatrics** 2004, 113 No. 1:133-134. 6. AHA, AAP. **Pediatrics** 2006, 117 No. 5:e1029-e1038. 7. Kinsella JP, **Pediatrics in Review** 1999,20:e110-e113. 8. Shankaran S, et al. New England J Med

**TRANSLATING HUMAN NEWBORN INTENSIVE CARE TO APPLICATIONS IN VETERINARY MEDICINE**

**Michael R. Gomez, M.D.**
**Tulsa, OK**

Neonatology has advanced tremendously during my twenty years in the field. The early years of Neonatology resulted in treatment methodologies that dramatically improved survival among term and preterm infants. This transfer of medical knowledge has traditionally been from animal to human. With refinements in medical technology for use in infants, it has become increasingly apparent that the technology transfer can be bidirectional. As certain treatments and technologies are refined, miniaturized and amortized, they become increasingly more feasible to use in the treatment of analogous conditions that afflict animals.

Surfactant replacement therapy (RDS) has been the "holy grail" of treatments for respiratory distress syndrome (RDS) in preterm infants. Born just 5.5 weeks premature, Patrick Kennedy died of RDS in 1963 despite the tremendous resources available to him at Boston Children's Hospital. Surfactants were postulated scientifically in 1929 by von Neegard, but almost twenty-five years passed before it was recognized that "anti-foams" present in the amniotic fluid of mature infants was the key to transitional respiration success in infants, and that its congenital deficiency was the cause for the gasping, hypoxic syndrome noted in infants born four or more weeks premature.[1] By the early 1960's, the structure of surfactant had been described, but it took another twenty years for Fujiwara to isolated bovine surfactant by lung washout and was able to administer it successfully to ten human infants near death from RDS.[1] Today, SRT is commonplace in all Neonatal intensive care units and there are three to four formulations available on the market, from the animal derived to purely synthetic. As a result, there has been a significant decline in annual deaths from RDS and in permanent impairment that results from the abnormal lung development that occurs when premature babies are placed on chronic mechanical ventilators and oxygen. From work with surfactants has come profound work on how mechanical ventilators and less invasive methods of applying positive pressure ventilation that has continued to improve survival rates and lower complications rates among preterm infants.[2] With the advent of NICU care for animals via research, it is now feasible to care for these infants in these environments with readily available equipment and proven therapies. As the cost of surfactants has steadily declined, it has become more feasible to utilize this therapy, first tested in animal models to be utilized for animals with surfactant deficiency RDS. SRT can also be used to modulate and prevent the development of bronchopulmonary dysplasia, the typical lung abnormality seen in when an immature lung is exposed to higher than tolerated pressure and oxygen related toxicity.

Another perplexing dilemma of the newborn was the development of persistent pulmonary hypertension of the newborn (PPHN) that caused profound hypoxemia and cardiorespiratory death in term newborn infants who were not typically suspected of having RDS. These term infants were, for the most part, born with an intact pool of surfactant, but their lung development, and normal post-natal decline in pulmonary artery pressure were adversely affected. Unlike the premature infant who had a predictable rate of mortality and complications, these term infants were expected to be normal. Various processes such as congenital infections, meconium aspiration, in utero maladaptation from placental abnormalities could result in the failure of the normal decline in post-natal pulmonary artery pressure. Sometimes, and for no apparent cause, pulmonary hypertension would develop unexpectedly (idiopathic) and result in profound hypoxemic respiratory failure and death. The proposed mechanism for this fall in pulmonary vascular pressure was felt to be self-determined by an intrinsic factor first called endothelial derived relaxing factor. Nobel prize winning research by Furchgott[3] and others determined that this EDRF was in fact a soluble compound derived from the metabolism of arginine called nitric oxide. Nitric oxide was the fabled second messenger that crossed cell membranes and dissolved into the lung vasculature causing a relaxation of blood vessels and a perfusion of the infant pulmonary vascular bed. After being described in ----, reverse engineering led to the theory that if the lung was deficient in soluble nitric oxide intrinsically, it could be delivered extrinsically in small quantities enough to pass from the alveolus into pulmonary blood vessels. Kinsella[4] and others described the first human experiments using inhaled nitric oxide and its use has profoundly affected the management of PPHN. While the delivery system in humans is quite complex because of safety regulations, the delivery of nitric oxide for animal use at the recommended five to twenty parts per million can be done safely and effectively with relatively simple and inexpensive equipment.[5]

Sometimes despite optimal medical treatment, cardiorespiratory failure ensues. In 1972, Hill et al.[6] described the use of cardiopulmonary bypass techniques to artificially support the heart and lungs of a motorcycle crash victim. Shortly thereafter, Bartlett[7] described the use of extracorporeal membrane oxygenation (ECMO) methods for the treatment of cardiorespiratory treatment in human newborns. Some twelve thousand infants have undergone ECMO therapy since then, and although its use has declined somewhat from improvements in conventional medical therapies, ECMO remains the best treatment of last resort with a potential for a normal outcome in over 85% of infants who had a predicted mortality of over 80% with medical treatment. Like with other treatments, the principles of ECMO techniques were developed in animal models. After considerable refinements, ECMO devices are more compact, less expensive and can be done with minimal surgical techniques. It is now feasible to reverse the process and use ECMO methods for short-term support of animals with reversible cardiorespiratory failure and expect a good long-term outcome despite a different lifespan expectation.

Perinatal asphyxia is a problem that affects humans and animals alike. The mammalian placenta, irrespective of its structure or physiology, is near the end of its useful life at the end of its gestational period. It is supposed to separate shortly after the birth, so its senescence is physiologic. It is this senescence of the placenta that makes mammals vulnerable to abnormalities in parturition that can result in the lack of oxygen and blood flow at delivery. This vulnerability manifests itself as perinatal asphyxia and can affect all mammalian species resulting in profound end organ injury most significantly for humans, the brain, resulting in a type of muscle spasticity called cerebral palsy, and can also result in cognitive injury as well. In its most severe forms, it can harm multi-organ systems and result in death. The development of whole body[8] and selective[9] hypothermia techniques in infants has its genesis in long term animal modeling including pigs, rats, sheep, and baboons among many others. Despite some doubt as to its utility in humans[10], it has been shown to be neuro-protective in some circumstances, and while not a panacea, can mitigate the most devastating types of neurological injuries in human infants while not adversely affecting survival.

Few areas in human medicine have yielded the statistical successes in survival and the recovery of meaningful years of life as have neonatal intensive care[11]. Neonatal medicine is an area of innovation and relies heavily on animal modeling to develop successful therapies that benefit newborn infants. High expectations parents and society have on a health pregnancy continues to drive innovation in this area. Nearly every hospital with at least 1500 annual deliveries (about 4 deliveries a day) has some neonatal intensive care capability. Neonatology remains the most intensive area of training in the 35 month training program for general Pediatricians, and is the single largest sub-specialty in Pediatrics with more board certified Neonatologist than the next three sub-boards combined (Pediatric Cardiology, Hematology-Oncology, Pediatric Critical Care). With the success seen in NICU care extrapolated from animal models, it makes sense to take these successes in the opposite direction and apply them to animals. As technologies and therapies are developed, refined and standardized, their cost decline and benefits increase. Most Americans expect the same level of expertise, if not commitment, from their Veterinarian than from their Pediatrician and Neonatologist, which will allow for a free flow of information between the medical disciplines for the benefit of all.

## REFERENCES

[1.] Wrobel S. **FASEB Journal** 2004; 18:1624e. [2.] Vitali SH and J Arnold**. Critical Care** 2005; 9:177-183. [3.] Furchgott R. **JAMA** 1996; 276(14):1186-88. [4.] Kinsella JP, et al. **Lancet** 1992;340:819-820. [5.] Gomez MR, et al. **IEEE Annual Conference** 1995; v.2:1675-6. [6.] Hill JD, et al. NEJM 1972; 286:629-634. [7.] Bartlett RH, et al. **Trans Am Soc Artif Intern Organ** 1976;22:80–88S. [8.] Shankaran S, et al. **Pediatrics** 2002:110(2):377-385. [9.] Jacobs S, et al. **Cochrane Database Syst Rev**. 2007 Oct 17;(4):CD003311. [10.] Cabanac M. **FASEB Journal** 1993; vol. 7:1143-6. [11.] Cutler DM, et al. NEJM 2006; 355(9):920-7.

# TEAT LESIONS & DIAGNOSTIC APPROACH

**Sylvain Nichols, DMV, MS, DACVS**
**St-Hyacinthe, QC**

## INTRODUCTION

Teat injuries are frequent in dairy cattle.[1] In many cases, they result in premature culling of the animal. Veterinarians are frequently asked to evaluate malfunctioning teat. With palpation alone, it can be difficult to diagnosis precisely the origin and the severity of the lesion. Without a precise diagnosis, the treatment plan is difficult to elaborate and frequently fails to re-establish the normal function of the teat. This session will focus on diagnosis of teat injuries. Older and newer diagnostic tools will be reviewed. Treatment option will be given but will not be elaborated on.

## SPECIFIC ANATOMY

From the lactocyte to the external sphincter, there are 2 areas where the milk has to go through a narrowing: the annular ring and the rosette of Furstenberg/streak canal (teat's distal end). Most of the problems are encountered in those areas. The annular ring is a large vascular plexus at the base of the teat where the vascularisation of the teat starts and ends. Internal lesions at the annular ring or enlargement of the vascular plexus impede the milk outflow from the gland cistern to the teat cistern.

The rosette of Furstenberg is the junction of the teat mucosa at the distal end of the teat cistern. It is often referred to as the internal sphincter of the teat. This structure marks the transition between the teat cistern and the streak canal. Lesions in this area act as a one way valve decreasing the milk flow during milking.

## LESIONS AT THE ANNULAR RING

The lesions at the annular ring will be divided in 2 categories: internal lesions and external lesions

### Internal lesions

These lesions are generally secondary to mastitis. In the younger animal, the lesion is often extensive and can involve the entire length of the teat and the gland cistern. Many heifers are infected with mastitis at calving.[2] Depending on the agent involved, an increase in the somatic cell count (SCC) can be the only clinical signs. On other occasions, the result of the mastitis is dramatic and the heifer has a partially or a totally blind quarter.

After calving, the owner complains about a quarter that is not letting its milk down. He is convinced that the animal should produce milk since the quarter is enlarged. Most of the time, this enlargement is caused by edema. In this case, he wants his veterinarian to determine the extent of the damage to the teat and hopefully re-establish the normal milk flow. The first thing that has to be determined is the productivity of the quarter. If the quarter is producing milk, the teat can be evaluated. By only palpating the structures, it is difficult to evaluate the milk production of the quarter and the integrity of the teat. Those structures might be firmer than usual. No milk or only a few drops will be obtained by hand milking. To obtain a diagnosis, an ultrasound evaluation of the teat and quarter should be performed. A few descriptive papers of the normal and abnormal anatomy of the teat have been published in the past.[3,4] The ideal probe to use is a linear 7.5 mHz transrectal probe. First the probe is placed on the affected quarter. Normally, the gland cistern should be filled with milk. If no milk is detected and the animal is only a few days post-calving, it is right to assume that the quarter has been compromised. The evaluation can end at this point since no treatment options are available to treat a fibrotic quarter. If milk is present in the teat cistern, the teat is evaluated longitudinally and transversally. The evaluator is trying to find an intact portion of the teat cistern distally. The quarter is massaged to stimulate the milk to go down which should fill the teat cistern if present. If a teat cistern is identified, it is possible to approach the lesion through a thelotomy to create an artificial teat cistern with the help of an implant.[5]

If no milk is detected distally, a small canula is introduced through the streak canal and contrast product can be injected into the cistern. 50% sodium diatrizonate solution (Hypaque) can be safely infused into the teat and gland cistern after appropriate cleaning of the teat's end. A lateral radiograph is then taken. If a cistern is detected, the lesion can be approached surgically. If no cistern is detected, the lesion can still be approached surgically. However, the prognosis for regaining a functional teat decreases dramatically.

When contrast radiography was compared to other imaging modalities (ultrasound and theloscopy), the sensitivity of the technique was lower for the detection of proximal obstructions.[6] However, in regard of the prognosis for return to normal milking, the radiographic evaluation provides important information.

Similar lesions can be found in adult cattle. Generally, they have a history of mastitis in the affected quarter at the dry period. The clinical presentation is similar. At calving, no milk or only a few drops of milk come out of the affected quarter. The lesion can be palpated at the base of the teat. The ultrasound is used to identify the location and the thickness of the lesion. In some cases, a teat endoscopy (theloscopy) can be performed to confirm the diagnosis. In that case, a specialized endoscope is introduced through the streak canal. The teat cistern is distended with saline or air and the lesion is visualized. Frequently, the lesion has a cauliflower shape. The treatment is surgical. The success rate varies according to the thickness of the mass at the ultrasound evaluation.[7]

### External lesions

The most frequent external lesion, at the base of the teat, causing milking problems is a varicose vein. This condition is generally found in older cattle. The animal is presented for milking difficulty. On palpation, the teat appears normal. However, a large vein is usually found at the base of the teat. In some case, this vein will protrude one third of the way down on the teat. It is difficult, with palpation only, to determine if this enlarged vein is the cause of the milking difficulties. To confirm the diagnosis, an ultrasound evaluation of the vein is performed using a linear 7.5 mHz transrectal probe. The base of the teat is evaluated. The large vessel is visualized compressing the junction between the gland and the teat cistern. In some cases, the compression is not easily visualized. However, when the milking machine is in place, the vein becomes engorged and creates an obstruction. Since an ultrasound can't be done during milking, the teat is completely evaluated with the ultrasound to exclude any other possible problems.

In most of the cases, the varicose vein can be followed on the udder. A small gauge catheter can be placed in the vein away from the teat. An angiogram is then performed by injecting contrast product while a lateral radiograph is taken. This procedure is not essential for the diagnosis. However, it is useful to elaborate the surgical plan. It helps in the localization of the collateral vessels. Ideally, the varicose vein is double ligated and removed.[8]

**DISTAL END INJURIES**

Seventy percent of the cows culled for milking problems have a lesion at the distal end of the teat.[9] Those lesions are traumatic in nature. They occur when the animal steps on its teat or when a neighbouring cow steps on the teat. The trauma can be inflicted by the inappropriate use of teat inserts and canulas. Shortly after the injury, the teat can be bruised and swollen. However, at the time of the evaluation there is no sign of external trauma. On palpation, the distal portion of the teat may feel firmer. It is assumed that the lesion is located at the distal end of the teat. In general, no further diagnostics are performed and a blind surgical procedure is performed. In this situation, the recovery rate is very poor.

To improve the recovery rate of distal end injuries it is important to choose the appropriate treatment for each condition. To choose the right treatment it is important to have the correct diagnosis. For example, a lesion at the rosette of Furstenberg does not need the streak canal to be opened and a ruptured streak canal does not need to have the rosette curetted.

To evaluate the streak canal, a graduated probe has been developed. This probe evaluates the length of the canal. Streak canals over 8 mm are assumed to be abnormal.[10] However, there is normal variation between individual cattle making the interpretation of the measurement difficult. It is recommended to compare the measure with the contralateral teat. A ruptured or traumatized streak canal will be longer than normal.[10]

An ultrasound evaluation of the distal end of the teat should be performed using a linear 7.5 mHz transrectal probe. To improve the image, the evaluation is performed with the teat in a plastic cup filled with water. The acoustic gel and probe are placed on the plastic cup. The rosette of Furstenberg and the streak canal are evaluated and compared with the contralateral teat. The evaluator is looking for a hyperechoic mass at the rosette of Furstenberg, an irregular streak canal and ruptured streak canal. If those lesions are found, blind surgical treatments are not recommended. If no lesions are found, the remainder of the teat is evaluated. If the ultrasound evaluation is normal, an exclusion diagnosis of fibrosis of the streak canal can be made. Only in this case should a blind surgical technique be used to open the streak canal.

Ideally, the diagnosis of distal end injuries should be confirmed by theloscopy. The axial theloscopy described previously is not appropriate for these cases. The theloscope should be placed through the teat wall. This is performed by using the sharp theloscopic trocar. This trocar is entered through the streak canal into the teat cistern. Then, it is directed laterally and pushed through the teat wall. The theloscopic canula is slid over the trocar from the outside in. The trocar is removed and the theloscope is introduced in the canula. The teat cistern is filled with air or saline and the distal end is evaluated. If a lesion is found, it can be treated immediately under theloscopic guidance. The instruments are inserted through the streak canal. This technique provides better results for the treatment of distal end injuries when compared to blind surgery and surgery performed through a thelotomy.[11-12]

**OTHER TEAT CONDITIONS**

**Webbed teats**

Supranumerary teats are cut at a young age. When they are in close proximity with another teat, they can be missed or left in situ deliberately to avoid traumatizing the principal teat. At calving, the supranumerary teat (webbed teat) may or may not have a distinct gland cistern. If a gland cistern is present, it may or may not empty itself during milking. If the webbed teat is not impinging milking and is not leaking milk between milkings, it can be left in place. However, if it is obstructing the normal milking of the principal or the accessory gland, it will leak between milkings and will need to be removed. In show cattle, the webbed teat will be removed for cosmetic purposes even if it does not cause any milking problems.

The webbed teat may look like a fistula or a normal teat attached to the principal teat. The diagnosis is made by visual exam and palpation. However, prior to elaborating a surgical plan, it is important to investigate the gland associated with the webbed teat. In some cases, the accessory gland is bigger than the principal gland.[13] If this is the case, the conjoined teat should be anastomosed to the principal teat.

The milk production of the accessory gland can be compared to the principal gland by measuring the milk production of each gland. A canula is inserted in each teat during milking. The milk of the 2 glands is caught separately and the quantities compared. An ultrasound evaluation can also be performed. A linear 7.5 mHz transrectal probe is used. However, it is difficult to evaluate each gland independently with this technique. Ideally, a contrast radiograph of the accessory and principal glands should be performed. A large volume (20-30ml) of contrast solution is injected first in the accessory gland. A lateral radiograph is taken. The contrast solution is milked out as much as possible and the same procedure is done on the principal gland. The 2 gland cisterns are compared. If the webbed teat has a significant gland, it can be worthwhile to anastomose both teats. If the webbed teat has an insignificant gland, it can be removed and the accessory gland is closed at the base of the teat.

Contrast radiography provides important information about the thickness of the septum between each teat. This information is important if the structures are anastomosed. The thicker the septum, the less likely it is that the anastomosis will be functional in the long term.

## REFERENCES

[1]Michigan Dairy Herd Improvement Association. **Animal summary**. Lansing (MI): Michigan DHIA Inc; 1981. [2]Roy JP, et al. **Can J Vet Res** 2007;71(4):283. [3]Takeda T. **Jpn J Vet Res** 1989;37:133. [4]Stocker H, et al. **Tierarztl Prax** 1989;17:251. [5]Ducharme NG, et al. **Can Vet J** 1987;28:757. [6]Frémont A, et al. **Rev Med Vet** 2002;153:40. [7]Ducharme NG, et al. **Can Vet J** 1987;28:763. [8]Couture Y,et al. **Vet Clin North Am Food Anim** 2005;21:201. [9]Geishauser T,et al. **Bov Pract** 2001;35:156. [10]Geishauser T, et al. **J Dairy Sci** 2000;83:1976. [11]Bleul UT, et al. **J Am Vet Med Assoc** 2005;226(7) :1119. [12]Hirsbrunner G, et al. **Vet Rec**, 2001;148:803. [13]Shappell KK, et al. **J Am Vet Med Assoc** 1989;195:623.

# CHALLENGING UMBILICAL PROBLEMS IN NEONATES

**André Desrochers DMV, MS, DACVS, DECBHM**
**St-Hyacinthe, Québec, Canada**

## INTRODUCTION

The umbilical rupture represents a step between uterine and air-breathing life. Although very simple in appearance, it is a stressful experience, with important metabolic and respiratory changes. The most important role of the umbilicus in utero is oxygen delivery but on the ground it becomes an open door to micro-organisms. There is no sterile environment therefore the calf has to fight back bacterial invasion rapidly. One study reported the incidence of umbilical infection and condemnation in a slaughterhouse to be around 1% in bob veal calves.[1] A retrospective study on umbilical pathology of Charolais calves, reported a frequency of umbilical infection of 7%.[2] Umbilical surgery are frequently performed on the farm. Challenging umbilical conditions or situations are often related to the size of the infected structures, the defect or the animal. The further the abscess is from the ventral abdominal wall the more difficult will be the dissection and resection. Adhesions of the greater omentum to infected umbilical structures are a constant surgical finding. However, the jejunum or the abomasum can be adhered too. Finally, some calves are poor candidates for surgery because of concomitant diseases related or not to the umbilical condition.

## SURGICAL PLANNING & EXAMINATION

After performing a complete physical examination, the clinician has to focus on certain aspects of the animal. Even if there is a large umbilical mass compatible with an abscess, the calf could certainly be bright and alert with a normal body temperature. Hathaway reported that amongst young slaughtered calves condemned for navel ill, 29.9% had lesions localized to the umbilicus.[3] Calves with umbilical infection have a time bomb ready to explode.

### Physical examination

With a meticulous palpation of the umbilical structures, an experienced clinician can certainly give an accurate diagnosis without using ultrasound. Palpation can be performed with the animal standing if the calf is not a ruminant yet and the abdomen is small. Otherwise, palpation can be done in dorsal recumbency with the animal being off feed for 24 hours. A soft reducible mass is compatible with a hernia. A hard non-reducible mass is most likely an abscess. Strangulated hernias may have similar features but it is rare. Small diameter omphalitis lesions can be the tip of the iceberg. Deep palpations of the abdomen and ultrasound examination are then performed for further evaluation. Rectal palpation can be performed on older animals if the size of the animal or your arm allows it. Older animal with urachal infection will have a bladder under tension that is difficult to palpate or a cord that can be palpated just on the rim of the pelvis. Sometimes abscesses are located at the dorsal and cranial aspect of the bladder. The thorax should be auscultated for any abnormal lung sounds especially if you plan a long surgery like a marsupialization of the umbilical vein. Every palpable joint must be examined on a calf with umbilical infection. They are highly susceptible to septic arthritis. The most common joints involved are carpus, stifle and tarsus. This is especially true with omphalophlebitis. We have also seen septic coxofemoral joint, atlantooccipal joint, elbow and shoulder joints.

### Ultrasound

Ultrasound is routinely used for any umbilical abnormalities. It helps to refine your diagnosis and evaluate the size of the structure involved and the location an abscess. Visualizing the diameter of the structures infected helps in surgical planning and to decide what anesthesia protocol to use. Ultrasound examination starts from the umbilicus and progress toward the bladder or the liver. There is a good correlation between ultrasound and surgical findings.[4,5]

### Off Feed Period

This is a critical point and the duration should be based on the following criteria: weight, size of the hernia and expected incision length and type of feeding. The rule of thumb is 12 hours for bottle fed or suckling calves (< 6 week-old or < 75 kg) and 36 hours for roughage, 24 hours for grain, and 12 hours for water in larger animals. If the defect is large or the animal weighs more than 350 kg, we use the following schedule for roughage, grain and water, respectively: 48-24-12. We will not hesitate to postpone a surgery if abdominal distension during the surgery is a risk.

### Abscess Drainage

The abscess should be drained if it is large or ready to rupture and let to drain for a week or two. We usually flush it well with a large volume of iodine diluted with tap water for about a week. Hydrotherapy helps decrease the swelling after you drain the abscess. The calf is started on Pen G (22 000 IU/kg q12) for 5 days or until the swelling is down and incision edges are not bleeding when manipulated. At this point, you have two choices based on the size of the abscess:

- Let it heal and reschedule the surgery in 3 weeks
- If the swelling is gone and drainage is small after one week of medical therapy, then the incision is sutured with a purse string and the surgery is performed.

**Antibiotics**

Antibiotics are administered just prior to surgery and immediately after if the pharmacodynamic of the drug allows it. Most of the surgical infections are related to intraoperative contamination. Antibiotics are continued for 2-3 days for clean and clean contaminated surgery, 5-7 days for contaminated surgery and up to 3 weeks if there is peritonitis or liver abscesses. Pen G is routinely used for umbilical surgery. The ntibiotic regimen will differ if concomitant disease has to be treated.

**Anesthesia protocols**

Duration and difficulty level of the surgery will guide the anesthetic protocol. We always perform a local anesthesia with lidocaine 2%. Avoid large amount of anesthetic solution and anticipate your surgical incision length. Flunixin or ketoprofen are routinely administered just before the surgery. We use the following protocol for complicated umbilical surgeries:

- Sedation: 0.1 mg/kg of Diazepam
- Epidural (coccygeal): lidocaine 2% (0.15 ml /kg) and xylazine (0.05mg/kg)
- Local anesthesia: lidocaine 2%

A coccygeal epidural (needle 20G or 18G on older calves) combines lidocaine 2% (without epinephrine) and 0.05 mg/kg of xylazine to an injected volume of 0.15 ml/kg. The solution is given slowly over 1-2 minutes and the calf is left to rest on sternal recumbency for 10-15 minutes before the surgery. Depending of the behavior of the animal and his general status, Glycerol Guiacolate could be given. A nasotracheal tube (7mm) is often passed while the animal is on dorsal recumbency while prepared for the surgery. Oxygen or an anesthetic machine can be hooked up rapidly as needed.

**LARGE UMBILICAL DEFECT**

Large umbilical defects on heavy animal are challenging because of the excessive tension on the sutured repair. Whether it is a large hernia or large defect resulting from extensive dissection (urachal infection or incision hernia), the approach is the same. We do not use mesh. Instead, the animal might be off feed up to 72 hours and the defect will be sutured by tissue apposition with PDS USP 2 in simple interrupted or cruciate fashion. Vicryl USP 3 might also be used in some critical areas under tension. An abdominal wrap is kept for 48 hours to avoid wound contamination and to decrease pressure on the suture line. Feeding restriction postoperatively is very important. Roughage is increased gradually to full ration over 5 to 7 days depending of the incision length.

**INFECTION OF THE URACHUS**

Urachus is the most common umbilical structure infected. Patent urachus is rarely seen in calves as it is in foals. Medical treatment could be tried for a while but recurrences are frequent because antibiotics do not penetrate well through a 1 cm fibrous capsule. Surgical approach is fairly similar to umbilical hernia but the incision is longer. Surgical preparation should include the area just cranial to the teats. Before preparation and local anesthesia, verify that the abscess is not leaking. Abdominal invasion is more delicate because of a possible incision through the abscess. The caudal aspect of the incision is dissected carefully to avoid cutting the urachus. The urachus is always adhered to the peritoneum and the omentum and is dissected with scissors until the apex of the bladder can be exteriorized. Do not hesitate to extend your incision more caudally if dissection or exteriorization is difficult. Abscesses at the dorsal aspect of the bladder are difficult to expose. Two assistants are necessary to retract the viscera while the surgeon can concentrate on the apical resection. Both arteries are double ligated as proximal as possible. The bladder is stabilized outside the incision with Doyen forceps across the apex. The apex of the bladder and the abscess are resected together. The bladder is sutured with double inverted pattern with 2-0 absorbable material (Vicryl and Dexon). The omentum is sutured and the abdomen is closed in a routine fashion

**OMPHALOPHLEBITIS**

Omphalophlebitis is the most challenging surgical procedure of all umbilical pathologies. The surgery itself is complicated and the calf has often concomitant pathologies like failure of passive transfer and septic arthritis. Once diagnosed, it should be operated rapidly to avoid spreading bacteria. Patient preparation should include clipping, scrubbing and local anesthesia of the right para-costal area in case you have to perform a marsupialization of the vein. The approach is the same as any umbilical infection. The vein is manipulated and dissected very carefully. The wall is thinner than an infected urachus. The vein is always adhered to the peritoneum. Careful dissection with scissors and fingers until the liver is reached. If the vein is tapering off near the liver, en bloc resection is possible with double ligatures. If not, marsupialization has to be done. A circular skin incision is performed at the right para-costal area. Muscles are bluntly dissected. Marsupialization through the abdominal incision has been associated with wound infection, dehiscence and hernias. Verify that the umbilical stump is not leaking before taking it into the abdomen and through the marsupialization site. The vein is exteriorized as much as possible without excessive tension. The fibrous wall of the vein is sutured to the muscle layers with interrupted horizontal mattress. A second suture is done between the vein and the skin. Abdominal wall, subcutaneous tissue and skin are sutured in a routine fashion.

Post-operative cares are demanding. The purse string on the umbilical stump is removed when the calf is standing for the first time. The vein is irrigated everyday with a diluted iodine solution until it comes up clear. Avoid excessive pressure because rupture of the umbilical vein has been reported as well as rupture of an abscess into the bloodstream. Complications are more frequent than other umbilical surgical procedures: post surgical pain, excessive granulation tissue at the marsupialization site, herniation at the marsupialization site, rupture of the vein if too much pressure is applied during irrigation, septic arthritis.

**INCISIONAL HERNIA**

Surgical wound infection is the primary cause of incisional hernia. Depending of the suture used to close the abdominal wall, the hernia becomes obvious 3 to 4 weeks after the previous surgery. Small abscesses surrounding previous knots can be seen with ultrasound imaging of the wound. The connective tissue forming the new hernia has to be mature before revising it. A waiting period of 60 days is essential otherwise the strength of the connective tissue will not hold sutures.

**ABOMASAL AND RUMEN INCARCERATION THROUGH UMBILICAL HERNIA**

The greater omentum is the most common structure herniated in the umbilicus in cattle. Sometimes a part of the abomasum or the rumen are herniated and incarcerated. Animal with incarceration might have clinical signs: weight loss, anorexia. Diagnosis is made with ultrasound imaging and tapping of the mass. The surgical approach is the same. Depending of the severity of the incarceration, the adhesions or the damage to the wall, simple reduction or partial resection of the organ might be necessary.

**CONCLUSIONS**

Complete physical examination and ultrasound imaging of the umbilicus are essential in complicated case of infection. It allows the surgeon to choose appropriately the anesthesia, plan the off feed schedule and give an appropriate prognosis to the owner. Under adequate anesthesia, the surgeon will perform delicate dissection and avoid unwanted traction on certain organs. The abdominal incision must be lengthened if necessary to gain adequate exposure. With appropriate antibiotic regimen and attention to Halsted principles, most umbilical surgeries will be successful.

**REFERENCES**

[1] Biss ME et al. **Br Vet J** 1994;150:377. [2] Bohy A et al. **Point Vétérinaire** 1990;22:542. [3] Hathaway SC et al. **New Zealand Vet J** 1993;41:166. [4] Staller GS et al. **J Am Vet Med Assoc** 1995; 206:77. [5] Lischer CJ et al. Schweiz Arch Tierheilk 1993;135:221.

# EFFECTS OF DISBUDDING ON HEART RATE, PHYSICAL ACTIVITY, AND SALIVARY CORTISOL CONCENTRATION IN NEONATAL HOLSTEIN-FRIESIAN CALVES.

**D Morin[1], L Brown[1], V Jarrell[1], R Wallace[1], P Constable[2].**
**[1]College of Veterinary Medicine, University of Illinois, Urbana, IL; [2]School of Veterinary Medicine, Purdue University, West Lafayette, IN.**

The effects of early disbudding on heart rate (HR), physical activity (PA), and salivary cortisol concentration (SCC) were studied in neonatal Holstein-Friesian calves. Seven calves between 6 and 11 d of age were placed in individual indoor pens and fitted with a self-contained, battery-operated, continuous-logging combined HR and movement sensor monitor (Actiheart®; Mini Mitter Company Inc, OR, USA). Monitors were attached to the thorax and mean HR and PA were recorded for each 15 second epoch, with activity being detected by accelerometry. Data from the Actiheart® were downloaded to a laptop computer at the conclusion of each study, and epochs with heart rates <60 beats per minute (bpm) or >180 bpm were considered to reflect noise from muscle movement and discarded from analysis of HR data. After 24 h of acclimation, calves were manually restrained in their pens for sham disbudding with a cold butane dehorner. Actual disbudding with a hot butane dehorner was conducted 24 h later. Milk replacer was fed at 12-h intervals, with sham or actual disbudding occurring 4 h after feeding. Saliva was collected 60, 30, and 5 min before sham and actual disbudding and every 30 min for 4 h thereafter. Free SCC was measured using an ELISA. Data were expressed as least squares means ± SE and analyzed using PROC MIXED with repeated measures; P<0.05 was considered significant.

Physiologic HR and PA data were obtained for 91% of the 7.5 h observation period. Sham disbudding caused a transient mean increase in HR of 30 bpm, which was similar to the increase seen at feeding time (8 and 20 h after sham disbudding). Heart rate in disbudded calves averaged 5.5 bpm higher (104.9±1.6 bpm; P=0.042) over the observation period compared with sham-disbudded calves (99.4±1.5 bpm), but there was no treatment x time interaction (P=0.89). Physical activity in disbudded calves (1.29±0.41 counts per minute [cpm]) was decreased (P=0.025) over the 7.5 h observation period compared with sham-disbudded calves (2.99±0.40 cpm), and there was a significant treatment x time interaction (P=0.0004). Although there was no effect of time (P=0.60), SCC appeared to peak between 90 and 150 min after sham or actual disbudding and mean SCC was higher (P=0.041) in disbudded (0.64±0.06 ng/mL) than sham-disbudded (0.56±0.06 ng/mL) calves.

In summary, physiologic responses (increased HR and SCC, decreased PA) were minimal when calves were disbudded in the first two weeks of life. A larger scale study is needed to characterize behavioral responses to disbudding at this age, and determine whether administration of analgesic agents mitigates physiologic responses to disbudding. We believe this is the first study to use a non-invasive and continuous method to monitor HR and PA in calves undergoing disbudding.

**ECHOCARDIOGRAPHIC PARAMETERS IN CATTLE KEPT AT ALTITUDE- CORRELATES WITH PULMONARY ARTERY PRESSURES (PAP).**

**GD Hallowell[1], TJ Potter[2] and TJ Holt[3]**
**1. 14, Threadcutters Way, Shepshed, Leicestershire, UK 2. Royal Veterinary College, Herts, UK. 3. Colorado State University, Fort Collins, CO.**

High altitude disease is a common condition found in cattle raised at >5000Ft and can lead to significant losses in susceptible herds. The aims of this study were to assess whether previously validated views for assessing right ventricular (RV) size correlated with PAP in cattle kept at altitude. Thirty-eight black Angus cattle (20 heifers and 18 bulls, 7months±19 days and weighed 233±35 Kg) were examined. All cattle underwent cardiac auscultation and pulmonary artery pressure (PAP) testing. Cardiac troponin I concentrations (cTnI) were measured in 25 animals. The following standard echocardiographic images were analysed using five cardiac cycles: right parasternal long axis projections optimised to show the pulmonary artery (PA), left ventricular outflow tract (LVOT), left ventricle (LV) and a short axis projection at the level of the left ventricle (LV-SA). Cardiac size was determined using a standard M-mode of the left ventricle. RV free wall thickness, RV and right atrial (RA) areas from all views and aortic (Ao) and pulmonary artery (PA) were interrogated. Data were analysed using Friedman test, linear regression and discriminant analyses. Mean PAP scores were 56±25mmHg and distributed as follows: 30-40mmHg (n=13; 34%); 41-70mmHg (n=13; 34%) and >70mmHg (n=12; 32%). Murmurs were detected in 29% (82% systolic and 18% diastolic) of cases but there was no correlation with mean PAP nor was there any correlation between presence of tricuspid (TR) or pulmonic regurgitation (PR) using colour-flow Doppler. cTnI were within normal limits in all but 1 animal and thus did not correlate with mean PAP. Bodyweight was correlated with Ao and PA but not with mean PAP scores. When parameters were corrected for bodyweight using Ao diameter the following were significantly correlated with mean PAP scores ($p<0.05$ and $r^2=60\text{-}85$): RV free wall diameter in systole, total RV area (on all 4 projections), RV area between moderator band (MB) and PA (PA projection), RV area between tricuspid valve (TV) and MB (LV projection), right atrial (RA) diameter and area (LVOT and LV projections), MB length (LV-SA and LV projections) and all RV diameters (LV-SA projection). The following echocardiocardiographic findings are ranked using discrimant analyses to demonstrate the best correlation with mean PAP: 1. RV total area (LVOT projection), 2. RV diameter (measure 2; LV-SA projection), 3. RV free wall in systole and equal 4. and 5. MB length (LVOT and RV-SA projections). In conclusion, there are several echocardiographic parameters on multiple projections that correlate with mean PAP in cattle with RV area, RV diameter and RV free wall thickness being the most useful. Reference ranges as an index of Ao diameter have been established for this age group of cattle. Further work would be required to ascertain if these ratios and correlates were applicable to other age ranges and breeds.

**EXPERIMENTAL INFECTION OF COLOSTRUM-DEPRIVED CALVES WITH BOVINE VIRAL DIARRHEA VIRUS TYPE-1 ISOLATED FROM FREE-RANGING WHITE-TAILED DEER (*ODOCOILEUS VIRGINIANUS*).**

**EA Raizman, RM Pogranichniy, M Lévy, M Negron, WG Van Alstine.**
**School of Veterinary Medicine, Purdue University, West Lafayette, IN.**

Bovine viral diarrhea virus (BVDV) is a pestivirus associated with important economic losses to the cattle industry in the United States. The role of wildlife as a reservoir for the virus is unclear. The objective of this study was to determine if calves could be experimentally infected with a BVD type 1 virus isolated from free ranging White-tailed deer (WTD), develop lesions and shed the virus.

Twelve colostrum-deprived male Holstein calves (8 infected and 4 controls) were used in this study. Both groups were housed in an isolation unit. After a 7 day period of acclimation, the experimental group was challenged intranasally with a 2 ml inoculum containing $10^6 TCID_{50}$/ml BVDV type 1 isolated from free ranging WTD in Indiana, while the control group was inoculated with a virus-free culture medium. Daily physical examinations were performed. Serology for BVD (virus neutralization and ELISA) was performed on days -5, 0, 3, 7, 10, 14 and 21 post inoculation (DPI). Whole blood, saliva, nasal and rectal swabs were collected on days 0, 3, 7, 10, 14 and 21 DPI for virus isolation, virus neutralization, ELISA and Real Time PCR (RT-PCR) assays. On day 14 DPI, 4 calves from the infected group and 2 controls were humanely euthanized; multiple tissue samples were collected for histopathology. The same protocol was used on day 21 for the remaining 4 infected and 2 control calves. The experimental protocol was approved by the Purdue Animal Care and Use Committee.

Several calves develop intermittent diarrhea before inoculation with BVDV. The most important microscopic lesions were thymic atrophy and lymphoid depletion of the Peyer's patches in all 8 infected calves. No significant lesions were identified in the control calves. RT-PCR was positive in the buffy coat of 8/8, in the nasal samples of 7/8 and in the saliva samples of 2/8 infected calves. The rectal samples were negative at all times for the 12 calves. On virus neutralization assay, 4/8 animals developed antibodies against BVDV and on ELISA assay 3/8 animals were positive for BVDV. All samples from the control calves remained negative.

These results support our hypothesis that colostrum-deprived calves can be successfully infected with a BVDV type 1 isolated from WTD and that infection results in the development of typical lesions in lymphoid tissues. Furthermore, positive identification of BVDV in nasal and saliva samples suggests that shedding of the virus in the environment is possible.

**EFFICACY AND PHARMACOKINETICS OF PANTOPRAZOLE IN ALPACAS.**

**GW Smith, JL Davis, SM Smith, DM Foster, MP Gerard, NB Campbell.**
**College of Veterinary Medicine, North Carolina State University, Raleigh, NC.**

Third compartment (stomach) ulcers are a common cause of morbidity and mortality in camelids of all ages. Although the exact cause and pathogenesis of stomach ulceration in llamas is not completely understood, stress appears to be the most common predisposing factor. Unfortunately, many of the accepted medications for treatment of gastroduodenal ulcers in other species have proven ineffective in New World camelids. Intravenous administration of omeprazole has been shown to significantly increase third compartment pH in llamas, however IV formulations are not readily available in North America and this route of administration is impractical for routine use by owners. Oral administration of omeprazole has been shown to be an ineffective method of increased third compartment pH in camelids. Pantoprazole is a proton pump inhibitor that is frequently used in humans and animals for the treatment of gastric ulcers. It is widely available and has been shown to effectively inhibit gastric acid production in several species. The purpose of this study was to determine whether intravenous and/or subcutaneous administration of pantoprazole would effectively increase third compartment pH in alpacas.

In this study, 5 adult male alpacas were anesthetized and fitted with a third compartment cannula for measuring gastric pH. Following recovery, alpacas received 1 mg/kg pantoprazole IV, q 24h for 3 days or 2 mg/kg SQ q 24h for 3 days. All alpacas received both IV and SQ pantoprazole, with a minimum of 3 weeks between treatments. Third compartment pH was recorded at regular intervals and plasma samples were taken for pharmacokinetic analysis. Following completion of the study, third compartment cannulas were removed and all alpacas were returned to their farm of origin.

Pantoprazole induced a slow but sustained increase in third compartment pH when given by both the IV and SQ routes. Baseline third compartment pH (1.81 $\pm$ 0.7) increased to 2.47 $\pm$ 0.8, 3.53 $\pm$ 1.0, and 4.03 $\pm$ 1.3 at 24, 48, and 72 hours following daily IV administration of pantoprazole. Third compartment pH increased from 1.73 $\pm$ 0.6 at baseline to 3.05 $\pm$ 1.1, 4.02 $\pm$ 1.4, and 3.61 $\pm$ 1.6 at 24, 48, and 72 hours following SQ administration.

This study showed that pantoprazole represents a safe and effective drug for increased third compartment pH in camelids. It is likely an effective treatment for third compartment ulcers and might be useful for prophylactic administration in stressed camelids at high risk for developing ulcers.

# EFFECTS OF MILKING FREQUENCY ON PHARMACOKINETICS OF CEPHAPIRIN ADMINISTERED BY INTRAMAMMARY INFUSION IN LACTATING DAIRY COWS.

**R Stockler[1], D Morin[2], R Lantz[3], P Constable[4].**

**[1]College of Veterinary Medicine, Mississippi State University, Mississippi State, MS; [2]College of Veterinary Medicine, University of Illinois, Urbana, IL; [3]Rocky Mountain Instrumental Laboratories, Fort Collins, CO; [4]School of Veterinary Medicine, Purdue University, West Lafayette, IN.**

Many dairy producers milk cows three times daily (3X) rather than twice daily (2X). However, little has been published on the effect of milking frequency on pharmacokinetics of intramammary (IMM) antimicrobial agents, making dosing recommendations in 3X herds largely empirical. We determined the effects of milking frequency (2X or 3X) on the pharmacokinetics of IMM cephapirin (CEPH) in healthy cows. Six multiparous Holstein-Friesians were milked 2X and dosed twice at a 12 h interval or milked 3X and dosed twice at an 8 or 16 h interval. Three cows had low milk production (21-24 kg/day) and three had high milk production (40-42 kg/day). A 200 mg dose of CEPH was infused into the designated rear mammary gland at each dosing time. A duplicated Latin square design allowed each cow to receive all three frequency-dose treatments, with intervening washout periods. Milk was collected from the infused gland for 96 h after treatment and concentrations of CEPH and its active metabolite desacetylcephapirin (DAC) determined using HPLC/MS/MS. Data were fitted using 1- and 2-compartment pharmacokinetic models, as well as a noncompartmental model. PROC MIXED was used for data analysis, and $P<0.05$ was considered significant.

Cephapirin was rapidly metabolized to DAC in the mammary gland, with DAC predominating in milk until 48 h after infusion. Pharmacokinetics of CEPH and DAC were similar regardless of milking frequency, with a 1-compartment model providing a better fit than a 2-compartment model in most cases. Mean residence time (MRT) for total CEPH equivalents (CEPH + DAC) was negatively and linearly correlated with milk production ($R^2=0.22$; $P=0.049$). Milk production also tended ($P=0.056$) to affect the percent of infused drug recovered in milk, with lower percent recovery in low producing cows (46±21%) than high producing cows (86±16%). The relationship between percent recovery and milk production was positive and curvilinear ($R^2=0.78$; $P<0.0001$). Neither milking frequency nor milk production significantly affected the time that milk CEPH concentration exceeded $MIC_{50}$ or $MIC_{90}$ values for common mastitis pathogens,but a 16-h dosing interval extended inhibitory concentrations approximately 8 h longer than an 8-h interval in cows milked 3X. Low milk production extended the time for milk CEPH concentration to reach the FDA tolerance level (0.020 µg/mL), but milking frequency had no effect. Findings suggest that this CEPH preparation, which is labeled for two doses 12 h apart, could be administered at a 16-h interval in herds milking 3X, without affecting inhibitory concentrations in milk or withdrawal time; extended withdrawal times may be prudent for cows with low milk production. However, more investigation is needed to determine the effects of milking frequency and milk production on CEPH pharmacokinetics in cows with mastitis.

## BLUETONGUE IN NEW WORLD CAMELIDS: SEROPREVALENCE AND RESULTS OF A VACCINATION TRIAL IN SWITZERLAND.

**P. Zanolari[1], V. Chaignat[2], C. Griot[2], C. Kaufmann[3], M. Mudry[1], M. Meylan[1].**
**1.Vetsuisse Faculty, University of Berne. 2.Institute of Viroroly and Immunoprophylaxis, Mittelhäusern. 3.Veterinary Practice Mondo a, Riehen, Switzerland.**

Bluetongue (BT) disease has been traditionally observed only in the milder climate of Southern Europe where the vector of the virus, midges of the genus Culicoides, was found. In the summer of 2006, however, first infections of cattle and sheep with the serotype 8 of BT virus (BTV-8) have been reported in Northern Europe and first clinical cases were observed in Switzerland in the fall of 2007. As BT incidence had increased dramatically in Northern Europe during 2007, a vaccination campaign in domestic ruminants was decided upon for 2008 in Switzerland. The susceptibility of New World Camelids (NWC) to BTV infection and their role in the epidemiology of the disease have not been clearly defined to date. Therefore, a study was undertaken to evaluate the seroprevalence of BT in the Swiss NWC population and to investigate the response of llamas and alpacas to commercial vaccines for use in domestic ruminants.

Blood samples were collected prior to the vaccination campaign from 354 NWC (159 llamas and 195 alpacas, 1 month to 17 years old) from 38 herds throughout the country, corresponding to approx. 10% of the Swiss NWC population. Each serum sample was analyzed with three serological methods (Bluetongue Virus Antibody Test Kit, VMRD Inc., USA; Ingezim BTV DR, Ingenasa, Spain; ID Screen Bluetongue Competition ELISA Kit, ID Vet, France). Forty-two healthy animals (17 llamas and 25 alpacas) were included in the vaccination trial, of which 9 were pregnant females. An inactivated vaccine (Bovilis BTV8, Intervet, 22 animals, or BTVPUR AlSap, Merial, 20 animals) was injected subcutaneously twice at an interval of 21 days. The animals were monitored clinically for local and systemic reactions to vaccination and blood samples were collected for serology (Bluetongue Virus Antibody Test Kit) on days 0, 14, 21, and 43.

All 354 animals tested prior to the vaccination campaign were negative for BTV-8 antibodies in all 3 tests. No adverse effects were observed after vaccination and all vaccinated animals developed antibodies to BTV-8 after the second application of the vaccines (day 43). After the first injection however, only 35% of the animals had detectable antibody levels. Seroconversion tended to be faster in alpacas than in llamas.

The Swiss NWC population was free of BT infection prior to the vaccination campaign in 2008. Two injections of inactivated vaccines led to the formation of antibodies against BTV-8. The degree of protection against infection with BTV-8 conferred by these antibodies should be assessed in the future through challenge trials or by epidemiological analysis in areas where BT is endemic in the domestic ruminant population.

**DEVELOPMENT OF PRIMARY BOVINE BRONCHIAL EPITHELIAL CELL (BBEC) CULTURES FROM LIVE CALVES, AND EVALUATION OF CELL RESPONSES TO VIRAL INFECTION.**

**AR Woolums, TM Krunkosky, CL Jarrett, MB Ard, LJ Berghaus, CL Bettis, KAE Hurley. College of Veterinary Medicine, University of Georgia, Athens, GA.**

The bovine respiratory disease complex (BRDC) impacts the health and productivity of millions of cattle annually. Primary infection of respiratory epithelium by viral pathogens initiates most cases of BRDC. Research in other species indicates that cytokines and cell surface molecules expressed by infected respiratory epithelial cells affect the progression and outcome of respiratory disease; however, little is known about how respiratory epithelial cell responses influence the BRDC. Bovine bronchial epithelial cells (BBEC) were isolated by bronchial brushing of secondary or tertiary bronchi of 2-week-old calves. Cells were established in culture, then induced to de-differentiate by growth media modification. De-differentiated BBEC were frozen as passage 1, then thawed and grown as passage 2 in an air-liquid interface culture system to a differentiated phenotype. Differentiated primary BBEC were evaluated by electron microscopy, confirming the presence of confluent monolayers of ciliated and nonciliated cells, including goblet cells. Prior to full differentiation cells were infected with bovine respiratory syncytial virus (BRSV). Expression of mRNA for the chemokines IL-8 and RANTES and the adhesion molecule ICAM-1 by infected cells was evaluated by real time RT-PCR; at 2 hours postinfection (PI), IL-8 and ICAM-1 mRNA were significantly increased over baseline. Levels returned to baseline at 4 and 24 hours PI. *In vitro* culture of passaged BBEC from live calves will enable evaluation of BBEC responses at various stages of viral infection. Information obtained will support the development of strategies to prevent or treat respiratory disease by targeting the earliest events in respiratory infection.

## *MYCOPLASMA BOVIS* COLONIZATION OF CALVES IN AN ENDEMICALLY-INFECTED HERD.

**FP Maunsell, JN Coan, GA Donovan, MB Brown.**
**College of Veterinary Medicine, University of Florida, Gainesville, FL.**

Our objective was to characterize patterns of nasal colonization by *M. bovis* in calves. In a prospective observational study, 359 heifer calves that were healthy at 4 days (d) of age were followed for 5 months. Calves were individually housed for the first 8 weeks (wk) of life in 3 barns in elevated stalls that permitted direct contact with calves in adjacent stalls. Stall design and calf management were identical for the 3 pre-weaning barns. The clinical status of calves was monitored daily. Nasopharyngeal swabs were collected for mycoplasma culture and PCR at 4 d and at 3, 6.5, 10, 14, and 23 wk of age, as well as prior to treatment for calves that developed clinical signs of respiratory disease or otitis media. Overall, *M. bovis* was isolated from 65% of calves. *M. bovis* was first isolated at 4 d and 3 wk of age in 0.8% and 15% of calves, respectively, and 41% of calves had one or more culture positive samples during the pre-weaning period. Calves reared in one barn had higher odds of being culture positive for *M. bovis* during the pre-weaning period than calves reared in two smaller barns (148/263 and 7/92, respectively; odds ratio [OR]=15.6, 95% confidence interval (CI)=8.0-30.4, $P<0.001$). The odds of a calf being culture positive for *M. bovis* during the pre-weaning period were higher if a neighboring calf was infected (OR=7.3, CI=4.6-11.6, $P<0.001$). There were 40 stalls that consecutively housed 2 study calves; the odds of a calf being culture positive during the pre-weaning period were higher if the calf previously occupying that stall had been culture positive (OR=7.9, CI=1.8-34, $P=0.007$). The age at which calves became colonized by *M. bovis* affected the risk of respiratory disease and otitis media; calves that were *M. bovis* positive prior to weaning had greater odds of disease than calves in which infection was first detected post-weaning (OR=2.4, CI=1.3-4.3, $P=0.004$). In summary, pre-weaning environment has a major effect on *M. bovis* infection and clinical disease in calves. Opportunities for improved control and prevention strategies for mycoplasmal disease were identified.

# ABDOMINAL DISCOMFORT IN LLAMAS & ALPACAS: CAUSES & CLINICAL CHARACTERISTICS

**Christopher Cebra VMD, MA, MS, DACVIM**
**Corvallis, OR**

The subject of abdominal pain in camelids has been addressed a number of times over the years, with the discussion becoming progressively more detailed as we gain experience with these species. We were truly learning from scratch, as the South American approach (hands off and deal with the consequences) was generally unsuitable to North American tastes. The tone has varied over the years, from extremes such as "we had better cut it, because it's worth so much," to, "why cut it, because they all die." The bases for these extreme views were the high proportion of underlying lesions that were a major morphologic defect, such as a ruptured ulcer, and they general uncertainty on how to proper assess the likely severity and nature of the disease in a sick camelid. As we have become more knowledgeable about disease states and more proficient and comfortable in dealing with them, we have improved in both those regards.

## CAUSES OF COLIC

Camelids develop abdominal pain for the same reasons as other animals. The major mechanisms are activation of stretch-pain receptors in the mesentery, various ligaments, or organ capsules, muscle spasms, inflammation, and ischemia. Similar to cattle, the capacity and location of the gastric compartments combined with the ability to regurgitate excess fluid decrease the likelihood of gastric distention or is tug on the mesentery contributing to pain signs. Additionally, relatively complete forestomach fermentation in adult camelids decreases the likelihood of aberrant intestinal fermentation, and hence gas colics are uncommon. However, aberrant fermentation can occur in younger camelids, or adults with depletion of their normal gastric microbial population. In crias, this "milk colic" may be severe, but usually only lasts a few hours. As in other species, poorly digestible milk replacers or viral damage to the intestinal brush border may contribute to aberrant fermentation. In adults, it is downright rare. In all ages, infection by gas-producing bacteria may contribute to signs with some forms of enteritis.

Capsular stretch is a more important source of pain when it involves non-digestive abdominal organs. Colic signs may be severe with distention of the kidney, ureters, or to a lesser degree, bladder. Milder signs are evident with enlargement of the liver due to lipidosis, hepatitis, or fluke infection.

Anatomically, the camelid stomach and intestines bear more similarity to cattle than horses. Aside from thrombosis due to systemic events, ischemia is most likely in regions with long mesenteric attachments and relative mobility. In camelids, this includes the distal phlange of the small intestine and the whole of the ascending colon, including its spiral loop. Rotation or displacement of the third compartment have not been described. Described ischemic events occur as a results of strangulation involving the epiploic foramen (rare), body wall defects (intermediate frequency), diaphragmatic defects (rare), mesenteric defects (rare), rotation around the mesenteric root (most common), intraabdominal adhesions (intermediate), or intussusception (rare). Non-strangulating entrapments can occur with less severe involvement of most of these structures as well. Intussusception has been described involving the jejunum, cecum, or ascending colon. Uterine torsion is among the more common causes of colic. Liver lobe or splenic torsion is extremely rare. Risk factors for entrapments obviously include the presences of body wall, mesenteric, or diaphragmatic defects. Enteritis or coccidial infection may promote mesenteric volvulus or intussusception, particularly when they involve the colon.

Intraluminal obstruction is usually caused by hair or plant fiber balls. In most cases, these appear to form in the gastric compartment, where multiple may be found. Some, smaller balls may form in the proximal loop of the ascending colon. Hairballs are most common in animals less than one year old. Ectoparasites, boredom, and post-weaning suckling behavior are thought to be risk factors. Phytobezoars may develop in animals with poor chewing or on excessively fibrous feed. Long-stem spring pasture appears to increase risk in the Pacific Northwest. Obstruction tends to occur near the pylorus or in the beginning of the duodenum in alpacas and smaller llamas, and in the centrifugal loops of the spiral colon in adult llamas. With time, bezoars lodged in the duodenum may migrate further along the intestine, but it is rare for them to make it into the colon or pass entirely. Tapeworm impaction with ulceration throughout the jejunum has been described in one alpaca. Foreign body obstruction is very rare.

Over recent years, a number of important causes of non-surgical colic have been identified. The earliest work on this identified that many camelids with colic signs that abated with time or medical treatment had serum biochemical abnormalities that reflected either obstruction (hypochloremic metabolic alkalosis) or enteritis (hyponatremia, hypochloremia, and often hypoalbuminemia). Our subsequent experiences suggest that "obstruction-like" episodes are rare. Small bezoars may pass, but that is uncommon and may result in mucosal abrasion. Hypochloremic metabolic alkalosis appears uncommon with ileus in sick camelids, which generally trend toward hyperchloremia. Whether ruptured ulcers with omental bursitis can cause this biochemical abnormality is unclear.

Camelids with "enteritis-like" blood work and colic signs may have a variety of conditions. The single most common in many different geographic areas appears to be Eimeria macusaniensis infection. These large coccidia invade the gut epithelium and lamina propia, in contrast to most GI parasites, which remain in the epithelium. It is possible that this deeper invasion has a more profound effect on GI motility or is simply more inflammatory, and hence more likely to cause colic. Other clinical evidence, including gut fluid distention and ileus detected ultrasonographically, by succussion, or at surgery, suggest that pooling of intraluminal fluid resulting in mesenteric tug is the main contributor to colic signs. This ileus can confound diagnosis, because it inhibits or delays the onset of diarrhea. Supportive evidence for a diagnosis of Eimeria macusaniensis infection includes hypoalbuminemia disproportionately severe for the level of anemia, occasionally hyponatremia, and identification of oocysts on fecal examination, if the infection has achieved patency.

Other common causes of enteritis without severe inflammation include coronavirus and rotavirus, Giardia, Trichuris, and Cryptosporidia. Salmonellosis is fairly uncommonly reported in camelids, and may or may not cause severe inflammation. Other bacterial enteritides include Clostridium perfringens and occasionally coliform organisms or Bacteroides-like organisms. In North America, the bacterial enteritides, except for Salmonellosis, are rarely considered primary diseases, and usually occur secondary to parasitic or viral infections, physical or chemical damage to the epithelium, or gut thrombosis. The less inflammatory enteritides are rarely associated with colic, and it usually resolves rapidly with the onset of diarrhea. Affected camelids may have hyponatremia and hypochloremia, relatively specific indicators of intestinal diseases in camelids, and hypoalbuminemia. Acidosis is generally less common or milder than with comparable disorders in cattle. Camelids with bacterial enteritis may additionally have fever, tachycardia, abnormal mucus membranes, high liver enzymes, lactic acidosis, and an abnormal leukogram. All abnormalities need not be present simultaneously to support this diagnosis. Diarrhea tends to be scant and less aqueous with bacterial enteritis or Eimeria macusaniensis infection, as opposed to profuse and watery with viral or other parasitic enteritides.

Camelids with colic due to bacterial enteritis or Eimeria macusaniensis infection can be very difficult to differentiate from camelids with gut strangulations, entrapments, or intraluminal obstructions, especially if the surgical lesion has been allowed to progress. Both types of disorder result in a camelid with low fecal output and also in declining systemic condition. For this reason, combined with a historical lack of rapid identification of surgical lesions, and the reluctance to watch hospital patients deteriorate without trying every possible treatment modality, many camelids with inflammatory conditions in the abdomen were surgically explored in the past. Exploration is helpful in confirming the lesion and establishing the prognosis in camelids with inflammatory diseases, and gives the opportunity to obtain biopsy specimens, which is especially helpful in diagnosing prepatent Eimeria macusaniensis infection, but is unlikely to lead to clinical improvement.

Slaframine poisoning is an uncommon, but important cause of non-surgical colic. It presumably has the same pathogenesis as in other species, and is associated with ingestion of the indolizidine alkaloid produced by fungus Rhizoctonia leguminicola. This fungus colonizes legumes, particularly red clover in damp areas and subterranean clover in dry areas, causing discoloration of the clover known as 'black patch'. Camelid exposure appears to be most common in the damp seasons, particularly Spring and Fall, which is conducive to both the clover and the fungus. Multiple animals may be affected simultaneously, which is uncommon with all other causes of colic. Signs occur within hours of ingestion in other species. In camelids, they last from 1 to 3 days, usually with a steady clinical improvement after the first 24 hours. In addition to colic, affected camelids may have profuse salivation ('Slobbers'), other SLUD signs, tremors, and a stiff gait. Treatment is not usually necessary, and the specific combination of signs, especially if multiple animals are affected or the discolored clover can be found, helps give the attending veterinarian confidence of the diagnosis. Clinical pathologic abnormalities have not been reported.

## CLINICAL SIGNS & PHYSICAL EXAMINATION

To the practitioner most acquainted with colic in equids, tachycardia, abdominal distension, absence of appetite, borborygmal sounds, and fecal passage as well as activities including pawing, restlessness, and stretching the abdomen are frequently taken as signs of colic. Although all of these signs may be present in some colicky camelids, most have neither tachycardia nor active signs of pain. Camelids have been described as being stoic animals that try to hide signs of illness and pain. While it is true that signs in camelids are less visible and violent than in horses, their behavior is restive and abnormal enough to be apparent to the careful observer. Specific signs of colic include lying in an abnormal position: sternal recumbency with one or more legs extended to the side or lateral recumbency; repeatedly getting up and lying down; kicking at the abdomen; and frequently changing position while recumbent. Lying quietly in sternal or lateral recumbency is not specific for colic.

Level of pain and presence or absence of other signs can be used to estimate the likelihood of the various categories of lesions. Small intestinal obstruction often leads to more subtle signs than other types of lesions, because fluid is refluxed into the gastric compartments and inflammation is minimal. This is most true for obstructions near the pylorus; the further into the jejunum the bezoar is pushed, the more tension there is on the mesentery and the more painful the camelid appears. Thus, in addition to the importance of recognizing pain signs, it is important to remember that camelids with mild to subtle signs still may have life-threatening obstructive lesions.

Very cranial obstructions can cause several other abnormalities that help differentiate them from other lesions. Fluid accumulation in the forestomach leads to bilateral, ventral abdominal distension, and a pot-bellied appearance. Gas may be eructated until the gastric fluid level reaches the cardiac sphincter, causing muscular spasms. After that time, bloat may be noticed. Other signs include dehydration, depression, and in rare cases, vomiting. Abdominal distension is difficult to assess in full-fleeced camelids; direct external palpation ("bear-hug") may be necessary. Other conditions, including advanced pregnancy, effusive peritonitis, venous congestion (as with strangulation), and enteritis can lead to abdominal distension without gastric engorgement. These conditions may be distinguished by other characteristics. It is also appropriate to mention here that subjective assessment of cranial abdominal pain by palpation of the flanks is extremely difficult to interpret; most camelids do not like being touched in that area and it is natural for them to flinch, jump, or lie down.

Dehydration and depression with small intestinal obstruction are due to sequestration of ingested and secreted fluids, with minimal functional intestinal absorptive surface area. Signs worsen with proximity to the pylorus, completeness of the obstruction, and time. Thus, a camelid with duodenal phytobezoar obstruction is likely to have subtle colic signs that abate after a few days due to progressive dehydration and depression. In contrast, a camelid with a colonic bezoar obstruction that leaves most of its intestinal absorptive epithelium functional is likely to have more obvious pain signs and develop minimal dehydration.

Camelids with strangulating obstructions or sepsis also develop signs of dehydration and depression (and also distension, as previously mentioned) due to venous pooling, peritoneal effusion, and the inflammatory response. Such camelids usually appear more painful and differ in other aspects of the diagnostic evaluation, but may be extremely difficult to differentiate from camelids with jejunal obstruction or any small intestinal obstruction with inflammatory complications.

In addition to the more obvious colic signs and possible dehydration and depression, camelids with inflammatory or cecocolic diseases may also show tenesmus. Camelids usually defecate and urinate at a communal dung pile several times a day, but rarely spend more than a minute or two at the pile or assume the posture when away from the pile. Tenesmus with obstruction is usually non-productive, whereas camelids with enteritis or peritonitis may have diarrhea. Thus, tenesmus with diarrhea and colic signs suggests a severe inflammatory condition. Tenesmus is also a common sign with urogenital disorders.

Anorexia, lack of feces, and gastric hypomotility are common findings in camelids with gastrointestinal disease, but lack specificity because they are also seen with diseases of other organ systems. However, finding appetence, normal fecal passage, and normal ruminations in a camelid with colic strongly supports a mild, transient problem, a non-inflammatory, non-gastrointestinal condition (i.e. uterine torsion), or the early stages of more severe disease. These signs should be monitored for resolution or progression.

Fecal passage may provide other insights into the nature of the disease. Camelids with obstructions may pass normal feces for 1 to 3 days (longer with small intestinal than cecocolic obstruction), but eventually change to dry, mucus-covered feces, and later stop defecating altogether: camelids with entrapments may continue to pass small quantities of normal or blood-tinged feces, whereas camelids with strangulations and necrotic enteritis and peritonitis may pass diarrhea in decreasing quantities. Consistent passage of small amounts of diarrhea in spite of ongoing pain signs for up to 4 days is common with *E. macusaniensis* infection. Continued, profuse diarrhea or normal feces are rare in camelids with colic due to gastrointestinal lesions requiring surgical repair.

Vital signs have been of little value in determining severity of colic and the presence of surgical lesions. Camelids with all types of lesions have been recorded as having low, normal, and high pulse and respiratory rates. It is our general bias that the more normal the vital signs, the better the prognosis, but this is not always true. High rectal temperatures are most common with peritonitis, but are not specific to the cause of inflammation. Low rectal temperatures can occur with most disorders and are usually a sign of circulatory compromise and more severe disease. When interpreting heart rate, it is imperative to remember that most healthy adult camelids have heart rates of 72 beats/minute or less, and that values between 72 and the published high end of 90 beats/minute may reflect pain or shock.

**FREQUENCY OF COLIC**

The types of lesions described here have been blamed for up to 19% of camelid deaths in camelids. It is possible that the percentages are even higher because colic signs may not have been recognized in all camelids, even if they were present. In our practice, colic is responsible for approximately 5% of camelid admissions. The percentage is likely to be lower in general practice, though poor recognition of signs may lead to underdiagnosis. The prevalence of individual diseases is likely to vary based on geographic, dietary, breed prevalence, and other factors. With many of these diseases, colic signs are transient, subtle, and may not be noticed without careful, constant observation.

**SELECTED REFERENCES**

[1.] Garry F. Proceedings. **11th Annu ACVIM Forum** 1993;817. [2.] Cebra CK, et al. **Vet Surg** 1998;27:112. [3.] Cebra CK. Proceedings. **15th Annu ACVIM Forum** 1997;447. [4.] Navarre CB, et al. Analysis **J Am Vet Med Assoc** 1999;214:812. [5.] Pearson EB. Proceedings. **13th Annu ACVIM Forum** 1995;671. [6.] Cebra CK, et al. **J Am Vet Med Assoc** 1997;211:600. [7.] Sullivan EK, et al. **Vet Surg**. 2005;34(5):524. [8.] Cebra CK, et al. **J Am Vet Med Assoc**. 2008;232(9):1357.

**ABDOMINAL DISCOMFORT IN LLAMAS & ALPACAS: DIAGNOSIS & TREATMENT**

**Christopher Cebra VMD, MA, MS, DACVIM**
**Corvallis, OR**

## RECTAL PALPATION

Although juvenile llamas and most alpacas may be too small, rectal examination of most adult llamas is possible with good restraint (preferably in a camelid chute), adequate lubrication and if necessary, chemical sedation. We find that adding 20 ml of a local anesthetic agent (2% lidocaine HCl) to 100ml of lubricant (for adults) and injecting the mixture into the rectum by plastic dose syringe or insemination pipette relaxes the anal sphincters and caudal rectum sufficiently to perform a good rectal examination. Individuals with large hands may not be able to do this and should stop if they encounter too much resistance. Rectal injuries and bladder rupture have been reported.

Rectal examination of the llama is similar to that of the cow; the bladder, female reproductive tract, left kidney and caudal first gastric compartment usually can be palpated. The right abdomen usually has no palpable viscera. Rectal palpation is most useful for abnormalities in the caudal abdomen, especially urogenital lesions. Obstruction, inflammation, entrapment, or strangulation of the terminal small intestine and ascending colon are likely to be detected by the presence of distended bowel or tight mesenteric roots, whereas generalized inflammatory conditions and small intestinal obstructions generally cause no palpable abnormalities.

Except for diagnosing uterine torsion, rectal examination of colicky camelids has been largely supplanted by flank ultrasonography in my practice.

## CLINICAL PATHOLOGY

A thorough clinicopathologic evaluation helps in identification of the lesions and selection of treatment. In general, the more normal the evaluation, the better the health of the bowel and the prognosis for survival. Camelids with hematologic abnormalities are likely to have inflammatory lesions and camelids with small intestinal obstruction are likely to have metabolic abnormalties.

### Complete Blood Count

Because of the frequency of inflammatory lesions in camelids with colic, differential blood counts are frequently abnormal. These changes can be obscured by stress neutrophilia, a common finding in sick camelids. More specific evidence of inflammation includes immature granulocytes, toxic changes in granulocytes, and hyperfibrinogenemia. Inflammatory changes are to be expected in camelids with peritonitis or enteritis; in camelids with obstructive bowel diseases they are an undesirable sign of advanced disease.

### Acid-base balance

Camelids with small intestinal obstruction frequently develop metabolic alkalosis due to the sequestration of gastric HCl within the gut lumen. Small intestinal entrapments and strangulations and long-standing cecocolic obstruction occasionally cause mild metabolic alkalosis, but such lesions usually cause so much pain that veterinary attention precedes development of this abnormality. Camelids with inflammatory or ischemic conditions have a tendency towards metabolic acidosis. This is more likely to be due to lactate production than intestinal bicarbonate loss, and is often accompanied by hyperlactemia and a high anion gap. Metabolic acidosis may also be seen with long-standing intestinal obstructive lesions as they become more inflammatory; hence, it is a sign of more severe disease.

### Electrolytes

Serum chloride is the most important electrolyte for diagnosing gastrointestinal causes of colic. Chloride ion is sequestered in the gastric compartments with cranial obstructions, but to a much lesser degree with more aboral obstructions. Thus, hypochloremia (< 106 mEq/L) is usually present under similar conditions as metabolic alkalosis.[2,6] Hypochloremia may also be found with metabolic acidosis in camelids with enteritis; hyponatremia and hypoproteinemia usually accompany it in these cases. *E. macusaniensis* infection is the most common cause associated with colic.

Sequestered chloride can also be detected by analyzing gastric fluid. This fluid may be retrieved by stomach tube or paracentesis. When analyzed correctly, forestomach fluid supernatant appears to have a similar chloride concentration as cattle (normal: < 20 mEq/L; camelids with anorexia: < 30 mEq/L). Higher values can be expected in camelids with obstructions that are more cranial, complete, and long-standing. Values >40 mEq/L almost certainly reflect an obstruction. Smaller increases are seen with strangulating lesions due to the rapid decline in clinical parameters.

Hypokalemia is almost ubiquitous in camelids with gastrointestinal disease, and worsens with duration and severity of anorexia and tendency towards metabolic alkalosis. Thus camelids with small intestinal obstructions have some of the lowest serum potassium concentrations, but this finding should be expected in camelids with other gastrointestinal disorders.

### Other blood constituents

Measures of dehydration and renal function are also likely to be most severe in camelids with small intestinal obstruction and are also in camelids in the shock phase which occurs with many of the inflammatory conditions. Creatinine is of more value than urea nitrogen in some cases, because urea nitrogen is affected by the rate of ruminal utilization. Albumin and serum total protein also tend to increase with dehydration, but camelids appear to lose or catabolize protein with many gastrointestinal disorders. Therefore, high and low albumin concentrations are associated with more severe disease. Additionally, protein is lost due to enteritis. Moderate to severe hypoproteinemia combined with continued passage of soft feces is suggestive of *E. macusaniensis* infection.

### Abdominal paracentesis

Using the right paracostal approach, abdominal fluid can be obtained from most sick camelids. Ultrasound guidance may improve yield but is usually not necessary. Protein concentrations and nucleated cell counts of peritoneal fluid are lowest in camelids with the least damage to the bowel and without peritonitis. Thus, they are most normal with early intraluminal obstructions and non-strangulating entrapments. Abnormal peritoneal fluid with these lesions (often accompanied by abnormal CBC results) indicates progression of the lesion and inflammatory complications. Strangulating obstructions with venous occlusion may increase the volume and red cell count of the peritoneal fluid before they increase protein and white cell counts. Very high (< 4.0 g/dL) protein concentrations and nucleated cell counts (< 10,000 cells/μL) are indicative of inflammatory diseases, as are low cell counts with toxic changes to neutrophils. Bacteria and plant material may be seen with sepsis and visceral rupture. Peritoneal fluid activities of amylase and lipase higher than serum activities may be associated with pancreatitis.

Peritoneal lactate higher than systemic lactate tends to reflect compromise of bowel in general, strangulation, or effusive cellular peritonitis. The greater the divergence, the more severe the lesion. Uroperitoneum is similar in camelids to other species. In general, the more normal the peritoneal fluid, the less severe the lesion. Exceptions to this rule include camelids with mature adhesions, which may have uncorrectable lesions but normal fluid and camelids with certain inflammatory conditions (pancreatitis, streptococcal peritonitis), which may have correctable lesions in spite of very abnormal fluid.

## ABDOMINAL IMAGING

### Radiographs

Plain abdominal imaging studies are of limited usefulness in diagnosing the cause of colic. In adult camelids, most abdominal detail is obscured by feed material in the gastric compartments, and few lesions lead to gas accumulation within the bowel. Neonatal llamas have a higher incidence of abnormal intestinal fermentative disorders and smaller gastric compartments; thus imaging studies of colicky crias may reveal elaborate intestinal gas patterns, or potentially impactions of the forestomach or colon.

Contrast studies have been performed on few occasions. They may be diagnostic with some obstructive lesions, and are useful in distinguishing whether a distended viscus is first compartment or not. Unfortunately, intestinal motility is often impaired, making the interpretation of contrast studies difficult.

### Ultrasound

Ultrasonographic imaging of the abdomen and its viscera may be performed transrectally or transcutaneously. It is ideal for detecting fluid accumulation within the bowel or abdomen. Transrectal imaging is best for the same lesions as are palpated per rectum, namely those in the caudal abdomen. This technique has the advantage of being possible in many camelids that are too small for palpation. Transcutaneous imaging through the right paralumbar fossa shows the same structures, as well as revealing the size and contents of the gastric compartments, and the amount, location, and nature of free fluid.

Ultrasonography allows assessment of bowel contents and motility. Changes in wall thickness are usually subtle. Normal intestines should have intermittent pockets of hypoechoic fluid and good evidence of peristalsis. Obstructed or damaged bowel has larger, stable pockets of fluid with minimal evidence of progressive motility: this is easier to find with lesions of the colon or jejunum due to their accessibility. Cranial obstruction may be inferred on finding fluid distension or atony of the third gastric compartment. Distinguishing obstructed from severely inflamed bowel is difficult; on occasion, the lumen of the inflamed bowel may have gas shadowing and that of obstructed bowel may contain a hyperechoic mass of fiber or intussusception. Inflamed, distended bowel may also show evidence of progressive motility.

Large amount of abdominal fluid are present with effusive peritonitis, severely inflamed bowel, uroperitoneum, venous occlusion with strangulation, and other ascites-causing conditions. The more inflammatory the cause, the more hyperechoic flecks are likely to be present. As mentioned previously, finding large pockets of fluid ultrasonographically aids in successful paracentesis. Ultrasonography may also be used to detect uterine torsion, urolithiasis, renal disease, liver disease, mass lesions, and rarely pancreatitis.

## TREATMENT

### Surgery vs medical therapy

Making the decision to surgically explore a camelid with an acute abdomen can be difficult. Historically, there has been some reluctance to perform surgery, mainly because of the perception that most camelids with colic (including many with intraluminal obstructions) eventually have uncorrectable lesions. This perception has been reinforced by poor surgical survival rates. However, improvements in diagnostic modalities, anesthetic management and clinical decision-making have led to many successful surgeries.

Our experiences with camelids are far less extensive than those with horses and cattle, but using similar criteria to those used in these species appears to give us reasonable guidelines for when to do surgery in camelids with colic. This includes clinical signs that worsen or do not improve with medical treatment. Deciding how long to attempt medical treatment before resorting to surgery is difficult. Although there is evidence that some intraluminal obstructions resolve with medical treatment, there is also ample evidence that delaying surgery in camelids with intraluminal obstruction can be detrimental; many intestinal ruptures found at surgery or necropsy occur at the site of a chronic intraluminal obstruction. Based on our best estimates, gut friability becomes life-threatening within 4 days of obstruction. Because cranial obstructions usually cause less overt signs than caudal obstructions, damage to the bowel wall may be more advanced on first examination of camelids with cranial obstructions. Abnormalities in the leukogram or peritoneal fluid and signs of sepsis often accompany bowel compromise.

Rising awareness of *E. macusaniensis* infection as a cause of colic has complicated the use of pain as the major determinant for surgical exploration. Camelids with parasitic enteritis tend to pass some abnormal feces, have blood work suggestive of enteritis, and have ultrasonographic evidence of progressive motility.

## MEDICAL TREATMENT

Medical treatment of colic focuses on the principles of patient stabilization and maintenance, pain control, treatment of infectious and inflammatory conditions, and occasionally attempts to resolve the obstruction.

Patient stabilization involves shock treatment and restoration of hydration, electrolyte, acid-base, and protein norms. Shock treatment is most vital with strangulation and inflammatory lesions. Intravenous fluids are key. An initial bolus of up to 5% of body weight (10% in neonates) may be administered, if the patient will tolerate it, and is not too hypoproteinemic. With alkalosis, 0.9% sodium chloride may be used, but with most disorders, a polyionic, buffered fluid is adequate. Sodium bicarbonate may be added to correct severe acidosis (pH <7.25), but this treatment is not particularly effective in the face of lactic acidosis. Supplemental potassium chloride, administered at <0.5 mEq/kg/h may also be helpful. Because of the small patient size and rarity of severe hyponatremia or hypochloremia, hypertonic saline is rarely necessary. Plasma transfusion or administration of a synthetic colloid may be necessary to treat hypoproteinemia. I typically administer plasma if the blood albumin concentration is <2.0 g/dL or total protein is <4.0 g/dL.

After the initial fluid bolus, I typically continue intravenous fluids at 2 ml/kg/h in adults and up to 4 mg/kg/h in neonates until the camelid is able to obtain its own water. Stabilization also involved common sense procedures such as protection from the elements, removal from stressful circumstances, provision of heat, water, and possibly feed, and protection from other animals. A bedded, sheltered stall with a controlled temperature serves nicely.

Pain control can be difficult to achieve and has the temporary danger of masking a surgical lesion. Flunixin meglumine (1.1 mg/kg, IV, q12h) or butorphanol tartrate (up to 0.05 mg/kg, IV or SC, q 4 to 6h) offer relief from mild pain, but are usually inadequate to control colic. In one case under my care, butorphanol appeared to allow sufficient patient relaxation for an alpaca to pass an enormous (>5cm in diameter) bezoar, but that is rare. More aggressive pain management includes Lidocaine (1.3 mg/kg IV over 15 minutes as a loading dose, followed by 0.05 mg/kg/min), ketamine (40mcg/kg/min), or fentanyl as constant rate infusions. Other options include epidural morphine (0.1 to 0.3 mg/kg of preservative-free morphine diluted to 12 ml), Buprenorphine (3 mcg/kg IM q4-6h), or transdermal fentanyl (300 mcg/h in an adult llama). Alpha-2 adrenergic agonists generally should be avoided due to their contribution to cardiovascular compromise and displacement of the soft palate.

Flunixin meglumine is also useful as an anti-inflammatory. I typically use 0.5 mg/kg, IV, q 12h, unless circumstances justify a higher dose. Antimicrobials round out the medical treatment plan. Broad spectrum antibiotics, such as pencillin-aminoglycoside combinations may be augmented with metronidazole to increase the anerobic coverage. Ceftiofur sodium may not have sufficient distribution to the peritoneal cavity to be effective there. If coccidia are suspected, the appropriate medication should be used.

Oral and parenteral agents to help move on obstruction have shown little promise in my hands. Oral agents tend to sit in the atonic, distended stomach, and parenteral agents cannot move the mass.

Determining how long to attempt medical treatment requires careful assessment of the patient. If I cannot control the pain, I am reluctant to delay surgery more than 24 hours in any camelid showing evidence of obstruction and diminishing fecal passage. To treat even that long, the camelid must show no evidence of bowel degradation on physical, blood, or peritoneal fluid examination. If the camelid continues to pass diarrhea or soft feces, possibly has a fever, and ultrasonography reveals evidence of progressive motility, the high prevalence of *E. macusaniensis* infection in our area often warrants extending medical treatment for 2 to 3 days.

Criteria for surgery include persistent pain and lack of feces, finding evidence strongly suggestive of an obstructive lesion, entrapment, or strangulation, and sufficient confidence in ruling out inflammatory conditions such as peritonitis, pancreatitis, enteritis, and hepatitis. Evidence for a small intestinal obstruction includes hypochloremic metabolic alkalosis, a high forestomach chloride concentration, abdominal distension, and possibly regurgitation or vomiting. Evidence for a distal small intestinal or colonic obstruction includes rectal or ultrasonographic identification of fluid-distended loops of bowel without good progressive motility in camelid showing overt pain signs. Characteristics of volvulus include initial high levels of pain, seroanguinous peritoneal fluid, and high concentrations of lactate in blood and especially peritoneal fluid.

## SURGICAL APPROACHES

Ventral midline, paracostal, and paralumbar celiotomy all have been performed successfully to diagnose and treat gastrointestinal disorders in camelids. Left paralumbar laparotomy offers the best exposure to the first gastric compartment, but poor exposure to the remaining abdominal viscera, and thus is not recommended unless the lesion is thought to affect those compartments. The choice between ventral midline and right paralumbar exploratory laparotomy often is based on the preference of the surgeon, as both approaches allow for palpation, visualization, and exteriorization of most abdominal viscera, including the spiral colon, cecum, most of the small intestine, and the pyloric antrum of the third gastric compartment. The pyloric antrum may be difficult to exteriorize through a ventral midline incision when the third gastric compartment is distended with fluid. Therefore, a right paracostal or paralumbar incision may be preferable when a gastric or duodenal lesion is suspected. Herniation of gastrointestinal viscera through the body wall incision may occur after ventral midline laparotomy, although we have noted this to occur more frequently after caesarean section in large female llamas than after general surgical exploration of the abdomen of camelids.

## REFERENCES

[1.] Garry F. Proceedings. **11th Annu ACVIM Forum** 1993;817. [2.] Cebra CK, et al. **Vet Surg** 1998;27:112. [3.] Cebra CK. Proceedings. **15th Annu ACVIM Forum** 1997;447. [4.] Navarre CB, et al. **Analysis J Am Vet Med Assoc** 1999;214:812. [5.] Pearson EB. Proceedings. **13th Annu ACVIM Forum** 1995;671. [6.] Cebra CK, et al. **J Am Vet Med Assoc** 1997;211:600. [7.] Sullivan EK, et al. **Vet Surg**. 2005;34(5):524. [8.] Cebra CK, et al. **J Am Vet Med Assoc**. 2008;232(9):1357.

# GASTROINTESTINAL DISORDERS IN CAMELIDS – CLINICAL EVALUATION & TREATMENT

**J. Lakritz DVM, PhD, DACVIM, C. Gerspach, Dr.med.vet, MS.**
**Columbus, OH**

## INTRODUCTION

Gastrointestinal (GI) disorders and parasitism are major health concerns of South American Camelid (SAC) producers in the USA.(Cebra *et al* 1998h,Cebra and Stang 2008b) Most commonly observed GI disorders in SAC are gastro-intestinal parasitism caused by nematodes (*Trichostrongyles*, *Trichuroids*, *Nematodirus spp.*) and coccidia (*Eimeria spp.*, *Cryptosporidium spp.*).(Cebra and Stang 2008a) Intestinal cestodiasis is only of significant concern in European camelid populations.(Hertzberg and Kohler 2006)

Commonly diagnosed disorders in SAC evaluated at our hospital include gastrointestinal parasitism, ulceration, intestinal obstruction due to plant or fiber impactions, fiber impactions of C1 or C3, displacement or volvulus of the colon or small intestine, megaesophagus and neoplasia.

### Gastrointestinal anatomy and physiology

The camelid digestive tract has some functional similarities with true ruminants; however, few anatomic similarities exist. The 1$^{st}$ compartment lacks papillae and is non-keratinized. There are distinct areas of sacculation containing glandular epithelium. Compartmental contractions are not palpable, and are more frequent (3-4 contractions/minute) than ruminants.(von Engelhardt 1998a) C1 epithelium is absorptive, removing water and other components leaving the ingesta homogenous and relatively dry.(Fowler ME 1998) The second compartment has a ventricular groove on the lesser curvature which functions as the esophageal groove. The glandular areas are divided by crests forming a net.(Fowler ME 1998) The mucosa of C3 is entirely glandular, with a net-like pattern on the lesser curvature of proximal C3.(Fowler ME 1998,Vallenas *et al* 1971b) From proximal C3 to distal, 40-50 longitudinal mucosal folds project into the lumen.(Fowler ME 1998,Vallenas *et al* 1971a) The distal 3$^{rd}$ compartment contains true gastric glands and the pH is $<2.5$ in the distal aspect of C3.(Fowler ME 1998) Rumination is initiated with contraction of the cranial sac of C1, inspiration of air, closure of epiglottis, a pressure drop within the thorax propelling feed into the esophagus that is transmitted aborally.(von Engelhardt 1998b) Eructation of gases occurs with contraction of the caudal sac of C1 and relaxation of the cranial region of C1.(von Engelhardt 1998c) As camelids possess C1 flora similar to ruminants, analysis of gastric fluids can be performed.(Belknap 1994) Measurement of pH (normal pH 6-7), evaluation of bacterial numbers and types (gram stain), methylene blue reduction time (normal 3-6 minutes) are similar to that observed in ruminants on a roughage diet.(Belknap 1994)

The duodenum starts as an enlargement (ampulla) extending cranially and dorsally to the upper quadrant of abdomen immediately caudal to liver. At the junction of the duodenum-jejunum, the jejunum curves medially and cranially around the mesenteric root and becomes the ileum coursing caudally and dorsally to join the cecum and colon at the cecocolic orifice. The cecum is present mainly on midline and its apex is directed caudally. Cranially to the cecum, the ascending colon moves forward and ventrally entering the proximal loop of the spiral colon. After 5 centripetal coils, there are 4 centrifugal coils. The diameter of the loops diminishes from approximately 3-5 cm in the proximal loop to 1-2 cm in the distal loops.(Fowler ME 1998) The smallest diameter is found at the central flexure which is associated with an 180$^{o}$ reflection of ingesta flow centrifugally out of the spiral colon. The small colon is approximately 2.5 cm in diameter.(Fowler ME 1998)

### Examination of the South American Camelid with GI problems

Two of the most important aspects of the clinical evaluation of the patient revolve around the clinical history provided and the clinical signs the animal demonstrates. Historical information should include signalment, management, purpose (breeding, fiber animal, etc), clinical signs and prior treatment.

Recent transport of breeding females with/without their cria, and treatments provided prior to examination are of significant importance to the history involving GI disease. Historical information regarding altered milk production of the dam is equally important for the history of a cria with depression, anorexia and colic. Evidence of failure to gain weight, reduced rate of gain or recent loss is commonly observed in association digestive disorders (nutritional, parasitism, enteritis). Common clinical signs include: anorexia and depression (change of behavior), straining over the dung pile, increased recumbence, lateral recumbency, rolling, kicking at abdomen, loss of weight or failure to gain weight, abnormal feces (clumped, soft pasty or diarrhea, no feces).(Cebra *et al* 1998g)

The presence of anorexia and altered behavior with or without colic in a febrile camelid is of significant importance to the physical examination. Animals with clinical signs of GI disease associated with tachycardia and septic peritonitis (not associated with rupture or leakage) may appear to be colicky without obstructive lesions.(Cebra *et al* 1998f)

Evaluation of the animals' behavior from a distance while observing the body contour for distension, urination or defecation and general attitude. Observing the animal assume a recumbent positions, especially if not "cushed" is associated with discomfort. Animals repeatedly standing and assuming recumbency, stand and assume lateral recumbency, rolling or both should be evaluated for gastrointestinal disorders. Abdominal distension and palpation of a tense and painful abdomen should be investigated by careful palpation of flanks to determine the presence of palpable loops of intestine or colon as undulations of the abdominal contour. These undulations become visible after clipping the fiber.

Auscultation or percussion of the abdomen in camelids with reduced or absent fecal output or evidence of colic will assist in the determination of abnormal GI content, as is observed with grain engorgement or C1 atony.(Fowler ME 1998) Rectal examination is a valuable tool in those camelids whose body size and rectal diameter allow this. Care should be taken to prevent rectal trauma. Use of liberal amounts of lubricant, sedation, caudal epidural, or rectal infusion of lubricant mixed with 2% lidocaine is advised. Abdominal distension in the SAC is commonly observed as pregnant females attain 9-11 months gestation. However, abdominal pain and excessive abdominal distension in pregnant females should be evaluated for torsion of the uterus. Further, young males which strain should be evaluated for difficulty urinating as a cause of GI signs.

Evaluation of the animals cardiovascular, respiratory and body temperature are obtained initially to evaluate potential for abnormalities in other body systems as a cause of discomfort. Evaluation of mentation and/or neurologic status is also important as camelids often present with aberrant parasite migration (*P. tenuis),* brain abscesses (*Listeriosis*), meningitis or increased intra-cranial pressure (Polio- encephalomalacia).

Depending upon the age of the animal, whether it is a weaned or nursing cria, both the dam and the cria should be carefully examined. In camelids nursing a cria, the presence of mammary inflammation (glands swollen, enlarged and painful) with normal milk, examination of the cria should be performed. Lack of fecal output and distended and taught abdominal contents in the cria may suggest digestive disorders. In other cases, crias from dams not producing enough milk may develop pica and nurse on fiber and develop obstructions.(Fowler ME 1998, Sullivan *et al* 2005a) Furthermore, in crias requiring bottle or tube feeding, GI disturbances associated with putrefaction of milk in C1 may lead to digestive disturbances. Chronically tube feeding crias may present with abdominal distension associated with C1 putrefaction GI obstructions (bezoar).(Cebra *et al* 1998e)

**C1 fluid analysis**

Because of the drier, more homogenous ingesta in C1, fluid collection from camelids is less reliable; however, if obtained, conducting gram staining number and motility of ciliates is possible.(Belknap 1994,Fowler ME 1998) Methylene blue reduction time is also helpful in animals where abnormal types and numbers of organisms are observed. Transfaunation of camelids with GI disturbances is routine in animals receiving antimicrobials or in animals with reduced GI function.

**Fecal examination**

All camelids examined for GI disturbances should have a fecal examination performed for evidence of gastrointestinal parasitism. Using sucrose (specific gravity >1.25) floatation incorporating centrifugation (Modified Stolls method) is appropriate for camelid parasite evaluation.(Cebra and Stang 2008c) Finding even a few *Trichuroid, Capillarid* or *Nematodirus* are significant even if there are few *Trichostrongylid* type eggs. Finding *Trichuroid (eggs passed intermittently, specific gravity 1.15),* or *Nematodirus* eggs is significant in camelids with digestive disorders.

**Imaging of the GI tract – Ultrasound**

Ultrasound imaging of the camelid GI tract has become a useful tool for imaging abdominal viscera of sick camelids. The evaluation should be conducted in systematic fashion, with liver ventral to the right diaphragmatic lung lobe, compartments 2 and 3, right kidney and intestines.(Cebra *et al* 2002b) If one is diligent, compartmental contractions and fluid movement is observed. Imaging of C1 is possible ventrally or on the left side.(Cebra *et al* 2002a) Visualization of sacculated and non-sacculated regions of this C1. The left kidney is superficial to the 1st compartment and immediately caudal to the spleen. There are generally intestinal loops or spiral colon present on the left side of the abdomen caudal to spleen, left kidney and C1. Small intestinal loops are most easily imaged on the right side of the abdomen dorsal and caudal to C3, extending caudally and dorsally to the paralumbar fossa. Intestines are commonly fluid filled, with fluid movement and some segmentation observed. Colon is often imaged as larger diameter loops with more echo-dense fluid in comparison to the small intestine.(Cebra *et al* 2002c) Spiral colon is not normally easily visualized, however it is suggested to be present further caudally near the pelvic inlet.(Cebra *et al* 2002d)

Imaging fluid and visceral edema is possible ventrally between the 1st and 3rd compartments, right paralumbar fossae, and left paralumbar fossae near the left kidney, spleen and C1.(Cebra *et al* 2002e) Collection of abdominal fluid is simplified with US guidance using an established method for camelids.(Cebra *et al* 2008c) Excess fluid or visceral edema is often present in hypoproteinemic animals, animals with intestinal obstructions or septic patients. The presence of a distended abdomen with repeated reverberation lines is suggestive of gas accumulation. Distended bowel or compartments on the right is generally more obvious than on the left side of the abdomen. Ultrasonographic imaging of both kidneys is important to document renal agenesis, congenital cystic kidneys or renal neoplasia which if present, could contribute to clinical signs associated with GI distension and discomfort.

**Imaging of the GI tract – Radiology**

Imaging of the camelid abdomen in a standing position is preferred to lateral recumbency. **Always compare lateral views to an orthogonal view (DV).** In animals with C1 distension, marked cranial displacement of the diaphragm and compression of the thorax or displacement of intestinal loops into pelvic inlet is expected. C1 distension may be the result of distal obstructions (C3, spiral colon, colon) or associated with consumption of poorly digestable feedstuffs or fiber (ectoparasitism and pruritis). In lateral images of C1, small gas pockets or fluid lines superimposed on the C1 compartment are suggestive of abnormalities distally and orthogonal views become critical. The presence of gas distended intestines or colon is associated with obstructive lesions which may be caused by intestinal volvulus, torsion, incarceration, intraluminal obstruction (bezoar), spiral colon impaction or colonic impaction. Determination of whether the distension is associated with enteritis or a surgical lesion should include ultrasound, abdominal fluid and clinico-pathologic evaluation.(Cebra *et al* 1998d)

**Clinical pathology**

Hematologic and serum biochemical profiles and blood gas analyses of healthy camelids and camelids with digestive disorders have been described.(Fowler ME 1998,Hawkey and Gulland 1988,Lassen *et al* 1986,Lewis 1976) **It is notable that camelids with proximal obstructive lesions are metabolically alkalotic, hypochloremic and hypokalemic.**(Cebra *et al*

1998c,Sullivan *et al* 2005b) In contrast, camelids with obstructions of the distal GI tract or GI perforations, these changes are not evident.(Cebra *et al* 1998b) Abdominal fluid analysis of animals with proximal or distal obstructions associated with intraluminal feed or hair impactions are less commonly associated with elevated cell counts or protein concentrations.(Cebra *et al* 1998a) Whereas, animals presenting for enteritis, peritonitis or longstanding obstructions, increased cell count, protein and feed particles or micro-organisms are observed.

In animals with clinical signs attributable to GI dysfunction, US evaluation and paracentesis is valuable in the assessment of the cause or nature of the patients situation and prognosis.(Cebra *et al* 2008b) In some animals, low plasma protein associated with intestinal damage (parasites), renal dysfunction, or neoplasia, large volume effusions may be easily sampled. Using the lateral, paracostal approach fluid recovery is rather easy.(Cebra *et al* 2008a) Cytologic evaluation of peritoneal effusions from animals in association with abnormalities of abdominal viscera to rule out neoplasia is prudent.(Cebra *et al* 1995,Fowler ME 1998,Sartin *et al* 2004,Valentine and Martin 2007) Fine needle aspirates or biopsy of abdominal organs, primarily liver and periodically kidney or lymph nodes is routine in adequately sedated/restrained animals.

## REFERENCES

[1.] Belknap EB (1994), **Vet.Clin.North Am.Food Anim Pract**. 10: 291-307; [2.] Cebra CK, et al (1998a), **Vet.Surg.** 27: 112-121; [3.] Cebra CK, et al (1995), **J.Vet.Intern.Med.** 9: 381-385; [4] Cebra CK, Stang BV (2008c), **J.Am.Vet.Med.Assoc**. 232: 733-741; [5] Cebra CK, et al (2008a), **J.Am.Vet.Med.Assoc.** 232: 1357-1361; [6] Cebra CK, et al., (2002e), **Vet.Radiol.Ultrasound** 43: 359-366; [7] Fowler ME (1998), **Medicine and Surgery of South American Camelids**, Iowa State University Press, Ames Iowa; [8] Hawkey CM, Gulland FM (1988), **Vet.Rec.** 122: 232-234; [9.] Hertzberg H, Kohler L (2006), **Berl Munch.Tierarztl.Wochenschr.** 119: 291-294; [10.] Lassen ED, et al., (1986), **Am.J.Vet.Res.** 47: 2278-2280; [11.] Lewis JH (1976), **Comp Biochem.Physiol A Comp Physiol** 55: 367-371; [12.] Sartin EA, et al., (2004), **J.Vet.Diagn.Invest** 16: 226-229; [13.] Sullivan EK, et al., (2005b), **Vet.Surg.** 34: 524-529; [14.] Valentine BA, Martin JM (2007), **J.Vet.Diagn.Invest** 19: 202-204; [15.] Vallenas A, et al., (1971a), **J.Morphol.** 134: 399-423; [16.] von Engelhardt W (1998), **Dtsch.Tierarztl.Wochenschr**. 105: 472-474

# DIFFERENTIAL DIAGNOSIS OF DIARRHEA IN CAMELID CRIAS

**Pamela G. Walker, DVM, MS, DACVIM**
**Findlay, Ohio**

Diarrhea in camelid crias can be a very frustrating situation for both owners and veterinarians. Crias can lose significant body weight (10 – 14%) in 24 hours and be in critical condition prior to owner recognition. Many causes of diarrhea in crias have been reported including *Cryptosporidium* spp,[1] *Giardia* spp,[2] Coronavirus/Rotavirus,[3] bacterial,[2] and *Eimeria* spp.[4] *Cryptosporidium parvum* occurs in calves greater than 7 days of age but in crias it can be as early as 5 days of age. *Giardia* spp can occur in crias as early as 14 days of age and is correlated most often with periods of high rainfall or problems with run-off water. *Eimeria* spp can be seen as early as 21 days of age[4] with *Eimeria macusaniensis* occurring as early as 36 to 40 days of age.[5] Diarrhea in crias less than 5 days of age is most likely to be nutritional (especially if bottle fed) or Gram negative infections.

## Cryptosporidium

Diarrhea in crias caused by *Cryptosporidium* spp can cause profuse, liquid, green, uniquely foul smelling diarrhea with some crias rapidly losing body weight (10 to 12%) requiring aggressive treatment with IV fluids. There are several ways to diagnosis *Cryptosporidium* in feces; acid fast stain and fecal floatation with Sheathers solution are two methods used with variable results due to the small size of the oocyst, variable staining characteristics and possible low numbers shed at the time the sample is taken. As a result, immunoassays have been developed to aid in diagnosis. One such test is a commercially available lateral flow immunoassay test called ImmunoCard STAT! for Crypto/Giardia distributed by Meridian Bioscience in Cincinnati Ohio. This test is very easy to use with results in 10 minutes at a cost of about $20 per card. When compared to Merifluor DFA (Meridian Bioscience, Inc), considered the Gold Standard by the Centers for Disease Control, the sensitivity for detection of *Cryptosporidium* was 68% with specificity equal to or greater than 99%.[6] This study determined that the ImmunoCard STAT! needed >175 organisms/10 µl for a positive result, and often reported false negative results with small numbers of parasites, especially with *Cryptosporidium.*[6] I have been providing the company with positive samples for several years. Those samples are retested with the Merifluor DFA at the lab with all the samples testing positive.

There are no labeled treatments for *Cryptosporidium* but many different drugs have been proposed, some with more success and others more toxicity. Paromomycin (Humatin®) is a nonabsorbable aminoglycoside that is used in immunocompromised humans,[7] cats[8] and calves[9] in an attempt to treat the infection. The human trial had a response rate of 67% and even though there were few cures, stated the diarrhea improved more with the drug than with the placebo.[7] Grinberg et al (2002) showed that neonatal calves started to shed oocysts and develop diarrhea only after the drug was withdrawn, suggesting strong anticryptosporidial effect.[9] Viu et al (2000) demonstrated a significant reduction of oocysts in lambs after being treated with paromomycin at 100 mg/kg, PO for 3 days, plus resolution of diarrhea immediately after the medication stopped.[10] In the author's experience, in a clinical setting, with natural occurring cases of diarrhea caused by *Cryptosporidium,* use of **Paromomycin at 50 mg/kg, PO for 5 to 10 days** has been very successful in shortening the duration and severity of diarrhea in crias. (personal communication) Other drugs used to treat diarrhea caused by *Cryptosporidium* (lasalocid, nitazoxanide, decoquinate or halofuginone) have a narrow margin of safety when used with camelids or access to the drug is limited.

## Giardia

*Giardia* causes soft, runny, slimy, slightly green feces that may have a frothy appearance, on occasion it can result in profuse, watery diarrhea. Some crias will exhibit consistent straining to defecate before the onset of diarrhea. *Giardia* cysts can be difficult to find in fecal floatations depending on the skill of the microscopist, so immunoassays such as the ImmunoCard STAT! above can be more reliable with quick results. In the Merifluor DFA comparison study, sensitivity for the detection of Giardia was 81% with specificity equal to or greater than 99%.[6] Giardia is seen on farms where run off water is a problem or when water accumulates from use of hoses and sprinklers to cool off the animals in the summer. The dams will lay in the puddles, contaminating their udder, then the cria will nurse and ingest the oocysts if present; also crias seem to have an affinity for water puddles! Diarrhea from Giardia is usually seen in the fall and spring, but the oocysts can overwinter and survive at 4C for several months. Once ingested, the trophozoite is released from the oocyst and invades the small intestines causing villous atrophy and subsequent malabsorptive diarrhea. Giardia diarrhea is an unusual finding in adults and is generally considered present on farms with overcrowded barns and pastures with a large population of young, immunologically naïve animals.[11] (Cebra 2003). While this is true, the curious and very oral nature of an active cria also contributes to exposure to many parasites. Camelid crias can be treated with **Fenbendazole at 50 mg/kg, PO, SID for 5 days** for diarrhea caused by Giardia. The author has also used **Metronidazole at 50 mg/kg, PO, BID, 5 – 7 days** without any problems, but only use in crias less than 2 months of age.

Diarrhea caused by *Cryptosporidium* and *Giardia* can be very frustrating and costly on some camelid farms. Most camelid owners clean their barns and even pastures daily, sometimes twice daily and are frustrated by still having crias develop diarrhea. Rulofson et al (2001) looked at 354 llamas from 33 different farms in California, ages from 3 weeks to 23 years. Using the Merifluor DFA test, 12 (3.4%) of the specimens were positive for *Giardia* and no specimens were positive for *Cryptosporidium.*[2] His conclusions were that farms with larger numbers of crias born the previous year would result in larger numbers of yearlings, which provided the source of animals shedding the parasites, keeping continual exposure for young crias.[2] A study in Spain looked at cows, sheep and goats, aged 3 to 13 years old. The study found *Cryptosporidium* in 8.4% of cows, 5.3% sheep and 7.7% goats and *Giardia* in 26.6% of cows, 19.2% sheep and 19.8% of goats.[12] Supporting the idea that neonates of many ruminant species are exposed to these parasites by subclinical carriers. There is no reason to conclude alpaca farms would be different.

**Viral diarrhea**

Both coronavirus and rotavirus have been recognized as causes of diarrhea in crias for many years.[13,14] Parreño et al (2001) determined that crias as young as 2 days of age can have severe diarrhea caused by rotavirus. In the same survey, it was determined that all the adult guanacos sampled were positive for antibodies to rotavirus.[3] We are also seeing diarrhea caused by both coronavirus and rotavirus on alpaca farms in North America.[11] Recently Jin et al (2007) sequenced the genome of an alpaca coronavirus (ACoV). He determined that the sequence of ACoV was closely related to bovine coronavirus.[15] In an attempt to prevent viral diarrhea alpaca owners are using oral vaccines made for calves; these vaccines are given orally to the newborn cria before nursing. Based on the findings of Jin et al (2007), and clinical experience, it appears as if these vaccines should be effective. The author has been using Pfizer's Calf-Guard® 3 mL, PO at birth before the cria nurses for 5 years without problems with a decrease of apparent viral diarrhea. In the past two years in the United States, alpaca owners have become very familiar with viral diarrhea. There have been several post-show outbreaks of both coronaviral and rotaviral diarrhea 3 to 5 days after returning from different shows. Most of these cases were self-limiting and could be controlled with common anti-diarrheal drugs (Kaolin, Imodium); however several fatalities also occurred (personal communication). In neonates, vial diarrhea tends to be high volume and high frequency and frequently requires intravenous fluids to maintain hydration in addition to other supportive care. It is important to rule out other pathogens that may also be present as is the case in calves.

**Bacteria diarrhea**

*E. coli* and *Clostridium perfringens* spp can cause diarrhea in crias, however *Salmonella* spp is not common.[2,10] *E. coli* diarrhea is often associated with neonatal sepsis and can be secondary to failure of passive transfer.[16] *E. coli* diarrhea can also occur secondary to viral diarrhea in older neonates as happens with calves. *E. coli* diarrhea typically occurs between 3 to 7 days of age, usually before any of the other more common causes of diarrhea can start. The diarrhea may be profuse and liquid, with the cria very lethargic and may have abdominal distension. Crias can be leukopenic, but unlike calves are usually hypernatremic.[17] If the cria is also septic, blood glucose is usually low, but glucose concentration should always be verified before any dextrose is administered as the cria may be hyperglycemic. Treatment should be aimed at correcting the dehydration and electrolyte imbalances and broad spectrum antibiotics administered. The author has had two cases of *Clostridium perfringens* type A secondary to Corona virus diarrhea in weanling crias. Both crias were given Procaine Penicillin G 44,000 IU/kg, SC, SID until the diarrhea resolved. Neither cria seemed to have any secondary side effects and both fully recovered. Many owners use oral vaccines made for calves to prevent diarrhea caused by *E. coli*; The author has used Boehringer Ingelheim's Bar-guard 99®, 5 mL, PO 3 to 5 hours after birth for years without problems.

**Coccidiosis**

There are six species of *Eimeria* currently recognized that can infect camelids. Diarrhea from "regular" coccidia (*E. alpacae, E. lamae, E. punoensis, E. peruviana*)[18] can occur as early as 21 days of age. *E. macusaniensis* (E. mac) or Large coccidia has a longer life cycle, with a prepatent period of 32 to 36 days.[5] Concurrent, fatal infection with *E. ivitaensis* has also been known to occur.[19] Diarrhea from coccidia is rare in adults, with the exception of E. mac that can cause debilitating, even fatal disease in naïve animals even before the development of diarrhea. "Regular" coccidia can be easily diagnosed with fecal floatation in any clinic. E. mac can be more difficult to find unless proper fecal analysis is used, sometimes serial analysis of fecal samples are required. A recent article comparing several techniques and floatation times concluded that the centrifugation-floatation technique, using concentrated sugar (specific gravity = 1.27) and a 60 minute floatation time was superior in detecting parasites that have the potential for causing problems in Camelids.[20] Treatment of "regular" coccidia is **Sulfadimethoxine 15 mg/kg, PO, BID X 5 days or Amprolium 10 mg/kg, PO, SID X 5 days.** With either treatment, caution the owner to monitor for signs of Polioencephalmalacia. Both "regular" coccidia and E. mac can produce a dark brown/black diarrhea. Usually it is a low volume diarrhea, but with E. mac there may be significant diarrhea, especially in adults. Which drugs to use when attempting to treat E. mac is controversial. Ponazuril (Toltrazuril-sulfone - Marquis®) has been used for several years to treat E. mac infections. No direct research has been done on camelids to date; Kritzner et al (2002) looked at absorption of Ponazuril in calves experimentally infected with Neospora. Absorption of Ponazuril was rapid with serum concentration apparent at day two of treatment with the half life of the active substance 9.8 days. When comparing 1 day versus 6 days of treatment at 20 mg/kg, PO the maximal peak was approximately 3 to 4 times higher in the 6 day treatment than in 1 day medicated calves.[21] Ponazuril was used to successfully treat *Isospora amphiboluri* infection in lizards.[22] This information suggest that Ponazuril would be absorbed in camelids and could be used to treat E. mac infections. The current recommendation is to treat with **Ponazuril 20 mg/kg, PO, SID X 3 days.** Because E. mac can cause clinical disease, even death before oocysts are present in feces,[23] prophylactic treatment with Ponazuril should be considered in camelids that have unexplained weight loss with concurrent hypoproteinemia and without severe anemia. Marque® is too concentrated to be given to camelids and should be diluted before administered. The pharmacist at Ohio State University recommends diluting the paste to 100 mg/mL by taking 40 grams of paste, q.s with distilled water to 60 grams total and mixing well. It is very water soluble and can be easily syringed into the animal.

Owners should be encouraged to weigh their crias daily until about 23 kg, then weekly until weaning. Weight should also be monitored weekly after weaning and other times of stress (shows, transport, etc). Initially crias with mild to moderate diarrhea may not show lethargy and unless the owners are weighing the crias daily they may not be aware of a significant loss in weight (0.45 kg to 0.9 kg). In some situations the owners may be unaware the cria has diarrhea unless they are closely monitoring the crias as the projectile nature of some diarrhea will not initially cause any staining of the perineal area. The characteristic of the feces can range from soft formed pellets to watery, projectile diarrhea. Due to the severity and rapidity of onset, it is important to attempt an "on the farm" diagnosis. It is also important to assess the degree of dehydration. It can be hard to determine dehydration in crias as they can have dry mucous membranes even if healthy and you can not tent their skin as in other ruminants. Capillary refill time and analysis of daily weights are the best way to determine dehydration on the farm. Careful consideration of the crias age is important when attempting to determine the cause of the diarrhea. In addition to the drugs mentioned above for specific causes of diarrhea, other anti-diarrheal drugs can be safely used in camelid crias. For low volume, normal frequency diarrhea Kaolin at 10 – 15 mL can be given SID to BID. For higher volume, increased frequency diarrhea liquid Imodium at 3 to 5 mL can be given SID to BID in addition to the Kaolin. For diarrhea that is so severe that the cria is still losing weight, despite IV fluids (probably viral in nature) conservative use of Deliver® electrolytes (4 – 5 oz, PO, SID via esophageal tube) can be used to prevent further dehydration. For "on the farm" diagnosis the chart below is based on the age of the cria to work through diagnosing the type of diarrhea. Early detection and aggressive treatment is essential for success.

**Assessment of Cria Diarrhea ≥ 5 days old**

- Check current and recent weights
- If ≥ 5 days old, run Crypto/Giardia test
  1. If negative, less than 3 weeks old, and diarrhea not severe:
     a. do nothing **OR**
     b. give active culture (plain) yogurt **AND/OR**
     c. give Kaolin (5 – 10 mL) and/or Imodium (3 mL) BID
  2. Monitor weight SID to BID
  3. Recheck if diarrhea continues
  4. If negative, less than 3 weeks old, and diarrhea moderately severe:
     a. give oral electrolytes (Pediolyte)
     b. treat with Metronidazole (for enteritis)
     c. give Kaolin (5 – 10 mL) and/or Imodium (3 mL) BID
  5. Submit sample for viral EM screening
  6. If negative and greater than 3 weeks old, check for Regular coccidia and treat with Sulfa if coccidia present
  7. If negative and greater than 1.5 months old, check for regular parasites and E. mac and treat accordingly
- If non – stop, liquid diarrhea, start on IV fluids and appropriate medications.
- If diarrhea persists and cria is still losing weight, use Deliver® electrolytes (4 – 5 oz, PO, SID via esophageal tube).
- If fecal is negative and cria greater than 1.5 months old, and there is E. mac on the farm, treat with Ponazuril anyway.
- Early treatment is KEY, crias can lose 10 – 14% of body weight overnight and will die if not given fluids (IV preferred for severely dehydrated crias (> 8 – 10%).

## REFERENCES

[1]Hovda,L.,et.al. **JAVMA**,1990;196(2)319; [2]Rulofson,F.,et.al. **AJVR**,2001;62(4)637; [3]Parreño,V., et.al. **J.Vet.Med.B**.,2001;48,713; [4]Rickard,L.,et.al. **Protozool**,1988;35(3)335; [5]Rohbeck,S., et.al., **Biology** E.mac in Llamas (abstr)Proc.4th **Eur Symp** SA Camelids,2004; [6]Johnston,S., et.al. **J Clin Micro**,2003;41(2)623; [7]Hashmey,R., et.al. Medicine,1997; 76(2)118; [8]Barr,S.,et.al. **JAVMA** 1944;205(12)1742; [9]Grinberg,A., et.al. Vet Rec,2002;151,606; [10]Viu,M.,et.al., **Vet Parasitol**,2000;90,163; [11]Cebra, C., et.al. **JAVMA**,2003 223(12)1806; [12]Castro-Hermida,J.,et.al. Parasitol Res,2007;101,1443; [13]Mattson,D., et.al. **VCNA-Food Animal**,1994,10,345; [14]Puntel,M.,et.al., **J.Vet.Med**,1999; 46,157; [15]Jin,L.,et.al.**Virology**,2007;365,198; [16]Adams,R.,et.al. **JAVMA**,1992;201(9)1419; [17]Whitehead,C., et.al. **Small Rumin Res**,2006; 61,207; [18]Fowler,M.,1998 **Med Surg** SA Camelids; [19]Palacios,C.,et.al. **Vet Rec**,2006;158,344; [20]Cebra,C.,et.al. **JAVMA**,2008;232 (5):733; [21]Kritzner,S.,et.al. **Ann Clin Microb Antimicrobials**,2002,1:4; [22]Bogoslavsky,B., **Proc Assoc Reptil Amphib Vet**, NewOrleans,LA,2007,8; [23]Cebra, C., et.al. **JAVMA**,2007;230(1)94.

# IMMUNE EVASION BY PATHOGENS OF BOVINE RESPIRATORY DISEASE COMPLEX

**Subramaniam Srikumaran, BVSc, PhD, Clayton L. Kelling, DVM, PhD, and Aruna PN. Ambagala, BVSc, PhD**
**Pullman, WA**

## INTRODUCTION

During evolution, the vertebrates have acquired several sophisticated immune mechanisms to combat the multitude of pathogens that they encounter during their life time. Such mechanisms include components of both innate and adaptive immune systems. Simultaneously, the pathogens too have evolved equally sophisticated mechanisms to suppress or evade the host immune responses. Thus the hosts and the pathogens have co-evolved over millions of years of evolution. The pathogens of bovine respiratory disease complex (BRDC) are no exception to this phenomenon. BRDC is a multifactorial disease with the involvement of several viruses and bacteria. The viral pathogens include bovine herpesvirus 1 (BHV-1), bovine respiratory syncytial virus (BRSV), bovine viral diarrhea virus (BVDV), and parinfluenza-3 (PI-3) virus. The bacterial pathogens involved in this disease complex are *Mannheimia haemolytica, Pasteurella multocida, Histophilus somni* and *Mycoplasma bovis*. Concurrent infection with two or more of these pathogens occurs frequently, and an evasion mechanism developed by one pathogen often benefits another. The same strategy can be developed by more than one pathogen as well. Hence we will discuss the different strategies under the respective categories. This presentation will focus on immune evasion mechanisms developed by BRDC pathogens that have broad impact on the pathogenesis of this complex disease. For a comprehensive description of immune evasion mechanisms of BRDC pathogens, the reader is referred to a recent review article on this subject[1]. Both humoral and cell-mediated immune responses are driven by cytokines secreted by the $CD4^+$ T-helper ($T_h$) cells, and hence it is appropriate to begin with evasion mechanisms evolved by pathogens to interfere with $T_h$ cell function.

## INTERFERENCE WITH T-HELPER CELL FUNCTION

The $CD4^+$ $T_h$ cells recognize peptides presented by MHC class II molecules expressed on antigen presenting cells[2]. This recognition induces the production by $T_h$ cells of cytokines required by macrophages, NK cells, B cells, $CD8^+$ cytotoxic T-lymphocytes (CTLs) and other proliferating T cells. Thus, the recognition of class II-presented peptides by $T_h$ cells would have dire consequences to the pathogens. Therefore, several pathogens have evolved strategies to interfere with $T_h$ cell function. BHV-1 induces the loss of CD4 molecules in cultured $CD4^+$ $T_h$ cells, and subsequently induces apoptosis of these cells[3]. In vivo experiments have validated these in vitro observations[4]. Apoptosis of $CD4^+$ $T_h$ cells occurs in peripheral blood and lymph nodes during acute BHV-1 infection. Down-regulation of expression of $CD4^+$ molecules or apoptosis of $CD4^+$ $T_h$ cells by BHV-1 would impair $T_h$ function and cytokine signaling which in turn would compromise the development of humoral as well as cell-mediated immune responses against not only BHV-1 but also the other pathogens of BRDC.

## DOWN-REGULATION OF EXPRESSION OF MHC CLASS II MOLECULES

Leukotoxin secreted by *M. haemolytica* has been reported to down-regulate the expression of class II molecules on bovine monocyte-derived macrophages[5]. Unfortunately, these important observations regarding the down-regulation of MHC class II expression by leukotoxin have not been extended by these authors or confirmed by any others up to now. It is possible that the observed effects are unique to the class II alleles from the two animals tested. If the expression of class II molecules is in fact down-regulated, activation of $T_h$ cells would be affected which in turn would affect the development of humoral as well as cell-mediated immune responses. In this regard, not only the antibody response to leukotoxin but also to all other antigens from other pathogens of BRDC would be affected. It would also affect the development of CTL response to the viral pathogens of BRDC.

## DOWN-REGULATION OF EXPRESSION OF MHC CLASS I MOLECULES

Down-regulation of expression of MHC class I molecules on the surface of infected cells is perhaps the most important immune evasion mechanism evolved by viruses. Neutralizing antibodies at the portal of entry constitute an important component of anti-viral defense mechanisms of the host, since they could prevent the entry of viruses into the host by neutralizing them. However, once the viruses breach this line of defense and enter the host cells, anti-viral antibodies are not very useful since they cannot get into the cells. Although the antibodies can mediate antibody-dependent cell cytotoxicity, this process is considered an inefficient process[6]. CTLs are the most effective component of immunity against most viral infections, particularly against herpesviruses which spread from cell to cell earlier than via the hematogenous route[7]. The CTLs recognize intracellular viral antigens presented by the self MHC class I molecules on the cell surface[8]. Efficient CTL response to a viral infection is orchestrated by a highly intricate multiple-step pathway of antigen processing and presentation which involves several intracellular organelles and chaperones. Therefore, several viruses including BHV-1 have evolved strategies to interfere at different steps of this pathway to down-regulate expression of MHC class I molecules as a means to escape CTL recognition[9].

The assembled class I molecules (heavy chain and $\beta_2$ microglobulin) present peptides originating from proteins synthesized in the cytoplasm. In a normal cell, the proteins which have completed their lifespan, as well as the abnormal proteins are degraded by the proteasomes, following ubiquitination of the target proteins[10]. The peptides, which are usually 8-10 amino acids long, are transported into the ER by the heterodimeric transporter associated with antigen processing (TAP) consisting of TAP1 and TAP2 subunits[11]. The chaperon, tapasin, bridges the TAP with the class I molecule, which facilitates the transfer of peptides onto the peptide binding groove of class I molecule. The trimolecular complex ($H/\beta_2m$/peptide) egresses from the ER, and translocates to the cell membrane via the secretory pathway. In cells infected with a virus, the viral proteins are processed in the same manner as the cellular peptides, and the viral peptides are presented by the class I molecules on the surface of the infected cells. The class I molecules presenting the viral peptides signal the virus-specific CTLs to recognize and lyse the virus-infected cell.

Studies in our laboratory revealed that BHV-1 down-regulates the expression of MHC class I molecules on infected cells[12,13]. Synthesis of the class I H chain as well as assembly/transport of class I molecules are affected by BHV-1 infection. Experiments with metabolic inhibitors revealed that an immediate-early/early protein of BHV-1 may be responsible for the down-regulation. Subsequent studies employed peptide transport assays, and confirmed that BHV-1 interferes with the TAP[14]. Results from studies conducted with BHV-1 in a human melanoma cell line[15] confirmed the findings with bovine cells described above. Subsequent studies have identified the protein encoded by BHV-1 UL49.5 as the protein responsible for TAP inhibition[16]. Inactivation of TAP by UL49.5 is by a two-stage process: inhibition of peptide transport, and degradation of TAP by proteasomes. The current model proposes that the interaction of ER luminal domain and transmembrane domain of UL49.5 with the TAP complex inhibits the conformational change that is necessary for the peptide-induced ATP-dependent translocation. The cytosolic tail of UL49.5 links the TAP heterodimer-UL49.5 complex to the ubiquitin proteasome pathway for degradation[16]. UL49.5-induced TAP inhibition and the consequent down-regulation of cell surface class I expression strongly reduces antigen presentation to $CD8^+$ CTLs[16]. Down-regulation of MHC class I expression by BHV-1 would compromise the development of CTL response against not only BHV-1 but also BRSV, BVDV, PI-3, and possibly any other hitherto un-identified viral pathogen of BRDC.

**SHUT-OFF OF HOST CELL PROTEIN SYNTHESIS**

BHV-1, like other alphaherpesviruses, encodes a structural protein which induces the rapid shut-off of host cell protein synthesis[17]. This so-called 'virion host shutoff' (vhs) protein inhibits host cell protein synthesis by inducing degradation of cellular mRNAs. BHV-1 UL41 gene product is necessary and sufficient for shut-off of host cell protein synthesis at early times post-infection[18]. Significant down-regulation of mRNA for class I molecules also occurs, which raises the question as to whether the BHV-1-induced down-regulation of class I molecules is due to the vhs activity. Studies with vhs-deletion mutants of BHV-1 by us[19] and others[18] revealed that the reduction in the synthesis of class I molecules is due to the vhs activity, but the TAP inhibition and retention of class I molecules in ER by BHV-1 are mediated by a protein other than the vhs, which has been shown to be UL49.5[16].

**INTERFERENCE WITH INTERFERON SYSTEM**

IFNs induce an anti-viral state and up-regulate immune functions by binding to type I or type II receptors on the cells, and activating the JAK/STAT signal transduction pathway[20]. IFN-induced anti-viral state and up-regulation of immune functions are detrimental for the replication and dissemination of the viruses, and hence several viruses have developed mechanisms to interfere with IFN function.

BHV-1 inhibits IFN-β-dependent transcription[21]. The bovine infected cell protein 0 (bICP0) induces proteosome-dependent degradation of interferon response factor 3 (IRF3), a transcription factor essential for interferon promotor activation. The zinc RING finger and the sequences near the C-terminus of bICP0 are necessary for the inhibition of IRF-3-induced IFN-β promoter activity and IRF3 degradation[22].

BRSV non-structural proteins NS1 and NS2 mediate virus escape from cellular anti-viral mechanisms induced by type I IFN[23]. NS1 and NS2 also prevent induction of IFN-β after virus infection or double stranded RNA stimulation. NS protein inhibition of IFN induction is mediated by inhibition of IRF-3 activation. Block of induction of type I IFN by NS proteins correlates with increased virulence of BRSV.

Neither the noncytopathic (ncp) BVDV[24] nor the cytopathic (cp) BVDV[25] infections induce detectable type I IFN responses in cell cultures, to avoid stimulation of innate immune responses in cell culture. BVDV $N^{PRO}$ inhibits production of type I IFN by specifically blocking the activity of IRF-3, preventing IRF-3 binding to DNA, enabling proteosomal degradation.

**INFECTION OF HOST CELLS**

Perhaps, BHV-1 infection of respiratory epithelial cells (rhinotracheitis), which constitute the first line of defense against the pathogens of BRDC, is more important than infection of other cell types. Lesions include loss of cilia and goblet cells leading to epithelial erosions, which could progress to necrosis of epithelium and adjacent lymphoid tissue[26]. This sequence of events favors the migration and colonization of the lower respiratory tract by bacterial pathogens such as *M. haemolytica* which occur as commensals of the nasopharynx of healthy animals. BHV-1 can infect many other cell types in cattle, but infection of lymphocytes contributes to immunosuppression more than the other cell types.

Experimental BRSV infection of calves is associated with reduction in proliferation of lymphocytes stimulated with mitogens[27]. Depression of lymphocyte responses to the mitogen phytohaemagglutinin in lambs experimentally infected with BRSV has been shown to be associated with $CD4^+$ lymphocytes. BRSV F protein mediates inhibition of mitogen-induced T-cell proliferation in vitro.

Immunosuppression due to BVDV infection could be related to tropism of virus for APCs. ncp BVDV infection of monocytes causes reduction in the proliferation of memory $CD4^+$ T cells when mixed with these APCs[28]. cp BVDV infection of monocytes in vitro results in cell death, while dendritic cell viability is not affected by cp BVDV infection. A difference in kinetics of the development of a specific T cell response after ncp and cp BVDV infection is evident. T cell proliferative responses are not reproducibly detectable until about 6-8 weeks after ncp BVDV infection, but are detectable by 3-4 weeks after infection with cp BVDV. BVDV infection invariably results in leukopenia leading to decreased CTLs and Th cells[29]. Both cp and ncp BVDV isolates induce apoptosis of T and B cells in vivo[30]. Those lymphotropic effects are marked in calves experimentally infected with ncp BVDV type 2, strain 890, resulting in cavitation in the Peyer's patches and moderate to marked thymic atrophy.

The immunosuppressive effects of BVDV infection are evident when calves experimentally co-infected with BRSV and NY-1 BVDV develop more severe respiratory tract disease compared to calves infected with either virus alone. Co-infected calves develop more extensive pneumonia characterized by caudodorsal as well as cranioventral interlobular edema, emphysema and bronchopneumonia in caudal lung lobes. In contrast, calves infected with BRSV alone develop bronchopneumonia limited to cranioventral distribution[31]. Subpopulations of lymphocytes are more markedly altered in peripheral blood and lymphoid tissues of co-infected calves. Co-infected calves have reduced percentage of $CD8^+$ lymphocytes and $CD4^+$ lymphocytes in the thymus and Peyer's patches[29].

## INTERFERENCE WITH PHAGOCYTOSIS AND INTRA-CELLULAR KILLING

Phagocytosis is a critical innate defense mechanism, but it is also an event where innate and adaptive immune responses come together. Macrophages and neutrophils play a key role in innate immunity since they can recognize, ingest and destroy many pathogens without the help of the adaptive immune response. Nevertheless, the internalization and killing of pathogens is an essential pre-requisite for microbial antigen processing and presentation for the induction of a specific adaptive immune response against the pathogen. Because of their importance in the induction of innate and adaptive immune responses, pathogens have evolved strategies to combat the phagocytes[32].

*M. haemolytica* is commonly found as a commensal in the tonsillar crypts and the upper respiratory tract of healthy cattle. In conjunction with active viral infection and stress factors, it reaches the lungs where it multiplies rapidly. Alveolar macrophages, in response to LPS stimulation, secrete the pro-inflammatory cytokines IL-8, IL-1β and TNF-α[33-35]. The chemoattractant effects of IL-8 results in a massive influx of neutrophils from the blood into the alveolar spaces of the lung. Both the cytokines IL-1β and TNF-α enhance the expression of $\beta_2$ integrins on the surface of macrophages, neutrophils, lymphocytes and other leukocyte subsets[36]. This sequence of events result in a situation that is very conducive to the immunosppressive effects of *M. haemolytica*.

*M. haemolytica* produces an exotoxin which is cytolytic to macrophages, neutrophils and all other leukocyte subsets[37] and hence referred to as leukotoxin. Leukotoxin binds to $\beta_2$ integrins which are leukocyte-specific integrins[38, 39]. Leukotoxin utilizes CD18, the β subunit of $\beta_2$ integrins, as its receptor which enables it to target not only the phagocytes, but the lymphocytes and every other leukocyte subset as well[40-42] which explains the earlier finding that leukotoxin–deletion mutant of *M. haemolytica* does not cause severe mortality and does not induce severe lesions in the lungs as the wild-type does[43]. It is an irony that this pathogen encodes a toxin that utilizes the very same host cell proteins that the host utilizes to orchestrate the immune response against the pathogens.

Macrophages and neutrophils produce several toxic products that aid in the killing of the phagocytosed bacteria. The superoxide anion, hydrogen peroxide and nitric oxide are the most potent toxic products against bacteria. PI-3 virus inhibits the generation of superoxide anion in alveolar macrophages without killing the cells[44]. Inhibition of superoxide anion generation is likely to be due to inhibition of a cytosolic, calcium-independent, phosphatidyl serine/diglyceride-dependent protein kinase C. It has been proposed that PI-3-induced inhibition of oxygen-dependent bactericidal function of alveolar macrophages and disturbances in signal transduction contributes to immunosuppression and secondary bacterial infection in BRDC.

In summary, the BRDC pathogens have collectively evolved an extensive array of evasion mechanisms which help each other and exacerbate the disease, which is likely responsible for the failure of the vaccines to confer satisfactory protection against this complex disease.

## REFERENCES

[1]Srikumaran S, et.al. **Anim Health Res Rev** 2008;8(2):215. [2]Weenink SM, et.al. **Immunol Cell Biol** 1997;75:69. [3]Eskra L, et.al. J Gen Virol 1997;78:2159. [4]Winkler MT, et.al. **J Virol** 1999;73:8657. [5]Hughes HP, et.al. **Infect Immun** 1994;62:1609. [6]Oldstone MBA. **Curr Top Microbiol Immunol** 1994;189:1. [7]Lodmell DL, et.al. **J Exp Med** 1973;137:706. [8]Zinkernagel RM, et.al. **Nature** 1974;248:701. [9]Yewdell JW et.al. **Nature Immunol** 2002;3:1019. [10]Tanaka K, et.al. **Adv Immunol** 1997; 64:1. [11]Momburg F, et.al. **Adv Immunol** 1998;68:191. [12]Hariharan MJ, et.al. **Viral Immunol** 1993;6:273. [13]Nataraj C, et.al. **Viral Immunol** 1997;10:21. [14]Hinkley S, et.al. **Virus Res** 1998;53:91. [15]Koppers-Lalic D, et.al. **Arch Virol** 2003;148:2023. [16]Koppers-Lalic D, et.al. **Proc Natl Acad Sci USA** 2005;102:5144. [17]Hinkley S, et.al. **Arch Virol** 2000;145:2027. [18]Koppers-Lalic D, et.al. **J Gen Virol** 2001;82:2071. [19]Gopinath RS, et.al. **Viral Immunol** 2002;15:595. [20]Darnell JE et.al. **Science** 1994;264:1415. [21]Henderson G, et.al. **J Gen Virol** 2005; 86:2697. [22]Saira K, et.al. **J Virol** 2007;81:3077. [23]Schlender J, et.al. **J Virol** 2000;74:8234. [24]Baigent SJ, et.al. **J Virol**;76:8979. [25]Gil LH, et.al. **J Virol** 2006;80:900. [26]Schuh JC, et.al. **J Comp Pathol** 1992;106:243. [27]Keles I, et.al. **J Comp Pathol** 1998;118:337. [28]Glew EJ, et.al. **J Gen Virol** 2003;84:1771. [29]Brodersen BW et.al. **Viral Immunol** 1999;12:323. [30]Chase CC, et.al. **Vet Clin North Am Food Anim Pract** 2004;20:95. [31]Brodersen BW, et.al. **Am J Vet Res** 1998;59:1423. [32]Coombes BK, et.al. **Curr Biol** 2004;14:R856. [33]Yoo HS, et.al. **Infect Immun** 1995;63:381. [34]Stevens P, et.al. **Can J Vet Res** 1995;59:110. [35]Malazdrewich C, et.al. **Vet Pathol** 2001;38:297. [36]Leite F, et.al. **Infect Immun** 2000;68:5581. [37]Shewen PE et.al. **Infect Immun** 1982;35:91. [38]Wang JF, et.al. **Microb Pathog** 1998; 25(6)317. [39]Ambagala TC, et.al. **FEMS Microbiol Lett** 1999;179:161. [40]Li J, et.al. **Vet Microbiol** 1999;67:91. [41]Deshpande MS, et.al. **Infect Immun** 2002;70:5058. [42]Dassanayake RP, **Infect Immun** 2007;75:5004. [43]Petras SF, et.al. **Infect Immun** 1995;63:1033. [44]Dyer RM, et.al. **J Immunol** 1994;153:1171.

**MOLECULAR AND CELLULAR PATHOGENIC MECHANISMS OF *MANNHEIMIA HAEMOLYTICA* LEUKOTOXIN**

**S. Srikumaran, BVSc, PhD, R.P. Dassanayake, BVSc, PhD, S. Shanthalingam, BVSc, P.K. Lawrence, PhD**
**Pullman, WA**

## INTRODUCTION

Bovine pneumonic pasteurellosis, more commonly known as shipping fever, is an economically important disease of both beef and dairy industry in North America and Western Europe. The annual losses to the US cattle industry has been estimated to be around $1 billion[1]. Although bovine respiratory disease ensues as a result of complex interaction between several viruses and bacterial species in conjunction with management associated stress, *Mannheimia haemolytica* serotype 1 is the primary bacterial agent responsible for the acute fibrinonecrotic pneumonia which kills the animals[2]. This bacterium is a normal inhabitant of the upper respiratory tract of healthy cattle. When the animals experience stress due to shipping and/or acute viral infection, it multiplies rapidly, reaches the lungs and causes acute pneumonia.

## LEUKOTOXIN IS THE MAIN VIRULENCE FACTOR OF *M. HAEMOLYTICA*

*M. haemolytica*, a Gram-negative coccobacillus, produces several virulence factors which include the capsule, outer membrane proteins, adhesins, neuraminidase, lipopolysaccharide and leukotoxin (Lkt)[3]. While the wild-type organisms cause severe lung lesions and mortality, Lkt-deletion mutants induce much milder lung lesions and reduced mortality[4,5]. Hence, Lkt has been accepted as the most important virulence factor of this organism. The Lkt operon contains four genes: C, A, B, and D[6]. The *lkt*A encodes the structural Lkt protein (LktA). The products of *lkt*C is responsible for activation of the LktA by acylation. The *lkt*B and *lkt*D encode proteins LktB/D which are responsible for the transport and secretion of LktA (Figure 1). Lkt is a 102kD protein formed by 953 amino acids. It is a member of the family of Gram-negative bacterial exotoxins, referred to as RTX, so called because of the tandemly repeating nonapeptide domains rich in glycine and aspartic acid[7]. Members of this family of toxins exhibit similar mechanisms of production, secretion, and cytotoxicity[8].

Comparative analysis of the structural domains of the wild type toxins, deletion mutants and hybrid molecules between different toxins by several investigators have resulted in the identification of certain domains common to the RTX toxins[1]. The N-terminal half of the toxin consists of four hydrophobic domains which are believed to be involved in pore formation in target cell membrane. The other domain that is common to all RTX toxins is composed of a number of glycine-rich nonapeptide repeats found in the C-terminal third of the toxin. These domains are involved in binding of $Ca^{2+}$ which is necessary for the biological activity of the toxin. A third domain composed of about 50-60 amino acids at the extreme C-terminal end is involved in the transport of the toxin across the bacterial envelope.

Deletions in the N-terminal half of the toxin involving the hydrophobic domains does not affect binding of the toxin to the target cells, but affects the lytic activity of the toxin. The binding of one Lkt-neutralizing monoclonal antibody has been mapped to the C-terminal half of the Lkt between amino acid residues 450 and 939[9,10]. The epitope recognized by another Lkt-neutralizing monoclonal antibody (MM601) developed in our laboratory lies between amino acid residues 841 and 873[10,11] providing evidence that the Lkt domain that binds to the target cells lies in the C-terminal third of the toxin. The epitope recognized by a Lkt-non-neutralizing monoclonal antibody (MM605) lies between amino acid residues 874 and 906 (Figure 1).

A number of studies have shown that Lkt has profound effects on normal cell functions at very low concentrations. The Lkt stimulates the neutrophil respiratory burst as measured by the generation of reactive oxygen intermediates such as superoxide anion and hydrogen peroxide, and the generation of secondary granules[12,13]. Cytoskeletal alterations which are normally seen in activated neutrophils also occur. Lkt at sublytic concentrations also induces the release of inflammatory mediators, leukotrienes[14,15], histamine[16], tumor necrosis factor $\alpha$ (TNF-$\alpha$) and interleukin $1\beta$ (IL-$1\beta$)[17]. These findings indicate an important role for the Lkt in the pathogenesis of bovine pneumonic pasteurellosis. As the concentration of Lkt increases, target cells are stimulated to undergo apoptosis. At high concentrations, necrosis of target cells occurs consequent to the membrane damage due to pore formation[15-17]. Pore formation results in the efflux of $K^+$, influx of $Ca^{2+}$, colloidal osmotic swelling, and eventual cell lysis.

The cytotoxicity of this toxin is specific to ruminant leukocytes. While all subsets of leukocytes are susceptible to Lkt-induced cytolysis, polymorphonuclear leukocytes (PMNs) are the most susceptible subset. Moreover, PMN depletion in calves has been shown to drastically reduce the lung lesions[18]. Therefore, Lkt-induced PMN lysis and de-granulation is the primary cause of acute inflammation and lung injury characteristic of pneumonia caused by *M. haemolytica.* Hence, PMN and Lkt are the key elements in the pathogenesis of this disease. The inhibition of leukocyte and Lkt interaction will prevent neutrophil degranulation and lysis which in turn should mitigate lung injury. Inhibition of Lkt-Leukocyte interaction necessitates the identification and characterization of the receptor for the Lkt on leukocytes.

## LEUKOTOXIN BINDS TO $\beta_2$-INTEGRINS EXPRESSED ON THE CELL SURFACE OF BOVINE LEUKOCYTES

Early studies in our laboratory[19] and that of others[20,21] have identified $\beta_2$ integrins as the receptors for Lkt on the leukocytes of cattle. Integrins are a large family of integral cell surface glycoproteins that are expressed on most of the cell types. They mediate cell-cell, and cell-extracellular matrix interactions. Integrins are heterodimeric molecules composed of two non-covalently associated subunits $\alpha$ and $\beta$. The integrins have been classified into subfamilies depending upon which $\beta$ subunit they contain. $\beta_2$ integrins are integrins that carry $\beta_2$ chain as their $\beta$ subunit. $\beta_2$ integrins are expressed exclusively on leukocytes[22].

The $\beta$ subunit of $\beta_2$ integrins is known as CD18, and the $\alpha$ subunit is known as CD11. Each subunit consists of a large extracellular domain, a single transmembrane domain and a small cytoplasmic domain. The $\beta$ subunit of $\beta_2$ integrins, CD18, associates with four different α chains CD11a, CD11b, CD11c and CD11d, giving rise to four different $\beta_2$ integrins: CD11a/CD18, known as lymphocyte function-associated antigen 1 (LFA 1); CD11b/CD18, known as Mac-1 and CR3; CD11c/CD18, known as CR4; CD11d/CD18 is a recently identified $\beta_2$ integrin that has not been well characterized in ruminants yet.

Each subunit consists of a large extracellular domain of several hundred amino acids, a single transmembrane domain and a small cytoplasmic domain of 20-70 amino acid residues. The extracellular domain binds a wide variety of ligands, whereas the intracellular cytoplasmic domain anchors to cytoskeletal proteins. In this manner the exterior and interior of a cell are physically linked, which allows for bidirectional transmission of mechanical and biochemical signals across the plasma membrane, and leads to a cooperative regulation of cell functions, including adhesion, migration, growth, and differentiation. Both α and β chains are required for the expression of the heterodimers on the cell surface. These molecules facilitate migration into areas of inflammation, phagocytosis, antigen presentation and cytotoxicity. They perform these functions by interacting with various intercellular adhesion molecules (ICAMs) which are expressed by leukocytes, endothelial cells and many other cell types. It is ironical that *M. haemolytica* Lkt would use the molecules evolved by the host for orchestration of immune response against pathogens, as its receptor on the host cells to kill the host cells and cause damage to lung tissue. As stated earlier, at sublytic concentrations, Lkt stimulates the production of pro-inflammatory cytokines IL-1β and TNF-α[23]. Furthermore, LPS of *M. haemolytica* also stimulates production of these cytokines by alveolar macrophages[24]. These cytokines enhance the expression of CD18 on the surface of PMNs[25] which in turn would enhance Lkt binding and lysis of PMNs. Thus, *M. haemolytica* mounts a multi-pronged attack on the host cells, and Lkt plays the most important role in this process.

**BOVINE CD18, THE β SUBUNIT OF $\beta_2$ INTEGRINS, SERVES AS A FUNCTIONAL RECEPTOR FOR LEUKOTOXIN**

The finding that Lkt binds to $\beta_2$ integrins raises the question as to whether the α or β subunit or both serve(s) as the functional receptor(s) for Lkt. To address this question, we transfected a Lkt-non-susceptible murine cell-line (P815) with bovine CD18 cDNA, and demonstrated that bovine CD18 is necessary and sufficient to mediate Lkt-induced cytolysis of bovine leukocytes[26]. However, in this study, we could not rule out any role for the α subunit, CD11a, because the murine cells that we used for transfection express murine CD11a, and the transfected bovine CD18 was co-expressed with the murine CD11a as a heterodimer. Therefore, we designed another study in which we developed transfectants exhibiting monomeric expression of CD18 or CD11a by transfecting the respective cDNA into Lkt-non-susceptible human cell-line HEK-293. Lkt bound and lysed the transfectants expressing bovine CD18, but not bovine CD11a confirming CD18 as the functional receptor for Lkt[27].

**CD18 OF ALL BOVINE $\beta_2$ INTEGRINS SERVES AS THE RECEPTOR FOR LEUKOTOXIN**

Earlier studies revealed that CD18 of the $\beta_2$ integrin LFA-1 serves as a receptor for Lkt[26,28]. However it was not clear whether CD18 of other $\beta_2$ integrins Mac-1 and CR4 also serve as a receptor for Lkt. Therefore we cloned and sequenced cDNA for bovine CD11b and co-transfected it along with bovine CD18 into the HEK-293 cell-line. Transfectants expressing bovine Mac-1 (CD11b/CD18) were bound and lysed by Lkt indicating that CD18 of Mac-1 also serves as a functional receptor for Lkt[29]. The fact that CD18 of both LFA-1 and Mac-1 serves as functional receptors for Lkt suggests that CD18 of CR4 is also likely to serve as a receptor for Lkt. Experiments to confirm the role of CR4 are currently underway in our laboratory.

**THE BINDING DOMAIN OF LEUKOTOXIN ON BOVINE CD18 LIES WITHIN AMINO ACIDS 1-291**

In order to locate Lkt binding domain on bovine CD18, we designed a study with bovine-murine CD18 chimeras[30]. In this study, nucleotide sequences of bovine and murine CD18 were compared to identify unique restriction sites present in both sequences at identical positions. Two unique restriction sites were identified. Cleavage at these sites divided CD18 into three fragments A, B, and C. A set of 3 chimeric CD18 cDNA was constructed by deleting fragment A/B/C from murine CD18 and inserting the corresponding fragment of bovine CD18. cDNA encoding the chimeric CD18 was transfected into the murine cell line, P815. Stable transfectants were tested for Lkt-binding and cytolysis. Of these 3 transfectant cell lines, Lkt bound and lysed the transfectant cell line 6S1 which had fragment A of bovine CD18. This indicated that the Lkt binding domain lies within amino acids 1 to 291 of bovine CD18.

In summary, *M. haemolytica* Lkt is the most important virulence factor secreted by this organism. It utilizes as its receptor a leukocyte cell surface protein, the β subunit of $\beta_2$ integrins, encoded by the host for orchestration of immune response against this and other pathogens. Lkt-induced lysis and degranulation of PMNs is the primary cause of acute inflammation and lung injury characteristic of pneumonia caused by *M. haemolytica.*

**REFERENCES**

[1]Jeyaseelan S, et.al. **Anim Health Res Rev** 2002;3(2):69. [2]Frank GH, et.al. **Am J Vet Res** 1987;48(12):1674. [3]Confer AW, et.al. **Can J Vet Res** 1990;54(suppl.):S48. [4]Tatum FM, et.al. **Microb Pathog** 1998;24(1):37 [5]Petras SF, et.al. **Infect Immun** 1995;63(3):1033. [6]Narayanan SK, et.al. **Vet Microbiol** 2002;84(4):337.
[7]Welch RA, **Mol Microbiol** 1991;5(3):521. [8]Coote JG, **FEMS Microbial Rev** 1992;8(2):137. [9]Gerbig DG, et.al. **Infect Immun** 1992;60(5):1734. [10]Lainson FA, et.al. **Micorbiology** 1996;142(Pt9):2499. [11]Gentry MJ, et.al. **Microb Pathog** 1991;10(5):411. [12]Czuprynski CJ, et.al. **Infect Immun** 1991;59(9):3126. [13]Maheswaran SK, **Vet Immunol Immunopathol** 1992;33(1-2):51. [14]Henricks PA, et.al. **Infect Immun** 1992;60(8):3238. [15]Clinkenbeard KD, et.al. **J Leukoc Biol** 1994;56(5):644. [16]Adusu TE, et.al. **Can J Vet Res** 1994;58(1):1. [17]Stevens PK, et.al. **Infect Immun** 1996;64(7)2687. [18]Breider MA, et.al. **Can J Vet Res** 1988;52(2):205. [19]Ambagala TC, et.al. **FEMS Microbiol Lett** 1999;1791(1):161. [20]Li J, et.al. **Vet Microbiol** 1999;67(2):91. [21]Wang JF, et.al. **Microb Pathog** 1998; 25(6)317. [22]Harris ES, et.al. **J Biol Chem** 2000;275(31)23409. [23]Leite F, et.al. **Infect Immun** 2002;70(8):4336. [24]Lafleur RL, et.al. **Microb Pathog** 2001;30(6):347. [25]Leite F, et.al. **Microb Pathog** 2003;34(6):267. [26]Deshpande MS, et.al. **Infect Immun** 2002;70(9):5058. [27]Dassanayake RP, et.al. **Infect Immun** 2007;75(10):5004. [28]Jeyaseelan S, et.al. **Infect Immun** 2000;68(1):72. [29]Lawrence PK, et.al. **Vet Immunol Immunopathol** 2008;122(3-4):285. [30]Gopinath RS, et.al. **Infect Immun** 2005;73(9):6179.

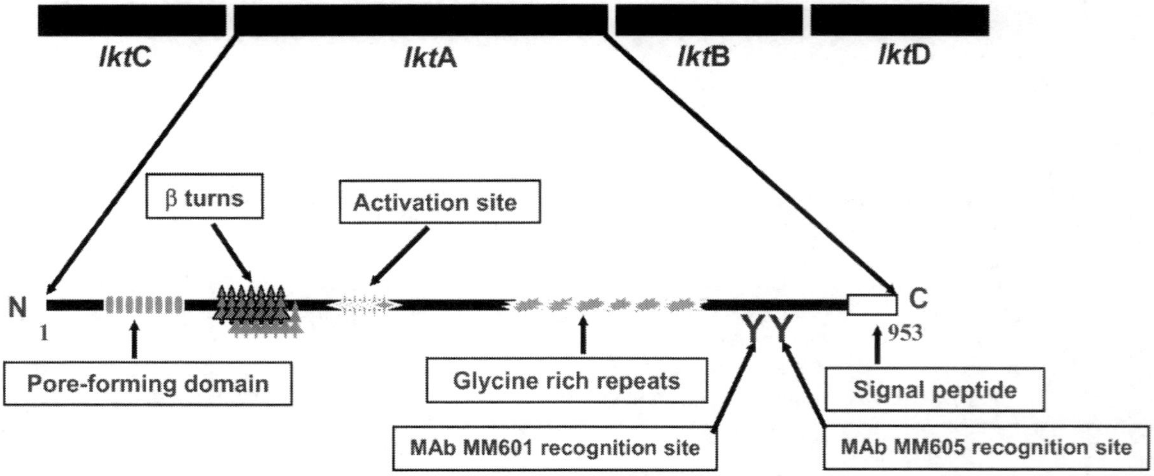


**Figure 1.** Functional domains of *Mannheimia haemolytica* leukotoxin

## COMPARATIVE PATHOLOGY OF AND PATHOGENIC MECHANISMS IN BOVINE BACTERIAL PNEUMONIA

**Anthony W. Confer, DVM, PhD, Diplomate ACVP, Regents Professor & Sitlington Endowed Chair**
**Stillwater, OK**

In North American cattle, *Mannheimia haemolytica, Pasteurella multocida, Histophilus somni, Arcanobacterium pyogenes,* and *Mycoplasma bovis* are isolated from severe bacterial pneumonia frequently seen in dairy calves (enzootic pneumonia) and in feedlot cattle (shipping fever) in what is known as the Bovine Respiratory Disease (BRD) Complex. All of these bacteria are ubiquitous in the cattle population as commensals in the nasopharynx, and following stress or viral infection, each can proliferate and be inhaled into the lungs. Each has its own cadre of known virulence factors such as adhesins, toxins, and enzymes that enhance their ability to colonize, cause tissue destruction, and incite an intense inflammatory response. The purpose of this talk is to review the virulence factors, pathogenesis, and lesions associated with these various bacteria.

### *Mannheimia haemolytica* (formerly *Pasteurella haemolytica* A)

Of the pathogenic bacteria discussed herein, *M. haemolytica* is arguably the most important of the group and has traditionally been associated with the acute fulminating pleuropneumonia seen within the first weeks of cattle feeding. This Gram-negative bacterium has 12 representative capsular serotypes, but serotype 1 (S1) and S6 are the most prevalent in bovine pneumonia. *M. haemolytica* is acquired by calves early in life through contact with their dams and other cattle. The bacterium resides in the nasopharynx and tonsil with the latter considered being the major reservoir. (Rice et al., 2007)

Virulence factors are considered to be many and include protein adhesins, capsular polysaccharide, lipopolysaccharide (LPS), iron-binding proteins, secreted enzymes, and a ruminant-specific RTX toxin – leukotoxin (LKT). Specific adhesins include a glycoprotein, N-acetyl-D-glucosamine, that mediates adherence to tracheal epithelial cells and activates the oxidative burst of bovine neutrophils. Heat-modifiable outer membrane protein A (OmpA) and the surface lipoprotein 1 (Lpp1) mediate *M. haemolytica* binding to bronchial epithelial cells. In addition, the *M. haemolytica* capsule may function as an adhesin in addition to its antiphagocytic properties. Neuraminidase and sialoglycoprotease modify cell surfaces and may enhance bacterial adhesion. *M. haemolytica* LPS has typical endotoxemic and proinflammatory properties, causes vasculitis, and complexes with LKT enhancing LKT-receptor production and augmenting LKT activity. LKT induces dose-related changes in bovine leukocytes. At high LKT concentrations, cells undergo rapid osmotic swelling, membrane pore formation, and necrosis. At reduced doses, LKT can induce leukocyte apoptosis or release of proinflammatory cytokines, oxygen free radicals, and cellular proteases. The sum result of these virulence factors is alveolar and vascular damage and inflammation (Confer et al., 1995; Czuprynski and Welch, 1995; Czuprynski et al., 2004)

### *M. haemolytica*

Induced pneumonia is characterized by acute cranioventral fulminating fibrinous to fibrinopurulent pleuropneumonia. On cut surface, there is characteristic "marbling" of the lung with lobules varying from normal, gray, or red with whole lobules undergoing hemorrhage or coagulation necrosis. Interlobular septa are usually distended with fibrin-rich edema fluid and interlobular lymphatics may contain fibrin thrombi. Histologically, alveoli are flooded with fibrin-rich exudate, a variable neutrophil and macrophage infiltrate, and the traditional "oat cells" that are flattened and streaming necrotic macrophages within the affected alveoli. Lobules undergoing coagulation necrosis appear as infarcts with thick neutrophilic infiltrates around the periphery of the necrotic lobule. Vasculitis with fibrin thrombi is frequently seen. Often bronchioles are spared or contain fibrin-rich exudate that may have extended from the alveoli.

### *Pasteurella multocida*

*P. multocida* has been isolated from up to 40% of the cases of enzootic and shipping fever pneumonia. This Gram-negative bacterium is acquired at an early age from the dam and is a common nasopharyngeal commensal. Serotype A:3 is the most common *P. multocida* isolate from bovine pneumonia, and pneumonia associated with *P. multocida* infections whether in young dairy calves, in weaned and shipped beef cattle, or in calves experimentally challenged is often difficult to discern from pneumonia associated with other bovine bacterial pathogens such as *M. haemolytica* and *Histophilus somni.* Dungworth (Dungworth, 1985) indicated that "less fulminating…bronchopneumonias tend to be more often caused by *P. multocida* than by *P.* [*Mannheimia*] *haemolytica.*" In fact, isolation of more than one of these pathogens from pneumonic lungs in conjunction with respiratory viruses and/or *Mycoplasma spp.* occurs frequently (Gagea *et al.*, 2006).

*P. multocida* A:3 virulence factors are less numerous than those identified in *M. haemolytica.* Several adhesins, a thick polysaccharide capsule, and LPS are the major factors responsible for pneumonia. Adhesins responsible for adherence of the bacterium to cell surfaces include type IV fimbriae, OmpA, neuraminidase, and filamentous hemagglutinin (FHA)- similar to FHA found in *Bordetella spp.* In addition, OmpA, hemoglobin-binding protein A, and transferrin - binding protein A bind fibronectin and other extracellular matrix proteins suggesting that these aid in invasion. *P. multocida* LPSs have similar properties (biological and chemical) and structure to R-type LPS (with no polymeric O-antigen) found in many Gram-negative bacteria and induce classic signs of endotoxic shock (Harper *et al.*, 2006). *P. multocida* LPS is a potent stimulator of inflammatory cytokines and a predominate inciter of pulmonary inflammation. The importance of capsule as a virulence determinant in the pathogenesis of *P. multocida* infection is due to its antiphagocytic properties (Dabo et al., 2007).

Various authors have described the gross pulmonary pathological changes differently, and those designations probably reflect the age of the lesion, whether or not other infectious agents were involved but not identified, and descriptive preferences of the various authors. The lesion is a typical cranioventral bronchopneumonia and has been characterized simply as bronchopneumonia or as a bronchopneumonia with various descriptive modifiers including acute fibrinosuppurative, subacute to chronic fibrinopurulent, fibrinous to fibrinopurulent, suppurative , and fibrino-necrotizing. The presence of fibrinous to fibrinopurulent pleuritis, distended interlobular septa with edema or fibrin, and/or abscesses is variable with *P. multocida* infection.

***Histophilus somni* (formerly *Haemophilus somnus*)**

*H. somni*, a Gram-negative bacterium, is associated with numerous pathological processes including pneumonia, septicemia, myocarditis, abortion, thrombotic meningoencephalomyelitis, and synovitis. The bacterium is associated with enzootic or shipping fever pneumonia with epidemiological studies demonstrating marked variation in its prevalence compared to *M. haemolytica* or *P. multocida* in shipping fever pneumonia.

*H. somni* are non - encapsulated, and the virulence factors include lipooligosaccharide (LOS), various outer membrane proteins – especially transferrin binding proteins, and immunoglobulin-binding proteins (IgBPs). Studies of adhesion proteins are generally lacking. Of particular interest with respect to virulence are LOS and IgBPs. LOS can mediate endothelial cell apoptosis and through antigenic phase variation can assist the bacterium to escape the host immune response. IgBPs are surface-exposed fibrillar networks of proteins that bind the Fc domain of bovine IgG2 and are responsible for resistance to complement-mediated serum killing, a characteristic of pathogenic strains of *H. somni*. In addition, *H. somni* produces histamine and can form biofilm within the host (Corbeit, 2007).

Gross lesions of acute fulminating pneumonia due to *H. somni* are similar to those seen with *M. haemolytica* infection, i.e. a cranioventral fibrinous pleuropneumonia with coagulation necrosis that can involve entire lobules, and hemorrhage. Pulmonary vasculitis is often seen and those vessels may contain fibrin thrombi. In several studies, *H. somni* was associated with pneumonia appeared later in the feeding period than was *M. haemolytica,* and in those cases the pneumonic lesions included fibrosis, fibrous pleural adhesions, and bronchiolitis obliterans.

***Mycoplasma bovis***

This bacterium has been subject of considerable investigation; however, its primary role in bovine bacterial pneumonia is controversial. Much of the controversy is because many healthy feedlot cattle shed *M. bovis* from the nasal passages with approximately 50% of cattle positive upon feedlot entry and nearly 100% positive by day 12 (Allen et al. 1992). In addition, *M. bovis* has been isolated from up to 45% of grossly and histologically normal bovine lungs (Gagea et al., 2006). The bacterium is predominately an extracellular pathogen that is present on respiratory epithelial surfaces. *M. bovis* antigen is present within macrophages of infected lung, and that antigen is most likely phagocytized debris (Caswell and Archambault, 2007).

Virulence factors include variable surface proteins (VSPs) that function as adhesins, are responsible for phenotypic variation among *M. bovis* strains, and allow for evasion of host immune responses. Several other surface proteins have been demonstrated that assist in colonization. A polysaccharide toxin has been described; however, the role or even existence of that toxin remains controversial. Hydrogen peroxide is produced by some *M. bovis* strains and forms oxygen free radicals resulting in host lipid peroxidation. Heat shock proteins are produced, but their role in virulence is not known. The formation of biofilm has been associated with numerous *M. bovis* strains, and that trait enhances immune and antimicrobial resistance and colonization.

Transmission of *M. bovis* is through nasal shedding and fomites; however, direct contact with others is the most common because survival outside the host is limited. Lesions that are attributable to *M. bovis* are a caseonecrotic bronchopneumonia that may have abscesses, bronchiectasis, and sequestration. Arthritis may accompany respiratory disease. The chronic nature of the pneumonia is apparent at the gross and microscopic levels. In addition, the highest percentages of *M. bovis* pneumonia are in cattle previously treated for BRD. Death often occurs after 4-6 weeks in the feedlot. Histologically, *M. bovis* antigen is present within and surrounding the caseous foci.

***Arcanobacterium pyogenes* (formerly *Actinomyces pyogenes*)**

*A. pyogenes* is a Gram-positive, rod-shaped bacterium. It is the most common bacterium isolated from bovine abscesses in all locations, including the lungs. It is frequently isolated from chronic abscessing bovine pneumonia. *A. pyogenes* is thought to be a secondary invader into a lung already pneumonic due to other infectious agents. *A. pyogenes* is an inhabitant of many mucosal surfaces in cattle including nasopharynx. Factors precipitating its proliferation and inhalation are not known; however, Catry et al. (2008) recently demonstrated that fluoroquinolone administration to healthy cattle diminished upper respiratory *P. multocida*, and *A. pyogenes* became the predominant nasopharyngeal isolate. Therefore, it may survive and proliferate in cattle that are treated for BRD.

*A. pyogenes* has several important virulence factors. These include a cholesterol-depend cytolysin (pyolysin) that is a pore-forming cytolysin for immune cells and is also a hemolysin. Several adhesins have been identifiedincluding two neuraminidases, which cleave sialic acids and expose cell receptors, and type 2 fimbrae. Several extracellular matrix binding proteins that bind to collagen or fibronectin and exoenzymes (DNAse and proteases) assist in invasion of tissue and degradation of proteins and nucleic acids. In addition, *A. pyogenes* can evade host defenses by invasion of epithelium, intracellular survival in macrophages, and formation of biofilm (Jost and Billington, 2005).

Transmission of *A. pyogenes* is thought to occur at a young age with calves acquiring the bacterium from their dams. Pulmonary lesions ascribed to *A. pyogenes* are primarily severe abscesses within areas of chronic bronchopneumonia or chronic pleuropneumonia. These abscesses are typically characterized by liquefactive necrosis surrounded by thick fibrous connective tissue band, whereas they are often larger and less caseous than *M. bovis*-induced lesions.

## REFERENCES

[1.] Allen JW, Viel L, Bateman KG, Nagy E, Røsendal S, Shewen PE, 1992. Serological titers to bovine herpesvirus 1, bovine viral diarrhea virus, parainfluenza 3 virus, bovine respiratory syncytial virus and *Pasteurella haemolytica* in feedlot calves with respiratory disease: associations with bacteriological and pulmonary cytological variables. **Can J Vet Res**. 156:281-288. [2.] Caswell JL, Archambault M, 2007. *Mycoplasma bovis* pneumonia in cattle. **Anim Health Res Rev**. 8:161-186 [3.] Catry B, Croubels S, Schwarz S, Deprez P, Cox B, Kehrenberg C, Opsomer G, Decostere A, Haesebrouck F, 2008. Influence of systemic fluoroquinolone administration on the presence of *Pasteurella multocida* in the upper respiratory tract of clinically healthy calves. Acta Vet Scand. 50:36-39. [4.] Confer AW, Clinkenbeard KD, Murphy GL: Pathogenesis and virulence of *Pasteurella haemolytica* in cattle: an analysis of current knowledge and future approaches. In Donachie W, Lainson FA, Hodgson JC (eds) *Haemophilus, Actinobacillus* and *Pasteurella*. **Plenum, London**, pp 51-62, 1995. [5.] Corbeil LB , 2007. *Histophilus somni* host-parasite relationships. **Anim Health Res Rev**. 8:151-160. [6] Czuprynski CJ, Leite F, Sylte M, Kuckleburg C, Schultz R, Inzana T, Behling-Kelly E, Corbeil L, 2004. Complexities of the pathogenesis of Mannheimia haemolytica and Haemophilus somnus infections: challenges and potential opportunities for prevention? **Anim Health Res Rev**. 5:277-282. [7.] Czuprynski CJ, Welch RA, 1995. Biological effects of RTX toxins: the possible role of lipopolysaccharide. **Trends Microbiol.** 3:480-483. [8.] Dabo SM, Taylor JD, Confer AW, 2007. *Pasteurella multocida* and bovine respiratory disease. **Animal Health Research Review** 8: 129-150. [9.] Dungworth, D.C., 1985, The respiratory system, In: Jubb, K.V.F., Kennedy, P.C., Palmer, N. (Eds.) **Pathology of Domestic Animals. Academic** Press, Orlando, pp. 448-489. [10.] Gagea MI, Bateman KG, van Dreumel T, McEwen BJ, Carman S, Archambault M, Shanahan RA, Caswell JL, 2006. Diseases and pathogens associated with mortality in Ontario beef feedlots. **J Vet Diagn Invest**. 18:18-28 [11.] Harper M, Boyce JD, Adler B, 2005. *Pasteurella multocida* pathogenesis: 125 years after Pasteur. **FEMS Microbiol Lett.** 265:1-10 [12.] Jost BH, Billinton SJ, 2005. *Arcanobacterium pyogenes*: molecular pathogenesis of an animal opportunist. **Antonie Van Leeuwenhoek**. 88:87-102. [13.] Rice JA, Carrasco-Medina L, Hodgins DC, Shewen PE, 2007. *Mannheimia haemolytica* and bovine respiratory disease. **Anim Health Res Rev**.8:117-128.

# IMMUNITY AGAINST *MANNHEIMIA HAEMOLYTICA*: CURRENT KNOWLEDGE AND NEW APPROACHES

**Anthony W. Confer, DVM, PhD, Diplomate ACVP, Regents Professor & Sitlington Endowed Chair**
**Stillwater, OK**

*Mannheimia haemolytica* serotype 1 (S1) is the main bacterium responsible for the clinical signs and lesions of severe fibrinous pleuropneumonia in stressed beef cattle (shipping fever).[4,8,13,20] The bacterium is also isolated from neonatal calf enzootic pneumonia. Although less commonly isolated, *M. haemolytica* S6 was identified in approximately 25% of *M. haemolytica* – associated cases of bovine pneumonia. *M. haemolytica* is a Gram-negative commensal in the bovine nasopharynx. Following stress or viral infections, the bacterium proliferates and is inhaled into the lungs stimulating a series of pathological events that lead to acute, severe fulminating fibrinopurulent inflammation with necrosis and hemorrhage. The bacterium was formerly classified as *Pasteurella haemolytica* and reclassified as *Mannheimia haemolytica* due to substantial genomic differences between it and other members of the *Pasteurella* genus. Despite that reclassification, the bacterium is still sometimes referred to in commercial vaccines by its previous name, *P. haemolytica*.

Specific *M. haemolytica* S1 immunogens that are important for stimulating immunity in cattle are surface antigens, leukotoxin (LKT), and sialoglycoprotease; the latter two are present in bacterial culture supernatant.[20] Immunity to experimental *M. haemolytica* pneumonia requires stimulation of LKT-neutralizing antibodies and antibodies to surface antigens that likely enhance phagocytosis and complement-mediated bacterilysis.[22,23] Potentially important surface antigens are capsular polysaccharide, lipopolysaccharide (LPS), outer membrane proteins, and iron-regulated outer membrane proteins. Experimentally, the most important *M. haemolytica* surface antigens are likely outer membrane proteins. Studies have demonstrated positive correlation between high antibody responses to OMPs such as PlpE, GS60 and others.[15,18,19] Although capsular polysaccharide and LPS are immunogenic, antibodies to these surface antigens do not appear to play major roles in stimulating protective immunity.[10,12]

Various formulations of *M. haemolytica* vaccines are currently available including a live mutant and various killed vaccines.[9] Efficacy has been demonstrated primarily with experimental challenge methods using direct *M. haemolytica* challenge via intratracheal, intrabronchial, or transthoracic routes or using a combination respiratory viral/*M. haemolytica* challenge. In several field trials, *M. haemolytica* vaccination had positive effects in feedlot cattle.[25] Differences exist among the various studies with respect to efficacy of different vaccines that were compared in controlled field trials. Numerous commercially available bovine biologicals contain *M. haemolytica* antigens and are often in combination with antigens of other infectious agents. These include combinations with respiratory viral vaccines, *H. somni* and/or *P. multocida* bacterins, and occasionally *Clostridium spp.* biologicals. The various licensed and available *M. haemolytica* biologicals fall into one of eight formulation categories, seven of which are non-living vaccines. These are described by their manufacturers as follows: 1) Bacterin with aluminum hydroxide adjuvant, 2) Bacterin with water-in-oil adjuvant, 3) Outer membrane extract, 4) Bacterin-toxoid (LKT toxoid), 5) Toxoid-cell-associated antigen, 6) Adjuvanted toxoid (culture supernatant), 7) Autogenous (herd-specific) bacterins commercially produced from isolates submitted by practicing veterinarians, and 8) Live streptomycin-dependent mutant. The latter vaccine is the only currently licensed live *M. haemolytica* biological. In the past, several live *M. haemolytica* vaccines were commercially available, and those vaccines showed potential efficacy; however, untoward side effects such as severe local and systemic reactions were documented after vaccination with these products.

In recent years, new approaches have been studied for improving *M. haemolytica* vaccine efficacy. Some of these include genetically modified live vaccines derived from AroA- or LKT- mutated *M. haemolytica*.[14,24] Those mutant *M. haemolytica* vaccine candidates have not commercialized in North America. Bacterial ghosts, which are empty cell envelopes, are derived through a temperature-controlled, bacteriophage-derived lysis gene. Adjuvanted *M. haemolytica* bacterial ghost vaccine enhanced protection against experimental challenge.[17] Other studies demonstrated that while vaccination with commercial vaccines can enhance resistance against experimental *M. haemolytica* challenge, addition of one of several *M. haemolytica* recombinant proteins, including recombinant LKT, sialoglycoprotease, or outer membrane lipoprotein PlpE, enhanced efficacy of the commercial *M. haemolytica* products. [5,7,11,22] Lipoprotein PlpE is identical in S1 and S6 strains, and addition of S1 PlpE to an S1 commercial vaccines significantly enhanced resistance against a *M. haemolytica* S6 challenge.[1,2,7] We have also developed an experimental *M. haemolytica* vaccine composed of a recombinant chimeric protein containing the LKT-neutralizing epitope and the surface-exposed, complement-fixing epitope of the major outer membrane lipoprotein, PlpE.[3] Addition of that protein to a *M. haemolytica* bacterin markedly enhanced protection against experimental challenge.[6] Mucosal delivery of *M. haemolytica* vaccines also under investigation.[21] One approach under intense study is the use of edible vaccines, wherein alfalfa expressing various *M. haemolytica* antigens including truncated LKT and the GS60 outer membrane lipoprotein are fed to cattle, thus stimulating mucosal and systemic antibodies to these immunogens.[15,16] Intranasal vaccination has also been explored.[21] One approach used ISCOMs carrying soluble antigens of *M. haemolytica* including native LKT. A recently published study demonstrated that intranasal vaccination of young calves with that vaccine stimulated LKT-specific IgA antibodies.[21] In addition, we have used the recombinant LKT-PlpE chimeric protein plus cholera toxin (CT) to intranasally immunize young dairy calves and have developed and immunized calves with another recombinant LKT-PlpE chimeric protein vaccine that also expresses CT subunit B. Intranasal vaccination with the chimeric proteins stimulated variable mucosal and systemic antibody responses depending on the age of the calf (Confer et al., submitted for publication, 2008).

## REFERENCES


[1] Ayalew S, Blackwood ER, Confer AW. Sequence diversity of the immunogenic outer membrane lipoprotein PlpE from *Mannheimia haemolytica* serotypes 1, 2, and 6. **Vet Microbiol** 114: 260–268, 2006. [2] Ayalew S, Confer AW, Payton ME, Garrels KD, Shrestha B, Ingram K, Montelongo M, Taylor JD. *Mannheimia haemolytica* Chimeric Protein Vaccine Composed of the Major Surface-Exposed Epitope of Outer Membrane Lipoprotein PlpE and the Neutralizing Epitope of Leukotoxin. **Vaccine** 26: 4955–4961, 2008 [3] Ayalew S, Confer AW, Blackwood ER. Characterization of Immunodominant and Potentially Protective Epitopes of *Mannheimia haemolytica* Serotype 1 Outer Membrane Lipoprotein PlpE. **Infect and Immun** 72: 7265-7274, 2004 [4] Booker CW, Abutarbush SM, Morley PS, Jim GK, Pittman TJ, Schunicht OC, Perrett T, Wildman BK, Fenton RK, Guichon PT, Janzen ED. Microbiological and histopathological findings in cases of fatal bovine respiratory disease of feedlot cattle in Western Canada. **Can Vet J.** 49473-481, 2008 [5] Confer AW, Ayalew S, Montelongo M, Panciera RJ, Whitworth LC, Hammer, JD. Immunogenicity of *Mannheimia haemolytica* serotype 1 recombinant outer membrane protein PlpE and augmentation of a commercial vaccine. **Vaccine** 21: 2821-2829: 2003 [6] Confer AW, Ayalew S, Montelongo M, Step DL, Wray JH, Hansen RD, Panciera RJ. Immunity of cattle following vaccination with a *Mannheimia haemolytica* Chimeric PlpE-LKT (SAC89) protein. **Vaccine**, in press, available online – Nov. 2008 [7] Confer AW, Ayalew S, Panciera RJ, Montelongo M, Wray JH. Recombinant *Mannheimia haemolytica* serotype 1 outer membrane protein PlpE enhances commercial *M. haemolytica* vaccine-induced resistance against serotype 6 challenge. **Vaccine** 24: 2248-2255, 2006 [8] Confer AW, Clinkenbeard KD, Murphy GL: Pathogenesis and virulence of *Pasteurella haemolytica* in cattle: an analysis of current knowledge and future approaches. In Donachie W, Lainson FA, Hodgson JC (eds) **Haemophilus, Actinobacillus and Pasteurella. Plenum**, London, pp 51-62, 1995. [9] Confer AW, Montelongo M, Brown MJ, Fergen BJ, Clement JC. Onset of Serum Antibodies to *Pasteurella* (*Mannheimia*) *haemolytica* Following Vaccination with Five Commercial Vaccines. **Bov Practitioner** 35: 141-148, 2001. [10] Confer AW, Simons KR, Panciera RJ, Mort AJ, Mosier DA: Serum antibody response to carbohydrate antigens of *Pasteurella haemolytica* serotype 1: relation to experimental bovine pneumonic pasteurellosis. **Am J Vet Res** 50:98-105, 1989. [11] Conlon JA, Shewen PE, Lo RY. Efficacy of recombinant leukotoxin in protection against pneumonic challenge with live *Pasteurella haemolytica* A1. **Infect Immun**. 59:587-591, 1991 [12] Conlon JA, Shewen PE. Clinical and serological evaluation of a *Pasteurella haemolytica* A1 capsular polysaccharide vaccine. **Vaccine.**11:767-772, 1993 [13] Gagea MI, Bateman KG, van Dreumel T, McEwen BJ, Carman S, Archambault M, Shanahan RA, Caswell JL, 2006. Diseases and pathogens associated with mortality in Ontario beef feedlots. **J Vet Diagn Invest**. 18:18-28 [14] Homchampa P, Strugnell RA, Adler B. Construction and vaccine potential of an aroA mutant of *Pasteurella haemolytica*. **Vet Microbiol**. 42:35-44, 1994 [15] Lee RW, Cornelisse M, Ziauddin A, Slack PJ, Hodgins DC, Strommer JN, Shewen PE, Lo RY. Expression of a modified *Mannheimia haemolytica* GS60 outer membrane lipoprotein in transgenic alfalfa for the development of an edible vaccine against bovine pneumonic pasteurellosis. **J Biotechnol.** 135:224-231, 2008 [16] Lee RW, Strommer J, Hodgins D, Shewen PE, Niu Y, Lo RY. Towards development of an edible vaccine against bovine pneumonic pasteurellosis using transgenic white clover expressing a *Mannheimia haemolytica* A1 leukotoxin 50 fusion protein. **Infect Immun**. 69:5786-5793, 2001 [17] Marchart J, Dropmann G, Lechleitner S, Schlapp T, Wanner G, Szostak MP, Lubitz W. *Pasteurella multocida*- and *Pasteurella haemolytica*-ghosts: new vaccine candidates. **Vaccine.** 21:3988-3997, 2003 [18] Pandher K, Confer AW, Murphy GL. Genetic and immunologic analysis of PlpE, a lipoprotein important in complement-mediated killing of *Pasteurella haemolytica* Serotype 1. **Infect Immun** 66:5613-5619, 1998. [19] Pandher K, Murphy GL, Confer AW. Identification of immunogenic, surface-exposed outer membrane proteins of *Pasteurella haemolytica* serotype 1. **Vet Microbiol** 65:215-226, 1999. [20] Rice JA, Carrasco-Medina L, Hodgins DC, Shewen PE, 2007. *Mannheimia haemolytica* and bovine respiratory disease. **Anim Health Res Rev**.8:117-128. [21] Shewen PE, Carrasco-Medina L, McBey BA, Hodgins DC. Challenges in mucosal vaccination of cattle. **Vet Immunol Immunopathol**. 2008 Oct 17. [Epub ahead of print] [22] Shewen PE, Lee CW, Perets A, Hodgins DC, Baldwin K, Lo RY. Efficacy of recombinant sialoglycoprotease in protection of cattle against pneumonic challenge with *Mannheimia (Pasteurella) haemolytica* A1. **Vaccine.** 21:1901-1906, 2003 [23] Shewen PE, Wilkie BN. Vaccination of calves with leukotoxic culture supernatant from *Pasteurella haemolytica*. **Can J Vet Res.** 52:30-36, 1988 [24] Thumbikat P, Briggs RE, Kannan MS, Maheswaran SK. Biological effects of two genetically defined leukotoxin mutants of *Mannheimia haemolytica*. **Microb Pathog**. 34:217-226, 2003 [25] Wildman BK, Perrett T, Abutarbush SM, Guichon PT, Pittman TJ, Booker CW, Schunicht OC, Fenton RK, Jim GK. A comparison of 2 vaccination programs in feedlot calves at ultra-high risk of developing undifferentiated fever/bovine respiratory disease. **Can Vet J.** 49:463-472, 2008

# COMPUTED TOMOGRAPHIC EVALUATION OF OTITIS IN THE ALPACA

**Silver Tawni I., BSA, DVM, MS**
**Saskatoon, SK**

## INTRODUCTION

Most ear infections in South American camelids are not diagnosed until the disease process has advanced to chronic, severe otitis externa or media.[2,3,4,5,7] Consequently the etiology of otitis in the new world camelid is not always apparent. Foreign bodies, ticks, mites, Listeria spp, and bacterial infections have all been implicated in individual cases.[2,3,4,5,7,8] The elongated, thin, bent, ear canal anatomy, surrounded with hair, may also predispose these animals to the condition.

The osseous anatomy of the camelid ear is complex. The long ear canal has a maximum diameter of 6 mm but narrows to only 4 mm in width.[4] Approximately 1.8 cm into the canal there is a ventral bend and then the canal is a further 1 cm in depth before the tympanic membrane is reached.[4] This complex anatomy makes external otoscopic examination of the ear canal difficult. In the larger llama the tympanic membrane and complete external ear canal can only be visualized in 50% of the adult animals without the aid of an endoscope.[2,5] In the alpaca it is usually impossible to visualize the tympanic membrane unless endoscopy is employed. Combined with the narrow, bent canal, the camelid tympanic bulla is large, lacks a fundic cavity and has an extensive honeycomb shape.[1,2,4] It has been theorized that the multi-chambered middle ear cavities provide more than one resonant frequency when sound is transmitted allowing them to distinguish between resonant frequencies.[20,21] In theory therefore the alpaca has an enhanced auditory reception of both high and low frequency sounds than a single chambered middle ear cavity.[20,21] This enhanced auditory feature however does not aid in the timely diagnosis and effective treatment of advanced otitis as the alpaca is often reluctant to have their ears handled and because of the complex anatomy.[3,4,5]

The difficulty in thoroughly examining otic structures did prompt the use of radiography to describe the normal osseous anatomy of the camelid ear. The lack of a true fundic cavity combined with the complexity of the otic and skull structures however creates middle and inner ear superimposition. Added to the superimposition is the difficulty in positioning for dorsoventral or oblique views and it has been found that plain films are of little diagnostic value. In order to further define the ear canal and determine tympanic membrane integrity, contrast radiography has also been described.[1,2,3] Contrast radiography is also difficult because contrast must fill the entire canal to be diagnostic. Loss of the contrast from the ear canal, difficult animal positioning during imaging, and lack of complete filling of the ear canal limit the value of the information gained from a contrast study without improving the usefulness of radiographic imaging.[2,22]

Superimposition of other structures and poor soft tissue contrast makes conventional radiography inferior to computed tomography in its investigation of many disease processes, including ear disease.[6] Computed tomography has been used to diagnose, stage, and treat middle ear disease in the dog.[6] The skull anatomy of the llama has been described using computed tomography.[4] The pathological changes associated with otitis externa, media, and interna have not been described by computed tomography in the camelid. Given the superior imaging ability, computed tomography more accurately describes, diagnoses and assists in the treatment of otitis in the camelid.[4,6]

A study was performed to describe the normal anatomy of the alpaca ear as imaged by computed tomography. This normal reference was then used to determine if the anatomical changes associated with otitis would specifically correlate with clinical signs demonstrated by alpacas affected with otitis.

### Normal Study

Three alpacas, euthanized for reasons unrelated to ear or vestibular disease, had CT scans for the purposes of describing normal ear anatomy. There was no history of ear disease and no gross clinical evidence of otitis. The CT scans were performed within one hour of euthanasia. The skull and neck were positioned in sternal recumbency with the hard palate parallel to the scanning table. Contiguous, 1 mm slice thickness, transverse images were made through the entire skull. Images were reconstructed with a high resolution algorithm and displayed using both a bone window and a soft tissue window. Histopathological evaluation of each ear revealed normal ears that were free of any disease process. Anatomical descriptions of the ear were described using published references on the anatomy of the new world camelids, gross anatomy of skulls and comparison to CT imaging descriptions of canine ear structures.[1,11]

## CLINICAL CASE STUDIES

Twelve alpacas presented to Colorado State University Veterinary Teaching Hospital or Ohio State University Veterinary Teaching Hospital between September 2004 and January 2009 for clinical signs associated with otitis externa, media, or interna. Animals were included in the study if a computed tomography examination of the entire skull or ear structures had been performed to evaluate their presenting complaint. Only initial clinical and CT examinations of these animals were used for data collection. Animals were male or female, were between 1 month and 3 years of age, and weighed between 15.0 and 147.0 pounds. All alpacas had results confirmed by histopathology, culture or necropsy evaluation of the affected ear(s).

### CT Scan Technique

Under general anesthesia all the alpacas were positioned either in sternal or dorsal recumbency with their forelimbs positioned caudally. The head was positioned to ensure the hard palate was parallel to the scanning table to allow symmetrical comparison. A pre-scan lateral image was made to ensure correct positioning. Transverse, contiguous, slices of a mean thickness of 3.0 mm (range 1.0-5.0 mm), were imaged through the skulls. Pre- and post contrast studies were obtained of the ear structures for each animal. Specific areas of interest were reimaged with 1 mm slice thicknesses. Images were reconstructed with a high resolution algorithm to minimize pixel size and maximize image resolution allowing display using both bone window and soft tissue windows.

**CT Scan Analysis**

Specific pathologic characteristics, differing from the normal anatomy, were looked for in the clinical cases. Each ear was evaluated for the presence of fluid or soft tissue attenuation of the tympanic bullae, thickening of the tympanic bulla wall, the presence of osteolysis or expansile osseous proliferation of otic structures, and/or the presence of peri-otic soft tissue disease. Regions of interest were placed over the bullae cavities and used to obtain Hounsfield unit measurements. Due to the complexity of the bullae these regions of interest had to include both the septae and the cavity portion of the middle ear. The thickness of the tympanic bulla wall was also measured in the normal and abnormal animals. All of the characteristics were then correlated with the animal's presenting clinical signs.

**RESULTS**

**Normal Study**

In the normal ear the external acoustic meatus, internal acoustic meatus, and septated tympanic bulla, were air filled and hypoattenuating. The osseous horizontal canal measured 6-7 mm in width then narrowed to a mere 3 mm wide near the tympanic membrane. The tympanic membrane was a soft tissue-attenuating line across the base of the osseous canal. Positioning and slice thickness gave some tympanic membranes the appearance of being mildly thickened and triangulated while in others the membrane could not be visualized.

The tympanic bulla is complex in that it is diffusely septated with thin bone. The entire bulla "drapes" over the proximal portions of the stylohyoid bone, creating the stylohyoid fossa. In order to obtain Hounsfield unit measurements, region-of-interests that included the septae were drawn over normal air-filled bullae. The normal HU values ranged between -500 to -300. The normal tympanic bulla wall measured a maximum of 1 mm in thickness. The density of the petrous temporal bone was highly bone attenuating with HU of +2100 to +2300. The osseous inner ear structures discernable included the spiral cochlea, incus, malleolus, vestibular aquaduct and vestibular window.

**Clinical Case Study**

Fourteen alpacas presented with a history of chronic non-responsive otitis externa in one or both ears, for a total of 19 clinical ears. Eleven of these animals presented with unilateral neurological signs that ranged in severity from a mild chronic head tilt to facial paralysis or Horner's syndrome and ataxia. Chronic facial paralysis was evident in some cases as oral mucosal ulceration, corneal ulceration, muscle wasting, mandibular deviation, or weight loss was present.

Otitis externa was evident on CT in thirteen of the nineteen clinically diagnosed ears. Six clinical ears were completely normal on CT examination. Of the eleven neurological animals ten alpacas had ipsilateral evidence of tympanic bulla attenuation (HU +100 to +750). Nine of these eleven alpacas had ipsilateral bulla wall thickening (2 to 4 mm) and osteolysis. Eight of these animals had severe osteoproliferative osteolysis which invaded or deviated surrounding soft tissues, bone, or the brain case. Peri-otic tissue involvement was evident in nine animals one of which was non-neurologic.

Upon evaluation of the ears that were contralateral to the neurologically affected ears CT revealed that four had otitis externa and seven were normal. None of the non-neurological ears had evidence of bulla attenuation, bulla wall thickening, osteolysis, or peri-otic soft tissue involvement. Of the number of ears with clinical otitis externa twelve had attenuation of the bulla; two of these animals were not neurological.

These findings were statistically significant in that for those animals with any cranial nerve deficits there is likely the presence of middle ear attenuation and osseous involvement. The severity of this involvement however can vary remarkably further indicating the need for CT evaluation in these cases. In addition 10% of the animals with clinical otitis externa had evidence of middle ear disease on CT that was not clinically detectable.

**TREATMENT**

The medical treatment of otitis externa can be difficult and requires a certain amount of dedication of both the owner and clinician. Initial determination of the inciting cause though difficult may be paramount in resolving the disease process. Removal of foreign material, diagnosing and treating systemic disease, or killing of ticks is necessary before successful treatment of otitis can occur. Ivermectin in mineral oil, permethrin (catron fly spray) and fipronil (Frontline®) can be used to kill ticks if they are detected. Long term antibiotics are used to treat the infections with the common choices including either Procaine Penicillin G or NuFlor®.

The combined use of endoscopy and CT are vital in diagnosis and therapeutic progress evaluation of otitis in most chronic or neurological cases. Ear lavages can be helpful to clean ear canals. The bent canal anatomy however can prevent thorough lavage of the horizontal canal in the standing animal especially if significant debris or inflammation is present. An endoscopic ear canal flush can be performed while the animal is under anesthesia for the CT examination and both modalities can be used to confirm successful lavage of the horizontal canal.

When chronic or severe otitis media, otitis interna or severe peri-otic disease is detected on CT a bulla osteotomy is usually recommended as treatment. During an osteotomy surgeons will often require intra-operative CT guidance to determine and maintain orientation in the complex otic anatomy. This is especially true if severe peri-otic disease distorts normal tissue structures and landmarks. CT is also used to drain abscesses, follow draining tracts, and confirm appropriate antibiotic bead placement. In difficult cases repeat CT evaluation may be vital in assessing treatment progression and determine prognosis.

## DISCUSSION

Computed tomography provides disease description information that could not be otherwise determined clinically. The factors were measurable in that their presence indicated severe or aggressive disease and therefore warranted more aggressive therapy. Interestingly some of the clinical cases that presented with severe clinical signs did not have as severe CT changes which may redirect the treatment options of a particular case.

Attenuation of the bulla is an important indication of middle ear disease. Hounsfield Unit measurement can help detect fluid or tissue in an affected bulla. The normal aerated bullae and its thin septae have a HU measurement between -500 to -300. Ears affected with otitis media have HU values that range between +100 to +750. Correlation between HU values and bullae fluid cytology was not possible in this study although higher values could infer more cellular material and be a severity or chronicity indicator.

The osseous bulla wall thickness was compared between normal and abnormal animals. Wall thickness did not exceed 1 mm in normal ears. Animals with middle ear disease that was either chronic or severe had thickened bulla walls. These thickened walls measured between 2 and 4 mm. The more severe the osteolytic osteoproductive changes the more abnormal the shape and thickness of the bulla wall and likely the more severe or aggressive the disease process.

The presence of neurological signs in the alpaca with clinical otitis has a significant correlation to the presence of middle ear inflammation or destruction. Importantly the study demonstrated that the severity of the middle ear disease can vary markedly between cases and although most neurologic animals had obvious osseous destruction this was not the case in all animals. One animal with unilateral vestibular disease, Horner's syndrome and ataxia had evidence only of otitis externa on both clinical and CT examinations. This demonstrates the importance of a CT evaluation on all neurologically affected alpacas with otitis to ensure appropriate therapy is undertaken.

Two clinical animals without neurological signs had evidence of mild to moderate attenuation of the tympanic bulla. There was no evidence of osseous lysis or peri-otic disease in either of these animals. The chronicity of the middle ear involvement could be inferred in one animal in that the bulla wall was mildly thickened to 2-3 mm. This animal had historical clinical otitis externa for at least 9 months. Interestingly the second alpaca went onto develop progressively worsening seizure activity. Consecutive CT examinations of this animal did indicate a change in bulla attenuation that correlated with the increase in seizure activity. Despite the presence of the seizures however there was no CT evidence of post-contrast meningeal disease, inner ear disease, periotic disease or the osseous change.

In the dog, signs of peripheral vestibular disease do not occur until the middle ear inflammation reaches the inner ear.[9] The progression of the disease into the inner ear structures is not necessarily visible on any imaging modality.[9] The current recommendation for this dilemma in the dog is that if there is visible middle ear disease, and there are clinical signs indicating peripheral vestibular involvement, otitis interna is likely and the animal should be treated accordingly.[9] The increased incidence of seizures in the initially non-neurologic alpaca would suggest meningeal involvement that had spread from the middle ear through the inner ear. Importantly this would suggest that if there is evidence of visible middle ear disease and there are clinical signs supporting neurological disease then inner ear or meningeal disease is likely and the animal should be treated accordingly.[9] This also suggests that for those non-neurologic animals with clinical otitis externa, which have tomographically confirmed middle ear disease, repeat CT evaluations and careful clinical monitoring should be undertaken to try to prevent disease progression with timely aggressive surgical intervention.

CT plays a vital role in the staging and treating otitis in alpacas because although there is a significant relationship between the presence of neurological signs and osseous involvement of the middle ear there are cases when this middle ear destruction is not present. In addition CT is required to diagnose middle ear disease in the non-neurological cases to ensure appropriately aggressive treatment. Computed tomography has also proven to be required for directional guidance during surgical and treatment procedures. Finally, repeat CT evaluations are useful in monitoring healing and prognosis especially in challenging cases.

## REFERENCES

[1]Hathcock JT, et al. **Vet Rad & Ultrasound** 1995; 36: 290.[2]Koenig JB, et al. **JAMVA** 2001; 218:1619. [3]Fowler ME, et al. **J Zoo Anim Med** 1985; 16:9. [4]Fowler ME. Medicine and Surgery of South American Camelids. Ames, Iowa: Iowa State University Press, 1989; 441. [5]Marriott MR, et al. Aust Vet J 1999; 77:301. [6]Rohleder JJ, et al.. **Vet Rad & Ultrasound** 2006; 47: 45. [7]Van Metre DC, et al. **JAVMA** 1991;199: 236. [8]Foreyt WJ, et al. **J Parasit** 2006; 78:153. [9]Dvir E, et al. **Vet Rad & Ultrasound** 2000; 41: 46. [10]Barthez PY, et al. **Vet Rad & Ultrasound** 1996; 37: 95.[11]Russo M, et al. **Vet Rad & Ultrasound** 2002; 43:22. [12]Garosi LS, et al. **Vet Rad & Ultrasound** 2003; 44:137. [13]Little CJL, et al. **Vet Rec** 1991; 128:293.[14]Bischoff MG, et al. **Vet Clin Small Anim** 2004; 34: 437. [15]Allgoewer I, et al. **Vet Rad & Ultrasound** 2000; 41: 413. [16]Sargent SJ, et al. **Eur Soc Vet Derm** 2006; 17: 175. [17]Stern-Bertholtz W, et al. **J Sm Anim Prac** 2003; 44: 253.[18]Detweiler DA, et al. **J Vet Intern Med** 2006;20:1080. [19]Shell LG. **Vet Clin Small Anim** 1988; 18:885. [20]Coleman MN, et al. **Anat Rec Part A** 2004; 281A:1123. [21]Mason MJ. **J Morph** 2006; 267:678. [22]Eom KD, et al. **Vet Rad & Ultrasound** 2000;41:231. [23]Defalque VE, et al. **Vet Rad & Ultrasound** 2005; 46:490. [24]Spangler EA, et al. **J Am Anim Hosp Assoc** 2000;36:239.

# DELIVERING CLONED CALVES: CHALLENGES & OPPORTUNITIES

**Gilles Fecteau, DMV, DACVIM, AC Brisville, DMV, S Boysen, DVM, DACVECC, S Buczinski, DMV, DACVIM, P Dorval, DMV, R Lefebvre, DMV, PhD, DACT, P Hélie, DMV, DACVP, A Desrochers, DMV, MSC, DACVS & LC Smith, DMV, PhD**
**Saint-Hyacinthe, Québec**

## INTRODUCTION & BACKGROUND

Somatic cell nuclear transfer (SCNT) is a reproductive biotechnology, which has a broad spectrum of applications. However the cloning process remains relatively inefficient. In cattle, a success rate of 5 to 7% of reconstructed embryos is reported (1, 2). The success rate is variable depending of the laboratory technique, cell-line or genotype. The low success rate is attributable to: early losses during implantation period, late pregnancy losses which encompass placental and foetal abnormalities, and peri and post-natal mortalities.

Due to client concerns and the high risks associated with clone pregnancies, the food animal and the emergency and critical care services from the teaching hospital at the Université de Montréal collaborated in the development of a special intensive care unit to manage, follow and care for the carrier heifers and their newborn clones. Industry demand projected 2 cohorts of clones per year. Each cohort could potentially (depending on pregnancy losses) produce 8 to 10 newborn clones. Until that particular request, the food animal service has been challenged sporadically with one clone birth at a time (mainly animals developed for research purposes). The uneventful birth of Starbuck II (January 2001) created great expectations. At that time, there was no clinically oriented research project related to cloning.

Several teams around the world have previously published their findings regarding newborn cloned calves. The objective of this paper is to summarize our experience delivering 26 cloned calves (2004-2007) and how we tried to create opportunities from the challenges that newborn cloned calves represent (1-8).

## BUILDING THE EXPERT TEAM: A MULTIDISCIPLINARY APPROACH

A multidisciplinary team was build over time. As the management of the initial cohort faced numerous diversified challenges with newborn clones, different specialists were consulted. It became routine procedure to meet following each cohort delivery to review and discuss what went right and most often what went wrong. How could we improve success rates and develop a more efficient standardised approach. Which experts could be consulted to try and better our understanding of the problem? Possible mistakes or misinterpretation of clinical signs and/or laboratory data were discussed. Different hypotheses were developed to explain failure. Over time, the standardised approach was modified based on outcome. The disciplines involved were: critical and intensive care, theriogenology, surgery, neurology, pathology and clinical pathology. Fundamental researchers were also involved and a grant proposal was funded.

## THE SUPPORT TEAM

Animal health technicians and veterinary students were trained to insure 24 hours monitoring. Special patient record forms were developed to facilitate data collection. Veterinary students were recruited on a voluntary basis with one being appointed as the coordinator. The coordinator was responsible of organising schedules and ensuring there was sufficient number of students involved to meet demand (8 to 10 students were involved for each cohort). Students were paid per hour of work according to regular hospital policy. Informal training sessions were scheduled but most students training took place in clinics working with the animal health technician. The regular animal health technicians working in the bovine clinics were also involved and contributed greatly to the development of techniques and/or material that would better meet patient needs (e.g. auricular arterial catheter placement technique, O2 delivery systems).

## THE PHYSICAL CHALLENGES

A critical care unit was developed to accommodate the presence of several newborns (n=8) at the same time. It involved temporarily remodelling a surgery suite into a neonatal intensive care unit (NICU). The room was disinfected and designated strictly for the care and management of cloned calves (typically for a period of 2 to 4 weeks per cohort). This area was physically separated from the regular veterinary hospital and designated animal caretakers responsible for NICU room sanitation were instructed to change clothes when working in the unit. Biosecurity protocols were implemented to reduce the risk of neonatal enteric disease outbreaks as well as transmission of hospital pathogens to the clone NICU.

## STANDARDIZING THE MANAGEMENT OF DELIVERY & MONITORING OF THE NEWBORN

The standard management included several aspects briefly summarized here. Dexamethasone (25 mg, IM) was administered 24 to 36 hours prior to the day of cesarean section. The dams were injected over 2 to 3 days to limit cesarean sections to a maximum of 3 per day. The cesarean and the first few moments of life of each calf was video recorded to allow subsequent analysis. During the cesarean section, macroscopic abnormalities of the placentomes were recorded and photographied. Cloned calves were examined promptly after birth to assess respiratory function. Nasal and pharyngeal secretions were removed by suction, and cold water was poured into the ear canal. The cardiorespiratory function was assessed with particular attention paid to respiratory effort, heart rate and rhythm. Apnea, persistent bradycardia, bradyarrhythmia or bradypnea would justify interventional support. Orotracheal intubation, followed by manual or mechanical ventilation was performed when considered necessary. Clinical assessment of the calf was repeated every 15 min if cardiorespiratory parameters were outside reference ranges or respiratory assistance was required. If the calf was stable, clinical assessment was performed every hour for the first 24 hours, then twice daily. Complete blood count (CBC), serum biochemistry profile (SBP), thoracic radiographs (first 24 hours) and arterial and/or venous blood gas analysis was included in the minimal dataset recorded.

Placement of an auricular arterial catheter (medial branch) was attempted in all calves (24G, 0,75 inch). When successfully placed, the catheter was flushed every hour with 1 mL of diluted heparin solution (10 IU/mL). Catheters were left in place as long as there was no local irritation or until no longer required. Catheters remained functional for an average of 3.5 days (1.5 to 6.5 days).

Oxygen administration was performed by intranasal route. A soft plastic canula was inserted in a nostril, up to the level of the medial canthus, and was fixed on a modified small halter. The oxygen released was humidified and at room temperature. The maximal oxygen rate in one canula was 15 L/min. If greater concentration was needed, a second canula could be placed in the other nostril. Rates of oxygen administration were variable (2 L/min to 30 L/min) and adjusted based on the results of blood gas analysis. The lowest oxygen rate required to maintain the PaO2 above 85 to 90 mm of Hg was used.

If respiratory failure was diagnosed (persistent hypoxemia combined with hypercapnia) then assisted mechanical ventilation was initiated using an Esprit ventilator (Respironics Inc.). No sedation was used. A nasotracheal tube was placed and changed every 24 hours or as needed). The technique used is described in Buczinski *et al* (9).

Umbilical care included disinfection with a 5% iodine tincture solution. Frozen colostrum from healthy donor cows was offered to the calves within 2 h of birth. The colostrum was provided by the client and originated from a herd with a known status against important infectious diseases (BVD, BLV, paratuberculosis and IBR). Colostrum quality (IgG concentration) was verified. A sample of colostrum was submitted for bacteriological culture at the moment of freezing and only low bacterial count colostrum was used. The feeding schedule after colostrum feeding (day 2 and further) consisted of administering 2 L of whole milk, by bottle, every 8 h. The whole milk was provided by the client and originated from the herd described above. Orogastric intubation was used if the calf failed to nurse on its own.

**OVERALL RESULTS**

Twenty-six Holstein calves were delivered by caesarean between August 2004 and June 2007 (21 males and 5 females). Births were grouped in 5 different cohorts (2 to 8 calves per cohort). Ten different cell lines were cloned. At least one survivor from each cell line was discharged from the VTH. The cloned calves were delivered by cesarean section between 271 and 279 days of gestation (mean and median at 274 days). Birth weights varied from 35 to 82 kg (median; 55kg, mean; 56 kg). The overall survival rate upon discharge was 21 of 26 (81%). The 3 month survival rate was 17 of 26 (65%). Three calves were euthanized without any attempt at treatment (severe arthrogryposis). Some degree of respiratory problem was identified in 20 of the 23 remaining calves. Thirteen calves were treated with intranasal O2, all of which were discharged alive and in good physical condition. Seven calves were mechanically ventilated, 5 of which were discharged alive.

**MEDICAL CHALLENGES ENCOUNTERED**

**Ante Partum**

The heifers carrying the cloned embryos were monitored closely by the owner and in our experience 2 major problems were faced during pregnancy; hydrops and hepatic lipidosis. Progressive and variable abdominal distension caused by uterine fluid accumulation was a common observation in the last trimester. It was subtle in some cases while it became a reason for veterinary consultation and intervention in other cases. The precise incidence was not known as the heifers were only referred to our VTH on the day of delivery or if severely affected. In the first cohort in particular, several cases of hepatic lipidosis were diagnosed ante partum (pregnancy toxemia). The animals were presented to the VTH several days prior to the planned date of delivery. They were anorectic and severely ketotic. Some animals were unable to stand. All animals responded poorly to initial therapy and the foetus they were carrying aborted. Our hypothesis was that the large fluid filled uterus combined with the overweight foetus was occupying most of the abdominal space, thus preventing the dam from obtaining its nutritional requirements. In the subsequent cohorts, nutrition planning was adjusted, particularly in the last trimester. To characterize the anomalies developing during the pregnancy, ultrasound monitoring was used (10). Transabdominal ultrasonography can also be a valuable technique for the evaluation of the fetuses and the fetal adnexa in the late pregnancy (11). In general, the examination technique was easy to perform and well tolerated by the dams. The information obtained may be limited by the size of the fetus and the abdominal depth of the dam, however fetal integrity, morphology of the placentomes and uterine fluids can be assessed (12). Ultrasonography allowed the early identification of fetal anasarca, which indicates the potential of this technique for early detection of fetal abnormalities (12).

**At Birth**

The newborn clones were extremely challenging. Medical problems were frequent and diverse. One common problem was excessive birth weight (only 2 calves weighed less than 45 kgs). Heavy newborns cannot stand on their own, and seem to loose considerable weight in the first few days of life. A second common problem was the enlarged umbilical vessels (virtually 100% of animals) and concurrent excessive umbilical haemorrhage (very frequent). The umbilical vessels were grossly abnormal in size. Being delivered by C-section, the natural stretching and subsequent rupture of umbilical vessels does not take place. Despite our initial effort to manually stretch the umbilicus, the spontaneous rupture was frequently in close proximity to the body wall. As a consequence, persistent bleeding was inevitable and subsequent infection occurred frequently. In an effort to reduce this problem, we used emasculators and sterile tie-raps applied to the umbilical vessels during the c-section (as a sterile procedure prior to any contamination). While this tended to control most of the bleeding, other significant problems likely arose from these procedures (see below). We are hoping to develop a novel approach to address this problem. A third common and important problem is the cardiorespiratory maladaptation (respiratory failure) at birth. Some newborns were totally apneic (n=5) or severely bradypneic/tachypneic (n=9). They lacked efficient respiratory effort and could not initiate normal respiratory function without assistance. Using assisted ventilatory support, we were able to successfully resuscitate several neonates (n=7). Finally, macroscopic musculoskeletal abnormalities were frequent and often resulted in euthanasia of calves. Severe arthrogryposis and bulldog calf were the most common anomalies we encountered (n=3).

**After Birth**

Even if cardiorespiratory adaptation occurs uneventfully, newborn clones require constant attentive vigilance. Hypoxemia without hypercapnia was commonly diagnosed (n=13). It is our impression that while this hypoxemia is common and marginal, it is associated with clinical signs of weakness and lost of suckling reflex (loss of interest in nursing). We observed clinical improvement when O2 supplementation was provided (13/13 discharged). O2 supplementation was needed for an average of 111 hours and a median of 70 hours. One calf was supplemented for 340 hours (+/- 2 weeks). Other newborns developped hypoxemia and hypercapnia (hypoventilation) (n=7). While less frequent, the affected calves cannot survive without ventilatory support (5/7 were discharged). The precise aetiology of respiratory dysfunction remains unclear and further investigations are in progress. Chest radiographs revealed abnormal findings of various severity and nature in 22 animals in which radiographs were taken. Post mortem examination of 5 animals revealed amniotic aspiration (n=2), interstitial pneumonia (n=2), suspected hyaline membrane disease (n=3) and pulmonary atelectasia (n=3). Respiratory difficulties were the most prevalent but not the only clinical anomalies identified in cloned calves examined (20/26). Associated with respiratory problems, several neonates became extremely weak and unable to stand. The generalised weakness is so severe that calves could not elevate their heads or maintain them in an elevated position. It is not possible to conclude if the weakness resulted from respiratory failure (tissue hypoxemia) or contributed to respiratory failure if the diaphragm also became weakened. In association with the generalised weakness, the nursing ability was often impaired. Twenty-one calves missed at least one meal and 18 missed more than one feeding. The subsequent insufficient feed intake was addressed by esophageal tubing. Providing required nutrient via esophageal feedings can pose particular challenges to the GI tract (a large quantity of milk is required for a 75 kg calf). This is even more concerning when the animal is weak, lacks physical exercise and may or may not have proper GI motility. As a consequence, this problem was approached with a combination of partial parenteral nutrition combined with esophageal feedings. The parenteral nutrition, even if partial, allowed a reduction in the quantity of milk administered per day while still meeting nutritional requirements and preventing a malnourished state from developing. A very common and important problem we faced was umbilical infection and related complications (13/23). Umbilical vessels were large and normal physiological rupture did not occur at the time of parturition. The external structures became infected despite careful management. In our experience, the infection was not limited to the external umbilical structures but also involved the umbilical vein (n=3), arteries (n=5) or the urachus (n=9) (or a combination). Antimicrobial therapy combined with surgical resection was often necessary. For unknown reasons, a high incidence of wound infection and/or dehiscence was observed. In many cases, a second surgery was necessary to correct the hernia that developed following the first surgery. While the use of emasculators and tie raps seemed to reduce umbilical bleeding after birth, it would appear that this procedure may have increased the risk of infection by preventing drainage of the umbilical vessels. In addition to the frequent problems listed above, multiple others conditions were encountered at a lower incidence: infection, hyperthermia, abomasal ulcers, bloat (vagal indigestion), panosteitis and bone marrow aplasia. In addition to the above clinical problems, several haematological or biochemical abnormalities were also identified: anemia, hyperglycemia/hypoglycaemia and azotemia are some examples.

**CONCLUSION**

Although morbidity rate was high as a result of respiratory dysfunction, the mortality rate in this study was low. Intranasal oxygen supplementation seemed to be an effective means of supporting calves suffering from hypoxemia without significant hypercapnia (success rate of 100%). When hypoventilation was associated with hypoxia, mechanical ventilation also appeared to be a viable option, although the number of calves treated was small (success rate; 5/7). Finally 65% of the calves discharged from the hospital were considered healthy at 3 months of age.

In a teaching hospital situation, it is possible to accommodate the birth of a relatively large number of clones if a special team and a standardized protocol is developed. A multidisciplinary approach and a complete commitment of the client are extremely important. Fundamental and clinical research opportunities arise from the challenges encountered. Multiple techniques and material adapted or developed for cloned calves becomes available for the regular NICU.

**REFERENCES**

[1]Chavatte-Palmer P, et. al. **Cloning Stem Cells** 2004;6(2):94. [2]Wells DN, et. al. **Cloning Stem Cells** 2004;6(2):101. [3]Chavatte-Palmer P, et. al. **Biol Reprod** 2002;66(6):1596. [4]Fecteau ME, et. al. **Vet Clin North Am Food Anim Pract** 2005;21(3):637. [5]Hill JR, et. al. **Theriogenology** 1999;51(8):1451. [6]Garry FB, et. al. **Theriogenology** 1996;45:141. [7]Batchelder CA, et. al. **Cloning Stem Cells** 2007;9(1):63. [8]Batchelder CA, et. al. **Cloning Stem Cells** 2007;9(1):83. [9] Buczinski SMC, et. al. **J Vet Emerg Crit Care** 2007;17:179. [10]Kohan-Ghadr HR, et. al. **Therio** 2008;69:218. [11]Buczinski SMC, et. al.**Can Vet J** 2007;48:178. [12] Buczinski S, et. al. **Can Vet J** In press.

**CHALLENGING CASES IN SMALL RUMINANT MEDICINE**

**Mélanie Boileau, DVM, MS, DACVIM**
**Stillwater, OK**

## INTRODUCTION

Small ruminant population in North America is on the rise with a total of 6 million sheep and 3 millions goats, just in the United States alone. Food animal veterinarians are increasingly in demand to provide advice on management, herd health, and disease prevention in these species. Herd investigation of disease outbreak, uncommon illness or simply, unusual presentation of a common disease can be challenging. Obtaining a thorough history and performing a complete physical examination is essential to appropriate diagnosis. Treatment and preventive measures can only be instituted once the problems are identified.

## CASE 1

### Signalment & History

The time of year is October. A valuable 18 month old LaMancha buck weighing 78 kg was presented to our teaching hospital for lethargy, recumbency and anorexia of 4 days duration. The buck was kept on pasture and fed supplemental grain and hay, had free access to goat mineral mix with added monensin, and was drinking out of an automated water through. He was vaccinated with *Clostridium perfringes* type C, D and tetanus toxoid (CDT), and received two 12.5g copper oxide boluses orally, 1.5 month prior to presentation to control gastrointestinal nematodes. All other goats present in the flock appeared healthy.

### Physical exam

Initially, the animal was mentally depressed, sternally recumbent, and tachypneic (80 breath/min). Rumen contractions were 1/2min. Oral mucous membranes (MM) were pink but tacky. When stimulated, hyperexcitability and hypermetria was evident. When left alone, "head pressing" was exhibited. Pupillary light reflex was present but menace response was absent.

### Problem list

History of administration of copper boluses labeled for Cu supplementation in 100-500lb beef or dairy calves, subacute episode of lethargy and anorexia, tachypnea, rumen hypomotility, 5-6% dehydration, mental dullness, hyperexcitability, hypermetria, head pressing and cortical blindness.

### Differential diagnosis

Copper, monensin, lead, salt or sulfur toxicity; thiamine deficiency, pyrrolizidine alkaloid (*Senecio*) or kale (*Brassica*) poisoning, bacterial hepatitis, *Clostridium perfringes* type D, grain overload, cyanide or nitrate toxicity, and rabies.

### Diagnostic plan

Diagnostic work up included CBC, serum chemistry, rumen fluid analysis, urinalysis, blood ammonia levels, ultrasound of the liver, and serum copper levels. During its hospitalization, the most significant changes on CBC included a neutrophilia (d1) followed by profound neutropenia and leucopenia (d4-7), and decreasing hematocrit, from 32% (d1) to 22% (d4-12). Liver enzymes: AST (993; N: 46-161 IU/L) and GGT (291; N: 34-65 IU/L); total bilirubin (2.5; N: 0-0.1 mg/dL), and blood ammonia (99; N: 11-35 umol/L) were initially increased. Rumen fluid analysis was normal (pH=7). Urinalysis was unremarkable except for trace protein, S.G.=1.012, and 1+ bilirubinuria (d2-11). On ultrasound, liver parenchyma was diffusely hyperechoic. Serum copper level was elevated (1.4; toxic levels: 1.2-7.5 ppm)

### Treatment plan

With confirmed diagnosis of copper toxicosis, the animal was treated with sodium thiosulfate (1000 mg PO q24h x 24d), ammonium molybdate (200 mg PO q24h x 24d), D-penicillamine (26 mg/kg PO q12h x 6d), thiamine (10 mg/kg SC q12h x 10d), vitamin E (3200 mg PO q24h x 10d), activated charcoal (1g/kg PO once (d1)), vitamin K1 (1mg/kg SC once (d1)), polyonic fluids with added KCl, and antibiotics.

### Outcome

The animal was discharged 12 days after presentation and reported to be doing very well 3 months later.

## CASE 2

### Signalment & History

The time of year is December. Two female Boer goats, 2 and 4-years old respectively, were admitted to our teaching hospital with presenting complaint of weakness which progressed to paralysis over a period of 6-7 days. The owner reported seeing similar clinical signs in 12 of 30 herd mates, over a period of 2 weeks. Treatment instituted by the RDVM included procaine penicillin and thiamine. With this therapy, 3 affected goats improved partially, 3 did not respond, and 6 eventually died or were euthanized. Post-mortem examinations were not performed. All affected goats were housed on the same 20 acre pasture, fed supplemental prairie hay, had free access to goat mineral mix, and were drinking out of an automated water trough. All other goats present on the premise and housed on different pastures, appeared healthy. Two months prior to presentation, the entire herd was vaccinated with CDT, dewormed with levamisole, and sprayed with a topical insecticide (unknown product).

### Physical examination

*Goat #1* was bright, alert and responsive (BAR) but recumbent. Approximately 10 ticks were found, attached at various locations on the animal's body. Neurological exam showed normal mentation, and no evidence of cranial nerve deficits. Front limbs were normal. Severe ataxia was exhibited when assisted to stand and walk. Paresis was evident in both hind limbs (left > right), with conscious proprioceptive (CP) deficits, increased muscle tone and exaggerated patellar reflexes. Tail and anal tone were normal. The lesion was localized to T2-L3 level of spinal cord white matter, worse on the left. *Goat #2* was also mentally alert with no cranial nerve deficits, and carried low number of ticks. Paresis and CP deficits were present in both front limbs (right > left). Withdrawal reflexes were absent in the right, and delayed in the left front limb. Hind limbs, tail and anal tone were normal. The lesion was localized to the C6-T2 level of spinal cord grey matter, worse on the right.

**Problem list**

History of weakness developing to paralysis within 1 week in BAR animals. Only 2 goats presented for diagnostic investigation but several animals (12/30) also affected with similar clinical signs. Neurologic deficits exhibited relates to multifocal, asymmetrical spinal cord disease. Ticks were present on both goats.

**Differential diagnosis**

Tick paralysis, aberrant migration of *Parelaphostrongylus tenuis* or other parasites, protozoal myelitis (*Toxoplasma*), listeriosis, delayed organophosphate neurotoxicosis, vertebral body osteomyelitis or fracture, spinal abscess, copper deficiency, caprine-arthritis and encephalitis virus, and rabies.

**Diagnostic plan**

Initial diagnostic work up included an electrolyte panel, tick removal and identification, and collection and analysis of CSF. All electrolytes were within normal limits. Ticks were identified as *Dermacentor albipictus*. Lumbosacral CSF revealed increased WBC (100; N: 0-7/uL), red blood cells (5355/uL), protein (123; N: < 40 mg/dL), and CPK (1046; N: ≤ 20 IU/dL). WBC differential was as follow: 53% eosinophils, 43% mononuclear cells, and 4% plasma cells.

**Treatment plan**

With the eosinophilic pleocytosis present on the CSF, a presumptive diagnosis of *P. tenuis*-induced cerebrospinal nematodiasis was made. Both goats were treated with fenbendazole (50 mg/kg PO q24h x 5d), ivermectin (0.2 mg/kg SC), dexamethasone (3 mg/kg IV q3d x 2Tx), and thiamine (10 mg/kg IM q12hr x 5d). Cyfluthrin was administered topically to kill any remaining ticks.

**Outcome**

Goat #1's condition continued to deteriorate over the course of 7days; was euthanized and necropsied. No gross lesions were observed on postmortem examination. On microscopic exam of the spinal cord, several nematodes were present on cross-section. They were identified as *Parelaphostrongylus tenuis*. Goat #2 recovered completely within 7 days of treatment.

**CASE 3**

**Signalment & History**

The time of year is May. A 3-year-old Suffolk ram was referred to our teaching hospital with a presenting complaint of weight loss and reduced appetite of 3 months duration. The ram responded poorly to treatment (tilmicosin, dexamethasone, B-vitamins) administered 2.5 month prior to presentation. He was housed in a dry lot, fed Bermuda grass hay ad lib and drinking from a water trough. Vaccination (CDT) and deworming (fenbendazole) were current.

**Physical examination**

The ram was moderately depressed, had a body condition score of 1/5, prolonged skin turgor, and delayed CRT (~3sec). Rumen contractions were counted at 5 to 6 per 2 min. A firm watermelon sized mass could be palpated in the right ventral abdomen which was distended.

**Problem list**

Chronic weight loss, poor appetite and body condition score, 5-6 % dehydration, hypermotile rumen, right-sided ventral abdominal distension with a firm palpable mass.

**Diagnostic plan**

Initial diagnostic work up consisted of abdominal ultrasound, CBC, serum chemistry, and rumen chloride concentration. Ultrasonographic evaluation was consistent with the firm mass being an enlarged abomasum. The animal was leucopenic (WBC: $3.5 \times 10^3$; N: 4-12 $\times 10^3$/µL). Serum electrolytes were within normal limits, except for calcium (8.8; N: 11.5-12.8 mEq/L). TCO2 was elevated (43.2; N: 21-28 mEq/L). Rumen chloride concentration was above published reference range for Suffolk sheep (24; N: 8-15 mEq/L).[1] Liver enzymes (GGT, AST, LDH) were elevated. Exploratory celiotomy was scheduled for the following morning.

**Differential diagnosis**

Abomasal emptying defect, abomasal impaction, obstruction of reticulo-omasal orifice or pylorus with foreign body or trichobezoars, and peritonitis.

**Treatment plan**

A jugular catheter was placed and polyonic fluids, supplemented with calcium, were administered over night. The animal was fasted for 24 hours. Following routine presurgical preparation while under general anesthesia, the abdomen was entered via a right paracostal approach. The abomasum was approximately 6 times normal size and filled with abnormally fibrous material. There was no evidence of peritonitis. The abomasum was opened and no obstructions were detected in the reticulorumen or abomasum.

**Outcome**

Due to poor prognosis associated with abomasal emptying defect, the ram was euthanized on the surgical table with the owner's prior consent.

**COPPER TOXICITY**

Copper (Cu) is an essential nutrient for domestic animals. However, accidental or unintentional ingestion of abnormally high doses of Cu or prolonged excessive supplementation can result in acute or chronic toxicity. Sources of Cu include concentrated feed mixtures or mineral supplements designed for horses/cattle/pigs, Cu sulfate foot bath, overdoses of parenterally administered Cu salts, Cu piping or plumbing, and access to slurry from pig units and poultry litter[2] used as fertilizer. Sheep are the most susceptible species to the effects of Cu,[3,4] but poisoning has been documented in calves,[5] and in goats,[6] the latter species being the least susceptible. Because daily hepatic biliary Cu excretion is less than 1% of the amount ingested,[7] Cu accumulates in the liver over periods of time ranging from a few weeks to several months. During that time, the animals remain asymptomatic. Cu buildup is followed by a sudden release of the accumulated Cu into the blood, initiating an acute hemolytic crisis and clinical signs. Stress factors such as transportation, lactation, and diet modification typically precipitates the hemolytic crisis. The toxic dose for each species is variable and depends on the duration of exposure, and the amount of Mo, Zn, Ca, and sulfate being fed.

In sheep, clinical signs include anorexia, dehydration, tachycardia, icterus or pale mucous membranes, red serum, dark-red urine, bruxism, diarrhea, and death. In goats, intravascular hemolysis appears to be uncommon to absent.[6] In chronic cases, dullness, weakness, hypothermia and abortion may also be present. Clinicopathological abnormalities include severe anemia with Heinz bodies, hemoglobinemia, methemoglobinemia (<5%); increased serum bilirubin, liver enzymes (GGT, AST, SDH, LDH) and blood ammonia; hemoglobinuria, and proteinuria. Liver biopsy sample is the most sensitive indicator of Cu poisoning, however plasma or serum can also be submitted for Cu level determination. Normal blood and liver Cu levels in goats range from 0.6-1.5 ppm, and from 12-115 ppm, respectively. On postmortem, the presence of generalized icterus, watery blood, enlarged friable yellow liver combined with swollen, dark brown to black metallic kidneys is strongly suggestive of Cu toxicosis. Treatment is directed at removing the animals from the offending source and promoting elimination of Cu from blood and liver. In acute hemolytic syndrome where PCV < 10%, nasal oxygen, vitamin E (2000 IU PO q24h), and packed red blood cells or whole blood transfusion is indicated. Chelation therapy includes D-penicillamine (26-52 mg/kg PO q24h x 6d), ammonium molybdate (300 mg PO q24h), and sodium thiosulfate (300 mg/kg PO q24h). Ammonium tetrathiomolybdate (2-15 mg/kg IV q24h x 3-6d) can also be used to increase hepatic biliary Cu excretion. Recommended withdrawal times in meat and milk for these antidotes in food animals have been published.[8] Control of acute Cu poisoning depends on common sense to avoid accidental exposure to and ingestion of high doses of Cu.

## PARELAPHOSTRONGYLUS TENUIS (MENINGEAL WORM)

Migration of *Parelaphostrongylus tenuis* (PT) in the central nervous system of white-tailed deer (WTD), definitive host, is relatively inoffensive, but aberrant migration does occur in domestic and wild ungulates. Cerebrospinal nematodiasis caused the migrating third-stage larvae of PT (L3) has been reported in camelids (llama, alpaca), sheep, goat, cattle, antelope, moose, wapiti, caribou, reindeer, fallow and mule deer. The disease is endemic in North America, especially in the northeastern United States and western Canada. Highest incidence of clinical disease occurs during cool and moist weather, typically in late fall or winter. The parasite has an indirect life cycle that involves terrestrial slugs and snails as intermediate hosts.[9] Host susceptibility to PT varies between species. Fatal infection can be experimentally induced in llamas with as little as five L3 larvae.[10] Moderate signs of ataxia and paresis have been reported in Bighorn sheep given 300 larvae.[11] In goats, recovery has been observed in experimentally challenged animals and in natural outbreaks. The wide variety of neurologic signs observed in clinical cases reflect the frequently multifocal or asymmetrical nature of the parasite distribution in the central nervous system. Typically, clinical signs of disease reflect spinal cord deficits such as paresis or paralysis of one or more limbs, ataxia, CP deficits, toe dragging, dog-sitting position or inability to rise. In most cases, affected animals are mentally alert and continue to eat and drink.

Presence of eosinophilic pleocytosis on cytologic examination of the cerebrospinal fluid is suggestive but not pathognomonic of PT infection.[9]A serum antigen-capture ELISA has been developed for the diagnosis of PT in elk, WTD[12] and moose, but it has not been validated for other domestic species. Definitive diagnosis depends on identifying the parasite in nervous tissue at time of necropsy. To be successful in treating clinical parelaphostrongylosis, diagnosis and aggressive treatment must be instituted early. Published treatment regimens include multiple high doses of fenbendazole, ivermectin, moxidectin, and diethylcarbamine in combination with corticosteroids. Treatment success is variable. Prevention should be based on keeping deer out of the pastures with proper fencing, raising domestic species in confinement, or avoid grazing the edge of the woods or the water where deer drink.[13] Deworming monthly with ivermectin as been recommended in llamas and alpacas in areas cohabitated with WTD,[9] however this practice may eventually lead to resistance of more common gastrointestinal parasites (ex: *Haemonchus contortus* in small ruminants).

## ABOMASAL EMPTYING DEFECT

Abomasal emptying defect (AED) has been described as functional pyloric stenosis, abomasal impaction, and abomasal outflow defect. It presents similarly to abomasal impaction in cattle, but is recognized mainly in the Suffolk breed.[1] AED in Dorset, Hampshire, Texel, and recently British Toggenburg goats, has also been also documented.[14] Reported age of onset is from 2 years-old and greater. The precise etiology and pathogenesis of AED is unknown. One author has suggested that AED may be an acquired form of dysautonomia.[15] This disease is associated with concentrate feeding and often occurs in the post lambing period although pregnant ewes are also susceptible.[16]

Clinical signs include chronic weight loss, anorexia, abdominal distention, and palpation of a doughy mass in the right ventral abdomen. Tachycardia and atrial fibrillation may be present.[17] Elevated rumen chloride is indicative of disease. Hypochloremic metabolic alkalosis is uncommon. Liver enzymes (AST, GGT, SDH) may be increased. On postmortem, the abomasum is greatly distended (2 to 6 times normal size), and the contents may be liquid or dry. Treatment with laxatives (mineral oil, magnesium sulfate), motility modifiers (metoclopramide, erythromycin), and abomasotomy have been mostly unsuccessful.[16]

## REFERENCES

[1] Rings DM et al. **J Am Vet Med Assoc** 1984;185(12):1520-1522. [2] Christodoulopoulos G et al. **Aust Vet J** 2007;85(11):451-453. [3] Ortolani EL et al. **Vet Hum Toxicol** 2004;46(6):315-318. [4] Roubies N et al. **J Vet Pharmacol Therap** 2008;31(2):181-183. [5] Dwayne WH et al. J Vet Diagn Invest 1997;9:441-443. [6] Cornish J et al. **J Am Vet Med Assoc** 2007;231(4):586-589. [7] Smith BP. In **Large animal internal medicine, 4th ed**, Mosby Elsevier 2009;1166-1169. [8] Haskell SRR. **J Am Vet Med Assoc** 2005;226(6):884-887. [9] Nagy DW. **Vet Clin Food Anim Pract** 2004;20:393-412. [10] Rickard LG et al. **J Zoo Wild Med** 1994;25(3):390-402. [11] Pybus MJ et al. **J Wildlife diseases** 1996;32(4): 614-618. [12] Ogunremi O et al. **J Vet Diagn Invest** 1999;11:515-521. [13] Smith MC. In: **Proceedings Am Assoc Bov Pract** 2005;38:123-127. [14] Edwards GT, et al. **Vet Rec** 2008;162:418-419. [15] Pruden SJ, et al. **Vet Pathol** 2004;41:164-169. [16] Ruegg PL, et al. **J Am Vet Med Assoc** 1988;193(12):1534-1536. [17] Moresco A, et al. **J Am Vet Med Assoc** 2001;218(8):1264-1266.

# JOHNE'S DISEASE: SUPERSHEDDERS, PASSIVE SHEDDING & ADULT INFECTION

**Robert H. Whitlock, DVM, PhD, DACVIM**
**Kennett Square, PA**
**IA Gardner, Y Schukken, BL Mangold, RW Sweeney, J Van Kessel, J Karns, J Smith, E Hovingh, D Wolfgang**

## INTRODUCTION

Johne's disease continues to emerge as an important infectious disease of many ruminant species having it's greatest impact on dairy cattle. Both the 1996 and the 2002 Dairy NAHMS study reported Johne's as important economical disease of dairy cattle. Isolation of *Mycobacterium avium* Subsp. *paratuberculosis* (MAP) bacteria by fecal culture is widely recognized as the ante-mortem "gold-standard" to identify cattle with Johne's disease (Collins). Current laboratory practice standards in the U.S. does not include enumeration of MAP colony-forming unit (CFU) counts beyond 50 visible CFU per tube on solid media such as Herrold's Egg Yolk Medium (HEYM). Culture tubes with CFU exceeding this level are commonly designated as *Too Numerous to Count* (*TNTC*) and the sample is reported as being a "high" or "heavy shedder". Serial dilution of samples collected in a study has established that samples classified as TNTC can have a wide range of CFU counts (100,000 to more than 4 million cfu/gram). We propose the term "**super-shedders**" to describe a subset of shedding cows based on further enumeration of MAP CFU counts. This so-called super-shedder phenomenon has been reported for other bacteria including verotoxigenic *Escherichia coli* O157 in cattle (Bach; Cray) and MAP infection in sheep (Whittington). We believe that super-shedder dairy cows have always existed but that this phenomenon has not been adequately recognized or accounted for in control programs such as the Voluntary Bovine Johne's Disease Control Program (VBJDCP).

Preliminary observations suggest that most super-shedder cows do not have clinical signs of Johne's disease (weight loss & Diarrhea), although they shed as much or significantly more MAP into the environment than a typical cow with clinical Johne's disease. Since these cows are not readily recognized by dairymen, super-shedder cows may be responsible for massive environmental contamination that results in a disproportionate percentage of the new MAP infections that occur in a herd. In addition, they may contribute to the fecal 'pass-through' phenomenon previously described by Sweeney(1992), where positive fecal culture results are obtained from non-infected cows for several days after they have consumed feces from infected cattle.

Traditionally, cattle were classed as either positive or negative for Johne's disease based on growth of MAP from a fecal culture. Culture-positive cattle were judged to be infected, shedding MAP, and potential sources of infection for herd-mates. The standard practice was to recommend that all positive animals be culled as quickly as possible because a positive fecal culture has been considered the "gold standard" reference test with 100% specificity (Collins). Over time, however, it has been recognized that differences exist in MAP shedding levels among culture-positive cattle, and the higher the number of visible colony forming units (CFU) on the surface of solid media, typically Herrold's egg yolk medium (HEYM), the greater the risk of spreading disease. Cattle are classified as low (1 to 9 CFU/tube), moderate (10 to 50 CFU/tube) or high/heavy (>50 CFU/ tube) shedders. One recent study (Crossley) showed that most infected cows were in the low (51.4%) or high (30.8%) CFU categories. Although the categorization of CFU may vary slightly among laboratories, other published studies (Whitlock) have yielded similar findings.

Most U.S. laboratories now report the number of visible MAP colonies on each tube of HEYM. However, most laboratories do not enumerate above 50 to 70 CFU/tube. Samples with CFU > 50 are classed as TNTC or heavy shedders. During 2004, the laboratory committee of the National Johne's Working Group (NJWG) discussed the need to enumerate heavy shedders. Based on an overwhelming consensus opinion of experienced Johne's investigators, little justification could be provided to count > 50 CFU/tube. As a result, national policy was developed whereby Johne's testing laboratories doing solid media culture are required to report any fecal sample with > 50 CFU of MAP/tube as a heavy shedder. Most members of the laboratory committee concurred at that time that some cattle classified as heavy shedders had >100 CFU/tube but there was **NO biological reason** to justify enumerating MAP CFU above 50 CFU/tube.

We now realize that this decision was flawed because there is a vast range of shedding levels greater than 50 CFU. Some cows with TNTC culture results are super-shedders, and if they are not identified, the consequences could be disastrous. Current recommendations have generally evolved to identification of the most infectious cows by culture or ELISA with subsequent culling of the most heavily infected animals first. Theoretically, removal of all TNTC cows should prevent any further contamination of the dairy environment with MAP. However, not all dairymen expeditiously cull cows that have fecal MAP counts that are reported as TNTC. If a super-shedder cow is producing well and has no clinical signs of Johne's disease, the animal is likely to stay in the herd even though it is shedding massive numbers of MAP. Prioritization of cows for culling on the basis of more categories of shedding levels (with less emphasis on low and moderate CFU) would have practical utility just as ELISA interval-based likelihood ratios (Collins) have had for serologic results.

Fecal samples for RT-PCR (qrt-PCR) were evaluated using the Tetracore quantitative real-time PCR, according to manufacturer directions. The target sequence of the MAP PCR is the hspX gene, which contains MAP-specific sequences capable of distinguishing MAP from other organisms including other Mycobacteria. Use of the *hsp*X gene for diagnostic detection of MAP is covered by US Patent No. 5,985,576, under exclusive license to Tetracore, Inc., Rockville, MD from the USDA. The extraction procedure involves bead-beater, and chaotropic separation of DNA.

Retrospectively, fecal samples classified as heavy shedders (>70 cfu/tube) from three demonstration herds (herd-A, 312 cows; herd B, 103 cows and herd C, 125 cows respectively) were serially diluted at 1:100 and 1:1,000 to determine MAP cfu/gm of manure. Many of these samples were also tested by RT-PCR to determine Ct. Lower cfu samples (both low and moderate shedders) were cultured repeated to determine which culture positive fecal samples may be the result of "passive shedding" due to manure contamination of the environment. Cattle that were both ELISA negative and culture negative on two or more sampling times were classified as "passive shedders".

**Supershedders**

The initial pilot dilutions from 1:1 to 1:50,000 suggested that a 1:1 dilution would be valuable to re-confirm the original MAP cfu prior to freezing at -70$^0$C for a period of time. Dilutions of 1:100 and 1:1,000 for routine processing of samples suspected to be super-shedders provided an acceptable range to determine final MAP cfu/gm of manure. The estimated range of mycobacteria in these five samples previously recorded as heavy shedders was 63,000 to 1,470,000 MAP cfu/gram. In a similar manner the cycles to positive (Ct) ranged from 33.3 for the lowest cfu/gm to 19.4.

The RT-PCR cycles to positive (Ct) was closely related to the MAP cfu/gm in the original five serially diluted samples. The Ct range was 33.3 to 19.4 for those fecal samples with 63,000 to 1,470,000 cfu MAP/gm. Subsequent comparisons of RT-PCR to serial dilutions of fecal samples has provided an excellent correlation. Our preliminary estimates are that 10% of heavy shedders (or about 2 to 3% of all culture-positive cattle at a single time-point) may be super-shedders, excreting >10 billion CFU MAP per cow per day.

Initially, super-shedders were thought to represent unique biological events, developing only in the occasional infected animal. It was theorized that some trigger was required to cause a MAP-shedding cow to become a super-shedder, or that perhaps the super-shedder was infected either with a unique strain that grew faster in the laboratory or a more pathogenic strain resulting in heavier tissue infection and thus higher shedding concentrations of MAP. The biological significance remained elusive, but these animals represent a new fundamental understanding of Johne's disease. Contrary to our initial supposition that it would be a rarity, super-shedders have been detected in nearly every herd with Johne's disease our laboratory has tested since the start of the study in 2006. Hence, we now believe that any active MAP shedder has the potential to become a SS, if the animal remains in the herd for an adequate period.

**MAP SS are superspreaders**

Super-shedder cattle represent the Trojan horse for Johne's disease. Cows at this later progressive stage of the disease, which is not necessarily clinical, represent the greatest threat to spread MAP to other cows in several ways including:

- Cows with > $1 \times 10^6$ CFU MAP/gram disseminate MAP so effectively on the farm, that <1g ingested by another cow in the ration or on pasture will likely result in transient fecal shedding of MAP. It may also induce adult infection, if the cow is not already infected
- Footwear, vehicle wheels, and other fomites contaminated with just a few grams of manure from these cows are potential sources of infection for newborn calves
- Calves born to these cows are likely have massive tissue concentrations of MAP soon after birth.
- Milk and colostrum from super-shedder cows are more likely to contain MAP and infect the newborn calf. The skin of the udder is easily contaminated with MAP and may be positive, even though fecal culture from the same cow is negative.
- Super-shedders serve to contaminate pasture extensively with MAP resulting in exposure of grazing cohorts (as in a leader-follower system) to large numbers of MAP

**Passive shedding**

The initial concept of "pass-through" was first reported from this laboratory 16 yrs ago (Sweeney). Feces from a cow with clinical Johne's disease was administered via oral-gastric intubation to 6 two-year-old dairy heifers. Two heifers received 240,000,000 CFU, two heifers received 60,000,000 CFU and two heifers received 15,000,000 CFU of MAP. A second dose was administered to each heifer 24 hours later. The dosing was repeated two weeks later with feces obtained from another cow with clinical Johne's disease. Fecal samples were collected from each of the heifers three times per day for two weeks. MAP was first detected in the heifers' fecal samples approximately 24 hours after initial dosing. In this experiment, not only did passive shedding occur after oral ingestion of MAP, but the concentration of MAP in the fecal samples was dose-dependent. Extrapolated results from the above data suggest that ingestion of 1,000,000 CFU would result in 2 CFU per each of four tubes and 100,000 CFU MAP would yield 1 CFU on one of four tubes. This in turn supports the conjecture that detectable passive shedding could result from ingestion of feed containing 100,000 CFU MAP (the equivalent of 1 small drop of feces from a super-shedder).

One herd with a super-shedder resulted in extensive environmental MAP contamination and we suspect that 14/17 (80%) of the culture positive cattle in that herd were the result of pass-through. It now is apparent that most low shedders and some moderate shedders may be the result of super-shedder environmental contamination. Our current estimate suggests that 30%, and in some extreme circumstances up to 90% of fecal culture positive cattle may be the result of "pass-through."

Since the presence of a super-shedder in a herd may result in many cows with pass-through culture positive fecal samples, newer diagnostic efforts need to focus on identification of the super-shedder and less attention given to low or moderate shedders (in terms of necessity to cull these cows). More sensitive organism detection techniques will only detect more pass-through positive cows and not necessarily cattle with potential infectious disease. Research efforts need to be re-directed to detect potential super-shedders and not the low-shedding, potentially "pass through" false positive cattle in the herd.

**Adult infections**

Over the past 4 years our laboratory has followed three dairy herds with Johne's disease to better understand the biology of the natural disease. Super-shedders identified in each herd have been implicated in passive MAP-shedding by herd mates. Passive-shedder status in herd mates has been confirmed by (1) the detection of MAP in their fecal samples with subsequent fecal samples yielding negative results, (2) negative Johne's serologic tests and (3) MAP-negative or low-positive tissue culture results (the most sensitive method to detect MAP in tissues) from tissues obtained at slaughter.

Over time in these three RDQMA herds more than 50% of all cows with culture-positive fecal samples were confirmed to be passive shedders. Passive shedders were typically serologically negative and culture positive on one fecal sample at one point in time when a super-shedder was present. The great majority of passive shedders were low or moderate shedders, ostensibly because the small amount of contaminating feces was diluted out by the huge rumen capacity (120 liters) and the total amount of feed materials consumed each day.

Approximately 30% of the cows culled from each herd were followed to slaughter to collect four tissue samples plus an additional fecal sample. Most (70%) cattle characterized as passive shedders had no positive tissue culture results, and 30% were found to have low-concentration MAP-positive tissue culture results. By comparison, cattle characterized as active shedders had massive concentrations of MAP in their tissues, often exceeding 10 million CFU/gram of tissue; nearly a thousand fold greater MAP concentrations than tissues from passive shedders. Based on earlier data from this laboratory, these cows are more likely to have disseminated MAP infection throughout their body tissues and in their milk. Thus MAP super-shedders efficient MAP superspreaders disseminating infection to many herd-mates. This infection of adult herd-mates is generally unrecognized by the scientific community. A recent national survey reported than more than 68% of herds were infected and with the recognition of MAP SS, a much lager proportion of cows within herds are likely to be infected than heretofore recognized, perhaps 50% of cows in the herds.

**SUMMARY**

This report further documents the critical importance to identify super-shedders since they contribute billions of MAP cfu daily to the environment serving to disseminate MAP to susceptible cattle and in many herds result in significant number of culture positive fecal samples through passive shedding. The frequency of super-shedders among culture-positive cattle in infected herds is unknown. However, our preliminary investigations have shown that each of 3 current herds has at least 1 super-shedder cow, with 5 super-shedder cows having been identified among the 7.1% (41/575) culture-positive cattle in the herds.

Two types of fecal MAP shedding exist; 1) passive shedding due to pass-through of MAP in the intestinal tract and 2) active shedding due to more extensive MAP infection of the host intestinal tissues. In some herds passive shedding or false positive fecal cultures may account for more than 50% of the culture positive samples. The challenge will be to differentiate active shedders from passive shedders.

**Acknowledgements**

Financial support for this work was provided by the USDA Agricultural Research Service (58-1265-3-115) by USDA-APHIS-VS field studies funding, JDIP funds (Q6286224122) and by the USDA-VS-National Demonstration Herd Project.

**REFERENCES**

Crossley BM, et al. Vet Micro 2005; 107:257 ; Collins MT, et al. Clin Diagn Lab Immunol 2005;12:685; Bach SJ, et al. FEMS Microbiol Lett 2005;252:25; Cray WC. Et al. Appl Environ Microbiol 1995;61:1586; Pradhan A, et al. J Dairy Sci Accepted 2009; Sweeney RW, et al. Am J Vet Res 1992;53:1312; Whitlock RH, et al. Vet Micro 2000; 77:387; Whittington RJ, et al. Aust Vet J 2000;78:34.

## PARATUBERCULOSIS IN CAMELIDS: CLINICAL PRESENTATION, DIAGNOSIS, & CONTROL

**Marie-Eve Fecteau, DVM, DACVIM**
**Kennett Square, PA**

### INTRODUCTION

Despite the fact that paratuberculosis, or Johne's disease has long been reported in camelid species[1-2], the body of knowledge regarding most aspects of the disease in camelids remains poor. For instance, disease prevalence in North American herds is still unknown. For this reason, many large animal clinicians will rely on the abundant bovine literature on the topic to guide their clinical approach in the diagnosis and control of the disease in their camelid patients. Some aspects of the clinical presentation may be very similar between species, but some other aspects may also differ greatly.

### EPIDEMIOLOGY

The bacterium responsible for paratuberculosis, *Mycobacterium avium* subsp. *paratuberculosis* (MAP) causes granulomatous inflammation of the intestines and mesenteric lymph nodes. Several strains have now been identified in most domestic and some wild ruminant species such as cattle, sheep, goats, camelids, bison, and deer. Although most strains appear to be species specific, other strains have also been reported to infect ruminants between species[1, 3-4]. In all 3 previously referenced studies, DNA fingerprinting of several alpaca or llama MAP isolates were identified as cattle strains[3-4]. Therefore, most strains should be regarded as infectious to ruminant species other than the species of origin. The type of strain responsible for an infection in a given animal may also have an influence on the diagnosis as some strains are regarded as more fastidious to grow in laboratory[5].

As in cattle, transmission of MAP in camelids is assumed primarily fecal-oral. The possibility of transfer of infection *in utero*, or via milk or colostrum is unknown in camelids. As in bovine herds, when paratuberculosis is introduced into a previously unexposed herd of camelids, many individuals can become infected prior to any animals showing clinical signs of disease. The subclinically infected animals may be contaminating the premises with the organism, thus exposing susceptible herd mates to infection. By the time an animal shows clinical signs of disease, several herd mates may be in the subclinical phase of the disease. No studies have documented age-related resistance to infection in camelids, however, in cattle, neonates and younger animals are assumed more susceptible.

### CLINICAL PRESENTATION

Clinical signs of paratuberculosis in camelids are similar to those reported in cattle and small ruminant species and may include lethargy, chronic weight loss, and dependent edema. In contrast to cattle, but similar to small ruminants, not all camelids in advanced stage of disease will develop diarrhea or changes in fecal consistency. Both llamas from a previously mentioned report showed signs of weight loss associated with diarrhea[1]. In another report including 10 alpacas, 5 animals had experienced diarrhea before death[2]. The difference in the occurrence of diarrhea between cattle and camelid (and small ruminants) may be related to the latter's intestinal tract physiology and their ability in reabsorbing water. Also in contrast with cattle, camelids and small ruminants affected with paratuberculosis can present with anemia of variable severity. The exact pathophysiology for the development of anemia in camelids and small ruminants with paratuberculosis is unclear, but could be secondary to the chronicity of the disorder, especially when the anemia is classified as normocytic, normochromic, and nonregenerative. In addition, concurrent disease processes such as internal parasitism or infection with *Mycoplasma haemolamae* may contribute to the degree of anemia. The reported age at onset of clinical signs tend to be younger in camelids than in cattle. Three of the 10 alpacas reported before were between 11 and 14 months of age[2], and one of the 2 llamas in the other report was 16 months-old[1]. A possible explanation for this finding is that the incubation period for certain MAP strains is shorter in camelids that in other species; however, this has not been proven.

### DIAGNOSIS

#### Culture

Fecal culture is considered the standard method of diagnosis. However, it requires 8-16 weeks incubation, which may be considered as a downside when attempting to obtain a diagnosis on a sick animal, or when a quick turnaround is needed during a herd screening. Both fecal and tissue samples can be cultured for MAP using standard Herrold's Egg Yolk Medium (HEYM) according to methods previously described[6-7]. Whenever possible, tissue of euthanized animals should be submitted for culture, in order to increase detection of positive animals that may not yet be shedding, or that may have intermittent shedding. Tissue submitted should include samples of ileum, ileocecal junction, and ileocecal lymph node. Also whenever possible, quantitative fecal cultures should be performed in order to identify the heavier shedders.

#### Serology

Serologic testing of clinically affected camelids has been reported using both AGID[1-2] and ELISA[1] techniques. The AGID detected one of four clinically affected alpacas tested in the case report by Ridge et al[2], and both of the infected llamas in the Belknap[1] study, giving an overall sensitivity of 50% with both groups combined. A commercial ELISA identified both cases as positive in the Belknap[1] study. Later, the use of a modified bovine ELISA was reported to have more promising results for the detection of camelid antibodies to MAP[8]. The increased sensitivity of the modified ELISA allowed for a more rapid, and cost effective detection of the disease in clinical animals. In another publication, the same group also compared the specificity of 3 modified bovine ELISAs and one AGID kit in a culture-negative herd of llamas and alpacas[9]. Specificities were 48%, 92%, and 98% for the 3 ELISA tests respectively; and 100% for the AGID test. The authors' conclusions were that these essays are best used for rapid presumptive diagnoses of llamas and alpacas with diarrhea and weight loss, and as a screening tool for herds known to be exposed to infection. They also recommended that all seropositive animals be confirmed by culture.

**RT-PCR**

More recently, most laboratory specializing in the diagnosis of paratuberculosis in cattle have started using bovine RT-PCR kits for the detection of MAP in the feces of other ruminants, including llamas and alpacas. RT-PCR allows for very fast results and is considered by the author to be quite reliable in the detection of animals that are shedding MAP in their feces.

**Necropsy**

Necropsy findings of clinically affected animals are very similar to cattle and include a thickened and corrugated mucosal surface of the distal small intestine and colon, and enlarged mesenteric lymph nodes. Histologic findings include infiltration of the lamina propria by epithelioid macrophages that may contain acid-fast staining organisms, lymphocytes, and plasmocytes. It is the author's opinion that not all clinical cases will be positive for acid-fast stain organisms, and that tissues should always be harvested for culture in addition to histology.

## CONTROL

Management of paratuberculosis in camelid herds relies on the same principles described in cattle. Controlling fecal-oral transmission in home-reared replacements, buying replacement from herds with no history of disease, frequent testing, and culling or segregation of test-positive animals are key to any good control program[10]. When suspecting a herd problem, testing RT-PCR testing of individual fecal samples or pooled fecal samples may help ruling out the presence of MAP within the herd. Animals from positive pools should be re-tested individually. RT-PCR-positive individuals should then be confirmed via standard culture prior to an animal being culled. All positive animals should be identified and segregated from the rest of the herd and should be pastured on separate fields.

## REFERENCES

[1]Belknap EB, et al. **JAVMA** 1994; 204(11): 1805. [2]Ridge SE, et al. **Aust Vet J** 1995; 72(4): 150. [3]Cousins DV, et al. **Aust Vet J** 2000; 78(3): 184. [4]Feizabadi MM, et al. **Aust Vet J** 1997; 75(12): 887. [5]Juste RA, et al. **Vet Microbiol** 1991; 28 : 385. [6]Shin SJ. ***Proceedings.* 93rd US** Animal Health Association Meeting 1989; 381. [7]Sweeney RW, et al. **J Clin Microbiol** 1992; 30: 166. [8]Kramsky JA, et al. **Vet Microbiol** 2000 ; 77 : 333. [9]Miller DS, et al. **J Vet Diagn Invest** 2000 ; 12 : 345. [10]Stehman SM. **Vet Clin NA** : Food Animal Practice 1996 ; 12 : 441.

# RENAL DISEASE IN CAMELIDS

**Christian Gerspach, Dr. med. vet., MS, & J. Lakritz DVM, PhD, DACVIM**
**Zurich, Switzerland & Columbus, OH**

## INTRODUCTION

Renal disease can occur primary or secondary to an underlying disease. Diseases involving the urinary tract appear not to be very frequent, according to the literature. It is uncertain if they are infrequent or under diagnosed in camelids. Most commonly reported have been congenital abnormalities[1-3], systemic infections including the kidneys[4-7] primary renal neoplasia[8-11] and neoplasia including the kidneys.[12,13] Clinical signs of renal disease are often not specific. Therefore a careful work up should include hematologic and biochemical profiles as well as urinalysis. Nephrotoxic drugs should be used carefully and avoided in azotemic camelids.

## HISTORY & EXAMINATION

Renal disease usually is not a primary complaint in camelids referred to a veterinary clinic. Therefore a good history is crucial to obtain evidence of clinical signs accompanying renal disease or nephrotoxic drugs given to the patient previously.
A clinical history should include information about, weight loss, medications (dose, duration), supplementations (Vitamin D), acute or chronic progress of the disease, deworming and recent fecal examinations for parasites.

Common historical information in camelids with renal disease includes weight loss or failure to gain weight, weakness, anorexia, soft feces and anemia. A careful physical examination should include a urinalysis.

### Clinical Pathology

A complete blood cell count (CBC) and a biochemical profile will reveal evidence of anemia, hypoproteinemia, electrolyte abnormalities and azotemia. Anemia can be associated with hematuria. Causes of hemorrhage from the proximal portion of the urinary tract may include infection, trauma, toxicosis, nephrolithiasis, neoplasia, and vascular anomalies.
Hypoproteinemia is a common problem in camelids and can be associated with parasitism, blood loss, protein losing enteropathy or protein losing nephropathy. Poor healing of wounds has been observed in camelids with hypoproteinemia in our clinic and might be an indicator. Azotemia is a very common finding in sick camelids. Most of the camelid patients are dehydrated and azotemia is prerenal. Prerenal azotemia can only be assumed if renal and post renal azotemia has been ruled out.

### Urinalysis

A urinalysis is a very important part of the evaluation of the urinary tract and renal function. First the ability to pass urine should be evaluated. Specific gravity of the urine is a simple test for the ability of the kidneys to excrete solute and conserve fluid.
In patients with proteinuria or glucosuria measurement of the specific gravity can be inaccurate. In patients with pigmenturia the urine should be centrifuged to differentiate between hematuria and hemoglobinuria/myoglobinuria.

### Ultrasound

Ultrasound of the kidneys provides important information to investigate size and texture of the kidneys. This technique is important to detect renal neoplasia, renal agenesis, hydronephrosis and parenchymal abnormalities. Both kidneys have a size of 5 x 9 cm. The left kidney can be approached from the left paralumbar fossa, ventral to the transverse processes 5-7 of the lumbar vertebrae. The abdominal organs adjacent to the left kidney are compartment 1 cranially and the spleen caudally. The right kidney can be imaged ventral to the transverse processes 4-6 of the lumbar vertebrae. On the right side the liver is adjacent to the kidney.

## RENAL DISEASES

Congenital abnormalities of the kidneys have been described in camelids. Unilateral[1] and bilateral[2] renal agenesis of the kidneys were reported in crias, leading to azotemia and death. Both cases were diagnosed by abdominal ultrasound. Ureteral duplication was reported in a cria resulting in hydronephrosis.[3]

Systemic infections with *Candida albicans*[4,5] and *Cryptococcus neoformans*[6] affected different organs but also the kidneys. All camelids reported above were adult.

Renal neoplasms have been reported in camelids, such as unilateral renal papilloma[8], multicentric urinary papilloma involving both kidneys and a ureter[10], both cases associated with anemia and hematuria. Other renal neoplasms reported were renal teratoma in a llama[9] , nephroblastoma in a guanaco[11] and renal lymphosarcoma.[12,13]

## CASE-BASED APPROACH

### Case One

A two year old alpaca was presented for severe edema, diarrhea and depression of 4 month duration. The animal has been treated with lincomycin and sulfamethoxazole previously. The alpaca has been dewormed repeatedly and fecal examinations were positive for coccidian and trichostrongyles on each occasion.

On physical examination the vital parameters were normal. The animal showed severe weight loss with a body condition score of 1/10. There was a severe pitting edema of eyes, head, face and ventrum. On succussion of the abdomen a fluid wave was observed. The alpaca had profuse, watery (cow pie) diarrhea. The initial diagnostic procedures included a complete blood count (CBC), serum biochemical profile, urinalysis and abdominal ultrasound.

The CBC revealed hypoproteinemia (plasma protein 2.5 g/dl; reference range 5.4-7.2 g/dl) and slight anemia (PCV 20%, reference range 24-35%). The biochemical profile revealed panhypoproteinemia with hypoalbuminemia (albumin 0.9 g/dl, reference range 2.6-4.7 g/dl) and hypoglobulinemia (globulin 1.1 g/dl; reference range 2.7-2.9 g/dl). There was no azotemia. The specific gravity of the urine was 1.051 and there was proteinuria (dipstick 3+). On an abdominal ultrasound examination, voluminous ascites was observed. The cortex of both kidneys appeared normal. However, on both kidneys, a cortico-medullary hyperechogenicity was evident.

A renal biopsy was performed in order to further examine the kidneys. Histopathologic diagnosis was membranous glomerulopathy with protein casts in tubules and loops of Henle . A congo red stain was negative suggesting that this was not associated with amyloid, however immunohistochemical studies were not undertaken.

The owner elected to euthanize the alpaca and a necropsy was performed. There were 2 liters of free fluid in the abdomen and 0.5 liters of fluid in the thorax and marked mesenteric edema. Based on the necropsy finding the diagnoses were severe diffuse membaranous glomerulopathy, mineralization of collecting duct epithelia and acute tubular necrosis resulting in protein losing nephropathy and end stage renal disease. The glomerular lesions were most likely due to immune complexes although congo red stain was negative. However, immunohistochemistry was not performed in this case. Immunohistochemistry has been used to demonstrate amyloid deposits using anti-AA amyloid, in tissues that were negative for congo red staining.[14]

In this case the cause for hypoproteinemia was protein losing nephropathy, diagnosed by urinalysis, abdominal ultrasound and renal biopsy.

**Case Two**

An eight year old female alpaca was referred for weight loss and diarrhea of 2 month duration. A fecal examination revealed trichostrongyles. Physical examination revealed normal vital parameters with marked depression, increased time spent in recumbence, severe weight loss with a body condition score of 1/10 and submandibular edema.

A CBC revealed hypoproteinemia (plasma protein 4.0 g/dl; reference range 5.4-7.2 g/dl). A biochemical profile revealed azotemia (BUN 69 mg/dl; reference range 12-33 mg/dl; Creatinine 2.8 mg/dl; reference range 1.3-2.7 mg/dl). There was also a panhypoproteinemia with hypoalbuminemia (albumin 1.4 g/dl, reference range 2.6-4.7 g/dl) and hypoglobulinemia (globulin 2.2 g/dl; reference range 2.7-2.9 g/dl). Electrolyte abnormalities included hyponatremia (126 mEq/l; reference range 142-156 mEq/l) and hyperphosphatemia (8.9 mg/dl; reference range 2.5-8.6 mg/dl). Liver enzymes were elevated (GGT 112 IU/l; reference range 6-36 IU/l; SDH 16.5 IU/l; reference range 3-20 IU/l). The urine specific gravity was 1,016 (reference range 1,013-1,048) and there was trace proteinemia as well as pigmenturia (dipstick 2+).

On ultrasound of the kidneys the cortex appeared normal. There was cortico-medullary hyperechogenicity and mineralization of the renal pelvices. The liver appeared normal with the exception that multiple hyperechoic densities causing shadowing were present throughout the parenchyma. There was also marked ascites.

The owner elected to euthanize the alpaca and a necropsy was performed. Based on the necropsy findings the diagnoses were multifocal interstitial fibrosis with interstitial nephritis, granulomatous hepatitis with multifocal fibrosis, granulomatous lymphadenitis and submucosal eosinophilic colitis.

The low sodium, high phosphorus and azotemia were suggestive for renal disease. The elevated liver enzymes as well as the ultrasound findings were suggestive for hepatitis. In this case the clinical signs and the abnormal biochemical profile suggestive for renal disease need to be considered in light of the hepatitis.

Hypoproteinemia and weight loss are common in camelids. The two cases above represent important differential diagnoses that need to be taken into consideration for a clinical work up.

**Case Three**

A eight day old, 7 kg, male alpaca cria was presented for lethargy anorexia and anuria. The owner gave 100,000 Units of vitamin D once daily for 5 days. Over 5 days the cria received 71,428.0 IU/kg Vitamin D (recommended dose: 1000 IU/kg).

Physical examination revealed lethargy, tachycardia (159 beats per minute) bilateral epiphora and no urination.

A CBC revealed hypoproteinemia (plasma protein 5.2 g/dl; reference range 5.4-7.2 g/dl). A biochemical profile revealed azotemia (BUN 120 mg/dl; reference range 12-33 mg/dl; Creatinine 6.2 mg/dl; reference range 1.3-2.7 mg/dl). There was also a panhypoproteinemia with hypoalbuminemia (albumin 4.2 g/dl, reference range 2.6-4.7 g/dl) and hypoglobulinemia (globulin 1.8 g/dl; reference range 2.7-2.9 g/dl). Electrolyte abnormalities included hypercalcemia (15.2 mg/dl; reference range 8.0-10.4 mg/dl) and hyperphosphatemia (10.2 mg/dl; reference range 2.5-8.6 mg/dl).

Based upon the historical information and the results of serum biochemical analyses, the possibility of Vitamin D toxicity was suspected. To detect possible tissue mineralization transabdominal ultrasonography and radiographic evaluation of thorax and abdomen were performed. Abdominal radiographs showed a thin mineral opaque rim outlining the kidneys and mineralization of the wall of the first compartment.

To determine whether the hypervitaminosis D was likely, serum samples were submitted to the Diagnostic Center for Population and Animal Health, Michigan State University. The results of the evaluation indicated the 25-hydroxy-vitamin D3 concentration was 602 nmol/L (range provided by the laboratory 50-200 nmol/L).

Therapy was initiated with lactated ringers supplemented with 2% dextrose at a rate of 1ml/pound at a constant rate infusion.

After rehydration furosmide was given in escalating doses every 30 minutes for 2 hours (1.1 to 4.4 mg/kg). Urine output was monitored using cage pads and ultrasound. No urine production has been observed. A Tenckoff dialysis catheter was placed into the abdomen in order to perform peritoneal dialysis. The cria received 60 ml of warmed Dianeal PD-2 peritoneal dialysis solution with 1% Dextrose intraperitoneally every 45 minutes. Before application of more fluid the dialysis fluid was aspirated from the abdomen. The treatment was monitored with repeated evaluation of serum blood urea nitrogen, creatinine and potassium levels, clinical assessment and ultrasonographic evaluation of the accumulation of urine in the bladder. On ultrasound the bladder did not increase in size over the entire course of treatment. By the fourth day, the cria was unable to rise, did not attempt to nurse and it developed increased respiratory effort. With the owner's permission the cria was euthanized.

Necropsy revealed multifocal glomerular mineralization, glomerular necrosis, tubular basement membrane mineralization and tubular epithelial necrosis. The bladder contained less than 1 ml urine.

This case demonstrates that vitamin D hypervitaminosis occurs in crias due to over supplementation resulting in tissue mineralization.

**Case Four**

An eleven year old, 72 kg, female alpaca was presented with a history of hematuria of 4 months' duration.
On Physical examination the vital parameters were normal. Urine dipstick analysis revealed 3+ blood, 2+ protein, and pH of 9.0; specific gravity was 1.022. After centrifugation, the urine was clear yellow with red sediment. Evaluation of a urine sediment smear revealed large numbers of RBCs and squamous epithelial cells.

The CBC revealed hypoproteinemia (plasma protein 2.5 g/dl; reference range 5.4-7.2 g/dl) and anemia (PCV 17%, reference range 24-35%). The biochemical profile revealed panhypoproteinemia with hypoalbuminemia (albumin 2.4 g/dl, reference range 2.6-4.7 g/dl) and hypoglobulinemia (globulin 2.5 g/dl; reference range 2.7-2.9 g/dl).

In order to identify the source of blood loss a vaginal examination was performed. The mucosa appeared normal and the cervix was closed. On ultrasound examination the uterus, bladder and kidneys appeared normal. A positive and double contrast cystogram did not reveal abnormalities of the bladder. These findings decreased the likelihood that urinary bladder neoplasia, urolithiasis, or cystitis was a cause of hematuria. Cystoscopy revealed pulses of red urine originating from the left ureteral opening. Since there was no evidence of azotemia a unilateral nephrectomy was carried out through a left paralumbar laparotomy incision. The approach was chosen on the basis of the presence of the left kidney in the retroperitoneal space, adjacent to the spleen on the left side, and the reduced need to manipulate the intestines with this approach.

The pelvis of the left kidney was filled with a 1.5 cm mass, which was connected to the epithelium but not invading. Hemorrhage was detected within the papillary mass. The final diagnosis, based on histopathology, was a transitional cell papilloma.

After surgery hematuria and anemia resolved and there were no signs of disease or renal failure. Within one year following surgery the alpaca did well, repeated CBC and serum biochemical profile continued to be normal and the alpaca gave birth to a healthy cria. This alpaca continues to do well and is a productive breeding female with a 2nd pregnancy carried to term since the unilateral nephrectomy.
This case demonstrates a primary renal disease with no signs of renal failure. Hematuria is an uncommon cause of anemia in camelids. A thorough evaluation of the urinary tract should be performed to locate the side of hemorrhage.

## CONCLUSION

Renal problems are an important differential diagnosis in the work up of anemia, hypoproteinemia, weight loss, and hematuria. The risks of chronic drug therapy should be considered. Nephrotoxic drugs should be used carefully and avoided in azotemic camelids. One should consider renal disorders in any animal with clinical signs including anemia, hypoproteinemia, diarrhea and weight loss, especially when these changes occur chronically.

## REFERENCES

[1]Hardefeldt L, et. al. **Aust Vet J** 2007;85(5):185; [2]Poulsen KP, et al. **Can Vet J** 2006; 47: 159; [3]Cardwell JM, et al. **Vet Rec** 1999; 145: 104; [4]Kramer K, et al. **J Comp Path** 2008; 139: 141; [5]Smith SH, et al. **Vet Rec** 2008; 162: 485; [6]Bildfell RJ, et al. **J Vet Diagn Invest** 2002; 14: 337; [7]Severo LC et al. J Med Vet Mycol 1988; 25(4): 329; [8]Gerspach et al. JAVMA 2008; 232(8): 1206; [9]Patel JH, et al. Can Vet J 2004; 45: 938; [10]Peauroi JR, et al. J Zoo Wildl Med 1995; 26: 315; [11]Dietrich JG, et al. Vet Pathol 1988; 25: 329; [12]Cebra CK, et al J Vet Intern Med 1995; 9: 381; [13]Irwin JA, et al. Can Vet J 2001; 42: 805; [14]Yamada M, et al. J Vet Med Sci 2006; 68(7): 725.

## BIOCHEMICAL & HISTOLOGICAL DIAGNOSIS OF LIVER DISEASES IN DAIRY CATTLE

**Elizabeth Doré, DMV, DACIM**
**Saint-Hyacinthe, Québec, Canada**

### INTRODUCTION

Liver diseases are a challenge to diagnose in the bovine specie. Presenting complaints and physical examination findings are rarely specific enough to indicate liver disorder. Diagnostic tests such as serum biochemistry profile and liver biopsy are helpful to diagnose and characterize liver disease. The objective of this presentation is to review the different parameters of the serum biochemistry profile that relate to liver integrity and liver biopsy technique. Finally, cases of bovine affected by liver disease diagnosed by histopathology will be discussed.

### SERUM BIOCHEMISTRY PROFILE

#### Markers of liver damage (hepatocellular injury)

Aspartate aminotransferase (AST) is an enzyme found in the cytoplasm and the mitochondrial membranes of hepatocytes. This enzyme is not specific to the liver since it is also found in the skeletal and cardiac muscles. The half-life is relatively long and increased of AST can persist for over a week after initial liver injury. More specific enzymes can be used to assess hepatic damage in cattle such as sorbitol dehydrogenase (SDH) and glutamate dehydrogenase (GLDH). SDH is found in the cytoplasm of the hepatocytes. GLDH is a more stable mitochondrial enzyme. These 2 enzymes have a short half-life (few hours) and their elevation can be interpreted as an active an outgoing liver damage.

#### Markers of cholestasis & hepatobiliary disorders

Gamma-glutamyltransferase (GGT) is found in the cell membranes of the hepatocytes, mostly the one boarding the canalicular surfaces and the bile duct epithelium. It is not a liver specific enzyme since it is also found in pancreatic and kidney cells. Damage to the renal tubular cells will result in an increased GGT in the urine and likely not in the serum. Pancreatitis will cause an increase in serum GGT and has been previously reported as an incidental finding at necropsy in 2 cattle.[1,2] For neonates, increased serum GGT is associated with colostrum absorption. Bilirubin is the product of haemoglobin degradation. Total bilirubin included indirect or unconjugated bilirubin which is bound to albumin and direct or conjugated bilirubin. Total bilirubin may increase in serum following a haemolytic crisis, inadequate uptake of unconjugated bilirubin by the liver, decrease conjugation of bilirubin and biliary stasis. Moderate hyperbilirubinemia may be present in sick cattle without primary liver disease.[3] This could be consecutive to an impaired liver blood flow. Bilirubin is responsible for the yellow discoloration of the mucous membranes termed jaundice or icterus. In a review of sick cattle with hyperbilirubinemia, McSherry et al. found that icterus was clinically visible in cases were the total bilirubin was higher than 65μmol/L.[3]

### LIVER BIOPSY TECHNIQUE

Percutaneous liver biopsy can be easily performed on a standing cattle at the farm. The technique is safe and does not cause significant liver injury.[4] Rogers et al. studied the growth performance of cattle after a percutaneous liver biopsy and found no difference in average daily gain in the group that was biopsied and the control group.[5] The site to biopsy is in the tenth or eleventh intercostal space at the level of a line drawn from the middle of the right paralumbar fossa. A TrueCut trocar (Cardinal Health, McGaw Park, IL) or other biopsy instrument is inserted cranioventrally and to the left side or aiming for the left elbow. In conditions where the liver is diffusely affected there is a greater chance to get a representative sample. Ultrasonographic examination of the liver can be done prior to the procedure to identify the proprer puncture site. In cases where liver abscesses are suspected (inflammatory process with or without increase in liver enzymes[6]) liver biopsy is contraindicated since there is a risk of rupturing the abscess and secondary peritonitis. Liver sample can be preserved in 10% formalin for histophatological evaluation. Due to the small sample size, it is recommended to place the liver tissue in a histology cassette. For bacteriologic culture, the sample should be submitted in a sterile container. If the sample needs to be shipped to a laboratory, it should be place in a special medium transport to preserve anaerobic conditions. Liver samples can also be used to diagnose trace mineral deficiencies. For that purpose, a larger biopsy instrument should be used. Care should be taken to send the samples as fresh as possible to the laboratory. If the moisture content of the liver sample differs from the liver moisture *in situ*, concentrations of trace minerals reported might be affected. Normal reference ranges are often reported on a wet weight. Recently, a correlation of potassium content and moisture in bovine liver samples was studied and this allows to better predict mineral concentration regardless of sample moisture.[7]

### CHOLANGIOHEPATITIS

Histopathological changes seen with cholangiohepatitis are accumulation of inflammatory cells (lymphocytes, plasma cells, neutrophils) in the bile ducts and the surrounding hepatocytes. In the feline species, inflammatory liver diseases are the most common liver affectation after hepatic lipidosis. The feline cholangiohepatitis complex is a well described condition.[8,9] Cholangiohepatitis in horses is often associated with cholelithiasis and is thought to be secondary to ascending bacterial infection from the proximal small intestine. Horses with cholangiohepatitis have been treated successfully with supportive care and long term antibiotic therapy.[10] Only 2 cases of cholangiohepatitis in the bovine species have been reported in the literature. The first case, published in 1969, describe the manifestations of chronic cholangiohepatitis in a 3 year-old Holstein dairy cow.[11] The cow presented for anorexia and reduced milk production. Icterus became apparent only 1 week after initial presentation. A blood sample revealed elevated bilirubin, SGOT (Serum Glutamooxaloacetate Transferase) and OCT (Ornithine Carbamoyl Transferase). The liver was found to be enlarged and adhered to the body wall during an exploratory laparotomy. A diagnosis of diffuse chronic cholangiohepatitis with biliary fibrosis was made at necropsy. The second case, reported in 2002, was a 2 month-old Holstein heifer calf presented with diarrhea and fever.[12] The calf had elevated GGT, AST and bilirubin. The diagnosis of cholangiohepatitis was done from a liver biopsy and *Escherichia coli* was cultured form the liver sample. Treatment consisted of intravenous fluid therapy and antibiotic. The calf was discharged alive form the clinic.

**Cholangiohepatitis in Holstein dairy cattle: 15 cases (1992-2006)**
**E. Doré, G. Fecteau, N. Queney, P. Hélie, D. Francoz**

The purpose of this study was to characterize the clinical manifestations and the clinicopathological changes in bovine diagnosed with cholangiohepatitis. The liver samples were taken using a percutaneous liver biopsy in 14 cases and 1 case was diagnosed post mortem. Age ranged from 7 months to 7.3 years (median: 3.9 years). Six cases presented during the 6 weeks after calving and 7 cases were first calf heifers. Out of the 15 cases, 9 cases presented concomitant diseases: surgical digestive problem (n=4), salmonellosis (n=3) and cranial peritonitis (n=2). Clinical presentation was anorexia (15/15), reduced milk production (10/10), scant feces (14/15), ruminal hypomotility (13/15), tachycardia (12/15), fever (11/15), icterus (7/15), diarrhea (4/11) and colic (3/15). For 6 cases with an exclusive diagnosis of cholangiohepatitis, the following parameters were all above reference range: total bilirubin (median: 74 μmol/L), GGT (median: 487 U/L) and AST (median: 254 U/L). Bacteriologic culture of 10 liver samples was positive in 5 cases. *Escherichia coli* was isolated from 3 of these cases. Treatment consisted of antibiotic therapy and intravenous or *per os* fluids. Seven cases out of the 15 were discharged from the clinic with a good prognosis. Cholangiohepatitis exists in bovine and should be included in the differential diagnosis when an animal present with vague clinical signs combined to elevated liver enzymes and bilirubin. Diagnostic should be confirmed by a liver biopsy submitted for histopathologic examination and bacterial culture.

## OTHER HISTOLOGICAL CHANGES

Records of bovine older than 6 months that presented to the food animal clinic (University of Montréal) over a 10 year period with liver changes diagnosed at histopathology were reviewed. All cases reviewed were Holstein dairy cattle. Fifteen cases were diagnosed with cholangiohepatitis (see above), 23 with hepatic necrosis and 18 with portal hepatitis.

### Hepatic necrosis

Out of 23 cases, the diagnosis of hepatic necrosis was done with a percutaneous liver biopsy in 12 cases and at necropsy for 11 cases. Five cases presented icterus of their mucous membranes. Concomitant diseases were: abomasal volvulus (n=7), cardiomyopathy (n=4), copper intoxication (n=3), salmonellosis (n=2). The median values for the bilirubin was 16.2 μmol/L (normal values: < 14 μmol/L), for the AST 428 U/L (normal values: 30-104 U/L)) and for GGT 86 U/L (normal values: 9.5-39 U/L). Hepatic necrosis is not a primary liver disease but a secondary change. It is interesting to note that almost 1/3 of the cases had an abomasal volvulus which might have compromise blood supply to the liver and therefore results in necrosis of the hepatocytes and elevated liver enzymes. A similar process have been reported in horses with a right dorsal displacement of the large colon where the GGT was often elevated and was attributed to possible bile duct compression.[13]

### Portal hepatitis

All cases (n=18) with portal hepatitis were diagnosed using a percutaneous liver biopsy. The histopathological changes were accumulation of inflammatory cells at the portal space without involvement of the bile ducts. Five of the cases presented with a left displaced abomasum. Two of the cases had icterus. The median values for the bilirubin was 25.7 μmol/L (normal values: < 14 μmol/L), for the AST 132 U/L (normal values: 30-104 U/L)) and for GGT 57.5 U/L (normal values: 9.5-39 U/L). This histologic change appeared to be more an incidental finding and does not seems to have a clinical relevance.

## CONCLUSION

To diagnose a liver disease in the bovine specie, the clinician should rely on a thorough physical examination and changes in hepatic parameters on the serum biochemistry profile. A liver biopsy should be submitted for histological examination to confirm the diagnosis and for bacteriologic culture to better aim antibiotic therapy in case of cholangiohepatitis.

## REFERENCES

[1]Braun U, et al. **Schweiz Arch Tierheilk** 2001;143:99. [2]Doherty ML, et al. **Vet Rec** 1998;142(18):493. [3]McSherry BJ, et al. **Can J Comp Med** 1984;48:237. [4]Amorim RM, et al. **Ciencia Rural** 2003;33(3):519. [5]Rogers GM, et al. **Bovine Pract** 2001;35(2):177. [6]Doré E, et al. **J Vet Intern Med** 2007;21:853. [7]Ludwick TP, et al. **J Vet Diagn Invest** 2008;20:314. [8]Day DG, et al **Vet Clin North Am Small Anim Pract** 1995;25(2):375. [9]Gagne JM, et al. **Vet Pathol** 1996;33:521. [10]Peek SF, et al. **Equine Vet J** 2000;32(4) :301. [11] Whitlock RH, et al. **Cornell Vet** 1969;59(4):515. [12]Coombs DK, et al. **Vet Rec** 2002;150(17):551. [13]Gardner RB, et al. **J Vet Intern Med** 2005;19 :761.

# VERTEBRAL COLUMN MRI

**Jonathan M. Levine, DVM, DACVIM (Neurology)**
**College Station, TX**

## INTRODUCTION

Rapid advances in neurodiagnostics have occurred since the early 1970s, in large part due to the development and refinement of magnetic resonance imaging (MRI). In human medicine, MRI is typically the modality of choice for evaluating the vertebral column due to its accuracy, non-invasive nature, and ability to provide prognostic information. Limited data also support the existence of these advantages in companion animals. The use of MRI for vertebral column imaging in veterinary medicine is rising, but is limited by access to this technology, perceptions concerning cost, and clinician familiarity amongst generalists and specialists.

## TRADITIONAL IMAGING

### Radiography

Vertebral column radiography has been used in veterinary medicine since the early part of the 20$^{th}$ century. Radiography is advantageous in that it is readily available, easy to perform, and inexpensive. It is probably best used as a screening diagnostic for diseases where bony structures are primarily involved such as vertebral fracture, diskospondylitis, atlantoaxial subluxation, disk herniation, and lytic bone tumors. Even in these diseases, the accuracy of vertebral column radiography is not ideal. The reported sensitivity (true positives/true positives + false negatives) of radiography for detecting vertebral column fracture and luxation in dogs is 72% and 77.5% respectively compared to the gold standard of CT, although specificity (true negatives/true negatives + false positives) is high.[1] In a population of dogs with various thoracolumbar vertebral column disorders, the accuracy (sensitivity*prevalence + specificity*[1-prevalence]) of radiographs for determining the presence and site of disk herniation was 51-61% using myelography and surgery as gold standard.[2] In another report, the overall accuracy for correct identification of cervical disk herniation in dogs was 35%, using myelography as gold standard.[3]

### Myelography

Myelography was first described in humans by Sicard and Forestier in 1923.[4] Widespread use did not occur in veterinary medicine until the technique was refined and popularized by Hoerlein.[5] Lesions can usually be classified as extradural, intradural-extramedullary, and intramedullary. Myelogram is believed to be an accurate means of diagnosing disk herniation in veterinary species; its ability to detect non-disk associated lesions is likely more limited. It has the disadvantage of being invasive (iodinated contrast is injected into the sub-arachnoid space). Adverse events related to myelographic contrast delivery can include seizures, allergic reactions, myelopathy, cardiac arrhythmia, hemorrhage, and death.[6-9] Myelography does not allow direct visualization of the spinal cord parenchyma and investigators have expressed concerns regarding the possibility of lesion mis-classification, even in the setting of extradural compression.[10] Prognostic information that can be derived from myelography is limited. Intraparencymal contrast uptake in the setting of severe spinal cord injury may indicate myelomalacia. Dogs lacking deep nociception due to disk herniation and with an intramedullary pattern:L2 ratio ≥5 have a 26% recovery (defined as voluntary motor function at discharge) compared to a 61% recovery for dogs with ratios <5.[11]

### Computed tomography (CT)

Computed tomography is a non-invasive, rapid, x-ray based technique for generating multi-planar images of the vertebral column. Bone detail is excellent and this is probably the advanced technique of choice for imaging dogs with vertebral fractures or osseous neoplasia. The relative sensitivity for detecting thoracolumbar disk herniation in dogs is comparable to myelography (81.8% versus 83.6%, respectively).[12] Limited information suggests that non-mineralized disk herniation may be challenging to detect.[12] Computed tomography does offer improved spinal cord soft tissue detail compared to myelogram, but soft tissue detail is not nearly as impressive as MRI.

## HOW MRI WORKS

The science behind MRI is complex and went through several evolutionary phases in the 1950s, 1960s and 1970s. The importance of this early work resulted in 6 Nobel prizes. MRI detects subtle changes in the magnetism of atomic nuclei in water molecules, which make up approximately 70% of most tissues. The water content of tissues is often altered in disease, which makes MRI a very sensitive diagnostic technique.

The simplified sequence of events during image acquisition is as follows. A patient is placed on a gantry within the bore of a high field magnet. A receiver coil has already been attached to the body part of interest. Once the patient is in the magnet, protons within molecules in the body will align in the longitudinal direction (along the longitudinal axis of the patient; referred to as "z" or "Bo") in high energy and low energy orientations. Magnetic gradients are applied in other planes to allow the machine to identify image slices and "encode" protons in frequency and phase. A radiofrequent (RF) pulse is then delivered and flips the protons 90 degrees from the longitudinal plane (z) to the transverse plane (x). Typically an additional 180 degree RF pulse is used to "re-phase" protons and eliminate field inhomogeneity. Protons are next allowed to return to their original energy state, which is termed "relaxation". Relaxation occurs in the longitudinal-transverse plane (so called spin-lattice) and axial (y) – transverse plane (so called spin-spin). The receiver coil registers the energy that is released during relaxation.

Repetition time (TR) is the time between RF pulses and echo time (TE) is the time when the receiver coil "listens" for energy release. Combinations of TR and TE are referred to as pulse sequences and determine the image weighting.

**IMAGE WEIGHTINGS**

Image weighting is akin to the contrast characteristics of an MRI, which are in part determined by the pulse sequence selected. T1 weighted images (T1WI) have a higher signal to noise ratio than T2 weighted images (T2WI) and thus provide excellent anatomic detail. Structures that appear hyperintense (bright) on T1 include fat, gadolinium contrast, methemoglobin (intra- and extra-cellular), melanin, calcium in dystrophic mineralized lesions, manganese, slow flowing blood, and protein. Hypointensity on T1WI is often due to the presence of fluid, bone, fibrous tissue, air, flowing blood, or certain phases of hemorrhage. T2 weighted images are often more helpful than T1WI at identifying abnormalities in tissue. Regions of hyperintensity may result from edema, necrosis, cerebrospinal fluid, fat, demyelination, glial scar, inflammatory infiltrates, and certain phases of hemorrhage. T2W hypointensity can be caused by bone, air, flowing blood, and certain phases of hemorrhage.

**DIAGNOSTIC ACCURACY AND PROGNOSTIC DATA**

In the era of evidence-based decision making, information concerning the accuracy as measured by sensitivity and specificity of various imaging techniques must be used to select an appropriate modality. Unfortunately, data in humans and especially veterinary patients are limited. Defining gold standards for many disease processes can be challenging. In some instances, in vivo diagnosis is likely to be presumptive (eg, infarction). In the case of disk herniation, surgical confirmation is usually viewed as gold standard, but is biased by the results of pre-surgical imaging in the first place. Reliance on necropsy confirmation may affect outcome data and bias the population towards those individuals that were difficult to diagnose. Finally, there may be some instances when imaging is more sensitive for detecting lesion burden than necropsy such as in humans with multiple sclerosis.[13,14] In addition to challenges associated with defining gold standards, all studies have limitations associated with design. For example, the assignment of patients to imaging groups may be non-random or sufficient power may not be present to detect important adverse events or differences in accuracy. The following is a list of available information, organized by etiology and species, comparing various imaging techniques.

**Vertebral column trauma**

No veterinary data exist. In humans, CT and MRI are roughly equivalent for detecting vetertebral column fracture or subluxation.[15] In patients with vertebral canal compromise >25% based on CT, 100% of individuals had MR evidence of spinal cord compression.[15] In patients with <25% canal compromise based on CT, 78% had evidence on MRI of spinal cord compression.[15] MRI does provide prognostic data not available from CT or myelogram. The presence of T1W hyperintensity, T2W hyperintensity, and hypointensity on gradient echo images within the spinal cord has been associated with poorer standardized functional outcome at follow-up.[16-18] The extent of spinal cord compression as well as the degree of vertebral canal compromise has also been associated with standardized functional outcome at follow-up.[16]

**Disk herniation**

Data are not currently available that compare the sensitivity or specificity of MRI with other modalities for detecting disk herniation in veterinary species. In humans, most studies concerning this are older and do not reflect recent improvements in MRI field strength, coils, protocols, and software. In one report, MRI was roughly equivalent for detecting lumbar disk herniation compared to CT-myelogram (82.6% vs. 83%, respectively) when surgical exploration was used as the gold standard.[19] In another study on human lumbar disk herniation, MRI was more specific but less sensitive than CT-myelogram (86.5% versus 78.9% and 64.3% versus 72.8%, respectively) when surgical exploration was used as the gold standard.[20] In a single center, prospective veterinary study T2W hyperintensity within the spinal cord in dogs with paraplegia due to thoracolumbar disk herniation was negatively associated with functional outcome (voluntary ambulation at follow-up), especially if >3 L2 vertebral articulations in length.[21]

**Diskospondylitis**

Data comparing various modalities are not available in veterinary medicine. In humans, MRI has a reported sensitivity of 96%, compared to 82% for radiography.[22] Radiography is believed to be inaccurate in some cases of early diskospondylitis and in humans false positives have been reported to be associated with severe spondylosis deformans.[22,23] The accuracy of MRI was 94% in one study compared to 73% for radiography.[22]

**Fibrocartilagenous embolic myelopathy (FCEM)**

Myelography is reported to disclose an intramedullary pattern in 25.9% of dogs with presumptive FCEM.[24] MRI, in a recent study, detected lesions in 78.8% of dogs.[25] The sensitivity and specificity of MRI in discerning FCEM from other etiologies is believed to be higher than myelogram as myelogram discloses fewer lesions and the intramedullary pattern seen is non-specific.[26] In one report, if MR lesion length within the spinal cord was >2 vertebrae or lesion cross sectional area was >67% successful outcome (defined as normal neurological exam or ability to perform daily activities without assistance and fecal/urinary continence) was adversely affected.[26]

### Cervical spondylomyelopathy (CSM)

Recent data in humans suggest that MRI is likely to be roughly equivalent to CT-myelogram for detecting lesions associated with cervical spondylomyelopathy.[27,28] There are concerns that MRI may over estimate vertebral canal narrowing in severe cases due to chemical shift artifact or CSF motion artifact.[27,28] In dogs, information is available from a single center, prospective study evaluating 18 dogs with myelogram compared to MRI.[10] Both MRI and myelogram appeared to have reasonable inter-rater agreement as to the location of the main lesion (100% vs. 83%, respectively). With MRI, 10/18 dogs have signal changes within the spinal cord. Adverse events associated with myelography included seizures (28%) and worsening of neurological signs (33%).

### Neoplasia

Data comparing various imaging modalities is not available in veterinary medicine. In humans, MRI is more sensitive for detecting spinal cord metastasis compared to CT or CT-myelogram. In a study of 91 patients with spinal cord neoplasia where MRI was judged to not be mandatory, MRI changed the radiotherapy plan in 53% of cases and 21% of cases radiotherapy plan changes were deemed "major".[29] In 33 humans with cerebrospinal fluid metastasis, MRI shows more lesions (n=24) compared to CT-myelogram (15) and all lesions seen on CT-myelogram were visualized by MRI.[30]

## CONCLUSION

MRI appears to be an excellent means of imaging the vertebral column in domestic species. It is non-invasive, provides excellent soft tissue detail, and produces multi-planar images. Preliminary data is available in some veterinary diseases, which suggests MRI is a sensitive means of detecting lesions and may assist in determining prognosis for functional recovery.

## REFERENCES

[1]Kinns J et al. **Vet Radiol Ultrasound** 2006;47:563. [2]Lamb CR et al. **Vet Radiol Ultrasound** 2002;43:222. [3]Somerville ME et al. **J Am Anim Hosp Assoc** 2001;37:563. [4] Sicard JA et al. **Rev Neurol** 1923;6:676. [5]Hoerlein BF. **Am J Vet Res** 1953;14:275.
[6] Barone G et al. **J Am Vet Med Assoc** 2002;220:1499. [7]Carroll GL et al. **Vet Radiol Ultrasound** 1997;38:284. [8]Lewis DD, Hosgood G. **J Am Vet Med Assoc** 1992;200:1381. [9]Packer RA et al. **Vet Radiol Ultrasound** 2007;48:323. [10]da Costa RC et al. **Vet Radiol Ultrasound** 2006;47:523. [11]Duval J et al. **Vet Surg** 1996;25:6. [12]Israel SK et al. **Vet Radiol Ultrasound** 2009 (In Press). [13]Bo L et al. **Neuropathol Appl Neurobiol** 2004;30:106. [14]De Groot CJ et al. **Brain** 2001;124:1635. [15]Fehlings MG et al. **Spine** 1999;24:605. [16]Miyanji F et al. **Radiology** 2007;243:820. [17]Mascalchi M et al. **Clin Radiol** 1993;48:100. [18]Yamashita Y et al. **Br J Radiol** 1991;64:201. [19]Modic MT et al. **Am J Roentgenol** 1986;147:757. [20]Jackson RP et al. **Spine** 1989;14:1362. [21]Ito D et al. **J Am Vet Med Assoc** 2005;227:1454. [22]Modic MT et al. **Radiology** 1985;157:157. [23]Honan M et al. **Am J Med** 1996;100:85. [24]Gandini G et al. **J Small Anim Pract** 2003;44:76. [25]De Risio L et al. **J Vet Intern Med** 2007;21:1290. [26]De Risio L et al. **J Am Vet Med Assoc** 2008;233:129. [27]Reul J et al. **Neuroradiology** 1995;37:187. [28]Stafira JS et al. **AJNR Am J Neuroradiol** 2003;24:766. [29]Husband DJ et al. **Br J Radiol** 2001;74:15. [30]Heinz R et al. **AM J Neuroradiol** 1995;16:1147.

## NEUROGENIC LAMENESS - THE CHALLENGING LAME DOG

**Ronaldo C. da Costa, DMV, MSc, PhD, Diplomate ACVIM – Neurology**
**Columbus, OH**

### INTRODUCTION

Lameness is a common problem in small animal practice. The majority of cases of lameness are caused by orthopedic diseases. However, neurological diseases also cause lameness. It is important to always consider neurological causes, especially in patients with chronic and/or recurrent episodes of lameness.

The goal of this overview is to present the clinical approach, ancillary tests and the main characteristics of the primary differential diagnosis of neurogenic lameness. Emphasis will be placed on the clinical presentations and diseases where lameness can be the main clinical sign.

### CLINICAL APPROACH

The diagnostic approach should initially involve thorough physical and orthopedic examinations to clearly establish the absence of orthopedic disease. Neurogenic lameness should be strongly considered in chronically lame patients with no evidence of musculoskeletal disease that respond to neither conservative treatment nor rest.

A complete neurological examination (evaluation of mental status, gait and posture, cranial nerves, postural reactions, spinal reflexes and pain perception) should be always carried out to assess the extent, if any, of neurological involvement. In a patient with lameness, neurological signs that are highly suggestive of nervous system disease are ataxia, weakness, postural reaction deficits and obvious spinal pain. If any of these signs are associated with lameness, the index of suspicion for an underlying neurogenic cause is raised; however, many, and perhaps the majority of patients with neurogenic lameness do not present with clear neurological signs associated with lameness. These are the challenging cases. In these patients, clinical signs suggestive of neurogenic involvement are: severe muscle atrophy, Horner's syndrome (miosis, ptosis, third eyelid protrusion and enophtalmus), decreased or absent cutaneous trunci (formerly panniculus) reflex, decreased muscle tone and spinal reflexes (lower motor neuron signs), signs of weakness/hypertonicity in the ipsilateral pelvic limb in cases of lameness involving a thoracic limb, and pain on spinal, axillary or limb palpation/manipulation. Clinical signs that may be seen and are highly suggestive of neurogenic involvement are monoparesis or monoplegia and self-mutilation. It is important to emphasize that any dog with chronic lameness is expected to develop some degree of muscle atrophy from disuse alone. However, neurogenic muscle muscle atrophy is almost always much more severe than disuse atrophy. Severe muscle atrophy, either affecting one, or several muscle groups is, in this author's opinion, the most important feature of neurogenic lameness in patients lacking other neurological signs and by itself warrants pursuit of additional diagnostic testing.

### DIFFERENTIAL DIAGNOSIS

The primary differential diagnoses of neurogenic lameness are: malignant peripheral nerve sheath tumor (MPNST), intervertebral disk disease, spinal neoplasia, brachial plexus avulsion, traumatic neuropathy, degenerative lumbosacral stenosis, cervical spondylomyelopathy and diskospondylitis.[1, 2] Not all of these differentials affect the thoracic and pelvic limbs equally and their clinical presentation in relation to lameness and other neurological signs may be quite different.

### ANCILLARY TESTS

Several ancillary tests can be used to aid the diagnosis of neurogenic lameness. The selection of the each test should be guided by the differential diagnosis being considered.

**Electrodiagnostic tests**[3]

- <u>Electromyography (EMG)</u> – EMG is commonly used to differentiate neurogenic from disuse muscle atrophy. It is a highly sensitive technique. Fibrillation potential and positive sharp waves are suggestive of neurogenic muscle atrophy.
- <u>Nerve conduction velocity</u> – Motor or sensory conduction velocities may be prolonged and the amplitude reduced in cases with peripheral nerve involvement.
- <u>F-wave, H-wave, cord dorsum potentials</u> – Evaluate the root portion of the nerves and can be useful to determine whether a lesion affects predominantly the sensory or motor root.

These techniques are invaluable to confirm neurogenic involvement and are very sensitive. However, they are not specific to any of the differential diagnosis, and require specialized equipment and the use of general anesthesia.

**Imaging**

- <u>Survey radiographs</u> – Radiographs can demonstrate evidence of intervertebral disk disease, osteolytic/osteoproliferative lesions, and occasionally may allow visualization of an enlarged intervertebral foramen in cases of proximal peripheral nerve sheath tumors. Radiographs are usually utilized as a screening test to rule out major abnormalities.
- <u>Ultrasonography (US)</u> – Axillary ultrasonography is an extremely useful technique due to its simplicity and sensitivity in cases of distal nerve sheath tumors. US can often identify masses in cases where axillary palpation is unrewarding. The contralateral axillary or inguinal area should be examined for comparison. Most patients tolerate the examination well without sedation, although it is occasionally necessary. If a mass is identified, US guided fine needle aspiration biopsy should be attempted. Core biopsies can also be performed under US guidance but requires heavier sedation or anesthesia. The echogenicity of MPNSTs is variable. One limitation of US is that it only identifies neoplasms at the level of the brachial/lumbosacral plexus or distal to it, therefore a normal US exam can not rule out primary or secondary nerve tumors.

- Myelography – Can establish a definitive diagnosis if the disease causes compression of the subarachnoid space surrounding the spinal cord. Most diseases that cause neurogenic lameness involve the nerve roots, the spinal nerves or peripheral nerves, thus myelography may be nondiagnostic. If signs of spinal pain or spinal cord involvement (proprioceptive ataxia, paresis, upper motor neuron signs – increased muscle tone and reflexes) are absent, other tests such as ultrasonography, computed tomography or magnetic resonance imaging should be selected.
- Computed tomography (CT) and magnetic resonance imaging (MRI) – these advanced imaging modalities allow visualization of the vertebral canal, plexus area and distal nerve regions. Therefore they offer significant advantages over radiographs, myelography or ultrasonography. It is important to select a field of view large enough to allow visualization of the plexus area. MRI is the diagnostic modality of choice because it provides the greatest soft tissue characterization.

**Laboratory analysis**

- Cerebrospinal fluid analysis – Usually only shows albuminocytologic dissociation, which is a non-specific finding.
- Cytology –Cytology is extremely useful in the diagnosis of neoplastic masses because the treatment of primary nerve sheath tumors is very different from that of secondary nerve tumors such as lymphoma. Other tumors like osteosarcomas may also be in the plexus region. The cytological features of MPNSTs are characteristic of a soft tissue sarcoma.[4]

## CAUSES OF NEUROGENIC LAMENESS

### Malignant peripheral nerve sheath tumor

Malignant peripheral nerve sheath tumors, previously called schwannomas or neurofibromas, are the main causes of chronic neurogenic lameness in dogs.[5] The clinical presentation of MPNST is indistinguishable from other causes of lameness in the early stages of the disease. The majority of MPNSTs (80%) occur in the thoracic limbs. Approximately 55% of MPNSTs are located in the plexus or peripheral nerve regions (distal tumors), while 45% are located in the nerve root region (proximal tumors).[6]

There is no breed predisposition, but large breed dogs are over-represented. The mean age of affected dogs is 7.9 years, and it is uncommon to see MPNSTs in dogs younger than 3 years.[1] The most common sign is chronic progressive thoracic limb lameness; however, acute cases have been reported. Muscle atrophy (neurogenic atrophy) is often present and is more severe than that observed with orthopedic disorders. Decreased muscle tone and decreased flexor reflexes may also be observed. A palpable axillary mass is present in less than 30% of cases.[4] Axillary pain is occasionally elicited upon palpation. Horner's syndrome may be seen if the tumor involves the T1-T3 nerve roots. Ataxia, paresis and asymmetric proprioceptive deficits, may be seen if the tumor compresses the spinal cord.[2] The cutaneous trunci reflex may be asymmetric with involvement of the spinal cord or the lateral thoracic nerve. Self-mutilation may also be observed in some cases of nerve sheath tumors. Rectal palpation should always be performed when a lumbosacral nerve sheath tumor is suspected.

Diagnosis of MPNST is established by imaging accompanied by cytology or histopathology. Ultrasonography is the initial diagnostic test of choice because it can be readily performed and does not require general anesthesia.[7,4] Axillary ultrasonography should allow detection of at least 50% of tumors. Proximal tumors will not be visible with US, therefore myelography, CT or MRI would be required. If a tumor is detected via US and surgical resection is planned, advanced imaging is strongly recommended. MRI is the most sensitive test for the diagnosis of MPNST. The CT and MRI features of MPNSTs have been reported.[8, 9]

The treatment of choice for MPNSTs is surgical resection following the principles of oncologic surgery. Amputation is usually required to minimize the chance of local recurrence. Hemilaminectomy or dorsal laminectomy may be required to allow resection of the nerve root region. Nerve sheath tumors are locally invasive and rarely metastasize; however, postoperative local recurrence is common (up to 72% of cases). Anedoctal evidence suggests that some cases may respond to radiation therapy. Recently, a post-operatory chemotherapy protocol using metronomic therapy with cyclophosphamide and piroxicam was shown to significantly delay tumor recurrence in dogs with incompletely resected soft tissue sarcomas, including MPNSTs.[10] Oral corticosteroids are used as a palliative measure in nonsurgical cases.

### Intervertebral disk disease (IVDD)

Both intervertebral disk protrusion and extrusion can cause neurogenic lameness. The direction of the disk herniation is usually lateralized, causing compression of the nerve roots. Intervertebral disk herniation tends to causes lameness when the herniation occurs in the regions of the cervicothoracic (C4-5, C5-6, C6-C7, C7-T1) or lumbosacral (L4-5, L5-6, L-6-7) intumescences.[11] However, cranial cervical disk herniations (C2-3, C3-4) have also been reported to cause lameness.[12,13] Usually the lameness in these cases is associated with spinal pain and it is called "nerve root signature". It is much more common to see this presentation with cervical disk herniations, compared to caudal lumbar disk herniations. Pain is an important feature of lateral or foraminal disk extrusions. Dogs may vocalize periodically or cry as if they were experiencing sharp pain. Pain may be elicited by limb manipulation or spinal palpation.[11]

Diagnosis of lateral or foraminal disk extrusions is easily achieved by advanced imaging with CT or MRI. Oblique radiographs may be able to show asymmetric opacity in the foraminal region.[13] In cases of lateral and foraminal extrusions, myelography is often unrewarding.

Surgical and medical treatment options exist for the management of IVDD. Surgical treatment is usually recommended because the pain often does not subside until the offending disk is removed and the nerve freed from entrapment.[11]

**Brachial Plexus Avulsion (BPA)**

This is a common sequel of major traumas, usually traffic accidents. The majority of dogs have avulsion of all roots of the brachial plexus (C6-T2) causing dysfunction of the suprascapular, musculocutaneous, radial, median, ulnar, lateral thoracic nerve and sympathetic nerves. The second most common type of avulsion involves only the caudal roots (C8-T1-T2) affecting radial, median, ulnar, lateral thoracic nerve and sympathetic nerves. In both instances, dogs present with thoracic limb monoparesis or monoplegia. Other signs commonly seen are analgesia of the limb and absence of flexor reflex. Horner's syndrome and unilateral absence of the cutaneous trunci reflex are seen in approximately 70% of cases.[14] Less commonly, dogs with BPA can have avulsion of only one or two nerve roots. In these cases of focal brachial plexus avulsion, instead of monoparesis or monoplegia, the clinical picture is dominated by a non-weight bearing lameness. Diagnosis of BPA is based on a history of trauma and electrodiagnostic abnormalities. CT and MRI may delineate the areas of avulsion. Unfortunately the only specific treatment is reimplantation of the avulsed nerve roots and it has only been done experimentally in dogs.[15] Physical therapy and limb protection are recommended for at least 2-3 months to allow resolution of edema, and remyelination of partially affected axons. The only reliable negative prognostic indicator is complete analgesia distal to the elbow.[16]

**Spinal Neoplasia**

Spinal neoplasias can be extradural, intradural/extramedullary or intramedullary. Extradural tumors (e.g. osteosarcoma, chondrosarcoma) are the most common. The osseous proliferation associated with some of these tumors can cause nerve root compression or entrapment, and lead to lameness. Spinal pain is often present. Spinal radiographs may show osteolytic and/or osteoproliferative changes. Computed tomography, CT myelography and MRI allow assessment of spinal cord involvement.[17, 18] Biopsy and histopathology is often necessary to select the most appropriate treatment.

**Cervical spondylomyelopathy**

Some dogs with cervical spondylomyelopathy (CSM) have an asymmetric tetraparesis with more severe weakness of one of the thoracic limbs and difficulty bearing weight with frequent collapse. This is sometimes interpreted as "lameness" by owners, but a careful examination readily reveals other neurological signs. This presentation can be observed with either disk-associated asymmetric compressions, or osseous compressions causing pedicular malformation and foraminal stenosis. The compressions are often located at C5-6 and C6-7 regions. Clinical signs of myelopathy (proprioceptive ataxia and paresis) are often observed and tend to dominate the clinical picture. Cervical hyperesthesia may or may not be present. Diagnosis of CSM requires myelography, CT or MRI. Treatment can be surgical or non-surgical.

**Degenerative lumbosacral stenosis (DLS)**

This is a disease typical of older large breed dogs. Male dogs and German Shepherds are over-represented. DLS is often caused by a combination of intervertebral disk protrusion with vertebral canal stenosis. Compression of the nerve roots or spinal nerves is often caused by lateralized disk protrusion and/or foraminal stenosis at the L6-L7 or L7-S1 regions. Lameness is usually associated with other historical findings, such as reluctance to jump, difficulty in rising and climbing stairs.[2] Caudal lumbar and lumbosacral pain is a very consistent finding in dogs with DLS. Pain can be elicited by digital pressure over the L6-L7 or L7-S1 regions, or by lifting the tail and pushing it dorsally. Proper support in the pelvic region is important when applying digital pressure during palpation because many patients have concomitant hip dysplasia. Diagnosis is best established by advanced imaging, preferably by MRI.[19] Treatment of DLS can be surgical or non-surgical. Recently, descriptions of surgical techniques using a lateral approach have been reported to allow more direct decompression of foraminal stenosis.[20, 21]

**References**

[1.] da Costa RC. Nerve sheath tumors. In: Tilley LP, Smith FWK, eds. **Blackwell's Five-minute veterinary consult: canine and feline. 4th ed.** Philadelphia: Blackwell; 2007:972. [2.] McDonnell JJ, Platt SR, Clayton LA. **Neurologic Conditions Causing Lameness in Companion Animals**. Vet Clin North Am Small Anim Pract 2001;31:17-38. [3.] Cuddon PA. **Electrophysiology in Neuromuscular Disease**. Vet Clin North Am Small Anim Pract 2002;32:31-62. [4.] da Costa RC, Parent JM, Dobson H, et al. Ultrasound-guided fine needle aspiration in the diagnosis of peripheral nerve sheath tumors in 4 dogs. **Can Vet J** 2008;49:77-81. [5.] Summers BA, Cummings JF, de Lahunta A. Neoplasia and the peripheral nervous system. In: Summers BA, Cummings JF, de Lahunta A, eds. V**eterinary neuropathology**. St. Louis: Mosby; 1995:472-481. [6.] Brehm DM, Vite CH, Steinberg HS, et al. A Retrospective Evaluation of 51 Cases of Peripheral Nerve Sheath Tumors in the Dog. **J Am Anim Hosp Assoc** 1995;31:349-359. [7.] Rose S, Long C, Knipe M, et al. Ultrasonographic Evaluation of Brachial Plexus Tumors in Five Dogs. **Vet Radiol Ultrasound** 2005;46:514-517. [8.] Rudich SR, Feeney DA, Anderson KL, et al. Computed Tomography of Masses of the Brachial Plexus and Contributing Nerve Roots in Dogs. **Vet Radiol Ultrasound** 2004;45:46-50. [9.] Kraft S, Ehrhart EJ, Gall D, et al. Magnetic Resonance Imaging Characteristics of Peripheral Nerve Sheath Tumors of the Canine Brachial Plexus in 18 Dogs. **Vet Radiol Ultrasound** 2007;48:1-7. [10.] Elmslie RE, Glawe P, Dow SW. Metronomic Therapy with Cyclophosphamide and Piroxicam Effectively Delays Tumor Recurrence in Dogs with Incompletely Resected Soft Tissue Sarcomas. **J Vet Int Med** 2008;22:1373-1379. [11.] Bagley RS, Tucker R, Harrington ML. Lateral and Foraminal Disk Extrusion in Dogs. **Compend Contin Educ Vet** 1996;18:795-804. [12.] Morgan PW, Parent J, Holmberg DL. Cervical Pain Secondary to Intervertebral Disc Disease in Dogs; Radiographic Findings and Surgical Implications. **Prog Vet Neurol** 1993;4:76-80. [13.] Felts JF, Prata RG. Cervical Disk Disease in the Dog: Intraforaminal and Lateral Extrusions. **J Am Anim Hosp Assoc** 1983;19:755-760. [14.] Wheeler SJ, Clayton Jones DG, Wright JA. The Diagnosis of Brachial Plexus Disorders in Dogs: A Review of Twenty-Two Cases. **J Small Anim Pract** 1986;27:147-157. [15.] Moissonnier P, Lavieille S, Duchossoy Y, Blot S, Horvat JC. Treatment of Avulsion of the Plexus Brachial in Dogs using Root Reimplantation: Experimental Study. **Bull Acad Vet Fr** 1998;71:175-181. [16.] Faissler D, Cizinauskas S, Jaggy A. Prognostic Factors for Functional Recovery in Dogs with Suspected Brachial Plexus Avulsion. **J Vet Intern Med** 2002;16:370. [17.] Kippenes H, Gavin PR, Bagley RS, et al. Magnetic Resonance Imaging Features of Tumors of the Spine and Spinal Cord in Dogs. **Vet Radiol Ultrasound** 1999;40:627-633. [18.] Drost WT, Love NE, Berry CR. Comparison of Radiography, Myelography and Computed Tomography in the Evaluation of Canine Vertebral and Spinal Cord Tumors in Sixteen Dogs. **Vet Radiol Ultrasound** 1996;37:28-33. [19.] Mayhew PD, Kapatkin AS, Wortman JA, Vite CH. Association of Cauda Equina Compression on Magnetic Resonance Images and Clinical Signs in Dogs with Degenerative Lumbosacral Stenosis. **J Am Anim Hosp Assoc** 2002;38:555-562. [20.] Gödde T, Steffen F. Surgical Treatment of Lumbosacral Foraminal Stenosis using a Lateral Approach in Twenty Dogs with Degenerative Lumbosacral Stenoss. **Vet Surg** 2007;36:705-713. [21.] Carozzo C, Cachon T, Genevois JP, et al. Transiliac Approach for Exposure of Lumbosacral Intervertebral Disk and Foramen: Technique Description. **Vet Surg** 2008;37:27-31.

# CANINE LUMBOSACRAL SPINAL FUSION TECHNIQUES

Amy E. Fauber, DVM, MS, Diplomate ACVS
West Lafayette, IN & Charlotte, NC

## BACKGROUND

The severity of clinical signs for dogs presenting with lumbosacral instability can vary. The patient's disease must be distinguished from other factors causing pain or neurologic impairment that localizes to the lumbosacral spine or pelvis. The diagnosis and treatment of lumbosacral instability can be challenging. Many dogs are successfully treated by surgical decompression alone.[1] However, as lumbosacral disease varies between dogs, decompression alone may not be adequate in some patients. Specific guidelines for the definitive diagnosis of instability do not exist. Often following imaging, the neurosurgeon is left asking, "To fuse or not to fuse?" It is the author's contention that in some patients, continued movement of the lumbosacral spine despite decompression will result in a failure of resolution of clinical signs. Of particular concern is that narrowing of the intervertebral foramen at L7-S1 will continue to cause nerve compression and subsequent pain. Fusion of the lumbosacral spine eliminates this dynamic source of nerve compression. Another mechanism for spinal nerve decompression is a foramintomy.[2] This procedure may provide adequate nerve decompression, but may be technically challenging and may contribute to instability.

Furthermore, lumbosacral fusion can be difficult to accomplish surgically due to anatomical constraints. There is a limited amount of bone available. The lumbosacral vertebral segment has unique characteristics that make implant placement challenging. Limited reports exist regarding techniques used to produce fusion at the lumbosacral space.[3,4,5,6]

## DIAGNOSIS

The diagnosis of lumbosacral instability is best accomplished with dynamic imaging of the lumbosacral spine. Flexed and extended lateral survey radiographs may serve as an important screening tool for lumbosacral instability. However, dynamic imaging of the lumbosacral spine with MRI or CT is our preference for the diagnosis of instability. Imaging of the LS spine has been well described.[7,8,9] The authors prefer CT for evaluation of the vertebrae and foramen. This type of imaging tends to be easier to perform compared to MRI. Positioning of the patient is less challenging. CT also provides excellent bone resolution. The patient's lumbosacral junction is imaged in flexion and extension. Axial images are reconstructed to evaluate and compare the position of L7 and S1 between extension and flexion. Compression of the nerve roots as they exit the foramen is evaluated on the images as well as signs of subluxation of the sacrum in relation to L7.

## SURGICAL TECHNIQUE

Several techniques have been described for stabilization of the lumbosacral spine.[4,5,6,10] These procedures may be biomechanically inadequate, technically challenging, or require expensive implants.[11,12] Most cases of lumbosacral instability have a protrusion of the degenerate lumbosacral intervertebral disk. Many patients also have circumferential compression from soft tissues and bony compression of the nerves as they exit the foramen. Because disk protrusion is often encountered, most surgical procedures for lumbosacral fusion involve a dorsal laminectomy followed by a partial diskectomy with placement of a graft into the LS space once the disk has been removed. Implants are applied to L7 and S1 to maintain distraction and stabilization so bony fusion can occur. One reported technique for LS fusion is placement of screws through the articular processes of L7-S1.[4] Articular cartilage is removed from the facets and the articular processes are then stabilized using a positional screw. Cancellous bone graft should be placed around the articular processes to encourage fusion. A variation of this has also been reported using negatively threaded pins instead of screws through the articular processes.[5,6] These implants are under significant cyclical strain while the fusion is occurring and complications of these implants include bending, breakage and migration. In order to avoid cyclic fatigue of the implants, some surgeons place additional implants in the body of the L7 vertebra, sacral body and sometimes the ilium.

## NOVEL SURGICAL TECHNIQUE

The author's preference is to place the implants in locations not previously described. Implants are placed directly into the vertebral body, maximizing available bone purchase. This lessens the risk of implant failure due to loosening or breaking. For this technique the patient is placed on the table in sternal recumbency with the legs in a neutral position. A standard dorsal laminectomy is performed at L7-S1. The intervertebral disk is then partially removed and the endplates are curetted. Cancellous bone graft is placed into the intervertebral disk space to promote fusion. The laminectomy is extended to allow for clear identification of the sciatic nerve roots and the endplates of L7 and S1. Screws or threaded pins are placed across the L7-S1 articular processes. Positive profile threaded pins are then placed within the spinal canal. The nerve roots are retracted while these pins are being placed. Two pins are placed in the caudal endplate of L7 and two pins are placed in the cranial endplate of S1. The nerve roots are avoided and protected using a blunt Freer periosteal elevator or a nerve root retractor during the placement of the pins. The pins are cut so that they do not extend higher than the dorsal aspect of the spinous processes and epaxial musclulature. Once the implants are in place, gelfoam is placed over the nerve roots to protect them during the application and hardening of the PMMA. Careful technique is used when mixing the PMMA to minimize air bubbles. The PMMA is then applied over the pins dorsally to incorporate the implants. Following lavage, the incision is closed routinely. Radiographs are made most-operatively to evaluate implant and PMMA placement.

**Potential complications of this technique include**

- Iatrogenic injury to the nerve roots or nerves as a result of implant placement and secondary to heat generated by PMMA
- Nerve compression secondary to implant placement within the spinal canal
- Implant loosening or breakage
- Infection
- Lack of fusion of the vertebrae
- Adjacent segment disease

To date the author has not experienced any significant complications with the use of this procedure. Fusion of the vertebrae may be difficult to definitively identify radiographically although clinically the dog shows no signs of complications. It has also been reported that in while some dogs have fusion of the LS space as a result of surgery, they may not have significant improvement of their clinical signs.[12] Lumbosacral stenosis continues to present both diagnostic and therapeutic challenges. Undoubtedly refinements in our understanding of this disease will contribute to improvements in treatment.

## REFERENCES

[1]Danielsson F, et.al. **Vet Surg** 1999;28:91. [2]Goedde T, et.al. **Vet Surg** 2007;36:705. [3]Auger J, et.al. **Vet Surg** 2000;29:70. [4]Bagley R. In Slatter, D(ed) **Textbook of Small Animal Surgery** (ed3) 2003;1238. [5]Slocum B, et.al. **J Am Vet Med Assoc** 1986;188:31. [6]Slocum B, et.al. In Bojrab, M(ed) **Current Techniques in Small Animal Surgery** (ed4) 1998;861. [7]Jones J, et.al. **Vet Radiol Ultrasound** 1995;36:91. [8]Jones J, et. al. **Vet Radiol Ultrasound** 1999;40:108. [9]Goedde, T. **ACVIM Forum Proceedings** 2007. [10]Sharp N. In Sharp and Wheeler: **Small Animal Spinal Disorders** (ed 2) 2005;203. [11]Carpenter L. **ACVIM Forum Proceedings** 2003. [12]Carpenter L. **ACVIM Forum Proceedings** 2006.

## INCLUSIONS IN CANINE NEURODEGENERATIVE DISEASES: VILLAIN, HERO OR INNOCENT BYSTANDER?

**Dennis O'Brien DVM PhD DACVIM Neurology**
**Columbia, Missouri**

Inclusion bodies are nuclear or cytoplasmic aggregates of stainable material, typically proteins. Most commonly, they represent sites of viral replication and consist of viral capsids. The resulting inclusions can be pathognomonic for the infection such as Negri bodies found in rabies. In this article, however, we will limit our review to inclusion bodies in developmental or degenerative diseases of the nervous system. The appearance of the inclusion may provide an important clue to the diagnosis in those diseases as well. More recently, however, there has been a shift in paradigm from viewing these protein aggregates as merely makers of disease to recognizing that they provide important clues to the underlying pathogenesis. Particularly when considering therapeutic approaches, this raises the question of whether the inclusions themselves are the cause of the cell dysfunction and death, an attempt of the cell to defend itself against disease, or innocent bystanders caught up in a process that they do not contribute to. We will review some mechanisms that are known to produce inclusion bodies and the current thinking about the contribution of the inclusions to the pathogenesis of disease.

### Axonal transport

In all cells, proteins are synthesized and folded in proximity to the nucleus and transported through the cytoplasm. Neurons are the only cells where this transportation extends for long distances down the axon, and thus they are more susceptible to diseases affecting transport. Proteins, other macromolecules, neurotransmitters and mitochondria are carried away from the nucleus anterograde by kinesin (Figure 1). Microtubules serve as the main "rail" along which the kinesin molecule "walks" with its attached cargo. Fast and slow axonal transport are now thought to be mediated by the same mechanisms with the difference being greater time spent paused in transit with slow axonal transport. Damaged macromolecules are transported retrograde back to the cell body in a similar fashion by dynein. In the cell body, they can be degraded and recycled by the processes discussed below.

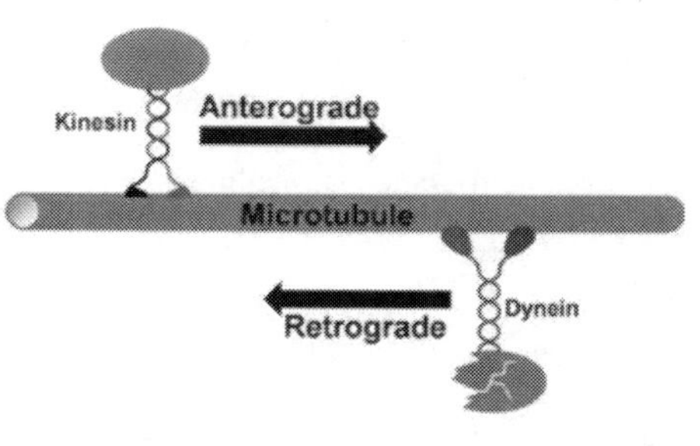


**Figure 1:** Proteins & other macromolecules manufactured in the cell body are transported anterograde down the axon along microtubules by kinesin. Dynein transports damaged proteins retrograde to the cell body where they are recycled.

Disruption of axonal transport can result in an accumulation of material in the axon called a spheroid. The longer the axon, the more likely it is to be affected by transport difficulties. Thus mutations in transport motors for example have been associated with motor diseases such as spastic paraplegia and with peripheral neuropathies such as Charcot-Marie-Tooth disease in humans. Spheroids will accumulate proximal near the cell body, or distal in the nerve terminal depending on which system is most affected. The latter can be observed in the nucleus gracilis where proprioception afferents from the hind limbs terminate. In these disorders, the spheroids appear to be mainly innocent bystanders. The failure of transportation of the needed materials to the axon results in axonopathy or the failure of retrograde transport of neurotrophic factors may lead to neuronal dysfunction or apoptosis.

### Lysosomal storage disease

The complex macromolecules that make up neurons and myelin will wear out and must be replaced. Some enzymes involved in the degradation process must be separated from the rest of the cytoplasm so that the appropriate pH for their activity can be achieved and the normal proteins in the cell protected from degradation. The lysosome is a double membrane bound organelle that accomplishes this process. The enzymes involved will break the complex molecules down into reusable or excretable fragments through a series of steps. Each step depends on the preceding step to efficiently catabolize the substrate. A mutation in the gene for one of these enzymes can lead to decreased or no enzyme activity (Figure 2). The catabolism cannot proceed beyond this roadblock and the substrate immediately upstream accumulates within the lysosome producing a membrane bound storage body.

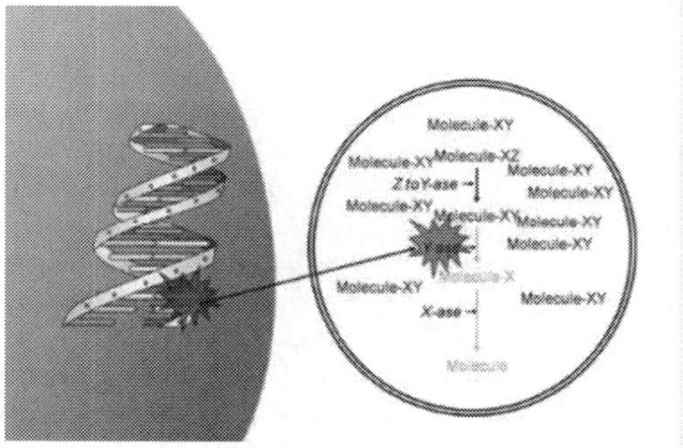

**Figure 2:** Mutations that lead to a deficiency of a lysosomal enzyme can block a step in the degradation of complex macromolecules. The molecule upstream from the deficit then builds up within the lysosome.

Most storage diseases are characterized by a delay in the onset of clinical signs. One form of neuronal ceroid lipofuscinosis (NCL) in dachshunds is caused by a mutation in the TPP1 gene that leads to a premature stop codon and no detectable enzyme activity. Yet these dogs do not develop clinical signs of blindness, ataxia, and myoclonic seizures until over 6 months of age. The macromolecules that are degraded by TPP1 are synthesized normally, and thus it is not until those molecules need to be turned over that problems arise. As the undegraded molecule builds up in the lysosomes, this eventually interferes with cell function. In some cases, a delay in onset may be due to a residual enzyme activity. For example, sheep or humans with NCL due to a mutation in the cathepsin D gene die as neonates whereas American bulldogs with the disease do not show the progressive dysmetria and paresis until 1-3 years of age. While sheep with NCL have no cathepsin D activity, dogs have 35% of normal activity in their brains. This residual enzyme activity may delay the onset of signs. In Tibetan terriers, onset of clinical signs of behavior changes, ataxia, and blindness do not develop until late in life. The gene responsible and the reason for the very late onset in this breed are not known.

Exactly how the deficiency of lysosomal enzyme leads to neuronal dysfunction and death is not clear. Therapies have focused on preventing the formation of lysosomal storage bodies through gene or enzyme replacement therapies. Such therapies are currently being investigated in a number of the canine diseases.

**The ubiquitin proteasome system**

The ubiquitin proteasome system is another mechanism for degrading proteins with cells. This includes misfolded or damaged proteins as well as some short-lived regulatory proteins. The unwanted protein is tagged for degradation with ubiquitin through a three step process (Figure 3). The E1 and E2 enzymes are relative non-specific while there are many E3 ligases that are fairly substrate specific. The tagged protein is targeted for degradation by proteasome (Figure 4). The regulatory cap of the proteasome acts as a reverse chaperone (see below) to unfold the protein in an energy dependant process. The proteolytic core of the proteasome contains sites which possess chymotrypsin-, trypsin- and peptidylglutamyl-peptide hydrolytic activity. The peptide fragments or amino acids produced can then be reutilized by the cell. Meanwhile, a ubiquitin C-terminal hydrolase cleaves off the ubiquitin molecules which are also recycled. In contrast to lysosomal enzymes, proteasomes are found free in cytoplasm or the nucleus.

The protein parkin is an E3 ubiquitin ligase. Mutations in the gene for parkin cause a common form a familial Parkinson's disease and the locus for canine multiple system degeneration (a movement disorder of Kerry blue terriers and Chinese crested dogs) includes this gene. Mutations in ubiquitin C-terminal hydrolase and α synuclein (a substrate for ubiquitination by parkin) also produce familial parkinsonism. Thus this system appears to play a key role in the disease. The inclusion bodies seen in sporadic Parkinson's disease (Lewy bodies) contain parkin, ubiquitin and α synuclein suggesting that proteasome dysfunction contributes to the acquired as well as the hereditary disease. One theory suggests that sporadic Parkinson's disease arises when proteins subjected to oxidative damage with aging overwhelm the ability of the ubiquitin proteasome system to degrade them. Dopaminergic cells are subject to a high level of oxidative stress due to the ability of dopamine itself to produce radials. This increased level of protein damage in a cell with reduced ability to deal with the damaged proteins could explain the selective degeneration of the dopaminergic neurons.

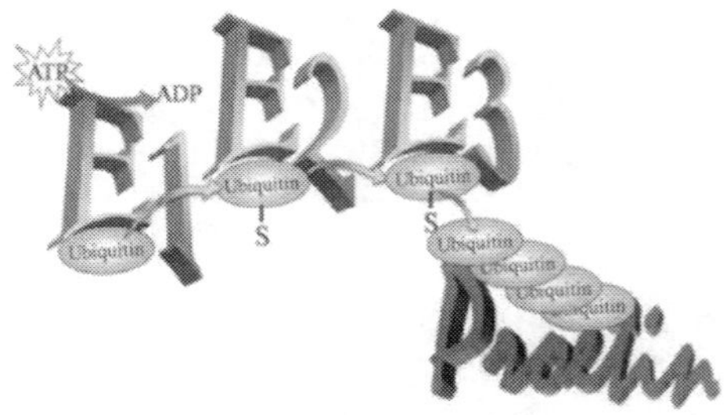


**Figure 3:** Misfolded proteins are tagged for degradation with a chain of ubiquitin molecules through a 3 enzyme system. E1 activates ubiquitin in an energy dependant process. E2 transfers the activated ubiquitin to E3 (ubiquitin ligase) which covalently binds it to the targeted protein.

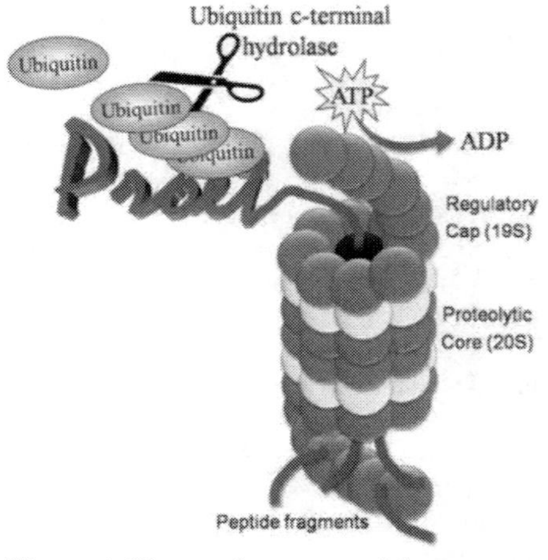


**Figure 4:** The regulatory cap of the proteasome unfolds the protein in an energy dependant process and feeds it into the core where proteolytic sites break it down into animo acids or peptide fragments. Ubiquitin is cleaved off the protein by ubiquitin c-terminal hydrolases and recycled.

**Misfolded proteins**

Proteins must be folded into the proper conformation to be functional. This folding process is typically accomplished by chaperone molecules as the protein is synthesized. Proteins that are misfolded are targets of the proteasome system but tend to form insoluble aggregates. These aggregates can defy the action of the proteasome by resisting unfolding which is necessary to feed the protein into the proteasome. Other misfolded proteins will tend to polymerize with the aggregate leading to its growth into an inclusion body.

The best know example of protein misfolding in veterinary medicine is prion diseases. The prion protein is a constituent of neurons which normally forms a triple helix structure: the cellular isoform. In scrapie and bovine spongiform encephalopathy an altered isoform of the prion protein is found: the scrapie isoform. The scrapie isoform is misfolded into a β-pleated sheet which resists proteases. These abnormal isoforms accumulate in the cell as scarpie-associated fibrils. Originally considered merely a marker of the disease, Stanley Pruisner proposed that these abnormal proteins actually were the infectious agent of the transmissible spongiform encephalopathies. The idea that there could be an infectious agent without nucleic acid was heresy at the time, but it is now widely accepted. The current theory is that the scrapie isoform of the prion protein serves as a molecular chaperone and induces the misfolding of the normal isoform into the protease resistant β-pleated conformation. Thus prion disease harkens back to the idea of inclusions as components of infectious agents, only in this case gene transcription does not play the same role.

In other diseases, a mutation in the protein leads to abnormal folding or function of the protein which then accumulates within the cell. In familial amyotrophic lateral sclerosis (ALS) in humans and degenerative myelopathy (DM) in dogs, a mutation in the SOD1 gene causes a progressive paralysis in aged patients. Inclusions which bind anti-SOD1 antibodies occur in neurons in both diseases. In ALS these inclusions are sometimes organized into a Lewy-body like aggregate with a dense core and hallow structure, while in DM they are more diffuse granular structures throughout the cytoplasm. Exactly how the mutant SOD1 leads to the loss of neurons is still unclear, and the composition of the inclusion bodies is being explored for clues.

A common reason for protein misfolding in human neurodegenerative disease is repeat expansions. Some genes contain repeated sequences which typically code for chains of glutamine or alanine. These areas of the gene tend to be unstable, and during meiosis the repeated sequences can become expanded. Too much expansion of the repeat leads to a protein that becomes misfolded and resistant to degradation. The only repeat expansion recognized thus far in dogs is in Lafora disease, a progressive myoclonic epilepsy in miniature wire-haired dachshunds. In this case an 11-12 base pair long repeat is expanded. Lafora disease is characterized by a PAS positive polyglycosan inclusions (Lafora bodies) found in neurons and muscle. Polyglycosan differs from normal glycogen in being short-chained, densely packed, insoluble, and heavily phosphorylated. As a result, it is relatively resistant to degradation by amylase. Laforin is a phosphatase that appears to be involved in glycogen regulation by dephosphorylating complex carbohydrates. The other gene associated with Lafora disease in humans, malin, is a ubiquitin ligase again implicating the proteasome.

Inclusions can be randomly located, but in cells such as neurons with a microtubule system, the aggregates can be localized near the centrosome. This has led to the concept that these inclusions are actively transported to this area by the cell. Such aggregates are referred to as aggresomes (Figure 5). Aggresomes may serve as staging areas for degrading the proteins. In some diseases, such as Parkinson's disease, ubiquitin and other components of the proteasome are present in the inclusions suggesting involvement of the proteasome as discussed above. Proteasomes are distributed throughout the cytoplasm and nucleus, however, so there should be little need for consolidation of the aggregates. More likely, the consolidation of abnormal proteins would allow for more efficient engulfing of the proteins into autophagasomes which mature into lysosomes. Alternatively, the aggresomes may simply represent an attempt to wall off the proteins from the cytoplasm as suggested by the fact that many are coated with a "cage" of intermediate filaments. The distinction between aggresome as cage to protect the cell from the abnormal protein versus staging ground for degradation has important implication for therapy.

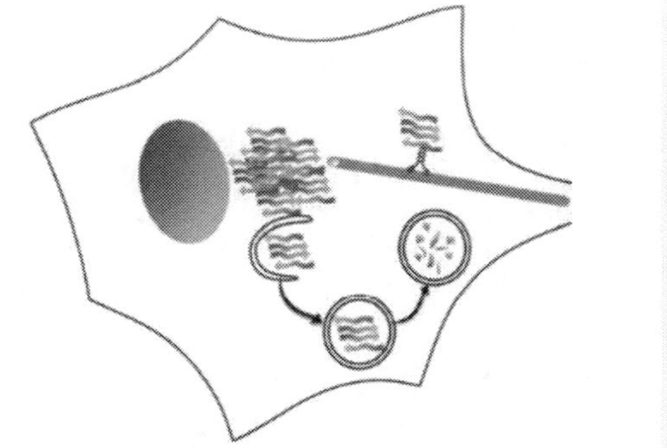

**Figure 5:** In one model of inclusion body formation, small aggregates of misfolded proteins are transported along microtubules to a perinuclear location where they form an aggresome. This may enable easier engulfing of the proteins into an autophagosome that matures into a lysosome and degrades the protein. Alternatively, it may represent an attempt of the cell to wall off the protein preventing it from interfering with cell function.

Huntington's disease (HD) is one of the glutamine expansion diseases of humans characterized by degeneration of the basal ganglia and inclusions. Recent studies of tissue cultures expressing the mutant gene for HD have shown that cells that have inclusions have less mutant HD protein in their cytoplasm and survive longer than cells that do not develop the inclusions. This implies that therapeutic strategies to prevent the aggregation of the abnormal protein could actually be detrimental.

How the abnormal proteins lead to cell death is another important question when considering therapy. One possible mechanism is through inhibition of the proteasome. Misfolded proteins may obstruct the proteasome preventing it for breaking down other proteins. Since the proteasome is also responsible for breaking down regulatory elements in the cell, a dysregulation of cell function could result in triggering of the apoptotic pathway. Alternatively, the misfolded protein may monopolize chaperones that are attempting to refold the protein into proper conformation. These chaperones would then not be available to fold newly synthesized proteins into their proper conformation.

**Conclusion**

The character of the inclusions in a disease can provide an important clue to diagnosis, but the role they play in the pathogenesis will be critical in devising therapeutic strategies. In some, like axonal transport diseases, they are simply innocent bystanders stuck in obstructed traffic. In the case of lysosomal storage diseases, the accumulated proteins appear to be a significant contributor to disease. In others such as Huntington's disease, the aggregation of the abnormal proteins appears to be an important protective strategy by the cell.

**REFERENCES**

[1.] M. Arrasate et al. **Nature** 431 (7010):805-810, 2004. [2.] T. Awano et al. **Molecular Genetics and Metabolism** 87 (4):341-8, 2006. [3.] T. Awano et al. **Molecular Genetics and Metabolism** 87 (4):341-8, 2006. [4.] J.R. Coates et al. **PNAS** in press 2009. [5.] F. Fornai et al. **Brain Research Bulletin** 65 (4):275-290, 2005. [6.] S. Ganesh et al. **Journal of Human Genetics** 51 (1):1-8, 2006. [7.] R. R. Kopito. **Trends in Cell Biology** 10 (12):524-530, 2000. [8.] H. Lohi et al. **Science** 307 (5706):81, 2005.S. Mittal, **Human Molecular Genetics** 16 (7):753-762, 2007. [9.] D. P. O'Brien et al**. Journal of Heredity** 96 (7):727-34, 2005. [10.] S. Roy et al. **Acta Neuropathologica**.*109(1):5-13,* 2005. [11.] J. M. Woulfe. **Neuropathology & Applied Neurobiology** 33 (1):2-42, 2007.

# BROADLY REACTIVE PAN VIRAL, BACTERIAL, & RICKETTSIAL PCR FOR PATHOGEN DETECTION IN CANINE MENINGOENCEPHALITIS OF UNKNOWN ETIOLOGY (MUE)

**SJ Schatzberg[1], Q Li[1], SR Platt[1], M Kent[1], JM Levine[2], B Porter[2], GJ Levine[2], K Chandler[3], EB Breitschwerdt[4], AJ Birkenheuer[4], LJ Anderson[5], S Tong[5]. 1. University of Georgia, College of Veterinary Medicine (CVM), Athens, GA. 2. Texas A&M University, CVM, College Station, TX. 3. Royal Veterinary College, Hatfield, UK. 4. North Carolina State University, CVM, Raleigh, NC. 5. Centers for Disease Control and Prevention, Division of Viral Diseases, Atlanta, GA.**

Meningoencephalitis comprises a group of complex, debilitating, and often fatal disorders affecting both people and dogs. The approaches to patient evaluation, differential diagnosis, diagnostic testing, and treatment are evolving rapidly in both human and veterinary neurology. Despite many advances, definitive etiologies for human and canine meningoencephalitis still remain elusive in approximately 50%-70% and 75%-95% of cases, respectively. This largely is due to limitations in the spectrum of diagnostic tests utilized routinely in the clinical setting.

In this investigation, broadly reactive PCR tests for a diverse group of viruses, bacteria and rickettsia have been utilized to evaluate approximately 200 cases of canine meningoencephalitis of unknown etiology (MUE). The ability to detect known and potentially novel pathogens should help to determine the etiology of fatal disorders such as meningoencephalitis, may identify emerging and zoonotic pathogens, and in a clinical setting should improve survival with the subsequent implementation of more directed therapies in animals and people. Our broader objective is to improve standard pathogen detection methods in veterinary and human medicine.

## RESEARCH DESCRIPTION.

### Problem Statement and Specific Aims

Human and canine meningoencephalitis are life threatening diseases that include a broad spectrum of pathological entities, for which etiologic agents are lacking in 50%-90% of cases.(*1-3*) A diverse and sensitive pathogen screening strategy is needed desperately in a clinical setting for canine and human encephalitis for several key reasons. 1.) It is critical to discriminate among infectious (and noninfectious) causes of meningoencephalitis, as the various meningoencephalitides overlap in clinical features yet require different therapies. 2.) A changing pattern for the etiologies of meningoencephalitis has been demonstrated in multiple prospective studies.(*2, 4*) 3.) The principal causes for meningoencephalitis vary by geographic region.(*5*) 4.) Specific PCR and serological tests typically are selected based on pathogens that are indigenous to a geographic region, and such tests may fail to identify pathogens unknowingly (or maliciously) brought into a region. 5.) Novel causes for meningoencephalitis are likely to be missed with conventional PCR and serological tests.

At CDC and UGA-CVM we have developed broad-based, consensus, degenerate PCR primer sets for viral families associated with encephalitis including adeno-, herpes-, alpha-, flavi-, bunya-, corona-, borna-, rhabdo-, paramyxo-, orthomyxo-, polyoma-, and picornaviruses. In collaboration with NCSU-CVM, we also have developed and routinely utilize degenerate primer sets for the detection of bacterial, rickettsia, *Bartonella* and *Borrelia spp.* In this investigation, broadly reactive PCR tests have been used screen for this extremely diverse group of pathogens in canine meningoencephalitis. The work has been based on the following hypothesis and objectives:

| |
|---|
| ***Hypothesis*:** **Broadly reactive PCR will identify infectious agents in the CNS of dogs with meningoencephalitis of unknown etiology.** |
| ***Objective 1:*** *To determine whether or not nucleic acids from infectious agents can be identified in CSF by applying broadly reactive pan viral, bacterial and rickettsial PCR assays to 200 CSF samples, collected pre- and/or postmortem, from dogs with meningoencephalitis of unknown origin.* |
| ***Objective 2:*** *To determine whether or not nucleic acids from infectious agents can be identified in fresh brain samples by applying broadly reactive pan viral, bacterial and rickettsial PCR assays to 50 fresh brains, collected postmortem from dogs with meningoencephalitis of unknown origin.* |

### Background & Significance

The inciting causes for canine meningoencephalitis are unknown in up to 75% of published cases.(*6-8*) In 2002, a 10 year literature review disclosed only 23 cases of bacterial meningoencephalitis published in dogs.(*9*) Viral meningoencephalitis also is reported rarely in dogs apart from regional outbreaks of canine distemper and rabies virus infections. These prevalence data are similar to what the investigators have observed in clinical practice. Herpes-, parvo-, and parainfluenza viruses may cause encephalitis in puppies, but confirmed reports are exceptionally rare and essentially not described in mature dogs.(*7, 10-13*) Recently, there have been sporadic reports of flavivirus (tick-borne encephalitis virus, west Nile virus),(*14-18*) bornavirus,(*19, 20*) and Eastern equine encephalitis infection in dogs,(*21*) suggesting that canine viral meningoencephalitis is more prevalent than currently accepted. However, the etiologies for the majority of cases of canine meningoencephalitis remain elusive because of current limitations for pathogen detection in routine diagnostic testing.

Several methods can be used to determine the presence or absence of infectious agents in clinical specimens. The classical approaches to identifying pathogens, such as tissue culture isolation, electron microscopy, immunologic assays (e.g., immunofluorescence and enzyme immunoassays) and serologic assays, have identified the etiology of many infectious diseases. However, conventional laboratory diagnostics such as serum (or CSF) antigen and antibody testing and CSF culture are accepted to be non-specific and insensitive, respectively, for the identification of CNS pathogens.(*22, 23*) Moreover, conventional diagnostic methods utilizing agent-specific known protein or nucleic acid signatures are likely to miss unsuspected pathogens and importantly fail to identify novel or emerging microbial agents.

Genome-based assays including PCR-based assays are powerful tools to detect pathogens, but are limited to individual pathogen species. Broader molecular methods have been developed more recently to identify novel and/or unsuspected pathogens. For example, consensus, degenerate PCR is designed to target highly conserved genomic regions that are present in all members of pathogen family and/or genera.(*24*) This methodology not only amplifies nucleic acids from known members of a microbial family and/or genera, but critically also detects novel microbes due to the evolutionary relationships of the organisms.

Consensus, degenerate PCR has been used successfully to identify and characterize a number of novel human viruses including Severe Acute Respiratory Syndrome Coronavirus (SARS-CoV),(*25*) hepatitis G,(*26*) Sin Nombre virus,(*27*) Human Retrovirus-5 (HRV-5),(*28*) and novel animal viruses such as the macaque gamma herpesvirus and pig endogenous retrovirus (PERV).(*29, 30*) Recently, CDC has applied pan viral PCR to patient specimens from selected disease outbreaks, individual cases of severe illness, and prospective etiologic studies of encephalitis. CDC has detected a human-related metapneumovirus, as a potential causative agent for a 2006 respiratory disease outbreak in wild chimpanzees in western Tanzania. More recently, CDC's degenerate pan viral PCR assays have identified several known pathogens from patients with encephalitis in Bangladesh and Thailand including HSV 1-2, EBV, adenovirus, Japanese encephalitis virus, and Saint Louis encephalitis virus. These methodologies have been further evaluated in the investigation of canine meningoencephalitis of unknown etiology.

The most common post-mortem diagnoses for canine encephalitis include necrotizing meningoencephalitis (NME), necrotizing leukoencephalitis (NLE), and granulomatous meningoencephalitis (GME).(*3, 31, 32*) Without aggressive immunotherapy these disorders are uniformly fatal. The etiologies for these canine meningoencephalitides are unknown. On an antemortem basis, when histopathological confirmation is lacking, the terminology meningoencephalitis of unknown etiology (MUE) may be preferable when these differentials are prioritized. A recent landmark study using paraffinized canine brains disclosed the presence of viral antigens in 14 of 53 cases (26%) of encephalitis, previously of unknown origin.(*7*) Broadly reactive consensus, degenerate PCR assays have even greater potential for identifying causative viruses in canine meningoencephalitis, on both a retrospective and prospective basis. Our current methods should detect nucleic acids from hundreds of common, but perhaps more importantly, emerging, novel, and zoonotic pathogens that may be present in neural tissues from canine meningoencephalitis patients.

**Preliminary Results**

Consensus *Bartonella spp.*, *Borrelia spp.*, and *Rickettsia spp.* PCR has been negative on all samples tested to date. Eubacterial and specific bacterial PCR has been predominantly negative, but has identified sporadic positive cases including *Staphylococcus spp* (2 out of 140 CSF samples) and *Mycoplasma spp.* (10 out of 30 MUE brains). These results have been confirmed by sequence analysis and bacterial culture of CSF or brain tissues.

Viral PCR has disclosed positive results for bunya- (6 out of 140 CSF samples), polyoma (5 out of 140 CSF samples), and paramyxoviruses (1 out of 30 brains) in MUE. Sequence analysis of PCR amplicons thus far has confirmed: LaCrosse virus, Merkel cell carcinoma virus (MCV, a recently identified human polyomavirus) and canine distemper virus in sporadic CSF and brain samples.

At the time of the research report submission, PCR testing was ongoing for 140 cases of MUE for adeno-, herpes-, alpha-, flavi-, bunya-, corona-, borna-, rhabdo-, paramyxo-, orthomyxo-, polyoma-, and picornaviruses. In addition, Bartonella, rickettsia, and Ehrlichia results from 200 cases were in the process of a being repeated as standard operating procedure. These results will be presented at the 2009 ACVIM forum.

**Conclusions**

Both negative and positive results must be interpreted cautiously. For example, the demonstration of nucleic acids from a pathogen does not definitively establish causality, and simply may be a reflection of a patient's immune-compromised state. Conversely, negative results do not exclude the possibility of an autoimmune response secondary to prior exposure (eg post-infectious meningoencephalitis), since pathogens can silently prime for and trigger meningoencephalitis via molecular mimicry (*33*). Moreover, causative agent(s) may be present but at undetectable levels in the presence of a self-perpetuating immune response, a phenomenon that has been well described for flavivirus infections.(*34*) Work is ongoing to investigate para- and post-infectious considerations in the idiopathic meningoencephalitides. The specific roles, if any, of recently identified viruses and *Mycoplasma spp.* are being further investigated in GME, NME, and NLE.

The ability to detect known and potentially novel pathogens should help to determine the etiology of a subset of cases of canine meningoencephalitis, may identify emerging (eg MCV) and zoonotic (eg LaCrosse) pathogens, and in a clinical setting should improve survival with the subsequent implementation of more directed anti-microbial therapies. In addition, these broadly reactive PCR assays should prove useful for the identification of numerous veterinary and human disorders of unknown etiology.

## REFERENCES

[1]C. A. Glaser et al., **Clin Infect Dis** 43, 1565 (Dec 15, 2006). [2.] M. Koskiniemi et al., **European journal of pediatrics** 156, 541 (Jul, 1997). [3.]A. Tipold, R. Fatzer, A. Jaggy, A. Zurbriggen, M. Vandevelde, **Journal of Small Animal Practice** 34, 623 (1993). [4.]E. Beghi et al., **Ann Neurol** 16, 283 (Sep, 1984). [5.]H. Rantala, M. Uhari, **The Pediatric infectious disease journal** 8, 426 (Jul, 1989). [6.] S. J. Schatzberg, N. J. Haley, S. C. Barr, A. de Lahunta, N. J. Sharp, **J Vet Intern Med** 19, 553 (Jul-Aug, 2005); [7.] S. Schwab et al., **J Comp Pathol,** (Jan 31, 2007); [8.] A. Tipold, **Journal of Veterinary Internal Medicine** 9, 304 (1995). [9.]S. T. Radaelli, S. R. Platt, **J Vet Intern Med** 16, 159 (Mar-Apr, 2002). [10.]D. R. Agungpriyono, K. Uchida, H. Tabaru, R. Yamaguchi, S. Tateyama, **Vet Pathol** 36, 77 (Jan, 1999).[11.]W. K. Baumgartner, S. Krakowka, A. Koestner, J. Evermann, **Vet Pathol** 19, 79 (Jan, 1982).[12.]B. J. Johnson, A. E. Castro, **J Am Vet Med Assoc** 184, 1398 (Jun 1, 1984).[13.]D. H. Percy, H. J. Olander, L. E. Carmichael, **Pathol Vet** 5, 135 (1968).[14.]S. Buckweitz et al., **J Vet Diagn Invest** 15, 324 (Jul, 2003).[15.]J. Klimes, Z. Juricova, I. Literak, P.

Schanilec, E. Trachta e Silva, **The Veterinary record** 148, 17 (Jan 6, 2001).[16]M. W. Leschnik, G. C. Kirtz, J. G. Thalhammer, **Int J Med Microbiol** 291 Suppl 33, 66 (Jun, 2002).[17]R. W. Read, D. B. Rodriguez, B. A. Summers, **Vet Pathol** 42, 219 (Mar, 2005).[18]H. Weissenbock, A. Suchy, H. Holzmann, **Acta neuropathologica** 95, 361 (Apr, 1998).[19]H. Weissenbock, N. Nowotny, P. Caplazi, J. Kolodziejek, F. Ehrensperger, **J Clin Microbiol** 36, 2127 (Jul, 1998).[20]M. Okamoto et al., **J Comp Pathol** 126, 312 (May, 2002).[21]M. D. Farrar, D. L. Miller, C. A. Baldwin, S. L. Stiver, C. L. Hall, **J Vet Diagn Invest** 17, 614 (Nov, 2005).[22]J. Dong, J. P. Olano, J. W. McBride, D. H. Walker, **J Mol Diagn** 10, 185 (May 1, 2008, 2008).[23]R. J. Whitley, J. W. Gnann, **Lancet** 359, 507 (Feb 9, 2002).[24]T. Kaur et al., **American journal of primatology** 70, 755 (Aug, 2008).[25]T. G. Ksiazek et al., **The New England journal of medicine** 348, 1953 (May 15, 2003).[26]J. N. Simons et al., **Nature medicine** 1, 564 (Jun, 1995).[27]S. T. Nichol et al., Science 262, 914 (Nov 5, 1993).[28]D. J. Griffiths, P. J. Venables, R. A. Weiss, M. T. Boyd, **J Virol** 71, 2866 (Apr, 1997).[29]C. Patience, Y. Takeuchi, R. A. Weiss, **Nature medicine** 3, 282 (Mar, 1997).[30]T. M. Rose et al., **J Virol** 71, 4138 (May, 1997).[31]D. R. Cordy, Vet Pathol 16, 325 (May, 1979).[32]D. R. Cordy, T. A. Holliday, **Vet Pathol** 26, 191 (May, 1989).[33]D. J. Theil, I. Tsunoda, F. Rodriguez, J. L. Whitton, R. S. Fujinami, **J Neurovirol** 7, 220 (Jun, 2001).[34]N. Krueger, H. W. Reid, **The Veterinary record** 135, 224 (Sep 3, 1994).

# TRANSLATION OF CANINE DEGENERATIVE MYELOPATHY TO HUMAN ALS

**Joan R. Coates, DVM, MS, Diplomate ACVIM (Neurology)**
**Columbia, MO**

## INTRODUCTION

### Degenerative Myelopathy

Canine degenerative myelopathy (DM) is a spontaneously occurring, adult-onset, progressive spinal cord disease.[1] This disease is not uncommon in some pure bred dogs with an overall prevalence rate of 0.19%.[2] Most dogs are at least 8 years old before they show clinical signs.[1,2-5] The initial upper motor neuron (UMN) and dorsal column degeneration produces a spastic and general proprioceptive ataxia in the pelvic limbs which progresses to paraplegia. Dog owners usually elect euthanasia within a year of diagnosis; however, when euthanasia is delayed, lower motor neuron (LMN) signs emerge with ascending tetraparesis, flaccid paralysis and widespread muscle atrophy.[2,6] Because a variety of common acquired compressive spinal cord diseases can mimic DM by compromising the UMN pathways, a definitive diagnosis of DM can only be accomplished postmortem by the histopathologic observation of axonal and myelin degeneration which occur in all spinal cord funiculi, but are consistently most severe in the dorsal portion of the lateral funiculus within the middle to lower thoracic region.[1,2,4,5,7] The progression to LMN signs was not realized until the disease was more thoroughly studied in DM affected Pembroke Welsh Corgi dogs.[2] We have additionally documented neuromuscular pathology as evidenced by denervation atrophy in muscle, and nerve fiber loss, axonal degeneration and demyelination of myelinated fibers in peripheral nerves.

DNA samples from 38 DM-affected Pembroke Welsh corgi cases and 17 related clinically normal controls were used for genome-wide association mapping[8-10] which produced the strongest associations with markers on CFA31 in a region containing the canine superoxide dismutase 1 (*SOD1*) gene.[11] *SOD1* was considered a regional candidate gene because mutations in human *SOD1* can cause amyotrophic lateral sclerosis (ALS), an adult-onset fatal paralytic neurodegenerative disease with both upper and lower motor neuron involvement. The resequencing of *SOD1* in normal and affected dogs revealed a G to A transition, resulting in an E40K missense mutation.[11] Homozygosity for the A allele was associated with degenerative myelopathy in five dog breeds: Pembroke Welsh corgi, Boxer, Rhodesian ridgeback, German Shepherd dog and Chesapeake Bay retriever. Microscopic examination of spinal cords from affected dogs revealed myelin and axon loss affecting the lateral white matter and neuronal cytoplasmic inclusions that bind anti-superoxide dismutase 1 antibodies. These inclusions are similar to those seen in spinal cord sections from ALS patients with *SOD1* mutations.[11] Our findings identify canine DM to be the first recognized spontaneously occurring animal model for ALS.

SOD1 functions as a homodimer which converts superoxide radicals to hydrogen peroxide and molecular oxygen. Active sites in each subunit contain one copper ion and one zinc ion within an eight-stranded antiparallel β-barrel. In canine SOD1, amino acid position 40 lies within the short "Greek key" connecting loop that stabilizes the alignment between the two sets of β-strands which comprise the β-barrel.[17,18] The codon for amino acid position 40 of human SOD1 lies within a region that contains a cluster of missense mutations that are associated with human ALS, including E40G at the position homologous to the canine E40K mutation.[19,20] The canine E40K mutation, like the human E40G mutation and several other ALS-associated *SOD1* mutations, reduces the net negative charge of the predicted protein product. The SOD1 isoforms with reduced net negative charge may be prone to aggregation because of reduced repulsive Coulombic forces or because of increased interaction with anionic membrane surfaces.[20,21]

All of the strictly diagnosed DM-affected dogs were A/A homozygotes; however, several of the aged A/A homozygotes were symptom free. Thus, DM appears to be an incompletely penetrant autosomal recessive disease. The discovery that homozygosity for E40K is a major genetic risk factor underlying DM should lead to marker-based long-term dog-breeding strategies for avoiding future generations of dogs at risk for developing DM. The disease incidence is unknown but 1,000's of dogs are homozygous for the mutant allele, thus at risk for DM. The allelic frequency may be as high as 75% in breeds prone to DM (unpublished data, Dr. Gary S. Johnson – Animal Molecular Genetics Laboratory, University of Missouri). The high frequency of the mutation in some breeds such as the Boxer and Pembroke Welsh corgi suggests that strict avoidance of this mutation could severely reduce the effective population sizes of these breeds. The potential of identifying modifier loci and altering their frequency may offer an alternative route to breeding improvement.

### Amyotrophic Lateral Sclerosis

Amyotrophic lateral sclerosis (Lou Gehrig's disease) refers to a heterogeneous group of adult onset human diseases in which progressive degeneration of the motor neuron system causes advancing weakness and muscle atrophy, culminating in paralysis and death. It is a syndrome of upper and lower motor neuron dysfunction at several levels of the neural axis without involvement of other systems. Amyotrophic lateral sclerosis is the most common adult motor neuron disease and was first described in 1869 by Jean-Martin Charcot. The disease affects people in their fourth to sixth decade of life. Most people die within 5 years after disease onset. Approximately 5 to10% of ALS cases are familial; the rest appear to be sporadic.[22-25] A landmark discovery by Rosen et al. found 11 missense mutations in the SOD1 gene in 13 families with ALS.[25] Mutations in *SOD1* account for approximately 20% of the familial ALS cases and 1% to 5% of the cases of sporadic ALS.[23-25] Over 150 different *SOD1* mutations have been identified in ALS patients (http://alsod.iop.kcl.ac.uk/Als/index.aspx). All SOD1 mutations associate with dominant ALS with the exception of N86S and D90A substitution that follows a recessive inheritance pattern.[26]

Elucidation of mechanisms underlying ALS has been hampered by a paucity of biological material from affected individuals in early stages of the disease.[27] At present, there exists no prophylactic or curative treatment for ALS. The prescribed anti-glutamatergic drug Rilutek (riluzole) may have a marginal but unsatisfactory effect in extending life by a few months.[28] Thus the need for significant disease modifying therapies is paramount.

**TRANSGENIC RODENTS AS ANIMAL MODELS**

Animal models permit testing of hypotheses about the pathogenesis of a disease and evaluation of therapies prior to clinical trials in humans. To be valuable for study of ALS, an animal model must mimic the major clinical signs and pathologic changes found in the human disease.[28] The discovery of SOD1 mutations in familial ALS was followed by the generation of transgenic rodents expressing human *SOD1* mutations (h*SOD1*$^{m}$).[29,30] In contrast, nullizygous *SOD1* knockout mice develop normally,[31] suggesting that the neurodegeneration in h*SOD1*$^{m}$ mice and in ALS patients results from a toxic gain of function.[22,30] Thus, ALS research has relied heavily on transgenic rodents expressing human *SOD1* mutations to produce a motor neuron disease which recapitulates many features of ALS.[28,30] At present, 15 different human SOD1 mutations have been expressed in mice.[32] The transgenic SOD1$^{G93A}$ mice (glycine substituted to alanine at position 93), carrying 23 transcopies of the human *SOD1* gene are principally used in ALS research. The essential clinical features of these mice include: initial hindlimb tremor and weakness around 3 months detected by locomotor deficits, progressing to paralysis and premature death after 4 months.[30] Hallmark pathologic features include concomitant spinal motor neuron loss and gliosis.

These transgenic rodent models have served as an important platform for evaluating pathogenic hypotheses of ALS that are now having important implications for disease intervention and treatment of ALS. Although the nature of the toxin is unclear, several experiments suggest that the neurodegeneration occurs because conformational changes in the mutant SOD1 protein alter the biological activity and/or promote the formation of intracellular SOD1 aggregates.[21,22,33] Transgenic mutant SOD1 mice have confirmed that experimental ALS is a disease of peripheral denervation and synaptic dysfunction long before motor neuron loss.[34,35] ALS mice have identified potential nonneuronal influences in the pathogenesis, one being that disease progression rate triggered in motor neurons may also be governed by its neighboring glial cells.[22,36]

The h*SOD1*$^{m}$ rodent models also have been used as a tool for evaluation of 'proof of concept' and preclinical studies in over 150 potential therapeutic anti-oxidant, anit-aggregation, anti-glutaminergic, anti-inflammatory, anti-apoptotic and neurotrophic pharmacologic agents.[32] A proof of concept study has the goal to elucidate the mechanism of disease which may use a drug as an investigational tool. A preclinical study has the primary goal to develop a drug for use in humans. Treatment approaches are also being broadened to include those models that focus on causes other than *SOD1* mutations – target validation.[37] Histone deacetylase inhibitors, anti-apoptotics, anti-inflammatories and novel anti-oxidants appear effective at extending survival in genetic ALS models.[32] Progression in therapeutic advances beyond riluzole that serve as rescue protocols to motor neurons of ALS mice include viral mediated delivery of growth factors suggesting that targeted therapy to the CNS is key for neurotrophic factor therapy in ALS. Delivery of small molecule therapies to the spinal cord of ALS mice with viral vectors has yielded the greatest rescue to date.[38] Other therapies on the horizon include targeted stem cell therapy[39] and immunotherapy[40] and consideration of combination therapies that affect multiple mechanisms.

**Problems with Translational Approaches from Mice to Humans**

The choice of therapeutic agents in many clinical trials of human ALS has been predicated on the efficacy of these drugs when studied in the h*SOD1*$^{m}$ mouse. One of the disappointments of the current mouse and rat models is that many drugs that showed an effect in rodents have failed to be efficacious in human ALS.[41,42] One reason for failure is that transgenic rodents better represent a model for familial ALS rather than sporadic ALS, the most common form of ALS. Differences in pathophysiology between SOD1-linked ALS and those forms of ALS not associated with the *SOD1* mutation are still incompletely understood; thus, biologic differences between these forms of ALS may also limit treatment approaches. Another reason is that preclinical therapies are tested in rodents prior to the symptomatic phase of disease which has limited strategy in treatment of sporadic ALS. Lastly, drug dosage and bioavailability also may not translate to humans based on experimental data in rodents. A recent and the only meta-analysis study of treatment trials in the h*SOD1*$^{m}$ mouse explored specific reasons for failure of translational research.[41] This study revealed significant publication bias in small studies reporting beneficial results with a paucity of studies reporting no treatment effect. An over optimistic impression is generated in the utility of the model and bias toward treatment efficacy in animals that may lead to unnecessary clinical trials in people. The study also brought forth that most treatment trials in mice are of limited methodological quality in randomization, blinding, differentiation between clinical and statistical significance and relevance of the mouse model to human disease (i.e. selection of agents for clinical trials in familial ALS). Blinding observers is essential in preclinical studies and recommended in proof of concept studies.[42,44] Due to its lethality and rapid progression, ALS is appropriate for 'proof of concept' testing of potential therapies.[42] It will be important to make a distinction in future studies between those that delay or prevent onset of disease and those that aim to identify therapeutic agents that may prolong survival after the onset of disease.

Attempts have been made to establish guidelines for preclinical *in vivo* evaluations of pharmacotherapeutics with a goal to establish standard methods for drug testing in ALS.[43] Guidelines were established for model characterization, symptomatic versus asymptomatic treatment phases, endpoints, environment, dose/response and methods of drug application. Efforts are being made to standardize the G93A animals according to genetic background, sex and litter matching, number of transgenes and inbreeding.[44] Standardization of these transgenic models will provide the advantage for reproducibility of preclinical studies and for establishing adequate power analysis. In the vast majority of therapeutic approaches tested in rodent models compounds are administered months before disease onset – leading to positive therapeutic effects. Thus, it is predictable that a drug efficient in the preclinical phase may not have a similar chance of efficacy at clinical disease.[43] Study endpoints used for motor neuron disease studies in rodent models include motor function or survival.[43] Survival is often the primary endpoint in preclinical studies. Survival is measured in the ability of the mouse to right itself within 30 seconds. Animal weight is considered a secondary endpoint for survival. Motor function is assessed using methods that include rotarod, grip strength, gait analysis, and electrophysiologic techniques. Translation of test results in animal models further requires the demonstration of dose-response curves and an understanding of the pharmacokinetics. For drug delivery assessments, measurements of brain and spinal cord levels of the compounds are necessary and may differ between CNS compartments. This is important when considering routes of administration such as oral, parenteral, and more advance delivery techniques using intrathecal and intraventricular routes.

## CAN DOGS BRIDGE THE GAP BETWEEN MICE AND HUMANS?

Until the discovery of an *SOD1* mutation in canine degenerative myelopathy, there have been no previous reports of spontaneously occurring animal models of *SOD1*-linked ALS. A canine model of disease can form a key step in the translation of basic science knowledge to clinical applications. Because of the small size, primitive nervous system and limited cognitive capacity of mice, their usefulness is restricted in studies of the pathophysiologic mechanisms and therapeutic interventions utilizing gene-targeted strategies with respect to future human applications. Treatments that are effective in rodent trials often will fail in human patients and may vice versa and thus, may be discarded when potential exists for benefiting humans.[45]

The ***rationale*** for the dog as an animal model for study of human ALS is that canine DM will mimic the UMN onset form of human ALS and that responses to therapy in the dog will predict similar responses in humans. Clinical signs for canine DM have similarities to the El Escorial criteria developed to standardize diagnosis of ALS for research purposes.[46] The major hallmarks of the El Escorial criteria include signs of upper and lower motor neuron degeneration, and progressive spread of these signs within a region or to other regions. Moreover, there must be an absence of electrophysiological or neuroimaging evidence of other disease processes that could explain the combination of progressive upper and lower motor neuron dysfunction.

The development of therapeutic interventions for ALS has been complicated by the diverse disease pathophysiology and lack of technology to deliver some therapeutics to the nervous system. The canine DM model caused by an *SOD1* mutation may offer several advantages over the current mouse and rat models that are typically used to further understand human ALS. Compared to the h*SOD1*$^{m}$ rodent models, dogs with DM are more similar to people in size, in the structure and complexity of their nervous systems, and in the duration of the disease. In addition, they are unlikely to possess the very high levels of mutant *SOD1* expression which occur in many of the h*SOD1*$^{m}$ rodent models[28,47] and which may induce pathologic processes distinct from those affecting ALS patients. The larger size of the dog will facilitate some studies and provide a closer model for treating humans. For example, intraventricular or intrathecal injections of growth factors, stem cells, or small molecules may be more relevant in the dog model rather the mouse or rat, especially regarding CSF dynamics, and pharmacologic properties of drugs penetrating into a larger brain and spinal cord. It is likely that this spontaneous mutation in dogs will more accurately mimic ALS in humans than the current mouse and rat models. Thus, the results from clinical trials conducted with DM-affected dogs may better predict the efficacies of therapeutic interventions for treating ALS.

### Establishment of Outcome Measures

With the discovery that an *SOD1* mutation is associated with canine DM, we now can better focus our efforts in studies of clinical disease progression and neuropathology in order to establish outcome measures. An outcome measure is any measurement that is used for assessing the design phase and used to calculate sample size for sufficient power of the study. Since there are no established biomarkers of disease progression in ALS, researchers use potential surrogate markers. Outcomes in human ALS trials evaluate for symptomatic improvement, reduced rate of deterioration and prolonged survival time.[48] Measurements of muscle strength, motor function, respiratory function and survival are used as primary outcomes in ALS patients. The ALS Functional Rating Scale (ALSFRS) which measures activities of daily living was first designed to augment the standard primary outcomes.[49,50] Electrophysiologic testing procedures using motor unit number estimate (MUNE) along with conventional electromyography and nerve conduction studies can be applicable to the dog. MUNE provides a quantitative assessment of LMN survival and loss by estimating the number of functional motor units and is an established technique for monitoring ALS progression in ALS rodent models and human patients.[51] We have the potential to apply some of these techniques to dogs and expand upon an already validated gait scale used to monitor recovery of pelvic limb function in dogs.[52] If we can better characterize the clinical disease spectrum and the UMN and LMN disease distributions in DM affected dogs, these dogs could serve as a valuable animal model with which to evaluate the efficacy and safety of therapeutic interventions.

## REFERENCES

[1]Averill DR. **J Am Vet Med Assoc** 1978;162:1045. [2]Coates JR, et.al. **J Vet Int Med** 2007;21:1323. [3]Griffiths IR, et.al. **J Small Anim Pract** 1975;16:461. [4]Braund KG, et.al. **Am J Vet Res** 39:1309. [5]Johnston PEJ, et.al. **Vet Rec** 2000;146:629. [6]Matthews NS, et.al. **J Am Vet Med Assoc** 1985;186:1213. [7]March PA, et al. **Vet Path** 2009, in press. [8]Sutter NB, et.al. **Genome Res** 2004;12:2388. [9]Lindblad-Toh K, et.al. **Nature** 2005;438:803. [10]Karlsson EK, et.al. **Nature Genetics** 2007;39(11):1321. [11]Awano T, et al. **Proceed Nat Acad Sci** 2009, in press. [17]Green SL, et. al. (2002) **J Hered** 2002;93:119. [18]Boissinot M, et al. **EMBO J** 1997;16:2171. [19]Deng HX, et al. **Science** 1993;261:1047. [20]Sandelin E, et. al. **J Biol Chem** 2007;282:21230. [21]Shaw BF, et. al. **Trends Biochem Sci** 2007;32:78. [22]Boillee S, et. al. **Neuron** 2006;52:39. [23]Schymick JC, et. al. **Hum Mol Genet.** 2008;17:768. [24]Rosen DR, et. al. **Nature** 1993;362:59. [25]Pasinelli P, et. al. **Nat Rev Neurosci** 2006;7:710. [26]Andersen PM, et. al. **Brain** 1996;119:1153. [27]Kato S. **Acta Neuropathol** 2008;115:97-114. [28]Bensimon G, et. al. **N Eng J Med** 1994;330:585. Rakhit R. et. al. **J Biol Chem** 2002;277:47551. [30]Gurney ME, et. al. **Science** 1994;264:1772. [31]Reaume A, et. al. **Nat Genet** 1996;13:43-47. [32]Turner BJ, et. al. **Prog Neurobiol** 2008;85:94. [33]Rakhit R, et. al. **Biochimica et Biophysica Acta** 2006;1762:1025. [34]Fischer LR, et al. **Exp Neurol** 2004;185:232. [35]Gould TW, et. al. **J Neurosci** 2006;26:8774. [36]Clement AM, et. al. **Science** 2003;302:113. [37]Bruijn LI, et. al. **Annu Rev Neurosci** 2004;27:723. [38]Smith RA, et. al. **J Clin Invest 2006**;116:2290. [39]Nayak MS, et. al. **Biochimica et Biophysica ACTA** 2006;1762:1128. [40]Urushitani M, et al. **PNAS** 2007;104:2495. [41]Benatar M. **Neurobiol Dis** 2007;26:1. [42]DiBernardo AB et. al. **Biochimica et Biophysica Acta** 2006;1762:1139. [43]Ludolph AC. **Amyotroph Lat Scler** 2007;8:217. [44]Scott S, et al. **Amyotroph Lat Scler** 2008;9:4. [45]Ellinwood NM, et. al. **J Gene Med** 2004;6:481. [46]Brooks BR, et. al. Amyotroph Lat Scler and Other Motor Neuron Dis 2000;1:293-299. [47]Dal Canto MC, Gurney ME. **Brain Res** 1995; 676:25. [48]Bromberg MB. **Adv Neurol** 202;88:53. [49]Gordon PH, et. al. **ALS and Other Motor Neuron Dis** 2004;5:90. [50]The ALS CNTF Treatment Study Phase 1-11 Study Group. **Arch Neurol** 1996;53:141. [51]Shefner JM, et. al. **Muscle Nerve** 2006;34:603. [52]Olby NJ, et. al. **J Neurotrauma** 2004;21:49-59.

## THE ROLE OF CHOLESTEROL IN NEURODEGENERATIVE DISEASES OF SMALL ANIMALS

**Charles H. Vite, DVM, PhD, Dipl ACVIM (Neurology)**
**Philadelphia, PA**

Sterols in the brain are predominantly in the form of unesterified cholesterol with the majority present in myelin and lesser amounts present in neurons and glia. Most if not all of the cholesterol used in the development of the brain and spinal cord is synthesized within the central nervous system (CNS) with little or no cholesterol crossing the blood-brain barrier. Endogenous CNS cholesterol leaves the brain following conversion to 24(S)-hydroxycholesterol that can cross the blood-brain barrier to enter the serum.[1]

Proteins transport cholesterol in lipoproteins in the CNS as they do in the blood. The major CNS apolipoproteins are apoE, synthesized by astrocytes and microglia, and apoJ, synthesized by astrocytes. ApoE-lipoproteins bind to neuronal receptors of the LDL family which mediate their uptake.[1,2]

### ALZHEIMER DISEASE

Alzheimer disease (AD) is a progressive neurodegenerative disorder resulting in cognitive impairment and memory loss. Histologically, beta-amyloid is deposited in the extracellular space and neurofibrillary tangles occur in the intracellular space. A well-known risk factor of Alzheimer disease in the inheritance of the apo-E4 allele.[1] An increase in the cholesterol content of the brain, possibly due to an increase in beta-amyloid deposition, is also postulated to correlate with an increased risk of developing Alzheimer disease. Data is inconclusive as to whether increasing serum cholesterol concentrations are associated with a risk of developing Alzheimer disease. In spite of this inconclusive data, the use of cholesterol lowering drugs to decrease the incidence of Alzheimer disease is being considered and may decrease the risk of developing Alzheimer disease.[1] What is known regarding the effect of cholesterol on canine cognitive dysfunction will be discussed.

### NIEMANN-PICK TYPE C DISEASE

Niemann-Pick type C (NPC) disease is a neurovisceral lysosomal storage disease characterized biochemically by the endosomal/lysosomal accumulation of unesterified cholesterol, sphingomyelin, and other glycolipids, and clinically by hepatosplenomegaly, progressive neurologic dysfunction, and early death. Mutations in either the *NPC1* or *NPC2* gene can result in NPC disease with *NPC1* mutations accounting for 95% of the cases. The proteins NPC1 and NPC2 are critical for the movement of unesterified cholesterol and glycosphingolipids from the endosomal/lysosomal compartment to the Golgi apparatus, plasma membrane, and endoplasmic reticulum and the molecular mechanism by which this transfer occurs is becoming clearer. The mechanisms by which deficient activity of these proteins results in clinical disease are not understood. NPC disease causes motor dysfunction, dementia, and death of children. Histologically, Purkinje cell loss, axonal spheroid formation, and cell swelling due to intralysosomal accumulation of cholesterol, gangliosides, and other lipids occurs.[3,4]

Cats with NPC disease have a spontaneously-occurring missense mutation in *NPC1* (2864G-C) and clinical, neuropathological and biochemical abnormalities similar to those present in juvenile-onset patients making this model homologous to the most common form seen in human patients. Affected cats have specific signs of cerebellar and vestibular system dysfunction with progressive severity of dysfunction that can be quantified by post-rotatory nystagmus and brain stem auditory evoked response measures. NPC disease cats also show increasing serum activity of alanine aminotransferase, asparate aminotransferase, and cholesterol with advancing age. The feline model has been critical for identifying the late endosomal/lysosomal accumulation of gangliosides (GM2 and GM3) and unesterified cholesterol, for evaluating the association of GM2 storage with meganeurite formation and abnormal dendritogenesis, and for correlating neuroaxonal dystrophy with neurologic dysfunction. The feline model has also been used to evaluate the efficacy of experimental therapy of NPC disease.[5]

There are two promising experimental therapies currently being investigated for NPC disease, β-cyclodextrins and miglustat.[6,7,8] 2-hydroxypropyl-β-cyclodextrins (HPBCD) are cyclic oligosaccharides consisting of seven β -(1-4)-glucopyranose units. HPBCDs have a hydrophilic exterior and a hydrophilic interior making them useful for increasing the aqueous solubility of hydrophobic molecules such as cholesterol, steroids, and vitamins. *In vitro* studies using β -cyclodextrins have shown a marked removal of cholesterol from cultured neuronal and non-neuronal cell lines. *In vivo* studies in 7 day old mice have shown that HPBCD improved hepatic function testing, decreased neurodegeneration and prolonged lifespan presumably due to the mobilization of stored cholesterol from the lysosome.[6]

The second treatment being investigated is N-butyldeoxynojirimycin (NB-DNJ; miglustat), an inhibitor of all glucosylceramide-based glycosphingolipids including lactosylceramide and gangliosides.[8] Preliminary studies performed in two affected cats treated at 7 and 13 weeks of age with daily miglustat for 3 weeks and 8 weeks, respectively, showed a delay in the onset of intention tremor and ataxia, diminished parvalbumin-positive axonal spheroids, increased numbers of calbindin-positive Purkinje cells, and a reduction in brain ganglioside accumulation. This study underscored the importance of glycosphingolipid storage on disease progression and was the first therapy to effectively ameliorate disease progression in any species with NPC disease.

The mechanism for the neurodegeneration seen in NPC disease is still undetermined. Activated macrophage/microglia populations have been identified in the brains of mice with several lysosomal storage diseases including NPC disease. The increase and activation of these cells occurred before detectable apoptosis and it was hypothesized that they may cause a cascade of neuronal apoptosis and resultant neurological dysfunction. Microglia, the resident macrophages of the brain, can be activated to release pro-inflammatory mediators resulting in amplification of the inflammatory response;[9] activated microglia *in vitro* can cause neuronal cell death.[10] Additionally, since microglia also carry the *NPC1* mutation, their inflammatory response may be more aggressive and inappropriate. Indeed, in two other genetic diseases, Sandhoff disease and metachromatic leukodystrophy, bone marrow transplantation reduced neurodegeneration and improved neurological functional. In both instances, improvement was attributed to normalization of microglial function. Since preliminary studies in the cat have shown that miglustat is effective in improving neuropathology in treated cats, we are evaluating how therapy affects markers of microglial activation in treated cats.

**Similarities between AD & NPC disease**

Neurofibrillary tangles are present in NPC disease and are indistinguishable from the tangles present in the brain of patients with Alzheimer disease. In contrast, amyloid plaques do not form in NPC disease. Microglial activation is proposed to be involved in the pathogenesis of neurodegeneration in both diseases.

## PRION DISEASES

Prion infections of neurons increases the free cholesterol content, an decrease esterified cholesterol, in cell membranes, and this increase in membrane cholesterol may results in neurodegeneration possibly by decreasing membrane fluidity.[11]

## SUMMARY

Cholesterol storage is postulated to be at the root of several diseases characterized by neurodegeneration. Understanding the effect of therapies which modulate the cholesterol storage will be at the heart of understanding how cholesterol storage results in neurodegeneration.

## REFERENCES

[1]Vance JE et al. **Sem Cell Dev Biol** 2005; 16: 193-212; [2] Li H et al. **J Neuropathol Exp Neurol** 2005; 64: 323-333; [3] Liu B et al. **J Lipid Res** 2007; 48: 1710-1723; [4] Infante RE et al. **Proc Natl Acad Sci** 2008; 105;15223; [5] Vite CH et al. **Ped Res** 2008; 64: 544-549; [6] Liu B et al. **Proc Natl Acad Sci** (early edition); [7] Platt FM et al. J **Biol Chem** 1994; 269: 27108; [8] Zervas M et al. **Curr Biol** 2001; 11: 1283. [9] Culbert AA et al. **J Biol Chem** 2006; 281:23658; [10] Skaper SD et al. **Glia** 2006; 54:234; [11.] Bate C et al. **BMC Biology** 2008; 6.

**FIRST EVIDENCE THAT SPORADIC INCLUSION BODY MYOSITIS OCCURS IN AGED DOGS.**

**GD Shelton[1], J King[2], RA LeCouteur[2], BK Sturges[2], DC Williams[2], M Aleman[2], LT Guo[1].**
**1. School of Medicine, University of California San Diego, La Jolla, CA. 2. School of Veterinary Medicine, University of California Davis, Davis, CA.**

Sporadic inclusion body myositis (sIBM) is the most common, progressive myopathy in people over the age of 50 years, leading to severe disability. While other immune-mediated inflammatory myopathies such as masticatory myositis, polymyositis and dermatomyositis are well documented in dogs, sIBM has not been identified to date. In this study, muscle biopsies from 5 dogs >10 years of age showing clinical signs of chronic (evident for >6 months) and progressive neuromuscular dysfunction, and 5 clinically normal aged dogs, were screened for evidence of IBM using current pathologic, immunohistochemical and electron microscopic diagnostic criteria. Red-rimmed vacuoles and congophilic intracellular inclusions, canonical markers for IBM, were identified in biopsies from 2 dogs with myopathy but not in biopsies from any of the control dogs. Intracellular amyloid deposits were detected by fluorescence microscopy using rhodamine optics of Congo red stained sections. Such deposits were also labeled with antibodies against amyloid-β peptide, amyloid-β precursor protein and ubiquitin. Endomysial inflammation, with or without invasion of non-necrotic fibers by mononuclear cells, was highlighted using antibodies against CD4 and CD8 T cell antigens. Ultrastructural studies revealed aggregates of cytoplasmic filamentous inclusions, membranous structures and myeloid bodies in a proportion of vacuolated muscle fibers. In conclusion, muscles from 2 of 5 aged dogs evaluated in this study with chronic myopathy fulfilled the requirements for a diagnosis of sIBM. This study provides the first evidence of sIBM in dogs showing both the inflammatory and degenerative aspects of human sIBM. These findings should alert clinicians to this chronic and progressive inflammatory myopathy affecting muscles of aged dogs.

## THORACOLUMBAR PARTIAL LATERAL CORPECTOMY IN 50 DOGS: SLOT SIZE; ACHIEVED DECOMPRESSION AND COMPLICATIONS.

**T Flegel, IC Boettcher, E. Ludewig, P. Boettcher.**
**Department of Small Animal Medicine, University of Leipzig, Leipzig, Germany.**

Intervertebral disc disease (IVDD) has traditionally been treated using a dorsolateral or lateral approach to the spinal canal. However, removal of ventrally located disc material can be challenging using those techniques often resulting in iatrogenic spinal cord damage. Therefore, thoracolumbar partial lateral corpectomy (TLPLC) has recently been introduced for ventral disc herniation. During TLPLC a lateral slot is created in the vertebral bodies adjacent to the affected disc allowing access to ventrally located disc material with minimal spinal cord manipulation.

We investigated efficacy and safety TLPLC in dogs with IVDD causing ventral spinal cord compression. The objectives were 1. to determine degree of spinal cord decompression, 2. determine actual size and angle of the created slot, 3. determine type and rate of complications. Inclusion criteria were: diagnosed IVDD between Th10 and L5 with the majority of disc material located ventrally, presurgical spinal imaging using myelography, computed tomography-myelography (CT-myelography) or magnetic resonance imaging (MRI), as well as postsurgical spinal computed tomography (CT). 50 dogs treated with 59 corpectomies met the inclusion criteria. 51 dogs had a single TLPLC performed, whereas 6 dogs had 2 and 1 dog 3 corpectomies performed during the same surgical procedure. Surgery was performed as initially described by Moissonnier et al. 2004, except that after creating the slot the dorsal longitudinal ligament was partially resected in order to allow visualization of the spinal canal and the decompressed spinal cord after removal of disc material. After surgery, dogs underwent control CT-myelography using the still visible presurgically injected contrast agent in 46 dogs. In one dog, where the diagnosis was based on MRI, subarachnoidal contrast was injected after surgery for CT-myelography, and in 3 more dogs plain CT was performed without contrast injection.

Decompression of the spinal cord was considered excellent in 59% and good in 32% of dogs. However, immediate revision was performed in 16% in order to achieve a better result. The actual mean slot dimensions were: 64% of vertebral body width for laterolateral extension, 43% of the vertebral body height for dorsoventral extension, 30% and 25% of the cranial and caudal vertebral body length respectively for craniocaudal extension. In 68% of cases, the slot was angled ventrally more than 10% off the horizontal, making decompression more difficult. The proximal part of a rib had to be resected for adequate drilling in 15%, whereas rib exarticulation without resection was performed in another 5% of cases. Surgical complications included: significant hemorrhage (25%), inadvertent transection of the spinal nerve (12%), pneumothorax (32%), and laceration of the dura mater (5%).

Lateral partial corpectomy is a surgical technique allowing for complete spinal cord decompression in ventral IVDD in the majority of cases, but surgeons must be aware of potential complications.

**EVALUATION OF A MAGNETIC RESONANCE IMAGING COMPATIBLE, FRAMELESS STEREOTACTIC BRAIN BIOPSY DEVICE IN THE DOG.**

**AV Chen, FA Wininger, RS Bagley, RL Tucker, AR Schneider, JM Gay.**
**Washington State University College of Veterinary Medicine, Pullman, WA.**

A stereotactic brain biopsy system that is MRI compatible has not been validated in dogs. The purpose of this study was to determine the mean needle placement error in canine cadaver brains using the modified Brainsight frameless stereotactic system.

Cadaver heads were disarticulated at C2-3, stored at 10◦C and used within 24 hours of euthanasia. Relocatable reference points (fiducial markers) were attached to the cadaver heads using a dental bite block. A T1-weighted gradient echo 3D sequence was acquired using set parameters. Fiducial markers were used to register the head to the acquired MR images in reference to a 3D position sensor. This allowed the planning of trajectory path to brain targets in real time. Coordinates (X,Y,Z) were established for each target and 0.5ul of diluted gadolinium was injected at each target using a 26 gauge needle to create a lesion. The center of the gadolinium lesion was identified on the post-operative MR images and coordinates (X',Y',Z') were established. The precision of this system in bringing the needle to target (needle placement error) was calculated using the formula: error = $\sqrt{[(X-X')^2+(Y-Y')^2+(Z-Z')^2]}$.

Seventeen sites were targeted in the brain. Mean needle placement error for the caudate nucleus(n=8), thalamus(n=6), and midbrain(n=3) was 1.52+/-0.87(SD), 1.70+/-0.77, and 2.67+/- 0.65mm, respectively. The overall mean needle placement error for all target sites(n=17) was 1.79+/-0.87mm. There was no statistically significant relationship between target depth and degree of error(p=0.54). We conclude this stereotactic system has acceptable precision and may be utilized clinically for biopsying brain lesions in dogs.

# CHIARI–LIKE MALFORMATION IN THE GRIFFON BRUXELLOIS.

C Rusbridge[1], SP Knowler[1], L Pieterse[2] AK McFadyen[3]
1 Stone Lion Veterinary Centre, Wimbledon, UK. 2 Statuesque Griffon Bruxellois, Sydney, Australia. 3. Glasgow Caledonian University, UK.

The Griffon Bruxellois (GB) is a brachycephalic toy with terrier characteristics which, like the Cavalier King Charles Spaniel (CKCS), has ancestral origins from the Toy Spaniel. Both the CKCS and the GB have predisposition for Chiari-like malformation (CM) and syringomyelia (SM). The purpose of this study was to describe CM and SM in the GB and to investigate the (null) hypothesis that CM alters the rostrocaudal and ventrodorsal dimensions of the skull.

MRI from 56 GB dogs were assessed for CM and CSF pathway abnormalities. Skull radiographs were obtained in 33 of 56 dogs. Two rostrocaudal and two ventrodorsal measurements were made and ratios of one length to another were calculated. Four factor unbalanced general linear ANOVA models were then used to analyse each of the ratio measures for each category of sex, CM, SM and ventricular dilatation. Age was not found to be highly correlated with any of the ratios and hence was not used as a co-factor. A multiple discriminant analysis model was developed to ascertain which of the ratios best discriminated between those dogs with and without CM. The level of significance was set at 1% and single factor tests were used to ascertain if there was a significant differences in the means of the length ratios between 1) male and female dogs; 2) dogs with and without CM: 3) dogs with and without SM.

In this selected sample, 60.7 % had CM. SM occurred with and without CM (37.5% and 8.9% study population respectively) however there was a significant association ($p = 0.004$) between having CM or not and having SM. The radiographic study demonstrated that one measurement ratio could be used to predict CM (sensitivity 87%, specificity of 78%) and that there were significant interaction factors between sex and SM for two measurement ratios. Four of 56 dogs had clinical signs of CM/SM; all had an asymptomatic CM/SM affected dam. Pedigree analysis suggested a hereditary predisposition to both CM and SM and that there may be a "carrier" state for CM as normal dogs may have affected offspring.

The study suggested that CM in the GB is characterised by a shortening of the basicranium, a deficiency of the supraoccipital bone and a compensatory lengthening of other bones in the skull especially the parietal bone. It is hypothesized that this could be due to insufficiency of the bone and /or craniosynostosis of the lambdoid (occipitoparietal) and cranial base sutures. SM in the GB occurs with and without CM. Dogs with SM but without CM tended to have dilatation of the entire ventricular system suggesting that there may be obstruction of the foramen magnum other than CM - for example arachnoiditis. In addition it is further hypothesized that the bony changes in brachycephalism and/or CM could result in venous narrowing at the jugular foramina, elevated venous pressure and impaired CSF absorption predisposing to ventriculomegaly and SM.

# CLINICAL AND NEUROPATHOLOGICAL FINDINGS IN AMERICAN STAFFORDSHIRE TERRIERS WITH CEREBELLAR CORTICAL DEGENERATION.

**NJ Olby[1], PM Mehta[1], T Flegel[2], S Blot[3], M Abtibol[3]**
**[1] North Carolina State University, Raleigh, NC; [2]Universität Leipzig, Leipzig, Germany; [3]Ecole Nationale Vétérinaire d'Alfort, Maison d'Alfort, France.**

Cerebellar cortical degeneration has emerged as a serious and common hereditary problem in the American Staffordshire Terrier and increasingly in the Pit Bull Terrier. The mutation associated with this autosomal recessive, adult onset neurodegenerative disease has recently been discovered. A large cohort of affected dogs has been monitored for several years and a wider range of signs than originally described has emerged. The aim of this research is to describe the clinical, magnetic resonance imaging and histopathological findings in a large group of affected dogs.

The clinical signs of over 100 American Staffordshire Terriers with cerebellar cortical degeneration were described by evaluation of the dog, videotapes of the dog or owner description, and the clinical course was followed by e mail contact with the owner every 3 months. When possible, MRI's of the brain were performed and if available for necropsy, histopathological evaluation of the brain was completed.

This large overview of affected dogs confirmed the previous description of cerebellar and central vestibular signs but the severity of signs ranged widely from late onset of mild signs that did not affect lifespan to onset within the first two years of life and progression to euthanasia within 18 months. In addition, a small proportion of dogs had generalized seizures, some owners described 'cerebellar seizures', and many dogs developed difficulty in initiating movement and coarse body tremors at rest. These signs suggest that the clinical effects of this disease are not restricted to the cerebellum and while the overwhelming pathology identified affected the cerebellar cortex, more subtle neuronal pathology was identified in the thalamus and brainstem. In summary, this genetic disease produces signs of cerebellar, central vestibular and forebrain dysfunction due to widespread neuronal involvement.

**IDENTIFICATION OF A SINGLE POINT MUTATION STRONGLY ASSOCIATED WITH THE CEREBELLAR CORTICAL ABIOTROPHY OF THE AMERICAN STAFFORDSHIRE TERRIER.**

**Marie Abitbol[1], Jean-Laurent Thibaud[1], Natasha Olby[2], Laurent Tiret[1], Catherine André[3], Delphine Delattre[4], Stéphane Blot[1]**
**[1]Alfort School of Veterinary Medicine, Maisons-Alfort, FRANCE; [2]College of Veterinary Medicine, North Carolina State University, Raleigh, NC, USA; [3]UMR6061 CNRS/Université Rennes1, Rennes, FRANCE;**
**[4] ANTAGENE, Limonest, FRANCE.**

A late onset hereditary cerebellar cortical abiotrophy has been described in the American Staffordshire Terrier (AST) breed. Affected ASTs show gait abnormalities, cerebellar ataxia with hypermetria, stiffness and loss of balance often beginning between 3 and 5 years of age. The disease is characterized by loss of Purkinje cells, a thinning of the granular and molecular cell layers and a concomitant accumulation of lipofuscin pigments in neurons. This hereditary cerebellar abiotrophy segregates as an autosomal recessive trait.

Using 77 French ASTs, we performed a microsatellites-based whole genome scan analysis and identified a region on CFA09 that was strongly associated with the disease. The region overlapped with the critical region found by Olby and collaborators (ACVIM 2008). Fine mapping of the region led to the identification of a single point mutation in a candidate gene that was strongly associated with the disease. Then, one hundred eighty four healthy ASTs and 150 clinically diagnosed affected AST from France and the US were tested and confirmed a strong genotype-phenotype correlation between the mutation and a predisposition to the disease. A genetic test is available for AST owners and breeders and we anticipate it will help reduce the high prevalence of this incurable disease in the breed. Also, the identification of the mutated gene will allow to better understand the pathophysiology of the hereditary cerebellar cortical abiotrophy affecting US and French ASTs and will shed light on the corresponding human condition.

## NEONATAL NEUROLOGY: ASSESSING THE RUNT

**Dennis O'Brien DVM PhD DACVIM Neurology**
**Columbia, Missouri**

Neonatal diseases have received relatively little attention in veterinary medicine. Most ill neonates are dismissed as runts or "fading" and little effort put into identifying the underlying cause. With the advent of genomics, the genetics of many hereditary diseases are being elucidated permitting DNA testing to prevent the disease. Thus there is added incentive to recognize neonatal disease. Breeders want to decrease the incidence of these diseases both to improve the health of their breed and to improve their bottom line. Precise characterization of neonatal diseases will define the syndrome and can provide clues to the underlying genes responsible if the disease is genetic. In this review we will focus on the normal development and neurologic exam of the neonatal dog, and how molecular genetic approaches can be applied to these problems.

**Development & the neurologic examination of the neonate**

Recognizing and localizing neurologic lesions in the neonate is difficult because the nervous system is undergoing rapid change during development. For example, crossed extension which is a sign of upper motor neuron disease in the adult, is normal during the first few weeks of life. Thus the exact age of the animal should be recorded and the stage of development taken into consideration when interpreting behavior and the neurologic exam. Table 1 outlines the different stages of development and their accompanying behaviors and neurologic development. Approximate ages for the developmental milestones are given based on experimental observations with limited breeds represented. It should be noted that breed variations are the rule rather than the exception, but these provide some guidance as to whether an individual is delayed.

As mentioned above, components of the neurologic examination will change with development. Figure 1 illustrates the appearance and disappearance with development of components of the examination. See O'Brien 1994 for details of the neurologic examination of the neonate.

**Table 1:** Stages of Development in the Dog

- Neonatal period (0-13 days; ends when eyes first open)
  - Focus on feeding and warmth
  - Behavior
    - Quiet if warm and fed even in strange environment
    - If cold or hungry – scanning, crying, crawling forward or circles but no direction unless contacts something warm
    - Rooting and suckling. Poor feeding from artificial nurser
    - No social interactions
    - Reflexive urination & defecation
  - Neurologic development
    - Eyes and ears closed
    - Trigeminal and vestibular only cranial nerves myelinated
    - Initially lie flat; begin to crawl at 2-3 days
    - Righting from birth but character different from adult
    - EEG unchanged with state: high amplitude spindles
- Transitional period (13-21 days; ends when ears open & develops acoustic startle)
  - Transition between neonatal and socialization
  - Behavior
    - Takes artificial nurser more readily; begins to lap at 3 weeks
    - Decreased crying with cold but begin to recognize strange environment
    - Begin orienting in space. Can move backwards and will withdraw from visual stimuli
    - Lose need for reflex elimination but continue to respond to stimulation
    - Beginnings of agonistic behavior; paw and mouth clumsily. Corresponds with tooth eruption.
  - Neurologic development
    - By 3 weeks can stand and eat
    - Balance improves to where can maintain on a balance board
    - EEG spindles disappear and slow wave sleep appears
- Socialization period (3-12 weeks; ends when critical socialization period ends)
  - Establish social interactions with littermates and mother
  - Behavior
    - Lap and eat while mother less tolerant of nursing
    - Leave nest to urinate & defecate. Specific spot by 8.5 weeks
    - Play fighting prominent
  - Neurologic development
    - Adult locomotion though clumsy and low endurance
    - EEG REM sleep & low voltage fast activity by 5 weeks
- Juvenile period (3-~24 months; ends when sexually mature)
  - Matures into an adult
  - Behavior
    - No longer cries in strange environment
    - Definite dominance hierarchies established
  - Neurologic development
    - EEG adult by 20 weeks

**Diagnostic approach**

In general, the diagnostic approach to a neonate will not differ from that in an adult except that the small size of the patient may preclude some tests and changes with development may confound interpretation of others. While we focus on the nervous system in this discussion, it is important to remember that congenital defects can affect multiple organ systems. Thus a thorough evaluations is warranted to detect concurrent deficits affecting other systems such as cardiac, urogenital, or musculoskeletal.

At a minimum a PCV and blood glucose should be evaluated, and if sufficient samples are available, a CBC and serum chemistry profile should be performed. A blood smear should be examined for evidence of vacoulations or inclusions since some lysosomal storage disease will affect leukocytes. Electrolyte abnormalities are common in other neonates and should be investigated in our patients as well. Thyroid can be measure to rule out congenital hypothyroidism. Blood ammonia values are easier to interpret when searching for evidence of porto-systemic shunts and will detect other congenital diseases such are urea cycle enzyme deficiencies. The test must be run immediately after the sample is taken, however, which limits its usefulness in most practices. Blood gas should be considered if there is a high anion gap or low bicarbonate on serum chemistries. Ketonuria in a neonate that is not anorexic could indicate an inborn error of metabolism. Many inborn errors of metabolism produce abnormal byproducts which are excreted in the urine where they can be readily detected through special assays. In contrast to blood, where care must be taken with the volume collected from a neonate, urine can be collected with impunity. Thought should be given to collecting whole blood for DNA extraction if a genetic disease is suspected (see below).

**Figure 1: Changes in the neurologic exam with development in the neonate.**

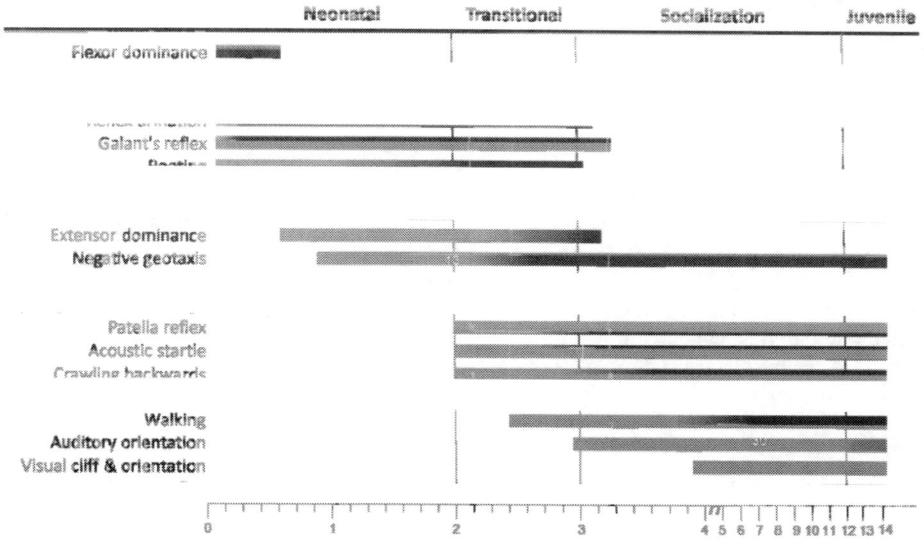


Shading indicates development or disappearance of a response. Number is the average day of onset when known.

Imaging of the brain can reveal congenital malformations of the brain such as polymicrogyria, Dandy-Walker malformation, or hydrocephalus. Ultrasound may be able to detect the cerebellar atrophy of Dandy-Walker malformation through the foramen magnum and dilated ventricles of hydrocephalus through the fontanels if open. MRI will be the most sensitive modality and due to the small size of the neonate, a high field magnet will provide the best images. Myelination will be incomplete so the gray-white matter distinction will not be as great as in an adult. If electrodiagnostics are performed to diagnose neuromuscular disease or seizure disorders, the changes with development need to be taken into account.

Often post mortem examination offers the best opportunity for a definitive diagnosis, but it is important to collect data such as serum chemistry profiles or electrodiagnostics antemortem. Ideally a pathologist should perform the post-mortem, but this is sometimes impractical. In addition to routine liver, kidney, lung, etc.; brain, spinal cord, peripheral nerve, and/or muscle should be taken depending upon the nature of the condition. If myopathy is suspected, submit fresh muscle overnight to a lab that can process it for immunohistochemistry. The skull of a neonate is often thin and easily removed with rongeurs or even a side-cutting wire cutter. The neonatal brain is very soft, however, and care must be taken not to damage it in removal. Thought should be given to collecting frozen tissues for biochemical assays and/or DNA studies. Liver and kidney are very metabolically active tissues that are often used for biochemical studies. Some studies must be done on brain tissues, but cutting the brain before it is fixed can create artifact. If more than one pup in a litter is affected, one brain could be fixed *in toto* while another split sagittally and half frozen for potential biochemical analysis.

**Gene discovery strategies**

Many mutations will be manifest as neonatal disease. Development of the nervous system is a complex process and many genes are only expressed for a brief period during development. Thus there is a bias for mutations to interfere with that process. Inbreeding and founder effects can make genetic diseases more common in pure breed animals, but genetic disease can arise even in randomly breeding populations. History of littermates and the extended family can help determine if genetics appears to be playing a role in the disease. Breed tables can be consulted to see what familial diseases have been reported in the breed; the signs can then be compared to the case at hand. Breed club websites can often be a link to information about genetic diseases in a breed and DNA tests available, though as with all information on the internet, a healthy degree of skepticism is needed when evaluating the information. Modern molecular genetics has provided us with the canine genome map and the tools to identify the genes responsible for hereditary diseases. Breeders are aware of the value of DNA testing, but veterinarians have the background to interpret the results of such tests as well as take the lead in developing new test. Identifying the specific mutation responsible for a genetic disease can improve therapy and provide the breeds the tools to decrease the problem in the breed. When facing an unknown disease, gene discovery strategies can be used to identify the gene responsible. Current approaches to gene discovery being used include the candidate gene approach, linkage mapping, and SNP association.

The candidate gene approach involves identifying the comparable disease in humans, rodents or other species for which genes have been identified. The more accurate and molecular the diagnosis in the case in question, the more refined the list of candidates can be. For example, when two dachshund littermates with a syndrome of blindness, progressive ataxia and terminal myoclonic seizures were necropsied, fluorescent microscopy of brain tissue showed fluorescent material characteristic of neuronal ceroid lipofuscinosis within lysosomes. Mutations in eight genes have been identified in humans, mice and sheep with NCL, thus these were candidates for cause of the disease in dachshunds. Electron microscopy showed curvilinear bodies in the dog similar to those found in humans with one form NCL. The gene associated with this form of NCL codes for the lysosomal enzyme TPP1 and the affected dog showed no activity for that enzyme in biochemical assays using frozen brain tissue. Thus based on EM and biochemical data, there was one highly likely candidate. Sequencing the comparable canine gene demonstrated a missense mutation that led to a premature stop codon truncating the protein prior to the active site. The candidate gene approach can identify the mutation with a single affected case as was done here, allowing for quick development of a DNA test to identify carriers as well as definitively diagnose affected dogs. For many diseases, however, there are too many candidates, such as seizure disorders where hundreds of mutations are known to produce epilepsy in mice. Alternatively, the gene associated with a particular disease may not have been identified yet in any species. Then alternative approaches are necessary.

Linkage mapping uses genetic markers to follow the inheritance of chromosomes within a family. The disease causing gene will be in the vicinity of some of these markers on a single chromosome. If a set of markers are consistently inherited in affected dogs, then those markers are "linked" to the disease. Neonatal diseases are ideally suited for mapping studies because the entire family, except maybe the sire, is available for DNA sampling at the time the disease is recognized. Neonatal encephalopathy with seizures (NEWS) in standard poodles is characterized by developmental delay, ataxia, seizures and death before weaning age. We demonstrated linkage between markers on chromosome 36 and NEWS. Recombinations occur when a portion of one chromosome is traded with its pair during meiosis. The resulting shuffle of markers allows more precise localization of the portion of that chromosome where the disease gene resides. This area of a chromosome where the responsible gene is located is called the locus. In NEWS, recombinations within the families studied further narrowed the locus to an area containing 26 genes (Figure 2). Of these, three were clearly involved in CNS development, and thus became the prime candidates to be sequenced. We found a missense mutation in one of these genes *ATF2,* which codes for a major transcription factor. DNA testing now allows standard poodle breeders to use wise breeding strategies to avoid producing affected dogs while still being able to use carriers in their breeding program.

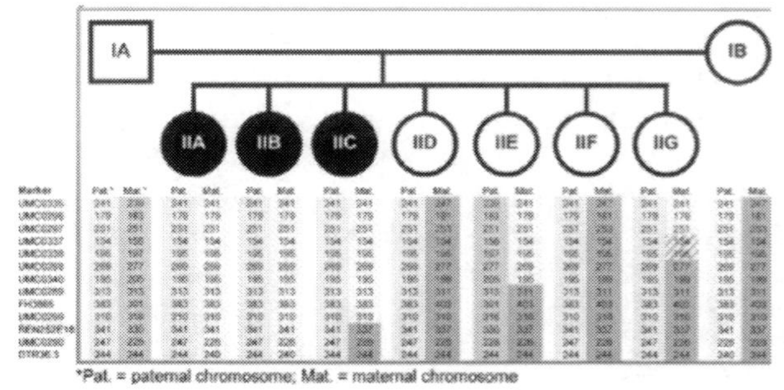


*Figure 2: In this pedigree, solid symbols are pups affected with NEWS and the combinations of markers in the locus for each pup are listed below the symbol. Markers in each parent are distinguished by shade. Recombinations that narrowed the locus are seen as mixed shading of the column in IIC, IIE, and IIG. The cross-hatched area in IIG was not informative.*

SNP association is similar to linkage mapping in that marker genes are used. In this case, the markers are single nucleotide polymorphisms (SNPs). SNPs are much more numerous than the type of markers used in linkage analysis, thus providing more dense coverage of the genome. This increased density allows us to look for a statistical association between having a disease and having a particular set of SNP. The pattern of SNPs in dogs with the disease is compared to that of healthy dogs of the same breed. Affected dogs will inherit the particular group of SNPs which reside close to the mutant gene more commonly than normal dogs will. This can be done without having samples from the extended family. SNP association mapping has been utilized to identify the mutations responsible for exercise induced collapse in Labrador retrievers and degenerative myelopathy in a number of breeds.

## Conclusions

Quality breeders understand the role genetic testing will play in their attempts to improve the health of their breed. We as veterinarians should offer the diagnostic tests necessary to definitively diagnose a neonate with neurologic signs. In some cases, we may find a treatable condition such as congenital hypothyroidism. We need to be aware of the genetic tests available and how to interpret them, so that we can counsel owners and breeders appropriately. In cases where the gene has not been identified, the diagnostic tests can provide the clues necessary to identify the gene responsible. Then wise breeding strategies can ensure that affected dogs are not produced in the future and the incidence of the mutant allele in the population decreased over time.

## REFERENCES

[1.] Chen X et al, **Neurogenetics** 9:41-49, 2008. [2.] Coates JR et al **PNAS** in press 2009. [3.] Coates JR & Kline KL. Congenital and inherited neurologic disorders in dogs and cats. In: **Current Veterinary Therapy XII,** edited by J. D. Bonagura, Philadelphia:WB Saunders Co., 1995, p. 1111-1120; [4.] Coates JR & O'Brien DP **Vet Clin North Am (Small Anim Pract)** 34 (6):1361-1401, 2004; [5.] Fox MW: **Integrative Development of Brain and Behavior in the Dog**. Chicago, University of Chicago Press, 1971. [6.] Fyfe JC et al, **J Vet Int Med** 17 (1):50-57, 2003. [7.] Jurney C et al **J Vet Int Med in press** 2009. [8.] M. D. Lorenz and J. N. Kornegay. **Handbook of Veterinary Neurology**, Philadelphia:W.B. Saunders Company, 2004; [9.] O'Brien DP: **Seminars in Veterinary Medicine and Surgery (Small Animal)** 9:62-67, 1994. O'Brien DP & Packer RA: [10.] Organic acidurias in August J ed **Consultations in Feline Internal Medicine**, 5th ed 2009. D. P. [11.] O'Brien. Organic Acidurias. **Proceedings of the 23rd ACVIM Forum**. Baltimore, MD pp346-8, 2005; [12.] Patterson EE et al: **Nature Genetics** 40:1235-9, 2008. [13.] Scott JP, Fuller JL: **Genetics and the Social Behavior of Dogs**. Chicago, University of Chicago Press, 1965.

**THERAPEUTIC OPTIONS FOR CANINE GLIOMAS**

**Andy Shores, DVM, MS, PhD, DACVIM (Neurology)**
**Auburn, AL**

## INTRODUCTION

Advancements in treatment options for complex disorders of the central nervous system in veterinary medicine usually shadows the development of techniques and instrumentation in the human neurology and neurosurgery fields. In most instances, the success rates in the human fields have steadily increased; however, this has not been the case in the treatment of gliomas. All medical disciplines continue to struggle with this tumor of the central nervous system with consistently low mean survival times. The advent of applicable chemotherapy as an adjunct to the treatment of intracranial gliomas has added additional weeks to the mean survival times and has been the only definitive breakthrough in a number of years. In the veterinary neurosurgical field, data concerning the surgical therapy for canine gliomas is almost non-existent partially because of the relatively low case numbers and a lack of a centralized approach to specific therapies. A wider availability of advanced imaging techniques is now allowing for a larger number of cases diagnosed and for the opportunity to treat many of the patients at less advanced stage of the disease. Seemingly, we are at a point in our development where a collection of a larger number of cases is feasible and the use of some of the newer multi-modality techniques can yield clinically applicable data on its effectiveness in the treatment of canine gliomas. This presentation will review the diagnosis and treatment protocol for canine gliomas at the Auburn University Small Animal Teaching Hospital (AUSATH)and briefly review or identify some other available treatment options.

## CANINE GLIOMAS

In dogs, the brachycephalic breeds have the highest incidence of brain tumors among domestic animals. Glial tumors (astrocytomas, oligodendrigliomas, glioblastoma multiforme, unclassified gliomas) are the most common. Patients are usually middle aged or older with a median age of 9 years in previous studies. It is likely the known incidences have increased and the median age has decreased since the last comprehensive study because of the availability of advanced imaging in the last decade. The glial tumors arise from the glial cells within the brain parenchyma and infiltrate the surrounding tissue. The majority are located in the cerebrum, but also are found in the brainstem and cerebellum. Clinical signs are related to the location and size of the mass. Most canine patients with cerebral glial cell tumors present for seizure episodes, although some have more advanced changes to include alterations in mental status, contralateral visual deficits, gait abnormalities (circling, hemiparesis), and head pressing. Because of the slow progressive nature of many intracranial neoplasias, owners will not always note the minor changes until a major event (such as a seizure) triggers a recollection of differences in behavior over a period of several weeks or months.[1,2] Unfortunately, many patients are treated palliatively for a period of weeks or even months before referral is suggested or recommended. This delay in the diagnosis may result in a lost opportunity to deliver an effective therapy that will provide for a substantial period of a quality life style.

## DIAGNOSTIC PROTOCOL

Patients admitted to the AUSATH for evaluation of an intracranial disease all receive a standard evaluation that includes physical and neurologic evaluations, routine blood and urine analyses, three-view thoracic radiographs (in patients $\geq$ 5 years of age), and any additional testing if deemed warranted by results of the aforementioned studies. After the initial assessment determines a low anesthetic risk, patients are anesthetized using a protocol designed for patients with a potential for increased intracranial pressure. Next, an MRI is performed on-site, and includes T1 and T2 weighted studies, FLAIR and STIR sequences, and a contrast enhanced T1 study. The identification of an intraaxial intracranial mass usually precludes CSF collection as a part of the patient evaluation unless an inflammatory disease is highly suspected. Following the identification of a suspected glial tumor, based on location and imaging characteristics, and following consultation with the owners, additional tests may be ordered to include abdominal ultrasonography. In most instances, pre-surgical biopsy is not pursued with the thought that the definitive diagnosis will be made based on the tissue samples at the time of surgical excision. At the time of surgery tissue samples for culture and microscopic evaluation are obtained directly from the tumor and from the sediment (fragments) of the tumor tissue available from the CUSA tissue trap. These fragments retain good preservation of histological detail and tissue culture viability. The quantity has been consistently greater than that obtained by standard biopsy methods.[3]

## TREATMENT PROTOCOL

The initial step in providing an effective treatment regime is confirmation of the diagnosis by microscopic evaluation. Samples are submitted in formalin for histopathology and impression smears are submitted for cytological evaluation. Additional samples are placed in specialized growth media for tissue culture; however, this process requires considerable time and is not our standard for a definitive diagnosis. A diagnosis from the formalized tissue is usually rendered within 48 hours.

### Surgical Approach

For the purposes of this presentation, the surgical approach and therapy of cerebral glial cell tumors will be described. A standard lateral rostrotentorial craniotomy-craniectomy approach is used for tumors of the cerebral hemispheres. Tumors involving the frontal lobe are usually approached using a unilateral or bilateral frontal or transfrontal craniotomy-craniectomy. An extended rostrotentorial craniotomy-craniectomy is used when the mass involves the more lateral aspects of the cerebral hemisphere.[4] On occasion, a sphenotemporal craniectomy is used to approach masses involving the most ventral aspect of the temporal lobe. In all cases, the extent of the craniotomy-craniectomy is made to provide more than adequate exposure and to allow the placement of an intraoperative ultrasound probe on the dura to both identify the mass and monitor the completeness of removal with the Cavitron Ultrasonic Surgical Aspirator (CUSA). It is often difficult to distinguish the borders of glial tumors from normal brain parenchyma intraoperatively; however, with the use of the intraoperative ultrasonography, the borders are quite distinct. Obviously on a microscopic basis, the ultrasound image cannot make this distinction, but its use has proven very helpful in the removal of the macroscopically obvious mass in total.

Following satisfactory removal of the tumor, the exposed cerebral tissue is covered in swine intestinal submucosa (SIS) material and sometimes an additional layer of gelatin foam. The boney defect is spanned with a polypropylene mesh. A mesh is covered using the temporalis muscle and subcutaneous tissue and the skin is closed in a routine fashion with a nylon suture material.

An immediate postoperative MRI screen is used to further asses the surgical removal of the mass. This image also serves as a reference for follow-up, post therapy images are made at 3, 6, and 12 months post-surgery.

**Postoperative Care**

Patients are placed in the intensive care ward at the AUSATH for the duration of their initial hospitalization. Following confirmation of the diagnosis of a glial tumor by the pathology report, consultation with the oncology section is made. A postoperative therapeutic plan to include radiation therapy using the on-site linear accelerator and a chemotherapy regime is designed. Patients can be discharged to the owners care following successful recovery from the surgical procedure, with the follow-up radiation and chemotherapy beginning approximately 2 weeks postoperatively.

**Radiation Therapy**

Radiation therapy is initiated following CT planning images and consists of 16 fractions. Specific details of the therapy guidelines for the radiation protocol at the AUSATH will be presented.

**Chemotherapy**

Temozolomide is a recently developed imidazotetrazine derivative that is essentially an oral formulation of dacarbazine. Whereas dacarbazine must be activated by the cytochrome P-450 system in the liver, temozolomide is metabolized to its active form by a nonenzymatic pathway in the blood. Temozolomide is absorbed rapidly, exhibits near 100% p.o. bioavailability within 1–2 h of administration, and has demonstrated antineoplastic activity in high-grade gliomas.[5,6] We start at 80 mg/m2. The dose is given for 5 consecutive days. A CBC is drawn before treatment and 2 weeks after the 5 day course. In the absence of myelosupression, 5 additional courses at 21 day intervals are given at a dose of 100 mg/m2. A CBC is evaluated 1 week before each cycle to determine the degree or severity of myelosupression before initiating the next course. The drug is given on an empty stomach due to absorption issues but it is a gastric irritant so we feed promptly at 20 minutes after pilling to prevent immediate vomition. Also, it is very caustic to mucous membranes so care must be taken that the dog does not bite the capsule and it is essential that the person giving the drug wear gloves because of the caustic nature of the powder.

**Results**

A small number of cases and limited long-term follow-up times make a full discussion of results unwarranted at this time. A collaborative study would seem appropriate to answer many of the questions on the efficacy of our therapeutic protocol. In humans, this therapeutic regime has quickly become a standard for most institutions treating glioblastoma multiforme and many other glial cell tumors. This is the first breakthrough in many years that has significantly increased the mean survival time in humans. Studies to institute therapies that will reduce tumor cell resistance to temozolomide are ongoing. Overall, the general impression at the AUSATH is that the surgical resections have been more than adequate and the postoperative complications have generally been limited to continued seizure activity in some patients. The radiation therapy and the chemotherapy have been well tolerated and no severe toxic effects have been noted with the temozolomide.

## OTHER AVAILABLE THERAPIES

Gama-knife therapy is available in the veterinary realm and results have been mixed.[7] Therapies being evaluated in humans and are possibly undergoing clinical trails in animals include interstitial therapy with biodegradable wafers of chemotherapeutic agents and convection-enhanced delivery of chemotherapy and radiation. The use of fluorescing compounds (5-ALA) may assist in a more complete surgical removal of glial cell tumors.[8] The use of viral vectors is being evaluated in both human and veterinary studies. A more readily available conduit for assessing ongoing trails and protocols will help delineate currently available treatments and may stimulate collaborative efforts.

## CONCLUSIONS

Successful surgical management of canine gliomas is a developing strategy in the treatment of this malady and its sophistication only needs the addition of more experience and surgical tools to make it a widely successful and generally accepted practice. Continued concentration on the development of a multi-modality approach to canine gliomas is a very important step in the quest for long-term, enhanced quality of life outcomes. Additional development of therapeutic approaches can only be further assessed by large collaborative studies. An increased awareness of the possibility of success in the treatment of canine brain tumors in general may facilitate more referrals, earlier diagnosis, and ultimately an enhanced opportunity to successful treatment.

## REFERENCES

[1]Dewey CW. **A Practical Guide to Canine and Feline Neurology**. 127-137, 2003. [2]Long S. Neoplasia of the Nervous Sys, **in Braund's Clinical Neurology in Small Animals: Localization, Diagnosis and Treatment**, Vite C.H. (ed.). International Veterinary Information Service, 2004. [3]Richmond IL, et al. **Neurosurgery.** 1983 Oct;13(4):415-9. [4]Bagley RS. **Textbook of SA Surgery**, ed 3. 1163-1173, 2003. [5]Dervisis NG, et al. **JAVMA** 231(4):563-9, 2007. [6]Hart MG, et al. *Cochrane* **Database of Systematic Reviews** 2008, Issue 4. Art. No.: CD007415. DOI: 10.1002/14651858.CD007415. [7]Dickinson P. Proced 11th **Can Med Symposium**, May 10-11, 2008. [8]Strummer W, et al. **Lancet Oncol** 7:392-401, 2006.

# EVIDENCE BASED MEDICINE – A TOOL TO CONTAIN THE COSTS OF DIAGNOSIS & TREATMENT OF NEUROLOGICAL DISEASES

**John Speciale DVM, DACVIM, DABVP**
**Webster, NY**

## INTRODUCTION

Cost effective medicine is not just the least expensive medicine. It is the best medicine. Convincing evidence, compiled from cohort studies of Medicare patients, documented that treatment regimens varied between geographical regions. The most expensive regimens had lower patient satisfaction and worse clinical outcomes. These areas employed more technology and specialists than areas with lower expenditure and better results. Apparently, over-saturation of specialists in a geographic region led to more expensive but less beneficial medical care.[1,2]

Cost effective, efficient, and high quality medical care requires a conscientious cognitive process that includes both evidence-based (EBM) and problem oriented (POMR) approaches. Both strategies necessarily require clinicians to address the unique qualities of each individual patient (N=1). As practicing physicians and veterinarians migrate away from individualized care, referral to specialists and injudicious use of technology can undermine care and escalate expense. [1-6]

## EVIDENCE-BASED & PROBLEM ORIENTED MEDICINE

Evidence-based medicine is defined as "the integration of current best evidence with clinical expertise and patient values"[7]. It is "the conscientious, explicit and judicious use of current best evidence in making decisions about the care of ***individual patients***. The practice of evidence-based medicine means integrating ***individual clinical expertise*** with the best available external clinical evidence from systematic research" (emphasis added). The "best available external clinical evidence" is rarely the randomized controlled clinical trial, or even a systemic review of randomized trials. Randomized clinical trials provide the most reliable (i.e. repeatable) information, but qualitative, non-numerical reports offer greater validity.[4] This apparent paradox that underlies the nature of EBM can be appreciated by consideration of EBM's short and interesting history.

### History of EBM [8]

EBM arose from two traditions, each founded by remarkable and very different individuals – Archibald Cochrane from the UK and American Alvan Feinstein.

Archibald Cochrane was an egalitarian and a humanist who emphasized the potential for population studies to benefit community health, not individual patient care. As a prisoner of war for 4 years, Cochrane studied disease within his community of prisoners. He meticulously recorded symptoms in his patients and linked them to environment, e.g. better nutrition increased resistance to TB. Cochrane emphasized randomized clinical trials as the most efficient tool to intervene in public health, but his efforts were directed towards environmental conditions and studies that could benefit the population at large, not a few individuals. He emphasized direct and detailed examination of patients and precise notation of their clinical signs. Cochrane's proponents founded, and named in his honor, the Cochrane Collaboration, an organization of unpaid analysts dedicated to formally collecting, evaluating, and interpreting clinical trials.

Alvan Feinstein's first career was as a mathematician. He claimed to pursue a career in medicine after he concluded that he was only a mediocre mathematician. In contrast to the humanistic Cochrane, Feinstein, a realist, entered medicine when he recognized that, unlike mathematics, medicine allowed a mediocre practitioner the opportunity to become prosperous.

Feinstein became frustrated by the medical profession's inability to interpret clinical signs. He recognized that more complete interpretation of signs was a potentially powerful and underused clinical tool. To Feinstein, important evidence arose from direct observation of patients and their specific signs. He discouraged "pigeonholing" patients and encouraged the scientific study of each individual patient (N=1). To Feinstein, the randomized clinical trial was a useful, but limited tool. Feinstein's evidence-based clinical trials were necessarily performed at the bedside. He deemphasized numerical data and emphasized symptomatology. He found laboratory research ("rat turd grinding") too far removed from patients to be clinically useful. Feinstein was the founding spirit behind McMaster University's effort to formalize the study of clinical evidence.

The simple term "scientific medicine" was proposed as a title for this science of bedside evaluation, but it was unacceptable to most doctors who were offended by the implication that prior efforts were unscientific. McMaster University first institutionalized what is now known as EBM, under David Sackett, in a discipline termed "clinical epidemiology." However, this title too was poorly received both by medical students, who had little interest in epidemiology, and epidemiologists, who were not inclined to dwell at patient bedsides. In 1982 Suzanne and Robert Fletcher published the first effort at formalizing EBM with "Clinical Epidemiology, the Essentials." The term evidence-based medicine was socially more acceptable (although, perhaps, not particularly descriptive) and first appeared in 1990, in a guide that Sackett had prepared for residents. Sackett went on to publish the first textbook of EBM (Evidence-based medicine: How to practice and teach EBM) in 1997.

### History of POMR

LL Weed published the first example of a problem-oriented medical record in 1966. Weed's original manuscript is useful for neurologists to review because it can help to organize case management when, like so many neurology cases, pathophysiology is poorly understood. Weed presented an example of an organized clinical approach to patients when diagnosis was uncertain. To Weed, problems were not diagnoses, but, rather, characterizations of signs and/or disease. Weed's definition of problem-oriented medicine as "perpetual muddleheadness" recognized the need to redefine problems as new facts arose.[9]

In current medical records Weed's system is hardly recognizable. Problems are often stated as diagnoses, which lead to treatment of a disease, not a patient (disease-oriented evidence, DOE). The need to reexamine the approach to POMR has been studied and, perhaps, resonant with Feinstein's emphasis on bedside observations, greater patient-involvement in problem definition facilitated the process of patient evaluation and improved care measureably.[10-13]

**History of POEM**[14,15]

"Patient-oriented evidence that matters" (POEM) was introduced in 1994. POEM emphasizes information that is important to the patient. Generally, reviews authored by experts and manuscripts that measure surrogate data (e.g. cardiac arrhythmia and not patient mortality) tend to be less useful or even misleading. POEM is more likely to report direct measures of patient outcome (e.g. function or survival) and is opposed to DOE.

**Relationship of EBM & POMR to health care costs**

Competing interests complicate the implementation of evidence-based interventions. Certainly, in human health care, managers have a vested interest in minimizing expenditures for patient care. There is concern that managers interpret data to favor economy over care.[4,8,16] In contrast, excessive costs arise when ineffective diagnostics are promoted for political reasons or data is contorted to promote expenditures for minimally effective (or, worse, harmful) pharmaceuticals and procedures.[8]

The "Cochrane Test" has been adopted by some health care managers to help avoid ethical conflicts and provide those services that most effectively deal with health problems. Free from competing influences, the guidelines should serve to deliver cost-effective care: 1, consider anything that works; 2, make effective treatments available to all; 3, minimize ill-timed interventions; 4, treat patients in the most cost-effective place; 5, prevent only what is preventable; 6, diagnose only if treatable.[8]

Often verbalized maxims of companion animal practitioners contrast with the Cochrane Tests. These maxims can be undermined with a minimum of scrutiny, e.g.:

**"Owners want to know what's wrong with their pet."**

Pet owners who have been assured that specific diagnosis is unnecessary for recovery are often disinclined to subject the pet to additional tests or to incur substantial expense for such testing.

**"Owners want the best care for their pet."**

Survey information showed a clear and direct relationship between the likelihood of successful treatment and the willingness to pay for an intervention. Nevertheless, a majority of clients, even in a fairly high income community, were unable or unwilling to spend more than $3000 even when success of treatment (99%) was nearly guaranteed.

**"The best care is the most expensive."**

Objective, well-controlled studies have clearly shown that this has not been the case for humans.[1,2] The inverse relationship between very costly care and treatment is likely to be more pronounced in veterinary medicine.

**"More information will allow better treatment."**

Identification of subclinical diseases leads to unnecessary treatment, morbidity, and sometimes mortality. Also, the more data accumulated, the greater the likelihood that some data will be incorrect and, possibly, lead to misdiagnosis. The Russell Soundex Code[17], a strategy used in patient data bases, serves as a metaphor for the hazard of excessive data. This code limits all patients' surnames to only 4 characters. This stark abbreviation ignores most of the information in a surname, but the Code consistently minimizes misidentification of patients. Clearly, in this instance, *less* information (used wisely) is more reliable.

## COST ASSESSMENT

A variety of measures are used to balance quality of care with cost.[4]

**Cost minimization** is appropriate when two treatments are essentially identical, e.g. choice between a generic and brand name drug. There is no outcome measure and the choice is solely based upon cost.

**Cost effectiveness** is based upon a single measurable variable, e.g. life-years gained by treatment of an otherwise fatal condition. Cost-effectiveness is difficult to assess in veterinary medicine because client perception of quality of life needs to be considered. *Quality- of- life years* is a formal assessment for human patients that, for better or worse, ascribes a numerical value on the benefit of survival.

**Cost utility** is used to assess 2 dimensions, e.g. therapeutic benefit of a cosmetic procedure vs. the risk of surgery.

**Cost benefit** is used to make institutional decisions, e.g. whether to invest in an MRI or additional nursing staff.

Cost-effectiveness is a concern that should be addressed for each client and patient. The value of relief from pain, function and, for that matter, survival, is subjective and cannot be addressed competently if we unaware of a particular client's perspective and a pet's normal activities. In my patient population, nearly all deaths result from euthanasia, a condition that is exceedingly rare in human medicine. Therefore, the transference of survival data from human to veterinary medicine is inappropriate. Because the selection of euthanasia is based on subjective, and, usually, non-profession opinions, communication with the client is critical. Communication must be more than dissemination of information. When a veterinarian understands the individual preferences of a client, and their physical and financial abilities, treatment is more likely to be successful, because treatments can be modified to meet the clients' needs and expectations.[18]

**Undermined quality care with technology and excessive specialty interventions.**

Lewis Thomas, former dean of the Yale College of Medicine,[19] identified 3 levels of technology. "**High technology"** reflects a genuine understanding of disease mechanisms; vaccinations for distemper and rabies are examples of the highest level of applied medical technology. Opposed to high technology are conditions that are "**non-technical**," conditions for which we can offer only palliative care (e.g. degenerative myelopathy).

In between these two extremes is, what Thomas termed "**half technology"**; medical or surgical interventions to forestall death, e.g. organ transplants. Thomas noted that these interventions are "at once highly sophisticated and profoundly primitive." Half-technology obscures the boundary between science and technology. It attracts headlines and public awareness. Nevertheless, few, if any, of these interventions will ever influence the overall health of a population, whereas simple interventions, which are much more likely to be important (e.g. life style changes, or aspirin therapy to prevent heart attacks), often attract less attention. The proliferation of "half technology" has escalated medical expenditures and defies EBM's emphasis on population health.

**Strategy to approach cost-effective medicine**

Application of the Cochrane Tests would be a useful intervention to control costs and avoid overaggressive diagnostic and therapeutic interventions in veterinary medicine: 1, investment in "cutting edge" technology must be balanced with affordability; 2, inexpensive and benign interventions should require less evidence of efficacy; 3, interventions may be considered ill-timed for patients that are not uncomfortable and are recovering spontaneously; 4, home treatment or treatment at the referring practice may be most cost-effective; 5, diagnostic efforts should be encouraged only if the result is likely to change the treatment; 6, specific diagnosis is necessary only for specifically treatable problems or those that would clarify the suitability of euthanasia.

Some prevalent veterinary diagnostics and treatments warrant objective scrutiny because they may undermine veterinary clinical science and patient care, e.g. routine MRI imaging. There has been little or no objective evidence to document clinically important advantages for MRI.[20,21] Furthermore, MRI may needlessly hasten euthanasia, or call inappropriate attention to findings not directly related to the patients symptoms.

Surgical intervention for spinal disc disease should be recognized as "half technology." Preliminary survey data from clients in a group of upper New York State practices suggested that surgery was unaffordable as a treatment for a large majority of dogs with spinal disc disease. Surprisingly, clients who could afford their pet's spinal surgery, when shown comparison data, nearly all preferred medical treatment both for their pet and for themselves! Prevention of spinal disease may be more important than treatment.

The incidence of spinal disc disease is greatest in dachshunds. However, I have never examined a dachshund for signs of disc disease if the dog was engaged in the rigorous sport of "flyball," and rarely have I encountered signs of disc disease in agility trained dachshunds. A formal survey of athletic dachshunds is anticipated, but an inverse relationship between exercise and clinical disc disease was noted long ago. Hoerlein recognized that the incidence of clinical disc disease was lower for dogs of all breeds (and people) subjected to routine prolonged exercise.[22] If physical activity deters spinal disc disease, then the intervention of exercise should have a much greater impact on pet health than spinal surgery. Such a universally affordable intervention would be in the spirit of Cochrane and Feinstein. EBM deserves more attention from the veterinary community than the continual effort to study and employ ever more advanced and expensive, technologies.

**Cultural change in veterinary neurology**

Unfortunately, veterinary medicine may be following in the footsteps of the seriously encumbered U.S. human health care. Analysts proposed that the "rescue" of human health care lies in the empowerment of primary providers. Such a shift is logical in veterinary medicine, but it cannot be applied until the training of veterinary students begins to emphasize "bed side" EBM and techniques for acquisition of reliable clinical information, and to deemphasize expert opinion. As many as 1 in 3 patients may warrant a literature search, so use of appropriate data bases should be incorporated routinely into training and practice. Information technology is no less valuable than biomedical technology and, in the last several years, databases have become so accessible that they should be used as a screening "test" at least as routinely as thoracic radiography. This technology requires some statistical knowledge, but statistical expertise is unnecessary. An excellent published review of clinical statistics that stressed ***natural frequencies*** and ***number needed to treat*** demonstrated that clinicians can readily grasp all relevant statistical concepts.[23-27]

Medical technology should not be allowed to diminish clinical skills. Even in excellent human medical facilities the skills of physical examination have waned and inexpensive and potentially life-saving interventions (e.g. aspirin) are often overlooked.[28-33] In veterinary neurology too, we should deemphasize technology that has little proven utility, lest we lose skills, the opportunity for more practical treatments, and, eventually the public trust.[34]Veterinary neurologists' most cost effective and least ambiguous contribution to clinical practice should be exceptionable ability to recognize and interpret clinical signs, not facility with technology.

If fewer expensive and undocumented treatments are performed, the business model of practice must change to maintain income. As has been suggested for human medicine, a doctor's greatest compensation should arise from patient evaluation and interpretation of clinical signs. Income generated from tests and procedures that are capital intensive (and therefore expensive both to clients and the profession) should be limited. Fewer facilities should be burdened with the overhead of extremely expensive equipment and facilities.

## REFERENCES

[1]Fisher ES et al. **Ann In Med** 2003;138:273. [2] Fisher ES, et al: **Ann In Med** 2003;138:288.[3] Sackett D **BMJ** 1996; 312:71). [4] Greenhalgh How to read a paper. **The basics of evidence based medicine**. [5] Choudry NK **Ann Int med** 2005; 142:260. [6] Gitlin JN **Acad Radiol** 20048: 843. [7] Staunton M: Radiology 2007; 242: 23. [8]**Daly J evidence-based medicine** 2005. [9] Weed LL **NEngJMed** 1966; 80: 593. [10] Donnelly WJ et al.**Acad Med** 2005; 80: 205. [11] Masi AT et al. **Semin Arthritis Rheum** 2002; 32: 71. [12] Speciale, **Proceed ACVIM** 2006; 19 . [13] Garland AD, et al. **Obstet & Gynec** 2005;105: 205. [14]Shaughnessy AF, **BMJ** 2003; 327:266. [15] White B fam **Prac Med** 2004;11: 51. [16] Loewy EH, **MedGenMed** 2007; p:30. [17] McKibbin CW, **J Med Syst** 1977;1: 283. 17 [18] Speciale **proceed ACVIM** 2005; 18. [19] O'Brien, Bad Medicine. [20]Suwankon N, **JAVMA** 2006; 229: 1924. [21] Jones JC, **JAVMA** 2000; 216:1769. [22] Hoerlein BF: **Canine Neurology** 1978: 486. [23] Barratt A, et al., **CMAJ** 2004; 171: 353. [24] Montori VM et al., **CMAJ** 2004; 171: 611. [25] McGinn T et al.. **CMAJ** 2004; 171: 1369. [26] Hatala R et al., **CMAJ** 2005; 172: 661. [27] Montori VM, **CMAJ** 2005; 173: 385. . [28]Krumholz HM **Ann Int Med** 1996; 124:292. [29] Hirsch AT, et al.: **JAMA** 2001; 286: 717-722. [30] Mangione S, et al. **JAMA** 1997; 278: 717. [31] Mangione **S Am J Med** 2001;110:210-216. [32]: Mangione **S Am J Pulm & Crit Care** 1999; 159: 1119. [33] Zoneraich S Circulation 1995; 91: 2089. [34] Greenhalgh et al. **Brit Med Bull** 2004; 69: 197.

# SURGICAL TREATMENT OF CEREBRAL HEMORRHAGES IN DOGS. NEW HORIZONS

**Filippo Adamo, DVM, DECVN**
**San Leandro, CA**

## INTRODUCTION

Cerebrovascular accidents (CVAs) are the third leading cause of death following neoplasia and cardiovascular disease in humans.[1, 2] Cerebrovascular accidents were previously considered to be uncommon in dogs and cats, but with the more recent advances in neuroimaging in veterinary medicine, including Computed Tomography (CT) and Magnetic Resonance Imaging (MRI), CVAs are now recognized more frequently. Given recent increases in reporting and awareness, CVAs should be included in the list of differential diagnoses in an animal with an acute onset of progressive or non-progressive brain disease. Cerebrovascular accidents include focal or global ischemic encephalopaties. These involve interruption of perfusion to the brain parenchyma, resulting in neuronal injury, which may lead to transient or permanent neurologic dysfunction.

## FOCAL ISCHEMIC ENCEPHALOPATHYY

A "stroke" or brain infarction is a focal encephalopathy secondary to a CVA and resulting in neurological deficit of sudden onset. [3] Brain infarction can result from arterial or venous disease, however because of abundant venous anastomoses, venous infarction is uncommon in dogs.[4]

**The causes of strokes can be divided into basic groups**

1) Obstruction of the blood vessels, leading to infarction. The obstruction may result from thrombosis and embolism. Thrombosis is defined as vascular obstruction developing within the occluded vessels. Embolism is defined as occlusion of a vessel by a fragment of blood clot or other substance that has flowed to the site of obstruction from a distant location.[5]

2) Rupture of blood vessels walls, leading to hemorrhage. [4] Intracranial hemorrhage is classified as intraparenchymal, epidural, subdural, and intraventricular.[4, 5]

An additional form of focal ischemic encephalopathy is the transient ischemic attack (TIA), which is a well-recognized vascular disorder in humans. [6] TIA is a symptomatic episode of a brief focal, neurologic deficit secondary to embolism, vascular constriction or spasm that resolves within 24 hours, with 90% of spasms resolve usually within the first 4 hours.[5,6] This form of vascular event is also believed to occur in animals but is seldom recognized and difficult to document.[4]

## GLOBAL ISCHEMIC ENCEPHALOPATHY

Hemodynamic compromise can result in decreased global cerebral perfusion. The most common potential causes of global ischemic brain damage include severe hypotension, advanced pulmonary disease, cardiopulmonary arrest, diffuse cerebral edema, or an anesthetic accident with insufficient oxygen delivery to the brain.[5] Areas of the brain most susceptible to global ischemia include cerebral and cerebellar cortex, thalamus, hippocampus, and some basal nuclei.[5]

## CEREBRAL HEMORRHAGE

### Intraparenchymal hemorrhage

In the author's experience, intraparenchymal supratentorial hemorrhage is the most common form of intraparenchymal hemorrhage occurring in dogs and cats. Hemorrhagic infarction results in immediate extravasation of proteins with both vasogenic and eventual cytotoxic formation. In addition, ongoing hemorrhage may lead to focal space-occupying lesions, resulting in a mass effect with compression and ischemic necrosis of surrounding brain parenchyma.[5] When the hemorrhage and the consequent hematoma becomes severe enough to cause a mass effect within the brain parenchyma, the outcome may be fatal. The brain hematoma initiates edema and neuronal damage the surrounding parenchyma.[4, 7] Fluid begins to collect immediately around the hematoma, and edema usually persists for 5 days to 2 weeks. [4,11, 8] The release and accumulation of osmotically active serum proteins from the clot are responsible for the early edema around the hematoma.[4,11] This is followed by vasogenic and cytotoxic edema, which in turn contribute to the further disruption of the blood-brain barrier, the failure of the potassium/sodium pump, and the death of neural tissue.[4, 9] The delay in the breakdown of the blood-brain barrier and the development of cerebral edema after intracerebral hemorrhage suggest that there may be secondary mediators of both neural injury and edema.[4] It is commonly believed that both blood and plasma products are the mediators of most of the secondary processes initiated after intracerebral hemorrhage.[4, 10] Several observations have suggested that the genesis of a local hyperfibrinolytic state leads to the development of an expanding hematoma. [11, 12, 13] Elevated levels of fibrinogen degradation products have been demonstrated.[16,17, 14] Immature blood vessels in the capsule of the hematoma are poorly formed. These vessels may leak erythrocytes into the hematoma or be more likely to bleed secondary to minimal trauma. [15, 16] This low grade persistent hemorrhage is perpetuated by the anticoagulant, fibrinolytic environment; this process is the primary mechanism for hematoma expansion.[2,21] Daily hemorrhage of up to 10% of an existing hematoma volume has been demonstrated.[16,17]

### Subdural Hematoma

Subdural space is the virtual space located between the inner dural layer and the arachnoid. [17] Traumatic laceration of veins crossing this potential space is the usual cause of acute subdural hematoma (SDH) formation in people.[17] The time between a traumatic event and the development of clinical signs of intracranial dysfunction is inversely related to the severity of cerebral and vascular damage sustained at the time of injury.[16] Acute SDHs are often accompanied by cerebral contusion and laceration, and can be rapidly life threatening.[16, 18] Chronic SDHs are often associated with a minor traumatic episode that may have been forgotten by the time the patient develops clinical signs.[16] Subacute SDHs have an intermediate pattern of development of clinical signs.[16] Small

SDHs in people usually completely resolve, leaving only a residual membrane, which represents the residual capsule. Larger hematomas tend to undergo clot lysis, become encysted, and may subsequently show a tendency to expand.[16,18] In people, SDHs are classified into four categories according to the time elapsed between the traumatic incident and the onset of clinical signs: hyperacute, acute, subacute, and chronic.[20] SDHs in dogs have been rarely reported.[6] SDHs may expand following the same mechanism described for the intraparenchymal hemorrhage.

**Epidural & Intraventricular Hematoma.**

Epidural hematoma is usually associated with traumatic brain injury in which the build up of blood occurs between the dura mater and the skull. Hematoma expansion may occur similar to the same mechanism described above. Intraventricular epidural hematoma tend to be spontaneous and are events rarely repoirted in animals.

## IINCIDENCE & RISK FACTORS

**Incidence**

The true incidence of CVA in animals remains still unknown at this time, and there is no reported breed predilection. Risk Factors: Hyperfibrinogenemia, polycythemia, and multiple myeloma have been identified as cause of increased blood viscosity in both humans and dogs.[5] In one study, several dogs diagnosed with idiopathic focal cerebral or cerebellar infarction were treated long term with phnylpropanolamine. [19] Underlying conditions such as chronic renal failure and hyperadrenocorticism may predispose dogs to brain infarction or hemorrhage and should be screened for in animals with a suspected vascular event.[5] Suspected risk factors for the development of global ischemia during anesthesia include brachicephalic conformation and the use of ketamine for induction and maintenance of anesthesia. [20] Most cases of global ischemia may be the result of predisposing factors for the patients to cardiopulmonary arrest, idiosyncratic anesthetic reactions, insufficient anesthetic monitoring, or excessive anesthetic dosing.

## ETIOLOGY

The cause of primary intraparenchymal hemorrhage is not completely understood, but in human's patients is frequently associated with hypertension with concurrent fibrinoid degeneration of cerebral arteries. [21]

**Cerebral infarction**

Brain infarction in dogs and cats is typically nonhemorrhagic; [22] however, in the author's experience hemorrhagic infarction is not uncommon. The most common causes of arterial cerebral infarction are atherosclerosis (usually secondary to hypothyroidism, hyperlipoproteinemia or idiopathic), sepsis, neoplasia, *Dirofilaria immitis*, heart disease, and vasculitis.

**Cerebral Hemorrhage**

Hemorrhagic infarction has been associated with many underlying conditions including hypertension (usually secondary to renal diseases, hyperadrenocorticism, or idiopathic), coagulopathies, metastatic or primary central nervous system neoplasia, vascular malformation, amyloid angiopathy and idiopathic causes, and less commonly to arterial thrombosis and embolism.[5]

## CLINICAL SIGNS

Clinical signs are variable and are related to the specific cerebral area affected (forebrain: cerebral cortex and thalamic structures; brainstem: midbrain, pons or medulla oblongata; or cerebellum), the type (global or focal, ischemic or hemorrhagic) and the extension of the lesion. Clinical signs are usually acute in onset focal, asymmetric and non progressive. However, progression of the clinical signs and worsening of the neurologic deficits can be seen up to 24 to 72 hours following the initial infarction. Cerebral hemorrhage may be an exception to this description and can present with a more progressive course.[4] Clinical signs in ischemic encephalopathy usually regress after 24-72 hours; this is most likely due to edema resorption. Clinical signs secondary to cerebral hemorrhage may take longer to regress and are secondary to hematoma, hemorrhage and edema resorption. Seizures are reported to be very common in association with CVA in dogs.[4] With brain hematomas, seizures, dementia and comas are common signs.

## DIAGNOSIS

Imaging studies of the brain including CT and MRI are necessary to confirm the suspicion of a CVA and to rule out other intracranial diseases. Computed tomography is very sensitive for acute hemorrhage, with a linear relationship demonstrated between CT attenuation and hematoma hematocrit. The periphery of the lesion may enhance on CT imaging from approximately 6 days to 6 weeks after onset.[4, 23] MRI findings correlate well with the age of a hematoma and the onset of related symptoms. [24] Characteristic MRI intensities allow for determination of the age of hemorrhage based on biochemical transformation of oxyhemoglobin (hyperacute stage) to deoxyhemoglobin (acute stage) to methemoglobin (subacute stage) and to hemosiderin (chronic stage). [25, 26] The MRI appearance of intraparenchymal and SDH brain hematomas in dogs has been previously reported.[28,29]

A minimum data base including: CBC, serum biochemistry panel, urinalysis, thyroid function testing (TT4, fT4, TSH), coagulation profile (PT, PTT, fibrinogen degradation products), and multiple systolic blood pressure measurements should be evaluated in any animal suspected of having a CVA. A fecal analysis should also be performed to rule out parasitic infestation and blood and urine should be cultured if sepsis is suspected. Cerebrospinal fluid (CSF) analysis is seldom indicated to confirm the diagnosis but may help to rule out infectious or inflammatory diseases. CSF abnormalities associated with CVAs may include xanthochromia, increased protein, and a mild neutrophilic or mononuclear pleocytosis.[4,10]

## TREATMENT & PROGNOSIS

There is no specific treatment for ischemic encephalopathy or for most type of intraparenchymal and subarachnoid hemorrhage. In the acute phase of a CVA particularly when the clinical signs are severe, the goal of treatment is to maintain good cerebral perfusion pressure (cerebral blood flow). This is achieved via the maintenance of systemic blood pressure, providing good oxygenation, controlling seizures if present, and removing the underlying cause when detected. In the immediate period following a suspected stroke, including cerebral hemorrhage, mannitol therapy (0.5-1 g/kg body weight) is warranted to combat brain edema. There is no evidence that glucocorticoid therapy provides any beneficial effect to human stroke victims. [27] The outcome of patients affected by CVA depends on the extension and the localization of the lesion, the severity of clinical signs, and the underlying etiology.

Most cases of cerebral infarction improve over a few days to weeks. Intraparenchymal, intraventricular and subarachnoid hemorrhage may also cause reversible signs, but the severity of the clinical signs are often more severe. In large intraparenchymal, subdural or subarachnoid hematoma, if the clinical signs do not improve despite medical therapy, surgical evacuation of the hematoma should be employed, this may be a life saving treatment and may reverse the clinical signs.

## REFERENCES

[1] Starkman S, Dobkin B: Cerebral vascular emergencies, in Shoemaker WC, Ayers SM (eds): **Texbook of critical care,** ed 4. Philadelphia, WB Saunders, 2000, pp 1539-1545 [1] Mortel KF: Prospective study of vascular events and cerebral perfusional changes following transient ischemic attack. **Angiology** 47(3):215-223, 1996 [1] Garcia JH. The evolution of brain infarcts: a review. **J Neuropatholo Exp Neurol** 1992;51:387-393 [1] Platt SR, Garosi L: Canine Cerebrovascular disease: Do Dogs have strokes? **Pearl of Veterinary Medicine. JAAHA** (39)337-342, 2003 [1] Thomas WB. Cerebrovascular disease. **Vet Clinic North Amer: Sm An Pract** (26) 925-943, 1996 [1] Mortel KF: Prospective study of vascular events and cerebral perfusional changes following transient ischemic attacks. **Angiology** 47(3):215-233, 1996 [1] Wagner KR, Xi G, Hua Y, et al. Lobar intracerebral hemorrhage model in pigs: rapid edema development in perihematomal white matter stroke. **Stroke** (27)490-497 1996 [1] Zazulian AR, Derdeyn CP, Powers WJ. Progression of mass effect after intracerebral hemorrhage. **Stroke** (30)1167-1173, 1999 [1] Wagner KR, Xi G, Hua Y, Kleinholtz M, de Courten-Myers GM, Myers RE. Early metabolic alterations in edematous perihematomal brain regions following experimental intracerebral hemorrhage. **J Neurosurg** (88)1058-1065, 1998 [1] Qureshi AI, Tuhrim S, Broderick JP, Batjer HH, Hondo H, Hanley DF: Spontaneous intracerebral hemorrhage. **N Englis J Med** (344)1450-1460, 2001 [1] Hopkins A, Wheeler S. Subdural hematoma in a dog. **Vet Surg** 1991;20:413-417. [1] Ito H, Yamamoto S. Fibrinolitic enzyme in the lining walls of chronic subdural hematoma. **J Neurosurg** 1978:197-200. [1] Sciarra D. Head injury In: Rowland LP, Merritt's **Texbook of neurology**. Philadelphia, Lea and Febiger, 1984;277-299. [1] Weir B, Gordon P. Factor affecting coagulation: fibrinolysis in chronic subdural fluid collection. **J Neurosurg** 1983:242-245. [1] Evans S, Gean A. Craniocerebral trauma In: **Corra E, Magnetic Resonance Imaging**. 3rd ed. China, Mosby Inc., 1999;1347-1360. [1] Yamashima T, Yamamoto S, Friede R. The role of endothelial gap junctions in the enlargement of chronic subdural hematomas. **J Neurosurg** 1983:298-303. [1] Fletcher T. Spinal cord and Meninges In: 3rd ed. Miller's **Anatomy of the Dog**. Philadelphia, W.B. Saunders Co, 1993;800-828. [1] McKissock W, Richardson A, Bloom W. Subdural hematoma. A review of 389 cases. ***Lancet*** 1960:1365-1369. [1] Irwin DC, Dewy CW, Stefanacci JD: Suspected cerebellar infarcts in four dogs. **J Vet Emerg Crit** Care, 2005. [1] Panarello GL, Dewey CW, Barone G, Stefanacci JD: Magnetic resonance imaging of two suspected cases of global ischemia. **J Vet Emerg Crit Care** 14(4):269, 2004 [1] Castel JP, Kissel P. Spontaneous intracerebral and infratentorial hemorrhage. In: Youmans JR, ed. **Neurological surgery.** Phyladelphia: WB Saunders 1890-1917, 1990 [1] Garosi L, McConnel JF, Platt SR, Barone G, Baron JC, de Lahunta A, Schatzberg SJ: Results of diagnostic investigations and long term outcome of 33 dogs with brain infarction (2000-2004) **J Vet Intern Med** (19) 725-731, 2005 [1] Grossman RI. Intracranial hemorrhage. In: Latchaw RE, ed MR and CT imaging of the head, neck, and spine. St Louis: **Mosby Yearbook**, 171-202, 1991[1] Kaminogo M, Moroki J, Ochi A, et al. Characteristic of symptomatic chronic subdural hematomas on high-field MRI. **Neurorad** 1999:109-116. [1] Thomas W, Sorjonen D, Scheuler R, *"et al"*. Magnetic resonance imaging of brain infarction in seven dogs. **Vet Radiol Ultrasound** 1996;37:345-350 [1] Thomas W, Adams W, McGavin M, *"et al"*. Magnetic resonance imaging appearance of intracranial hemorrhage secondary to cerebral vascular malformation in a dog. **Vet Radiol Ultrasound** 1997;38:371-375.

## PATHOPHYSIOLOGY OF SPINAL CORD INJURY

**Nick Jeffery BVSc PhD DECVS DECVN FRCVS**
**Cambridge, UK**

### MECHANISMS OF INJURY

The spinal cord can be injured through three major pathological mechanisms: i) ***contusion***, in which there is a rapid on/off deformation of the spinal cord (similar to that resulting from a hammer blow); ii) ***compression***, in which the spinal cord undergoes persistent deformation; and, iii) direct impairment of ***vascular*** supply, such as that caused by fibrocartilaginous embolisation. Clinical spinal cord injury (SCI) most commonly results from a combination of compression and contusion, for instance the injury caused by vertebral fracture-luxation or resulting from extrusion of the nucleus of an intervertebral disc. From a clinician's viewpoint, it is important to attempt to estimate the 'proportion' of damage resulting from each of these components, since it provides a means realistically to estimate the likely effect of surgical intervention. Studies in many species have suggested that the severity of injury is largely determined at the instant of the injury, especially in those in which contusion predominates.

Most of what is known about the pathogenesis of SCI is derived from experimental studies on laboratory rodents. For ease of interpreting results laboratory studies have predominantly examined the effects of contusion and comparatively little is known about the effects of chronic cord compression; it is generally assumed that the mechanisms of injury are similar but occur at a slower rate. This presentation will focus on the effects of contusive injury.

### PRIMARY VERSUS SECONDARY INJURY

The immediate effects of a contusive injury to the spinal cord are relatively minor, as defined by what can be observed histologically. Crucially however, the primary injury initiates a cascade of secondary events, collectively termed the secondary injury[1,2], that lead to progressive tissue disruption and which occur over the following hours, days and weeks. The primary injury results from direct mechanical impact which disrupts axons, blood vessels, and cell membranes. The secondary injury results from vascular dysfunction, oedema, ischaemia, excitotoxicity, electrolyte shifts, free radical production, inflammation, and delayed apoptotic cell death. The recognition of these separate phases, and the constituent events of the secondary injury, has been a key achievement of spinal cord injury research during the last two decades.

Immediately after injury, damage to the microvasculature leads to haemorrhage (most commonly in the central grey matter) and generalised swelling of the spinal cord extending for many segments rostral and caudal to the epicentre. Although histopathological changes are minimal in the first few hours, injurious pathophysiological processes, such as microglial activation and glutamate release, are already initiated. During the following hours to days, the secondary injury processes gain momentum, through persistent vascular disruption, hypoperfusion and ischaemia, and include free radical production, ionic dysregulation, glutamate–mediated excitotoxicity, and immune-associated neurotoxicity that all contribute to further axonal injury and cell death.

### COMPONENTS OF THE SECONDARY INJURY

Poor blood flow through the damaged region leads to failure of energy supply and therefore loss of the ability to maintain the normal ionic balance across the cell membrane. Depolarisation in itself will lead to release of excitotoxic amino acids (most notably glutamate), which then allow supra-physiological accumulation of calcium in both neurons and glial cells. In axons, the influx of sodium that results from depolarization will then activate the sodium calcium exchange mechanism in reverse, meaning that intra-axonal calcium concentration rapidly increases. Dysregulation of calcium ion concentration is a central event in mediating cell death since it initiates many damaging processes, including the activation of calpains, mitochondrial dysfunction, promotion of free radical and vasoactive amine production, many of which culminate in cytokeletal or membrane disruption, axonal and cell death.

Free radical mediated injury, caused by both nitrogen and oxygen species, is an important final common pathway of injury to the spinal cord following SCI, since it induces lipid peroxidation and nitrosylation of all cell membranes and consequent cell death. Oxygen free radicals are found at highest concentration at about twelve hours following injury but remain high for ~ 1 week, only returning to normal levels by 4–5 weeks after injury. Other free radical species (*e.g.* peroxynitrite) can also directly induce apoptosis.

Inflammation[3] is initiated immediately through activation of microglia, but the inflammatory process following SCI is highly complex and involves numerous cellular populations (including astrocytes, microglia, T cells, neutrophils, and invading monocyte) and many chemical mediators (including TNFα, interferons, and interleukins). Because the blood-spinal cord barrier is disrupted after SCI inflammatory cells are free to enter and leave the damaged region, thus allowing a systemic (body-wide) immune response to occur. However, the role of inflammation is double-edged, since there is evidence for both deleterious and beneficial effects of the evoked inflammatory response.

It is important to remember that loss of glial cells is also an important contributor to the loss of function after SCI. Oligodendrocytes are very susceptible to ischaemia and energy deprivation and are also to glutamate-mediated excitotoxicity. Oligodendrocyte death will lead to loss of associated myelin sheaths and the consequent axonal demyelination leads to loss of conduction. Although demyelination is a widespread phenomenon in the early stages after SCI, the extent to which it persists for weeks or months afterwards is controversial[4], since it is known that both invading Schwann cells and newly-generated oligodendrocytes can both contribute to remyelination.

**EARLY REPARATIVE RESPONSES**

Phagocytic activity, mainly involving removal of cell and myelin debris by macrophages, commences early in the course of SCI but persists for many weeks[5]. There is evidence that macrophage activity is a key event in mediating subsequent reparative processes (as it is in other body tissues). Although initially astrocytes at the epicenter will be killed by SCI, during the subacute phase astrocytes at the periphery of the lesion become hypertrophic and proliferative and increase expression of the astrocytic intermediate filament glial fibrillary acidic protein (GFAP). These reactive astrocytes grow multiple, large cytoplasmic processes that interweave to form the astrocytic scar. This scar is of critical importance for its effects as both a physical and chemical barrier to axonal regeneration. The astrocytic reaction should not be viewed as wholly detrimental to eventual recovery, since it also aids in reestablishment of ionic homeostasis and the integrity of the blood-spinal cord barrier, which is important for the resolution of oedema and in limiting the infiltration of immune cells.

**SUBSEQUENT PROGRESS**

In some patients there is evidence of cavitation at the epicentre that may progress to form syrinxes. Syringomyelia in itself can have detrimental consequences for the patient, such as neuropathic pain, and may progress. This development is apparently relatively unusual in canine patients although few severely affected animals have been monitored for many years.

**END RESULT**

After contusive injury spinal cord tissue is lost in a stereotypical manner[6] – the centre of the cord is destroyed first – implying loss of grey matter – and tends to spare the sub-pial axons. It has also been found that there is relative sparing of the smaller diameter axons, perhaps because of their smaller surface area to volume ratio.

The astrocytic scar (sometimes compounded by meningeal cell infiltration) defines a boundary around the damaged region of spinal cord and characterized by expression of GFAP and high levels of other inhibitory molecules, notably chondroitin sulphate proteoglycans of various types.

**REFERENCES**

[1] Tator CH & Fehlings MG (1991) **J Neurosurg** 75:15–26. [2] Kwon BK et al. (2004) **Spine J** 4:451–464. [3] Fleming JC et al. **Brain** (2006)129:3249–3269. [4] Lasiene J et al. (2008) **J Neurosci** 28:3887–3896. [5] Hagg T & Oudega M (2006) **J Neurotrauma** 23:264–280. [6] Smith PM & Jeffery ND (2006) **Brain Pathol** 16: 99-109.

# MECHANISMS OF SPINAL CORD REGENERATION & REPAIR

**Nick Jeffery BVSc PhD DECVS DECVN FRCVS**
**Cambridge, UK**

## RESTORATION OF FUNCTION WITHOUT REGENERATION

In mild-moderate injuries the restoration of physical integrity through scarring, combined with the well-known mechanisms of nervous system 'plasticity' are sufficient for the restoration of useful function[1,2]. In experimental cats the ability to walk remains intact when as few as 5% of the original number of axons in the thoracolumbar spinal cord survive following SCI. Currently available rehabilitative therapy in humans has also been shown effective to make best use of surviving spinal cord tissue. In future, pharmacological manipulation of the spinal pattern generators or development of novel methods to activate the correct patterning of limb movements may also be able to mediate restoration of specific functions.

## SPONTANEOUS REGENERATIVE RESPONSES

The processes that can be considered to contribute towards spinal cord regeneration and repair commence simultaneously with the processes that cause progressive damage during the early phases of SCI. Thus, removal of debris, activation of inflammatory cells and astrocytes all have important effects in initiating a regenerative and reparative response. However, many of the reparative events – notably astrocytic hypertrophy and proliferation - can probably best be considered as a means of excluding pathogens, rather than promoting recovery of function. On the other hand, at variance with previous dogma, there is the potential for many cell types to be replaced. For instance, oligodendrocyte precursor cells are widely distributed throughout the spinal cord and there is evidence that there are neural stem cells within the ependymal lining of the central canal. Newly-generated oligodendrocytes are able to remyelinate previously demyelinated axons and thus restore normal conduction, providing an example of successful regeneration within the CNS. Spontaneous replacement of lost neurons by neural stem cells has rarely been observed, the major problem being that the environment of the damaged CNS encourages differentiation towards astrocytes.

In general the CNS can be viewed as lacking in regenerative responses and, from the point of view of recovery of function, the most notable obstacle is the lack of significant axonal regeneration across the astrocytic scar. Much research effort has been focused on developing methods to overcome this problem.

## UNDERSTANDING THE LACK OF AXONAL REGENERATION IN THE SPINAL CORD

Observations made about a century ago – and repeated frequently since! – demonstrated that there was at least some, albeit poor, ability for CNS neurons to regenerate damaged axons[3], an ability that varies greatly depending on neuronal location and phenotype. This is apparent in the spinal cord in that abortive regenerative sprouting can be observed in the weeks after acute SCI. Repeated observations have implicated the astrocytic scar as a prime culprit in the inhibition of axonal regeneration, and recent work has also incriminated CNS myelin.

## METHODS TO OVERCOME THE INHIBITORY NATURE OF THE DAMAGED CNS

Since there are two distinct mechanisms underlying failure of axonal regeneration there are two approaches to overcoming the problem. In order to overcome the *intrinsically* poor regenerative response of CNS neurons, efforts have been made to identify molecular signals that stimulate axonal growth following injury. Several regeneration-associated genes have been found to be upregulated after SCI, such as GAP-43. However, even in the face of up-regulation of appropriate genes the regenerative response in the CNS is not as robust as that in the PNS. By carrying out experiments on neurons in the dorsal root ganglia a decrease in cAMP levels has been implicated in limiting the regenerative response in CNS axons[4]. Therefore, a recently proposed method (currently undergoing clinical testing) to improve outcome after SCI is to elevate intracellular cAMP levels using the phosphodiesterase inhibitor rolipram[5].

On the other hand there are the '*extrinsic*' factors – environmental factors that inhibit the ability of the axon to regenerate. These molecules are divided into inhibitors associated with CNS myelin and inhibitors associated with the astrocytic scar. Numerous growth inhibitory molecules have been found associated with CNS myelin, including Nogo, MAG, OMgp, ephrin B3 and Netrin-1. Importantly, many (though not all) of these inhibitory factors function through interaction with the Nogo receptor (NgR), meaning that this forms a target for therapeutic blocking agents[6]. Numerous experiments have documented the axon growth-promoting effects of agents that block the action of this receptor. The second category of inhibitory signal are those that arise from the glial scar, which results from reactive astrocytosis, and forms a source of many growth inhibitory extracellular matrix components known as chondroitin sulphate proteoglycans (CSPGs). Recently a strategy has been developed to degrade these molecules, using a bacterial enzyme, ChABC. A series of experiments has clearly demonstrated improvements in axonal regeneration and both electrophysiological and behavioural function[7]. Further experiments are underway to determine whether it might be possible to genetically manipulate either host or transplant cells to express this enzyme.

## CELL-BASED STRATEGIES

Many cell types could potentially play a role in improving outcome after SCI, and they can be used to fulfill a variety of functions: 1) direct cell replacement (*e.g.* oligodendrocytes, neurons); and 2) altering the environment in such a way as to either enhance or support axonal regeneration, provide neuroprotection, or both. Although there is clearly considerable potential for multipotent (*e.g.* stem) cells to fulfill both these functions, there are many current obstacles to their direct therapeutic use. For instance, a multipotent cell transplanted into a region in which the environmental cues are driving production of reactive astrocytes will likely become a reactive astrocyte! Nevertheless, it is now possible to manipulate stem cells ex-vivo so as to direct their maturation when transplanted[8]. On the other hand, using a cell that has already differentiated has the advantage that its function and phenotype is clearly defined and can be selected based on those properties. It is with this prospect in mind that the properties of properties of Schwann cells and, more recently, olfactory ensheathing cells, have been investigated in detail. The interest in Schwann cells is that, if transplanted, they are able to reproduce the PNS environment – which is permissive to axonal growth - within the damaged region of the CNS. The widespread interest in olfactory ensheathing cells (OEC) was originally derived from observations of their function within their normal environment, in which they promote the growth of axons from the newly-generated neurons that are continuously being formed within the olfactory mucosa. Furthermore, they are able to bridge the divide between the PNS and CNS in this location. Subsequent experiments demonstrated that these properties were reproduced when they were transplanted into the damaged CNS[9]. Evidence of their ability to promote both functional and histological improvement after various types of spinal cord injury has now been produced in many laboratories throughout the world. The possibilities for translating OEC transplants into a clinically applicable therapy will be discussed in the next presentation in this series.

**REMYELINATION**

Following SCI there is widespread demyelination, owing to the death of oligodendrocytes. Evidence from most experimental paradigms has suggested that remyelination follows quickly (within ~four weeks) and uneventfully, mediated by both oligodendrocytes (derived from progenitors within the CNS) and Schwann cells (derived from the PNS and CNS), but there are some contradictory data. Post mortem examination of human patients with chronic SCI has suggested that some axons may remain demyelinated for years after the injury[10] and there is indirect supporting evidence from electrophysiological investigations. Since demyelinated axons are unable faithfully to transmit impulses and may be at risk of future degeneration there is interest in promoting remyelination as another method to improve outcome after SCI in humans. There is abundant evidence that transplanted glial cells or their progenitors can effectively remyelinate demyelinated CNS axons. The remyelinatiing ability of OEC and Schwann cells provides another reason for them to be good candidates for cell-based transplantation therapy of SCI. However, from post mortem examinations that we have carried out there may be reason to doubt whether long-term widespread demyelination is a common consequence of SCI in dogs[11].

**REFERENCES**

[1] Jeffery ND & Blakemore WF (1999) Vet Rec 145: 183-90. [2] Bareyre FM et al. (2004) Nat Neurosci. 7: 269-77. [3] David S & Aguayo AJ (1981) *Science* 214:931–933. [4] Cai D et al. *J Neurosci* (2001) 21:4731–4739. [5] McKerracher L et al. (1994) *Neuron* 13:805–811. [6] Freund P et al. (2006) *Nat Med* 12:790–792. [7] Bradbury EJ et al. (2002) *Nature* 416:636–640. [8] Hofstetter CP et al. (2005) *Nat Neurosci* 8:346–353. [9] Ramon-Cueto A et al.(2000) *Neuron* 25:425–435. [10] Bunge RP et al. (1993) *Adv Neurol* 59:75–89. [11] Smith PM & Jeffery ND (2006) Brain Pathol 16: 99-109.

## TRANSLATION OF LABORATORY RESEARCH INTO CLINICAL PRACTICE

**Nick Jeffery BVSc PhD DECVS DECVN FRCVS**
**Cambridge, UK**

### ISSUES IN TRANSLATION

There are very many interventions that have been shown to improve histological and / or functional outcome after SCI in experimental rodents. These findings have of course engendered a great deal of optimism that it might be possible to translate the experimental interventions into clinical therapies. However, there are many steps that must be made to fill this gap[1], but the key data relate to four issues:

1. Will the interventions used successfully in rodents also promote recovery in 'higher' mammals (notably primates)? This question centres on the differences in anatomy between species, especially the dominance of corticospinal tract function over motor control in humans.
2. There is a size difference between experimental rodents and other species; the necessary 'scaling up' may vary in complexity depending on whether dealing with an administered drug or cell transplantation.
3. Experimental lesions are designed to be homogenous (see also point 4 below), whereas all clinical classification schemes will inherently permit a degree of (often unknowable) heterogeneity.
4. There is a difference between defining a statistical difference to prove a principle and finding an effect that is medically relevant.

To address these issues several lines of research are required[2]. Firstly, it is necessary – partly from a licensing point of view in human medicine – to demonstrate efficacy in another species. In human SCI research there is a strong viewpoint that therapies require testing in primates because of the differences in moor control compared to that in rodents. Secondly, there is a need to devise a 'clinical model' of disease to address points 3 & 4. A very large number of interventions that are statistically useful in laboratory animals for treatment of many lesions will not prove sufficiently efficacious for application as a medical therapy. There are many examples of failed translation of therapeutics in CNS trauma, most notably MK-801, 21-aminosteroids and, arguably, methylprednisolone sodium succinate. Thirdly, the issue of applicability to patients may be critical. Interventions that could prove highly efficacious cannot be used if they have high toxicity (*e.g.* tetrodotoxin), if successful use requires intervention within a non-achievable time window (*e.g.* vitamin E) or if application carries too many technical risks.

### CLINICAL TRIAL SEQUENCE

Translation of experimental therapeutics into the clinic depends upon successful completion of a series of clinical trial phases[3]. These are designed so as to test in sequence the safety and efficacy in selected groups of patients[4]. Clinical trials are conventionally divided into three phases – i) Phase I to determine safety; ii) Phase II to determine efficacy in a small closely defined group of patients; and iii) Phase III are randomized controlled multicentre trials on large patient groups.

In Phase I trials, conventionally the best case selection is of patients that have a poor chance of recovery with conventional therapy, for instance in SCI research perhaps those that have lost all function below the lesion. The aim is to search for evidence of detrimental effects, such as neurological worsening (although for patients with extremely poor function this might be very difficult) or more pertinently the development of neuropathic pain (for instance). A Phase I trial would also aim to record evidence suggesting efficacy, especially as a means to aid in design of methods formally to test efficacy in subsequent phases[5].

Phase II trials consist of carefully designed studies (that may be randomised controlled trials) to detect efficacy. In SCI research the NIH considers testing surgical interventions against 'placebo' essential since any treatment not so tested could be erroneously thought to be effective, thereby resulting in large numbers of future patients potentially receiving a non-efficacious, and perhaps dangerous, treatment. Testing the results in terms of relevant outcome measures is an important aspect of this process; these require careful selection and should be relevant to everyday life. Conventionally there is one primary outcome measure for which the power of the study is defined and then observations are made on other outcome measures for which further evidence can be sought in other, subsequent, trials if necessary.

Phase III trials aim to examine the wider applicability of the results obtained in a wider population than that used in the Phase II trial and commonly compare the novel treatment against a 'gold standard'.

Finally, in humans there are many licensing issues, which are not so severe in animal patients but still exist. There are ethical issues with all such interventions, most notably with surgery - and especially 'sham' surgery.

### THE CAMBRIDGE PHASE II TRIAL ON TRANSPLANTATION OF OLFACTORY ENSHEATHING CELLS (OEC) FOR SEVERE SCI IN DOGS

Our clinical laboratory has the essential requisites for carrying out trials on OEC transplantation[6], since we are linked with a basic science laboratory that has been investigating the in vivo and in vitro properties of OEC for many years. OECs can be regarded as front-runners for application in human and animal SCI because of the wealth of background data supporting their efficacy in rodents[7]. The primary purpose of our study is to examine the effects of OEC transplantation into *clinical* lesions (thus addressing points 3 & 4 above); primate experiments on these cells are also taking place because there is also the need to show that the observed recovery in rodents can be duplicated in primate species.

During the past few years we have been establishing quantitative outcome measures for SCI dogs[8,9], to permit quantitative analysis of outcome following any intervention, including OEC transplants. Thus, we are using a battery of tests including:

1. Gait analysis – to examine the coordination between fore and hind limb movements (testing the TL cord segment) and coordination of the paw placement with the centre of mass (thus examining the function of brainstem control mechanisms)
2. Electrophysiological outcomes – MEP and SEP
3. Urodynamic assessment using bladder cystometry (outcome in bladder care is a high priority for many human patients)

We have powered the study to examine the difference between group of dogs treated with and without cell transplants on the coordination between fore and hind limbs.

**The protocol**

Olfactory ensheathing cells are taken from the nasal mucosa for autotransplantation, then grown in culture for ~three weeks and then transplanted percutaneously into the damaged region of the spinal cord; both these techniques have been developed in our laboratory specifically for this purpose. Cells are pre-labelled by BrDU (so that they can be identified at subsequent post mortem examination). The dogs are randomized to receive either a cell transplant or a non-cell transplant; it is a double-blinded trial – both the client and the researcher carrying out the analysis of function are blinded to treatment category. .

The animals are then re-examined at monthly intervals for functional assessments. We have requested that the owners return the dogs to us for histological assessment when they are euthanased.

**REFERENCES**

[1] Kleitman N (2004) J Spinal Cord Med 27: 311-8. [2] Jeffery ND et al. (2006) Spinal Cord 44: 584-593. [3] Lammertse D et al (2007) Spinal Cord 45: 232-42. [4] Tuszynski MH et al (2007) Spinal Cord 45: 222-31. [5] Steeves JD et al (2007) Spinal Cord 45: 206-21. [6] Jeffery ND et al. (2005) J Neurotrauma 22: 1282-1293. [7] Raisman G & Li Y (2007) *Nat Rev Neurosci* 8:312–319. [8] Hamilton L et al (2007) BMC Neurosci 8:77. [9] Hamilton L et al. (2008) BMC Vet Res 4:47.

## MATRIX METALLOPROTEINASES: THERAPEUTIC TARGET IN DOGS WITH SPINAL CORD INJURY

**Jonathan M. Levine, DVM, DACVIM (Neurology)**
**College Station, TX**

### INTRODUCTION

According to recent studies, approximately 230,000 people in the United States are living with spinal cord injury (SCI) with 10,000 new injuries occurring per year.[1] The physical disabilities associated with SCI in humans often impact quality of life due to their affects on emotional wellbeing, family relationships, and lost productivity. In the United States, the yearly total healthcare expenses associated with human SCI are 9.7 billion dollars with approximately $1,000,000 spent on each affected patient over a lifetime.[2] Years lost due to disability (YLD), a measure of the impact of disease on society, is as high in SCI as for almost any other form of trauma requiring hospitalization.[3]

In dogs, the incidence of SCI is higher than in humans with the majority of injuries resulting from disk herniation. Estimates suggest that 2.3% of all veterinary hospitalizations in dogs relate to disk herniation and in certain breeds, such as the Dachshund, lifetime incidence may be as high as 19-24%.[4,5] As in humans with SCI, disability impacts owner-perceived quality of life in dogs and results in secondary impairments such as decubital ulcers and urinary tract infection.[6]

In SCI, the mechanisms underlying tissue damage are often classified as primary and secondary. Primary injury refers to the initial mechanical insult sustained by the spinal cord and can be sub-classified into components such as compression, contusion, concussion, and laceration. Secondary injury refers to the biochemical cascade of events that follows primary injury. Processes involved in secondary injury, such as oxidative stress, vasculopathy, inflammation, and excitotoxicity, are complex and often over-lapping. Oxidative stress occurs in SCI due to energy depletion, which leads to free radical generation, and as a direct result of neutrophil degranulation. Free radicals cause lipid peroxidation as well as altered nucleic acid and amino acid structure. Traditionally, inflammation has been seen as a key component of secondary SCI. Innate responses occur within hours after SCI and are typified by local cytokine release from injured vessel walls and resident microglia, neutrophil and macrophage diapedesis, and local tissue destruction which is predominantly neutrophil-associated.[7,8] Adaptive responses occur hours to days after SCI and are mediated through local antigen presenting cells (microglia and macrophages) that direct T-cell responses.[8] More recently, beneficial aspects of inflammation in the sub-acute phases of SCI have been recognized. Macrophages may improve axonal regeneration at lesion margins and CD4+ T-cell recognition of self antigens such as Nogo-A and myelin basic protein derived peptide may boost neurite growth.[9,10]

### MATRIX METALLOPROTEINASES IN SCI

Matrix metalloproteinases (MMPs) are zinc-dependent endopeptidases involved in the breakdown of extra-cellular matrix (ECM). They have a well defined role in physiologic processes such as wound healing, angiogenesis, and embryonic development. Metalloproteinases are also associated with lesion progression in diseases including multiple sclerosis, glial neoplasia, traumatic brain injury, and SCI. MMP-2 and MMP-9, both of which are gelatinases, have been most closely studied in the context of SCI.[11,12] MMP-2 is constitutively expressed, whereas MMP-9 is induced within hours of SCI and persists for only 72 hours in rodent models.[11,12]

The sources of MMP-9 are believed to include macrophages, neutrophils, and endothelium which are critical players in the innate inflammatory response.[11,12] Early MMP-9 activity has been shown to disrupt laminin and collagen IV, reduce cell-cell contact, increase pro-inflammatory cytokine levels, and directly induce apoptosis.[11-14] In combination, these events result in blood-spinal cord barrier breakdown, spinal cord inflammation, necrosis, and apoptosis of glia and neurons in rodent models. MMP-9 has been shown to be expressed in the cerebrospinal fluid of a proportion of dogs with thoracolumbar disk herniation.[15] All dogs expressing MMP-9 had clinical signs of <24 hours duration at admission and expression was associated with more severe neurological injury (paraplegia).

In sub-acute SCI, MMPs likely play divergent roles. This is perhaps best exemplified with the gelatinases. Whereas, MMP-2 supports white matter sparing and limits the formation of an inhibitory glial scar, MMP-9 is a determinant of glial scarring in both in vitro models as well as in a murine model of SCI.[16,17] MMP-9 facilitates glial scarring by directing the early migration of reactive astrocytes into injured parenchyma, leading to the formation of immature gliotic meshwork.[17] Release of pro-inflammatory cytokines associated with MMP-9 expression may also induce glial scarring.[17]

### MMP INHIBITION

Early, short-term inhibition of MMPs improves outcome across several disease models. In traumatic brain injury, animals that received inhibition within hours of injury with GM-6001 (Illomostat) had improved functional recovery compared to controls.[18] In SCI, inhibition of MMPs reduces apoptosis and improves locomotor functional recovery.[1,12] In all models of injury, early inhibition appears crucial, although a therapeutic window for initiation has not been defined. Traditionally, animals have been provided GM 6001 for 72 hours after SCI; longer periods of drug administration (7 days) do not improve functional outcome compared to vehicle.[19] This finding may be explained by loss of neuro-protection associated with MMP-2 in the sub-acute phases of SCI.[19]

GM-6001 is a broad spectrum MMP inhibitor that interferes with the "cysteine switch" needed for MMP activation. It is highly liophilic and is detectable in the CSF of rodents 30 minutes after intraperitoneal (IP) delivery with a Cmax of 120 minutes.[20] GM-6001 has been the drug used in the vast majority of MMP inhibition studies for these reasons and the fact that it has been well tolerated in small mammals. It has been shown to reverse clinical signs of experimental allergic encephalitis (EAE), vascular stent-associated hyperplasia, and scarring of scleral incisions.[20-22] In addition, GM 6001 has been used in rodents to treat experimental SCI at 100 mg/kg IP BID for 3 days or in traumatic brain injury as a single 100 mg/kg IP bolus.[12,18] A 4 dog GM 6001 tolerance study was performed according to FDA guidelines at Texas A&M University in December 2007. Dogs were administered vehicle, 100 mg/kg

GM 6001, 150 mg/kg GM 6001, or 300 mg/kg GM 6001 subcutaneously BID for 72 hours. Transient regional discomfort and dose dependent transient hyperthermia were the only physical adverse effects seen. Complete blood counts, serum biochemistries, and coagulation profiles were within normal limits for all dogs prior to, 1 day after, and 7 days after drug delivery. Histopathology was within normal limits for the control and 300 mg/kg experimental dog, the only dogs assessed. Serum levels of GM-6001 were detected within 30 minutes with apparent Cmax at 30 minutes.

## RANDOMIZED, DOUBLE BLIND, PLACEBO-CONTROLLED STUDY

Naturally occurring veterinary disease offers advantages over experimental models when trying to determine the efficacy of a therapy and the potential for expanded human clinical trials. For example, SCI associated with canine disk herniation mimics in vivo human injuries as the primary mechanisms tend to be diverse, lesions occur in dogs with varied genetic backgrounds, and affected animals have intact immune responses. This is in contrast to experimental models.[23] The incidence of canine disk herniation is relatively high and pre-referral treatment is often purely supportive.

An NIH-funded, clinical trial has been initiated at Texas A&M University to examine the outcome effects of GM 6001 in dogs with thoracolumbar disk herniation. Injury duration prior to diagnostic imaging is limited to <48 hours, dogs cannot receive corticosteroids prior to referral, and all dogs have to receive surgical decompression. Various clinical outcome measures and adverse events are recorded. The trial is still on-going.

## REFERENCES

[1] Dang AB et al. **Spine** 2008;33:E576. [2]Ackery A et al. **J Neurotrauma** 2004;21:1355. [3]Polinder S et al. **Bull World Health Organ** 2007;85:27. [4]Ball MU et al. **J Am Vet Med Assoc** 1982;180:519. [5]Hoerlein BF. *Canine Neurology: Diagnosis and Treatment*. Philadelphia: WB Saunders, 1978. [6]Levine JM et al. **J Am Vet Med Assoc** 2008;233:931. [7]Trivedi A et al. **Clin Neurosci Res** 2006;6:283. [8]Keane RW et al. **J Neurotrauma** 2006;23:335. [9]Baptiste DC, Fehlings MG. **J Neurotrauma** 2006;23:318. [10]Rossignol S et al. **J Neurosci** 2007;27:11782. [11]Duchossoy Y et al. **Mol Cell Neruosci** 2001;17:945. [12]Noble LJ et al. **J Neurosci** 2002;22:7526. [13]Schonbeck U et al. **J Immunol** 1998;161:3340. [14]Mannello F et al. **Apoptosis** 2005;10:19. [15]Levine JM et al. **Am J Vet Res** 2006;67:283. [16]Hsu JY et al. **J Neurosci** 2006;26:984. [17]Hsu JY et al. **J Neurosci** 2008;28:13467. [18]Sifringer M et al. **Neurobiol Dis** 2007;25:526. [19]Trivedi A et al. **Int J Neuroprotect Neuroregen** 2005;2:30. [20]Gijbels K et al **J Clin Invest** 1994;94:2177. [21]Li C et al. **J Am Coll Cardiol** 2002;39:1852. [22]Wong TL et al. **Invest Ophthalmol Vis Sci** 2005;46:2018. [23]Faden AI, Stoica B. **Arch Neurol** 2007;64:794.

# BONE-SEEKING NANOPARTICLES ENCAPSULATING THERAPEUTICS FOR TREATING FOCAL MALIGNANT OSTEOLYSIS.

**Timothy Fan[1], Laura Garrett[1], Jackie Wypij[1], Robert O'Brien[1], and Jianjun Cheng[2];**
**Department of Veterinary Clinical Medicine, University of Illinois, Urbana, IL;**
**Department of Materials Science and Engineering, University of Illinois, Urbana, IL.**

Neoplastic bone destruction is a painful condition in dogs, and requires the development of new and effective treatment options. Nanotechnology drug delivery systems hold promise for treating various pathologic conditions, including cancer. The selective targeting of nanoparticles to the tumor microenvironment would have the added benefit of not only enhanced localized effects, but also reduced systemic toxicity.

The purposes of this study were 1) to fabricate surface-coated pamidronate (bone-seeking) nanoparticles encapsulating doxorubicin, 2) to determine the hematologic and biochemical safety of bone-seeking nanoparticles in research dogs, 3) to demonstrate minimal reticuloendothelial system retention of bone-seeking nanoparticles in normal dogs, and 4) to demonstrate targeting to areas of malignant osteolysis in dogs with osteosarcoma.

Bone-seeking nanoparticles encapsulating doxorubicin were fabricated and characterized *in vitro* for their release and binding kinetics. Three research dogs were administered bone-seeking nanoparticles encapsulating a doxorubicin (dose equivalent $30mg/m^2$) and monitored for hematologic and biochemical toxicity weekly for 4 weeks. *In vivo* the biodistribution of bone-seeking nanoparticles conjugated with Tc-99m was studied in normal dogs and in dogs with appendicular osteosarcoma.

*In vitro*, bone-seeking nanoparticles demonstrated controlled release of doxorubicin over 28 days. Administration of bone-seeking nanoparticles encapsulating doxorubicin appears safe in research dogs, and intravenous Tc-99m-labelled bone-seeking nanoparticles concentrate in areas of malignant osteolysis with minimal retention in the reticuloendothelial system as demonstrated by nuclear scintigraphy scans in normal dogs and dogs with spontaneously-arising appendicular osteosarcoma.

Based upon *in vitro* and *in vivo* studies, bone-seeking nanoparticles encapsulating doxorubicin appears safe and home to areas of malignant bone destruction. Additional studies are warranted for investigating this novel drug delivery system in dogs with appendicular osteosarcoma to determine direct anticancer effects within the primary tumor, with future studies assessing the efficacy of bone-seeking nanoparticles in the management of malignant bone pain.

# GENISTEIN AS A NOVEL TREATMENT OF CANINE LYPMHOMA

**Steven E. Suter VMD, MS, PhD DACVIM (Oncology)**
**Raleigh, NC**

## INTRODUCTION

Genistein (4, 5, 7-trihydroxyisoflavone) is a readily available isoflavone found in soy-based products. Epidemiological studies indicate that consumption of soy-containing diets is associated with a lower incidence of many human tumors[1,2]. Genistein's effects on various human solid cancer cell lines have been extensively studied[3-9]. Genistein is also active against human lymphoid neoplasia[10,11]. When included into a CHOP (cyclophosphamide, doxorubicin, vincristine, prednisone) lymphoma protocol, genistein has growth modulatory effects, via G2-M arrest and decreased NF-kb binding, in a DLBCL xenograft setting[12].

Genistein Combined Polysacharride (GCP™), a commercially available form of genistein (26), is a complex mixture produced by fermentation of soybean extracts with a mushroom mycelia (*Ganoderma lucidum*) culture that contains approximately 40% isoflavones. Fermentation deglycosylates soy isoflavones, producing aglycone isoflavones that are highly absorbable across the gut lumen[13]. GCP™ has documented *in vitro* and *in vivo* effects on a wide variety of human cancers[14,15].

Canine high-grade B-cell lymphoma (LSA), which is similar phenotypically and biologically to the most common variant of human non-Hodgkin's lymphoma (NHL) called diffuse large B-cell lymphoma (DLBCL), is the most common hematopoietic tumor of the dog (83%)[16]. Dogs are evolutionarily more closely related to humans than rodents and are large, long-lived animals; therefore, canine B-cell LSA represents an excellent model to investigate novel therapeutics and treatment strategies which may have direct applicability to the human disease[17].

Based on the similarity between canine B-cell LSA and human DLBCL, we hypothesized that genistein would inhibit the cell growth of two well-established canine lymphoid B-cell lines. We show that genistein induces cell death via apoptosis at concentrations similar to those reported in the human literature. We also present our preliminary findings of the oral absorption of GCP™ in domestic dogs.

## RESULTS

This study analyzed the effect of pure genistein and GCP™ on the cell growth of two well characterized canine lymphoid cell lines, GL-1[18] and 17-71[19]. These cell lines were exposed to increasing concentrations of genistein (5-100 mg/ml).

**FIGURE 1**

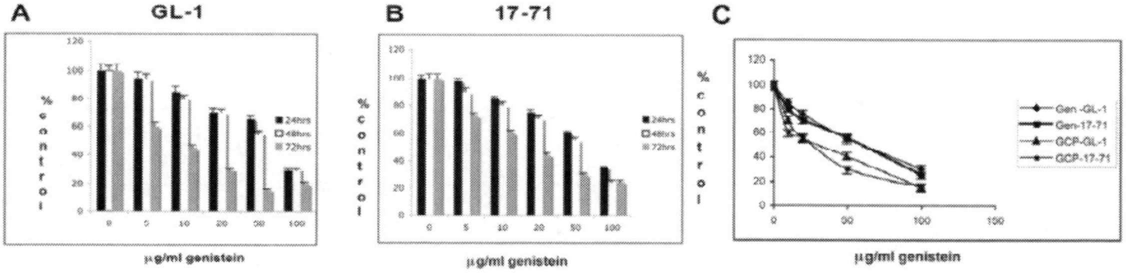


Inhibition of proliferation of two canine B cell lines, *(A)* GL-1 and *(B)* 17-71, by genistein and GCP™. Proliferation of both cell lines was assessed in the presence of various concentrations of either compound for 72 h in a 96-well format using an MTT cell proliferation kit. Absorbances were measured at 570 nm and expressed in relation to absorbances of control (carrier only) cells. Values are expressed as means $\pm$ SD (n=3). $P \leq 0.005$ (compared with control cells in both GL-1 and 17-71 cells). *C*, comparison of purified genistein to amounts of GCP™ extract with equivalent concentrations of genistein on cell growth of both GL-1 and 17-71 cells over 72 h.

Based on these experiments, we calculated the 72 h *in vitro* $IC_{50}$ of both genistein and GCP™ against the GL-1 and 17-71 cells to be 10 µg/ml and 20 µg/ml, respectively. For the remainder of the experiments, only GCP™ was used.

Using a number of assays that measure apoptosis, we next demonstrated that GCP™ initiates the process of apoptosis rather than acting as a cytostatic agent in the canine lymphoid cell lines, which may account for the growth inhibition observed earlier in the cell proliferation studies. The translocation of membrane phospholipid phosphatidylserine (PS) from the inner to the outer leaflet of the plasma membrane, an early event in the apoptotic cascade, can be detected via staining with Annexin V.

**FIGURE 2**

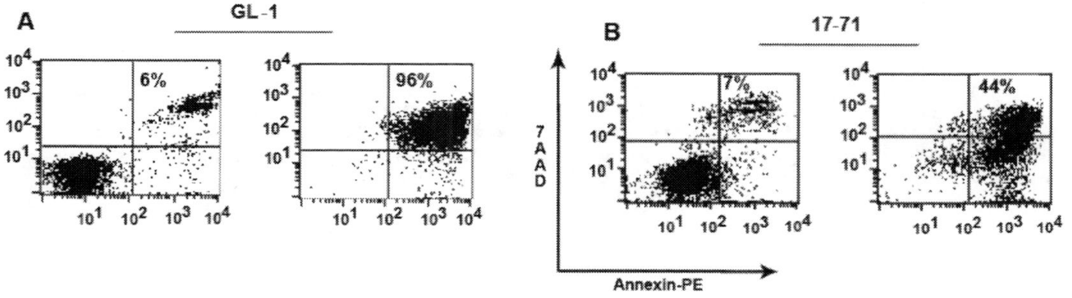

GCP™ induces apoptosis in both *(A)* GL-1 and *(B)* 17-71 cells. Cells were exposed to the $IC_{50}$ of GCP™ for 72 h. Adherent and floating cells were pooled, washed, stained with an anti-Annexin antibody and 7-AAD and analyzed by flow cytometry to detect early signs of apoptosis. The left had scatter plot for both cell lines is representative (n=3) of cells treated with carrier only, while the right scatter plot is representative (n=3) of GCP™ treated cells.

We also examined the induction of apoptosis of GCP™ treated cells by examining the activation of the cytsteine proteases, caspase 9 and 3.

**FIGURE 3**

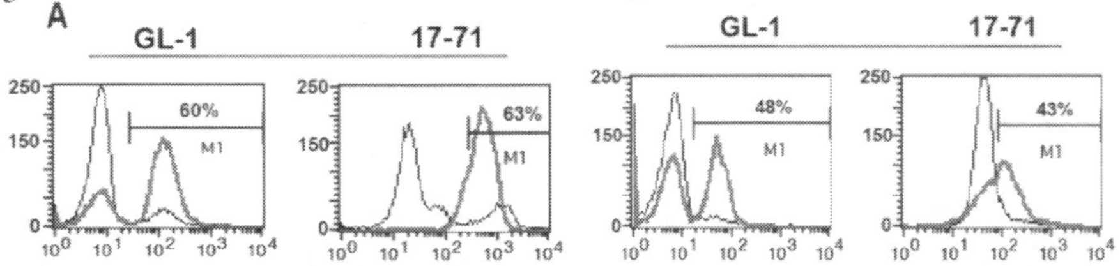


GCP™ induces caspase activation leading to DNA degradation in both GL-1 and 17-71 cells. A, anti-caspase 9 antibody and B, anti-caspase 3 antibody. In all histograms, the thin black tracings are representative (n = 3) of cells treated with carrier only, while the thick grey tracings are representative (n = 3) of GCP™ treated cells.

This experiment showed that both caspase 3 and 9 were activated in both cell lines after exposure to GCP™, further documenting that apoptosis was the cause of growth inhibition.

Isoflavone glycosides, such a genistein, are not readily absorbed across enterocytes in most species, which limits their therapeutic use to intravenous administration. GCP™, which is an aglycone isoflavone, is highly bioactive due to unimpeded intestinal absorption. We performed a dose escalation pharmacokinetic study to determine if GCP™ fed to normal dogs led to plasma genistein levels that approached the *in vitro* $IC_{50}$ of the compound and evaluate any toxicities associated with oral dosing. We fed 1.5, 3.5, and 5 grams (0.45%, 1.0%, and 1.45% GCP-supplemented diet, respectively) GCP™ to normal dogs with a dose level ranging from 0.05 mg/kg/day to 0.19 mg/kg/day. Plasma genistein levels were determined as described for 48 h post-feeding while CBCs (complete blood counts) and serum biochemistries were analyzed before and 48h post-feeding. In addition, the dogs were monitored clinically throughout the study and for two weeks thereafter for signs of GCP™ toxicity.

**FIGURE 4**

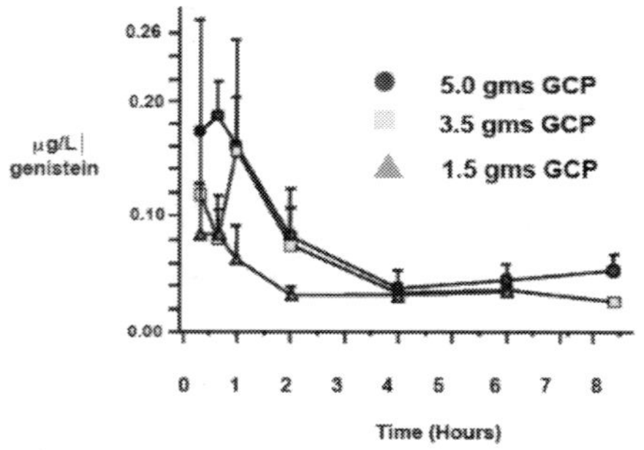


**TABLE 1**

| Parameter | Dog 1 | | | Dog 2 | | | Dog 3 | | |
|---|---|---|---|---|---|---|---|---|---|
| | 1.5 | 3.5 | 5.0 | 1.5 | 3.5 | 5.0 | 1.5 | 3.5 | 5.0 |
| Tmax (hr) | 0.67 | 1.0 | 0.67 | 0.33 | 0.33 | 0.33 | 0.33 | 1.0 | 1.0 |
| Cmax (µg/ml) | 0.0547 | 0.0777 | 0.01412 | 0.106 | 0.076 | 0.252 | 0.055 | 0.2396 | 0.1554 |
| AUC (hr*µg/ml) | 0.2096 | 0.2743 | 0.7195 | 0.1955 | 0.2125 | 0.372 | 0.1378 | 0.4377 | 0.5527 |
| Half-life (hr) | 5.1997 | 2.9417 | 7.0874 | 3.6433 | 4.2442 | 2.5027 | 6.0987 | 2.8577 | 5.7273 |

Orally administered GCP™ leads to detectable genistein plasma levels in normal dogs (n=3). In Fig 6, genistein plasma levels were obtained using HPLC and plotted against time (h) after 1.5, 3.5, and 5.0 gms GCP™ were added to the dogs food. In Table 1, the pharmacokinetic parameters, Tmax, Cmax, AUC, and Half-life, were calculated for each dog when fed increasing amounts of GCP™.

These results support the notion that GCP™ has significant in vitro anti-proliferative effects against canine B-cell lymphoid cell lines. Our initial pharmacokinetic data emphasizes the importance of bioavailability and disposition kinetics studies when attempting to evaluate the *in vivo* effects of compounds in different species. Based on our data showing orally administered GCP™ leads to detectable genistein plasma levels, we hypothesize that genistein may be useful in both chemopreventive and treatment regimes for canine B-cell lymphoma. Based on this promising preliminary data, additional studies in our laboratory are ongoing in an effort to determine if canine B-cell lymphoma will be a relevant large animal model for examining the *in vivo* efficacy of genistein for human DLBCL treatment regimes.

## REFERENCES


[1]Adlercreutz H, et al. **Am J Clin Nutr** 1991; 54:1093. [2]Messina MJ, et al. **Nutr Cancer** 1994; 21:113. [3]Park OJ, et al. **Toxicol Lett** 2004; 150:43. [4]Dean NM, et al. **Biochem Biophys Res Commun** 1989; 165:795. [5]Markovits J, et al. **Cancer Res** 1989; 49:5111. [6]Shao ZM, et al. **Anticancer Res** 1998; 18:1435. [7]Day JK, et al. **J Nutr** 2002; 132:2419S. [8]Cabanes A, et al. **Carcinogenesis** 2004; 25:741. [9]Fang MZ, et al. **Clin Cancer Res** 2005; 11:7033. [10]Raynal NJ, et al. **J Nat Prod** 2008; 71:3. [11]Yamasaki M, et al. **Cancer Sci** 2007; 98:1740. [12]Mohammad RM, et al. **Mol Cancer Ther** 2003; 2:1361. [13]Izumi T, et al. **J Nutr** 2000; 130:1695. [14]Bemis DL, et al. **Clin Cancer Res** 2004; 10:5282. [15]Yuan L, et al. **Ann N Y Acad Sci** 2003; 1010:347. [16]Moulton J. Tumors of Lymphoid and Hematopoietic Tissue. Berkeley: Univeristy of California Press; 1990. [17]Khanna C, et al. **Nat Biotechnol** 2006; 24:1065. [18]Nakaichi M, et al. **J Vet Med Sci** 1996; 58:469. [19]Steplewski Z, et al. **Cancer Immunol Immunother** 1987; 24:197.

## PEGYLATED ARGININE DEIMINASE IN CANINE CANCER

**CO Rodriguez, Jr., J Zhang, A Fascetti, Z Yu, RK Bold, K Skorupski, H-J Kung, and X Chen.**
**University of California, Davis, CA.**

Arginine is a nonessential amino acid in both healthy adult mice and men[1-3] that undergoes both synthesis and degradation in complex series compartmentalized steps at the cellular and systemic level[3]. The urea cycle enzymes responsible for this axis are argininosuccinate synthetase (ASS1) and argininosuccinate lyase (ASL), which converts this metabolite to fumarate and L-arginine[1-3]. Catabolism may occur via arginase, another urea cycle enzyme, which degrades L-arginine to L-ornithine[3]. L-ornithine can be converted back to L-citrulline by ornithine carbamoyltransferase (OCT) and thus close a recycling biochemical loop in the synthesis and degradation of arginine. This closure is dependent on the levels of intracellular ASS1 and ASL[4, 5]. It has been shown that many tumor types are auxotrophic for L-arginine due to very low or no specific activities of one or both of the L-arginine synthetic enzymes, ASS1 and ASL or to corruption of the citrulline-arginine pathway and are thus sensitive to depravation of this single nutrient[6]. The purpose of this study was to determine the sensitivity of canine cancer cell lines to arginine deprivation by pegylated arginine deiminase (PEGADI).

The kinetics of arginine depletion was determined by HPLC. Cell growth inhibition was determined by counting cells, MTS assay and clonogenic assay. The presence or absence of arginine metabolic enzymes (ASS, OCT) and detection of apoptosis (PARP) were determined by western blot. Student's two-tailed *t*-test was used to assess for statistical differences in paired samples. ANOVA was utilized to assess for differences between groups. Dunnett's t ad hoc test was used to confirm significance.

PEGADI depleted arginine from medium in a time- and concentration-dependent manner. Arginine levels in medium were halved at 24h with 0.03uM but not detectable within 2h at 0.1uM. When quantitated by counting, MTS or clonogenic assay, osteosarcoma (D17 and two others), and three canine melanoma cell lines had an IC50 ranging between 0.03uM and 0.1uM PEGADI at 24h. All cell lines evaluated expressed detectable ASS1 and OCT. Death in arginine-deprived cells occurred by apoptosis in a p53-independent manner.

These data support the notion that PEGADI-mediated arginine deprivation may possibly be a viable treatment strategy in the management of canine melanoma and osteosarcoma. Treatment of osteosarcoma-, melanoma- and lymphoma-bearing dogs with PEGADI is planned.

[1]Rogers Q. In: Symposium Honoring Willard J. Visek 1994; 1994. 9-21.[2]Yu YM, Am J Physiol 1996;271(6 Pt 1):E1098.[3]Wu G Biochem J 1998;336 ( Pt 1):1.[4]Shen LJ, Cancer Lett 2003;191(2):165.[5]Shen LJ, Cancer Lett 2006;231(1):30. [6]Wheatley DN, Mol Cell Biochem 2003;244(1-2):177.

**PHASE I/II TRIAL OF SATRAPLATIN IN TUMOR-BEARING DOGS.**

**KA Selting,[1] CJ Henry,[1,2] DL Gustafson,[3] X Wang,[4] JD Robertson[4].**
**University of Missouri, Departments of. Veterinary Medicine and Surgery, Internal Medicine, and Chemistry and MU Research Reactor, Columbia, MO; and The Animal Cancer Center, Colorado State University, Fort Collins, CO.**

Satraplatin is the first orally bioavailable platinum anticancer chemotherapeutic. Platinum agents are effective in the treatment of many cancers. The objectives of this dose-escalation study were to determine the maximally tolerated dose (MTD) of satraplatin in tumor-bearing dogs, to identify the dose-limiting and other toxicities in dogs, and to characterize the myelosuppression that occurs secondary to satraplatin in this species.

Dogs were entered in standard 3+3 cohorts. When dose-limiting toxicity was encountered in 2/3 or 2/6 dogs, the preceding dosage was considered the MTD. Twice weekly blood counts were used to determine the timing and severity of neutrophil and platelet nadirs.

The MTD was determined to be 35 $mg/m^2$/day for 5 days. Pharmacokinetics at the MTD are summarized in the table below. Twenty-one dogs were treated, with pharmacokinetic evaluation performed for 14 dogs. Platinum levels were determined using ICP-MS. Ultrafiltrate showed an accumulation factor for Cmax = 1.19 (0.85) and for AUC = 1.30 (0.47), and total platinum showed an accumulation factor for Cmax = 2.01 (0.83) and for AUC = 2.99 (0.69).

Interestingly, the platelets consistently reached nadir sooner than the neutrophils (day 15 vs. 18). Myelosuppression was the dose-limiting toxicity and gastrointestinal toxicity was mild. No neuro- or nephrotoxicity was noted.

A metastatic nodule resolved in one dog, and 4 of 8 dogs treated in the adjuvant setting for osteosarcoma had prolonged disease-free intervals. Satraplatin was well tolerated in tumor-bearing dogs and showed preliminary efficacy, thus warranting further investigation.

| **PK parameter** | **$Pt_{TOT}$ (day 1)** | **PUF (day 1)** | **$Pt_{TOT}$ (day 5)** | **PUF (day 5)** |
|---|---|---|---|---|
| **Tmax (hr)** | 2.1 (1.1) | 1.8 (0.7) | 2.1 (1.2) | 1.9 (1.1) |
| **Cmax (ng/ml)** | 115 (38) | 71.7 (26.1) | 206.2 (60) | 70.8 (25.1) |
| **AUC (hr*ng/ml)** | 2492 (931) | 316 (63) | 12921 (6174) | 451 (221) |
| **Cl (ml/hr/$m^2$)** | 15345 (4150) | 115092 (26139) | 3156 (1221) | 91663 (36366) |
| **MRT (hr)** | 36 (14.2) | 7 (1.3) | 81 (25) | 14 (14.3) |

Table 1: PUF=plasma ultrafiltrate (free Pt). $Pt_{TOT}$=Total platinum. T max = time to maximal plasma concentration; Cmax = maximal plasma concentration; AUC = area under the curve (time-concentration curve); Cl = clearance; MRT = mean residence time. Values are expressed as mean (standard deviation).

**EVALUATION OF CISPLATIN (CDDP) AS A ELECTROCHEMOTHERAPY AGENT FOR THE TREATMENT OF INCOMPLETELY EXCISED MAST CELL TUMORS IN DOGS.**

**Spugnini EP,[1] Vincenzi B,[2] Baldi A,[3] Citro G.[1]**
**S.A.F.U. Department, Regina Elena Cancer Institute – Rome – Italy [2]Medical Oncology, University Campus Bio-Medico, Rome – Italy [3]Department of Biochemistry, section of Pathology, II University of Naples- Italy**

Electrochemotherapy (ECT) couples the administration of anticancer drugs with the delivery of electric pulses that increase the drug uptake through the cell membranes, thus resulting in an improved efficacy. Aim of the study was the evaluation of tolerability and efficacy of cisplatin as ECT agent for incompletely excised canine mast cell tumors.

Dogs with incompletely excised MCTs received two sessions of ECT under sedation. Explicitly, after marginal resection, confirmed by the surgeon and the pathologist reports, the tumor bed and margins were infiltrated with cisplatin (0.5 mg/ml), and then exposed to trains of permeabilizing biphasic electric pulses. A second ECT session was performed one or two weeks later based on clinical considerations. More precisely, five minutes after the injection of the chemotherapy agent, sequences of 8 biphasic pulses lasting 50+50 microseconds each, with 1 ms interpulse intervals, were delivered in bursts of 1300 V/cm for sclerosed cutaneous fields, and of 800 V/cm for softer lesions, using caliper or needle array electrodes respectively. Adherence of the external electrodes to the lesion was maximized using an electroconductive gel.

Thirty-seven dogs entered the study so far. The treatment was well tolerated and side effects were minimal. Twenty-nine dogs are still in remission at variable times, seven had tumor recurrence, one died of multiple cutaneous mast cell tumors and one died of unrelated causes. The estimated median time to recurrence was 1200 days. Remarkably, no case of recurrence was observed among the long term (>1 year) treated patients.

Cisplatin seems a effective agent for ECT and this combined treatment could be a useful addition to the current options for incompletely excised MCT in consideration of its low cost, limited toxicity, and ease of administration.

**CARBOPLATIN DOSAGE IN CATS BASED ON A TARGETED AREA UNDER THE PLATINUM CONCENTRATION-VERSUS-TIME CURVE AND PATIENT GLOMERULAR FILTRATION RATE.**

**DB Bailey[1], KM Rassnick[2], NL Dykes[2], L Pendyala[3], J Prey[3].**
**Oradell Animal Hospital, Paramus NJ., Cornell University, Ithaca NY. Roswell Park Cancer Institute, Buffalo NY.**

Carboplatin-associated neutropenia is strongly correlated with area under the platinum concentration-versus-time curve (AUC). We prospectively evaluated a novel dosing strategy in tumor-bearing cats based on a targeted AUC and an estimation of platinum clearance ($CL_{Pt}$) derived from patient glomerular filtration rate (GFR). We also evaluated two different GFR assays to predict $CL_{Pt}$.

First, GFR was measured using plasma clearance of technetium 99m-labeled diethylenetriaminepentaacetic acid ($^{99m}$Tc-DTPA). Carboplatin was administered using the equation: Dose = $AUC_{Target}$ x (2.6 x $GFR_{99mTc\text{-}DTPA}$) x Body Weight. Initial $AUC_{Target}$ was set at 2.0 min·mg·mL$^{-1}$ and escalated in cohorts of 3 cats. Myelotoxic effects were assessed with weekly CBCs. The maximum tolerated $AUC_{Target}$ was defined as the highest $AUC_{Target}$ where less than one-third of treated cats experienced unacceptable toxicity. Next, GFR was measured simultaneously using plasma $^{99m}$Tc-DTPA and serum iohexol clearance. Each cat received a dose of carboplatin using the above equation and the previously-identified maximum tolerated $AUC_{Target}$. Plasma platinum concentrations were measured using atomic absorption spectrophotometry.

$AUC_{Target}$ ranged from 2.0 to 3.0 min·mg·mL$^{-1}$; the maximum tolerated $AUC_{Target}$ was 2.75 min·mg·mL$^{-1}$. Nineteen cats were treated at the maximum tolerated AUC. Thirteen became neutropenic, but only 1 developed grade 4 neutropenia. $GFR_{99mTc\text{-}DTPA}$ and $GFR_{Iohexol}$ were strongly correlated ($r = 0.90$), but $GFR_{99mTc\text{-}DTPA}$ measurements were significantly smaller by a factor of approximately 1.4. $CL_{Pt}$ was linearly related to $GFR_{99mTc\text{-}DTPA}$: $CL_{Pt} = 2.5 \times GFR_{99mTc\text{-}DTPA}$ ($P < 0.0001$). It also was linearly related to $GFR_{Iohexol}$: $CL_{Pt} = 1.3 \times GFR_{Iohexol} + 1.4$ ($r^2 = 0.83$, $P = 0.0002$).

Carboplatin-associated myelotoxicity is accurately predicted with this dosing strategy. The maximum tolerated $AUC_{Target}$ is 2.75 min·mg·mL$^{-1}$. Both $^{99m}$Tc-DTPA and iohexol clearance measurements can be used to estimate $CL_{Pt}$, but they are not interchangeable. Consequently, while our initial dosing equation is appropriate when using $^{99m}$Tc-DTPA clearance, it must be modified when using iohexol clearance: Dose = $AUC_{Target} \times [(1.3 \times GFR_{Iohexol}) + 1.4) \times$ Body Weight.

**INVESTIGATING A PROCASPASE-3 ACTIVATOR FOR THE TREATMENT OF CANINE LYMPHOMA.**

**PW Lucas, LD Garrett, AM Barger, J Wypij, D Deorge, PJ Hergenrother, TM Fan. Comparative Oncology Research Laboratory and Cancer Care Clinic, University of Illinois at Urbana-Champaign, Urbana IL 61802-4714.**

Procaspase-3 leads to DNA degradation and apoptosis of cells when activated. A mechanism for tumor immortality may be sequestration of procaspase-3, thereby preventing its dimerization to active caspase-3. The first procaspase-3 activating compound is PAC-1, which has been shown to induce cancer cell apoptosis *in vitro* and anticancer effects *in vivo*. The purposes of the study were 1) to determine procaspase-3 concentrations in two canine lymphoma (cLSA) cell lines, 2) to investigate the *in vitro* cytotoxic capabilities of PAC-1, and 3) to determine the bioavailability and safety of PAC-1 in dogs.

Two cLSA cell lines, 17-71 and OSW, were evaluated for basal procaspase-3 concentrations via flow cytometry using a commercially-available antibody (Epitomics). Cell lines were incubated with increasing concentrations of PAC-1 for varying times, and cytotoxicity was analyzed with flow cytometry. Six research dogs were administered PAC-1 (oral and intravenous) to determine its pharmacokinetic and safety profiles.

Immortalized cLSA lines expressed measurable concentrations of procaspase-3, and incubation with PAC-1 caused a dose- and time-dependent cytotoxicity. The pharmacokinetic profile of oral PAC-1 was variable with bioavailability ranging from 2.7-38%. Conversely, intravenous PAC-1 demonstrated a consistent profile with a $T_{1/2}$ of 3.6 hours. At serum target concentrations of PAC-1 predicted to exert anticancer activity, no biochemical or hematologic toxicity was observed in research dogs.

Based upon *in vitro* cytotoxicity data conducted in cLSA cell lines, in conjunction with favorable *in vivo* pharmacokinetic and safety profiles in research dogs, further investigation of PAC-1 in spontaneously-arising canine LSA is warranted.

# CYSTOSONOGRAPHIC MEASUREMENTS OF CANINE BLADDER TUMORS

**Dr. Carrie Tupper Hume, G Seiler, A Cáceres, Y Porat-Mosenco, F Shofer, A LaBlanc, K Sorenmo**
**Philadelphia, PA**

## INTRODUCTION

Cystosonographic measurements of bladder tumors are used to monitor treatment response. The purpose of this study was to evaluate the difference in cystosonographic tumor measurements between operators and with changes in bladder size. Hypothesis: Inter-operator variations, as well as changes in bladder volume, result in clinically significant differences.

## METHODS

Ten dogs with bladder tumors were included. A urinary catheter was passed, the bladder emptied and then filled with saline. Three operators, who were blinded to each other's measurements, determined the length, width, and height of the tumors at three different bladder volumes. RECIST criteria was used for comparison. Intraclass correlation coefficient was used to assess operator reliability. The percent difference between measurements was calculated by dividing the difference of the measurements by the smallest measurement.

## RESULTS

A poor correlation between operators was found ($r_I$=0.6). The percent difference of tumor length between operators, evaluating the same bladder volume, was as high as 121%. The difference in length was greater than 20% in 57% of the measurements and greater than 30% in 36% of the measurements with changes in operator. Inconsistent measurements were also found with tumor length and changes in bladder volume. The average percent difference of tumor length for different bladder volumes, without changing operators, ranged from 1.2% to 41%. The difference in length was greater than 20% in 17% of the measurements and greater than 30% in 11% of the measurements with changes in bladder volume. Discrepancies between operators and bladder volumes were significant enough to change response categories and alter treatment.

## CONCLUSIONS

Cystosonographic measurements of bladder tumors are affected by both operator and bladder volume. These results may have implications for studies using cystosonography to evaluate tumor progression.

## REFERENCES

[1] Phase II clinical trial of carboplatin in canine transitional cell carcinoma of the urinary bladder. Chun, et al. ***JVIM*** 5:279-283, 1997. [2] Effects of the Cyclooxygenase inhibitor, piroxicam, in combination with chemotherapy on tumor response, apoptosis, and angiogenesis in a canine model of human invasive urinary bladder cancer. **Mohommed, et al. Molecular Cancer Therapeutics**. Vol 2, 183-188, February 2003. [3] Cisplatin versus cisplatin combined with piroxicam in a canine model of human invasive urinary bladder cancer. Knapp, et al. **Cancer Chemother Pharmacol** 46: 221-226, 2000. [4] Evaluation of cisplatin administered with piroxicam in dogs with transitional cell carcinoma of the urinary bladder. Greene, et al. **JAVMA** Vol 231, No 7, October 1, 2007. [5] Clinical evaluation of mitoxantrone and piroxicam in a canine model of human invasive urinary bladder carcinoma. Henry, et al. **Clinical Cancer Research**. Vol. 9, 906-911, February 2003

**APPLICATION OF IMMUNOPHENOTYPING IN THE ONCOLOGY PATIENT USING FLOW CYTOMETRY**

**Melinda J. Wilkerson, DVM, MS, PhD, DACVP**
**Manhattan, KS**

## INTRODUCTION

Flow cytometry is broadly defined as a technology that measures multiple parameters of cells or particles as they pass through a light source in a fluid stream. Although application of flow cytometry is new to many in the field of veterinary medicine, the technology has been available since 1969. The first fluorescence detection cytometer that utilized an argon laser as a light source was developed by Marvin Van Dilla in at the Los Alamos Laboratory. Most bench top flow cytometers are dedicated to analytical methods that measures light scatter and emitted fluorescent light. The clinical importance of analytical flow cytometry was brought to light in the 1980's when monoclonal antibodies became available and physicians used flow cytometers to monitor the numbers of CD4+ and CD8+ T lymphocytes in patients infected with the human immumodeficiency virus. Today, the laser technology of flow cytometers is incorporated into routine clinical hematology analyzers and in research facilities. Analytical flow cytometers use combined systems of fluidics, optics, electronics, and computer. I will provide an overview of the principals of analytical bench top flow cytometers and briefly discuss the use of the technology in patients with hemic neopasia.

Classification of canine hemic neoplasias can be difficult based on cytologic or histopathologic assessment alone. With the advent of flow cytometry and monoclonal antibodies to canine cluster of differentiation (CD) molecules, identification of leukocyte subsets are now possible and may help in lineage determination of hemic neoplasias. Determination of B or T cell lineage in lymphomas is an important prognostic tool because T cell lymphomas have a poorer prognosis than B cell lymphomas. However, human oncology studies demonstrate that surface antigen receptor and gene rearrangements can be aberrantly expressed in lymphoma/leukemic cases, requiring simultaneous immunophenotyping, polymerase chain reaction (PCR), and chromosomal assays to fully understand the cell lineage. Little is known about these issues in canine hemic neoplasia, however, new data is beginning to emerge and these caveats need to be discussed by the veterinary medical community.

## LINEAGE DETERMINATION OF HEMIC CELLS BY CD MOLECULES

Cluster of differentiation (CD) nomenclature was first introduced at an international conference in Paris (1982) during the boom in monoclonal antibodies. This nomenclature was proposed as a way to standardize the nomenclature used to define the specificity of monoclonal antibodies to surface molecules and to define the biological function of the molecule expressed by various hematopoietic lineages. Today there are over 339 CD molecules defined; this explosion in CD molecules is the result of the use of molecular biology techniques that identify new molecules (http://prow.nci.nih.gov).

## CANINE HEMIC NEOPLASIA AND CLASSIFICATION BASED ON CD MOLECULES

### Overview of lineage-associated markers

Cellular markers commonly used in immunophenotyping of hematologic neoplasias were initially determined by differentiation and maturation studies of hematopoietic cells. Lineage associated markers can be broadly classified into the groups: B-cell, T-cell, natural killer (NK)-cell, myeloid/monocytic, erythroid, and activation makers (i.e. MHC II, CD25, CD69). Noncommitted hematopoietic stem cells express CD34, a glycoslyated surface glycoprotein. This marker is often used to differentiate acute immature leukemias of lymphoid or myelogenous origin from chronic lymphocytic leukemia or a leukemic state of lymphoma.

Stages of B cell development can be identified by the sequence of expression of cell membrane markers. Early B cell precursors express CD34, CD19, CD45, CD45R, and intracytoplasmic CD79a and CD79b followed by the loss of CD34 and orderly expression of CD10, CD20, CD21, and surface IgM. [1,2] CD79a, the signal transduction portion of the B cell receptor spans the transmembrane and intracytoplasmic regions of the cell and is expressed in all immature and mature B cells and in the majority of B cell neoplasms, including precursor B cells and plasma cell tumors. [3] Human antibodies to CD79a are cross reactive and have been used in immunohistochemistry and flow cytometry format in the dog. The appearance of immunoglobulin M (IgM) and CD21on the surface defines mature young B cells that exit the bone marrow. Antibodies specific for CD19, CD10 and CD20 have not been developed to canine B cells. Major molecules of importance for B cell maturation in the dog are illustrated in Figure 1.

Cells that enter the thymus are capable of giving rise to T cell subsets, NK cells, and dendritic cells. Markers that are present on early T-cell precursors (thymocytes) include cytoplasmic CD3, CD2, CD5, CD7, CD1a. During thymic maturation, the T-cell precursors undergo rearrangement of the TCR and begin to express the accessory molecules CD5, CD4, and/or CD8. Mature T cells that enter the circulation and lymph nodes loose early markers while retaining surface CD3, CD4 or CD8 [1,4] (Figure 1).

During development and early maturation lymphoid cells undergo genetic rearrangements of genes encoding the immunoglobulin (Ig) and the T-cell receptor (TCR) that lead to sequential expression of the receptors of B and T lineage. In most lymphoid neoplasms, antigen receptor gene rearrangement precedes transformation; therefore the daughter cells derived from the malignant progenitor have the same antigen receptor gene, representing a monoclonal population. Using PCR assays that identify these surface receptors and genes that encode for them, it is possible to differentiate monoclonal neoplasms from polyclonal reactive processes and to screen for minimal residual disease after treatment or recurrences before they become clinically apparent.

CD3 is a complex molecule associated with the T cell receptor (TCR) and is expressed in early thymocytes through maturation into mature T cells. B cell lineages are typically defined by the expression of cytoplasmic CD79a and others described above to determine the maturation state of the lymphoma cells. For example, early pre-B cells express only CD79a, whereas mature B cells also express CD21 (Epstein Barr virus receptor) and surface molecules of MHC class II, and IgM (Figure 1).

Cells of the myeloid lineage are typically identified by the expression of CD13 (aminopeptidase), CD33 (sialoadhesion), and CD117 (c-kit). Canine antibodies to these molecules are not available. The maturation of myeloblasts into promyelocytes is accompanied by a loss of CD34 followed by expression of the beta integrin CD11b and the respiratory burst enzyme myeloperoxidase. CD14 expession (LPS receptor) is a characteristic of monocyte maturation.

Erythroid precursors are identified by the absence of CD45, common to all leukocytes, and high expression of CD71 (transferrin receptor). Natural killer cells express the cell markers CD56 or CD16.

Although a set of 20 or more antibodies to CD molecules are routinely used to define lineages of hematopoietic neoplasias in humans [5], a recent study of 81 patients with lymphocytic leukemia identified CD79a and CD3 as the most sensitive and specific B and T markers that were expressed in virtually all cases of B and T cell lymphoma/leukemia. [6] Based on this observation similar immunophenotyping panels can be adapted for use in canine lymphoproliferative disorders. In fact, chronic lymphocytic leukemia in dogs, once thought to be of B cell lineage, has been redefined as a T cell phenotype through the recent availability of a comprehensive set of monoclonal antibodies.[7]

**Determination of clonality by detection of receptor rearrangements**

Applications of a polymerase chain reaction (PCR) to amplify the variable regions of immunoglobulin genes and T-cell receptor genes have been used to confirm lineage specificity and detect the presence of clonal populations in canine samples. In both B and T cells, immunoglobulin and T cell receptor (TCR) gamma gene encodes for the antigen binding region and contains much of the unique sequence of DNA that is expressed by a clonal population of cells. This strategy uses primers to a single variable and two joining regions for amplification of immunoglobulin genes and TCR-gamma genes.[8] Amplified products identified PCR primers are resolved by agarose gel electrophoresis and the test is identifying antigen receptor rearrangements by PCR (PARR). Single amplicons indicate a clonal population of cells, whereas multiple products indicate a polyclonal response. The PCR technique has been reported to correlate well with immunophenotyping for 78% of the T cell types and 96% of the B cell types.[8]

**Caveats of diagnostic interpretation**

Diagnostic interpretation of the immunophenotyping results becomes difficult when one or more of the T-cell antigens or B-cell antigens co-express with the other. For example, CD3 and cytoplasmic CD79a expression has been reported in 10% of human cases of T cell lymphoma/leukemias.[9] Using additional CD markers and PCR assays to identify receptor rearrangements many of the CD79a/CD3 positive cases were further defined to be of T cell lineage.[10] Early studies in the dog noted co-expression of T-cell antigen and surface immunoglobulin at a high frequency (up to one third) interpreted as B-cell lymphoma.[11,12] Later studies surveying a total of 164 canine cases of lymphoma showed no co-expression of B and T-cell markers.[13,14] and attributed the earlier findings by to a lack of specificity of the pan-T cell markers. Although further studies need to be done in dogs, T and B cell co-expression in humans have been associated with reduced survival studies [9,15] an angioblastic T cell lymphomas.[16,17]A single canine case reported with CD79a and CD3 antigen co-expression was defined as a high grade T-cell lymphoma.[18] We have noted aberrant expression of various molecules (B or T) in canine cases of lymphomas and leukemia.[(19)] We and others have noted Ig rearrangements in a few canine leukemia cases that have myeloid or myelomonocytic morphology. Mixed lineage leukemias reported in humans have poorer prognosis [20] and are proposed to be very early hematopoietic progenitor cells that have a transient expression of myeloid or lymphoid markers which later can lead to down regulation of CD molecules and differentiation toward either myeloid, T-lymphoid or B-lymphoid differentiation pathways.[21]

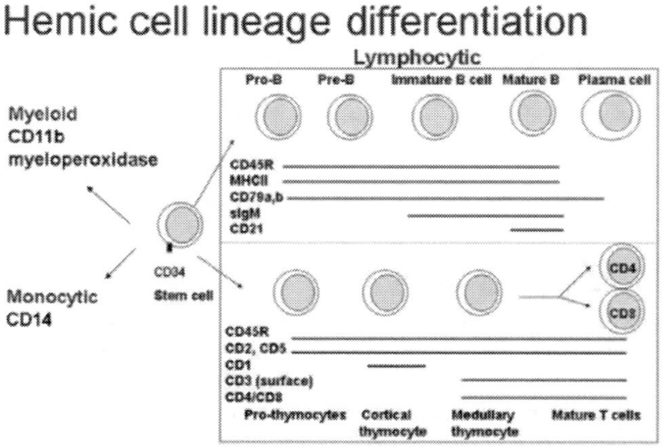


**Figure 1.** Sequential expression of major CD molecules associated with differentiation of myeloid and B and T lymphocytes.

**SUMMARY**

Identification of cellular lineage and clonality can be accomplished by a combination of immunophenotyping by flow cytometry and PCR methods. These diagnostics improve our ability to classify hemic neoplastic syndromes. However, appropriate and aberrant CD molecule expression and gene rearrangements by neoplastic cells can be quite variable and complex leading to difficulties in interpretation. Aberrant expression of CD molecules occurs in human cases of angioblastic T cell lymphomas and large cell lymphomas indicating that no marker has absolute lineage specificity and that immunophenotyping studies should always be performed with panels of monoclonal antibodies. Features of malignancy not only include aberrant antigen expression but also the loss or markedly dimmed expression of CD45.[5, 22] Cases with ambiguous phenotypes may require genotypic studies for precise lineage assignment. As a result of these issues, there is a good basis to consider other avenues such as gene arrays that would define our oncology patients into gene groups rather than morphologic syndromes. In this manner we may target upregulated genes or gene products that function in angiogenesis, cellular differentiation, apoptosis, and signal regulators as alternative therapeutic targets.[23,24]

## REFERENCES

[1] Moore P, Vernau W. Lymphocytes: Differentiation molecules in diagnosis and prognosis. In: Feldman B, Zinkl J, Jain N, eds. **Schalm's Veterinary Hematology.**5th Philadelphia: Lippincott Williams & Wlkins;2004:247-255[2] Aster J, Kumar V. White cells and lymph nodes. In: Coltran R, Kumar V, eds. **Robbin's Pathologic Basis of Disease.**6th Philadelphia: W.B. Saunders;1999:652-657[3] Chu PG, Arber DA. CD79: a review. **Appl Immunohistochem Mol Morphol.** 2001;9:97-106.[4] Rothenberg EV, Yui MA, Telfer KC. T-cell developmental biology. In: William E.Paul, ed. **Fundamental Immunology.**5th Philadelphia: Lippincott Williams & Wilkins;2003:259-301[5] Nguyen D, Diamond L, Braylan R. **Flow cytometry in Hematopathology. A visual approach to data analysis and interpretation.** 2003; Totowa: Humana Press.[6] Paredes-Aguilera R, Romero-Guzman L, Lopez-Santiago N, et al. Flow cytometric analysis of cell-surface and intracellular antigens in the diagnosis of acute leukemia. **Am J Hematol.** 2001;68:69-74.[7] Vernau W, Moore PF. An immunophenotypic study of canine leukemias and preliminary assessment of clonality by polymerase chain reaction. **Vet Immunol Immunopathol.** 1999;69:145-164.[8] Burnett RC, Vernau W, Modiano J, et al. Diagnosis of canine lymphoid neoplasia using clonal rearrangements of antigen receptor genes. **Veterinary Pathology.** 2003;40:32-41.[9] Pilozzi E, Pulford K, Jones M, et al. Co-expression of CD79a (JCB117) and CD3 by lymphoblastic lymphoma. **J Pathol.** 1998;186:140-143.[10] Pilozzi E, Muller-Hermelink HK, Falini B, et al. Gene rearrangements in T-cell lymphoblastic lymphoma. **J Pathol.** 1999;188:267-270.[11] Appelbaum FR, Sale GE, Storb R, et al. Phenotyping of canine lymphoma with monoclonal antibodies directed at cell surface antigens: classification, morphology, clinical presentation and response to chemotherapy. **Hematol Oncol.** 1984;2:151-168.[12] Greenlee PG, Filippa DA, Quimby FW, et al. Lymphomas in dogs. A morphologic, immunologic, and clinical study. **Cancer.** 1990;66:480-490.[13] Fournel-Fleury C, Ponce F, Felman P, et al. Canine T-cell lymphomas: a morphological, immunological, and clinical study of 46 new cases. **Vet Pathol.** 2002;39:92-109.[14] Culmsee K, Simon D, Mischke R, Nolte I. Possibilities of flow cytometric analysis for immunophenotypic characterization of canine lymphoma. **J Vet Med A Physiol Pathol Clin Med.** 2001;48:199-206.[15] Lai R, Juco J, Lee SF, Nahirniak S, Etches WS. Flow cytometric detection of CD79a expression in T-cell acute lymphoblastic leukemias. **Am J Clin Pathol.** 2000;113:823-830.[16] Attygalle AD, Diss TC, Munson P, et al. CD10 expression in extranodal dissemination of angioimmunoblastic T-cell lymphoma. **Am J Surg Pathol.** 2004;28:54-61.[17] Yao X, Teruya-Feldstein J, Raffeld M, Sorbara L, Jaffe ES. Peripheral T-cell lymphoma with aberrant expression of CD79a and CD20: a diagnostic pitfall. **Mod Pathol.** 2001;14:105-110.[18] Thomas R, Smith KC, Gould R, et al. Molecular cytogenetic analysis of a novel high-grade canine T-lymphoblastic lymphoma demonstrating co-expression of CD3 and CD79a cell markers. **Chromosome Res.** 2001;9:649-657.[19] Wilkerson, M. J., K. Dolce, T. Koopman, W. Shuman, R. Chun, L. Garrett, L. Barber, and A. Avery. 2005. Lineage differentiation of canine lymphoma/leukemias and aberrant expression of CD molecules. *Vet Immunol Immunopathol* 106:179-196.[20].Kyoda K, Nakamura S, Matano S, Ohtake S, Matsuda T. Prognostic significance of immunoglobulin heavy chain gene rearrangement in patients with acute myelogenous leukemia. **Leukemia.** 1997;11:803-806.[21]. Schmidt CA, Przybylski GK. What can we learn from leukemia as for the process of lineage commitment in hematopoiesis? **Int Rev Immunol.** 2001;20:107-115.[22] Gorczyca W, Weisberger J, Liu Z, et al. An approach to diagnosis of T-cell lymphoproliferative disorders by flow cytometry. **Cytometry.** 2002;50:177-190.[23] Villalobos A. Recent VCS focus on lymphoma. **Veterinary Practice News.** 2002.[24] Yang L, Zhao H, Li SW, et al. Gene expression profiling during all-trans retinoic acid-induced cell differentiation of acute promyelocytic leukemia cells. **J Mol Diagn.** 2003;5:212-221.

VETERINARY ONCOLOGY SCREENING TESTS & BIOMARKERS-WHERE'S THE EVIDENCE?

Carolyn J. Henry, DVM, MS, DACVIM (Oncology)
Columbia, MO

## INTRODUCTION

A discussion of screening tests available in veterinary oncology is predicated upon an understanding of the very term, "screening test". According to the National Cancer Institute (http://cancernet.nci.nih.gov), screening refers to "checking for disease when there are no symptoms" By this definition, the benefits of cancer screening are evident. Cancer detection in the asymptomatic individual would be anticipated to prevent cancer deaths and reduce the morbidity associated with treating patients with advanced disease. In human oncology, in order for a screening test to be considered effective, it must meet at least two requirements: 1) it must permit detection of the cancer prior to development of symptoms and 2) there must be evidence that treatment initiated earlier as a consequence of screening results in an improved outcome.[1] By these standards, veterinary oncology currently has very little to offer in terms of cancer screening. However, advances in the field of biomarker discovery have led to development of new candidate screening tests that are currently under evaluation. Some of the most promising biomarker discovery tools include metabolomics (the study of metabolite expression), transciptomics (the study of RNA transcript expression), and proteomics (the study of protein expression patterns).[2] The latter has been the subject of intense interest in veterinary medicine in the last decade, with recent publications suggesting that proteomics will provide a means to distinguish cancer-bearing dogs from those without disease.[2-7] This session is intended to review the scientific evidence supporting the use of screening tests in veterinary oncology, discuss the limitations and correct application of such tests, and encourage dialogue and collaboration amongst veterinary oncologists and others in the field that will ulitmately lead to clinically relevant advances in veterinary cancer biomarker discovery.

## CURRENT STATE OF VETERINARY CANCER SCREENING TESTS

A review of the US and European veterinary medical literature reveals only a handful of manuscripts related to cancer screening tests. Most of these address proteomic technologies to identify biomarkers that may differentiate dogs with lymphoma from unaffected dogs. The first, initially presented at the 2006 Veterinary Cancer Society Annual Conference, described a study in which Gaines, et al. utilized surface-enhanced laser desorption-ionization time-of-flight mass spectrometry (SELDI-TOF-MS) to identify unique biomarker peaks in serum from 29 dogs with lymphoma compared to 87 control dogs.[2] A classification tree was then built on the basis of three of these protein peaks. The best individual serum biomarker peak had a reported 75% sensitivity and 86% specificity. The classification tree had a reported 97% sensitivity and 91 % specificity for the classification of B-cell lymphoma.

In a 2009 report[3], investigators from PetScreen, Ltd. describe the use of similar techniques to develop their canine lymphoma screening test that is now commercially available. The investigators tested serum samples from 92 control dogs and 87 dogs with lymphoma collected at veterinary practices in the US. They subjected the serum to ion exchange chromatography and analyzed the samples via SELDI-TOF-MS. They identified 19 serum protein peaks that differed significantly ($P<0.05$) between the two groups based upon their normalized ion intensities. From these peaks, two biomarkers were identified that were capable of differentiating lymphoma patients from non-lymphoma patients. Data analysis indicated a positive predictive value (PPV) of 82%. Next, a clinical follow-up study was conducted, enrolling 96 canine patients suspected of having lymphoma. Data from the latter study indicate a specificity value of 91%, sensitivity of 75%, PPV of 80% and negative predictive value of 88%. The specific proteins responsible for the biomarker peaks were not identified in either of the above studies; thus, it is not known if the unique peaks associated with lymphoma patients were related to the same proteins in both studies. A proteomic evaluation of lymph node tissue from dogs with lymphoma (n=11) was described by McCaw, et al. in 2007.[6] In this study, protein expression in lymph node extracts of dogs with multicentric B-cell lymphoma was compared to that of control lymph nodes (n=13)from unaffected dogs. Twenty-one proteins were found to be differentially expressed between the lymphoma and control samples. Using matrix-assisted laser desorption-ionization time-of-flight mass spectrometry MALDI-TOF-MS, the investigators characterized differentially expressed proteins and found that the expression of prolidase (proline dipeptidase), triosephosphate isomerase and glutathione S-transferase was decreased in lymphoma samples, whereas macrophage capping protein was up-regulated. Prolidase and triosephosphate isomerase have been previously implicated in breast and pancreatic cancer.

An optical cancer screening and monitoring technique was recently patented for commercialization based on a pilot study of canine serum samples.[8] The technique involves measuring serum concentrations of albumin, globulins, and hemoglobin by fitting the near UV absorbance and photoluminescence spectra of diluted serum as a linear combination of component reference spectra. Serum protein levels were also tracked over the course of treatment to determine if they could help monitor patient immune response to carcinoma and therapy. When protein assessment was applied as a cancer screening test, 70% of affected dogs had both suppressed serum hemoglobin levels (below 20 mg/dL) and atypical serum protein compositions (i.e, albumin and globulin levels outside reference range). Of the dogs that met the criteria established by the investigators for hemoglobin, albumin, and globulin levels, 20% had a false positive result with this cancer screening test.

Other veterinary cancer screening tests include the Veterinary Bladder Tumor Antigen test (V-BTA, Bion Diagnostics) discussed in more detail below, as well as more general health screens including the ERD-HealthScreen Canine Urine Test (Heska) to detect microalbuminuria and the alphafetoprotein (AFP) and cancer antigen 15/3 tests that differ between dogs with neoplastic disease and healthy dogs or those with non-neoplastic disease. In order to accurately assess the value of any of these tests in clinical practice, it is important to understand the intended application of each test and to choose the appropriate population in which to prospectively evaluate test performance.

**APPLICATION OF SCREENING TESTS TO CLINICAL PRACTICE**

In the context of a screening test being defined as a method to facilitate disease detection in the asymptomatic patient, the veterinary community has had a decidely poor track record in correctly applying such tests to clinical practice. A classic example is the V-BTA test first evaluated and reported in 1999.[9] The initial report cited 90% sensitivity (likelihood of detecting TCC in affected dogs) and 78% specificity (likelihood of a negative test result in a dog that is free of TCC). In a subsequent multi-institutional study, the author and others evaluated over 250 samples with the V-BTA and reported similar results.[10] The latter study included two control populations: one group of healthy controls and one with non-neoplastic urologic disease. The test performed poorly in the population with non-neoplastic urologic disease (specificity = 41%), with false positive results occurring when samples contained blood, protein, or glucose. However, false negative results were unlikely. As such, the test has a high negative predictive value; ie, a high proportion of the dogs that test negative in the screen truly do not have the disease. The term, "**SnNOut**" has been coined to help one remember that with a **Sen**sitive test, a **N**egative result can rule **Out** a diagnosis.[11] Therein lies what is perhaps the most valuable aspect of the V-BTA test. Unfortunately, the response of many veterinarians to the reported results was to declare the test to be of no clinical value because hematuria and proteinuria are often what prompts one to assess patients for TCC. In such cases, a positive test result would not discriminate dogs with neoplastic disease from those with other urinary tract disease. However, if applied in the strictest sense of the word, "screening", the V-BTA test would be used to evaluate an at-risk population <u>before</u> they develop clinical signs of disease. For example, using the test as part of an annual wellness exam to screen geriatric Scottish terriers for TCC is an entirely reasonable application. Unfortunately, the inappropriate use of the V-BTA test in young animals that present with hematuria has lead to disappointing results (high rate of false positives) and the misconception that the test has no value. Because the strengths and weakness of this test have been misunderstood, we have not taken full advantage of its value as a screening test for high risk breeds. This is but one example of the veterinary community rejecting a screening test based upon a misinterpretation of how it should be used clinically. The term, "**SpPIn**" can be used as a reminder that for tests with high **Sp**ecificity, a **P**ositive result can rule **In** a diagnosis.[11] With its specificity of only 41% for dogs with signs of urologic disease, the V-BTA test is clearly not useful for ruling <u>in</u> disease. Again, it is important to recognize that a screening test is ideally used to test a population with a high prevalence of the disease. Furthermore, if a positive test result will prompt invasive diagnostic procedures, then the test should have high sensitivity. Positive predictive value and, to a lesser degree, negative predictive value of a test depend upon the test specificity and the prevalence of the disease for which one is testing. It is ill-advised to take sensitivity and specificity values out of context, as they are not intrinsic properties of a given test. In order to interpret sensitivity and specificity data and to honestly evaluate test performance, the nature of the test, the population tested, and the disease itself must be well-defined.[12]

Other factors required for the critical evaluation of a new screening test are biases that cause the test to appear more promising than it actually is. These biases include length bias, selection bias, and lead-time bias. ***Length bias*** refers to when a screening test facilitates detection of cancers that are less aggressive or more slowly growing.[1] Patients with slow-growing occult cancers are likely to have a more favorable outcome than those patients with overt clinical signs that caused their cancer to be detected between planned screenings. Thus, the improved outcome in the former group may be incorrectly attributed to the early detection afforded by the screening test when it is actually due to a difference in tumor behavior. An example would be if a screening test detected indolent lymphoma in a dog without clinical signs and one compared outcome in this dog to a dog presenting with T-cell lymphoma that was detected due to clinical lymphadenomegaly and signs of hypercalcemia. One might attribute the improved survival for the dog with indolent lymphoma to the fact that the disease was detected early via cancer screening. However, the true reason for the improved outcome is more likely the inherent differences in the disease, rather than a benefit of early detection. ***Selection bias*** occurs when the screened population is not representative of the overall population. For example, very conscientious clients may elect to have screening tests done as part of their pets' annual wellness exams. If survival rates are then compared between this population and one that has not been routinely screened for disease, the population undergoing regular screening is likely to fare better. They may have a decreased overall mortality rate compared to the non-screened population, but the underlying reason for the difference may be the better overall health care of the screened population, rather than to the impact of the screening test. Finally, ***lead–time bias*** is a confounding factor that has been reported for human screening tests. Lead-time bias occurs when an outcome such as survival time is compared between those individuals with disease that was detected via screening and that of individuals with disease that was detected only after the development of clinical signs.[1] The time from diagnosis to death is increased in the screened patients, but this is not due to an actual prolongation in life. Screened patients, then, do not really live longer, but are simply diagnosed earlier and have their "survival clock" started sooner. This, by definition, does not fulfil the previously stated requirement of a screening test that earlier treatment equates to longer survival. While it is important to consider these potential biases, it is also essential not to allow anecdotal reports or assessments based on insufficient patient numbers to bias our decision of whether or not to promote the development and use of screening tests in veterinary medicine.

**POTENTIAL PITFALLS OF CANCER SCREENING**

A variety of screening tests are used routinely in human medicine, including mammography to detect breast cancer and sigmoidoscopy for detection of colon cancer. In both instances, the utility of the screening depends upon the population being screened, especially with regard to patient age and risk factors for the disease. The same should be considered when applying screening tests to the companion animal population. This cannot be overemphasized because there are a number of potential pitfalls relative to inappropriate cancer screening. First and foremost is when a false positive result leads to unnecessary testing. The aforementioned example of using the V-BTA test in an inappropriate screening population could result in unnecessary and expensive testing in a patient that is much more likely to have non-neoplastic urinary tract disease. Another potential pitfall of cancer screening is when a false negative test result leads to inappropriate reassurance that cancer is not present and, thus a delay in eventual diagnosis.

This is of most concern for cases in which delayed diagnosis will have a serious negative impact on case outcome. A third potential pitfall of screening tests is when the testing procedure itself poses a significant risk to the patient. An example of this in human medicine is when complications occur secondary to endoscopic biopsy in a patient who was not at significant risk for having occult intestinal malignancy. Obviously, the risk of any invasive procedure must be considered in light of the likelihood that the procedure will detect the disease—especially with diseases for which early detection will result in significant improvement in outcome. As cancer screening methods evolve to include less invasive and more refined techniques, this pitfall of cancer screening may be diminished. Lastly, the risk of overdiagnosis is a very real concern related to cancer screening. One relevant example of this in human medicine is the overdiagnosis of prostate cancer in men since the implementation of regular screening via PSA testing and digital rectal examination. It has been estimated that over 30% of localized prostate cancer diagnosed as a result of screening tests is clinically insignificant, resulting in overtreatment of a disease that is unlikely to be the reason for death in affected men.[13] A veterinary corollary to this example is when ultrasound evaluation of the abdomen is used to screen for occult metastatic disease in an animal with a known primary malignancy. If such imaging reveals an 'incidentalonoma', the clinician and client are then faced with the difficulty of deciding whether this ancillary finding warrants separate diagnostic evaluation. Given the large proportion of dogs that are noted on necropsy to have had occult malignancy in the abdomen at the time of death due to another cancer, one might question the clinical value of detecting such lesions.[14,15]

## BEYOND SCREENING—OTHER APPLICATIONS FOR BIOMARKER TESTING

In addition to the potential value of biomarkers for early detection of cancer, patient monitoring and case management decisions may be facilitated by identification of tumor-specific biomarkers. An example is the use of tests such the V-BTA test and the PetScreen lymphoma test to monitor remissions status and determine when re-initiation of therapy is warranted. If found to be accurate predictors of remission status, such tests may negate the need for more expensive diagnostics such as abdominal ultrasound or thoracic radiography to assess remission status. Likewise, the ability to predict loss of remission before it is clinically evident would facilitate earlier medical intervention that my improve outcome. There is inherent difficulty in assessing tests for this purpose in that a gold standard by which to validate each test may not exist. For example, there is currently no gold standard by which to confirm complete remission in dogs with lymphoma. Thus, if the PetScreen test is utilized and results are positive, yet one is of the clinical impression that the dog is in remission, which assessment is accurate and how can we determine this? Polymerase chain reaction for antigen receptor rearrangements (PARR) testing is one candidate test that has the potential to "make the call" in such instances. The test is currently available at Colorado State University and was evaluated as a means to detect canine lymphoma cells in a study reported by Lana, et al. in 2006.[16] Although the purpose of their study was to evaluate the utility of PARR for molecular staging and predicting prognosis in canine lymphoma patients, the data provides useful information regarding the sensitivity of PARR testing for canine lymphoma. The study included 86 dogs with histologically or cytologically confirmed lymphoma. PARR assays of a representative tumor-infiltrated lymph node and peripheral whole blood were performed in all cases. Sixty-two patients had clonal rearrangements that were detected in the lymph node and these dogs were subsequently staged by PARR. Of these, 45 (73%) had detectable tumor cells in the blood by PARR assay. Eighty percent of the dogs with Stage 3 lymphoma were found to be blood PARR-positive, suggesting that the PARR is more sensitive for detecting circulating tumor cells than visual assessment of blood smears or bone marrow. However, these results also suggest that PARR testing is not necessarily a "gold standard" in that 17 of 65 dogs with confirmed lymphoma were PARR-negative.[16] It is likely that we will continue to move away from the concept of applying one gold standard across the board for screening and patient monitoring, but rather will customize testing by identifying so-called "molecular fingerprints" that can discriminate between affected an unaffected individuals.

## FUTURE DIRECTIONS

In order to determine the clinical utility and appropriate application of screening tests in veterinary oncology, prospective and unbiased evaluation is essential. Unfortunately, the skepticism present amongst the veterinary community, along with a tendency to incorrectly apply screening tests to the companion animal population has historically created a veterinary market that has been unattractive to commercial organizations considering development of these tests. An uncooperative and, at times, even adversarial posture toward those hoping to develop cancer screening tests for companion animals has left us with little to offer our clients in terms of early diagnosis of many cancers. Given that the late stage of disease recognition is a major impediment to curative therapy, it would seem prudent to promote the development and testing of effective cancer screening tools for veterinary medicine.

As we look to the future of cancer screening tests, it is evident that comparisons of molecular fingerprints between affected and unaffected individuals will be key to discovering appropriate biomarkers of disease. Whereas biomarkers have in the past been used as individual tests to provide a yes/no answer with regard to disease presence, it is now apparent that single biomarkers are unlikely to effectively discriminate disease state. Simply put, most biomarkers are not unique to cancer tissue and, thus, cannot reliably distinguish normal from diseased states. Rather, multiple biomarkers that create a molecular phenotype may be identified through bioinformatics and used to recognize patterns unique to patients with occult cancer.[5] We are on the cusp of an important era that will change our approach to diagnosis and case management of veterinary cancer patients. It is essential that we work collectively to identify, refine, and validate biomarkers so as to maximize their positive impact on the field of veterinary oncology.

## REFERENCES

[1]Brawley OW, et al. **J Clin Oncol** 2005;23:293; [2]Gaines PJ, et al..**AJVR** 2007;68:405 [3]Ratcliffe L, et al. **Vet Comp Oncol** 2009 (in press); [4]Matharoo-Ball B, et al **Vet Comp Oncol** 2008;6:224; [5]Mian S, et al. **EJCAP** 2006;16:63; [6]McCaw DL, et al. **Clin Cancer Res** 2007;13:2496; [7]Doherty MK, et al, **Proteomics-Clinical Appl** 2008;2:135; [8]Dickerson BD, et al. **Applied Optics** 2007;46:8080; [9]Borjesson DL, et al. **Vet Clin Pathol** 1999;28:33; [10]Henry, CJ, et al. **AJVR** 2003;64:1017; [11]Lamb CR, **Vet Radiol & US** 2007;48:486; [12]Lamb CR, **Vet Radiol & US** 2007;48:189; [13]Etzsioni R, et al. **J Natl Cancer Inst** 2002;94:981; [14]Snyder JM, et al. **JVIM** 2006; 20:669;[15]Vos JH, et al. **Vet Quarterly** 2005;27:2; [16]Lana SE, et al. **JVIM** 2006;20:329.

# AUTOLOGOUS PERIPHERAL BLOOD STEM CELL TRANSPLANTATION FOR THE TREATMENT OF CANINE LYMPHOMA

**Steven E. Suter VMD, MS, PhD DACVIM (Oncology)**
**Raleigh, NC**

## INTRODUCTION

Canine lymphoma is one of the most common malignant tumors of dogs (up to 25% of all cancers) and it is the most common hematopoietic tumor of the dog (83%)[1,2]. The incidence of lymphoma has been reported to approach 2.4% in susceptible older animals, with an annual incidence rate of 1.5 per 100,000 for dogs less than 1 year of age and 84 per 100,000 dogs in the 10 to 11 year old group[3]. The notion that the incidence of canine lymphoma is increasing is supported by a report documenting an increasing caseload of canine lymphoma patients presenting to 20 U.S. veterinary institutions from 1987 to 1997[4]. In the veterinary literature, canine high grade B-cell lymphoma reportedly accounts for approximately 70% to 75% of canine cases[5-7]. The treatment of choice typically involves sequential multi-drug chemotherapy using CHOP or L-VCAP based protocols. The median survival time of T and B-cell canine lymphoma patients is 7-9 months and 12-14 months, respectively, with an expected 2-yr. survival of ~25% and an overall cure rate of less than 5%[8,9] when considering B-cell lymphoma. Death is uniformly the result of recrudescent disease that is resistant to conventional chemotherapeutic drugs. Although not definitively proven, it is now generally accepted that canine nodal B-cell lymphoma is a heterogeneous population of B-cell neoplasms, mirroring the human non-Hodgkin's lymphoma (NHL) experience[5,10].

The same drugs have been used to treat lymphoma for approximately 30-40 years with no increase in either cure rates or overall survival since the introduction of Adriamycin in the early 1970s, and there are no new drugs currently in the developmental pipeline. Additionally, the toxicity of most currently used chemotherapeutics severely limits our ability to cure this disease. Although lymphoma is extremely common, there is a surprising paucity of veterinary literature describing the development of novel treatments for this disease, which clearly underscores the need for relevant research in this area.

Bone marrow transplantation (BMT) has become an important component in the treatment of human lymphoma, leukemia, and other hematologic malignancies, as well as some other non-malignant bone marrow diseases. Originally, bone marrow transplantation was developed to allow dose intensification of myelosuppressive anticancer therapies including high-dose chemotherapy and lethal total-body radiation (TBI). In these initial studies, normal bone marrow cells were harvested from a patient before intensive myelosuppressive therapy was given. The harvested cells, which contain a population of pluripotent stem cells, were then reinfused into the patient in an effort to "rescue" the patient from complete bone marrow collapse and ultimate death. There is a variety of human literature showing that these dose intense schemes followed by BMT have a significant anticancer effect and an overall increase in patient survival.

For close to fifty years, dogs have served as an excellent, large-animal model for human hematopoietic cell transplantation (HCT).[11] Based on a huge body of literature, dogs have directly contributed to the development and optimization of HCT protocols for human patients with a variety of hematologic and nonhematologic malignancies. As early as 1979, it was shown that dogs with lymphoma who were treated with an aggressive chemotherapy protocol followed by bone marrow harvesting, lethal TBI, and re-infusion of the harvested marrow enjoyed a significantly longer median disease-free survival than dogs receiving chemotherapy alone.[12] In fact, over 20% of these dogs remained alive in an unmaintatined complete remission 1-2 years after TBI. In a subsequent paper by the same group, 30% of the dogs maintained complete clinical remission 200-663 days after TBI.[13] This same group, mostly working at the Fred Hutchinson Cancer Research Center (FHCRC) in Seattle, WA, have gone on to publish a wide variety of reports using various chemotherapy protocols and radiation schemes showing that at least 20-30% of dogs with lymphoma can be long-term survivals when treated with a combination of chemotherapy, TBI, and bone marrow transplantation.

With continuing development and refinement of BMT protocols, it became clear that hematopoietic stem cells (HSC) and other committed hematopoietic progenitors are present at very low numbers in the peripheral blood of a variety of mammalian species, including people, mice, baboons, and dogs.[14-17] These cells, called peripheral blood stem cells (PBSC), were originally shown to contribute to bone marrow reconstitution of all the blood cell lineages when infused into myeloablated mice. However, the large-scale collection of enough PBSC to transplant a human patient was not possible during this time. It wasn't until the 1990s that many human hematopoietic growth factors were cloned and produced in quantities large enough for clinical use. These powerful factors facilitated the recruitment of large numbers of hematopoietic stem and progenitor cells from the bone marrow into the peripheral blood. This, in conjunction with the production of monoclonal antibodies against hematopoietic progenitor cells, namely CD34 antibodies, permitted the development of a technology whereby sufficient numbers of peripheral blood stem cells could be isolated for human transplantation. A number of canine cytokines, including recombinant canine G-CSF (rcG-CSF)[18,19], stem cell factor (rc-SCF)[18,19], and granulocyte macrophage colony-stimulating factor (rc-GM-CSF)[20] were also produced, along with an anti-canine CD34 monoclonal antibody.[21] Again, dogs were instrumental in developing the technology needed for the large-scale collection of these cells for what is now known as peripheral blood stem cell transplantation (PBSCT).

Leukaphoresis, the technique used to harvest stem cells from the peripheral blood, was originally reported in dogs in 1967 whereby circulating stem cells were transplanted to evaluate granulocyte therapeutic support.[22] A variety of leukaphoresis techniques were developed and many preclinical studies showed that both autologous[23] (stem cells collected from the patient) and allogeneic[24] (stem cells collected from a matched donor) PBSCT led to the reconstitution of normal hematopoiesis in dogs who received otherwise lethal TBI. It was also established during this time that approximately 10 times more PBSC were needed to induce hematopoietic cell lineage regeneration and subsequent stable engraftment.[25] With the development of highly automated microprocessor-controlled blood separators (leukaphoresis machines), which provide continuous-flow peripheral blood removal and separation into component parts via centrifugation on the basis of specific densities, a variety of cells, including platelets, granulocytes, and mononuclear cells, can be isolated from the peripheral blood, with the remainder of the cells being returned to the circulation.[26] These leukaphoresis machines are now routinely used in humans for platelet collections, whose product is usually given to thrombocytopenic patients who have received chemotherapy, and peripheral mononuclear cell collections, whose product is used for PBSCT. Leukaphoresis machines can also be used with canine patients to harvest both platelets[27] and PBSC.[28]

Based on the extensive data describing the cure of dogs with lymphoma using chemotherapy, TBI, and BMT, and the finding that canine CD34+ stem cells can be harvested using human leukaphoresis machines, the stage has been set to begin to develop treatment protocols based on PBSCT. In one report, a 7-year-old golden retriever with T-cell lymphoma was given an allogeneic PBSCT using cells harvested from a dog leukocyte antigen (DLA)-matched littermate.[29] In this report, the donor dog received rcG-CSF for 6 days prior to leukaphoresis and the harvested mononuclear cells, which contained ~0.4% CD34+ hematopoietic stem cells, were infused into the recipient immediately after TBI at a total dose of 3.6 x $10^6$ CD34+ cells/kg. Complete hematopoietic engraftment, as manifested by sustained increases in granulocyte and platelet counts after the post radiation nadir, was documented. This study was reported in 2006, and the dog is still alive at this writing. Approximately 6 other dogs have undergone PBSCT using this protocol (Dr. Ed Sullivan, Bellingham Veterinary & Critical Care, Seattle, WA, personal communication), with complications only arising in one dog post-transplant. All of these dogs are currently clinically normal and living normal lives at home.

Through a generous donation from the Mayo Clinic in Rochester, MN, NC State College of Veterinary Medicine was able to obtain 3 Baxter-Fenwal leukaphoresis machines in February, 2007. Training and certification was completed shortly thereafter using cows blood. Next, using live dogs, we were also able to show that Neupogen, which is recombinant human G-CSF, cross-reacted with canine bone marrow cells and dramatically increased peripheral white blood cell counts. We were then able harvest peripheral blood CD34+ stem cells in sufficient quantities to consider using PBSCT for the treatment of dogs with hematologic malignancies. Based on these initial studies, the Canine Bone Marrow Transplant Unit began accepting cases in October, 2008 (http://www.cvm.ncsu.edu/vth/clinical_services/onco/BoneMarrowTransplant.html) for the treatment of lymphoma, and possibly leukemia and multiple myeloma. At this writing, we have treated 3 dogs with lymphoma and 1 dog with acute lymphoblastic leukemia. Based on extensive human data which describes an ~50%-66% cure rate of patients with non-Hodgkin's lymphoma using autologous transplantation, we expect canine lymphoma cure rates to be similar. The treatment protocol we use for these patients, which includes chemotherapy, TBI, and autologous BMT, is based on the work of Dr. Ranier Storb at the FHCRC and Dr. Ed Sullivan, who has performed these transplants in a private practice setting while working closely with Dr. Storb. We are currently accepting dogs with both B and T-cell lymphoma that weigh more than 17 kgs. It is preferable that these patients are not ill, and are in complete remission.

## BMT PROTOCOL

1. Induction chemotherapy using any multi-agent chemotherapy protocol
2. High-dose Cytoxan (500mg/$m^2$) IV with Mesna and Lasix
3. Neupogen @ 5ug/kg SQ BID staring 1 week after Cytoxan for 6 days
4. Leukaphoresis to harvest peripheral blood CD34+ stem cells on d6 of Neupogen
5. CD34+ cell enumeration-target is 2 x $10^6$ CD34+ cells/kg-cells refrigerated overnight
6. Total body radiation @ 2 x 5Gy with a 3 hour interfraction interval the following day
7. Infusion of the harvest product immediately after TBI
8. Supportive care in a semi-isolation ward for ~2 weeks after TBI-
    a. GI side-effects typically 3-5 days post-TBI
    b. Neutrophil nadir typically 7-10 days post-TBI
    c. Platelet nadir typically 10-12 days post-TBI
    d. Patients are discharged when neutrophils are > 1,000/ul for 3 consecutive days and platelets are >25,000/ul.

This talk will emphasize the major points of PBSCT in dogs and discuss the various toxicities we have seen with TBI. Our preliminary results, including patient management, hematopoietic engraftment kinetics, and owner's perception of the procedure for our first six patients will also be discussed.

## REFERENCES

[1]Kaiser H. **Animal Neoplasms: A systemic review** 1981;747. [2]Moulton J. **Tumors of lymphoid and hematopoietic tissue.** Berkeley: Univeristy of California Press; 1990. [3]Dorn CR, et al. **Bibl Haematol** 1970;403. [4] Withrow SJ, et al. **Small Animal Clinical Oncology**, 3rd ed., Philadelphia, W.B. Saunders, 2001. [5]Fournel-Fleury, et al. **J Comp Pathol** 1997;117:35. [6]Wilkerson MJ, et al. **Vet Immuno Immunopathol** 2005;106:179. [7]Culmsee K, et al. **J Vet Med A Physiol Pathol Clin Med** 2001;48:199. [8]Piek CJ, **Vet Q** 1999;21:44. [9] Valerius KD, et al. **J Am Vet Med Assoc** 1997;210, 512. [10]Valli VE, et al. **Vet Pathol** 2006;43:241. [11]Lupu M, et al. **Vet Comp Oncol** 2007;5(1):14. [12]Weiden PL, et al. **Blood** 1979;54(5):1039. [13]Weiden PL, et al. **Exp Hematol** 1979;Suppl 5:160. [14]Goodman JW, et al. **Blood** 1962;42:19.
[15]McCredie KB, et al. Science 1971;171:93. [16]Debelak-Fehir KM, et al. **Exp Hematol** 1975;3:109. [17]Storb R, et al. **Blood** 1977;50:537. [18]de Revel T, et al. **Blood** 1994;83:3795. [19]Sandmaier BM, et al. **Blood** 1996;87:3508. [20]Nash RA, et al. **Blood** 1994;83:1963. [21]McSweeney PA, et al. **Blood** 1998;91:1977. [22]Storb R, et al. **Blood** 1967;30:806. [23]Cavins JA, et al. Blood 1964:23:38. [24] Storb R, et al. **Blood** 1967;30:805. [25]Appelbaum FR, et al. **Exp Hematol** 1979;7(Suppl 5):7. [26]McLeod BC, et al. Apheresis: Principles and Practice. Bethesda, MD: AABB Press;2003. 27Callan B, et al. [28]Lupu M, et al. **J Vet Intern Med** 2008;22:74. [29]Lupu M, et al. **JAVMA** 2006;5:728.

**PROSPECTIVE EVALUATION OF CLINICALLY RELEVANT TYPE B HYPERLACTATEMIA IN DOGS WITH CANCER.**

**M Touret, S Boysen, ME Nadeau.**
**University of Montreal, St Hyacinthe, Canada.**

Cancer is considered a cause of type B hyperlactatemia in dogs. However, studies evaluating cancer as a cause of clinically relevant type B hyperlactatemia (>2.5 mmol/L) are lacking. It is well accepted that cancer cells have a higher lactate production due to increased aerobic glycolysis, known as the "Warburg effect". The mechanisms through which aerobic glycolysis occurs are not well elucidated but it has been suggested that neoplasia may cause type B hyperlactatemia via this process.

The aim of our study was to determine if canine malignant tumors could be associated with a clinically relevant type B hyperlactatemia (> 2.5 mmol/L).

Thirty-three dogs with malignant tumors were included: 20 with hematopoietic and 13 with solid tumors. Histology was used to confirm the diagnosis (cytology was considered appropriate for lymphoma). Confounding factors associated with hyperlactatemia were excluded. Lactate measurements were obtained from a free flow jugular whole blood sample and immediately analysed using the LactatePro®.

All dogs had lactate values less than 2.5 mmol/L. The mean blood lactate concentration was 1.16 mmol/L ± 0.06 (mean ± SD). The mean blood lactate concentration for solid and hematopoietic tumor was 1.24 mmol/L and 1.02 mmol/L respectively. Dogs with lymphoma (n = 17) had a mean blood lactate concentration of 1.23 mmol/L.

Malignant tumors were not considered a cause of clinically relevant type B hyperlactatemia. Therefore, hyperlactatemia in cancer patients should be carefully investigated for causes other than cancer.

# A COMPARATIVE APPROACH TO METASTASIS BIOLOGY AND THERAPY.

**L Ren[1], S-H Hong[1], Joseph Briggs[1], M Paoloni[2], C Mazcko[2], A Mendoza, T Osborne[1], M Mendoza[1], M Ahluwalia[1], K Nuckolls[1,2] C Khanna[1,2],**
**Tumor and Metastasis Biology Section[1], Pediatric Oncology Branch, and Comparative Oncology Program[2], Center for Cancer Research, National Cancer Institute, Bethesda Maryland.**

The study of cancer requires in vivo modeling. It is unlikely, however, that this complexity can be entirely captured by one model alone. By understanding the strengths and weaknesses of a set of models, it becomes possible understand the limits of a model and to choose the appropriate model(s) for particular problems or questions. With this modeling approach in mind, our laboratory has focused its efforts on the problem of metastasis in the pediatric solid tumor, osteosarcoma. Despite successful control of the primary tumor and adjuvant chemotherapy, death from pulmonary metastases occurs in over 30% of ostetosarcoma patients within 5 years. A more aggressive course of disease is known to exist in pet dogs afflicted with the same cancer.

To define genetic determinants of pulmonary metastasis in osteosarcoma we have utilized a cross-species comparative approach (murine, canine, and human) for gene identification and evaluation. Using this comparative approach we have identified the cytoskeleton linker protein, ezrin, to be associated with metastasis. Our work suggests that ezrin is necessary for metastasis in murine transplantable osteosarcoma and genetically engineered rhabdomyosarcoma models, that it is relevant in human-murine sarcoma xenograft models, and that its expression is associated with metastatic progression in pet dogs with naturally occurring osteosarcoma, finally we have found an association between ezrin expression and risk of relapse in pediatric osteosarcoma patients. Through both in vitro and in vivo modeling we have recently determined that ezrin acts a biological node or nexus in the metastatic phenotype of cancer by facilitating signaling of many vital metastasis-associated pathways and by enabling the translational machinery of the cell. We hypothesize that the efficient delivery of proteins to specific locations in a cell at specific times is necessary for metastatic cells to overcome the numerous stresses experienced by cells during metastasis.

A significant hurdle exists in the translation of this information from the laboratory to the clinic. This hurdle is in part the result of shortcomings in conventional preclinical models of cancer therapy. To provide a unique and informative prospective on the translation of biology to therapy we have recently launched the Comparative Oncology Program (COP), of the National Cancer Institute. Its mission is to *provide an integrated mechanism by which naturally occurring cancer models can be used to generate new information about cancer, translate biological concepts towards clinical application and bring novel therapeutic options to the management of human cancers.* The goal of his approach will be improved outcomes for both canine and human cancer patients.

**RETROSPECTIVE EVALUATION OF PROGNOSTIC FACTORS IN DOGS WITH HISTIOCYTIC DISEASE.**

**Moore, AS, Frimberger AE, Sullivan N, Reppas G**
**Veterinary Oncology Consultants, NSW, Australia. Idexx Laboratories, NSW. Symbion Vetnostics, NSW.**

Splenic fibrohistiocytic nodules (SFHN) form a continuum between splenic lymphoid nodular hyperplasia, lesions of >70% lymphocytes with interspersed fibrohistiocytic cells (grade 1); to malignant splenic stromal neoplasms (histiocytic sarcoma), primarily fibrohistiocytic cells with <40% lymphocytes (grade 3). Histiocytic sarcoma is an uncommon disease. One report characterized tumors as localized histiocytic sarcoma if there was a history of initially single site involvement and metastasis to the regional lymph node; whereas systemic histiocytic sarcoma is described as widespread system involvement including lung and other organs.

Our study retrospectively evaluated the clinical course in 53 dogs with histiocytic disease to attempt to identify prognostic factors. 28 dogs were diagnosed with SFHN (2 grade 1, 8 grade 2, 18 grade 3) following splenectomy. An additional 25 dogs with histiocytic sarcoma affecting the subcutaneous tissues (18 dogs), or with systemic dissemination (7 dogs) were also evaluated.

Factors evaluated for prognostic significance by Kaplan-Meier and Cox-regression on survival time included; histopathologic criteria, abnormalities in hematology and serum biochemistry analysis (graded) and chemotherapy used. Additional factors examined included, for dogs with SFHN, lymphoid : fibrohistiocytic ratio and for dogs with histiocytic sarcoma, nodal metastases, systemic involvement and CD-18 staining.

With SFHN, there were 9 males (7 castrated) and 19 spayed females. Median age was 9 years and median weight was 26 kg. Median overall survival was 387 days. Those factors found to negatively alter survival were: lymphoid ratio (<40% = grade 3), % necrosis (<50%), male gender, and body weight (>20kg). Although thrombocytopenia and hypoalbuminemia were significant predictors of poor prognosis in another study of histiocytic sarcoma, they did not appear to affect survival in this study.

For histiocytic sarcoma, there were 12 males (9 castrated) and 13 spayed females. Median age was 8 years (both local and systemic forms) and median weight was 29 kg (localized) and 35kg (systemic). Of the 18 dogs with localized histiocytic sarcoma, 8 had normal sized regional nodes, and 7 had biopsy-confirmed metastasis at diagnosis.

The prognosis for dogs with SFHN is confirmed to be largely determined by the proportion of lymphocytes in the tumor; dogs with fewer than 40% lymphocytes have a poor prognosis, and chemotherapy is probably warranted, although the response rate appears to be low in this small group of dogs.

**RETROSPECTIVE EVALUATION OF CHEMOTHERAPY OUTCOMES IN DOGS WITH ATYPICAL SITE LYMPHOMAS.**

**Frimberger AE[1], Moore, AS[1]. Veterinary Oncology Consultants, NSW, Australia.**

Lymphoma is the most common hematologic malignancy in dogs and in general is reasonably well understood in terms of its natural history, prognosis, and responsiveness to treatment. Occasional dogs present with unusual anatomic variants ("atypical sites") and the impact of this presentation on prognosis with and without therapy is not well understood. Our goal is to better characterize the course of disease and responsiveness to treatment in these dogs. Cutaneous lymphoma, is divided into epitheliotropic (also called mycosis fungoides) and non-epitheliotropic forms. Epitheliotropic lymphoma is generally considered a difficult disease to treat, with variable response rates; a number of different treatments have been reported to have some success.

A search of the records of Veterinary Oncology Consultants and the Animal Referral Hospital from 2003-2008 yielded 28 cases of canine primary cutaneous lymphoma for which follow up information could be obtained. 14 dogs each were managed at the Referral Hospital and in a primary care setting. Factors evaluated for prognostic significance by Kaplan-Meier and Cox-regression on survival time included; sex, age, weight, mycosis fungoides stage (premycotic / mycotic / tumor), lymphoma substage (a or b), mucosal / mucocutaneous location versus cutaneous location, solitary or multiple lesions, and treatment used.

22 dogs were considered to have epitheliotropic lymphoma, whereas 6 had non-epitheliotropic disease. Of the non-epitheliotropic lymphomas, 4 were CD3+, 1 was CD79a+, and 1 was CD3-, but CD79a was not performed. There were 14 males (8 castrated) and 14 spayed females. Median age was 11 years (4 - 18) and median weight was 20 kg (6 - 56). Lesions had been present a median of 4 weeks (0 – 104). 4 dogs had premycotic lesions, 9 had mycotic, and 13 had tumor lesions; and 2 could not be determined from the records. 23 dogs were considered in lymphoma substage a and 5 dogs were considered in substage b. 19 dogs had cutaneous disease and 9 dogs had mucosal / mucocutaneous junction disease. 8 dogs had solitary lesions, 19 had multiple lesions, and 1 could not be determined from the record. 20 dogs received chemotherapy.

Median overall survival was 331 days. The only factor found to significantly affect survival on multivariate analysis was location, with mucosal / mucocutaneous location being more favorable (median survival 1070 days) than cutaneous (median survival 274 days). A trend was apparent for improved outcomes with solitary lesion compared to multiple lesions, although it was not statistically significant.

The prognosis for dogs with epitheliotropic lymphoma may not be as grave as has generally been thought. Mucosal location is a strong predictor of favorable outcome, and the presence of only a solitary lesion may be somewhat favorable. The best treatment option for this condition has yet to be determined.

## CHEMOTHERAPY & THE IMMUNE SYSTEM: OLD MYTHS & NEW DIRECTIONS

**Barb Biller, DVM, PhD, DACVIM**
**Fort Collins, CO.**

### INTRODUCTION

Chemotherapy has historically been viewed as immunosuppressive, yet a growing body of evidence clearly demonstrates multiple immunostimulatory effects of a number of commonly used chemotherapy drugs. Studies in mice and humans have found that administration of chemotherapeutic agents has variable effects on different components of the immune system. For example, administration of high doses of the drug cyclophosphamide (CYC) results in severe myelosuppression whereas low repeated doses of CYC augments immunity through a number of mechanisms (1-3). The immunomodulatory effects of other chemotherapeutic agents, such as doxorubicin (DOX) and the vinca alkaloids, are less clearly dose-dependent. These agents have both suppressive and stimulatory effects at equivalent doses, suggesting that factors other than dose determine their immunomodulatory potential (4-5). For example, the administration of DOX at doses known to cause neutropenia can also activate macrophages and natural killer (NK) cells. Interestingly, DOX and other anthracyclines such as mitoxantrone have recently been shown to kill tumor cells in such a way that the dying (or apoptotic) tumor cells themselves further stimulate a potent antitumor immune response.

Despite the promise offered by new treatment modalities for cancer such as tumor vaccines and other forms of immunotherapy, novel therapies appear most effective when used in combination with traditional treatments such as chemotherapy. Therefore, there is considerable interest in the identification of chemotherapy agents that can be combined with cancer immunotherapy and an implicit need to better understand the effects of chemotherapy on the immune system.

### THE EFFECTS OF CHEMOTHERAPY ON IMMUNE RESPONSES IN DOGS WITH CANCER

There have been a limited number of studies evaluating the impact of chemotherapy on immune responses in dogs with cancer; the majority of these studies have focused on assessment of lymphocyte numbers and antibody titers in dogs with lymphoma (6-8). In a recent study of dogs undergoing chemotherapy for lymphoma, for example, a > 50% reduction in numbers of total lymphocytes was observed. However, lymphocyte numbers and subsets were not evaluated in these dogs prior to the onset of chemotherapy making differentiation of the effects of chemotherapy on lymphocyte numbers from those of lymphoma impossible (6). To better understand how chemotherapy affects immunity in dogs with either lymphoma or osteosarcoma, we recently conducted a prospective study evaluating T and B cell numbers and humoral immune responses in dogs being treated with one of two different chemotherapy protocols (9). Numbers and subsets of T and B lymphocytes were assessed by flow cytometry and antibody responses to *de novo* vaccination with a novel antigen were determined before, during, and after chemotherapy. When compared to healthy, age-matched control dogs, dogs with either lymphoma or osteosarcoma were found to have significantly fewer CD4+ and CD8+ T cells <u>prior</u> to administration of chemotherapy. This data is presented below in **Figure 1**.

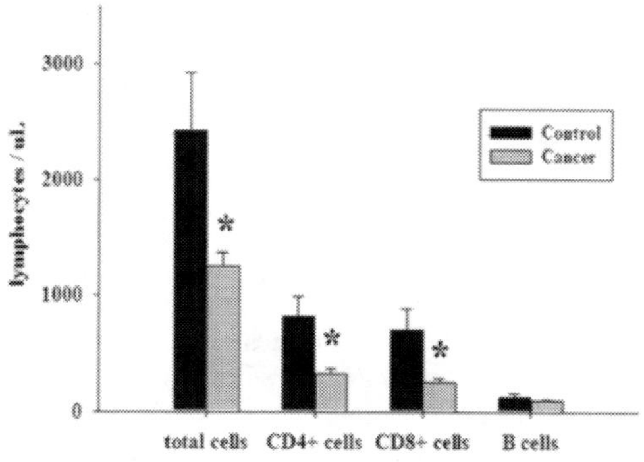


**Figure 1:** Total number of CD4+ and CD8+ T-cells and B-cells in healthy dogs and in dogs with cancer

Evaluation of lymphocyte subsets over time showed that treatment with single-agent DOX did not cause a significant decline in T or B cell numbers compared to pre-treatment values. This was true whether the dogs had lymphoma or osteosarcoma. For dogs with lymphoma treated with combination chemotherapy (using a CHOP-based protocol), we found that although CD8+ T cell numbers decreased one week after the first treatment, they rebounded within one week and were not significantly different at any other time point. However, in dogs treated with CHOP, B cell numbers declined significantly after the first treatment and remained decreased throughout the 6-month period of study. Despite this change, antibody titers following vaccination with the KLH antigen were not significantly different between control and chemotherapy-treated dogs. These findings suggest that chemotherapy may have less impact on T cell numbers and the ability to mount an antibody response than previously appreciated.

**COMBINING CHEMOTHERAPY WITH IMMUNOTHERAPY IN VETERINARY CANCER PATIENTS**

One of the primary areas of interest in our laboratory is the synergistic combination of immunotherapy and chemotherapy in companion animals with cancer. We recently reported the investigation of a novel immune stimulant containing cationic liposomes and DNA complexes (LDC) for use as an allogeneic tumor cell lysate vaccine in dogs with splenic hemangiosarcoma (HSA) (10). Since this was an open phase I study, its purpose was to assess safety and immunologic responses in dogs receiving concurrent DOX chemotherapy. The tumor vaccine was found to be safe in dogs with HSA causing mild to moderate vomiting and/or diarrhea in up to 18% of patients (5 of 28 dogs). All of these dogs were also being treated concurrently with chemotherapy therefore separation of side effects induced by chemotherapy versus those caused by the vaccine was not possible. The tumor vaccine was found to elicit humoral immune responses against control antigens and HSA tumor cells in most dogs that were vaccinated while receiving concurrent DOX chemotherapy. The median disease-free interval for dogs treated with the vaccine in combination with DOX chemotherapy was 146 days with a median overall survival time of 182 days (see **Figure 2**, below). This compared favorably to the disease-free interval and overall survival time for control dogs treated with surgery followed by DOX chemotherapy alone (126 days and 133 days, respectively). Although addition of the LDC vaccine appeared to confer a survival advantage compared to dogs treated with chemotherapy alone, a larger prospective clinical trial will be necessary to better assess the vaccine's efficacy.

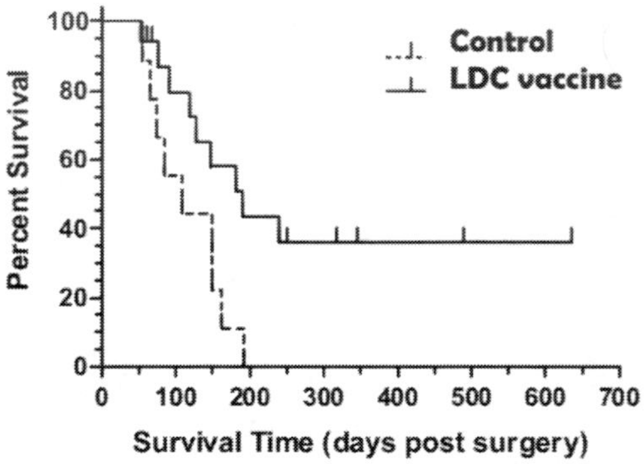


**Figure 2:** Survival times for 17 dogs receiving a HSA tumor lysate vaccine in combination with surgery and doxorubicin chemotherapy

A number of other cell-based vaccination strategies are also being explored in veterinary patients including xenogeneic DNA vaccines, viral vector-mediated delivery of immunostimulatory cytokines and dendritic cell (DC) vaccines transfected with tumor-associated antigens (11-13). An intriguing alternative approach to the generation of canine DC vaccines was recently described by Mason and colleagues at the University of Pennsylvania (14). These investigators found that large numbers of CD40-activated B cells could be produced from the peripheral blood of canine lymphoma patients and used to stimulate effective T-cell mediated immune responses. This technique overcame many of the technical hurdles inherent in the generation of DC vaccines and was found to be effective whether B cells were obtained from dogs that had already undergone chemotherapy or were collected from naïve lymphoma patients. The therapeutic benefit of this approach is currently being investigated in combination with chemotherapy in a larger clinical trial.

**CONCLUSIONS & FUTURE DIRECTIONS**

Tumor immunotherapy is an exciting and rapidly expanding field. Many studies suggest, however, that cancer immunotherapy is likely to be most effective when used in the setting of minimal residual disease or as an adjuvant to traditional treatments such as surgery, radiation therapy and chemotherapy. Fortunately, the long-held view that immunotherapy and chemotherapy are antagonistic therapeutic modalities is beginning to crumble in the face of studies that document the significant immunostimulatory effects of commonly used chemotherapy drugs.

In addition to evaluating the effects of chemotherapy on adaptive immune responses, our laboratory is also exploring the impact of chemotherapy on innate immunity. Because dendritic cells can strongly influence the outcome of tumor vaccine therapy, we are currently investigating the effects of chemotherapy on dendritic cells in several rodent tumor models. Although preliminary, our results suggest that administration of DOX to mice with large, established tumors results in a significant increase in dendritic cell numbers and function. This observation is of particular importance because dendritic cells are often significantly depleted and dysfunctional in mice (and in humans) with cancer. Using this information, we hope to be able to determine whether DOX chemotherapy can be synergistically combined with tumor vaccination to improve the minimal tumor control seen with either treatment modality alone. The results of these studies will then be used to design a clinical trial investigating a novel tumor vaccine combined with chemotherapy in dogs with lymphoma.

**REFERENCES**

[1.] Ghiringhelli F, et al. **Cancer Immunol Immunother** 2007;56:641-8. [2.] Lake RA, et al. **Nat Rev Cancer** 2005;5:397-405. [3.] Schiavoni G, et al. **Blood** 2000;95:2024-30. [4.] Mihich E, et al. **Transplant Proc** 1984;16:499-501. [5.] Ujhazy P, et al. **Cancer Immunol Immunother** 2003;52:463-72. [6.] Gauthier MJ, et al. **J Vet Intern Med** 2005;19:193-9. [7.] Henry CJ, et al. **J Am Vet Med Assoc** 2001;219:1238-41. [8.] Winnicka A, et al. **J Vet Med A Physiol Pathol Clin Med** 2002;49:303-6. [9.] Walter CU, et al. **J Vet Intern Med** 2006;20:342-347. [10.] U'Ren LU, et al. **J Vet Intern Med** 2007;21:113-120. [11.] Bergman PJ, et al. **Vaccine** 2006;24:4582-5. [12.] Jourdier TM, et al. **Gene Ther** 2003;10:2126-32. [13.] Gyorffy S, et al. **J Vet Intern Med** 2005;19:56-63. [14.] Mason NJ, et al. **Gene Ther** 2008;15:955-65.

# MICRORNAS: A REVIEW & CURRENT STUDIES OF THEIR EXPRESSION IN CANINE CANCERS

William C. Kisseberth, DVM, PhD, DACVIM (Oncology)
Columbus, OH

## INTRODUCTION

The sequencing of the canine genome has presented new and unique opportunities to better understand the molecular basis of disease, which ultimately is necessary for making significant advances in disease diagnosis and treatment. Although the initial goal of sequencing the genomes of the dog and other species was to sequence the "entire" genome, it is that portion of the genome that is ultimately expressed as a protein (approximately <1.5% of the mammalian genome) that has been the primary focus of subsequent investigation, i.e. the post-genomic era. Recently, a major advance occurred in our understanding of the "other 99%" of the genome that does not code for a protein. This was the identification and recognition of microRNAs (miRNAs), short 19-25 nucleotide non-protein coding RNAs.[1,2] It is now clear that miRNAs play an important role in gene regulation and perhaps maintenance of the genome's integrity. Importantly, dysregulation of miRNA expression has been implicated in a wide variety of diseases, including cancer. Dysregulated miRNAs are also potential targets for therapy and biomarkers of disease. MiRNA-based therapies already have been proposed.[3] In this session, we will review miRNA structure and function, their roles in cancer, methods of measuring their expression, miRNA expression profiling, and their potential uses as therapeutics and biomarkers.

## MICRORNA STRUCTURE & FUNCTION

MiRNA genes are a recently recognized large class of genes that do not encode proteins. MiRNAs are initially transcribed in the nucleus as long, capped and polyadenylated precursors, called pri-miRNAs, which are then cleaved into 70- to 100-nt hairpin precursors, called pre-miRNAs, by a ribonuclease called Drosha and the double-stranded DNA binding protein DGCR8/Pasha. Pre-miRNAs are exported from the nucleus and translocated into the cytoplasm, where the ribonuclease Dicer, in association with the double-stranded RNA-binding protein TRBP (transactivation-responsive RNA-binding protein) processes the pre-miRNAs into 19- to 25-nt miRNA duplexes. The duplexes are then incorporated into the RISC (RNA-induced silencing complex). The mature miRNA drives the complex to the target mRNA, and translation inhibition can occur either through cleavage and degradation of the messenger (for perfect complementarity to the 3' untranslated region of the target) or through blocking of the translation (without mRNA degradation), if the complementarity is partial. MiRNAs are highly evolutionarily conserved, and have been identified in a wide variety of organisms.[1,2] Hundreds of miRNAs have been identified in the genomes of the worm, fly, human, and others. It is believed that miRNAs act to regulate gene expression during development and differentiation at the post-transcriptional level, although the most relevant targets for specific biologic processes are in many cases still elusive. The first members of this gene family were identified almost 20 years ago; however, it has been over only the past few years that the size and importance of this large gene family has been fully appreciated. It is estimated that the genomes of complex organisms, such as the dog, contain up to a thousand miRNAs, representing approximately 1% of currently predicted genes. This may be a conservative estimate. Most mammalian mRNAs contain conserved targets for miRNAs and are thus potentially regulated by miRNAs.[4]

Although miRNAs may regulate gene expression by other mechanisms and the details of these mechanisms remain somewhat controversial, miRNAs are generally thought to down-regulate gene expression by two primary post-transcriptional mechanisms, mRNA cleavage and translational repression. According to the prevailing model, the choice of mechanism is determined by the identity of the mRNA target. Once incorporated into a cytoplasmic RISC, the miRNA will specify cleavage of the target mRNA if there is sufficient complementarity to the miRNA, or it will repress productive translation if the mRNA does not have sufficient complementarity to be cleaved but does have a suitable number of partially complementary sights.[1,2,4] The complementary sites for known targets reside in the 3' UTRs of mRNAs. Computational approaches to finding miRNA targets suggest thousands of potential targets, with most miRNAs having multiple potential targets. The apparent pervasiveness of this previously undetected mechanism of gene regulation has led some to declare that miRNAs are "the genome's guiding hand".[5]

## MICRORNAS ARE INVOLVED IN DIVERSE BIOLOGICAL PROCESSES

MiRNAs play key roles in a wide variety of biological processes, including regulating cell fate specification, development, cell death, proliferation, immune response, cardiovascular biology, and others. Similarly, their roles in various disease processes are beginning to be elucidated. A role for altered miRNA expression has been demonstrated in heart failure, lung disease, rheumatic disease, diabetes, and neurological disease. It should not be surprising that miRNA dysregulation also has a significant role in carcinogenesis.

## MICRORNAS & CANCER

Cancer is a complex genetic disease characterized by alterations in oncogenes and tumor suppressor genes. One of the first observations that miRNAs might have an important role in cancer was that miRNAs are frequently located in cancer-associated genomic regions, which include minimal regions of amplification, loss of heterozygosity, fragile sites, and common breakpoint regions in or near oncogenes or tumor suppressor genes. It has since been shown that miRNAs can contribute to oncogenesis by functioning as tumor suppressors or as oncogenes through their interactions with protein coding genes.[6,7] Because miRNAs have multiple mRNA targets, it is postulated that two or more protein coding genes from different interacting molecular pathways may be disturbed when a miRNA is dysregulated. Just like protein coding oncogenes and tumor suppressor genes, miRNA genes may undergo chromosomal rearrangements, genomic amplifications or deletions, mutations, or epigenetic modifications. Because miRNAs act indirectly, by interacting with their target mRNAs, activation of oncogenic miRNAs and inactivation of tumor suppressor protein coding genes having the same molecular consequences: reduced levels of proteins blocking proliferation and activating apoptosis. Similarly, inactivation of suppressor miRNAs and activation of oncogenic protein coding genes are followed by accumulation of proteins that stimulate proliferation and inhibit apoptosis.

## MICRORNA EXPRESSION PROFILING

Just as mRNA transcriptional profiling of tumors has been done for the past decade, high throughput miRNA expression profiling can be done too. In general, all existing methods can be separated into one of two broad categories: direct hybridization to probes without RNA amplification and RNA amplification and quantification. We have used a combination of oligonucleotide microarrays for miRNA expression profiling and quantitative RT-PCR for array validation to investigate miRNA expression in canine cancers.[8,9]

MiRNA expression profiling has been done and reported for most human cancer histologies.[10,11] In general; different types of cancers can be distinguished from one another based on their miRNA expression profiles. Surprisingly, miRNA expression profiling of hundreds of miRNAs appears to be more robust than mRNA transcriptional profiling of thousands of mRNAs for correctly classifying tumors.[10] This may be a reflection of the important role of miRNAs in differentiation and development. Overall miRNA expression appears to be down-regulated in cancer. Whereas dysregulated expression of specific miRNAs is associated in some cases with specific cancers (e.g. miR-15, miR-16; chronic lymphocytic leukemia), dysregulation of other miRNAs is commonly found in multiple different cancer types (e.g. miR-21, miR-17-5p, miR-191).[11]

## AN OLIGONUCLEOTIDE MICROARRAY FOR GENOME-WIDE PROFILING OF CANINE CANCERS

In order to do miRNA profiling of canine tissues, we have constructed a custom canine miRNA oligonucleotide microarray.[12] These microarrays utilize nucleic acid hybridization of 5' biotin-labeled complementary cDNA target with DNA oligonucleotide probes attached to a gel matrix. The biotin labeled cDNA targets are prepared by reverse transcription into first strand cDNA. Total RNA is primed for reverse transcription by a random octomer conjugated with two biotins and a 5' poly(A) tail. This procedure results in an equal number biotin-cDNA target to templates of miRNA. To design the oligo probes, we identified putative canine miRNAs based on homology to previously identified miRNAs in other species. The first-generation canine miRNA microarray we constructed consisted of over 600 40-mer oligonucleotide probes complementary to previously identified canine miRNAs, miRNAs highly homologous to those in other species, and predicted miRNAs. The oligos were modified at the 5' end with an amine-C6 linker. When this microarray was used to determine miRNA expression in a panel of normal canine tissues, unique miRNA expression profiles were identified for each tissue. Many of the miRNAs that were differentially expressed in specific canine tissues have been identified previously as being differentially expressed in the same tissues in other species. A second generation canine miRNA microarray currently is under development. MiRNA gene expression profiling using these arrays is being done in a variety of canine tumor types including lymphoma, osteosarcoma, and mast cell tumors.

## CLINICAL IMPLICATIONS OF MICRORNA EXPRESSION IN CANCER

The recognition of miRNAs as important players in cancer pathogenesis has tremendous significance for our understanding of cancer biology and for the diagnosis and treatment of cancer. As previously stated, miRNA profiles can be used to correctly classify tumors. Thus, tests can be developed to diagnose tumors of uncertain origin based on their miRNA expression profiles.[13] MiRNAs may also have a use as biomarkers for cancer and other diseases. Free (non-cell associated) miRNAs can be detected in the blood of healthy and (some) cancer-bearing human patients.[14] Furthermore, the blood miRNA expression profiles from these populations may be different. The extent to which these differences exist and whether such differences can be exploited as biomarkers is an area of intense ongoing investigation. MiRNAs also represent a potential new therapeutic target.[3] Therapeutic replacement of down-regulated tumor suppressor miRNAs, or antagonizing the effects of up-regulated oncogenic miRNAs offer new approaches to cancer therapy. MiRNA-based therapies have begun to be investigated in experimental animals, but have not been applied yet to humans or dogs with spontaneously occurring tumors.

## REFERENCES

[1]Bartel DP. **Cell** 2004;116(2):281. [2]Bartel DP, Chen CZ. **Nat Rev Genet** 2004;5(5):396. [3]Stenvang J, et al. **Biochem Soc Trans** 2008;36(Pt 6):1197. [4]Bartel DP. **Cell** 2009;136(2):215. [5]Dennis C. **Nature** 2002;420(6917):732. [6]Calin GA, Croce CM. **Nat Rev Cancer** 2006;6(11):857. [7]Calin GA, Croce CM. **Cancer Res** 2006;66(15):7390. [8]Liu CG, et al. **Methods** 2008;44(1):22. [9]Schmittgen TD, et al. **Methods** 2008;44(1):31. [10]Lu J,et al. **Nature** 2005;435(7043):834. [11]Volinia S, et al. **Proc Natl Acad Sci U S A** 2006;103(7):2257. [12]Liu CG, et al. **Proc Natl Acad Sci U S A** 2004;101(26):9740. [13]Tam W. **J Mol Diagn**. 2008;10(5):411. [14]Chen X,et al. **Cell Res**. 2008;18(10):997.

## CANINE ONCOGENOMICS: WHAT IS IN OUR GENES?

**Melissa C. Paoloni, DVM, DACVIM (Oncology)**
**Bethesda, MD**

### INTRODUCTION

Cancer can be defined as a genetic disease. Many different mutations lead to the development of a tumor and others to its subsequent metastatic spread, these include the activation of oncogenes and inactivation of tumor suppressor genes. The multi-step pathogenesis of cancer has been described by many, although most notably by Hanahan and Weinberg. These include manipulation of growth factors, loss of growth inhibition, evading programmed cell death, inhibition of cellular senescence/aging and the ability to metastasize. Mutations described in canine cancers are a more recent development, with non-candidate approaches made possible by the publishing of the canine genome in 2005. A variety of high throughput methods are now available for interrogating canine cancers including RT-PCR, PARR, oligonucleotide and SNP microarrays and comparative genomic hybridization (CGH) technologies. Applications of each technology will be briefly explored in the following review. Lastly, the clinical utility of gene discovery will be emphasized with descriptions of canine clinical trials utilizing novel targeted therapies.

### ONCOGENES & TUMOR SUPPRESSOR GENES

Two classes of genes have been broadly defined as contributing to cancer development and progression: (1) oncogenes and (2) tumor suppressor genes. These two classes may contribute to one or many of the credentials of cancer in a given tumor. As such, a brief review of tumor oncogene and tumor suppressor gene biology guides the discussion of the credentials of cancer.

### ONCOGENE FORMATION

Oncogenes, first recognized in the study of RNA viruses, are integral genes which when inappropriately regulated can contribute to the development of cancer. When expressed under the control of intact and physiological regulatory pathways, these genes are referred to as proto-oncogenes. Proto-oncogenes are involved in several physiological functions including, cellular proliferation, differentiation, modulation of gene expression, progression through the cell cycle, cell adhesion and motility, as well as the co-ordination of DNA synthesis. Beyond their role as effectors in these diverse cellular processes, their expression also facilitates signal transduction and therein are responsible for a cell's ability to respond to external signals. Mechanisms of onogene formation include:

- **Gene Mutation:** Constitutive activation of a gene
- **Gene Amplification:** Over-expression of a otherwise normal gene product
- **Chromosome Translocation:** Expression of an aberrant fusion protein or activates a latent oncogene in a new location

Examples of oncogenes important in human and canine cancers include:

| CATEGORY | MECHANISM | ASSOCIATED HUMAN TUMORS | ASSOCIATED CANINE TUMORS |
|---|---|---|---|
| **Growth factors** | | | |
| FGF | Amplification | Bladder, breast CA | Bladder CA (TCC)? |
| PDGF | Overexpression | Glioma | Unknown |
| IGF-1 | Overexpression | Osteosarcoma | Osteosarcoma |
| **Growth factor receptors** | | | |
| c-kit | Mutation | GIST, mastocytsosis | Mast cell tumors, GIST |
| EGF-R | Mutation | Breast, lung CA | Osteosarcoma, mammary CA |
| c-met | Overexpression | Many | Osteosarcoma |
| **Cytoplasmic kinases** | | | |
| abl | Translocation | CML | Unknown |
| **G proteins** | | | |
| ras | Mutation | Lung carcinoma | Lung carcinoma |
| **Nuclear proteins** | | | |
| myc | Translocation | Burkitt's lymphoma | unknown |
| MDM2 | Overexpression | Soft tissue sarcoma | Soft tissue sarcoma |

### TUMOR SUPPRESSOR GENE INACTIVATION

Tumor suppressor genes act to control the progress of cells through the cell cycle, control messenger pathways, survey damage to the genome, activate DNA repair and initiate cell death when irreversible damages occurs. Tumor suppressor genes are recessively regulated. As such the loss of function of both alleles is required for the abnormal function of a tumor suppressor gene.

Mutations in one allele of a tumor suppressor gene can be carried asymptomatically as a germ-line mutation and passed silently into successive generations. This explains the association of tumor suppressor genes with familial or inherited cancers. According to the Knudson's two hit model of carcinogenesis a "second hit" or subsequent somatic mutation in the remaining wild-type (normal) tumor suppressor allele is necessary for loss of tumor suppressor gene function and for increased cancer risk. This "second hit" is more often categorized as a duplication of the mutant allele and subsequent loss of heterozygosity, rather than somatic mutation in the wild type allele. Regardless, it is this dual loss in a tumor suppressor gene that contributes to tumor formation

Examples of tumor suppressor genes important in human and canine cancers include:

| SUBCELLULAR LOCATION | FUNCTION | ASSOCIATED HUMAN TUMORS | ASSOCIATED CANINE TUMORS |
|---|---|---|---|
| **Cell Surface** | | | |
| E-cadherin | Cell adhesion | Gastric, breast CA | Unknown |
| **Cytoskeleton** | | | |
| *NF-2* | Unknown | Meningioma, schwanoma | Unknown |
| **Cytosol** | | | |
| APC | Inhibits signal transduction | Colon, pancreas CA | Unknown |
| **Nucleus** | | | |
| *p53* | Regulates cell cycle and apoptosis | Many | Osteosarcoma, lymphoma, mammary CA, soft tissue and hemangiosarcoma |
| Rb | Regulates cell cycle | Retinoblastoma, osteosarcoma | Osteosarcoma? |
| p21<br>p16 | Regulates cell cycle | Melanoma | Melanoma |
| PTEN | Controls survival signaling | Breast, Prostate CA | Osteosarcoma |

## OK, IT IS NOT THAT SIMPLE….

There are many other elements integral in the development and progression of cancer, many of which work in cohort with the genetic mutations described above. Other driving factors involve epigenetic modifications that regulate of gene expression. Mechanisms include post-translational modification of histones and commonly, DNA methylation at CpG sites. Genetic and epigenetic modifications can occur in tumor cells as well as neighboring "normal" cell such as endothelial or stromal cells. All of these modifications can affect tumor growth, response to therapies and propensity to metastasize.

## CURRENT & FUTURE MANAGEMENT OF CANCER IN THE DOG

The opportunity for these strategies to be included in the treatment of pet dogs with cancer is increasingly likely as a result of the release of the canine genome sequence and the development of technologies that have lowered the costs of production of these resources. It is now the case that genomic and proteomic platforms, antibodies and other biological reagents are available for use in the dog. Examples of expression microarray studies in osteosarcoma and cytogenetics in lymphoma will be presented, illustrating both the application of the canine genome release and non-candidate approaches to discovery of possible genetic targets for therapeutic development. As our understanding of the biology of cancer has expanded a new list of cancer management strategies, including diagnostic tests, prognostic markers, and targeted therapies has emerged.

## CLINICAL TRIALS UTILIZING TARGETED THERAPIES

Studies in dogs will not only result in the development of new drugs for dogs with cancer but will likely inform the development of novel treatment strategies in human patients. Some examples of targeted therapies evaluated in canine patients advanced by our enriched understanding of candidate genes will be described. These include c-kit inhibition in canine mast cell tumors, and both IGF and mTOR blockade in canine osteosarcoma.

## REFERENCES

[1.] Paoloni, M and Khanna C. **Nature Rev Cancer** 2008; 8 (2): 147-156. [2.] London et al. **Clin Cancer Res** 2003; 9 (7): 2755-2768. [3.] Hanahan, D and Weinberg, R. **Cell** 2000; 100 (1): 57-70.

# UNCOVERING HERITABLE INFLUENCES ON DOG CANCER

**Jaime F. Modiano, VMD, PhD**
**Minneapolis, MN**

Familial cancer syndromes associated with single gene defects have been characterized in humans, rats, and dogs [1-3], and myriad cancers have been modeled in laboratory mice using genetic engineering [4]. It is curious that cancer syndromes with spontaneous mutation of the BHD gene have been identified in each of these species [5,6]. Nevertheless, most cancers occur sporadically as complex diseases where multiple genes with variable penetrance interact with environmental factors to progressively endow a single "tumor initiating cell" with malignant properties. At present, the role individual genetic backgrounds play on phenotypes and biological behavior of sporadic (non-heritable) tumors remains to be determined in any species.

Recent studies have touched on how race and ethnicity might influence gene expression and in turn contribute to disease susceptibility in humans, but few differences have been found [7-9]. Dog breeds may provide a close parallel to human ethnic groups. While dogs retain individual (outbred) traits, the derivation and maintenance of unique breeds has led to restricted gene pools. This can be used to study heritable contributions to cancer susceptibility in animals that develop tumors spontaneously and share our environment, but with the benefit of less "noise" from other phenotypic variation [10-13]. Recent work has emphasized this utility of spontaneous canine tumors as robust, non-redundant models that complement studies in humans and laboratory animals to understand cancer genetics [14-16]. For example, the degree of medical surveillance in dogs is second only to that in humans [17]; diseases such as cancer, where traits are genetically complex and whose prevalence increases with inbreeding, are well documented in dogs [14,17]; and dog populations are structured into >400 partially inbred isolates (breeds) and a heterogeneous population of mixed-breed dogs. Gene flow between breeds is restricted by pedigree barriers and dogs of different breeds are often more (or less) susceptible to different diseases [14]. Equally important, the canine genome closely resembles the human genome, pet dogs share the human environment, and the lifetime cancer risk in dogs and humans is similar [18,19]. Indeed, some cancers occur more frequently in dogs and the incidence of most cancers varies according to breed, providing opportunities to study tumors that are difficult to replicate in humans or in inducible rodent models.

For example, cancer is reported as the cause of death for 40-60% of Golden Retrievers and Rottweilers in the United States in contemporary breed health surveys [20,21]. In these dogs, there appears to be reciprocal relationships among breed, lifetime risk, and age of non-Hodgkin lymphoma (greater risk in Golden Retrievers, earlier age of onset in Rottweilers) and osteosarcoma (greater risk in Rottweilers, earlier age of onset in Golden Retrievers) [11]. These associations suggest that heritable risk factors for cancer segregate with breed-specific traits; however, the association of breed and tumor is not restricted to Golden Retrievers and Rottweilers. A number of dog breeds, including Boxers, Golden Retrievers, Labrador Retrievers, Cocker Spaniels, Bassett Hounds, and others are at increased risk for lymphoid cancer, while other breeds, such as Chihuahuas, Dachshunds and Pomeranians show a lower incidence of the disease [22]. Similarly large and giant breeds are at greater risk for osteosarcoma [23], although additional factors, including heritable traits that are independent of size, clearly play a role [24,25].

Recently, we showed that breed predisposition for non-Hodgkin lymphoma and other lymphoproliferative diseases could be extended to specific ontogenic and functional phenotypes ("B-cell" and "T-cell") [26]. We showed that Spitz breeds and Asian "lap dogs" belonging to the oldest domestic dog groups [27] almost exclusively develop T-cell tumors, and some European breeds like Cocker Spaniels and Bassett Hounds almost exclusively develop B-cell tumors. Boxers and Golden Retrievers also show more T-cell disease than what is seen as an average for the population, and lymphoproliferative diseases occur earlier than the average for all dogs [28], but in these breeds, the frequency of B-cell and T-cell phenotypes is more balanced [26]. The prevalence of B-cell and T-cell lymphoma and leukemia in most other breeds is not significantly different from the average of all dogs or from that seen in dogs of mixed breeding (*i.e.*, ~2:1 B-cell to T-cell).

Environmental factors cannot account for the observed breed predilections for lymphoid tumors [29]. There is, therefore, a basis for the hypothesis that heritable risk factors predispose dogs to develop B-cell or T-cell malignancies. Such risk factors are presumably mutations or epigenetic changes in genes that regulate lymphocyte development, although the involvement of genes that regulate the fidelity of the genome cannot be ruled out, as familial non-Hodgkin lymphomas in people are largely associated with conditions such as the p53 mutation in Li-Fraumeni syndrome [30,31]. Whether these factors arose ancestrally, or developed more recently during the process of breed derivation, they are now firmly embedded in the genome, and their identification will provide information that will be valuable for prevention and treatment. While this task may seem difficult, we already have documented the occurrence of breed-specific cytogenetic changes in lymphoid tumors from Golden Retrievers [26], and analyses of pedigrees from Golden Retriever families suggest there is shared susceptibility to specific forms of lymphoma or leukemia among family members. The use of genomic and expression arrays provide contemporary tools that will allow us to refine familial data by tumor type and clinical response, thus providing greater insight and statistical power to find specific genes that are important in disease susceptibility and progression. For example, preliminary analyses of gene expression signatures of lymphoma samples from Golden Retrievers and non-Golden Retrievers shows a ~1.2 to 1.4-fold decrease in expression of genes that localize to canine chromosome (CFA) 14, and a corresponding increase in expression of genes that localize to CFA 15 in the Golden Retrievers samples, with $p$ values <0.05 [11]. The data suggest there is a significant difference in expression levels, but the extent of the reduction is small. Several non-mutually exclusive possibilities can account for the small magnitude of these changes, including low levels of expression for genes along these chromosomes, loss of a single copy (or portion thereof) of CFA 14 in tumors from Golden Retrievers, leading to haploinsufficiency rather than absolute deletion of genes encoded therein, or epigenetic modification (including by losses of microRNAs). Intriguingly, complementary experiments suggest that loss of CFA 14 may be a more widespread occurrence in multiple types of tumors from Golden Retrievers [32,33]. This is the subject of additional investigation in our lab and those of our collaborators.

New molecular and phenotypic classification of lymphoma and leukemia also may assist in predictions regarding the course of disease and outcomes in response to treatment. A series of genomic microarrays have been developed to analyze DNA copy number aberrations, and are being used now to investigate the cytogenetics of numerous canine cancers, including lymphoma and leukemia, brain tumors, osteosarcoma, and soft tissue sarcomas [33-37]. The genome assembly-integrated nature of the most recent iterations of these genomic arrays facilitates rapid transition from the identification of a region within a chromosome associated with cancer presentation to the precise location with the canine genome assembly, and thus to the identification of specific genes within those regions [33,36]. Simultaneously, studies are underway to assess the predictive ability of gene expression profiles for lymphoma and leukemia. We are using chromosome-specific FISH to directly define and catalogue altered genome organization within individual cells comprising the malignant mass [38-41]. We also are using canine comparative genomic hybridization (CGH) analysis to determine DNA copy number status across the entire genome [33,36-39,42]. FISH analysis also has enabled the precise definition of structural chromosome aberrations that show a convincing evolutionary history between homologous cancers of dogs and people [41]. These analyses can be used to predict both tumor origin and response to therapy. For instance, gain of dog chromosome 13 (CFA 13), particularly in a region syntenic to human chromosome 4q (HSA 4q), occurs in ~70% of canine diffuse B-cell lymphoma [39]. This suggests that this region of the genome contains heretofore-unidentified genes that are etiologically and prognostically significant for this disease. There are also indications that gain of CFA 13 also is predictive for chemotherapy response [39,43], perhaps because amplification of the c-*myc* and c-*kit* oncogenes, both of which are encoded in this region of CFA 13, increases the proliferative rate of the malignant cells and consequently their susceptibility to anti-mitotic compounds. These approaches also can be used to define specific chromosomal regions that are associated with heritable risk by identifying unique tumor genomes that segregate with selected breeds or groups, and that can pinpoint new regions for gene discovery [26].

Although they are at earlier stages of development, tools such as arrays that can be used to analyze global gene expression are making rapid progress toward clinical application. The ability to analyze thousands of genes at once maximizes the efficiency with which we can identify genetic alterations associated with tumor pathogenesis, as well as prognostic "gene signatures" [44]. Importantly, even simple tests that can be used as surrogates for cytogenetic changes such as the deletion of CFA11 in high-grade T-cell lymphomas have predictive value for this disease [28]. Our data using gene expression profiles is providing new first steps to document not only the potential to identify prognostically useful markers for canine tumors [45], but also ontogenetically significant relationships (for example, underlying the biology of multiple different subtypes of lymphoma) and heritable associations defined by unique profiles that segregate according to breed [32].

A contemporary area of emphasis is to define whether these properties are inherent to cancer cells or whether they arise by natural selection and clonal evolution. We can use molecular tools to predict risk, prognosis, and response to therapy in some cancers of companion animals, and we believe the availability and usefulness of such tools in clinical practice will expand rapidly. Therefore, as we improve our understanding of basic mechanisms that account for malignant transformation and tumor progression, we will be able to design even better strategies for cancer prevention and therapy.

## REFERENCES


[1]. Fearon ER. **Science** 1997;278:1043. [2]. Hino O. **Genes Chromosomes Cancer** 2003;38:357. [3]. Lingaas F, et al. **Hum Mol Genet** 2003;12:3043. [4]. Sharpless NE, Depinho RA. **Nat Rev Drug Discov** 2006;5:741. [5]. Comstock KE, et al. **Mamm Genome** 2004;15:544. [6]. Okimoto K, et al. **Proc Natl Acad Sci U S A** 2004;101:2023. [7]. Ferguson SE, et al.. **Gynecol Oncol** 2006;101:209. [8]. Spielman RS, et al. **Nat Genet** 2007;39:226. [9]. Wallace TA, et al. **Cancer Res** 2008;68:927. [10]. Sutter NB, et al. **Genome Res** 2004;14:2388. [11]. Modiano JF, Breen M. **Cancer Therapy** 2008;6:239. [12]. Thomas R, et al. **Chromosome Res** 2009;in press. [13]. Breen M. Cytogenet **Genome Res** 2008;120:50. [14]. Sutter NB, Ostrander EA. **Nat Rev Genet** 2004;5:900. [15]. Khanna C, et al. **Nat Biotechnol** 2006;24:1065. [16]. Paoloni M, Khanna C. **Nat Rev Cancer** 2008;8:147. [17]. Ostrander EA, et al. **Trends Genet** 2000;16:117. [18]. Peterson MR, et al. **J Vet Intern Med** 2000;14:140. [19]. Jemal A, et al. **CA Cancer J Clin** 2007;57:43. [20]. Glickman LT, et al. The Golden Retriever Club of America National Health Survey (2000). Available at http://www.grca.org/healthsurvey.pdf. [21]. Slater M. RHF Health Survey Results (2008). Available at http://www.rottweilerhealth.org/RHF_surveyresults.html. [22]. Priester WA, McKay FW. **Natl Cancer Inst Monogr** 1980;54:1. [23]. Ru G, et al. **Vet J** 1998;156:31. [24]. Cooley DM, et al. **Cancer Epidemiol Biomarkers Prev** 2002;11:1434. [25]. Phillips JC, et al. **Genomics** 2007;90:354. [26]. Modiano JF, et al. **Cancer Res** 2005;65:5654. [27]. Parker HG, et al. **Science** 2004;304:1160. [28]. Modiano JF, et al. **Leukemia** 2007;21:184. [29]. Ritt MG. In: Schalm's Veterinary Hematology, 6th ed. Ames, IA: Blackwell Publishing; 2009. [30]. Siddiqui R, et al. **Curr Oncol Rep** 2004;6:380. [31]. Segel GB, Lichtman MA. **Blood Cells Mol Dis** 2004;32:246. [32]. Tamburini BA, et al. **PLoSONE** 2009 (in revision). [33]. Thomas R, et al. **Chromosome Res** 2009;in press. [34]. Thomas R, et al. **Cytogenet Genome Res** 2003;102:254. [35]. Thomas R, et al. **Cytogenet Genome Res** 2003;102:249. [36]. Thomas R, et al. **J Hered** 2007;98:474. [37]. Thomas R, et al. **Genome Res** 2005;15:1831. [38]. Thomas R, et al. **Chromosome Res** 2001;9:649. [39]. Thomas R, et al. **Br J Cancer** 2003;89:1530. [40]. Milne BS, et al. **Chromosome Res** 2004;12:825. [41]. Breen M, Modiano JF. **Chromosome Res** 2008;16:145. [42]. Dunn KA, et al. **Vet J** 2000;160:77. [43]. Hahn KA, et al. **Vet Pathol** 1994;31:528. [44]. Rosenwald A, et al. **N Engl J Med** 2002;346:1937. [45]. Gavin K, et al. In: Genes Dogs & Cancer V. Orlando, FL: International Veterinary Information Services; 2009.

## MANIPULATING THE IMMUNE RESPONSE TO TARGET CANCER

**Nicola Mason, BVetMed, PhD, DACVIM**
**Philadelphia, PA**

### INTRODUCTION

Since William Coley's epiphany in the late 1800s that the immune system when appropriately awakened can prevent tumor progression and induce established tumor regression, increasingly sophisticated strategies have been developed to harness cell mediated immunity to target and eliminate tumors. With our ever increasing understanding of innate and adaptive immune cell activation, trafficking and effector function together with the realization that strategically combined immunological/biological therapies may provide greater therapeutic efficacy than individual therapies alone, immunotherapy is set to become a key asset in our armory against cancer.

In order for strategies aimed at augmenting tumor-specific T cell immunity to achieve their full therapeutic potential, they must incorporate mechanisms to achieve all of the following:

1) Increase the numbers of activated, fit tumor-specific T cells
2) Generate tumor-specific memory T cells
3) Optimize trafficking of tumor-specific T cells into the tumor site
4) Counteract the immunosuppressive effects of the tumor microenvironment

### STRATEGIES TO GENERATE CLINICALLY RELEVANT, TUMOR-SPECIFIC T CELL-MEDIATED IMMUNITY

#### Antigen-specific T cell activation

In the veterinary arena, the majority of efforts to stimulate anti-tumor immunity have focused on active immunization strategies to increase the number of tumor-specific T cells in vivo. These approaches have largely utilized cell based vaccines intended to prime functional, tumor-specific cytotoxic T cell responses and generate immunological memory against multiple tumor associated antigens (TAA) either directly (e.g. antigen-presenting cell (APC) vaccines such as autologous dendritic cell or CD40-activated B cell vaccines loaded with whole tumor RNA) or indirectly (e.g. genetically-modified autologous and allogeneic whole tumor cells or lysates).[1-6] In each case the antigenic payload presented to peripheral T cells should comprise a multitude of TAA that are aberrantly expressed or over-expressed by the tumor. These approaches aim to induce polyvalent T cell responses against the tumor and minimize antigen variant loss that can lead to immune editing and loss of therapeutic anti-tumor immunity. Furthermore, they are particularly attractive in the veterinary arena where few bona fide TAA have been identified. Whole tumor cell vaccines that have been tested to date in the canine system have been genetically modified to express immunostimulatory molecules (B7-1 and B7-2) and cytokines such as GM-CSF and IL-2 that promote APC recruitment, tumor uptake and antigen processing as well as T cell activation, growth and proliferation.[4, 5] While the targeting of multiple tumor-derived antigens is important to promote polyvalent anti-tumor responses and reduce the risk of antigen variant loss, the use of whole tumor cell vaccines or APCs loaded with whole tumor lysates or whole tumor RNA carries with it the risk of inducing autoimmunity to antigens that are not specific for malignant cells. However, no serious adverse side effects have been documented to date in either human or veterinary trials evaluating different strategies using whole tumor antigens as the antigenic payload.[4, 6-10]

In a wide range of human clinical trials designed to evaluate safety and therapeutic efficacy, cell based vaccines have been shown to be capable of augmenting tumor-specific T cell responses and in mediating clinical responses in up to 50% of trial patients. These promising results have lead to efforts to enhance the immunogenicity particularly of APC based cancer vaccines by utilizing combined activating stimuli (e.g. CD40L plus TLR agonists) together with therapeutic strategies to break immunological tolerance to TAA such as the concurrent depletion of regulatory T cells.[11-13] While these additional effects may augment anti-tumor immunity they may also increase the risk of autoimmunity, when whole tumor lysates or genetic material is used as the antigenic payload.

Peptide-based vaccines and DNA vaccines that encode antigens that are specifically expressed or over-expressed by tumor cells reduce the risk of inducing systemic autoimmunity, however, these approaches have been limited in veterinary medicine by our current restricted knowledge of canine MHC haplotypes (required for the design of peptide vaccines) together with the paucity of identified bona fide TAA in the dog. One exception to this is tyrosinase, a melanocyte specific gene expressed in both human and canine malignant melanoma. The first, and to date only DNA vaccine licensed for use in veterinary oncology encodes a xenogenic tyrosinase gene coupled with huGM-CSF which has shown safety, immunogenicity and clinical efficacy in phase I/II clinical trials in dogs with malignant melanoma.[14,15] A recombinant human VEGF vaccine has also been evaluated in dogs with soft tissue sarcomas and has been shown to reduce circulating VEGF concentrations and induce clinical responses in a proportion of treated dogs.[16] Taken together, these are exciting results and it is likely that over the next few years as TAA are identified in the dog and common MHC haplotypes are determined, peptide and DNA vaccines targeting TAA will be developed, optimized and marketed.

#### Non-specific T cell activation

Immunomodulatory agents including Toll Like Receptor (TLR) agonists such as oligodeoxynucleotides (ODNs) with unmethylated CpG dinucleotide motifs and Bacillus Calmette-Guerin (BCG) have been used either alone or in combination with DNA vaccines, peptide vaccines and whole tumor cell vaccines to augment their immunogenicity and therapeutic efficacy in a wide variety of different tumor types in humans. These agents exert their immunostimulatory effects via TLRs expressed on the surface of APCs such as macrophages and dendritic cells. Following TLR engagement, APCs become activated and up-regulate cell surface expression of MHC class I and II molecules and co-stimulatory molecules, empowering them to prime tumor-specific T cells. TLR stimulated APCs also secrete chemokines that attract T cells increasing the likelihood of cognate T cell interaction and activation. However, despite promising results in augmenting T cell priming and activation in human clinical trials and evidence of immunomodulatory effects of CpG and BCG in the dog their evaluation as potent immunomodulatory agents in the canine cancer patient remains understudied.[17, 18] Cationic liposome-DNA complexes have been evaluated alone and together with whole tumor cell lysates in an adjuvant setting to elicit non-specific tumor activity in the dog.[7, 19] While the cationic liposomes have been used successfully as a transfection agent to delivery a transgene encoded by plasmid DNA in to canine cells, it is apparent that the liposome complexes themselves exert anti-tumor and anti-angiogenic effects.[16] Adenoviral gene delivery of immune stimulatory agents such as CD40L and Flt3L have also been evaluated in dogs and have shown promising effects on immune function and tumor regression.[20, 21]

**Cancer vaccines and the generation of immunological memory**

While cell-based immune therapies increase the number of tumor-specific, IFN-γ secreting CD8+ T cells in some human cancer patients, these immunological readouts often fail to correlate with clinical response.[22, 23] These findings may be attributed in part to the lack of induction of tumor-specific memory responses that act to maintain a steady stream of "fit," high quality effector T cells that can mediate anti-tumor immunity.[24] Successful cancer vaccine strategies must not only act to boost the numbers of potent tumor-specific CD8+ effector T cells but also generate high quality memory T cell responses that mediate long-lasting anti-tumor immunity. CD4+ T cell help is required for the maintenance and secondary expansion of long-lived memory CD8+ T cells and therefore cancer vaccines should aim to stimulate both CD4+ and CD8+ tumor-specific T cells. Although the optimal ways to stimulate central and effector memory T cell responses via vaccination are currently unknown, recent studies in non-human primates have indicated that the use of TLR agonists together with peptide vaccines not only enhances the magnitude and quality of effector T cell responses but it also promotes the generation of memory T cell responses in prime-boost anti-viral vaccine schedules.[25] It is likely that similar strategies will be evaluated in human and veterinary cancer vaccine strategies to promote the generation of tumor-specific memory T cells in the near future.

**Tumor-specific T cell trafficking**

While much attention in both the human and veterinary cancer arenas has focused on ways to generate large numbers of tumor-specific T cells, it is becoming increasingly apparent that significant barriers exist to these cells entering into solid tumor parenchyma and exerting their cytotoxic effects. As such, efforts to promote anti-tumor immunity should incorporate strategies to promote T cell extravasation from the microvasculature into the tumor itself. Understanding the mechanisms responsible for limited T cell infiltration into tumors is likely to yield important information that may be used to optimize tumor-specific T cell trafficking. For example, recent elegant studies using laser microdissection techniques and gene expression array analysis have identified over-expression of the endothelin B receptor in tumor microvasculature as an impediment to T cell infiltration into tumor tissue. Blockade of this receptor with a specific peptide leads to enhanced effector T cell infiltration into the tumor itself, suggesting that this agent may be a useful adjunct to increase the therapeutic efficacy of our current cancer vaccines.[26] Other investigators have found that tumor microvasculature has a sub-threshold level of expression of adhesion molecules that impedes extravasation of tumor-specific T cells into the tumor itself and that systemic inhibition of TGF-β increases the number of tumor-specific T cells that gain entry into the tumor parenchyma.[27, 28] Furthermore, immunomodulatory agents that promote inflammation within the tumor itself have also been shown to enhance the trafficking of effector T cells into the tumor.

**Immunosuppression within the tumor microenvironment**

Although many different vaccination strategies lead to the activation of tumor-specific T cells in cancer bearing patients, the ability of these cells to mediate therapeutic benefit is limited in many cases. These findings have been attributed not only to the failure of tumor–specific T cells to traffic into the tumor but also to the immunosuppressive effects of the tumor microenvironment. Immunosuppressive factors such as regulatory CD4+ T cells, myeloid suppressor cells, CD8 suppressor T cells, and TGF-β, secreted by tumor or stromal cells all act to promote a state of immunological tolerance within the tumor microenvironment.[29, 30] The presence of CD4+, FoxP3+ regulatory T cells in the tumor draining lymph nodes and peripheral blood of canine cancer patients has been shown to correlate with a poor prognosis in a variety of different tumor types.[31] Regulatory T cells inhibit T cell priming, proliferation, effector function and memory recall responses through direct cell contact and the release of inhibitory cytokines and as such represent a major obstacle to the success of cancer vaccine strategies.[32] In particular, regulatory T cells have been shown to reduce the number and function of memory CD8+ T cells and contribute to the dysfunctional CD8 T cell memory responses reported in tumor-bearing hosts. Regulatory T cells also inhibit DC maturation and antigen-presenting function and lead to the induction of tolerogenic DC. Strategies to deplete regulatory T cells have lead to tumor regression in mouse cancer models and have shown beneficial effects in human clinical cancer vaccine trials.[12, 13, 32] The benefit of regulatory T cell depletion in canine cancer patients has yet to be determined but is likely to enhance therapeutic efficacy when used in combination with cancer vaccines.

## ASSESSMENT OF VACCINE-INDUCED, TUMOR-SPECIFIC IMMUNE RESPONSES

A significant impediment to progress in the field of veterinary cancer immunotherapy is our limited ability to measure vaccine-induced tumor-specific immune responses coupled with a general lack of understanding of which measurable immune parameters correlate with clinical outcome. In human clinical trials, the immunogenicity of cancer vaccines is frequently assessed by evaluating the increase in antigen-specific $CD8^+$ T cell numbers post vaccination using flow cytometry and tetramer staining. The lack of knowledge of specific TAA and available reagents in dogs means that tetramer staining is currently unavailable to evaluate antigen-specific T cells in dogs. However, vaccine induced tumor antigen-specific T cell responses can be monitored effectively both in vitro and ex vivo using commercially available ELISpot kits.[3] In addition, monoclonal antibodies specific for important effector cytokines such as IFN-$\gamma$ and TNF-$\alpha$ that cross react with canine cytokines can be used in combination with ex vivo stimulation/amplification techniques to evaluate the numbers of functional, fit antigen-specific CD4+ and CD8+ T cells induced by cancer vaccines in canine patients on clinical immunotherapy trials. Cytotoxic T lymphocyte (CTL) assays have also been used in the dog and these are likely to provide important insights into the ability of cancer vaccines to induce cytotoxic T cell responses in veterinary patients.[33] In vivo assessments of the immunogenicity of trial vaccines include the performance of DTH responses. These tests have been used in both human and veterinary clinical trials and aim to provide a more accurate assessment of antigen-specific T cell function than the previously described in vitro assays.[2, 6] However, despite the ability to evaluate these parameters and demonstrate that cancer vaccines increase tumor-specific immune responses in some cancer patients, these immunological readouts often fail to correlate with clinical response. Thus the ability to determine whether vaccine induced, antigen-specific T cells traffic into the tumor and exert an effective cytotoxic response there must be performed concurrently with these in vitro and in vivo evaluations of immunogenicity to provide a complete picture that may more accurately explain the presence or absence of observed therapeutic effects and direct future efforts to improve therapeutic efficacy. To this end, repeat biopsies of solid tumors pre and post vaccination, coupled with advanced imaging strategies will aid in the evaluation of T cell trafficking and intra-tumor T cell function, the results of which will direct the way to optimize current cancer vaccines in both human and veterinary oncology.

## SUMMARY

In summary, our ever increasing understanding of the basic biology of cell mediated immune responses both in health and disease coupled with an increased ability to manipulate the size, specificity, longevity and location of these responses predicts that combination immunotherapies for the treatment of both human and veterinary cancer patients will becoming increasingly sophisticated and therapeutically efficacious in many malignancies in the near future and will become a common fixture in our armory against cancer.

## REFERENCES

[1]Tamura K, et al. J Vet Med Sci. 2007; 69 (9):925. [2]Tamura K, et al. Vet J. 2008; 175 (1):126. [3]Mason N, et al. Gene Therapy 2008; 15 (13):955. [4]Finocchiaro L, et al. Gene Therapy 2008; 15 (4):267. [5]Whitley E, et al. Anticancer Res. 2002; 22 (5):2567. [6]Alexander A, et al. Cancer Immuno Immunother. 2006; 55 (4):433. [7]U'Ren L, et al. J Vet Intern Med. 2007; 21 (1):113. [8]Milazzo, C, et al. Blood 2003; 101:977. [9]Soiffer, R, et al. 1998; Proc Natl Acad Sci U S A 95:13141. [10]Jaffee, E. M, et al. 2001; J Clin Oncol 19:145, [11]Lapteva N, et al. Cancer Res. 2007; Nov 1;67 (21):10528. [12]Morse M, et al. Blood. 2008;112 (3):610. [13]Dannull J, et al. J Clin Invest. 2005; 115 (12):3623. [14]Bergman P, et al. Vaccine 2006; 24 (21):4582. [15]Liao J, et al. Cancer Immun. 2006; 21 (6):8. [16]Kamstock D, et al. Cancer Immunol Immunother. 2007; 56 (8):1299. [17]Milner R, et al. Vet Immuno Immunopathol. 2006; 114 (3-4):273 . [18]Henry CJ, et al. Am J Vet Res. 2007; 68 (11):1246. [19]Dow S, et al. Hum Gene Ther. 2005; 16 (8):937. [20]Von Euler H, et al J Immunother. 2008; 31 (4):377. [21]Candolfi M, et al. Neurosurgery 2007 60 (1):178. [22]Hersey, P, et al. Cancer Immunol Immunother 2004; 53:125. [23]Kyte J, et al. Cancer Gene Ther. 2006; 13 (10):905. [24]Klebanoff, C. A, et al. 2005; Proc Natl Acad Sci U S A 102:9571. [25]Wille-Reece, U, et al. 2005; Proc Natl Acad Sci U S A 102:15190. [26]Buckanovich RJ, et al. Nat Med. 2008; 14 (1):28. [27]Fisher DT, et al. Immunol Invest. 2006; 35 (3-4):251 [28]Wallace A, et al. Clin Cancer Res. 2008; 14 (12):3966. [29]Serafini P, et al. Semin Cancer Biol. 2006;16 (1):53 . [30]Kaminska B, et al. Acta Biochim Pol. 2005; 52 (2):329. [31]Biller B, et al. Vet Immuno Immunopathol 2007; 116 (1-2):69. [32]Curiel TJ, et al. Curr Opin Immunol. 2008; 20 (2):241. [33]Gyorffy S, et al. J Vet Intern Med. 2005; 19 (1):56.

# LESS MAY BE MORE: METRONOMIC (LOW DOSE CONTINUOUS) CHEMOTHERAPY

**Anthony J. Mutsaers, DVM, DACVIM (Oncology) & J. Paul Woods, DVM, DACVIM (Internal Medicine, Oncology)**
**Toronto, ON, Canada & Guelph, ON, Canada**

## INTRODUCTION

The use of continuous schedules of chemotherapy, delivered at lower doses by necessity compared to traditional dosing schemes, has become an increasingly popular addition to patient management in veterinary medical oncology. While the concept itself is not new, as low dose, maintenance chemotherapy protocols have been important components of treatment for certain cancers (such as acute lymphoblastic leukemia) for decades, research into the reasons for potential benefit have evolved as more knowledge is acquired in the areas of cancer biology and the tumor microenvironment. There is still much to be learned, however the low cost, ease of administration, and acceptable toxicity profiles, make metronomic chemotherapy protocols both attractive and suitable to veterinary applications. Preliminary clinical trial results have now been reported in both human and veterinary medicine, and further, more powerful studies are ongoing.

## MECHANISMS

The most logical assumption regarding the mechanism(s) responsible for clinical benefit with metronomic chemotherapy would be the cytotoxic effect of the drugs on the tumor cells themselves. This is, after all, how most of these drugs were discovered and proven to be effective anti-cancer therapeutics. The question becomes whether or not a chemotherapeutic drug that is effective in conventional protocols might also be beneficial at lower dosages, provided the dose chosen enables the drug to reach susceptible tumor cells in a sufficient concentration. Conventionally, the goal has been to raise the dose to the highest level that is tolerated by the patient, the so-called "maximum tolerated dose" (MTD). The evidence to support this has generally been attributed to the work of Skipper,[1] who demonstrated a logarithmic tumor cell kill with increasing chemotherapeutic dose. The rebuttal has generally consisted of an acknowledgement of efficacy in certain "liquid tumors", such as leukemias, but an inapplicability of such experimental results obtained *in vitro*, to the majority of solid tumors that exist within a complex microenvironment. By eliminating the unavoidable break period that accompanies MTD chemotherapy, tumor cells may be more continuously exposed to drug, which would permit more efficient cell kill as cells continue to cycle[2] – an argument also used in the justification of low dose, continuous radiation therapy schedules.

### Tumor Angiogenesis

The area of the tumor microenvironment that has perhaps been most studied with respect to the effects of metronomic chemotherapy has been angiogenesis.[3] It was likely the preclinical studies by Browder[4] and Klement[5] that focused attention on this potential mechanism (as well as adoption of the term "metronomic chemotherapy"). Since then, 3 aspects of angiogenesis have been shown to be impacted by continuous chemotherapeutic schedules. The first is a direct cytotoxic effect on the endothelial cells of the blood vessels themselves. Based primarily on cell culture experiments, it has been demonstrated that a variety of chemotherapeutic agents are cytotoxic to endothelial cells at concentrations in the picomolar range, that are well below what was required to produce a similar toxicity to tumor cells or other microenvironmental cells, such as fibroblasts.[6,7] The second means of decreasing angiogenesis appears to be an indirect effect, brought on by tipping the relative balance of angiogenic growth factors and inhibitors toward the latter, thereby preventing further vessel expansion. The endogenous inhibitor thrombospondin-1 appears to be a critical molecule in this process, and one that is stimulated by metronomic chemotherapy schedules.[8,9] Finally, the expansion of tumor blood vessels may also require, or at least be influenced by, recruitment of bone marrow-derived endothelial progenitor cells. Continuous scheduling of certain chemotherapy drugs appears to suppress this effect, while MTD schedules appear to promote bone marrow cell recruitment, as part of a vasculogenic rebound effect.[10,11]

### Tumor Immunology

An interesting alternative, and perhaps complementary mechanism, involves alterations in immune cells, also present in the tumor microenvironment. These observations focus mainly on a decrease in the number of regulatory T cells, which are normally involved in autoimmune disease prevention as well as tumor-induced tolerance. Low doses of cyclophosphamide are able to decrease $CD4^+CD25^+$ regulatory T cells in human cancer patients,[12] and recently these cells have been identified and quantified in canine cancer patients as well.[13] Interestingly, a recent study of immune responses in canine lymphoma and osteosarcoma cases during chemotherapy treatment also revealed a lower impact of chemotherapeutic drugs on T-cell numbers and antibody responses than one would expect based on an assumption that chemotherapy has generally been considered immunosuppressive.[14]

## CLINICAL TRIAL RESULTS

To date, there have not been any randomized, controlled, phase III clinical trial reports for the use of metronomic chemotherapy, other than perhaps the established maintenance regimens in human oncology, such as with acute lymphoblastic leukemia mentioned above. Conversely, in fact, a veterinary study of the utility of maintenance chemotherapy for canine lymphoma showed no statistical benefit to extension of the UW-Madison protocol beyond 25 weeks.[15] In addition, a study involving the administration of doxorubicin at 10mg/m$^2$ IV weekly, rather than 30mg/m$^2$ IV every 3 weeks showed no response benefit in canine lymphoma,[16] and weekly administration of lower doses of cisplatin (20mg/m$^2$ IV) with the intent to utilize the drug as a radiation sensitizer, resulted in progressive leucopenia beyond 5 consecutive doses.[17]

There has been reasonable speculation as to whether or not certain chemotherapy drugs will be more or less useful for a metronomic scheduling approach, and indeed, most single-agent and doublet metronomic chemotherapy clinical trials published have utilized cyclophosphamide (or trophosphamide). The study by Colleoni et al,[18] was one of the first to utilize cyclophosphamide in this way. This trial in metastatic breast cancer patients treated with metronomic cyclophosphamide and methotrexate resulted in an overall benefit rate of 31.7%, with a median response duration of 6.8 months. Since that time, numerous other trials have been conducted for treatment of tumors such as prostate[19] and non-small cell lung carcinoma,[20] among others. Interestingly, there have also been beneficial results obtained with tumors of particular veterinary relevance, such as non-hodgkin's lymphoma,[21] melanoma,[22] and sarcomas, including human angiosarcoma.[23]

Veterinary trials published to date include an investigation of metronomic cyclophosphamide and etoposide with piroxicam for the treatment of stage II hemangiosarcoma.[24] This trial evaluated 9 dogs and compared the results to a historical control population of 24 dogs treated with doxorubicin. The median disease free interval and median survival were comparable (178/178 days vs 133/126 days) between the 2 populations, and little toxicity was observed with the metronomic protocol. Another study, utilizing metronomic cyclophosphamide and piroxicam for incompletely resected canine soft tissue sarcomas showed a significant improvement in disease-free interval in the 30 dogs that underwent this adjuvant treatment, when compared to 55 historical control dogs treated with surgery alone.[25]

## FUTURE DIRECTIONS

Clinical trial evaluation of low dose, continuous chemotherapy regimens continues in both human and veterinary oncology. As more trial results are reported, we will gain an increased understanding of those diseases, drug combinations, and schedules that will be most relevant for this type of treatment strategy. Possible new applications of continuous dosing schedules may be found in the combination with standard drug dosing, or long term use with newer targeted oncology drugs, that are also meant to be given on a continuous basis at the optimal biologic dose.

## REFERENCES

1. Skipper H.E. et al. **Cancer Chemother Rep** 1970;54:431-50. 2. Kamen B.A. et al. **J Pediatr Hematol Oncol** 2006;28:325-7. 3. Kerbel R.S. and Kamen B.A. **Nature Rev Cancer** 2004;4:423-436. 4. Browder T. et al. **Cancer Res** 2000;60:1878-1886. 5. Klement G. et al. **J Clin Invest** 2000;105:R15-R24. 6. Bocci G. et al. **Cancer Res** 2002;62:6938-43. 7. Miller K.D. et al. **J Clin Oncol** 2001;19:1195-1206. 8. Bocci G. et al. **Proc Natl Acad Sci** USA 2003;100:12917-12922. 9. Hamano Y. et al. Cancer Res 2004;64:1570-1574. 10. Bertolini F. et al. **Cancer Res** 2003;63:4342-4346. 11. Shaked Y. et al. **Cancer Cell** 2008;14:263-73. 12. Ghiringhelli et al. **Cancer Immunol Immunother** 2007;56:641-8. 13. Biller BJ et al. **Vet Immunol Immunopathol** 2007;116:69-78. 14. Walter CU et al. **J Vet Intern Med** 2006;20:342-7. 15. Garret L.D. et al. **J Vet Intern Med** 2002;16(6):704-709. 16. Ogilvie G.K. et al. **J Am Vet Med Assoc** 1991;198:1762-4. 17. Hahn K.A. et al **Vet Clin Pathol** 1997;26:29-31. 18. Colleoni M. et al. **Ann Oncol** 2002;13:73-80. 19. Glode L.M. et al. Cancer 2003;98:1643-1648. 20. Kato H. et al. New Engl J Med 2004;350:1713-1721. 21. Buckstein R. et al. **Clin Cancer Res** 2006;12:5190-8. 22. Reichle A. et al. **Cancer** 2004;101:2247-55. 23. Vogt T. et al. **Cancer** 2003;98:2251-2256. 24. Lana S. et al. **J Vet Intern Med** 2007;21:764-9. 25. Elmslie R.E. et al. **J Vet Intern Med** 2008;22:1373-9.

## TYROSINE KINASE INHIBITOR (TKI) THERAPY IN COMPANION ANIMALS: A PRACTICAL APPROACH

**Cheryl London , DVM, PhD, ACVIM (Oncology), Associate Professor**
**Columbus, OH**

### INTRODUCTION

Advances in molecular biology have permitted the identification of key abnormalities in cancer cells involving proteins that regulate signal transduction, cell survival and cell proliferation. These often involve a class of proteins called tyrosine kinases that phosphorylate and activate other proteins in the cell, regulating a number of cellular processes. Several inhibitors of these kinases (known as tyrosine kinase inhibitors, TKIs) are approved for the treatment of human cancers, and many have significant clinical efficacy.

### TKIS IN NORMAL CELLS & CANCER CELLS

Tyrosine kinases are proteins that phosphorylate other proteins on tyrosine residues thereby regulating cell growth and differentiation. They bind ATP and use this to add phosphate groups to key residues on themselves (termed autophosphorylation) and on other molecules, resulting in intracellular signaling and alterations in gene transcription that affect cell proliferation and survival(*1*). Those tyrosine kinases expressed on the cell surface (receptor tyrosine kinases, RTKs) bind growth factors that regulate their activation. Examples include Kit, Met, EGFR, all of which are known to be dysregulated in particular forms of cancer (*2-5*). Certain RTKs such as VEGFR, PDGFR, and FGFR are also important in promoting the growth of tumor blood vessels known as tumor angiogenesis. Tyrosine kinases are also present in the cytoplasm and nucleus of cells, participating in signal transduction and regulation of gene transcription.

An extensive characterization of tyrosine kinase dysregulation has been undertaken in human cancers. Evidence suggests that tyrosine kinases are often abnormally activated in malignant tumors through mutation, overexpression, the generation of fusion proteins from chromosomal translocation, or autocrine loops of activation(*6*). The consequence of this is persistent cell signaling thereby inducing uncontrolled cell proliferation and survival. The most well-characterized example of tyrosine kinase dysregulation is the Bcr-Abl fusion protein found in human patients with chronic myelogenous leukemia (CML) that results in chronic activation of the cytoplasmic kinase Abl(*7, 8*). An RTK frequently altered in cancer is Kit, normally expressed on hematopoietic stem cells, melanocytes, and on mast cells (*9*). Dysregulation of Kit occurs in several human cancers including systemic mastocytosis (*10*), acute myleogenous leukemia (AML) (*11*) and gastrointestinal stromal tumors (GISTs)(*12, 13*) through a variety of different mutations that induce constitutive activation of Kit. Another example is the presence of a point mutation in EGFR in a subset of human primary lung tumors that induces prolonged signal transduction following stimulation by its ligand EGF(*14*).

Tyrosine kinase dysfunction has been far less characterized in veterinary oncology. Several authors have identified the presence of Kit mutations in dog mast cell tumors (MCTs) and these also result in uncontrolled signaling(*15-18*). Up to 30% of all dog MCTs may carry Kit mutations and these have been shown to be significantly associated with tumor grade, local recurrence and decreased survival. Mutations in Kit have also been found in canine GISTs and are nearly identical to those present in the human disease. Other kinases currently under investigation for their potential role in canine cancers include EGFR and Met, among others. With respect to feline cancers, vaccine associated sarcoma (VAS) cell lines were shown to express PDGFRβ that was phosphorylated following PDGF exposure and this induced a protective effect on VAS cells after treatment with chemotherapy(*19*). Therefore, PDGFR may promote the growth and survival of VAS *in vivo*. Recently, activating mutations in Kit have also been identified in feline MCTs(*20*).

While many strategies exist for targeting protein kinases, the most successful approach has been the use of small molecule tyrosine kinase inhibitors (TKIs) that work by blocking the ATP binding site of kinases, essentially acting as competitive inhibitors(*21-23*). In the absence of ATP binding, the kinase is not able to phosphorylate itself or initiate downstream signaling. Such inhibitors can be made in large quantities, are usually orally bioavailable, and can readily enter cells to gain access to the intended target.

### REVIEW OF THE HUMAN EXPERIENCE

Perhaps the most successful small molecule kinase inhibitor developed to date is Gleevec (imatinib mesylate, Novartis), an orally administered drug that blocks the activity of the cytoplasmic kinase Abl. This drug was designed specifically to target the constitutively active Bcr-Abl fusion protein found in human patients with CML(*7, 8*). For those individuals in the chronic phase of CML, Gleevec induces a remission rate close to 95% and most patients remain in remission for longer than 1 year (*24-29*). Gleevec also binds to the ATP binding pocket of the RTK Kit. Gleevec induces responses in 50-70% of patients with GISTs, far better than the 5% response rate treatment with a variety of chemotherapeutics (*30, 31*). Based on the high response rate, Gleevec has become standard of care therapy for patients with GISTs.

The TKI Sutent (SU11248, sunitinib, Pfizer), was developed as an anti-angiogenic agent and blocks the activity of the split kinase RTK family including VEGFR, PDGFR, FGFR, and Kit(*32*). Sutent has activity against several cancers including neuroendocrine, colon and breast carcinomas, and is currently under investigation in a variety of combination regimens(*32*). Definitive efficacy was further demonstrated in patients with Gleevec resistant GIST(*33*) and renal cell carcinoma that had failed IL-2 and/or interferon therapy(*32*). Another orally active small molecule TKI that has been successful in treating human cancers is Gefitinib (Iressa, Astra Zeneca) which inhibits EGFR signaling(*34, 35*). In human patients, Iressa has demonstrated clinical activity in non-small-cell lung cancer (NSCLC), with 12-20% of patients experiencing complete or partial responses and an additional 30-40% of patients experiencing stable disease(*3, 34*)

## CLINICAL TRIALS OF TKIS IN VETERINARY ONCOLOGY

Limited data exists on the clinical efficacy of small molecule inhibitors in veterinary medicine. In part, this is due to the fact that targets for therapeutic intervention are not clearly defined for most canine or feline cancers. Additionally, many of the human TKIs are currently cost-prohibitive, preventing their widespread use. Recently, Gleevec has been used to treat cancers in dogs and cats, and two TKIs (Palladia and Kinavet) have undergone registrational studies in dogs with mast cell tumors. These studies are discussed below and their general use is summarized in Table 1.

### Palladia (toceranib phosphate, SU11654)

The first evaluation of TKIs in veterinary patients was a phase I clinical trial exploring activity of the multi-targeted TKI Palladia (SU11654, Pfizer)(*36*). This compound is similar to Sutent, exhibiting activity against members of the split-kinase RTK family (VEGFR, PDGFR and Kit), and was therefore predicted to have both anti-angiogenic and anti-tumor activity. In 57 dogs with a variety of cancers (carcinomas, sarcomas, MCTs, melanomas, and lymphomas) objective responses occurred in 16 dogs (RR=28%) with stable disease in an additional 15 dogs for an overall biological activity of 54%. Responding tumors included sarcomas, carcinomas, melanomas, myeloma, and MCTs. The highest RR was in MCTs, with 10/11 dogs with Kit mutations exhibiting responses (n=9) or stable disease (n=1). A placebo controlled randomized registrational study of Palladia was then performed in dogs with non-resectable Grade II and III MCTs (*37*). During the blinded phase, the RR in Palladia treated (n=86) dogs was 37.2% (7 CR, 25 PR) versus 7.9% (5 PR) in placebo-treated (n=63) dogs. Of 58 dogs that received Palladia following placebo-escape, 41.4% (8 CR, 16 PR) experienced an objective response. The overall RR for all 145 dogs receiving Palladia was 42.8% (21 CR, 41 PR). Dogs with Kit mutations were much more likely to respond (SD≥10 weeks, CR, or PR) to Palladia than those without (82% vs 55%) and dogs without lymph node metastasis had a higher response rate than those with involvement (67% vs. 46%).

### Kinavet (masitinib)

Kinavet (AB Science) is another TKI that primarily targets Kit. In a phase II study of a Kinavet in 13 dogs with grade II and III MCTs there were 2 CRs, 2 PRs, and stable disease in an additional 2 dogs (Axiak et al, VCS 2006). A randomized double blind placebo controlled phase III clinical trial of Kinavet was then performed in over 200 dogs with non-metastatic Grade II or III MCTs(*38*). While the overall RR was not significantly different between placebo and Kinavet treated dogs (15% vs. 16%), there was a significant difference in time to progression between the two groups (75 vs 118 days), suggesting that Kinavet has biologic activity in MCTs. Although dogs with MCTs possessing Kit mutations did not experience a significantly greater response to therapy when treated with Kinavet (20%) compared to placebo (10%), they did experience a significantly longer time to progression in dogs with MCTs possessing Kit mutations.

### Gleevec (imatinib mesylate)

Gleevec has been used in dogs to treat MCTs. However, it has idiosyncratic hepatotoxicity in proportion of dogs necessitating discontinuation of therapy. A recent study demonstrated response to therapy in 10/21 dogs treated with Gleevec; the RR was 100% in dogs whose MCTs possessed a Kit ITD (n=5) (*39*). No dogs exhibited hepatotoxicity, although the duration of treatment was relatively short in most patients. Another study reported partial responses to therapy in 3 dogs with systemic mast cell disease treated with Gleevec(*40*). A phase I clinical trial evaluating the toxicity of Gleevec was performed in 9 cats with a variety of tumors(*41*). Doses of 10-15 mg/kg were well tolerated with no evidence of hematologic/hepatic toxicity and only mild gastrointestinal toxicity. Recently, a cat with systemic mastocytosis was treated with Gleevec and experienced a complete response to therapy(*20*). As previously mentioned, VAS cell lines were shown to express PDGFRβ; Gleevec was shown to block PDGF induced phosphorylation in these cells, inhibit the growth of VAS tumors in murine xenografts and reverse the protective effect of PDGF on chemotherapy induced growth inhibition(*19*).

## PRACTICAL ASPECTS OF TKI THERAPY IN COMPANION ANIMALS

The use of TKIs should be considered when standard therapies have not been of benefit, or have failed, or if no good therapy exists for a particular cancer. With respect to Palladia, tumor types shown to benefit from therapy include MCTs, as well as potentially metastatic mammary carcinoma, metastatic soft tissue sarcoma, and multiple myeloma, while Kinavet has shown some utility in the treatment of MCTs. The use of TKIs in combination with standard therapies such as radiation and chemotherapy is only beginning to be explored in human medicine, and has not yet been evaluated in veterinary medicine. However, preliminary evidence suggests that the addition of TKI therapy to standard therapeutics may improve outcome in some human cancers. Given the potential for additive toxicities, until clinical trials of combinations have been performed, it is probably advisable to use TKIs after standard treatments have been completed.

Similar to chemotherapeutics, TKIs often induce toxicities that target normal tissues, likely due to the effects of inhibition of receptors expressed on normal cells that require these for survival/proliferation under homeostatic conditions. Therefore, in many respects, TKIs should be viewed as similar to chemotherapeutics with respect to prevention and management of side effects. Both Palladia and Kinavet can induce anorexia, vomiting, diarrhea and gastrointestinal (GI) bleeding in treated patients (*36-38*). These are magnified in the setting of MCTs in which subclinical GI ulceration may be present. Treatment for mild GI toxicities primarily involves the initiation of supportive care (antacids/proton pump inhibitors, anti-diarrhea drugs). Should the toxicities be significant, a treatment break is warranted, and an alteration in dose or schedule may be indicated. Both Palladia and Kinavet also have the capacity to cause unique side effects. Palladia can induce a mild, non-life threatening neutropenia in a subset of treated patients. This does not seem to predispose dogs to bacterial infection, and often resolves over time. Kinavet can induce a protein losing nephropathy and a syndrome of hemolytic anemia in a small subset of patients, the origins of which are not clear.

## SUMMARY

Dysfunction of tyrosine kinases occurs frequently in human cancers and recent work indicates that a similar pattern of dysfunction is present in dog and cat cancers. Several inhibitors of tyrosine kinases are currently used in human cancer therapy and many of these have significant clinical activity. TKIs have only recently entered veterinary oncology, but the success of studies evaluating Palladia and Kinavet indicate that TKIs will soon become available to treat veterinary cancer patients. Several challenges remain for TKIs in veterinary oncology including defining cancers in which they are most likely to be effective, establishing regimens that reduce their toxicities, evaluating their biologic activity in the microscopic disease setting, and investigating how to combine them with standard therapeutics such as radiation therapy and chemotherapy.

**Table 1**

| TKI | Targets | Tumor Types | Species | Dose | References |
|---|---|---|---|---|---|
| Palladia® (toceranib) | Kit<br>VEGFR<br>PDGFR<br>Flt-3 | MCTs<br>Sarcomas<br>Carcinomas<br>Melanoma<br>Myeloma | Dogs | 3.25 mg/kg EOD | 36, 37 |
| Kinavet® (masitinib) | Kit<br>(PDGFR) | MCTs | Dogs | 12.5 mg/kg SID | 38 |
| Gleevec® (imatinib) | Kit<br>Abl<br>PDGFR | MCTs<br>Sarcomas | Dogs<br>Cats | 5-10 mg/kg SID | 19, 39, 40, 41 |

## REFERENCES

[1] C. A. London, **Vet Comp Onc** 2, 177 (2004). [2] P. C. Ma, G. Maulik, J. Christensen, R. Salgia, **Cancer Metastasis Rev** 22, 309 (Dec, 2003). [3] J. J. Laskin, A. B. Sandler, **Cancer Treat Rev** 30, 1 (Feb, 2004). [4] J. A. Fletcher, **Semin Oncol** 31, 4 (Apr, 2004). [5] A. Neubauer et al., **Leuk Lymphoma** 25, 91 (Mar, 1997). [6] I. Shchemelinin, L. Sefc, E. Necas, Folia Biol (Praha) 52, 81 (2006). [7] R. H. Alvarez, H. Kantarjian, J. E. Cortes, **Semin Hematol** 44, S4 (Jan, 2007). [8] M. D. Moen, K. McKeage, G. L. Plosker, M. A. Siddiqui, **Drugs** 67, 299 (2007). [9] S. J. Galli, K. M. Zsebo, E. N. Geissler, **Advances in Immunology** 55, 1 (1994). [10] J. Gotlib, Immunol Allergy **Clin North Am** 26, 575 (Aug, 2006). [11] A. S. Advani, Curr Hematol Rep 4, 51 (Jan, 2005). [12] J. Lasota, M. Miettinen, Semin Diagn Pathol 23, 91 (May, 2006). [13] M. A. Siddiqui, L. J. Scott, **Drugs** 67, 805 (2007). [14] T. J. Lynch et al., **N Engl J Med** 350, 2129 (May 20, 2004). [15] D. Zemke, B. Yamini, V. Yuzbasiyan-Gurkan, **J Vet Diagn Invest** 13, 341 (2001). [16] C. A. London et al., **Exp Hematol** 27, 689 (1999). [17] Y. Ma et al., **J Invest Dermatol** 112, 165 (1999). [18] S. Downing, M. B. Chien, P. H. Kass, P. E. Moore, C. A. London, **Am J Vet Res** 63, 1718 (Dec, 2002). [19] R. Katayama et al., **Cancer Chemother Pharmacol** 54, 25 (Jul, 2004). [20] M. Isotani et al., **Vet Immunol Immunopathol** 114, 168 (Nov 15, 2006). [21] H. J. Wanebo, A. Argiris, E. Bergsland, S. Agarwala, H. Rugo, **Cancer Metastasis Rev** 25, 279 (Jun, 2006). [22] I. Shchemelinin, L. Sefc, E. Necas, Folia Biol (Praha) 52, 137 (2006). [23] A. E. Wakeling, **Endocr Relat Cancer 12 Suppl** 1, S183 (Jul, 2005). [24] M. J. Mauro, B. J. Druker, **Oncologist** 6, 233 (2001). [25] H. Kantarjian et al., **N Engl J Med** 346, 645 (2002). [26] C. Beham-Schmid et al., Blood 99, 381 (2002). [27] B. J. Druker et al., **N Engl J Med** 344, 1031 (2001). [28] B. J. Druker et al., **N Engl J Med** 344, 1038 (2001). [29] C. L. Sawyers, Curr Opin Genet Dev 12, 111 (Feb, 2002). [30] M. Miettinen, M. Sarlomo-Rikala, J. Lasota, Hum Pathol 30, 1213 (1999). [31] M. Miettinen, M. Sarlomo-Rikala, J. Lasota, **Ann Chir Gynaecol** 87, 278 (1998). [32] L. Q. Chow, S. G. Eckhardt, J Clin Oncol 25, 884 (Mar 1, 2007). [33] T. G. Hopkins, M. Marples, D. Stark, **Eur J Surg Oncol** 34, 844 (Aug, 2008). [34] M. Ranson, S. Wardell, **J Clin Pharm Ther** 29, 95 (Apr, 2004). [35] R. Roskoski, Jr., **Biochem Biophys Res Commun** 319, 1 (Jun 18, 2004). [36] C. A. London et al., **Clin Cancer Res** 9, 2755 (Jul, 2003). [37] C. A. London et al., **Clin Cancer Res** In Press (2009). [38] K. A. Hahn et al., **J Vet Intern Med** 22, 1301 (Nov-Dec, 2008). [39] M. Isotani et al., **J Vet Intern Med** 22, 985 (Jul-Aug, 2008). [40] L. Marconato et al., **J Vet Intern Med** (Jun 18, 2008). [41] J. L. Lachowicz, G. S. Post, E. Brodsky, **J Vet Intern Med** 19, 860 (Nov-Dec, 2005).

**UNDERSTANDING & TREATING BONE CANCER PAIN IN COMPANION ANIMALS**

**Timothy M. Fan, DVM, PhD, DACVIM**
**Urbana, IL**

## INTRODUCTION

Bone is a living organ, rich in blood supply and nerves. Like any other organ, sensations of pain affecting the skeleton can dramatically decrease quality-of-life scores in companion animals. Given its principal anatomic function for bearing weight, any significant compromise in bone mineral density and quality, poses a risk for severe pain and pathologic fracture. Neoplasms which involve the skeleton can arise primarily from the bone, or secondarily invade or metastasize to the skeleton. In dogs, osteosarcoma is the most common tumor histology associated with focal skeletal pain; however, several additional tumor types have a predilection to involve bone including metastatic carcinoma and hematopoietic neoplasms such as multiple myeloma. In cats, primary bone tumors are much less frequent in comparison to dogs; however, involvement of bone from secondary invasion is common for oral squamous cell carcinoma. Despite the diverse tumor histologies which may pathologically affect bone, the mechanisms for how tumor cells invade and cause skeletal pain are universally shared between dogs and cats. Understanding how bone cancer pain is generated and sustained remains paramount for developing newer and more effective treatment strategies to improve quality-of-life scores in companion animals suffering from skeletal tumors.

## PATHOPHYSIOLOGY OF PAIN

Although pain is a universal sensation, its categorization can be differentiated based upon temporal aspects (acute, chronic, or intermittent), intensity (mild, moderate, severe, or excruciating), and anatomic origin (somatic, visceral, or neuropathic). Often in the cancer patient, all three types of pain may occur alone or in combination with one another. Nociception is the transduction, conduction and processing of afferent nerve signals generated by stimulated nociceptors, resulting in the perception of pain.[1-3] Signals from noxious stimuli (mechanical, thermal, or chemical) are transmitted mainly via two types of nerve fibers: the small myelinated Aδ fibers, and the small unmyelinated C fibers.[1-4] The larger myelinated Aβ fibers normally transmit non-noxious stimuli, such as touch, vibration, pressure, movement and proprioception.[2] Based upon their large diameter and heavy myelination, electrical transmission is rapidly conducted along these fibers. In comparison, electrical transmission conducted by C fibers is slower than Aδ, given their smaller diameter and unmyelinated nature which does not permit for rapid saltatory conduction. The Aδ and C fibers enter dorsal horn of the spinal cord via the dorsal root ganglia, where they synapse with second order neurons of the gray matter.[1-4] Some integration and modulation may take place in the dorsal horn, through interactions with excitatory and inhibitory interneurons. The resulting nociceptive information is carried to the brain via the spinothalamic tracts, where it can be integrated, processed, and recognized in multiple areas of the brain.[1,2]

Pain associated with skeletal neoplasms is attributed to 2 specific host responses. First, the presence of cancer cells in the bone microenvironment results in the release of chemical mediators by non-neoplastic stromal cells which in turn stimulate periosteal nociceptors, leading to the generation of painful sensations.[5,6] Second, the genesis, maintenance, and exacerbation of bone cancer pain may be directly attributed to dysregulated and pathologic osteoclastic bone resorption.[7-9] Based upon these mechanisms of bone pain generation, the optimal management of malignant osteolytic pain would combine eradication of cancer cells in bone matrix with inhibition of tumor-induced osteoclastic bone resorption. While some veterinary cancer patients' pain may be relieved adequately by treating the underlying malignancy, others will require symptomatic therapy to achieve improved quality of life, the primary objective of veterinary oncology.

## RECOGNITION OF PAIN

One of the biggest problems in controlling bone cancer pain with veterinary patients is that of its recognition by the clinician. Unlike human beings, companion animals tend to be more stoic when it comes to the outward expression of painful sensations. As such, pet owner and veterinarians are often oblivious to painful conditions affecting companion animals until skeletal diseases become very advanced with corresponding extreme pain levels. Although subjective pain recognition is useful in a clinical setting, certain physiological parameters, such as heart and respiratory rates, temperature, pupil size, and blood pressure, have been used as surrogate measures of pain in veterinary patients, should also be integrated into determining the extent and intensity of pain. Importantly, changes in clinical behavior provides a better estimate of painful conditions, and pet owner or veterinarian observations regarding changes in movement, posture, grooming, appetite and thirst, licking of the painful area, vocalization, growling when approached, among others overt changes in behavior may provide a semi-objective information on the degree of pain experienced by a given dog or cat suffering from malignant osteolytic diseases.

## OSTEOSARCOMA- A COMMON MODEL FOR EVALUATING ANALGESIC STRATEGIES

Osteosarcoma (OSA) is by far the most common primary bone tumor in dogs, and it is estimated that over 8,000 dogs per year will be diagnosed with OSA in the United States. Canine patients diagnosed with OSA are often middle-aged to older dogs of large or giant breeds, with definite overrepresentation of certain breeds such as the Rottweiler, Saint-Bernard, Golden Retriever, Irish Wolfhound, and Greyhound. The standard therapy for canine OSA includes radical surgical resection, adjuvant chemotherapy, and palliative radiation therapy. The median survival time of appendicular OSA treated with amputation followed with adjuvant systemic chemotherapy (doxorubicin, platinum compounds) is approximately one year.

The most common and striking presenting sign of dogs presented with OSA is lameness associated with severe pain of the affected bone. The mechanisms of bone pain have been well characterized. The link between osteolytic bone metastasis and pain is provided by the concomitant activation of osteoclasts and nociceptors by prostaglandins and cytokines. In fact, osteolytic destruction of bone is mediated by several humoral mechanisms in addition to prostaglandins and includes parathormone, cytokines, and growth factors that stimulate osteoclast activity. It is suspected that similar mechanisms take place in pain associated with primary bone tumors such as canine osteosarcoma. Since osteoclasts appear to play a preponderant role in the genesis of bone pain associated with malignant osteolysis, therapies that inhibit pathologic osteoclastic activities could be effective in significantly decreasing bone pain in canine osteosarcoma patients that are not candidate for radical surgery or do not have access to megavoltage radiation therapy.

## DOCUMENTED EFFECTIVE PALLIATIVE THERAPIES

### Palliative radiation therapy

Radiation therapy is considered the most effective treatment modality in human cancer patients for the management of osteolytic bone pain. Unfortunately, in human cancer patients, the incidence of multifocal or diffuse skeletal metastases secondary to primary carcinoma progression often precludes the efficient implementation of megavoltage radiation therapy. Luckily, this is the not case in dogs with OSA, as most often these patients only experience severe bone pain associated with a focal primary tumor. In this clinical setting when only one or few lesions are involved, the effective management of malignant bone pain is typically achieved by administering coarse fractions of 8 to 10 Gy of megavoltage irradiation, in 3 treatments at 0, 7 and 21 days. Palliative RT reportedly improves limb function and quality of life in about 75% of patients, and for a median of 2-3 months duration.[10] Although megavoltage radiation therapy appears effective when used as a single agent treatment option, recent studies suggest that the concurrent administration of systemic chemotherapy along with palliative radiation appears to enhance analgesic response rates and durations.[11] Given that apparent superior response rates can be achieved with combining radiation therapy with systemic chemotherapy, the dual use of these treatment modalities should always be recommended to pet owners willing to pursue the most effective palliative treatment options.

### $^{153}$Samarium-EDTMP

The use of a therapeutic radionuclide called $^{153}$Samarium-EDTMP has been described for both appendicular and axial OSA in dogs, and provides pain relief in many treated patients. $^{153}$Samarium-EDTMP is commercially available for the treatment of skeletal metastases in human cancer patient, and is sold by the name Quadramet™. In human cancer patients, Quadramet™ is effective for alleviating bone pain associated with diffuse or multifocal skeletal lesions, and therefore has the advantage over palliative radiation therapy which is limited for treating only a few lesions. By means of delivery concentrated radiation doses to the site of active bone remodeling, $^{153}$Samarium-EDTMP administration is capable of providing significant and meaningful palliation of bone pain in dogs suffering from appendicular OSA. $^{153}$Samarium-EDTMP therapy is well tolerated and alleviates osteolytic bone pain in the majority of dogs treated.[12] Side effects associated with treatment include transient decreases in platelet and white blood cell counts. The combination of $^{153}$Samarium-EDTMP with systemic carboplatin therapy has been recently investigated for the management of bone pain associated with appendicular OSA, and appears well-tolerated but does carry the risk for grade III or IV neutrophil nadirs.

### Stereotactic radiosurgery (SRS)

Although the delivery of radiation therapy is effective for managing a variety of tumor histologies, there always remains the concern for excessive normal tissue damage. For the treatment of brain tumors in human cancer patients, excessive normal tissue damage is unacceptable, and prompted the development of stereotactic radiosurgery, as coined the term "gamma knife". Stereotactic radiosurgery involves the precise delivery of a single large dose of radiation to a designated tumor target; achieved by multiple arrays of overlapping radiation beams. Similar to its utility for treating brain tumors in people, stereotactic radiosurgery has also been evaluated for alleviating bone pain associated with appendicular OSA. Recent results suggest that the use of stereotactic radiosurgery in dogs with OSA can provide pain alleviation, long-term local tumor control and improvement in limb function.[13] Similar to palliative radiation therapy, combining systemic chemotherapy with stereotactic radiosurgery appears to enhance response rates and durations.

## INVESTIGATIONAL PALLIATIVE THERAPIES- THE AMINOBISPHOPHONATES

The pharmaceutical use of bisphosphonates is accepted for the treatment of neoplastic bone disorders in human cancer patients. At low concentrations bisphosphonates inhibit bone resorption without inhibiting the process of bone mineralization. This results in stabilization and even enhancement of bone mineral density. Bisphosphonates directly inhibit bone resorption by binding to hydroxyapatite crystals, as well as inducing osteoclast apoptosis. In part, pain associated with bone cancers is a direct consequence of malignant bone resorption. Therefore, inhibiting pathologic bone resorption with bisphosphonates would theoretically decrease the likelihood of pathologic fracture, as well as alleviate intense bone pain.

### Aminobisphosphonate Background

Bisphosphonates are synthetic analogs of inorganic pyrophosphate ($PP_i$) that were initially utilized in the detergent industry as demineralizing agents, and then for diagnostic purposes in bone scanning, based on their ability to adsorb to bone mineral. The pharmaceutical use of bisphosphonates has now gained wide acceptance in human non-neoplastic bone disorders such as osteoporosis and Paget's disease. In the last decade, bisphosphonates have been intensely investigated as novel antineoplastic agents. Currently, several bisphosphonates have demonstrated efficacy for treatment of tumor-induced hypercalcemia, multiple myeloma, and metastatic bone diseases.

The effective treatment of bone disorders by bisphosphonates is attributed to their differential effect on bone resorption and bone mineralization. At low concentrations bisphosphonates inhibit bone resorption without inhibiting the process of bone mineralization. This results in stabilization and even enhancement of bone mineral density. Bisphosphonates directly inhibit bone resorption by binding to hydroxyapatite crystals. Once incorporated into the hydroxyapatite matrix of bone, bisphosphonates inhibit further calcium and phosphorous mineral dissolution. Perhaps more importantly, bisphosphonates impede osteoclast activity and induce osteoclast apoptosis; both mechanisms result in inhibition of bone resorption.

Several bisphosphonates of varying potency have been designed to optimize selective inhibition of bone resorption, in conjunction with minimal inhibition of bone mineralization. The relative potency of bisphosphonates is attributed to the chemical structure of their side chain ($R^2$ group). Bisphosphonates with a nitrogen atom within the $R^2$ group are extremely powerful inhibitors of bone resorption and are classified as aminobisphosphonates (Table 1). Although several aminobisphosphonates are commercially available, pamidronate has been used most extensively in people for the management of tumor-induced hypercalcemia, Paget's disease, osteolytic bone metastases of breast cancer and osteolytic lesions of multiple myeloma. One important difference between aminobisphosphonates and non-nitrogen containing bisphosphonates is that the latter cause osteoclast apoptosis by taking the place of phosphate groups in the adenosine triphosphate (ATP) molecule, thus yielding a cytotoxic compound, while aminobisphosphonates act on the mevalonate pathway by inhibiting the geranylgeranylation of certain proteins involved in cell signaling.

**Table 1:** Bisphosphonate Structure and Relative Antiresorptive Potencies

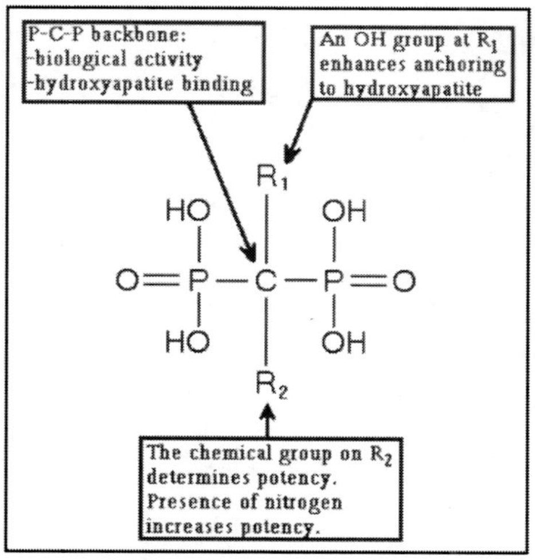


| Bisphosphonate | $R_1$ | $R_2$ | Relative Antiresorptive Potency (RAP) |
|---|---|---|---|
| Etidronate (Didronel©) | -OH | - $CH_3$ | 1 |
| Clodronate (Ostac©) | -Cl | -Cl | 10 |
| Pamidronate (Aredia©) | -OH | $-(CH_2)_2NH_2$ | 100 |
| Alendronate (Fosamax©) | -OH | $-(CH_2)_3NH_2$ | 1,000 |
| Zoledronate (Zometa©) | -OH | 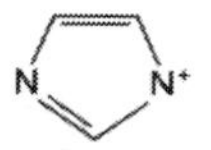 | 10,000 |

## REFERENCES

[1]Bruera ED, Portenoy RK (eds). **Cancer Pain: Assessment and Management.** Cambridge, UK: Cambridge University Press, 2003;485pp., [2]Gaynor JS, Muir WW. **Handbook of Veterinary Pain Management**. St. Louis, Mo: Mosby, 2002; 479pp., [3]Tranquilli WJ, Grimm KA, Lamont LA.*Pain* **Management for the Small Animal Practitioner**, 2nd Ed. Jackson, WY: Teton NewMedia, 2004; 152pp., [4]O'Brien MG. Cancer pain management: A review of recent considerations and advancements. *Vet* **Cancer Soc Newsletter** 1998;22:1-6.,[5]Clohisy DR, Mantyh PW. Bone cancer pain. Cancer 2003;97:866-873.,[6]Mantyh PW, Clohisy DR, Koltzenburg M, et al. Molecular mechanisms of cancer pain. **Nat Rev Cancer** 2002;2:201-209.,[7]Clohisy DR, Mantyh PW. Bone cancer pain and the role of RANKL/OPG. **J Musculoskelet Neuronal Interact** 2004;4:293-300.,[8]Goblirsch M, Mathews W, Lynch C, et al. Radiation treatment decreases bone cancer pain, osteolysis and tumor size. **Radiat Res** 2004;161:228-234.,[9]Luger NM, Honore P, Sabino MA, et al. Osteoprotegerin diminishes advanced bone cancer pain. **Cancer Res** 2001;61:4038-4047.,[10]Ramirez O, 3rd, Dodge RK, Page RL, et al. Palliative radiotherapy of appendicular osteosarcoma in 95 dogs. **Vet Radiol Ultrasound** 1999;40:517-522.,[11]Green EM, Adams WM, Forrest LJ. Four fraction palliative radiotherapy for osteosarcoma in 24 dogs. **J Am Anim Hosp Assoc** 2002;38:445-451.,[12]Milner RJ, Dormehl I, Louw WK, et al. Targeted radiotherapy with Sm-153-EDTMP in nine cases of canine primary bone tumours. **J S Afr Vet Assoc** 1998;69:12-17.,[13]Farese JP, Milner R, Thompson MS, et al. Stereotactic radiosurgery for treatment of osteosarcomas involving the distal portions of the limbs in dogs. **J Am Vet Med Assoc** 2004;225:1567-1572

# GETTING THE MOST FROM YOUR TUMOR HISTOPATHOLOGY REPORT

**Antony S Moore MVSc, BVSc, Diplomate ACVIM (Onc) and Angela E Frimberger VMD, Diplomate ACVIM (Onc)**
**Veterinary Oncology Consultants (www.vetoncologyconsults.com) & Wauchope, NSW Australia**

## INTRODUCTION

Being able to obtain diagnostic samples and establish definitive diagnoses are crucial to developing an effective treatment plan for a pet with cancer. Developing a comfortable relationship with a veterinary pathologist is essential to good oncology practice. Such a relationship begins with sample submission; remember that the pathologist can only provide limited information about the biopsy if he or she is provided with a poor sample or limited clinical information. It is very important to ensure that the pathologist has a complete, detailed, written physical description of the lesion and a history of the condition. If the histopathologic description does not fit the clinical picture or observations about the biological behaviour of the disease, the clinician should have further discussions with the pathologist. In such cases, it is in the patient's best interest to talk to the pathologist and for them to obtain additional evaluations and even a second opinion. Histopathologists, like all other specialists, are very comfortable obtaining second opinions to ensure that the patient receives the best treatment possible. Additionally, remember to give the pathologist feedback on the clinical outcome for the patient, which helps them to determine the value and accuracy of their work.

In addition to providing clinical information the veterinarian should ensure that the sample is obtained in an appropriate manner, from an appropriate site, and that the tissue is handled gently and prepared appropriately. Specific attention should be focused on obtaining appropriate samples and appropriate fixation.

This presentation will focus on the role of the histopathologist. The veterinary histopathologist has additional impact on the management of a patient with cancer beyond establishing a diagnosis.

## WHAT A VETERINARY HISTOPATHOLOGIST CAN PROVIDE

The selection and success of treatment of a cancer patient is directly dependent on the information obtained from a veterinary pathologist. While most clinicians have a good understanding of the basic terms used in a pathology report, if the report contains terms that are unfamiliar, the clinician should feel comfortable in discussing such terms with the histopathologist.

As long as there is adequate tissue available for review, a histopathologist should be able to answer such questions as:

- Is it a tumor?
- What type of tumor is it?
- Has the tumor metastasized (depending on the tissue sampled) or is there blood vessel and lymphatic invasion that could mean it is likely to metastasize?
- Is the tumor likely to have been completely excised (if the entire excised tissue is submitted)?
- If the tumor is not completely excised, where the margins are closest (this may depend on accurate marking by the clinician) and what margin (in mm) has been obtained in close resections.
- Would additional stains or tests allow better characterization of this tumor?

## TUMOR GRADING & STAGING

The grading of tumors is perhaps one of the most important tasks of a veterinary pathologist. There is sometimes confusion between definitions of the term grade, which is primarily determined by histopathology; and staging, which is primarily determined clinically. The biological behaviour of tumors are not always predicted by the histologic grade, however there is a long list of tumors where grade is extremely important in predicting behaviour and hence in directing treatment options. Grading is performed by either objective or subjective evaluation of the histopathologic characteristics of cancer and is often based on mitotic index (number of mitotic figures per 10 high power fields (hpf)), the degree of differentiation, degree of cellular or nuclear pleomorphism, proportion of necrosis, invasiveness, stromal reaction, and inflammatory response. These characteristics can be used to grade the tumor as a low grade (well differentiated, grade 1), intermediate grade (moderately differentiated, grade 2), or high grade (poorly differentiated, grade 3) tumor. Grading does include some subjective criteria; therefore variation may occur between pathologists when an individual tumor is graded.

Examples where grade is an important prognostic factor include mast cell tumors, soft tissue sarcomas melanoma, mammary gland carcinoma, synovial cell sarcoma, multilobular osteochondrosarcoma, hemangiosarcoma, splenic sarcomas, transitional cell carcinoma, squamous cell carcinoma, pulmonary carcinoma, osteosarcoma, histiocytic sarcomas and. lymphoma.

For example, the histologic scoring system shown in Table 1 was devised for primary lung carcinomas, and has been shown to carry prognostic significance. The histologic grade affected survival in one study of dogs with pulmonary carcinomas; dogs with grade 1 tumors had a median survival time of 26 months; dogs with grade 2 tumors had a median survival of 8 months, and dogs with grade 3 tumors had a median survival of 5 days.

**Table 1.** Histologic grading of primary pulmonary carcinomas

| Score | Parameter | | | | | | |
|---|---|---|---|---|---|---|---|
| | Differentiation | Nuclear pleomorphism | Mitotic index | Nucleolar size | Tumor necrosis | Fibrosis | Demarcation |
| 0 | | | | | none | none | |
| 1 | Well | Mild | 1-10 | Small | 1-20% | 0.5 (1-20%) | Well |
| 2 | Moderately | Moderate | 11-20 | Medium | 21-50% | 1.0 (21-50%) | Moderate |
| 3 | Poorly | Severe | 21-30 | Large | >50% | 1.5 (>50%) | Invasive |
| 4 | | | 31 + | | | | |

| Grade | Description |
|---|---|
| 1 | Cumulative score of 8.5 or less |
| 2 | Cumulative score of 9 to 14 |
| 3 | Cumulative score of 14.5 or greater |

The histopathologist may further assist in *staging* the tumor by determining the size of the tumor (measurement of the tumor specimen submitted), depth of invasion into surrounding normal tissues and by identification of metastasis to regional lymph nodes or other distant sites such as the lung. Just like grading, staging is essential to direct therapy and to help render an accurate prognosis.

**MITOTIC INDEX**

Mitotic index is an independent predictor of tumor behaviour for many cancers in pets; osteosarcomas, melanomas and mast cell tumors in dogs are tumors for which higher mitotic index means a much worse prognosis for the patient. Mitotic index is defined by convention as number of mitoses counted in 10 high power (40 times magnification) fields. Selection of fields for counting of mitoses should be random, avoiding areas of necrosis and abundant matrix. A meaningful index cannot be calculated from a rate of "0 to 3 per high power field", as this would give a mitotic index of anywhere from 3 to 27 which will cover the gamut of low grade to high grade for many tumors. Some tumor scoring schemes characterize mitotic rate as a mitosis score (0-4; 0 = 0-10 mitotic figures per 10 high power fields; 1 = 11-20; 2 = 21-50; 3 = 51-74; 4 = 75 or more) but the most useful to the clinician is the **number of mitoses per 10 high power fields**.

**EVALUATION OF TUMOR MARGINS**

For surgery to be successful, complete margins of normal tissue around the tumor are essential whether the tumor is benign or malignant. Margins should be checked in all directions including those lateral and deep to the tumor, and this can only be done if all resected tissue is submitted for histopathology. While a biopsy result of apparent complete excision is encouraging, the histopathological evaluation of margins is limited by the pathologist's inability to examine the entire deep and lateral margins; however the pathologist *can* examine a number of sites around the sample to increase the confidence that it was removed completely. This process can be made more helpful to the veterinarian if different colored inks or sutures are placed to guide the pathologist to the particular areas that the surgeon feels the least "comfortable" about having removed with adequate margins.

The pathology report should always be interpreted in light of your clinical impression of the surgery. For example, for a grade 2 mast cell tumor, most investigators feel a clear *histologic* margin of 5mm is adequate; the published reports requiring 2-3cm margins are *clinical* margins, not histologic. If the deep margin includes fascia this may be encouraging; both muscle and fascia may act as a partial barrier to invasion by some tumors such as mast cell tumors and soft tissue sarcomas, particularly early on. Ideally, if the margins were less than 5mm, the best option is a second, "scar revision" surgery to remove margins around the remaining scar. Any tissue removed at a scar revision should, again, be labeled and submitted for histologic assessment.

Thus, if the clinical impression was that the tumor was *not* completely resected, then a histopathology report of clear margins should not be relied on. On the other hand, if the histopathologist finds evidence of tumor invasion to the margins of the excision, then this should take precedence even if the impression at surgery was of a clear excision.

**SPECIAL STAINS**

While the diagnosis of cancer still relies heavily on morphologic assessment of specimens under light microscopy with routine staining (hematoxylin and eosin; H & E), in some situations, the tumors may be too undifferentiated to determine their cell of origin. The use of special stains can be helpful for determining the tissue of origin which can be pivotal for treatment and hence outcome.

Deciding on the specific diagnosis for an undifferentiated round cell tumor (which could be a mast cell tumor or a lymphoma) would be very important as their treatment is dramatically different. In this situation a poorly granulated mast cell tumor may be identified with toluidine blue and lymphoma would stain with immunohistochemical staining for CD-79a (B-lymphocyte marker) or CD-3 (T-lymphocyte marker).

**Some examples of special stains (histochemistry) include:**

- Toluidine blue is a marker of mast cell granules and helpful in distinguishing poorly differentiated (and poorly granulated)

mast cell tumors.

- Melanin bleach or iron stains are helpful in distinguishing melanin and haemosiderin to determine if the patient has a melanoma
- Alcian blue stains mucin and acid-mucous substances and is helpful for determining if the tumor is a neurofibrosarcoma or myxosarcoma.
- Mucicarmine or periodic acid Schiff (PAS) is used to identify glycogen that is in mucus of poorly differentiated carcinomas.

## IMMUNOHISTOCHEMISTRY

Immunohistochemistry utilizes antibodies directed toward specific epitopes such as intermediate filaments, secretory substances, hormones, and proteins of various types. Immunohistochemistry takes advantage of the preservation of certain molecules that can demonstrate the source of an otherwise undifferentiated neoplasm. Some examples of commonly used immunohistochemical stains for the more specific diagnosis of certain tumors include:

### Mesenchymal marker

*Vimentin* is an intermediate filament staining positive for all tumors of mesenchymal origin. It is also a good internal control as a marker of technique.

### Epithelial marker

*Cytokeratin* is expressed in epithelial cells and is a good marker for carcinomas. Cytokeratin may be co-expressed in some mesenchymal tumors (such as synovial sarcomas) and may be prognostic for those tumors. Mesotheliomas may stain positive for cytokeratin.

### Muscle markers

*Desmin* is an intermediate filament of skeletal, cardiac and smooth muscle. It is expressed in most rhabdomyosarcomas, variably in smooth muscle tumors, and rarely in histiocytic sarcomas.

*Actins: muscle specific actin* stains skeletal, smooth and cardiac muscle and their tumors and may also be found in histiocytic sarcomas; *smooth muscle actin*, as the name implies, is more restricted to leiomyomas and leiomyosarcomas.

Positive staining for *myoglobin* may be a marker of skeletal muscle tumors (rhabdomyosarcomas).

### Fibrohistiocytic markers

*CD-18* (canine or feline derived) is a histiocytic (monocyte and macrophage) marker. If an undifferentiated "round-cell" tumor is a histiocytoma or a spindyloid tumor is a histiocytic sarcoma then tumor cells would be CD-18 positive, CD-3 negative, and CD-79a negative. Other stains such as alpha-1-antitrypsin or lysozyme have been used to identify histiocytes, but are not as specific as CD-18.

### Lymphoid markers

T lymphocyte markers (e.g.: *CD-3*) and B lymphocyte markers (e.g.: *CD-79a*) stain T and B cell lymphomas, respectively. *CD-34* is felt to be a stem (progenitor) cell marker, and appears to be valuable in differentiating acute lymphoblastic leukemia (ALL) from stage 5 lymphoma that is infiltrating bone marrow. Chronic lymphoid leukemia (CLL) should not express CD-34.

### Endothelial marker

*Factor VIII* can be used to identify tumors of endothelial cells and is therefore useful in diagnosis of haemangiosarcoma. In humans *CD-31* (platelet endothelial cell adhesion molecule) and *CD-34* (endothelial as well as progenitor cells) are also useful markers for haemangiosarcomas

### Neuronal, nerve sheath and melanocyte markers

*S-100* has a wide range of activity in the peripheral and central nervous systems being expressed on astrocytes, oligodendrocytes and Schwann cells; and so is positive in tumors deriving from those tissues as well as many peripheral nerve sheath tumors. In addition, S-100 is expressed on melanocytes and so is a very powerful marker for unpigmented or spindyloid melanomas.

*HMB-45 (melan-A)* is a marker of pre-melanosomal vesicles and is a useful marker to stain most melanomas. *Chromogranin* is a marker for some neuroendocrine tumors.

*Neuron-specific enolase* may be used to stain uncommon tumors of the nervous system such as neuroblastoma and paraganglioma.

*Glial fibrillary acidic protein (GFAP)* is an intermediate filament found in astrocytic tumors and expressed in some nerve sheath tumors.

### Markers of cellular proliferation

*Ki-67* is a cell proliferation rate marker used to determine the speed of tumor progression. For most tumor types, increased staining for Ki-67 correlates with a worse prognosis. Ki-67 staining appears to be more helpful in establishing prognosis than other markers such as *PCNA* (proliferating cell nuclear antigen) and *AgNOR* (silver staining nucleolar organizing regions). AgNOR staining has proven prognostically significant in canine mammary gland tumors, lymphoma, and mast cell tumors, but is often technically challenging.

**Endocrine markers**

*Insulin, glucagon, calcitonin, parathormone and thyroglobulin* can be used to identify tumors of hormone producing tumors of these substances, such as pancreatic, parathyroid or thyroid tumors.

**Miscellaneous markers and tests**

Immunostaining for *C-219* (and other markers) can detect *p-glycoprotein (multidrug resistance; MDR)*, a protein associated with resistance to naturally occurring chemotherapeutic agents (such as doxorubicin and vincristine).

**Table 2.** Special stains / immunohistochemistry for undifferentiated tumors

<table>
<tr><td rowspan="2">CK+</td><td>Vim +</td><td colspan="5">Synovial sarcoma</td></tr>
<tr><td>Vim -</td><td colspan="5">Carcinoma</td></tr>
<tr><td rowspan="8">CK-</td><td rowspan="6">Vim+</td><td>Muscle# markers +</td><td colspan="4">Leiomyosarcoma</td></tr>
<tr><td rowspan="5">Muscle markers -</td><td rowspan="2">S100 +</td><td>Melan A +</td><td colspan="2">Melanoma/ clear cell sarcoma</td></tr>
<tr><td>Melan A -</td><td colspan="2">PNST, liposarcoma</td></tr>
<tr><td rowspan="3">S100 -</td><td>Factor VIII +</td><td colspan="2">Haemangiosarcoma</td></tr>
<tr><td rowspan="2">Factor VIII -</td><td>CD18 +</td><td>Histiocytic sarcoma</td></tr>
<tr><td>CD18 -</td><td>Other sarcoma, fibrosarcoma</td></tr>
<tr><td rowspan="2">Vim -</td><td>Internal control +</td><td colspan="4">Consider other markers (neuroendocrine etc)*</td></tr>
<tr><td>Internal control -</td><td colspan="4">Consider technical problem</td></tr>
</table>

# desmin, actin (muscle-specific or smooth muscle)
* neuron-specific enolase

**Table 3.** Special stains / immunohistochemistry for round cell tumors

<table>
<tr><td>Toludine blue +</td><td colspan="5">Mast cell tumor</td></tr>
<tr><td rowspan="7">Toludine blue -</td><td>CD 79a +</td><td>CD 3 -</td><td colspan="3">B-cell lymphoma</td></tr>
<tr><td rowspan="6">CD 79a -</td><td>CD 3 +</td><td colspan="3">T-cell lymphoma</td></tr>
<tr><td rowspan="5">CD 3 -</td><td>Melan A +</td><td colspan="2">Melanoma</td></tr>
<tr><td rowspan="4">Melan A -</td><td>CD18 +</td><td>Histiocytoma, histiocytic sarcoma</td></tr>
<tr><td rowspan="3">CD18 -</td><td>TVT in dogs</td></tr>
<tr><td>Plasma cell markers #</td></tr>
<tr><td>Consider other markers * and re-evaluate</td></tr>
</table>

# Kappa chain; Lambda chain, CD 138 (human).
* S-100, cytokeratin, vimentin, desmin, actin.

## REFERENCES


1. Kuntz CA, et al. **JAVMA** 1997; 211:1147. 2. Kirpensteijn JK, et al. **Vet.Pathol** 2002;39: 240. 3. Dernell WS, et al. **JAAHA** 1998;34:11. 4. McNiel EA, et al. **JAVMA** 1997;211:1422. 5. Ruslander DA, et al. **In vivo** 1997;11:169. 6. Ponce F, et al. **Vet J** 2004;167:158. 7. Murphy S, et al. **Vet Rec** 2004;154:743. 8. Spangler WL, et al. **Vet Pathol** 2006;43:136. 9. Romansik EM et al. **Vet Pathol**. 2007;44:335.

# HEMATOLOGY MONITORING FOR SMALL ANIMAL CANCER PATIENTS

**Angela E Frimberger VMD, Diplomate ACVIM (Onc) & Antony S Moore MVSc, BVSc, Diplomate ACVIM (Onc)**
**Veterinary Oncology Consultants (www.vetoncologyconsults.com) & Wauchope, NSW Australia**

## INTRODUCTION

Hematologic evaluation is an important part of initial diagnostics and staging in cancer patients, as well as a vital part of ongoing monitoring for patients undergoing treatment. The cancer process itself may cause hematologic abnormalities by direct involvement of the bone marrow or paraneoplastic effects that may aid the diagnostic process or require management in themselves. During cancer treatment, some of the most common side effects of chemotherapy are hematologic.

## HEMATOLOGIC ABNORMALITIES ASSOCIATED WITH MALIGNANT DISEASE

A variety of hematologic abnormalities are seen in patients with cancer, either as a direct physical result of the tumor, or as paraneoplastic syndromes. In general these resolve with tumor control.

### Abnormal red blood cell numbers

Many cancer patients have anemia detected on baseline CBC evaluations. Potential causes can include anemia of chronic disease, blood loss, immune-mediated hemolytic anemia, and marrow infiltration by disease.

Anemia of chronic disease occurs when red blood cell lifespan is shortened, compounded by disordered iron metabolism and blunted eyrthopoietic response. The anemia seen here is usually of low grade and normocytic, normochromic, and nonregenerative. It is not usually a significant clinical problem, and usually resolves (sometimes slowly) with tumor control.

Acute blood loss can result in a regenerative anemia; if peracute, the anemia may appear nonregenerative. Chronic blood loss results in a microcytic, hypochromic anemia. Common sites of bleeding include splenic or hepatic tumors, nasal tumors, and GI tumors. Even if not involved in tumor, the GI tract may also be a site of blood loss as a paraneoplastic effect in mast cell tumors and gastrinomas.

Immune-mediated hemolytic anemia can occur as a paraneoplastic effect in a number of cancers, particularly lymphoma. This anemia varies in severity, is usually regenerative, and can be acute or chronic.

Primary erythrocytosis, or polycythemia vera, is a myeloproliferative disorder of the erythroid lines. Secondary erythrocytosis can be seen as a rare paraneoplastic syndrome associated mostly with renal tumors, but other tumors have also been reported.

### Abnormal white blood cell numbers

Leukemias can result in increased numbers of any line of circulating cells, the most common in dogs and cats being lymphoid leukemias. Increased circulating lymphoid cells can also occur with stage 5 lymphoma. In chronic leukemias the circulating cells may resemble normal cells cytologically. Increased circulating numbers of white blood cells can also occur as paraneoplastic syndromes and it is necessary to differentiate a chronic leukemia from a paraneoplastic syndrome.

Decreased numbers of white blood cells usually raise a suspicion of bone marrow infiltration with disease (myelophthisis), particularly if other blood cell lines are affected. Pancytopenia can also occur as a paraneoplastic syndrome with estrogen-producing tumors (Sertoli cell tumors), which may not resolve with removal of the tumor.

### Abnormal platelet numbers

Thrombocytopenia can occur as a result of several mechanisms, including myelophthisis, immune-mediated destruction, and consumption. Thrombocytosis can occur as a rare paraneoplastic syndrome.

## HEMATOLOGIC ABNORMALITIES ASSOCIATED WITH CANCER TREATMENT

The cancer treatment most likely to cause hematologic changes is chemotherapy. All anticancer chemotherapy drugs have the potential to produce adverse side effects. However, the side effects that can occur in pets are usually not as severe or as long lasting as those which occur in humans due to the lower doses used. Most clients are pleasantly surprised at how well their pets feel during chemotherapy. Despite the lower doses and lower risks of toxicity, some toxicity may occur, primarily following the first time a patient receives a chemotherapy drug and before the dose has been "individualized" for that patient. Of these toxicities, neutropenia is the one that is most likely to be life-threatening, due to the increased risk of sepsis.

*Myelosuppression* is a general term applied to the toxic effects of chemotherapy on the bone marrow. The most chemosensitive cells in the bone marrow are the proliferating progenitors and precursors, which are starting to commit to a particular lineage but are still immature. The more differentiated cells form a non-proliferating pool of maturing cells that will be unaffected by chemotherapy and will provide mature cells for a period of days. Like the maturing cells, haematopoietic stem cells are largely non-proliferating and so are relatively resistant to chemotherapy toxicity. However, they are stimulated to divide by the loss of proliferating precursor cells, and rapidly replace the lost cells so that cytopenias following chemotherapy rarely last more than several days. This also has implications for the interval between administrations of myelosuppressive drugs. If these drugs are given when the stem cell pool is dividing (i.e., soon after the previous administration), then severe prolonged myelosuppression due to stem cell destruction may occur. The usual interval between for myelosuppressive drug administrations is every 2 to 3 weeks. Some drugs (such as lomustine and carboplatin) may have delayed or prolonged nadirs, and dosing intervals are longer for these drugs. The relative likelihood of causing myelosuppression varies between drugs (Table 1) and between species.

**Table 1.** Myelosuppressive effects of chemotherapeutic agents used in veterinary medicine

| *Highly Myelosuppressive* | *Moderately Myelosuppressive* | *Mildly Myelosuppressive* |
|---|---|---|
| Doxorubicin | Melphalan | L-asparaginase[a] |
| Vinblastine | Chlorambucil | Vincristine[a] |
| Cyclophosphamide | 5-Fluorouracil | Bleomycin |
| Lomustine (CCNU) | Methotrexate | Corticosteroids |
| Actinomycin-D | Procarbazine | |
| Mitoxantrone | Cisplatin | |
| Carboplatin | | |

[a] More pronounced myelosuppression can occur if these two drugs are administered concurrently.

### Anemia

Although myelosuppression does disrupt erythropoiesis, anemia as a direct chemotherapy side effect is rarely of clinical significance in veterinary oncology, because of long lifespan of red blood cells. When it does occur it is usually only in long term chemotherapy patients, and tends to be a low grade anemia with a gradual onset.

### Thrombocytopenia

Thrombocytopenia occasionally may occur as a direct chemotherapy side effect, although usually it is not dose-limiting. Although it can follow treatment with any myelosuppressive drug, it is most likely to be significant after treatment with carboplatin or lomustine. Chemotherapy-related thrombocytopenia rarely causes clinical signs, however at counts of less than 50,000/uL the risk of bleeding increases and the veterinarian should be alert to petechiation, ecchymoses or mucosal bleeding. Myelosuppressive chemotherapy should not be administered if the platelet count is less than 100,000/uL.

### Neutropenia

Sepsis due to chemotherapy or cancer-related neutropenia is one of the more common emergencies in veterinary cancer medicine but is usually preventable through judicious monitoring and appropriate supportive care during cancer therapy. In addition, caregivers should be educated about the early clinical signs of neutropenia and thrombocytopenia induced by cancer treatment, so that they can assist in early detection and seek immediate treatment. The risk of developing neutropenic sepsis is probably less than 5% in veterinary patients, but because dogs may hide their symptoms until late in the disease, the condition of sepsis may be quite advanced when first recognized and requires prompt intervention by the clinician. Timely treatment can make the difference between life and death for these patients.

Knowing the myelosuppressive potential for chemotherapy drugs being administered, and having an estimate of the time that the neutrophil (not just total white cell count) nadir - or low point - will occur, allows the veterinarian to plan monitoring blood counts and to administer prophylactic antibiotics, if that approach is chosen. Chemotherapy agents vary as to how myelosuppressive they are and therefore how likely they are to cause neutropenia and hence increase the risk of sepsis. Table 1 outlines the myelosuppressive potential for a number of commonly used chemotherapeutics.

The timing of the neutrophil nadir also varies from drug to drug. The neutrophil nadir occurs 7 to 10 days after administration for most chemotherapeutics, and for consistency a CBC is collected 7 days after administration in order to assess the risk of sepsis, and whether the dosage of that drug needs to be changed. The aim is to have a nadir between 1,000 and 3,000 segmented neutrophils/uL. The nadir is usually of short duration at the doses used in veterinary medicine, recovering to normal within 48-72 hours. Some drugs may cause a "delayed" neutrophil nadir, for example carboplatin may cause a nadir 2 or even 3 weeks after administration in cats and in some dogs, and recovery to normal may be further delayed, particularly in cats. The clinician is advised to check the expected nadir time and severity for each drug they are not familiar with, and in each species, and be familiar with factors that could cause alterations in chemotherapy drug metabolism that could increase the risk of myelosuppression (e.g. reduced renal clearance may exacerbate the myelosuppressive effects of carboplatin).

### Prophylaxis

Prophylactic antibiotic therapy is often recommended for patients receiving myelosuppressive chemotherapy as a method of reducing the risk of sepsis. The costs of prophylactic treatment may be weighed against the decreased risk of morbidity due to sepsis, as well as decreased risk of hospitalization and costs associated with it. Some oncologists recommend use of trimethoprim sulfa as a prophylactic measure in any dog receiving a myelosuppressive chemotherapy agent for the first time (until the nadir neutrophil count is known) as this drug is relatively inexpensive, orally administered, and has little negative effect on gastrointestinal flora. This approach has been shown to reduce morbidity and hospitalization in dogs treated with doxorubicin chemotherapy for osteosarcoma or lymphoma.

**How to manage mild (grade 1) neutropenia**

For most chemotherapy patients the neutrophil count at the expected nadir is between normal and 1,500/uL. At this level the risk of complicating illness is low, but because the time of the true nadir is not known for an individual patient prophylactic antibiotics should be continued on an outpatient basis. A good broad spectrum orally administered antibiotic choice for most dogs is trimethoprim-sulpha combinations such as Tribrissen (see more below under the heading Prevention). In cats enrofloxacin (or orbifloxacin; which appears to cause less nausea) can be given. Prior to the next dose of myelosuppressive chemotherapy another CBC should be collected and the next drug should not be administered unless the neutrophil count is above 3,000/uL.

For animals experiencing mild neutropenia, the dosage of the chemotherapy drug causing that effect should **not** be changed the next time the drug is administered unless recovery is delayed requiring the next treatment to be postponed.

**How to manage moderate (grade 2) neutropenia**

With a neutrophil count between 1,500/uL and 1,000/uL following chemotherapy, the risk of complicating illness is still fairly low. However because the time of the true nadir is not known for an individual patient, prophylactic antibiotics should be continued on an outpatient basis, or started in pets that are not already receiving them. Prior to the next dose of myelosuppressive chemotherapy another CBC should be collected and the next drug should not be administered unless the neutrophil count is above 3,000/uL.

For animals experiencing moderate (grade 2) neutropenia, the dosage of the drug causing that effect does **not** need to be changed the next time the drug is administered, although veterinarians less experienced in giving chemotherapy may prefer to make a 25% dose reduction for all subsequent doses of **that drug**.

**How to manage moderate (grade 3) neutropenia**

For some animals the myelosuppressive effects of chemotherapy may be more pronounced; in these pets the neutrophil count may be between 1,000/uL and 500/uL following chemotherapy. At this level the risk of complicating illness is much higher, although most animals still will not experience sepsis as long as they are receiving prophylactic antibiotics. These should be continued, or started in pets that are not receiving them. A full examination of the patient should be performed, but the pet should **not** be hospitalized unless parenteral care is needed, as this may increase the risk for sepsis (see below for risk factors). Owners should be instructed to check the temperature (preferably rectal) at least twice a day, and contact the veterinarian if it is above normal on 2 occasions 30 minutes apart. Additionally, any signs of illness (vomiting, diarrhoea, anorexia) should be reported to the veterinarian promptly, and the patient treated as an emergency. Prior to the next dose of myelosuppressive chemotherapy another CBC should be collected and the next drug should not be administered unless the neutrophil count is above 3 ,000/uL.

For animals experiencing moderate (grade 3) neutropenia, the dosage of the drug causing that effect should be reduced by 25% for all subsequent doses of **that drug**; and a CBC should be collected at the expected nadir after administering the reduced dose of that drug.

**How to manage severe (grade 4) neutropenia**

Rarely the myelosuppressive effects of chemotherapy may be marked; in these pets the neutrophil count is below 500/uL following chemotherapy. At this level the risk of complicating illness is very high, and some animals will become septic even if they are receiving prophylactic antibiotics. A full examination of the patient should be performed, but, again, the pet should not be hospitalized unless there is a fever, or unless parenteral supportive care is needed, as this may increase the risk for sepsis (see below for risk factors). Owners should be instructed to check the temperature (preferably rectal) at least twice a day, and contact the veterinarian if it is above normal on 2 occasions 30 minutes apart. Additionally, any signs of illness (vomiting, diarrhoea, anorexia) should be reported to the veterinarian promptly, and the patient treated as an emergency. Prior to the next dose of any myelosuppressive chemotherapy another CBC should be collected and the next drug should not be administered unless the neutrophil count is above 3,000/uL.

For animals experiencing severe (grade 4) neutropenia, the dosage of the drug causing that effect should be reduced by 25% for all subsequent doses of **that drug**; a CBC should be collected at the expected nadir after administering the drug at a dose reduction.

## CONCLUSION

Careful reading of the hematology report, both at baseline pre-treatment evaluation, and during treatment, can provide key information that will allow the clinician to modify treatment for optimal management of each individual patient, and direct further diagnostics when needed. While many anticancer chemotherapeutics have myelosuppressive potential, at the doses used in veterinary medicine most neutropenia caused by cancer treatment can be managed resulting in minimal impact on the patient's quality of life. It is important to be aware of the variation in myelosuppressive potential and timing of the individual drugs being used. Careful education of caregivers in combination with careful monitoring and prompt appropriate therapy when indicated will result in good outcomes for almost all neutropenic chemotherapy patients.

## REFERENCES

[1] Lara-Garcia A, et al. **JAVMA.** 2008;232:1488-95. [2] Chretin JD, et al. **JVIM** 2007;21:141-148. [3] Grindem CB et al. **JVIM** 8:400, 1994.

# SERUM BIOCEHMISTRY ANALYSIS & URINALYSIS IN SMALL ANIMAL CANCER PATIENTS

**Angela E Frimberger VMD, Diplomate ACVIM (Onc) & Antony S Moore MVSc, BVSc, Diplomate ACVIM (Onc)**
**Veterinary Oncology Consultants (www.vetoncologyconsults.com) & Wauchope, NSW Australia**

## INTRODUCTION

Cancer patients are often geriatric, and may have the potential for a number of underlying conditions that may adversely affect their health and the potential success of therapy. In addition, any neoplastic process may result in a number of paraneoplastic conditions that affect the well being of the cancer patient. In addition, other unrelated but serious concerns may be identified that may bear on the patient's overall prognosis and could affect decisions regarding treatment options.

In many instances, correcting or supporting underlying problems such as renal failure, urinary tract infections, heart disease, and metabolic disturbances may significantly improve the general health of the patient and thus improve the potential for successful cancer care, and the overall quality of life.

When complicated surgical procedures or multiple radiation therapies that require repeated or prolonged anesthesia are planned, adequate renal and hepatic functions are vital.

The likelihood of a successful outcome for a patient treated with chemotherapy is as dependent on drug metabolism and elimination (and drug absorption for orally administered chemotherapy) as it is on the sensitivity of the tumour. Consequently, information gained during staging may identify problems that will impact the type and dosage of chemotherapy to optimize efficacy while limiting toxicity.

For example, hepatic dysfunction may lead to delayed drug elimination (examples; vinca alkaloids, doxorubicin) and therefore to greater toxicity such as myelosuppression. In contrast, cyclophosphamide is activated in the liver so hepatic dysfunction may result in poor efficacy.

Similarly, renal dysfunction may worsen toxicity for some drugs that are themselves nephrotoxic. For example, cisplatin and streptozotocin should not be used in patients with renal azotemia. Reduced renal excretion of carboplatin will exacerbate myelosuppression.

## SERUM BIOCHEMISTRY EVALUATION

A full serum chemistry panel is always recommended for baseline staging of any pet with cancer, particularly if surgery, chemotherapy or multiple anaesthesias (such as for radiotherapy) are anticipated. This is even more critical for geriatric patients or those at high risk for paraneoplastic syndromes (i.e. dogs with lymphoma). Information gained from these evaluations can also be helpful in planning for supportive and palliative care. These tests are easy to perform, relatively reliable, and minimally invasive. Findings that may be relevant in evaluation of a cancer patient and treatment planning can include the following.

### Alanine aminotransferase (ALT)

Some chemotherapeutics are hepatotoxic. CCNU may cause elevations in serum ALT prior to causing functional hepatopathy, and prompt discontinuation will resolve the elevations in most dogs. Dogs with elevations in ALT prior to receiving a dose of CCNU are at increased risk of toxicity. Therefore monitoring of serum ALT prior to patients receiving each dose of CCNU is a good screening test for toxicity. In dogs with elevations over baseline, subsequent measurement of bile acids will be a more specific test for hepatic dysfunction.

### Alkaline phosphatase (ALP)

Multiple studies have demonstrated an association between elevated serum alkaline phosphatase (ALP) levels at the time of diagnosis and poor survival after amputation for osteosarcoma. In one of these, elevated total ALP levels were associated with shorter survival times, 5.5 months versus 12.5 months for dogs with normal serum ALP. Bone ALP levels may be even more powerful in their ability to predict survival in dogs with osteosarcoma.

### Bilirubin

As above for ALT. Elevations in bilirubin are of major concern if hepatotoxic agents are to be used. Additionally, chemotherapy drugs that are metabolized by the liver (vincristine, vinblastine, doxorubicin) will cause exaggerated toxicity if administered to a patient with liver dysfunction; therefore doses of these drugs should be reduced for patients with liver dysfunction.

### Creatinine

Elevated serum creatinine, if accompanied by isosthenuria or hyposthenuria, is indicative of renal dysfunction. When the urine specific gravity is high, an elevated creatinine level suggests dehydration. In older cats, serum creatinine should be interpreted in light of serum T4 level; i.e. hyperthyroidism may mask renal insufficiency. In animals with renal dysfunction, doses of renally excreted chemotherapy drugs (e.g. carboplatin) should be reduced; and nephrotoxic drugs (e.g. cisplatin in dogs and doxorubicin in cats) should be avoided altogether.

### Urea

Elevated urea with a normal creatinine and urine specific gravity can occur with a high protein diet, but in other patients it may be suggestive of GI tract hemorrhage, so this possibility should be considered in light of the clinical history (mast cell tumour, NSAID therapy, etc.).

**Calcium**

Hypercalcemia is one of the most common paraneoplastic syndromes seen in veterinary oncology. When it occurs it is most commonly associated with lymphoid malignancies or anal sac adenocarcinoma. Occasionally the PU/PD that results from hypercalcemia is the first presenting sign of one of these cancers. When severe, this paraneoplastic syndrome requires immediate management to avoid the development of hypercalcemia nephropathy. The serum calcium level should always be interpreted in light of the serum albumin level.

**Albumin**

Increased albumin concentration should prompt consideration of dehydration. Decreased albumin can occur with various metabolic abnormalities including liver failure, renal disease, hemorrhage, and GI tract disease (maldigestion, malabsorption). Decreased albumin concentration can affect other tests (i.e. hypoalbuminemia artifactually decreases the calcium level) and drug binding.

**Globulin**

Hyperglobulinemia may be polyclonal or monoclonal, and serum protein electrophoresis is necessary to make this vital distinction. When polyclonal the most likely cause is chronic inflammation / infectious disease, although less commonly a polyclonal gammopathy can be associated with cancer. When monoclonal (or biclonal) gammopathy is identified, the most common cause is plasma cell dyscrasia; however infectious diseases can also be a cause (e.g. Ehrlichiosis).

**Glucose**

Hypoglycemia is seen as a paraneoplastic syndrome with a variety of neoplasms, including large hepatic tumors and insulinomas. Abdominal ultrasonography and measurement of serum insulin level should be considered.

## URINE EVALUATION

Urinalysis is a commonly overlooked test in patients not displaying signs referable to the urinary tract. However, it is an important part of pre-treatment baseline evaluation in cancer patients. The urine specific gravity is important in interpreting changes in urea and creatinine (see above). Proteinuria may be useful in interpretation of hypoalbuminemia. Bence-Jones proteinuria (as seen in myeloma-related disorders), however, is not detectable on routine urinalysis.

A microscopic examination should also be performed, as the risk of occult bacterial infection (UTI) is high in patients with cancer and geriatric patients in general. Secondary urinary tract infections can particularly occur in animals with urinary tract tumors or hemorrhagic cystitis as a chemotherapy toxicity, and the diagnosis of an infection does not rule out these possibilities as an underlying cause. Occult UTI can act as a source of bacteremia leading to sepsis in myelosuppressed patients. Detection of active sediment should always prompt culture and sensitivity testing. For some animals, it may be necessary to start chemotherapy while awaiting culture results; in these cases broad spectrum antibiotics (cephalosporins) should be used until the result is known.

## REFERENCES

[1.] Garzotto CK, et al. **JVIM** 2000;14:587. [2.] Moore AS, et al. **JVIM** 2007;21:783. [3.] Webster CRL. In Ettinger SJ et al. (eds): **Textbook of Veterinary Internal Medicine**. St. Louis, Elsevier, 2005:1422. [4.] Litster A, et al.**Vet Microbiol**. 2008. [Epub ahead of print]

## ANCILLARY LABORATORY TESTING IN SMALL ANIMAL CANCER PATIENTS

**Antony S Moore MVSc, BVSc, Diplomate ACVIM (Onc) & Angela E Frimberger VMD, Diplomate ACVIM (Onc)**
**Veterinary Oncology Consultants (www.vetoncologyconsults.com) Wauchope, NSW Australia**

### INTRODUCTION

In cancer patients, basic testing such as CBC biochemistry panel and urinalysis are a vital component of diagnostics and the best possible management. However, these tests may not always give enough information for optimal patient care. In these cases, ancillary testing may be used to determine supplementary information about individual patients.

### MULTIPLE DRUG RESISTANCE & THE MDR-1 (ABCB1-1Δ) GENE

When using combination chemotherapy, cross-resistance between different classes of agents is a serious threat to successful treatment. In particular, multiple drug resistance (mdr) due to a transmembrane efflux pump protein (p-glycoprotein; gp 170) is able to reduce intracellular levels of vinca alkaloids, anthracyclines and other "naturally" derived chemotherapeutics to non-toxic levels. As these drugs are often used in the treatment of lymphoma it is wise not to overly rely on these drugs in the "rescue" setting, but rather to use them in combination in first line therapy.

Studies have shown that the mdr pump protein is present at low levels at the outset of chemotherapy for canine lymphoma, but increases in approximately 75% of cases after chemotherapy is instituted. Studies in both dogs and cats with lymphoma have found that if mdr is detected in pretreatment biopsies, the remission duration and survival times are shorter. Dogs with lymphoma that expressed mdr lived a median of 7.4 months compared to 12 months if mdr was not present. Interestingly, drugs classified as alkylating agents are not susceptible to the mdr mechanisms. There is very little cross-resistance even between different alkylating agents. This may explain the success of the strategy of using alkylating agents as rescue therapy for animals with lymphoma. It is also possible that the use of alkylating agents in combination with mdr-type drugs during the early phases of treatment (induction) may slow the onset of resistance. Trials of agents that competitively "block" mdr have not yet led to practical veterinary treatment strategies.

Immunostaining for C-219 (and other markers) can detect p-glycoprotein in biopsies. While not widely used clinically, such testing could provide prognostic value and assist in treatment selection for dogs undergoing chemotherapy.

A further implication of MDR-1 (ABCB1-1Δ) gene status in dogs, is when the gene is mutated and the protein product is abnormally functioning. Dogs that have a mutated MDR1 gene can have reduced drug elimination compared to others, resulting in elevated plasma drug levels and an increased tendency to toxicity. This is well recognized in collies, where MDR-1 mutations lead to increased sensitivity to a number of drugs, including ivermectin, but also those chemotherapeutics listed above. Breeds that have been shown to be affected in the USA are Collies, Australian shepherd, English shepherd, Shetland sheepdog, McNab, old English sheepdog, longhaired whippet and silken windhound. In Germany additional affected breeds include the border collie and Wäller. Homozygosity for the wild-type allele means the dog is not at undue risk of toxicity when mdr excreted chemotherapy drugs are used; on the other hand both heterozygotes (at high risk) and those homozygous (at very high risk) for the mutant allele show undue toxicity from these drugs. This has been shown clinically, in a recent study, where affected dogs were significantly more likely to develop myelosuppression (neutropenia and thrombocytopenia) after vincristine chemotherapy than dogs without mutations.

A patented test for the mutated gene is commercially available and is recommended when considering using chemotherapy to treat dogs that are at risk (any of the breeds listed above). Dose reductions in doxorubicin, vincristine or vinblastine (as the most commonly used drugs) are recommended in dogs that are heterozygotes, or homozygous for the mutation.

### PCR FOR ANTIGEN RECEPTOR-GENE REARRANGEMENTS (PARR) ASSAY & FLOW CYTOMETRY

PARR: By this technique DNA is extracted from a sample of either a lymph node or circulating lymphocytes (in an animal with lymphocytosis). Amplification of immunoglobulin and T-cell receptor sequences is performed. If the proliferation is due to a neoplastic cause, then the population of lymphocytes would be expected to have derived from a single "clone" of cells, and there should be a large accumulation of the specific receptor (immunoglobulin for B-cell) seen as a band on gel electrophoresis. If the lymphocytosis or nodal enlargement is due to a heterogeneous population of lymphocytes (as would be expected in a patient with a reactive, inflammatory, infectious cause), there is no distinct band. This is a powerful tool in dogs with a slowly progressive lymphoid disease, where it is important prognostically, and to decide whether to institute therapy or to look for another cause.

Flow cytometry can identify and quantify lymphocyte subsets using cluster of differentiation (CD) markers. The full prognostic significance of panels of these markers has yet to be determined in cats and dogs with lymphoma, but it could be a powerful tool in directing therapy and making a prognosis for our patients.

However, the use of flow cytometry to detect the presence of CD-34 in leukemic blood samples has enormous prognostic significance. CD-34 is a putative stem cell marker and the finding of tumor cells with CD-34 in a patient with lymphocytosis implies that patient has acute lymphoid leukemia (ALL), rather than chronic lymphoid leukemia (CLL). This can be of value in patients where the lymphocytosis is neoplastic (see PARR above), but the lymphoid cells are "reactive" or "medium sized" and a true distinction between CLL and ALL cannot be made. In one study, dogs that were CD-34 positive had a median survival of 16 days; if CD-34 negative, regardless of other markers dogs with all other subtypes lived considerably longer (between a median of 289 and 474 days).

Submission forms are obtained from the Veterinary Clinical Immunology Laboratory, Colorado State University. Instructions for collection of samples for PARR and for Flow Cytometry include:

PARR: Aspirate the mass or node, and squirt the sample into the bottom of an EDTA tube with no additional saline. Slides, stained or unstained, can also be submitted, but they should be cellular. Biopsies can also be put into an EDTA tube. NO FORMALIN FIXED SAMPLES! Marrow and cavity fluids can be placed directly into EDTA tubes.

Flow cytometry: Draw blood or cavity fluid into EDTA, and then immediately refrigerate. For aspirates, RPMI, DMEM or similar cell culture media with 10% fetal bovine serum is ideal. This can usually be found in a University setting. In practices where this is not available, put 0.5 to 1 ml of saline (LRS, Norm R) into a red top tube, then aspirate the mass or node several times, squirting the contents into the saline. Rinse the syringe with the saline to increase yield. It is essential to add protein to these samples. This can be accomplished by adding serum from a patient of the same species to the tube such that the serum comprises approximately 10% of the volume. Protein should also be added to cavity fluids.

**VETERINARY BLADDER TUMOR ANTIGEN (V-BTA) TEST**

This non-invasive urine test was developed for human patients and directed against a glycoprotein-complex found in the urine of patients with transitional cell carcinoma (TCC). It has been used in humans to detect early relapse in patients treated for TCC. The test can be run using as little as 0.5ml of urine, and it may be more accurate on centrifuged samples. If the glycoprotein complex is present it will agglutinate causing a color change.

A veterinary version of the test appears to work similarly in dogs (not yet evaluated in cats). Limitations in both dogs and people appear to be the presence of inflammatory sediment (proteinuria or pyuria), aciduria, glucosuria or hematuria; all increase the risk of a false positive reaction.

In studies the sensitivity (true positive) rate is high at around 85%, and the specificity (true negative) rate is similar. For dog with urinary tract disease the specificity drops to less than 50%. Positive tests were seen for dogs with urinary bladder, urethral, prostatic and even renal cancer.

The high rate of false positives means this is not a test that can be used to detect TCC in the general population. On the other hand, the high rate of true positives means that the likelihood of missing a TCC (false negative) is very low. For this reason, the V-BTA could be used in dogs considered at high risk of developing TCC (for example; Scottish terriers, Shetland sheepdogs) as part of their geriatric wellness testing. A positive result would not tell you that the patient has TCC, but would be a reason to investigate further. Urinalysis, as well as ultrasonography may be justified in these patients, as early detection could allow more successful treatment. Most importantly if the test was negative, the chances of that patient having a "silent" TCC would be very low; and the owners could feel reasonably confident that their dog was free of cancer at that time and avoid any more invasive or expensive testing.

**CREATININE CLEARANCE**

Some chemotherapy agents are nephrotoxic (cisplatin and streptozotocin in dogs, doxorubicin in cats), and some are renally cleared. Carboplatin has anecdotal reports of efficacy for a wide range of carcinomas in both dogs and cats. Although carboplatin is not a major nephrotoxin, it is renally cleared, and so any renal insufficiency necessitates a dose reduction in order to avoid unexpectedly severe myelosuppression. Although azotemic chronic renal failure in cats is uncommon in the general feline pet population, many of our older oncology patients have renal insufficiency that may not be causing azotemia (renal function must decrease by 75% before azotemia is detected). One option for older feline patients with renal azotaemia for whom carboplatin chemotherapy is indicated, is to perform an arbitrary dose reduction (25-50%), after which dosages can be increased following each cycle that occurs without toxicity. This approach may leave some animals under-treated with a resistant tumor cell population, and other that still show marked myelotoxicosis. Even in cats where there is no renal azotemia, there is commonly individual variability for dosing interval with carboplatin. Some cats may take 5 or more weeks to recover their hematologic parameters to a level where a subsequent dose can be delivered safely.

In humans, carboplatin is also excreted primarily by the kidney via glomerular filtration, clearance of carboplatin is directly correlated with creatinine clearance and carboplatin dosage is based on glomerular filtration rate (GFR) rather than on body weight, BSA or other parameters. In studies in humans, as the area under the concentration-versus-time curve (AUC) for carboplatin increases so does the severity of myelosuppression. The same appears to be true for cats.

Recent studies at Cornell University have shown that basing carboplatin treatment on creatinine clearance in cats is able to accurately direct the dose that should be used that will cause predictable and manageable myelosuppression. These studies showed that in cats, the appropriate target AUC should be 2.75. Based on that study, the following formula is used to calculate a carboplatin dose (in mg) = 2.75 x 2.6 x GFR (ml/min based on tech scan) x body weight every 4 weeks. Using this formula, cats that were normal, had suboptimal renal function or were azotemic received doses equivalent to 85-307mg/m2; however this wide range of doses caused a predictable and non-life-threatening myelosuppression. Similar strategies could be helpful in selecting carboplatin doses in dogs.

**PRE- & POST-PRANDIAL BILE ACID LEVELS**

Hepatic metabolism is important in the detoxification and excretion of many commonly used chemotherapy drugs (vincristine, vinblastine, doxorubicin). In addition, for drugs that have hepatotoxic potential such as lomustine, patients with any pre-existing hepatic disease are at greatly increase risk of toxicity. Increases in serum hepatocellular enzyme levels should raise suspicion of hepatobiliary disease but are not specific, and other causes of such elevations may occur in both geriatric and cancer patients. In these patients, further evaluation of the liver is warranted before chemotherapy administration.

Total serum bile acid levels are increased when the enterohepatic circulation is disturbed. An increased level of fasting serum bile acid is more specific, but not more sensitive, than serum ALP and hepatocellular enzymes levels for hepatobiliary disease. An increased level of postprandial serum bile acid is more sensitive than is an increased level of fasting bile acids. Thus, combining fasting and postprandial bile acid levels improves both sensitivity and specificity for suspected hepatobiliary disease. However, it should be noted that the type of hepatobiliary disease cannot be determined by serum bile acid levels. In addition, the degree of increase in serum bile acid levels does not correlate with the severity of dysfunction or histologic change; therefore in evaluating progression or improvement of disease, only the difference between normal and abnormal, not the actual level, is relevant.

## REFERENCES

**MDR :** Mealey KL, et al. JVIM 2008;22:996. Martinez M, et al. J Vet Pharmacol Therap 2008;31:285. Brenn SH, et al. Vet Compar Oncol 2008;6:201. Bergman PJ, et al. JVIM 1996;10:354. Moore AS, et al. Cancer Invest 1995;13:475. **PARR :** Williams MJ, et al. JVIM 2008;22:596. **V-BTA:** Billet J-P HG, et al. AJVR 2002;63:370. Henry CJ, et al. AJVR 2003;64:1017. **Creatinine clearance :** Bailey DB, et al. AJVR 2004;65:1502. **Bile acids:** Kristal O, et al. JVIM 2004;18:75. Center SA, et al. Semin Vet Med Surg 1990; 5:83-93. Webster CRL. In Ettinger SJ et al. (eds): Textbook of Veterinary Internal Medicine. St. Louis, Elsevier, 2005:1422.

## CANINE PARVOVIRUS VARIANTS CIRCULATING IN USA (2006-2008)

**Sanjay Kapil, DVM, MS, PhD, ACVM, Professor, OADDL**
**Stillwater, OK**

### INTRODUCTION

Canine parvovirus (CPV) emerged as a new pathogen of dogs in the 1970's. Today, CPV is the most common viral disease of dogs. There are two canine parvoviruses: canine parvovirus-1(Minute virus of canines) and canine parvovirus-2, the major cause of puppy enteritis, morbidity and mortality. CPV-1 and CPV-2 are antigenically and genetically different. This review deals only with CPV-2.

#### Canine Parvovirus

Canine parvovirus contains a single-stranded, DNA of about 5000 bases. This small virus has only major proteins, viral protein 1 (VP1) and viral protein 2 (VP2). CPV is a non-enveloped virus that is resistant to lipid solvents, temperature, and pH changes. Due to a compact DNA genome, parvovirus has high physical resistance. CPV-2 can survive in the kennel environment for months or years. If the kennel is not properly cleaned, the virus can be transmitted to the next batch of naïve puppies. Canine parvovirus genome tends to mutate at critical amino acids (about 10 residues) of VP-2 capsid protein.

#### Evolution of CPV-2 in the Last 3 Decades

Canine parvovirus-2 (CPV-2) pandemic started in 1977. It was associated with high mortality due to hemorrhagic diarrhea and myocarditis. The virus probably originated from a carnivore parvovirus (Feline panleukopaenia, FPV; or mink parvovirus (MPV); or blue fox parvovirus). These viruses have high genetic similarity (99%). However, they have biological differences, such as differences in the suitable pH for maximum hemagglutination activity. For example, based on a recent study, feline parvovirus interacts with transferrin receptor at pH of 6.5. However, a recent CPV-2c can interact with transferrin from pH (3-8). We could not test CPV-2c and transferrin interactions outside this range of pH because of erythrocyte hemolysis. These genetically closely related viruses have acquired host-range differences due to single-point mutations on the VP-2 protein. The prototype CPV-2 isolates did not infect cats but could grow in feline cell lines such as Crandall Reese feline kidney (CRFK) cell line. CPV-2 soon acquired the ability to infect cats by point mutations that broadened the host range.

By 1982, a genetic variant emerged that had amino acid changes at positions: 87, 101, 300, 305, and 555 of the VP2 protein. This genetic variant was designated CPV-2a to distinguish it from the original CPV-2. CPV-2a had acquired the ability to infect cats and it had isoleucine instead of valine at position 555 of the VP-2 protein. In 1985, another genetic variant designated CPV-2b emerged spreading worldwide. This genetic variant is the major genotype of CPV-2 in the USA. In 2000, Dr. Buonavoglia from Bari, Italy, reported a new genotype of CPV-2 that had glutamic acid at position 426 of the VP-2. This virus is the second major genotype of CPV-2 in the USA. CPV-2c was first detected in the USA by my laboratory[1]. CPV-2c has now been documented on all continents except Australia. In addition to natural evolution of CPV-2, there are other mechanisms of antigenic and genetic changes in the current CPV-2 isolates.

### DIAGNOSIS

Canine parvovirus is the most common cause of enteric disease in puppies. After oral infection followed by systemic infection, canine parvovirus is shed for 5-7 days. However, it has been shown that dogs infected with CPV-2c may shed the virus up to 51 days. There are a variety of tests available for laboratory-based diagnosis of CPV-2 infection. For ante-mortem diagnosis, fecal samples (2-5 grams) are preferred. For post-mortem diagnosis, pieces of unfixed, small intestine and tongue are preferred. Canine parvovirus can cause focal infection of the intestines. Three discontinuous loops of the small intestine should be submitted for direct fluorescent antibody testing. Anti-CPV-2 monoclonal antibody 3B10 from VMRD, Inc. (WA, USA) has been found to react with all circulating genotypes of CPV-2 in the USA[1]. This antibody is also suitable for immunohistochemistry on formalin-fixed tissues. For genotyping, nucleic acid-based techniques are needed. At OADDL, we use CPV-PCR followed by sequencing and analysis of the sequence. Amino-acid position 426 determines the genotype of the CPV-2. For CPV-2 and CPV-2a, the amino acid at position 426 is asparagine (Asn); for CPV-2b, the 426 is aspartic acid (Asp); and for CPV-2c the amino acid at position 426 is glutamine (Glu).

In the clinic, ante-mortem diagnosis can be performed using ELISA and SNAP tests. A variety of factors including the time of collection of the samples after infection can affect the outcome of the diagnostic tests.

All CPV-2 viruses hemagglutinate erythrocytes. We have used this universal property of the virus to develop an instant, easy, robust animal-side diagnostic tests for CPV-2 antigen (SAT) and antibody diagnosis (SIT)[2]. These instant tests do not require any instrumentation and are suitable for application in kennels by owners because they require minimal training. The procedure requires mixing a drop of fecal sample and erythrocytes. We used porcine erythrocytes in our experiments; however, canine and feline erythrocytes should also perform well. The agglutination reaction is quick/instant. The test performs well at room temperature, also. The complete details of the protocol and the performance characteristics are described elsewhere[2]. We found that the test performs well with all circulating genotypes of CPV-2 in the USA.

### VACCINES

Current CPV-2 vaccines can be classified according to the amount of CPV-2 antigen: high- and low-antigen dose CPV-2 vaccines. High antigen dose vaccines claim to immunize the puppies in the presence of maternal antibodies. However, I could not find the scientific criteria that can define the amount of virus in high antigen vaccines in the literature. Higher amounts of CPV-2 vaccine will lead to higher and longer shedding by the puppies leading to second-hand exposure to the virus. Most major commercial vaccines contain modified live virus (MLV) of CPV-2 or CPV-2b genotypes. There is no commercial vaccine that contains CPV-2a. Moreover, CPV-2a has not been found in the nature in the USA based on our results at OADDL in the last 3 years. Commercial vaccines containing CPV-2 isolates may not provide robust serum antibody responses against CPV-2b and CPV-2c isolates circulating in the USA. It has been recently published that all commercial vaccines provide protection against current circulating genotypes of CPV-2 based on experimental studies just presented recently. However, other published studies contradict these claims of CPV-2 vaccine efficacy under field conditions.

Clearly, the observations made by diagnostic virologists/pathologists, canine breeders, field veterinarians and shelters do not agree with the universal efficacy of CPV-2 vaccines, worldwide. After the introduction of commercial CPV-2 vaccines, the problem has significantly reduced but CPV-2 continues to be the number-one infectious problem of puppies, worldwide.
CPV-2b vaccines have been found to have higher efficacy compared to some available CPV-2 based on the published literature[3]. Currently, there is no commercial vaccine that contains CPV-2c antigen. Commercial companies are currently evaluating the efficacy of MLV CPV-2c vaccines.

**Antigenic Cross-Reactivity among CPV-2 Isolates**

Serum antibody titers are a good measure of immune status of the dog against CPV-2. Antibodies against one genotype will cross react with other genotypes of CPV-2. Traditionally, hemagglutination-inhibition test using porcine erythrocytes is used to assess the level of antibodies in serum samples. A titer of $\geq$ 1:80 is considered protective. Most commercial vaccines induce this titer after a single vaccination in puppies that are serologically negative. The take of the MLV vaccine have been reported to be excellent in the experimental vaccination experiments.

Serum neutralization testing is slightly cumbersome but it is the preferred method for determination of the animal's immune status. Differences in SN titers among CPV-2 isolates are considered insignificant if it is $\leq$ 4-fold. SN titers among CPV-2 isolates that differ by 4-fold or more are considered significant. We have found that some CPV-2c isolates react 10-fold less with anti-CPV-2b serum compared to homologous reaction with anti-CPV-2b antiserum using a quick SIT method. Anti-CPV-2c antiserum reacts better with CPV-2c and CPV-2b isolates compared with CPV-2b antiserum [4].

**Vaccine Failure**

Under field conditions, vaccine failures happen. The most common reason for vaccine failure is the incomplete take of the MLV vaccine due to interference from maternal antibodies. Moreover, the problem is exacerbated due to less efficient neutralization of the variant CPV-2 isolates that have emerged recently. Other factors, such as biological variation among CPV-2 isolates may be responsible for the recent problems due to CPV-2c. We have noticed that in the last 3 years, the frequency of GCA at position 440 has increased in the CPV-2c isolates. Moreover, the CPV-2c and CPV-2b isolates recovered from puppies over 5months of age have a higher frequency of the GCA at position 440 compared to ACA among CPV-2c and CPV-2b isolates recovered from puppies less than 5 months of age.

**Disinfection**

Canine parvovirus is resistant to most of the common disinfectants. Both household bleach (1:30 dilution in water) and potassium peroxide (Trifectant or Virkon) is suitable for all CPV-2 outbreaks. Ten minutes of exposure to these chemicals will destroy CPV-2 virus. All canine parvovirus isolates have similar physical properties. Items that cannot be treated with either of these chemicals can be steam disinfected.

## ECOLOGY

**Geographic Distribution of CPV-2c in the USA**

To date, OADDL has detected CPV-2c in samples from a total of 15 states in USA. Unrestricted movement of infected dogs has been responsible for transatlantic transmission of CPV-2c from Italy. Highest numbers of CPV-2c cases have been detected in Oklahoma.

**Global Distribution of Canine Parvovirus Genotypes**

As of January, 2009, CPV-2c has been documented in Asia (Japan) in cats, Africa (Tunisia), Europe (Italy, Germany, Portugal, and Spain), CPV-2c has not been detected in England; South America (Uruguay); and North America (USA) but not in Canada. Australia is free of CPV-2c. The geographic and regional monitoring is important so that proper genotypes can be selected for vaccine production in different countries of the world. Moreover, it will be important to understand how to introduce the vaccines with different genotypes in different states in the USA. It will be prudent to continue using the current MLV CPV-2 vaccines for most of the states in the USA and introduce CPV-2c as soon as it becomes available.

**Effects of Canine Parvovirus on Wild Life**

Canine parvovirus is a major threat to conservation of wolves. About half of the wolf puppies in Minnesota have succumbed to canine parvovirus. Carnivore parvovirus isolates have caused disease in Lynx, bob cats, and raccoons. There are few studies in which the wild life parvovirus isolates have been characterized. Most of the studies have used serology to detect the exposure to parvovirus. We have recently found that raccoons can be infected with canine parvovirus. Thus, raccoons can play a role in the epidemiology of CPV-2.

## CONCLUSIONS

Canine parvovirus emerged about 30 years ago. Thus, it is a relatively new pathogen of dogs. The virus spread rapidly across continents as most dog populations were naïve. The virus has undergone rapid evolution in adapting to the new host. Unlike CPV-2, the feline panleukopaenia (FPLV) is a relatively genetically stable virus. FPLV has been a pathogen of cats for over 100 years. Due to movement of dogs, CPV-2c has spread world wide and CPV-2c has become the major genetic variant in Europe. In USA, CPV-2c accounts for about 50% of the cases of parvovirus in samples submitted to OADDL. CPV-2c cannot be clinically distinguished from other CPV-2 genotypes. There is no published report of histopathology in cases affected with CPV-2c. At OADDL have examined cases of CPV-2c in the USA and found them to be similar to other genotypes of CPV-2. More puppies from smaller breeds of dogs were submitted to OADDL with a diagnosis of CPV-2c; however, we have not found any enhanced susceptibility for any breed of dogs. Several of cases have had a history of yellow or hemorrhagic diarrhea. CPV-2c is mostly a disease of young puppies. However, we have had at least 9 cases of CPV-2c infections in adult dogs.

Currently, there is a debate if the canine parvovirus-2 vaccines need to be updated. Several groups, including vaccine companies have shown that current vaccines provide protection against the Italian CPV-2c isolate under experimental conditions in sero-negative dogs. Based on large numbers of independent reports from dog breeders in South-Central USA, I suggest that current vaccines need to be carefully evaluated for their efficacy against current circulating CPV-2 isolates. These studies are in progress at OADDL using about 40 CPV-2 isolates in cell culture from incidence reports/large outbreaks from kennels with some history of use of commercial vaccines. The vaccination histories can be unreliable and frequently incomplete. However, there are at least three independent published reports that suggest that CPV-2 isolates can differ in their biological properties and CPV-2 isolates (about 30%) can have significant antigenic differences (about ten-fold)[2,4]. This level of antigenic distance is significant and will make a difference in puppies that are maternal antibody positive at 6 weeks of age. The passive antibodies will block vaccination and will leave the puppy susceptible to CPV-2c exposure and could lead to vaccine failure[5]. Even if there is a need for updating the vaccine to contain CPV-2c, the introduction of the new vaccine should be based on the circulating CPV-2 genotypes. There is an obvious need to know the genotypes of CPV-2 circulating in various parts of the world. This critical information is unavailable. In mean time, it is critical that veterinarians use the MLV vaccines containing the updated genotypes of CPV-2 according to the instructions on the label of vaccine[6]. There has been a recent recommendation to give a booster shot to puppies at 4 months of age to cover the emerging CPV-2c isolates. This additional booster vaccine will help increasing the number of puppies (population immunity) that have actively seroconverted before they are put in the CPV-2 contaminated environment.

## REFERENCES

[1]Kapil, S. et al., **J. Clin. Microbiol.** 2007;45:4044. [2]Marrulappa S, et al. **Clin. Vac. Immunol.** 2009; 16:127. [3]Greene.**Inf. Dis. Dogs Cats**2006. 3rd edition. [4]Cavalli et.al. **Clin. Vacc. Immunol.** 2008. 15:534. [5]Decaro et al. **New Microbiol.** 2008.31:125. [6]Larson et al. **Vet. Ther.** 2008. 9:94.

# CANINE DISTEMPER VARIANTS CIRCULATING IN USA (2007-2008)

**Sanjay Kapil, DVM, PhD, ACVM Professor, OADDL, CVHS**
**Stillwater, OK**

## INTRODUCTION

Canine distemper virus is highly contagious. Canine distemper is endemic in most parts of the world. However, introduction of high-take vaccines in 1960's has reduced the incidence of the disease. Canine distemper virus is a negative strand RNA virus that belongs to family *Paramyxoviridae* under genus *Morbillivirus*. This family of viruses has properties that make them emerging. Canine distemper virus shares some similarities with human measles virus. Canine distemper virus has emerged recently with first case described in 1905 [1]. CDV is responsible for multi-systemic infection involving ocular, respiratory, gastro-intestinal and nervous systems.

### Structure of the CDV

The virus has two glycosylated envelope proteins: Fusion (F) and Hemagglutinin (H). The H protein is responsible for binding to the host cell receptor (SLAM). These two proteins are responsible for induction of neutralizing antibodies and protection from reinfection. After proteolytic degradation, the viral proteins can bind to major histocompatibility antigens and serve as targets for T cell responses.

### Host range

Canine distemper virus has a broader host range and can infect members of Felidae (cats), Canidae (dogs and foxes), Mustelidae (weasles) and Proconidae (raccoons). Phocine distemper virus (PDV) can cause disease in dogs after experimental infection. These viruses are transmitted by horizontal transmission and no vectors are known. CDV in domestic dogs is considered as a reservoir of infection for endangered carnivores.

### Transmission

The F and H surface proteins are critical immunogens in dogs for antibody-mediated protection against CDV. Respiratory route via aerosols is the most common means of transmission. However, transplacental infections have been rarely reported.

### Genotypes of CDV

On phylogenetic analysis, the CDV shows temporal and spatial differentiation among isolates[2] . Reemergence of CDV has been linked to a limited number of CDV lineages that are unique to the area. CDV isolates form 7 unique lineages: American-1 (vaccines), American-2, Arctic-like, Asia-1, Asia-2, Europe, and European wildlife[3]. The majority of CDV isolates currently circulating in the USA belong to European-wildlife lineage.

## DIAGNOSIS

Presumptive diagnosis of CDV can be made in an unvaccinated dog of 5-6 months of age with clinical signs compatible with CDV. The blood examination will reveal lymphopenia and thrombocytopenia. Serological examination of serum and CSF can be supportive of the diagnosis. Virus isolation in a recombinant cell line (Vero with canine SLAM) provides a very sensitive method for isolation of wild-type CDV isolates. In acute cases, RT-PCR using the nucleocapsid gene of CDV as a target is diagnostic for the presence of virus. However, for genotyping the extracted RNA sample can be processed for H gene sequencing and phylogenetic analysis. For antemortem diagnosis, conjunctiva, blood, CSF and urine sediment can be examined by FAT with FITC-labeled direct-conjugate (VMRD, Pullman, WA). Detection of CDV antigen 1 week after infection can be difficult due to clearance by virus-specific antibodies produced by dogs. Chronic cases of CDV can be diagnostically challenging and RT-PCR on fresh (not stored) CSF may be used in the cases with nervous symptoms.

## VACCINES

Immunity to CDV after natural infection is very long term (about 8 years). In vaccinated bitches, about 97% of the antibody is transferred in colostrum and 3% is transplacental. Serum neutralizing titers are a good measure of the immune status of an animal.

### Types of CDV vaccines

Vaccines are the primary preventive measure against CDV and current vaccines provide good immunity.

### Chemically Inactivated CDV Vaccines

These vaccines are not available in the USA. However, they are safer and are recommended only for ferrets and pandas that succumb to CDV with MLV vaccines.

### Heterologous Measles Vaccine

Heterologous immunity induced by antigenically related human measles virus has been used for initial vaccination of puppies. However, this vaccine does not induce high antibody responses compared to MLV CDV vaccines[4] .

**Modified Virus CDV Vaccines**

These commercial vaccines contain the CDV genotype American-1. This genetic variant is extinct in the USA and has not been detected in nature for the last 2 decades. In our studies over the last 2 years, we have not detected the American-1 CDV genotype in dogs that were naturally infected with CDV that were collected from 5 states in the USA (OK, KS, MO, CA, and FL).

**Genotypes of Canine Distemper Virus Circulating in the USA**

It is well documented that various genetic variants of CDV are circulating worldwide, including USA[5] . Despite this genetic variation, it is widely accepted that outbreaks due to newer variants will not occur in MLV-vaccinated dogs[4] . This is due to good cross-reactivity among CDV variants. However, efficacy of heterologous protection has not been experimentally and scientifically confirmed. Based on the genotyping of CDV isolates collected during the last 2 years, European wildlife CDV is the most predominant CDV genotype in the USA [5].

**Safety of MLV CDV Vaccines**

Modified live CDV vaccines are safe in general. However, when there is a familial history of problems with MLV CDV vaccines, the recombinant CDV vaccine is recommended. However, the recombinant subunit vaccine induces lower immune responses. The risk of potential vaccine-induced encephalitis should be weighed against relative level of risk of the dog contacting CDV from natural exposure in the population.

**Population Immunity**

The incidence of CDV in an area is affected by the vaccine coverage in the dog population. Based on measles virus vaccine coverage of 90 % in an area is needed to stop the transmission of the Morbillivirus infections and break the cycle of an outbreak due to CDV. In Finland, reappeance of CDV coincided with decrease in population immunity[6]. Based on the experience in Finland outbreaks, a level of herd immunity above 70% is sufficient to control the CDV outbreak.

**Vaccine Failure**

Outbreaks of canine distemper can occur in vaccinated dogs[6]. CDV is an extremely fragile virus, thus, the lyophilized vaccine should be used immediately after reconstitution so that environmental conditions do not decrease the efficacy of the vaccine. CDV vaccine should be administered to dogs before exposure to shelter environment. An SN titer to CDV of 1:100 is presumed to be protective.

**Disinfectants**

CDV is an enveloped virus. CDV is extremely easy to disinfect with common disinfectants.

**Zoonosis**

Role of canine distemper virus in Multiple Sclerosis and Paget's disease in humans is still questionable.

## CASE REPORT # 1: CANINE DISTEMPER IN A AN ADULT VACCINATED DOG

**Clinical History**

A 28 months-old, spayed female Scottish terrier from Oklahoma was presented to the Boren Veterinary Teaching Hospital, CVHS, Stillwater, OK, in Dec. 2008, with 2 week's history of fever with no response to treatment with antibiotics. The dog had grand mal seizures that started 2 days ago. Bilirubin values were increased and possible cholangiohepatitis was suspected based on ultrasound examination. WBC count was increased and large granular lymphocytes were present in the CSF. Euthanasia was performed at the request of the owners. The dog had received 3 puppy shots of a commercial MLV CDV vaccine at a veterinary hospital in Oklahoma. The last vaccination was a booster shot in September 2007 at 14 months of age. This dog was presented to VTH in December 2008 with neurological problems.

**Antemortem Examination**

A sample of cerebrospinal fluid and serum were submitted. On SN testing for CDV, a titer of 1:256 was recorded for the serum and a titer of 1:6 was recorded for the CSF sample. A fresh sample of urine (pellet) was found to be negative for CDV RNA by RT-PCR.

**Postmortem Examination**

External examination revealed mild to moderate, multifocal erythema of the skin. Histologically, the cerebrum showed neuronal cell necrosis that was severe and locally extensive with intracytoplasmic inclusion bodies consistent with CDV infection. Intracytoplasmic inclusion bodies were observed in the uroepithelium and these inclusions were compatible with the diagnosis of CDV infection. Fluorescent antibody examination on kidney, tonsil and brain sections was performed. Tissues tested were negative by FAT for CDV except for two of the brain sections which contained intensely fluorescing, small foci (n=3) of positive signal compatible with the presence of CDV antigen. Based on the focal, small areas involved an interpretation of "suspect for CDV" was recorded.

**Diagnosis**

Chronic Canine Distemper Virus Encephalitis

## CASE REPORT #2: CANINE DISTEMPER IN A TWO-DAY OLD PUPPY

**Case History**

A two–day old Dachshund puppy was submitted to OADDL with history of extensor rigidity and sawhorse stance. The kennel owner had lost a total of about ten puppies from different litters exhibiting similar clinical signs. At least 5 different litters were affected out of a total of 30 breeding females (at least 10%). The owner had vaccinated the dogs with a commercial MLV vaccine purchased at a feed store in Oklahoma. Due to CDV problems, the owner has switched to measles CDV vaccine. The CDV problems in this kennel had continued.

**Gross Examination**

Lissencephaly characterized by absence of cerebral gyri was noted in this puppy. There were severe focal hemorrhages of the cerebrum. Kidneys showed cortical petechial hemorrhages. Soft palate had multifocal hemorrhages.

**Histologically**

Diffuse, moderate interstitial pneumonia was noted with lymphocytic and histiocytic infiltration in the lung.

**Fluorescent Antibody Test**

Canine distemper virus FAT was strongly positive in all tissues examined. Canine adenovirus and canine herpes virus were negative.

**Virus Isolation:** Positive

**Serological Examination of 8 Adult Animals**

High CDV titers detected in bitches.

**Final Diagnosis**

Transplacental CDV infection in a newborn puppy.

## CONCLUSIONS

Majority of the current CDV isolates circulating in the USA belong to European wildlife. A few CDV isolates belonging to other lineages are also circulating in the USA. However, further studies are required to assess the antigenic distance between the vaccine CDV isolates (American-1) and the CDV isolates especially collected from cases with a history of CDV vaccine failure. These *in vitro* cross-reactive studies are in progress at OADDL. MLV vaccines should be used because vaccination is the only dependable method of CDV prevention. Whole virus CDV vaccines are safe in dogs. Recombinant CDV vaccines are useful for exotic species of carnivores (ferrets, pandas, and dogs with a family history of adverse reaction to CDV vaccines) that do not tolerate MLV vaccines. The vaccines should be used according to the label on the vaccine to prevent unknown side effects. It is critical that MLV vaccines are properly stored and the cold chain is maintained due to lability of CDV in the vaccine. Serological monitoring by SN testing at a diagnostic laboratory after vaccination should provide assurance that CDV vaccine has been properly administered and taken by the animal.

## REFERENCES

[1]Pomeroy et. al. **J. Mol. Evol.** 2008; 66:98-106; [2]Mochizuki et. al. **J. Clin. Microbiol.** 1999; 37:2936-2942; [3]McCarthy et. al.. **Proc. Biol. Sci.** 2007; 274:3165-3174; [4]Greene **Inf. Dis Dogs Cats.** 2006. 3rd edition; [5]Kapil et al. **Clin. Vacc. Immunol.**2008; 15:707-712; [6]Ek-Kommonen et al., 1997. **Vet. Rec.**:380-383.

## IMMUNOMODULATORS IN VETERINARY MEDICINE – IS THERE EVIDENCE OF EFFICACY?

**Katrin Hartmann, Prof., Dr. med. vet., Dr. habil., DECVIM**
**Munich, Germany**

### INTRODUCTION

Immunomodulators (or immunostimulatory agents or biological response modifiers) are commonly used drugs in veterinary medicine, mainly for prophylaxis and treatment of infectious diseases and tumors in dogs and cats. In oncology, various cytokines are with defined mechanism of action are used effectively in immunotherapy, but use of "unspecific immunomodulators" in animals with malignancies is also very popular. There are only few controlled studies published in animals with tumors. Liposome-encapsulated muramyl tripeptide phosphatidylethanolamine (L-MTP-PE) has been used to treat oral melanomas in dogs,[1] hemangiosarcomas of the spleen in dogs,[2] osteosarcomas in dogs,[3] and mammary adenocarcinomas in cats.[4] Another drug use to treat tumor patients is bacillus Calmette-Guérin that has been use in different mamma tumors[5] and in Sticker sarcoma (TVT) in dogs.[6] *Propionibacterium acnes* been use in different mamma tumors in dogs[5] as well as mammary carcinomas in cats, and hemangiopericytomas and mast cell tumors in dogs.[7] More commonly, unspecific immunomodulators are use to prevent and especially treat infectious diseases, mainly virus infections. Thus, the talk will mainly focus on immunomodulator therapy in viral diseases in dogs and cats. It has to be realized that an unspecific stimulation of the immune system might be contraindicated in some infections (e.g., in feline immunodeficiency virus (FIV) because these drugs can lead to an increased virus replication caused by activation of latently infected lymphocytes and macrophages, and therefore can effect a progression of disease, or in cats with feline infectious peritonitis (FIP) as clinical signs of FIP develop as a result of an overwhelming immune-mediated response). Immunomodulators modify the responses of immunocompetent cells through cytokines or other mechanisms. It has been suggested that these agents benefit infected animals by restoring compromised immune function, thereby allowing the patient to control viral burden and recover from associated clinical syndromes. It is difficult to test the efficacy of these drugs *in vitro*. Most of the *in vivo* reports in the veterinary literature are difficult to interpret due to unclear diagnostic criteria, lack of clinical staging or follow-up, the natural variability of the course of disease, the lack of placebo control groups, small numbers of cats used, and additional supportive treatments given. Although reports of uncontrolled studies frequently suggest dramatic clinical improvement, these effects are usually not observed when followed by controlled studies.

### INTERFERON INDUCERS

So-called "interferon inducers" are compounds that induce synthesis of interferons and other cytokines. Several macromolecules and microorganisms are known to produce antiviral and other antimicrobial activities in the host. Most of the substances known to have this effect have a structure similar to double-stranded DNA that can be of microbial origin or synthetic.

**Polyriboinosinic-polyribocytidylic acid**, poly-IC, is one of the most common chemical interferon inducer used in cats and dogs in the United States. Although poly-IC acid has only induced minimal interferon induction in dogs,[8] it was reported to be effective in protecting dogs from infectious canine hepatitis (ICH). In a controlled experimental study, however, mortality in dogs with ICH was not decreased and survival time was only prolonged for a few days.[9] Side effects in cats and dogs include lymphopenia, lymphoid necrosis, CNS depression, hemorrhagic gastroenteritis, and coagulation disorders.

**PIND-AVI** (parapoxvirus avis) and **PIND-ORF** (parapoxvirus ovis) are attenuated and by treatment with gamma rays inactivated poxviruses that can be classified as interferon inducers as their suggested mode of action is through e.g., induction of interferons and colony stimulating factors, and activation of natural killer cells. Inducers of paramunity are on the market Germany and some other European countries and have been used against a variety of virus infections prophylactically and therapeutically in many animal species. However, most of the reports on their efficacy are anecdotal or lack control groups. In all placebo-controlled double-blind studies published so far, no beneficial effect could be documented. The drug caused a sensation in Germany when it was published that they would cure 80 to 100 % FeLV-infected even those in moribund condition.[10] However, 2 placebo-controlled double-blind trials using the same treatment protocol (1 ml subcutaneously 1 to 3 times a week for 4 to 30 weeks) were performed to determine the therapeutic efficacy of these compounds in naturally FeLV-infected cats under controlled conditions and were not able to repeat these striking results, as there was no significant difference in the number of cats that became FeLV-negative during the 6 week treatment period, and no statistically significant differences could be demonstrated application in any investigated parameters between paramunity inducer and placebo.[11,12]

**Acemannan** is a water-soluble long-chained complex carbohydrate (mannan) polymer derived from aloe vera plant that is licenced for veterinary medicine. Acemannan is thought to be taken up by macrophages stimulating them to release cytokines, which induce cell-mediated immune responses including cytotoxicity. In murine tissue culture, acemannan stimulates macrophages to produce tumor necrosis factor-α and interleukin-1. Acemannan has been used as an adjuvant in vaccines or by itself to enhance regression of tumors, e.g., in the management of post-vaccinal injection-site sarcomas in cats.[13] Acemannan has been used in FIV-infected cats in a study including 49 FIV-infected cats receiving acemannan IV or SC once weekly for 12 weeks or by daily PO administration for 12 weeks. The beneficial effects observed in this study are difficult to interpret because cats were concurrently treated with antibiotics and other symptomatic therapies, and the study did not include a control group.[14] In one non-controlled open-label trial, 50 cats with natural FeLV infection were treated with acemannan (2 mg/kg intraperitoneally q 7 days for 6 weeks). At the end of the 12 week study period, 71 % of the cats were known to be alive. All cats remained FeLV antigen-positive, and there was no significant change in clinical signs or hematologic parameters. The study did not include a control group, and clinical and laboratory evaluations failed to show improvement; this makes it difficult to determine whether the use of acemannan improved the outcome of infection.[15] In dogs, intralesional injection of acemannan might offer a treatment for dogs with papillomatosis but data are limited. SC injection can cause hypersensitivity reactions and necrosis at the injection site.[16] Intralesional injections may cause discomfort or bleeding from the injection sites. High dosages administered intraperitoneally can cause an atypical microgranulomatous peritonitis with mononuclear macrophage infiltrates in liver, spleen, and peritoneal surface leading to abdominal pain, vomiting, and diarrhea.

***Staphylococcus* protein A**, SPA, is a bacterial polypeptide product purified from cell walls of *Staphylococcus aureus* Cowan I that has the ability to bind to the Fc portion of certain IgG subclasses by a non-immune mechanism, without disturbing antigen binding. SPA may combine with immune complexes and stimulate complement activation, as well as induce T-cell activation, stimulation of natural killer cells, and interferon-α production. Anti-tumor activity has been described. A variety of SPA sources and treatments have been used in FeLV-infected cats. Interest was first generated when plasma from FeLV-infected lymphoma-bearing cats was passed over SPA or *Staphylococcus aureus* columns to remove circulating immune complexes and then returned to the cats. More than 100 cats were treated in this manner, generally undergoing treatments twice per week for 10 to 20 weeks. In some studies, a high rate of tumor remission and conversion to FeLV-negative status was observed; in others, responses were less dramatic and short-lived. Subsequently, it was determined that SPA may have leached from the filters and columns used for immunosorption and been returned to the cats as contaminants in the treated plasma. Following that observation, cats were treated with small doses of SPA. In such a study including kittens with experimental FeLV infection, treatment with SPA neither reversed anemia nor improved humoral immune function. In a more recent placebo-controlled study, treatment of ill client-owned FeLV-infected cats with SPA (10 µg/kg twice per week for up to 10 weeks) did not cause a statistically significant difference in the FeLV status, survival time, or clinical and haematological parameters when compared to a placebo group but it caused a significant improvement in the owners' subjective impression on the health of their pets.[17] Hypersensitivity, anaphylactic reactions, and low grade peritonitis can occur as side effects.

***Propionibacterium acnes***, formerly *Corynebacterium parvum*, is available for veterinary use and consists of a killed bacterial product that has been shown to stimulate macrophages resulting in release of various cytokines and interferons and to enhance T-cell, and natural killer cell activity in mice. In mice, it was effective in preventing development of rabies in some animals after challenge.[18] In cats and dogs, it has been used for the treatment of certain tumors, e.g., malignant melanoma.[1] As antiviral, it has been applied to cats with FeLV infection, FHV-1 infection, or FIP. *Propionibacterium acnes* has been used in FeLV-infected cats, but no prospective studies have been performed. Veterinarians have described their clinical experience in anecdotal reports and have observed some beneficial effect. *Propionibacterium acnes* treatment was used in 74 cats with experimentally induced FIP that received either *Propionibacterium acnes*, high dose SC human interferon-α, a combination, or placebo. Although *Propionibacterium acnes* alone was ineffective, there was some indication that a combination of *Propionibacterium acnes* and high doses of interferon-α was more effective than interferon-α alone in prolonging the mean survival time in the treated cats for a few days.[19] Adverse reactions include fever and lethargy within 24 hours of administration. Localized pain and swelling can occur in cases of extravasation during intravenous injection. Proliferative glomerulonephritis and liver cirrhosis in a dog has been associated with *Propionibacterium acnes*.

**Bacille Calmette-Guérin**, BCG, is a cell wall extract of a nonpathogenic strain of *Mycobacterium bovis* that was originally developed in 1908 by Calmette and Guerin as "vaccine" against tuberculosis in humans. Facultative intracellular organisms such as mycobacteria are immunostimulators that have a marked affinity for localizing in and stimulating mononuclear-phagocyte clearance mechanisms. BCG has been utilized in treating neoplasia in cats and dogs.[5,6]. Kittens experimentally infected with feline sarcoma virus (FeSV), which is a recombinant virus that can develop in a cat with FeLV infection, were inoculated with BCG SC at the same time and site as the FeSV inoculation but 1 week later, or with a mixture of viable autochthonous neoplastic cells approximately 35 days after FeSV inoculation, but BCG was not able to prevent tumor development or increase the survival rate.[20]

A commercially available biological extract of ***Serratia marcescens*** containing DNA and membrane components of *Serratia marcescens* is used as immunomodulatory substance. *Serratia marcescens* is a motile facultative anaerobe gram-negative bacillus that occurs naturally in soil and water as well as the intestines and produces a red pigment at room temperature. It can cause nocosomial infections associated with urinary and respiratory tract infections, endocarditis, osteomyelitis, septis, wound infections, eye infections, and meningitis in humans; systemic infections in cats and dogs are also described. It has been shown that *Serratia marcescens* preparations stimulate normal feline bone marrow-derived macrophages to release maximum concentrations of interleukin-6, tumor necrosis factor-α, and interleukin-1 leading to elevations in rectal temperature and neutrophil counts.[23] In a study with FeLV-infected cats, weekly treatment with *Serratia marcescens* preparations, however, failed to prevent or reverse viremia in cats when initiated prior to or 6 weeks after inoculation with FeLV.[21]

## OTHER DRUGS WITH IMMUNOMODULATORY ACTIVITY

A number of drugs used for different other reasons in animals also have immunomodulatory activity. Some of them have been used to treat virus infections in dogs and cats.

**Levamisole** is a broad-spectrum anthelmintic, e.g., used for example as heartworm preventive in cats and dogs, that nonspecifically stimulates cell-mediated immunity and was detected as immunomodulatory substance when treatment applied against nematode infestation did not only kill the parasites but also improved clinical signs of other infections. Levamisole influences phosphodiesterase activity leading to increase of cyclic guanosin-3',5'-monophosphate and decrease of cyclic adenosine monophosphate. Increase in cyclic guanosin-3',5'-monophosphate in lymphocyes stimulates proliferative and secretory responses. It also potentiates mononuclear cells in phagocytosis, chemotaxis, and intracellular destruction of bacteria. Levamisole has been used in cats with mammary tumors but had no significant effect. It has been given to FIV- and in FeLV-infected cats but its effect has never been substantiated by controlled studies. Toxicity of levimasole is relatively high; hypersalivation, vomiting, diarrhea, and CNS signs (similar to signs observed in nicotine poisoning) have been observed. Morphologic lesions, characterized by perivascular, nonsuppurative, or granulomatous meningo-encephalitis were described in the CNS of treated dogs. Potential induction of the cell-mediated immune system may produce the lesions in the CNS by causing the body to react against latent agents (e.g., CDV).

**Diethycarbamazine**, N,N-diethyl-4-methyl-1-piperazine carboxamide, DEC, is an antiparasitic agent widely used, especially in tropical regions, to prevent and treat filariasis in humans. It also has been used as heartworm preventative in cats and dogs. The antifilarial effect of this drug has been attributed to immunomodulation. There was some evidence indicating that DEC may mitigate the course of FeLV infection in cats. Uncontrolled studies have suggested that continuous PO DEC treatment given shortly after FeLV infection prevents or delays FeLV-assoviated lymphopenia and prolongs survival. In one controlled study, its therapeutic effect against FeLV infection was investigated in 24 specific pathogen-free kittens experimentally infected with a lymphoma-causing strain of FeLV. The kittens were divided into 4 groups and were administered either a high dose of DEC (12 mg/kg PO q 24 h), a low dose of DEC (3 mg/kg PO q 24 h), zidovudine or a placebo for 10 weeks. While zidovudine was effective in preventing persistent viremia, DEC in either dose was not effective; however, both doses of DEC as well as zidovudin could prevent lymphoma development.[22] DEC has severe side effects including mainly liver toxicity.

**Lactoferrin** is a protein of bovine origin that binds iron, reducing its availability for bacteria. It has been used for its local immunomodulatory effect in the oral cavity, and it has been shown to increase phagocytic activity of neutrophils.[23] A protective effect of lactoferrin during lethal bacteremia has been reported in mice. In a study in healthy humans, orally administered lactoferrin increased the phagocytic activity of peripheral mononuclear cells in some but not all individuals. Lactoferrin has been used in cats with FIV infection and stomatitis. It was applied topically (40 mg/kg q 24 h for 14 days) to the oral mucosa of the cats with intractable stomatitis (FIV-infected cats and FIV-negative controls) and improved clinical signs in all cats independent of their FIV status.[23] Recently, a cats with stomatitis and FCV infection was treated with lactoferrin powder (200 mg) applied directly to the lesions. Clinical signs began to resolve after 11 months and resolution of clinical signs coincided with the cessation of FCV shedding.[24] Side effects are not described.

**Table 1:** Immunomodulators used in virus-infected cats and dogs (n.d. = not determined)

| drug | infection | efficacy *in vitro*? | controlled studies *in vivo*? | efficacy *in vivo*? | Forecast |
|---|---|---|---|---|---|
| **interferon inducers** | | | | | |
| poly-IC | FIV, FIP | n.d. | no | n.d. | Contraindicated |
| | FeLV, FCV | n.d. | no | n.d. | likely ineffective |
| | FHV-1, CHV | n.d. | no | n.d. | possibly effective |
| | ICH | n.d. | yes | no | ineffective |
| PIND-AVI/PIND-ORF | FeLV | n.d. | yes | no | ineffective |
| | FHV-1, FCV | n.d. | no | n.d. | likely ineffective |
| | FIV, FIP | n.d. | no | n.d. | contraindicated |
| Acemannan | FIV, FIP | n.d. | no | n.d. | contraindicated |
| | FeLV, papillomatosis | n.d. | no | n.d. | likely ineffective |
| SPA | FIV, FIP | n.d. | no | n.d. | contraindicated |
| | FeLV | n.d. | yes | +/- | possibly effective |
| *Propionibacterium acnes* | FIV | n.d. | no | n.d. | contraindicated |
| | FeLV | n.d. | no | n.d. | likely ineffective |
| | FIP | n.d. | yes | no | contraindicated |
| BCG | FIV, FIP | n.d. | no | n.d. | contraindicated |
| | FeLV | n.d. | yes | no | ineffective |
| *Serratia marcescens* | FIV, FIP | n.d. | no | n.d. | contraindicated |
| | FeLV | n.d. | yes | no | ineffective |
| **other drugs with immunomodulatory activity** | | | | | |
| levamisole | FIV, FIP | n.d. | no | n.d. | contraindicated |
| | FeLV | n.d. | no | n.d. | likely ineffective |
| DEC | FIV, FIP | n.d. | no | n.d. | contraindicated |
| | FeLV | n.d. | yes | no | ineffective |
| Lactoferrin | FIV | n.d. | yes | +/- | possibly in stomatitis |
| | FeLV | n.d. | no | n.d. | possibly in stomatitis |
| | FCV | n.d. | no | +/- | possibly in stomatitis |

## REFERENCES

[1]MacEwen EG, et al. **Clin Cancer Res** 1999;5:4249. [2]Vail DM, et al. **Clin Cancer Res** 1995; 1:1165. [3]Kurzman ID, et al. **Clin Cancer Res** 1995;1:1595. [4]Fox LE, et al. **Cancer Biother** 1995;10:125. [5]Parodi AL, et al. **Cancer Immunol Immunother** 1983;15:172. [6]Hess AD, et al. **Cancer Res** 1977;37:3990. [7]Misdorp W, **Vet Quarterly** 1987;9:279. [8]Smith et al., 1980. [9]Wooley RE, et al. **Am J Vet Res** 1974;35:1217. [10]Hörber D, et al. **Tierarztl Prax** 1991;19:311. [11]Hartmann K, et al. **Vet Immunol Immunopathol** 1998;65:267. [12]Hartmann K, et al. **Vet Microbiol** 1999;69:111. [13]King GK, et al. **J Am Anim Hosp Assoc** 1995;31:439. [14]Yates KM, et al. **Vet Immunol Immunopathol** 1992;35:177. [15]Sheets MA, et al. **Mol Biother** 1991;3:41. [16]Fogleman RW, et al. **Vet Hum Toxicol** 1992;34:201. [17]McCaw DL, et al. **J Am Anim Hosp Assoc** 2001;37:356. [18]Megid J, et al. **Comp Immunol Microbiol Infect Dis** 2000;23:91. [19]Weiss RC, et al. **Am J Vet Res** 1990;51:726. [20]Bachman D, et al. **Am J Vet Res** 1982;43:475. [21]Elmslie RE, et al. **Mol Biother** 1991;3:231. [22]Nelson P, et al. **Vet Immunol Immunopathol** 1995;46:181. [23]Sato R, et al. **Am J Vet Res** 1996;57:1443. [24]Addie DD, et al. **J Small Anim Pract** 2003;44:172.

# USE OF MOLECULAR ASSAYS FOR THE DIAGNOSIS OF FELINE INFECTIOUS DISEASES

**Michael R. Lappin, DVM, PhD, DACVIM**
**Fort Collins, CO**

Clinical syndromes induced by infectious agents are common in cats. Documenting that the infectious agent is still present is the best way to make a definitive diagnosis. Commonly used techniques vary by the body system but include fecal flotation, cytology, histopathology, immunohistochemistry, culture, antigen tests, and molecular diagnostic assays. For some agents, antibody test results are also used to help make a clinical diagnosis. However, presence of antibodies may only document prior exposure, not current infection.

Sensitivity is the ability of an assay to detect a positive sample; specificity is the ability of an assay to detect a negative sample. Sensitivity and specificity vary with each assay. Positive predictive value is the ability of a test result to predict presence of disease; negative predictive value is the ability of a test result to predict absence of disease. Many of the infectious agents encountered in feline practice infect a large percentage of the population, resulting in positive organism detection techniques or serum antibody production. However, they only induce disease in a small number of cats in the infected group. Examples include coronaviruses, *Toxoplasma gondii*, and *Bartonella* spp.. For these examples, even though assays with good sensitivity and specificity are available, the predictive value of a positive test is actually very low.

One of the newest organism demonstration techniques used in practice is the polymerase chain reaction (PCR) assay. This reaction amplifies small quantities of DNA to detectable levels. By use of a reverse transcriptase step, RNA is converted to DNA; therefore the technique can also be used to detect RNA (RT-PCR). PCR assays are of great value for documentation of infections, particularly if the organism in question is difficult to culture (e.g., *Mycoplasma* spp.) or cannot be cultured (e.g., hemoplasmas). Specificity can be very high, depending on the primers used in the reaction. For example, primers can be designed to detect one genus but not others. Primers can also be designed to identify only one species. For example a PCR assay can be developed to detect all haemoplasmas or just one species such as *Mycoplasma hemofelis*.

Because of the inherent sensitivity of the reaction, PCR can give false-positive results if sample contamination occurs during collection or at the laboratory performing the procedure. False-negative results can occur if the sample is handled inappropriately. Results may also be affected by treatment. For example, many cats with hemoplasmosis become negative by PCR assay of blood while on doxycycline or fluoroquinolone therapy and so should be sampled prior to treatment. Other potential problems are that minimal standardization exists among commercial laboratories offering PCR assays and minimal external quality control exists.

Although PCR assays can be one of the most sensitive for documentation of infections, positive test results do not always prove that the infection is resulting in clinical illness. For example, because the technique detects DNA of both live and dead organisms, positive test results may be achieved even if the infection has been controlled. When the organism being tested for commonly infects the background population of healthy cats, interpretation of results for a single animal can be difficult. For example, '*Candidatus* M. haemominutum' infects up to 20% of healthy cats in some environments and so detection of a positive test result in a clinically ill cat does not prove a disease association. In addition, for some agents the currently available PCR assays cannot discriminate between vaccine strains and field strains. For example, currently available PCR assays for FHV-1 also amplify modified live vaccine strains, so a positive result does not even indicate presence of a pathogenic strain. Real-time PCR can be used to determine the amount of microbial DNA in a sample. It is possible that the DNA load will correlate to the presence of disease for some agents. However, some agents are very host adapted and can have large amounts of DNA present in samples from healthy carrier cats. For example, the number of '*Candidatus* M. haemominutum' copy numbers/μl of blood does not correlate to the PCV. Based on these findings, it is very important that small animal practitioners carefully assess the predictive values of currently available PCR and the expertise and reliability of the laboratory that will be performing the assays. New PCR assays are being developed almost daily. The purpose of this lecture is to use several common infectious disease agents to emphasize important points concerning molecular assays in the diagnosis of feline infectious diseases.

**Respiratory agents**

Feline calicivirus (FCV) is a common differential diagnosis for cats with clinical evidence of rhinitis, stomatitis, and conjunctivitis. Less commonly, FCV is associated with polyarthritis and lower airway disease in kittens. Virus isolation can be used to document current infection but takes at least several days for results to return. Because of wide-spread exposure and vaccination, the positive predictive value of serological tests is poor. Reverse transcriptase (RT) PCR assays can be used to amplify the RNA of FCV and results can be returned quickly. However, these assays also amplify vaccine strains of FCV. FCV RNA can be amplified from samples collected from normal carrier cats as well as clinically ill cats and so have poor positive predictive value. For example, in one study in our laboratory, presence of FCV RNA failed to correlate to the presence or absence of stomatitis in cats (1). In addition, amplification of FCV RNA cannot be used to prove virulent systemic calicivirus infection. Results of FCV RT PCR can also be falsely negative and so can have poor negative predictive value.

Feline herpesvirus 1 (FHV-1) is a common differential diagnosis for cats with clinical evidence of rhinitis, stomatitis, conjunctivitis, keratitis, and facial dermatitis. Because of wide-spread exposure and vaccination, the positive predictive value of serological tests is poor. FHV-1 can be documented by direct fluorescent staining of conjunctival scrapings, virus isolation, or PCR. FHV-1 DNA can be amplified from conjunctiva, nasal discharges, and pharynx of healthy cats and so the positive predictive value of conventional PCR assays is low (2). Currently used PCR assays also detect vaccine strains of FHV-1, further lessening the positive predictive value of the assays (3). In one study in our laboratory, presence of FHV-1 RNA failed to correlate to the presence or absence of stomatitis in cats. Quantitative PCR may ultimately prove to correlate to the presence or absence of disease but failed to correlate to presence of conjunctivitis in one study (4). The negative predictive value of FHV-1 PCR assays is also in question because many cats that are likely to have FHV-1 associated disease are negative. This may relate to clearance of FHV-1 DNA from tissues by a hypersensitivity reaction. Tissue biopsies have greater sensitivity than conjunctival swabs but do not necessarily have greater predictive value. FHV-1 DNA can be amplified from aqueous humor of some cats but whether this indicates FHV-1 associated uveitis is unknown.

*Mycoplasma* spp. *Chlamydophila felis*, and *Bordetella bronchiseptica* are other common respiratory pathogens in cats. As for FHV-1 and FCV, PCR positive test results for these organisms cannot be used to distinguish a carrier from a clinically ill cat. In addition, PCR assays do not provide antimicrobial drug susceptibility testing and so for cats with potential bordetellosis, culture and sensitivity is the optimal diagnostic technique, especially if an outbreak is occurring. *Toxoplasma gondii* DNA has been amplified from airway washings of some cats with lower respiratory tract disease and so PCR is an option for evaluation of samples from diseased animals from which the organism is not identified cytologically.

**Gastrointestinal agents**

The diagnosis of *Giardia* spp. infection is easy to make in cats with small bowel diarrhea with the combination of fecal flotation techniques and wet mount examination. Fecal antigen tests are also accurate and there are several assays available for point of care use, included one labeled for veterinary use (5). Fecal PCR assays are often falsely negative because of PCR inhibitors in stool. However, *Giardia* spp. PCR can be used to determine whether the infective species is a zoonotic assemblage which is the primary indication for this technique.

While *Cryptosporidium* spp. infection is common, it is unusual to find *C. felis* oocysts after fecal flotation in cats. Acid-fast staining of a thin fecal smear is cumbersome and insensitive. Antigen assays titrated for use with human feces are inaccurate when used with cat feces. Thus, PCR may be aid in the diagnosis of cryptosporidiosis in dogs and cats and has been shown to be more sensitive than IFA in cats (6). *Cryptosporidium* spp. PCR assays are indicated in IFA negative cats with unexplained small bowel diarrhea and when the genotype of *Cryptosporidium* is to be determined. However, *C. felis* infection in cats is common and so positive tests results do not always prove that the agent is the cause of the clinical disease. No drug is known to eliminate *Cryptosporidium* spp. infections and small animal strains are not considered significant zoonotic agents so PCR is never indicated in healthy animals.

PCR assays are also available for detection of DNA of *Tritrichomonas foetus, Salmonella* spp., *Campylobacter* spp., parvoviruses, and *T. gondii* and a RT-PCR assay is available for coronaviruses. Trophozoites of *T. foetus* can often be detected on wet mount examination of fresh feces which can be completed as an in clinic test. DNA of *T. foetus* can be detected in healthy carrier cats and so positive results do not always prove illness from the organism (7). Cases with suspected salmonellosis or campylobacteriosis should be cultured to determine the anti-microbial susceptibility patterns. There is no current evidence that parvovirus PCR on feces is superior to currently available antigen assays. *Toxoplasma gondii* is only shed for about 7-10 days and millions of oocysts are generally shed during this time making the organism very easy to identify. Thus, PCR assays are usually not needed to diagnosis this infection. Because virus isolation is not practical clinically, RT-PCR is used most frequently to detect coronaviruses RNA in feces. However, positive test results do not differentiate FIPV from FECV.

**Blood borne agents**

The new names for *Haemobartonella felis* are *Mycoplasma haemofelis* (Mhf), '*Candidatus* Mycoplasma haemominutum' (Mhm), and '*Candidatus* M. turicensis' (Mtc). In at least two studies of experimentally infected cats, Mhf is apparently more pathogenic than Mhm; all Mhf inoculated cats became clinical ill whereas Mhm inoculated cats were generally subclinically infected. Cats with chronic Mhm infection had more severe anemia and longer duration of anemia when experimentally infected with Mhf when compared to cats infected with Mhf alone. It appears that Mtc has intermediate pathogenicity. Diagnosis is based on demonstration of the organism on the surface of erythrocytes on examination of a thin blood film or PCR assay. Organism numbers fluctuate and so blood film examination can be falsely negative up to 50% of the time. The organism may be difficult to find cytologically, particularly in the chronic phase. Thus, PCR assays are the tests of choice due to sensitivity (8). Primers are available that can amplify all three hemoplasmas. Real time PCR assays can be used to monitor copy numbers during and after treatment but do not have greater sensitivity, specificity, or predictive value than conventional PCR assays. PCR assays should be considered in the evaluation of cats with unexplained fever or anemia that are cytologically negative. In addition, the ACVIM recommends screening cats for use as blood donors by PCR assays for haemoplasmas (9). Many cats are carriers of the relatively non-pathogenic 'Candidatus M. haemominutum' and so positive test results may not always correlate to the presence of disease.

Cats can be infected by *E. canis* like organism (10) and *Anaplasma phagocytophilum* (11). Little is known about the other agents in these genera in regards to cats. As the organisms are in different genera, serological cross reactivity is variable. Thus, while the clinical syndromes can be similar, there is no one serological test to document infection and there is currently no standardized serology for cats. In addition, some cats with *E. canis* infection do not seroconvert and so PCR assay is superior to serology in cats. PCR assays can be designed to amplify each organism. Alternately, primers are available to amplify all of the organisms in a single reaction and then sequencing can be used to determine the infective species.

Cats can be infected by *Rickettsia felis* and have been shown to have antibodies against *R. rickettsii*. Fever, headache, myalgia, and macular rash in humans have been attributed to *R. felis* infection in several countries around the world. In recent study in our laboratory, we assayed 92 pairs of cat blood and flea extracts from Alabama, Maryland and Texas, using PCR assays that amplify a region of the citrate synthase gene (gltA) and the outer membrane protein B gene (ompB). Of the 92 pairs, 62 of 92 (67.4%) flea extracts and none of the cat blood samples were positive for *R. felis* DNA (12). In another study, we showed *R. felis* and *R. rickettsii* antibody prevalence rates in cats with fever to be 5.6% and 6.6%, respectively but neither organism was amplified from blood (13).

These results prove that cats are sometimes exposed but further data are needed to determine significance of diseases associations.

Blood culture, PCR assay on blood, and serologic testing can be used to assess individual cats for *Bartonella* spp. infection. Serological methods include IFA, ELISA, and Western blot immunoassay. Cats that are culture-negative or PCR-negative and antibody-negative and cats that are culture-negative or PCR-negative and antibody-positive are probably not a source of flea, cat, or human infection. However, bacteremia can be intermittent and false-negative culture or PCR results can occur, limiting the predictive value of a single battery of tests. While serologic testing can be used to determine whether an individual cat has been exposed, both seropositive and seronegative cats can be bacteremic, limiting the diagnostic utility of serologic testing. Thus, testing healthy cats for *Bartonella* species infection is not currently recommended (14). Testing should be reserved for cats with suspected clinical bartonellosis. However, because *Bartonella* spp. infection in cats is so common in healthy cats, even culture or PCR positive results does not prove clinical bartonellosis. For example, while we detected *Bartonella* spp. DNA in more cats with fever than pair matched cats without fever, the healthy cats were still commonly positive (15).

*Cytauxzoon felis* in clinically affected cats is usually easily identified on cytological examination of blood smears or splenic aspirates. Serologic testing is not commercially available. PCR can be used to amplify organism DNA from blood from cats that are cytologically negative (16).

Antibodies against feline immunodeficiency virus (FIV) are detected in serum in clinical practice most frequently by enzyme-linked immunosorbent assay (ELISA). Comparisons between different tests have shown the results of most assays are comparable (17). Clinical signs can occur before seroconversion in some cats, and some infected cats never seroconvert; thus false-negative reactions can occur. Results of virus isolation or PCR on blood are positive in some antibody-negative cats. False-positive reactions can occur using ELISA; hence, positive ELISA results in healthy or low-risk cats should be confirmed using Western blot immunoassay. Kittens can have detectable, colostrum-derived antibodies for several months. Kittens less than 6 months of age that are FIV seropositive should be tested every 60 days until the result is negative. If antibodies persist at 6 months of age, the kitten is likely infected. Virus isolation or RT-PCR on blood can also be performed to confirm infection. However, FIV is not present in the blood in high levels and so false negative results are common. Thus, the assay is not very accurate for distinguishing a vaccinated cat from a naturally exposed cat (18).

Most cats with feline leukemia virus infection are viremic and so molecular diagnostic assays are not usually needed in clinical practice. However, use of newer sensitive real time PCR assays have been used to accurately characterize the stages of infection (). However, these assays are not commonly available commercially.

RNA of both FIPV and FECV can be amplified from the blood of cats and so positive test results do not always correlate with the development of FIP. Amplification of the mRNA of the M gene by RT-PCR has had mixed results in two studies performed to date. In the one study, 13 of 26 apparently normal cats were positive for FECV mRNA in blood suggesting that the positive predictive value of this assay for the diagnosis of FIP was low (20).

**Ocular agents**

*Toxoplasma gondii*, *Bartonella* spp., FHV-1 and coronavirus are the organisms for which DNA or RNA has been amplified most frequently from the aqueous humor of cats with endogenous uveitis. While little is know about the predictive value of these assays when used with aqueous humor, the combination of molecular assays with local antibody production indices may aid in the diagnosis of some cases.

## REFERENCES

[1]Quimby JM, et. al. **J Feline Med Surg**. 2008;10:66. [2]Veir JK, et. al. **J Feline Med Surg**. 2008;10:551. [3]Maggs DJ et. al. **Am J Vet Res.** 2005;66:1550. [4]Low HC, et. al. **Am J Vet Res.** 2007;68:643. [5]Mekaru SR, et. al. **J Vet Int Med** 2007;21:959. [6]Scorza AV, et. al. **J Parasitol** 2003;89:423-426. [7]Gookin JL, et. al. **J Clin Microbiol** 2004;42:2707. [8]Jensen WA, et. al. **Am J Vet Res** 2001;62:604. [9]Wardrop KJ, et. al. **J Vet Intern Med**. 2005;19:135. [10]Breitschwert EB, et. al. **J Vet Intern Med** 2002;16:642. [11]Lappin MR, et. al. **J Am Vet Med Assoc** 2004; 225:893. [12]Hawley JR, et. al. **J Feline Med Surg**. 2007 9:258-562. [13]Bayliss DB et. al. **J Feline Med Surg**. 2008 Sep 9. [Epub ahead of print]. [14]Brunt J, et. al. **J Feline Med Surg**. 2006;8:213. [15]**Lappin MR** et. al. **J Fel Med Surg** 2009;11:141. [16]Haber MD, et. al. **Vet Parasitol.** 2007;146:316. [17]Hartmann K et. al: **J Feline Med Surg** 2007;9:439. [18]Crawford PC et. al. **J Am Vet Med Assoc** 2005;226:1503. [19]Levy J, et. al. **J Feline Med Surg**. 2008;10:300. [20]Can-S Ahna K et. Al. **J Feline Med Surg** 2007;9:369..

## APPROACH TO IDENTIFYING INFECTIOUS CAUSES OF FELINE DIARRHEA


**Jody L. Gookin, DVM, Ph.D., DACVIM**
**Raleigh, NC**


### INTRODUCTION

Feline diarrhea can be a frustrating clinical problem. In developing a diagnostic plan, there is often a level of uncertainty as to whether or not all of the potential infectious causes have been ruled out – particularly prior to embarking on more invasive diagnostic tests or therapeutics. This problem is compounded by the large number of infectious agents that could be present, lack of highly sensitive means to detect these agents, and high prevalence of many of these "pathogens" in healthy animals. Whether or not a given infectious agent is responsible for disease must always be considered in light of the patients signalment, history and clinical signs.

### LARGE VERSUS SMALL BOWEL DIARRHEA

Distinguishing between whether or not the diarrhea is small or large bowel in origin from the onset will be important in terms of prioritizing your differential diagnoses and diagnostic plan.

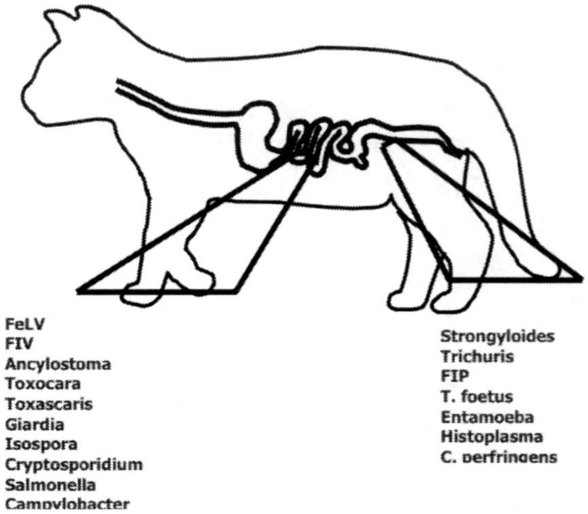


**Table 1**. Distinguishing clinical features between small and large bowel diarrhea.

| Clinical Sign | Small Bowel | Large Bowel |
|---|---|---|
| Weight loss/polyphagia | + | – |
| Volume | + | – |
| Frequency | +/– | + |
| Melena | + | – |
| Frank blood/mucus | – | + |
| Tenesmus | – | + |

### DIAGNOSTIC PLAN

There are many diagnostic tests that are useful for identifying infectious causes of diarrhea and for many of these agents there are multiple diagnostic tests. Routine complete blood cell count, serum biochemistry profile, and urinalysis sometimes provide insight into the probability of an infectious agent but rarely identify a specific cause.

**Table 2.** Prioritized database for ruling out infectious causes of chronic diarrhea in the cat.

| First Tier Diagnostic Tests | Small Bowel | Large Bowel | Second Tier Diagnostic Tests | Small Bowel | Large Bowel |
|---|---|---|---|---|---|
| Complete physical examination | + | + | Fecal cytology | + | + |
| CBC and FeLV/FIV test | + | + | Rectal scraping | – | + |
| Fecal wet mount(s) | + | + | CP enterotoxin | – | + |
| Fecal centrifugation flotation | + | + | Fecal culture | + | – |
| *Giardia* fecal antigen test | + | + | *Cryptosporidium* testing | + | – |
| Specific test for *Tritrichomonas foetus* | – | + | | | |
| Response to treatment with Fenbendazole | + | + | | | |

#### Complete blood cell count and FeLV/FIV test

Potentially helpful findings on a CBC include the following. Blood-loss anemia (either regenerative or consistent with iron deficiency) may indicate the presence of blood-sucking endoparasites. Neutrophilia (*Salmonella*, *Campylobacter*), neutropenia (*Panleukopenia*), and eosinophila (helminthes or fungi). High or low plasma protein concentrations should be investigated further by determination of serum albumin and globulin concentrations. Panhypoproteinemia in the absence of anemia suggests protein-losing enteropathy. Hyperglobulinemia may suggest the presence of FIP. FeLV/FIV testing is recommended for all cats presenting with clinical signs of chronic diarrhea. Positive results dictate a search for secondary or opportunistic infections that may be responsible for signs, however sometimes no other etiologic agent can be identified.

#### Fecal wet mount

A wet mount is performed by obtaining *fresh* feces directly from the colon. A *small* amount of the feces (less than a rice-grain) is mixed immediately with a drop of saline on a microscope slide. A coverslip is applied and the slide examined using a light microscope (40X objective). Feces obtained using a loop inserted into the proximal colon or by means of a saline "flush" of the proximal colon with a red-rubber catheter is preferable. Agents to look for include: *Tritrichomonas foetus* trophozoites (LB), *Entamoeba histolytica* (LB), and *Giardia* trophozoites (SB), characteristic "darting" motility of *Campylobacter* (SB). *T. foetus* and *Giardia* trophozoites can be easily confused. *T. foetus* has an undulating membrane and *Giardia* has motility resembling a falling leaf.

A video of their motility can be viewed at the following website http://www.JodyGookin.com. None of these infectious agents can be ruled out by their absence on a wet mount.

**Zinc sulfate centrifugation-flotation**

A centrifugation flotation is performed by mixing 1-2 grams of feces in 33% zinc sulfate solution and straining into a centrifuge tube. The tube is centrifuged at 1,500 rpm for 3-5 min. If the tubes tilt during centrifugation, then each is topped with a coverslip which is examined for parasites. If the tubes are fixed upright during centrifugation, then a drop from the surface of the liquid is examined. Agents to look for include: *Strongyloides* ova (LB), *Trichuris* ova (LB), *Entamoeba* cysts (LB), *Ancylostoma* ova (SB), *Toxocara* and *Toxascaris* ova (SB), *Giardia* cysts (SB), *Isospora* spp (SB). Testing of 3 separate fecal samples is recommended, especially for detection of *Giardia*. Cats may also be tested for *Giardia* using a ***Giardia* antigen ELISA**. Due to intermittent shedding of ova and oocysts, all cats should be empirically treated with fenbendazole as this drug has broad-spectrum efficacy against enteric helminths and *Giardia*.

**Fecal cytology/rectal scraping**

Cytology performed on samples of mucus and epithelium of the rectal mucosa can give valuable insight into possible underlying infectious diseases, particularly of the large bowel. These samples can be obtained by scraping the mucosa with a gloved finger, with a wet (saline) cotton-tipped swab, or conjunctival spatula. In each case the sample is smeared onto a glass slide and stained with either Diff-Quick or Wright's-Giemsa stain. Agents to look for include: *Histoplasma capsulatum* (organisms inside mononuclear cells), *Clostridium perfringens* endospores (large bacterial rods containing a clear center with dark staining ends resembling "safety pins"), *Campylobacter* (suggested by the presence of gull-shaped bacteria). An acid fast or monoclonal antibody stain (Meridian MeriFluor) can be requested to look for *Cryptosporidium* spp. oocysts. If neutrophils are present consider fecal culture for *Campylobacter* or *Salmonella*. If there are ≥3-5 *Clostridium* spores per hpf consider a fecal assay for enterotoxin or a therapeutic trial of amoxicillin for treatment of *C. perfringens*.

**Culture or PCR for *Tritrichomonas foetus***

Recommended for any cat with large bowel diarrhea, particularly young purebred cats, those developing diarrhea after introduction of a kitten, or any cat raised in a densely-housed facility[1,2]. If repeated wet mount results are negative for *T. foetus*, feces may be cultured in commercially available pouches (In Pouch™ TF; www.BiomedDiagnostics.com). Pouches should be inoculated with 0.05 g (about the size of a rice grain) of freshly voided, loop-collected, or colonic flush fecal specimens and incubated either at 37°C for 48 hrs or at room temperature (25°C) for up to 12 days[3]. Pouches can be placed within a manufacturer-provided clamp that allows the pouch to be mounted onto the stage of a light microscope. Although feces may be shipped to some laboratories for cultivation, given the fragile nature of the organisms it is strongly recommended that this user-friendly culture technique be performed in-house for best results. A sensitive and specific PCR for detection of *T. foetus* is commercially available through the author's laboratory. PCR is superior to fecal culture for diagnosis of infected cats. A lima-bean sized amount of fresh feces is placed in a red-top vacutainer tube which is then filled with rubbing alcohol (isopropyl alcohol). Refrigeration or freezing is not necessary. Feces must be diarrheic and free of litter. Formed samples rarely test positive even if infection is present. Samples collected by flush or loop are preferable. Cats should not be receiving any antibiotics within several days prior to or at the time of testing. Submission forms and instructions can be found on our website (http://www.JodyGookin.com). Questions about the PCR can be directed to Maria Stone. Maria_Stone@ncsu.edu (919) 513-6365.

**Fecal culture for *Salmonella* or *Campylobacter***

Should be considered in kittens with a history consistent with stressful housing conditions, exposure to contaminated foods (including raw or improperly cooked meat) or facilities, and clinical signs of acute diarrhea and systemic illness (fever, leukocytosis, dehydration). The clinical and pathologic features of these infections may be indistinguishable from feline panleukopenia. Because of the high prevalence of asymptomatic carriage of these bacteria in cats, demonstration of the organisms does not prove that they are responsible for disease.

**Electron microscopy of feces**

Should be considered if looking for unusual viral etiologies of diarrhea (e.g. torovirus)

**Intestinal mucosal biopsy**

Surgical or endoscopic intestinal mucosal biopsies are often appropriate in cases of diarrhea where less invasive diagnostics fail to identify a cause for diarrhea and appropriate empirical medical and dietary treatment trials have failed to result in a resolution of signs. When looking for infectious causes of diarrhea, it is important to indicate what agents are suspected so that appropriate special stains are likely to be performed. When performing endoscopy for diagnosis of small bowel diarrhea a duodenal aspirate for *Giardia* can be performed. Approximately 10 cc of saline can be flushed into the duodenum, aspirated, centrifuged and pellet examined for the presence of motile trophozoites.

**Table 3.** Infectious causes of small or mixed bowel diarrhea

| Helminths | Diagnostic test(s) | Diagnostic finding(s) | Treatment |
|---|---|---|---|
| *Ancylostoma tubeforme* | | | |
| *Toxocara cati* | Flotation | Ova | Fenbendazole |
| *Toxascaris leonina* | | | |
| **Viruses** | | | |
| Feline leukemia virus | FeLV ELISA | Antigen positive | |
| Feline immunodeficiency virus | FIV ELISA | Antibody positive | Supportive |
| Feline panleukopenia virus | ELISA for CPV antigen | Antigen positive | |
| **Protozoa** | | | |
| *Giardia* | Flotation | Oocysts | Fenbendazole, metronidazole |
| | ELISA | Antigen positive | |
| *Cryptosporidium* | Fecal cytology | Acid-fast positive | None proven |
| | ELISA | Antigen positive | |
| *Isospora* | Flotation | Ova | Sulfonamides |
| **Bacteria** | | | |
| *Salmonella* | Culture or PCR | *Salmonella* | Supportive<br>± antibiotics based on sensitivity |
| *Campylobacter* | Wet mount | | Erythromycin |
| | Fecal culture | | |

**Table 4.** Infectious causes of large bowel diarrhea

| Helminths | Diagnostic test(s) | Diagnostic finding(s) | Treatment |
|---|---|---|---|
| *Strongyloides* | Flotation | Embryonated ova | |
| | Baermann | Larvae | Fenbendazole |
| *Trichuris* spp. | Flotation | Bipolar operculated unembryonated ova | |
| **Viral** | | | |
| Feline infectious peritonitis | Histopathology | Pyogranulomas with intralesional FCoV | None |
| **Protozoa** | | | |
| *Tritrichomonas foetus* | Direct smear | Trichomonads | |
| | Fecal culture | Trichomonads | Ronidazole |
| | PCR | Positive | |
| *Entamoeba histolytica* | Wet mount | Amoebas | Metronidazole |
| | Flotation | Cysts | |
| **Fungi** | | | |
| *Histoplasma capsulatum* | Rectal scraping or mucosal biopsy | *Histoplasma* organisms in mononuclear cells | Itraconazole |
| **Bacteria** | | | |
| *Clostridium perfringens* | Fecal culture | *C. perfringens* | Amoxicillin, |
| | Fecal cytology | ≥ 3 endospores per hpf | metronidazole, |
| | Immunodetection | Enterotoxin (CPE) | or tylosin |

## REFERENCES

[1]Gookin JL, et al. **J Am Vet Med Assoc** 1999;215:1450. [2]Gookin JL, et al. **J Clin Microbiol** 2004;42:2707.[3]Gookin JL, et al. **J Am Vet Med Assoc** 2003;222:1376.

## CAUSES OF FELINE ANEMIA: OLD AND NEW?

**Kristy Dowers, DVM, MS, DACVIM**
**Fort Collins, CO**

### INTRODUCTION

Anemia is a common blood work abnormality in many species, including cats. Determining the regenerative nature of the anemia guides the workup of the case. Regenerative anemias suggest blood loss or red cell lysis. Red cell lysis can be due to toxins, infectious agents, neoplasia (as a secondary immune-mediated phenomenon) or primary immune-mediated hemolytic anemia. Non-regenerative causes of anemia include iron-deficiency, anemia of inflammatory disease, renal disease, bone marrow disorders, neoplasia, infectious disease, and immune-mediated disorders directed at erythrocyte precursors.

### CLINICAL PRESENTATION & HISTORICAL FINDINGS

Classic signs of anemia include lethargy, weakness, inappetence, pale mucous membranes or icteric membranes if hemolysis is occurring. Given that infectious disease and neoplasia are two major rule-outs for anemia, fever may be present. Since many of the infectious agents are vector-borne, flea and tick exposure should be determined. Outdoor status increases the chance of contact with other cats and thereby exposure to the retroviral agents. Exposure to drugs or toxins such as onions and acetaminophen is important to ascertain. Other systemic signs, such as PU/PD with renal disease, can indicate the presence of other chronic diseases.

### DIAGNOSTIC EVALUATION

A complete blood count, including reticulocyte count, will determine whether the anemia is regenerative or non-regenerative. Acute blood loss and red blood cell lysis will not be regenerative for 2-4 days and so clinical and clinicopathologic parameters must be taken into account when making an assessment of the regenerative nature of the anemia. Clinical evidence of bleeding and a low plasma protein should differentiate acute blood loss from hemolytic anemia. Spherocytes are difficult to identify in feline blood smears and are therefore not a reliable means to diagnose hemolytic anemia. Presence of reticulocytes and/or nucleated red blood cells and macrocytosis with or without autoagglutination is diagnostic for regenerative anemia. An iron-deficiency anemia will be characterized as microcytic and FeLV-induced non-regenerative anemia will often be macrocytic. A careful evaluation of blood smears can reveal intra- or extracellular organisms. Presence of Heinz bodies may indicate an exposure to toxins that cause oxidative damage such as onions, propylene glycol, etc. A direct Coombs' test may be performed, but a negative test does not rule out an immune-mediated anemia since its sensitivity and specificity are low in cats. An osmotic fragility test is indicated in breeds such as Abyssinians and Somalis which can have a hereditary erythrocyte membrane defect. A biochemistry panel is necessary to rule in or out other causes of anemia, such as renal disease and other systemic illnesses. Hyperbilirubinemia may be present if hemolysis is severe. Other abnormalities in the biochemistry panel depend on the specific chronic disease, infectious agent or neoplastic process. FeLV antigen/FIV antibody testing should be done and thoracic and abdominal radiographs (or abdominal ultrasound) should be performed to help rule-out neoplasia.

### TOXINS

A variety of toxins and drugs are associated with hemolytic anemia in cats. Zinc, onions, propylene glycol, acetaminophen, methylene blue, methionine, methimazole and benzocaine preparations are few of the most common agents. The majority of these toxins induce Heinz body formation and some, such as acetaminophen toxicity, cause methemaglobinemia as well. Supportive care is the primary therapy for Heinz body anemia.

### INFECTIOUS AGENTS

#### Hemoplasmosis

*Mycoplasma haemofelis*, 'Candidatus *M. haemominutum'* and 'Candidatus *M. turicensis*' are the organisms responsible for hemoplasmosis. Virulence is attributed primarily to *M. haemofelis* and *M. turicensis*. The disease has worldwide distribution with rates of prevalence varying from 4%-23% for *M. haemofelis*, depending on the region.[1] Since the flea is the likely vector, prevalence rates vary according to climate. Other modes of transmission include blood transfusion (and by extension between cats if blood contamination occurs) and from the queen to her kittens (in utero or via lactation). Clinical signs are those associated with anemia and fever is often present. Identification of the gram-negative, epierythrocytic organisms on a blood smear is an unreliable means of diagnosis because of cyclic parasitemia, staining artifacts and sample handling. PCR for organismal DNA is the most sensitive and specific test for the hemoplasmas. The standard treatment is doxycycline (5 mg BID x 14 days), but there is substantial evidence that the fluoroquinolones are equally efficacious. In several studies, enrofloxacin (5-10 mg PO q 24 hrs x 14 days),[2] marbofloxacin (2.75 mg PO q24 hrs)[3] and pradofloxacin[4] (a novel fluoroquinolone not yet available in the U.S.) have demonstrated clearance of the organism by PCR weeks to months after therapy. In cats in which organism clearance has not occurred, recrudescence during times of stress is possible.

#### Cytauxzoonosis

*Cytauxzoon felis* is a tick-borne protozoal organism that is predominantly found in the south central, southeastern and mid-Atlantic states.[5] The bobcat is the reservoir host and *Dermacenter variabilis* is the presumed tick vector. These signet ring-shaped organisms are located within the red blood cell during the "erythrocytic phase" of the infection. *C. felis* invades monocytes/macrophages which are distributed to the tissue for the "tissue phase" or "leukocytic phase" of the disease. Schizontengorged monocytes clog vessels and account for the multi-organ failure and rapid death, usually within 1 week, associated with cytauxzoonosis. Considered to be 100% fatal in cats, recent reports of healthy carriers and cats which survive the infection have suggested that aggressive therapy may improve survival rates.[6] A presumptive diagnosis is often made on acute onset of clinical signs (severe lethargy, anorexia, icterus, dyspnea) in an endemic region with possible exposure to ticks. A definitive diagnosis can be made if the parasite is visualized within the RBCs; PCR is also now available.[7] A promising combination of the anti-malarial drug atovaquone (15mg/kg PO TID) and azithromycin (10mg/kg PO Q24), in addition to aggressive supportive care and heparin, is currently being recommended. With this combination, a survival rate of 64% was reported.[8]

**Bartonellosis?**

*Bartonella* spp. have been implicated in multiple diseases in humans, dogs and cats. In cats, the primary infective species are *B. henselae* and *B. clarridgeiae*. In humans, *B. bacilliformis* causes a hemolytic anemia in which nearly 100% of RBCs are infected.[9] Although this organism does not infect cats, a case of human hemolytic anemia caused by *Bartonella henselae* has been documented.[10] Studies have also demonstrated the intraerythrocytic location of *B. henselae* in naturally-infected cats using electron microscopy, making this organism a candidate for unexplained cases of feline anemia. Recently, however, a retrospective study attempting to correlate Bartonella spp. with cases of hemolytic anemia found no significant prevalence differences in the healthy cats versus anemic cats with a positive *Bartonella* PCR assay.[11] As with many high-prevalence diseases, it will be difficult to make the association between *Bartonella* spp. and anemia.

**Ehrlichiosis?**

*Ehrlichia* and *Anaplasma*-like DNA have been amplified from naturally-infected cats and have been demonstrated to cause clinical signs in cats, including fever, lethargy, inappetence and, in some cases, anemia.[12,13] Though these organisms are listed under hemolytic causes, the anemias in one case report were non-regenerative.[14] In the retrospective study referenced above, however, no sample from any of the healthy or anemic cats was PCR-positive for *Ehrlichia* or *Anaplasma* spp.[11] Given the fact that ehrlichiosis in cats is rare and that only some of these cats had anemia, it is unnecessary to routinely test for these organisms in cats with anemia.

**FeLV/FIV**

FeLV can cause both non-regenerative and regenerative anemias. The latter may be due to concurrent infectious with the hemoplasmas, but an immune-mediated process is also possible. FeLV antigen testing is therefore recommended for any cat presenting with hemolytic anemia. FIV infections are nearly always associated with a non-regenerative anemia. There are only a few controlled studies evaluating use of antiviral agents in these diseases and most show either no effect or are toxic. Feline interferon-ω and human interferon-α showed some effect for FeLV and FIV, respectively.[15]

## PRIMARY IMMUNE-MEDIATED HEMOLYTIC ANEMIA

There is a subset of cats with hemolytic anemia for which neoplasia, toxins and the majority of infectious diseases have been ruled out, either by infectious disease assays or trial therapies with anti-bacterial agents. In one study of 19 cats, no identifiable infectious, neoplastic or toxic could be found as a cause of their anemia.[12] A majority of these cats were Coombs' positive suggesting that anti-erythrocyte antibodies were present. In another retrospective study of anemic cats whose blood was submitted for extensive infectious disease testing, approximately 75% of cats with either non-regenerative or regenerative anemia and 60% of cats with non-regenerative anemia were negative for all organisms tested.[11] This would suggest that primary IMHA may be more common than previously thought. Unfortunately, a definitive diagnosis may be difficult to obtain. The direct Coombs' test appears to be neither specific nor sensitive in cats. New techniques, such as flow cytometry and anti-erythrocyte antibody assays, may be helpful in identifying anti-erythrocyte antibodies in cats with primary hemolytic anemia but are currently not commercially available. Therefore, the diagnosis of primary IMHA cats is usually made by ruling out all other causes of anemia with an exhaustive workup and trial therapy. Treatment with immunosuppressive doses of steroids is the mainstay of therapy.

## REFERENCES

[1)] Tasker S. Consult Feline **Vet Int Med**, 2005 [2)] Dowers KL et al. **J Amer Vet Med Assoc** 221(2):250-3, 2002. [3)] Ishak AM et al. **J Vet Int Med** 22(2):288-92, 2008. [4)] Dowers KL et al. **Am J Vet Res**. *2009* 70(1):105-111, 2009. [5)] Birkenheuer AJ et al. **J Amer Vet Med Assoc** 228(4):568-71, 2006. [6)] Haber MD et al. **Vet Parasit**. 146(3-4):316-20, 2007. [7)] Birkenheuer AJ et al. **Vet Parasit** 137(1-2):144-9, 2006. [8)] Birkenheuer AJ. ACVIM Proceedings 2008 (on-line). [9)] Spach DH et al. **Infect Disease Clin North Amer** 12(1):137-55, 1998. [10)] Van Audenhove A et al. **Brit J Haematology** 115, 924-925, 2001. [11)] Ishak AM et al. *J Fel Med & Surg* 9, 1-7, 2007. [12)] Buoro IB et al. *Vet Record* 125(17):434-6, 1989. [13)] Stubbs CJ et al. **Compend Cont Educ Pract Vet** 22, 307–317, 2000. [14)] Kordick DL et al. **J Clin Microbio** 33, 1655-1656, 1995. [15)] Hartmann, **ACVIM Proceedings**, 2009 (on-line). [16)] Kohn B et al. **J Vet Int Med**20, 159-166, 2005.

## FELINE SINO-ORBITAL ASPERGILLOSIS: AN EMERGING CLINICAL SYNDROME

**Vanessa R. Barrs, BVSc(hons), MVetClinStud, FACVSc (Feline Medicine) GradCertEd (Higher Ed)**
**Sydney, NSW, Australia**

### INTRODUCTION

Until recently, little information has been available regarding aspergillosis affecting the upper respiratory tract (URT) of cats. URT aspergillosis in cats is a more aggressive disease than canine sinonasal aspergillosis (SNA). Whilst some cats present with SNA characterised by sneezing and nasal discharge, there is a propensity for extension of infection beyond the sinonasal cavity to involve adjacent structures including the orbit, palate, nasopharynx and cribriform plate. Thus, the first presenting signs of URT aspergillosis in cats may be referable to orbital invasion resulting in sino-orbital aspergillosis (SOA), which can be considered a progression from SNA. The fungal pathogens identified in many feline cases, *Neosartorya* spp, are related to but distinct from the major canine pathogen, *Aspergillus fumigatus*.

### ETIOLOGY

**Classification of fungal pathogens**

Both *Aspergillus* spp and *Neosartorya* spp are implicated as fungal pathogens in feline URT aspergillosis.[1,2] These two closely related genera are classified within *Aspergillus* section *Fumigati*. Both are ubiquitous, filamentous, saprophytic, ascomycetes distributed primarily in soil and decaying vegetation. Confusion may arise over nomenclature because fungi can exist in different physical states corresponding to different stages of the life cycle; the anamorphic, asexual stage is typically mould-like and bears conidia (spores) whereas the teleomorphic, sexual stage, produces fruiting bodies (cleistothecia) containing ascospores. The total organism is referred to as the holomorph. Some *Aspergillus* species are strictly anamorphic, and therefore reproduce by asexual means only. Teleomorphic species within *Aspergillus* section *Fumigati* have been assigned to the genus *Neosartorya*. *Aspergillus* section *Fumigati* currently includes 9 strictly anamorphic *Aspergillus* spp and 24 *Neosartorya* spp.[3,4] *Aspergillus fumigatus*, previously thought to be strictly anamorphic, was recently shown to have a fully functional sexual reproductive cycle that leads to the production of cleistothecia and ascospores. Its teleomorph is *Neosartorya fumigata*.[4]

**Identification of fungal pathogens in cats**

Prior to 2007, complete reports of feline URT aspergillosis have been limited to case reports describing infections in 13 cats.[5-13] Affected cats presented with signs referable to sinonasal cavity infection (SNA) or sino-orbital infection (SOA). The fungal pathogen was often not identified either because fungal culture was negative or was not performed. In 5 cases where fungal culture was positive, isolates were proposed to be *Aspergillus* spp in 4 cases and *Penicillium* spp in another.[5,6,11,13] Identification to species level based on fungal culture morphology was performed in only 2 cases where *A fumigatus* and *A niger* were documented.[11,13]

In 2006, 3 cases of feline URT aspergillosis were diagnosed at the author's institution. This prompted a medical record search, which identified only 3 cases in the preceding 19 year period (January 1986 to December 2005). Following publication of a description of the syndrome of SOA and a call for cases, a further 15 cases were recruited for study between 2006 and 2008. Clinicopathological data from the resulting cohort of 21 cats from Eastern Australia provides new information on this disease in cats.[1,2] Fungal pathogens were readily isolated; culture on Sabouraud's dextrose agar was attempted in all 21 cases and was positive in 20. On the basis of fungal colony morphology, *A fumigatus* was the most common isolate identified in this cohort. However, it is not possibly to correctly identify some species belonging to *Aspergillus* section *Fumigati* using morphological and cultural characteristics alone. Thus, molecular diagnostics were performed on the fungal cultures and/or on formalin-fixed paraffin embedded (PE) or fresh tissues from affected cats to confirm the identity of the isolates. A panfungal PCR that amplifies the internal transcribed spacer 1 (ITS1) region of the ribosomal DNA gene cluster between the 18S and 5.8S rRNA genes was used for PE or fresh tissues. A second panfungal PCR that targets a larger region of the rDNA gene cluster including the ITS1, 5.8S gene and ITS2 regions was performed on fungal cultures. The ITS1 regions was targeted because it is multicopy ($\geq$ 100 copies in the fungal genome), universal fungal primers are available and it contains highly variable regions for species identification.[14] PCR products were sequenced and compared with those in the GenBank database. Results of these molecular analyses indicated that the majority of feline isolates that had been identified as *A fumigatus* based on culture morphology were actually *Neosartorya* spp.[1,2] In several cats, *A fumigatus* was identified using both fungal colony morphology and PCR of culture material or tissue. PCR of fungal culture material and/or tissue obtained from canine SNA cases was also carried out at the same time and identified *A fumigatus* in all cases.[1] For *Neosartorya* spp isolates, in most cases it was not possible to distinguish between several possible species of *Neosartorya* using the panfungal PCR because of limited sequence variation in the ITS1 region. In one case, *N pseudofischeri* was identified from PCR of both culture material and fresh tissue with 100 percent identity to sequences in the GenBank database.

Because of limitations for species identification using morphological criteria alone, a polyphasic taxonomic approach is recommended for definitive identification of species within *Aspergillus* section *Fumigati*. This involves a combination of macro- and micromorphology, extrolite (fungal metabolite) profiles and PCR determination of targeted genes including β-tubulin, calmodulin, ITS and actin.[3]

### EPIDEMOLOGY

No age or sex predilection for feline URT aspergillosis is apparent. Disease has been reported in 15 females and 18 males; most were desexed.[1,2,5-13] The median and mean age at diagnosis was 6 years with a range from 1 and ½ to 13 years. An intriguing finding is the preponderance of brachycephalic cats affected; 16 of 34 cases were brachycephalic, including 8 Persians and 6

Himalayans. This contrasts starkly with canine SNA where dolicocephalic or mesocephalic breeds are typically affected. The basis for this potential brachycephalic breed association in cats is not clear. Impaired drainage of URT secretions due to brachycephalic conformation may be important. In humans, decreased sinus aeration and drainage of respiratory secretions secondary to infection, polyps and allergic rhinosinusitis have been identified as risk factors for invasive SNA aspergillosis. It is likely that additional risk factors are present in brachycephalic cats, since brachycephalic dogs are under-represented for SNA. These may include heritable defects in mucosal immunity or common environmental factors such as antecedent viral URT infection.[11] Chronic FHV infection markedly alters sinonasal cavity architecture because of turbinate lysis secondary to intense inflammation, resulting in altered local mucosal defense mechanisms. One third of the Australian cohort of cats had a history of chronic recurrent rhinosinusitis before aspergillosis was diagnosed.[1,2] So far there is no evidence of a retroviral association with feline URT aspergillosis; of 19 affected cats tested for feline leukaemia virus antigen, only 1 was positive and all 19 cats tested for feline immunodeficiency virus were seronegative. Therefore, it appears that URT aspergillosis in cats occurs in apparently systemically immunocompetent individuals, some of which have identifiable breaches in local defence mechanisms. The apparent increased prevalence at our institution is likely a consequence of increased recognition of cases. Cases have been reported mostly in Eastern Australia and the USA with individual cases from Switzerland and the United Kingdom.[1,2,5-13]

## PATHOGENESIS

*Neosartorya* spp have only rarely been associated with disease; isolated cases of invasive pulmonary infection, osteomyelitis, endocarditis, peritonitis and mycotic keratitis have been described in immunocompromised humans.[15] Mistaken identification of *Neosartorya* spp as *A fumigatus* is considered likely to have resulted in underestimation of disease caused by *Neosartorya* spp in humans.[15] Compared to SNA in dogs where fungal infection is usually confined to the sinonasal cavity, infections in cats have a tendency to become locally invasive with extension to involve the orbit.[1,2,5,6,8-10,13] In canine SNA fungal hyphae do not invade the nasal mucosa and are typically present within overlying superficial necrotic plaques. Like SNA in dogs, infection in cats starts in the sinonasal cavity. It is likely that SNA and SOA in cats represent a spectrum of disease, with SOA being a manifestation of more chronic, invasive infection. Progression from SNA to SOA has been documented in individual cases.[10] Furthermore, all cases of feline SOA to date have shown evidence of sinonasal cavity involvement on diagnostic imaging or at necropsy.[1,2,5,6,8-10,13] The orbital lamina, situated between the orbit and frontal sinus, is the most common anatomic region where extension of infection from the sinonasal cavity to the orbit has been identified.[2] However, only some cases of SNA progress to SOA. The factors that determine whether infections will progress from SNA to SOA in individual cats have not yet been identified.

## CLINICAL SIGNS

In dogs the triad of muzzle pain, profuse mucoid to haemorrhagic chronic nasal discharge and depigmentation, crusting or ulceration of one or both nares is highly suggestive of SNA. In cats, presenting signs of SNA are less specific. Most cats will have a history of sneezing and a unilateral or bilateral serous to mucopurulent nasal discharge.[1,2,7,11,12] Intermittent epistaxis occurs in 40%. In contrast to dogs, neither nasal depigmentation nor ulceration has been documented in cats with SNA. A discharging sinus or soft-tissue mass may be identified overlying the frontal sinus or the nasal bone as a result of bony lysis and fungal proliferation. Stertor is variably present due to excessive nasal secretions and/or a mass lesion in the caudal nasal cavity and/or nasopharynx.

Combining data from previously reported cases and the Australian cohort, 21 of 39 cats had SOA.[1,2,5-13,16] Although most cats with SOA have a history of sneezing or nasal discharge in the 6 months prior to presentation, at the time of presentation, nasal signs may be subtle or absent. Rather, cats with SOA typically present with a constellation of clinical signs referable to invasive expansion of a fungal granuloma in the ventromedial orbit. These include unilateral exophthalmos with third eyelid prolapse, exposure keratitis and conjunctival hyperemia and a mass or ulcer in the ipsilateral pterygopalatine fossa behind the last molar tooth.[1,2,13] Invasion through the palatine bone causes ulceration of the hard palate. Stertor is also common. In severe, chronic infections, exophthalmos can be bilateral.[1,2,5,6] Pain on opening the mouth is not usually present. Extension of infection outside the sinonasal cavity can also result in facial distortion, including swelling of the nasal bridge, periorbital tissues and soft tissues adjacent the maxilla. Neurological signs are uncommon but develop subsequent to CNS invasion through the cribriform plate or sphenoid sinus. Submandibular lymph node enlargement and pyrexia are also common. In contrast, cats with SNA are not usually pyrexic.

## DIAGNOSIS

Definitive diagnosis of URT aspergillosis in cats is achieved by cytological or histological identification of fungal hyphae in affected tissue together with identification of the fungal pathogen. Obtaining a definitive diagnosis may require various combinations of diagnostic tests including radiography, CT, endoscopy, cytology, serology, fungal culture and molecular techniques. Serum anti-*Aspergillus* antibodies can be detected by counter-immunoelectrophoresis (CIE), agar gel immunodiffusion (AGID) or ELISA. The diagnostic utility of serum anti-*Aspergillus* antibody detection in cats with URT aspergillosis has not been evaluated. Two of 3 cats with SNA tested positive for serum antibodies against *A fumigatus* using CIE and another cat with SNA tested positive using AGID.[7,11] Galactomannan (GM), a water-soluble polysaccharide cell wall component of *Aspergillus* and some other fungal species, is released in variable quantities during hyphal growth in tissues. This antigen can be detected in the serum of human patients with invasive *Aspergillus* infections using a commercially available sandwich ELISA (Platelia™ *Aspergillus*). GM antigen levels correlate with the tissue fungal burden and can be used to monitor clinical outcomes and response to therapy.[17] Rising GM levels are associated with treatment failures in humans. The use of GM assays for diagnosis and monitoring of feline URT aspergillosis is currently under investigation in our laboratory.

For investigation of suspected sinonasal cavity or sino-orbital disease, CT is the imaging modality of choice. Reported changes in mycotic rhinosinusitis in cats include lysis of turbinates, palatine bone, nasal septum, frontal bone and/or cribriform plate.[16] Nasal cavity involvement is usually bilateral. CT findings on cats with SOA within the Australian cohort included punctate lysis of the orbital lamina, ventromedial soft tissue orbital mass, opacification of the sphenoid and frontal sinuses due to fluid or soft-tissue, and a soft-tissue mass in the choanae or nasopharynx.[2] Sinonasal cavity involvement in SOA is usually bilateral. In cats with URT aspergillosis differentiation of invasive mycotic disease from malignant neoplasia may not be possible from imaging findings alone.[16]

## TREATMENT & PROGNOSIS

The typical antifungal susceptibility profile of *Neosartorya* spp isolated from cats with aspergillosis includes susceptibility to amphotericin B (AMB) and posaconazole (PCZ), intermediate or dose-dependent susceptibility to itraconazole (ICZ) and voriconazole (VCZ), and resistance to 5-flucytosine, ketoconazole and fluconazole. Elevated minimum inhibitory concentrations of VCZ to *Neosartorya* spp have been documented elsewhere.[15] In contrast, both VCZ and PCZ typically have greater activity than ICZ *in vitro* against *A fumigatus*.[18] However, the correlation between *in vitro* antifungal susceptibility test results and *in vivo* response to therapy is only modest at best. For treatment of invasive aspergillosis in humans, VCZ has replaced AMB as the drug of choice for first-line therapy. PCZ is used for antifungal prophylaxis in high-risk patients and for salvage therapy. The echinocandins, including caspofungin, are reserved for salvage therapy or combination antifungal therapy for refractory invasive aspergillosis.[19] Combination antifungal therapy may provide enhanced efficacy compared to single drug therapy for treatment of invasive aspergillosis.[20,21]

URT aspergillosis in cats is more difficult to treat than SNA in dogs. Treatment of SNA carries a better prognosis than for SOA.[2] The prognosis for resolution of SOA is generally poor. Antifungal triazole drugs are the mainstay of therapy for SNA, given alone or in combination with terbinafine or AMB. [1,2,11,12] AMB can be administered as amphotericin deoxycholate (AMB-D) or as liposomal AMB (AMB-L).[2] ICZ or PCZ are the triazoles of choice, although the pharmacokinetics and pharmacodynamics of the latter are yet to be studied in cats. Single topical sinonasal clotrimazole (CLT) infusions have been reported in 3 cats.[10,11,13] In one cat where infection was confined to the sinonasal cavity treatment with oral ICZ and a 1 per cent CLT infusion was successful.[11] In 2 other cats with SOA, a CLT infusion was part of a multimodal therapeutic approach. One cat was euthanased because of suspected progression of infection despite treatment with ICZ and radical surgical debridement of the orbit, including extenteration of the globe.[10] In another cat treatment with ICZ, topical CLT and AMB-D did not resolve infection. However, the cat was cured after treatment with oral PCZ administered for 16 weeks.[13] The use of topical CLT infusions should be restricted to treatment of cats with SNA where there is an intact cribriform plate, no evidence of orbital involvement on CT, and no histological evidence of submucosal fungal invasion.[2] Concomitant systemic antifungal therapy with ICZ or PCZ is also recommended.

The optimal treatment regimen for SOA in cats has not been identified. Of 14 cases where treatment was attempted, infection resolved in only 4.[2,13] Three of those were treated with medical therapy alone, including 1 which resolved after PCZ administration for 16 weeks.[13] In another cat, infection resolved after 7 months of combination therapy using ICZ and AMB-D (cumulative dose 11 mg/kg) and terbinafine. A third cat responded well to 8 months of therapy with ICZ and AMB-D (cumulative dose 14 mg/kg). However, clinical and CT evidence of infection recurred 3 weeks after ICZ was stopped. Treatment was reinstituted with terbinafine and PCZ for 4 months, followed by PCZ alone for a further 3 months. Signs resolved and the cat remained asymptomatic 1 year later. The fourth case was treated with surgical debridement of the frontal sinuses during diagnostic investigation followed by AMB-D (cumulative dose 10 mg/kg) and ICZ for 7 months. Clinical signs of infection resolved and the cat was asymptomatic 2 years later. Treatment failures occurred in 4 cats which were treated with medical therapy alone and in 6 cases treated with medical therapy and radical surgical debridement of orbital granulomas, including exenteration in 5 cases.[2] Ongoing studies of the pathogenesis of URT aspergillosis in cats, combined with pharmacokinetic/pharmacodynamic data for newer antifungals in this species will assist in identifying the optimal therapeutic approach to treatment of this disease.

## REFERENCES

[1] Barrs VR, et al. **J Vet Intern Med** 2007; 21:579 (abstr). [2]Barrs VR, Beatty JA. Feline URT aspergillosis. In, August, JR (Ed). **Consultation in Feline Medicine**, vol 6, Elselver, St Louis, MO, In press. [3] Samson RA, et al. **Stud Mycol** 2007; 59:147. [4]O'Gorman CM et al. **Nature** 2008; 457:471. [Epub ahead of print]. [5]Peiffer RL, et al. **J Am Vet Med Assoc** 1980; 176:449. [6]Wilkinson GT, et al. **J Small Anim Pract** 1982; 23:127. [7]Goodall SA, et al. J Small Anim Pract 1984;25:627. [8]Davies C, et al. **J Am Anim Hosp Assoc** 1996;32:380. [9]Halenda RM, et al. **Vet Radiol Ultrasoun** 1998;38:208. [10]Hamilton HL, et al. **J Am Anim Hosp Assoc** 2000;36:343. [11]Tomsa K, et al. **J Am Vet Med Assoc** 2003; 222:1380. [12]Whitney BL, et al. **J Feline Med Surg** 2005;7:53. [13]McLellan GJ, et al. **J Am Anim Hosp Assoc** 2006;42:302. [14]Lau A, et al. **J Clin Microbiol** 2007;45:380. [15]Balajee SA, et al. **J Clin Microbiol** 2005; 43:5996. [16]Karink K, et al. **Vet Radiol Ultrasoun** 2009;50:65. [17]Husain S, et al. **Am J Transplant** 2004;4:796. [18]Arikan S, et al. **Med Mcyol** 2008;46:567. [19]Patterson TF. **Med Mcyol** 2006;44:5357. [20]Navjar LK. **Antimicrob Agents Ch** 2004;48:758. [21]Clemons KV, et al. **Antimicrob Agents Ch** 2005;49:4867.

**APPROACH TO METABOLIC ACIDOSIS: A NEED FOR 2 MAJOR AMENDMENTS**

**Mitchell L Halperin MD & Kamel S Kamel MD**
**Toronto, Ontario, Canada**

## INTRODUCTION

Metabolic acidosis is a disease process or diagnostic category rather than a final diagnosis [1]. There are two main causes, gain of acids or loss of sodium bicarbonate ($NaHCO_3$). Therefore there can be a deficit of bicarbonate ($HCO_3^-$) without a low concentration of $HCO_3^-$ in plasma ($P_{HCO3}$) if the ECF volume is appreciably contracted. Hence the $P_{HCO3}$ may not be low in this setting.

If clinical approaches do not take this into consideration, errors in diagnosis and therapy may place patients at undue risk. In addition, there may be complications due to the anions that have accumulated. For example, when a large amount of citric acid accumulates in plasma, the concentration of ionized calcium may fall sufficiently to impair myocardial contractility and hyperkalemia is also present [2]. Hence it is very important to consider factors in addition to the $H^+$ load and the blood pH.

On the one hand, the anion gap in plasma ($P_{Anion\ gap}$) is a lynch pin in the diagnostic approach to patients metabolic acidosis when nephrologists make acid-base diagnoses and when they design therapy [3]. On the other hand, there are claims that the strong ion difference (SID) [4] offers important advantages over the anion gap in this setting [5], but proof of this contention is lacking. Accordingly, we shall compare both methods in a patient where the basis for metabolic acidosis is a very large loss of $NaHCO_3$ in diarrhea fluid over 24-hr. This patient will have a very contracted ECF volume.

It is important to ensure that $H^+$ are removed by the bicarbonate ($HCO_3^-$) buffer system (BBS) during acidemia. Accordingly, the $PCO_2$ in blood that is in diffusion equilibrium with the vast majority of the bicarbonate buffer system (BBS) is the critical $PCO_2$ to examine [6]. The organ that has this buffer system is in skeletal muscles.

### Test case

A healthy 50 kg subject develops massive diarrhea today. There was no input or urine output in hospital. The ECF volume was obviously contracted on physical examination. The arterial blood pH was 7.36, $HCO_3^-$ 22 mM, and $PCO_2$ 38 mm Hg, indicating that if metabolic acidosis is present, it is mild in degree. Nevertheless, the diarrhea fluid contained 40 mM $HCO_3^-$, and its volume was ~ 5 L. The additional relevant laboratory data include the concentration of potassium ($K^+$) in plasma ($P_K$), which was 4.5 mM, the hematocrit 0.60 and the concentration of albumin in plasma ($P_{Albumin}$) 8.0 g/dL. When the latter was adjusted for this high $P_{Albumin}$, the $P_{Anion\ gap}$ was not increased due to added acids.

## THE DIAGNOSIS OF A MILD DEGREE OF METABOLIC ACIDOSIS IS INCORRECT

There are two ways to reveal that there was a severe degree of metabolic acidosis due to a very large deficit of $HCO_3^-$ in the body.

### Direct measurements

Close to 200 mmol of $HCO_3^-$ would be lost in 5 L of diarrhea fluid (5 L X 40 mM = 200 mmol); however, the volume of diarrhea fluid is not known clinically [7, 8].

### Calculated deficit of $HCO_3$

The amount of $HCO_3^-$ in the ECF compartment is the product of the $P_{HCO3}$ and the ECF volume, but a quantitative assessment of the ECF volume is not possible using a clinical assessment or hemodynamic measurements [9, 10]. The reason why these methods fail is that the bulk of the vascular volume is in venous capacitance vessels, and when these vessels contract, this has a major impact on the caliber of large central veins, which affects filling pressure and thereby clinical and hemodynamic measurements. Said another way, these assessments are affected by pressure, which is not synonymous with volume.

To obtain a quantitative assessment of the ECF volume on admission, the hematocrit can be used, because when it is high in an acute illness, this rise is due to a lower plasma volume and not an increased red blood cell (RBC) volume (2 L in our 50 kg subject), as an increased RBC mass (polycythemia) is a rare condition. In addition, changes in the hematocrit during therapy will provide a reasonable estimate of changes in the ECF volume [8, 11].

The hematocrit (0.40) is the ratio of the RBC volume (2 L) to the blood volume (5 L, 2 L RBC + 3 L plasma). In the test case, the hematocrit was 0.60 (see **equation 1**).

0.60 = 2 L RBC/(X L blood volume); X = 3.33 L, which contains 2 L RBC

∴ **Plasma volume = 1.33 L versus the usual 3.0 L** **(1)**

### Estimate of the deficit of $HCO_3^-$ in the ECF compartment

Before the diarrhea, 250 mmol of $HCO_3^-$ were present the ECF compartment (ECF volume 10 L X $P_{HCO3}$ of 25 mM). On admission, the ECF contains 110 mmol $HCO_3^-$ (~ 5 L X 22 mM). Hence there is a deficit of ~ 140 mmol of $HCO_3^-$, and this is the definition of metabolic acidosis due to a loss of $NaHCO_3$ together with a second factor that raised the $P_{HCO3}$, the very contracted ECF volume.

**FAILURE TO ASSESS BUFFERING OF $H^+$ BY THE BBS**

From a chemical perspective, buffering of $H^+$ minimizes the change in the $[H^+]$ when an acid is added to the body. From a physiological perspective, however, buffering of $H^+$ is '**good**' if it occurs by the BBS and '**bad**' if it occurs when $H^+$ bind to proteins [6] (**Fig 1**). In more detail, removal of $H^+$ via binding to intracellular proteins is potentially dangerous, because these proteins will now have a more positive net charge and this can affect their shape and possibly their function. Recall that proteins may be enzymes, transporters or contractile elements.

A low $PCO_2$ in cells is critical for removal of $H^+$ by $HCO_3^-$. The vast majority of the BBS is in the ECF and ICF of skeletal muscles. Since the $PCO_2$ in capillaries is similar to the venous $PCO_2$ draining these muscles, the brachial or femoral venous $PCO_2$ reflects the ICF $PCO_2$ in this location. Three factors lead to an elevated venous $PCO_2$, a high arterial $PCO_2$, a higher production of $CO_2$, and a very reduced rate of blood flow (i.e., a very contracted ECF volume). Since the $P_{Anion\ gap}$ and SID methods rely only on the arterial $PCO_2$, they fail to assess the BBS.

**Figure 1:**

Failure of the BBS in a patient with metabolic acidosis

Failure of the BBS to remove $H^+$ occurs if the arterial $Pco_2$ is not low enough (left side) and/or when the blood flow rate is very slow (right side).

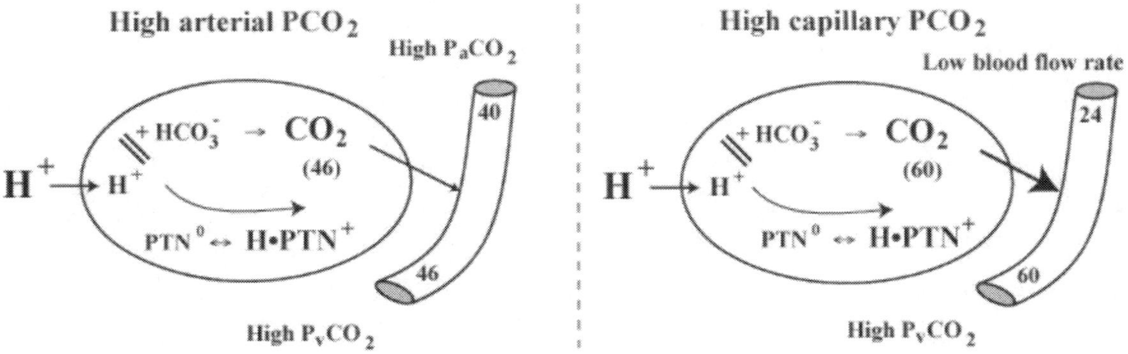


**Guideline for intravenous therapy**

A high venous $PCO_2$ on admission would provide an objective measure for rapid infusion rates. Once the large deficit of HC is recognized, the infusate should contain 25 to 45 mM $HCO_3^-$ or L-lactate. Once the venous $PCO_2$ falls, tissue perfusion has improved, and the infusion rate should be decreased. The target for the venous $PCO_2$ is 6 to 8 mm Hg higher than the arterial $PCO_2$.

**CONCLUDING REMARKS**

All methods that rely solely on concentrations, do not include a quantitative analysis of the ECF volume, and fail to measure the brachial or femoral venous $PCO_2$ need amendments. To lower the $PCO_2$ near the bulk of the BBS, tissue perfusion should be improved by re-expanding the ECF volume with a solution that contains a source of $HCO_3^-$.

**REFERENCES**


1. Kamel KS, Halperin ML: **Clin Nephrology** 65:000-000, 2006. 2. DeMars C, et al. **Ann Emerg Med** 38:588-591, 2001. 3. Emmett M, Narins RG: **Medicine** 56:38-54, 1977. 4. Stewart PA: **Can J Physiol Pharmacol** 61:1444-1461, 1983. 5. Corey HE: **Kidney Int** 64:777-787, 2003. 6. Gowrishankar M, et al: Buffering of a H+ load; A 'brain-protein-centered' view. **J Amer Soc Nephrol** 18:2278-2280, 2007. 7. Watten RH, Morgan FM, Songkhla YN, Vanikiati B, Phillips RA: Water and electrolyte studies in cholera. **J Clin Invest** 38:1879-1889, 1959. 8. Zalunardo N, Lemaire M, Davids MR, Halperin ML: Acidosis in a patient with cholera: A need to redefine concepts. **Quart J Med** 97:681-696, 2004. 9. Chung HM, et al: **Am J Med** 83:905-908, 1987. 10. McGee S, et al: Is this patient hypovolemic? **J Am Med Assn** 17:1022-1029, 1999. 11. Napolova O, et al: **Nephrol Dial Trans** 18:2674-2677, 2003

## WHY LIFE-THREATENING HYPERKALEMIA DOES NOT OCCUR WHEN WE EAT OR SPRINT: LESSONS FOR THE APPROACH TO EMERGENCY HYPOKALEMIA

**Mitchell L Halperin MD, Surinder Cheema-Dhadli PhD, and Kamel S Kamel MD**
**Toronto, Ontario, Canada**

### INTRODUCTION

The major reasons why a patient may develop a severe degree of hypokalemia with extreme weakness include a prolonged and very low intake of potassium ($K^+$), a shift of $K^+$ into cells, and/or a very large loss of $K^+$ in the urine (or rarely via the intestinal tract) [1]. If the time course for the fall in the concentration of $K^+$ in plasma ($P_K$) is very short time, the major component of the pathophysiology will involve a shift of $K^+$ into cells. There are two more issues to evaluate in this setting, first, the process must obey the law of electroneutrality; second, the mechanism will usually be related to the physiological basis for a shift of $K^+$ into cells.

As shown in the test case, the cause may not be obvious, but careful attention to the clinical story and the laboratory results should provide the clues to reach the correct diagnosis and to develop a plan for future investigations and therapy [2]. In addition, the diagnosis for the shift becomes more certain if the rate of excretion of $K^+$ is very low (< 2 mmol of $K^+$/10 mmol or per gm of creatinine in the urine), there are no sites for a large non-renal loss of $K^+$, there are no acid-base disorders, and if the $P_K$ returns to the normal range with only a small input of $K^+$. Since virtually all of these features were consistent with acute hypokalemia in the test case, the diagnostic category, but not the final diagnostic entity is known with confidence.

Before presenting the test case, we shall review the principles of the physiology of a shift of $K^+$ into cells [3], as this can help reveal to discover why our patient had an acute shift of $K^+$ into cells.

### Review of the physiology of the movement of $K^+$ across cell membranes

$K^+$ will shift into cells when there is an increased negative voltage in their ICF compartment. This voltage is generated by an increase in ion flux through the Na-K-ATPase because it is an electrogenic pump (i.e., it exports 3 $Na^+$ while importing only 2 $K^+$) [4].

### When high levels of $\beta_2$-adrenergic hormones activate the Na-K-ATPase in the cell membrane, flux through the Na-K-ATPase should be increased.

Should this be an important reason for the hypokalemia, the clinician should seek other signs of an adrenergic surge including signs of anxiety, a high systolic blood pressure and/or a wide pulse pressure, and tachycardia [2]. These signs will be present in the test case. In addition, the adrenergic stimulus must last for many hours. Hence drugs or adrenergic mimics (e.g., caffeine) with a long half-life are more likely to be part of the pathophysiology.

### There is an increased number of Na-K-ATPase pump units in cell membranes

This occurs primarily with high levels of thyroid hormone and possibly insulin. Therefore it is important to seek signs and symptoms consistent with hyperthyroidism and to do the appropriate lab tests.

### Increase ion flux through the Na-K-ATPase by having a high concentration of $Na^+$ in cells

In this pathophysiology, the rise in the concentration of $Na^+$ is due to its entry into cells in an electroneutral fashion, which occurs when $Na^+$ enters cells in exchange for $H^+$ on the $Na^+/H^+$ exchanger (NHE) [5]. There are two activators of NHE, insulin and a rise in the concentration of $H^+$ in cells. The signs for a persistently high insulin level are those related to hypoglycemia (which are usually obvious, but not in all cases) plus an adrenergic surge, which almost always plays an important role. The commonest cause for a persistently high concentration of insulin in plasma ($P_{Insulin}$) is when insulin is given to a patient with diabetic ketoacidosis (DKA). On the other hand, this role of insulin may also explain why a high carbohydrate load is associated with an acute attack in patients with hypokalemic periodic paralysis.

There is experimental evidence to suggest that the NHE in cell membranes can be activated by a rise in the concentration of $H^+$ in cells. Hence we shall examine several possible mechanisms for this acute shift of $K^+$ in acid-base disorders.

### Role of academia

This does not seem to play an important role because an infusion of HCl into experimental animals leads to the exit of $K^+$ and a rise in the $P_K$. Accordingly, HCl causes the voltage in cells to be less negative (**Figure 1**).

**Figure 1:**

Shift of $K^+$ out of cells when HCl is infused

The circles depict cell membranes. The transport mechanisms for the entry of $H^+$ into and the exit of $HCO_3^-$ from cells are shown in the left portion of the Figure. The two transport mechanisms for net gain of $H^+$ in cells are shown above the dashed horizontal line, the monocarboxylic acid transporter (MCT) and the $Cl^-/HCO_3^-$ anion exchanger (AE). Shown below the dashed horizontal line is the $Na^+/H^+$ exchanger (NHE), which is activated by a high intracellular [$H^+$] and can only catalyze the exit of $H^+$ from cells. As shown in the right portion of the Figure, when there is a $H^+$ load with anions that cannot enter cells (e.g., HCl), the fall in the $P_{HCO3}$ leads to more flux through the $HCO_3^-/Cl^-$ anion exchanger, which accelerates the electroneutral exit of $HCO_3^-$ from, and the entry of $Cl^-$ into cells. The subsequent exit of $Cl^-$ via $Cl^-$ ion channels decreases the net negative intracellular voltage, leading to the exit of $K^+$ from cells via $K^+$ channels.

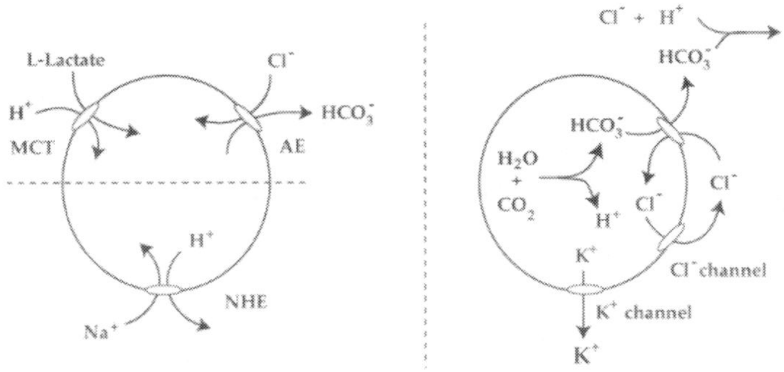


**Role of monocarboxylic acids**

An infusion of L-lactic acid or ketoacids into rats caused a similar degree of acidemia, but this had the opposite effect to HCl, the entry of $K^+$ into cells and an acute fall in the $P_K$ [6]. Thus acidemia seems to be an insufficient stimulus to activate NHE. Our hypothesis to explain these results is that L-lactic acid uniquely raises the local $H^+$ concentration near NHE (**Figure 2**), whereas HCl cannot do this [7]. Note that the test case had a fall in the $P_{HCO3}$ and a rise in the L-lactate level in plasma in the 3 to 4 mM range, and this could help explain the acute shift of $K^+$ into cells in this person.

**Figure 2:** Role of l-lactic acid on a shift of $K^+$ into cells

The circle represents the cell membrane. The initial event is the entry of L-lactic acid into cells, which leads to the release of H+ in cells immediately adjacent to the $Na^+/H^+$ exchanger (NHE). These $H^+$ bind to the $H^+$ modifier site in NHE and this activates this cation exchanger. As a result, $Na^+$ enters cells and this drives flux through the Na-K-ATPase, which increases the negative voltage in these cells. Thus only monocarboxylic acids lead to a shift of $K^+$ into cells.

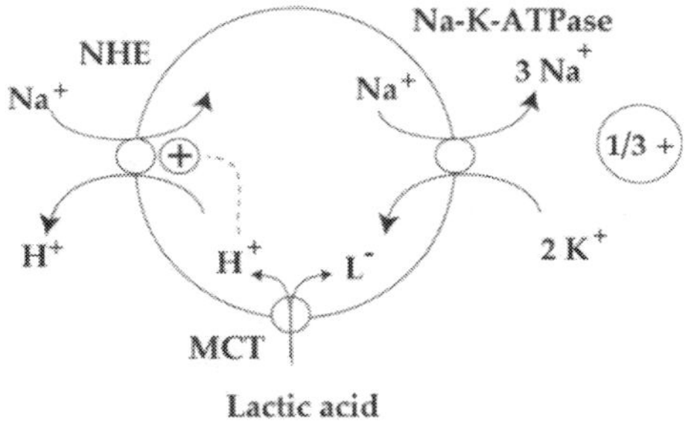


**Rise in the $P_K$ due to sudden anabolism**

Patients with catabolic states such as DKA will have a shift of $K^+$ into cells when they are treated with insulin while phosphate is provided in their diet. A similar caution is merited for patients with pernicious anemia who are initially treated with vitamin $B_{12}$ and in cachectic patients who are treated with parenteral nutrition. This type of acute hypokalemia was not present in our test case.

**Test case**

A 25 year old very fit female was perfectly healthy until 6 months ago. Since that time, she has had 3 admissions for an unusually severe degree of weakness after she ate a large breakfast and performed exercise. The weakness became very severe over a few hours and she was brought to hospital. She did suffer from asthma, and had taken a bronchodilator on her first admission, but she did not take this drug on the past 2 admissions. She denied the intake of all medications with possible adrenergic side effects including cocaine, and she did not drink caffeine after the first admission. On physical examination, her blood pressure was close to 125/60 mm Hg and her pulse rate was 120 beats/min, but after she recovered, her blood pressure was 105/ 75 mm Hg and her pulse rate was 60/min. These findings suggested that she had an adrenergic response during her illness (she was very worried about the weakness). The key laboratory finding was hypokalemia (1.9-2.3 mM) and there were no other abnormalities (her blood pH was 7.38, $HCO_3^-$, 22 mmol/L, $PCO_2$ was 37 mm Hg, glucose 114 to 120 mg/dL. Therapy consisted of an infusion of KCl and her $P_K$ rose to 3.8 mmol/L after she received 60 to 80 mmol of $K^+$ (i.e., in 2 hr). Her urine contained very little $K^+$.

**Clinical approach to the patient with acute hypokalemia**

When faced with a patient with hypokalemia, the first step is to recognize whether an emergency is present. If an emergency is present, therapy to raise the $P_K$ must begin without delay. In this context, the most serious emergency due to hypokalemia is a cardiac arrhythmia; this threat and the absolute $P_K$, however, are not tightly correlated. In the test case, there is a normal EKG, so we can proceed with diagnostic issues.

The major symptom directly related to hypokalemia is weakness, and this can be serious if respiratory muscles are affected, especially if there is a need for deep and frequent respirations (i.e., if metabolic acidosis is present). Since there is no acid-base disorder, again we can proceed with diagnostic issues.

## DISCUSSION OF THE CASE

### The sustained adrenergic surge

There were several factors highlighted in the above discussion. First there was evidence for a sustained $ß_2$-adrenergic surge, but its basis was not yet clear, as the only suspect was the bronchodiator, and this was not used before all subsequent admissions. Moreover, she denied drug intake that could result in an adrenergic surge including cocaine and amphetamines. Although she was worried about her illness, this should not produce a repetitive acute illness.

### An insulinoma with a secondary adrenergic response to hypoglycemia

Since her blood sugars were never low during and between bouts of the illness, the concentration of insulin in plasma was consistently in the normal range, and the C-peptide level in plasma, a sign of excessive insulin release from an insulinoma, was in the normal range, this diagnosis is unlikely.

### Pheochromocytoma

This can cause a $ß_2$-adrengic surge, but it is a very rare condition. Moreover, all subsequent tests for this condition were negative. Hence we are extremely uncomfortable with this diagnosis.

### The suspicious slightly low $P_{HCO3}$ and high $P_{L\text{-}lactate}$

This may be the only remaining clue for the hypokalemia and the $ß_2$-adrengic surge, but its likelihood to be the primary cause of the disorder is unclear.

### Improving the laboratory testing

Blood tests reveal what is present at one small point in time. On the other hand, examination of urine that was present in the bladder did reveal the final diagnosis.
We ask the reader to think of what should be measured to reveal the final diagnosis.
As a hint, this is extremely important, as there is a cure in store for the patient.

### Integrative physiology

We shall also reveal how and where L-lactate plays an important role in this pathophysiology.

## REFERENCES

[1] Halperin ML, Kamel KS: **Lancet** 352:135-142, 1998. [2] Lin SH, et al: **Quart J Med** 94:133-139, 2001. [3] Lin S-H, Halperin ML: **Clinical medicinal chemistry** 14:1551-1565, 2007. [4] Clausen T, Everts ME: **Kidney Int** 35:1-13, 1989. [5] Counillon LL, Pouyssegur RJ: in **The Kidney: Physiology & Pathophysiology** (vol 1), edited by Seldin DW, Giebisch G, Philadelphia PA, Lippincott Williams & Wilkins, 2000, pp 223-234. [6] Cheema-Dhadli S, et al. **J Am Soc Nephrol** 19:Abstract, 2008. [7] Halperin ML, et al: **Fluid, Electrolyte and Acid-Base physiology; a problem-based approach.** Philadelphia, W.B. Saunders, 2009.

# PATHOPHYSIOLOGY OF HYPOADRENOCORTICISM IN DOGS & IN PEOPLE

**Patty Lathan, VMD, MS, DACVIM**
**Mississippi State, MS**

## INTRODUCTION

Hypoadrenocorticism is a rare disease in dogs and in people; roughly 1/20,000 dogs and 1/25,000 people are estimated to have the disease.[1] In both species, naturally-occurring Addison's disease usually results from destruction of 90% of the adrenal cortices, leading to primary adrenocortical failure. All three layers of the adrenal cortices (zona glomerulosa, zona fasciculata, and zona reticularis) are generally affected, resulting in a combined deficiency of glucocorticoid and mineralocorticoid activity. However, recent reports from referral institutions revealed isolated cortisol deficiency in 11 and 24% of dogs diagnosed with primary Addison's disease.[2,3]

Secondary hypoadrenocorticism, resulting from insufficient ACTH secretion from the pituitary gland, causes clinical cortisol deficiency only. While natural causes are rare, iatrogenic secondary hypoadrenocorticism from long-term glucocorticoid use is much more common in dogs and people. Hypoadrenocorticism resulting from deficient CRH secretion has been reported in people, but not in dogs.

## ETIOLOGY OF PRIMARY HYPOADRENOCORTICISM

### Humans

When Thomas Addison first described hypoadrenocorticism in 1855, six of eleven reported cases suffered from tuberculosis, four had adrenal neoplasia, one had adrenal, hemorrhage, and one had idiopathic adrenal fibrosis. Addison's description of adrenal fibrosis was later found to be consistent with findings in autoimmune adrenalitis.[4] Today, tuberculosis remains the most common world-wide cause of Addison's disease in people. In the United States and other developed countries, however, autoimmune adrenalitis results in 65-80% of the cases. Other causes of Addison's disease include histoplasmosis and other fungal diseases, neoplasia, and hemorrhage.[5,6]

Histopathologic findings in people with active autoimmune adrenalitis include generalized adrenocortical infiltration of mononuclear cells (lymphocytes, plasma cells, and macrophages). Adrenocortical necrosis ensues, ending with the replacement of the cortex with fibrous tissue. Following complete destruction of the cortex, inflammatory cells are no longer present.[5] Autoantibodies against the adrenal cortex and 21-hydroxylase, an enzyme involved in aldosterone and cortisol synthesis, are present in the majority of people with naturally-occurring Addison's disease, and are markers for autoimmune adrenalitis. However, there is currently no evidence *in vivo* to support a direct pathogenic role for these antibodies.[7]

Approximately half the patients with Addison's disease have adrenocortical insufficiency alone. The other half suffer from an autoimmune polyendocrinopathy syndrome (APS). Adrenocortical failure is present in approximately 70% of the cases of APS type 1, in conjunction with hypoparathyroidism and mucocutaneous candidiasis; various extra-endocrine organs may also be targeted in APS-1. All cases of APS type 2a (Schmidt's syndrome) have hypoadrenocorticism, accompanied by thyroiditis and/or diabetes mellitus [1]

In patients that have autoimmune adrenalitis, but not APS, Addison's disease appears to be inherited as a complex genetic trait, caused by the interaction of several genes and environmental factors. However, only 2% of these patients have another family member with the disease. APS-1, on the other hand, has been shown to be caused by mutations in the AIRE (autoimmune regulator) gene. Inheritance appears to be autosomal recessive; people with two copies of the mutated gene are generally affected with APS-1, while those with only one copy are not. APS-2 occurs with an autosomal dominant, incomplete penetrance, pattern of inheritance. Various HLA class 2 genes (human leukocyte antigens) have been implicated in disease causation. Whereas various ethnic groups may have similar manifestation patterns of APS-1, the order of occurrence of component diseases in APS-2 varies by individual.[5,8]

### Dogs

Although canine hypoadrenocorticism is occasionally caused by fungal disease, neoplasia, and hemorrhage, most cases are labeled as "idiopathic" in etiology. Necropsy findings typically reveal bilateral adrenocortical atrophy, with or without fibrosis—similar to Addison's initial description of idiopathic adrenal insufficiency in people. However, histological examination of adrenal glands from a dog with recently-diagnosed hypoadrenocorticism revealed adrenocortical atrophy with foci of lymphocytes and plasma cells scattered throughout the cortex, consistent with findings in people with active autoimmune adrenalitis.[9] These findings suggest an immune-mediated mechanism for canine hypoadrenocorticism. Although anti-adrenal antibodies have been reported in a few dogs with primary hypoadrenocorticism, identification of antibodies against canine adrenocortical enzymes has not been achieved.[9,10]

In a combined series of 187 dogs with primary hypoadrenocorticism, 28 dogs had a concurrent endocrinopathy; hypothyroidism and diabetes mellitus were most commonly represented.[11] Additionally, two out of four related leonbergers with hypoadrenocorticism were concurrently diagnosed with hypothyroidism, suggesting a syndrome similar to APS-2 in people.[12] Familial predispositions to hypoadrenocorticism have also been reported in the Portuguese water dog, Nova Scotia duck tolling retriever, standard poodle, and bearded collie.

Isolated cortisol deficiency in patients with primary hypoadrenocorticism appears to be more common in dogs than in people. Familial glucocorticoid deficiency (FGD) is a rare autosomal-recessive disease in humans, caused by one of two mutations in the ACTH receptor.[6] Thus, the mechanism is not immune-mediated in people. Interestingly, in two of the most recent case series assessing atypical hypoadrenocorticism in dogs, females represented 55% of the cases.[2,3] In contrast, most of the larger case series have reported approximately 70% of all canine Addison's cases to be female. Also of note is that there appear to be similar breed predispositions in combined and glucocorticoid-deficient hypoadrenocorticism. Poodles and Great Danes were over-represented in both groups in the recent reports. Given that one of the Addisonian leonbergers with concurrent hypothyroidism was glucocorticoid-deficient only, one might suspect a similar autoimmune etiology for both typical and atypical presentations of hypoadrenocorticism. However, multiple etiologies may exist.

**MANIFESTATION OF DISEASE**

As described by Thomas Addison in 1855, adrenal insufficiency in humans is a condition characterized by "anemia, general languor and debility, remarkable feebleness of the heart's action, irritability of the stomach and a peculiar change of skin color…"[4]
With the exception of the "peculiar change of skin color," these findings are similar in dogs. People with Addison's often exhibit darkening of their skin, hair, nails, and scars. Patients with FGD also have skin hyperpigementation. The most likely cause is long-term overproduction of ACTH due to lack of negative feedback from cortisol. ACTH is a peptide composed of 39 amino acids. Melanocyte-stimulating hormone (MSH-α) consists of the first 13 amino acids in the ACTH peptide, and is released following cleavage . Skin pigmentation in Addison's disease is proposed to be caused by binding of ACTH, via its MSH sequence, to melanocortin receptors in the skin. Whether this binding occurs following cleavage of ACTH to MSH peptides is unknown.[13] The reason that this occurs in people with hypoadrenocorticism, but not dogs, is also unclear.

**REFERENCES**

[1]Laureti S, et.al. **J Clin Endo & Metab** 1999;84(5):1762. [2]Lifton SJ, et.al. **J Am Vet Med Assoc** 1996;209(12):2076. [3]Thompson AL, et.al. **J Am Vet Med Assoc** 2007;230(8):1190. [4] Addison T. Published collection of his works; 1868. [5]Betterle C, et.al. **Endo Rev** 2002;23(3):327. [6]Clark AJL, et.al. in DeGroot, et.al. (eds). **Endocrinology ed 5**; 2006;1:819. [7]Husebye ES, et.al. **Endo** 2006;147(5):2411. [8]Maclaren N, in DeGroot, et.al. (eds). **Endocrinology ed 5**; 2006;1:819. [9]Schaer M, et al. **J Am An Hosp Assoc** 1986;22:789. [10]Kooistra HS, et.al. **Vet Q** 1995;17(2):59. [11]Feldman EC, et.al. **Canine & Feline Endocrinology and Reproduction ed 3**; 2004:394. [12]Smallwood LJ, et.al. **J Am An Hosp Assoc** 1995;31:301. [13]White A, in DeGroot, et.al. (eds). **Endocrinology ed 5**; 2006;1:323.

ENDOCRINE CHANGES DURING THE PROGRESSION OF CRITICAL ILLNESS

Johan P. Schoeman BVSc, MMedVet, DSAM, DECVIM
Pretoria, South Africa

## INTRODUCTION

The integration of endocrinology into critical illness generated important new insights into the complex and constantly changing metabolic/endocrine environment accompanying severe illness. These studies uncovered endocrine biomarkers of illness severity, identified harbingers of incipient recovery and documented the effects of hormonal interventions.

Developments in intensive care medicine have enabled patients to survive conditions for which natural pathophysiological coping mechanisms do not exist. These highly technological interventions in the natural process of dying have unmasked previously unknown conditions such as non-specific wasting syndrome, hyperglycemia and insulin resistance, atrophy of intestinal mucosa and disturbed motility of the gastrointestinal tract, all of which contribute to the protracted dependency on intensive care support. This review will provide an up to date overview of the current knowledge. It is hoped that this would stimulate further interdisciplinary research into this important and exciting field.

### Human Studies

Several studies in human critical illness have shown a positive association between high basal serum cortisol and mortality[1]. In addition, human critical illness is characterized by multiple and complex changes in the thyroid axis[2], where low concentrations of T3 and T4 have been shown to indicate prolonged illness and a poor prognosis[3], while an increase in TSH heralds the onset of recovery from severe illness[4]. Prolonged critical illness is also associated with decreased testosterone levels and attenuated LH release in men[5], and a reversible reduction in LH, FSH and estradiol in women[6]. Relative hyperprolactinemia during the initial phase of critical illness is thought to contribute to the initial activation of the inflammatory cascade[7]. In a human study testing several endocrine predictors, high basal cortisol and low thyroxine correlated best with outcome[8]. After many years of debate regarding the incidence and importance of "relative adrenal insufficiency" (RAI), consensus is emerging. Initial studies originating from the Glasgow Royal Infirmary in Scotland in 1982, which highlighted the phenomenon of adrenal failure in critically ill patients, seems to have been fundamentally flawed[9]. It was later shown that most of these patients received the sedative etomidate, which suppresses adrenal function[10]. It was therefore not surprising that this study obtained such a high success rate with corticosteroid replacement therapy, because these patients had been iatrogenically deprived of endogenous cortisol in the first place. Their excellent response to steroid replacement therapy has not been rivalled by any other study in human critical illness. A landmark study conducted in the 1990's, in which a positive response to physiological replacement doses of corticosteroids had been shown, has recently been questioned[11]. Firstly, the study performed *post hoc* subgroup analysis and showed a positive response only in a subset of patients. Importantly, these were septic shock patients which had RAI (based on a delta cortisol < 9 ug/dl [250 nmol/l]) and remained hypotensive after at least 1 hour of vasopressor therapy following prior adequate fluid replacement. There is convincing evidence that a subset of patients exist in which physiological glucocorticoid replacement therapy is beneficial – the problem lies in prospectively identifying these patients.

After many conflicting studies, the most recent multi-centre *Corticus* study has shown that the ACTH stimulation test with serum total cortisol measurements is not predictive of response to corticosteroid therapy. Moreover, this study showed no survival benefit in administering replacement doses of corticosteroids to septic shock patients; on the contrary, it showed a higher incidence of super-infection, including repeated septic shock in treated patients[12]. However, critics of the *Corticus* study rightly pointed out that patient recruitment had been slow and biased towards the selection of patients unlikely to benefit from glucocorticoid therapy. A major reason for this was that attending clinicians believed at the time that glucocorticoid therapy was beneficial and therefore refused to allow their patients to be randomized. In addition, the final 500-strong study cohort was much less than the pre-calculated and adequately powered sample size of 800 patients; furthermore, the cohort contained far more surgical than medical ICU patients – a group less likely to benefit from glucocorticoid replacement therapy. In the author's opinion of the current human critical care literature, *absolute* adrenal insufficiency is a rare event (except when caused by etomidate), while true *relative* adrenal insufficiency is a nebulous and poorly characterized entity, reserved for a subset of patients with vasopressor-resistant hypotensive septic shock.

Reasons for the poor characterization of relative adrenal insufficiency based on total cortisol measurements are manifold: First, it is generally accepted that free cortisol, rather than total cortisol, is responsible for biological effects; second, glucocorticoid receptor function, rather than total cortisol, determines the effective intracellular glucocorticoid activity; third, disparate patient populations have been assessed at varying times from their onset of illness; fourth, most studies are of a cross-sectional nature, not accounting for changes in RAI status during the progression of illness and are confounded by the use of intensive care drugs known to modify endocrine function such as dopamine and etomidate.

### Canine studies

Studies performed 60 years ago on adrenalectomized dogs showed that the adrenal response to shock and critical illness was essential for survival[13]. A study conducted in critically ill dogs with a variety of diseases failed to show any significant difference in basal cortisol concentrations between survivors and non-survivors and also failed to detect adrenal insufficiency[14]. Non-thyroidal illness in dogs is a well-known cause of euthyroid sick syndrome and the effects of such illness on lowering thyroid hormones have been well-described[15].

A recently published study performed on 33 critically ill dogs has purported to document RAI in septic patients[16]. It calculated delta cortisol values for each patient by subtracting basal cortisol from ACTH-stimulated cortisol. This delta cortisol value was tested for its association with the following outcome variables; mortality by the time of hospital discharge, 28 day mortality and hypotension. No significant association was demonstrated between delta cortisol and outcome at the time of hospital discharge, but lower delta cortisol was significantly associated with hypotension. When delta cortisol was dichotomized into non-responders — a value below 3 ug/dl (83 nmol/l) and responders — a value above 3 ug/dl, it was shown that non-responders were much more likely to die. Potential critique on this particular study is: Firstly, the study population had only one ACTH stimulation test performed, which was temporally separated from the outcome variables (i.e. discharge mortality and especially 28 day mortality) by many days. Secondly, dogs that received a single dose of corticosteroids more than 72 hours before the ACTH stimulation test could potentially have been included in the study and thirdly ACTH was administered intramuscularly. The lowest delta cortisol concentrations were demonstrated in hypotensive dogs. Both adrenal blood flow and the systemic uptake of ACTH could have been markedly impeded by the systemic hypotension – the poorer the adrenal blood flow, the worse the uptake of synthetic ACTH and hence the poorer the adrenal response; potentially explaining the association between non-response and hypotension that was found. Finally and most importantly, the basal cortisol concentrations of the survivors in comparison to the non-survivors were not reported. The concept behind this important omission has been addressed in a recent human publication and concerns the definition of relative adrenal insufficiency[17]. If we accept that the adrenal gland has a limit to the amount of cortisol it can produce, then it follows that dogs with the highest basal cortisol concentrations (the sickest dogs) would have the lowest delta cortisol values – and hence be classified as non-responders. The association of low delta cortisol with increased mortality could therefore be confounded by high basal cortisol.

In two different, comparatively homogenous models of acute critical illness and sepsis viz. virulent canine babesiosis and canine parvoviral diarrhoea, we found the following: In 95 dogs with canine babesiosis, admission median basal cortisol concentration was significantly higher in non-survivors than survivors; 482 nmol/L (IQR 456-562) versus 115 nmol/L (IQR 64-208)[18]. Within the group of more severely ill, admitted dogs, the median cortisol of the non-survivors was still significantly higher than that of the admitted dogs that survived; 482 nmol/L (IQR 456-562) versus 151 nmol/L (IQR 70-274)(Fig 1). Basal plasma ACTH concentrations were also significantly higher in dogs that died compared to dogs that survived. ACTH stimulation tests performed in these dogs demonstrated lower delta cortisol in the non-survivors than in the survivors, but a strong inverse correlation between basal cortisol and delta cortisol. Interestingly, babesiosis dogs categorized as non-responders (delta cortisol < 83 nmol/l) had significantly higher cortisol to ACTH ratios than responder dogs, providing conflicting information regarding adrenal function compared to the delta cortisol parameter[19]. Although basal T4 and fT4 was significantly lower in non-survivors than survivors of babesiosis[18] (Fig 2), TSH concentrations were similarly low in both groups.

In 57 dogs with parvoviral diarrhoea, median basal day 1, day 2 and day 3 cortisol concentrations were significantly higher in non-survivors versus survivors[20] (Fig 3.). The above finding was confirmed in a follow-up study on a different group of 63 dogs with parvoviral diarrhea on which daily intravenous ACTH stimulation tests were performed[21]. In keeping with many human studies and the Burkitt canine study, these 63 dogs had significantly lower delta cortisol concentrations in the non-survivors than survivors on day 1 and day 2. However, on day 3 there was no statistical difference in delta cortisol between the survivors and non-survivors, mainly because of a relative reduction in basal cortisol (and therefore increased delta cortisol values) in the non-survivors. In addition, although 44% of surviving dogs were classified as non-responders to the ACTH stimulation test on day 1, only 3% of them were still non-responders on day 3; thereby confirming the marked changes occurring in RAI status during the progression of critical illness. Furthermore, all these so-called non-responder dogs survived their illness without receiving glucocorticoid treatment, providing novel, albeit circumstantial, canine evidence supporting recent human findings of no survival benefit for corticosteroid supplementation in critical care[12]. There was a significant negative correlation between basal and delta cortisol, in part explaining the apparent adrenal insufficiency in very sick patients with high basal cortisol concentrations on days 1 and 2.

These findings demonstrate the limitations of extrapolating and according patients non-responder status based on one-off ACTH stimulation tests. Basal hormone concentrations changed dramatically in these dogs during the progression of their illness, while their ACTH-stimulated cortisol concentrations remained relatively stable, calling for a re-assessment of the time-honored ACTH-stimulated cortisol value in the determination of adrenal reserve, as is the case in conventional endocrine screening. Very different neuroendocrine paradigms prevail in acute versus chronic human critical illness and indications are that the same is true for dogs[22].

Another canine study that found relative adrenal insufficiency in canine sepsis, could not demonstrate a significant association between delta cortisol and mortality, but found that dogs with delta cortisol < 83 nmol/l were 5.7 times more likely to receive vasopressors. Small sample size and a one-off ACTH stimulation test were also cited as limitations. Baseline cortisol differences between survivors and non-survivors were not compared. Cases were included up to 48 hours after the onset of illness, during which time basal cortisol could have decreased as alluded to in the above discussion[23].

Both groups of parvovirus infected puppies also demonstrated significantly lower T4 and fT4 in the non-survivors versus the survivors on all three days[20,24] (Fig 4.). Interestingly, the endogenous TSH concentrations were higher in non-survivors on day 1 and 2, before also decreasing to below the limit of detection in the non-survivors as critical illness progressed. T4 and fT4 are thus more sensitive than TSH in predicting mortality and decrease earlier than TSH in canine critical illness[24]. A study conducted in a mixed population of sick dogs also found significantly lower T4, fT4 and T3 in dogs that died compared to dogs that made a full recovery[25]. Low T3 concentrations have also been found in 56% of critically ill dogs and non survivors had significantly lower T3 than survivors[26].

In conclusion, it would appear that (similar to humans with critical illness) dogs with the most severe acute illness have the highest concentrations of the pituitary-adrenal-axis hormones ACTH and cortisol, providing substrates for survival and protecting the host against the deleterious effects of the immune system. In contrast, the pituitary-thyroid-axis hormones are markedly suppressed in these very sick patients, with the putative beneficial effect of inactivating anabolic pathways. RAI has been described in dogs, but as in humans, its definition warrants clarification. Data on the somatotropic and gonadal axes are lacking in small animals. We need large prospective, longitudinal studies, containing enough cases with adverse outcomes, so that multivariate analysis can identify the combined value of various hormone parameters in the prediction of mortality. In addition, the role of inflammatory cytokines as mediators of the aforementioned processes needs to be investigated.

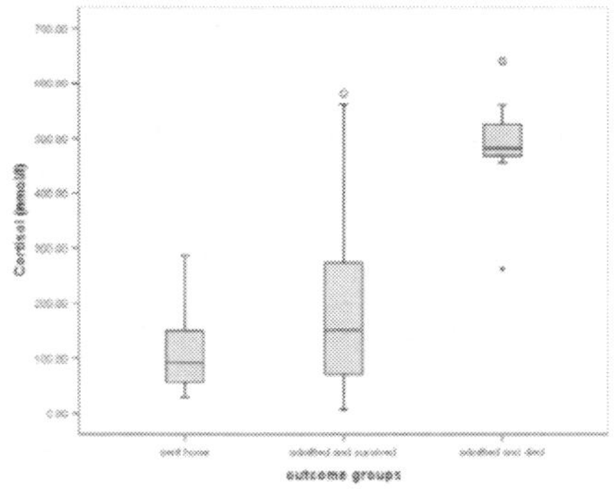


Figure 1.

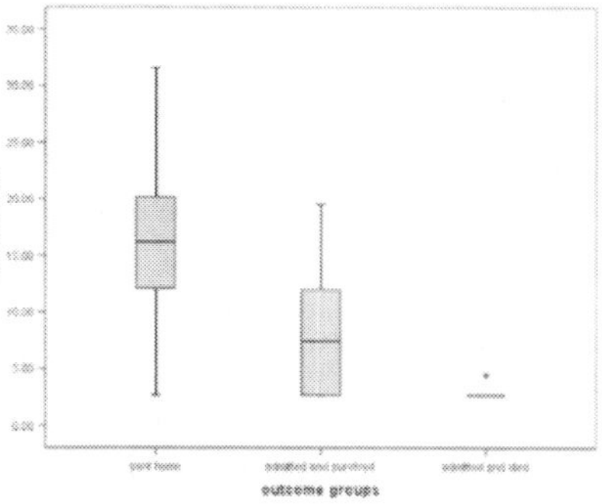


Figure 2.

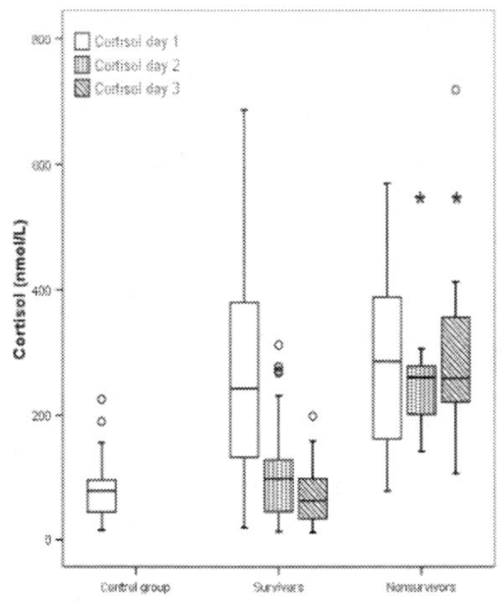


Figure 3.

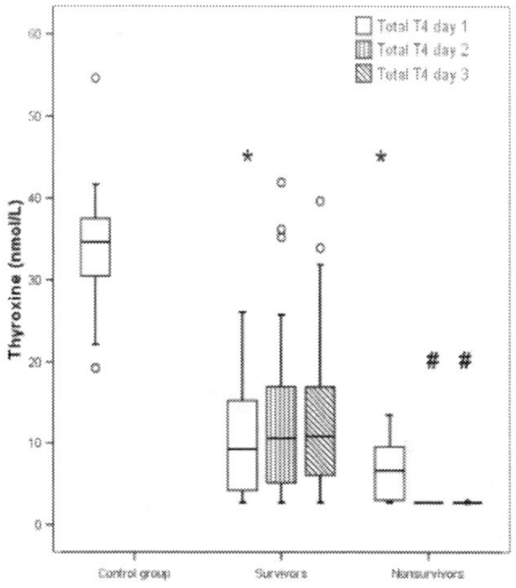


Figure 4.

## REFERENCES

[1]Annane D, et al. **J Am Med Assoc** 2000;283(8):1038 [2]Wartofsky L, et al. **Endocr Rev** 1982;3(2):164 [3]Peeters RP, et al. **J Clin Endocr Metab** 2005;90(8):4559 [4]Bacci V, et al. **J Clin Endocr Metab** 1982;54(6):1229 [5]van den Berghe G, et al. **Clin Endocrinol** 1994;41(5):563 [6]van Steenbergen W, et al. **Neuroendocrinol** 1994;60(2):165 [7]Devins SS, et al. **Crit Care Med** 1992;20(12):1644 [8]Ray DC, et al. **Intensive Care Med** 2002;28(9):1301 [9]Finlay WEI, et al. **Lancet** 1982;319(8286):1414 [10]Ledingham IM, et al. **Lancet** 1983;321(8336):1270 [11]Annane D, et al. **J Am Med Assoc** 2002;288(7):862 [12]Sprung CL, et al. **N Engl J Med** 2008;358(2):111 [13]Swingle WW, et al. **Proc Soc Exp Biol Med** 1957;96(2):446 [14]Prittie JE, et al. **J Am Vet Med Assoc** 2002;220(5):615 [15]Kantrowitz LB, et al. **J Am Vet Med Assoc** 2001;219(6):765 [16]Burkitt JM, et al. **J Vet Intern Med** 2007;21(2):226 [17]Lipiner-Friedman D, et al. **Crit Care Med** 2007;35(4):1012 [18]Schoeman JP, et al. **Vet Parasitol** 2007;148:75 [19]Schoeman JP, et al. **Vet Parasitol** 2008;154:205 [20]Schoeman JP, et al. **J Am Vet Med Assoc** 2007;231(10):1534 [21]Schoeman,JP, et al. Proceedings of the American College of Veterinary Internal Medicine 2008. [22]van den Berghe G, et al. **J Clin Endocr Metab** 1998;83(6):1827 [23]Martin LG, et al. **J Am Vet Med Assoc** 2008;233(1):87 [24]Schoeman JP, et al. **Microbes Infect** 2008;10(2):203 [25]Mooney CT, et al. **J Small Anim Pract** 2008;49(1):11 [26]Elliott DA, et al. **J Vet Emerg Crit Care** 1995;5(1):17

# THE INTESTINAL MICROBIOME IN DOGS & CATS

**Jan S. Suchodolski, med. vet., Dr. med. vet, PhD**
**College Station, TX**

## INTRODUCTION

The intestinal microbiota plays an important role in gastrointestinal health and disease, yet our understanding of the composition and the dynamics of the intestinal ecosystem is very rudimentary. Past studies characterizing the canine and feline intestinal microbiota have used traditional microbiological culture techniques. It is now well recognized that culture dependent methods fail to characterize the tremendous microbial diversity of the intestine, as the majority of microbial species escape identification.

In humans, based on molecular analysis of microbial composition of intestinal samples it has been estimated that the intestine harbors up to 1,800 bacterial genera and between 15,000 and 36,000 individual species.[1] The total microbial load in the intestine is estimated to contain $10^{12}$ to $10^{14}$ organisms, containing at least 100 times as many genes as the human genome. This mutually interacting system comprising the host cells and the resident microbial community is called the **intestinal microbiome**. Similar molecular studies, although on a more limited scale, have now also been reported for dogs and cats, and have allowed the identification of many previously unidentified bacterial and fungal phylotypes in the dog and cat intestine.[2-6]

In this presentation we will discuss new molecular approaches for describing the microbiota in dogs and cats, and describe how these methods have furthered our understanding of microbe-host interactions in GI disease.

## ASSESSMENT OF INTESTINAL BACTERIA

### Bacterial culture

The canine and feline intestinal microbiota has been mainly evaluated by direct culture techniques. The duodenum harbors total bacterial counts ranging from $10^2$ to $10^9$ cfu/g, considerably higher than the human duodenum ($<10^5$ cfu/g). Cats appear to have higher counts of anaerobic bacteria. The jejunum is quantitatively similar to the duodenum whereas the ileal microbiota is increased in numbers. The total bacterial count in the colon ranges between approximately $10^9$ and $10^{11}$ cfu/g. *Bacteroides, Clostridium, Lactobacillus, Bifidobacterium,* and *Enterobacteriaceae* spp. are the predominant bacterial groups cultivable from the canine and feline intestine.

Bacterial culture has limitations for assessing the complex bacterial diversity found in the gastrointestinal tract of mammals. Samples must be processed immediately to preserve aerobic and anaerobic species, which are prone to damage during sample handling. Selective culture media followed by phenotypic and biochemical identification systems often fail to accurately classify all microorganisms. For example, in one study it was demonstrated that despite using Beerens agar, a medium specifically designed for the isolation of *Bifidobacterium* spp., a mixture of various organisms other than *Bifidobacterium* spp. was isolated. However, the major limitation of bacterial culture is that only very minor proportions (estimated 5-20%) of intestinal microbiota are cultivable. Therefore, a culture dependent approach exhibits bias, allowing only the characterization of cultivable members of the microbiome.

### Molecular tools

Several molecular techniques have been developed that facilitate either direct identification of bacterial species from intestinal samples or yield a "fingerprint" of bacterial diversity in a given sample. Bacteria can be directly identified by PCR amplification using universal, group- or species-specific primers targeting a known gene, with subsequent sequencing of the PCR product and comparing the obtained sequence to existing public databases (e.g., GenBank, Ribosomal Database Project). The 16S rRNA gene is most commonly targeted, since more than 400,000 unique 16S rRNA sequences are available for comparison in public databases. Other genomic targets include the 16S-23S internal transcribed spacer (ITS) region, the 65-kDa heat shock protein (HSP) or the chaperonin (cpn60) sequences.

The amplification of the 16S rRNA gene using universal bacterial primers that target conserved regions in the 16S rRNA gene allows amplification of theoretically all bacteria present in the sample. For identification purposes, 16S rRNA gene clone libraries are constructed. This approach is very informative, but also a laborious and expensive procedure that is not well suited for high-throughput assessment of a large number of samples. New approaches utilize 454-pyrosequencing technology (Roche®). This method has revolutionized studies of microbial ecology, allowing several thousand sequences to be analyzed within a few hours. Furthermore, this approach allows detection of low abundant sequences (<0.1% of total). An alternative and rapid method is the use of molecular fingerprinting techniques, (e.g., denaturing gradient gel electrophoresis or analysis of terminal restriction fragment length polymorphisms). These allow rapid assessment of the predominant bacterial species present in a sample by yielding a fingerprinting patterns representative of the bacterial diversity found in the sample. However, these latter techniques yield more superficial information about the microbial composition.

## MOLECULAR BASED CHARACTERIZATION OF INTESTINAL MICROBIOTA

Based on molecular fingerprinting it has been shown that each individual animal harbors very unique microbial communities, with a pronounced spatial variation along the gastrointestinal tract.[7] Using 16S rDNA clone libraries and/or 454-pyrosequencing, we have identified a much greater bacterial diversity in the intestinal tracts of cats and dogs than previously reported.[4] For example, up to 40% of our sequences originated from previously uncharacterized phylotypes. We have identified an average of 110 different genera, and up to 450 unique phylotypes in the canine jejunum alone. The canine and feline intestine harbors 10 different bacterial phyla. Of these, *Proteobacteria, Firmicutes, Actinobacteria, Bacteroidetes, Spirochaetes, and Fusobacteria* make up 99% of all gut microbiota

(Table 1). The remaining 1% is represented by members of the bacterial phyla *Tenericutes, Verrucomicrobia, Cyanobacteria, and Chloroflexi*. The *Firmicutes* are the most diverse bacterial phylum and consists mainly of *Clostridiales* and *Lactobacillales*. *Clostridiales* is the most abundant and diverse bacterial order in all segments of the canine and feline intestine. Based on phylogenetic analysis, the order *Clostridiales* is comprised of many different phylotypes, which are organized into so called *Clostridium* clusters.[3,4] This various clusters differ in their locations within the intestine: sequences affiliated with *Clostridium* cluster XI and *Clostridium* cluster XIVa dominate in the proximal small intestine and colon, respectively. *Bifidobacterium* spp. are considered to be beneficial bacteria and are commonly used a probiotic agents. Using group-specific primers, we have characterized the most common *Bifidobacterium* spp. in fecal samples from dogs and cats: *B. longum, B. pseudolongum, B. subtile, B. adolescentis, B. catenalatum, B. saeculare*, and *B. pullorum*.

**Molecular characterization of fungal organisms**

Molecular tools have also allowed the characterization of fungal diversity in the intestine. We have recently reported the prevalence and identification of fungal organisms in the small intestine of healthy dogs and dogs with chronic enteropathies.[5] The results indicated a high prevalence (up to 76.1% of dogs) and high diversity of fungal organisms in the canine duodenum. Furthermore, dogs with gastrointestinal disease harbored opportunistic fungal pathogens. We have identified a total of 51 different phylotypes. The most commonly observed phylotypes were *Pichia* spp., *Cryptococcus* spp., *Candida* spp., and *Trichosporon* spp.

**Molecular characterization of Archea & viruses**

While much recent work has focused on the molecular characterization of bacterial communities, only very limited information is available about Archea and viruses in the canine and feline gut. Recent *in vitro* studies suggest that, similar to humans, methanogenic Archea are present in the canine intestinal tract, but have yet to be characterized.[8] Methanogenic Archea have been associated with oral and intestinal disease. The knowledge about the presence of viral communities in the gastrointestinal tract is limited to a few families (including rotavirus, coronavirus, parvovirus). Recent studies in humans revealed that the viral community in the gastrointestinal tract is more diverse than the bacterial, with several hundred different phylotypes.[9] It is likely that a similar diverse viral community is present in dogs and cats, but needs yet to be characterized.

## MICROBIOTA IN HEALTH & DISEASE

The resident intestinal microbiota offers nutritional advantage to the host by production of short chain fatty acids (e.g., butyrate, propionate, and acetate), which stimulate mucosal growth and epithelial cell proliferation. Lactate, produced by microbial fermentation, serves as an energy source for the host. In addition, the normal intestinal microbiota plays a major role in protection of the host from invasion by harmful bacteria through exclusion of potentially pathogenic organisms. The implicated defense mechanisms against these harmful microbes include competition for oxygen and nutrient substrates, competition for mucosal adhesion sites, creation of a physiologically restrictive environment for non-resident bacterial species (e.g., production of substances that are toxic to other bacteria, changes in pH and redox potential, or hydrogen sulfide production), and secretion of antimicrobial substances (e.g., bacteriocins).[10]

On the other side various microbes may be detrimental to gastrointestinal health. Specific pathogens (i.e, *Salmonella*, enterotoxigenic *Clostridium perfringens, Campylobacter* spp, and others) may disturb the architecture of the gastrointestinal mucosa, penetrate into the submucosa and Peyer's patches, or produce exotoxins that interfere with enterocyte function. Enterotoxins often stimulate fluid secretion by the mucosa, while mucosal absorptive capacity may be decreased due to villus effacement and loss of surface area, resulting in diarrhea.

**Microbiota & chronic enteropathies**

There is mounting evidence that the residential gut microbiota plays an integral part in the development of inflammatory bowel diseases (IBD) in several species including humans, dogs and cats.[1,2] Studies in humans have shown an association between IBD and altered microbial communities in the intestine.[1,11,12] Patients with IBD respond also to a modification of the enteric microbiota via antibiotic therapy. In these studies, a decrease in the bacterial phyla *Firmicutes* and *Bacteroidetes*, and an increase in *Proteobacteria* and *Actinobacteria* was associated with IBD.[1] It is important to note, that some bacterial groups that traditionally were considered to be non-beneficial (i.e., *Clostridia*) are usually the most diverse bacterial group in the intestine, forming several phylogenetic *Clostridium* clusters. In most studies evaluating the microbiota between controls and IBD patients, a reduction in the diversity of *Clostridium* clusters XIVa and IV (i.e., *Lachnospiraceae* and *C. coccoides* subgroups) was associated with IBD, suggesting that this bacterial group may play (possibly due to production of butyrate) a greater role in maintenance of gastrointestinal health than previously believed.

Gastrointestinal disease due to alterations in the microbiota is also believed to occur commonly in dogs. In a recent study evaluating histiocytic ulcerative colitis, a specific form of canine IBD that affects predominantly Boxer dogs it was observed that these dogs respond favorably to antimicrobial therapy with fluroquinolones (i.e, Baytril), and an association with the presence of adherent and invasive *Escherichia coli* (AIEC) and intestinal inflammation was observed.[13] Interestingly, these AIEC isolates obtained from Boxer dogs with histiocytic ulcerative colitis share similarities to AIEC isolates that were obtained from ileal tissues of humans with Crohn's disease.[13]

We have recently also demonstrated imbalances in bacterial communities in dogs with IBD.[2] Similar to findings in humans, the species richness was significantly lower in IBD dogs compared to controls. IBD dogs were also significantly enriched in *Enterobacteriaceae*. Similar findings were also observed in cats with IBD. The composition of the mucosa-associated microbiota was distinctly different between healthy cats and cats with IBD, and IBD cats had a significantly higher number of mucosa-associated organisms belonging to the *Enterobacteriaceae* family than healthy cats.[14] It is hypothesized, that such alterations in commensal microbial communities may perpetuate intestinal inflammation in predisposed animals. Therefore, treatment strategies geared towards modulating microbial populations (i.e., nutritional intervention or probiotics) may be of therapeutic benefit in such patients.

**Table 1.** Percentages of 16S rDNA clones belonging to the major phylogenetic lineages in different compartments of the canine intestinal tract.

| | Duodenum | Jejunum | Ileum | Colon |
|---|---|---|---|---|
| Clostridiales | 40.0% | 38.8% | 24.8% | 26.1% |
| Lactobacillales | 24.7% | 12.6% | 1.4% | 13.4% |
| Fusobacteriales | 3.3% | 14.2% | 32.6% | 28.9% |
| Enterobacteriales | 32.0% | 27.3% | 18.4% | 1.4% |
| Bacteroidales | 0.0% | 7.1% | 22.7% | 30.3% |

## REFERENCES

[1] Frank DN et al. Molecular-phylogenetic characterization of microbial community imbalances in human inflammatory bowel diseases. **PNAS** 2007; 104:13780. [2] Xenoulis P et al. Molecular-phylogenetic characterization of microbial communities imbalances in the small intestine of dogs with inflammatory bowel disease. **FEMS Microbiol Ecol** 2008; 66:579. [3]Ritchie LE et al. Assessment of microbial diversity along the feline intestinal tract using 16S rRNA gene analysis. **FEMS Microbiol Ecol** 2008; 66:590. [4]Suchodolski JS et al. Analysis of bacterial diversity in the canine duodenum, jejunum, ileum, and colon by comparative 16S rRNA gene analysis. **FEMS Microbiol Ecol** 2008;66:569. [5]Suchodolski JS et al. Prevalence and identification of fungal DNA in the small intestine of healthy dogs and dogs with chronic enteropathies. **Vet Microbiol** 2008; 132:379. [6]Desai AR et al. Characterization and quantification of feline fecal microbiota using cpn60 sequence-based methods and investigation of animal-to-animal variation in microbial population structure. **Vet Microbiol** 2008 (epup ahead of print). [7]Suchodolski JS et al. Assessment of the qualitative variation in bacterial microflora among compartments of the intestinal tract of dogs by use of a molecular fingerprinting technique. **Am J Vet Res** 2005; 66:1556.[8]Middelbos IS et al.. In vitro evaluation of methanogenesis in the dog. **FASEB J.** 2008;22:444. [9]Breitbart M et al. Metagenomic analyses of an uncultured viral community from human feces. **J Bacteriol** 2003; 185:6220. [10]Kanauchi O et al. The beneficial effects of microflora, especially obligate anaerobes, and their products on the colonic environment in inflammatory bowel disease. **Curr Pharm Des** 2005; 11:1047. [11]Darfeuille-Michaud A et al. High prevalence of adherent-invasive Escherichia coli associated with ileal mucosa in Crohn's disease. **Gastroenterology** 2004; 127:412. [12]Kotlowski R et al. High prevalence of Escherichia coli belonging to the B2+D phylogenetic group in inflammatory bowel disease. **Gut** 2007; 56:669. [13]Simpson KW et al. Adherent and invasive **Escherichia coli** is associated with granulomatous colitis in boxer dogs. **Inf and Imm** 2006; 74:4778. [14]Janeczko S et al. The relationship of mucosal bacteria to duodenal histopathology, cytokine mRNA, and clinical disease activity in cats with inflammatory bowel disease. **Vet Microbiol** 2008; 128:178.

## CANINE CONGENITAL NEUTROPENIA SYNDROMES : DIAGNOSIS AND TREATMENT

**Clinton D. Lothrop Jr. PhD, DVM, DACVIM**
**Birmingham, AL**

### INTRODUCTION

Although there is no definition for how low the neutrophil (PMN) count must be to be classified as neutropenic most hematologists would classify an absolute PMN count less than 1500/ul as neutropenic and less than 500/ul as severe neutropenia. Causes of neutropenia in dogs include overwhelming infection, drug toxicity, parvovirus infection, myelopthitic syndromes, aplastic anemia, ricketsial infection and congenital neutropenic disorders. Genetic causes of neutropenia include cyclic hematopoiesis (CH), better known as gray collie syndrome , and trapped neutrophil syndrome of Australian and New Zealand Border collie dogs. This presentation will review the congenital neutropenia disorders of man and dogs with emphasis on the molecular basis of severe congenital neutropenia (SCN ) and cyclic neutropenia(CN) in humans and CH in dogs. The hallmark of the of the congenital neutropenia disorders is maturation arrest of myeloid precursors at the promyelocyte stage with almost complete absence of mature PMNs in the bone marrow.

### Human congenital neutropenia disorders

Severe congenital neutropenia, Kostmann's syndrome, was first described as an autosomal recessive disorder in members of a Swedish family with consanguinity by Rolf Kostmann in his PhD thesis in 1956 . We now know that autosomal recessive SCN is a relatively rare disorder that can be caused by mutations in 1 of several distinct genes (table 1). Autosomal dominant SCN is a much more common disease and is most commonly,50-60% of SCN patients, caused by a mutation in the ELA2 gene which encodes neutrophil elastase. All cases of CN are caused by ELA2 mutations,some identical to those reported to cause SCN. Elastase is a serine protease that is stored in the primary granules along with myeloperoxidase, aurocidin and proteinase 3. The genes for the primary granule proteins are coordinately regulated and maximal synthesis occurs in promyelocytes . More than 100 different ELA2 mutations have now been identified to cause SCN .Surprisingly none of the mutations result in a substantial loss or gain in enzyme activity or change in substrate specificity. Other genes in which mutations have been demonstrated to cause SCN include HAX1,the original kindred indentified Kostmann, p14 ,Gfi1, AP3B1, WAS protein,CSF3R and G6PC3. Apparently, a common mechanism for neutropenia in all of these disorders involves induction of the unfolded protein response (UPR) and increased apoptosis of myeloid progenitor cells.

**Genetic Variants of Severe Congenital Neutropenia.**

| Gene | Gene Function | Incidence of Variant | Inheritance | Associated Features |
|---|---|---|---|---|
| *ELA2* | Serine protease | 50–60% | Autosomal dominant, sporadic | Isolated neutropenia |
| *HAX1* | Mitochondrial function | Unknown | Autosomal recessive | Neurologic and neuropsychological abnormalities in some cases |
| *GFI1* | Transcription factor | Rare | Autosomal dominant | Monocytosis and defects in lymphocyte number and function |
| *WAS* | Cytoskeleton function | Rare | X-linked recessive | Monocytopenia and T-lymphocyte activation |
| *CSF3R* | G-CSF receptor | Rare | Autosomal dominant | Severe myeloid hypoplasia in the bone marrow; resistant to G-CSF treatment |
| *G6PC3* | Glucose metabolism | Unknown | Autosomal recessive | Cardiac defects, thrombocytopenia, and urogenital abnormalities |

NEJM 2009; 360: 3

### Canine congenital neutropenia disorders

Canine cyclic hematopoiesis (CH), first described by Pat Lund in 1967 , is an autosomal recessive disorder of collie dogs that has been referred to as "lethal gray trait" or "gray collie syndrome" which refers to the coat color dilution that is apparent at birth. It has not been reported in other breeds of dogs . The hallmark of CH is the 12-14 day PMN cycles with 2-4 days of neutropenia each cycle. Monocytes, reticulocytes and platelets also cycle with a similar periodicity but out of phase with the PMN cycles (fig1). Canine CH is caused by an insertional mutation in exon 21 of the AP3B1 gene. Clinical features of CH in addition to the hematopoietic cycles include melanocyte dysfunction causing the coat color dilution and a qualitative platelet disorder associated with a dense granule abnormality. The disease has been induced and cured by bone marrow transplantation demonstrating a stem cell defect. Elastase is thought to be an AP3 cargo protein which shuttles nascent elastase from the ER/TGN to the primary granules where it is stored.. Presumptively, in the absence of AP3 elastase accumulates in the ER/TGN and induces the UPR and apoptosis of myeloid precursor cells similar to that observed in SCN/CN patients with ELA2 mutations. Administration of granulocyte colony stimulating factor (G-CSF) eliminates the neutropenic cycles but the cycles resume once daily administration is stopped. Long term administration of human G-CSF is not possible because of development of neutralizing antibodies after 2-3 weeks of daily administration. Continued administration of human G-CSF will result in persistent severe neutropenia due to development of high titer antibodies that cross react with endogenous G-CSF. Long term administration of canine G-CSF doesn't result in the development of neutralizing antibodies but recombinant canine G-CSF is not routinely available.

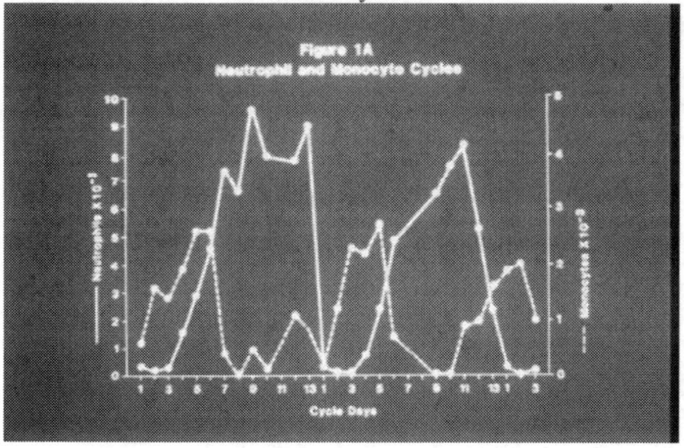


**Trapped Neutrophil Syndrome**

A related disorder with autosomal recessive inheritance, known as trapped neutrophil syndrome (TNS) has been recognized in Australian and New Zealand Border collie dogs. The genetic basis for TNS is not known but mutations in the ELA2 and AP3B1 genes have been eliminated by linkage analysis and autozygosity mapping. A linkage test has been described to assist breeders in the identification of heterozygous animals.

**SUMMARY**

Recent advances in our understanding of the genetic causes of SCN/CN and canine CH has provided key insight into critical steps in myelopoiesis and PMN production. Mutations that result in increased ER stress and apoptosis of myeloid progenitor cells provide a common mechanism to explain how mutations in seemingly unrelated genes all result in neutopenia. Current treatment options are limited to G-CSF administration and bone marrow transplantation. Future treatment modalities include gene therapy made possible by our understanding of the genetic basis of these disorders and use of small molecule agents to modify the UPR.

**REFERENCES**

[1.] Lund JE et al **Blood** 1967;29:452 [2.] Benson KF et al **Nat Genet** 2003;35: 90 [3.] Shearman JR and Wilton AN **Animal Genetics** 2007;38:188 [4.] Meng RH et al **Expt Hematol** 2009,Submitted [5.] Horwitz MS et al **Blood** 2007; 109:1817 [6.] Grenda DS et al **Blood** 2007; 110:4179 [7.] Dale DJ and Link DC **NEJM** 2009; 360:3 [8.] Pacheco JM et al **AM J Hematol** 2008; [9.] Berliner N **Blood** 2008;111:5427 [10.] Boztug K et al **NEJM** 2009; 360:32

## CANINE INDUCED PLURIPOTENT STEM CELLS (iPS) APPLICATIONS FOR REGENERATIVE MEDICINE

**Clinton D. Lothrop Jr. PhD, DVM , DACVIM**
**Birmingham, AL**

### INTRODUCTION

Clinical applications for regenerative medicine using embryonic stem (ES) cells offer considerable promise for treatment of hematopoietic, cardiac, neuromuscular and autoimmune diseases. In spite of the broad potential applications for ES cells in regenerative medicine significant ethical concerns for human embryo derived ES cells and potential concerns for tissue rejection following transplantation may limit the acceptance of using human ES cells in regenerative medicine. Recently, it has been shown that these concerns can be bypassed by generating pluripotent stem cells directly from somatic cells. These cells, termed induced pluripotent stem (iPS) cells, can be maintained in vitro in an undifferentiated state indefinitely and can be differentiated into the 3 germ cell layers . Injection of iPS cells into immunocompromised mice results in the development of teratomas similar to ES cells. The potential to use iPS cells to treat a genetic disease has been demonstrated in a mouse model of sickle cell anemia. In this presentation we will review the current state of the art as it relates to the development of canine iPS and ES cells and their application in regenerative medicine..

### Canine ES Cells

Embryonic stem cells are cell lines that can be maintained in vitro in an undifferentiated state permanently but can be differentiated under appropriate conditions to generate the three embryonic germ cell layers endoderm ,mesoderm and ectoderm. Much of our understanding of ES biology comes from work with murine ES cells and to a lesser extent with human and domestic animal ES cells. Long term maintenance of pluripotency of murine ES cells requires specific expression of transcription factors such as *Oct 3 / 4,Sox 2 and Nanog* and induction of widely expressed tumor –related genes such as *Stat3,Myc,Klf4* and *B-catenin.* Recognition of this fact formed the basis for current strategies to transform somatic cells into iPS cells. Similar to human and murine ES cells, canine ES cells express unique markers such as *Nanog,Sox2, SSEA-3,SSEA-4, TRA-1-60,TRA-1-81, Oct ¾ and alkaline phosphatase.* The cells can be passaged indefinetively in vitro , and can be differentiated into hematopoietic, cardiac ,neuronal and endothelial cells.

### iPS Cells

In landmark studies by Thomson and Yamanaka it was shown that over expression of a cocktail of just four transcription factors, Oct 4, c-myc, Sox2, and Klf4 was sufficient to reprogram mouse fibroblasts into pluripotent stem cells. Since that initial report there have been multiple reports confirming and extending the initial findings to demonstrate that it is possible to generate iPS with just 2 factors and that small molecule chromatin modulators can improve the efficiency of generating iPS cells. It has been shown that under certain conditions a variety of differientiated somatic cell lineages can be transformed into iPS cells . Cells already undergoing cell division such as keratinocyte stem cells or that have transcriptional profiles in which ES like transcription factors are already upregulated can be more efficiently transformed into iPS cells than skin fibroblasts Generation of iPS cells requires a sequential reprogramming of the cellular transcriptional profile which takes 2-3 weeks while the cells are maintained on embryonic feeder cells. Subsequently , the cells can be induced into distinct cell lineages by culturing the cells in the appropriate inductive microenviroment. Crucial steps in the clinical application of iPS cells to regenerative medicine is the differentiation of the pluripotent stem cells to specific cell lineages and scale up to generate adequate numbers of cells for clinical therapy. Our laboratory has demonstrated that a similar cocktail of retroviral or lentiviral vectors encoding Oct4, Sox2, c-myc and Klf4 can be used to induce canine iPS cells from skin fibroblasts or keritinocytes. The cells formed neural tumors when injected into nude mice. We are currently trying to differentiate the canine iPS cells into hematopoietic stem cells for bone marrow transplantation and gene therapy applications.

**Table 1. Reprogramming of different mouse cell types into iPS cells**

| Germ layer | Cell type | Efficiency | Selection type | References |
|---|---|---|---|---|
| Mesoderm | MEFs | 0.01-0.1% | Fbx15/Oct4/Nanog | Takahashi and Yamanaka, 2006; Maherali et al., 2007; Okita et al., 2007; Wernig et al., 2007 |
| | MEFs* | 2-4% | Nanog | Wernig et al., 2008 |
| | Mature B cell*,† | 3% | Nanog | Hanna et al., 2008 |
| Endoderm | Stomach cells | ND | Fbx15 | Aoi et al., 2008 |
| | Liver cells | ND | Fbx15 | Aoi et al., 2008 |
| | Intestinal epithelium* | ND | Nanog | Wernig et al., 2008 |
| | β cells | 0.1-0.2% | Morphology | Stadtfeld et al., 2008 |
| Ectoderm | Neural progenitors | <3.6% | Oct4 | Kim et al., 2008; Shi et al., 2008; Eminli et al., 2008 |
| | Keratinocytes* | ND | Nanog | Wernig et al., 2008 |

*Secondary cells. Differentiated cells derived from iPS cells carrying the four viral transgenes under the control of a doxycycline-inducible promoter.
†Note that mature B cells required ectopic expression of Cebpα in addition to the four reprogramming factors to produce iPS cells.
MEFs, murine embryonic fibroblasts. ND, not determined.

Development 136, 509-523, 2009

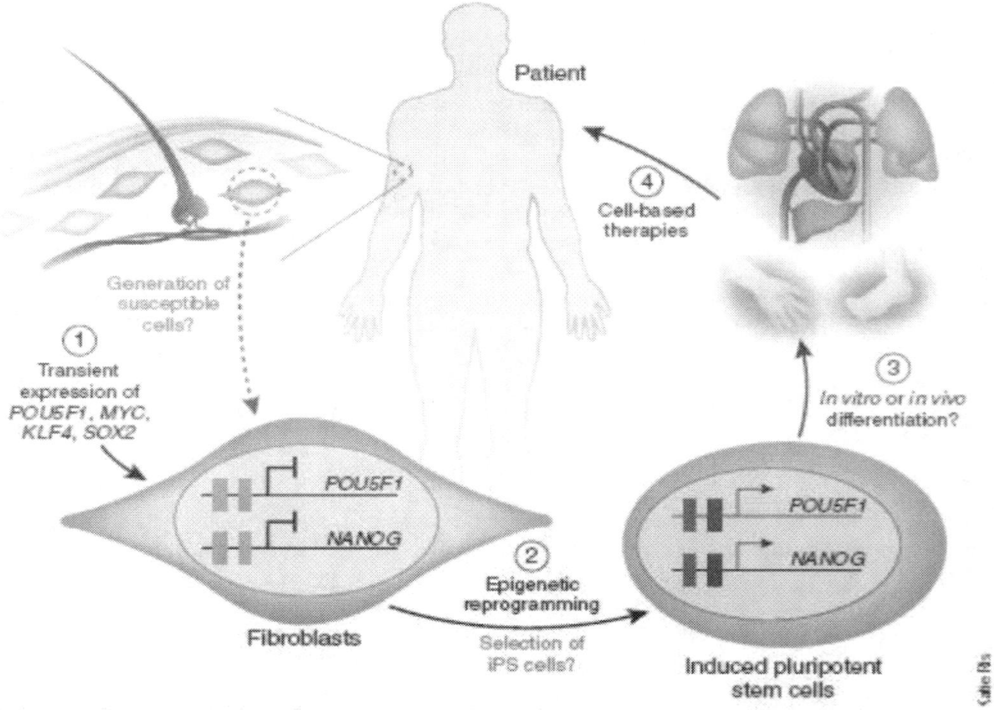


**Figure 1** Hypothetical strategy for using iPS cells in cell-based therapies. Fibroblasts, which are ubiquitous but differentiated skin cells[12], are isolated from the patient by skin biopsy. Next, the four core genes are ectopically and transiently expressed, causing epigenetic changes in the genome (boxes in front of genes) that reprogram the cell and activate endogenous *POU5F1* and *NANOG*, two genes that drive ES fate. The iPS cells are then differentiated into the desired adult cell types or tissues, although conditions for directed differentiation need to be determined. The desired cell type or tissue, genetically identical with the patient, can be use to replace damaged or diseased organs. Because iPS cells are susceptible to environmental cues for differentiation, it may not be necessary to fully differentiate them before reintroduction to the patient. Alternatively, it may be advantageous to directly reprogram them to the lineage of interest for tissue engineering purposes.

Nature Medicine, Volume 13, Number 7, July 2007

## REFERENCEs

1. Welsttead GG et al **Jove** 2008:1 2. Hatoya S et al **Mol Repro and Develop** 2005;73:298 3. Vaags AA et al **Stem Cells** 2008:1 4. Jang G et al **Therio** 2006:1 5. Wernig M et al **Nature** 2007:1 6. Takahashi K et al **Nature Protocols** 2007;2: 3081 7. Takahashi K and Yamanaka S **Cell** 2006;126:663 8. Takahashi K et al **Cell** 2007;131:861 9. Hanna J et al **Science** 2007;318:1920 10. Chang HY and Cotsarelis G **Nat Med** 2007; 13:783

**CANINE EOSINOPHILIC BRONCHOPNEUMOPATHY: IMMUNOPATHOGENESIS**

**Dr. Dominique Peeters**
**Belgium**

Eosinophilic lower airway diseases are a heterogeneous group of disorders in which an increased number of eosinophils is present in the airways and/or lung parenchyma. In some cases, eosinophils are merely a part of the inflammatory process and may even be present to protect host tissues against parasites or other organisms. In other cases, eosinophils appear to be directly responsible for tissue damage. In the dog, eosinophilic lower airway diseases are broadly separated into airway, parenchymal and mixed disorders [1]. Most **eosinophilic parenchymal disorders** are associated with a known underlying condition like parasitic infection (occult heartworm disease, *Angiostrongylus vasorum*…), chronic bacterial pneumonia, idiopathic hypereosinophilic syndrome or lung cancer. **Eosinophilic diseases** that are **strictly confined to the airways** are not commonly encountered in the dog and include idiopathic eosinophilic (tracheo)-bronchitis and parasitic tracheobronchitis caused by *Oslerus osleri*. Finally, eosinophilic bronchopneumopathy (EBP) is a condition in which **eosinophilic infiltration involves both the bronchial wall and the pulmonary parenchyma**. A cause is rarely identified and most cases of EBP are considered idiopathic [2].

Although the etiology of EBP is still unknown, part of the pathogenesis has been elucidated. A selective increase in $CD4^+$ T-cells and a selective decrease in $CD8^+$ T-cells has been demonstrated in bronchoalveolar lavage fluid (BALF), bronchial mucosa and pulmonary interstitium of affected dogs [3, 4]. This is similar to the situation in human bronchial asthma, eosinophilic bronchitis and idiopathic chronic eosinophilic pneumonia, where the ratio of $CD4^+$ T-cells/$CD8^+$ T-cells increases and activated Th2 cells accumulate at sites of inflammation [5-7].

Eosinophilic infiltration and a predominance of $CD4^+$ T-cells in BALF support the role for a dominant Th2 immune response in the lower airways in dogs with EBP. Despite this, real-time RT-PCR has not confirmed a significant difference in bronchial Th2 cytokine expression (IL-4, IL-5 and IL-13) in dogs with EBP compared to control animals [8]. The lack of significant difference between control and diseased dogs is thought to be related to methodology used. First of all, by using mucosal biopsies, the actual amount of mRNA produced in mucosal T cells could have been diluted in the total amount of mRNA found in other mucosal cell types. Secondly, RT-PCR methods have recently been improved by assessing RNA quantity and quality before doing the PCR and by using multiple internal control genes for calculation of gene expression. These changes should be used to assess the cytokine mRNA expression in T cells isolated from BAL collected from EBP dogs before any conclusion can be drawn regarding the Th1/Th2 balance in canine EBP. Another way to evaluate the cytokine profile in lower airways of dogs with EBP would be to determine the cytokine protein concentrations in BALF by using capture ELISA methods.

Quantification of mRNA encoding for several CC-chemokines and one of their receptors (CCR3) [8] has not revealed a significant difference in expression of monocyte chemoattractant protein (MCP)-1, -2, -4 and CCR3 between control and EBP dogs. However, expression of transcript for MCP-3, eotaxin-2 and -3 was significantly greater in bronchial biopsies from dogs with EBP than in samples from control dogs, and significantly less mRNA encoding for regulated on activation normal T cell expressed and secreted protein (RANTES) was found in the mucosa of dogs with EBP [8]. Eotaxins are the strongest chemoattractants for eosinophils and basophils [9]. MCP-3 attract eosinophils but also other cell types such as monocytes, dendritic cells, basophils and T cells [10]. Increased mRNA levels for MCP-3, eotaxin-2 and -3 in bronchial biopsies from dogs with EBP suggest that these chemokines drive the recruitment of eosinophils and mononuclear cells into the airways in EBP.

Hypersensitivity to aeroallergens is suspected in canine EBP [2]. In one study, an intradermal skin test using a panel of 48 standardized allergens including house dust mite, *D. pteronyssinus, D. farinae*, *Tyrophagus*, human dander, mixed feathers, moulds, pollens of grasses, trees, and weeds and mixed insects, was positive in 4 of 12 dogs with untreated EBP [3]. In another study, 3 dogs with EBP were tested with various antigens and all three were negative [11]. The relationship between positive intradermal skin testing and documentation of aeroallergens responsible for EBP is difficult to establish. A positive intradermal skin test does not necessarily indicate that the allergen identified is responsible for the pulmonary response. This may be explained by factors such as a difference in mast cell distribution between the lungs and skin or in the route of allergen exposure leading to hypersensitivity. Indeed, there is a discrepancy between localized and systemic immune responses after antigen challenge in the lung [12]. Measurement of serum allergen-specific IgE might provide additional insight into the role of aeroallergens in eosinophilic lung disease. A study comparing serum and BALF allergen-specific IgE in dogs with EBP and dogs with chronic bronchitis is conducted in our laboratory. Preliminary results show that allergen-specific IgE are found in the serum of most dogs with either EBP or chronic bronchitis. However, allergen-specific IgE are found in BALF from most dogs with EBP in greater quantity than in the BALF collected from dogs with chronic bronchitis. These IgE are specific mostly of indoor allergens such as flea saliva, *D. farinae* or *Tyrophagus*.

Lower airway and parenchymal destruction and remodelling observed in canine EBP is at least in part related to up-regulation of collagenolysis and proteolysis. Indeed, collagenase activity of matrix metalloproteinases (MMP) is increased in BALF from dogs with EBP as compared with that found in BALF from control animals [13, 14]. This increased collagenolytic activity is due in part to increased activity of MMP-8, MMP-9, and MMP-13 [13, 14]. In EBP, these MMPs seem to be produced by macrophages and epithelial cells and not by eosinophils [13, 14]. Epithelial laminins are among the proteins that are degraded by MMPs in canine EBP [15], and increased laminin-5$\gamma$2-chain degradation products in BALF from these dogs indicates epithelial injury. Epithelial sloughing leading to temporary denudation of the basement membrane is evident histologically at both the bronchial and alveolar levels in canine eosinophilic lung disease [15].

Procollagen type III amino terminal propeptide (PIIINP) is a marker of extracellular matrix turnover [16]. A quantitative test to identify PIIINP has been developed in an attempt to evaluate organ fibrosis. This test is a sensitive but non-specific marker for assessment of tissue collagen type III turnover [17]. High BALF PIIINP concentrations have been found in dogs with EBP [18]. Although further investigations in large populations of dogs with varying bronchopulmonary pathologies are warranted, this study suggests that BALF PIIINP could be a promising marker of lung disease in the dog [18].

## REFERENCES

[1] Clercx C and Peeters D. **Vet Clin North Am Small Anim Pract** 2007;37(5):917-35, vi. [2] Clercx C, et al. **J Vet Intern Med** 2000;14(3):282-91. [3] Clercx C, et al. **J Vet Intern Med** 2002;16(3):229-37. [4] Peeters D, et al. **J Comp Pathol** 2005;133(2-3):128-35. 5. Albera C, et al. **Eur Respir J** 1995;8(8):1281-5. [6] Brightling CE, et al. **J Allergy Clin Immunol** 2002;110(6):899-905. 7 Robinson DS, et al. **Thorax** 1993;48(1):26-32. [8] Peeters D, et al. **Vet Immunol Immunopathol** 2006;110(1-2):65-77. [9] Lloyd C. **Immunology** 2002;105(2):144-54. [10] Lukacs NW. **Nat Rev Immunol** 2001;1(2):108-16. [11] Corcoran BM, et al. **J Small Anim Pract** 1991;32(10):494-502. [12] Bice DE, et al. **Am J Respir Cell Mol Biol** 1993;8(6):662-7. [13] RajamakiMM, et al. **J Vet Intern Med** 2002;16(6):658-64. [14] Rajamaki MM, et al. **Vet J** 2002;163(2):168-81. [15] Rajamaki MM, et al. **Vet J** 2006;171(3):562-5. [16] Risteli L and Risteli **J. Methods Enzymol** 1987;145:391-411. [17] Jensen LT. **Dan Med Bull** 1997;44(1):70-8. [18] Schuller S, et al. **Am J Vet Res** 2006;67(5):749-55.

# CANINE SINONASAL ASPERGILLOSIS: IMMUNOPATHOGENESIS

**Dr. Dominique Peeters**
**Belgium**

Sino-nasal aspergillosis (SNA) is the second most common cause of nasal discharge in dogs after nasal neoplasia. The disease is most often caused by *Aspergillus fumigatus* although occasional cases are caused by other *Aspergillus* species such as *A. flavus or A. niger*. Typical clinical signs are those of a severe rhinosinusitis and include profuse mucopurulent to purulent nasal discharge, nasal pain, and ulceration or depigmentation of the nostrils. This latter sign is almost exclusively seen in SNA and is thought to be caused by fungal toxins in the nasal discharge. Other clinical signs that are often observed are sneezing, reverse sneezing, epistaxis, depression, decreased appetite and increased nasal airflow.

Most studies have aimed to determine the best diagnostic method and the most appropriate therapeutic approach for canine SNA. Nevertheless, although the causative agent of the disease is clearly identified, the pathogenesis of SNA is still poorly understood.

*Aspergillus* is an ubiquitous fungus and the reason that *A. fumigatus* causes disease in only a small proportion of exposed dogs remains unclear. In rare cases only, a predisposing factor such as a facial trauma, a nasal foreign body, nasal carcinoma, or an impacted tooth is present. Many potential virulence factors for *A. fumigatus* have been studied *in vitro*. For example, gliotoxin, aflatoxins and ribotoxins may interfere with mucociliary clearance, opsonization and neutrophilic phagocytosis [1]. Whether these toxins play a role in canine SNA remains to be determined.

Dogs with SNA are usually **not systemically immunocompromised**. Affected dogs do not generally have any immune suppressive condition. Moreover, fungal infection does not spread throughout the body and is restricted to the nose and/or frontal sinus. Although impaired peripheral blood lymphocyte proliferative responses have been shown in dogs with SNA [2], the significance of this finding is unclear because *A. fumigatus* has been demonstrated to inhibit lymphocyte proliferation *in vitro*, suggesting that the abnormal lymphocyte function is secondary to the infection rather than a primary immunodeficiency. The absence of systemic immune suppression in dogs with SNA is also suggested by the lack of invasiveness of respiratory mucosa by the fungus. In Grocott-stained sections obtained from nasal biopsies of 15 dogs with SNA, fungal hyphae were seen at the surface of the mucosa in only six cases and were not identified within or beneath the mucosal epithelium in any case [3]. In addition, examination of Grocott-stained sections from the frontal bone of five SNA dogs with frontal bone osteomyelitis did not contain fungal elements in bony tissue [4]. Despite the fact that SNA is a non-invasive disease in dogs, marked destruction of nasal turbinates is present in almost all patients. Thus, bony destruction is not caused by the fungus itself but is thought to be due to the host inflammatory response.

As *A. fumigatus* is ubiquitous and a systemic immune suppression is not present in affected dogs, **a local immune dysfunction** is suspected to be involved in the pathogenesis of canine SNA. Generation of a dominant Th1-cell response, characterized by the activation of $CD4^+$ Th1 cells producing interferon (IFN)-γ and macrophages producing pro-inflammatory cytokines, is required for the expression of protective acquired immunity to fungi [5]. Cytokine analysis (using mRNA quantification of mucosal biopsies by real-time RT-PCR) has confirmed that the nasal mucosa of dogs with SNA is dominated by a Th1 response, with up-regulation of mRNA encoding IFN-γ, IL-6, IL12p35, IL-12p40, IL-18 and TNF-α [6, 7]. This strong Th1 and pro-inflammatory immune response may help confine *Aspergillus* infection to the nasal cavities and frontal sinuses.

However, mRNA encoding the immunomodulatory cytokine IL-10 is also up-regulated in nasal tissue from dogs with SNA [6, 7]. The increase in IL-10 transcripts is suspected to be implicated in the failure to clear the *Aspergillus* infection despite a strong mucosal Th1 immune response. Indeed, IL-10 exerts detrimental effects on host responses to fungi [5]. However, high-level production of IL-10 in the nasal mucosa of dogs with SNA might prove beneficial by limiting of the inflammatory response that extends local tissue damage initiated by the fungal infection. In experimental rodent models, animals with chronic infectious disease are permitted to have persistent infection when Th1 cells switch from an IFN-γ producing phenotype to one dominated by IL-10 production [8]. Similar mechanisms involving T regulatory (Treg) cells are proposed in experimental aspergillosis [9]. The cause and the population of cells (Th1 or Treg?) implicated in the increase in IL-10 mRNA expression in dogs with SNA is currently unknown but this increased expression seems particular to SNA as the expression of mRNA encoding for IL-10 is not increased in the nasal mucosa of dogs with idiopathic lymphoplasmacytic rhinitis [7].

Another factor potentially involved in the failure to clear the fungus despite a strong Th1 immune response resides in the possible involvement of Th17 cells in the pathogenesis of canine SNA. Interleukin-12 family cytokines are heterodimeric cytokines that are mostly produced by macrophages and dendritic cells. IL-12 is composed of a light chain (IL-12p35) and a heavy chain (IL-12p40). The IL-12p40 component of IL-12 can also dimerize with IL-12p19 to form IL-23 [10]. In nasal tissue from dogs with SNA, the level of IL-12p19 and IL-12p40 transcripts is greater than in tissue from control animals, suggesting that IL-23 mRNA expression is increased in dogs with SNA [7]. The role of IL-23 in the protection against fungal infection is less clear than that of IL-12. Interleukin-23 stimulates IL-17 production by Th17 cells [10]. Interleukin-17 in turn stimulates IL-8 production and promotes neutrophilic inflammation [11]. Nevertheless, both IL-23 and IL-17 have been shown to subvert the inflammatory program of neutrophils, which results in severe tissue inflammatory pathology associated with *Aspergillus* infection in a murine model of aspergillosis [12]. Thus, the IL-23/IL-17 pathway promotes neutrophilic inflammation but inhibits the fungicidal inflammation of these cells, thereby perpetuating infection and the associated inflammatory response and counteracting the effects of protective Th1 immunity. As Th17 cells have not yet been defined in the dog and the expression of IL-17 has not been reported in canine SNA, it is difficult to assert that Th17 cells exert similar roles in canine SNA but the increased expression of IL-23 is suggestive of activation of this T cell subset in this disease.

Clearly, further studies are required to understand the implication of IL-10, IL-17 and IL-23 in the pathogenesis of canine SNA. Isolation and characterization of T cells from mucosal biopsies by flow cytometry and/or RT-PCR are needed to determine whether Treg or Th1 cells are producing IL-10 and whether Th17 cells are found in the mucosa of affected dogs.

We have also studied the expression of some chemokines in the nasal mucosa of dogs with SNA. The inflammatory infiltrate found in the nasal/sinusal mucosa of dogs with SNA is dominated by lymphocytes and plasma cells, although neutrophils are present in great numbers in some cases [3]. Eosinophils and mast cells are only rarely observed. The nature of the inflammatory infiltrate can be explained by the upregulation of mRNA encoding IL-8, monocyte chemoattractant protein (MCP)-1, MCP-2, MCP-3 and MCP-4 in the nasal mucosa of dogs with SNA compared with control animals [6]. Indeed, IL-8 is the primary neutrophil chemoattractant and MCPs attract mostly mononuclear cells.

Innate and adaptive immune responses act to generate the most effective form of immunity for protection against *A. fumigatus*. Although the actions of T cells feed back into this dynamic equilibrium to regulate the balance between pro-inflammatory and anti-inflammatory signals, the decision of how to respond is still primarily determined by interactions between fungi and cells of the innate immune system. Therefore, studying the innate immune system in the nasal mucosa of dogs with SNA is of the upmost importance. In particular, the expression of Toll-like receptors (TLRs), which are known to play pivotal roles in the immunity to fungi, is under investigation in our laboratory.

In conclusion, despite a strong cell-mediated mucosal immune response mediated mostly by IL-12, IL-18, IFN-γ and MCPs in canine SNA, complete elimination of *A. fumigatus* does not occur, probably in part due to local production of IL-10, IL-23 and possibly IL-17. Further studies determining the T cell subsets producing these cytokines and the expression of TLRs and other components of the innate immune system are needed to better understand the immunopathogenesis underlying canine SNA. These studies will hopefully offer new therapeutic targets for this condition in dogs.

## REFERENCES

[1.] Tomee JF and Kauffman HF. **Clin Exp Allergy** 2000;30(4):476-84. [2.] Sharp NJH et al. **Comp Cont Educ** 1991;13:41-7. [3.] Peeters D et al. **J Comp Pathol** 2005;132(4):283-8. [4.] Peeters D and Clercx C. **Vet Clin North Am Small Anim Pract** 2007;37(5):901-16. [5.] Romani L. **Nat Rev Immunol** 2004;4(1):1-23. [6.] Peeters D et al. **Vet Microbiol** 2006;114(3-4):318-26. [7.] Peeters D et al. **Vet Immunol Immunopathol** 2007;117(1-2):95-105. [8.] Trinchieri G. **J Exp Med** 2007;204(2):239-43. [9.] Romani L and Puccetti P. **Trends Microbiol** 2006;14(4):183-9. [10.] Hunter CA. **Nat Rev Immunol** 2005;5(7):521-31. [11.] Kolls JK and Linden A. **Immunity** 2004;21(4):467-76. [12.] Zelante T et al. **Eur J Immunol** 2007;37(10):2695-706.

# THE ROLE OF PETS AS A RESERVOIR FOR HUMAN CAMPYLOBACTERIOSIS

Els Acke VetSurg, PhD, DECVIM-CA, CertSAM, MRCVS; Kevina McGill BSc; Boyd R. Jones BVSc, FACVSc, DECVIM-CA, MRCVS; Seamus Fanning BSc, PhD; Paul Whyte MSc, PhD
Dublin, Ireland

## INTROCUCTION

Campylobacteriosis is the most common bacterial cause of human enteritis in the developed world. The epidemiology of *Campylobacter* spp. infections is highly complex and knowledge on different sources and transmission routes is lacking. Cats and dogs can be asymptomatic carriers of *Campylobacter* spp. and contact with pets has been recognized as a risk factor for human campylobacteriosis.[1, 2] With reported *Campylobacter* spp. isolation rates of up to 87.0% and 75.0% in dogs and cats in Ireland[3], pets may play a role in the epidemiology of campylobacteriosis.

## STUDY DESIGN

The aim of our study was to assess the role of pets as a reservoir for human *C. jejuni* infections. A database was assembled with genetic fingerprints of *C. jejuni* isolates obtained from pets. The profiles of each pet isolate were compared to those in a database containing Irish human clinical and retail food *Campylobacter* genotypes. Samples were obtained from pets by rectal swab from cats and dogs in shelters and private households in Ireland. A range of specialised culture media was used to isolate *Campylobacter* spp.[3, 4] Two molecular methods were applied for the genotyping of *C. jejuni* isolates: flagellin (*fla*) typing using *Dde*I and *Hinf*I restriction enzymes[5, 6] and pulsed-field gel electrophoresis (PFGE) using *Sma*I restriction enzyme.[7] Genoprofiling and cluster analysis was carried out using Fingerprinting II software, Bionumerics Version 3.0 (Applied Maths, Kortrijk, Belgium). Isolates clustering at 90% or over were considered 'indistinguishable' isolates.

## RESULTS

A total of 20.6% of profiles in the *fla Dde*I comparison of 164 human strains and 79 pet strains were detected in clusters including at least one human clinical and one pet *C. jejuni* strain. A total of 63.6% of profiles in the *fla Hinf*I comparison containing 164 human strains and 78 pet strains were in clusters. Clustering of 60 pet strains compared to a total of 109 human clinical strains based on PFGE profiles revealed 4.1% of profiles were in clusters with at least one human clinical and one pet strain. A total of 12.9% of profiles in the *fla Dde*I comparisons of 269 food strains and 79 pet strains were detected in clusters with at least one food and one pet *C. jejuni* strain. A total of 54.9% in the *fla Hinf*I comparison of 332 food strains and 78 pet strains were in clusters with at least one food and one pet *C. jejuni* strain. Clustering of 60 pet strains compared to a total of 310 food isolates based on PFGE profiles revealed 1.6% of profiles were in clusters with at least one food and one pet *C. jejuni* strain.

## CONCLUSION

The link detected between Irish pet *C. jejuni* strains and those isolated from humans with enteritis suggests that *C. jejuni* strains isolated from pets are capable of causing clinical infection in humans. In addition, contaminated food may be a potential source of *Campylobacter* infection in pets, which may consequently result in human infection. As asymptomatic animals can act as reservoirs, especially pets owned by those more susceptible to infection may represent an important health risk and adequate hygiene practices in households with pets are essential.

## REFERENCES

1. Tenkate et al. (2001) Risk factors for campylobacter infection in infants and young children: a matched case-control study **Epidemiol. Infect.** 127, 399-404. 2. Carrique-Mas et al. (2005) Risk factors for domestic sporadic campylobacteriosis among young children in Sweden. **Scand. J. Infect. Dis.** 37, 101-110. 3. Acke et al. (2006) Prevalence of thermophilic Campylobacter species in cats and dogs in two animal shelters in Ireland. **Vet. Rec.** 158, 51-54. 4. Acke et al. (2009) Prevalence of *Campylobacter* species in household pets in Ireland. **Vet. Rec.** 164, 44-47. 5. Nachamkin et al. (1993) *Fla*gellin gene typing of Campylobacter jejuni by restriction fragment length polymorphism analysis. **J. Clin. Microbiol.** 31, 1531-1536. 6. Nachamkin et al. (1996) Analysis of HL and O serotypes of Campylobacter strains by the *fla*gellin gene typing system. **J. Clin. Microbiol.** 34, 277-281. 7. Campynet protocol: http://campynet.vetinst.dk/

**GENOMIC LINKAGE OF OBESITY AND DISEASE.**

**WO Carter, RM Yamka, MS Abrahamsen,**
**Hill's Pet Nutrition Center, Topeka, KS.**

Obesity has been linked to many diseases, including diabetes, pancreatitis, cruciate ligament rupture, hypothyroidism, hyperadrenocorticism, lower urinary tract disease, oral disease, neoplasia, dyslipidemia, osteoarthritis, hypertension, and altered kidney function. Application of new research tools such as gene expression, proteomics and metabolomics, provides the opportunity to begin to understand the dysregulation of molecular processes in obese animals that may contribute to the development of obesity-related diseases. Genomic analysis of adipose tissue and lymphocyte samples from lean and obese dogs revealed that key genes important in beta-oxidation of fats such as PPAR-gamma, carnitine O-palmitoyltransferase 1 A and acyl-CoA synthetase are down regulated in obese animals. Similarly, adipose tissue from obese dogs display a decreased expression of genes involved in glucose metabolism. Additional analysis of circulating lymphocytes in obese and lean dogs identified differences in expression of numerous pathways involved in complex biological processes, including immune activity, inflammation, wound healing and cell replication, that may be contributing factors to the development of future disease states.

## THE EFFECT OF VARYING TIDAL VOLUME ON PCO2 AND pH IN HEALTHY BEAGLES

**TJ Oura, EA Rozanski D Bedenice,**
**Cummings School of Veterinary Medicine at Tufts University, North Grafton, MA.**

Acute respiratory distress syndrome (ARDS) is a common condition in critically ill people and associated with a high mortality. In people, no pharmacological therapy has been proven beneficial; however, low tidal volume (Lo $V_T$) (6 ml/kg) has been related to increased survival. In some patients, Lo $V_T$ may be associated with severe hypercarbia and subsequent respiratory acidosis. Critically ill dogs also develop ARDS and the appropriate ventilatory strategy is unknown in this species. The purpose of this study was, therefore, to evaluate the effect of varying tidal volume ($V_T$) on $PCO_2$ and pH in healthy dogs, thus determining if healthy dogs develop hypercarbia and/or respiratory acidosis when ventilated at Lo $V_{T,}$. These data are necessary to allow subsequent validation of a Lo $V_T$ approach in critically ill dogs

Five healthy beagles were enrolled. Dogs were pre-medicated with hydromorphone and diazepam, and induced with propofol. Anesthesia was maintained with a propofol continuous rate infusion. Dogs were intubated and ventilated using a Puritan-Bennett 840 ventilator at 6 ml/kg, 8 ml/kg, 10 ml/kg, 12 ml/kg and 15 ml/kg. The minute ventilation was held consistent at 200 ml/kg. The inspiratory oxygen concentration was 25%, without use of positive end-expiratory pressure. Routine anesthetic monitoring was provided. Arterial blood gases for determination of $PCO_2$ and pH were collected after 15 minutes equilibration at each tidal volume. Using 12 ml/kg as the standard, t-tests were used to compare the effects of different tidal volumes on pH and $pCO_2$. Additionally, incidences of 1) mild acidosis (pH 7.21- 7.36) and 2) Severe acidosis (ph< 7.20) were recorded.

The tidal volume had a significant effect on pH and PCO2 as outlined below.

| Tidal volume | 6ml/kg (mean ± stdev) | 8 ml/kg | 10 ml/kg | 12 ml/kg | 15 ml/kg |
|---|---|---|---|---|---|
| $PCO_2$(mmHg) | 40.4 ± 2.2 (p=0.001) | 38.6 ± 2.2 (p=0.07) | 36 ± 2.7 NS | 32.8 ± 3.5 | 29.1 ±1.7 (p=0.03) |
| pH | 7.325 ± .021 (p=0.002) | 7.336 ± .02 (p=0.005) | 7.368 ± .03 NS | 7.39 ±0.04 | 7.431 ± .03 NS |

Out of 25 analyses, mild acidosis was present in 12 samples. However; no severe disturbances were identified in any dog.

This study demonstrates that Lo $V_T$ is well-tolerated in healthy dogs anesthetized using a clinically relevant anesthetic protocol and a critical Care ventilator. Alterations in tidal volume resulted in predictable effect on PCO2 and pH. Further evaluation in dogs with naturally-occurring ARDS is warranted to determine if respiratory acidosis occurs in these animals.

## LAPAROSCOPIC SPLENIC BIOPSY IN DOGS AND CATS: 15 CASES.

**Radhakrishnan, Anant,**
**Bluegrass Veterinary Specialists, Lexington, KY.**

Laparoscopic biopsy of the spleen has not been described in the veterinary literature, although biopsy of the spleen is well documented in human medicine. Laparoscopy is routinely performed for exploratory evaluation and biopsy of internal organs. Histopathology may be preferred to cytology for assessing disease conditions due to potential inadequacy of samples obtained by fine needle aspiration. Negative cytologic evaluation for neoplasia does not preclude the diagnosis of neoplasia. The ability to obtain tissue samples of the spleen minimally invasively and safely allows for greater confidence in diagnosis and can decrease the incidence of unnecessary splenectomy.

15 client-owned patients (10 dogs, 5 cats) underwent laparoscopy to obtain biopsy of the spleen. A two port technique was used. Samples were obtained using a 5 mm cup biopsy forceps. Hemostasis was achieved with an absorbable gelatin sponge placed at the biopsy site using 5 mm Babcock grasping forceps. All patients recovered uneventfully and survived to discharge.

Average survival time was 453 days (range 2-960) with 10/15 patients alive at the time of submission. Three patients were diagnosed with neoplasia (2 lymphoma, 1 mastocytosis). Of the twelve patients with non-neoplastic disease, nine were still alive at the time of this writing (average 550 days; range 120-960).

Laparoscopic splenic biopsy can be accomplished safely and effectively. Greater confidence in non-neoplastic diagnoses is achieved when there is histopathologic confirmation compared to cytologic evaluation.

**MOLECULAR ANALYSIS OF *GIARDIA* SPP. INFECTIONS IN DOGS IN THE UNITED STATES.**

**MR Lappin, AV Scorza.**
**Colorado State University, Fort Collins, CO.**

*Giardia* spp. are common in many animals including dogs, cats, and people, thus, there is potential for zoonotic transmission. However, there are host-adapted assemblages as well as zoonotic assemblages. In addition to routine fecal examination techniques, direct fluorescent antibody staining (DFA), fecal antigen assays, and polymerase chain reaction assays are commonly used to aid in the diagnosis of *Giardia* infections. However, most commercially available assays were initially developed with isolates from humans and so whether the assays detect dog and cat assemblages is essentially unknown.

Fecal samples from dogs positive for *Giardia* spp. by fecal flotation, DFA, or fecal antigen assay were selected. DNA was extracted from each sample and the *Giardia* assemblage determined by genetic sequencing of two gene targets (β giardin and GDH).

*Giardia* spp. DNA was amplified from 83 samples. Assemblage A (zoonotic) was amplified from one sample and the remaining samples were assemblage C or D. Some *Giardia* positive samples were negative by PCR assay. The assemblage results varied between the two genes for some samples but no assemblage C or D samples were assemblage A in the alternate assay. The commercially available DFA and antigen assays assessed detected *Giardia* assemblages C and D.

Most dogs were infected with host adapted genotypes and commercially available DFA and antigen assays appear to detect dog assemblages. *Giardia* PCR assays can give falsely negative results and so should not be used alone to screen for *Giardia* infection. *Giardia* genotyping studies should always compare the results from at least two genes.

**EVIDENCE FOR A ROLE OF INNATE IMMUNITY IN THE PATHOGENESIS OF INFLAMMATORY BOWEL DISEASE IN GERMAN SHEPHERD DOGS (GSD)**

**K Allenspach[1], A Kathrani[1], JS Suchodolski[2], A House[1], JM Steiner[2], N Gruetzner[2], D Werling[3].**
**Department of Veterinary Clinical Sciences[1] and Department of Pathology and Infectious Diseases[3], Royal Veterinary College, University of London, UK, and the Gastrointestinal Laboratory[2], Department of Small Animal Clinical Sciences, College of Veterinary Medicine, Texas A&M University, College Station, Texas.**

There is mounting evidence that inflammatory bowel disease (IBD) in people is the consequence of a deranged immune response to luminal bacteria in genetically susceptible individuals. The hypothesis of this study was that a primary innate immune defect involving pattern recognition receptor (PRR) function contributes to the pathogenesis of IBD in GSD and that similar disturbances in PRR expression profiles and the composition of the microbiota can be found in GSD with IBD as have been identified in people.

mRNA expression profiles of Toll-like receptor (TLR)2, TLR4 and TLR5 in duodenal biopsies from 13 GSD with IBD and 11 healthy ex-racing greyhounds were evaluated by quantitative real-time PCR. Simultaneously, the mucosa-adherent bacterial microbiota in the same dogs was assessed by obtaining cytology brush samples subjected to DNA extraction and PCR amplification using universal 16 rDNA primers. This was followed by construction of clone libraries. Genotyping for single nucleotide polymorphisms of canine TLR5 was performed in a total of 422 dogs, including 192 healthy GSD and 22 GSD with IBD. The exon of canine TLR5 was sequenced in 10 GSD with IBD and allele frequency between the groups was assessed using the SNPShot technique.

TLR2, TLR4 and TLR5 mRNA expression was significantly elevated in the GSD compared to healthy controls ($p<0.02$). Phylogenetic analysis of the mucosa-adherent microbiota revealed significant differences in microbial populations between both groups of dogs ($p<0.001$). Four non-synonymous SNP were identified in the canine TLR5 gene. The C100T SNP was present in 85% of the population in 96 different canine breeds, but in only 40% of the GSD population. So far, no difference in SNP frequency was identified between GSD with IBD and healthy GSD.

In conclusion, this study provides evidence that there is a similar dysregulation of PRR expression and the bacterial microbiota in the mucosa of GSD with IBD as has been described in people. In addition, the allele frequency of certian TLR5 SNP is different in the GSD population versus other breeds. Polymorphisms in PRR could in part be contributing to the pathogenesis of IBD in this breed.

**CANINE INTESTINAL INTRAEPITHELIAL LYMPHOCYTES – GOOD OR BAD GUYS?**

**N. Luckschander , N. Pfammatter, D. Sidler, I.A. Burgener, S. Jakob, A. Zurbriggen, N. Corazza, T. Brunner**
**Departement of Clinical Veterinary Medicine, Vetsuisse Faculty, University of Berne, Switzerland;**
**Division of Immunopathology, Institute of Pathology, University of Berne, Switzerland**

Little is currently known about the lymphocyte populations in the normal and diseased canine gut. The aims of this project were thus the phenotypical and functional characterization of canine intraepithelial lymphocytes (IEL), and to assess changes in dogs with chronic enteropathies (CE).
For the phenotypical characterization multi-parameter flow cytometry analysis (FACS) techniques were used. IEL were either isolated from full-thickness biopsies or endoscopical biopsies of adult Swiss Beagle dogs (n = 10 each group) and compared to lymph node cells. In addition, we compared endoscopical biopsies form healthy and dogs with CE (n = 6 per group). FACS analysis revealed that canine IEL consist of various unconventional lymphocyte subsets, unique to mucosal surfaces. These include $\gamma\delta$ T cells, $CD4^-CD8^-$ and $CD8\alpha\alpha^+$ T cells. While no statistical difference was observed between full-thickness and endoscopical biopsies, validating the sampling methodology, significant changes were found between healthy and diseased dogs. These included a significant increase in $CD4^-CD8^-$ ($p<0.05$) and decrease in $CD8\alpha\alpha$ and $CD4^+$ $TCR\alpha\beta^+$ IEL ($p<0.05$) in CE dogs.
As IEL in other species have been previously suggest to exhibit regulatory functions we investigated the role of IEL on lymph node T cell proliferation. While IEL alone did not show an activation-induced proliferation ($^3$H-thymidine incorporation), they significantly inhibited the proliferation of activated lymph node T cells in a cell number-dependent manner. These findings demonstrate for the first time that canine IEL have an immunoregulatory phenotype, which may contribute to the maintenance of the intestinal immune homeostasis and its loss in canine CE.

# CANINE $CD4^{+}CD25^{HIGH}$ T CELLS EXHIBIT SUPPRESSIVE PROPERTIES *IN VITRO* — FURTHER EVIDENCE OF REGULATORY T CELLS IN THE DOG

**DY Pinheiro, Y Singh, F Cunningham and OA Garden**
**Regulatory T Cell Laboratory, Department of Veterinary Clinical Sciences, The Royal Veterinary College, Camden Campus, Department of Veterinary Basic Sciences, The Royal Veterinary College, Hawkshead Campus, Department of Immunology, Imperial College London**

Naturally occurring regulatory T cells (T*regs*) play a key role in maintaining immune homeostasis. They are identified in mice and human beings on the basis of high constitutive expression of both the IL-2 receptor $\alpha$ chain (CD25) and the transcription factor FoxP3. Preliminary studies conducted by us and others have demonstrated FoxP3 expression by $CD4^{+}$ and $CD8^{+}$ $CD5^{+}$ T cells isolated from the blood and peripheral lymph nodes of systemically healthy dogs. Neutrophils and B cells showed negligible FoxP3 expression under the conditions examined.

When lymphocytes isolated from the popliteal lymph nodes were activated with concanavalin A (ConA) for 72 hours *in vitro*, there was an increase in proportional FoxP3 expression by $CD5^{+}$ T cells of both the $CD4^{+}$ and $CD8^{+}$ subsets (mean values: $CD4^{+}FoxP3^{+}$ – from 17.7% to 32.2% of all $CD5^{+}CD4^{+}$ T cells, n=2; $CD8^{+}FoxP3^{+}$ – from 1.4% to 8.2% of all $CD5^{+}CD8^{+}$ T cells, n=3). The removal of Con A stimulation and continued culture disclosed populations that were $FoxP3^{high}$ (median fluorescence intensity [MFI] = 787.9 in one representative experiment; n=3) and $FoxP3^{intermediate}$ (MFI = 90.6) after 72 hours – the former IFN-$\gamma^{-}$ and the latter IFN-$\gamma^{+}$.

FoxP3–pre-enriched lymphocytes were sorted on the basis of both $CD4^{+}$ and $CD25^{high}$ expression by FACS™ and placed in an assay system in which magnetically-isolated $CD4^{+}$ T cells were used as an effector population. Titratable suppression of $CD4^{+}$ T cell proliferation by the sorted cells was demonstrable within 48 hours, with enhanced suppression at 72 hours.

In summary, healthy dogs possess T cells sharing the phenotypic and functional characteristics of naturally occurring T*regs* of man and mouse. Further work is ongoing to interrogate more fully the functional properties of these cells *in vitro*, within both allogeneic and xenogeneic contexts.

# FELINE OBESITY: INFLAMMATORY, HEMATOLOGIC & METABOLOGIC ALTERATIONS

**Charlotte R. Bjornvad, DVM, PhD, Resident ESVCN**
**Copenhagen, Denmark**

## INTRODUCTION

Obesity is an increasing problem in the companion animal population. In cats, Bodyweight is generally not sufficient for estimating body condition. Several clinical systems for estimating body composition has been proposed and of these, the feline BMI, the 5 and 9 point body condition systems has been most extensively used. They all show good correlation to dual energy X-ray measurements of body fat mass percentage, good repeatability and limited interobserver bias (German et al., 2006; LaFlamme, 1997). Despite that these systems are being promoted by nutritionists, researchers and petfood industry they are not yet an integrated part of the common clinical examination in most practices (German and Morgan, 2008).

## FACTS OF FELINE OBESITY

Recent reports from the United States, Australia and France support the contention that feline obesity is a global problem (Colliard et al., 2009; Laflamme et al., 2008; Lund et al., 2005). The prevalence of overweight cats with a body condition score (BCS) of 6-7 out of 9 was 19-29% and obese cats (BCS 8-9/9) was 6.5- 7.8%. Compared with lean cats, overweight cats have an increased prevalence of urinary tract infections and oral disease while obese cats have an increased prevalence of diabetes mellitus, neoplasms, dermatopathy and oral disease (Lund et al., 2005). Factors supporting a sedentary lifestyle such as neutering and indoor confinement predispose to obesity in cats (Colliard et al., 2009; Slingerland et al., 2007; Lund et al., 2005). Neutering results in both a decreased energy expenditure due to lower physical activity and an altered appetite regulation resulting in an increased food intake (Belsito et al., 2009; Kanchuk et al., 2003; Martin et al., 2001). Furthermore, being male, middle age (5-11 years) and a domestic shorthaired or mixed breed increases the risk of overweight. Feeding premium and therapeutic foods has, in some studies, correlated with obesity, but this may be due to the confounding factor that obese cats may be fed prescribed weight loss diets. Furthermore, dry food and high levels of carbohydrates have been theorized to increase the prevalence of obesity, but this connection has not been verified (Buffington, 2008; Slingerland et al., 2007).

## FAT AS AN ENDOCRINE ORGAN

Until recently, body fat was believed to be an inert repository for lipids. In addition to energy storage, white adipose tissue is now recognised as being an important endocrine organ. Adipocytes produce hormones (adipokines) that are important mediators for metabolism and inflammation (Rasouli and Kern, 2008). Adipokines are continuously being discovered and only a few will be discussed in this abstract. Leptin is a 16 kDa peptide produced primarily by adipose tissue (Lazar, 2005). It acts centrally to regulate appetite and energy expenditure and has peripheral effects to coordinate whole body energy metabolism (Houseknecht and Portocarrero, 1998). The plasma leptin concentration is proportional to white adipose tissue volume and it has been shown to influence growth hormone levels, immunology homeostasis, hematopoesis, angiogenesis, wound healing, osteogenesis and gastrointestinal function. Decreased leptin signalling or receptor function increases energy intake and lowers energy expenditure. Nevertheless, for reasons not fully understood, high levels of leptin do not suppress appetite in obese individuals. Retinol-binding protein 4 is produced primarily in adipocytes, serum levels increase with obesity and it may enhance insulin resistance (Graham et al., 2006). Resistin is a 12 kDa peptide produced in mononuclear cells, adipocytes, muscle and pancreatic cells. It acts primarily as a proinflammatory mediator and may be involved in insulin resistance (Graham et al., 2006). Adiponectin is produced in adipocytes but interestingly, serum concentrations are decreased in morbidly obese individuals. Adiponectin increases insulin sensitivity, augments endothelial nitrous oxide production and inhibits macrophage activation (Rasouli and Kern, 2008).

## OBESITY & METABOLIC ALTERATIONS IN HUMANS & RODENT MODELS

In humans, obesity is considered an epidemic disease, decreasing life longevity and quality due to adverse consequences such as type 2 diabetes and resulting complications, cardiovascular disease and certain cancers. Abdominal obesity, insulin resistance, dyslipidemia (hypercholesterolemia), and/or high blood pressure together referred to as the metabolic syndrome predispose to development of type 2 diabetes and cardiovascular diseases. These factors may seem unrelated, but increasing evidence suggests that a common pathogenesis is closely linked to low grade inflammation, partly relayed by adipose derived cytokines such as leptin, resistin, adiponectin, TNFα and interleukin 6 (Dandona et al., 2005). This low grade inflammation results from chronic activation of the innate immune system and can subsequently lead to insulin resistance, impaired glucose tolerance and eventually diabetes (Bastard et al., 2006). In obesity, white adipose tissue is characterized by an increased production and secretion of a wide range of inflammatory molecules including leptin, TNFα, interleukin 6 (IL-6) and resistin compared to the adipose tissue in the non-obese individual. Conversely, the secretion of adiponectin, an insulin sensitizing effector is down regulated. Recent data indicate that the white adipose tissue in obese subjects is infiltrated with macrophages that may be a contributing source of the locally produced proinflammatory cytokines (Bastard et al., 2006). Leptin resistance is a hall mark in rodent and possibly human obesity and appears to be due to defects in leptin-receptor signalling as well as impaired transport into the brain (Schmidt et al., 2006). Leptin could modulate TNFα production and macrophage activation resulting in increased levels of TNFα and IL-6 (Bastard et al., 2006; Lazar, 2005). TNFα and IL-6 may then alter insulin sensitivity by triggering different key steps in the insulin signalling pathway (Bastard et al., 2006). In addition TNFα and IL-6 activates the acute phase reactants and the haemostatic system through tissue factor resulting in a procoagulant stage (Aso et al., 2003). The high level of circulating free fatty acids as well as retinol binding protein 4 (RBP4) and other recently discovered adipokines may all be additional contributors to the development of insulin resistance (Lazar, 2005). In rodents, the interaction between inflammation and insulin signalling is supported by the ability of aspirin to improve insulin resistance, in part by preventing the antagonistic effects of fatty acids and cytokines (Yuan et al., 2001).

## OBESITY & METABOLIC ALTERATIONS IN CATS

Pathologically, feline obesity seems to share similarities with human obesity. Feline obesity is characterized by dyslipidemia (Jordan et al., 2008), insulin resistance (Hoenig et al., 2007) and increased insulin secretion (Lutz and Rand, 1996) that may result in islet amyloid formation and β-cell loss (Lutz and Rand, 1997) leading to inadequate insulin production (Lutz and Rand, 1996) and overt type 2 diabetes. Serum leptin and TNFα concentrations are increased in obese cats compared with lean while adiponectin levels are decreased (Hoenig et al., 2007; Appleton et al., 2000) and the insulin receptor GLUT4 but not GLUT1 expression is decreased in both muscle and fat tissue in obese cats (Brennan et al., 2004). Another strong link between feline and human obesity is the fact that the domestic cat develop diabetes mellitus that is clinically and pathologically similar to human type 2 diabetes (Lutz and Rand, 1993). Furthermore, obesity and physical inactivity are the main predisposing factors for diabetes development in both species (Henson and O'Brien, 2006). Nevertheless, important differences also exist. Obesity does not decrease HDL cholesterol in cats as in humans and obesity does not cause atherosclerosis and cardiovascular diseases in cats.

## CONCLUSION

Feline obesity is an increasing global problem where neutering and physical inactivity are some of the major predisposing factors. Obesity in cats leads to metabolic alterations and possibly low grade inflammation that ultimately increases the risk of developing type 2 diabetes and certain cancers. Understanding these metabolic alterations may lead to improved preventive and therapeutic strategies.

## REFERENCES

[1]German AJ, et al. **J Nutr** 2006;136: 2031S-2033S. [2]LaFlamme DP, **Feline Practice** 2007;25:13-18. [3]German AJ & Morgan LE, **Vet Record** 2008;163:503-5. [4]Colliard L et al. **J Feline Med Surg** 2009;11:135-40. [5]Lund EM et al. **Intern. J Appl Res Vet Med** 2005; 3:88-96. [6]Laflamme DP et al. **J Am Vet Med Assoc** 2008;232:687-694. [7]Slingerland LI et al.**Vet J** 2007;179:247-53. [8]Kanchuk ML et al. **J Nutr** 2003;133:1866-74. [9]Martin L et al. **J Anim Physiol Anim Nutr** 2001;85:195-199. [10]Belsito KR et al. **J Anim Sci** 2009;87:594-602. [11]Buffington CA **Can Vet J** 2008;49:561-63. [12]Rasouli N & Kern PA **J Clin Endocrinol Metab** 2008;93: S64-S73.[13]Lazar MA **Science** 2006,307:373-375. [14]Houseknecht KL & Portocarrero CP **Domest Anim Endocrinol** 1998;15:457-75. [15]Graham TE et al. **N Engl J Med** 2006;354:2552-2563. [16]Dandona P et al. **Circulation** 2005;111:1448-54.[17]Bastard JP et al. **Eur Cytokine Netw** 2006;17: 4-12. [18]Schmidt MI et al. **Diabetologia** 2006;49:2086-96. [19]Aso Y et al. **Diabet Med** 2003;20:930-34.[20]Yuan M et al. **Science** 2001;293:1673-77. [21]Jordan E et al. **Domest Anim Endocrinol** 2008;35:290-99. [21]Hoenig M et al. **Am J Physiol Regul Integr Comp Physiol** 2007;292:R227-34. [22]Lutz TA & Rand JS **Can Vet J** 1996;37:27-34. [23]Lutz TA & Rand JS **J Comp Pathol** 1997;116:157-70. [24]Appleton DJ et al. **J Feline Med Surg** 2000;2:191-9. [25]Brennan CL et al. **Domest Anim Endocrinol** 2004;26:291-301. [26]Lutz TA & Rand JS **Br Vet J** 1993;149:527-536. [27]Henson MS & O'Brien TD **Ilar Journal** 2006;47:234-242.

# CATS & CALCIUM: THE CAUSES & CONSEQUENCES OF HYPERCALCEMIA

Audrey K. Cook, BVM&S, MRCVS, DACVIM, DECVIM-CA
College Station, TX

## INTRODUCTION

There are three forms of circulating calcium: *bound* to serum proteins (approximately 40%), *complexed* with various anions such as citrate and phosphate (approximately 8%), and *ionized* (iCa; approximately 52%). The latter is the biologically active form and the term 'hypercalcemia' should only be used when this fraction is elevated. The calcium value reported on a standard serum biochemical profile reflects the total calcium (tCa) value (i.e., bound, complexed and ionized); this is a useful indication of calcium homeostasis but is influenced by serum protein concentrations and acid-base status. Formulas designed to 'correct' for serum protein concentrations have not been validated in cats.

## CALCIUM HOMEOSTASIS

Over 99% of the body's calcium is stored in bone; less than 0.1% in the extracellular fluid. However, iCa must be tightly regulated as it impacts numerous metabolic functions. Under normal circumstances, calcium intake and loss are carefully balanced, and about 90% of ingested calcium is excreted through the gastrointestinal (GI) tract. If necessary, the GI tract can excrete >100% of intake, as calcium is added to secretory fluids. The kidneys make small adjustments to calcium homeostasis; active renal secretion does not occur, so a balance exists between glomerular filtration and tubular uptake. Usually, 99% of filtered iCa is reabsorbed, primarily by the proximal tubules, where it essentially mirrors the handling of sodium. Hormonal factors impact reabsorption in the later sections of the nephron, but 24-hour urinary calcium excretion is principally determined by serum concentration.

Parathyroid hormone, calcitonin and vitamin D are the major hormonal regulators of calcium uptake, excretion, and storage.

### Parathyroid Hormone (PTH)

The essential role of PTH is to increase serum calcium concentration. This is achieved primarily through stimulation of osteoclast activity and release of calcium from bone. Osteoclasts do not have receptors for PTH, so this response occurs through an unidentified intermediary. Bone reabsorption falls dramatically in the absence of PTH. PTH also increases renal reabsorption of calcium in the loop of Henle and distal tubules, whilst promoting the excretion of phosphorus. There is a feedback loop between PTH and 1,25-hydroxycholecalciferol (calcitriol) in which PTH promotes the renal hydroxylation of 25-hydroxycholecalciferol, and calcitriol deficiency triggers the release of PTH, independent of serum calcium concentrations. The release of PTH is stimulated by both decreased iCa and increased serum phosphate levels.

### Calcitonin

This hormone is released in response to increased iCa. It decreases serum calcium concentrations through inhibition of osteoclast activity and increased renal excretion of calcium. However, calcitonin has limited biological potency and maximal secretion cannot counterbalance a potent hypercalcemic drive.

### Vitamin D

Vitamin D promotes the intestinal uptake of calcium and phosphorus; absorption of ingested calcium is highly dependent on adequate amounts of this vitamin. Cats obtain vitamin D from food as cholecalciferol. This form has little biological activity and must undergo hydroxylation, first in the liver and then in the kidney. Hepatic hydroxylation is influenced by serum concentrations of its product, but the second hydroxylation step is limited by renal $\alpha$ 1 hydroxylase activity. This is stimulated by PTH but inhibited by hyperphosphatemia and (to a modest extent) by hypercalcemia. At physiological concentrations, 1,25-hydroxycholecalciferol promotes bone calcification; excessive amounts cause bone reabsorption.

Numerous other factors also influence calcium homeostasis, including hepatic, renal and GI function, adrenocortical hormones (cortisol and aldosterone), thyroid hormone, and serum sodium, phosphate and magnesium concentrations. Bone contains exchangeable calcium for immediate buffering, along with providing vast storage capabilities. The interplay between these hormones, organs and ions is complex, and our understanding of the full intricacies of calcium homeostasis is still incomplete.

## THE CAUSES OF HYPERCALCEMIA

In general, hypercalcemia is the result of increased gastrointestinal uptake and/or excessive release of calcium from bone. Decreased renal excretion may play a minor role. The most important considerations in feline medicine include the following:[1]

### Hyperparathyroidism

Primary hyperparathyroidism is usually due to a benign adenoma of one of the 4 parathyroid glands. A distinct cervical mass may be palpable. Siamese cats have been over-represented in several case studies, but a clear genetic trait has not been identified. The diagnosis is simple if the PTH concentration is above the upper limit of normal; however, in some patients the PTH is still "normal" but is inappropriate in light of the concurrent hypercalcemia. If the result is unclear, the tests should be repeated after 6 weeks to determine if iCa and PTH are trending upwards. As excessive PTH promotes phosphate excretion, hypophosphatemia is often noted; this appears to protect against renal compromise. Removal of the affected gland results in a rapid normalization of iCa.

Primary hyperparathyroidism in human beings may be managed successfully with calcimimetics. These novel drugs bind specifically to calcium receptors in parathyroid tissue and are a potent inhibitor of PTH release. In one study, 16 human patients with solitary adenomas had normalization of iCa within 24 hours.[2] This approach may prove to be a useful therapeutic option for our patients. It also provides some insight into the biological behavior of PTH-secreting adenomas, as (at least in people) the secretory activity of these tumors is still influenced by iCa. Human clinical trials are underway, but these agents are not yet available.

Increased iCa and PTH have been described in a subset of cats with chronic kidney disease (CKD); this is referred to as tertiary hyperparathyroidism (see "Renal"). These cats differ from patients with primary hyperparathyroidism as they are substantially azotemic (IRIS stage 3 or 4) and hyperphosphatemic. In addition, hypertrophy of all 4 parathyroid glands will occur and may be identified ultrasonographically.

**Idiopathic (IHC)**

This term describes cats with (generally) mild to moderate hypercalcemia (i.e., tCa < 15 mg/dl) for which no underlying cause can be identified.[3] PTH concentrations are subnormal or in the lower half of the reference range, vitamin D levels (25-hydroxycholecalciferol and calcitriol) are normal and parathyroid hormone-related protein (PTH-rp) is undetectable. In most cases, serum phosphate concentrations are normal, but may be increased if renal function is compromised. Over the last decade, this syndrome has become the most common cause of hypercalcemia in cats n the USA. Some of these cats appear clinically normal whilst others have non-specific signs such as mild weight loss, vomiting and constipation. Affected cats are predisposed to calcium oxalate urolithiasis due to increased calciuresis.

It appears that IHC cats are a heterogeneous group and there is presently no consensus about how to manage these patients or when action is necessary. As a broad guideline, cats with tCa > 13 mg/dl, with urolithiasis, or with evidence of CKD merit intervention. Although several therapeutic strategies have been proposed, consistent responses have not been reported. Ideally, iCa would be maintained at a normal level through modification of net GI uptake or inhibition of osteolysis. Any therapy that lowers iCa through increased calciuresis may promote urolithiasis in susceptible patients.

As the early descriptions of cats with IHC reported the frequent use of acidifying diets, dietary modification has been recommended.[4] Some cats do improve on alkalinizing diets; however, these diets generally contain less calcium and phosphorus than maintenance foods and these factors (rather than simply pH) may be beneficial. As an added consideration, alkalinizing diets decrease calciuresis and inhibit formation of calcium oxalate uroliths. High-fiber diets also have been suggested, on the basis that rapid GI transit may limit calcium absorption. Once again, some cats show a positive response. It is interesting to note however that many high-fiber diets are calcium-rich, in a deliberate attempt to counteract this effect. Information about the vitamin D content of commercial cat foods is limited, so the impact of this dietary variable is not easily evaluated.

Some patients with IHC achieve a normal iCa with glucocorticoid therapy, although high doses (prednisolone at 10 mg/cat/day) may be necessary. This probably reflects decreased intestinal absorption of calcium. Increased calcium loss in the urine also may occur, although a short term study indicated that prednisolone did not promote calciuresis in normal cats.[5] However, the consequences of chronic steroid administration are substantial and the putative benefits should be weighed carefully against the risks.

Bisphosphonates have been proposed for cats with IHC.[4,6] These drugs block specific enzyme pathways within osteoclasts, thereby inhibiting bone reabsorption. They are used more commonly for patients with acute and severe hypercalcemia mediated by PTH-rp or due to vitamin D toxicity. Both oral and injectable forms are available; the oral forms are associated with severe esophageal damage and should be administered very carefully. In theory, a sustained positive response to bisphosphonates seems unlikely if PTH is suppressed, as osteoclast activity in normal individuals is minimal when PTH is scarce.

**Renal**

CKD is listed as a common cause of feline hypercalcemia.[7] However, many studies documenting this association relied upon tCa concentrations; iCa measurements were not reported. As complexed calcium concentrations are often increased in cats with CKD, it is likely that many CKD patients with elevated tCa are not truly hypercalcemic. However, some cats with CKD do have elevated iCa, in combination with increased PTH concentrations. This is often referred to as tertiary hyperparathyroidism. The reason for PTH hypersecretion in this patient subset is unclear. It has been suggested that prolonged stimulation alters the set-point for PTH release and/or that calcium receptor expression within the parathyroid tissue is compromised.[8] In essence, serum calcium concentration becomes dependent on PTH secretion; it is no longer the determinant. Interestingly, this syndrome sometimes persists in human patients despite successful renal transplantation, suggesting that affected parathyroid glands become truly and permanently autonomous.

Calcitriol appears to inhibit PTH synthesis in normal circumstances and has been proposed as a therapeutic option in cats with tertiary hyperparathyroidism. However, calcitriol receptor density in parathyroid tissue is decreased in uremic human patients, thereby mitigating the response to calcitriol supplementation. The effect of calcitriol in cats with this syndrome has not been reported, although one study in cats with renal (secondary) hyperparathyroidism did not show a positive response.[9] It is essential to control hyperphosphatemia before considering calcitriol, because there is a substantial risk of tissue mineralization and further compromise to renal function. Phosphate, iCa and PTH concentrations should be monitored regularly and the calcitriol dose titrated slowly upwards as appropriate. If iCa increases, the drug must be discontinued immediately.

Parathyroidectomy is a recognized treatment for tertiary hyperparathyroidism in human beings who fail medical therapy.[10] Intraoperative PTH measurements are performed so that the appropriate amount of tissue is removed and a portion of a gland is cryopreserved for implantation later if hypoparathyroidism occurs. Symptomatic improvement (reduced fatigue and less musculoskeletal pain) often occurs within 24 hours of surgery. The technical requirements of this procedure certainly limit its use in feline patients.

Although they are still under investigation, calcimimetics warrant consideration in patients with tertiary hyperparathyroidism. Aberrant PTH release may be regulated effectively without the possible complications associated with calcitriol.

**Tumors**

Neoplasia is an important cause of hypercalcemia in cats. A number of tumor types are listed, with lymphoma and squamous cell carcinoma being most common. The mechanism usually is due to release of a fetal protein called PTH-related protein (PTH-rp), which mimics the effect of PTH. PTH-rp can be assayed in feline serum; a positive result confirms hypercalcemia of malignancy.[11] However, an undetectable value does not exclude an underlying tumor as other paraneoplastic responses may be present. In most patients, a thorough physical examination, abdominal ultrasonography and thoracic radiographs are adequate to locate the malignancy. Bone marrow aspiration, bone scans, and organ/lymph node aspiration may be necessary in the more challenging cases.

In the short term, hypercalcemia should be managed with saline diuresis, loop diuretics and injectable bisphosphonates. Although glucocorticoids may decrease serum calcium concentrations, they should not be administered until a definitive diagnosis is made. Hypercalcemia usually resolves rapidly with chemotherapy in cats with lymphoma. Patients with inoperable carcinomas or other non-lymphoid neoplasms may be managed with glucocorticoids and subcutaneous fluids. Chronic bisphosphonate therapy should be considered as these drugs may inhibit tumor angiogenesis, whilst controlling hypercalcemia.[12]

## CONSEQUENCES OF HYPERCALCEMIA

The clinical signs associated with hypercalcemia are often vague and non-specific, and include lethargy, vomiting, anorexia and constipation. There are many explanations for GI dysfunction. Decreased motility may be a factor and might in turn increase calcium uptake by the intestine. Hypercalcemia in human beings is associated with pancreatitis and experimental studies in cats indicate that acute hypercalcemia destroys pancreatic acinar cells and increases the permeability of the pancreatic duct.[13,14] However, a clinical correlation between spontaneous hypercalcemia and pancreatitis has not been demonstrated. Similarly, peptic ulcers are associated with hypercalcemia in people, probably as a result excessive acid secretion, but this has not been reported in cats.

Even small increases in iCa result in marked calciuresis. Consequently, many cats present with signs related to calcium oxalate urolithiasis.[15,16] Dysuria and hematuria may be reported if stones form in the urinary bladder. Abdominal pain and renal compromise may occur with an obstruction in the pelvis or ureter. Because of the close connection between calcium disorders and stone formation, all hypercalcemic cats should be screened for stones with abdominal imaging, and iCa should be measured in any cat with urolithiasis.

Independent of the etiology of hypercalcemia, renal damage is a major cause of patient morbidity. This is somewhat dependent on serum phosphate concentrations, but can occur regardless of the calcium-phosphorus product if hypercalcemia is severe. The classic assumption that renal damage occurs due to aberrant mineralization is certainly valid, but excessive calcium probably has many subtle and complex toxic effects on tubular cell function.

Severe hypercalcemia is associated with abnormal neuromuscular function, CNS compromise, and cardiac arrhythmias. Rapid increases in serum calcium can be fatal.

## EVALUATING THE HYPERCALCEMIC CAT

The initial approach to the hypercalcemic cat is influenced by several factors. These include the magnitude of the hypercalcemia, the serum phosphorus concentration, renal status, and the severity of clinical signs. If hypercalcemia is severe (i.e., tCa > 15 mg/dl), if the calcium-phosphorus product is > 70, or the patient is azotemic, it is appropriate to start supportive care (i.e., saline fluid therapy, +/-loop diuretics, +/-bisphosphonates) whilst diagnostic tests are performed.[17]

The diagnostic approach should be guided by patient history and physical examination findings, along with the results of the complete blood count, biochemical profile, and urine analysis. If the serum phosphate is increased, vitamin D toxicity, osteolytic disease, and renal failure are considered likely. Commercial assays for vitamin D are readily available and should be considered if imaging studies are inconclusive. However, it is important to remember that the standard assay measures 25-hydroxycholecalciferol (i.e., 25-hydroxyvitamin D); synthetic vitamin D analogues (e.g., calcipotriene) and 1,25-hydroxycholecalciferol are not detected.

If the serum phosphate level is low or subnormal, hypercalcemia of malignancy or primary hyperparathyroidism should be considered. In general, thoracic and abdominal imaging studies will identify most cats with neoplasia, and should be performed early in the diagnostic evaluation. The next step is the measurement of PTH and PTH-rp. Depending on these findings, cervical ultrasonography may be appropriate.

A diagnosis of idiopathic hypercalcemia can only be made when all other possibilities have been satisfactorily excluded.

## REFERENCES

[1] Schenck PA, et al. In: August J. **Consultations in Feline Internal Medicine**. 5th ed. St. Louis, Mo.: Elsevier, 2005;157. [2]Shoback DM, et al. **J Clin Endo Met** 2003;88(12):5644. [3]Midkiff AM, et al. **J Vet Intern Med** 2000;14(6):619. [4]Chew DJ, et al. In: **Proceedings of NAVC** 2007;732. [5]Geyer N, et al. **Vet Ther** 2007;8(4):239. [6]Hostulter RA, et al. **J Vet Intern Med** 2005;19(1):29. [7]Savary KC, et al. **J Vet Intern Med** 2000;14(2):184. [8] Parfitt AM. **Kid Int** 1997;52:3. [9]Hostulter RA, et al. **J Vet Intern Med** 2006;20(6):1307. [10] Milas M, et al. **Surgery** 2004;136(6):1252. [11]Bollinger AP, et al. **Vet Clin Pathol** 2002;31(1):3. [12]Wypij JM, et al. **J Vet Intern Med** 2008;22(1):158. [13]Frick TW, et al. **Eur J Surg** 1992;158(5):289. [14]Cates MC, et al. **Surgery** 1988;104(2):137. [15]Kyles AE, et al. **J Am Vet Med Assoc** 2005;226(6):932. [16]Aronson LR, et al. **J Am Vet Med Assoc** 2006;228(5):743. [17]Shaer M. **Vet Clin North Am Small Anim Pract** 2008;38(3):513.

## WHAT CAN VETERINARIANS LEARN FROM NEW FINDINGS IN HUMAN TYPE 1 DIABETES?

**Lucy J. Davison MA VetMB PhD DSAM DipECVIM-CA MRCVS**
**Cambridge, UK**

### SUMMARY

In answering the question 'What can veterinarians learn from new findings in human type 1 diabetes?', first it is necessary to review briefly what is known of the pathogenesis of human type 1 diabetes (T1D), and then compare it to the disease thought by many to be its veterinary counterpart, canine diabetes mellitus (DM). Following the comparison of the latest knowledge of disease pathogenesis in both species, the lecture will focus on the recent 'genetic revolution', which is taking place in many complex human diseases, including T1D. This will involve discussion of how new genes have recently been associated with T1D in genome-wide association studies (GWAs), as well as the potential implications for our knowledge about mechanisms of pancreatic islet cell destruction. The final part of the lecture will be devoted to the specific lessons that veterinarians are learning from this GWA approach to understanding complex diseases such as T1D.

### THE PATHOGENESIS & GENETICS OF HUMAN T1D

It is estimated that around 20 million people worldwide suffer from type 1 diabetes (T1D), mostly young adults and children, and the incidence of the disease is increasing[1]. The autoimmune destruction of pancreatic islets in T1D has been described as a multi-step process : genetic susceptibility, triggering event (e.g. drugs, virus infection), active autoimmunity, progressive loss of insulin response to glucose, overt diabetes with residual insulin, and finally complete beta cell destruction[2]. Convincing evidence for autoimmunity in humans and rodents includes the presence of circulating autoantibodies to pancreatic proteins such as insulin, proinsulin GAD-65 (glutamic acid decarboxylase 65) and IA-2 (insulinoma antigen-2) in the serum of diabetic patients. These autoantibodies can precede clinical signs of diabetes by months to years and are suspected to be a marker of disease rather than directly involved in beta cell destruction[3].

Genetic susceptibility to T1D is particularly associated with the genes of the Major Histocompatibility Complex (MHC), known as the *HLA (Human Leukocyte Antigen)* genes[4]. The association between *HLA* and T1D in humans was first reported almost 30 years ago and certain *HLA* alleles can confer susceptibility to or protection from T1D. The critical role of the MHC genes, which are involved in antigen presentation to T cells, underlines the importance of the adaptive immune response in T1D. Polymorphisms in other immune function and pancreatic genes such as *CTLA-4 (cytotoxic lymphocyte antigen-4)* and *INS (insulin)* have also been associated with an individual's T1D risk.

It must also be remembered that certain non-genetic factors are thought to be important in triggering immune-mediated diseases. Since the concordance rate for T1D in human identical twins is less than 50%, environmental factors are believed to play an important role in human diabetes. Infectious organisms such as Coxsackie B virus, Epstein Barr virus, enterovirus and rotavirus are thought by some authors to play a role in triggering autoimmunity as well as possible dietary and geographical influences[5].

### THE PATHOGENESIS & GENETICS OF CANINE DM

It has been suggested that dogs suffer from a disease equivalent to human T1D, implying that all diabetic dogs are insulin deficient as the result of autoimmune pancreatic beta cell destruction. This is because clinically, T1D shares many similarities with canine DM and recent research has demonstrated some underlying similarities in pathogenesis between the two species[6]. In reality, however, canine DM is a heterogeneous disease, with multiple pathological processes leading to similar clinical signs. For example, as well autoimmune pancreatic destruction, congenital anomalies resulting in DM have been reported in dogs less than 6 months of age, (although such cases are rare). More commonly, acute or chronic pancreatitis can result in canine DM following damage to the pancreatic endocrine tissue. Additionally, canine DM can occur secondary to hormonal antagonism and insulin resistance during dioestrus, when growth hormone is synthesised in the canine mammary glands or in canine hyperadrenocorticism.

In canine DM, the evidence for autoimmunity is less convincing than that in humans and rodents, although only a small number of studies have been undertaken. A proportion of canine diabetic sera will show autoreactivity by immunofluorescence to pancreatic islet sections and are capable of stimulating complement-mediated beta cell lysis. Evidence of antigen specific reactivity to insulin, proinsulin, GAD65 and IA-2 has also been demonstrated in the sera of a small proportion of canine DM patients[7].

Certain breeds of dog seem to be predisposed to developing DM, suggesting that the canine disease has a genetic component. In the UK, studies have shown that the Samoyed, Tibetan terrier and Cairn terrier have the highest odds ratio for developing disease, whereas certain breeds such as the Boxer and Golden retriever seem to be at decreased risk[8]. The *Dog Leukocyte Antigen* (*DLA*) genes are the canine equivalent of the *HLA* system and certain canine *DLA* haplotypes (combinantions of alleles) have been associated with canine DM risk[9]. It is important to note that any studies undertaken to assess the genetics of human T1D rely on accurate classification of the patient as a 'Type 1' diabetic patient by assessment of serum autoantibodies and a young age at presentation. In canine studies, however, this classification is more difficult and currently relies mainly on exclusion of other factors such as pancreatitis, dioestrus and other forms of hormonal antagonism.

### WHAT IS THE 'GENETIC REVOLUTION'?

The study of genetics of a complex disease allows further understanding of its pathogenesis and, most importantly, has the potential to reveal new therapeutic targets. Previously, investigation of T1D genetics has relied on methods such as linkage analysis within families or the investigation of individual 'candidate genes' for testing differences between patient and control samples. The

MHC proved to be an excellent candidate, since polymorphisms in the region encoding the antigen binding / peptide presenting pocket were found to have a direct influence on the adaptive immune system, the development of autoimmunity and hence T1D susceptibility. Similarly, disease-associated polymorphisms within the insulin gene have been identified which affect insulin transcription within the thymus and hence the ontogeny of insulin-autoreactive T lymphocytes.

This approach, however, is labour intensive and has little chance of discovering genes of novel function. The classical model of the genetics of a complex disease suggests that a small number of genes have a large effect on disease susceptibility (e.g. the *HLA* genes account for just under 50% of a person's genetic T1D risk), whereas a larger number of genes each make a smaller contribution. Within the last 3 years, since the start of the so-called 'genetic revolution', advances in technology have made it possible to identify new T1D genes with small but significant effects on disease susceptibility, using genome-wide association (GWA) studies in large sample cohorts. The difference between diabetes risk and susceptibility within a gene can sometimes be associated with a polymorphism in only a single nucleotide of the gene, known as a SNP. It is also interesting to note that these disease-associated SNPs can be intronic as well as exonic and hence do not always alter the amino acid sequence of the protein. Disease-associated SNPs can affect downstream pathways in many ways, including control of gene transcription or splicing.

## WHAT FACTORS WERE NECESSARY TO ALLOW THIS 'GENETIC REVOLUTION' TO COME ABOUT?

A GWA performed by the Wellcome Trust Case Control Consortium (WTCCC), published in 2007, genotyped 2000 human T1D cases and 3000 controls for 500,000 SNPs, throughout the whole genome[10]. GWAs require a 'gene-chip', which can identify the genotyope of each individual sample at thousands of SNPs. Genotypes are then compared in patients and controls and statistical analysis performed to identify any significant association with disease. This gene-chip technology was the critical factor in the WTCCC being able to test such a large number of T1D case and control samples in a cost-effective manner. In addition, design of a gene-chip requires knowledge of which base pairs within a gene are polymorphic between individuals (i.e. contain SNPs), and this effort was therefore underpinned by on-line bioinformatic resources such as 'Hap-Map' and 'dbSNP'[11]. These databases contain information generated by consortia attempting to map the variation in the human genome. They act as a reference map for the human genome and are crucial in the selection of which SNPs to test. They provide detailed information such as which SNPs are the most variable within populations, the frequencies of each allele, and the degree of linkage disequilibrium (LD) between SNPs. The final necessary factor in a GWA study is the availability of a large number of well-phenotyped DNA samples from patients and controls, to provide adequate power to detect smaller genetic effects. In human T1D, and the other 6 diseases investigated by the WTCCC, this involved collaborative efforts between many laboratories, involving both clinicians and researchers.

## WHAT NEW FINDINGS HAVE COME FROM GWA STUDIES IN HUMAN T1D?

The initial analysis of WTCCC data reassuringly revealed genetic associations with T1D which had previously been identified by other techniques, such as the *HLA* genes. Importantly, however, several other previously unidentified genetic regions were shown to contain SNPs that significantly correlated with T1D risk, such as *CD69* (involved in lymphocyte proliferation and signalling), *KIAA0350* (of unknown function) and *PTPN2* (a protein tyrosine phosphatase). The second phase of the WTCCC analysis is currently underway, involving re-sequencing of the associated regions in an attempt to fine-map the T1D associated signals. A meta-analysis of 3 recent GWA in T1D, involving a total of 3561 cases and 4646 controls, undertaken by the Type 1 Diabetes Genetics Consortium (T1DCG) has revealed further candidates for T1D susceptibility such as *BACH2* (a transcription factor found in B lymphocytes)[12]. Following GWA studies, there are now more than 40 genetic regions known to be associated with T1D, compared to only 5 discovered by more traditional methods.

## WHAT ARE THE CLINICAL IMPLICATIONS OF THESE FINDINGS?

Some of these newly discovered regions encode proteins with defined roles within the immune system or pancreas, whereas others are of unknown function, and therefore investigation of their biology may offer new insights into autoimmune disease. The challenge following a GWA study is to identify exactly how an increased risk genotype translates to an increased risk phenotype and particularly whether it is possible to intervene clinically at any point in this process. Some such genotype-phenotype correlations have already been made in T1D. For example, a SNP in the T cell regulatory gene *CTLA-4* has been shown to affect the relative levels of the soluble versus membrane bound version of this molecule, which acts as a negative regulator of the immune system[13].

As well as providing new insights into disease pathogenesis and potential therapeutic targets, the results of GWAs will improve the identification of genetically 'at risk' or 'protected' individuals for recruitment into T1D preventation clinical trials, since ideally it would be best to prevent the disease rather than try to stop pancreatic autoimmunity once it has been initiated. Good candidates for preventative strategies would be those individuals carrying multiple SNPs associated with high genetic T1D risk, who could be recruited even before they develop autoantibodies.

Finally, genes which are found to be associated with one autoimmune disease by GWA, might also be important in other autoimmune or immune mediated diseases. For example, since their discovery by GWA study, some novel T1D genes have also been shown to be associated with multiple sclerosis[14] and celiac disease[15].

## WHAT LESSONS CAN VETERINARIANS LEARN FROM THESE STUDIES?

Studies of human T1D remind us that as in human medicine, canine diseases with a genetic basis could be the result of a few major genes with large effects plus many other genes with smaller effects. We already know that the canine MHC genes and some cytokine genes are involved in canine DM susceptibility within certain breeds, but there are many further genes waiting to be discovered. As veterinarians, we are in the unique position of being able to study individual breeds, in some cases with relatively small gene pools. The population structure of purebred dogs lends itself very well to the identification of disease-associated genes by GWA with fewer markers and fewer individuals needed compared to human studies[16]. The fact that most specific dog breeds are <200 years old means that their genome contains very large LD blocks of alleles that tend to be inherited together. There are also many examples of canine genes in which there is a lot of inter-breed variation but minimal intra-breed variation, meaning that a large case-control GWA study of T1D could be very powerful, especially if performed in pure breed dogs. A further argument for breed-specific analysis in case-control studies is the fact that we do not yet know whether canine DM is caused by the same mechanism within or between breeds, and is certainly possible that canine DM genes might differ between breeds[17-19]. It must be remembered, however, that a disadvantage of single breed studies lies not only in the difficulty of collecting enough 'case' samples, but also the fact that any SNP found to be associated with disease might lie within a very large LD block. Investigation of the functional significance of the SNP in this case could involve re-sequencing of a very large region (up to 2 megabases long), potentially containing several candidate genes, only one of which might be associated with disease[16].

The lessons we can learn from these recent studies are not simply related to molecular mechanisms in T1D itself, although undoubtedly some of the genes which are important in T1D could be investigated as candidate genes in canine DM. Indeed, this work has already begun, with evidence for SNPs in several cytokine genes (*IL-6, IL-4, IL-10*) being associated with canine DM susceptibility or protection in certain breeds[18,19]. A major lesson from these studies relates to the collaborative way in which they were conducted. Successful GWAs teach us that a large amount of data relating to genetic susceptibility can only be generated and analysed if successful multi-centre collaborations are established for sample collection, phenotyping and analysis, and a good genome polymorphism 'reference-map' is available for SNP assay design. Progress is already being made in all these areas in canine medicine, including the recent publication of papers relating to canine GWA studies - for example the analysis of breed phenotypic traits[20]. The dog genome assembly was published in 2005, and the UK Canine Diabetes Database and Archive Initiative was started in 2001, meaning that there are exciting times ahead.

## REFERENCES

[1]Adeghate E *et al.* 2006, **Ann N Y Acad Sci**. 1084: 1-29. [2]Eisenbarth,GS 1986, **N.Engl.J.Med**. 314, 1360-1368. [3]Taplin CE & Barker JM 2008, **Autoimmunity** 41(1):11-8. [4]Redondo MJ & Eisenbarth GS 2002,. **Diabetologia** 45, 605-622. [5]Peng H, Hagopian W. 2006, **Rev Endocr Metab Disord**. 7(3):149-62. [6]Catchpole B *et al.* 2005, **Diabetologia.** Oct;48(10):1948-56. [7]Davison LJ *et al.* 2008, **Vet Immunol Immunopathol**. Nov 15;126(1-2):83-90. [8]Davison LJ *et al.* 2005, **Vet Rec**. Apr 9;156(15):467-471. [9]Kennedy LJ *et al.* 2006, **Tissue Antigens**. 2006 Dec;68(6):467-76. [10]Wellcome Trust Case Control Consortium. 2007, **Nature.** 447(7145):661-78. [11]http://www.hapmap.org/ and http://www.ncbi.nlm.nih.gov/projects/SNP/ [12]Cooper JD *et al.* 2008, **Nat Genet**. 40(12):1399-401. [13]Ueda H *et al.* 2003, **Nature** 423(6939):506-11. [14]International Multiple Sclerosis Genetics Consortium (IMSGC) **Genes Immun**. 2008, Nov 6. Epub. [15]Smyth DJ *et al.* 2008, **N Engl J Med**. 2008 Dec 25;359(26):2767-77. [16]Karlsson EK & Lindblad-Toh K. 2008, **Nat Rev Genet**. Sep;9(9):713-25. [17]Catchpole B *et al.* 2008, **J Small Anim Pract**.;49(1):4-10. [18]Kennedy LJ *et al.* 2007, **J Hered**. 98(5):518-25. [19]Short AD *et al.* 2008, **Vet Immunol Immunopathol**. 2008 Oct 17. Epub. [20]Andersson L 2008, **Genetica**. 2008 Aug 15. Epub.

## CONTINUOUS GLUCOSE MONITORING IN CATS & DOGS

**Cynthia R.Ward, VMD, PhD, DACVIM & Alexis M. Cistola, DVM, DACVIM**
**Athens, GA**

Diabetes mellitus is a common endocrinopathy in dogs and cats, affecting as many as 1 in 100 to 1 in 500 animals.[1] Monitoring of the diabetic patient can be a challenge for both owners and veterinarians. The 12-hour glucose curve has been the gold standard method for monitoring insulin therapy in dogs[2] and cats.[3] This technique involves hospitalizing patients to have blood glucose measurements typically every 2-4 hours, but up to every 1 hour if necessary. This traditional glucose curve has many limitations, including disruption of the patient's normal activity and eating routine, the introduction of stress-related hyperglycemia,[4,5] and labor intensiveness of the procedure. The use of long-acting insulins such as Glargine and PZI may also necessitate a 24-hour glucose curve to fully evaluate their efficacy in some animals.[6] Furthermore, both diabetic dogs and cats have been shown to have significant variations in their day-to-day glycemic control.[7,8] Intermittent blood sampling over only a 12-hour period may grossly over or underestimate a patient's glycemic control, and glucose peaks and nadirs may be missed if they occur between samplings.

Continuous glucose monitoring systems (CGMS) have been approved by the United States Food and Drug Administration for use in human diabetics and provide a minimally invasive method for continuously evaluating glycemic control for up to 72 hours. Interstitial glucose has been shown to correlate well with plasma glucose levels,[9] and the relationship between plasma glucose and interstitial glucose is not affected by insulin.[10]

Continuous glucose monitoring systems have been utilized in human medicine since 2000.[11] The systems are comprised of an external sensor with a flexible electrode that reacts with glucose when it is inserted into the subcutaneous tissue. This sensor then communicates to a pager-sized monitor, which is capable of storing up to 288 daily measurements. Many CGMS have been developed in recent years, including the CGMS Gold, Medtronic Guardian-RT, and Medtronic Guardian-Real Time. The Gold system transmits glucose measurements from sensor to monitor via a cable, where the Guardian-RT and Guardian-Real Time models send data from sensor to monitor wirelessly via a transmitter connected to the sensor. Both the Guardian-RT and the Guardian-Real Time models also provide real-time glucose values, which can be read on the monitor without disruption of the recording. The CGMS Gold requires removal of the device from the patient and downloading of the entire monitor content in order to review glucose measurements. The Guardian-Real Time is the most recent model. It has the added features of decreased transmitter size and the ability to display 3, 6, 12 or 24 hours of glucose measurements as a curve on the monitor.

The sensor used in the CGMS Gold and Guardian systems was designed and developed by Minimed for three consecutive days of subcutaneous glucose monitoring.[12] Each sensor contains a glucose oxidase-coated platinum working electrode, which is kept in constant potential to a reference electrode. The sensor is encased in flexible polymer tubing with a small window of polyurethane that covers the active area of the electrode. This polyurethane window acts as a glucose diffusion limiting membrane so a linear relationship between interstitial glucose and sensor current is maintained.[10,13] Hydrogen peroxide, which is generated from the oxidation of glucose by glucose oxidase, is then oxidized on the electrode to produce a current with intensity that is proportional to the surrounding glucose concentration.[10,12] The sensor is then calibrated with blood glucose levels to produce the interstitial glucose readings in accordance with the correlating reaction intensity. As such, it is important to interpret absolute values generated by the CGMS with caution.[11] The manufacturer has set a working range of 40-400 mg/dl for the device. Glucose measurements outside of this range will not be recorded.

Interstitial glucose measurements are obtained every 10 seconds with an average value recorded every 5 minutes. The recorded values are then downloaded via ComLink™ connector to a computer and a curve is generated for analysis. Current versions of analysis software are web-based, and each patient's information can be saved under an individual username and password. This allows easy information sharing with colleagues and owners.

The Guardian-Real Time can be purchased for approximately $1500. This includes a monitor, transmitter, transmitter charger, Sen-serter insertion device, ComLink™ and 4 single-use sensors. Additional sensors can be purchased for $35 each and are supplied in packs of 4 or 10. Each box of sensors is marked with an expiration date, but sensors are usually good for 1 year from the date of purchase. Analysis software is web-based and provided at no additional charge. The CGMS Gold and Guardian-RT can no longer be purchased from the manufacturer.

CGMS placement may vary slightly. At the authors' institution, a 3 cm by 3 cm patch of hair over the interscapular region is clipped and aseptically prepared. Alternatively, a similar sized region can be prepared over the lateral thoracic region.[12-17] The sensor with an attached 22-gauge stylet is loaded into an insertion device (Sen-serter) provided by the manufacturer. The insertion device is spring-loaded and allows quick implantation of the sensor tip into the subcutaneous tissue. Quick implantation and the small stylet size preclude the need for sedation or anesthesia (general or local) for monitor placement. The stylet is then removed and discarded. The external portion of the sensor is attached to the patient by means of the sensor's own self-adhesive.

The sensor must undergo a five-minute "wetting" period as the electrode becomes immersed in the interstitial fluid. After this five-minute period, the monitor is turned on and then the transmitter or cable is connected to the sensor. A 1 hour (CGMS Gold) or 2 hour (wireless models) initializing "warm-up" phase occurs before the first calibration can be entered. The glucose curve begins with the first calibration and ends automatically 72 hours later.

The sensor and attached transmitter are then covered with a protective dressing. The monitor is then attached to the collar of medium and large dogs, or in the case of cats and small dogs, the monitor is placed on or within the cage. The monitor may also be secured to the animal's dorsum by means of an elastic bandage and tape harness, or by a recording vest.[13]

The wireless Guardian-RT and Guardian-Real Time models must have the transmitter and monitor remain within 5 feet of each other or data will fail to transmit and record. If the Guardian-RT transmitter is separated from its monitor for more than five minutes, then data after this time period will not be recorded. The Guardian-Real Time transmitter has the capacity to store 45 minutes of data if it is separated from its monitor. Once the monitor and transmitter are replaced in proximity, the stored data will be recorded and continued.

Each CGMS requires calibration after the 1 or 2 hour initialization period. Calibrations are performed by drawing a blood sample and then entering the glucose level obtained into the monitor. Glucose measurements must be entered within 5 minutes of acquisition to ensure proper calibration. A reliable portable glucometer [18-20] may be used for the purposes of obtaining quick glucose measurements. The patient's glucose must fall within the working range of 40-400 mg/dl to be used for CGMS calibration or to be recorded. Animals residing in the hospital during their recordings may have measurements taken via a central catheter or by peripheral venipuncture. Alternatively, owners may be educated on techniques for collection of a capillary blood sample from the ear pinna.[21] Good correlation has been documented between glucose concentrations in capillary blood and venous blood,[22] and this collection method has been shown to be well tolerated by pets and owners.[21]

The first calibration must occur as soon as possible after this initialization period or improper interstitial glucose measurements may occur.[13] Three separate calibrations are required per each 24 hour period the monitor is recording. It is not necessary to evenly space the timing of calibrations, but the interval between calibrations may not exceed 12 hours. It is recommended by the manufacturer to avoid performance of calibrations in the 2 hours following insulin administration or consumption of a meal.

The use of continuous glucose monitoring systems has been evaluated in veterinary medicine.[12-17] Results of a study of 10 diabetic dogs showed good correlation when a CGMS was compared to a standard venipuncture generated glucose curve.[12] In general, glucose concentrations recorded by the CGMS were similar to, but lower than, glucometer readings. The authors did note in 6 of 10 dogs that a postprandial rise in blood glucose concentration was detected with measurement by a glucometer, but the CGMS failed to capture such an event. This phenomenon has not been demonstrated by results from other studies involving diabetic dogs,[13,14,16] but their results do confirm good correlation between blood glucose curves and CGMS recordings.

Researchers in another study evaluated 16 CGMS[a] recordings in 14 diabetic cats.[15] Good correlation was noted between CMGS values and blood glucose concentrations. In 7 of 16 tracings, the 40-400 mg/dl working range was a limiting factor in the use of the monitor. Four cats experienced hyperglycemia greater than 400 mg/dl, 3 were hypoglycemic less than 40 mg/dl, and another was both hyper and hypoglycemic during the attempted curves. All cats tolerated the device well, although removal and kinking of the sensor required sensor replacement in one and two cats, respectively.

A CGMS has been used successfully in a diabetic horse.[14] Also, good correlation between glucose measurements obtained from a CGMS and standard venipuncture collected samples has been documented in horses and a cow undergoing intravenous glucose tolerance testing.[14,16] A CGMS has also been used at the authors' institution to evaluate stress-induced hyperglycemia in a white-faced capuchin monkey and sepsis-induced hypoglycemia in a cria.

At the time of writing, research is currently ongoing at the authors' institution to validate the Guardian-Real Time's use for monitoring diabetic dogs. As this model is wireless and provides real time glucose trends, it is ideal for glucose monitoring of hospitalized animals with minimal handling of the patient. Studies evaluating the use of CGMS in ketoacidotic patients, septic and juvenile patients prone to hypoglycemia, and other patients with abnormal glycemic control have yet to be performed. The continuous 72 hour curves generated also may help to evaluate day-to-day variations in glycemic control. Variations in treatment recommendations for the same animal may be made when only 12- or 24-hour curves are examined versus a continuous 3 day curve.

Continuous glucose monitoring systems are continually undergoing improvement as they advance in the human market. Improved sensor technology, upgraded analysis software, decreased equipment size and real-time glucose measurements have served to increase the utility of these monitors.

## REFERENCES

[1]Feldman EC, Nelson RW. **Canine and feline endocrinology and reproduction**. 3rd ed. St. Louis: W.B. Saunders, 2004:486-579. [2]Fleeman LM, Rand JS.**Vet Clin North Am Small Anim Pract** 2001;31:855-880 [3]Rand JS, Martin GJ. **Vet Clin North Am Small Anim Pract** 2001;31:881-913.[4]Rand JS, Kinnaird E, Baglioni A, et al. **J Vet Intern Med** 2002;16:123-132.[5]Nelson R.**J Vet Intern Med** 2002;16:121-122.[6]Cook AK. **Vet Med** 2007;102:600-614.[7]Fleeman LM, Rand JS. **J Am Vet Med Assoc** 2003;222:317-321.[8].Alt N, Kley S, Haessig M, et al. **J Am Vet Med Assoc** 2007;230:1011-1017.[9]Rebrin K, Steil GM. **Diabetes Technol Ther** 2000;2:461-472. [10]Rebrin K, Steil GM, van Antwerp WP, et al. **Am J Of Physiol** 1999;277:E561-571.[11]Melki V, Ayon F, Fernandez M, et al. **Diabetes Metab** 2006;32:123-129.[12]Davison LJ, Slater LA, Herrtage ME, et al. **J Small Anim Pract** 2003;44:435-442.[13]Wiedmeyer CE, Declue AE. **J Vet Intern Med** 2008;22:2-8.[14]Wiedmeyer CE, Johnson PJ, Cohn LA, et al. **J Am Vet Med Assoc** 2003;223:987-992.[15]Ristic JM, Herrtage ME, Walti-Lauger SM, et al. **J Feline Med Surg** 2005;7:153-162.[16]Wiedmeyer CE, Johnson PJ, Cohn LA, et al. **Diabetes Technol Ther** 2005;7:885-895.[17]DeClue AE, Cohn LA, Kerl ME, et al. **J Am Anim Hosp Assoc** 2004;40:171-173.[18]Wess G, Reusch C. **Am J Vet Res** 2000;61:1587-1592.[19]Cohn LA, McCaw DL, Tate DJ, et al. **J Am Vet Med Assoc** 2000;216:198-202.[20]Wess G, Reusch C. **J Am Vet Med Assoc** 2000;216:203-209.[21]Casella M, Wess G, Reusch CE. **J Am Anim Hosp Assoc** 2002;38:239-245.[22]Wess G, Reusch C. **J Small Anim Pract** 2000;41:60-66.

## HOST FLORAL INTERACTIONS IN THE GASTROINTESTINAL TRACT

**Kenneth W. Simpson BVM&S, PhD, DipACVIM, DipECVIM-CA.**
**Ithaca, NY.**

Inflammatory bowel disease (IBD) is the term applied to a group of poorly understood enteropathies that commonly affect animals and people. IBD in both people and animals is increasingly considered a consequence of uncontrolled intestinal inflammation in response to a combination of elusive environmental, enteric luminal constituents (principally microbial and dietary) and immunoregulatory factors in genetically susceptible individuals.

In people, genetic susceptibility is linked increasingly to defects in innate immunity exemplified by mutations in the innate immune receptor NOD2/CARD15, that in the presence of the enteric microflora may lead to up-regulated mucosal cytokine production, delayed bacterial clearance and increased bacterial translocation, thereby promoting and perpetuating intestinal inflammation (Packey 2008). While the mucosa-associated flora is implicated frequently as a pivotal factor in the development of IBD in people and animals, the specific bacterial characteristics that drive the inflammatory response have remained elusive.

In dogs, the clinical responses of Boxer dogs with granulomatous colitis to enrofloxacin, and a subset of dogs (particularly German Shepherd) with idiopathic chronic diarrhea to antibiotics such as tylosin or oxyteracycline, points to a similar interaction of host susceptibility and microflora (Batt 1988, German 2003, Hostutler 2004, Westermarck 2005,). The *E. coli* strains isolated from the colon of Boxer dogs with GC/HUC are broadly similar to those in healthy dogs, and the numbers of cultivable aerobic and anaerobic bacteria in the duodenal juice of dogs with antibiotic responsive enteropathy are similar to dogs that respond to food or immunosuppression suggesting that affected individuals are susceptible to their resident microflora rather than obligate pathogens (German 2003, Simpson 1994, *Craven 2009), but this remains to be determined.

Recent advances in molecular microbiology have enabled the analysis of complex bacterial communities without bacterial culture. Culture-independent analyses of bacterial 16S rDNA libraries in people reveal that only 30% of the fecal flora appears cultivable, and there is significant variation in the flora in different gastrointestinal segments and luminal contents versus the mucosa of healthy individuals (Eckburg 2005). The application of these culture independent techniques to people, dogs and cats with gastrointestinal disease has revealed that intestinal inflammation is associated with a floral shift from Gram-positive to Gram-negative bacteria, predominantly Enterobacteriaceae. It is noteworthy that increased numbers of Enteobacteriaceae have been found to correlate with mucosal inflammation and clinical signs in cats with signs of gastrointestinal disease, and a novel group of adherent and invasive E. coli (AIEC) have been associated with intestinal inflammation in people and Boxer dogs with granulomatous/histiocytic ulcerative colitis. In many patients it remains to be determined if these floral alterations are a cause or a consequence of the inflammation, but their discovery has provided new opportunities for therapeutic intervention. This is exemplified by intramucosal E. coli in Boxer dogs with granulomatous colitis, where eradication of invasive E. coli correlates with long term clinical and histological remission.

This presentation will present an in depth overview of the mucosal microflora associated with chronic intestinal inflammation in dogs, cats and people, and data from mouse models that reveal acute inflammation can also perturb the enteric microbiota, and trigger the proliferation of adherent and invasive *E. coli* similar to those described in IBD.

The results of the principal studies covered in this presentation are summarized as follows:

### MUCOSAL MICROFLORA AND CHRONIC INFLAMMATORY BOWEL DISEASE

#### Granulomatous Colitis in Boxer dogs

Granulomatous colitis (GC) of Boxer dogs, also known as Histiocytic Ulcerative Colitis (HUC), was first described by Van Kruiningen in a kennel of Boxer dogs in 1965. The clinical hallmarks of the disease are severe large bowel diarrhea that is often accompanied by profound weight loss, anemia and hypoalbuminemia. GC/HUC, although rare, occurs world-wide with reported cases originating from Australia, Japan, North America and Europe. Boxer dogs are most commonly affected although other breeds such as Mastiff, Alaskan Malamute, Doberman Pinscher, and French Bulldogs, are sporadically affected. The pathognomonic lesion of GC/HUC in Boxer dogs is mucosal infiltration with large numbers of macrophages staining positively with period-acid-Schiff (PAS), and is usually accompanied by mucosal ulceration and loss of goblet cells.

The poor response of GC/HUC to immunosuppression led to a re-appraisal of antibiotic therapy, and there are now three independent studies that describe dramatic improvements in clinical signs and histological lesions of Boxer dogs treated with antibiotic protocols containing enrofloxacin. Colonic biopsies from affected Boxer dogs ($n$=13) and controls ($n$=38) were examined by fluorescent in situ hybridization (FISH) with a eubacterial16S rDNA probe. Culture, 16S ribosomal DNA sequencing, and histochemistry were used to guide subsequent FISH. Intramucosal Gram-negative coccobacilli were present in 100% of GC/HUC samples from Boxer dogs but not controls, and invasive bacteria hybridized with FISH probes to *E. coli*. Independent support for these findings is provided by the immunolocalization of *E. coli* to macrophages within the colons of 10/10 Boxer dogs with HUC. *E. coli* strains isolated from affected Boxer dogs are novel in phylogeny, and have an adherent and invasive phenotype (AIEC) similar to strains isolated from people with Crohn's disease.

A recent study has examined further the relationship of intramucosal E. coli to clinical and histological outcome in 7 Boxers with GC/HUC treated with enrofloxacin. Clinical response was noted in 7/7 dogs within 2 wks of enrofloxacin given at a mean dose of 7 mg/kg/d for a mean duration of 9.5 weeks, and was sustained in 6 dogs (mean disease-free interval to date of 45 months). Post-enrofloxacin FISH was negative for *E. coli* in 4/5 dogs. *E. coli* resistant to enrofloxacin were present in the FISH positive dog that relapsed clinically. These findings suggest that *E. coli* invasion plays a critical role in the initiation and/or progression of GC/HUC in Boxer dogs. Antibiotic resistance is an increasing problem and we have documented enrofloxacin resistant *E.coli* in 6/6 Boxers that were empirically treated with enrofloxacin prior to definitive diagnosis.

Since GC/HUC in dogs is remarkably breed-specific it may be due to a heritable anomaly in Boxer dogs that confers susceptibility to invasion and persistence of AIEC within the colonic mucosa. As Boxer dogs are also prone to other granulomatous diseases such as intestinal and systemic protothecosis and cutaneous intracellular mycobacteria (canine leproid granuloma syndrome), it is possible that Boxers have a genetic defect in how they respond to granuloma-inducing microorganisms in general, e.g. defective bacterial killing. Further studies are ongoing to identify the host and bacterial factors related to invasion and persistence of *E. coli* in GC/HUC susceptible Boxers.

**Mucosal flora associated with non-specific colitis in dogs**

The colonic mucosa of 29 dogs undergoing evaluation for signs of gastrointestinal disease (endoscopic biopsies) and 12 dogs without signs of intestinal disease (full thickness biopsies) was evaluated using a standardized histopathological scoring system. The number and spatial distribution of mucosal bacteria was determined by FISH with probes to 16S rDNA and an antibody to *Brachyspira* spp.

Colonic histopathology ranged from normal (18), through mild (10), moderate (11), and severe (2) colitis. The mucosal flora of healthy and inflamed colon hybridized with probes to *Bacteroides/Prevotella*, *Clostridium*, Enterobacteriaceae, *Streptococcus* and *Helicobacter* spp. 16S rDNA sequence analysis showed *Helicobacter* spp had highest homology to *H. canis, F. rappini,H. hepaticus* and *H. cinaedi*. *Brachyspira* were not detected in any samples. The number of mucosa-associated bacteria and Enterobacteriaceae was lower in dogs with colitis and was inversely correlated with severity. Colitis was associated with a significant reduction in the proportion of *Clostridium* spp relative to total bacteria and Enterobacteriaceae. These data establish that the density and composition of the colonic mucosal flora is related to the presence and severity of colonic inflammation in dogs. The relationship of the abnormal flora to the etiopathogenesis of "lymphoplasmacytic" colitis remains to be established.

**Mucosal flora associated with duodenal inflammation in dogs**

In a study by Xenoulis et al. DNA was extracted and 16S rRNA gene was amplified using universal bacterial primers from endoscopic duodenal brush cytology samples from 10 dogs with IBD and nine healthy control dogs (6 Beagles, 3 Greyhounds). Constructed 16S rRNA gene clone libraries were compared between groups. From a total of 1240 selected clones, 156 unique 16S rRNA gene sequences were identified, belonging to six phyla: Firmicutes (53.4%), Proteobacteria (28.4%), Bacteroidetes (7.0%), Spirochaetes (5.2%), Fusobacteria (3.4%), Actinobacteria (1.1%), and Incertae sedis (1.5%). Species richness was significantly lower in the IBD group. PCA showed IBD and control dogs are composed of distinct microbial communities. The most profound difference involved enrichment of the IBD dogs with members of the Enterobacteriaceae family. However, differences involving members of other families, such as Clostridiaceae, Bacteroidetes and Spirochaetes, were also identified.

A recent study has sought to gain insight into the relationship of the mucosal flora to clinical response to diet, antibiotics, or immunosuppression. A culture-independent bacterial tag-encoded FLX 16S rDNA amplicon pyrosequencing (bTEFAP) approach was used to evaluate the microbiome of endoscopic duodenal biopsies from 16 dogs with a histopathological diagnosis of IBD (7MN 7FN 2M, median age 5 yrs) and 14 clinically healthy (H) dogs (5FN, 6MN, 3F, median age 7.5 yrs). 10 dogs were food responsive (FR), 3 steroid responsive (SR), and 3 were partially responsive to a combination of food and antibiotics (FA). A total of 85,065 sequences (54776 from H, 30289 from IBD) were analyzed. The number of sequences per individual (median) was higher in H (3,912) than IBD (1,893) ($P=0.007$). Dogs with IBD had lower overall species diversity than H at all primary levels of taxonomy based upon rarefaction ($p<0.05$). Analysis by clinical response indicated differences in numbers of sequence between groups (H, FR, SR and SA: Kruskal Wallis, $P<0.003$,), and lower numbers in FR than H ($p=0.007$, see graph).

These studies indicate that duodenal inflammation is associated with decreased diversity of the mucosal flora and altered relative proportions of dominant floral groups. Further work is required to enumerate the different bacterial subgroups, and to determine whether differences in the mucosal flora impact the clinical response.

**Feline Inflammatory Bowel disease**

Duodenal biopsies were collected from 27 cats: 17 undergoing diagnostic investigation of signs of gastrointestinal disease,and 10 healthy controls. Subjective duodenal histopathology ranged from normal (10), through mild (6), moderate (8), ands evere (3) IBD. The number and spatial distribution of mucosal bacteria was determined by fluorescence in situ hybridizationwith probes to 16S rDNA. Mucosal inflammation was evaluated by objective histopathology and cytokine profiles of duodenal biopsies.The number of mucosa-associated Enterobacteriaceae was higher in cats with signs of gastrointestinal disease than healthy cats ($P < 0.001$). Total numbers of mucosal bacteria were strongly associated with changes in mucosal architecture ($P < 0.001$) and the density of cellular infiltrates, particularly macrophages ($P < 0.002$) and CD3+lymphocytes ($P < 0.05$). The number of Enterobacteriaceae, E. coli, and Clostridium spp. correlated with abnormalities in mucosal architecture (principally atrophy andfusion), upregulation of cytokine mRNA (particularly IL-1, -8 and -12), and the number of clinical signs exhibited by theaffected cats.

**Crohn's disease**

Analysis of 16S rDNA libraries constructed from endoscopic biopsies of the ileum from patients with Crohn's ileitis (a disease phenotype that occurs in approximately 70% of individuals with CD), demonstrates selective enrichment of ileal mucosa in *E. coli* and relative depletion in Clostridiales compared to both patients with CD restricted to the colon, and to healthy individuals. PCR of ileal DNA yielded no evidence of MAP, *Listeria* or *Shigella*, and confirmed the increase in *E. coli*. Similarly, 16S rDNA library-based evaluation of surgically resected CD ileum shows an increase in Proteobacteria (this phylum includes Enterobacteriaceae), decrease in Firmicutes (this phylum contains Clostridiales), and the absence of DNA from MAP in CD mucosa. The relative decrease in Clostridiales in CD ileal mucosa echoes findings in studies of feces and colonic mucosa from patients with CD, and suggests that this clostridial dysbiosis may be a feature of CD in general rather than a marker of CD ileitis.

The in depth knowledge of microfloral composition provided by 16S sequence analysis can be used to guide *in situ* examination of the spatial distribution of the mucosal flora interacting most closely with the innate immune system. FISH with a probe restricted to *E. coli/Shigella* reveals that the ileal mucosa of CD patients harbors significantly more *E. coli* than normal ileum, and mucosally invasive *E. coli* are only detected in CD mucosa. In patients with Crohn's ileitis, *E. coli* represent approximately 3.5% of the mucosal flora labeled with the panbacterial FISH probe, EUB338 and approximately 40% of the Enterobacteriaceae recognized by the Enterobacterial FISH probe, 1531. This contrasts with culture-based studies reporting that *E. coli* represent >50% of the cultivable mucosal flora in 40 to 58% of patients with ileal CD. However, this discrepancy is readily explained by the inability to culture the vast majority of enteric mucosal flora, and supports the use of culture-independent analysis. Furthermore, the number of *E. coli* visualized in ileal biopsies by Baumgart *et al.*, but not bacterial colonization in general, correlates with histological and endoscopic disease activity, which suggests *E. coli* could be specifically involved in the inflammatory process. These findings complement previous studies that have documented *E. coli* antigens and DNA in granulomas and peri-ulcerative lesions in CD, and provide a rational explanation for the higher prevalence of antibodies directed against *E. coli* outer membrane porin C (OmpC) and flagellin in patients with CD involving the ileum versus colon restricted CD.

Culture based analysis of *E. coli* isolated from the mucosa of patients with CD has revealed that many strains display pathogen-like behavior in cultured cells. The best-characterized CD associated *E. coli* strain LF82 (isolated by Dr. Arlette Darfeuille-Michaud from the ileum of a patient from Lille France), adheres to, invades, persists, and replicates in cultured intestinal epithelial cells. Invasion is associated with trigger-type endocytosis and utilizes host cell microfilaments and microtubules. The predominance of AIEC in the inflamed ileum has recently been addressed by Barnich *et al.* who found that AIEC strains adhere to the brush border of primary ileal enterocytes isolated from CD patients but not controls without inflammatory bowel disease. AIEC adhesion is dependent on both bacterial surface expression of type 1 pili and on expression of carcinoembryonic antigen–related cell adhesion molecule 6 (CEACAM6) on the apical surface of ileal epithelial cells, an interaction that can be inhibited by mannose. These *in vitro* findings are highly consistent with recent *in situ* observations that *E. coli* colonization correlates with disease activity.

The increased numbers of mucosa-associated and invasive *E. coli* relative to other bacteria observed in patients with Crohn's ileitis suggest that AIEC, with the appropriate stimulus, are most likely opportunistic pathogens that can exploit the mucosal environment of a Crohn's susceptible individual, who has, for example, reduced defensin production and delayed bacterial clearance associated with functionally important polymorphisms in NOD2/CARD15. Alternatively, AIEC may permissively proliferate subsequent to a depletion of normal flora such as Faecalibacterium, or alterations in bacterial products like butyrate, which is important for colonic health and ileal regeneration, or a combination of these possibilities. In this respect it has recently been shown that a decrease in Faecalibacterium can lead to upregulated mucosal inflammation.

## MUCOSAL MICROFLORA AND ACUTE INFLAMMATION

We examined the effects of acute intestinal inflammation induced by infectious agents and NSAID on the composition and spatial distribution of the ileal mucosal flora. Wildtype C57BL/6 mice were gavaged with *Toxoplasma gondii* (n=10 mice), *Giardia muris* (n=10), or indomethacin at doses of 0.1 mg/mouse for 5 days (LDI, n=5), or 1mg/mouse for 3 days (HDI, n=5). Ileal tissue was harvested from controls (n=8) and affected mice: 4 and 8 days after *T.gondii*, 7 and 14 days after *Giardia*, and 3 and 7 days after HDI and LDI respectively. Inflammation was assessed histopathologically by an investigator masked to treatment. Composition and spatial distribution of the mucosal flora was evaluated by 16S rDNA sequencing, quantitative PCR, fluorescence in situ hybridization (FISH) and culture. Mucosal *E. coli* isolates were characterized by phylogroup, genotype (RAPD-PCR, virulence determinants) and pathotype (invasion of Caco-2, persistence in J774 macrophages).

Infection with *T. gondii* (8 days post-infection,Toxo8) and HDI resulted in severe ileitis, that was fatal in 2 of 5 HDI mice. The proportion of clones in 16S rDNA libraries corresponding to Enterobacteriales was significantly greater in Toxo8 (84%) and HDI (90%) than in controls (0%) and other groups (≤2%). The number of *E. coli* (median copies *uidA*:18S.$10^6$) was higher in Toxo8 ($3x10^6$) and HDI ($1.7x10^6$) compared to controls ($2.6x10^4$). FISH positive mucosally invasive *E. coli* were observed solely in the Toxo8 and HDI groups. *E. coli* isolates from Toxo8 and HDI were phylogroup B1 and had identical RAPD-PCR profiles. A representative strain (CUMT8) was serogroup O8:H21, contained few known virulence genes (*hcp,gsp,lpfA*), and displayed an AIEC pathotype. We conclude that acute ileitis perturbs the enteric microbiota and triggers the proliferation of adherent and invasive *E. coli* similar to those described in IBD. The microbial alterations induced by acute inflammation may promote and perpetuate intestinal inflammation in an individual susceptible to IBD.

## REFERENCES

Barnich N J Clin Invest. 2007 Jun;117(6):1566-74, Butt RM, Gut. 1988 Apr;29(4):473-82, Baumgart M, ISME J. 2007 Sep;1(5):403-18, Darfeuille-Michaud A, *Gastroenterology* 2004;127:412-421, Eckburg PB, Science. 2005 Jun 10;308(5728):1635-8, Frank DN Proc Natl Acad Sci U S A. 2007 Aug 21;104(34):13780-5, German AJJ Vet Intern Med. 2003 Jan-Feb;17(1):33-43, Hostutler RA, *J Vet Int Med* 2004;18:499-504, Janeczko S, Vet Microbiol. 2008 Apr 1;128(1-2):178-93. Packey CDJ Intern Med. 2008 Jun;263(6):597-606, Rhodes JM. *Gut* 2007;56:610-2, Simpson KW, *Infect Immunity* 2006;74:4778-4792, Simpson KW. J Am Vet Med Assoc. 1994 Aug 1;205(3):405-7, Sokol H. Proc Natl Acad Sci U S A. 2008 Oct 28;105(43), Van Kruiningen HJ, *APMIS* 2005;113:420–425, Van Kruiningen HM*Path Vet* 1965;2:521-544, Westermarck E, J Vet Intern Med. 2005 Mar-Apr;19(2):177-86, Xenoulis PG, FEMS Microbiol Ecol. 2008 Dec;66(3):579-89.

## NATURALLY OCCURING LIVER DISEASE OF ANIMALS AS MODELS IN BIOMEDICAL RESEARCH

**Bud C. Tennant, DVM, DACVIM**
**Ithaca, NY**

The idea of "One Medicine" is not new. In ancient times, priests and other healers treated the illnesses of both humans and other animals (Cornelius and Arias, 1969). During emergence from the Dark Ages, veterinary medicine separated from human medicine in a manner that was not necessarily beneficial to either or to the public health. During the Renaissance, the fields of comparative anatomy and comparative physiology originated as a result of animal experimentation conducted by early medical scholars such as William Harvey. Later, comparative pathology would develop as a result of the work of Virchow who was an early and strong advocate of animal disease research to enhance understanding of human disease and is considered to be the founder of modern medicine (Cardiff, et al., 2008). The classic work of veterinarians followed by that of Jenner in the control of small pox represents an early contributions to comparative medicine. During the 20th Century, "One Medicine" was advocated by Sir William Osler, a physician (Cornelius and Arias, 1969), and by Calvin Schwabe, a veterinarian (Cardiff, et al., 2008).

In recent years, leading medical and veterinary associations in the United States (Enserink, 2007) and Europe (Favre, 2005; Michell, 2005) have initiated constructive rapprochements that have the objective, under the rubric of "One Medicine" or "One Health," to enhance professional and scientific collaboration. The current symposium is directly related to these recent initiatives and is focused on recognizing the importance of research on naturally occurring animal diseases to biomedical knowledge and ultimately to the public health. The objective of the following presentation is to demonstrate 1.) how comparative research on naturally occurring hepatitis B like virus infections of animals has contributed to knowledge of the pathogenesis of infection in affected species including humans and 2.) how development of the woodchuck animal model has been useful in the preclinical development of the drugs now used to treat HBV infection in humans and in the development of new drugs for treatment and prevention of viral induced hepatocellular carcinoma (HCC).

The Eastern woodchuck (*Marmota monax*) harbors a DNA virus (Woodchuck hepatitis virus [WHV]) that is similar in structure and replicative life cycle to the human hepatitis B virus (HBV). Like HBV, WHV infects the liver and can cause acute and chronic hepatitis. Furthermore, chronic WHV infection in woodchucks may lead to development of HCC as it may in humans. The woodchuck model has been important in studies of the pathogenesis of hepadnavirus infections and in the preclinical evaluation of safety and efficacy of the antiviral drugs now in use for treatment of HBV infection. The woodchuck continues to serve as an important, predictive model for innovative forms of therapy of hepatitis B using antiviral nucleosides and immune response modifiers alone or in combination. Almost all woodchucks that become chronic WHV carriers after experimental neonatal infection develop HCC with a median HCC-free survival of 24 months and a median life expectancy of 30–32 months. The woodchuck model of viral-induced HCC has been used effectively for the development of new imaging agents for enhancement of detection of hepatic neoplasms by ultrasound and magnetic resonance imaging. The chemoprevention of HCC using long-term antiviral nucleoside therapy has been shown in the woodchuck (Colonno et al.; 2001Tennant et al., 2004), and "proof of principal" has been established for other innovative, molecular and immunotherapeutic methods for treatment of chronic HBV infection and HCC. The model is available for fundamental investigation of viral and molecular mechanisms of hepatocarcinogenesis and should have substantial value for future development of chemoprevention and gene therapy of human HCC (Tennant, et al. 2004).

Cardiff RD, Ward JM, Barthold SW. "One medicine-one pathology': are veterinary and human pathology prepared? Laboratory Investigation 2008; 88:18-26

### REFERENCES

[1.] Colonno RJ, Genovesi EV, Medina L, et al. Long-term entecavir treatment results in sustained antiviral efficacy and prolonged life span in the woodchuck model of chronic hepatitis infection. **J Infect Dis** 2001;184:1236-1245 [2.] Cornelius CE and Arias IM. Biomedial Models in Veterinary Medicine. **Am J Med** 1969; 40; 165-169 [3.] Enserink M. Initiative aims to merge animal and human health science to benefit both. **Science** 2007; 316:1553 [4.] Favre J. Editorial. Human and veterinary medicine. **Brit Med J** 2005; 330: 858-859 [5.] Michell AR. Comparative clinical science: The medicine of the future **The Veterinary Journal** 2005; 170: 153-163 [6.] Tennant BC, Toshkov IA, Peek SF, et al. Hepatocellular carcinoma in the woodchuck model of hepatitis B virus infection. **Gastroenterology** 2004; 127: S283-S293

## TRANSLATIONAL RESEARCH IN OSTEOSARCOMA: DOGS TO KIDS & BACK AGAIN!

**Stephen J. Withrow, DVM, DACVIM (Oncology), DACVS**
**Fort Collins, CO**

### INTRODUCTION

Osteosarcoma is one of the most common primary bone cancers in humans and companion animals (**Table 1**).[1,2] Until the 1960's, the disease was characterized in all species by rapid progression of the primary tumor and early metastasis making even amputation at best a minimally palliative procedure soon followed by death.

**Table 1:** Comparison of Canine and Human Osteosarcoma[1]

| *Variable* | *Dog* | *Human* |
|---|---|---|
| Mean age | 7 years | 14 years |
| Race/breed | Large or giant purebreds | None |
| Body weight | 90% > 20 kg | Heavy |
| Site | 77% long bones<br>Metaphyseal<br>Distal radius > proximal humerus > Distal femur > tibia | 90% long bones<br>Metaphyseal<br>Distal femur > proximal tibia > Proximal humerus |
| Etiology | Generally unknown | Generally unknown |
| % clinically confined to the limb at presentation | 80-90% | 80-90% |
| % histologically high grade | 95% | 85-90% |
| DNA index | 75% aneuploid | 75% aneuploid |
| *Molecular and genetic alterations* | *See Table 2* | *See table 2* |
| *Prognostic indicators* | *Young age, alkaline phosphatase* | *Alkaline phosphatase, MDR1* |
| Metastatic rate without chemotherapy | 90% before 1 year | 80% before 2 years |
| Metastatic sites | Lung > bone > soft tissue | Lung > bone > soft tissue |
| Improved survival with chemotherapy | Yes | Yes |
| Regional Lymph node metastasis | poor prognosis, < 5% incidence | Poor prognosis |

Early work showed remarkable similarities between human and canine osteosarcoma which formed the basis for translational research.[3,4] Similarities include slight prevalence in males over females, large patient size, limb location > axial, metaphyseal site, rare lymph node metastasis, and common hematogenous lung metastasis. Two differences exist between humans and dogs: canine osteosarcoma is ten times more prevalent than osteosarcoma in people, and the median age of affected dogs is eight to ten years while osteosarcoma affects humans predominantly in the teenage years.

Specific etiologic factors are rare but often shared in both species: exposure to ionizing radiation (experimental radionuclides and secondary to therapeutic radiation), and malignant transformation around metallic implants.[5,6,7,8] Both of these factors were first discovered or clarified in dogs. As is true in humans, canine osteosarcoma is characterized by bizarre and irregular karyotypes and marked anuploidy.[9] Studies of high risk breeds should add some clarity to heritable and genetic risk factors.[10]

### TREATMENT OPTIONS

Local control of osteosarcoma in dogs is traditionally achieved by amputation, whereas up to 80% of people may now undergo limb sparing procedures. The dog model has contributed to technology in support of limb sparing including proof that intra-arterial cisplatin can be safely delivered and that preoperative chemotherapy (with or without moderate radiation) can increase the percentage tumor necrosis, leading to a higher number of and safer limb sparing resections.[11] Allograft and metal implant technologies have often been developed and studied in canines before implementation in humans. Local biodegradable chemotherapy with cisplatin has modestly reduced local recurrence rates but has not yet attained Food and Drug Administration approval.[12] Just as important as the potential efficacy is the toxicity profile for severe neurotoxicity of cisplatin polymers when implanted in or on brain and spinal cord.

Osteosarcoma has long been considered radioresistant to traditional doses of external beam radiation in both dogs and humans. A dose response for percent necrosis (with or without cisplatin) exists in dogs which gives hope to future studies of dose intensification by altering fraction size, treatment duration, total dose, custom delivery (SRS, IMRT), or systemic radionuclide delivery.[13,14] It is now possible to kill osteosarcoma invivo with radiation and recovery of secondary bone necrosis to full weight bearing capacity is the new challenge.

**SURVIVAL**

Local control alone does not significantly improve survival. Human (and subsequently canine) osteosarcoma was one of the first solid tumors to show definite benefit from adjuvant chemotherapy (doxorubicin and cisplatin). Five year disease free intervals in humans have improved from 10% to 70% or more. Canine adjuvant trials followed the human experience with parallel positive albeit not curative results.[1]

Immunologic approaches to therapy have been proposed. A fascinating event, first demonstrated in dogs with osteosarcoma, is that dogs with an infected limb spare and adjuvant therapy live longer than dogs without infection.[15] This phenomenon has since been proven in people as well.[16] Further evidence of immunomodulatory effects on survival were demonstrated with MTP-PE in dogs and more recently in pediatric osteosarcoma patients.[17,18]

**CONCLUSION**

Dogs with osteosarcoma offer a unique and relavent model for studies of etiology, local invasion, and metastasis investigations via various gene and protein expression pathways (**Table 2**).[2] The recent public release of the canine genome and clarification of molecular pathways of cellular initiation, progression, and metastasis will offer unique prevention and targeted treatment pathways.[19] Several federal initiatives under the auspices of the National Cancer Institute and Children's Oncology Group recognize this unique model and have supported infrastructure for tissue archiving (Pfizer Canine Comparative Oncology and Genomics Consortium) and clinical trials (Comparative Oncology Trials Consortium). In addition, the Food and Drug Administration has recognized the value of comparative models in accelerating drug discovery to the human patient. The future of comparative and translational research is bright.

**Table 2:** Molecular and Genetic Factors associated with OS in Dogs[1]

| ***Factor*** | ***Comments*** |
|---|---|
| p53 | Mutated and/or overexpressed in several investigations |
| IGF-1/IGF-1R | May contribute to the malignant phenotype |
| HGF/c-Met | May contribute to the malignant phenotype |
| *erb*B-2/HER-2 | Overexpressed in several canine OS cell lines |
| PTEN | Mutated or down-regulated in high percentage of canine OS cell lines |
| *sis*/PDGF | Overexpressed in some canine cell lines |
| Matrix metalloproteinases | Overexpressed in canine OS cell lines |
| Ezrin | A membrane-cytoskeletin linker associated with the metastatic phenotype in canine OS |
| COX-2 | Expression upregulated in some canine OS; prognostic in some investigations, not in others |
| Angiogenic factors | VEGF measurable in plasma of dogs with OS; angiostatin present in urine of dogs with OS |
| Telomerase reverse transcriptase gene | Upregulated in some canine OS |

**REFERNCES**

[1]Dernell WS, et. al. **Withrow and MacEwen's Small An Clin Oncol** 4th ed. 2007:540. [2]Withrow SJ, et. al. **Clin Orthop** 1991;270:159. [3]Brodey RS, et al. **J Am Vet Med Assoc** 1959;134(2):53. [4]Misdorp W, et al. **J Natl Cancer Inst** 1979;62:537. [5]Stevenson S, et al. **J Am Vet Med Assoc** 1982;180:1189. [6]Miller SC, et al. **Radiat Res** 2003;160(5):517. [7]White RG, et al. **Radiat Res** 1994;137(3):361. [8]McChesney-Gillette S, et al. **Cancer Res** 1990;50:54. [9]Paoloni M., et al. **Nature Reviews Cancer** 2008;8:147. [10]Phillips JC, et al. **Genomics** 2007;90(3):354. [11]Withrow SJ, et al. **Cancer** 1993;71:2484. [12]Withrow SJ, et al. **Ann Surg Oncol** 2004;11(7):705. [13]Farese JP, et al. **J Am Vet Med Assoc** 2004;225(10):1567. [14]Thrall DE, et al. **Int J Rad Onc Biol Phys** 1990;18:1354. [15]Lascelles BDX, et al. **Ann Surg Oncol** 2005;12(12):1073. [16]Jeys LM, et al. **Ann Surg Oncol** 2007;14(10):2887. [17]Kurzman ID, et al. **Clin Cancer Res** 1995;1:1595. [18]Meyers PA, et al. **J Clin Oncol** 2008;26(4):633. [19]Khanna C, et al. **Nature Medicine** 10(2): 182-186, 2004

# VETERINARY MEDICINE & COMPARATIVE GENOMICS: NEW TECHNOLOGIES, NEW OPPORTUNITIES

**Matthew Breen PhD, C.Biol, M.I.Biol**
**Raleigh, NC**

## SUMMARY

Comparative genomics provides great insight into the ancestrally retained pathogenic nature of numerous diseases, including cancer. Veterinarians need to be aware and prepared for the plethora of new diagnostic opportunities that will be demanded by clients. This presentation will highlight some of the major advances and how these may impact veterinary medicine.

## INTRODUCTION

Comparative genomics may be defined broadly as any area of research in which the sequence and function of genomes of different species are compared. Since the first draft of the human genome sequence was reported [1] and subsequently refined [2] there has been increasing interest in comparing our genome sequence with those of the other organisms. Using this approach we are able to identify genome regions that we share with other species and those that are different. This approach informs us of how our own genome has evolved. While the reference sequence is generally that of the human genome, comparative genomics involves the study of genome organization between any species. As data from an increasing number of species are collated we are able to identify genes that have been retained (conserved) among species, suggesting their importance across species, as well as those genes that are unique to each species. Even prior to genome sequence technology it was a plausible hypothesis that the human genome would be very similar to that of our most closely related species, other primates, and studies have shown that there is more than 95% to 98% similarity between related genes in humans and other primates. As the divergence between species lengthens the similarity of corresponding genomes declines. For example when the genome of the mouse was analyzed, the average similarity between genes in human and mouse was estimated to be 85% and there was considerable variation from gene to gene [3]. Perhaps one of the major findings of comparative genomics is that even in organisms that are separated by hundreds of millions years of divergent evolution there remains a common set of shared genes. For example, studies of the genome of the fruit fly revealed that 60% of genes within this insect are shared with the human genome[4-6].

### Animal models of genetic disease – One Medicine

The core similarity of genes defining distantly related animals suggests that we can learn a great deal about the structure and function of our own genome by a close consideration of the genomes of other animals. In this process identification of models of human genetic disease have been identified in organisms as distantly related as insects. For example researchers have found that two-thirds of human genes known to be involved in cancer have counterparts in the fruit fly[7]. Even more surprisingly, when scientists inserted a human gene associated with early-onset Parkinson's disease into fruit flies, the insects developed Parkinson-like symptoms suggesting that these insects could serve as a new model for testing therapies aimed at Parkinson's [8].

Even with a complete human genome sequence to hand, there is still much to understand about genome organization and function as we pursue the biomedical goal of finding treatments and cures for diseases. As genome sequencing technology advanced the major genome centers turned their attention to developing complete genome sequences of up to 26 additional vertebrate species, selected on the basis that each may reveal features of our own genome associated with health. Among these species the domestic dog was the first companion animal to hit the headlines. In July 2005, the 7.6x sequence of the 2,400,000,000 bases comprising the genome of a Boxer named *Tasha* was released [9]. Soon after, a first generation 'light' sequence of a domestic cat, *Cinnamon*, was reported [10] along with a detailed sequence of a thoroughbred horse, *Twilight* (http://www.genome.gov/20519480). With additional genome sequences becoming available for cattle, pigs, chicken etc, it is inevitable that our knowledge of the genome organization of animal species will provide new opportunities for veterinary medicine. Further, since we know that dogs, cats and horses each possess traits and diseases that are shared with humans, the availability of genome sequences for each species now provides an essential resource to thoroughly evaluate these shared phenotypes. The comparative value of biomedical research is widely accepted and there is every expectation that data generated from studies of animal diseases will have a major impact on human health

### Clinical genomics in veterinary medicine

In the early stages of veterinary clinical genetics, screening tests for specific genetic diseases were developed for what one would consider the low hanging fruit. Examples include PCR screening for the presence of male-specific DNA in heifers as means to identify intersex cattle. This molecular approach offers a considerably faster throughout then earlier cytogenetic screening, thus providing significant cost saving [11-13]. The development of molecular based assays to detect the presence of the mutant allele associated with equine hyperkalemic periodic paralysis (HYPP) allows for to accurately reliably identification of affected as well as carriers individuals of this disease. To this end, as of as of 2007, the American Quarter Horse Association has mandated that all AQH foals shown to be homozygous for the mutant allele will not be eligible for AQHA registration (http://www.aqha.com/association/registration/hypp.html). By adopting this policy the AQHA are hoping to minimize the transmission of the disease allele in subsequent generations through a process of controlled breeding. Of all companion and livestock species, the domestic dog has perhaps received the most attention with regards to the prevalence of genetic diseases and the need to develop diagnostic tests to effectively identify individuals affected by or susceptible to such diseases. There are numerous resources that provide detailed information and links to describe the vast range of inherited disease in dogs, one such being the IDID (Inherited disease in dogs) based at the University of Cambridge UK (http://www.vet.cam.ac.uk/idid)[14] as well as components of the Online

Mendelian Inheritance in Animals (OMIA) database (http://omia.angis.org.au). Within the IDID database there are data for over 1,000 canine genetic diseases, the majority of which are reported to be autosomal recessive monogenic disorders.

With advances in technology the emerging field of veterinary clinical genomics can be defined as the use of data derived from genomics activities for veterinary medical purposes. Genomics activities in this context include, but are certainly not limited to areas of research based on DNA sequencing, gene discovery, DNA copy number, gene expression and haplotype association. It is perhaps the ability to be able to most appropriately associate relevant clinical/medical information to the emerging molecular information that provides us with the single largest challenge over the coming decades. In tandem with the generation of laboratory based data, sophisticated bioinformatics tools are being developed to tease out those genomic features that represent the key signatures and thus translate to delivery of appropriate information to medical decision makers.

**Cancer cytogenetics**

The presence of chromosome aberrations have been described in over 25,000 human neoplasms that representing 75 different types of cancer (http://cgap.nci.nih.gov/Chromosomes/Mitelman). Medical genomics has embraced these findings and led to the development of sophisticated molecular cytogenetic tests that are now of increasing importance in the detection, prevention, and management of cancer. In human medicine, the identity of cytogenetic aberrations has been shown also to assist in the localization of cancer-associated genes and selection of the most appropriate therapeutic approach.

**Comparative cytogenetic signatures in canine and human cancers**

The most well known cytogenetic aberration observed in human cancer was the Philadelphia chromosome ($Ph^1$). This cytogenetic aberration is observed in over 80% of adults diagnosed with chronic myelogenous leukemia (CML)[15]. Using classical cytogenetics this rearrangement was shown to involve an exchange of chromosomal material between the distal ends of human chromosomes (HSA) 9 and 22, generating a derivative chromosome 22 described as t(9;22)(q34;q11) [16]. Using locus specific fluorescence *in situ* hybridization (FISH), the Philadelphia chromosome may be observed in affected cells in which the gene *ABL* (usually located at HSA 9q34) is adjacent to the gene *BCR* (usually located at HSA 22q11). The chimeric activity of the *BCR-ABL* fusion gene results in elevated tyrosine kinase activity and it is this that is crucial to the oncogenic potential. In simple terms, the BCR-ABL protein sends signals that result in proliferation of abnormal white blood cells. Progression of the disease may be halted by a therapy that blocks this inappropriate signal. The signal blocker used is the BCR-ABL kinase antagonist, imatinib mesylate (Gleevec®) [15] which acts to stop the formation of new abnormal cells. While this effect does not apply to all patients with CML it is sufficiently effective that Gleevec therapy is now considered standard of care for patients shown to present with the Philadelphia chromosome. The presence of the Philadelphia chromosome is generally determined with molecular cytogenetics, though PCR based tests are also available. Since Imatinib mesylate actually inhibits several protein tyrosine kinases including platelet-derived growth factor receptor (PDGFR) and c-kit, which are preferentially expressed in tumor cells, this therapy is in ongoing studies to evaluate its effects in other cancers (eg [17-20]

From a veterinary perspective, CML is rare and those cases reported in the dog (eg [21-24]) faced a poor prognosis. A recent study aimed at demonstrating the evolutionarily conserved cytogenetic changes in corresponding human and dog cancers showed that some dogs diagnosed with CML also present with a BCR-ABL translocation (the 'Raleigh' chromosome) [24]. These data suggest that, cost aside, treatment with Gleevec (using careful monitoring for liver toxicity) could be an option for therapy of these canine CML cases. In the same study we showed that canine CLL and Burkitt lymhoma similarly presented with the diagnostic cytogenetic signatures of the corresponding human cancers.

In broader studies of a variety of canine cancers we have demonstrated the presence of numerous conserved cytogenetic signatures associated with cancer subtypes that are evolutionarily conserved with the corresponding cancer in humans. These data provide strong evidence that if the pathogenetic etiology of cancers are shared between human and dog, there is opportunity for persevering with human derived therapies in canine oncology. Moreover, from a comparative angle, genome wide assessment of genomic changes in canine cancers is leading to the identification of key regions of the canine genome, and hence the human genome. Within regions we will be able to identify a key subset of evolutionarily-conserved genomic changes that are intimately associated with the disease process, isolating these from the multitude of aberrations that are secondary effects of tumor-associated genomic instability. For example the development of genome-integrated canine molecular cytogenetics resources allow the interrogation of tumor DNA samples and tumor cells to determine the presence of genome-wide aberrations at increasing resolutions [25-28]. Using these reagents and resources we have evaluated genome wide DNA copy number changes in canine osteosarcoma [29]and intracranial tumors (ICTs) [30]and compared these to what is known in the corresponding human cancers. Analyses of threes data revealed that progress towards development of more sophisticated molecular subclassification of cancers and ensuing tailored therapies will come from evaluating and comparing genomic changes in both species.

**Emerging approaches to clinical genomics**

The development of DNA sequence-based microarrays allow the genotype of an individual to be surveyed at ever increasing resolution. In addition, gene expression microarrays (gene chips) are available to allow simultaneous assessment of the level of activity for several thousand genes within a cell population. Currently, gene expression microarrays are commercially available for numerous veterinary species, including dog, horse, cattle, pig and chicken. Similarly, microarrays are being developed for veterinary species that genotype simultaneously tens of thousands of discrete regions within a single genome, evaluating the genome of an individual for variation at the level of single nucleotides. These single nucleotide polymorphisms, (SNPs) may be grouped together to develop specific haplotypes that are associated with particular complex disease. Using these approaches, it is now possible to study several hundred individuals, grouped according to a shared 'phenotype' in a relatively short timeframe. Through the use of complex statistical algorithms and bioinformatics, significant phenotype-genotype associations may be revealed, involving small regions of the genome. Subsequent study of these regions will then reveal the underlying genes involved and hence provide the means to develop appropriate diagnostic processes.

## IN CONCLUSION

Currently the application of advanced genomic technologies for routine health is cost prohibitive, even for human medicine. However, technology is becoming faster and les expensive. The human genome was estimated to have cost approximately \$3-5,000,00,000 and took almost 15 years to complete, the mouse genome cost ~\$100,000,000 and took approx. three years, while the dog genome cost ~ \$40,000,000 and took about a year and the horse genome cost ~ \$15,000,000 and took approx. six months. The race is on for the \$1,000 genome sequence [31, 32]. It is conceivable that, as with human health, the next decade may bring genotyping and/or whole genome sequencing/analysis and/or proteinomic evaluation of animal patients for the purposes of health management. Such analysis of patients will begin to play a central role in routine veterinary care as we approach the ultimate goal in clinical veterinary genomics, individualized medicine based on the genetic health status of our animals. While private practitioners may not be required to perform these emerging tests, it is vital that as with a CBC, or chest radiograph, the clinician be able to interpret the results of these new assays and articulate the results to their clients. Accessibility to new information is becoming ever easier and faster, and the demands of the public are increasing. The veterinary profession needs to be prepared to adapt to these changing and challenging times.

## REFERENCES


[1]Lander, E.S., et al. **Nature** 2001;409(6822):860, [2]**Nature** 2004;431(7011):931, [3]Waterston, R.H., et al. **Nature** 2002;420(6915):520, [4]Adams, M.D., et al. **Science**2000;287(5461):2185, [5]Kornberg, T.B., et al. **Science** 2000;287(5461):2218, [6]Myers, E.W., et al. **Science** 2000;287(5461):2196, [7]Tenenbaum, D. **J Natl Cancer Inst** 2003;95(23):1742, [8]Feany, M.B., et al. **Nature** 2000;404(6776):394, [9]Lindblad-Toh, K., et al. **Nature** 2005;438(7069):803, [10]Pontius, J.U., et al. **Genome Res** 2007;17(11):1675, [11]Schellander, K., et al. **Anim Genet** 1992;23(6):549, [12]Olsaker, I., et al. **Anim Genet** 1993;24(4):311, [13]Ennis, S., et al. **Anim Genet** 1994;25(6):425, [14]Sargan, D.R. **Mamm Genome** 2004;15(6):503, [15]Kurzrock, R., et al. **Ann Intern Med** 2003;138(10):819, [16]Rowley, J.D. **Nature**1973;243(5405):290, [17]Cristofanilli, M., et al. **Ann Oncol** 2008;May 29, [18]Caudell, J.J., et al. **Appl Immunohistochem Mol Morphol** 2005;13(2):167, [19]Scheinfeld, N. **J Drugs Dermatol** 2006;5(2):117, [20]Vuky, J., et al. **Invest New Drugs** 2006;24(1):85, [21]Tarrant, J.M., et al. **Vet Clin Pathol** 2001;30(1):19, [22]Leifer, C.E., et al. **J Am Vet Med Assoc** 1983;183(6):686, [23]Pollet, L., et al. **J Small Anim Pract** 1978;19(8):469, [24]Breen, M., et al. **Chromosome Res** 2008;16(1):145, [25]Thomas, R., et al. **Cytogenet Genome Res** 2008;122(2):110, [26]Thomas, R., et al. **Genome Res** 2005;15(12):1831, [27]Young, A.C., et al. **Chromosome Res** 2008;16(8):1193, [28] Thomas, R., et al. **J Hered** 2007;98(5):474, [29]Thomas, R., et al. **Chromosome Res** 2009;in press, [30]Thomas, R., et al. **Journal of NeuroOncology** 2009;in press, [31]Service, R.F. **Science** 2006;311(5767):1544, [32]Wise, J. **BMJ** 2008;336(7638):237

**DISPELLING THE MYTHS ABOUT DIAGNOSIS & TREATMENT OF FELINE RESPIRATORY INFECTIONS**

**Leah A. Cohn, DVM, PhD, DACVIM (SAIM)**
**Columbia, MO**

## INTRODUCTION

In an ideal world, diagnosis and treatment would be based on scientific evidence. Unfortunately, we often have poor quality or no real evidence to guide clinical practice. This leads to practice based on anecdote and opinion. Even when evidence exists, there is such a huge body of medical literature that we as practitioners may not be familiar with what has been scientifically substantiated or countered. This talk aims to evaluate some of the common "myths" associated with the diagnosis and treatment of feline respiratory infections and provide scientific evidence when available to support or refute common beliefs.

## MYTHS

### Domestic cats are not susceptible to influenza virus infection

Although as yet undocumented in North America, natural infection of cats with H5N1 avian influenza (not canine influenza H3N8) has been recognized in both Asia and Europe.[1,2] The first reports of the highly pathogenic avian influenza H5N1 virus' ability to infect cats came from Thailand, where tigers and leopards housed in a zoo became infected after consuming tainted poultry.

In addition to natural infections, experimental infections have been transmitted by direct inoculation of the virus parenterally or through the respiratory tract in domestic cats.[1-9] It is believed that initial viral replication occurs in the respiratory and GI tract with subsequent viremia. It is apparent that some cats can remain well despite infection; these cats most likely have exposure to lesser viral loads.[9] In those with greater exposure, extensive pulmonary damage and multifocal organ hemorrhage and necrosis are responsible for mortality. Neurologic signs including ataxia and seizure in naturally infected cats likely results from non-suppurative encephalitis. More common signs are non-specific, and include fever, depression, elevation of the third eye lid, conjunctivitis, increased respiratory effort, nasal discharge, and icterus. Sometimes, sudden death is observed within days of infection.

There are more likely causes for an acute onset of respiratory signs in cats (eg, herpes virus, calicivirus, bacterial pneumonia) but veterinarians must be vigilant to the possibility of influenza infection. Documented infections in birds in the area, high fever, and access to the outdoors all increase the index of suspicion for influenza. Viral isolation or RT-PCR from oropharyngeal or rectal swabs or necropsy specimens are the typical methods of confirmation. Immunohistochemistry can be used on infected organs. Serologic diagnosis using hemagglutination inhibition is also possible. Treatment would be largely supportive, although the wisdom of treating a cat with confirmed infection is uncertain considering the possible zoonotic potential of this disease. Transmission of virulent avian influenza from infected cats to humans has not been documented.[10] However, the virulent H5N1 virus has been transmitted from infected poultry to humans with resultant high morbidity and mortality rates.

### Chronic nasal discharge in cats is due to persistent/recurrent herpes viral infection

Chronic nasal discharge in adult cats (eg. chronic rhinosinusitis or CRS) is a common and frustrating condition. Often, veterinarians tend to reach for antibiotics to treat this condition without pursuing a specific diagnosis in a belief that all or nearly all CRS is the result of persistent or recurrent viral infection with secondary bacterial infection. In fact, many non-viral disease conditions may result in similar clinical signs of chronic nasal discharge. In two separate studies of chronic discharge in adult cats, the most common single etiologic diagnosis was nasal neoplasia, including both carcinoma and lymphoma. [11,12] Neoplasia accounted for between 25 and 39% of specific causes of chronic nasal discharge. These findings emphasize the importance of a diagnostic investigation rather than simply assuming that CRS is the result of viral infection. Other important differential diagnosis include inflammatory rhinitis, mycotic rhinitis (e.g., cryptococcosis, aspergillosis), dental disease, nasal polyps, and foreign bodies.

Feline herpes virus (FHV-1) is a common cause of acute upper respiratory infection in cats. After clinical recovery, cats remain latent carriers of the virus. The latent virus can be re-activated later in life and result in pathology, especially ocular pathology. Many veterinarians therefore believe that CRS for which no other specific cause (eg, neoplasia, fungal rhinitis, foreign body) is found is likely due to reactivation of latent FHV-1. In a study of 10 cats with CRS and 7 control cats without nasal disease, Johnson et al failed to isolate FHV-1 from any cat in either group as would have been expected had reactivation of virus been the cause of clinical signs.[13] Additionally, the same study documented FHV-1 by PCR assay from both CRS affected cats and control cats lacking nasal signs in equal proportions. Although tests used in this study certainly may have missed some FHV-1 infected cats, there is no reason to expect that more sensitive test methodologies would have resulted in a different proportion of cats with FHV-1 in either group.

Just because active virus was not detected in cats with CRS more often than in controls does not negate a possible role for FHV-1 in causation of CRS. FHV-1 may have a role without viral reactivation. The virus can cause cytolysis of nasal turbinates which might impair local physical defenses to the extent that secondary bacterial infections would result.[14] Further, FHV-1 infection might cause CRS either by triggering an immune mediated or inflammatory response in the nasal passages. In fact, nasal samples from cats with detectable FHV-1 mRNA displayed increased transcription for a variety of cytokines.[15]

### Lung worms can be ruled out by fecal floatation

In certain geographic regions, lung worms account for a substantial proportion of respiratory disease in cats. Important parasitic respiratory pathogens include the trematode fluke *Paragonimus kellicotti*, the lung worm *Aelurostrongylus abstrusus*, and *Eucoleus aerophilia* (aka, *Capillaria aerophilia*). A common misconception holds that these parasites can be ruled out through fecal floatation. This belief is incorrect on two fronts. Firstly, pulmonary parasites are shed through the feces only an intermittent basis. Unlike intestinal worms, immature pulmonary parasites (ova or larvae) must be expectorated and swallowed prior to fecal shedding. For this reason, shedding is only intermittent at best. Even when using the most sensitive, most appropriate fecal examination techniques, intermittent shedding means that infections may be missed from a single sample. Therefore, it is recommended that fresh feces be examined on at least 3 different occasions prior to "ruling out" lungworms as a cause of respiratory disease. Alternatively, a course of anthelmintic sufficient for the treatment of lung worms can be employed as a therapeutic trial. The second reason that the belief that fecal flotation can rule out pulmonary parasites is just a myth is that floatation is not always the most sensitive test for detection of immature parasites. Fecal flotation is the test of choice for *E. aerophilia*, but not for the other lung worms. Simple sedimentation techniques are preferred for the detection of *A. abstrusus* and *P. kellicotti*.

**Azithromycin is the "best" antibiotic for cats with upper respiratory infection**

Acute upper respiratory infections (URI) are common in cats. The organisms most often detected in shelter or cattery environments included feline herpesvirus-1 (FHV-1), feline calicivirus (FCV), *Pasteurella* species, *Chlamydophila felis*, *Mycoplasma* species, *Bordetella bronchiseptica*, and coagulase-negative *Staphylococcus* species.[16] Veterinarians often treat cats with URI with antimicrobials both because of the possibility of primary bacterial infection (eg, *C. felis*, *B. bronchiseptica*, +/- *Mycoplasma*), and because of the potential for secondary bacterial infections (eg, *Pasteurella*, *Staph*) in association with other pathogens, including viral pathogens. Anaerobic bacteria too may have a role in feline URI.[13] A variety of antimicrobials have been used for empiric treatment of URI. Considerations for choosing an antimicrobial include cost, ease of administration, and efficacy. Efficacy is determined by several factors including the susceptibility of involved pathogens as well as pharmacokinetics and pharmacodynamics of the drug. Traditionally, amoxicillin has been commonly used for the treatment of cats with URI. It is inexpensive, safe, available in convenient dosage strengths, and has activity against most Gram + and anaerobic pathogens. However, it would not be efficacious for the treatment of either *B. bronchiseptica*, *Mycoplasma*, or *C. felis*. Traditional treatment of *C. felis* (which is predominantly associated with conjunctivitis rather than rhinitis) involves administration of doxycycline, a drug with likely efficacy in the treatment of *B. bronchiseptica* and *Mycoplasma* as well.[17]

Recently, many veterinarians have switched from these "traditional" treatments with amoxicillin or doxycycline to azithromycin. A macrolide antimicrobial commonly used for the treatment of respiratory infection in people, azithromycin has a long tissue half-life in cats allowing it to be dosed infrequently.[16,18] The drug would be expected to be efficacious for the treatment of *Mycoplasma* as well as many Gram + and Gram – organisms. In fact, many veterinarians now believe that azithromycin has been demonstrated to be the "best" antibiotic for the treatment of URI in cats. Although only a couple of comparative studies have been completed, azithromycin has not been found to be superior to more traditional antibiotics. Cats with URI suspected to be related to bacterial infection (primary or secondary) in a shelter population were given a clinical disease score and then treated with either azithromycin or amoxicillin.[16] Cats that continued to display clinical signs after therapy were crossed over to receive the other drug. Over half of the cats had FHV-1 detected by PCR, and a large variety of bacterial organisms were isolated for the nose or pharynx, with a large proportion growing either *Pasteurella* or *Mycoplasma* species. Despite the fact that some infections would be predicted to not respond to amoxicillin based on in vitro culture results, there was no difference in outcome between cats treated with amoxicillin and those treated with azithromycin. In another study, cats with experimental *C. felis* infection were treated with either azithromycin or doxycycline.[17] While both drugs resulted in rapid resolution of clinical signs, azithromycin failed to eliminate the pathogen. Because infected but non-clinical cats could remain contagious to cats (or humans), failure to eliminate the pathogen is a serious concern.

Azithromycin remains a reasonable antimicrobial choice for the treatment of URI in cats and does have the advantage of infrequent dosing. However, it is considerably more expensive than some more traditionally used medications and may not offer any substantial advantage in regards to efficacy of treatment.

**Pyothorax is caused by cat fight or puncture wounds**

Pyothorax is a cause of substantial morbidity and mortality in cats. Traditionally, veterinarians have been taught that pyothorax follows cat fight wounds. This seemed to fit with the predominance of infections caused by mixed bacterial populations of anaerobes and facultative anaerobes plus oral flora.[18] Recently, it has been proposed that infections are more likely the sequelae of parapneumonic spread of infection after colonization and invasion of lung tissue by oropharyngeal flora related to upper respiratory infection.[19] This observation is not only supported by the findings of the retrospective study which proposes this theory, but also by prior observations that pyothorax is most often identified in young cats from multiple cat households regardless of exposure to the outdoors.[20] Cats within the same household seldom fight but often transmit respiratory infections.

**Identification of an organism from nasal culture or PCR in a cat with upper respiratory signs proves causation**

Unfortunately, a variety of both bacterial and viral organisms can be found as normal flora in healthy cats, after recovery from infection of previously ill cats, or as pathogens involved in current infections. Simply finding an organism doesn't prove cause and effect. This is true even if the organism is a known pathogen, such as FHV-1. Cats infected with FHV-1 remain latently infected long after clinical recovery from any illness; cats previously vaccinated for FHV-1 may be infected but never develop any clinical illness yet remain latently infected. Therefore, FHV-1 may be identified in either perfectly healthy cats, or in cats with upper respiratory signs due to some altogether different cause. To even make matters more complicated, commonly used assays such as PCR cannot distinguish naturally occurring virulent FHV-1 infection from modified live vaccine strains.[21] So as an example of the frustrations of organism identification confirming disease causation, one could be presented with a cat suffering from chronic nasal discharge as a result of nasal lymphoma with a positive FHV-1 PCR test. Despite test results, the cat is ill because of lymphoma, not herpes viral infection!

A similar problem occurs in relation to calici virus. Cats may be healthy carriers of the virus for prolonged periods of time. One of these carrier cats might easily develop upper respiratory signs for some reason completely independent of calici virus infection and yet "test positive" for calici. Simply finding calici virus doesn't prove causation.

Similar issues arise with bacterial infections. Most organisms which cause bacterial rhinitis as secondary, opportunistic pathogens are found as part of the normal nasal or oropharyngeal flora. Simply growing these organism from nasal culture does not prove that they are the primary cause of nasal/upper respiratory signs, nor does it even prove that they are a secondary causative factor in disease. Even the organisms considered to be primary pathogens (e.g., *B. bronchiseptica* and *C. felis*) can be identified in cats lacking any clinical signs of illness. The lack of correlation between results of culture and sensitivity from nasal samples and disease is even supported by some treatment trials. When cats with upper respiratory infections are treated with antimicrobials which would not be expected to be efficacious for a given bacterial isolate, they often recover quickly anyway.[16]

**Chronic nasal discharge should prompt culture of the discharge**

As mentioned above, the nasal passages and nasopharynx have a rich flora. This flora is not only normal but may help protect the airways from infection by occupying an ecologic niche in a non-harmful manner. For this reason, nasal culture is apt to grow a variety of organism which may or may not be in any way related to clinical signs of nasal disease. Interestingly, different organisms may be retrieved when the same cat is cultured at the same time by different means or at different sites (e.g., lavage, swab, or biopsy; nasal or oropharyngeal samples).[13,22] It seems reasonable that culture growth from tissue obtained by nasal biopsy would be more meaningful that growth from lavage or swab samples, but this has not been investigated. In the author's opinion, culture of nasal discharge is rarely useful, and culture of nasal biopsies or deep tissue swabs is only occasionally useful in the management of chronic feline nasal discharge. However, this is an opinion only, with little of merit to support or refute the assumption! It is rare for specific antibiotic therapy guided by culture results produces a sustained resolution of chronic nasal discharge, while acute nasal discharge often resolves with any of a variety of antibiotics (even those not predicted to be effective based on in vitro testing).[16] Veterinarians who wish to obtain nasal cultures should request anaerobic and *Mycoplasma* culture since these types of pathogens may have relevance and are not detected by routine aerobic culture.

**Testing by PCR is always sensitive & specific**

PCR can only detect what is present in the area sampled at the moment it is sampled. Even when a given pathogen causes disease, it may not be continuously present in the sampled tissue. A good example is PCR for detection of influenza virus in dogs; if the nasopharynx is swabbed when there is no viral shedding, false negative results will be had. On a given day PCR results may be positive, while they would be negative the next day in the same animal. In a recent comparison of results when cats with URI were sampled by PCR from both the nasal passages and oropharynx, results of PCR for *Mycoplasma* and FHV-1 were in agreement in only 19 or 39 cats tested (48.7%).[22]

In other cases, PCR results may be positive when a given organism is not responsible for disease. This has already been discussed in regards to FHV-1. Another relevant example is PCR testing for feline infectious peritonitis (FIP). The dry form of FIP may result in fatal pulmonary granulomas and viral pneumonia in cats. However, the diagnosis of dry form FIP is extremely difficult. A mRNA RT-PCR test of feline mononuclear cells has been touted as a useful new tool in this difficult diagnosis.[23] Unfortunately, this relatively sensitive tool may not be as specific as first believed with false positive results in 54% of healthy cats tested.[24]

Additionally, not all PCR tests nor laboratories are created equally. Some tests are able to detect fewer organisms than others. Some laboratories have more stringent quality control than others. When it comes to PCR testing, the moto "buyer beware" comes to mind.

## OTHER MYTHS

**FVRCP vaccination prevents infection with feline herpes and calici virus**

Vaccination minimizes clinical disease but prevents neither infection nor spread of infection by vaccinated cats.

**Conjunctivitis is always caused by *Chlamydophila felis***

There are many potential causes, include FHV-1.

**Cats don't get mycotic pneumonia/rhinitis**

Cats are susceptible to both mycotic pneumonia and rhinitis.

**Calici virus is the predominant cause of feline gingivitis/stomatitis**

Gingivitis/stomatitis is often unassociated with any infection .

**Virulent calici virus can be confirmed by PCR**

There is no single test to distinguish hyper-virulent strains.

***Mycoplasma* is only a commensal organism and not a feline pathogen**

As we learn more, it appears that at least some species of *Mycoplasma* do contribute to both lower and upper airway disease in cats.[13,22]

## REFERENCES

[1.] Starick E, et al. Vet Microbiol 2008,128:243; [2.] Buranathai C, et al. Avian Dis 2007, 51:194; [3.] van Riel D, et al.: **Science** 2006,312:399; [4.] Rimmelzwaan GF, et al.: **Am J Pathol** 2006,168:176; [5.] Vahlenkamp TW, et al. **J Gen Virol** 2008,89:968; [6.] Klopfleisch R, et al.: **Vet Pathol** 2007,44:261; [7.] Songsermn T, et al.: **Emerg Infect Dis** 2006,12:681; [8.] Kuiken T, et al.: **Science** 2004,306:241; [9.] Leschnik M, et al.: **Emerg Infect Dis** 2007,13:243; [10.] Butler D: Nature 2006,440:135; [11] Demko JL et al. **J Am Vet Med Assoc.** April 2007;230:1032. [12)] Henderson SM, et al. **J Feline Med Surg**. 2004;6:245. [13.] Johnson LR, et al. **J Am Vet Med Assoc**. 2005;227:579. [14.] Hoover EA, et al. **Lab Invest** 1971;25:457. [15.] Johnson LR et al. **Vet Micro** 2005 ;108 :225; [16.] Ruch-Gallie RA et al. **J Feline Med Surg** 2008, 10: 542; [17.] Owen WMA et al. **J Feline Med Surg**. 2003; 5:305; [18.] Hunter R P et al. **J Vet Pharmacol Ther**. 1995;18:38. [19.] Walker AL, et al. **J Am Vet Med Assoc**. 2000, 216:359-63. [20.] Barrs VR, et al. **J Feline Med Surg**. 2005;7:211. [21]. Waddell LS, et al. **J Am Vet Med Assoc**. 2002;221:819. [22.] Maggs DJ, et al. **Am J Vet Res**. 2005, 66:1550. [23.] Vier JK, et al. **J Feline Med Surg**, 2008, 10, 551.[24.] Simons, FA et al. .2005 **J Virol Methods** 124:111. [25] Can-Sahna K, et al. **J Feline Med Surg** 2007, 9:369.

**DISPELLING THE MYTHS ABOUT DIAGNOSIS & TREATMENT OF FELINE ASTHMA**

**Carol Norris Reinero, DVM, DACVIM, PhD**
**Columbia, MO**

## INTRODUCTION

Misconceptions about the definition, causation, diagnostics and therapeutics for feline asthma are commonplace. This lecture will present the evidence or discuss the lack of evidence associated with each of these, with a focus on therapeutics.

## DEFINITION

Asthma is a chronic inflammatory disorder of the lower airways and also has airway hyperresponsiveness and airflow limitation (leading to clinical respiratory signs), and airway remodeling as prominent features [1]. Both genetics and environmental exposures are thought to play an important role in the development of asthma. In humans, the genetic predisposition to forming aeroallergen-specific IgE is the strongest identifiable underlying cause of asthma [1].

## CAUSATION: TRUTH OR MYTH?

Asthma in cats is also believed to be induced by inhalation of aeroallergens. In sharing the same environment, humans and their pet cats are exposed to the same types of aeroallergens and develop a remarkably similar syndrome of asthma. Evidence to support that these inhaled allergens are clinically important contributors to allergy in both species include the following: (1). Over the past two decades there has been an increase in the incidence of human allergic asthma, especially in developed countries [2,3]. Similarly, one abstract on pet cats with asthma in a large urban city showed a parallel increase in diagnosed cases of feline asthma over the past 20 years [4]. (2). A large study showed concurrent diagnosis of allergy in humans with a veterinary diagnosed allergy in dogs or cats and this concomitant occurrence of allergies in both species was postulated to be due to shared environmental factors [5]. Similar environmental allergens have been identified in both species. (3). Pet cats with clinicopathologic features of asthma naturally develop allergen-specific IgE antibodies [6-9]. When the most common allergens identified in serum IgE panels were used to create an experimental model of asthma, characteristic features of allergic asthma including allergen-specific airway hyperreactivity, eosinophilic inflammation, an early Thelper2 mediated cytokine profile in blood and bronchoalveolar lavage fluid (BALF) and airway remodeling changes were induced [8]. While allergen avoidance has been suggested to ameliorate clinical signs of asthma [7,10,11] evaluation of airway inflammation or hyperreactivity were not performed in these studies and this can only be considered weak support of an allergic etiology. Similarly, allergen-specific hyposensitization ("allergy shots") also reportedly resulted in clinical improvement, although documentation of resolution of pathologic abnormalities was not done [6,7].

Asthma in cats has been suggested to be related to dietary hypersensitivity. With the waxing and waning signs that normally occur with this disease in many cats, it might appear that a change of diet could alter clinical signs. However, there is no scientific evidence supporting a dietary causation of feline allergic airway disease. Definitively proving this association would be challenging, as even in cats with chronic gastrointestinal disease speculated to be related to food sensitivity, assaying for food allergen-specific IgE in serum or with gastroscopic food sensitivity testing has been unrewarding [12]. In humans, food allergy manifesting as asthma alone is both rare and atypical and avoidance of specific foods has not been shown to improve asthma [13]. Studies on nutritional factors in the development of atopic disease in infants and children has been performed and demonstrate: there is no role for maternal dietary restrictions during pregnancy or lactation, breastfeeding for at least 4 months may prevent or delay wheezing in early childhood, and there is no protective effect of any dietary intervention beyond 4-6 months of age for the development of atopic disease [14].

## DIAGNOSTICS: TRUTH OR MYTH?

Allergic asthma is associated with an eosinophilic inflammatory response. However, it is not true that the absence of a peripheral eosinophilia can either make the diagnosis of feline asthma or rule it out. Depending on the study, a peripheral eosinophilia was noted in only 17-46% of cases [10,11,15,16]. Importantly, the presence of a peripheral eosinophilia has not been correlated with the degree of airway eosinophilia [10,15,16]. It is also a common misconception that thoracic radiographs can be used to make a definitive diagnosis of asthma. First, normal thoracic radiographs have been reported in up to 23% of cases [11]. Second a more characteristic bronchial or bronchointerstitial pattern can be seen with other respiratory diseases of the cat including chronic bronchitis, heartworm associated respiratory disease (HARD) and lungworm infection (*Aleurostrongylus abstrussus*). Other clinical and diagnostic features must be considered, including collection of bronchoalveolar lavage fluid (BALF). While there is still ongoing debate on what should be considered a "reference range" for the various cell types in BALF and how to best discriminate feline asthma from feline chronic bronchitis, eosinophilic airway inflammation is supportive of a diagnosis of feline asthma (especially once parasitic diseases have been ruled out). Unfortunately in cats, one of the key tests performed in humans to definitively diagnose asthma (pulmonary mechanics) is not readily available [1]. Spirometry to measure forced expiratory volume in one second (FEV-1) and forced vital capacity (FVC) is routinely performed in humans over 5 years of age. Demonstration of reversibility of airway obstruction (at least in part) with bronchodilators is fundamental in the diagnosis of asthma. In cats, direct and indirect measures of airway resistance have been showed to be of value, although these are rarely performed due to the requirements for specialized training and equipment [8,15,17]. The aforementioned means of measure airway resistance can also be performed in tandem with specific allergen-challenge or a non-specific bronchoprovocant (e.g., carbachol or methacholine). Other diagnostic approaches (eg, evaluation of exhaled nitric oxide, various mediators in exhaled breath condensate, and other biomarkers in blood or BALF) still need further investigation in cats [1,18,19].

**TREATMENT: TRUTH OR MYTH?**

Feline asthma is routinely treated with injectable or oral glucocorticoids. In healthy cats, inhaled glucocorticoids have minimal systemic effects on the adaptive immune system, although they can still suppress the hypothalamic-pituitary-adrenocortical axis [20]. Inhaled glucocorticoids have been shown to reduce eosinophilic airway inflammation in experimental feline asthma [21]. There are many different types and strengths of inhaled glucocorticoids, and there is a lack of peer-reviewed studies that have compared all these different inhalant preparations in cats with lower airway disease. One study suggests that (as in humans) more is not always better, and at least in experimentally asthmatic cats a dose of fluticasone at 44ug/actuation BID is as effective in reducing eosinophilic airway inflammation as 220ug/actuation BID [22].

It is commonly accepted that glucocorticoids effectively control lower airway inflammation and should be titrated to the lowest effective dose to control clinical signs of disease. While evidence is still preliminary, a series of cases was previously reported documenting subclinical airway inflammation from BALF specimens in pet cats with lower airway disease [23]. Even though cats were placed on high doses of prednisolone and owners reported improvement or complete resolution of clinical signs, a high percentage of inflammatory airway cells persisted in BALF. This brings into question whether clinical signs should be used to guide response to therapy, or if repeated collection of BALF will be the gold standard to document resolution of inflammation. Peer reviewed studies still need to be performed to examine this further.

Bronchodilators are important in medical management of asthma. Albuterol (also called salbutamol in Europe), a short acting beta-2 agonist, is frequently given by metered dose inhalation using a spacer in pet cats. Chronic use of inhalant racemic albuterol has been shown to induce neutrophilic airway inflammation de novo in healthy cats and exacerbate eosinophilic airway inflammation in experimentally asthmatic cats (the so called "beta agonist paradox") [24]. Inhalant racemic albuterol should thus be administered as "rescue" therapy and not be given as monotherapy in the daily treatment for feline asthmatics. Other evaluated bronchodilators in cats include ipratropium (a anticholinergic) and salmeterol (a long acting beta 2 agonist). Specifically, a combination of inhaled albuterol and ipratropium has been shown to reduce bronchoscopy-induced airflow limitation in experimentally asthmatic cats [25]. In that same model of asthma, albuterol, ipratropium and salmeterol were administered by a metered dose inhalant/spacer or nebulizer and airway hyperreactivity in response to carbachol challenge was measured [26]. Results showed salmeterol was associated with the weakest degree of bronchoprotection and the MDI/spacer had relatively comparable efficacy to the nebulizer.

A subpopulation of humans with allergic asthma responds to cysteinyl leukotriene (cysLT) antagonists and given similarities between human and feline asthma, leukotriene modifiers have been advocated in cats with this disease. Urinary excretion of cysLT (a reflection of endogenous total body production) did not differ between healthy cats and pet cats with lower airway disease [27]. Using an experimental model of feline asthma, there was no difference in urinary excretion of cysLT or in BALF cysLT compared with control non-asthmatic cats or in asthmatic cats before and after allergen-challenge[28]. In an in vitro model using tracheal and bronchial smooth muscle strips from experimentally asthmatic cats, incubation with a 5-lipoxygenase inhibitor to block cysLT production prior to allergen challenge did not blunt the type I immediate hypersensitivity reaction (ie, smooth muscle contraction)[29]. Finally, in a placebo controlled drug trial using zafirlukast (a cysLT receptor antagonist) there was no significant difference in BALF eosinophilia or in airway hyperreactivity after methacholine challenge[21]. In humans, genetic variation in key proteins in the leukotriene pathway influences response to the leukotriene modifiers and it is speculated that a genetic signature might be able to be developed to predict which patient are likely to respond to this type of treatment[30]. While this may theoretically be similar in cats, the lack of an ability to detect differences in whole body (urinary) or BALF cysLT production between non-asthmatic and asthmatic cats supports the lack of importance of this mediator in the pathogenesis of feline asthma. Thus, at this point, there is no compelling evidence to recommend use of leukotriene modifiers in cats with asthma.

Serotonin is an important mediator in the immediate type I hypersensitivity reaction in cats with experimentally induced asthma, at least in vitro[29]. A drug trial using a common dose of the serotonin antagonist cyproheptadine (2 mg BID) did not reduce airway eosinophilia compared with placebo; although overall airway hyperreactivity was not significantly reduced with treatment, individual cats (2 of 6) did have decreased airway hyperresponsiveness[21]. When a higher dose of cyproheptadine was used (8 mg BID) [31], no significant effect on eosinophilic airway inflammation was noted[32]. This study did not evaluate airway hyperreactivity. While serotonin can not be advocated as monotherapy for the treatment of feline asthma since it does not blunt airway inflammation, studies evaluating its effect (at a higher dose of 8 mg BID) on dampening bronchoconstriction are still warranted.

Like serotonin, histamine is speculated to play a role in the acute phase response of allergic asthma. Histamine content in BALF (reflective of mast cell degranulation) was documented to increase in experimentally asthmatic cats after allergen challenge [33]. Use of the 2nd generation histamine 1 receptor antagonist cetirizine has some protective effects in humans with mild asthma after histamine challenge and may decrease pro-inflammatory mediators; however, in experimentally asthmatic cats, it did not significantly reduce eosinophilic airway inflammation or other measured immunologic variables [32]. Like serotonin, it can not be advocated as monotherapy for feline asthma.

Dietary omega-3 fatty acid supplementation has been reported to be of therapeutic use for a variety of pro-inflammatory diseases including lower airway disease. In an experimental model of feline asthma, no significant anti-inflammatory effect (evaluated by analysis of BALF total cell numbers and percentage of eosinophils) was noted with a 4 week course of oral omega-3 fatty acid supplementation and luteolin [34]. However, in this pilot study, decreases in airway hyperreactivity in response to carbachol challenge were noted with treatment using an indirect measurement (barometric whole body plethysmography).

If an underlying allergic trigger can be identified as a cause for asthma, allergen avoidance or allergen-specific immunotherapy can be tried. It is inherently challenging to identify clinically relevant allergens in pet cats, as timing and amount of allergen exposure, different numbers and types of allergens, concurrent medications, and other environmental factors may influence test results. Both intradermal skin testing and serum allergen-specific IgE determination can be used; in experimentally asthmatic cats with known/controlled aeroallergen exposure the former is sensitive and the latter (by the high affinity Fc epsilon receptor based ELISA) is highly specific [35]. Abbreviated protocols for allergen-specific immunotherapy called rush immunotherapy (RIT) have shown great promise in experimental feline asthma [36,37]. Ongoing studies are focusing on "cross-protection" with RIT using one allergen unrelated to the allergen used for experimental sensitization and challenge.

## FUTURE DIRECTIONS

Much work needs to be done to understand the pathogenesis of naturally developing feline asthma and to develop better diagnostics and therapeutics. In particular, future efforts should focus on: 1. Coming to a consensus on reference ranges for BALF cytology, ideally not using research or shelter cats. 2. Determining the best ways to discriminate between feline asthma and chronic bronchitis (ie, what should be a standard work up for these cases); this should include identification and evaluation of less invasive "biomarkers" and ideally finding a way to include pulmonary mechanics (especially with bronchoprovocation testing) 3. Multicenter studies looking at novel treatments for pet cats with naturally developing asthma.

## REFERENCES

1. Busse W. Expert panel report 3: Guidelines for the diagnosis and management of asthma. Bethesda, MD: US Department of Healthy and Human Services; NHLBI, 2007. 2.Redd S. Asthma in the United States: burden and current theories. **Environ Health Perspect** 2002;110 Suppl 4:557-560. 3.Tattersfield A, Knox A, Britton J, et al. Asthma. **Lancet** 2002;360:1313-1322. 4.Ranivand L, Otto C. Feline asthma trends in Philadelphia, Pennsylvania 1896-2007. 26th Annual Symposium of the **VCRS** 2008. 5.Schafer T, Merkl J, Klemm E, et al. We and our pets: allergic together? **Acta Vet Hung** 2008;56:153-161.6.Halliwell R. Efficacy of hyposensitization in feline allergic diseases based on upon results of in vitro testing for allergen-specific immunoglobulin E. **J Am Anim Hosp** 1997;33:282-288. 7.Prost C. Treatment of feline asthma with allergen avoidance and specific immunotherapy: experience with 20 cats. **Revue Francaise d'Allergologie et d'Immunologie Clinique** 2008;48:409-413.8.Norris Reinero C, Decile K, Berghaus R, et al. An experimental model of allergic asthma in cats sensitized to house dust mite or bermuda grass allergen. **Int Arch Allergy Immunol** 2004;135:117-131.9.Moriello K, Stepien R, Henik R, et al. Pilot study: prevalence of positive aeroallergen reactions in 10 cats with small-airway disease without concurrent skin disease. **Vet Dermatol** 2007;18:94-100.10.Corcoran B, et al. Feline asthma syndrome: A retrospective study of the clinical presentation in 29 cats. **J Small Anim Pract** 1995;36:481-488.11.Adamama-Moraitou K, Patsikas M, Koutinas A. Feline lower airway disease: a retrospective study of 22 naturally occurring cases from Greece. **J Fel Med Surg** 2004;6:227-233.12.Guilford W, Jones B, Markwell P, et al. Food sensitivity in cats with chronic idiopathic gastrointestinal problems. **J Vet Intern Med** 2001;15:7-13.13.Beausoleil J, Fiedler J, Spergel J. Food intolerance and childhood asthma: what is the link? **Paediatr Drugs** 2007;9:157-163.14.Greer F, Sicherer S, Burks A, et al. Effects of early nutritional interventions on the development of atopic disease in infants and children: the role of maternal dietary restriction, breastfeeding, timing of introduction of complementary foods, and hydrolyzed formulas. **Pediatrics** 2008;121:183-191.15.Dye J, McKiernan B, Rozanski E, et al. Bronchopulmonary disease in the cat: historical, physical, radiographic, clinicopathologic, and pulmonary functional evaluation of 24 affected and 15 healthy cats. **J Vet Intern Med** 1996;10:385-400.16.Moise N, Wiedenkeller D, Yeager A. Clinical, radiographic, and bronchial cytologic features of cats with bronchial disease: 65 cases (1980-1986). **J Am Vet Med Assoc** 1989;194:1467-1473.17.Kirschvink N, Leemans J, Delvaux F, et al. Non-invasive assessment of airway responsiveness in healthy and allergen-sensitized cats by use of barometric whole body plethysmography. **Vet J** 2007;173:343-352.18.Sparkes A, Mardell E, Deaton C, et al. Exhaled breath condensate (EBC) collection in cats--description on a non-invasive technique to investigate airway disease. **J Fel Med Surg** 2004;6:335-338.19.Reinero C, Nafe L, DeClue A. Bronchoalveolar lavage fluid inflammatory markers in cats with asthma and chronic bronchitis. 18th Annual **ECVIM Congress** 2008;Abstract.20.Reinero CR, Brownlee L, Decile KC, et al. Inhaled flunisolide suppresses the hypothalamic-pituitary-adrenocortical axis, but has minimal systemic immune effects in healthy cats. **J Vet Intern Med** 2006;20:57-64.21.Reinero C, Decile K, Byerly J, et al. Effects of drug treatment on inflammation and hyperreactivity of airways and on immune variables in cats with experimentally induced asthma. **Am J Vet Res** 2005;66:1121-1127.22.Cohn L, DeClue A, Cohen R, et al. Dose effects of fluticasone propionate in an experimental model of feline asthma. **ACVIM Forum 2008**.23.Reinero C. The importance of subclinical inflammation in canine and feline lower airway disease. **ACVIM Forum 2008**.24.Reinero C, Delgado C, Spinka C, et al. Enantiomer specific effects of albuterol on airway inflammation in healthy and asthmatic cats. **Int Arch Allergy Immunol** In press.25.Kirschvink N, Leemans J, Delvaux F, et al. Bronchodilators in bronchoscopy-inuced airflow limitation in allergen-sensitized cats. **J Vet Intern Med** 2005;19:161-167.26.Leemans J, Kirschvink N, Bernaerts F, et al. A pilot study comparing the antispasmodic effects of inhaled salmeterol, salbutamol and ipratropium bromide using different aerosol devices on muscarinic bronchoconstriction in healthy cats. **Vet J** in press.27.Mellema M, Gershwin L, Norris C. Urinary leukotriene levels in cats with allergic bronchitis. **ACVIM Forum 1999**;724.28.Norris C, Decile K, Berghaus L, et al. Concentrations of cysteinyl leukotrienes in urine and bronchoalveolar lavage fluid of cats with experimentally induced asthma. **Am J Vet Res** 2003;64:1449-1453.29.Padrid P, Mitchell R, Ndukwu I, et al. Cyproheptadine-induced attenuation of type-I immediate-hypersensitivity reactions of airway smooth muscle from immune-sensitized cats. **Am J Vet Res** 1995;56:109-115.30.Lima J. Treatment heterogeneity in asthma: genetics of response to leukotriene modifiers. **Mol Diagn Ther** 2007;11:97-104.31.Norris C, Boothe D, Esparza T, et al. Disposition of cyproheptadine in cats after intravenous or oral administration of a single dose. **Am J Vet Res** 1998;59:79-81.32.Schooley E, McGee Turner J, Jiji R, et al. Effects of cyproheptadine and cetirizine on eosinophilic airway inflammation in cats with experimentally induced asthma. **Am J Vet Res** 2007;68:1265-1271.33.Mitchell R, Cozzi P, Ndukwu I, et al. Differential effects of cyclosporine A after acute antigen challenge in sensitized cats in vivo and ex vivo. **Br J Pharmacol** 1998;123:1198-1204.34.Leemans J, Cambier C, Chandler T, et al. Dietary supplementation with an omega-3 polyunsaturated fatty acid and luteolin combination in cats with experimentally-induced asthma: a preliminary study. 26th Annual Symposium of the VCRS 2008.35.Lee-Fowler T, Cohn L, Reinero C. Comparison of intradermal skin testing and allergen-specific serum IgE in experimental feline asthma. 18th annual ECVIM Congress 2008.36.Reinero CR, Byerly JR, Berghaus RD, et al. Rush immunotherapy in an experimental model of feline allergic asthma. **Vet Immunol Immunopathol** 2006;110:141-153.37.Reinero C, Cohn L, Delgado C, et al. Adjuvanted rush immunotherapy using CpG oligodeoxynucleotides in experimental feline allergic asthma. **Vet Immunol Immunopath** 2008;121(3-4):241-250.

# DIAGNOSIS & TREATMENT OF FELINE PULMONARY VASCULAR DISEASE: WHAT DO WE KNOW?

**Deborah M. Fine, DVM, MS, DACVIM (cardiology)**
**Columbia, MO**

## INTRODUCTION

The two most common forms of pulmonary vascular disease in cats are pulmonary thromboembolism[1-6], and right to left shunting congenital cardiac disease.[7,8] Of these two general categories, pulmonary thromboembolic disease is the more common, though both of these disorders are relatively rare.

## PULMONARY THROMBOEMBOLISM

As in dogs and humans, pulmonary thromboembolism (PTE) in cats is almost always associated with some form of serious underlying abnormality. The pathogenesis of PTE includes venous stasis, exposure of sub-endothelial collagen, and hypercoaguability (Virchow's triad). One or more of these factors is almost always found to be present in cats that have been diagnosed with PTE. In two relatively large retrospective studies in cats, a number of disorders were found to be associated with PTE.[1,2] These included cardiac disease, neoplasia, disseminated intravascular coagulation, pancreatitis, protein losing nephropathies and enteropathies, immune-mediated hemolytic anemia, sepsis, and in one instance there was no underlying abnormality found.

The age of affected cats at the time of death or euthanasia tends to be middle-age to older; however the age ranges reported have been from 10 months to 18 years. There does not appear to be any breed or gender predisposition associated with feline PTE. Most of the cats diagnosed with PTE were domestic shorthairs, and this most likely reflects their general prevalence.

### Diagnosis

The most common complaints reported by the cats' owners were very nonspecific and included lethargy, anorexia, weight loss, difficulty breathing, vomiting, and coughing. Abnormalities detected during initial physical examination included lethargy, tachypnea or dyspnea, dehydration, hypothermia, and icterus.

Clinical pathologic abnormalities of PTE are nonspecific. The most common abnormalities reported have included increased neutrophils, BUN, glucose, and bilirubin. Decreases were reported in hematocrit, platelet count, sodium, calcium, potassium, and albumin. Results of coagulation testing were only reported in a very few cats (n = 7).[1,5] One-stage prothrombin time was high in three cats, and APTT was high in seven cats. Fibrin degradation product concentrations were only measured in two cats, and were within the reference ranges.

One study specifically looked at the association with indwelling intravenous catheters during hospitalization.[1] In that study the majority of cats had short-term peripheral venous catheters placed. However, a number of cats had intermediate to long-duration placement of catheters. This included two long-term hemodialysis catheters. It is likely in the majority of these cats that the intravenous catheters were simply associated with the treatment of the underlying disease process, and not the inciting event. None the less, in some cases this may represent an important risk factor.

Definitive diagnosis of PTE requires the use of pulmonary scintigraphy or spiral computed tomography.[4,6] Pulmonary perfusion scintigraphy is carried out by obtaining images with a gamma camera after intravenous injection of biodegradable radioactive particles which are temporarily held up in the pre-capillary level of pulmonary circulation. Poorly or non-perfused sectors of pulmonary parenchyma thus exhibit less radioactivity than normally perfused sectors. In humans, the introduction of the multi-detector row spiral CT has made high-resolution image acquisition possible. This allows identification of small thrombi. This form of CT has replaced selective pulmonary angiography and ventilation perfusion scanning as the image technique of choice in humans. CT can also assess for parenchymal changes associated with pulmonary thromboembolism such as hemorrhage or infarction and can identify or rule-out alternative diagnoses. Unfortunately, there is limited availability of this equipment in veterinary medicine, and both modalities require general anesthesia for the patients.

A radiographic angiogram is less sensitive for detection of small emboli than the above modalities due to the large overlap of blood vessels within the lung fields. However, larger filling defects indicative of the mass or thromboembolism can be detected. This form of imaging also requires general anesthesia or at least sedation. Plain film radiographs can be helpful for assisting in the diagnosis of PTE, particularly when there is a high index of suspicion for the disorder. There is a wide variety of radiographic abnormalities that have been associated with PTE in cats. Pulmonary vascular abnormalities are among the most common findings and descriptions include uneven vascular diameter and uneven distribution of blood flow between lung lobes. Also seen are larger than normal central portions of pulmonary arteries with disproportionate tapering of the diameters in the middle and peripheral lung fields. Evidence of hypovolemia is manifested as a lucent appearance to the periphery of the lung lobes in the absence of hyperinflation. Also reported are abnormalities of the pulmonary parenchyma, including peripheral consolidations usually involving the caudal lobes, isolated demarcated thoracic masses, or atelectasis of one or more lung lobes. Additional findings noted fairly frequently included varying degrees of pleural effusion, and mild to moderate cardiomegaly. In some instances, radiographs were interpreted as normal or showing only mild vascular or parenchymal changes. The presence of severe respiratory distress in the patient coupled with normal or near-normal radiographs should highly increase the index of suspicion for PTE.

Echocardiography can sometimes be used to visualize a thrombus in the main pulmonary artery or large right and left branches. Pulmonary thromboembolus causes pulmonary hypertension, and this manifests as dilation of the right atrium and ventricle. In severe cases the interventricular septum may actually deviate toward the left heart.

A variety of pulmonary function tests were performed in one case report.[4] Results in this cat documented a normal $PaCO_2$ with markedly decreased $Pa0_2$ on arterial blood gas analysis. The alveolar-arterial gradient was also markedly increased in comparison to normal. An increased A-a gradient is a sensitive measurement of pulmonary impairment. Other tests included measurements of total lung capacity, functional residual capacity, alveolar volume, and calculation of physiologic dead space. Most of these measures do not have normal reference ranges established in cats; however those that did were abnormal.

**Treatment**

In most of the reported cases of PTE in cats, the diagnosis was not suspected antemortem, and therefore specific therapy was not directed at this problem. However, there are case reports of successful treatment of feline PTE.[4,6] In these instances therapy included anticoagulants and supportive measures. Cage rest to avoid additional mobilization of any thrombi is important. Supplemental oxygen should be given in most cases. However, depending upon the extensiveness of thromboembolic disease, these patients may be non-responsive or only partially responsive to oxygen therapy.[2] The ideal anticoagulant therapy is unknown in these patients. Treatment strategies that have been utilized include low molecular weight heparin or unfractionated heparin, aspirin, and warfarin. In one cat whose PTE was associated with heartworm disease, treatment with corticosteroids was also utilized.[4]

## PULMONARY HYPERTENSION & RIGHT TO LEFT SHUNTING CARDIAC DEFECTS

The pathophysiology of right to left shunting cardiac defects is uncertain. There are two commonly hypothesized explanations for this abnormality. The first hypothesis proposes that there is a persistence of fetal-type vasculature. That is, the fetal pulmonary circulation fails to remodel at birth and remains a high pressure/ resistance system. The other hypothesis suggests that the increased flow through a large left to right shunting defect stimulates vasoconstriction of the precapillary arterioles in an attempt to protect the microcirculation from flooding. Over time this results in progressive morphologic changes including medial hypertrophy an intimal hyperplasia. Regardless of which hypothesis is true, the end result is an animal with varying degrees of exercise intolerance, respiratory difficulty, cyanosis, and sometimes syncope. The cause of death in these cases is often progressive right heart failure, but in some instances sudden-death may occur due to arrhythmias or pulmonary thromboembolism.

Although right to left shunting cardiac defects in the cat are rarely described, there are numerous case reports in the literature.[7,8] Most of the cases reported were in relatively young cats, however it is not uncommon for cats with congenital heart disease to survive well into adulthood.

**Diagnosis**

Echocardiography is generally a very sensitive modality for the diagnosis of congenital cardiac defects. The most common features seen on echocardiography with right to left shunting defects are a dilated right atrium and right ventricle. The right ventricular free wall is often markedly hypertrophied, and the interventricular septum is usually flattened and deviated toward the left ventricle. These findings are very indicative of markedly increased right-sided pressures. In the absence of pulmonic stenosis, the only other cause of increased right ventricular pressure is pulmonary hypertension. If tricuspid valve regurgitation is present, continuous-wave Doppler can be used to determine the severity of pulmonary hypertension. The velocity of the tricuspid regurgitated jet is measured, and the modified Bernoulli equation is applied to determine the pressure gradient (pressure gradient = $4 * velocity^2$). Anything over 50 mmHg is considered indicative of severe pulmonary hypertension.

A saline-contrast echocardiogram ("bubble study") is often helpful for confirming the diagnosis of right to left shunting defect. To perform this procedure a peripheral venous catheters placed. A small volume of saline (2 to 3 mL) is drawn into a syringe with a few drops of the patient's blood (this helps to suspend the micro bubbles). The syringe is vigorously agitated, then all of the large bubbles are ejected from the syringe. The agitated saline is then injected rapidly into the catheter while imaging the heart. In normal patients, the bubbles will appear in the right atrium and then the right ventricle and then will be taken up by the pulmonary circulation. No bubbles should appear in the left heart. In a patient with a right to left shunting ventricular septal defect, bubbles will be visualized crossing the defect into the left ventricle and aorta. In a patient with of right to left shunting patent ductus arteriosus, the abdominal aorta should be imaged for ultrasonographic appearance of bubbles. Because the shunt is distal to the heart, no bubbles will appear in the left ventricle.

**Treatment**

The presence of pulmonary hypertension makes surgery difficult to impossible in these cases, however many of the reported cats appear to live a reasonable quality of life with just conservative management.

## REFERENCES

[1]Norris CR., et.al. **JAVMA** 1999;215(11):1650. [2]Schermerhorn T., et.al. **JVIM** 2004;18:533. [3]Sottiaux J., et.al. **JSAP** 1999;40:88. [4]Davidson BL., et.al. **JVIM** 2006;20:1037. [5]Whitley NT., et.al. **Aust Vet J** 2001;79(6):403. [6]Pouchelon JL., et.al. **JSAP** 1997;38:306. [7]Connolly DJ., et.al. **JSAP** 2003;44:184. [8]Hearon K., et.al. **JAVMA** 2006;228(10);1495.

# DISPELLING THE MYTHS ABOUT FELINE HEARTWORM ASSOCIATED RESPIRATORY DISEASE

**Ray Dillon, DVM, MS, MBA, DACVIM**
**Auburn, AL**

## CURRENT BASICS OF HEARTWORM INFECTION IN CATS

The last 30 years have increased the basic understanding of heartworm disease in the cat and emphasized the clinical importance of this disease. Several basic concepts have been confirmed by data collected in clinical practices, shelter cat populations, and experimental studies:

- Cats develop bronchial disease as a consequence of immature adult heartworms which never become fully adult heartworms and initiate Heartworm Associated Respiratory Disease (HARD) which can persist for 12 months after a single short lived infection
- Cats which get only immature adults that live for 2-3 months and cats which develop adult heartworms both develop HARD associated with death of immature adults
- Cats get immature heartworms as early as 75-90 days after infection and immature heartworms die creating lung and arterial disease which persist for up to 12 months
- Cats which develop lung disease from immature heartworms which die may be antibody positive for only a few months although the bronchial disease persists
- Cats with only immature adult heartworms will be antigen negative and echocardiographically worm cannot be visualized although bronchial disease is evident radiographically
- The bronchial disease of heartworms in cats has: 1.)a bronchial epithelium component, 2.)a smooth muscle proliferation remodeling, and 3.) alteration of bronchial smooth muscle reactivity
- Some cats do develop adult heartworms which may live for up to 3-4 years in client cats
- After adult heartworm develop, the inflammatory reaction of the lung is reduced in cats and symptoms may be limited for long periods of time although bronchial pattern continues to be evident radiographically
- The lung recovery from heartworm infections are not uniform in parenchymal distribution leaving severely diseased parenchymal adjacent to relatively normal lung.
- Fully mature adult heartworms have the ability to down regulate the pulmonary intravascular macrophages in the lung
- The death of immature as well as adult heartworms may be associated with acute lung injury and severe Type 1 cell injury of alveolar capillary beds
- Some cats which develop fully mature adult heartworms demonstrate minimal clinical signs prior to acute crisis typically associated with adult worm death
- Most cats which develop adult mature heartworms survive the infection as the worms die over the next 2-4 years

### Implications in Determining Prevalence of Disease

The weight of the evidence has defined that where ever there is heartworm disease of dogs, there is heartworm disease of cats, it just does not present as heartworm disease of dogs. Prevalence of a disease (HARD) which is caused by a transient infection which eludes detection by physical and immunologic testing, tends to under report the disease.

If cats infected with mature adult heartworms were combined with cats with HARDs from active immature heartworm death, and then were combined with cats that have residual lung disease from heartworms have died (and antibody is not detectable), then heartworm disease in cat and dogs would probably be of equal incidence in most locations, assuming a suitable vector in the area to feed on both dogs and cats.

### Response of the Cat Lung to Adult Heartworm

If the course of a D immitis infection is evaluated chronologically, the changing nature of the disease is evident. As the parasite first arrives in the lungs as early as 75 days after being infected by a mosquito, the lung responses with intense inflammation and "asthma- like" symptoms may develop. The cat has specialized macrophages (Pulmonary Intravascular Macrophages- PIMs) in the capillary beds of the lung that are not present in the dog. If the mature parasite develops, the clinical signs may be intermittent or absent.

The adult parasite seems to be able to suppress the macrophage function in the lung. The cat will have classic radiographic and histologic findings of feline heartworms, but may not show clinical signs. After the adult parasite develops, the pulmonary parenchymal changes and even enlarged caudal pulmonary arteries on VD radiographs may decrease. However, at the time of worm death, the suppression of macrophage function is decreased and the lungs become extremely inflamed and the specialized macrophages may become important in the intense reaction. The result is a non-functioning lung and an acute respiratory distress syndrome. Although this reaction can occur as the result of even a single worm burden, spontaneous death of cats from heartworm infection is uncommon compared to clinical disease associated with the early 3-6 month disease.

Usually severe dyspnea associated with heartworm disease and especially that which results in death is the consequence of an adult heartworm dying. After the removal of dead heartworms, there may be continued inflammatory lung disease in some cats. Following a clinical group of cats in Italy (Venco), client cats were positive on antigen tests and echocardiographically for over 4 years before all cats cleared the infection.

**Response of the Cat to the Early Infection**

The disease associated with feline heartworm infection is a moving target, with the pathology, and resulting clinical signs, dependent on the stage of the life cycle involved. Early arrival of L5 results in classic asthma-like radiographic and clinical signs. In cats during this early part of the infection, coughing and dyspnea can be intermittent. A peripheral eosinophilia may or may not be present. Typically, the cytology of BAL reveals an eosinophilic reaction which continues even if all the immature heartworms die. The radiographic pattern can be dynamic and right-sided cardiac changes are not present. The inflammatory pattern in the lung parenchyma is peribronchial but may be severe enough to be a diffuse alveolar pattern. Pulmonary arterial patterns may be normal although if the periarterial inflammation is severe, the right and/or caudal pulmonary arteries may appear enlarged. Often, the inflammatory lung pattern is severe enough that the pulmonary arteries cannot be visualized. The arrival of immature L5s in the 75-90 post infection period represent a time frame where many immature heartworms are dying, causing inflammatory lung disease, even in cats which later develop adult heartworms. Even in experimental cats with adult heartworms, the finding of heartworm fragments 6 and 8 month after L3 infection would confirm that immature heartworms have a higher mortality rate that in dogs where almost all become mature adult heartworms.

**Three Month Disease Cycle- Diagnostic Considerations**

With the initiation of pulmonary disease at 3 months after the infection, pathology and clinical signs may be present 3 months before antigen or microfilaria would be present in the blood and 3 months before heartworms could be visualized echocardiographically. Thus, dogs and cats with lesions during this early phase are "heartworm negative" by the typically applied screening tests used in practice. Although the radiographic and histopathologic lesions are present during the early phases (months 3-6 post L3 infection) in the dog, rarely are dogs presented for clinical signs during this time frame. However, because of the intense reaction of the cat lung to the arrival of these 2-3 cm L5 at 3 months and/or the reaction of the lung to the higher mortality of these early young adults, clinical signs are frequently noted during this stage of the infection.

Some cats will develop these early lesions and become antibody positive, but over time (3-6 months) the young adult parasites in the distal pulmonary arteries all die. Although this form of heartworm infection could be considered as "self-curing," lesions were induced, clinical signs may have develop, and long term consequences of this infection often persist for at least a year in some cats. The alveolar and bronchial changes associated with this early form of the disease will clinically and radiographically mimic "feline asthma" but the concept that these bronchiole with smooth muscle proliferation are "hypercontractile" cannot be assumed. Cats will develop an eosinophilic BAL cytology, radiographic bronchial lesions without pulmonary arterial changes, and clinical signs of coughing and/or dyspnea will respond to typical corticosteroid therapies. Cats that are heartworm antibody positive have been successfully infected with the L3 which have molted to the L4 and L5, typically (depending on the test methods) lived 2 ½ to 3 months, and may or may not have gone on to arrive in the distal pulmonary arteries as immature adults.

**Clinical Evidence of the HARDs**

In a study of cats presented to practicing veterinarians with a history of coughing and/or dyspnea, 42% were heartworm initially antibody positive. When cats with radiographic lesions and/or positive serology were examined over a 3-month time frame, the changes demonstrated that this infection is dynamic and a diagnostic challenge. Discordant serologic results between test methods can occur because each identifies a different antibody during the initial phases of the infection. Over the course of the 3 month observation period of client cats, 60% of cats presented for the first time for coughing were heartworm antibody positive at least on one test and had radiographic disease consistent with heartworm disease. Of cats initially antibody positive at presentation with coughing and lung disease, 30% were antibody negative 3 months later.

**HEARTWORM ASSOCIATED LUNG DISEASE (HARDs)**

The HARDs term has been coined to describe the lesions associated with the arrival and death of immature heartworms (2-3 cm) as early as 75-90 days after infection. It is important to point out that this pathology develops also in cats which then develop adult heartworm infections as well. The radiographic lesions are identical in cats which have only immature adult heartworms and those which develop adult heartworms. Thus the HARDs term can be used in cats which develop adult heartworms as well.

With the new understanding of the effects of immature heartworms on cat bronchial disease, the most likely time to identify this syndrome would be 3-6 months after the peak mosquito season in heartworm areas. Testing of chronic bronchial cats would prove to be unlikely to identify cause and affect relationships because even if the lung disease were induced by immature heartworms, the antibody test would most likely be negative 3-4 months after the initial insult.

**HARDs may not be "asthma like"**

The bronchial disease of heartworms is a combination of 1.) epithelial disease and early eosinophilic cytology on BAL, 2.) proliferation of smooth muscle around the bronchioles and dissecting through parenchymal areas (remodeling) 3.) altered reactivity of bronchial smooth muscle to stimuli (reversible hypercontracitility) or inhibited reactivity in the inability to respond to a bronchodilator. The dynamics of bronchial disease should dissuade veterinarians from using the term "asthma like" in feline lung disease characterized as a radiographic "bronchial pattern". The finding of a bronchial pattern on radiographs and the clinical response of a coughing cat to corticosteroids has resulted in the assumption that all of these cats are "asthma like". Intermittent reversible bronchoconstriction may NOT correlate with bronchial remodeling (histopathologic lesions or radiographic bronchial patterns). This is in the same clinical concept that response to corticosteroids does not always implicate a histologically evident inflammatory disease. Likewise, the reversal of remodeling and resolution of smooth muscle proliferation after the single insult of a single HARD event, does not lessen the possibility of clinical disease as the smooth muscle left may be "twitchy" and potentially hyper responsive to stimuli. Evaluation of isolated bronchial rings for contractility and relaxation mediators, and the 3D CT studies of feline lung before and after provocholine will assist in this evaluation. Until these are determined, the clinical signs of dyspnea and coughing in heartworm cats should be characterized as bronchial disease and the term "asthma like" used only as it is a convenient "lay" term to use with cat owners.

The clinical and experimental evidence continues to support the basic fact that immature heartworms often infect cats, cause bronchial and lung disease, and no physical evidence (worm fragments) or antibody to heartworms can be identified as quickly as 2-3 months after worm death. The author used this phrase to describe heartworm disease in cats to clients: "The juvenile delinquent form of heartworms is a hit and run driver in cats. The immature forms come in, create disease, and any evidence that they were the cause is not present by the time we arrive on the scene."

### Three Month Disease Cycle- Preventative Considerations

The obvious goal for veterinary medicine is to prevent disease. Further, given the nature of immunity to repeated infections, in climates where the presence of mosquitoes' is seasonal, the risk for lung disease created by immature adults could be seasonal each year. The profession has been timid in recommending heartworm preventation to cat owners. The supporting evidence of the clinical studies of coughing client cats, shelter cat studies, and experimental studies have all confirmed that bronchial disease of cats can be caused by heartworms, even if adult heartworms never develop. Veterinarians should be reminded that cats placed on monthly heartworm prophylaxis may become heartworm antibody positive if infected by L3 from mosquitoes, as the worm may survive long enough to stimulate an antibody response but heartworms never reach the heart or lungs. Prevention of heartworm disease should be recommended to both dogs and cats where heartworm disease is present.

### REFERENCES

[1] Brawner B, Dillon AR, Robertson C. Guerrero H. Radiographic diagnosis of feline heartworm disease and correlation to clinical criteria: results of a multicenter clinical case study. Veterinary Therapeutic: Research in Applied Vet Med. 1,2: 81-87, 2000. Robertson C, Dillon AR, Brawner B, Guerrero H. Prevalence of feline heartworm infections among cats with respiratory and gastrointestinal signs. **Veterinary Therapeutic: Research in Applied Vet Med**. 1,2: 88-95, 2000. [2] Dillon AR, Robertson C, Brawner B, Guerrero H. Relationship of serology to clinical disease in feline heartworm disease. **Veterinary Therapeutics: Research in Applied Vet Med.** 1,3: 176-181, 2000. [3] L Browne, TD Carter JK Levy, P Synder, C Johnson, Pulmonar arterial disease in cats seroprositve of DI but lacking adult heartworms in the heart or lung. **AJVR** 9:1544-1549, 2005. [4] Dillon AR, Blagburn B, Tillson DM, Brawner W, Spencer J, Welles, Boothe D. Immature heartworm infection produces pulmonary parenchymal, airway, and vascular disease in cats. **J Amer Vet Int Med** 21: 608-609, 2007.[5] Buzhart L, Blagburn B, Cousins M, Dillon R, Levy J, Nelson T. Feline Heartworm Disease. **Compend Cont Educ** 30:8A,1-13, 2008. [6] Dillon AR, Warner AE, Brawner W, Hudson J, Tillson DM. Activity of pulmonary intravascular macrophages in cats and dogs with and without adult Dirofilaria immitis. **Vet Parasitol**, 158:171-176, 2008. [7] Venco L et al, **Amer Heartworm Society Symposium**, 2004.

# PATHOPHYSIOLOGY OF URINARY TRACT INFECTIONS IN DOGS & CATS

**Jeanne Barsanti, DVM, MS, DACVIM (Specialty of Internal Medicine)**
**Athens, GA**

## INTRODUCTION

UTI refers to microbial colonization of the urine or of any urinary tract organ, except the distal urethra, which has a normal bacterial flora. Infection of the urinary tract may affect more than one organ or may be localized to the upper tract (kidney and adjacent ureter; bacterial pyelonephritis) or to the lower tract (bladder and adjacent urethra; bacterial cystitis; bacterial urethrocystitis). Infections of the lower urinary tract in intact male dogs are considered to concomitantly affect the prostate gland (bacterial prostatitis). Infection of one part of the urinary tract increases the likelihood that the rest of the tract will become infected.

## ETIOLOGY

Bacterial UTIs are most common in female spayed dogs, followed by male castrated dogs and intact female dogs.[1] In a large study over a long period of time, 3.9% of female dogs and 2.9% of male dogs examined in one teaching hospital had positive urine cultures, although some dogs had their urine cultured only if the urinalysis was abnormal.[1] This lowers the infection rate since some dogs with UTI documented by culture have normal urinalyses. Infection rate in 85 asymptomatic, adult, intact male dogs was 9%.[1] Half of these dogs had the same organism isolated by direct aspiration of small prostatic cysts.

Over 70% of UTI in dogs are caused by a single bacterial species.[1] In complicated infections secondary to anatomic or functional abnormalities of the urinary tract, infection with multiple organisms becomes more likely. The most common gram-negative organisms are *E. coli*, Proteus, Klebsiella, Pseudomonas, and Enterobacter. *E. coli* is by far the most common urinary tract pathogen causing 37-45% of UTI.[1] In comparison with human urinary isolates, canine *E. coli* are more likely to be antibiotic resistant and to have a higher R-plasmid transmissibility rate.[1] Gram-positive organisms (Staphylococcus, Streptococcus, Enterococcus) account for 25-30% of naturally occurring UTIs.[1] Infection with Staphylococcus or Proteus is often associated with struvite calculi because of alkalinization of the urine by the organisms' metabolism of urea. The most common 8 species listed above account for 93% of all UTI.[1] Mycoplasma species have been reported as infrequent causes of UTI in dogs.[2] In the largest study of UTI in dogs, 35 bacterial genera and 4 species of fungi were identified as causing UTI.[1]

UTI are less common in cats than in dogs. Numerous studies have shown that young adult cats having signs of lower urinary tract problems (dysuria with hematuria) rarely have bacterial UTIs.[1] However, UTIs may develop secondary to urinary catheterization or to urinary tract surgery, such as perineal urethrostomy, in cats with urethral obstruction.[1] UTIs are more common in cats older than 10 years.[1] When UTI occurs in cats, the organisms most frequently involved are the same as those in dogs, except *Pasteurella* organisms are more commonly reported in cats.[1] *Mycoplasma felis* and *M. gateae* could not survive in osmotic conditions present in normal feline urine, although a *Ureaplasma* sp. could.[1] However, ureaplasmas were not isolated from any cats with lower urinary tract signs.[1]

Yeast and fungi in urine often represent contamination of the sample. However, fungi in properly collected (cystocentesis) and promptly examined urine samples are abnormal. Fungal UTI with *Trichosporon domesticum, Candida spp, or Aspergillus spp.* usually have been associated with abnormalities in host defenses, antibiotic usage, and urogenital diseases.[1, 3]

## PATHOGENESIS

UTIs are usually caused by bacterial organisms that are constituents of the microflora of the intestinal or lower urogenital tracts. Dogs with UTI associated with *E. coli* or *Proteus mirabilis* were likely to carry the same organism in their intestinal tracts and prepuce or vagina.[1] The usual method of infection of the bladder is by organisms ascending the urethra. Pyelonephritis is believed to be most commonly caused by ascending infection from bladder urine. Inflammation from cystitis can cause edema and deformation of the ureterovesicular junction, allowing bacteria to reach the ureters and then the kidneys.[4] Renal infections also occur in the absence of signs of cystitis. The renal cortex is much more resistant to infection than the renal medulla, decreasing the likelihood of hematogenous infection.

Development of UTI indicates an alteration in the host and bacterial flora inter-relationship.[1] To accomplish infection, bacteria must attach to and colonize the mucosa of the urethral orifice and transport themselves up the urethra, adhering to the uroepithelium. Both host defense mechanisms and bacterial virulence properties are important in determining whether infection occurs as well as what part of the urinary tract is affected.

### Host Factors

The inflammatory response in UTI involves three steps: the bacteria stimulate uroepithelial cells to produce inflammatory mediators, these mediators direct inflammatory cells to the site of infection, and the local inflammatory response determines if the infection is eliminated or if tissue damage results. Thus, host response is important in determining the degree of clinical signs and severity of tissue injury.[1] Many UTI are asymptomatic. Asymptomatic infections are associated with diabetes mellitus, hyperadrenalcorticism, and treatment with glucocorticoids in dogs, but occur in dogs and cats without these factors as well. The reasons are not understood, but one idea is that affected people are genetically different from those that develop symptomatic infections.[1] Genetic characteristics have also been investigated in relation to susceptibility to UTI.[1] For example, vaginal epithelial cells of women with recurrent UTI demonstrate increased bacterial adherence *in vitro* as compared to epithelial cells from women without UTI.[1,4]

Important host factors in resistance to infection are normal, frequent emptying of the urinary bladder, the presence of the normal flora, the age and sex of the host, characteristics of urine, and normal urinary tract anatomy, including normal urinary tract epithelium with its glycoprotein coat, and immunologic responses.[1] The importance of host factors is shown by the fact that many *E. coli* isolated from canine UTI have no detectable urovirulence factors.[1]

Structural, functional, metabolic abnormalities and treatment with certain classes of drugs predispose to UTI. Abnormalities include obstructive lesions, metabolic diseases such as diabetes mellitus, hyperadrenalcorticism, and renal failure, instrumentation or catheterization or abnormal voiding, particularly with neurologic diseases[5], and abnormal anatomy. UTI has been found in 40% of dogs with hyperadrenalcorticism[1], 20-40% of dogs with diabetes mellitus.[1,6] and approximately 20% of dogs treated with glucocorticoids.[7] Treatment with dexamethasone in association with surgery for acute thoracolumbar intervertebral disk herniation increased the risk for UTI 11 times.[8] In people, antibiotic use up to 4 weeks before onset of UTI increased risk of UTI by 3 to 6 times.[4] In cats, 17% with chronic renal failure, 13% with diabetes mellitus, and 22% with uncontrolled hyperthyroidism were found to have UTI.[9,10] Lower urinary tract obstruction with secondary vesicoureteral reflux of infected urine is a contributing factor to the development of pyelonephritis, as are anatomic abnormalities such as ectopic ureters. Approximately 17% of dogs with ectopic ureters were found to have pyelonephritis.[1] Other anatomic abnormalities associated with chronic UTI include persistent urachus, other bladder diverticuli, perivulvar dermatitis, and recessed vulva.[1] Although vestibulovaginal stenosis was associated with chronic UTI in one paper[1], a more recent study found no difference in vestibulovaginal ratios between normal spayed female dogs and dogs with lower urinary tract diseases, 10 of 19 of which had UTI.[11]

Bacteria adhere poorly to normal bladder epithelium owing to the presence of a glycosaminoglycan coating. This coating, which can be replaced within 24 hours if injured, is extremely hydrophilic, so that a layer of water forms at the surface. This aqueous layer provides a barrier between the transitional epithelium and the urine, explaining in part why bladder epithelium can tolerate constant exposure to a substance as irritating as urine. Infection is more likely to occur if this surface coating is damaged as by uroliths, neoplastic transformation, or exposures to chemical irritants such as cytoxan. There is no natural barrier against bacterial adherence in the kidney.[1]

Bladder epithelial cells exfoliate in response to infection and are cleared with the flow of urine, an important defense mechanism. In humans with UTI, large numbers of exfoliated bladder epithelial cells are often found.[1] Although this exfoliation is considered a host defense mechanism, it is also a way for bacteria to be spread into the environment and may expose undrelying epithelial cells to infection.

The antibacterial effects of urine include osmolality, urea concentration, and organic acid concentration.[1] Variations in pH (markedly acidic with a pH of 5) and specific gravity (markedly concentrated) have an inhibitory effect on bacterial growth. The ability of cats to highly concentrate their urine may be one explanation for the low incidence of bacterial UTI in young cats. However, one large study found that decreasing urine specific gravity was not correlated with UTI in cats, although increasing age, chronic renal failure, diabetes mellitus, and uncontrolled hyperthyroidism were.[9] Urine which is less acidic and not well concentrated supports multiplication of urinary tract pathogens almost as well as nutrient broth.[1]

It is clear from recent research in UTI, that minor weaknesses in host defenses necessitate more virulent organisms to induce disease. With normal host defenses, some infections will clear spontaneously or require minimal antimicrobial therapy. The more severe the abnormality in host defense, the more prone to infection even with less virulent organisms, and the more prone to severe infections that require precise and extensive therapy.[1]

**Bacterial Virulence Factors**

Uropathogenic *E. coli* (UPEC) are a genetically heterogeneous group of *E. coli* that differ from non-pathogenic *E. coli* by the presence of virulence factor genes (VFG). UPEC typically carry large blocks of genes, called pathogenicity-associated islands (PAIs) that are not found in fecal isolates.[1] These PAIs code for the virulence factors such as hemolysin, adhesins, iron acquisition systems, fimbriae, and toxins.[1] The presence of PAIs allows virulence genes to be easily spread between bacterial populations by horizontal gene transfer. People and pets can share *E. coli* strains, including those that cause UTI;[12] however, this is not common.[13] Virulence factor genotypes of feline UPEC were found to vary considerably with geographic location, New Zealand versus Great Brittain.[14]

The type of virulence genes present in a certain strain appears to determine the site of infection (lower versus upper tract) and the severity of that infection.[1] One genetic type of UPEC can change virulence properties depending on host response during infection.[1] Studies continue in regard to determining whether *E. coli* isolates that cause cystitis are distinct from those that cause pyelonephritis in genetic profiles or whether the host response determines how the genes operate.[1] *E. coli* from UTI in dogs have been found to have varying propensities to cause pyelonephritis and renal damage in mice based on urovirulence factors.[1] Even though bacterial virulence factors are important, the severity of the tissue injury is also dependent on the inflammatory reaction of the host.

UPEC are usually thought of as extracellular pathogens as they are cultured from the urine. Recent observations have found that these organisms are not always extracellular.[1] Type 1-piliated *E. coli* are able to enter bladder epithelial cells. Usually, this triggers the epithelial cell to undergo apoptosis and exfoliate. However, some *E. coli* seem to be able to persist intracellularly in some hosts, by moving to other superficial cells or to deeper epithelial cells.[1] Thus, infection can persist even though bladder urine is sterilized by antimicrobial therapy. *Fim*H, the adhesin for type 1 fimbriae which can trigger the uptake of *E. coli* into murine bladder cells, has been found in UPEC from cats.[14] Many recurrent infections are relapses with the same bacterial strains. The ability of *E. coli* to persist intracellularly is a possible explanation for some cases of relapsing UTI.[1]

**Re-infection Versus Relapse**

A re-infection is defined as a UTI due to a different organism that recurs at variable intervals after therapy is discontinued. A relapse is a UTI due to the same organism that recurs after therapy is discontinued. Although most relapsing UTI recur quickly, some do not relapse for months.[15] When recurrence of UTI with the same organism does not occur for months, it is difficult to distinguish a relapsing infection from a re-infection due to the same organism.
*E. coli* is the most common cause of UTI and there are many varieties of *E. coli*. Clinicians have relied on the sensitivity pattern of the organism to help determine whether the current infection is due to a new or the same *E. coli*. However, the same *E. coli* can shift in

regard to antibiotic sensitivity so this method is not very accurate.[15,16] Unfortunately, more accurate methods, such as pulsed-field gel electrophoresis, are not readily available.

**Relapsing Infections**

In relapsing infections, the problem is usually that the organism has found a deep seated niche in the urinary tract, protected from antimicrobial action. The antimicrobial kills the bacteria that are in the urine, but cannot reach those deep in tissue sites. The likely tissue sites are the kidney (chronic pyelonephritis) and the prostate gland in intact males. Struvite uroliths are another potential site of infection. Other sites are the submucosa of the bladder and any location associated with a partially obstructive process, such as a ureterolith.

**Re-Infections**

With re-infections, the problem is not curing the infection. The problem is that the pet becomes re-infected with a different organism at a variable interval. The causes include poor systemic immune function (such as hyperadrenalcorticism or immunosuppressive drug therapy), loss of one of the antimicrobial properties of urine (such as glucosuria associated with diabetes mellitus or poor urine concentrating ability due to any one of many causes), an anatomic predisposition to infection (such as a stricture, a urolith, a prior urethrostomy, or a neoplasm), or a physiologic predisposition to infection (such as urine retention from a neurologic problem or decreased urethral sphincter tone).

**Catheter Induced UTI**

One important iatrogenic cause for UTI is urinary tract catheterization. Bacteria, especially *E. coli* and Klebsiella, establish periurethral colonization, and these bacteria are inoculated into the bladder during intermittent catheterization. Even a single catheterization can result in UTI in female dogs.[1] In people, the risk of UTI with intermittent catheterization is 1-3% per insertion so that if the catheter must be inserted several times a day, most people become infected after a few weeks.[1] Risk of infection is greater if the catheter is left in place (indwelling). Bacteria can readily ascend either around or through the catheter. Closed, sterile systems can prevent bacterial access within the catheter lumen, which is the most rapid route of entry, but no method to date has been effective to prevent access up the extraluminal side of the catheter. In dogs, the risk of UTI with an indwelling catheter with a closed system increased by 20% for each year increase in dog age, 27% for each day increase in duration of catheterization, and 454% with administration of antimicrobial drugs.[17] The shorter the duration of indwelling catheterization, the less risk of UTI and the less risk of antibiotic resistant UTI.[18] Three days or more of indwelling urinary catheterization increases risk of UTI in both dogs and cats.[18,19]

Biofilm formation on the catheter is important in the development of infection when catheters are indwelling. Certain uropathogenic *E. coli* can promote biofilm formation on abiotic surfaces.[1] Biofilms consist of host materials including Tamm-Horsfall protein and crystals and bacterial products including bacterial polysaccharides and glycocalyces, and living bacteria.[1] A biofilm can be deposited within hours of catheter insertion.[1] Bacteria embedded in the biofilm are less susceptible to antimicrobial agents. Once infections become established on a urinary catheter, they are difficult to treat successfully, requiring catheter removal.

**REFERENCES**

[1]Barsanti JA. **Infectious Diseases of the Dog and Cat**, 3rd Edition, 2006, pp 935-961. [2]L'Abee-Lund TM, et al. **Vet Rec** 2003; 153:231. [3]Jin Y, et al. **J Am** Anim Hosp Assoc 2005; 41:373. [4]McLaughlin SP, et al. **Med Clin** N Am 2004; 88:417. [5]Stiffler KS, et al. **Vet Surg** 2006; 35:330. [6]Hume DZ, et al. **J Vet Intern Med** 2005;19:426. [7]Torres SMF, et al. **J Am Vet Med Assoc** 2005; 227:239. [8]Levine JM, et al. **J Am Vet Med Assoc** 2008; 232:411. [9]Bailiff N, et al. **J Vet Intern Med** 2007; 21:649. [10]Bailiff N, et al. **J Vet Intern Med** 2006; 20:850. [11]Wang KY, et al. J Vet Intern Med 2006; 20:1065. [12]Murray AC, et al. Ann Intern Med 2004; 140:848. [13]Stenske KA et al. **J Vet Intern Med** 2008; 22:785. [14]Freitag T, et al. **Vet Microbiol** 2005; 106:79. [15]Freitag T, et al. **J Vet Intern Med** 2006; 20:245; [16]Drazenovich N, et al. **J Vet Intern Med** 2004; 18:301. [17]Bubenik LJ, et al. **J Am Vet Med Assoc** 2007; 231:893. [18]Smarick SD, et al. **J Am Vet Med Assoc** 2004; 224:1936. [19]Ogeer-Gyles J, et al. **J Am Vet Med Assoc** 2006; 229:1584.

# FELINE URINARY TRACT INFECTIONS: NEW PATHOGENS, PRESENTATIONS, & TREATMENTS

**Annette Litster, BVSc, PhD, FACVSc (Feline Med), MMedSci (Clin Epi)**
**West Lafayette, IN**

## INTRODUCTION

The lower urinary tract has a variety of host defense mechanisms to prevent bacterial urinary tract infection (UTI). Physical barriers such as the length of the urethra; high pressure zones within the urethra; bacteria-trapping longitudinal folds in the proximal urethra; and urethral peristalsis resulting in a unidirectional urine flow, form the first line of defense. These are supported by mucosal defense barriers, including a glycosaminoglycan layer and intrinsic mucosal antimicrobial properties, to prevent bacterial migration and colonization, and by the composition of the urine (Blanco et al, 2001; Bartges, 2005).

Normal cat urine is highly concentrated, with urine specific gravity often exceeding 1.045 and an associated high osmolality (Lees et al., 1979). High concentrations of urea and organic acids and secreted antimicrobial peptides that inhibit bacterial colonisation work together with the acquired cell-mediated and antibody-mediated immune response (Blanco et al, 2001; Bartges, 2005), making it a remarkably hostile environment for bacterial growth. It is not surprising therefore, that bacterial urinary tract infections in cats are relatively rare events (Bartges et al, 2000). The reported prevalence is variable, depending on the inclusion criteria of investigating studies. Studies of cats with clinical signs of lower urinary tract disease (dysuria, stranguria, pollakiuria) have consistently shown that the overall prevalence of positive bacterial urine cultures is less than 3% (Kruger et al., 1991; Buffington et al., 1997; Lekcharoensuk et al., 2001). However, some studies have reported much higher prevalences (15-43%) in cats that have normal urinary tract defence mechanisms against infection compromised by the effects of other diseases and/or the treatment (e.g. urinary catheterization, perineal urethrostomy) of such conditions (Lees, 1996; Mayer-Roenne et al., 2007).

## NEWLY REPORTED BACTERIAL PATHOGENS OF THE FELINE URINARY TRACT

Our group recently published a study detailing bacterial isolates which caused bacterial urinary tract infection in 107 cats, their prevalence, and their antimicrobial susceptibility data (Litster et al, 2007a). The majority of Gram-negative bacterial isolates were *E. coli*, echoing the results of other published studies in cats (Davidson et al., 1992; Bartges et al., 2000;), while the most common Gram-positive bacterial isolate was *E. faecalis* (Table 1). In 84.1% (90/107) of specimens there was a pure growth of a single bacterial isolate; in 14.9% (16/107) of specimens two bacterial isolates were cultured; and in one specimen (0.93%) four different isolates including a multi-drug resistant *E. coli* were identified.

**Table 1.** Bacterial Isolates (n=126) from 107 Cats with Lower Urinary Tract Infections.

| Bacterial Isolate | Total |
|---|---|
| *E. coli* | 47 |
| *Enterococcus* species [1] | 36 |
| *Staphylococcus felis* | 25 |
| *Proteus species* [2] | 6 |
| *Enterobacter/Klebsiella* species [3] | 4 |
| *Pseudomonas aeruginosa* | 2 |
| *Staphylococcus aureus* | 2 |
| *Staphylococcus intermedius* | 2 |
| *Streptococcus bovis* | 2 |
| Total number of isolates | 126 |

[1] - *Enterococcus faecalis* (n=34) and *Enterococcus faecium* (n=2).
[2] - *Proteus mirabilis* (n=5) and one isolate identified as *Proteus vulgaris.*
[3] - One isolate identified as *Enterobacter aerogenes,* two isolates identified as *Enterobacter cloacae* and one *Klebsiella pneumoniae subspecies oxytoca* isolate.

We attempted to identify all Staphylococcal isolates to species level using standard biochemical protocols, but in the case of the coagulase-negative Staphylococci (CNS), the current commercial phenotypic identification systems were not discriminatory and could not differentiate between *Staphylococcus felis* and the other CNS species, in particular *Staphylococcus simulans*. This necessitated the use of partial 16S rDNA sequencing, which identified all the coagulase-negative isolates as *S. felis*. *S. felis* was first recognized from feline clinical specimens in 1989 (Igimi et al., 1989) and is regarded as a normal commensal organism present on the skin (Lilenbaum et al., 1998), the conjunctival sac and eyelid margins (Espinolaet al., 1996) and in the saliva of normal healthy cats (Lilenbaum et al., 1999). Igimi (1989, 1994) describes strains of *S.felis* isolated from clinical specimens from cats with cystitis, but there is no description of the method of specimen collection, the clinical history of the cats, or the degree or purity of growth. CNS are not widely known to be feline pathogens and individual species are difficult to identify (Takahashi et al., 1997). However, in our study, all *S. felis*-positive specimens were obtained aseptically by cystocentesis from cats with clinical signs of lower urinary tract disease, thereby minimizing the possibility of sample contamination by commensal organisms from the skin or lower urethra. The prevalence of *S. felis*-positive UTIs reported in our study suggests that *S. felis* is a common urinary tract pathogen of cats. However, further work, such as analysis of the isolates for virulence factors, would be required to confirm this. As *S. felis* is not found on the databases of most of the commercial Staphylococcal identification systems currently used by veterinary pathology laboratories, it is likely to be reported as a CNS species. The majority of *S. felis* isolates were susceptible to all antimicrobials tested, with only two strains resistant to penicillin/ampicillin and only one strain resistant to tetracycline, erythromycin, clindamycin and penicillin/ampicillin.

There have recently been case reports of *Corynebacterium urealyticum* (previously known as *Corynebacterium* group D2) in association with lower urinary tract infections in cats (Bailiff et al., 2005; Cavana et al., 2008). These non-hemolytic Gram-positive, rapid urea-splitting bacilli are uncommon causes of urinary tract infections in cats and in dogs and they present diagnostic and therapeutic challenges because of slow in vitro growth and the multidrug resistant nature of the pathogen (Bailiff et al., 2005). Risk factors for this type of infection include urological procedures, foreign bodies, bladder mucosa abnormalities, immuno-suppressed states and antibiotic treatment (Cavana et al., 2008). In particular, because of the ability of this organism to hydrolyze urea, infection may be associated with encrusting cystitis, a condition causing precipitation of struvite or calcium phosphate crystals on the bladder mucosa. Treatment should be based on the results of the most recent antimicrobial susceptibility patterns available and treatment of any predisposing factors (Bailiff et al., 2005).

**FELINE BACTERIAL UTIs: A NEWLY REPORTED CLINICAL PRESENTATION**

Although clinical signs of lower urinary tract disease (e.g. dysuria, stranguria, pollakiuria, hematuria) are common in feline practice (Buffington et al., 2006), bacterial urinary tract infections (UTI) in cats are relatively rare (Bartges et al., 2000). However, numerous studies have shown that risk factors such as age ($\geq$ 10 years), urinary catheterization and perineal urethrostomy can increase this rate markedly (Gregory et al., 1983; Barsanti et al., 1985; Davidson et al., 1992; Lekcharoensuk et al., 2001). Interestingly, a recent study which examined whether decreasing urine specific gravity was associated with positive urine culture reported that no such association existed, but pyuria, bacteriuria and hematuria were all associated with increased likelihood of positive urine culture (Bailiff et al., 2008).

Our group recently published a study that specifically examined bacterial lower UTI in cats in the absence of lower urinary tract clinical signs, a history of inappropriate urination, or previous UTI (including pyelonephritis) (Litster et al., 2009). Asymptomatic bacteriuria is a common but usually benign finding in women, and risk factors include pregnancy, diabetes mellitus, spinal cord injury, indwelling urinary catheterization and being an elderly resident of a nursing home (Nicolle, 2006). Occult bacterial UTIs have also been reported in diabetic dogs, but urinalysis results could reliably predict urine culture status and the presence of bacteriuria was inconsistent (McGuire et al., 2002). Care was taken in our study design to include only urine specimens that were collected by cystocentesis and yielded moderate or heavy growth on bacterial culture, so that iatrogenic bacterial contamination was not a confounding factor. All of our specimens met the widely used interpretive criteria for significant bacteriuria in feline urine ($\geq$1,000 CFU/mL of bacteria grown from a quantitative culture of urine collected by cystocentesis; Osborne et al 1999).

Culture positive urine specimens in our study of occult lower UTIs were more likely to be from older female cats (Litster et al., 2009), confirming the results of a previous large epidemiologic study of symptomatic cats (Lekcharoensuk et al., 2001). This finding is also echoed in studies of asymptomatic bacteriuria in humans (Nicholle, 2006) and occult canine UTIs (McGuire et al., 2002) and is not surprising given the relative ease with which resident gastrointestinal flora can ascend the relatively short and wide female urethra from the perineum and establish an infection. The bacteria cultured from urine specimens in asymptomatic cats also appeared to have initiated an inflammatory response in the urinary tract as positive cultures were accompanied by high urine erythrocyte and leukocyte counts (Litster et al., 2009). Asymptomatic bacteriuria in humans is also usually accompanied by pyuria (Nicolle, 2006).

Determination of urine bacterial culture status is increasingly becoming part of a minimum database of laboratory information collected for a wide variety of clinical presentations. In our study, approximately 10-15% of cats that presented for hyperthyroidism, diabetes mellitus or chronic renal disease had a bacterial lower UTI, confirming the findings of other studies (Bailiff et al., 2006; Elliott, 2007; Mayer-Roenne et al., 2007; Bailiff et al., 2008). Also, some cats that presented for routine geriatric checkups had culture positive urine, but these cats had no clinical signs of illness of any kind. The positive relationship between age and positive urine culture status reported in our study and another other study (Lekcharoensuk et al., 2001) may, at least in part, explain this finding.

While healthy women identified with asymptomatic bacteriuria on population screening subsequently experience more frequent episodes of symptomatic infection, antimicrobial treatment of asymptomatic bacteriuria does not decrease the occurrence of these episodes (Nicolle, 2006). Clinical trials have consistently found no benefits with antimicrobial treatment of human asymptomatic bacteriuria and negative outcomes can occur, such as adverse drug effects and re-infection with organisms of increasing antimicrobial resistance (Nicolle, 2006). Prospective clinical trials comparing clinical outcomes in cats with occult UTIs treated with antimicrobials with an untreated control group would be necessary to decide the optimal management recommendations for these cats.

**NEW ANTIMICROBIAL TREATMENTS FOR BACTERIAL URINARY TRACT INFECTIONS IN CATS**

An important concept in modern antimicrobial chemotherapeutics is the mutant prevention concentration (MPC), which is the drug concentration required to prevent selection of first-step resistant variants naturally present in large bacterial populations. Such mutants may be amplified by consecutive steps of clonal enrichment under selective drug pressure during therapy (Wetzstein et al., 2005). Our group recently performed a controlled clinical trial to assess the efficacy and palatability of pradofloxacin 2.5% oral suspension compared to two other commonly prescribed antimicrobials (doxycycline and amoxicillin/clavulanic acid) for the treatment of feline bacterial lower UTI (Litster et al., 2007b). Pradofloxacin is an 8-cyano-fluoroquinolone developed to treat bacterial infections in dogs and cats (Körber B, et al., 2002). Molecular substitutions at positions C-7 and C-8 have greatly enhanced the bactericidal activity of pradofloxacin compared to earlier fluoroquinolones, especially for pathogens with reduced fluoroquinolone susceptibility, as its molecular structure restricts bacterial selection for antimicrobial resistance. Pradofloxacin has been shown in vitro to be highly active against a range of canine and feline urinary tract pathogens, including *E. coli* and *Staphylococcus spp.* (Körber B, et al., 2002). In our clinical trial, the post-treatment urine specimen in all cats in the pradofloxacin group (n=27) was negative for bacteria on urine culture. In each of the other 2 treatment groups (doxycycline, n=23; amoxicillin/clavulanic acid, n=28), there were 3 cats (doxycycline, 13.0%; amoxicillin/clavulanic acid, 10.7%) in which the post-treatment urine specimen was positive for bacteria on urine culture. The difference in proportions of treatment failure among groups was not statistically significant ($p>0.05$). Adverse events that were attributed to drug exposure by the attending veterinarian were not reported in the pradofloxacin or doxycycline groups. Three cats initially allocated to the amoxicillin/clavulanic acid treatment group developed gastrointestinal signs that were attributed by the attending veterinarian to drug exposure, resulting in their exclusion from the trial. None of the cats in the pradofloxacin group had pupillary light reflex test abnormalities during therapy (Litster et al., 2007b).

There is an emerging need for palatable oral pharmaceutical formulations for companion animals, and this factor has a major impact on issues such as owner convenience and compliance, particularly for chronically administered medications (Thombre et al., 2004). Our clinical trial also assessed the ease of administration by owners of each of the three antimicrobials used, all of which were presented in palatable liquid form. All three antimicrobials tested performed equally well, and none of the participating owners reported inability to medicate their cat during the trial. Owners' perceptions of the difficulty of administering oral medication to their cats was more positive post-treatment than pre-treatment ($P=0.001$) (Litster et al., 2007b).

Cefovecin, a new extended spectrum semi-synthetic cephalosporin, has a 14-day dosing interval after a single subcutaneous injection and combines rapid absorption with slow elimination due to the protein bound fraction of the drug acting as a reservoir with slow release of free (unbound) drug (Stegemann et al., 2006a). Concentrations of cefovecin in urine collected from cats with normal renal function are maintained above the $MIC_{90}$ of *E coli* for at least 10 days (Stegemann et al., 2006a). A recent in vitro study reported that cefovecin exhibited in vitro activity against all major aerobic and anaerobic bacterial pathogens associated with skin, urinary tract, and periodontal infections in dogs and cats (Stegemann et al., 2006b). Subsequently, multi-center, masked, randomized study was performed to compare the efficacy and safety of cefovecin with cephalexin for the treatment of feline bacterial UTIs (Passmore et al., 2008). Post-treatment urine cultures revealed that cefovecin eliminated 75.9% of all pathogens (41/54); 76.7% of single and mixed infections with *E. coli* (23/30); and 72.7% of single infections with *E. coli* (16/22) - a lower efficacy than that reported for cefovecin in canine UTIs (Passmore et al., 2007). There were no suspected adverse drug reactions attributable to treatment with cefovecin or cephalexin UTIs (Passmore et al., 2008).

In both studies performed by our group to examine bacterial lower UTI in cats, *Enterococcus spp.* and *E. coli* were the most common isolates (Litster et al, 2007a; Litster et al., 2009). These were also the most common bacteria isolated from cats with clinical signs of lower urinary tract disease in other published studies (Davidson et al., 1992; Bartges et al., 2000; Gottlieb et al, 2008). In our studies, the *E. coli* isolates were sensitive to the majority of the antimicrobials tested, whereas the *Enterococcus* isolates were uniformly sensitive to amoxycillin/clavulanic acid and penicillin/ampicillin and mostly resistant to cephalothin and clindamycin. Based on these findings, in a clinical situation, if high numbers of bacteria are visualized on microscopic examination of a cystocentesis-obtained urine specimen from a cat without lower urinary tract signs, and it is decided that immediate antimicrobial treatment is warranted, it would seem reasonable to prescribe amoxycillin/clavulanic acid for Gram-negative infections and amoxicillin for Gram-positive infections. Of course, regional variations in bacterial prevalence, bacterial sensitivity patterns and registered drug availability and palatability could affect this recommendation and quantitative urine culture and antimicrobial susceptibility testing is still the recommended "gold standard" to confirm a bacterial UTI and to determine the appropriate antimicrobial treatment.

## REFERENCES

Bailiff NL, et al., **JAVMA** 2005;226(10):1676-80. Bailiff NL, et al. **JVIM** 2006;20(4):850-5. Bailiff NL, et al. **Vet Clin Path** 2008;37(3):317-322. Barsanti J, et al., **JAVMA** 1985;187(4):384-388. Bartges J, et al. In: Current Veterinary Therapy XIII, 2000 WB Saunders Company, Philadelphia, pp 880-882. Bartges J, **J Vet Med** 2005;100(3):224-230. Blanco L, et al. **J Vet Med** 2001;96(10):776-790. Buffington C, et al. **JAVMA** 1997;210(1):46-50. Buffington C, et al. **JAVMA** 2006;228:722-725. Cavana P, et al., **J Feline Med Surg** 2008;10(3):269-73. Davidson A, et al. **California Vet** 1992:46(5):32-34. Elliott J, *25th Annual Scientific Meeting of the American College Veterinary Medicine*, 2007 Seattle, WA. Espinola MB, et al. **J Small Anim Prac** 1996;37:364-366. Gottlieb S, et al. **Aust Vet J** 2008;86(4):147-152. Gregory C, et al. **Vet Surg** 1983;12:210-212. Igimi S, et al. **Int J Syst Bacteriol** 1989;39(4):373-377. Igimi S, et al. **Vet Micro** 1994;39: 255-260. Körber B, et al. *42nd Interscience Conference on Antimicrobial Agents and Chemotherapy*. American Society for Microbiology. 2002 San Diego, CA. Kruger J, et al. **JAVMA** 1991;199(2):211-216. Lekcharoensuk C, et al. **JAVMA** 2001;218(9):1429-1435. Lees GE, et al. **JAVMA** 1986;189(6):648-652. Lees GE, **Vet Clin North Am Small Anim Pract** 1196;26(2):297-304. Lilenbaum W, et al. **Let Appl Micro** 1998;27(4):224-228. Lilenbaum W, et al. **Let Appl Micro** 1999; 28:448-452. Litster AL, et al. **Vet Micro** 2007a;121:182-188. Litster AL, et al. **JVIM** 2007b;21(5):990-995. Litster AL, et al. **Vet Micro** 2009 ***In Press, Corrected Proof***, *Available online 28 October 2008.* Mayer-Roenne B, et al. **J Feline Med Surg** 2007;9(2):124-32. McGuire NC, et al., **J Am Anim Hosp Assoc** 2002;38(6):541-4. Nicolle LE, **Int J Antimicrob Agents** 2006;28 Suppl 1:S42-8. Osborne CA, et al., Urinalysis: A Clinical Guide to Compassionate Patient Care. 1999 Veterinary Learning Systems. USA pp17-30. Passmore CA, et al., **J Small Anim Prac** 2007;48(3):139-144. Passmore CA, et al., **J Small Anim Prac** 2008;49(6):295-301. Stegemann MR, et al., **J Vet Pharmacol Ther** 2006a;29(6):513-24. Stegemann MR, et al., **Antimicrob Agents Chemother** 2006b;50(7):2286-2292. Takahashi T, et al. **J Vet Med Sci** 1997;59(9):775-783. Thombre AG, **Adv Drug Deliv Rev** 2004;56:1399-413. Wetzstein HG, **Antimicrob Agents Chemother** 2005;49:4166-73.

# FELINE INTERSTITIAL CYSTITIS: BLADDER DISEASE OR "DEVELOPMENTAL DISORDER"?

**T. Buffington**
**Columbus, OH**

Retrospective studies suggest that the majority of non-obstructed cats with lower urinary tract signs (LUTS) have an idiopathic disorder, and that this percentage has not changed appreciably during the past 4 decades. The importance of lower urinary tract (LUT) disorders to feline health is emphasized by the finding that elimination disorders (the vast majority of which are urinary) result in destruction of millions of cats in animal shelters in the United States every year. Feline interstitial, or idiopathic, cystitis has been described as an acute or chronic disease of waxing and waning LUTS, sterile urine, absence of cellular abnormalities suggesting neoplasia, and failure to identify an alternative cause for these signs after appropriate imaging procedures. In the absence of cystoscopic evaluation, the term idiopathic is preferred.[1] Although the terminology surrounding this syndrome in cats and humans has been the topic of spirited[2,3] discussion, the plain fact is that we currently argue from ignorance. The term interstitial cystitis, coined in 1887,[4] suggests the etiology of the problem to be inflammation of the bladder interstitium, despite rather compelling evidence to the contrary in most cases.[5] Unfortunately, diseases ascribed to the organ associated with some presenting clinical signs tend to be categorized by medical specialty, with the resulting organ-specific focus. Like most organs, however, the lower urinary tract has only a limited repertoire of responses to insult, regardless of the source. For example, LUTS may result from insults external to the bladder, such as microbiological infection, those intrinsic to the bladder, such as cancer, and by disorders, such as viral disease, spinal cord injury or diabetes mellitus, of other body systems that secondarily affect the bladder.

Despite the "organocentric" focus of some specialists, cats with FIC[6,7] appear to suffer from variable combinations of co-morbid symptoms. For example, LUTS has been reported to be co-morbid with separation anxiety syndrome, hypertrophic cardiomyopathy, and obesity. While it is conceivable that a bladder disease could generate symptoms in other organs, one might predict that the LUTS would precede these symptoms – Hill's criterion of temporality.[8] This often is not the case, although this topic has not been adequately studied.

Based on my current understanding of this problem, I believe that research in cats contains enough contrary instances of sufficient weight to support consideration of the hypothesis that *some cases of* FIC might represent a syndrome affecting the bladder rather than a bladder syndrome.[9,10] According to this view, FIC might represent a "central sensitivity" syndrome.[9] Commonalities across these syndromes include history of adverse early experience(s), sudden onset (often after a precipitating event such as an injury or infection), multiple comorbid syndromes in the same individual with no obvious pattern or order of onset, waxing and waning course, altered stress response system (SRS) function (generally including variably altered neuro-endocrine-immune function, particularly during stressful periods), and resistance to current therapeutic approaches.

Recent research suggests that one potential mechanism underlying the sensitization of the SRS involves a process called epigenetic modulation of gene expression.[10] Epigenetic modulation of gene expression is a prominent candidate mechanism for the identified differences in stress responsiveness found in patients with MUS because it has been shown to be an ancient process that occurs in the offspring of pregnant females exposed to stressors, potentially resulting in long term neuro-endocrine-immune abnormalities. Evolutionarily conserved developmental processes appear to interact with environmental cues (often transmitted from the mother via the placenta) to attempt to match the physiology of the developing organism to its post-natal environment. The available evidence suggests that FIC also may be a consequence of early life experiences.[11] A threatening or nutrient limiting environment that is perceived from placental and/or maternal signals may alter the developmental trajectory of the fetus in response to the available information to enhance reproductive fitness in the predicted environment, potentially resulting in durable changes in brain, autonomic, endocrine and immune function.[12] While likely to enhance the probability of survival, these changes also appear to increase vulnerability to stressors, putting these individuals at greater risk to develop disorders characterized by pain and discomfort in some environments. Genetic variability in the pattern and magnitude of each of these responses occurs among individuals, so the final pattern of the stress response will vary in outbred populations.

One important challenge to any hypothesis for a non-bladder etiology for FIC is, "Why are symptoms related to the bladder?". A more illuminating question might be, "how can so many organ systems be affected in patients with FIC?" One possibility is that increased outflow of the SRS, especially in the absence of adrenocortical restraint, can adversely affect a variety of important functions. Events external to the central nervous system, from both within and without the body, are conveyed by the sensory neurons to the thalamus, where they are evaluated and forwarded to the cortex for further processing prior to activation of the appropriate motor program.[13] Potentially threatening events can activate the hypothalamic SRS directly via the amygdala, bypassing cortical inhibitory control.[14]

Peripherally, activation of the SRS can increase epithelial permeability, permitting environmental agents greater access to sensory neurons,[15] which could result in increased neuron firing and local inflammation. Sympathetic neural-epithelial interactions appear to play an important role in permeability. For example, Birder, et al., have shown that application of norepinephrine (NE) to urinary bladder strips induces release of nitric oxide (NO) from the urothelium. Application of capsaicin results in release of NO from urothelium and neurons in the bladder. In light of reports that NO may increase urothelial permeability[16,17] these results suggest that some of the sympathetically mediated alterations in permeability may be mediated by NE via this mechanism. Moreover, increased permeability related to increased SRS activation does not require direct interaction with epithelial cells, nor is it restricted to the urinary bladder.

Neural release of NE is one of a variety of mechanisms whereby SRS-mediated activation of efferent nerves can activate mast cells, which can in turn increase epithelial permeability.[18] Sensory neurons, too may increase epithelial permeability by releasing neurotransmitters at the peripheral process of the nerve via sympathetic-sensory coupling, dorsal root reflexes, and axon reflexes.[19]

Moreover, it should be noted that the presence of inflammation is not well correlated with pain; in the bladder, presence of submucosal petechial hemorrhages has been observed both in cats[20] and humans with no signs referable to the lower urinary tract, and others have identified urothelial disruption (and presumably increased permeability) in painless bladder conditions.[21] This may be because emotional and environmental factors such as fear or threat can modulate the experience of pain through descending pathways from the midbrain.[22] Thus, even the heightened firing of afferent nerves noted in FIC cats[23] could result in differences in perceived bladder sensations depending on the effects of the emotional state of the animal on the balance between descending inhibition and facilitation of afferent transmission.

One also must recognize that an important limitation of many experimental studies of disease is that their focus is limited to the organ of interest. For example, although we have identified a number of abnormalities related to the urinary bladder, we know far less about whether these are specific to the bladder, or occur in other organs as well. Based on my experiences, I consider cats more likely to have a syndrome affecting the bladder when they have:

1. A history of early adverse experience (orphaned, bottle fed, rescued);
2. Evidence of problems in other organs, particularly preceding the presenting LUTS; and
3. Evidence of familial involvement.

This information often has not been obtained, (and may be unavailable). By taking the time to obtain a careful review of the cat's life history and conduct a thorough physical examination *before* assuming that the cat has a bladder disease, I have found that some cats appear to have a disease affecting the bladder, which helpfully informs my therapeutic recommendations. I urge others not to believe me (and my biases), but to test this hypothesis for themselves.

## REFERENCES

[1.] Buffington CAT, Chew DJ, Woodworth BE. Feline Interstitial Cystitis. **Journal of the American Veterinary Medical Association** 1999;215:682-687. [2.] Osborne CA, Kruger JM, Lulich JP, et al. Feline urologic syndrome, feline lower urinary tract disease, feline interstitial cystitis: what's in a name? **Journal of the American Veterinary Medical Association** 1999;214:1470-1480. [3] Baranowski AP, Abrams P, Berger RE, et al. Urogenital pain--time to accept a new approach to phenotyping and, as a consequence, management. **Eur Urol** 2008;53:33-36. [4.] Skene AJC. **Diseases of Bladder and Urethra in Women**. New York: William Wood, 1887. [5.] Tomaszewski JE, Landis JR, Russack V, et al. Biopsy features are associated with primary symptoms in interstitial cystitis: results from the interstitial cystitis database study. **Urology** 2001;57:67-81. [6.] Buffington CAT, Westropp JL, Chew DJ, et al. Clinical evaluation of multimodal environmental modification in the management of cats with lower urinary tract signs. **Journal of Feline Medicine and Surgery** 2006;8:261-268. [7.] Buffington CAT, Westropp JL, Chew DJ, et al. A case-control study of indoor-housed cats with lower urinary tract signs. **Journal of the American Veterinary Medical Association** 2006;228:722-725. [8.] Hill AB. The Environment and Disease: Association or Causation? **Proc R Soc Med** 1965;58:295-300. [9.] Buffington CAT. Bladder Pain Syndrome/Interstitial Cystitis In: Baranowski AP, Abrams P,Fall M, eds. **Urogenital Pain in Clinical Practice**. New York: Informa Healthcare, USA, 2008;169-183. [10.] Buffington CAT. Developmental Influences on Medically Unexplained Symptoms. **Psychotherapy and Psychosomatics** 2009;In Press. [11.] Gluckman PD, Hanson MA. **Mismatch: Why our World no Longer Fits our Bodies.** Oxford: Oxford University Press, 2006. [12.] Van den Bergh BR, Mulder EJ, Mennes M, et al. Antenatal maternal anxiety and stress and the neurobehavioural development of the fetus and child: links and possible mechanisms. A review. **Neurosci Biobehav Rev** 2005;29:237-258. [13.] Craig AD. How do you feel? Interoception: the sense of the physiological condition of the body. **Nat Rev Neurosci** 2002;3:655-666. [14.] LeDoux J. **The Synaptic Self: How our brains become who we are.** New York: Penguin Books, 2002. [15.] Veranic P, Jezernik K. The response of junctional complexes to induced desquamation in mouse bladder urothelium. **Biology of the Cell** 2000;92:105-113. [16.] Jezernik K, Romih R, Mannherz HG, et al. Immunohistochemical detection of apoptosis, proliferation and inducible nitric oxide synthase in rat urothelium damaged by cyclophosphamide treatment. **Cell Biol Int** 2003;27:863-869. [17.] Oter S, Korkmaz A, Oztas E, et al. Inducible nitric oxide synthase inhibition in cyclophosphamide induced hemorrhagic cystitis in rats. **Urological Research** 2004;32:185-189. [18.] Theoharides TC, Cochrane DE. Critical role of mast cells in inflammatory diseases and the effect of acute stress. **Journal of Neuroimmunology** 2004;146:1-12. [19.] Wang J, Ren Y, Zou XJ, et al. Sympathetic influence on capsaicin-evoked enhancement of dorsal root reflexes in rats. **Journal of Neurophysiology** 2004;92:2017-2026. [20.] Chew DJ, Buffington CA, Kendell MS, et al. Amitriptyline treatment for severe recurrent idiopathic cystitis in cats. **Journal of the American Veterinary Medical Association** 1998;213:1282-1286. [21.] Romih R, Korosec P, Jezernik K, et al. Inverse expression of uroplakins and inducible nitric oxide synthase in the urothelium of patients with bladder outlet obstruction . **BJU Int** 2003;91:507-512. [22.] Fields H. State dependent opioid control of pain. **Nature Reviews, Neuroscience** 2004;5:565-575. [23.] Roppolo JR, Tai C, Booth AM, et al. Bladder A-delta afferent nerve activity in normal cats and cats with feline interstitial cystitis. **Journal of Urology** 2005;173:1011-1015.

## FELINE INTERSTITIAL (IDIOPATHIC) CYSTITIS (FIC): EXPANDING TREATMENT APPROACHES1

**T. Buffington**
**Columbus, OH**

FIC can have a variable outcome. Lower urinary tract signs (LUTS) resolve spontaneously in many cats within 2-3 days of an initial episode or subsequent recurrence, with or without treatment. As many as 50% of these cats will have another episode within 12 months and 39% recurred in a more recent study of cats consuming dry food. It is not yet possible to predict which cats with FIC will relapse; some cats have multiple recurrences, while clinical signs never resolve in a small population of severely affected cats. In my experience, the most important consideration for a successful outcome is good communication with the owner. After the initial appointment and diagnostic evaluation of the cat, I commonly schedule an appointment with only the owners present to thoroughly review the disease process, discuss short and long-term treatment options, and to ensure that the owners have a clear understanding of this problem in their cat. Some common client questions and my responses are presented in Table 1.

**Table 1.** Common Client questions: (adapted from[3])

| | |
|---|---|
| Are the signs caused by a clearly understood medical entity? | No, although current research suggests that they may be the result of a problem that occurred during development |
| Is FIC life threatening? | No, FIC is definitely not life threatening in itself (although can be life- altering) |
| Does a definitive cure exist? | Not yet |
| Can symptoms be ameliorated? | Usually yes, and often quite effectively |
| Is FIC likely to plateau or progress? If so, over what period? | FIC usually has reached its plateau by the time the veterinarian sees the patient, although symptoms may wax and wane around this plateau. |
| Are there attendant complications? If so, can these be mitigated, or better still, prevented? | Stress-exacerbated symptoms, which may be reduced, can occur. |
| If that is a possibility, at what compromise in living? | Some work to learn to identify and cope with sources of stress may be necessary. |
| Will a change in lifestyle make a substantial difference in outcome in relation to the cat's well-being? | Absolutely! |

Once the diagnosis of FIC is made, the veterinarian can explain to the client that FIC is not a syndrome commonly treatable by antibiotics or surgery, is manageable with a combination of analgesic medications (acutely) and environmental modification (chronically), and that most care can be provided in collaboration with their technician. The clinician also should listen to the client's (often frustrated) story of the effects of having a cat with FIC, provide a satisfactory explanation for the sources of the signs (I often liken it to a "bladder migraine"), express care and concern for the situation, and enhance the client's sense of control. Effective doctor-patient interactions appear to enhance patient adherence to treatments, and quality of life outcomes of therapy in humans.[4] In some cases, email can provide an effective communication tool, permitting clients to send finite messages regarding their cat's condition for the clinician or technician's consideration and reply.

### Acute therapy

Antibiotics are rarely indicated for most cats that present for LUTS, and unless a bacterial urinary tract infection is documented, empiric use of antibiotics is not warranted in patients with FIC. We provide short term analgesic therapy in cases of an acute flare of signs as appropriate, and sometimes recommend pharmacotherapy in refractory cases after all previous recommendations have been instituted to the extent possible, because we have not found a drug that is more effective than multimodal environmental enrichment (MEMO). We most commonly use buprenorphine at 5 to 10 micrograms/kg orally BID to QID for 3 to 5 days. We have administered the injectable preparation orally in many cats successfully; it appears to be odorless and tasteless to cats. Butorphanol has been recommended, but its effects are not as potent, and do not last as long as those of buprenorphine. It is not known whether or not providing analgesia during acute episodes impacts the development of future episodes. To my knowledge, no controlled studies of the safety or effectiveness of NSAIDS for FIC have been reported. Although anecdotal evidence of their use exists, chronic use can predispose cats to the development of acute intrinsic renal failure, particularly with the overzealous use of Metacam®. Also, NSAIDS have been ineffective in pain management of interstitial cystitis in humans.

### CHRONIC THERAPY

### Environmental Enrichment

Multimodal environmental enrichment provides the primary therapy for prevention of recurrence of FIC. This opinion is based on the documented neuroendocrine abnormalities suffered by cats with FIC, and on our published and clinical experience. We define MEMO for indoor-housed cats to mean provision of all "necessary" resources, refinement of interactions with owners, a tolerable intensity of conflict, and thoughtful institution of change(s). Although we are not aware that a particular resource list has been validated for indoor housed cats, some recommendations are available in the many excellent publications about cat housing and behavior that currently are available. We also recommend extending the "1+1" rule traditionally applied to litter boxes (1 for each cat in the home, plus 1 more) to all pertinent resources (particularly food water and litter containers) in the household.

**Food**

Cats prefer to eat individually in a quiet location where they will not be startled by other animals, sudden movement, or activity of an air duct or appliance that may begin operation unexpectedly. Although canned food may be preferable for cats with FIC due to the increased water content or a more natural "mouth feel", some cats may prefer dry foods. If a diet change is recommended, offering the new diet in a separate, adjacent container rather than removing the usual food and replacing it with the new food permits the cat to express its preference. Natural cat feeding behavior also includes predatory activities such as stalking and pouncing. These may be simulated by hiding small amounts of food around the house, or by putting dry food in a container from which the cat has to extract individual pieces or move to release the food pieces (if such interventions appeal to the cat).

**Water**

Cats also seem to have preferences for water that can be investigated. Water-related factors to consider include freshness, taste, movement (water fountains, dripping faucets or aquarium pump-bubbled air into a bowl), and shape of container (some cats seem to resent having their vibrissae touch the sides of the container when drinking). As with foods, changes in water-related factors should be offered in such a way that permits the cat to express its preferences. Additionally, food and water bowls should be cleaned regularly unless individual preference suggests otherwise.

**Litter Boxes**

Litter boxes should be provided in different locations throughout the house to the extent possible, particularly in multiple cat households. Placing litter boxes in quiet, convenient locations that provide an escape route if necessary for the cat could help improve conditions for normal elimination behaviors. If different litters are offered, it may be preferable to test the cat's preferences by providing them in separate boxes, since individual preferences for litter type have been documented. For cats with a history of urinary problems, unscented clumping litter should be considered. Litter boxes should be cleaned regularly and replaced; some cats seem quite sensitive to dirty litter boxes. Litter box size and whether or not it is open or covered also may be important to some cats.

**Space**

Cats interact with both the physical structures and other animals, including humans, in their environment. The physical environment should include opportunities for scratching (both horizontal and vertical may be necessary), climbing, hiding and resting. Cats seem to prefer to monitor their surroundings from elevated vantage points, so climbing frames, hammocks, platforms, raised walkways, shelves or window seats may appeal to them. Playing a radio to habituate cats to sudden changes in sound and human voices also has been recommended and videotapes to provide visual stimulation are available.

**Play**

Some cats seem to prefer to be petted and groomed, whereas others may prefer play interactions with owners. Cats also can be easily trained to perform behaviors ("tricks"); owners just need to understand that cats respond much better to praise than to force, and seem to be more amenable to learning if the behavior is shaped *before* feeding. Cats also may enjoy playing with toys, particularly those that are small, move, and that mimic prey characteristics. Many cats also prefer novelty, so a variety of toys should be provided, and rotated or replaced regularly to sustain their interest. Also, some cats seem to have specific prey preferences. For example, some cats prefer to catch birds, while others may prefer to chase mice or bugs. Identifying a cat's "prey preference" allows owners to buy or make toys that the cat will be more likely to play with.

**Conflict**

When cats' perception of safety becomes threatened, they appear to respond in an attempt to restore their "perception of control". During such responses, some cats become aggressive, some become withdrawn, and some become ill. In our experience, inter-cat conflict commonly is present when multiple cats are housed indoors together and health problems are present. Conflict among cats can develop because of threats to their perception of their overall status or rank in the home, from other animals in the home, or from outside cats. With a little practice, one can recognize the signs of conflict and estimate its potential role in exacerbation of signs of FIC. If it is, owners usually can identify the causes after the signs of conflict are explained to them. Once this has been done, clients often are well on their way to reducing the intensity of conflict. Of course, some conflict between housemates is normal, regardless of species. Our goal is to reduce unhealthy conflict to a more manageable level for the cats involved. Signs of conflict between cats can be open or silent. Signs of open conflict are easy to recognize; the cats may stalk each other, hiss, and turn sideways with legs straight and hair standing on end up to make themselves look larger. If neither cat backs down, the displays may increase to swatting, wrestling, and biting. The signs of silent conflict can be so subtle they are easily missed. Because cats do not appear to possess distinct dominance hierarchies or conflict resolution strategies, threatened cats may attempt to circumvent agonistic encounters by avoiding other cats, by decreasing their activity, or both. Threatened cats often spend increasingly large amounts of time away from the family, staying in areas of the house that others do not use, or they attempt to interact with family members only when the assertive cat is elsewhere. Two types of conflict can occur; offensive and defensive. In offensive conflict, the assertive cat moves closer to the other cats, and to control the interaction. In defensive conflict situations, a threatened cat will attempt to increase the distance between itself and the thing it perceives as a threat. Although cats engaged in either type of conflict may spray or eliminate outside the litter box, we find that threatened cats are more likely to develop elimination problems.

The most common cause of conflict between indoor-housed cats that we have been able to identify is competition for resources. Cats may engage in open or silent conflict over space, food, water, litter boxes, perches, sunny areas, safe places where the cat can watch its environment, or attention from people. There may be no obvious limitation to access to these resources for conflict to develop. The change may only be the cat's perceptions of how much control it wants over the environment or its housemates' behaviors. Open conflict is most likely to occur when a new cat is introduced into the house, and when cats that have known each other since kittenhood reach social maturity. Conflict occurring when a new cat is introduced is easy to understand, and good directions are available from many sources for introducing the new cat to the current residents. Clients may be puzzled by conflict that starts when one of their cats becomes socially mature, or when a socially mature cat perceives that one of its housemates is becoming socially mature. Cats become socially mature between 2 and 5 years of age, and start to take some control of the social groups and their activities. This may lead to open conflict between males, between females, or between males and females. The cats involved in the conflict may never be "best friends", but they usually can live together without showing signs of conflict or conflict-related disease. In severe cases, a behaviorist may be consulted for assistance in desensitizing and counter conditioning of cats in conflict so they can share the same spaces more comfortably if this is desired.

Treatment for conflict between cats involves providing a separate set of resources for each cat, preferably in locations where the cats can use them without being seen by other cats. This lets the cats avoid each other if they choose to without being deprived of an essential resource. Conflict also can be reduced by neutering all of the cats, and by keeping all nails trimmed as short as practicable. Whenever the cats involved in the conflict can not be directly supervised, they may need to be separated. This may mean that some of the cats in the household can stay together, but that the threatened cat is provided a refuge from the other cats. This room should contain all necessary resources for the cat staying in it. Cats generally require and use more space than the average house or apartment affords them. The addition of elevated spaces such as shelves, "kitty condos", cardboard boxes, beds, or crates may provide enough space to reduce conflict to a tolerable level. In severe situations, some cats may benefit from behavior-modifying medications. In our experience, however, medication can help only after environmental enrichment has occurred, it can not replace it.

Conflict with other animals, dogs, children, or adults is relatively straightforward. In addition to being solitary hunters of small prey, cats are small prey themselves for other carnivores, including dogs. Regardless of how sure the client is that their dog will not hurt the cat, to the cat the dog represents a predator. If the cat does not assert dominance over the dog, as often happens, it must be provided ways to escape at any time. For humans, it usually suffices to explain that cats may not understand rough treatment as play, but as a predatory threat.

Most cats in urban areas in the United States are housed indoors and neutered, so conflict with outside cats can occur when a new cat enters the area around the house the affected cat lives in. To cats, windows are no protection from a threatening cat outside. If outside cats are the source of the problem, a variety of strategies to make ones garden less desirable to them are available. Because of the dearth of controlled trials, it currently is not possible to prioritize the importance of any of these suggestions, or to predict which would be most appropriate in any particular situation. Appropriately designed epidemiological studies might be able to identify particularly important factors, after which intervention trials could be conducted to determine their efficacy in circumstances where owners successfully implemented the suggested changes.

## ADDITIONAL APPROACHES

### Pheromones

A novel aspect of environmental enrichment that recently has become available is application of pheromones to the living space. Pheromones are chemical substances that seem to transmit highly specific information between animals of the same species. Although the exact mechanism of action is unknown at this time, pheromones appear to effect changes in the function of both the limbic system and the hypothalamus to alter the emotional status of the animals. Five facial pheromones have been isolated from cats; cats deposit the F3 fraction on prominent objects (including humans) by rubbing against the object when the cat feels safe and at ease. The function of this secretion is not only to mark objects, but also as an antagonist for urine marking and scratching. Feliway® a synthetic analogue of this naturally occurring feline facial pheromone, was developed to decrease anxiety-related behaviors of cats. Although a statistically significant difference was not found when Feliway® was compared to a placebo in cats with FIC, cats exposed to Feliway® exhibited a trend for fewer bouts of cystitis, and reduced negative behavioral traits.[2] Decreased spraying in multi-cat households, decreased marking, and a significant decrease in scratching behavior also has been reported subsequent to its use. Although, Feliway is not a panacea for unwanted cat behaviors or FIC, we have sometimes used it successfully in combination with environmental enrichment, and/or drug therapies. Feliway® is sold as both a spray and room diffuser. The spray can be used to treat areas of the house where the cat is urinating by use of a single spray to the affected spot for 30 days. We also have found Feliway® to be beneficial to decrease anxiety associated with traveling. Clients can spray the cat carrier at least 15 minutes prior to the trip and then place the cat in the carrier to help decrease the stress and anxiety most cats associate with travel. The treated areas should be sprayed at least 15 minutes prior to the cat encountering the area because the vehicle (ethanol) the pheromone is carried in is offensive to most cats. The room diffuser can be placed in a room where the cat inappropriately urinates. One room diffuser is reported to cover approximately 650 square feet and last for 30 days.

### Pharmacologic Therapy

Therapy with the tricyclic antidepressants (TCAs) may be helpful in chronic cases if MEMO and phermonotherapy cannot adequately manage clinical signs. TCAs should not be used for acute treatment of FIC, since they may increase the risk of recurrence. Amitriptyline has been reported to successfully decrease LUTS in cats with severe, recurrent FIC during a 12 month period in an uncontrolled trial. I usually begin by prescribing 5 mg once daily, administered in the evening. The dosage can be slowly adjusted up (to not more than 10 mg/day) or down until the desired effects are achieved. If a favorable response is not achieved, the drug should be tapered slowly and then discontinued. Side effects of amitriptyline include lethargy, weight gain, and urine retention because its anticholinergic properties.

Clomipramine (Clomicalm®, veterinary; Anafranil®, human) is another tertiary amine like amitriptyline, with more selectivity for blocking serotonin reuptake. Clomipramine (initial dose used was 0.25-0.5mg/kg orally once daily) significantly decreased the number of episodes of urine spraying in cats while the cats were given the drug. Clomipramine has milder anticholinergic effects than does amitriptyline, although sedation is a common side effect of this drug as well. Clomipramine, in conjunction with environmental modification, has also been described to successfully decrease anxiety related and "obsessive compulsive" behaviors in cats. Although anecdotal improvements in some patients with recurrent FIC have been observed, to my knowledge no controlled data are available. Fluoxetine (Prozac®), used to help decrease the rate of urine marking after environmental (litterbox-focused) alterations, was reported to help cats with periuria with variable success.

Glycosaminoglycan replacement drugs such as pentosan polysulfate (Elmiron®) and glucosamine also have been used to treat cats with FIC. IN both cases, placebo controlled trials testing the oral effectiveness of these compounds for the management of cats with FIC found no significant differences between the two groups, although cats in both treatment groups improved significantly over the course of the study.

## CONCLUSIONS

Many indoor housed cats appear to survive perfectly well by accommodating to less than perfect surroundings. The neuro-endocrine-immune abnormalities in the cats we treat, however, do not seem to permit the adaptive capacity that healthy cats enjoy, so these cats may be considered a separate population with greater needs. Moreover, veterinarians are concerned more with optimizing the environments of indoor cats than with identifying minimum requirements for indoor survival. My current approach is to let the client choose the most appropriate intervention *for their particular situation,* and to let technicians do the enrichment implementation and follow-up (under veterinary supervision as appropriate). Finally, the question of the relative merits of indoor housing to promote the welfare of cats (and the different opinions on what constitutes animal welfare in general) is beyond the present scope, and is a subject of controversy among experts. I hope to encourage extension of the welfare efforts of individuals working in zoos, who have recognized the effects of the quality of housing on the health on animals in their care and worked to enrich the environments of these animals, to all "captive animals" in our care. I believe that FIC and a variety of related disorders may be better prevented than treated, and that we have a huge opportunity to encourage this husbandry approach in veterinary clinical practice.[5] Further information about environmental enrichment for indoor housed cats is available at: http://www.indoorcat.org/.

## REFERENCES

[1] Westropp J, Buffington CAT. Feline Lower Urinary Tract Diseases. In: Ettinger SJ, Feldman EC, eds. **Textbook of Veterinary Internal Medicine**, 7th ed. St. Louis: Elsevier-Saunders; 2009:In Press. [2] Gunn-Moore DA, Cameron ME. A pilot study using synthetic feline facial pheromone for the management of feline idiopathic cystitis. **J Feline Med Surg** 2004;6:133-138. [3] Lown B. The Lost Art of Healing Practicing Compassion in Medicine. New York: **Ballantine Books**; 1999;344. [4] Abood SK. Increasing adherence in practice: making your clients partners in care. **Vet Clin North Am Small Anim Pract** 2007;37:151-164; abstract ix-x. [5] Buffington CAT. External and Internal Influences on Disease Risk in Cats. **Journal of the American Veterinary Medical Association** 2002;220:994-1002.

# CURRENT UNDERSTANDING OF THE PATHOGENESIS OF FELINE DIABETES MELLITUS & PRINCIPLES OF THERAPY

**Jacquie Rand BVSc, DVSc, MACVS, Dip ACVIM**
**Brisbane Australia**

## CLASSIFICATION

The current classification divides diabetes mellitus into type 1, type 2, and other specific types of diabetes.

Type 1 diabetes is an uncommon cause of diabetes in cats based on histological studies and absence of islet cell antibodies.

Type 2 diabetes appears to be the most common form of diabetes in cats, based on islet histology, risk factors, and clinical behavior of the disease. Type 2 diabetes is characterized by inadequate insulin secretion and impaired insulin action, that is, insulin resistance. The relative severity of these two defects varies between patients, both feline and human. The defect in beta cell function is usually progressive, and in some cats and humans results in complete loss of insulin secretion

A substantial minority of diabetic cats have other specific types of diabetes, previously called secondary or type 3 diabetes. Diseases in this category that have been reported in cats include diseases causing insulin resistance, and those resulting from non-specific destruction of pancreatic tissue. The most frequent is pancreatic adenocarcinoma, accounting for up to 19% of feline diabetes presented to referral institutions. Pancreatitis is a common histological finding in diabetic cats, but whether it is a cause or a result of diabetes is unclear. A few cats have diabetes as a result of extensive pancreatitis, and may or may not have concurrent signs of exocrine pancreatic insufficiency. Occasional causes of naturally-occurring insulin resistance in cats include growth-hormone producing tumors resulting in acromegaly, and hyperadrenocorticism. Iatrogenic administration of megestrol acetate or long-acting steroids is commonly associated with the development of diabetes in cats.

### Type 2 diabetes

Type 2 diabetes is characterized by insulin resistance and decreased insulin secretion. In humans it has been debated whether reduced insulin secretion or insulin resistance is the predominant initial defect in type 2 diabetes, and it appears that the initial defect varies between individuals. Insulin's ability to control blood glucose is dependent on both adequate insulin concentrations and adequate tissue sensitivity, and decreases in either interact to reduce the capacity to maintain blood glucose. Both insulin resistance and decreased insulin secretion are present once clinical signs of diabetes occur. Genotype, obesity, physical inactivity, and some drugs are all associated with insulin resistance.

### Insulin resistance

Insulin resistance is the second major abnormality in type 2 diabetes. Diabetic cats are approximately 6 times less sensitive to insulin than normal cats, and humans with type 2 diabetes have similar magnitudes of insulin resistance.

### Insulin resistance & obesity

Obesity is a recognized risk factor for the development of diabetes in cats and humans, although not all cats or humans with type 2 diabetes are overweight. In cats and humans, obesity causes insulin resistance.

### Genetics of type 2 diabetes

In humans with type 2 diabetes and in some cats, there is strong evidence for a genetic basis for the disease. Population studies have demonstrated a higher incidence of diabetes within some families of cats and ethnic groups of people.
In humans, it is considered likely that the multiple genes involved in predisposition to type 2 diabetes are both primary genes controlling insulin secretion and action, as well as secondary genes, influencing factors such as propensity to obesity.

Breed is a recognised as a risk factor in cats. The incidence of diabetes is over-represented in Burmese cats in Australia, New Zealand, and the United Kingdom. The frequency of diabetes in Burmese cats is approximately 4 times higher than that of Domestic cats in Australia, with 1 in 50 Burmese affected compared to less than 1 in 200 Domestic long- and short-hair cats. In some Burmese families, more than 10% of the offspring are affected.

### Age

Increasing age is also a risk factor for type 2 diabetes, and most cats are older than 8 years of age when diagnosed, with a peak incidence between 10 and 13 years of age. Although 1 in 50 Burmese cats have diabetes, the incidence increases to 1 in 10 for Burmese cats 8 years or older.

### Obesity & physical inactivity

Environmental influences interact with genetic influences, and play an important role in the development of diabetes in humans, and most likely in cats. Environmental or lifestyle factors shown to be important in humans and probably in cats, include obesity, physical inactivity, dietary factors, and urban rather than rural residence. The lifestyle of many domestic cats has changed similarly to humans, with inactivity and obesity increasing in urban cats. Exclusively indoor cats are usually less active than outdoor domestic cats that hunt and defend territory, and significantly less active than feral cats, which have to hunt to obtain all their nutrition. In humans and rats, exercise has been shown to increase insulin sensitivity. Lack of exercise impairs insulin action, and adds to the underlying level of genetically-determined insulin resistance. Less physically active cats and exclusively indoor cats are at increased risk of diabetes.

Obesity associated with insulin resistance is an important risk factor in the development of type 2 diabetes in humans, and even small increases in body mass index are associated with an increased risk of diabetes. Obesity has also been identified as a significant risk factor for diabetes in cats. In a study where cats were allowed free-access to a highly palatable energy dense diet over 10 months and increased their bodyweight by 44%, their insulin sensitivity decreased by more than half. Following weight gain 25% of the cats in the study had insulin sensitivity values within the range reported for diabetic cats.

**Diet & type 2 diabetes**

In both cats and humans, obesity has been shown to be a major risk factor for diabetes. Overfeeding of highly palatable, calorie-dense food in cats with reduced physical activity, likely contributes to obesity, and hence diabetes. Recent evidence in cats suggests that a high carbohydrate diet increases the demand for insulin secretion when compared to a low carbohydrate, high protein diet. In susceptible cats, this long-term demand for increased insulin secretion may lead to beta cell apoptosis and a decline is insulin secretory capacity, precipitating impaired glucose tolerance and diabetes, as hypothesized by the Carnivore Connection Theory. *Ad libitum* feeding of cats is not recommended except in cats which self-regulate food intake and maintain an ideal body condition. Most cats fed commercial food *ad libitum* ingest excess energy, resulting in obesity and increased risk of diabetes.

**Amyloid**

Many, but not all cats and humans with diabetes, have amyloid deposition replacing islets cells. Amyloid deposition does not appear to be an essential component of type 2 diabetes in cats or humans, but contributes to beta cell loss and failure of insulin. In a model of induced diabetes in cats, cats with high amylin concentrations tended to have more profound amyloid deposition. Obesity likely contributes to amyloid deposition in susceptible cats by stimulating hyperamylinemia and hyperinsulinemia, secondary to insulin resistance.

**Glucose & lipid toxicity**

Once persistent hyperglycemia occurs, insulin secretion is reduced through a phenomenon termed glucose toxicity. There is evidence that initially, suppression of insulin secretion is functional and reversible, and is not associated with visible lesions in beta cells. With hyperglycemia of longer than 2 weeks duration, histological abnormalities are evident, including glycogen deposition and cell death. The severity of the glucose toxicity effect is dependent on the degree of hyperglycemia and the duration.

Increased fatty acids produce a similar effect to glucose toxicity, called lipotoxicity. The clinical implications of glucose and lipid toxicity are very important. It is vital that effective therapy be instituted as soon as possible to reduce hyperglycemia in diabetic cats, if beta cell function is to be preserved. This is important, because data in humans indicates that patients with residual beta cell function have better glycemic control when treated with insulin, than patients with no significant endogenous insulin secretion. Secondly, it is important because the majority of cats will undergo remission of their diabetes, if the effects of glucose toxicity are minimized.

**Insulin resistance & chronic hyperglycemia**

Chronically elevated blood glucose also causes insulin resistance. Once overt diabetes mellitus with persistent hyperglycemia occurs, the added insulin resistance further compounds the problem of inadequate insulin secretion. This has implications for therapy, because once glucose concentrations are decreased with treatment, insulin sensitivity may improve.

**Diabetic remission or transient diabetes**

Diabetic remission occurs in up to 90% of newly diagnosed cats if treated appropriately. Remission occurs most commonly after 1 to 4 months of insulin therapy, and in some cats, glucose tolerance is normal in remission. Based on experience, remission is more likely if glycemic control is optimum, so beta cells can recover from glucose toxicity. Therefore insulin therapy is recommended as the initial therapy to maximize control of blood glucose and increase the probability of remission.

**Clinical Signs and Diagnosis**

There are no internationally agreed criteria for diagnosis of diabetes mellitus in cats. Clinical signs such as polydipsia/polyuria, weight loss, or polyphagia are non-specific, and diagnosis cannot be confirmed by clinical examination. Diagnosis in cats is often complicated by stress hyperglycemia, which in sick non-diabetic cats may lead to glycosuria or blood glucose levels in excess of 360 mg/dL (20 mmol/L). Blood glucose in non-diabetic unstressed client-owned cats is usually less than 171 mg/dL (9.5 mmol/L). When sampling blood, it is very important to avoid struggling. as this has been shown to be associated with transient hyperglycemia as high as 288 mg/dL (16 mmol/L) in normal cats. Blood glucose should be measured several hours after the first sample to confirm persistent hyperglycemia, especially if the blood glucose is less than 360 mg/dL (20 mmol/L). Signs of diabetes occur once blood glucose concentration exceeds the renal threshold, which is approximately 250-288 mg/dL (14-16 mmol/L) for normal cats.

Fructosamine may be useful in assisting diagnosis, especially when typical clinical signs of diabetes have not been observed or reported. However, some cats that do not have diabetes have elevated fructosamine levels similar to those of diabetic cats (false positive), and occasionally untreated diabetic cats may have fructosamine levels similar to that of normal cats (false negative). A fructosamine level of greater than 400 µmol/L in a cat strongly supports a diagnosis of diabetes. In experimentally-induced hyperglycemia in cats, although fructosamine concentration increased significantly from baseline within 3 days when blood glucose was maintained at 306 or 540 mg/dL (17 or 30 mmol/L), mean fructosamine did not increase above 350µmol/L when glucose was 306 mg/dL (17 mmol/L) for 6 weeks.

Measurement of water intake is inexpensive and useful for confirming polydipsia once blood glucose is above the renal threshold. In normal cats, total water intake including water in food ranges from 60 to 100 mL/kg/24h, but water drunk is much less, especially if the cat is consuming canned food, with average values of approximately 20 mL/kg or less. Measurement of water intake to document polydipsia may only be practical in a cat which is clinically healthy, and the decision to treat can be delayed by a few days.

If there is doubt whether the hyperglycemia is transient and associated with stress or is from diabetes, in sick cats it is prudent to begin insulin therapy and monitor glucose concentrations carefully for the next few days. Reducing glucose concentrations with exogenous insulin reduces the suppressive effect of glucose toxicity and makes recovery of beta cells more likely. Because glucose toxicity can reduce insulin secretion in normal cats to levels of insulin-dependent diabetic cats within 3 to 7 days, do not wait to begin insulin therapy if blood glucose concentration is 270 mg/dL (15 mmol/L) or higher. Likewise, in cats with iatrogenic or spontaneous hyperadrenocorticism or acromegaly, begin insulin therapy immediately to preserve remaining beta cells. Therapy for the underlying disease can then be instituted and glucose concentrations monitored to adjust insulin dose. Do not wait to see if the diabetes resolves once the underlying disease process is treated or the exogenous hormone eliminated, because this makes permanent diabetes more likely.

**Ketoacidosis**

Ketoacidosis occurs in approximately 12 – 37% of diabetic cats at the time of diagnosis, and a smaller percent are ketotic without acidosis. Ketoacidosis results in depression, vomiting and anorexia. Ketoacidosis can be precipitated by concomitant disease, especially infection. It is also important to realize that once cats become markedly insulinopenic, even previously normal cats progress to ketosis within 10 – 30 days, without evidence of precipitating disease. If untreated, acidosis quickly ensues. Small doses of insulin will usually correct ketonemia and prevent life-threatening acidosis, even if marked hyperglycemia persists. In experimentally induced diabetes, ketonemia occurred on average 5 days before ketonuria was detected using a urine test strip, and acetone odor to the breath was also detected before ketonuria. Although ketotic cats usually have low insulin concentrations, with appropriate therapy to overcome glucose and lipotoxicity, some beta cell function may return, and many of these cats achieve remission.

**Principles of therapy**

Therapy for diabetes should be instituted as soon as possible after diagnosis. The main aim of therapy is to achieve exemplary glycemic control to facilitate remission. In cats in which remission is not possible, resolution of clinical signs and avoidance of clinical hypoglycemia are the goals. Administration of insulin and dietary modification are the principal therapies used for management of diabetic cats. Oral hypoglycemic drugs may be useful in some cats depending on the loss of function from glucose toxicity, residual beta cell mass, and other concurrent therapy. A recent study has shown that if good glycemic control is achieved early in newly diagnosed diabetic cats, very high remission rates occur within 4 weeks of treatment. Good glycemic control reverses the glucose toxicity suppressing beta cells, and maximizes the chance of preserving beta cell function and achieving diabetic remission.

**SUMMARY**

Understanding the features of feline diabetes is important for good patient management, and may help to decrease the incidence of diabetes.

**RECOMMENDED READING**

Selected articles are available our website: www.uq.edu.au/vetschool/centrecah

[1] Rand JS, Fleeman LM, Farrow HA, Appleton DJ & Lederer R. (2004). Canine and Feline Diabetes Mellitus: Nature or Nurture? ***Journal of Nutrition***, 134(8S), p 2072S - 2080S. [2] Rand JS, Marshall R. Diabetes mellitus in cats. **Vet Clin North Am Small Anim** Pract 35[1]:211-24 2005 [3] Nelson RW (2005) Diabetes mellitus. *In* Ettinger SJ, Feldman EC (Eds.), ***Textbook of veterinary internal medicine*** (Vol. 2, pp. 1563): Elsevier Saunders

# MANAGEMENT OF FELINE DIABETES MELLITUS: PART 1. WHICH INSULIN DO I CHOOSE & HOW DO I ADJUST THE DOSE?

**Jacquie Rand BVSc, DVSc, MACVS, Dip ACVIM; Rhett Marshall BVSc, MACVS**
**Brisbane Australia**

## MANAGEMENT OF FELINE DIABETES

Therapy for diabetes should be instituted as soon as possible after diagnosis. The main goal of therapy is to achieve euglycemia without the need for insulin therapy, commonly termed diabetic remission. For those cats not achieving remission the goals are to resolve clinical signs and avoid hypoglycemia. Administration of insulin and dietary modification are the principal therapies used for management of diabetic cats. Oral hypoglycemic drugs have limited usefulness in the initial management of diabetic cats, and if used as a sole agent without insulin, will substantially decrease the probability of remission.

Cats that are not substantially dehydrated and are still eating can be treated with subcutaneous insulin. Cats initially presented with diabetic ketoacidosis can be treated with subcutaneous insulin after stabilization. When used intramuscularly or intravenously, glargine acts like regular insulin and can be used instead of regular insulin for initial stabilization of diabetic cats. The use of long-acting insulins, particularly glargine and detemir, and low carbohydrate diets facilitates achieving better glycemic control and increases the probability of remission, while minimizing the chance of hypoglycemia. If using lente insulin, because clinical hypoglycemia is more common and can be life-threatening, avoid aiming for perfect glycemic control.

### Insulin therapy

Long-acting insulin therapy remains the preferred initial and long-term treatment of choice for diabetes mellitus in cats. Its effectiveness and safety is be enhanced when combined with a low carbohydrate diet. Oral hypoglycemic agents such as acarbose may be effective as an additional therapy in cats which require a high carbohydrate diet, provided they are not *ad libitum* feeding.
Many types of insulin are available and have been used in cats. Achieving good glycemic control with intermediate acting potent insulins such as NPH and lente is often difficult, and increases the risk of clinical hypoglycemia. Recent data suggests that the long-acting insulins glargine and detemir provide better glycemic control and reduced risk of clinical hypoglycemia when given twice daily and combined with a low carbohydrate diet. More importantly, their use results in a significantly higher probability for remission.

### Oral hypoglycemic drugs

The use of oral hypoglycemic drugs to treat feline diabetes has been limited for a number of reasons. Many owners find administering tablets more difficult than injecting with insulin. Drugs which stimulate insulin secretion (eg. sulphonylureas) require adequate beta cell function to be effective, and if there is inadequate glucose-lowering effect, persistent hyperglycemia can lead to continued beta cell loss through glucose and lipid toxicity. These drugs may also stimulate accelerated islet amyloid deposition exacerbating beta cell loss. Acarbose acts by reducing glucose absorption from the gastrointestinal tract and is indicated as an adjunct to insulin therapy in cats which are meal fed a high carbohydrate diet. The greatest indication for using them is in cats with advanced renal failure which require a restricted protein diet. However, these cats usually have a reduced appetite and need to be fed *ad libitum*, which greatly reduces the efficacy of acarbose.

### Monitoring therapeutic efficacy

Response to treatment can be evaluated in a number of ways, and no individual modality should be used as the sole parameter for adjusting therapy. A combination of owner assessment, clinical signs, and changes in body weight and water intake are often the best indicators of glycemic control. Dosage changes can be made based on a number of different blood glucose parameters, and optimally should include more than one parameter. The pre-insulin glucose concentration is important when using glargine, detemir and PZI, as there is often a persisting effect from the previous injection. Nadir (lowest) glucose concentration is important for dosage adjustment. It limits the dose increase that can be made when nadir glucose concentration is in the lower end of the normal reference range. Water drunk and urine glucose concentration are probably more important when using glargine than for other types of insulin, and is discussed below.

When using other insulins (eg lente or NPH), dosage changes are usually based on nadir blood glucose. Pre-insulin glucose, time to nadir and the time to return to baseline are also used where appropriate, and recommendations for their use are listed in table 2. Water drunk, urine glucose and clinical assessment may be less important for making dosage changes with lente and NPH insulins, but should still receive consideration when adjusting dosage.

**Table 1:** Parameters for changing insulin dosage and frequency based on blood glucose measurements when using lente or NPH insulin in diabetic cats.

| Blood Glucose Variable | Recommendation |
|---|---|
| Use an initial dose 0.5 U/kg of lean body weight BID if blood glucose is ≥360 mg/dL (> 20 mmol/L) and 0.25 U/Kg BID if glucose < 360 mg/dL (> 20 mmol/L). | |
| If pre-insulin blood glucose concentration is <210mg/dL (< 12mmol/L) | With-hold insulin and check for diabetic remission provided cat treated for a minimum of 2 weeks; otherwise reduce dose |
| If pre-insulin blood glucose concentration is 211-250mg/dL (13 - 16mmol/L) | Total dose should be no more than 1U/cat bid |
| If nadir blood glucose concentration is < 54mg/dL (<3mmol/L) | Dose should be reduced by 50 % |

| | |
|---|---|
| If nadir blood glucose concentration is 54-90mg/dL (3 – 5mmol/L) | Dose should be reduced by 1U if poor control of clinical signs of diabetes; dose should remain the same if exemplary control of clinical signs |
| If nadir blood glucose concentration is 91-180mg/dL(6 – 9mmol/L) | Dose should remain the same |
| If nadir blood glucose concentration is >180mg/dL(> 10mmol/L) | Dose should be increased by 1U |
| If nadir blood glucose concentration occurs within 3 hours of insulin administration, or blood glucose returns to baseline within 8 hours | Change to longer acting insulin (ie. Glargine, detemir or PZI) |
| If the nadir blood glucose concentration occurs at 8 hours or later | Once daily administration may be used, although twice daily administration at a reduced dose is preferred |

**Glargine**

Glargine is a new human synthetic insulin analogue produced by recombinant DNA technology utilizing *E.Coli*. It differs from human insulin in that the amino acid asparagine at position A21 is replaced by glycine and two arginines are added to the B-chain terminus. Glargine is a clear aqueous solution in 100U/ml vials with pH=4 until injected subcutaneously. The interaction of the acidic insulin and the relatively neutral pH of the subcutaneous tissues forms micro-precipitates, and has a relatively constant systemic absorption profile. The formation of micro-precipitates and slow absorption are dependant on the acidity of glargine, hence glargine cannot be mixed or diluted.

Glargine is marketed for human patients as a very long-acting "peak less" insulin, with regard to its glucose lowering effects. It is designed to provide a basal or background insulin concentration, with the intention that a shorter-acting insulin be administered at meal times to achieve optimal glycemic control. Insulin glargine gained approval from the United States Food and Drug Administration in June 2000, for use in treating type 1 and type 2 diabetes in humans. The expected benefits in diabetic cats of an insulin preparation with a longer duration of action include improved glycemic control resulting in increased rates of diabetic remission, reduced rates of euthanasia, and decreased cost to clients.

The pharmacokinetics and pharmacodynamics of once daily administration of glargine compared with two of the most commonly used insulin preparations, porcine lente and protamine zinc insulin (PZI), has been reported in healthy cats. Once daily administration of glargine was found to have a similar mean daily glucose concentration and area under the 24hr glucose curve to PZI, and both were significantly lower than lente insulin. Glargine produced a glucose nadir later than PZI or lente, and had longer duration of action than lente. The duration of action for glargine was 22hrs, and 5 of the 9 cats had significantly decreased blood glucose concentration at 24hrs.

The administration of glargine once daily versus twice daily has been compared in healthy cats. A longer effect was achieved by administering glargine BID compared to once daily. Once daily administration of glargine has been shown to produce similar remission rates to twice daily dosing of lente insulin. Because excellent glycemic control facilitates remission and superior glycemic control is achieved if glargine is injected twice daily, twice daily dosing should be used for at least the first 4 months of treatment after diagnosis.

The usefulness of glargine for treating newly diagnosed diabetic cats has been evaluated. Twenty-four newly diagnosed diabetic cats (17m,7f) were treated with either glargine, PZI or lente (n=8 for each group) and fed a very low carbohydrate-high protein diet (Purina DM canned). Insulin was initially given at 0.5U/kg BID S/C if blood glucose was >360mg/dl, and 0.25U/kg BID S/C if blood glucose was <360mg/dl. Insulin dose was then adjusted based on serial blood glucose curves and water intake. Cats were defined as achieving diabetic remission if normoglycemia was maintained without insulin therapy for more than 2 weeks.

At diagnosis, there was no statistical difference between treatment groups for age, body weight, body condition score, or concentrations of fructosamine, blood glucose, B-hydroxybutyrate or bicarbonate.

There was a non-significant trend for glargine treated cats to have lower 12hr glucose concentrations after 10 and 17 days, than those treated with PZI or lente. Mean 12hr blood glucose at 4 weeks was significantly lower for glargine than PZI and lente treated cats. Fructosamine concentration after 4 weeks of treatment was significantly lower than at diagnosis for glargine treated cats but not for PZI or lente.

All 8 cats treated with glargine went into diabetic remission within 4 months of beginning treatment, while 3 cats treated with PZI and 2 cats treated with lente achieved diabetic remission.

Only 1 cat treated with glargine required an increase in insulin dose above 0.5U/kg BID, and 7 of 8 cats had their insulin dose reduced in the first 3 days of treatment. This is an important factor when initiating treatment with glargine, as there is usually a carry-over effect from the previous dose that may take several days to become apparent.

A significant finding in this trial was that no cat treated with glargine showed clinical hypoglycemia despite having biochemical hypoglycemia, while 2 cats treated with lente and 1 cat treated with PZI insulin exhibited signs of clinical hypoglycemia.

Glargine can be safely instituted at 0.5U/kg bid and serial blood glucose curves should be obtained daily for 3 days either in hospital or at home. When evaluating the blood glucose curve using glargine, it is often more useful to assess pre-insulin glucose concentration rather than the nadir glucose. We have found it often takes 3-5 days for a good glucose-lowering effect to be seen in the glucose curves, possibly because of the long duration of action and carry-over effect of glargine. Almost all cats will need to have their initial dose reduced within 2 weeks and many will achieve remission within 4 weeks.

Detemir is a newer synthetic insulin analogue with long duration of action that is produced using recombinant DNA technology in yeast (*Saccharomyces cerevisiae*). The insulin molecule is modified by the addition of an acylated fatty acid chain that enables reversible binding to plasma proteins, especially albumin, from where it is released slowly into plasma. This modification leads not only to a prolonged absorption from the subcutaneous tissue, but slow release from plasma proteins. Detemir results in similar remission rates and time to remission as glargine, but the median maximum dose used (1.75 IU/cat BID) is about 30% less than with glargine (2.5 IU/cat BID)

Monitoring and adjusting insulin dose when using glargine or detemir should be based on a number of parameters including; pre-insulin and nadir glucose concentration, water intake, urine glucose concentration and clinical assessment as shown in Table 2. We have found pre-insulin glucose concentrations measured at home an excellent tool for well-educated owners to safely modify daily doses of glargine or detemir. Cats treated with glargine should have a negative, 1+ or 2+ urine glucose (scale 0-4+) and a value of 3+ or 4+ likely indicates that a dose increase is required.

The good glycemic control achieved when using glargine or detemir likely reverses glucose toxicity of the B-cells, which facilitates endogenous insulin production and a reduced requirement for exogenous administration. Insulin dose may be reduced sequentially as indicated by blood glucose concentration, urine glucose and water intake until the dose is ½-1U/cat SID. Even if normoglycemic, it is recommended that insulin is not withdrawn within 2 weeks of commencement of therapy. Sequential deduction of insulin dose to ½-1 U/cat SID is recommended before insulin is withdrawn, and the cat carefully monitored afterwards to ensure remission has continued. It is also imperative that cats remain on a low-carbohydrate diet with calorie control to prolong the remission period. Newly diagnosed diabetic cats that have good glycemic control within the first few weeks of therapy, are very likely to go into diabetic remission. Cats that have been long-term diabetics are less likely to go into remission probably because of progressive B-cell loss associated with glucose toxicity. Remission rates are very significantly reduced if good glycemic control is instituted longer than 6 months after diagnosis.

In conclusion, glargine and determir are safe and effective in treating feline diabetes and are the preferred insulins in newly diagnosed diabetic cats. Long-term diabetic cats should be changed to glargine or detemir if there is poor glycemic control or owners wish to pursue once daily injections. High remission rates are expected in newly diagnosed cats when combined with a low-carbohydrate diet and twice daily injections.

**Table 2.** Parameters for changing insulin dosage when using insulin glargine or detemir in diabetic cats.

| **Parameter used for dosage adjustment** | **Change in dose** |
|---|---|
| Begin with 0.5 U/kg if blood glucose (≥360 mg/dL (≥ 20 mmol/L) or 0.25/kg of <u>ideal weight</u> if blood glucose is lower.<br>Do not increase in first week unless minimum response to insulin occurs, but decrease if necessary. Monitor response to therapy for first 3 days<br>If no monitoring is occurring in first week, begin with 1 U/cat BID | |
| If pre-insulin blood glucose concentration >216mg/dL (>12mmol/L) provided nadir is not in hypoglycemic range<br>***or***<br>If nadir blood glucose concentration >180mg/dL (>10mmol/L) | Increase by 0.25-1U |
| If pre-insulin blood glucose concentration 180<216mg/dL (≥ 10 - ≤12mmol/L)<br>***or***<br>Nadir blood glucose concentration is 90-160mg/dL (5-9mmol/L) | Same dose |
| If pre-insulin blood glucose concentration is 198-252mg/dL (11-14mmol/L).<br>***or***<br>If nadir glucose concentration is 54-72 mg/dL (3-4 mmol/L). | Use nadir glucose, water drunk, urine glucose and next pre-insulin glucose concentration to determine if insulin dose is decreased or maintained. |
| If pre-insulin blood glucose concentration <180mg/dL (10 mmol/l)<br>***or***<br>If nadir blood glucose concentration < 54mg/dL (<3 mmol/l) | Reduce by 0.5- 1UU ***or*** if total dose is 0.5-1U SID, stop insulin and check for diabetic remission |
| If clinical signs of hypoglycemia are observed | Reduce by 50% |
| If blood glucose measurements are not available: | |
| If water intake is ≤20mls/kg on wet food or ≤60mls/kg on dry food | Same dose |
| If water intake is >20mls/kg on wet food or >60mls/kg on dry food | Increase dose by 0.5-1U |
| If urine glucose is > 3+ (scale 0 - 4+) | Increase dose by 0.5-1U |
| If urine glucose is negative | Decrease dose until 0.5-1 U SID and then check for diabetic remission |

**SUGGESTED READING**

Selected articles are available our website: www.uq.edu.au/vetschool/centrecah. [1.] Rand JS and Marshall RD Diabetes mellitus in cats. **Vet Clin North Am Small Anim Pract** 35[1]:211-24 2005

## MANAGEMENT OF FELINE DIABETES MELLITUS: PART 2: WHAT DIET SHOULD I CHOOSE & HOW DO I MANAGE PROBLEM CATS

**Jacquie Rand BVSc DVSc Dip ACVIM; Rhett Marshall BVSc, MACVS,**
**Brisbane Australia**

### FEEDING

While the ideal combination of macronutrients (protein, fat and carbohydrate) to feed diabetic cats is not known, diets low in carbohydrates and high in protein reduce post-prandial hyperglycemia and insulin concentrations in healthy cats. Initial data from diabetic cats also suggest that low carbohydrate-high protein diets result in better clinical control, reduced insulin requirements and increased rates of diabetic remission. Thus a commercial low carbohydrate should be used in diabetic cats, unless contraindicated by other disease. During the first few weeks of treatment, diabetic cats may have a reduced appetite, and if they refuse these low carbohydrate diets, they should be offered any palatable food.

Care should be taken with cats diagnosed with advanced renal disease, as diets high in protein may have a deleterious effect. For these cats, dietary management of renal disease using a restricted protein diet should take precedence over dietary management of diabetes, and acarbose used to reduce glucose absorption from the GIT.

Obesity in cats markedly reduces insulin sensitivity and hence energy fed should be restricted so weight loss occurs in obese cats at a rate of 1-2% loss of body weight per week. Because of the decreased postprandial hyperglycemia with a low carbohydrate diet, it is suggested that diets with less than 20% of energy from carbohydrate (eg.Hills m/d, Purina dm) should be used for obese diabetic cats during the calorie restriction phase. Currently, most feline weight loss diets are low fat, high carbohydrate diets. Weight loss improves insulin sensitivity, and may reduce insulin requirements. In some cats, diabetic remission is obtained after weight loss and short-term insulin or oral hypoglycemic therapy.

### Acarbose

The α-glucosidase inhibitors (eg acarbose) reduce intestinal glucose absorption (Greco 1999) and are generally not effective in the treatment of feline diabetes alone, but can be used in conjunction with insulin and/or other oral agents to gain better glycaemic control. Cats given acarbose and meal fed a high carbohydrate diet have significantly reduced blood glucose concentrations but the same effect can be achieved by using a low carbohydrate diet.

### Problem cats

When cats treated with insulin fail to stabilise, a number of underlying causes and approaches to treatment should be considered. It is important to remember that some cats take 1 to 4 months to stabilize. Lente and NPH insulins provide inferior control of blood glucose and are more difficult to adjust to get good glycemic control, and are more likely to be associated with problem cats. These insulins are more often associated with marked changes in blood glucose concentration and clinical hypoglycemia than longer acting insulin such as glargine, detemir and PZI,.

The most common problems resulting in poor control are excessive dose, miscalculation of dose, too short duration of insulin action, or poor absorption of insulin. When using lente, NPH or ultralente, some cats are mistakenly labelled problem cats when the clinical signs are well controlled, but blood glucose measurements are less than ideal. This usually occurs when there are unrealistic goals for glycemic control using these insulins and lack of understanding that for 4 hours twice daily, cats treated with these insulins have negligible blood glucose lowering effect from the insulin. If the glucose nadir is below 10 mmol/L (182 mg/dL) after each insulin injection, peak action occurs >3 hours after administration, and hypoglycaemia is not occurring, glycemic control is usually adequate. These cats usually have good clinical control (stable body weight, good coat condition, active, alert, water drunk <100ml/kg/24h). Swapping to a longer acting insulin such as glargine and detemir will usually improve glycemic control and improve the probability of remission.

Problem cats have persistent clinical signs including polydipsia (water drunk > 100 mL/kg/24h), low body condition score, polyphagia, lethargy, and a poor hair coat; an insulin dose higher than normal (1.5 – 2 or more IU/kg/injection); and either a nadir glucose> 180 mg/dL (10 mmol/L) or hypoglycemia. For problem-solving in problem cats, it is important to first rule-out administration problems. Expired insulin, heat affected insulin (eg. left in a car in summer), poor mixing of suspensions (eg. lente insulin), failure of administration (eg. injecting through the skin pinch onto the hair-coat), and the presence of air bubbles in the syringe causing a lower administered dose, all occur regularly in practice. Insulin syringes can be difficult to manage for elderly owners with arthritic hands and poor vision. These owners are often better able to cope with insulin pens. Misunderstandings between the owner and veterinarian regarding the number of units to be administered can occur when using 40-IU/mL insulin in a 100-IU/mL syringe, because the cat is only getting 40% of the dose indicated by the markings on the syringe. Watch the owner administer the insulin. If it is an old bottle of insulin, change to a new one. If the cat has been treated for at least 8 to 12 weeks and insulin is being correctly administered but poor control is still evident, measure water intake over consecutive days at home (measure fructosamine concentration if water intake cannot be measured), and obtain a blood glucose curve.

Poor control may result from an excessive dose of intermediate acting insulin, which may cause apparent insulin resistance (dose >1.5 – 2 IU/injection with persistent hyperglycaemia), or short duration of insulin action. In many cats treated with the intermediate-acting insulins such as lente and NPH (isophane), these potent insulins rapidly lower blood glucose. This stimulates counter-regulatory responses, even when blood glucose concentration is not in the hypoglycemic range. The resulting counter-regulatory response increases blood glucose concentration, and causes an apparent short duration of insulin action and insulin resistance. This can be very frustrating for veterinarians when managing diabetic cats. This response happens because of the action of the hypothalamic neurons that sense blood glucose concentration and initiate counter-regulation. The hypothalamic neurons control entry of glucose into their cytoplasm, and actively maintain a large concentration gradient with plasma glucose when blood glucose concentrations are high. When potent insulin such as lente insulin is given and blood glucose concentration decreases rapidly, the intracellular glucose concentration of the hypothalamic neurons decreases more quickly into the range perceived by the neurons as hypoglycemic and a counter-regulatory response is triggered, even before hypoglycemia develops. The resultant secretion of glucagon, epinephrine, cortisol and growth hormone increases blood glucose concentration, and causes an apparent short duration of insulin action. Because the glucose lowering effect of lente and NPH in cats is less than 8 hours, most diabetic cats have blood glucose concentrations of 360 to 430 mg/dl (20 to 24 mmol/l) at the time of the next insulin dose, predisposing them to premature counter-regulation. The result is that in some cats, lente and NPH insulins may only lower blood glucose for 2 to 3 hours. This inherent short duration of action of lente, NPH and ultralente insulins, coupled with the response of the hypothalamic neurons can be very frustrating for practitioners. It also is dangerous for diabetic cats, because their insulin dosage often is wrongly increased. The end result is that the effect of lente, NPH and ultralente is often too short to achieve good glycemic control, and insulin resistance and signs of hyperglycemia and hypoglycemia ensue.

For cats on potent insulins such as lente or NPH, if the cat is polydipsic and insulin seems to have little effect, especially when previously it caused substantial lowering of glucose, or the duration of action seems to be short, there are two options. The preferred option is to swap to glargine or detemir. Alternatively try lowering the dose of insulin to 0.3 to 0.5 IU/kg for 10-14 days to see if blood glucose or water intake improve towards the end of the period. If clinical control is not improved with a lower dose, check the glucose response to a standard dose of 0.5 IU/kg of insulin, to determine the duration of effect. If the glucose nadir occurs 2 – 3 hours after injection, switch to a longer acting insulin (glargine or detemir), or increase the frequency of administration to TID. With PZI or ultralente, if there is little response to insulin, try swapping to glargine or detemir and slowly increasing the dose until glycemic control is achieved, which may require a dose of 5-10U/cat twice daily in some problem cats. If there is still polydipsia (water drunk>100ml/kg/24 h) and little glycemic response after 1-2 months at a dose of 5-6U/cat twice daily, check the cat for hyperthyroidism, hyperadrenocorticism, acromegaly or other systemic disease such as renal failure. We have seen improved glycemic control in cats with periodontal disease, following dental surgery in combination with short-term antibiotics. In the meantime, increase the dose by 1 IU every 1-2 weeks until some glycemic response is achieved. Control is achieved in most difficult cats, with the exception of cats with acromegaly, once glargine dose is 5-10 U/cat twice daily (4-5 U/cat with detemir). Warn the owner that a severe hypoglycemic episode can occur with this protocol, and to be particularly vigilant regarding the early signs of hypoglycaemia (lethargy, mental dullness, wobbliness, trembling and dilated pupils). In some cats which appear insulin resistant, but no cause can be found, admitting them to hospital for carefully observed intensive insulin therapy to normalize blood glucose for several days, may substantially reduce the subsequent insulin doses that achieve control. Most cats are eventually controlled on 1-3 U/cat BID of glargine or 1-2 U/cat BID of detemir, even if they required a dose as high as 5-6 U/cat BID in the first 1-3 months to control blood glucose.

## SUGGESTED READING

Selected articles are available our website: www.uq.edu.au/vetschool/centrecah [1.] Mazzaferro EM, Greco DS, Turner AS, Fettman MJ (2003) Treatment of feline diabetes mellitus using an α-glucosidase inhibitor and a low-carbohydrate diet, **Journal of Feline Medicine and Surgery 5**(3), 183-189 [2.] Nelson RW (2005) Diabetes mellitus. *In* Ettinger SJ, Feldman EC (Eds.), **Textbook of veterinary internal medicine** (Vol. 2, pp. 1563): Elsevier Saunders [3.] Marshall RD, Rand JS. Comparison of the pharmacokinetics and pharmacodynamics of glargine, protamine zinc and porcine lente insulins in normal cats, **Journal of Veterinary Internal Medicine** 2002, 16(3):358. [4.] Marshall RD, Rand JS. Comparison of the pharmacokinetics and pharmacodynamics of once versus twice daily administration of insulin glargine in normal cats, **Journal of Veterinary Internal Medicine** 2002, 16(3):373. [5.] Marshall RD and Rand JS. Treatment with glargine results in higher remission rates than lente or protamine zinc insulins in newly diagnosed diabetic cats.**Proceedings 24th ACVIM Forum** *2005*. [6.] Rand JS and Marshall RD Diabetes mellitus in cats. **Vet Clin North Am Small Anim Pract** 35[1]:211-24 2005

# MANAGEMENT OF FELINE DIABETES: HOME MONITORING & PROTOCOLS

Jacquie Rand BVSc[1] DVSc Dip ACVIM; [2]Kirsten Rump BSc, MSc
[1]Centre for Companion Animal Health, School of Veterinary Science, The University of Queensland, Queensland, Australia
[2]MPI for Informatics, Saarbruecken, Germany

## INTRODUCTION

The primary goal of management of feline diabetes is to achieve diabetic remission. Secondary goals are to control clinical signs and avoid clinical hypoglycemia. Achieving exemplary control of blood glucose concentrations allows the beta cells to recover from the effect of glucose toxicity which suppresses insulin secretion. With appropriate management instituted early after diagnosis of diabetes, approximately 90% of cats achieve remission. Three factors are crucial in increasing the probability of remission. Firstly, use of a long-acting insulin, secondly, feeding a low carbohydrate diet and thirdly, frequent monitoring of blood glucose concentration and appropriate dose adjustments aimed at achieving excellent glycemic control. Blood glucose concentrations vary from day to day, even when diet and insulin dose are not changed. In addition, cats are prone to stress hyperglycemia which can increase blood glucose concentration by as much as 180 mg/dL (10 mmol/L) if struggling occurs. Maximizing the quantity and quality of information on which to base insulin dose changes is important, and home monitoring provides a number of advantages.

### Advantages of home monitoring

Home monitoring has many advantages for both the pet and its owner. Spuriously high blood glucose concentrations caused by stress hyperglycemia are less likely if a cat is tested in its home environment compared to a veterinary hospital. Eating affects blood glucose concentrations, and therefore it is important when measuring blood glucose concentrations as a basis for making insulin dose adjustments, that the amount and composition of food eaten represent the cat's normal diet. In the home environment, cats are likely eating and behaving normally, therefore giving more accurate readings than those obtained in a veterinary clinic. Home monitoring also provides the opportunity for blood glucose concentrations to be assessed immediately by the owner, should the cat begin to behave abnormally. This is helpful for identifying and managing clinical hypoglycemia. Early optimization of blood glucose concentration is facilitated with home monitoring, because protocols which involve daily home testing can be used. This may increase the probability of remission compared to testing blood glucose concentrations weekly or every two weeks in a veterinary hospital because home monitoring provides more data on which to make more frequent insulin dose adjustments. Hospitalization of a pet is relatively expensive and inconvenient for the client, and stressful for the cat. It is time-consuming for veterinarians and their staff, and because of the labor involved, is often not very profitable. Home monitoring reduces the cost and inconvenience to the owner, and can improve the profitability of managing diabetic patients by reducing management practices that are labor intensive. Clients still should have frequent veterinary visits particularly in the initial stabilization phase, so that their home-log of glucose concentrations and insulin dose can be reviewed, along with review of blood collection and blood glucose measurement techniques. In addition, visits provide the opportunity to examine and weigh the pet, and check accuracy of the glucose meter.

### Blood glucose meters

It is critical owners are made aware of differences in blood glucose meters. All the reputable companies such as Abbott, Bayer and Roche make precise meters that have close correlation (r=>0.90) with feline blood glucose concentration measured by a serum chemistry analyzer. This means that the results they provide are very repeatable. However, both accuracy and precision are important. Overall, the minimum accuracy levels required by regulating authorities internationally for human-use meters lie between +/-15-20%. There are no standards for use of these meters in veterinary medicine, and the accuracy of human-use meters decrease substantially when used for feline (or canine) blood. This is assumed to be because of the different distribution of glucose between plasma and red blood cells in feline and human blood.

Unpublished data indicates that Ascensia Contour (Bayer) measures on average 35 mg/dL (Bayer) lower and Accu-Chek Aviva (Roche Diagnostics) 25mg/dL (Roche) lower in feline blood compared to an automated serum analyzer for samples in the normal range. In contrast, Abbott AlphaTRAK reads on average 0.06 mg/dL higher. In dogs, similar differences were found with these meters, and in the normal range of blood glucose concentrations, the percent difference from the actual value was 73%, 45% and 18% respectively (Cohen et al, ACVIM abstract 2008). Therefore, it is critical that the owner be aware that if they are using a meter calibrated for feline blood that they should not change to a meter calibrated for human blood without also changing the target blood glucose concentrations in the dosing protocol. A greater danger is the change from a meter calibrated for human-use to one calibrated for feline-use when using a protocol aimed at tight glycemic control and with target blood glucose values designed for a whole blood human-use meter. In this scenario, there is an increased risk of unintentional hypoglycemia. It is very important that the client be made aware that the target glucose concentration is dependent on the meter used. Control solutions are available for glucometers from the individual manufacturers and allow owners to test for proper functioning of their glucometer. The meters should be regularly checked for accuracy and this can also be done during client visits. Blood glucose concentration can be measured both on the owner's meter and with a serum chemistry analyzer. This will also allow some estimation of how much the cat is affected by the stress of visiting the veterinary clinic.

For human-use meters in some countries, blood glucose concentrations are measured from whole blood, while in others, measurements are taken from whole blood but the glucometer calculates a plasma equivalent. Measurements from meters calibrated for human blood which provide plasma-equivalent values are approximately 11% higher.

Important features in appropriate blood glucose meters are a small sample size ($\leq$0.6 µl), ease of use and reliability. Additionally, the cost of test strips should be reasonable. For owners of diabetic cats who follow a protocol which involves daily home testing, test strips are often the most expensive component of the therapy. The availability of test strips is also worth considering, as test strips may run out unexpectedly. Consideration of the long-term benefits to the client of utilizing veterinary products from reputable companies that support research and education of veterinarians and clients should be made when considering the cost of test strips and making recommendations to the client regarding veterinary or human-use meters.

**Practical methods for obtaining blood in cats**

Typically, blood samples are taken from capillaries of a cat's ear using a lancing device for human diabetics. There are many types of lancing devices available, including devices which create vacuum to facilitate collection of sufficient blood volume. The depth of penetration is often adjustable for lancet devices and the lancet needles are disposable. Some owners prefer not to use a lancing device, but draw blood "free hand" with just a lancet needle or syringe needle. Some lancets work better in some cats and with some owners. Often a successful approach is for the veterinarian and client to test sites and lances with the pet in the consulting room, to determine the best device and site to maximize the likelihood of obtaining a sample of adequate volume. It is vital that when introducing home monitoring, the owner is shown how to set up the meter and gets practice using it under guidance of a veterinarian or technician before going home.

Warming the ear sufficiently prior to taking blood is a crucial factor in being able to easily draw blood in the initial phases of introducing home monitoring. Warm and rub the ear for several minutes prior to drawing blood. Once a cat has been home monitored for a period of time, extensive warming is often no longer necessary.

Some owners prefer to draw blood on inner surface of the pinna, as this is hairless. Alternatively, the outer surface of the pinna can also be used. Obtaining an adequate sample can be facilitated if the hair is clipped and/or a thin layer of Vaseline or generic petroleum jelly is first applied to the surface to encourage beading of the blood. When the ear has been well warmed, the marginal ear vein often produces too much blood, and sites near the vein will become preferable as they produce sufficient amounts of blood. With repeated use of an area for obtaining blood, samples are more readily obtained, probably because of increased vascularization.

Using paw pad pricks is less commonly used by owners as a technique for obtaining capillary blood. The carpal or pisiform pad at the back of the leg may be more reliable for obtaining blood than the relatively thick paw pad. For some fractious cats, or cats with ear problems, this method can be effective.

**Adjusting Insulin Doses**

Parameters for changing insulin dosage when using insulin glargine or detemir together with home monitoring of blood glucose concentrations have been developed and tested in more than 70 diabetic cats (Roomp and Rand, ACVIM abstract, 2008). In cats which had been previously treated for a median of 4 months with intermediate-acting insulin and had failed to achieve remission, 81-84% achieved remission with detemir or glargine and home monitoring using the intensive protocol, provided the changes were instituted within 6 months of diagnosis. Higher remission rates are achieved if the protocol is instituted immediately after diagnosis. These protocols are available on www.uq.edu.au/ccah. The protocol aimed to achieve intensive blood glucose control. Blood glucose was measured at least 3 times daily with a glucometer.

Until blood glucose concentration is well controlled, frequent monitoring of blood glucose concentration should be encouraged. It is recommended that as a minimum, owners measure blood glucose just before the morning and evening insulin injection and again before they go to bed, 6 days out of 7 and on one day in the week, measure blood glucose every 3-4 hrs throughout the day until they go to bed. If they are able to measure blood glucose more frequently than 3 times a day on more than one day a week, that is preferable, but it requires a committed owner with the time available. Even 3 blood glucose measurements per day and one full blood glucose curve a week provides more information to make insulin dose changes than blood glucose measurements obtained at the veterinary clinic once every 7-14 days.

In cats, dose adjustments for intermediate-acting (lente or NPH) insulin rely on measurement of nadir glucose concentration which occurs typically 3-6 hours after insulin administration. Pre-insulin blood glucose concentration is only used as an indicator to decrease the dose in cats on intermediate-acting insulin. For cats on intermediate-acting insulin, blood glucose concentration should be measured every 2 hours until a nadir (usually 3-6 hrs after insulin) is evident. Use of intermediate acting insulin in cats significantly reduces remission rates and is not recommended in newly diagnosed diabetic cats. For long-acting insulin, dose is adjusted based on pre-insulin blood glucose concentration and nadir glucose concentration.

Single measurements of blood glucose concentration have limited value and can lead to inappropriate dose adjustments unless they are pre-insulin measurements or taken when an animal has signs consistent with hypoglycemia. The time of nadir glucose concentration typically varies considerably between days and is usually missed if a single measurement of blood glucose is being made, although in cats treated twice daily with intermediate-acting insulin, the nadir is usually between 3 and 6 hours, with 4 hours typical. For long-acting insulin in cats, nadir glucose concentration may occur just prior to the next insulin dose.

Because normoglycemia is being aimed for, choice of glucometer is important. To avoid clinical hypoglycemia or unnecessarily high blood glucose concentrations, it is important to recognize and understand whether the meter is calibrated for human or cat blood. Meters calibrated for human diabetic patients that report glucose concentrations for whole blood (not plasma-equivalent) read cats' blood glucose on average 25-35 mg/dL (1.5 to 2 mmol/L) lower than actual blood glucose concentration. The protocol for intensive blood glucose control utilizing home monitoring recommends dose increases to initially achieve a nadir (lowest glucose concentration) close to the normal range and then later to increase dose so the majority of maximum concentrations are also in this range. For meters calibrated for feline blood, aim for a range of 83-133 mg/dL (4.6-7.4 mmol/L) and for meters which give readings in whole blood and calibrated for human patients, use a target range of 50-100 mg/dL (2.8-5.6 mmol/L). Details of protocols are available on www.uq.edu.au/ccah

**Ketosis**

Prior to treatment with insulin or before regulation is achieved, hyperglycemic cats may develop ketosis. Both high and moderate levels of hyperglycemia can be associated with the development of ketones in cats, and a few cats with blood glucose concentrations 200 to 300 mg/dL (11-15 mmol/L) are also ketotic.

Monitoring for ketones can be achieved using urine test strips or in the blood using the Precision Xtra (Abbott). The new models of the Precision Xtra only require 1.5 μl of blood and samples are taking the same way as samples for blood glucose measurements. Test strips for this device are relatively expensive.

The type of ketone (ketone body) detected by commonly used testing devices in urine and blood differs. While urine test strips measure the concentration of acetoacetate, the Precision Xtra measures beta-hydroxybutyrate (BHBA) in whole blood. There are two further important practical differences between blood and urine ketones: blood ketones appear days earlier than urine ketones, and lower yet significantly elevated blood ketone levels can already be detectable while urine ketones still remain negative i.e. undetectable. For example in one cat being followed by the owner, the cat tested negative for urinary ketones up to when blood ketones were 1.8 mmol/l and only began producing (trace) urine ketones once blood ketones exceeded 18.5 mg/dL (1.8 mmol/l). A study in healthy cats infused with glucose showed that beta-hydroxybutyrate was present in urine approximately 5 days before acetoacetate were detectable in urine using urine dipsticks. Well regulated cats have blood ketone concentrations from 0- 2.1 mg/dL (0.0 – 0.2 mmol/l). Therefore, cats with a tendency to develop ketones or previously had diabetic ketoacidosis may benefit from blood ketone testing.

**Cats requiring high doses of insulin**

Cats with blood glucose concentrations that do not seem to respond to typical doses of glargine or detemir should be tested for acromegaly. Recent studies have shown that acromegaly is more common than previously thought and cats that require >1.5 IU/kg should have their IGF-1 concentrations tested. Hyperadrenocorticism (Cushings Syndrome) is very rare in comparison to acromegaly.

**Avoiding and dealing with relapse in remission cats**

Once remission is achieved, it is likely that the pancreas retains decreased function because of the underlying disease process and damage caused by hyperglycemia. Therefore, situations which reduce insulin sensitivity and require the pancreas to increase insulin secretion to maintain normoglycemia can lead to diabetic relapse, and should be avoided. Medications that produce insulin resistance and hyperglycemia (in particular corticosteroids and megestrol acetate) should be avoided and the cats should continue receiving a strict low carbohydrate diet. Continued once monthly home testing is recommended, to ensure that owners have an up-to-date set of equipment and remain practiced in home testing should a relapse occur. It is also vitally important to detect early increases in blood glucose concentration before ensuing damage to the pancreatic beta cells occurs. Six monthly veterinary exams including comprehensive blood panels and, when necessary, dental prophylaxis is also recommended.

Approximately one quarter of cats in remission will relapse and require insulin again. Cats which relapse should be investigated to determine the cause of the relapse. One possible reason for relapse is the development of hyperthyroidism, which is common in older cats. Euthyroid older cats' T4 will gradually drop with age and therefore older cats with rising T4 concentrations or T4 concentrations at the upper end of the normal range should be carefully evaluated for hyperthyroidism.

# PROTEIN-LOSING ENTEROPATHIES IN DOGS & CATS: DIAGNOSIS, MANAGEMENT & COMMON MISTAKES

**Michael D. Willard, DVM, Dipl ACVIM**
**College Station, Texas**

## HYPOALBUMINEMIA IN GENERAL

When concerned with protein loss of any cause, one should measure serum albumin concentrations (NOT serum total protein concentrations). Do not use human clinical pathology laboratories as their technology typically does not detect canine albumin (meaning that they routinely report serum albumin concentrations of < 1.5 gm/dl in clinically normal dogs). If the patient has substantial hypoalbuminemia, the first step is to examine the skin for obvious lesions which can cause protein loss. Cutaneous lesions sufficient to cause such hypoalbuminemia are obvious – you should be able to just look at the patient and know. Next, hepatic function testing (e.g., resting and post-prandial serum bile acid concentrations) and a urinalysis are requested. If there is any doubt on the urinalysis, then a urine protein:creatinine ratio will quantify the magnitude of urinary protein loss. Severe hypoalbuminemia (i.e., < 2 gm/dl) in an animal with diarrhea suggests a protein-losing enteropathy (PLE). If severe, exudative cutaneous disease, protein-losing nephropathy, and hepatic insufficiency are eliminated, then PLE is a reasonable tentative diagnosis in patients with a serum albumin < 2.0 gm/dl. Contrary to what the textbooks say, PLE may be associated with a low, normal or increased serum globulin concentration.

## PROTEIN-LOSING ENTEROPATHY

Perhaps the most important point of this discussion is that while hypoalbuminemia has repeatedly been reported to be a poor prognostic sign in patients with chronic GI disease[1,2], there may be one or more subset(s) of patients that respond well to appropriate therapy. Therefore, diagnosing PLE is not necessarily cause for despair. However, since many of these animals have severe alimentary tract disease that needs to be diagnosed promptly to maximize the chance for successful therapy, aggressive diagnostics are typically an appropriate recommendation. Although therapeutic trials can be chosen in place of classic diagnostic tests in many of the more common alimentary tract diseases (e.g., dietary allergy, dietary intolerance, antibiotic-responsive enteropathy, parasites), such an approach is generally ill-advised if the serum albumin concentration is less than 2.0 g/dl. This is true because it may be necessary to perform an antibiotic and/or dietary therapeutic trial for 3-6 weeks in order to ascertain if it is being effective, and a patient with severe PLE can become markedly worse in that time, especially if the serum albumin concentration is falling rapidly.

Any GI disease can cause PLE if it is severe enough. Many acute GI diseases cause PLE (e.g., parvoviral enteritis); however, these diseases typically are comparatively easier to treat than the chronic GI disease causing PLE. Therefore, the focus in this lecture is PLE in animals with chronic GI disease. The major causes of PLE in adult dogs tend to be intestinal lymphangiectasia, inflammatory bowel disease (IBD), alimentary tract lymphoma (LSA), and fungal infections (i.e., histoplasmosis and pythiosis). Other causes include alimentary tract ulceration/erosion, severe disease of intestinal crypts, antibiotic-responsive enteropathy, and parasites. The major causes of PLE in juvenile dogs tend to be parasites and chronic intussusception. Cats with PLE usually have IBD or alimentary tract lymphoma.

Many dogs with PLE have hypocholesterolemia. Pets with protein-losing nephropathies usually have hypercholesterolemia, while those with hepatic insufficiency often have hypocholesterolemia. Fecal examinations for parasites are appropriate. Although parasites are an uncommon cause of PLE in adult animals, pets in select environments (e.g., confined areas where patients can reinfect themselves) may incur substantial parasitic loads.

A substantial number of dogs and cats with PLE do not have vomiting or diarrhea, just like a substantial number of dogs and cats with severe hepatic disease do not have an increased ALT or SAP. This may be especially true of dogs with primary intestinal lymphangiectasia. Fecal concentrations of alpha-1 protease inhibitor can be used as a means of confirming PLE. The major use for this test seems to be the hypoalbuminemic patient in which you suspect PLE, but which also has PLN and/or hepatic disease. However, there are several nuances about this test, especially collecting samples ,that make it potentially difficult to interpret. We seldom need this test in clinical practice.

## DIAGNOSTIC APPROACH TO PLE PATIENT

Once PLE has been diagnosed, intestinal biopsy is usually the ultimate means of establishing a diagnosis. Biopsy can be done via laparotomy, laparoscopy, or endoscopy. Feeding a small, fatty meal (use canned food, not dry, and add in cream or corn oil) the night before the procedure might make it easier to diagnose lymphangiectasia. Flexible endoscopy, when done by someone who is trained in how to take diagnostic tissue samples and submit them is usually more than adequate to obtain diagnostic samples. However, if endoscopy will be used to biopsy the small intestines, it is preferable to first ultrasound the abdomen to make sure that there are no focal infiltrates that are out of reach of the endoscope, or which might be more easily diagnosed by ultrasound-guided fine needle aspiration. Radiographs and barium series are seldom as sensitive as ultrasound. If flexible endoscopy will be done, one should biopsy both the duodenum and ileum. There have been numerous cases in which lymphangiectasia, IBD or LSA were obvious in the ileum but not in the duodenum. It is not necessary to enter the ileum with the endoscope to obtain a good tissue sample of the ileal mucosa.

Laparotomy and laparoscopy are good means of obtaining diagnostic samples, but it is surprisingly easy to procure non-diagnostic samples with these techniques (i.e., "full-thickness sample" is not synonymous with "diagnostic sample"). Endoscopy does have the advantage of allowing one to visualize mucosal lesions that are "invisible" when looking at the serosa. In some cases, the diagnosis can only be obtained by biopsying these focal lesions. If full-thickness biopsies are obtained in severely hypoalbuminemic animals, then serosal patch grafting will minimize the risk of suture line leakage. A nonabsorbable or a poorly absorbable suture (PDS) should be used.

## INTESTINAL LYMPHANGIECTASIA

Intestinal lymphangiectasia seems particularly common in Yorkshire terriers and Soft-Coated Wheaten terriers[3-5], but may occur in any breed. Sometimes dogs with lymphangiectasia have distinct ultrasonographic findings: "streaks" in the mucosa that represent dilated lymphatics.[6] While histopathology is obviously the desired means of diagnosis, one can sometimes make a definitive diagnosis based upon gross endoscopic findings (i.e., numerous, erratic, grossly engorged lacteals seen as large white blebs on the mucosa). These lesions are "fragile" and apparently may be destroyed by biopsying them (both endoscopically and surgically) if the endoscopist or surgeon is not careful. It is important to note that lymphangiectasia can be a relatively localized disease in the intestines, being present in only the ileum or only the jejunum or only the duodenum; therefore, it is important to biopsy as much of the intestinal tract as possible. Furthermore, if one biopsies the intestines and cannot find a cause of PLE, sometimes lymphangiectasia can be diagnosed by default (i.e., by eliminating IBD, lymphoma, parasites, intussusception, fungal infections, etc).

Diagnosis by means of endoscopic biopsy is certainly possible, but recent work has shown that poor quality biopsies (e.g., primarily villus tips) makes is much more difficult to find the lesions. If one is taking high quality tissue samples (i.e., total length of the villi plus subvillus mucosa down to the border of the mucosa and muscularis mucosa), it typically takes 3-6 samples to have 90-99% confidence in finding lymphangiectasia. However, it can take 5-7 times as many tissue samples to have the same assurance if you are obtaining poor quality tissue samples that primarily consist of villus tips.[7]

Therapy for intestinal lymphangiectasia revolves around an ultra-low fat diet, preferably with anti-inflammatory therapy designed to alleviate the lipogranuloma formation that commonly occurs within the intestinal wall and/or mesentery. Supplementation with medium chain triglyceride oil (MCT) used to be recommended because MCT oil supposedly bypasses intestinal lymphatics thus preventing further rupturing of the lacteals. Pancreatic enzymes were often added to the diet to ensure digestion of the medium chain triglyceride oil. MCT oil is seldom used anymore, probably because appropriate dietary therapy is usually more than sufficient. Feeding homemade diets that are highly digestible and ultra-low in fat (e.g., white turkey meat plus potato or rice) or feeding commercial diets is often very helpful in these patients. Such a diet can be so successful that it might occasionally be appropriate to use it as a therapeutic trial. Dogs with lymphangiectasia often show a marked increase in serum albumin concentration within 7-14 days of starting such a diet.

The important of lipogranulomas in the intestinal wall and mesentery is uncertain. However, it is hypothesized that some patients fail to respond to appropriate dietary therapy because of formation of very large or excessive numbers of lipogranulomas that so completely obstruct the intestinal lymphatics that even an ultra-low fat diet cannot prevent lacteal rupture. Therefore, once a diagnosis of lymphangiectasia is made (either by histology, grossly at endoscopy, or tentatively by response to an ultra-low fat diet), it may be appropriate to use anti-inflammatory therapy designed to prevent granuloma formation/enlargement. Prednisolone, azathioprine, and/or cyclosporine are commonly used for this purpose.

If the serum albumin is very low (e.g., $\leq$ 1.3 gm/dl), one is often tempted to administer a plasma transfusion while waiting to see what effect the diet will have. However, it is exceedingly difficult to increase the serum albumin concentration by transfusing PLE patients with plasma because so much of the albumin is quickly lost out the gut. You would probably have to give at least two units of plasma to a 15 lb dog in order to raise the serum albumin from 1.0 gm/dl to 1.8 gm/dl, and sometimes you would have to give 3 or 4 units. If it is critical to raise the plasma oncotic pressure, then administering hetastarch may be preferred because it costs less than plasma and it stays in the intravascular compartment longer than albumin.

These patients may be a increased risk for hypomagnesemia which may potentiate the problem of hypocalcemia.[8] At this time, we do not know how important it is to supplement magnesium to patients, but severe hypomagnesemia can be resolved by a constant rate infusion of maganesium sulfate.

## CRYPT LESIONS

Lesions of the intestinal crypts have been recognized as being associated with PLE in dogs.[9,10] We have identified two different lesions of the small intestinal crypts that can cause PLE. One type is characterized by crypts (usually duodenal) that are filled and somewhat distended with proteinaceous fluid and necrotic inflammatory cells. While such dilated crypts can be found in many animals, including clinically normal dogs, finding large numbers of them in multiple tissue samples seems to be consistently associated with PLE. We do not know if this is a cause-and-effect relationship, or if the dilated crypts are simply a marker for some other process but are not causing the protein loss themselves. Several of these patients have responded to therapy with elemental diets, total parenteral nutrition, prednisolone, azathioprine, and/or metronidazole. We have seen this lesion associated with IBD as well as lymphangiectasia.

A second type of crypt lesion that appears to be less common than the first, is characterized by focal accumulations of mucus causing massive distention of the intestinal crypts. This has been reported once before, and we have seen two such cases. The most important aspect of diagnosis seems to be the fact that the lesion may be very focal, almost appearing as ulcers when looking at the intestinal mucosa through an endoscope. Therapy similar to that used on animals with the other form of crypt lesion may be helpful. We have used cyclosporine, but do not know if it is helpful, or if the clinical response is due to the other drugs that the patient is receiving.

These lesions have not been commonly reported. Recent work has shown that these lesions are extremely easy to miss if poor quality endoscopic biopsies are performed. While 7-12 high quality tissue samples (i.e., full length of villi plus subvillus mucosa down to the level of the muscularis mucosa) will find these lesions 90-99% of the time, about 7 times as many tissue samples will be needed if poor quality samples primarily consisting of villus tips are submitted.[7]

## CHRONIC INTUSSUSCEPTION

Chronic intussusception is a relatively important, and often missed cause of PLE in juvenile animals. The classic history is one of acute enteritis (e.g., parvoviral enteritis) which does not resolve as expected. The patient feels somewhat better, but continues to have diarrhea, and the serum albumin concentration gradually diminishes. It can be very hard to palpate an ileo-colic intussusception; abdominal ultrasound is clearly the preferred way to diagnose intussusception. Therapy is surgical.

## PARASITES

Although uncommon, nematodes may cause PLE in adult animals if there are large numbers of them. Whipworms and hookworms in particular may occasionally be responsible for PLE in older dogs. However, giardiasis has been reported to cause PLE in people.

## ANTIBIOTIC-RESPONSIVE ENTEROPATHY (ARE)

We believe that we are starting to recognize ARE as a cause of PLE in dogs. We now have several patients that appeared to have marked increases in their serum albumin concentration associated with antibiotic therapy. However, because dietary change is often performed simultaneously with antibiotic therapy in these patients, cause-and-effect is not clearly established.

## REFERENCES

[1]Allenspach K, et al. **J Vet Int Med** 2007, 21: 700; [2]Craven M, et al. **J Small Anim Pract** 2004, 45: 336; [3]Littman MP, et al. **J Vet Int Med** 2000, 14: 68; [4]Peterson PB, et al. **Vet Clin N Am** 2003, 33: 1061; [5]Kull PA, et al. **J Am Vet Med Assoc** 2001, 219: 197; [6]Gaschen L, et al, **Vet Radio Ultrasound** 2008, 49: 56; [7]Willard M, et al. **J Vet Int Med** 2008, 22: 1084. [8]Bush WW, et al. **J Am Vet Med Assoc** 2001, 219: 1732; [9]Willard MD, et al. **J Vet Int Med** 2000, 14: 298; [10]Willard MD, et al. **J Am Anim Hosp Assoc** 2003, 39: 187.

## HEMATEMESIS/GI BLOOD LOSS: CAUSES, DIAGNOSIS & MANAGEMENT OF OCCULT ULCERATION & COAGULOPATHIES IN DOGS & CATS

**Michael D. Willard, DVM, Dipl ACVIM**
**College Station, Texas**

Hematemesis and melena necessitate a slightly different approach because some rule-outs become more likely while others become much less likely. We will be including upper gastrointestinal bleeding in this discussion. For starters, we will not be discussing vomiting that produces "flecks" of blood because this can be seen in any dog (and perhaps cat) with vigorous vomiting in which the gastric mucosa is traumatized by the physical act of vomiting. It is easy to identify fresh blood in the vomited material as long as the patient is not eating something that is red or that produces a pink color to the vomited material simple secondary to pigment leaching out of the food material. Most of the time, hematemesis is denoted by a "coffee-grounds"-like material that most clients (and some veterinarians) do not recognize as blood. In like manner, we will not discuss hematochezia; rather, we will discuss melena and occult GI bleeding. A common mistake is being concerned over "dark stools". Noting that a patient has dark stools is generally useless. Lots of dogs have dark stools and no problems or GI blood loss at all. The color of the stool is not an issue until the stool is pitch, tar, coal, asphalt black. Then it may be melena (if it is not due to Bismuth or a lot of green bile giving it a near-black appearance). If in doubt, place some of the fresh feces on white, absorbent paper. If it is melena, there will be a red color diffusing from the feces. Melena is only seen if there is acute loss of a lot of blood. Most dogs losing blood in the upper GI tract do not have any important changes in the color of the feces. Rather, you will see anemia and hypoalbuminemia. Also remember, you may or may not see hypoglobulinemia; it all depends upon what the serum globulin concentration was before you started losing blood. Fecal occult blood tests are seldom that helpful or necessary, but can occasionally be informative in confusing cases. However, you need to use a test for which the laboratory has substantial experience in dogs so that the results can be meaningfully interpreted. Some fecal blood tests will routinely give a positive reaction when used on canine feces.

When there is a substantial amount of blood being ejected from the mouth, there tend to be 3 major reasons: coagulopathy, swallowing blood from elsewhere and gastrointestinal ulceration/erosion (GUE).

### COAGULOPATHIES

Most coagulopathies cause concomitant bleeding from the nose or accumulation of blood in body cavities or petechia. However, there are many cases in which the only sign of a systemic coagulopathy is GI bleeding. Therefore, it is always appropriate to check the platelet count and some measure of clotting factor adequacy in animals with hematemesis or GI blood loss. While coagulopathies are a relatively uncommon cause of GI bleeding compared to ulceration/erosion, they can have devastating consequences if not diagnosed promptly. In particular, remember that some cats with intestinal disease will malabsorb vitamin K to the point of having vitamin K-responsive coagulopathies. Also remember that just because the patient did well during a surgical procedure a few days before the current bleeding episodes does not mean that a coagulopathy is impossible; sometimes the prior surgical procedure apparently depletes the limited amounts of coagulation factors in patients with subclinical coagulation defects, causing them to become clinical after the procedure.

### ULCERS AND EROSIONS

#### Gastritis

Gastritis due to any number of causes typically has some degree of GUE present. We will not discuss this cause too much since most of these cases quickly and spontaneously resolve. *Helicobacter* is an extremely important cause of GUE in people, but to date has not been shown to have a cause-and-effect relationship with GUE in dogs and cats[1]. There are often incredible numbers of *Helicobacter* found in ulcer craters; however, *Helicobacter* can be found in ulcers caused by almost anything. Therefore, it seems likely that the *Helicobacter* are there because a) they are found in most dogs and cats and b) the ulcer crater is an opportune place for them to grow. Despite many people's opinions, the gastritis associated with renal disease seldom causes ulcers.[2]

#### Drugs

NSAIDs are still a very important cause of GUE in the dog, despite all the newer, "safe" NSAIDs. High doses of dexamethasone also have substantial potential for significant GUE. Prednisolone by itself is generally not ulcerogenic unless it is used in very high doses (e.g., > 2-3 mg/lb/day) or is administered to a patient with other "ulcerogenic" risk factors (hypoxia, poor perfusion), and even then it is not particularly bad. Combining steroids and non-steroidal drugs can be devastating. You can use ultra-low dose aspirin (0.5 mg/kg once daily)[3] when treating IMHA dogs with steroids, but this only works because the NSAID is being used at such a low dose.

There continues to be a substantial problem with the use of non-steroidal anti-inflammatory drugs (NSAIDs) in dogs. All NSAIDs have the potential to cause devastating GUE, and some of these non-steroidal drugs are renowned for their toxic effects (i.e., indomethacin, naproxen, flunixin). While able to cause significant ulceration and bleeding all by themselves, the ulcerogenic potential of NSAIDs is particularly augmented by other factors, especially concurrent administration of another NSAID or a corticosteroid, and hypoperfusion of the alimentary tract. Even though many dogs tolerate such combination therapy, you need to realize that you are "walking on thin ice" (see comments above on use of ultra-low dose aspirin). Many to most of the dogs treated with NSAIDs have endoscopically visible erosions, hemorrhages and/or ulcers, depending upon the drug used and the dose administered. It is important to note that most dogs with GUE due to NSAIDs may be completely asymptomatic. Finally, there is tremendous between-dog variation regarding the alimentary tract response to NSAID's; some dogs may almost bleed to death because of a small dose of aspirin while most dogs would tolerate a much larger dose with relative impunity.

While the newer Cox-2 NSAIDs (e.g., carprofen, etogesic, deracoxib, meloxicam, etc) have much less potential for causing GUE than the older NSAIDs[4,5], you can still see GUE due to these drugs. Part of the problem is that these "safe" drugs are being used so extensively and casually that it is hard to imagine side effects not occurring. In particular, it appears that the problem often revolves around using inappropriately high doses (after all, the drug is so safe that ...), using the drug at the wrong time (e.g., when the patient is experiencing shock or has poor perfusion to the alimentary tract), and possible using the drug too soon after stopping some other NSAID. While controversial, it appears that it might be necessary to have a washout period between stopping one NSAID and starting one of the newer Cox-2 NSAIDs. There is published literature to the contrary, but this is an area that bears additional work.

**Other causes**

Mast cell tumors may look like any skin lesion. In particular, they can perfectly mimic the appearance and feel of lipomas, such that the only way to distinguish them from lipomas is by aspirate cytology. When these tumors degranulate, they release histamine which if of sufficient magnitude can cause gastric acid hypersecretion. This can result in severe ulceration, especially just inside or just beyond the pylorus.

"Stress" in the context of this lecture refers to poor perfusion. This poor perfusion can be caused by hypovolemic shock, Systemic inflammatory response syndrome, or perhaps neurogenic shock[6] (although it seems more likely that the mucosal defects are due to the attendant steroid therapy that commonly occurs in these patients). It can also refer to major, extreme exertion (e.g., endurance sled dogs racing in subzero weather).[7] The history is usually diagnostic.

Hepatic failure seems to be another important cause of GUE in the dog. Anytime a dog with hepatic disease suddenly becomes clinically worse (especially if it becomes obviously encephalopathic), you should consider the possibility of GUE. Bleeding into the intestine counts as a high protein meal and predisposes to hepatic encephalopathy in these patients. In ill patients that cannot be scoped, aggressive therapy with famotidine and/or carafate is reasonable. Hepatic disease may cause disseminated intravascular coagulopathy which can cloud the picture when trying to determine the cause of hematemesis.

Gastric tumors may cause bleeding. The leiomyoma and leiomyosarcoma in particular may cause especially dramatic bleeding due to their propensity to ulcerate. This is especially important because this tumor is often curable with surgery, as opposed to lymphomas and carcinomas that are more common, have less dramatic signs, seldom cause GI blood loss and yet have a much worse prognosis. Unfortunately, it can be hard to adequately image the stomach with ultrasound, and these masses can be missed if there is blood, ingesta and/or air in the stomach.

Surgery of the stomach can be responsible for GI bleeding. If the closure is done improperly and the mucosa does not cover the defect, then bleeding can easily result.

Hypoadrenocorticism may be responsible for severe hematemesis that can produce life-threatening shock. Such severe hematemesis appears to be a rare complication of hypoadrenocorticism, but should be considered in cases with appropriate history, CBC and/or serum biochemistry changes, as well as patients that do are not readily diagnosed with other causes.

Inflammatory bowel disease has been reported to be associated with GUE, especially in the dog. Right now, the definition of IBD is under scrutiny, but at any rate it appears IBD is not commonly associated with GUE. However, when idiopathic GUE is found during endoscopy, the stomach and duodenum should be biopsied.

Heavy metal intoxication causes hematemesis, but this is fortunately uncommon at this time.

Gastrinomas are typically small pancreatic tumors which produce large quantities of gastrin, a hormone which causes gastric acid secretion. Multifocal duodenal ulceration/erosion is very suggestive of this tumor, as is a large ulcer just past the pylorus (as for mast cell tumors). Gastric erosions are sometimes seen, but gastric ulcers appear to be rare in this disorder. GI bleeding is not particularly common in these patients although it is possible. Esophageal ulceration may also occur if there is gastroesophageal reflux of the highly acidic gastric contents. Measurement of serum gastrin concentrations may be diagnostic. However, anything which causes gastric distention or renal failure can produce increased fasting serum gastrin concentrations.

Foreign objects get a lot of press as causes of GUE, but in fact they are relatively uncommon causes. However, they are particularly important in patients that have GUE because even the most innocuous of GI foreign objects (e.g., paper, small piece of soft cloth) can sometimes prevent a pre-existing ulcer from healing. They typically need to be removed in patients with GUE.

Cats seldom have gastric ulcers, and when they do, lymphoma is a somewhat common cause. However, there are ulcers in feline stomachs that are idiopathic.

**INGESTING BLOOD**

This is a possibility that is typically forgotten. However, it is surprisingly easy to have bleeding pulmonary lesions in which the blood is coughed up, swallowed, and later vomited. In most of these cases, the client does not report coughing (perhaps because the hematemesis "wipes" all else from their mind). In like manner, we have seen cases in which blood was trickling from the choanae into the pharynx and being swallowed, and yet the patient had no history of sneezing or coughing or nasal discharge.

**GENERAL DIAGNOSTIC APPROACH TO THE PATIENT WITH HEMATEMESIS OR UPPER GI BLEEDING**

There is often something in the history that is suggestive of the cause of the bleeding (e.g., use of NSAIDs, shock, etc). If that is the case, then it is often reasonable to begin appropriate therapy after requesting basic laboratory testing (e.g., CBC, serum chemistry panel) to determine the severity of the bleeding and if there are other diseases (e.g., hepatic disease, renal disease) present.

Imaging (especially ultrasound) is typically appropriate but not necessarily imperative at this time. If the cause of the GI bleeding or hematemesis is not obvious, if the patient has not responded to 5-7 days of appropriate therapy, or if the bleeding is severe, then additional diagnostics are important and should be performed promptly.

First, as stated above, it is wise to first eliminate coagulopathy with a platelet count and some measure of clotting factor adequacy. I typically request PT and PTT, but a mucosal bleeding time is a very useful screening test in these patients. Sometimes there is both a mucosal defect and a coagulopathy. In particular, if ehrlichiosis is possible, one must consider the possibility that the patient has what would normally be an insignificant mucosal defect but which is bleeding because of the effects of *Ehrlichia* spp. upon platelet numbers and their function. After coagulopathy has been eliminated, then imaging should be done if it has not already occurred, and ultrasound is especially important as it may reveal masses that can be aspirated percutaneously, thus avoiding the need for endoscopy/surgery.

If these tests have not revealed the diagnosis, then gastroduodenoscopy is generally performed next. The specific reasons to do gastroduodenoscopy in a patient with GI bleeding are to:

a) determine if this is a case in which surgery can remove a defined number of ulcers (this is for cases that are bleeding and have not responded to medical management or cases that are bleeding so badly that one cannot wait on medical management). In these cases, it is important to be sure that bleeding is not due to widespread erosions that cannot be cured surgically. There is no relationship between the size of the mucosal defect and the amount of bleeding; patients with lots of small erosions often bleed as bad or worse than patients with ulcers. It is also important to determine the number and location of such ulcers as they may be hard to find during a gastrotomy.

b) determine if there is a gastric tumor or some other infiltrate in a patient with GUE that is non-responsive to appropriate therapy.

c) determine the cause in a patient with GUE and no apparent cause on the history, physical examination, or routine blood work.

d) look for a cause of bleeding in a patient with GI blood loss of unknown cause.

It is important to note that endoscopy will not generally allow one to determine if an ulcer will or will not respond to medical management. Only treating and observing the patient will accomplish that.

It is important to not give carafate within 24-36 hr before endoscopy because it will cover the lesions and make evaluation more difficult. It is best if food has been withheld for at least 24 hours. Avoid prokinetics (e.g., metoclopramide). Endoscopy of these patients may be difficult is there is substantial blood present in the stomach. Patience is necessary to flush in water and then aspirate it and the blood over and over again. It is important to be able to view as much of the gastric mucosa as possible. In some cases, there are lots of huge blood clots that cannot be removed endoscopically, in which case surgery might be necessary. It can be especially easy to miss ulcers that are in the pylorus. The pylorus is typically infiltrated and inflamed, making it more difficult to pass the tip of the scope through that area. Therefore, the endoscopist often does not obtain a good view of this area. One may have to go in and out of the pylorus multiple times to be sure that there is or is not a lesion present. If a cause of upper GI bleeding cannot be found in the stomach or duodenum, strong consideration should be given to examining the trachea, bronchi, and choanae while the patient is anesthetized. Patients with hematochezia may benefit from colonoileoscopy, but patients with hematemesis or melena rarely benefit from lower endoscopy.

If there is substantial upper GI blood loss and these tests do not allow diagnosis, then exploratory surgery is the next step. However, it is very easy to not be able to find the cause of the bleeding in these patients, so warn the owner before doing the procedure.

## REFERENCES

[1]Wiinberg B, et al. **J Vet Int Med** 2005, 19: 4; [2]Peters R, et al. **J Vet Int Med** 2005, 19: 315; [3]Weinkle T, et al. **J Am Vet Med Assoc** 2005, 226: 1869; [4]Reimer ME, et al. **J Vet Int Med** 1999, 13: 472; [5]Boston SE, et al. **Am J Vet Res** 2003; 64: 1369; [6]Neiger R, et al. **J Vet Int Med** 2000, 14: 33; [7]Davis M, et al. **J Vet Int Med** 2006, 20: 835.

**Due to unavailable paper there is a gap in pagination.**

**Pages 486-488**

**CANINE & FELINE CHRONIC DIARRHEAS: DIAGNOSIS/MANAGEMENT OF INFILTRATIVE DISORDERS**

**Michael D. Willard, DVM, Dipl ACVIM**
**College Station, Texas**

The first diagnostic steps in essentially all animals with chronic diarrhea cases are typically history, physical examination, basic laboratory data (serum biochemistry profile, + CBC), and fecal examination. Then, if the patient is so sick that you cannot chance a 2-6 week therapeutic trial that may fail; or, if the patient is substantially hypoalbuminemic; or, if the owners insist upon obtaining a definitive diagnosis now, then more invasive tests (e.g., histopathology) are typically the next step. The main purpose of biopsy, whether it is of the small or large intestine, is to look for infiltrative diseases (i.e., inflammatory diseases such as IBD or fungal infection; or, neoplastic disease such as lymphoma).

**ABDOMINAL ULTRASOUND**

Ultrasound is useful to look for enlarged mesenteric lymph nodes, focal intestinal/gastric lesions, and loss of mucosal layering. Focal enlargements may suggest a tumor (e.g., alimentary lymphoma or carcinoma), as may lymphadenopathy. However, animals with severe IBD may also have mesenteric lymphadenopathy (as may dogs with histoplasmosis or pythiosis). If the lymph nodes are enlarged, it is reasonable to aspirate them percutaneously with ultrasound guidance. Mesenteric lymph nodes are typically reactive, making it more difficult to interpret cytology from them. However, finding obvious sheets of lymphoblasts or fungal organisms (e.g., histoplasmosis) allows diagnosis. Sonographic examination of the intestines is important (i.e., you may make a diagnosis), but it does not detect intestinal mucosal disease in many patients that are afflicted with such disease. Thickness of the mucosa is not as important a finding as loss of mucosal layering.[1,2] Loss of layering suggests that severe infiltration is likely (either inflammatory or neoplastic), but normal-appearing mucosa may have marked disease present. Most of the time, ultrasound's major use is to help you decide whether to perform intestinal biopsy using endoscopy or laparotomy. If there is an obvious lesion where an endoscope cannot reach, it is best to perform laparotomy instead of endoscopy. In contrast, abdominal radiographs (plain or contrast) are rarely helpful or cost-effective in these patients.

**INTESTINAL BIOPSIES**

Endoscopic biopsy of the alimentary tract can utilize either a rigid scope or a flexible scope, depending upon what organ will be biopsied. In general rigid scopes routinely retrieve much better tissue samples than flexible scopes; however, they can only be used for obtaining tissue samples of the descending colon. There is almost no reason to take surgical biopsies of the descending colon; rigid forceps can obtain diagnostic samples of almost any mucosal or submucosal lesion in the descending colon. There are exceptions, but they are rare. The advantages of rigid colonoscopy are a) it is easy to do, b) the equipment is relatively inexpensive, and c) one may almost always obtain diagnostic samples. The only disadvantages of rigid colonoscopy are a) it is possible to perforate the colon (something that is very rare) and b) you cannot examine or sample the ascending or transverse colon.

Flexible endoscopic biopsy of the alimentary tract is a) quick (i.e., usually < 20 min to obtain gastric and duodenal samples), b) safe, c) of minimal stress to very ill animals and d) is typically diagnostic, if done correctly. However, flexible endoscopy has gotten a bad reputation because of all the inaccurate diagnoses that have occurred in animals that were sampled using this technique. This centers around the comment that flexible endoscopy is typically diagnostic *if done correctly*. Unfortunately, a lot of endoscopy is done incorrectly, especially in regards to obtaining tissue samples. The problem is that while the smaller diameter endoscopes are very easy to use (i.e., it is easy to put them into small orifices and pass them through small pylori), these are the scopes with the smaller biopsy channels. It is easy to obtain inadequate tissue samples with biopsy instruments designed for scopes with a 2.2 mm biopsy channel. Scopes with a biopsy channel of 2.8 mm typically allow excellent tissue samples to be consistently obtained; however, these scopes are often larger in diameter (e.g., usually ≥ 9.0 mm), making them harder to use in cats and dogs < 7 lbs. If you are practiced, you can usually get a 9.0 mm scope through the pylorus of cats > 6 lbs, but even accomplished endoscopists cannot always do this. Therefore, the clinician must make a decision based upon how often he/she will be performing endoscopy, because that usually determines how accomplished they will become and what they can afford. One should use fenestrated, ellipsoidal forceps with a serrated edge. Then, you keep taking sample after sample until you obtain good samples that encompass the full thickness of the mucosa down to and preferentially including the submucosa. If the sample fragments easily, then it is primarily villus tips and is inadequate. You may submit it, but do not count it. At least 6-8 good tissue samples should be obtained.[3] Immediately after obtaining the sample, it is carefully removed from the biopsy forceps with a 25 gauge needle and gently spread out on a tissue cassette sponge such that the luminal side (i.e., the side with the villi) is oriented upwards and the sample is no longer folded. The samples are allowed to adhere to the sponge, but are not allowed to dry out. The sponge with the tissue samples is placed in formalin promptly.[4]

It is becoming apparent that biopsying both the duodenum and ileum is useful in many patients (especially cats). It is desirable to use a flexible scope when obtaining colonic samples so that the ileo-colic valve area can be inspected and one may obtain ileal mucosal samples. An experienced clinician can usually enter the ileum of dogs weighing > 10-15 lbs, but it rare that the tip of the scope can be advanced into the ileum of cats. However, even in small dogs and cats, the biopsy forceps can usually be passed through the ileocolic valve and ileal tissue obtained by performing blind biopsies of the mucosa. As alluded to, there are some disadvantages to flexible endoscopy of the alimentary tract: a) it requires expensive equipment, b) it is easy to take inadequate samples, c) some lesions are out of reach of the endoscope, and d) some lesions are so dense that you cannot obtain a diagnostic biopsy of them with this equipment. Scirrhous carcinomas and pythiosis are the two major examples of such lesions.

Full-thickness biopsies of the intestines are performed during exploratory laparotomy or laparoscopy. The main advantage is that these samples have excellent morphology and include the entire submucosa *if they are properly taken*. The main disadvantages of surgical biopsies are a) you seldom need full thickness biopsies to obtain a diagnosis, b) there is substantially more risk for morbidity and mortality with laparotomy compared to endoscopy (e.g., dehiscence), c) you cannot obtain as many full thickness biopsies as you can endoscopic biopsies, which is important because some lesions are very spotty and are only found in 3 out of 7 endoscopic samples, d) you cannot see mucosal changes during surgery and hence it is easy to biopsy the wrong area of the stomach or intestines, and e) it is easy to take a "wedge" of intestine in which you have a lot of serosa and very little mucosa, making the sample of poor diagnostic usefulness. It is worth remembering that it is easy to miss the lesion or take a nondiagnostic intestinal biopsy during laparotomy. Regardless of whether you are performing endoscopy or surgery, you should usually investigate the entire alimentary tract unless there is an obvious foreign body or mass.

## INFLAMMATORY BOWEL DISEASE

Inflammatory bowel disease (IBD) is defined as "idiopathic inflammation of the intestines" and as such is probably not the most common cause of chronic small or large bowel diarrhea in dogs and may not be as common in cats as was once believed. This definition means that you cannot diagnose IBD solely by histopathology. You need the biopsy, but you ultimately diagnose IBD by finding intestinal inflammation and showing that it is idiopathic by eliminating diet, parasites, bacteria and fungal agents as the cause. You cannot eliminate dietary causes and bacterial causes by histopathology or blood tests; therapeutic trials are necessary. This is very important because diagnosing IBD generally results in anti-inflammatory or immunosuppressive drugs being used. However, if the patient has dietary-responsive or antibiotic-responsive disease, then appropriate therapy directed at the cause usually renders these drugs unnecessary. I stress this point because many patients have been erroneously diagnosed with IBD, improperly treated, and significantly harmed because IBD is a "fashionable" diagnosis. While IBD is a real syndrome and is important for the veterinary practitioner to understand, it can easily degenerate into an excuse and a diagnosis of convenience.

The cause of IBD is unknown. The cells found in most patients with IBD primarily consist of lymphocytes and plasma cells, and less often eosinophils and/or neutrophils. Mononuclear infiltrates appear to be a generic inflammatory response of the intestines to any number of causes. Current thought about the pathogenesis of IBD is that the intestines experience an insult from any number of causes, but then the inflammatory response becomes self-perpetuating. Because the intestines are replete with dietary, bacterial, and parasitic antigens, it has been hypothesized that these are the antigens that maintain the inflammatory response in these patients. Bacteria in particular are starting to look especially important as a means of perpetuating the intestinal inflammation in these patients.[5] Clinically normal dogs and cats routinely harbor large numbers of bacteria (i.e., > $10^5$ bacteria/ml of fasting intestinal fluid) in the upper small intestine. Therefore, it seems reasonable that any patient diagnosed with IBD should receive dietary and/or antibacterial therapy in addition to the typical symptomatic anti-inflammatory drugs. Even if these adjunctive therapies do not cure the patient, they may at least allow lower doses of the anti-inflammatory and immunosuppressive drugs to be effective. This may be the reason why metronidazole has long been considered a useful adjunct in the therapy of IBD; it controls anaerobic bacterial populations (as opposed to altering cytokines, as has been suggested).

## DOGS

### Lymphocytic-plasmacytic enteritis (LPE)

LPE is primarily a disease of middle-aged to older dogs. It is rarely seen in young dogs. Some canine LPE patients respond well to elimination diets and antibiotics; therefore, they should be used in addition to anti-inflammatory drugs. However, high dosages of corticosteroids (i.e., 2.2-4.4 mg/kg/day) may be necessary. Stronger immunosuppressives (e.g., azathioprine and cyclosporine) may also be needed. Even when you are using effective therapy, you may have to wait for 2-4 weeks before seeing results.

### Eosinophilic gastroenteritis (EGE)

Many cases respond well to a strict elimination diet. Because high quality hypoallergenic diets are commonly used, we diagnose this form of IBD much less commonly than in years past. Occasionally dietary therapy by itself is ineffective, and then adding 2.2-4.4 mg prednisolone/kg body weight/day will be needed.

### Lymphoma (LSA)

LSA can obliterate the mucosa and cause malabsorption and protein-losing enteropathy. There may be peripheral involvement (e.g., lymphadenopathy), but the lesion is often limited to the abdomen and does not cause palpable masses. The prognosis is poor, but rare patients live for years after therapy (e.g., surgery and/or chemotherapy). Dogs are more likely to have lymphoblastic or prolymphocytic alimentary LSA, versus cats which may have well differentiated, small cell lymphoma. It is common to misdiagnose lymphoma as IBD in poor, superficial endoscopic biopsies.

### Pharmacologic options

Azathioprine is an excellent immunosuppressive, but occasionally causes hepatic failure or pancreatitis. Although administering it daily gives faster results, giving it every other day is safer.

Cyclosporine is a useful therapy for canine IBD, and it may be helpful in dogs with lymphangiectasia.[6] It does not have the side effects of steroids or azathioprine, but it is substantially expensive. It is critical to measure blood levels of the drug (can start at 7 days after starting the drug) to be sure that a particular animal is absorbing sufficient amounts of the orally-administered drug. The typical starting dose is 3-5 mg/kg bid. This can be very expensive in larger dogs; therefore, some people simultaneously treat with ketoconazole (3-7 mg/kg bid) to inhibit metabolism of the cyclosporine and thereby achieve higher blood levels at less cost to the client.

## CATS

### Lymphoplasmacytic enteritis (LPE)

The average age of affected cats is about 7-9 years old, but cats $\leq$ 1 year old have been affected. Vomiting and weight loss are often the primary signs. In general, severe weight loss tends to correlate with more severe inflammatory infiltrates in the intestines. Anorexia and/or diarrhea seen about 50% of the time. Hypoalbuminemia tends to be much less common in cats with LPE than in dogs with LPE. Increased serum activities of hepatic enzymes are relatively commonly. Rarely there can be bleeding due to vitamin K malabsorption. Prednisolone is typically successful in resolving or significantly controlling the disease.

There has been the concern that LPE can transform into LSA. In one study which examined 60 cases, there were 6 instances in which an initial diagnosis of LPE was later changed to lymphosarcoma. This appeared to be due to an initial misdiagnosis rather than transformation of the IBD into lymphoma.

**Lymphoma (LSA)**

LSA is the most common gastrointestinal malignancy of cats, and alimentary LSA is usually the most common form of LSA identified in cats. Most alimentary LSA's are FeLV negative. The most common signs of intestinal LSA are similar to that found for IBD. Thickened intestinal loops occur but are not invariable. There are no CBC or serum chemistry profile findings diagnostic or even strongly suggestive of alimentary LSA. Abdominal ultrasound may reveal thickened intestines and/or enlarged mesenteric lymph nodes. Neither of these findings diagnose LSA or eliminate IBD; however, marked mesenteric lymphadenopathy seems to be more common in LSA than in IBD. Alimentary LSA may be solitary or diffuse. The former may or may not have enlarged mesenteric lymph nodes while the latter usually does. LSA involving large granular lymphocytes has been reported; these animals typically have intestinal masses (usually jejunal) and are FeLV negative.[7] However, cats with large granular lymphocytes usually are hypoproteinemic, a finding that is seems less consistent with other forms of alimentary LSA. It is becoming increasingly apparent that ileal biopsies can be crucial in diagnosing LSA[8]; therefore, colonoileoscopy is crucial in cats suspected of having infiltrative bowel disease. While much has been made of the need for full-thickness biopsies to reliably diagnose alimentary LSA, this is not clear cut. Biopsying sites other than just the duodenum (i.e., the ileum) and obtaining adequate tissue samples with the endoscope are probably much more important than having full-thickness tissue samples.

There are some instances in which an apparently correct diagnosis of IBD is latter followed by a diagnosis of intestinal LSA. It is suspected that in some patients, LPE may be a preneoplastic disorder. The frequency of this occurrence is unknown; the point is that one should not hesitate to rebiopsy a patient if the disease (e.g., IBD) does not respond to apparently appropriate therapy. This concept works in "reverse" also.[9] We have documented a cat in which LSA was diagnosed twice over 9 months, and the patient was ultimately found to have a dietary-responsive enteropathy. The prognosis for cats with alimentary LSA is often poor; however, some cats with well-differentiated alimentary LSA will respond for months or years to combination chemotherapy.

**Pharmacologic options**

Budesonide is largely eliminated by first pass metabolism in the liver. It is typically administered orally at 1 mg/cat, once to twice daily. The main reason for using this drug is the patient that responds to steroids but cannot tolerate the side effects of the steroids (e.g., the diabetic cat whose intestinal disease responds best to prednisolone). Hyperadrenocorticism is still possible.[10]

Azathioprine should never be used; chlorambucil is safer and just as effective. There are at least two methods of administering chlorambucil to cats. In the first method, the initial dose is 2 mg of chlorambucil/$M^2$ of body surface area given daily for 4-7 days. The dose is then decreased to 1 mg/$M^2$ daily for 7 days. If the clinical signs are lessening, one then starts to administer the drug daily, but only every other week. The second method is to give large cats (i.e., > 7 lbs) 2 mg twice weekly and smaller cats (i.e., < 7 lbs) 1 mg twice weekly.

**REFERENCES**

[1]Rudorf H, et al. **J Small Anim Prac** 2005, 46: 322; [2]Gaschen L, et al, **Vet Radio Ultrasound** 2008, 49: 56; [3]Willard M, et al. **J Vet Int Med** 2008, 22: 1084; [4]Mansell J, et al. **Vet Clin N Am** 2003, 33: 1099; [5]Janeczko S, et al. **Vet Micro** 2008, 128: 178; [6]Allenspach K, et al. **J Vet Int Med** 2006, 20: 239; [7]Roccabianca P, et al. **Vet Path** 2006, 43: 15; [8]Evans S, et al. **J Am Vet Med Assoc** 2006, 229: 1447; [9]Ragaini L, et al. **Vet Res Comm** 2003, 27; 791; [10]Tumulty J, et al. **J Am Anim Hosp Assoc** 2004, 40: 120.

**FELINE OBESITY: UNDERSTANDING THE PROBLEM & AFFECTING THE SOLUTION**

**Debra L. Zoran, DVM, PhD, DACVIM**
**College Station, TX**

Cats are obligate carnivores. While this is not a news bulletin, we often fail to consider this in our nutritional approach for them. While cats can use carbohydrates (CHO) as a source of metabolic energy, they have no requirement for them (nor do dogs for that matter). But, more importantly, because cats evolved consuming prey (e.g. high protein, low to moderate fat, minimal carbohydrate), they are metabolically adapted for higher protein metabolism and lower CHO utilization. What does that mean metabolically and nutritionally? There are a number of specific metabolic and biochemical differences in feline physiology that are important. For those who are interested in the specific details of these metabolic and physiologic differences in the nutritional biochemistry of cats, the reader is referred to several recent reviews on this subject. This paper will discuss why obesity has become such a significant problem in cats, the approaches we can (and must) take to prevent it, the important medical aspects of obesity and the keys to management of it.

## DEVELOPMENT OF FELINE OBESITY

The primary reason for development of obesity in any animal is that they are consuming more energy than they are expending. This can occur when a cat has excessive dietary intake of calories (food or treats) or when there is a reduction in energy expenditure (reduced activity, illness or injury resulting in less exercise, etc). However, it is very important to recognize that obesity is not just a simple matter or intake or output – as we all recognize, there are many other factors that may control or play significant roles in the development of this disease. These factors include genetic predisposition, sex, neuter status, hormonal disturbances, and other conditions that can influence or control appetite, metabolism, and homeostasis. In indoor cats, reduced energy expenditure is a very important problem, and this is compounded by the fact that it is not easy to increase energy expenditure in cats like dogs with directed exercise. Additionally, because many cat owners prefer to feed free choice dry food, the risk of overfeeding, even in very small amounts is very high. In either case, the primary reason that weight gain occurs in cats is that they have a positive energy balance and this must be changed to affect weight loss. In both male and female cats, neutering is an important risk factor for obesity, due to the hormonal changes that occur that result in changes in levels of leptin, progestins, prolactin, insulin like growth factor (IGF-1) and other hormones that result in increased appetite, and reduced energy metabolism and metabolic rate. The key factors for prevention of obesity in neutered animals appears to be careful control of intake immediately after neutering (no free choice feeding, reduction of intake by 25% to account for the hormonal changes resulting in reduced energy needs), and close monitoring of body weight and BCS to allow adjustments in intake if needed. In addition, because feeding cats is often an important social interaction, feeding and food interaction with the cat can become a daily ritual that can become a problem: resulting in overfeeding and inappropriate food intake patterns. In households where the owners are health conscious (conscious of diet and nutrition, who exercise regularly, and watch their own weight) the pets tend not to be obese (except where free choice feeding is used). Thus, there are clearly human behavioral and "food is love" issues that have to be considered in the development of obesity, and these must be addressed for successful weight control to be achieved as well.

## THE IMPORTANCE OF OBESITY AS A DISEASE

Obese humans generally do not live as long as their lean counterparts, and are much more likely to suffer from obesity-related diseases such as type II diabetes, coronary artery disease, osteoarthritis, hypertension, and some cancers. Cats are susceptible to the same detrimental effects, including decreased longevity, and development of a variety of disorders that are associated with being obese. In a lifetime study in 24 Labrador dogs, dietary calorie restriction over their lifespan (to maintain a lean body condition) was clearly shown to increase longevity and reduce chronic disease conditions common to older dogs. In particular, the lean dogs lived an average of two years longer and had a reduced incidence of hip dysplasia, osteoarthritis and glucose intolerance. A similar study has not been performed in cats, but there is a clear difference in morbidity and risk for specific diseases (i.e. 4x greater risk of diabetes, 5x greater risk of lower urinary tract disease) in obese cats compared to their lean counterparts.

In order to understand the role of obesity in disease development, it is necessary to first understand the role of adipose tissue in energy balance and metabolism. The expansion of adipose tissue was long thought to be simply a depot for the deposition of fatty acids (triglycerides) that occurred because of the excess energy intake. However, research in the past decade has revealed that adipose tissue is not just a storage site, but also is responsible for production of many key hormones (e.g. leptin, adiponectin, and many others) involved in energy balance and a variety of other processes. Leptin was first identified in 1994 and its discovery opened the door to a whole new world of understanding of white adipose tissue that continues to this day. Leptin is an inhibitor of food intake – primarily through suppression of neuropeptide Y and orexin in the brain, and plays a role in the neural regulation of resting energy rate. In obese mice that lack leptin or have leptin resistance, appetite suppression mechanisms are non-functional and energy metabolism is abnormal. However, despite the importance of leptin in appetite suppression and energy expenditure, it is not the "silver bullet" of obesity management: simply replacing leptin does not result in weight loss or changes in the metabolism of obese animals. Further research in recent years has elucidated an entire list of adipokines secreted by fat cells. These adipokines are involved in a wide variety of metabolic and physiologic, as well as inflammatory changes that are implicated in the pathophysiology of obesity. In addition to the hormones produced by adipose tissue, it has also been discovered to be a major secretory organ – and does not just release fatty acids during fasting – the tissue also releases a wide variety of lipid based moieties: cholesterol, retinol, steroid hormones and prostaglandins. Finally, and possibly more ominously, fat cells in obese animals secrete a large number of pro-inflammatory cytokines (e.g. tumor necrosis factor alpha was the first to be discovered), including interleukins (IL-1, IL-6, IL-10), chemokines (such as monocyte chemoattractant protein -1), proteins involved in hemostasis (platelet activator inhibitor -1) and blood pressure (angiotensinogen). The concentrations of pro-inflammatory cytokines are greatly increased in obese animals and decreases with normalization of body weight. What does this mean for the obese cat? Obesity is a metabolic disease that results in major changes in appetite control, energy expenditure, and induces a chronic, low-grade, pro-inflammatory state that may be responsible for many of the diseases associated with increased body weight. There are a number of diseases in cats are reported to be associated with obesity, including orthopedic diseases, diabetes, abnormal circulating lipids, certain cancers, lower urinary tract disease, dyspnea due to compromised ability to breathe (pickwickian syndrome) and possibly increased risk of allergic airway disease, decreased immune function, and pancreatitis to name just a few. The interested reader is directed to large number of recent reviews on this subject for additional details, as this area of research is rapidly changing as we seek to better understand obesity as a disease.

**OBESITY & DIET**

While figures vary, recent studies indicate that greater than 35% of cats in the United States are overweight or obese. There are a large number of factors that contribute to this problem, however, one factor that is that is important to consider, both in the development of and treatment of obesity, is the role of the type and amount of diet fed. First, it is important to remember that cats utilize protein as an energy source, even under conditions of availability of other energy sources such as fat or carbohydrates (CHO). The second major consideration is that CHO in the diet that are not immediately used for energy (e.g. via exercise or other utilization for energy) will be stored as fat. Thus, in sedentary (indoor) cats consuming diets with a large amount of CHO (typical dry foods) often are consuming more calories from CHO than they will burn (remember in low energy usage states, cats will use protein for energy), and thus, even if they are seemingly not consuming very much, they will gain weight. Finally, for any weight loss plan to be successful, lost weight must be in the form of fat mass, not lean body mass (muscle). In the past, traditional weight loss plans include feeding an energy restricted (e.g. low fat, high CHO, high fiber, moderate protein) diet. However, while these diets may result in weight loss, they do so to the detriment of lean body mass – a fact that has two negative implications: loss of lean body mass is detrimental to overall health, and loss of muscle mass can initiate a starvation response that results in unsuccessful weight loss. Healthy weight loss requires loss of adipose tissue along with maintenance of lean body mass, as lean body mass is the driver of basal energy metabolism (loss of lean body mass is a major contributor to weight regain as appetite is not reduced and satiety not reached). Several recent studies have shown that use of a high protein, low CHO diet (at least 45% protein or higher, less than 15% CHO) in cats undergoing weight loss resulted in preservation of lean body mass, reduction of body fat, and successful weight loss. Importantly, high protein, low carbohydrate diets not only result in sustained weight loss in these cats, but also in normalization of appetite (reduced urge to eat constantly because they are satiated). Conversely, cats fed low fat, high CHO, high fiber diets not only lost muscle mass, but also had issues of glucose tolerance. Because dry foods must be extruded (i.e. made into a biscuit), CHO are required in the production process, and thus, it is difficult to achieve a very low CHO diet that is dry. Further, many of the commercially available high protein, low CHO dry foods are NOT low calorie, so it is extremely easy to feed too much. For example, high protein, low CHO diabetic diets are typically 500 kcal/cup or higher. This is a statement worth repeating: Too many calories of any kind, including protein calories, will cause weight gain or failure to lose weight. Thus, at this time, the best commercial diets for achieving a high protein, low CHO profile, along with controlled calories (portion control), are by feeding canned cat foods. Typical diabetic diet canned foods have anywhere from 165-190 kcal/5.5 oz can. While there are many canned foods that have a high protein/low CHO profile, it is also important to recognize that it is not universal – canned foods can be high CHO, low protein, or have poor quality ingredients resulting in ineffective or unhealthy weight loss too. The key is to remember that "one size does not fit all" in cat foods – and careful reading of the label can help determine the protein, CHO and fat levels, which is the start of the process.

**Calculating calories**

Most indoor cats do not need more than their resting energy requirement (RER) to meet their daily nutritional requirements. In other words, most cats do not require more than approximately 200 kcal/day. Cats that are very obese may require reductions in intake to levels significantly less than 200 kcal/day, however, this must be done slowly (weight loss should proceed at no faster than 1-2%/week), and decreases should not be implemented more often than every 3-4 weeks (to allow metabolic adjustment). Because reduced intake is essential to achieve weight loss, and preservation of lean body mass is essential to long term success and health, feeding cats greatly reduced amounts of food during weight loss requires that the protein quantity and quality be increased. The key point is this: set a target calorie intake, then weigh the cat monthly, and adjust the amount of food based on weight loss. The goal is for 1% weight loss per week or 3-4% per month – and if this goal is not being achieved, a reduction in calories (10-20%) must be done to meet the weight loss goals. But, equally important is that protein not become limiting, so a diet containing 45% protein or higher is essential.

**Key Points about diet selection in obesity management in cats**

The commercially available diets lowest in CHO are canned foods. It is important to remember that just because you are feeding a canned food, it does mean that you are feeding a high protein, low CHO diet (you must read the label) and it also doesn't mean you are feeding a high quality protein (low quality protein in foods can cause fecal odor and diarrhea due to poor digestibility and will also lose muscle mass). Most cats should be fed some (50% is a starting point) canned food as part of their diet throughout their life – both to reduce the CHO in their diet, but also to better control calories (dry foods are very calorie dense), and to increase the amount of water consumed daily. Furthermore, eating canned food is a learned behavior – if canned food is part of a kittens' diet, they will more readily eat canned food as an adult (e.g. when they need canned food for urinary disease or renal disease later in life). An important follow up point to remember about all diets is that calories count. You cannot free choice feed most indoor cats – even with high protein, low carb diets – because if they consume too many calories (and the diabetes diets are very calorie dense) they will become or remain obese. Also, calorie control must be started when they are kittens as obesity starts in young adults due to the above mentioned issues with neutering and energy intake.

The key point for obesity prevention (or correction) is balancing the energy intake / energy expenditure equation. In indoor cats, where exercise is reduced by the nature of the lifestyle, energy restriction becomes paramount to obesity prevention or correction. Energy restriction can be achieved by low fat high fiber diets, but many of these diets are not high enough in protein to preserve muscle and result in loss of muscle mass first, then eventually they will lose fat mass, resulting in an unhealthy weight loss and a strong tendency to regain weight (muscle mass loss will always increase the likelihood of yo-yo weight effects or failure of weight loss). High protein, low carb, low fat diets are ideal for weight loss in cats because they preserve muscle mass while restricting energy sources that will induce fat loss. However, portion control is ultimately the key to controlling energy intake – and the easiest way to achieve portion control is to feed canned food. In cats that won't eat canned food, the only options are a few high protein, low fat, low carb dry foods – but these should be selected for use in weight loss programs if at all possible.

**Table 1** Cats and Nutrition: Some Key Nutritional Differences

- Cats have an obligate need for protein and amino acids in their daily diet because they are unable to down regulate their urea cycle or transaminases (protein conversion to energy) as other species can in times of starvation.
- Cats utilize protein for energy, even in the face of large amounts of CHO in the diet
- Taurine, arginine, methionine, cysteiine, and possibly carnitine requirements for cats are greater than non-carnivores
- Arachidonic acid is also an essential fatty acid in cats (it is not in dogs), and is found only in fats from animal tissue
- Cats require vitamin A and D to be present in the active form in their diet as they are unable to synthesize adequate amounts from other dietary precursors (e.g. carotenoids or vitamin D precursors in skin)
- Cats have an increased need for many B vitamins in their diet (e.g. thiamin, pyridoxine, niacin, pantothenic acid) as they have greater metabolic needs for these vitamins and cannot synthesize or get them from other sources.
- Salivary amylase is absent in cats, and they have greatly reduced levels of intestinal and pancreatic amylases – so CHO digestion is much less efficient.
- Cats have fewer disaccharidases and other brush border enzymes in their small intestine designed to digest and absorb starches.
- The small intestine of cats is much shorter than that of an equally sized omnivore – longer GI tracts are necessary for handling of complex carbohydrates.
- Cats have greatly reduced activities of hepatic enzymes (e.g. glucokinase) designed to convert a post prandial glucose load to glycogen and thus are less able to handle this glucose load.
- There are no fructokinases in cats – they are unable to utilize fructose and other simple sugars.

## REFERENCES

[1.] Hoenig, et al. **Dom Anim Endocrin** 2007. [2.] Jeusette et al. **J Anim Phys Anim Nutr** 2007. [3.] German, AJ, **J Nutr.** 2006. [3.] Zoran DL **JAVMA**, 2002.[4.] Backus et al., **J Nutr**, 2004.[5.] Hiskett et al. **Vet Res Comm** 2008. [6.] Martin et al. **J Fel Med Surg** 2006. [7.] Scott et al. **J app Anim Welfare Sci** 2002. [8.] Trayhurn et al **Proc Nutr Soc** 2001.

## THE ENIGMA OF FELINE CARBOHYDRATE METABOLISM

**Thomas Schermerhorn VMD, ACVIM(SAIM)**
**Manhattan, Kansas**

As obligate carnivores, cats must have evolved specialized mechanisms suited for consumption of a diet containing large amounts of protein but relatively little carbohydrate (reviewed by Zoran 2002). The idea that feline metabolism is adaptive, along with seminal findings from early feline studies, is the conceptual framework on which current ideas about feline glucose metabolism are hung. Nevertheless, much more remains to be discovered about the physiologic, cellular, and molecular regulation of feline glucose metabolism, and controversy persists in many areas where conflicting or incomplete data are available. From the perspective of clinical disease, understanding normal is essential for understanding abnormal. Feline diabetes mellitus, in particular, is a disorder where detailed knowledge of normal glucose metabolism is needed to facilitate advances in our understanding of etiology, diagnosis, and treatment. Most of our knowledge to date has come from physiological and whole animal studies and comparatively little is known about the cellular and molecular basis for feline dietary adaptations.

### CARBOHYDRATE INTAKE & DIGESTION

All *felidae*, including domestic cats, are classified as obligate carnivores, meaning that meat is an essential component of their diet. Carbohydrate content in the diet of carnivores is typically low compared to protein and fat but it is not lacking entirely. Carbohydrate in the carnivore diet tends to be present in the form of complex carbohydrates (e.g. starches) rather than simple sugars (e.g. fructose and glucose). In the wild, carnivores ingest plant-based carbohydrates when they consume the entrails of their prey, which is a reported behavior in non-domestic large felids. The intestinal contents of prey animals serve as a source of micronutrients, vitamins, and fiber for carnivores. However, a species requirement for a meat based diet does not necessarily translate into carbohydrate intolerance. In fact, the absolute physiologic need for glucose as a nutrient in nervous tissue and its overall role in cellular energy balance secures a crucial role for glucose in feline metabolism.

### OVERVIEW OF CARBOHYDRATE ADAPTATIONS

Cats appear to lack salivary amylase, which is necessary for initiating digestion of some forms of carbohydrate, and have reduced activities of pancreatic amylase and intestinal disaccharidases when compared to other species (Zoran 2002). The glycemic response to feeding is somewhat unusual in cats with a few studies reporting a lack of significant postprandial hyperglycemia. Recent work by de-Oliveira and colleagues (2008) suggests a diminished insulin and glucose responses after carbohydrate feeding in cats. Certain important features of feline intestinal carbohydrate digestion and absorption, such as carbohydrate transport and disaccharidase activity, are non-adaptable to changes in dietary carbohydrate content.

Hepatic carbohydrate metabolism of cats has several features not found in non-carnivores. A principal feature of hepatic glucose metabolism in cats is the predominance of low Km hexokinase (HK) activities in the feline liver compared with other species and low activity of glucokinase (GK) activity (Washizu et al. 1999; Tanaka et al. 2005). Unlike non-carnivores, feline hepatic gluconeogenesis is reported to be active even in the fed state and glucose phosphorylation capacity is not subject to dietary regulation (i.e. it is non-adaptive) (Rogers et al. 1977). Consistent with the need for a high gluconeogenic capacity, the activity of the glucose-6-phosphatase (G6Pase) enzyme system is comparatively high in feline liver compared to canine liver (Washizu et al. 1999; Tanaka et al. 2005). The gluconeogenic capacity of the feline liver is not inhibited by glucose, unlike the situation in rodents and humans[9].

The lack of hepatic glucokinase expression in cats presents a paradox because the requirement for glucokinase to maintain normal glucose metabolism in other species has been established beyond doubt by transgenic (Postic et al., 1999) and gene knockout studies (Nordlie et al., 1999), as well as study of naturally-occurring glucokinase gene mutations in humans (Zelent et al, 2005). Thus, while reduced capacity for metabolism of simple sugars might be consistent with consumption of a low carbohydrate diet, the lack of glucokinase activity remains difficult to reconcile from a physiological perspective. The situation in cats is even more puzzling now that dietary induction of hepatic glucokinase activity has been shown to occur in several non-feline species previously reported to lack glucokinase gene expression in liver (Soengas JL, et al, 2006; Rideau N, et al., 2008).

Despite the unique adaptations observed in cats, it is clear that cats can efficiently digest carbohydrates. Diets containing >40% digestible carbohydrate can be tolerated by cats. Digestibility of most sugars is >90%, including simple dietary carbohydrates such as glucose and fructose (Morris et al., 1977; de-Oliveira, 2008).

### FELINE GLUCOSE-SENSING MECHANISMS

Blood glucose concentration is maintained within narrow physiologic limits and represents a balance between glucose entry into the circulation and its removal by the tissues. The endocrine pancreas and liver have major roles in the regulation of blood glucose concentration. Precise minute-to-minute regulation of insulin secretion from the pancreatic beta cells requires a glucose-sensing mechanism to modulate beta cell function according to the prevailing glucose concentration, while glucose-sensing pathways in hepatocytes ensure a proper balance between hepatic glucose usage and production. Central to both glucose-sensing mechanisms is the concept of a molecular "glucose sensor", which acts as the principal regulator of glucose flux through the cell and, thus, the generation of downstream signals that adjust cellular function. Outside of liver and beta cells, islet alpha cells, endocrine cells in the intestinal epithelium, and specialized neurons in the brain also have a glucose-sensing function. Together these cells comprise a glucose-sensing network that coordinates the physiologic response to changes in glucose concentration.

Glucokinase (GK), the 52 kD protein product of the *GCK* gene, is the "glucose sensor" found in hepatocytes and pancreatic islet cells of most species. *GCK* expression in the liver and endocrine pancreas conveys on these tissues the ability to respond metabolically to acute changes in blood glucose concentration (Nordlie RC et al., 1999). Depending on the tissue, hormones, such as insulin and glucagon, and substrate availability acutely regulate GK activity (Nordlie RC et al., 1999). *GCK* gene expression in beta cells and liver is regulated by separate promoters, which permits differential regulation by nutrients and hormones in these tissues (Zelent et al., 2005). The hepatic *GCK* promoter is strongly activated by insulin and repressed by glucagon while the beta cell promoter is not regulated by insulin or glucagon (Nordlie RC et al., 1999).

**GLUCOSE-SENSING IN THE ENDOCRINE PANCREAS**

Maintenance of blood glucose within tightly controlled limits requires beta cells to respond to changes in blood glucose with an appropriate amount of insulin release. The cellular pathways involved in the beta cell response have been well studied (Bratanova-Tachova et al., 2002). Briefly, glucose enters the beta cell via GLUT 2 transporters. Once inside the cell, it is metabolized to glucose-6-phosphate (G6P) by GK. G6P metabolism via glycolysis and the Krebs cycle generates ATP and other metabolic signals. The change in the intracellular ATP/ADP ratio that results from glucose metabolism is the major stimulus for insulin secretion. ATP generated by glucose metabolism binds to ATP-sensitive potassium channels ($K_{ATP}$ channels) in the beta cell membrane. After ATP is bound, the $K_{ATP}$ channels close, leading to membrane depolarization. As the membrane potential becomes more positive, voltage-dependent calcium channels (VDCC) open allowing an influx of extracellular calcium. The resultant rise in intracellular free calcium activates calcium-sensitive exocytotic proteins, which leads to insulin granule secretion.

Feline pancreatic islets possess an intact glucose sensing mechanism. RT-PCR studies have demonstrated GK expression in the feline pancreas (Tanaka et al, 2005), although GK activity in feline pancreas has yet to be confirmed. Our laboratory has also documented GK, Kir6.2 and SUR1 expression in the feline pancreas (unpublished). Feline islets also express numerous other genes with roles in insulin signaling (Zini et al, 2009). Thus, the available molecular data showing GK and $K_{ATP}$ channel subunit expression is consistent with roles for these proteins in stimulus-secretion coupling in feline beta cells, as in other species. When viewed in light of results from *in vivo* clinical and experimental studies of glucose-stimulated insulin secretion in cats, the importance of glucose as the principal physiological stimulator of insulin secretion in cats is affirmed.

**GLUCOSE-SENSING IN THE LIVER**

Hepatic GK activity is influenced by nutritional status and acute changes are observed soon after feeding. Dietary nutrients, especially glucose and fructose, acutely stimulate hepatic GK activity. Within individual hepatocytes, GK activity is also regulated by the glucokinase regulatory protein (GKRP), which is localized in the hepatocyte nucleus. GK undergoes a well-described nuclear-cytosolic translocation cycle and its subcellular localization varies with the nutritional state (Fernandez-Novell et al., 1999). In the fasted state, GK is bound to GKRP and localized within the heptocyte nucleus, where it is functionally inactive (Agius L, 1998). After feeding, GK translocates to the cytosol and becomes active (Agius L, 1998). GK entry into the nucleus requires GKRP but exit from the nucleus is mediated by a nuclear export signal in the GK amino acid sequence (Shiota C. et al, 1999).

Although feline genes involved in glucose sensing and the proteins they encode are nearly identical to homologous genes in other mammals (Lindbloom et al, 2008; Hiskett et al, 2008; Springer et al, 2008), work in cats has clearly demonstrated that the molecular basis of feline hepatic glucose-sensing differs from the model outlined above. Hepatic GK gene expression is low and GKRP is completely absent in feline liver (Hiskett et al, 2008). Lack of GKRP expression means that cats do not possess a mechanism for sequestering GK protein in the hepatocyte nucleus (Hiskett et al, 2008). In other mammal species, GKRP expression has effects on GK gene and protein levels distinct from its action as a regulator of glucokinase activity. Thus, the lack of GCKR expression in cats may contribute to reduced GK mRNA expression and low enzyme activity observed in feline liver. Interestingly, the fructokinase pathway upstream of GKRP appears intact in feline liver (Springer et al, 2008)

**GLUCOSE-SENSING AS A PHARMACOLOGIC TARGET**

The discovery of a family of organic compounds that act as glucokinase activators opened new options for diabetes treatment. Several mechanisms of GK activation have been described for this group of drugs including increasing GK $V_{max}$ and enhancing GK affinity for glucose (Grimsby et al, 2003). GK activators bind directly to GK at a site that is distinct from the catalytic site and exert their action independent of GKRP (in the liver) (Grimsby et al, 2003). Studies of various GK activators (GKA1, GKA2, RO-28-1675, and compound A) in rodents have shown enhanced hepatocyte glucose metabolism, reduced threshold for glucose-stimulated insulin secretion from isolated rodent islets, and decreased blood glucose concentrations after oral glucose administration in various animal DM models (Grimsby et al, 2003). The dual actions of GK activators to enhance hepatic glucose metabolism and stimulate pancreatic insulin secretion have great therapeutic potential for diabetes mellitus.

**WHY STUDY FELINE GLUCOSE METABOLISM?**

Several key questions remain about feline glucose metabolism that, if answered, might shed additional light on the molecular mechanisms that facilitate dietary adaptation of felids and carnivores in general.

1. How can efficient hepatic glucose metabolism occur in the absence of a complete hepatic glucose-sensing mechanism and glucokinase? In species that normally express hepatic glucokinase, any disruption of hepatic glucose sensing or glucokinase activity results in impaired glucose intolerance. There is also evidence to suggest that glucose acted on by GK is a more suitable substrate for subsequent metabolism, including glycogen synthesis, than is glucose acted on by hexokinase. The selectivity of metabolic pathways may arise from the specific subcellular location of the metabolic enzyme.
2. What mechanisms underlie the failure of postprandial signals to initiate the metabolic switch that is so prominent in non-carnivores? In non-carnivores, expression of GK, G6Pase and related genes are regulated in a reciprocal fashion by

postparandial signals. This level of coordinated gene regulation requires highly integrated signal pathways that include a myriad of transcriptional factors that interact with DNA. Presumably, a fundamental difference in one or more of these pathways underlies feline hepatic gene regulation.

3. Can nutritional or pharmacological interventions be used to modulate the feline hepatic glucose-sensing gene network? Enhancing GK gene expression or activity improves glucose tolerance in experimental models, including diabetic rodents. Further, several recent studies have demonstrated dietary induction of hepatic GK in carnivorous species previously found to lack hepatic GK activity. Understanding the regulation of relevant feline genes involved in glucose metabolism is essential if it is to be a useful clinical target.

## REFERENCES

[1.] Agius L. **Adv Enzyme Regul**. 1998;38:303-31. [2.] Bratanova-Tochkova TK, et al. **Diabetes.** 2002 Feb;51 Suppl 1:S83-90. [3.] de-Oliveira LD, et al. **J Anim Sci**. 2008 Sep;86(9):2237-46. [4.] Fernandez-Novell JM, et al. **FEBS Lett**. 1999 Oct 8; 459(2):211-4. [5.]Grimsby J, et al. **Science**. 2003 Jul 18;301(5631):370-3. [6.] Hiskett E, et al. **Vet Res Comm.** 2008 Sep 9. [Epub ahead of print] [7.] Lindbloom S, et al. **DNA Seq**. 2008 Jun;19(3):256-63. [8.] Nordlie RC, et al. **Annu Rev Nutr** 1999;19:379-406.
[9.] Postic C, et al. **J Biol Chem**. 1999;274(1):305-15. [10.] Rideau N, et al. **Gen Comp Endocrinol**. 2008;158(2):173-7. [11.] Rogers QR, et.al. **Enzyme**. 1977; 22(5):348-56 [12.] Shiota C. et al. **J Biol Chem**. 1999 Dec 24;274(52):37125-30. [13.] Soengas JL, et al. **Am J Physiol Regul Integr Comp Physiol**. 2006;291(3):R810-21. [14.] Springer N, et al. **Res Vet Sci.** 2008 Dec 22. [Epub ahead of print] [15.] Tanaka A, et al. **Vet Res Commun**. 2005 Aug;29(6):477-485. [16.] Washizu T, et.al.. **Res Vet Sci** 1999 Oct;67(2):205-6. [17.] Zelent D, et al. **Biochem Soc Trans**. 2005;33(Pt 1):306-10. [18.] Zini E, et al. **Vet J.** 2008 Dec 30. [Epub ahead of print] [19.] Zoran DL. **J Am Vet Med Assoc**. 2002 Dec 1; 221(11):1559-67.

# CANINE OBESITY MANAGEMENT: TRADITIONAL & NEW APPROACHES

**Sherry Sanderson, BS, DVM, PhD, Dipl ACVIM, Dipl ACVN**
**Athens, GA**

Obesity remains the most common nutritional health problem in dogs, and obesity-associated risks continue to increase. Adipose tissue was long considered metabolically inert, and its primary role in disease was attributed to stress on the joints caused by increased weight bearing and to increased workload on the heart. However, the understanding of how obesity contributes to health problems has recently grown. Adipose tissue is now known to be metabolically active and is a major endocrine organ in the body. In addition to secreting hormones, adipose tissue produces protein signals and factors called adipokines, and many adipokines are linked to inflammation and the inflammatory response.[1-5] Thus, obesity is considered a chronic inflammatory disease, and many obesity-related risks can be attributed to this inflammatory state. However, obesity is treatable, and our therapeutic options continue to increase.

The number of overweight or obese dogs in the United States has reached epidemic proportions, and the incidence of obesity continues to increase despite attempts to control weight with diet, exercise, and drugs.

- it is estimated that 25-44% of dogs in the United States are overweight or obese, and about 50% of dogs age 5–10 years are overweight or obese.
- obesity is the excessive accumulation of adipose tissue and occurs when dogs are >20% above ideal weight. Dogs 10%–20% above ideal weight are considered overweight.[6]

**Genetic implications**

Obesity genes have not been found in dogs, but a genetic component to obesity in certain breeds is likely.

- some breeds are predisposed to obesity, including Labrador retrievers, miniature schnauzers, dachshunds, Shelties, cocker spaniels, beagles, basset hounds, and Cairn terriers.
- in a 1999 study,[7] beagles had the highest incidence of obesity, followed by cocker spaniels, golden retrievers, and Labrador retrievers.
- obesity is more common in females than males, and neutered females are almost twice as likely to be overweight or obese as intact female dogs.

**Risk Factors**

- Overeating plus inadequate exercise: the most common risk factor for obesity.
- Breed (see breed predilection section above).
- Age: Metabolic rate decreases with advancing aging as lean muscle mass decreases and fat mass increases.
- Neutering: Gonadectomy increases the incidence of obesity in dogs.
- Endocrine disorders: Hypothyroidism and hyperadrenocorticism are the most common endocrine causes of obesity in dogs.
- Drugs: Corticosteroids and phenobarbital increase appetite in dogs and contribute to obesity.

**Pathophysiology**

Obesity is characterized by chronic inflammation, and inflammation increases as adipose tissue expands.
Adipose tissue is a major endocrine organ that produces hormones and protein factors and signals called *adipokines*. The expression, production, and release of many adipokines are increased with obesity.[1-5,8]

- Cytokine-like adipokines (leptin and adipsin):
  - ❖ In normally responsive individuals, leptin triggers the hypothalamus to decrease appetite and reduce food intake and stimulates diet-induced thermogenesis.
  - ❖ Obese individuals can become leptin resistant (leptin is no longer effective at downregulating food intake or increasing metabolic energy expenditure).
  - ❖ Leptin is elevated in obese dogs, and levels increase with increasing body condition score.[9-13]
- Inflammatory cytokines (interleukin [IL]-1β, IL-6, IL-10, tumor necrosis factor-α [TNFα])
  - ❖ TNFα promotes inflammation and causes insulin resistance by blocking activation of insulin receptors.
- Chemokines (IL-8, monocyte chemotactic protein-1 [MCP-1])
  - ❖ MCP-1 contributes to inflammation by attracting macrophages.
- Acute phase proteins (haptoglobin, plasminogen activator inhibitor-1, serum amyloid A)

- Adiponectin, an antiinflammatory adipocyte hormone, is decreased in obese humans and dogs.[14]
- Obesity-associated inflammation extends beyond adipose tissue. The liver is affected by fatty infiltration and adipose-derived inflammatory mediators that cause activation of inflammatory signaling pathways in the liver, including Kupffer cell activation. Activation of Kupffer cells helps further produce inflammatory mediators and recruit other immune cells, and may play a role in hepatic insulin resistance.[15]
- Persistent, low-grade inflammation and increased oxidative stress secondary to obesity may also play a role in many chronic diseases, such as osteoarthritis and diabetes mellitus.

**<u>Issues or concerns veterinarians face when trying to manage an obese patient</u>**

1. Failure of veterinarian to acknowledge obesity as a disorder in the patient's record even when noted on PE.
2. Uncertainty of veterinarian on how to manage obesity.
3. Fear by the veterinarian of offending client if they tell a client that their pet is overweight.
4. Failure of clients to recognize when their pet is overweight.

5. Failure of clients to perceive risks associated with obesity in their pet.
- it is essential that the client buys in to the need for their pet to lose weight.
- successful obesity management should be viewed as a partnership between the client and the veterinarian.
- it is important for the client to realize some of the risks associated with obesity in their pet.

**Obese dogs & cats have increased risk of chronic health problems**

✓**conditions caused or complicated by obesity include:**
- decreased life expectancy
- chronic inflammation
- pulmonary and cardiovascular disease
- exercise and heat intolerance
- joint/musculoskeletal problems and arthritis
- compromised immune function
- pancreatitis
- hyperinsulinemia, glucose intolerance, and diabetes mellitus
- hepatic lipidosis
- increased morbidity and mortality during and following anesthesia (many anesthetics are fat soluble, so it may take longer for recovery)
- increased dystocia

**Diagnosing obesity**

✓many sophisticated methods, such as duel x-ray absorption (DEXA), are available for diagnosing obesity.

✓the most practical ways to diagnose obesity in a clinical setting include:
- body weight
- body condition score (BCS)

BODY CONDITION SCORE SCALES*

| | 1 – 5 scale | 1 – 9 scale |
|---|---|---|
| very thin | 1 | 1 |
| ideal | 3 | 4-5 (dogs); 5 (cats) |
| obese | 5 | 9 |

*it doesn't matter which scale you use as long as it is noted in the record (i.e. BCS = 4/5 or BCS = 4/9)

**Treatment**

**Obesity management programs should have short- and long-term goals.**
- Short-term goals: Lose weight and reach an ideal BCS.
- Long-term goals: Maintain ideal BCS once the weight loss program is completed.

**Traditional Management of Obesity**

Traditional Obesity Management Program Consists of Caloric Restriction in Combination with Exercise

**1. Obtain thorough diet history and correct or control any underlying diseases (i.e. endocrine diseases).**

**2. Induce a negative energy balance**

✓restrict caloric intake

✓increase energy expenditure (exercise)

✓combination of the two – **most successful**

**3. Calculate Caloric Restriction [See calculation sheet at end of notes]**

**4. Divide total daily caloric intake into at least 2 meals rather than one large meal**

✓may help with satiety in some animals

✓there is an obligatory cost in energy while digesting and absorbing food called thermogenesis.
- eating multiple smaller meals rather than one large meal induces more thermogenesis, and therefore burns more calories

**5. Choose an appropriate diet**

✓therapeutic weight loss diets have modified the levels of nutrients such that they contain adequate levels of nutrients while supplying less calories.
- larger quantities of food can also be fed which may help with satiety.
- some therapeutic weight loss diets are also supplemented with carnitine. Carnitine helps spare lean body mass during a weight loss program, which is beneficial.

✓"light" diets can be used if the patient does not have a large amount of weight to lose, but it may not work as well in patients with large amount of weight to lose.

***What about total fasting (starvation) as a weight loss strategy?***

✓absolutely not recommended in dogs because fasting has both immediate and long-term negative consequences, such as:
- starvation restricts not only calories but also every nutrient, and animals, like people, have daily requirements for nutrients.
- excess loss of lean body tissue
- decreased intestinal mass and surface area - increases risk to animal of infections and absorption problems

➢studies comparing the outcome of "starvation" weight loss programs *vs* restricted caloric programs reveal that the net weight loss with starvation is roughly equivalent to that on restricted caloric intake

➢increases the likelihood of **weight "rebound"** when starvation ceases

➢starvation is perceived as "inhumane" by most clients (and veterinarians) and the "partnership" between the client and the veterinarian that is needed for successful weight loss could be adversely affected

***What about treats?***

✓the human-animal bond is important to maintain during a weight loss program, and treats play an important role in some of those relationships

✓if clients that are used to giving treats are told to discontinue treats "cold turkey", they may tell you that they are not giving any treats, but likely may still be giving treats on the side.

✓it is better to provide the client with a list of <u>options</u> or <u>alternatives</u> that they can use for treats rather than approaching giving treats with a militant view that completely outlaws treats

➢one very important goal is to change the habits of the owner that may have contributed to their pets weight gain. Providing the client with a list of options that they can use for treats gets this process started.

**It is important to restrict intake of treats to <10% of total daily caloric intake**

✓most treats are not nutritionally balanced

**6. Decide on a rate of weight loss**

✓due to lack or "uniform" sizes of dogs and risk of hepatic lipidosis in cats, it is better to recommend **a percent loss of body weight per week** rather than a standard generic quantity

✓a reasonable goal for rate of weight loss is 1% of body weight/week and weigh animal every 2 weeks.

***What should you do if an animal's weight loss plateaus during a weight loss program?***

➢decrease the quantity of food

➢recalculate caloric needs based on patient's new "leaner" body weight

➢**increase exercise**—this is an important but often overlooked option

-benefits of exercise include:

(a) inducing positive changes in body composition (increase muscle while decreasing fat)

(b) raising metabolic rate

## ALTERNATIVE OPTION FOR WEIGHT LOSS IN DOGS THAT DO NOT RESPOND TO TRADITIONAL METHODS OF WEIGHT MANAGEMENT

**Traditional Methods of Weight Loss in Dogs**

• induce a negative energy balance using caloric restriction and exercise

• unfortunately not all dogs respond to traditional approaches of weight loss

• a new, innovative approach for weight loss in dogs is now available

**<u>DIRLOTAPIDE (SLENTROL)</u>**

• Dirlotapide (Slentrol, Pfizer Animal Health;www.pfizerah.com) has recently been approved by the FDA for management of canine obesity. Slentrol is an additional tool for the veterinarian's armamentarium that provides effective and reliable weight loss.

• Slentrol is a selective microsomal triglyceride transfer protein (MTP) inhibitor that blocks the assembly and release of some of the lipoproteins from enterocytes into the bloodstream.

• Microsomal transfer proteins are found in the intestines and in the liver; in vitro studies show that Slentrol selectively inhibits MTP in intestinal cells over hepatic MTP at a ratio 37.5:1.

• Slentrol does not completely block the release of all triglycerides into the bloodstream. Some triglycerides get trapped in enterocytes and end up in the feces when cells are sloughed.

• Slentrol works primarily by decreasing food intake. Acting locally in the small intestine to decrease lipid absorption, dirlotapide increases levels of chemical mediators of satiety such as gut hormone PYY in dogs. The hormone is released from the GI tract and travels in the blood to the level of the hypothalamus where it stimulates satiety.

• Dirlotapide is not a pancreatic enzyme inhibitor and therefore does not interfere with fat digestion in the gastrointestinal tract. Fat is broken down normally in the gastrointestinal tract and undergoes normal absorption into the enterocytes. Thus, the drug does not cause problems associated with pancreatic enzyme inhibitors such as undigested fat passing through the gastrointestinal tract resulting in an oily anal discharge.

• ~ 90% of dirotapide's effect in promoting weight loss is due to a reduction in food intake; the remaining 10% of the effect is due to loss of fat trapped in enterocytes.

• Dirlotapide is administered orally q 24 H.

• The most commonly reported side effects with Slentrol are vomiting, diarrhea and lethargy. Mild to moderate elevations in hepatic transaminases (ALT/AST) that return to normal after the discontinuation of drug therapy have been observed in dogs treated with Slentrol.

• Dirlotapide is <u>contraindicated</u> in cats, as well as in dogs with liver disease or in dogs receiving long-term corticosteroid therapy.

• The safe use of Slentrol has been studied for one year.

• Dirlotapide does not promote weight loss in every dog. Weight rebound is a problem in some patients once therapy is stopped. Successful long-term management of weight loss achieved with Slentrol requires on-going management in the form of caloric restriction and exercise after dirlotapide therapy has been completed.

**REFERENCES**

[1] Kershaw EE, Flier JS. Adipose tissue as an endocrine organ. **J Clin Endocrin Metab** 2004;89:2548-2556.[2] Trayhurn P, Bing C, Wood IS. Adipose tissue and adipokines-energy regulation from the human perspective. **J Nutr** 2006;136:1935S-1939S.[3] Miner JL. The adipocytes as an endocrine cell. **J Anim Sci** 2004;82:935-941.[4] Greenberg AS, Obin MS. Obesity and the role of adipose tissue inflammation and metabolism. **Am J Clin Nutr** 2006;83(suppl):461S-465S.[5] Inadera H. The usefulness of circulating adipokine levels for the assessment of obesity-related health problems. **Int J Med Sci** 2008;5:248-262.[6] Obesity. Burkholder WJ, Toll PW: in Hand MS, Thatcher CD, Remillard RL, Roudebush P (ed): **Small Animal Clinical Nutrition**, 4th ed. Marceline, Missouri: Walsworth Publishing Company, 2000, pp 401-430.[7] Lund EM, Armstrong PJ, Kirk CA, et al. Health status and population characteristics of dogs and cats examined at private veterinary practices in the United States. **J Am Vet Med Assoc** 199:214:1336-1341.[8] Trayhurn P. Endocrine and signaling role of adipose tissue: new perspectives on fat. **Acta Physiol Scand** 2005;184:285-293.[9] Yamka RM, Frantz NZ, Friesen KG. Effects of 3 canine weight loss foods on body composition and obesity markers. **Intern J Appl Res Vet Med** 2007;5:125-132.[10] Yamka RM, Friesen KG, Frantz NZ. Identification of canine markers related to obesity and the effects of weight loss on the markers of interest. **Intern J Appl Res Vet Med** 2006;4:282-292.[11] Effects of chronic obesity and weight loss on plasma ghrelin and leptin concentrations in dogs. **Res Vet Sci** 2004;79:169-171.[12] Ishiola K, Soliman MM, Sagawa M, et al. Experimental and clinical studies on plasma leptin in obese dogs**J Vet Med Sci** 2002;64:349-353.[13] Ishioka K, Hosoya K, Kitagawa H, et al. Plasma leptin concentrations in dogs: Effects of body condition score, age, gender and breeds. **Res Vet Sci** 2007;82:11-15. [14] Ishioka K, Omachi A, Sagawa M, et al. Canine adiponectin:cDNA structure, mRNA expression in adipose tissues and reduced plasma levels in obesity. **Res Vet Sci** 2006;80:127-132. [15] Shoelson SE, Lee J, Goldfine AB. Inflammation and insulin resistance. **J Clin Invest 2006**;116:1793-1801.

# THE USE OF NMR TO UNRAVEL GLUCOSE & LIPID METABOLISM OF LEAN & OBESE CATS

**Margarethe Hoenig, Dr.med.vet., Ph.D.**
**Urbana, IL**

## INTRODUCTION

Nuclear magnetic resonance ("NMR") was discovered in 1946 by Felix Bloch and Edward Purcell and NMR systems have become a powerful diagnostic tool. We used NMR technology in the study of physiology of cats to unravel metabolic changes associated with obesity.

"The NMR phenomenon occurs because atomic nuclei behave like tiny magnets under an external magnetic field. When the small magnetic fields generated by the spinning nuclei are stimulated with a specific frequency of electro-magnetic energy, the nuclide reacts by emitting radio waves. By measuring the radio waves, one can create an image of the molecular structure and environment of the sample under study.In NMR spectroscopy, a solid, liquid, or gas sample is subjected to a strong uniform magnetic field. Simultaneously, the NMR spectrometer applies RT (radio frequency) pulses at right angles to the magnetic field. The nuclei in the sample absorb the energy, causing the nuclei to go from parallel to nonparallel orientation (spin) in the magnetic field. Maximum absorption occurs when the characteristic frequency of the nucleus is the same as the frequency of the RT pulse. The intensity of the absorption is directly proportional to the number of nuclei changing orientation at the specific frequency. So peak areas correspond to the numbers of nuclei in a sample and also the absolute concentrations. NMR is a non-destructive technique for mapping molecular structures and investigating molecular function" (1).

We have used NMR technology to study lipid and glucose metabolism in lean and obese cats. The proton NMR method of lipoprotein measurement takes advantage of the natural spectroscopic "fingerprints" of lipoprotein particles of varying sizes. A patient's plasma sample is subjected to a short pulse of radio energy within a strong magnetic field. The resonant sound that is broadcast by the lipid molecules is recorded and analyzed to determine the number and size of lipoproteins present. Lipids associated with larger lipoproteins broadcast a signal that is characteristically distinct from those of smaller lipoproteins. A computer algorithm separates the signals into groups and then quantifies the number of particles in that group (2,3). We found that obese cats had higher plasma non-esterified fatty acids and triglycerides, as well as very-low-density-lipoproteins (VLDL) consisting primarily of medium-sized particles which has been associated with cardiovascular disease in people. The concentration of low-density-lipoproteins (LDL) was comparable between lean and obese cats , although obese cats had mostly very small, whereas lean cats had mostly large particles. The concentration of high-density-lipoproteins (HDL) was lower in obese cats and consisted primarily of small particles. Very small and small LDL and HDL partiles are also associated with cardiovascular disease in people. Our results showed that long-term obese cats develop similar lipoprotein changes to obese humans, yet, hypertension and atherosclerosis have not been described in obese cats. This suggests that dyslipidemia alone is not sufficient to induce hypertension and atherosclerosis and that other anti-atherogenic factors may be present in the obese, dyslipidemic cat (4).

We recently used a triple tracer method to study glucose metabolism in lean and obese cats . Applying this method, $^{13}$C-labeled glucose was used to measure glucose turnover by conventional indicator dilution; deuterium was used to measure the fractional contribution of glycogen, glycerol and the TCA cycle to endogenous glucose production, and $^{13}$C–labeled propionate was used as a gluconeogenic tracer to measure fluxes through pathways associated with the TCA cycle (5). We found that lean and obese cats have significant differences in their regulation of key components of hepatic glucose metabolism.

## CONCLUSION

Nuclear magnetic resonance technology allows the detection of key steps in glucose and lipid metabolism metabolism non-invasively with a single blood sample.

## REFERENCES

1. http://www.wangnmr.com/NMR_technology.htm. 2. Otvos JD, Jeyarajah EJ, Bennett DW, et al. Development of a proton nuclear magnetic resonance spectroscopic method for determining plasma lipoprotein concentrations and subspecies distributions from a single, rapid measurement. **Clin Chem** 1992;38:1632-1638. 3. Otvos JD, Jeyarajah EJ, Bennett DW. Quantification of plasma lipoproteins by proton nuclear magnetic resonance spectroscopy. **Clin Chem** 1991;37:377-386. 4. Jordan E, Kley S, Le N-A, Waldron M, Hoenig M. Dyslipidemia in obese cats. **Dom. Animal. Endocrinol**., 35: 290-9, 2008. 5. Kley S, Hoenig M, Glushka J, Jin ES, Burgess SC, Waldron M, Jordan ET , Prestegard JH, Ferguson DC, Wu S, Olson D. The impact of obesity, gender and diet on hepatic glucose production in cats. **Am. J. Physio**l., in press

# MANAGING CONSTIPATION IN CATS

**Anthony P. Carr, Dr. med. vet. DACVIM**
**Saskatoon, Saskatchewan**

Constipation is a frequent complaint middle aged to older cats. In some cases the disease becomes refractory enough to treatment that either subtotal colectomy or euthanasia have to be considered. The problem is thought to be caused by underlying metabolic problems in some patients such as kidney disease or other issues that generally result in dehydration. These are however relatively rare and do not generally cause clinical signs. Most clinical cases of recurrent constipation/obstipation are idiopathic in nature. Pelvic abnormalities and strictures represent some of the occasional causes of this problem that can be identified with work up as is nerve trauma to the sacral region. Megacolon represents the extreme manifestation of obstipation/constipation. In cats with megacolon abnormal smooth muscle cell function of the colon has been detected though this was in cats with advanced disease so it is uncertain if this was truly the cause or a manifestation of chronic constipation.

The consequences of constipation/obstipation usually are metabolic derangements. With prolonged problems endotoxemia and even death can occur. Long term this can also lead to megacolon, though in many cases megacolon can occur without a clear history of constipation/obstipation.

## TREATMENT

A variety of treatments have been recommended for the constipated/obstipated cat as well as the cat with megacolon.

### Initial management

When initially presented relieving the constipation is indicated. This can be done with a variety of ways, whereby manual disimpaction is the least "nice" of the options and should be reserved for refractory cases. Enemas can often be helpful to help to moisten dried out feces. In general 5 to 10 ml/kg of warm water can be given as an enema. Alternatively smaller volumes of DSS (5 to 10 ml total dose) can be given, though this is more irritating. This can be supplemented with oral lactulose and fluid therapy to maximize efficacy.

Recently we have adapted the use of PEG solutions administered via NE tube to help relieve obstipated/constipated cats. This is similar to methods used in humans. We give PEG solution as a slow trickle via NE tube (4 to 18 hours). This generally results in defecation within 6 to 12 hours. Obviously before embarking on this therapy it is wise to rule out obstructions of the GI tract that would make passing feces difficult or impossible. To date we have not had any significant adverse side effects and have not had to resort to manual disimpaction. In some cases enemas were given concurrently, thought this does not appear to be necessary. There is always concern about giving PEG solutions to cats as oxidative injury to RBCs could occur, though we have not documented this to date.

### Long term management

Ultimately in those cats where the problem constantly recurs, surgical intervention may be needed. Medical therapy (life-time) can in many cases avert the need for surgery or significantly delay the need for surgery.

Diet is an important part of management and it is difficult to be sure which diet is best in each individual case. Increased fiber and low residue diets are the most popular.

Fiber has been recommended for many years. This can be a psyllium product (Metamucil 1-4 tsp per meal), pumpkin pie filling or wheat bran. Fiber has been shown in humans to be only moderately effective as a laxative.

Lactulose is also a very good option for maintaining soft stools. The dosage is 0.5 ml/kg two to three times daily. Dosage is adjusted to obtain the stool quality desired. In humans this product is known to cause flatulence and GI cramping. Recently PEG containing laxatives (Miralax) have been recommended for use in cats. In humans PEG laxatives have been shown to be safe and effective with few adverse side effects. Rarely severe hyponatremia has been reported because of SIADH. These products have been recommended in cats though there is no published data showing efficacy or safety.

Prokinetic medications are vital to managing the chronically impacted cat. Cisapride (2.5 to 5 mg/cat q8 to 12 hours) has helped many cats avoid surgery for megacolon. It can be obtained from many compounding pharmacies. Other medications are available as well, however in most cases Cisapride is adequate.

## REFERENCES

[1.] Candy DCA, Edwards D, Geraint M. Treatment of faecal impaction with polyethylene glycol plus electrolytes (PGE + E) followed by double-blind comparison of PEG + E versus lactulose as maintenance therapy. **Journal Pediatric Gastroenterology and Nutrition** 2001; 43:65. [2.] Ramkumar D, Rao SSC. Efficacy and safety of traditional medical therapies for chronic constipation: systematic review. **Am J Gastrorenterol** 2005;100:936. [3.] Washabau RJ, Holt DE. Diseases of the large intestine. In Ettinger SJ, Feldman EC (eds): **Textbook of Veterinary Internal Medicine**, 6th ed. St. Louis, Elsevier, 2005.

## COBALAMIN DEFICIENCY IN DOGS & CATS: WHY SHOULD YOU CARE?

**Jörg M. Steiner, med.vet., Dr.med.vet., PhD, DACVIM, DECVIM-CA**
**College Station, TX**

### INTRODUCTION

Cobalamin (vitamin B12) is a cyclic tetrapyrrol that contains a corrin ring with a cobalt atom in the center. Cobalamin is actually made up of a group of compounds and is exclusively derived from bacterial sources.[1] The biologically active forms of this vitamin are methylcobalamin (required for methyl-group transfers) and adenosylcobalamin (required for adenosyl-group transfers), but there are other molecules that belong to this group of vitamins, such as hydroxycobalamin or cyanocobalamin.[2] Cyanocobalamin, does not occur naturally, but is manufactured by bacterial fermentation and cyanide incorporation for treatment of cobalamin deficiency. Cobalamin has important functions in amino acid metabolism and DNA synthesis.

### COBALAMIN FUNCTION

Cobalamin is an essential cofactor for several enzyme systems in mammalian species. The first enzyme system, methylmalonyl-CoA mutase, is located in the mitochondria and plays a crucial role in the transformation of propionyl-CoA to succinyl-CoA. Propionyl-CoA is formed as a product of beta-oxidation of odd-chain fatty acids, as a product of isoleucine or valine metabolism, or as a product of threonine or methionine metabolism. Briefly, propionyl-CoA ($C_3H_6O_2$-CoA) and bicarbonate ($HCO_3$) are transformed to D-methylmalonyl-CoA ($C_4H_6O_4$-CoA) by propionyl-CoA carboxylase.[2] D-methylmalonyl-CoA is either transformed to methylmalonic acid ($C_4H_6O_4$) by D-methylmalonyl-CoA hydrolase or to L-methylmalonyl-CoA ($C_4H_6O_4$-CoA) by D,L-methylmalonyl-CoA racemase.[2] Finally, L-methylmalonyl-CoA is transformed into succinyl-CoA ($C_4H_6O_4$-CoA) by the cobalamin-dependent enzyme methylmalonyl-CoA mutase.[2] Succinyl-CoA in turn enters directly into the citric acid cycle. Thus, cobalamin plays a major role in the metabolism of several amino acids. Cobalamin is also important in the transformation of the sulfur-containing amino acids methionine and cysteine. Homocysteine is an intermediary amino acid that is being formed from methionine and is not found in the diet. It can be transformed back into methionine by the cobalamin-dependent enzyme methionine synthase, which acts by incorporating an additional methyl group. There is an alternative non cobalamin-dependent pathway for this transformation, the betaine pathway. Homocysteine can also be transformed to cysteine through cystathionine by transsulfuration, which is not a cobalamin-dependent process that removes a different methyl group from homocysteine. The transformation of homocysteine to methionine is linked to another metabolically crucial process, the generation of the biologically active tetrahydrofolate from $N^5$-methyltetrahydrofolate. Simplistically, the cobalamin-dependant enzyme methionine synthase transfers a methyl group from $N^5$-methyltetrahydrofolate to homocysteine, which results in tetrahydrofolate and methionine. Therefore, this enzyme is also known as 5-methyltetrahydrofolate-homocysteine methyltransferase. Thus, this enzyme not only plays a role in the transformation of sulfur-containing amino acids, but may be even more important in the generation of the biologically active tetrahydrofolate, which is involved in the synthesis of both purines and pyrimidines.

### COBALAMIN ABSORPTION

Dietary cobalamin is tightly bound to dietary animal-derived protein. In the stomach, dietary protein is partially digested by pepsin and HCl and cobalamin is being released. However, cobalamin immediately binds to a transporter protein called haptocorrin, R-protein, or R-binder.[3] Haptocorrin is mostly synthesized and secreted by the gastric mucosa. Haptocorrin in turn is digested by pancreatic proteases in the small intestine. Free cobalamin binds to intrinsic factor (also known as S-binder).[3] In humans, intrinsic factor is mostly synthesized and secreted by the parietal cells of the gastric mucosa, but there is good evidence that in dogs and cats most of intrinsic factor is synthesized and secreted by pancreatic acinar cells.[4] Cobalamin/intrinsic factor complexes are being absorbed by a complex of receptors in the microvillus pits of the apical brush border membrane of the ileal enterocytes. The receptor complex consists of cubilin and amnionless.[3] This receptor complex appears to be stabilized by another protein, megalin. Thus, the absorption of cobalamin is an extremely complex system that relies on a multitude of factors and processes. As cobalamin is being absorbed into the intestinal epithelial cells, it dissociates from intrinsic factor and free cobalamin is released into the circulation, where most of it binds to yet another protein, transcobalamin II. One of the main storage compartments for cobalamin is the kidney, which maintains serum cobalamin concentrations by releasing cobalamin when needed.

### COBALAMIN DEFICIENCY

Cobalamin deficiency is quite common in humans. Several causes of cobalamin deficiency have been described in humans, including hereditary causes that are due to mutations of the genes encoding carrier proteins, dietary insufficiency, postsurgical malabsorption after gastrectomy or ileal resections, food-cobalamin malabsorption, and idiopathic cobalamin deficiency.[3] Dietary insufficiency is quite uncommon as a cause of cobalamin deficiency and mainly occurs in elderly people, who are already malnourished and consume an all vegetarian diet. Food-cobalamin malabsorption is a term that combines any cause of cobalamin deficiency that is not due to dietary deficiency, such as exocrine pancreatic insufficiency (EPI), intestinal lymphoma or tuberculosis, celiac disease or Crohn's disease.

In dogs and cats the most common causes of cobalamin deficiency are chronic and severe small intestinal disease and EPI. In addition, hereditary cobalamin deficiency has been described in various dog breeds. A recent study has shown that 82% of dogs with EPI were cobalamin deficient.[5] A similar but smaller study in cats has suggested that most, if not all, cats with EPI are cobalamin deficient.[6] While cobalamin deficiency can occur in humans with EPI it is certainly not as prominent of a feature as it is in dogs and cats, which can be explained by the differences in cobalamin absorption between species. As discussed above, intrinsic factor in humans is mostly supplied through the gastric mucosa, while in dogs and cats it is mostly supplied by pancreatic acinar cells. The lack of pancreatic proteases and the alteration of the small intestinal microbiota may also play a role, but appear to be less important than the lack of intrinsic factor.

Dogs and cats with severe and long-standing small intestinal disease involving the ileum may also show cobalamin deficiency. In one study, 49 of 80 cats (61%) with chronic signs of gastrointestinal disease had cobalamin deficiency, as evidenced by a subnormal serum cobalamin concentration.[7] It is interesting to note that there is one study from the UK that would suggest that cobalamin deficiency is much less common in cats in the UK than in the USA.[8] However, there are other reports from the UK that would suggest that cobalamin deficiency does occur frequently in cats with gastrointestinal disease in the UK.[9] These differences are interesting as they point to difficulties of measuring serum cobalamin concentrations in dogs and cats. As in humans, there appear to be considerable cobalamin stores. It is unclear where most of the cobalamin stores are located in dogs or cats, but as in humans it takes a considerable amount of time for these body stores to be depleted if an insufficient amount of cobalamin is being absorbed. In one study the half-life for exogenously administered cobalamin in healthy cats was estimated to be between 11 and 14 days, while the half-life in cats with gastrointestinal disease was being estimated at 4.5 to 5.5 days in the same study. However, it should be noted that serum concentrations may not accurately reflect the true half life of cobalamin in cobalamin-deficient patients as body stores may get replenished with the exogenously administered cobalamin. Regardless, this study would suggest that similarly to humans, cobalamin stores are large and cobalamin malabsorption has to be long-standing before cobalamin deficiency occurs in dogs and cats.

Hereditary cobalamin deficiency has been recorded in a few dog breeds, including the Giant Schnauzer, Beagle, Border Collie, Australian Shepherd, and Chinese Shar Pei.[10-13] Recently, a region of chromosome 13 has been identified that cosegregates with cobalamin deficiency in the Chinese Shar Pei, but the actual gene causing the disease has not yet been identified.[14]

Clinical signs or complications of cobalamin deficiency have been reported in humans and veterinary species. The most important clinical signs observed in humans with cobalamin deficiency are hematological (i.e., macrocytosis, neutrophil hypersegmentation, non-regenerative macrocytic anemia, thrombocytopenia, neutropenia, pancytopenia) and neuro-psychiatric (i.e., polyneuritis, dementia, others). In addition, digestive abnormalities (i.e., villous blunting, inflammatory infiltration of the GI mucosa, cobalamin malabsorption, malabsorption of other nutrients), immunodeficiencies, and other signs (i.e., tiredness, loss of appetite) have been reported, but are still under debate.[3,15,16]

Clinical signs or complications of cobalamin deficiency have been much less well described in dogs and cats. Most dogs and cats with cobalamin deficiency only show clinical signs of gastrointestinal disease, which could either be a cause or the effect of cobalamin deficiency. In a recent case report a Border Collie with selective cobalamin deficiency was described.[12] The dog presented with hyperammonemic encephalopathy and fully responded to cobalamin supplementation.[12] In another case report a juvenile Beagle presented with failure to gain weight, lethargy, intermittent vomiting, seizures, anemia, and leucopenia.[17] This dog also fully responded to treatment with cobalamin supplementation. In a separate case report a 4-year old cat presented with severe encephalopathy and was diagnosed with an organic acidemia and cobalamin deficiency.[13] Interestingly, in contrast to the Border Collie mentioned above, this cat had a normal plasma ammonia concentration.[12,13]

## DIAGNOSIS OF COBALAMIN DEFICIENCY

A definitive diagnosis of cobalamin deficiency can be challenging. Clinical signs are ultimately caused by cobalamin deficiency on a cellular level. However, the cellular cobalamin status is difficult to assess. Serum cobalamin concentration has been traditionally measured to help assess cobalamin status, but some patients with cobalamin deficiency on a cellular level do not always have severely decreased serum cobalamin concentrations. Thus, in order to avoid missing patients with cobalamin deficiency, cobalamin supplementation should be considered even when serum cobalamin concentration is low normal. Several assays for the measurement of serum concentrations of cobalamin in humans are available. In order to be used in dogs and cats, these assays designed for use in humans must be validated for use in dogs and cats. The GI Lab at Texas A&M University has recently analytically validated an automated immunochemiluminescent assay for the measurement of cobalamin concentration in humans for use in dogs and cats. A reference range for serum cobalamin concentration in dogs and cats was established using this assay. It should be noted that reference ranges are not transferrable between labs and each lab should have their own reference range established.

Serum or urine methylmalonic acid (MMA) concentration can also be used as an indicator of cobalamin status. Cobalamin deficiency leads to accumulation of MMA and thus concentrations of MMA are often dramatically increased in the serum or urine of patients with cobalamin deficiency. Serum MMA concentrations have been shown to be increased in cats with cobalamin deficiency and have been shown to decrease with cobalamin supplementation.[18] Also, recently dogs with severely decreased serum cobalamin concentrations were shown to have increased serum MMA concentrations (unpublished data). Interestingly, several dogs with low-normal serum cobalamin concentrations were also shown to have increased serum MMA concentrations, demonstrating that a severely decreased serum cobalamin concentration is not optimally sensitive for the diagnosis of cobalamin deficiency on a cellular level and that a cut-off value for cobalamin supplementation should be chosen that is in the low-normal reference range. This is especially true if one considers that cobalamin supplementation is minimally invasive, safe, and relatively cheap. As suggested by these data, measurement of serum MMA concentration may be a better diagnostic test for cobalamin deficiency than serum cobalamin concentration. However, measurement of MMA concentration in serum or urine is technically involved and expensive. Thus, MMA is currently not routinely assessed in patients evaluated for cobalamin deficiency.

Humans with cobalamin deficiency have been shown to have increases in serum 2-methylcitric acids I and II, homocysteine, and cystathionine. To the authors knowledge, serum concentrations of 2-methylcitric acids I or II have not yet been systematically assessed in dogs or cats with cobalamin deficiency. However, in contrast to human patients with cobalamin deficiency, who almost always have increased serum concentrations of homocysteine and cystathionine, cats with cobalamin deficiency appear to have an unaltered serum homocysteine concentration and a decreased serum cystathionine concentration.[3,19] However, it is unknown what leads to these species-specific differences or whether these differences have any clinical implications.

Another way to assess cobalamin status is indirectly by use of a Schilling test.[20] The Schilling test involves the oral administration of radioactively marked cobalamin that has been bound to an animal protein. However, to the author's knowledge the Schilling test has never been systematically evaluated in dogs or cats and is not practical for routine clinical practice.

Thus, the only routinely available diagnostic tool to assess cobalamin status in dogs and cats is serum cobalamin concentration, which should be evaluated in every dog and cat with chronic signs of gastrointestinal disease or with clinical signs compatible with cobalamin deficiency that cannot be attributed to other conditions (i.e., unexplained immunodeficiencies, anemias, neuropathies).

## COBALAMIN SUPPLEMENTATION

Patients with severe cobalamin deficiency often do not respond to therapy of the underlying gastrointestinal disorder until cobalamin is supplemented. Unfortunately, only empirical suggestions are available concerning protocols for cobalamin supplementation in dogs or cats. However, there is no indication that over-supplementation of cobalamin leads to clinical disease. In humans, the standard route of cobalamin application is by parenteral administration. This is because cobalamin deficiency has been shown to lead to cobalamin malabsorption in the ileum.[15] However, there are recent data that would suggest that with certain forms of cobalamin deficiency, oral or nasal supplementation may be efficacious. There are no such data for dogs or cats and the data in humans is controversial. Thus, currently, the recommendation is to supplement cobalamin in veterinary patients by subcutaneous injection. The most common form of cobalamin used for supplementation is cyanocobalamin, but hydroxycobalamin can also be used. The author empirically uses the following dosing schedule: 150-250 µg per injection in cats, 250-1200 µg per injection in dogs; SC q 7 days for 6 weeks, then q 14 days for 6 weeks, then q 30 days for one injection, then reevaluate serum cobalamin concentration one month later. If the underlying disease process has resolved and cobalamin body stores have been replenished, serum cobalamin concentration should be supranormal at the time of re-evaluation. However, if serum cobalamin concentration is in the normal range, treatment should be continued at least monthly and the owner should be warned that clinical signs may recur sometime in the future. Finally, if serum cobalamin concentration at the time of recheck is subnormal, further work-up is required to definitively diagnose the underlying disease process and cobalamin supplementation should be continued weekly or bi-weekly. In a recent study cats with chronic gastrointestinal disease that had failed therapy and were shown to have cobalamin deficiency quickly gained weight after initiation of cobalamin supplementation.[18]

It is also interesting to note that the recommended dose for cobalamin supplementation in humans by far exceeds what is believed to be the physiologic cobalamin requirement in humans. Thus, there is speculation that the beneficial responses of cobalamin supplementation in human patients with cobalamin deficiency may at least in part be due to pharmacologic effects of cobalamin.[16]

## CONCLUSION

Cobalamin is a water-soluble vitamin with several essential functions in mammalian organisms. The absorption of cobalamin is complex. Longstanding and severe distal small intestinal disease or EPI can cause cobalamin deficiency. Because of the important metabolic role of cobalamin, deficiency of this vitamin leads to clinical signs (i.e., gastrointestinal, hematological, immunological, and neurological) and must be corrected. Serum cobalamin concentration is currently the most practical diagnostic tool for cobalamin deficiency available. Dogs and cats with cobalamin deficiency must be supplemented parenterally.

## REFERENCES

1. Markle HV. **Crit Rev Clin Lab Sci** 1996; 33:247-356. 2. Allen RH, et al. **FASEB J** 1993; 7:1344-1353. 3. Dali-Youcef N, et al. **QJM** 2009; 102:17-28. 4. Fyfe JC. **J Vet Int Med** 1993;7:133 (abstract). 5. Batchelor DJ, et al. **J Vet Intern Med** 2007; 21:54-60. 6. Steiner JM, Williams DA. **J Vet Int Med** 1995;9:193 (abstract). 7. Simpson KW, et al. **J Vet Int Med** 2001; 15:26-32. 8. Ibarrola P, et al. J Feline **Med Surg** 2005; 7:341-348. 9. Reed N, Gunn-Moore DA, Simpson KE. Scientific Proceedings of the 50th annual congress of the **BSAVA** 2007;476 (abstract). 10. Fyfe JC, et al. **Jour Am An Hosp Assoc** 1989; 50:533-539. 11. Bishop MA, et al. **J Vet Int Med** 2007; 21:613-614 (abstract). 12. Battersby IA, et al. **J Sm Anim Pract** 2005; 46:339-344. 13. Kelmer E, et al. **J Vet Emerg Crit Care** 2007; 17:299-304. 14. Gruetzner N, et al. **J Vet Int Med** 2007; 21:653-654 (abstract). 15. Arvanitakis C. **Acta Hepato-Gastroenterol** 1978; 25:313-318. 16. Solomon LR, **Blood Rev** 2007; 21:113-130. 17. Fordyce HH, et al. **J Sm Anim Pract** 2000; 41:407-410. 18. Ruaux CG, et al. **J Vet Int Med** 2005; 19:155-160. 19. Ruaux CG, et al. **Am J Vet Res** 2001; 62:1852-1858. 20. Brugge WR, et al. **Gastroenterol** 1980; 78:937-949.

## HIGHLIGHTS FROM THE IVECCS 2008 MEETING PART I & II

**Sheila McCullough, DACVIM, DACVECC**
**Yorkville, IL**

The lofty goal of these two sessions is to present the highlights of many excellent presentations from the International Emergency and Critical Care Society meeting held in Phoenix, Arizona in September, 2008. IVECCS is currently celebrating our $20^{th}$ year as a specialty college and hosted an excellent set of speakers.

Both in these proceedings and at both presentations, I will be presenting the work of many excellent speakers and refer you to the original proceedings from IVECCS 2008. In these two sessions, I will discuss:

- key journal articles from both human and veterinary medical journals for 2008
- the use of thromboelastography (current IVECCS abstracts included)
- the controversial use of albumin in critically ill and healthy patients
- the applications of interventional radiology in veterinary medicine
- imaging of the microcirculation

Key articles chosen for discussion will be summarized below.

**The Corticus Trial**

[1]Sprung CL, Annane D, Keh D, et al. Hydrocortisone therapy for patients with septic shock. *N Engl J Med* 2008; 358:111-24

This randomized, double blind, placebo-controlled clinical trial included 499 patients in ICU's across Europe and the UK and investigated the effects of hydrocortisone (HC) in patients with septic shock. The trial was designed to validate the results of a similar study published in 2002 (Annane, et al, JAMA, 2002) using slightly different methods (no etomidate allowed) and a larger patient population from a more diverse geographical area. The results of this study call into question the link between relative adrenal insufficiency (RAI) and blood pressure responsiveness to hydrocortisone in septic shock. In this trial, forty seven percent of patients were diagnosed with RAI, and hydrocortisone administration did not change the outcome in RAI patients compared to patients with normal adrenal responsiveness. Hydrocortisone administration was associated with more rapid shock reversal and increased incidence of infectious complications, regardless of adrenal responsiveness at baseline.

**Cardiopulmonary-Cerebral Resuscitation:[1] Vasopressin in CPCR**

Gueugniaud PY, David JS, Chanzy E, et al. Vasopressin and epinephrine vs. epinephrine alone in cardiopulmonary resuscitation. *N Engl J Med* 2008; 359:21-30.

This paper was undertaken after another paper (Wenzel, et al., NEJM 2004) found that patients with refractory cardiopulmonary arrest (CPA) had improved outcome with repeated administration of vasopressin and epinephrine than with epinephrine alone.

Data from 2894 patients with CPA were evaluated. Those patients with asystole or pulseless electrical activity were randomized immediately, while those with ventricular fibrillation were included only if three defibrillation attempts failed. There was no difference in any outcome measure between the group that received epinephrine and vasopressin and the group that received epinephrine alone. Post-hoc analysis suggested that epinephrine alone may be superior to combination therapy in patients with pulseless electrical activity.

**Tight Glycemic Control[1]**

Brunkhorst FM, Engel C, Bloos F, et al. Intensive insulin therapy and pentastarch resuscitation in severe sepsis. *N Engl J Med* 2008;35:125-39.

Van de Berghe, et al (NEJM, 2001) showed improved outcome when blood glucose concentration was maintained < 110 mg/dl in critically ill post-operative patients. Additional studies since that time have not supported these results. In this study, 488 patients were randomized to receive either conventional or tight glycemic control. Intensive insulin therapy to maintain BG within 80-110 mg/dl range resulted in hypoglycemic events in 12.1% of patients compared to 2.1% of patients treated with conventional insulin therapy ($P < 0.001$). No difference in morbidity outcome was measured or the rate of death between the conventional and the insulin therapy groups at 28 days. Patients receiving intensive insulin therapy had a longer length of stay.

Veterinary studies such as Chan et al found an association between the degree of hyperglycemia and mortality (JVIM 2007).

**REFERENCES**

1. Burkitt J Veterinary Journal in Review, **IVECCS Proceedings**, handout given at time of discussion, September, 2008.

**Update on Coagulation: TEG: Is it worth it in private practice or just an academic dream?**

Although thromboelastography (TEG) was developed in the 1940's, it is over the past decade that TEG has been heralded as a global test of coagulation that has value in both academia and private practice. TEG is designed to measure the viscoelastic properties of blood clot formation. Currently, TEG is the only coagulation test that can evaluate all phases of coagulation by measuring how fast a clot forms, the speed of the clot growth and overall clot strength. This ability to evaluate both hypercoagulable and hypocoagulable states provides a distinct advantage for the clinician. In addition, TEG may also evaluate fibrinolysis and the interaction of platelets and leukocytes with the proteins of the coagulation cascade.

One advantage for the clinician is the simplicity of TEG instrumentation. A sample of citrated blood (with calcium added) or fresh whole blood is placed in a sampling cup and placed into the instrument. A disposable pin, which is connected to a torsion wire inside the instrument, is slowly oscillated at 4 degrees in the blood sample. As the clot begins to form the resultant tension is recorded by the computer and the tracing is produced. From this graphical representation, a number of parameters are measured:

**R value or R time:** defined as the time (in minutes) from initiating analysis to the beginning of clot development where the tracing just begins to split or diverge

**K value or K time:** defined as the time from the end of R until it reaches a standard deviation (20 mm) of the tracing

**Alpha ( $\alpha$ ) or alpha-angle:** defined as the slope of the line formed by R and K

**MA = maximum amplitude:** defined as the maximal divergence in the tracing and is thought to reflect overall clot strength.

Overall, the standard TEG illustration of 'normal' coagulation may be represented as the following graph:

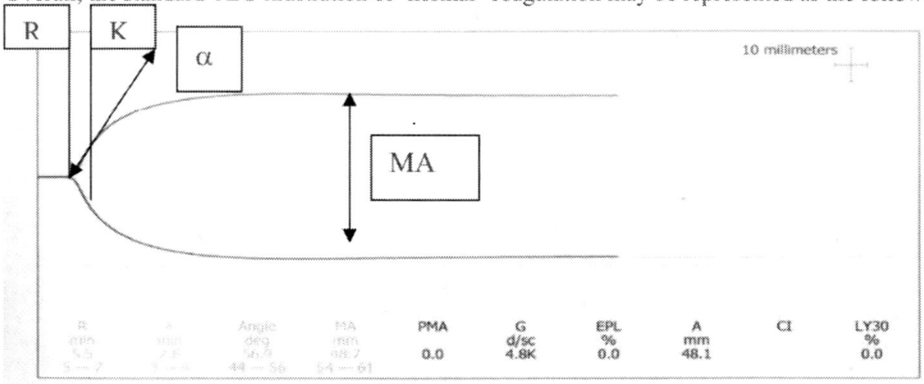


Based on the above definitions, a TEG in a hypercoagulable state ((R and K are decreased; alpha and MA values are increased) and a hypocoagulable state ((R and K values are increased; alpha and MA values are decreased) are depicted below.

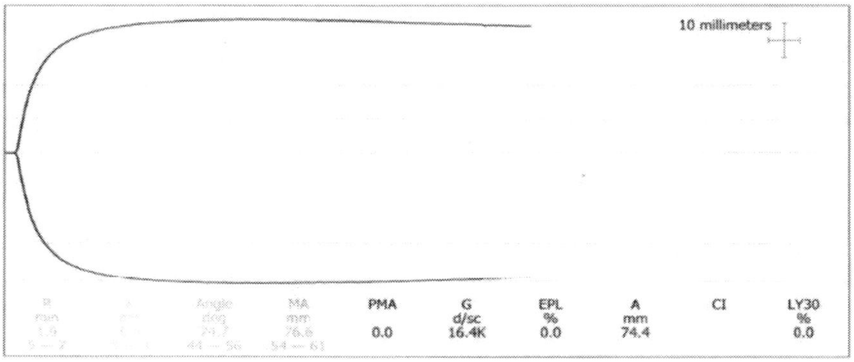


**Figure 1:** Hypercoagulable Patient (L.Waddell)

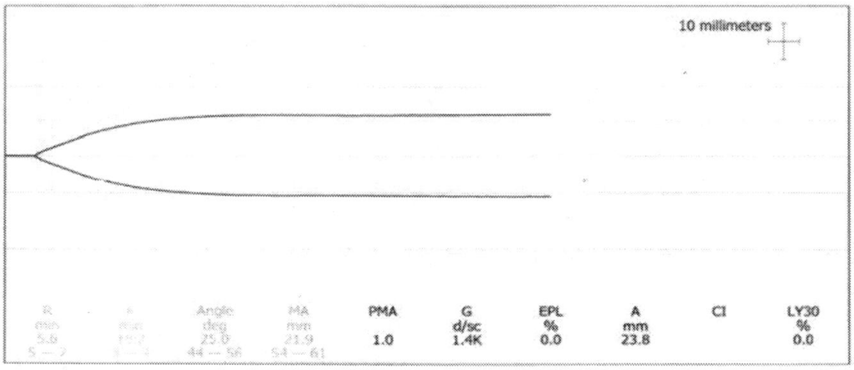


**Figure 2.** Hypocoagulable Patient (L. Waddell)

The benefit of TEG in both academia and private practice is the ability to identify hypercoagulable states in patients earlier as well as to correctly identify dogs with bleeding disorders prior to conventional testing .

Reference: D Chan Thromboelastography: Academic toy or indispensable tool?, IVECCS Proceedings, September, 2008. References from this proceedings available on request.

The most recent IVECCS 2008 small animal abstracts on TEG veterinary research are summarized below.

**Identification of hypercoagulability in dogs with primary immune mediated hemolytic anemia utilizing thromboelastography**

Fenty RK, deLafrocade AM, Shaw SP, O'Toole TE (Tufts University Cummings School of Veterinary Medicine, North Grafton, MA, USA)

In this study, citrated whole blood samples taken from dogs with primary IMHA prior to any treatment modality were evaluated against control dogs. Thromboelastography variables were determined using a computerized coagulation analyzer. Five dogs had met the inclusion criteria. Compared with controls, dogs with primary IMHA have significantly lower K values and higher mean MA values. Mean values for R and a were not significantly different between IMHA and control dogs. The conclusion made by the authors is that TEG tracings in dogs with primary IMHA document hypercoagulability.

**The effect of unfractionated heparin on TEG analysis in healthy dogs.**

Pittman JR, Koenig A, Brainard BM College of Veterinary Medicine, University of Georgia, Athens, GA

Baseline coagulation parameters (CBC, plt count, PT, PTT and AT) on six female dogs (with jugular catheters) were taken and heparin (200 IU/kg SQ every 8 hours) was administered for 3 days. PTT and TEG were performed hourly for 12 hours after the first 200 IU/kg SQ heparin dose was given and Anti-Xa was assayed at 0,3,6 and 8 hours. Day 4 protocol was similar to the Day 1 protocol. This study concluded that administration of SQ heparin resulted in consistent progressive changes in the TEG tracings, with maximal changes occurring 3-5 hours after dosing. These changes do not correlate with the PTT values. The authors suggest that abrupt discontinuation of heparin therapy may result in transient hypercoagulability.

**Retrospective evaluation of thromboelastrography in critically ill dogs with IMHA**

Sinnott VB, Otto CM School of Veterinary Medicine, University of Pennsylvania, Philadelphia, PA, USA

The goal of this study was to evaluate TEG as a prognostic indicator in dogs with IMHA.

Dogs diagnosed with IMHA over an 8 year period that had at least one TEG were reviewed. Forty dogs met the inclusion criteria with 31 found to be hypercoagulable (coagulation index (CI) > 4), 6 normocoagulable (-4 <CI> 4) and none were hypocoagulable (CI < 4). No dog with a normal CI survived however 19/34 hypercoagulable patients survived to discharge. This study concluded that 'a normal coagulation index in critically ill dogs with IMHA appears to portend a worse prognosis than a hypercoagulable CI. A normocoagulable TEG is associated with a prolonged PT and PTT in dogs with IMHA.

**The albumin controversy continues....**

Over the past decade, human serum albumin (HAS) has played an increasing role in the treatment of critically ill veterinary patients though it's role remains controversial. Benefits include improving albumin levels and COP with minimal volume infusion however there is increasing concern regarding both immediate and delayed sensitivity reactions. Veterinary literature is limited but it is growing and spans from 2004 through the present.

1) Bell, et al (ECVIM-CA Congress abstract, 2004) reported the use of HAS in 4 dogs and 1 cat with 3/5 surviving to discharge. There were no reported side effects or signs of acute immunologic reactions.

2) Chan, et al (IVECCS abstract, 2004) reported the use of HAS in 37 dogs. /37 of the dogs were surgical cases and /37 were diagnosed with septic peritonitis. Overall, 57% of the dogs survived and had higher post HAS albumin levels, higher net COP increases as well as receiving higher doses of HAS. Complications included prolonged clotting times (3), increased breathing effort (3), vomiting (2) and fever (1). There were no reports of immediate anaphylactic reactions.

3) Matthews and Barry (JVECC, 2005) retrospectively evaluated the use of HAS in 64 dogs and 2 cats with multiple diseases (PLE/PLN, hepatic disease, peritonitis, IBD + peritonitis and miscellaneous disease). Significant elevations in albumin were noted after HAS treatment and beyond. No significant side effects were noted with the exception of 2 dogs with facial edema for the initial transfusion. Two dogs were given HAS months later then euthanized (no side effects reported).

4) Torrente (IVECCS abstract, 2006) reported the use of HAS in 25 critically ill patients (23 dogs, 2 cats). The duration of HAS therapy lasted from 1- 11 days with a 68% survival rate (6 patients euthanized and 2 deaths). One acute allergic reaction was reported (urticaria and facial swelling/edema). In this group, 5 patients received repeated transfusions several weeks or months later without reactions.

5) Cohn, et al (AJVR, 2007) reported the presence of anti-human albumin antibodies in healthy dogs infused with higher doses (50gr) of HAS. These antibodies were noted within a week of the transfusion, peaking at 2 and 3 weeks post transfusion. Of the 9 dogs in this study, 2 developed an allergic reaction. Another two dogs were given another transfusion 5 weeks later and they developed anaphylactic shock. The authors recommended that all dogs given HAS transfusions be carefully monitored for anaphylactic shock. Given the discovery that all 9 dogs developed anti HAS antibodies, there is an increased risk for anaphylaxis and delayed hypersensitivity.

6) Martin, et al (JAVMA, 2008) reported the presence of IgG antibodies to HAS in 14 critically ill dogs. The antibodies peaked at 4-6 weeks after administration. Important to note, 5 of 68 negative control dogs (47 healthy dogs and 21 critically ill dogs did not receive HAS) also had detectable antibodies to HAS.

Recent studies have reported a significant number of immediate and delayed sensitivity reactions. Delayed reactions appear to be more significant. Recent reports of HAS infusion into healthy dogs have resulted in fatal complications. Since these reactions are mediated by an immune response (type III hypersensitivity), it is theorized that healthy individuals are able to mount a more intense immunological reaction than in potentially immunocompromised critically ill animals. Current recommendations are still controversial as there are no optimal protocols for the use and administration of human albumin to dogs. Both the benefits and risks of human albumin transfusions should be carefully reviewed for each critical patient.

Reference: Chan, D Revisiting the Human Albumin Controversy, IVECCS Proceedings, 2008.
Powell, L Use of Albumin in Dogs with Septic Peritonitis, IVECCS Proceedings, 2008.

**Interventional Radiology (IR)**

Presentations by Chick Weisse, VMD, DACVS and Matthew Beal, DVM, DACVECC

Growing interest in interventional radiology is fueled by our need to develop minimally invasive procedures to aid our critical patients.

***Advantages of IR techniques include:***

- Lower morbidity and mortality rates
- Shorter anesthetic time
- Minimal hospital stay
- Application of novel treatments over std therapy

***Disadvantages of IR techniques include:***

- Need for specialized, expensive equipment with advanced imaging (fluoroscopy)
- Limited formal training programs (one) and few specialists with such expertise

Current applications for veterinary patients are listed below. If you have patients that would benefit from these procedures, please contact these clinicians directly (Chick Weisse, weissec@vet.upenn.edu or Matt Beal, bealmatt@cvm.mus.edu ).

***Current applications for veterinary patients :***

- Stent placement for treatment of tracheal collapse
- Percutaneous transvenous coil embolization of congenital intra-hepatic PSS
- Palliative stenting for treatment of stenosis(nasopharyng/airway/esophagus/colon)
- Malignant urethral obstruction
- Transarterial embolization for treatment of hemorrhage (epistaxis) or vascular malformations (AV fistulas or AV malformations)
- Foreign body removal – vascular (catheter fragments) and tracheobronchial
- Antegrade urethral catheterization – performed percutaneously; useful in small female dogs or cats with urethral tears in which standard retrograde catheterization has failed
- Placement of percutaneous drainage tubes (peritoneal, cystostomy, nephrostomy)
- Ureteral stenting for ureterolithiasis or obstruction due to ureteral or trigonal neoplasia
- Percutaneous feeding tube placement – fluoro guided nasojejunal/esophojejunal
- Intra-arterial chemotherapy – allows for intra-arterial administration of chemotherapeutic drugs directly to the tumor, reducing adverse systemic effects.(osteosarcoma, and transitional cell carcinoma)
- Chemoembolization – combines intra-arterial administration of chemotherapeutic drugs with particle embolization. Used primarily on hepatic tumors.
- Palliative stenting of malignant nonvascular obstructions – treatment of obstructions of upper or lower urinary tract, respiratory tract, cardiovascular system, and GIT

The following procedures have been used less commonly and when traditional treatments have failed.

- Cyanoacrylate embolization for treatment of idiopathic chylothorax
- Percutaneous nephrolithotomy for nephroliths or proximal ueretoliths
- Trans-arterial embolization for treatment of metastatic disease
- Palliative stenting of malignant venous obstructions – used when perivascular tumors such as vascular sarcomas or adrenal gland tumors invade a vessel lumen, resulting in edema, ascites, or both.
- Thrombectomy – rheolytic thrombectomy has been used for feline saddle thrombi

For additional information on these procedures, please see the following reference:

Weisse CW, Berent AC, Todd KL, Solomon JA. Potential applications of interventional radiology in veterinary medicine. ***JAVMA***, 2008; 233:1564-1574.

**Imaging the Microcirculation**

Presentation by Deborah Silverstein, DVM, DACVECC

In sepsis, normal auto regulatory mechanisms of the microcirculation do not respond appropriately, leading to heterogeneous microcirculatory perfusion. The ability to measure a critical patient's microcirculation has led to the development of side stream dark field (SDF) technology that goes beyond standard measurements of poor perfusion such as lactate and metabolic acidosis.

Side stream dark field (SDF) technology utilizes a novel method of reflectance avoidance in which the illuminated light and reflected light travel via independent pathways. A light guide is surrounded by green light (530 nm) emitting diodes (LEDs). The light from the LEDs is absorbed by the hemoglobin of erythrocytes, resulting in the ability to observe the moving cells. The concentrically placed LEDs at the tip of the probe, which are protected by a disposable cap, send the penetrating green light deep into the tissue, illuminating the microcirculation. A 5X magnifying lens in used to project the image onto a video camera, providing clear images of the capillaries and allowing for computer automatic analysis of the images. The model currently being used in veterinary medicine is the MicroScan, produced by Cytometrics, Philadelphia, PA. The tissue bed that is used in humans is the sublingual and in veterinary patients is the mucogingival junction at base of canine tooth. Once video has been obtained, the microcirculatory flow index, perfused vessel density, and proportion of perfused vessels of various sizes can be determined.

In humans, research in septic patients has shown that microcirculatory dysfunction is associated with decreased capillary density, increased perfusion heterogeneity, and an increased proportion of intermittently perfused or non-perfused capillaries. Stagnant blood flow in the smallest capillaries with normal to almost normal blood flow in larger capillaries can often be seen. In multiple human studies, differences were seen in microcirculatory function in normal controls versus septic patients, and microcirculatory indices (flow velocity score and perfused vessel density) were lower in nonsurvivors versus survivors.

The MicroScan has been used experimentally in septic sheep and pigs and similar results have been obtained. Sepsis causes a dramatic reduction in perfusion scores in septic pigs, with survivors showing increasing perfusion scores while non-survivors' perfusion scores did not show significant change. This occurs without any correlation between cardiac output and microcirculatory indices, and despite normalization of hemodynamic variables with resuscitation, perfusion scores remain decreased. The MicroScan is currently being used to gather data on dogs with naturally occurring sepsis.

In humans, research in septic patients has shown that microcirculatory dysfunction is associated with decreased capillary density, increased perfusion heterogeneity, and an increased proportion of intermittently perfused or non-perfused capillaries. Stagnant blood flow in the smallest capillaries with normal to almost normal blood flow in larger capillaries can often be seen. In multiple human studies, differences were seen in microcirculatory function in normal controls versus septic patients, and microcirculatory indices (flow velocity score and perfused vessel density) were lower in nonsurvivors versus survivors.

Continued research into the use of this exciting new technique is warranted. Areas of interest include not only sepsis, but also cardiac disease, head trauma, burns, and organ transplantation. For our veterinary septic patients in particular, microcirculatory analysis may help identify patients at risk for development of MODS and MOF, and allow us to prognosticate, monitor these patients, and potentially guide our therapy.

I would like to thank all the excellent speakers that contributed to the IVECCS 2008 meeting and Lori Waddell, DVM, DACVECC for her help in preparation of this manuscript.

# EMERGING STREPTOCOCCAL DISEASES OF DOGS & CATS

**Dr. Kate Hurley & Patricia Pesavento**
**Davis, CA**

## BACTERIOLOGY

The genus Streptococcus is made up of non-motile, Gram-positive cocci that grow in pairs or chains because they divide perpendicular to the long axis of a developing chain. This can be seen in smears of a lesion, and often in histologic sections.

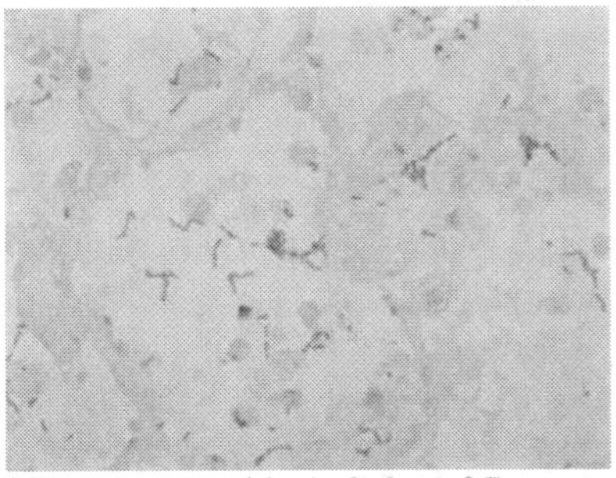

**Photo inset:** *K9 lung:* alveolae are laden with blue (Gram-positive) chains of *Streptococcus zooepidemicus* in a dog that died acutely with hemorrhagic pneumonia. Brown and Brenn's (modified Gram's) stain.

Streptococci are more acid-tolerant than most bacteria, which is important in their ecology: in general the skin surface is acidic, an environment that is discouraging to non-commensal bacteria. Streptococci have nutritional requirements that include vitamins, amino acids, and nucleic acids, and are therefore not "free-living" in the environment (which is not to say that they cannot persist in the environment).

Speciation of members of the Streptococcus genus can be daunting, as can understanding the significance of the findings. The Streptococcus genus has considerable genetic and serologic diversity. Streptococci are first designated as exhibiting either alpha, beta or gamma **hemolytic activity**. In ß-hemolysis , the sheep RBCs embedded in a blood agar plate and surrounding the bacterial colony are completely lysed. Specific protein "hemolysin" toxins are responsible for this phenotype. In the case of alpha-hemolysis, RBCs remain intact but their pigment is changed from red to green. Gamma hemolysis is, ironically, lack of hemolysis.

One early and common step in classification includes a "**Lancefield**" serologic grouping, which is determined by capsular polysaccharides. S. pneumoniae (pneumococcus), a human pathogen producing pneumonia and meningitis, and one species of Strep. suis lack a group specific antigen. Streptococci from many Lancefield groups are recognized as veterinary pathogens. Perhaps the most important reason for this grouping is to quickly narrow down the Streptococcus of interest in a lesion which can predict the odds ratio of causing infectious disease; however these serogroups do not provide information about *genetic relatedness* of strains. Even when the bacteria contain a similar capsular antigen, researchers have shown differences in up to 40 other genes that are not uncovered by this system. Genetic relatedness becomes most important in analyzing epidemiology or pathogenesis in outbreak situations (see section below). Molecular epidemiological methods are routinely used in human medicine, and include multilocus sequence typing (MLST), pulsed-field gel electrophoresis (PFGE), microarray, or even whole genome sequencing.

A discussion of **surface virulence factors** for a genus as genetically diverse as Streptococcus is well beyond the scope of these presenters, however we will at least mention the M protein. M protein is the single most studied and appreciated virulence feature of the Streptococci. The M protein has also been used to differentiate strains, and its expression is not limited to the genus Streptococci. There are over 80 immunologically distinct M proteins, with antigenic variation due primarily to single amino acid substitutions. In humans, streptococcal toxic shock syndrome has been associated predominantly with M types 1 and 3. Also in humans, induction of anti-M protein antibodies or cytotoxic T-cells which cross-react with cardiac myosin or basal ganglia neurons are felt to play a role in the pathogenesis of myocarditis and a variant reaction of rheumatic fever. Similar sequelae of Streptococcal infection in veterinary medicine have been postulated (purpura hemorrhagica, glomerulonephritis), but these are difficult associations to prove.

## STREPTOCOCCAL DISEASES

### Streptococci of veterinary interest

The spectrum of diseases attributable to Streptococcus infection in animals is a fair reflection of the breadth of disease patterns seen in humans. The higher rate of isolation of these bacteria in recent studies may be due to improved detection methods, an increase in virulence, or an expanding population of compromised hosts. In part, increase in isolation and in virulence in human medicine has been attributed to acquisition of antibiotic resistance. In our experience at the VMTH with outbreaks or individual isolates in dogs and cats, antibiotic resistant strains are uncommon (Barbara Byrne, Chief of Microbiology Service, VMTH, UC Davis). However, the capacity to acquire antibiotic resistance should not be in question, as capacity of some members of the Streptococcus genus to evolve decreased susceptibility to antibiotics is notorious. Most notable is S. pneumoniae, isolates of which are now >90% penicillin resistant worldwide (compared to <10% of isolates 40 years ago). The large capacity and frequency of horizontal gene transfer among strains would make this theoretically possible for any member of the genus.

**STREPTOCOCCUS CANIS**

**Case**

Several outbreaks of necrotizing Streptococcus canis have been reported in shelter cats. In two confirmed outbreaks, S. canis infection followed or accompanied upper respiratory infection, manifesting clinically as severe sinusitis and/or abscesses and affecting between 1/3 and 2/3rds of cats present in the shelter. These outbreaks resolved after treating all affected and exposed cats with antibiotics to which the S. canis was susceptible.

The case described in detail in this lecture occurred amongst feral cats at a shelter in Southern California between November of 2005 and February of 2006. The outbreak was confined to one room containing 37 feral cats at the time of the investigation. A "rabies pole" was used to remove cats from traps and place them into cages with slatted floors. The same pole was used to remove cats for euthanasia. Clinical signs included ulceration and oozing of one or more feet, and death in some cases within 4-6 days of exposure. Approximately 45/325 cats were found dead in cages over the course of the suspected outbreak; based on the time of death, approximately 25 of these died within 4-7 days of admission and are believed to be related to Streptococcal infection. Twelve of the 37 cats in the room at the time of the outbreak investigation were affected to varying degrees. Four cases were confirmed with histopathology of necrotizing cellulites/fasciitis involving the feet of the cats.

All cats were euthanized due to the inability to handle the cats for treatment and because this was the shelter's standard policy for feral cats that were not claimed by care-takers during the legal holding period. Initial control measures included breaking down all cages, thorough cleaning and spraying with bleach. All newly admitted feral cats were moved to another room. However, the outbreak persisted in spite of these measures. All affected and exposed cats were euthanized again, and newly admitted cats were placed in the original room in the freshly disinfected cages. The rabies pole was reportedly disinfected between every use. Additional cases were again identified within days. A second investigator went to the shelter to observe cat handling and noted that the rabies pole was heavily contaminated with hair, saliva and blood. "Disinfection" consisted of dipping the pole briefly in a bucket of potassium peroxymonosulfate, which failed to remove gross organic debris on the pole. The pole was removed immediately and the cages again carefully disinfected as cats were euthanized. The outbreak resolved with this action.

**DISCUSSION**

S. canis, although typically considered a commensal and extracellular pathogen in cats, is capable of causing severe life-threatening invasive infections, such as necrotising fasciitis, sinusitis, bacteremia, and toxic-shock-like syndrome. S. canis possesses the Lancefield group G antigen, and is present in up to 10% of cultures from the nasal cavity of cats with chronic upper respiratory infections (URI) (M. Lappin, personal communication). Reports of S. canis in cats have associated the bacteria with epidemics of arthritis, cervical lymphadenitis, urogenital infections, upper respiratory infections, and neonatal septicemia. Most of these reports reflect time-limited outbreaks of disease in young, intensively housed, closed-colony animals.

**STREPTOCOCCUS ZOOEPIDEMICUS**

**Case**

The presenters are aware of confirmed outbreaks of *S. zooepidemicus* causing hemorrhagic pneumonia at shelters in Nevada, Florida, Wisconsin and New York, and at a boarding kennel in Colorado. The first two recognized outbreaks occurred at very large municipal shelters at which dogs were not vaccinated on intake and other canine respiratory viral and bacterial infectious agents were frequently isolated. However, subsequent outbreaks have been documented at facilities lacking these obvious risk factors. In the outbreak at the Colorado boarding kennel, at least three otherwise healthy, fully vaccinated adult dogs died of hemorrhagic pneumonia, including one dog belonging to the kennel owner that was not housed about 20 yards from the affected facility.

The outbreak described in this presentation occurred over a period of a little more than a year (2005 – 2/2007) at an open-intake animal shelter admitting over 50,000 animals/year. Shelter staff estimated that over 1000 cases occurred, with a case defined as a dog found dead in its run with blood coming out the nose and mouth. Most cases were in adult dogs and occurred within 7-14 days of intake, but some occurred in dogs that had been in the shelter for six weeks or more. Death often occurred within hours of clinical signs first being observed. S. zooepidemicus was isolated from lungs of affected dogs, from oropharyngeal swabs of neighboring dogs and from the environment, including a sample from a puddle of supposedly clean water in a newly-disinfected run. Other respiratory pathogens were also isolated frequently but inconsistently: of 46 dogs tested (via oropharyngeal swab), 57% were positive for *Bordetella*, 54% for canine herpesvirus, 15% for canine adenovirus, and 9% for canine distemper. Parvo was also frequent in this population.

In this case, because of the frequency of severe disease in this population and the scant knowledge about community risk posed by the infection, the decision was made to euthanize all dogs with clinical signs of respiratory disease. Healthy dogs were grouped in specific wards and treated with a two week course of Cephalexin. Wards were emptied and thoroughly cleaned, disinfected with quarternary ammonium followed by bleach, and dried. Newly admitted dogs were kept in separate wards from exposed dogs. No further cases were detected after this strategy was implemented.

**DISCUSSION**

In horses, Strep. zooepidemicus is a commonly isolated mucosal commensal that infrequently invades and usually as a sequela to virus infection, high ambient temperature, or prolonged transportation. Infections are typically individual and endogenous in that pre-existing clones in tonsilar tissue are selected, amplified and eventually dominate in the lower respiratory tract. Isolates found in pneumonic lungs of any two affected horses vary greatly and reflect the varied genotypes resident in the normal tonsil. In canine shelter outbreaks that we have investigated, however, and similar to the molecular genetic findings of the S. canis isolates described above, dominance of a single clone occurred.

**WHY ARE WE SEEING OUTBREAKS OF STREPTOCOCCUS IN SHELTERS?**

Shelters are uniquely suited for emergence of virulent strains of Streptococci. We will discuss the factors that might contribute to emergence and clonality.

- Facilitated by transmission – carriers and non-carriers in confined areas (child care centers, shelters)
- Likelihood of transmission is dependent on
    - number of individuals (level of crowdedness)
    - host susceptibility (stress)
    - persistence
    - sensitization (viral induction of interferon-gamma, which decreases clearing response by macrophages)
    - ability to exchange virulence factors: some investigators have attributed virulence, and more specifically necrotizing fasciitis ("flesh eaters") to expression of bacterial enzymes such as hyaluronidase, which degrades the fascia.

**WHAT MORE DO WE KNOW? IS CLONALITY IMPORTANT?**

**Basics of outbreak control**

The incubation, likelihood of transmission outside of shelter/kennel situations, and frequency/manner of shedding following recovery is unknown for highly pathogenic streptococcal infections in dogs and cats. This makes sweeping generalizations for management impossible. That said, the outbreaks of which we are aware have been successfully managed by the following steps:

- Isolation and treatment of clinically affected *and exposed* animals with a two week course of an antibiotic documented or likely to be effective against this gram positive bacteria (or euthanasia if humane care and safe housing can not be provided for the duration of treatment)
    - Treatment must be started early; by the time severe signs are detected, it is often too late for antibiotics to have any beneficial effect.
    - Clinically affected animals should be housed separately from exposed but apparently healthy animals
    - Treatment of newly admitted animals may also be considered, particularly if complete decontamination of premises or isolation of exposed animals is not possible
    - If animals are sent off-site to private homes (e.g. rescue or foster groups) for treatment, individuals should be made aware of the possible risks to pet animals
- Quarantine of affected animals until two weeks following complete resolution of clinical signs
- Complete cleaning, disinfection and drying of all locations where affected animals were housed

Although there have been no reports to these presenters' knowledge of transmission by adopted or rescued animals within pet homes, the possibility can not be excluded.

# MECHANISMS OF ESOPHAGEAL MOTILITY IN DOGS AND CATS

**Robert J. Washabau, VMD, PhD, Dipl. ACVIM**
**St. Paul, MN**

## ANATOMY

The transport of ingested liquids and solids from the oral cavity to the stomach is the major function of the esophagus. Anatomic structures that permit this function are the striated muscle of the cranial esophageal sphincter (cricopharyngeus), the striated and smooth muscle of the esophageal body, and the smooth muscle of the caudal esophageal (gastro-esophageal) sphincter. An important species difference between the dog and cat is in the musculature of the esophageal body. The full length of the canine esophageal body is composed of striated muscle, whereas the distal one-third to one-half of the feline esophageal body is composed of smooth muscle. The striated muscle of the cranial esophageal sphincter and esophageal body is innervated by somatic branches (glossopharyngeal, pharyngeal, and recurrent laryngeal) of the vagus nerve arising from the brainstem nucleus ambiguus. The smooth muscle of the esophageal body and caudal esophageal sphincter is innervated by autonomic branches (esophageal) of the vagus nerve, arising from the dorsal motor nucleus of the vagus.[1]

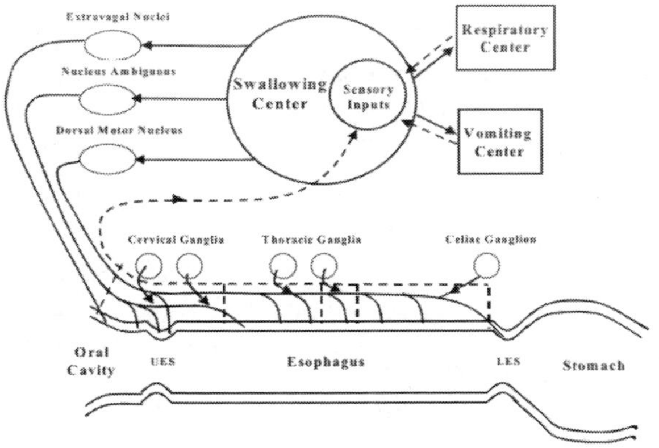


**Neural Regulation of Swallowing**
**UES = Upper or Cranial esophageal sphincter**
**LES = Lower or Caudal esophageal sphincter**

## PHYSIOLOGY

During fasting, the elevated pressures of the cranial and caudal esophageal sphincters prevent movement of food and chyme into the esophageal body from the oral cavity and stomach, respectively. When an animal swallows, the cranial esophageal sphincter relaxes to permit the movement of liquids and solids into the proximal esophageal body. Swallowing also initiates a wave of peristaltic contractions (primary peristalsis) in the esophagus that transports food into the distal esophageal body. Primary peristaltic contractions are reinforced by a secondary wave of contraction (secondary peristalsis) physiologically mediated by intraluminal distension. The caudal esophageal sphincter relaxes in advance of the propagated pressure wave to permit food to empty into the stomach. Once the bolus of food has passed into the stomach, the caudal esophageal sphincter resumes its high resting pressure.[2,3,4,5]

## PATHOPHYSIOLOGY

Inflammation (esophagitis, gastro-esophageal reflux, esophageal fistulae), hypomotility (idiopathic megaesophagus, dysautonomia, diverticula), and obstruction (stricture, hiatal hernia, neoplasia, intussusception, foreign bodies, vascular ring anomalies) are the major pathophysiologic processes involving the esophagus of dogs and cats.[6] With mild esophagitis and hypomotility lesions, the esophagus may undergo healing without further complication. In more severe cases, the esophagus may respond with extensive fibrosis, muscular hypertrophy, esophageal narrowing, and/or loss of neural regulation (i.e., flaccidity). Luminal flow is directly related to the fourth power of the esophageal radius so that even small diminutions result in significant reductions in esophageal transit.

## HISTORY

Signs consistent with esophageal disease include regurgitation, odynophagia (painful swallowing), dysphagia (difficulty in swallowing), multiple swallowing attempts, excessive salivation (particularly with inflammatory disorders), and changes in appetite (ravenous appetite with motility disorders or reduced appetite with inflammatory disorders). Some breeds are at increased risk for swallowing disorders. Regurgitation is the most important clinical sign of esophageal disease and should be differentiated from vomiting, gagging, and dysphagia. Regurgitation differs from vomiting in that is characterized by the passive retrograde evacuation of undigested food from the esophagus. Vomiting is characterized by coordinated activities of the gastrointestinal, musculoskeletal, and nervous systems culminating in the active evacuation of digested or partially digested food from the gastrointestinal tract.[7] Vomiting usually signifies disease caudal to the gastro-esophageal sphincter. The severity of clinical signs with esophageal disease is somewhat dependent upon the pathogenesis of the disease. Animals with vascular ring anomaly, for example, may have severe regurgitation, but the appetite is usually excellent because of secondary malnutrition. Animals with inflammatory esophageal disease (e.g. esophagitis), on the other hand, may have anorexia, dysphagia, odynophagia, and salivation, without much evidence of regurgitation. The latter group clearly presents a diagnostic challenge because of the differing clinical signs. Finally, signs referrable to aspiration pneumonia, e.g. coughing and dyspnea, may be the major presenting complaint in some animals.

## PHYSICAL EXAMINATION

The physical examination findings may be minimal in some animals with primary esophageal disease. Severe regurgitation and malnutrition may result in mild to moderate cachexia. Fever and pulmonary crackles or wheezes occur in association with aspiration pneumonia. An occasional foreign body or esophageal dilation may be detected during the examination. The physical examination is important from the standpoint of excluding other gastrointestinal or systemic disease.

## DIAGNOSTIC TESTS

Initial laboratory testing should include routine hematology, serum biochemistry, urinalysis, and fecal parasitologic examination. This data base will be useful in excluding systemic or metabolic disease as a cause of secondary esophageal signs. In the absence of systemic or metabolic disease, hypoproteinemia (associated with malnutrition) and leukocytosis (associated with esophageal inflammation or aspiration pneumonia) are the only laboratory abnormalities that are sometimes encountered. Hyponatremia and hyperkalemia may be evident in animals with Addison's disease (hypoadrenocorticism) and secondary mega-esophagus.

Survey radiography, contrast radiography, fluoroscopy, and esophageal endoscopy are the diagnostic methods currently available in modern veterinary practice. Survey radiography of the neck and thorax should be performed in all animals suspected of having esophageal disease. Definitive diagnosis or evidence in support of a diagnosis may be obtained with survey radiographs in many cases, including esophageal foreign body, mega-esophagus, neoplasia, hiatal hernia, and gastro-esophageal intussusception. Thoracic radiographs will also identify some of the complications of esophageal disease, including aspiration pneumonia, pleural effusion, mediastinitis, and pneumothorax.

Contrast radiography and fluoroscopy can be performed to identify esophageal lesions or to confirm a tentative diagnosis. Disorders not readily diagnosed by survey radiography (e.g. radiolucent foreign body, esophagobronchial fistula, esophagitis, diverticula, and stricture) may be more readily diagnosed by contrast radiography. Dynamic contrast studies, e.g. videofluoroscopy, should be used instead of static barium radiographs whenever possible. In addition to structural information, dynamic studies will provide some information about esophageal motility. Contrast studies may be performed using barium paste (80-100 per cent weight/volume), barium suspension (30 per cent weight/volume), barium-coated meals, or iodinated contrast agents. The choice of a specific contrast agent will depend upon which esophageal disease is suspected.

Esophageal endoscopy has become a very useful tool in the diagnosis and treatment of esophageal disease. Usually performed after survey radiographic assessment,[8] endoscopy may also be used therapeutically to remove foreign bodies, dilate esophageal strictures, or to place gastrostomy feeding tubes.

Ultrasonography has proved useful in the diagnosis of peri-esophageal masses or other mediastinal disease. Esophageal manometry is useful for diagnosing cricopharyngeal achalasia, generalized esophageal motility disorders, and caudal esophageal sphincter incompetence, but the technique is currently only performed at major referral centers and university teaching hospitals. Nuclear scintigraphy[9] and esophageal pH monitoring have also been used to diagnose esophageal motility disorders and gastro-esophageal reflux, respectively.

## REFERENCES

[1.] Venker-van Haagen AJ (1986). Contributions of the glossopharyngeal nerve and the branch of the pharyngeal branch of the vagus nerve to the swallowing process in dogs. **American Journal of Veterinary Research** 47, 1300-1305. [2.] Blank EL (1989). Cholinergic control of smooth muscle peristalsis in the cat esophagus. **American Journal of Physiology** 257 (Gastrointestinal Liver Physiology 20), G517-G523 [3.] Lang IM, Dantas RO, Cook IJ, and Dodds WJ (1991). Videoradiographic, manometric, and electromyographic analysis of canine upper esophageal sphincter. **American Journal of Physiology** 260 (Gastrointestinal and Liver Physiology 23), G911-G919 [4.] Mayrand S (1994). In vivo measurement of feline esophageal tone. **American Journal of Physiology** 267 (Gastrointestinal Liver Physiology 30), G914-G221 [5.] Washabau RJ, Fudge M, Ormsbee HR, Barone F (1995). GABA receptors in the dorsal motor nucleus of the vagus influence feline lower esophageal sphincter and gastric function. **Brain Research Bulletin** 38, 587-598. [6.] Washabau RJ, Holt DE (2003). Pathophysiology of Gastrointestinal Disease. In: **Textbook of Veterinary Surgery**. 3rd edition. Ed. D Slatter, WB Saunders Company, Philadelphia, pp 530-555 [7.] Lang IM, Sarna SK, and Dodds WJ (1993). Pharyngeal, esophageal, and proximal gastric responses associated with vomiting. **American Journal of Physiology** 265, G963-972 [8.] Gualtieri M. (1995). Upper gastrointestinal tract videoendoscopy. **Compendium on Continuing Education for the Practicing Veterinarian** volume 7, issue 2 [9.] Iwanaga Y, Wen J, Thollander MS, Kost LJ, Thomforde GM, Allen RG, and Phillips SF (1998). Scintigraphic measurement of regional gastrointestinal transit in the dog. **American Journal of Physiology** 275, G904-910

**THE EMERGENCE OF A BAD BUG: TREATMENT, INFECTION CONTROL & PUBLIC HEALTH CHALLENGES OF METHICILLIN-RESISTANT STAPHYLOCOCCUS AUREUS (MRSA)**

**Jeffrey B. Bender, DVM, MS, DACVPM,**
**St. Paul, Minnesota**

## INTRODUCTION

*S. aureus* has long been recognized as an important human pathogen and is the leading cause of suppurative infections in humans, including superficial skin infections such as boils and furuncles as well as more serious infections such as bloodstream infections, pneumonia, osteomyelitis and endocarditis. *S. aureus* is also a major cause of nosocomial infections, including surgical site infections and infections associated with indwelling medical devices. Currently, the proportion of *S. aureus* isolates that are MRSA is more than 55 percent in some U.S. human hospitals. Established risk factors for MRSA include current or recent hospitalization or surgery, residence in a long-term care facility, dialysis, and indwelling percutaneous medical devices and catheters[1,2]. Recently, community-associated cases of MRSA have been recognized[3].

MRSA has recently been a concern in equine hospitals[4,5]. This includes wound infections and transmission between horses to humans and humans to horses[6]. As a result of these concerns, Veterinary Teaching Hospitals have developed surveillance and control programs, including periodic human surveillance, glove use when handling any horse and isolation of MRSA culture positive horses. MRSA has been reported in dogs [7-9] and rarely in cats. In 1988, an outbreak of MRSA was reported from a human hospital geriatric ward[10]. Thirty-eight percent of the nursing staff was positive for MRSA. It was assumed that the high rate of colonization was due to a ward cat. Infection control measures and removal of the cat led to resolution of the outbreak.

Recent MRSA infections in animals seen by clinical staff in the University of Minnesota, Veterinary Medical Center has created diagnostic and treatment challenges[11]. From October 2003 through December 2008, MRSA has been identified from 29 refractory cases presented to the veterinary medical clinic (Table 1). There has been a steady increase in the number of cases detected from 2003 through 2008. Ten (34%) of cases were identified in 2008. Cases included 20 dogs, 7 cats, and 2 horses. Twenty-four isolates were sent to the Minnesota Department of Health for confirmation, antimicrobial susceptibility testing, and DNA fingerprinting. Seventeen of the 24 isolates have been characterized as clonal group USA100, representing strains commonly identified from recently hospitalized human patients. Other clonal groups include USA300, USA400, and USA500 representing community associated. With the identification of community associated strains, there is a detection of both USA100 and USA300 strains in human hospitals. This likely will be the case in veterinary hospitals. Veterinary isolates have all been multidrug resistant. Nosocomial spread is suspected from several case clusters (see Table 1). This has been further supported with indistinguishable MRSA strains isolated from environmental surfaces and from clinicians.

Fifteen (68%) of 22 interviewed family members of the infected pet were recently hospitalized, diagnosed with MRSA, had on-going severe illnesses (i.e. chemotherapy), or were healthcare providers. No identifiable risk factors were reported in 6 instances and clustered temporally suggesting nosocomial events.

Pet owners were informed that a multidrug resistant organism was isolated from their pet. Client oriented fact sheets are presented to clients (http://www.cvm.umn.edu/cahfs/sheets/H-M/home.html). Clients were also contacted individually by a public health veterinarian to address potential concerns.

**Table 1.** MRSA clinical cases at the University of Minnesota, Veterinary Medical Center, 2003-2008

| Date | Species | Source | PFGE subtype | Clonal Group |
|---|---|---|---|---|
| 10/30/03 | Feline | Abscess | MR11 | USA100 |
| 12/4/04* | Feline | Skin | MR191 | USA100 |
| 3/21/05* | Feline | Skin | MR191 | USA100 |
| 6/8/05 | Canine | Skin | MR191 | USA100 |
| 6/8/05 | Canine | Skin | MR360 | USA100 |
| 7/27/05 | Feline | Joint | MR4 | USA100 |
| 10/13/05 | Feline | Skin | MR391 | USA300 |
| 2/21/06* | Canine | Skin | MR107 | USA100 |
| 3/13/06* | Canine | Trachea | MR107 | USA100 |
| 8/5/06 | Canine | Chest tube | MR340 | USA300 |
| 8/21/06 | Canine | Pleural fluid | MR340 | USA300 |
| 11/16/06 | Canine | Wound | MR107 | USA100 |
| 12/2/06 | Canine | Skin | MR4 | USA100 |
| 12/5/06 | Canine | Bone | MR120 | USA100 |
| 4/11/07 | Canine | Skin | MR771 | USA100 |
| 5/22/07 | Canine | Wound | MR107 | USA100 |
| 7/9/07 | Canine | Wound | -- | -- |
| 7/19/07 | Equine | Abscess | MR882 | USA500 |
| 8/14/07 | Canine | Joint | MR391 | USA100 |
| 1/4/08 | Canine | Wound | MR391 | USA100 |
| 1/7/08* | Canine | Incision site | MR307 | USA300 |

| | | | | |
|---|---|---|---|---|
| 1/11/08* | Canine | Wound | MR307 | USA300 |
| 2/22/08 | Feline | Urine | -- | -- |
| 3/18/08 | Canine | Skin | | |
| 6/17/08 | Canine | Incision site | | |
| 6/26/08 | Canine | Incision site | MR340 | USA300 |
| 8/21/08 | Canine | Urine | | |
| 8/30/08 | Equine | Incision site | MR959 | USA400 |
| 9/19/08 | Feline | Urine | MR958 | USA100 |

*temporal clusters of indistinguishable subtypes

**Community Studies**

To characterize the potential for interspecies transmission of MRSA, several community studies were conducted. This included the collection of nasal and rectal swabs from asymptomatic animals in a long-term care facility and pets of human patients recently diagnosed with MRSA. Two of 11 resident cats from the long-term care facility were identified with MRSA. Isolates were genotype USA100 (Healthcare-associated strains). MRSA was isolated from 2 of 28 asymptomatic pets of pet owners previously diagnosed with community-associated MRSA. Isolates from the 2 animals were genotype USA300 (Community-associated strains). These studies demonstrate the potential for interspecies transmission in community settings and in a long-term care facility. Colonized animals in these studies were asymptomatic and appeared to be transiently colonized[12, 13].

**Infection Control Strategies**

The single most important prevention measure is appropriate hand hygiene. Personnel should wash their hands between handling each animal. This will decrease the risk of pathogen transmission to other animals and to staff. It is also important to incorporate "standard precautions" to protect staff and prevent movement of MRSA. This may include appropriate isolation and disinfection protocols. Staff should be aware of the potential risks and notify appropriate clinic personnel to prevent nosocomial and zoonotic spread. Immune deficiencies may put personnel at an increased risk. Individuals with an immune dysfunction should discuss their status with their employer, so that appropriate preventative measures can be taken. Ideally, veterinary clinics should provide "right to know" materials and have in place procedures to reduce zoonotic disease transmission.

**SUMMARY**

Pets with MRSA likely acquire their infection from their owners as demonstrated by the presence of common genotypes among the various populations. There is a need to re-enforce precautionary measures and hand hygiene to pet owners diagnosed with MRSA infection. Owner education should describe the potential risk of transmission from and/or to pets. Further research to quantify this household risk, the length of carriage in pets, and the potential treatment options is needed. In addition, stringent precautions are necessary to prevent environmental contamination and spread by staff within veterinary clinics.

The emergence of MRSA in companion animals highlights the importance of astute clinical and microbiologic monitoring for multi-drug resistant infections. In addition, there is a need to train veterinary practitioners and students about these new pathogens and their potential for nosocomial spread. The development of integrated human/animal surveillance systems and continued communication between human and veterinary professionals can help to identify new infections and treat those infections crossing species.

**REFERENCES**

[1] Brumfitt W, Hamilton-Miller J. Methicillin-resistant **Staphylococcus aureus. N Engl J Med** 1989;320:1188-1196.[2] Lowy FD. *Staphylococcus aureus* infections. **N Engl J Med** 1998;339:520-532.[3] Naimi TS, LeDell KH, Boxrud DJ, Groom AV, Steward CD, Johnson SK, Besser JM, O'Boyle C, Danila RN, Cheek JE, Osterholm MT, Moore KA, Smith KE. Epidemiology and clonality of community-acquired methicillin-resistant **Staphylococcus aureus** in Minnesota, 1996-1998. *Clin Infect Dis* 2001;33:990-996.[4] Hartmann FA, Trostle SS, Klohnen AA. Isolation of methicillin-resistant **Staphylococcus aureus** from a postoperative wound infection in a horse. **J Am Vet Med Assoc** 1997;211:590-592.[5] Seguin JC, Walker RD, Caron JP, Kloos WE, George CG, Hollis RJ, Jones RN, Pfaller MA. Methicillin-resistant **Staphylococcus aureus** outbreak in a veterinary teaching hospital: potential human-to-animal transmission. **J Clin Microbiol** 1999;37:1459-1463.[6] Weese JS, Archambault M, Willey BM, et al. Methicillin-resistant **Staphylococcus aureus** in horses and horse personnel, 2000-2002. **Emerg Infect Dis** 2005;11(3):430-5.[7] Guardabassi L, Loeber ME, Jacobson A. Transmission of multiple antimicrobial-resistant **Staphylococcus intermedius** between dogs affected by deep pyoderma and their owners. **Vet Microbiol** 2004;98:23-27.[8] Tomlin J, Pead MJ, Lloyd DH, Howell S, Hartmann F, Jackson HA, Muir P. Methicillin-resistant **Staphylococcus aureus** infections in 11 dogs. **Vet Rec** 1999;144:60-64.[9] Baptiste KE, Williams K, Willams NJ, et al. Methicillin-resistant **Staphylococci** in companion animals. **Emerg Infect Dis** 2005;11(12):1942-4.[10] Scott G, Thomson R, Malone-Lee J, Ridgeway G. Cross-infection between animals and man: possible feline transmission of **Staphylococcus aureus** infections in humans? **J Hosp Infect** 1988;12:29-34.[11] Bender JB, Torres SM, Gilbert SM, Olsen KE, LeDell KH. Isolation of methicillin-resistant **Staphylococcus aureus** from a non-healing abscess in a cat. **Vet Rec** 2005;157:388-389.[12] Umber J, Bender JB. Pets and antimicrobial resistance. **Vet Clinics of North America, Small Animal**. March 2009.[13] Bender J, Coughlan K , Water K, Boxrud D, Peterson K, Buck J. Methicillin-resistant **Staphylococcus aureus** (MRSA) infections among pets in Minnesota. Proceedings of the **International Emerging Infectious Diseases Conference**, 2008. Atlanta, GA March 16-20. p.87

## EFFECT OF CO-INFECTION WITH CANINE VECTOR-BORNE PROTZOA

**Gad Baneth, DVM, PhD, Dipl. ECVCP**
**Rehovot, Israel**

### INTRODUCTION

Concurrent infection of a host by several pathogenic microorganisms is frequently detected and managed in clinical situations. The pathogenesis and pathological consequences of co-infection are commonly not well understood and involve multiple mechanisms related to transmission, host and cell invasion, immune responses, pathogen multiplication, and dissemination mechanisms. Recent studies have shown that co-infection can enhance pathogenic processes, parasite transmission, and augment disease severity (Graham et al., 2007). Much of the knowledge available on simultaneous infection has been learned from HIV co-infection in humans infected with protozoa, fungi, bacteria and viruses. The synergistic effects of co-infection with other vector-borne pathogens such as *Anaplasma phagocytophilum* and *Borrelia burgdorferi* have been described and their mechanisms are being elucidated (Grab et al., 2007). The purpose of this review is to discuss co-infection involving canine vector-borne protozoa and our current understanding of the possible relationships between co-infecting agents.

Co-infection in the host can be studied at different levels starting from a single cell model, through evaluation in an isolated tissue, the organ level and the whole organism or animal. The degree of host response complexity increases and becomes more multifaceted as more types of host cells, immune mechanisms and body systems become involved. Protection mechanisms involving signal molecules and proteins including cytokines, chemokines and T cells interact in the tissue or whole organism level and produce complicated immune responses. Additional considerations are involved in the case of vector-borne pathogens transmitted by arthropods such as ticks, fleas, mosquitoes, and flies. Transmission and introduction into the host of several pathogenic agents simultaneously elicits more diverse host responses which may conflict or override each other allowing the pathogens to synergistically colonize the infected host more successfully.

### LESSONS FROM HIV CO-INFECTION & TICK-BORNE BACTERIA CO-INFECTION IN HUMANS

HIV depresses the immune system, decreases T cell counts and exposes infected individuals to severe manifestations of infection with protozoal diseases including malaria, visceral leishmaniasis, babesiosis, American and African trypanosomiases, amoebiasis, toxoplasmosis, and cryptosporidiasis. HIV is a risk factor for severe malaria in both young children and adults and co-infection is extremely common in areas where the endemicity of these infections overlap especially in Africa (Grimwade et al., 2004; Cohen et al., 2005). Severe babesiosis usually occurs in splenectomized humans, however severe, persistent, and/or recrudescent babesiosis with *Babesia microti* is prevalent in immunosupressed HIV-infected patients with intact spleen (Karp and Auwaerter, 2007). After toxoplasmosis, leishmaniasis is the most common tissue protozoan opportunistic infection in patients with AIDS. HIV Patients with low CD4+ cell counts (<200 cells/mm$^3$) may suffer from primary new infection or reactivation of clinically latent infection. Although the basic clinical features of visceral leishmaniasis in HIV patients are similar to those in immunocompetent hosts, aberrant manifestations of disease are frequently observed including detectable peripheral parasitemia found in more than 50% of patients (Riera et al., 2005). An inverse relationship was found between the CD4$^+$ T-cell count in HIV+ humans and their infectiousness *to Phlebotomus perniciosus* sand flies (Molina et al., 1999). A similar relationship between infectiousness to sand flies and CD4+ cell counts was found also in dogs with naturally occurring *Leishmania infantum* infection (Guarga et al., 2000). The control on blood parasitemia with *Leishmania* appears to be lifted as CD4+ lymphocyte numbers decrease resulting in higher parasitemia and infectiousness to vector sand flies in both dogs and humans.

Borreliosis caused by *B. burgdorferi* is the most common human arthropod-borne infection in North America. It is transmitted by the bites of *Ixodes* spp. ticks and infects both humans and dogs in the northern hemisphere. *B. burgdorferi* spirochetes can spread to the skin, joints, heart, eyes and the nervous system causing a multitude of clinical manifestations in humans (Grab et al., 2007). Some studies have suggested that concurrent infection by other tick-borne pathogens could influence the course of the disease leading to more severe infection, persistence and decreased response to therapy. *A. phagocytophilum* is, the causative agent of human granulocytic anaplasmosis (HGA), is transmitted by the same tick spp. and has been implicated as a frequent co-infecting agent accompanying infection with *B. burgdoerferi*. Co-infection that results in simultaneous clinical manifestations of Lyme disease and HGA has been documented in humans and studies have provided evidence that co-infections contribute to more severe morbidity and a longer duration of clinical manifestations than reported with either disease alone (Krause et al. 2002; Ramsey et al., 2002; Steere et al., 2003). Likewise, experimental infections in rodent models have shown that co-infection is associated with higher bacterial loads, prolonged persistence and more severe disease (Thomas et al., 2001; Holden et al., 2005).

Recently, it has been shown that *A. phagocytophilum*-infected neutrophils release endothelial cell-derived matrix metalloproteases, cytokines and chemokines which increase vascular permeability and facilitate trans-endothelial cell migration of *B. burgdorferi* and its invasion into tissues including traversing the blood-brain barrier (Nyarko et al., 2006). *A. phagocytophilum-B. burgdorferi* co-infection has also been shown to enhance chemokine, cytokine and endothelial cell-derived matrix metalloproteases expression by human brain microvascular endothelial cells, thus potentially increasing vascular permeability and augmenting the inflammatory response in affected brain tissues (Grab et al., 2007). These studies have shown that tick-borne pathogens which commonly also affect animals including dogs have a complex synergistic relationship that can provide insights into the reasons for the severe and sometime fatal disease reported in cases of co-infection.

### CO-INFECTION WITH CANINE VECTOR-BORNE PROTOZOA

Dogs are commonly infected with multiple vector-borne protozoal and bacterial organisms (Baneth et al., 1998; Kordick et al, 1999). Viral diseases such as parvoviral enteritis or canine distemper may augment and increase the severity of these infections (Baneth et al, 1997). Fatal cases of severe co-infection have been reported (Kodick et al, 1999; Gal et al. 2007) but direct relationship between co-infecting agents is yet to be studied further at the basic level to reveal mechanisms that trigger more severe clinical manifestations of co-infection with comparison to infection with a single pathogen. Canine momocytic ehrlichiosis and canine leishmaniasis are major vector-borne diseases of dogs with a worldwide distribution. The causative agents of these diseases, *Ehrlichia canis* and *L. infantum* are intracellular pathogens that use the monocyte-macrophage as their host cell. The epidemiology of canine ehrlichiosis and leishmaniasis overlaps in many areas of the world because the vector activity and transmission periods of these pathogens are similar. While *E. canis* is transmitted by ticks, *L. infantum* is transmitted by phlebotomine sand flies. Both infectious agents disseminate from the skin to the spleen, liver and bone marrow (BM) and activate host immune mechanisms that induce a range of immunopathological responses. A recent study described forty three dogs introduced into an endemic area of canine ehrlichiosis and leishmaniasis in southern Italy and followed for 2 transmission periods. Thirty four dogs that acquired *Ehrlichia canis* infection also developed leishmaniasis. Evidence of *E. canis* infection preceded *L. infantum* ($p<0.05$) in 28 (82%) dogs. Furthermore, clinical signs were more frequent ($p<0.01$) in dogs with dual infection than in those with single infection and it was concluded that ehrlichiosis could have been a contributing factor to establishment of leishmaniasis (Mekuzas et al., 2009).

**MODELS OF CONCOMMITANT INFECTION**

Several modes of concomitant infection with two pathogens leading to disease may be visualized. These include:

1) Synergism between two pathogens that normally cause mild to moderate disease if present alone, leading to more severe disease.
2) Induction of a severe disease by a pathogen that normally causes mild infection provoked by a concomitant highly pathogenic or immunosuppressive disease agent
3) Induction of extremely severe disease by synergistic action of co-infecting highly pathogenic agents.

Examples for these modes of concomitant infections include cerebral toxoplasmosis in HIV patients which complies with the second mode describing a relatively mildly pathogenic agent which becomes extremely virulent in the face of immunosupression induced by HIV. The combination of *Plasmodium falciparum* and HIV in severe acute malaria may fit the third mode of described.

**TIMING OF INFECTION**

Although co-infections by a vector such as a tick may be transmitted simultaneously during the same blood meal, concurrent infections are frequently associated with one infection preceding the other. The presence of one infection may facilitate the spread and dissemination of a latent infection already present in the animal, or allow the establishment of a new infection which would normally be eliminated by immune defense mechanisms. The sequence of events leading to a disease presented as a co-infection may frequently not be traceable unless specific tests have been taken prior to the appearance of the second infecting agent.

**METHODS FOR SIMULTANEOUS THE DETECTION OF INFECTION**

Correct diagnosis and management of co-infection patients may often be complicated because one infection frequently stands out while the presence of other infective agents may be less obvious at first. Different infections may present with similar and often non-specific clinical signs. An effort must be made to think of the possibility of co-infection and to make the diagnostic effort timely and efficient. Detection of co-infection may be carried out by submitting samples for several different tests. Some diagnostic techniques allow the simultaneous assessment of infection with several pathogens. These include commercial serological kits that evaluate the presence of antibodies reactive with several agents in the same sample. Other test for the simultaneous detection of pathogen DNA include multiplex PCR assays and the reverse line blot (Sirigireddy et al, 2005; Matjila et al, 2008). These assays are based on molecular biological techniques for the detection of pathogen-specific DNA sequences and typically have very high levels of sensitivity and specificity.

**TREATMENT AND PREVENTION**

Management of co-infections frequently requires special therapeutic considerations which could include different drugs aimed at each infecting agent. Treatment may be of different durations and follow up tests could be variable. For instance, in the case of *L. infantum-E. canis* co-infection, treatment of leishmaniasis will be prolonged and may include allopurinol for the dog's entire life while *E. canis* is usually treated with doxycyline for 21days. Prevention is carried out with vaccination if available, or by minimizing exposure to pathogens, for instance by using insecticides against ticks or other hematopahgeous arthropods.

**CONCLUSIONS**

Co-infections frequently present with more severe clinical signs than infections with one pathogen only. The possible lack of specific clinical signs characterizing the different co-infecting agents may be confusing and hinder the detection of all the infecting agents involved in the disease. Appropriate treatments should be selected to cover all of the pathogens involved in the concomitant infection. Better understanding of the pathogenic mechanisms involved with canine co-infections will facilitate improved future management of these conditions.

## REFERENCES

[1] Baneth G, et al, **Veterinary Parasitology** 1998;74: 133-142. [2] Baneth G, et al, _**Veterinary Parasitology** 1997;70: 201-206. [3] Cohen C, et al., **Clinical Infectious Diseases** 2005;41:1631-1637. [4] Gal A, et al, **Canadian Veterinary Journal** 2007;48: 619-622. [5] Grab DJ, et al., **Clinical and Vaccine Immunology** 2007;14:1420-1424. [6] Graham AL, et al., **Trends in Parasitology** 2007;23:284-291. [7] Grimwade K, et al., **AIDS** 2004;18:547-554. [8] Guarga JL, et al., **Research in Veterinary Science** 2000;69:249-253. [9] Holden et al, **Infection and Immunity** 2005;73:3440-3444. [10] Karp CL, and Auwaerter PG, **Clinical Infectious Diseases** 2007;45:1208-1213. [11] Kordick SK, et al, **Journal of Clinical Microbiology** 1999;37:2631-2638. [12] Krause et al, **Clinical Infectious Diseases** 2002;34:1184-1191. [13] Matjila PT, et al, **Veterinary Parasitology** 2008;155,152-157.
Mekuzas Y, et al, **Clinical Microbiology and Infection**. In press. [14] Molina R, et al, **American Journal of Tropical Medicine and Hygiene** 1999; 60:51-53. [15] Nyarko et al, **International Journal for Parasitology** 2006;36:601-605. [16] Ramsey et al, **Emerging Infectious Diseases** 2002;8:398-401. [17] Riera C, et al., **American Journal of Tropical Medicine and Hygiene** 2005;73:1012-1015. [18] Sirigireddy KR and Ganta RR. **Journal of Molecular Diagnostics** 2005;7:308-316. [19] Steere et al, **Clinical Infectious Diseases** 2003;36:1078-1081. [20] Thomas et al, **Infection and Immunity** 2001;69:3359-3371.

# UPDATE ON THE EPIDEMIOLOGY & TREATMENT OF FELINE CYTAUXZOONOSIS

**Leah A. Cohn, DVM, PhD, DACVIM (small animal internal medicine)**
**Columbia, MO**

## INTRODUCTION

Infection of the domestic cat with the protozoan *Cytauxzoon felis*, transmitted by the bite of an infected tick, is usually fatal. The disease follows a rapid course culminating in death within days of initial illness. From 12-15 days after being bitten by an infected tick, the cat develops non-specific signs of illness, including anorexia, fever, and pallor. Pre-mortem diagnosis is usually made by identifying erythrocytic organisms on a blood smear. Despite attempted therapy, most cats succumb within 5 days of the onset of illness. Although *C. felis* infection is uncommon nationwide, it is quite common in certain areas of the Midwestern and Southern United States and the endemic geographic region may be expanding with cases reported along the Atlantic coast.

## EPIDEMIOLOGY

Because of the rapidly fatal course of disease in this species, the domestic cat is a presumed dead-end host for *C. felis*. The reservoir host is the bobcat, *Lynx rufus*. Infection of the bobcat with *C. felis* usually leads to a brief, non-lethal schizogonous phase, followed by life-long maintenance of the intraerythrocytic parasites. Infection has been documented in several species of cat, including multiple big cat species. Those with limited schizogony seem to experience limited morbidity and mortality, while those cat species with more pronounce schizogony succumb to infection. Experimental studies have failed to document infection in any type of non-felid species, including immunodeficient mice.

Close contact between ill, infected cats and healthy cats does not transmit infection. Transmission of the disease to domestic cats can occur in one of two ways. Transmission requires either the bite of a tick that carries the organism, or inoculation of schizonts from an infected animal. Interestingly, transfusion of blood containing erythrocytic piroplasms to the domestic cat results in apparent erythrocytic infection of the recipient without either clinical disease or protection from clinical disease following inoculation of schizonts. Only ticks are believed to transmit the disease naturally. Although *C. felis* has been identified in multiple species of tick, only *Dermacentor variabilis* is documented as a competent vector of infection.

Feline cytauxzoonosis can occur in cats of any age, although most reported cases have been in young adult cats. Immune suppression is not necessary for cats to become infected, and there is no evidence that infection is more likely in retroviral infected cats than in healthy cats. No sex or breed predilections have been identified for cytauxzoonosis. Outdoor cats are more likely to acquire this disease, presumably due to an increased exposure to the tick vectors. The greatest risk of infection seems to occur in the spring and summer, presumably during times the tick vector is most active. In Oklahoma, peak incidence occurred in May, while infections were very rarely identified from November through March.[1] Similarly, 75% of clinical cases identified by the University of Missouri Veterinary Medical Diagnostic laboratory over a several year period were presented between May and July.

For many years after its original description during the mid 1970's in cats from Missouri, feline cytauxzoonosis was recognized only in states in the south central USA.[2] In recent years, the geographic range of infections has expanded greatly, with infections now reported in cats throughout the Southeastern and Mid-Atlantic as well as the South central USA (Figure 1).[3,4] Although not yet demonstrated in domestic cats, *C. felis* infection has been demonstrated in bobcats living as far north as Pennsylvania.[5] Because the diagnosis is often based on visualization of RBC inclusions, and because necropsy diagnosis of cytauxzoonosis is even more straightforward than antemortem diagnosis, it seems unlikely that infections were simply "missed" in these areas prior to the last decade. Instead, it appears that the infection is now occurring in areas where it did not previously occur.

Because the reservoir host for the pathogen is the bobcat, infected domestic cats are typically from rural or suburban rather than urban areas. Cats living close to wooded areas or less intensely managed land are more likely to become infected.[1] It is common for multiple cats in a neighborhood or even a house-hold to become infected within a single season, likely reflecting the presence of infected reservoir hosts in the area.

There are no large published epidemiologic studies of *C. felis* infection in domestic cats. In a single study of healthy cats presented to trap-neuter-return programs in Tennessee (n=75), Florida (n=494), and North Carolina (n=392), prevalence of *C. felis* as detected by PCR ranged from 1.3% to 0.4% to 0%, respectively.[6] It is likely that prevalence rates are higher in regions traditionally found to have enzootic infection; the author is aware of several colonies of cats in Missouri with prevalence rates of up to 45%. Prevalence of infection in cats from Brazil with a hematoprotozoan parasite morphologically indistinguishable from *C. felis* is similarly high, up to 48.5%.[7] Prevalence of infection in bobcats again depends on geography, but estimates range from 7% in Pennsylvania to 33% in North Carolina to 62% in Oklahoma.[5,8,9] There are no published studies on incidence of infection. In a retrospective review, cytauxzoonosis accounted for approximately 1% of feline submissions to the Oklahoma Animal Disease Diagnostic Laboratory and 1.5% of feline admissions to the Boren Veterinary Medical Teaching Hospital between 1995-2006 and 1998-2006, respectively.[1] Personal communications with the author suggest that veterinarians in enzootic "hot spots" see as many as 5 cases of cytauxzoonosis a week during the peak season, with many more veterinarians in surrounding regions reporting that they see 5-10 cases per summer.

## TREATMENT OPTIONS

Once described as a uniformly fatal infection, over the last decade there have been several reports of surviving cats. In some cases cats survived with aggressive supportive care, in other cases cats survived with supportive care plus antiprotozoal therapy, and yet others cats have been identified during the carrier piroplasm stage of infection with no known history of clinical illness. Unfortunately, no controlled prospective studies have been completed to demonstrate efficacy of any treatment regimen.

For ill cats, supportive care is certainly indicated although there is no evidence as to which aspects of supportive care are important. It is reasonable to assume that fluid therapy is indicated to correct dehydration, preserve vascular volume, and maintain tissue perfusion. For those cats which become anemic and demonstrate tachypnea or tachycardia, provision of oxygen carrying capacity via transfusion of whole blood, packed red blood cells, or hemoglobin based oxygen carrying solutions is indicated. Cytauxzoonosis is often accompanied by DIC; prophylactic (100-200 U/kg unfractionated heparin SC q 8 hrs) or therapeutic (goal, 2X prolongation of aPTT) heparin administration may prove useful. In order to replenish coagulation factors when treating active DIC, whole blood or plasma transfusion typically accompanies heparin therapy. Non-steroidal anti-inflammatory drugs such as meloxicam have been used to provide analgesia and reduce fever. Some veterinarians treat ill cats with low dose prednisolone. The potential benefit (or harm) of NSAIDs and corticosteroids has not been evaluated. A variety of antimicrobial drugs (e.g., sodium ampicillin, enrofloxacin, and doxycycline) have been used in infected cats alone or in combination.[10-12] None of these drugs has demonstrated efficacy against protozoal agents such as *C. felis*.

Definitive treatment of cytauxzoonosis would involve antiprotozoal drug administration. The anti-protozoal hydroxynapthoquinolone agents parvaquone and buparvaquone are the treatment of choice for Theileria infection of African cattle. Despite the close relationship of *C. felis* to *Theileria* spp., these drugs demonstrate a complete lack of efficacy for treatment of cytauxzoonosis in trials to date.[13] The aromatic diamidine compounds imidocarb dipropionate (Imizol®, Schering-Plough Animal Health, Kenilworth, NJ) or diminazene aceturate are used for the treatment of various protozoal agents, including *Babesia* spp and African trypanosomiasis. In a retrospective study, 5 of 6 cats treated with diminazene plus aggressive supportive care survived infection.[11] Unfortunately, diminazene is not approved for use in the USA. Although requests can be made to the Food and Drug Administration for importation of diminazene under compassionate drug use importation rules (http://www.fda.gov/ora/import/ora_import_system.html), the time required for importation prevents its practical utility in treatment of *C. felis* since each request must be made on behalf of a specific animal (i.e., the drug cannot be stockpiled for future use). Imidocarb is readily available in the USA, and has been widely used for the treatment of cytauxzoonosis at a variety of different dose regimens (e.g., two doses at 2 mg/kg given intramuscularly 7 days apart, two doses at 5 mg/kg given intramuscularly 7 days apart, two doses of 5 mg/kg given 4 days apart). Although a prospective trial evaluating the use of imidocarb at 3-4 mg/kg IM, twice 7 days apart is ongoing, results are not yet available. Included in a retrospective report of 18 surviving infected cats (only one of which was treated with imidocarb) was the anecdotal mention that imidocarb lacked efficacy in the treatment of experimentally induced cytauxzoonosis.[14] In the author's personal communications with veterinarians from endemic regions, some believe treatment with imidocarb results in no improvement in clinical outcome (rare survival) while others claim as much as a 75% clinical recovery in cats treated with imidocarb and supportive care. When used, pre-treatment with atropine may minimize the potential adverse cholinergic effects of imidocarb treatment. Recently, treatment has been attempted with a combination of the antiprotozoal drug atovaquone and the antibiotic azithromycin. This combination has been used successfully to treat both human and canine babesiosis.[15,16] Although a prospective clinical trial is underway, results are not yet available to either demonstrate or refute the efficacy of this combination therapy. Survival of up to 62% of cats treated with this combination has been reported, although the report was not based on a controlled or comparative study.[17]

It is unclear if treatment should be administered to healthy cats with persistent erythroparasitemia. In the few cats followed for months or years, clinical disease has not occurred as a result of the piroplasm infection. However, the possibility exists that these cats may experience hemolysis at some future date. Of additional importance, it is theoretically possible that these carrier cats may serve to infect naïve ticks which could then go on to inoculate pathogenic sporozoites when they next feed. Treatment with imidocarb is apparently not effective in eliminating parasitemia in healthy cats.[18] The combination of atovaquone and azithromycin was unable to uniformly cure infection, but treated cats were often cytologically and PCR negative for parasites 8 weeks after treatment.[18]

Because no therapy has been proven effective and the disease remains largely fatal, and ounce of prevention is worth a pound of cure. Since the only natural means of transmission seems to be via the tick vector, measures to control these ectoparasites are indicated. Keeping cats indoors minimizes tick exposure. No chemical is completely effective to prevent tick attachment, but the author prefers topical fipronil spray.

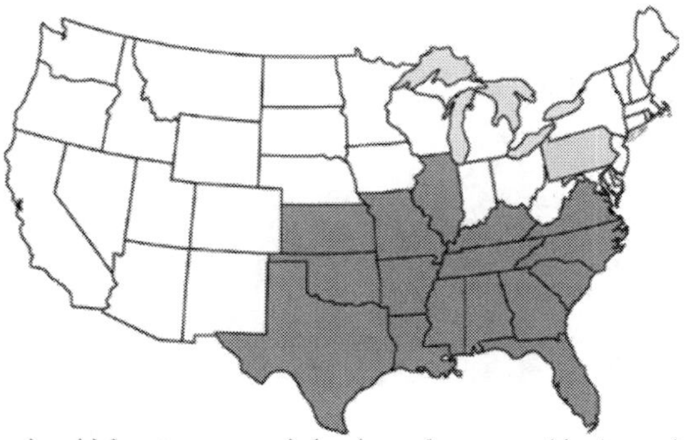

**Figure 1:** Map of the USA indicating states in which cytauxzoonosis has been documented in domestic cats (green). In Pennsylvania (grey), the pathogen has been detected in bobcats but disease has not been documented in domestic cats.

## REFERENCES

[1.] Reichard MV, et al. **Vet Parasit** 2008;152:314.; [2.] Wagner JE. **J Am Vet Med Assoc** 1976;168:585.[3] Birkenheuer AJ, et al. **J Am Vet Med Assoc** 2006;228:568. [4.] Jackson CB, et al. **Vet Parasit** 2006;139:192.[5] Birkenheuer AJ, et al. **Vet Parasit** 2008;153:126.[6.] Haber MD, et al. **Vet Parasit** 2007;146:316.[7.] Mendes-de-Almeida F, et al. **Vet Parasit** 2007;147:9.[8.] Glenn BL, et al. **J Am Vet Med Assoc** 1982;181:1251.[9.] Kocan AA, et al. **J Wildl Dis** 1985;21:190.[10.] Walker DB, et al. **J Am Vet Med Assoc** 1995;206:1363.[11.] Greene CE, et al. **J Am Vet Med Assoc** 1999;215:497.[12.] Hoover JP, et al. **J Am Vety Med Assoc** 1994;205:455.[13.] Motzel SL, et al. **Vet Parasit** 1990;35:131.[14.] Meinkoth J, et al. **J Vet Intern Med** 2000;14:521.[15.] Birkenheuer AJ, et al. **J Vet Intern Med** 2004;18:494.[16.] Raju M, et al. **Ped Infect Dis** *J* 2007;26:181.[17.] Birkenheuer AJ, et al. **J Vet Intern Med** 2008;22:703 (abstract).[18.] Cohn LA, et al. **J Vet Intern Med** 2008;22:704 (abstract).

**RECOGNITION & MANAGEMENT OF ATYPICAL CANINE BABESIOSIS**

**Adam J. Birkenheuer, DVM, PhD, DACVIM**
**Raleigh, NC**

## INTRODUCTION

Canine babesiosis was first described in South Africa in the late 1800's and was originally presumed to be a "biliary form" of canine distemper. Most researchers assume that this was actually a case of babesiosis caused by *Babesia canis rossi*. The first report of canine babesiosis in the United States was made in 1934 and is presumed to have been caused by *Babesia canis vogeli*. The first report a small *Babesia* infection in a dog was in 1968. The first case of endemic babesiosis caused by a small piroplasm was in 1979. Since the early 1990's the number of cases of babesiosis in the US and around the world has increased dramatically. Classical babesiosis is characterized by fever, hemolytic anemia and splenomegaly. Over the past twenty years there has been a dramatic increase in the number of as cases reported as well as the geographic distribution of canine babesiosis around the world.

## EMERGING INFECTIONS OR EMERGING INFORMATION

The number of publications on PubMed that include the term *Babesia* has increased steadily and the total number of publications has nearly doubled in the past ten years. The number of piroplasms known to infect dogs has increased from just 4 in 1989 t oat least 10 genetically distinct piroplasms at the time of this writing. While some of these reports have documented the true emergence of an infectious agent such as the spread of *Babesia gibsoni* across North America, South America and Europe, other reports are more likely to have simply shed light on an a genotype of *Babesia* that has been present for years.

The most important reason for the "sudden" appearance of so many "new" species of piroplasms is the advent of the polymerase chain reaction and improvements in DNA sequencing technology that have made it widely available and affordable. Prior to PCR and DNA sequencing, species determinations were made almost exclusively based on parasite morphology and presumed host and vector specificity. We now clearly know that these characteristics alone are not sufficient to definitively identify a piroplasm to the species level.

## SMALL PIROPLASMS OF DOGS

### *Babesia gibsoni*

First described in 1910 in Asia, *Babesia gibsoni* was believed to be the only small piroplasm of dogs for nearly a century. The first molecularly confirmed cases of *B. gibsoni* in North American dogs were reported in 1999. Since that time, *B. gibsoni* has been recognized in dogs throughout the United States. Similarly, this infection has been identified in Australia, Europe, South America and South Africa all of which were believed to be free of this parasite. *Babesia gibsoni* is now the most commonly identified cause of canine babesiosis in the US. This parasite has emerged and spread rapidly throughout the world in the absence of known competent tick vectors in many regions. Dog to dog transmission rather than tick-transmission is believed to be the primary route of infection outside of Asia. Epidemiological evidence suggests that dog to dog transmission via bite wounds is a common route of infection. Additionally this parasite can be spread via transplacental transmission. Breeds associated with dog-fighting, American Pit Bull Terriers and the Tosa Inu, are the most common breeds in which *B. gibsoni* is identified. They are not believed to have any increased genetic susceptibility to infection. They are believed to have increased infection rates due to increased chance of being used in illegal dog-fighting as well as through transplacental transmission.

### *Babesia conradae*

*Babesia conradae* was first described in 1991 from dogs in Southern California. Due to the lack of complete genetic database at the time, this parasite was believed to be *B. gibsoni*. *Babesia conradae* appears to be geographically limited to Southern California and very little is known about the life-cycle or risk factors for this infection.

### *Babesia microti*-like parasite (AKA *Theileria* annae, *Babesia* annae)

A *Babesia microti*-like parasite was first described in 1999 in a German dog that had traveled to Northwest Spain. Since that time several large case series have described babesiosis associated with this parasite. To date, these cases have been limited to dogs in NW Spain or dogs that have traveled to that region. A large proportion of dogs infected with the *B. microti*-like parasite have suffered from renal failure (presumably due to advanced glomerulonephritis). A closely related parasite has been identified in the blood of North American foxes.

### *Theileria* species

Several molecular surveys have identified *Theileria* sp. DNA in the blood of dogs. *Theileria* equi DNA has been amplified by PCR from the blood of European dogs. More recently a large molecular survey of dogs from South Africa identified a genetically unique *Theileria* sp. in nearly 40% of the dogs from one region. Clinical data from these dogs is sparse so few conclusions can be drawn regarding the clinical significance of *Theileria* infections in dogs.

## LARGE PIROPLASMS OF DOGS

### *Babesia canis vogeli*

*Babesia canis vogeli* or *Babesia vogeli* has a nearly worldwide distribution as its primary vector is the brown dog tick *Rhipicephalus sanguineus*. *Babesia canis vogeli* has long been considered to be a puny pathogen only affecting puppies and splenectomized dogs. With increased vigilance and testing *B.* c. *vogeli* is being recognized more frequently causing clinical illness in adult dogs.

***Babesia canis canis***

*Babesia canis canis* has long been recognized as the most common cause of canine babesiosis in Europe. More recently this genotype has been identified in both North and South America. It is generally considered a moderately pathogenic parasite. A vaccine against *Babesia canis canis* reduces morbidity and mortality, but does not induce protective immunity is available in Europe. This vaccine does not confer cross-protection against other species of *Babesia*.

***Babesia canis rossi***

*Babesia canis rossi* or *Babesia rossi* is a highly virulent parasite that is endemic to Africa and has not yet been described in other parts of the world. Dogs with *B.* c. *rossi* typically present with either "uncomplicated" babesiosis characterized by anemia, fever and splenomegaly or "complicated" babesiosis which can include organ failure in one or more systems including: CNS signs, renal failure, hepatic failure, myocardial dysfunction, ARDs and pancreatitis.

**Other large piroplasms**

Recently two "new" *Babesia* spp. have been identified in dogs. One species has been identified in North America and appears to be a cause of thrombocytopenia and mild anemia in dogs that have been splenectomized or are undergoing chemotherapy. The second species has been described in a dog from the United Kingdom that had no history of travel. *Rangelia vitali* is an apicomplexan parasite that is believed to occasionally infect erythrocytes and appear morphologically similar to *Babesia* spp. has received renewed attention from Brazilian veterinary researchers. *Rangelia* primarily infects endothelial cells and causes hemorrhagic disease.

**CLINICAL DISEASE**

Hemolytic anemia and its accompanying signs are what most practitioners associate with canine babesiosis. While these remain common and important signs of canine babesiosis, we are recognizing an ever widening spectrum of signs associated with canine piroplasmosis.

The first form of "atypical" babesiosis that we need to be on the alert for is "typical" babesiosis caused by an "atypical" species. Unfortunately the available tests for babesiosis are far from perfect. Serological assays can have both false positive and false negative results due to differences in parasite strains, laboratory technique and variations in immune responses by individual animals. Molecular tests such as PCR can result in false positive tests due to laboratory error. Additionally PCR tests may actually be too specific. For example an assay that is designed to detect *Babesia gibsoni* may fail to detect EVERY other piroplasm and result in a "false negative" result if the dog is infected with *B. canis*. Broad-range assays designed to screen for all piroplasms with follow-up species specific assays or DNA sequencing can minimize these types of errors. As it is very easy to import dogs into the US it is crucial that diagnostic assays are able to detect all species.

Another common form of "atypical" babesiosis is *Babesia* infection in a non-anemic or mildly anemic dog. A close scrutiny of the literature reveals that although many dogs with *Babesia* infections are anemia, many only have mild anemia (i.e. PCV of 30-35%) or have packed cell volumes well into the normal range. Many of these dogs have a moderate to severe thrombocytopenia as their sole hematologic abnormality. In most large studies, thrombocytopenia is the most common hematological abnormality associated with canine *Babesia* infection. The practitioner that only tests for *Babesia* when anemia is present will miss some cases.

Lastly we are recognized *Babesia* infections in dogs where neither anemia or thrombocytopenia appear to be the most prominent clinicopathological abnormality. Some cases of babesiosis have presented solely for evaluation of hyperglobulinemia. These have been polyclonal in nature and resolved with anti-protozoal therapy. Additionally protein losing nephropathies have been identified in dogs with *Babesia* infections caused by species other than the *B. microti*-like parasite from Spain.

**TREATMENT**

Note: Since the incidence of canine babesiosis is increasing in the US, specific anti-*Babesial* treatment is appropriate and indicated in dogs highly suspected of having babesiosis even without a definitive diagnosis. However since all treatments are not effective against all piroplasms and cost may vary considerably it is preferable to correctly identify which genotype is present in stable patients.

**Antibiotics**

Antibiotics are not the treatment of choice, but are reasonable to consider while waiting for specific anti-protozoal therapy. Doxycycline will reduce clinical signs and is associated with decreased morbidity and mortality. The dose is 10mg/kg/day. Administered IV or PO once or twice daily. I prefer once daily dosing unless the patient is vomiting after the medications. Clindamycin has been used to treat babesiosis. The dose in a recent study was 25mg/kg PO BID for 14 days, and was associated with a reduction of clinical, hematological and biochemical abnormalities. Some recent work has shown that a combination of doxycycline (5mg/kg PO BID), clindamycin (25mg/kg PO BID) and metronidazole (15mg/kg PO BID) may be effective in clearing *B. gibsoni* infections however the duration of therapy is not well defined with some dogs receiving treatment for several months. Administration of antibiotics alone without specific testing has resulted in a number of dogs being tentatively diagnosed with ehrlichiosis based on a partial or complete recovery. Cost: $10-50

**Imidocarb dipropionate**

Imidocarb dipropionate (Imizol®, Schering-Plough) is the only FDA approved drug for the treatment of canine babesiosis. Dose 6.6mg/kg IM once. Repeat dose in 2 weeks. I will usually re-dose in one week if the dog is not responding well. Likely to cure *B. canis vogeli*. Good to reduce morbidity, mortality and parasitemia in *B. gibsoni* (Asian genotype), but is not curative. Side-effects include pain at injection site and cholinergic effects (SLUD). Pre-treatment with atropine reduces cholinergic effects. Cost: $50-100

**Atovaquone and Azithromycin**

An atovaquone and azithromycin drug combination is the only treatment shown to reduce *B. gibsoni* parasitemia below the limit of detection by a sensitive PCR test. Was effective in 83% of treated dogs in a controlled trial. Atovaquone is an anti-protozoal drug believed to inhibit the action of cytochrome *B*. It is available in two formulations in the US, Mepron® (GlaxoSmithKline) and Malarone® (GlaxoSmithKline). Mepron is a single drug formulation and is available in a liquid suspension. Malarone is a multi-drug combination containing atovaquone and proguanil hydrochloride. In our experience Mepron is more expensive, but is very well tolerated, while Malarone has been associated with severe GI side-effects in some cases. The atovaquone dose is 13.5mg/kg PO TID administered with a fatty meal for 10 consecutive days in combination with azithromycin. The azithromycin dose is 10mg/kg PO Q24 for 10 consecutive days. We have not used this combination to treat *B. canis* or *B. gibsoni* (USA/California genotype). Cost: $300-900

**Diminazene aceturate**

Not FDA approved. Dose 3-7mg/kg IM once. Diminazene is likely to be curative for *B. canis*. It does not cure *B. gibsoni* (Asian genotype), but reduces morbidity, mortality and parasitemia. Anecdotal evidence suggests that *B. gibsoni* infected dogs recover more quickly than those treated with Imidocarb. Side-effects are similar to Imidocarb. Mortality has been noted at doses greater than 10mg/kg. Cost: ?

**Supportive care**

Intravenous fluids are indicated for hypovolemic patients. Whole-blood or Packed RBC transfusions (from screened donors of course) for dogs that are clinical for their anemia. We have also used a hemoglobin based oxygen carrying solution in some dogs with babesiosis.

**Immune-suppressive drugs**

The use of immune-suppressive drugs for the immune-mediated components of the disease (anemia and thrombocytopenia) is controversial. If a dog is stable (i.e., eating and drinking and does not require ICU), regardless of PCV, I will treat with anti-protozoal drugs alone. I have treated dogs with PCV's as low as 8% with imidocarb alone and they have done very well. If a dog is crashing despite anti-protozoal therapy, I will treat like an IMHA case with 2mg/kg/day of prednisone. I will taper the prednisone faster in these cases than I do with idiopathic IMHA cases. Cases that have been treated with long-term immune-suppressive drugs prior to specific anti-protozoal treatment tend to have a poor response to treatment.

**REFERENCES**

[1.] Birkenheuer AJ, et al. **Vet Parasitol**. 2004 Oct 5;124(3-4):151-60. [2.] Holm LP, et al. **Vet Rec**. 2006 Aug 5;159(6):179-80. [3.] Matjila PT et al. **Vet Parasitol**. 2008 Oct 20;157(1-2):34-40. [4.] Garcia AT. **Vet Parasitol**. 2006 May 31;138(1-2):97-102. [5.] Kjemtrup AM, Conrad PA. **Vet Parasitol**. 2006 May 31;138(1-2):112-7.

# CANINE HEPATOZOONOSIS: UPDATE ON PREVALENCE, DIAGNOSIS AND TREATMENT

**Douglass K. Macintire, DVM, MS Diplomate ACVIM and ACVECC**
**Auburn, AL**

Over 300 distinct species of the genus Hepatozoon have been recognized. Of these, over 120 are found in snakes and 46 are in mammals.

Two species of Hepatozoon organisms have been reported to infect domestic dogs worldwide. The first species described was *Hepatozoon canis*, initially reported in India in 1905. Since then, *H. canis* has been reported in dogs from Japan, Africa, Israel, Spain, France, Italy, Portugal, Greece, the Philippines, Brazil, Argentina, and the United States.

In 1978, hepatozoonosis was first reported in dogs in the USA along the Texas Gulf Coast region. Initially, the causative agent was thought to be *Hepatozoon canis*, but differences in clinical features of disease, parasite stages, parasite location, and gametocyte structure resulted in the designation of a new species, *Hepatozoon Americanum*. *H. Americanum* now appears to be a serious emerging tick-borne disease in the USA.

Of 614 samples submitted to the Molecular Diagnostic Laboratory at Auburn University, 32% were positive for *H. americanum*. Positive samples originated from 13 states, including Washington, California, Nebraska, Vermont, and Virginia as well as the Southeastern US. In addition, *H. canis* was found in 14% of the samples, mostly in dogs from Alabama and Mississippi. Before 2008, it was believed that the only species that naturally infected dogs in the US was *H. americanum*, and *H. canis* only occurred in Europe, the Middle East, Africa, and Asia. Recent reports have also described *H. canis* in dogs from Argentina and Brazil. Researchers in Spain found 90% of foxes tested and 27% of symptomatic dogs to be positive for Hepatozoonosis. Two species, separate from *H. canis* and *H. americanum*, were also described in the Spanish report. Another distinct species of hepatozoon has been found in Japanese dogs as well.

## LIFE CYCLE

The definitive host of *H. canis* is the brown dog tick, *Rhipicephalus sanguineus*. Ticks become infected by ingesting circulating gametocytes (infected neutrophils) from the bloodstream of a dog as nymphs. The gametocytes fuse in the tick to form an ookinete. Each ookinete becomes an oocyst which contains numerous sporocysts. Within each sporocyst are 12-24 sporozoites. In order to become infected, a dog must ingest the adult tick. Following ingestion, sporozoites are released when exposed to bile and penetrate the dog's intestinal tract to be carried by blood or lymph to mononuclear phagocytic cells in various organs (spleen, lymph nodes, liver, lung, pancreas, and bone marrow). Asexual reproduction of the organism occurs through merogony. Merozoites are formed within the meront and released to either become circulating gametocytes in white blood cells, or to become additional tissue meronts. A single tick can contain hundreds to thousands of infective organisms.

The life cycle of *Hepatozoon americanum* is essentially the same, except that the definitive tick host is *Amblyoma maculatum*, the Gulf Coast Tick, and the host cell is the monocyte or macrophage rather than the neutrophil. All stages of the tick – larva, nymph, and adult – can become infected. The gametocyte is identical in appearance to H. canis, but parastemia generally affects < 0.1% of leukocytes. In contrast, parasitemia is a common finding in dogs with *H. canis*, and 1-90% of leukocytes can be infected. Another difference in the *H. americanum* life cycle is the presence of a cystic tissue stage that is not seen in *H. canis* infections.

## TISSUE STAGES

*Hepatozoon canis* infection is associated with the occurrence of meronts in the spleen, liver, lymph nodes, kidney, pancreas, lungs, and bone marrow. Schizonts have a characteristic "wheel-spoke" pattern and are not associated with tissue inflammation, and do not occur in muscle tissue.

The most common tissue stage of *H. americanum* is a large "onion-skin" cyst found most often in skeletal muscle. At the center of the "cyst" is the host cell with the intracellular parasite. The infected cell is surrounded by lamellar muco-polysaccharide secretions, resulting in the characteristic "onion-skin" appearance. Merogony occurs and merozoites are released resulting in an intense pyogranulommatous inflammatory reaction. Cysts can be found in skeletal muscle, cardiac muscle, pancreas, kidneys, lymph nodes, and spleen. Micromerozoites become gamonts that circulate in the blood and are infective to *Amblyoma maculatum* ticks. Macromerozoites can also enter macrophages and form more tissue cysts, thus propagating the continued infection in the canine host without further tick exposure.

## TRANSMISSION

Until recently, the only known method for transmitting American hepatozoonosis has been through the ingestion of an infected tick. Domestic dogs were thought to become infected during grooming and by ingesting tick-infested wild life. Coyotes are thought to be important reservoirs in endemic areas, as over 50% of randomly collected coyotes in Oklahoma were found to be positive for *H. americanum*. Adult Gulf Coast ticks can be found in large numbers on coyotes, but the larval and nymphal stages of *A maculatum* prefer to feed on ground-dwelling birds and small rodents. Fifty-eight percent of cotton rats and 33% of white footed mice from endemic areas were found to be positive for a hepatozoon species distinct from *H. americanum*. However, when they were experimentally infected with sporulated oocysts of *H. americanum*, they developed inflammatory lesions of skeletal muscle, cardiac muscle, lung, and kidney that contained a previously undescribed tissue stage called a cystozoite. These small cysts are similar to cystozoites found in snakes with hepatozoonosis that cause infection in predatory reptiles that ingest them. Recently, infected rodent carcasses were fed to a dog and induced clinical signs of fever, stiffness, lethargy and ocular discharge 28 days post ingestion. Gametocytes were observed on Day 42, the same time the PCR test became positive. A muscle biopsy taken at 90 days revealed the classic "onion skin" cysts. We now know that wild and domestic canids can be infected by either grooming or predation, and that ticks can be infected as larvae, nymphs, or adults. Once infected, the organism is transmitted through both transtadial and transovarial transmission and remains in the tick population. The large wildlife reservoir and the persistence of the organism in endemic tick populations explain why *H. americanum* has been recognized as an emerging disease that is increasing in number of cases as well as geographic distribution in the USA.

**CLINICAL SIGNS**

The pathogenicity of *H. canis* has been questioned because most dogs with circulating gametocytes are asymptomatic. With severe parasitemia or concurrent infection, however, dogs may exhibit fever, cachexia, splenomegaly, anemia, and lethargy.

*H. americanum* infection commonly results in severe disease characterized by fever, muscle pain, stiffness, ocular discharge, and weight loss. Gait abnormalities range from stiffness to total recumbency. The disease has a chronic course, and clinical signs often wax and wane. Without treatment, death usually occurs within 12 months. Chronic infection may result in amyloid deposition, vasculitis, and glomerulonephritis with protein-losing nephropathy.

**DIAGNOSIS**

*H. canis* is most commonly diagnosed by finding circulating gametocytes on the CBC or buffy coat. *H. americanum* is most commonly diagnosed by finding cysts or pyogranulomas on a muscle biopsy specimen or obtaining a positive PCR test. The gametocyte appearance in WBC's is similar, but can affect 20-60% of the cells in dogs with *H. canis* and is rarely found in more than 0.1% of WBC's in dogs with *H. americanum*. A PCR test for hepatozoonosis has recently become available through the Molecular Diagnostics Lab at Auburn University. The test can differentiate species and provides a quantitative assessment of infection. The test is very effective in identifying early infections, but false negatives sometimes occur in chronic disease. In suspect cases with a negative PCR, a muscle biopsy should be done to confirm the negative result.

**LABORATORY & RADIOGRAPHIC FINDINGS**

The most common findings in dogs with *H. canis* are mild anemia and mild thrombocytopenia. Dogs with marked parasitemia may have elevated CPK, elevated SAP, hypoalbuminemia, and leukocytosis, but dogs with low parasitemia are usually subclinical.

*H. americanum* infection generally causes severe disease and is associated with marked leukocytosis (typically in the range of 25,000-150,000/uL), non-regenerative anemia, elevated SAP, hypoalbuminemia, low BUN, and hypoglycemia. The anemia is generally associated with chronic disease. The elevation in SAP is thought to be secondary to periosteal inflammation associated with myositis. Hypoalbuminemia and low BUN are often the result of altered protein metabolism associated with the production of acute phase proteins and macroglobulins as a consequence of inflammation. Hypoglycemia may be spurious because of the increased white blood cell count. The clinicopathologic findings of H. americanum are so consistent that, when they are noted, they should prompt consideration of a muscle biopsy or PCR blood test to confirm the diagnosis. Radiography frequently demonstrates periosteal proliferation at sites of muscle attachment as a result of pyogranulomatous myositis and is also considered a good "screening test" for the disease.

**TREATMENT**

For *H. canis,* the recommended treatment is imidocarb dipropionate, 5 mg/kg SC or IM every 14 days until parasitemia clears. One or two injections are usually sufficient. Dogs with concurrent tick borne infections should receive doxycycline, and heavily parasitized dogs may require supportive care.

Treatment of *H. americanum* has been frustrating because no therapy is effective in eliminating the tissue stage of the organism. Palliative therapy with NSAIDs is often helpful in controlling fever and muscle pain, but definitive treatment requires anti-protozoal therapy. Beneficial response has been obtained with a 14 day course of trimethoprim sulfa (15 mg/kg q 12 h), clindamycin (10-15 mg/kg q 8 h), and pyrimethamine (0.25 mg/kg q 24 h) given concurrently. Unfortunately, relapse will occur within 1-6 months as merozoites are released from tissue schizonts and re-establish infection in the dog. To prevent maturation of merozoites that are being released, decoquinate, a quinolone anticoccidial agent, is used to prevent continued development of tissue stages of the organism following the combination treatment. Doses of 10-20 mg/kg (¼ - ½ teaspoon per 10 pounds) mixed in the food twice daily have been effective in preventing relapse and extending survival time in infected dogs. Long term treatment of 18 months to 2 years is recommended as well as aggressive tick control measures to prevent re-infection of susceptible dogs. Treatment should not be discontinued until the dog is asymptomatic and the PCR test is negative. Long term cure (>8 years with no clinical signs) has been achieved in dogs using combination therapy followed by long-term decoquinate.

## REFERENCES

[1] Ewing SA, et al. American canine hepatozoonosis. **Clin Microbiol Rev** 2003.16(4):688-697. [2] Macintire DK, et al. Hepatozoonosis in dogs: 22 cases (1989-94). **J Amer Vet Med Assoc**, 1997. 210:916-922 [3] Mathew JS, et al. Experimental transmission of *Hepatozoon americanum* Vincent-Johnson et al., 1997 to dogs by the Gulf Coast Tick, *Amblyoma maculatum* Koch. **Vet Parasitol** 1998.474:1-14. [4] Macintire DK, et al. Treatment of dogs infected with *Hepatozoon americanum:* 53 cases (1989-1998). **J Am Vet Med Assoc**, 2003. 218:77-82. [5] Cummings CA, et al. Characterization of stages of *Hepatozoon americanum* and parasitized canine host cells **Vet Path** 2005.42:788-796. [6] Vincent-Johnson NA, et al. A new Hepatozoon species from dogs: Description of the causative agent of canine hepatozoonosis in North America. **J Parasitol** 1997. 83:1165-1172. [7] Allen KE, et al. Diversity of Hepatozoon species in naturally infected dogs in the southern United States. **Vet Parasitol**, 2008. 154:220-225. [8] Baneth G, et al. Life cycle of *Hepatozoon canis* (Apicomplexa: adeleorina: Hepatozoidae) in the tick *Rhipicephelus sanguineus* and domestic dog (*Canis familiaris*) **J Parasitol 2007.** 93:283-299. [9] Li Y, et al. Diagnosis of canine *Hepatozoon spp* infections by quantitative PCR. **Vet Parasitol** 2008. 157:50-58. [10. Johnson] E, et al. Experimental transmission *of Hepatozoon americanum* to rodents. **Vet Parasitol** 2008. 151:164-169. [11] Johnson EM, et al. Field survey of rodents for Hepatozoon infections in an endemic focus of American canine hepatozoonosis. **Vet Parasitol**. 2007. 150:27-32. [12] Johnson EM, et al. Infectivity of *Hepatozoon americanum* cystozoites for a dog. **Vet Parasitol** 2008.154:148-150.

# THE COMPLETE BLOOD COUNT & NON-NEOPLASTIC HEMATOPATHOLOGY

**Joanne B. Messick, VMD, PhD, Diplomate ACVP**
**Ithaca, NY**

The complete blood count (CBC) is the most commonly ordered test in veterinary medicine. Mistakes are made interpreting hematologic data that cost time and money and could potentially jeopardize the health of your patient. In this presentation, several problem and/or less usual cases in hematology will be presented and the interpretation of the CBC findings discussed. The pathologic and clinical features, differential diagnoses for these non-neoplastic conditions, and diagnostic pitfalls of each case will be discussed.

In approaching the CBC concentrate on the following 5 parameters: 1) The Hemogram: Red Blood Cell Count (RBC), Hematocrit (HCT), and Hemoglobin (Hmb); 2) The Indices: Mean Corpuscular Volume (MCV) and Mean Corpuscular Hemoglobin Concentration (MCHC); 3) The Total White Blood Cell Count (tWBC) and Absolute Numbers of each Leukocyte (tWBC X % cell); 4) The Platelet count; and 5) Comments on the report! If all 5 of these parameters are normal, then significant blood disease is unlikely to be present. Nonetheless, scan the blood smear for "things that go bump in the night" and may not be associated with any CBC changes. The presence of microfilaria, low numbers of mast cells, blast cells or other cells not normally found in the peripheral circulation are often detected while scanning the smear. In addition, platelet clumps that may cause the automated platelet count to be falsely decreased can detected by scanning.

Anemia is present when a patient has a reduced amount of hemoglobin per unit volume of blood when compared with the reference population for that animal. Overhydration with parenteral fluids can cause reduced Hct, Hmb, RBC values, simulating anemia. Various factors can cause increases in the Hct, Hmb, and RBC count. Increased RBC mass (absolute polycythemia) or decreased plasma volume as seen with dehydration (relative polycythemia) cause high values. Splenic contraction may cause spurious high values. Absolute polycythemia may be primary, polycythemia vera (erythropoietin independent) or secondary to conditions such as reduced oxygen saturation, renal cyst, hydronephroma, and renal carcinoma (erythropoietin dependent).

There are several systems for classification of anemias and none of these is perfect. For pathophysiologic classification, all anemias are due either reduced or ineffective production of erythrocytes or excessive loss or destruction of erythrocytes. A third possibility, that of dilution due to an increased plasma volume, may also occur. Morphologic classification of anemia is made according to size (MCV) and Hmb concentration (MCHC). For a morphologic classification of anemia, the following diagnostic scheme is followed: 1) Anemias may be normocytic, macrocytic, or microcytic based on MCV values; 2) Anemias may be normochromic or hypochromic based on MCHC values; an absolute increase in MCHC does not occur except for the occasional elevations noted with spherocytosis. This system of classification is of greatest value for detecting an iron deficiency anemia (microcytic hypochromic), megaloblastic anemia (macrocytic normochromic), and a regenerative anemia (macrocytic hypochromic). A third system for classification of anemia depends on whether or not there is evidence of bone marrow response. Using this system, evidence of regeneration is based on the presence of polychromasia and reticulocytosis in the peripheral blood. To calculate the absolute number of reticulocytes the RBC count is multiplied by % reticulocytes; for dogs = >80,000 reticulocytes per µl and for cats = > 60,000 reticulocytes per µl are minimum number. These absolute values are the bare-bones minimums for indicating regeneration! Another calculation, which can provide information about regeneration is the corrected reticulocyte %. The % of reticulocytes for the patient is multiplied by the patient's PCV/45 for dogs or by PCV/38 for cats. A value >1.5 indicates regeneration, whereas a value between 1.0 and 1.5 is equivocal and <1% means NO regeneration.

The duration of the anemia will alter your interpretation of whether or not it's regenerative – remember, it takes 3 to 5 days before a regenerative response will be seen in the peripheral blood. With an acute bleed or acute hemolysis, the expected regenerative response may not yet be present. The interpretation of a non-regenerative anemia is based on the absence of polychromasia, reticulocytosis, and decreased or ineffective production of erythrocytes in the bone marrow. The absence of regeneration suggests either a production problems or insufficient time has elapsed for a response to be seen

In normal individuals, there are a majority of neutrophils, followed by lymphocytes, then monocytes, with a few eosinophils and basophils. The degree to which leukocyte numbers are increased and cell type(s) involved are an indication of the underlying etiology. General etiologies of leukocytosis include physiologic leukocytosis, reactive leukocytosis and leukemia. The stress leukon may have a leukocytosis, neutrophilia, lymphopenia, eosinopenia, and monocytosis, whereas the leukocytosis due to fear is associated with a neutrophilia and lymphocytosis. Dogs often display a stress leukon whereas fear is more common seen in cats and horses. Usually the leukocytosis due to stress in dog is less than 30,000 cells/ul and in the cat, less than 25,000 cells/ul. Bands must not exceed 1,000/ul and toxic changes should not be observed. Reactive leukocytosis is secondary to infection, neoplasia (not hematopoietic in origin), inflammation, etc. Acute or chronic inflammation may be associated with increased numbers of neutrophils and/or monocytes. An inflammatory leukon is present when leukocyte numbers exceed 30,000 cell/ul in the dog and 25,000 cell/ul in the cat, and/or when bands exceed 1,000 cell/ul in either species, and/or when comments indicated significant toxicity of the granulocytes. Hypersensitivity or allergic reactive leukocytosis associated with increased numbers of eosinophils. Basophils may also be increased. Viral or prolonged antigenic response may be associated with lymphocytosis. Leukemoid response is a severe leukocytosis that mimics a leukemia but is just reactive (you must find the underlying cause to distinguish whether it is a leukemoid response or a chronic leukemia!!). The following are characteristics of leukemoid response: 1) leukocytosis does not persistent over months; 2) response to appropriate therapy; and 3) able to find the underlying cause for the massive cell elevations (ie. If neutrophils are markedly elevated, you are looking for a nidus of inflammation such as pyelonephritis, pneumonia, etc., whereas if eosinophils are massively elevated, you are looking for a reason that might explain why these cells are increased such as mast cell tumor, lymphoma, or other tumor type (autoimmune, infectious, and hypersensitivity or allergic disorders usually don't cause massive eosinophilias).

When peripheral consumption or destruction exceeds the bone marrows ability to respond, a leukopenia may develop. Overwhelming infection and marrow failure or toxicity are mechanisms associated with leukopenia.

Primary thrombocytosis is due to a neoplastic proliferation (essential thrombocythemia), whereas secondary or reactive thrombocytosis has an underly cause such as bleeding, infection or neoplasia etc. The mechanisms of thrombocytopenia may include destructive/consumption thrombocytopenia- e.g. ITP and DIC; loss/sequestration thrombocytopeia – e.g. blood loss, endotoxemia, hypothermia, hypersplenism or sequestration in tumor; marrow failure or decreased production thrombocytopenia - usually involves other blood cells as well (ie there is a pancytopenia) and may be due to etiologies such as myelophthisis, myelofibrosis, drugs, and ehrlichiosis. The presence of macroplatelets may indicate a "regenerative" thrombocytopenia. Platelet estimate – for every platelet in a 100X/oil immersion field multiply by 20,000. (ie. 6 platelets/100X represents approximately 120,000 platelets/ul). This is NOT an automated count, it's an estimate! Make your estimate in 5 to 10 100X fields and take an average and remember platelet clumping will also falsely decrease estimates.

## DIAGNOSIS & DIFFERENTIAL DIAGNOSIS OF ABNORMAL BLOOD SMEARS

**Anne Barger, DVM, MS, Diplomate ACVP**
**Urbana, IL**

Examination of the peripheral blood smear begins on low power. The feathered edge should be scanned for platelet clumps, large cells and microfilaria. The body of the smear should be scanned for proteinaceous background material, identification of the monolayer and again, large cells and microfilaria. A white blood cell estimate can also be performed at low power. Getting an accurate number at this level is challenging so the emphasis should be placed on whether the white blood cell count is increased, decreased or adequate. A rough estimate can be obtained by counting the cells per field and multiplying by 100-150. Within the monolayer, examination of the red blood cell density and red blood cell arrangement is important. Both agglutination and rouleaux can be observed at low power and may actually be missed at higher magnification. The red blood cell density is important because a decreased density is consistent with anemia. Red blood cells should be evenly spaced from each other and almost touching. Increased space between the red blood cells indicates anemia.

At higher power (40x, 50x or 100x) more specific information about red blood cells, white blood cells and platelets can be evaluated. Red blood cell morphology should be evaluated at higher power. Different morphologic changes can indicate very different disease states in the patient or even artifact. Polychromasia indicates erythrocytes which are grayish-blue in color due to the presence of RNA, ribosomes and other organelles. Reticulocytes are often polychromatophilic. All polychromatophils are reticulocytes but not all reticulocytes are polychromatophilic. Anisocytosis indicates variation in cell size and may indicate the presence of small or large cells. Reticulocytes are often larger than mature erythrocytes so they are described as macrocytes. Other causes of macrocytosis include feline leukemia virus, folic acid deficiency (uncommon in animals) and macrocytosis of toy and miniature poodles. Microcytosis or small red blood cells can be observed with several different disease processes, the most common of which is iron deficiency. Due to the impaired heme synthesis, the red blood cells undergo an additional division. Microcytosis will often precede hypochromasia. Diseases associated with iron deficiency or iron misuse include chronic blood loss, dietary deficiency in neonates and portosystemic shunts. Other causes of microcytes include hereditary microcytosis (Akita and Shiba Inu) and familial dyserythropoiesis of English Springer Spaniels[1]. Hypochromasia can also be observed in patients with iron deficiency and indicates a decrease in hemoglobin content. If severe enough, this may result in a decrease in the mean corpuscular hemoglobin concentration or MCHC. Membrane abnormalities or poikilocytosis can result in various morphological abnormalities. Spherocytes are red blood cells which have lost their biconcave shape. On a blood smear, they lack central pallor and they can appear a little smaller. They are commonly associated with immune-mediated hemolytic anemias but have also been reported in dogs with spectrin deficiency[2]. Acanthocytes are spiculated red cells with irregularly spaced spicules. These cells can be observed in patients with liver disease, portosystemic shunts or high-cholesterol diets. In contrast echinocytes are also spiculated red blood cells but have regularly spaced spicules. These are generally considered an artifactual change however they have been observed in dogs with glomerulonephritis, lymphoma, hemangiosarcoma and other neoplasms, immune mediated hemolytic anemia, pyruvate kinase deficiency, rattlesnake envenomation and doxorubicin toxicosis among others[3]. Cats likely have echinocytes with many of these diseases as well but are specifically reported in the literature associated with chronic doxorubicin administration. Schistocytes are red blood cell fragments usually resulting from direct physical damage to the erythrocyte secondary to vascular abnormalities or turbulent blood flow. The shape of the fragments may vary from pointed or triangular to spiculated. Microangiopathic fragmentation has been described in dogs in a number of different disorders including disseminated intravascular coagulation (DIC), glomerulonephritis, hemangiosarcoma, myelofibrosis, dyserythropoiesis and chronic doxorubicin toxicosis. Other poikilocytes include target cells or codocytes, which can be observed with iron deficiency, liver disease or as an artifact or dacryocytes (tear-drop shaped red blood cells) which have been reported in dogs with myelodysplastic disease and myelofibrosis. Keratocytes have one or two spicules and are thought to represent blister cells which have opened, whereas eccentrocytes have a peripheralization of their hemoglobin.

The erythrocytes should also be examined for inclusions. Nucleated erythrocytes are commonly observed in anemic patients as part of the regenerative response. However in the absence of anemia or polychromasia, bone marrow and splenic disorders should be considered. Howell-Jolly bodies are nuclear remnants and appear as a small basophilic inclusion. These structures can accompany a regenerative response. Basophilic stippling are part of the regenerative response to anemia in some species such as cattle, however in cats and dogs, this observation is less common. In these species, lead toxicity should be considered especially if the patient is not anemic or has increased numbers of nucleated red blood cells. Oxidative damage can result in a number of morphologic changes to the red blood cells including the formation of Heinz bodies. Heinz bodies are small aggregates of oxidized hemoglobin which can occur as hemoglobinized projections extending from the cell or as pale areas within the cell but close to the cell membrane. Staining the sample with new methylene blue or other vital stains, stains the Heinz body pale blue, allowing them to become much more obvious. Cats are more susceptible to Heinz body formation because of their hemoglobin structure; therefore a healthy cat can have low numbers of Heinz bodies in circulation. Many toxins have been reported to cause Heinz bodies. Hemic parasites also appear as red blood cell inclusions. In dogs two forms of babesial organisms can be identified; *Babesia gibsoni* is a small, signet ring shaped organism with an eccentrically placed nucleus and *Babesia canis* is a piroplasmic organism, much larger than *B. gibsoni*, generally with two organisms/cell. In cats, *Mycoplasma hemofelis* and *M. hemominutum* are common extracellular parasites. These organisms appear as small cocci or chains of cocci on the erythrocyte membrane. *Cytauxzoon felis* is an intraerythrocytic parasite that is regionally specific and appears as a small, signet ring shaped organism, similar to *B. gibsoni*.

A differential count of 100-200 cells should be performed at high power, in the monolayer of the blood smear. The white blood cells observed in the peripheral blood include neutrophils, eosinophils, basophils, monocytes, lymphocytes and less commonly, plasma cells, mast cells and blasts. In dogs and cats, the neutrophil is the predominant white blood cell. Neutrophils are granulocytes and originate from the same precursor cell as eosinophils and basophils. Neutrophils have a multilobulated nucleus with pale, basophilic cytoplasm. During an inflammatory response, toxic changes may occur which can change the appearance of the cytoplasm. Basophilic cytoplasm, foamy cytoplasm, döhle bodies and azurophilic granules are common changes that can occur with inflammation. Additionally, if there is increased tissue demand for white blood cells, particularly neutrophils, bands and metamyelocytes or hyposegmented neutrophils may be observed in circulation. Bands have a U shaped nucleus with no indentation and metamyelocytes have a reniform shaped nucleus. Neutrophils can also have a hypersegmented nucleus. Hypersegmented neutrophils are generally thought of as older cells and can be observed in patients with chronic inflammation, myeloproliferative disorders or receiving glucocorticoids. Eosinophils are also granulocytes and have a lobulated nucleus and contain numerous eosinophilic granules. In dogs these granules are round and in cats they are rod shaped. Eosinophilia can be observed with parasitic infestation, allergic or hypersensitivity responses and marked eosinophilia (>10,000/µl) can be seen in paraneoplastic responses, hypereosinophilic syndrome, eosinophilic pulmonary granulomas, hepatozoonosis and eosinophilic leukemia[4]. Basophils are not commonly observed in peripheral blood but can accompany eosinophils. In cats with eosinophilia and basophilia, heartworm infestation should be considered. Basophils have a lobulated nucleus with grayish-blue cytoplasm. In dogs, few basophilic granules are observed. In cats, basophils are filled with pale, lavender granules which are difficult to visualize with Dif-Quik type stains.

Lymphocytes are not granulocytes. These cells are small and round with a scant rim of basophilic cytoplasm and a round nucleus which fills the cytoplasm. Variable lymphocyte morphology can be observed in peripheral blood. Reactive lymphocytes are seen with antigenic stimulation. These cells are larger and have increased amounts of deeply basophilic cytoplasm. Observation of "atypical" lymphocytes is a common notation on CBCs. These cells have an increased amount of basophilic cytoplasm and are slightly larger than classic lymphocytes. The clinical significance of these lymphocytes is uncertain but if they are present in low numbers, they likely represent a reactive population. However if the patient has a significant lymphocytosis (>20,000/µl) and most or all of the lymphocytes appear atypical, a neoplastic proliferation should be considered. Monocytes are larger than lymphocytes and have a moderate rim of basophilic cytoplasm, often containing discrete cytoplasmic vacuoles and variably shaped nucleus from round to reniform to lobulated. These cells are often part of an inflammatory response but in dogs, increased monocytes can be observed as part of a stress response.

Other white blood cells that can be observed in peripheral blood include mast cells, plasma cells and blasts. Mast cells are round cells with a round nucleus containing numerous metachromatic granules. These cells can be seen in peripheral blood with a number of different diseases including mast cell neoplasia, parvoviral enteritis, fibrinous pericarditis, bacterial peritonitis, aspiration pneumonia, acute pancreatic necrosis, IMHA, allergic dermatitis and gastric torsion[5]. Presence of significant mastocytemia (>10,000/µl) is more consistent with a neoplastic proliferation of mast cells such as mast cell tumor, systemic mastocytosis or mast cell leukemia. Plasma cells are infrequently observed in the peripheral blood but can be observed as part of an immune response, particularly post-vaccination. Blasts in the peripheral blood are a fairly common finding especially in patients with lymphoma or other myeloproliferative or lymphoproliferative disorders disorders. Blasts are larger cells with varying amounts of basophilic cytoplasm, round nuclei with a stippled chromatin pattern and prominent nucleoli. Morphologically, we cannot consistently distinguish a large lymphocyte from myeloblast or myelomonocytic precursor[6]. Immunophenotyping of these cells is essential. Acute leukemias can have a marked leukocytosis, often >100,000 cells/µl with a predominance of neoplastic cells. In patients with lymphoma, there may only low numbers of neoplastic cells in circulation, sometimes representing 1-2% of the nucleated cell population. Identification of these cells in the peripheral blood does not necessarily indicate neoplastic infiltration of the marrow however the recognition of the cells in circulation however is very important.

Platelets are cytoplasmic fragments of megakaryocytes. They can be estimated on high power (100x). In the monolayer, platelets should be counted in 10 fields and the average is calculated. This value is multiplied by 15,000-20,000. The blood smear should also be evaluated at low power for platelet clumps. The presence of platelet clumps will decrease automated platelet counts as well as the estimate. Generally, platelets are smaller than red blood cells. They are a basophilic disc of cytoplasm with varying numbers of granules. However, in thrombocytopenic patients, giant platelets can be observed and indicate that the marrow is responding or attempting to respond to the thrombocytopenia.

In conclusion, examination of the peripheral blood smear is an essential component of a CBC. Abundant information can be gathered from the blood smear, including subtle changes such as toxic changes in the neutrophils, hemic parasites in the erythrocytes or low numbers of neoplastic cells or mast cells that may help to make a diagnosis.

## REFERENCES

[1]Allen L,et al.**J Am Vet Med Assoc** 1999;214:218-221. [2]Slappendel RJ, et al. **J Vet Intern Med** 2005;19:187-192.[3]Weiss DJ, et al.**Vet Clin Pathol** 1990;19:114-118.[4]Marchetti v, et al. **Vet Clin Pathol** 2005 Sep;34(3):259-63. [5]McManus PM. **J Am Vet Med Assoc** 1999;215:355-357.[6]Gauthier MJ. **J Vet Intern Med** 2005;19:193-199.

# DISTINGUISHING LEUKEMOID REACTION FROM CHRONIC LEUKEMIA

**Anne Barger, DVM, MS Diplomate ACVP**
Urbana, IL

## INTRODUCTION

In human medicine the definition of a leukemoid reaction is a persistent neutrophilic leukocytosis above 50,000/µl when the cause is something other than leukemia. We use a similar definition in veterinary medicine however we often use this term to also include extreme lymphocytosis and eosinophilia in addition to neutrophils. The most challenging task is to differentiate this extreme leukocytosis from a chronic leukemia, most often a chronic myelogenous leukemia (CML).

### WHO Classification of Chronic Myeloproliferative Diseases

The 2001 WHO classification scheme recognizes four separate groups of CML: chronic myeloproliferative diseases (CMPDs), myelodysplastic syndrome (MDS), MDS/MPD and mast cell disease. The focus of this review will be on the CMPDs group because it includes CML, chronic neutrophilic leukemia (CNL) and chronic eosinophilic leukemia/hypereosinophilic syndrome. The shared feature of these diseases is the clonal myeloproliferation that is devoid of dyserythropoiesis, granulocytic dysplasia or monocytosis[1]. Diagnosis of CML requires detection of either the Philadelphia chromosome or the bcr/abl fusion gene. There are three phases of this disease: chronic, aggressive or accelerated and finally, the blast phase. The chronic phase is the most challenging to diagnose because the patients are often asymptomatic. A neutrophilia, often >170,000/µl is found on routine blood work. Interestingly, the neutrophilia is always accompanied by a basophilia and occasionally an eosinophilia. Bone marrow aspiration reveals marked granulopoiesis with hyperplasia of all three granulocytic cell lines[2]. The aggressive or accelerated phase consists of >10% blasts in the bone marrow and may result in few blasts observed in circulation. The blast phase consists of >20% blasts, more consistent with a typical acute leukemia. CNL is a genetically distinct disease considered a diagnosis of exclusion and is challenging to differentiate from a leukemoid response (LR). These patients have a mature neutrophilia with a left shift limited to bands as well as extramedullary hematopoiesis of the liver and spleen[3]. However, these cells are recognized as a clonal expansion of neutrophils resistant to apoptosis. Diagnosis of this neoplasm relies on the exclusion of CML and leukemoid reactions. There is no definite molecular marker like the bcr/abl translocation to distinguish CNL so clonality studies are used to differentiate CNL from leukemoid reactions. In people, the human androgen receptor gene assay (HUMARA) is used to identify clonality[4]. A combination of cytogenetics, karyotyping, RT-PCR, cytochemistry and ICC are used to definitively diagnose chronic myeloproliferative disorders[2].

### Diagnosis of Leukemoid Reactions in People

Major causes of LR include infection (*Clostridium difficile*, tuberculosis, shigella), intoxication (ethylene glycol), paraneoplastic (carcinoma of lung, GI, bladder, Hodgkin's lymphoma, melanoma), severe hemorrhage and acute hemolysis[5]. The primary method used to differentiate LR from CMPDs is identification of the underlying cause which is often obvious, so clear history, thorough physical exam and diagnostic imaging are also very helpful. However, additional diagnostic tests may be utilized and include a white blood cell count with differential, blood smear, leukocyte alkaline phosphatase (LAP) score, vitamin $B_{12}$ levels, bone marrow aspiration, cytogenetic studies, immunophenotyping and serum levels of growth factors (table 1)[5]. The differential count is useful because CNL patients often have a left shift with bands but rarely have metamyelocytes and myelocytes. Human neutrophils normally have leukocyte alkaline phosphatase and activated neutrophils have even higher levels, so LR results in a markedly increase in the LAP score. Vitamin $B_{12}$ is elevated in CNL and CML however they can elevate in LR so this test is not consistently diagnostic. LR patients have a hyperplastic bone marrow with marked myeloid hyperplasia with orderly and complete maturation and normal morphology. Patients with CML will also have a basophilia, eosinophilia, sometimes megakaryocyte hyperplasia and monocytosis and reticulin fibrosis. Marrow from patients with CNL appears very similar to LR marrow with a mild increase in reticulin fibrosis. Immunophenotyping of these three processes is not helpful because they have a very similar phenotype. There does not appear to be the expression of a particular protein which will allow differentiation. However there are distinct cytogenetic differences. As mentioned earlier, the Philadelphia chromosome and bcr/abl translocation are both reported with CML. As expected CNL and CML are both clonal and LR is polyclonal. In differentiating CNL and LR, there is no one test that is definitive. It takes a panel of tests, thorough history and physical exam to differentiate these two processes. In people, CML is easy to differentiate due to the distinguishing cytogenetic abnormalities.

**Table 1.** Diagnostic testing to differentiate LR, CNL and CML

| | LR | CNL | CML |
|---|---|---|---|
| Blood smear examination | Left shift with bands, metamyelocytes and myelocytes | Left shift but bands only | Immature cells with basophils and eosinophils |
| LAP score | High | High | Low |
| Serum vitamin $B_{12}$ | Variable | High | High |
| Bone marrow | Myeloid hyperplasia with orderly maturation and normal morphology | Similar to LR with slight increase in reticulin | Basophilia, eosinophilia, monocytosis, reticulin fibrosis |
| Cytogenetic analysis | No cytogenetic abnormalities | 37% of cases have cytogenetic abnormalities | bcr/abl translocation |
| Clonality | Polyclonal | Monoclonal | Monoclonal |

Adapted from Sakka V, et al. European J of Int Med 2006;17:394-398

**CML in Dogs & Cats**

CML in veterinary species include chronic myelogenous or granulocytic leukemia, chronic myelomonocytic leukemia, chronic monocytic leukemia, chronic eosinophilic leukemia/hypereosinophilic syndrome, polycythemia vera and essential thrombocytosis. Bone marrow evaluation of these patients reveals <30% blasts however a blast-phase for these leukemias is likely. The M:E ratio in these patients can be as high as 36:1[6]. Dysplastic changes may not be present in the peripheral blood, though they have been reported[6]. Clinical signs are often nonspecific but include lethargy, pyrexia, hepatosplenomegaly if the neoplastic cells infiltrate liver and spleen and mild to moderate lymphadenopathy. Chronic leukemias are thought of as a diagnosis of exclusion. There is not one diagnostic test but more a panel of tests including examination of the peripheral blood smear, immunophenotyping, cytochemical staining, bone marrow aspiration and bone marrow core biopsy, and biopsy or aspiration of abdominal organs and lymph nodes.

**Leukemoid Reaction in Dogs & Cats**

Extreme leukocytosis can observed with many different diseases including infectious and inflammatory diseases, paraneoplastic disorders, genetic disorders such as canine leukocyte adhesion deficiency, immune-mediated diseases and tissue necrosis. Specific diseases reported to cause extreme leukocytosis in dogs, include immune-mediated diseases such as IMHA and ITP, inflammatory disease such as cholangiohepatitis, pyelonephritis, pyometra and pneumonia and tumors including lymphoma, mammary adenocarcinoma and bronchoalveolar carcinoma[7]. In cats, inflammatory diseases such as pyothorax, abscess, pyelonephritis and enteritis, immune mediated disease including IMHA and glomerulonephritis, tissue necrosis from pancreatitis and thrombosis and neoplasms including lymphoma, mast cell tumors, and squamous cell carcinoma all have been reported to cause extreme leukocytosis or LR[8]. Leukocyte adhesion deficiency is an autosomal recessive disorder that has been reported in Irish Setters. These dogs often have recurrent bacterial infections. Signs manifest shortly after birth and include omphalitis and severe bacterial infections resulting from a single point mutation in the CD18 gene[9]. These dogs have neutrophil counts which often exceed 100,000/μl. Paraneoplastic neutrophilia and eosinophilia have been reported with lymphoma[10] and can be difficult to differentiate from hypereosinophilic syndrome or eosinophilic leukemia. Similar to people, a thorough history and physical exam can be helpful in identifying a source of inflammation. In many cases, the source of inflammation is obvious and can be easily distinguished however there are elusive cases which will be difficult to diagnose. Cytogenetic abnormalities are not common in animals and the Philadelphia chromosome has not been identified making diagnosis of CML vs. LR even more challenging[2]. Toxic changes in neutrophils are more common in LR and cytochemical staining can be useful as well. Canine neutrophils are normally negative for alkaline phosphatase activity, however neoplastic myeloid precursors have been reported to stain positive for ALP. This test may also be useful in differentiating LR from CML. The best diagnostic test for differentiating LR from CNL appears to be clonality and may be an area of future work in diagnosis of chronic leukemias in veterinary medicine.

**CONCLUSION**

At this time there is no clear answer however, now that the canine genome has been mapped, cytogenetic testing can become more readily available. The panel of tests mentioned above combined with thorough physical examination currently appear to be the best diagnostic tools.

**REFERENCES**

[1]Tefferi A, et al. **Leukemia** 2008;22:14-22. [2]McManus PM. **Vet Clin Pathol** 2005;34:189-212. 3Reilly JT. Br J Haematol 2002;116:10-18. [4]Böhm J, et al. **J Clin Pathol** 2003;56:292-295. [5]Sakka V, et al. **Eur j of Int med** 2006;17:394-98. [6]Fine DM, et al. **J Am Vet Med Assoc** 1999;214:1809-1812. [7]Lucroy MD, et al. **J Am Vet Med Assoc** 1999;214:805-807. [8]Lucroy MD, et al. **J Am Vet Med Assoc** 2001;218:736-739. [9]Creevy KE, et al. **Vet Immunol Immunopath** 2003;94:11-22. [10]Marchetti V, et al. **Vet Clin Pathol** 2005;34:259-263.

# INTERPRETATION & EXAMPLES OF ATYPICAL NON-LYMPHOID ONCO-CYTOLOGY

**Craig A. Thompson, DVM, DACVP**
**West Lafayette, IN**

## INTRODUCTION

As cytology becomes more a more available, more veterinarians are become proficient at obtaining, processing and evaluating samples. With this rise in interest and practice, it has become necessary to be able to deal with a variety of tissues and samples. Furthermore, with CT scans, ultrasound and other imaging modalities becoming readily available, the variety of samples available to interpret has likewise become more diverse. Therefore, today's cytopathologist can and will be faced with a number of samples previously unheard of and as such, should be prepared to deal with samples from essentially any tissue in the body. This session will look at samples that in the author's experience are rare, unusual or tricky to interpret (often with undesirable results when misinterpreted). There are dozens if not hundreds of samples that fit these criteria; therefore, only a few will be chosen to cover, based on experience as well as the experience of several oncologists. Each example will be described, with case-based examples used to illustrate the points raised in this manuscript. Unfortunately, a picture is indeed with a thousand words, so attendance of the session or use of references and cytology atlases will be paramount to gain the full experience of this session

### Primative Neuroectodermal Tumors (PNET)

These tumors arise from germinal neuroepithelial cells and can present as embryonal / congenital to juvenile to adult neoplasms.[1] Included in this category are neoplasms such as medulloblastomas, neoroblastomas, esthesioneuroblastomas and nephroblastomas, amongst others. The sample presented to the cytopathologist depends on the site of the neoplasm; however, aspirates, impression smears and fluid samples (e.g. cerebrospinal fluid) are all possible sources of cells from these tumors. Although varied in clinical presentation, the cytomorphology of the tissues are surprisingly similar. The cells seen are generally small to medium in size and round in shape. They often contain a scant to small amount of cytoplasm surrounding a round nucleus. Cell margins are often discrete; however, can be somewhat wispy. In general, these cells are often mistaken for small lymphocytes if a PNET is not on the differential diagnosis list. Features that can be useful in distinguishing a PNET from lymphoma include the arrangement of the cells, clinical features and immunocytochemical staining properties. Depending on the specific PNET, there will often be a specific arrangement noted, including crude glomerular formations (nephroblastoma), sheets and chords, and especially rosette formation. All of these cohesive arrangements are never seen with lymphoma, as it is exclusively a discrete cell and does not form cohesive arrangements. If additional verification is needed to confirm a PNET, immunocytochemical stains can be utilized. Lymphoma should stain with either CD3 or CD29a, a T-cell and B-cell marker, respectively; whereas PNETs should not stain positively for these stains. Other specific staining characteristics vary depending on the specific tumor. An additional clue that the author looks for, but puts minimal weight on is the presence of lymphoglandular (LG) bodies. LG bodies are cytoplasmic fragments from lysed and/or divided cells.[3] Although commonly associated with lymphocytes, these structures can be seen with any rapidly dividing tissues. In the author's experience, they can be used as a crude indicator that the cells present are more likely lymphocytes.

### Endocrine - Neuroendocrine / Naked Nuclei Tumors

The naked nuclei / neuroendocrine category of tissue and thus neoplasms is a wide group that includes many different tissue Types. Included in this category of cells includes: adrenal gland (cortex and medulla), beta cells from the Islets of Langerhans, cells from chemodectomas, thyroid, parathyroid, pituitary and from carcinoids. One of the cardinal rules employed when evaluating cytology is to never evaluate cells that do not have an intact nucleus or intact cytoplasmic borders. This tissue type stretches this cardinal rule to the limit and makes identifying these neoplasms a challenge. The cytomorphology of these tissue types is a typically monotonous population of round nuclei seen in a pool of cytoplasm with rare, if any, intact cells.[3] The challenge lies not only in be astute enough to identify these lysed cells are pertinent, but also to recognize them as malignant, as they show very few criteria of malignancy. This challenge is particularly steep in lymph nodes, where these cells are often surrounded by other cells with round nuclei. Tricks that can be used to identify these lesions include using the clinical presentation, lesion location, as well as a few key cytologic features. Being epithelial, these cells should be found in close proximity to one another, even when lysed. In addition, when other cells, such as lymphocytes, lyse, the nuclei often begin to break apart, with their chromatin patterns becoming open. Although lysed, the nuclei of these cells remain relatively intact. Finally, occasionally these cells will be arranged into sheets or acini (e.g. thyroid tissue), which can also declare the tissue of origin. As mentioned before, the monotony of these cells presents problems in identifying a lesion as malignant and as such, sometimes cytomorphology alone is insufficient to make this determination.

### Apocrine Gland (Anal Sac) Adenocarcinoma vs. Hepatoid Tumors

A common quandary that practitioners face is how to deal with masses in the peri-anal region. Two commonly seen causes of masses are the apocrine gland (Anal Sac) and perianal glands. Cytologic samples from these two lesions are often distinguishable; however, on occasion there can be considerable cross-over.[4] In addition to the signalment, clinical findings and presence or absence of a humoral hypercalcemia of malignancy, the cytomorphology of the cells present can be used to distinguish the two. The key feature of the anal sac adenocarcinoma is the tight clusters of small round cells that have a high nucleus to cytoplasm (N:C) ratio. These cells generally have a uniformly high N:C; however, are extremely monomorphic and monotonous. Conversely, cells from hepatoid tumors are generally larger, have angular margins and a low N:C ratio. In addition to the hepatoid cells, there are frequently many reserve cells present around the clusters of hepatoid cells and scattered about the smear. The confusion arises when the reserve cells become more numerous or predominate the sample. Features that the author finds useful to distinguish these two biologically dichotomous lesions include the arrangement of the small cells. Reserve cells tend to have distinct margins and a small amount of blue cytoplasm. Cells from apocrine gland adenocarcinomas typically have indistinct margins, may arrange themselves into acini and are packed much tighter than reserve cells from hepatoid tumors. In addition, using other clinical findings as support for cytologic observations can be useful to build a more definitive case.

**Granulomatous / Pyogranulomatous Inflammation vs. Malignant Histiocytic Diseases**

Another common problem facing today's cytopathologist is discerning inflammatory lesions involving macrophages with malignant neoplasia involving histiocytes. Being able to confidently define the lesion can clearly be helpful in choosing treatment. The primary feature used by the author to distinguish the two is the presence of other inflammatory component other than macrophages. In mammals it is uncommon to seen an inflammatory lesion composed solely of macrophages. Neutrophils, lymphocytes, plasma cells, etc. are often present, depending on the etiologic agent at work. When the neutrophils are degenerate (i.e. exhibit karyolysis, karyorrhexis and/or pyknosis), this lends even more support to an inflammatory lesion. In addition, the identification of specialized macrophages such as epithelioid macrophages and/or multinucleated giant cells helps identify a lesion as inflammatory rather than neoplastic. Finally, if an etiologic agent can be identified, a diagnosis can be made. Common agents seen with this type of inflammation include: *Mycobacteria sp.*, fungal agents (e.g. Blastomycosis, *Coccidioides immitis*, opportunistic infections, etc.), foreign bodies

**Plasma Cell Neoplasia vs. Osteosarcoma (OSA)**

Differentiation of these two somewhat common neoplasms can be extremely difficult, especially since OSA can present in an extraskeletal fashion, and plasma cells neoplasms can certainly present with bony lesions (i.e. multiple myeloma). These cells share many common characteristics; however, a few features may help declare the true tissue of origin. If a plasma cell tumor is suspected, a serum (and/or urine) protein electrophoresis can be extremely useful in quickly ending the debate. This test can and should be run, even without a hyperglobulinemia. Cytologic features to look for include the presence of immunoglobulin / amyloid associated with the neoplastic cells. This material is a grainy pink material, whereas osteoid is often smooth to somewhat fibrillar in nature. Another useful, rapid test is to use alkaline phosphatase staining to help differentiate the two.[5] Although somewhat slower, immunocytochemical staining using B-cell markers and/or immunoglobulin can also be employed.6

**REFERENCES**

[1]Koestner A, et al. **Tumors in Domestic Animals** 4th ed.2002; 714. [2]Raskin RE. **Atlas of Canine and Feline Cytology.** 2001; 109. [3]Raskin RE. **Atlas of Canine and Feline Cytology**. 2001; 31. [4]Raskin RE. **Atlas of Canine and Feline Cytology**. 2001; 65. [5]Barger A, et al. **Vet pathol** 2005;42(2): 161. [6]Ramos-Vara, et al. **Vet Pathol** 2007 ; 44(6) : 875.

## NON-LYMPHOID ONCO-CYTOLOGY; THE NEW NOMENCLATURE FOR OLD LESIONS

**Craig A. Thompson, DVM, DACVP**
**West Lafayette, IN**

### INTRODUCTION

Over the years, diagnostic modalities have evolved. Initially, gross observations were adequate to describe lesions and formulate names and assign biologic activity to these names. As technology advanced, so did the resolution of these lesions. Histologic examination became available to supplement gross observations. In its beginning, these microscopic descriptions were based on polychromatic stains, with hematoxylin and eosin (H&E) being the primary stain used, which is still the workhorse today. Technology progressed and transmission electron microscopy was utilized to examine the ultrastructure of cells and tissue, further revealing differences. More recently, immunohistochemical stains have been developed to identify markers to a number of features that can be used to classify and sub-classify tissues. As all of these and other methods have been employed, tissues and lesions have been classified and reclassified. This session will discuss a few of the more recent additions / changes to the nomenclature of lesions. Unfortunately, a picture is indeed with a thousand words, therefore attendance of the session or use of references and cytology atlases will be paramount to gain the full experience of this session

### Hemangiopericytoma (HEP) / Peripheral Nerve Sheath Tumor (PNST) vs. Perivascular Wall Tumors (PWT)

In years past, lesions that were evaluated cytologically and described as HPCs or PNSTs presented as highly cellular samples with numerous mesenchymal cells that had wispy blue cytoplasm, were often associated with capillaries and showed moderate to marked criteria of malignancy. The lesion location, and presence or absence of capillaries often helped distinguish these lesions, although biologic behavior was described as similar. Recently, these lesions have come under scrutiny and re-examined. Cytologic specimens were compared to histologic counterparts supplemented with immunohistochemical stains. It was discovered that many lesions previously described as HEPs were in fact other lesions such as myopericytomas, angioleimyomas and angioleiomyosarcomas in addition to true HEPs.[1] Armed with this new knowledge, the cytopathologist should bear in mind a broader range of differential diagnoses to offer the clinician and consider the use of histologic examination in addition to immunohistochemical stains.

### Histiocytic Diseases

Few diseases have generated more debate and controversy than the histiocytic diseases. This section will cover the different name changes and cytologic features of these lesions in a case based format.

### Follicular Neoplasms

Similar to the histiocytic diseases, the follicular tumors have likewise undergone some suggested name changes including basal cell tumors. This section will cover the different name changes and cytologic features of these lesions in a case based format.

### REFERENCES

[1]Avallone G, et al. **Vet Pathol** 2007; 44(5) : 607. [2]Valli et al. **Veterinary Comparative Hematopathology** 2007. 505.

# OCULAR CYTOLOGY OF INFLAMMATORY, CYSTIC, TRAUMATIC & NEOPLASTIC LESIONS IN CATS & DOGS

**Rose E. Raskin, DVM, PhD, DACVP**
**West Lafayette, IN**

## INTRODUCTION

Cytologic examination of the eye and surrounding structures is frequently helpful in determining general categories of pathology prior to performing more invasive or expensive procedures. Diagnoses may involve one or more of the following cytodiagnostic groups. These categories can apply to the various anatomic sites of the eye.

**General Cytodiagnostic Groups for Ocular Cytology**

- Normal tissue
- Cystic or hyperplastic tissue
- Inflammation
- Neoplasia
- Response to tissue injury

The presentation will discuss the cytologic features of the selected abnormalities listed below plus others as time permits.

## EYELID

### Cyst

*Chalazion* or meibomian cyst refers to the granuloma formed from the irritating effects of ruptured sebaceous secretions caused by inflammation or tumors. The leakage of meibomian cyst material incites an inflammatory response similar to a foreign body reaction. In this situation, numerous foamy macrophages, few giant cells, neutrophils, lymphocytes, amorphous debris, and sebaceous epithelium are present.

## CONJUNCTIVAE

The normal palpebral conjunctiva is lined by a pseudostratified columnar epithelium containing goblet cells. Goblet cells are characterized by a columnar shape, eccentric nucleus, and pale foamy cytoplasm. The bulbar conjunctiva consists of sheets of stratified nonkeratinized squamous epithelium, mostly intermediate and superficial with less than 1% goblet cells as detected by impression cytology.[1] There are no significant numbers of inflammatory cells; however, mucosa-associated lymphoid tissue is located below the squamous layer of the palpebral conjunctiva near the fornix where goblet cells are absent.

### Goblet cell hyperplasia

Conditions such as keratoconjunctivitis sicca, vitamin A deficiency, chronic disease, and trauma from mechanical irritants result in increased cell numbers sometimes with evidence of metaplasia. These specimens contain many keratinized epithelial cells and goblet cells. Increased pigmentation of the epithelium may accompany the lesion displayed as numerous fine black-green melanin cytoplasmic granules. Hyperplastic epithelium should be recognized by an increased nuclear to cytoplasmic ratio.

### Herpes conjunctivitis

Neutrophilic conjunctivitis may be associated with infectious agents such as bacteria and viruses as well as noninfectious causes. One should suspect a bacterial origin if degenerative changes are present. With viruses such as feline herpesvirus and chronic canine distemper virus, nondegenerate neutrophils predominate along with reactive epithelium that often appear as small deeply basophilic basal cells. Multinucleated epithelial cells were found in approximately half of the cases with feline herpesvirus (FHV-1) infection diagnosed by culture or immunofluorescent assay.[2] Additionally, eosinophils have been associated with FHV-1 infection .[3] Intranuclear inclusions in FHV-1 infection are occasionally encountered in conjunctival smears but are more readily seen in histologic sections. A polymerase chain reaction (PCR) test has greater sensitivity for detection of the virus than other tests.[4]

### Chlamydiosis

The infectious agent, *Chlamydophila felis,* affects conjunctival tissue producing epithelial inclusions which present as variably sized discrete basophilic to bright magenta bodies in the cytoplasm of epithelium within first 2 weeks of infection. Nondegenerate neutrophils and small lymphocytes may accompany the lesion. Small elementary bodies (0.3 $\mu$m) are shed which infect new cells and grow into larger (0.5 to 1.5 $\mu$m) initial bodies. The initial bodies proliferate to become a large, membrane-bound reticulate body containing many elementary bodies. Visible epithelial inclusions are only present in a third of the cases positively identified by immunofluorescent assay. Besides fluorescent antibody detection, cell inoculation, enzyme-linked immunosorbent assay (ELISA), PCR procedures, or immunochemistry may be used.[3,5] In another study, of 226 conjunctival samples, 39% were positive for non-*C. felis* chlamydial DNA, identified as *Neochlamydia hartmannellae*, in cytologic samples with eosinophilic inflammation.[5]

**Mycoplasmosis**

Mycoplasmal infection may have an incidence of 25% in cats with conjunctivitis. Infection with feline *Mycoplasma* sp presents as small basophilic granules similar to the elementary bodies of chlamydial infections with the exception that they are adherent to the surface membrane. These small gray granules may be found extracellularly in the background. Nondegenerate neutrophils are frequent.

## NICTITATING MEMBRANE

The third eyelid, or nictitating membrane, is a large fold of conjunctiva protruding from the medial canthus. It contains a T-shaped cartilaginous plate surrounded by glandular epithelium that is serous in the cat and seromucoid in the dog.[6] The palpebral and bulbar surfaces are covered by nonkeratinized stratified squamous epithelium. The free margin of the membrane is pigmented and cytologically, fine green-black melanin granules can be found within the epithelium. Numerous variably sized lymphoid aggregates with overlying epithelium having microvilli or microfolds (M cells) on the apical surface are associated with the bulbar surface of the nictitating membrane.[7] The stroma also contains fibrous connective tissue.

**Follicular Hyperplasia**

With follicular conjunctivitis there is a mixed population of lymphoid cells that resembles a hyperplastic lymph node. Plasma cells and intermediate lymphocytes are often present in addition to the small lymphocytes. Reactive epithelium also accompanies the lesion.

**Squamous Cell Carcinoma**

Squamous cell carcinoma when present on the nictitating membrane is considered an extension from the eyelid. It appears similar to those found in the skin often accompanied by nonseptic purulent inflammation. Sheets of epithelium appear with some features of malignancy, including anisokaryosis, multinucleation, variable nucleocytoplasmic ratio, and coarse chromatin. In addition to the previously mentioned malignant features there is perinuclear vacuolation, a feature associated frequently with malignant squamous epithelium. Caution may be warranted when severe suppurative inflammation is present as the epithelial reaction may appear dysplastic mimicking neoplasia. The presence of very bizarre morphologic changes and the absence of sepsis further supports a malignant rather than dysplastic epithelium.

## CORNEA

The outer surface is composed of nonkeratinized stratified squamous epithelium. Below the surface epithelium is a thick layer of parallel bundles of collagenous stroma with infrequent intermixed fibrocytes called keratocytes. Deeper to the stroma is Descemet's membrane, composed of fine collagen fibrils. The deepest layer consists of a single layer of flattened endothelium. Cytologically, basal and intermediate squamous epithelial cells without pigmentation predominate in scrapings.

**Eosinophilic Keratitis**

Eosinophilic keratitis is a raised vascular lesion with white granular surface seen in cats. It is composed of squamous epithelium, cellular debris, including eosinophil granules, and numerous mast cells with lower numbers of intact eosinophils. Deeper scrapings contain predominantly eosinophils and lymphoid cells. There appears to be a significant association between feline herpevirus infection and epithelial keratitis.[3]

## ANTERIOR UVEA

The anterior uvea consists of the iris and ciliary body.. The uvea is highly vascular and often pigmented. The melanocytes in the dog and cat contain rod to oval brown melanin granules. Large round melanin granules compose the surface of the ciliary body.

**Neoplasia**

Melanomas are the most common primary intraocular tumor in the dog, with the anterior uvea more often affected. When melanoma occurs in the cat, it involves more often the iris and ciliary bodythan oral and dermal melanomas in that species. Ocular melanoma is also more malignant than dermal melanoma, with higher rates of mortality and metastasis. Aspiration of the neoplasm may be accomplished through the anterior chamber using a 25-gauge or smaller needle. Cytologic features of mitotic index, nuclear size and pleomorphism, and degree of pigmentation are helpful for prognostic purposes in the dog but not in the cat. The presence of anisokaryosis, variable nucleocytoplasmic ratio, and prominent nucleoli should help to distinguish malignant melanocytes from melanophages.

Ciliary neoplasms such as iridociliary adenoma and adenocarcinoma are the second most common primary intraocular tumors in the dog. These tumors are rare in the cat. Ciliary body adenocarcinoma appears as a tight cluster of epithelial cells with high nucleocytoplasmic ratios, coarse chromatin, anisokaryosis, and mitotic figures. The cytoplasm is foamy with numerous discrete vacuoles.

Lymphoma has been recognized as the most frequent intraocular tumor in cats. It occurs as a diffuse or nodular iris lesion. Ocular lymphoma often occurs as a metastatic site so a primary site should be investigated. Cytologic changes are similar to lymphoid neoplasms in other sites. B-cell forms are common and may have sizes up to 25 microns with large prominent nucleoli and deeply basophilic cytoplasm.

## AQUEOUS HUMOR

Normal aqueous humor in the dog has a direct cell count mean of 8/μl (range 0 to 37) and protein mean of 36 mg/dl (range 21 to 65). Normal aqueous humor in the cat has a direct cell count mean of 2/μl (range 0 to 15) and protein mean of 44 mg/dl (range 22 to 75).[8] Cytologically, the low-protein fluid is acellular with only occasional free melanin granules or melanin-containing cells.

### Inflammation and Hemorrhage

Anterior uveitis most often produces a neutrophilic infiltrate with infectious causes such as bacteria, *Blastomyces* sp, *Prototheca* sp, and *Leishmania* sp. Mononuclear uveitis may occur in the presence of hemorrhage. Evidence of damage to the ciliary body and/or iris can result in frequent rod to round melanin granules engulfed by macrophages.

Hyphema may present as acute or chronic hemorrhage. Acute hemorrhage may be indicated by the presence of erythrophagocytosis by macrophages. Platelets may be present if ongoing hemorrhage is occurring but the presence of platelets alone may indicate blood contamination only. Chronic hemorrhage is indicated by the presence of hemosiderin-laden macrophages.

## VITREOUS BODY

Fluid obtained by vitreocentesis is acellular except for hyalocytes, a type of histiocyte. In addition to hyalocytes, fibrocytes and glial cells make up a small portion of the vitreal cells found in the fluid. Occasional lanceolate or caraway seed-shaped melanin granules may be present that are likely retinal in origin.

### Inflammation

Infectious causes of endophthmalitis include bacteria, aspergillosis, blastomycosis, cryptococcosis, histoplasmosis, and protothecosis.[9]

### Tissue Injury

Lens fibers may be seen with primary lens disease or secondary to accidental puncture during sample collection. They appear as uniform amorphous basophilic strands or ribbon structures. Their degeneration may induce a neutrophilic inflammatory response. Retinal cells, when present, are usually the result of inadvertent aspiration or may reflect detachment secondary to disease e.g., uveitis or glaucoma. Photoreceptor and ganglion cells have been rarely identified on cytology; photoreceptor nuclei have marginated heterochromatin with central pale euchromatin producing the appearance of a divided or segmented nucleus.

## ORBIT MASSES

Neoplasms involving the orbit include lymphoma, melanoma, mast cell tumor, histiocytic sarcoma, squamous cell carcinoma, adenocarcinoma of the lacrimal or salivary glands, chondrosarcoma, and other sarcomas. The classification of feline intraocular neoplasms by histochemistry and immunohistochemistry was published recently.[10] Histiocytic sarcoma must be considered in the differential diagnosis of dogs with intraocular masses, especially in Rottweilers and Retriever breeds. It may be distinguished from melanoma by finding CD18-positive cells and no reactivity using Melan-A. The ocular manifestation of histiocytic sarcoma arises from the systemic form of the disease.[11]

## REFERENCES

[1] Bolzan et al. **Vet Ophthalmol** 2005;8:401. [2] Naisse et al. **J Am Vet Med Assoc** 1993;203:834. [3] Volopich et al. **Vet Ophthalmol** 2005;8:25. [4] Stiles et al. **Am J Vet Res** 1997;58:338. [5] von Bomhard et al. **Am J Vet Res** 2003;64:1421. [6] Samuelson. In: Gelatt, **Veterinary Opthalmology, 3rd ed.** 1999 pp. 31-150 [7] Guiliano et al. **Graefe's Arch Clin Exp Ophthalmol** 2002;240:220. [8] Hazel et al. **Am J Vet Res** 1985;46:657. [9] Stenner et al. **Med Mycol** 2007;45:249. [10] Grahn et al. **Vet Ophthalmol** 2006;9:395. [11] Naranjo et al **Vet Ophthalmol** 2007;10:179.

# NEURO-CYTOLOGY OF CEREBROSPINAL FLUID, IMPRESSION SMEARS, & FINE NEEDLE BIOPSIES

**Rose E. Raskin, DVM, PhD, DACVP**
**West Lafayette, IN**

## INTRODUCTION

Cytologic examination of cerebrospinal fluid and solid neural tissues is frequently helpful in determining general categories of pathology prior to performing more invasive or expensive procedures. Diagnoses may involve one or more of the following cytodiagnostic groups. The presentation will discuss the cytologic features of the selected abnormalities described below plus others as time permits.

**General Cytodiagnostic Groups for Neuro-Cytology**

- Normal tissue
- Cystic or hyperplastic tissue
- Inflammation
- Neoplasia
- Response to tissue injury

## CEREBROSPINAL FLUID

### Inflammation

### Lymphocytic Pleocytosis

Necrotizing encephalitis in small breed dogs is one disease presenting with lymphocytic pleocytosis. Pugs, Malteses, Shih Tzus, French bulldogs, and Yorkshire terriers are reported to demonstrate multifocal to massive necrosis and non-suppurative inflammation of the cerebrum and meninges that is fatal or leads to euthanasia.[1] These dogs are usually less than 4 years of age and present frequently with seizures, depression, and ataxia; they do not respond significantly to glucocorticoids. The CSF presents with mild to moderate pleocytosis, generally > 200 cells/$\mu$l, and these are predominantly lymphocytes, generally greater than 70%. Mononuclear pleocytosis is evident and large mononuclear cells may display marked cytoplasmic vacuolization consistent with demyelination. CSF protein concentration is often greater than 50 mg/dl. The cause is considered unknown but an immune-mediated syndrome is suspected in those dogs. A similar population of cells may be found in the CSF in GME necessitating a histologic examination of the brain to detect the necrotizing lesions.

### Mixed Cell Pleocytosis

Pleocytosis with a mixture of cell types may be seen with a variety of underlying causes including GME, FIP, canine distemper, steroid-responsive meningoencephalomyelitis, toxoplasmosis, neosporosis, sarcocystis, encephalitozoonosis, cryptococcosis, blastomycosis, aspergillosis, histoplasmosis, degenerative disc disease, ischemia, and neoplasia. In the case of sarcocystosis, the animal may present with paraparesis and pain on palpation. *Sarcocystis* sp. infected animals may have merozites within the cytoplasm of large mononuclear cells and to a lesser extent within neutrophils as well as extracellularly. The pear-shaped merozoite measures approximately 2-3 x 5 µm. Neutrophils appear mostly non-degenerate and compose the majority of the nucleated cell population along with lymphocytes and large mononuclear cells. PCR and gene sequencing are used to establish the specific diagnosis.

### Neural Tissue Injury

### Myelomalacia

Homogeneous "ribbons" of basophilic material when present in CSF may represent degenerated myelin or fragments of phospholipids from damaged cytomembranes. Spinal cord infarction with diffuse myelomalacia may produce foamy macrophages in the CSF. Luxol fast blue staining of the amorphous eosinophilic material found within the macrophages can support the product being myelin. Similar myelin-like extracellular material was found in a dog with spinal subdural hemorrhage secondary to an intervertebral disk protrusion.[2] Other demyelinating conditions such as degenerative myelopathy may present with free myelin. Evaluations of five cats with myelomalacia due to fibrocartilagineous embolism noted abnormal CSF in four animals with three cases having neutrophilic pleocytosis.[3]

### Neoplasia

### Lymphoma

Elevated protein concentration in approximately 30% of neoplastic cases is generally the only abnormality in CSF with occasional neoplastic cells being present. Finding cells will depend on the location of the mass with its proximity to the ventricle, its involvement with the meninges, or its communication with the subarachnoid space. The presence of mitotic cells in the CSF is unusual and often indicates a proliferative population such as a neoplasm. The presence of immature lymphocytes is highly diagnostic for the presence of CNS lymphoma. In a study of CSF collected from 51 dogs with primary intracranial neoplasia, two lymphoma cases had detectable neoplastic cells or cellular atypia in the fluid supporting the use of CSF analysis to help achieve a diagnosis.[4] Similar results were found in a study of CSF from 28 cats with intracranial neoplasia in which the CSF in one cat with lymphoid blast cells was diagnostic.[5] Well-differentiated lymphoid malignancies may not be readily distinguished from a lymphocytic pleocytosis involving granular lymphocytes. As a sequel to a splenic granular cell lymphocyte lymphoma with circulating cells, these cells may appear within normal limits when only occasionally noted. However, a majority of granular lymphocytes in a case with lymphocytic pleocytosis would support the spread through altered circulation.

**Other Round Cell Tumors**

Plasma cell tumors have occasionally been found that include encephalic and spinal sites.[6] Use of MUM-1 immunochemistry can be used to confirm the origin of the cells as plasma cells in less differentiated cases. Histiocytic neoplasms when benign appearing may be difficult to distinguish from GME.[7] Even non-hematopoietic neoplasms such as medulloblastoma should be considered as a differential diagnosis for atypical round cells in the CSF of young dogs.[8]

## NORMAL CENTRAL NERVOUS SYSTEM TISSUE

**Brain Cortex**

Normal cerebral tissue has low cellular density. In general, the grey matter contains neurons and a few non-myelinated fibers, while the white matter displays myelinated axonal fibers. Neuropil is the term used to define the dense network of fine glial processes, neuronal processes (axons and dendrites), and fibrils in the gray matter of the CNS. The characteristic and almost distinctive blue-purple staining of normal neuropil exhibits foaminess and may contain blood vessels.

Neurons are the principal components of the CNS, which can vary in size from location to location. Most neurons are very large cells, measuring up to 40 µm in diameter but their size can vary from 5 µm (granular layer of the cerebellum) to 100 µm (motor cortex). Despite this variation in size, neurons share common morphologic features in both dogs and cats that have an angulated shape with multiple and branching cytoplasmic processes composed of dendrites and a single axon. All neurons have a very large, centrally placed nucleus and frequently a single, prominent nucleolus. The cytoplasm is usually abundant and granular, due to the presence of abundant rough endoplasmic reticulum (Nissl substance) that may obscure the nucleus. In some areas of the brain, the cytoplasm may contain melanin pigments (neuromelanin) and microvacuoles (neuromediators).

**Cerebellum**

Smears from the cerebellar cortex have a highly characteristic appearance as the cellularity is usually higher than in the brain cortex with sheets of small, hyperchromatic granular cells from the inner, granular layer that are occasionally intermixed with large Purkinje cells. Due to their small size and near absence of cytoplasm, the small granular cells can be confused with lymphoid cells. Purkinji cells are a large distinctive flask-shaped neuron with single, central nucleus and characteristic single large extended axon. The numerous highly branched dendrites for this cell are usually not evident with routine cytologic stains. The cytoplasm surrounding the nucleus also contains basophilic granular material known as Nissl bodies or substance.

**Oligodendrocytes**

Oligodendrocytes are the myelin-forming cells of the central nervous system. Their nuclei are smaller (5-7 µm) and rounder than astrocytes and, like astrocytes, their cytoplasm is not well defined. Due to their size and shape, oligodendrocytes can be mistaken for lymphocytes. Oligodendrocytes may surround neurons in a process called satellitosis.

**Ependymal and choroid plexus cells**

Choroid plexus cells can be considered as specialized ependymal cells, lining the brain ventricles and the central canal of the spinal cord. These neuroepithelial cells show similar cytomorphologic features as they are usually organized in small loose clusters and small sheets of cuboidal to columnar cells with a single, small, round and centrally placed nucleus. Choroid cells are cytologically indistinguishable from ependymal cells.

**Mesenchymal Cells**

Meningeal cells are only rarely seen and recognized in brain smears. The cells are usually organized in sheets or in loose clusters showing a rather pleomorphic storiform pattern. Cell borders are poorly defined and the nuclear shape ranges from round to oval to elongate. Occasionally, cells are organized in pseudo-acinar structures mimicking a glandular origin, which are more common in meningiomas, but can also be identified in normal meningeal tissue. Also seen are cell whorls, which may or may not be associated with psammoma bodies.

Microglial are neuroglial cells derived from bone marrow elements, likely macrophages. They have small and elongated nuclei. Microglia when reactive often show lipophagocytosis producing a foamy appearance.

**Hyperplasia**

**Astrocytosis**

Astrocytes have a supportive function to neurons and are distributed throughout the nervous system. The characteristic star-shaped appearance cannot be evaluated with routine cytologic stains. The cells appear as small, oval, naked nuclei, which measure between 7-10 µm, surrounded by neuropil. In response to neural tissue injury in the brain and spinal cord there is a proliferation and hypertrophy of the resident astrocytes which will display an increased nucleocytoplasmic ratio and a single small prominent nucleolus.

**Neoplasia**

**Meningioma**

Non-neuroepithelial tumors that arise from the arachnoid meningeal layer are termed meningiomas. These are the most common intracranial tumor in dogs and cats. The tumors are derived from leptomeningocytes that associate with neural crest tissue and have both epithelial and fibroblastic ultrastructural characteristics. As a result, these tumors have several variant forms that are found both in cervical and lumbar regions of the spinal cord as well as intracranially and within the retrobulbar region.[9,10] Spinal cord meningiomas are mostly extramedullary. Those with a meningotheliomatous appearance may present as cell clusters with plump cells having an oval to round eccentric nuclei and occasional prominent nucleoli. The cytoplasm contains an eosinophilic secretory material. Meningiomas with the more common spindle cell cytologic appearance have a round to oval nucleus with finely granular chromatin, small nucleoli, and lightly granular and basophilic cytoplasm that forms wispy tails. A cohesive aggregate of meningeal cells may be encountered with a distinctive whorl formation characteristic of a psammatous meningioma. Fine-needle aspirations, crush preparations, and incisional cutting needles have been used to obtain cytologic and tissue samples for biopsy.[11-13]

The characteristics of cisternal CSF in a series of 56 dogs[14] with confirmed intracranial meningioma found neutrophilic pleocytosis in only 19% of dogs; significantly different than expected based on earlier reports. Neutrophilic pleocytosis may not be detected in CSF samples from dogs with meningiomas located within the middle or rostral portion of the cranial fossae.

**Granular Cell Tumor**

Another tumor associated with the meninges is the granular cell tumor, an uncommon tumor of animals and humans that occurs both within and outside the nervous system. The cell of origin is unclear but their morphologic appearance is thought to be the result of the accumulation of lysosomes as demonstrated by electron microscopy that reflect metabolic derangements in the cell. The cytologic features include large round cells with eccentric nuclei and cytoplasm distended by many variably stained eosinophilic granules. Cells stain strongly positive for PAS as well as for the immunocytochemical stain for ubiquitin. Other staining reactions for these cells include a variably positive reaction for S-100, α-1-antichymotrypsin, α-1-antitrypsin, and vimentin as well as a negative reaction for glial fibrillary acidic protein, pancytokeratins, and markers for subpopulations of leukocytes and macrophages.[15, 16]

## PERIPHERAL NERVOUS SYSTEM TISSUE

### Neoplasia

**Peripheral Nerve Sheath Tumor**

Peripheral nerve sheath tumors (PNST) are most often associated with peripheral nerve roots and include those of the neural crest-derived Schwann cell that assists in myelination as well as fibroblastic connective tissue cells that surround nerve bundles. A cytological distinction among benign PNST (Schwannoma and neurofibroma) can be difficult or impossible. Moreover, sometimes benign Schwannomas and neurofibromas can show cytological features of atypia so that differentiation between benign and malignant PNST becomes even more difficult. Cytological features include moderate to high cellularity with cells mainly grouped in thick fragments with characteristically spindle shapes, elongate plump nuclei, and inconspicuous nucleoli. An unusual cytological presentation of a feline PNST displayed a pleomorphic population of individual round cells resembling histiocytes or plasma-cells with round, central to eccentric nuclei, basophilic cytoplasm, numerous mitotic figures, and large, multinucleated cells.[17]

**Gliomas**

Gliomas refer to neoplasms of specific neuroglial cells that include oligodendrocytes and astrocytes. Tumors from these cells most often produce a normal CSF related to their deep parenchymal location. Oligodendroglioma tumor cells that are rarely found in the CSF appear large with nuclei four to six times the size of red cells. Nuclei were eccentric within a densely basophilic moderately abundant cytoplasm. An aspirate of the mass displays a monomorphic population of large mononuclear cells arranged in loose sheets or small clusters. Nuclei are round to oval with fine chromatin and indistinct nucleoli. Astrocytomas examined by crush preparation preserve the cytoplasmic features of an eccentrically placed nucleus within a moderately abundant basophilic cytoplasm resembling histiocytes. A positive immunohistochemical stain for glial fibrillary acid protein (GFAP) is a marker used to distinguish astrocytes from other neuroglial and meningeal cells but may sometimes yield a positive reaction in neuroepithelial tumors, such as ependymoma and choroid plexus tumor, and meningeal tumors.[9]

**Choroid Plexus Tumors**

Neuroepithelial cells also include ependymal cells and choroid plexus cells. The ependymal cells line the ventricular system of the brain and central canal of the spinal cord. The choroid plexus cells represent a highly vascular portion of the pia mater that projects into the ventricles of the brain and is thought to secrete the CSF. Positive cytokeratin staining is expected in choroid plexus tumors. The choroid plexus tumor cells appear similar to mesothelial cells in the CSF, having a large round nucleus with abundant well-defined deeply basophilic, finely granular cytoplasm having surface projections. A papillary appearance with epithelial features can be seen in cytologic preparations of both choroid plexus papilloma and choroid plexus carcinoma. Tight cohesion between the cells is seen along with a high nucleocytoplasmic ratio. The nucleus has finely granular chromatin and large prominent nucleoli.

## REFERENCES

[1]Timmann et al. **J Small Anim Pract** 2007;48:339. [2]Bauer et al. **Vet Clin Pathol** 2006;35:123. [3]Mikszewski et al. **J Am Anim Hosp Assoc** 2006; 42:226. [4]Snyder et al. **J Vet Intern Med** 2006; 20: 669. [5] Troxel et al. **J Vet Intern Med** 2003; 17:850. [6]Greenberg et al. **J Vet Intern Med** 2004; 18: 581. [7]Zimmerman et al. **Vet Clin Pathol** 2006; 35:127. [8]Thompson et al. **Vet Clin Pathol** 2003; 32:143. [9]Montoliu et al. **J Comp Path** 2006; 135: 200. [10]Zimmerman et al. **Vet Clin Pathol** 2000; 29: 29. [11]DeLorenzi, et al. **Vet Clin Pathol** 2006; 35: 208. [12]Moissonier et al. **J Small Anim Prac** 2002; 43:115. [13]Platt et al. **Vet Surg** 2002; 31: 65. [14]Dickinson et al. **J Am Vet Med Assoc** 2006; 228: 564. [15]Higgins et al. **Vet Pathol** 2001; 38: 620. [16]Sharkey et al.**Vet Clin Pathol** 2004; 33: 111. [17]Tremblay N, Lanevschi A, Doré M et al: Of all the nerve! A subcutaneous forelimb mass in a cat. Vet Clin Pathol 2005; 34: 417-420.

# UPDATE ON HUMAN IMMUNE-MEDIATED HEMATOLOGICAL DISEASE FROM THE AMERICAN SOCIETY OF HEMATOLOGY

**Stephanie A. Smith DVM, MS, Diplomate ACVIM**
**Urbana, IL**

## INTRODUCTION

In December 2008 over 24,000 participants gathered in San Francisco for the 50th Annual Meeting of the American Society of Hematology. The program included 4 days of scientific sessions, a special symposium on thrombosis and hemostasis, multiple satellite symposia, and over 6,300 submitted abstracts. What follows are some selected topics from the meeting that may be of interest to veterinarians hematologists, or have potential relevance to improving the understanding and treatment of hematologic disease in veterinary patients.

## PLATELETS

### Normal Platelets[1, 2]

The mechanisms by which platelets are generated from megakaryocytes remain poorly defined. Real-time imaging techniques have provided a fascinating window into the processes by which platelets are born. Recent work has suggested that megakaryocytes are sessile cells that extend dynamic proplatelet-like protrusions into the microvasculature. The shear forces generated by flowing blood then result in removal of the protrusions, resulting in the appearance of platelets within the flowing blood. Other recent studies have suggested, in contrast to the current concept of platelets as mature differentiated cells, that proplatelets or platelets themselves may continue to divide into smaller platelets within the peripheral vasculature.

### Immune Mediated Thrombocytopenia (ITP) and *Helicobacter pylori*[3, 4]

Recent studies have linked infection with *H. pylori* to ITP, although the exact nature of the relationship is controversial. The prevalence of *H. pylori* infection in humans is high (>50% worldwide) but highly variable between populations. The prevalence does not appear to differ between the general adult population and ITP patients, which argues against a cause and effect relationship.

The strongest evidence for the relationship between ITP and *H. pylori* is that eradication of *H. pylori* has been reported to result in an increase in platelet count in patients with chronic ITP. Eradication with triple therapy (amoxicillin, clarithromycin, and a proton pump inhibitor), as described in 25 different case series including 792 patients, has reported an increase in platelet count following therapy in 84.7% of patients. In a randomized prospective trail, 46% of patients receiving eradication therapy had a platelet response while no patients in the non-treated group had a platelet response. In other studies the response rate has varied form 0 to 100%, with wide variation between countries. Patients with shorter ITP duration appear more likely to respond to eradication therapy.

Possible mechanisms for the relationship between infection and ITP include:

1) Molecular mimicry due to cross-reaction between *H. pylori* antigens (in particular CagA) and platelet glycoproteins resulting in cross-reacting anti-platelet antibodies.
2) Since, lipopolysaccharide from gram-negative bacteria significantly enhances Fc-dependent platelet phagocytosis, infection with *H. pylori* may lead to increased platelet destruction in the presence of anti-platelet antibodies.
3) *H. pylori* binds to vWF and may induce GPIb and FcRIIa- dependent platelet aggregation. Activation would then promote platelet clearance and antigen exposure.

Could immune-mediated thrombocytopenia in dogs have an association with Helicobacter infection, and might eradication therapy improve outcome in these patients? These questions await study in veterinary populations.

### Thrombopoiesis in ITP[5]

Historically, therapy for ITP has focused on the immune-destruction of platelets through the use of immunosuppressive therapy, immunomodulation, and removal of the spleen. Recent work has indicated that abnormal levels of thrombopoietin (TPO) and impaired megakaryocyte maturation and production of platelets also play roles in the pathogenesis of ITP.

The role of TPO in thrombopoiesis was first described only 15 years ago. This protein is produced by the liver, and binds to a receptor on the megakaryocyte, resulting in stimulation of platelet production. It is thought that circulating levels are regulated via platelet uptake of TPO. However, patients with ITP do not have the expected elevations in TPO levels. Anti-TPO antibodies are only rarely identified.

The new understanding of the role of TPO in the pathogenesis of ITP has led to development of new pharmaceuticals. Several platelet growth factors are in the development pipe-line. Romiplostim (Nplate™), a TPO peptide mimetic, and Eltrombopag (Promacta®), a small molecule mimetic, were both recently approved for use in humans with chronic ITP that has not responded to traditional therapy. Nplate is administered as a weekly subcutaneous injection. Data suggests that many patients achieve good platelet responses, and half of treated patients were able to discontinue immunosuppressive therapy. Drug availability is limited through a restricted distribution program. Promacta is orally administered. More than 80% of ITP patients achieved a platelet response and about 1/3 were able to discontinue immunosuppressive therapy. Whether or not the TPO receptor agonists (either those currently available or those in development) will be effective in dogs with immune-mediated thrombocytopenia remains to be seen.

**Platelets in Malignancy**[6, 7]

Emerging data supports a major role for platelets in metastatic disease. Platelets express a glycoprotein receptor on their surface, GP 1b-1X. Binding of its ligand (vWF) to this receptor is a major initiator of thrombus formation in the arterial circulation. Recently published data from knockout and transgenic mouse models now suggests that GP 1b-1X is a major contributor to tumorigenesis, and that functional absence of GP 1b-1X significantly reduces primary tumor burden. The role of platelets in the metastatic potential of various neoplastic disorders is being further investigated. Platelets may represent important targets in the therapeutic management of neoplasia.

**Platelet Transfusion**[8]

Platelet concentrates have long been available in human medicine. They are produced from platelet-rich plasma, from buffy coat, or through apheresis. Platelet concentrates are not often available in veterinary medicine due to cost issues and the labor and specialized equipment needed for production. However, cryopreserved and lyophilized platelets have recently become available. The clinical utility of such products is not entirely certain at this point, but with availability use is likely to increase.

The issue of appropriate use of platelet transfusion is far from settled in human medicine. The issue of prophylactic (aimed at increasing the platelet count to prevent future bleeding) vs therapeutic (aimed at stopping bleeding when associated with severe thrombocytopenia) is a controversial one. Many clinical practice guidelines recommend a prophylactic platelet transfusion trigger of 10,000/uL while common practice is to use a trigger of 20,000/uL. A variety of randomized and non-randomized studies comparing the two thresholds have demonstrated no increase in bleeding risk with the lower trigger. More recent studies have focused on whether a transfusion trigger of 5,000/uL may be acceptable. Three ongoing multicenter randomized controlled trails (PLADO, SToP, and TOPPS) are in process to address the questions of triggers and the use of prophylactic vs therapeutic platelet transfusion.

## ERYTHROCYTES

**Anemia of Inflammation and Malignancy**[9]

It is widely recognized that a normochromic/normocytic anemia commonly develops with inflammatory and neoplastic disorders. The mechanisms by which these anemias develop have recently been better delineated.

Cytokines, in particular IL-1B, play significant roles in the development of this anemia. Cytokines interfere with the appropriate hypoxia-induced upregulation of erythropoietin-gene expression. This results in low endogenous erythropoietin levels with respect to the degree of anemia. Cytokines also impact the response of the erythroid components of the bone marrow to erythropoietin. *In vitro* and *in vivo* studies have indicated profound cytokine-mediated suppression of erythropoiesis that is sometimes (but not always) able to be overcome with exogenous erythropoietin.

The anemia of inflammation is generally associated with a reduction in serum iron concentrations. This change is mediated by the hormone hepcidin. Under normal circumstances, hepcidin is produced by the liver. It travels through the blood to interact with ferroportin on cell surfaces. Hepcidin binding to ferroportin prevents release of iron from the cells. Hepcidin also downregulates expression of ferroportin. The net effect of hepcidin is to trap iron within storage cells and prevent iron uptake from the gastrointestinal tract. In disease states, hepcidin is also produced by monocytes. Hepcidin production in monocytes and the liver is upregulated by IL-6. As a consequence, hypercytokinemia results in excess hepcidin acitivity, and therefore a reduction in serum iron level and the percent of transferrin saturation. The end-product is iron-restricted erythropoiesis. If the iron-restriction is profound, the anemia may be characterized as microcytic and hypochromic.

The ideal, but often not possible, method of treating anemia of inflammation and malignancy is to eliminate the underlying cause. This anemia, if clinically relevant such as that associated with chemotherapeutic management of malignancy, may benefit from exogenous erythropoietin. Early studies of erythropoietin therapy were mixed as to improvement in quality of life or survival factors. Some studies in a variety of human cancers have actually suggested decreased survival in patients receiving erythropoietin, leading to a "black box" warning on drug labels. Recent data has suggested that providing additional iron supplementation can improve hematologic response to exogenous erythropoietin. It is not yet clear whether iron in conjunction with erythropoietin provides significant clinical benefit versus transfusion support.

**Interactions between Reticulocytes and Other Cells**[10, 11]

Recent work by Parise and others has focused on mechanisms of vaso-occlusive disease in sickle cell anemia. These studies have led to a better understanding of interactions between reticulocytes and other cells. Reticulocytes express the integrin α4B1 on their surface. This integrin can bind via its ligand (the plasma protein fibrinectin) to a variety of other cell types, including monocytes, platelets, and endothelial cells. It was proposed, in part because reticulocytes are enriched in the peripheral circulation in sickle-cell disease, that reticulocyte aggregates with other cells could contribute to the vascular obstruction that is a common occurrence in sickle cell crisis.

While canine immune-mediated hemolytic anemia (IMHA) shares many clinical features with human sickle cell disease, the pathophysiologies are quite different. However, reticulocytosis is common and quite often profound in canine IMHA, as is the leukocytosis. It is possible that one of the mechanisms contributing to thrombotic disease could be cellular aggregates of reticulocytes, leukocytes, and platelets.

**Eculizumab**[12, 13]

Eculizumab (Soliris™) is a humanized monoclonal antibody that specifically binds human C5 complement protein. It prevents cleavage of C5 to C5a and C5b, which consequently prevents formation of the membrane attack complex. It was initially targeted to prevent complement activation in autoimmune disorders (ie rhuematoid arthritis, systemic lupus erythematosis, glomerulonephritis) but is FDA approved for use in paroxysmal nocturnal hemoglobinuria (PNH). Because the genetic defect in PNH results in extreme senstivity of RBCs to lysis by complement, eculizumab directly impacts hemolysis in this disease. In several trials eculizumab improved anemia and markedly decreased degree of hemolysis and transfusion needs of affected patients.

Because complement fixation by IgM (and possibly IgG) may play a role in hemolysis in canine immune-mediated hemolytic anemia, a therapy that prevents complement-mediated RBC destruction could have a role in this disease. Whether or not this human monoclonal antibody would be effective against canine complement C5 remains to be determined.

## REFERENCES

1. Junt T, Schulze H, Chen Z et al. Dynamic visualization of thrombopoiesis within bone marrow. **Science.** 2007;317:1767-1770. 2. Elizabeth Cramer-Borde. Platelet Birth. from Scientific Committee on Platelets. **50th ASH Annual Meeting**. 2008. 3. Stasi R, Provan D. Helicobacter pylori and Chronic ITP. **Hematology Am Soc Hematol Educ Program**. 2008;2008:206-211. 4. Arnold DM, Stasi R. Does Helicobacter pylori Eradication Therapy Result in a Platelet Count Improvement in Adults with Immune Thrombocytopenic Purpura Regardless of H pylori Infection? **Hematology Am Soc Hematol Educ Program**. 2008;2008:31-32. 5. Gernsheimer TB. The Pathophysiology of ITP Revisited: Ineffective Thrombopoiesis and the Emerging Role of Thrombopoietin Receptor Agonists in the Management of Chronic Immune Thrombocytopenic Purpura. **Hematology Am Soc Hematol Educ Program**. 2008;2008:219-226. 6. Jain S, Zuka M, Liu J et al. Platelet glycoprotein Ib alpha supports experimental lung metastasis. **Proc Natl Acad Sci U S A.** 2007;104:9024-9028. 7. Ware, J. and Jain S. Platelet glycoprotein 1b-1X in hemostasis and malignancy. Scientific Committee on Hemostasis. **50th ASH Annual Meeting.** 2008. 8. Blajchman MA, Slichter SJ, Heddle NM, Murphy MF. New strategies for the optimal use of platelet transfusions. **Hematology Am Soc Hematol Educ Program**. 2008;2008:198-204. 9. Adamson JW. The anemia of inflammation/malignancy: mechanisms and management. **Hematology Am Soc Hematol Educ Program.** 2008;2008:159-165. 10. Brittain JE, Knoll CM, Ataga KI, Orringer EP, Parise LV. Fibronectin bridges monocytes and reticulocytes via integrin α4B1. **Br J Haematol.** 2008;141:872-881. 11. Parise, L. Platelet Activation and Inflammation. Special Symposium on Basic Science of Hemostasis and Thrombosis. **50th ASH Annual Meeting**. 2008. 12. Hillmen P. The Role of Complement Inhibition in PNH. **Hematology Am Soc Hematol Educ Program**. 2008;2008:116-123. 13. Haspel RL, Hillmen P. Which Patients with Paroxysmal Nocturnal Hemoglobinuria (PNH) Should Be Treated with Eculizumab? **Hematology Am Soc Hematol Educ Program**. 2008;2008:35.

## INNOVATIVE THERAPIES FOR IMMUNE-MEDIATED HEMOLYSIS IN DOGS

Katharine F. Lunn, BVMS, MS, PhD, MRCVS, DACVIM
Fort Collins, CO

### INTRODUCTION

Immune-mediated hemolytic anemia (IMHA) is the most common cause of hemolytic anemia in dogs.[1,2] This may be a primary or idiopathic process in which an underlying trigger is not identified, or IMHA may be secondary to other diseases such as neoplasia, infection, or a drug reaction.[1] One study has shown a significant temporal relationship between vaccination and the onset of IMHA,[3] but this observation has not been supported by other case series and remains controversial.[4,5] In addition, some case series have suggested a seasonal occurrence of IMHA,[1,6,7] but this finding is not consistent across all retrospective studies.[5] Idiopathic or primary IMHA appears to account for 60-75% of cases of IMHA in dogs.[2] These patients may have a true autoimmune hemolytic anemia in which autoantibody is produced against self-antigen on the red cell membrane.[1] Red blood cell (RBC) destruction in IMHA is the result of a type II hypersensitivity reaction in which RBCs are phagocytosed or lysed after the binding of antibody and/or complement to the cell surface.[1] Antibodies of the IgG class are most commonly involved, and IgM autoantibodies appear to be less frequently detected.[8,9] Binding of IgG to the surface of RBCs triggers removal of the cells by macrophages that carry receptors for the constant fragment (Fc) portion of the IgG molecule. These macrophages are found primarily in the spleen, but also in the liver. Antibody coated RBCs may be also destroyed intravascularly after complement fixation and activation of a membrane attack complex that leads to cell lysis. Complement activation also enhances extravascular lysis of RBCs, which leads to the production of spherocytes in IMHA.

Dogs with IMHA typically have signs of weakness, lethargy, exercise intolerance, and anorexia. Physical examination findings may include pale mucus membranes, tachycardia, tachypnea, fever, hepatosplenomegaly, and icterus. In some cases the signs of IMHA may be complicated by the development of disseminated intravascular coagulation or thromboembolic disease.[4,7,10-12] Most patients with IMHA have a regenerative anemia, characterized by an increased absolute reticulocyte count (> 60,000 /µl), together with anisocytosis, polychromasia and sometimes nucleated RBCs. A proportion of patients with IMHA demonstrate a non-regenerative anemia, either due to detection of IMHA before enough time has elapsed to initiate a bone marrow response, or due to immune-mediated disease directed against RBC precursors in the bone marrow.[13] The presence of spherocytosis (usually marked) and in-saline microscopic or macroscopic agglutination of RBCs indicate the presence of immune-mediated disease. Additional findings that support the diagnosis of IMHA are the presence of hemoglobinuria or hemoglobinemia, and elevated bilirubin levels. In the absence of RBC autoagglutination, the presence of antibody or complement on the surface of red cells can be demonstrated by the direct antibody (Coombs') test,[14] or by flow cytometry. The latter technique appears to be a more sensitive test for the detection of RBC-bound antibody.[8,9] The diagnosis of IMHA also requires the ruling out of other causes of regenerative anemia, including blood loss and non-immune-mediated RBC damage. In addition, underlying infection, neoplasia, or drug reactions must be ruled out in order to diagnose a primary, or idiopathic, IMHA.

Reported mortality rates for canine IMHA range from 20% to 70%, even in the face of aggressive therapy.[3-7,15-17] Factors associated with increased mortality include the presence of autoagglutination,[3,6] elevated serum bilirubin concentrations,[4,6] increased plasma urea concentration,[18] and the presence of thrombocytopenia.[4,18] Dogs that die of IMHA may succumb to tissue hypoxia due to severe anemia, or may have hemostatic disorders.[4,10,12] Patients with IMHA may also be euthanized due to poor prognosis, disease severity, the expense of medical therapy, or side effects and complications of medical management. Survival times for dogs with IMHA are infrequently reported, but it is apparent that the most severely affected patients often die within 1-2 weeks of diagnosis, and that patients that survive 10-14 days after initiation of treatment are likely to have longer survival times.[3,5,18] Thus there is a clear need for rapidly effective therapies that improve short-term survival and allow patients the opportunity to respond to longer-term immunosuppression.

### TREATMENT OF IMMUNE-MEDIATED HEMOLYTIC ANEMIA

The mainstay of treatment for IMHA is the use of high doses of glucocorticoids; usually prednisone.[2] The effects of glucocorticoids include decreased immunoglobulin production, reduced binding of immunoglobulin to RBCs, inhibition of the complement cascade, and decreased production of cytokines.[2] Many published case series have documented the use of additional immunosuppressive therapies in the management of IMHA. Unfortunately many of these reports involve small numbers of patients and non-uniform treatment protocols, therefore there is still no strong consensus regarding the safest and most effective medical therapy for these patients. Azathioprine is a purine analogue antimetabolite that is commonly combined with prednisone therapy.[7] There are no studies that directly compare the use of prednisone with or without azathioprine, however many authors consider it to be beneficial.[5,7,18] Cyclophosphamide is a potent cytotoxic agent that has been used in the management of IMHA. One prospective study suggested that cyclophosphamide provides no advantage over prednisone alone in the treatment of IMHA,[17] and 2 retrospective studies have also cast doubt on the benefits of this therapy.[5,16] Cyclosporine, danazol, leflunomide, and mycophenolate mofetil have also been used as immunosuppressive agents in the management of IMHA, however there are no prospective studies documenting their benefits over glucocorticoid therapy alone.[16,19-22]

### ALTERNATIVE APPROACHES TO THERAPY FOR IMMUNE-MEDIATED HEMOLYTIC ANEMIA

Despite the reported use of a variety of immunosuppressive therapies, mortality rates for canine IMHA have not improved over the past 20 years.[15] It is apparent that many patients that die from IMHA do so within 1-2 weeks of diagnosis. This is most likely because glucocorticoids and immunosuppressive therapies such as azathioprine take several days to weeks to become effective. Thus there is a clear need for safe and effective new strategies to more rapidly gain control of the disease process. Potential approaches to this problem include splenectomy, the administration of human intravenous immunoglobulin, and the use of liposomal clodronate.

**Splenectomy**

The spleen plays an important role in the pathogenesis of IMHA as it is the major site of removal of immunoglobulin coated RBCs from the blood. Splenectomy might therefore be predicted to be beneficial in decreasing RBC destruction in patients with IMHA. This approach was first described in dogs in 1985,[23] but has not been widely adopted. A more recent preliminary report of a small number of dogs suggested that early splenectomy may be associated with increased patient survival and more rapid recovery of RBC mass.[24] Potential drawbacks of this approach include the risks of anesthesia and surgery in acutely ill dogs with IMHA, an increased susceptibility to certain infectious diseases, and the possibility of disease relapse due to activity of the hepatic mononuclear phagocytic system. There is a need for larger prospective long-term studies of the effects of splenectomy in dogs with IMHA.

**Human Intravenous Immunoglobulin**

Intravenous immunoglobulin G (IVIG) is used in human medicine for replacement therapy in immunodeficiencies, and for immunomodulatory therapy in autoimmune disease, particularly immune thrombocytopenic purpura.[25] Human IVIG is derived from pooled plasma from 1000-100,000 donors, and contains the entire spectrum of antibodies produced by the population. The mechanism of action of IVIG is not completely understood, and it is likely that it affects the immune system at many levels. Immunomodulatory effects of IVIG may include inhibition of B cells, neutralization of autoantibodies, interference with the complement system, inhibition of T cell activation and antigen presentation, effects on leukocyte migration, and modulation of the cytokine network.[26,27] Of particular significance in the acute treatment of IMHA is the proposed decrease in RBC clearance through blockade by IVIG of Fc receptors on mononuclear phagocytic cells.[28] It has been demonstrated that human IVIG binds to canine lymphocytes and monocytes, and that IVIG inhibits Fc-mediated phagocytosis of antibody-coated canine RBCs.[28] In a prospective clinical trial, 10 dogs with IMHA that did not respond to conventional immunosuppressive therapy were treated with IVIG. This therapy appeared to be useful for short-term stabilization of some patients, but an improvement in long-term survival was not demonstrated.[29] A retrospective study of the use of IVIG in 13 dogs with IMHA has also been reported, and the authors suggested that this therapy may be useful for dogs that do not respond within 7 days to more conventional immunosuppressive therapy.[30] Disadvantages of the use of IVIG in canine IMHA include expense, periodic limited availability, a potential risk of hypersensitivity reactions to this human product, and a potential, but unproven, increased risk of thromboembolic disease.[29]

**Liposomal Clodronate**

Numerous studies in mice and rats have demonstrated the impressive macrophage-depleting properties of liposomal clodronate (LC).[31-33] Clodronate is a bisphosphonate which, when incorporated into liposomes, is preferentially phagocytosed by macrophages and some dendritic cells, leading to rapid apoptosis of these cells.[34,35] Systemic injection of LC can induce very efficient depletion of splenic and hepatic macrophages within 24 hours. [34] Most significantly, several studies have demonstrated the effectiveness of systemic administration of LC for treatment of autoimmune diseases in mouse models, including thrombocytopenia and hemolytic anemia.[36,37] These findings in mice lead to preliminary studies at Colorado State University in dogs with IMHA. [38]

Significant findings of these studies were:

1. Intravenous infusion of LC is well-tolerated in normal dogs
2. Intravenous infusion of LC induces a dose-dependent blockade of clearance of opsonized autologous RBCs
3. Studies performed *in vivo* in normal dogs indicated that splenic macrophages and dendritic cells are the primary antigen presenting cells responsible for the uptake of LC
4. Studies performed *in vitro* showed that LC can induce rapid killing of canine splenic macrophages and dendritic cells
5. Seven dogs with spontaneous IMHA were treated with LC and the effects on RBC clearance and patient survival were assessed. Clearance of RBC was found to be decreased in 2 of the 7 dogs following LC treatment
6. Overall survival of LC treated dogs was significantly ($p = 0.035$) increased compared to historical control animals that were not treated with LC.[38]

A prospective randomized placebo-controlled clinical trial of LC treatment of dogs with IMHA is currently underway at Colorado State University. Dogs are enrolled in the study if they have clinicopathological findings consistent with idiopathic IMHA. Specifically, patients must be anemic, with evidence of regeneration (absolute reticulocyte count > 60,000/μl) and hemolysis (hyperbilirubinemia, hemoglobinemia, hemoglobinuria or presence of spherocytosis), and a positive Coombs' test, flow cytometric evidence of RBC-bound immunoglobulin, or in-saline autoagglutination of RBCs. Initial evaluation of patients includes a complete blood count (CBC), serum chemistry profile, urinalysis, thoracic radiographs, abdominal radiographs or ultrasound, and infectious disease testing according to geographical location and the discretion of the attending clinician. Patients with blood loss and with non-immune-mediated causes of hemolytic anemia are excluded. Patients are also excluded if serious concurrent diseases are present (such as congestive heart failure, renal or hepatic disease) and if IMHA is found in association with an underlying disorder such as neoplasia, infectious disease, or known drug reaction.

Eligible patients are randomly assigned to receive either LC or placebo (liposomes containing phosphate buffered saline). The clinician and veterinary staff responsible for the care of each patient are masked as to whether the patient received LC or placebo. All patients entered into the study receive prednisone and azathioprine at standard therapeutic doses. Supportive fluid therapy, blood or packed RBC transfusions, anticoagulant therapy, and gastrointestinal protectants are used at the discretion of the attending clinician. Patients are ineligible for the study if they undergo splenectomy or receive intravenous human or equine immunoglobulin or a hemoglobin based oxygen carrying solution. Other immunosuppressive agents, such as cyclophosphamide, danazol, or cyclosporine, cannot be used during the initial treatment period, unless strongly indicated for medical reasons, such as azathioprine intolerance. Patients that receive these agents during the first 2 weeks are removed from the study.

Liposomes are administered intravenously by continuous infusion over 90 minutes at a dose of 0.5 ml/kg. Preliminary studies showed that this dose is effective and well tolerated. Dogs are monitored throughout the infusion period for clinical signs such as nausea, vomiting, diarrhea or fever. Adverse reactions to the liposomes appear to be minor, and extremely rare.

Parameters evaluated in the study include survival rates at 7, 14, and 90 days, duration of hospitalization, cost of therapy, requirement for blood transfusion, the presence of surface-bound immunoglobulin on RBCs (assessed by flow cytometry), and the time to resolution of spherocytosis and stabilization of PCV.

## REFERENCES

[1]Day MJ. In: **Schalm's Veterinary Hematology** (2000) Feldman BF, Zinkl JG, Jain NC, eds. Lippincott, Williams and Wilkins, Philadelphia, PA. [2]McCullough S. **Vet Clin North Am Small Anim Pract** 2003;33:1295. [3]Duval D, et al. **J Vet Intern Med** 1996;10:290. [4]Carr AP, et al. **J Vet Intern Med** 2002;16:504. [5]Burgess K, et al. **J Vet Intern Med** 2000;14:456. [6]Klag AR, et al. **J Am Vet Med Assoc** 1993;202:783. [7]Weinkle TK, et al. **J Am Vet Med Assoc** 2005;226:1869. [8]Wilkerson MJ, et al. **J Vet Intern Med** 2000;14:190. [9]Quigley KA, et al. **J Vet Diagn Invest** 2001;13:297. [10]Klein MK, et al. **J Am Vet Med Assoc** 1989;195:246. [11]McManus PM, et al. **J Am Vet Med Assoc** 2001;218:1308. [12]Scott-Moncrieff JC, et al. **J Am Anim Hosp Assoc** 2001;37:220. [13]Stokol T, et al. **J Am Vet Med Assoc** 2000;216:1429. [14]Wardrop KJ. **Vet Clin Path** 2005;34:325. [15]Reimer ME, et al. **J Am Anim Hosp Assoc** 1999;35:384. [16]Grundy SA, et al. **J Am Vet Med Assoc** 2001;218:543. [17]Mason N, et al. **J Vet Intern Med** 2003;17:206. [18]Piek CJ, et al. **J Vet Intern Med** 2008;22:366. [19]Holloway SA, et al. **J Am Vet Med Assoc** 1990;197:1045. [20]Miller E. **Sem Vet Med Surg** 1997;12:167. [21]Gregory CR, et al. **Transp Proc** 1998;30:4143. [22]Nielsen L, et al. **ECVIM Proceedings**, 2005. [23]Feldman BF, et al. **J Am Vet Med Assoc** 1985;187:617. [24]Toll J, et al. **J Vet Intern Med** 2003;17:383. [25]Looney RJ, et al. **Best Pract Res Clin Haem** 2006;19:3. [26]Scott-Moncrieff JC, et al. **Sem Vet Med Surg** 1997;12:178. [27]Stangel M, et al. **J Neurol** 2006;253(suppl):V/18. [28]Reagan WJ, et al. **Am J Vet Res** 1998;59:1568. [29]Scott-Moncrieff JC, et al. **J Am Vet Med Assoc** 1997;210:1623. [30]Kellerman DL, et al. **J Vet Intern Med** 1997;11:327. [31]van Rooijen N, et al. **Cell Tissue Res** 1984;238:355. [32]Claassen I, et al. **J Immunol Methods** 1990;134:153. [33]van Rooijen N, et al. **J Immunol Methods** 1994;174:83. [34]van Rooijen N. **J Immunol Methods** 1989;124:1. [35]Naito M, et al. **J Leukoc Biol** 1996;60:337. [36]Alves-Rosa F, et al. **Blood** 2000;96:2834. [37]Jordan MB, et al. **Blood** 2003;101:594. [38]Mathes M, et al. **Exp Hem** 2006;34:1393.

**IMMUNO-MAGNETIC ISOLATION OF CANINE CIRCULATING ENDOTHELIAL CELLS.** **TB Wills, A Heaney, KJ Wardrop, GJ Haldorson**
**Washington State University College of Veterinary Medicine, Pullman, WA.**

This study was performed to determine if immunomagnetic isolation could be used to obtain and enumerate circulating endothelial cells (CEC) from canine whole blood.

Five mls of blood was obtained from ten healthy client owned dogs. Magnetic beads were conjugated with an antibody against an endothelial cell marker, CD146, and added to canine EDTA anticoagulated whole blood to bind and isolate CD146-positive cells. Because low numbers of activated T lymphocytes have surface CD146, CD146-positive cells were additionally stained with a fluorescein conjugated endothelial marker, UEA-1 (ulex europaeus agglutinin-I), to confirm endothelial origin. The cells were then manually counted using a fluorescent microscope and the results were reported as the number of CECs per mL.

The numbers of CECs in ten healthy dogs ranged from 24 to 76 (average 43) per mL of whole blood.

Increased numbers of circulating endothelial cells are thought to be a biomarker of vascular injury and have been observed in human patients with cardiovascular disease, neoplasia, vasculitis, sickle cell anemia, shock, and sepsis. Procoagulant proteins and receptors have been found on CECs and increased numbers of CECs in peripheral blood may also be a potential marker for thromboembolic disease. Immunomagnetic isolation can detect low numbers of cells and is one technique currently used to enumerate human CECs. This study determined that canine CECs can be enumerated with immunomagnetic isolation, however further studies are needed to evaluate the clinical impact of CEC enumeration in different canine diseases.

**MARKERS OF PLATELET REACTIVITY: COATED PLATELETS AND PLATELET -DERIVED MICROPARTICLES.**

**MB Brooks, JL Catalfamo.**
**College of Veterinary Medicine, Cornell University, Ithaca, NY.**

Stimulated platelets promote thrombin generation through a change in composition of their outer membrane surface. The tenase and prothrombinase coagulation complexes assemble on a catalytic surface provided by externalized phosphatidylserine (PS). The term "coated" platelets was coined to describe a subpopulation of highly procoagulant platelets that not only externalize PS, but shed membrane microparticles (MP), and are covered by a stable meshwork of serotonin-derivatized alpha granule proteins (e.g. Factor V, fibrinogen). The physiologic role of coated platelets and platelet-derived MP, and their pathologic significance in human thrombotic syndromes remains controversial. We therefore undertook comparative studies in dogs to determine whether coated platelet formation is a conserved feature of platelet procoagulant response and to assess the thrombin generating potential of MP.

Platelet rich plasma (PRP) was isolated from healthy dogs, diluted in a buffer containing calcium, and stimulated with a series of agonists: thrombin (0.5 U/mL); thrombin/convulxin (0.5 U/mL/500 ng/mL); ionophore A23187 (3 uM); tetracaine (2 mM). The reaction mixtures were labeled with fluorescent probes for flow cytometric detection of 5 parameters of coated platelet formation: Mitochondrial potential (TMRE); PS externalization (AnnexinV-FITC); calcein permeability (calcein-AM); MP-release (CD61-PE); bound fibrinogen (PE-streptavidin-biotinylated fibrinogen). In addition, a fluorimetric assay (Technothrombin TGA, Technoclone) was used to measure thrombin generation in MP-enriched and MP-free plasma fractions. For these experiments, PRP was spun sequentially at 20,000 g to sediment MP pellets and at 100,000 g to isolate particle free plasma (PFP) supernatants.

Thrombin/convulxin induced a canine mean coated platelet response (30% platelets with bound fibrinogen; n=12 dogs) similar to that of healthy people. All agonists induced loss of mitochondrial potential, with response ranging from approximately 8% of platelets (thrombin) to 100% of platelets (ionophore). The remaining response parameters varied among agonists. Ionophore-induced increase in intra-platelet calcium produced the most profound calcein loss, PS externalization, and MP release. Bound fibrinogen was displayed exclusively by thrombin/convulxin treated platelets. Results of TGA revealed that peak thrombin production and total thrombin generating potential of MP were approximately 300 fold greater than PFP.

Canine platelets display a complex activation response, however all of the characteristic features of human coated platelets can be induced with physiologic agonists. Moreover, platelet-derived MP are the primary mediators of thrombin generation in canine plasma. These markers of platelet reactivity provide new tools to characterize the role of procoagulant platelets in canine hypercoagulable syndromes.

# FUNCTIONAL ASSESSMENT OF THE UPPER DIGESTIVE TRACT

**Stanley L. Marks, BVSc, PhD, DACVIM (Internal Medicine, Oncology), DACVN**
**Davis, CA**

## INTRODUCTION

Disorders affecting the oropharynx and esophagus are relatively common in dogs, and are typically associated with dysphagia. Despite the frequency of this problem, the diagnosis of disorders affecting the oropharyngeal, esophageal or gastroesophageal phase of swallowing can be extremely challenging, and warrant a thorough physical examination, followed by clinical observations of the animal eating and drinking. The causes of dysphagia may be secondary to either a neurologic or muscular disturbance of the swallowing reflex (**functional**) or secondary to strictures, traumatic injury, foreign bodies, or neoplastic processes (**structural**) involving the oropharyngeal region or esophagus. The swallowing reflex involves a coordinated process involving the tongue, hard and soft palate, pharyngeal muscles, esophagus, and gastroesophageal junction. The swallowing reflex is also dependent on normally functioning striated muscle and neuromuscular transmission, the integrity of the trigeminal, facial, glossopharyngeal, vagus, and hypoglossal nerves, their nuclei in the brainstem, and the swallowing center in the reticular formation of the brain.

Dogs with an abnormal oral phase of swallowing typically have difficulty with prehension and/or abnormal transport of a bolus to the tongue base. These disorders can often be diagnosed on physical or neurological examination, or by watching the animal eat. Oropharyngeal dysphagias affecting the pharyngeal phase of swallowing can be more challenging to diagnose, and often present with non-specific signs such as gagging, retching and the necessity for multiple swallowing attempts prior to the successful movement of a bolus into the proximal esophagus.[1] These patients have abnormal transport of bolus from the oropharynx to the hypopharynx or hypopharynx to the esophagus. Cricopharyngeal dysphagia is associated with the abnormal transport of a bolus through the proximal esophageal sphincter (PES). Signs are similar to those seen with pharyngeal disorders. Cricopharyngeal dysphagia may result from incomplete or lack of opening of the PES (true cricopharyngeal achalasia) or from abnormal timing of PES opening (cricopharyngeal asynchrony) such that PES opening lags behind bolus presentation from pharyngeal contraction.[1,2] In addition, a subset of dogs have combined disorders affecting both the pharyngeal and cricopharyngeal phases of swallowing, necessitating the implementation of videofluoroscopic swallow studies to accurately diagnose this disorder.

## ASSESSMENT PROCESS

Assessment of dogs with signs and symptoms of feeding and/or swallowing disorders encompass multiple dimensions that include, but may not be limited to: a) review of the signalment; b) review of drug history and history of recent anesthesia; c) physical examination (prefeeding assessment); d) clinical feeding and swallowing evaluation; and e) instrumental evaluation of swallowing.

### Signalment

The importance of the animal's signalment cannot be overemphasized. Puppies can be diagnosed with a variety of congenital functional disorders affecting swallowing, including pharyngeal weakness, cricopharyngeal dysphagia, esophageal dysmotility, and sliding hiatal hernias. Breeds that have a hereditary predisposition or a high incidence of oropharyngeal or esophageal dysphagia include the Golden Retriever (cricopharyngeal dysphagia), Cocker and Springer spaniels (cricopharyngeal dysphagia), English Bulldog (sliding hiatal hernia), Chinese Char Pei (sliding hiatal hernia), Bouvier des Flandres (muscular dystrophy), Boxer (inflammatory myopathy), and German Shepherd (vascular ring anomaly). In addition, large breed dogs are predisposed to masticatory muscle disorders.

### Physical Examination

Physical examination of the animal must include careful examination of the pharynx using sedation or anesthesia if necessary. The pharynx and neck should be carefully palpated for masses, asymmetry, or pain. The chest should be carefully auscultated for evidence of aspiration pneumonia. Evaluation of cranial nerves should be performed including assessment of tongue and jaw tone, and abduction of the arytenoid cartilages with inspiration. Complete physical and neurological examination may identify clinical signs supportive of a generalized neuromuscular disorder, including muscle atrophy, stiffness, or decreased/absent spinal reflexes. Contrary to popular belief, the presence or absence of a gag reflex does not correlate with the efficacy of the pharyngeal swallow nor adequacy of deglutitive airway protection.[3]

### Cervical and Thoracic Radiographs

The pharynx of normal animals is evident on radiographs because it is air filled. The size of the air-filled space can be decreased by local inflammation or neoplasia, laryngeal edema, or elongation of the soft palate. Pharyngeal size can also appear increased with dysfunction of the pharynx or upper esophageal sphincter, chronic respiratory (inspiratory) disease, and chronic severe megaesophagus. The normal esophagus is not visible on survey radiographs. An exception occurs following aerophagia due to excitement, nausea, dyspnea, or anesthesia.

### Other Diagnostic Modalities

Confirmation of disorders affecting cricopharyngeal function, esophageal motility, and gastroesophageal function require additional diagnostic modalities, including videofluoroscopic assessment of swallowing, esophagoscopy, esophageal pH testing with/without concurrent impedance testing, esophageal manometry, and electromyographic analysis. These modalities are complimentary to one another, and their implementation is based on the results of clinical observation of the animal drinking and eating, and clinical suspicion for disorders affecting the oropharynx or esophageal motility. Although esophagoscopy and survey radiography provide anatomic information about the structures involved with the swallowing reflex, neither provides information about esophageal function. This is an important limitation of these diagnostic techniques, particularly in animals that are dysphagic secondary to dynamic disorders such as cricopharyngeal disease or esophageal dysmotility.

**Contrast Videofluoroscopy**

Contrast videofluoroscopy involves real time image capture of the dog as it is swallowing liquid barium or barium-soaked kibble. Videofluoroscopy is used to determine the normal sequence of events that make up a swallow and to measure the timing of these events in relation to one another.[4] Additionally, the movement of certain anatomic structures is measured in relation to a fixed point to further assess function. Swallowing events that occur out of sequence, at inappropriate times or with reduced vigor can cause significant morbidity. For purposes of the radiographic examination of swallowing, it must be remembered that videofluoroscopy captures only a brief window in time and it does not simulate a real meal. A normal study should not be used to conclusively rule out a diagnosis of hiatal hernia in a dog. More recently, a contrast videofluoroscopy method for quantifying pharyngeal contractility has been described in the dog.[5] The pharyngeal constriction ratio (PCR) is calculated by dividing the pharyngeal area at maximum contraction by the pharyngeal area at rest.[5] As pharyngeal contractility diminishes, the ratio approaches 1.0. This simple procedure provides important information regarding the strength of pharyngeal contraction in dysphagic dogs, and facilitates the improved selection of dogs diagnosed with cricopharyngeal dysphagia for surgical intervention.

**Esophageal Manometry**

Esophageal manometry measures pressure within the esophageal lumen and sphincters, and provides an assessment of the neuromuscular activity that dictates function in health and disease. Manometric techniques have improved in a step-wise fashion from a single pressure channel to the development of high-resolution manometry (HRM) with up to 36 pressure sensors.[6] Advances in computer processing allow pressure data to be presented in real time as a compact, visually intuitive "spatiotemporal plot" of esophageal pressure activity. This spatiotemporal plot provides objective measurements of the forces that drive food and fluid from the pharynx to the stomach. This diagnostic modality is limited to academic institutions conducting clinical research studies on esophageal motility and reflux.

**Esophageal pH/Impedance Testing**

Esophageal pH/impedance testing is a useful diagnostic tool that is used to diagnose acid and non-acid reflux in animals with suspected gastroesophageal reflux (GER), unexplained esophagitis, or hiatal hernias. The technology of esophageal pH testing has advanced tremendously in recent years, and clinicians have several choices when selecting esophageal pH probes. The catheter-free Bravo pH Monitoring System from Medtronic is the first catheter-free system used to measure esophageal pH levels in human patients who are suspected of having GER, and is revolutionizing the way esophageal pH testing is done, because it is allows people to maintain their regular diet and activities during pH testing. The Bravo system is an alternative to the traditional pH trans-nasal pH catheter that can cause patient discomfort, and is easily dislodged by dogs and cats if the animal is not closely monitored. The Bravo System consists of two primary components: a) A small pH capsule the size of a gelcap that is attached to the wall of the esophagus and transmits data to a receiver; and b) a pager-sized receiver worn by the patient or animal that receives pH data from the Bravo capsule. After the test is completed, data from the receiver is downloaded to pH analysis software (Polygram™ Net pH Analysis Software) using infrared technology. The main disadvantage of the Bravo system is that one can only record esophageal pH, and the system does not utilize impedance technology that allows one to measure both acid and non-acid reflux. Esophageal pH testing has been extensively utilized in anesthetized dogs in an effort to identify risk factors for GER, and the effects of body position and prokinetic agents on GER.[7,8]

The veterinary profession has made tremendous strides in our ability to evaluate and diagnose disorders of the upper gastrointestinal tract in dogs. The diagnosis of cricopharyngeal and esophageal functional disorders is challenging, but can be facilitated with a comprehensive and structured approach, including observation of the animal eating and drinking, a thorough physical and neurological examination, evaluation of survey radiographs of the chest and neck, and selection of optimal advanced diagnostic modalities based on initial diagnostic findings, clinical suspicion, and availability of these tools.

**REFERENCES**

[1]Davidson AP et al. **Am J Vet Res** 2004;65(3):344-349; [2]Pollard RE, et al. **Vet Radiol Ultrasound** 2000;41(5):409-412; [3]Cook IJ. **Dig Dis** 1998;16:125-133; [4]Kendall KA, et al. **Dysphagia** 2000;15(2):74-83; [5]Pollard RE, et al. **Vet Radiol Ultrasound** 2007;48(3):221-6; [6]Fox MR, et al. **Gut** 2008;57:405-423; [7]Wilson DV, et al. **Am J Vet Res** 2005;66:386-390; [8]Wilson DV, et al. **Am J Vet Res** 2006; 67:16-31.

# NEUROMUSCULAR DISORDERS AFFECTING THE PROXIMAL DIGESTIVE TRACT

G. Diane Shelton, DVM, PhD, DACVIM
La Jolla, CA

## INTRODUCTION

Dysphagia, defined as an abnormality in swallowing, can be one of the most difficult diagnostic challenges encountered in clinical veterinary practice. Normal swallowing is a well-coordinated process involving the tongue, hard and soft palate, pharyngeal muscles, esophagus, and gastroesophageal junction. As the muscles of the proximal digestive tract are predominantly striated, any disease that affects skeletal muscle is a possible cause of dysphagia. In addition to normally functioning striated muscle, the integrity of neuromuscular transmission and of several cranial nerves, including sensory and motor fibers of the trigeminal (CN V) and facial (CN VII) nerves, the glossopharyngeal (CN IX), vagus (CN X), and hypoglossal (CN XII) nerves, their nuclei in the brainstem, and the swallowing center in the brain reticular formation, are critical to the swallowing process. Difficulties in swallowing may be found in very young animals associated with congenital abnormalities, or as an acquired condition in mature animals. This discussion will focus on essential neuromuscular diagnostic procedures and on the most common neuromuscular disorders in which dysphagia may be the sole presenting clinical sign or associated with multiple clinical abnormalities.

## PHYSICAL & NEUROLOGICAL EXAMINATION

The accurate identification of a swallowing disorder is critical to reaching a specific diagnosis and devising a therapeutic plan. Careful physical and neurological examinations, including evaluation of cranial nerves, should be performed including assessment of tongue and jaw tone, pharyngeal tone and visualization of movement of the arytenoid cartilages with inspiration. Complete physical and neurological examinations may identify clinical signs of a generalized neuromuscular disorder including muscle atrophy, stiffness, or depressed or absent spinal reflexes.

Observation of the animal eating and drinking may demonstrate repeated unsuccessful attempts to swallow, excessive head movements, dropping food from the mouth, reingestion of dropped food, and gagging. The gag reflex should be evaluated by placing a finger in the pharynx. The normal response is to repel the finger from the pharyngeal area. An animal with a depressed reflex or absent sensory response may move its head away without actively pushing away the stimulus. Clinical signs of aspiration pneumonia, a common complication, and of esophageal dilation and regurgitation may occur concurrently with pharyngeal dysphagia.

## DIAGNOSTIC TESTING

### Neuromuscular Minimum Database

Once dysphagia has been identified on physical and neurological examination, whether as the only clinical sign or associated with signs of generalized neuromuscular disease, the following minimum database is suggested: a complete blood count, serum chemistries including creatine kinase (CK) activity and electrolyte concentrations, urinalysis, evaluation of thyroid function, and the acetylcholine receptor (AChR) antibody titer for acquired myasthenia gravis (MG). A persistently elevated CK, even if only mildly elevated, could be an indication of an underlying inflammatory myopathy. Markedly elevated CK concentrations may suggest a necrotizing or dystrophic myopathy. Electrolyte abnormalities may be indicative of an endocrine disorder such as Addison's disease. Neuromuscular complications of hypothyroidism have been described.

### Imaging Modalities

Survey neck and thoracic radiographs should be performed and may demonstrate megaesophagus, a cranial mediastinal mass, or metastatic lung disease. Since swallowing is not a static process, motility should be evaluated fluoroscopically using both liquids and solids. While all three phases of swallowing can be evaluated, the three stages of the oropharyngeal phase can be separately evaluated.[1,2] Most recently, magnetic resonance imaging (MRI) has been used to identify inflammatory lesions in difficult to diagnose cases of dysphagia.[3] In inflammatory myopathies, cellular infiltrates can have a patchy or multifocal distribution and be missed on individual biopsy specimens. Thus, MRI examination can aid in locating lesions and guiding a diagnostic biopsy, or for follow-up examinations when repeat biopsies are not available. Areas of hyperintensity may also be identified on screening examinations when the diagnosis is in question. Since dysphagia may be the primary clinical abnormality in dogs with inflammatory myopathies [4] without other clinically evident signs of generalized weakness, imaging studies should be performed as an early diagnostic procedure. Neoplasia involving cranial nerves may also be evident on MRI evaluations. If a central cause of pharyngeal dysphagia is suspected based on the clinical signs and neurological examination, evaluation of the brain by MRI should be performed along with concurrent evaluation of the cerebrospinal fluid. Evaluation of intracranial diseases is described in standard textbooks of veterinary neurology.

### Electrodiagnostic Testing

Electrodiagnostic evaluation, including electromyography and measurement of motor and sensory nerve conduction velocities, can provide important information as to the severity, distribution, and character of a myopathic or neuropathic disease process. In addition to evaluation of limb muscles, testing should also include evaluation of the pharyngeal and laryngeal muscles, and the tongue. Results of these studies can also help guide the direction of biopsy procedures. Electrodiagnostics are performed by specialists trained in the procedures, are only available at specialized centers, and performed under general anesthesia. As a lengthy anesthesia is required for these procedures, the health status of the animal must be taken into consideration, weighing the risks and benefits. In cases of suspected MG, evaluation of the serum AChR antibody titer should be performed prior to anesthesia. If the antibody titer is negative on at least 2 separate occasions, and a diagnosis of MG is still suspected, electrodiagnostic testing can then be performed to measure the decremental response of the compound muscle action potential in response to repetitive nerve stimulation if the dog is stable and aspiration pneumonia under control. In most cases electrodiagnostics will not provide a specific diagnosis, but can give an indication as to whether the underlying disease process is myopathic or neuropathic, and to the pattern of muscle and peripheral nerve involvement. A complete discussion of electrodiagnostic testing is described elsewhere.[5]

**Muscle & Nerve Biopsies**

The goal of diagnostic testing is to identify a specific disease so an optimal treatment plan can be formulated. With the exception of MG which is best diagnosed by the serum AChR antibody titer, and the congenital ion channelopathies including myotonias which are best identified by electromyography, the muscle biopsy, and in some cases the peripheral nerve biopsy, is the only way to reach a specific diagnosis. If the problem is in the muscle or nerve, go to the source of the problem. As with most treatable neuromuscular diseases, it is critical to reach an accurate diagnosis early in the course of the disease, before fibrosis and myofiber loss is extensive and treatment options are few. In cases of pharyngeal dysphagia, a biopsy should be obtained from a pharyngeal muscle if possible, in addition to a biopsy from a large proximal pelvic limb muscle such as the vastus lateralis or a thoracic limb muscle such as the triceps muscle. If a pharyngeal muscle biopsy is not possible, a limb muscle biopsy may provide the diagnosis. A complete discussion of how to correctly collect and prepare muscle and peripheral nerve biopsy specimens is described elsewhere.[6]

## NEUROMUSCULAR DISEASES COMMONLY ASSOCIATED WITH DYSPHAGIA

**Acquired Myasthenia Gravis**

Acquired MG is the most commonly diagnosed neuromuscular disorder in this author's laboratory, and should be high on the list of differential diagnoses in dogs with acquired dysphagia, whether alone or as part of a generalized neuromuscular disorder. A broad spectrum of presenting clinical signs can be associated with MG and clinicians should be aware of the many presentations. The gold standard for the diagnosis of acquired MG continues to be the demonstration of acetylcholine receptor antibodies by immunoprecipitation radioimmunoassay. In a study of a large group of dogs with acquired MG,[7] focal signs including pharyngeal, esophageal, and laryngeal weakness, without clinically detectable limb muscle weakness, were described in 43% of dogs. Pharyngeal weakness, as the only clinical sign of MG, was described in 1% of the myasthenic dogs. Focal clinical signs, including megaesophagus and dysphagia, were also described in 14% of cats in a similar study.[8] Myasthenia gravis is a treatable disease if diagnosed early and treated appropriately. Although information is not yet available for cats, most dogs that survive an aspiration pneumonia early in the course of the disease and do not have a thymoma, should go into spontaneous remission with resolution of clinical signs.[9]

**Generalized or Focal Inflammatory Myopathies/Myositis**

The most common inflammatory myopathies (IMs) in dogs include the highly specialized and focal immune-mediated masticatory muscle myositis; immune-mediated polymyositis with variable involvement of many muscle groups including the pharyngeal, laryngeal, masticatory and limb muscles, and the tongue and heart; and generalized myositis from infectious causes.[10] Extraocular myositis is a less common focal form of immune-mediated myositis. Although dysphagia can occur as the predominant clinical sign in IMs, most commonly it is associated with other muscle groups as part of a polymyositis syndrome. Involvement of other muscle groups may be clinically evident or mild and subclinical. The serum CK activity may be mildly to moderately elevated. However, a normal CK activity should not rule out a diagnosis of myositis. Chronic dysphagia was described as the primary clinical abnormality in two Boxer dogs with IM.[4] Recently, a syndrome has been identified in Vizsla dogs in which dysphagia, excessive salivation, masticatory muscle atrophy and esophageal dilatation were described (September 2007 Case of the Month on http://vetneuromuscular.ucsd.edu, http://vizslamyositis.blogspot.com/). Evaluation of muscle biopsy specimens is critical for the diagnosis of an inflammatory myopathy. Immunohistochemical demonstration of CD8+ T cells and upregulation of MHC Class I in biopsy specimens is proving to be a valuable adjunct to the diagnosis of immune-mediated myositis when the diagnosis is in question.[11] Reaching an early and accurate diagnosis in IMs is important to a good clinical outcome. Once muscle mass is lost and fibrosis significant, the chance of regaining normal muscle mass and function is poor.

**Muscular Dystrophies & other Congenital (Inherited) Myopathies**

Muscular dystrophies (MD) are a diverse group of inherited myopathies that have been reported in several species including humans, dogs, cats, mice, hamsters and chickens. The most common form of human MD (and probably of dogs and cats) results from mutations in the gene encoding for dystrophin. Other relatively common forms of MD result from mutations in genes encoding dystrophin-associated glycoproteins within the dystrophin-glycoprotein complex (DAG). The DAG complex is a transmembrane protein complex that links the muscle cytoskeleton to the extracellular matrix and is thought to confer structural stability to the sarcolemma during contraction. There are over 30 molecularly characterized forms of muscular dystrophy (MD) in humans with some still unclassified. Only a few different forms of MD have been identified to date in dogs and cats. Dystrophin deficient MD has been described in many breeds of dogs and the mutation identified in the Golden Retriever, Rottweiler, Labrador Retriever, and most recently, in the Cavalier King Charles Spaniel (Piercy 2009, personal communication). Sarcoglycan deficient MD has been identified in the Boston Terrier,[12] laminin alpha-2 deficiency in cats and a dog,[13,14] and recently, MD associated with deficiency of alpha-dystroglycan in Sphynx and Devon Rex cats was reported.[15] While the clinical presentation in these dystrophies is an early age of onset and a diffuse neuromuscular disease, dysphagia and regurgitation can occur as a result of hypertrophy of the pharyngeal, esophageal, or lingual musculature. A degenerative myopathy presenting with dysphagia has been reported in Bouvier des Flandres dogs.[16] Bouvier dogs were described as having a muscular dystrophy, although results of immunostaining or immunoblotting for dystrophy related proteins, or identification of a specific mutation, have not yet been reported. Older onset forms of MD, such as oculopharyngeal MD, have been described in humans but not in yet in dogs or cats. Clinical signs are slowly progressive with prominent dysphagia. As more cases are studied in dogs and cats, and as the spectrum of dystrophic diseases is expanded, it is likely that older onset forms of MD will be identified.

Other non-inflammatory and congenital myopathies have been reported in dogs and cats[17] including nemaline rod myopathy, centronuclear myopathy, myotubular myopathy, and myofibrillar myopathy with desmin accumulations. Where tissues have been evaluated, abnormal cytoarchitectural changes have been identified in the esophageal and pharyngeal musculature.

### Cranial & Peripheral Neuropathies

Pharyngeal weakness and laryngeal paralysis are commonly associated with generalized peripheral neuropathy syndromes in many breeds of dogs. Clinical signs of dysphagia and dysphonia may occur early in the course of the disease or later with chronicity and progression. In many cases, a specific underlying cause cannot be identified and most are considered degenerative in nature. Cranial neuropathies can present with clinical signs of abnormal prehension or tongue paralysis if there is involvement of cranial nerves V, XII or the caudal brainstem; with dysphagia if involvement of cranial nerves IX, X and the caudal brainstem; or with megaesophagus if there is involvement of cranial nerve X.

## PROGNOSIS

The prognosis is dependent on early recognition of dysphagia, identification of an underlying cause, and the presence of a treatable disease. While the most common causes of acquired dysphagia such as MG, myositis, and possibly the endocrine myopathies associated with hypothyroidism and Addison's disease are treatable, they must be diagnosed early in the course of the disease for the best clinical outcome. Many neuromuscular diseases are not treatable, such as the muscular dystrophies and most of the peripheral neuropathies, and the prognosis for recovery is poor in these cases. Aspiration pneumonia and severe debilitation may be fatal complications.

## REFERENCES

[1]Suter PF, et al. **Vet Radiol** 1980;21(1):24, [2]Watrous BJ, et al. **Vet Radiol** 1983;24(1):11, [3]Platt SR, et al. **Vet Radiol Ultrasound** 2006;47(6):532, [4]Ryckman LR, et al. **J Am Vet Med Assoc** 2005;226(9):1519, [5]Cuddon PA. **Vet Clin North Am Small Anim Pract** 2002;32(1):31, [6]Dickinson PJ , et al. **Vet Clin North Am Small Anim Pract** 2004;34(6):63, [7]Shelton GD, et al. **J Am Vet Med Assoc** 1997;211(11):1428, [8]Shelton GD, et al. **J Am Vet Med Assoc** 2000;216(1):66, [9]Shelton GD, et al. **Neurology** 2001;57(11):2139, [10]Shelton GD. **Neuromuscul Disord** 2007;17(9-10):663, [11]Van der pas J, et al. **J Neurol Neurosurg Psychiatry** 2004;75(1):136, [12]Dietz K, et al. **J Vet Intern Med** 2008;22(2):476, [13]O'Brien DP, et al. **J Neurol Sci** 2001;189(1-2):37, [14]Poncelet L, et al. **J Small Anim Pract** 2003;44(12):550, [15]Martin PT, et al. **Neuromuscul Disord** 2008;18(12):942, [16]Peeters ME, et al. **Vet Rec** 1994;134(17):444, [17]Shelton GD, et al. **Neuromuscul Disord** 2005;15(2):127.

## REVISITING TREATMENT OF DYSPHAGIA

Stanley L. Marks, BVSc, PhD, DACVIM (Internal Medicine, Oncology), DACVN
Davis, CA

### INTRODUCTION

Dysphagia, or difficulty swallowing, can result from an abnormality in the oropharyngeal, esophageal or gastroesophageal phase of swallowing. Oropharyngeal dysphagias include abnormalities affecting the oral, pharyngeal or cricopharyngeal phases of swallowing.[1] Cricopharyngeal dysphagia is a swallowing disorder of the upper esophageal sphincter (UES) characterized by either cricopharyngeal asynchrony or cricopharyngeal achalasia.[1,2] Cricopharyngeal asynchrony is the incoordination between the contraction of the dorsal cranial and middle pharyngeal contractor muscles (hyopharyngeus, pterygopharyngeus, and palatopharyngeus mm) and opening of the upper esophageal sphincter (cricopharyngeus and thyropharyngeus mm).[2] Cricopharyngeal achalasia is the inability of the cricopharyngeus muscle to open during the cricopharyngeal phase of swallowing.[2] The exact underlying etiologies have not been determined, although the disorder can be reproduced by transection of the pharyngeal branch of cranial nerve X.[1] Cricopharyngeal dysphagia and achalasia are diagnosed via contrast videofluoroscopy.

In contrast, pharyngeal dysphagia is the interruption of the transport of the food bolus from the oropharynx to the hyopharynx, and impairment of the initiation of the involuntary portion of the swallowing reflex.[1] Morphologic causes for interference with the pharyngeal phase of swallowing include inflammation, penetrating foreign bodies, cleft palate, and iatrogenically shortened soft palate. Clinical signs of pharyngeal dysphagia are similar to cricopharyngeal dysphagia. Fluoroscopy demonstrates the absence of a strong aboral pharyngeal contraction with an incomplete bolus transport. Differentiation between cricopharyngeal and pharyngeal dysphagia is crucial as the treatment for cricopharyngeal dysphagia (cricopharyngeal myotomy or myectomy) can exacerbate signs caused by pharyngeal dysphagia.[3]

### Treatment of Cricopharyngeal Dysphagia

Definitive treatment for cricopharyngeal dysphagia has been reported with myotomy or myectomy of the cricopharyngeal muscles; however, a recent study documented a high failure rate in 14 dogs with cricopharyngeal dysphagia following cricopharyngeal myotomy and myectomy.[4] Clinically, six dogs showed no improvement after surgery, three of which worsened postoperatively. Eight of the 14 dogs were euthanized due to problems related to cricopharyngeal dysphagia, including persistent dysphagia (8 dogs) and aspiration pneumonia (5 dogs). This study underscored the importance of critically assessing dogs for surgical intervention, and has revised our therapeutic approach to dogs with cricopharyngeal dysphagia. Likely reasons for the high failure rate observed in the dogs of the study following myotomy or myectomy of the cricopharyngeus muscle included incomplete work-up of dogs prior to surgical intervention, resulting in failure to accurately diagnose dogs with focal myasthenia gravis, laryngeal paralysis, and other underlying polyneuropathies/myopathies. In addition, a few dogs with severe aspiration pneumonia had exacerbation of their pneumonia following the surgical procedure, and these dogs should have been better stabilized prior to surgery. Finally, dogs with concurrent pharyngeal weakness and cricopharyngeal dysphagia likely had exacerbation of their dsyphagia following the myotomy or myectomy procedure. For these reasons, all dogs with cricopharyngeal dysphagia are currently worked up comprehensively to assess pharyngeal function via fluoroscopy, rule out concurrent or systemic polyneuropathies/myopathies. Dogs that are diagnosed with underlying neuropathies are managed conservatively with alterations of feeding practice or low-profile gastrostomy devices if the underlying neuropathy or myopathy cannot be specifically managed.

In veterinary medicine, the standard surgical approach for myotomy or myectomy has remained constant over the years. The cricopharyngeal and thyropharyngeal muscles can be approached by either a standard ventral midline approach, with 180° rotation of the larynx on its longitudinal axis[5] or via a lateral approach with 90° rotation of the larynx.[6] Cricopharyngeal myotomy involves transecting the cricopharyngeal muscle to the level of the pharyngeal mucosa.[5] Endoscopic $CO_2$ laser cricopharyngeal myotomy is being increasingly utilized in people, and this procedure has been shown to be safe and effective, with decreased anesthesia time and morbidity compared with the more traditional transcervical cricopharyngeal myotomy.[7] This procedure has not been performed in the dog to date, although it may well have application in appropriately selected animals.

Dogs with cricopharyngeal achalasia which is less common than cricopharyngeal dysphagia, can be managed surgically with myotomy or myectomy of the cricopharyngeus muscle. An alternative and less invasive procedure involves the injection of botulinum toxin injection into the cricopharyngeus muscle.[8] Botulinum toxin A (BTA) is a neurotoxin synthesized from the bacillus *Clostridium botulinum*. It acts at the presynaptic cholinergic nerve terminals to block the release of acetylcholine at the myoneuronal junction. In a dose-related manner, it weakens contraction when injected into the target muscle.[8] The toxin has been used successfully in people for the treatment of esophageal achalasia, a condition characterized by hypertonicity of the lower esophageal sphincter. It has also been reported to be of benefit for the management of people with cricopharyngeal achalasia. For clinical use, botulinum toxin activity is defined in units such that one unit represents the median lethal dose in mice. Because of its short half-life (4 h), the toxin is reconstituted shortly before injection with 0.9% sterile saline to a concentration of 25 units/cc. The average injection dose in people ranges between 25-50 units.[8] In people with cricophayryngeal achalasia, physicians frequently balloon dilate the proximal esophageal sphincter during the botulinum injection procedure. We have successfully managed a dog with severe cricopharyngeal achalasia with injection of botulinum toxin, although the benefits of the toxin wore off after approximately 4 months. This is an expected and well-documented phenomenon in people, and can be easily reapplied if warranted. Alternatively, a more permanent myotomy or myectomy of the cricopharyngeus muscle can be performed at this time. The limited duration of botulinum toxin's effect is a benefit, as animals that respond favorably to the toxin should do well following surgical myotomy or myectomy. In contrast, animals that do poorly following botulinum injection can be supported with an enteral feeding device until the effects of the toxin have worn off. These animals should not undergo surgical correction of their cricopharyngeal disease, as the disorder will be exacerbated. Dogs diagnosed with concurrent cricopharyngeal dysphagia associated with dyssynchrony should not receive botox or surgical intervention. A comprehensive search for an underlying neuropathy/myopathy should be performed, and if unable to be documented, these dogs should be managed with a low-profile gastrostomy device.

**Treatment of Gastroesophageal Reflux Disease**

Gastroesophageal reflux disease (GERD) is a common chronic disorder in the Western world. The basic cause of GERD has been well characterized—the fundamental defect is a loss of integrity of the gastroesophageal barrier. What is less clear is the most appropriate means of addressing this reflux. GERD has a variety of symptoms in people, ranging from typical presentations of heartburn and regurgitation (without esophagitis) to atypical presentations, such as severe erosive esophagitis and its associated complications. Because of its symptomatic diversity, physicians may select from a variety of therapeutic approaches.

In dogs, most cases of GERD occur secondary to hiatal hernia; however, the author has documented cases of severe flaccidity of the lower esophageal sphincter in the absence of a hiatal hernia. Medical therapy aims at decreasing acidity by suppressing proton secretion and has been well established. Available medications include antacids and alginates, H2-receptor antagonists, motility agents, and proton pump inhibitors (PPIs). All of these medications do little to prevent the reflux of gastric contents, and antireflux surgery is usually warranted. In people, antireflux surgery is commonly performed laparoscopically, and is aimed at reinforcing and repairing the defective barrier through plication of the gastric fundus.[9] The earliest performed successful procedures were the Nissen and Toupet fundoplications, to which several modifications have since been made. It has been demonstrated in preliminary studies and long-term outcomes of such open surgery and preliminary studies of such laparoscopic surgery that antireflux surgery is an effective approach, with overall outcomes superior to those achieved with medications.[9]

**Endoscopic Antireflux Therapies**

Endoscopic implantation of bovine dermal collagen and Teflon paste injected into the lower esophageal sphincter of dogs with surgically induced GERD was first reported in the late 1980's.[10] In recent years, endoscopic intraluminal antireflux approaches have attracted the attention of physicians, surgeons, and commercial companies, especially after the approval of two endoscopic intraluminal methods by the United States FDA in 2000. The common element is prevention of acid reflux by construction of a functional or controlled barrier in the lower esophageal sphincter zone. Three main methods are currently employed: endoscopic intraluminal valvuloplasty,[11] endoscopic radiofrequency therapy,[12] and endoscopic injection or implantation of foreign material.[13] The endoluminal suturing method is highly demanding technically, and its short-term results are encouraging, although largely dependent on the experience of the endoscopist. Several prospective cohort studies have shown that the radiofrequency procedure (Stretta®) significantly improves GERD symptoms and quality of life while reducing esophageal acid exposure and eliminating the need for antisecretory medications in the majority of patients within 6–12 months. Most recently, some researchers have studied the endoluminal implantation of polymers, such as Plexiglas™ (polymethyl-methylacrylate), Gatekeeper® hydrogel, and Enteryx® (ethylene vinyl alcohol copolymer).[14] The preliminary results of these studies showed that the implantation method was feasible and safe; however, the only multicenter trial related to outcome that has been published has included just 1 year of follow-up.

It is clear that there are a plethora of research studies using animal models to evaluate less-invasive approaches to prevent reflux in people. These models provide an ideal template to further our investigations in an effort to provide additional options for the surgical and endoscopic management of congenital or acquired hiatal hernias and reflux disease in the dog.

**REFERENCES**

[1]Watrous BJ. **Vet Clin North Am Small An Pract** 1983;3: 437-459; [2]Suter PF, et al. **Vet Radiol** 1980;21:24-39; [3]Allen SW. **Prob Vet Med Head Neck Surg** 1991; 3:290-297; [4]Warnock JJ. **J Am Vet Med Assoc.** 2003;223(10):1462-8; [5]Shelton, GD. The Esophagus. In: Slatter D, ed. **Textbook of Small Animal Surgery, 2nd Ed.** WB Saunders Company: Philadelphia, 1993; 553-557; [6]Niles JD, et al. **J Small Anim Pract** 2001; 42: 32-35; [7]Pitman M, et al. **Laryngoscope** 2009;119:45-53; [8]Haapaniemi JJ, **Dysphagia** 2001;16(3):171-5; [9]Peters JH. **Ann Surg** 1998;228:40-50; [10]O'Connor KW. **Gastrointest Endosc** 1988;34:106-112; [11]Mason RJ. **Gastrointest Endosc** 1997;45(3):283-90; [12]Comay D. **Can J Gastroenterol** 2008 Jun;22(6):552-8; [13]Domagk D. **Am J Gastroenterol** 2006 Mar;101(3):422-30; [14]Zhi XT. **J Long Term Eff Med Implants** 2005;15(4):375-88.

**ESOPHAGEAL STRICTURES – UPDATE ON THERAPEUTIC OPTIONS**

**Keith Richter, DVM, DACVIM**
**San Diego, CA**

## INTRODUCTION

Esophageal strictures are being recognized more commonly in dogs and cats. Despite the fact that we continue to learn more about possible causes, risk factors, and potential preventative measures of esophageal strictures, there remains a lack of evidence based medicine assessing optimal treatment strategies. Thus, treatment decisions are often based on individual experience and reports from the human literature. This lecture will summarize some of the current thoughts on esophageal strictures in dogs and cats, and give the author's opinions on some unresolved issues regarding balloon dilatation and esophageal intraluminal stent placement.

## ETIOLOGY & CLINICAL SIGNS

The causes of esophageal strictures may be divided into benign and malignant causes. Malignant strictures are rare in the esophagus, and are beyond the scope of this discussion. The most common cause of benign esophageal stricture is secondary to reflux during anesthesia. Two separate studies reported that this occurred in 64% and 65% of cases. The median onset to clinical signs was 7.5 days post-anesthesia, with a range of one to 32 days. It is presumed that decreased lower esophageal sphincter tone occurring during anesthesia allows for gastroesophageal reflux, which results in subsequent acid injury (peptic injury) to the esophageal mucosa. It is estimated that over 270-degrees of mucosal injury is required to result in stricture formation.

The next most common cause is pill-induced esophagitis with subsequent stricture. Doxycycline has been the drug most frequently reported in association with the formation of benign strictures, though clindamycin hydrochloride and aspirin have also been implicated. The proposed mechanism of tablet-induced focal esophagitis and subsequent esophageal stricture formation is tablet retention in the esophagus (resulting in esophagitis) due to poor esophageal clearance with a "dry" swallow (defined as swallowing the pill without following with water or food). In human beings, less than 10% of tablets rapidly pass into the stomach after a "dry" swallow. This number increases to >95% after 6-ml of water are ingested after the pill. Similar findings have been described in cats. In one study, 63% of tablets and 83% of capsules did not pass into the stomach after five minutes following a "dry" swallow, whereas 100% reached the stomach within two minutes if 6-ml of water were subsequently ingested. Additional causes of benign esophageal stricture include vomiting of gastric contents, secondary to damage following foreign body ingestion and retention in the esophagus, and secondary to swallowing of caustic substances.

Regardless of the initiating event, once esophagitis develops, there is a reflex decrease in lower esophageal sphincter tone. This results in more acid reflux and subsequent worsening of the esophagitis. This vicious cycle can eventually result in potentially more damage to the esophagus. Once severe esophagitis is present, damage can extend to the lamina propria and muscularis layers. This incites a fibroblastic proliferation and contraction, leading to stricture formation. Strategies to prevent stricture formation once esophagitis develops include administration of cisapride (to increase lower esophageal sphincter tone) and administration of a proton pump inhibitor or histamine H2 receptor antagonist (to decrease gastric acid formation). Administration of sucralfate is of questionable efficacy because of the lack of acidic environment in the esophagus (though gastroesophageal reflux may reduce esophageal pH enough to make it effective).

Clinical signs of esophageal stricture include pain during swallowing, dysphagia, increased salivation, regurgitation, anorexia, and weight loss. These signs tend to be progressive as the stricture gets progressively more narrowed.

## TREATMENT OF BENIGN ESOPHAGEAL STRICTURES

The first line treatment of benign esophageal strictures is mechanical dilatation of the stricture. Techniques have evolved from whale bones (for achalasia), to rigid bougienage, to flexile bougies, to balloon dilation. The latter technique is generally thought to be superior to bougienage, but objective evidence for this is lacking in both human and veterinary patients. The theoretical advantage of balloon dilatation is that the forces applied to the stricture are a radial stretch, whereas some of the forces applied with bougienage also are longitudinal, thus resulting in greater potential for esophageal perforation.

The technique of balloon dilatation varies depending on available equipment and balloon styles. Esophageal dilatation balloons are made of special plastic that makes the balloon extremely rigid when maximally inflated at high pressures, generally up to 45 psi (unlike the balloon of a Foley catheter or endotracheal tube). These balloons are readily available in various diameters, up to 20-mm. Balloons larger than this however, are necessary for large breed dogs, and can be custom made by certain manufacturers. Commercially available achalasia balloons are rated at too low of an inflation pressure to dilate very firm strictures. If these balloons are used, fluoroscopy should be used to image the dilation to confirm stricture dilatation. Balloons are made to be passed through the channel of an endoscope (TTS™ meaning through the scope) or passed guided by the use of a guide wire. The author generally prefers the latter system. If a guide wire is used, the soft end is passed through the channel of the endoscope (or imaged with fluoroscopy) and advanced beyond the stricture into the stomach or caudal esophagus. The scope is removed as the guide wire is advanced through the channel, thus leaving the guide wire near its original position. Fluoroscopy is helpful for this, but not necessary. The balloon catheter is then passed over the guide wire (with the balloon deflated) until it is positioned within the stricture. Alternatively, if a TTS™ balloon is used, it is advanced through the cannel of the scope and positioned within the stricture under direct endoscopic visualization. The position of the balloon is verified by visualizing radiopaque markers defining the ends of the balloon, either with direct visualization through the endoscope or via fluoroscopy. The balloon is slowly inflated until it reaches the manufacturer's rated pressure for that particular balloon. This is accomplished with an inflating device attached to a pressure transducer. Typically the balloon is kept in an inflated state for one to two minutes before it is deflated. Sequentially larger balloons are used until the clinician subjectively judges the degree of mucosal tearing to be acceptable.

The technique is very subjective in veterinary medicine, with many variables not defined by controlled studies. These include the sequence of dilatation, the optimal final dilation diameter, use of corticosteroids, post-dilation drugs, post-dilation feeding regime, the elective use of percutaneous endoscopic gastrostomy (PEG) tubes, and repeat elective dilatations (and at what interval). The following discussion describes the regime used at the author's hospital, but it must be emphasized that these opinions have not been rigorously studied in dogs and cats. Typically in our hospital, an initial balloon diameter is selected that is 50-100% larger than the estimated stricture diameter. If there is doubt, selection of a smaller balloon with gradually larger subsequent dilations is safest. The sequence of subsequent larger dilations is then determined by the degree of mucosal tearing. The final dilation diameter is usually chosen by the degree of mucosal tearing and the size of the patient. As a general guideline, the end diameter is as follows: cats and dogs <8 kg: 12-mm; dogs 8 to 15 kg: 16-mm; dogs 15-30 kg: 20-mm; dogs 30-50 kg: 25-mm; and dogs >50 kg: 30-mm. This is very subjective, and is a balance between the desire to achieve a large enough opening at the completion of procedure and the risk of perforation when too large a balloon is used.

The use of corticosteroids has been studied in human patients with esophageal strictures, and allows guidelines to be considered for veterinary patients. Oral corticosteroids are not considered effective. Intralesional corticosteroids are considered likely effective. One randomized trial of human patients (n = 30) with recurrent peptic stricture involved injection of intralesional triamcinolone in a four quadrant technique (or sham injection). The group that received the triamcinolone injections had longer dilation-free time intervals. Only 13% of the patients who received triamcinolone required repeat dilations compared with 60% of the sham-injected patients. The author uses a 21-gauge flexible injection needle through the channel of the endoscope, and prefers to inject the triamcinolone prior to dilation. This is because it is easier to visualize the needle entering the submucosa and subsequent injection bleb (whereas after the dilation, the mucosal damage and hemorrhage makes visualization more difficult). Approximately 0.5-cc is injected into each site in a four quadrant pattern.

Following the procedure, the author does not use a PEG tube and allows oral feeding with a canned food consistency. There is no evidence that the esophageal passage of food with normal bolus size results in exacerbation of esophagitis and increases the risk of re-stricture. Medications for esophagitis are used, including cisapride (0.5-0.75 mg/kg tid) and omeprazole (0.7-1 mg/kg sid). As mentioned above, the use of liquid sucralfate is of uncertain benefit, but unlikely of any detriment. Some authors recommend elective dilations at one week intervals to decrease the likelihood of re-stricture formation. The author does not routinely recommend this since the published median number of dilations required for successful treatment is two (range 1-5). If three dilations are required, then the author often recommends an elective dilation one week later. The overall successful treatment rate is reported to be between 70% and 88% in dogs and cats.

**Using Intraluminal Stents for Treatment Failures**

There is an increasing interest in the placement of intraluminal stents for treatment failures rather than surgical correction. Stents are used in human patients for malignant strictures or complicated benign strictures. The latter includes recurrent strictures (>5 to 10 dilations), long strictures (> 2-cm), tortuous, or narrow strictures. Very little information is available for esophageal stent placement in veterinary patients, and much of the information must be extrapolated from experience in human patients. Stents are available both covered (usually with a polypropylene covering around the outer surface of the stent) and uncovered. This covering prevents ingrowth of tissue within the stent. Available stent composition materials include Nitinol (nickel plus titanium), Elgiloy (cobalt, nickel, plus chromium), stainless steel, polyester plastic/silicone, or a biodegradable material (such as PDS™). These differ in radial force, buckling characteristics, degree of shortening upon deployment, or releasing proximally versus distally. Selection is influenced by stricture characteristics, including straight or angulated, fully or semi-circumferential, intrinsic or extrinsic, location, and benign versus malignant. Currently the most common indication for stent placement in veterinary patients is failure of balloon dilatation (usually > 5 dilations). Success has been described for biodegradable stents (and thus eliminating the need for stent removal) and Nitinol stents. The size of the deployed stent is usually chosen to be slightly smaller than the esophagus after balloon dilatation. Once the stent is deployed, it must be anchored in place or it will rapidly migrate into the stomach. If the cranial aspect of the stent is in the cervical esophagus, it is sutured in place through a small surgical incision. If the stent is completely within the thoracic esophagus, it can be sutured in place using a suturing device (GI Stitch™, Pare Surgical) that can be used through a double channel endoscope. The stent can also be sutured via a limited thoracotomy (or thoracoscopy). Alternatively, selection of a stent that is longer than needed will allow suturing to the cervical esophagus, thus avoiding a thoracotomy.

Complications of stent placement include stent migration, tissue ingrowth into the stent resulting in stent occlusion, food obstruction, hemorrhage, perforation, airway compression, pressure necrosis/fistula formation, dysphagia, and pain.

## REFERENCES

[1] Adamama-Moraitou KK et al. **Can J Vet Res** 2002; 66:55. [2] Leib MS et al. **J Vet Intern Med** 2001; 15:547. [3] Ramage JI et al. **Am J Gastroenterol** 2005; 100:2419. [4] Saito Y et al. **Dig Dis Sci** 2008; 53:330. [5] Shah JN. **Gastrointest Endoscopy** 2006; 63:164. [6] Siersema PD. **Nat Clin Pract Gastro Hepatol** 2008; 5: 142. [7] Tierney W et al. **Gastrointest Endoscopy** 2006; 63:920. [8] Westfall DS et al. **J Vet Intern Med** 2001; 15:467.

# ANIMAL WELFARE REGULATION IN CANADA & THE UK

**Michael S. Cockram BVetMed, PhD, MRCVS**
**Charlottetown, PEI**

## INTRODUCTION

Animal welfare regulation in Canada and the UK will be compared. Some of the reasons for differences between the countries will be speculated and the elements required for comprehensive and effective animal welfare regulation will be discussed.

## ANIMAL WELFARE REGULATION

"Regulation, in its broadest sense, is a principle, rule or condition that governs the behaviour of citizens and organizations."[1] Governments use regulation to achieve policy objectives. The extent to which legal regulation is used for the regulation of animal welfare issues differs between Canada and the UK. In Canada, there is federal enforcement of regulations on transport and slaughter, but provinces have the primary responsibility for the protection of animals in other circumstances, with minimal national co-ordination. The Criminal Code of Canada provides some anti-cruelty protection, but is deficient in the protection that it offers for the neglect of animals.[2] All provinces have legislation on animal cruelty, but in Quebec this only applies to dogs and cats, and some provinces have specific regulations for farm animals. Provincial animal welfare legislation is enforced by either police officers, officers working for the provincial society for the prevention of cruelty to animals, or provincially appointed veterinarians. Many provinces provide funding for enforcement, but in some cases this is left to self-funded or special interest groups [3,4]. In the UK, Government policies are framed within the context of European Union (EU) law. The EU lays down minimum welfare standards for farmed animals by means of Directives on the welfare requirements of specific categories of farmed animals (e.g. poultry, calves and pigs), transport and slaughter. The EU also undertakes regular monitoring of member states, such as the UK, to ensure compliance with the directives. Other international institutional arrangements are likely to be increasingly important to both the UK and Canada. The World Organisation for Animal Health (OIE) is responsible for the oversight of animal diseases and provides recommendations and guidance covering animal health and animal welfare. OIE health standards are recognised by the World Trade Organisation (WTO), which governs international trade agreements. In the UK, animal welfare is a devolved responsibility. There is a separate Animal Welfare Act in England and Wales from that in Scotland, and the secondary legislation is different in England from that in Wales. There are in general two types of animal welfare legislation; anti-cruelty legislation and legislation that is proactive in requiring a certain level of care for the animals, to provide a minimal standard of animal welfare provision and to prevent unnecessary suffering.[4] Although in the UK both types of legislation were in place for farmed and research animals, recent changes have extended a duty on the person responsible for a domesticated vertebrate animal to take reasonable steps to ensure that the needs of the animal are met to the extent required by good practice. These include its need for: a suitable environment; a suitable diet; to exhibit normal behaviour patterns; any need to be housed with, or apart from, other animals; and to be protected from pain, suffering, injury and disease. Many of the regulations either ban or place restrictions on several types of intensive animal production systems. Inspections are performed by government veterinarians, animal health officers, local authority inspectors and Government agencies (e.g. the Meat Hygiene Service) to enforce animal welfare regulations that affect animals kept by man, including livestock on-farms, in-transit, at markets and at slaughter; companion animals, horses, zoos, research laboratories.[5]

### What are the differences and why do they exist?

Compared to the UK, Canada appears to rely mainly on self-regulation. Although there is some provincial legislation, a national infra-structure for comprehensive and effective government regulation to ensure the welfare of animals is not present in Canada. However, the need for more comprehensive and/or additional farm animal welfare standards or regulations in Canada has been recognised.[6]

### Public views

A major factor influencing government policy is public opinion. In the UK, animal welfare is a topic of public concern that is expressed through political activity, at times direct action and increasingly by consumer action, directly through consumer choice, but often through supermarket driven quality assurance schemes. In the UK, regulation to protect animal welfare is viewed by the public to be important in avoiding cruelty[7]. Animal welfare organisations are well funded by public donations and are active in producing campaign materials, influencing the media and in lobbying politicians at national and European level. This has resulted in detailed legislation. Animal welfare as an issue has less public profile in Canada than in the UK. In a recent survey, less than 8% of Canadian consumers considered that the humane treatment of animals used to produce food in Canada was poor.[8]

### Government policy

Agriculture and Agri-Food Canada operates to ensure that Canada's agriculture and agri-food sector is able to compete in international and domestic markets to derive economic returns to the sector and the Canadian economy as a whole. Although it recognizes animal welfare as a consumer issue, animal welfare is not mentioned in their latest report on plans and priorities.[9] The responsibility for animal welfare is left mainly to industry. However, the potential benefit of the federal government in providing leadership and an appropriate infra-structure has been discussed.[6] In the UK, an independent advisory body; the Farm Animal Welfare Council (FAWC) was established by the Government to provide advice for Ministers on animal welfare issues. An Animal Health and Welfare Strategy has also been developed as a partnership between stakeholders and Government. In the UK, Government intervention on animal welfare has been justified for a number of reasons: (1) to deal with potentially negative aspects of market forces, such as animal suffering and/or to encourage positive policies to improve animal welfare; (2) to provide 'public goods' – things that are shared by and benefit many in society but which are not directly linked to any payment, e.g. high levels of animal welfare, (3) to supply information, about animal welfare and (4) in pursuit of social goals to achieve a 'caring society'. [5]

**Market forces**

In both Canada and the UK there is an underlying belief that 'the market' is often more efficient and effective in supplying the needs of society than government intervention. Therefore governments seek to promote (or ensure the operation of) private markets in the provision of goods and services required by society rather than intervene directly to supply them itself. "For example, increased consumer demand for food with a high animal welfare provenance could have a more rapid and widespread effect in improving farm animal welfare than government legislation but only providing that appropriate markets exist for this to happen." [5].

Regulation is essentially interference in the market process. Business often argues that they are capable of self regulating to a sufficiently high standard and have a natural aversion to legislation mandating changes in business practices.[7,10] Canadian industry operates in a very competitive marketplace and is affected by practices and the associated costs of livestock production in the USA. Two major drivers of change in the area of farm animal welfare in North America are consumer and public attitudes and developments affecting international trade in livestock products. [10] When making purchasing decisions, consumers primarily consider prices and product attributes, but wealthy consumers are beginning to consider animal welfare. [10] Unless low animal welfare standards are associated with animal production practices that result in lower productivity and efficiency or pose a threat to human health, they do not impose an economic cost on society. It could be argued that society as a whole gains economically from current production practices that produce low prices for food products. [10] In some cases, higher welfare standards are likely to increase production costs for animal products. For example, changes in confinement operations, particularly increased space requirements, may require additional capital costs through the modification of existing facilities or construction of new or larger facilities. Extensive production systems, such as free-range eggs, require more land. Operating costs may increase, e.g. due to greater use of labour in caring for animals. "Increases in production costs due to higher animal welfare standards will not be borne solely by producers in the form of lower profits. At least part of the costs will be passed through to consumers in the form of higher prices."[10]

"Producers will be able to absorb the higher costs associated with higher welfare standards if consumers actively demand conforming product. For this to happen, consumers must not only want to purchase the product and be willing and able to pay a price premium, but they also must be able to identify it, i.e., it must be differentiated from non-conforming products. This is the rationale behind the use of labeling to identify welfare-friendly products.".[10] "Because consumers may find it difficult and time consuming to ensure that each product they buy meets all their requirements, they may rely on food retailers or restaurants." [10]

As animal welfare itself has no explicit financial value it cannot be left solely to market forces. In the UK, government intervention is considered to be necessary to protect animal welfare and legal minimum standards are considered to be a requirement to ensure that animals are not treated cruelly or in ways that the majority of society finds unacceptable. These legal minimum standards are consistently and rigorously enforced, and this requires dedicated enforcement activity with appropriate institutional arrangements. [5]

**Self-regulation**

Self regulation is where society, sections of society or institutions are allowed to regulate themselves. This form of regulation needs to be carefully monitored, e.g. by independent auditing. An example of self-regulation is the Canadian Council on Animal Care that sets and maintains standards for the welfare of animals used in research, teaching and testing. It conducts assessments of institutions to ensure compliance with guidelines and practices. Farm assurance schemes that have expanded greatly over recent years are another example of self-regulation and have a potentially important role in promoting animal welfare. In Canada, the National Recommended Codes of Practice for the Care and Handling of Farm Animals are now the responsibility of the National Farm Animal Care Council that is composed of representatives from various stakeholder groups. In some provinces there are farm animal industry organised farm animal care councils. At the national level, the codes represent voluntary guidelines and include various minimum standards for producers and others. Some provinces reference the codes in provincial regulations. In court cases involving enforcement of federal or provincial statutes, the codes have been used as a basis for determining acceptable operating practices. However a "low level of visibility and awareness of the existing codes" is considered to be a problem. [6]

In the UK, farm assurance has had a beneficial influence on animal welfare. However, when most farmers in a particular sector are members of a scheme, the standards might only indicate compliance with legal minimal standards. The influence of farm assurance schemes on animal welfare depend on their animal welfare assessment and auditing procedures. They can help to ensure that current legislation is implemented and can incorporate requirements above legislative standards. The UK Codes of Recommendation for the welfare of farmed animals are examples of enforced self-regulation. These lay down recommendations concerning the treatment of specific farm species kept under different systems. In the UK, welfare codes do not lay down statutory requirements, but livestock farmers and employers are required by law to ensure that all those attending to their livestock are familiar with, and have access to, the relevant codes. Although the main aim of the welfare codes is to encourage farmers to adopt high standards of husbandry, they may also be used to back up legislative requirements. Where a person is charged with a welfare offence, failure to comply with the provisions of a welfare code may be relied on by the prosecution to establish guilt. Farm assurance schemes usually incorporate elements of the welfare codes into their requirements. [5]

### Understanding of the concept of animal welfare

Although in Canada the concepts of animal welfare have been articulated extremely well, [11] there still appears to be some confusion between animal welfare and issues such as animal rights and organic farming and there is a fear of the impact of improved welfare conditions on the productivity of livestock farming. Animal welfare issues are raised by a range of interest groups that have different agendas. Animal rights groups that are opposed to the use of animals for food, clothing and research are more likely to engender an antagonist reaction from industry than animal welfare organisations that wish to work with industry to improve the conditions under which animals are kept and treated. The anti-intensive and naturalistic views of those that promote organic farming can also cause confusion to industry and government who might not be aware of more rational and scientific approaches to animal welfare that have been developed in Europe. Whether the inclusion of irrelevant and disproportionate arguments against recent proposals to reform the cruelty to animals section of the Criminal Code was part of a deliberate attempt by vested interests to oppose change or was genuine confusion over the issues involved is not clear.[12] Animal welfare is considered by the Federal Government to be a food quality issue,[6] but this narrow view is not shared by Canadian consumers. [8]

### Issues of openness, transparency and confidentiality

In the UK, the Government places emphasis on openness, transparency and engagement with stakeholders. It is considered that policy can only be effectively implemented with effective communication and information. Readily available information on the effectiveness of animal welfare regulation is vital for democratic debate. In contrast, Canada seems to place considerable emphasis on issues of confidentiality and privacy.

### Media influence

In the UK, media exposure of animal welfare issues, e.g. discussion of free range poultry systems on cooking programmes influence consumer buying habits.[5] In Canada, the media is sometimes accused of been only interested in sensationalizing animal welfare issues[6] and the opportunities for the provision of information and for rational discussion of animal welfare is more limited.

### Education

In the UK, education is considered to be beneficial to engender an awareness of animal welfare in (future) citizens and consumers. There is no mandatory school education of animal welfare issues, but it is used in several parts of the curriculum. Various organisations have produced educational materials, but there is concern as to whether they provide appropriate and balanced information for teachers to use.[5]

### Research

In the UK, research is used as a significant instrument to develop animal welfare policy. There is a clear role for government to fund or stimulate research in areas of importance that the private sector is failing to address. Such research provides evidence for decision-making, policy solutions, identifying and tackling future issues and for animal welfare monitoring and surveillance activities. Whether Canada has the research and development capability and sufficient funding from non-industry sources to fulfill this role has been questioned. [6] Research is required to identify what animal welfare issues need to be addressed by regulation and how to identify animal welfare issues e.g. pain and distress and their associated risk factors.

## CONCLUSION

Legislation will always be necessary to effectively protect animals from unnecessary suffering and distress. Society in both countries has an expectation that government should intervene to ensure that animal welfare is appropriately protected. Legislation should be based on scientific and technical information, established experience and society's expectations. It is essential that compliance with regulations is rigorously enforced. Enforcement requires well drafted legislation, inspection to identify an offence, the collection of evidence, an evaluation of whether prosecution is the most appropriate course of action, where appropriate, the prosecution of the case through the judicial system and effective deterrent penalties if the offender is found guilty. In the UK, there is a reasonably effective infrastructure for this process. However, the advisability of leaving many aspects of animal welfare regulation and enforcement in Canada to self-funded or special interests groups has been questioned.[4] The effectiveness of animal welfare regulation needs to be evaluated against specified targets or animal welfare outcomes and this information made readily available to the public. The veterinary profession has an important role to play both in the processes of government and industry regulation and in the provision of information and expert advice on animal welfare.

## REFERENCES

[1] Government of Canada 2006 **http://www.regulation.gc.ca/index-eng.asp**, [2]IFAW. 2008. **http://www.ifaw.org/Publications/Program_Publications/Regional_National_Efforts/North_America/Canada/asset_upload_file751_15788.pdf**, [3] Wepruk, J. 2004. http://www.afac.ab.ca/lawsregs/awlcanada.pdf, [4]Whiting,T.L., et al. **Can. Vet. J.** 2006. 47 1065, [5] Farm Animal Welfare Council 2008. **http://www.fawc.org.uk/pdf/policy-instr-081212.pdf.**, [6] Agriculture and Agri-Food Canada 2007. **http://www4.agr.gc.ca/AAFC-AAC/display-afficher.do?id=1184935828879&lang=eng**, [7] Pettigrew, N et al. 2007. **http://www.berr.gov.uk/files/file44591.pdf.**, [8] Ipsos-Reid Corporation. 2006. **http://www4.agr.gc.ca/resources/prod/doc/agr/pdf/canadian_perceptions_e.pdf**, [9] Agriculture and Agri-Food Canada 2009. **http://www.tbs-sct.gc.ca/rpp/2008-2009/inst/agr/agr00-eng.asp**, [10] Farm Foundation 2006. **http://www.farmfoundation.org/news/articlefiles/373-finalenglishreport1156973539.pdf**
[11] Fraser, D.; **Anim. Welfare** 1997. 6 187., [12] Sorenson, J. **Social & Legal Studies** 2003. 12. 377.

# CASE STUDIES ILLUSTRATING ANIMAL WELFARE REGULATION IN CANADA &0 THE UK

**Michael S. Cockram, BVetMed, PhD, MRCVS**
**Charlottetown, PEI**

## INTRODUCTION

Case studies will be used to provide examples of animal welfare regulation in Canada and the UK. Although a range of different topics will be reviewed, this summary deals only with the topic of neglect.

### Case Studies of Neglect

Example: failure of a person responsible for the care of animals to provide adequate feed and water to their animals resulting in starvation of some of the animals.

### The Criminal Code of Canada

In the UK, conviction rates for animal welfare offences are high, but in Canada, the prosecution of cases of neglect is problematic.[1] If the Criminal Code of Canada [2] is used, the requirement to prove that neglect is a wilful action requires proof of a person's intent. The requirement to prove that a person intended to neglect their animals makes the use of the Criminal Code to prosecute cases of neglect practically redundant. [1] Ill-treatment of animals that leads to suffering can be classified into: (1) Abuse that is deliberate cruelty, (2) Neglect that results from factors such as idleness, ignorance or overwork and (3) Deprivation that is inherent within certain management systems that do not provide animals with an opportunity to fulfil their behavioural and/or physiological needs. [3] Although UK legislation does, The Criminal Code does not appear to adequately differentiate between (a) deliberate acts of cruelty that lead to animal suffering and (b) neglect that also leads to suffering, but is a result of a lack of adequate care, where there was a duty of care for the animals, regardless of whether the neglect was intentional or not.

There is legislation in all Canadian provinces that make inadequate provision of food and water an offence. However, in Quebec, this only applies to dogs and cats. The capacity to enforce the legislation and provide an effective deterrent for this form of neglect is variable between provinces. At first sight, the Criminal Code should also offer animals protection from this form of neglect, but this is not the case. The relevant sections of the Criminal Code in relation to this example of neglect are listed below.

### Wilfully causing event to occur

429. (1) Every one who causes the occurrence of an event by doing an act or by omitting to do an act that it is his duty to do, knowing that the act or omission will probably cause the occurrence of the event and being reckless whether the event occurs or not, shall be deemed, for the purposes of this Part, wilfully to have caused the occurrence of the event.

### Cruelty to Animals

Causing unnecessary suffering 445.1 (1) Every one commits an offence who (a) wilfully causes or, being the owner, wilfully permits to be caused unnecessary pain, suffering or injury to an animal or a bird; Failure to exercise reasonable care as evidence (3) For the purposes of proceedings under paragraph (1)(a), evidence that a person failed to exercise reasonable care or supervision of an animal or a bird thereby causing it pain, suffering or injury is, in the absence of any evidence to the contrary, proof that the pain, suffering or injury was caused or was permitted to be caused wilfully, as the case may be.

### Causing damage or injury

446. (1) Every one commits an offence who
b) being the owner or the person having the custody or control of a domestic animal or a bird or an animal or a bird wild by nature that is in captivity, abandons it in distress or wilfully neglects or fails to provide suitable and adequate food, water, shelter and care for it.

### Canadian Case Studies

### Criminal Code

Mr Guy Russell, Mr Warren Russell and Mr Clayton Russell were charged under section 446(1) of the Criminal Code at the Provincial Court of Saskatchewan in 2000 with failure to provide suitable and adequate food for their cattle. This case was cited by the Canadian Veterinary Medical association in evidence to the Parliamentary Standing Committee on Justice and Human Rights as an example of the necessity to reform the Criminal Code. [4]

The accused had care and control of the cattle for several months under an agreement with the owners of the cattle and had asserted that they were knowledgeable about livestock. The cattle were seized by the SPCA and RCMP in December 1998 because they were thin and a number of calves had died of starvation and malnutrition. On seizure several aspects of poor animal husbandry were observed:

1. although there was sufficient feed available at the time of seizure, because of dominance by the older animals, failure to segregate young and mature animals resulted in reduced access to feed by the younger animals;
2. minimal bedding and cold weather contributed to the effects of inadequate feed; and
3. drinking water supplies were frozen.

Although the cattle were Bos indicus and the breed was Brahman, the court concluded that if they had been provided with feed and shelter they would have survived the winter in Saskatchewan (however, the calves may have been susceptible to cold conditions).

Expert evidence suggested that the cattle would have been in a poor body condition for several months and had not been adequately fed before the onset of winter.

The court decision highlighted that it was necessary to show that the accused in failing to provide adequate feed either:

1. "intended to produce the result of the cattle dying" or
2. "that they did do so with reckless disregard for the obvious consequences of what was going to happen."

The court concluded (a) that the accused were responsible for the cattle, (b) that they had inadequately cared for them and (c) that the result of the inadequate care was that some cattle died of starvation. The court considered whether indifference or incompetence in looking after the cattle amounted to wilful negligence by causing the death of the cattle. It concluded that the care of the cattle was highly negligent. However, it considered that there was still a requirement to prove intent beyond reasonable doubt in that the accused understood that what they were doing was likely to have a result that was illegal. Although the court considered that the accused had been incompetent to look after cattle and that an experienced and knowledgeable cattleman would have been aware that the cattle were suffering from malnutrition and were about to die, it did not consider that the accused had the knowledge that they were causing the result that they did and they did not intend to cause the death of the cattle. The charges were dismissed.

**Provincial Legislation**

Mr Jim Jenkins, Sherwood Park, Alberta was charged in 2007 with depriving his cattle of adequate food. Alberta SPCA Peace Officers responded to a complaint that there were dead cattle on a property.[5] They obtained a search warrant and accompanied by a local veterinarian and two RCMP officers, searched the premises. Sixty-nine dead cattle were found at various locations on the premises. In addition, 174 extremely emaciated cattle, including pregnant and calving cows were found. There was no evidence of feed on the property. Following veterinarian recommendations, the cattle were rounded up and provided with food. The following day the cattle were relocated, and later sold at an auction. The owner was found guilty under Section 2(1) of the Alberta Animal Protection Act [6] and was fined $1500. He was also prohibited from owning more than 20 head of cattle, or up to 30 cattle upon approval from a veterinarian.

**UK Case Study**

Mr John Kilford, Oaktree Farm, Cadnam, Southampton SO40 2UU, UK was charged at New Forest Magistrates' Court, in 2007 [7], with: six cases of unreasonably omitting to provide long term feeding, caused unnecessary suffering to a cow; one case of unreasonably omitting to clean out and ventilate a shed containing pigs, allowing a build up of strong ammonia fumes such as to cause unnecessary suffering; and one case of keeping livestock which did not have permanent access to a sufficient quantity of fresh drinking water. He pleaded guilty. The penalty was a Community Punishment Order that required 180 hours of work in the community and was disqualified from keeping cattle for life. He was also ordered to pay the court costs associated with the prosecution of £1,544.56.

Although the legislation under which the above charges were made has now been replaced, similar charges could have been laid under the new Animal Welfare Act 2006 and associated Welfare of Farmed Animals Regulations.

**Animal Welfare Act, 2006**[8]

Prevention of harm

4 Unnecessary suffering

(1) A person commits an offence if—
- (a) an act of his, or a failure of his to act, causes an animal to suffer,
- (b) he knew, or ought reasonably to have known, that the act, or failure to act, would have that effect or be likely to do so,
- (c) the animal is a protected animal, and
- (d) the suffering is unnecessary.

(2) A person commits an offence if—
- (a) he is responsible for an animal,
- (b) an act, or failure to act, of another person causes the animal to suffer,
- (c) he permitted that to happen or failed to take such steps (whether by way of supervising the other person or otherwise) as were reasonable in all the circumstances to prevent that happening, and
- (d) the suffering is unnecessary.

(3) The considerations to which it is relevant to have regard when determining for the purposes of this section whether suffering is unnecessary include—
- (a) whether the suffering could reasonably have been avoided or reduced,
- (b) whether the conduct which caused the suffering was in compliance with any relevant enactment or any relevant provisions of a licence or code of practice issued under an enactment,
- (c) whether the conduct which caused the suffering was for a legitimate purpose, such as—
  - (i) the purpose of benefiting the animal, or
  - (ii) the purpose of protecting a person, property or another animal,
- (d) whether the suffering was proportionate to the purpose of the conduct concerned,
- (e) whether the conduct concerned was in all the circumstances that of a reasonably competent and humane person.

Promotion of welfare

9 Duty of person responsible for animal to ensure welfare
(1) A person commits an offence if he does not take such steps as are reasonable in all the circumstances to ensure that the needs of an animal for which he is responsible are met to the extent required by good practice.
(2) For the purposes of this Act, an animal's needs shall be taken to include—
(a) its need for a suitable environment,
(b) its need for a suitable diet,
(c) its need to be able to exhibit normal behaviour patterns,
(d) any need it has to be housed with, or apart from, other animals, and
(e) its need to be protected from pain, suffering, injury and disease.
(3) The circumstances to which it is relevant to have regard when applying subsection (1) include, in particular—
(a) any lawful purpose for which the animal is kept, and
(b) any lawful activity undertaken in relation to the animal.

**The Welfare of Farmed Animals (England) Regulations 2007**[9]
Feed, water and other substances
22. Animals must be fed a wholesome diet which is appropriate to their age and species and which is fed to them in sufficient quantity to maintain them in good health, to satisfy their nutritional needs and to promote a positive state of well-being
24. All animals must have access to feed at intervals appropriate to their physiological needs (and, in any case, at least once a day), except where a veterinary surgeon acting in the exercise of his profession otherwise directs.
25 All animals must either have access to a suitable water supply and be provided with an adequate supply of fresh drinking water each day, or be able to satisfy their fluid intake needs by other means.

**Penalties for neglect to provide feed and water**
UK, Animal Welfare Act
Potential penalties for neglect to provide feed and water
- Imprisonment for up to 51 weeks, or a fine not exceeding £20,000, or both.
- Deprivation of ownership of an animal and disposal of an animal
- Disqualification from owning or keeping animals
- Reimbursement of court expenses

**Criminal Code**
- Imprisonment for up to six months or a fine not exceeding $5,000 or both.

**Provincial Legislation**
In Quebec there is no relevant provincial legislation and therefore no penalty for the neglect of farm animals. The penalties in other provinces vary from either no penalty of imprisonment to imprisonment for up to 2 years. Maximum fines vary from $500 to $60,000. Only some provinces have the sanction of disqualification from owning or keeping animals for a period of time. [10]

**REFERENCES**
[1] IFAW. 2008. http://www.ifaw.org/Publications/Program_Publications/Regional_National_Efforts/North_America/Canada/asset_upload_file751_15788.pdf. [2] Criminal Code. http://laws.justice.gc.ca/en/ShowTdm/cs/C-46///en, [3] Ewbank, R. In Moberg G.P. Animal Stress. 1985. 71., [4] 39th Parliament, 2nd Session. Standing Committee on Justice and Human Rights. 2008. http://www2.parl.gc.ca/HousePublications/Publication.aspx?DocId=3239999&Language=E&Mode=1&Parl=39&Ses=2, [5] AlbertaSPCA. **AnimalKind** 2008. 28 5. http://www.albertaspca.org/PDFs/AnimalKind_Spring2008.pdf, [6] Government of Alberta. 2009. http://www.qp.gov.ab.ca/documents/Acts/A41.cfm?frm_isbn=0779743571, [7] DEFRA. 2008. http://www.defra.gov.uk/animalh/ahws/deliver/health_act-1981_year2007.pdf, [8] Great Britain Parliament. 2006. http://www.opsi.gov.uk/ACTS/acts2006/pdf/ukpga_20060045_en.pdf, [9] Great Britain Parliament. 2007. http://www.opsi.gov.uk/si/si2007/uksi_20072078_en_1, [10] Canadian Federation of Humane Societies. 2009. http://cfhs.ca/law/provincial_legislation.

# WHAT YOU NEED TO KNOW TO INTRODUCE NEW SERVICES TO CLIENTS

**Dr. Louise S. Dunn**
**Greensboro, NC**

Call it a tired cliché, but it's true: It's a dog-eat-dog world. And veterinarians just may know this better than anyone.

Staying competitive in the animal health field depends on you and your hospital team's ability and success in remaining on the cutting-edge of veterinary medicine. Besides routinely investing in continuing education and seeking specialized training, another way to stay ahead of the pack is to continually invest in technology. Having the latest, greatest technical tools available in the animal health field will boost your hospital's medical capabilities and, in turn, increase client compliance for advanced care.

But here's the deal: Shelling out thousands of dollars for a fancy new ultrasound, digital X-ray or another piece of high-tech equipment isn't the simple solution to this challenge. No, you'll first need to "sell" that new toy to your hospital team so that they can in turn "sell" clients on the new advanced services that you have to offer.

How do you do this? By coaching, teaching and getting buy-in from your hospital team.

Through this session, you'll learn how to educate your team about the equipment or service and its capabilities and help them understand why it's in the patient's best interest to utilize these new advanced services you're offering. Language matters, and you'll learn what key phrases will help you better educate your hospital team members and gain their buy-in.

To help you succeed, keep this phrase in mind: Everything is local. This means that every member of your team must be committed to educating your clients. While some veterinarians are prone to throwing up their hands in a "I'm a veterinarian, not a salesman!" manner, it's the wrong attitude to adopt. It's absolutely critical that you get your team on board and learn to educate. Besides ensuring their ability to sell additional services to clients, it will also boost overall team morale and help them feel empowered.

For example, if a team member questions your $40,000 outlay for a new ultrasound rather than giving staff raises, you can show her how additional revenue from these services affects the practice's bottom line — and, in turn, her paycheck. Plus, providing advanced patient care will help bond clients further to your practice, ensuring long-term loyalty and compliance — another key factor impacting your team's profit potential.

Not sure you can communicate the nitty-gritty numbers to get your team on board? Don't worry. It's amazing what better communication, the correct talking points and excitement can do. The way you approach your team, the language you use and the energy you devote to this endeavor plays a major role in your success.

And as an added bonus to keep you on track, you'll get a simple spreadsheet that will do the math for you. You can use this tool to determine how many times you'll need to use the equipment to pay it off and start making a profit. As they say, seeing is believing, and this will help you further convince your team the new equipment or service is a winning addition to your practice.

# OPEN BOOK MANAGEMENT: HOW & WHY IT WILL TAKE YOUR PRACTICE TO THE NEXT LEVEL

**Louise S. Dunn**
**Greensboro, NC**

How many times have you heard a business management consultant proclaim, "Numbers talk"? It might seem like the buzz phrase du jour, but it's been proven time and time again by experts in the business world.

Open Book Management (OBM) is at the heart of this phrase. It's a management technique created in the 1990s that was taken mainstream by Jack Stack and John Case (see references). In a nutshell, OBM allows employees access to all relevant financial information and key performance indicators (such as profit and loss statements, balance sheets, expenses, etc.) so that they can make better decisions on the job. These numbers are compiled and calculated on a "scorecard" that gives owners and employees an at-a-glance view of the business' current performance and financial state.

Harvard Business School leaders have said OBM, or Scorecard Management, has had the greatest impact of any business idea in the last 75 years. OBM gives practice owners and the management team a step-by-step way to measure performance on every level of the company. Through OBM, employees have a better understanding of a business' financial health and they have a stake in how it performs. It helps them understand how a business works and conditions them to think like business owners. It's exactly what every veterinary practice needs to create a winning culture.

However, to create this practice culture, you as the leader has to believe in OBM and get your team to believe in it too. In doing so, this culture will transcend to all parts of the business: hiring, firing, equipment investment, profit building, and so on.

Your team members will be positively impacted by the OBM process as well. The skills they learn will help them in their professional and personal lives. In addition, they will begin to feel like more a part of the practice as they directly share in its successes and failures. OBM empowers and encourages your team members to share information that will, in turn, create a happy, dynamic, successful team and business.

But there are challenges. Let's face it: We see lots of turnover in veterinary medicine. People get involved because they love animals but they don't realize it's a business too. Practices are constantly losing their knowledge base. Veterinarians need to be more savvy with their hiring protocols to land dedicated people from the start. OBM will help you with performance incentives and employee retention.

In order to implement OBM into your practice, you must follow certain steps. These include meetings, teaching, tracking performance with scorecards, and building trust in your team.

Teamwork is essential to OBM. As Bob Levoy says, "You can't lead the band and play all the instruments." (see references) You need a team to be successful.

People are not as committed to work *on* the practice as they are to work *in* the practice. Think of your scorecards as bowling pins at the end of your bowling lane. OBM yields in a better bottom line, creates team loyalty, and boosts compliance when your team knows what you expect. With OBM, you get the best results when your team knows what you're doing, why you're doing it, and how you do it.

## REFERENCES

[1] Open-Book Management: **The Coming Business Revolution by John Case** (Collins Business, 1996) [2] The Open-Book Experience: **Lessons from Over 100 Companies Who Successfully Transformed Themselves** by John Case (Nicholas Brealey Publishing, 1998) [3] **The Great Game of Business** by Jack Stack and Bo Burlingham (Broadway Books, 1994) [4] **101 Secrets of a High-Performance Veterinary Practice** by Bob Levoy (Veterinary Medicine Publishing Company, 1996)

## THE STEPS INVOLVED IN THE IMPLEMENTATION OF OPEN BOOK MANAGEMENT & SCORECARDS

**Louise S. Dunn**
**Greensboro, NC**

Being a successful practice owner takes more than just practicing good medicine and celebrating the human-animal bond. Your success depends on your ability to utilize viable business strategies and make enough money to retain a successful veterinary team and provide the top-quality medicine and surgery that clients demand.

Open Book Management (OBM) will help you achieve your business and clinical goals. It's a proven fact in the business world that what you measure and track affects your business' bottom line. Specifically, benchmarking and scorecards will help your team members understand how their job performance affects the practice's bottom line and its overall success.

The fact is that, in most practices, compliance goes up when there is a financial incentive involved for the team. A motivated hospital team will work together to ensure that patients are treated and clients are educated about all medical options, which will appropriately result in the practice running smoothly.

In his best-selling book, *Getting Employees To Fall In Love With Your Company*, Jim Harris outlines methods to implement OBM. (See references.) But you have to understand the process as well as the importance of the bottom line and trends. This is what we'll cover today.

### NUMBERS TALK

In the previous session (032), we looked at training team members in basic business skills that will allow them to use that knowledge and help the practice achieve its goals. OBM teaches that making money is not only OK but it's essential in enabling you and your team to provide quality medicine, facilitate the human-animal bond, as well as provide viable compensations packages and ROI to the practice owner(s).

Let's talk about industry standards and how you compare. If you're not gauging how you size up to competitors (locally, regionally, and nationally), you're missing the boat. Not sure where to start? Look no further than the veterinary industry organizations — AVMA, AAHA, NCVEI, etc. — for data on comparable practices. These are the stats that you'll want to include on your scorecard.

Next, consider the niches that big-name brands fill. Nordstrom prides itself on superior customer service, while Marriott has a strong foothold in pricing. These companies know exactly how much they can charge at each of their locations bases on what areas they excel in. Likewise, Walmart and Dell use supply/inventory as one of their benchmarks. Once you identify your niche you must measure how you stack up to others.

For example, are you tracking how many of your clients are returning or how many other veterinarians are continuing to refer cases to you? Once you know this data, you can continue to build on the good things that you're doing and celebrate this success with your team. And when trends show you're not stacking up, you can quickly take action and come up with a different game plan.

In one practice, the team found that their accounts receivable were out of control. But once they began tracking numbers, they were able to identify the root cause of the problem. They found that the front desk needed to improve on their ability to enforce the no-billing policy and collect payment at the time of service.

As you establish benchmarks, you will also have to address the fact that your team members will have different levels of business training and confidence and comfort when dealing with numbers. This is where training is key so that everyone understand how the time commitment required for OBM will pay off for everyone in the end.

As Ross Clark, DVM, has identified, there are actually four stakeholders in the business of veterinary medicine. When determining what to track, consider patient care, client service, team knowledge and performance, and ROI for the four key stake holders:

#### The Pet

This is all about the **patient**. Your team's job is to understand that the pet fears being at your practice, make the pet as comfortable as possible, and relieve the pet's pain whenever needed.

#### The Client

This is all about **superior service**. Your team's job is to focus on the hours of operation, exceeding client expectations, and developing "the bond."

#### The Team

This is all about **rewarding careers**. Your job is to ensure your associates work as a team, they feel like they've done a good job, they have the ability to progress, they have an opportunity for continuing education.

#### The Investor

This is all about the **profitability that grows veterinary hospitals.** Your job is to control expenses while striving for professional excellence (utilizing great equipment, continuing to train doctors and team members, maintaining a professional demeanor at all times, etc.).

Remember accountability is an absolute in a today's economic climate for success. OBM provides a tool for leadership to maintain focus and be accountable to the patient, client, team and business outcomes. Every member of the healthcare team must see themselves as a leader and buy in to the implementation of OBM.

All winning hospital teams know their purpose and live their culture and unique brand as they work toward consistently achieving their goals. Relationships internal to the practice team and external with business partners must support the purpose of educating clients and providing passionate care.

The desire to change when necessary and understand the desired outcome of change will lead a successful practice. The human-animal bond continues to drive our profession, even in the tough times. OBM provides the basis for success.

**REFERENCES**

[1.] Open-Book Management: **The Coming Business Revolution** by John Case (Collins Business, 1996) [2.] The Open-Book Experience: **Lessons from Over 100 Companies Who Successfully Transformed Themselves** by John Case (Nicholas Brealey Publishing, 1998) [3] **The Great Game of Business** by Jack Stack and Bo Burlingham (Broadway Books, 1994) [4] **Getting Employees To Fall In Love With Your Company**, Jim Harris (AMACOM, 1996)

**THE HUMAN RESOURCE PROCESS: HOW TO FIND, HIRE & RETAIN SUPPORT TEAM MEMBERS**

**Louise S. Dunn**
**Greensboro, NC**

It doesn't matter what industry you work in, hiring the right team is No. 1 on every business owner's wish list. Success derives from a synergistic team of a committed, focused leader and a team that believes in the practice vision. A dedicated, proficient team will always achieve extraordinary goals, but simply being a good clinician or surgeon is no longer enough. As the practice leader, you must go beyond being a doctor or practice manager into becoming an effective coach and motivator.

But your success in veterinary medicine doesn't stop with hiring a great team. Once you've attracted the best, it's up to you to train these folks, keep them energized to do their very best and retain them for the long haul to keep your practice thriving.

Team building is a continuous process filled with fluctuating performances and stress levels. With each new hire or change, you'll likely see a nosedive in one and a peak in the other. To ease these moments, use proven managerial tools:

**Clearly defined written job descriptions**

Most employees support the practice's vision and want to please the team. Specific, detailed job descriptions outline exactly what they need to do to fulfill their roles. And written job descriptions are crucial for smaller practices in which one employee may assume several roles during a week or even a day. Be sure to include in the employee manual job descriptions for all positions as a ready, accessible reference and encourage your team to use them when they must step outside of their own roles. Effective written job descriptions must include a detailed summary of the position and how it affects the practice, performance expectations, required and desired qualifications, and a thorough list of specific duties as well as others the employee might assume during unique circumstances. Use job descriptions as a training outline.

**Ongoing training and individual opportunities for self-improvement**

Remember this rule of thumb: Hire for attitude and train for skills. Find and match a team member's niche with the practice's needs. Once the match is achieved, nurture and encourage each employee to grow and contribute to the hospital's goals. Proactively cash in on teachable moments and create them whenever possible. Take advantage of regular team meetings to incorporate training. Encourage doctors and team members to share information from CE courses or conferences attended during team meetings. Compensate employees who take on extra responsibilities and complete them outside normal business hours. And pay for or reimburse self-improvement expenses.

**Constructive, specific, and recurring performance appraisals**

Traditional performance appraisals (formerly called evaluations) try to motivate a person based on historical data that is often subjective or non-specific. The wave of the future is to provide positive evaluations based on goal setting and accountability. A two-part review is most effective: The first part counts for 40 percent to 50 percent of the total review and is filled out by the team member in advance to review goals and contributions and set the stage for his or her continued progress; the second part is filled out by the evaluator and measures operational tasks and counts toward 50 percent to 60 percent of the total review. Just as you base wages on skills, base raises on skills enhancement. Evaluate individual raises on how well the employee has improved patient care, customer care and the financial health of the practice. This is how you can position yourself as a mentor.

Combined, these provide your team with accountability and direction and further define your expectations.

Another proven strategy that has worked for many of my clients is implementing the four principles of enduring success. In other words:

1. Remember your mistakes and understand how you can avoid repeating them.
2. Be conservative about change.
3. Know your customers and what they want.
4. Manage your growth and leverage your assets.

During this session, we'll explore these four principles and also tried-and-true tips from the trenches on how to take the "TPR" of your practice. What's that? Think:

- **T**emperature — is your practice culture warm and fuzzy or more like ice?
- **P**ulse — do things hum or is everything a rat race in your practice?
- **R**espiration — are you taking deep breaths or constantly gasping for air?

The team looks to you for direction on how to behave. As the practice leader, you must hold yourself to a higher standard for patient care, customer service, performance, and self-mannerisms. Together, you'll achieve the goals of providing exceptional patient care and customer service and ensuring a successful business.

# TOP HUMAN RESOURCE MISTAKES & HOW TO FIX THEM

**Louise S. Dunn**
**Greensboro, NC**

Have you ever been asked, "Why re-create the wheel?" It's a valid question and one that you should always keep in mind when it comes to business. Far too many practice owners overlook the tried-and-true solutions — tips learned in the trenches — to common human resource challenges.

In this session, we're going to focus on these four strategies that all practice owners should be utilizing to run a more successful, effective business:

## TURNING A PASSIVE AGGRESSIVE ORGANIZATION AROUND INTO A POSITIVE WORK ENVIRONMENT

Management experts have long counseled business owners throughout a range of industries to utilize the Appreciative Inquiry (AI) process or theory in evaluating and implementing organizational change. It's a great tool that you can use to help you identify strengths among your hospital team members and build upon them.

What is AI? In a nutshell, it's an idea that encourages you to build upon your business around what works rather than trying to fix what doesn't. Through the AI process, you'll empower your team members by recognizing each individual's strengths and unique skills. This helps you identify key areas in which they excel so that you can make personnel adjustments accordingly.

For example, let's say that your lead receptionist is great with details and is an organizational whiz. However, she may lack in interpersonal skills — or not come across as a "people person." This really isn't the type of person you want meeting and greeting clients as they come through your door. So maybe you should consider moving this person into a practice manager role where she's still minding all the key business details but not interacting with the public so much.

## TEACHING & TRAINING EVERYONE THE SAME WAY

For this one, you need to throw out that one-size-fits-all mindset. Researchers like educator Howard Gardner have found that people learn in various styles. So when you're training your team members, it's important to teach in a manner that each individual understands.

To properly train your team, you must address all of these styles:

- Active Learner (Tactile)
- Reflective Learner (Thinker)
- Sensing Learner (Factual)
- Visual Learner
- Intuitive Learner (Investigative)
- Verbal Learner (Auditory)
- Sequential Learner (Step by Step)
- Global Learner (Big Ideas)

### Identifying & working with different conflict resolution styles

Let's talk about conflict. Unfortunately, conflict is a fact of life: Everybody experiences it, and it's always present. And if you think that it's not present in your practice, then your head is in the sand.

Conflict isn't something that you can just ignore and hope it goes away. I won't. Instead, you must deal with it because it's a huge drain on your practice and energy. The best course of action? Have an action plan to reduce the different types of conflict in your practice:

- Operational
- Ethical
- Personality

Just as there are different communication styles, there are different styles or approaches of conflict resolution. (See references.) These include:

- Smoothing (accommodating style)
- Forcing or win-lose negotiations (competitive style)
- Withdrawing (avoidant style)
- Problem-solving negotiations (collaborative style)
- Compromising (compromising style)

It's important that you identify these various styles and know how to use each one to implement change and resolutions.

Being proactive and having standard operating procedures (SOPs) in place to address conflict in your practice will help you be successful and help you turn conflict into teachable moments or learning opportunities. Standard operating procedures tell your team exactly what their jobs entail, the performance level and professionalism you expect, and the reasons behind the business decisions you make. With everyone one the same page and working toward the same practice goals, you become a unified force for excellence.

## RUNNING EFFECTIVE PRODUCTIVE MEETINGS

Don't forget the power of open communication at meetings. Effective team meetings are one of the most powerful practice builders. They will bond and excite your team. Meetings will raise your practices benchmarks, provide TQM (total quality management) increase profits and improve compliance.

In some cases, a short team huddle is all that is needed. But this huddle should still be a structured meeting designed to keep the players focused — it usually lasts about five minutes. (See references.)

While some people consider a meeting successful if a desired goal was met, that's not the definition of success. What's important is that all parties were able to communicate, a resolution occurred, and a trackable plan of action was set up. That's the definition of a successful, effective meeting.

## REFERENCES

[1.] **Getting Employees To Fall In Love With Your Company**, Jim Harris (AMACOM, 1996) [2] **"Teaching Students to be Peacemakers," conflict strategies game**, DK Johnson, 3rd edition

# THE BIGGEST MANAGEMENT CHALLENGE: A SYSTEM TO HOLD YOUR TEAM ACCOUNTABLE

**Louise S. Dunn**
**Greensboro, NC**

A common mantra that I continually preach to my consulting clients is "Make sure your practice is centered around the Circle of Care." The Circle of Care consists of three key components:

- Patient care
- The client experience
- The business

The Circle of Care should be the nucleus of your practice and be your guiding force when making all business decisions. It's the key component that ensures consistency in successful veterinary practices.

We know the No. 1 reason for turnover in businesses is a feeling of inequality among team members. It's often easy for team members to feel slighted and presume that others are being treated more favorably. But in successful practices, owners create internal systems and processes that ensure that this doesn't happen. For example, an important component of these systems and processes includes accountability.

One way that you can ensure that your team focuses on the Circle of Care is by establishing an accountability structure that works. To do so, you should have clear set goals, incentives for team members to achieve goals, and defined consequences when these goals aren't met.

I like to remind my clients of this sage advice from business guru Jim Collins: "First, you need the right people, in the correct seats on your bus." (see references) In other words, do you have the correct team members fulfilling the appropriate roles in your practice?

Once you've done this, the secret to establishing a solid accountability structure is to implement the nine key elements, which include trust, training, team buy-in, and leading with direction. These elements all work together to form the solid base that you'll need for a fool-proof accountability structure. During this session, we'll discuss each of these nine elements and talk about how they all play into the big picture.

Successful veterinary practices have adopted a mantra to keep their teams committed to excellence, consistency and minding details: "Always take care of every patient, every client, every record, every time." Once your team is on board and everyone knows the standards you expect and the consequences they'll face when expectations aren't met, you and your team will work smarter not harder. And that's the key to successful businesses.

## REFERENCES

1. *Good to Great:* **Why Some Companies Make the Leap…and Others Don't** by Jim Collins (Collins Business, 2001)

## SUCCESSFUL PRACTICES REQUIRE NUMEROUS LEADERSHIP & MANAGEMENT ROLES: WHAT ARE THEY?

**Louise S. Dunn**
**Greensboro, NC**

Talk about the potential for an identity crisis. As the head of a veterinary practice, you're required to wear many hats. At any given time, you're likely to find yourself fulfilling the role of:

- Practice owner
- Veterinarian
- Leader/CEO
- Medical specialist
- Manager
- Chief communicator
- Teacher/researcher
- Head cheerleader/motivator

Whatever hat you find yourself wearing, you have one overarching goal: to get your team members to buy into the practice's mission of providing the best care possible to patients.

Through good leadership, team members better appreciate how their actions contribute to a larger good of providing exceptional patient care and client service. So what's the difference between a manager and a leader?

A leader sets the practice's vision and goals. A manager communicates the vision to the team so they can accomplish the practice's goals. A leader defines values, establishes the culture, fosters professionalism, and sets benchmarks. A manager handles interpersonal conflicts and provides hands-on direction. A practice's leader and manager isn't necessarily the same person, but every practice needs someone performing each role.

Leadership styles vary. An inspirational leader generates enthusiasm and excitement. An organizational leader plots the path to achieving goals. The strongest leaders meld both qualities and also can recognize strengths and weaknesses in themselves and others. They then match complementary qualities to further benefit the practice while encouraging personal growth.

Your culture and goals must support every team member's commitment to raising the practice's standards of care:

- Patient care
- Client experience
- The business (human resources, operations, marketing, finance and strategic planning)

Here's how this can benefit your practice. Say you have a ward attendant or animal care attendant who is committed to improving senior pet care. He may suggest buying measuring cups to ensure that the animals are not over fed. This will decrease your food costs.

Is leadership easy? It is both an art and a science. But it can — and must — be learned.

The team looks to you for direction on how to behave. As the practice leader, you must hold yourself to a higher standard for patient care, customer service, performance, and self-mannerisms. Together, you'll achieve the goals of providing exceptional patient care and customer service and ensuring a successful business.

# MECHANISMS OF HEPATIC FIBROSIS, THE COMMON PATHWAY OF CHRONIC LIVER INJURY

**Scott L. Friedman, M.D.**
**New York, NY**

The clarification of mechanisms underlying the fibrotic response in hepatic injury and repair has been a significant turning point in understanding the basis of hepatic fibrosis. In particular, the identification of *stellate cell activation* as a key event in fibrogenesis has provided an important framework for conceptualizing the liver's response to injury. This review is not intended to focus primarily on mechanisms of hepatic fibrosis, but rather on broader aspects of stellate cell behavior. The reader is referred to many recent reviews for more comprehensive updates on fibrosis pathogenesis (1-6).

Stellate cell 'activation' refers to the conversion of a resting vitamin A-rich cell to one that is proliferating, fibrogenic and contractile. While it is increasingly clear that other mesenchymal cell populations also contribute to extracellular matrix accumulation, stellate cell activation remains the most dominant pathway leading to hepatic fibrosis. Moreover, stellate cell activation represents a continuum, such that early changes in cellular phenotype may be distinct from those occurring with progressive injury and activation in terms of growth characteristics, response to soluble mediators, inflammatory signaling and apoptotic potential (7-12).

Activation consists of two major phases, *Initiation* and *Perpetuation*, followed by *Resolution* of fibrosis if injury subsides.

**Initiation** (also called a "preinflammatory stage"), refers to early changes in gene expression and phenotype that render the cells responsive to other cytokines and stimuli. Initiation results mostly from paracrine stimulation, primarily due to changes in surrounding extracellular matrix, as well as exposure to lipid peroxides and products of damaged hepatocytes.

**Perpetuation** results from the effects of these stimuli on maintaining the activated phenotype and generating fibrosis. Perpetuation involves autocrine as well as paracrine loops. It is comprised of several discrete responses including proliferation, contractility, fibrogenesis, matrix degradation, retinoid loss, and inflammatory cell infiltration.

**Resolution** of fibrosis refers to pathways that either drive the stellate cell to apoptosis, or contribute to their reversion to a more quiescent phenotype.

## INITIATION

The earliest changes observed during stellate activation result from paracrine stimulation by all neighboring cell types, including sinusoidal endothelium, Kupffer cells, hepatocytes, and platelets. As noted above, early injury to endothelial cells stimulates production of cellular fibronectin, which has an activating effect on stellate cells (13). Endothelial cells are also likely to participate in conversion of TGFβeta from the latent to active, profibrogenic form. Platelets are another important source of paracrine stimuli, including PDGF, TGFβeta$_1$, and EGF (14).

Kupffer cell infiltration and activation also contribute to stellate cell activation. Kupffer cells stimulate matrix synthesis, cell proliferation, and release of retinoids by stellate cells through the actions of cytokines (especially TGFβeta$_1$) and reactive oxygen intermediates/lipid peroxides (15).

Hepatocytes are a potent source of fibrogenic lipid peroxides, although effects on stellate cell collagen synthesis and proliferation may be dose-dependent (16). Hepatocyte apoptosis following injury also promotes stellate cell initiation through a process mediated by Fas (17, 18). This process may involve the tumor necrosis factor-related apoptosis-inducing ligand (TRAIL) (17, 18). Whereas hepatocyte necrosis associated with lipid peroxidation is considered a classical inflammatory and fibrogenic stimulus, recent findings also implicate apoptosis, or programmed cell death, in the fibrogenic response. Apoptotic fragments released from hepatocytes are fibrogenic towards cultured stellate cells (19), and activate Kupffer cells (20). Also, Fas-mediated hepatocyte apoptosis is fibrogenic *in vivo* in experimental animals (18).

The cytochrome CYP2E1 plays an important role in the generation of reactive oxygen species that stimulate hepatic stellate cells (21). Cultured hepatic stellate cells grown in the presence of the HepG2 cell line expressing CYP2E1 (E47 cells) leads to increased production of collagen, an effect prevented in the presence of antioxidants or a CYP2E1 inhibitor (21). These data suggest that the CYP2E1-derived reactive oxygen species are responsible for the increased collagen production. In similar experiments using co-cultured hepatic stellate and E47 cells, the addition of arachidonic acid plus ferric nitrilotriacetate (agents that potentiate oxidative stress) further induced collagen synthesis (22). These findings may help to explain the pathogenesis of liver injury in alcoholic liver disease since CYP2E1 is alcohol-inducible.

## PERPETUATION

Perpetuation of stellate cell activation involves at least seven discrete changes in cell behavior: proliferation, chemotaxis, fibrogenesis, contractility, matrix degradation, retinoid loss, and WBC chemoattractant / cytokine release. The net effect of these changes is to increase accumulation of extracellular matrix. As an example, proliferation and chemotaxis lead to increased numbers of collagen-producing cells. Cytokine release by stellate cells can amplify the inflammatory and fibrogenic tissue responses, and matrix proteases may hasten the replacement of normal matrix with one typical of the wound "scar."

### Proliferation

PDGF is the most potent stellate cell mitogen identified (23). Induction of PDGF receptors early in stellate cell activation increases responsiveness to this potent mitogen (24). Downstream pathways of PDGF signaling have been carefully characterized in stellate cells and include PI3 kinase, among others (25) (26). In addition to proliferation, PDGF stimulates $Na^+/H^+$ exchange, providing a potential site for therapeutic intervention by blocking ion transport (27). Other compounds with mitogenic activity in stellate cells and a potential role in fibrogenesis include vascular endothelial cell growth factor (28), thrombin and its receptor (29, 30), EGF, TGFβ, keratinocyte growth factor (31) and bFGF (32). Signaling pathways for these and other mitogens have been greatly clarified in stellate cells, offering many potential sites for therapeutic intervention (see (25)).

**Chemotaxis**

Stellate cells can migrate towards cytokine chemoattractants (25, 33, 34) explaining in part why stellate cells align within inflammatory septae *in vivo*. A number of chemoattractants have been identified, prominent among which are PDGF (35, 36), MCP-1 (37) and CXCR3 (38). In contrast, adenosine (39), blunts chemotaxis and may immobilize cells once they reach the site of injury. The mechanical features of stellate cell chemotaxis have recently been explored, revealing that PDGF-stimluated chemotaxis is associated with cell spreading at the tip, movement of the cell body towards the stimulant, and retraction of trailing protrusions, associated with transient myosin phosphorylation (40).

**Fibrogenesis**

Stellate cells generate fibrosis not only by increased cell numbers, but also by increasing matrix production per cell. The best-studied component of hepatic scar is collagen type I, the expression of which is regulated both transcriptionally and post-transcriptionally in hepatic stellate cells by a growing number of stimuli and pathways. A detailed review of collagen gene regulation in stellate cells is beyond the focus of this review, but several recent references are recommended (41-51).

The most potent stimulus for production collagen I and other matrix constituents by stellate cells is TGFβetaı, which is derived from both paracrine and autocrine sources (see (4, 52, 53) for reviews). Signals downstream of TGFβ include a family of bifunctional molecules known as Smads, upon which many extracellular and intracellular signals converge to fine-tune and enhance TGFβ's effects during fibrogenesis (52) TGFβetaı also stimulates the production of other matrix components including cellular fibronectin and proteoglycans (54). In addition to a major role for Smad proteins, TGFβeta1 stimulates collagen in stellate cells through a hydrogen peroxide and C/EBPβeta-dependent mechanism (55). The response of Smads in stellate cells differs between acute and chronic injury to further favor matrix production (10, 56, 57).

As mentioned above, lipid peroxidation products are emerging as important stimuli to extracellular matrix production (58). Their effects may be amplified by loss of antioxidant capacity of stellate cells as they activate (59). These important insights have provided the rationale for the study of antioxidants in the treatment of a variety of liver diseases.

Connective tissue growth factor (CTGF/CCN2) is also a potent fibrogenic signal towards stellate cells (60-63) and may be specifically upregulated by hyperglycemia and hyperinsulinemia (64). While stimulation of CTGF production has traditionally been considered TGFβeta-dependent (65), the possibility of TGFβeta-independent regulation is increasingly likely (66).

**Contractility**

Contractility of stellate cells may be a major determinant of early and late increases in portal resistance during liver fibrosis. The collagenous bands typical of end-stage cirrhosis contain large numbers of activated stellate cells (67-69). These impede portal blood flow by constricting individual sinusoids and by contracting the cirrhotic liver. The acquisition of a contractile phenotype during stellate cell activation has been documented in culture and *in vivo*, and is mediated in part by receptors that interact with the extracellular matrix and are driven by calcium signaling (70). Endothelin-1 and nitric oxide are major counter-regulators controlling stellate cell contractility, in addition to a growing list of additional mediators including angiotensinogen II, eicosanoids, atrial natriueretic peptide, somatostatin and carbon monoxide, among others (see (67-69, 71) for reviews). While most studies implicate calcium signaling in response to endothelin-1 induced contractility (72), a recent study contradicts that view by demonstrating calcium-independent contractile force in stellate cells (70).

As stellate cells activate, the expression of the cytoskeletal protein alpha-SMA is increased (73, 74), which confers increased contractile potential. More extensive descriptions of βlphaSMA and other contractile filaments are reviewed above and in several reports (71, 75-78). While the conventional wisdom is that alpha actin expression enhances tissue fibrosis, mice lacking this protein in myofibroblasts have increased renal fibrosis in experimental glomerulonephritis (79), suggesting that alpha actin induction may be a counter-regulatory response to enhanced fibrogenesis by myofibroblasts, including activated stellate cells.

**Matrix degradation**

Fibrosis reflects a balance between matrix production and degradation. The degradation of extracellular matrix is a key event in hepatic fibrosis. Early disruption of the normal hepatic matrix by matrix-degrading proteases hastens its replacement by scar matrix, which has deleterious effects on cell function. Disruption of the normal liver matrix is also a requirement for tumor invasion and desmoplasia (80) and may be particularly relevant to pancreatic stellate cells (81). Degradation in these contexts is referred to as being "pathologic." On the other hand, resorption of excess matrix in patients with chronic liver disease provides the opportunity to reverse hepatic dysfunction and portal hypertension.

An understanding of mechanisms involved in matrix remodeling has evolved significantly in the past several years. A critical element in matrix remodeling is a family of matrix-metalloproteinases (also known as matrixins). These are calcium-dependent enzymes that specifically degrade collagens and non-collagenous substrates (82-84). As a general rule, the matrix-metalloproteinases fall into five categories based upon their substrate specificity:

- Interstitial collagenases (MMP-1, -8, -13)
- Gelatinases (MMP-2,-9 and fibroblast activation protein)
- Stromelysins (MMP-3, -7, -10, 11)
- Membrane type (MMP-14, 15, -16, -17, -24, -25)
- Metalloelastase (MMP-12)

Stellate cells are the principal source of MMP-2 (85, 86), MMP-13 (the rodent equivalent of MMP-1) (87) and stromelysin (88). Activation of latent MMP-2 may require interaction with hepatocytes (89, 90). Markedly increased expression of MMP-2 is characteristic of cirrhosis (91). MMP-9 may also be secreted by stellate cells (92, 93).

A major determinant of progressive fibrosis is failure to degrade the increased interstitial, or scar matrix. Matrix metalloproteinase-1 (MMP-1) is the main protease that can degrade type I collagen, the principal collagen in fibrotic liver. Sources of this enzyme are not as clearly established as for the type IV collagenases. Stellate cells express MMP-1 mRNA but little enzyme can be detected (86).

Regulation of matrix metalloproteinase activity occurs at many levels, among which is their inactivation by binding to tissue inhibitors of metalloproteinases (TIMPs) (82). Stellate cells also produce functional TIMP-1 and TIMP-2 (94), and sustained production of these proteins during liver injury could inhibit the activity of interstitial collageneases, leading to reduced degradation of the accumulating matrix during liver injury. TIMP-1 also is anti-apoptotic towards stellate cells (95), and thus its sustained expression in liver injury will enlarge the population of activated stellate cells by preventing their clearance. In support of TIMP's role *in vivo*, transgenic overexpression of TIMP-1 in liver, or administration of TIMP neutralizing antibodies both delay regression of liver fibrosis in experimental animals (96).

Stellate cells express uroplasminogen activator receptor (uPA-R) and its inhibitor (PAI-1), as well as other components of the plasmin system (97-102). These findings suggest that stellate cells contain most, if not all of the molecules necessary to either activate or inhibit metalloproteinases.

Most recently, increased proteolytic activity ascribed to *A Disintegrin and Metalloproteinase with ThromboSpondin type repeats* (ADAMS)-13 has been reported in activated stellate cells (103). However, the native substrate(s) and biological role of this protease is not known.

**RESOLUTION**

As attention has turned to the treatment of liver fibrosis, the issue of how stellate cell activation resolves has become quite critical (82, 104-106). Two potential pathways account for reduction in activated stellate cells, either reversion to a quiescent phenotype or clearance through apoptosis. Although reversion can be accomplished in cultured stellate cells through transfer of activated cells to a basement membrane matrix (107, 108), this has not been validated *in vivo*. To do so, genetic lineage tracing would be required to confirm that cells once activated have resumed a quiescent phenotype, however such studies have not been reported yet.

In contrast, a large amount of evidence supports the importance of stellate cell apoptosis during regression of liver fibrosis (82, 109 ). In culture, stellate cells are sensitive to CD95-L and TRAIL-mediated apoptosis, and NK cells can induce apoptosis of stellate cells by a TRAIL mediated mechanism (110) (see Section 5C, "Immunoregulation"). Nerve growth factor (NGF) derived from hepatocytes is also apoptotic towards stellate cells (111), and is antagonized by serotonin receptor signaling (112). Apoptosis requires an intact proteomomal degradation pathway, since its inhibition prevents stellate cell apoptosis (113). In a recent study (110), an antifibrotic effect of NK cells was indicated by the presence of increased fibrosis in mice depleted of NK cells by anti- asialo-GM1 antibody, and by decreased fibrosis after NK cell activation by a TLR3 ligand poly I:C. The NK cell induced stellate cell apoptosis was specific for activated stellate cells that expressed the NK cell activating receptor NKG2D. The activated NK cells deliver a lethal blow to stellate cells by inducing apoptosis with TRAIL. In this study NK cell function was dependent on IFNβ, and provided an explanation for earlier experiments demonstrating an important antifibrotic role for IFN□ (114). The antifibrotic role of NK cells was further supported by evidence of their direct adhesion to stellate cells in mouse livers, and by the development of greater fibrosis in mice genetically deficient in NK cells (115). Most recently, these findings have been reinforced by studies in humans with HCV (116). In addition to NK cells, activated Kupffer cells can also provoke stellate cell apoptosis by a unique caspase 9- and a receptor-interacting protein-dependent mechanism (117).

The antifibrotic role of NK cells is also consistent with the clinical data of increased liver fibrosis in the setting of therapeutic immunosuppression. The effect of single immunosuppressive agents on NK cell function is minimal, but the combination of cyclosporine and corticosteroids results in significant loss of NK cell cytotoxicity (118). In addition, cyclosporine renders some cells resistant to NK cell-mediated cytotoxicity. The effect of HIV infection on NK cell number and function is more complex. Some NK cell subsets express co-express CD4 and HIV co-receptors, and are targets for infection with HIV. NK cells from HIV infected patients have reduced cytolytic activity and decreased production of cytokines (119). The hypothesis that NK cells limit liver fibrosis by inducing stellate cell apoptosis predicts that NK cell function will be relatively impaired in individuals with rapid progression of fibrosis compared to those in whom liver fibrosis progresses slowly. It may also explain why fibrosis accelerates with aging (120), since NK cell function declines with age.

**REFERENCES**

[1.] Bataller R, Brenner DA. Liver fibrosis. **J Clin Invest** 2005;115:209-218. [2.] Lotersztajn S, Julien B, Teixeira-Clerc F, Grenard P, Mallat A. Hepatic fibrosis: molecular mechanisms and drug targets. **Annu Rev Pharmacol Toxicol** 2005;45:605-628. [3.]Friedman SL. Transcriptional regulation of stellate cell activation. **J Gastroenterol Hepatol** 2006;21 Suppl 3:S79-83. [4.]Gressner AM, Weiskirchen R, Breitkopf K, Dooley S. Roles of tgf-Beta in hepatic fibrosis. Front **Biosci** 2002;7:D793-807. [5.]Friedman SL. Molecular regulation of hepatic fibrosis, an integrated cellular response to tissue injury. **J Biol Chem** 2000;275:2247-2250. [6.]Kisseleva T, Brenner DA. Hepatic stellate cells and the reversal of fibrosis. **J Gastroenterol Hepatol** 2006;21 Suppl 3:S84-87. [7.]Gressner AM. Transdifferentiation of hepatic stellate cells (Ito cells) to myofibroblasts: a key event in hepatic fibrogenesis. **Kidney Int Suppl** 1996;54:S39-45. [8.]Schnabl B, Purbeck CA, Choi YH, Hagedorn CH, Brenner D. Replicative senescence of activated human hepatic stellate cells is accompanied by a pronounced inflammatory but less fibrogenic phenotype. **Hepatology** 2003;37:653-664. [9.]Roth-Eichhorn S, Eberheim A, Bode HP, Gressner AM. Transformation-dependent calcium influx by voltage-operated calcium channels in stellate cells of rat liver. **J Hepatol** 1999;30:612-620. [10.] Dooley S, Delvoux B, Lahme B, Mangasser-Stephan K, Gressner AM. Modulation of transforming growth factor beta response and signaling during transdifferentiation of rat hepatic stellate cells to myofibroblasts. **Hepatology** 2000;31:1094-1106. [11.]Gong W, Pecci A, Roth S, Lahme B, Beato M, Gressner AM. Transformation-dependent susceptibility of rat hepatic stellate cells to apoptosis induced by soluble Fas ligand. **Hepatology** 1998;28:492-502. [12.] Purps O, Lahme B, Gressner AM, Meindl-Beinker NM, Dooley S. Loss of TGF-beta dependent growth control during HSC transdifferentiation. Biochem Biophys **Res Commun** 2007;353:841-847. [13.] Jarnagin WR, Rockey DC, Koteliansky VE, Wang SS, Bissell DM. Expression of variant fibronectins in wound healing: cellular source and biological activity of the EIIIA segment in rat hepatic fibrogenesis. **J Cell Biol** 1994;127:2037-2048. [14] Bachem MG, Melchior R, Gressner AM. The role of thrombocytes in liver fibrogenesis: effects of platelet lysate and thrombocyte-derived growth factors on the mitogenic activity and glycosaminoglycan synthesis of cultured rat liver fat storing cells. **J Clin Chem Clin Biochem** 1989;27:555-565. [15] Bilzer M, Roggel F, Gerbes AL. Role of Kupffer cells in host defense and liver disease. **Liver Int** 2006;26:1175-1186. [16] Novo E, Marra F, Zamara E, Valfre di Bonzo L, Caligiuri A, Cannito S, Antonaci C, et al. Dose dependent and divergent effects of superoxide anion on cell death, proliferation, and migration of activated human hepatic stellate cells. **Gut** 2006;55:90-97. [17] Canbay A, Friedman S, Gores GJ. Apoptosis: the nexus of liver injury and fibrosis. **Hepatology** 2004;39:273-278. [18] Canbay A, Higuchi H, Bronk SF, Taniai M, Sebo TJ, Gores GJ. Fas enhances fibrogenesis in the bile duct ligated mouse: a link between apoptosis and fibrosis. **Gastroenterology** 2002;123:1323-1330. [19] Canbay A, Taimr P, Torok N, Higuchi H, Friedman S, Gores GJ. Apoptotic body engulfment by a human stellate cell line is profibrogenic. **Lab Invest** 2003;83:655-663. [20] Canbay A, Feldstein AE, Higuchi H, Werneburg N, Grambihler A, Bronk SF, Gores GJ. Kupffer cell engulfment of apoptotic bodies stimulates death ligand and cytokine expression. **Hepatology** 2003;38:1188-1198. [21] Nieto N, Friedman SL, Cederbaum AI. Cytochrome P450 2E1-derived reactive oxygen species mediate paracrine stimulation of collagen I protein synthesis by hepatic stellate cells. **J Biol Chem** 2002;277:9853-9864. [22] Nieto N, Friedman SL, Cederbaum AI. Stimulation and proliferation of primary rat hepatic stellate cells by cytochrome P450 2E1-derived reactive oxygen species. **Hepatology** 2002;35:62-73. [23] Pinzani M. PDGF and signal transduction in hepatic stellate cells. **Front Biosci** 2002;7:d1720-1726. [24] Wong L, Yamasaki G, Johnson RJ, Friedman SL. Induction of beta-platelet-derived growth factor receptor in rat hepatic lipocytes during cellular activation in vivo and in culture. **J Clin Invest** 1994;94:1563-1569. [25] Pinzani M, Marra F. Cytokine receptors and signaling in hepatic stellate cells. **Semin Liver Dis** 2001;21:397-416. [26] Lechuga CG, Hernandez-Nazara ZH, Hernandez E, Bustamante M, Desierto G, Cotty A, Dharker N, et al. PI3K is involved in PDGF-beta receptor upregulation post-PDGF-BB treatment in mouse HSC. **Am J Physiol Gastrointest Liver Physiol** 2006;291:G1051-1061. [27] Di Sario A, Bendia E, Taffetani S, Marzioni M, Candelaresi C, Pigini P, Schindler U, et al. Selective Na+/H+ exchange inhibition by cariporide reduces liver fibrosis in the rat. **Hepatology** 2003;37:256-266. [28] Yoshiji H, Kuriyama S, Yoshii J, Ikenaka Y, Noguchi R, Hicklin DJ, Wu Y, et al. Vascular endothelial growth factor and receptor interaction is a prerequisite for murine hepatic fibrogenesis. **Gut** 2003;52:1347-1354. [29] Marra F, Grandaliano G, Valente AJ, Abboud HE. Thrombin stimulates proliferation of liver fat-storing cells and expression of monocyte chemotactic protein-1: potential role in liver injury. **Hepatology** 1995;22:780-787. [30] Marra F, DeFranco R, Grappone C, Milani S, Pinzani M, Pellegrini G, Laffi G, et al. Expression of the thrombin receptor in human liver: up-regulation during acute and chronic injury. **Hepatology** 1998;27:462-471. [31] Steiling H, Muhlbauer M, Bataille F, Scholmerich J, Werner S, Hellerbrand C. Activated hepatic stellate cells express keratinocyte growth factor in chronic liver disease. **Am J Pathol** 2004;165:1233-1241. [32] Yu C, Wang F, Jin C, Huang X, Miller DL, Basilico C, McKeehan WL. Role of fibroblast growth factor type 1 and 2 in carbon tetrachloride-induced hepatic injury and fibrogenesis. **Am J Pathol** 2003;163:1653-1662. [33] Maher JJ. Interactions between hepatic stellate cells and the immune system. **Semin Liver Dis** 2001;21:417-426. [34] Marra F. Chemokines in liver inflammation and fibrosis. **Front Biosci** 2002;7:d1899-1914. [35] Ikeda K, Wakahara T, Wang YQ, Kadoya H, Kawada N, Kaneda K. In vitro migratory potential of rat quiescent hepatic stellate cells and its augmentation by cell activation. **Hepatology** 1999;29:1760-1767. [36] Kinnman N, Hultcrantz R, Barbu V, Rey C, Wendum D, Poupon R, Housset C. PDGF-mediated chemoattraction of hepatic stellate cells by bile duct segments in cholestatic liver injury. **Lab Invest** 2000;80:697-707. [37] Marra F, Romanelli RG, Giannini C, Failli P, Pastacaldi S, Arrighi MC, Pinzani M, et al. Monocyte chemotactic protein-1 as a chemoattractant for human hepatic stellate cells. **Hepatology** 1999;29:140-148. [38] Bonacchi A, Romagnani P, Romanelli RG, Efsen E, Annunziato F, Lasagni L, Francalanci M, et al. Signal transduction by the chemokine receptor CXCR3: activation of Ras/ERK, Src, and phosphatidylinositol 3-kinase/Akt controls cell migration and proliferation in human vascular pericytes. **J Biol Chem** 2001;276:9945-9954. [39] Hashmi AZ, Hakim W, Kruglov EA, Watanabe A, Watkins W, Dranoff JA, Mehal WZ. Adenosine inhibits cytosolic calcium signals and chemotaxis in hepatic stellate cells. Am J Physiol Gastrointest **Liver Physiol** 2007;292:G395-401. [40] Melton AC, Yee HF. Hepatic stellate cell protrusions couple platelet-derived growth factor-BB to chemotaxis. **Hepatology** 2007;45:1446-1453. [41] Stefanovic B, Hellerbrand C, Holcik M, Briendl M, Aliebhaber S, Brenner DA. Posttranscriptional regulation of collagen alpha1(I) mRNA in hepatic stellate cells. **Mol Cell Biol** 1997;17:5201-5209. [42] Garcia-Ruiz I, de la Torre P, Diaz T, Esteban E, Fernandez I, Munoz-Yague T, Solis-Herruzo JA. Sp1 and Sp3 transcription factors mediate malondialdehyde-induced collagen alpha 1(I) gene expression in cultured hepatic stellate cells. **J Biol Chem** 2002;277:30551-30558. [43] Anania FA, Womack L, Jiang M, Saxena NK. Aldehydes potentiate alpha(2)(I) collagen gene activity by JNK in hepatic stellate cells. **Free Radic Biol Med** 2001;30:846-857. [44] Lindquist J, Stefanovic B, Brenner D. Regulation of collagen alpha1(I) expression in hepatic stellate cells. **J Gastroenterol** 2000;35 Suppl 12:80-83. [45] Stefanovic B, Hellerbrand C, Brenner DA. Regulatory role of the conserved stem-loop structure at the 5' end of collagen alpha1(I) mRNA. **Mol Cell Biol** 1999;19:4334-4342. [46] Rippe RA. Role of transcriptional factors in stellate cell activation. **Alcohol Clin Exp Res** 1999;23:926-929. [47] Tsukada S, Parsons CJ, Rippe RA. Mechanisms of liver fibrosis. **Clin Chim Acta** 2006;364:33-60. [48] Inagaki Y, Truter S, Greenwel P, Rojkind M, Unoura M, Kobayashi K, Ramirez F. Regulation of the alpha 2(I) collagen gene transcription in fat-storing cells derived from a cirrhotic liver. **Hepatology** 1995;22:573-579. [49] Inagaki Y, Mamura M, Kanamaru Y, Greenwel P, Nemoto T, Takehara K, Ten Dijke P, et al. Constitutive phosphorylation and nuclear localization of Smad3 are correlated with increased collagen gene transcription in activated hepatic stellate cells. **J Cell Physiol** 2001;187:117-123. [50] Inagaki Y, Nemoto T, Kushida M, Sheng Y, Higashi K, Ikeda K, Kawada N, et al. Interferon alfa down-regulates collagen gene transcription and suppresses experimental hepatic fibrosis in mice. **Hepatology** 2003;38:890-899. [51] Stefanovic B, Stefanovic L, Schnabl B, Bataller R, Brenner DA. TRAM2 protein interacts with endoplasmic reticulum Ca2+ pump Serca2b and is necessary for collagen type I synthesis. **Mol Cell Biol** 2004;24:1758-1768. [52] Inagaki Y, Okazaki I. Emerging insights into Transforming growth factor beta Smad signal in hepatic fibrogenesis. **Gut** 2007;56:284-292. [53] Breitkopf K, Godoy P, Ciuclan L, Singer MV, Dooley S. TGF-beta/Smad signaling in the injured liver. Z **Gastroenterol** 2006;44:57-66. [54] George J, Wang SS, Sevcsik AM, Sanicola M, Cate RL, Koteliansky VE, Bissell DM. Transforming growth factor-beta initiates wound repair in rat liver through induction of the EIIIA-fibronectin splice isoform. **Am J Pathol** 2000;156:115-124. [55] Garcia-Trevijano ER, Iraburu MJ, Fontana L, Dominguez-Rosales JA, Auster A, Covarrubias-Pinedo A, Rojkind M. Transforming growth factor beta1 induces the expression of alpha1(I) procollagen mRNA by a hydrogen peroxide-C/EBPbeta-dependent mechanism in rat hepatic stellate cells. **Hepatology** 1999;29:960-970. [56] Tahashi Y, Matsuzaki K, Date M, Yoshida K, Furukawa F, Sugano Y, Matsushita M, et al. Differential regulation of TGF-beta signal in hepatic stellate cells between acute and chronic rat liver injury. **Hepatology** 2002;35:49-61. [57] Liu C, Gaca MD, Swenson ES, Vellucci VF, Reiss M, Wells RG. Smads 2 and 3 are differentially activated by transforming growth factor-beta (TGF-beta ) in quiescent and activated hepatic stellate cells. Constitutive nuclear localization of Smads in activated cells is TGF-beta-independent. **J Biol Chem** 2003;278:11721-11728. [58] Svegliati Baroni G, D'Ambrosio L, Ferretti G, Casini A, Di Sario A, Salzano R, Ridolfi F, et al. Fibrogenic effect of oxidative stress on rat hepatic stellate cells. **Hepatology** 1998;27:720-726. [59] Whalen R, Rockey DC, Friedman SL, Boyer TD. Activation of rat hepatic stellate cells leads to loss of glutathione S- transferases and their enzymatic activity against products of oxidative stress. **Hepatology** 1999;30:927-933. [60] Paradis V, Dargere D, Vidaud M, De Gouville AC, Huet S, Martinez V, Gauthier JM, et al. Expression of connective tissue growth factor in experimental rat and human liver fibrosis. **Hepatology** 1999;30:968-976. [61] Paradis V, Dargere D, Bonvoust F, Vidaud M, Segarini P, Bedossa P. Effects and regulation of connective tissue growth factor on hepatic stellate cells. **Lab Invest** 2002;82:767-774. [62] Rachfal AW, Brigstock DR. Connective tissue growth factor (CTGF/CCN2) in hepatic fibrosis. **Hepatol Res** 2003;26:1-9. [63] Gao R, Brigstock DR. Connective tissue growth factor (CCN2) induces adhesion of rat activated hepatic stellate cells by binding of its C-terminal domain to integrin alpha(v)beta(3) and heparan sulfate proteoglycan. **J Biol Chem** 2004;279:8848-8855. [64] Paradis V, Perlemuter G, Bonvoust F, Dargere D, Parfait B, Vidaud M, Conti M, et al. High glucose and hyperinsulinemia stimulate connective tissue growth factor expression: a potential mechanism involved in progression to fibrosis in nonalcoholic steatohepatitis. **Hepatology** 2001;34:738-744. [65] Grotendorst GR. Connective tissue growth factor: a mediator of TGF-beta action on fibroblasts. **Cytokine Growth Factor Rev** 1997;8:171-179. [66] Brigstock DR. The CCN family: a new stimulus package. **J Endocrinol** 2003;178:169-175. [67] Rockey DC. Hepatic blood flow regulation by stellate cells in normal and injured liver. **Seminars in Liver Disease** 2001;21:337-350. [68] Rockey DC. Cellular pathophysiology of portal hypertension and prospects for management with gene therapy. **Clin Liver Dis** 2001;5:851-865. [69] Rockey DC. Vascular mediators in the injured liver. **Hepatology** 2003;37:4-12. [70] Melton AC, Datta A, Yee HF, Jr. [Ca2+]i-independent contractile force generation by rat hepatic stellate cells in response to endothelin-1. Am J Physiol Gastrointest **Liver Physiol** 2006;290:G7-13. [71] Reynaert H, Thompson MG, Thomas T, Geerts A. Hepatic stellate cells: role in microcirculation and pathophysiology of portal hypertension. **Gut** 2002;50:571-581. [72] Pinzani M, Failli P, Ruocco C, Casini A, Milani S, Baldi E, Giotti A, et al. Fat-storing cells as liver-specific pericytes. Spatial dynamics of agonist-stimulated intracellular calcium transients. **J Clin Invest** 1992;90:642-646. [73] Rockey DC, Boyles JK, Gabbiani G, Friedman SL. Rat hepatic lipocytes express smooth muscle actin upon activation in vivo and in culture. **J Submicrosc Cytol Pathol** 1992;24:193-203. [74] Ramadori G, Veit T, Schwogler S, Dienes HP, Knittel T, Rieder H, Meyer zum Buschenfelde KH. Expression of the gene of the alpha-smooth muscle-actin isoform in rat liver and in rat fat-storing (ITO) cells. Virchows **Arch B Cell Pathol Incl Mol Pathol** 1990;59:349-357. [75] Geerts A. History, heterogeneity, developmental biology, and functions of quiescent hepatic stellate cells. **Semin Liver Dis** 2001;21:311-335. [76] Uyama N, Zhao L, Van Rossen E, Hirako Y, Reynaert H, Adams DH, Xue Z, et al. Hepatic stellate cells express synemin, a protein bridging intermediate filaments to focal adhesions. **Gut** 2006. [77] Niki T, Pekny M, Hellemans K, Bleser PD, Berg KV, Vaeyens F, Quartier E, et al. Class VI intermediate filament protein nestin is induced during activation of rat hepatic stellate cells. **Hepatology** 1999;29:520-527. [78] Rombouts K, Knittel T, Machesky L, Braet F, Wielant A, Hellemans K, De Bleser P, et al. Actin filament formation, reorganization and migration are impaired in hepatic stellate cells under influence of trichostatin A, a histone deacetylase inhibitor. **J Hepatol** 2002;37:788-796. [79] Takeji M, Moriyama T, Oseto S, Kawada N, Hori M, Imai E, Miwa T. Smooth muscle alpha-actin deficiency in myofibroblasts leads to enhanced renal tissue fibrosis. **J Biol Chem** 2006;281:40193-40200. [80] Arii S, Mise M, Harada T, Furutani M, Ishigami S, Niwano M, Mizumoto M, et al. Overexpression of matrix metalloproteinase 9 gene in hepatocellular carcinoma with invasive potential. **Hepatology** 1996;24:316-322. [81] Omary MB, Lugea A, Lowe AW, Pandol SJ. The pancreatic stellate cell: a star on the rise in pancreatic diseases. **J Clin Invest** 2007;117:50-59. [82] Iredale JP. Hepatic stellate cell behavior during resolution of liver injury. **Semin Liver Dis** 2001;21:427-436. [83] Arthur MJ. Collagenases and liver fibrosis. **J Hepatol** 1995;22:43-48. [84] Benyon RC, Arthur MJ. Extracellular matrix degradation and the role of hepatic stellate cells. **Semin Liver Dis** 2001;21:373-384. [85] Arthur MJ, Stanley A, Iredale JP, Rafferty JA, Hembry RM, Friedman SL. Secretion of 72 kDa type IV collagenase/gelatinase by cultured human lipocytes. Analysis of gene expression, protein synthesis and proteinase activity. **Biochem J** 1992;287:701-707. [86] Milani S, Herbst H, Schuppan D, Grappone C, Pellegrini G, Pinzani M, Casini A, et al. Differential expression of matrix-metalloproteinase-1 and -2 genes in normal and fibrotic human liver. **Am J Pathol** 1994;144:528-537. [87] Schaefer B, Rivas-Estilla AM, Meraz-Cruz N, Reyes-Romero MA, Hernandez-Nazara ZH, Dominguez-Rosales JA, Schuppan D, et al. Reciprocal modulation of matrix metalloproteinase-13 and type I collagen genes in rat hepatic stellate cells. **Am J Pathol** 2003;162:1771-1780. [88] Vyas SK, Leyland H, Gentry J, Arthur MJ. Rat hepatic lipocytes synthesize and secrete transin (stromelysin) in early primary culture. **Gastroenterology** 1995;109:889-898. [89] Theret N, Lehti K, Musso O, Clement B. MMP2 activation by collagen I and concanavalin A in cultured human hepatic stellate cells. **Hepatology** 1999;30:462-468. [90] Theret N, Musso O, L'Helgoualc'h A, Clement B. Activation of matrix metalloproteinase-2 from hepatic stellate cells requires interactions with hepatocytes. **Am J Pathol** 1997;150:51-58. [91] Benyon RC, Iredale JP, Goddard S, Winwood PJ, Arthur MJ. Expression of tissue inhibitor of metalloproteinases 1 and 2 is increased in fibrotic human liver. **Gastroenterology** 1996;110:821-831. [92] Han YP. Matrix metalloproteinases, the pros and cons, in liver fibrosis. **J Gastroenterol Hepatol** 2006;21 Suppl 3:S88-91. [93] Han YP, Zhou L, Wang J, Xiong S, Garner WL, French SW, Tsukamoto H. Essential role of matrix metalloproteinases in interleukin-1-induced myofibroblastic activation of hepatic stellate cell in collagen. **J Biol Chem** 2004;279:4820-4828. [94] Arthur MJ. Fibrogenesis II. Metalloproteinases and their inhibitors in liver fibrosis. **Am J Physiol Gastrointest Liver Physiol** 2000;279:G245-249. [95] Murphy FR, Issa R, Zhou X, Ratnarajah S, Nagase H, Arthur MJ, Benyon C, et al. Inhibition of apoptosis of activated hepatic stellate cells by tissue inhibitor of metalloproteinase-1 is mediated via effects on matrix metalloproteinase inhibition: implications for reversibility of liver fibrosis. **J Biol Chem** 2002;277:11069-11076. [96] Yoshiji H, Kuriyama S, Yoshii J, Ikenaka Y, Noguchi R, Nakatani T, Tsujinoue H, et al. Tissue inhibitor of metalloproteinases-1 attenuates spontaneous liver fibrosis resolution in the transgenic mouse. **Hepatology** 2002;36:850-860. [97] Knittel T, Fellmer P, Ramadori G. Gene expression and regulation of plasminogen activator inhibitor type I in hepatic stellate cells of rat liver. **Gastroenterology** 1996;111:745-754. [98] Leyland H, Gentry J, Arthur MJ, Benyon RC. The plasminogen-activating system in hepatic stellate cells. **Hepatology** 1996;24:1172-1178. [99] Imai S, Okuno M, Moriwaki H, Muto Y, Murakami K, Shudo K, Suzuki Y, et al. 9,13-di-cis-Retinoic acid induces the production of tPA and activation of latent TGF-beta via RAR alpha in a human liver stellate cell line, LI90. **FEBS Lett** 1997;411:102-106. [100] Zhang LP, Takahara T, Yata Y, Furui K, Jin B, Kawada N, Watanabe A. Increased expression of plasminogen activator and plasminogen activator inhibitor during liver fibrogenesis of rats: role of stellate cells. **J Hepatol** 1999;31:703-711. [101] Fibbi G, Pucci M, Grappone C, Pellegrini G, Salzano R, Casini A, Milani S, et al. Functions of the fibrinolytic system in human Ito cells and its control by basic fibroblast and platelet-derived growth factor. **Hepatology** 1999;29:868-878. [102] Fiorucci S, Antonelli E, Distrutti E, Severino B, Fiorentina R, Baldoni M, Caliendo G, et al. PAR1 antagonism protects against experimental liver fibrosis. Role of proteinase receptors in stellate cell activation. **Hepatology** 2004;39:365-375. [103] Niiya M, Uemura M, Zheng XW, Pollak ES, Dockal M, Scheiflinger F, Wells RG, et al. Increased ADAMTS-13 proteolytic activity in rat hepatic stellate cells upon activation in vitro and in vivo. **J Thromb Haemost** 2006;4:1063-1070. [104] Fallowfield JA, Iredale JP. Reversal of liver fibrosis and cirrhosis--an emerging reality. **Scott Med J** 2004;49:3-6. [105] Friedman SL, Bansal MB. Reversal of hepatic fibrosis -- fact or fantasy? **Hepatology** 2006;43:S82-88. [106] Friedman SL, Arthur MJ. Reversing hepatic fibrosis. **Science & Medicine** 2002;8:194-205. [107] Olaso E, Eng F, Lin C, Yancopoulous G, Friedman SL. The discoidin domain receptor 2 (DDR2) is induced during stellate cell activation, mediates cell growth and MMP2 expression, and is regulated by extracellular matrix. **Hepatology** 1999;30:413A. [108] Gaca MD, Zhou X, Issa R, Kiriella K, Iredale JP, Benyon RC. Basement membrane-like matrix inhibits proliferation and collagen synthesis by activated rat hepatic stellate cells: evidence for matrix-dependent deactivation of stellate cells. **Matrix Biol** 2003;22:229-239. [109] Issa R, Williams E, Trim N, Kendall T, Arthur MJ, Reichen J, Benyon RC, et al. Apoptosis of hepatic stellate cells: involvement in resolution of biliary fibrosis and regulation by soluble growth factors. **Gut** 2001;48:548-557. [110] Radaeva S, Sun R, Jaruga B, Nguyen VT, Tian Z, Gao B. Natural killer cells ameliorate liver fibrosis by killing activated stellate cells in NKG2D-dependent and tumor necrosis factor-related apoptosis-inducing ligand-dependent manners. **Gastroenterology** 2006;130:435-452. [111] Oakley F, Trim N, Constandinou CM, Ye W, Gray AM, Frantz G, Hillan K, et al. Hepatocytes express nerve growth factor during liver injury: evidence for paracrine regulation of hepatic stellate cell apoptosis. **Am J Pathol** 2003;163:1849-1858. [112] Ruddell RG, Oakley F, Hussain Z, Yeung I, Bryan-Lluka LJ, Ramm GA, Mann DA. A role for serotonin (5-HT) in hepatic stellate cell function and liver fibrosis. **Am J Pathol** 2006;169:861-876. [113] Anan A, Baskin-Bey ES, Bronk SF, Werneburg NW, Shah VH, Gores GJ. Proteasome inhibition induces hepatic stellate cell apoptosis. **Hepatology** 2006;43:335-344. [114] Rockey DC, Chung JJ. Interferon gamma inhibits lipocyte activation and extracellular matrix mRNA expression during experimental liver injury: implications for treatment of hepatic fibrosis. **J Investig Med** 1994;42:660-670. [115] Melhem A, Muhanna N, Bishara A, Alvarez CE, Ilan Y, Bishara T, Horani A, et al. Anti-fibrotic activity of NK cells in experimental liver injury through killing of activated HSC. **J Hepatol** 2006;45:60-71. [116] Morishima C, Paschal DM, Wang CC, Yoshihara CS, Wood BL, Yeo AE, Emerson SS, et al. Decreased NK cell frequency in chronic hepatitis C does not affect ex vivo cytolytic killing. **Hepatology** 2006;43:573-580. [117] Fischer R, Cariers A, Reinehr R, Haussinger D. Caspase 9-dependent killing of hepatic stellate cells by activated Kupffer cells. **Gastroenterology** 2002;123:845-861. [118] Hudnall SD. Cyclosporin A renders target cells resistant to immune cytolysis. **Eur J Immunol** 1991;21:221-226. [119] Fauci AS, Mavilio D, Kottilil S. NK cells in HIV infection: paradigm for protection or targets for ambush. **Nat Rev Immunol** 2005;5:835-843. [120] Poynard T, Mathurin P, Lai CL, Guyader D, Poupon R, Tainturier MH, Myers RP, et al. A comparison of fibrosis progression in chronic liver diseases. **J Hepatol** 2003;38:257-265.

# DIAGNOSIS & TREATMENT OF HEPATIC FIBROSIS – CHALLENGES & EMERGING STRATEGIES

**Scott L. Friedman M.D.,**
**New York, NY**

The clarification of the mechanisms of fibrosis has led to a surge in enthusiasm for treating hepatic fibrosis, and animal models strongly support the rationale of this approach. In fact, much like the paradigm of cancer treatment, success to date in animal models has been far easier to achieve than in human trials, possibly because the duration of injury and extent of matrix deposition and cross-linking is greater in humans. Key obstacles remain to be overcome before antifibrotic therapies can reach widespread clinical use.

## DIAGNOSTIC APPROACHES

The foremost challenge is to find the optimal method of diagnosing fibrosis. While liver biopsy combined with connective tissue stains has been a mainstay of diagnosis, it is prone to sampling error and inter-observer variability. This may account for a one-stage difference in up to one-third of patients when using conventional 5- or 6-point scoring scales (e.g., Metavir, Ishak). A potentially more quantitative and feasible application of liver biopsy might include the measurement of key fibrogenic mRNAs by real time PCR. For example, changes in the expression of fibrogenic genes, including those correlating with stellate cell activation (e.g. PDGFb receptor) may reveal early evidence of regression even before the matrix content has changed. Efforts to make this application a reality are currently under development.

There has been significant progress in the development of non-invasive markers, which will be essential to obtain early biomarkers of efficacy for clinical trials and to guide clinical usage. Current clinical trials of antifibrotic drugs anticipate that at least 6-12 months will be required to demonstrate efficacy by biopsy, based on observed rates of regression in patients cured of their underling liver disease (e.g. HBV, HCV); however, lengthy trials are costly and may discourage drug development.

The ideal non-invasive fibrosis marker must be reproducible and linear, reflecting fibrosis in all types of chronic liver disease, and correlating with matrix content and clinical outcome (1). It must also be sensitive enough in individual patients to discriminate between different stages of fibrosis, and reflect the response to antifibrotic therapy over time. There are several approaches currently being developed. First is the serum measurement of one or more circulating matrix proteins and/or serum biochemistries in combination. Second are serum proteomics or glycomics, which assess patterns of protein peaks or glycoprotein branching, respectively. A recent study combined glycomics with a serum marker panel(2). It achieved a high level of discrimination in diagnosing compensated and decompensated cirrhosis, however, these patients are also readily diagnosed by standard clinical means. Third are imaging methods (CT, MRI, PET, elastography, radionuclide receptor scanning), which can assess either intrahepatic blood flow patterns, organ texture, or activated stellate cell mass using cell-specific reagents to bind membrane receptors. Currently, several of these approaches can already successfully distinguish patients with little or no fibrosis, as well as those with advanced disease, but they less reliably discriminate intermediate stages over time. Nonetheless, evidence of either minimal or advanced fibrosis can be extremely important in making treatment decisions, especially in patients with HCV, for which current antiviral therapies are often poorly tolerated or ineffective in a substantial fraction of patients.

### Clinical trials

The choice of patients and endpoints for clinical trials of antifibrotic drugs presents a conundrum. Current efforts focus mainly on HCV patients with non-cirrhotic fibrosis who have failed antiviral therapies, in part because of the large number of such patients and their well characterized natural history. On one hand, antifibrotics are more urgently needed in patients with HCV cirrhosis and are more likely, if successful, to yield improved survival and reduced morbidity. On the other hand, cirrhosis is probably less reversible than earlier fibrosis stages and will require longer treatment, and these patients will be less able to tolerate any idiosyncratic hepatotoxicity from an investigational agent. They are also at higher risk for hepatocellular cancer. Thus, initial trials should be aimed a demonstrating the capacity for matrix resorption without insisting that complete reversal of fibrosis or improved morbidity and mortality be documented; findings from such studies will support the rationale for longer, larger trials with 'hard' clinical endpoints including survival and decreased complications. Other potential target populations include either patients with HCV post-liver transplantation or children, since fibrosis is accelerated in these groups, as well as those with cholestatic liver disease or NASH.

### Current treatment approaches

There is tremendous activity in the field of drug development and testing for hepatic fibrosis, with realistic hopes for significant progress in the near future. To date, however, no drugs are approved as antifibrotic agents in humans. Such therapies need to be safe and without long-term toxicity or increasing the risk of hepatocellular cancer.

The liver offers a unique targeting advantage since orally administrated agents will be metabolized first in liver and need not reach extrahepatic tissues, possibly allowing for lower, safer dosages. Combination therapies may prove synergistic rather than additive, but there is concern that a drug that is not effective as a single agent may never be included in a multi-drug combination. Antifibrotic therapies might require only intermittent rather than continuous administration. Finally, drugs that are most likely to reach clinical trial first are those that have already been approved for other indications, have a clear rational basis for use in liver fibrosis, and have amassed extensive safety data. Once proof-of-principle for any antifibrotic is achieved, it is likely that interest in the development of other agents will accelerate further.

Because no antifibrotic drugs have been approved yet, this article does not review ongoing clinical trials. However, a framework for attacking the pathways of injury and fibrosis is clearly established and summarized in Table 1, and detailed in recent reviews(3, 4). Multiple targets have been identified and many antagonists are being tested in cell culture and animal models. Clearly, the best 'antifibrotic' is to eradicate the underlying cause of liver disease wherever possible, for example by clearing HCV or HBV. Other approaches include reducing hepatic injury and inflammation, attenuating stellate cell activation by manipulating intracellular signaling and transcriptional pathways, antagonizing cytokines and growth factors that drive stellate cell activation, provoking apoptosis of activated stellate cells, or accelerating degradation of extracellular matrix. Several agents have overlapping activities and many are nearing clinical trials.

Finally, it is important to realize that for an antifibrotic drug to be successful, it need not eradicate hepatic fibrosis entirely, since the liver has an enormous functional reserve and most patients with fibrosis have normal liver function. Instead, any drug that sufficiently attenuates fibrosis progression to prevent the development of cirrhosis and/or hepatocellular carcinoma will be viewed as a success, and will fuel what is already a dynamic and rapidly advancing field.

## CONCLUSIONS

We are on the cusp of a new era, in which the significant progress in basic research is moving the field towards novel approaches for the diagnosis and therapy of hepatic fibrosis. Accelerating progress is anticipated once validated non-invasive biomarkers of fibrosis are developed, as there are ample therapeutic targets, and many compounds in the drug development pipeline will soon be available for clinical study. If successful, antifibrotic therapy could transform the management of patients with progressive liver disease, having a significant beneficial public health impact.

## REFERENCES

[1.] Albanis E, Friedman SL. Non-invasive markers of hepatic fibrosis. **Clinical perspectives in Gastroenterology** 2002;5:182-187. [2.] Callewaert N, Van Vlierberghe H, Van Hecke A, Laroy W, Delanghe J, Contreras R. Noninvasive diagnosis of liver cirrhosis using DNA sequencer-based total serum protein glycomics. **Nat Med** 2004;10:429-434. [3.] Pinzani M, Rombouts K. Liver fibrosis: from the bench to clinical targets. **Dig Liver Dis** 2004;36:231-242. [4.] Schuppan D, Porov Y. Hepatic fibrosis: From bench to bedside. **J Gastroenterol Hepatol** 2002;17 Suppl 3:S300-S305.

# ADVANCES IN CHARACTERIZATION OF FELINE NONSUPPURATIVE CHOLANGITIS/CHOLANGIOHEPATITIS SYNDROME

**Amy L. Warren, BVSc. (hons.), DACVP (anatomic & clinical)**
**Calgary, AB**

## INTRODUCTION

Cholangitis and cholangiohepatitis syndrome (CCHS) is the second most common acquired liver disorder of the domestic cat, second only in prevalence to the syndrome of hepatic lipidosis[1, 2]. Historically, CCHS has been histologically subcategorized on the basis of inflammatory infiltrates (suppurative or non-suppurative), subjective assessment of portal fibrosis (acute having minimal fibrosis or chronic having notable fibrosis), presence of bile ductule destruction, and association with metazoan parasites (liver fluke infestation, e.g. Opisthorchiidae)[3].

Clinical differentiation of the subcategories of the CCHS is unreliable without a liver biopsy because clinical features are widely divergent; some cats show no signs and others present with severe acute illness. Typically, non-suppurative CCHS is slowly progressive, clinically silent during its initial stages, and is usually chronically established at the time of definitive diagnosis by liver biopsy[1, 4]. A common association between inflammatory bowel disease, pancreatic inflammation, and non-suppurative CCHS was first suggested in 1984 by Zawie and Garvey and has subsequently been confirmed clinically[5].

We are currently conducting a study comparing liver biopsies from 42 cats with non-suppurative CCHS (n=37) and hepatic lymphoma (n=5). The aims of this study are four-fold. Firstly, we are seeking to refine the histopathologic features of non-suppurative CCHS, particularly with respect to the inflammatory infiltrate, apparent targeting of bile ductules, bile ductule proliferation, and portal fibrosis. Secondly, we are seeking to compare the infiltrative lymphocyte populations (immunophenotyping) between cats with advanced lymphocytic CCHS, cats with modest portal inflammation, and cats lacking inflammatory liver lesions. Thirdly, we are investigating the potential role of bacteria in non-suppurative CCHS. Fourthly, we are evaluating the utility of lymphocyte immunophenotyping and PCR detection of T-Cell receptor (TCR) gene rearrangement in differentiating inflammatory reactive or lymphoproliferative responses from true malignant lymphoma. Some preliminary findings from this study, as well as a how these fit with current concepts, will be discussed.

## HISTOLOGIC CHARACTERIZATION OF CCHS

While non-suppurative CCHS (synonyms: lymphocytic cholangiohepatitis, lymphocytic portal hepatitis) is the most common subcategory of feline CCHS, histologic features vary, reflecting disease chronicity and perhaps immune-mediated cell targeting. Well recognized features include small lymphocytes infiltrating within and restricted to the portal region (zone 1), a variable degree of portal fibrosis, and biliary ductular or oval cell proliferation. However, in some cats, lymphocytes seemingly target bile ductules and infiltrate into biliary epithelium causing progressive ductopenia. In others, expansion of an infiltrating mantle of lymphocytes confuses differentiation from a well-differentiated malignant lymphoma or lymphoproliferative disorder. While it has been speculated that suppurative CCHS precedes development of non-suppurative CCHS, proof of this transition has never been documented nor modeled in the cat.

### Portal Fibrosis

Similar to previous studies, cats with non-suppurative CCHS had a high degree of fibrosis and bile ductular proliferation. However, the degree of fibrosis was highly variable among cats, and not correlated to the degree of inflammatory infiltrate. Unexpectedly, the degree of fibrosis also was substantial in cats with lymphoma such that there was no significant difference when these cats were compared to cats with non-suppurative CCHS. However, cats with lymphoma were unique in that they did not demonstrate severe bridging fibrosis. The discovery that cats with lymphoma also develop portal fibrosis suggests the presence of an inflammatory milieu. The routine application of Masson trichrome allowed objective scoring of hepatic fibrosis that was often inapparent on routine HE stained sections.

While the concentric rings of peribiliary fibrosis observed in some cats with non-suppurative CCHS suggested chronic cholangiolar damage, we generally observed peribiliary fibrosis in all groups of cats examined. Consequently, we were unable to demonstrate a significant difference in the extent of peribiliary fibrosis between cats with CCHS, cats with lymphoma, and control cats. We conclude that chronic but silent development of peribiliary fibrosis may be common in cats and is not unique to lymphocytic CCHS.

### Bile ductule hyperplasia, targeting and ductopenia

Bile ductule hyperplasia was moderate to severe in cats with non-suppurative CCHS as well as in cats with lymphoma. Our findings in cats with non-suppurative CCHS were consistent with previous descriptions of lymphocytic cholangitis[1, 2,7,8,9]. However, finding that cats with lymphoma also demonstrate bile ductule hyperplasia precludes use of this feature for discrimination of lymphoma from non-suppurative CCHS. It is possible that this finding reflects inflammation associated with non-suppurative CCHS antecedent to developing lymphoma. Alternately, increased biliary pressure, aberrations of the peribiliary vascular plexus or stimulation of cholinergic or dopaminergic nerves may also have influenced development of biliary hyperplasia pursuant to portal lymphoma[10]. We were interested to note bile ductule hyperplasia in control cats, where cross sections of bile ducts ranged from 2-6 per portal region. While no studies to date have specifically quantified the average number of bile ductules in portal areas of cats, it is generally assumed that normal cats usually have one to two bile ductule profiles per portal region. The increased bile duct profiles in control cats may represent anatomic variation, or may indicate a degree of bile duct tortuosity rather than an absolute increase in bile ductule number.

Cats with non-suppurative CCHS and bile ductule proliferation often demonstrated a peripheralization of bile ductules to the margins of portal regions. This seemingly reflects a pathophysiologic response to bile ductule destruction and induction of oval cell hyperplasia[11]. While this histologic feature has not been previously described in cats with non-suppurative CCHS, this may serve as a marker of bile duct immunoinjury as lymphocytes target or "chase" bile ductules. Ductopenia (defined as the absence of bile ductules in >30% of portal areas) was unique to cats with non-suppurative CCHS (n=6). Loss of bile ducts has only been briefly mentioned in a single prior report[12]. We found that ductopenia was often patchy in distribution. Ductopenic portal triads often had a relatively reduced lymphocytic infiltrate compared to intact portal triads. In humans, ductopenia characterizes a number of diverse syndromes[13, 14]. Common to these syndromes is that ductopenia represents the consequence of immune-mediated necroinflammatory bile duct destruction[13]. Cholestasis, a sequel of ductopenia, is usually associated with classic bile accumulation within canaliculi and in the cytoplasm of hepatocytes. However, cholestasis can also subtly manifest as hepatocellular swelling and cytoplasmic accumulation of copper and copper-associated protein[13]. Though overt canalicular cholestasis was noted in only one case with ductopenia, a number of cases had notable copper accumulation within the cytoplasm of hepatocytes.

Bile duct targeting, characterized by lymphocytic infiltration within or immediately adjacent to the biliary epithelium and association with concurrent bile ductule epithelial swelling, necrosis or attenuation, was detected in approximately one-half of cats with non-suppurative CCHS. This feature appears to be diagnostically useful as none of the control cases or cats with lymphoma demonstrated similar bile duct targeting. This feature has been inconsistently recognized in other reports of feline lymphocytic cholangitis[1, 2,9,12]. In one report, bile duct targeting occurred only in sections demonstrating a suppurative inflammatory reaction[2]. Subsequently, this feature was interpreted to represent a chronic response in ascending cholangitis. In the present study, a suppurative inflammatory process was never observed in cats with the bile duct targeting lesion. Our findings suggest that lymphocytic bile duct targeting reflects an immune-mediated pathogenesis.

Aggregates of foamy macrophages (lipogranulomas) found in portal regions was specific to non-suppurative CCHS. Thus, this feature may be useful in discriminating between non-suppurative CCHS and lymphoma. Lipogranulomas, more commonly observed in liver tissue from older dogs compared to cats, represent a residual lesion marking cell death. As such, lipogranulomas within the adventitia of the portal triad may reflect cholangiolar destruction. This is consistent with the finding of residual cytokeratin positive cellular debris within some lipogranulomas.

## CHARACTERIZATION OF CCHS INFLAMMATORY INFILTRATES USING IMMUNOHISTOCHEMISTRY & TCR CLONALITY

Immunohistochemical characterization of infiltrating lymphocytes in feline lymphocytic CCHS was described in 20 cats with a progressive disease syndrome[9]. Because malignant lymphoma, the most frequently observed hematopoietic neoplasm in feline livers, often shows predilection for portal areas, it can be confused with non-suppurative CCHS. In such cases, immunohistochemical characterization of infiltrating lymphocytes and inspection for T-cell receptor (TCR) gene rearrangements might assist in the differentiation of inflammation from neoplasia[15].

### Immunohistochemistry

Our data corroborates previous findings that non-suppurative CCHS is predominantly a T-cell mediated inflammatory disorder[9]. In approximately 70% of cats with non-suppurative CCHS, T-cells intimately encircled biliary ductules and infiltrated biliary epithelium. In most cats, B-cells were randomly dispersed among dense T-cell populations with occasional formation of aggregates or early follicles. Similar B-cell aggregates have previously been described in cats with progressive lymphocytic cholangitis[9]. While B-cell follicle formation is known to follow chronic immune stimulation, none of the aggregates identified in the present study formed germinal centers. Because B-cell aggregates were also observed in one cat with lymphoma, their presence cannot be used to rule out neoplasia. However, this observation deserves further scrutiny in larger study. Some cats maintained a mixed B-cell T-cell infiltrate with approximately equal proportions of each cell type. Finding such a mixed but balanced population of lymphocytes may indicate differences in etiopathogenesis or disease stage, relative to cats with T-cell predominant inflammation. Continued study of cats with non-suppurative CCHS will be necessary to more fully investigate the relevance of such pleotrophic lymphocyte subsets.

### TCR clonality

We hypothesized that using PCR for the detection of TCR clonality would aid in differentiating neoplastic (likely clonal) populations of lymphocytes from inflammatory (likely polyclonal) populations. Unexpectedly, 27% of the cats that were histologically diagnosed with non-suppurative CCHS had clonal or pseudoclonal products. Of the 3 cats histologically diagnosed with T-cell lymphoma, one cat had a clonal product, one had an oligoclonal product and one had a polyclonal product. When the clonal and oligoclonal cases were combined in a single group, there was no significant difference in TCR clonality between hepatic lymphoma and control cats, non-suppurative CCHS and control cats or non-suppurative CCHS and lymphoma cats. Several considerations may explain our unexpected TCR findings. Firstly, a restricted diversity of responding inflammatory T-cells could represent a clonal population. Secondly, it is known that the liver harbors a large number of NKT-cells and that these cells have a restricted TCR diversity[16]. Similarly, since enteric intraepithelial T-cells have oligoclonal TCR beta-chain repertoires, it is possible that similar cells comprise the intraepithelial lymphocytes in bile ductules and that these dominate the inflammatory infiltrate in non-suppurative CCHS[17]. Finally, it is also possible that cats with non-suppurative CCHS may have clonal or oligoclonal T-cell populations that truly reflect an emerging lymphoma, as recognized in enteropathy-type intestinal T-cell lymphoma associated with refractory sprue in humans[18].

## ETIOPATHOGENESIS OF CCHS

To date, the etiopathogenesis of lymphocytic CCHS remains speculative. While an immune-mediated pathogenesis is suspected[8, 9], some authors have alternatively proposed that the syndrome represents a chronic manifestation of ascending bacterial infection[1, 2]. An immune mediated pathogenesis is consistent with the predominant CD3+ T-lymphocytes, B-lymphocyte aggregates and follicle formation, bile duct targeting and ductopenia, and membranous and cytoplasmic MHC class-II expression on bile duct epithelia documented in one report[9]. However, evidence implicating an infectious primary cause is sparse, with conflicting conclusions regarding the potential for *Helicobacter* generated lesions and a single study demonstrating zone 1 inflammation in cats with chronic *Bartonella* infections[19, 20, 21]. Thus, non-suppurative CCHS remains enigmatic with respect to etiopathogenesis.

We were specifically interested in the potential role of bacteria in the etiopathogenesis of non-suppurative CCHS. To address this, we performed fluorescence in situ hybridization using eubacterial probes on some of the liver biopsies (n=31). Most cases (>90%) of non-suppurative CCHS had no detectable organisms and overall, our results argue against an active role of bacteria in the etiopathogenesis of this disease. However, the possible role of bacteria in inciting this disease cannot be ruled out.

Evidence provided by the present study, particularly the histopathologic and immunohistochemical findings, supports that an immune-mediated process is a major mechanism of feline non-suppurative CCHS. Our findings do not implicate a primary bacterial pathogenesis. Further studies to investigate the possible role of autoantigens (as occur in similar human diseases) are necessary to further dissect the pathogenesis of this disease.

## REFERENCES

[1]Day DG. **Vet Clin North Am Small Anim Pract** 1995; 25:375. [2]Gagne JM, et. al. **Vet Pathol** 1996; 33:521. [3]van den Ingh TSGAM, et. al. *In:* **WSAVA Standards for Clinical and Histological Diagnosis of Canine and Feline Liver disease 1st ed.**, 2006; 68. [4]Center SA. *In:* **Strombeck's Small Animal Gastroenterology 3rd ed.** 1996; 860. [5]Zawie DA, et. al. **Vet Clin North Am Small Anim Pract** 1984 2:1201. [6]Gagne JM, et. al., **J Am Vet Med Assoc** 1999; 214:513. [7]Weiss DJ, et. al. **J Am Vet Med Assoc** 1996; 209:1114. [8]Prasse KW, et. al. **Vet Pathol** 1982; 19:99. [9]Day MJ. **J Comp Pathol** 1998; 119:135. [10]LeSage G, et. al. **Liver** 2001; 21:73. [11]Theise ND. **Hepatology** 1999; 30:1425. [12]Lucke VM, et. al. **J Small Anim. Pract.** 1984; 25:249. [13]Burt AD. **Clin Liver Dis** 2002; 6:363. [14]Hubscher SG, et. al. **Hepatology** 1993; 17:70. [15]Moore PF, et. al. **Vet Immunol Immunopathol** 2005; 106:167. [16]Knolle PA. *In:* **Zakim and Boyer's Hepatology: A Textbook of Liver Disease, 5th ed**. 2006; 149. [17]Helgeland L, et.al. **J Immunol** 1999; 162:2683. [18]Daum S, et. al. **Gut** 2001; 49:804. [19]Greiter-Wilke A, et. al. **J Vet Intern Med** 2006; 20:822. [20]Boomkens SY, et. al. **FEMS Immunol and Med Microbiol** 2004; 42:307. [21]Kordick DL, et. al. **J Clin Microbiol** 1999; 37:1536.

**CLINICAL CHARACTERISTICS & ROLE OF BACTERIA IN FELINE NEUTROPHILIC CHOLANGITIS**

**David C. Twedt, DVM, DACVIM**
**Fort Collins, CO**

Liver disease is common in the cat with the most frequent disorders described being lipidosis, inflammatory liver disease, neoplasia and hepatocellular necrosis. A 10-year study of feline liver biopsy data found that inflammatory liver disease is the second most common pathology after hepatic lipidosis.[1] Due to the lack or uniform liver histology descriptions and nomenclature in small animal liver disease the WSAVA sponsored an international group to attempt to standardize the histological classifications.[2] Collectively this group referred to the predominate inflammatory biliary associated diseases in cats as cholangitis with three distinct forms: neutrophilic, lymphocytic and cholangitis associated with liver flukes. The neutrophilic form of cholangitis can also be clinically and pathologically further subdivided into an acute neutrophilic form (ANF) and a chronic (and less suppurative) neutrophilic form (CNF).[3] These two sub-classifications are thought to likely represent different stages of a single disease process. Lymphocytic cholangitis (LC) appears to be less common occurrence in a review of 175 consecutive feline liver biopsies at Colorado State University (unpublished observations). Some authors suggest LC may have an immune mediated mechanism of disease.[4] There is also a separate entity called lymphocytic portal hepatitis (LPH), which is usually mild and may represent a non-specific reactive or aging change occurring in older cats.[2]

## NEUTROPHILIC CHOLANGITIS

Neutrophilic cholangitis was also previously referred to as suppurative, exudative or lymphocytic plasmacytic cholangitis/cholangiohepatitis. Neutrophilic cholangitis is thought by some authors to be the result of a biliary tract infection ascending from the gastrointestinal tract. The lesions are histologically characterized by the presence of neutrophils in the lumen and or epithelium of bile ducts. In the early stage neutrophils are also found in the portal areas. The inflammation may subsequently extend into the hepatic parenchyma (ie, cholangiohepatitis). As the disease progresses in chronicity the inflammatory infiltrate becomes more mixed with the portal areas containing neutrophils, lymphocytes and plasma cells. Fibrosis and bile ductular proliferation may also occur.

In many cases there appears to be a relationship of neutrophilic cholangitis with pancreatitis and inflammatory bowel disease (IBD). This is supported by a study finding 83% of cats with chronic biliary tract disease to have IBD and 50% having concurrent pancreatitis.[5] It is speculated that the common channel, the junction of the pancreatic duct and the common bile duct before they enter the duodenum, may explain the association of cholangitis and pancreatitis whereby ascending bacteria from the intestine entering the common channel can ascend into both organs. This anatomical relationship in conjunction with normally high concentrations of enteric bacteria in the proximal intestine of cats as compared to dogs and the frequent concurrent association with IBD may play a part in the role of the etiopathogenesis of feline inflammatory liver disease. Some refer to this multiple organ involvement as the feline triaditis syndrome; the association of liver, pancreatic and intestinal disease.

### Evidence For Bacteria In Neutrophilic Cholangitis

It is not uncommon to culture bacteria from cats having cholangitis however cultures are more often positive in the ANF than CNF. In most cases the organisms identified are enteric pathogens. We sought to determine the presence and distribution of bacteria within the livers of cats with neutrophilic cholangitis (ANF and CNF) by use of culture-independent methods and to compare those findings to culture results. A total of 40 archived liver biopsy samples were obtained at Colorado State University of cats with inflammatory hepatic disease and 14 cases with histologically normal livers. Histopathology was classified according to WSAVA guidelines. We found 16/40 (40%) cases had neutrophilic cholangitis with 3 cases of ANF and 13 cases of CNF. Other histological groups included LC, LPH and reactive hepatopathies (RH). The unstained hepatic tissue sections were subjected to fluorescence in situ hybridization (FISH) using a 16S rDNA probe that recognizes bacteria in the genera (EUB338) and a non-EUB probe as a control.[6] Antibodies against cytokeratins (clones AE1, AE3) and factor VIIIa were used to enable unequivocal distinction between bile ducts and vascular structures respectively. Tissues were examined with a fluorescence microscope for the presence and location of bacteria (KW Simpson, Cornell University). Bacteria (EUB338) were observed in 3/14 normal livers (2 had bacteria on their liver surface and one in a venous sinusoid) that likely represented contamination. In the 3 cases of ANF all had a positive culture (2 *Enterococcus spp*, 1 *E. coli*). All samples were FISH negative using EUB338 however the EUB338 probe does not detect *Enterococcus spp*. The one cat that was *E. coli* culture positive had histological evidence of cholestasis without hepatic biliary inflammation but had a positive bile culture. Two of the ANF cases had acute pancreatitis and 1 had chronic pancreatitis and IBD. 10/13 CNF cases were cultured and 4/10 cases had positive cultures (3 *E. coli* and 1 anaerobe) but 7 were FISH positive for EUB338 giving a total of 8/13 (61%) cases having bacteria by culture, FISH analysis or both. Both FISH and culture detected *E. coli* in 3/3 CNF having invasive bacteria. 6/13 CNF also had either concurrent pancreatic or intestinal disease or both. From this data and the expanded data presented on all 40 liver samples in an abstract presented at this Forum we believe there is evidence to support the role of bacteria involving the majority of cases of neutrophilic cholangitis with far less bacteria observed in other forms of feline inflammatory liver disease and the rare presence of bacteria in normal feline liver tissue that probable reflects surface contamination. From these studies and previous reports in the literature the information presented below represents our current understanding of neutrophilic cholangitis in the cat.

**Acute Neutrophilic Form**

ANF of cholangitis is associated with neutrophilic inflammation centered on the bile ducts. *E. coli* is most often cultured from either the bile or liver followed then by *Enterococcus spp.* Other bacterial isolates reported in affected cases include *Bacteroides spp, Enterobacter, Staphylococcus spp, Streptococcus spp, Klebsiella, Pseudomonas, Salmonella spp,* and occasionally anaerobes such as *Clostridium spp.*[7,8] ANF is observed primarily in younger to middle-aged cats (generally 3 to 7 years). Cats typically present with acute illness that is usually only a week or several weeks in duration. Clinical signs include acute vomiting, anorexia, and lethargy. Physical examination often reveals a fever, dehydration, icterus and abdominal pain on cranial abdominal palpation. Laboratory findings with all types of cholangitis are quite variable but usually a mild to moderate leukocytosis and icterus are identified. Liver enzymes are variable however most have GGT values higher than ALP elevations reflecting bile duct involvement. Morphologic changes in bile ducts may sometimes be observed using ultrasound. The bile ducts and gall bladder can be dilated or thickened due to edema and inflammation. Occasionally biliary obstruction may occur and is suggested by a dilated gall bladder and tortuous bile ducts. Definitive diagnosis requires a liver biopsy but a tentative diagnosis is often made based on a response to appropriate antibiotic therapy. Cytological examination from fine needle liver or bile aspirates may show suppurative inflammation. Bile aspirates for cytology and culture are relative safe if done correctly and should be considered in suspected cases. All cats having a liver biopsy should have a liver culture for aerobes and anaerobes. The ANF cases may also have concurrent acute suppurative pancreatitis and clinical investigation of the pancreas is always warranted (i.e. ultrasound and fPLI).

ANF of cholangitis requires prompt initiation of antibiotic therapy, ideally based on culture and sensitivity results. Without a culture I would suggest selecting an antibiotic that is effective against most enteric gram-negative aerobes and *Enterococcus spp.* and one that has good hepatic and biliary penetration. I prefer a cephalosporin, amoxicillin or amoxicillin-clavulanic acid as an empirical antibiotic choices. Antibiotics therapy should extend well after clinical recovery (possibly 3 to 4 weeks if not longer). Additional therapy includes nutritional support, pain management (buprenorphine), ursodeoxycholic acid (Ursodiol) and surgery if a bile duct obstruction is identified (see below).

**Chronic Neutrophilic Form**

It is possible that untreated or poorly treated cases having ANF of cholangitis may progress to a more chronic form. In the CNF there is a variable infiltration in portal areas by plasma cells, lymphocytes and neutrophils. The inflammation is centered on the bile duct walls but the process may also extend through the limiting plate to involve parenchyma. Biliary hyperplasia, periductal (sclerosing) fibrosis and bridging fibrosis may develop. The etiology for CNF is unknown but it is postulated that persistent presence of the enteric microflora may lead to upregulated cytokine production thereby promoting and perpetuating the biliary inflammation.

Cats tend to be older than those with the ANF at the time of the diagnosis. The clinical signs are often intermittent and can include vomiting, lethargy, anorexia and weight loss. The signs often tend to wax and wane over months with intermittent flare ups. A leukocytosis is uncommon and liver enzymes are variable but reflect predominately cholestasis. Icterus is common. Ultrasound evaluation in CNF may have similar changes as seen with the ANF. As fibrosis and bile duct replication take place, the portal areas become more hyperechoic and the bile ducts can become more torturous. Changes suggesting chronic pancreatitis may be seen on ultrasound showing a mottled nodular appearance with ductular changes. Feline pancreatic lipase immunoreactivity (fPLI) concentration may also be abnormal. A definitive diagnosis of the CNF of cholangitis requires a liver biopsy and cultures.

The etiology of CNF is unknown and specific therapeutic recommendations are difficult to provide. I believe an appropriate course of antibiotic therapy as described above for ANF is indicated with or without a positive culture based on our findings of bacteria in 60% of the cats having the CNF. Corticosteroids have traditionally been recommended for the inflammatory component of the disease but again no controlled studies are available. With a failure to respond initially to antibiotics and having negative liver cultures corticosteroids may be a reasonable consideration. I also always prescribe ursodeoxycholic acid and employ nutritional and/or antioxidant (i.e., vitamin E, SAMe and milk thistle [silibin]) support as well. In some cats with either ANF or CNF the bile may become an inspissated sludge or result in cholelith formation and possibly causing biliary obstruction. This may be the result of deconjugation of bile salts and bilirubin by bacteria in the biliary tree. Surgical by-pass or diversion procedures are reported to be associated with a poor prognosis and the mortality rate is high and is not recommended if other options are available.[9] I recommend surgical flushing of the extrahepatic biliary tree followed by appropriate therapy for the neutrophilic cholangitis. Mortality for non-diversionary techniques such as biliary stent placement is reported to be much lower than diversion surgeries.

**REFERENCES**

[1]Gagne JM, et.al. **Vet Pathol** 1996;33:521. [2]van den Ingh TSGAM, et.al. **WSAVA Standards for Histological and Clinical Diagnosis for Canine and Feline Liver Disease** 2006, p61. [3]Twedt DC, et.al. **Kirk's Current Veterinary Therapy XIV** 2008, p576. [4]Day DG, **Vet Clin North Am Small Anim Pract** 1995;25(2):375. [5]Gagne JM et.al. **J Am Vet Med Assoc** 1999;214(4):513. [6]Simpson KW, et.al. **Infection and Immunity** 2006, 74(8); 4778. [7]Hiesch VM, et.al. **J Am Vet Med Assoc** 1983,182(11):1223. [8]Brain PH, et.al. **J Feline Med and Surg** 2006, 8(2); 91. [9]Bacon NJ, et. al. **J Small An Pract** 2003,44; 231.

## WSAVA CLASSIFICATION & ROLE OF BACTERIA IN FELINE INFLAMMATORY HEPATOBILIARY DISEASE

**Mark P. Rondeau, DVM, DACVIM (SAIM)**
**Philadelphia, PA**

### INTRODUCTION

Historically, the veterinary literature regarding inflammatory disease of the feline liver and biliary tract has utilized differing and non-standardized classification systems to describe the complex of conditions that may occur.[1-3] Consequently, it has been difficult to compare results between studies. A standardized classification system for this disease complex has recently been proposed by the World Small Animal Veterinary Association (WSAVA) Liver Standardization Group.[4] This group stratifies cholangitis in dogs and cats into the following four categories: neutrophilic cholangitis (NC), lymphocytic cholangitis (LC), destructive cholangitis and chronic cholangitis associated with liver fluke infestation. NC and LC represent the common inflammatory diseases that have been described previously in the cat. NC is characterized histologically by the presence of neutrophils within the lumen or epithelium of bile ducts. This disease can be seen in acute or chronic stages. During the acute stage, neutrophilic inflammation and edema are often present in portal areas. In chronic NC, mixed inflammation is often present in portal regions, possibly coexistent with bile duct hyperplasia and fibrosis. In the previous literature, NC may have been referred to as suppurative cholangitis/cholangiohepatitis[1,2] or just cholangiohepatitis[3]. LC is a disease process characterized histologically by lymphocytic inflammation restricted to portal areas with varying degrees of fibrosis and bile duct hyperplasia. In addition to small lymphocytes, solitary plasma cells and eosinophils may be present. Previous nomenclature for this syndrome may have included lymphocytic cholangitis/cholangiohepatitis[1], lymphocytic-plasmacytic cholangitis/cholangiohepatitis[1], non-suppurative cholangitis/cholangiohepatitis[2], or lymphocytic portal hepatitis[3].

A single pathogenesis for cholangitis in cats remains unproven to date. It has been hypothesized that some, perhaps many, cases of cholangitis are associated with ascending bacterial infection from the intestine to the biliary tract, resulting in acute neutrophilic inflammation of bile ducts and ductules.[1-4] Absent appropriate diagnosis and therapy, the acute neutrophilic form of cholangitis may then be followed by the chronic, lymphocytic form of the disease.[2] Evidence in support of this hypothesis derives from the association reported between cholangitis, pancreatitis, and inflammatory bowel disease.[5] Despite an obvious association, studies documenting the prevalence of bacterial infection of the biliary tract in affected cats have yielded inconsistent results. Infection rates as low as 14% with NC and 5% with LC have been reported.[1] However, other studies have documented higher infection rates, including positive bile cultures in 5 of 6 cats with NC in a recent report[6]. Two studies describing cats with extra-hepatic bile duct obstruction, the majority of which had cholangitis, reported high rates of infection (53% and 55%).[7,8] Explanations for negative cultures in some cases may include timing of culture (some patients may have been receiving antibiotics at the time of culture), infection with organisms that are difficult to grow by traditional methods, or inadequate sampling. Evidence of *Helicobacter* infection has been reported in some cats with cholangitis.[9,10] A survey of cats with all types of hepatobiliary disease at one institution revealed that positive bacterial cultures were more common when sampling bile (36%, excluding contaminants) than hepatic tissue (14%).[11]

This lecture will present recent and ongoing research regarding cholangitis in cats and the possible role of bacteria in its pathogenesis.

### RESEARCH

**A Survey of Feline Cholangitis Using the WSAVA Liver Standardization Group Classification System.**
**Co-Investigators: Morgan MJ, Rankin SC, Shofer FS, Berent A, VanWinkle T, Washabau RJ.**

The primary goal of this study was to characterize a population of cats with cholangitis using the classification system developed by the WSAVA Liver Standardization Group, and to determine whether this classification system was applicable to our population of clinically affected cats. The biopsy database at the University of Pennsylvania was searched for cases of feline cholangitis, hepatitis or cholangiohepatitis diagnosed between January 1990 and December 2005. Histopathology was reviewed by a single pathologist (TVW) and all cases having cholangitis as the primary lesion and a complete medical record were included. Cases were stratified into groups using the WSAVA classification system for feline cholangitis. For the purposes of data analysis and reporting, cases of NC and LC were further characterized based on the histologic presence or absence of fibrosis and bile duct hyperplasia. Cases in which the neutrophil was the predominant inflammatory cell type and in which fibrosis and bile duct hyperplasia were characterized as minimal or mild were classified as acute neutrophilic cholangitis (ANC). Cases with mixed inflammatory cells and a component of neutrophilic inflammation and in which moderate to severe fibrosis or bile duct hyperplasia were present were classified as chronic neutrophilic cholangitis (CNC). Cases in which there was predominantly lymphocytic or lymphoplasmacytic inflammatory infiltrate with minimal to no neutrophilic infiltrate and minimal to mild fibrosis or bile duct hyperplasia were classified as acute lymphocytic cholangitis (ALC). Cases in which there was predominantly lymphocytic or lymphoplasmacytic inflammation, minimal to no neutrophilic infiltrate, and moderate to severe fibrosis or bile duct hyperplasia were classified as chronic lymphocytic cholangitis (CLC). Medical records were reviewed by a single author (MJM).

Seventy-four cases were included in the study. Of these, 13 had ANC, 31 had CNC, 14 had ALC, and 16 had CLC. The median age for all cases was 11 years, with no significant difference between groups. 63.5% of cats in the study were male. No breed predilection was found. There were no significant differences between groups in the frequency of any clinical signs or physical examination findings. Cats with CNC were significantly more likely than cats in any other group to have increased ALT, GGT, total bilirubin, total protein, and globulin measurements. Ultrasonographic abnormalities associated with the liver were seen frequently in all groups, with no significant differences detected between groups. Pancreatic inflammation was common in the few cases that had pancreatic biopsies performed (11/14, 78.5%). Similarly, intestinal inflammation was common in the cases that had intestinal biopsies (16/21, 76.2%). However, there were no differences between groups and the small number of cases that had both intestinal and pancreatic biopsies performed makes interpretation of a relationship between cholangitis, pancreatitis and inflammatory bowel disease difficult. A significantly higher percentage of cats with CNC had bile duct obstructions (76.2%) compared to cats with other forms of cholangitis (14.3%). Excluding probable contaminants from analysis, culture yielded bacterial growth in 41% of the cases in which it was performed, though no significant differences were identified between groups. Overall, bile cultures were more likely to be positive and less likely to be contaminated than were hepatic cultures. This suggests that bile culture may be preferable to hepatic tissue culture for documentation of bacterial infection in cases of cholangitis. There was a trend towards an increased frequency of bacterial growth in cats with EHBDO compared to those without EHBDO, although the trend did not reach statistical significance. The overall survival to discharge was 72%, with no differences between groups.

As a secondary goal of the study, real-time PCR using 16S rRNA gene primers was performed on paraffin-embedded formalin-fixed hepatic tissue from 40 cases and 50 control cats in an effort to amplify bacterial DNA. Results were positive in 6/40 cases and 5/50 controls. DNA sequencing revealed contaminant organisms in all positive samples, likely a result of non-sterile handling and prolonged storage of the paraffin blocks. This does not appear to be a useful diagnostic tool for identifying bacterial infections in patients with cholangitis when the samples are not handled in a sterile manner and are stored for extended periods of time

**Prospective Evaluation for Bacterial Infection in Hepatic Tissue and Bile of Cats with Diffuse Hepatobiliary Disease. Co-Investigators: Morgan MJ, Rankin SC, Berent A, Weinstein N, VanWinkle T.**

The goals of this ongoing study are to determine the prevalence of bacterial infection in cats with diffuse hepatobiliary disease (including cholangitis), to evaluate the diagnostic utility of hepatic tissue versus bile for the identification of bacterial pathogens in these cats, and to assess the utility of real-time PCR (RT-PCR) for detection of bacterial organisms in frozen hepatic tissue and bile. Cases at the University of Pennsylvania in which hepatic biopsy is performed as part of a complete diagnostic evaluation of diffuse hepatobiliary disease are being enrolled consecutively. Cats receiving antibiotic therapy within 2 days prior to sampling are excluded from the study. For all cases, hepatic tissue is acquired using the method desired by the attending clinician, and bile is collected via cholecystocentesis. Control samples were obtained at the time of euthanasia from cats without liver disease that were part of a breeding colony involved in an unrelated study. Control hepatic tissue was obtained via wedge biopsy and bile via cholecystocentesis within 10 minutes of euthanasia.

To date, 22 cases and 8 control cats have been included since April 2006. Aerobic and anaerobic bacterial culture was performed on all samples (hepatic tissue and bile) at the time of acquisition. Real-time PCR using 16S rRNA gene primers was performed on batched, frozen samples (hepatic tissue and bile) within 6 months of their acquisition. Routine histopathology using light microscopy was performed on all cases and controls. Cases with cholangitis were categorized as having acute neutrophilic cholangitis (ANC), chronic neutrophilic cholangitis (CNC), acute lymphocytic cholangitis (ALC), or chronic lymphocytic cholangitis (CLC) as described above. Cases without cholangitis were categorized as having hepatic lipidosis (HL) or other hepatic disease. All histopathology was reviewed by a single pathologist (TVW).

Twelve cases with cholangitis (7 with CNC, 4 with CLC, 1 with ALC), 4 cats with HL, 6 cats with other types of hepatic disease, and 8 control cats have been evaluated to date. Liver histopathology was normal in all controls. Bile culture was positive in 5 of 22 cases, including 4/7 (57%) cats with CNC, and 1 cat with concurrent HL and biliary carcinoma. The following bacterial organisms were isolated from bile: *E.coli* (3), α-hemolytic *Streptococcus* (1), *P. aeruginosa* (1) and an unidentified anaerobic gram-positive rod. RT-PCR of the bile was positive in 4 cases, all of which had positive bile cultures. In 3 of the 4 cases in which RT-PCR was positive, DNA sequencing identified the same organism that had been found using traditional bacterial culture of the bile. In one case, DNA sequencing identified *B.fragilis*, which had not been identified using bacterial culture. Three of the 22 cases had extra-hepatic bile duct obstruction and all 3 had evidence of bacterial infection of the bile. Hepatic tissue culture was positive in 2 of 22 cases. In both cases *E.coli* was isolated from both hepatic tissue and bile of cats with CNC. RT-PCR of hepatic tissue was negative in all cases. Culture and RT-PCR revealed no evidence of bacterial infection in hepatic tissue or bile of any control samples.

Due to the small case numbers it is difficult to draw significant conclusions from the data obtained to date from this ongoing study. However, it appears that bacterial infection is more common in cats with CNC than in cats with other forms of cholangitis and other types of hepatobiliary disease. It also appears that bile may be more useful than hepatic tissue for the purpose of bacterial identification. RT-PCR may be a viable method to rapidly and accurately detect bacterial DNA in bile, but perhaps not in hepatic tissue. All 3 cases with extra-hepatic bile duct obstruction had bacterial infection of bile, further suggesting a strong relationship between these 2 conditions. Further data is required to better define the relationships between cholangitis, extra-hepatic bile duct obstruction and bacterial infection in cats and to determine the optimal diagnostic and therapeutic interventions in these cases.

## REFERENCES

[1]Center SA et.al. **Proceedings 12th ACVIM Forum** 1994:766. [2]Day DG. **Vet Clin North Am Small Anim Pract** 1995;25:375. [3]Gagne JM et.al. **Vet Pathol** 1996;33:521. [4]Rothuizen et.al. **WSAVA Standards for Clinical and Histologic Diagnosis of Canine and Feline Liver Disease** 2006:68. [5]Weiss DJ et.al. **J Am Vet Med Assoc** 1996;209:1114. [6]Brain PH et.al. **J Feline Med Surg** 2006;8:91. [7]Mayhew PD et.al. **J Small Anim Pract** 2002;43:247. [8]Eich CS et.al. **J Am Anim Hosp Assoc** 2002;38:290. [9]Boomkens SY et.al. **FEMS Immunol Med Microbiol** 2004;42:307. [10]Greiter-Wilke A, et.al. **J Vet Int Med** 2006;20:822. [11]Wagner et.al. **J Vet Int Med** 2007;21:417.

# ROLE OF AMPK IN SKELETAL MUSCLE: FROM METABOLIC SIGNALING TO IMMUNE REGULATION

**Emmanuelle St-Amand, B.Sc, Geneviève Pilon, Ph.D, and André Marette, Ph.D.**
**Québec City, QC**

## INTRODUCTION

Skeletal muscle is the major tissue responsible for whole-body glucose disposal when insulin levels rise after a meal and during physical exercise. It is now well established that the translocation of GLUT4 glucose transporters from intracellular compartments to the cell surface represents an important mechanism by which both insulin and exercise stimulate glucose transport in muscle cells. Furthermore, there is now substantial evidence that impaired insulin stimulation of glucose transport in skeletal muscle of type 2 diabetic individuals is mainly caused by a defect in insulin receptor signaling to the PI3K/Akt pathway, leading to impaired GLUT4 translocation and/or activation of the transporter protein at the cell surface. Importantly, the effect of exercise/contraction on glucose transport appears to be relatively well preserved in diabetic muscle. It is therefore critical to fully understand the signaling mechanisms that induce glucose transport in skeletal muscle of exercised/contracted muscles, with the hope of discovering novel targets for the treatment of muscle insulin resistance in type 2 diabetic subjects. Muscular contraction activates 5'-AMP-activated protein kinase (AMPK) and there is accumulating evidence that this energy-sensing enzyme mediates at least part of the metabolic effects of exercise. Because activation of AMPK also increases lipid oxidation in muscle, it is considered to be largely responsible for the health benefits of exercise and is thought to be a promising target for the development of new treatments for obesity, insulin resistance and type 2 diabetes.

## AMPK

### A Metabolic Master Switch

AMPK is a member of a metabolite-sensing protein kinase family that acts as a fuel gauge monitoring cellular energy level. When AMPK "senses" decreased energy stores, it acts to switch off ATP-consuming pathways and switch on alternative pathways for ATP regeneration. AMPK is active as a heterotrimer consisting of one catalytic subunit ($\alpha$) and two noncatalytic subunits ($\beta,\gamma$). The noncatalytic subunits are essential for optimum enzyme activity and may participate in substrate targeting. Several isoforms for each subunit have been identified, and the contribution of each isoform to the AMPK heterotrimer varies among different tissues. All subunits of AMPK seem to be expressed in human and rodent skeletal muscle tissue, but only three AMPK heterotrimers ($\alpha1\beta2\gamma1$, $\alpha2\beta2\gamma1$, and $\alpha2\beta2\gamma3$) appear to account for the AMPK activity present in this tissue. LKB1 is still considered the most likely (albeit not the only) candidate for the upstream kinase in the AMPK cascade. Indeed, purified recombinant LKB1 phosphorylates and activates AMPK whereas cells lacking LKB1 do not respond to AICAR, a pharmacological activator of AMPK, or to other stimuli that increase the AMP:ATP ratio. Genetic studies in animals also support a role for LKB1 as an upstream AMPK kinase. LKB1 deletion in skeletal or cardiac muscle severely abolishes AMPK$\alpha$2, but not AMPK$\alpha$1, activity in response to AICAR and electrical stimulation-induced AMPK phosphorylation/activity is blunted to a large extent in the absence of LKB1[1,2].

### Role of AMPK in Contraction-Induced Glucose Uptake in Skeletal Muscle

Glucose transport in skeletal muscle is mediated by facilitated diffusion through a family of membrane-embedded proteins known as the glucose transporters (GLUTs). At least 5 isoforms (GLUT1, 4, 5, 8, and 11) of the GLUT family are expressed in adult muscle but GLUT4 i is by far the predominant isoform expressed in muscle. Under basal or resting conditions, GLUT4 is sequestered in intracellular vesicles. When muscle is exercised or stimulated by insulin, GLUT4 is translocated from the intracellular vesicles to the cell surface. Using subcellular fractionation, we have previously shown that both insulin and contraction increase glucose transport by mobilizing GLUT4 to the plasma membrane but also to the T-tubules[3,4,5], which are deep inward extensions of the sarcolemma that interact with the terminal cisternae of the sarcoplasmic reticulum. The nature and number of intracellular GLUT4 storage compartment(s) from which GLUT4 is recruited is still debated. We previously reported that contraction stimulates GLUT4 translocation from a "contraction-responsive" recycling endosomal pool distinct from that of insulin[6]. These results lend support to earlier studies suggesting that insulin and contraction recruit distinct but undefined GLUT4 storage pools in muscle.

The stimulatory effects of moderate exercise on muscle glucose disposal are relatively well preserved in insulin-resistant type 2 diabetic subjects. It is therefore critical to delineate the proximal signaling mechanisms leading to contraction-induced muscle glucose uptake. AMPK was rapidly considered an attractive candidate for mediating the beneficial actions of exercise. Indeed, early studies showed that exercise and electrically-stimulated contraction increase the activity of AMPK in muscle. Moreover, several studies have shown that AICAR stimulates glucose transport in fast-twitch muscles and that its effect is not additive to that of contraction. Acute infusion of AICAR was found to activate skeletal muscle AMPK and increase glucose transport in obese insulin-resistant Zucker rats[7] and defective activation of $\alpha$1-AMPK in response to contraction in obese Zucker rats did not affect contraction-stimulated glucose transport, which was paralleled by normal activation of $\alpha$2-AMPK in muscle of these animals[8]. Notably, protein expression of all $\alpha$ -$\beta$, and $\gamma$ -subunits of AMPK is maintained in skeletal muscle of obese type 2 diabetic subjects. Furthermore, exercise-induced $\alpha$2-AMPK activity is preserved or only slightly attenuated in type 2 diabetic subjects[9,10]. Accordingly, AICAR increases phosphorylation of AMPK and its downstream target acetyl-CoA carboxylase (ACC) similarly in skeletal muscle from type 2 diabetic subjects and nondiabetic controls[11]. Thus AMPK activation may be an attractive strategy to improve glucose transport in insulin-resistant skeletal muscle.

Further genetic evidence for a role of AMPK in contraction-induced glucose uptake came from the work of Mu et al.[12] who showed that expression of a dominant inhibitory mutant of AMPK in transgenic mouse muscle reduced contraction-induced glucose uptake by 30-40% whereas it fully abrogated the stimulatory effect of AICAR or hypoxia on glucose transport. This study not only supports a role for AMPK in contraction-induced glucose uptake but also indicates that other AMPK-independent mechanism(s) are likely to explain the full effect of contraction on glucose uptake. Knockout (KO) mouse models for individual AMPKα subunits were recently generated. Whereas α1-AMPK KO mice exhibited normal glucose homeostasis, α2-AMPK-KO showed impaired insulin secretion and in vivo insulin resistance. However, α2-AMPK KO mice also exhibited increased adrenergic tone which could explain both decreased insulin secretion and insulin resistance in vivo. Surprisingly, isolated muscles from α2-AMPK KO mice failed to respond to AICAR but exhibited normal response to contraction-induced glucose uptake. However, this is likely explained by a compensatory increase in expression and activity of the α1-AMPK subunit. This compensatory response of the α1-AMPK subunit was not observed in muscle-specific expression of a dominant inhibitory mutant of α2-AMPK.

Our own studies support the growing concept that contraction-induced glucose transport is at least partly AMPK-independent. Indeed, we found that the AMPK activator AICAR increases GLUT4 translocation to the plasma membrane but fails to induce recruitment of the transporter to T-tubules[13], which represent 60% of the total cell surface area in skeletal muscle fibers. Since we have reported that exercise and contraction stimulate GLUT4 translocation to both the plasma membrane and the T-tubules[4,5], it is conceivable that contraction induces GLUT4 mobilization to the tubular elements through an AMPK-independent mechanism. This observation is consistent with the finding that contraction-induced glucose uptake is only partially inhibited in muscle expressing a dominant-negative AMPK[12].

Despite the accumulated evidence indicating a partial role for AMPK in exercise dependent glucose metabolism, Fujii *et al.*[14] recently suggested that contraction-mediated glucose transport is fully independent of AMPKα2. This is based on observations made in another mouse model expressing an inactive form of AMPKα2 in skeletal muscle (α2i TG). Their conclusion relies on the premise that when force generation was matched between muscles of α2i TG mice and their WT littermates, no deficit in contraction-induced glucose transport was observed. Indeed, mouse models lacking an active α2AMPK have a reduced capacity for maximal force production, a finding we have confirmed in transgenic mice expressing a dominant inhibitory mutant of AMPK (AMPKα2-KD mice)[15]. Since, in skeletal muscle, contraction mediated glucose transport is proportional to force development, it is key that this parameter be controlled. In their study, Fujii and colleagues opted to use a non supramaximal voltage (8.4-21V) in the WT group in order to match force production to the level of the α2i TG animals that were stimulated with a supramaximal voltage (*i.e.* >30V). However, during electrical stimulation of isolated muscles, the voltage must remain supramaximal in order to ensure stimulation of all deep muscle fibres. It is likely that the option of reducing the stimulation voltage in the WT mice would inevitably lead to a lesser recruitment of the deepest muscle fibres, and hence reduced glucose transport, which ultimately cannot be compared with muscles of α2i TG animals that were stimulated by a supramaximal voltage in which all muscle fibres are solicited. Thus the conclusion that no differences in contraction-induced glucose transport is observed between WT and α2i TG mice is questionable based on the fact that a greater number of fibres were likely recruited during the electrical stimulation of α2i TG muscles even though force production was equal to that of WT mice.

To circumvent these limitations, we designed a contraction stimulation protocol to match force production between WT and AMPKα2-KD Tg mice without compromising the voltage stimulation parameters in the WT mice and thus fibre recruitment remained similar between groups. This was made possible by stimulating both WT and AMPKα2-KD Tg EDL muscles at a lower frequency (50Hz), where genotypic differences in absolute force production were no longer observed[15]. This allowed us to investigate the role of AMPK in contraction-mediated glucose transport under conditions of equivalent muscle force responses. When force development between AMPKα2-KD and WT EDL muscles is matched, using identical stimulation voltage and contraction protocols, glucose transport is diminished by 50% in EDL of AMPKα2-KD mice. Our data are therefore entirely consistent with a significant role for AMPK in submaximal contraction-induced glucose transport, but they also suggest that AMPK-independent mechanisms are also required for stimulation of hexose uptake.

**Role of AMPK in the Regulation of Inflammation by Anti-Diabetic Drugs**

Over the last 6 years, my laboratory has also focused on the potential role of NO in the modulation of glucose metabolism and in the pathogenesis of insulin resistance in acute and chronic inflammatory settings such as sepsis and obesity-linked diabetes. A major finding of this project was that iNOS expression is increased in skeletal muscle and adipose tissue of dietary and genetic models of obesity-linked diabetes[16]. More importantly, we confirmed the causative role of iNOS in obesity-linked insulin resistance by showing that mice with a targeted disruption of iNOS (iNOS KO) do not develop muscle insulin resistance when made obese by the consumption of a high-fat diet[16]. We also made the striking observation that clinically-used anti-diabetic drugs such as PPARγ agonists and metformin inhibit iNOS induction in myocytes, adipocytes and macrophages, all major sites of iNOS expression associated with insulin resistance in inflammatory settings[17]. These drugs mediate their inhibitory action on iNOS via stimulation of AMPK. Indeed, we have confirmed that AMPK is the mediator of iNOS inhibition by AICAR and anti-diabetic drugs by knocking down AMPK using RNA interference[17]. This was the first demonstration that AMPK inhibits iNOS and thus suggest that PPARγ agonists and metformin exert their anti-diabetic actions not only via their well known lipid and glucose-related metabolic effects but also via inhibiting iNOS, a key inflammatory mediator of insulin resistance.

**Reciprocal Regulation of Glucose Metabolism and Insulin Action by PPARγ/AMPK and iNOS/NO pathways**

However, while PPARγ activation reduces iNOS expression in several cell types, there is also evidence for reciprocal down-regulation of PPARγ by iNOS in macrophages[18]. This suggests that the iNOS/NO pathway modulates PPARγ activity, at least in myeloid cells. Supporting this concept, we recently found that high-fat-fed obese mice lacking iNOS are sensitize to the beneficial effect of a low dose of the PPARγ agonist Rosiglitazone on glucose tolerance[19]. This genotype-specific effect of Rosiglitazone was linked to a marked enhancement in the ability of the PPARγ agonist to raise oligomeric complexes of adiponectin in the plasma and activate the AMPK pathway in target tissues of obese iNOS KO mice. Our results therefore identify iNOS as a critical inflammatory modulator of PPARγ activation and circulating adiponectin levels, and show for the first time that invalidation of a key inflammatory mediator can improve the therapeutic efficacy of PPARγ agonism and AMPK activity in an animal model of obesity.

## CONCLUSIONS

AMPK is clearly a fascinating protein and joins the growing number of molecules that are known to modulate both metabolic processes and immune function. Its ability to promote lipid and glucose utilization while exerting anti-inflammatory actions makes it a very promising target for the development of new anti-obesity and insulin sensitizing drugs. However, further studies are warranted to uncover the molecular mechanisms underlying AMPK dual actions on metabolic and immune pathways.

## REFERENCES

[1]Sakamoto K, et.al. **Embo J** 2005; 24, 1810-20. [2]Thomson DM, et.al. **Am J Physiol Endocrinol Metab** 2007; 292, E196-202. [3]Dombrowski,L, et.al. **Am J Physiol Endocrinol Metab** 1996; 270, E667-76. [4]Roy D, et.al. **Am J Physiol Endocrinol Metab** 1997; 273, E688-94. [5]Roy D, et.al. **Biochem Biophys Res Comm** 1996; 223, 147-152. [6]Lemieux K, et.al. **Diabetes** 2000; 49, 183-9. [7]Bergeron R, et.al. **Diabetes** 2001; 50, 1076-82. [8]Barnes BR, et.al. **Diabetes** 2002; 51, 2703-8. [9]Hojlund K, et.al. **Am J Physiol Endocrinol Metab** 2004; 286, E239-44. [10]Sriwijitkamol A, et.al. **Diabetes** 2007; 56, 836-48. [11]Koistinen HA, et.al. **Diabetes** 2003; 52, 1066-72. [12] Mu J, et.al. **Mol Cell** 2001; 7, 1085-94. [13]Lemieux K, et.al. **FASEB J** 2003; 17, 1658-65. [14]Fujii N, et.al. **J Biol Chem** 2005; 280, 39033-41. [15]Lefort N, et.al. **Am J Physiol Endocrinol Metab**, 2008; 295: E1447 - E1454. [16]Perreault M & Marette A. **Nat Med** 2001; 7, 1138-43. [17]Pilon G, et.al. **J Biol Chem** 2004; 279, 20767-74. [18]Shibuya A, et.al. **FEBS Lett** 2002; 525, 43-7. [19] Dallaire P, et.al. **Diabetes** 2008; 57: 1999-2011.

**GENETICS OF TYPE 1 DIABETES: CURRENT STATUS & THE FUTURE**

**Constantin Polychronakos M.D.**
**Montreal, Quebec, Canada**

## INTRODUCTION

Type 1 diabetes (T1D) is one of the most severe of the common metabolic disorders, uniformly resulting in death without insulin treatment. Even with the most sophisticated treatment approaches, long-term complications are common and lead to severe disability and premature death. Diabetes costs the Canadian economy 14 billion / year and, although T1D cases are a minority, having a much earlier onset and greater severity, they account for a large proportion of this estimate. T1D is due to autoimmune destruction of the insulin-producing pancreatic β-cells and, by the time clinical symptoms appear, most insulin secretory capacity has been lost. This makes prediction and prevention a high priority but both require a better understanding of aetiology. The cause of loss of immune tolerance with such high specificity for the β-cell is unknown but it includes an important component of genetic predisposition. Knowledge of the genetic basis of the disease is expected to reveal valuable clues about pathogenesis, permit the early identification of individuals at risk and aid in the development of effective preventive interventions. Moreover, genetic heterogeneity turns out to be an important factor in the disease, genetic profiling is likely to aid in subclassification of modes of loss of tolerance that may be different in different individuals and respond differentially to specific treatments (personalized medicine).

## THE GENETICS OF T1D

### Genetic epidemiology

T1D predisposition is highly heritable. Disease concordance in monozygotic twins exceeds 50% and concordance for the presence of T1D-related autoantibodies is even higher (1), indicating that genetic predisposition has an important role in determining why one out of 200-400 individuals of European descent is affected. Concordance in dizygotic twins falls by an order of magnitude to ~5%, no different from that of siblings and still much higher than in the general population, which is 0.3%. This indicates a relative sibling risk ($\lambda_s$) of >15, one of the highest in common complex traits. The order-of-magnitude difference between monozygotic and dizygotic twins also suggests that little, if any, of this familial clustering is due to *shared* environment. The rapid fall with increasing genetic distance also points to the polygenic nature of the trait.

In addition, epidemiologic evidence suggests strong environmental factors but none have been identified to date with certainty. Knowledge of the genetic underpinnings of T1D will undoubtedly make their identification easier, both by increas-ing statistical power with the removal of genetic "noise" and by providing mechanistic insights. Both genetics and environment appear to play an important role in large discrepancies in the geographic distribution of the disease. Europe has the highest incidence and, within the continent, a strong north-to-south gradient is detectable, with Scandinavian countries having as much as four-fold higher incidence than in the south. T1D risk is considerably lower in individuals of African descent and much lower in East Asians.

As noted above, there is no compelling reason to assume that tolerance to β-cells is lost through the same mechanism in all human patients. Evidence for genetic heterogeneity comes from the animal models. The inbred strains NOD mouse and BB rat have autoimmune diabetes that, like human T1D, is T-cell mediated and dependent on the major histocompatiblity locus (MCH, see below). However, in the two inbred strains (genetically, the equivalent of two human patients in multiple copies) the same autoimmune end-point is the result of entirely different immune dysregulation phenotypes (and, correspondingly, genotypes), neither of which has been convincingly reproduced in human T1D patients, taken as one group (Reviewed in Ref. 2). I have proposed (2) that to the extent that rodents can tell us anything about human disease, there are at least two different ways of losing immune tolerance to the β-cells. We do not know how many additional immune dysregulation models are represented in the genetically diverse human population. Indeed, it is logical to expect that in any human or animal population, disease risk will depend on the genetic diversity among individuals or inbred strains. A complete picture of the genetics of human T1D will help clarify this.

### Molecular Genetics

Until recently, identification of individual loci in complex traits was almost exclusively the result of candidate-gene searches. On this basis, it has been known for three decades that the major single contributor to T1D risk maps to the MHC complex on Chr.6p21, a family of adjacent genes encoding the human leukocyte antigen (HLA) proteins, each of which is highly polymorphic. In the human, they can be subdivided into class I (HLA-A, B, C) and class II (HLA-DP, DR, DQ). Their importance lies in their function (antigen presentation to T-lymphocytes) and their high degree of protein polymorphism, unique in the genome and driven by the evolutionary pressure to present diverse antigens from emerging pathogens. HLA is estimated to account for ~45% of the familial clustering. At least one of the two common and strongly predisposing haplotypes (DR3-DQ2 or DR4-DQ8) can be found in ~20% of individuals of European descent and in 85-90% of T1D patients, making HLA virtually necessary, although far from sufficient. Individuals who do not share a single HLA haplotype with their affected sibling still have a five-fold higher risk than the general population, indicating the existence of additional loci of smaller effect (3).

Of these non-HLA loci the strongest known effect comes from the insulin gene, mapping to a repeat polymorphism upstream of the insulin promoter (Fig. 1). Short alleles (class I) predispose to T1D, while presence of at least one copy of the long alleles (class III) reduces disease risk to less than half. Interestingly these protective class III alleles are associated with three-fold higher expression of insulin in the thymus (4), where the peptide is ectopically expressed for purposes of immune tolerance.

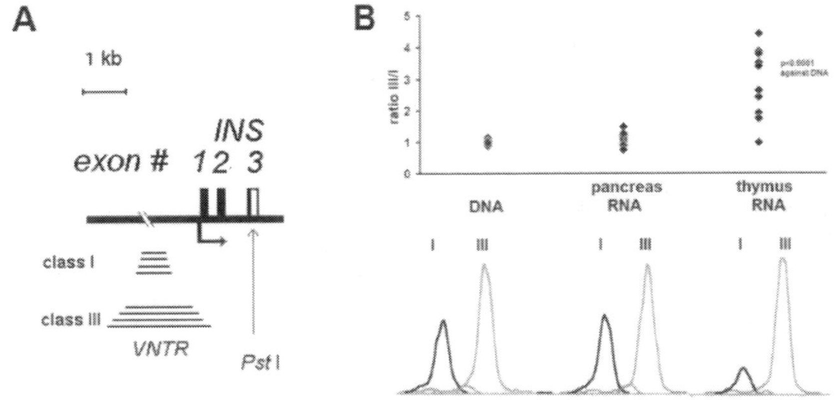


**Fig. 1A** schematic representation of the Variable Number of Tandem Repeats (VNTR) alleles upstream of the insulin promoter, composed of from 36 to 150 repeats of variants of the consensus unit ACA GGGG TG T GGGG.
**B** a PstI polymorphism at the 3' UTR of the gene allows distinction of mRNA transcripts derived from the chromosome carrying each allele in heterozygous samples. By single-nucleotide primer extension, the ratio of the two fluorochromes representing each allele in the RT-PCR product from pancreas is identical to that in genomic DNA, indicating no transcriptional effect. However, the same ratio in thymus RNA clearly favours higher expression from the T1D-protective class III allele. Higher expression in the thymus would promote self tolerance to insulin, a major autoantigen in T1D.

The third most important T1D locus maps to the *PTPN22* gene, where the common Arg620Trp substitution alters the function of the protein product LYP, a phosphatase crucial in the signalling of the T-cell receptor (TCR). Of the very large number of other published T1D associations, only a handful has been convincingly replicated. These include polymorphisms affecting IL2RA (encoding the $\alpha$–subunit of the IL2 receptor), *IFIH1* (interferon-induced with helicase domain, involved in innate anti-viral response) and CTLA4 (Cytotoxic T-lymphocyte Antigen 4, a co-receptor modulating T-cell activation).

Such loci represent the "low-hanging fruit," genes whose functional importance is already known. Linkage analysis, so successful in monogenic disorders, has not resulted in any convincing locus discovery. Loci *IDDM3-IDDM19* have not been replicated and *IDDM1* and *IDDM2* were discovered through candidate genes, HLA and INS respectively.

**THE GENOME-WIDE ASSOCIATION (GWA) ERA**

The past two years have seen the discovery of additional loci, thanks to two recent breakthroughs. The conceptual breakthrough is the genome-wide definition of human linkage disequilibrium (LD) patterns by the HapMap project. It was shown that chromosomal recombination is concentrated in defined "hot spots," separated by long blocks of sequence (up to hundreds of kb) where little or no recombination has happened. Within those blocks, polymorphic alleles do not reach equilibrium with each other through recombination and thus can provide information about each other, such that most of the variation contained in millions of common human polymorphisms can be captured by genotyping less than one million (considerably less if confined to European-descent subjects). The technical breakthrough was the ability to genotype thousands of individuals for this number of single-nucleotide polymorphisms (SNPs) through high-density genotyping arrays that became available in the past 2 years. To date, two published genome-wide association (GWA) studies have interrogated the entire genome for T1D association:

**The McGill-CHOP Study.**

In our collaboration with the Children's Hospital of Philadelphia (CHOP), we used the *Illumina* Hap550 array to genotype one thousand paediatric T1D cases for 550,000 SNPs, in Stage 1 of a two-stage design (5,6).Two novel loci reached the genome-wide (GW) significance threshold of $10^{-8}$ (calculated to correct for testing 550,000 hypotheses).

The first involves several SNPs on a 250 kb LD block on Chr16p that contains CLEC16A (C-Lectin 16A, formerly KIAA0350), a gene of unknown function that contains a C-type lectin domain and is expressed preferentially on cells of the immune system. The second locus involves a 210 kb LD block on Chr. 12q, containing nine RefSeq genes, some of which have known immune functions. Identification of the gene (and variant) responsible will require fine mapping by a combination of genetic and functional studies (in progress).

Subsequent stages have identified three additional loci, involving genes with interesting immune functions: *UBASH3A*, *BACH2* and *RASGRP1*.

**The Wellcome Trust Case-Control Consortium (WTCCC)**

Testing 2,000 T1D cases on the *Affymetrix* GeneChip for 400k SNPs, this group have identified four novel loci (7,8). The two with the highest statistical significance and effect sizes coincide with the two discovered in our effort. In addition, results from our Stage 2 confirm the remaining 2. This simultaneous independent discovery of non-candidate loci constitutes an important validation of the WGA approach

**The Type 1 Diabetes Genetics Consortium (T1DGC)**

An additional GWA on 2,500 cases and 2,000 controls using the Hap550 array has been performed and discovered several additional loci which have not been published yet and remain to be replicated.

**TOWARD THE FUTURE**

As described above, the number of known T1D loci now approaches twenty. Together, however, they are far from entirely explaining the genetic component of the disease. This can be illustrated by the Receiver Operator Characteristic (ROC) curve (Fig. 2),

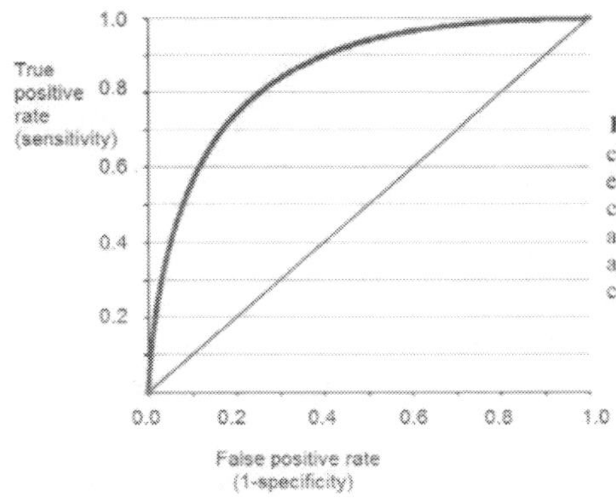


**Fig. 2.** The Receiver Operator Characteristic curve for classifying affected vs. unaffected on the basis of risk in each of 39,368 risk strata resulting from all possible combinations of genotypes in 10 T1D loci. For example, a relatively benign intervention to prevent T1D could be applied to 25% of the general population in order to capture 85% of T1D cases.

constructed by plotting the number of true positives as a function of false positives that would be predicted by placing the risk-level cut-off for prediction in each 39,368 risk strata. The strata result from all possible genotype combinations in the first 10 known loci, based on their corresponding genotype frequencies and relative risk (RR) conferred. The area under the ROC curve is ~0.8, in a range from 0.5 (random call) to 1 (perfect diagnostic test). Adding newly discovered loci of ever decreasing effect sizes will only marginally improve this performance.

**What are we Missing?**

It is clear, therefore, that GWA studies by the currently existing microarrays have failed to capture a significant portion of the genetic determinants of the familial clustering of T1D. The genotype arrays on which they are based, are designed to capture relatively common alleles, even when they confer very small RRs (i.e. $<1.2$). The most plausible explanation for the yet undiscovered portion of T1D familial clustering is a relatively large number of alleles with very low frequencies each, conferring very high RRs. The genetic signal from these alleles is weak only because of their rarity. For the few individuals that carry them, they are much more important than the common variants.

**The Next Steps.**

Detection of rare alleles entails sequencing of the large number of individuals necessary for the sample sizes required for statistical significance with rare alleles. Fortunately, the large cohorts necessary for such studies have now been assembled and next-generation, ultra-high throughput sequencing methods have been validated and up-and-running. It is an ideal juncture of important questions arising out of recent information (the GWA results) and recently developed technologies.

If rare but highly penetrant variants are indeed as important as predicted, very substantial genetic heterogeneity will be the result, implying functional heterogeneity in the immune dysregulation causing T1D. Genetic profiling for these variants will then be the basis of meaningful personalized medicine.

**Copy-Number Variation**

The examination of rare variants with strong biological effects has gained impetus by the discovery, over the past 2-3 years, of the widespread existence of structural variation in the human genome. Deletions and duplications ranging in size from kilobases to megabases are not uncommon in chromosomes of normal individuals. These copy-number variations (CNVs) often encompass entire genes, or even clusters of adjacent genes, generating gene-dosage variation and, occasionally, even homozygous deletions. Numerous examples of their involvement in complex disorders have been recently discovered, mostly in neuropsychiatric disorders.

Our Hap550 GWA raw intensity data were mined for possible CNV signals (lower intensity and higher-than-expected "homozygosity" for deletions and higher intensity with imbalance of allele signals in duplications). We statistically detected 30,453 CNV signals, 11 of which had significant frequency differences between 1,000 cases and 4,000 controls. The WTCCC data similarly revealed many additional loci. The development of high-throughput methods for physical genotyping to confirm these signals is in progress.

**CONCLUSION**

Much is known about the genetic basis of T1D as a result of recent breakthroughs, but much remains to be known. Fine-mapping of the existing loci, and discovery of the remaining ones, especially the rare but highly penetrant variants theoretically predicted, will be an arduous task. The results, however, are likely to revolutionize our understanding of the disease.

**REFERENCES**


[1.] Redondo et al., **N Engl J Med** 2008 359:2849-50. [2.] Polychronakos C. **Curr Diab Rep**. 4:151-4, 2004. [3.] Risch N. **Am J Hum Genet** 40:1-14 1987 [4.] Vafiadis et al., **Nat Genet** 1997 15:289-92. [5.] Hakonarson et al. **Nature** 2007 448(7153):591-4 [6.] Hakonarson et al., **Diabetes** 2008 57:1143-6 [7.] WTCCC et al., **Nature** 2007 447(7145):661-78 [8.] Todd et al., **Nat Genet** 2007 39:857-64.

# DIAGNOSTIC ASSESSMENT OF PROTEINURIC RENAL DISEASE: THE 21ST CENTURY VISION

**Larry D. Cowgill, DVM, PhD, DACVIM**
**Davis & San Diego, CA**

## INTRODUCTION

Pathologic glomerular proteinuric diseases have become recognized increasingly as a major cause of acute and chronic renal injury in dogs and cats. They are easily detected on the basis of routine laboratory parameters and screening and quantitative urine tests. Proteinuria can be recognized with tremendous sensitivity with microalbumin screening and quantitated effectively with urine protein/creatinine measurements. The clinical consequences of proteinuric renal diseases are readily recognizable admixtures of weight loss, hypoalbuminemia, hyperlipidemia, progressive renal dysfunction, hypertension, and fluid disorders. What has remained a diagnostic mystery is the spectrum of the pathologic disorders which predispose this "common denominator" clinical sign and how to approach this spectrum of conditions therapeutically. Technological advances in digital pathology, distance conferencing, and global data sharing portend revolutionary opportunities for the clinical and pathologic assessment of proteinuric renal disease in the 21st century.

## THE PAST & THE FUTURE

Nephrology and nephropathology have advanced to subspecialty niches within each of their respective disciplines of internal medicine and anatomic pathology. The internist and diagnostic pathologist have little difficulty recognizing overt injury to the glomerulus, but the internist is limited by a lack of therapeutic strategies with defined clinical and pathophysiologic correlations; and the diagnostic pathologist is constrained by an overly subjective and imprecise morphologic classification system based hematoxaylin and eosin staining and light microscopy. Over the past decade nephrologists have begun to appreciate the existence of a broad spectrum of distinct glomerular entities which heretofore have been loosely grouped into a limited collection of pathologic categories. Nephropathology also is poised (and mandated) to abandon its simplistic, subjective, and outdated classification scheme for glomerular disease that has been generally adopted from human pathology.

Consider a vision for the 21st century: Dr. Domingo in rural Spain has a canine patient with proteinuria but no regional expertise to aid or guide clinical decisions for the evaluation and management of his patient. An international team of expert nephrologists with specialty interest and 150+ years of collective expertise in glomerular disease convene at a a routinely scheduled meeting to discuss their favorite topic in nephrology (the glomerulus)—some are enjoying morning coffee, some are contemplating lunch, some are close to catching a train for home, and some sip an evening cocktail. An equally impressive international team of veterinary nephropathologists (and a few human nephropathologists) assemble in the comfort of their familiar surroundings to join the nephrologists to review, discuss, and advise Dr. Domingo about his patient. This is not a fantasy vision for the future of the 21st century but one that exists today. It is not just a vision to serves nephrology but a model for other disciplines in which advances in digital pathology, telemedicine, distance conferencing, data sharing, and international patient registries can generate new knowledge and facilitate rapid and worldwide access to authoritative consensus and expert opinion to any and all who seek it.

The vision (reality) so described is the culmination of an idea conceived collectively (see Acknowledgements, below) nearly 5 years ago as the International Veterinary Renal Pathology Initiative. Subsequently, the "initiative" has come under the auspices of the WSAVA as the International Renal Standardization Project with generous support from Hill's Pet Nutrition and Bayer Animal Health. The overarching purpose of the initiative was to more accurately characterize proteinuric glomerular diseases in dogs employing light, transmission electron, and immunofluorescent microscopy as diagnostic modalities, and to relate the specific pathologic findings to the clinicopathologic presentations and outcomes of the patients. Within this overview, a variety of specific achievements were envisioned which have established a foundation and scaffolding for the future pathologic assessment of proteinuria in dogs.

### Vision/Achievement #1

A series of regional Diagnostic Renal Pathology Centers was developed as a global network with identical, quality-controlled procedures to process and share high-resolution digital micrographic images of kidney specimens for light, immunofluorescent, and transmission electron microscopic evaluation. On the foundations of a prototype center developed by Dr. George Lees, 2 of 4 projected Centers have been established at Texas A&M University in the USA and at Utrecht University in the Netherlands. At the core of each Center is an Aperio digital slide scanning system for the production of virtual whole-slide images that can be reviewed simultaneously throughout the world and a robust digital imaging and data management platform to facilitate sharing and analysis of the digital imagery.

### Vision/Achievement #2

An extensive electronic relational medical database was envisioned to archive and integrate pathologic and clinical information, to facilitate data analysis, and to provide a platform for international data sharing. Such a database has been created collectively by the Initiative Committee and Prelude Dynamics which enables compilation of all historical, clinical, laboratory, therapeutic, follow-up, and outcome data for the entire course of the patient's disease and management. The clinical database has been merged with the pathologic database to establish a single seamless interface to all clinical and pathologic information which is accessible throughout the world. The same medical database is positioned also to serve as a renal disease registry to facilitate the collection and analysis of clinical information on animals with renal disease worldwide. The interface to the database can be viewed at www.http://ivrpi.com/vision.

**Vision/Achievement #3**

The Renal Pathology Initiative and Standardization Project sought to establish a new paradigm for the diagnostic assessment and classification of canine glomerular disease. The paradigm would transition pathologic diagnoses from subjective opinion based on qualitative light assessment by a single anatomic pathologist to a system of diagnosis and classification based on quantitative lesion assessment, multiple microscopic modalities, and consensus opinion of a consortium of nephrologists and nephropathologists. To provide greater objectivity, independence, flexibility, and quantitative evaluation to the pathologic description, a scoring system has been established to characterize the extent and magnitude of identifiable lesions of the renal glomeruli, tubular epithelium, interstitium, and vasculature. The digital microscopy can be extended further by processing with sophisticated image analysis algorithms that can both focus the pathologist's evaluation and quantitate discrete pathologic entities. This fusion of morphologic assessment and quantitative annotation and scoring will serve as a basis for an objective and unbiased classification system for proteinuric dogs. The resulting classification system will enable more precise and statistical comparison of future therapeutic strategies subjected to clinical trial.

**Vision/Achievement #4**

We further imagined a forum for international case discussion to integrate clinical nephrology with nephropathology. Technological advances in WEB-based conferencing provided solution to an extremely effective and high quality platform for the simultaneous, interactive, and fully participatory review of the clinical presentation/course and digital imagery by the combined nephrology and nephropatholgy consortium. From these case conferences, a consensus-based, clinical-pathological diagnosis can be rendered which conforms to an internationally standardized glomerular disease classification scheme. The case review platform serves secondarily as an effective venue to promote advanced specialty training in nephropathology and to facilitate impromptu scientific exchange and insightful opinion by invitation of noted authorities to the review sessions.

The International Veterinary Renal Pathology Initiative and Standardization Project have put into practice a quantum shift vision for the clinical and pathologic assessment of proteinuria in dogs. The strategic and implementable components of this vision hold great promise to improve the recognition and management of glomerular disease and to promote the exchange of knowledge to all segments of the veterinary profession.

## REFERENCES

**Original IVRPI Committee:** Claudio Brovida (Italy), Cathy Brown (USA), Larry Cowgill (USA), Jonathan Elliott (UK), Roel Goldschmeding (Netherlands), Reidun Heiene (Norway), Johan Jansen (Norway), George Lees (USA), David Polzin (USA), Kinji Shirota (Japan), Bill Spangler (USA), Shelly Vaden (USA), Astrid van Dongen (Netherlands). **Current Pathology Team:** Brian Berridge (USA), Cathy Brown (USA), Rachel Cianciolo (USA), Fred Clubb, Jr (USA), Roel Goldschmeding (Netherlands), Johan Jansen (Norway), George Lees (USA), Chuck Mohr (USA), Kinji Shirota (Japan), Bill Spangler (USA), Jaco van der Lugt (Netherlands),. **Current Nephrology Team:** Claudio Brovida (Italy), Larry Cowgill (USA), Jonathan Elliott (UK), Reidun Heiene (Norway), George Lees (USA), David Polzin (USA), Shelly Vaden (USA), Astrid van Dongen (Netherlands).

# SPECTRUM OF PATHOLOGIC LESIONS IN PROTEINURIC DOGS: INITIAL DATA FROM THE WSAVA PROJECT

**George E. Lees, DVM, MS, DACVIM**
**College Station, TX**

## INTRODUCTION

The purpose of the Renal Standardization Project, which was launched under the auspices of the World Small Animal Veterinary Association (WSAVA) in 2008, is to routinely use three diagnostic modalities (light, electron, and immunofluorescent microscopy) as is done in human nephropathology to accurately characterize glomerular disease in proteinuric dogs, and to relate these findings to clinicopathologic presentation and outcome. The long-term goal is to better understand and evaluate and thus optimize the medical management of dogs with proteinuric renal disorders by identifying distinct glomerular diseases for which specific prognoses and therapeutic guidelines can be provided.

To achieve this goal, an international network of Diagnostic Renal Pathology Centers linked to a shared database of clinical and pathologic information about all cases evaluated has been created. The first center that was established was the Texas Veterinary Renal Pathology Service (TVRPS), which is located at Texas A&M University and began accepting case submissions in 2005. Cases submitted to the TVRPS have provided valuable preliminary data about the spectrum of pathologic lesions in proteinuric dogs.

## TVRPS CASE ACCESSIONS: 2005 – 2008

From March, 2005, through December, 2008, the TVRPS received submissions for 294 cases, of which 269 were dogs and 25 were cats. Most but not all of the 269 dogs were being evaluated for renal diseases with clinical features that included proteinuria, and glomerular disease was diagnosed in about two-thirds of the cases examined (Table 1).

**Table 1-** Categories of Renal Disease Found in Canine Cases Submitted to the TVRPS: 2005 – 2008

| | All Submitted Cases | | | Glomerular Categories Alone | |
|---|---|---|---|---|---|
| **Patterns of Injury** | Number | % | | Number | % |
| Glomerular categories | | | | | |
| Amyloidosis | 30 | 11.2 | | 30 | 17.6 |
| Membranous GN | 32 | 11.9 | | 32 | 18.8 |
| Membranoproliferative GN | 50 | 18.6 | | 50 | 29.4 |
| Mesangial glomerulopathy | | | | | |
| Mild mesangial changes | 4 | 1.5 | | 4 | 2.4 |
| Mesangioproliferative GN | 7 | 2.6 | | 7 | 4.1 |
| Proliferative GN | 4 | 1.5 | | 4 | 2.4 |
| Glomerulosclerosis | | | | | |
| Focal and segmental | 16 | 5.9 | | 16 | 9.4 |
| Diffuse and segmental to global | 18 | 6.7 | | 18 | 10.6 |
| Miscellaneous glomerulopathies | 9 | 3.3 | | 9 | 5.3 |
| Subtotal | 170 | 63.2 | | **170** | **100.0** |
| Other categories | | | | | |
| Acute tubular injury | 11 | 4.1 | | | |
| Nephrosclerosis | 26 | 9.7 | | | |
| Miscellaneous others | 34 | 12.6 | | | |
| Normal kidney / negligible changes | 15 | 5.6 | | | |
| Non-diagnostic specimens | 13 | 4.8 | | | |
| Subtotal | 99 | 36.8 | | | |
| **Total** | **269** | **100.0** | | | |

The "big three" patterns of glomerular injury are amyloidosis, membranous glomerulopathy, and membranoproliferative glomerulonephritis, which is consistent with the findings of other previous studies of canine glomerular disease. In the current case series, these 3 categories account for 41.6 percent of all cases, and 65.9 percent of the cases in which glomerular patterns of injury were identified. Nonetheless, although less common, a variety of other glomerulopathies have been found in the dogs evaluated to date.

## CLINICOPATHOLOGIC OBSERVATIONS

### Amyloidosis

The 30 dogs with amyloidosis included slightly more males (n, 19) than females (n, 11), and were a variety of different breeds (8 Mixed breeds, 3 English Bulldogs, 2 Chinese Shar Peis, and 17 other breeds with 1 dog each). The dogs generally were middle aged (mean age, 7.4; median age, 6.8 years), but the age of affected dogs ranged from 3 to 13 years. Half of the dogs in which amyloidosis was diagnosed were not yet azotemic (15/30 with SCr < 1.5 mg/dl). Most exhibited marked proteinuria, the mean and median UPC in 28 dogs for which it was known were 12.7 and 12.9 respectively; however, UPC values in these dogs ranged from 1.5

to 45. The proteinuria was associated with severe hypoalbuminemia (SAlb ≤ 1.4 g/dl) in 6 (20.7%), moderate hypoalbuminemia (SAlb = 1.5 to 1.9 g/dl) in 13 (44.8%), mild hypoalbuminemia (SAlb, = 2.0 to 2.4 g/dl) in 5 (17.2%), and normal serum albumin levels (SAlb ≥ 2.5 g/dl) in 5 (17.2%) of the 29 dogs for which serum albumin concentration was known. Hypertension was uncommon in dogs with amyloidosis; only 3 (13.6%) of 22 dogs in which blood pressure was known had systolic BP ≥ 180 mmHg.

**Membranous GN**

Two-thirds (22) of the 32 dogs with membranous GN were males; only 10 were females. Many different breeds were included (7 Mixed breeds, 2 Golden retrievers, 2 Labrador retrievers, 2 Dachshunds, and 19 breeds with 1 dog each). These dogs also generally were middle aged (mean age, 6.7 years; median age, 7.0 years), but their ages ranged from 1 to 11.5 years. Dogs with membranous GN had magnitudes of proteinuria (mean UPC, 12.9; median UPC, 9.0) similar to what was observed in dogs with amyloidosis, but dogs with membranous GN were less likely to be azotemic. Only 10 (31.2%) of 32 dogs with membranous GN had SCr ≥ 1.6 mg/dl. Additionally, despite similar magnitudes of proteinuria, dogs with membranous GN tended to have somewhat lower serum albumin concentrations than dogs with amyloidosis; there was severe hypoalbuminemia (SAlb ≤ 1.4 g/dl) in 12 (37.5%), moderate hypoalbuminemia (SAlb = 1.5 to 1.9 g/dl) in 10 (31.2%), mild hypoalbuminemia (SAlb, = 2.0 to 2.4 g/dl) in 3 (9.4%), and normal serum albumin levels (SAlb ≥ 2.5 g/dl) in 7 (21.9%) of the 32 dogs. As in dogs with amyloidosis, dogs with membranous GN were not often hypertensive; only 4 (13.8%) of 29 dogs in which blood pressure was known had systolic BP ≥ 180 mmHg.

**Membranoproliferative GN**

Similar numbers of males (n, 28) and females (n, 22) exhibited a membranoproliferative GN pattern of injury. Many breeds were represented, but Labrador retrievers (n, 11) and Golden retrievers (n, 7) were more common in this group than in the amyloidosis or membranous GN groups. Other breeds included Mixed (n, 5), Boxers (n, 3), Dachshunds (n, 2), Miniature Schnauzers (n, 2), Australian shepherds (n, 2), Soft-coated Wheaten terriers (n, 2), and 16 breeds with 1 dog each. These dogs usually were azotemic; only 10 (20.4%) of 49 dogs with MPGN in which the value was know had SCr ≤ 1.5 mg/dl. Dogs with MPGN had magnitudes of proteinuria (mean UPC, 10.4; median UPC, 10.0; with UPC values ranging from 1.4 to 34.2) similar to those of dogs with amyloidosis or membranous GN, but the degree of associated hypoalbuminemia was less frequently severe than was the case in dogs with membranous GN. In dogs with MPGN, there was severe hypoalbuminemia (SAlb ≤ 1.4 g/dl) in 9 (18.4%), moderate hypoalbuminemia (SAlb = 1.5 to 1.9 g/dl) in 16 (32.7%), mild hypoalbuminemia (SAlb, = 2.0 to 2.4 g/dl) in 12 (24.5%), and normal serum albumin levels (SAlb ≥ 2.5 g/dl) in 12 (24.5%) of the 49 dogs in which serum albumin concentration was known. Hypertension was more common in dogs with MPGN than in dogs with amyloidosis or membranous GN; 17 (37.8%) of 45 dogs with MPGN in which blood pressure was known had systolic BP ≥ 180 mmHg.

**Glomerulosclerosis**

Dogs with glomerulosclerosis that was focal and segmental (n, 16) were grouped separately from dogs with glomerulosclerosis that was diffuse and segmental to global (n, 18), but at least some dogs in these two groups might represent a single pattern of injury merely observed during different stages in its pathogenesis and progression. In both groups, and in contrast to the other groups cited previously, females were more common than males; 11 (68.8%) of 16 dogs with focal and segmental lesions were females, and 13 (72.2%) of 18 dogs with diffuse and segmental to global lesions were females. Dogs with these lesions also were middle aged (FSGS: mean age, 8.2 years; median age, 7.8 years, and DGGS: mean age, 8.3 years; median age, 8.0 years), but tended to be slightly older than dogs with amyloidosis, membranous GN, or MPGN.

As a rule, dogs with focal and segmental glomerulosclerosis were either non-azotemic or only mildly azotemic; 9 (56.2%) of 16 dogs had SCr ≤ 1.5, and 7 (43.8%) of 16 had SCr = 1.6 to 2.5 mg/dl; no dog with this lesion had SCr > 2.5 mg/dl. Dogs with FSGS (mean UPC, 6.0; median UPC, 5.0, with UPC values ranging from 2.4 to 12.7) had magnitudes of proteinuria that were more moderate than observed with amyloidosis, membranous GN, or MPGN. Additionally, hypoalbuminemia was infrequent and mild in this group; only 2 dogs had SAlb values < 2.0 g/dl (1 was 1.8, and 1 was 1.9), while 4 (26.7%) of 15 had mild hypoalbuminemia (SAlb = 2.0 to 2.4 g/dl), and 9 (60.0%) of 15 had normal (SAlb ≥ 2.5 g/dl) serum albumin concentrations. Hypertension was observed in some dogs with FSGS; 4 (26.7%) of 15 dogs in which blood pressure was known had systolic BP ≥ 180 mmHg.

Dogs with diffuse and segmental to global glomerulosclerosis were more often azotemic than the dogs with FSGS; only 7 (38.9%) of 18 dogs had SCr values ≤ 1.5 mg/dl, while 11 (61.1%) of 18 dogs had SCr values > 1.5 mg/dl. However, magnitudes of proteinuria (mean UPC, 6.4; median UPC 6.2) and severities of hypoalbuminemia (2 with SAlb slightly < 2.0 g/dl; 4 with SAlb = 2.0 to 2.4 g/dl; and 10 with SAlb ≥ 2.5 g/dl among the 16 dogs in which SAlb was known) were similar to what was observed in the dogs with FSGS. The frequency of hypertension also was similar; 4 (23.5%) of 17 dogs for which blood pressure was known had systolic BP ≥ 180 mmHg.

**Other glomerulopathies**

Taken together, other glomerulopathies accounted for 8.9 percent of all TVRPS case accessions and thus 14.2 percent of the cases of glomerular disease that were identified. This is a mixed group of uncommon disorders, and the accumulated data are not yet sufficient for us to formulate generalities about any of them.

**REFERENCES**

[1.] Jennette JC, et al. In Jennette CJ, Olson JL, Schwartz MM, Silva FG (eds): **Heptinstall's Pathology of the Kidney**, ed 6, 2007. [2.] Walker PD, et al. **Mod Pathol** 2004;17(12):1555-1563.

# CLINICAL PATHOLOGIC CORRELATES OBSERVED IN CANINE PROTEINURIC RENAL DISEASES

**David J. Polzin, DVM, PhD DACVIM**
**St. Paul, Minnesota**

## CLINICOPATHOLIGICAL SYNDROMES

### Overview

Proteinuric kidney diseases (glomerulopathies) are an important cause of kidney disease in dogs, but their diagnostic evaluation, classification, and management remain poorly characterized compared to other renal pathology. Accurate identification and appropriate treatment of canine glomerular diseases can be expected to yield consequential improvements in clinical outcomes by providing a basis for clinical trials that may provide a basis for more specific management of these diverse conditions. In humans, specific glomerular diseases are characterized and identified by their light, immunopathologic, and electron microscopic features which are correlated with the particular clinical and clinicopathologic findings that are exhibited by affected patients. The clinical characteristics and the morphologic features of the renal biopsy are used to provide a final diagnosis for patients. This occurs as the result of a collaborative effort engaging both the clinical specialist (nephrologist) and renal pathologist. Diagnosis of a patient's glomerular disease provides insights about the disorder's pathogenesis and prognosis, as well as guidelines for appropriate treatment.

Glomerulopathies as a group may be described in terms of their clinical presentation (clinical diagnosis), their structural pattern of injury (pathologic diagnosis – see table 1), or both. It is important to appreciate the difference between clinical manifestations and structural expression of the disease. In other words, recognizing clinical pattern in a patient with glomerular disease may or may not allow an understanding of the underlying pathology. However, in humans where thousands of renal biopsies and their clinical presentations have been studied, it is often possible to limit the possible morphologic diagnoses to a limited number of options. The obvious benefit of recognizing such associations is that they may guide the clinician in choosing both diagnostic and therapeutic options for the patient.

The two major glomerular syndromes are the nephrotic syndrome and the nephritic syndrome. While these syndromes may be described as discrete entities (as below), most patients and most glomerular diseases manifest as a combination of these two syndromes in various combinations (table 1).

## NEPHROTIC SYNDROME

The major clinical finding in patients with nephrotic syndrome (NS) is proteinuria with urine sediment that is relatively inactive in that it contains few cells or casts. The clinical picture in patients presenting with this syndrome can range from asymptomatic proteinuria detected incidentally on routine examination through full-blown nephrotic syndrome including heavy proteinuria, hypoalbuminemia, edema, lipiduria and hyperlipidemia. The hypoalbuminemia in these patients results, in part, from urinary protein losses that exceed the increased rate of hepatic albumin synthesis. Edema results primarily from renal sodium retention rather than hypoalbuminemia. Hyperlipidemia is a consequence of increased hepatic lipoprotein synthesis induced through an unknown mechanism by the fall in plasma oncotic pressure (a function of serum albumin concentration).

The nephrotic syndrome generally reflects noninflammatory injury to the glomerular capillary wall that can be induced by one of three basic mechanisms: 1) injury to podocytes (e.g. minimal change disease or primary focal glomerulosclerosis), 2) immune complex formation with activation of complement in the subepithelial space (e.g. membranous nephropathy), or 3) deposition disorders affecting the glomerular capillary walls (e.g. amyloidosis or diabetic nephropathy). The common structural lesion in nephrotic disorders is podocyte injury characterized by effacement of foot processes (also called: diffuse simplification or "fusion" of the foot processes). Because these processes are essentially non-inflammatory and there is little influx of cells into the glomerulus (in part because the site of injury is outside the glomerular basement membrane, away from the systemic immune system), the urine sediment is typically devoid of inflammation and the plasma creatinine concentration remains normal or only mildly increased.

## NEPHRITIC SYNDROME

In contrast to the nephrotic syndrome, the nephritic syndrome reflects an inflammatory reaction typified by cellular infiltration into the glomerulus including neutrophils, macrophages, monocytes, and lymphocytes. As a consequence, the nephritic syndrome is typically recognized clinically as hematuria, white blood cells, and cellular and granular casts present in the urine sediment. The type and extend of glomerular injury determines the extent of renal dysfunction that may accompany the condition. Severe glomerular injury affecting most or all glomeruli may exhibit variable and usually sudden development of azotemia. This reduction in glomerular filtration rate (GFR) reflects a decreased surface area available for filtration occurring as a consequence of inflammatory infiltration and/or proliferation of intrinsic glomerular cells inducing partial or complete closure of some glomerular capillaries. This reducing in GFR is typically associated with salt and water which promotes extracellular fluid volume expansion, edema and arterial hypertension. This clinical pattern is typical of the "acute nephritis syndrome."

Other inflammatory glomerulopathies may result in focal necrotizing and/or crescentic lesions; however, these processes typically develop over a longer period of time and are characterized by inflammatory urine sediment, but slowed development of azotemia. This clinical presentation is less characterized by fluid retention and its associated signs and is called "rapidly progressive glomerulonephritis." Another type of glomerulopathy recognized in humans presents primarily as "asymptomatic hematuria" with mild or no proteinuria, normal plasma creatinine and blood pressure. The hematuria in this syndrome reflects defective glomerular basement membrane structure rather than inflammation.

There are 3 primary mechanisms underlying glomerular inflammation and the nephritic syndrome: 1) immune complex formation and complement activation in the subendothelial space and mesangium (e.g. post-infectious glomerulonephritides, IgA nephropathy or some patients with SLE), 2) antibodies directed against the glomerular basement membrane itself, or 3) circulating antibodies directed against neutrophil cytoplasmic antigens (ANCA). The common structural lesion of nephritic disorders is active inflammatory injury of the endothelium, usually due to immune complex deposition or formation in subendothelial or mesangial areas or antibody-mediated binding in close proximity to the endothelium.

## OTHER SYNDROMES

As can be seen by examining table 1, many glomerular disorders present with a mixture of features characteristic of the nephrotic and nephritic syndromes, and glomerulopathies cannot easily be categorization based solely on their clinical presentation.

Similarities and dissimilarities between canine and human clinical syndromes in patients with glomerulopathies has not been well developed, in part due to the absence of adequate identification of the structural lesions present in proteinuric dogs. The WSAVA Renal Pathology study Group has developed a database which is being used in an attempt to define these associations in dogs and ascertain similarities and dissimilarities that may exist between the two species. This study began in 2009 and will continue over the next four years to prospectively develop this database and study the clinicopathologic correlations in proteinuric kidney diseases of dogs.

**Table 1:** Tendencies of Human Glomerular Diseases to Manifest Nephrotic and Nephritic Features

| | Nephrotic Features | Nephritic Features |
|---|---|---|
| Minimal change glomerulopathy | ++++ | - |
| Membranous glomerulopathy | ++++ | + |
| Diabetic glomerulosclerosis | ++++ | + |
| Amyloidosis | ++++ | + |
| Focal segmental glomerulosclerosis (FSGS) | +++ | ++ |
| Fibrillary glomerulonephritis | +++ | ++ |
| Mesangioproliferative glomerulopathy | ++ | ++ |
| Membranoproliferative glomerulonephritis | ++ | +++ |
| Proliferative glomerulonephritis | ++ | +++ |
| Acute diffuse proliferative glomerulonephritis | + | ++++ |
| Crescentic glomerulonephritis | + | ++++ |

## REFERENCES

[1.] Jennette JC and Falk RJ: Glomerular Clinicopathologic Syndromes. p. 150. In, Greenberg A (ed), **Primer on Kidney Diseases**. 4th Ed. Elsevier Saunders. Philadelphia. 2005. [2.] Rennke HG and Denker BM: Renal Pathophysiology: **The Essentials**. 2nd edition. Lippincott, Williams & Wilkins. Philadelphia, 2007.

## WHAT'S ON THE HORIZON FOR MANAGEMENT OF PROTEINURIA RENAL DISEASES?

**Shelly L. Vaden, DVM, PhD, DACVIM**
**Raleigh, NC 27606**

### INTRODUCTION

Proteinuria is an important medical problem that should not be ignored. Unfortunately, proteinuria has frequently gone unnoticed in dogs and cats because of many factors. In recent years, it has become apparent that renal proteinuria is associated with a greater risk of renal morbidity, renal mortality and all cause mortality in dogs and cats. In dogs with chronic renal failure, an initial urine protein: creatinine ratio (UPC) of >1.0 was associated with a greater risk of developing a uremic crisis and death.[8] In cats with chronic renal failure, a UPC ratio in excess of 0.4 at the initial evaluation was associated with and increased risk of mortality due to all causes.[14] In nonazotemic cats, proteinuria has also been associated with reduced survival (UPC ratio >0.3).[15] Moreover, a reduction in renal proteinuria may be associated with more favorable patient outcomes.[6]

Much more is known about the impact of proteinuria in people with kidney disease. Large clinical trials in people have demonstrated that proteinuria is an independent risk factor for progressive renal disease. People with chronic kidney disease (CKD) who excrete more than 3 g of albumin per day have a 8.1x higher risk of progressing to end stage renal disease (ESRD).[7] Reducing proteinuria by >30% of baseline within 6 to 12 months of treatment was predictive of long term renal and cardiovascular outcomes in people with type 2 diabetes.[4] Furthermore, the more protein excretion can be reduced, down to about 500 mg/day, the better the renal outcome.[7] In order to achieve such reductions in urine protein excretion, so called "renoprotective" drugs must be used. These agents lower protein excretion more than would be expected from antihypertensive effects alone.

This purpose of this presentation is to discuss the therapeutic management of proteinuria beyond the use of monotherapy with angiotensin-converting enzyme inhibitors. Because there is a wealth of information on this topic from studies in people, and very little information from studies in dogs and cats, much of the information presented herein is from people. Much more study is needed to verify that these concepts can be transferred safely to dogs and cat.

### RENAL HANDLING OF PROTEIN

The glomerulus is a modified capillary bed that functions as a filter, across which a plasma ultrafiltrate is formed. The filtration barrier is composed of three layers: the fenestrated endothelium, the glomerular basement membrane (GBM), and the visceral epithelial cells, or podocytes. Slit diaphragms are specialized cell junctions between the podocytes that bridge the filtration slits. The filtrate passes through the endothelial fenestrae, permeates the GBM and then passes through the filtration slits into the urinary space. This complex structure is freely permeable to water and small dissolved solutes but retains cells and most macromolecules, such as proteins. Albumin, a negatively charged protein with a molecular weight of 69,000 daltons, is normally largely excluded from the filtrate. Hemodynamic forces also influence the transglomerular movement of proteins. Decreases in renal plasma flow accompanied by increases in the transcapillary hydraulic pressure difference (capillary lumen minus Bowman's space), as may occur following increased angiotensin II concentrations, will increase the protein concentration in Bowman's space relative to the plasma concentration. Despite this complex system, small amounts of albumin and other proteins are normally found in the filtrate. Substantial degradation of these proteins occurs; proteins and peptide fragments may undergo resorption in the distal nephron.

### INHIBITION OF THE RENIN-ANGIOTENSIN-ALDOSTERONE SYSTEM (RAAS)

Because hemodynamic forces influence the transglomerular movement of proteins, it follows that altering renal hemodynamics would be effective in reducing proteinuria. RAAS has been the major target system for this approach to reducing proteinuria. Although the mechanisms of this effect have not been fully elucidated, it appears that many of these agents reduce proteinuria greater than would be expected on the basis of their antihypertensive effects alone.

#### Angiotensin Converting Enzyme Inhibitors (ACEi)

ACEi may reduce proteinuria and preserve renal function by several possible mechanisms in addition to decreased efferent glomerular arteriolar resistance leading to decreased glomerular transcapillary hydraulic pressure. Enalapril significantly reduced proteinuria and delayed the onset or the progression of azotemia in dogs with glomerulonephritis.[6] Treatment of dogs with glomerular diseases with ACEi is considered a standard of care. Typically an ACEi is given once daily, and if no reduction in proteinuria is seen after 2 to 4 weeks, the frequency is increased to twice daily (approximately half of the dogs may eventually need twice daily administration).[6] Benazepril has been shown to reduce proteinuria in cats with CKD, with greater reductions occurring in the face of higher UPCs.[9] Although the serum creatinine concentration should be monitored, it is uncommon for dogs and cats to have worsening of azotemia due to enalapril administration alone. Hyperkalemia appears to be a common side effect in dogs with renal disease. There is no evidence to support that one ACEi is superior to another.

ACEi have been shown to slow renal disease progression in people with mild renal failure and high proteinuria and reduce the risk of ESRD in people with chronic kidney disease and protienuria.[3] However, in the absence of angiotensin converting enzyme, angiotensin II is produced by other kinases and is therefore, not completely suppressed by an ACEi alone, even when given at maximally effective doses. In recent years, there has been increased interest in the use of other agents known to inhibit RAAS.

#### Angiotensin Receptor Blockers (ARB)

ARB block the angiotensin II type 1receptor. Losartan resulted in an average reduction in proteinuria of 35% from baseline during a 3.4 year follow up period; much of this reduction was in the first 6 months of therapy.[2] In irbesartan treated patients, every

50% reduction in proteinuria during the first 12 months of therapy reduced the risk of a negative renal outcome by more than half.[2] Telmisartan, an ARB that is more lipophilic and has a longer half-life than losartan, was shown to be more effective in reducing proteinuria than was losartan in patients with diabetic nephropathy.[3] Telmisartan also has a higher affinity for the angiotensin-1 receptor and a greater percentage of binding compared with losartan. In addition to its antiproteinuric effects, losartan was associated with a decrease in urine concentrations of NAG, a marker of tubular damage that correlates with proteinuria, and scores for glomerular obsolescence and interstitial fibrosis.[13] Unfortunately, blocking the angiotensin II type 1 receptor with an ARB, may give rise to a compensatory increase in renin activity, and therefore and incomplete block of the RAAS.[10]

**ACEi & ARB combination therapy**

In theory, there may be an added benefit to combined therapy with an ACEi and an ARB because of the inability of monotherapy with either class of drug to provide complete RAAS blockade.[2] Blockade may be more complete when ACEi is combined with an ARB, although it is still only 75-80% complete. Studies in people have suggested that these drugs are additive or perhaps even synergistic in reducing proteinuria, although a benefit has not been demonstrated in every study.[12] Because the dose of each individual drug can be reduced during combined therapy, adverse effects may be reduced.

**Aldosterone Receptor Antagonists**

Serum aldosterone increases over time (i.e., aldosterone escape) in people treated even with maximal doses of ACEi and ARB. Prolonged hyperaldosteronism may have adverse effects on the heart, systemic blood vessels and glomeruli. Aldosterone-receptor antagonists have been shown to reduce proteinuria and stabilize kidney function in an additive fashion to ACEi and ARB in people. Spironolactone has been used most commonly in veterinary medicine however there is little published data supporting efficacy of this drug in dogs and cats. When compared to spironolactone, eplerenone is more selective at mineralocorticoid receptor blockade but less potent. Eplerenone may be the drug of choice in people because the relative lack of binding to androgen and progesterone receptors produces fewer endocrine side effects. However, endocrine side effects of spironolactone in dogs and cats are less problematic and the preference is unclear in veterinary medicine.

**Renin Inhibitors**

Aliskiren blocks the catalytic site of renin, thereby reducing the formation of angiotensin I. Preliminary data suggests that the drug provides renoprotection similar to that of the ARB. As previously mentioned, ACEi may lead to suboptimal blockade of RAAS, because of the production of angiotensin II through alternate kinases. Renin inhibitors overcome the increases in renin activity and angiotensin II that occurs during therapy with ARB and ACEi, respectively, and also reduce urine aldosterone excretion.

**Optimizing RAAS Blockade**

An individual titration study of losartan and lisinopril demonstrated that increasing doses did result in greater reductions of proteinuria and that the optimal antiproteinuric dose differed between individuals. Combination therapy had the greatest antiproteinuric effect when the most effective dosages of the individual drugs for the individual patient were used.[10] High sodium intake could blunt the antiproteinuric effects of ACEi. In some people, giving a diuretic improved the antiproteinuric effects of ACEi.[12] Animals fed renal diets with reduced salt content should not need an added diuretic.

## CONTROL OF HYPERTENSION

Blood pressure and the transglomerular movement of protein are undeniably linked. Adequate control of blood pressure will often lead to a reduction in the magnitude of proteinuria in a previously hypertensive patient but the degree of reduction differs with the different antihypertensive drugs. Nearly 40% of people do not have decreased protein excretion with lowering of their blood pressure.[7] In this case, a renoprotective effect may not be realized with antihypertensive therapy. In one study, a 15-mmHg decrease in blood pressure with a concomitant > 30% increase in proteinuria was associated with a 4x increased risk of ESRD.[5]

Patients with renal disease and hypertension should be treated with antihypertensive drugs that also reduce proteinuria. Reduction of proteinuiria may occur with amlodipine therapy but it is less than that which is seen with ACEi or ARB therapy. The risk of a negative renal end point in people treated with amlodipine after adustment for mean arterial pressure was greater than in those treated with an ARB.[11] Amlodipine has been shown to active RAAS in dogs and this may also occur in cats.[1] Some believe that amlodipine should not be administered to a dog or cat with CKD without simultaneous administration of an agent that blocks RAAS.

## DISEASE SPECIFIC TREATMENT

As we move towards improving the evaluation of renal biopsy specimens, we will be better able to make accurate histologic diagnoses. Disease specific treatment can then be explored more thoroughly. It is clear that certain glomerular diseases in people are responsive to immunosuppressive treatment (e.g., membranous glomerulonephritis, minimal change disease) but this needs further study in dogs and cats.

## REFERENCES

1. Atkins CE, et al. **J Vet Pharm Therap** 2007;30:394. 2. Bakris GL, **Clin JASN** 2008;3:S3. 3. Bakris GL, et al. **KI** 2008;74:364. 4. De Zeeuw D, et al. **Circulation** 2004;110:921. 5. Eijkelkamp WB, et al. **JASN** 2007;18:1540. 6. Grauer GF, et al. **JVIM** 2000;14:526. 7. Hirsch S, et al. Cleveland **Clin J Med** 2008;75:705. 8. Jacob F, et al. **JAVMA** 2005;226:393. 9. King JN, et al. **JVIM** 2006:20:1054. 10. Laverman et al. **KI** 2002;62:1020. 11. Lewis EJ, et al. **NEJM** 2001;345:851. 12. Linas SL, **Clin JASN** 2008;3:S17. 13. Shimizu A, et al. **Hypertens Res** 2008;31:1711. 14. Syme HM, et al. **JVIM** 2006;20:528. 15. Walker D, et al. **JAMA** 2002;288:2421.

## CALCIUM OXALATE: BIOLOGIC BEHAVIOR & RISK FACTOR MANAGEMENT

**Jody Lulich, DVM, DACVIM**
**Saint Paul, MN**

### BIOLOGIC BEHAVIOR

Epidemiological studies have revealed that the risk of calcium oxalate (CaOx) uroliths is approximately 7 times greater in Bichon Frise dogs admitted to veterinary hospitals than dogs admitted without urinary tract disease. However, the frequency and time of recurrence of CaOx uroliths in Bichons has not been evaluated in a systematic fashion. This information is needed to determine whether therapy designed to prevent recurrence of CaOx uroliths is warranted.

Records of 50 Bichons (34 males and 16 females) with CaOx uroliths admitted to our hospital between 1990 and 2002 were reviewed. Mean age of Bichons at initial urolith detection was 7.6±2.4 years. Urinalyses (n=27) revealed that specific gravity was 1.027±1.012 and pH was 7.3±0.8; only 37% had CaOx crystalluria. Thirty-four Bichons were evaluated following surgery. CaOx recurrence occurred in 25. Recurrence rate increased with length of time that Bichons were evaluated. After 1 year, 37% had their first recurrence; after 2 years, 64% had their first recurrence and 8% had their second recurrence; after 3 years, 90% had their first recurrence, 15% had their second recurrence, and 4% had their third recurrence. Urolith recurrence was detected in 100% of dogs evaluated ≥ 4 years. At first recurrence; neither age, gender, urine specific gravity, urine pH, CaOx crystalluria, nor serum calcium concentration was associated with the propensity for urolith formation. These observations indicate that following urolith removal CaOx recurrence is imminent. Therefore, long-term monitoring and therapy is needed to prevent repeat urolith removal.

### SELECTING AN APPROPRIATE DIET

Results of experimental and clinical investigations have confirmed the importance of dietary modifications in medical protocols to prevent urolith recurrence. However for calcium oxalate, selection of an appropriate diet is challenging because: 1) the exact mechanisms underlying calcium oxalate urolith formation are not completely known, 2) results of epidemiological studies do not always match physiology response of diet characteristics, and 3) diet efficacy has not been evaluated using clinically relevant endpoints.

#### The Exact Mechanisms Underlying Calcium Oxalate Urolith Formation Are Not Completely Known.

Although formation of CaOx uroliths is associated with a complex and incompletely understood sequence of events, it is accepted that initial crystal formation and subsequent crystal growth are at least partly a reflection of urine supersaturation. However, unlike other stone types, the urinary concentrations of calculogenic minerals may not be the predominant driving force for CaOx urolith formation. For example, 68 proteins were recently identified in matrix of CaOx uroliths from humans (Canales 2008). We have identified over 30 proteins in CaOx uroliths from dogs. The function of many of these proteins is unknown. Those proteins known to influence crystal nucleation, growth, and aggregation include osteopontin, nephrocalcin, Tamm-Horsfall, bikunin, and α1-antitripsin. In experimentally induced hyperoxaluria, intratubular CaOx crystalluria was detected in ostoeopontin and Tamm-Horsfall knockout mice, but did not occur in wild-type controls. These findings indicate the importance of other substances besides the concentrations of minerals in urine in the generation of CaOx uroliths.

#### Results of Epidemiological Studies Do Not Always Match Physiology Response of Diet Characteristics

Few studies have evaluated the risk of dietary ingredients on CaOx urolith formation. In these studies, the most influential dietary attributes are urine acidifying potential and the concentrations of calcium, protein, and sodium. (Table 1). Of the many dietary attributes only two results of epidemiologic studies are consistently corroborated by results of experimental studies: diets formulated to promote more acidic urine and diets with low moisture content (i.e. kibble) increase the risk of CaOx formation. Therefore, canned diets designed to promote formation of neutral or alkaline urine should be considered to minimize CaOx urolith recurrence.

**Table 1.** Evaluation of dietary risk factors for calcium oxalate formation in dogs

| Dietary component | Epidemiologic Result | Comments | Conclusion |
|---|---|---|---|
| Urine Acidifying potential | Diets promoting urine pH values less than 6.6 had the highest association with calcium oxalate formation in dogs consuming dry formulations of food (Lekcharoensuk March 2002). | Acidic urine was also the most important risk factor for calcium oxalate urolith formation in cats (Lekcharoensuk 2001, Kirk 1995). Urine pH values greater than 6.5 were associated with decreased risk. This is physiologically plausible because in dogs alkalosis reduced urine calcium excretion (Marone 1983) | To reduce calcium oxalate urolith recurrence select diets promoting alkalosis and urine pH values greater than 6.5. |
| Protein | Lower protein diets (1.9 to 5.8g/100kcal) were associated with increased risk. Higher protein diets | Epidemiological studies do not demonstrate cause and effect but associations to investigate further. In contrast to the studies by Lekarchoensuk and colleagues, hypercalciuia | Until experimental studies evaluating different protein contents are performed in dogs with calcium oxalate stones, |

| | | | |
|---|---|---|---|
| | (>6.25g/100kcal) were associated with fewer stones. Lekcharoensuk Feb & Mar 2002) | from increased animal protein consumption (>77g/day) increased the risk of calcium oxalate stones in man compared to the lowest intake (<50g/day) (Curhan 1993). Several mechanisms are responsible for the hypercalciuria: dietary protein increases glomerular filtration and thus the filtered load of calcium, renal fractional reabsorption of calcium is depressed attributed to protein's acid load or due to the hyperinsulinemic effects of protein. | we recommend that protein be minimized to prevent calcium oxalate recurrence. The exact quantity has not been determined. |
| Calcium | Ironically, epidemiologic studies indicate that diets with lower quantities of calcium (<2.2mg/kcal) are a risk factor for urolith formation. | In an experimental study in dogs, lower both calcium (1.8mg/kcal) and oxalate (0.1mg/kcal) were associated with the lowest relative supersaturation (RSS) for CaOx (Stevenson 2003). Diets with high calcium (7.5mg/kcal) and low oxalate resulted in the highest CaOx RSS. | Diets to reduce urolith recurrence should have lower quantities of calcium and oxalate |

**Diet Efficacy Has Not Been Evaluated Using Clinically Relevant Endpoints.**

Ideally, studies to prevent CaOx urolith formation should be performed in dogs with a history of forming calcium oxalate stones. In addition, investigators and clients should be masked as to which dogs are randomly assigned the therapeutic food and the control food. Lastly, studies should report reliable endpoints of urolith recurrence (i.e. medical imaging). We are not aware of any reported studies meeting these basic criteria. So how can veterinarians make recommendations? The two most commonly prescribed diets to prevent CaOx urolith recurrence in dogs are Prescription Diet u/d® and Urinary SO™. How do these diets compare (table 2)?

**Table 2.** Comparing the Dietary Characteristic of Diets Commonly Used to Prevent CaOx urolith Recurrence

| | Prescription diet u/d Canned | Prescription diet u/d Dry | Urinary SO Canned | Urinary SO Dry |
|---|---|---|---|---|
| Protein g/100kcal | 2.9 | 2.3 | 3.7 | 4.0 |
| Fat g/100kcal | 5.8 | 4.7 | 5.9 | 4.0 |
| Carbohydrate g/100kcal | 12.2 | 14.8 | 7.9 | 13.0 |
| Calcium g/100kcal | 0.076 | 0.075 | 0.2 | 0.19 |
| Phosphorus g/100kcal | 0.038 | 0.045 | 0.17 | 0.15 |
| Sodium g/100kcal | 0.061 | 0.050 | 0.29 | 0.33 |
| Potassium g/100kcal | 0.098 | 0.125 | 0.16 | 0.23 |
| Magnesium g/100kcal | 0.011 | 0.011 | 0.012 | 0.016 |
| Urine pH | 7.1-7.7 | | 5.5-6.0 | |
| Urine Ca mmol/L | 1.22±0.18* | | 1.29±0.90† | |
| Urine Ox mmol/L | 0.744±0.014* | | 0.91±0.78† | |
| CaOx RSS | ND | | 7.76±7.1† | |

* Lulich 2001, †Stevenson 2004

**REFERENCES**


[1] Canales BK, et al..**Journal of Endourology**. June 1, 2008, 22(6): 1161-1168. [2] Curhan GC. et al. **New England Journal of Medicine**. 1993;328(12):833-8. [3] Kirk C. et. al. **JAVMA** 1995;207:1429. [4] Marone CC, et.al. **C.V.** Mosby Co. 1983, pp264 [5] Lekcharoensuk C et al **American J of Veterinary research**, 2002, 63:163-169. [6] Lekcharoensuk C. et.al. **American Journal of Veterinary Research**. 2002; 63:330-337. [7] Lekcharoensuk C., et. al. **JAVMA** 2001; 219(9):1228-1237. [8] Lulich JP, et. al. **JAVMA** 2001;218:1583 [9] Stevenson AE, et. al. **Research in Veterinary Science** 2003;75:33-41. [10] Stevenson AE. et. al. **Veterinary Therapeutics** 2004;5:218-231.

# CALCIUM OXALATE: MANAGEMENT BY ESTIMATING URINARY SATURATION

Joe Bartges, DVM, PhD, DACVIM, DACVN
Knoxville, TN

## INTRODUCTION

Urolith formation, dissolution, and prevention involves complex physical processes. Major factors include: 1) supersaturation resulting in crystal formation, 2) effects of inhibitors of crystallization and inhibitors of crystal aggregation and growth, 3) crystalloid complexors, 4) effects of promoters of crystal aggregation and growth, and 5) effects of noncrystalline matrix.[1,2]

## URINARY SATURATION

An important driving force behind stone formation is saturation state of urine with calculogenic substances. When a solution such as urine is saturated, it refers to the maximal amount of a substance, such as calcium oxalate, that can be completed dissolved. This point is termed the thermodynamic solubility product. When calcium oxalate is present in urine at a concentration less than the solubility point, the urine is undersaturated with calcium oxalate and calcium oxalate completely dissociates and dissolves. When calcium oxalate is present in urine at a concentration that is equal to the solubility point, the urine is saturated with calcium oxalate and calcium oxalate begins to precipitate. When calcium oxalate is present in urine at a concentration above the solubility point, the urine is supersaturated with calcium oxalate and calcium oxalate precipitates.

Urine contains ions and proteins that interact and/or complex with calcium and oxalic acid so as to allow them to remain in solution. This explains why calcium and oxalic acid in urine do not normally precipitate to form calcium oxalate crystals. Urine is normally supersaturated with respect to calcium and oxalic acid. But energy is required to maintain this state of calcium and oxalic acid solubility, and, therefore, the urine must constantly "struggle" to maintain calcium and oxalic acid in solution. Thus, urine is described as being metastable, implying varying degrees of instability with respect to the potential for calcium oxalate crystals to form. In this metastable state, new calcium oxalate crystals will not precipitate, but if already present, crystals can be maintained and even grown in size. If the concentration of calcium and oxalic acid is increased, a threshold is eventually reached at which urine can not hold more calcium and oxalic acid in solution. The urine concentration at which this occurs is the formation point of calcium oxalate. Above the thermodynamic formation product, urine is oversaturated and unstable with respect to calcium and oxalic acid. Thus, calcium oxalate crystals will spontaneously precipitate, grow in size, and aggregate together.

## ESTIMATES OF URINARY SATURATION FOR CALCIUM OXALATE

**Relative supersaturation.** Determining the relative supersaturation (RSS) of a stone-forming substance such as calcium oxalate in a patient's urine is one technique used to assess risk of stone formation.[3,4] RSS is determined by measuring urine concentrations of several analytes including sodium, potassium, calcium, oxalate, magnesium, etc., in the patient's urine. These values are then entered into a computer program (EQUIL or SUPERSAT), which calculates the saturation of various minerals in the patient's urine compared to the concentration of the same minerals in predetermined samples of urine obtained from healthy human beings.

Relative supersaturation is equal to the quotient between the current ion activity product of a salt(s) (AP) in solution and its solubility product (SP).

$$\text{RSS} = \text{ion AP of the patient's urine / ion SP}$$

Relative supersaturation is related to the energy available for crystal nucleation and growth; however, RSS values are limited by the fact that the thermodynamic solubility products used for these calculations has not been measured in the patient's urine. This is of concern because it is probable that different macromolecules, including inhibitors and promoters of crystal formation, growth, and aggregation, in the patient's urine have a pronounced influence on free ion concentrations. By using calculations measured in urine from healthy human beings, RSS may overestimate solubility products and activity products of different minerals, and thus tend to underestimate the risk of urolith formation. Another technical problem is evaluating canine patients is that the EQUIL computer program used to calculate RSS involves comparison of the canine patient's urine values to standardized values based on the composition of human urine.

In summary, compared to activity product ratios, RSS values are less sensitive in predicting the likelihood of stone formation; however, RSS is usually a more sensitive index of risk for urolith formation than measurements of urine pH or mineral concentrations.

**Activity product ratios.** Activity product ratios (APR) also are designed to express the degree of supersaturation of solutions with calculogenic minerals such as calcium oxalate. APR's are obtained by calculating the ion activity products in the patient's urine samples before and after equilibrium with various seed crystals such as calcium oxalate.

$$\text{APR} = \frac{\text{ion AP of patient's urine before incubation with seed crystals}}{\text{ion AP of patient's urine after incubation with seed crystals}}$$

Compared to RSS values, APR values are a more sensitive index of the degree of urine supersaturation with various calculogenic minerals and therefore are more reliable in predicting the likelihood of urolith formation. To obtain APR values, first the activity product (not the activity product ratio) of calcium oxalate is calculated as the mathematical product of the activity of calcium and activity of oxalic acid. The term "activity" of a mineral is an index of the likelihood that the mineral will combine with other substances in urine. The "activity" of a mineral is dependent on several factors including: 1) the urine concentration of that mineral, 2) the urine concentrations of other substances such as sodium, potassium, calcium, etc., 3) the quantity and functional state of nonmineral inhibitors and promoters of crystal formation, growth, and aggregation, 4) urine pH, and 5) temperature of urine.

In summary, the first step in obtaining the APR is to determine the activity product (AP) of calcium oxalate in a patient's urine, which encompasses measurement of the urine concentration of these analytes. The numerically expressed concentrations of analytes are then entered into the EQUIL computer program, which calculates the "pre-incubation" AP.

The next step in determining the APR is to incubate an aliquot of the same urine sample with preformed seed crystals, composed of pure calcium oxalate. Following incubation with the seed crystals, the urine concentration of the same analytes are again measured in the incubated urine sample to determine if their concentrations have increased as result of dissolution of the seed crystals, or to see if their concentrations have decreased as a result of precipitating out of solution. The post-incubation urine concentrations of analytes are then entered into the EQUIL computer program to calculate a second "post-incubation" activity product (AP). Dividing the "pre-incubation" AP by the "post-incubation" AP gives the activity product ratio for that patient's urine sample.

An exact measurement of supersaturation is not obtained by determining APR, but the method provides useful information about the relative increase or decrease of the ion activity products in the patient's urine that results from seed crystal growth or seed crystal dissolution. An APR less than one represents undersaturation of urine with the mineral being evaluated (calcium oxalate in our example). An APR equal to one represents saturation of the patient's urine samples with calcium oxalate. An APR greater than one indicates that the patient's urine sample is supersaturated with calcium oxalate. The actual numerical APR value above one provides information about the degree of supersaturation that is present.

Activity product ratios can be calculated for any calculogenic mineral as long as pure seed crystals for that type of mineral are available. The APR is a more sensitive index of urine saturation for stone-forming compounds than RSS. Use of APR methodology will not eliminate errors associated with the effect of unknown factors such as crystallization inhibitors or promoters on ion activities; however, since the same urine sample obtained from the patient is analyzed before and then after equilibration with seed crystals (such as calcium oxalate), the same type of error occurs in evaluation of both analyses and therefore the errors cancel. Whereas calculation of RSS can overestimate supersaturation, saturation, and undersaturation, the APR method overestimates undersaturation, underestimates supersaturation, and correctly measures saturation, provided that a sufficient amount of seed crystals have been used.

In general, APR is a more sensitive index of urine saturation for stone-forming compounds than RSS. APR's can be used to investigate the efficacy of special diets and/or specific dietary components on the degree of supersaturation of a patient's urine or a healthy dog's urine. Calculation of the APR may be used in animals that have had uroliths to determine whether they are at risk for formation of new stones before they develop.

## RESULTS OF CALCIUM OXALATE URINARY SATURATION STUDIES IN DOGS AND CATS

Studies utilizing urine saturation testing has been performed in veterinary medicine, but despite the number of studies, very few have been performed in dogs or cats that are urolith-formers and no studies exist that compares estimates of urinary saturation with recurrence rates of uroliths. In dogs, calcium oxalate urolith formation typically occurs when urinary relative supersaturation for calcium oxalate is greater than 10; the metastable zone lies between a relative supersaturation value for calcium oxalate of 1 and 10 to 14.[5] In cats, calcium oxalate urolith formation typically occurs when urinary relative supersaturation for calcium oxalate is greater than 12; the metastable zone lies between a relative supersaturation value for calcium oxalate of 1 and approximately 12.[6] Sterile struvite urolith formation in cats typically occurs when urinary relative supersaturation for struvite is greater than 2.5; the metastable zone lies between a relative supersaturation value for struvite of 1 and approximately 2.5.[6] Urinary supersaturation represents a risk for urolith formation, but as in human beings, there is overlap in values between urolith-forming animals and healthy, non-urolith-forming animals.[7,8]; therefore, other factors are important. Use of urinary saturation studies can provide further information as to mechanisms of urolith formation, screening of animals at risk for urolith formation, and monitoring efficacy of urolith management.

## REFERENCES

[1]Bartges JW, Osborne CA, Lulich JP, et al. Methods for evaluating treatment of uroliths. **Vet Clin North Am Small Anim Pract** 1999;29:45-57. [2]Coe F, Parks J. *Nephrolithiasis:* **Pathogenesis and Treatment.** 2nd ed. Chicago: Year Book Medical Publishers Inc., 1988. [3]Brown C, Purich D. Physical-chemical processes in kidney stone formation In: Coe F,Favus M, eds. **Disorders of Bone and Mineral Metabolism.** New York: Raven Press, 1992;613-624.[4]Pak CYC, Hayashi Y, Finlayson B, et al. Estimation of the state of saturation of brushite and calcium oxalate in urine: A comparison of three methods. **J Lab Clin Med** 1977;89:891-901.[5]Stevenson AE, Rutgers C. Nutritional management of canine urolithiasis In: Pibot P, Biourge V,Elliott DA, eds. **Encyclopedia of Canine Clinical Nutrition.** Aimargues, France: Aniwa SAS-Royal Canin, 2006;284-315.[6]Houston DM, Elliott DA. Nutritional management of feline lower urinary tract disorders In: Pibot P, Biourge V,Elliott DA, eds. **Encyclopedia of Feline Clinical Nutrition.** Aimargues, France: Anawa SAS-Royal Canin, 2008;285-321.[7]Robertson WG, Peacock M, Nordin BEC. Activity products in stone-forming and non-stone forming urine. **Clin Sci** 1968;34:579-594.[8]Kavanagh JP. Supersaturation and renal precipitation: the key to stone formation? **Urol Res** 2006;34:81-85.

# MANAGEMENT OF IDIOPATHIC HYPERCALCEMIA IN CATS WITH CALCIUM OXALATE STONES

**Dennis J. Chew[1], DVM, DACVIM (Internal Medicine), Patricia A. Schenck[2], DVM, PhD, Brian Hardy[3] DVM**
**Columbus, Ohio; East Lansing, Michigan; Saint Paul-Minneapolis, Minnesota**

## INTRODUCTION

Idiopathic hypercalcemia (IHC) has been recognized in cats over the past 15 years with increasing frequency. It now appears to be the most common cause of ionized hypercalcemia in cats in the USA, and is becoming more common in other parts of the world. Unfortunately, the underlying cause remains unknown. IHC may only develop in genetically susceptible cats in a provocative environment (diet, toxins, environment, stress, other unknown factors), and it is likely that IHC may have multiple etiologies. The lack of a definitive cause makes specific recommendations for treatment difficult.

## PATHOPHYSIOLOGY & DIAGNOSIS OF IDIOPATHIC HYPERCALCEMIA

Calcium status is usually initially determined based on serum total calcium (tCa) concentration, even though it is the ionized calcium (iCa) fraction that is physiologically active. This reliance on tCa measurement is based on the assumption that the iCa concentration is directly proportional to tCa. This assumption is incorrect, and can lead to errors in interpretation of laboratory data. In a study of over 400 cats, a 40% diagnostic disagreement between serum iCa and tCa was noted. Hypercalcemia and normocalcemia were underestimated, and hypocalcemia was overestimated when using tCa to predict iCa status. In addition, over 11% of cats classified as normal by serum tCa concentration were found to be hypercalcemic when serum iCa was measured[1]. Thus, without serum iCa measurement, some truly hypercalcemic cats are not being diagnosed as hypercalcemic, which is probably underestimating the potential number of cats with IHC. It would be beneficial to assess the serum iCa status in all cats with urolithiasis, or vague gastrointestinal signs such as weight loss, constipation, and vomiting. Serum iCa is preferably measured at the same time that parathyroid hormone (PTH) concentration is measured.

Many potential causes of hypercalcemia can be ruled out after analysis of the history, and findings from the physical examination. In cases where the cause of hypercalcemia is not obvious, then hematology, serum biochemistry, body cavity imaging, cytology, or histopathology may be necessary, along with evaluation of serum iCa and PTH concentrations. Cats with IHC will have increased serum iCa, with low or normal range PTH concentrations (a parathyroid-independent hypercalcemia). Parathyroid hormone related polypeptide (PTHrP) should be negative, and concentrations of 25-hydroxyvitamin D and 1,25-dihydroxyvitamin D (calcitriol) will be normal to low. Serum phosphorus concentration is usually within normal limits unless chronic renal failure is also present.

In a review of over 400 cats with IHC, the age at presentation ranged from 0.5 to 20 years old (mean 10 years), and 27% of the cases were longhaired cats (compared to a 14% submission rate from longhaired cats); thus longhaired cats appear to be over-represented. There was no affect of gender on incidence. Almost half of the cats had no obvious clinical signs at the time of presentation. The most common clinical sign was mild weight loss with no other clinical signs (noted in 18% of cases); 15% had uroliths or renoliths, 6% had inflammatory bowel disease, 5% had chronic constipation, 4% exhibited vomiting, and 1% were anorectic. Of those with uroliths or renoliths, calcium oxalate stones were specifically noted in 10% of the cases[2]. Polyuria and polydipsia are common clinical signs of hypercalcemia in dogs, but not in cats. Unlike dogs, most cats with hypercalcemia can still concentrate urine to a high degree as long as they do not have concurrent renal failure.

## CALCIUM OXALATE STONE FORMATION & HYPERCALCEMIA

Calcium oxalate stones form in patients following urinary supersaturation with calcium and oxalate precursor minerals in a susceptible individual. The presence of a promoter or absence of an inhibitor will enhance the chances for stone genesis. Sufficient time with retention of crystals within the urinary tract is required to allow aggregation of crystals and stone growth. Supersaturation with precursor minerals can occur due to hypercalciuria or hyperoxaluria as primary events, or their concentration can be increased due to inadequate intake and subsequent urinary excretion of water.

When calcium enters the blood stream following excessive intestinal absorption or bone resorption, the higher concentration of circulating calcium undergoes glomerular filtration, some tubular reabsorption, and hypercalciuria as the response expected from normal kidneys. In some situations, so-called "renal leak" hypercalciuria occurs in the absence of an increased filtered load of calcium. An increased urinary calcium excretion expressed as mg/min or total mg per day does not necessarily correlate to the concentration of urinary calcium expressed as mg/dl as this also depends on the amount of water excreted along with the calcium. The urinary concentration of calcium may be more important than its total daily excretion in the pathophysiology of stone genesis.

Approximately 35% of cats with calcium oxalate urinary stones identified from the Minnesota Urolith Center in the mid 1990's were found to have hypercalcemia – most of these cats had IHC[3]. Cats with calcium oxalate ureterolithiasis have been described with increasing frequency since the early 1990's. From 14 to 36% of cats with ureterolithiasis were described to have hypercalcemia by two different groups[4,5].

## POSSIBLE TREATMENT STRATEGIES

Aggressive treatment of hypercalcemia is warranted in cats with calcium-containing urinary stones, since continuing hypercalcemia poses a risk for development of new stones and enlargement of existing stones. The major goal of treatment is to lower the filtered load of calcium through the kidneys by restoring normocalcemia – excessive bone resorption and GI absorption of calcium are targeted.

**Increased Dietary Fiber**

There is the potential that switching to a diet that is higher in dietary fiber could have benefit. Feeding of increased dietary fiber has been reported to restore normocalcemia in some cats with IHC and calcium-oxalate urolithiasis[3,6]. In another study[7], the feeding of a high fiber diet did not restore normocalcemia in any cat with IHC. Intestinal absorption is complex, and the effect of fiber is dependent on dietary fiber type, the amount of fiber, and also other nutrients present in the diet. Intestinal absorption of calcium may be decreased with the feeding of high fiber diets by decreasing the gastrointestinal transit time. Most pet food manufacturers have taken this potential decrease in absorption of calcium into account, and have increased the quantity of dietary calcium in high fiber diets to enhance calcium absorption.

**Less Acidifying Diets**

Serum ionized calcium concentration and the degree of calciuria is increased in cats that eat acidifying and Mg restricted diets[8]. It is likely that these changes result from buffering effects with bone that release calcium into the circulation. It is possible that in some cats hypercalcemia will resolve when fed a less acidifying diet. Though renal diets are less acidifying and possibly alkalinizing, their degree of phosphorus restriction may enhance calcitriol synthesis that could blunt the salutary effects of the alkalinization

**Vitamin D Restriction**

It is possible that IHC could in part be a consequence of excessive dietary vitamin D, even though the serum concentration of 25-hydroxyvitamin D is within normal limits. The minimum requirement for dietary vitamin D is debatable, since 'normal' values for vitamin D have been determined using cats that consume vitamin D supplemented diets. While cats with IHC typically have 25-hydroxyvitamin D and calcitriol concentrations within normal limits, some cats may be particularly sensitive to the effects of dietary vitamin D. There are minimum and maximum allowable limits for vitamin D content in cat foods; however, there is a 20-fold difference between these values. Since the vitamin D content is not listed on the label, it is difficult to determine how much vitamin D is being ingested, and difficult to choose a diet that may have a lower level of vitamin D supplementation. The effects of Vitamin D may be additive to those of Vitamin A on bone resorption.

**Glucocorticosteroids**

Glucocorticosteroid administration can result in a significant drop in serum iCa concentration in some cats with IHC. Steroids act by decreasing intestinal calcium absorption, reducing bone resorption, and increasing calcium excretion[9]. Treatment with glucocorticoids should be considered in those cats that do not become normocalcemic within 6 to 8 weeks following a dietary change. Prednisone is given at 5 mg/cat/day for one month, and then serum iCa concentration is reassessed. If the iCa concentration has returned to normal, this dose of prednisone is continued for several months. If the cat is still hypercalcemic, the dose is increased to 10 mg/cat/day. Some cats may require 15 to 20 mg prednisone/day to maintain normocalcemia. Even with maximal doses of prednisone, some cats will exhibit an escape from the effects, with a return of hypercalcemia. Conventional wisdom based on data in other species is that steroids reduce tubular reabsorption of calcium thereby enhancing calciuria which could increase the risk for calcium oxalate stone growth. When treatment lowers serum calcium concentration, the filtered load of calcium is decreased which likely offsets any tendency to increase calciuria from any effects on tubular function. A recent short-term study of normal cats given 10 mg prednisolone once daily for 2 weeks did not induce diuresis, enhance calcium excretion, or increase urinary saturation for calcium oxalate[10].

**Bisphosphonates**

Once weekly oral alendronate reduced serum ionized calcium concentration in most cats with IHC in a recent pilot study conducted at The Ohio State University VMTH (Dr. Brian Hardy unpublished observations). No side effects were documented in the 6 months of this study with an average weekly dose of 10 mg per cat. Care must be taken to ensure that tablet medication does not stick in the esophagus as this is a known risk for erosive esophagitis in humans. Tap water PO following pilling is recommended to help lessen this possibility. Oral bisphosphonate treatment can restore normocalcemia in some cats that fail to do so during diet and prednisolone treatment. Some cats have required up to 30 mg weekly per cat to achieve normocalcemia. Oral bisphosphonate treament has failed to achieve normocalcemia in a small number of cats with IHC. The salutary response during bisphosphonate treatment indicates that blunting of osteoclastic bone resorption can be effective to decrease serum ionized calcium but it does not prove that accelerated bone resorption is the underlying cause of the hypercalcemia.

**Diuretics**

Thiazides should not be used since they can result in hypocalciuria and further hypercalcemia. Oral furosemide is of potential value during chronic IHC due to its calciuretic effects though there are no reports of its use chronically for this purpose. Though furosemide may initially lower serum calcium, some cats develop dehydration that is not fully corrected by increased water intake during treatment. Dehydration can increase serum calcium concentration by hemoconcentration and by stimulation of the renal tubules to increase sodium and calcium reabsorption.

## REFERENCES

[1]Schenck PA. **Proceedings of the 10th Congress of the International Society of Animal Clinical Biochemistry**. Gainesville, FL, 2002.[2] Schenck PA. **J Vet Intern Med** 2004; 18:442.[3] Osborne CA. **Vet Clin North Am Small Anim Pract 26**(2):217-32, 1996.[4] Kyles AE. **J Am Vet Med Assoc** 226(6): 932-936, 2005[5]Aronson LR. **J Am Vet Med** *Assoc* 228(5): 743-749, 2006.[6]McClain HM. **J Am Anim Hosp Assoc** 35(4):297-301, 1999.[7]Midkiff AM. **J Vet Intern Med** 14(6):619-26, 2000.[8]Ching SV. J Nutr 119(6): 902-915, 1989.[9]Schenck PA. In: SP DiBartola, ed. **Fluid therapy in small animal practice**. 3rd ed. St. Louis: Elsevier, 2005;122-194. [10]Geyer N. **Vet Ther** 8(4): 239-246, 2007.

# CALCIUM OXALATE: MANAGEMENT BY LASER LITHOTRIPSY

**Larry G. Adams, DVM, PhD, DACVIM (SAIM)**
**West Lafayette, IN**

Calcium oxalate uroliths are not amenable to medical dissolution; therefore they must be mechanically removed from the urinary tract. Although most uroliths are removed by open cystotomy, lithotripsy is a minimally invasive option for removal of uroliths. Lithotripsy is crushing or fragmenting uroliths by high energy shock waves or by laser energy. Types of lithotripsy include extracorporeal shock wave lithotripsy, electrohydraulic lithotripsy and laser lithotripsy. Shock wave lithotripsy (SWL) uses repeated shock-waves generated outside the body to fragment the uroliths into many small fragments that can pass spontaneously through the excretory system. Laser lithotripsy is an option for fragmentation and removal of lower tract uroliths that can be visualized using a rigid or flexible cystoscope or ureteroscope.[1] Although electrohydraulic lithotripsy can also be performed in dogs,[2] this technique has been largely replaced by laser lithotripsy because of greater efficiency and efficacy of laser lithotripsy compared to electrohydraulic lithotripsy.[1-3] In humans, laser lithotripsy using the holmium: YAG laser is the procedure of choice for intracorporeal fragmentation of large bladder calculi, ureteroliths and nephroliths.[4,5] The mechanism of urolith fragmentation with the holmium: YAG laser is mainly photothermal, and involves a thermal drilling process rather than a shock wave effect.[6,7] The 2100 nm wavelength of the holmium: YAG laser is absorbed in 0.5 mm of fluid. Therefore it can be used to safely fragment uroliths in tight locations such as within the urethra or ureter without damage to the urothelium.[4,5]

## RESULTS OF LASER LITHOTRIPSY IN DOGS

Laser lithotripsy of bladder and urethral stones is possible in most dogs. Because of the small diameter of the laser fibers, laser lithotripsy can be performed through small diameter flexible ureteroscopes used for urethrocystoscopy in male dogs. *In vitro* studies confirm that the holmium laser can efficiently fragment uroliths from dogs regardless of chemical composition.[8] Implanted urethroliths in male dogs were safely fragmented by laser lithotripsy using a holmium laser, although no attempt was made to remove urolith fragments in this research study.[9] There are several recent clinical studies on effectiveness of laser lithotripsy for removal of uroliths from the lower urinary tract in dogs.[1,3,10,11]

We recently reported the results from 73 dogs treated with laser lithotripsy at Purdue University and the University of Pennsylvania.[1] Laser lithotripsy resulted in complete removal of all uroliths in all female dogs (28/28; 100%) and a majority of male dogs (39/45; 86.7%).[1] Dogs with urethroliths had shorter laser time than dogs with cystic uroliths. Once uroliths were sufficiently fragmented such that they were small enough to pass through the urethra, larger fragments were initially removed with a stone basket to verify that the fragments were small enough to allow removal. Once the larger fragments were safely removed through the urethra via basket extraction, voiding urohydropropulsion was then performed to evacuate remaining fragments.

Considering recent publications on laser lithotripsy in dogs, the following observations can be made about the efficacy of laser lithotripsy. All urolith fragments were removed more frequently in female dogs (83 to 100%) than in male dogs (68% to 87%).[1,3,10,11] Complete urolith removal was achieved in 100% of dogs with urethroliths.[1,3,11] Procedure time was also shorter for removal of urethroliths than urocystoliths.[1] One case-controlled study showed that laser lithotripsy was comparable in efficacy to open cystotomy, although procedure time was slightly shorter for surgery than laser lithotripsy.[10] Two studies showed that procedure times were significantly shorter once the operator gained experience with the technique.[3,11] In a recent study comparing laser lithotripsy to cystotomy for removal of lower urinary tract uroliths, residual urolith fragments were detected in approximately 10% of dogs treated by either method.[10] Therefore cystoscopic examination (female dogs) or post-procedural radiographs (male dogs) are indicated to confirm complete urolith removal.

## RESULTS OF LASER LITHOTRIPSY IN CATS

Laser lithotripsy of bladder and urethral stones is also possible in female cats. Cystoscopes that may be utilized for laser lithotripsy and basket extraction of uroliths in female cats include a 9 French rigid pediatric cystoscope and a 7.5 Fr flexible ureteroscope. Laser lithotripsy is not possible in male cats because the available flexible cystoscopes for male cats do not have a working channel large enough for passage of a laser fiber. There are no publications on laser lithotripsy for removal of urocystoliths in cats. The author has successfully performed laser lithotripsy in 4 female cats. Cats with uroliths have more friable bladder mucosa than dogs with uroliths, and bleeding limited visualization during the procedure in some cases. All urolith fragments were successfully removed from each cat during the procedure or within 2 weeks. Two cats had small urolith fragments trapped in intravesical blood clots at the end of the laser lithotripsy procedure, although the cats spontaneously voided the fragments within 2 weeks.

## CASE SELECTION FOR LASER LITHOTRIPSY

As veterinarians consider the relative roles of different minimally invasive techniques for stone removal, the relative stone burden and location of the uroliths affects which techniques should be considered. Case selection is important for laser lithotripsy because not all dogs with urocystoliths can be treated by this technique. Male dogs less than 5-7 kg may be too small for transurethral passage of flexible ureteroscopes to the urinary bladder for fragmentation and removal of urocystoliths. Laser lithotripsy via transurethral approach may not be the best option for removal of very large numbers of urocystoliths, especially in male dogs. One study showed that in male dogs, the overall stone burden was inversely related to successful removal of all uroliths by laser lithotripsy whereas stone burden was less likely to affect outcome in female dogs.[11] As holmium: YAG lasers become more widely available in veterinary medicine, this technique may replace open cystotomy for removal of urocystoliths. Laser lithotripsy is especially effective and efficient from removal of urethroliths and, if available, laser lithotripsy should be utilized to avoid urethrotomy. Alternative minimally invasive approaches for removal of urocystoliths in small male dogs include percutaneous cystoscopy, laparoscopic assisted cystoscopy, and SWL fragmentation of bladder stones followed by voiding urohydropropulsion.[12,13]

## LASER LITHOTRIPSY TECHNIQUE

Positioning of dogs for cystoscopy and laser lithotripsy is a matter of clinician preference. The author positions female cats and dogs in dorsal recumbency and male dogs in left lateral recumbency. The hair around the prepuce or vulva is clipped and a sterile prep of the surrounding skin should be performed. In male dogs, the penis is extruded and cleansed prior to cystoscopy. The prepuce or vaginal vestibule is flushed several times with dilute betadine saline solution. The animal is draped with a small opening for access to the urethra. Strict attention to aseptic technique is important in handling the cystoscope or ureteroscope and stone baskets. By positioning female dogs in dorsal recumbency and using sterile drapes, it is easier to maintain aseptic technique during repeated passage of the cystoscope for laser lithotripsy and basket extraction of urolith fragments. Laser lithotripsy requires that the veterinarian has good fundamental cystoscopy skills in both male and female dogs.

When working through the biopsy channel of rigid cystoscope, the author uses a 550 micron fiber (or 365 micron fiber in female cats) to deliver the laser energy to the surface of the urolith. Although both 200 and 365 micron fibers pass through the 3-French working channel of a flexible ureteroscope, the 200 micron fiber is preferable because it allow for greater degree of deflection of the tip of the ureteroscope than the 365 micron fiber. The increased flexibility of the ureteroscope during laser lithotripsy allows for greater precision in targeting uroliths, especially in the urethra. The author uses an initial setting of 0.6 to 0.8 joules per pulse and 6 to 10 Hz for laser lithotripsy. Either the joules (energy per pulse) or Hz (pulses per second) may be increased to facilitate fragmentation of the uroliths. Increasing the Hz tends to increase the speed of lithotripsy provided the urolith shows some fragmentation, whereas increased joules per pulse is utilized if the urolith does not fragment at lower power settings. During laser lithotripsy, the laser fiber should be in contact with the urolith or within <0.4 mm of the urolith surface. The laser fiber should be parallel to the urethral or bladder mucosa or the laser fiber must be positioned > 0.5 mm from the mucosa. Working parallel to the mucosa is especially important in the urethra to avoid iatrogenic trauma to the urethral mucosa.

During laser lithotripsy fluid irrigation using sterile saline is needed to maintain visualization of uroliths and provide fluid to dissipate heat near the laser fiber tip. In female dogs and cats, gravity fluid irrigation is sufficient with the fluid bag positioned no more than 70 cm above the patient. In male dogs, a 60 cc syringe is connected to the flexible ureteroscopes via an urolock adaptor (Boston Scientific) to allow pressurized fluid infusion during laser lithotripsy. Careful attention to the total infused volume is necessary to prevent overdistension or rupture of the urinary bladder. The approximate maximum infusion volume should be ≤10 ml/kg of body weight assuming the bladder was empty prior to the procedure. Periodic palpation of the bladder is advisable to monitor the degree of bladder distention.

## CONCLUSIONS

Transurethral cystoscope-guided laser lithotripsy is successful in female dogs and cats and most male dogs for fragmentation and subsequent removal of cystic and urethral uroliths. Laser lithotripsy is especially effective for fragmentation and removal of urethroliths, and if available, laser lithotripsy is the procedure of choice for removal of urethroliths.

## REFERENCES

[1.] Adams LG et al. Use of laser lithotripsy for fragmentation of uroliths in dogs: 73 cases (2005–2006), **J.Am.Vet.Med.Assoc**. 2008,232: 1680-1687. [2.] Defarges A et al. Use of electrohydraulic lithotripsy in 28 dogs with bladder and urethral calculi, **J.Vet.Intern.Med**. 2008,22: 1267-1273.[3.] Grant DC et al. Holmium: YAG laser lithotripsy for urolithiasis in dogs, **J.Vet.Intern.Med**. 2008,22: 534-539.[4.] Lingeman JE et al. Surgical management of urinary lithiasis, in **Campbell's Urology**, ed. Retik AB, Vaughan ED, Jr., Wein AJ, W.B.Saunders, 2002, p 3361-3451.[5.] Wollin TA et al. The holmium laser in urology, **J.Clin.Laser.Med.Surg**. 1998,16: 13-20.[6.] Chan KF et al. Holmium:YAG laser lithotripsy: A dominant photothermal ablative mechanism with chemical decomposition of urinary calculi, **Lasers in Surg Med** 1999,25: 22-37.[7.] Vassar GJ et al. Holmium: YAG lithotripsy: photothermal mechanism, **J.Endourol**. 1999,13: 181-190.[8.] Wynn VM et al. In vitro effects of pulsed holmium laser energy on canine uroliths and porcine cadaveric urethra, **Lasers.Surg.Med**. 2003,33: 243-246.[9.] [D]avidson EB et al. Laser lithotripsy for treatment of canine uroliths, **Vet.Surg**. 2004,33: 56-61. [10] Bevan JM et al. Comparison of laser lithotripsy and cystotomy for the management of canine urolithiasis, **J.Am.Vet.Med.Assoc**. In Press.[11.] Lulich JP et al. Efficacy and safety of laser lithotripsy to remove urocystoliths and urethroliths in dogs: 100 consecutive cases, **J.Am.Vet.Med.Assoc**. In Press.[12.] Rawlings CA et al. Use of laparoscopic-assisted cystoscopy for removal of urinary calculi in dogs, **J.Am.Vet.Med.Assoc**. 2003,222: 759-761.[13.] Goldman CK et al. Extracorporeal shock-wave lithotripsy for the treatment of urocystoliths in dogs. **Urol.Res**. 36, 179. 2008.

# INDOOR AIR QUALITY & DISEASE

**Janice A. Dye, DVM, MS, PhD, DACVIM**
**Research Triangle Park, NC**

## INTRODUCTION

Concern over the quality of indoor (i.e., residential) as well as outdoor (i.e., environmental) air is increasing. Accordingly, owners of companion animals may approach their veterinarian about the potential for airborne irritants, allergens, pollutants, or infectious agents to negatively impact their pet's health. Furthermore, if the pet has an underlying respiratory condition, an owner may inquire as to ways to improve the quality or purity of the air within the home. This session will review commonly encountered indoor air contaminants and related home conditions that contribute to adverse respiratory health outcome in small animals. Owing to the fact that many dogs, cats, and birds spend the majority of their lives within the home residence, the focus of the review will be on indoor air factors. This will include several outdoor air pollutants that readily permeate indoor spaces. Similar principles and precautions may apply to horses and camelids that winter in barns with dusty hay, minimal ventilation, and excessive dampness.

## CHEMICAL CONTAMINANTS

### Combustion products

*Second-hand smoke* (SHS), also known as environmental tobacco smoke, is simply the smoke from someone else's cigarette. SHS contains many toxins and ciliotoxins that — particularly in poorly ventilated air spaces, cause ocular and respiratory tract irritation. This is especially problematic for young children in whom exposure is associated with more severe asthma, reduced resistance to respiratory tract infections, and increased incidence of middle-ear infections, bronchiolitis, and pneumonia.[1] Based on urinary cotinine levels, house pets are similarly exposed to SHS, with levels in dogs increasing proportionately with the number of cigarettes smoked in the household.[2] Predisposition toward analogous conditions in exposed pets is therefore likely. The Surgeon General has concluded that there is no risk-free level of exposure to SHS and thus, a smoke-free environment is the only way to fully protect nonsmokers from the hazards of SHS.[1] By extension, cats and dogs with pre-existing upper or lower respiratory disease may be especially sensitive to the irritating effects of SHS.

*Nitrogen dioxide* ($NO_2$), a poisonous, brownish-colored gas, is a major indoor and outdoor air pollutant. Importantly, levels in the home often exceed (by as much as fives times) ambient concentrations, well in excess of levels considered safe in outdoor air by the USEPA. The major sources of indoor $NO_2$ are poorly ventilated or maintained gas appliances (e.g., water heaters, clothes dryers, kitchen ranges), fireplaces, and coal- or wood-burning stoves. Adverse health effects include impairment of immune defense mechanisms. Data further suggest that infants and children with atopy or asthma are more sensitive to the respiratory effects of $NO_2$, in part due to enhanced asthmatic reactions to inhaled allergens.[3] Thus, in small animals, $NO_2$ exposure may result in chronic bronchial disease or increased frequency / severity of respiratory infection. Long-term (near-lifetime) exposure to $NO_2$ is also purported to contribute to chronic emphysematous lung changes.

*Ozone* ($O_3$), primarily an outdoor air pollutant, is produced in the lower atmosphere by photochemical reactions involving sunlight, oxygen, and combustion products from gasoline and other solvents. Hot temperatures and stagnant air masses contribute to increased ambient concentrations. As significant accumulations occur near ground level, the phenomenon of "smog" is observed. Indoor $O_3$ levels arise mainly from outdoor levels permeating the home during normal air exchange. Indoor levels are usually less than half that of outdoor levels, but during *smog* episodes, they may be sufficient to impact health. Ozone can be emitted directly into the indoors by, forexample, electric generators, office equipment (e.g., photocopy and fax machines, laser printers), and ironically, ionizing "air-purifiers" that actually generate $O_3$ to react with and rid air of odors. During the latter instance, indoor $O_3$ concentrations can exceed levels found in smog.[4] As an oxidant gas, $O_3$ is among the most injurious of the atmospheric pollutants. When inhaled, it can damage airway cells and cause lung inflammation. In healthy dogs and humans, acute $O_3$ exposure can also increase airway responsiveness. In fact, exposure to relatively low $O_3$ concentrations (comparable to that of many urban areas) can increase airway responsiveness during allergen challenge in humans with atopic asthma. Moreover, infant rhesus monkeys repeatedly exposed to house dust mite antigen alternating with $O_3$ developed severe, long-lasting asthma-like disease.[5] In humans and laboratory animals, $O_3$ has been shown to have variable effects on mucociliary clearance, a primary mechanism for clearing inhaled particles or offending microorganisms from the airways, thereby raising concerns about host resistance to infectious agents.[3]

*Sulfur dioxide* ($SO_2$), another outdoor air pollutant, is the major precursor to acidic deposition ("acid rain"). Veterinary concerns regarding $SO_2$ pollution are mainly over aquatic and wildlife health. One important exception relates to use of unvented kerosene space heaters. Such heaters emit significant $SO_2$ and associated acid aerosols. Total airborne concentrations can be exorbitant in closed-space situations. When $SO_2$ gas is combined with other organic acid emissions, space heater emissions can cause ocular and respiratory irritation in animals and humans alike. In addition, acid aerosols exposure from ambient $SO_2$ pollution may occur in pets that are either housed outdoors or that are exercised for significant periods outdoors. Again, asthmatic humans are at particular risk, thus dogs and cats with chronic bronchial disease may be at similar peril to this irritant gas-particulate complex.[3]

*Indoor particulate matter* (PM) can be classified according to the source of origin, including SHS, cooking, heating (fireplaces), candle burning, house dust, and intrusion of outdoor PM (e.g., road traffic or diesel exhaust) with associated secondary organic aerosols. Decrements in peak expiratory flow measurements of asthmatics were more significantly associated with increases in indoor $PM_{2.5}$ levels than with outdoor $PM_{2.5}$ levels.[6] Importantly, combined exposures of PM with allergens or endotoxin may lead to enhanced pulmonary inflammation, thus further exacerbating allergic symptoms and asthma.[7]

*Carbon monoxide* (CO) is a colorless, odorless gas. A product of incomplete fuel combustion, sources of CO are similar to those of $NO_2$, the principal source being motor vehicle exhaust. Classic cases of human poisonings, accidental or intentional, arise from car engines left running in attached garages or sheds. Significant exposure may also result during use of charcoal grills or unvented kerosene heaters in relatively closed air spaces, and when chimneys or venting ducts of central heating systems are obstructed. Birds are especially sensitive to this air toxicant, as evidenced by their classic use as sentinels in coal mine shafts. Carbon monoxide poisoning is also an important cause of the morbidity and mortality in pets managing to escape from burning homes. Owing to its overt affinity for hemoglobin, CO prevents cellular oxidation, resulting in severe cellular anoxia and death. Unlike most air pollutants where problems occur due to low level, chronic exposures, a single (likely to be encountered) exposure to CO may prove lethal. Clinical signs range from vague lethargy and fatigue, to headache and depression, and progress to confusion, nausea, and ultimately, unconsciousness and death. Fisher *et al*[8] demonstrated that isocapnic hyperpnea with 100% $O_2$ in dogs more than doubled the rate of CO elimination compared with normal ventilation with 100% $O_2$; a simple but effective means to hasten recovery of poisoned pets.

**Off-gassing emissions**

*Volatile organic compounds* (VOCs). Home exposure to various chemical compounds is often significant. It is estimated that the average home contains at least 45 aerosol products, ranging from cleaning supplies [e.g., citrus-scented (limonene) air fresheners, furniture polish], to personal care items (hair spray, deodorant, static guard, perfume), to solvents, paints, and pesticides (insect and roach sprays, fumigants, and flea control products). In turn, these compounds are composed of active and inert ingredients including a numerous VOCs. When moving into a new home or during extensive remodeling, both owners and pets are frequently exposed to fumes from new carpeting, fresh paint or varnish, and new furniture or draperies. If used in well-ventilated areas, as per label instructions, most people (and pets) experience minimal adverse effects from use of these products. However, many VOCs are potential sensory irritants. Accordingly, exposure can be associated with ocular, dermal, and respiratory tract irritation. During excessive or prolonged contact, dizziness, headache, nausea, and even allergic reactions may arise. Retrospective epidemiologic studies have associated chronic topical insecticide application, especially in obese dogs, with increased risk of developing TCC of the bladder.[9]

Perhaps most importantly, many owners fail to use products as per the warning label instructions. They insist on using these products carelessly or excessively. VOC exposure is therefore a realistic cause of illness in pets. It should be considered if signs reflect relatively acute onset of ocular, dermal, or respiratory system effects. Avian veterinarians should be especially attuned to this possibility. As with carbon monoxide, pet birds may be exceedingly sensitive to aerosolized VOCs. Even the fumes from over-heated cooking pans lined with polytetrafluoroethylene (Teflon® or Silverstone®) have been associated with acute and fatal toxicosis.[10]

*Formaldehyde* is a pungent, highly water soluble, reactive gas. Indoor exposures may occur during off-gassing of formaldehyde from resins present in particleboard, fiberboard, and wood paneling. Other sources may include foam insulation, carpeting, upholstery, and drapery fabrics. Mobile homes, being relatively air-tight and constructed from many of these materials, are especially problematic. Formaldehyde gas is an ocular and upper respiratory irritant. Exposures may cause headaches, nausea, and dizziness.

**CARCINOGENS**

*Asbestos* fibers are naturally-occurring mineral (hydrated silicate) materials that were used extensively as a noncombustible insulation for walls and pipes, in flooring and roofing materials, in cement and spackling compounds, and in friction products (e.g., brake shoes and clutch pads). Of concern, asbestos fibers are small and when abraded or disturbed, readily become air-borne and are easily inhaled. However, pulmonary defense mechanisms have great difficulty removing or destroying these fibers. Asbestos-associated diseases have long latency periods (up to 30 years in humans), and include development of pleural plaques, asbestosis (i.e., interstitial fibrosis), bronchogenic carcinoma, and the rare cancer, malignant mesothelioma. Asbestos removal and/or containment remain costly and controversial issues. In dogs, mesothelioma was associated with occupational exposure of owners (e.g., shipyard workers or auto mechanics) presumably due to exposure of re-entrained asbestos dust from clothing. Male dogs were at a higher risk for developing mesothelioma.[11] Mesothelioma is also more common in human males.

*Radon* is a type of radioactive gas emitted by soil or rock containing trace amounts of radium or uranium. Radon may enter the home via the underlying soil through cracks in the foundation or holes in the geologic deposits beneath the home. Ionized atoms adsorb to dust particles and may be inhaled. The amount of radon present in soil varies widely depending on one's geographic location. As an alpha-emitter, it primarily targets the lung. Despite availability of home detection kits and numerous lay press reports on radon risk, current scientific evidence for an association between residential radon exposure and lung cancer is weak.

Conversely, exposure to *SHS* has been identified as a risk factor for lung cancer for over three decades. Because side stream smoke is generated at lower temperatures and under different conditions than mainstream smoke, it contains even higher concentrations of several key toxins. At least 250 chemicals in SHS are known toxic or carcinogenic compounds; hence SHS is designated both as *a known human carcinogen* and as an *occupational carcinogen*. SHS smoke exposure during childhood and adolescence can increase the risk of developing lung cancer as much or more than exposure in adulthood.[12,13] Studies associating SHS exposure with cancer in pet animals are rather limited. One study showed increased risk for lung cancer in dogs was restricted to brachycephalic and mesocephalic breeds.[14] Conversely, reports of risk in dogs showed that dolichocephalic breeds were at increased risk for developing nasal cancer.[15,16] It was postulated that these trends reflect the tendency for carcinogens to be deposited in the nasal regions of dolichocephalic breeds, thereby preventing deposition into the lower respiratory tract. Unfortunately, increased nasal carcinogen deposition may in turn be associated with increased risk of nasal neoplasia.

## BIOLOGIC CONTAMINANTS

### Allergens

Ironically, for the purpose of this review, pets in general and specifically the airborne feline saliva allergen (Fel d I) are important indoor air allergens for humans. Although pet exposure can undoubtedly trigger or worsen allergy symptoms in sensitized individuals, recent data also suggests that pet exposure in early childhood may actually prevent the development of allergic sensitization and allergic diseases — including allergic rhinitis, asthma, and atopic dermatitis. The "protective pet effect" may be influenced by multiple factors including type of pet; timing, duration, and intensity of exposure; and genetic factors. Mechanisms underlying these observations remain under investigation but appear to include altered immune development, pet-specific tolerance, and stimulation of the innate immune system.[17,18] Data defining the role of allergens in feline or canine inflammatory airways disease has been limited, in large part by lack of commercially available cat- or dog-specific antibody assays. Still, a recent pilot study of 10 cats with or without feline bronchial disease demonstrated that in affected cats, intradermal skin testing (although not serum IgE levels) had significantly more positive reactions to house dust mites.[19]

### Dampness & mold

Due to the increasing in home-heating costs in the 1970's, residential dwellings are more efficiently insulated and "air-tight" than ever before. Decreased air flow through a dwelling, possibly combined with extensive use of wall-to-wall carpeting, promotes accumulation of select microbes, bacteria, fungi, and dust (including dust mites and their excreta) within the indoor air at concentrations higher than previously encountered. While the role of micro-organisms or their by-products in the worsening of atopic conditions is uncertain, their potential for inflicting infectious disease is epitomized by "Legionnaire's Disease" (which arose from microbial overgrowth within air conditioning systems). Moisture damage and consequent mold contamination (or mold-related volatile emission products) have been commonly reported in homes. Numerous studies have reported associations between dampness or mold with adverse health effects, implicating a variety of biologic mechanisms ranging from actual infections, to hypersensitivity or irritant reactions.[20,21] A recent report describing acute pulmonary hemorrhage in two cats exposed to toxic black mold[22] was remarkably similar to a cluster of cases in infants of the Cleveland area that also developed acute pulmonary hemorrhage. Most studies have not objectively measured mold growth or health outcomes. Furthermore, home dampness may be a proxy for mold growth; alternatively, dampness may also be a "marker" for increased levels of dust mites, endotoxin, or other factors that arise during conditions of reduced ventilation. A causal relation is biologically plausible given that indoor fungi produce allergens, toxins, and irritants. The US Institute of Medicine concluded that there was sufficient evidence of an association between dampness and adverse health outcomes; however, causality had not yet been established.[23]

## IMPROVING INDOOR AIR QUALITY

Simple preventative measures can help to minimize owner and pet exposure to air irritants and pollutants. Owners should be cautioned about use of ionizing air purifiers that increase indoor ozone concentrations well in excess of levels considered safe. CO monitors should be installed. Gas appliances should function properly. Gas stoves and heaters should be vented to the outside of the house. Gas ranges should *never* be used as heat sources. Kerosene heaters should be used carefully and only as per the manufacturer's instructions. Heaters should be refueled outdoors and only specified low-sulfur fuel should be used. Cars or lawn mower engines should never be left running inside a garage or shed. This is especially critical if the garage is ever used to house a pet.

Aerosol household products should be used only in well-ventilated areas. If cleaning products, remodeling materials, new paint or varnish, new or refinished furniture, new carpeting or draperies, etc., are associated with an offensive odor or seem to cause headaches and ocular irritation to the owners, the owners should ensure that neither they, nor their pet, maintain prolonged exposure with the fumes until the odors dissipate. If this is not possible, efforts to improve air ventilation are vital. Similarly, flea control dips, shampoos, and spray products should only be applied in well-ventilated areas and as directed. Water leaks and conditions contributing to excess moisture levels should be repaired and eliminated. Improved ventilation not only helps decrease humidity, it may also decrease accumulation of gases or chemicals arising from materials in the living space.

The above precautions may be especially important for puppies and kittens, geriatric dogs and cats, and pets with known respiratory or cardiac disease. Analogous to the human situation, is it reasonable to assume that dogs with chronic bronchitis or cats with bronchial disease are at increased risk of exacerbating their conditions with prolonged exposure to air pollutants. This is consistent with a recent study that noted increased feline asthma prevalence over the last 20 year in a large urban city.[24] Avoidance of exercising with pets outdoors during peak ozone or PM concentrations is deemed prudent. Perhaps the one precaution that would result in the most benefit to the pet as well as the owner(s) is for the owner to stop cigarette smoking altogether or at least restrict smoking to outside the home. Again, the Surgeon General has concluded that eliminating smoking in indoor spaces is the only way to fully protect nonsmokers from secondhand smoke exposure. (This information does not reflect USEPA policy.)

## REFERENCES

[1]US Department of HHS. The Health consequences of involuntary exposure to tobacco smoke. 2006. **www.surgeongeneral.gov/library/secondharndsmoke/.** [2]Bertone-Johnson ER, et.al. **Environ Res** 2008;106(3):361. [3]Bernstein JA, et.al. **J Allergy Clin Immunol** 2008;121(3):585. [4]Britigan N et.al. **J Air Waste Manag Assoc** 2006; 5:601. [5]Schelegle ES, et.al. **Toxicol Appl Pharmacol** 2003;191(1):74. [6]Ma L, et.al. **J Epidemiol** 2008;18(3): 97. [7]Ormstad H. **Toxicology** 2000;152:52. [8]Fisher JA, et.al. **Am J Respir Crit Care Med** 1999;159(4 Pt 1):1289. [9]Glickman LT, et.al. **J Tox Environ Health** 1989;28:407. [10]Wells RE. **J Am Vet Med Assoc** 1983;182:1248. [11]Glickman LT, et.al. **Environ Res** 1983;32:305. [12]Janerich DT, et.al. **N Eng J Med** 1990;323:632. [13]Asomaning K, et.al. **Lung Cancer** 2008;61(1):13 [14]Reif JS, et.al. **Am J Epid** 1992;135:234. [15]Hayes HM et.al. **Cornell Vet** 1982;72:168. [16]Reif JS, et.al. **Am J Epidemiol**. 1998;147(5):488. [17]Gern JE, et.al. **J Allergy Clin Immunol** 2004;113(2):307-14. [18]Bufford JD, et.al. **Curr Allergy Asthma Rep** 2007;7(5):37. [19]Moriello KA, et.al. **Vet Dermatol** 2007;18(2):94. [20]Hope AP, et.al. **Allergy Asthma Proc** 2007;28:262-70. [21]Institute of Medicine. Damp Indoor Spaces and Health. Washington DC: National **Acadamy of Sciences**: 2004. [22]Mader DR, et.al **J Am Vet Med Assoc** 2007;231:731-5. [23]The US Institute of Medicine's Committee on Damp Indoor Spaces and Health Committee Report, 2004. [24] Ranivand L and Otto C. **VCRS Proceedings** 2008.

# EFFECTS OF SECOND HAND CIGARETTE SMOKE ON RESPIRATORY HEALTH OF COMPANION ANIMALS.

**EA Rozanski, J Abrams,D Bedenice,**
**AM Hoffman Cummings School of Veterinary Medicine at Tufts University, North Grafton, MA.**

Exposure to environmental tobacco smoke [ETS] ("second-hand" smoking or passive smoking") represents a significant human health risk to people. As pets live in close proximity, the potential effects on their health is of interest as well. Several studies have been performed to evaluate the incidence of neoplasia in pets as well as to evaluate methods of monitoring objectively the degree of smoke exposure. Additionally, our laboratory sought to quantify the pulmonary functional changes in outwardly healthy dogs living with smokers.

Neoplasia risk: A weak relationship to passive exposure and lung cancer was found in dogs (Reif et al 1992). A convincing relationship, particularly in dolichiocephalic dogs, was found between ETS and nasal and paranasal cancers in dogs. (Reif et al 1998). In cats, lymphoma was increased in cats exposed to ETS, and there was a trend toward an increase in the rate of squamous cell carcinoma (Bertone et al).

Cotinine: Cotinine is a metabolite of nicotine and has been successfully measured in the urine of dogs and exposed to ETS. This is a useful objective method of quantifying the degree of exposure.

Lung function: Intra-cellular debris has been identified in the BAL of a cat from a smoking household. (Dye et al) Our laboratory, in testing dogs smoke exposed and non-smoke exposed, identified that following abnormalities Expiratory $S_{raw}$ was significantly higher in smoking dogs ($p=0.02$) and there was a trend ($p=0.06$) for a higher inspiratory Sraw in smoking dogs as well. The FRC/kg was significantly higher ($p < 0.01$) in NS dogs (75.7 ml/kg ± 18.5 ml/kg) in comparison to S dogs (48.2 ml/kg ± 9.5 ml/kg). In contrast, DLCO did not differ between groups ($DLCO_S$ = .22.44 ± 11.5 mL/min/mm Hg, $DLCO_{NS}$ = 22.08± 10.6 mL/min/mm Hg) . Additionally, the DL/VA was 6.7 ± 2.9 in NS and 7.0 ± 1.9 in S dogs, which was not statistically different. Arterial blood gas analysis documented a trivial reduction in PaO2 from 87 to 83 mmHg in the NS compared to the S group. However, neither the A-a gradient (27 in NS versus 29 in S), nor alveolar dead space (19% in S and 15% in NS) differed significantly between groups

These studies support there is a significant effect of ETS in companion animals and future investigation is warranted.

References Reif JS Am J Epidemiol 1992 135: 234-239. Reif JS et al. AM J Epidemiol 1998 147: 488-492. Bertone et al. Am J Epidemiol 2002 156: 268-273. Bertone et al. JVIM 2003 17: 557-562. McNiel et al AJVR 2007 68: 349-352, Bertone-Johnson ER et al Envirn Res 2008 106: 361-364. Dye et al. JVIM 1996 10: 385-400.

**COLD AIR AIRWAY DISEASE IN DOGS.**

**MS Davis.**
**Oklahoma State University Center for Veterinary Health Sciences, Stillwater**

Human athletes who perform strenuous exercise under cold conditions, such as cross-country skiers, hockey players, and figure skaters, have been shown to be at increased risk for chronic lower airway disease commonly referred to as "ski asthma". This disease syndrome shares many features with asthma, including acute bronchoconstriction in response to specific stimuli (in this case, exercise while breathing cold air), airway hyperreactivity to non-specific stimuli such as aerosolized histamine or methacholine, granulocytic airway inflammation, and airway wall remodeling. Despite demonstrably abnormal airway function, there appears to be minimal impact on exercise capacity between acute episodes of bronchoconstriction, with extremely high reported prevalence in elite, Olympic-caliber winter athletes. Laboratory models of "ski asthma", including a canine model employing segmental airway challenge through a wedge bronchoscope, have reproduced virtually all of the elements of "ski asthma", suggesting that these responses to cold air are conserved throughout athletic mammals. Therefore, we conducted a series of studies to test the hypothesis that dogs that exercise strenuously in cold conditions also had an asthma-like condition analogous to "ski asthma".

Studies were conducted using multiple populations of ultra-endurance racing sled dogs. Ultra-endurance events, such as Iditarod and the Yukon Quest, consist of teams of dogs competing over 1000 miles over 10-14 days, typically at subfreezing temperatures and frequently at temperatures as cold as -20 to -40°C. Sixty-nine dogs were examined within 48 hr after completing an 1100 mile exercise challenge in 10-12 days, and compared to a group of 12 control dogs which had received routine training, but had not exercised for at least 2 weeks. Evaluation consisted of physical examinations and bronchoscopy and, in 10 exercised dogs and all control dogs, bronchoalveolar lavage (BAL) was performed. A second set of studies was performed in which pulmonary mechanical properties and BAL was assessed in a group of dogs receiving routine ultra-endurance training after 4 months of detraining, and compared to untrained laboratory hounds and to age-matched cohorts housed in the same kennel but not receiving any training.

Most dogs completing the 1100 mile exercise challenge had abnormal accumulations of mucus in the trachea, and exercised dogs had higher BAL fluid concentrations of leukocytes compared to trained, unexercised dogs. Sled dogs had higher concentrations of BAL leukocytes and greater pulmonary reactivity to aerosolized histamine compared to laboratory hounds, but these values were unaffected by routine endurance training.

These data support the hypothesis that chronic exposure of airways to cold air during exercise can create an asthma-like syndrome, without an overt impact on exercise capacity. Once established, this syndrome does not completely resolve with 4 months of rest in temperate conditions.

**IMMUNOMODULATION WIITH ALLERGEN-SPECIFIC IMMUNOTHERAPY IN FELINE ASTHMA.**

**CR Reinero, T Lee-Fowler, H Liu, AE DeClue, LA Cohn. University of Missouri, College of Veterinary Medicine, Columbia, MO.**

Allergen-specific immunotherapy (ASIT) diminishes eosinophilic airway inflammation in experimental feline allergic asthma. It is thought that ASIT induces immune tolerance to the sensitizing allergen which may normalize (increase) IL-10 production and suppress allergen-specific T cell responses. The purpose of this study was to evaluate alterations in IL-10 producing cells and allergen-specific proliferation of lymphocytes with ASIT in experimental feline asthma.

Cats were randomly assigned to have an asthmatic phenotype induced by Bermuda grass allergen (BGA), house dust mite allergen (HDMA) or both; or they were mock sensitized. Abbreviated ASIT (=rush, RIT) using BGA was performed in all cats. Aerosol challenges with BGA, HDMA, both or saline (according to group assignment) were administered weekly for 6 months. Peripheral blood mononuclear cells (PBMCs) were harvested prior to and after 6 months of BGA RIT for use in an IL-10 ELISPOT (RnD Systems). Also after 6 months of RIT, PBMCs were incubated in vitro in media alone or media with BGA or HDMA for use in a lymphocyte blastogenesis assay (Roche Diagnostics). Lymphocytic proliferative responses were reported as stimulation indices (SI) for BGA and HDMA. Biologically significant values were considered when the SI>2. As this study is ongoing, not all data are available for each cat at each time point.

In asthmatic cats prior to BGA RIT, the group mean±SEM number of IL-10 producing cells in cats sensitized to a single or to dual allergens was 138±23 and 115±39, respectively. After BGA RIT, the number of IL-10 producing cells increased only in cats sensitized with the same allergen used in RIT: 255±52, BGA sensitized cats; 122±29 HDMA sensitized cats (for comparison, control non-asthmatic cats had 273±70 IL-10 producing cells). Results of the blastogenesis assay demonstrated that prior to RIT, the group mean $SI_{BGA} > 2$ only in BGA sensitized cats ($SI_{BGA} = 2.1$; supportive of lymphocyte recognition of this allergen only with prior sensitization). At months 3 and 6 of BGA RIT, the group mean $SI_{BGA} < 2$ for BGA sensitized and mock sensitized cats (implying lymphocyte hyporesponsiveness to BGA); however, the $SI_{BGA}$ in the HDMA sensitized group was 3.1 and 2.1 at these time points (suggesting BGA was now recognized by these lymphocytes). Additionally, the $SI_{HDMA}$ was >2 at all time points in only the HDMA sensitized cats but decreased with RIT (pre-BGA RIT, 2.5; month 3, 2.0; month 6, 2.0).

In conclusion, RIT appeared to increase IL-10 producing cells and dampen T cell responsiveness in an allergen-specific manner. Cats sensitized with an allergen (HDMA) unrelated to the allergen used in RIT (BGA) did not have increases in IL-10 producing cells, but did respond by proliferating to BGA in vitro; interestingly these cats also had decreases in HDMA-specific proliferation over time. Thus, there may be "cross-talk" within the immune system with unrelated allergens in ASIT.

## PATHOPHYSIOLOGY OF NON-INFECTIOUS LOWER AIRWAY INFLAMMATION

**Nathalie Kirschvink, DVM, PhD**
**University of Namur, Belgium**

### INTRODUCTION

Acute, but most often chronic non-infectious lower airway inflammation occurs in human patients (asthma, COPD, …) as well as in domestic animal species. Among the most investigated diseases figure asthma and chronic bronchitis in cats, eosinophilic bronchopulmonary disease and "ski asthma" in dogs and recurrent airway obstruction (RAO) or heaves in horses.

Independently of the origin of lower airway inflammation, the pathophysiologic mechanisms leading to acute and chronic lung dysfunction are similar. Clinical repercussions however depend on species-related particularities of lung function and the ideal methods of assessment of inflammatory changes also depend on the animal species. As chronic airway inflammation leads to airway remodeling and irreversible lung dysfunction, early detection of inflammatory changes is of major importance (Figure 1).

Figure 1

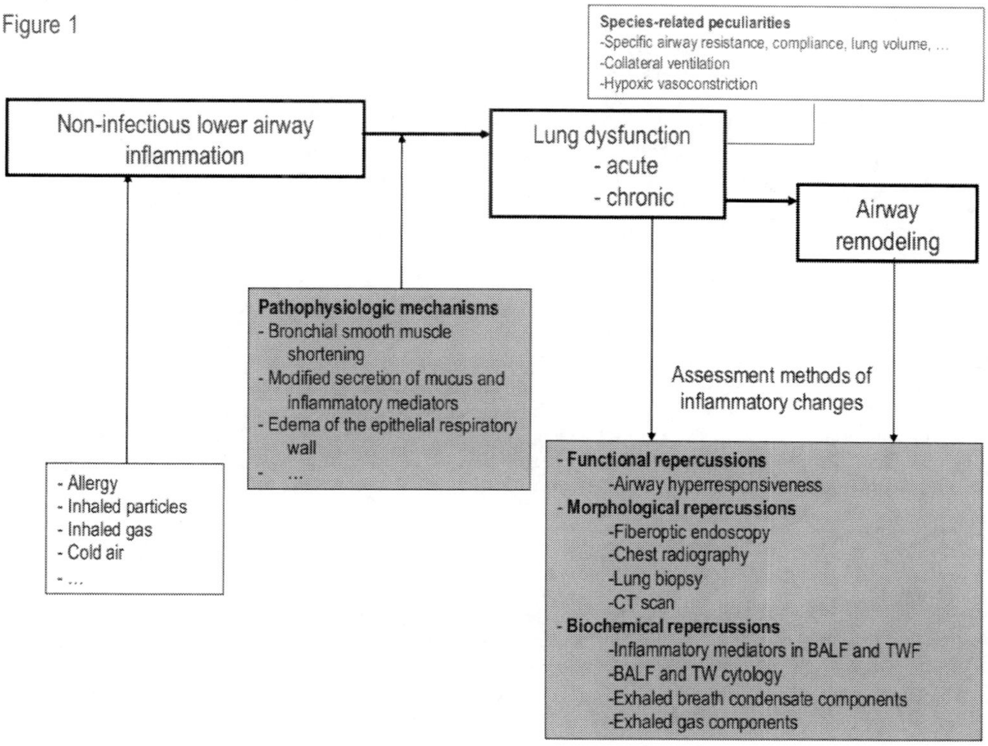


### NON INFECTIOUS LOWER AIRWAY INFLAMMATION AND ITS IMPACT ON LUNG FUNCTION

**Pathophysiologic mechanisms leading to lung dysfunction**

Inflammation of the lower airway may induce lung dysfunction by three essential pathways: the most important factor leading to airway narrowing is bronchial smooth muscle shortening, followed by modified secretion of mucus and modification of the epithelial lining fluid of the airways as well as edema of the mucosa and submucosa of the bronchial walls. The functional impact of each component may depend on the severity of the inflammatory process and on the relative importance of each pathway. Identification of the main pathway(s) and origins of lower airway inflammation will influence the choice of medical treatment[1].

**Species-specific peculiarities of lung (dys)function**

Among species-specific peculiarities of lung function figure the differences of pulmonary variables such as specific tidal volume, compliance, resistance etc (Table 1), the differences regarding collateral ventilation and the capacity of hypoxic vasoconstriction (for review see [2]).

**Specificities of pulmonary function values**

**Table 1 :** Comparison of specific pulmonary function values between adult Belgian White and Blue (BWB) cattle, adult Friesian cattle, adult horse, man, dog and cat[3].

| Variable | Unit | BWB | Friesian | Horse | Man | Dog | Cat |
|---|---|---|---|---|---|---|---|
| $sV_t$ | ml/kg | 6.9 | 8.9 | 11 | 9.2 | 8.7 | 8.1 |
| $sV_e$ | l.m²/min | 25 | 25 | 23 | 6.6 | 5.3 | 4.3 |
| $sR_L$ | kPa | 8.5 | 7 | 3.6 | 1.5 | 1.3 | 1.2 |
| $sC_{Dyn}$ | l.kPa/kg | 2.2 | 2.4 | 4.3 | 2.8 | 2.4 | 2.1 |
| sWvis | kPa | 0.74 | 0.47 | 0.46 | 0.34 | 0.24 | 0.24 |

$sV_t$ : specific value of tidal volume; $sV_e$ : specific value of minute volume; $sR_L$ : specific value of total pulmonary resistance; $sC_{Dyn}$ : specific value of dynamic lung compliance; sWvis : specific value for minute work of breathing.

**Collateral ventilation**

Within an anatomical region of lung parenchyma, ventilation normally occurs through a given bronchus *via* the standard airway branching pattern from larger to smaller airways. In species with collateral airways, collateral ventilation refers to any ventilation to this given region that arrives from a neighboring airway. In species lacking for collateral ventilation, atelectases frequently occur in any case of airway obstruction. Three possible pathways for collateral ventilation have been considered in the human lung: (i) epithelialized tubular communications through the bronchiolar wall to adjacent alveoli (i.e. bronchiole-alveolar communications or 'channels of Lambert'), (ii) alveolar pores ('pores of Kohn'), and (iii) interbronchial connections that occur at the level of the respiratory bronchioli and/or connections near the alveolar duct ('channels of Martin')[4]. Bronchiole-alveolar communications are present in lungs of cats and sheep, but also in rabbits. Alveolar pores may be found in all the common laboratory mammals and man. Channels of Martin have been found in canine lungs first[4]. Due to their size of 120 to 200 µm, they are thought to be the primary pathways for collateral ventilation connecting normal airways at the level of the alveolar duct[3].

The presence of collateral ventilation has been proven in dogs, cats, rabbit, ferret, sheep and horses[4]. Consequently, collateral ventilation may serve partially to bypass airway obstruction in these animal species. However, differences exist among species and those differences have clear physiological significance. In dogs, cats, rabbits and ferrets, collateral ventilation is greater compared to sheep or horses. Cat lungs are probably similar to dog lungs. In sheep, a significant decrease in collateral resistance (i.e. improvement of collateral ventilation) was observed with maturation indicating that collateral resistance may change with age. Also in dogs, the number of interalveolar pores increases with age. Despite the presence of alveolar pores the pig has no collateral ventilation[4]. The same is true for the bovine lung. Since pigs and cattle lack normal pathways for collateral airflow, atelectases frequently occur in bovine and porcine lungs in any obstructive condition. In consequence (since atelectatic lobes cannot be inflated by air through collateral airways), pigs and cattle easily develop ventilatory asynchronisms or regional inhomogeneities in alveolar ventilation. On the other hand, both the lack of collateral ventilation and the presence of connective tissue septa between lobules often limit inflammatory processes to a lobule in these species. Consequently, consolidated lobules may be adjacent to healthy lobules within the same lobe.

**Hypoxic vasoconstriction**

A consequence of alveolar hypoventilation may be hypoxic vasoconstriction, aiming to improve the ventilation/perfusion ratio in case of atelectases. The vascular smooth muscle of pulmonary arterioles is particularly well developed in bovines and swine, which can lead to important vascular responses and pulmonary hypertension during persistent hypoxia. Although this mechanism of blood redistribution is beneficial in case of localized alveolar hypoxia, it may have serious consequences on right heart work when hypoxia is generalized, such as in animals with severe diffuse lung disease or in animals living at high altitude (i.e. "brisket disease" in cattle)[5].

## ASSESSMENT OF LOWER AIRWAY INFLAMMATION: FUNCTIONAL, MORPHOLOGICAL & BIOCHEMICAL APPROACH

It is important to consider the different consequences of airway inflammation if an appropriate treatment has to be chosen. As airway inflammation can lead to functional, morphological and biochemical changes within the respiratory tract, its quantification might include several approaches.

**Functional approach of inflammation: hyperresponsiveness**

Bronchial hyperresponsiveness is defined as an abnormal sensitivity and exaggerated response of the airways to pharmacological, chemical, physical or physiological stimuli. Airway luminal diameter, smooth muscle mass, elastic recoil, airway inflammation, epithelial injury and neural activity may all play a role in hyperresponsiveness[6]. Remodeling-associated hyperresponsiveness can be suspected in case of treatment-resistant and irreversible hyperresponsiveness. Numerous lung function tests are available and might be adapted to different animal species, allowing assessment of bronchial reactivity in clinical situations as well as in respiratory research.

**Morphological approach of inflammation by diagnostic imaging**

Diagnostic imaging can be performed using different techniques and is complementary to functional or biochemical evaluation, especially when chronic inflammation is leading to airway remodeling. A classical but not very sensitive method is thoracic radiography. This technique is mainly used for clinical purposes and bears the advantage of being applicable to large and small animal species. Pulmonary computed tomography (CT) provides qualitative information about large lung fields and offers

the possibility of quantifying structural changes. CT is widely used in human respiratory medicine and research, whereas its application in veterinary medicine or animal respiratory research is still limited to some animal species.
The evaluation of the large airways' aspect by use of bronchoscopy provides valuable information about mucosal aspect and mucus production. However, this technique can only be used in animals of a certain size. Bronchoalveolar lavage fluid, bronchial biopsies or brushings might be sampled using a bronchoscope, which provides insights about cytological and histological changes of the airways. In horses, histological changes of peripheral airways and lung tissue might be evaluated using thoracoscopy-guided biopsies.

**Biochemical approach of inflammation: Assessment of biomarkers**

Inflammation- and remodeling-associated biomarkers can be determined in bronchoalveolar lavage fluid, tracheal washes, biopsy samples, induced sputum (in humans), exhaled air and exhaled breath condensate. In some particular cases, blood markers associated with a pulmonary process might also be used.
The choice of the matrix used for biomarker determination depends on the animal species and the marker selected. All sampling techniques are potentially applicable in large animals, whereas the reduced size might become a limiting factor in small animals. The need for anesthesia in small animals is a further limiting factor, especially with regard to repeated investigations. The sampling technique used further depends on the chemical properties (molecular weight, stability, solubility,…) and the synthesis and secretion of the biomarker.

Among the biomarkers that are frequently assessed figure cytokines, prostaglandins, leukotriens, growth factors, transcription factors, volatile molecules (NO, ethane), H2O2, pH, metalloproteinases,…. [1,6].The equilibrium between matrix metalloproteinases (MMP) and their endogenous inhibitors (TIMP) seems to play a particularly important role for the development of airway inflammation and remodeling and is therefore investigated in human and animal respiratory research[7].

## CONCLUSION

Appropriate treatment of non-infectious lower airway inflammation is based on the identification of the major components leading to inflammation and lung dysfunction. Species-dependent peculiarities of lung function also impacts on the manifestations and clinical repercussions of airway inflammation. Assessment of airway inflammation should consider more than a biochemical or cytological approach.

## REFERENCES

[1]Bousquet J, et al. **Am J Respir Crit Care Med** 2000;161:1720 [2]Kirschvink N and Reinhold P.**Curr Drug Targets** 2008; 9(6):470 [3]Gustin P et al., **Res.Vet.Sci**. 1988; 45: 405 [4]Mitzner, W. (1991) THE LUNG: Scientific Foundations, (Crystal, R.G.; West, J.B. et al., Eds.) Raven Press, Ltd., New York, Chapter 5.1.2.7., pp. 1053 [5]Reinhold and Hoechel **Berl Munch.Tierarztl.Wochenschr.** 2005; 118: 52 [6]Rutgers S.R. et al.. **Clin Exp Allergy** 2001;31:193 [7]Lagente, V., et al. **Brazilian J Med Biol Res** 2005:38:1521.

## COLD AIR AIRWAY DISEASE IN HORSES.

**MS Davis.**
**Oklahoma State University Center for Veterinary Health Sciences, Stillwater**

Human athletes who perform strenuous exercise under cold conditions, such as cross-country skiers, hockey players, and figure skaters, have been shown to be at increased risk for chronic lower airway disease commonly referred to as "ski asthma". This disease syndrome shares many features with asthma, including acute bronchoconstriction in response to specific stimuli (in this case, exercise while breathing cold air), airway hyperreactivity to non-specific stimuli such as aerosolized histamine or methacholine, granulocytic airway inflammation, and airway wall remodeling. Despite demonstrably abnormal airway function, there appears to be minimal impact on exercise capacity between acute episodes of bronchoconstriction, with extremely high reported prevalence in elite, Olympic-caliber winter athletes. The mechanism of initiation of "ski asthma" is excessive water loss from the lower airway surfaces due to penetration of unconditioned air, and is related to the combination of high inspiratory airflows and the ambient air temperatures. Therefore, we conducted a series of studies to test the hypothesis that a condition analogous to "ski asthma" could be triggered in horses by a single exercise challenge while breathing cold air.

Pilot studies were conducted to establish treadmill exercise conditions that would result in the exposure of equine lower airways to unconditioned air without the confounding effects of exercise-induced pulmonary hemorrhage. All study designs consisted of a controlled exercise session (5 min canter at a heart rate of 170 bpm) while breathing either room temperature (25oC) or chilled air (-5oC), administered in a randomized cross-over design with a 2 week washout between exercise sessions. Bronchoalveolar lavage (BAL) was collected 5, 24, and 48 hr after separate exercise sessions, and pulmonary mechanical properties were measured at the same timepoints after a single exercise session. In addition to standard BAL fluid cellular evaluation, selected cytokines in the homogenized BAL fluid cell pellet were measured using RT-PCR. In all instances, the results from the cold air exercise were compared to those obtained following warm air exercise.

BAL fluid collected 5 hr after exercise while breathing cold air had increased concentration of airway epithelial cells, demonstrating mild airway injury, and increased expression of cytokines representative of a Th2 phenotype. The Th2 cytokine phenotype persisted through 24 hr post-exercise, at which time there was increased expression of pro-inflammatory cytokines and increased BAL fluid concentration of neutrophils. Total pulmonary resistance gradually increased following exercise while breathing cold air, reaching statistical significance by 48 hr post-exercise.

These data support the hypothesis that a single episode of airway exposure to cold air during exercise can create a mild asthma-like syndrome, and may alter airway immune function for up to 48 hr. Further studies are needed to determine whether REPEATED exposure, typical of routine training, can result in clinically-detectable airway abnormalities.

## AIR QUALITY IN BARNS.

**Melissa R.Mazan, DVM, Diplomate ACVIM, Jessica Svatek, DVM, Louise Maranda, DVM, PhD, David Christiani, MD, Andrew Ghio, MD, Jenifer Nadeau PhD, Andrew M. Hoffman, DVM, DVSc, DACVIM.**

Background: People working in cattle, swine and poultry barns have a higher prevalence of respiratory symptoms and decreased lung function. There is scant evidence regarding the respiratory health of humans working in horse barns, although it iswell documented that stabled horses have a high prevalence of airway disease.

Aim: To determine whether people spending time in horse barns have a higher prevalence of self-reported respiratory symptoms than non-exposed controls.

Methods: A cross-sectional questionnaire study was conducted from May 2005 to January 2006 to investigate the prevalence of self-reported respiratory symptoms in 82 barn-exposed subjects and 74 control subjects. Logistic regression and the chi-square test were used to analyse the data.

Results: There was a significantly higher prevalence of self-reported respiratory symptoms in the barn-exposed group (50%) versus the control group (15%). Exposure to horse barns, smoking and family history of asthma or allergies was independent risk factors for respiratory symptoms. High exposure to the horse
barn yielded a higher odds ratio for self-reported respiratory symptoms (8.9).

Conclusions: Exposure to the equine barn is a risk factor for respiratory symptoms. Investigation of organic dust exposures, lung function and horse dander allergies in the barn-exposed group will be necessary to determine how best to protect the health of this group.

# EVIDENCE-BASED TREATMENT OF NON-INFECTIOUS AIRWAY DISEASE

**Melissa R. Mazan, DVM, DACVIM**
**Grafton, MA**

Inflammatory airway disease (IAD) is reported to be the second most common cause of wastage in racehorses and affects up to 80% of stabled horses. Although it has been well-established that IAD is not associated with bacterial infection, many horses continue to treated with antimicrobials despite lack of evidence for bacterial infection such as pyrexia, fever, or depression. Indeed, a plethora of treatments are reported for IAD, with varying degrees of scientific evidence supporting their use. The purpose of this paper is to review this evidence in order to determine a rational basis for treatment of IAD.

The first and most important treatment – environmental remediation – is also the most long lasting and free of side effects, but many owners are reluctant to fully pursue this path. The following evidence may serve as greater impetus to do so. Multiple studies have shown that environmental remediation, usually meaning pasture with pelleted complete feed and complete cessation of hay feeding, results in almost complete remission from RAO. If pasture is not available, then cardboard bedding accompanied by complete pelleted feed is a suitable alternative. A randomized controlled study demonstrated that a combination of complete pelleted feed and outdoor living resulted in remission in heavey horses, using PFTs and BAL to confirm. Only in severe cases was the addition of inhaled fluticasone propionate necessary. Subsequently, this group has shown that RAO horses kept in low dust environments with no medical management for up to 5-6 years are indistinguishable from a control group without probing the lower airways via measurements of FEV.

Standard pharmacologic treatment of non-infectious airway disease in horses usually includes corticosteroids, which may be administered parenterally or via inhalation. Efficacy of inhaled medication depends on both the drug and the inhalational device. The newest device, the Aerohippus, has greater efficiency than others, with up to 21% of dose delivered to the lower lung. Inhaled fluticasone propionate (2200 ug BID) results in complete resolution of clinical signs, normalization of PFTs, and decrease in BAL neutrophilia, along with a significant decrease in mRNA expression of IL-4 and increase in IFN-gamma. Using BAL but not PFTs, the efficacy of fluticasone propionate has also been demonstrated in a large-scale randomized trial in young racing horses over a 14-month period. The addition of the anti-cholinergic inhaled drug ipratropium bromide did not improve the outcome. Finally, a double blind placebo-controlled, cross-over design in heavey horses that FP twice daily normalized response to histamine challenge. One study has shown that parenteral dexamethasone was more effective in heavey horses than was inhaled beclomethasone dipropionate. Although 3750 ug BID of inhaled beclomethasone caused suppression of adrenocortical function similar to dexamethasone, a lower dose (500 ug) did not, and had similar effect. Parenteral corticosteroids may be preferable for short-term treatment for reasons of cost or owner compliance. The most commonly used drugs are dexamethasone and prednisolone. A recent study showed using BAL and endoscopy that dexamethasone (0.1 mg/kg IM) worked more rapidly than prednisolone (1.0 mg/kg), but that prednisolone effects were longer lasting. Occasionally, triamcinolone is used for its longer duration of action, and its ability to improve PFTs and BAL cytologic features has been shown.

Bronchodilators are frequently used for rapid effect; the majority are inhaled, however, the only FDA approved bronchodilator is the parenterally administered beta-2 agonist, clenbuterol. Although one study has shown that a mid-level dose of clenbuterol given chronically is deleterious to aerobic performance in the horse, and has adverse effects on cardiac performance, recent work shows that the combination of a corticosteroid and IV clenbuterol causes a decrease in BAL neutrophils, and IL-1beta, although only FP was able to cause a decrease in IL-8. Beta-2 agonists are more commonly given via metered dose inhaler. One of the earliest-studied of these drugs was pirbuterol, which was shown to give 30 minutes of bronchodilation at a dose of 800 ug; higher doses resulted in trembling and excitement. Albuterol sulfate has been shown to have a slightly longer duration of action (30 min to 2 hours) at a dose of 360 ug. There have been questions raised in the human literature concerning the safety of chronic administration of albuterol. A blinded, placebo-controlled, crossover study showed that chronic (2-week) treatment with inhaled albuterol does not result in diminished bronchodilatory response. Multiple studies have shown that albuterol does not affect aerobic performance.

Anticholinergics are also used to affect bronchodilation. The most commonly used is ipratropium bromide, a quaternary amine of atropine, which has very little crossover into systemic circulation. Multiple studies have shown the efficacy of ipratropium, lasting approximately 6 hours, both by nebulization and by dry powder. A less well-supported treatment is the mast cell stabilizer, such as sodium cromoglycate or nodocromil sodium. Although an initial study showed that heaves-affected horses, when nebulized once daily for 1-4 days, remained free of symptoms for up to three weeks, a previous study contradicted this in heaves horses nebulized with up to 500 mg for 2 days before barn exposure.

Xanthine derivatives have received some attention over the years, However, most recently the combination of theophylline and dexamethasone was shown NOT to be useful in treating heaves.

Acute human asthma is sometimes treated with intravenously administered magnesium, and oral supplementation is given for prevention of further attacks. Recently one group has found that mid-level magnesium (640 mg) given with inhaled albuterol resulted in longer lasting improvement both in PFTs and clinical signs.

Many owners are anxious to try herbal remedies with the thought that they may have fewer side effects than other drugs. There has been very little research into the efficacy of herbal remedies; most is anecdotal. However, a recent report showed no effect of an herbal composite in alleviating respiratory dysfunction in RAO. Similarly, owners frequently would like to use acupuncture to alleviate clinical signs of heaves. However, a recent study has shown that there is no specific effect of a single acupuncture treatment in horses with severe RAO.

Although there is good rationale for phosphodiesterase inhibitors to be efficacious in treating RAO or IAD in horses, no improvement was seen in PFT or inflammation after 14 days treatment.

There is some debate as to whether antioxidants are of use in treating lower airway disease in the horse. Heavey horses treated with vitamins E and C and selenium had improved exercise tolerance using standard treadmill tests, as well as improved inflammatory score on endoscopy. In contrast, although control horses had an increase after exercise in systemic ascorbic acid after nutritional supplementation, heavey horses did not.

Still categorized as experimental, a MAPkinase –p38 inhibitor has recently been shown to help delay onset of symptoms and reduction in pulmonary neutrophilia in RAO horses without improving lung function.

Hyposensitization has long had anecdotal support as treatment for IAD and RAO, but the evidence remains purely anecdotal. RAO is not an immediate hypersensitivity, thus not entirely suitable theoretically for hyposensitization.

There is some good evidence, on the other hand, that immunomodulation may have limited efficacy. Although there appears to be no systemic absorption of low dose human interferon alpha, it is reasonable that it may work through interactions with the oropharyngeal lymphoid tissue. Using a low dose PO (50-150 IU per horse), one study reported less exudate in the respiratory tract, lower cell counts in BAL, and conversion to a non-inflammatory BAL cytologic picture. In support of these findings, a double-blind randomized clinical trial with 34 horses found that there was a much lower relapse rate (17/22 did not relapse v 8/12 treated with saline did relapse) when horses were treated with human interferon alpha. Finally, caprine serum has been used for immunomodulation of lower airway disease. This study, unlike those cited above, used neither BAL nor pulmonary function tests for assessment, so must be viewed as delivering less robust outcomes measurements.

In the end, however, the majority of studies, albeit with excellent and robust measures of disease such as pulmonary function tests, BAL, endoscopy, and clinical scoring, involve few horses. Most crossover studies involve no more than 5-6 horses. Studies seen as large scale seldom involve more than 30 horses. Horse studies are, unfortunately, expensive and cumbersome; these factors limit the findings. The best data support the use of environmental control (i.e., low dust surroundings), corticosteroids, both inhaled and parenteral, and limited use of bronchodilators.

## REFERENCES


[1.] Weese JS et al **Can Vet J.** 2005 May;46(5):438-9. [2.] Derksen F, et al **Am J Vet Res**. 1987 Mar;48(3):423-6. [3.] Vandenput S, et al **Vet Q.** 1997 Nov;19(4):154-8. [4.] Vandenput 1997, [5.] Kirschvink N, et al **Vet J.** 2002 May;163(3):319-25.[6.] Couetil LL, et al **Am J Vet Res**. 2005 Oct;66(10):1665-74[7.] Miskovic M, et al **J Vet Intern Med**. 2007 Sep-Oct;21(5):1060-6.[8.] Giguere S, et al **Vet Immunol Immunopathol**. 2002 Mar;85(3-4):147-58.[9.] Spendlove PJ **Proceedings of the VCRS**, 2006, Jena Germany[10.] Viel L et al, **Proceedings of the AAEP** 1999 Albuquerque New Mexico[11.] Rush BR et al **Am J Vet Res**. 1998 Aug;59(8):1039-43.[12.] Courouce-Malblanc A, et al **Vet J.** 2008 Feb;175(2):227-33. Epub 2007 Mar 23.[13.] Lapointe JM, et al **Am J Vet Res**. 1993 Aug;54(8):1310-6.[14.] Kearns CF et al, **Vet J.** 2008 Oct 15. [Epub ahead of print][15.] Laan TT, et al **Vet Res Commun**. 2006 Aug;30(6):623-35.[16.] Derksen FJ, et al **Am J Vet Res**. 1999 Jun;60(6):689-93.[17.] Mazan MR, et al, **Proceedings of the Annual Convention of the AAEP**, 2005, San Antonio, Texas[18.] Mazan MR, et al **Am J Vet Res.** 1998 Feb;59(2):176-81[19.] Bailey, et al **Bailey J, Colahan P, Kubilis** P, Pablo L.[20.] Bayly WM, et al **Equine Vet J.** 2001 May;33(3):302-10.[21.] Robinson NE , et al **Equine Vet J**. 1993 Jul;25(4):299-303.[22.] Duvivier DH, et al **Equine Vet J.** 1999 Jan;31(1):20-4.[23.] Hare JE et al, **J Vet Pharmacol Ther.** 1994 Jun;17(3):237-44.[24.] Soma LR, et al **Vet Res Commun**. 1987;11(4):339-51.[25.] Cesarine C, Hamilton E, Pcandet V, Lavoie JP. Efficacy of theophylline associated to a low dose of dexamethasone in the treatment of horses with heaves. **VCRS** 2004 Montreal[26.] V., Beuchner Maxwell U christmann, S Witonsky Nebulized mag and albuterol: a novel treatment for RAO. **VCRS** 2007}[27.] Wilson DV et al, **Equine Vet J.** 2004 Sep;36(6):489-94[28.] Lavoie JP et al. **JVIM** 2006 175-181[29.] Kirschvink N et al. **Equine Vet J.** 2002 Nov;34(7):705-12.[30.] Cuff TL et al **Proceedings of the VCRS** 2003[31.] Lavoie JP et al. **Proceedings of the VCRS** 2004[32.] Polikepahad S, et al **Can J Vet Res**. 2008 Jul;72(4):340-9.[33.] McGorum BC, et al **Vet Immunol Immunopathol**, 1993;36:223-237[34.] Hamm D, et al. **Eq Vet J** 2002, 71-75[35.] Rush MJoore BR, et al. **AJVR** 1996, 49, 347-358.[36.] Rush Moore BR et al **AJVR** 1995 1995;56:562-567.[37.] Moore ,I et al,. **Can Vet J** 2004 594-60.

# EFFECT OF ANTIGEN CHALLENGE AND WITHDRAWAL ON AIRWAY SMOOTH MUSCLE REMODELING IN HEAVES.

M. Leclère[1*], A. L.-Lamoureaux, J. G. Martin[2], J.-P. Lavoie[1]
Faculty of Veterinary Medicine, Université de Montréal, St-Hyacinthe, Qc.
Meakins Christie Laboratories, Department of Medicine, McGill University, Montréal, Qc.
Faculty of Veterinary Medicine, Université de Montréal, 3200 Sicotte, Saint-Hyacinthe, Québec, Canada.

Increased airway smooth muscle (ASM) mass may contribute to airway obstruction in heaves. It is unknown if this remodeling is reversible under natural conditions once it is established. In this study, peripheral ASM was measured in heaves-affected horses after a 3-month period of antigen withdrawal and following a 30-day challenge.

Eleven adult horses (6 heaves-affected and 5 controls) were kept at pasture for a minimum of 3 months. They were then stabled and fed hay for 30 days. Diseased horses had been intermittently symptomatic for at least 3 years prior to the study. Lung biopsies were obtained under thoracoscopic guidance after antigen withdrawal (remission) and after antigen challenge (exacerbation). Smooth muscle was identified by immunohistochemical staining for smooth muscle specific alpha-actin and its area in the airway wall was measured and corrected for airway size using standard morphometric techniques.

All horses had normal lung function at pasture and only heaves-affected horses developed airway obstruction after antigen exposure. Compared to controls, diseased horses had more than twice the amount of ASM in their peripheral airways while in remission ($p \leq 0.01$). There was no further significant increase after the 30-day challenge period.

Results of this study indicate that ASM remodeling, once established, may not be reversible by antigen avoidance. Furthermore, no notable increase in the ASM mass was observed after a month of continuous challenge, suggesting that increase in ASM may reach a plateau, possibly through a balance between proliferation and apoptosis. These results also provide the basis for a long-term longitudinal study of smooth muscle remodeling reversibility.

**COMPARATIVE EFFICACY OF THREE COMMON TREATMENTS FOR EQUINE RECURRENT AIRWAY OBSTRUCTION.**

**L C Lee, VA Buechner-Maxwell, U Christmann, S. Witonsky, K. Zimmermann, S Werre. Virginia-Maryland Regional College of Veterinary Medicine, Blacksburg, Virginia.**

Equine Recurrent Airway Obstruction (RAO) is characterized by airway inflammation, bronchoconstriction, and mucus plugging. Acute RAO is often treated with a combination of environmental management and tapering doses of glucocorticoids. While empirical evidence suggests this is an efficacious therapy, there is little scientific data to support this conclusion. Using a three-way cross-over design, we evaluated the response of 6 acutely affected RAO horses to three treatment regimes: tapering doses of the glucocorticoid, dexamethasone (dex), environmental modification (env), and a combination of both treatments (dex-env). Clinical parameters, pulmonary function and bronchoalveolar cytology were measured before initiation of treatment and throughout the 21 day treatment period. A mixed-effects repeated measure of analysis of variance was used to test the main effects of severity of RAO and treatment as well as their interactions. All treatments initially had a beneficial effect on clinical parameters and pulmonary function in association with a reduction in airway inflammation. However, as frequency and amount of medication was reduced, the horses in the dex group experienced a significant rise in airway inflammation in the absence of clinical disease. In contrast, horses in the env group as well as the dex-env group demonstrated a reduction in clinical compromise and airway inflammation that that was maintained throughout the 21 day experimental period. These results verify the importance of environmental modification in the management of RAO. When environmental modification is not included as part of the treatment regime, horses are prone to recrudescence of airway inflammation in the absence of clinical disease.

**ANATOMICAL LOCALIZATION OF RESPIRATORY DISTRESS**

**Dr. Kate Hopper BVSc, PhD, DACVECC**
**Davis, CA**

## INTRODUCTION

Respiratory distress is a common reason for emergency presentation and is characterized by an increased respiratory rate and effort. Animals are often uncomfortable and restless; they may extend their neck and have increased respiratory noise. Respiratory distress patients can be extremely challenging as without the correct, specific therapy they may rapidly deteriorate and die but the very nature of their disease limits the feasibility of many diagnostic options. This discussion presents eight major mechanisms for increased respiratory rate and effort, the approach to differentiation of each problem in clinical patients and their specific therapeutic principles.

The causes of increased respiratory rate and effort can be divided into eight broad categories;

1. Upper airway disease
2. Lower airway disease
3. Pulmonary parenchymal disease
4. Pleural space disease
5. Chest wall disease
6. Pulmonary thromboembolism
7. Abdominal distension
8. "Look alike" diseases

### General Principles

All animals believed to have respiratory difficulties should receive oxygen therapy immediately. This should be continued until it is established that the patient no longer requires it. Emergency stabilization principles of A-B-C should be adhered to. Limiting stress in the respiratory distress patient is essential. Diagnostic procedures such as radiographs should be delayed until the animal is considered stable and in many cases therapeutic intervention may need to be performed without a confirmed diagnosis.

### Upper Airway Disease

The upper airways include the nasal passages, larynx, pharynx, trachea and mainstem bronchi. Upper airway obstructions will cause increased respiratory noise and effort. The noise can be described as stertor (low, snoring sounds) or stridor (high pitched, musical sounds). Because the diameter of the extrathoracic airway reduces on inspiration, extrathoracic airway obstructions will cause respiratory distress on inspiration.[1] Although fixed extrathoracic airway obstructions commonly cause respiratory distress on both inspiration and exhalation. In contrast, intra-thoracic airway obstructions will cause noise and difficulty on exhalation. Complete large airway obstructions will be associated with an absence of respiratory noise and marked respiratory efforts which will quickly progress to respiratory arrest. Causes of obstruction include foreign bodies, mass lesions, laryngeal paralysis, laryngeal edema, inflammatory lesions, abscesses, tracheal collapse and brachycephalic syndrome.

Initial diagnosis of an upper airway problem is based on the clinical signs and response to treatment. The loud respiratory signs are often easily recognizable as upper airway in origin. Specific diagnosis generally requires direct visualization under anesthesia using laryngoscopy, bronchoscopy, etc. Thoracic radiographs can be normal or may show signs of pulmonary parenchymal disease such as aspiration pneumonia or non-cardiogenic pulmonary edema. Radiographs of the extra-thoracic airway can be helpful but are prone to artifacts secondary to positioning and may be challenging to interpret.

Treatment is based on oxygen therapy and reducing the degree of respiratory effort. Patients with extra-thoracic airway disease have fewer difficulties if they will breathe with slow and shallow breaths. If the temperature is elevated the animal should be actively cooled with water and fans. Sedation with low doses of acepromazine and/or butorphanol is often beneficial. An anti-inflammatory dose of corticosteroids maybe beneficial to reduce airway swelling and edema. It is important to weigh the benefits of corticosteroid therapy with the potential adverse effects of these drugs. Animals in severe distress demonstrated by cyanosis, dramatic respiratory efforts, gasping, etc. or animals failing to respond to stabilization efforts require establishment of a patent airway. Rapid induction of anesthesia and orotracheal intubation is ideal. If routine intubation is unsuccessful, intubation with a smaller endotracheal tube is attempted. If this cannot be performed intubation with a catheter (red rubber tube, urinary catheter) is attempted, this may be used as a stylet to guide an endotracheal tube or for oxygen administration into the trachea as a temporary measure. When orotracheal intubation cannot be achieved, an emergency tracheostomy is required. Definitive therapy will require resolution of the cause of the airway obstruction.

### Lower Airway Disease

Lower airway disease causes wheezes and respiratory distress on exhalation. Lower airway disease is most commonly due to bronchoconstriction, often of an allergic nature. The classic example of lower airway disease is feline asthma (allergic bronchitis) although anaphylaxis can also present with severe bronchoconstriction. The wheezing may be audible on initial evaluation or it may require auscultation. Initial diagnosis is based on visualization of marked expiratory efforts and detection of wheezes. Cats present far more commonly for lower airway disease than dogs. A history of coughing in cats is also suggestive of allergic bronchitis.[2] Thoracic radiographs are often diagnostic but rarely possible before stabilization. In acute disease thoracic radiographs may be unremarkable.[2]

Initial treatment includes oxygen therapy and bronchodilators. The most potent bronchodilators are the beta-2 agonists such as terbutaline, weaker bronchodilators such as aminophylline should only be used in less severe cases. As most animals are believed to have an allergic component to their bronchoconstriction, corticosteroids may also be beneficial. In the case of life threatening bronchoconstriction IV or IM epinephrine (0.1 - 0.2 mg/kg for dogs, 0.1mg/cat) should be given.

Allergic bronchitis is a common cause of acute respiratory distress in cats. As stress can significantly worsen their state minimal handling is often essential. Administering medications intramuscularly and leaving these cats alone in an oxygen cage maybe the safest approach.

**Pulmonary parenchymal disease**

Pulmonary parenchymal disease can be due to trauma, hemorrhage, infection, edema, neoplasia or inflammatory disease. All these diseases result in variable degrees of alveolar fluid accumulation and alveolar collapse. This produces ventilation perfusion inequalities within the lung that impairs the oxygenating ability. Clinically patients will have respiratory distress; auscultation may reveal increased breath sounds and moist crackles or may be normal. The initial, emergency room diagnosis maybe based on history, ausclutation, evidence of hypoxemia without other cause evident and response to treatment. Diagnosis is confirmed with thoracic radiographs where pulmonary infiltrates will be present. In some cases bronchoscopy and bronchoalveolar lavage will be required to obtain a definitive diagnosis.

Inadequate oxygenation is a consistent finding in patients with pulmonary parenchymal disease.[1] Arterial blood gases are the most accurate measurement of oxygenating ability. The normal $PaO_2$ in an animal breathing room air (at sea level) is 80 to 120 mmHg. A $PaO_2$ of less than 80 mmHg ($SpO_2 < 95\%$) is considered hypoxemia and less than 60 mmHg ($SpO_2 < 90\%$) is severe hypoxemia. Visible cyanosis is a sign of severe hypoxemia ($PaO_2$ of ~ 40 mmHg).[1,3] If the patient is on supplemental oxygen the $PaO_2$ should be equal to 5 times the $FIO_2$ (measured in percent). If the patients $PaO_2$ is significantly less than this predicted number it suggests pulmonary parenchymal disease. Pulse oximetry is very insensitive to oxygenation abnormalities in patients receiving supplemental oxygen.

Treatment is always based on oxygen therapy to resolve hypoxemia coupled with specific therapy for the primary pulmonary disease process. Oxygen therapy is aimed at maintaining the $PaO_2$ in the normal range ($SpO_2$ greater than 95%). If this cannot be achieved with oxygen therapy alone, mechanical ventilation is indicated.[4]

**Pleural Space Disease**

Accumulation of air, fluid or soft tissue in the pleural space prevents lung expansion, this reduces tidal volume and predisposes to atelectasis. Animals with pleural space disease may present with acute respiratory signs or they may have a history of chronic progressive respiratory disease. Auscultation maybe abnormal, fluid and soft tissue accumulations will dampen lung sounds making them difficult to hear. When there is fluid accumulation the loss of lung sounds is most distinct ventrally. Initial diagnosis is based on history and ausculation findings. If an ultrasound machine is nearby is can be used to quickly identify fluid or soft tissue in the pleural space. Pneumothorax can also be identified by loss of the glide sign.[5] In patients with severe respiratory distress, if there is any uncertainty as to the presence of pleural space disease a thoracocentesis should be performed immediately. If fluid or air is present it should be drained maximally. In more stable patients thoracic radiographs are diagnostic. Fluid analysis will differentiate the numerous causes of pleural effusion.

Pneumothorax is defined as air accumulated in the pleural space and can be considered as closed, tension or open. A closed or simple pneumothorax has an intact thoracic cavity and can lead to mild, moderate or severe dsypnea. Treatment consists of evacuating the pleural space via thoracocentesis or a chest tube. Commonly needle thoracocentesis is used initially. If this does not allow complete evacuation or if repeated thoracocentesis is required in a relatively short time period the placement of a chest tube is indicated.

Diaphragmatic hernia can be traumatic or congenital in origin. The degree of associated respiratory distress is determined by the volume of organs herniated into the thoracic cavity and the chronicity of the disease. Often diaphragmatic hernia is an incidental finding possibly many months or years following the initial traumatic event. In these stable patients surgical repair can be considered an elective procedure. In animals with an acute herniation, surgical repair is performed as soon as the patient is otherwise stable. If there is evidence that the stomach is displaced into the thoracic cavity surgery should be considered a priority and not delayed.

**Chest Wall Disease**

Chest wall defects are primarily the result of trauma and are clearly evident at the time of physical examination. A flail chest is caused by multiple rib fractures leading to a freely mobile segment of the thoracic wall, a pseudoflail refers to a segment of the chest wall that is still attached but able to move independently of the rest of the thorax. Flail segments result in paradoxical movement of the segment – moving inward on inhalation and outward on exhalation. It may compromise a patient's ability to oxygenate and ventilate but in most cases it is the associated pulmonary contusions that are the greater cause of respiratory distress. Treatment consists of oxygen therapy to ensure the animal is not hypoxemic, evaluation for the presence of a pneumothorax and appropriate analgesia. Rarely does the flail segment require specific intervention. Surgical stabilization of the flail segment maybe required is severe cases.

An open pneumothorax results from a defect in the thoracic wall, the 'sucking chest wound'. It may or may not be associated with underlying pulmonary trauma. In this situation pleural pressure equilibrates with atmospheric pressure and pulmonary collapse will occur. Immediate treatment consists of covering the defect with a sterile dressing that still allows air to escape on inspiration. Chest tube(s) are placed in a location remote from the defect and then complete occlusion of the defect by the dressing can be achieved. As soon as the patient is sufficiently stable surgical intervention is required to repair the defect and possibly address associated pulmonary trauma.

**Pulmonary Thromboembolism**

A pulmonary thromboembolism (PTE) occludes blood flow to a section of the lung. This section of the lung is no longer contributing to the oxygenation of arterial blood and theoretically the rest of the lung should be capable of effectively oxygenating the patient in most situations. But PTE often causes hypoxemia; the cause of this hypoxemia is ventilation perfusion abnormalities in other (non-embolized) regions of the lung.[1,6] Patients with PTE have dyspnea and blood gas abnormalities. If the PTE is large there may be right sided heart dysfunction and hemodynamic compromise. Thoracic radiographs may reveal pulmonary infiltrates but they are often normal. Dyspnea and hypoxemia in a patient with normal thoracic radiographs is suggestive of PTE. Definitive diagnosis requires angiography, computer tomography or ventilation / perfusion scanning.[6] In mild to moderate cases oxygen therapy and anticoagulant treatment to prevent further emboli formation is recommended. In cases of severe respiratory distress and/or cardiovascular compromise fibrinolysis with tissue plasminogen activator or streptokinase is indicated. This therapy is associated with a significant risk of hemorrhage.

**Abdominal distension**

Abdominal distension causes anterior displacement of the diaphragm leading to respiratory compromise. Common causes of severe abdominal distension include gastric dilation volvulus, heavy pregnancy, ascites and intra-abdominal masses. Significant abdominal distension in a patient with respiratory distress should be decompressed as possible. In cases of severe abdominal distension pulmonary atelectasis can occur.

**"Look alikes"**

There are several causes of increased respiratory rate and/or effort that are due to stimulation of the central respiratory center in animals with no primary respiratory tract disease. These patients will have normal oxygenation and no abnormalities on auscultation or thoracic radiographs. They require no treatment of their respiratory signs but may benefit from treatment of their primary disease process.

Many brain diseases can cause stimulation of the medullary respiratory center. Diseases such as neoplasia, inflammatory brain disease and trauma can all lead to respiratory changes. These patients usually have other signs of brain disease such as obtundation, cranial nerve defects and behavioral changes. They may also have abnormal respiratory patterns such as apneustic breathing and cheyne stokes breathing.

Hyperthermia is also a potent stimulus of respiratory rate and effort, especially in dogs. Hyperthermia is the consequence of high ambient temperatures and/or exertion and should be differentiated from fever. Fever describes an elevation in body temperature due to endogenous pyrogens and will not trigger increases in respiratory rate and effort. Measurement of body temperature in animals with respiratory distress is very important. Hyperthermia may be the primary cause or a contributor to increased respiratory rate and effort and cooling will tend to improve these clinical signs.

Drugs such as opioids will directly stimulate the central respiratory centre producing rapid panting. On examination these animals will have a normal or lower than normal body temperature and a normal or higher than normal $PCO_2$. Metabolic acidosis will also stimulate increases in respiratory rate and depth of breathing in an effort to lower $PCO_2$ and return pH towards normal.

**REFERENCES**

[1] Lumb AB, **Nunn's Applied Respiratory Physiology**, 5th Ed. Boston: Butterworth-Heinemann; 2000. [2] Adamama-Moraitou KK, et al. **J Feline Med Surg** 2004;6:227-233. [3] King LG, **Respiratory Disease in Dogs and Cats**. St Louis: Saunders, 2004. [4] Hopper K, et al. **J Am Vet Med Assoc** 2007;230:64-75. [5] Sartori S, et al. **AJR Am J Roentgenol** 2007;188:37-41. [6] Dalen JE. **Chest** 2002;122:1440-1456

## CLINICAL CONSIDERATIONS OF HYPOXEMIA


**Dr. Kate Hopper BVSc, PhD, DACVECC**
**Davis, CA**


## OXYGENATION

Maintaining the delivery of oxygen to the tissues is the cornerstone of emergency medicine and critical care. The delivery of oxygen ($DO_2$) is the product of cardiac output (CO) and the content of oxygen in arterial blood ($CaO_2$). The lungs are responsible for the movement of oxygen from alveoli into the blood. Once in the blood, oxygen is carried in two forms. A very small proportion of the total oxygen content of blood is present dissolved in physical solution in plasma. This is measured as the partial pressure of oxygen or $PO_2$. The majority (>98%) of oxygen is carried bound to hemoglobin and is measured as a saturation or $SO_2$. This is reflected in the equation for oxygen content.[1]

$$CaO_2 = ([Hb] \times SaO_2 \times 1.34) + (0.003 \times PaO_2) \text{ ml } O_2/100\text{ml blood}$$

Where [Hb] is the concentration of hemoglobin, $SaO_2$ is saturation of hemoglobin with oxygen and $PaO_2$ is the partial pressure of oxygen dissolved in plasma. Each gram of Hb will bind 1.34 ml of $O_2$ when fully saturated. 0.003 is the solubility coefficient of oxygen in plasma at body temperature. From this equation it can be seen that the three factors contributing to oxygen content are hemoglobin concentration, the saturation of this hemoglobin with oxygen and the partial pressure of oxygen dissolved in plasma. Although hemoglobin saturation with oxygen clearly contributes the majority of the oxygen content, it is the $PaO_2$ in the plasma that will determine the degree of saturation and hence $PaO_2$ has a vital functional role in maintaining oxygenation even though it contributes minimally quantitatively. The relationship between $PaO_2$ and $SaO_2$ is described by the oxygen-hemoglobin dissociation curve. The term hypoxemia is used to describe inadequate blood oxygen levels and is defined by the $PaO_2$. For patients at sea level (barometric pressure ~ 760mmHg), hypoxemia is commonly defined as a $PaO_2$ less than 80 mmHg., a $PaO_2$ less than 60 mmHg is considered severe hypoxemia and may require immediate therapeutic intervention.[1,2]

### Mechanisms of Hypoxemia

The $PaO_2$ is primarily dependent on the quantity of oxygen inspired which is measured as the partial pressure of alveolar oxygen ($P_AO_2$) and pulmonary function. The $P_AO_2$ can be determined from the alveolar air equation.[1,3]

$$\mathbf{P_AO_2 = (F_IO_2 \times (Barometric\ Pressure - Water\ Vapor\ Pressure)) - PaCO_2 / RQ}$$

Where RQ = respiratory quotient ($CO_2$ Production / $O_2$ Consumption). RQ ~ 0.9 (0.7 – 1.0). Water vapor pressure = 50 mmHg and barometric pressure is ~ 760 mmHg at sea level.

As the alveolar air equation describes, the quantity of oxygen in the alveolus is determined by the fraction of inspired oxygen ($F_IO_2$), barometric pressure and the animal's $PaCO_2$. The fraction of inspired oxygen can be measured as a percentage or a decimal. For example the $F_IO_2$ of room air can be considered 21% or 0.21. At sea level the barometric pressure is ~ 760 mmHg, with increasing altitude this pressure falls. So even though the $F_IO_2$ remains 21%, animals' breathing room air in Denver where barometric pressure is ~ 600 mmHg will have a lower $P_AO_2$ than those at sea level. As a result the normal blood oxygen level will change with changes in altitude.[2]

The alveolar air equation describes the quantity of oxygen in the alveolus available for gas exchange. The amount of this oxygen that is successfully taken up by the arterial blood depends on pulmonary function. In patients with normal lungs the $PaO_2$ is expected to almost equilibrate with $P_AO_2$. The normal $PaO_2$ of patients at seal level, breathing room air is 80-120 mmHg. With increasing severity of lung disease there will be an increasing disparity between $P_AO_2$ and $PaO_2$.

As previously mentioned a $PaO_2$ of less than 80mmHg is considered hypoxemia. The causes of hypoxemia can be divided into three broad categories.

1. Low fraction of inspired oxygen ($F_IO_2$)
2. Hypoventilation (only if breathing room air)
3. Venous Admixture
    a. Ventilation/perfusion mismatch
        i. Low ventilation/perfusion alveoli
        ii. No ventilation/perfusion alveoli
    b. Anatomical vascular right to left shunts
    c. Diffusion defects

Low inspired oxygen occurs at altitude, in the clinical setting it is a possible cause of hypoxemia in patients connected to a breathing circuit. The oxygen supply should be evaluated immediately when hypoxemia is identified in anesthetized animals. Hypoventilation causes an elevated $PCO_2$ which dilutes out the oxygen in the alveolus causing a low $P_AO_2$. The impact of $PaCO_2$ on $P_AO_2$ is limited and will only cause hypoxemia in animal's breathing room air. If the $F_IO_2$ is increased to 30% or more, hypoxemia due to hypoventilation will be resolved.[1] Consequently any patients with significant hypoventilation should be placed on supplemental oxygen immediately to prevent against hypoxemia.

Venous admixture describes all the ways blood can pass from the right side of the heart to the left side of the heart without being adequately oxygenated. The most common cause of this is ventilation/perfusion (V/Q) abnormalities subsequent to pulmonary disease such as pulmonary edema, pneumonia and pulmonary hemorrhage. The clinical relevance of considering lung disease associated more with low V/Q issues (ie. Alveoli that are underventilated but still receiving gas flow) separately from lung disease associated more with no V/Q issues (ie collapsed alveoli receiving no portion of the tidal volume), is their response to oxygen.

Patients with mainly low V/Q lung disease will usually respond adequately to oxygen therapy while patients with more no V/Q alveoli tend not to be oxygen responsive and will likely require positive pressure ventilation. Right to left anatomical shunts are uncommon and are usually accompanied by other clinical signs such as an audible murmur and polycythemia. Pulmonary diffusion defects are uncommon and occur in association with diseases such as emphysema, oxygen toxicity and smoke inhalation.[1,2]

**Identification of Inadequate Oxygenation**

Clinically, accurate assessment of a patient's oxygenation status provides diagnostic and therapeutic information. Monitoring oxygenation is important in all patients with evidence of respiratory compromise, all critically ill patients, all anesthetized patients and any patient receiving oxygen therapy.

There are several oxygen related parameters that can be evaluated. In the clinical setting the choice of which parameter to assess will be largely dictated by the clinical scenario and equipment availability.

What can we measure in the clinical setting?

- Mucous membrane color
- Partial pressure of oxygen in the arterial blood - $PaO_2$
- Oxygen saturation of hemoglobin - $SpO_2$

**Mucous Membrane Color:**

Assessment of mucous membrane color is part of any physical examination. Although the presence of pink mucous membranes is a normal and somewhat reassuring finding it is important to appreciate that the presence of pink mucous membranes does not imply adequate arterial oxygenation. The presence of cyanosis, a blue discoloration of mucous membranes, is however a sign of severe arterial hypoxemia. In animals with normal hematocrits cyanosis occurs at PaO2 levels of ~40 mmHg. Hence, animals can be mildly to moderately hypoxemic yet their mucous membranes will remain pink in color. Factors such as anemia and the type of ambient lighting can impair the ability to identify cyanosis. In summary mucous membrane color is not a reliable measure of blood oxygenation. Cyanosis is a near-terminal sign and emergency therapy is indicated.

**$PaO_2$**

Measurement of $PaO_2$ requires an arterial blood sample and a blood gas machine. Appropriate sample handling is important to avoid pre-analytical errors. An effort needs to be made to ensure excessive heparin dilution or exposure of the sample to the air does not occur. The sample should be run as soon as possible. If there is a delay prior to analysis the sample should be capped and stored in ice water.[4]

The normal $PaO_2$ in a patient depends on the $F_IO_2$, the altitude and the $PCO_2$. The normal $PaO_2$ of animals with normal $PCO_2$ levels breathing room air ($F_IO_2 = 21\%$) at sea level is 80-120 mmHg. A $PaO_2$ of less than 80 mmHg is considered hypoxemic and less than 60 mmHg is considered severe hypoxemia. It is important to note that animals on oxygen therapy can have an inadequate or 'less than expected' ability to oxygenate, but as long as the $PaO_2$ is greater than 80 mmHg these animals are not considered hypoxemic. Less than expected ability to oxygenate is due to venous admixture as discussed above.

**Pulse Oximetry**

Pulse oximetry is non-invasive and the technology has become widely available so it is a very valuable tool in monitoring the oxygenation status of a patient. It relies on spectrophotometry to detect patterns of reflected light. Unfortunately accuracy can be a problem with pulse oximetry. It was designed for cooperative, hairless human patients and so its adaptation to veterinary medicine brings some inherent issues. The standard pulse oximeters emit only 2 wavelengths of light and assume that there is only oxyhemoglobin and deoxyhemoglobin present. If a significant amount of carboxyhemoglobin is present this will be read as oxyhemoglobin and a falsely high $SpO_2$ will result. Methemoglobin will tend to read as an $SpO_2$ of 85% despite the real $SpO_2$ being lower or higher than that.[1]

Because pulse oximetry requires pulsatile blood flow, hypotension or peripheral vasoconstriction may affect the accuracy of the readings. Movement of the patient can confuse the probe and it may mistakenly take readings of tissue or veins. Haired skin or pigmented skin can make the readings spuriously low or prevent the equipment from making any reading at all.[5] In veterinary patients the easiest, most reliable readings tend to come from unpigmented mucous membranes in anesthetized patients.

Ways to maximize the accuracy of the pulse oximetry reading:

- Chose a site that has minimal pigmentation and is warm and thin. Mucous membranes are ideal.
- Protect the probe from ambient light (e.g. cover it with a cloth)
- Encourage the patient to be still for the reading
- Take continuous readings or multiple readings
- Ensure the pulse rate displayed by the pulse oximeter is the same as the patient's pulse rate at that time
- If the reading does not correlate with the animal's clinical picture take an arterial blood gas or persist with more readings before believing it

Animal's breathing room air should have maximally oxygenated hemoglobin if they have normal lung function (99-100%) corresponding to a $PaO_2 \sim 100$ mmHg. Any fall in $PaO_2$ below this will see a corresponding drop in $SpO_2$. An $SpO_2$ of $< 95\%$ represents hypoxemia ($PaO_2 < 80$ mmHg) and an $SpO_2 < 90\%$ ($PaO_2 < 60$ mmHg) represents severe hypoxemia, so there is a very narrow range of acceptable readings allowing very little room for error. Animals on supplemental oxygen should have $PaO_2$ levels of

110-500 mmHg but the $SpO_2$ will still read 99-100%, hence the pulse oximeter is insensitive to changes in oxygenation until the $PaO_2$ falls to 100mmHg or less. Consequently, if an animal is on supplemental oxygen and has a true $SpO_2$ reading of < 99% it is of immediate concern given that it should be 99% on room air.

**Venous Blood Gases**

The partial pressure of oxygen in venous blood ($PvO_2$) cannot be interpreted as a measure of pulmonary function. $PvO_2$ is dependent on the arterial oxygen content, cardiac output and oxygen consumption by the tissues. Arterial oxygen content in turn is dependent on $PaO_2$ and hemoglobin concentration while cardiac output is a product of preload, afterload, contractility and heart rate. Abnormalities in any one of these factors can lead to a low $PvO_2$; inadequate oxygenation of arterial blood is only one of them. For this reason $PvO_2$ cannot be used as a measure of lung function or oxygenation.

## THERAPEUTIC CONSIDERATIONS OF HYPOXEMIA

The first priority of oxygen administration should be the resolution of life threatening hypoxemia. During initial stabilization a high $F_IO_2$ is often used as it is considered safe for short term administration. When patients require ongoing oxygen therapy (more than 8-12 hours) it is important to decrease the $F_IO_2$ to the lowest level required to attain the oxygenation goals. The goal of oxygen therapy is a $PaO_2$ of 80 – 120mmHg ($SpO_2$ of 95 – 99%). A higher $PaO_2$ is of no clinical benefit assuming the patient has an adequate hematocrit and stable cardiovascular status. A lower $PaO_2$ (~60 mmHg) may need to be tolerated in patients if the $F_IO_2$ cannot be decreased sufficiently. Indirect measures of tissue oxygenation such as lactate and venous $PO_2$ can be used to aid in titration of $F_IO_2$ to lower $PaO_2$ levels. Once significant hypoxemia has been resolved further diagnostic tests can be performed and specific therapies can be instituted. If severe hypoxemia ($PaO_2$ < 60 mmHg) cannot be resolved with oxygen therapy, positive pressure ventilation (PPV) is indicated. Such cases of severe hypoxemia that are not sufficiently responsive to oxygen therapy are almost always due to pulmonary parenchymal disease such as pneumonia, pulmonary edema or pulmonary hemorrhage. Positive pressure ventilation is effective as it can re-expand previously collapsed alveoli and better ventilate poorly functioning alveoli to improve the efficiency of the lung. In addition it takes over the work of breathing preventing patient exhaustion and subsequent cardiopulmonary arrest. Another indication for PPV is when adequate oxygenation can only be achieved with unacceptably high levels of oxygen.

## OXYGEN TOXICITY

Pulmonary oxygen toxicity is related to the partial pressure of oxygen of the inspired gas ($P_IO_2$) rather than the concentration or fraction of inspired oxygen ($F_IO_2$) of the inspired gas. A $P_IO_2$ of 255 mmHg is considered the safe level for continuous, long-term exposure in people.[1] This is equivalent to an $F_IO_2$ of ~33% at 1ATM (sea level). Higher levels of $P_IO_2$ maybe tolerated for short periods of time. In addition to $P_IO_2$, oxygen toxicity is also a function of the duration of exposure. The lower the level of inspired oxygen the longer animals will survive. For example dogs exposed to 100% oxygen survive ~2 to 5 days while some dogs exposed to 75-80% oxygen were still alive after 14 days. Interruption in the exposure to oxygen can also reduce its toxicity. In a study in dogs exposed to 95-100% oxygen the mean survival time was 39 hours for continuous exposure compared to 117 hours if oxygen exposure was interrupted for 45 minutes every day.[7]

Susceptibility to oxygen toxicity varies between species. For example breathing 100% oxygen is lethal in 1.5 to 7 days in rats while some frogs and sea turtles can breathe 100% oxygen indefinitely. Initial experimental evidence suggested that resistance to oxygen toxicity increases with body size. This concept is somewhat controversial at present.

**Dogs & Cats**

There is quite a range of reported survival times for dogs exposed to 95-100% oxygen. This variation is not just between studies, there also appears to be a marked variation in susceptibility between individual animals. Mean reported survival times range from 39 hours to ~5 days.[6,7] In addition the progression of disease is variable with some dogs dieing within hours of first developing dyspnea while others will have severe dyspnea for as long as 48 hours before death.

The sole publication the author could find reported a mean survival time of 9 healthy cats exposed to 98-100% oxygen of 83±22 hours (3.5 days).[7] It is tempting to suggest that cats maybe more susceptible to oxygen toxicity than dogs but its impossible to be definitive without further information.

## REFERENCES


[1] Lumb AB, **Nunn's Applied Respiratory Physiology**, 5th Ed. Boston: Butterworth-Heinemann; 2000. [2] King LG, **Respiratory Disease in Dogs and Cats**. St Louis: Saunders, 2004. [3] Riley RL, et al. **Am J Physiol** 1946;147:191-198. [4] Rezende ML, et al. **J Vet Emerg Crit Care** 2007;17:67-71. [5] Tobin MJ. **Principles and Practice of Intensive Care Monitoring**. New York, McGraw-Hill, 1998. [6] Ashbaugh DG. **J Appl Physiol** 1971;31:664-668. [7] Clark JM, et al. ***Pharmacol Rev*** 1971;33:37-133

**INDICATIONS FOR INITIATION OF POSITIVE PRESSURE VENTILATION**

**Dr Kate Hopper BVSc, PhD, DACVECC**
**Davis, CA**

## INTRODUCTION

Positive pressure ventilation (PPV) describes the procedure of performing some or all of the breathing for a patient via delivery of positive airway pressure. It can be performed manually using a bag-valve or breathing circuit. For therapeutic applications a mechanical ventilator is used. Positive pressure ventilation is a standard of care in human intensive care units and is becoming more widely available in veterinary medicine.

## INDICATIONS

Often the most challenging aspect of PPV is making the decision when an animal requires ventilation. In short, there are three main indications for positive pressure ventilation (PPV);[1,2]

1. Severe hypoxemia despite oxygen therapy; as indicated by cyanosis or $PaO_2$<60mmHg or $SpO_2$ < 90%, despite administration of oxygen.*
2. Severe hypercapnia; $PaCO_2$ > 60 mmHg despite therapy
3. Excessive respiratory effort

* Animals that require high levels of oxygen therapy (>60%) for prolonged periods (>24-36 hours) to prevent hypoxemia maybe considered candidates for mechanical ventilation in an effort to manage them on a lower $F_iO_2$ and avoid oxygen toxicity.

### Hypoxemia

Hypoxemia is defined as a $PaO_2$ of < 80mmHg ($SpO_2$ < 95%). Patients with a $PaO_2$ < 60 mmHg ($SpO_2$ < 90%) have severe hypoxemia and may be candidates for mechanical ventilation if they do not respond to other therapy. Supplemental oxygen is the initial therapy for hypoxemia. Inspired oxygen concentrations of > 60% can be given for short periods of time (<24 hours) but oxygen toxicity can develop if animals are exposed to inspired oxygen concentrations of > 60% for prolonged periods. If the $PaO_2$ does not increase to > 60mmHg with oxygen therapy, or if it takes > 60% inspired oxygen to maintain an acceptable $PaO_2$ for a prolonged period, PPV is indicated. Overall animals that will require PPV for hypoxemia will be animals with pulmonary parenchymal disease. Causes include pneumonia, pulmonary edema and pulmonary hemorrhage.

It is important to avoid PPV of animals with a closed pneumothorax as it can rapidly become a fatal tension pneumothorax. If an animal has a closed pneumothorax and also requires PPV, one or more chest tubes should be placed. If there is not sufficient time for chest tube placement or if chest tube drainage is insufficient to keep the pleural space evacuated, a small thoracotomy incision may be required. This will convert the process to an open pneumothorax and the patient can now be safely ventilated.

### Hypoventilation:

An arterial $PaCO_2$ of > 45 mmHg defines hypoventilation, a $PaCO_2$ of > 60 mmHg is severe hypoventilation and requires therapeutic intervention. Venous carbon dioxide measurements are slightly higher than arterial levels (4-5mmHg higher) but are a reliable indicator of ventilation. Central venous samples such as jugular venous samples are ideal. End tidal $CO_2$ estimates $PaCO_2$ although it will tend to be slightly lower than arterial $CO_2$. In animals with pulmonary disease, $ETCO_2$ correlation with arterial carbon dioxide can vary and should be interpreted with caution. It is very difficult to predict if an animal has hypoventilation from clinical observation of its respiratory pattern. Hypoventilation should always be confirmed prior to instigating therapy.

Hypoventilation is a result of a reduction in alveolar minute ventilation. General causes include brain disease, cervical spinal cord disease (C1-C2), peripheral nerve disease, neuromuscular junction disorders and myopathies effecting the respiratory muscles. It is interesting to note that these patients will not have the classic signs of respiratory distress. Instead they will tend to have decreased respiratory rate and effort and the severity of their respiratory compromise maybe missed without carbon dioxide monitoring. Patients with intracranial disease cannot tolerate an elevated $CO_2$ and may require PPV to control an arterial $CO_2$ that would be acceptable in a normal patient.

### Excess work of breathing:

Animals that are working very hard to breath are at risk of muscle fatigue and acute respiratory arrest. These patients are often very hard to resuscitate. The increased muscle activity associated with an increased work of breathing increases oxygen consumption and can lead to hyperthermia which further stimulates the respiratory effort. The evaluation of an increased work of breathing is based on clinical judgment. These animals may be maintaining acceptable blood gases but the concern of impending fatigue is a compelling reason to place the animal on PPV. If an animal has abnormal blood gases despite an excessive work of breathing, the situation requires urgent attention. In one more recent study excessive respiratory effort was reported as a reason for PPV in 46 of 148 cats and dogs evaluated. In the same study 22% of dogs and 50% of cats had no pulse oximetry or blood gas evaluation prior to beginning PPV.

### Unable to decide??

If considerable time is spent debating if PPV is indicated or not – there is a strong likelihood it is! If the problem is significant enough for you to ask the question, the problem is nearly always significant enough to warrant intervention.

## PATIENT SET UP

In order to begin PPV it is necessary to secure an artificial airway, set up the machine and organize the patient monitoring equipment. In an emergency the patient maybe manually ventilated until the ventilator can be organized. Patients can be ventilated via endotracheal tubes or tracheostomy tubes. As neurologically intact animals require full general anesthesia in order to tolerate PPV, it is most common to use an endotracheal tube for ventilation. In comatose or paralyzed animals, a tracheostomy tube can provide the option to ventilate the patient with minimal if any sedatives. This allows effective monitoring of the neurological status and minimizes the cost and many of the adverse effects associated with anesthetic drugs.

The anesthetic induction protocol of patients that require PPV should include preoxygenation and ensure rapid control of the airway. Consideration of the animals' cardiovascular status and underlying disease processes should also effect the choice of induction drugs. Once anesthetized and on the ventilator a maintenance anesthetic protocol will be necessary. This is usually achieved by the use of constant rate infusion of injectable sedatives. It is common to use two to three drugs in combination in an effort to minimize the dose and hence the adverse effects of any one drug. Common maintenance sedative drug combinations used by the author are a benzodiazepine with propofol ± an opioid, a benzodiazepam with pentobarbital ± an opioid, a benzodiazepam with an opioid. Long term use (> 48 hrs) of propofol should be avoided as hyperlipidemia usually develops. Pentobarbital infusion (1-2 mg/kg/hr) is the main anesthetic drug used for long term ventilation of dogs and cats at at the authors institution.[3]

Effective patient monitoring is essential during both initiation of PPV and maintenance of PPV. A continuous ECG, pulse oximeter, end tidal carbon dioxide, blood pressure and continuous temperature monitoring are all ideal. As much of this monitoring that can be attached just prior, or immediately after initiation of PPV as possible the more likely it is that adverse patient events (including death) can be avoided.

## GOALS OF PPV

The goal of ventilator therapy is to maintain acceptable blood gas values with minimally aggressive ventilator settings. Even if more aggressive machine settings can produce improved lung oxygenating efficiency, there is no therapeutic advantage to the patient and the increased risk of ventilator induced lung injury makes this approach detrimental. The ideal ventilator settings for each individual animal cannot be predicted and are determined through trial and error. The patient should be fully evaluated after every change in ventilator settings, including blood gas analysis if possible. It is advisable to make only one change in ventilator settings at a time in order to accurately evaluate what effect each change has on the patient.

Blood gas goals of PPV[1,2]

- $PaO_2$ of 80 – 120 mmHg ($SpO_2$ > 95%)
- $F_IO_2$ < 60%
- $PaCO_2$ of 35 – 55 mmHg (35 - 40mmHg in patients with brain disease)
- $PvCO_2$ of 40-60 mmHg (38 – 45 mmHg in patients with brain disease)

Ideally all ventilator patients are monitored with arterial blood gases. When this is not possible pulse oximetry, venous blood gases and end tidal carbon dioxide monitoring can be used to guide therapy.

## INITIAL STABILIZATION ON THE VENTILATOR

As previously stated the ideal ventilator settings for a given patient cannot be predicted. It is likely that animals with lung disease will need more aggressive settings than those with neurological/neuromuscular disease processes. This is because pulmonary parenchymal disease will reduce the compliance of the lung. This will mean that higher pressures will be needed to achieve the same tidal volume. It is necessary to choose some initial settings prior to connecting the patient to the machine, this can be done using guidelines such as those shown in Table 1. It is best to have the ventilator turned on and confirm it is working appropriately before connecting the patient. It is imperative to always have a method by which to perform manual ventilation close at hand at all times during PPV in case of machine malfunction or operator error.

When the patient is first connected to the machine an $F_IO_2$ of 100% is advised as a safety measure. When ventilating patients with lung disease it can beneficial to keep them in sternal recumbency for the initial stabilization period. These animals almost always oxygenate better in sternal compared to lateral recumbency. Hopefully the $F_IO_2$ can be reduced once the stability of the patient is verified. After connection the patient's chest is observed for appropriate movement – if there is insufficient or over aggressive chest inflation the ventilator settings should be adjusted appropriately. The chest is then auscultated bilaterally to be sure there is ventilation of both lungs and all monitoring is evaluated (blood pressure, ECG, pulse oximeter, $ETCO_2$ etc.). Any concerning changes should be addressed immediately.

**Table 1:** Suggested initial ventilator settings

| Ventilator Parameter | Initial Setting |
|---|---|
| Fraction of inspired oxygen | 100% |
| Tidal volume | 5-15 ml/kg |
| Respiratory rate | 10-30 breaths per minute |
| Minute ventilation | 150-250 ml/kg |
| Peak inspiratory pressure | 10 - 15 $cmH_2O$ |
| Positive end expiratory pressure | 0 - 5 $cmH_2O$ |
| Inspiratory time | ~1 second |
| Inspiratory : Expiratory ratio | 1:2 |
| Inspiratory trigger | -1 to -2 $cmH_2O$ |

### $PaO_2$

Once the patient is apparently stable on the ventilator, assessment of arterial blood gases is performed. The first priority is to lower the $F_IO_2$ to < 60% while maintaining an acceptable $PaO_2$, although 100% $F_IO_2$ can be maintained for 12-24 hours in most patients without major concern. Positive end expiratory pressure (PEEP) can increase the oxygenating efficiency of the lung by preventing alveolar collapse. Patients with lung disease will need some degree of PEEP; initially it can be set at 4 – 6 $cmH_2O$. In animals with significant lung disease, higher levels of PEEP are often required to allow adequate reductions in $F_IO_2$.

### $PaCO_2$

An elevated $PCO_2$ is due to reduced alveolar minute ventilation. Alveolar minute ventilation is the product of tidal volume (effective) and respiratory rate.

Possible causes of hypercapnia in ventilated patients are:

- Increased apparatus dead space – excess tubing / connectors between the patient and the ventilator circuit Y-piece
- Endotracheal tube kink or obstruction
- Incorrect assembly of the ventilator circuit
- Pneumothorax
- Large pulmonary embolism (uncommon)

If evaluation of the machine and patient rules out these issues, then it is to be assumed that there is insufficient alveolar minute ventilation (MV = TV x RR). The ventilator settings should be adjusted to increase TV or RR or both and the patient re-evaluated.

## PROGNOSIS

Prognosis for successful weaning from mechanical ventilation is largely dependent on the primary disease process. A study of 148 dogs and cats receiving long term PPV (>24 hours) reported that an average of 36% of patients requiring ventilation for pulmonary parenchymal disease (hypoxemia) were weaned from the ventilator and 22% left the hospital, compared to a weaning rate of 50% for patients with neuromuscular disease processes (inability to ventilate) and 39% of these animals survived to hospital discharge.[3] The outcome of animals within each group varied according to the underlying disease. For example 50% (10/20) of dogs with aspiration pneumonia were successfully weaned from the ventilator while only one of 12 dogs with acute respiratory distress syndrome was weaned successfully.

From the small amount of literature available it appears that cats have a lower success of weaning from PPV. In one study of 53 cats ventilated for varying lengths of time only 15% survived.[4] In the long term ventilation study mentioned above (>24 hours) 25% of cats were weaned from PPV (compared to 44% for dogs) and 21% survived to hospital discharge (compared to 29%).[3] From evaluation of this study it appears that the underlying diseases of cats that require PPV tend to be ones associated with less success than dogs. For example one of the more common reasons for PPV of cats in these two studies was post-cardiopulmonary arrest.

## REFERENCES

[1.] Hess DR, et al. **Essentials of mechanical ventilation**. 2nd ed. McGraw-Hill, New York, 2002; [2.] King LG. **Textbook of Respiratory Disease in Dogs and Cats**. St Louis, Saunders, 2004. [3.] Hopper K, et al. **JAVMA** 2007;230:64-75. [4.] Lee JA, et al. **J Am Vet Med Assoc** 2005;226:924-931

# MANAGEMENT & WEANING OF VENTILATED PATIENTS

Dr. Kate Hopper BVSc, PhD, DACVECC
Davis, CA

## INTRODUCTION

Maintaining animals on mechanical ventilation requires intensive patient care, constant evaluation of ventilator settings and a focus on weaning. In almost all cases a trained, dedicated care giver is required 24 hours a day, 7 days a week. This presentation provides some suggestions and guidelines for management and weaning of ventilated patients and a review of ventilator related complications.

## PATIENT CARE

Appropriate patient care will make the difference between successful positive pressure ventilation (PPV) and unsuccessful. Most of the complications reported in ventilator patients can be prevented by adequate care. In order to provide this level of care, staff training will be required in addition to developoing clearly defined patient care protocols and detailed record keeping.

Unless the patient has significant neuromuscular or intracranial disease anesthesia will be necessary to allow positive pressure ventilation and maintain an airway. Balanced anesthesia incorporating 2-3 drugs is advised to reduce the adverse effects of any one agent. Benzodiazepines ± opioids are commonly used in conjunction with propofol (if short term ventilation of 1-2 days) or pentobarbital (if longer term). The aim is to keep the patient anesthetized sufficiently that it will tolerate the endotracheal tube and the PPV but not so deeply anesthetized that more severe adverse cardiovascular effects of the anesthetic drugs is likely. This requires some experience in anesthetic monitoring and constant tinkering with the anesthetic drug infusion rates as drug requirements will change with time. It is important to have drug dose charts that correctly convert the mg/kg/hr drug doses to ml/hr for the drug concentration you are using. In addition it is recommended to have pre-calculated some 'rescue doses' of drugs in case animals suddenly become light and try to extubate themselves.

The breathing circuit should be humidified appropriately without excessive accumulation of liquid in the circuit. There are two basic options for airway humidification, active heated water humidifiers or passive heat moisture exchangers.[1,2] The heated water humidifiers are very effective, are ideal in patients with substantial airway secretions and help prevent hypothermia which can be an issue in small patients. They are expensive and unless they are used in conjunction with heated wire circuits (which are also expensive) they are associated with substantial 'rain out' in the circuit. This refers to water accumulating in the limbs of the breathing circuit which in large volumes can interfere with ventilator function. More concerning, it greatly increases the risk of ventilator associated pneumonia (VAP). The heat moisture exchangers (HME) are inexpensive devices that fit on the end of the endotracheal tube and passively take heat and moisture from the exhaled gas and return it to the inhaled gas. They have been found to be equally effective at humidifying the airways as compared to the active humidifiers and may reduce the incidence of VAP by reducing moisture in the circuit. They can occlude which has the potential to be a life threatening event and for this reason should be avoided in patients with lots of airway secretions.

The endotracheal or tracheostomy tube should be suctioned every 4 hours maintaining strict sterile technique. In addition the tube cuff should be deflated, repositioned and reinflated at regular intervals. It is commonly recommended to replace endotracheal (and tracheostomy tubes if they don't have a removable inner cannula) with a sterile tube every 24 hours to avoid occlusion with concretions of airway secretions. It is important to thoroughly suction the oropharynx prior to cuff deflation and tube replacement. The practice of instilling sterile saline down the trachea during suctioning to help loosen airway secretions is controversial.[1,2] Although many feel it makes suctioning more effective there is concern that the saline can wash bacteria from the upper airway or colonized endotracheal tube down into the lung. Interestingly a recent randomized clinical trial in human ventilator patients reported a reduced incidence of VAP when saline was instilled prior to suctioning.[3]

Care of the oral cavity is of utmost importance in these patients. Oral ulceration is a common problem and with poor oral care may occur in 90% or more of our patients.[4] Oral ulceration is a primary source of bacterial proliferation and has been reported as a major cause of VAP in human PPV patients.[1,2] Prevention of oral ulceration with careful positioning of mouth gags, keeping the tongue and mucous membranes moist and frequent repositioning of the tongue and ET tube is essential. The mouth and the oropharynx should be carefully rinsed and suctioned every 4 hours. A chlorhexidine rinse of the mouth should be performed every 8 hours.[5] In addition to oral care, frequent application (every 2 hours) of ocular lubricant is essential to reduce the incidence of corneal ulcers.

Recumbent patient care including repositioning the animal every 2 hours and passive range of motion exercises must be performed. Well padded bedding and careful positioning of the limbs is essential to avoid pressure sores. Soiling should always be immediately cleaned.

Ventilator patients are critically ill and will require careful management of their fluid and nutritional requirements. Most ventilator patients will develop fluid retention as a result of the abnormal neurohormonal signals triggered by the positive intrathoracic pressure. The development of a low urine output should prompt careful patient evaluation for pre-renal causes.[1] In many patients diuretic administration is required to maintain an adequate urine output. Other medications will be dictated by the patients varying underlying disease processes.

Providing nutritional support is always challenging in these patients if they require long term PPV. Although enteral nutrition is always desirable the risk of aspiration pneumonia is very high and commonly is a terminal event in the ventilator patient. Consequently enteral nutrition is frequently avoided in the anesthetized ventilator patient. Some long term ventilator patients do very well with gastrotomy tube feedings and some neuromuscular patients ventilated via a tracheostomy tube can take oral food. When enteral feeding is not used, parenteral nutrition is indicated.

## COMPLICATIONS

### Bucking the Ventilator

When patients are fighting or 'bucking' the ventilator they will not be able to be ventilated effectively and its common to note desaturation and hypercapnia. When patients begin to buck the ventilator a systematic approach should be taken to determine the nature of the problem. Possible causes include:

- Hypoxemia
  - Loss of oxygen supply. Worsening of underlying disease or development of new pulmonary disease such as pneumothorax, VAP, aspiration pneumonia or acute respiratory distress syndrome
- Hypercapnia
  - Circuit disconnect/leak, tube obstruction or kink, pneumothorax etc.
- Pneumothorax
  - Typified by a rapidly climbing $PCO_2$ and a plummeting $PaO_2$. Need auscultation and potentially thoracocentesis to diagnose
- Hyperthermia
  - Anesthetized animals like to have relatively low temperatures, even 102F may cause dogs to pant on the ventilator
- Patient-ventilator dysynchrony (inappropriate ventilator settings)
  - Observe when the patient is trying to inhale and exhale, see if its possible to match that pattern with your ventilator settings
- Inadequate depth of anesthesia
  - Rely on the routine clinical signs of anesthetic depth. This is the most common problem but care should be taken not to blindly increase the anesthetic drug dose without fully assessing the patient.

### Hemodynamic compromise

Positive pressure ventilation can reduce cardiac output in some patients. The increases in intrathoracic pressure from ventilation can reduce systemic venous return. The increase in lung volume can also cause decrease ventricular compliance. This can shift the interventricular septum and impair left ventricular filling.[1] This is mainly a problem for animals with hypovolemia or vasodilation.[1,2] Although animals with more aggressive ventilator settings are of more concern for developing negative cardiovascular effects, the compliance of their lungs if generally very poor such that little of the airway pressure provided is transmitted to the intrathoracic structures. Constant heart rate and blood pressure monitoring is important in PPV patients, fluid and drug therapy should be given as appropriate if evidence of hypovolemia and/or hypotension develop.

### Ventilator Associated Pneumonia

In human medicine ventilator associated pneumonia has been reported to occur in 20-30% of patients and leads to an increase in mortality.[1,2] Endotracheal intubation compromises the normal upper airway defenses and bacteria from the oral cavity can migrate down the airway to the lungs. This problem can be exacerbated by aspiration of gastric contents. VAP is recognized as developing new pulmonary infiltrates 48 hours or longer after beginning PPV. Decreases in oxygenation, changes on the cbc and a fever are also common clinical signs. Prevention of VAP includes maintaining careful oral hygiene and using a sterile technique when handling the circuit or endotracheal/tracheostomy tube. The less the circuit is disconnected the better. It is now recommended to replace ventilator circuits only every 5 days (earlier if gross contamination).[1] Many VAP infections have been found to come from the hands of caregivers, for this reason wearing gloves and frequent hand washing is strongly recommended for all ventilator patients.[1,2] Wearing gloves should be considered mandatory when dealing with the oral cavity.

### Ventilator Induced Lung Injury

Ventilator induced lung injury (VILI) is a significant risk to any patient receiving PPV. In addition to increasing pulmonary pathology, in human medicine VILI has been found to increase the incidence of multiple organ dysfunction and patient mortality. It is believed that pulmonary inflammation resulting from VILI leads to the elaboration of inflammatory mediators which can enter the systemic circulation amplifying the systemic inflammatory response and subsequent organ damage.

The main mechanism of VILI is 'stretch-induced' trauma. For many years excessive airway pressure or barotrauma was blamed for VILI. Numerous experimental and clinical investigations have since led to the conclusion that volutrauma and the associated over distension of alveoli is the major cause of VILI.[1,2] If normal lungs are ventilated with high tidal volumes progressive lung injury occurs. This is typified by a decrease in pulmonary compliance, increases in lung protein and cytokine concentrations and hyaline membrane formation and is morphologically indistinguishable from the acute respiratory distress syndrome (ARDS). Alveolar epithelial and vascular endothelial cells are stretched on inspiration. If this stretching is extreme it can cause cellular disruption and alveolar rupture resulting in interstitial emphysema and/or pneumothorax. Pneumothorax has been reported to occur in 10-20% of human and veterinary ventilator patients. More moderate degrees of deformation triggers mechano-sensitive cellular responses that can initiate cytokine mediated pulmonary inflammation. Inflamed vascular endothelial cells recruit activated leukocytes to adhere and transmigrate into the pulmonary interstitium. Accumulation of polymorphonucleur cells is largely credited with causing increases in alveolar-endothelial barrier permeability and hyaline membrane formation, characteristics of VILI. This mechanical injury is most prominent in alveoli that are overdistended, alveoli on the border between aerated and non-aerated lung regions and alveoli that open and collapse with each breath.[1,2]

**Lung Protective Ventilation**

It is now recognized that PPV triggers an inflammatory response in the lungs and the degree of this response is determined by the ventilation strategy employed. It has been well demonstrated that hyperinflation of the lung is injurious and should always be avoided. In many patients this can be achieved by maintaining recommended tidal volumes when designing a ventilation protocol. In patients with severe lung disease, most commonly acute respiratory distress syndrome (ARDS) the usual recommended tidal volume may be excessive. ARDS causes collapse and consolidation of alveoli leaving fewer aerated lung regions. These regions would be vulnerable to overdistension if a regular tidal volume was delivered. The ARDS Network reported a significant reduction in mortality of ARDS patients ventilated with a tidal volume of 6ml/kg versus 12 ml/kg.[6] This lung protective ventilation strategy also includes the use of high PEEP (5-12 $cmH_2O$ in human patients) and limits peak inspiratory pressures to no higher than 30 $cmH_2O$. High PEEP increases the proportion of aerated lung and reduces cyclic opening and closing of alveoli on each breath. A consequence of low tidal volume ventilation is elevations in $PCO_2$ a situation referred to as 'permissive hypercapnia'. There are approaches to controlling $PCO_2$ during this ventilatory strategy but in many cases it is well tolerated by the patient. Given the evidence from experimental animal studies and human clinical trials it would be reasonable to assume that lung protective ventilation has a valid role in veterinary patients. It is important to appreciate that this ventilation strategy is designed for the ARDS lung and should be applied with caution to other disease states.

**Weaning from Ventilation**

Weaning from the ventilator is generally a continuous process of titrating down the ventilator settings. Often as the patient improves the mode of ventilation may be changed to one that requires the animal to perform a greater proportion of the work of breathing. Such modes include synchronized intermittent mandatory ventilation (SIMV) and pressure support.

Prior to disconnection from the machine the patient must have obtained certain physiologic goals.[1,2,7] These include:

- Original disease process stable or improving
- Normal respiratory drive
- No longer requiring significant ventilator support for adequate gas exchange
- Adequate oxygenation ($PaO_2/F_IO_2$ ratio of at least 150-200 is recommended)

Reasons an animal cannot be weaned:

- Hemodynamically unstable
- Requirement of high $F_IO_2$ (greater than 50-60%)
- High peak inspired airway pressures (>25 $cmH_2O$)
- High PEEP levels (>4-5 $cmH_2O$)

Disconnection from the ventilator should be considered a 'spontaneous breathing trial' during which the patient is intensely monitored. The development of hypoxemia, hypercapnia, hyperthermia, tachycardia, hypotension and/or tachypnea are all indications that the trial has failed and positive pressure ventilation should be reinstituted. Such spontaneous breathing trials should be practiced daily once the animal qualifies according to the criteria listed above.

**REFERENCES**

1. Tobin MJ. **Principles and Practice of Mechanical Ventilation**. New York, McGraw-Hill, 2006. 2. MacIntyre NR, et al. **Mechanial Ventilation,** 2nd ed. St Louis, Saunders, 2009. 3. Caruso P, et al. **Crit Care Med** 2009;37:32-38. 4. Fudge M, et al. **J Vet Emerg Crit** Care 1997;7:79-87. 5. Genuit T, et al. **Surgical Infect** 2001;2:5-18 6. The Acute Respiratory Distress Syndrome Network. **N Engl J Med** 2000;342:1301-1308. 7. Hess DR, et al. **Essentials of mechanical ventilation**. 2nd ed. New York McGraw-Hill, 2002.

# SEDATION & ANESTHESIA FOR THE CRITICALLY ILL PATIENT COMPREHENSIVE REVIEW

**Jane Quandt DVM, MS, DACVA, DACVECC**
**St. Paul, MN**

The critically ill patient that requires general anesthesia presents a challenging situation. The patient needs a thorough pre-operative assessment to define what type of trauma or compromise the patient is undergoing. The nature of their disease will alter physiology and decrease reserves which affects the pharmacokinetic and pharmacodynamic behavior of sedative and anesthetic drugs. Stabilization of the critical patient, including minimizing stress levels and optimizing oxygen delivery, prior to sedative or anesthetic drug exposure is essential as the risk of sedation and anesthesia in an unstable patient increases the potential for complications. Appropriate monitoring should be performed at all times to ensure that these delicate patients survive their sedation or anesthetic episode. Intensive care may include vasopressor and inotropic support, aggressive colloid and/or crystalloid therapy, analgesic support, antibiotic therapy, oxygen therapy, and nursing care to improve the survivability in this critically ill patient population.

Prior to anesthesia, diagnostics, including physical examination, radiographs, blood chemistry, complete blood count, coagulation profile, acid-base status, blood glucose and lactate levels as necessary should be performed to determine the severity of the illness. Dehydration or hypovolemia along with fluid, acid-base, and electrolyte deficits should be corrected if possible before anesthesia.

Venous access is imperative in managing and anesthetizing the critical patient. Intravenous (IV) administration of drugs is preferred as intramuscular (IM) or subcutaneous (SC) may have delayed absorption, particularly when the patient is dehydrated, hypovolemic, poorly perfused or hypothermic. IV access also allows for ease of anesthetic drug titration. The critically ill patient of may benefit from having more than one IV catheter to provide a port for not only drug administration, but antibiotic delivery, vasopressor and inotropic support, and fluid therapy. Because of different fluid rates and possible incompatibilities of various agents, such as vasopressors, sodium bicarbonate, and blood transfusion products, a minimum of two IV catheters should be placed prior to anesthesia in unstable patients. It is not uncommon for the critically ill animal to experience hypotension during the anesthetic period necessitating the need for inotropic support. Blood products should have a dedicated catheter and no other fluids or drugs should be administered in the same line during the transfusion due to concerns for possible contamination and potential for bacterial growth. This is also true for those patients that are receiving total parenteral nutrition (TPN). The catheter for TPN is a dedicated line and should never be disconnected or have any other fluid running through the same line due to the risk of sepsis. Total parenteral nutrition is considered a hyperosmotic crystalloid and should be accounted for as part of the crystalloid fluid volume.

A packed cell volume (PCV) greater than 25% is recommended for adequate oxygen carrying capacity and oxygen delivery. During anesthesia the PCV can decrease by 3 to 5 %, therefore even a small volume of blood loss may be significant in the anesthetized animal. Similarly, hypoproteinemic patients, total protein (TP) $\leq$ 3.5 g/dl and/or an albumin $\leq$ 2 g/dl, may benefit from the use of colloids to help maintain normal colloid osmotic pressure (COP) (n=18-24 mmHg) and to prevent edema formation or vascular leak. Measurement of COP prior to anesthesia is helpful in determining the need for colloid support and to help determine when to terminate colloid therapy. If patients are hypoproteinemic, colloid options include Hetastarch™, Dextran™, 25% human serum albumin or even Oxyglobin™. If the patient is hypocoagulable, FFP should be considered.

Lactate plasma concentrations can be a useful prognostic indicator. The normal plasma lactate concentration is less than 2.5 mmol/L. It has been shown that lactate levels are increased in ill and injured dogs, and the values were significantly higher in nonsurvivors compared to survivors. The most common and usually most severe cause of increased lactate levels is systemic hypoperfusion, which requires prompt treatment and identification of the cause. Therapy includes fluid resuscitation to treat hypoperfusion, correction of the cause, and a PCV >20% to help maintain arterial oxygenation.

Patients with renal insufficiency may require a higher fluid rate to maintain renal perfusion, and should have urine output (UOP) carefully monitored during anesthesia. Monitor blood pressure closely; if possible a pre-anesthesia value is helpful. The mean arterial blood pressure (MAP) should be greater than 60 mm Hg to maintain renal perfusion. Because critically ill patients are often hypotensive during anesthesia a mean blood pressure (MAP) of <60 mm Hg or a systolic < 90 mm Hg requires prompt treatment to maintain appropriate organ perfusion. A fluid bolus should be initiated, if appropriate, of either a crystalloid at a rate of 10 to 20 ml/kg IV over 15 to 20 minutes or a colloid bolus of 5 to 10 ml/kg IV over 10 to 20 minutes. If the hypotension persists despite fluid therapy there may be a need for inotropic support. Drugs excreted by the kidney may have delayed excretion, and should be used cautiously (i.e. ketamine in cats). Urine output (UOP) should be monitored carefully, and goals to achieve normal UOP of 1 to 2 ml/kg/hr should be achieved.

Patients with heart murmurs may have decreased ability to compensate under anesthesia, and fluid overload should be avoided. Consider the use of colloids which requires less volume to be administered. Blood pressure should also be carefully monitored, as anesthesia-induced hypotension may result in decompensation. Fluid overload can be assessed by measuring the body weight both pre and post operatively. The measurement of central venous pressure (CVP) will help in the assessment of fluid therapy, (normal 0 to 10 cm $H_2O$), and volume overload, but only assesses the right-side of the heart.

In patients with liver disease, anesthetic protocols and monitoring may be affected due to decreased glucose production, albumin production, altered drug metabolism, via cytochrome p-450 enzymes, and the decreased production of clotting factors. Consider the use of reversible injectable agents such as the opioids. It may be tempting to use just inhalant agents such as isoflurane or sevoflurane. These two agents are the most commonly used, but both agents cause cardiovascular and respiratory depression. The cardiovascular depression can lead to hypotension.

Blood glucose levels should be closely monitored in those animals that have severe liver disease, and in patients that are pediatric, septic, or diabetic. Dextrose supplementation may be necessary.

One should always consider if the patient is on any pre-existing drug therapy, such as non-steroidal anti-inflammatories (NSAIDS), diuretics, anticonvulsants, and cardiac medications. These agents may predispose to possible complications such as GI bleeding, renal impairment, electrolyte abnormalities, hypotension, and added depressant effects with anesthesia.

Protocols should be implemented to minimize the amount of time the animal is under anesthesia; therefore, options such as pre-clipping the surgical site with the animal awake should be performed if possible. Preoxygenation of the animal prior to induction will allow for additional time that may be needed to intubate the animal; this is especially helpful with those animals in respiratory distress or with an airway which may be difficult to intubate. Electrocardiography (ECG) and blood pressure monitoring should be in place prior to induction to detect evidence of arrhythmias, hypotension, or cardiovascular collapse that may occur during induction of the critically ill animal. The ECG should be monitored closely for changes in heart rate, rhythm, and for the presence of malignant arrhythmias, which may be more prevalent with trauma, splenic disease, septic peritonitis, hypoxic patients, or gastric dilatation volvulus patients.

Physical examination findings for perfusion, such as the capillary refill time (CRT), mucous membrane color and pulse quality should also be monitored closely. The use of pulse oximetry will add information on hemoglobin saturation (Hb) and oxygenation.

Body temperature should be continuously monitored as anesthetic drugs disrupt normal thermoregulatory mechanisms and hypothermia leads to prolonged recovery. The use of forced warm air heating blankets will help in the treatment of hypothermia.

The critically ill patient may not exhibit classic pain response symptoms due to their debilitated state, they should be carefully but appropriately treated with analgesic. Analgesia promotes well-being and has a positive effect on the speed and quality of recovery. Pain can lead to catabolism and complications such as delayed wound healing, sepsis, and nosocomial disease. Analgesics are imperative in these painful, critically ill patients, and should be administered as soon as possible following patient assessment to provide a significant benefit. It is vital however, that the underlying disease process be addressed while pain relief is provided, as analgesia therapy may mask the underlying disease processes or the hemodynamic stability of a patient. Ideally analgesics are administered before pain develops, preemptive analgesia, as less drug therapy may then be necessary to control pain. This is especially important prior to surgery or other invasive procedures; however this is not always feasible in trauma cases or in emergent cases.

Because of complex pain pathways, it is often unlikely that one agent alone will completely alleviate pain, regardless of how high the dose is. The use of more than one drug class can improve analgesia by acting at multiple receptor types and may also overcome the problem of varying onset times and durations between drug classes. Patients should be frequently reassessed to ensure that the analgesic regimen is adequate and appropriate. Hemodynamics, respiration, and body temperature should be continuously monitored as sedative drugs can disrupt normal hemodynamics, ventilation and thermoregulatory mechanisms leading to hypothermia, hypotension and hypoventilation. The use of analagesics is incorporated into the sedative and anesthetic protocol. Commonly used agents are opioids, tranquilizers, dissociatives, alpha 2 agonists, lidocaine, and inhalants.

Opioids act centrally to limit the input of nociceptive information to the central nervous system (CNS), which will reduce central hypersensitivity. Opioids are commonly used in critically ill patients as they have a rapid onset of action, are safe, reversible, and are potent analgesics.

Partial mu agonist, buprenorphine, 5 to 20 μg/kg IV, IM, or SC q6–8h, and mixed mu agonist–kappa antagonist, butorphanol, 0.1 to 0.4 mg/kg IV q4h, reach maximal effect at the upper end of the dose range. If the pain is severe or the analgesia is inadequate, additional doses of partial or mixed agonists–antagonists are unlikely to be effective. Using a pure μ agonist, morphine, hydromorphone, fentanyl, or oxymorphone, would be more effective because there is no upper limit to the analgesia provided by a pure μ agonist. Opioid antagonist, naloxone, bind to the same receptor as agonists but cause no effect, and can competitively displace the agonist from the receptor and therefore reverse the agonist effect.

Potent side effects such as respiratory depression and bradycardia may be seen at the higher dose range with a pure mu agonist; therefore, the higher dose range should be used cautiously in critically ill patients. Anticholingerics are not routinely used unless there is a need to treat the bradycardia. An additional side effect of the μ agonists morphine and meperidine is histamine release, particularly when given rapidly intravenously, this can lead to severe hypotension due to vasodilation. Opioids can create gastroparesis and ileus, which may result in vomition, regurgitation, and aspiration of gastrointestinal contents, particularly in depressed, sedated, weak, or critically ill patients. Vomition is more likely to occur when the mu agonist is given IM or SC.

Opioids can be safely administered to cats to provide analgesia. Morphine or oxymorphone, 0.05 mg/kg titrated slowly IV, can be administered for analgesia; however, side effects such as hyperexcitability or agitation may occur. It has been shown that the onset of mydriasis following administration of opioids correlates with adequate analgesia in cats; continual dosing after achieving mydriasis may result in adverse side effects such as dysphoria and agitation. The mixed partial mu agonist buprenorphine,10 to 20 μg/kg transmucosally-oral, IV, IM q6–8h, has been shown to be an effective analgesic in cats.

One advantage of opioid administration in critically ill patients is that their effects can be reversed if necessary with a pure antagonist such as naloxone, 0.02 to 0.1 mg/kg IV, IM, or SC. Naloxone can reverse CNS depression, respiratory depression, and bradycardia, but the reversal of sedation and analgesia can cause pain, excitement, emergence delirium, aggression, and hyperalgesia Low-dose naloxone, 0.004 mg/kg titrated slowly IV, has been recommended to reverse CNS depression without affecting analgesia. The duration of effect for naloxone is relatively short, 20 to 30 minutes, because of its rapid metabolism in dogs and cats, which may predispose patients to re-narcotization when the drug is used to reverse long-acting opioids. Agonist–antagonists such as butorphanol, 0.05 to 0.1 mg/kg IV, may also be used to reverse sedation and respiratory depression from μ agonists. The benefit of using butorphanol as a reversal agent is that complete reversal of analgesia does not occur because of the κ agonist effects of butorphanol. In contrast, buprenorphine is not as easily reversed as butorphanol or the mu agonists because it is difficult to displace from the receptor.

$\alpha_2$-Adrenergic agonists bind to receptors in the CNS, leading to sedation, peripheral vasoconstriction, bradycardia, respiratory depression, diuresis, muscle relaxation, and analgesia. Medetomidine and dexmedetomidine are the most common $\alpha_2$-adrenergic agonist administered in small animals. The sedative effects of medetomidine have a longer duration of action than do the analgesic effects, which last approximately 30 to 90 minutes. Low dose medetomidine, 1 to 10 µg/kg IV, can be safely used in stable patients, or administered in conjunction with opioids to produce analgesic synergism and increase the analgesic duration to as long as 4 hours. In the extremely painful or vicious animal the mu agonist narcotic, morphine, can be combined with the alpha 2 agonist, medetomidine, 5 to 10 ug/kg, IM for enhanced analgesia, sedation, and restraint. At higher doses, medetomidine can be used to sedate distressed animals and for minor procedures such as restraint and analgesia for radiographic positioning. The effects of $\alpha_2$-adrenergic agonists can be reversed with atipamezole, 0.05 to 0.2 mg/kg IM, SC, or IV, a specific $\alpha_2$-adrenergic antagonist which reverses analgesia, sedation, and respiratory depression. Intramuscular or subcutaneous routes of administration are preferred for reversal because intravenous administration can lead to abrupt hypotension and/or aggression.

NMDA receptor antagonists block multiple binding sites at this receptor resulting in analgesic, amnestic, psychomimetic, and neuroprotection. Ketamine, a noncompetitive NMDA receptor antagonist, can reverse central hypersensitivity due to prevention of wind-up activity and central sensitization of wide dynamic-range neurons in the dorsal horn of the spinal cord. Ketamine can be administered PO, SQ, IM, and IV. It causes minimal cardiovascular depression, does not depress laryngeal protective reflexes, and produces less ventilatory depression compared with opioids. Side effects include tremors and sedative effects along with increased cardiac output via increased sympathetic Oral ketamine can also be used, dogs: 8 to 12 mg/kg PO q6h, to provide pain relief following burn injuries. Subanesthetic or low doses in dogs and cats, 0.1 to 1 mg/kg IV, followed by a CRI, 2 µg/kg/min, may produce analgesic effects without causing anesthesia or profound sedation.

Benzodiazepines are minor tranquilizers, and are most often used in combination with other agents, such as opioids and ketamine as they may cause dysphoria if used alone. The mechanism of action is via potentiation of the inhibitory neurotransmitter GABA. The effects are anti-anxiety, euphoria, dysphoria, anti-convulsant activity, muscle relaxation, and ataxia. They have a wide safety margin and are commonly used as part of an IV induction technique. There may mild respiratory depression. Benzodiazepines can be reversed.

Midazolam, Versed[R], has a water soluble base which makes for excellent absorption following IM or SC injections. Duration of action is 1 to 2 hours.

Diazepam, Valium[R] has a propylene glycol base which makes IM injection painful and absorption unpredicatable. Therefore it is best injected IV. The propylene glycol is implicated in causing hypotension following rapid IV bolus injection, when giving IV give slowly and flush with fluids to decrease the risk of phlebitis. The rate of metabolism is variable depending on the age and species.

Flumazenil, Romazicon[R], is a benzodiazepine antagonist, and will reverse all the effects of a benzodiazepine. It is not a controlled drug, but it is expensive. Dose is 0.01 to 0.02 mg/kg IV.

Acetylpromazine, 0.01 to 0.05 mg/kg IV, not to exceed a total dose of 2 mg, may be used in combination with opioids as an anxiolytic and to enhance sedative effects. Acetylpromazine, a major tranquilizer, and should be used with caution due the potential for vasodilation and resultant profound hypotension and hypothermia. Acetylpromazine does not provide analgesia and should not be administered as a single agent if analgesia is desired. It may take up to 15 minutes before the effect of an intravenous dose of acetylpromazine is clinically observed; therefore, repeated doses should be avoided until the full effect is evident. Acetylpromazine can be safely administered to ICU patients if given at low doses, 0.005 to 0.01 mg/kg, in hemodynamically stable animals with adequate respiratory function.

Sedation may be needed for procedures such as ultrasound and biopsy, feeding tube placement, urinary catheter placement, bandage or cast application or removal, bone marrow aspiration, and to facilitate taking radiographs.

These drug techniques can also be used as part of a sedative- premed or induction protocol leading to general anesthesia.

**Possible protocols would include**

For a minor procedure such as ultrasound with fine needle aspiration or biopsy of visceral organ:

Butorphanol 0.2 to 0.3 mg/kg IV

Diazepam or midazolam 0.2 mg/kg IV

**For a more painful procedure but not requiring general anesthesia**

Hydromorphone 0.05 to 0.1 mg/kg IV

Diazepam or midazolam 0.2 mg/kg IV

If the above are not quite sufficient, titrate propofol 0.5 to 1 mg/kg IV closely watching for decreased respiratory rate, recommend monitoring with pulse oximeter, providing oxygen supplementation, have ETT and laryngoscope ready if need to intubate.

If you wish to add additional analgesia, instead of propofol substitute ketamine 1 to 2 mg/kg IV with the same recommendations as for propofol.

When general anesthesia is required a balanced anesthetic technique will help to minimize the side effects from the use of one single agent. Endotracheal intubation is used to control the airway, to provide the ability to ventilate the patient, and to protect the airway from aspiration. A laryngoscope and a variety of endotracheal tube sizes, and a breathing circuit that matches the patient's size should all be readily available. All supplies and machinery should be checked thoroughly prior to induction and intubation. One should be ready to implement intermittant positive pressure ventilation (IPPV), if the patient hypoventilates, becomes apneic or is to undergo a thoracic procedure.

Induction is usually via IV injection. Commonly used agents include a neuroleptanalgesic combination such as hydromorphone, 0.05 to 0.1 mg/kg IV, oxymorphone,0.05 to 0.1 mg/kg IV, or fentanyl, 5 to 10 ug/kg IV, and diazepam, 0.2 mg/kg IV, or midazolam, 0.2 mg/kg IV, with the addition of ketamine, 0.5 to 4 mg/kg IV, and possibly lidocaine, 1 to 2 mg/kg IV. Lidocaine can be used with those patients that have ventricular arrhythmias, to enhance analgesia, and to help with reperfusion injury as may be seen with a GDV. Propofol at 2 to 4 mg/kg IV can also be used for induction. Propofol is short acting, 5 to 10 minutes, has no analgesic properties, and can lead to hypotension and apnea, therefore it can be used in combination with opioids and tranquilizers. Propofol is useful for a rapid induction, 30 seconds, for those patients that need rapid control of their airway. In some patients following the neuroleptanalgesic administration, mask induction with sevoflurane or isoflurane may be accomplished. The use of multiple agents allows for a lower level of inhalant anesthesia and this may lessen the potential for hypotension.

If a patient undergoing general anesthesia remains hypotensive even following appropriate therapy it may be necessary to consider discontinuing the inhalant anesthetic agent to eliminate the hypotensive effects of the inhalant and continuing the anesthetic maintenance via an injectable drug regimen. It may consist of the use of a constant rate infusion, CRI, of a mu agonist such as fentanyl or morphine in combination with ketamine and lidocaine. Some patients may only need fentanyl as an intermittent IV bolus or as a CRI. Constant rate infusion of analgesics has the advantage of maintaining effective plasma concentrations for continued pain relief. They can be delivered by infusion from a fluid bag or via a syringe pump. The use of a syringe pump is preferred as it will allow for accurate dosing.

If it is suspected that the patient about to undergo general anesthesia will be prone to develop hypotension based on the pre-operative assessment the induction plan can be modified to include agents that can also be incorporated into a CRI. The dose of those agents used for induction can also act as the loading dose for a CRI. Following induction the CRI can then be started immediately. A loading dose is important in order to establish plasma drug levels. Inhalants can be used initially at a lower concentration than is normally used due to the presence of the CRI. If it is deemed necessary to discontinue the inhalant the CRI is already in place. The IV induction may be accomplished with the use of a neuroleptanalgesic combination such as hydromorphone, 0.05 to 0.1 mg/kg IV, oxymorphone, 0.05 to 0.1 mg/kg IV, or fentanyl, 5 to 10 ug/kg IV, and diazepam or midazolam, 0.2 mg/kg IV, with the addition of ketamine, 0.5 to 4 mg/kg IV, and possibly lidocaine, 1 to 2 mg/kg IV. The use of multiple agents, hydromorphone, or fentanyl,diazepam, ketamine, and lidocaine, will have a slower onset, but will provide analgesia and is more cardiovascularly sparing.

Ketamine may be used to enhance analgesia and will increase heart rate and blood pressure. Ketamine given perioperatively may prevent "wind-up" pain from occurring and thereby reduce postoperative pain. Low doses of ketamine do not create the undesirable side effects such as dysphoria or hallucination and can be used for intra- and post-operative analgesia in dogs. A loading dose of ketamine, 0.5 mg/kg IV, should be immediately followed by a CRI of 10 to 20 μg/kg/min for the duration of the surgery. This is then reduced to 2 μg/kg/min during the recovery phase and post-operative phase.

To avoid histamine release which may occur with rapid intravenous morphine administration), a morphine CRI, 0.1 to 1 mg/kg/hr, should be started after administration of an initial loading dose of 0.15 to 0.5 mg/kg IV, diluted given slowly over 5 to 10 minutes. A CRI of morphine, 0.12 mg/kg/hr, reportedly induces effects similar to intramuscular morphine,1 mg/kg q4h, in dogs undergoing laparotomy. Regardless of how morphine is administered, its use may result in bradycardia, hypothermia, and panting. Other opioids instead of morphine can be administered if undesirable side effects of morphine occur.

Butorphanol has been administered at a loading dose of 0.1 mg/kg followed by an infusion of 0.03 to 0.4 mg/kg/hr. This agent does not provide sufficient analgesia for severe pain.

Fentanyl can also be given to enhance analgesia, starting with a loading dose of 2 to 10 μg/kg IV immediately followed by a 0.3 to 1 ug/kg/min CRI in the dog during surgery. A CRI of 2 μg/kg/hr IV, increased as needed up to 5 μg/kg/hr IV is used the post-operative and recovery phase. Fentanyl is also a safe analgesic for cats when administered at a loading dose of 1 μg/kg IV followed by 0.1 to 0.3 ug/kg/min CRI in the cat. In some critical cases a CRI is not used, instead intermittent boluses of fentanyl IV are given, 2 to 10 ug/kg. A single dose of fentanyl has a duration of approximately 20 to 30 minutes.

In some patients, such as those with head trauma, a fentanyl CRI, in conjunction with a propofol CRI, can provided adequate, safe, cardiovascular-sparing anesthesia therefore reducing or minimizing the amount of inhalant anesthesia necessary. Side effects such as bradycardia may require treatment with an anticholinergic. The propofol is used as part of the induction technique at a dose of 2 to 4 mg/kg IV with a CRI of 0.1 to 0.4 mg/kg/min. Propofol will help decrease intracranial pressure, but as it is not an analgesic, fentanyl is added. This type of patient will benefit from intermittent positive pressure ventilation.

Lidocaine can also be administered for pain control, and it may retard the effects of compromised viscera, reperfusion injury or ventricular arrhythmias, due its free radical scavenging abilities, analgesic effects, and anti-arrhythmic properties. It's use as CRI is with an initial loading dose of 1 to 2 mg/kg followed by 0.025 mg/kg/min. Lidocaine doses as high as 2 to 3 mg/kg/hr IV have been reported. Recently published research in the cat has shown that CRI s of lidocaine should be used cautiously (if at all) in this species due to cardiopulmonary depression; this would be an especially important consideration in the critically ill animal.

Morphine, 3.3 μg/kg/min, lidocaine, 50 μg/kg/min, and ketamine, 10 μg/kg/min, can be administered as a CRI analgesic combination in dogs. These agents can be given separately or mixed together in a single bag or syringe. Using a combination of agents may result in enhanced analgesia through synergism and multiple receptor activation.

In the cardiovascularly stable patient, medetomidine can be used as a constant rate infusion for analgesia with an initial loading dose of 1 ug/kg IV then CRI of 1 to 3 ug/kg/hr. It has been used alone and in combination with morphine, lidocaine, and ketamine. Medetomidine will enhance the analgesia and sedative effects of the other agents. Bear in mind that medetomidine and the opioids are reversible agents, but lidocaine and ketamine are not.

One can always start with a single agent and add others if deemed necessary.

The use of CRI agents for general anesthesia has the advantage of continuing the CRI in the recovery phase. The dose can be decreased as the surgical stimulus is no longer present. There are a variety of combinations that can be used and it may depend on the duration of analgesia and sedation needed as well at the level of pain that is expected. As the patient improves the number of agents as well as the dose can be decreased.

When sedating or anesthetizing the critically ill animal drug doses used should begin at the lower dose range and if possible titrated to effect. In addition vigilant monitoring is necessary to minimize possible risks and complications that may occur.

**THE USE OF VASOPRESSIN & POSITIVE INOTROPES FOR THE TREATMENT OF HYPOTENSION**

**Jane Quandt DVM, MS, DACVA, DACVECC**
**St. Paul, MN**

Early goal directed therapy is recommended in the treatment of shock, sepsis, and persistent hypotenstion. This involves the use of early, aggressive fluid resuscitation, antibiotics, and vasoactive agents if needed. The most commonly used vasoactive agents are dopamine, dobutamine, noreprinephrine, ephedrine, phenylephrine, epinephrine, and vasopressin. Hypotension is defned as a mean blood pressure, MAP, of less than 60 mmHg or a systolic blood pressure, SAP, of less than 90 mmHg. If hypotension persists despite aggressive IV fluid therapy, antibiotics, blood products and other treatments appropriate for the disease process then the use of vasoactive agents should be considered.

The splanchnic vasculature serves as an important blood reservoir for the circulatory system. Catecholamines play a major role in regulating the tone of the arterial resistance vessels and venous capacitance vessels. The major effect of catecholamines on splanchinc capacitance vessels is venoconstriction. Active venoconstriction and passive elastic recoil mechanisms work in concert to shift splanchnic blood volume to the systemic circulation. The effects of catecholamines depend on the volume of blood in the splanchnic reservoir, therefore fluid therapy is vital. Vasopressors and inotropes will not be effective without adequate volume resuscitation. Cardiac output is regulated by preload, contractility, heart rate, and afterload. Unless the heart fails, it will pump out all of the blood that returns to it. Mobilization of splanchnic blood volume is almost entirely the result of beta 2 and alpha adrenergic receptor activation. Activation of beta 2 adrenoceptors almost invariably enhances venous return by increasing arterial flow and decreasing hepatic venous resistance. In the cardiovascular system alpha 1receptors are responsible for vasoconstriction. The predominant receptor mediating venoconstriction is the alpha 2 receptor. Generally alpha agonists increase venous return, but they decrease it when used at high doses or in the presence of severe hypovolemia. This is most likely due to removing a portion of the vasculature from the systemic circulation by arterial vasoconstriction. Once cardiovascular function has been maximized further catecholamine exposure will cause a reduction in cardiac output. There can be cardiovascular collapse due to excessive vasoconstriction during acute sympathetic over-stimulation.

It can be difficult to choose the appropriate catecholamine for cardiovascular support due to the inability to know which adrenergic stimulation would be most beneficial. The choice to use catecholamine support is after the patient has failed to respond to adequate fluid resuscitation. Start with the lowest dose and titrate up as needed. Typically refractory systemic hypotension is managed with a beta agonist, if the hypotension does not improve or if there are signs of vasodilation despite the hypotension, then alpha agonist therapy is considered. If one agent is insufficient consider layering or using more than one agent simultaneously to activate multiple receptors. The goal is to use the lowest dose possible for the shortest period of time. If the patient is dependent on catecholamine support or has been on this support for some time, the catecholamines should be weaned slowly to determine that the patient is capable of sustaining a normal blood pressure. If the patient can not sustain a normal blood pressure then catecholamine support must be re-instituted.

## RECEPTORS

***Alpha 1***

Post-synaptic vascular smooth muscle vasoconstriction, increased resistance.

***Alpha 2***

Pre-synaptic stimulation inhibits norepinephrine release negative feedback. Post-synaptic there is vascular smooth muscle vasoconstriction.

***Beta 1***

Post-synaptic effect on heart is positive inotropy and chronotropy. Increased automaticity and conduction velocity.

***Beta 2***

Post-synaptic there is vascular smooth muscle vasodilation and bronchial smooth muscle bronchodilation. Pre-synaptic beta 2 facilitates norepinephrine release.

Beta receptors mediate vasodilation of vascular beds to help maintain blood flow to vital organs and relax bronchial smooth muscle to facilitate ventilation and oxygenation. Beta mediated vasodilation is the predominant effect in the vascular supply to the heart.

***Dopaminergic 1***

Post-synaptic there is renal and mesenteric vascular smooth muscle vasodilation.

***Dopaminergic 2***

Pre-synaptic stimulation inhibits norepinephrine and Ach release.

When a dose is given as a CRI, the agent is commonly diluted in 0.9% saline to a concentration that makes dosing easier for patient size, such as 25 ug/ml for small patients and 100 ug/ml for larger animals. These agents are given as a CRI due to their short duration of approximately 1 to 5 minutes.

## NATURAL CATECHOLAMINES

***Epinephrine***

Epinephrine has a high affinity for all splanchnic adrenergic receptors but seems to exert greater effects on alpha 1 and beta receptors, is direct acting at these receptors. Epinephrine is the primary physiologic agonist for the pre-synaptic beta 2 receptors. Increases cardiac output, heart rate and mean arterial blood pressure. Can increase myocardial oxygen consumption. Low dose 0.02

mg/kg, high dose 0.2 mg/kg IV. CRI low dose 0.005 to 0.05 ug/kg/min IV, high dose 0.05 to 1 ug/kg/min IV. It can lead to severe vasoconstriction.

***Norepinephrine***

Direct acting at alpha 1 and 2 and beta 1. Increases peripheral vascular resistance and mean arterial blood pressure. Can increase myocardial oxygen consumption. Seems to increase blood flow to the heart and kidneys without significant ischemia to other tissues. Low dose is 0.05 to 0.3 ug/kg/min IV, high dose is 1 to 2 ug/kg/min IV.

***Dopamine***

Is the precursor to norepinephrine, and acts at dopaminergic 1 and 2, alpha and beta. Increases cardiac output, heart rate, renal blood flow, and mean arterial blood flow. Dopamine at low doses stimulates dopaminergic 1 and 2 receptors and produces vasodilation, natriuresis, diuresis and variable alterations in renal and GI blood flow. At relatively low doses there is an increase in portal blood flow related to stimulation of beta 2 receptors. At mid-range doses there is stimulation of beta receptors leading to an increase in cardiac contractility and heart rate. At higher doses dopamine directly and via conversion to norepinephrine activates alpha 1 and alpha 2 receptors. Low dose 1 to 4 ug/kg/min IV, medium dose 5 to 10 ug/kg/min, high dose > 10 ug/kg/min IV. The dose-response of dopamine can be variable in the critically ill patient.

**SYNTHETIC CATECHOLAMINES**:

***Dobutamine***

Acts primarily at beta 1. Increase cardiac contractility, cardiac output, heart rate, and mean arterial blood pressure. The preferred inotropic agent if there is poor myocardial contractility. Dose 2 to10 ug/kg/min IV, high doses may lead to tachycardia. In cats doses > 5 ug/kg/min may induce CNS signs such as tremors or seizures.

**SYNTHETIC NON-CATECHOLAMINES**:

**Indirect acting**:

***Ephedrine***

Acts at alpha and beta. Increases cardiac output, heart rate and blood pressure. Given as a single IV dose, duration is approximately 20 minutes. Can get tachyphylaxis with repeated doses. Most commonly used in large animal species. Low dose 5 to 10 ug/kg, high dose 100 to 200 ug/kg.

**DIRECT ACTING**

***Phenylephrine***

Acts at alpha. Increases peripheral vascular resistance and mean arterial blood pressure. May get a reflex bradycardia in response to significant vasoconstriction. The vasoconstriction can lead to excessive decreases in blood flow to liver, GI tract, and kidneys, although coronary blood flow is increase. Low dose 1 ug/kg/min IV, high dose 3 ug/kg/min IV.

**Vasopressin**

Acts through 5 different receptors: three subtypes, $V_1$, $V_2$, $V_3$, the oxytocine receptor, and the purinergic $P_2$ receptor. The pressor effects are achieved through stimulation of $V_{1A}$ receptors found in the vascular smooth muscle, causing intense, nonadrenergic vasoconstriction. Vasopressin preferentially constricts arterioles in extracerebral tissues, with relatively less constriction in renal and coronary blood vessels, and may dilate the cerebral vasculature thereby improving cerebral perfusion.

The osmotic effects of vasopressin occur at a serum concentration of 1-7 pg/ml and the smooth muscle regulation occurs at 10-200 pg/ml. The vasopressin concentration increases 20 to 200 fold in hypotensive states and acute septic shock to help maintain vasoconstriction. Hypotension is a potent stimulus for vasopressin release. In septic shock, plasma vasopressin levels are initially elevated in order to maintain adequate blood pressure and organ perfusion. But vasoactive levels of vasopressin become rapidly depleted with continued stimulation. Vasopressin deficiency may contribute to the pathogenesis of irreversible shock, and it's use has been found to helpful in the treatment of this shock state. Autonomic dysfunction with the loss of baroreceptor function, decreased endogenous vasopressin production, and exhausted vasopressin stores may account for this observed dysfunction of the vasopressinergic system in critically ill patients. It has been suggested that an adequate initial endogenous stress response may correlate with a rapid recovery or even subsequent survival.

Plasma levels of catecholamines are markedly increased during septic shock, therefore fewer receptors are available for activation. Giving escalating doses of catecholamines results in the alpha adrenergic receptors becoming supersaturated leading to increased adverse effects and little if any hemodynamic benefit. This outlines one of the primary causes of catecholamine-resistant shock in the septic shock patient. The early administration of vasopressin can help restore the physiologic levels that are needed to maintain blood pressure in septic shock.

It has been shown that the use of vasopressin will decrease the doses of catecholamines, which will decrease the adverse effects seen with high dose catecholamine administration. Vasopressin action is independent of adrenergic stimulation, which suggests that the drug has an additive effect. This ability to cause vasoconstriction nonadrenergically allows for lower doses of the other vasopressors, which may be one of the keys to successfully treating hypotension and poor tissue perfusion and minimize adverse drug reactions.

In addition to IV fluids, antibiotics, ventilator support, and stress-dose corticosteroid, vasopressin can be used with other vasopressors for restoring hemodynamics and minimizing adverse effects and the need for exogenous catecholamines. It should be emphasized that it is important to use vasopressin and norepinephrine in combination to optimize the pressor response, while minimizing the potential deleterious effects of higher doses of either agent.

There is debate whether vasopressin activity at intestinal $V_1$ receptors jeopardizes or ameliorates the splanchnic circulation. A cecal ligation-perforation sepsis model in rats indicated there may be microvascular injury to the intestinal villi. There needs to be the realization that where as systemic blood pressure may increase, the splanchnic blood flow may decrease.

During cardiac arrest and resuscitation, progressive ischemia and acidosis are present. Epinephrine has routinely been the vasopressor of choice, however epinephrine and other catecholamines lose much of their effectiveness as vasopressors in the hypoxic and acidotic body state. Successful resuscitation depends on coronary perfusion pressure being adequate. Coronary perfusion pressure above 15 mm Hg is thought to be a predictor of return of spontaneous circulation. The addition of vasopressin along with epinephrine may improve rates of restoration of spontaneous circulation and survival. Vasopressin continues to function in the acidotic, whether metabolic or respirator, environment, as the response of the $V_1$ receptoress remains intact. Epinephrine consumes oxygen, where as vasopressin increases coronary blood flow and the availability of oxygen to the myocardium. Vasopressin therefore avoids the myocardial hypoxia and tachycardia that may occur with epinephrine. When used during CPCR, vasopressin causes vasoconstriction of peripheral arterioles and preferential shunting of blood toward the heart and CNS.

Doses: this drug is dosed in units. For vasodilatory shock 0.5 to 2 mU/kg/min IV, or 0.01 to 0.04 U/kg/hr IV, for CPCR 0.4 to 0.8 IU/kg IV. Duration is approximately 20 to 30 minutes.

Terlipressin is a synthetic long-acting analogue of vasopressin. Terlipressin has a higher affinity for vascular receptors and a longer duration of action, more than 2 hours. Terlipressin in intermittent IV dosing has been used for the treatment of hypotension with shock. It does have powerful vasoconstrictive effects that may be of concern in relation to ischemia. It is currently being used as a rescue therapy in human patients with refractory septic shock. There is no current information on it's use in the veterinary patient.

I recommend Long KM, Kirby R. An update on cardiovascular adrenergic receptor physiology and potential pharmacological applications in veterinary medicine. **J Vet Emerg Crit Care** 2008;18(1):2-25. This in an excellent review.

**RECOGNITION & CLASSIFICATION OF ACUTE KIDNEY INJURY IN THE ICU**

**Marie E. Kerl DVM, DACVIM (SAIM), DACVECC**
**Columbia, MO**

Acute, intrinsic renal failure (ARF) is the abrupt, sustained decline in glomerular filtration resulting in azotemia. The kidneys are uniquely susceptible to acute damage secondary to ischemic events or nephrotoxin exposure due to the large quantity of blood filtered for each cardiac cycle, regional differences in blood supply within the kidney, and a high rate of metabolic activity of renal tubular cells. [1] In critical illness, ARF occurs secondary to nephrotoxic therapeutic agents, hypovolemia, drugs that impair renal autoregulatory mechanisms, or secondary to the systemic inflammatory response. In physician-based medicine, the concept of hospital-acquired renal failure (HARF) has gained recognition for the negative effects on morbidity and mortality for critically ill people. [2-4] Acute renal failure is easier to prevent than to cure, and early recognition of situations that are potentially harmful to renal function is paramount to developing preventative strategies.

The kidney is the only organ possessing one capillary bed interposed between two arterioles in the glomerulus, and a second capillary bed supplying the renal interstitium. The glomerular capillary bed lies between the afferent and efferent arterioles. Smooth muscle in the arteriolar walls causes arteriolar constriction or dilation secondary to a variety of causes. (Figure 1) Changes in arteriolar diameter may occur from influence of locally produced or circulating substances, and will increase or decrease glomerular filtration rate and renal blood flow. Renal tubules have regional differences in solute transport functions, with some tubular regions having proportionally higher metabolic energy consumption and oxygen demands than others. The kidneys are also responsible for biotransformation of some drugs and toxins.

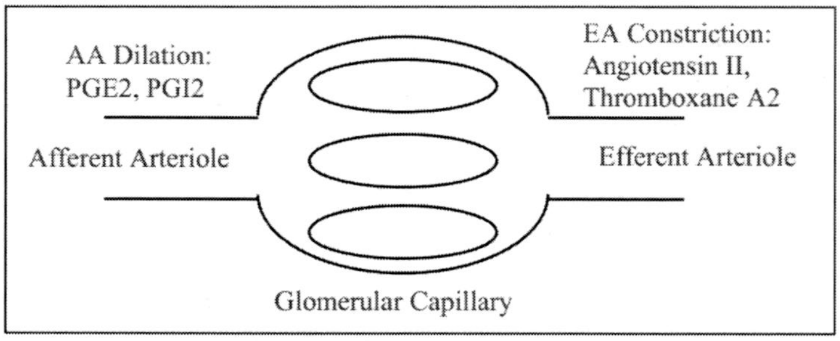


**Figure 1:** Afferent and Efferent Arteriolar Influencing Factors

Two major mechanisms predispose to establishment of ARF: injury from exposure to a nephrotoxin, or reduced renal blood flow causing ischemia. [5] Common events predisposing to ARF are listed in Table 1. The renal cortex receives 90% of total renal blood flow, making it uniquely susceptible to toxin-induced renal injury. The S3 segment of the proximal tubule and the thick ascending limb of the loop of Henle have high solute transport rates and metabolic rates which increase susceptibility to concentration of toxicants and depletion of cellular energy stores during ischemia. [1] Toxicants also accumulate within the renal medulla due to the countercurrent multiplier system. Certain toxicants are secreted or absorbed by tubular epithelial cells to cause damage. Toxins with renal biotransformation form metabolites that are more toxic than the parent compound, such as metabolism of ethylene glycol into glycolic acid. The renal medulla only receives 10% of total renal blood flow, which exacerbates medullary hypoxic renal injury. As cellular adenosine triphosphate (ATP) is depleted, sodium-potassium ATP-ase pumps are paralyzed resulting in cell swelling and death. Cellular debris may cause obstruction and loss of function of individual tubules. [1, 6]

Commonly recognized phases of ARF include initiation, extension, maintenance, and recovery. The initiation phase occurs with exposure to the toxicant or the ischemic event. Extension occurs during the response to injury and the subsequent renal damage that exacerbates the response to the initial insult. Recognizing the possibility of renal injury and administering appropriate therapy during the initiation phase prevents the establishment of ARF or lessens the severity of injury. The maintenance phase consists of established tubular lesions and nephron dysfunction. The recovery phase occurs with nephron repair and return to function, with compensatory hypertrophy of the remaining functional nephrons. In general, renal recovery occurs over 6-8 weeks following establishment of renal failure. [7]

In physician-based medicine, there are more than 30 definitions of ARF in the literature. [8] Over the past 10-15 years, nephrologists and criticalists have sought to reach consensus on the definition of ARF and develop mechanisms for earlier identification of ARF in critically ill patients. [8] The term “hospital acquired acute renal failure” was coined, and has been studied in various human and veterinary settings. [4, 9-12]. While only 5% of a general hospital population develops ARF, 30% of patients hospitalized in the intensive care unit (ICU) develop ARF. [13] Mortality is significantly worse for septic patients in the ICU, with a 45% mortality rate for non-septic patients and 75% mortality rate in septic patients. [14] As a result of a consensus panel on ARF which published findings in 2004, Bellomo and colleagues from the Acute Dialysis Quality Initiative Group published a set of guidelines known as the RIFLE criteria, an acronym which stands for “Risk of renal dysfunction, Injury to the kidneys, Failure of kidney function, Loss of kidney function, End-stage kidney disease”. These guidelines were established to identify both early, mild injury as well as severe dysfunction, and also to take into account the contribution of chronic renal injury. The guidelines take into account both changes in glomerular filtration rate or serum creatinine, as well as urine output. Specific criteria are outlined in Table 2.

HARF has also been described in the veterinary literature, although specific references to HARF compared to ARF are limited.[9, 12, 15 16] Vaden et al. reported that intact male dogs and non-sporting dogs were more likely to experience HARF, and that dogs with severe azotemia were less likely to survive.[12] Worwag et al. reported on 32 cases of ARF in cats. Fifty-six percent of cats developed ARF following exposure to a known or suspected nephrotoxin, including administration of medications with known nephrotoxic properties. The remainder of the cats reported was grouped into non-nephrotoxic ARF, and included cats with post-anesthetic ARF, and ARF following hypotensive-hypovolemic episodes.[16] Of the common causes of ARF listed in Table 1, HARF would be more likely in critical illness causing ischemic events. In addition, drug therapies with known nephrotoxic properties such as aminoglycoside antibiotics, amphotericin, and cisplatin can induce renal failure during hospitalization. Therapeutic agents also have the potential to cause renal ischemia; however a direct association between ARF in dogs receiving non-steroidal anti-inflammatories has not been proven.[17-19]

**Table 1:** Common Causes of Acute Renal Failure

| Common Nephrotoxic Agents | Common Ischemic Events | Other Events |
|---|---|---|
| Therapeutics | Shock | Leptospirosis |
| Aminoglycosides | Hypovolemic | Pyelonephritis |
| Cephalosporins | Cardiogenic | Glomerulonephritis |
| Tetracycline | Distributive | Hypercalcemia |
| Amphotericin B | Obstructive | Ureteral obstruction |
| Cisplatin | Decreased cardiac output | Urethral obstruction |
| Acetaminophen | Trauma | Systemic immune- mediated disease |
| Aspirin | Hypotension | |
| Ibuprofen | Hyperthermia/hypothermia | |
| Radiographic contrast agents | Extensive cutaneous burns | |
| Nontherapeutics | Renal vessel thrombosis | |
| Heavy metals | Hyperviscosity syndromes | |
| Ethylene glycol | Nonsteroidal anti-inflammatory agents inhibiting renal prostaglandins | |
| Pesticides | | |
| Herbicides | | |
| Envenomnation | | |
| Hemoglobin/Myoglobin | | |
| Raisins / Grapes | | |
| Easter lily | | |

**Table 2:** The RIFLE Classification Scheme for Identification of Acute Renal Failure [8]

| RIFLE Classification | Glomerular Filtration Rate (GFR) Criteria | | Urine Output Criteria |
|---|---|---|---|
| | Serum Creatinine increase from baseline | GFR Measurement | |
| **R** - Risk of renal dysfunction | > 1.5 X | < 25% | < 0.5 ml/kg/hr for 6 hours |
| **I** - Injury to the kidney | > 2 X | < 50% | < 0.5 ml/kg/hr for 12 hours |
| **F** - Failure of kidney function | > 3 X, or > 4 ug/dl | < 75% | < 0.3 ml/kg/hr for 24 hrs, or anuria |
| **L** - Loss of kidney function | Persistent ARF – complete loss of kidney function for > 4 weeks | | |
| **E** - End-stage kidney disease | End-stage Kidney Disease > 3 months | | |

Of the causes of ARF, some might be preventable with appropriate patient management.[20] Recognizing the predisposing factors for ARF is the first step in prevention. The initiation phase of ARF occurs in a matter of minutes to hours. The extension phase lasts for hours to days. These two phases represent the time during which damage might be prevented, discontinued, or minimized to alter the outcome of ARF. Once the maintenance phase is reached, significant time is required for any recovery to occur.[6] In the human literature, HARF includes prerenal-, renal, or post-renal etiologies. Prerenal azotemia is caused by poor renal blood flow secondary to hypovolemia or cardiac failure, and is characterized by azotemia with concentrated urine specific gravity. Left unresolved, prerenal azotemia becomes ARF, so every effort should be made to re-establish euvolemia. Likewise, prolonged urinary obstruction (bilateral ureteral or urethral) can cause significant renal injury from back-pressure on tubules and failure of solute excretion and should be resolved in a timely manner.

Although HARF cannot be prevented entirely, maintenance of fluid balance is recommended as a preventative measure.[20] Fluid therapy should be tailored to each patient and clinical situation. For initial fluid resuscitation, either crystalloid or colloid fluids (or a combination) can be utilized, and there is no clear indication in the literature of superiority of either. Since crystalloid fluids largely redistribute to the interstitial space within an hour of administration, these should be considered for fluid resuscitation in the dehydrated patient. Fluids containing synthetic colloids (hydroxyethyl starch, dextran) can be administered for shock therapy concurrent with or instead of crystalloids depending upon the specific situation. Following initial resuscitation, fluid therapy should be continued to resolve dehydration and address maintenance therapy and ongoing losses. To prevent hospital-acquired acute renal failure caused by inadequate circulating volume, parameters of volume status should be carefully evaluated at least twice daily. These include maintenance of clinical parameters of hydration (skin turgor, mucous membrane moistness), maintaining arterial blood pressure at a mean arterial pressure of > 80 mm Hg, increasing central venous pressure to 6-8 cm H2O or > 5 cm H2O above baseline. In addition urine output should be quantified, with an attempt to maintain output at greater than 1 ml/kg/hour in normally hydrated animals receiving intravenous fluid therapy. Visual assessment and quantification provides a rough measure of urine output; however urethral catheterization should be considered if exact urine volume needs to be quantified. Drawbacks of urinary catheterization include iatrogenic ascending urinary infection, urethritis from urethral trauma. Alternatively, for small dogs or cats in which urethral catheterization is impractical, the cage may be lined with absorbent pads that are weighed when placed in the cage, and weighed again when soiled to measure urine output. One-gram weight equals one-milliliter volume.

As is shown in Table 1, there are a number of therapeutic agents that increase risk for acute renal failure. If at all possible, avoidance of specific drugs known to cause ARF should be practiced in the critically ill, volume-compromised patient with risk factors known to increase likelihood of ARF. When there is no reasonable substitute for a nephrotoxic drug (e.g. antimicrobial resistance to all but an aminoglycoside for gram negative infection in a compromised patient), drug levels should be monitored and either dosing frequency or interval should be altered based on drug pharmacokinetics. Fresh urine sediment can be evaluated for presence of tubular casts as an early indicator of renal injury. Nonsteroidal anti-inflammatory medications (NSAIDS) have both a direct toxic effect on the kidneys as well as predisposing to an ischemic injury. The direct toxic effect occurs at high doses to cause a renal papillary necrosis. At therapeutic doses, the anti-prostaglandin effect of NSAIDS prevents afferent arteriolar dilation that occurs during low volume states to maintain glomerular filtration rate and renal blood flow. In studies on healthy, normovolemic animals undergoing general anesthesia for routine surgical procedures, NSAIDS that are currently available as analgesics in dogs and cats have not shown any evidence of changes in biochemical parameters for renal failure. [18, 21] Likewise, ARF did not occur in dogs receiving carprofen and experiencing hypotension while under general anesthesia. [17] Despite these studies, animals with critical illness or geriatric animals can have multiple reasons for development of ARF and it would be prudent to select analgesics in animals at risk for renal injury that do not alter renal hemodynamics.

Immediate short-term prognosis with acute renal failure is poor. [22, 23 2, 24] Greater than 60% of patients die from complications of renal failure from all etiologies. In the patients that respond to initial therapy, some percentage will develop complete resolution of azotemia, some will have chronic renal failure and stabilize, and some may need ongoing dialysis therapy or renal replacement at the end of the renal recovery period. Renal biopsy may sometimes offer information as to the underlying etiology of acute renal failure, however does not generally provide any prognostic information for patient recovery.

## REFERENCES

[1]Racusen LC. **Acute Renal Failure: A Companion to Brenner & Rector's The Kidney**. Philadelphia: Saunders, 2001;1.[2]Hoste EA, et al. **Crit Care** 2006;10(3):R73.[3]Nolan CR, et al. **J Am Soc Nephrol** 1998;9(4):710.[4]Ostermann M, et al. **Crit Care Med** 2007;35(8):1837.[5]Lameire N, et al. **Lancet** 2005;365(9457):417.[6]Bonventre JV, et al. **J Am Soc Nephrol** 2003;14(8):2199.[7]Anderson RJ. **Acute Renal Failure: A Companion to Brenner & Rector's The Kidney**. Philadelphia: Saunders, 2001;157.[8]Bellomo R, et al. **Crit Care** 2004;8(4):R204.[9]Behrend EN, et al. **J Am Vet Med Assoc** 1996;208(4):537.[10]Network VNARFT, et al. **N Engl J Med** 2008;359(1):7.[11]Uchino S, et al. **J Am Med Assoc** 2005;294(7):813.[12]Vaden SL, et al. **J Vet Intern Med** 1997;11(2):58.[13]Edelstein CL, et al. **Kidney Int** 1997;51(5):1341.[14]Neveu H, et al. **Nephrol Dial Transplant** 1996;11(2):293.[15]Rieser TM. **Vet Clin North Am Small Anim Pract** 05;35(2):359.[16]Worwag S, et al. **J Am Vet Med Assoc** 2008;232(5):728.[17]Bostrom IM, et al. **Am J Vet Res** 2002;63(5):712.[18]Forrester SD, et al. **Compend Contin Educ Pract Vet** 1999;21(10):910.[19]Ko JC, et al. **J Am Vet Med Assoc** 000;217(3):346.[20]Venkataraman R. **Crit Care Med** 2008;36(4 Suppl):S166.[21]Lobetti RG, et al. **Am J Vet Res** 2000;61(12):1501.[22]Brivet FG, et al. **Crit Care Med** 1996;24(2):192.[23]Morgera S, et al. **Crit Care Med** 2008;36(4 Suppl):S193.[24]Barrantes F, et al. **Crit Care Med** 2008;36(5):1397.

# PATHOPHYSIOLOGY & MANAGEMENT OF ACUTE RENAL FAILURE ASSOCIATED WITH CANINE ENDOCRINE DISEASE

**Marie E. Kerl DVM, Diplomate, ACVIM, ACVECC**
**Columbia MO**

The kidneys are both at risk of injury from a wide variety of endocrine diseases, and are endocrine organs as well. Renal response to endocrine disease can be recognized as failure to adequately concentrate urine, as is seen in diabetes insipidus and hyperadrenocorticism, while other diseases such as hypoadrenocorticism and hypercalcemia result in renal failure that can be acute or chronic. Treatment for some endocrine disorders like feline hyperthyroidism is confounded in the presence of chronic kidney disease. This presentation will review the mechanisms of acute renal failure, and highlight specific occurrences of renal disease with concurrent endocrine disease.

**Adrenal Disease: Hypoadrenocorticism**

Hypoadrenocorticism mimics many other disease presentations. In young dogs, hypoadrenocorticism is frequently misdiagnosed as acute renal failure. Azotemia and isosthenuria are commonly found on laboratory testing in dogs with hypoadrenocorticism.[1] Hyponatremia and hyperkalemia are also frequently present on the same chemistry profiles in laboratory settings where electrolytes are measured; although not all in-house chemistry analyzers used in private practice routinely measure electrolytes. Although azotemia and isosthenuria occur concurrently, the azotemia might still be pre-renal. In hypoadrenocorticism, the medullary concentration gradient is lost due to chronic sodium wasting, so it is impossible for an affected animal to demonstrate adequate urine concentration. Additionally, absence of glucocorticoid attenuates the ability of the hypothalamus to make and release antidiuretic hormone. Due to absence of aldosterone and chronic sodium wasting, hypovolemia and secondary prerenal azotemia occurs. If the hypovolemia is not corrected, permanent renal tubular injury occurs causing nephron damage and establishment of acute renal failure. The long-term prognosis of hypoadrenocorticism is generally guarded to poor once renal failure has been established.

Signalment can be very helpful to increase suspicion of hypoadrenocorticism causing azotemia. The typical age of onset of hypoadrenocorticism is 1-5 years of age, and the primary etiology is immune-mediated adrenal gland destruction. Hypoadrenocorticism is rare in cats and in older dogs. Additionally, the azotemia will sometimes resolve when intravenous fluid therapy is administered, and reoccur once the dog is released from the hospital. Any breed and either gender can be affected, although familial hypoadrenocorticism has been reported in various breeds (e.g. standard poodles, Rottweilers).

Establishing a diagnosis coincides with recognizing the concurrent electrolyte abnormalities and performing adrenocorticotropic hormone (ACTH) stimulation testing. Confirming the diagnosis is imperative since hyponatremia and hyperkalemia can occur from acute renal failure as well as hypoadrenocorticism. Acute renal failure can cause hyperkalemia from poor perfusion and hyponatremia either from renal salt wasting or from antidiuretic hormone-mediated dilution with free water in response to severe hypovolemia.[1]

If hypoadrenocorticism is diagnosed, definitive therapy with glucocorticoid and mineralocorticoid replacement should begin immediately. Any life-threatening fluid or electrolyte disorders should be corrected at the outset of treatment. Adequate circulating volume should be maintained with isotonic crystalloids. If the azotemia resolves completely and does not return when fluid therapy is discontinued, the azotemia is considered to have been pre-renal. Tubular damage might have occurred, but have been below the threshold to result in renal failure, so measures should be taken to protect the kidneys from additional insult for at least 6 weeks which is the minimum recovery period following renal injury. Further insult within this time period has the potential of causing a subclinical renal injury to become clinical renal failure, and make the prognosis worse.[1]

**Adrenal Disease: Hyperadrenocorticism**

Although hyperadrenocorticism is a slowly progressive endocrinopathy and is not typically associated with renal injury, there are some distinct possibilities for chronic renal injury as a result of this disease. Dogs with hyperadrenocorticism are predisposed to ascending urinary tract infections. These infections are associated with inactive urine sediment in 50% of cases. If the infection is localized to the lower urinary tract, then lower urinary signs might be seen. If the infection has ascended to the renal pelvis to cause pyelonephritis, the most common clinical sign is polyuria and polydipsia which is also the most common presenting sign of hyperadrenocorticism. This presentation of renal failure would typically be of chronic renal failure. Therefore, urine culture is indicated prior to antibiotic therapy in all cases of hyperadrenocorticism.

Additionally, hyperadrenocorticism is associated with an increased incidence of calcium oxalate urolithiasis. Nephroliths could cause ureteral obstruction which can cause unilateral renal death if not diagnosed. Animals seemingly show minimal clinical signs of ureteral obstruction, and sometimes obstructing ureteroliths are not diagnosed until the remaining kidney becomes obstructed.

Finally, if hyperadrenocorticism causes hypertension, chronic hypertension can cause glomerular injury and chronic renal failure

**Parathyroid Disease and Hypercalcemia**

Hypercalcemia can be caused by an increase in parathyroid hormone production that is not responsive to negative feedback (primary hyperparathyroidism or parathyroid hyperplasia), by production of a parathyroid hormone-related polypeptide (PTHrP) by malignant cells, or by excess production of a vitamin D analogue by granulomatous cells or ingestion of a toxin or pharmaceutical that is an active vitamin D compound. Idiopathic hypercalcemia has been recognized in cats.

There are several mechanisms that contribute to renal failure with hypercalcemia. Initially, hypercalcemia causes nephrogenic diabetes insipidus and polyuria. Anorexia or nausea might prevent adequate volume replacement, resulting in dehydration. This loss of circulating fluid volume is one method by which hypercalcemia impairs glomerular filtration. Other methods include altered glomerular capillary permeability and sustained renal vasoconstriction causing ischemic injury. The ascending loop of Henle, distal tubule, and collecting tubule basement membranes become sites of soft tissue mineralization. Hypercalcemia impairs mitochondrial function in renal epithelial cells, and cell death and tubular cast formation cause loss of tubular flow. Additionally, chronic hypercalcemia can be associated with formation of calcium-containing urinary stones in the upper or lower urinary tract.

Not all causes of hypercalcemia result in permanent renal injury. Disorders that cause slow, gradual increase of calcium over months to years (e.g. primary hyperparathyroidism) are frequently associated with normal blood urea nitrogen and creatinine. [2] Rapidly progressive hypercalcemia from disorders like vitamin D intoxication or neoplasia is more commonly associated with renal injury and failure.

Prevention of renal failure in hypercalcemia involved rapidly establishing a diagnosis in order to begin definitive therapy, and instituting renal protective measures pending diagnosis. With mild ($Ca^{++}$ < 14-15 mg/dl) hypercalcemia that is not rapidly progressive, insuring appropriate fluid intake via oral replacement for urinary losses or intermittent subcutaneous crystalloid therapy may be the only steps needed. In animals that are more severely hypercalcemic, dehydrated, nauseous or vomiting, intravenous fluid therapy should be instituted. Normal saline (0.9% NaCl) is the treatment of choice since this fluid competes with calcium absorption sites in the kidneys to promote calciuresis. Once the animal is normally hydrated, calcium measurement should be repeated. If hypercalcemia is persistent, furosemide therapy can be initiated at 2-4 mg/kg BID to TID. Furosemide can exacerbate renal injury in the dehydrated patient. If calcium remains elevated and definitive diagnosis has not yet been established, then salmon calcitonin can be instituted at 4-6 IU/kg SQ BID to TID. This drug reduces serum calcium by stopping osteoclast activity. A blunting effect occurs over time to attenuate the effect. This drug is also expensive. Finally, for hypercalcemia with no definitive cure such as vitamin D containing rodenticide intoxication or certain cancers, bisphosphonates can be used to lower serum calcium. Bisphosphonates lower calcium by reducing osteoclast activity and function. Side effects in people include renal impairment, and nausea and esophageal reflux with the oral form. Hydration status should be normalized prior to initiating this therapy. Intravenous injection is typically used to lower calcium initially; whereas oral treatment can be used long term. Oral therapy should not be associated with feeding. Bisphosphonates used either experimentally or clinically in dogs include pamidronate, clodronate, and etidronate.

**Diabetes Mellitus**

Azotemia is an atypical finding in dogs with diabetes mellitus, and might represent concurrent renal failure or prerenal azotemia. In people, diabetic nephropathy is well-described; however this is an unusual occurrence in dogs with diabetes. [3] Diabetic nephropathy is a glomerular disease that results in glomerulosclerosis over time. Pathogenesis is unknown, but is probably multifactorial. [4] [3] Chronic hyperglycemia causes chronic renal hyperperfusion and intraglomerular hypertension. Protein deposition occurs in the mesangium, and thickening of the glomerular basement membrane. These factors cause proteinuria, reduce glomerular bloodflow, and lead to glomerulosclerosis and nephron loss. [4] Proteinuria is inflammatory for renal tubules, and can exacerbate nephron damage.

Urinalysis should be performed regularly in the diabetic patient to monitor for proteinuria. In the absence of overt proteinuria, microalbuminuria can be monitored. Microalbuminuria indicates loss of albumin in the urine that is greater than normal (normal < 1 mg/dl albumin in the urine), but less than would be identified on a urine dipstick for protein (> 30 mg/dl). Strict glycemic control helps to control progression in people, and should be considered in dogs and cats with proteinuria or microalbuminuria. Additional treatments for protein losing nephropathy can also be considered. Angiotensin-converting enzyme inhibitors have been proven to reduce proteinuria, and are used routinely in people to delay progression of diabetic nephropathy.

Animals with concurrent renal failure and diabetes mellitus are extremely challenging to treat. Sometimes an animal that has been managed as a stable diabetic for several years develops chronic renal failure, and becomes extremely difficult to manage. Insulin is renally excreted, therefore insulin kinetics are altered with renal failure. Careful monitoring is necessary to avoid complications of hypoglycemia or hyperglycemia. [3]

**Diabetes Insipidus**

Diabetes insipidus (DI) is an inability to appropriately conserve free water in the kidneys, resulting in polyuria and, consequently, polydipsia. Because of ongoing free water loss, serious metabolic consequences can occur with DI including hyperosmolarity, hypernatremia, and hypovolemia. For clinical understanding, DI is typically divided into central DI, in which ADH is produced in insufficient quantities, and nephrogenic DI, in which the kidneys are not able to respond appropriately to ADH. Renal failure is not necessarily a presenting complaint when DI is diagnosed; however if water is accidentally withheld in an animal with obligate diuresis from DI, renal failure can be the end result. Also, renal failure can be precipitated following water deprivation testing if the clinician does not exercise great care in administering this test. [5-7]

**REFERENCES**

[1]Kintzer PP, et al. **Kirk's Current Veterinary Therapy XIV**. St. Louis: Saunders Elsevier, 2009;231.[2]Feldman EC, et al. **Kirk's Current Veterinary Therapy XIV**. St. Louis: Saunders Elsevier, 2009;247.[3]Greco DS. **Kirk's Current Veterinary Therapy XIV**. 214-218 ed. St. Louis: Saunders Elsevier, 2009.[4]Stokes MB, et al. **Am J Kidney Dis** 2003;41(6):1327.[5]Rose BD. **Clinical Physiology of Acid-Base and Electrolyte Disorders**. 4 ed. New York: McGraw-Hill, 1994;235.[6]Rose BD. **Clinical Physiology of Acid-Base and Electrolyte Disorders**. 4 ed. New York: McGraw-Hill, 1994;261.[7]DiBartola SP. **Fluid, Electrolyte, and Acid-Base Disorders in Small Animal Practice**. 3 ed. St. Louis, MO: Saunders Elsivier, 2006;47.

## ULTRASOUND PERFUSION IMAGING AS A TOOL TO DIRECT PATIENT MANAGEMENT: HUMAN CASE EXAMPLES

**Robert F. Mattrey, M.D.**

The availability of contrast agents has allowed x-ray computed tomography (CT) and magnetic resonance imaging (MRI) to expand the role imaging from defining and depicting structure of normal and diseased tissues to assessing function. Using intravenous contrast, which is common practice with these two techniques, provides information on the relative vascularity and perfusion of a region within a solid organ or a suspected lesion. With repeated observations the wash-in and washout patterns of the contrast agent allows for gross assessment of the temporal patterns of tissue perfusion and with kinetic modeling the extraction of blood flow, fractional blood volume and the capillary-surface product that describes the rate of equilibration between the vascular and interstitial fluid. The major challenge of CT and MRI is the extremely rapid equilibration between these two compartments. In fact, the intravascular contrast concentration on the venule side of the capillary becomes nearly equal to the interstitial concentration during each pass particularly when capillary flow is slow as may be the case in tumors. This extremely rapid equilibration makes it difficult to model the kinetics because of the limited temporal resolution. Attempts at quantifying tumor response to anti-angiogenesis therapy have had mixed results [1].

Three major recent advances have significantly impacted the field of contrast sonography: microbubbles have become potent agents affecting tissue echogenicity at miniscule doses; digital systems with broadband transducers and powerful on-board computers have increased imaging flexibility and improved both contrast and spatial resolution; and greater understanding of bubble interaction with sound has allowed for manipulation of transmitted ultrasound pulses and signal processing to promote image contrast by better suppressing of non-contrast-containing tissues. Several microbubble-based contrast agents are available for non-cardiac applications worldwide, but none are available in the U.S.A. The worldwide experience has reproduced early predictions and solidly positioned contrast-enhanced ultrasound (CEUS) as a reliable, reproducible and powerful imaging technique. While we utilize the approved agents in the U.S.A. for several possible indications off-label, the experience worldwide has mainly focused on liver tumor detection and more so characterization. This presentation will discuss contrast specific imaging that to date has relied on two specific characteristics unique to microbubbles, sensitivity to ultrasound pressure and the generation of harmonic frequencies because of non-linear oscillation of microbubbles or when microbubbles are destroyed when exposed to the compressive and rarefactive forces of the ultrasound wave. These microbubble characteristics and the emerging imaging techniques that utilize them to promote image contrast will be highlighted and the impact of this technology on clinical applications will be presented.

Aside from the extreme sensitivity of current ultrasound imaging to microbubbles, the 1-3μm diameter of microbubbles defines their clinical applications. Because of their relatively large size (little smaller than a red blood cell), they are limited to the vascular compartment. Therefore, vascular enhancement is limited to the filling of the vascular lumen. This allows the assessment of arterial wall disease, luminal patency, detection and definition of extent of thrombi and more important the distinction of tumor from bland thromboses. Because microbubbles only fill the vascular space and because ultrasound acquires images in real time – 10 to 30 frames/sec, ultrasound allows the tracking of microbubbles from the feeding vessels to the capillary bed providing tissue angiography within solid organs that cannot be achieved with CT or MRI. Further, microbubbles reach resonant oscillations in an ultrasound field that cause their collapse and loss. While this fragility limits their survival when insonated at sufficient power, it allows the elimination of tissue enhancement and the monitoring of the rate of tissue refill within a region of interest at high temporal resolution to extract physiologic data. The time intensity curve has an exponential rate of rise and a plateau that define blood flow and fractional blood volume, respectively. While CEUS cannot assess the capillary leak constant (k-trans), it can more accurately assess relative tumor blood flow and fractional blood volume [2, 3], two important tumor indices, where the latter has been well correlated with microvascular density [2], a reliable index of angiogenesis. Further, the extreme sensitivity of US to microbubbles where a single moving microbubble can be seen makes the detection of miniscule residual tumor blood flow possible. For example, CEUS was able to assess and predict hepatocellular carcinoma response to trans-arterial chemo-embolization few days after treatment while DCE-CT and DCE-MRI required three months to achieve sufficient specificity [4]. Early data on quantification of relative blood flow and fractional blood volume and analysis of time-intensity curves in cancer patient on antiagniogenesis therapy suggest that CEUS can detect tumor response to therapy [3]. These tumor indices changed at 7 days after therapy and the change was predictive of tumor response on CT three months later, the standard technique used by the FDA to assess therapeutic response.

For the assessment of liver lesions, the combination of enhancement patterns of tumors and tissue angiography made possible with CEUS has proved to be the most specific imaging tool in recognizing not only benign from malignant tumors but also specific tumor types. Hemangiomas, even those with extremely rapid fill rates that just appear bright on CT and MRI can be seen filling from the periphery to allow their recognition. Focal nodular hyperplasia that is induced by a central vascular lesion can be seen fill from the center outward, the exact opposite of the way hemangiomas fill. These lesions are the only lesions that fill from the center outward. Tortuous irregular vessels that fill tumors heterogeneously and completely are only seen in hepato cellular carcinomas. Metastases develop a complete rim of enhancement rapidly during the arterial phase and fail to fill completely. Because the liver is 30% by volume blood, and because of the irregular sinusoids, liver enhance markedly on delayed scans. This degree of enhancement can only be achieved or exceeded by hempangiomas and other benign lesions. Malignant lesions, even though with arterial flow, have less blood volume than the liver to become hypoechoic relative to the liver on delayed scans. Simply assessing whether lesions are iso- or hyperechoic versus hypoechoic to liver on delayed scans, carries over 90% accuracy in recognizing benign from malignant lesions.

CEUS therefore provides new data not available with CT and MRI that can be acquired at the bedside repeatedly to increase confidence or show temporal changes that increases the statistical power of observations or quantifications. The ability to assess vascular patency can affect management of acute vs. chronic vascular occlusion. Filling the vascular lumen can allow the direct measurement of stenoses. Enhancing perfused tissues allows the assessment solid organ injury or infarction. More, important, tissue enhancement during the arterial phase aids in the detection of tumors and the assessment of tumor response to therapy. Worldwide experience has shown that CEUS is the most specific imaging technique at characterizing liver tumors and classifying benign from malignant lesions. These capabilities that are made available at the bedside or in operating rooms have the potential to significantly impact patient care in a more timely and less costly manner than is currently possible.

## REFERENCES

[1]Ashton E, Raunig D, Ng C, Kelcz F, McShane T, Evelhoch J. Scan-rescan variability in perfusion assessment of tumors in MRI using both model and data-derived arterial input functions. **J Magn Reson Imaging** 2008; 28 (3): 791-6. [2]Lucidarme O, Kono Y, Corbeil J, Choi SH, Golmard JL, Varner J, Mattrey RF. Angiogenesis: noninvasive quantitative assessment with contrast-enhanced functional US in murine model. **Radiology** 2006; 239 (3): 730-9.[3]Lassau N, Lamuraglia M, Chami L, Leclere J, Bonvalot S, Terrier P, Roche A, Le Cesne A. Gastrointestinal stromal tumors treated with imatinib: monitoring response with contrast-enhanced sonography. **AJR Am J Roentgenol** 2006; 187 (5): 1267-73.[4]Kono Y, Lucidarme O, Choi SH, Rose SC, Hassanein TI, Alpert E, Mattrey RF. Contrast-enhanced ultrasound as a predictor of treatment efficacy within 2 weeks after transarterial chemoembolization of hepatocellular carcinoma. **J Vasc Interv Radiol** 2007; 18 (1 Pt 1): 57-65.

## EXPERIMENTAL & POTENTIAL CLINICAL APPLICATIONS OF MOLECULAR ULTRASOUND IMAGING & THERAPY

**Robert F. Mattrey, M.D.**

In vivo cellular and molecular imaging (ICMI) became possible because of the convergence of higher resolution and higher sensitivity imaging and the rapid advance in our understanding of the molecular biology of cell function and signaling in health and in disease. With a wealth of animal models and sophisticated gene manipulations possible, the potential now exists to image cellular mechanisms, including gene expression, in support of new paradigms in therapeutic interventions. As an imager, I view ICMI as a tool development effort to enable non-invasive probing of tissue and tumor function at their basic molecular level. Essentially ICMI is non-invasive "immunohistochemistry" of living systems or a "virtual tissue biopsy". I also believe that imaging must be directly integrated with therapy (see and treat) or must appropriately guide patient management to positively affect patient health. Discovering small tumors at early stages must be accompanied by the ability to distinguish killer from indolent cancers to avoid the risks associated with unnecessary therapy. For instance, the high prevalence of prostate cancer, the high probability that a man will die with rather than from their cancer, and the deterioration of the quality of life associated with all prostate cancer therapeutic techniques, require that the effort should be prioritized for tumor characterization rather than early detection. Finally, cancers kill patients because the treated primary tumor recurs or metastases develop. Chemotherapy offers benefits by its ability to treat invisible tumor deposits, and limitations, because not all immortal cancer cells are sensitive to the agent allowing a resistant clone to immerge. Tissue ablation via minimally invasive means such as radiofrequency, freezing, etc. or non-invasive means such as photoablation or high-intensity focused ultrasound, etc. induce effective mechanical tissue killing regardless of the tissue's genetic or molecular biologic make-up, however, ablative therapy requires tumor visualization and image-guided intervention. It is our belief that using molecular imaging principles to guide non-invasive tissue killing would be an effective method to control cancer. We are working on the potential for sound-guided intervention, a new concept with reasonable probability for success. Should it be successfully accomplished it could allow for tumor ablation without the need to visualize tumors. The concept and goals of molecular imaging and the potential contributions that ultrasound brings to this new field will be presented. for

Molecular imaging does not aim to alter imaging instruments to push their spatial resolution limits to the molecular level. Rather, it aims to develop smart reporter systems that when detected with in vivo imaging equipment implies the presence of specific signaling pathways, receptor systems, or enzymes. Should a gene be introduced into the cell that turns on when a promoter of interest is stimulated to produce a receptor or an enzyme, then molecular imaging tools could be used to recognize the expression of the gene. To be able to recognize areas of enhancement in vivo, a high target to background concentration ratio must be achieved. Therefore, the contrast agent must accumulate in the target at sufficient concentrations to be detected and must be eliminated from the background to increase image contrast. This sensitivity requirement has essentially eliminated the use of x-ray CT agents from the molecular imaging goals. Unfortunately, imaging tools with highest spatial resolution have limited detection sensitivity, while those with highest sensitivity have poor spatial resolution. The decreasing spatial resolution from 10s of μm for CT to mm for MR to cm for PET and optical imaging is accompanied by increasing sensitivity to the reporter from 0.1 moles of an agent for CT to nm concentration for optical imaging. Ultrasound is in a unique position that offers not only high spatial resolution (<1mm) and high sensitivity (single microbubble [1], but does so at very high temporal resolution of many frames per second.

Since a single microbubble can now be detected by the efficient ultrasound techniques [1], the opportunity to target and detect microbubbles generated interest, with several successful in-vivo attempts [2-8] targeting microbubbles to blood clots [2] areas of inflammation [3], tissue rejection [4] and angiogenesis [5, 6]. All these targets are intravascular because the 1-3 μm microbubble-size prevents them from exiting the vascular space. When the 100-200nm droplets of non-targeted PFC emulsions are injected, they accumulate in tumors due to leak across tumor capillaries [9-10]. When these particles are targeted, intravascular [7] as well as extravascular sites can be reached [8] particularly since tumor vessels may even lack endothelial-cell lining [11 ]. However, liquid PFC droplets are minimally reflective (7 times more reflective than RBCs [12]) compared to microbubbles ($10^{12}$ times more reflective than RBCs [13]). Further, liquid PFC droplets are neither destroyed by ultrasound to take advantage of decorrelation imaging [14], nor produce harmonics to take advantage of contrast specific imaging, however, when aggregated liquid PFC droplets become visible in animals [15, 16] and human subjects [9, 17] using clinical scanners.

While current microbubbles can only interact with endothelial cells, these cells offer a plethora of receptors that reflect the status of the surrounding tissues. Inflamed and, ischemic tissues as well as tumors have developed a sophisticated means to interact with blood elements via the endothelial surface. Local and hormonal factors affect endothelial cell function that is induced to attract cells and other effectors to these sites for subsequent trapping and transport across the capillaries to support the needs of the surrounding tissues. The most studied of these changes is the over-expression of integrins, notably alpha-v-beta-3 and others [5, 6]. These receptors have been exploited to target imaging as well as therapeutic agents. Microbubbles can be made to carry a variety of molecules including ligands, antibodies, DNA and/or drugs. These microbubbles have been effectively targeted to activated endothelial cells to detect tumors and potentially deliver a payload.

The microbubble size of 2-3μm is ideal for imaging since these bubbles reach resonance at 2 to 5MHz, the frequencies used for clinical imaging [13]. Further, at these sizes, microbubbles survive plasma surface tension and ambient pressure as they travel through the circulation. Although it would be advantageous to target smaller bubbles that are in the range of 100nm to detect extravascular targets, it is not clear whether smaller bubbles can survive plasma surface tension that increases with the square of the radius. We and others have recently been successful at producing bubble suspensions with a size distribution between 0.1-1μm where the dominant population is ~100-200nm. We anticipate that the nanobubbles will be less echogenic than their larger counterparts despite their much larger number, and will be more fragile when exposed to ultrasound. However, the nanobubbles that now have a comparable diameter to the emulsion particles could be used to reach the tumor matrix [15, 16] as well as target extravascular sites [8].

## REFERENCES

[1] Klibanov AL, Rasche PT, Hughes MS, Wojdyla JK, Galen KP, Wible JH, Brandenburger GH. Detection of Individual Microbubbles of Ultrasound Contrast Age nts Imaging of Free-Floating and Targeted Bubbles. **Invest Radiol** 2004;39: 187–195. [2] Wu Y, Unger EC, McCreery TP, Sweitzer RH, Shen D, Wu G, Vielhauer MD. Binding and lysing of blood clots using MRX-408. **Invest Radiol.** 1998;33:880-5.[3] Lindner JR, Song J, Christiansen J, Klibanov AL, Xu F, Ley K. Ultrasound assessment of inflammation and renal tissue injury with microbubbles targeted to P-selectin. **Circulation**. 2001;104:2107-12. [4] Weller GER, Lu E, Csikari MM, Klibanov AL, Fischer D, Wagner WR, Villanueva FS. Ultrasound Imaging of Acute Cardiac Transplant Rejection With Microbubbles Targeted to Intercellular Adhesion Molecule-1. **Circulation** 2003;108:218-224.[5] Leong-Poi H, Christiansen J, Klibanov AL, Kaul S, Lindner JR. Noninvasive assessment of angiogenesis by ultrasound and microbubbles targeted to alpha(v)-integrins.**Circulation** 2003; 107:455-460[6] Ellegala DB, Leong-Poi H, Carpenter JE, Klibanov AL, Kaul S, Shaffrey ME, Sklenar J, Lindner JR. Imaging Tumor Angiogenesis With Contrast Ultrasound and Microbubbles Targeted to Alpha-v Beta-3. **Circulation** 2003;108:336-341.[7] Hall CS, Marsh JN, Scott MJ, Gaffney PJ, Wickline SA, Lanza GM. Time evolution of enhanced ultrasonic reflection using a fibrin-targeted nanoparticulate contrast agent. **J Acoust Soc Am**. 2000; 108:3049-57. [8] Lanza GM, Abendschein DR, Hall CS, Scott MJ, Scherrer DE, Houseman A, Miller JG, Wickline SA. In vivo molecular imaging of stretch-induced tissue factor in carotid arteries with ligand-targeted nanoparticles. J Am Soc Echocardiogr. 2000; 13:608-14.[9] Behan, M. O'Connell D, Mattrey RF, Carney DN. Perfluorooctyl bromide as a Contrast Agent For Sonography and Computed Tomography. AJR 1993;160:399-405.[10] Mattrey RF, Long DM, Multer F, Mitten R, Higgins CB: Perfluorooctyl bromide: A Reticuloendothelial-Specific and A Tumor Imaging Agent for Computed Tomography. **Radiology**. 1982; 145:755-758.[11] Folberg R, Hendrix MJ, Maniotis AJ, Vasculogenic mimicry and tumor angiogenesis. Am J Pathol, 2000. 156(2): p. 361-81.[12] Mattrey RF, Steinbach GC: Ultrasound Contrast Agents: State of the Art. **Invest Radiol**. 1991; 26:S5-S11.[13] Ophir J, Parker KJ. Contrast agents in diagnostic ultrasound [published erratum appears in Ultrasound Med Biol 1990;16:209] **Ultrasound Med Biol** 1989;15:319-33[14] Tiemann K, Becher H, Bimmel D, Schlief R, Nanda NC. Stimulated Acoustic Emission Nonbackscatter Contrast Effect of Microbubbles Seen with Harmonic Power Doppler Imaging. **Echocardiography**. 1997;14:65-70[15] Mattrey RF, Scheible FW, Gosink BB, Leopold GR, Long DM, Higgins CB: Perfluorooctyl bromide: A Liver\Spleen-Specific and a Tumor-Imaging Ultrasound Contrast Material. **Radiology**. 1982; 145:759-762.[16] Mattrey RF, Leopold GR, vanSonnenberg E, Gosink BB, Scheible FW, Long DM: Perfluorochemicals as Liver- and Spleen-seeking Ultrasound Contrast Agents. **J Ultrasound Med**. 1983; 2:173-176.[17] Mattrey RF, Strich G, Shelton RE, Gosink BB, Leopold GR, Lee T, Forsythe J: Perfluorochemicals as US Contrast Agents for Tumor Imaging and Hepatosplenography: Preliminary Clinical Results. **Radiology**. 1987; 163:339-343.

## CONTRAST-ENHANCED CT IMAGING IMPROVES LESION DETECTION & CHARACTERIZATION: CASE EXAMPLES

**Dr. Gabriela Seiler, Dr.Med.Vet. DACVR, DECVDI**
**Philadelphia, PA**

### INTRODUCTION

The use of contrast media is an integral part of Computed Tomographic (CT) examinations in veterinary medicine. Intravenous, intrathecal and oral applications are available, although as opposed to human medicine, oral contrast media such as barium sulfate are rarely used in veterinary CT imaging. This may change in the future as abdominal CT is likely to become more frequently used with the increasing availability of fast scanning multislice CT units.

### Types of contrast used & dosages

For intravenous contrast-enhanced CT, the use of non-ionic iodinated contrast is the standard of care in most institutions. Non-ionic iodinated contrast media are well tolerated by most patients and are associated with a low incidence of adverse effects, most of which are transient and mild, whereas ionic contrast media have been shown to be associated with higher incidence and potentially serious side effects[1-3]. Dosages for intravenous contrast injections are 600-800mg I/kg. Intrathecal contrast-enhanced CT (CT-myelogram) is often performed following a conventional myelogram and therefore the normal dose for myelography is applied (0.45ml/kg of 200-300mgI/ml)[4]. If CT-myelography is performed without preceding conventional myelography, the dose can likely be reduced; however there are no detailed studies available on the optimal dose for this application.

### Applications of contrast-enhanced CT and imaging protocols

At our institution, about 90% of all small animal CT studies include the administration of water-soluble iodinated contrast agents. While there are specific indications for which contrast application is not necessary such as joint or fracture evaluation, contrast enhanced CT is advantageous if not essential for most other applications for various reasons. First of all, the normal anatomy is better delineated after contrast administration because of differential perfusion of tissues, and large blood vessels can be identified. Pathology is more easily detectable and better outlined, as most inflammatory and neoplastic lesions enhance after contrast administration. In some instances, the type and degree of enhancement even allows to diagnose specific disorders and lesions. Surgical planning is greatly facilitated by determining the extent and margination of a lesion and potential involvement or invasion of the adjacent vessels.

Imaging protocols always include a pre-contrast image series followed by a post-contrast series with thick slice thickness. Thicker slices (3-5mm) are required after contrast administration as the images have to be displayed in a narrow window in order to see the differences in contrast uptake, and thin slices result in a high degree of noise when viewed in a narrow window. If the vessels need to be visualized, the image series is acquired immediately after injection of a bolus of contrast, typically using an automatic power injector. Timing of the acquisition has to be adjusted to the vasculature of interest, the exact timing can be established with a test injection of a small volume of contrast followed by continuous scanning of a single slice containing the vasculature of interest. Arrival of the contrast in the arterial and venous or portal system can then be measured exactly using CT software. Typically the arterial phase starts about 10 seconds, the venous phase starts between 25 and 35 seconds after i.v. injection into a cephalic vein. Following angiography a second post-contrast image series is often obtained to allow any lesions to be fully saturated with contrast.

## SELECTED APPLICATIONS OF CONTRAST-ENHANCED COMPUTED TOMOGRAPHY

### Head and Neck imaging

Especially in head and neck imaging where many small soft tissue structures have to be differentiated, contrast enhancement is necessary to outline the normal anatomy. Tissues with normally very strong contrast enhancement include the soft palate, the salivary glands, lymph nodes, and the esophageal mucosa. Assessment of the origin of a lesion, involvement of other tissues, and lymph node size and morphology in patients with head and neck cancer is difficult without contrast administration.

One of the more common indications for CT in our practice is nasal disease. Presence and extent of bone destruction as well as soft tissue proliferation are the main CT criteria used to distinguish between common diseases such as nasal neoplasia, fungal rhinitis or non-destructive rhinitis[5]. Nevertheless, comparison of pre- and post-contrast images is essential to determine if focal lesions are perfused or if they just represent accumulations of fluid, mucus or necrotic tissues. The frontal sinuses are a typically filled with soft tissue attenuating material in the presence of nasal disease and only after contrast application it can be determined if it solely represents fluid or mucus accumulation or if there is extension of the disease process into the sinus. Similarly, while erosion of the cribriform plate can be diagnosed on pre-contrast images, confirmation of brain involvement requires contrast administration[5]. Once a lesion is detected, an attempt can be made to target the well perfused areas when obtaining biopsy samples, thus increasing the diagnostic quality of the sample.

A characteristic pattern of contrast enhancement has been described in dogs with masticatory myositis[6]. These patients often present with reluctance to eat, weight loss, drooling, and swelling of the masticatory muscles, or if the disease is chronic, with muscle atrophy. On clinical examination, most dogs show pain during muscle palpation or when opening their mouth. Other diseases with similar clinical presentation include temporomandibular joint disease, bone and soft tissue neoplasia retro-orbital space-occupying lesions and polymyositis. CT is a means to rule out any space-occupying lesions, assess the temporomandibular joint and characterize the changes within the masticatory muscles. Changes in size of the masticatory muscles is a common finding on CT, typically the digastricus muscle is spared though. Additionally, marked contrast enhancement with an inhomogeneous pattern is usually present, mainly in the temporal and pterygoid muscles. The enhancement can be linear or geographic (focal) or can present as rim enhancement around a central hypoattenuating lesion. The inhomogeneous enhancement pattern was interpreted as areas of inflammation with increased vascularity and interspersed hypoattenuating areas of necrosis[6].

**Thoracic imaging**

CT is the imaging method of choice for many thoracic conditions in human medicine and is increasingly used in veterinary medicine as well. It has been shown to add valuable In our clinic contrast is routinely used for most indications, exceptions are thoracic CT to search for metastatic pulmonary disease, pulmonary bullae or for diffuse pulmonary parenchymal diseases such as pulmonary fibrosis, in these cases contrast application does not significantly add to the evaluation. Indications that always warrant contrast-enhanced imaging protocols include evaluation of mediastinal masses, lymph nodes, and heart base masses[7]. The perfusion of the mass itself is not predictive of the origin, except for mediastinal cysts which have a thin wall and contain non-enhancing material. A large percentage of thymomas are cystic, however other tumors such as lymphoma may comprise necrotic areas that are hypoperfused as well. Angiography protocols are frequently used in combination with a delayed post contrast image series to determine the degree of vascular involvement or invasion, which in turn decides if surgical removal of the mass can be attempted[8].

Hypoperfusion of a lung lobe after contrast administration, in conjunction with other findings such as abrupt ending bronchi, abnormal position of a lung lobe, and vesicular gas pattern within the lobe is a sign of lung lobe torsion[9]. In the presence of pleural effusion intravenous contrast application will allow detection of pleural masses (mesothelioma, carcinoma) and facilitate distinction between the fluid, mediastinal structures and collapsed lung lobes.

**Abdominal imaging**

In recent years, ultrasound has been the principal imaging method to evaluate abdominal disease in small animals. CT has already established itself as the imaging method of choice for nasal disease and many thoracic diseases, and it may very well be the imaging modality that will challenge the leading position of ultrasound for abdominal imaging in the future.

A relatively frequent indication to perform contrast-enhanced CT is the pre-surgical evaluation of adrenal tumors. In the veterinary literature the CT appearance of the adrenal glands of normal dogs and dogs with pituitary-dependent hyperadrenocorticism has recently been described[10], but there is no information available on differentiation between adrenal adenomas and malignant adrenal masses such as carcinomas, pheochromocytomas and metastatic disease. A different arterial enhancement and contrast washout phase has been described in the human literature[11]. In our clinic CT is a very valuable method though to detect adrenal masses in patients with retroperitoneal hemorrhage and to determine the presence or absence of vascular invasion of adrenal masses. For this purpose, CT angiography is performed.

Other abdominal masses that can be assessed with contrast-enhanced CT are hepatic and splenic masses. Again, in the veterinary literature there is not much information available on CT characteristics of those lesions, whereas in human medicine they are very well described[12,13]. In dogs, one study describes the use of contrast enhancement for determination if a splenic lesion is benign or malignant. The authors found that on post-contrast CT images there was a significant difference in attenuation characteristics (measured in Hounsfield Units, HU) among malignant splenic masses, hematoma, and hyperplasia. A cutoff value of post-contrast medium HU values of 55 was determined to be the best threshold value to distinguish malignant from nonmalignant masses, with those less than 55 HU being malignant[14]. There is not much information available for differentiation between benign and malignant liver masses, even though CT is widely used in human medicine to characterize liver lesions. More work has been done in veterinary medicine using contrast harmonic ultrasound (CHUS) which has been shown to be able to differentiate between benign and malignant liver nodules based on their perfusion characteristics[15].

In human medicine, dual-phase CT angiography is used to differentiate between pancreatic carcinoma, nodular hyperplasia and insulinoma, based on their perfusion characteristics during the arterial, portal and late venous phase of contrast enhancement[16]. Insulinomas tend to have a very strong arterial blood supply and therefore enhance very strongly immediately after injection of a bolus of contrast. Carcinomas on the other hand are poorly perfused and are hypoattenuating compared with the surrounding pancreatic parenchyma during the arterial phase, whereas nodular hyperplasia has very similar contrast enhancement patterns compared with the surrounding pancreas and therefore blend in with the remaining pancreatic tissue after contrast administration[16]. It is not known if the same criteria apply to our canine and feline patients. However, recently a few cases with pancreatic insulinoma were described which also had a strong arterial phase enhancement[17].

Other applications where CT angiography is used for lesion characterization are masses of unknown origin (pancreatic vs. liver for example); in those cases the vascular supply of the mass can be determined using CT angiography. For advanced therapy options such as chemoembolization, determination of the arterial vascular supply of a mass is commonly performed using CT angiography.

**Soft tissue tumor imaging**

Assessment of the exact size, location and margination of a lesion is especially important for surgical and radiation therapy planning. Important examples are feline fibrosarcomas. They present as a distinct mass lesion in the subcutaneous tissues which can be easily recognized without contrast administration. However, fibrosarcomas have a tendency to produce long tendrils extending in the deeper tissues and those tendrils become only apparent after contrast administration[18]. Soft tissue sarcomas in general are commonly evaluated with contrast-enhanced CT to determine if they are resectable and to what extent surrounding normal structures have to be removed to achieve acceptable margins.

Large lipomas can clinically not be distinguished from infiltrative lipomas, and extent and involvement of adjacent structures often needs to be assessed before surgical removal is attempted. Infiltrative lipomas Due to the superb contrast resolution of CT imaging the fat-attenuating mass can readily be distinguished from the surrounding musculature[19]. Preliminary information though suggests that the distinction between infiltrative lipoma, which is locally aggressive with infiltration of surrounding tissues but does not metastasize, and liposarcoma which can metastasize can be made using contrast-enhanced CT. Liposarcomas appear to have a soft tissue attenuating, contrast enhancing component whereas infiltrative lipomas are generally not contrast enhancing.

**Neuroimaging**

Magnetic resonance imaging (MRI) is the method of choice for neuroimaging in general as it is very sensitive to tissue alterations, and CT is only used if MRI is not available. In acute head trauma contrast administration is usually not necessary as acute hemorrhage can be readily identified based on the increased attenuation. The majority of brain lesions though is isoattenuating to the surrounding brain tissue on pre-contrast images and can only be detected if they are associated with a significant mass effect seen as shift of the midline structures or compression of the ventricular system, or if they are surrounded by hypoattenuating edema[20]. Most inflammatory and neoplastic lesions however enhance after contrast administration. In the central nervous system the blood-brain barrier normally prevents any contrast enhancement of neural tissues. Tumor vessels on the other hand have an abnormal wall structure resulting in leakage of contrast into the extravascular space and enhancement of the neoplastic tissue. Slow-growing brain tumors such as gliomas may remain non-enhancing and therefore undetected because of a relatively normally developed vasculature with an intact blood-brain barrier.

Evaluation of the spinal cord with CT is mainly performed if there is a suspicion of an aggressive vertebral lesion. Even though the bone destruction is readily seen on pre-contrast images, the increased contrast uptake of the soft tissue portion of the mass will allow better evaluation of size and degree of spinal cord compression. For diagnosis of extradural spinal cord compression caused by intervertebral disc herniation, a myelogram is ideally performed in conjunction with the CT. If for any reason intrathecal contrast application is not feasible then intravenous contrast administration is helpful in delineating the epidural vascular structures from abnormal material in the vertebral canal such as herniated disc material.

**REFERENCES**

[1]Pollard et al. **J Am Vet Med Assoc** 2008;233:274-278. [2]Pollard RE et al. **Am J Vet Res** 2008;69:1274-1278. [3]Pollard RE et al. **Am J Vet Res** 2008;69:1268-1273.
[4]Wallack ST. In: **The handbook of veterinary contrast radiography**. 2003:95. [5]Tromblee TC et al. **Vet Radiol Ultrasound** 2006;47(3):241-248. [6]Schwarz T et al. **J Am Vet Med Assoc** 2007;231:924-930. [7]Yoon J et al. **Vet Radiol Ultrasound** 2004;45(6):542-546. [8]Scherrer W et al. **N Z Vet J** 2008;56(6):330-333. [9]Seiler G et al. **Vet Radiol Ultrasound** 2008;49(6):504-508. [10]Bertolini G et al. **Vet Radiol Ultrasound** 2008;49(5):449-453. [11]Szolar DH et al. **Radiology** 2005 ;234:479-485. [12]Morana G et al. **Cancer Imaging** 2008;8:S132-135. [13]Takayasu K et al. **Oncology** 2007;72 Suppl 1:83-91. [14]Fife WD et al. **Vet Radiol Ultrasound** 2004; 45: 289-297. [15]Ivancic M et al. **J Am Vet Med Assoc** 2009;234(1):88-94. [16]Tamm EP et al. **Haematol Oncol Clin N Am** 2007;21:409-432. [17]Mai W et al **Vet Radiol Ultrasound** 2008;49:141-148. [18]Davis KM et al. **Compend Contin Educ Vet** 2007;29(12):712-714. [19]McEntee MC et al. **Vet Radiol Ultrasound** 2001;42(3):221-225. [20]Kraft SL et al. **Clin Tech Sm Anim Pract** 1999;14(2):112-123.

## CASE EXAMPLES OF CONTRAST ENHANCED & DIFFUSION-WEIGHTED MR IMAGING

**Shannon P. Holmes, DVM, MSc, DACVR**
**Gainesville, Florida**

The definition of image contrast is the *relative* difference in echogenicities/attenuations/intensities between two adjacent regions within an object on a gray (or color) scale. In medical imaging, the process of contrast enhancement most commonly refers to the intravenous administration of a contrast agent to increase this relative difference between normal and disease states. In ultrasound imaging, contrast is providing by the intravenous administration of gas-containing bubbles. Iodine based contrast agents are used in x-ray based imaging modalities, like CT. MRI differs from the other modalities because a variety of contrast agents are available which result in different contrast characteristics. The gadolinium-based contrast agents are the most widely used in clinical studies, but other are continuously being developed and explored. MR imaging also differs from other imaging modalities because contrast can be created through manipulation of MR physics. Examining the different intrinsic contrasts of tissue and define the chemical nature of tissues. Intrinsic contrast includes T1 and T2 relaxation, T2-star (T2*), proton density (PD), diffusion, perfusion and BOLD imaging. This session will cover, through case examples, the use and benefit of the various contrast enhancing MR imaging techniques of veterinary clinical imaging.

### INTRAVENOUS CONTRAST AGENTS

The contrast agents of other imaging modalities increase contrast by their visible presence; that is, the contrast agent is visualized. All MR contrast agents used to increase contrast differ because the agent is not visible. It is the effect of the contrast agent on the regional protons that is visualized as increased image contrast. There are 2 main groups of the intravenous MR contrast agents. The paramagnetic contrast agents increase signal intensity, where as the effect of supraparamagnetic contrast agents decreases the signal intensity of tissue.

Another manner that MR contrast agents differ from the contrast agents of the other imaging modality is that their effect is no only determined by the concentration in the tissue. Proton density is important since MR contrast agents act indirectly through the protons within the body. The resultant effect of MR contrast agents is also influence by proton diffusion and the regional chemical environment of the protons.

#### Gadolinium-Based Contrast Enhanced Imaging

In clinical MR imaging, gadolinium-enhanced studies are routine, included in almost all studies. The appearance of lesions on different pulse sequences, as well as on gadolinium-contrast enhanced images, facilitates more definitive diagnoses than achievable with other imaging modalities.

Gadolinium-based contrast agents are the most common paramagnetic contrast agents. Clinical gadolinium contrast agents are considered extracellular compartment agents, because they rapidly accumulate in blood plasma and the extracellular space following administration. Tissue enhancement, therefore, is indicative increased vascularity and/or vascular permeability. The time to peak contrast enhancement is approximately 7-10 minutes with a half life of 15-19 minutes. This contrast agent is relatively rapidly excreted by the renal system. T1-weighted spin echo or turbo/fast spin echo images are most commonly acquired following the administration of gadolinium-based contrast. These agents shorten T1 relaxation and therefore as relatively hyperintense areas on the T1 images. Recently in veterinary medicine, consistent with human MR techniques, contrast enhanced FLAIR images have been found to have superior lesion conspicuity to conventional T1-weighted contrast enhanced imaging. Gradient-recalled echo sequences made prior to and following contrast administration are also being explored. The advantage of these sequences is the rapid acquisition times and higher resolution images.

The benefit of gadolinium contrast enhancement was initially realized in the detection of intracranial neoplasms. The identification of dural tails on post-contrast enhanced MR images facilitates definitive meningioma diagnoses. The pattern and amount of contrast enhancement of an intracranial neoplasm can be an identifying quality when considered with the lesions signal intensity characteristics on the other routine brain images. The largest report on the appearance of histologically diagnosed intracranial neoplasms demonstrated that size, shape, location, relative signal intensity and contrast enhancement pattern were repeatable within tumor type. Continued development of our knowledge of the MR appearance of lesions and their contrast enhancement patterns in other body systems is needed.

There are also endogenous compounds that produce paramagnetic contrast enhancement. These include melanin and methemoglobin. When present, the tissue occupying these compounds will appear hyperintense on T1-weighted images. Melanin tumors can be definitively diagnosed on MR images based on this signal intensity characteristic. An extension of this process is the appearance of multifocal hyperintense round lesions in the brain parenchyma, representing metastatic melanomas. The blood breakdown product, methemoglobin, is known to evolve at a specific time post-hemorrhage. Approximately 3 days following the hemorrhagic event, methemoglobin will produce hyperintense areas on T1-weighted images. Gradient recalled sequences are useful for confirming the presence of hemorrhage and reaching a very specific MR diagnosis.

#### Dynamic Contrast Enhanced MR Imaging (DCE-MRI) using Gadolinium

In DCE-MRI, a series of images are acquired before and repeatedly after the injection of a paramagnetic contrast agent. The images produced are a temporal representation of the MR signal intensity change for each image voxel, following the influx and efflux of contrast into a tissue. It is considered a noninvasive imaging technique for tissue microvasculature qualitative and quantitative evaluation.

In veterinary medicine, qualitative DCE-MRI is used for the detection of pituitary microadenomas in dogs. The micoradenoma has a similar enhancement pattern to that described for most malignancies. The microadenoma will enhance more greatly in the arterial phase than the remaining pituitary parenchyma, which is followed by a relative rapid decrease in enhancement relative to the surrounding tissue. As a result, the microadenoma is hypointense to the surrounding parenchyma in the images acquired towards the end of the study which correlate to the venous phase.

Clifford et al. (2004) used the DCE-MRI to correlate lesions detected on MR images with the hepatic and splenic histology/cytology. The lesion characterization was made in conjunction their appearance on T2-weighted images, but their enhancement pattern tended to better distinguish between benign and malignant lesions. Of the 35 focal hepatic or splenic lesions, the appearance on the MR images classified a lesion as malignant or benign with an overall accuracy 94% (33 of 35 lesions). The overall sensitivity was 100%, where as the overall specificity was 90%, which represents unparallel diagnostic accuracy compared to other imaging modalities. At this time, abdominal MR in veterinary practice is limited, which is most commonly attributed to the cost of MR procedures and the added expense and risks associated with the need for general anesthesia. With results as accurate as these, better diagnostic and therapeutic plans could translate into better cost-benefit outcomes. This is a commonly recognized phenomenon in human medicine.

DCE-MRI is considered a very exciting area of oncologic research and clinical trials in drug development. The images have been shown to have strong correlation to tumor grade, microvessel density/perfusion, and vascular endothelial growth factor [VEGF] expression. Since these are important prognostic factors, DCE-MRI has the potential to function as a predictive tool for recurrence and survival. Numerous animal models have been used to investigate different applications of DCE-MRI to oncologic studies. There is one study that looked at canine spontaneous nasal and oral tumors and changes in the DCE-MRI tumor patterns observed during the course of radiation therapy. Dramatic changes were seen in these tumors on the DCE-MRI studies which were correlated to radiation-induced changes in tissue contrast enhancement, as well as reduction in viable tumor tissue and tumor size. Intratumoral heterogeneity is a recognized challenge to radiation therapy protocols. With DCE-MRI, the radiosensitivity may be better spatially defined. Voxel-based dose prescriptions, based on voxel signal intensity, could be employed and refined during the course of treatment. This is referred to as "biological image-guided radiotherapy."

**Nephrogenic Systemic Fibrosis (NSF)**

The major reported advantage of MR imaging, compared to CT imaging as an example, has been that it is truly a non-invasive procedure, since it does not use ionizing radiation. NSF, previously known as nephrogenic fibrosing dermopathy (NFD), represents the first major adverse effect of MR imaging. While it was first reported in 1997, it has only been reported in humans with severe renal impairment (glomerular filtration rate <30 mL/min/1.73 $m^2$) and with acute renal insufficiency. It has not been reported in the veterinary literature. It is a late occurring reaction, characterized by progressive thickening of the skin, intermittent pigment changes, joint stiffness, extremity swelling, cutaneous redness, pruritus, and a burning sensation. The etiology of NSF is suspected to be multifactorial; etiologies currently being investigated as triggers included cytokine induction, procoagulable status and vascular injury, and gadolinium (Gd)-based contrast agents.

NSF attributed to contrast-enhanced MR imaging is most commonly seen in patients undergoing MR angiography. This association is suspected to be related to higher Gd doses administered during these procedures. In response to this side effect, new methods of MR angiography are being explored in which contrast administration is unnecessary. More MR angiograms are being performed in veterinary medicine and clinicians should consider patient's renal status prior to procedures involving Gd contrast.

**Supraparamagnetic Iron Oxide (SPIO) Contrast Enhanced Imaging**

SPIO is referred to as a negative contrast agent, because it produces a darkening of the tissue. The alteration in signal intensity is due to shortening T2* and T2 relaxation. These contrast agents consist of suspended superparamagnetic nanoparticles with an iron oxide moiety (including magnetite, $Fe_3O_4$, maghemite, γFe2O3, or other ferrites). These possess a relatively large net magnetic dipole, when compared to its magnetic dipole due to its individual unpaired electrons. In tissue, the contrast agent produces large regional magnetic-field heterogeneity. As the protons relax in this environment, there is greater dephasing of their magnetic moments which equates to shortening of T2 relaxation. Because of their effect on the magnetic field, these are also referred to susceptibility contrast agents.

They are also considered tissue-specific contrast agents. The liver extracts the majority of the contrast, via the reticolendothelial system. Liver contrast imaging is the extent to which SPIO has been explored in veterinary medicine. The one study was a descriptive study on the MR appearance of the liver following SPIO administration. The other study, performed by Holmes et al. (unpublished data), investigated differences in hepatic SPIO uptake in patients with normal livers and in the livers of patients diagnosed with microscopic and/or macroscopic portosystemic vascular anomalies. In human MR imaging, this contrast agent has been extensively research for distinction of hepatic tumors types. Because malignant hepatic neoplasms lack normal a RES, they appear hyperintense to the surrounding markedly hypointense normal liver parenchyma.

SPIO is actually not truly a tissue specific contrast agent. It is almost exclusively phagocytosed by activated macrophages. In the liver, the high population of Kupffer cells is the reason it was initially thought to be tissue specific. It is now thought to be an excellent contrast agent for imaging inflammation. Investigation of multiple sclerosis have used SPIO contrast enhanced images to better understand the waxing and waning intraneural inflammation and its deleterious effects on myelin. MR microscopy has detected SPIO uptake in damaged spinal cords of rats and the perivascular cuffs of multiple sclerosis. The level of resolution of MR microscopy and SPIO contrast enhancement has reached single cell imaging.

## DIFFUSION WEIGHTED IMAGING *(DWI)*

Diffusion images represent the *in vivo* characteristics of water diffusion in each designated voxel (volume of tissue pictured in each pixel of an image). Regular MR techniques are modified to display the Brownian motion of molecules and quantify different chemical properties of tissues. In DWI, the diffusion of water molecules along a magnetic field gradient reduces the MR signal; with less diffusion, the decrease in signal intense is less intense and is thus displayed in the image as a relatively brighter area.

The Echo Planar Imaging (EPI) pulse sequence is used to produce DW images. EPI involves the application of 2 temporally separated RF pulses (called a bipolar gradient pulses) that produce a gradient in the magnetic field strength. The protons precess at different rates, which equates to dephasing. When refocused with the second pulse, the non-uniformity of the precessing protons results in signal loss. This loss of signal is related to water diffusion. These signal values are translated as apparent diffusion coefficients (ADC) and are mapped as an image (called an ADC map). Like other "weighted" MR sequences, the degree of diffusion weighting is a selectable parameter. The weighting of a DWI sequences is determined by the strength of the diffusion gradients, which are determined the selected b-value.

DWI has revolutionized MR imaging of cerebral vascular accidents in both human and veterinary imaging. It is one disease process in which reduced diffusion results and can therefore be non-invasively examined with DWI. Demyelination processes are another. At this time, the delineation of cerebrovascular accidents ("strokes") is the only clinical application of DWI routinely used in veterinary MR studies. The addition of this sequence to a study facilitates definitive distinction of infarcts and intracranial neoplasms. DWI continues to be extensively researched in other areas of the human body, including differentiation of malignant breast and prostatic neoplasms. Similar potential can be realized in veterinary MR imaging as progression is made outside the cranial vault and the value of DWI is better understood.

### Diffusion Tensor Imaging

DTI is a more intricate version of DWI because it maps the magnitude and direction of diffusion. In the world of neurology, it is referred to as "Tractography." This terminology originates from the visual portrayal of the white matter fibers in the brain . DTI is extremely sensitive to small changes in the white matter which has lead to extensive MR research in multiple sclerosis and epilepsy. Water diffusion in white matter regions of the nervous system are usually more anisotropic, where as the gray matter and CSF diffusion is isotropic. With DTI, the fractional anisotropy (FA) gives information about the shape of the diffusion tensor at each voxel. The fractional anisotropy reflects differences between an isotropic diffusion and a linear diffusion. The direction of diffusion is normally parallel to the local white matter fascicle. Reduction or loss of anisotropy is seen in white matter diseases and the severity of the disease can be estimated. Maps of entire white matter tracts can be elaborated from the voxel diffusivity information. These maps have proven beneficial in pre-surgical planning of intracranial tumor resection.

Numerous veterinary institutes have installed high field MR units (>2.0 Tesla), which will ease the investigation of DTI applications in veterinary patients, due to the increase in available MR signal. DT images can be produced on clinical MR units (1.5 Tesla), but this process if more challenging in the small patients examined in veterinary imaging.

## REFERENCES

CE-MRI: Burtea C et al. 2008;(185 Pt 1):135-65. Falzone C et al. **Vet Radiol Ultrasound**. 2008 Jul-Aug;49(4):333-8. Søvik A et al. **Acta Oncol**. 2008;47(7):1249-56. Schultz RM et al. **Vet Radiol Ultrasound**. 2007 Mar-Apr;48(2):101-4. Kraft S et al. **Vet Radiol Ultrasound**. 2007 Jan-Feb;48(1):1-7. Varejão AS et al. **Vet Radiol Ultrasound**. 2006 Jul-Aug;47(4):328-33. Chang D et al. **Vet Radiol Ultrasound**. 2002 Jan-Feb;43(1):37-42. DWI: DGarosi L et al. J Vet Intern Med. 2006; 20(2):311-21. McConnell JF et al. **Vet Radiol Ultrasound**. 2005 Jan-Feb;46(1):1-10. Alexander, AL et al. **Neurotherapeutics**. 2007 July; 4(3): 316–329. Mukherjee P, et al. **AJNR** 2008 Apr;29(4):632-41.

# KEYNOTE ADDRESS

## "Creating *Buy-in* to Avoid a *Bail-out* During Challenging Economic Times"

## Presented by:
## Dr. Pat LeBlanc, DVM, MS, Diplomate ACVA

In a difficult economic environment, maintaining employee engagement and positive morale has the potential to be inversely related to a maintaining the organizations bottom line. In his Keynote address, Dr. Pat LeBlanc, Director of the Veterinary Teaching Hospital at Michigan State University, will talk about this link and how to forge a positive work atmosphere without losing focus on the importance of financial success in your practice or clinic.

# ABSTRACT DISCLOSURE INDEX

| Abstract # | Disclosure |
|---|---|
| 36 | All authors are employees of Boehringer Ingelheim; Funding by Boehringer Ingelheim |
| 43 | Previously presented at ECVIM in Belgium, September 2008 |
| 58 | Previously presented at Conference of Research Workers in Animal Diseases |
| 93 | Previously presented at ACVS, October 2008 |
| 99 | The study features a diet which is manufactured by Royal Canin. The control dogs were seen at the *Royal Canin Weight Management Clinic*, University of Liverpool, which is supported by a grant from WALTHAM. Funding for the study was kindly provided by Royal Canin US, IDEXX and Platinum Performance. AJ German: Senior lectureship funded by Royal Canin. SL Holden: Nursing post supported by Royal Canin. D. Elliott: employee of Royal Canin US. |
| 100 | Previously presented at European Society of Veterinary Comparative Nutrition in Vienna, September 2008 |
| 119 | Previously presented at Dorothy Havermeyer Foundation Workshop in Neonatal Septicemia, November 2008 |
| 141 | Previously presented at World Buiatrics Congress, Budapest 2008 |
| 145 | Previously presented at ECVIM in Belgium, September 2008 |
| 146 | Previously presented at ECVIM in Belgium, September 2008 |
| 173 | Previously presented at ECVIM in Belgium, September 2008 |
| 190 | The abstract has data about the canine exercise induced collapse mutation for which there is genetic testing available at the University of Minnesota Veterinary Diagnostic Lab. EE Patterson, KM Minor, SM Taylor, and JR Mickelson - receive a portion of testing royalties |
| 237 | Previously presented at InnLab 2009, Berlin, Germany |
| 333 | Previously presented at IVECCS, September 2008 |
| 334 | Previously presented at The Second Havemeyer EHV-1 Workshop, Steamboat Springs, CO, September 2008 |
| 350 | H Kaukinen, J Aspegrén, JS Salonen PhD, LTamm are employed by the manufacturer of the study drugs, Orion Corporation Orion Pharma, Espoo, Finland |

## Research Abstract Program of the
## 2009ACVIM Forum/Canadian Veterinary Medical Association Convention
## Montréal, Québec, June 3 – June 6, 2009

## Index of Abstracts

## ORAL PRESENTATIONS – Thursday, June 4

| Time | # | Presenting Author | Abstract Title |
|---|---|---|---|
| **SMALL ANIMAL – CARDIOLOGY**** | | | |
| 9:00 am | 1 | Allison Heaney | Use of Polymerase Chain Reaction to Detect Bacteremia in Dogs with Bacterial Endocarditis |
| **9:15 am** | **2** | **Andrea Lantis** | **The Effect of Furosemide and Pimobendan on the Renin-Angiotensin-Aldosterone System (RAAS) in Dogs** |
| **9:30 am** | **3** | **Lisa Keller** | **Assessment of Change in Left Ventricular Systolic Function by Tissue Doppler Imaging Methods after Closure of Patent Ductus Arteriosus in Dogs** |
| 9:45 am | 4 | Giosi Farace | Correlation of N-Terminal Prohormone Brain Natriuretic Peptide with Left Ventricular Outflow Tract in Dogs with Sub Aortic Stenosis |
| *BREAK* | | | |
| 10:30 am | 5 | Ashley Saunders | NT-proBNP Concentrations in Canine Congenital Heart Disease |
| 10:45 am | 6 | Gretchen Singletary | Utility of NT-proBNP Assay to Detect Occult Dilated Cardiomyopathy in Doberman Pinschers |
| 11:00 am | 7 | Gerhard Wess | Evaluation of NT-proBNP in the Diagnosis of Various Stages of Dilated Cardiomyopathy in Doberman Pinschers |
| **11:15 am** | **8** | **Sarah Achen** | **Serial Evaluation of NT-proBNP in Dogs with CHF Predicts Clinical Score and the Presence or Absence of Radiographic Pulmonary Edema** |
| 11:30 am | 9 | Gerhard Wess | The Utility of NT-proBNP to Detect Early Stages of Hypertrophic Cardiomyopathy in Cats and to Differentiate Disease Stages |
| 11:45 am | 10 | Giosi Farace | Pulmonary Hypertension and N-Terminal Prohormone Brain Natriuretic Peptide in Dogs |
| 12:00 pm | 11 | Kathryn Meurs | Genome-Wide Association Identifies a Mutation for Arrhythmogenic Right Ventricular Cardiomyopathy in the Boxer Dog |
| 12:15 pm | 12 | Eva Oxford | Phenotypic Differences in the Ultrastructure of Cardiomyocytes from Boxer Dogs Afflicted with Arrhythmogenic Right Ventricular Cardiomyopathy (ARVC) |

*****Also see Cardiology abstracts 61 - 73 (Thursday, June 4, 2:15 pm – 6:00 pm)***

| Time | # | Presenting Author | Abstract Title |
|---|---|---|---|
| **SMALL ANIMAL – NEPHROLOGY/UROLOGY** | | | |
| 9:00 am | 13 | Barrak Pressler | *In vivo* Determination of Canine Cystolith Mineral Composition Using Computed Tomography-Generated Hounsfield Units |
| 9:15 am | 14 | Allyson Berent | Ureteral Stenting for Feline Ureterolithiasis: Techinical and Clincial Outcomes |
| 9:30 am | 15 | Allyson Berent | The Use of a Percutaneously Controlled Hydraulic Occluder for the Treatment of Urethral Sphincter Mechanism Incompetence in 11 Dogs and 1 Cat |
| 9:45 am | 16 | Michael Wood | Development of a Ussing Chamber Model for Studying the Canine Urothelial Response to Infection |
| *BREAK* | | | |
| **10:30 am** | **17** | **Christine Wu** | **Urodynamic Evaluation of Cats with Feline Interstitial Cystitis (FIC)** |
| 10:45 am | 18 | Dennis Chew | Randomized, Placebo-Controlled Clinical Trial of Pentosan Polysulfate Sodium for Treatment of Feline Interstitial (Idiopathic) Cystitis |
| 11:00 am | 19 | Herve Lefebvre | Comparison of Plasma Clearance of Exogenous Creatinine and Iohexol in a Canine Population |
| 11:15 am | 20 | Andrea Zatelli | Effects of Supplementation with Amino Acids in Dogs with Glomerulonephritis |
| **11:30 am** | **21** | **Karen Hilling** | **Documentation of Hypercoagulability in Protein-Losing Nephropathy via Thromboelastography in 10 Dogs** |
| 11:45 am | 22 | Mark Acierno | Measuring Level of Agreement between Directly Measured Blood Pressure and Values Produced by Oscillometric Units in Cats |
| **12:00 pm** | **23** | **Carly Bloom** | **Pharmacokinetics and Cardiovascular Effects of Intravenous Fenoldopam in Healthy Awake Dogs** |

**Boldface type** indicates presentation by award eligible resident

Presentation times are subject to change

| | | | |
|---|---|---|---|
| 12:15 pm | 24 | Bart Maddens | Urinary Immunoglobulin G, C-Reactive Protein and Retinol Binding Protein as Candidate Early Biomarkers for Renal Dysfunction in Dogs with Pyometra ***(ECVIM Award Winner)*** |

**SMALL ANIMAL – HEMATOLOGY**

| | | | |
|---|---|---|---|
| 9:00 am | 25 | Stephanie Smith | Whole Bloodthromboelastometry (TEM) is Related to Cell Counts and Plasma Coagulation Tests in Healthy Dogs |
| **9:15 am** | **26** | **Lara Rose** | **Effect of Canine Hyperadrenocorticism on Thrombelastography Parameters** |
| **9:30 am** | **27** | **John Thomason** | **Effects of Cyclosporine on Canine Platelet Procoagulant Activity** |
| **9:45 am** | **28** | **Sarah Helmond** | **Treatment of Canine Immune-Mediated Hemolytic Anemia with Individually Adjusted Heparin Dosing: A Pilot Study** |
| *BREAK* | | | |
| **10:30 am** | **29** | **Elizabeth Orcutt** | **Comparison of Individually Monitored Unfractionated Heparin Versus Low-Dose Aspirin on Survival of Dogs with Immune Mediated Hemolytic Anemia** |
| **10:45 am** | **30** | **Kristopher Sharpe** | **Platelet Impedance Aggregometry in Clinically Healthy Dogs with and without Ultralow-Dose Aspirin** |
| 11:00 am | 31 | Paulo Vilar Saavedra | Hemostatic Effects of Cryoprecipitate in Dogs with Disseminated Intravascular Coagulation |
| 11:15 am | 32 | Rebecca Kessler | Extended Canine Blood Typing by Gel Column Technique |
| 11:30 am | 33 | Rebecca Kessler | *Pseudomonas fluorescens* Contamination of Canine and Feline Packed Red Blood Cell Units |

**SMALL ANIMAL – OTHER**

| | | | |
|---|---|---|---|
| 11:45 am | 34 | Mads Kjelgaard-Hansen | Combination of Inflammatory and Hemostatic Markers in Mortality Model for Critically Ill Canine ICU Patients Significantly Improves Efficacy |
| 12:00 pm | 35 | Bianka Schulz | Investigation of Doxycycline-Related Side Effects in Dogs |

**SMALL ANIMAL – PHARMACOLOGY**

| | | | |
|---|---|---|---|
| 12:15 pm | 36 | Fabrice Thoulon | Metabolism and Excretion of Oral Meloxicam in the Cat |

**SMALL ANIMAL – ENDOCRINOLOGY**

| | | | |
|---|---|---|---|
| **9:00 am** | **37** | **Arnon Gal** | **Cloning of the Calcium-Sensing Receptor from the Feline Parathyroid Gland** |
| 9:15 am | 38 | Eric Zini | Hypertriglyceridemia does not Impair Whole Body Insulin Sensitivity in Cats, Possibly Through Increased Circulating Adiponectin |
| 9:30 am | 39 | Eric Zini | Role of Hyperglycemia in the Pathogenesis of Pancreatitis in Cats |
| **9:45 am** | **40** | **Kirk Sears** | **Use of Lispro Insulin for Treatment of Dogs with Diabetic Ketoacidosis** |
| *BREAK* | | | |
| 10:30 am | 41 | Kirsten Roomp | Evaluation of Detemir in Diabetic Cats Managed with a Protocol for Intensive Blood Glucose Control |
| **10:45 am** | **42** | **Andrew McGraw** | **Determination of the Concentrations of Trilostane and Ketotrilostane that Inhibit *Ex Vivo* Synthesis of Cortisol in Canine Adrenal Tissues** |
| 11:00 am | 43 | Ghita Benchekroun | Plasma ACTH Precursors (Pro-OpioMelanoCortin and Pro-Adrenocorticotropin) in Cats with Hyperadrenocorticism |
| 11:15 am | 44 | Katharine Lunn | Endocrine Function in Cats after Stereotactic Radiosurgery Treatment for Acromegaly |
| 11:30 am | 45 | John Rossmeisl | Blood-Brain-Barrier Disruption in Chronic Canine Experimental Hypothyroidism |
| 11:45 am | 46 | Jennifer Wakeling | Induction of Aryl Hydrocarbon Receptors in Cats |
| 12:00 pm | 47 | Audrey Cook | Prevalence of Hypocobalaminemia in Cats with Hyperthyroidism |
| 12:15 pm | 48 | Lisa Morrow | Hypertension in Hyperthyroid Cats: Prevalence, Incidence and Predictors of its Development |

**EQUINE****

| | | | |
|---|---|---|---|
| 9:00 am | 49 | Natália Rodrigues | Expression of Cyclooxygenase Isoforms in Equine Gastric Ulcers |
| **9:15 am** | **50** | **Kati Vainio** | **Primary Gastric Impaction in Horses: A Retrospective Study of 22 Cases (2005-2008)** |
| 9:30 am | 51 | Frank Andrews | Effect of Volatile Fatty Acid Mixes on Equine Nonglandular Mucosa Bioelectric Properties: Pathogenesis of Gastric Ulcers in Horses |
| **9:45 am** | **52** | **Peggy Moreau** | **Study of the Variability of Endoscopic Duodenal Biopsies in Healthy Horses and in Heaves-Affected Horses Fed with Different Diets** |

*BREAK*

| | | | |
|---|---|---|---|
| 10:30 am | 53 | Emilie Setlakwe | Sub-Epithelial Fibrosis is Present in the Peripheral Airways of Heaves-Affected Horses |
| 10:45 am | 54 | Modest Vengust | Furosemide does not Affect Transvascular Fluid Fluxes across the Lung in Exercising Horses |
| **11:00 am** | **55** | **James Nicol** | **Partitioning of Pulmonary Resistance in Horses** |
| 11:15 am | 56 | Julie Dauvillier | Effects of Prolonged Inhaled Corticosteroid Treatment on Cell-Mediated Immunity in Horses |
| 11:30 am | 57 | Lutz Goehring | EHV-1 Positive PBMC Fractions During Cell-Associated Viremia following Experimental Infection |
| **11:45 am** | **58** | **Sonya Wilsterman** | **The Efficacy of Nitazoxanide as an Antiviral Agent for the Treatment of EHV-1 Infection in Horses** |
| 12:00 pm | 59 | Alexandra Draper | Repeatability and Reproducibility of Transabdominal Ultrasonographic Renal Loci and Dimensions in Thoroughbred Horses |
| **12:15 pm** | **60** | **Gemma Tyner** | **Renal Biopsies in Horses: A Retrospective Study of 71 Cases** |

***Also see Equine abstracts 116 - 129 (Thursday, June 4, 2:15 pm – 6:15 pm) and Equine abstracts 142 – 143 (Thursday, June 4, 5:45 pm – 6:15 pm)***

## SMALL ANIMAL – CARDIOLOGY**

| | | | |
|---|---|---|---|
| **2:15 pm** | **61** | **Caryn Reynolds** | **Owner Perceptions and Priorities Regarding Quality of Life and Quantity of Life in Cats with Heart Disease** |
| **2:30 pm** | **62** | **Sara Bordelon** | **Validation of Platelet Factor 4, Soluble P-Selectin, β-thromboglobulin, and 11-dehydro-thromboxane B2 ELISAs in the Feline Population** |
| 2:45 pm | 63 | Karl Jandrey | Plasma PECAM-1 and Endothelin-1 Concentrations are Elevated in Cats with Hypertrophic Cardiomyopathy |
| 3:00 pm | 64 | Francesco Porciello | Association of Hypertrophic Cardiomyopathy Phenotype and Genotype in Italian Maine Coon Cats |
| 3:15 pm | 65 | Denise Schwartz | Pulsed Tissue Doppler Identifies Preclinical Left Ventricular Myocardial Dysfunction in Obese Dogs |
| **3:30 pm** | **66** | **Sarah Scruggs** | **Serotonin Transporter Expression Is Locally Down-Regulated in Canine Degenerative Mitral Valves** |
| 3:45 pm | 67 | Sophia Moesgaard | B-Type Natriuretic Peptide and Cyclic Guanosine Monophosphate Levels Increase whereas Nitric Oxide Bioavailability Decreases with Increasing Degree of Myxomatous Mitral Valve Disease |
| *BREAK* | | | |
| **4:30 pm** | **68** | **Adonia Hsu** | **Trans-Thoracic Three-Dimensional Echocardiographic Evaluation of Normal Canine Heart Valves** |
| 4:45 pm | 69 | Ingrid Ljungvall | Cardiac Troponin-1 is Associated with Severity of Myxomatous Mitral Valve Disease, Age, Heart Rate and C-Reactive Protein in Dogs |
| 5:00 pm | 70 | Carl Kirker-Head | Evaluation of a Minimally Invasive Treatment for Mitral Valve Regurgitation |
| 5:15 pm | 71 | Sonya Gordon | Caval Syndrome; Outcome in 42 Dogs |
| 5:30 pm | 72 | Brian Maran | Modulation of Inflammatory and Oxidative Stress Markers in Coronary Artery Smooth Muscle Cell Model |
| 5:45 pm | 73 | Barthel Schmelting | High Definition Oscillometry: A Novel, Non-Invasive Approach for Precise Blood Pressure Determination and Cardiovascular Assessment in Old World and New World Monkeys |

***Also see Cardiology abstracts 1 - 12 (Thursday, June 4, 9:00 am – 12:30 pm)***

## SMALL ANIMAL – NEUROLOGY

| | | | |
|---|---|---|---|
| 2:15 pm | 74 | Luisa De Risio | Prevalence, Heritability and Genetic Correlations of Congenital Sensorineural Deafness and Pigmentation Phenotypes in 4143 Border Collies |
| **2:30 pm** | **75** | **Fiona James** | **Investigation of a New Electroencephalography (EEG) Electrode in Sedated and Awake Dogs** |
| 2:45 pm | 76 | Simon Platt | Cerebrospinal Fluid Neurotransmitter Concentrations in Dogs with Ischemic Infarction of the Brain |
| 3:00 pm | 77 | Ronaldo da Costa | Risk Factors for Post-Myelographic Seizures in Dogs – 503 Cases |
| 3:15 pm | 78 | John Rossmeisl | Effects of Palliative Pharmacotherapy on Survival and Quality of Life in 50 Dogs with Primary Intracranial Neoplasms |
| **3:30 pm** | **79** | **Sarah Moore** | **The Pharmacokinetics of Levetiracetam in Dogs Concurrently Receiving Phenobarbital** |
| **3:45 pm** | **80** | **Liran Tzipory** | **Metabolism and Action of Nelarabine in Glioma Cell Lines** |

*BREAK*

| | | | |
|---|---|---|---|
| 4:30 pm | 81 | Veronika Stein | Expression of Matrix Metalloproteinases and Reck in Microglial Cells in Different Intracranial Diseases |
| 4:45 pm | 82 | Steven De Decker | Low Field Magnetic Resonance Imaging (MRI) Findings of the Caudal Cervical Region in Clinically Normal Doberman Pinschers and Foxhounds |
| 5:00 pm | 83 | Ronaldo da Costa | Computed Tomographic Findings In Large and Giant Breed Dogs with Cervical Spondylomyelopathy: 58 Cases |
| 5:15 pm | 84 | Filippo Adamo | Cervical Arthroplasty in Dogs with Disc Associated Caudal Cervical Spondylomyelopathy and Cervical Disc Herniation: Preliminary Study in Two Cases |
| **5:30 pm** | **85** | **Jenny Scarano** | **F Ratio of the Sciatic and Ulnar Nerve in Dogs** |
| 5:45 pm | 86 | G. Diane Shelton | Peripheral Nerve Pathology in Canine Degenerative Myelopathy with Mutation in Superoxide Dismutase 1 Gene |
| 6:00 pm | 87 | Joan Coates | An SOD1 Mutation Associated with Degenerative Myelopathy Occurs in Many Dog Breeds |

**SMALL ANIMAL – ONCOLOGY**

| | | | |
|---|---|---|---|
| 2:15 pm | 88 | Karen Jackson | Lymphocyte Counts and Circulating Lymphoblasts in Canine Lymphoma Patients |
| 2:30 pm | 89 | Gina Michels | Pharmacokinetic Properties of Toceranib Phosphate (PalladiaTM, SU11654), A Novel Tyrosine Kinase Inhibitor, in Laboratory Dogs and Dogs with Mast Cell Tumors |
| 2:45 pm | 90 | Rachel Dean | Feline Injection Site Sarcomas in the United Kingdom: One Hundred and Fifty Seven Cases |
| 3:00 pm | 91 | Christine Kempf | Nonviral Gene Transfer of the Feline Cytokine Genes IL-2, IFNγ and GM-CSF as Adjuvant Immunotherapy of the Feline Fibrosarcoma |
| 3:15 pm | 92 | Melani Fork | Immunhistochemical Differentiation of Canine Prostatic Carcinoma |
| 3:30 pm | 93 | Chick Weisse | Feasibility and Safety Associated with Selective and Superselective Intra-Arterial Carboplatin +/- Meloxicam Delivery for Urothelial Tumors in Dogs |
| 3:45 pm | 94 | Marisa Ames | Effect of Intracavitary Chemotherapy as Treatment of Recurrent Pleural Effusion Following Pericardiectomy: Eight Cases |

**SMALL ANIMAL – NUTRITION/METABOLISM**

| | | | |
|---|---|---|---|
| 4:30 pm | 95 | Matthew Beal | Technique for Fluoroscopic Nasojejunal Tube Placement in Dogs |
| 4:45 pm | 96 | Matthew Beal | Technique for Percutaneous Radiologic Gastrojejunostomy in the Dog |
| 5:00 pm | 97 | John Bauer | Dietary γ-linolenic Acid Supports Arachidonic Acid Enrichment of Feline Red Blood Cell Membranes |
| 5:15 pm | 98 | Alexander German | Adipokine Expression and Secretion by Canine Adipocytes: Stimulation of Inflammatory Adipokine Production by LPS and TNFα |
| 5:30 pm | 99 | Anne Chauvet | Exercise and Active Client Motivation Improve Rate of Weight Loss in Obese Dogs |
| 5:45 pm | 100 | Vincent Biourge | Comparaison of Dietary Strategies on the Perception of Hunger During a Field Feline Weight Loss Program |
| 6:00 pm | 101 | Joana Nery | Influence of Dietary Protein Source and Content on Fecal Quality and Gene Expression of Water and Electrolyte Transporters of Colon in Miniature Poodles and German Shepherds |

**SMALL ANIMAL – INFECTIOUS DISEASE**

| | | | |
|---|---|---|---|
| 2:15 pm | 102 | Iain Keir | Influence of Empirical Antimicrobial Choice on Outcome following Septic Peritonitis in Dogs |
| 2:30 pm | 103 | Dawn Boothe | Detection of Fluoroquinolone Resistance Level in Clinical Canine and Feline *Escherichia coli* Pathogens Using Rapid and Early Real-Time PCR Assay |
| **2:45 pm** | **104** | **Ashley Cruse** | **Evaluation of Potential Biomarkers for Canine Sepsis** |
| 3:00 pm | 105 | Barbara Kohn | Pulmonary Manifestation in Canine Leptospirosis |
| 3:15 pm | 106 | Susan Little | Seroprevalence of FeLV and FIV in Canada |
| 3:30 pm | 107 | Desiree Kruse | Prognostic Factors in Cats with Feline Panleucopenia |
| **3:45 pm** | **108** | **Elizabeth Lechner** | **Prevalence of Protective Antibody Titers for Distemper Virus and Parvovirus in Dogs Entering a Florida Animal Shelter** |
| *BREAK* | | | |
| 4:30 pm | 109 | Melissa Beall | Detection of Canine Parvovirus Type 2c by a Commercially Available In-House Rapid Test |
| 4:45 pm | 110 | Stephanie Janeczko | Prevalence of, Risk Factors for, and Assemblage Types of *Giardia* Infection in Cats Housed in an Animal Shelter |

| | | | |
|---|---|---|---|
| 5:00 pm | 111 | Jody Gookin | Assessment of Reproductive Tract Disease in Cats at Risk for Enteric *Tritrichomonas foetus* Infection |
| 5:15 pm | 112 | Matt Eberts | Use of Snap®4Dx® Assay to Monitor Tick-Borne Infection in Minnesota: A 4-Year Study |
| 5:30 pm | 113 | Michael Lappin | Prevalence of Hemoplasma DNA in Field-Caught Mosquitoes in Colorado |
| **5:45 pm** | **114** | **Christina Bradbury** | **Topical Imidacloprid and Moxidectin Prevents Flea Transmission of *Bartonella henselae* in Cats** |
| 6:00 pm | 115 | Katrin Hartmann | Outcome and Prognostic Factors in Dogs with Canine Leishmaniasis in a Non-Endemic Country |

**EQUINE****

| | | | |
|---|---|---|---|
| 2:15 pm | 116 | Mary Rose Paradis | The Use of Computer Tomography in the Diagnosis of Septic Arthritis/Osteomyelitis in the Neonatal Foal |
| **2:30 pm** | **117** | **Rosa Barsnick** | **Endocrine Energy Response in Septic Foals: Insulin, Leptin and Adiponectin** |
| **2:45 pm** | **118** | **Dominic Dawson** | **Opsonization of *Rhodococcus equi* with High Antibody Plasma Decreases Bacterial Viability and Promotes Phagocyte Activation** |
| 3:00 pm | 119 | Lais Costa | Plasma Endothelin Concentrations in Septic and Non-Septic Neonatal Foals |
| 3:15 pm | 120 | Jose L. Mendez | Thromboelastography (TEG): An Innovative Technique for Early Detection of Hemostatic Abnormalities in Foals with Septicemia |
| 3:30 pm | 121 | Gayle Hallowell | Effects of Dobutamine and Atropine and Acetylpromazine on Aortic Valve Function in the Horse |
| 3:45 pm | 122 | Andrew van Eps | Intravenous 99m-technetium Liposomes in Horses: A Safety and Biodistribution Study |
| *BREAK* | | | |
| 4:30 pm | 123 | Jill Muno | Prevalence and Risk Factors for Hyperinsulinemia in Clinically Normal Horses in Central Ohio |
| 4:45 pm | 124 | Jill Muno | Seasonal Variation in Plasma Insulin and Glucose Concentration in Normal Horses in Central Ohio |
| 5:00 pm | 125 | Nicholas Frank | Effects of a Supplement Containing Chromium and Magnesium on Insulin Sensitivity in Horses with Equine Metabolic Syndrome |
| **5:15 pm** | **126** | **Rebecca Funk** | **Seasonal Changes in the Combined Glucose Insulin Tolerance Test in Normal Aged Horses** |
| 5:30 pm | 127 | Nicholas Frank | Effects of Sampling Time and Hay Feeding on Blood Glucose, Insulin, and Adrenocorticotropin Hormone (ACTH) Concentrations in Horses |
| 5:45 pm | 128 | Nicholas Frank | Effects of Endotoxemia and Carbohydrate Overload on Glucose and Insulin Dynamics and the Development of Laminitis in Horses |
| **6:00 pm** | **129** | **Teresa Burns** | **Pro-Inflammatory Cytokine and Chemokine Expression Profiles of Various Adipose Tissue Depots of Insulin-Resistant and Insulin-Sensitive Light Breed Horses** |

*****Also see Equine abstracts 49 - 60 (Thursday, June 4, 9:00 am – 12:30 pm) and Equine abstracts 142 – 143 (Thursday, June 4, 5:45 pm – 6:15 pm)***

**FOOD ANIMAL**

| | | | |
|---|---|---|---|
| 2:15 pm | 130 | Jody Gookin | An Animated Model of Reticulorumen Motility |
| 2:30 pm | 131 | Alexandra Burton | A Retrospective Study of 106 Cases of Bovine Lymphosarcoma |
| 2:45 pm | 132 | David Francoz | Determination of Dairy Cows Hematological Reference Interval by 3 Methods and Evaluation of Confounding Factors |
| **3:00 pm** | **133** | **Anita Varga** | **Correlation of Serum Cardiac Troponin I and Myocardial Damage in Cattle with Monensin Toxicosis** |
| **3:15 pm** | **134** | **Anita Varga** | **Validation of the I-STAT Stallside Immunoassay for the Measurement of Bovine Cardiac Troponin I and its Clinical Relevance** |
| 3:30 pm | 135 | Baljit Singh | RGD-Conjugated Helical Rosette Nanotubes Inhibit Chemotaxis and Cell Signaling of Cattle Neutrophils |
| 3:45 pm | 136 | Michel Levy | Experimental Infection of Colostrum-Deprived Calves with Bovine Viral Diarrhea Virus Type-1 Isolated from Free-Ranging White-Tailed Deer (*Odocoileus virginianus*) |
| *BREAK* | | | |
| 4:30 pm | 137 | Nathalie Kirschvink | Evidence for Transplacental Passage of Bluetongue Virus Seroytpe 8 in Sheep |
| 4:45 pm | 138 | Nathalie Kirschvink | Long Term Follow up of Rams` Semen Quality after Natural Infection with Bluetongue Virus Serotype 8 |
| **5:00 pm** | **139** | **Teresa Buchheit** | **Prevalence of BVDV on Alpaca Farms in Eastern and Middle Tennessee** |

| | | | |
|---|---|---|---|
| 5:15 pm | 140 | Mireille Meylan | Seroprevalence of Bovine Viral Diarrhea in New World Camelids in Switzerland and Implications for the Eradication Campaign in the Bovine Population |
| 5:30 pm | 141 | Mireille Meylan | Tuberculosis Caused by *Mycobacterium microti* in New World Camelids |
| **EQUINE**** | | | |
| 5:45 pm | 142 | Jennifer Davis | Pharmacokinetics of Intravenous Enrofloxacin and its Active Metabolite Ciprofloxacin in Hospitalized Horses |
| 6:00 pm | 143 | James Belknap | Laminar Leukocyte Accumulation and Epithelial Stress in Horses with Carbohydrate Overload-Induced Laminitis |

*** Also see Equine abstracts 49 - 60 (Thursday, June 4, 9:00 am – 12:30 pm) and Equine abstracts 116 - 129 (Thursday, June 4, 2:15 pm – 6:15 pm)*

---

## ORAL PRESENTATIONS – Friday, June 5

| | | | |
|---|---|---|---|
| **SMALL ANIMAL – RESPIRATORY** | | | |
| 8:00 am | 144 | Elizabeth Rozanski | The Effect of Tracheal Lavage Administration on Lung Function in Healthy Kittens |
| 8:15 am | 145 | Laura Nafe | Utility of Biomarkers in Bronchoalveolar Lavage Fluid for Discrimination of Asthma and Chronic Bronchitis in Cats |
| **8:30 am** | **146** | **Tekla Lee-Fowler** | **Comparison of Intradermal Skin Testing and Allergen-Specific Serum Immunoglobulin E (IgE) in Experimental Feline Asthma** |
| **8:45 am** | **147** | **Jason Eberhardt** | **Chronic Administration of feG-COOH in Experimental Feline Asthma** |
| 9:00 am | 148 | Ray Dillon | Progression of Circulating Antigen and Antibody Responses in Cats Experimentally Infected with *Dirofilaria immitis* |
| 9:15 am | 149 | Anne Wooldridge | Isometric Responses of Isolated Intrapulmonary Bronchioles from Cats with and without Adult Heartworm Infection |
| 9:30 am | 150 | Henna Heikkilä | Matrix Metalloproteinase -2 and -9 in Bronchoalveolar Lavage Fluid of Dogs with Idiopathic Pulmonary Fibrosis |
| 9:45 am | 151 | Baljit Singh | Recruitment of Pulmonary Intravascular Macrophages in Immune-Mediated Hemolytic Anemia In Dogs |
| **SMALL ANIMAL – HEPATOLOGY** | | | |
| 8:00 am | 152 | Cynthia Webster | Cyclic AMP Attenuates Fatty Acid Induced Apoptosis in an In vitro Model of Hepatic Lipotoxicity |
| 8:15 am | 153 | Kenny Simpson | Culture-Independent Detection of Bacteria in Feline Inflammatory Liver Disease |
| **SMALL ANIMAL – IMMUNOLOGY** | | | |
| **8:30 am** | **154** | **Todd Archer** | **Development of a Flow Cytometric Panel of T-Lymphocyte Biomarkers to Evaluate the Immunosuppressive Effects of Cyclosporine in Dogs** |
| **SMALL ANIMAL – GASTROENTEROLOGY** | | | |
| 8:45 am | 155 | Niels Gruetzner | Evaluation of Serum Concentrations of Immunoglobulin A and Immunoglobulin M in Chinese Shar Peis with Cobalamin Deficiency |
| 9:00 am | 156 | Aarti Kathrani | Genetical Analysis of Toll-Like Receptor Genes in German Shepherd Dogs Reveals Potential Polymorphism Associated with Inflammatory Bowel Disease |
| 9:15 am | 157 | Iwan Burgener | Urinary Leukotriene E4 Excretion is Increased in Dogs with Chronic Enteropathies |
| 9:30 am | 158 | Melanie Craven | High Throughput Pyrosequencing Reveals Reduced Bacterial Diversity in the Duodenal Mucosa of Dogs with IBD |
| 9:45 am | 159 | Melanie Craven | *E. Coli* Associated with Granulomatous Colitis of Boxer Dogs Frequently Manifest Resistance to Antibiotics |
| **2:15 pm** | **160** | **Susan Simmerson** | **Description of Protein-Losing Enteropathy in Yorkshire Terrier Dogs using the W.S.A.V.A. Gastrointestinal Classification System** |
| 2:30 pm | 161 | Joshua Hobbs | Doppler Ultrasound Analysis of Gastrointestinal Blood Flow for Differentiating Food Allergic from Non-Food Allergic Pruritic Dogs |
| **2:45 pm** | **162** | **Heather Graham** | **Effects of Prednisone or Prednisone with Ultralow-Dose Aspirin on the Gastroduodenal Mucosa of Healthy Dogs** |
| 3:00 pm | 163 | Panagiotis Xenoulis | Serum Triglyceride Concentrations in Miniature Schnauzers with and without a History of Pancreatitis |
| 3:15 pm | 164 | Lisa Prior | Serial Evaluation of Canine Pancreatitic Lipase (Spec cPL™) in Dogs with Clinical Signs of Pancreatitis |

| | | | |
|---|---|---|---|
| 3:30 pm | 165 | Marnin Forman | Evaluation of Feline Pancreas-Specific Lipase (Spec fPL™) for the Diagnosis of Feline Pancreatitis |
| **3:45 pm** | **166** | **Kelly McCord** | **A Multi-Institutional Study Evaluating Diagnostic Utility of Spec cPL™ in the Diagnosis of Acute Pancreatitis in Dogs** |
| *BREAK* | | | |
| 4:30 pm | 167 | Suzanne Bailey | Clinical Significance of Increased Serum Feline Pancreatic Lipase Immunoreactivity Concentrations in Cats with Inflammatory Bowel Disease |
| 4:45 pm | 168 | Dennis O`Brien | Serum D-Lactate Concentrations in Cats with Gastrointestinal Disease |
| 5:00 pm | 169 | Nora Berghoff | Association of Serum Cobalamin and Methylmalonic Acid Concentrations in Dogs |
| **5:15 pm** | **170** | **Kelly McCord** | **Comparison of Gastrointestinal Motility in Dogs Treated with Metoclopramide, Cisapride, Erythromycin or Maropitant Using the SmartPill™** |
| 5:30 pm | 171 | Frederic Gaschen | Variability Associated with Repeated Measurements of Gastric Emptying using the SmartPill pH.p Wireless Capsule and Scintigraphy in Dogs |
| 5:45 pm | 172 | Michael Lappin | The Esophageal Transit Time of Tablets or Capsules in Cats following Administration with FlavoRx® Pill Glide or Pill Delivery Treats |
| 6:00 pm | 173 | Diane Frank | Gastrointestinal Disease in Dogs with Excessive Licking of Surfaces |

## POSTER PRESENTATIONS

**On Display:** Thursday, June 4, 7:00 am - 10:00 pm  Friday, June 5, 7:00 am – 10:00 pm
Saturday, June 6, 7:00 am – 12:30 pm

**Attended by Author:** Thursday, June 4, 7:00 am – 9:00 am; Friday, June 5, 9:50 am – 10:30 am &
Friday, June 5, 6:00 pm – 7:30 pm (Wine & Cheese Reception)

---

| # | Presenting Author | Abstract Title |
|---|---|---|
| **SMALL ANIMAL – NEUROLOGY** | | |
| 174 | Byeong-Teck Kang | High Resolution Magnetic Resonance Imaging of the Canine Brain at 7 Tesla: A Comparison Study with 0.2 and 1.5 Tesla |
| 175 | Alberta de Stefani | Clinical Signs, Magnetic Resonance Imaging Findings and Survival in Dogs with Intracranial Meningiomas and Glial Cell Tumours |
| **176** | **Michaela Cautela** | **Oral Hydroxyurea Therapy for Dogs with Suspected Intracranial Meningioma: A Retrospective Cohort Study (2004-2009)** |
| 177 | Christopher Mariani | Immune Cell Infiltration into Canine Meningiomas |
| 178 | Byeong-Teck Kang | Fluorodeoxyglucose Positron Emission Tomography and Magnetic Resonance Imaging Findings of Non-Suppurative Meningoencephalitis in Seven Dogs |
| 179 | Arianna Negrin | Steroid Responsive Meningitis-Arteritis Treatment with Three Potential Different Protocols: Clinical Signs, Laboratory and Long Term Follow Up in 48 Dogs |
| 180 | Byeong-Teck Kang | Three-dimensional Time-of-Flight MR Angiography of Intracranial Vessels in a Canine Model of Ischemic Stroke with Permanent Middle Cerebral Artery Occlusion |
| 181 | Maria Luisa Suarez | Prevalence of Cerebral Microbleeds in Aged Dogs |
| **182** | **Fred Wininger** | **Preliminary Observations between Cox-2 Inhibitor Administration and Acute Cerebrovascular Disease in Dogs** |
| 183 | Renee Barber | Multiplex Analysis of Cytokines in the Cerebrospinal Fluid of Dogs after Stroke |
| 184 | N. Matthew Ellinwood | Validation of R&D Systems Kits to Detect Cytokine Levels in Canine Cerebral Spinal Fluid |
| 185 | Ronaldo da Costa | Magnetic Resonance Imaging Findings In 60 Dogs with Cervical Spondylomyelopathy |
| 186 | Hiroaki Kamishina | Phosphorylated Neurofilament NF-H as a Biomarker of Spinal Cord Injury in Dogs |
| 187 | Natasha Olby | Epidural and Intradural Scarring Post Laminectomy and Durotomy: A Comparison of Gelfoam and SentrX Film |
| **188** | **Allison Haley** | **Breed Specific Polymyositis in the Hungarian Vizsla Dog** |
| 189 | Richard Piercy | Mutational Analysis of Dystrophin-Deficient Muscular Dystrophy in Cavalier King Charles Spaniels |
| 190 | Edward (Ned) Patterson | Frequency of the Canine Exercise Induced Collapse Gene in Diverse Breeds |
| 191 | Charles Vite | BRDU Labeling Pattern of the Rostral Migratory Stream in Normal Canine and Feline Brains |
| 192 | Charles Vite | Intrathecal Enzyme Therapy In Mucopolysaccharidosis I Cats Reduces Storage Throughout The Brain |
| 193 | Ji-hey Lim | Neural Stem Cell Sources in Adult Dogs |
| 194 | German Santamarina | Dogs with Cortical Beta-Amyloid Pathology Have Less Serotonergic Neurons in the Rostral Raphe Nuclei than Aged Matched Healthy Controls |
| **SMALL ANIMAL – ONCOLOGY** | | |
| 195 | Clair Oliveira | Evaluation of the Carcinoembrionic (CEA) and Carbohydrate 15-3 (CA 15-3) Antigen in Malignant Mammary Tumors in Bitches |
| 196 | Clair Oliveira | Expression of Her-2/Neu Protein in Mammary Tumors of Bitches |
| 197 | Sílvia Regina Ricci Lucas | Ceruloplasmin Concentrations in Dogs with Multicentric Lymphoma Undergoing Chemotherapy |
| 198 | Emily Marcus | Flow Cytometric Prognostic Factors in Canine B Cell Lymphoma |
| 199 | Genevieve Hammond | Denamarin® for the Prevention of Lomustine (CCNU) Induced Hepatotoxicity in Tumor Bearing Dogs: Interim Analysis |
| 200 | Timothy Stein | Effects of Mutant Beta-Catenin on Hepatocyte Growth Homeostasis ***(VCS Award Winner)*** |

**Boldface type** indicates presentation by award eligible resident  Presentation times are subject to change

**SMALL ANIMAL – HEMATOLOGY**

| | | |
|---|---|---|
| 201 | Jacqueline Whittemore | Comparison of Complete Blood Count Values for Samples Collected Using Traditional and Micro-sampling Tubes |
| 202 | Shiguang Yu | Aging Associated Changes in Blood Cells of Cats |
| 203 | Rebecca Haviland | Clopidogrel Therapy in Dogs with Immune-Mediated Hemolytic Anemia |
| 204 | Sharon Center | Platelet P-Selectin (CD62P) Expression and Platelet-Leukocyte Interactions in Clinically Healthy Dogs with and without Ultralow-Dose Aspirin |
| **205** | **Lisa Shearer** | **Effects of Aspirin and Clopidogrel on Platelet Function in Normal Dogs** |
| 206 | Reinhard Mischke | Influence of rhFVIIa on Haemostatic Variables in Normal and Factor VII Deficient Dogs |

**SMALL ANIMAL – CARDIOLOGY**

| | | |
|---|---|---|
| 207 | Rebecca Manley | Exercise-Associated Heart Rate Recovery Time in Normal Dogs |
| 208 | Elise Mercier | Influence of Age on Pulmonary Arterial Pressure in Healthy Beagle Dogs: Analysis by Right Ventricular Catheterization, Pulsed Doppler Echocardiography and Pulsed Doppler Tissue Imaging |
| 209 | Beate Egner | Evaluation of HDO® (High Definition Oscillometry) - A New NIBP Technique - In Comparison to Invasive Measurement (HSE - Hydro Sachs Electronics) in Anesthetised Dogs |
| 210 | Yasutomo Hori | Action Duration-Related Diuretic Effects and Rennin-Angiotensin-Aldosterone System in Healthy Dogs with a Furosemide |
| 211 | Yasutomo Hori | Changes in the Plasma ANP and NT-proBNP Concentrations in Healthy Dogs with an Acute Volume Overload |
| 212 | Alora Carrier | Analytical Validation of a Commercially Available Feline N-Terminal Prohormone Brain Natriuretic Peptide ELISA |
| 213 | Katie Cyr | Measurement of Canine B-Type Natriuretic Peptide Using a Human Analyzer |
| **214** | **Kristine Yee** | **Intraday Variation In Feline N-Terminal Prohormone Brain Natriuretic Peptide Concentration** |
| 215 | Nobuyuki Kanno | Plasma Adrenomedullin as a Biomarker for Severity of Pulmonary Hypertension in Dogs |
| 216 | Maria Helena Matiko Akao Larsson | Pulsed Tissue Doppler Evaluation in Dogs with Idiopathic Dilated Cardiomyopathy |
| 217 | Maria Helena Matiko Akao Larsson | Evaluation of Left Ventricular Systolic and Diastolic Function Using Pulsed Tissue Doppler in Dogs with Chronic Mitral Valve Insufficiency |
| 218 | Miki Nishida | Pathological Findings in Dogs with Mitral Valve Insufficiency: Biopsies during Open Heart Surgery with Cardiopulmonary Bypass |
| 219 | Miki Nishida | Long-Term Survival of Two Dogs after Mitral Valve Plasty using Expanded Polytetrafluoroethylene Chords: Pathological Study |
| 220 | Miki Nishida | Mitral Valve Plasty in 11 Cavalier King Charles Spaniels |
| 221 | Takeshi Mizuno | Intra- and Post-Operative Complications in 47 Dogs that Underwent Mitral Valve Plasty |
| 222 | Takeshi Mizuno | Efficacy of Methylprednisolone Therapy for Inflammation after Cardiac Surgery |
| 223 | Megumi Fujiwara | Recovery of Atrial Fibrillation in Three Dogs after Mitral Annuloplasty Under Cardiopulmonary Bypass for Severe Mitral Regurgitation |
| 224 | Yoko Fujii | LV 2D Strain Analysys Pre- and Post- Balloon Valvuloplasty in Dogs with Pulmonary Stenosis |
| 225 | Yoshinori Nishijima | Effects of Bisoprolol on Ventricular Myocyte Electrophysiology During Chronic Heart Failure |
| 226 | Megan Rector | Heart Rate in Irish Wolfhounds with Lone Atrial Fibrillation & Assessment of Digoxin as Sole Method of Rate Control |
| 227 | Jonathan Abbott | Prevalence of the Myosin-Binding Protein C Mutation in Apparently Healthy Cats of Diverse Lineage |
| 228 | Ann-Sabin Berli | The Use of Strain, Strain Rate, and Displacement by 2D Speckle Tracking for Assessment of Systolic Left Ventricular Function in Goats: Applicability and Influence of General Anesthesia |

**SMALL ANIMAL – RESPIRATORY**

| | | |
|---|---|---|
| 229 | Miquel Laporta | Barometric Whole-Body Plethysmography Reference Values in Different Age-Range Cats with Natural Bronchial Disease |
| 230 | Yan-Ann Jhuo | Clinical Application of Barometric Whole-Body Plethysmography and Bronchoprovocation Test in Feline Bronchial Disease |
| **231** | **Michelle Patrick** | **The Quality of Canine Bronchoalveolar Lavage Fluid Cytology is Preserved for Up to 72 Hours with or without Autologous Serum** |

| | | |
|---|---|---|
| 232 | Stephanie Pumphrey | Clinical and Laboratory Characteristics of 35 Dogs with Pulmonary Thromboembolism (2002-2008) |
| 233 | M Josefa Fernandez Del Palacio | Comparative Study of the Tolerance and Efficiency of Four Different Techniques for Nebulisation Delivery in Cats |
| 234 | Carol Reinero | Utility of Tepoxalin, a Dual Lipoxygenase and Cyclooxygenase Inhibitor, in Canine Chronic and/or Eosinophilic Bronchitis |
| 235 | Eric de Madron | Treatment of Tracheal Collapse with Vet-Stents® in 36 Dogs: Complications and Long-Term Results |

**SMALL ANIMAL – ENDOCRINOLOGY**

| | | |
|---|---|---|
| 236 | Melania Osto | Subacute Endotoxemia Temporarily Impairs Insulin Sensitivity and β-cell Function in Cats |
| 237 | Johanna Rieder | Comparison of Wireless and Conventional Continuous Glucose Monitoring Systems (CGMS) in 4 Dogs |
| 238 | Richard Nelson | Extended Use of Protamine Zinc Recombinant Insulin for Treating Diabetes Mellitus in Cats |
| 239 | Pascale Smets | Urinary Markers and Plasma Iohexol Clearance in Healthy Dogs and Dogs with Pituitary-Dependent Hyperadrenocorticism |
| 240 | Kellie Fecteau | Effect of Phytoestrogens and Melatonin on Adrenal Steroidogenesis Using Human Adrenal Cells as a Model for Canine Hyperadrenocorticism |
| 241 | Linda Martin | Effects of Low Doses of Cosyntropin on Cortisol Concentrations in Clinically Healthy Cats |
| 242 | Linda Martin | Dose Response of Aldosterone to ACTH Stimulation in Clinically Healthy Cats |
| **243** | **Bronwyn Rutland** | **Determination of the Optimal Time for Testing Thyroid Hormones During Treatment with Methimazole in Healthy and Hyperthyroid Cats** |
| 244 | Kate Hill | The Efficacy and Safety of a Novel Lipophilic Formulation Methimazole for the Once Daily Transdermal Treatment of Cats with Hyperthyroidism |

**SMALL ANIMAL – GASTROENTEROLOGY**

| | | |
|---|---|---|
| 245 | Jan Suchodolski | The Effect of a Multi-Strain Probiotic Formulation on Markers of Gastrointestinal Function in Healthy Dogs and Cats |
| 246 | Debbie Minikhiem | Effect of *Bifidobacterium animalis* AHC7 on Resolution of Acute Diarrhea in the Canine |
| 247 | Jose F Garcia | Evaluation of Intra-Stool Variability of Three Lactic Acid Bacterial Genera in Dogs by Quantitative Real-Time PCR |
| **248** | **Jodi Nicklas** | **Molecular and Histopathological Characterization of Enteroadherent Bacteria in Failure to Thrive Kittens** |
| 249 | Melanie Craven | Absence of a Bacterial Association in Yorkshire Terriers with Protein-Losing Enteropathy and Cystic Intestinal Crypts |
| 250 | Panagiotis Xenoulis | Serum fPLI and Spec fPL Concentrations in Cats with Experimentally Induced Chronic Renal Failure |
| **251** | **Brier Bostrom** | **Immunohistochemical Characterization of Canine Pancreata with Lymphocytic Infiltration** |
| 252 | Romy Heilmann | Purification and Partial Characterization of Canine S100A12 |
| 253 | Niels Gruetzner | Association Study of Single Nucleotide Polymorphisms with Cobalamin Deficiency in the Chinese Shar Pei |
| 254 | Albert Jergens | Apoptosis in the Pathogenesis and Treatment of Canine Inflammatory Bowel Disease |
| 255 | Isabelle Ruhnke | Feasibility of a Modified Using Air Suction Chamber for the Evaluation of Intestinal Mucosal Function in Endoscopic Biopsies from Dogs and Cats |
| 256 | Allison Bradley | Comparison of Gastrointestinal Motility in Dogs Treated with Iberogast™ and Ondansetron using the Smartpill™ |
| **257** | **Bettina Mayer-Roenne** | **Successful Treatment of Benign Esophageal Strictures with Balloon Dilation and Submucosal Triamcinolone Injection in Five Dogs and One Cat** |
| **258** | **Adam Ginman** | **Small Animal Gastrointestinal Endoscopy: A Computer-Aided Instructional Program** |

## SMALL ANIMAL – HEPATOLOGY

| | | |
|---|---|---|
| 259 | Kensuke Nakamura | Usefulness of Contrast-Enhanced Ultrasonography for Characterization of Focal Liver Lesions |
| **260** | **Andrea Johnston** | **Hepatic Copper Concentrations in Labrador Retrievers with and without Chronic Hepatitis (1980-2008): An Emerging Syndrome or Oversupplementation?** |
| 261 | Angela Au | Reduction of Cytokine-Induced Prostaglandin E2 and Chemokine Production in Cultured Canine Hepatocytes by Natural Products |
| 262 | Melinda Camus | Serum Protein Electrophoresis Beta-Gamma Bridging is a Poor Predictor for Hepatic Disease |

## SMALL ANIMAL – NEPHROLOGY/UROLOGY

| | | |
|---|---|---|
| **263** | **Kevin Byl** | **In Vitro Comparison of Plain Radiography, Double Contrast Cystography, Ultrasonography, and Computed Tomography for Measuring Urocystolith Size Using a Bladder Phantom** |
| 264 | Ben Young | Influence of Food Consumption on Variation of Urinary Volume, pH, Analytes, and Relative Supersaturation for Calcium Oxalate and Struvite in Healthy Adult Male Hound Dogs |
| 265 | Bernhard Gerber | Alpha 1-Acid Glycoprotein, Serum Amyloid A, and Monocte Chemoattractant Protein-1 in Cats with Idiopathic Cystitis |
| 266 | Ryan Keegan | Oxidative Stress in Cats with Chronic Renal Failure |
| 267 | Marcia Kogika | Erythrocyte Zinc Protoporphyrin Evaluation in Cats with Chronic Renal Disease |
| 268 | Christiane Prosser | Sequential Evaluation of Erythrocyte Zinc Protoporphyrin in Cats with Chronic Renal Disease |
| 269 | Shigeki Yamano | Endogenous Erythropoietin Levels and Iron Utility in Cats with Chronic Kidney Disease |
| 270 | Natalie Finch | Is Cadmium Exposure Associated with Chronic Kidney Disease in Cats? |
| 271 | Jerome Seguela | Glomerular Filtration Rate in Healthy English Pointer, English Setter and German Shepherd Dogs |
| 272 | David Grant | 24-Hour Water Consumption in Healthy Cats using a Fountain versus a Bowl |
| 273 | Courtney North | Clinical and Genetic Characterization of a Congenital Entlebucher Mountain Dog Urinary Tract Syndrome |
| 274 | Mary Nabity | Proteomic Analysis of Urine from Male Dogs in Early Stages of Progressive Tubulointerstitial Injury Due to X-linked Hereditary Nephropathy (XLHN) |

## SMALL ANIMAL – NUTRITION/METABOLISM

| | | |
|---|---|---|
| 275 | Nolan Frantz | Metabolomic Profiling of Plasma in Arthritic vs Non-Arthritic Dogs |
| 276 | Nolan Frantz | Metabolomic Profiling of Plasma in Arthritic vs Non-Arthritic Cats |
| 277 | Ryan Yamka | Metabolomic Profiling of Plasma in Lean vs Obese Dogs |
| 278 | Yuka Mitsuhashi | Determination of Maintenance Energy Requirement of Cats after Spaying |
| 279 | Márcia Jericó | Assessment of Metabolic Syndrome Occurrence in Obese Client-Owned Dogs |
| 280 | Amy Chamberlin | A Vegetarian Dietary Oil Source of $\gamma$-linolenic Acid Results in Arachidonic Acid Accumulation in Feline Uterine Tissues: Evidence of Delta-5 Desaturation |
| 281 | Ji-Houn Kang | Effect of Parenteral L-alanyl-L-glutamine Supplementation on Phagocytic Responses of Canine Peripheral Blood Polymorphonuclear Neutrophilic Leukocytes Exposed to Methylprednisolone Sodium Succinate |

## SMALL ANIMAL – PHARMACOLOGY

| | | |
|---|---|---|
| **282** | **Jessica Quimby** | **The Pharmacokinetics of Mirtazapine in Healthy Cats** |
| 283 | Robin Butchin | Bioavailability of a Novel, Bioenhanced Preparation of Curcumin in Dogs |
| 284 | Meaghan Whealen | Bioavailability of a Novel Formulation of S-Adenosylmethionine in Beagle Dogs |
| 285 | Steven Budsberg | In vivo Effects of Firocoxib, Meloxicam and Tepoxalin Adminstration on Eicosanoid Production in Target Tissues of Normal Cats |

## SMALL ANIMAL – INFECTIOUS DISEASE

| | | |
|---|---|---|
| 286 | Elise Gingrich | Intestinal Parasites of Dogs on the Galapagos Islands |
| 287 | Merle Olson | Canadian Prevalence and Diagnosis of Giardia Infection in Dogs and Cats using a Fecal Antigen Test and Fecal Smear |
| 288 | Sahatchai Tangtrongsup | Prevalence and Molecular Analysis of *Giardia* spp. Infections in Dogs in Chiang Mai, Thailand: Preliminary Findings |
| 289 | Nipattra Debavalyi | Serotypes and Pathogenicity Associated with Antimicrobial-Induced Resistance in Fecal *Escherichia coli* of Healthy Dogs |

| | | |
|---|---|---|
| 290 | Julia Veir | Comparison of Quantitative PCR and Conventional Endpoint PCR for Amplification of Parvovirus DNA in Blood from Naturally Infected and Recently Vaccinated Dogs |
| **291** | **E. Freya Kruger** | **The Epidemiology of *Streptococcus canis* Infections in Dogs and Cats** |
| 292 | Jean-Pierre Pages | Bacterial and Fungal Colonization of Peripheral Intravenous Catheters in Dogs and Cats |
| 293 | Mary Marcondes | Determination of Four IgG Subclasses in Asymptomatic Dogs with Visceral Leishmaniasis and in Animals Vaccinated Against the Disease |
| 294 | Regis Krah | Performance Evaluation of Species-Specific Peptide-Based Assays for Detection of Canine Antibodies to *E. canis* and *E. chaffeensis* |
| 295 | Ramaswamy Chandrashekar | *Ehrlichia/Anaplasma* spp. Infection and Exposure Rates in Dogs from Oklahoma and Arkansas |
| 296 | Pedro Paulo Diniz | Efficacy of Doxycycline Treatment in Dogs Naturally Infected with *Anaplasma phagocytophilum* |
| 297 | Audra Fenimore | *Bartonella* spp. Associated Endocarditis in Dogs in Colorado and Wyoming |
| **298** | **Jeanne Ficociello** | **Detection of *Bartonella henselae* IgM in Cats with Experimentally Induced Bartonellosis** |
| 299 | James Eucher | Prevalence of *Bartonella* and Haemoplasma DNA in Samples from Non-Owned Free-Roaming Cats Admitted to Animal Shelters in Colorado and Florida |
| 300 | Christian Leutenegger | Association of Pathogenicity and Rate of Coinfection of Three Feline Hemotropic Mycoplasma Strains with Presence and Severity of Anemia |
| 301 | Christian Leutenegger | Analytical Sensitivity and Specificity of a Real-Time PCR Assay Detecting Pathogenic Leptospira in Dogs Based on the hap-1 Gene |
| 302 | Kristy Dowers | Prevalence of Infectious Agents and Anti-Erythrocyte Antibodies in Cats with Anemia |
| 303 | Michael Lappin | *Bartonella henselae* Antigen Recognition Patterns in the Serum and Aqueous Humor of Cats with and without Uveitis |
| 304 | Michael Lappin | Effect of Orbifloxacin or Doxycycline Administration on Bartonella spp. and Hemoplasma Test Results from Naturally-Exposed Cats |
| 305 | | Abstract Withdrawn |
| 306 | Jacqueline Whittemore | Alpha Enolase and Annexin A2 Antibodies in Cats |
| 307 | Kristopher Curtis | Prevalence of Feline Immunodefficiency Virus Antibody and Feline Leukemia Virus Antigen in Samples Submitted to IDEXX Reference Laboratories for Feline Hemotropic Mycoplasma Testing |
| 308 | Yasuyuki Endo | An Updated Epidemiological Survey of Feline Immunodeficiency Virus Infection in Japan |
| 309 | Mitika Hagiwara | Evaluation of CD4+ and CD8+ T Lymphocytes in Cats Naturally Infected with Feline Leukemia Virus |
| 310 | Archivaldo Reche Junior | Cutaneous Mycoflora and CD4+:CD8+ Ratio of Cats Infected with Feline Immunodeficiency Virus |
| 311 | Archivaldo Reche Junior | Prevalence of Feline Calicivirus and Herpesvirus Type 1 in Maine Coon Cats with Chronic Gingivitis |

**SMALL ANIMAL – OTHER**

| | | |
|---|---|---|
| 312 | Sara Zaldivar-Lopez | Disease Prevalence and Causes of Death in American Kennel Club Registered Greyhounds |
| 313 | Paulo Vilar Saavedra | Thromboelastographic Changes after Gonadectomy in Retired Racing Greyhounds |

**FOOD ANIMAL – IMMUNOLOGY**

| | | |
|---|---|---|
| 314 | Daniela Bedenice | Immunologial Responses in Persistently BVDV Infected Alpacas |
| 315 | Claire Whitehead | Humoral Response to Bluetongue Vaccination in Alpacas |

**FOOD ANIMAL – INFECTIOUS DISEASE**

| | | |
|---|---|---|
| **316** | **Marcio Costa** | **Incidence of *Clostridium difficile* in Feces of Male Holstein Veal Calves** |

**FOOD ANIMAL – OTHER**

| | | |
|---|---|---|
| **317** | **Jennifer Sonis** | **Prairie Rattlesnake Envenomation in 25 New World Camelids** |

**EQUINE – CARDIOLOGY**

| | | |
|---|---|---|
| 318 | Marc Kraus | Concentrations of Cardiac Tropinin I Measured with an I-Stat® I Analyzer in Normal Horses Undergoing a Standard Treadmill Performance Examination |
| 319 | Rose Nolen-Walston | The Effects of Dehydration on Central Venous Pressure in Adult Horses |
| **320** | **Joan Norton** | **Comparison of Water Manometry to Two Electronic Pressure Monitors for Central Venous Pressure Measurement in Horses** |

| | | |
|---|---|---|
| **321** | **Joan Norton** | **Repeatability and Effect of Head Position on Central Venous Pressure Measurement in Standing Adult Horses** |
| 322 | Olga Seco | Cardiac Troponin I Concentrations in Horses Referred for Colic |

**EQUINE – ENDOCRINOLOGY**

| | | |
|---|---|---|
| **323** | **Rosa Barsnick** | **Association of Growth Hormone and Ghrelin with the Energy Metabolism and Mortality in Septic Foals** |
| **324** | **Rebecca Funk** | **Seasonal Changes in Plasma Alpha-Melanocyte-Stimulating Hormone and Adrenocorticotropic Hormone in Response to Thyrotropin-Releasing Hormone Administration in Normal Aged Horses** |
| 325 | John Haffner | Effect of Insulin Infusion into the Digital Artery on Glucose Concentrations in the Digital Vein in Healthy Horses |
| 326 | Johan Brojer | Validation of a Species Specific Enzyme-Linked Immunosorbent Assay for Measurement of Serum Insulin in Horses |
| 327 | Harold Schott II | Hyperinsulinemia in Horses with Pituitary Pars Intermedia Dysfunction and Effects of Housing on Insulin Response During a Glucose Tolerance Test |

**EQUINE – HEMATOLOGY**

| | | |
|---|---|---|
| 328 | Susana Macieira | Characterization of the cDNA Sequence Encoding for the Platelet Integrin AlphaIIb and Beta3 in a Horse with Glanzmann thrombastenia |
| **329** | **Maeva May** | **Comparison of Hematologic and Biochemical Values Obtained Via Intravenous Catheter Versus Venipuncture in Hospitalized Horses** |

**EQUINE – IMMUNOLOGY**

| | | |
|---|---|---|
| 330 | Jennifer Day | Effects of Lipopolysaccharide, Lipoteichoic Acid, and Peptidoglycan on Tumor Necrosis Factor Production in Equine Whole Blood Culture |
| 331 | Britta Leise | Laminar Inflammatory Gene Expression in the Carbohydrate Overload Model of Equine Laminitis |
| 332 | Brett Sponseller | IL-10 Downregulates Production of Peroxynitrite in Equine Monocyte-Derived Macrophages: Implications for Rhodococcal Infection |
| 333 | Jarred Williams | Effect of Intravenous Lidocaine Administration on Leukocyte Emigration in the Black Walnut Extract Model of Laminitis |

**EQUINE – INFECTIOUS DISEASE**

| | | |
|---|---|---|
| 334 | Stephanie Bell | The Immune Response of Foals to Natural Infection with Equid Herpesvirus-2 and its Association with Febrile Illness |
| 335 | Emir Hodzic | Use of Quantitative Real-Time PCR for the Detection of *Salmonella* Sp. in Fecal Samples of Horses Admitted to a Veterinary Medical Teaching Hospital |
| 336 | Alma Roy | Comparison of PCR Testing for Management of an Outbreak of EHVI at a Racetrack |
| **337** | **Katharyn Mitchell** | **Salmonella Oranienburg in a Large Animal Veterinary Teaching Hospital** |

**EQUINE – NEPHROLOGY/UROLOGY**

| | | |
|---|---|---|
| 338 | Harold Schott II | Paradoxical Aciduria in Horses: 37 cases (2000-2008) |

**EQUINE – NEUROLOGY**

| | | |
|---|---|---|
| 339 | Cristina Sogaro-Robinson | Diagnostic Utility of Computed Tomography of the Head in Horses Affected by Neurologic Disorders |

**EQUINE – NUTRITION/METABOLISM**

| | | |
|---|---|---|
| 340 | Anna Firshman | The Effect of Metformin on Insulin Sensitivity in Horses |
| 341 | David Lewis | The Positive Interaction of Propionate, Insulin, and Cortisol in the Horse |
| 342 | Jessica Suagee | Acute Hyperinsulinemia Modulates Expression of Genes Involved in Glucose and Fat Metabolism in Subcutaneous Adipose Tissue of Horses |
| 343 | Kibby Treiber | Insulin Sensitivity in Ponies is Influenced by Season and Pasture |

**EQUINE – OTHER**

| | | |
|---|---|---|
| 344 | James Belknap | Mononuclear Leukocytes in the Laminae of Normal Horses and Those with Black Walnut Extract-Induced Laminitis |
| 345 | | Abstract Withdrawn |
| 346 | Lee Ann Fugler | Evaluation of Various Matrix Metalloproteinase Inhibitors (MMPIs) in the Horse |
| **347** | **Cristobal Navas de Solis** | **Ultrasonographic Measurement of Spleen Volume in Horses** |

**EQUINE – PHARMACOLOGY**

| | | |
|---|---|---|
| 348 | Grace Forbes | Effects of Flunixin Meglumine on Leukocyte Activation in a Low-dose Endotoxemia Model in Horses |
| 349 | Jennifer Davis | The Pharmacokinetics of Intravenous and Intramuscular Buprenorphine in the Horse |
| 350 | Cynthia North | Bioavailability of Detomidine Administered to Horses as an Oromucosal (Sublingual) Gel and Comparison of Absorption of Detomidine by the Sublingual and Intramuscular Routes |
| 351 | Elizabeth Davis | Pharmacokinetics of Orally Administered Terbinafine in Horses |
| **352** | **Jessica Thompson** | **Assessment of Cardiac Function, Skeletal Muscle Morphology, and Muscle Enzymes in Horses on Long-Term Clenbuterol Therapy** |
| **353** | **Dorothy Whelchel** | **Effects of Clopidogrel on Platelet Function in the Horse** |

**EQUINE – RESPIRATORY**

| | | |
|---|---|---|
| 354 | Renaud Leguillette | Upper and Lower Airway Endoscopic Findings and Bronchoalveolar Lavage-Induced Lower Airway Collapse in Equine Respiratory Inflammatory Diseases |

**ABSTRACT #1**

USE OF POLYMERASE CHAIN REACTION TO DETECT BACTEREMIA IN DOGS WITH BACTERIAL ENDOCARDITIS. AM Heaney[1], KM Meurs[1], JL Oaks[1], PR Fox[2], TC DeFrancesco[3]. 1. Washington State University, College of Veterinary Medicine, Pullman, WA, 2. The Animal Medical Center, Caspary Institute, New York, NY, 3. North Carolina State University, Raleigh, NC.

Infectious endocarditis (IE) is associated with substantial morbidity and mortality. Outcome is influenced by effectiveness of antimicrobial therapy. The reliability of blood cultures to identify inciting pathogens has been disappointing. Panbacterial polymerase chain reaction (PCR) used on DNA extracted directly from human valve tissue has resulted in a threefold increase in sensitivity compared to blood cultures. Application of panbacterial PCR primers to canine blood samples holds promise to identify the offending bacteria and improve of antimicrobial therapy. Panbacterial PCR amplification (using primers to amplify the conserved 16s ribosomal bacterial DNA) with direct sequencing to identify the organism would be beneficial if it allows earlier targeted therapy.

Our hypothesis was that PCR can be used to identify circulating bacteria in canine IE patients.

Our objective was to perform PCR with primers targeted to amplify the 16s ribosomal bacterial DNA from IE suspects' blood samples and compare these findings with those obtained from blood cultures.

Thirteen dogs with IE were studied. Diagnosis was based upon history, physical examination, and echocardiographic examination. All had PCR perfomed and 9 of the 13 also had blood cultures. Two cases were positive using standard blood culture techniques yielding a sensitivity of 22.2%. PCR was positive in 4 of 13 cases yielding a sensitivity of 30.8%

In conclusion, bacteremia was detected by PCR in 4 out of 13 dogs with IE. Panbacterial PCR is a viable and potentially useful test to detect bacteremia in dogs with IE and warrants further investigation. Case enrollment is ongoing.

**ABSTRACT #2**

THE EFFECT OF FUROSEMIDE AND PIMOBENDAN ON THE RENIN-ANGIOTENSIN-ALDOSTERONE SYSTEM (RAAS) IN DOGS. AC Lantis, CE Atkins, TC DeFrancesco, BW Keene. North Carolina State University College of Veterinary Medicine, Raleigh NC.

We hypothesized that the labeled dosage of pimobendan would not have an additive effect on furosemide-induced RAAS activation. Furthermore, we asked whether furosemide's effect on RAAS would plateau with subacute administration.

Twelve healthy hounds were used. Group 1 (control) received furosemide (2 mg/kg PO, BID) for 10 days. Group 2 received furosemide (2 mg/kg PO, BID) and pimobendan (0.25 mg/kg PO, BID) for 10 days. To measure the response of the RAAS, urinary aldosterone:creatinine ratio (A:Cr) was determined on days 0, 5 and 10.

Urinary aldosterone secretion, as indicated by A:Cr, increased 3-fold with both furosemide monotherapy and the combination of pimobendan and furosemide at both days 5 and 10. There was no evidence of synergism between pimobendan and furosemide in terms of RAAS activation. Furosemide's effect on RAAS appears to plateau at 5 days or sooner but is maintained for at least 10 days.

We have shown that pimobendan, at the labeled dosage, does not accentuate furosemide- induced RAAS activation. We observed a three-fold increase in RAAS activity with furosemide alone and in combination with pimobendan. Therefore, furosemide, with or without pimobendan, is not recommended for chronic use in the absence of concurrent therapy to blunt RAAS activity, such as ACEI, aldosterone receptor blockers, or angiotensin II type I receptor blockers.

**ABSTRACT #3**

ASSESSMENT OF CHANGE IN LEFT VENTRICULAR SYSTOLIC FUNCTION BY TISSUE DOPPLER IMAGING METHODS AFTER CLOSURE OF PATENT DUCTUS ARTERIOSUS IN DOGS. LJM Keller, M Killich, K Hartmann, G Wess. Clinic of Small Animal Internal Medicine, Ludwig-Maximilians-University, Munich, Germany.

Patent ductus arterious (PDA) initially causes left ventricular (LV) volume overload; systolic dysfunction may develop later. Studies in veterinary medicine showed minimal or no improvement of left ventricular systolic volume indices after PDA closure indicating that myocardial remodeling in PDA is at most only partially reversible. Human long term follow-up studies using LV endsystolic volume measurements and evaluating ejection fraction (EF) showed, however, that systolic function deficits are reversible after closure. The aim of the present study was to determine systolic function in dogs with PDA before and after closure using systolic parameters of Tissue Doppler Imaging (TDI) and comparing them to endsystolic volume indices and EF in dogs.

In 31 dogs, echocardiographic studies including volume measurements and TDI parameters were performed before, one day after, and $\geq 3$ months after interventional or surgical PDA closure. To evaluate LV volume indices and EF, the modified Simpson's method was used. Systolic Tissue Velocity, systolic Strain Rate, and Strain using Echo Pac 2D Strain® analysis (General Electric Medical Systems, Horten, Norway) were evaluated in the basal, midventricular, and apical segment of the interventricular septum. Systolic TDI parameters at $\geq 3$ months after closure were compared to the TDI parameters of 49 age-matched healthy control dogs.

Systolic Tissue Velocity in all segments as well as systolic Strain Rate and Strain in the apical segment significantly decreased immediately after closure. At the $\geq 3$ month follow-up, these parameters had reversed to normal, and there was no significant difference compared to preclosure values. In addition, TDI parameters at the $\geq 3$ month follow-up did not show decreased values in comparison to healthy control dogs. Systolic volume indices did not change over the study period, whereas EF decreased and stayed low at long term follow-up.

PDA with a relevant amount of left-to-right shunting, left-to-right shunting causes LV remodeling. This is the first study in dogs that showed, however, that LV remodeling can be reversible after PDA closure using TDI methods. TDI in the present study showed that systolic function deteriorates immediately after closure but recovers within 3 months. Systolic volumes did not show any change over the study period, and EF decrease immediately after closure and stayed low at long-term follow-up. TDI seems to be a more sensitive tool and, therefore, should be preferred over volume measurements for the assessment of LV systolic function in congenital heart diseases.

**ABSTRACT #4**

CORRELATION OF N-TERMINAL PROHORMONE BRAIN NATRIURETIC PEPTIDE WITH LEFT VENTRICULAR OUTFLOW TRACT IN DOGS WITH SUB AORTIC STENOSIS. G Farace[1], SJ Ettinger[2], S Forney[3], A Beardow[1], A Carrier[1], K Yeung[1], [1]IDEXX Laboratories, Inc., Westbrook, ME, [2]California Animal Hospital, Los Angeles, CA, [3]Las Vegas Veterinary Referral Center, Las Vegas, NV.

In human medicine aortic stenosis has been shown to correlate with N-terminal prohormone brain natriuretic peptide (NTproBNP). This is to be expected because an obstruction in the aorta is likely to cause increased ventricular systolic pressures which in turn will increase the ventricular wall stress and thus cause a release of BNP and NTproBNP. Dogs with sub aortic stenosis (SAS) are likely to undergo similar pathophysioloic changes and so it is probable that there is a correlation between the left ventricular outflow tract (LVOT) peak flow velocity, severity of stenosis, and NTproBNP concentration

To test this hypothesis we retrospectively analyzed samples from 169 normal dogs of various breeds and ages and 69 dogs with varying degrees of SAS. Each dog was given a full cardiac exam including a radiograph and echocardiogram with Doppler flow analysis prior to collecting plasma for NTproBNP testing.

There is a moderate correlation (0.56) between LVOT and NTproBNP and this does not change if only dogs with a diagnosis of SAS are considered.

Grouping the population according to the severity of SAS (none, mild, moderate and severe) results in mean LVOT velocities of 1.46 (95% CI: 1.40-1.52), 2.62 (2.49-2.75), 3.58 (2.99-4.17) and 6.39 (5.83-6.96) m/s respectively and NTproBNP mean values of 361 (307-415), 344 (239-448), 1011 (86-1936) and 2529 (1841-3217) pmol/L. Clearly, there are significant differences in LVOT across the groups while NTproBNP is only significantly different from the normal population once the severity of the SAS reaches a moderate level. This would appear to suggest that while there is a change in the outflow tract velocity in those patients with mild SAS it is not large enough to cause hemodynamic changes resulting in BNP (and NTproBNP) release. However the data does suggest that NTproBNP can differentiate patients with moderate or severe disease from those with no disease or only mild disease. Indeed an ROC analysis gives an AUC of 0.86 and a negative predictive value of 96% at a cut-off of 713pmol/L.

Mean NTproBNP levels in the moderate and severe groups are extremely high. Interpretive criteria would suggest these dogs were in congestive heart failure. This would make SAS a confounding factor for detecting heart failure associated with acquired cardiac disease were it not for the fact that the 2/3 of the dogs in this study with SAS are under the age of three.

In conclusion, NTproBNP correlates with LVOT and both correlate with severity of SAS.

**ABSTRACT #5**

NT-PROBNP CONCENTRATIONS IN CANINE CONGENITAL HEART DISEASE. AB Saunders[1], SG Gordon[1], MW Miller[1], SE Achen[1], RM Roland[1], LT Drourr[1], CD Hariu[1] and MA Oyama[2]. [1]The Michael E. DeBakey Institute, College of Veterinary Medicine and Biomedical Sciences, Texas A&M University, College Station, TX, [2]University of Pennsylvania, Philadelphia, PA.

Many canine congenital heart defects (CHD) such as patent ductus arteriosus (PDA) and pulmonic stenosis (PS) can be repaired or palliated with interventional or surgical procedures. Puppies with CHD may have murmurs that are undetected by primary care veterinarians, and CHD such as atrial septal defects (ASD) may go undetected in adult dogs. Echocardiography is indicated for definitive diagnosis, but is not always available to primary care veterinarians who may need to encourage clients to obtain a referral for definitive diagnosis.

The objective of this study was to report NT-proBNP concentrations in dogs with a variety of CHD. Thirty nine dogs with CHD (PDA=13, PS=20, ASD=6) and 6 young, healthy dogs were evaluated. All dogs had echocardiograms. Median NT-proBNP for PDA was 742 pmol/L (range, 50-3000), PS was 746 pmol/L (278-3000), ASD was 833 pmol/L (388-2194) and for normal dogs was 333 pmol/L (161-826). Dogs with PS had Doppler estimated right ventricular systolic pressures consistent with moderate (1/20; 1024 pmol/L) to severe (19/20) PS. Dogs with ASD had hemodynamically significant defects with moderate to severe right atrial and ventricular dilation. Dogs with PDA, PS and ASD had elevated NT-proBNP concentrations when compared to the normal dogs. NT-proBNP >354 and >826 pmol/L possessed sensitivity and specificity for detecting CHD (PDA, PS, ASD) of 89.7% and 66.7% and 46.5% and 100% respectively (area under curve = 77.9%).

Further studies are needed to better characterize the clinical utility of NT-proBNP as a screening test for various canine congenital heart diseases.

**ABSTRACT #6**

UTILITY OF NT-PROBNP ASSAY TO DETECT OCCULT DILATED CARDIOMYOPATHY IN DOBERMAN PINSCHERS. N Morris[1], MA Oyama[2], ML O'Sullivan[3], GE Singletary[2], SG Gordon[4]. 1. Massachusetts Veterinary Cardiology Services, Pembroke MA. 2. University of Pennsylvania, Philadelphia, PA. 3. Ontario Veterinary College, Guelph ONT. 4. Texas A&M University, College Station, TX.

Detection of preclinical (occult) DCM is challenging. Blood testing for occult DCM represents a potential means of screening large numbers of dogs. We sought to determine ability of NT-proBNP to detect occult DCM in Doberman pinschers undergoing screening with Holter and echocardiography.

One-hundred Doberman pinschers were prospectively screened using Holter monitoring, echocardiography, and NT-proBNP assay. Occult disease was defined as LV systolic dimension greater than 38.8, 39.5, 40.2, 40.9, 41.6, 42.3, and 43.0mm in dogs weighing up to 25, 30, 35, 40, 45, 50, and 55kg respectively, or >50 VPCs on 24-hour Holter. Median

NT-proBNP in dogs with occult DCM (n=33) was significantly higher than normal dogs (n=67) (occult, 552 [IQR, 318-1154] pmol/L; normal, 220 [IQR, 171-315]; P<0.0001). NT-proBNP >194, >315, and >478 pmol/L possessed sensitivity and specificity for detecting occult DCM of 90.9% and 40.3%, 78.8% and 76.1%, and 57.6% and 91.4%, respectively (area under curve=82.3%). With regards to how diagnosis of occult DCM was achieved, 5 dogs fulfilled echo criteria alone, 16 dogs fulfilled Holter criteria alone, and 12 dogs fulfilled both. NT-proBNP detected dogs with abnormal echo (AUC=91.6%) or abnormal echo and Holter (AUC=95.3%) better than dogs with only abnormal Holter (AUC= 69.7%). NT-proBNP >478 pmol/L had 88.2% sensitivity and 91.4% specificity for detecting the 17 dogs with abnormal echo. The use of abnormal Holter or NT-proBNP >478 pmol/L possessed 86.4% sensitivity, 96.8% specificity, and 93% accuracy for detecting occult cases, suggesting that use of these two diagnostics could potentially be used to detect occult DCM.

**ABSTRACT #7**

EVALUATION OF NT-PROBNP IN THE DIAGNOSIS OF VARIOUS STAGES OF DILATED CARDIOMYOPATHY IN DOBERMAN PINSCHERS. G. Wess, V. Butz, M. Killich, A. Schulze, K. Hartmann. Clinic of Small Animal Internal Medicine, Ludwig Maximilian University of Munich, Germany.

Dilated cardiomyopathy (DCM) in Doberman pinschers is an autosomal inherited diseases with a very high prevalence in this breed. The disease is usually divided into three disease stages. The first stage consists of a genetic defect that cannot be diagnosed with current diagnostic tools. This stage is followed by an occult phase, during which dogs appear to be clinically healthy. The occult phase is characterized by ventricular arrhythmias, leading to sudden death in about 1/3 of the dogs. In the occult phase, dogs may or may not show abnormalities on echocardiography. This stage is followed by the clinical phase with typical echocardiographic findings of DCM. Previous studies have shown elevated NT-proBNP in dogs with congestive heart failure and in dogs with heart diseases without heart failure. The purpose of this study was to evaluate NT-proBNP in the diagnosis of DCM in Doberman pinschers in various disease stages.

The study included 484 NT-proBNP measurements of 324 dogs (187 male and 137 female Doberman pinschers, mean age 4,7 years). Samples were collected from dogs at various rechecks in different disease stages. Based upon echocardiography and 24-hour-ECG, the following groups were compared: healthy controls with normal echocardiographic exam and <50 VPCs in 24-hours-ECG (n=329), only ventricular premature contractions (VPCs) (n=64), only echocardiographic changes (n=26), VPCs and echocardiographic changes (n=31), decompensated stage (n=13) and finally the stage "still normal" consisting of dogs that were considered to be free of disease at the first exam, but that developed the disease later (n=17). Nt-proBNP was measured in plasma samples using an ELISA (VETSIGN Canine CardioSCREEN Nt-proBNP, Guildhay Ltd, UK).

NT-proBNP levels were significant different the between the control group and the group with only VPCs, the group with only echocardiographic changes, the group with both VPCs and echocardiographic changes and the decompensated group. Between the control group and the "still healthy" group only a trend, but no significant difference was detected. NT-proBNP had a sensitivity of 76,1% and a specificity of 76,9% using a cut-off value of 400 pmol/l to detect the occult phase of cardiomyopathy. Using the same cut-off value, the sensitivity reached 90%, the specificity 79% to detect dogs with echocardiographic changes. NT-proBNP increased with progressing disease stages. Even VPCs without echocardiographic changes led to increased NT-proBNP values. NT-pro-BNP is a useful additional test to screen for Doberman cardiomyopathy, but should not replace conventional tests.

**ABSTRACT #8**

SERIAL EVALUATION OF NT-proBNP IN DOGS WITH CHF PREDICTS CLINICAL SCORE AND THE PRESENCE OR ABSENCE OF RADIOGRAPHIC PULMONARY EDEMA. SE Achen[1], SG Gordon[1], RM Roland[1], AB Saunders[1], MM Boggess[2] and MW Miller[1]. The Michael E. DeBakey Institute, [1]College of Veterinary Medicine and Biomedical Sciences & [2]College of Science, Texas A&M University, College Station, TX.

Human studies have found NT-proBNP useful in clinical decision making in CHF. The objective of this prospective study was to evaluate serial NT-proBNP in dogs with CHF due to chronic valve disease (CVD). Evaluations included thoracic radiographs, blood work, BP, echocardiogram, NT-proBNP, and therapy changes.

Statistical analysis evaluated the predictive value of the change in NT-proBNP (ΔNT-proBNP) against clinical score (CS; -1 = ↓ furosemide ± ↑ pimobendan ± fluid therapy, 0 = no therapy change, 1 = ↑ furosemide ± ↑ pimobendan ± parenteral furosemide) and radiographic cardiogenic infiltrate score (RIS; 0 = none, 1 = cardiogenic infiltrate). Eighteen dogs had 58 total serial evaluations. ΔNT-proBNP between successive visits was calculated, giving 38 observations on 14 dogs. Median ΔNT-proBNP was -170 pmol/L (range -2,000 - 3,700 pmol/L).

Using logistic and multinomial logistic models, with Huber-White standard errors to account for correlation of observations on the same dog, ΔNT-proBNP is significantly predictive of both CS and RIS ($p<0.05$). For a ↓ in ΔNT-proBNP ≥ 2,000 pmol/L, a CS of -1 was likely, with an 80% probability. For ΔNT-proBNP between -1,000 and 2,000 pmol/L, a CS of 0 was likely, with a probability > 70%. For ΔNT-proBNP >2,000 pmol/L, a CS of 1 was likely, with a 60% probability. A ΔNT-proBNP >2,000 pmol/L, predicted an RIS = 1 while a ↓ in ΔNTproBNP, or an ↑ <2,000 pmol/L, predicted an RIS = 0.

These results suggest serial evaluation of ΔNTproBNP may facilitate management of CHF due to CVD.

**ABSTRACT #9**

THE UTILITY OF NT-PROBNP TO DETECT EARLY STAGES OF HYPERTROPHIC CARDIOMYOPATHY IN CATS AND TO DIFFERENTIATE DISEASE STAGES. G. Wess, P. Daisenberger, J. Hirschberger, K. Hartmann. Clinic of Small Animal Internal Medicine, Ludwig-Maximilians-University, Munich, Germany.

Hypertrophic cardiomyopathy (HCM) is the most common cardiac disease in cats. In practice, it is currently necessary to use echocardiography to confirm the diagnosis and to assess disease severity. Early stages of the disease often show no or only subtle clinical symptoms and might therefore not be detected if echocardiography is not available. N-terminal proBNP (Nt-proBNP) has no physiological activity, but is more stable than BNP and therefore easier to measure. Measurement of NT-proBNP concentration is helpful to distinguish cardiac from non-cardiac causes of dyspnea in cats, dogs, and humans. Previous studies have shown elevated NT-proBNP concentrations even in the occult phase of cardiomyopathy.

The purpose of this prospective study was to evaluate the utility of Nt-proBNP to detect early disease stages of HCM in cats and to assess whether a differentiation between mild, moderate, and severe disease stages is possible using NT-proBNP measurements.

Nt-proBNP was measured in plasma samples from 159 cats using an ELISA (VETSIGN Feline CardioSCREEN Nt-proBNP, Guildhay Ltd, UK). The cats were classified according to echocardiography into one of the following groups: (1) clinical healthy (control) group (n = 33), (2) mild HCM with focal hypertrophy of the left ventricular (LV) free wall or septal hypertrophy between 5.5 and 5.9 mm (n = 13), (3) moderate HCM with focal or generalized LV free wall or septal wall between 6.0 and 7.0 mm and normal left atrial size (LA) (n=12), and (4) severe HCM with LV free wall or septal wall > 7.0 mm and enlarged LA (n = 43) and clinical symptoms.

Nt-proBNP was significantly higher in all HCM groups compared to healthy control cats. Mean NT-proBNP was 58 +/- 65 pmol/l in the control group (1), 333 +/- 244 pmol/l in the mild group (2), 433 +/- 299 pmol/l in the moderate group (3) and 835 +/- 314 pmol/l in the severe group (4). Values of the mild (2) and moderate (3) group cats were significantly lower then those of the severe (4) group cats, but there was no difference between the values of the mild (2) and moderate (3) group. It was shown that the recommended cut-off value of 49 pmol/l had a high specificity of 97.1% but only a specificity of 56% to differentiate healthy cats from cats with HCM. This low cut-off value revealed too many false positive results. Using a cut-off value of 100 pmol/l, Nt-proBNP had a sensitivity of 94.2% and a specificity of 81.3% for the differentiation of control and HCM cats and might therefore be preferable for use in practice. In conclusion, this feline ELISA Nt-proBNP assay is helpful in the diagnosis of even early cases of HCM in cats. Cats with elevated levels, therefore, should be further worked up with echocardiography.

**ABSTRACT #10**

PLUMONARY HYPERTENSION AND N-TERMINAL PROHORMONE BRAIN NATRIURETIC PEPTIDE IN DOGS. G Farace[1], SJ Ettinger[2], S Forney[3], A Beardow[1], A Carrier[1], K Yeung[1] [1]IDEXX Laboratories, Inc., Westbrook, ME, [2]California Animal Hospital, Los Angeles, CA, [3]Las Vegas Veterinary Referral Center, Las Vegas, NV.

N-terminal prohormone brain natriuretic peptide (NTproBNP) is usually viewed as a marker of left ventricular dysfunction but there is a significant body of literature in human medicine that shows a correlation between pulmonary hypertension (PH), right ventricular dysfunction and NTproBNP. In dogs it has been reported that B-type natriuretic peptide (BNP) is elevated in cases of PH but the degree of elevation was not correlated to the ISACHC class. PH may, therefore be a confounder if NTproBNP alone is used to diagnose disease. If the NTproBNP level correlates with the severity of PH it might be useful in staging disease.

The aim of this retrospective study was two-fold: firstly to determine if NTproBNP levels are elevated in dogs with PH independent of disease severity and to determine if NTproBNP correlates to PH severity.

374 dogs had their maximum tricuspid regurgitation pressure gradient (max TRPG) measured together with a blood draw upon entry into the California Animal Hospital Veterinary Specialty Group (CAHVSG)/IDEXX Cardiac Study. Those with congenital disease were excluded. (10) The remainder was divided according the magnitude of the max TRPG. Dogs with a max TRPG <30mmHg were classified as normal; 31-50mmHg were considered to have mild PH; 51-75mmHg moderate and >75mmHg had severe PH.

Plotting max TRPG against NTproBNP concentration gives a weak but significant correlation (0.17, p=0.001) thus indicating a relationship between these two measures. Comparing those dogs with PH to those with no PH shows a significant difference (1289 vs. 1076pmol/L, p=0.02). However controlling for disease severity (according to the CAHVSG) there is no significant difference between the two groups at any individual grade.

The mean NTproBNP levels for the four groups of PH were: none – 1076pmol/L (n=193), mild – 1185pmol/L (n=132), moderate - 1686pmol/L (n=24) and severe – 1573 (n=15). While this suggests that there is an increase in NTproBNP as PH increases significant overlap between groups occurs and only those dogs with moderate PH are significant elevated (p=0.04) over the dogs with no PH. Controlling for the disease severity fails to identify any significant difference between the four groups.

In conclusion, this study concurs with the study on BNP. NTproBNP is elevated in the presence of PH but that there is no correlation to the severity of cardiac disease or severity of PH when the evaluation is controlled for disease severity.

**ABSTRACT #11**

GENOME-WIDE ASSOCIATION IDENTIFIES A MUTATION FOR ARRHYTHMOGENIC RIGHT VENTRICULAR CARDIOMYOPATHY IN THE BOXER DOG. KM Meurs[1], E Mauceli[2], G Acland[3], K Lindblad-Toh[2,4] 1.Washington State University College of Veterinary Medicine, Pullman, WA, 2.Broad Institute of MIT and Harvard, Cambridge, MA 3.Cornell University, Ithaca, NY, 4. Uppsala University, Uppsala, Sweden.

Arrhythmogenic right ventricular cardiomyopathy (ARVC) in the Boxer dog may result in syncope, sudden cardiac death and the development of congestive heart failure. ARVC is more common in the boxer than in other breeds, suggesting a genetic predisposition in this breed. There is a familial inheritance, possibly suggesting a major dominant gene. Based on the new tools available with the canine genome project, it is now possible to perform genome wide association with a canine 50k SNP array. The objective of this study was to identify the genetic alteration(s) associated with the disease.

A 10 Mb region of genome-wide significant association was identified on chromosome 17 (p genome< 0.02) using DNA samples from 48 adult boxer dogs with ARVC (> 500 ventricular premature complexes (VPC)/24 hours) and 48 unaffected boxers (<100 VPC/24 hours). Fine-mapping of 24 additional SNPs within the region localized a signal to between two candidate genes of potential cardiac importance. All exons of the candidate genes and non-coding conserved elements within this region were re-sequenced in affected and unaffected dogs.

A 7 base pair deletion within a non-coding conserved element in a regulatory region of a calcium modulating gene was identified and observed to be highly associated with the disease status (p<0.0001). The deletion was not observed in unaffected Boxer dogs or in any of 31 dogs from other breeds tested. Affected dogs were either homozygous or heterozygous for the deletion. Homozygous dogs (N=15) had a statistically greater number of VPC/24 hours than heterozygous (mean of 7860 and 2845, respectively) (p<0.05).

We conclude that this 7 base pair deletion is associated with the development of Boxer ARVC and that dogs that are homozygous for the mutation appear to have a more severe degree of arrhythmias.

**ABSTRACT #12**

PHENOTYPIC DIFFERENCES IN THE ULTRASTRUCTURE OF CARDIOMYOCYTES FROM BOXER DOGS AFFLICTED WITH ARRHYTHMOGENIC RIGHT VENTRICULAR CARDIOMYOPATHY (ARVC).

Eva M. Oxford[1], Karen Maass[2], Philip R. Fox[3], Mario Delmar[4], and N. Sydney Moise[1]. 1. Cornell University, College of Veterinary Medicine, Ithaca, NY. 2. NYU Langone Medical Center, NY, NY. 3. Animal Medical Center, NY, NY 4. Center for Arrhythmia Research, U. of Michigan, Ann Arbor, MI.

ARVC is an inherited myocardial disease associated with sustained monomorphic ventricular tachycardia, sudden cardiac death (SCD), and replacement of myocardium with fatty or fibro-fatty tissue. Our previous work demonstrated altered localization of intercalated disc (ID) proteins in afflicted boxers. In this study we used transmission electron microscopy (TEM) to assess the ultrastructure of the ID and sarcomere in ARVC-afflicted boxers.

Afflicted boxers were identified by 24-hour Holter monitoring and histopathological examination of the heart. Samples from the right and left ventricles (RV and LV) of 2 unafflicted dogs (1 beagle, 1 German shepherd), and 7 ARVC afflicted boxers were collected and examined by TEM (data from 3 boxers reported here). Samples from control tissue showed normal morphology, normal alignment of the sarcomeres and myofibrils, and attachment of myofibrils to the ID. Samples from ARVC-afflicted boxers displayed myofibral disorganization, with fewer attachments to ID. Furthermore, electron dense areas up to one sarcomere in length were observed radiating from Z bands (Z-band streaming) in ARVC afflicted samples. Within RV and LV longitudinal sections, the following parameters were measured from an average of 120 µm of ID per ventricle: (1) average length of desmosomes (DSM), adherens junctions (AJ), and gap junctions (GJ), and (2) number of DSM, AJ, and GJ per 10 µm of ID. No significant differences were found in the length (all measurements in µm) of AJ or GJ from RV or LV samples of ARVC versus control. AJ length was 0.27 +/- 0.01 (RV) and 0.25 +/-0.02 (LV) in control samples versus 0.26+/-.04 (RV) and 0.23 +/- 0.02 (LV) in afflicted animals (pNS). GJ length in control samples was 0.48+/-0.03 (RV) and 0.48 +/- 0.02 (LV). In afflicted animals, GJ length was 0.40 +/- 0.02 and 0.47 +/-0.13, for RV and LV respectively (pNS). In contrast, the average length of DSM was significantly reduced in afflicted animals. DSM length was 0.22 +/- 0.05 (RV) and 0.20 +/- 0.02 (LV) in control. DSM length in afflicted samples was 0.16 +/- 0.003 (RV) and 0.15 +/-0.03 (LV) ($p< 0.05$). In addition, our data indicated a significant ($p< 0.05$) decrease in the number of GJ per 10 µm ID in the ARVC samples (1.8 +/-0.12 RV, 1.5 +/- 0.27 LV in control; 0.4 +/- 0.02 RV, 0.6 +/-0.24 LV in ARVC). No decrease in the number of DSM or AJ was observed. Our results indicate that ARVC in the boxer is associated with reduced length of DSM and reduction of GJ at the intercalated disc; the latter which may act as a substrate for ventricular arrhythmias. Furthermore, Z band streaming suggests a novel concept: genes associated with the cardiomyocyte contractile apparatus may play an etiological role in the development of ARVC in the boxer dog.

**ABSTRACT #13**

*IN VIVO* DETERMINATION OF CANINE CYSTOLITH MINERAL COMPOSITION USING COMPUTED TOMOGRAPHY-GENERATED HOUNSFIELD UNITS.

BM Pressler, LG Adams, HG Heng, JJ Rohleder, GE Moore. Purdue University, West Lafayette, IN.

Determination of mineral composition is essential for effective management of urolithiasis in dogs. Composition of uroliths retrieved from people has been determined *in vitro* by computed tomographic peak attenuation measurements (Hounsfield units, HU), and HU allow *in vivo* differentiation of the most common types of uroliths. In a previous study the three most common cystolith mineral types recovered from dogs could be differentiated *in vitro* using CT-generated HU with >90% accuracy. This study was designed to determine the accuracy of CT-generated HU for differentiation of canine cystolith mineral composition *in vivo*.

Forty-eight cystoliths of known composition were placed in a phantom array (15 struvite, S; 15 calcium oxalate, O; 11 urate, U; 7 mixed, M) and scanned at 80 kVp/400 mAs using two different CT scanners. Mean HU were calculated for each cystolith on each CT scanner by two radiologists; region-of-interest (ROI) for HU calculation was determined using the largest cystolith cross-section, with a manually-centered oval ROI encompassing two-thirds to three-fourths of the cystolith. Reference ranges were constructed using receiver operating curves to optimize sensitivity and specificity of mineral identification. To determine *in vivo* applicability of these HU reference ranges, 20 dogs with confirmed cystolithiasis were anesthetized and CT scans (1 mm slices) were obtained of their bladders immediately prior to removal of cystoliths. Each radiologist calculated mean HU using the largest urolith cross-section, and mineral composition was predicted. True cystolith mineral composition was determined using standard techniques by a veterinary urolith center.

The 20 enrolled dogs were determined to have 12 calcium-containing cystoliths (9 pure O, 1 pure calcium phosphate, 2 cystoliths with 80-85% O and 15-20% silica), 6 S-containing cystoliths (5 pure S, 1 85% S/15% calcium phosphate), and 2 U cystoliths. 9 of 12 (75%) of calcium-containing cystoliths were correctly identified by both radiologists; 2 were misidentified as S by one radiologist, and one was misidentified as S by both radiologists. 2 of 6 (33%) S cystoliths were correctly identified by both radiologists; 3 were incorrectly identified as U by both radiologists, 1 was identified as O by both radiologists. 2 of 2 (100%) U cystoliths were correctly identified by both radiologists. In total, the two radiologists were in agreement in their analysis of likely mineral composition in 18 of the 20 (90%) cystoliths. Overall, 75% and 88% of non-dissolvable and dissolvable uroliths, respectively, were correctly identified.

These preliminary results suggest that CT-generated HU may be used to differentiate canine cystolith mineral composition *in vivo*. Accuracy is highest for calcium-containing and urate cystoliths. Ongoing studies will increase the number of enrolled dogs and determine if alternative methods of reference range calculation may improve correct identification of cystolith mineral composition.

**ABSTRACT #14**

URETERAL STENTING FOR FELINE URETEROLITHIASIS: TECHNICAL AND CLINICAL OUTCOMES.

A Berent[1], C Weisse[1], D Bagley[2], C Adan[1], K Todd[1], J Solomon[1]. 1. Matthew J. Ryan Veterinary Hospital of the University of Pennsylvania, Philadelphia, PA. 2.Thomas Jefferson University Hospital, Philadelphia, PA.

Ureteral stenting has been performed for a variety of benign and malignant disorders, bypassing an obstruction while directing urine from the renal pelvis to the urinary bladder. The most common ureteral dilemma in feline patients is ureterolith-induced obstruction, of which traditional therapy is associated with excessive morbidity/mortality. The authors' objective is to describe the technical and clinical outcome of ureteral stenting in feline patients with ureterolith induced obstructions.

Twenty-two cats (25 ureters) had stent placement (3 french double pigtail) attempted for nephro-ureterolithiasis. Access was attempted endoscopically (12) and/or surgically (17). 14 female and 8 male cats, weighing 2.1-6.2 kg, were included. Surgical placement was possible in 14/17 (82.4%) and endoscopic in 4/12 (30%), with an overall success in 18/25 ureters (72%). 18/18 stents (100%) resulted in ureteral/pelvis decompression and azotemia stabilization. Procedure associated complications included: temporary stranguria (5), imperfect stent location (2) and a ureteral tear/trauma (1). One of two displaced stents required adjustment 3 days later. All other procedure associated complications resolved spontaneously without further intervention.

Seventeen of 18 stents remained in place and patent long-term (range: 2->780 days). There was no reported need for pre-mature removal. 4/16 cats with stents developed urinary tract infections (2/4 while a nephrostomy tube was present). All infections were successfully cleared. One cat had evidence of intermittent hematuria without associated dysuria, infection, azotemia, or pyelectasia. Further investigation was declined.

Overall, ureteral stenting is possible, safe, and effective for the treatment of feline ureterolithiasis, maintaining ureteral patency regardless of stone number or size.

**ABSTRACT #15**

THE USE OF A PERCUTANEOUSLY CONTROLLED HYDRAULIC OCCLUDER FOR THE TREATMENT OF URETHRAL SPHINCTER MECHANISM INCOMPETENCE IN 11 DOGS 1 CAT.

A Berent[1], C Weisse[1], C Adan[1], K Todd[1]. 1. Matthew J. Ryan Veterinary Hospital of the University of Pennsylvania, Philadelphia, PA.

Refractory urinary incontinence, where traditional medical/surgical techniques fail to resolve urinary leakage, is a major problem in animals with urethral sphincter mechanism incompetence (USMI), particularly after fixation of ureteral ectopia. The authors' objective is to describe the technical and clinical outcome using a percutaneously controlled hydraulic occluder (HO) for refractory incontinence secondary to USMI in small animal patients.

Eleven dogs and 1 cat were included. 8/11 dogs had a history of ectopic ureters. All patients were females ranging from 2.3 kg-37.3 kg, and 6-55 months. All patients failed medical management with phenylpropanolamine (12), diethylstilbestrol (6) and/or collagen injections (6). An uninflated silicone ring (HO) was surgically placed around the proximal urethra and connected to a subcutaneous injection port. Cystopexy was performed when necessary (6/12). Saline injections, providing extraluminal urethral compression, were performed if necessary. Median follow-up time was 7 months (range 2-10 months).

Seven of 11 dogs and 1/1 cat (66.7%) were continent after HO placement. 2/4 incontinent dogs were not yet injected. The two remaining dogs had severe anatomical abnormalities (intrapelvic/hypoplastic bladder [2], urethral dysplasia [1], short/wide urethra [2]). All incontinent dogs were reported to have dramatic improvements (50-80%). Two urethral stenoses occurred (1 extraluminal stricture; 1 intraluminal webbing) 3 weeks after placement requiring removal (n=2) and replacement (n=1) with a larger HO and intraluminal urethral stent. One dog returned to incontinence after removal.

Overall, the use of a HO for refractory USMI may result dramatic improvement toward urinary continence. The risk of urethral narrowing should be considered.

**ABSTRACT #16**

DEVELOPMENT OF A USSING CHAMBER MODEL FOR STUDYING THE CANINE UROTHELIAL RESPONSE TO INFECTION.

M. Wood, E. Breitschwerdt, M. Suyemoto, M. Stone, J. Gookin. North Carolina State University College of Veterinary Medicine, Raleigh, NC.

The urothelium prevents pathogen access to the body. Tight junctions linking umbrella cells of the urothelium serve as a physical blockade passively preventing infectious agents from accessing the underlying submucosa. These junctions assist in maintaining a measurable transepithelial resistance (TER) across viable tissue. The urothelium also serves active roles. During the acute phase of urinary tract infections the epithelium can produce cytokines including IL-6. In-vitro studies have demonstrated that cultured urothelial cells will produce IL-6 when challenged with bacteria. To date however, a precise understanding of the role of IL-6 during the acute phase response to urinary pathogens has eluded researchers. While cell culture systems identify the response of individual cells, the mucosal response of the tissue may be more complex. With the ultimate goal of characterizing the global urothelial response to infection, the described work sought to develop an in-vitro model that preserves the microenvironment constructed by the urothelium and lamina propria. To achieve this aim bladder mucosa from recently euthanized dogs was stripped of its muscularis layer and mounted within Ussing chambers. This construct created independent luminal and submucosal compartments. Each side of the tissue was bathed with an oxygenated, 37°C, physiological saline solution. A 10mM glucose substrate added to the submucosal side provided nutrient support. Model success was defined as a demonstration of intact barrier function and cytokine production over a minimum course of 5h. Barrier function was measured using voltage clamps to record the potential difference and short circuit currents traversing the tissue over time. These values were used to calculate the TER over a 10h period for tissue treated on the mucosal side with *E. coli* J96 or saline control. Additional chambers received PMA/ionomycin treatment for 3h as a positive control for cell death. Tissue from each of these time course treatments were formalin fixed and stained with H&E to visually identify evidence of cell death. The second criterion of model success was evidence of active urothelial function as evidenced by cytokine production. Fluid from both the serosal and mucosal reservoirs was sampled at 3, 5, 7 and 10h after mucosal *E. coli* and saline treatments. IL-6 concentrations were determined via a validated canine specific IL-6 ELISA. Results indicated that both saline and *E coli* treated tissue remained viable beyond 5h as the TER increased over this time period and there were no indicators of cell death via histology. This contrasted with the PMA/ionomycin treated tissue in which the TER decreased and there was evidence of apoptosis within 3h. *E. coli* treatment significantly increased submucosal IL-6 concentrations at 3 and 5h when compared to the saline control. These data provide evidence that canine bladder mucosa sustains normal barrier function, remains histologically intact, and responds biologically to *E. coli* infection by synthesis of IL-6 in an ex-vivo model system. The model generated represents a powerful tool for the study of molecular mechanisms of the urothelial response to urinary tract infection.

**ABSTRACT #17**

URODYNAMIC EVALUATION OF CATS WITH FELINE INTERSTITIAL CYSTITIS (FIC). Wu CH[1], Westropp JL[1], Buffington CAT[2]. School of Veterinary Medicine, University of California, Davis, CA[1]. and The Ohio State University College of Veterinary Medicine[2].

Although urodynamic studies of humans with interstitial cystitis have been reported, similar evaluations of cats with FIC have not been reported to our knowledge. The purpose of this study was to compare urodynamic parameters from 12 female cats with FIC with previously published data from six healthy female cats using the same anesthetic regimen.

All cats were anesthetized with propofol for cystometrogram (CMG) and urethral pressure profile (UPP) measurements. Threshold pressure (TP) and volume were recorded for CMGs. If a detrusor reflex was not evident, TP was recorded when urine passively leaked, or intravesical pressure exceeded 85 $cmH_2O$. The maximal urethral closure pressure (MUCP) was evaluated during the UPP.

Four cats had obvious detrusor reflexes, but none had evidence of detrusor overactivity (DO). Bladder compliance during the first CMG was significantly lower in FIC than in healthy cats (0.3±0.2 vs. 0.6±0.2; p=0.04). Compliance of cats with FIC increased significantly on the second CMG compared to the first (0.4±0.3 vs. 0.3±0.2 mls/cm H20; p=0.007). The maximal urethral closure pressure (MUCP) was significantly higher in FIC than in healthy cats on the first and second UPP evaluated (first MUCP obtained: 158±48 vs. 91±23 cm H20; p=0.003, second: 152±45 vs. 86±29).

Although a decreased compliance was documented, it was not a result of DO; therefore drugs aimed at this problem do not seem warranted in cats with FIC. Because higher MUCP values were noted, selective alpha adrenoceptor antagonism may be warranted, particularly if these problems are found in male cats.

**ABSTRACT #18**

RANDOMIZED, PLACEBO-CONTROLLED CLINICAL TRIAL OF PENTOSAN POLYSULFATE SODIUM FOR TREATMENT OF FELINE INTERSTITIAL (IDIOPATHIC) CYSTITIS. DJ Chew[1], JW Bartges[2], LG Adams[3], JM Kruger[4], CAT Buffington[1] 1. The Ohio State University, Columbus, OH. 2. University of Tennessee, Knoxville, TN. 3. Purdue University West Lafayette, IN. 4. Michigan State University, East Lansing, MI.

The objective of this study was to evaluate the treatment effect of pentosan polysulfate (PPS) compared to placebo on lower urinary tract signs (LUTS) in cats with feline interstitial (idiopathic) cystitis (FIC). This study was a multicenter, double-blinded, placebo-controlled, randomized clinical trial that involved 107 cats with at least two episodes of LUTS within the past six months, cystoscopic findings of glomerulations, and absence of an alternative diagnosis.

Cats were randomly assigned to 0.0 (vehicle placebo), 2.0, 8.0 or 16.0 mg/kg PPS orally twice daily for 26 weeks. Owners rated hematuria, stranguria, pollakiuria, periuria and vocalization during voiding attempts on a scale of 0-3 (none, mild, moderate, severe) weekly. Glomerulations were counted at the end of the study, and 3 of the 4 sites also recorded urothelial friability, edema and vascularization. Data were analyzed within and between groups using repeated measures; ANOVA; $P \leq .05$ was considered significant.

All treatments were well tolerated. Owner-recorded average symptom scores decreased by 75% in both groups; recurrent episodes occurred in approximately 40% of cats in each group. No statistically significant differences were observed between any of the groups based on the owner's evaluation of clinical signs or overall improvement in cystoscopic score. A statistically significant decrease in friability was observed at the 16.0 mg/kg dose.

Clinical improvement occurred in all cats, regardless of the dose of PPS administration or changes in cystoscopic appearance of the bladder. These results indicate that PPS is equivalent to placebo for treatment of FIC, and demonstrate the importance of placebo controlled clinical trials for evaluation of therapy for FIC.

**ABSTRACT #19**

COMPARISON OF PLASMA CLEARANCE OF EXOGENOUS CREATININE AND IOHEXOL IN A CANINE POPULATION. C Collignon[1], R Heiene[2], N. Harran[1], U Risoen[2], D. Balouka[1], C Germain[1], M Faucher[1], K Eliassen[2], Y Queau[1], B Reynolds1, HP Lefebvre[1] 1. Clinical Sciences, Ecole Nationale Vétérinaire de Toulouse, Toulouse, France. 2. Companion Animal Clinical Sciences, Norwegian School of Veterinary Science, Oslo, Norway.

Plasma clearance of exogenous creatinine (PCEC) and iohexol (PCI) have been validated for measurement of glomerular filtration rate (GFR) in dogs. The aim of this prospective study was to compare PCEC and PCI in a canine population.

Simultaneous intravenous administration of iohexol (300 mg I/kg) and creatinine (40 mg/kg) was performed in 50 healthy and renal-impaired Anglo Francais dogs. Blood samples were then obtained at 5, 10, 30, 60, 120, 240, 360, and 480 min. Plasma creatinine and iohexol were assayed by enzymatic method and HPLC, respectively. Exo-iohexol concentrations were used. The method effect on the GFR value was tested using ANOVA.

Plasma clearances (mean±SD) of PCEC and PCI were 3.0±0.7 and 3.4±0.8 mL/min/kg. Their ranges were 1.1-5.7 and 1.2-5.1 mL/min/kg, respectively. A significant effect of the method on GFR value was evident ($P=0.001$; $R^2=0.767$). Nevertheless, both methods identified the same dogs with abnormally low GFR value, discrepancies being observed for higher GFR values. 34/50 dogs showed a difference between both clearances <0.6 mL/min/kg and only 5/50 dogs a difference >1 mL/min/kg.

In conclusion, both methods provide similar results for screening patients with low GFR values but may differ for higher GFR values.

**ABSTRACT #20**

EFFECTS OF SUPPLEMENTATION WITH AMINO ACIDS IN DOGS WITH GLOMERULONEPHRITIS. A Zatelli[1], F Nizi[1], E Zini[2]. 1 Clinica Veterinaria Pirani, Reggio Emilia, Italy; 2 Clinic for Small Animal Internal Medicine, Vetsuisse Faculty, University of Zurich, Switzerland.

Severe malnutrition is frequent in dogs with renal proteinuria and is likely associated with increased morbidity and mortality. Diets with reduced protein content are routinely used in the management of dogs with kidney disorders, including glomerulonephritis. They were shown to reduce disease progression, delay uremia onset and increase survival. However, in dogs with glomerulonephritis renal diets may not adequately meet protein requirements, thus leading to malnutrition. Aim of this study was to investigate the effects of oral amino acid supplementation on body weight (BW), serum albumin, creatinine and urea levels, and urine protein-to-creatinine ratio (UPC) in dogs with glomerulonephritis treated with enalapril and a renal diet.

Forty-seven dogs with glomerulonephritis were allocated to receive the amino acid supplement (29 dogs, Group A), or not (18 dogs, Group B). The daily amount (mg) of amino acids (IT IS, ACME, Reggio Emilia, Italy) was calculated with the following formula: BW (kg) × UPC × 20. All dogs were treated with enalapril (0.5 mg/kg, q24h) and received a commercial renal diet. BW, serum albumin, creatinine and urea, and UPC were determined at baseline and after 4-8 weeks in all dogs. In Group A the same parameters were also recorded at 16-20 weeks. Statistical differences in each group were determined with paired t-tests.

At 4-8 weeks, BW of dogs in Group A increased on average 3.1% ($P<0.01$) whereas in Group B was stable. Albumin levels increased of 0.25 g/dl ($P<0.05$) in Group A and did not differ in the other. Urea levels and UPC significantly decreased only in Group B (average decrease of urea: 16.0 mg/dl; and of UPC: 1.91). In both groups creatinine levels did not vary. At 16-20 weeks in Group A, compared to 4-8 weeks, creatinine levels increased of 0.45 mg/dl ($P<0.05$) and UPC decreased of 0.84 ($P<0.05$). BW and levels of albumin and urea remained stable.

Treating dogs affected by glomerulonephritis with enelapril and a renal diet, in addition to lowering proteinuria, is beneficial because maintains stable BW and albumin levels. Even though supplementation with amino acids increases BW and albumin levels, it delays the decrease of proteinuria and prevents lowering of urea, one of the major toxins in chronic renal failure. In light of these findings, supplementation with amino acids should be regarded with caution in dogs with glomerulonephritis. Whether amino acids may have a place in the treatment of dogs with more severe hypoalbuminemia is currently explored.

**ABSTRACT #21**

DOCUMENTATION OF HYPERCOAGULABILITY IN PROTEIN-LOSING NEPHROPATHY VIA THROMBOELASTOGRAPHY IN 10 DOGS. KM Hilling, MA Labato, AM de Laforcade, S Shaw. Cummings School of Veterinary Medicine at Tufts University, North Grafton, MA.

Protein-losing nephropathy (PLN) has been previously associated with hypercoagulable states, including the presence of fatal pulmonary thromboemboli. A sensitive method for detecting Hypercoagulablility (HC) in affected dogs is warranted, as the appropriate use of either anti-platelet or anti-thrombotic therapies may improve patient survival and limit complications of PLN. Conventional tests of coagulation are insensitive at detecting HC. Serum albumin has previously been shown to correlate with antithrombin (AT) activity in PLN; reduced AT is associated with thrombotic disease. The thromboelastography (TEG) assay has recently been re-introduced as a point-of-care test of the global clotting function. The purpose of this pilot study was to evaluate the TEG assay as a reliable and early predictor of HC in canine PLN patients, including those with evidence of thromboembolism.

Kaolin-activated TEG was performed in accordance with manufacturer's recommendations, using the TEG 5000 Haemostasis Analyzer (Haemoscope Co). The results of TEG variables *R, K, α, MA* and *G* for study dogs were compared to established local reference ranges. Based on the TEG *G* value, dogs were categorized as hypo- (<5.7K), normo- (G=5.7-6.9K), and hypercoagulable (G>6.9K); absolute numbers of the TEG variables *R, K, α, MA* and *G* were also evaluated as was serum albumin. All dogs with PLN were eligible for study inclusion; dogs were divided into uncomplicated PLN (PLN-uc) and PLN with thromboembolic disease (PLN-TED). Results between groups were compared using a t-test with a p <0.05 considered significant.

Ten dogs were enrolled; all dogs were hypercoagulable. There was no correlation between G and albumin values. Three dogs had TED, albumin was significantly higher in dogs with PLN-TED then PLN-uc (p=0.02), but there was no other difference in any TEG variables between groups.

| Variable | R | K | α | MA | G | Albumin |
|---|---|---|---|---|---|---|
| Ref range | 3-4 | 1-2 | 61-66 | 54-58 | 5.7-6.9 | 2.8-4.0 |
| PLN | 5.4 ± 2.2 | 1.5 ± 0.6 | 69.9± 5.6 | 68.7± 7.3 | 11.8± 3.9 | 2.5 ± 0.7 |
| PLN-TED | 6.8 ± 3.7 | 2.0 ± 1.0 | 66.3± 8.4 | 71.4±10.0 | 13.8±5.6 | 3.1± 0.1 |
| PLN-uc | 4.9±1.2 | 1.3± 0.4 | 71.4±3.8 | 67.6±6.4 | 10.9±3.1 | 2.2±0.6 |

These data confirm that dogs with PLN are hypercoagulable as assessed by TEG. However, in contrast to antithrombin activity, the degree of HC does not correlate to serum albumin. Further studies are warranted to delineate coagulation changes in PLN, as well as define optimal anticoagulation strategies.

## ABSTRACT #22

MEASURING LEVEL OF AGREEMENT BETWEEN DIRECTLY MEASURED BLOOD PRESSURE AND VALUES PRODUCED BY OSCILLOMETRIC UNITS IN CATS. MJ Acierno[1], DC Seaton[1], MA Mitchell[2], A da Cunha[1]. 1. The Louisiana State University, Baton Rouge, LA. 2. University of Illinois, Urbana, IL.

While the Doppler detector has been validated for use in dogs and cats, studies involving the use of oscillometric units have shown that these devices consistently underestimate feline blood pressure. Recently, several oscillometric devices that claim to have been specifically optimized for use in cats have been introduced The goal of this study is to compare three of these newly designed oscillometric devices.[a,b,c] to directly measured blood pressure in cats undergoing anesthesia.

A 22 gauge catheter was placed in the dorsal pedal artery of 15 client owned cats receiving anesthesia in preparation for surgery. The arterial catheter was connected to a continuous multifunction monitor[c] via a disposable pressure transducer system.[d] The directly obtained blood pressure readings were checked for stability and consistency while each of the oscillometric devices were setup and calibrated in accordance with the manufacturers' directions. A series of four paired blood pressure measurements (directly vs. oscillometric) was then made for each device.

Agreement between the directly and indirectly obtained systolic, diastolic and mean arterial pressure measurements were determined by use of the Bland-Altman method for multiple observations. Good agreement was defined as a bias and limit of agreement (LOA) within 10% of the expected mean. Results for 15 cats were available for units 1[a] and 2[b], and 13 for unit 3[c]. Unit 1[a] demonstrated a systolic bias of -7.0 mmHg (LOA -52.2 to 38.2), a diastolic bias of 9.2 mmHg (LOA -36.47 to 54.87), and a MAP bias 7.0 mmHg (LOA -35.5 to 49.5). Unit 2[b] demonstrated a systolic bias of -13.18 mmHg (LOA -53.75 to 27.39), a diastolic bias of 0.03 mmHg (LOA -26.8 to 26.9), and a MAP bias -4.1 mmHg (LOA -28.2 to 19.98). Unit 3[b] demonstrated a systolic bias of -9.7 mmHg (LOA -41.5 to 22.1), a diastolic bias of 5.6 mmHg (LOA -35.5 to 41.7), and a MAP bias -4.6 mmHg (LOA -36.6 to 27.4).

Results suggest that there is significant disagreement between all of the oscillometric devices and the directly measured blood pressure. At this time, use of these oscillometric devices in cats cannot be recommended.

a- VET HDO, Vetline LLC, Saint Kitts and Nevis, WI
b-PetMap, Ramsey medical, Tampa FL
c- Cardell Max-1, Sharn Veterinary, Tampa, Fla
d - DTX Plus, Becton Dickinson, Sandy, Utah.

## ABSTRACT #23

PHARMACOKINETICS AND CARDIOVASCULAR EFFECTS OF INTRAVENOUS FENOLDOPAM IN HEALTHY AWAKE DOGS. CA Bloom[1], MA Labato[1], MH Court[2], S Hazarika[2]. 1. Tufts University Cummings School of Veterinary Medicine, North Grafton, MA. 2. Tufts University School of Medicine, Boston, MA.

The purpose of this study was to collect pharmacokinetic data on fenoldopam in healthy awake dogs. Fenoldopam is a selective dopamine-1 receptor agonist that causes peripheral arterial vasodilation, increased renal blood flow, and diuresis. Main uses in human medicine are acute hypertensive crisis and non-polyuric renal failure. Enthusiasm exists for the use of fenoldopam in non-polyuric renal failure in dogs, and while pharmacodynamic studies have been performed, there is no pharmacokinetic data.

Six healthy, awake beagles were given a 180-minute fenoldopam CRI at 0.8 mcg/kg/min, followed by a 120-minute washout period. Citrated blood was collected at minutes 0, 5, 10, 20, 30, 60, 90, 120, 150, and 180 during infusion, and minutes 182, 185, 190, 210, 225, 240, 270, and 300 following infusion. HR, RR, and BP via Doppler were concurrently measured.

All dogs completed the study uneventfully. There was no significant change in HR, RR or BP.

Steady state fenoldopam concentrations averaged 19.9 ng/ml and were achieved within 10 minutes of starting infusion. Area under the curve averaged 3590 ng/ml x minute. Clearance averaged 40 ml/min/kg. Elimination was rapidly achieved in all dogs. Three dogs had no fenoldopam in the plasma at time 180, within seconds of terminating the infusion; therefore, elimination data were not determined for these dogs. Elimination half-life ranged from 1.1 – 6.7 minutes in 3 dogs, and volume of distribution ranged from 30-727 ml/kg in 3 dogs. Clinical studies of fenoldopam in dogs with non-polyuric renal failure using a fenoldopam CRI at 0.8 mcg/kg/min are warranted.

**ABSTRACT #24**

URINARY IMMUNOGLOBULIN G, C-REACTIVE PROTEIN AND RETINOL BINDING PROTEIN AS CANDIDATE EARLY BIOMARKERS FOR RENAL DYSFUNCTION IN DOGS WITH PYOMETRA. B.E.J. Maddens[1], S. Daminet[2], P. Smets[2], K. Demeyere[1], H. de Rooster[2], E. Meyer[1]; [1]Department of Pharmacology, Toxicology, Biochemistry and Organ Physiology [2]Department of Medicine and Clinical Biology of Small Animals, both Faculty of Veterinary Medicine, Ghent University, Belgium.

Renal dysfunction in dogs has been associated with pyometra. Nevertheless, the mechanism and type of renal injury remain controversial. Routine measurement of serum urea nitrogen and creatinine detects renal injury only in a late irreversible stage and does not indicate the localisation of the damage. We hypothesize that urinary immunoglobulin G (uIgG), C-reactive protein (uCRP) and retinol binding protein (uRBP) may serve as superior renal biomarkers. Both IgG and CRP are high molecular weight proteins indicating glomerular lesions, whereas RBP may serve as a tubular damage marker.

In this study, 14 dogs with *Escherichia coli* (*E. coli*) pyometra (P) without concurrent other diseases were included. Age-matched clinically healthy bitches (H) served as controls (n=14). At ovariohysterectomy (P, H) or ovariectomy (H) blood, urine and uterine swabs were taken. Serum biochemical analysis, CBC, urinanalysis and uterine bacteriological culture were performed. Commercial canine ELISAs were validated and used to quantify uIgG and uCRP; uRBP concentrations were determined with a human RBP ELISA kit validated for use in the dog. All concentrations were related to urinary creatinine concentration (uC) and expressed as ratios.

In P dogs, uIgG/C (196 ± 45.2 mg/g) and uCRP/C (706 ± 298 µg/g) (mean ± SEM) ratios were significantly increased compared to those in H bitches (2.1 ± 0.4 mg/g respectively 0.1 ± 0.1 µg/g) ($P<0.01$). Furthermore, uRBP/C ratios were significantly higher in P (43.7 ± 10.1 µg/g) than in H dogs (16.7 ± 5.5 µg/g) ($P<0.05$). A positive correlation at the 0.05 level was found between the urinary concentrations of all three proteins. Urinary total protein/C ratios (uTP/C) were significantly higher in P (0.77 ± 0.19) than in H dogs (0.16 ± 0.06) ($P<0.05$). A highly significant positive correlation was found between uTP/C and uIgG/C ($R = 0.79$) or uCRP/C ($R = 0.72$), but not between uTP/C and uRBP/C.

Concentrations of uIgG and uCRP, the latter for the first time examined in the dog, are strongly elevated in P bitches indicating glomerular dysfunction immediately after *E. coli* pyometra. The milder increase in uRBP is suggestive for tubular dysfunction, possible secondary to the proteinuria related to immune-mediated glomerulonephritis. We suggest that the observed increases in the candidate renal markers uIgG, uCRP and uRBP might be indicative of renal dysfunction at the glomerular and tubular level, associated with *E. coli* pyometra.

*Previously presented at ECVIM in Belgium, September 2008.*

**ABSTRACT #25**

WHOLE BLOODTHROMBOELASTOMETRY (TEM) IS RELATED TO CELL COUNTS AND PLASMA COAGULATION TESTS IN HEALTHY DOGS. SA Smith[1], MA McMichael,[2] AJ Galligan[2], S Gilor,[2] C Hoh[2] College of Medicine[1] and College of Veterinary Medicine,[2] University of Illinois, Urbana IL.

TEM use is increasing. An understanding of the relationships between TEM and other hematologic values is needed. Blood samples were collected from 78 apparently healthy dogs. Hematocrit (Hct), Platelet count (Plat), prothrombin time (PT), activated partial thromboplastin time (aPTT), and Fibrinogen (Fib) were performed using standard methods. Citrated whole blood was evaluated using Rotem (Pentapharm) and either the manufacturer supplied Extem (tissue factor) or Intem (contact activator). TEM parameters included clot time (CT), clot formation time (CFT), alpha angle (α), and maximum clot firmness (MCF). Strength of correlations was determined using Pearson Product Moment Correlation (r) calculated with Sigma Stat.

Comparisons between Extem (Ex) and Intem (In) parameters indicated that there was no significant relationship between ExCT and InCT. r values for CFT (0.77***), α (0.69***), and MCF (0.52*) indicated that Ex values and In values were highly correlated. Additional r values:

| | ExCT | ExCFT | Exα | ExMCF | InCT | InCFT | Inα | InMCF |
|---|---|---|---|---|---|---|---|---|
| PT | **0.34*** | 0.19 | -0.18 | -0.23 | | | | |
| aPTT | | | | | **0.78*** | 0.08 | -0.07 | -0.03 |
| Plat | -0.16 | **-0.36**** | **0.44**** | **0.35*** | -0.20 | **-0.36*** | **0.29*** | **0.58**** |
| Fib | -0.21 | **-0.51**** | **0.50***** | **0.50***** | -0.29 | **-0.53*** | **0.53*** | **0.62**** |
| Hct | **0.34*** | **0.49***** | **0.45***** | **-0.39**** | 0.28 | **0.66**** | **-0.66**** | **-0.53**** |

*p<0.05, **p<0.005, ***p<0.0005

CT is a function of time to initial fibrin polymerization, and is primarily dependent on thrombin formed in the initiation phase of coagulation. CT is related to the type and strength of trigger used and consequently correlates with plasma coagulation testing utilizing a comparable trigger. In contrast, CFT, α, and MCF are related to thrombin generated (and fibrin polymerized) during propagation. These parameters are correlated with platelet count and fibrinogen concentration, and independent of the type of clot trigger used. Interestingly, Hct was correlated with all parameters, with the exception of In-CT, despite the presence of normal red cell mass in all dogs (Hct range 41-60%). Dogs with higher Hct were relatively hypocoagulable, while those with lower Hct were relatively hypercoagulable. The significant impact of Hct on TEM results may be an *in vitro* artifact associated with relative under- or over-citration of blood samples, or may be an indication of interference of red cell mass with clot development. TEM parameters obtained from anemic or polycythemic dogs may need to be assessed in reference to HCT adjusted reference ranges.

**ABSTRACT #26**

EFFECT OF CANINE HYPERADRENOCORTICISM ON THROMBELASTOGRAPHY PARAMETERS. L Rose, M Dunn, C Bédard. University of Montreal Faculty of Veterinary Medicine, Saint-Hyacinthe, Quebec.

Hyperadrenocorticism (HAC) is associated with an increased risk of thromboembolism in dogs. The purpose of this prospective clinical study was to use thrombelastography (TEG®) to detect and qualify the hypercoagulable state in canine patients with HAC. We hypothesized that the presence of HAC in dogs would result in a hypercoagulable profile on

TEG® tracings that is reversible with management of the disease.

Seventeen dogs suffering from HAC as diagnosed by ACTH stimulation test, low dose dexamethasone suppression test or complete adrenal profile were included. TEG® tracings were obtained before treatment in all dogs and after normalisation of ACTH stimulation and improvement of clinical signs in 4 dogs. TEG® analyses were performed in duplicate on citrated whole blood 30 minutes after collection using recombinant human tissue factor at a final concentration of 1:3600. The reaction time (R), the alpha angle ($\alpha$), the kinetic time (K) and the maximum amplitude (MA) were recorded.

When individual dogs were compared to the normal reference range, no dog was considered hypercoagulable. Furthermore, a paired t test showed no significant difference in TEG® values before and after treatment. However, a t test for values with unequal variance was used to compare the population of HAC dogs with a population of healthy beagles and showed that the mean values of healthy dogs were significantly higher for R ($p < 0.0001$) and K ($p < 0.0001$) and significantly lower for alpha ($p < 0.0001$) and MA ($p < 0.0001$) suggesting that HAC dogs tend toward hypercoagulable TEG® tracings.

**ABSTRACT #27**

EFFECTS OF CYCLOSPORINE ON CANINE PLATELET PROCOAGULANT ACTIVITY. <u>JM Thomason</u>, K Lunsford, A Mackin, L Pinchuk, S Pruett, C Langston. Mississippi State University College of Veterinary Medicine, Starkville, MS.

Cyclosporine has become increasingly popular for treating canine immune-mediated diseases such as immune-mediated hemolytic anemia (IMHA). Cyclosporine has recently been shown to increase platelet procoagulant activity in humans, potentially predisposing to thrombosis. Our study was designed to evaluate the effects of cyclosporine on canine platelet procoagulant activity.

Platelet reactivity was quantified by flow-cytometric evaluation of activated platelet surface markers P-Selectin and phosphatidylserine, platelet function was assessed via a point-of-care analyzer (PFA-100®), and clot formation was evaluated using viscoelastometry (SonoClot®). Evaluation of platelet reactivity in two healthy dogs after cyclosporine PO BID for 1 week, and in a third dog after a cyclosporine IV CRI for 1 day (dose adjusted to attain target blood levels) revealed increased P-Selectin expression (mean increase 95%, range 37 - 138%), decreased phosphatidylserine expression (mean decrease 35%, range 5-64%), and variable Sonoclot® signatures. *In vitro* platelet exposure to cyclosporine resulted in decreased PFA-100® closure times (mean change 96 seconds, range 41-170 seconds), and no consistent change in P-Selectin or phosphatidylserine expression or Sonoclot® signatures.

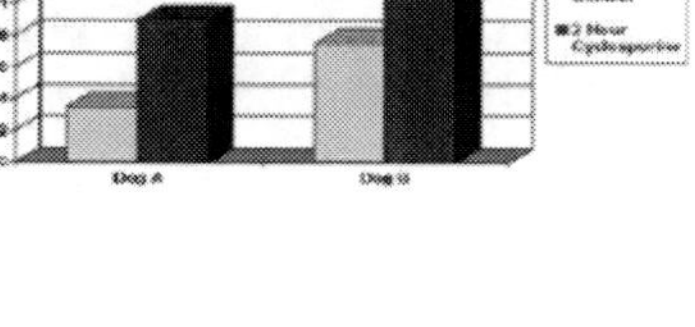


Our results suggest that cyclosporine may increase platelet procoagulant activity in some dogs. Individual variations in canine platelet thromboxane receptor function, thromboxane-mediated degranulation, and P-selectin expression may explain variable responsiveness to cyclosporine.

Cyclosporine may potentially predispose some dogs to thrombotic complications such as pulmonary thromboembolism, possibly negating the benefits of the drug in patients with prothrombotic conditions such as IMHA. Further investigation of the potential effects of cyclosporine on platelet procoagulant activity in both normal dogs and clinic patients is warranted.

**ABSTRACT #28**

TREATMENT OF CANINE IMMUNE-MEDIATED HEMOLYTIC ANEMIA WITH INDIVIDUALLY ADJUSTED HEPARIN DOSING: A PILOT STUDY. <u>SE Helmond</u>[1], DJ Polzin[1], PJ Armstrong[1], M Finke[1], SA Smith[2]. 1. University of Minnesota College of Veterinary Medicine, St Paul, MN. 2. University of Illinois College of Medicine, Urbana, IL.

A major cause of mortality and morbidity in canine Immune-Mediated Hemolytic Anemia (IMHA) is thrombotic disease. Previous studies have suggested that unfractionated heparin (UH) may not be effective at increasing survival in IMHA; however, subtherapeutic dosing may explain the seeming lack of efficacy. We hypothesized that providing appropriate therapeutic plasma levels of UH by individually adjusting doses based on anti-factor Xa activity would improve survival in dogs with IMHA.

To investigate this hypothesis, a randomized, prospective, masked, placebo-controlled clinical trial was conducted using 15 dogs with newly diagnosed primary IMHA. Dogs were recruited following diagnosis and received standardized therapy for IMHA. In addition, dogs were treated with either a Constant Dose (CD) (150U/kg SQ q 6h) or an Individually Adjusted Dose (IAD) of UH. UH dosing was monitored via an Anti-Xa chromogenic assay modified for use in dogs and adjusted according to a nomogram developed for dogs. UH dose was adjusted daily until day 7, and then once weekly until day 28, after which heparin dosing was tapered over 1 week. Dogs were followed for 6 months (180 days).

The primary endpoint was 6 month survival. Secondary endpoints were: 1) frequency of thromboembolism (TE), 2) frequency of hemorrhage, 3) the dose of UH required to maintain a therapeutic plasma level in the IAD group.

At Day 180, 7 out of 8 dogs in the IAD group were alive; 1 dog out of 7 in the CD group was alive ($p = 0.01$). Median survival time for the IAD group was 180 days, and 68 days for the CD group. Five dogs in the CD group had confirmed TE events, three of which were fatal. Two dogs in the IAD group had suspected TE events; one of these dogs died of TE. There were no hemorrhagic complications in either group. The dose of UH required to maintain therapeutic plasma levels (0.3-0.7 IU/mL) in dogs in the IAD group ranged from 180 IU/kg to 712 IU/kg q6-8h. There were marked inter- and intra-individual variations in UH doses, with no correlations between dose and anti-Xa activity. Anti-Xa activity in the CD group ranged from <0.1IU/mL to 0.33 IU/mL (median 0.1 IU/mL).

Results of this pilot study suggest that IAD UH therapy, using anti-Xa monitoring, reduces mortality in canine IMHA. This is the first study to demonstrate successful adjustment of therapeutic heparin levels in dogs with IMHA. In addition, we found that fixed, low dose UH therapy resulted in poor survival rates, as has been suggested in other studies.

**ABSTRACT #29**

COMPARISON OF INDIVIDUALLY MONITORED UNFRACTIONATED HEPARIN VERSUS LOW-DOSE ASPIRIN ON SURVIVAL OF DOGS WITH IMMUNE MEDIATED HEMOLYTIC ANEMIA. ES Orcutt[1], DJ Polzin[1], PJ Armstrong[1], SE Helmond[1], S Smith[2]. 1. University of Minnesota College of Veterinary Medicine, St Paul, MN. 2. University of Illinois, Urbana, IL.

A previous retrospective study on low-dose aspirin therapy (LDA) in dogs with immune mediated hemolytic anemia (IMHA) reported a survival rate of 75% in LDA treated dogs. As a result, LDA has become the standard anticoagulant therapy for IMHA because of its perceived effectiveness, lack of need for special monitoring, minimal cost and availability. However, multiple studies in human patients comparing the TE prophylactic effect of aspirin versus heparin have shown marginal to no benefit with aspirin therapy compared to a significant reduction in TE or vascular events in patients treated with heparin. We hypothesized that subcutaneous administration of unfractionated heparin with dosage individually adjusted using anti-Xa monitoring (IAH) is superior to LDA in limiting mortality in dogs with IMHA managed with glucocorticoids. Study objectives were: (1) to evaluate 6 month survival in dogs treated with LDA versus IAH and (2) to evaluate the incidence of TE in both groups.

Medical records from the U of MN Veterinary Medical Center (VMC) were searched for dogs with IMHA treated with low-dose aspirin therapy (0.5 mg/kg/day). Inclusion criteria were: 1) regenerative anemia, 2) one or more of the following: a) autoagglutination, b) spherocytosis, or c) positive Coomb's test; 3) platelet count > 40,000/µl; and 4) glucocorticoid therapy for IMHA. Dogs from a study at the VMC evaluating IAD and selected using the same inclusion criteria were used as the comparison group. Data collected included signalment, biochemical and hematological result, treatment regimes, TE complications and outcome.

Twenty five dogs met inclusion criteria for the LDA group and 8 dogs from the IAD study served as the comparison group. Ten of 25 dogs treated with LDA survived and 7 of 8 dogs treated with IAD survived (p=0.084). Seven of 25 LDA treated dogs and 2 of 8 IAD treated dogs developed TE. Six of 7 LDA dogs developing TE died; 1 of 2 IAD dogs developing TE died.

Results of this retrospective study suggest that IAD may be superior to LDA in reducing mortality in canine IMHA. However, dogs enrolled in the IAD study may have fared better due to closer monitoring occurring in the prospective clinical trial.

**ABSTRACT #30**

PLATELET IMPEDANCE AGGREGOMETRY IN CLINICALLY HEALTHY DOGS WITH AND WITHOUT ULTRALOW-DOSE ASPIRIN. KS Sharpe, SA Center, JF Randolph, MB Brooks, KL Warner. College of Veterinary Medicine, Cornell University, Ithaca, NY.

Improved long-term survival in dogs with IMHA has been achieved when ultralow-dose aspirin (ULDAsp, 0.5 mg/kg/day PO) is combined with glucocorticoids and azathioprine. Whether this survival benefit solely reflects antithrombotic influences on platelets, or modulatory effects on vascular endothelium, systemic inflammation, or drug protein-binding, remains unclear.

This study was conducted to evaluate the efficacy of ULDAsp as an antithrombotic agent in the dog using a relevant *in vitro* platelet aggregation methodology. Our hypothesis was that ULDAsp would inhibit *in vitro* platelet aggregation in clinically healthy dogs.

Impedance platelet aggregation was studied in 19 dogs without and with ULDAsp administered orally once daily for 2 days. Blood was collected into 3.2% sodium citrate (1:9 dilution) 24-hrs after the last treatment. Each dog had a platelet count $\geq$ 250,000/µL. Blood was maintained at 37°C during aggregation assessments. Agonists evaluated included 20, 10, and 2 µmol/L ADP and 10, 5, and 2 µg/mL bovine collagen. Onset (seconds) and rate (ohms/second) of the aggregation response and maximum aggregation (ohms) were recorded. Aggregation parameters were normally distributed. Differences with and without ULDAsp were determined using a paired T-test with $\alpha = 0.05$.

ULDAsp significantly delayed aggregation-onset elicited with ADP by 54-104%, but no change occurred with collagen. ULDAsp significantly slowed and attenuated the maximum aggregation response by 23-25% and 25-29%, respectively, with ADP and higher concentrations of collagen.

Findings substantiate that ULDAsp impairs platelet aggregation in healthy dogs. This data will serve as a platform for future functional assessment of platelet reactivity in dogs with IMHA.

**ABSTRACT #31**

HEMOSTATIC EFFECTS OF CRYOPRECIPITATE IN DOGS WITH DISSEMINATED INTRAVASCULAR COAGULATION. P Vilar, R Ball, N Westendorf, MC Iazbik, L Marín, and CG Couto. The Ohio State University, Veterinary Clinical Sciences, Columbus, OH.

Hemorrhagic thrombocytopenic complications (e.g disseminated intravascular coagulation-DIC, severe acute hemorrhage) are frequently observed in trauma patients and in those undergoing prolonged surgery, even in those with no previous hemostatic abnormalities. The severity of bleeding and prognosis are not well correlated with the prothrombin time (PT) or activated partial thromboplastin time (aPTT) (Wiinberg B et al; 2007). Thromboelastography (TEG®) is able to assess *ex vivo* hemostasis using a whole blood sample, since it evaluates most of the components that play a role in the formation of the hemostatic plug *in vivo* (i.e blood cells, platelets, clotting factors). TEG has been proposed as the hemostasis test of choice for monitoring patients receiving plasma components for coagulopathies (Johansson P; 2007). Based on data from previous studies (Fries D et al; 2005, 2006) that showed improvement in clot formation during thrombocytopenia after administration of fibrinogen concentrates, we evaluated the use of canine cryoprecipitate (CRYO) in 3 dogs with postoperative bleeding due to DIC in an attempt to reestablish normocoagulable conditions. Patients were monitored using TEG®. Blood samples were collected via a central venous catheter into a 2.7 ml Vacutainer tube containing 3.2% buffered sodium citrate, and TEG® was done using citrated native technique, as we previously described (Vilar P et al; 2008). Three hypocoagulable ($TEG_{R,K}$>mean+2SD, and/or $TEG_{angle,MA,G}$<mean-2SD) dogs with postoperative bleeding associated with DIC were transfused with 50-70 ml/10kg of CRYO (containing approximately 500 mg/dl fibrinogen). Two dogs had been splenectomized for hemangiosarcoma and 1 had gastric dilation-volvulus. Postransfusion TEGs revealed normocoagulable tracings and increased hematocrits in all cases less than 24h post-CRYO; no additional blood component transfusions were needed. No clinical thromboembolic complications were detected in any dog. TEG® showed excellent clinical correlation with the bleeding status in all dogs. TEG® is a useful test for monitoring transfusion therapy in patients with coagulopathies; CRYO infusion should be considered in dogs in DIC.

**ABSTRACT #32**

EXTENDED CANINE BLOOD TYPING BY GEL COLUMN TECHNIQUE. RJ Kessler[1], J Reese[1], D Chang[1], A Hale[2], U Giger[1]. 1. Section of Medical Genetics, University of Pennsylvania (PennVet), Philadelphia, PA 2. Animal Blood Resources International (ABRI), Stockbridge, MI.

To ensure blood compatibility, safety, and efficacy of transfusions in dogs, typing for the major antigenic blood type DEA 1.1 is recommended by using one of several well established methods. The role of extended typing for other canine blood types is more controversial regarding their clinical importance, and these typing procedures have been hampered by the limited availability of typing reagents and difficulties in both performing the tube typing protocol and interpreting those results due to weak agglutination reactions. Our goal was to develop and standardize a laboratory method of extended typing using polyclonal reagents that minimizes the use of reagent while maximizing sensitivity, specificity, interpretability, and reproducibility.

We utilized available polyclonal typing reagents (ABRI) at optimized concentrations with the saline or canine antiglobulin gel columns (novel GEL; DiaMed, Switzerland) similar to the commercially available gel column DEA 1.1 typing technique (standard GEL). All dogs were also typed using the conventional tube method according to the manufacturer's instructions (TUBE). A total of 54 dogs including 22 patients and 32 blood donors at PennVet were typed for DEA 1.1, 1.2, 3, 4, and 7, as well as the *dal* red cell antigen. Agglutination reactions were graded on a scale of 0 to 4+ and reactions $\geq$2+ were considered positive.

Of the 43 dogs typed for DEA 1.1, 23 were positive for DEA 1.1 using the standard GEL as well as with the novel GEL method (with antiglobulin) using polyclonal DEA 1.1 and 1.X reagents. Twenty of those 23 were also DEA 1.1 positive with the TUBE. Two of the 3 remaining were positive with the TUBE for 1.X but only 1+ for 1.1 (suggesting a DEA 1.2 type). In addition, 2 dogs tested DEA 1.1 positive using the novel GEL method, but not the standard GEL or TUBE method. All samples tested negative for autoagglutination (in saline $\pm$ antiglobulin).

All 54 dogs were typed for DEA 3, 4, and 7 using the standard TUBE and novel GEL (saline) method. Only 4 dogs were found to be DEA 3 positive with the TUBE, 3 of which were also positive with the novel GEL method. All dogs tested 3+ or 4+ for DEA 4 via both methods. Twelve dogs tested positive for DEA 7 with the TUBE, half of which were also positive with the novel GEL; an additional 2 dogs were positive by the novel GEL method alone. All 36 dogs typed for the *dal* antigen were positive (no Dalmatians were tested). Only 1 dog was positive for all tested red cell antigens.

In conclusion, this gel column technique was able to detect the DEA 1.1, 3, 4, and 7 as well as the *dal* antigen using available polyclonal antibodies. The agglutination reactions for extended typing were stronger and more readily interpreted with the novel GEL than the TUBE method. Few discrepancies between the TUBE and GEL methodology remain and require further investigation.

**ABSTRACT #33**

*PSEUDOMONAS FLUORESCENS* CONTAMINATION OF CANINE AND FELINE PACKED RED BLOOD CELL UNITS. RJ Kessler, S Young, DA Oakley, S Rankin, U Giger. Departments of Clinical Studies and Pathobiology, School of Veterinary Medicine, University of Pennsylvania, Philadelphia, PA.

While rigorous screening programs have drastically reduced the risk of infectious disease transmission in human and veterinary blood banking, bacterial contamination of blood products has emerged as a major cause of morbidity and mortality in human transfusion medicine. In particular, packed red blood cells (pRBCs) and platelet concentrates stored at 4°C and 20°C, respectively, have been found to be contaminated with *Streptococcus spp* (platelets), *Yersinia spp*, and *Pseudomonas spp* (pRBCs). In veterinary medicine, there has been only sparse mention of bacterial contamination of canine and feline blood products. We describe here the discovery of a *P. fluorescens* (*Pf*) contamination of a unit of feline pRBCs which led to further investigations of experimentally *Pf*-contaminated units and utilization of a semi-quantitative real-time 16S bacterial ribosomal DNA PCR test (16S PCR) for blood product screening.

A unit of feline pRBCs that turned black on the 22nd day of storage at 4°C was removed from the blood bank and further examined. There was red cell lysis and cytology showed many free bacteria, however, aerobic culture of the blood at 37°C was negative. Real-Time 16S PCR testing was strongly positive, and isolated DNA was sequenced and identified as *Pf*. An aliquot of the unit grew when cultured at 20°C (room temperature). Extensive evaluations failed to identify a source of this contamination, nor has a similar contamination or color change been observed with any other stored pRBCs in our blood bank in the year prior to or in the 4 months after this incident.

Units of canine pRBCs (25ml each) were inoculated with 0, 5, and 25µl of *Pf*-rich pRBCs from the sentinel feline unit and stored at either 4°C or 20°C for 48-72hr. Some units were then switched from 4°C to 20°C after 48hr. Prior to and immediately post-inoculation all 16S PCR results were negative; separately spiked 25ml pRBCs only became positive after adding ≥100µl *Pf*-rich pRBCs. Only *Pf*-spiked pRBC units became 16S PCR positive (≥8hr [25µl at 20°C]; 48hr [25µl at 4°C and 5µl at 20°C]) and showed a color change as early as 24hr; cultures confirmed the presence of a pure culture of *Pf*. One spiked unit [5µl at 4°C] that was 16S PCR negative at 48hr became positive within 4hr at 20°C.

In conclusion, *Pf* appears to have the unique capacity to grow in feline and canine pRBCs slowly when stored cold and rapidly at room temperature. Screening of blood products with 16S PCR is a simple, rapid, and sensitive test method to detect bacterial contamination before a gross color change is noted or bacterial culture results are positive, and thus may be useful for routine screening of pRBCs. While a source of contamination and potential accidental exposure to 20°C for our sentinel unit was not determined here, aseptic collection and processing methods and temperature controlled storage along with regular visual evaluation of units is recommended.

## ABSTRACT #34

COMBINATION OF INFLAMMATORY AND HEMOSTATIC MARKERS IN MORTALITY MODEL FOR CRITICALLY ILL CANINE ICU PATIENTS SIGNIFICANTLY IMPROVES EFFICACY. M Kjelgaard-Hansen[1], B Wiinberg[1], AL Jensen[1], E Rozanski[2], AT Kristensen[1]. [1]Department of Small Animal Clinical Sciences, University of Copenhagen, Copenhagen, Denmark. [2]Cummings School of Veterinary Medicine at TUFTS University, North Grafton, MA.

Key inflammatory and hemostatic markers have been identified as possible prognostic markers in critically ill canine patients; serial measurements of Protein C and Antithrombin (AT) was reported to correlate to survival of septic dogs (deLaforcade et al., 2008) and the level of IL-6 correlated to survival time in dogs with SIRS and sepsis (Rau et al., 2007). However, studies combining inflammatory and hemostatic markers are rare. Recent human studies have reported significant improvement in prognostic efficacy by addition of simple hemostatic and/or inflammatory markers to traditional risk assessment scores (e.g. SOFA and SAPS-II). The objective of the present study was to investigate the effect of combining key inflammatory and hemostatic markers in prognostic models for critically ill canine ICU patients.

Fifty critically ill dogs admitted to the ICU at Cummings School of Veterinary Medicine at Tufts University, North Grafton, MA (April to July 2006) were included prospectively. Citrated whole-blood, serum and citrated plasma were obtained by standardized procedures. Tissue-factor Thromboelastography (TEG [MA, G, R, K and α]) and platelet count was performed within 30 minutes, while serum and plasma was stored and analyzed in batches for AT, D-dimer, Protein C, Protein S, PT, aPTT, Plasminogen, Plasminogen inhibitor, C-reactive protein (CRP) and IL-10. All methods were validated for use in dogs. Twenty-eight day survival was recorded.

Multiple logistic regression analysis was performed with survival as outcome variable based on A) Inflammatory markers, B) Hemostatic markers and C) all markers, using backwards exclusion and forward inclusion (P>0.1). Area under Receiver operation characteristic curves (ROC-AUC) was used to evaluate and compare model discriminative efficacies, and optimized sensitivity and specificity were assessed.

Two dogs were excluded (missing data). Mortality was 42% (20/48). Final models and results were (significant parameters included, (ROC-AUC), [sensitivity and specificity], respectively: A) CRP, (0.71), [0.75, 0.68], B) MA, G, AT (0.80), [0.70, 0.89] and C) MA, G, AT, PT, aPTT, CRP, (0.91), [0.75, 0.93]. Model C) had significantly better discriminative efficacy than A) [P=0.019] and B) [P=0.046], while no difference was found between A) and B) [P=0.32].

In conclusion, prognostic models for 28-day mortality was developed by means of key inflammatory and hemostatic markers, available and validated for use in dogs. Significantly improved discriminative efficacy was observed when both inflammatory and hemostatic markers were included in the model. Future studies refining prognostic modeling in critically ill dogs are recommended to include key markers of both inflammation and hemostasis.

## ABSTRACT #35

INVESTIGATION OF DOXYCYCLINE-RELATED SIDE EFFECTS IN DOGS. B. Schulz, S. Hupfauer, K. Hartmann. Clinic for Small Animal Internal Medicine, Ludwig-Maximilians-University, Munich, Germany.

Doxycycline is a commonly used antibiotic in small animal practice with a broad spectrum of activity against bacteria, *Rickettsia*, *Mycoplasma*, and *Chlamydophila* species. Although there are many data published on pharmacokinetics and toxicology in humans and laboratory animals, little information is available on side effects in dogs. Side effects described in single case reports include gastrointestinal problems, hepatotoxicity, and injection site reactions. In young animals bone and dental problems can occur.

Aim of the retrospective study was to assess the incidence of doxycycline-related side effects (anorexia, vomiting, diarrhea, pyrexia, and increased liver enzyme activities) in dogs and to investigate correlations between side effects and signalement, dose, duration of treatment, frequency of application, doxycycline preparation, and use of additional drugs. Statistical comparison was performed using likelihood ratio tests and logistic regression analyses.

Three hundred eighty six dogs, that had received doxycycline, were included in the study. Of these, 2.5 % (8/314) developed anorexia during therapy. The longer dogs were treated, the less frequently anorexia was observed (reduced by factor 0.83; p = 0.050). Vomiting developed in 18.3 % (63/344) of dogs, and the risk increased with age by factor 1.06 per year (p = 0.030). Diarrhea was documented in 7.0 % (24/342) of dogs. None of the dogs developed fever. An increase in ALT was observed in 39.4 % (26/66), an increase in ALP in 36.4 % (16/44) of patients receiving doxycycline therapy. The factors sex, dose, frequency of application, doxycycline preparation, and application of additional drugs did not have an influence on occurrence of side effects.

In conclusion, vomiting and increase in liver enzyme activities represent the most frequently observed doxycycline-related side effects.

**ABSTRACT #36**

METABOLISM AND EXCRETION OF ORAL MELOXICAM IN THE CAT. Thoulon F[1], Narbe R[2], Johnston L[2], Ingwersen W[3], Watson P[2]. Boehringer Ingelheim Animal Health. 1. Reims, France 2. Ingelheim, Germany 3. Burlington, Ontario, Canada.

The objective of this study was to investigate the metabolic pathways and routes of excretion of oral meloxicam in the cat. Many nonsteroidal antiinflammatory drugs (NSAIDs) are metabolised by the hepatic glucuronyl transferase enzyme pathway (i.e. glucuronidated), before excretion. Cats are relatively and variably deficient in this enzyme activity, making prolongation of NSAID half life and toxic accumulation a possibility. Meloxicam is metabolised via oxidation in all other species previously studied.

A mean oral dose of 0.75 mg/kg or 397.55 kBq/kg of oral [$^{14}$C]-meloxicam was administered to 3 fasted male cats. The cats were fed 4 hours after administration. No concurrent medication was administered. Urine, faeces, vomit and cage washes were collected over the following 144 hour period. Blood was collected pre-dosing and at 3 and 12 hours post-dosing. Metabolites were identified by HPLC/MSMS. Where possible a metabolic structure was proposed for each metabolite detected.

Only the unchanged parent compound was identified in plasma. In urine 5 major metabolite peaks were detected and in faeces 4 major metabolite peaks were detected, which were identified by HPLC/MSMS as products of oxidative metabolism. No conjugated metabolite was detected.
Elimination occurred early (34% during the first 24 hours, 61% during the first 48 hours).

| % recovered dose in first 48 hours | Meloxicam | Metabolites | Total |
|---|---|---|---|
| Urine | 2 | 19 | 21 |
| Faeces | 49 | 30 | 79 |

The results indicate that the major route of excretion of meloxicam in the cat is faecal. The findings indicate that, as with other species investigated, the main pathway of biotransformation of meloxicam in the cat is oxidation. This results in predictable pharmacokinetics, making meloxicam suitable for long-term administration in the cat.

**ABSTRACT #37**

CLONING OF THE CALCIUM-SENSING RECEPTOR FROM THE FELINE PARATHYROID GLAND. A Gal, T Ridge, TK Graves. University of Illinois College of Veterinary Medicine, Urbana, IL.

Gain- or loss-of-function mutations in the plasma membrane-bound calcium-sensing receptor (CaSR), which is responsible for detecting extracellular concentraions of calcium ion, can result in aberrant calcium regulation. Such mutations cause familial benign hypocalciuric hypercalcaemia, neonatal severe primary hyperparathyroidisim, and autosomal dominant hypoparathyroidisim in people. To enable future study of these disorders in cats, we cloned and sequenced CaSR mRNA from the feline parathyroid gland.

Total RNA was extracted from cat parathyroid gland, and reverse-transcribed cDNA was used for PCR amplification of the CaSR. Plasmids containing CaSR amplicons were transferred to *E. coli,* which were grown on selective media. Sequenced inserts were used to design specific exonic CaSR PCR primers. Peripheral blood leukocytes of 5 healthy normocalcemic cats were used for amplification and sequencing of the six CaSR exons.

The feline CaSR has one positive reading frame and consists of 3243 base pairs. Its overall homology to homology to canine, bovine and human CaSR is 92%, 90% and 91%, respectively. The amino acid sequence is as follows:
MAFYSCCLILLAITWCTSAYGPDQRAQKKGDIILGGLFP
IHFGVAAKDQDLKSRPESVECIRYNFRGFRWLQAMIFAI
EEINSSPVLLPNMTLGYRIFDTCNTVSKALEATLSFVAQ
NKIDSLNLDEFCNCSEHIPSTIAVVGATGSGISTAVANLL
GLFYIPQVSYASSSRLLSNKNQFKSFLRTIPNDEHQATA
MADIIEYFRWNWVGTIAADDDYGRPGIEKFREEAEERD
ICIDFSELISQYSDEEEIQQVVEVIQNSTAKVIVVFSSGPD
LEPLIKEIVRRNITGRIWLASEAWASSSLIAMPEYFHVV
GGTIGFALKAGQIPGFREFLQKVHPRKSVHNGFAKEFW
EETFNCHLQEGAKGPLALDTFLRGHEEGGGRISNSSTAL
RPLCTGDENISSVETPYMDYTHLRISYNVYLAVYSIAHA
LQDIYTCLPGRGLFTNGSCADIKKVEAWQVLKHLRHLN
FTNNMGEQVTFDECGDLVGNYSIINWHLSPEDGSIVFK
EVGYYNVYAKKGERLFINEEKILWSGFSREVPFSNCSR
DCLAGTRKGIIEGEPTCCFECVECPDGEYSDETDASACD
KCPDDFWSNENHTSCIAKEIEFLSWTEPFGIALTLFAVL
GIFLTAFVLGVFLKFRNTPIVKATNRELSYLLLFSLLCCF
SSSLFFIGEPQDWTCRLRQPAFGISFVLCISCILVKTNRV
LLVFEAKIPTSFHRKWWGLNLQFLLVFLCTFMQIVICVI
WLYTAPPSSYRNHELEDEIIFITCHEGSLMALGFLIGYTC
LLAAICFFFAFKSRKLPENFNEAKFITFSMLIFFIVWISFIP
AYASTYGKFVSAVEVIAILAASFGLLACIFFNKVYIILFK
PSRNTIEEVRCSTAAHAFKVAARATLRRSNVSRKRSSSL
GGSTGSTPSSSISSKSNSEDPFPQPERQKQQQPLALTQQE
QQPQPQQPSSLQQQPQPQPQPRCKQKVIFGSGTVTFSLS
FDEPQKSAMAHRNSMHQNSLEAQKSNETLTRHQALLP
LQCGETDSELSAQERGLQGPVDGDFXPEMEDPEEMSPA
LVVSSSQSFVISGGGSTVTENILHS

This is the first report of the feline CaSR sequence. This sequence may serve to elucidate mechanisms of disorders of calcium regulation in cats.

**ABSTRACT #38**

HYPERTRIGLYCERIDEMIA DOES NOT IMPAIR WHOLE BODY INSULIN SENSITIVITY IN CATS, POSSIBLY THROUGH INCREASED CIRCULATING ADIPONECTIN. E Zini[1], D Konrad[2], M Osto[3], M Franchini[4], M Ackermann[4], TA Lutz[3], CE Reusch[1]. Clinic for Small Animal Internal Medicine[1], Vetsuisse Faculty, University of Zurich, Switzerland. University Children's Hospital[2], Zurich, Switzerland. Institute of Veterinary Physiology[3], Institute of Virology[4], Vetsuisse Faculty, University of Zurich, Switzerland.

Obesity and hyperlipidemia impair insulin sensitivity in human type 2 diabetes mellitus, partly due to activation of a mild inflammatory response. Because obesity-induced insulin resistance predisposes cats to diabetes and hyperlipidemia is a frequent concurrent finding, we propose that excess lipids impair insulin sensitivity in this species. We hypothesize that hyperlipidemia may be accompanied by decreased circulating adiponectin and increased pro-inflammatory markers, such as $\alpha_1$-acid glycoprotein (AGP) and monocyte chemoattractant protein-1 (MCP-1) and, in insulin sensitive tissues, by decreased mRNA of adiponectin, glucose transporter-4 (GLUT4), peroxisome proliferative activated receptor-γ1 (PPARγ1) and PPARγ2, and increased resistin and GLUT1.

Healthy cats (n=6) were infused with lipids (Lipovenoes 10%, Fresenius Kabi, Switzerland) for 10 days to clamp their blood triglycerides at the approximate level of untreated feline diabetes (265-620 mg/dl). Controls received saline (n=5). On day 10, adiponectin and pro-inflammatory markers were measured in plasma. An intravenous glucose tolerance test was performed and whole body insulin sensitivity was calculated. Specimens were collected from subcutaneous and omental fat, liver and skeletal muscles. Tissue mRNAs of glucose metabolism-related genes were quantified. Statistical differences were determined with non parametric tests.

Whole body insulin sensitivity did not differ between lipid- and saline-infused cats. Compared to cats on saline, cats infused with lipids had 50% higher plasma adiponectin ($P<0.05$) and 2-3 times higher AGP and MCP-1 ($P<0.01$ and $P<0.05$, respectively). Lipid-infused cats had decreased resistin and GLUT1 mRNA and increased GLUT4 in the omental fat. GLUT1 mRNA was decreased and PPARγ2 increased in subcutaneous fat. Adiponectin and PPARγ1 expression did not differ in any tissue.

Although hyperlipidemia induced systemic inflammation in cats, whole body insulin sensitivity was not impaired. Increased circulating adiponectin levels may have contributed to preventing insulin resistance, possibly by increasing GLUT4 and PPARγ2 transcripts and decreasing resistin expression in subcutaneous or omental fat. The source of increased plasma adiponectin is unknown.

**ABSTRACT #39**

ROLE OF HYPERGLYCEMIA IN THE PATHOGENESIS OF PANCREATITIS IN CATS. E Zini[1], M Osto[2], M Franchini[3], F Guscetti[4], M Ackermann[3], TA Lutz[2], CE Reusch[1]. Clinic for Small Animal Internal Medicine[1], Institute of Veterinary Physiology[2], Institute of Virology[3], Institute of Veterinary Pathology[4], Vetsuisse Faculty, University of Zurich, Switzerland.

Pancreatitis is commonly diagnosed in cats and humans with diabetes mellitus. It is characterized by infiltration with neutrophils and macrophages and by varying degrees of fibrosis. Based on studies in diabetic humans, various hypotheses have been put forward to explain concurrence of exocrine and endocrine disease, such as the spread of an inflammatory disorder from the former to the latter, a common immune-mediated inflammation or a local vascular damage. We propose that diabetic hyperglycemia or hyperlipidemia promotes inflammation of the exocrine pancreas in cats.

Healthy cats were infused with glucose (n=5) or lipids (n=6) for 10 days to clamp their blood concentrations at the approximate level of untreated feline diabetes (glucose: 450-540 mg/dl; triglycerides: 265-620 mg/dl). Controls received saline (n=10). On day 10, blood samples and tissue specimens were collected. Levels of amylase, lipase, and feline pancreatic lipase and trypsin-like immunoreactivity (fPLI and fTLI, respectively) were measured in serum samples. To quantify neutrophils in the exocrine pancreas, pancreatic sections were immunostained with insulin and myeloperoxidase. Myeloperoxidase was also used to quantify neutrophils in liver and omental fat sections. Statistical differences between groups were determined with non parametric tests.

Plasma amylase was lower than the reference range in 4 hyperglycemic cats. Control and hyperlipidemic cats had normal amylase. Lipase, fPLI and fTLI concentrations were similar between groups. Compared to controls, hyperglycemic cats had three times more neutrophils in the exocrine pancreas per high-power field ($P<0.001$). The number of neutrophils did not differ in the liver and omental fat. Control and hyperlipidemic cats had similar neutrophil counts in all examined tissues.

Hyperglycemia selectively increases the number of neutrophils in the exocrine pancreas of cats, possibly predicting the development of pancreatitis. As shown in experimental pancreatits in cats, low amylase may represent an early marker of pancreatic inflammation. Based on the present feline model, hyperlipidemia does not seem to affect the exocrine pancreas.

**ABSTRACT #40**

USE OF LISPRO INSULIN FOR TREATMENT OF DOGS WITH DIABETIC KETOACIDOSIS. KW Sears, KJ Drobatz, RS Hess. University of Pennsylvania School of Veterinary Medicine, Philadelphia, PA.

The goals of this study were to characterize the use of lispro insulin (LI) in dogs with diabetic ketoacidosis (DKA) and to compare the rate of resolution of hyperglycemia and ketoacidosis in dogs treated with LI to these rates in dogs treated with regular insulin (RI).

A prospective randomized study of dogs with DKA (blood glucose (BG) concentration >250 mg/dl, venous pH <7.35, and beta hydroxybutyrate concentration >2.0 mmol/L) was performed. Dogs received either LI or RI as an intravenous (IV) continuous rate infusion according to a previously published protocol. IV fluid therapy, adjustment of the insulin infusion, antibiotic therapy, and electrolyte supplementation were also standardized. Insulin therapy was begun after 6 hours of IV fluid therapy. BG was measured q2 hours and electrolytes and venous pH were measured q6 hours.

Ten dogs have been enrolled in the study: four received LI and six received RI. Median BG concentrations at the time of admission and at the time that insulin treatment was begun were not significantly different between LI (392 mg/dl and 311 mg/dl) and RI (500 mg/dl and 369 mg/dl) groups; however, median BG in all 10 dogs at the time insulin treatment began (328 mg/dl) was significantly lower than at admission (400 mg/dl, p=0.025). There was no significant difference between LI and RI groups in the median time from initiation of insulin therapy to the time BG dropped below 250 mg/dl (3 and 7 hours, respectively). Venous pH was not significantly different between LI (7.26) and RI (7.23) groups at the time of admission and there was no significant difference in the time it took for venous pH to reach 7.35 (28 and 44 hours, respectively). Similarly, beta hydroxybutyrate concentration was not significantly different between LI (4.4 mmol/L) and RI (4.7 mmol/L) groups at the time of admission and there was no significant difference in the time it took for beta hydroxybutyrate concentration to reach 2.0 mmol/L (29 and 40 hours, respectively). Length of hospitalization was not different between LI (85 hours) and RI (108 hours) groups. Adverse effects to insulin therapy were not identified in any of the dogs.

It is concluded that LI is a safe and effective alternative to RI treatment in dogs with DKA.

**ABSTRACT #41**

EVALUATION OF DETEMIR IN DIABETIC CATS MANAGED WITH A PROTOCOL FOR INTENSIVE BLOOD GLUCOSE CONTROL. K Roomp[1], JS Rand[2]. 1. Max Planck Institute for Informatics, Germany. 2. Centre for Companion Animal Health, Uni of Queensland, Australia.

There are no reported studies of long-term use of detemir in diabetic cats. The aim of this study was to report outcomes using detemir and a protocol aimed at intensive blood glucose control with home monitoring in diabetic cats, and to compare the results to a previous study using the same protocol with glargine.

Eighteen cats diagnosed with diabetes mellitus were included in the study. Cats diagnosed with acromegaly were excluded. Data were provided by owners who joined the online German Diabetes-Katzen Forum, and followed an intensive blood glucose regulation protocol for a minimum of 5 months or until remission was achieved. Detemir was administered twice daily and a low carbohydrate wet food diet was fed. The insulin dose was adjusted aiming to achieve euglycemia (50-100 mg/dL as measured using a portable blood glucose monitor calibrated for human blood). Owners performed an average of 5 $\pm$ 2 blood glucose measurements per day in the stabilization period, and supplied spreadsheets recording daily insulin dosages, blood glucose concentration and clinical information.

Seventeen cats in the cohort were initially treated with another insulin type (16 with porcine lente insulin) for a median of 9 weeks, but failed to achieve remission prior to switching to detemir. Most (15/17) of these cats were fed a low carbohydrate diet while on the other insulin.

The overall remission rate was 67% (12/18). For cats that began the protocol within 6 months of diagnosis, the remission rate was 81% (9/11) and for those that began 6 months after diagnosis, the remission rate was 42% (3/7). The median time to remission was 1.7 months after beginning the intensive protocol (range = 10 days to 5.3 months). Nine cats of 12 cats (75%) achieving remission remained off insulin, and the median duration of remission was 12.3 months (range = 6.4 months to 2 years). Three cats (25% of remission cats) relapsed and required insulin again. Only one of these relapsed cats achieved a second remission.

Six of 18 cats (33%) in the cohort required insulin throughout the study to control blood glucose concentrations and did not achieve remission. The median length of time on the protocol was 10.3 months (range = 5.4 months to 1.2 years). The majority (83%; 5/6) of long-term diabetics were considered well regulated with a median blood glucose concentration of $\leq$150 mg/dL and 17% (1/6) were moderately well regulated (median blood glucose $\leq$200 mg/dL). Clinical hypoglycemia was rare, with only a single event in one cat which had mild signs. The median maximum insulin dose administered to cats in the study was 1.75 IU twice daily.

These results are comparable to those of the glargine study. No significant differences were identified between outcomes for glargine and detemir, with the exception of a lower maximal dose for detemir (p-value = 0.045). The median maximum glargine dose was 2.5 IU (range = 1.0 to 9.0 IU) compared with a median detemir dose of 1.75 IU (range = 0.5 to 4.0 IU).

**ABSTRACT #42**

DETERMINATION OF THE CONCENTRATIONS OF TRILOSTANE AND KETOTRILOSTANE THAT INHIBIT *EX VIVO* SYNTHESIS OF CORTISOL IN CANINE ADRENAL TISSUES. AL McGraw[1], EM Whitley[2], DM Boothe[3], EN Behrend[1], HP Lee[1] [1]Departments of Clinical Sciences, [2]Pathobiology and [3]Anatomy, Physiology and Pharmacology, Auburn University, Auburn, AL.

Interest has been stimulated recently regarding use of trilostane in the treatment of canine hyperadrenocorticism. Trilostane competitively inhibits 3-β-hydroxysteroid dehydrogenase, an essential enzyme in the cortisol production pathway. Pharmacodynamic data for trilostane in dogs is lacking, and the potency of ketotrilostane, a metabolite of trilostane, relative to trilostane in dogs is unknown. The purpose of this study was to evaluate the *ex vivo* ability of trilostane and ketotrilostane to inhibit cortisol secretion as a model for future study of the drug and to determine whether trilostane or ketotrilostane is more potent in dogs.

For each trial (n=4), adrenal gland tissues were sliced, placed in culture and stimulated with ACTH for the first 3 hrs. From 3-7 hrs of incubation, slices were incubated with ACTH alone (none) or with one of 5 different concentrations of trilostane or ketotrilostane (T1-T5 = 140, 700, 1,400, 4,200 and 7,000 ng/mL and K1-K5 = 40, 200, 400, 1,200 and 2,000 ng/mL). In each trial, a single concentration of each compound was used in 6 tubes (total 24 tubes each for trilostane and ketotrilostane). At the end of 0, 1, 2, 3, 5 and 7 hr(s), tubes were harvested, and media and tissue slices were divided for different assays (cortisol and potassium concentrations, respectively). An adrenal slice exposed to each drug concentration was submitted for histopathology to judge tissue health. For each drug concentration, % inhibition between 3 (maximal secretion) and 7 hrs (maximal inhibition) in secreted cortisol concentration per mg tissue was calculated and compared using an ANOVA. Significance was set at $p<0.05$.

All tissue slices were judged to be healthy. The mean % inhibition when no drug was present was 2.8%. The mean % inhibition ranged from 19.5%-77.3% for T1-T5 and from 36.1%-65.2% for K1 through K5. When comparing % inhibition between the 5 trilostane concentrations and no drug, a significant difference was detected ($p<0.0001$). The % inhibition was significantly less when no drug or the lowest concentration of trilostane was present compared to T2 through T5. When comparing % inhibition between the 5 ketotrilostane concentrations and no drug, a significant difference was detected ($p<0.0056$). The % inhibition was significantly less when no drug was present compared to all ketotrilostane concentrations.

Both trilostane and ketotrilostane inhibit cortisol secretion in the *ex vivo* model. Ketotrilostane appears to be more potent than trilostane.

**ABSTRACT #43**

PLASMA ACTH PRECURSORS (PRO-OPIOMELANOCORTIN AND PRO-ADRENOCORTICOTROPIN) IN CATS WITH HYPERADRENOCORTICISM. G. Benchekroun[1], P. de Fornel Thibaud[1-2], M. Dubord[3], M. Le Chevoir[1], C. Petit[4], O. Dossin[5], F. Fracassi[6], F. Garnier[7], C. Maurey-Guenec[1] and D. Rosenberg[1]. [1] Internal Medicine Unit, [2] Veterinary Anticancer Center, Maisons Alfort, France [3] Veterinary Biochemistry Unit, [4] Parasitology-Dermatology Unit, [5] Internal Medicine Unit, [6] Veterinary Clinical Dpt, Faculty of Veterinary Medicine, Bologna, Italia and [7] Biochemistry Unit, National Veterinary School of Lyon, Marcy l'Étoile, France. [1, 3] National Veterinary School of Alfort, Maisons Alfort, France. [4, 5] National Veterinary School of Toulouse, Toulouse, France.

Feline hyperadrenocorticism (FH) is a rare condition. Pituitary-dependant hyperadrenocorticism (PDH) is observed in approximately 80% of cases. Most of the cases are associated with diabetes mellitus. Diagnosis of FH is quite difficult. First, clinical signs are non specific, except for the non systematic feline skin fragility syndrome. Secondly, the specificity of all tests validated for cats is questionable as well. Recently, an ACTH precursor (POMC/pro-ACTH) assay has been validated in cats (Benchekroun *et al.*, J Vet Intern Med 2008; 22, 794 (abstract). The aim of this preliminary study is to evaluate prospectively the plasma concentration of ACTH precursors in a small cohort of cats with PDH and estimate its usefulness in its diagnosis.

Three groups of cats were defined. Group 1 included cats with PDH. The diagnosis of FH was based on clinical data and low-dose dexamethasone suppression test (LDDST). PDH was demonstrated by adrenal and pituitary gland CT scan. Group 2 and 3 included diabetic and apparently healthy cats respectively. For the two groups, FH was excluded by LDDST or Urine Cortisol:Creatinine Ratio (UCCR).

Whole blood was collected from the cats on EDTA-tubes; after immediate centrifugation at 4°C, plasma was promptly frozen (-80°C) until assay.

Six cats were included in group 1. 5 cats had a large pituitary tumor with a height ranging from 6 mm to 27 mm. No pituitary tumor was visualised in the remaining cat. Plasma ACTH precursor concentrations ranging from 229 to 1412 pmol/L were measured in PDH cats with large tumors; the remaining cat of group 1 had an ACTH precursor concentration <53 pmol/L. Group 2 and 3 included 8 and 13 cats respectively. Plasma ACTH precursors ranged respectively from <53 to 96 pmol/L and from <53 to 99 pmol/L.

Although being found in a small number of cats, these results suggest that large corticotropic tumors are associated in that species with large plasma concentration of ACTH precursors like in dogs. No plasma POMC/pro-ACTH concentration above 100 pmol/L was found in cats free of PDH. The specificity of high plasma ACTH precursor concentration in cats with PDH has to be confirmed on a larger cohort. If confirmed, given the high prevalence of large pituitary tumors, the introduction of this tool could offer a gain of specificity in the general approach of feline PDH.

*Previously presented at ECVIM in Belgium, September 2008.*

**ABSTRACT #44**

ENDOCRINE FUNCTION IN CATS AFTER STEREOTACTIC RADIOSURGERY TREATMENT FOR ACROMEGALY. KF Lunn, SM LaRue, College of Veterinary Medicine and Biomedical Sciences, Colorado State University, Fort Collins, CO.

Four cats received stereotactic radiosurgery (SRS) for treatment of acromegaly. Endocrine testing was performed for up to 26 weeks after SRS, to document changes in pituitary function. All cats had typical clinical signs of acromegaly, including diabetes mellitus with insulin resistance. A macroscopic pituitary tumor was detected on CT in 3 of 4 cats. SRS was administered in 2-4 fractions over 4 days using a Varian Trilogy™. Dose was based on normal tissue constraints. Serum IGF-1, thyroid hormones, endogenous ACTH, and cortisols before and after ACTH stimulation were measured before and 2, 6, 12, and 26 weeks after SRS.

All cats had elevated IGF-1 levels prior to SRS. Insulin resistance decreased markedly by 26 weeks in 2 cats, and in 1 of these, IGF-1 levels fell to 50% of baseline. At all time points evaluated, thyroid panels were unremarkable, and ACTH stimulation tests showed elevated pre- and post-ACTH cortisol values in all cats, however there were no clinical signs of hyperadrenocorticism. Endogenous ACTH levels did not indicate SRS-induced pituitary hypofunction. There was no evidence of diabetes insipidus and electrolyte values remained normal in all cases. No adverse effects of SRS were noted in any patient.

These results indicate that SRS is well-tolerated in cats with acromegaly, and may lead to improved clinical signs within 6 months. Pituitary function remains normal at least for 26 weeks post-SRS, but IGF-1 levels may improve within this time. These patients should be monitored beyond 26 weeks, as continued endocrine changes are likely.

**ABSTRACT #45**

BLOOD-BRAIN-BARRIER DISRUPTION IN CANINE CHRONIC EXPERIMENTAL HYPOTHYROIDISM. Rossmeisl JH, Panciera DL. VA-MD Regional College of Veterinary Medicine, Blacksburg VA.

Hypothyroidism has been documented to cause central nervous system (CNS) abnormalities, most commonly central vestibular disease. The purpose of this study was to evaluate the effects of chronic hypothyroidism on CNS vasculature. It was hypothesized that canine hypothyroidism would cause blood-brain-barrier (BBB) abnormalities that would be detectable using indirect peripheral biomarkers.

Eighteen clinically normal, euthyroid, mixed breed females were studied. Hypothyroidism was induced by $^{131}$Iodine administration in nine dogs; nine served as untreated controls.

Clinical and routine cerebrospinal fluid (CSF) analyses, blood and CSF protein/lipoprotein electrophoresis, vascular endothelial growth factor (VEGF) and S-100beta concentrations, and CSF albumin quota (AQ) were performed at baseline and 6, 12, and 18 months after induction of hypothyroidism. Data were analyzed using repeated measures ANOVA.

At baseline, no differences between groups were detected for any variable. Throughout the study, controls dogs remained free of neurologic disease, and had test variables that remained within reference ranges. Two hypothyroid dogs developed CNS signs during the study, associated with necropsy evidence of brain infarction. At 12 and 18 months, the CSF total protein, VEGF, and S-100beta, and fractional albumin concentrations as well as AQ were significantly higher ($p < 0.04$) in hypothyroid dogs than controls. Among test variables assayed in blood the only significant difference was a higher S-100beta concentration in hypothyroid dogs ($p = 0.003$) compared to controls at 18 months.

Chronic hypothyroidism results in multiple alterations compatible with BBB disruption that may be asymptomatic, which may have clinical implications for pharmacologic treatment of coincident conditions. Further studies are needed to identify risk factors for the development of clinically apparent CNS dysfunction in hypothyroid dogs with cerebrovascular disease.

**ABSTRACT #46**

INDUCTION OF ARYL HYDROCARBON RECEPTORS IN CATS. J. Wakeling, J. Elliott, H. Syme. Royal Veterinary College, London, UK.

Hyperthyroidism is a common disease of older cats and is insidious in onset. Recent studies have suggested that persistent organic pollutants (POPs) may be implicated in the development of this disease. POPs are ubiquitous halogenated aromatic industrial pollutants that have many effects on reproductive and thyroid function. The most toxic POP is 2,3,7,8-tetracholodibenzo-p-dioxin (2,3,7,8-TCDD). When given to pregnant rats 2,3,7,8-TCDD decreases tT4 concentrations leading to a sustained excessive secretion of TSH followed by hyperplasia of the thyroid follicular cells in the offspring. POPs are thought to induce many of their endocrine disrupting effects by binding to the cytoplasmic aryl hydrocarbon receptor (AhR). It is possible to quantify POP binding to the AhR in a patented cell-culture system (CALUX®; Chemical Activated Luciferase Gene Expression), allowing an estimation of the total burden of POPs in a sample. This retrospective pilot study tested the hypothesis that cats accumulate POPs throughout their lifetime and that this accumulation is related to development of hyperthyroidism.

AhR induction was measured in fat samples collected from cats either at post mortem (senior cats, >8 years; n=17) or at the time of ovariohysterectomy (young cats; n=6). The luminescence detected in the CALUX® assay was used to generate results expressed as a toxic equivalence (TEQ) benchmarked to a 2,3,7,8-TCDD standard curve. Total thyroxine and TSH concentration were measured in all cats; blood samples were collected within 6 months of death or at the time of neutering, depending on the group. Senior cats were further divided into three groups; euthyroid (tT4<55nmol/l) with TSH concentration below the limit of quantification of the assay (<0.03 ng/ml; Undetectable TSH group; n=6), euthyroid cats with detectable TSH concentrations (≥0.03 ng/ml; Detectable TSH group; n=5) and cats with hyperthyroidism confirmed ante mortem (Hyperthyroid group; n=6). Thyroid histopathology was performed in all senior cats and nodular hyperplastic changes were graded as previously described. Where data were available, the percentage daily weight loss was calculated for each senior cat in the months prior to death.

Percentage weight loss and TEQ pg /g fat were compared between groups using the Kruskal Wallis test. Normal distribution was assessed using the Kolmogorov Smirnov test. The TEQ pg /g fat was normally distributed when logarithmically transformed. Correlations between normally distributed continuous variables were tested using the Pearson's Correlation coefficient.

Median [IQ range] TEQ for all cats was 17 [8-30] pg/g fat, similar to median values reported in humans using the same assay. There was no significant difference in the induction of the AhR (TEQ pg/g fat) between the four groups of cats ($p=0.72$) or in the amount of weight loss between the three senior cat groups ($p=0.21$). There was no significant correlation of Log [TEQ] pg/g fat with histopathological thyroid grade ($p=0.25$); TSH concentration ($p=0.64$) or age ($p=0.71$).

These limited data do not provide evidence that POPs accumulate with age in cats, or that accumulation of POPs is a risk factor for the development of hyperthyroidism in this species.

**ABSTRACT #47**

PREVALANCE OF HYPOCOBALAMINEMIA IN CATS WITH HYPERTHYROIDISM. AK Cook[1], JM Steiner[1], JS Suchodolski[1], JE Robertson[2]. 1. College of Veterinary Medicine and Biomedical Sciences, Texas A&M University, College Station, TX. 2. IDEXX Laboratories, West Sacramento, CA.

Serum cobalamin concentrations are routinely measured in cats with suspected gastrointestinal disorders, and hypocobalaminemia is commonly associated with exocrine pancreatic insufficiency and diffuse distal small intestinal disease. In human medicine, abnormalities in serum cobalamin concentrations are commonly noted in geriatric individuals and those with thyroidal disease, but similar associations have not been reported in feline patients. The goal of this study was to determine the prevalence of hypocobalaminemia in cats with spontaneous hyperthyroidism.

Serum cobalamin concentrations were measured using an automated chemiluminescence assay in 76 cats with total thyroxine (T4) concentration ≥ 8.0 µg/dl (range: 8.0-30.2 µg/dl; median 11.9 µg/dl).

Cobalamin concentrations ranged from < 150 ng/L to >1000 ng/L, with a median value of 409 ng/L. Thirty one cats (40%) had a serum cobalamin concentration below the reference range of 290-1,500 ng/L. There was no correlation between serum cobalamin concentrations and T4 concentrations. In addition, the mean T4 concentration for hypocobalaminemic cats and those with serum cobalamin concentrations within the reference range were not significantly different.

This study indicates that hypocobalaminemia is a common finding in cats with hyperthyroidism. The mechanism is unknown, but possible explanations include compromised cobalamin uptake due to changes in gastrointestinal transit time or changes in cobalamin requirements or metabolism. It is unknown at this time if serum cobalamin concentrations normalize when a euthyroid state is achieved, or if some cats remain cobalamin deficient following successful therapy for their hyperthyroid state.

**ABSTRACT #48**

HYPERTENSION IN HYPERTHYROID CATS: PREVALENCE, INCIDENCE, AND PREDICTORS OF ITS DEVELOPMENT.

LD Morrow[1], VJ Adams[2], J Elliott[1], HM Syme[1]. 1. Royal Veterinary College, University of London, UK. 2. Animal Health Trust, Newmarket, UK.

Hypertension and azotemic chronic kidney disease (aCKD) can be diagnosed concurrently with hyperthyroidism and have also been detected after initiation of treatment. The aims of this study were to estimate the prevalence of hypertension in cats at first diagnosis of hyperthyroidism, to estimate its incidence after treatment for hyperthyroidism and to identify predictors of the development of hypertension in hyperthyroid cats.

Hyperthyroidism was diagnosed when total plasma thyroxine concentration was >55 nmol/l. Systolic blood pressure (SBP) was measured using the indirect Doppler method and the mean of five consecutive measurements was calculated. Hypertension was diagnosed when a cat had SBP greater than 170 mm Hg on two separate occasions or when it had SBP greater than 170 mm Hg on one occasion and compatible ocular signs. Hyperthyroid cats were treated medically, surgically or with a combination of these methods. A cat was diagnosed as having aCKD if it had a plasma creatinine concentration of >2.0 mg/dl and a urine specific gravity of <1.035 on the same visit. Three hundred and twenty four cats newly diagnosed as hyperthyroid at two first opinion geriatric cat clinics in London, UK between April 1, 1999 and June 18, 2008 were included in this retrospective analysis of longitudinal data using Kaplan-Meier survival analysis and Cox regression with time-varying covariates. The time-varying data that were analysed represented a total of 2205 visits to the clinic for 215 cats.

Of 324 cats diagnosed with hyperthyroidism, 21 were already receiving amlodipine for treatment of hypertension and were excluded from further study. Thirty nine of the remaining 303 cats were hypertensive at time of diagnosis of hyperthyroidism; a prevalence of 12.9% (95% CI: 9.4-17.3). The incidence with which hypertension developed in hyperthyroid cats following treatment was 22.8% (49/215, 95% CI: 17.5-29.1). Hypertension developed a median of 5.3 months (95% CI: 3.2-9.9) after treatment began. Of the newly hypertensive cats for which renal status data were available, 15 out of 42 (35.5%, 95% CI: 22.0-52.0) had a creatinine concentration of >2.0 mg/dl. Both urine and blood samples had been taken at diagnosis of hypertension in 31 of the cats; of these, 11 (35.5%, 95% CI: 19.8-54.6) had aCKD. Overall survival did not differ between hypertensive and normotensive cats ($p=0.4$). Age and treatment for hyperthyroidism were predictive of the development of hypertension.

We conclude that although hypertension can present concurrently with hyperthyroidism, in many cats hypertension develops after anti-thyroid treatment. In addition, hypertension does not always develop in association with the unmasking of underlying renal disease. These findings suggest that it would be prudent to assess hypertensive status in both newly diagnosed and treated hyperthyroid cats.

**ABSTRACT #49**

EXPRESSION OF CYCLOOXYGENASE ISOFORMS IN EQUINE GASTRIC ULCERS.

NATÁLIA L. F. RODRIGUES, MONIQUE DORÉ and MICHÈLE Y. DOUCET. Faculté de Médecine Vétérinaire, Université de Montréal, Saint-Hyacinthe, Québec, Canada.

In order to characterize cyclooxygenase-1 and cyclooxygenase-2 isoforms expression in naturally occurring equine gastric ulcers, 38 ulcerated and 10 healthy equine stomachs were collected from a local abattoir. Two gastric squamous mucosa samples were taken from each stomach. When ulcers were present, samples were collected in the immediate periphery of the lesion. The first sample from each site was fixed in 10% formalin for immunohistochemical preparation and the other was frozen at -70 °C for immunoblotting analysis. The inflammatory reaction present in ulcerated tissues was evaluated on standard histology sections using a scoring system from 0 to 2, where 2 characterized the most severe reaction. Immunnoreactivity to two antibodies, MF-241 (selective for COX-1) and MF-243 (selective for COX-2) , was evaluated using a scoring system from 0 to 3, where 0 denoted the absence of COX expression and 3 corresponded to the maximum expression of either COX isoform. COX-1 and COX-2 characterizations were confirmed by immunoblotting analyses. Fifty-five percent (55%) of the ulcerated gastric mucosa samples presented an inflammation score of 2 while 37% were scored 1. All normal gastric mucosa samples strongly expressed COX-1. The expression of COX-1 was significantly lower and the expression of COX-2 was significantly higher in ulcerated compared to normal mucosae ($p < 0.0001$). An increased expression of COX-2 in equine squamous gastric ulcers suggests a role for this enzyme in gastric ulcer healing as observed in other species. The use of specific COX-2 inhibitors should therefore be used with caution in horses with EGUS.

**ABSTRACT #50**

PRIMARY GASTRIC IMPACTION IN HORSES: A RETROSPECTIVE STUDY OF 22 CASES (2005-2008).

KME Vainio, BW Sykes. Hyvinkaa Horse Hospital, Hyvinkaa, Finland.

Gastric impaction is a disease characterized by excessive accumulation of ingesta in the stomach after appropriate fasting. The objective of this retrospective study was to describe the clinical findings and outcome of horses with a primary gastric impaction in a referral population. Diagnosis of a primary gastric impaction was made if the horse had been fasted for a minimum of 16 hours, a ball of ingesta precluded visualization of the margo plicatus, and there was no evidence of concurrent intestinal pathology.

Hospital records were examined, and 22 horses with a primary gastric impaction treated between 2005 and 2008 were identified. The median age of horses was 8 years (range 8 months to 25 years). Twelve of 22 horses were geldings, 9 of 21 were mares, and 1 of 21 was a stallion. Half (50%) of the patients were Warmblood riding horses. This is comparable to the hospital population (54%). Finnhorses (27%) appeared to be overrepresented compared to the hospital population (17%). Fourteen of 22 horses presented on an emergency basis. The most common complaint was inappetence (59%) followed by acute colic (32%) and recurrent colic (32%). On admission, 36% of horses were tachycardic and 36% had subjectively decreased gastrointestinal borborygmi. The most common laboratory findings included leucopenia (29%), granulocytosis (29%) and hyperfibrinogenemia (29%).

All horses were treated with enteral fluid therapy with or without analgesia and laxatives. The median dose of fluids administered per day was 5 doses of 2-10 liters of isotonic electrolyte solution (range 1 to 8 doses). The mean length of treatment until resolution was 2.1 days (range 1 to 5 days). The median duration of hospitalization was 5 days (range 2 to 9 days). The majority of horses (91%) survived to discharge. One horse was euthanized due to gastrointestinal tract rupture and one due to a non-resolving impaction. The condition reoccurred in two cases; one of these horses was euthanized and the other one was treated successfully a second time. Primary gastric impaction appears to be a condition with clinical signs of inappetence and mild abdominal discomfort. It has a good prognosis. Enteral fluid therapy may be of value in treating gastric impactions.

**ABSTRACT #51**

EFFECT OF VOLATILE FATTY ACID MIXES ON EQUINE NONGLANDULAR MUCOSA BIOELECTRIC PROPERTIES: PATHOGENESIS OF GASTRIC ULCERS IN HORSES. RE Reese[1] , FM Andrews[2], SB Elliott[1], AM Saxton[3]. 1. University of Tennessee College of Veterinary Medicine, Knoxville, TN. 2. Louisiana State University School of Veterinary Medicine, Baton Rouge, LA. 3. University of Tennessee, College of Agriculture, Animal Science, and Natural Resources, Knoxville, TN.

Gastric ulcers are common in horses and volatile fatty acids (VFAs), fermentation products from the action of gastric bacteria on high grain diets, have been implicated in their cause. The purpose of this study was to measure nonglandular (NG) tissue bioelectric properties in an Ussing Chamber system following exposure to normal Ringers solution (NRS), a VFA mixture containing sub-threshold VFA concentrations found 2 hours after feeding a sweet feed diet (0.5 Kg/Kg BW) and 3 VFA mixtures containing high acetic (40 mM), high butyric (40 mM) and high propionic (40 mM) acids in NRS, at pH 1.5, 4, or 7. After NRS or VFA exposure, tissues were examined under light microscopy for histopathologic evidence of cell swelling and necrosis.

Horses (n=10) were euthanized and NG gastric tissues obtained within 30 min. Tissues were then mounted in an Ussing Chamber system and NRS or VFA mixtures, at pH 1.5, 4 or 7 added to the mucosal side of the chamber and NRS, at pH 7 was added to the serial side. Tissues were allowed to acclimate for 30 mins, then pH on the mucosal side was adjusted to pH 1.5, 4 or 7 and exposed for 330 mins. Tissue short circuit current (Isc) and potential difference (PD) were measured ever 15 mins and resistance (R) and conductance (G) calculated. After exposure, tissues were removed from the Ussing Chamber and placed in 10% formalin and stained with hematoxylin and eosin for histopathologic examination.

NG mucosa exposed to HCl alone (pH 4 and 1.5) in NRS caused an immediate significant decrease in PD, followed by a decrease in Isc and R. NG mucosa, when exposed to the sub-threshold VFA mixture ($\leq$ 20 mM), showed no significant change in PD or Isc, when compared to tissues exposed to NRS at the same pH. However, NG mucosa exposed to VFA mixtures containing high acetic, propionic and butyric acids, caused an immediate significant decrease in PD and Isc. Results suggest that a sub-threshold mixture of VFAs, found after feeding a grain diet (0.5 Kg/Kg BW) do not cause bioelectric changes in the equine NG mucosa other than that caused by HCl alone. However, higher concentrations of VFAs in gastric juice from excessive grain feeding, at pH $\leq$ 4.0 results in changes in barrier function and may lead to gastric ulcers. Histopathologic changes of cell swelling were noted only in tissues exposed to the VFA mixtures containing high VFA concentrations. This study confirms that when stomach VFA concentrations are at or below threshold ($\leq$ 20 mM), tissue barrier function is maintained; whereas if gastric juice VFA concentrations exceed threshold concentrations (> 20 mM), as found after feeding grain in excess of 0.5 Kg/Kg BW, damage may occur in NG mucosa resulting in ulcers.

**ABSTRACT #52**

STUDY OF THE VARIABILITY OF ENDOSCOPIC DUODENAL BIOPSIES IN HEALTHY HORSES AND IN HEAVES-AFFECTED HORSES FED WITH DIFFERENT DIETS. P. Moreau, P. Hélie, T. Munoz, J.-P. Lavoie, D. Jean. Faculté de médecine vétérinaire, Université de Montréal, Saint-Hyacinthe, QC, Canada.

Endoscopic duodenal biopsies are currently used in equine medicine to evaluate small intestinal diseases. However, there is limited information concerning the normal duodenal histology of horses. A recent study reported minimal to moderate infiltration of inflammatory cells in the duodenal lamina propria of healthy horses. Diet, anti-inflammatory drugs or systemic inflammatory diseases are factors possibly contributing to mucosal inflammatory cell infiltration. The objectives of the study were: 1) to evaluate duodenal histology with healthy horses fed with different diets and 2) to compare duodenal histology in healthy horses with that of horses with heaves, a condition associated with systemic inflammation.

Glucose absorption tests, endoscopic duodenal biopsies and rectal biopsies were performed on six healthy horses fed with three different diets (sweet feed/hay, fat and fibre/hay, sweet feed/alfalfa pellets). Ten horses with heaves were evaluated during symptomatic and asymptomatic periods, when fed with different diets (sweet feed/hay or sweet feed/alfalfa pellets), and with or without inhaled corticosteroids administration. Morphology (villous stunting, epithelial injury, crypt distention, lacteal dilatation, mucosal fibrosis, lymphoid follicles) and inflammation (lymphocytes, eosinophils, plasma cells, macrophages and neutrophils in epithelium and/or in lamina propria) were assessed in a blind fashion by a board certified pathologist. Inflammatory cell infiltration was graded as minimal, mild, moderate or severe.

Glucose absorption tests and rectal biopsies were considered normal in all horses and were not different with types of food, diseases (heaves) or inhaled corticosteroids. Duodenal biopsies showed minimal to moderate infiltration with lymphocytes/plasma cells (43% of biopsies), eosinophils (11%), mixed eosinophils and lymphocytes/plasma cells (27%) with variability between horses and for a same horse, but not associated with different types of diets, heaves or corticosteroids.

In conclusion, variability of duodenal histology in horses was confirmed, but neither types of diets, heaves or inhaled corticosteroids were identified as significant factors of variability.

**ABSTRACT #53**

SUB-EPITHELIAL FIBROSIS IS PRESENT IN THE PERIPHERAL AIRWAYS OF HEAVES-AFFECTED HORSES. E L Setlakwe, M Leclère, JP Lavoie. Université de Montréal, St-Hyacinthe, QC.

It has been proposed that extra-cellular matrix remodelling, when present, could either serve as a protective mechanism or a contributing factor to airway obstruction or airway hyperresponsiveness. The aim of this study was to evaluate, using morphometric techniques, whether sub-epithelial fibrosis is present in the peripheral airways of heaves-affected horses.

Peripheral lung biopsies acquired under thoracoscopic guidance were obtained from 6 heaves-affected horses and 5 controls, both before and after a 30 day antigenic exposure. All horses had been kept in an antigen poor environment for at least 3 months prior to the study. Lung function parameters were derived from an esophageal ballon catheter and a heated pneumotachograph. Lung biopsies were microscopically examined using picrosirius red, a collagen specific histological staining technique. Collagen area in the sub-epithelial layer, e.g. the region between the airway smooth muscle and the epithelial layer, was measured and corrected for airway size using standard morphometric techniques.

At baseline, all horses had a normal lung function, while only heaves-affected horses developed airway obstruction after antigenic challenge. In comparison with controls, horses with heaves had a nearly doubled collagen deposition in the airway sub-epithelial layer before challenge. No further increase in sub-epithelial fibrosis was observed in diseased horses after the 30 day antigenic challenge.

Sub-epithelial fibrosis is present in the peripheral airways of heaves-affected horses and is unchanged by a 30 day continuous antigenic challenge. Further studies will be required to determine the role of fibrosis in equine obstructive lung disease.

**ABSTRACT #54**

FUROSEMIDE DOES NOT AFFECT TRANSVASCULAR FLUID FLUXES ACROSS THE LUNG IN EXERCISING HORSES. M. Vengust[1], C. Kerr[2], H. Staempfli[2], J. Pringle[3], G. Heigenhauser[4], L. Viel[2]. 1. University of Ljubljana, Veterinary Faculty, Slovenia; 2. University of Guelph, Ontario Veterinary College, Canada; 3. Swedish University of Agricultural Sciences, Faculty of Veterinary Medicine and Animal science, Uppsala, Sweden; 4. McMaster University Medical Centre Hamilton, Ontario, Canada.

Use of intravenous furosemide is used in some racing jurisdictions for purported effect in racehorses experiencing Exercise-Induced Pulmonary Hemorrhage. The effects of furosemide appear to be a reduction in pulmonary transcapillary pressures. In addition furosemide inhibits $Cl^-$ and $HCO_3^-$ exchange, which will effect erythrocyte volume regulation. Erythrocyte volume regulation and increased pulmonary transcapillary pressures during exercise are major contributors to pulmonary transvascular fluid fluxes. For this study it is hypothesised that furosemide treatment would attenuate transpulmonary fluid fluxes in horses during high intensity exercise.

Six Standardbred horses were exercised twice on a high-speed treadmill (Säto Sweden) at 80% $VO_2$ peak until fatigue. Horses were randomly assigned treatment with 250mg of furosemide (FurTr) or placebo (Con) intravenously 4 hours prior to exercise, with cross over treatment used at the repeated exercise test (8 days later). Resting arterial and mixed venous blood, as well as $CO_2$ elimination and $O_2$ uptake, were sampled simultaneously 5 minutes apart. During exercise, the sampling was performed at 60 sec intervals until fatigue. Erythrocyte and plasma volume changes across the lung were calculated from changes in hemoglobin and hematocrit values in venous and arterial blood. Cardiac output (Q) was calculated using Fick equation. Erythrocyte and plasma fluid fluxes, and transvascular fluid fluxes across the lung were calculated using Q and erythrocyte, plasma, and whole blood volume changes across the lung. Variables were analyzed using two-way repeated-measures ANOVA ($P<0.05$).

At rest furosemide had no effect on Q, erythrocyte and plasma fluid flux, and transvascular fluid fluxes across the lung. During exercise Q increased in Con and in FurTr, with Q being lower in FurTr at fatigue (301.8 (mean) ±8.5 (SE) L/min) compared to Con (336.5 ±15.6 L/min) ($P < 0.006$). Fluid flux from erythrocytes increased during exercise in Con and FurTr and at fatigue reached (14.6 ±2.3 L/min) and (11.8 ±2.2 L/min), respectively ($P = 0.2$). There were no plasma fluid fluxes observed in either group. Transvascular fluid flux from the pulmonary circulation increased and at fatigue reached 14.1 ±3.5 L/min in Con and 11.8 ±3.4 L/min FurTr ($P = 0.7$).

At the standard dose furosemide is given as premedication to race horses, it appears that changes in pulmonary transcapillary pressures and $Cl^-$ and $HCO_3^-$ exchange blockage do not have significant effects on erythrocyte and plasma volume changes across the lung, and transpulmonary fluid fluxes.

**ABSTRACT #55**

PARTITIONING OF PULMONARY RESISTANCE IN HORSES. JA Nicol, R. Léguillette. Faculty of Veterinary Medicine, University of Calgary, Calgary, Alberta.

The relative contribution of larger and smaller airways to total lung resistance is unknown in horses. The purpose of this study is to determine the relative contribution of central airway (bronchi binomial numbering: #1.1), large bronchi (first generation), small bronchi (third generation), and smaller airways embedded in the lung tissue to the total lung or lobar resistance.

To address this question we measured lung mechanics ex-vivo using 5 horses' lungs. The lungs were obtained from an abattoir, then only their cranial lobes were used and their central airway (#1.1) was cannulated. Two retrograde catheters were then blindly inserted respectively in a larger first generation bronchus, as well as a smaller third generation bronchus. One alveolar capsule was also glued to the surface of each lobe. The lobes were then suspended in a sealed box and subjected to quasi-sinusoidal pressure changes. A pneumotachograph was inserted at the outlet of the main bronchus and used to measure airflow and calculate volumes. Five pressure transducers were used to measure transpulmonary, central airway, peripheral airway, and alveolar pressures. The lungs were ventilated in the box at 3 different transpulmonary pressures (Ptp) (15, 20, 30cm$H_2O$) and frequencies (10, 20, 30 breaths/min). Total lobar resistance (RL) and elastance (EL) were calculated. Individual resistances to airflow from the central airway (Rc), the two different size catheterized airways (RpLarge and RpSmall) and lower tissue elastance (Et) and resistance (Rt) were calculated and expressed as relative values of RL. The diameter of the catheterized airways was measured using a standard method after the lung mechanics.

The measured diameter of the catheterized airways was 5.7mm (±0.9 SD) for the larger airways and 2.5mm (±1.1 SD) for the smaller airways.

Effect of an increase in Ptp: The following results were obtained when Ptp was increased from 15cm $H_2O$ to 30cm $H_2O$ at the different respiratory rates mentioned above:

RL and EL are increasing significantly for all respiratory rates. The relative contribution of Rc, RpLarge and RpSmall to RL significantly decreased for all respiratory rates. Conversely, the relative contribution of Rt to RL significantly increased for all respiratory rates. Furthermore, the relative contribution of first and third generation peripheral airway to RL was not significantly different for all respiratory rates and Ptp values.

Effect of an increase in respiratory rate: The change in respiratory rate from 10 breath/ min to 25 breaths/min had no significant affect on RL, Rc, Rp and Rt. The increase in EL with increasing respiratory rate was not significant.

The effect of Ptp on lung mechanics was predominant on the change in respiratory rate. The most important contributor to RL is Rt. The relative contributions of Rc, RpSmall and RpLarge to RL decrease when Ptp increases, whereas the relative contribution Rt to RL does the opposite.

**ABSTRACT #56**

EFFECTS OF PROLONGED INHALED CORTICOSTEROID TREATMENT ON CELL-MEDIATED IMMUNITY IN HORSES. Julie Dauvillier[1], M. Julia B.F. Flaminio[2], Jean-Pierre Lavoie[1]. 1. Faculté de Médecine Vétérinaire, Université de Montréal, Saint Hyacinthe, QC, Canada. 2. College of Veterinary Medicine, Cornell University, Ithaca, NY.

A single therapeutic dose of systemically administered corticosteroids significantly modifies cell-mediated immunity in horses. Thus, inhaled medications have been used in an attempt to attenuate systemic side effects when the prolonged administration of corticosteroids is necessary for the treatment of chronic inflammatory pulmonary diseases. The objective of this study was to determine whether a prolonged administration of inhaled corticosteroids altered the cell-mediated immunity in horses.

Ten adult horses with heaves were divided into 2 groups: 5 horses received therapeutic dosages of inhaled fluticasone while 5 others received no medication. Nine months after the beginning of treatment, the following parameters were measured in peripheral blood using flow cytometry: lymphocyte subpopulation distribution (CD4+ T lymphocytes, CD8+ T lymphocytes, B lymphocytes), lymphocyte function-associated antigen-1 (LFA-1) and major histocompatibility complex (MHC) class II expression on lymphocytes and monocytes, and *in vitro* lymphocyte proliferation when stimulated by pokeweed mitogen or concanavalin A. Values of treated and untreated horses were compared using a Student t-test. A complete blood count was performed to evaluate total leukocyte subpopulations before (T0), and after 2, 6 and 11 months of treatment. A repeated-measures linear model, with group as between-subject factor and months of treatment as within-subject factor, was used and comparisons of means were conducted with the sequential Bonferroni procedure.

Results of the different immunologic tests were comparable between groups ($p>0.05$). Moreover, no significant differences were observed in the leukocytes subpopulations, neither between groups nor within groups between different time points. In addition, none of the horses developed signs of infection that could have resulted from immunosuppression. These results suggest that prolonged administration of inhaled fluticasone is not associated with major alteration in cell-mediated immunity in horses.

**ABSTRACT #57**

EHV-1 POSITIVE PBMC FRACTIONS DURING CELL-ASSOCIATED VIREMIA FOLLOWING EXPERIMENTAL INFECTION. S Wilsterman, LV Ashton, RJ Callan, DP Lunn, S Rao, LS Goehring. College of Veterinary Medicine and Biomedical Sciences, Colorado State University, Fort Collins, CO.

Viremia during Equid Herpesvirus type 1 (EHV-1) infections is a prerequisite for horses to develop neurological disease. Cell-associated viremia occurs in peripheral blood mononuclear cells (PBMC), which consist of 4 cell subsets: CD4, CD8, B lymphocytes and monocytes. The goal of this study was to identify the virus-infected PBMC subsets during EHV-1 viremia.

PBMC were collected from 2 independent infection studies between days 4 – 9 *post infection*. The first study included 8 cross-bred yearling horses, infected with EHV-1, strain Ohio 03. The second study used 7 cross-bred yearling ponies, and EHV-1 strain Ab4. PBMC were isolated, labeled using subset-specific monoclonal antibodies, and separated into 4 fractions by using a magnetic bead technique. Fractions from each animal on each collection day were submitted for quantitative PCR analysis using 2 PCR assays. PCR EHV1 gB provided viral genome copy numbers detected in each sample, and PCR equine β-actin estimated the number of nucleated cells in a sample. Results were expressed as qualitative data, absence or presence of viremia in a fraction, and as quantitative data as viral genome copy numbers/ 2.5x $10^5$ cells. Multivariable linear regression analysis was used to determine the association of 'cell subset' and 'day post infection' with 'EHV-1 gB copy number'. The viral load of the monocyte cell population, and 'day 4' were used as references for 'cell subset' and 'day post infection', respectively. Poisson regression was used to analyze the average number of viremic days in each fraction.

Neurological disease was mild and transient in 3 horses, and absent in all ponies. Duration of viremia in horses (days 5 - 9) was longer than in the ponies (days 5 - 7), and was undetectable in 2 horses, and in 1 pony. Total horse viremic days for each subset were 17 (CD8), 7 (B lymphocyte), 3 (monocyte) and 2 (CD4). Virus quantity was significantly higher in CD8 when compared to the reference. Viral load in CD4 and B lymphocytes was not significantly different from the monocyte reference. Total pony viremic days were 12 (CD8), 10 (B lymphocytes), 7 (monocytes) and 5 (CD4). In the pony study virus quantities were not different between CD8, B lymphocytes and monocytes. CD4 lymphocytes consistently contained fewer viral genome copies than the reference.

Differences in number of viremic days, virus preference and quantity within a PBMC subset may depend either on the strain that was used for infection (Ohio 03 *versus* AB4), or on the host (horse *versus* pony). Both experiments show that the CD8 subset contained the highest cumulative viral load followed by B lymphocytes, monocytes and CD4. The results on subset preferences do not suggest a mechanism on how endothelial cells become infected. Future studies will be necessary to clarify this mechanism.

**ABSTRACT #58**

THE EFFICACY OF NITAZOXANIDE AS AN ANTIVIRAL AGENT FOR THE TREATMENT OF EHV-1 INFECTION IN HORSES. S Wilsterman, RJ Callan, LS Goehring, L Ashton, S Rao. College of Veterinary Medicine and Biomedical Sciences, Colorado State University, Fort Collins, CO.

There is currently no specific antiviral medication for the treatment of EHV-1 myeloencephalopathy. The purpose of this study was to determine whether or not treating horses with nitazoxanide (Navigator®) paste prior to and during infection with EHV-1 was efficacious in protecting the treated horses against clinical disease, and whether or not this treatment diminished the nasal shed of virus or decreased the extent of viremia.

16 yearling horses were randomly assigned to a treatment (n=8) or control group (n=8). Horses were verified EHV-1 seronegative prior to the experiment. The treatment group was treated with 25 mg/kg of nitazoxanide by mouth twice per day starting two days prior to EHV-1 infection for a total of 25 days. The control group received a placebo. All horses were challenged with a 5x $10^7$ PFU of EHV-1 (Ohio 03). Horses were evaluated daily by a veterinarian who did not have knowledge of the treatment groups, a clinical score was determined (0= no abnormalities). Viremia and the nasal shed of virus were determined using quantitative PCR for the viral glycoprotein B gene. Clinical score data was analyzed using a Wilcoxon non-parametric test. Physical exam parameters, viremia, and nasal shedding were analyzed by linear regression.

There were no significant differences in the clinical scores or physical exam parameters between the treatment and control groups following EHV-1 infection. The log peak titer of viremia was lower in the treatment group but not significantly different. There were no significant differences noted between the treatment and control groups in regard to the peak nasal titer, the duration of nasal shed of virus or the duration of viremia.

The treatment of horses with nitazoxanide does not significantly decrease viremia, nasal viral shed or ameliorate the clinical signs associated with EHV-1 infection.

*Previously presented at Conference of Research Workers in Animal Diseases*

**ABSTRACT #59**

REPEATABILITY AND REPRODUCIBILITY OF TRANSABDOMINAL ULTRASONOGRAHIC RENAL LOCI AND DIMENSIONS IN THOROUGHBRED HORSES. ACE Draper[1] and GD Hallowell[2]. 1. Wendover Heights Veterinary Centre, Halton, Buckinghamshire, UK 2. 14, Threadcutters Way, Shepshed, Leicestershire, UK.

Transabdominal renal ultrasonography is the most widely utilised imaging modality for the investigation and diagnosis of renal disease but to the authors' knowledge no reference ranges or reports of repeatability and reproducibility have previously been published in horses. The aims of this study were to report intercostal space (ICS) visibility of Thoroughbred kidneys, normal references ranges for various kidney dimensions in longitudinal (Long) and transverse (Trans) section which include total kidney, cortex, medulla and pelvic length and width dimensions and to calculate repeatability and reproducibility of these dimensions using transabdominal ultrasonography. Seven Thoroughbred horses (500±30kg) were scanned in triplicate by two scanning technicians over three consecutive days. Images were analysed on three separate occasions by one observer (to assess repeatability) and once by a second observer (to assess reproducibility). Reproducibility was also assessed using a questionnaire sent to 10 clinicians; this consisted of 10 ultrasound images (5 normal and 5 abnormal kidneys) from clinical cases (diagnosis was confirmed by post-mortem examination). Statistical analysis included ANOVA, repeated measures ANOVA and *post-hoc* Bonferoni test and paired T-test. Both kidneys were visible in the paralumbar fossa (PLF) and $17^{th}$, $16^{th}$ and $15^{th}$ ICS, with the right kidney rarely visible in 14th ICS. Mean values (±SD) were obtained for all renal dimensions and renal length and width are displayed in table 1. There were significant differences between renal dimensions in different ICS on both left and right sides ($p=0.03$) except between the 17th and 16th ICS on both the left and right sides. Good repeatability and reproducibility was demonstrated from the horses examined with an intra-class correlation coefficient (ICC) of >0.9 for all renal dimensions. Reproducibility of normal and abnormal renal dimensions from the questionnaire was also deemed to be excellent (ICC>0.93) as was identification of normal versus abnormal. In conclusion a normal reference range for various renal dimensions in Thoroughbred horses has been established. Excellent repeatability and reproducibility of renal measurements and identification of abnormal versus normal mean that this is a valuable clinical modality for examination of this organ.

| | | PLF | | 17th ICS | | 16th ICS | | 15th ICS | |
|---|---|---|---|---|---|---|---|---|---|
| | | Long | Trans | Long | Trans | Long | Trans | Long | Trans |
| L | Length (cm) | 12.3(1.2) | 13.1(2.1) | 10.9(2.9) | 11.4(3.1) | 13.9(1.3) | 12.5(1.4) | 12.7(2.2) | 12.1(1.6) |
| | Width (cm) | 7.7(1.1) | 8.8(1.4) | 7.6(2.2) | 8.0(1.6) | 6.6(0.7) | 7.5(1.1) | 7.2(1.2) | 7.7(2.1) |
| R | Length (cm) | 14.7(0.8) | | 13.7(2.1) | 11.6(1.0) | 14.8(1.8) | 10.3(0.6) | 16.0(0.4) | 11.8(1.2) |
| | Width (cm) | 7.5(0.6) | | 7.5(0.8) | 9.6(0.9) | 7.4(0.7) | 4.3(0.3) | 7.9(0.5) | 8.5(0.9) |

Table 1. Dimensions for the left (L) and right (R) kidneys displayed as mean (±SD).

**ABSTRACT #60**

RENAL BIOPSIES IN HORSES: A RETROSPECTIVE STUDY OF 71 CASES.
GA Tyner[1], S Lindborg[1], A Gruntman[2], L Javsicas[3], C Sommardahl[4], LC Fennell[5], N Menzies-Gow[6], L Coutëil[7], T Norman[8], L Hardefeldt[9], RD Nolen-Walston[1] . 1. New Bolton Center, University of Pennsylvania, Kennett Square, PA. 2. Tufts University, Grafton, MA. 3. University of Florida, Gainesville, FL. 4. University of Tennessee, Knoxville, TN. 5.University of Melbourne, Melbourne, Aust. 6. Royal Veterinary College, London, UK. 7. Purdue University, West Lafayette, IN. 7. Texas A&M University, College Station, TX. 9. University of Wisconsin, Madison, WI.

Renal biopsies are infrequently performed in horses due to concerns regarding complication rate and poor diagnostic yield. This multi-center retrospective analysis was performed to determine the utility and safety of renal biopsies in equids. Nine institutions identified a total of 71 horses that had undergone renal biopsy. The median age was 8 years (range 11 days-26 years), consisting of 25 mares, 33 geldings, and 13 stallions. Presenting complaints included chronic anorexia (28% of cases), azotemia (27%), weight loss (27%), colic (15%), hematuria (15%), and polyuria/polydipsia (15%). Azotemia was present in 69% of cases (median creatinine 3.0mg/dl, range 0.4-29.0), 44% were anemic, and 30% were hypoproteinemic. Electrolyte abnormalities included hypochloremia (59%), hypocalcemia (42%), hyponatremia (32%) and hypercalcemia (19%). Urinalysis results revealed isosthenuria in 35%, and an increased fractional excretion of sodium in 26%. Diagnostic ultrasonography was performed in 94% of cases prior to biopsy. The principal sonographic findings were increased echogenicity (23%), renomegaly (15%), loss of corticomedullary distinction (15%), renal atrophy (8%), and renal mass (9%). The biopsy was taken from the right kidney in 77% of cases, the left in 15%, and both kidneys in 7%. A median of 2 samples (range 1-4) were taken per kidney, and one specimen was obtained laparoscopically.

A histologic diagnosis was obtained for 79% of samples, 12% revealed no lesion, and 9% of samples were inadequate for diagnosis. Complications occurred in 11% (8/71) cases; most were mild and self-limiting, and included colic (2/8), hematuria or hematoma not requiring transfusion (4/8), pain (1/8), and abortion (1/8). No cases received a blood transfusion. One case died as a result of the biopsy procedure (1/71), resulting in a mortality rate of 1.4%. This outcome occurred in an adult gelding that underwent biopsy of a left kidney mass and died of exsanguination associated with arterial disruption. The majority of cases were discharged alive (77%), although 10% were euthanized at home within one month. A total of 22% of cases were euthanized in hospital; 13% due to poor prognosis, 3% from worsening of clinical signs, 3% for various co-morbidities, and 2% due to financial constraints. In conclusion, renal biopsies in the horses were found to be associated with a low rate of serious complications, and provided a definitive histopathologic diagnosis in most cases. Further investigation is warranted to establish if data generated by renal biopsy significantly alters therapeutic plan or ability to provide accurate prognostic information.

**ABSTRACT #61**

OWNER PERCEPTIONS AND PRIORITIES REGARDING QUALITY AND QUANTITY OF LIFE IN CATS WITH HEART DISEASE.
CA Reynolds[1], MA Oyama[1], D Brown[1], J Rush[2], E Rozanski[2], P Fox[3], D Adin[4], R Williams[5], K MacDonald[6], R Malakoff[7], KE Schober[8], ML O'Sullivan[9], GE Singletary[1]. 1. University of Pennsylvania, Philadelphia, PA. 2. Tufts University, North Grafton, MA. 3. The Animal Medical Center, New York, NY. 4. Veterinary Specialists of Rochester, Rochester, NY. 5. Veterinary Emergency Clinic, Toronto, ON. 6. The Animal Care Center, Rohnert Park, CA. 7. Angell Animal Medical Center, Boston, MA. 8. The Ohio State University, Columbus, OH. 9. Ontario Veterinary College, University of Guelph, Guelph, ONT.

Owners' perceptions and priorities regarding quality of life (QoL) are important considerations given the unknown efficacy of many commonly administered medications, stress of hospital visits, difficulties providing home care, and personal choices involving euthanasia. We sought to better understand these issues through questionnaire-based study of owners of 182 cats with a variety of heart diseases from 8 sites. Appetite, owner interaction, sleep patterns, and litterbox habits were deemed significantly more important to QoL than pet-pet interaction, play habits, and desire to go outdoors. Concern over pet suffering was significantly greater than life expectancy. Using an 11-point scale ranging from 0 (low QoL, long lifespan) to 10 (high QoL, short lifespan), owners desired a median balance between these priorities of 8. 94% of owners were willing to trade survival time for comfortable QoL; 54% of these were willing to trade up to 6mo. On univariate analysis, factors associated with willingness to trade 6mo included owner age, study site, clinician, ISACHC class, stress of hospital visit, number of cats in the household, and current medications. On multivariate analysis, factors associated with increased willingness to trade were study site, cats not receiving medications, and older age of owners. Owner concern whether stress of administering medications at home would decrease QoL increased with number and frequency of medications. These results indicated that QoL is more important to owners of cats with heart disease than longevity. The various priorities and concerns of cat owners should be taken into account in order to provide optimal care.

**ABSTRACT #62**

VALIDATION OF PLATELET FACTOR 4, SOLUBLE P-SELECTIN, β-THROMBOGLOBULIN, AND 11-DEHYDRO-THROMBOXANE B2 ELISAS IN THE FELINE POPULATION. SB Bordelon, AM Heaney, OL Nelson, TB Wills, KM Meurs. Washington State University – College of Veterinary Medicine, Pullman, WA.

Arterial thromboembolism (ATE) is a common complication of feline cardiac disease and identification of cats at risk for ATE is important. In human patients traditional markers of coagulation with one or more of the following; platelet factor 4 (PF4), soluble P-selectin (sP-selectin), beta-thromboglobulin (β-TG), or 11-dehydro-thromboxane B2 (11-dehydro-TXB2) have been used to identify patients at risk for thromboembolism.

We hypothesized that human ELISAs for PF4, sP-selectin, β-TG and 11-dehydro-TXB2 could be used to establish normal values for these markers in cats.

The objective of this study was to validate human ELISAs for PF4, sP-selectin, β-TG and 11-dehydro-TXB2 in the feline population.

Twenty-five cats were determined to be free of systemic and cardiac disease via physical examination, echocardiography, electrocardiography, blood pressure, complete blood counts, biochemical profiles, and total T4 levels. Three milliliters of blood was collected from each cat and placed into CTAD tubes. Samples were centrifuged to obtain citrated (sP-selectin) and platelet poor plasma (PF4 and β-TG). Urine samples for 11-dehydro-TXB2 were collected into C&S tubes. Intra-and inter-assay variability was determined for each ELISA.

The β-TG ELISA did not detect β-TG in any of the cats. Intra-assay variability for PF4, sP-selectin and 11-dehyro-TXB2 was 1-33%, 3- > 1000%, and 14-18%, respectively. Inter-assay variability for PF4, sP-selectin and 11-dehyro-TXB2 was 6-110%, 3-89%, and 10-44% respectively.

We conclude that these ELISAs for PF4, sP-selectin, β-TG, and 11-dehydro-TXB2 are not useful in cats. However, the devastating nature of feline ATE warrants further investigation into markers that may identify cats at risk for thromboembolism.

**ABSTRACT #63**

PLASMA PECAM-1 AND ENDOTHELIN-1 CONCENTRATIONS ARE ELEVATED IN CATS WITH HYPERTROPHIC CARDIOMYOPATHY. KE Jandrey, JW Norris, K Huang, MD Kittleson, F Tablin; University of California, School of Veterinary Medicine, Davis, CA.

The purpose of this study was to determine the plasma concentrations of platelet/endothelial cell adhesion molecule-1 (PECAM-1) and endothelin-1 (ET-1) in normal cats and cats with hypertrophic cardiomyopathy (HCM). Our hypothesis was that cats with HCM have increased concentrations of both variables when compared with normal cats.

Platelet rich plasma and platelet poor plasma (10ug/ml PGE1, 1% aprotinin, and 2mg/ml leupeptin) from normal cats and cats from a research colony with familial HCM were prepared from anticoagulated whole blood. All cats were evaluated by echocardiography.

ET-1 concentration was determined in triplicate using a commercially available ET-1 ELISA assay previously validated for use in cats. Recombinant human ET-1 was used to generate a standard curve for assay calibration. Plasma PECAM-1 concentration was measured by Western blot analysis using a polyclonal antibody and densitometry (arbitrary units) was quantified by Image J analysis (NIH Shareware). Paired t-tests were performed to compare groups (Sigma Stat, Systat, San Jose CA) and statistical significance was set at $p<0.05$.

The mean normal cat plasma PECAM-1 concentration ($n=5$) was 1276 + 684, and was 2939.9 + 768.9 ($p=0.015$) in the group of cats with HCM ($n=7$). The mean normal cat plasma ET-1 concentration ($n=9$) was 3.56 + 0.98 fmol/ml and was 6.55 + 1.25 fmol/ml in the cats with HCM ($n=4$; $p<0.001$).

This increase in PECAM-1 and ET-1 likely indicate that cats with HCM have platelet and endothelial up-regulation. An increase in the concentration of these plasma proteins may be an indication of increased risk of thromboembolism.

**ABSTRACT #64**

ASSOCIATION OF HYPERTROPHIC CARDIO-MYOPATHY PHENOTYPE AND GENOTYPE IN ITALIAN MAINE COON CATS. F Porciello[1], P Ferrari[2], F Birettoni[1], M Rishniw[3], G Pertica[4], M Polli[4], M Longeri[4]. 1. University of Perugia, Italy. 2. Observatory for Feline HCM, Italy. 3. Cornell University, Ithaca, NY. 4. University of Milano, Italy.

The Italian Observatory for Feline Hypertrophic Cardiomyopathy (HCM), established March 2008, aims to monitor the natural history of HCM in Maine Coons via an integrated echocardiographic and genetic database and provide advice to breeder groups. We examined data from 75 Maine Coon cats (3 months to 6 years) enrolled during the first 7 months.

Standard echocardiographic methods were used to define the disease status in all cats: HCM echo-positive or HCM echo-negative (wall thickness ≥ or < 6mm respectively). Any echo-positive cats with concurrent hypertension, nephropathy or hyperthyroidism, were excluded from analysis. All cats were screened for MYBPC3 mutation G91C by sequencing both strands.

Sixty-three cats were echo-negative; 12 were echo-positive for HCM. Genotype frequencies were 64% N/N (Negative/Negative), 28% P/N (Positive/Negative) and 8% P/P (Positive/Positive). Average discrepancy between genetic and echocardiographic results was 35% with most discrepancy (86%) associated with the P/N genotype. In cats >3 years, 50% of echo-negative cats had a P/N genotype and no echo-negative cats had a P/P genotype. However, the incidence of echo-positive cats with N/N genotype increased from 0% to 50% with increasing age.

Our study demonstrates substantial discordance between genotype and phenotype in heterozygous Italian Maine Coons <3-4 years of age. Construction of age-specific detection curves for P/N and P/P genotypes is necessary to calculate the probability that asymptomatic subjects will develop the disease. The presence of some echo-positive cats with an N/N genotype indicates that factors other than the MYBPC3 G91C mutation are likely involved in HCM in Maine Coons.

**ABSTRACT #65**

PULSED TISSUE DOPPLER IDENTIFIES PRECLINICAL LEFT VENTRICULAR MYOCARDIAL DYSFUNCTION IN OBESE DOGS. DS Schwartz, VMC de Oliveira, PRR Melo, AM Mazini, P Claus; GT Goldfeder; AC Ferreira, MHMA Larsson. Faculdade de Medicina Veterinária e Zootecnia, USP, São Paulo, SP, Brazil.

Human overweight subjects without overt heart disease show changes on left ventricular (LV) function, detected by tissue Doppler imaging (TDI), even in the absence of coronary heart disease or other comorbidities, but this effect on dogs is not well established.

We sought to investigate preclinical effects of obesity on left ventricular function of dogs, using pulsed TDI.

LV myocardial Doppler derived systolic (Sm), early (Em) and late (Am) diastolic velocities were obtained by transthoracic echocardiography from 16 obese dogs and 30 non-obese, classified by body condition scoring. Other comorbidites were ruled out. Student´s t test (mean±SD) and Mann-Whitney test (median, $P_{25}$ and $P_{75}$) were applied (Sigma Stat 3.5), significance was considered if $P < 0.05$.

Obese presented significantly lower Em at anular interventricular septum (IVS) (0.0719±0.0229 vs 0.0915±0.0247; P=0.012); Em/Am at IVS (1.114±0.35 vs 1.384±0.370, P = 0.020); Em at anular left ventricular free wall (LVFW) (0.103 ± 0.0256 vs 1.20±0.0267, P= 0.048); Em/Am at anular LVFW (1.320±0.303 vs 1.576±0.336, P=0.115) and increased Am at mid LVFW (median=0.085, $P_{25}$=0.07, $P_{75}$=0.0917 vs median=0.0685, $P_{25}$=0.06, $P_{75}$ =0.077, P=0.029), a decreased Sm at anular LVFW (0.126±0.0378 vs 0.154±0.0473, P=0.049), and an increased E/Em (10.398±2.915 vs 8.648±2.267, P=0.029), suggesting an increased LV filling pressure. Obese and controls were comparable regarding other longitudinal and radial velocities.

Overweight dogs without overt heart disease present preclinical left myocardial systolic and diastolic function changes detected by pulsed TDI. Even without a predisposition of dogs to ischemic heart disease, obesity has an impact on ventricular function.

**ABSTRACT #66**

SEROTONIN TRANSPORTER EXPRESSION IS LOCALLY DOWN-REGULATED IN CANINE DEGENERATIVE MITRAL VALVES. SM Scruggs, S Disatian, EC Orton. Department of Clinical Sciences, Colorado State University, Fort Collins, CO.

Tryptophan hydroxylase 1 (THP1) is the rate limiting enzyme necessary for peripheral serotonin synthesis and the serotonin transporter (SERT) is a key component of serotonin metabolism. SERT transports serotonin across the cell membrane for intracellular metabolism and competes with serotonin receptors for interaction with serotonin. We have recently reported an autocrine serotonin signaling mechanism in canine degenerative mitral valves based on up-regulation of THP1 and down-regulation of SERT in late-stage disease. It is unknown whether SERT down-regulation occurs early or late in degenerative mitral valve disease (DMVD) or whether its down-regulation is a local or systemic phenomenon. SERT expression was determined in canine normal (n=4), early-(n=4), and late-stage (n=4) degenerative mitral valves by immunohistochemical (IHC) and immunoblot (IB) analyses. SERT expression was also evaluated in circulating platelets from normal dogs and dogs with DMVD. SERT was expressed in all layers of normal and early-stage mitral valves based on IHC. SERT was markedly down-regulated in valve interstitial cells, but not valve endothelial cells, of late-stage mitral valves. SERT expression was significantly decreased ($p<0.05$) in late-stage, but not early-stage mitral valves on IB. Up-regulation of TPH1 in late-stage degenerative mitral valves was confirmed. No difference in SERT expression was apparent in circulating platelets from normal and DMVD dogs. We conclude that SERT down-regulation is a local and late-stage phenomenon in canine DMVD. SERT down-regulation likely enhances autocrine serotonin signaling in late-stage DMVD by decreasing local serotonin metabolism and increasing interaction of serotonin with its receptor.

**ABSTRACT #67**

B-TYPE NATRIURETIC PEPTIDE AND CYCLIC GUANOSINE MONOPHOSPHATE LEVELS INCREASE WHEREAS NITRIC OXIDE BIOAVAILABILITY DECREASES WITH INCREASING DEGREE OF MYXOMATOUS MITRAL VALVE DISEASE. SG Moesgaard[1], T Falk[1], T Teerlink[2], HH Guðmundsdóttir[1], S Sigurðardóttir[1], CE Rasmussen[1], LH Olsen[1]. 1. University of Copenhagen, Frederiksberg, Denmark, 2. VU University Medical Center, Amsterdam, The Netherlands.

Myxomatous mitral valve disease (MMVD) is a widespread heart valve disease in dogs, and a lot of research has focused on identifying biomarkers in asymptomatic stages which could be beneficial in the future diagnosis, prognosis and treatment of MMVD. Among several potential biomarkers the natriuretic peptides have so far been the only ones showing significant changes in several studies of both asymptomatic and symptomatic dogs. The aim of this study was to measure plasma cyclic guanosine monophosphate (cGMP) concentration in dogs with different degrees of MMVD. cGMP is a second messenger for both natriuretic peptides and nitric oxide (NO). We therefore studied the association between cGMP and both N-terminal pro-B-type natriuretic peptide (BNP) and the L-arginine/asymmetric dimethylarginine (ADMA) ratio (a measure of NO bioavailability).

Five groups consisting of 1) Cavalier King Charles Spaniels (CKCS) with no or minimal mitral regurgitation (MR), 2) CKCS with mild MR 3) CKCS with moderate to severe MR,

4) dogs in heart failure due to MMVD (different small breeds) and 5) Cairn Terriers with no or minimal MR were subjected to a clinical examination, blood sampling and echocardiography.

Plasma BNP increased with increasing MR and it was significantly elevated in the CKCS with moderate to severe asymptomatic MR ($P<0.05$) and further elevated in the heart failure dogs ($P<0.0001$). Plasma cGMP concentration also increased with increasing MR and was highly correlated with BNP ($P<0.0001$). However, cGMP was only significantly increased in the heart failure dogs ($P<0.01$). The cGMP/BNP ratio decreased with increasing MR and the decrease was significant in the heart failure dogs ($P<0.05$). The L-arginine/ADMA ratio decreased with increasing MR which was also significant in the heart failure dogs ($P<0.005$). Furthermore, both cGMP and L-arginine/ADMA ratio levels differed between the CKCS and the Cairn terrier breeds ($P<0.05$).

These results confirm that BNP increases already in the moderate to severe asymptomatic stages of MMVD. The increase in BNP is closely correlated to an increase in cGMP, however, the cGMP/BNP ratio decreases with heart failure, which indicates that the dogs may have developed natriuretic peptide resistance. Finally, the decrease in the L-arginine/ADMA ratio suggests that there is a decrease in NO bioavailability in the heart failure dogs which could then again have a negative influence on the cGMP concentration. These results are interesting considering the promising treatment results with Pimobendan, which is a phosphodiesterase inhibitor that may restore cGMP and improve endothelial function.

**ABSTRACT #68**

TRANS-THORACIC THREE-DIMENSIONAL ECHOCARDIOGRAPHIC EVALUATION OF NORMAL CANINE HEART VALVES. A Hsu, WP Thomas. William R. Pritchard Veterinary Medical Teaching Hospital, University of California, Davis, Davis, CA.

With its unique ability to display and spatially orientate the heart's structures in three dimensions, we sought to determine if three-dimensional echocardiographic (3DE) views of the four canine heart valves (mitral = MV, tricuspid = TV, aortic = AV, pulmonic = PV) could be obtained consistently, and to determine the quality of the 3DE images and whether they provided views and information not obtainable by the standard two-dimensional echocardiographic (2DE) examination.

Both 2DE and 3DE examinations (Philips iE33; 2D: 5MHz or 8MHz transducer; 3D: X3 or X7 transducer) were performed in 12 normal dogs which were restrained in lateral recumbency on a table designed to allow ventral transducer placement. A continuous electrocardiogram (ECG) was obtained during all examinations. Standard 2DE views were used to assess the four valves, including a right parasternal long-axis view (MV, TV, AV), a right parasternal short axis view at the level of the mitral valve (MV), a right parasternal short axis view at the level of the heart base (AV, PV), a left apical four chamber and/or five chamber view (MV, TV, AV), and left cranial long axis and short axis views (TV, AV, PV). 3DE datasets included full volume 4 beat acquisitions as well as live 3DE images of the valves. 3DE images of the valves were obtained and then evaluated and manipulated on a separate dedicated workstation. Valves were evaluated in each view and quality was graded subjectively using the following scale: 1=poor; 2=fair; 3=good; 4=excellent. Images of the mitral valve were most consistently obtained (12/12), with a quality grade of 3 or 4 in 10/12 dogs. The tricuspid valve was imaged in most dogs (10/12), with 3 dogs receiving a quality grade > 2. The aortic valve was imaged in every dog (12/12), with 5 dogs receiving a quality grade > 2. The pulmonic valve was the most difficult to view, with only 3/12 dogs with a valve that could be evaluated, all of which received a grade of 2.

These preliminary results indicate that trans-thoracic 3DE images of the canine heart valves can be obtained and evaluated within the context of the routine cardiology evaluation and provide unique views complimentary to the 2DE examination. Good to excellent images of the MV were consistently recorded. High quality images of TV, AV, and PV were more difficult to obtain.

**ABSTRACT #69**

CARDIAC TROPONIN-I IS ASSOCIATED WITH SEVERITY OF MYXOMATOUS MITRAL VALVE DISEASE, AGE, HEART RATE AND C-REACTIVE PROTEIN IN DOGS. I. Ljungvall[1], K. Höglund[1], A. Tidholm[2], L. H. Olsen[3], M. Borgarelli[4], P. Venge[5], J. Häggström[1]. [1]Swedish University of Agricultural Sciences, Uppsala, Sweden, [2]Albano Animal Hospital, Stockholm, Sweden, [3]University of Copenhagen, Denmark, [4]Kansas State University, USA, [5]University of Uppsala, Sweden.

Analysis of cardiac troponin I (cTnI) may increase the knowledge of cardiac remodeling in dogs with myxomatous mitral valve disease (MMVD). An age- and gender dependent variation in cTnI has been shown in rats, dogs and humans, and cTnI is known to be related to systemic inflammatory response. The role of inflammatory mediators in the progression of cardiac remodeling in MMVD is unknown. The aims of the study were to investigate whether plasma levels of cTnI and C-reactive protein (CRP) were associated with severity of MMVD, and to investigate potential associations between cTnI and signalment variables in dogs with MMVD.

Prospective evaluation by physical examination, electrocardiography, blood pressure measurement, and echocardiography was performed in 81 dogs. Plasma cTnI was analysed by a high sensitivity assay (Beckman Coulter AccuTnI) with a lower limit of detection of 0.001 ng/ml (10% CV at 0.014ng/ml), and plasma CRP was analysed by a canine CRP ELISA (TriDelta Development Ltd).

Healthy dogs (n=11) had a median cTnI of <0.001 ng/ml (interquartile range (IQR) <0.001-0.004), dogs with mild MMVD (n=39) had a median cTnI of 0.003 ng/ml (IQR 0.001-0.024), dogs with moderate MMVD (n=13) had a median cTnI of 0.014 ng/ml (IQR 0.008-0.029), and dogs with severe MMVD (n=18) had a median cTnI of 0.043 ng/ml(IQR 0.031-0.087). Cardiac troponin-I concentrations differed significantly between healthy dogs and dogs with moderate MMVD ($P<0.001$), between healthy dogs and dogs with severe MMVD ($P<0.001$), between dogs with mild and severe MMVD ($P<0.001$), and between dogs with moderate and severe MMVD ($p<0.002$). Multiple regression analysis indicated a major effect of age, heart rate, CRP concentration, and left ventricular end diastolic diameter corrected for body weight on cTnI concentration. Age was the variable that contributed most to the model $R^2$. C-reactive protein levels did not differ significantly between severity groups, but increased significantly with increasing age.

In conclusion, plasma cTnI concentration was found to increase with increasing severity of MMVD, and cTnI concentration increased with increasing age, heart rate, left ventricular end diastolic diameter, and CRP concentration in dogs. C-reactive protein concentration was not associated with severity of MMVD.

**ABSTRACT #70**

EVALUATION OF A MINIMALLY INVASIVE TREATMENT FOR MITRAL VALVE REGURGITATION.

C Kirker-Head[1], T Piemonte[2], L Svensson[3], J Wilson[4], A Walsh[1], M Stark[1], J Rush[1]. 1. Cummings School of Veterinary Medicine at Tufts University, North Grafton, MA. 2. Lahey Clinic, Burlington, MA. 3. Cleveland Clinic, Cleveland, OH. 4. Cardiosolutions Inc, Stoughton, MA.

Mitral valve regurgitation (MR), due to myxomatous degeneration of the valve leaflets or secondary to dilated cardiomyopathy, is common in older people and older dogs. Medical management can provide some relief, but mitral valve surgery via cardiopulmonary bypass has been required to improve long term outcome. Percutaneous (transcatheter) beating heart mitral valve repair offers an alternative to open heart surgery. This study was designed to test the feasibility of a novel intracardiac implant system designed to ameliorate MR by enhancing functional valve coaptation (valve-valve or valve-device) during systole.

Following IACUC approval, mature healthy female sheep (65-75 kg; n=6) and pigs (50-65 kg; n=17) were anesthetized to permit fluoroscopic and intracardiac echocardiographic (ICE) guidance of a transcatheter (femoral vein) delivery system. Transseptal atrial puncture was employed to gain access to the left atrium and left ventricle. A steerable 14-French delivery catheter was used to deploy the implant, which consisted of a cylindrical (9 x 40-mm) collapsible Nitinol wire scaffold enveloped within a silicone polyurethane sheath. The implant was suspended within the mitral annulus by a single Nitinol support wire anchored to the ventricular apical wall using either a screw-in or an umbrella design. Procedural success was recorded and animal health, cardiac function and implant stability were assessed post-implantation using physical exam, echocardiography, fluoroscopy, and necropsy (range 23-328 days post-implantation).

System design modifications and increased operator experience culminated in successful, uncomplicated, and repeatable implantations of the device. When compared with sheep, the pigs' cardiac anatomy and physiology (eg, fewer moderator bands, easier transseptal approach, vigorous recovery) facilitated the implantation procedure. Procedure duration declined from 240 minutes to 45 minutes. Intraoperative and postoperative fluoroscopy and echocardiography documented optimal positioning of the implant within the mitral annulus and coaptation with the mitral leaflets with minimal mitral regurgitation. Seven animals succumbed to device perforation of the myocardium, excessive anticoagulation or cardiac failure. Necropsy confirmed accurate device placement in most cases, but post-implantation anchor detachment occurred in 2 animals and metal fatigue resulted in fracture of some umbrella-style anchor tines in 3 animals. A mild focal inflammatory response was noted where the support struts contacted the left ventricular wall and mild myxomatous changes were occasionally noted where the valve leaflets contacted the implant.

This novel implant, documented to achieve functional valve coaptation (valve-valve or valve-device) in systole with minimal regurgitation and cardiac pathology offers a potential percutaneous treatment for human and canine MR patients. Efficacy studies are warranted.

**ABSTRACT #71**

CAVAL SYNDROME; OUTCOME IN 42 DOGS.

SG Gordon, C Bové, AB Saunders, MW. Miller, RM Roland, SE Achen, LT Drourr and MM Boggess. The Michael E. DeBakey Institute, Colleges of Veterinary Medicine and Biomedical Sciences and Science, Texas A&M University, College Station, TX.

The objective of this retrospective study is to report the clinical outcome of canine caval syndrome (CS). Medical records of 42 client owned dogs diagnosed with CS at Texas A&M University were reviewed. Information including signalment, history, physical examination, clinicopathologic testing, diagnostic imaging, treatment and outcome were obtained.

Twenty-one of 42 dogs had transvenous heartworm extraction (TVHWE) attempted. Two of the 21 dogs died during the procedure, 4 died post-operatively, and following anesthesia induction in one dog, worms migrated into the distal pulmonary artery and therefore extraction was not attempted. Fourteen dogs underwent successful TVHWE and were discharged from the hospital. Mean follow up time in these 14 dogs was 24.4 ± 17.7 months with a range of 2-56 months. To date 10 (71.4%) of these dogs have survived at least 18 months, and 7 (50.0%) greater than 24 months. Of the remaining 4 dogs, 3 were lost to follow-up and one was euthanized due to intervertebral disc disease. A parametric proportional hazards model, with a Weibull distribution, was run to see if any of the pre-operative clinicopathologic parameters or diagnostic imaging results had an effect on survival time. Dogs with elevated alanine transferase (ALT) or worms documented in the pulmonary artery prior to TVHWE had significantly ($p<0.05$) shorter overall survival.

These results suggest that dogs with CS who undergo TVHWE and survive to discharge have a good mid-term prognosis. Dogs with elevations in ALT or pre-operative evidence of worms in the pulmonary artery may have increased mortality rates.

**ABSTRACT #72**

MODULATION OF INFLAMMATORY AND OXIDATIVE STRESS MARKERS IN CORONARY ARTERY SMOOTH MUSCLE CELL MODEL.

BA Maran[1]; GE Snow [2,3]; JM Robinson[2,5]; RY Au[2]; AY Au,[2,4]; CG Frondoza[1,2,5]. 1. Mississippi State University, Mississippi State, MS. 2.

Nutramax Laboratories, Inc., Edgewood, MD. 3. Dartmouth College, Hanover, NH. 4. Syracuse University, Syracuse, NY. 5. Johns Hopkins University, Baltimore, MD.

Dilated cardiomyopathy (DCM) is the leading cause of heart failure in certain breeds of dogs. DCM is associated with inflammation and oxidative stress, but the pathogenesis of the disease is not well understood. Mediators implicated in the pathogenesis of DCM include cytokines, prostaglandins, nitric oxide, and reactive oxygen species. To help clarify the molecular events involved, we evaluated coronary artery smooth muscle cells (CASMCs) cultured *in vitro*. Using this model, we determined whether cytokine-induced inflammation and oxidative stress can be inhibited by silybin phosphatidylcholine complex (SPC) and epigallocatechin gallate (EGCG). These natural products are known to have anti-inflammatory and anti-oxidant properties. SPC has been shown to be bioavailable in dogs and cats and is commercially available (Marin® and Denamarin®). Induction of inflammation was assessed by gene expression analysis of cytokines and COX-2, as well as production of prostaglandin E2 ($PGE_2$). Induction of oxidative stress was monitored by analysis of superoxide dismutase (SOD) gene expression.

Phenotype expression was monitored by immunostaining for α-actin. CASMCs were pre-incubated with clinically relevant concentrations of SPC (298 ng/ml), EGCG (0.4 μg/ml) or control media alone for 24 hours. Cells were then incubated for 24 hours with the combination of IL-1β (10ng/ml) and TNF- α (1ng/ml) or control media alone. Expression of cytokines (IL-1β and TNF- α), COX-2, and SOD was determined by semi-quantitative RT-PCR while $PGE_2$ production was measured by ELISA. Statistical significance was set at $p<0.05$ using one-way ANOVA and the Tukey post-hoc test.

CASMCs proliferated in monolayer culture and maintained their phenotype, as indicated by intense immunostaining for α-actin. Cells responded to cytokine activation by more than a 40% increase in expression of IL-1β, TNF- α, COX-2, and SOD. Activation significantly increased production of $PGE_2$. Pretreatment of CASMCs with SPC or EGCG reduced gene expression of cytokines, COX-2, and SOD. Pre-treatment also reduced $PGE_2$ production to near baseline levels. Results from this study demonstrate that SPC and EGCG inhibit the induction of inflammation and oxidative stress markers *in vitro*. The CASMC model serves as a tool to identify agents that may be beneficial in reducing inflammation and associated oxidative stress in veterinary practice.

**ABSTRACT #73**

HIGH DEFINITION OSCILLOMETRY: A NOVEL, NON-INVASIVE APPROACH FOR PRECISE BLOOD PRESSURE DETERMINATION AND CARDIOVASCULAR ASSESSMENT IN OLD WORLD AND NEW WORLD MONKEYS. Barthel Schmelting[1], Marc Niehoff[1], Beate Egner[2], Gerhard F Weinbauer[1]. [1]Covance Laboratories, Muenster, Germany, [2] Clinical Center for Small Animals, Hoerstein, Germany.

Direct invasive techniques are regarded as gold standard for blood pressure determination. However, these are not feasible for day-by-day routines and large animal numbers as frequently demanded during preclinical safety assessment. Non-invasive cuff methods are used to overcome these limitations with variable success. Small body size and high heart rates (> 300 bpm), e.g. in marmosets (*Callithrix jacchus)*, often obviate further cardiovascular investigation of conscious animals. High Definition Oscillometry (HDO) was evaluated as a potential non-invasive approach for blood pressure recordings in cynomolgus monkeys (*Macaca fascicularis*) and marmoset monkeys. In conscious animals, systolic, diastolic, mean arterial blood pressure (MAP) and pulse/min were determined 15 times within approx. 9 min per individual. This session was performed during three consecutive days. After procedural habituation the MAP (in mmHg, mean ± SD) in *C. jacchus* was 94.9 ± 4.3, in *M. fascicularis* males 96.2 ± 3.4 and 86.9 ± 3.0 in females. Importantly, these data correspond to those reported for invasive techniques. Mean intraindividual coefficient of variation for systolic, diastolic, MAP and bpm ranged between 2.6 and 10.1 %, demonstrating the high precision and feasibility even in conscious animals. Direct comparison to telemetry revealed that HDO accurately detected drug related cardiovascular changes. Under anaesthesia, a significant decrease in pulse rate and blood pressure could be determined. Due to its accuracy and precision, HDO may be regarded as an alternative approach to invasive surgeries in animal models used for cardiovascular research. HDO is applicable to large number of animals and day-by-day routines.

**ABSTRACT #74**

PREVALENCE, HERITABILITY AND GENETIC CORRELATIONS OF CONGENITAL SENSORINEURAL DEAFNESS AND PIGMENTATION PHENOTYPES IN 4143 BORDER COLLIES. Luisa De Risio, Julia Freeman, Sarah Blott, Alberta de Stefani, Lara Matiasek, Tom Lewis. The Animal Health Trust, Newmarket, UK.

The objectives of the present study were to estimate prevalence, heritability and genetic correlations of congenital sensorineural deafness (CSD) and pigmentation phenotypes in Border Collies.

Full litters of purebred Border Collies that presented to the Animal Health Trust (1994–2008) for assessment of their hearing status by brain stem auditory evoked response (BAER) at 4-10 weeks of age were included. Heritability and genetic correlations were estimated using restricted estimate maximum likelihood.

4143 puppies (1951 female, 2192 male) met the inclusion criteria. Of these, 4043 (97.59%) had normal hearing status, 84 (2.03%) were unilaterally deaf (UD) and 16 (0.39%) were bilaterally deaf (BD). Iris color was brown bilaterally in 3890 puppies (1.57% UD, 0.18% BD), blue unilaterally in 158 puppies (8.86% UD, 1.27% BD), and blue bilaterally in 95 puppies (9.47% UD, 7.37% BD). White pigmentation of the head was estimated visually at > 50% in 111 dogs (19.82% UD, 7.21% BD) and < 50% in 4032 dogs (1.54% UD, 0.20% BD). 366 puppies (5.19% UD, 2.19% BD) had merle coat color. The sire and dam hearing status was known (based on BAER results) in 3491 and 3801 dogs respectively (3482 and 3776 normal, 5 and 21 UD, 4 and 4 BD respectively). Heritability of deafness (normal/unilateral/bilateral) was estimated as 0.37. The genetic correlations of deafness with iris color, percentage of white pigmentation of the head, and merle coat color were 0.63, 0.12, and 0.24, respectively.

These results indicate that there is significant genetic effect on CSD in border collies, and that some of the genetic regions determining deafness also influence other pigmentation phenotypes.

**ABSTRACT #75**

INVESTIGATION OF A NEW ELECTROENCEPHALOGRAPHY (EEG) ELECTRODE IN SEDATED AND AWAKE DOGS. FMK James, C Kerr, A Bersenas, J Parent, L Grovum, D Allen, R Poma; Ontario Veterinary College, University of Guelph, Guelph, ON.

The new subdermal wire electrode (SWE) was compared with the traditional gold cup electrode (GCE) and needle electrode (NE), in six sedated and awake dogs.

Eight EEG channels were recorded during twenty-minute video-EEG recording sessions (at 0.5, 2, 4, 6 and 8 hours) in this prospective, randomized, and blinded trial. Non-physiological artifacts were identified. Duration of non-physiological artifact was summed per channel. The number of unaffected channels (NUC) was also noted.

After multivariate analysis of NUC, there were main effects of electrode type and sedation over time, with median SWE (NUC = 2.8, 95% confidence interval: 0.84-5.7) significantly different than GCE (7.87, 7.44-7.94) and NE (7.6, 6.61-7.89) medians ($p < 0.05$). After 4 hours, and regardless of the electrode type, the NUC decreased in awake dogs, but not in sedated dogs. The duration of non-physiological artifact was significantly different between SWE and the other two electrodes, the medians for SWE, GCE, and NE at the 8-hour recording sessions being 61.55 (95% confidence interval: 21.81-173.65), 1.33 (0.47-3.75), and 21.01 (6.85-64.42) seconds respectively ($p < 0.05$).

In conclusion, choices of EEG electrode and level of sedation do not affect the overall recording quality in EEG sessions lasting up to 4 hours. With longer EEG recording sessions, caution should be used in selecting EEG electrodes and sedation state, although the difference between electrodes may not be clinically significant.

**ABSTRACT #76**

CEREBROSPINAL FLUID NEUROTRANSMITTER CONCENTRATIONS IN DOGS WITH ISCHEMIC INFARCTION OF THE BRAIN. SR Platt,[1] R Barber,[1] L De Risio,[2] M Kent,[1] J Eagleson,[1] SJ Schatzberg.[1] 1. University of Georgia, College of Veterinary Medicine, Athens, GA USA. 2. Animal Health Trust, Newmarket, UK.

Cerebrospinal fluid (CSF) concentrations of excitatory (glutamate) and inhibitory (GABA) neurotransmitters may reflect central nervous system excitotoxicity. Excitotoxicity is a component of the purported pathophysiology of cerebrovascular accidents, specifically ischemic infarction. Although CSF glutamate levels have been evaluated in dogs with epilepsy, spinal disease and brain tumors, neurotransmitter concentrations have not been documented in dogs with ischemic infarction (stroke) of the brain. Such information could provide therapeutic targets for clinically affected dogs. The objective of this study was to compare CSF neurotransmitter levels in dogs with clinically confirmed ischemic infarction to those of normal dogs.

Cerebellomedullary cisternal CSF was collected from 10 healthy beagles with normal brain MRI and from 12 various dog breeds which had clinical signs and cerebral MRI consistent with ischemic infarction. The medical records of the clinical dogs were searched to determine duration of disease at time of CSF tap and prior use of anti-inflammatory medications. The CSF samples (20μl) were analyzed for glutamate, GABA and glycine concentrations with electrochemical high performance liquid chromatography following derivitisation with o-phtaldialdehyde/2-mercaptoethanol. A student's t-test was used to test for differences between the levels of the 3 neurotransmitters in the following groups: clinical dogs vs. normal; duration of signs <=5 days vs. >5 days; anti-inflammatory medicines administered vs. none given. Student's t-test was implemented in PROC GLM in SAS and significance was set at $P \leq 0.05$.

There were no significant differences between the mean glutamate concentrations of all clinically affected dogs and normal dogs. However, there were significant differences between the mean CSF glutamate levels for dogs clinically affected for $\leq 5$ days (0.28μmo/l) when compared to those affected > 5 days (0.09μmol/l) and to normal dogs (0.1μmol/l) ($P=0.03$). There were also significant differences in the mean glycine concentrations detected between normal dogs (1.64μmol/l) and those which had experienced ischemic infarction (1.09μmol/l) $P=0.0099$. There were no differences detected between the GABA concentrations of any of the groups assessed. There was no effect of the use of anti-inflammatory medications on any of CSF neurotransmitters.

The results of this preliminary retrospective study suggest that glutamate –mediated excitotoxicity may be involved in the pathogenesis of canine cerebral ischemic infarction within the first 5 days of disease. The reduction of glycine concentrations may also play an important role. Glutamate elevation is a potential future treatment target for dogs with cerebral ischemic infarction.

**ABSTRACT #77**

RISK FACTORS FOR POST-MYELOGRAPHIC SEIZURES IN DOGS – 503 CASES. Ronaldo C. da Costa[1], Howard Dobson[2], Aleisha Lusk[2], Joane Parent[3]. [1]The Ohio State University, Columbus, OH, [2] Ontario Veterinary College, University of Guelph, Guelph, ON, Canada, [3]Université de Montréal, St-Hyacinthe, QC, Canada.

The reported incidence of post-myelographic seizures in dogs using iohexol has been highly variable, reaching up to 21.4% on a recent report. Most studies, however, investigated relatively small populations of dogs. Since our experience appeared different, we retrospectively investigate the incidence and risk factors of post-myelographic seizures in a large patient population.

Medical records of dogs that had a myelogram performed between March 2002 and December 2004 were reviewed. Data relating to breed, sex, age, weight, volume and dose per kilogram of iohexol (240mg/ml), site and number of injections, location of the lesion, duration of anesthesia, time from injection to recovery, whether or not surgery immediately followed the myelogram, whether acepromazine was used immediately before or after anesthesia, and presence or absence of seizures were collected. Multiple statistical analyses were performed (univariate exact conditional analysis, Fisher's exact tests, and logistic regression with conditional tests).

The files of 503 dogs were evaluated. Only 15 dogs (2.98%) suffered a post-myelographic seizure. Although several factors influenced on seizure occurrence, due to space constraints only a few will be described here. The size of the dog was an important factor. Thirteen out of 140 large (>20 kg) dogs (9.29%) and two out of 124 medium (>9<20kg) dogs (1.61%) had a seizure. No small (<9kg) dogs (out of 239) had seizures. Large dogs were 35.35 times more likely to have a post-myelographic seizure ($P$ <0.001) than small dogs. The location of contrast injection was also important. Out of the 424 dogs that were injected into the lumbar cistern, 6 suffered a post-myelographic seizure (1.42%), while this occurred in 6 out of 62 (9.68%) dogs injected into the cerebellomedullary (CBM) cistern. Dogs that were injected into the CBM cistern had a significantly ($P$ =0.001) higher risk of seizures and were 7.46 times more likely to develop seizures than those injected into the lumbar cistern. Multiple injections in the lumbar region increased the seizure risk ($P$=0.002), but not in the CBM cistern ($P$=0.36). The total volume of iohexol was highly significant in increasing the seizure risk ($P$ <0.0001). The mean total volume of iohexol for dogs that had a seizure was 11.73 ± 5.52 ml, while for dogs that did not have a seizure it was 4.57 ± 4.13 ml. For each 1 ml increase in iohexol volume, the seizure risk increased by 1.203 (95% CI, 1.11 to 1.32). The location of the spinal lesion was also important. Dogs with cervical lesions had a higher seizure risk ($P$=0.004) and were 4.65 times more likely to develop seizures than dogs with lesions in other spinal regions.

We conclude that the overall seizure risk is low in small and medium breed dogs with lumbar iohexol injections. Large breed dogs with cervical spinal disease and injection of large volumes of iohexol in the CBM cistern have a significantly higher seizure risk.

**ABSTRACT #78**

EFFECTS OF PALLIATIVE PHARMACOTHERAPY ON SURVIVAL AND QUALITY OF LIFE IN 50 DOGS WITH PRIMARY INTRACRANIAL NEOPLASMS. JH Rossmeisl, JL Robertson, JC Jones, KL Zimmerman. VA-MD Regional College of Veterinary Medicine, Blacksburg, VA.

This prospective study intended to evaluate effects of palliative therapies on survival and quality of life (QOL) in dogs with brain tumors. We hypothesized that survival and QOL for forebrain tumors would be superior to tumors elsewhere.

All dogs had clinical and CT/MRI evidence of a brain mass, were discharged from the hospital alive, and had histopathologic confirmation of a brain tumor upon death. Animal- , treatment-, and tumor-related factors were examined for survival effects, with treatments standardized based on clinical and tumor-related factors. Therapies included phenobarbital (Group 1), prednisone (Group 2), and phenobarbital and prednisone (Group 3). The impacts of tumors on behavior, physical status, and human-animal interactions were assessed by serial client QOL questionnaires.

Fifty dogs with 19 meningiomas, 10 gliomas, 9 pituitary macroadenomas, 8 choroid plexus tumors, and 4 neuronal tumors were included. Multivariate analysis identified tumor location, clinical signs/severity, and pre-treatment QOL scores as significant prognostic variables. The rates of mortality in Groups 2 (hazard ratio 16.9, 95% CI 1.3 - 225.1; p =0.03) and 3 (hazard ratio 12.9, 95% CI 1.7 - 99.3; p =0.01) were significantly greater compared to Group 1, and Group 1 dogs survived longer (median 212 days, 95% CI 178 – 343 days). No differences existed between groups in QOL scores for days 14 - 90 post-treatment, although these QOL scores improved in all groups compared to pre-treatment values.

Palliative treatments resulted in short-term QOL improvement in all groups. Dogs with mild neurologic dysfunction due to forebrain tumors can survive for prolonged periods.

**ABSTRACT #79**

THE PHARMACOKINETICS OF LEVETIRACETAM IN DOGS CONCURRENTLY RECEIVING PHENOBARBITAL. SA Moore, KR Muñana, MG Papich, JA Nettifee-Osborne. North Carolina State University College of Veterinary Medicine, Raleigh, NC.

Levetiracetam (LEV) is commonly used as an add-on medication in dogs with refractory epilepsy, although the pharmacokinetics of LEV when given in conjunction with other antiepileptic drugs has not been evaluated in the dog. The objective of this study was to determine if the pharmacokinetics of LEV change when administered in conjunction with phenobarbital (PB). Six healthy dogs were administered a single oral dose of LEV (16.7–27.8 mg/kg). Blood samples were collected at baseline and intermittently for 24 hours . The study was repeated after the same dogs received oral PB (2.0-3.3 mg/kg) twice daily for 21 days. All dogs achieved steady state concentrations of PB. Plasma LEV levels were evaluated by high pressure liquid chromatography. Pharmacokinetic data was analyzed using a compartmental model. Peak concentration ($C_{MAX}$) for LEV administered alone was 32.39 ug/ml ± 6.76, compared to18.22 µg/ml ± 8.97 when given with PB. The elimination $T_{1/2}$ in dogs given LEV alone and dogs given LEV and PB concurrently was 3.43h ± 0.47 and 1.73 h ± 0.22, respectively. Clearance increased from 124.93 ml/hr/kg ± 26.93 for LEV alone to 252.99 ml/hr/kg + 35.43 for LEV and PB in combination. Volume of distribution was 611.59 ml/kg ± 76.74 for LEV alone compared to 697.75 ml/kg + 85.43 for LEV with PB. Concurrent PB administration significantly alters the pharmacokinetics of LEV in the dog, indicating that dosage adjustments may be necessary when he drug is administered to dogs in conjunction with PB.

**ABSTRACT #80**

METABOLISM AND ACTION OF NELARABINE IN GLIOMA CELL LINES. L Tzipory, PJ Dickinson, CO Rodriguez Jr. University of California-Davis, Veterinary Medical Teaching Hospital, Davis, CA.

Nelarabine, pro-drug of ara-G, is phosphorylated in the initial rate-limiting step by cytosolic deoxycytidine kinase (dCK) and mitochondrial deoxyguanosine kinase (dGK) to ara-GMP. This moiety is then quickly converted to the active 5'-triphosphate, ara-GTP, which competes with the natural endogenous nucleotide, dGTP for incorporation into the replicating DNA strand. Incorporation of ara-GTP into the DNA results in replication block and subsequently inhibition of DNA synthesis leading to apoptosis.

In phase I and II human clinical trials, the most common adverse effects were non-hematological, neurological toxicities, suggesting ara-G penetration the blood brain barrier. These data, combined with the knowledge that tissues from the central nervous system have the highest specific activity of dGK in the body, led us to hypothesize that ara-G would have efficacy in the treatment of tumors of the central nervous system.

The purpose of this study is to evaluate and compare the intracellular pharmacokinetics and pharmacodynamics of Nelarabine in glioma cell lines of canine, murine and human origin.

Short term growth inhibition was assessed with MTT assay. Accumulation of intracellular ara-GTP was quantitated by HPLC. Repeated measure ANOVA and Student' t test were used to assess for significance.

The IC50 for murine (4C8) was 100uM at 72h. The IC50 for canine (JT3-BG) and human (U87) was >100uM at 72h. The accumulation of intracellular ara-GTP was similar between the cell lines.

Nelarabine appears to have efficacy against glioma cell lines of canine, murine and human origin. Its efficacy *in vivo* remains to be determined.

**ABSTRACT #81**

EXPRESSION OF MATRIX METALLOPROTEINASES AND RECK IN MICROGLIAL CELLS IN DIFFERENT INTRACRANIAL DISEASES. V.M. Stein[1], C. Puff[2], W. Baumgärtner[2], A. Tipold[1]. Department of Small Animal Medicine and Surgery[1], Institute of Pathology[2], University of Veterinary Medicine, Hannover, Germany.

Microglial cells are the brain-resident tissue macrophages and are closely associated with active central nervous system (CNS) pathology. They are known be activated by a large variety of stimuli and subsequently show de novo expression of numerous immunologically relevant molecules. As such they can induce the synthesis and secretion of MMPs and TIMPs. The question whether microglial reaction repertoire to a given stimulus is stereotyped or adapted to the underlying pathology is still a matter of debate.

Matrix metalloproteinases (MMPs) are proteolytic enzymes involved in the degradation of the extracellular matrix. They are closely implicated in the pathogenesis of inflammation in the CNS, demyelination and tumor pathogenesis. MMP inhibitors (tissue inhibitors of matrix metalloproteinases, TIMPs) are crucial in the regulation of MMP activities, and are therefore important candidates for the therapy of different diseases. RECK (reversion-inducing cysteine-rich protein with Kazal motifs) is a membrane-anchored glycoprotein that regulates extracellular matrix integrity and angiogenesis. To address the question whether microglia shows a specific repertoire of MMP expression, MMP-2, MMP-9, MMP-12, MMP-13, MMP-14, TIMP-1, TIMP-2, and RECK (reversion-inducing cysteine-rich protein with Kazal motifs) mRNA was evaluated in dogs with different intracranial diseases.

Microglial cells of twenty four dogs were isolated ex vivo with density gradient centrifugation. The dogs suffered from different intracranial diseases which were categorized according to histopathological examination. Disease categories comprised intracranial tumors, intracranial inflammation, idiopathic epilepsy, other intracranial diseases such as trauma and malformations, and extracranial diseases which served as a negative control group. The microglial cell pellets were examined via quantitative reverse transcriptase polymerase chain reaction (qRT-PCR) with SYBR-green using the Mx3005P Quantitative PCR System™ (Stratagene, Amsterdam, The Netherlands). Statistical analysis was conducted using the Kruskal Wallis test and the Wilcoxon rank sum test with the significance level set at $p \leq 0.05$.

MMP-2, MMP-9, MMP-12, MMP-13, MMP-14, TIMP-1, TIMP-2, and RECK mRNA was expressed by microglial cells, although no significant differences could be found between the different intracranial disease categories. However, in single dogs elevated MMP expression was found, such as MMP-2 in a dog with canine distemper virus infection. TIMPs and RECK were elevated in single dogs of each disease category. In conclusion, microglial expression of MMPs and TIMPs reflects their potential to respond to various disease conditions in the CNS rather than to indicate a specific disease etiology. However, the pathogenesis of a given disease might be crucially influenced by microglial MMP and TIMP expression.

**ABSTRACT #82**

LOW FIELD MAGNETIC RESONANCE IMAGING (MRI) FINDINGS OF THE CAUDAL CERVICAL SPINAL CORD IN CLINICALLY NORMAL DOBERMAN PINSCHERS AND FOXHOUNDS. S. De Decker, IMVL. Gielen, L. Duchateau, I. Van Soens, V. Bavegems, T. Bosmans, H. Van Bree, LML Van Ham. Faculty of Veterinary Medicine, Ghent University, Belgium.

Little is known about the spectrum and incidence of MRI abnormalities in the cervical vertebral column and spinal cord of clinically normal dogs.

Clinically normal Doberman Pinschers (n=20) and Foxhounds (n=17) were studied. The dogs were divided in 2 age categories: $\leq 5$ years (n=20) and $> 5$ years (n=17) and were considered clinically normal on the basis of a complete physical, neurological, and blood examination and an echocardiographic examination in all Doberman Pinschers. MRI was performed with a 0.2 Tesla magnet. Sagittal and dorsal T1 and T2 weighted images (WI) of the entire cervical spine and T1 and T2 WI transverse images from C4 to T1 were obtained. On sagittal T2 WI, disc degeneration was classified based on signal intensity as normal, partially or completely degenerated. On sagittal and when available (C4-T1) transverse T2 WI, ventral and dorsal compression were classified as normal, partial or complete subarachnoidal space compression and spinal cord compression. Vertebral body abnormalities were assessed on sagittal T1 WI and were defined as a flattening of the ventrocranial border of the vertebral body. Abnormal intraparenchymal signal intensity (ISI) changes were classified based on the surrounding spinal cord parenchyma on T1 and T2 WI.

Only 1 dog showed no single abnormality on MRI examination. Severe MRI abnormalities were seen in 17 dogs with the occurrence or combination of complete disc degeneration (n=14) and/or spinal cord compression (n=11). Vertebral body abnormalities were seen in 7 dogs and hyperintense ISI changes were seen in 2 dogs. The occurrence of vertebral body abnormalities was significantly higher in the Doberman Pinscher (P=0.04). The severity of disc degeneration (P=0.005) and ventral compression (P=0.05) were significantly higher in the higher age category. There was a significant difference in disc degeneration, ventral and dorsal compression (P<0.0001; P=0.004; P<0.0001, respectively) according to the location of the different intervertebral disc spaces, with increasing abnormalities for the more caudal intervertebral disc spaces (P<0.0001; P=0.0192; P<0.0001, respectively). The correlation coefficient between severity of disc degeneration and spinal cord compression was 0.52 (P=0.0003).

The results of this study indicate that MRI abnormalities are commonly seen in the caudal cervical vertebral column and spinal cord of clinically normal Doberman Pinchers and Foxhounds. This suggests that such lesions are part of the normal, or at least, common aging process of spinal degeneration.

**ABSTRACT #83**

COMPUTED TOMOGRAPHIC FINDINGS IN LARGE AND GIANT BREED DOGS WITH CERVICAL SPONDYLOMYELOPATHY: 58 CASES. Ronaldo C. da Costa, Rita Echandi, Dustin Beauchamp. Department of Veterinary Clinical Sciences, College of Veterinary Medicine, The Ohio State University, Columbus, OH.

Computed tomography (CT) is routinely used in the diagnosis of cervical spondylomyelopathy (CSM); however, no report presents CT findings in a large population of dogs with CSM. The purpose of this study was to compare the CT findings in large and giant dogs with CSM.

Medical and radiology records were searched for cases of canine CSM between 2003 and 2008. Inclusion criteria included a definitive diagnosis of CSM achieved by myelography and CT myelography (CTM) in large or giant dogs with clinical signs of cervical spinal disease. Fifty eight dogs met the inclusion criteria. Twenty-three were large breed dogs and 35 giant breed dogs. Among the large breed dogs, 11 were Dobermans, 3 Weimaraners, 3 mixed large breeds, 2 Dalmatians, 2 Labrador retrievers, 1 Greyhound, and 1 German shorthaired pointer. Their median age was 7 years (mean 7.1 years). There were 11 males and 12 females. Among the giant breed dogs, 18 were Great Danes, 8 Rottweilers, 7 Mastiffs, 1 Great Pyrenees, and 1 Bernese mountain dog. Their median age was 2.5 years (mean 4.1 years). There were 21 males and 14 females.

CTM findings in large breed dogs revealed that 11 dogs had a single site of spinal cord compression, while 12 dogs had multiple sites. The main site of compression was at C6-7 in 14 dogs, C5-6 in 7 dogs, and C4-5 and C2-3 with one dog each. The main cause of compression was disk-associated in 19 dogs, and osseous with or without ligamentous hypertrophy or synovial cysts in 4 dogs. The direction of compression was ventral in 19 dogs, dorsolateral in 2 dogs, dorsal in 1 dog and lateral in 1 dog. The severity of the spinal cord compression was graded as severe (>50% cord diameter), moderate (between 25-50%) and mild (<25%). The main compression was severe in 2 dogs, moderate in 14, and mild in 7 dogs. Four of the 23 dogs had spinal cord atrophy.

CTM findings in giant breed dogs showed that 30 dogs had multiple sites of compression, while 5 dogs had a single site. The main compressive site was at C6-7 in 13 dogs, C5-6 in 8 dogs, C4-5 in 5 dogs, C2-3 in 2 dogs, and C3-4 in 1 dog. Six dogs had 2 or 3 compressive sites of similar severity. Osseous compressions were also observed at C7-T1 in 8 dogs, and T1-T2 or only T2 in 5 dogs. The primary cause of compression was osseous in 27 dogs (2 dogs had concurrent synovial cysts), disk-associated in 4 dogs, and a combination of disk and osseous compression in 4 dogs. The direction of compression was lateralized in 18 dogs, dorsolateral in 5 dogs, dorsal in 3 dogs, ventral in 5 dogs, and a combination in 4 dogs. The main spinal cord compression was graded as severe in 9 dogs, moderate in 19 dogs, and mild in 7 dogs. Seven dogs had spinal cord atrophy.

Results indicate that giant dogs often have multiple compressions, usually caused by malformed or osteoarthritic articular processes that cause lateralized compression. Additionally, an important number of giant dogs had secondary osseous compressions at C7-T1 and T1-2 sites.

**ABSTRACT #84**

CERVICAL ARTHROPLASTY IN DOGS WITH DISC ASSOCIATED CAUDAL CERVICAL SPONDYLOMYELOPATHY AND CERVICAL DISC HERNIATION: PRELIMINARY STUDY IN TWO CASES. PF Adamo, G Burns. Bay Area Veterinary Specialists, San Leandro, CA.

The purpose of this preliminary study was to determine whether cervical arthroplasty is tolerated in dogs with disc associated Caudal Cervical Spondylomyelopathy (CCSM) or cervical disc herniation.

Two dogs diagnosed with disc associated cervical myelopathy using MR imaging were included in this study. Dog 1: a 4 year old MN Doberman Pincher 31 kg, with acute onset of cervical pain and ataxia. Dog 2: an 8 year old MN mixed breed 23 kg, with a 4 month history of progressive ataxia/tetraparesis. MRI findings showed acute myelopathy secondary to C6-C7 disc herniation in dog 1; and chronic myelopathy secondary to C5-C6 disc disease in dog 2.

Both dogs were surgically treated with standard ventral slot and spinal cord decompression followed by the insertion of the Titanium alloy version (Ti-6AI-4V-ELI) of the canine cervical artificial disc, previously tested by Adamo et al. - *in vitro* study. Post-operatively cervical radiographs were taken, a cast was applied for 6 weeks and serial neurologic examinations with cervical radiographs were taken at 2 and 6 weeks (both dogs) and at 12 weeks and 6 months (dog 1).

Dog 1 had a complete neurologic recovery; dog 2 showed significant improvement at 2 months post-op. On immediate post-operative radiographs the implant was well seated in the slot providing good distraction. In all post-operative cervical radiographs, distraction was maintained with no evidence of ventral bridging spondylosis at the treated site.

Cervical arthroplasty was well tolerated in both dogs, maintained distraction and prevented bridging spondylosis. This procedure warrants further study.

**ABSTRACT #85**

F RATIO OF THE SCIATIC AND ULNAR NERVE IN DOGS. JM Scarano, C Massicotte. Animal Emergency and Referral Associates, Fairfield, NJ.

We propose a reference value for the F ratio for the sciatic and ulnar nerve of dogs adjusted for the length of their limbs. Eleven healthy dogs with no clinical signs of peripheral nerve or spinal cord diseases were studied thus far.

F ratios were calculated for the sciatic nerve with supramaximal stimulation at the hock, popliteal fossa and hip. F ratios were calculated for the ulnar nerve with supramaximal stimulation at the carpus and the elbow. The formula $^{F-M-1}/_{2M}$ was used to calculate the F ratio. Dogs were divided into three groups based on total limb length which was measured from the hip to the fourth digit in hindlimbs and from the proximal scapular spine to the fourth digit in forelimbs . Small dogs had a total limb length less than 400mm, medium dogs greater than 400mm and less than 600mm and large dogs greater than 600mm. The mean value for each group was calculated and the following reference values proposed (+/- standard deviation): small hock = 4.7+/-2.6, small stifle = 2.7+/-1.0, small hip=1.7+/-0.9, medium hock = 3.7+/-0.9, medium stifle = 3.7+/-0.6, medium hip=1.9+/-1.9, large hock =1.9+/-0.1, large stifle=0.9+/-0.1, large hip =0.9, small carpus = 5.3+/-1.2, small elbow = 3.3+/-1.6, medium carpus = 4.7+/-0.8, medium elbow = 1.8+/-0.4, large carpus = 4.2+/-0.4, large elbow 1.6+/-1.1. These preliminary results suggest the necessity to consider limb length when calculating the F ratio. This study is ongoing and will include one hundred dogs to improve the reliability of these reference ranges.

**ABSTRACT #86**

PERIPHERAL NERVE PATHOLOGY IN CANINE DEGENERATIVE MYELOPATHY WITH MUTATION IN SUPEROXIDE DISMUTASE 1 GENE. GD Shelton[1], GC Johnson[2], GS Johnson[2], DP O'Brien[2], ML Katz[3], JR Coates[2]. 1. School of Medicine, University of California, San Diego, La Jolla, CA. 2.College of Veterinary Medicine, 3. School of Medicine, University of Missouri, Columbia, MO.

Canine degenerative myelopathy (DM) is a progressive neurodegenerative disease which has recently been shown in several canine breeds to be a result of a missense mutation in the superoxide dismutase 1 (*SOD1*) protein (Awano T et al, PNAS,2009). Based on this mutation and similarities in disease phenotype, canine DM appears to be a spontaneously occurring model for human amyotrophic lateral sclerosis (ALS). In canine DM, spasticity and general proprioceptive pelvic limb ataxia occurs in dogs older than 8 years of age, and if euthanasia is delayed, clinical signs will ascend causing flaccid tetraparesis and other lower motor neuron signs. A similar course of clinical progression has been described in upper motor neuron onset human ALS. To determine if pathologic changes typical of denervation are present in muscle (biceps femoris and gastrocnemius), and nerve (peroneal) of dogs with advanced DM, specimens were collected either as biopsies or following euthanasia from Pembroke Welsh Corgi (5), Boxer (2), German Shepherd Dog (1), Kerry Blue Terrier (1) and Chesapeake Bay Retriever (1) dogs homozygous for the *SOD1* mutation and confirmed as DM with microscopic demonstration of DM-specific lesions in spinal cords. Muscle specimens were evaluated in both frozen and paraffin sections, and peripheral nerve specimens were evaluated in resin sections. Abnormalities were generally classified as denervation atrophy, nerve fiber loss, and axonal degeneration or demyelination, and were scored 0 to +++ (normal to marked, respectively). Similar specimens were collected and processed from age-matched control dogs including Boxer (3), German Shepherd Dog (1) and English Cocker Spaniel (1) shown to be clear or heterozygous for the mutation. The most dramatic and consistent abnormalities were found in the Pembroke Welsh Corgi in which denervation atrophy (+++) was present in all muscle specimens and nerve fiber loss (++ to +++), myelin ovoids (++), myelin splitting and ballooning (++ to +++) and inappropriately thinly myelinated fibers (++ to +++) consistent with mixed axonal degeneration and demyelination was found in peroneal nerves. Similar but milder changes (+ to ++) were present in peripheral nerves in the Boxer, German Shepherd Dog, Kerry Blue Terrier and Chesapeake Bay Retriever breeds with variable muscle atrophy (0 to ++). A longer disease duration in the Pembroke Welsh Corgi (24-48 months) versus other affected breeds (7-12 months) may explain the more extensive disease severity. In conclusion, this study provides pathologic evidence for peripheral nerve involvement in canine DM consistent with the clinical signs of lower motor neuron disease.

**ABSTRACT #87**

AN SOD1 MUTATION ASSOCIATED WITH DEGENERATIVE MYELOPATHY OCCURS IN MANY DOG BREEDS. JR Coates[1], R Zeng[1], T Awano[1], LA Hansen[1], S Khan[1], GC Johnson[1], JF Taylor[2], SN Long[3], DP O'Brien[1], CM Wade[4], ML Katz[1], K Lindblad-Toh[4,5], GS Johnson[1]. 1. University of Missouri 1. College of Veterinary Medicine and 2. Division of Animal Sciences, Columbia, MO. 3. University of Pennsylvania School of Veterinary Medicine, Philadelphia, PA. 4. Broad Institute of Harvard and MIT, Cambridge, MA. 5. Uppsala University, Sweden.

Canine degenerative myelopathy (DM) is a fatal neurodegenerative disease prevalent in several dog breeds. Typically, the initial progressive upper motor neuron spastic and general proprioceptive ataxia in the pelvic limbs occurs at 8 years of age or older. If euthanasia is delayed, the clinical signs will ascend causing flaccid tetraparesis and other lower motor neuron signs. We previously identified a missense mutation in the canine superoxide dismutase 1 (*SOD1*) gene and showed that homozygosity for the A allele of this SOD1:c.118G>A mutation is a risk factor for DM in five dog breeds: Boxer, Chesapeake Bay retriever, German Shepherd dog, Pembroke Welsh corgi and Rhodesian ridgeback (Awano *et al,* PNAS, in press, 2009). We here report the results of genotyping 6407 individual dogs at SOD1:c.118.

The genotyped dogs were from 105 distinct breeds and at least 25 dogs represented each breed. A TaqMan® SNP allele discrimination assay was used for genotyping. The PCR primer sequences were GTGGGCCTGTTGTGGTATCA and CAAACTGATGGACGTGGAATCC and the probe sequences were VIC-CTCGCCTTTAGTCAGC for the mutant allele and FAM-CGCCTTCAGTCAGC for the wild type allele. The frequencies of the *SOD1* A allele ranged from 90% (Wire fox terrier) to zero (54 different breeds). Seventeen breeds had A-allele frequencies over 20%. DM is a well-recognized problem in 6 of these breeds with high A-allele frequencies: Pembroke Welsh corgi (A allele frequency = 85%), Boxer (71%), Chesapeake Bay Retriever (45%), Rhodesian ridgeback (33%), German Shepherd dog (32%) and Cardigan Welsh corgi (32%). Clinical histories resembling DM are common among dogs in 7 other breeds with high A-allele frequencies: American water spaniel (46%), Bernese mountain dog (41%), Kerry blue terrier (37%), Canaan dog (33%), Bloodhound (28%), French bulldog (23%) and Kuvasz (22%). Conversely, clinical histories consistent with DM were rare in four breeds with high A-allele frequencies: Wire fox terrier (90%), Tibetan terrier (29%), Welsh terrier (27%), and Chow chow (20%).

The putative existence of dog breeds that have high A-allele frequencies but rarely suffer from DM could be explained by genetic modifiers that alter susceptibility to DM.

**ABSTRACT #88**

LYMPHOCYTE COUNTS AND CIRCULATING LYMPHOBLASTS IN CANINE LYMPHOMA PATIENTS. <u>Karen V. Jackson[1],</u> Beth Overley[2], Reema Patel[1]. 1. Department of Pathobiology, University of Pennsylvania, Philadelphia, PA. 2. Langhorne, PA.

The purpose of this study was to evaluate the absolute lymphocyte and lymphoblast count in pretreatment complete blood counts (CBCs) from dogs with lymphoma with the suspicion that non-lymphopenic animals likely had circulating lymphoblasts.

Fifty-six dogs diagnosed with lymphoma were prospectively enrolled if they had no previous treatment (including glucocorticoids) and if they had intention-to-treat with a CHOP based protocol. All dogs had pre-treatment CBCs with review by a board-certified clinical pathologist. One hundred lymphoid cells were counted and the percentage lymphoblasts recorded.

The mean absolute lymphocyte count was 2690/µL and mean absolute lymphoblast count was 1350/µL. Twenty-six dogs were lymphopenic (<1000 total lymphocytes/µL) and thirty dogs were non-lymphopenic (range: 1120-26230/µL). The presence of measurable numbers of lymphoblasts (>100/µL) was significantly correlated with a lymphocyte count over 1000/µL ($p<0.001$). The sensitivity and specificity, of a lymphocyte count over 1000/µL predicting the presence of circulating lymphoblasts, are 0.73 and 0.77 respectively.

These results indicate that if the total absolute lymphocyte count is over 1000/µL (i.e. non-lymphopenic) there are likely circulating lymphoblasts. In this situation, a blood smear review by a clinical pathologist is warranted.

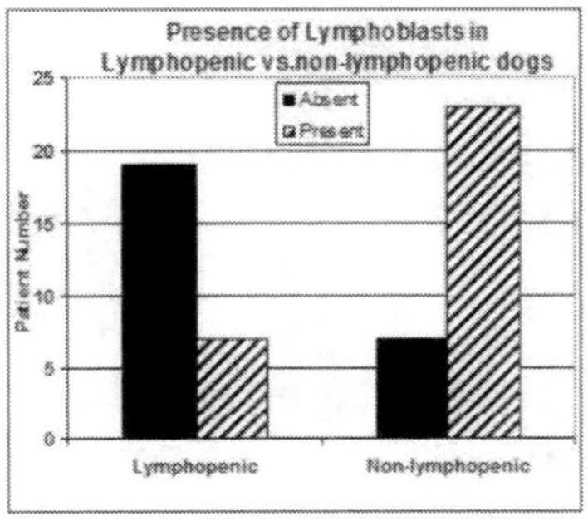


**ABSTRACT #89**

PHARMACOKINETIC PROPERTIES OF TOCERANIB PHOSPHATE (Palladia™, SU11654), A NOVEL TYROSINE KINASE INHIBITOR, IN LABORATORY DOGS AND DOGS WITH MAST CELL TUMORS. MF Yancey, <u>GM Michels,</u> DA Merritt, SP Lesman, JF Boucher. Pfizer Animal Health, Kalamazoo, MI.

Toceranib phosphate (Palladia™, SU11654), an oral multi-kinase inhibitor, is under investigation for treatment of mast cell tumors (MCT) in dogs. Several studies were performed to characterize the pharmacokinetics (PK) of toceranib in dogs. Means of the following PK parameters were estimated following a 1.0 mg (free base equivalents, fbe)/kg IV dose administered to laboratory beagles: plasma clearance, 1.45 L/kg/h; volume of distribution at steady state, 29.7 L/kg; terminal half-life, 17.5 h. Following single oral doses of 3.25 mg (fbe)/kg administered to laboratory beagles, mean Cmax estimates ranged from 67.8 - 112 ng/mL and occurred between 5.5 and 7.0 h post-dose. Terminal half-life was 31 h. Oral bioavailability was estimated to be 76.9%. There were no statistically significant ($p>0.05$) changes to any PK parameter due to the fed/fasted state or between days 0, 28, and 86 during 13 weeks of every-other-day dosing (3.25 mg/kg) to laboratory beagles. Toceranib concentrations were proportional with dose over the range studied here (2.0 to 6.0 mg/kg). The PK of toceranib in client-owned dogs of a variety of pure and mixed breeds with MCTs was similar to that in healthy laboratory dogs. In summary, toceranib exhibited moderate clearance, high volume of distribution, and a moderate elimination half-life. After a single oral dose at 3.25 mg/kg, the concentration vs. time curve showed broad, sustained exposure with measurable concentrations for more than 48 h. The PK parameters of toceranib described here support every-other-day administration at an initial dose of 3.25 mg/kg for the treatment of MCTs in dogs.

**ABSTRACT #90**

FELINE INJECTION SITE SARCOMAS IN THE UNITED KINGDOM: ONE HUNDRED AND FIFTY SEVEN CASES. <u>RS Dean[1]</u>, VJ Adams[1], DU Pfeiffer[2]. 1. Epidemiology Unit, Centre for Preventive Medicine, Animal Health Trust, Lanwades Park, Kentford, Newmarket, Suffolk, UK. 2. Veterinary Epidemiology & Public Health Group, Royal Veterinary College, Hawkshead Lane, Hatfield, Hertfordshire, UK.

Feline injection site sarcomas (FISS) are rare, clinically aggressive tumours of cats. Vaccinations, other injections and local trauma have been implicated in the pathogenesis of the tumour.

A consensus on the histopathological features of FISS was developed in a separate study and specific criteria for diagnosis were identified. These criteria were used to provide a clear case definition. Cases were prospectively recruited and a questionnaire was completed by the owner about the history of each cat.

A total of 259 cases were identified over a 14 month period and questionnaires were returned for 184 cases (71% response rate). When reviewed, 157 (85%) cases met the histopathology consensus. The most common location of the tumour was the interscapular space (148), other sites included the thoracic wall, flank and hindlimb. The mean age of the cats was 10.6 years (sd 3.6) with a mode of 13 and 17% were under 7 years of age. Both sexes were similarly represented with 83 males and 74 females, 99% of cats were neutered and only 8% were pedigree.

The majority of cats (98%) had been vaccinated, 67% in the last 12 months. Of the 154 vaccinated cats 120 (78%) were reported to have received leukaemia vaccinations, 66% (79/120) in the last 12months. Only 3 cats had been vaccinated against rabies. Microchips had been placed in 44% of cats with 11% of these in the last 12 months. Only 12 of the cats were reported to have received an injection of lufenuron, 5 within the last 12 months. Fighting was reported in 43% of cats.

This large case series of FISS did not use previous exposure to vaccination/injection and/or anatomic location as part of the case definition. The tumours were located in places other than the interscapular space. A wide age range was represented in this study including several younger cats. Nearly all the cats had been vaccinated, though rabies vaccines were rarely used and approximately half of the cats had a microchip.

## ABSTRACT #91

NONVIRAL GENE TRANSFER OF THE FELINE CYTOKINE GENES IL-2, IFNγ AND GM-CSF AS ADJUVANT IMMUNOTHERAPY OF THE FELINE FIBROSARCOMA. CK Kempf [1], BC Schwarz [1], T Brill [2], C Plank [2], R Köstlin [3], J Hirschberger [1], U Schillinger [2]. 1. Clinic of Small Animal Medicine, Faculty of Veterinary Medicine, Ludwig-Maximilian University Munich, Germany. 2. Institute of Experimental Oncology, Technical University Munich, Germany. 3. Clinic of Small Animal Surgery and Reproduction, Ludwig-Maximilian University Munich, Germany.

This phase I dose-escalating study was performed to determine toxicity and feasibility of a non viral gene-transfer system transmitting a combination of three different feline cytokines. Fifteen cats with primary or recurrent fibrosarcoma were included in the study. All cats had to be healthy based on blood work and physical examination and were staged by ultrasound and radiographs for signs of metastasis. A prior treatment with chemo-, radio- or an immunosuppressive therapy was not allowed. Following a wide surgical resection of the tumor an absorbable collagen sponge was implanted into the wound cavity. Each collagen sponge was loaded with equal amounts of plasmid DNA (pDNA), polyethylenimin and a protecting copolymer (P6YE5C). The plasmids were coding for feIL-2, feIFNγ and feGM-CSF, respectively. The dose of each plasmid was escalated in 4 different treatment groups starting with 75µg to 600µg per plasmid. Each cat was evaluated for treatment-related toxicity within the first 90 days after surgery and all occurring adverse events were registered and graded according to the VOCG-CTCAE scale. The most striking adverse event observed during the examination period was a drop in the lymphocyte count. Lymphocytes in cats treated with a dose of 600µg/plasmid dropped by 75% compared to initial values. This was considered to be therapy-related but not life-threatening. In conclusion, a dose of 600µg pDNA per cytokine is safe for the patient and feasible for a phase II clinical trial. The collagen sponge proved to be an easy to handle transfer medium for this nonviral gene therapy.

## ABSTRACT #92

IMMUNNOHISTOCHEMICAL DIFFERENTIATION OF CANINE PROSTATIC CARCINOMA. M. Fork[1], P. Bock[2], U. Kaim [2], W. Baumgärtner[2], I. Nolte[1]. Small Animal Clinic, University of Veterinary Medicine Hanover, Germany; Department of Pathology, University of Veterinary Medicine, Hanover, Germany.

In distinction from human prostate carcinoma, the disease in the dog is thought to be mostly androgen independent. Therefore basal cells, androgen refractory ductal cells, and transitional cells are a possible origin of canine prostate carcinoma. In human medicine Prostate Specific Antigen (PSA) and Prostatic Acid Phosphatase (PSAP) are markers commonly used for screening for prostate carcinoma.

The aim of this study was to identify the cells of origin of canine prostate carcinoma and to evaluate the use of PSA and PSAP in the diagnosis of the disease.

Nine samples of canine prostate carcinoma, nine samples of benign prostate hyperplasia and thirteen samples of transitional cell carcinoma (TCC) were evaluated using immunohistochemistry for PSA, PSAP, High Molecular Weight Cytokeratin (HMW-CK), staining for basal cells, Cytokeratin 7 (CK7), staining for urothelial cells and ductal cells, and Uroplakin III (UPIII), being specific for urothelial cells. Castration status and in the case of TCC also gender were recorded.

In benign prostate hyperplasia there was a strong positivity for PSA and PSAP, canine prostatic carcinoma was less likely to be positive for PSA and PSAP. In addition the staining was weaker in cells of malignant compared to cells of benign prostatic tissue. No TCC was positive for PSAP, but there was positive staining for PSA in five samples of transitional cell carcinoma. No sample of prostatic origin, benign or malignant, stained positive for UPIII, and only five samples of TCC (38%) showed a positive staining for UPIII. The staining for CK7 was significant more likely to be positive in samples of transitional cell carcinoma than in cells of prostatic origin. There was no significant difference in the staining for HMW-CK between either group.

We conclude that PSAP is a more useful marker for canine prostate carcinoma than PSA. In addition the secretion of PSAP and PSA is reduced in prostate carcinoma of the dog. For differentiation between transitional cell carcinoma and prostate carcinoma, CK7 and PSAP are useful markers. Basal cells do not seem to play a major role in the development of canine prostatic carcinoma. Uroplakin III is not useful for detecting transitional cell carcinoma.

**ABSTRACT #93**

FEASIBILITY AND SAFETY ASSOICATED WITH SELECTIVE AND SUPERSELECTIVE INTRA-ARTERIAL CARBOPLATIN +/- MELOXICAM DELIVERY FOR UROTHELIAL TUMORS IN DOGS. C Weisse, A Berent, Sorenmo K, Todd K, Soulen M, Solomon JA. Veterinary Hospital of the University of Pennsylvania, Philadelphia, PA.

Dogs with transitional cell carcinoma (TCC) are typically diagnosed with locally advanced tumors and up to 50% have distant metastasis at the time of diagnosis. Despite treatment most dogs will die within the first 12 months of diagnosis, often from progression of local disease. More effective local tumor control is necessary in order to improve outcome. Chemotherapeutic agents have activity in dogs with TCC, however these agents are associated with systemic toxicity, therefore preventing significant dose escalation. Intra-arterial delivery of chemotherapy and NSAIDs can provide increased intra-tumoral drug concentration and may therefore improve the anti-tumor activity. The investigators hypothesized that these treatments would be well tolerated and not result in any excessive morbidity or toxicity when compared with intravenous administration.

Twenty-two dogs received a total of 53 treatments (1-6 treatments/dog). All patients were placed under general anesthesia and vascular access to the femoral or carotid artery was obtained. Interventional radiology techniques were used under fluoroscopic guidance to gain selective or superselective access to the internal iliac or prostatic/caudal vesical arteries, respectively. Systemic doses of chemotherapy with or without meloxicam were administered and the animals recovered. The treatments were repeated following standard chemotherapy protocols.

Complications were mostly minor and identified following eleven (11/53) procedures. One dog developed presumptive sepsis after the first procedure and was euthanized one week later. Non-clinical myelotoxicity was identified following five (5/53) procedures.

Selective and superselective intra-arterial administration of chemotherapy with or without meloxicam is safe in dogs with naturally-occurring urothelial tumors.

*Previously presented at ACVS, October 2008*

**ABSTRACT #94**

EFFECT OF INTRACAVITARY CHEMOTHERAPY AS TREAMENT OF RECURRENT PLEURAL EFFUSION FOLLOWING PERICARDIECTOMY: EIGHT CASES. M Ames, EA Rozanski, LM Freeman, LBarber, SM Cunningham, JE Rush. Cummings School of Veterinary Medicine at Tufts University, North Grafton, MA.

Recurrent pleural effusion in dogs following surgical pericardiectomy for idiopathic or malignant pericardial effusion is often a therapeutic challenge. The purpose of this study was to evaluate the survival time and side effects in dogs treated with intracavitary platinum-based chemotherapy after pericardiectomy for treatment of pericardial effusion.

All dogs that developed recurrent pleural effusion following pericardiectomy that were treated with intracavitary cisplatin or carboplatin were retrospectively evaluated. Dogs that had pericardiectomy for treatment of either hemangiosarcoma or chemodectoma were excluded. Data collected from the medical record included underlying diagnosis, survival time from the initial diagnosis of pericardial effusion, survival time from pericardiectomy and evidence of side effects of chemotherapy. Survival times were compared with Kaplan-Meier survival curves and Cox proportional hazards analysis.

Eight dogs met the inclusion criteria. The median age at first diagnosis of pericardial effusion was 8.5 years (range 5.9-10.9 years). Six were large breed dogs (> 25 kg) and 2 were small dogs (< 10 kg). All dogs had pericardiectomies; 4 via lateral thoracotomy, 2 thoroscopically, and 2 via median sternotomy. The pericardial biopsy was definitive for mesothelioma in 3 dogs and for unclassified neoplasia (likely carcinoma) in 1, while 4 dogs had severe inflammatory pericarditis with notation by the pathologist that neoplasia could not be excluded. No dog had hypoalbuminemia. Six dogs received intracavitary cisplatin (50 mg/m$^2$) following a therapeutic thoracentesis. Dogs were rolled to distribute the chemotherapy evenly. Dogs received intravenous saline diuresis (18.3 ml/kg/hr x 4 hours) in conjunction with cisplatin chemotherapy. Dogs received between 1-7 cisplatin treatments (median = 3); treatments were repeated when effusion returned, but no more often than every 4 weeks. Three dogs received intracavitary carboplatin (300 mg/m$^2$), one of which had first received 4 cycles of cisplatin. The median survival time from the first pericardial effusion was 11.5 months (range 2.5-96.0 months) and the median survival time from pericardiectomy was 9.0 months, (range 1.5- 60.0 months). Two dogs developed azotemia during treatment (creatinine values > 3.0 mg/dl) which resulted in discontinuation of chemotherapy. One dog, treated with carboplatin, developed severe neutropenia (32 PMN/ul) and associated sepsis; this dog recovered with supportive care. There was no difference in survival time from initial diagnosis (p=0.23) or from pericardectomy (p=0.29) based upon histopathological evidence of cancer.

Intracavitary chemotherapy was associated with few side effects, and represents a viable therapeutic option for dogs that have recurrent pleural effusion after pericardiectomy.

**ABSTRACT #95**

TECHNIQUE FOR FLUOROSCOPIC NASOJEJUNAL TUBE PLACEMENT IN DOGS. Beal MW, Mehler SJ, Staiger BA, Moore TW, Lam NK. College of Veterinary Medicine, Michigan State University, East Lansing, MI.

Nutritional support contributes to a positive outcome in the management of dogs with critical illness. The advantages of enteral nutritional support when compared to methods of parenteral nutritional support are well documented in humans. Dogs with critical illness are often in need of nutritional support, but are often intolerant of feedings into the stomach. Post-pyloric feeding ay obviate many of these tolerance problems. Surgical jejunostomy techniques are well established, however, the methods are invasive and complications are common. We hypothesized that fluoroscopic nasojejunal feeding tube (NJT) placement is a safe, rapid, and uncomplicated method for achieving sustained post-pyloric access for enteral nutritional support in normal dogs.

Six normal, male, beagle dogs (7-12Kg) were anesthetized using standard techniques common to those utilized during gastrointestinal endoscopic procedures. Dogs were placed in left lateral recumbency and lidocaine was administered into the right nostril. A 0.035in, 260cm, standard stiffness, hydrophilic guide wire with a straight, flexible tip (HGW) was advanced through the nostril and into the stomach. A 5F 100cm berenstein catheter (BC) was advanced over the HGW and into the stomach and used to guide it across the pylorus. The HGW was then advanced as far into the jejunum as possible. The BC was removed and an 8F end-hole and multi fenestrated feeding tube was advanced over the HGW into the jejunum. Gradual NJ feeding escalation was performed over the ensuing 48hr and NJT location was monitored radiographically (0hr, 48hr, 72hr) to assess for migration. At 72hr, emesis was induced to determine the effect of emesis on NJT location. Tube migration was determined by measuring the location of the tip of the tube in relation to the caudal duodenal flexure (CDF). Dogs were monitored for gastrointestinal complications, mechanical complications with the tube, and self-removal. A paired t-test was used to compare tube locations at different time points. Statistical significance was set at $p<0.05$.

NJT placement was successful in all dogs. Median time to crossing the pylorus with the HGW was 11.3min (range 2-28). Median total procedural time was 34.1min (range 26-49). Median initial location of the tip of the NJT was 20cm beyond the CDF (range 14-37). Complications of tube placement included mild epistaxis in two dogs. Three dogs removed the NJ tube within the first 48hr. In the remaining 3 dogs, no significant tube migration occurred between 0hr and 72hr. Emesis induction resulted in oral migration of the NJT into the descending duodenum in one dog.

Fluoroscopy guided NJT placement is a safe, rapid, uncomplicated, minimally invasive technique for gaining post-pyloric access for enteral nutritional support in dogs. Further evaluation of NJTs in clinical patients will be necessary to better assess NJT tolerance and tube migration.

**ABSTRACT #96**

TECHNIQUE FOR PERCUTANEOUS RADIOLOGIC GASTROJEJUNOSTOMY IN THE DOG. Beal MW, Mehler SJ, Staiger BA, Moore TW, Lam NK, Brown AJ. College of Veterinary Medicine, Michigan State University, East Lansing, MI.

Nutritional support contributes to a positive outcome in the management of dogs with critical illness. Post-pyloric feeding may obviate tolerance problems associated with gastric feeds in critically ill dogs. Surgical jejunostomy techniques are invasive and complications are common. We hypothesized that percutaneous radiologic gastrojejunostomy (PRGJ) is a safe and uncomplicated technique for gaining post-pyloric access for enteral nutritional support in dogs.

Six normal, male, beagle dogs (7-12Kg) were anesthetized using standard techniques common to those utilized during gastrointestinal endoscopic procedures. The dogs were positioned in right lateral recumbency and an orogastric tube allowed for gastric insufflation. Fluoroscopy was utilized to identify a location caudal to the left, 13$^{th}$ rib for placement of 3 gastrointestinal suture anchors (GSA) into the stomach in a triangular pattern, creating a gastropexy. An 18g puncture needle was advanced through the center of the triangle created by the GSA. A 0.035in, 150cm, standard stiffness, hydrophilic guide wire with a straight, flexible tip (HGW) was passed through the puncture needle and into the stomach and a 5F angled guiding catheter was then passed over the HGW and used to direct the HGW across the pylorus and duodenum, and into the jejunum. Serial over-the-wire dilation of the body wall facilitated placement of an 18F peel-away introducer through which an 18F/8F 58cm dual-lumen gastrojejunal feeding tube (GJT) was advanced over the HGW and into the jejunum. Gradual jejunal feeding escalation was performed over 48hr. Tube location was determined radiographically at 48hr, 96hr, after emesis induction at 96hr, and at the time of tube removal (days 16-18). Location of the distal tip of the jejunal portion of the tube was noted in reference to the caudal duodenal flexure (CDF). A paired t-test was used to compare tube locations at different time points during the study. Statistical significance was set at $p<0.05$, with Bonferroni correction for multiple comparisons. Dogs and GJT were monitored for complications.

PRGJ was successful in all dogs. Median time to traversing the pylorus with the HGW was 23.5min (range 14-93). Median total procedural time was 53min (range 42-126). The distal tip of the GJT was located median 35cm (range 27-40) past the CDF at initial placement (0hr). No procedural complications were noted. No significant tube migration was noted at any time point or after induction of emesis. 5/6 dogs completed the study period (96hr). 1/6 dogs removed the tube on day 3. Mild gastrostomy site redness and discharge was noted in all dogs.

PRGJ in the dog is a safe, minimally invasive technique that allows for gastric decompression and jejunal feeding without the need for surgical intervention.

**ABSTRACT #97**

DIETARY γ-LINOLENIC ACID SUPPORTS ARACHIDONIC ACID ENRICHMENT OF FELINE RED BLOOD CELL MEMBRANES. L Trevizan[1], M Hoernis[1], KE Bigley[1], JE Bauer[1,2]. 1. Companion Animal Nutrition Laboratory, 2. Intercollegiate Faculty of Nutrition, Texas A&M University, College Station, TX.

Conversion of linoleic acid (LA) to arachidonic acid (AA) in cats is limited due to low Δ6-desaturase activity. However, it is unknown whether this enzyme can be induced by feeding high dietary concentrations of its LA substrate. Also unknown is whether bypassing the Δ6-desaturase step by providing a dietary source of the enzymatic product, γ-linolenic acid (GLA), will enable AA synthesis. The objective of this study was to evaluate the *in vivo* accumulation of AA after feeding specific fatty acid precursors and measuring the incorporation of their fatty acid metabolites into feline red blood cell (RBC) membranes. It was hypothesized that feeding a GLA enriched diet would bypass the Δ6-desaturase step resulting in AA synthesis via chain elongation and Δ5-desaturation, with its subsequent incorporation into RBC membranes. It was also hypothesized that Δ6 desaturase may be directly induced by providing large dietary amounts of LA precursor. To test these hypotheses, adult female cats (n=29) were divided into three groups and fed complete and balanced diets containing high LA (HL, n=10), high GLA (GLA, n=10), or adequate LA (LL, control, n=9). The diets were similar in caloric density, total protein, total fat, carbohydrate, vitamins, and minerals and differed only in their fatty acid compositions. Diets were fed for 56 days according to metabolic body weight ($100\times W_{kg}^{0.67}$) to maintain body condition scores (BCS) of 5/9 with water provided *ad libitum*. Food consumption was recorded daily and body weights (BW) and BCS were determined weekly. After 56 days, blood samples were collected via saphenous vein into EDTA-containing tubes. Red blood cell ghosts were prepared from the samples and total lipids were extracted. Membrane phospholipids were fractionated via thin layer chromatography and fatty acid profiles were determined by gas chromatography. The Shapiro-Wilks test was used to determine normal distribution of fatty acid data and one-way ANOVA was performed followed by Tukey ($\alpha$=0.05) multiple comparisons. It was found that high dietary LA concentration did not result in RBC AA enrichment vs. control suggesting no induction of Δ-6 desaturase activity. However, chain elongation of LA was observed with the accumulation of 20:2n-6 and the presence of the Δ-5 desaturase product, 20:3n-6 (Δ 5, 11, 14) in the HL group. Most notably, feeding high dietary GLA was successful at bypassing the Δ-6 enzymatic step yielding increased amounts of AA. These findings support the presence of a functionally active Δ-5 desaturase enzyme in cats and that diets containing high amounts of GLA may be suitable for feline species.

## ABSTRACT #98

ADIPOKINE EXPRESSION AND SECRETION BY CANINE ADIPOCYTES: STIMULATION OF INFLAMMATORY ADIPOKINE PRODUCTION BY LPS AND TNFα. VH Ryan[1, 2], AJ German[2], IS Wood[1], L Hunter[1], PJ Morris[3], P Trayhurn[1]. [1]Obesity Biology Research Unit, School of Clinical Sciences, University of Liverpool, Liverpool, Merseyside, UK; [2]Department of Veterinary Clinical Sciences, University of Liverpool, Liverpool, Merseyside, UK; [3]WALTHAM Centre for Pet Nutrition, Mars UK, Freeby Lane, Melton Mowbray, Leicestershire UK.

The aim of the current study was to examine the expression of key inflammation-related adipokines in canine adipose tissue and adipokine expression and secretion determined in canine adipocytes.

Adipose tissue samples were collected from adult Staffordshire Bull Terriers of both genders immediately after euthanasia, which had been performed for reasons unrelated to the research. The expression of inflammatory adipokines was assessed in a previously validated canine primary adipocyte culture system. Gene expression was assessed using real-time polymerase chain reaction methodology, whilst ELISA techniques were used to examine adipokine protein secretion into culture medium. The study was performed according to the University of Liverpool Guidelines on Animal Ethics and all tissue used was otherwise to be discarded.

Adiponectin, leptin, IL-6, monocyte chemotactic protein 1 (MCP-1) and TNFα genes were expressed in all the major adipose tissues depots of dogs, but there were no major depot differences in mRNA levels. Each adipokine examined was expressed in adipocytes differentiated in culture and secreted into the medium (leptin was not detected). IL-6, MCP-1 and TNFα were also expressed and secreted by preadipocytes; adiponectin and leptin were, however, only expressed after the induction of adipocyte differentiation. The inflammatory mediators LPS and TNFα had a major stimulatory effect on the expression and secretion of IL-6, MCP-1 and TNFα; there was a >5000-fold increase in IL-6 mRNA level with LPS. IL-6 release into the medium was increased >50-fold over 24 h with both LPS and TNFα, while MCP-1 release was increased 23- and 40-fold by TNFα and LPS, respectively. In contrast, there was no effect or a small reduction in adiponectin and leptin mRNA level with the inflammatory mediators. Dexamethasone stimulated leptin gene expression, had no effect on adiponectin expression, but decreased the expression and secretion of IL-6 and MCP-1. The peroxisome proliferator-activated receptor gamma (PPARγ) agonist rosiglitazone stimulated the expression of both adiponectin and leptin, and inhibited the expression of IL-6, MCP-1 and TNFα; however, only MCP-1 secretion was reduced.

These results demonstrate that canine adipocytes express and secrete key adipokines. It is concluded that canine adipocytes are highly sensitive to inflammatory mediators, these inducing major increases in the production and release of inflammatory adipokines.

## ABSTRACT #99

EXERCISE AND ACTIVE CLIENT MOTIVATION IMPROVE RATE OF WEIGHT LOSS IN OBESE DOGS. A Chauvet[1], J Laclair[2], SL Holden[3], D Elliott[4], AJ German[3]. [1]Veterinary Neuro Services, Sarasota, FL, [2]Jim LaClair's Dog-Friendly Dog Training, Sarasota, FL, [3]Department of Veterinary Clinical Sciences, University of Liverpool, Liverpool, Merseyside, UK; [4]Royal Canin USA.

Obesity is a now a major concern in companion animals, and management strategies can either utilize dietary energy restriction or drug therapy. Lifestyle changes, involving increasing activity levels, are usually recommended as an adjunct to weight management, but there is little information on how this impacts success. The hypothesis of the current study was that including an organized exercise program using an underwater treadmill would improve weight loss on a conventional weight management regime.

Eight client-owned obese dogs (condition score 9/9), of various breeds and genders, were enrolled in a weight loss program using a high protein diet and lifestyle changes. An exercise regime was devised for each dog based on regular lead-walking and treadmill exercise. During the 3-month program, dogs were re-examined at 33, 54 and 86 days; at each stage, body weight, thoracic girth, and abdominal girth were recorded. The percentage weight loss over the program was compared with a matched control group enrolled at a weight management referral clinic and using an equivalent diet for weight loss.

Mean (±standard deviation, SD) percentage of starting weight lost was 7.4 ±2.20%, 10.8 ±1.88% and 18.9 ±5.44%, at

33, 54 and 86 days, respectively. The overall mean rate of weight loss for the whole 3-month trial was 1.5 ±0.43% per week. Both thoracic and abdominal girth also declined significantly over the course of the weight loss program (P<0.0001 for both). The median number of treadmill exercise sessions per dog was 13 (range 5 to 17). Median session duration was 30 minutes (range 8 to 37), median speed was 1.2 mph (0.2 to 3.5 mph), and median distance walked was 0.60 miles (range 0.03 miles to 1.70 miles). The session speed (P=0.0023), session duration (P=0.0032), and distance travelled per session (P=0.0010) all increased significantly over the course of the study. Cases lost significantly more weight than their respective controls (56 days: median 11.4% [range 8.1-13.1%] vs. 7.3% [range 3.0-12.8%], P=0.0043; 86 days: median 15.5% [range 14.3-25.3%] vs. 9.7% [range 4.6-18.5%], P=0.0047).

This study is the first to show that outcome of weight loss is improved when organized exercise sessions are included in conventional canine weight management programs.

**ABSTRACT #100**

COMPARISON OF DIETARY STRATEGIES ON THE PERCEPTION OF HUNGER DURING A FIELD FELINE WEIGHT LOSS STUDY. T. Bissot[1], E. Servet[1], S. Vidal[1], M. Deboise[1], G. Egron[2], M. Hugonnard[2], S Lardy[2], A.J. German[3], V. Biourge[1] [1]Royal Canin , Aimargues, France. [2] Ecole Nationale Vétérinaire de Lyon, [3] Veterinary Clinical Sciences, University of Liverpool, UK.

Various dietary strategies exist for inducing weight loss in obese cats, but success in field trials is worse than expected from colony-based studies. The purpose of the current study was to assess the efficacy of different dietary strategies on the perception of hunger by owners and the rate of weight loss in a controlled field weight loss program.

48 overweight (body condition score (BCS) ≥4/5), healthy adult neutered cats completed a 20-wks weight loss trial. Three dietary strategies, intended to promote satiety and weight loss, were evaluated. DietA[a] (ME: 2940 Kcal/1000 Kcal as fed, prot.: 116 fat: 31 , TDF: 78 , starch: 15 g/1000 Kcal ,respectively) and C[c] (ME: 3060 KCal/kg, prot.: 115 fat: 36 , TDF: 81 , starch: 62 g/MJ,respectively) were dry diets, whereas dietB was a mixed ration: dietA and a moist diet[b] (ME: 550 Kcal/kg as fed, prot.: 137 , fat: 36 , TDF: 27 , starch: 56 , water: 1512 g/100 Kcal). On day 0, cats underwent physical examination, BW and BCS measurement. Target body weight (TBW) was estimated from current BCS. Cats were then randomly-allocated to the 3 diet groups, and owners instructed to feed their cat 30 to 35 Kcal /kg TBW/day based on the BCS.. On weeks 4, 12 and 20, cats were weighed; if the rate of weight loss was <0.5 or >2 %/week, energy allocation was re-adjusted. Hunger was assessed with a questionnaire, enabling calculation of a begging behaviour score (from 1= no begging to 4 = permanent begging), begging manifestation scores (incessant vocalisation, excessive clinginess, food stealing, and aggression; scored as 0 or 1), and a global begging score (sum of both). Data were analysed using a repeated-measure ANOVA design, with cats, weeks and diets as main factors. All results are expressed as mean±sem.

Cats remained healthy and consumed all their food. Mean weight loss (%) over 20 weeks was similar amongst groups (11.0±1.3%, 10.9±1.2% and 11.9±1.7%, respectively). Mean rate of weight loss and energy allocation was not different amongst groups. Begging behaviour, the sum of begging manifestations, and the global begging score were significantly (p<0.001) higher with diet C than with diets A and B. At week 12, significantly more owners regarded food allowance as insufficient on diet C than on diets A and B (64% vs. 31%, 39%, respectively).

Diets A and B performed better than diet C, based on their lower begging scores and owner perception of excess hunger. This confirms that diets formulated to induce satiety can help limit hunger during weight loss program, and are therefore likely to improve chances of success.

[a] Royal Canin, Veterinary Diet, Feline Satiety Dry, Aimargues, France; [b]Royal Canin, Veterinary Diet, Feline Obesity Wet, Aimargues, France; [c] Hill's Prescription Diet, Feline R/D Dry, Topeka, KS, USA.

*Previously presented at European Society of Veterinary Comparative Nutrition in Vienna, September 2008*

**ABSTRACT #101**

INFLUENCE OF DIETARY PROTEIN SOURCE AND CONTENT ON FECAL QUALITY AND GENE EXPRESSION OF WATER AND ELECTROLYTE TRANSPORTERS OF COLON IN MINIATURE POODLES AND GERMAN SHEPHERDS. J Nery[1,2], V Leray[1], C Tournier[3], V Biourge[3], L Martin[1], H Dumon[1] and P Nguyen[1]. 1. Ecole Nationale Vétérinaire de Nantes, France, 2. University of Turin, Italy, 3. Royal Canin, Aimargues, France.

When compared to smaller dogs, large dogs are prone to producing poor quality feces, that could be partially attributed to higher colonic fermentations. Fermentation of undigested proteins could have deleterious effects on the colonic mucosa and affect colonic functions. One of them is water and electrolyte absorption. Highly digestible protein sources, such as wheat gluten, lead to a low protein residue in the colon. The aim of this study was to evaluate the effect of dietary protein source and amount on fecal quality and expression of proteins involved in colonic absorption, as water channel (aquaporin 8; AQP8), and electrolyte channel (Na+/K+ ATPase, as well as cystic fibrosis transmembrane conductance regulator (CFTR)) in the colon of large and small dogs.

Eleven female dogs (5 Miniature poodles: MP, 4.9±1.5 kg BW, and 6 German shepherds: GS, 26.6±2.4 kg BW) were used. Two diets varying in protein source and level were tested in a crossover study. Diets were isoenergetic with similar fat, TDF and ash content. Main protein sources and levels were wheat gluten meal at low protein level (LP; CP=21.4% as fed) and poultry meal at high protein level (HP; CP=34.8% as fed). The study design comprised two 14-d experimental periods and a 24-d washout period. Feces were scored (1= dry feces to 5= liquid diarrhea) daily for 7 days. Biopsies from proximal and distal segments of the colon were performed at the end of each experimental period. mRNA expression of two variants of AQP8 (AQP8v1 and AQP8v2), $Na^+/K^+$ ATPase and CFTR were measured by RT-real time PCR by comparison with GAPdH expression. Fecal score data were analyzed using Kruskal Wallis test and gene expression data were normalized and analyzed with repeated measures ANOVA.

Fecal score was higher in GS than MP dogs whatever the diet ($p<0.001$ for LP and $p<0.01$ for HP) and in HP fed dogs whatever the breed ($p<0.05$ for MP and $p<0.01$ for GS). A trend for higher mRNA expression of AQP8v1 was found in MP, compared with GS dogs, when fed HP diet ($p<0.09$). A trend for higher expression in the distal colon, associated with a breed effect was also observed ($p=0.15$). Similarly an effect of [breed*biopsy location] was observed in AQP8v2 expression in dogs fed HP ($p<0.09$); expression was higher in the distal colon of MP dogs. mRNA expression of AQP8v1 and AQP8v2 was constant in GS dogs. No effect of diet, breed or biopsy location was found for $Na^+/K^+$ ATPase nor CFTR expression.

Poorer fecal score in large compared to small dogs could be partially attributed to lower water channels mRNA expression in the distal colon. This divergence between the two breeds appeared to decrease when feeding highly digestible protein at low level. To improve fecal score in large sensitive dogs modulation of protein source and content could therefore be an adequate nutritional strategy.

**ABSTRACT #102**

INFLUENCE OF EMPIRICAL ANTIMICROBIAL CHOICE ON OUTCOME FOLLOWING SEPTIC PERITONITIS IN DOGS. I Keir, A Mateus, J Wignal, A Boag.

Despite advances in veterinary medical practice, canine sepsis is still associated with a mortality of greater than 50%. With any bacterial infection antimicrobial therapy is the cornerstone of treatment, along with drainage and debridement or removal of infected fluids and tissue. Studies in humans have shown that inappropriate initial empirical antimicrobial therapy is associated with a poorer outcome even when the antimicrobial therapy was subsequently changed based on culture results. Current human recommendations also advise the instigation of antimicrobial therapy within 1 hour of a diagnosis of sepsis.

Medical records of the Queen Mother Hosptial for Animals, Royal Veterinary College were reviewed retrospectively from 2003 to 2008 for dogs with a diagnosis of septic peritonitis. Data collected from the record included choice of empirical antimicrobial therapy, time from admission to first antibiotic therapy, bacterial culture and sensitivity results and any change in antimicrobial therapy made after culture results. Antimicrobial therapy was defined as appropriate if at least one of the drugs included in the empirical antibiotic treatment was effective against the pathogen(s) isolated. Survival was defined as alive at time of discharge. Chi squared analysis was used to compare the survival of patients receiving appropriate antimicrobial therapy with those receiving inappropriate therapy. Fisher's exact test was used to analyze the effect of the time from admission to commencing empirical antimicrobial therapy.

A total of 51 records were available with empirical antimicrobial and culture data available in 38 cases. The overall survival rate was 55%. Dogs that received appropriate empirical antimicrobial therapy had a survival to discharge rate of 61%; dogs that received inappropriate empirical antimicrobial therapy had a survival to discharge rate of 55%. No statistical difference was found between the rate of survival and appropriate empirical antimicrobial selection (p 0.191). The administration of antimicrobials within 1 hour of admission to the hospital did not impact on survival to discharge (p 0.437).

Factors affecting outcome in canine septic peritonitis are likely multifactorial. In our study the use of appropriate empirical antimicrobials and timing of antimicrobial administration were not significant factors, however considering the small number of cases, the possibility of a type II error should be considered.

**ABSTRACT #103**

DETECTION OF FLUOROQUINOLONE RESISTANCE LEVEL IN CLINICAL CANINE AND FELINE *ESCHERICHIA COLI* PATHOGENS USING RAPID AND EARLY REAL-TIME PCR ASSAY. BW Shaheen[1], DM Boothe[1], C Wang[2], CM Johnson[2], B Kaltenboeck[2]. 1. Department of Anatomy, Physiology and Pharmacology, 2. Department of Pathobiology, College of Veterinary Medicine, Auburn University, Auburn, AL.

Fluoroquinolones are used to treat infections caused by *Escherichia (E.) coli* in canine and feline veterinary patients, particularly those infecting the urinary tract. The *gyrA* gene is a primary target causing fluoroquinolone resistance in Gram negative coliforms, with mutations in codon 83 and 87 generally associated with high level of resistance *E. coli* clinical isolates. The purpose of this study was to develop a fluorescence resonance energy transfer (FRET) quantitative PCR that can precisely and specifically identify enrofloxacin-resistance in the canine and feline clinical *E. coli* isolates that carry mutations in codon 83 and 87 of *gyrA*. A total of 70 isolates were randomly selected that represent different levels of resistance to enrofloxacin between May and December 2008. Isolates were collected from 4 different regions: West, South, Midwest, and Northeast. We then tested 5 infected urine specimens to challenge the real-time PCR assay with urine specimens containing the following colony forming unites (CFU): $10^6$, $10^5$, $10^4$, $10^3$, $10^2$, 10, and 0 CFU/ml. The susceptibility to enrofloxacin was determined by a standard micro-dilution test and according to CLSI guidelines and interpretive standards. Quantitative determination of *E. coli* was performed by real-time FRET-PCR in a LightCycler (Roche Diagnostics). This real-time quantitative PCR assay is rapid, economical, and sensitive compared to cultured antimicrobial susceptibility testing. The assay identified as low as five genome copies per reaction. For the 70 isolates tested, the sensitivity (correctly predicting true positives, that is either intermediate or high level resistance) was 87.5% (95% CI = 75% to 95.3%) (n=42/48) whereas specificity (correctly identifying true negatives or susceptible) was 100% (95% CI = 87.3% to 100%) (n=22/22). Furthermore, we were able to accurately differentiate between the wild type and mutants *E. coli* directly from infected canine urine samples (n=5) within 1-2 hrs (as low as 10 CFU/ml corresponded to 19 genome copies). These results were confirmed by sequence alignments of the PCR products and comparison with the susceptibility testing. The FRET-PCR assay appears to have promising clinical application as an early diagnostic tool for rapid and sensitive detection and differentiation of the level of fluoroquinolone resistance among clinical *E. coli* isolates.

**ABSTRACT #104**

EVALUATION OF POTENTIAL BIOMARKERS FOR CANINE SEPSIS. AM Cruse, JA Yoder, SK Nordone, HS Marr, and AJ Birkenheuer. North Carolina State University, Raleigh, NC.

Sepsis is a complex clinical syndrome defined as the presence of systemic inflammatory response syndrome (SIRS) with documentation or suspicion of concurrent infection. Early recognition and treatment of sepsis has been shown to be crucial for the improvement of patient outcome. The purpose of this study was to evaluate the expression of three potential biomarkers of canine sepsis. Potential biomarkers were selected based upon increased expression in murine and zebrafish models.

Whole blood samples were collected from client-owned dogs presenting to the NCSU-VTH between February-October 2008. The patient populations included dogs with sepsis (n=12), systemic inflammatory response (SIRS without documented infection) (n=10), cancer (n=30), and healthy controls (n=25). Each sample was evaluated for FLT-4, KLF-1, and Tmem150 (TM6P1) mRNA expression using real-time polymerase chain reaction. Relative expression was assessed compared to an internal control gene (β-actin). Differences were compared using a one-way ANOVA.

FLT-4 expression was significantly up-regulated ($p<0.001$) in dogs with sepsis and SIRS compared to dogs with cancer. FLT-4 expression was significantly up-regulated ($p<0.001$) in dogs with sepsis compared to healthy dogs. However there was no difference in FLT-4 expression between dogs with sepsis and SIRS. Tmem150 expression was down regulated ($p<0.05$) in dogs with SIRS compared to healthy dogs. No other significant differences in gene expression were detected between groups.

FLT-4, also known vascular endothelial growth factor receptor 3, is involved in lymphangiogenesis and may be a useful marker for the diagnosis of sepsis and/or SIRS. Further studies with larger numbers of animals are needed to determine if FLT-4 can differentiate between sepsis and SIRS patients.

**ABSTRACT #105**

PULMONARY MANIFESTATION IN CANINE LEPTOSPIROSIS. K Radeke[1], B Kaser-Hotz[1], G Arndt[2], AD Gruber[3], B Guerra[4], A Jansen[5], F Lotz[2], E Luge[4], K Nöckler[4], B Kohn[1]. 1. Clinic of Small Animals, Faculty of Veterinary Medicine, Freie Universität Berlin; 2. Institute for Biometrics and Data Processing, FU Berlin; 3. Department of Veterinary Pathology, FU Berlin; 4. Federal Institute for Risk Assessment, Berlin; 5. Robert Koch Institute, Berlin, Germany.

Canine leptospirosis is a disease with multi-organ involvement with mostly renal and hepatic manifestations. Severe pulmonary hemorrhage syndrome" with respiratory signs (dyspnea, coughing, hemoptysis), radiological pulmonary changes and mortality rates of over 50% has been described in human medicine. However, thus far lung manifestation in dogs with leptospirosis has been reported infrequently.

Therefore, the medical records of dogs diagnosed with leptospirosis at the Small Animal Clinic, FU Berlin between 2006 and 2008 were reviewed with regard to: 1) clinical signs with special emphasis on respiratory signs and radiological pulmonary changes, 2) comparison of laboratory results among dogs with and without lung manifestation, 3) correlation between severity of pulmonary distress and outcome of disease.

The diagnosis of leptospirosis was based on results of MAT titers, urine PCR, and histopathology. Based on clinical and radiological signs patients were grouped in dogs with (group 1) and without (group 2) lung manifestation. Severity of pulmonary distress was graded as mild to moderate (grade 1) and severe (grade 2). Thoracic radiographs were graded based on pulmonary changes (interstitial, alveolar) and location (dorsal, generalized) in grade 1 (interstitial, dorsal), 2 (alveolar, dorsal), 3 (interstitial, generalized), and 4 (alveolar, generalized). The results of CBC and clinical biochemistry were compared between group 1 and 2.

37 dogs with leptospirosis were included in the study. 22 of 37 dogs (59.5%) suffered from pulmonary distress (15 dogs of grade 1, 7 of grade 2) either at presentation or during course of treatment. Radiological pulmonary changes of grade 1 were detected in 1 dog, of grade 3 in 11 dogs, and of grade 4 in 10 dogs with pulmonary distress. At admission or during course of disease 89% of the dogs were azotemic, 81% had increased liver enzymes and hyperbilirubinemia; no significant differences between groups were detected. There were significant differences between group 1 and 2 with regard to the severity of thrombocytopenia (median platelet count group 1 $68x10^3/\mu l$, group 2 $151x10^3/\mu l$, $p=0.01$) and anemia (median Hct group 1 29%, group 2 36%, $p=0.04$). All dogs with dyspnea grade 2 had to be euthanized due to respiratory distress. 33% of the dogs with dyspnea grade 1 and 27% of the dogs without respiratory signs were euthanized due to acute renal or hepatic failure.

A pulmonary form of leptospirosis was detected in 60% of the patients. Lung involvement represents a severe complication causing increased mortality depending on the severity of the respiratory signs.

**ABSTRACT #106**

SEROPREVALENCE OF FELV AND FIV IN CANADA. Susan Little, Bytown Cat Hospital, Ottawa, Ontario; Dorothee Bienzle, William Sears, University of Guelph, Guelph, Ontario; Jessica Lachtara, IDEXX Laboratories, Westbrook, ME.

FeLV and FIV are retroviruses that represent two of the most common and important infectious diseases of cats worldwide. Several published studies have evaluated the seroprevalence of FeLV and FIV in North American cats, but little of the available data is from Canada. The purposes of this study were to determine the seroprevalence of FeLV and FIV infection among cats in Canada and to identify risk factors for seropositivity.

Veterinary clinics, animal shelters, cat rescue programs, and feral cat programs in Canada participated in the study by testing cats and kittens for FeLV antigen and FIV antibody from August to November 2007. Cats were tested using a commercially available point-of-care ELISA test (IDEXX Snap Combo) or samples were submitted for ELISA testing to a diagnostic laboratory (IDEXX PetChek FIV Antibody,

PetChek FeLV Antigen). Confirmatory tests were not performed as part of the study. Seroprevalence was calculated as the percentage of positive tests in the study population for each virus.

A total of 11,144 cats were tested. Seroprevalence for FeLV was 3.4% and for FIV was 4.3%. Of these, 58 cats (0.5%) were co-infected with both viruses. The probability of a positive test for FIV was not significantly higher for cats tested at shelters or rescue programs than at veterinary clinics. The probability of a positive test result for FeLV for cats tested at veterinary clinics was 1.7x higher than for cats tested at shelters or rescue programs (p=0.023). Prevalence was significantly higher in sick cats (6.7% FIV, 6.6% FeLV) versus healthy cats (3.2% FIV, 2.0% FeLV), and in adult cats (5.9% FIV, 4.4% FeLV) versus juvenile cats (1.6% FIV, 1.7% FeLV). The probability of a positive test result for FIV was 3.4x higher (p<0.001) in cats with access to outdoors when compared to cats with no outdoor access. The probability of a positive test result for FeLV was 1.4x higher (p=0.027) in cats with access to outdoors when compared to cats with no outdoor access. The probability of a positive test result for FIV was highest in intact males (7.4%), and for FeLV was highest in intact females (7.3%) and intact males (7.1%).

The seroprevalence for FeLV (3.4%) and FIV (4.3%) reported here is higher than in a recent large study that evaluated samples predominantly from the U.S. (FeLV 2.3%, FIV 2.5%). These findings emphasize variability in prevalence rates, and suggest that despite advances in testing and vaccination, infection with FeLV and FIV remains common in Canada.

**ABSTRACT #107**

PROGNOSTIC FACTORS IN CATS WITH FELINE PANLEUCOPENIA.

D. Kruse[1], S. Unterer[1], K. Horbacher[1], C. Sauter-Louis[2], K. Hartmann[1]. [1]Medizinische Kleintierklinik, [2]Clinic for Ruminants, Ludwig Maximilian University Munich, Germany.

Feline panleucopenia is a highly contagious viral disease characterized by enteritis, panleucopenia, and high mortality. The purpose of this study was to identify prognostic factors for the survival of cats with panleucopenia.

Medical records of 244 cats diagnosed with panleucopenia between 1990 and 2007 were evaluated retrospectively with regard to history, signalment, as well as clinical and laboratory parameters. Cats that tested positive for feline panleucopenia virus (FPV) *via* electron microscopy, polymerase chain reaction (blood, feces), antigen ELISA (feces), or that had histopathological lesions consistent with panleucopenia at necropsy were included.

Survival rate was 51.1%. No significant correlation was found between outcome and living conditions (indoor *versus* outdoor cat), age, vaccination status (vaccinated *versus* unvaccinated cats) or severity of clinical signs. Cats that died from the disease had significantly lower leucocyte and thrombocyte counts at time of presentation when compared to surviving cats. The relative risk of death for patients with a leucocyte count of < 1000/µl was 1.77 times as high as in patients with a leucocyte count of 1000-2500/µl (p = 0.038), and 1.85 times as high as in patients with a leucocyte count of > 2500/µl (p = 0.001). Furthermore, the likelihood of a fatal outcome was higher when the serum albumin concentration was < 30 g/l or the serum potassium concentration was < 4 mmol/l.

In conclusion, leucocyte and thrombocyte counts as well as albumin and potassium serum concentrations at time of presentation can be used as prognostic markers in cats with panleucopenia.

**ABSTRACT #108**

PREVALENCE OF PROTECTIVE ANTIBODY TITERS FOR DISTEMPER VIRUS AND PARVOVIRUS IN DOGS ENTERING A FLORIDA ANIMAL SHELTER.

ES Lechner[1], PC Crawford[2], CH Edinboro[3], EJ Dubovi[4], R Caligiuri[5], and JK Levy[2]. 1. University of Florida College of Veterinary Medicine, Gainesville FL. 2. University of Florida Maddie's Shelter Medicine Program, Gainesville, FL. 3. Exponent®, Inc. Health Sciences Group, Menlo Park, CA. 4. the Animal Health Diagnostic Center, College of Veterinary Medicine, Cornell University, Ithaca, NY. 5. Alachua County Animal Services, Gainesville, FL.

Outbreaks of canine distemper virus (CDV) and canine parvovirus (CPV) are common in animal shelters, resulting in death of thousands of dogs each year. The purpose of this study was to determine the proportion of dogs entering an animal shelter with protective antibody titers (PAT) for CDV and CPV and to identify predictive factors associated with these titers.

CDV and CPV antibody titers were measured in 431 dogs on admission into a Florida animal shelter. Overall, 65% had insufficient CDV and/or CPV titers at admission. Only 35% of the dogs had PATs for both CDV and CPV, 8% had PATs for CDV but not CPV, 32% had PATs for CPV but not CDV, and 25% lacked PATs for either virus. In univariate analyses, dogs 1 year of age or older, neutered dogs, and dogs reclaimed by owners were more likely to have CDV and CPV PATs. In multivariate analyses, age and neutered status were significant predictors for CDV PAT, while age was a significant predictor for CPV PAT. Dogs with PATs for one virus were more likely to have PATs for the other virus.

Crowding, stress, and random co-mingling of dogs from different sources are common at shelters and contribute to risk for disease outbreaks. In this study, most dogs did not have protective antibody titers to CDV and/or CPV at the time of admission to the shelter. This supports current guidelines to vaccinate all dogs immediately upon admission to shelters, regardless of source, age, or physical condition.

**ABSTRACT #109**

DETECTION OF CANINE PARVOVIRUS TYPE 2C BY A COMMERCIALLY AVAILABLE IN-HOUSE RAPID TEST.

Nicola Decaro,[1] Costantina Desario,[1] Melissa J. Beall,[2] Marco Campolo,[1] Maria Loredana Colaianni,[1] Anthony A. Dimarco,[2] Canio Buonavoglia[1]. [1] Department of Public Health and Animal Sciences, Faculty of Veterinary Medicine, Valenzano (Bari), Italy, [2] IDEXX Laboratories, Inc., Westbrook, Maine, USA.

Canine parvovirus (CPV) is one of the main pathogens causing haemorrhagic gastroenteritis and mortality in pups. In addition to the recognized variants CPV-2a and CPV-2b, a third variant, CPV-2c, emerged in Italy in 2000 and quickly spread to the canine population worldwide. Recently, concerns have been expressed about the ability of the in-house tests to detect the new variant 2c at the same extent of CPV-2a/2b. In this study, we have evaluated for the first time the detection rates of the different CPV variants by using a commercially available in-house test. A total of 201 specimens, containing CPV DNA loads higher than $10^5$ DNA copies $mg^{-1}$ faeces, were tested with the SNAP® Canine Parvovirus Test kit (IDEXX Laboratories, Inc.). The sample distribution according to the virus type was determined by minor groove binder (MGB) probe PCR assays. The samples originated from Italy ($n$ = 125), United Kingdom ($n$ = 56), Spain ($n$ = 5), and Greece ($n$ = 15). Performance was assessed by calculating the confidence intervals about the means using exact binomial limits with an alpha of 5%. The in-house assay was able to detect 41/51 types 2a, 39/50 types 2b and 77/100 types 2c CPVs, and percent positives was 80.4%, 78.0% and 77.0% for CPV types 2a, 2b and 2c, relative to real-time PCR. The in-house test gave positive results for specimens ranging from $3.92x10^5$ to $1.5x10^{11}$ DNA copies $mg^{-1}$ faeces. The detection rate for CPV-2c was not significantly different from those of the other two CPV types, thus dissipating any previous concerns about the hypothesised but never demonstrated less efficiency of the test in detecting the new variant. At the same time, some important mutations identified more recently were proven not to affect test sensitivity.

**ABSTRACT #110**

PREVALENCE OF, RISK FACTORS FOR, AND ASSEMBLAGE TYPES OF *GIARDIA* INFECTION IN CATS HOUSED IN AN ANIMAL SHELTER. SD Janeczko, B Griffin, SC Barr, JM Scarlett. Cornell University College of Veterinary Medicine, Ithaca, NY.

*Giardia* spp. infection in cats housed in animal shelters is presumably common, with the reported prevalence ranging from 1-80%. Infected cats may suffer from small bowel diarrhea but are frequently asymptomatic, and risk factors for infection are not well understood. Data regarding the frequency of zoonotic transmission is lacking and is generally based on circumstantial rather than experimental evidence, making it difficult to assess the public health risk associated with *Giardia* in cats. Both host adapted and potentially zoonotic assemblages have been identified in the small number of feline *Giardia* isolates that have been genotyped to date. Thus, confusion exists regarding the significance of *Giardia* in cats and the need for surveillance and treatment in animal shelters. The purpose of this study was to estimate the prevalence of *Giardia* in cats available for adoption at an animal shelter in central New York, assess zoonotic risk by determining the assemblage type(s) found in cats shedding *Giardia* cysts, and determine risk factors for infection.

A single fecal sample was collected monthly for 5 months from all cats and kittens available for adoption. Samples were collected from cats housed in cages and in colony rooms. Cats consumed food coloring or glitter once daily in canned food during the sample collection. These served as fecal markers, allowing identification of each cat's feces in the litter box. Each cat was individually examined in order to determine age, sex, body weight, and body condition score. Additional information collected for each cat included: origin (stray vs. relinquished), length of time in the shelter, type of housing, number in the housing unit, square footage per cat, and total number of cats in the shelter at the time of sample collection. A fecal sample was identified for each cat, scored using a standardized fecal scoring system (Ralston Purina, St. Louis, MO), and tested for the presence of *Giardia* antigen using a commercially available ELISA test (SNAP Giardia, IDEXX Laboratories, Westbrook, ME). Magnesium sulfate centrifugation was performed on all positive samples to concentrate the cysts and assemblage type was determined. Risk factors for infection were analyzed.

Of 554 fecal samples from 302 different cats, 60 positive samples were obtained from 49 different cats during the 5 month study period. The monthly prevalence as determined by the SNAP ELISA ranged from 6.6% to 14.8%, with increasing rates of infection seen during months with higher population densities and correspondingly lower square footage per cat. Cats housed in group settings were more likely to be shedding *Giardia* cysts than those cats housed individually. Preliminary sequencing results have identified both the zoonotic assemblage A and host adapted assemblage F, suggesting the potential for zoonotic transmission from some of the shelter cats.

**ABSTRACT #111**

ASSESSMENT OF REPRODUCTIVE TRACT DISEASE IN CATS AT RISK FOR ENTERIC *TRITRICHOMONAS FOETUS* INFECTION. SG Gray, SA Hunter, MR Stone, JL Gookin. North Carolina State University, College of Veterinary Medicine, Raleigh, NC.

The purpose of this study was to determine if *T. foetus* infection can be demonstrated in reproductive tract tissue from cats housed under conditions of intense breeding and for which a high prevalence of intestinal *T. foetus* infection has been identified.

The sample population consisted of purebred, cattery-housed cats undergoing elective ovariohysterectomy or castration and for which reproductive tract tissue, feces, and a reproductive history were submitted. Reproductive tract tissue was examined for *T. foetus* by light microscopy, immunohistochemistry, and PCR. History of reproductive tract disease was evaluated for statistical association with identified or reported exposure to intestinal *T. foetus* infection.

Sixty-one cats from 36 catteries met the inclusion criteria. 25% (15/61) of cats and 67% (22/33) of catteries were identified with active or reported *T. foetus* infection. Light microscopic, immunohistochemical, or molecular evidence of *T. foetus* infection of the reproductive tract was not found in any cats including; 15 with intestinal *T. foetus* infection, 29 residing in a cattery in which *T. foetus* infected cats were identified, and 8 for which gross or light microscopic evidence of reproductive tract disease was identified. Despite intense breeding and a high prevalence of reported reproductive problems, there were no differences in total number of litters, number of litters per breeding, kitten mortality, or birth defects between cats or catteries infected with *T. foetus* and those for which *T. foetus* infection was not identified.

In this study, there was no evidence that the feline reproductive tract is colonized by *T. foetus*. Accordingly, it is unlikely that reproductive tract infection with *T. foetus* plays a significant role in overall disease transmission or response to treatment.

**ABSTRACT #112**

USE OF SNAP®4Dx® ASSAY TO MONITOR TICK-BORNE INFECTION IN MINNESOTA: A 4-YEAR STUDY. B. Thatcher[1], M. Beall[1], K. Cyr[1], B. Stillman[1], M. Eberts[2], R. Chandrashekar[1]. 1. IDEXX Laboratories, Inc, Westbrook, ME. 2. Lakeland Veterinary Hospital, Baxter, MN.

In an effort to study the prevalence as well as efficacy of treatment and prevention of canine tick-borne diseases, we have been conducting a prospective study in Baxter, Minnesota. A total of 1809 dogs were tested by SNAP® 4Dx® between July 2004 and June 2008. The majority of the dogs were sampled as part of routine wellness exams (1630, 90.1%), while 179 (9.9%) were suspected of anaplasmosis and/or borreliosis at time of presentation. Serological trends were assessed by calculating the percentage of positive results for each condition by year. Confidence intervals about the means were calculated using normal approximation with an alpha of 5%. Overall, there appeared to be a decline in the number of dogs seroreactive for *Borrelia burgdorferi* since 2005 (39.4% in 2005 vs. 30.3% in 2008), coincident with the adoption of rigorous vaccination and treatment protocols. Similar to earlier findings (Beall et al. 2008), dogs suspected of clinical anaplasmosis and/or borreliosis were more likely to be seroreactive to both organisms (40.6%) compared to healthy dogs (21.1%). However, the number of dogs seroreactive to both *A. phagocytophilum* and *B. burgdorferi* appeared to be on the decline from 30.7% in 2005 to 21.3% in 2008 with an overall decrease in the number of clinically ill dogs from 17.4% in 2005 to 3.1% in 2008. Twenty-four dogs suspected of anaplasmosis and/or borreliosis from the previous study were followed from 2005 to 2008 to assess health, risk of re-infection, and serologic reactivity. None of the dogs reported relapses that would be consistent with chronic anaplasmosis or borreliosis; 42% became seronegative by SNAP® 4Dx® for at least one of the infections. Data from this canine field population suggests that the awareness of tick-borne diseases has encouraged the use of rigorous preventive and treatment strategies resulting in fewer seroreactive dogs as well as fewer dogs with clinical disease.

**ABSTRACT #113**

PREVALENCE OF HEMOPLASMA DNA IN FIELD-CAUGHT MOSQUITOES IN COLORADO. PC Lin, JR Hawley, BG Bolling, LM Eisen, MR Lappin. College of Veterinary Medicine and Biomedical Sciences, Colorado State University, Fort Collins, CO.

*Mycoplasma haemofelis*, '*Candidatus* M. haemominutum', and '*Candidatus* M. turicensis' (haemoplasmas) are common blood-borne organisms in cats. In dogs, *M. haemocanis* and '*Candidatus* M. haematoparvum' also exist. Other mammalians also harbor hemoplasmas including *M. wenyonii* in cattle. Transmission of hemoplasmas by arthropods has been suspected in cats, however, results of flea transmission studies have been inconclusive. The purpose of this study was to determine whether canine or feline hemoplasma DNA could be amplified from mosquitoes.

*Aedes vexans* mosquitoes caught in 2006 and 2007 as part of a West Nile Virus surveillance study were used. 100 sample pools containing approximately 50 mosquitoes each were selected from one of four field sites in Fort Collins, Colorado that were chosen based on probable close proximity to dogs and cats. Mosquitoes were pulverized and DNA extracted using a modified Qiagen protocol. Polymerase chain reaction for hemoplasma DNA was performed using a previously published assay and positive amplicons were sequenced.

Of the 100 mosquito pools, 5 pools (5%) had a band on gel electrophoresis consistent with a haemoplasma and adequate DNA for sequencing. *Spiroplasma* spp. DNA (a bacterium that infects insects) were amplified from four pools and DNA most consistent with *M. wenyonii* were amplified from the fifth pool.

Amplification of *M. wenyonii* DNA suggests that the positive mosquitoes had ingested blood from a large animal, likely cattle. Further study is needed to determine whether mosquitoes could be a mechanical vector for feline or canine hemoplasmas.

**ABSTRACT #114**

TOPICAL IMIDACLOPRID AND MOXIDECTIN PREVENTS FLEA TRANSMISSION OF *BARTONELLA HENSELAE* IN CATS. CA Bradbury, MR Lappin. Colorado State University, School of Veterinary Medicine, Department of Clinical Sciences, Fort Collins, CO.

*Bartonella* species are common pathogens in cats and people and are considered significant zoonotic agents especially in immunocompromised people. The most common species isolated from cats are *Bartonella henselae* and *B. clarridgeae* which are both transmitted by fleas. The purpose of this study was to determine whether monthly topical administration of imidacloprid and moxidectin would lessen flea transmission of *B. henselae* among cats.

Eighteen specific pathogen free cats were housed in three groups of six. The three enclosures were separated by mesh so as to allow fleas to pass among groups yet prevent cats from contacting one another. One group was inoculated intravenously with *B. henselae* and after infection was confirmed in all cats based on positive polymerase chain reaction (PCR) assay results, the cats were housed in the middle enclosure. The *B. henselae* infected cat group was flanked by a group that was administered topical imidacloprid and moxidectin monthly for three months and by a group that was not treated. On days 0, 30, and 60, 100 fleas per cat were placed on each of the six cats in the *B. henselae* infected group. Blood was collected from all cats weekly for *Bartonella* spp. PCR, serology and culture.

While *B. henselae* infection was ultimately confirmed in all of the untreated cats by blood culture and PCR assay results, none of the treated cats became infected.

In this setting, monthly topical imidacloprid and moxidectin prevents flea transmission of *B. henselae* in cats.

**ABSTRACT #115**

OUTCOME AND PROGNOSTIC FACTORS IN DOGS WITH CANINE LEISHMANIASIS IN A NON-ENDEMIC COUNTRY. K. Hartmann[1], K. Geisweid, C. Sauter-Louis[2]. [1]Medizinische Kleintierklinik, [2]Clinic for Ruminants, Ludwig Maximilian University Munich, Germany.

Due to increasing numbers of dogs that are imported from or have been visiting Mediterranean countries, canine leihmaniasis has become a very important infectious disease in Germany. So far, the disease is not yet endemic, and autochthonic infections are rare. The aim of this study was to find out, (1) which clinical and laboratory parameters can be used as prognostic factors, (2) whether height of antibody titers is a prognostic factor in a country in which re-infection is extremely unlikely, and (3) whether treatment of asymptomatic dogs leads to a better prognosis.

Data of 93 dogs infected with *Leishmania infantum*, that had been presented to the Veterinary Teaching Hospital of the University of Munich and whose owners answered a detailed questionnaire, were evaluated retrospectively.

Most of the dogs had been imported from endemic countries (78%) and were mixed-breed (34%) with a median age of 7.5 years. Most common problems were lethargy (69%), skin lesions (63%), proteinuria (55%), lymphadenopathy (52%), and hypoalbuminemia (43%). There was no correlation between height of antibody titers and severity of clinical signs at presentation or survival time. However, there was a significant correlation between survival time and presence of proteinuria ($p < 0.001$) and presence of lymphopenia ($p = 0.016$). Also, treatment was correlated with a longer survival time ($p = 0.017$), even in asymptomatic dogs, and treated dogs lived about two years longer then non-treated.

In conclusion, height of antibody titers does not correlate with prognosis. More useful parameters to predict shorter survival are proteinuria and lymphopenia.

**ABSTRACT #116**

THE USE OF COMPUTER TOMOGRAPHY IN THE DIAGNOSIS OF SEPTIC ARTHRITIS/OSTEOMYELITIS IN THE NEONATAL FOAL. Mary Rose Paradis, Mauricio Solano, Amy Tidwell , Louise Maranda Cummings School of Veterinary Medicine at Tufts University, N. Grafton, MA.

Septic arthritis/osteomyelitis is a recognized sequela to bacteremia in the neonatal foal. It manifests as lameness with increase joint effusion. Historically diagnosis of osteomyelitis has been made by radiographic examination of suspected joint or growth plate but radiographic evidence of osteomyelitis often lags behind the clinical signs. The hypothesis of this study was that computer tomography (CT) would detect bone lesions earlier and more reliably than radiography in the neonatal foal. Twenty foals between the ages of 8 hours to 36 days who were presented to the hospital with lameness secondary to septic arthritis/osteomyelitis during the foaling seasons of 2002 -2006. Clinical information was collected for each foal. A total of 54 joints (34 clinically affected and 20 normal) were evaluated by radiographs and CT. Foals enrolled in the study had radiographs and CT of the affected joint and contralateral normal joint within 5 days of each procedure (mean =1 day). The images were evaluated by 2 board certified radiologists (individually and blinded) for the presence of osteomyelitis. Descriptions of the lesions were noted. It was found that investigators were more likely to agree on the presence of osteomyelitis from CT on normal and abnormal joints (Kappa 1 and 0.86 respectively) than normal and abnormal radiographs (Kappa 0.714 and 0.476.). CT was 1.7 times more likely to detect osteomyelitis than radiography. Our conclusion was that osteomyelitis lesions in foals with septic arthritis are more likely to be seen earlier with the use of CT than radiographs. This could have an effect on both the treatment choices and prognosis for affected foals.

**ABSTRACT #117**

ENDOCRINE ENERGY RESPONSE IN SEPTIC FOALS: INSULIN, LEPTIN AND ADIPONECTIN. RJIM Barsnick[1], SDA Hurcombe[1], NM Slovis[2], KA Sprayberry[2], PA Smith[1], RE Toribio[1]. 1. The Ohio State University College of Veterinary Medicine, Columbus, OH. 2. Hagyard Equine Medical Institute, Lexington, KY.

Anorexia and hypoglycemia are common clinical findings in critically ill septic foals. The therapeutic plan for hypoglycemic foals often requires the administration of parenteral glucose or parenteral nutrition. A number of septic foals develop glucose intolerance (hyperglycemia, glycosuria), and insulin is frequently administered to affected foals. These complications, we believe are the result of sepsis-mediated dysregulation of hormones involved in energy homeostasis (insulin, leptin and adiponectin).

The goal of this study was to investigate the association of insulin, leptin and adiponectin with sepsis, severity of disease, glycemia, triglyceridemia and mortality in foals.

We measured the plasma concentrations of aforementioned hormones using validated immunoassays in 44 septic, 62 sick non-septic (hospitalized for other diseases) and 19 healthy neonatal foals. Sepsis was defined as a sepsis score >11.

Insulin concentrations were higher in healthy foals than in sick non-septic and septic ($p<0.01$) foals. Septic foals had lower serum glucose and higher triglyceride concentrations than healthy foals ($p<0.01$). Among the septic foals, non-survivors had lower glucose and higher triglyceride values than survivors ($p<0.05$), but no difference was found with insulin concentrations. A number of septic foals had a poor insulin response to hyperglycemia (normal or low insulin concentrations).

Leptin and adiponectin concentrations were not different comparing septic, sick non-septic and healthy foals. However, plasma leptin and adiponectin concentrations were significantly lower in septic foals that died ($p<0.05$).

These results indicate that endocrine dysregulation of the energy metabolism in septic foals is frequent and likely important in the pathogenesis of sepsis. The lack of insulin response to hyperglycemia indicates sepsis-mediated impairment of insulin release. The low leptin and adiponectin concentrations in septic foals support energy impairment and are potential prognosticators of survival.

Our findings are clinically applicable in the evaluation, prognosis and treatment of critically ill foals.

**ABSTRACT #118**

OPSONIZATION OF *RHODOCOCCUS EQUI* WITH HIGH ANTIBODY PLASMA DECREASES BACTERIAL VIABILITY AND PROMOTES PHAGOCYTE ACTIVATION. DR Dawson[1], MJ Flaminio[1], D Nydam[1], JE Graham[2], M Cynamon[3], TJ Divers[1]. 1. Cornell University College of Veterinary Medicine, Ithaca, NY. 2. University of Louisville School of Medicine, Louisville, KY. 3. Department of Veterans Affairs Medical Center, Syracuse, NY.

Susceptibility to *Rhodococcus equi* pneumonia in foals is exclusive to the first few months of life. Once infection is established, virulent *R. equi* survive and replicate inside macrophages. There are conflicting published data about the actual effect of intravenous high antibody plasma transfusion in the prophylaxis of *R. equi* disease in foals. In this study, we evaluate the effect of commercially available high antibody plasma on bacterial viability and phagocyte activation *in vitro*. We hypothesized that opsonization of virulent *R. equi* would decrease bacterium viability inside macrophages and increase phagocyte activation. We used a virulent *R. equi* (strain 103S) construct that expressed luciferase (*R. equi*-LUX). Luciferase activity in live *R. equi* reflects both the number of viable organisms, and the metabolic activity of the bacteria. Opsonization of *R. equi*-LUX with plasma decreased bacterial luciferase activity in comparison to non-opsonized bacteria ($p<0.0001$). A similar suppression was also observed in bacteria phagocytized by monocyte-derived macrophages after 24hr culture ($p=0.006$). The opsonization of *R. equi*-LUX increased oxidative burst activity in phagocytes ($p<0.0001$). TNFα production increased in macrophages infected with opsonized versus non-opsonized bacteria ($p<0.0001$). The effect of opsonization in *R. equi*-LUX was associated with antibody and heat-resistant factors. Low opsonization (10%) significantly decreased bacterial viability and increased phagocytic activation, and greater levels of opsonization (40%) revealed further advantages. It is possible that passively transferred antibodies against *R. equi* affect bacterial viability before macrophage invasion *in vivo*. These data suggest a possible role of colostrum and plasma antibody in protection of foals.

**ABSTRACT #119**

PLASMA ENDOTHELIN CONCENTRATIONS IN SEPTIC AND NON-SEPTIC NEONATAL FOALS. L.R.R. Costa[1*], S.E. Eades[1], A. Polkes[2], R.J. MacKay[2]; R.M. Moore[1]. 1. Equine Health Studies Program, Department of Veterinary Clinical Sciences, School of Veterinary Medicine, Louisiana State University, Baton Rouge, LA.2.Large Animal Clinical Sciences, School of Veterinary Medicine, University of Florida, Gainesville, FL. * Currently at Department of Clinical Sciences, Cummings School of Veterinary Medicine, Tufts University, North Grafton, MA.

Endothelins are endogenous peptides known by their potent spasmogenic action of smooth muscles, but shown to have diverse roles in physiological and pathophysiological processes. The best studied isoform, endothelin (ET)-1, has been implicated in the pathophysiology of vascular, cardiac, respiratory and inflammatory conditions.

The purpose of this study was to compare plasma ET-1 concentration among neonatal foals considered healthy, sick but unlikely septic and sick and likely septic.

Plasma ET-1 concentrations were determined in samples collected at the University of Florida and Louisiana State University from a total of 137 foals including 119 ill foals (divided into two age groups: < 7-d-old and ≥ 7-d-old) and 18 normal foals. Foals were divided into 3 disease groups: Group 1 = healthy, Group 2 = sick but unlikely septic (i.e., sepsis score less than 11), and Group 3 = sick and likely septic (sepsis score greater or equal 11). Differences among groups were tested for significance using a Kruskal-Wallis test, followed by Dunn test to separate group medians.

Plasma ET-1 levels were significantly higher in Group 3 foals than in foals from Groups 1 or 2 ($P < 0.001$). Overall, younger foals (< 7-d-old) had significantly higher plasma ET-1 levels than older foals (≥ 7-d-old) ($P < 0.05$). Within Group 3, the median and interquartile range for ET-1 in plasma from foals < 7-d-old was 5.58 pg/ml (IQR = 3.52 to 8.96, n=39) and was 4.74 pg/ml (IQR = 1.67 to 8.77, n=10) for those ≥ 7-d-old. Within Group 2, the median and interquartile range for ET-1 in plasma from foals < 7-d-old was 2.16 pg/ml (IQR = 0.48 to 5.26, n=47) and 0.33 pg/ml (IQR = 0.10 to 0.43, n=23) for those ≥ 7-d-old. The median for ET-1 in plasma from foals in Group 1 was 1.48 pg/ml (IQR = 0.75 to 2.78; n=18). Moreover, plasma ET-1 concentrations appeared to be affected by the presence of enteritis, hypoxic-ischemic encephalopathy, and maturity (premature birth or term foal). These results suggest a pathophysiological role of ET-1 in equine neonatal sepsis.

*Previously presented at Dorothy Havermeyer Foundation Workshop in Neonatal Septicemia, November 2008*

**ABSTRACT #120**

THROMBOELASTOGRAPHY (TEG): AN INNOVATIVE TECHNIQUE FOR EARLY DETECTION OF HEMOSTATIC ABNORMALITIES IN FOALS WITH SEPTICEMIA. JL Mendez, P Vilar, MC Mudge, CG Couto. The Ohio State University College of Veterinary Medicine, Columbus, OH.

Hemostatic abnormalities are common in neonatal foals with severe septicemia, and frequently result in secondary complications such as disseminated intravascular coagulation and multiple organ dysfunction syndrome. Newborn foals with severe septicemia have recently been shown to have widespread intravascular fibrin deposits with evidence of hypercoagulability (Cotovio 2008). Conventional hemostasis profiles do not detect hypercoagulability reliably. Thromboelastography provides data about the entire hemostatic system, from clot formation to fibrinolysis. Early detection of hypercoagulability by TEG should allow for interventional therapy and may change the outcome in hypercoagulable septic foals. The purpose of this study was to evaluate and compare TEG parameters in sick septic and non-septic neonatal foals.

Blood samples were collected from 18 foals (<2 weeks) upon admission to the Emergency Service into citrated tubes; they were placed in the thrombelastograph (TEG) cup and recalcified by adding $CaCl_2$ 30-90 minutes after sampling. Group 1 included septic foals (n=9) with a positive blood culture and/or sepsis score >11 and a focus of infection, and Group 2 included sick non-septic foals (n=9) with sepsis score <10 and no evidence of systemic inflammation. CBC, fibrinogen, prothrombin time and activated partial thromboplastin time were performed. Specific TEG parameters including R-time, K-time, angle "α", maximum amplitude "MA" and G-value were compared between groups.

There were significant differences ($p<0.05$) in TEG parameters; foals with septicemia had shorter K-time, wider "α", increased "MA" and G-value, changes consistent with hypercoagulability.

In conclusion, TEG can be used to accurately detect early hemostatic changes such as hypercoagulability in septic foals.

**ABSTRACT #121**

EFFECTS OF DOBUTAMINE AND ATROPINE AND ACETYLPROMAZINE ON AORTIC VALVE FUNCTION IN THE HORSE. GD Hallowell[1], TJ Potter[2] and IM Bowen[1] 1.School of Veterinary Medicine and Science, University of Nottingham, UK.2.Royal Veterinary College, North Mymms, UK.

Echocardiograms are performed on horses after drug administration for stress echocardiography and sedation and therefore an understanding of the effects of these agents on valve function is important for image interpretation. The objective of this study was to evaluate the effects on aortic valve function of administration of the sedative acetylpromazine (ACP) and atropine and dobutamine in horses subjected to stress echocardiography, to ascertain if these agents induced or masked the echocardiographic appearance of aortic valve prolapse (AVP).

For evaluation of ACP, echocardiography was performed on eight healthy Thoroughbred (TB) research horses that were sedated for routine dental work. For evaluation of atropine and dobutamine, images from eight TB or TB cross horses that were presented for investigation of poor performance were examined. Following a detailed clinical investigation none of the horses with poor performance were subsequently found to have underlying cardiac dysfunction.

Two dimensional and Doppler echocardiographic views were obtained before and after acetylpromazine and atropine-dobutamine stress echocardiography. Changes in internal cardiac dimensions, presence and extent of AVP (and physiological aortic regurgitation (PAR)) were recorded.

Acetylpromazine administration did not affect the occurrence of AVP (25% to 33%; $p=0.46$). There was no effect upon occurrence of PAR ($p=0.75$). However although there were no significant differences in jet length or width, there was an increase in jet area ($p=0.02$) identified.

The occurrence of AVP increased from 25 to 75% with atropine and dobutamine administration ($p=0.02$). PAR was noted in 75% of horses prior to drug administration, in 63% post-atropine administration and in 100% post-dobutamine administration ($p=0.26$). There was no significant difference in the length ($p=0.73$), width ($p=0.75$) or area ($p=0.21$) of the regurgitant jet following drug administration.

In conclusion it appears that the administration of dobutamine and atropine changes aortic valve competency leading to an increased occurrence of AVP and PAR. Acetylpromazine may increase the degree of PAR. Changes due to the direct effects of these agents should be taken into account on interpretation of echocardiographic results.

**ABSTRACT #122**

INTRAVENOUS 99m-TECHNETIUM LIPOSOMES IN HORSES: A SAFETY AND BIODISTRIBUTION STUDY. C. Underwood[1], A.W. van Eps[1], P. Laverman[2], G. Storm[3], L. van Bloois[3], S.R. van Tomme[3], M.W. Ross[1], T.P. Schaer[1]. 1. University of Pennsylvania, New Bolton Center, Kennett Square, PA. 2. Radboud University Nijmegan Medical Center, Nijmegen. NL. 3. Utrecht Institute for pharmaceutical Sciences, University of Utrecht, Utrecht, NL.

Liposomes are phospholipid nanoparticles that can extravasate at sites of increased vascular permeability. They can be used for targeted drug delivery and diagnostic imaging. This study is the first to describe the intravenous administration of $^{99m}$Tc labeled polyethylene glycol (PEG) coated liposomes in normal horses.

The liposomes were prepared via the film hydration method. Labeling of the glutathione liposomes was performed using $^{99m}$Tc-hexamethyl-propylene-amine-oxime ($^{99m}$Tc-HMPAO). An assay was developed to establish the 50% serum hemolytic complement activity ($CH_{50}$) in horses, as complement-mediated reactions are a common adverse effect in other species. *In vitro* $CH_{50}$ assays were performed prior to liposome injection. Seven horses were administered 0.4 µmol/kg $^{99m}$Tc-HMPAO PEG-liposomes and 2.5µmol/kg unlabelled PEG-liposomes intravenously. Clinical parameters, hematology, plasma biochemistry and serum complement activity were monitored serially. Scintigraphic imaging was performed at 1, 12, and 21 hours post-infusion. Three of the horses were euthanized at 23 hours. The percentage of injected dose per kilogram of tissue (%ID/kg) was calculated for multiple organs. Results were analyzed using ANOVA and Friedman tests.

Complement assays required the addition of C3 depleted human serum to the standard hemolytic assay to obtain a $CH_{50}$ value. No significant complement activation was detected. No significant clinical, hematological or biochemical changes occurred following liposome administration, although one horse developed mild transient urticaria and mild agitation during the infusion. Scintigraphic studies revealed a prolonged vascular phase that lasted to 21 hours, with increased radiopharmaceutical uptake in the liver, lungs, spleen and kidney. There was a repeatable pattern of organ distribution. Evaluation of tissue samples revealed the highest concentrations within the lung, kidney, liver and spleen, with significantly increased $^{99m}$Tc activity in the lung and spleen compared with blood. The mean circulatory half life of the $^{99m}$Tc PEG liposomes was 11.51 +/- 2.2 hours.

There were minimal adverse effects associated with intravenous liposome administration in normal horses. This study establishes normal scintigraphic findings after administration of $^{99m}$Tc PEG liposomes. Liposomes have significant potential in horses, as a diagnostic tool and for targeted drug delivery, particularly in the management of refractory infections and neoplasia.

**ABSTRACT #123**

PREVALENCE AND RISK FACTORS FOR HYPERINSULINEMIA IN CLINICALLY NORMAL HORSES IN CENTRAL OHIO. J Muno[1], L Gallatin[1], RJ Geor[2], GA Anderson[3], KW Hinchcliff.[3] The Ohio State University, Columbus, OH.[1] Virginia Tech, Blacksburg, VA.[2] University of Melbourne, Australia.[3]

Obesity is common in horses and is a risk factor for insulin resistance. However, other risk factors for insulin resistance in horses have not been well documented. The aim of this study was to determine the prevalence and risk factors for hyperinsulinemia in clinically normal, adult horses.

A convenience sample of 300 horses (138 mares, 143 geldings and 19 stallions; 4 to 30 years) was drawn from 18 farms in central Ohio. Plasma insulin and glucose were measured after a 10-12 hour period of grain and grass withholding. Body condition score, height, weight tape, neck circumference and management practices were recorded. A univariable logistic regression and a multivariable logistic regression with a random effect of farm were used to estimate odds ratios for hyperinsulinemia.

The prevalence of hyperinsulinemia, (>15 μU/L) was 22.3% (67/300). A number of factors including breed, age, body condition score, girth circumference, not feeding grain and lack of access to pasture were associated with hyperinsulinemia in the univariable analysis. Multivariable analysis revealed that increasing age (odds ratio per year of increasing age 1.1, 95% CI 1.03-1.17) and body condition score (odds ratio 2.43, 95% CI 1.6-3.7) increased the risk for hyperinsulinemia while access to pasture (odds ratio 0.34, 95% CI 0.14-0.86) decreased the risk. Breed, sex, exercise, and measures of neck and girth circumference were not associated with risk of hyperinsulinemia.

This study demonstrates that when accounting for common management factors, only indirect measures of obesity and increasing age are independently associated with increased risk of hyperinsulinemia.

**ABSTRACT #124**

SEASONAL VARIATION IN PLASMA INSULIN AND GLUCOSE CONCENTRATION IN NORMAL HORSES IN CENTRAL OHIO. J Muno[1], L Gallatin[1], RJ Geor[2], GA Anderson[3], KW Hinchcliff[3]. The Ohio State University, Columbus, OH[1]. Virginia Tech, Blacksburg, VA[2]. University of Melbourne, Australia[3].

Some tests of endocrine function in horses, including the dexamethasone suppression test and plasma adrenocorticotrophin concentration have seasonal variation. The purpose of this study was to determine if plasma insulin and glucose concentrations vary with season in healthy horses in central Ohio.

Twenty-nine healthy horses, as determined by physical examination, were evaluated at three-month intervals over a one year period. Plasma insulin and glucose concentrations were measured after a 10-12 hour period of grain and grass withholding. Body condition score, height, weight tape and neck circumference were measured at the same time. Friedman's test and Wilcoxon's signed rank test were used to compare the overall and individual differences between the four seasons in the values of insulin and glucose. The type 1 error rate was 5%.

There was an effect of season on the concentration of both insulin and glucose (P<0.001). The concentration of insulin in summer was lower (median 4.6 μU/L, range 1.2 to 22.0) than any of the other three months (P<0.001). There was no significant difference between any of the other seasons (autumn 13.4, range 4.3 to 271.7; winter 13.6, range 5.3 to 401.7; spring 12.5, range 2.0 to 412.4). The concentration of glucose in autumn was less than in any of the other three seasons (P<0.001), and the concentration in summer was less than in spring (p=0.041).

This study demonstrates that an apparent seasonal variation exists in the plasma concentrations of both insulin and glucose in normal horses. This information may modify the interpretation of laboratory findings indicative of insulin resistance.

**ABSTRACT #125**

EFFECTS OF A SUPPLEMENT CONTAINING CHROMIUM AND MAGNESIUM ON INSULIN SENSITIVITY IN HORSES WITH EQUINE METABOLIC SYNDROME. K Chameroy, N Frank, SB Elliott, RC Boston. Department of Large Animal Clinical Sciences, College of Veterinary Medicine, University of Tennessee, Knoxville, TN 37996 (Chameroy, Frank, Elliott); and the Department of Clinical Studies, University of Pennsylvania, Kennett Square, PA (Boston).

Chromium and magnesium are commonly recommended as supplements for horses with equine metabolic syndrome (EMS), which is a clinical syndrome comprised of obesity, regional adiposity, insulin resistance, and laminitis. The aim of this study was to determine whether a combination supplement containing chromium, magnesium, and herbs would alter physical measurements and/or insulin sensitivity in horses affected by EMS. Fourteen horses with resting hyperinsulinemia (>30 μU/mL) and/or abnormal combined intravenous glucose-insulin tolerance test results were randomly allocated to treatment (n = 8) and control (n = 6) groups. Treated horses received the supplement once daily in 0.5 lb oats for a total of 16 weeks. The supplement evaluated contained 5 mg chromium as yeast and 8.8 g magnesium oxide/proteinate. Frequently sampled intravenous glucose tolerance tests and physical measurements were performed at 0, 8, and 16 weeks.

Body weight and neck circumference measurements remained unaffected by treatment in this study. Glucose and insulin data from one horse in the treatment group were excluded because pain associated with recurrent laminitis affected results. Mean ± SD (n = 13) insulin sensitivity was $0.64 \pm 0.62 \times 10^{-4}$ $L{\bullet}min^{-1}{\bullet}mU^{-1}$ for horses included in this study and no effect of treatment was detected over time. Other measures of glucose and insulin dynamics and resting blood glucose, insulin, and triglyceride concentrations also remained unaffected by treatment.

We conclude that the specific formulation evaluated, given once daily in the feed for 16 weeks did not alter physical measurements or glucose and insulin dynamics in horses with EMS.

**ABSTRACT #126**

SEASONAL CHANGES IN THE COMBINED GLUCOSE INSULIN TOLERANCE TEST IN NORMAL AGED HORSES.

RA Funk, AJ Stewart, AA Wooldridge, RJ Kemppainen, EN Behrend, AK Johnson. Auburn University College of Veterinary Medicine, Auburn, Alabama.

Insulin resistance (IR) is increasingly recognized in adult horses. Insulin resistant horses are often obese and predisposed to developing laminitis, especially in spring and summer. Insulin resistance can also occur secondary to pituitary pars intermedia dysfunction (PPID). Diagnostic testing for PPID is challenging in the fall due to seasonal increases in hormones in normal horses. The purpose of this study was to evaluate seasonal changes in the combined glucose insulin tolerance test (CGIT) for IR.

Ten healthy aged horses with normal dexamethasone suppression tests (indicating absence of PPID) were administered CGIT in February, May, June, August, September, and November. Horses were administered dextrose (150 mg/kg) and insulin (0.1 U/kg) intravenously. Glucose concentrations were measured at 0, 1, 5, 15, 30, 45, 60, 75, 90 and 150 minutes. Insulin concentrations were analyzed at 0, 5, and 75 minutes. Mean glucose and insulin concentrations for each time point and mean area under the curve for positive and negative glucose peaks were compared between months using 2-way analysis of variance. Significance was set at $p<0.05$.

Glucose returned to baseline levels by 45 minutes in all horses, except for two individuals in September. No significant differences in mean glucose and insulin concentrations at each time point nor in mean area under the curve for positive and negative glucose peaks between months were identified.

Therefore, the majority of horses were considered insulin sensitive throughout the study. Based on this initial analysis, seasonal changes do not appear to affect the results of the CGIT in the majority of normal aged horses.

**ABSTRACT #127**

EFFECTS OF SAMPLING TIME AND HAY FEEDING ON BLOOD GLUCOSE, INSULIN, AND ADRENOCORTICOTROPIN HORMONE (ACTH) CONCENTRATIONS IN HORSES.

NS Chumbler, Tóth F, Elliott SB, N Frank. Departments of Large Animal Clinical Sciences, College of Veterinary Medicine, University of Tennessee, Knoxville, TN.

Ideal sampling conditions for glucose, insulin, and adrenocorticotropin hormone (ACTH) measurements have not been established for horses, so this study was undertaken to measure interday and intraday variability, and to determine whether horses should be deprived of feed or fed hay prior to sampling.

Six healthy crossbred mares were compared with 6 horses with hyperinsulinemia (> 30 μU/mL). Blood samples were collected at 0800 for 4 days and then every hour from 0800 to 2000 on the fourth day. Sampling took place after feed deprivation the first week and after hay feeding the next week according to a crossover design.

Serum insulin concentrations remained below 30 μU/mL for 5 of 6 mares in the control group, but hyperinsulinemia (maximum: 43.3 μU/mL) was detected in the remaining mare when hay was fed. Overall mean interday coefficients of variation were 6%, 43%, and 20% for glucose, insulin, and ACTH, respectively. Hay feeding was associated with significantly ($P < 0.05$) higher serum insulin concentrations in both healthy (8.2 versus 2.2 μU/mL) and insulin resistant (56 versus 39 μU/mL) horses across the 4-day testing period. Significantly higher mean serum insulin concentrations were also detected when hay was fed to healthy (9.5 versus 2.3 μU/mL) and insulin resistant (78.8 versus 19.1 μU/mL) horses across 12h.

Feeding hay raised serum insulin concentrations, which suggests that horses should be deprived of feed before blood samples are collected. However, hyperinsulinemia developed after hay feeding in one mare, so a provocative test may be useful in these cases.

**ABSTRACT #128**

EFFECTS OF ENDOTOXEMIA AND CARBOHYDRATE OVERLOAD ON GLUCOSE AND INSULIN DYNAMICS AND THE DEVELOPMENT OF LAMINITIS IN HORSES.

F Tóth, N Frank, RC Boston. Department of Large Animal Clinical Sciences, College of Veterinary Medicine, University of Tennessee, Knoxville, TN (Tóth, Frank); and the Department of Clinical Studies, University of Pennsylvania, Kennett Square, PA (Boston).

Endotoxemia lowers insulin sensitivity and has been anecdotally associated with laminitis in horses, but a direct cause-and-effect relationship between endotoxemia and laminitis has not been established. The aims of this study were therefore to examine the effects of endotoxemia and alimentary carbohydrate overload on glucose and insulin dynamics and to determine whether administration of both treatments sequentially affected the incidence or severity of laminitis.

Horses (n = 24) were randomly allocated to three treatment groups, each containing 8 horses. A lipopolysaccharide (LPS) group received LPS intravenously for 8h, an oligofructose (OF) group received 5g/kg OF via nasogastric intubation, and a LPS/OF group received LPS for 8h followed by OF the next day. Glucose and insulin dynamics were assessed before and after treatment by frequently sampled intravenous glucose tolerance testing and minimal model analysis. Physical examinations and complete blood counts were performed and the severity of laminitis was determined using the Obel grading scale.

Horses receiving LPS exhibited clinical signs of a systemic inflammatory response and leucopenia. Mean insulin sensitivity decreased by 66%, 74%, and 49% in LPS, OF, and LPS/OF groups, but responses did not differ significantly among groups. In contrast, development of Obel grade ≥ 2 laminitis was significantly associated with treatment, with the highest incidence detected for the LPS/OF group.

Insulin sensitivity decreases after endotoxin and/or OF administration in horses. Endotoxemia predisposes horses to carbohydrate overload-induced laminitis and increases the severity of disease. Medical management of endotoxemia is warranted to prevent laminitis in horses with intestinal disease.

**ABSTRACT #129**

PRO-INFLAMMATORY CYTOKINE AND CHEMOKINE EXPRESSION PROFILES OF VARIOUS ADIPOSE TISSUE DEPOTS OF INSULIN-RESISTANT AND INSULIN-SENSITIVE LIGHT BREED HORSES.

Burns TA[1], Geor RJ[2], Mudge MC[1], McCutcheon LJ[2], Belknap JK[1]. 1. The Ohio State University, Columbus, OH. 2. Michigan State University, East Lansing, MI.

Insulin resistance has been associated with predisposition to laminitis in horses and ponies. In humans and rodents, omental adipose tissue (AT) expresses pro-inflammatory cytokines and adipokines at significantly higher levels than subcutaneous or retroperitoneal AT, in correlation with the degree of insulin resistance. While this has been postulated to also be a factor contributing to insulin resistance in horses, no published data currently support or refute this hypothesis. This study characterized the expression of pro-inflammatory cytokines and chemokines in several AT depots of insulin-resistant (IR) and insulin-sensitive (IS) horses. Eleven mares (8 Quarter horses, 2 Thoroughbreds, and 1 Standardbred), categorized as IR (insulin sensitivity [SI] = 0.58 ± 0.31 $L \cdot min^{-1} \cdot mU^{-1}$; n = 5) or IS (SI = 2.59 ± 1.21 $L \cdot min^{-1} \cdot mU^{-1}$; n = 6) based on results of an insulin-modified frequently-sampled intravenous glucose tolerance test were studied. No statistically significant differences in weight, body condition score, neck circumference, girth, or ultrasonographic retroperitoneal fat thickness were noted between groups. Omental, retroperitoneal, and mesocolonic AT was collected concurrently via ventral midline celiotomy; nuchal ligament and tail head AT biopsies were collected via skin incision. All tissues were snap-frozen and stored at -80°C. For each depot, total RNA was extracted and cDNA analyzed via real-time quantitative PCR to quantify expression of TNF-α, IL-1β, IL-6, plasminogen activator inhibitor-1 (PAI-1), and monocyte chemoattractant protein-1 (MCP-1). Data were analyzed using a Kruskal-Wallis test with a Dunnett's post test (p = 0.05). No differences in AT expression of TNF-α, IL-1β, IL-6, PAI-1, or MCP-1 were noted between IR and IS groups for each depot. However, when data from IR and IS groups were combined for each depot, the expression of IL-1β (p=0.009) and IL-6 (p=0.023) was significantly higher in nuchal ligament AT than in other depots, suggesting that this AT depot has different biological behavior in the horse and is more likely to adopt an inflammatory phenotype than other depots examined. Importantly, these data indicate that omental adipose tissue (and other visceral fat depots) may not be as important to the pathophysiology of obesity in the horse as in other species.

**ABSTRACT #130**

AN ANIMATED MODEL OF RETICULORUMEN MOTILITY.

JL Gookin, DM Foster, AM Harvey. North Carolina State University, College of Veterinary Medicine, Raleigh, NC.

Akin to the heart, faithful contraction of the ruminant forestomach is required to sustain life. Reticulorumen motility mediates the mixing of fiber with fermentative bacteria, regulates the rate of nutrient intake and delivery, permits offloading of ~60 liters of fermentative gases per hour, and enables these animals to eat in haste and re-masticate their food at leisure and in greater safety. An understanding of reticulorumen motility is important to the assessment of ruminant health, optimal production, and in recognition, diagnosis and treatment of disease. Teaching reticulorumen motility can be challenging because real-time images of the contraction events are largely unavailable. Didactic material is based mostly on written descriptions, line drawings, or pressure tracings obtained during a contraction sequence. These approaches make learning reticulorumen motility dry and tedious and fails to leave the student with an appreciation for the integrated nature and periodicity of the contraction events. Our goal was to develop an animated model of reticulorumen motility. We hypothesize that first-year veterinary students will prefer use of the animated model over traditional instructional methods for learning.

An animated model of the reticulorumen was created using Adobe®Flash. Illustration components were created in Adobe®Illustrator and exported into Flash; 4 different sequences of motility were then created by manipulating the Illustrator paths frame by frame within Flash. Each sequence was saved as a scripted movie clip and placed in an interactive interface with a main menu for user access and control of each animation. An interactive overview of the associated anatomy with labels that can be turned on/off is accessible for reference and self testing. The program contains 4 animated learning modules: primary reticulorumen contraction, secondary reticulorumen contraction with eructation, rumination, and rumination superimposed on primary and secondary contraction sequences. Contraction events are depicted with real-time periodicity or can be rate-controlled using a slider bar. Written description of the contraction events accompany each module. This project is the first to use animation as an alternative to traditional instructional methods for teaching reticulorumen motility. Use of this model may translate into an improved understanding of reticulorumen motility and, subsequently, improved problem-solving skills.

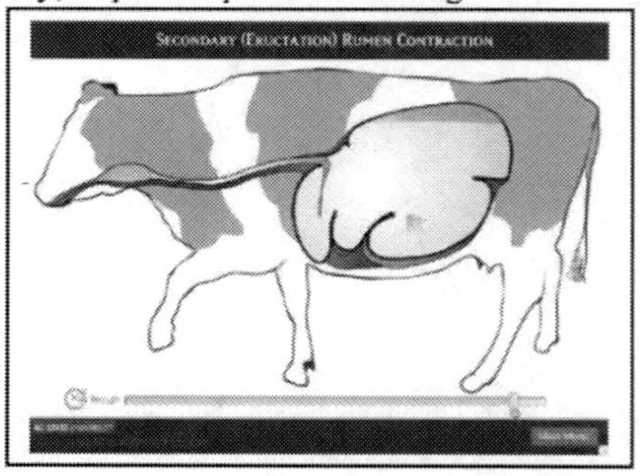


**ABSTRACT #131**

A RETROSPECTIVE STUDY OF 106 CASES OF BOVINE LYMPHOSARCOMA

AJ Burton[1], DV Nydam[1] and TJ Divers[2]. [1]Dept of Population Medicine and Diagnostic Sciences and [2]Dept of Clinical Sciences, College of Veterinary Medicine, Cornell University, Ithaca, NY.

Prevalence of infection with the Bovine Leukemia Virus (BLV) in the USA is high, with 83.9% of dairy operations testing positive for BLV (NAHMS, 2007). Less than 5% of cattle infected with BLV go on to develop neoplasia (lymphosarcoma). There are many publications discussing the transmission, epidemiology, testing and control of BLV.

However, there are no peer-reviewed publications reporting on large numbers of clinical cases of cattle affected with lymphosarcoma. The objective of this study was to describe the initial clinical presentation, ante mortem diagnostic tests and tumor location in cattle diagnosed with lymphosarcoma.

A retrospective study of all cattle admitted to the Cornell University Farm Animal Hospital between January 1980 and December 2008, that had a definitive diagnosis of lymphosarcoma was conducted. Data on signalment, presenting complaint, clinical pathology, ante mortem diagnostic tests and necropsy findings was collected. Categorical data were analyzed using and Chi-square tests with calculation of Exact binomial 95% confidence intervals for proportions. Alpha was set at 0.05.

Of 106 total animals, 94% were cows and 6% were bulls. The median age (n = 95) was 5 years. Enlarged peripheral lymph nodes (PLNs) were detected in 54% of animals (n = 102). Median lymphocyte count (n = 96) was 4,900 cells/L with 10.4% having circulating lymphoblasts. Ante mortem aspirate/biopsy sites were based upon localization of clinical disease. The proportions of false negatives for ante mortem diagnostic tests were as follows: PLN aspirate, 57% (C.I. 34 – 78), PLN wedge biopsy, 0% (C.I. 0 – 29), surgical abdominal exploration and biopsy, 0% (0 – 23), abdominocentesis, 64% (C.I. 35 – 87), pleurocentesis, 20% (C.I. 1 – 72), pericardiocentesis, 33% (C.I. 4 – 78) and cerebral spinal fluid tap, 81% (C.I. 54 – 96). In cows that had a full necropsy (n = 83) the most frequently identified locations of a lymphosarcoma tumor were the heart (66%), abomasum (59%), uterus (37%), kidney (31%) and epidural space (27%). Retrobulbar tumors were identified in only 10%.

This study has identified more kidney tumors and fewer retrobulbar tumors, in contrast to the frequently cited 5 'classical' predilection sites (abomasum, heart, uterus, epidural space and retrobular). A higher percentage of cows exhibited PLN enlargement (54%) than previously estimates (≈25%). Abdominocentesis, PLN aspirate and cerebral spinal fluid taps were found to be fairly insensitive tests while PLN biopsy and surgical exploration and biopsy found to be sensitive tests for lymphosarcoma.

**ABSTRACT #132**

DETERMINATION OF DAIRY COWS HEMATOLOGICAL REFERENCE INTERVAL BY 3 METHODS AND EVALUATION OF CONFOUNDING FACTORS. D Francoz, SM Buczinski, M Moreau, C Bédard. Faculté de Médecine Vétérinaire, Université de Montréal, Montréal, Canada.

The purpose of this study was to establish hematological reference intervals in dairy cows in Québec to compare them to those established by Schalm in 1965, and to evaluate the effect of herds and physiological stage on hematological values.

Blood samples were obtained from 45 healthy Holstein cows between 1 to 60 days in milk (DIM), 21 between 61 to 180 DIM, 21 between 181 and 300 DIM and 19 dry cows, randomly selected in 4 different farms. White blood cells counts were obtained using an Advia 120 analyzer. Fibrinogen concentration was determined by heat precipitation technique. Serologic testing for bovine leukosis virus (BLV) was performed by ELISA. A bootstrap resampling method (BRM) was performed yielding 1000 samples from the study cohort to validate reference intervals (95%) estimated by parametric (PM) and nonparametric (NPM) methods. Statistical inferences for between class comparisons were done under a probability value of 5%.

Farms of origin and DMI categories had significant effect on monocyte and fibrinogen levels, respectively. However, values remain within the interval references. BLV status had a significant effect on leukocyte (p=0.003), lymphocyte (p<0.001) and basophil (p=0.003) counts.

| Blood parameters | PM | | NPM | | BRM | |
|---|---|---|---|---|---|---|
| | Mean | | percentile | | percentile | |
| [Schalm's reference interval] | -1.96SD | +1.96SD | 2.5 | 97.5 | 2.5 | 97.5 |
| **Reference interval for all selected cows (n=106)** | | | | | | |
| **Fibrinogen** g/L [2.0-6.0] | 0.9 | 5.5 | 1.6 | 5.0 | 1.6 | 5.0 |
| **Leukocyte** $10^9$/L [4.0-12.0] | 0.5 | 15.7 | 4.2 | 16.8 | 4.2 | 16.8 |
| **Neutrophil segmented** $10^9$/L [0.6-4.0] | 1.0 | 4.6 | 1.3 | 4.6 | 1.3 | 4.6 |
| **Lymphocyte** $10^9$/L [2.5-7.5] | 0.0 | 11.2 | 1.9 | 12.3 | 1.9 | 12.3 |
| **Monocyte** $10^9$/L [0.0-0.8] | 0.0 | 0.9 | 0.1 | 1.1 | 0.1 | 1.1 |
| **Eosinophil** $10^9$/L [0.0-2.4] | 0.0 | 1.0 | 0.0 | 0.9 | 0.0 | 0.9 |
| **Basophil** $10^9$/L [0.0-2.2] | 0.0 | 0.3 | 0.0 | 0.3 | 0.0 | 0.2 |
| **Reference interval for BLV-negative cows (n=75)** | | | | | | |
| **Leukocyte** $10^9$/L [4.0-12.0] | 3.6 | 10.7 | 4.2 | 11.3 | 4.2 | 11.3 |
| **Lymphocyte** $10^9$/L [2.5-7.5] | 1.0 | 5.9 | 1.9 | 6.9 | 1.9 | 6.8 |
| **Basophil** $10^9$/L [0.0-2.2] | 0.0 | 0.3 | 0.0 | 0.1 | 0.0 | 0.1 |

Based on BRM, NPM appears more adequate for the determination of reference intervals. On the opposite to BLV status, farms of origin and DMI categories do not seem important confounding factors when establishing reference interval. As recently reported, reference intervals have changed over the time when comparing to those established in 1965.

**ABSTRACT #133**

CORRELATION OF SERUM CARDIAC TROPONIN I AND MYOCARDIAL DAMAGE IN CATTLE WITH MONENSIN TOXICOSIS. A Varga[1,2], KE Schober[2], CH Holloman[2], PC Stromberg[2], J Lakritz[2], DM Rings[2]. 1. William R. Pritchard Veterinary Medical Teaching Hospital, University of California, Davis, CA. 2.The Ohio State University, Columbus, OH.

Cardiac troponin I (cTnI) is a sensitive and specific biomarker of myocardial injury in people and small animals. Little is known about the diagnostic use of cTnI in cattle. We hypothesized that serum cTnI correlates to myocardial function and histopathologic lesions in cattle with monensin-induced myocardial injury.

Ten cows received one dose of monensin orally; 30 mg/kg ($n$ = 1) or 40 mg/kg BW ($n$ = 1) (Group A) or 50 mg/kg monensin ($n$ = 8, Group B). Repeated measurements of serum cTnI and echocardiographic variables were performed until study termination at 80 (Group A) and 144 hours (Group B) post dosing. Semi-quantitative histopathologic examinations of the heart (90 fields in 9 target areas of the myocardium) were performed in each cow. A scoring system with regard to the magnitude of myocardial injury was established and the total heart score per cow was compared to maximum cTnI concentration measured after monensin administration. Five hearts from healthy cows served as controls.

Increased cTnI concentration (> 0.07 ng/ml) was found in 9/10 cows. cTnI was significantly associated with left ventricular shortening fraction ($r^2$ = 0.51, p = 0.02) and myocardial histopathologic lesion score ($r^2$ = 0.49; p = 0.021).

All cows (n = 7) with evidence of myocardial necrosis had a cTnI concentration ≥ 1.04 ng/ml.

cTnI is related to myocardial necrosis and severity of myocardial damage in cattle with monensin toxicosis. cTnI may become a useful diagnostic tool in the non-invasive assessment of myocardial injury in cattle with naturally-occurring cardiac disease.

**ABSTRACT #134**

VALIDATION OF THE I-STAT® STALLSIDE IMMUNOASSAY FOR THE MEASUREMENT OF BOVINE CARDIAC TROPONIN I AND ITS CLINICAL RELEVANCE. A Varga, JA Angelos, JM Cornish, M Chigerwe. William R Pritchard Veterinary Medical Teaching Hospital, University of California, Davis, CA.

Measurement of cardiac troponin I (cTnI) is the reference method for non-invasive diagnosis of myocardial injury in humans and small animals. Increased circulating cTnI concentrations also have been associated with non-cardiac diseases. The objective of this study was to validate the I-STAT analyzer for the detection of bovine serum cTnI and to measure cTnI levels in cattle with cardiac and non –cardiac diseases.

Purified bovine cTnI was diluted in cTnI-free serum from a healthy cow. Precision, linearity and recovery were assessed using the I-STAT (HeskaCorporation,CO) immunoassay. At concentrations 0.5, 1, 2, 8 and 32 ng/ml the mean intra-assay precision was 11 ± 6.1 % and mean inter-assay precision 13 ± 4.9 % (coefficient of variation). The assay demonstrated good linearity of serial dilutions from a concentration of 2 to 32 ng/ml and mean cTnI recovery was 100.5 ± 26.7 %. Thirty healthy cattle had serum cTnI concentrations between 0 and 0.01 ng/ml.

In our study cows with non-cardiac primary diseases such as left displaced abomasum, had cTnI concentrations ≤ 0.29 ng/ml, while cows with cardiac diseases such as endocarditis, had significantly higher cTnI concentrations.

These preliminary data suggest that the I-STAT immunoassay has sufficient test performance to detect bovine serum cTnI concentrations. Cattle with myocardial diseases have higher cTnI concentrations as compared to cows without primary cardiac diseases, but those concentrations are still significantly higher than in healthy cattle. Further studies are required to evaluate the relationship between cTnI concentration and clinical outcomes in cattle.

**ABSTRACT #135**

RGD-CONJUGATED HELICAL ROSETTE NANOTUBES INHIBIT CHEMOTAXIS AND CELL SIGNALING OF CATTLE NEUTROPHILS. LMH Anh[1], H Fenniri[2], SS Suri[1] and B Singh[1]. 1. Western College of Veterinary Medicine, University of Saskatchewan, Saskatoon, Canada. 2. National Institute for Nanotechnology, Edmonton, Canada.

Bovine respiratory disease complex is the most common disease that causes significant economic loss, typically in feedlot cattle. Current treatment methods are focused on reducing inflammatory responses, control of airway reactivity and improvement of pulmonary functions without potential side effects. Neutrophils are the key contributors in acute lung inflammation. However, activated neutrophils live longer and cause excessive tissue damage upon migration into lungs. Therefore, modulation of their migration and lifespan are attractive approaches in treatment strategies of bovine respiratory disease. Nanotechnology holds significant potential to design new compounds by our ability to manipulate at the nanoscale. We used helical rosette nanotubes, which are a class of novel, biologically inspired, water soluble and metal-free tubes. We used helical rosette nanotubes (RNT) conjugated to arginine-glycine-aspartic acid (RGD-RNT) to study their effects on neutrophil chemotaxis and cell signaling. Bovine neutrophils exposed to 5% RGD-RNT reduced their migration in response to fMLP ((formyl-Methionyl-Leucyl-Phenylalanine), compared to the non-treated group ($P<0.001$). This inhibitory effect was the same as that of groups treated with ERK1/2 inhibitor (UO126) and p38 mitogen-activated protein kinase (p38 MAPK) inhibitor (SB239063). In addition, the phosphorylated ERK1/2 and p38 MAPK for the first time were quantified by sandwich ELISA to elucidate the mechanism of neutrophil migration. The phosphorylation of both the ERK1/2 and p38 was inhibited at 5 minutes by RGD-rosette nanotubes ($P<0.05$). Furthermore, integrin $\alpha v\beta 3$ is possibly involved in migration of bovine neutrophils. These experiments provide the first evidence that RGD-rosette nanotubes suppress phosphorylation of ERK1/2 and p38 MAPK and inhibit chemotaxis of bovine neutrophils.

**ABSTRACT #136**

EXPERIMENTAL INFECTION OF COLOSTRUM-DEPRIVED CALVES WITH BOVINE VIRAL DIARRHEA VIRUS TYPE-1 ISOLATED FROM FREE-RANGING WHITE-TAILED DEER (*ODOCOILEUS VIRGINIANUS*). EA Raizman, RM Pogranichniy, M Lévy, M Negron, WG Van Alstine. School of Veterinary Medicine, Purdue University, West Lafayette, IN.

Bovine viral diarrhea virus (BVDV) is a pestivirus associated with important economic losses to the cattle industry in the United States. The role of wildlife as a reservoir for the virus is unclear. The objective of this study was to determine if calves could be experimentally infected with a BVD type 1 virus isolated from free ranging White-tailed deer (WTD), develop lesions and shed the virus.

Twelve colostrum-deprived male Holstein calves (8 infected and 4 controls) were used in this study. Both groups were housed in an isolation unit. After a 7 day period of acclimation, the experimental group was challenged intranasally with a 2 ml inoculum containing $10^6$TCID$_{50}$/ml BVDV type 1 isolated from free ranging WTD in Indiana, while the control group was inoculated with a virus-free culture medium. Daily physical examinations were performed. Serology for BVD (virus neutralization and ELISA) was performed on days -5, 0, 3, 7, 10, 14 and 21 post inoculation (DPI). Whole blood, saliva, nasal and rectal swabs were collected on days 0, 3, 7, 10, 14 and 21 DPI for virus isolation, virus neutralization, ELISA and Real Time PCR (RT-PCR) assays. On day 14 DPI, 4 calves from the infected group and 2 controls were humanely euthanized; multiple tissue samples were collected for histopathology. The same protocol was used on day 21 for the remaining 4 infected and 2 control calves. The experimental protocol was approved by the Purdue Animal Care and Use Committee.

Several calves develop intermittent diarrhea before inoculation with BVDV. The most important microscopic lesions were thymic atrophy and lymphoid depletion of the Peyer's patches in all 8 infected calves. No significant lesions were identified in the control calves. RT-PCR was positive in the buffy coat of 8/8, in the nasal samples of 7/8 and in the saliva samples of 2/8 infected calves. The rectal samples were negative at all times for the 12 calves. On virus neutralization assay, 4/8 animals developed antibodies against BVDV and on ELISA assay 3/8 animals were positive for BVDV. All samples from the control calves remained negative.

These results support our hypothesis that colostrum-deprived calves can be successfully infected with a BVDV type 1 isolated from WTD and that infection results in the development of typical lesions in lymphoid tissues. Furthermore, positive identification of BVDV in nasal and saliva samples suggests that shedding of the virus in the environment is possible.

## ABSTRACT #137

EVIDENCE FOR TRANSPLACENTAL PASSAGE OF BLUTONGUE VIRUS SEROTYPE 8 IN SHEEP.

Kirschvink N.[1], Bolkaerts B.[1], Baricalla C.[1], Wiggers L.[1], de Leeuw I.[2], Vandenbussche F.[2], De Clercq K.[2], Saegerman C.[3] [1]Animal Physiology, Veterinary Department, University of Namur, Belgium; [2]Veterinary and Agrochemical Research Centre, Department of Virology, Brussels, Belgium; [3]Epidemiology and Risk Analysis applied to Veterinary Sciences, Faculty of Veterinary Medicine, University of Liège, Belgium.

Serotype 8 of bluetongue virus (BTV8) emerged in summer 2006 in Northern Europe and strongly spread in 2007. Transplacental passage of the BTV8 has recently been documented in naturally infected cattle, but not yet in sheep. This study aimed at assessing within one sheep flock whether a natural BTV8 infection passes the placental barrier.

309 pregnant ewes belonging to the flock of Namur University were followed since summer 2007 and pregnancy diagnosis, which was performed by ultrasonography 6-8 weeks after conception. 185 ewes had developed clinical signs of BTV8 during or shortly prior to gestation in summer 2007. Lambing occurred at 4 periods: November 2007 (70 ewes, 121 living lambs, 16 aborted fetuses), January 2008 (40 ewes, 65 living lambs, 0 aborted fetuses), March 2008 (141 ewes, 215 living lambs, 8 aborted fetuses) and May 2008 (54 ewes, 74 living lambs, 2 aborted fetuses). At lambing, blood was sampled and serum was prepared from all ewes and all lambs prior colostrum intake in order to perform BTV antibody detection (ELISA, Id-Vet France). At the age of 2-4 weeks red blood cells (RBC) were sampled from lambs when they were BTV-antibody positive at birth. Abortions occurring during the last 3 weeks of gestation were recorded. Aborted lambs underwent necropsy and if possible, their spleen was sampled. RBC and spleen were analyzed by RTqPCR for BTV8 detection.

All ewes were BTV-positive (ELISA) at lambing and 9 lambs had BTV antibodies at birth before colostrum intake. The RBC of 7 of those lambs were also positive for BTV8 by RTqPCR. The seven ELISA and RTqPCR positive lambs were all born in November 2007, 1 lamb born in January 2008 and 1 lamb born in March 2008 had BTV antibodies but were RTqPCR negative. Lambs born in May 2008 were all negative for BTV antibodies. Among the 26 aborted animals, the fetuses were most often in an advanced state of decomposition. Only 10 spleens could be analyzed by RTqPCR. 4 samples from fetuses aborted in November 2007 and March 2008 were positive for BTV8, the only 2 samples from abortions in May 2008 were negative, other samples did not provide conclusive results.

These data provide evidence that BTV8 passes the placental barrier in sheep. It has to be determined to what extent this way of virus transmission contributes to the overwintering phenomenon of BTV8.

## ABSTRACT #138

LONG TERM FOLLOW-UP OF RAMS' SEMEN QUALITY AFTER NATURAL INFECTION WITH BLUETONGUE VIRUS SEROTYPE 8.

Saegerman C.[1], Bolkaerts B.[2], Raes M.[2], Kirschvink N.[2] [1]Epidemiology and Risk Analysis applied to Veterinary Sciences, Faculty of Veterinary Medicine, University of Liège, Belgium; [2]Animal Physiology, Veterinary Department, University of Namur, Belgium.

Bluetongue virus (BTV) is reported to induce transient infertility in ruminants. Serotype 8 of BTV emerged in 2006 in Northern Europe and infected ~ 90% of the Belgian sheep population in 2007. Whilst recovery of normal semen quality in BTV8-infected rams occurs within 2-4 months after infection (Kirschvink et al., 2007; Vet. J), long term follow up of naturally infected rams has not yet been described. The aim of this field study was to assess in 2008 the semen quality of rams that underwent a natural BTV8 infection in 2007 and whose semen quality had been monitored in 2007 and 2008.

11 healthy rams tested prior to the BTV8 outbreak in Belgium were used as controls (C) and 20 rams with natural infection by BTV8 in 2007 were included in this study. BTV-infected rams were tested in 2007 between 5 days and 2 months after development of clinical signs of disease (Clinical BTV8) as well as 6 to 12 months later (BTV8 recovery). Semen was evaluated for motility (score ranging from 0-5), spermatozoa concentration and viability (% of normal living [NL]; normal death [ND]; abnormal death [AD] spermatozoa).

Results are shown as mean ± SD.

| | Controls (n=11) | Clinical BTV8 in 2007 (n=20) | BTV8 recovery in 2008 (n=20) |
|---|---|---|---|
| Motility score | 4.5 ± 0.2 | 1.2 ± 1.5 * | 4.3 ± 0.4 $ |
| Conc ($10^9$ sperm/ml) | 3.2 ± 0.7 | 1.2 ± 1.1 * | 3.2 ± 0.6 $ |
| % NL spermatozoa | 75 ± 6 | 21 ± 27 * | 75 ± 9 $ |
| % ND spermatozoa | 21 ± 7 | 49 ± 20 * | 23 ± 11 $ |
| % AD spermatozoa | 4 ± 3 | 30 ± 18 * | 4 ± 4 $ |

* significantly different from respective control-value ; $ significantly different from respective Clinical BTV8-value ; $p<0.05$

These results show (1) an important impact of serotype 8 BTV on rams' semen quality within 5 days and 3 months after development of clinical disease and (2) that BTV8-infected rams show a fully recovered normal semen quality within 6 to 12 month after infection.

**ABSTRACT #139**

PREVALENCE OF BVDV ON ALPACA FARMS IN EASTERN AND MIDDLE TENNESSEE. TM Buchheit, SR Van Amstel, KV Thomas, M Abd-Eldaim. The University of Tennessee, College of Veterinary Medicine, Knoxville, TN.

The purpose of this study was to survey the seroprevalence to BVDV and possible presence of persistently infected (PI) animals in alpaca herds. Blood was collected from 244 alpacas in 5 herds with a negative BVDV vaccination history. All animals in each herd were tested for the presence of antibodies to BVDV by use of immunofluorescence assay (IFA) and for the presence of viral RNA by conventional polymerase chain reaction assay using reverse transcriptase RT-PCR. Overall, antibodies to BVDV were detected in 1.6% (4/244) of the samples which is similar to most other BVDV prevalence studies. Two herds had at least 1 BVDV-positive animal with a titer ≥ 1:80, with 3 of the 4 seropositive alpacas from the same herd (Farm 1). All 244 samples were found virus-negative by PCR. No animals younger than 2 years of age were found to be seropositive and none of the BVDV-positive animals were pregnant at the time of testing. Only one of the herds tested had cattle on the farm; however, none of the alpacas in this herd tested BVDV-positive. No suspected PI animals were identified. The presence of 3 seropositive animals on Farm 1 may be due to unrestricted comingling of animals. In conclusion, although no animals with viremia were identified in the area, BVDV should still be considered as a cause of recurring illness, ill-thrift, low birth-weight or premature crias and abortion.

**ABSTRACT #140**

SEROPREVALENCE OF BOVINE VIRAL DIARRHEA IN NEW WORLD CAMELIDS IN SWITZERLAND AND IMPLICATIONS FOR THE ERADICATION CAMPAIGN IN THE BOVINE POPULATION. M. Mudry[1], M. Meylan[1], R. Zanoni[1], G. Schüpbach Regula[2], P. Zanolari[1]. 1.Vetsuisse Faculty, University of Berne. 2.Swiss Veterinary Office, Berne, Switzerland.

Infection with the Bovine Viral Diarrhea Virus (BVDV) is common in the Swiss cattle population, with seroprevalence rates of 60-90% and 0.6% of persistently infected (PI) animals. The economic losses resulting from BVDV infection (estimated at 4-6 Mio US$/year for a cattle population of 1.5 Mio) prompted the decision to eradicate the disease in Switzerland. The entire Swiss bovine population has been tested since the fall of 2008 and all calves are currently tested immediately after birth, PI animals are culled. The susceptibility of New World Camelids (NWC) to BVDV infection and their role in the epidemiology of the disease have not been clearly defined to date, seroprevalence rates of 2 to 11% in NWC can be found in the literature. As no data were available for Switzerland prior to the start of the eradication campaign, a study was undertaken to evaluate the seroprevalence of BVD in the Swiss NWC population.

Blood specimens were collected during the summer of 2008 from 349 NWC from 40 herds, a representative sample of the Swiss population. Information was obtained, among others, about possible contact of the NWC with other ruminants, especially with cattle. The serum samples were analyzed for antibodies to BVDV using an ELISA method (Institute of Veterinary Virology, University of Berne). For comparative purposes, 250 serum samples collected in 40 herds in 2000 and representative of the Swiss NWC population at that time were analyzed with the same serological method.

The individual seroprevalence rates were 3.6% (9 animals out of 250) in 2000 and 5.7% (20 animals out of 349) in 2008. Similarly, herd prevalence was higher in 2008 (30%) than in 2000 (20%). Seropositive animals were found in all regions of the country. Contact with cattle was possible in 42% of the herds with at least one seropositive animal whereas this was the case in 24% of herds without seropositive NWC.

The seroprevalence of BVD in the Swiss NWC population was higher than the 2-3% expected based on the fact that cattle and NWC are rarely kept in close contact to each other in Switzerland, even in herds where both species are present. The apparent seroprevalence rate increased from 3.6% to 5.7% between 2000 and 2008. Whether NWC only seroconvert after contact with the virus or if they serve as a reservoir by circulating the virus within the camelid population or through PI animals needs now to be determined. The possible persistence of BVDV in the NWC population is highly relevant in view of the efforts undertaken to eradicate the infection in cattle.

**ABSTRACT #141**

TUBERCULOSIS CAUSED BY *MYCOBACTERIUM MICROTI* IN NEW WORLD CAMELIDS. P. Zanolari[1], N. Robert[1], G. Pfyffer[2], K. Lyashchenko[3], M. Meylan[1]. 1.Vetsuisse Faculty, University of Berne, Berne, Switzerland. 2.Department of Medical Microbiology, Luzern, Switzerland. 3.Chembio Diagnostic Systems, Medford, NY.

Tuberculosis is one of the world's leading infectious causes of death among humans, with one third of the world's population currently being infected with *Mycobacterium tuberculosis*. In ruminants, tuberculosis is mainly caused by *M. bovis* and rarely by *M. tuberculosis*. The disease is significant not only because of economic losses but also because of its zoonotic potential. Infection with *Mycobacterium microti*, a member of the *M. tuberculosis* complex, has been reported in several species, including humans. The goal of the present study was to describe the clinical and pathological findings in 11 New World Camelids (NWC) infected with *Mycobacterium microti*.

The records of 11 animals (10 llamas and 1 alpaca, aged 4 to 18 years) from 6 different herds with a history of wasting and weakness admitted to the Clinic for Ruminants and/or the Institute for Animal Pathology within a timeframe of 7 years (2001-2008) were reviewed, and the clinical findings, clinicopathologic abnormalities, diagnostic imaging findings and necropsy results were described.

Clinical signs were limited to weight loss, recumbency and anorexia in the late stages of the disease. Two thirds of the animals showed increased breathing effort or respiratory distress. Palpable lymph nodes were enlarged in 4 animals. No consistent hematologic abnormalities were identified. The comparative intradermal tuberculin test with bovine protein purified derivate (PPD) and avian PPD was negative in all animals. Abdominal and/or thoracic ultrasound and thorax radiology revealed abnormal findings such as enlarged mediastinal, mesenteric and/or hepatic lymph nodes only in advanced cases. The presence of granulomatous lesions in various lymph nodes, in the lungs, pleura, mediastinum, peritoneum and mesenterium, liver, spleen, pancreas and/or kidneys was confirmed at necropsy in all animals. Infection with *M. microti* was confirmed by bacteriological culture and/or spoligotyping.

An infection caused by *M. microti* should be considered as a differential diagnosis in chronic debilitating diseases with or without respiratory signs in NWC. However, the clinical diagnosis remains challenging, particularly in the early stages of the infection. As cases of *M. microti* infection have been reported in immunocompromized human patients, the zoonotic potential of the organism should be kept in mind when dealing with the disease in NWC and the owners of affected animals should be informed about the potential health hazard related to clinical and subclinical infection in NWC.

*Previously presented at World Buiatrics Congress, Budapest 2008*

**ABSTRACT #142**

PHARMACOKINETICS OF INTRAVENOUS ENROFLOXACIN AND ITS ACTIVE METABOLITE CIPROFLOXACIN IN HOSPITALIZED HORSES.

JL Davis, MK Sheats, MG Papich. North Carolina State University, College of Veterinary Medicine, Raleigh, NC.

Enrofloxacin is a fluoroquinolone antibiotic with excellent activity against gram-negative enteric bacteria, as well as some staphylococci, and *Mycoplasma, Chlamydia,* and *Rickettsia* spp. Enrofloxacin is metabolized to ciprofloxacin, which is believed to produce an additive effect on antimicrobial activity. The pharmacokinetics of enrofloxacin administered via various routes of administration have been reported in multiple studies using healthy adult horses. However, there are no reports of the pharmacokinetics of this drug when used in sick, hospitalized horses that are frequently on multiple medications and may have hepatic or renal compromise. The purpose of this study is to examine the pharmacokinetics of enrofloxacin and its active metabolite ciprofloxacin in hospitalized horses using a population sampling method.

Data from nineteen horses was available for analysis. Horses were being treated for post-operative colic (n=5), other gastrointestinal diseases (n=5), liver disease (n=4), musculoskeletal diseases (n=4), and pneumonia (n=1). Heparinized plasma samples were collected from horses after 3 or more doses of 5 mg/kg intravenous (IV) enrofloxacin once daily, to approximate steady state concentrations. Horses were assigned to a sampling schedule so that each horse had blood drawn at 5 time points for analysis. Samples were frozen at -70°C until analysis by high pressure liquid chromatography with fluorescence detection. Data was analyzed using a naïve pooled approach with noncompartmental modelling.

Following intravenous administration, enrofloxacin had a mean ± SD clearance of 2.14 ± 1.62 mL/kg/min, and a plasma half-life ($t_{1/2}$) of 9.55 ± 5.14 hr. This represents a slower clearance and a longer half-life compared to other reports of IV enrofloxacin in horses. The volume of distribution was 1.25 ± 0.34 L/kg, which is lower than other reported values. The combined total area under the curve (AUC) for enrofloxacin and ciprofloxacin was 71.2 ± 48.5 hr*ug/mL and ciprofloxacin represented 12.4 ± 8.3% of the total AUC. In the subset of horses with liver disease, t ½ and clearance of enrofloxacin was 12.3 ± 6.4 hr and 1.5 ± 1.3 mL/kg/min, respectively, compared to 8.7 ± 4.7 hr and 2.3 ± 1.7 mL/kg/min, respectively, in horses without liver disease. These differences were not statistically significant ($P>0.05$).

The pharmacokinetic-pharmacodynamic indicator of success for fluoroquinolone antibiotics is based on a ratio of the plasma AUC to the minimum inhibitory concentration (MIC) of the bacteria of ≥ 100-125. Based on the results of this study and this AUC:MIC ratio, enrofloxacin administered at 5 mg/kg IV q24h in hospitalized horses should be successful in treating bacteria with a MIC of ≤ 0.5 μg/mL.

**ABSTRACT #143**

LAMINAR LEUKOCYTE ACCUMULATION AND EPITHELIAL STRESS IN HORSES WITH CARBOHYDRATE OVERLOAD-INDUCED LAMINITIS.

RR Faleiros[1,2], GJ Nuovo[1], PJ Johnson[3], SJ Black[4], JK Belknap[1] 1.Ohio State University, Columbus, Ohio. 2.Universidade Federal de Minas Gerais, Belo Horizonte, Brazil. 3. University of Missouri, Columbia, MO. 4. University of Massachusetts, Amherst, MA.

Laminar failure in laminitis results from a dysadhesion of the basal laminar epithelium from the underlying basement membrane. We recently reported prominent laminar leukocyte infiltration prior to the onset of laminar epithelial stress in the developmental stage in the Black Walnut Extract (BWE) laminitis model. The purpose of this study was to characterize laminar leukocyte populations and the temporal relationship of leukocyte infiltration to laminar epithelial stress in the CHO model of laminitis, a model that more commonly results in laminar failure observed in clinical laminitis cases. Immunohistochemistry for calprotectin (CP, marker for neutrophils, classically activated macrophages, and stressed [not normal] keratinocytes) and CD163 (antibody specific for resident macrophages and monocytes in alternatively activated phenotype) was performed on laminar paraffin sections obtained: 24 h following water administration (CON, n=8), and at onset of fever (10-20 hours, DTP group, n=6) and at the onset of lameness (20-46 hours, LAM group, n=6) following CHO administration. Minor histologic changes (slightly abnormal PAS staining of basement membrane [BM]) were present in 3/6 horses in the LAM group. Similar to the BWE model, laminar epithelial stress (increased CP signal) occurred at the onset of lameness. Increases ($P<0.05$) in laminar CP+ leukocytes and CD163+ monocytes/macrophages occurred in the LAM group (not DTP group) compared to CON group. The temporal pattern of laminar leukocyte infiltration and epithelial stress is markedly different in the two models, with leukocyte infiltration preceding epithelial stress in the BWE model, and both events occurring simultaneously in the CHO model. Importantly, the marked leukocyte infiltration in the CHO model prior to severe histologic changes in this model indicates that, in contradistinction to previous reports, leukocyte infiltration may be a cause of and not a reaction to BM degradation and structural failure.

**ABSTRACT #144**

THE EFFECT OF TRACHEAL LAVAGE ADMINISTRATION ON LUNG FUNCTION IN HEALTHY KITTENS. EA Rozanski, G Buckley, MR Mazan, D Bedenice, AH Hoffman,. Cummings School of Veterinary Medicine at Tufts University, North Grafton, MA.

The purpose of this study was to evaluate the effect of tracheal lavage on lung function in healthy kittens. While tracheal wash sampling for evaluation of lung cytology and bacterial culture is considered routine, there are reports of respiratory embarrassment or failure and even death following this procedure in cats. Seven healthy kittens were recruited from consenting owners for study enrollment. The study was performed prior to routine castration or ovariohysterectomy. All kittens were healthy on examination. Cats were pre-medicated with ketamine (7 mg/kg mg), butorphanol (0.75 mg/kg), and diazepam (0.3 mg/kg) intramuscularly. Anesthesia was induced with propofol 2-4 mg/kg intravenously to effect. Cats were intubated with a standard 3.5 mm ID sterile endotracheal tube and ventilated with a 10 ml/kg tidal volume at 15 breaths/min and 30% inspired oxygen, using a Puritan Bennett 840 critical care ventilator. A standard volume history was created by 2 breaths to total lung capacity (25 cmH20) prior to testing. Using commercial software, static compliance (Cstat) and plateau pressure (Pplat) were measured. Additionally, cats' oxygenation was assessed via pulse oximetry. All animals underwent a standard tracheal lavage with 3 aliquots of 2 ml/kg sterile saline, followed by measurements. Repeated measures ANOVA was used to compared the effect of lavage. Tracheal lavage caused a significant decrease in Cstat, significant increase in Pplat and desaturation.

| | Cstat (mean +/SD | Pplat | Pulse oximetry |
|---|---|---|---|
| Baseline | 6 ±1.1 ml/cmH20 | 6.1 ± 0.8 cmH20 | 99±1 % |
| Post lavage | 2.9 ± 0.9 ml/cmH20 [a] | 9.6 ± 2.1 cmH20 [a] | 93± 4% [b] |

a- $p = 0.001$ compared to baseline, b= $p = 0.01$ compared to baseline.

The results of this study demonstrated that tracheal lavage in healthy cats is associated with an acute 'restrictive effect' on lung function most likely associated with airway closure or derecruitment (atelectasis) as the likely causes of mechanical dysfunction and hypoxemia.

**ABSTRACT #145**

UTILITY OF BIOMARKERS IN BRONCHOALVEOLAR LAVAGE FLUID FOR DISCRIMINATION OF ASTHMA AND CHRONIC BRONCHITIS IN CATS. LA Nafe, AE DeClue, TM Lee-Fowler, JE Eberhardt, CR Reinero. University of Missouri, College of Veterinary Medicine, Columbia MO.

Feline allergic asthma and chronic bronchitis are lower airway diseases that are often grouped together as one syndrome. We postulate that as in humans, the molecular mechanisms which trigger and sustain these diseases are not identical and that differences in biomarker signatures in bronchoalveolar lavage fluid (BALF) exist. The study objective was to compare IL-4, IFN-γ, TNF-α and total nitric oxide (NO) in BALF from cats with naturally developing asthma (CLIN ASM) and chronic bronchitis (CLIN CB) and further compare them to research cats with experimentally induced asthma (EXP ASM), non-septic suppurative inflammation (RES SUPP) and healthy controls (RES NORM).

Ninety-seven cats were divided into five groups [CLIN ASM (n=13), CLIN CB (n=8), EXP ASM (n=23), RES SUPP (n=33), RES NORM (n=20)] based on predefined BALF cytological criteria. Unconcentrated BALF supernatant was assayed for IL-4 and IFN-γ concentrations, TNF-α bioactivity and total NO. Statistical analysis was performed using a one-way ANOVA and *post-hoc* Tukey test, with a $p<0.05$ considered significant.

Concentrations of BALF IL-4 and IFN-γ were below the limits of detection for most cats, precluding statistical analysis. No significant differences were detected between groups for TNF-α bioactivity. The group mean±SEM of BALF NO was significantly higher in CLIN CB (7.9±0.6 μM) compared to RES SUPP (5.0±0.4 μM; $P=0.002$) and EXP ASM cats (4.6±0.2 μM; $P=0.001$).

Using the currently available relatively insensitive methodologies, none of the measured cytokines or NO are useful for discriminating local immune changes in the airways of pet cats with naturally developing asthma versus chronic bronchitis.

*Previously presented at ECVIM in Belgium, September 2008.*

**ABSTRACT #146**

COMPARISON OF INTRADERMAL SKIN TESTING AND ALLERGEN-SPECIFIC SERUM IMMUNOGLOBULIN E (IgE) IN EXPERIMENTAL FELINE ASTHMA. TM Lee-Fowler[1], LA Cohn[1], AE DeClue[1], CM Spinka[2], RD Ellebracht[2], CR Reinero[1]. [1]Department of Veterinary Medicine and Surgery, College of Veterinary Medicine, [2]Department of Statistics, College of Arts and Sciences, University of Missouri, Columbia, MO.

Reliably identifying clinically relevant allergen(s) for allergen-specific immunotherapy as a treatment for feline allergic asthma may be challenging. Poor correlation has been found between intradermal skin testing (IDST) and serum allergen-specific IgE; however, these studies evaluated atopic pet cats with unknown type(s), duration, timing and quantity of allergen exposure. We proposed to control these factors by using a feline model of allergic asthma induced by specific allergens to compare IDST and serum IgE tests. We hypothesized that IDST results would more closely correlate with the known allergen exposure than would allergen-specific serum IgE determination.

Thirteen cats were randomly assigned to saline (placebo), bermuda grass allergen (BGA), or house dust mite allergen (HDMA) groups for sensitization and challenge. On day 0 (prior to sensitization) and days 28 and 50 (post-allergen sensitization) IDST was performed and serum collected. Serum from each cat at each time point as well as pooled serum samples (BGA or HDMA) had determination of allergen-specific IgE using an Fcepsilon receptor based ELISA (Heska, CO); pooled samples were also analyzed using an enzymoimmunometric assay (VARL, CA). Half of each pooled sample was heat inactivated (HI) to selectively destroy IgE. Sensitivity (SE), specificity (SP), and positive and negative predictive values (PPV and NPV, respectively) were calculated. Prausnitz-Kustner (PK) testing using the pooled and pooled HI samples was performed.

Combined results for IDST found SE=90.9%, SP=86.7%, PPV=83.3%, and NPV=92.9%. For ELISA-based serum IgE testing, the SE=22.7%, SP=100%, PPV=100% and NPV=63.8%. The enzymoimmunometric assay did not detect sensitizing IgE, but did detect IgE reactivity to a variety of irrelevant allergens (even in HI samples). Results from the heat inactivated samples (ELISA) and PK test suggest either incomplete destruction of IgE or the presence of homocytotropic IgG. In conclusion, both IDST and allergen-specific IgE determination via ELISA were specific; either test can be used to guide selection of allergens for immunotherapy. The enzymoimmunometric assay was unreliable and can not be recommended.

*Previously presented at ECVIM in Belgium, September 2008.*

**ABSTRACT #147**

CHRONIC ADMINISTRATION OF feG-COOH IN EXPERIMENTAL FELINE ASTHMA. JM Eberhardt, AE DeClue, CR Reinero. University of Missouri, College of Veterinary Medicine, Columbia MO.

Previously we demonstrated that a single dose of the immunomodulating tripeptide, feG-COOH, administered *prior* to allergen challenge partially dampens eosinophilic airway inflammation in experimental feline asthma. We hypothesized that chronic (2 weeks) administration of feG-COOH would decrease clinical signs and airway inflammation in experimentally asthmatic cats. Asthma was induced in 9 cats using Bermuda grass allergen (BGA). Cats were randomized to receive oral feG-COOH (1 mg/kg, q 24h) or placebo (saline q 24h) for 2 weeks followed by a 2 week washout before crossover to the other treatment. A clinical scoring system was employed to grade clinical signs after allergen challenge. Bronchoalveolar lavage fluid (BALF) and blood were collected 24 hours after allergen challenge. Differential cell counts were performed on BALF. Peripheral blood mononuclear cells (PBMCs) were cultured in vitro with BGA. Supernatant from BALF and PBMCs and plasma was harvested for determination of IL-4 and IFN-$\gamma$ concentrations using commercial ELISAs or TNF-$\alpha$ bioactivity using a cell killing bioassay. Repeated measures ANOVA was used to compare the treatment groups with a $p<0.05$ considered significant.

There was no significant difference between treatments in clinical scores (mean±SD: feG 1.0±1.1; placebo 1.2±1.1, p=0.386), % BALF eosinophils (feG 33±18%, placebo 29±16%, p=0.479) or neutrophils (feG, 4.7±4.7%, placebo, 5.0±4.5%; p=0.504). No differences in cytokine concentrations were detected between treatments; however many samples were below the assay's lower limit of detection. In conclusion, chronic administration of feG did not significantly decrease clinical signs or airway inflammation in cats with experimentally-induced asthma.

**ABSTRACT #148**

PROGRESSION OF CIRCULATING ANTIGEN AND ANTIBODY RESPONSES IN CATS EXPERIMENTALLY INFECTED WITH *DIROFILARIA IMMITIS*. K Curtis[1], L Lorentzen[1] R Chandrashekar[1], J McCall[2], DM Tillson[3], AR Dillon[3]. IDEXX Laboratories, Westbrook, ME. 2. TRS Labs, Athens, GA. 3. College of Veterinary Medicine, Auburn University, Auburn, AL.

Cats can suffer complications from adult heartworm infections (*Dirofilaria immitis*) and emerging evidence demonstrates that immature adult heartworms, reaching the pulmonary arteries at approximately 2.5-4 months post-infection, can also cause significant lung pathology (Heartworm Associated respiratory Disease or HARD). To better understand the humoral immune response to *D. immitis*, 12 cats were infected by SC injection of 100 third stage ($L_3$) larvae per cat. Pre- and post-infection serum samples were collected from each cat at selected time points (days -1, 7, 14, 21, 28, 42, 56, 68, 75, 84, 112, 140, 168, 189, 217, 245 and terminal). The cats were necropsied 278-299 days post-infection and adult worms were recovered in 11 of 12 cats. Other imaging and histopathology data were collected. .

Temporal serum samples obtained from each cat were tested for antibodies to two *D. immitis* recombinant antigens, HWAg-1 and HWAg-2. In addition, samples were tested for circulating antigenemia (polycolonal assay) using PetChek® HTWM PF Test Kit (IDEXX Laboratories). Eleven of 12 cats were HWAg-1 antibody positive 68 days post-infection and the remaining cat became HWAg-1 antibody positive on day 140 post-infection. 11 of 12 cats remained HWAg-1 antibody positive through end of the study. All but one cat was positive for HWAg-2 antibody, with the first positive tests occurring between 42-84 days post-infection. Ten of the 11 HWAg-2 antibody positive cats remained positive during the study except one cat that became negative by day 271. Both antibody assays became positive during the pre-cardiac stages of the heartworm life cycle. Circulating antigen was detected in 10 of the 12 cats and 1 cat died at day 140 with heartworms. Of the remaining 9 positive cats, 2 were first positive on day 140 (4.7 months) and 5 first positive on day 168 (5.6 months). One cat first became antigen positive on day 217. Using SoloStep® CH Heartworm Test (Heska), antigen was detected in 3 of the cats at day 168.

This study demonstrates that in cats with experimental infections, parasite antigenmia can be detected prior to heartworms reaching sexually maturity. Historically, the antigen has been detected only in the circulation of dogs and cats harboring mature female heartworms, but target antigens have been demonstrated in the cuticle of heartworms. These data suggest that disintegration of immature adult heartworms can result in detectible levels of antigen utilizing a PetChek® HTWM PF Test Kit , and less commonly using SoloStep® CH Heartworm Test. In summary, in this experimental model, *D. immitis* antigen was detectable in some cats at time points associated with dying immature adult heartworms and the initiation of HARD (days 140 and 168).

**ABSTRACT #149**

ISOMETRIC RESPONSES OF ISOLATED INTRAPULMONARY BRONCHIOLES FROM CATS WITH AND WITHOUT ADULT HEARTWORM INFECTION. AA Wooldridge, AR Dillon, DM Tillson, Q Zhong, S Barney, Auburn University College of Veterinary Medicine, Auburn, Alabama.

Adult heartworm (HW) infections (*Dirofilaria immitis*) in cats cause bronchiolar pathology resulting in Heartworm

Associated Respiratory Disease (HARD). Peribronchiolar smooth muscle (SM) proliferation and increased bronchial wall to lumen ratios are induced by HW infections in cats, but the contribution of bronchiolar SM reactivity to the pathophysiology of HARD is unknown.

To better understand the bronchiolar response to *D. immitis*, 13 purpose-bred cats were utilized. Heartworm infected cats (n=7) received 100 third stage larvae subcutaneously, and 6 cats served as age matched controls. The left caudal lung lobe was collected at necropsy 278-299 days post-infection in heartworm cats. Surrounding lung tissue was removed, and four 2mm rings of 4th generation intrapulmonary bronchioles (IPB) were mounted on a wire myograph (Danish Myotechnologies) interfaced with a physiograph. Three cycles of pre-contractions to 1μM acetylcholine (ACh) were performed, followed by concentration-response curves to the contractile agonists ACh, histamine (HIS), and 5-hydroxytryptamine (5-HT). To evaluate relaxation, IPB were constricted with 1 μM 5-HT and concentration-response curves were generated to sodium nitroprusside (SNP) and isoproterenol (ISO). Data were analyzed by 2-way analysis of variance, and $p<0.05$ was significant.

In HW cats compared to control cats, relaxation to ISO was significantly reduced at 2 concentrations. HW cats demonstrated significantly reduced responses to ACh for 2 cycles of pre-contractions and at the highest concentration of ACh in the curve. There were no differences between control and HW cats in responses to 5-HT, HIS, or SNP.

The slightly reduced response of IPB in HW cats to ISO may indicate some level of refractoriness to bronchiolar relaxation. These results suggest that despite peribronchiolar SM proliferation and increased bronchial wall to lumen ratios in cats with heartworms, a direct hyper-reactive response of the bronchiolar SM cannot be demonstrated as the major mechanism of HARD.

**ABSTRACT #150**

MATRIX METALLOPROTEINASE -2 AND -9 IN BRONCHOALVEOLAR LAVAGE FLUID OF DOGS WITH IDIOPATHIC PULMONARY FIBROSIS. HP Heikkilä, MM Rajamäki. Faculty of Veterinary Medicine, University of Helsinki, Helsinki, Finland.

Idiopathic pulmonary fibrosis (IPF) is a chronic, progressive, interstitial lung disease of unknown etiology that mainly affects West Highland white terriers (WHWTs). Matrix metalloproteinases (MMPs) are proteases participating in pathological processes of different pulmonary diseases. The aim of this study was to investigate the role of gelatinolytic MMP-2 and -9 in bronchoalveolar lavage fluid of WHWTs with IPF.

Seven WHWTs with IPF (mean age 12 years, range 10-13) and 13 clinically healthy beagles (mean age 9 years, range 7-13) were included in the study. IPF was verified by clinical examinations and additionally by histopathology in 5 WHWTs. MMP-2 and -9 activities were analyzed by gelatinase SDS-page zymography, and the zymograms were evaluated for MMP activity with AlphaEaseFC software.

ProMMP-2 and -9 activities were significantly elevated in WHWTs compared with controls. Median proMMP2 was 0.72 (interquartile range 0.64-1.05) in WHWTs vs. 0.23 (0.11-0.43) in controls ($P < .01$). The corresponding figures for median proMMP-9 were 0.39 (0.20-0.48) and 0.05 (0.03-0.06) ($P < .01$). Mean arterial PaO2 was significantly lower in WHWTs than in controls ($69.0 \pm 15.2$ mmHg vs. $92.5 \pm 7.6$ mmHg, $P < .05$). A trend towards a negative correlation between proMMP2 and PaO2 was detected.

Our results suggest that upregulated MMP-2 and -9 participate in the pathogenesis of canine IPF and that the level of MMP-2 increases during disease progression.

**ABSTRACT #151**

RECRUITMENT OF PULMONARY INTRAVASCULAR MACROPHAGES IN IMMUNE-MEDIATED HEMOLYTIC ANEMIA IN DOGS. P Hodgson, A Carr and B Singh. Western College of Veterinary Medicine, University of Saskatchewan, Saskatoon, Canada.

Immune-mediated hemolytic anemia (IMHA) is a devastating disease in which macrophages, complement and antibodies attack the circulating red blood cells leading to their premature destruction. Dogs with IMHA have an increased incidence of pulmonary thromboembolism. The mechanisms of IMHA and the contributions of pulmonary pathophysiology to IMHA-induced mortality are poorly understood and consequently there is no effective treatment and a high mortality rate of 20% to 80%. Pulmonary intravascular macrophages (PIMs) are not normally found in the dog and rat lungs but can be recruited in sepsis in rats. Recruited PIMs promote systemic pathophysiology and mortality. The objective of the experiments was to examine the recruitment of PIMs in dogs with IMHA and in a mouse model of IMHA.

First, we examined lung tissues from dogs (N=16) that had died of IMHA and the normal dogs (N=4). Lung sections stained with H&E showed large macrophage-like cells in pulmonary microvessels. Lung sections were stained with antibodies against macrophages (MAC387), platelet-endothelial cell adhesion molecule-1 (PECAM-1) and Toll-like receptor-4 (TLR4). While MAC-387 antigen is observed in granulocytes, monocytes and tissue macrophages while PECAM-1 is expressed on the monocytes, neutrophils, platelets and a subpopulation of T-cells and TLR4 is found in peripheral blood leukocytes, macrophages and endothelial cells. Our data showed increased numbers of PIMs along with increased expression of PECAM-1 in airway epithelium in the lungs of IMHA dogs compared to the lungs from normal dogs. The expression of TLR4, which regulates sensitivity to endotoxin, was also increased in macrophages and vascular endothelium in the lungs from the IMHA dogs. Second, we examined recruitment of PIMs in a mouse model of IMHA. The mice were treated with Ter-119, an antimurine red blood cell antibody, which induced significant anemia, splenomegaly and recruitment of PIMs compared to saline-treated mice. These data show recruitment of PIMs and increased expression of inflammatory molecules PECAM-1 and TLR-4 in lungs of IMHA dogs. Furthermore, evidence of recruitment of PIMs in the mice provides us with a model to study the role of PIMs in IMHA-associated pulmonary pathophysiology in dogs.

**ABSTRACT #152**

CYCLIC AMP ATTENUATES FATTY ACID INDUCED APOPTOSIS IN AN IN VITRO MODEL OF HEPATIC LIPOTOXICITY. K Ponzetti, CRL Webster, Tufts Cummings School of Veterinary Medicine, N Grafton, MA.

In the last several years, it is becoming increasingly evident that lipid accumulation within hepatocytes can lead to hepatic failure. In humans, lipidosis (or steatosis) contributes to the hepatic pathology seen in alcoholic hepatitis and nonalcoholic steatohepatitis (NASH). In cats, a syndrome of acute hepatic failure is accompanied by widespread lipidosis. Free fatty acids such as palpitate and oleate, which are increased in the liver of cats with the hepatic lipidosis syndrome, are important mediators of lipotoxicity. In rodent models, fatty acid sensitizes hepatocytes to apoptotic cell death. The aim of this study was to establish an in vitro model of fatty acid hepatotoxicity and to initiate investigations into potential therapeutic options. Rat hepatocytes were isolated by standard collagenase perfusion and placed in culture. Twenty-four hrs later they were treated with 200 uM palmitate or 300 uM oleate. After 5-6 hrs the cells were fixed and stained with Hoescht for morphologic determination of apoptosis by fluorescence microscopy. Whole cell lysates were prepared from treated hepatocytes and used for immunoblotting for the cleavage fragment of the effector caspase, caspase 3. Some cultures were pre-treated for 1 hr with one of two cAMP analogues either 100 uM chlorophenyl thio-cAMP (CPT-cAMP) or 20 uM of the protein kinase A (PKA) independent cAMP analogue, 4-(4-chloro-phenylthio)-2'-O-methyladenosine-3'-5'-cyclic monophosphate (2-Me-CPT-cAMP) and then processed for morphologic or biochemical assessment of apoptosis. There was a significant increase in apoptosis in palmitate and oleate treated hepatocytes, 2.51 +/- 0.38 and 3.31 +/- 0.56 fold increase over control, respectively ($p<0.05$). In addition, immunoblotting revealed an increase in the amount of the cleavage fragment of caspase 3 in palmitate treated cells. Pre-treatment with CPT-cAMP or 2-Me-CPT-cAMP reduced the morphologic evidence of palmitate induced apoptosis by 57+/-10% and 57+/- 8.6% ($p<0.05$), respectively and ameliorated biochemical evidence of apoptosis. In conclusion, exposure of cultured rat hepatocytes to palmitate or oleate induces apoptosis. Increases in intracellular cAMP prevent fatty acid lipotoxicity in a PKA independent manner.

**ABSTRACT #153**

CULTURE-INDEPENDENT DETECTION OF BACTERIA IN FELINE INFLAMMATORY LIVER DISEASE. D.C. Twedt[1], S.D. Janeczko[2], K.W. McCord[1], J.L. Dudak[1], J.M. Cullen[3], P. Fisher[3], K.W. Simpson[3]. [1]Colorado State University, Fort Collins, CO, [2]Cornell University, Ithaca, NY, [3]North Carolina State University, Raleigh, NC.

The etiopathogenesis of feline inflammatory liver disease (ILD) is unclear. We sought to determine the presence and distribution of bacteria within the livers of cats with ILD by use of culture-independent methods.

40 archived samples from cats with inflammatory hepatic changes and 14 histologically normal livers were evaluated. Histopathology was classified according to WSAVA guidelines. Unstained hepatic tissue sections were subjected to fluorescence in situ hybridization (FISH) with a 16S rDNA probe that recognizes bacteria in general (EUB338) and a non-EUB probe used as a control. Antibodies against cytokeratins (clones AE1, AE3) and factor VIIIa were used to enable unequivocal distinction between bile ducts and vascular structures respectively. Tissues were examined with a fluorescence microscope for the presence and location of bacteria. Histopathology was classified as non-specific reactive hepatitis (RH,12), neutrophilic cholangitis (NC,10), lymphocytic cholangitis (LC,7), cholestasis/ obstruction (CO,6), probable lymphoma (LSA, 3), and neutrophilic hepatitis (NH, 2). Bacteria (EUB338) were observed in sections from 22/40 diseased and 3/14 normal livers (N) ($P<0.05$). In 11 sections bacteria were restricted to the outer surface of the liver capsule (4RH, 3LC, 2N, 1NC, 1CO) and may represent contaminants. Intrahepatic bacteria were visualized within portal vessels or venous sinusoids of 9 (4NC, 3RH, 1NH, 1LSA, 1N), the bile duct of 1 (NC), and within the hepatic parenchyma near the capsule in 4 (1LC, 1NC, 1NH, 1RH). Bacterial colonization was highest in 3 sections with NC and 1 cat with RH. Concurrent disease, predominately pancreatic and GI disease, was present in 12 of 14 cats with intrahepatic bacteria. Bacterial culture was positive in 12/24 samples (5 *E.coli*, 3 *Enterococcus*, 2 *Staphylococcus*, 1 *Actinomyces*, 1 *Streptococcus*). FISH and culture concurred in 15/24 cases. FISH and culture revealed *E.coli* in 3/3 NC with invasive bacteria. In 3/6 sections with capsular bacteria FISH and culture concurred, and 3 FISH positive samples were culture negative.

The results of this study suggest that bacteria may play a role in the etiology of feline inflammatory liver disease. The localization of intrahepatic bacteria suggests translocation of enteric bacteria is a likely source of infection.

**ABSTRACT #154**

DEVELOPMENT OF A FLOW CYTOMETRIC PANEL OF T-LYMPHOCYTE BIOMARKERS TO EVALUATE THE IMMUNOSUPPRESSIVE EFFECTS OF CYCLOSPORINE IN DOGS. T Archer, K Lunsford, A Mackin, C Fellman, J Stokes, L Pinchuk, S Pruett, C Langston. Mississippi State University College of Veterinary Medicine, Starkville, MS.

Cyclosporine is a potent immunosuppressive agent used to treat many canine diseases. Optimal dosing regimens remain unclear, primarily because standard methods that monitor effectiveness of immunosuppression have not been established. Our study was designed to establish a comprehensive panel of biomarkers in the dog that could be used to monitor the immunosuppressive effects of drugs such as cyclosporine. For our panel, we used flow cytometry to measure expression of cytokines (IL-2, IL-4, & INF-γ) and surface antigens (CD25 & CD95) on T-lymphocytes activated with concanavalin A and PMA/Ionomycin.

In an initial *in vitro* study, lymphocytes from 3 healthy dogs were incubated for different time periods in varying concentrations of activator and cyclosporine. This study demonstrated significant cyclosporine-associated reduction in biomarker expression: IL-2 expression was reduced by 54%, IL-4 by 34%, and IFN-γ by 81% (maximal suppression 12 hours after exposure to activator), and CD25 was reduced by 25% and CD95 by 18% (2 days after activator exposure). In a subsequent *in vivo* study, biomarkers were evaluated in 2 healthy dogs after one week of cyclosporine PO BID, with dose adjusted to attain published target blood levels, compared to a control dog. Treated animals had decreased expression of IL-2 (reduced by 51% & 42%), IL-4 (33% & 26%), and IFN-γ (51% & 38%), with more variable results for CD25 & CD95.

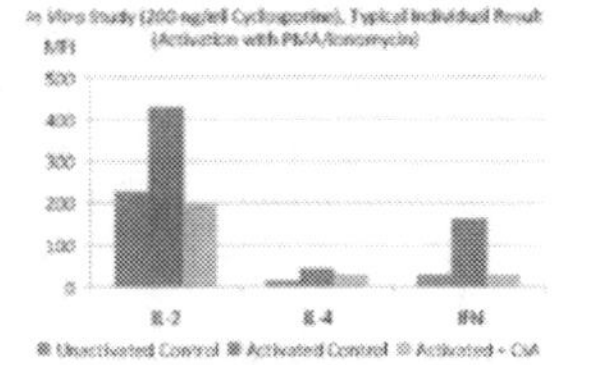


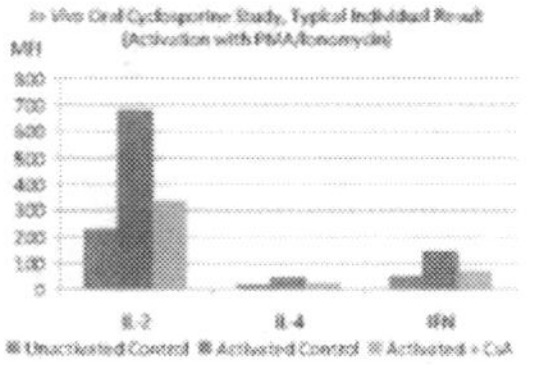


Our results confirm consistent cyclosporine-induced suppression of selected biomarkers of immunosuppression. Further investigation of the effects of cyclosporine on these biomarkers in both normal dogs and clinic patients is warranted.

## ABSTRACT #155

EVALUATION OF SERUM CONCENTRATIONS OF IMMUNOGLOBULIN A AND IMMUNOGLOBULIN M IN CHINESE SHAR PEIS WITH COBALAMIN DEFICIENCY.

N Grützner, JS Suchodolski, JM Steiner. Gastrointestinal Laboratory, Texas A&M University, College Station, TX.

Chronic small intestinal disease in Chinese Shar Peis (Shar Peis) has been associated with cobalamin deficiency. In humans, immunoglobulin A (IgA) has been suggested to play a role in the pathogenesis of cobalamin deficiency and may involve an anti-intrinsic factor-like activity or other mechanisms that interfere with normal cobalamin absorption. IgA producing cells in the small intestine belong to the B cell lineage. After an isoform switch from immunoglobulin M (IgM) to immunoglobulin A (IgA) expression, which is mediated through Th2 cytokine activity in the Peyer's patches of the lamina propria and the mesenteric lymph nodes, these cells return to the lamina propria of the intestinal tract by through the cardiovascular system (i.e., homing). Thus, serum IgA and IgM concentrations are the result of secretory function of the plasma cells. The aim of this study was to evaluate the serum concentrations of IgA and IgM in cobalamin deficient Shar Peis and Shar Peis with a normal serum cobalamin concentration.

Serum samples were collected from a total of 71 Shar Peis. Owners were asked to complete a questionnaire regarding the current health status of each dog. Serum cobalamin, folate, and canine trypsin-like immunoreactivity concentrations (reference ranges: 252-908 ng/L, 7.7-24.4 μg/L, and 5.7-45.2 μg/L, respectively) were measured by routine laboratory assays. Serum IgA and IgM concentrations were quantified by enzyme-linked immunosorbent assays using commercial kits (Bethyl Laboratories, Montgomery, TX). A Mann-Whitney test was used to compare the non-parametric data sets. Significance was set at $p<0.05$.

Undetectable serum cobalamin concentrations (<150 ng/L) were observed in 23 of 71 Shar Peis (32.4%), and these Shar Peis were considered to be severely cobalamin deficient. Serum cobalamin concentrations were within the reference range in the remaining 48 dogs. Serum trypsin-like immunoreactivity concentrations were within the reference range in all the dogs. Serum folate concentrations were within the reference range in all the control dogs. Three of the cobalamin deficient Shar Peis had slightly decreased serum folate concentrations (6.1, 6.4, and 7.2 μg/L). The median serum IgA concentration was significantly higher in cobalamin deficient Shar Peis (1.995 g/L) than in Shar Peis with a normal serum cobalamin concentration (0.812 g/L; $p<0.0001$). The median serum IgM concentration was significantly lower in cobalamin deficient Shar Peis (1.037 g/L) than in Shar Peis with a normal serum cobalamin concentration (2.071 g/L; $p<0.0001$).

In conclusion, cobalamin deficient Shar Peis had significantly higher median serum IgA and lower serum IgM concentrations compared to Shar Peis with a normal serum cobalamin concentration. These results warrant further investigation of gastrointestinal immunoglobulin secretion in Shar Peis with cobalamin deficiency.

## ABSTRACT #156

GENETICAL ANALYSIS OF TOLL-LIKE RECEPTOR GENES IN GERMAN SHEPHERD DOGS REVEALS POTENTIAL POLYMORPHISM ASSOCIATED WITH INFLAMMATORY BOWEL DISEASE.

A Kathrani[1], A House[1], D Werling[2], B Catchpole[2], A Murphy[3], H Trojer[3], N Gruetzner[4], K Allenspach[1]. Department of Veterinary Clinical Sciences[1], Department of Pathology and Infectious Diseases[2] and Department of Veterinary Basic Sciences[3], Royal Veterinary College, University of London, UK and the Gastrointestinal Laboratory[4], Department of Small Animal Clinical Science, College of Veterinary Medicine, Texas A&M University, College Station, Texas.

The identification of associations between specific polymorphisms in pattern recognition receptors such as Toll-like receptors (TLR) and human Inflammatory Bowel Disease (IBD) has provided data that demonstrates the importance of these genes and their role in innate immune function in the pathogenesis of this disease. To date no studies have been performed to evaluate polymorphisms in pattern recognition receptors in canine IBD. The aim of this study was to investigate whether polymorphisms in canine Toll-like receptor (TLR) 2, 4 and 5 genes are associated with IBD in German Shepherd dogs (GSDs).

Mutational analysis of TLR2, TLR4 and TLR5 was performed in 10 GSDs with IBD. Genomic DNA was extracted from blood stored in EDTA and the coding region of TLR2, TLR4 and TLR5 amplified using a polymerase chain reaction (PCR). PCR products were sequenced (Geneservice, Cambridge, UK) and full-length sequences obtained from each dog. Sequence data was compared to the canine genome (www.ensembl.org/Canis_familiaris). No non-synonymous single nucleotide polymorphisms (SNPs) were identified in the canine TLR2 gene. Four non-synonymous SNPs (T23C, G1039A, A1572T and G1807A) were identified in the TLR4 gene and three non-synonymous SNPs (G22A, C100T and T1844C) were identified in the TLR5 gene.

The three non-synonymous SNPs in TLR5 were evaluated further in a case-control study using a SNapSHOT multiplex reaction. Sequencing information from 21 GSDs with IBD from the UK were compared to two control groups consisting of 92 healthy GSDs recruited from the USA and 83 GSDs from patients treated for non-inflammatory disease at the Queen Mother Hospital for Animals referral hospital (UK). All three SNPs were found to be in Hardy-Weinberg equilibrium. When the case population was compared to the healthy controls from the USA the G22A SNP was found to be significantly associated with IBD ($p<0.001$). When the case population was compared to the GSDs with non-inflammatory disease from the UK the G22A SNP was found to be tending towards significance ($p=0.058$).

Our study suggests that the TLR5 SNP G22A could play a role in the pathogenesis of IBD in GSDs. Further studies are required to confirm the functional importance of this polymorphism in the pathogenesis of this disease.

**ABSTRACT #157**

URINARY LEUKOTRIENE E4 EXCRETION IS INCREASED IN DOGS WITH CHRONIC ENTEROPATHIES. M Im Hof[1], M Schnyder[1], S Hartnack[2], F Stanke-Labesque[3], N Luckschander[1], IA Burgener[1]. [1]Division of Small Animal Internal Medicine and [2]Department of Clinical Research and Veterinary Public Health, Vetsuisse Faculty of the University of Bern, Switzerland, and [3]Grenoble University Hospital, Grenoble, France.

Inflammatory bowel disease (IBD) and food-responsive diarrhea (FRD) are canine chronic enteropathies (CCE) that can only be differentiated by their response to treatment after exclusion of other diseases. Their severity can be scored with the canine IBD (CIBDAI) or the CCE clinical activity index (CCECAI). Cysteinyl leukotrienes (CysLTs) are proinflammatory 5-lipoxygenase-derived products that play a major role in the immune and inflammatory response. Leukotriene E4 (LTE4), the end product of CysLTs metabolism, is excreted in urine and is stable in this fluid. Its urinary quantification is therefore considered the best marker of in vivo CysLTs production in humans. The urinary LTE4 excretion is increased in human patients with IBD, and LTE4 levels are higher in patients with active IBD than in remission. The aims of this study were to evaluate the urinary LTE4 excretion in dogs with IBD, FRD and healthy controls, and to compare the LTE4 excretion with the activity of the disease using the CIBDAI scoring system.

Urine and CIBDAI were taken at initial presentation in dogs suffering from CCE. Due to their response to treatment, they were divided in FRD (responding to elimination diet alone) and IBD (need for supplementary immunosuppressive treatment). Quantification of LTE4 was performed by liquid chromatography-tandem mass spectrometry. The levels of LTE4 were corrected to the urinary level of creatinine measured with a Jaffe kinetic-based assay. Comparison between dogs suffering from CCE and healthy controls was performed with a Mann-Whitney U test, whereas comparison between IBD, FRD and control dogs were made with a Kruskal-Wallis one-way ANOVA with Dunn's test. Correlation between urinary LTE4 excretion and CIBDAI were analyzed using a Spearmann rank test. $P<0.05$ was considered significant.

15 IBD and 18 FRD patients were enrolled in the study together with 23 healthy control dogs. IBD dogs showed the highest urinary LTE4 excretion [median 79.9 pg/mg creatinine ($10^{th}$-$90^{th}$ percentiles 9.8-665.0)] followed by FRD [median 31.7 pg/mg creatinine ($10^{th}$-$90^{th}$ percentiles 6.2-169.4)] and control dogs [median 21.1 pg/mg creatinine ($10^{th}$-$90^{th}$ percentiles 9.1-86.5)]. The urinary LTE4 excretion was increased when comparing CCE (IBD + FRD) and control dogs ($P=0.024$) as well as IBD and control dogs ($P=0.018$). A ROC-analysis revealed a specificity of 95.7% for a LTE4 excretion of 90.6 pg/mg creatinine. No correlation was detected between LTE4 excretion and the clinical activity of the disease as measured with the CIBDAI.

In conclusion, dogs suffering from CCE have an increased urinary LTE4 excretion. Cysteinyl leukotriene pathway activation could contribute to the inflammation associated with IBD in dogs. The comparison of urinary LTE4 excretion before and after treatment is currently performed.

**ABSTRACT #158**

HIGH THROUGHPUT PYROSEQUENCING REVEALS REDUCED BACTERIAL DIVERSITY IN THE DUODENAL MUCOSA OF DOGS WITH IBD. M Craven, *SE Dowd, S McDonough, KW Simpson. College of Veterinary Medicine, Cornell University, Ithaca, NY and *Research and Testing Laboratory, Lubbock, Texas.

The enteric microflora is increasingly implicated in the etiopathogenesis of inflammatory bowel disease (IBD). Recent work has revealed reduced bacterial diversity and a shift towards a Gram -ve flora in the inflamed intestine across species, but the relationship of individual components of the enteric flora to different disease syndromes and clinical response is unclear. We sought to determine if the duodenal mucosal flora of dogs with IBD differs from healthy dogs, and to gain insight into the relationship of the mucosal flora to clinical response to diet, antibiotics, or immunosuppression.

A culture-independent bacterial tag-encoded FLX 16S rDNA amplicon pyrosequencing (bTEFAP) approach was used to evaluate the microbiome of endoscopic duodenal biopsies from 16 dogs with a histopathological diagnosis of IBD (7MN 7FN 2M, median age 5 yrs) and 14 clinically healthy (H) dogs (5FN, 6MN, 3F, median age 7.5 yrs). 10 dogs were food responsive (FR), 3 steroid responsive (SR), and 3 were partially responsive to a combination of food and antibiotics (FA). A total of 85,065 sequences (54776 from H, 30289 from IBD) were analyzed. The number of sequences per individual (median) was higher in H (3,912) than IBD (1,893) ($P=0.007$). Dogs with IBD had lower overall species diversity than H at all primary levels of taxonomy based upon rarefaction ($p<0.05$). When evaluated by Genus, we observed marked differences in the proportions of sequences in 16S rDNA libraries from H vs. IBD (**$p<0.001$, *$p<0.01$) for: Actinomycetales,* Bacteroidales,* Burkholderiales,* Campylobacteriales,* Clostridiales,* Coriobacteriales,* Enterobacteriales,** Fusobacteriales,* Lactobacillales* and Mycoplasmatales.** Analysis by clinical response indicated differences in numbers of sequence between groups (H, FR, SR and SA: Kruskal Wallis, $P<0.003$), and lower numbers in FR than H ($p=0.007$, see graph). Comparison of H and FR revealed most marked differences in proportions of Bacillales,** Erysipelotrichales,** Flavobacteriales,**

Pasteurellales,* Pseudomonadales,** and Sphingobacteriales.**

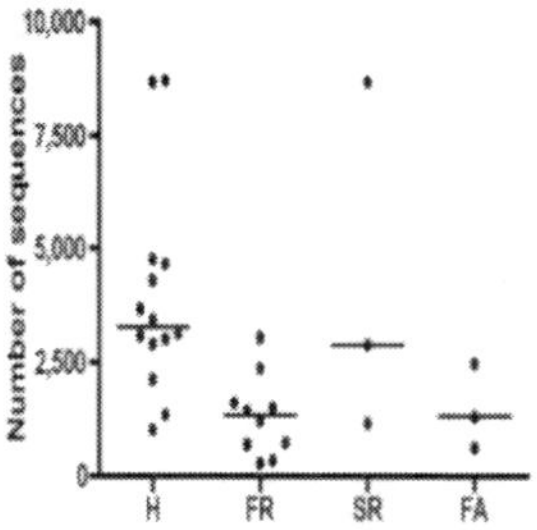


We conclude that duodenal inflammation is associated with decreased diversity of the mucosal flora, and altered relative proportions of dominant floral groups. Further work is required to determine whether differences in the mucosal flora impact the clinical response.

## ABSTRACT #159

*E. COLI* ASSOCIATED WITH GRANULOMATOUS COLITIS OF BOXER DOGS FREQUENTLY MANIFEST RESISTANCE TO ANTIBIOTICS. <u>M Craven</u>, B Dogan, A Schukken, M Volkman, A Chandler, PL McDonough, KW Simpson. College of Veterinary Medicine, Cornell University, Ithaca, NY.

Granulomatous colitis of Boxer dogs (GCB) is associated with the presence of *E. coli* within colonic mucosa and macrophages. A pivotal role for *E. coli* in the pathogenesis of GCB is supported by clinical and histological remission following the eradication of invasive *E. coli* with enrofloxacin. To optimize therapy against GCB associated *E. coli* we sought to determine antibiotic resistance profiles of *E. coli* strains isolated from GCB, and compared them with those from healthy dogs.

*E. coli* were isolated from colonic biopsies of 14 Boxers with a histological diagnosis of GC and mucosally invasive *E. coli* (8M, 6F, median age 12m), and from rectal mucosal swabs of 17 clinically healthy dogs (H, various breeds, 6M, 11F, median age 60m). 10 to 15 isolates from each sample were evaluated for differences in overall genotype and phylogroup (A, B1, B2, and D) by RAPD- and triplex PCR respectively. The antimicrobial resistance profile of representative strains from each sample was determined by broth microdilution MIC (NARMS 2004 Annual Report).

Culture yielded 23 different *E. coli* strains from GCB (1-3 / dog, median 2) and 34 strains from healthy dogs (1-3 /dog, median 2). The distribution of phylogroups was not significantly different (P=0.18) between GCB (5A, 7B1, 5B2, 6D) and H (2A, 10B1, 15B2, 7D), Resistance to antibiotics is shown in the table below (Fisher Exact, CGB vs H: *=P<0.05, **<0.01,***P<0.001).

| Antibiotic | Resistant (%) | | | | Antibiotic | Resistant (%) | | | |
|---|---|---|---|---|---|---|---|---|---|
| | strains | | patients | | | strains | | patients | |
| | GCB | H | GCB | H | | GCB | H | GCB | H |
| Clavamox | 35* | 8 | 57* | 12 | Trimethoprim-sulfa | 44*** | 6 | 57** | 7 |
| Ampicillin | 49** | 15 | 64** | 18 | Ciprofloxacin | 35*** | 0 | 43** | 0 |
| Cefoxitin | 30*** | 0 | 50*** | 0 | Gentamicin | 13 | 18 | 14 | 36 |
| Tetracycline | 48* | 18 | 64* | 24 | Chloramphenicol | 17* | 0 | 21 | 0 |

A high proportion of Boxers with GC harbored *E. coli* strains that were resistant to macrophage penetrating antibiotics ciprofloxacin (43%), trimethoprim (57%) and tetracycline (64%). Empirical treatment of GCB before colonic biopsy was associated with resistance against the antibiotic used in 6/6 Boxers receiving enrofloxacin, 3/8 receiving TMPS, 3/9 receiving Clavamox, 2/10 receiving ampicillin and 1/10 receiving tetracycline.

We conclude that antibiotic resistance is common in *E. coli* from GCB and suggest that antibiotic selection should be guided by sensitivity testing rather than empirical wisdom.

## ABSTRACT #160

DESCRIPTION OF PROTEIN-LOSING ENTEROPATHY IN YORKSHIRE TERRIER DOGS USING THE W.S.A.V.A. GASTROINTESTINAL CLASSIFICATION SYSTEM. <u>SM Simmerson</u>, A Wunschmann, L Crews, PJ Armstrong, M Fee, and R Washabau. University of Minnesota, College of Veterinary Medicine, St. Paul, Minnesota.

Yorkshire terrier dogs are reported to be at increased risk for the development of protein-losing enteropathy (PLE). Few reports have been published that document clinical features, histopathologic lesions, or outcome in this disorder, and the few cases that have been reported have been evaluated with differing non-standardized classification systems.

The study objectives were (1) to describe historical, physical examination, clinical pathologic, imaging, and histopathologic findings in a large group of Yorkshire terrier dogs with a diagnosis of PLE and (2) to use a standardized classification system for the evaluation of gut pathology.

Retrospective study. The diagnosis of PLE was made on the basis of gastrointestinal clinical signs, hypoalbuminemia and intestinal histopathology in Yorkshire terrier dogs evaluated between 2002-2007 in which this information was available. Dogs that were proteinuric, had abnormal bile acids, or received glucocorticoids within two weeks prior to diagnosis were excluded from the study. Intestinal biopsies were interpreted according to standards for cytoarchitectural change and cellularity using the W.S.A.V.A. G.I. Histopathology Classification system.

30 dogs met the inclusion criteria (mean age 6.7 yrs, range 1-12 yrs; 20 spayed females, 8 neutered males, 2 intact males). The most common presenting complaints were diarrhea (n=20), vomiting (n=11) and abdominal distention (n=11). Most common physical abnormalities were ascites (n=15) and pleural effusion (n=14). Hypoalbuminemia was often severe and ranged from 0.8-2.2 g/dL, mean 1.35 g/dL. Other abnormal laboratory results included hypoglobulinemia (n=14), hypomagnesemia (n=12), low serum creatinine (n=14), anemia (n=7) and thrombocytosis (n=7). Small intestinal mucosal speckling was present in 2/4 dogs that had abdominal sonography performed. Histopathologic lesions commonly included lymphatic dilation (n=24) and crypt abcessation (n=10) with lamina proprial inflammation ranging from mild to severe (n=30). Clinical course of disease was variable and ranged from resolution of clinical signs with glucocorticoid therapy (n=12) to unresponsive disease (n=9) and peracute death resulting from thromboembolism (n =2).

Yorkshire terrier dogs with PLE can be severely affected with ascites and pleural effusion. Clinical outcomes are variable and prognosis therefore is guarded. Intestinal abnormalities in this disorder include lymphatic dilation, intestinal crypt abscessation, and variable increases in the cellularity of the lamina propria.

**ABSTRACT #161**
DOPPLER ULTRASOUND ANALYSIS OF GASTROINTESTINAL BLOOD FLOW FOR DIFFERENTIATING FOOD ALLERGIC FROM NON-FOOD ALLERGIC PRURITIC DOGS. J Hobbs, L Gaschen, S Merchant, F Gaschen. Louisiana State University, Baton Rouge, LA.

Doppler ultrasound has been shown to be a non-invasive method of characterizing hemodynamic changes associated with food hypersensitivity in dogs with proven food allergies causing chronic enteropathy. The goal of this study was to identify whether or not dogs with cutaneous signs and documented food hypersensitivity also show changes in gastrointestinal hemodynamic responses in comparison to dogs with similar clinical signs but no food allergy.

Client-owned dogs with pruritus suspected of having a food allergy were included in this study. Transabdominal spectral Doppler analysis of the cranial mesenteric artery (CMA) was performed at time of initial presentation with the dog's regular diet and following a 10 week long hypoallergenic dietary trial with IVD® Rabbit and Potato. The following parameters were recorded at 0, 20, 40, 60, and 90 minutes post-prandially: resistive index (RI), pulsatile index (PI), mean diastolic flow (MDV) and the systolic/diastolic ratio (S/D ratio). Dogs were placed into either a food allergic (FA) or non-food allergic (NFA) group based on the dermatologic work up and response to the hypoallergenic diet. Dogs in the FA group were provoked with their original diet and re-examined at a third time point when clinical signs reoccurred. An ANOVA analysis of repeated measures was used to compare the Doppler parameters within groups between examinations and between groups at each examination.

Thirteen dogs were examined, 5 of which were confirmed to have skin disease due to food allergy. Eight dogs had pruritus due to other causes and did not respond to the dietary trial. At initial examination, there were no significant differences in the post-prandial blood flow parameters between the two groups. After a 10 week dietary trial neither the FA, nor the NFA group showed significant differences in the Doppler parameters compared to the initial examination. However, the FA group showed a significantly elevated ($p<0.05$) S/D ratio at t0 and t20 compared to the NFA group, representing a comparative decrease in diastolic blood flow (lack of vasodilation) at those time points. When provoked with their original diet with recurrence of clinical signs, the FA group showed a significant ($p<0.05$) increase in the MDV at t20 and t40 post-prandially compared to the exam following the dietary trial. The RI Doppler parameter never showed statistical differences in any test.

No significant differences in gastrointestinal blood flow could be identified when switching from a regular to a hypoallergenic diet within either group. This finding may limit the use of this technique as a diagnostic test for food allergies in dogs with cutaneous signs only. However, the FA group did show a statistically significantly more rapid increase in blood flow in the early post-prandial period (t20 and t40) when provoked with their original diet. This response may be explained by local immune reactions to food allergens at the gastrointestinal mucosa.

**ABSTRACT #162**
EFFECTS OF PREDNISONE ALONE OR PREDNISONE WITH ULTRALOW-DOSE ASPIRIN ON THE GASTRODUODENAL MUCOSA OF HEALTHY DOGS. AH Graham, MS Leib. Virginia-Maryland Regional College of Veterinary Medicine, Blacksburg, VA.

The co-administration of prednisone and ultralow-dose aspirin has been recommended for the management of various diseases, but the safety of this combination in dogs has not been studied. The authors hypothesized that the gastroduodenal lesions associated with prednisone and ultralow-dose aspirin would be similar to those caused by prednisone alone, but both treatments would result in more severe lesions than placebo.

This was a randomized, double-blinded, placebo-controlled study involving eighteen healthy adult purpose-bred dogs divided into three groups for 27 days of treatments: placebo, prednisone, and prednisone and ultralow-dose aspirin. Gastroduodenoscopy was performed prior to and days 5, 14, and 27 of treatment, and mucosal lesion scores assigned. Mucosal lesion scores were evaluated using a Kruskal-Wallis test. Clinical signs were compared using the Friedman's Chi-Square test (significance at $p<0.05$).

There were no significant differences in gastroduodenal lesion scores between groups at any time during the study, and there were no significant changes in gastroduodenal lesion scores over time within any group. Significantly more frequent diarrhea occurred in the prednisone and aspirin group during treatment, compared to baseline. No significant differences in clinical signs were found between any of the groups. Notably, although statistical significance was not reached, individual dogs in both the PN and PA group developed mucosal lesions consisting of hemorrhages (both), erosions (both), and a single small ulcer (PA), whereas the most severe lesions identified in the NN group were mucosal hemorrhages.

The concurrent use of prednisone and ultralow-dose aspirin did not increase the severity of gastroduodenal lesions compared to prednisone or placebo in healthy dogs. Co-administration of prednisone and ultralow-dose aspirin may increase the incidence of mild, self-limiting diarrhea in individual healthy dogs. The effect of this combination in older dogs with spontaneous disease was not determined from this study.

**ABSTRACT #163**
SERUM TRIGLYCERIDE CONCENTRATIONS IN MINIATURE SCHNAUZERS WITH AND WITHOUT A HISTORY OF PANCREATITIS. PG Xenoulis, MD Levinski, JS Suchodolski, JM Steiner. Gastrointestinal Laboratory, Texas A&M University, College Station, TX.

Miniature Schnauzers (MS) have been reported to be at increased risk for pancreatitis. One explanation for this might be the documented high prevalence of hypertriglyceridemia in this breed. In humans, severe hypertriglyceridemia has been described as risk factor for pancreatitis. An association between hypertriglyceridemia and pancreatitis has also been suggested for dogs, but a cause and effect relationship is less clear in this species. This is further complicated by the fact that hypertriglyceridemia can be the result rather than the cause of pancreatitis in some cases. The aim of this study was to compare serum triglyceride concentrations between MS with and without a history of pancreatitis.

Twenty-three MS with an initial diagnosis of pancreatitis and 44 age-matched healthy MS with no history of pancreatitis were included in this study. The diagnosis of pancreatitis was based on the presence of clinical signs associated with pancreatitis (e.g., vomiting, anorexia, abdominal pain) and a serum Spec cPL™ concentration >400 µg/L (the current cut-off value for pancreatitis). After recovery from the initial episode of pancreatitis, the owners of the dogs were asked to have a follow-up sample submitted. None of the 23 MS had any clinical signs of pancreatitis at the time the follow-up sample was collected, and none of the MS had any other known diseases that might affect lipid metabolism. Only the follow-up sample, which was obtained after clinical resolution of the initial episode of pancreatitis, was used for statistical analyses. This was done to exclude the possibility that hypertriglyceridemia in the initial sample had developed secondary to pancreatitis. Serum triglyceride concentrations were evaluated in the follow-up sample for all 23 MS and compared with those of the 44 age-matched healthy MS with no history of pancreatitis. Body weights were also compared between MS with and without a history of pancreatitis.

Follow-up samples from the 23 MS were collected within 2 to 41 weeks (median: 8.5 weeks) after the diagnosis of pancreatitis. There was no significant difference of the mean body weight between MS with and without a history of pancreatitis (p=0.438). Of the 23 MS with a history of pancreatitis, 18 (78.3%) had serum triglyceride concentrations above the reference range, while only 18 (40.9%) MS with no history of pancreatitis had serum triglyceride concentrations above the reference range (odds ratio: 5.2; 95% confidence interval: 1.7-16.6; p=0.004). The median serum triglyceride concentration was significantly higher in MS with a history of pancreatitis (median: 555.0 mg/dL) than in MS with no history of pancreatitis (median: 83.5 mg/dL; p=0.006).

In conclusion, MS with a history of pancreatitis were significantly more likely to have hypertriglyceridemia than age-matched MS with no history of pancreatitis. These findings provide further evidence that hypertriglyceridemia may be a risk factor for pancreatitis in MS. Additional studies are needed to further clarify the role of hypertriglyceridemia in the development of pancreatitis in MS as well as other dog breeds.

## ABSTRACT #164

SERIAL EVALUATION OF CANINE PANCREATIC LIPASE (SPEC cPL™) IN DOGS WITH CLINICAL SIGNS OF PANCREATITIS.

LM Prior[1], MA Forman[1], J Shiroma[1], JE Robertson[2], RA Hostutler[1], CD Brown[1], [1]MedVet, Medical and Cancer Center for Pets, Worthington, OH. IDEXX Laboratories, West Sacramento, CA.

Determining resolution of pancreatitis in dogs is currently based on a combination of improvement in clinical signs (ex. vomiting, inappetance, diarrhea), examination findings (ex. abdominal pain, fever), bloodwork parameters (ex. neutrophilia, amylase, lipase, hyperbilirubinemia), and abdominal ultrasound. These findings are not specific to the pancreas. The Spec cPL assay is a quantitative measurement of pancreatic lipase and is currently used in the diagnosis of pancreatitis. The aim of this study was to determine if elevated Spec cPL values (>201µg/L) decreased in dogs, as the clinical signs of pancreatitis (ex. vomiting) resolved.

Seventy-one dogs were prospectively evaluated and categorized as Spec cPL 201-399µg/L (Group 1), n=16 and ≥400µg/L (Group 2), n=55. Fifty-nine dogs with a Spec cPL ≤ 200µg/L (Group 3) were retrospectively evaluated and utilized as a baseline control group. All dogs were treated by board-certified internists and evaluations included CBC, biochemical panel, Spec cPL and an abdominal ultrasound performed by a board-certified radiologist. For all dogs with Spec cPL >201µg/L, multiple CBC, biochemical panel and Spec cPL tests (median 3, range 2-16) were performed during hospitalization and at recheck evaluations.

The median (range) age in Groups 1, 2 and 3 was 8.9 (4.6 - 14.1), 9 (2.5 - 15.1), and 7 (1.4 - 15.6) years, respectively. The median (range) Spec cPL at presentation in Groups 1, 2 and 3 was 306 (203 - 374), 661 (413 - >1000), and 37 (25 - 196) µg/L, respectively. Vomiting, reported at presentation, was present in Groups 1, 2, and 3 at 11/16 (69%), 37/55 (67%), and 49/59 (83%), respectively. In Groups 1 and 2, 9/11 (82%) and 30/37 (81%), had resolution of vomiting in a median (range) of 2 (1-83) and 2 (1-127) days, respectively. Median (range) Spec cPL, in Groups 1 and 2, at resolution of vomiting was 28 (<30 - 528) and 521 (<30 - >1000) µg/L, respectively. At last recheck evaluation, median (range) Spec cPL, in Groups 1 and 2, was 86 (<30 - 86) and 244 (<30 - >1000) µg/L, respectively. For dogs without resolution of vomiting, in Groups 1 and 2, median (range) of Spec cPL was 283 (77-489) and 364 (<30 - >1000) µg/L, respectively. Considering dogs with ≥3 samples, the Spec cPL did not decrease ≤ 200 µg/L, despite resolution of vomiting, in Group 1, 2/11 (18%) and in Group 2, 13/37 (35%).

In dogs with clinical sign of pancreatitis and elevated Spec cPL (>201 µg/L), a progressive decrease in Spec cPL was noted in dogs with resolution of vomiting and at final recheck evaluation. However, in a subpopulation of dogs the Spec cPL never returned to ≤ 200 µg/L, the established normal range, despite resolution of certain clinical signs of pancreatitis (vomiting).

## ABSTRACT #165

EVALUATION OF FELINE PANCREAS-SPECIFIC LIPASE (SPEC fPL™) FOR THE DIAGNOSIS OF FELINE PANCREATITIS.

MA Forman[1], J Shiroma[1], PJ Armstrong[2], JE Robertson[3], and J Buch[4]. [1]MedVet, Medical and Cancer Center for Pets, Worthington, OH. [2]University of Minnesota, College of Veterinary Medicine, St. Paul, MN. [3]IDEXX Laboratories, West Sacramento, CA. [4]IDEXX Laboratories, Westbrook, ME.

Pancreatitis appears to be a relatively common illness in cats but diagnosis is difficult due to vague clinical signs and clinicopathologic findings. Feline pancreatic lipase immunoreactivity (fPLI) has been shown to be moderately sensitive for the diagnosis of pancreatitis; however test availability has been limited and sensitivity was established using a limited number of cats with pancreatitis. The Spec fPL test is a monoclonal ELISA which has recently become commercially available. The aims of this study were 1) to establish the reference interval for the test in healthy cats and 2) to determine the sensitivity and specificity of the test in a large group of ill cats with and without pancreatitis.

Forty-one healthy cats and 141 cats with clinical signs consistent with pancreatitis were studied to determine the reference interval and the sensitivity and specificity of the Spec fPL test, respectively. The median ages of healthy and clinical cats were 5 (range 1-14) years and 11 (range 0.6-18) years, respectively. Based on a detailed history and examination, CBC, biochemistry panel, urinalysis, abdominal ultrasonography and clinical outcome, each clinical case was categorized by 2 board-certified internists, blinded to the Spec fPL results, into one of 6 categories (definitely not pancreatitis(DnP), n=26; probably not pancreatitis(PrnP), n=33; possibly not pancreatitis(PsnP), n=19; possibly pancreatitis(PsP), n=29; probably pancreatitis(PrP), n=25; and definitely pancreatitis(DP), n=9). Individual Spec fPL data points (dots) and median values (line) are displayed in the table. Statistical significant difference from healthy cats is indicated (*, ***).

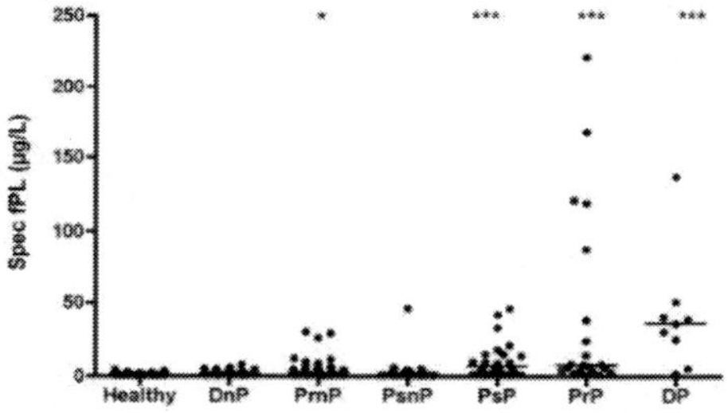


The reference interval, determined from the central 95 percentile of results from healthy cats was 0.7-3.5 µg/L. A concentration ≥5.4 µg/L was determined to be consistent with pancreatitis, using the results to maximize sensitivity and specificity. With 5.4 µg/L as the diagnostic cut off, the sensitivity of the Spec fPL assay for feline pancreatitis (DP and PrP) was 79% and the specificity for cats characterized as PrnP and NP was 82%.

These findings support the use of the Spec fPL as a screening test for feline pancreatitis.

## ABSTRACT #166

**A MULTI-INSTITUTIONAL STUDY EVALUATING DIAGNOSTIC UTILITY OF SPEC cPL™ IN THE DIAGNOSIS OF ACUTE PANCREATITIS IN DOGS.**

K McCord[1], J Davis[1], F Leyva[1], PJ Armstrong[2], KW Simpson[3], M. Rishniw[3], M.A. Forman[4], DS Biller[5], Comparative Gastroenterology Society Members, D Twedt[1]. [1]Colorado State University, Fort Collins, CO. [2]University of Minnesota, Saint Paul, MN. [3]Cornell University, Ithaca, NY. [4]MedVet Center for Pets, Worthington, OH. [5]Kansas State University, Manhattan, KS.

Measurement of pancreas-specific lipase (PL) may aid the diagnosis of acute pancreatitis in dogs. This study evaluated the performance of a commercial immunoassay for canine PL (Spec cPL) in dogs with a clinical diagnosis of acute pancreatitis.

Cases with an initial differential diagnosis that included acute pancreatitis (APS), or did not include acute pancreatitis (CO), were solicited from Comparative Gastroenterology Society (CGS) members. For each dog the history, physical examination, laboratory findings (including total amylase and lipase), abdominal ultrasound and the clinical course were evaluated by a panel of 4 board-certified internists blinded to the results of the Spec cPL. A board-certified radiologist reviewed and graded each ultrasound image. Cases were categorized to one of five pre-defined groups (0-4): 0-not pancreatitis, 1-not primary pancreatitis, 2-possibly pancreatitis, 3-probably pancreatitis or 4-pancreatitis.

The 84 cases enrolled were categorized as follows (group, n): APS (n=57), 0 (26), 1 (11), 2 (9), 3 (8) and 4 (3); CO (n=27), 0 (25) and 1 (2). There was a high level of agreement in the assignment of cases to different groups by the evaluators (kappa 0.87). Spec cPL differed according to group (1-wayANOVA, P<0.001) and was higher (P<0.001) in dogs with suspected pancreatitis (groups 2,3,4) than those without suspicion of pancreatitis (groups 0APS, and 0CO), but not group 1 (P>0.05) (Figure). Amylase and lipase activities did not differ between groups. Spec cPL sensitivity and specificity for cases with clinical score 0 (no pancreatitis), and 2,3,4 (pancreatitis), calculated using current cut off values of < 200µg/L as negative and >400 µg/L positive, yielded a 93% sensitivity and 78% specificity. The likelihood ratio of a negative test (0.029) was better than that for a positive test (1.3).

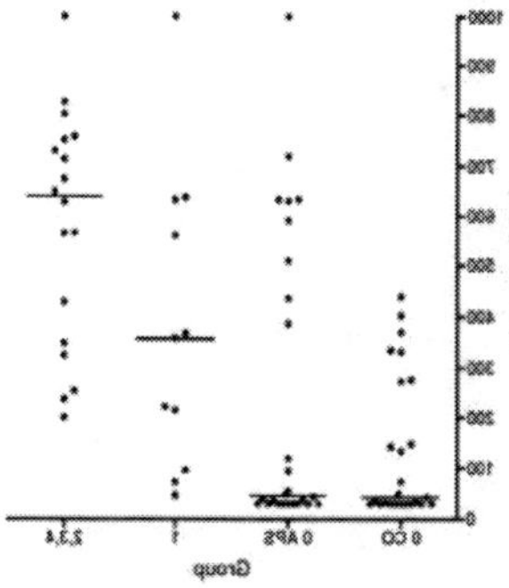

We conclude that Spec cPL is better able to discriminate dogs with suspected pancreatitis than amylase and lipase, and that dogs with a Spec cPL < 200µg/L are unlikely to have clinical acute pancreatitis.

## ABSTRACT #167

**CLINICAL SIGNIFICANCE OF INCREASED SERUM FELINE PANCREATIC LIPASE IMMUNOREACTIVITY CONCENTRATIONS IN CATS WITH INFLAMMATORY BOWEL DISEASE.**

S. Bailey[1], L. Benigni[1], J. Eastwood[1], O.A.Garden[1], L. McMahon[1], K. Smith[1], J.M. Steiner[2], J.S. Suchodolski[2], K. Allenspach[1]. [1]Department of Veterinary Clinical Sciences, Royal Veterinary College, University of London, UK, and [2]Gastrointestinal Laboratory, Department of Small Animal Clinical Sciences, College of Veterinary

Medicine and Biomedical Sciences, Texas A&M University, College Station, TX.

The aim of this study was to determine the prevalence of increased serum feline pancreatic lipase immunoreactivity concentrations (fPLI) in cats diagnosed with inflammatory bowel disease (IBD); and to investigate the clinical significance and compare the clinical characteristics between cats with IBD with serum fPLI concentrations within the reference range (2.0-6.8 μg/L), within the questionable range (6.9-11.9 μg/L) and above the diagnostic cut off for pancreatitis ($\geq$ 12 μg/L).

Medical records for 23 cats diagnosed with IBD and with serum fPLI concentrations available were retrospectively studied at the Royal Veterinary College, University of London. The cats were classified into three groups according to their serum fPLI concentration. A Kruskal Wallis test and Fisher's Exact test were used to analyze for any differences between the groups for continuous and categorical data, respectively.

Of the 23 cats, 7 cats had a serum fPLI concentration within the reference range, 9 cats within the questionable range, and 7 cats above the diagnostic cut off for pancreatitis. The clinical outcome of cats with increased serum fPLI concentrations was not significantly different from cats with normal serum fPLI concentrations. No significant differences were identified between the three groups for age, serum ALT and ALP activities, feline IBD clinical disease activity index, pancreatic ultrasound findings, intestinal histopathology scores, treatment, or clinical response. However, cats with serum fPLI concentrations $\geq$12.0 μg/L had a significantly lower median serum albumin concentration (27.1 g/L) compared to cats with serum fPLI concentrations within the questionable range (34.2 g/L; $p<0.05$), and a significantly lower median cobalamin concentration (204 ng/L) compared to cats with fPLI concentrations within the reference range (1142.5 ng/L; $p<0.05$).

A high number of cats with IBD had increased serum fPLI concentrations. However the outcome for cats with increased serum fPLI concentrations was not different from cats with normal serum fPLI concentrations. It is noteworthy that cats with serum fPLI concentrations $\geq$12.0 μg/L have lower serum albumin and cobalamin concentrations than cats with serum fPLI concentrations in the questionable and reference range, respectively.

**ABSTRACT #168**

SERUM D-LACTATE CONCENTRATIONS IN CATS WITH GASTROINTESTINAL DISEASE.

DP O'Brien[1], RA Packer[2], GE Moore[2], CY Chang[2], GA Zello[3], S Abeysekara[3], JN Naylor[3,4], JM Steiner[5], and JS Suchodolski[5]. 1. University of Missouri, Columbia, MO. 2. Purdue University, West Lafayette, IN. 3. University of Saskatchewan, Saskatchewan SK. 4. Ross School of Veterinary Medicine, St Kitts. 5. Texas A&M University, College Station, TX.

The D-enantiomer of lactic acid is not normally found in any appreciable quantities in serum from mammals. However, recently, we found markedly increased serum D-lactate concentrations in two cats with encephalopathy and pancreatic and gastrointestinal disease, presumably due to an alteration of the gastrointestinal microbiota and bacterial production of D-lactate. The purpose of this study was to determine if serum D-lactate concentration is increased in cats with gastrointestinal disease and if those elevations correlate with specific neurologic or gastrointestinal findings.

Serum D- and L-lactate concentrations were measured by stereospecific high-pressure liquid chromatography using a 3-mm octadecylsilane-packed analytical column coated with N,N-dioctyl-L-alanine as the chiral selector for the separation of lactic acid enantiomers. Serum samples from 30 normal cats were used as controls. Left-over serum samples from cats with abnormal gastrointestinal function test results, that were submitted to the GI Laboratory at Texas A&M University were randomly selected. Questionnaires were sent to veterinarians to survey for gastrointestinal and neurologic signs, and results of other laboratory tests. D-lactate and L-lactate concentrations were compared between normal cats and 100 cats with gastrointestinal disease and a completed questionnaire using a Wilcoxon rank sum test. Association of D-lactate concentration with tests of GI dysfunction and neurologic signs were evaluated with multiple regression and logistic regression analysis, respectively.

All 100 cats had a clinical history of gastrointestinal signs and/or abnormal gastrointestinal function test results. Only 3 cats had obvious neurological signs. D-lactate concentrations of cats with gastrointestinal disease (median and range: 0.41 and 0.01-8.33 mmol/l) were significantly higher than in healthy control cats (median and range: 0.22 and 0.05-0.87 mmol/l; $P=0.0069$). L-lactate concentrations were not significantly different between the two groups. D-lactate concentrations were not significantly associated with serum concentrations of fPLI, cobalamin, or folate or with the presence of neurologic abnormalities.

In this study, no association between D-lactate concentrations and neurologic signs were observed, although the number of cats with obvious neurologic signs may have been too small to detect such an association. The cat with the highest D-lactate concentrations (8.33 mmol/l) was reported as depressed, listless and weak. Serum D-lactate concentrations were also not associate with other tests of GI function. We conclude that D-lactate concentrations are increased in cats with gastrointestinal disease, possibly due to disarrangements of the intestinal microbiota. These findings warrant further investigations into the role of alterations of the intestinal microbiota in patients with gastrointestinal disease.

**ABSTRACT #169**

ASSOCIATION OF SERUM COBALAMIN AND METHYLMALONIC ACID CONCENTRATIONS IN DOGS.

N Berghoff, KC Stupka, JS Suchodolski, JM Steiner. Gastrointestinal Laboratory, Texas A&M University, College Station, TX.

Cobalamin (Cbl) is a co-factor for several enzymatic reactions in mammals, one of which leads to the conversion of methylmalonyl-coenzyme A (CoA) to succinyl-CoA. In human and veterinary patients with Cbl deficiency, the metabolite methylmalonic acid (MMA) accumulates and can thus be used as a marker for Cbl deficiency at the cellular level. The aims of this study were to calculate a reference range for MMA in dog serum and to determine the proportion of individuals with increased serum MMA concentrations in dogs with varying serum Cbl concentrations.

For determination of the MMA reference range, serum was collected from 43 healthy dogs. A total of 556 leftover canine serum samples were obtained from submissions to the GI Lab. The samples were divided into 10 groups according to their serum Cbl concentration: <150, 150-250, 250-350, 350-450, 450-550, 550-650, 650-750, 750-850, 850-950, and >950 ng/L ($n \geq 40$ for each group). Serum MMA was measured using an isotope-dilution GC/MS method. The reference range was calculated using the central 95$^{th}$ percentile. Groups were compared using a Kruskal-Wallis test with Dunn's post tests. A $\chi^2$-test for trend was calculated to evaluate the association of serum Cbl and MMA concentrations. Statistical significance was set at $p<0.05$.

The reference range for serum MMA was 414.7-1,192.5 nmol/L. Median and ranges for serum MMA concentrations for groups 1 through 10 were: 1,328.0 (492.0-24,795.0), 1,034.0 (469.1-3,003.0), 759.5 (378.9-4,064.0), 741.9 (397.2-2,447.0), 710.2 (405.0-1,396.0), 690.5 (398.6-1,733.0), 730.3 (391.7-1,807.0), 820.9 (332.0-1,523.0), 673.6 (343.1-1,765.0), and 733.0 (423.1-1,739.0) nmol/L, respectively. Median MMA concentrations were significantly different among the 10 groups of dogs ($p<0.0001$). In dogs with Cbl <150 ng/L, median MMA concentration was significantly different from that of all other groups ($p<0.05$ for all), except the "150-250" group. Median MMA concentration of the "150-250" group was significantly different from that of all other groups ($p<0.05$ for all), except the "750-850" group. No significant differences were found among the remaining groups. Percentages of dogs with MMA concentrations above the upper limit of the reference range in groups 1 to 10 were 62.5%, 31.7%, 19.2%, 16.0%, 10.0%, 14.3%, 6.0%, 9.6%, 10.2%, and 12.5%, respectively. The $\chi^2$-test for trend was significant ($p<0.0001$).

These data show that a large percentage of dogs with Cbl concentrations below the lower limit of the reference range (252-908 ng/L) have an increased serum MMA concentration. A significant trend for an increasing serum MMA concentration with a decreasing serum Cbl concentration was observed. Several dogs with Cbl of up to 450 ng/L had increased MMA concentrations, suggesting that a portion of dogs with Cbl concentrations in the low-normal range have evidence of Cbl deficiency. This may indicate a need to supplement these dogs with Cbl, despite a serum Cbl concentration within the reference range. It remains to be determined why some dogs with high-normal Cbl concentrations have an increased serum MMA concentration.

**ABSTRACT #170**

COMPARISON OF GASTROINTESTINAL MOTILITY IN DOGS TREATED WITH METOCLOPRAMIDE, CISAPRIDE, ERYTHROMYCIN OR MAROPITANT USING THE SMARTPILL™ KW McCord[1], P Boscan[1], K Dowers[1], A Bradley[1], S Cosgrove[2], K Dame[2], J Civil[2], D Holmberg[3], D Sacilowski[3] and DC Twedt[1]. [1]Colorado State University, Fort Collins, CO, [2]Pfizer Animal Health, Kalamazoo MI. [3]SmartPill Corporation, Buffalo, NY.

The effects of the drugs metoclopramide, cisapride, erythromycin and maropitant on gastrointestinal (GI) motility in the dog are poorly described. The aim of this study was to determine the effect of these drugs on motility in normal dogs using the SmartPill™. The SmartPill™ is a noninvasive wireless sensor capsule that is given orally and transmits data on pressures, transit time, luminal pH and temperature as it passes through the stomach, small and large bowel. This instrumentation has been validated to evaluate GI function in humans. A computer software displays the data and calculates the integral under the pressure curve, which is the area under the curve (AUC) and a motility index (MI=AUC/time).

Ten normal adult dogs were used in a randomized, crossover, blinded design. All dogs received metoclopramide 0.5 mg/kg SQ TID, cisapride 0.5 mg/kg PO TID, erythromycin 1 mg/kg PO TID, maropitant 1 mg/kg SID and a saline placebo 1 ml/dog SQ SID. Drugs were administered daily for three days to obtain a steady state and on day four the SmartPill™ was administered following a standard meal. All medications were continued until the SmartPill™ left the GI tract. There was at least 7-day washout period between treatments. Data was compared using a linear mixed model with 95% confidence interval. Significance was $p<0.05$.

The transit times for each treatment had large individual variation within and between dogs. Gastric and large bowel parameters were not significantly different between drugs and placebo. Small bowel mean pressures and MI values were higher for the cisapride treatment group when compared to control ($p<0.05$). The AUC and MI values for small bowel were higher for maropitant when compared to control ($p<0.05$).

Using the SmartPill™ technology there were no major differences in gastrointestinal motility measurements between metoclopramide, cisapride, erythromycin, maropitant and placebo. However, both cisapride and maropitant may decrease the contraction pressure pattern in the small bowel without altering the transit time.

**ABSTRACT #171**

VARIABILITY ASSOCIATED WITH REPEATED MEASUREMENTS OF GASTRIC EMPTYING USING THE SMARTPILL PH.P WIRELESS CAPSULE AND SCINTIGRAPHY IN DOGS. C.S. Boillat, F.P. Gaschen, L. Gaschen, R. Stout, G. Hosgood. School of Veterinary Medicine, Louisiana State University, Baton Rouge, LA.

This study evaluated and compared the repeatability of measurements of gastric emptying (GE) using the SmartPill pH.p® wireless capsule (SP) and scintigraphy in healthy dogs.

Six healthy adult dogs weighing $21.5 \pm 1.8$ kg were used. A $^{99m}$Tc-mebrofenin radiolabeled test meal was offered after *per os* administration of the SP. Serial abdominal images were acquired for 270 minutes. A dedicated remote receiver was used for data collection from the SP until SP was expelled in the feces. Each dog was examined 3 times 1 to 2 weeks apart with the above procedure.

The mean gastric emptying half-times measured by scintigraphy (T½-GES) for each dog ranged from 99.9 to 181.2 minutes. The mean gastric emptying times measured by the SP (GET-SP) for each dog ranged from 385.3 to 669.7 minutes. The mean coefficient of variation (CV) was 11.8% for T½-GES and 7.8% for GET-SP. The intra-class correlation coefficient was 69% for T½-GES and 71% for GET-SP. A nested analysis of covariance showed that changes in scintigraphy do not cause significant variation in GET-SP, and GET-SP and scintigraphy are alike, suggesting that both methods are comparable for GE evaluation.

Intra-individual variability of GE in healthy dogs is similar to values reported in people. These variations must be considered in interpreting individual test results. Repeatability of SP-measurements is at least as good as that of scintigraphy, the current gold standard for evaluation of GE. SP is a valuable, non-radioactive, out-patient alternative to scintigraphy for assessment of GE in dogs.

**ABSTRACT #172**

THE ESOPHAGEAL TRANSIT TIME OF TABLETS OR CAPSULES IN CATS FOLLOWING ADMINISTRATION WITH FLAVORx® PILL GLIDE OR PILL DELIVERY TREATS. MacPhail CM, Bennett AD, Gibbons DS, Lappin MR. Colorado State University, College of Veterinary Medicine and Biomedical Sciences, Fort Collins, CO.

Retention of tablets or capsules in the feline esophagus is known to be associated with local esophagitis and subsequent esophageal stricture formation. In previous work, tablets or capsules administered without water were found to be commonly lodged in the esophagus for greater than 5 minutes. Passage into the stomach can be facilitated by following dry-pilling with a 6 ml oral bolus of water or by coating medications with butter. The objective of this study was to evaluate the esophageal passage of tablets and capsules when administered with either a pill delivery treat (Pill Pockets®) or a one-step pill gun with flavored liquid (FlavoRx® Pill Glide).

Eight normal cats were enrolled in the study. Physical examination, serum biochemistry profile, and CBC were performed to ensure good health. Barium sulfate tablets, sized to 4 x 2mm, and barium sulfate-filled #4 gel capsules were used in this study. Cats were placed in a specially designed plexiglass box and fluoroscopy was used to evaluate tablet or capsule passage from 0 to 30 seconds, at 60, 90, and 120 seconds, and then at one-minute intervals until the tablet or capsule was in the stomach. Four separate medication administrations were performed in each cat on different days: tablets with pill delivery treats (T-PP), tablets with FlavoRx® pill glide (T-FG), capsules with pill delivery treats (C-PP), and capsules with FlavoRx® pill glide (C-FG).

The percentage of successful passages into the stomach at 30 seconds was 62.5% for T-PP, 62.5% for T-FG, 75% for C-PP, and 100% for C-FG. The percentage of successful passage at 60 seconds was 75% for T-PP, 87.5% for T-FG, 100% for C-PP, and 100% for C-FG. The average transit time was 59.3 sec for T-PP, 36.1 sec for T-FG, 24.5 sec for C-PP, and 15.9 sec for C-FG. The maximal time in the esophagus was 240 sec, 90 sec, 60 sec, and 29 sec for T-PP, T-FG, C-PP, and C-FG, respectively.

The results of this study demonstrate that either pill delivery method is acceptable for successful passage of tablets or capsules into the stomach of cats.

**ABSTRACT #173**

GASTROINTESTINAL DISEASE IN DOGS WITH EXCESSIVE LICKING OF SURFACES. V. Bécuwe, M.C. Bélanger, D. Frank, J. Parent, P.Hélie. Faculty of Veterinary Medicine, University of Montreal, Saint-Hyacinthe, QC, Canada.

Excessive licking of surfaces (ELS) is often considered as an obsessive-compulsive disorder and treated with antidepressant drugs. However, dogs with ELS respond poorly to this therapy. We hypothesized that ELS was a manifestation of nausea or discomfort, resulting from an underlying gastrointestinal (GI) disorder. The aims of the study were to 1) describe the ELS behavior, 2) perform a thorough clinical evaluation of the digestive system of dogs presented with ELS and 3) evaluate the outcome of this behavior after appropriate treatment of the underlying GI disorder.

Twenty dogs with ELS (L group) were divided in 2 subgroups of 10 dogs: L0 without and LD with digestive signs such as vomiting and diarrhea. Ten healthy dogs were assigned to a control group (C). Behavioral, physical and neurological exams were performed, prior to a complete evaluation of the GI system (CBC, serum biochemical profile, urinalysis, measurement of total serum bile acids and canine specific pancreatic lipase immunoreactivity, fecal flotation by zinc sulfate, fecal culture, abdominal ultrasonography and upper GI endoscopy with biopsies). Based on results, a board certified internist recommended an appropriate treatment. Dogs were monitored for a subsequent 90 days during which the licking behavior was recorded.

The underlying GI disorders found in the L group included eosinophilic infiltration, lymphoplasmacytic infiltration, delayed gastric emptying, irritable bowel syndrome, chronic pancreatitis, gastric foreign body, giardiasis. Complete resolution of ELS was observed in 9/18 dogs. At day 60, the LD dogs had a better outcome for ELS improvement than the L0 dogs ($P = 0.02$). There was no significant difference in behavior of L dogs and C dogs.

GI disorders should therefore be considered in the differential diagnosis of excessive licking of surfaces in dogs.

*Previously presented at ECVIM in Belgium, September 2008.*

**ABSTRACT #174**

HIGH RESOLUTION MAGNETIC RESONANCE IMAGING OF THE CANINE BRAIN AT 7 TESLA: A COMPARISON STUDY WITH 0.2 AND 1.5 TESLA. BT Kang[1], KJ Ko[1], DP Jang[2], JY Han[2], CY Lim[1], DI Jung[3], JH Yoo[1], C Park[4], YB Kim[2], EJ Woo[5], ZH Cho[2], HM Park[1]. 1. Konkuk University College of Veterinary Medicine, Seoul, South Korea. 2. Gachon University of Medicine and Science, Neuroscience Research Institute, Incheon, South Korea. 3. Gyeongsang National University College of Veterinary Medicine, Jinju, South Korea. 4. University of California Davis School of Veterinary Medicine, Davis, CA. 5. Kyung Hee University Department of Biomedical Engineering, Yongin, South Korea.

The purpose of this study was to describe the relevant canine brain structures using magnetic resonance imaging (MRI) at 7 T. T2-weighted imaging was performed on five healthy laboratory beagle dogs using 0.2, 1.5 and 7 T clinical scanners. The anatomic features were subjectively evaluated by direct comparison of the images obtained from the three different magnetic fields. The signal-to-nose ratio (SNR) and contrast-to-noise ratio (CNR) were calculated and compared between 1.5 and 7 T. The T2-weighted images at 7 T provided good spatial and contrast resolution for the identification of the clinically relevant brain anatomy; these images provided better delineation and conspicuity of the brain stem structures, which were difficult to unequivocally identify at 0.2 and 1.5 T. The SNR and CNR of the images at 7 T were significantly increased up to 218% and 615% compared to the 1.5 T images. If the image sequences could be optimized, 7 T clinical MRI could provide a good experimental and diagnostic tool for the evaluation of canine brain disorders.

**ABSTRACT #175**

CLINICAL SIGNS, MAGNETIC RESONANCE IMAGING FINDINGS AND SURVIVAL IN DOGS WITH INTRACRANIAL MENINGIOMAS AND GLIAL CELL TUMOURS. A de Stefani[1], A Sparkes[1], L S Garosi[2], L De Risio[1], F J Llabres[2], S R Platt[3]. [1]Animal Health Trust, Newmarket, Suffolk, UK. [2]Davies Veterinary Specialists, Higham Gobion, UK. [3]Department of Small Animal Medicine, College of Veterinary Medicine, University of Georgia, Athens, GA.

The aims of our study were to retrospectively assess possible associations between clinical signs, magnetic resonance imaging (MRI) findings and survival time in dogs with histopathologically confirmed meningiomas and glial cell tumours.

Our medical database was searched for dogs with intracranial neoplasia (1993-2004). Cases of meningioma (n=31) or glial cell tumour (n=27) with complete medical history, MRI and follow-up until time of death were included. Data were analysed using relevant non-parametric tests (chi-squared, Fishers Exact, Mann-Whitney, Spearman Rank, and Log-Rank Test). For all tests, significance was set at $P \leq 0.05$

Fifty-eight dogs satisfied the inclusion criteria. Association among clinical signs (presence of seizure activity, altered mentation, blindness and compulsive gait) and tumour site, tumour volume (tumour to total brain volume ratio) and MRI findings (peritumoural oedema, intra-tumoural haemorrhage and parenchymal shift) were assessed. In addition associations between tumour type and, tumour site, volume and MRI findings were evaluated. The influence of tumour site, tumour volume and tumour type on survival time was investigated. The tumour volume was significantly higher (P=0.0006) for gliomas (mean 5.3%) than meningiomas (mean 2.5%). There was no significant difference in the prevalence of oedema or parenchymal shift between the tumour types, but gliomas were significantly (P=0.03) more likely to have intratumoural haemorrhage (30% vs 6%). There was a significant (P=0.002) difference in anatomical distribution between meningiomas (most common in parietal and olfactory lobes, 58%) and gliomas (most common in frontal and temporal lobes, 78%). Tumour site significantly affected the likelihood of seizure activity being observed (P=0.004) or of a compulsive gait being present (P=0.04), but did not significantly affect the likelihood of altered mentation or visual deficits. Survival analysis was performed on 35 dogs that underwent surgery followed by radiation therapy. The median survival time from time of diagnosis of dogs with gliomas (327d) was not significantly different (P=0.26) from that of the dogs with meningiomas (486d) and no association between tumour site and survival (P=0.6) or between tumour volume and survival (P=0.14) was identified.

Association between tumour type, volume, MRI findings and anatomical distribution were identified in this study; however these variables did not influence survival in a subpopulation of dogs.

**ABSTRACT #176**

ORAL HYDROXYUREA THERAPY FOR DOGS WITH SUSPECTED INTRACRANIAL MENINGIOMA: A RETROSPECTIVE COHORT STUDY (2004-2009). MA Cautela[1], CW Dewey[1], S Cerda-Gonzalez[1], DJ Fletcher[1], G Barone[2]. [1]Cornell University College of Veterinary Medicine, Ithaca, NY; [2]Long Island Veterinary Specialists, Plainview, NY.

Meningioma is the most common canine brain tumor and it can be both difficult and expensive to treat. Supportive therapy typically consists of oral glucocorticoid (GC) therapy +/- antiseizure drugs. Survival times with supportive therapy are variable, but are typically between 1 and 4 mos. Standard modalities for definitive treatment of canine intracranial meningioma include surgical resection, megavoltage irradiation, or a combination of these options. Despite improved survival times with definitive therapies, owners often decide against these treatments due to both risk factors and cost. Hydroxyurea (HU) is an oral chemotherapeutic drug used with some success in human meningioma treatment. We have used oral HU in dogs (20 mg/kg SID) with suspected and confirmed meningiomas for several yrs. HU is inexpensive and has minimal adverse effects. Our clinical impression is that dogs treated with HU live longer than those treated with GC alone. The purpose of this investigation was to compare survival times between dogs with MRI-diagnosed intracranial meningioma treated with HU and GC and those treated with GC alone via a retrospective cohort study.

Case record inclusion was restricted to dogs from our 2 hospitals that had not received either surgical or radiation therapy. Diagnosis of intracranial meningioma was based on MRI interpretation by a board-certified radiologist. Data derived from each record included signalment, clinical signs, mass location, and survival time from diagnosis. 43 records were identified, 33 HU/GC dogs and 10 GC dogs. Survival times were estimated for each group using the Kaplan-Meier product limit method and compared using the log-rank test ($p<0.05$). A hazard ratio was also calculated. Survival was significantly longer in the HU/GC group (p=0.025) with these dogs living 2.97 times longer than dogs in the GC group (95% CI=1.15-7.68). At the time of data analysis, the median survival times for HU/GC and GC groups were 28 wks and 14 wks, respectively. 8 dogs (24%) were still alive in the HU/GC group and none in the GC group.

In this study, oral HU use in dogs with suspected intracranial meningioma was associated with longer survival times than those previously reported for dogs treated with GC alone and significantly increased survival likelihood compared with the cohort GC-only group. Our results support the use of oral HU for dogs with intracranial meningioma.

**ABSTRACT #177**

IMMUNE CELL INFILTRATION INTO CANINE MENINGIOMAS. CL Mariani, T Davis, NJ Olby. North Carolina State University, Raleigh, NC.

Meningiomas are the most common intracranial tumor in dogs. However, very little is known about the immune response to brain tumors in this species. Immune responses against tumor tissue may be beneficial for the host, resulting in delayed tumor growth or regression. Conversely, certain immune cell populations such as regulatory T cells (Tregs), may inhibit anti-tumor immune responses, facilitating tumor growth and progression. The immune system may also influence the expression of certain molecules contributing to tumor cell invasion, such as matrix metalloproteinases. Therefore, analysis of immune cell populations in tumor biopsy samples may provide useful prognostic information. In addition, the current therapy for canine meningiomas is inadequate, and novel treatments such as immunotherapy are needed to improve outcomes. Before exploring immunotherapeutic options, the populations of immune cells present in meningiomas must be characterized, which is the purpose of the current study.

Sections of meningioma tissue were obtained from archival formalin-fixed, paraffin-embedded blocks. Immunohistochemical evaluation of CD3 (T cells), CD79a (B cells), FoxP3 (Tregs), CD18 (microglia, macrophages), CD11d (dendritic cells), toll-like receptor 4 (TLR4) and TLR9 was performed. Consistent populations of CD3+ T cells and CD18+ cells were identified within and surrounding tumor tissue. In contrast, B cells and Tregs were rarely identified, although some samples showed clusters of positive cells. TLR4 and TLR9 positive cells were also noted. Further work to correlate these cell populations with outcome is underway. These observations may form the basis for intervention with immunostimulatory compounds such as TLR agonists.

**ABSTRACT #178**

FLUORODEOXYGLUCOSE POSITRON EMISSION TOMOGRAPHY AND MAGNETIC RESONANCE IMAGING FINDINGS OF NON-SUPPURATIVE MENINGOENCEPHALITIS IN SEVEN DOGS. BT Kang[1], DP Jang[2], CY Lim[1], SH Gu[1], DI Jung[3], JH Yoo[1], C Park[4], YB Kim[2], EJ Woo[5], ZH Cho[2], HM Park[1]. 1. Konkuk University College of Veterinary Medicine, Seoul, South Korea. 2. Gachon University of Medicine and Science, Neuroscience Research Institute, Incheon, South Korea. 3. Gyeongsang National University College of Veterinary Medicine, Jinju, South Korea. 4. University of California Davis School of Veterinary Medicine, Davis, CA. 5. Kyung Hee University Department of Biomedical Engineering, Yongin, South Korea.

The purpose of this study was to characterize the [18F]2-deoxy-2-fluoro-D-glucose positron emission tomography (FDG-PET) findings of encephalitis in dogs and to assess the role of the FDG-PET in the diagnosis of meningoencephalitis. The medical records, magnetic resonance (MR) and FDG-PET images of five dogs with necrotizing meningoencephalitis (NME), one dog with granulomatous meningoencephalitis (GME) and one dog with meningoencephalitis of unknown etiology (MUE) were reviewed. The T2-weighted images (WI) and fluid attenuated inversion recovery (FLAIR) images were characterized by hyperintensity, whereas the signal intensity of the lesions on the T1-WI images was variable. The metabolic changes on the brain FDG-PET corresponded well to the hyper- and hypointense lesions seen on the MR imaging. Glucose hypometabolism was identified in the NME, whereas hypermetabolism was noted in the GME. FDG-PET aided in the diagnosis of meningoencephalitis when the metabolic data was combined with clinical and MR findings. FDG-PET cannot be used alone to diagnose meningoencephalitis.

**ABSTRACT #179**

STEROID RESPONSIVE MENINGITIS-ARTERITIS TREATMENT WITH THREE POTENTIAL DIFFERENT PROTOCOLS: CLINICAL SIGNS, LABORATORY AND LONG TERM FOLLOW UP IN 48 DOGS. A Negrin[1], GB Cherubini[2], HA Volk[3], L Gaitero[1], S Añor[1] [1] Autonomous University of Barcelona, Barcelona, Spain, [2] Dick White Referrals, Newmarket, UK. [3] The Royal Veterinary College, University of London, UK.

The aims of this study were to describe the clinical signs, laboratory and follow up results, in particular the presence of paresis/ataxia and the CSF characteristics in 48 dogs affected by steroid-responsive meningitis-arteritis (SRMA) and treated with three different therapeutic protocols (Prednisolone, Prednisolone and Cytosine Arabinoside, Ciclosporine and Cytosine Arabinoside). In addition, the association between number of relapses and drug protocol (dose and duration), number of relapses and outcome, and side effects and survival were investigated. A retrospective search was performed for dogs fulfilling the following criteria: presence of cervical/diffuse spinal pain, diagnosis of aseptic suppurative meningitis based on neurological examination, CSF analysis, serum and CSF negative PCR results for *Toxoplasma gondii*, *Neospora caninum*, *Ehrlichia canis* and Canine Distemper virus, peripheral blood leukocytosis, normal spinal radiographs and complete/partial resolution of clinical signs with immunosupressive therapy. The same type and dose of each drug were used in all protocols. Long term follow up was obtained up to December 2008. A total of 48 dogs were included in the study. Boxers were the most prevalent breed (12), other breeds were not overrepresented. Mean age of onset of neurological signs was 14 months (4-84 months) and mean duration was 13 days (2-27 days). Clinical signs at presentation were: cervical pain (85%), pyrexia (73%), depression (56%), stiff gait and thoracolumbar pain (31%), ataxia (19%) and paresis (11%). On CSF analysis, median WBC count was 355/μL (31-8200/μL) with a median neutrophil percentage of 78% (49-97%). Median total protein concentration was 0.39 g/L (0.1-7.3 g/L). Dogs with multifocal spinal pain had a significant higher median WBC count than dogs with cervical pain only, while presence of paresis/ataxia was not statistically correlated with WBC count in CSF. Treatment protocols included: Prednisolone (Pred) (38 dogs), Pred and Cytosine Arabinoside (CA) (5 dogs) and Cyclosporine (Cyclo) and CA (5 dogs)..Remission of clinical signs was obtained in all cases at 1.5 months after treatment initiation, with a mean duration of 7.8 months. Eighty one percent (39/48) of cases were still alive and median life span since diagnosis was 16.7 months (8-83). Relapses of neurological signs were observed in 16/36 (44%) dogs treated with Pred, most of them (63%) at 5 months after treatment initiation. All 5 dogs treated with Cyclo-CA relapsed 1 month after treatment initiation, while none of the 5 dogs receiving Pred and CA had relapses during treatment. This study suggests that slower tapering of Pred alone may be more effective in avoiding relapses; moreover, the combination of Pred+CA could be a promising alternative as it seems to be effective to prevent from relapses and to decrease side effects, as well as to ensure good survival rates.

**ABSTRACT #180**

THREE-DIMENSIONAL TIME-OF-FLIGHT MAGNETIC RESONANCE ANGIOGRAPHY OF INTRACRANIAL VESSELS IN A CANINE MODEL OF ISCHEMIC STROKE WITH PERMANENT MIDDLE CEREBRAL ARTERY OCCLUSION. BT Kang[1], DP Jang[2], SH Gu[1], CY Lim[1], DI Jung[3], JH Yoo[1], C Park[4], YB Kim[2], EJ Woo[5], ZH Cho[2], HM Park[1]. 1. Konkuk University College of Veterinary Medicine, Seoul, South Korea. 2. Gachon University of Medicine and Science, Neuroscience Research Institute, Incheon, South Korea. 3. Gyeongsang National University College of Veterinary Medicine, Jinju, South Korea. 4. University of California Davis School of Veterinary Medicine, Davis, CA. 5. Kyung Hee University Department of Biomedical Engineering, Yongin, South Korea.

The purpose of this study was to evaluate the potential efficacy of 3-dimensional time-of-flight magnetic resonance angiography (TOF-MRA) to validate canine ischemic stroke model. Ischemic stroke was induced by permanent middle cerebral artery occlusion (MCAO) in five healthy beagle dogs. T2-turbo spin echo images and TOF-MRA were serially obtained three times with a 1.5-tesla magnetic resonance system: before, 3 and 10 days after the MCAO. TOF-MRA of the dogs showed the main cerebral arteries and their branches. In three dogs, angiograms of the brain obtained at 3 days after the MCAO showed complete occlusion of the MCA; in addition, T2 hyperintensity was found unilaterally in the striatocapsular and cerebral cortex lesions. However, partial occlusion of the proximal part of the MCA was identified in the other two dogs, with T2 hyperintensity found only in the striatocapsular lesions. Because improvement of the ischemic lesions was observed in all dogs, without reperfusion of the occluded vessels, recruitment of blood flow through collateral vessels was suggested. The occluded sites were confirmed at necropsy. The results of this study demonstrate the potential of TOF-MRA to provide a detailed description of intracranial arteries and aid in the diagnosis of flow impairment in canine ischemic stroke.

**ABSTRACT #181**

PREVALENCE OF CEREBRAL MICROBLEEDS IN AGED DOGS. D Insua[1], ML Suarez[2], G Santamarina[2], A González[2], M Sarasa[1], P Pesini[1]. 1. Araclon Biotech S.L. Zaragoza, Spain. 2. Department of Veterinary Clinical Sciences, University of Santiago, Lugo, Spain.

Cerebral microbleeds are focal lesions that can be visualized on magnetic resonance imaging (MRI). Hystopathological analysis shows that these are hemosiderin deposits from red blood cells that presumably have leaked out of small brain vessels. Little is known on microbleed prevalence, risk factors, and clinical correlates in the general population. Recent studies with gradient-echo T(2)* MRI technique to detect hemosiderin components in the brain have reported that latent microbleeds are more frequent in Alzheimer Disease (AD) patients than in normal controls. Furthermore, hystopathological studies have found that senile plaques are sites of microhaemorrhages. These co-localization raises the questions of whether microhaemorrhages are early events in plaque formation and whether therapies which stabilize cerebral microvessels can prevent the onset or slow the progress of dementias associated with plaque formation.

We have used the iron Perls staining to study the appearance and extension of microhaemorrhages in the brain of young (n = 6), aged dogs without cortical amyloid deposits (n = 7) and aged dogs with cortical amyloid deposits (n = 7). Five animals in this last group were diagnosed with canine counterpart of AD characterized by cognitive impairment, cortical amyloid deposits and neuronal loss. All the brains were obtained by necropsies carried out in our VTH with the explicit consent of the owners.

Extensive hemosiderin deposits were found in all the aged animals irrespective whether they did present or not cortical amyloid pathology. These deposits were intensely stained and particularly dense around the fiber tract of the internal capsule, the globus pallidum and the olfactory tubercle. The caudate nucleus was also considerably affected whereas the putamen used to be spared. Less dense deposits spread along the diagonal band, the septum, anterior commissure, and throughout the subcortical white matter. No appreciable hemosiderin deposits were seen in the cortical grey matter. A huge number of hemosiderin containing phagocytic microglia was seen within the stained areas. Furthermore hemosiderins deposits, although lighter and smaller, were found in the same brain regions of two young dogs.

These results showed that microbleeds have a great prevalence in the dog. In addition they do not give support to the existence of a mechanistic relation between this lesion and the presence of cortical amyloid-β pathology.

**ABSTRACT #182**

PRELIMINARY OBSERVATIONS BETWEEN COX-2 INHIBITOR ADMINSTRATION AND ACUTE CEREBROVASCULAR DISEASE IN DOGS. FA Wininger, RS Bagley, AV Chen. Veterinary Clinical Sciences-Washington State University, Pullman, WA.

Reasons for the increased recognition of acute cerebrovascular disease (ACVD) in dogs are currently being evaluated. ACVD in humans have been associated with the theoretical hypertensive, pro-thrombotic effect of new generation NSAIDs. This purpose of this study was test the hypothesis that the increased use of cyclooxygenase-2 inhibitor drugs in dogs may correlate with ACVDs.

A retrospective study of dogs with ACVD was performed to determine if a relationship with NSAID administration existed. 67 dogs met the inclusion criteria of a presumptive ACVD diagnosis based on clinical findings, cerebrospinal fluid cytology and MRI findings. All dogs had a minimum of one blood pressure measurement and a urinalysis. 33 of the 67 dogs had fibrin split products and D-dimers measured in serum. An age and breed matched cohort of an additional 67 dogs presenting for intracranial disease was used as a control population. To be included in the NSAID group, drug administration was within a year of presentation. Data was analyzed with chi-squared analysis. 4 of 67 dogs with ACVD had a NSAID administered within 5 days of presentation compared to 7 of 67 in the non-ACVD group (chi-square= 0.973, p-value=0.324). The type, frequency and duration of NSAID administration were highly variable. From the data evaluated no other risk factors have currently been identified. Based on this preliminary information, it is difficult at this time to determine a significant association between ACVD and the administration of COX-2 specific NSAIDs in dogs.

**ABSTRACT #183**

MULTIPLEX ANALYSIS OF CYTOKINES IN THE CEREBROSPINAL FLUID OF DOGS AFTER STROKE.

R Barber[1], S Platt[1], J Barber[1], L De Risio[2], J Eagleson[1], M Kent[1], MK Claiborne[1], S Schatzberg[1]. 1. University of Georgia College of Veterinary Medicine, Athens, GA. 2. Animal Health Trust, Newmarket, England.

Inflammation contributes to brain injury following human and animal strokes, and elevated inflammatory cytokines in CSF have been associated with clinical deterioration. As an initial characterization, pro- and anti-inflammatory CSF cytokines/chemokines were evaluated in dogs with stroke.

Cerebrospinal fluid was collected from 10 healthy beagles with normal brain MRI and 18 various dog breeds with clinical signs and MRI consistent with cerebrovascular infarction. Cerebrospinal fluid was collected in glass or polypropylene tubes, immediately frozen and stored at -80°C until use. Cerebrospinal fluid was evaluated with a canine-specific multiplex immunoassay for GM-CSF, IFN-γ, IL-2, IL-4, IL-6, IL-7, IL-8, IL-10, IL-15, IL-18, IFN-inducible protein (IP)-10, keratinocyte chemoattractant (KC), monocyte chemoattractant protein (MCP)-1 and TNF-α (Millipore) using a Luminex 200 instrument.

GM-CSF, IFN-γ, IL-6, IL-8, IL-10, IL-15, IL-18, IP-10 and TNF-α were not detected in the majority of dogs, however KC and MCP-1 concentrations were elevated significantly ($p < 0.01$ and $p < 0.0001$, respectively) in stroke dogs compared to normal dogs. Among stroke dogs, cytokine levels were not significantly changed by administration of anti-inflammatory medications (within two weeks of CSF collection) or by time of CSF collection after infarction. KC elevations occurred in stroke dogs compared to normal dogs when CSF was collected within 48 hours of stroke ($p < 0.05$) or greater than five days after stroke ($p < 0.01$). IL-2 and IL-4 levels were higher in CSF collected in polypropylene versus glass tubes in stroke dogs ($p < 0.005$ and $p < 0.05$, respectively).

While additional inflammatory mediators in canine stroke require evaluation, this preliminary analysis identified elevation of the chemokines KC and MCP-1 in stroke dogs consistent with post-infarct inflammation.

**ABSTRACT #184**

VALIDATION OF R&D SYSTEMS KITS TO DETECT CYTOKINE LEVELS IN CANINE CEREBRAL SPINAL FLUID.

AD Dierenfeld[1], PM Boggiatto[2], J Parkes[3], M Passage[4], JK Jens[2], EM Snella[2], KL Kline[3], CA Petersen[2], PI Dickson[4], NM Ellinwood[1]. Departments of 1) Animal Science, 2) Veterinary Pathology, 3) Veterinary Clinical Sciences, Iowa State University, Ames, IA; 4) Department of Pediatrics, LA Biomed at Harbor-UCLA, Torrance, CA.

Mucopolysaccharidosis (MPS) type I and type IIIB are lysosomal storage diseases caused by deficiency of lysosomal enzymes and subsequent lysosomal accumulation of dermatan and heparan sulfates (MPS I) and heparan sulfate (MPS IIIB). The phenotypes vary, although both forms are neuropathic in humans. Spontaneous canine models exist and have been critical in developing therapies for the corresponding human disease. There is interest in minimally invasive methods and techniques to monitor response to therapy that relies on authentic biomarkers and which can help to document any response to protein or gene therapy based treatments. Non-specific immune activation in the CNS is also hypothesized to play role in the pathology of these conditions. To investigate if CSF could be used as a source of biomarkers, we evaluated normal and MPS I and IIIB affected dog samples. Additionally, to confirm lack of immune response, we also assessed samples of animals treated as part of a previously reported intravenous and intrathecal based recombinant enzyme threatment for MPS I, in which animals showed tolerance as a result of a enzyme replacement protocol begun during the neonatal period.

ELISA Quantikine® IFNγ, IL-10, and DuoSet® IL-12/IL-23 p40 kits (R&D Systems) were validated for CSF by spiking CSF samples with serial dilutions of the standard, resulting in 90-95% of the spiked samples consisting of CSF. All spiked CSF samples showed expected values when assayed. CSF samples of normal, MPS IIIB affected, MPS I affected, and MPS I treated dogs were then evaluated using these validated kits. These samples were also assayed with TNF-α, a kit previously validated for canine CSF. Cytokine levels for all samples were below the level of detection for TNF-α, IFNγ, and IL-10. Efforts were made to extend the standard curve to include lower values. However this was unsuccessful. Slightly elevated levels were seen in 2 of 23 animals tested using the IL-12/IL-23 p40; values ranged from 45 to 85 pg/ml, slightly above the minimum detectible level of 31.25 pg/ml. The IL-12/IL-23 p40 positive CSF samples derived from a normal untreated dog and an MPS I enzyme treated dog, so no trend as to IL-12 levels between the various groups could be documented. The treated dog received quarterly intrathecal injections of rhIDU, and showed elevated levels of 85, 81, and 45 pg/ml over the first fifteen months of treatment, though the levels decreased below the limit of detection by the 18$^{th}$ month.

Validation of these kits for canine CSF will aid future research to determine immune response in the nervous system. The kits are limited to detecting elevated levels of cytokines in CSF. Normal dog CSF cytokine levels are absent or below the limit of detection with these assays. Although the levels for this particular study were not significantly elevated, these findings support our initial immune tolerance findings in that they do not document elevated CSF cytokine levels.

**ABSTRACT #185**

MAGNETIC RESONANCE IMAGING FINDINGS IN 60 DOGS WITH CERVICAL SPONDYLOMYELOPATHY.

Ronaldo C. da Costa[1], Joane Parent[2]. 1College of Veterinary Medicine, The Ohio State University, Columbus, OH, 2 Faculté de Médecine Vétérinaire, Université de Montréal, St-Hyacinthe, QC.

Medical and radiological records of dogs having cervical spondylomyelopathy (CSM) were searched between 2003 and 2006. Inclusion criteria was a definitive diagnosis of CSM achieved by magnetic resonance imaging (MRI) in large or giant breed dogs with clinical signs of cervical spinal disease. Our goal was to report the frequency of spinal changes seen on MRI of CSM cases.

Sixty dogs met the inclusion criteria. Forty three (71.7%) were large breed dogs and 17 (28.3%) were giant breed dogs. Thirty one of the 43 large dogs were Doberman pinschers. Their median age was 6.5 years (mean 6.8) with 18 males and 13 females. The 12 other large dogs were 3 Weimaraners, 2 Dalmatians, 3 mixed breed, 1 Labrador, 1 Boxer, 1 German shepherd and 1 Pit Bull. Their median age was 9 years (mean 8.7) with 8 males and 4 females. Among the giant dogs, 9 were Great Danes, 3 Bernese mountain dogs, 2 Rottweilers, 1 Irish Wolfhound, 1 Mastiff and 1 Swiss Mountain dog. Their median age was 2.8 years (mean 3.6) with 14 males and 3 females.

MRI findings in Dobermans revealed that 15 dogs had a single site of spinal cord compression while 15 dogs had 2 or more compressive sites (1 dog had only intervertebral foraminal stenosis). The main compression was located at C6-7 in 15 dogs, C5-6 in 14 dogs, and C4-5 in 1 dog. The compression was caused solely by intervertebral disk protrusion in 20 dogs, disk protrusion and ligament hypertrophy in 5 dogs, disk protrusion and osseous compressions (facet arthritis or malformation) in 3 dogs, and purely osseous in 2 dogs. Eighteen of the 30 dogs with cord compression (60%) had spinal cord increased signal intensity (ISI) on T2-weighted images.

MRI findings in other large breed dogs revealed that 5 dogs had a single compression, while 7 dogs had 2 or more compressions. The main compression was located at C6-7 in 5 dogs, C5-6 in 5 dogs and C4-5 in 2 dogs. The compression was caused by disk protrusion in 6 dogs, disk protrusion and osseous compressions in 4 dogs, purely osseous in 1 dog, and a combination of osseous and ligament hypertrophy in 1 dog. Six out of 12 dogs (50%) had spinal cord ISI.

MRI findings in giant breed dogs showed that 6 dogs had a single compression, while 11 dogs had two or more compressive sites. The main compression was at C6-7 in 7 dogs, C5-6 in 6 dogs, and C4-5 in 4 dogs. The compression was caused by osseous malformation and/or facet arthritis in 15 dogs, intervertebral disk protrusion in 1 dog, and a combination of disk protrusion and ligament flavum hypertrophy in 1 dog. Eight out of 17 dogs (47%) had ISI within the spinal cord.

The results show that large breed dogs (Dobermans and non-Dobermans) commonly have intervertebral disk-associated spinal cord compression (38 out of 42 dogs, 90%), with the compression most often located at C5-6 or C6-7 (92.8% of dogs). In contrast, most giant breed dogs (88%) had osseous compressions caused by lamina or pedicle malformation or facet arthritis. Increased signal intensity within the spinal cord was seen in approximately 50% of all dogs.

**ABSTRACT #186**

PHOSPHORYLATED NEUROFILAMENT NF-H AS A BIOMARKER OF SPINAL CORD INJURY IN DOGS.

Hiroaki Kamishina[1], Yui Kobatake[1], Masaaki Katayama[1], Yutaka Momota[2], Jun Yasuda[1], Reeko Sato[1], Takashi Uemura[3], Gerry Shaw[3]. 1.Iwate University, Morioka, Iwate, Japan. 2. Nippon Veterinary and Life Science University, Musashino, Tokyo, Japan. 3. University of Florida, Gainesville, FL.

Phosphorylated neurofilament NF-H is the major structural protein abundantly present in large-diameter axons of the spinal cord. Following injury, significant amounts of NF-H are expected to be released into the cerebrospinal fluid (CSF) and blood. The objectives of this study were to establish an ELISA assay that reliably quantifies canine NF-H in CSF and blood, and to evaluate the significance of NF-H as a biomarker of spinal cord injury (SCI) in dogs.

The specificity of the ELISA antibodies to canine NF-H was tested by immunohistochemistry and western blotting. Canine spinal cord NF-H was isolated and serially diluted samples were used to quantify the NF-H levels. CSF and plasma collected from 5 normal dogs and 20 dogs with SCI (acute thoracolumbar intervertebral disc disease) were used for NF-H measurements. In SCI dogs, plasma NF-H levels and hindlimb locomotor function were evaluated up to 3 months post-injury.

The assay was confirmed to specifically and quantitatively measure canine NF-H. NF-H was readily detectable both in CSF and plasma of dogs with SCI, but undetectable in normal dogs. In most cases, NF-H levels correlated with hindlimb functional scores; dogs with higher NF-H levels had lower functional scores (severe signs), however, some exceptions existed. The plasma NF-H level decreased by 4 weeks post-injury and returned to baseline by 3 months post-injury when maximum functional recovery was achieved. These findings suggest that measurement of NF-H arrows assessment of lesion severity and may aid in judging prognosis.

**ABSTRACT #187**

EPIDURAL AND INTRADURAL SCARRING POST LAMINECTOMY AND DUROTOMY: A COMPARISON OF GELFOAM AND SENTRX FILM.

NJ Olby, J-H Lim. North Carolina State University, Raleigh, NC.

Many neurosurgical procedures involve performing a laminectomy or craniotomy and durotomy, with the potential for post surgical scars to form both epidurally and intradurally. This scarring can cause disturbance of CSF flow and compression of the brain or spinal cord parenchyma. Many different materials have been developed to limit formation of scar tissue in the central nervous system but some of these materials are very expensive. Products based on cross-linked hyaluronic acid have been developed by SentrX Animal Care to promote healing while limiting scar formation. The aims of this study were to use magnetic resonance imaging (MRI) to compare epidural and intradural scar formation after hemilaminectomy and durotomy in dogs following application of either gelfoam or SentrX film.

Hemilaminectomies and durotomies were performed at L1/2 and T12/13 in six dogs. A piece of gelfoam or SentrX film was placed over each site before closure. MRI of the surgical sites was performed at 24 hours, 2, 6 and 12 weeks postoperatively. Axial and sagittal T2 weighted, T2 haste, T1 pre and post contrast and STIR images were obtained. Images of both hemilaminectomy sites were evaluated for the number of slices in which there was soft tissue within the epidural space, interruption of the subarachnoid space and distortion of the outline of the spinal cord. The identity of the film used at each site was revealed and the parameters measured compared between SentrX film and gelfoam using the students' t test with p values < 0.05 taken as significant.

All dogs recovered from surgery well with mild or no deficits in proprioceptive placing after two weeks. The combination of imaging sequences allowed evaluation of soft tissue within the epidural space (T2 and T1 weighted images), the integrity of the subarachnoid space (T2 haste images) and the outline of the spinal cord (T2 weighted images). At two weeks after surgery there was contrast enhancing soft tissue within the epidural space in all hemilaminectomy sites regardless of treatment. By six weeks, the contrast enhancement had disappeared but distortion of the spinal cord was evident with the cord pulled towards the dura mater at the durotomy site by intradural soft tissue. There was consistently less soft tissue within the epidural space and less interruption of the subarachnoid space and spinal cord distortion in the sites treated with SentrX film. This difference was statistically significant when the distance over which soft tissue was present within the epidural space was compared between treatments, but significance was not achieved for the number of slices in which there was distortion of the spinal cord or interruption of CSF flow. In conclusion, SentrX film is a safe and effective alternative to gelfoam for placement over a durotomy site and within a hemilaminectomy defect to limit post-operative scarring.

**ABSTRACT #188**

BREED SPECIFIC POLYMYOSITIS IN THE HUNGARIAN VIZSLA DOG. AC Haley[1], SR Platt, M Kent, SJ Schatzberg, A Durham, S Cochrane, GD Shelton. 1. University of Georgia, College of Veterinary Medicine, Athens GA. 2. University of Pennsylvania, School of Veterinary Medicine, Philadelphia, PA. 3. Veterinary Emergency Clinic and Referral Centre, Toronto, Ontario Canada, 4. University of California, San Diego, Department of Pathology, School of Medicine, La Jolla, CA.

Inflammatory myopathies are relatively common in dogs and may have a focal (masticatory muscle myositis) or generalized (polymyositis) distribution. Recently, a breed specific myositis presenting as pharyngeal dysphagia and masticatory muscle atrophy has been described in 14 Hungarian Vizsla dogs in the United Kingdom (Foale et al, BSAVA 2008). Here we report a similar syndrome in 5 Vizsla dogs from the North America. This retrospective study provides a preliminary clinicopathologic description and outcome in these 5 dogs.

Five Viszla dogs presented to 5 different veterinary hospitals (1 general practice, 2 referral practices, 2 university practices) with clinical signs of dysphagia (3/5), regurgitation (3/5), excessive salvation (3/5), masticatory muscle atrophy (4/5) and pain on opening the jaw (2/5). All dogs were male (4/5 castrated, 1/5 intact) and ranged in age from 1 to 9 years (mean 5.2 years). Creatine kinase activity was measured in 3 cases and was elevated (range 1061-9758 U/L; mean 5443). Antibodies against type 2M fibers and acetylcholine receptors were negative in those dogs tested (3/3 and 2/2 respectively). Three dogs had radiographically evident megaesophagus (ME). Two dogs had ME at the time of initial presentation while one dog developed ME 19 months after onset of dysphagia. Serum antibody titers against *Toxoplasma gondii* and *Neospora canis*, *Borrelia burgdorferi*, *Ehrlichia canis* and *Ehrlichia equi* were negative when tested. Electromyography was performed in 2 dogs with no abnormalities. Histopathologic examination of temporalis muscle biopsies was performed in 3 cases with multifocal areas of lymphocytic infiltration in 2 cases. Although cellular infiltrates were not evident in one case, bilateral multifocal hyperintensities in the temporalis muscle were observed on T2-weighted magnetic resonance images (MRI). A complete necropsy was performed on the fifth case and chronic lymphohistiocytic and plasmacytic myositis and fibrosis was evident in the esophagus, myocardium and skeletal muscle. Immunosuppression with prednisone and azathioprine has not resulted in clinical improvement at 2 and 3 months follow up in 2 dogs. Two dogs were lost to follow up. In conclusion this study, in combination with the previous report from the UK, should alert clinicians to the occurrence of a new breed associated polymyositis in the Vizsla dog which warrants further investigation to elucidate pathogenesis, genetic associations, response to treatment and prognosis.

**ABSTRACT #189**

MUTATIONAL ANALYSIS OF DYSTROPHIN-DEFICIENT MUSCULAR DYSTROPHY IN CAVALIER KING CHARLES SPANIELS. G. L. Walmsley[1], V.Arechavala-Gomeza[2], M. Fernandez-Fuente[1,2], N. Nagel[3], R. Stanley[1], K. Chandler[1], F. Muntoni[2], G. D. Shelton[4] and R.J. Piercy[1,2] 1. Department of Veterinary Clinical Sciences, Royal Veterinary College, London, UK; 2. Dubowitz Neuromuscular Centre, UCL Institute of Child Health London, UK; 3. Northdale Veterinary Practice, West Sussex, UK; 4. Department of Pathology, School of Medicine, University of California San Diego, La Jolla, CA.

Canine dystrophin-deficient muscular dystrophy, analogous to Duchenne muscular dystrophy of humans, is a severe inherited degenerative disorder of striated muscle. This debilitating and ultimately fatal condition results in a progressive destruction of skeletal and cardiac muscle due to mutations in the gene encoding dystrophin, a structural protein that links the contractile apparatus to the sarcolemma. The disorder is seen in several canine breeds but the genetic cause has only been reported in the Golden Retriever, German Short-haired Pointer and Rottweiler. Here we present the findings of clinical, histopathological and molecular characterisation of this condition in Cavalier King Charles Spaniels.

A 10 month old male neutered client-owned Cavalier King Charles Spaniel from the United Kingdom was presented with a chronic progressive history of lethargy, exercise intolerance and dysphagia. The dog was tetraparetic with poor skeletal muscle mass (body condition score = 2/9), reduced spinal reflexes, macroglossia and restricted jaw movement. Investigations documented a marked elevation in creatine kinase activity (33,695U/l; 61 – 394U/l) and electromyography revealed spontaneous activity indicative of a primary muscle disorder (complex repetitive discharges and pseudomyotonia). Dystrophin-deficient muscular dystrophy was diagnosed on the basis of skeletal muscle histopathology, immunohistochemistry and immunoblotting using monoclonal antibodies to the dystrophin rod and carboxy termini. Oligonucleotide primer pairs designed for RT-PCR to amplify overlapping regions of dystrophin cDNA identified, following sequencing, an exon deletion and a frame-shift not present in control cDNA, that is predicted to result in premature termination of the protein product. Sequencing the associated genomic DNA confirmed the causative (and novel) mutation. The ability of antisense oligonucleotide induced exon skipping to restore the reading frame was demonstrated *in vitro* in cultured myoblasts from the affected dog. Sequencing of amplified DNA from an additional Cavalier King Charles Spaniel with dystrophin-deficient muscular dystrophy from North America identified the same mutation.

In conclusion, dystrophin deficient muscular dystrophy in the Cavalier King Charles Spaniel may provide an excellent model for Duchenne muscular dystrophy due to the potential application for trials of antisense oligonucleotide-mediated exon skipping – one of the more promising research directions for genetic therapy in this fatal disorder.

**ABSTRACT #190**

FREQUENCY OF THE CANINE EXERCISE INDUCED COLLAPSE GENE IN DIVERSE BREEDS. KM Minor[1], EE Patterson[1], SD Gross[1], MK Keating[1], SM Taylor[2], GS Johnson[3], K Todd-Thompson[4], KJ Ekenstedt[1], JM Mickelson[1]. 1. University of Minnesota College of Veterinary Medicine, St. Paul, MN. 2. Western College of Veterinary Medicine, Saskatoon, Saskatchewan. 3. University of Missouri-Columbia College of Veterinary Medicine , Columbia MO. 4. University of Texas San Antonio School of Medicine, San Antonio TX.

A (G767T) dynamin 1 (*DNM1*) mutation in Labrador retrievers is highly associated with the syndrome of exercise-induced collapse (EIC). *DNM1* is essential for synaptic vesicle endocytosis. We previously reported a 37% carrier frequency with 3% homozygous affected dogs from a field-trial population of 400 dogs. The aim of this present study was to determine the frequency of the mutation in other Labrador sub-populations, as well as closely and distantly related breeds,

Breeds were chosen based on genetic marker clustering data, and/or breeds with readily available DNA samples with the goal of testing 200 or more dogs per breed. When possible, individuals selected were unrelated to the grandparent generation.

In all Labrador sub-populations (conformation, pet, and service), homozygous affected dogs and 30-40% carriers of the mutation were found. The mutation was also detected in Chesapeake Bay and Curly-coated retrievers (carrier frequency >10%) with clinically affected homozygous individuals in both breeds. A few carrier Welsh Pembroke Corgis (n=10) were also identified but we have not found any affected Corgis to date.

The mutation was not observed in Golden, Flat-coat or Duck tolling retrievers, nor in a number of other less related breeds (Newfoundlands, American water spaniels, Portugese water dogs, Jack russell terriers, Wheaton terriers, or Border collies). Ten Border collies (all mutation negative) had recurrent, unexplained collapse.

The *DNM1* mutation is prevalent in all Labrador sub-populations. A comprehensive program to carefully eliminate the mutant allele while still preserving genetic diversity and also accounting for reduction of other inherited diseases, could be implemented.

**ABSTRACT #191**

BRDU LABELING PATTERN OF THE ROSTRAL MIGRATORY STREAM IN NORMAL CANINE AND FELINE BRAINS. M Lewis,[1] S Malik,[2] T van Winkle,[1] DJ Watson,[2] CH Vite.[1] 1. School of Veterinary Medicine, University of Pennsylvania, Philadelphia, PA 2. Department of Neurosurgery, University of Pennsylvania, Philadelphia, PA.

In the human and rodent brain, a track of proliferative cells known as the rostral migratory stream extends from the subventricular zone (SVZ) of the lateral ventricle towards the olfactory bulb. This is the migratory pathway for SVZ neural progenitor cells which eventually differentiate into olfactory interneurons. We have characterized the pattern of proliferative cells in the normal canine and feline brain using 5-bromo-2-deoxyuridine (BrdU) labeling. All procedures were approved by the IACUC. BrdU was administered intravenously to dogs (n=2, 75 mg/kg) or cats (n=5, 25 mg/kg) every 24hr for 5 days. The dogs were perfused with saline followed by formalin and the brains were removed 1 or 5 days after the final BrdU infusion. Cats were perfused at 1 or 10 days followed the final BrdU infusion. Two additional dogs and one cat received a single dose of 75 mg/kg and were perfused 6 hr later. Animals that did not receive BrdU were used as controls. Frozen sagittal sections were prepared on a cryostat and stained with Nissl, hematoxylin and eosin, Luxol fast blue, anti-BrdU antibody, and anti Ki67 antibody. In canine and feline brains, a prominent, dense, continuous track of Nissl-stained cells was found, beginning in the subependymal layer at the base of the lateral ventricle, curving around the caudate nucleus and continuing ventrally to the olfactory peduncle. In brains from dogs and cats that received multiple doses of BrdU, we found BrdU-immunoreactive cells in the same track, continuously present from the ventral wall of the lateral ventricle to the rostral olfactory peduncle. The BrdU-labeled nuclei were single or in closely apposed small groups. These cells were not stained in control sections on which the primary (anti-BrdU) antibody was omitted. A similar pattern of staining was found in animals receiving one dose of 75 mg/kg, indicating cell division continuously along the track. Ki67 staining was identical to that seen in the BrdU stained sections. The distribution of dividing cells in the canine and feline brain is similar to the pattern of migrating subventricular zone neural progenitors in the rodent and human brains. The substantial size of this track in the dog and cats brain likely reflects the importance of the olfactory system in these animals. We will present our ongoing immunohistochemical studies in these and additional animals to describe the phenotype of these cells and the anatomical route and dimensions of the track.

**ABSTRACT #192**

INTRATHECAL ENZYME THERAPY IN MUCOPOLYSACCHARIDOSIS I CATS REDUCES STORAGE THROUGHOUT THE BRAIN. M Haskins,[1] P Wang,[1] S Walkley,[2] J Rhodes,[1] P O'Donnell,[1] C Bryan,[1] NM Ellinwood,[3] R Cahayag,[4] A Cheng,[4] C Henschel,[4] CA O'Neill,[4] J White, [4] CH Vite [1]. 1. School of Veterinary Medicine, University of Pennsylvania, Philadelphia, PA; 2. Albert Einstein College of Medicine, Bronx, NY; 3. Iowa State University, Ames, IA; 4. BioMarin Pharmaceutical Inc., Novato, CA.

Mucopolysaccharidosis (MPS) I is a lysosomal storage disease caused by deficient activity of the lysosomal hydrolase a-L-iduronidase (IDUA). The most common subtype in human patients has multisystemic disease including severe mental retardation associated with storage of glycosaminoglycans (GAGs) in the central nervous system (CNS). An orthologous cat model of MPS I has widespread storage of GAGs in the CNS. Approved clinical therapy for MPS I is weekly intravenous enzyme replacement with recombinant human IDUA (Aldurazyme; ALD) which is expensive and does not result in cure of the neurological disease. This study was designed to determine if intrathecal administration of ALD could alter the CNS lesions and how long the effect would persist.

Twelve MPS I cats between 18 and 27 months of age were divided into 4 groups. The cats were anesthetized with IV propofol, intubated, placed in lateral recumbancy, and ~2 cc of CSF collected. Two of the three cats in each group received intrathecal injections of 0.1 mg/kg of recombinant human α-L-iduronidase (Aldurazyme;ALD) in Elliots B solution on a Monday, Friday, and Tuesday. The remaining cat was treated with an equivalent volume/kg of Elliots B vehicle. Group 1 was sacrificed 2 days following the last injection; groups 2, 3, and 4 at 1 month, 2 months, and 4 months, respectively. The 4-month enzyme and GAG data are being analyzed together with 2- and 4-month immunocytochemistry and all the quantitative ganglioside data. Three cats given ALD and one given vehicle alone developed abnormal posture with lowered forelimbs, almost resting on their elbows. IDUA was not detected in CSF or blood from 2-28 days post-treatment. Brain IDUA activities were 4.2- and 3.8-fold normal at 2 days, dropping to 17% and 29% of normal at 1 month, and to affected levels by 2 months. Brain GAG levels were dramatically reduced at 2 days and 1 month, and were 30% and 48% below affected levels at 2 months. Blinded interpretation of brain ganglioside (GM2, GM3) immunohistochemistry and unesterified cholesterol showed a qualitative reduction at 2 days. These preliminary data indicate a need to repeat intrathecal injections at 2-3 month intervals to maintain suppression of GAG storage.

**ABSTRACT #193**

NEURAL STEM CELL SOURCES IN ADULT DOGS. JH Lim, NJ Olby, CL Mariani. North Carolina State University, Raleigh, NC.

Understanding adult neurogenesis is important to fully understand the potential for regeneration within the central nervous system (CNS), and to identify neural progenitors that could be used clinically for transplantation. We investigated whether neurospheres could be isolated from the adult canine CNS and whether adipose tissue-derived stromal cells (ADSCs) could be induced to differentiate into neural lineages.

Brain tissue was obtained from the dentate gyrus and the subventricular zone, and subcutaneous adipose tissue was collected from cadavers of adult dogs aged from 1 to 6 years. Following enzymatic dissociation, cells were cultured in appropriate media to produce neurospheres (brain tissue) or ADSCs (adipose tissue). Third passage ADSCs were induced to adipogenesis, osteogenesis and neurogenesis by addition of appropriate inductive agents. RT-PCR analyses were performed to identify neural progenitor cell surface markers in the induced ADSCs and neurospheres. Expression of neuronal lineage markers was evaluated immunohistochemically.

Neurospheres and ADSCs grew rapidly from the brain and adipose tissue respectively. Morphologically the ADSCs had fibroblast-like characteristics and could be induced to produce adipogenic and osteogenic lineages. Under neurogenesis culture conditions, ADSCs formed spherical cellular aggregates that resembled neurospheres. RT-PCR analyses revealed that these aggregates showed a similar phenotype to brain-derived neurospheres with expression of Sox2 and CD90. Neurospheres from both sources expressed Nestin, GFAP and Tuj1. In conclusion, we cultured neurospheres from adult canine brain tissue and generated neural progenitor cells from canine adipose tissue. The ADSCs have the potential for future autologous cell transplantation therapy for neurological disorders.

**ABSTRACT #194**

DOGS WITH CORTICAL BETA-AMYLOID PATHOLOGY HAVE LESS SEROTONERGIC NEURONS IN THE ROSTRAL RAPHE NUCLEI THAN AGED MATCHED HEALTHY CONTROLS. V Bernedo[1], D Insua[2], G Santamarina[1], ML Suárez[1], M Sarasa[2], P Pesini[2]. 1. Department of Veterinary Clinical Sciences, University of Santiago, Lugo, Spain. 2. Araclon Biotech Ltd. Zaragoza, Spain.

Dogs may naturally suffer an age-related cognitive dysfunction, which is sometimes described as the canine counterpart of Alzheimer disease (AD). This disorder reproduces key neuropsychological and hystopathological features of AD. Recently, we have found that dogs with the canine counterpart of AD lose noradrenergic neurons in the A6–A7 brainstem groups. However, no studies have previously investigated whether the serotonergic system is affected in this canine condition.

In the present study, we have used unbiased stereological procedures to estimate the number of the dorsal and median raphe nuclei (DRN and MRN, respectively) serotonergic neurons immunolabeled with an anti-tryptophan hydroxylase

(TrH) monoclonal antibody in young and aged dogs without beta-amyloid (Aβ) cortical deposits and in aged dogs with Aβ cortical deposits.

No significant variations were found between young and aged dogs without Aβ cortical deposits. In contrast, aged dogs with Aβ cortical pathology had 33% fewer serotonergic neurons in the DRN and MRN than aged dogs without Aβ cortical deposits ($p = 0.01$). However, in spite that the effect size of cortical beta-amyloid pathology on the number of serotonergic neurons could be considered large in statistical terms, these two variables did not correlated significantly.

These results suggest that degeneration of the serotonergic neurons could be involved in the cognitive damage that accompanies Aβ cortical pathology in the dog but it cannot be satisfactorily explained as secondary to cortical pathology in the way that the degeneration of the cholinergic and noradrenergic neurons in AD is generally explained. It should be considered that in the loss of serotonergic neurons other mechanisms might be implicated.

**ABSTRACT #195**

EVALUATION OF THE CARCINOEMBRIONIC (CEA) AND CARBOHYDRATE 15-3 (CA 15-3) ANTIGEN IN MALIGNANT MAMMARY TUMORS IN BITCHES. CM Oliveira, FHM Jardini. School of Veterinary Medicine – University of São Paulo, Brazil.

In last years, trough the use of biomarkers produced in cancerous tissue, diagnostics and therapeutics of human breast cancer changed significantly. These markers used in diagnosis, prognosis and monitoring of the tumor are able to signal biological behavior of cancers. Several substances are related to human breast cancer being CEA and CA 15-3 currently used. However in veterinary medicine these substances are not used regularly and there is a curiosity to investigate whether those biomarkers may be recommended in animals in the same manner as in humans.

Serum samples of 39 bitches, aging 5 to 13 year-old, with mammary tumors were collected immediately before and 2 weeks, 1, 3, 6 and 12 months after surgery, and frozen (-20°C). Fragments of the tumors were collected during surgery and fixated in 10% formalin and included in paraffin up to 48 hours after excision. The animals had radiographs and ultrasound scans done 3, 6 and 12 months after surgery in order to look for metastasis. The serum quantification of CEA and CA 15-3 was done by sequential immunometric chemoluminescence. For the detection of CEA in mammary- and lymph node tissue a commercial test kit (Dako®, LSAB2) was used; fragment of human breast cancer was used as control. Microscopic evaluation was based on coloration grades of Hercep test.

In none of the serum samples of the 39 bitches, with a primary complex and metastasis, CEA and CA 15-3 were found. Those findings may suggest that (1) none of the tumors produced the markers, (2) the markers were produced but not released, (3) biomarkers were in the blood but structural differences did not allow interaction with antibodies, or (4) other factors could interfere with biomarkers concentrations in the blood. There was no expression of CA in 1:100 and 1:200 dilutions on the tumoral fragments, independent of their histological type. The negative result may be explained by (1) no expression of CA, (2) no production of that protein, (3) production of the protein, but not detected by the antibodies, (4) the methodology did not allow an adequate interaction, or (5) the protein concentrations were inadequate to detect CA. Thus, findings observed in the present study showed that evaluation of CEA and CA 15-3, in serum and tumor tissue, were not adequate for dogs.

**ABSTRACT #196**

EXPRESSION OF HER-2/NEU PROTEIN IN MAMMARY TUMORS OF BITCHES. CM Oliveira. School of Veterinary Medicine – University of São Paulo, Brazil.

Mammary neoplasias in bitches make out an important portion of canine tumors. But different than in humans, early diagnosis has not been done and treatment usually has been limited to its excision. In humans knowledge of tumor markers, it is important to recognize markers for helping in diagnosis, in staging and as well as to evaluate prognosis, therapy and remission. Her-2 (human epidermal growth factor receptor-type 2) is one of these markers; it codifies synthesis of $p185^{erb}$ (Her-2/neu), a protein expressed in 30% to 40% of human breast cancers and is associated with a bad prognosis. Her-2 has raised some interest as an indicator to evaluate the sensibility to chemotherapy and it is associated to histology grade and survival rates. It is known that treatment with Herceptin R is only effective on tumors overexpressing Her-2 at their surface.

The objective of this study was to verify the expression of Her-2 in malignant and benign mammary tumors of bitches.

Fragments of tumors from 51 bitches where fixated in 10%formalin and included in paraffin, prepared and stained with Hematoxilin-eosin. Following tumors were identified: benign (n=10) and malignant (n=41). Detection of Her-2 was performed by immunohistochemistry (ABC) with streptavidine peroxidases. Results were classified in accordance to intensity of color and were denominated as negative (scores = 0 or 1+) and positive (scores = 2+ [weak] or 3+ [strong]). From those 51 tumors, Her-2/neu was present in 76.5% of the cases (n=39) and absent in 23.5% (n=12). The positive cases that expressed that protein, 79.5% were from malignant tumors and 20.5% were benign. Of the 51 analyzed tumors, the expression of Her-2/neu was positive (scores = 2+ or 3+) in 58.8% and negative (scores = 0 or 1+) in 41.2%. In conclusion benign and malignant tumors of bitches expressed Her-2/neu protein; however only malignant tumors presented strong positive expression of Her-2/neu.

**ABSTRACT #197**

CERULOPLASMIN CONCENTRATIONS IN DOGS WITH MULTICENTRIC LYMPHOMA UNDERGOING CHEMOTHERAPY. SRR Lucas, A. Merlo, RMS Mirandola, TP Gasparin. School of Veterinary Medicine and Animal Science - University of São Paulo, SP, Brazil.

Acute phase proteins (APPs) are synthesized under stimulation of cytokines such as IL-1, IL-6 and TNF-α. and represent a very fast and nonspecific response that occurs following inflammatory, infectious and neoplastic conditions in dogs. Ceruloplasmin (Cp) is a positive APP, responsible for the transport of copper and protection of cells and tissue against oxidant compounds. The aim of this study was to evaluate Cp concentrations at diagnosis and during chemotherapy in dogs with multicentric lymphoma (ML). Cp was measured using orto dianisine technique, in two groups of dogs: 10 healthy dogs (control) and 13 dogs with ML. All dogs were submitted to chemotherapy. (approved by the Bioethics Committee). Dogs with signs of concurrent disease or that were previously treated with prednisone were excluded from the study. Cp measurement was done before treatment and once a week, during the first month of chemotherapy, and each 3-week intervals until the relapse for dogs with ML, and until the 16th week in control dogs. ANOVA test followed by multiple Tukey´s tests were used to compare the groups. There was no difference between the mean of Cp concentration in dogs with ML at the diagnosis when compared to healthy dogs ($p>.05$). Levels of Cp decreased significantly at 4th week when compared to the first week, but Cp increase was not observed at the relapse. At all other times during the treatment, Cp concentrations for dogs with lymphoma were not significantly different from controls submitted to chemotherapy. As conclusion, Cp levels in ML at diagnosis were similar to healthy dogs, decreased when lymphoma remission was achieved and there was no change at the relapse.

**ABSTRACT #198**

FLOW CYTOMETRIC PROGNOSTIC FACTORS IN CANINE B CELL LYMPHOMA. EC Marcus, SE Lana, AC Avery. Colorado State University, College of Veterinary Medicine and Biomedical Sciences, Fort Collins, CO.

Recent studies suggest that length of survival can be predicted based on the type of non-neoplastic infiltrating cell populations in human B cell lymphoma. However, these trends have not yet been investigated in canine diffuse large B cell lymphoma. In addition, prognosis in canine B cell leukemia can be predicted by cell size, but similar information is not available for canine B cell lymphoma. The purpose of this study was to evaluate lymph node aspirates from confirmed cases of canine B cell lymphoma via flow cytometry to determine if correlation exists among the number macrophages and T cells present, B cells size, survival and disease free interval.

The study population included all dogs admitted to the Veterinary Teaching Hospital with B cell lymphoma within a two year period that had lymph node aspirates assessed via flow cytometry. Follow up data including disease free interval and survival duration from 43 cases of B cell lymphoma was collected and compared to immunophenotype data collected via flow cytometry. Relative size of B cells was determined by measuring geometric mean linear forward scatter.

Preliminary data analysis suggests a negative correlation between B cell size and disease free interval. B cell size may provide valuable prognostic information to clinicians and clients as they assess treatment plans for canine B cell lymphoma patients.

**ABSTRACT #199**

DENAMARIN® FOR THE PREVENTION OF LOMUSTINE (CCNU) INDUCED HEPATOTOXICITY IN TUMOR BEARING DOGS: INTERIM ANALYSIS. GM Hammond, MS Kent, AM Irish, TA Guerrero, CO Rodriguez, KA Skorupski. Veterinary Medical Teaching Hospital, University of California, Davis, Davis, CA.

Liver enzyme elevations occur in up to 86% of dogs treated with CCNU. Therapy may be prematurely discontinued, delayed, or the drug dose decreased due to liver enzyme changes or concerns about developing hepatotoxicity.

S-Adenosylmethionine (SAMe) and silybin are commonly used hepatoprotectants. Denamarin® incorporates both SAMe and silybin in a patented veterinary formulated product.

The goal of this clinical trial was to evaluate Denamarin as a hepatoprotectant in tumor bearing dogs undergoing CCNU therapy. Reported are the results of a planned interim analysis after the enrollment of 30 dogs.

A prospective non-blinded, randomized clinical trial was performed at the UC Davis VMTH. Dogs prescribed CCNU chemotherapy and that had a normal pre-treatment ALT were eligible. Dogs were randomly assigned to either Group 1, receiving Denamarin starting at the time CCNU therapy was initiated, or Group 2, receiving Denamarin only if they experienced a grade 4 hepatoxicity.

The two groups had no differences in pre-therapy ALT, bilirubin or tumor type. The mean post-therapy ALT elevation for Group 1 and Group 2 dogs was 119 IU/L ($p=0.10$) and 815 IU/L ($p=0.046$), respectively. Mean post therapy bilirubin elevation for Group 1 and Group 2 dogs were 0.4 mg/dl ($p=0.16$) and 0.15 mg/dl ($p=0.07$). Therapy alterations occurred in 3 dogs in Group 2, whereas one dog in Group 1 had chemotherapy altered due to liver enzyme elevations.

This data indicates that dogs receiving Denamarin may have less severe liver enzyme elevations after CCNU therapy and supports continued enrollment in this prospective clinical trial.

**ABSTRACT #200**

EFFECTS OF MUTANT BETA-CATENIN ON HEPATOCYTE GROWTH HOMEOSTASIS. Timothy J. Stein, DVM, PhD and Eric P. Sandgren, VMD, PhD; Department of Pathobiological Sciences, School of Veterinary Medicine, University of Wisconsin-Madison, Madison, WI.

Beta-catenin, a dual-function protein integral to cell-cell adhesion and intracellular transmission of Wnt signals, is altered in numerous human cancers including hepatocellular carcinoma. The comparative hepatocyte growth assay is a transplantation-based growth assay developed to quantify the effects of genetic alterations on hepatocyte growth homeostasis. For this study the comparative hepatocyte growth assay was used to assess the effects of a mutant beta-catenin on hepatocyte growth homeostasis.

The comparative hepatocyte growth assay was used to quantify the effects of mutant β-catenin on the growth kinetics of transplanted hepatocytes in growth permissive and restrictive environments. Two populations of transgenic mice were generated for donor hepatocytes; one harboring a liver-specific inducible mutant beta-catenin + marker transgene (hPAP), and the other carrying only a marker transgene

(LacZ). The growth of foci derived from hepatocytes expressing the mutant beta-catenin were compared to that of hepatocytes expressing only a marker transgene. The ability of mutant beta-catenin to give rise to clonal populations with extreme growth potential compared to normal hepatocytes was assessed through the determination of extreme outliers.

The expression of mutant β-catenin was unable to increase the growth of foci in growth permissive or restrictive environments. The production of foci with extreme growth potential was not different from control hepatocytes.

β-catenin does not affect measures of neoplasia assessed by the comparative hepatocyte growth assay. The contribution of beta-catenin to HCC is unique from that of other frequently identified genetic mutations in HCC, which have been shown to quantifiably effect hepatocyte growth homeostasis in this assay.

*Previously presented at the Veterinary Cancer Society Conference, 2008.*

**ABSTRACT #201**

COMPARISON OF COMPLETE BLOOD COUNT VALUES FOR SAMPLES COLLECTED USING TRADITIONAL AND MICRO-SAMPLING TUBES. JC Whittemore,[1] B Flatland.[2] From the Departments of Small Animal Clinical Sciences[1] and Pathobiology[2], University of Tennessee, College of Veterinary Medicine, Knoxville TN.

Anemia caused by diagnostic sampling is the main indication for blood transfusion in human neonates. Diagnostic sampling may pose the same risk to dogs and cats given their size. Microsampling tubes may be used to decrease sampling volumes; their accuracy under clinical conditions has not been reported.

Samples were collected from clinically healthy dogs (14) and cats (15). Three mL of blood were collected from each animal and transferred into EDTA Vacutainer™ (2.5 mL) and Microtainer™ tubes (0.5 mL) by a licensed veterinary technician blinded to the study purpose. Complete blood counts were performed within 12 hours on all samples. Parameters evaluated were RBC, HgB, Hct, MCV, MCH, MCHC, MPV, WBC, and plasma protein as well as neutrophil, lymphocyte, monocyte, eosinophil, basophil, and platelet counts. Results were compared by Pearson's correlation coefficients, Passing-Bablok regression analysis, Bland-Altman plots, and either paired t-tests or Wilcoxon Rank Sum (MCV, MCH) using MedCalc 10.0.1.

All data were normally distributed except MCV and MCH. Pearson's correlation coefficients were >0.9 for all analytes except monocytes (0.89), basophils (0.73) and eosinophils (0.77). There was no evidence of constant or proportional bias on Passing-Bablok regression analysis or Bland-Altman plots for any analyte. There were no significant differences on paired t-tests. Wilcoxon rank sum testing was significantly different (p=0.002) for MCV; however, this difference was not clinically significant.

These results suggest that Microtainer™ sampling tubes provide clinically equivalent complete blood count results to Vacutainers™ while minimizing total diagnostic sampling volumes.

**ABSTRACT #202**

AGING ASSOCIATED CHANGES IN BLOOD CELLS OF CATS. S Yu, MF Locniskar. Hill's Pet Nutrition, Inc., Topeka, KS.

The purpose of the present study was to investigate the aging effect on blood cells in healthy cats.

Complete blood cell count was measured with a CELL-DYN® 3700 (Abbott Dignostics) in 108 healthy American Domestic Shorthair cats. The measurements included RBC, Hb, PCV, MCV, MCH, MCHC, platelet, WBC, and WBC differential (neutrophils, lymphocytes, monocytes, eosinophils, and basophils). Cat age ranged from 1.5 to 16 years at the beginning of the study. Of these cats, 42 were males (6 intact and 36 castrated) and 66 were females (28 intact and 38 spayed). Cats were fed a dry cat food for four weeks before the blood cell measurements. The dry cat food met the AAFCO Cat Food Nutrient Profiles for adult maintenance.

RBC ($r = -0.62$, $p < 0.01$), Hb ($r = -0.57$, $p < 0.01$), PCV ($r = -0.61$, $p < 0.01$), platelets ($r = -0.40$, $p < 0.01$), lymphocytes ($r = -0.25$, $p = 0.010$), and eosinophils ($r = -0.36$, $p < 0.01$) were negatively while MCV ($r = 0.30$, $p = 0.002$), MCH ($r = 0.35$, $p < 0.01$), and neutrophils ($r = 0.28$, $p = 0.030$) were positively correlated with age in cats. Age had no effect ($p > 0.05$) on MCHC, WBC, monocytes, and basophils.

These data suggest that cats have a tendency of macrocytotic anemia when they get old. Although aging did not affect WBC, increased neutrophils and decreased lymphocytes and eosinophils may suggest altered immunity with aging of cats. The clinical implications of aging associated changes in blood cells need further investigation.

**ABSTRACT #203**

CLOPIDOGREL THERAPY IN DOGS WITH IMMUNE-MEDIATED HEMOLYTIC ANEMIA. RL Haviland, N Pacifico, D Bianco. Red Bank Veterinary Hospital, Tinton Falls, NJ.

Thromboembolic events (TE) are a common complication of canine immune-mediated hemolytic anemia (IMHA). Canine IMHA has a high mortality rate, especially during the first two weeks of therapy. Clopidogrel (Plavix®) is a second generation antiplatelet drug that may have a synergistic effect with aspirin. This study reports the use of clopidogrel with or without ultra-low dose aspirin therapy in client-owned dogs with primary IMHA.

Dogs were diagnosed with IMHA based on the presence of at least 4 of 5 criteria: spherocytosis, positive saline autoagglutination test, positive direct Coombs' test, reticulocytosis, and hyperbilirubinemia. Secondary causes of IMHA were excluded based on history, physical examination, complete blood count, coagulation studies, serum chemistry, urinalysis, thoracic radiographs, abdominal ultrasound and vector-borne disease testing. Dogs with severe thrombocytopenia or treated with heparin or unable to tolerate oral medications were excluded.

After informed owner consent was obtained, twelve dogs with primary IMHA were treated with clopidogrel using a single loading dose of 10 mg/kg PO on day 1, followed by a maintenance dose of 2 mg/kg PO q24hrs for 4 weeks. Five dogs received concurrently ultra-low dose aspirin therapy (0.5

mg/kg PO q24hrs for 4 weeks). All dogs received the same initial treatment, including corticosteroids, doxycycline, and gastroprotectant agents. Additional immunosuppressive therapy, such as azathioprine and cyclosporine, were used in few dogs after 1 week of therapy. All dogs were closely monitored for any evidence of TE, change in appetite, hemorrhagic, gastrointestinal or dermatologic adverse effects over a 6-month period. Physical examination, complete blood count and serum chemistry were performed at 1, 3, and 6 months.

All dogs survived to discharge. Two dogs relapsed within 6 months of therapy, and one of them was euthanized. No evidence of thrombosis or hemorrhage was identified at necropsy. The other eleven dogs were still alive at the conclusion of the study, and no evidence of TE was identified over a 6-month period. No hemorrhagic, gastrointestinal, or dermatologic adverse effects attributable to clopidogrel were noted in any dog. Clinicopathologic abnormalities were considered secondary to immunosuppressive drugs. Clopidogrel, either alone or in combination with ultra-low dose aspirin therapy, was well tolerated in this group of dogs with primary IMHA. Larger prospective controlled studies should be performed to evaluate the efficacy and safety of clopidogrel therapy for prevention of TE in dogs with primary IMHA.

**ABSTRACT #204**

PLATELET P-SELECTIN (CD62P) EXPRESSION AND PLATELET-LEUKOCYTE INTERACTIONS IN CLINICALLY HEALTHY DOGS WITH AND WITHOUT ULTRALOW-DOSE ASPIRIN. SA Center, JF Randolph, MB Brooks, KS Sharpe, T Stokol, KL Warner. College of Veterinary Medicine, Cornell University, Ithaca, NY.

Improved survival in canine IMHA has been achieved when ultralow-dose aspirin (ULDAsp, 0.5 mg/kg/day PO) is combined with glucocorticoids and azathioprine. However, it remains unclear whether this benefit reflects inhibition of thromboxane-induced platelet activation, modulatory effects on vascular endothelium and systemic inflammation, or altered drug-protein binding.

This study evaluated whether ULDAsp alters 2 flow cytometric parameters of platelet activation: P-selectin (CD62P) expression as a marker of alpha granule release and platelet-leukocyte aggregate (PLA) formation. We compared pre- and post-treatment basal platelet reactivity, and ex vivo platelet response to thrombin stimulation.

Whole EDTA-anticoagulated blood was collected from 10 clinically healthy dogs before and 24-hrs after 2 days of ULDAsp (0.5 mg/kg once daily). Ex vivo thrombin stimulation (1U/mL, 10 min, 37•C) or vehicle-control treatment was followed by paraformaldehyde fixation. Platelets were identified by labeling with an antibody to the constitutive membrane antigen GPIIIa (CD61-FITC). Platelet P-selectin expression was detected by dual labeling with CD61-FITC and CD62P-PE; PLA formation was measured in RBC-lysed samples by dual labeling with CD61-FITC and a pan-leukocyte antibody (CD18-PE). Differences were identified using the Wilcoxon Signed Rank test with a two-sided $P$=0.05.

We found no difference in basal CD62P expression or PLA formation between pre- and post- ULDAsp treatment samples. Although ex vivo thrombin stimulation significantly increased CD62P expression and PLA from basal levels, ULDAsp did not significantly influence either response. Seemingly, monitoring CD62P or PLA would have low utility in assessing a pathophysiologic benefit from ULDAsp in dogs with IMHA.

**ABSTRACT #205**

EFFECTS OF ASPIRIN AND CLOPIDOGREL ON PLATELET FUNCTION IN HEALTHY DOGS. L Shearer, SA Kruth, D Wood. Ontario Veterinary College, University of Guelph, Guelph, Ontario, Canada.

Aspirin has been widely used in both human and veterinary patients for its anti-thrombotic properties. Several studies have evaluated the effect of aspirin on platelet function in dogs, with conflicting results. Clopidogrel is an effective anti-thrombotic therapy in humans, but has been minimally evaluated in dogs. The objective of this study was to assess the effect of various doses of aspirin and clopidogrel on platelet function in healthy dogs. Results were correlated with plasma drug metabolite concentrations.

Six healthy dogs were randomized to six treatment groups (aspirin: 0.5, 1.0, 2.0 and 10.0 mg/kg/d; clopidogrel: 2.0 and 4.0 mg/kg/d). Blood samples were collected on days 0, 3, and 7 and evaluated using optical aggregometry (ADP and PAF as agonists) and the PFA-100 analyzer (collagen-ADP and collagen-EPI cartridges). Day 7 plasma drug metabolite levels were measured using high performance liquid chromatography.

No statistically significant effect on platelet function was identified for the currently recommended anti-thrombotic aspirin dose (0.5mg/kg/d) with either aggregometry or the PFA-100. ADP-induced aggregometry detected inhibition of platelet function by day 7 for the 1.0 and 2.0 mg/kg/d aspirin doses. No effect on platelet aggregation was detected for the 10 mg/kg/d aspirin dose. The PFA-100 detected significant inhibition of platelet function for all aspirin doses, except the 0.5mg/kg/d dose, on both day 3 and 7 using the collagen-EPI cartridge. Inhibition of platelet function was also detected with the PFA-100 for the 2 and 10 mg/kg/d aspirin doses using the collagen-ADP cartridge. Both doses of clopidogrel yielded significant inhibition of platelet function with both methodologies (ADP agonist) on day 3 and 7.

Plasma drug metabolite levels for aspirin yielded variable results (0.5mg/kg/d: 0.17-0.68ug/mL; 1.0mg/kg/d: 0.74-2.95ug/mL; 2.0mg/kg/d: 1.52-5.37ug/mL; 10mg/kg/d: 13.50-30.74). Some dogs had consistently low or high plasma metabolite levels regardless of the dose administered, suggesting variability in individual absorption. Measurement of drug metabolite levels for clopidogrel were less variable (2.0mg/kg/d: 0.043-0.519ug/mL; 4.0mg/kg/d: 0.303-0.489ug/mL), but displayed considerable overlap in the range of values between the two dose groups.

Currently recommended anti-thrombotic dosing of aspirin (0.5mg/kg/d) is ineffective as measured by optical aggregometry and the PFA-100 analyzer in healthy dogs. Treatment with 1mg/kg/d of aspirin was the lowest dosage to inhibit platelet function by both methodologies. Clopidogrel may be a reasonable anti-platelet therapy alternative. The anti-platelet effect of aspirin and clopidogrel was readily documented using the PFA-100. Ineffective anti-platelet therapy may be correlated with inadequate plasma drug metabolite concentrations. Plasma metabolite concentrations may subsequently be used to tailor anti-platelet therapy.

## ABSTRACT #206

INFLUENCE OF RHFVIIa ON HAEMOSTATIC VARIABLES IN NORMAL AND FACTOR VII DEFICIENT DOGS. R Mischke[1], B Dörsch[1], M Diedrich[1], H-J Schuberth[2], M. Kietzmann[3], M von Depka[4]. 1. Small Animal Clinic, Hannover School of Veterinary Medicine, Hannover, Germany. 2. Institute of Immunology, Hannover School of Veterinary Medicine, Hannover, Germany. 3. Institute of Pharmacology, Toxicology and Pharmacy, Hannover School of Veterinary Medicine Hannover, Germany. 4. Werlhof Institute for Haemostaseology, Hannover, Germany.

Recombinant human factor VIIa (rhFVIIa) is widely used as a haemostatic agent in human medicine. The objective of this study was to investigate the influence of rhFVIIa on different haemostastic parameters in normal and factor VII deficient dogs.

RhFVIIa was injected in 2 different dosages (100 and 500 µg/kg body mass [BM]) to either normal or factor VII deficient Beagle dogs (n=4–6). Factor VII:C activity, prothrombin time (standard test [$PT_{ST}$] and modified test [$PT_{MT}$]), activated partial thromboplastin time (APTT), platelet count and haematocrit as well as antibodies against rhFVIIa were measured at different time points.

Injections of rhFVIIa were well tolerated. The maximum values of factor VII:C activity which were measured 2 or 5 minutes after the rhFVIIa injection of 100 or 500 µg/kg were 4.5 ± 0.25 I.U./mL (1 I.U./mL is equivalent to 100 % activity) and 22.0 ± 2.0 I.U./mL in healthy dogs and 4.7 ± 0.28 I.U./mL and 19.1 ± 1.3 I.U./mL in factor-VII-deficient dogs. Terminal half-life values lay between 1 and 1.5 hours. The coagulation activity measured by $PT_{MT}$, which allowed a more differentiated monitoring of rhFVIIa treatment than $PT_{ST}$, increased to values of approximately 1.5times (after 100 µg/kg) or 3times (after 500 µg/kg) the initial values ($p<0.001$). A significant shortening of APTT from 14.7 ± 0.27 s to 13.0 ± 0.35 s was restricted to the group of healthy dogs receiving the high factor VIIa dosage and lasted for 1.5 hours. This group also showed the most remarkable effects on platetet count and haematocrit. Antibodies against rhFVIIa developed in nearly all the dogs.

In conclusion, $PT_{MT}$ is a useful screening test to monitor rhFVIIa treatment in dogs. Factor VII:C activity values and $PT_{MT}$ results measured after injection of 100 µg/kg rhFVIIa indicate that lower dosages should be effective for treatment of bleeding episodes in factor VII deficient dogs. High rhFVIIa concentrations have to be used to affect the APTT in dogs indicating a possible effect in haemophilic dogs.

## ABSTRACT #207

EXERCISE-ASSOCIATED HEART RATE RECOVERY TIME IN NORMAL DOGS. R Manley, E Côté, LA Pack. Atlantic Veterinary College, Charlottetown, PE, Canada.

Heart rate (HR) recovery time (HRRT) is the interval from peak HR during physical exertion to spontaneous return to resting HR. It indicates cardiovascular health and mortality risk in humans. We sought preliminary data that define normal HRRT in dogs.

Sixteen healthy dogs of various somatotypes, ages, and both sexes were outfitted with an electrocardiographic event recorder (King of Hearts Express, Alaris Medical). Each dog engaged in an exercise and cool-down protocol consisting of: 1 minute acclimation at the foot of a stairwell; climbing and immediately descending 6 flights of stairs (40 vertical feet; 12.25 m) at a brisk but comfortable pace, accompanied by a handler; and cool-down at the starting point. The procedures were repeated in triplicate with a minimum 30-minute rest period between each.

The procedure was tolerated by all dogs. The median resting HR prior to exercise was 110 beats/min. The median peak HR during exercise was 250 beats/min. Median duration of exercise was 52.5 seconds. The median HR decreased to 180 beats/min 40 seconds after onset of exercise; 145 beats/min 54 seconds after onset of exercise; and 135 beats/min 61 seconds after onset of exercise.

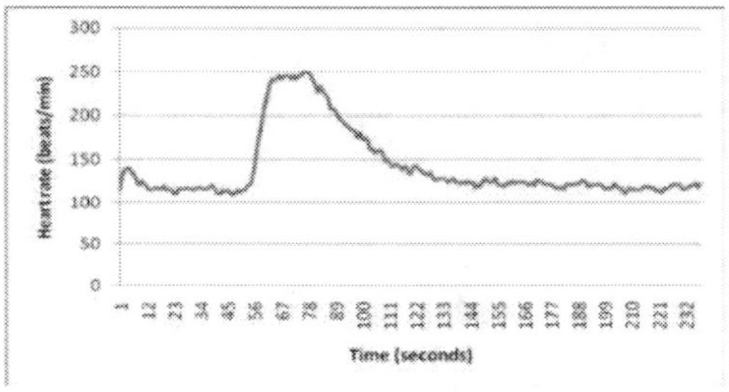


These results provide preliminary data that may be useful for assessing the cardiovascular function of dogs.

## ABSTRACT #208

INFLUENCE OF AGE ON PULMONARY ARTERIAL PRESSURE IN HEALTHY BEAGLE DOGS: ANALYSIS BY RIGHT VENTRICULAR CATHETERIZATION, PULSED DOPPLER ECHOCARDIOGRAPHY AND PULSED DOPPLER TISSUE IMAGING. E. Mercier[1], M. Mathieu[2], C. Clercx[1], K. Mc Entee[1,2]. 1 Department for Clinical Sciences B44, Faculty of Veterinary Medicine, University of Liège, Belgium 2 Laboratory of Physiology, Faculty of Medicine, Free University of Brussels, Belgium.

Pulmonary arterial pressure (PAP) increases with age in healthy human subjects and this has been attributed to vascular resistance; but no data exists in dogs. Pulmonary hemodynamics can be assessed directly by right heart catheterization, or indirectly by conventional Doppler-echocardiography and Doppler tissue imaging (TDI). The aims of this study were to assess the influence of aging on pulmonary hemodynamics and hemorheological properties in healthy dogs and to compare measurements obtained by cardiac catheterization, with systolic time intervals of pulmonary flow and myocardial velocities of the tricuspid lateral annulus.

Fourteen healthy experimental beagle dogs from two age groups were used in the study, 8 young dogs aged 10 months to 5 years (mean: 2.7 years) and 6 old dogs aged 8 to 15 years (mean: 12.1 years). Hematology, total protein and fibrinogen concentrations and whole blood viscosities were measured. Pulmonary flow, its maximal velocity, acceleration and ejection times were recorded by pulsed Doppler-echocardiography. Early (E') and late (A') diastolic myocardial velocities, isovolumic contraction velocity and systolic myocardial velocity of the free tricuspid annulus were measured by pulsed-wave TDI. Under general anesthesia, systolic (s), mean (m), diastolic (d) and occluded PAP, right atrial pressure and cardiac output were measured using a pediatric thermodilution Swan-Ganz catheter, while stroke volume, pulmonary vascular resistance (PVR), pulse pressure and compliance (C) were calculated. Values are given as mean ± SE.

Hemorheological measurements were not different between the 2 groups. Systolic PAP (30,6 ± 2,1mmHg), mPAP (19.3 ± 1.8mmHg) and dPAP (13.4 ± 1.4mmHg) were higher in old dogs compared with young dogs (respectively 21.4 ± 1,5mmHg, 13.7 ± 1,2mmHg and 9.5 ± 1.1mmHg). This increase was attributed to a higher PVR and lower C. Acceleration and ejection times of pulmonary flow were not modified. The A' wave was increased in old compared to young dogs (0.147 ± 0.003 m/s vs 0.109 ± 0.009 m/s, p = 0.02) and the E'/A' ratio was decreased (0.90 ± 0.05 vs 1.13 ± 0.06, p = 0.01). Systolic, m and dPAP were correlated with age; PVR and C were inversely related. The A' wave was correlated with age (r = 0.72, p<0.005) and with sPAP.

Results of the present study show that (1) in healthy Beagle dogs, PAP increases with aging, (2) factors implicated in this increase are PVR and C, (3) systolic time intervals of the pulmonary flow are not sensitive to detect this increase, (4) the tricuspid free annulus late diastolic velocity increases with aging and is correlated with sPAP.

**ABSTRACT #209**

EVALUATION OF HDO® (HIGH DEFINITION OSCILLOMETRY) – A NEW NIBP TECHNIQUE - IN COMPARISON TO INVASIVE MEASUREMENT (HSE - HYDRO SACHS ELECTRONICS) IN ANESTHETISED DOGS. Ch Baumgartner[1], W Erhardt[1], Ch Faltermeier[1], J Reinert[1], J Henke[1], B Egner[2], I Haas[3]. [1]ZPF (Center of Preclinical Research), Technical University of Munich, Germany, [2] Medical Center for Small Animals, Hoerstein, Germany, FZMB Bad Langensalza, Germany.

Precise measurement of high and low blood pressure situations is mandatory also in veterinary medicine. Even more if cardiac output situation is affected, like in arrhythmia or shock, accurate blood pressure information can be a vital parameter for treatment decision and survival. So far available NIBP technologies had clear limitations. Invasive techniques are regarded as gold standard however, these are not applicable for day-by-day routines.High Definition Oscillometry (HDO) was evaluated against direct line as a potential non-invasive approach. To investigate accuracy in the above situations, 7 beagle dogs where put under anesthesia (propofol) and blood pressure was affected by using either phenylephrin, dopamin or isufluran. Systolic, mean, diastolic arterial blood pressure (SAP/MAP/DAP) and pulse/min (bpm) were determined 75 times within 120 Minutes per individual.

Mean, standard deviation (SD) and standard error of the mean (SEM) were calculated. The agreement of non invasive SAP and DAP with the corresponding simultaneous invasive values was analyzed according to the Bland-Altman method. Mean difference (bias) and SD of the difference (precision) were calculated. A significant correlation for SAP was observed (r = 0.895; p<0.001). Mean difference between the two methods was 0.18 ± 9.15 mmHg. Also for DAP a significant correlation between the two methods was investigated (r = 0.875; p<0.001), with a mean difference of 5.6 ± 7.3 mmHg. As these results reflect all situations including severe arrhythmia, high and low heart rate (49 – 210 bpm) and blood pressure within 67-310 SAP, HDO may be regarded as a clear alternative for invasive methods.

**ABSTRACT #210**

ACTION DURATION-RELATED DIURETIC EFFECTS AND RENNIN-ANGIOTENSIN-ALDOSTERONE SYSTEM IN HEALTHY DOGS WITH A FUROSEMIDE. Y Hori*, N Ohshima, K Kanai, F Hoshi, N Itoh, S Higuchi. Kitasato University, Aomori, Japan.

This study investigated the difference in action duration-related diuretic effects and activity of rennin-angiotensin-aldosterone (RAA) system to a furosemide as a model of short-acting and long-acting loop diuretics.

Six dogs were anesthetized and were randomized to placebo, intravenous bolus administration (IB), and chronic rate infusion (CRI) groups. This study was conducted with crossover study. Furosemide, which was dissolved in sterile saline, was infused for 8 hours in CRI group, or was injected at 0- and 4-h in IB group. Total dosage of furosemide was 4 mg/kg and total amount of sterile saline was equal to each group at 18ml/head. Blood and urine samples were collected at baseline, 1, 2, 4, 5, 6, and 8-h.

Compared with baseline, the IB group increased significantly the urine output at 1- and 5-h. In contrast, the CRI group increased significantly the urine output compared to baseline, and the increase persisted for 4 hours. Compared to placebo, the 8-h urine output and the 8-h sodium excretion were increased significantly in IB and CRI groups. Similarly, those of CRI group were significantly higher than those of IB group. Plasma aldosterone concentration was elevated significantly in IB group but not significant in CRI group.

Our results indicate that action duration may be a predominant cause of loop diuretics-related diuretic differences. In addition, persistent diuresis may cause mild and long-acting effects with minimal activation of the RAA system.

**ABSTRACT #211**

CHANGES IN THE PLASMA ANP AND NT-PROBNP CONCENTRATIONS IN HEALTHY DOGS WITH AN ACUTE VOLUME OVERLOAD. Y Hori*, N Sano, K Kanai, F Hoshi, N Itoh, S Higuchi. Kitasato University, Aomori, Japan.

This study investigated the difference in the secretory responses of plasma ANP and NT-proBNP to an acute volume overload in dogs.

Six healthy dogs were anesthetized. Lactated Ringer's was infused intravenously at 90-100 ml/kg/h for 60 min as a volume overload. Subsequently, furosemide was administered as a preload reduction. The left ventricular (LV) pressures and pulmonary capillary wedge pressures (PCWP) were monitored at baseline and at 20, 40, 60, 80, 100, and 120 min. The plasma atrial natriuretic peptide (ANP) and N-terminal pro-B-type natriuretic peptide (NT-proBNP) concentrations were determined using a radioimmunoassay and enzyme immunoassay, respectively.

Volume overload significantly increased LV peak systolic pressure (LVPs), LV end-diastolic pressure (LVEDP), and PCWP. Preload reduction decreased the LVEDP and PCWP significantly. Volume overload increased the plasma ANP and NT-proBNP concentrations significantly. Preload reduction decreased the plasma ANP concentration significantly, but not the plasma NT-proBNP concentration. The plasma ANP concentration was strongly correlated with the heart rate, LVPs, LVEDP, and PCWP. The plasma NT-proBNP concentration was significantly correlated with LVPs and LVEDP.

The different responses in the natriuretic peptides may reflect different mechanisms regulating their secretion. The plasma ANP level may be a more useful parameter for the non-invasive estimation of the changes in acute volume overload than the plasma NT-proBNP level, especially when the hemodynamic state changes rapidly.

**ABSTRACT #212**

ANALYTICAL VALIDATION OF A COMMERCIALLY AVAILABLE FELINE N-TERMINAL PROHORMONE BRAIN NATRIURETIC PEPTIDE ELISA. A Carrier[1], A Beardow[1], G Farace[1], K Yeung[1], SJ Ettinger[2], K Yee[2]. [1]IDEXX Laboratories, Inc., Westbrook, ME, [2]California Animal Hospital, Los Angeles, CA.

An assay for feline N-Terminal Prohormone Brain Natriuretic Peptide (NT-proBNP) has been available since 2007 initially as CardioScreen™ NT-proBNP (Guildhay Ltd., UK) and now as Cardiopet™ proBNP (IDEXX Laboratories Inc, USA). The assay was initially commercialized by Guildhay with a time to result (the start to finish time of the assay) of around 20 hours. However in the US the assay has always been offered with a time to result of around 7 hours and to our knowledge a complete validation of this faster protocol has not been presented to date.

An analytical validation of assay covers a number of areas including accuracy - how do samples run on the faster assay correlate to the same samples run on the original assay; dynamic range – what is the lowest and highest concentrations of feline NTproBNP that can be reliably detected by the assay; precision – how consistent is the assay both within a single run (intra-assay precision) and across multiple runs on different days (inter-assay precision); interferences – do lipids, hemoglobin or bilirubin effect the performance of the assay and finally the sample types that can be analyzed, i.e. can NTproBNP be successfully detected in both serum and plasma.

The accuracy of the new method compared to the old method is very good with a correlation of $r^2$=0.95 and a linear slope of 0.91, both of these show that there is little difference between the two protocols.

The dynamic range of the assay is determined by serially diluting spiked samples to determine the minimum and maximum concentrations that can be reliably measured. The dynamic range was determined to be between ~20 and 1500 pmol/L.

Inter- and intra-assay precision was determined by running a set of spiked samples both in a single assay (minimum n=10) and in multiple assays (n=3). Inter- and intra assay precision was found to be between 10 and 20% across the dynamic range of the assay.

Spiking samples with lipids, hemoglobin and bilirubin were all found to have no effect on the assay performance.

A set of 20 matched serum and plasma samples did not show good agreement. No consistent bias or trend, that would have allowed a mathematical correction to be applied, was observed. We thus concluded that interpretive guidelines generated for plasma should not be applied to serum.

Overall, the kit is robust and the shortened protocol is acceptable as there appears to be no effect on the overall assay performance.

**ABSTRACT #213**

MEASUREMENT OF CANINE B-TYPE NATRIURETIC PEPTIDE USING A HUMAN ANALYZER. K Cyr, A Beardow, G Farace. IDEXX Laboratories, Inc., Westbrook, ME.

Canine and human B-type Natriuretic Peptide (BNP) are only moderately homologous (68%) and so it would seem unlikely that a human analyzer could be used to detect canine BNP. However there are regions of significant homology that might provide epitopes large enough to allow these assays to act as a semi-quantitative assay that could detect severe heart disease. To test this hypothesis we took an I-stat instrument (Abbot Laboratories, USA) and attempted to measure canine BNP either in spiked canine plasma or in real patient samples.

Samples spiked with synthetic human BNP-32 (Phoenix Pharmaceuticals, USA) showed expected results while samples spiked with synthetic canine BNP-32 (Phoenix Pharmaceuticals, USA) could not be detected even when spiked at levels of over 1000 pg/ml.

Samples from 4 dogs with cardiac disease (ISACHC class 1b or worse) were tested pet-side and once again no measurable BNP was detected in any of the samples.

This combination of failures to measure canine BNP would indicate that the regions homology of the two peptides are not sufficient to allow the antibodies used in the I-Stat assay to detect canine BNP. This confirms the view that the only avenue for a canine BNP assay is to have canine specific reagents.

**ABSTRACT #214**

INTRADAY VARIATION IN FELINE N-TERMINAL PROHORMONE BRAIN NATRIURETIC PEPTIDE CONCENTRATION. K Yee[1], J Allen[1], SJ Ettinger[1], A Beardow[2], A Carrier[2], G Farace[2], K Yeung[2]. [1]California Animal Hospital, Los Angeles, CA. [2]IDEXX Laboratories, Inc., Westbrock, ME.

N-Terminal Prohormone Brain Natriuretic Peptide (NT-proBNP) is a neurohormone and like other hormones it is possible that its circulating concentration in the blood of the patient could be cyclical in nature. If this is the case, then it is important to understand the nature and shape of the cycle since it could have a profound effect on the timing and interpretation of blood sample values.

Five normal cats were recruited for the study. Physical exam, thoracic radiographs, echocardiogram, systolic blood pressure, BUN, creatinine, TT4, and urine specific gravity were evaluated to determine health status of the cat and confirm there was no significant disease. Plasma samples were obtained over 24 hours at 9am, noon, 3pm, 6pm, midnight, 6am, and 9am. NT-proBNP concentrations were determined for each time point.

Four of the five cats showed no change throughout the time-course study; one cat did show some variability. However a one-way analysis of variance (ANOVA) repeated-measures design revealed no significant difference between the times of the day that samples were taken from the cats with $F(6,24)=1.42$; $p=0.25$.

The major limitation with the study is that the low levels on NT-proBNP mean that the assay is operating at the very limit of sensitivity; any minor well-to-well differences could mask some subtle variations.

In conclusion, this study suggests that there is no significant, measurable, cyclic (diurnal) variation of NT-proBNP in normal cats.

**ABSTRACT #215**

PLASMA ADRENOMEDULLIN AS A BIOMARKER FOR SEVERITY OF PULMONARY HYPERTENTION IN DOGS. N Kanno, K Asano, K Teshima, M Seki, K Edamura, S Tanaka. Nihon University, Kanagawa, Japan.

Plasma adrenomedullin (AM) concentration has been shown to increase in human patients with chronic heart failure (CHF). In addition, plasma AM level has been demonstrated to elevate in human patients with pulmonary hypertension (PH). We have reported that plasma AM is higher in dogs with CHF due to mitral regurgitation (MR) than in normal subjects. In this study, plasma AM was measured in dogs with MR and/or tricuspid regurgitation (TR) to evaluate whether canine AM has the potential for a useful biomarker as PH.

In this study, 76 canine patients including 52 with MR and 24 with MR and TR, and 10 healthy Beagles (control group) were used. The patients were assessed by physical examination, thoracic radiography and echocardiography, and categorized by International Small Animal Cardiac Health Council (ISACHC) classification. Plasma concentrations of AM were measured by a radioimmunometric assay. In the echocardiographic examination, TR velocity $\geq 2.8$ m/second was considered to indicate PH.

The plasma level of AM significantly increased in proportion to the severity of CHF in dogs with MR. Plasma AM level in dogs with PH ($52.83 \pm 29.93$ fmol/ml; n=12) significantly increased compared with that in control ($10.47 \pm 3.78$ fmol/ml). In ISACHC class II, plasma AM level in dogs with PH ($53.99 \pm 21.94$ fmol/ml; n=8) significantly elevated compared with that in dogs without PH ($37.79 \pm 14.58$ fmol/ml; n=24).

In conclusion, plasma AM has the potential to be a powerful indicator of PH in dogs with CHF.

**ABSTRACT #216**

PULSED TISSUE DOPPLER EVALUATION IN DOGS WITH IDIOPATHIC DILATED CARDIOMYOPATHY. GG Pereira, GT Goldfeder, VCM Oliveira, DG Prada, FL Yamaki, MHMA Larsson. Universidade de São Paulo, Faculdade de Medicina Veterinária e Zootecnia, São Paulo, SP, Brazil.

Idiopathic dilated cardiomyopathy (DCM) in dogs is recognized by myocardial lesions and dysfunction. We hypothesize that a pattern of regional myocardial dysfunction can be stablished by pulsed tissue Doppler in dogs with spontaneous DCM.

Fourty four dogs were evaluated, being 22 with DCM and 22 healthy, with similar weight, for control purpose. Longitudinal myocardial velocities were obtained in mitral annular (basal), papillary muscle level (medial) and apical segments of left ventricular free wall (LVFW) and interventricular septum (IVS). Radial myocardial velocities were assessed in subendocardial and subepicardial regions of medial LVFW. Peak myocardial velocities resulting from ventricular systole (Sm), early (Em) and late (Am) ventricular filling were determined. According to cardiac rhythm, recordings of three (sinus) or five (atrial fibrillation) consecutive cycles were averaged. Unpaired *t*-test was employed to investigate differences between groups.

Peak radial subendocardial and subepicardial Sm velocities were lower in DCM group compared to control ($0.068 \pm 0.019$ vs. $0.102 \pm 0.020$ m/s and $0.061 \pm 0.016$ vs. $0.099 \pm 0.027$ m/s, respectively; $p<0.0001$). Peak longitudinal Sm velocities were lower in basal and medial portions of LVFW ($0.093 \pm 0.033$ vs. $0.166 \pm 0.047$ m/s and $0.091 \pm 0.033$ vs. $0.141 \pm 0.044$ m/s, respectively; $p<0.0001$) and IVS ($0.063 \pm 0.021$ vs. $0.136 \pm 0.039$ and $0.066 \pm 0.026$ vs. $0.104 \pm 0.032$ m/s, respectively; $p<0.0001$). Peak diastolic waves were not significantly different.

Reduction in systolic longitudinal (basal and medial) and radial myocardial velocities represents the most significant pulsed tissue Doppler findings for DCM diagnosis in dogs.

**ABSTRACT #217**

EVALUATION OF LEFT VENTRICULAR SYSTOLIC AND DIASTOLIC FUNCTION USING PULSED TISSUE DOPPLER IN DOGS WITH CHRONIC MITRAL VALVE INSUFFICIENCY. LC Petrus; GT Goldfeder; VMC Oliveira, EC Soares, MHMA Larsson.

Thirty dogs with chronic mitral valve insufficiency, ISACHC class II and 30 healthy dogs were studied. The evaluation of the longitudinal myocardial function was assessed by means of the apical four chamber view, and the pulsed tissue Doppler was used to measure the velocity of six regions: apical (a), medial (m), and basilar or annular (b) of the interventricular septum (IVS) and of the left ventricular lateral wall (LVLW). From each studied segment the maximum myocardial velocity of the rapid filling (Em), atrial contraction (Am) and systolic (Sm) waves were obtained, as well as the ratio between the waves of the mitral inflow (E) and myocardial (Em), registered during the same cardiac cicle

(Ef/Em ratio). Mean, median and standard deviation of the echocardiographic indices were calculated. Student t- test was used to compare each group of dogs. Therefore, this study aimed for determining the abnormalities of dogs with mitral valve insufficiency and investigating the influence of the ventricular filling pressure over the studied indices. Sm LVLWb; Sm LVLWm; Sm IVSb were lower in dogs with CMVI than in healthy ones., and Am LVLWb; Am IVSb; Em IVSm; Am IVSm; Am IVSa and Ef/Em ratio were higher in dogs with CMVI then in healthy ones. Statistical significant difference was not observed between both groups when the other echocardiographic indices were evaluated.

In conclusion, dogs with chronic mitral valve disease, ISACHC class II, may show abnormalities of the left ventricular systolic function, and this can be observed by means of the pulsed tissue Doppler, even in dogs with normal shortening fraction. Moreover, the high Ef/Em ratio of dogs with CMVI compared to the value obtained from healthy dogs assures that this indice is very trustful and representative of the left ventricular filling pressure, useful to rule out other causes of dyspnea. The velocity of Em and Am were higher in dogs with valve disease than in healthy dogs probably because, despite heart insufficiency class II of dogs with CMVI, there was not myocardial remodeling able to change the left ventricular diastolic function, even though it is a chronic disorder in advanced stage.

**ABSTRACT #218**

PATHOLOGICAL FINDINGS IN DOGS WITH MITRAL VALVE INSUFFICIENCY: BIOPSIES DURING OPEN HEART SURGERY WITH CARDIOPULMONARY BYPASS. M Nishida, M Uechi, T Mizukoshi, T Ebisawa, M Mizuno, T Mizuno, K Harada, M Fujiwara, T Nakayama. Nihon University, Fujisawa, Kanagawa, Japan.

The aim of this study was to examine histopathological characteristics of myocardial and lung biopsy specimens taken from dogs with mitral valve insufficiency (MVI). Twenty-nine dogs with mild to severe MVI, aged 5–14 years, underwent mitral valve plasty with cardiopulmonary bypass between October 2007 and November 2008, and samples for biopsy were taken during surgery from the left atrial wall, the left ventricular free wall (between the anterior and posterior papillary muscles), ruptured chordae and/or the lung (at the apex of the left middle lobe). Specimens were fixed in 10% phosphate-buffered formalin before histopathological analysis. Histopathological examination of the left atrium (n=25) showed no abnormalities in 16 cases but revealed interstitial edema in 7, eosinophilic degeneration in 6, connective tissue hyperplasia in 3, inflammatory infiltration in 4, and myocardial fascicle atrophy in 3. The left ventricular wall (n=26) showed no abnormality in 24 cases, but interstitial edema was noted in 2. In the ruptured chords (n=16), 11 cases showed myxomatous degeneration, while 3 had eosinophilic degeneration with tissue disruption. In the lung specimens (n=12), no abnormality was found in 5 cases, but disrupted alveolar structure was seen in 1, alveolar septal thickening in 2 and heart failure cells in 5, indicating previous episodes of pulmonary edema. These results suggest that the heart and lung tissue structures are relatively well maintained in MVI until shortly before death.

**ABSTRACT #219**

LONG-TERM SURVIVAL OF TWO DOGS AFTER MITRAL VALVE PLASTY USING EXPANDED POLYTETRAFLUOROETHYLENE CHORDS: PATHOLOGICAL STUDY. M Nishida, M Uechi, T Mizukoshi, T Ebisawa, M Mizuno, T Mizuno, K Harada, M Fujiwara, N Nakayama. Nihon University, Fujisawa, Kanagawa, Japan.

Mitral valve plasty is one of the treatment options for mitral regurgitation (MR). Expanded polytetrafluoroethylene (e-PTFE) is a polymer, which has been widely used to make artificial chords. In this study, we report autopsy and histological findings in two dogs that underwent cardiopulmonary bypass for mitral annuloplasty using e-PTFE sheets and chordoplasty using e-PTFE sutures. Case 1 was a neutered, 7-year-and-5-month-old male Cavalier King Charles Spaniel with severe MR. Postoperative progress was favorable until 10 months later when the dog showed severe diastolic dysfunction with significantly decreased fractional shortening by echocardiography. The dog died unexpectedly at 23 months after surgery. Case 2 was a 10-year-and-3-month-old female Maltese with severe MR. Postoperative progress was also satisfactory, but the dog died at 26 months after surgery from respiratory failure caused by intrathoracic fibrosarcoma. By histopathological examination, the structural integrity of both atrial and ventricular muscle fibers was maintained in Case 1, with no evidence of degeneration, fibrosis or fiber disarray. In Case 2, mild fibrosis was noted at the base of the left ventricular papillary muscle, indicating an old myocardial infarct. In both cases, e-PTFE sheets and sutures were not damaged and well integrated in the surrounding, highly differentiated connective tissues. There was no evidence of reactive changes around e-PTFE. These results suggest that e-PTFE is excellent in tissue compatibility and durability and useful for canine mitral valve plasty.

**ABSTRACT #220**

MITRAL VALVE PLASTY IN 11 CAVALIER KING CHARLES SPANIELS. M Nishida, M Uechi, T Mizukoshi, T Ebisawa, M Mizuno, T Mizuno, K Harada, M Fujiwara. Nihon University, Fujisawa, Kanagawa, Japan.

This study aimed at retrospectively assessing the effectiveness of mitral valve plasty in Cavalier King Charles Spaniel (CKCS), a breed predisposed to mitral valve disease (MVD). Eleven CKCS (5 males and 6 females; bodyweight, 8.8 ± 1.4 kg; age, 110 ± 26.7 months) underwent cardiopulmonary bypass for mitral valve plasty between December 2006 and September 2008. Postoperative complications included 1 case of tricuspid valve insufficiency and 3 cases of left atrial thrombosis (one had a preexisting thrombus at the time of surgery). In 3 cases, neurological symptoms became evident after surgery due to preexisting syringomyelia. The mean survival time was 10.4 ± 6.8 months. One dog died from a suspected cardiac cause at 22 months after surgery, and another from possible thromboembolism at 4 months after surgery. Nine dogs were still alive at the time of the report. At 1 and 3 months after surgery, the left atrial to aortic root diameter ratio (1.76 ± 0.36 and 1.68 ± 0.33, respectively; n=7) and the plasma atrial natriuretic peptide level (117.9 ± 54.8 and 85.8 ± 38.2 pg/mL, respectively; n=4) were lower than those before surgery (2.60 ± 0.61 and 198.0 ± 109.9 pg/mL). There were also significant improvements in the number of prescribed cardiovascular drugs 1 month after surgery (1.6 ± 1.3 vs. 4.5 ± 1.6 preoperatively, $p<0.05$; n=11) and in the cardiac murmur grade (2.5 ± 0.8 vs. 5.1 ± 0.6 preoperatively, $p<0.05$; n=9). These results suggest that mitral valve plasty is beneficial in CKCS with MVD.

**ABSTRACT #221**

INTRA- AND POST-OPERATIVE COMPLICATIONS IN 47 DOGS THAT UNDERWENT MITRAL VALVE PLASTY. T Mizuno, M Uechi, T Ebisawa, M Mizuno, T Mizukoshi, K Harada, M Nishida, M Fujiwara, T Nakayama. Nihon University, Fujisawa, Kanagawa, Japan.

We previously reported that mitral valve plasty (MVP) under cardiopulmonary bypass (CPB) is an effective treatment for mitral regurgitation (MR) in dogs (2007 ACVIM Forum). To assess the incidence of intra- and post-operative complications, we retrospectively reviewed 47 cases that underwent MVP with CPB at the Nihon University Animal Medical Center between August 2006 and September 2008. The subjects were 47 dogs [22 males and 25 females, age: 62-175 (123 ± 25) months, bodyweight: 1.8-13.5 (5.7 ± 3.0) kg]. The mean age of the 10 dogs that died within 4 months after surgery was 140 ± 21 (range: 115-175) months, which was significantly higher compared to those that survived beyond 4 months postsurgery [119 ± 25 (range: 62-157) months]. The 4-month postoperative mortality was 29% for dogs aged 10 years or older and 11% for those younger than 10 years. The causes of death were surgical technical problems (2 cases), thrombosis (4 cases), pancreatitis (2 cases), pulmonary hypertension (1 case) and unknown (1 case). Of the 37 cases that survived for 4 months or longer, 4 cases had postoperative complications (thrombotic cerebral infarction, pulmonary infarction, cerebellar infarction and aspiration pneumonitis; 1 case each). Thrombus formation (including those in the left atrium) was observed in 12 cases and was the most frequent causes of postoperative complication and/or death. These results suggest that prevention of thrombosis is an important strategy for improving the surgical outcome of MVP. For dogs over 10 years old, in addition, preoperative stabilization and postoperative management are critical, and earlier surgery is recommended.

**ABSTRACT #222**

EFFICACY OF METHYLPREDNISOLONE THERAPY FOR INFLAMMATION AFTER CARDIAC SURGERY. T Mizuno, M Uechi, T Ebisawa, H Kamiyama, H Yamada, M Nishida, M Fujiwara, K Harada, M Mizuno, T Mizukoshi, T Nakayama. Nihon University, Fujisawa, Kanagawa, Japan.

Previously, we reported that dogs develop systemic inflammatory response syndrome (SIRS), with elevations of the IL-6 and IL-10 levels, after cardiac surgery with cardiopulmonary bypass (CPB) (2007 ACVIM Forum). SIRS can lead to serious complications such as multiple organ failure. In this study, the anti-inflammatory effect of methylprednisolone (MP) after cardiac surgery with CPB was evaluated. Fourteen dogs (bodyweight, 4.4-7.3 kg) were randomly assigned to a placebo (n=8) or a MP group (n=6) and underwent cardiac surgery with CPB between April 2007 and December 2008. Serum cytokine levels (IL-6, TNF-$\alpha$ and IL-10) were measured before surgery, at 5 minutes after heparinization, at 5 minutes after the onset of CPB, immediately after heart rebeating, at 15 minutes after protamine administration, and at 3, 6, 12, 24 and 48 hours after the end of CPB. Saline (placebo) or MP (10 mg/kg) was administered intravenously, concurrently with heparinization before the onset of CPB. Blood IL-6 peaked at 3 hours after the end of CPB in both groups, although its level was lower in the MP group than that in the placebo group (149 ± 30 vs. 345 ± 175 pg/mL). No intergroup difference was seen in IL-10 level at 3 hours after the end of CPB (placebo vs. MP, 57 ± 126.8 vs. 74 ± 128.1 pg/mL). There was no difference in TNF-$\alpha$ level between the groups. These results indicate that a single administration of methylprednisolone before CPB suppresses the elevation of IL-6 level following cardiac surgery.

**ABSTRACT #223**

RECOVERY OF ATRIAL FIBRILLATION IN THREE DOGS AFTER MITRAL ANNULOPLASTY UNDER CARDIOPULMONARY BYPASS FOR SEVERE MITRAL REGURGITATION. M Fujiwara, M Uechi, T Mizukoshi, T Ebisawa, M Mizuno, T Mizuno, K Harada, M Nishida. Nihon University, Fujisawa, Kanagawa, Japan.

In dogs with severe mitral regurgitation (MR), atrial dilatation and/or structural changes in atrial myocardium often lead to atrial fibrillation (AF). We report three cases of successful defibrillation following open heart surgery under cardiopulmonary bypass (CPB) performed for severe MR and cardiomegaly. Case 1 was a 7-year-old, male Cavalier King Charles Spaniel (CKCS) presenting with ascites. Body weight (BW) was 8.8 kg, systolic regurgitant murmur (SRM) was grade 5/6 (mitral area) and 4/6 (tricuspid area). Case 2 was a 12-year-old, female Miniature Schnauzer presenting with syncope. BW was 8.9 kg, SRM was grade 5/6 (mitral area). Case 3 was a 9-year-old, male CKCS with a history of ascites for 3 months. BW was 10.3 kg, SRM was grade 4/6 (mitral and tricuspid areas). AF was confirmed by ECG in all cases. All patients underwent CPB for mitral chordoplasty and annuloplasty. During surgery, significant enlargement in the left atrium and AF were confirmed. After return of spontaneous circulation, atrial and ventricular fibrillation occurred, but normal sinus rhythm was restored after defibrillation. AF was not observed for the following 3 months in all cases except for a transient recurrence in Case 1 at 1 week after operation. X-ray examination revealed reduced atrial and cardiac sizes in all cases. This study suggests that normal sinus rhythm can be restored by postoperative defibrillation in dogs with severe MR and AF, and that load reduction after mitral valve repair prevents the recurrence of AF.

**ABSTRACT #224**

LV 2D STRAIN ANALYSIS PRE- AND POST- BALLOON VALVULOPLASTY IN DOGS WITH PULMONARY STENOSIS. Y Fujii, H Takei, H Takano, Y Wakao. Azabu University School of Veterinary Medicine, Kanagawa, Japan.

Left ventricular (LV) deformation and paradoxical ventricular septal motion can alter coordination of LV regional dynamics in chronic right ventricular (RV) pressure overload. The purpose of this study was to assess the influence of right ventricular pressure overload on LV systolic motion using 2D strain parameters in dogs with spontaneous pulmonary stenosis (PS), and to see if these parameters change after successful balloon valvuloplasty.

Twenty-one clinically healthy beagles (control) and 9 dogs with PS were used. PS was severe in all PS dogs, based on Doppler pressure gradient (PG, 153.52±65 mmHg). Post-balloon PG was 54.05±30.9 mmHg. 2D strain parameters (radial strain [SR], strain rate [SrR], time to peak radial strain [SRT], radial displacement [DR], time to peak radial displacement time [DRT], radial displacement variation) were measured in the 6 segments of LV short-axis view. Eccentricity index (EI) and fractional shortening (FS) were also measured. FS, systolic and diastolic EI, DR, and SRT were significantly decreased in dogs with PS compared with control. In comparison of each segment, SR, SRT, DRT and radial variation in the septal segments were significantly decreased. No significant change was observed between pre- and post- balloon in 2D strain parameters except for segmental DRT. Based on these results, although LV dyssynchrony was improved, most of the 2D strain parameters did not change after successful balloon valvuloplasty. Impaired LV contraction dynamics would be due to not only pressure overload, but probably also other factors including right ventricular hypertrophy.

**ABSTRACT #225**

EFFECTS OF BISOPROLOL ON VENTRICULAR MYOCYTE ELECTROPHYSIOLOGY DURING CHRONIC HEART FAILURE. Y Nishijima[1], A Sridhar[1], RL Hamlin[1], G Beddies[2], CA. Carnes[1]. 1 The Ohio State University, Columbus, OH. 2. Bayer Healthcare LLC, Shawnee Mission, KS.

Beta-blockers are known to reduce the occurrence of sudden cardiac death in patients with chronic heart failure; however, the electrophysiologic mechanisms remain poorly defined. We hypothesized that chronic treatment with bisoprolol, a beta-1 selective adrenergic blocker, would attenuate heart failure (HF)-induced remodeling of ventricular myocyte electrophysiology. The canine RV tachypacing model of heart failure was used. After 12 weeks of RV tachypacing (LVFS 15 ± 2.5 %) dogs were randomized to 5 weeks of bisoprolol or placebo. A separate group of normal controls were used for all comparisons. Myocytes from the LV mid-myocardium were studied. Compared to normal controls, the placebo-treated HF myocytes exhibited significant prolongation of $APD_{50}$ and $APD_{90}$ at both 0.5 and 1 Hz ($p<0.05$). Bisoprolol treatment significantly attenuated the HF-induced changes in $APD_{50}$ and $APD_{90}$, to values that did not differ from normal controls. Heart failure did not alter the inward conductance of the inward rectifier current ($I_{K1}$) although HF significantly reduced peak outward $I_{K1}$ ($p<0.05$). Bisoprolol treatment restored ($p<0.05$) peak outward $I_{K1}$ to values which did not differ from control. Bisoprolol treatment did not alter the HF-induced reduction in the transient outward $K^+$ current ($I_{to}$). Chronic beta-1 selective adrenergic blockade normalizes the ventricular action potential in chronic HF. This effect may contribute to the reduced cardiac death observed with beta-adrenergic blockade during HF.

**ABSTRACT #226**

HEART RATE IN IRISH WOLFHOUNDS WITH LONE ATRIAL FIBRILLATION & ASSESSMENT OF DIGOXIN AS SOLE METHOD OF RATE CONTROL. M Rector[1], JM Bright[1], M Dentino[2]. [1]Colorado State University College of Veterinary Medicine & Biomedical Sciences, Ft. Collins, CO. [2]Nephrology Associates, Evansville, IN.

There is a high prevalence of lone atrial fibrillation (AF) in Irish wolfhounds (IWs), and it is possible that abnormally high heart rates over time may lead to dilated cardiomyopathy in affected dogs. Although digoxin is often prescribed for rate control in IWs with AF, it is not known whether this drug provides adequate rate control during typical daily activities. This study was done to determine whether average hourly heart rates (HRs) and peak hourly heart rates are higher in IWs with lone AF than in comparable IWs with normal sinus rhythm (NSR) and to determine whether administration of digoxin to IWs with lone AF would normalize HR. HR data was collected from 13 IWs with lone AF using 24 hour Holter monitor recordings obtained in the home environment. Comparable data was obtained from 13 healthy IWs of similar age and gender in NSR. All IWs with AF were then treated with digoxin (0.004-0.005 mg/kg q12h), and Holter studies repeated after verifying therapeutic serum concentration (1.0-2.0 ng/ml) in each treated dog.

Average and peak hourly HRs were compared using a mixed model repeated measures ANOVA to evaluate differences between groups (NSR vs AF) and between baseline and digoxin treatment within the AF dogs. The peak hourly HRs did not differ significantly in IWs with AF compared to normal IWs (AF 164.8 [SE=13.3] bpm vs NSR 159.0 [SE+13.8] bpm). However, the average hourly HRs were higher in AF dogs than normal dogs (AF 98.6 [SE=5.1] bpm vs NSR 72.5 [SE=5.3] bpm; $P=0.007$). Both the peak and the average hourly HRs were significantly reduced in IWs with AF by treatment with digoxin (peak baseline 164.8 [SE=13.3] bpm vs peak digoxin 138.9 [SE=14.1] bpm; $P=0.022$ and average baseline 98.6 [SE=5.1] bpm vs average digoxin 82.6 [SE=5.4] bpm; $P<0.001$). After treatment with digoxin the average hourly HRs of IWs with AF were not significantly different than those of normal dogs (AF digoxin 82.6 [SE=5.4] bpm vs NSR 72.5 [SE=5.3] bpm).

These results indicate that IWs with lone AF have significantly greater average hourly HRs than IWs with NSR and that these HR differences are resolved by administration digoxin. This study does not evaluate whether administration of digoxin alone will have the same HR lowering effect in IWs with AF when dilated cardiomyopathy and congestive heart failure are present.

**ABSTRACT #227**

PREVALENCE OF THE MYOSIN-BINDING PROTEIN C MUTATION IN APPARENTLY HEALTHY CATS OF DIVERSE LINEAGE. CF Paige[1], JA Abbott[1], KM Meurs[2]. 1. Virginia-Maryland Regional College of Veterinary Medicine, Virginia Tech, Blacksburg, VA. 2. College of Veterinary Medicine, Washington State University, Pullman, WA.

Hypertrophic cardiomyopathy (HCM) in Maine coon cats is associated with a mutation of the gene that codes for myosin-binding protein C; we sought to determine the prevalence of this mutation (MYBPC-MCC) in apparently healthy cats of diverse lineage.

Study subjects were identified during a community-based echocardiographic survey of apparently healthy cats. After DNA was extracted from buccal swabs, the region of the MYBPC gene containing the mutation was amplified by PCR and then sequenced. Based on identification of the constituent base pairs of codon 31 in exon 3, samples were classified as homozygous, heterozygous, or free of MYBPC-MCC.

One hundred three cats were echocardiographically examined. Within 10 months of echocardiographic examination, buccal swabs were obtained from 71 of these cats but it was not possible to isolate DNA from one of the samples. Of 70 cats for which results of DNA analysis were available, 66 (94%) were mix-breed, 2 (3%) were Siamese and 2 (3%) were Himalayan. Forty-two (60%) cats were neutered or sexually intact males. MYBPC-MCC was identified in DNA samples from 2 male and 1 female (4.3%; 95% Confidence Interval (CI): [0.9, 12.0]) young (< 5 years) mix-breed cats, all of which were heterozygous for the mutation. Two cases were domestic short-hair cats; one male was a domestic long-hair. The genetically affected cats were echocardiographically normal.

The MYBPC-MCC mutation is present not only in the Maine coon cat population but also in mix-breed cats. The prevalence of the mutation in apparently healthy cats is close to 4%.

**ABSTRACT #228**

THE USE OF STRAIN, STRAIN RATE, AND DISPLACEMENT BY 2D SPECKLE TRACKING FOR ASSESSMENT OF SYSTOLIC LEFT VENTRICULAR FUNCTION IN GOATS: APPLICABILITY AND INFLUENCE OF GENERAL ANESTHESIA. AJ Berli, R Jud, K Steininger, CC Schwarzwald. Vetsuisse Faculty, University of Zurich, Zurich, Switzerland.

Although cardiovascular disease is not commonly encountered in goats, this species often serves as an animal model for human cardiovascular disease. The goal of this study was to investigate the applicability of 2D speckle tracking (2DST) for assessment of segmental left ventricular (LV) function in goats, describe the techniques, and determine the influence of general anesthesia.

22 healthy, female Saanen-goats (age 3.7 ± 1.1 y, weight 59.8 ± 10.5 kg [mean ± SD]) were studied. The goats underwent a complete echocardiographic examination in an unsedated state in the standing position (control). All goats then underwent a second echocardiographic examination under general anesthesia and in sternal recumbency, 7.4 ± 3.4 days after the first examination (treatment). Echocardiography was performed using a GE Vivid 7 echocardiograph with a M4S probe (1.9/4.0 MHz). The LV was imaged in two-dimensional gray scale mode from a right-parasternal window using a long-axis, 4-chamber view (LAX) and a short-axis view at the level of the papillary muscles (SAX-PM). The recording frame rate ranged between 70.2 and 103.8 frames/second. Analyses were performed off-line (GE EchoPAC v6.1.2) and blinded to treatment. Measurements included longitudinal, circumferential, and radial peak systolic strain (SL, SC, SR), peak systolic strain rate (SrL-S, SrC-S, SrR-S) as well as longitudinal and radial peak systolic displacement (DL, DR). All measurements were reported by the software as average over each of 6 wall segments. Two-way RM ANOVA with a Holm-Sidak post-hoc test was used to assess the effects of segment and treatment on the variables. The level of significance was $p = 0.05$.

2DST analyses were feasible in 256 of 264 recordings (97%). In LAX view, SL and SrL-S decreased gradually and DL increased gradually from the apical to the basal segments. SL was unaffected by treatment, while SrL-S was significantly reduced during anesthesia. DL was significantly higher during anesthesia in the lateral, but not in the septal segments. In SAX-PM view, SR was higher in the LV free wall segments compared to the septal segments, but it was unaffected by treatment. The interacting effects of segment and treatment on SC, SrC-S, and SrR-S did not show any consistent, identifiable pattern. General anesthesia increased DR in the LV free wall segments.

We conclude that 2DST can be used to characterize segmental LV wall motion in awake and anesthetized goats. Anesthesia influences some of the 2DST variables, probably related to alterations in cardiac contractility and loading conditions. Assessment of LV diastolic function by 2DST, derivation of indices of global function, assessment of LV synchrony using time intervals, and evaluation of the clinical and experimental value of 2DST will require further investigations.

**ABSTRACT #229**

BAROMETRIC WHOLE-BODY PLETHYSMOGRAPHY REFERENCE VALUES IN DIFFERENT AGE-RANGE CATS WITH NATURAL BRONCHIAL DISEASE. L Garcia-Guasch[1], A Caro-Vadillo[2], M Laporta[1], J Manubens[1], JA Montoya-Alonso[3] [1] H.V.Molins, Barcelona, Spain. [2] Med. Cir. Anim., UCM, Madrid, Spain. [3] Int. Med., ULPGC, Las Palmas, Spain.

Feline bronchial disease (FBD) is characterized by inflammation of the lower airways without an obvious identifiable cause. Unfortunately there are no pathognomonic clinical signs or laboratory tests available in routine veterinary clinical practice for FBD. Barometric whole-body plethysmography (BWBP) is a noninvasive pulmonary function test (PFT) that allows a dynamic study of breathing patterns by placing the patient within an unrestrained Plexiglas chamber. BWBP is useful to study potential airway disease and the response to different treatments. The objectives of this preliminary study are (1) to establish a reference database for respiratory variables in cats with natural FBD using BWBP, (2) to evaluate if there are significant differences between healthy cats and cats with natural illness and (3) to evaluate if there are significant differences between healthy and affected cats in different age-range. The study was approved by the animal ethical committee of the University of Las Palmas (Spain). Twenty-three cats with natural FBD were included. Cats did not have a previous history of upper airway, cardiac or systemic diseases and had negative results when tested for FeLV/FIV diseases. Cats were arranged in three groups: young cats (n=5, from 0 to 2 years old), adult cats (n=11, from 2 to 8 years old), and old cats (n=7, more than 8 years old). Cats were placed in the chamber and after an adaptation period of time in a quiet and silent environment, four 3-minute periods were registered and data (tidal volume [TV], enhanced pause [Penh] and Pause) were shown as means with standard deviations. A $p$-value <0.05 was considered statistically significant.

| Age-range | TV (mL) | | Penh | | Pause | |
|---|---|---|---|---|---|---|
| | Healthy N=26 | FBD N=23 | Healthy N=26 | FBD N=23 | Healthy N=26 | FBD N=23 |
| 0-2 | 29,52±11,74 | 43,25±16,63 | 0,54±0,15 | 0,81±0,28 | 0,66±0,12 | 0,83±0,23 |
| 2-8 | 35,15±12,24 | 60,65±40,11 | 0,46±0,17 | 1,74±1,59* | 0,57±0,18 | 1,22±0,59* |
| >8 | 34,75±5,05 | 29,80±9,59 | 0,44±0,11 | 1,24±0,71* | 0,57±0,07 | 0,17±0,47* |
| *P<0,05 | P=0,037 | | P=0,02 | | P=0,001 | |

In conclusion: (1) measured results are firstly useful as baseline values for PFT by BWBP in cats with FBD and suggest that there are no significant differences in TV (p=0,120), Penh (p=0,349) and Pause (p=0,363) between different age-range groups of FBD cats, (2) there are significant differences for TV, Penh and Pause variables between healthy cats and cats with FBD (compared with reference values in healthy cats previously reported by authors), and (3) there are significant differences for Penh and for Pause between healthy cats and cats with FBD in adult ($p_{Penh}$=0,024 ; $p_{Pause}$=0,006) and old ($p_{Penh}$=0,05 ; $p_{Pause}$=0,03) cats.

**ABSTRACT #230**

CLINICAL APPLICATION OF BAROMETRIC WHOLE-BODY PLETHYSMOGRAPHY AND BRONCHOPROVOCATION TEST IN FELINE BRONCHIAL DISEASE. YA Jhuo, HP Huang. Department of Veterinary Medicine, National Taiwan University, Taipei, Taiwan.

The aim of this study was to evaluate the application of barometric whole-body plethysmography (BWBP) and bronchoprovocation tests in diagnosis of feline bronchial disease clinically.

Thirty nine client-owned cats were enrolled, 21 were clinically healthy and the other 18 were affected with bronchial disease. Diagnosis of feline bronchial disease was based on clinical manifestations (Feline Bronchial Disease Activity Scores), results of physical examinations, routine blood examinations, thoracic radiographs, electrocardiography and echocardiography in each cat. Baseline parameters of BWBP (respiratory rate, tidal volume, minute volume against body weight, peak inspiratory and expiratory flows, inspiratory and expiratory time, and relaxation time), and enhanced pause (Penh) were recorded over a 5-minute period. Bronchoprovocation tests were carried out using a jet nebulizer containing carbachol solutions at concentrations of 0.005, 0.01, 0.02, 0.04, 0.06, 0.08, 0.1, 0.2, and 0.5%. Saline (0.9% NaCl) and increased concentrations of carbachol solutions were nebulized for 1 minute and followed by 5 minutes recording until the carbachol concentration induced bronchoconstriction (Penh exceeded 300% of baseline, C-Penh300). Clinically healthy (n = 6) and FBD-affected cats (n = 4) were then further selected to determine the repeatability of BWBP on two independent occasions. Repeating tests were performed following the same procedure as described above at an interval of 2 weeks.

Baseline BWBP parameters between clinically healthy cats and cats with bronchial disease were not significantly different. C-Penh300 was significantly higher in clinically healthy cats (0.352 ± 0.196%) comparing to cats with bronchial disease (0.063 ± 0.044%, $P$ < 0.0001). No correlation was found between the baseline BWBP variables and C-Penh300 in cats with bronchial disease. In the test of repeatability of BWBP, the concentrations inducing a 300% increase of Penh were similar (mean C-Penh300 Test 1: 0.326 ± 0.225%; mean C-Penh300 Test 2: 0.258 ± 0.212%; $P$ = 0.50). Results of two repeating tests were positively correlated ($r$ = 0.76, $P$ = 0.00105).

These findings suggested that the results of BWBP assessment could be repeatable, and bronchoprovocation test could be applied as a diagnostic protocol for feline bronchial disease.

**ABSTRACT #231**

THE QUALITY OF CANINE BRONCHOALVEOLAR LAVAGE FLUID CYTOLOGY IS PRESERVED FOR UP TO 72 HOURS WITH OR WITHOUT AUTOLOGOUS SERUM. MH Patrick, T Wills, RK Sellon. Washington State University College of Veterinary Medicine, Pullman, WA.

Cytological examination of bronchoalveolar lavage (BAL) fluid is the gold standard for diagnosing canine lower respiratory tract disease. BAL fluid is typically evaluated almost immediately to limit changes in cytological quality due to low protein medium, cellular release of enzymes, or delayed cell preparation. The preservation of BAL sample quality over time has not been previously evaluated. We hypothesized that addition of autologous serum to BAL samples immediately after collection would preserve cytological quality for at least 72 hours. Serum was collected from each of 14 healthy dogs prior to BAL. BAL samples were divided into 13 aliquots of 250 µL; one aliquot was analyzed immediately. Cells were counted via hemocytometer and cytological quality assessed by scoring the degree of nuclear swelling and frequency of broken cells on cytospin preparations. Autologous serum (50 µl; 5%) was added to 6/12 aliquots, and all aliquots were stored at 4°C for 4, 24, 48, and 72 hours before analysis; the clinical pathologist was blinded to both serum treatment and storage time. Data were analyzed via ANOVA and paired t-test. There were no differences in cell count or sample quality at any of the time points between aliquots stored with or without serum. At 72 hours, the difference in the cell morphology between samples with serum versus those without serum approached significance. BAL samples remained of good diagnostic quality up to 48 hours. Adding serum may be useful for samples for which analysis will be delayed for more than 3 days.

**ABSTRACT #232**

CLINICAL AND LABORATORY CHARACTERISTICS OF 35 DOGS WITH PULMONARY THROMBOEMBOLISM. (2002-2008). SA Pumphrey, EA Rozanski, LM Freeman, SM Cunningham, JE Rush. Cummings School of Veterinary Medicine at Tufts University, North Grafton, MA.

Pulmonary thromboembolism (PTE) can be a life-threatening complication of critical illness. The purpose of this study was to evaluate clinical signs, underlying disease states, coagulation testing results, therapies, and outcome in dogs with confirmed PTE. The medical records and necropsy databases were searched for dogs with confirmed PTE between 2002 and 2008. Confirmation of PTE was made via necropsy, computed tomography (CT scan) with non-selective angiography, selective angiography, or visualization of a pulmonary thrombus on echocardiography (Echo). Dogs with suspected but not proven PTE were excluded as were dogs with PTE in association with heartworm infection. Information recorded from the medical record included history, presenting complaint, presence of key clinical signs, antemortem suspicion of PTE, therapies directed against PTE, co-morbidity, coagulation testing, and outcome.

Thirty-five dogs with confirmed PTE were identified. Twenty-six were confirmed at necropsy, while five had thrombi visualized on Echo, and 4 had PTE identified on CT scan. Thirty dogs presented with respiratory distress and 13 had cough, including 3 with hemoptysis. Seven dogs had right-sided congestive heart failure, 1 dog presented for collapse, and 1 experienced sudden death. Twenty dogs were treated with supplemental oxygen, including 4 who were mechanically ventilated. In 62% of cases, PTE was suspected antemortem. 12 of these dogs received anticoagulant or thrombolytic therapy, consisting of various combinations of unfractionated heparin, low molecular weight heparin, coumadin, clopidogrel, urokinase, tissue plasminogen activator and streptokinase. A variety of co-morbidities were identified, most commonly neoplasia, corticosteroid use, and protein-losing nephropathy. Prothrombin time (n=17) was elevated beyond the reference range in 8 dogs, and aPTT (n=19) was elevated in 7. Platelet count (n=27) was decreased in 15. D-dimer concentrations were measured in 13 dogs; 3 dogs had negative D-dimer concentration (< 250 ng/ml). Five dogs had D-dimer levels of 250-500 ng/ml, 2 dogs had concentrations 500-1000 ng/ml, 2 dogs had concentrations 1000-2000 ng/ml and 1 dog had a concentration > 2000 ng/ml. 30 dogs died or were euthanized, while 5 dogs were discharged. PTE was suspected antemortem in a higher percentage of dogs in this study than in previous reports. However, only 12 of 21 dogs with suspected PTE received anticoagulant or thrombolytic interventions. Additionally, in this population of dogs with confirmed PTE, D-dimers were not reliably elevated, with only 3/13 dogs having D-dimers ≥ 1000 ng/ml.

The clinical outcome associated with PTE remains poor. Future directions may include improving clinical suspicion for the disease in at-risk dogs, developing alternative means to aid antemortem diagnosis, and facilitating timely use of anticoagulant therapies.

**ABSTRACT #233**

COMPARATIVE STUDY OF THE TOLERANCE AND EFFICIENCY OF FOUR DIFFERENT TECHNIQUES FOR NEBULISATION DELIVERY IN CATS. Talavera J, Tudela A, Escobar M, Bayón A, Fernández del Palacio MJ. Departamento de Medicina y Cirugía Animal. 30100 Espinardo (Murcia) – Spain.

In humans, nebulisation/aerosolization is a well-established route to drug delivery to the respiratory tract and to provide airway humidification in patients with respiratory diseases. Protocols for nebulisation generation and delivery are well standardized. Because the inherent lack of cooperation of animals, most of these human protocols are not applicable to veterinary medicine. Successful therapy with nebulisation may depend more on adequate drug delivery than drug efficacy. The aim of this study was to compare the tolerance and efficiency of four different techniques for nebulisation delivery in cats.

Nine adult healthy cats were used. Tolerance to each nebulisation technique was subjectively evaluated. For objective evaluation of the efficiency of the nebulisation delivery, a cough challenge test was used. Capsaicine was nebulised (ultrasonic nebuliser) following a dose-response protocol with increasing concentrations (3, 10, 30, 60, 90, 100, 300 and 500 μM) during 1 minute each and lasting 1 minute between concentrations. When more than 5 coughs were induced the response was considered positive. The test was always performed with the cats lightly sedated (acepromacine, 0.06 mg/kg+buprenorfine, 0.01mg/kg, IM). In different days, the cough challenge was repeated in all cats once with each of the following techniques for nebulisation delivery: (1) direct approximation of the nebulisation outlet to the mouth/nose of the cat (flow-by method, FB); (2) using a face mask (FM); (3) using a plastic-covered Elizabethan collar (EC); and (4) using a plexiglas chamber (18.5 l) (PC). The number of positive responses (NPR) and the concentration of capsaicine that induced positive response (CCP) with the four techniques were statistically compared (Chi-square and Mann-Whitney test, respectively; $P<0.05$).

The PC technique was the best tolerated, although the FB and FM techniques showed also acceptable tolerance. However, the EC technique resulted very stressful and the test must be suspended in 4/9 cats. The technique that induced positive responses at lower CCP was the FM (13.8±6.0 μM), followed by the FB technique (20.0±9.3 μM), the PC technique (23.8±10.2 μM) and the EC technique (40.0±9.4 μM), although the differences only were statistically significant when compared FM and EC techniques ($P=0.02$). The FM technique induced positive responses in greater number of cats (NPR=8), followed by the FB technique (NPR=7), the PC technique (NPR=5) and the EC technique (NPR=3). Again, the differences only were statistically significant when compared FM and EC technique ($P=0.016$).

In conclusion, FM is the most efficient technique for nebulisation delivery in cats, offering good tolerance and the best quantitative results. The FB nebulisation is less efficient but it may become practical if the FM is not tolerated. The PC technique offers a minor efficiency but it may be useful in very stressful or aggressive cats. The EC technique presents low efficiency and tolerance.

**ABSTRACT #234**

UTILITY OF TEPOXALIN, A DUAL LIPOXYGENASE AND CYCLOOXYGENASE INHIBITOR, IN CANINE CHRONIC AND/OR EOSINOPHILIC BRONCHITIS. <u>CR Reinero</u>, CH Chang, AE DeClue; University of Missouri, College of Veterinary Medicine, Columbia, MO.

Eicosanoids like leukotriene B4 (LTB4), prostaglandin E2 (PGE2) and thromboxane B2 (TxB2) are inflammatory mediators in human chronic bronchitis. We hypothesized that canine chronic bronchitis is caused at least in part by products of the lipooxygenase and cyclooxygenase-2 pathways, and by selective inhibition with tepoxalin (Zubrin®), a parallel decrease in the number/percentage of airway inflammatory cells and concentrations of LTB4, PGE2 and TxB2 would be observed.

Twelve dogs with naturally developing chronic/eosinophilic bronchitis were enrolled. For inclusion, dogs needed baseline thoracic radiographs, bronchoalveolar lavage fluid (BALF) with >370 cells/μl, >5% neutrophils and/or >5% eosinophils, negative BALF culture and owner consent. Tepoxalin was administered at 20 mg/kg PO once followed by 10mg/kg daily for 2 weeks. At 2 weeks, thoracic radiographs and BALF collection were repeated. Concentrations of eicosanoid products were measured in BALF using commercial ELISAs. Data were analyzed using a paired t-test or a signed rank test with $p<0.05$ considered significant.

At baseline, mean±SD BALF total cell number, % neutrophils and % eosinophils were 992±652 cells/μl, 30±18% and 33±32%; no significant difference was noted after Tepoxalin (1190±1233 cells/μl, 40±28% and 36±26%, respectively). BALF LTB4 was significantly greater [median (q1, q3); pre 0.84 (0.3, 1.2); post 1.2 (0.8,2.2); p=0.02] post-tepoxalin treatment but there was no significant difference in mean±SD PGE2 (pre 1.17±0.38 ng/ml; post 2.40±1.00 ng/ml; p=0.22) or TxB2 (pre 0.05±0.02 ng/ml; post 0.07±0.03 ng/ml; p=0.54).

In conclusion, in dogs with chronic and/or eosinophilic bronchitis Tepoxalin does not significantly diminish airway inflammation or reduce concentrations of local airway eicosanoid products.

**ABSTRACT #235**

TREATMENT OF TRACHEAL COLLAPSE WITH VET-STENTS® IN 36 DOGS: COMPLICATIONS AND LONG-TERM RESULTS. <u>E De Madron</u>. Alta Vista Animal Hospital, Ottawa, ON, Canada.

Dyspnea due to severe tracheal collapse in dogs can be alleviated by the insertion of intra-tracheal stents specifically designed for dogs: the Vet-Stent® . The purpose of this retrospective study is to review the long-term results and complications of this approach.

Thirty six dogs with severe tracheal collapse (grade 3 or 4) and dyspneic symptoms underwent a Vet-Stent® placement. The entire trachea was collapsing (94% of dogs), as well as the bronchi (81%) . Duration of follow up was 2 to 822 days. Immediate post-operative complications included: ARDS (14%), death due to ARDS (11%), and pneumonia (17%). Cough was present in 69% of case weeks to months after stent placement. This cough proved intractable in 17% of dogs. The stent migrated in 8% of dogs due to improper sizing. Steroid responsive granulomas occurred in 6% of dogs. Stent rupture occured in 6% of dogs. A second stent had to be placed in 17% of dogs, and a third one in 6%.

Results were deemed very satisfactory in 25% of cases, acceptable but with lingering cough requiring medical treatment in 53% of dogs, poor due to severe persistent cough in 5% of dogs.

In conclusion, treatment of severe tracheal collapse with Vet-Stent® yields improvement in 88% of dogs. Medical control of cough is often necessary post stent placement. Proper measurement of the stent to place is primordial.

**ABSTRACT #236**

SUBACUTE ENDOTOXEMIA TEMPORARILY IMPAIRS INSULIN SENSITIVITY AND B-CELL FUNCTION IN CATS. <u>M Osto</u>[1], E Zini[2], M Franchini[3], M Ackermann[3], CE Reusch[2], TA Lutz[1]. 1. Institute of Veterinary Physiology; 2. Clinic for Small Animal Internal Medicine; 3. Institute of Virology; Vetsuisse Faculty, University of Zurich, Switzerland.

Systemic low-grade inflammation is a pathogenic component in chronic disease such as obesity and type 2 diabetes mellitus (T2DM). In humans and cats, chronic inflammatory conditions are associated with elevated proinflammatory cytokines. Activation of inflammatory signaling pathways are supposed to be linked to insulin resistance and to impairment of pancreatic β-cell function; the two most important features of T2DM. Because feline diabetes closely resembles human T2DM, we hypothesized that low-grade induced inflammation impairs insulin sensitivity and β-cell function in cats.

To test this hypothesis, 10 healthy cats were infused via the jugular vein with increasing doses (10, 200, 500 and 1000 ng/kg/h) of lipopolysaccharide (LPS; n=5) or saline (n=5) for 10 days. Body temperature, plasma glucose, insulin and $\alpha_1$-acid glycoprotein levels were assessed each day during the infusion period. Before and at the end of the infusion, intravenous glucose tolerance tests (ivGTT) were performed. On day 10, tissue specimens were collected. To quantify neutrophils, tissue sections were immunostained with myeloperoxidase. Statistical differences between groups were determined with non parametric tests.

Infusion of LPS was well tolerated. Body temperature and glucose levels increased during the first five days of LPS infusion and then progressively decreased to baseline levels. Insulin was not different from controls at any time point. Based on the homeostasis model assessment, β-cell function and insulin sensitivity were significantly decreased in the LPS group on day 1 and on day 2 and 3, respectively. From day 1 to 10, plasma levels of $\alpha_1$-acid glycoprotein were significantly increased in the LPS-infused cats as compared to controls. Based on the ivGTT at the end of the infusion, the insulin secretion pattern and insulin sensitivity were not different between groups. Compared to controls, on day 10 LPS-treated cats had increased number of neutrophils in the liver.

LPS induced an inflammatory response that lasted throughout the infusion. This was accompanied by reduced insulin sensitivity and β-cell function that were short-lasting. Because body temperature and glucose levels returned to normal, desensitization to LPS may have occurred. The underlying mechanisms are currently explored.

**ABSTRACT #237**

COMPARISON OF WIRELESS AND CONVENTIONAL CONTINUOUS GLUCOSE MONITORING SYSTEMS (CGMS) IN 4 DOGS.

J.Rieder[1], J.Seipel[1], B.Vaske[2], K.Rohn[3], D.Simon[1], I.Nolte[1]; [1]Small Animal Clinic and [3]Institute for Biometry and Epidemiology, Hanover School of Veterinary Medicine, Germany, [2]Institute for Biometry, Medical School Hanover, Germany.

Diabetes mellitus is a common endocrine disorder in dogs and its therapeutic approach is still a challenge for veterinarians. Conventional and wireless Continuous Glucose Monitoring Systems (CGMS) are useful tools for an optimal insulin treatment in dogs.

The aim of this study was the comparison of the conventional and the wireless CGMS concerning their diagnostic value.

Three diabetic dogs and one healthy dog underwent simultaneous continuous glucose monitoring with a conventional and, concurrently, a wireless device. The measurement of one diabetic dog was continued for 5 days. Blood glucose values were determined at least three times a day and were used for the calibration for the CGMS. A comparison of both devices was performed using the Wilcoxon test, differences of CGMS and blood glucose values were analyzed by the Friedmann-Test. p-values $<0,05$ were considered to be significant.

Diabetic dogs were represented by a Labrador Retriever, a Cocker Spaniel and a Dalmatian. The healthy dog was a Beagle. The glucose values measured by the conventional CGMS were significantly different compared to the wireless CGMS. Glucose values measured by the conventional CGMS were higher in comparison to the wireless type. Furthermore, significantly lower numbers of values were recorded by the wireless CGMS if compared to the conventional device. This is probably due to an increased rate of discontinuations during the measurement of the wireless device. There were no significant differences between blood glucose values and glucose values measured by the conventional and the wireless CGMS.

Because of the increased rate of discontinuations the results suggest that to date a continuous glucose monitoring is probably more reliable if it is accomplished with the conventional CGMS. Further studies are needed to confirm this conclusion by a more detailed comparison of blood glucose values and glucose values recorded by both CGMS devices.

*Previously presented at InnLab 2009, Berlin, Germany*

**ABSTRACT #238**

EXTENDED USE OF PROTAMINE ZINC RECOMBINANT INSULIN FOR TREATING DIABETES MELLITUS IN CATS.

RW Nelson[1], K Henley[2], S. Peterson[2] 1. School of Veterinary Medicine, University of California, Davis, CA. 2. Boehringer Ingelheim Vetmedica, Inc., St. Joseph, MO.

The study evaluated the long-term safety and efficacy of protamine zinc recombinant insulin (PZIR) on control of glycemia in diabetic cats. 145 diabetic cats that completed a 45 day field efficacy trial were enrolled for an additional 136 day period. All cats were treated with PZIR twice daily. Control of glycemia was assessed by physical examination, body weight and the investigator on Days 0 (entry into the extended study), 34, 68, 102, and 136, by serum fructosamine on Days 0, 68 and 136, and by owner assessment of clinical response weekly. Adjustments in PZIR dosage were made as needed to maintain control of glycemia.

113 cats completed the study. There was a significant ($p<0.0001$) decrease in serum fructosamine concentration and significant ($p<0.0001$) increase in body weight on Day 136, compared with Day 0. 95.6% of cats were in good clinical condition according to investigators and 95% of cats had the same or decreased water consumption and urination and 98% had the same or improved attitude according to owners at Day 136. Clinical signs of hypoglycemia were identified in 22 cats and resolved with appropriate therapy and insulin dosage adjustments.

Results of this study demonstrate that PZIR is safe and effective for control of glycemia in diabetic cats for at least 6 months.

**ABSTRACT #239**

URINARY MARKERS AND PLASMA IOHEXOL CLEARANCE IN HEALTHY DOGS AND DOGS WITH PITUITARY-DEPENDENT HYPERADRENOCORTICISM.

P.M.Y. Smets[1], E. Meyer[1], B.E.J. Maddens[1], S. Croubels[1], H.P. Lefebvre[2] and S. Daminet[1]. 1. Ghent University, Ghent, Belgium. 2. Ecole Nationale Vétérinaire, Toulouse, France.

Proteinuria and systemic hypertension have been associated with progression of chronic renal disease in dogs. Forty-six % of dogs with untreated pituitary-dependent hyperadrenocorticism (PDH) have proteinuria and 50 to 80 % have hypertension. Moreover, an increase in glomerular filtration rate (GFR) has been described in humans and several animal species receiving exogenous glucocorticoids. Still, the effect of PDH on renal function has not been documented.

Therefore, the aim of this study was to compare a selected set of glomerular and tubular markers, i.e. urinary protein to creatinine ratio (UPC), urinary albumin (uALB) and GFR, and urinary retinol binding protein (uRBP) and N-acetyl-β-D-glucosaminidase (uNAG), respectively, in healthy (H) dogs and in dogs with PDH.

Seven newly diagnosed PDH patients (age 6.7-11.6 years, bodyweight 20-42.8 kg) and eight H age-matched dogs (age 7.2-10.9 years, bodyweight 21.2-39 kg) were included. Urinalysis, bacterial culture and UPC were performed in all dogs. uALB was determined using a canine-specific ELISA, uRBP using a human RBP ELISA and uNAG activity using a colorimetric assay after validation for use in canine urine. Results were related to urinary creatinine (c) and expressed as ratios. In all dogs, GFR was calculated by means of plasma exo- ($Cl_{exo}$) and endo-iohexol clearance ($Cl_{endo}$) and expressed as ml/min/kg (WinNonlin® non-compartmental analysis).

In the H and the PDH dogs, the mean UPC ± standard deviation (SD) was 0.1 ± 0.05 and 2.1 ± 1.3, the uALB/c 33.4 ± 50.7 mg/g and 2385.8 ± 3236.7 mg/g, the $Cl_{exo}$1.8 ± 0.2 and

2.6 ± 0.7 and the $Cl_{endo}$ 2.2 ± 0.3 and 2.8 ± 1.1, the uRBP/c 18.6 ± 26.7 µg/g and 128.4 ± 93.3 µg/g, the uNAG/c 2.3 ± 1.3 U/g and 8.9 ± 7.7 U/g, , respectively. Based on a two sample t-test the UPC (p=0.005), uRBP/c (p=0.012), uNAG/c (p=0.047), $Cl_{exo}$ (p=0.015) and $Cl_{endo}$ (p=0.004) were all significantly higher in the PDH than in the H dogs. No significant difference was found for uALB/c, although increased ratios were observed in all PDH dogs.

In conclusion, results of this pilot study suggest the presence of glomerular and tubular renal dysfunction in PDH compared to H dogs, indicated by increases in UPC and GFR, and in uRBP/c and uNAG/c, respectively. Further research in a larger number of dogs is warranted to confirm these novel findings in order to gain a better insight into the effect of PDH on renal function.

**ABSTRACT #240**

EFFECT OF PHYTOESTROGENS AND MELATONIN ON ADRENAL STEROIDOGENESIS USING HUMAN ADRENAL CELLS AS A MODEL FOR CANINE HYPERADRENOCORTICISM.

KA Fecteau, JW Oliver, H Eiler. Department of Comparative Medicine, College of Veterinary Medicine, The University of Tennessee, Knoxville, TN.

Excessive secretion of adrenal sex hormones has been associated with clinical signs usually caused by cortisol in canine hyperadrenocorticism. Phytoestrogens and melatonin are effective in controlling hyperadrenocorticism in some dogs. The objectives of this study were to investigate: (1) the effectiveness of phytoestrogens, alone and in combination with melatonin, in blocking adrenal steroidogenesis at the cellular level and; (2) the effect phytoestrogens have on expression of aromatase which converts androgen to estrogen. Since human adrenal carcinoma cells but not canine adrenal cells are commercially available, human cells were used as a model for dog.

Cells were treated with 100 µM: cyclic AMP (cAMP), cAMP plus phytoestrogen enterolactone (ENL), cAMP plus phytoestrogen enterodiol (END), ENL plus melatonin, and END plus melatonin for 24 and 48 hours. Cell culture media were analyzed for estradiol, progesterone, 17-hydroxyprogesterone, androstenedione, aldosterone, and cortisol using radioimmunoassays. Treated-cell lysates were used for western blot immunoassay investigation of aromatase expression.

Results revealed a significant ($P < 0.05$) decrease in cAMP-stimulated estradiol (207 vs 430 pg/ml), androstenedione (29.5 vs 112.6 ng/ml), and cortisol (15.5 vs 57.3 ng/ml) concentrations when ENL was combined with melatonin and a decrease in estradiol (143.6 pg/ml), androstenedione (35.9 ng/ml), and cortisol (16.7 ng/ml) when END was combined with melatonin. Hormones were not significantly decreased when ENL or END was used alone. Aromatase expression was decreased in phytoestrogen and melatonin treated cells. Results suggest phytoestrogens plus melatonin act directly on adrenal cells lowering estradiol by decreasing aromatase as well as lowering cortisol and androstenedione.

**ABSTRACT #241**

EFFECT OF LOW DOSES OF COSYNTROPIN ON CORTISOL CONCENTRATIONS IN CLINICALLY HEALTHY CATS.

LG Martin,[1] AE DeClue,[2] EN Behrend,[1] LA Cohn,[2] DI Dismukes,[2] R Cohen,[2] HP Lee.[1] [1]Auburn University, College of Veterinary Medicine, Auburn, AL. [2]University of Missouri, College of Veterinary Medicine, Columbia, MO.

Use of low-dose ACTH stimulation testing may be important in diagnosing mild degrees of adrenocortical insufficiency and a new syndrome, relative adrenal insufficiency, which has been recognized in critically ill human and veterinary patients. In cats, the ACTH stimulation test utilizing a standard dose of cosyntropin (125 µg/cat, IV) is used for evaluation of adrenocortical function. A previous study has documented that lower doses of cosyntropin will stimulate maximal cortisol secretion in cats. However, this study used per cat dosing as opposed to per body weight dosing. The purpose of this study was to determine the lowest dose of cosyntropin on a per body weight basis that will produce maximal cortisol secretion in clinically healthy cats.

Six dose-response trials were performed in 7 clinically healthy cats (mean±SD weight 5.5 ± 1.1 kg) instrumented with vascular access ports (Norfolk Vet Products). Each cat was randomly given 5 doses (125 µg/cat, 10 µg/kg, 5 µg/kg, 2.5 µg/kg, 1 µg/kg) of cosyntropin (Cortrosyn®) and saline (control) IV with a 2-week wash out period between doses. Blood samples for determination of serum cortisol concentrations were obtained before and at 15, 30, 45, 60, 75, and 90 minutes after cosyntropin administration. Samples were centrifuged after clotting, and the serum was separated and stored at -20°C until analysis. Data were analyzed using a repeated measures ANOVA; post hoc comparisons were made using the Least Squared Means method. Significance was set at the $p \leq 0.05$ level.

Mean serum cortisol concentration increased significantly after administration of all 5 cosyntropin dosages when compared to baseline ($p < 0.0001$). After administration of 125 µg cosyntropin/cat, serum cortisol concentration peaked at 90 minutes. Mean serum cortisol concentrations peaked at 75, 60, 60, and 45 minutes following administration of 10, 5, 2.5, and 1 µg/kg cosyntropin, respectively. The peak serum cortisol concentrations subsequent to the 125 µg/cat, 10 µg/kg, and 5 µg/kg doses were not significantly different. Higher dosages of cosyntropin appeared to result in more sustained elevations in serum cortisol and a later time of peak response. No increase in cortisol concentration occurred after saline administration.

In conclusion, cosyntropin administered at a dose of 5 µg/kg IV produces maximal cortisol secretion in clinically healthy cats. Serum cortisol concentration was reliably increased in all cats after the administration of each dose of cosyntropin. These results can be used in subsequent studies to evaluate the hypothalamic-pituitary-adrenal axis in healthy and critically ill cats.

**ABSTRACT #242**

DOSE RESPONSE OF ALDOSTERONE TO ACTH STIMULATION IN CLINICALLY HEALTHY CATS. LG Martin,[1] AE DeClue,[2] EN Behrend,[1] LA Cohn,[2] DI Dismukes,[2] R Cohen,[2] HP Lee.[1] [1]Auburn University, College of Veterinary Medicine, Auburn, AL. [2]University of Missouri, College of Veterinary Medicine, Columbia, MO.

ACTH is involved in regulation of aldosterone secretion. Administration of exogenous ACTH to dogs causes a reliable increase in serum aldosterone secretion and can be used for evaluation of adrenal function. However, the aldosterone response to ACTH administration, the dose of ACTH which will stimulate maximal adrenocortical aldosterone secretion, and the time of maximal response has not been evaluated in cats. The purpose of this study was to determine the aldosterone response to 5 doses of cosyntropin in clinically healthy cats.

Six dose-response trials were performed in 7 clinically healthy cats (mean±SD weight 5.5 ± 1.1 kg) instrumented with vascular access ports (Norfolk Vet Products). Each cat was randomly given 5 doses (125 µg/cat, 10 µg/kg, 5 µg/kg, 2.5 µg/kg, 1 µg/kg) of cosyntropin (Cortrosyn®) and saline (control) IV with a 2-week wash out period between doses. Blood samples for determination of serum aldosterone concentrations were obtained before and at 15, 30, 45, 60, 75, and 90 minutes after cosyntropin administration. Samples were centrifuged after clotting, and the serum was separated and stored at -20°C until analysis. Data were analyzed using a repeated measures ANOVA; post hoc comparisons were made using the Least Squared Means method. Significance was set at the $p \leq 0.05$ level.

Mean serum aldosterone concentration increased significantly after administration of all 5 cosyntropin dosages when compared to baseline ($p < 0.0001$). After administration of 125 µg cosyntropin/cat, serum aldosterone concentration peaked at 60 minutes. Mean serum aldosterone concentrations peaked at 60, 45, 45, and 30 minutes following administration of 10, 5, 2.5, and 1 µg/kg cosyntropin, respectively. The peak serum aldosterone concentration subsequent to 125 µg/cat was similar to all other doses, except the 1 µg/kg dose. Higher dosages of cosyntropin appeared to result in more sustained elevations in serum aldosterone and a later time of peak response. No increase in aldosterone concentration occurred after saline administration.

In conclusion, 125 µg/cat and 10 µg/kg cosyntropin stimulated the highest serum aldosterone concentrations at 60 minutes post administration. Cosyntropin administered at doses of 5 and 2.5 µg/kg also produce maximal aldosterone secretion in clinically healthy cats, but time of peak concentration occurred at 45 minutes. Serum aldosterone concentration was reliably increased in all cats after the administration of each dose of cosyntropin. These results can be used in subsequent studies to evaluate the hypothalamic-pituitary-adrenal axis in healthy and adrenopathic cats, and cats with nonadrenal illnesses.

**ABSTRACT #243**

DETERMINATION OF THE OPTIMAL TIME FOR TESTING THYROID HORMONES DURING TREATMENT WITH METHIMAZOLE IN HEALTHY AND HYPERTHYROID CATS. Rutland BE[1], Nachreiner RF[2] & Kruger JM[1]; [1]College of Veterinary Medicine, [2]Diagnostic Center for Population and Animal Health (DCPAH); Michigan State University, East Lansing, MI.

Variation in the duration and magnitude of suppression of thyroid hormones, TT4 (total thyroxine), fT4 (free T4), TT3 (total triiodothyronine) and fT3 (free T3), over a 24hour period during methimazole treatment, has not been characterized in hyperthyroid cats. The purpose of this study was to determine the optimal sampling time to monitor therapeutic efficacy. Four healthy cats, in a crossover study, were treated with increasing doses of methimazole until a steady state of thyroid suppression was achieved. Additionally, thyroid fractions (including cTSH, canine thyroid stimulating hormone) were randomly and serially monitored for 24hrs post medication. Thyroid profiles from 461 hyperthyroid cats treated with methimazole were retrieved from the DCPAH database and retrospectively reviewed. Wilcoxin signed rank tests and two-way ANOVA were used to evaluate the difference between thyroid fractions in treated and untreated groups. Linear regression analysis evaluated relationships of all thyroid fractions to time post-pilling in the retrospective study. Cats were divided according to dosage (>0.5mg/kg/day), dosing interval (q24 vs q12hrs) and TT4<55nmol/L when correlations were performed.

All thyroid fractions were significantly suppressed and cTSH significantly elevated for the entire 24hours after once daily methimazole treatment in healthy cats (p-values<0.05). In hyperthyroid cats, there was no significant relationship between time post-pill and thyroid fractions, even when dosing interval, dosage and TT4<55nmol/L were taken into consideration. There was no correlation between dose rate and TT4. In conclusion, timing of blood sampling after methimazole administration does not appear to be a significant factor when assessing response to methimazole treatment.

**ABSTRACT #244**

THE EFFICACY AND SAFETY OF A NOVEL LIPOPHILIC FORMULATION METHIMAZOLE FOR THE ONCE DAILY TRANSDERMAL TREATMENT OF CATS WITH HYPERTHYROIDISM. KE Hill, MA Gieseg, DD Kingsbury, J Bridges, N Lopez-Villalobos & JP Chambers. Institute of Veterinary, Animal and Biomedical Sciences, Massey University, Palmerston North, NZ.

Hyperthyroidism is a common disease in cats over 6 years of age. There are three treatment options for hyperthyroidism: thyroidectomy, radioactive iodine, or anti-thyroid drugs such as carbimazole (Neomercazole) or methimazole (Tapazole). Most cats are notoriously hard to medicate with oral drugs, so the use of gel formulations of drugs applied to the inner surface of cats' pinnae are practical and popular. For good transdermal absorption the drug needs to be highly lipophilic and applied in the correct vehicle. Previous studies on transdermal methimazole in cats have used pleuronic lecithin organogel (PLO) as the vehicle which may not be the most suitable vehicle for a lipophilic drug such as methimazole. Our study formulated a novel vehicle, highly suitable for lipophilic drugs. The purpose of our study was to determine if the once daily transdermal administration of this novel lipophilic gel formulation of methimazole is as safe and effective as orally administered carbimazole in treating spontaneous feline hyperthyroidism.

Forty-five cats with newly diagnosed, untreated, naturally occurring hyperthyroidism were randomly assigned to receive either carbimazole (5mg PO, BID) or 10mg (0.1ml) of the lipophilic gel formulation of methimazole applied to the pinnae SID. Cats were examined on day 0 and after 1, 4, 8 and 12 weeks of treatment. The presence of clinical signs, body weight, systolic blood pressure, haematologic, serum biochemical and urine parameters, total serum thyroxine concentrations (TT4) and serum methimazole concentrations were recorded. At each visit owners were asked to complete a questionnaire on aspects of the health of their cat. Data were analysed with a linear mixed model that accounted for repeated measures on the same cat.

No significant differences between treatment groups were found at day 0. Both drugs were effective in treating hyperthyroidism as determined by a reduction in TT4, an increase in bodyweight and improvement in clinical signs. Repeated measurements of TT4, weight, blood pressure, heart rate, alkaline phosphatase, alanine aminotransferase, creatinine, urea and urine specific gravity showed no significant difference between treatment groups. The serum methimazole concentrations correlated poorly to the TT4 concentrations in both groups. No cats in the transdermal methimazole group developed pruritus or erythema of the ear.

The once daily application of a novel lipophilic formulation of transdermal methimazole applied to the pinnae is as effective as twice daily oral carbimazole in the treatment of feline hyperthyroidism. Once daily, transdermal application, compared to twice daily oral medications to cats has substantial, practical advantages to pet owners.

**ABSTRACT #245**

THE EFFECT OF A MULTI-STRAIN PROBIOTIC FORMULATION ON MARKERS OF GASTROINTESTINAL FUNCTION IN HEALTHY DOGS AND CATS. JS Suchodolski, JF Garcia, N Berghoff, N Grützner, CG Paddock, DJ Lanerie, JM Steiner. Gastrointestinal Laboratory, Texas A&M University, College Station, TX.

Probiotics are commonly recommended as an ancillary treatment for various gastrointestinal disorders. Administration of probiotics, due to their effect on intestinal microbial populations, could potentially lead to changes in concentrations of serum or fecal markers of gastrointestinal function. The aim of this study was to evaluate the effect of a commercially available multi-strain probiotic formulation (Proviable®, Nutramax Laboratories Inc.) on commonly used markers of gastrointestinal function in healthy dogs and cats.

Twelve healthy pet dogs and 12 healthy pet cats were enrolled in this study. The median age of the dogs was 3.5 years (range 1-10 years) and the median weight was 26.4 kg (range: 5.0-36.8 kg). The median age of the cats was 2.5 years (range 1-7 years) and the median weight was 4.9 kg (range 3.8-5.9 kg). Baseline serum and fecal samples were collected on day 0. A commercially available multi-strain probiotic formulation (1 capsule containing a total of $5x10^9$ cfu *Bifidobacterium bifidum, Lactobacillus acidophilus, L. plantarum, L. rhamnosus, L. delbrueckii, Streptococcus salivarius*, and *Enterococcus faecium*) was administered daily for 21 days. Additional serum samples were collected on days 21 (last day of probiotic administration) and 42 (21 days after the last dose of probiotics). Follow-up fecal samples were collected every 3-4 days until day 42. Serum concentrations of cobalamin, folate, TLI, and PLI and fecal $\alpha_1$-proteinase inhibitor concentrations were evaluated on days 0, 21, and 42. Probiotic species were detected by culture independent methods using denaturing gradient gel electrophoresis (DGGE) and genus specific 16S rDNA clone libraries. Quantitative changes in *Lactobacillus* spp., *Bifidobacterium* spp., and *Enterococcus* spp. were evaluated by quantitative real-time PCR using genus-specific 16S rDNA primers. Differences in gastrointestinal markers and bacterial groups between time points were determined using repeated measures ANOVA or Friedman's test where appropriate.

At least one of the probiotic species could be detected either by DGGE or by 16S clone libraries in 10 out of 12 dogs and in 11 out of 12 cats on day 21, but not on day 42. DGGE profiles revealed that probiotic bands appeared within 1-2 days of probiotic administration, but disappeared within a few days after discontinuation of administration. The number of *Lactobacillus* spp. (dogs and cats: $p<0.05$), *Bifidobacterium* spp. (dogs: $p=0.03$; cats: $p=0.07$), and *Enterococcus* spp. (dogs: $p=0.07$; cats: $p<0.05$) increased during probiotic administration, but returned to baseline values after administration of the probiotic was discontinued. None of the evaluated serum or fecal markers of gastrointestinal function changed significantly after treatment or discontinuation of treatment.

Probiotic species were detectable in the majority of treated dogs and cats. The results of this study would suggest that administration of this multi-strain probiotic for 21 days does not interfere with the interpretation of gastrointestinal function markers in healthy animals.

**ABSTRACT #246**

EFFECT OF *BIFIDOBACTERIUM ANIMALIS* AHC7 ON RESOLUTION OF ACUTE DIARRHEA IN THE CANINE. D. Minikhiem[1], R. Kelley[1], J. Park[1], T. Boileau[1], B. Kiely[2], L. O'Mahony[2, 3]. [1] Procter & Gamble Pet Care, Lewisburg, Ohio, USA, [2]Alimentary Health Ltd., Cork, Ireland, [3]Alimentary Pharmabiotic Centre, National University of Ireland, Cork, Ireland.

The effect of a canine-derived probiotic, *Bifidobacteria animalis AHC7,* on idiopathic diarrhea was evaluated in adult dogs. Forty-five dogs (n=24 placebo, n=21 probiotic) were recruited but only 31 (n=18 placebo, n=13 probiotic) met the inclusion criteria used for subsequent analysis. The most common reason for exclusion was an initial stool score < 4. Attending veterinarians prescribed metronidazole at their discretion. The mean age of dogs completing the trial was 1.58 years without difference between treatments. The probiotic *AHC7* ($1 \times 10^{10}$ CFU) and placebo were delivered twice daily for 14 days or until diarrhea resolved. Dogs were monitored daily for stool scores using a 4-point scale with 1 being ideal, and 4 being watery, liquid stool with little or no particulate matter. The mean time to diarrhea resolution was significantly reduced (P=0.008) for the *AHC7* group (3.9 days) compared to placebo (6.6 days). A subset analysis was performed for German Shepherd Dogs (n=10; n=6 placebo, n=4 AHC7) and for Labrador Retrievers (n=11; n=7 placebo, n=4 AHC7). In the German Shepherd Dogs, probiotic *AHC7* significantly (P=0.026) reduced the time to diarrhea resolution (3.3 days) compared to placebo (7.7 days). In Labrador Retrievers, probiotic *AHC7* also significantly (P=0.048) reduced the time to diarrhea resolution (3.3 days) compared to placebo (6.3 days). For all dogs, the percentage that were administered metronidazole was reduced in the probiotic group (38.5%) compared to placebo (50.0%). The results support that the probiotic, *B. animalis AHC7,* is an effective tool for the management of diarrhea in dogs.

**ABSTRACT #247**

EVALUATION OF INTRA-STOOL VARIABILITY OF THREE LACTIC ACID BACTERIAL GENERA IN DOGS BY QUANTITATIVE REAL-TIME PCR. JF Garcia, JM Steiner, JS Suchodolski. Gastrointestinal Laboratory, Texas A&M University, College Station, TX.

Lactic acid bacteria are commonly used in probiotic formulations as they are thought to have beneficial effects by modulating the intestinal microbiota. Quantitative real-time PCR (qPCR) allows detection and quantification of microbial genomic targets, and has been used for the quantification of bacterial groups in feces. To our knowledge, reproducibility of DNA extraction and intra-stool variation of lactic acid bacteria have not yet been described for qPCR from canine fecal samples. Therefore, the aims of this study were to evaluate the DNA extraction reproducibility and to determine intra-stool variability of three lactic acid bacterial genera assessed by qPCR in feces from dogs.

To test the reproducibility of the extraction method, DNA extraction was performed five times from four homogenized fecal samples. Additionally, feces from a total of eight dogs were used to evaluate intra-stool variability. Five sites were randomly sampled from each of the eight defecations (one from each dog). DNA was extracted from 100 mg of each sample using a phenol chloroform-based method. Separate qPCRs were performed for all samples using genus specific 16S rRNA gene primers for *Lactobacillus*, *Bifidobacterium*, and *Enterococcus* spp. Standard curves using serial 1:10 DNA dilutions (range 2 to 0.0002 ng) containing species of these three bacterial genera were used to calculate the unknown bacterial genomic targets. Samples and standards were run in duplicates using a commercial real-time PCR thermocycler (iCycler iQ, Biorad).

Twenty samples were evaluated for testing the reproducibility of the DNA extraction for each genus. Coefficients of variation (%CV) were calculated from the five extractions for each of the 3 genera: *Bifidobacterium*: median: 5.2 %CV, range: 2.7%CV to 6.2%CV; *Enterococcus*: median 7.1 %CV, range: 5.9%CV to 8.4%CV; and *Lactobacillus*: median: 4.7 %CV, range: 2.7%CV to 6.4%CV. Forty samples (eight defecations, five sites each) were analyzed to evaluate intra-stool variability for each genus. The %CV for the three genera: *Bifidobacterium*: median: 11.7 %CV, range: 2.2%CV to 32.2%CV; *Enterococcus*: median: 6.3 %CV, range: 1.8%CV to 21.1%CV; and *Lactobacillus*: median: 17.4%CV, range: 6.8%CV to 38.5%CV.

These results showed less than 10% variation among five different DNA extractions from the same fecal sample, thus showing that qPCR is a reproducible technique for the quantification of lactic acid bacteria in fecal samples from dogs. In this study we observed an intra-stool variability of up to 38.5%, which needs to be taken into account for future clinical studies.

**ABSTRACT #248**

MOLECULAR AND HISTOPATHOLOGICAL CHARACTERIZATION OF ENTEROADHERENT BACTERIA IN FAILURE TO THRIVE KITTENS. JL Nicklas[1], P Moisan[2], MR Stone[1], JL Gookin[1]. 1. North Carolina State University College of Veterinary Medicine, Raleigh, NC. 2. Rollins Animal Disease Diagnostic Laboratory, Raleigh, NC.

Diarrhea is a frequent cause of death or euthanasia of kittens, particularly those residing in rescue facilities. Bacterial culprits of diarrhea are particularly problematic to identify and it is likely that many have yet to be recognized. We sought to further characterize the molecular and histopathological features of failure-to-thrive kittens in which enteroadherent coccobacilli were observed on necropsy examination. Seven unrelated kittens aged 3-10 weeks from animal shelter or foster facilities died or were euthanized and submitted for necropsy to a state diagnostic laboratory. Clinical signs reported prior to death included: diarrhea (4), sudden death (1), upper respiratory infection (1), and unspecified (1). All kittens had histopathological examination findings of mild to moderate, subacute to acute, necrotizing enteritis with adherent colonies of coccobacilli. In each kitten, a diagnosis of attaching and effacing *E.coli* infection was presumed. Fecal bacterial culture results, performed in 4 kittens, were positive for *E.coli* in 3 cases, 2 of which were PCR positive for the attaching and effacing virulence factor. Concurrent infections (trichomonas, panleukopenia, and adenovirus) were identified in 3 kittens. Molecular characterization of the enteroadherent bacteria in each case was performed by fluorescence in-situ hybridization. Eubacterial (Eub338) and *E.coli*-specific fluorescence-labeled oligonucleotide probes were applied to formalin-fixed paraffin-embedded sections of intestinal tissue from each kitten. Only 2 of the 7 kittens demonstrated positive hybridization with Eub338 and *E.coli* probes. This was unexpected as all cases were presumed to be *E.coli* infections. Gram stains performed on fixed tissue from each kitten revealed Gram negative bacteria in the 2 cases positive for hybridization with Eub and *E.coli* probes, while the remaining 5 kittens demonstrated Gram positive infections. Hybridization conditions were optimized for lysozyme permeabilization of Gram positive bacteria and tissue sections re-probed using eubacterial (Eub338) and *Enterococcus* spp.-specific (Enc221) probes. Tissue sections from all 5 kittens with adherent Gram positive bacteria were positive for hybridization with eubacterial and *Enterococcus* spp. probes. Transmission electron microscopy of representative lesions from *E.coli* and *Enterococcus spp.* infected kittens revealed coccobacilli adherent to microvilli of intestinal epithelial cells.

Each of these kittens was histologically diagnosed as having attaching and effacing E.coli infection. In 2 of 7 kittens the adherent bacteria were molecularly identified in-situ as *E.coli.* In the majority of these cases however, adherent bacteria were identified as Gram positive Enterococci. Light microscopic and ultrastructural lesions resulting from these infections were highly similar. Enteroadherent *Enterococcus* spp. infection may be an unrecognized and important cause or co-morbid factor contributing to death or euthanasia in failure-to-thrive kittens.

**ABSTRACT #249**

ABSENCE OF A BACTERIAL ASSOCIATION IN YORKSHIRE TERRIERS WITH PROTEIN-LOSING ENTEROPATHY AND CYSTIC INTESTINAL CRYPTS. M Craven, GE Duhamel, NB Sutter, KW Simpson. College of Veterinary Medicine, Cornell University, Ithaca, NY.

Yorkshire Terriers' (YT) are predisposed (OR 4.2-10.1) to protein-losing enteropathy (PLE). Intestinal pathology of YT-PLE typically includes lymphangiectasia and mucosal lymphocytic plasmacytic infiltrates. Lesions described as "dilated intestinal crypts" (JVIM 14:298-307, 2000) or "mucoid cryptal ectasia" (JAAHA 39:187-191, 2003) consisting of cystic crypts filled with mucus and necrotic cellular debris, and occasional crypt abscessation have been reported in PLE. We sought to further describe the clinical and pathological features of YT-PLE and to explore a possible relationship between mucosal histopathology and mucosal bacteria

14 YT with PLE were identified between 1999-2008 (8M, 6F: 4 prospective, 10 retrospective). Clinical features and outcome were available for 12 dogs and intestinal biopsies for 14 (8 endoscopic, 6 surgical). Mucosal histopathology was examined by a blinded pathologist (GED) and inflammatory infiltrates, lymphangiectasia and crypt abnormalities were scored as normal-0, mild-1, moderate-2 or severe-3. Fluorescence in situ hybridization (FISH) with a eubacterial probe was used to ascertain the presence and distribution of bacteria in duodenal biopsies.

The median age and bodyweight at presentation were 96mo, and 3.1kg, respectively. Vomiting (7), diarrhea (6) and inappetance (6) were the most frequent clinical signs. Bicavity effusions were present in 5 dogs, and ascites alone in 3. Hypoalbuminemia (< 3.1g/dl) was present in all 12 dogs (median 1.6g/dl), and hypoglobulinemia (<1.9g/dl) in 7 (median 1.7g/dl). Additional biochemical abnormalities included hypocalcemia (12), hypocholesterolemia (11) hypomagnesemia (9), hypokalemia (5) and hypochloremia (5). Hematological abnormalities included mild anemia (5), thrombocytosis (8), mature neutrophilia (6), and neutrophilia with a left shift (n=3). Anti-thrombin III was low (<75%) in 4/6 dogs evaluated (mean 62%). Duodenal biopsies from all affected YT contained cystic intestinal crypts. Lymphangiectasia (median, range; 2,0-3), crypt hyperplasia (2,1-3) and villus blunting (4 dogs), were less consistent features. Mucosal infiltration of lymphocytes and plasma cells (villus 2,1-3; crypt 3,2-3) and eosinophils (1.5,1-2) was common. Empirical therapy with corticosteroids (11/12), azathioprine (2/12), antibiotics, plasma and diuretics was associated with a poor outcome. 7/12 cases died or were euthanased within 3m of diagnosis. Long-term survival occurred in 3 dogs, (36, 24, and 8m), and 2 are alive at 3m and 4m after diagnosis. FISH analysis showed no evidence of a bacterial association with crypt cysts or with mucosal inflammation.

We conclude that YT suffer a severe and often fatal form of PLE that is consistently associated with cystic intestinal crypts. The absence of a bacterial association suggests that this may be a primary morphogenetic disorder with or without a secondary environmental trigger.Further work is required to ascertain the etiopathogenesis of crypt lesions and their relationship to enteric protein loss.

**ABSTRACT #250**

SERUM fPLI AND SPEC fPL CONCENTRATIONS IN CATS WITH EXPERIMENTALLY INDUCED CHRONIC RENAL FAILURE. PG Xenoulis[1], DR Finco[2], JS Suchodolski[1], JM Steiner[1]. [1]Gastrointestinal Laboratory, Texas A&M University, College Station, TX. [2]College of Veterinary Medicine, University of Georgia, Athens, GA.

Although feline pancreatitis occurs frequently, its diagnosis can be challenging mainly because of its non-specific clinical presentation and the fact that until recently a sensitive and specific test for the diagnosis of this condition was not available. In the past decade, two immunoassays, feline trypsin-like immunoreactivity (fTLI) and feline pancreatic lipase immunoreactivity (fPLI) were developed and analytically validated as tests for pancreatic function and pathology. More recently, a new assay for the measurement of serum concentrations of feline pancreatic lipase, Spec fPL™, has been developed and validated. This assay shows the same performance characteristics as the original fPLI assay. Renal failure, a major differential diagnosis in cats with clinical evidence suggestive of pancreatitis, has previously been shown to be associated with significant increases of serum fTLI concentrations, therefore reducing the specificity of this assay for feline pancreatitis. The aim of the present study was to investigate the effect of experimentally induced chronic renal failure (CRF) on serum fPLI and Spec fPL concentrations.

Leftover serum samples (stored at -80°C) from 18 cats with CRF induced by subtotal nephrectomy for an unrelated project were used for the present study. Serum fPLI concentration was measured in samples from all 18 cats and compared with those in 41 healthy cats. Serum Spec fPL was measured in 16 cats with experimentally induced CRF. To test the stability of feline pancreatic lipase in serum samples stored at -80°C, nine unrelated samples were tested before and after 44 months of storage at -80°C.

There was no statistically significant difference of the mean serum fPLI concentrations for the 9 samples before and after 44 months of storage at -80°C ($p$=0.162). All cats with experimentally induced CRF had an increased serum creatinine concentration (>1.5 mg/dL; median: 2.9 mg/dL; range: 2.2–5.4 mg/dL) at the time of the study. Median serum fPLI concentration was not significantly different between cats with induced CRF (median: 8.6 µg/L, range: 5.4-9.9 µg/L) and healthy cats (median: 7.4 µg/L, range: 5.0-15.2 µg/L, p=0.124). All cats with induced CRF had serum fPLI and Spec fPL concentrations within their respective reference ranges. There was no significant correlation between serum creatinine concentrations and fPLI concentrations (Spearman r=0.037; p=0.884) or Spec fPL concentrations (Spearman r=0.127; p=0.639).

In conclusion, there was no significant difference in serum fPLI concentrations between cats with experimentally induced CRF and healthy cats. These results suggest that serum fPLI and Spec fPL concentrations are not affected by CRF and that serum fPLI and Spec fPL concentrations can be used for the diagnosis of pancreatitis in cats with renal failure. Further studies involving cats with naturally occurring CRF or with acute renal failure are warranted to verify these results for these groups of patients.

**ABSTRACT #251**

IMMUNOHISTOCHEMICAL CHARACTERIZATION OF CANINE PANCREATA WITH LYMPHOCYTIC INFILTRATION. BM Bostrom[1], SJ Newman[2], JS Suchodolski[1], JM Steiner[1]. 1. Gastrointestinal Laboratory, Texas A&M University, College Station, TX. 2. University of Tennessee, College of Veterinary Medicine, Knoxville, TN.

Lymphocytic infiltration of the canine pancreas is more prevalent than previously thought. However, the significance of such lymphocytic infiltration of the canine pancreas is unknown. In order to further characterize the lymphocytic infiltrate, immunohistochemistry (IHC) for CD3+ (suggesting T cell lineage) and CD79+ (suggesting B cell lineage) was performed.

Pancreatic samples were collected from dogs during necropsy performed at either the Animal Medical Center in Manhattan, NY or at the College of Veterinary Medicine at Texas A&M University. Hematoxylin and eosin stained slides of the pancreatic samples were graded using a previously published canine pancreatic histologic grading scheme (Newman et al. J Vet Diagn Invest, 2006). Fifteen fixed samples of pancreatic tissue representing 15 individual dogs were selected based on a histologic grade of 2 or higher for lymphocytic infiltration, which is defined as 10-40% of the section being affected (grade 2) and more than 40% of the section being affected (grade 3). The 15 pancreatic samples were recut and stained with monoclonal antibodies against CD3+ and CD79+ antigens using standard IHC techniques. A board certified veterinary pathologist (SJN) examined the stained slides and estimated the percentage of lymphocytes that were CD3+ or CD79+.

Lymphocytes of the sections analyzed were mostly CD3+ (median: 60%, range: <5% to 95%), but not very commonly CD79+ (median: <5%, range: 0% to 50%), suggesting mostly lymphocytes of the T cell lineage.

The predominance of T lymphocytes could indicate a cell-mediated immune response similar to that found in human patients with chronic pancreatitis. Additional characterization of the T lymphocyte population is warranted to further study the pathophysiology of pancreatic lymphocytic inflammation in the dog.

**ABSTRACT #252**

PURIFICATION AND PARTIAL CHARACTERIZATION OF CANINE S100A12. RM Heilmann, JS Suchodolski, JM Steiner. Gastrointestinal Laboratory, Texas A&M University, College Station, TX.

Canine S100A12 (cS100A12), also known as canine P6, MRP6, or calgranulin C, is a calcium-binding protein that is predominantly expressed in neutrophils and monocytes. Interaction of S100A12 with the receptor for advanced glycation end products (RAGE), a pattern recognition receptor with a central role in inflammation, has been shown to activate the NF-KB signal transduction pathway, leading to release of proinflammatory cytokines (e.g., TNF-$\alpha$). A positive feedback mechanism involving RAGE expression has been suggested to result in amplification and perpetuation of inflammation in conditions such as inflammatory bowel disease (IBD) in humans. Moreover, the concentration of S100A12 in feces has been shown to be a sensitive and specific marker for gastrointestinal inflammation in humans with IBD. To date, only human, porcine, bovine, and rabbit S100A12 has been purified, and an immunoassay for the quantification of S100A12 is only available for humans. Therefore, the aim of this study was to develop a protocol for the purification of cS100A12 from canine whole blood and to partially characterize this protein as a prelude to the development of an immunoassay for its quantification in canine serum and fecal specimens.

Leukocytes were isolated from canine whole blood by dextran sedimentation. Canine S100A12 was extracted from the cytosol fraction of the cells by three successive freeze-thaw-sonication cycles, and was further purified by ammonium sulfate precipitation, hydrophobic interaction chromatography, strong cation- and anion-exchange column chromatography, and preparative native polyacrylamide gel electrophoresis. Purified cS100A12 was partially characterized by determination of the molecular weight using automated fluorescence-based reducing gel electrophoresis (Protein 80 assay) on an Agilent Bioanalyzer 2100, estimation of isoelectric point by isoelectric focusing, estimation of specific absorbance, and N-terminal amino acid (AA) sequencing by Edman degradation.

Canine S100A12 was successfully purified from canine whole blood. Based on comparison to known molecular weight standards using the Protein 80 assay, the molecular weight of cS100A12 was estimated at 7,730. Isoelectric focusing revealed an isoelectric point of 6.0. The approximate specific absorbance of cS100A12 at 280 nm was determined to be 1.78 for a 1 mg/ml solution. The N-terminal AA sequence of the first 15 residues of cS100A12 was Thr-Lys-Leu-Glu-Asp-His-X-Glu-Gly-Ile-Val-Asp-Val-Phe-His, and showed 100% homology with the predicted protein sequence available through the canine genome project. Sequence homology for the 14 N-terminal residues of cS100A12 identified with those of feline, bovine, and human S100A12 was 78.6%.

We conclude that canine S100A12 can be successfully purified from canine whole blood using the method described. These findings will facilitate the development of an immunoassay for the quantification of cS100A12 in serum and feces from dogs.

**ABSTRACT #253**

ASSOCIATION STUDY OF SINGLE NUCLEOTIDE POLYMORPHISMS WITH COBALAMIN DEFICIENCY IN THE CHINESE SHAR PEI. N Grützner, MA Bishop, JS Suchodolski, JM Steiner. Gastrointestinal Laboratory, Texas A&M University, College Station, TX.

Based on a genome-wide scan using the canine minimal screening set-2 (MSS-2), cobalamin deficiency in the Chinese Shar Pei (Shar Pei) has recently been linked to a genomic locus in close proximity to two microsatellite markers on canine chromosome 13. The canine MSS-2 contains 327 microsatellite markers with an average marker spacing of 9 megabases (Mb), but no gaps larger than 17.1 Mb. However, these studies do not conclusively narrow down the region on chromosome 13 as the major locus for a gene or genes responsible for cobalamin deficiency in the Shar Pei. Also, not all genes that have been associated with cobalamin deficiency in humans or genes encoding for cobalamin binding proteins have been identified in the dog. The aim of this study was to scan the whole genome using single nucleotide polymorphisms (SNP) to identify genes or regions that may be associated with cobalamin deficiency in the Shar Pei.

Whole blood and serum samples were collected from a total of 42 unrelated Shar Peis. Serum cobalamin concentration (reference range: 252-908 ng/L) was measured and DNA was extracted from whole blood. A total of 49,633 SNPs were genotyped using the Affymetrix v2 Platinum canine SNP array. The analysis of genotype data was conducted by a whole genome association analysis toolset (Plink v1.05). A Bonferroni correction for multiple statistical comparisons was used to evaluate the significance of any potential association. Significance was set at a $p$-value <0.000001 ($p<1.0\times10^{-6}$).

Undetectable serum cobalamin concentrations (<150 ng/L) were observed in 14 of 42 Shar Peis (33.3%), and these Shar Peis were considered to be severely cobalamin deficient. The remaining 28 Shar Peis had serum cobalamin concentrations within the reference range. The SNP analysis revealed a total of 4 markers with a significant $p$-value of $<1.0\times10^{-6}$. These markers were located between 26,440,885-28,178,693 Mb on chromosome 13.

In conclusion, SNP analysis revealed a cluster of SNP markers on chromosome 13, which were significantly associated with cobalamin deficiency in this group of Shar Peis. Interestingly, previous results from linkage studies using the MSS-2 have pointed to the same region of chromosome 13. Thus, the findings of this study provide further evidence that this region of chromosome 13 contains the causative gene or genes for this condition.

**ABSTRACT #254**

APOPTOSIS IN THE PATHOGENESIS AND TREATMENT OF CANINE INFLAMMATORY BOWEL DISEASE. AE Jergens, L Flagg, D Moore, R Denadel, MJ Wannemuehler, College of Veterinary Medicine, Iowa State University, Ames, IA.

Idiopathic inflammatory bowel disease (IBD) is a common cause for chronic gastrointestinal disease in dogs. Abnormal cell death by apoptosis may result in the persistence of activated intestinal T cells that contribute to chronic mucosal inflammation and possibly promote impaired therapeutic responsiveness. The aim of this study was to compare rates of apoptosis in different intestinal compartments between healthy dogs and dogs with IBD, and to investigate whether lymphocyte apoptosis might predict the success of medical therapy. Fifty four dogs with IBD and 6 healthy control dogs were evaluated prospectively in the context of a controlled drug trial. Intestinal biopsies obtained by endoscopy were analyzed for their densities and distribution of apoptotic cells in the duodenum (apical villous [AV] and base [BV]) and in the colon (C). Intestinal lymphocytes expressing pro-apoptotic caspase 3 (Casp3) and anti-apoptotic B-cell leukemia/lymphoma 2 (Bcl-2) were assessed by immunohistochemistry and quantitated using light microscopy. TUNEL assays confirmed detection of fragmented DNA in tissue specimens. Correlation between Casp3/Bcl-2 expression and clinical response (CIBDAI score) to medical treatment was also determined. Results indicate that immunolabelling for Casp3 and Bcl-2 was predominantly (>90%) limited to the lamina proporia (LP). Bcl-2 positive cells were increased in the LP of IBD dogs as compared to control dogs, with the greatest numbers observed in AV and BV compartments of dogs having the greatest disease severity. Casp3 expression was greater in the AV and BV of healthy dogs versus IBD dogs. Most (17/20) moderate-to-severe IBD dogs responded to standardized medical therapy. In IBD dogs, Bcl-2 expression was greatest in a subset of dogs least responsive to drug therapy. We conclude that increased expression of Bcl-2 and decreased Casp3 expression in LP lymphocytes of dogs with IBD indicate a reduced susceptibility to apoptosis of lymphoid cells. Up-regulated Bcl-2 expression may be of value as a biomarker in predicting the response to therapy.

**ABSTRACT #255**

FEASIBILITY OF A MODIFIED USSING AIR SUCTION CHAMBER FOR THE EVALUATION OF INTESTINAL MUCOSAL FUNCTION IN ENDOSCOPIC BIOPSIES FROM DOGS AND CATS. I Ruhnke[1], JV DeBiasio[1], JS Suchodolski[1], MW Musch[2], JM Steiner[1]. [1]Gastrointestinal Laboratory, Texas A&M University, College Station, TX; [2]Department of Medicine, University of Chicago, IL.

Chronic intestinal diseases in dogs and cats, such as inflammatory bowel disease, food hypersensitivity, and intestinal lymphoma are poorly characterized, and new diagnostic modalities are needed to better understand these disorders. Ussing chambers have been used to study intestinal membrane transport in many species. The conventional circulating chamber requires collection of large tissue specimens, and thus, such experiments have rarely been performed in small animal patients with spontaneous gastrointestinal disease. A modified Ussing air suction chamber (MUAS) allows the use of small tissue pieces, which would allow the study of clinical patients with gastrointestinal diseases. The aim of this study was to evaluate the feasibility of a MUAS for evaluation of intestinal membrane function in endoscopic biopsies from dogs and cats.

Six duodenal biopsies from two dogs and three duodenal biopsies from one cat were collected during diagnostic gastroduodenoscopy and analyzed. Biopsies were transferred into the MUAS within 15 minutes of collection. To simulate physiological conditions, the water reservoir was heated to 37°C, oxygenated with 95% $O_2$ and 5% $CO_2$, and energy was provided by adding 10 mM glucose and 10 mM sorbitol. After 20 min of equilibration, the voltage applied through the tissue was clamped to zero. Voltage and the membrane-generated short circuit current (SCC) were constantly recorded. To determine absorptive function, the mucosal side of canine biopsies was exposed to 40 mM glucose, followed by 500 µM phloridizin and the feline biopsies were tested with 500 µM phloridizin. Secretory function was tested on feline biopsies with 200 µM histamine and 200 µM serotonin by adding those chemicals to the serosal side.

All biopsies showed reproducible responses to the applied stimuli. The mean ±SD conductance for canine and feline duodenal biopsies was 47.8 ±2.8 $mS \cdot cm^{-2}$ and 61.5 ±5.1 $mS \cdot cm^{-2}$, respectively. Addition of 40 mM glucose to the mucosal side of the dog tissue increased SCC by 55.0 ±36.2 $\mu A/cm^2$ in comparison to SCC before stimulation (ΔSCC). Application of 500 µM phloridizin was associated with a ΔSCC of -63.6 ±14.5 $\mu A/cm^2$ in canine biopsies and -22.0 ±8.7 $\mu A/cm^2$ in feline biopsies. Addition of 200 µM histamine caused a reproducible ΔSCC of 28.0 ±2.0 $\mu A/cm^2$ and that of 200 µM serotonin a ΔSCC of 30.6 ±1.1 $\mu A/cm^2$ in the cat duodenum.

These data demonstrate the feasibility of using a modified Ussing air suction chamber for evaluation of endoscopically collected canine and feline duodenal biopsies. Reproducible responses were observed after in-vitro exposure of biopsy specimens to glucose, phloridizin, histamine, and serotonin. Further studies to characterize tissue samples from small animal patients with spontaneous chronic intestinal disease with this technology are warranted.

## ABSTRACT #256

COMPARISON OF GASTROINTESTINAL MOTILITY IN DOGS TREATED WITH IBEROGAST™ AND ONDANSETRON USING THE SMARTPILL™. AM Bradley, P Boscan, K Dowers, KW McCord, M Marquez and DC Twedt. Colorado State University, Fort Collins, CO.

There are a number of drugs used for the treatment of gastrointestinal (GI) motility disorders in humans but few have adequately been evaluated in the dog. Iberogast (stw 5) is a combination of a number of herbs and is reported to affect GI motility in humans. Ondansetron is a 5-HT3 receptor antagonist used in dogs for antiemetic properties, but there are also a number of reports showing effects on GI motility in humans. The objective of this study was to determine the effects of Iberogast and ondansetron on GI motility using SmartPill technology. The SmartPill™ is a noninvasive wireless sensor capsule that is given orally and transmits data on gastric and small and large bowel pressures, transit time, luminal pH and temperature.

Ten healthy adult dogs were used in a cross-over placebo controlled design. All dogs received either ondansetron (0.5 mg/kg PO TID), Iberogast (1 ml PO TID) or placebo (saline 1 ml SQ TID) for three days prior to SmartPill administration and during motility recordings. On the fourth day a standard meal was fed, and the SmartPill was administered and motility data recorded until passage of the pill. There was a 7-day washout period between trials. Transit time (TT), pH, maximum pressure (MP), mean peak amplitude (MPA), contractions/minute (CPM) and motility index (MI, calculated AUC/time) for the stomach, small intestine (SI) and large intestine (LI). Data was analyzed using multivariable linear regression analysis with significance of $p<0.05$.

Total TT was significantly longer for Iberogast and ondansetron compared to placebo. Treatment with ondansetron led to significantly higher pH in each GI segment. Treatment had no significant effect on MP, MPA, CPM or MI.

Both Iberogast and ondansetron increase TT, but given the lack of individual segmental parameter differences, the mechanism of action is unclear. Iberogast and ondansetron may be beneficial in the management of disorders in dogs associated with decreased GI transit time.

## ABSTRACT #257

SUCCESSFUL TREATMENT OF BENIGN ESOPHAGEAL STRICTURES WITH BALLOON DILATION AND SUBMUCOSAL TRIAMCINOLONE INJECTION IN FIVE DOGS AND ONE CAT. B Mayer-Roenne, C Fraune, KA Ryan, FP Gaschen. From the Department of Veterinary Clinical Sciences, Louisiana State University, Baton Rouge, LA.

Development of benign esophageal strictures is an uncommon, but severe complication of general anesthesia, esophageal foreign bodies, severe prolonged vomiting and gastroesophageal reflux in small animals leading to regurgitation, weight loss and malnutrition. Successful treatment with balloon dilation has been described. Often multiple procedures are necessary for resolution of clinical signs.

Endoscopically guided submucosal triamcinolone injections at the stricture site were used in five dogs and one cat with esophageal strictures. In three dogs the injections were performed at the base of the stricture prior to balloon dilation. In one dog a small mucosal tear was noticed after injection. Subsequent dilation extended the tear both in length and depth, causing obvious bleeding. In two dogs and one cat triamcinolone was injected in the area of the stricture after balloon dilation without complications. All animals recovered well and were able to eat soft food without signs of regurgitation within twelve hours after the procedure. Three dogs had one ballooning procedure; two dogs had two procedures; the cat had four procedures with three triamcinolone injections. Two dogs continued to show regurgitation when eating regular food and were maintained on slurry. Four animals had recheck esophagoscopy revealing resolution of strictures; their signs had subsided completely.

Submucosal intralesional triamcinolone injections are a well tolerated adjunctive treatment for benign esophageal strictures in small animals. Controlled studies are necessary to assess whether this treatment improves outcome, and to evaluate the risks and benefits of performing injections prior to or after balloon dilation.

**ABSTRACT #258**

SMALL ANIMAL GASTROINTESTINAL ENDOSCOPY: A COMPUTER-AIDED INSTRUCTIONAL PROGRAM. AE Jergens, A Ginman, NS Rappa, T Leisen, A Staniger. College of Veterinary Medicine and Information Technology, Iowa State University, Ames, IA.

Gastrointestinal endoscopy is routinely performed in dogs and cats having clinical signs of chronic gastroenteritis. Endoscopy allows for direct visual inspection of mucosal surfaces of the alimentary tract and facilitates targeted biopsy for histopathologic interpretation. At present, experience with learning these advanced diagnostic techniques is sporadic and insufficient for most veterinary students to obtain prior to graduation. There is a real need for a concise, readily-accessible, and well illustrated teaching resource that can be quickly and conveniently reviewed by students prior to performing endoscopic procedures in the clinics. The aim of this project was to develop a novel, computer-aided instructional program for veterinary students which demonstrates performance of routine gastrointestinal endoscopic procedures in dogs and cats. The computer-aided instruction program (e.g., Flash, a multimedia authoring package) consisted of four individual lessons (e.g., esophagoscopy, gastroscopy, enteroscopy, colonoscopy) written for use on PC and MAC platform personal computers. Each lesson was designed using a standardized template to include clinical indications, instrumentation/patient preparation, normal endoscopic appearances, abnormal endoscopic appearances, biopsy technique with sample submission, and clinical case studies. Lessons were written for ease of learning and accompanied by digital endoscopic images, computer graphics, and short, real-time digital video-streams of actual gastrointestinal endoscopic procedures. Lesson integration with computer technology proved to be a seamless process. The completed lessons clearly and succinctly reviewed the key learning features of each endoscopic procedure. Performance assessment via WebCT is currently being developed to allow self-administration and immediate feedback by students in clinical rotations.

**ABSTRACT #259**

USEFULNESS OF CONTRAST-ENHANCED ULTRASONOGRAPHY FOR CHARACTERIZATION OF FOCAL LIVER LESIONS. K Nakamura, H Ohta, M Yamasaki, M Takiguchi. Department of Veterinary Clinical Sciences, Graduate School of Veterinary Medicine, Hokkaido University, Sapporo, Hokkaido, Japan.

In six normal beagles and twenty dogs with spontaneous focal or mutifocal hepatic lesions, contrast-enhanced ultrasonography was performed. Sonazoid, a newly developed second-generation contrast agent with the ability of real-time contrast imaging along with parenchymal imaging, was selected in this study. Appropriate protocol for the evaluation of all three phases (the arterial, portal and parenchymal phase) was established based on the results of normal beagles. The evaluation of echogenicity of hepatic nodules during the arterial, and parenchymal phase could differentiate malignant tumor from benign nodule with very high accuracy. In 17 of 18 dogs with malignant lesions, the nodule was hypoechoic to the surrounding normal liver parenchyma during the parenchymal phase. This finding was significantly ($P < 0.01$) correlated with malignancy with high sensitivity (100%) and specificity (94.4%). Moreover, the evaluation of the echogenicity during the arterial phase could differentiate nodular hyperplasia, hepatocellular carcinoma and hemangiosarcoma. Nodular hyperplasia, hepatocellular carcinoma and hemangiosarcoma showed characteristic echogenicity during the arterial phase (isoechoic, hyperechoic, and hypoechoic compared to the surrounding normal liver parenchyma, respectively). No characteristic finding was seen during the portal phase. Contrast-enhanced ultrasonography with Sonazoid appears to improve the characterization of canine focal or multifocal hepatic lesions.

**ABSTRACT #260**

HEPATIC COPPER CONCENTRATIONS IN LABRADOR RETRIEVERS WITH AND WITHOUT CHRONIC HEPATITIS (1980-2008): AN EMERGING SYNDROME OR OVERSUPPLEMENTATION? AN Johnston, SA Center, SP McDonough, J Wakshlag, KL Warner, College of Veterinary Medicine, Cornell University, Ithaca, NY.

Association between high hepatic copper and necroinflammatory liver injury is acknowledged. Transition metal status of copper imposes oxidant challenge during ordinarily innocuous hepatic responses and augments injury associated with chronic hepatitis (CH). Etiopathogenesis of hepatic copper accumulation remains enigmatic in breeds other than Bedlington terriers, while study of related Labrador Retrievers (LR) implicated breed predisposition. We have recognized increased incidence of pathologically elevated hepatic copper in dogs (diverse breeds) from 1998-2008, coincident with a pet food industry recommendation to replace cupric/cuprous oxide in feed formulations because of low bioavailability.[1]

Hepatic copper concentrations in LR (used as a sentinel breed) biopsies collected between 1980-2008 were segregated into 1980-1997 (T1) and 1998-2008 (T2). Copper quantification (μg/gm [dry matter basis], atomic absorption spectrometry) in formalin-fixed-deparaffinized tissue was validated against fresh liver. Cases were identified (institutional software), and tissue sections stained with H&E, Masson's Trichrome, Reticulin, Prussian Blue, and Rhodanine to categorize CH (n=36) vs control (CT, n=36). Copper concentrations and age (median [range]) were compared using Wilcoxon Rank Sum. Gender was compared using two-by-two tables. $P \leq 0.05$ determined significance. Relatedness of dogs or dietary management were not investigated.

Copper concentrations of CHT2 (1,150[191-3,950]μg/gm) were significantly higher than CHT1 (441[166-3,300]μg/gm); likewise, copper concentrations of CTT2 (347[101-981]μg/gm) were significantly higher than CTT1 (156[104-398]μg/gm); $P$=0.009, P=0.003, respectively. There were no differences in age or gender.

Increased hepatic copper in LR with and without CH during T2 may reflect increased copper bioavailability and consequently increased copper delivery in revised feed formulations.

[1]Czarnecki-Maulden G, et al: FASEBJ;1993:7:A305.

**ABSTRACT #261**

REDUCTION OF CYTOKINE-INDUCED PROSTAGLANDIN E2 AND CHEMOKINE PRODUCTION IN CULTURED CANINE HEPATOCYTES BY NATURAL PRODUCTS.

AY Au[1,2,3], JM Hasenwinkel[1], CG Frondoza[1,3,4]. 1. Nutramax Laboratories, Inc., Edgewood, MD. 2. Syracuse University, Syracuse, NY. 3. Johns Hopkins University, Baltimore, MD. 4. Mississippi State University, Mississippi State, MS.

Liver tissue is known to heal and regenerate with the critical participation of pro-inflammatory mediators and the growth factors that they induce. However, inflammation of the liver, which is a common ailment in dogs, can result in disease, fibrosis, and life threatening tissue damage when left uncontrolled. Hepatocytes that comprise the bulk of liver tissue have been documented to produce pro-inflammatory mediators and growth factors. How the production of pro-inflammatory mediators by hepatocytes is regulated is still unclear. Moreover, the mechanism by which hepatocytes play a role in liver repair and regeneration is also poorly understood. We have utilized a culture system which facilitates cellular adhesion, proliferation, and maintenance of the hepatocyte phenotype. Using this model, we tested the hypothesis that primary canine hepatocytes can be activated with interleukin 1-beta (IL-1β) to produce pro-inflammatory mediators. We also determined whether this activation can be counteracted by the hepatoprotective natural products Curcumin (NMXCC95™) and a melon-derived superoxide dismutase (MSOD) which is found in Oxstrin®.

Primary canine hepatocytes ($5x10^5$ cells) were plated in wells coated with 0.1% rat type I collagen. Cells were pre-treated for 48 hours with: (i) control media alone, (ii) NMXCC95 ™ (0.368, 3.68, or 36.8 μg/ml), or (iii) MSOD activity of 14000 IU/g (0.0143, 0.143, 1.43 μg/ml), and then activated for 72 hours with IL-1β10 ng/ml). Phenotype analysis was verified by immunostaining for albumin and cytokeratin 8 (CK8). Production of pro-inflammatory mediators prostaglandin E2 ($PGE_2$), interleukin-8 (IL-8), and macrophage chemotactic protein-1(MCP-1) were assayed by ELISA. Statistical analysis was performed using one-way analysis of variance (ANOVA) with significance set at $p<0.05$ using Student-Neuman-Keuls post-hoc analysis.

Hepatocyte cultures continued to produce liver phenotype markers albumin and CK8. Activation with IL-1β significantly induced pro-inflammatory $PGE_2$, IL-8, and MCP-1 production 2- to 6-fold. Pre-treatment with both Curcumin and melon-derived SOD significantly reduced $PGE_2$ production by as much as 50% to 60% of the activated control, respectively. These agents also reduced IL-8 and MCP-1 production. The present study shows that production of pro-inflammatory mediators by hepatocytes can be reduced by the hepatoprotective natural products NMXCC95™ and MSOD.

**ABSTRACT #262**

SERUM PROTEIN ELECTROPHORESIS BETA-GAMMA BRIDGING IS A POOR PREDICTOR FOR HEPATIC DISEASE.

MS Camus, PM Krimer, FS Almy, BE LeRoy. College of Veterinary Medicine, University of Georgia, Athens, GA.

Based on human studies conducted in the 1950's, beta-gamma bridging (β-γ bridging) on protein electrophoresis is presented in both the human and veterinary literature as virtually pathognomonic for hepatic disease. However, the criteria for β-γ bridging are not defined and very few veterinary publications exist to support a relationship between β-γ bridging and liver disease. The goal of this retrospective study was to confirm or repudiate an association between the two by measuring the positive predictive value of β-γ bridging for liver disease. All electrophoretograms generated at the University of Georgia between 1994 and 2008 were evaluated for the presence of β-γ bridging. β-γ bridging was identified if there were 1) an albumin to globulin ratio below the established reference interval; 2) indistinct separation/demarcation between all β and γ globulin fractions or between the $\beta_2$ and γ fractions, with a negative shoulder slope of <5%; and 3) predominance of γ proteins. There were 237 electrophoretograms examined, of which 25 cases (11 dogs, 11 cats, 3 horses) met the inclusion criteria for a β-γ bridge. Of these cases, 8/25 (32%) had hepatic disease based on biochemistry, cytology, histopathology, and/or necropsy findings, while 9/25 (36%) had a variety of infectious diseases. The positive predictive value of β-γ bridging for hepatic disease was determined to be 32.0% with a 95% confidence interval of 15.0-53.5% ($p<0.000$). The positive predictive value of this phenomenon for infectious disease was determined to be 36.0% with a 95% confidence interval of 18.0-57.5% ($p<0.000$). Though β-γ bridging can be found in some cases of hepatic pathology, it is not "pathognomonic" for liver diseases and is as frequently found with infectious diseases.

**ABSTRACT #263**

IN VITRO COMPARISON OF PLAIN RADIOGRAPHY, DOUBLE CONTRAST CYSTOGRAPHY, ULTRASONOGRAPHY, AND COMPUTED TOMOGRAPHY FOR MEASURING UROCYSTOLITH SIZE USING A BLADDER PHANTOM.

K Byl, JM Kruger, J Kinns, N Nelson, J Hauptman, CA Johnson. Michigan State University College of Veterinary Medicine, East Lansing, MI.

Accurate estimation of stone size is a critical factor in determining whether urocystoliths are amenable to removal by minimally invasive procedures. The purpose of this study was to compare the ability of plain radiography (PR), double-contrast cystography (DCC), ultrasonography (US), and computed tomography (CT) to accurately estimate the size a single urolith in an in vitro bladder phantom model.

The bladder phantom model consisted of a rubber balloon distended with a standard volume of 1% saline to simulate a urinary bladder. The balloon was positioned inside a 4-quart storage container on a 2 cm cushion of 2% gelatin and covered with 2 cm of water to approximate the soft tissue density present in a medium sized dog. Thirty canine urocystoliths of varying size (1-11 mm) and mineral composition (calcium oxalate, struvite, and ammonium urate) were individually placed in the bladder phantom and imaged by the four imaging modalities. Two radiologists, blinded to the stone size and mineral content, measured each stone at its greatest diameter using computerized calipers. Size estimates were then compared to actual stone size as determined by digital calipers.

Ultrasonography significantly overestimated urolith size by a mean of 2.95 mm +/- SEM 0.36 mm ($p<0.01$) compared to other imaging modalities, regardless of mineral composition. Size estimates obtained by PR, DCC, and CT were not significantly different. In conclusion, US consistently overestimated sizes of solitary urocystoliths; PR, DCC, or CT should be utilized whenever urolith size is a critical factor for determining treatment options.

**ABSTRACT #264**

INFLUENCE OF FOOD CONSUMPTION ON VARIATION OF URINARY VOLUME, PH, ANALYTES, AND RELATIVE SUPERSATURATION FOR CALCIUM OXALATE AND STRUVITE IN HEALTHY ADULT MALE HOUND DOGS. B Young, JW Bartges, S Cox, T Moyers. The University of Tennessee, College of Veterinary Medicine, Knoxville, TN.

In healthy human beings, there is diurnal variation to urinary excretion of calcium and pH; however, this circadian rhythm has been found to be absent in some human beings that form calcium oxalate uroliths. We hypothesized that healthy, adult, male hound dogs will exhibit a diurnal variation for urinary mineral excretion, pH, and relative supersaturation for calcium oxalate and struvite.

Six healthy adult male hound dogs weighing 18.6 to 26.8 kg were evaluated. They were fed an adult maintenance, dry-formulated dog food at 8:00AM each day. Dogs ate all food within 2 hours; fresh tap water was available at all times. Dogs were housed in individual metabolism cages for urine collection. The urinary bladder of each dog was emptied by transurethral catheterization to begin the study, and at 12 hours and 24 hours after food consumption. Urine from each 12-hour period (AM = 8AM to 8PM and PM = 8PM to 8AM) was measured, and pH and concentrations of sodium (Na), potassium (K), chloride (Cl), calcium (Ca), magnesium (Mg), phosphorous ($PO_4$), oxalic acid (Ox), citric acid (Cit), and ammonia ($NH_4$) were determined. Results were entered into a computer program (EQUIL 1.51b, College of Medicine, University of Florida) for estimation of relative supersaturation for calcium oxalate monohydrate (COMrss), calcium oxalate dihydrate (CODrss), and magnesium ammonium phosphate hexahydrate (struvite, MAPrss). Paired *t*-tests were used to compare results.

Differences were not found for urine volume, pH, or analytes between AM and PM: however, urine volume, pH, Na, K, Ca, $NH_4$, Ox, Cit, and Cl tended to be greater and urine Mg and $PO_4$, tended to be lower in PM when compared with AM. COMrss (AM: 5.25 vs PM: 3.27; $p = 0.025$) and CODrss (AM: 2.27 vs PM: 1.39; $p = 0.025$) were significantly greater and MAPrss (AM: 0.31 vs PM: 0.72; $p = 0.05$) was significantly lower in AM when compared with PM.

Results of our study are similar to previous studies in healthy beagles and calcium oxalate urolith-forming Miniature schnauzers (Lulich, et al 1991); however, those studies utilized pooled 24-hour urine samples. Although urinary concentrations of analytes were not significantly different between AM and PM, COMrss and CODrss were significantly higher after food consumption and MAPrss was significantly higher when food was withheld; therefore, these dogs exhibited diurnal variation in urine saturation for calcium oxalate and struvite. Our results suggest dietary conditions be considered when measuring urinary concentrations of analytes and urinary saturation for calculogenic minerals, and provide a basis for comparing results with calcium oxalate urolith-forming dogs.

**ABSTRACT #265**

ALPHA 1-ACID GLYCOPROTEIN, SERUM AMYLOID A, AND MONOCTE CHEMOATTRACTANT PROTEIN-1 IN CATS WITH IDIOPATHIC CYSTITIS. B Gerber, CE Reusch, E Zini. Clinic for Small Animal Internal Medicine, Vetsuisse Faculty, University of Zurich, Switzerland.

Feline idiopathic cystitis (FIC) is the most common diagnosis in cats with clinical signs of irritative voiding. The etiology of the disease is not known. It is suspected that FIC is a systemic disease rather than a disease restricted to the bladder. This is supported by findings like increased plasma catecholamine levels or an association with gastrointestinal tract signs in cats with FIC. In humans with interstitial cystitis, for which the cat is a model, similar associations were recognized. We hypothesized that in cats with non-obstructive FIC levels of the acute phase proteins alpha 1-acid glycoprotein (AGP) and serum amyloid A (SAA) and the chemokine monocyte chemoattractant protein-1 (MCP-1) are increased and show a similar pattern of increase as cats with obstructive lower urinary tract disease (LUTD). The aim of the study was to evaluate AGP, SAA and MCP-1 in serum of cats with FIC and to compare the values with healthy control cats and cats with obstructive LUTD.

FIC was diagnosed by excluding other causes of LUTD using urinalysis with culture, evaluation of the lower urinary tract by radiographs and/or ultrasound and hematology and serum biochemistry parameters. Cats were considered obstructed if the bladder was distended and they were unable to urinate. AGP was measured with a commercial radial immunodiffusion test. SAA was measured with a commercial multi-species ELISA assay and MCP-1 was measured with a commercial canine CCL2/MCP-1 ELISA kit which was validated for cats. Results were compared using the Mann–Whitney-U test. Differences were considered significant at $P< 0.05$.

Nine cats with FIC, 9 cats with obstructive LUTD of any cause and 9 control cats were included in the study. AGP was 360 ug/ml (200–600 ug/ml) (median and (range)) in control cats, 500 ug/ml (140-1400 ug/ml) in cats with FIC and 780 ug/ml (420-3200 ug/ml) in cats with obstructive LUTD. SAA was 0 ng/ml (0-5.2 ng/ml) in control cats, 0.5 ng/ml (0-23.4 ng/ml) in cats with FIC and 91.2 ng/ml (0-417.6 ng/ml) in cats with obstructive LUTD. MCP-1 was 56 pg/ml (21-73 pg/ml) in control cats, 204 pg/ml (54-806 pg/ml) in cats with FIC and 320 pg/ml (133-823 pg/ml) in cats with obstructive LUTD. All measured inflammatory proteins were significantly higher in cats with obstructive LUTD compared to control cats. Compared to control cats, AGP and SAA in cats with FIC were not significantly different, but MCP-1 was significantly higher. There was no significant difference in MCP-1 between FIC cats and cats with obstructive LUTD.

While AGP and SAA are not increased in cats with FIC, MCP-1 is increased similar to cats with obstructive LUTD, indicating that a systemic inflammatory reaction is present in FIC.

**ABSTRACT #266**

OXIDATIVE STRESS IN CATS WITH CHRONIC RENAL FAILURE. R Keegan & CB Webb, Colorado State University, Fort Collins, CO.

Oxidative stress is an important component in the progression of chronic renal failure (CRF) in humans, and neutrophil function may be impaired by oxidative stress in these patients. This study was designed to test the hypothesis that cats with CRF have increased oxidative stress and decreased neutrophil function when compared to Control cats.

A biochemical profile, complete blood count, urinalysis, blood pressure, plasma antioxidant capacity, erythrocyte lysate superoxide dismutase enzyme (SOD) level, whole blood reduced-to-oxidized glutathione ratio (GSH:GSSG), and neutrophil phagocytosis and oxidative burst were measured in 20 cats with CRF and 10 Control cats. Statistical comparisons (two-tailed t-test) are reported as mean ± standard deviation.

There was no difference in age or body weight between Control and CRF cats. The CRF cats had significantly greater serum BUN, creatinine, and phosphorus concentrations than Control cats, and significantly lower PCV and urine specific gravity than Control cats. The GSH:GSSG ratio was significantly greater in the CRF group (177.6 ± 197, 61.7 ± 33; $p < 0.02$), while the SOD level and antioxidant capacity were not significantly different between groups. Neutrophil oxidative burst following E. coli phagocytosis, measured flow cytometrically as an increase in mean fluorescence intensity (a unitless measure), was significantly greater in CRF cat then Control cats (732 ± 253, 524 ± 54; $p < 0.05$).

The significantly greater GSH:GSSG ratio in CRF cats is consistent with an activation of antioxidant defense mechanisms in this disease state. It remains to be determined if supplementation with antioxidants beyond the level of Control cats would be of benefit in cats with CRF.

**ABSTRACT #267**

ERYTHROCYTE ZINC PROTOPORPHYRIN EVALUATION IN CATS WITH CHRONIC RENAL DISEASE. CS Prosser, MM Kogika, MF Waki, AGT Daniel, DMN Simões, A Alves, MB Silva, BM Teixeira. School of Veterinary Medicine – University of São Paulo, Brazil.

Erythrocyte zinc protoporphyrin (ZPP) is used in human medicine as a screening method to detect iron deficiency. ZPP is produced instead of heme and, thus, increased concentrations of ZPP may be observed in cases of iron deficiency prior to detectable anemia. In cats, no studies concerning the evaluation of ZPP have been reported yet. Cats with chronic renal disease (CRD) may develop anemia as the disease progresses; iron deficiency may be one of the causes. Therefore, the aim of this study was to investigate whether ZPP may be altered in anemic and non-anemic cats with CRD.

Thirty-two client-owned domestic cats with CRD in stages II, III and IV were placed in one of two groups: group I (n = 12) – CRD with normocytic, normochronic anemia (PCV < 29%) and group II (n = 20) – CRD without anemia (PCV > 29%). Patients that received iron, vitamin B, blood transfusion or rHuEPO were excluded, as well as those with neoplasia or cardiac disease. Results were compared to a control group of 36 healthy cats (16 cats aging from 1 to 6 y-old and 20 cats older than 6 y-old). All cats of all groups were negative for FIV and FeLV, with normal serum levels of bilirubin (hyperbilirubinemia interferes with ZPP measurement). Blood was collected in EDTA tubes and hematofluorometric method was used to determine the ZPP (ProtoFluor® Z -Helena Laboratories); ZPP values were obtained three times/ sample prior to 6 hours post-collection.

The mean ZPP value of the control group was 38.39 μmol ZPP/mol heme (SEM ± 1.98; min = 18.67; max = 72.33), with no statistically significant difference between younger and older normal cats (P = 0.680). Mean ZPP measurements were 60.41 μmol ZPP/mol heme (SEM ± 6.76; min = 35.33; max = 117.33) and 45.62 μmol ZPP/mol heme (SEM ± 2.97; min = 30.00; max= 77.33) in groups I and II, respectively. Difference was detected between control group and group I (*t* test; P = 0.013), as well as between control group and group II (P = 0.008); however there was no difference between groups I and II (P = 0.11).

Anemic cats of group I showed significant increase of erythrocyte ZPP levels, suggesting that ZPP may have been produced instead of heme in consequence of iron deficiency. Similarly, non-anemic CRD cats (group II) showed higher level of ZPP than clinically normal cats; ZPP concentrations remained altered despite normal PCV and erythrocyte indices, indicating a degree of iron depletion. Development of iron-deficient erythropoiesis (microcytic anemia) may result only in late stages in cats, therefore ZPP may be an important tool to detect early iron depletion before anemia. Further studies are necessary to investigate the role of ZPP, serum iron and ferritin to confirm whether ZPP values correlate with the degree of iron deficiency in cats, and, therefore, recommended as a diagnostic tool to guide iron supplementation in clinical practice.

**ABSTRACT #268**

SEQUENTIAL EVALUATION OF ERYTHROCYTE ZINC PROTOPORPHYRIN IN CATS WITH CHRONIC RENAL DISEASE. CS Prosser, MM Kogika, MF Waki, BM Coelho, KK Kanayama, VABF Wirthl, MB Silva, PRE Mosko. School of Veterinary Medicine – University of São Paulo, Brazil.

Anemia may develop during the progression of chronic renal disease (CRD) in cats, and iron deficiency may be one of the causes. In the absence of iron, erythrocyte zinc protoporphyrin (ZPP) is formed during the final step of heme synthesis; in human medicine, ZPP has been used as a diagnostic indicator of iron deficiency. The aim of this study was to evaluate whether erythrocyte ZPP levels alter during the evolution of CRD in cats. Twelve domestic cats with CRD were evaluated for at least 50 days (range of 50 to 140) after the initial diagnosis. Animals were divided in three groups based on initial presentation: group I (#1-6; non-anemic CRD cats in stage II), group II (#7 and 8; anemic CRD cats, one in stage III and the other in IV, both receiving iron supplementation), and group III (#9-12; anemic CRD cats in stage III, treated with rHuEPO and iron). All cats were negative for FIV and FeLV, and had normal serum levels of bilirubin (hyperbilirubinemia interferes with ZPP measurement). Blood was collected in EDTA tubes and ZPP determinations were obtained three times/sample prior 6 hrs post-collection by hematofluorometry method (ProtoFluor® Z -Helena Laboratories).

Initial PCV values ranged from 30% to 39% and remained relatively consistent throughout the evaluation period for all group I cats, except cat #6 that developed progressive decrease in PCV, declining to 23% after 85 days. Regarding to ZPP levels of group I, it ranged consistently from 30.7 to 73.7 µmol ZPP/mol heme (previous study, ZPP in normal cats was 38.39 ± 1.98 [mean ± SEM; min=18.7 and max=72.3]), however cat #6 presented increase of ZPP levels (128.33 µmol ZPP/mol heme) as PCV was decreasing. The findings in cat #6 suggest that a degree of iron depletion may be also involved as one of the causes of anemia. Cat #7 (group II) developed a concomitant 16.5% decrease in ZPP value and 29.4% increase in PCV from initial levels following iron supplementation; similarly, cat #8 decreased ZPP levels by 26.5% and increased PCV values by 4.8%. These data may suggest that supplemental iron availability was adequate, since ZPP levels decreased. In group III, cat #9 responded during the treatment with increased PCV (120%) and ZPP (87.2%) values; ZPP probably increased due to increased iron demand secondary to rHuEPO stimulus, which may have resulted in iron depletion, and probably iron supplementation was not sufficient. In cat #10, ZPP values also increased and PCV values decreased, showing that there was not a good response for rHuEPO and other causes of anemia, other than iron deficiency, may have been also involved. Cats #11 and #12 showed decrease on ZPP (PCV increased in cat #11, but not in cat #12); probably iron supplementation may have been adequate. The detectable variation in ZPP levels along the progression of CRD suggests that iron may be involved in the mechanism; further studies are necessary to investigate the role of ZPP and iron depletion in the development of anemia in CRD cats.

**ABSTRACT #269**

ENDOGENOUS ERYTHROPOIETIN LEVELS AND IRON UTILITY IN CATS WITH CHRONIC KIDNEY DISEASE.

S Yamano[1,2], K Tanaka[2], M Nishida[1], K Harada[1], M Fujiwara[1], T Ebisawa[1], M Uechi[1]. 1. Nihon University, Fujisawa, Kanagawa. 2. Kanie Animal Clinic, Kanie, Aichi, Japan.

The aim of this study was to assess the changes in endogenous erythropoietin (EPO) and serum iron levels in cats with chronic kidney disease (CKD). The study included 14 healthy cats and 27 cats with CKD [stage 2 (n=8), stage 3 (n=7) and stage 4 (n=12), according to the International Renal Interest Society (IRIS) Staging]. Investigated parameters were complete blood count, serum creatinine, EPO, serum iron, total iron binding capacity, unsaturated iron binding capacity, and transferrin saturation (TSAT). The EPO level in healthy, stage-2, stage-3, and stage-4 cats was 15.6 ± 4.25, 14.6 ± 4.61, 17.6 ± 7.41 (n=4), and 10.6 ± 2.50 mIU/mL (n=9), respectively. There was a negative correlation between the logarithm of EPO ($Log_{10}$ EPO) and serum creatinine level ($r=-0.488$, $p=0.029$). Stage 3 and stage 4 cats showed significantly lower levels of hematocrit, hemoglobin and red blood cell count than healthy and stage 2 cats ($p<0.05$ for comparison with stage 3, $p<0.01$ for comparison with stage 4). The serum iron level in stage-4 (136.8 ± 63.2 µg/dL) was significantly higher than that in the healthy group (74.6 ± 20.4 µg/dL; $p<0.01$). TSAT in stage 4 was significantly higher than that in the healthy group ($p<0.05$). These results indicate that erythropoiesis is compromised, as indicated by reduced iron utility, in cats with advanced CKD.

**ABSTRACT #270**

IS CADMIUM EXPOSURE ASSOCIATED WITH CHRONIC KIDNEY DISEASE IN CATS? N.Finch, H.Syme, J.Elliott. Royal Veterinary College, London, UK.

Chronic cadmium exposure in human patients has been associated with kidney damage and the development of tubulointerstitial nephritis, the predominant pathological lesion associated with chronic kidney disease (CKD) in cats. There are no published studies investigating any association between chronic cadmium exposure and CKD in cats.

Urine samples were collected by cystocentesis from clinically healthy, non-azotaemic (plasma creatinine <2.0mg/dl), geriatric (<9 years old) cats and geriatric cats with previously confirmed CKD (persistent azotaemia and urine specific gravity <1.035). Any cats with concurrent diseases such as hyperthyroidism were excluded. Urinary cadmium concentrations were determined at a commercial laboratory using inductively coupled plasma mass spectrometry (ICP-MS) and urine creatinine concentration was measured using the kinetic jaffe method. The urinary cadmium: creatinine ratio was calculated to account for the concentration of the urine. Statistical analysis comparing the healthy and azotaemic groups was performed using the Mann-Whitney U test following assessment of the data for normality. Data are reported as median (range).

Thirty seven cats were recruited into the study (19 normal, 18 CKD). Twenty-five of the cats were classified as domestic shorthair, 5 as domestic longhair and 7 as purebred. The study included 17 male neutered cats, 19 female neutered cats and 1 female entire cat. Of the cats with CKD 11 were in IRIS stage II CKD and 7 in stage III CKD. The plasma creatinine concentration for the normal cats was 1.46 (1.06 – 1.73) mg/dl and for the cats with CKD was 2.63 (1.77 – 4.59) mg/dl. The urinary cadmium: creatinine ratio in normal cats and cats with CKD was 0.11 (0.02 – 0.63) $x10^{-6}$ and 0.14 (0.04 – 0.46) $x10^{-6}$ respectively. The urinary cadmium: creatinine ratio did not differ significantly between the groups of cats ($P = 0.261$).

This preliminary study does not provide any evidence to support the hypothesis that chronic exposure to cadmium from the environment plays a role in the pathogenesis of CKD in cats. However, the limitations of the study are that it involves spot urine samples, the content which may not reflect cadmium accumulation within the kidney tissue. Further work is required to determine whether cadmium accumulation in the kidney is associated with feline CKD.

**ABSTRACT #271**

GLOMERULAR FILTRATION RATE IN HEALTHY ENGLISH POINTER, ENGLISH SETTER AND GERMAN SHEPHERD DOGS. J Seguela[1], Y Queau[1], P Murgier[1], P Mimouni[2], D Concordet[1], C Duperron[1], HP Lefebvre[1]. 1. UMR 181 Physiopathologie et Toxicologie Expérimentales INRA-ENVT, National Veterinary School, Toulouse. 2. Clinique Vétérinaire, L'Isle Jourdain, France.

A breed effect on glomerular filtration rate (GFR) has been hypothetized in dogs (Lefebvre et al., JVIM, 2004;18:415). The aim of this study was to compare GFR in healthy populations from 3 different breeds.

44 English Pointer (EP), 50 English Setter (ES), and 55 German Shepherd (GS) healthy dogs were recruited. GFR was measured by plasma exogenous creatinine clearance test. Intravenous administration of exogenous creatinine (40 mg/kg) followed by repeated blood sampling was performed. Effect of age, body weight (BW) and breed was analyzed by ANOVA.

The mean±SD age and body weight (BW) were for 4.3±2.9 y. and 21.1±2.5 kg for EP, 3.5±2.7 y. and 18.4±3.2 kg for ES, 4.9±2.8 y. and 29.4±4.0 kg for GS. The age was similar between breeds and the BW was higher for GS ($P<0.001$). GFR was 3.5±0.6, 3.4±0.8 and 2.5±0.7 mL/min/kg for EP, ES and GS, respectively. GFR was lower ($P<0.001$) in GS than in EP and ES. The GFR distribution was normal for EP and ES, and after inverse transformation, for GS. Tentative reference intervals were 2.3-5.1, 1.8-5.0 and 1.7-3.8 mL/min/kg for EP, ES, and GS, respectively.

In conclusion, establishment of breed-specific reference intervals for GFR in dogs is needed, although BW could be a confounding factor.

**ABSTRACT #272**

24-HOUR WATER CONSUMPTION IN HEALTHY CATS USING A FOUNTAIN VERSUS A BOWL. DC Grant. Virginia-Maryland Regional College of Veterinary Medicine, Blacksburg, VA.

Increasing water consumption has been recommended to reduce urine concentration and recurrence of urolithiasis and idiopathic cystitis in cats. Water fountains are being advertised and some advocate their use to increase water consumption. This study evaluated the affect of a fountain versus a bowl on 24-hour water consumption and urine concentration.

Thirteen client-owned cats had a normal physical examination, hematocrit, biochemistry profile, and urinalysis. Cats were kept in their home environment and consumed their regular diet. Cats were allowed 7 days to acclimate to the fountain. Cats were randomized to be offered water from a fountain or their own bowl first, followed by the other, and 24-hour consumption was measured. The volume of water evaporated and spilled in 24 hours from the fountain or bowl was subtracted from the amount offered. Following each 24-hour period urine osmolality was measured. Analysis of variance was performed with significance set at a p-value <0.05.

One cat refused to drink from the fountain. For the remaining cats, there was no significant difference in water consumption from a bowl (22.9 ± 10.2 ml/kg/day) or from a fountain (31.6 ± 13.5 ml/kg/day), p=0.091. Similarly, there was no difference in urine osmolality when water was offered form a bowl (2469 ± 367.3 mOsm/kg) or a fountain (2538 ± 440.8 mOsm/kg), p= 0.682.

This refutes the theory that water fountains increase water consumption in cats.

**ABSTRACT #273**

CLINICAL AND GENETIC CHARACTERIZATION OF A CONGENITAL ENTLEBUCHER MOUNTAIN DOG URINARY TRACT SYNDROME. C North[1], JM Kruger[2], PJ Venta[2], JM Miller[3], DS Rosenstein[2], EK Randall[2], SD Fitzgerald[2]. 1.Estero, FL. 2.College of Veterinary Medicine Michigan State University, East Lansing, MI. 3. University of Michigan Health System, Ann Arbor, MI.

The "Entlebucher urinary syndrome" (EUS) was first recognized in North American Entlebucher mountain dogs in 1996. The purpose of this study was to characterize clinical and genetic features of EUS. Eleven female and 4 male client-owned Entlebucher mountain dogs were evaluated. Six female dogs had histories of urinary incontinence, recurrent urinary tract infection, or hydronephrosis; the remaining dogs were asymptomatic. Thirteen dogs were evaluated with a hemogram, serum biochemistries, urinalysis, urine culture, excretory urography, ultrasonography, and urethrocystoscopy. Two other dogs had partial evaluations that included necropsy. Five candidate genes (*PAX2, EYA1, FOXC2, HOXB7, and NPHP4*) were selected and evaluated by SINE insertion polymorphism or microsatellite marker-based exclusion analysis of DNA obtained from affected dogs. Clinical and pedigree histories from an additional 138 North American Entlebuchers were compiled for analysis.

Abnormalities identified in clinically affected dogs included bilateral ectopic ureters (6 dogs), hydroureter/hydronephrosis (4 dogs), and urinary tract infection (2 dogs). Both necropsied dogs had bilateral hydroureter/hydronephrosis with ureterovesical junction obstruction caused by chronic granulation tissue or lymphoplasmacytic inflammation. Bilateral intravesicular ectopic ureters also were identified in 3 asymptomatic female dogs. Probable EUS was identified in 12% of dogs in pedigree analyses; all affected dogs were female. Because of incomplete penetrance, mode of inheritance could not be determined. Exclusion analyses did not associate any candidate genes with EUS.

In conclusion, clinical features of EUS include ureteral ectopia (asymptomatic or with urinary incontinence), hydroureter/hydronephrosis, and urinary tract infection. Further analyses are necessary to confirm and characterize the hereditary nature of EUS.

**ABSTRACT #274**

PROTEOMIC ANALYSIS OF URINE FROM MALE DOGS IN EARLY STAGES OF PROGRESSIVE TUBULOINTERSTITIAL INJURY DUE TO X-LINKED HEREDITARY NEPHROPATHY (XLHN). MB Nabity, GE Lees, LJ Dangott, SK Ramaiah. Texas A&M University, College Station, TX.

Chronic kidney disease (CKD) is a major cause of morbidity and mortality in dogs, and methods to reliably identify tubulointerstitial disease in its early stages currently are lacking. The objective of this study was to identify predictive biomarkers of early tubulointerstitial injury in urine of male dogs with XLHN using 2-dimensional gel electrophoresis. Dogs with XLHN develop glomerular lesions due to a *COL4A5* mutation that causes the type IV collagen in their glomerular basement membranes to be abnormal. Affected males first develop glomerular proteinuria, but progressive tubulointerstitial injury is associated with the subsequent development of end-stage CKD during adolescence.

Urine samples from male dogs (n=6) were analyzed using 2-dimensional differential gel electrophoresis (DIGE), which uses 3 fluorescent dyes (Cy2, Cy3, and Cy5) for sample labeling. Samples were compared at two time points: 1) onset of urine protein:creatinine (UPC) persistently >2; and 2) onset of serum creatinine $\geq$ 1.2 (mildly azotemic). Albumin was removed from all samples by affinity chromatography, and samples were concentrated with ultrafiltration (10-kDa cut-off). Equal amounts of each sample were then precipitated with acetone, denatured, and fluorescently labeled. Proteins were focused on 13 cm IPG strips (pH 4-7) followed by separation on 12% SDS gels. Gels were scanned and spots analyzed using DeCyder software. Fifty-seven protein spots showed statistically significant changes between the time points. Of these, 20 were identified by mass spectrometry and included retinol binding protein (RBP), clusterin, alpha1-microglobulin, fetuin A, hemopexin, apolipoprotein A1, gelsolin, immunoglobulin, collagen type XXVII, complement C3, and haptoglobin.

RBP showed the greatest statistical significance and was subsequently evaluated by Western blot in the urine of affected male dogs (n=25) and their normal siblings (n=19). Urine samples were evaluated every 2-4 weeks starting two weeks prior to UPC persistently >0.5 in affected males and at several corresponding time points in their normal littermates. Samples were normalized to a creatinine of 20 mg/dl. RBP was detected approximately 2 months prior to the onset of azotemia (serum creatinine >1.2) in the urine of the affected male dogs and typically increased throughout the course of disease independently of the UPC value. RBP was not detected in the urine of the normal males.

In summary, these results suggest that 2-D gel electrophoresis can be successfully employed for biomarker identification in chronic kidney disease in the urine of dogs and that urine RBP may be a potential early marker of kidney disease and of CKD progression in dogs.

**ABSTRACT #275**

METABOLOMIC PROFILING OF PLASMA IN ARTHRITIC VS NON-ARTHRITIC DOGS. N.Z. Frantz, R.M. Yamka, and J. Brockman. Hill's Pet Nutrition, Inc., Topeka, KS.

Sixty-three neutered/spayed beagles (average age = 10.9 ± 2.3 years) were identified for this study. Dogs were radiographed to confirm changes consistent with degenerative joint disease and lameness in the arthritic dogs or absence in the non-arthritic dogs. Metabolomic profiling of plasma samples was performed by Metabolon (Durham, NC). T-test analysis was applied to log-transformed, day normalized data. Metabolites having a $P < 0.05$ (following a false discovery rate adjustment) were considered different among the two groups. KEGG and HMDB identifiers used were from commercial databases. Analysis of the plasma found differences in 10 metabolites between the two groups. Of the metabolites identified, 2 had increased fold-changes and 8 had decreased fold-changes in the arthritic group when compared to the non-arthritic group. The metabolites with increased fold changes in arthritic dogs were dimethylarginine and beta-sitosterol. The metabolites with decreased fold changes in arthritic dogs were glycine, trans 4-hydroxyproline, mannose, glucose, myo-inositol, threonate, 4-hydroxymandelate, and phenol sulfate. In summary, metabolites associated with increased oxidative stress were elevated while metabolites associated with glycosylation of proteins, Vitamin C metabolism, and collagen turnover were decreased in arthritic dogs.

**ABSTRACT #276**

METABOLOMIC PROFILING OF PLASMA IN ARTHRITIC VS NON-ARTHRITIC CATS. N.Z. Frantz, R.M. Yamka, and J. Brockman. Hill's Pet Nutrition, Inc., Topeka, KS.

Sixty American short-hair cats (average age = 11.3 ± 2.7 years) were identified for this study. Cats were radiographed to confirm changes consistent with degenerative joint disease and lameness in the arthritic cats or absence in the non-arthritic cats. A plasma sample was collected to determine metabolomic differences between the two populations. Of the sixty cats identified, 30 were classified as arthritic and 30 as non-arthritic based on radiographic evidence and visual appraisal. Metabolomic profiling of plasma samples was performed by Metabolon (Durham, NC). T-test analysis was applied to log-transformed, day normalized data. Metabolites having a $P < 0.05$ (following a false discovery rate adjustment) were considered different among the two groups. KEGG and HMDB identifiers used were from commercial databases. Analysis of the plasma found differences in 20 metabolites between the two groups. Of the metabolites identified, 11 had increased fold-changes and 9 had decreased fold-changes in the arthritic group when compared to the non-arthritic group. The metabolites with increased fold changes in arthritic cats were kynurenine, dimethylargnine, N-acetylornithine, homocitrulline, 2-aminobutyrate, 3-hydroxydeconate, 3-hydroxybutyrate, allantoin, 5-methylcytidine, pseudouridine, and trigonelline. The metabolites with decreased fold changes in arthritic cats were tryptophan, 3-indoleproprionate, creatinine, carnosine, gamma-glutamylfelinylglycine, glycerol 3-phosphate, campesterol, alpha-ketogluterate, and hippurate. In summary, plasma metabolomic analysis indicates differences of increased metabolites of oxidative stress and inflammation in arthritic compared non-arthritic cats.

**ABSTRACT #277**

METABOLOMIC PROFILING OF PLASMA IN LEAN VS. OBESE DOGS. RM Yamka, NZ Frantz and SC Zicker. Hill's Pet Nutrition, Inc., Topeka, KS.

Fifty-two neutered/spayed beagles (average age = 8.6 ± 3.1 years) were identified for this study. Dogs were weighed, given a body condition score (BCS; 1=lean, 3=ideal and 5=obese) and a plasma sample was collected. Of the fifty-two dogs identified, 24 were classified as lean (BCS = 2.8 ± 0.5; BW = 11.2 ± 1.9 kg) and 28 as overweight/obese (BCS = 3.9 ± 0.7; BW = 14.4 ± 3.6 kg). Metabolomic profiling of plasma samples was performed by Metabolon (Durham, NC). T-test analysis was applied to log-transformed, day normalized data. Metabolites having a $P < 0.05$ (following a false discovery rate adjustment value of 0.1) were considered different among the two groups. KEGG and HMDB identifiers were from commercial databases. Analysis of the plasma found differences in 20 metabolites between the two groups. Of the metabolites identified, 15 had increased fold-changes and 5 had decreased fold-changes in the overweight group when compared to the lean group. The metabolites with increased fold changes in overweight dogs were glutamine, 3-methylhistidine, lysine, isoleucine, valine, citrulline, oxidized glutathione, glutamylvaline, pyroglutamylvaline, gamma-glutamylleucine, gamma-glutamylmethionine, gamma-glutamylglutamine, stearoylglycerol (monostearin), 3-hydroxybutyrate and citrate. The metabolites with decreased fold changes in overweight dogs were p-cresol sulfate, 1,5-anhydroglucitol, glycerate, docosahexaenoate and threonate. There were minimal differences in metabolites for gender as all dogs were spayed or neutered. In summary, amino acid metabolism, lipid metabolism, carbohydrate metabolism and oxidative stress were different between overweight and lean dogs.

**ABSTRACT #278**

DETERMINATION OF MAINTENANCE ENERGY REQUIREMENT OF CATS AFTER SPAYING. Y Mitsuhashi[1], AJ Chamberlin[1], KE Bigley[1], JE Bauer[1, 2]. 1. Comp. An. Nutr. Lab., 2. Intercollegiate Faculty of Nutrition, Texas A&M Univ., Col Stn, TX.

Several cat studies have reported decreased energy expenditure after spaying. Spayed cats, therefore, tend to increase their body weights (BW) without changing calorie intake, which often results in obesity. The recently revised National Research Council (NRC) publication recommends a maintenance energy requirement (MER) for cats based on a 9-point body condition score (BCS) as follows: $100 \times BW^{0.67}$ for BCS≤5 and $130 \times BW^{0.40}$ for BCS>5. Although weight gain of spayed cats is widely recognized, the MER for these animals has not been clearly defined. Therefore, this study aimed to determine the MER for spayed cats whose BW began to increase shortly after surgery. Twenty-two short-haired adult female cats (2 yrs old, on average) were used. Three complete and balanced diets (diets A, B, and C) containing similar nutrient compositions but varying in fatty acid type had been fed in an amount to maintain their BW and BCS prior to the study. To begin the study, all cats were spayed and diet C was fed once daily for 9 weeks using the same MER before the study. The cats were then fed an amount of diet based on NRC recommendations for 2 additional weeks. During these 11 weeks, all cats gained weight. Beginning on wk 12 a weight loss regimen was initiated until each cat achieved a BCS of 5. Cats were fed approximately $65 \times BW^{0.67}$ to achieve a 1-2 % body weight loss per week during this period. After each cat obtained a BCS of 5, it was then fed an appropriate amount of diet to maintain its body weight for at least 4 weeks. BW and BCS were monitored weekly and bi-weekly before and during weight loss, respectively. Daily food consumption was recorded and used to determine an allometric factor (AF) calculated by dividing calories consumed by metabolic body weight ($BW^{0.67}$) based on the BW of cats at BCS of 5. Blood was collected at the time of surgery and after weight loss for plasma biochemistry profiles. Statistical analyses were performed on data both prior to the study and during weight gain. Weekly BW and AF were analyzed by repeated measures ANOVA using a general linear model. Weekly BCS data were analyzed by Friedman test followed by Wilcoxon signed ranks test with Bonferroni adjustment. All blood chemistries were within normal limits. BW and BCS did not change prior to spaying but showed significantly incremental increases thereafter such that a 16% increase in BW and nearly 1 point increase of BCS was observed ($p<0.01$). It should be noted that the AF was constant (AF=101.5±6.6 (SD) kcal/$BW^{0.67}$, $p=0.16$) both prior to surgery and thereafter during which time all cats significantly gained weight. The AF needed to maintain BW of each cat after obtaining a BCS of 5 was 75.0±5.6 kcal/$BW^{0.67}$. This AF is 25% lower than the current NRC recommendation and was statistically significantly lower than cats before surgery by paired t-test ($p<0.05$). In conclusion, spaying significantly increased body weight when using MER values for intact cats. Thus, $75 \times (\text{ideal BW})^{0.67}$ is proposed for the MER of spayed cats.

**ABSTRACT #279**

ASSESSMENT OF METABOLIC SYNDROME OCURRENCE IN OBESE CLIENT-OWNED DOGS. MM Jericó[1], FB Fusco[1], FC Chiquito[1], F Lorenzini[1], CF Pinto[2]; CB Ferreira[2]; RM Sousa[2]. 1. Anhembi Morumbi University, Laureate International Universities, SP, Brazil. 2. Santo Amaro University Veterinary Hospital, SP, Brazil.

The metabolic syndrome, a morbid condition well-described in humans, is defined as a set of risk factors such as obesity (mainly visceral obesity), arterial hypertension, dyslipidemia and insulin resistance; at least three of these events must occur simultaneously in the same individual for the diagnosis to be confirmed. It is known that the binding factor among these alterations is insulin resistance (characterized by hyperinsulinemia) and that the clinical relevance of the metabolic syndrome is related to its role in the development of atherosclerosis and onset of diabetes mellitus. In dogs, the metabolic syndrome has been well recognized in experimental conditions, where obesity is induced by the exaggerated administration of fat-rich diets. We aimed at assessing the metabolic syndrome occurrence in obese client-owned dogs (n=34), through the measurements of serum cholesterol, triacylglycerols, glycemia, amended insulin to glucose ratio (AIGR) and the homeostasis model assessment (HOMA). Normal values were established based on the results (mean value + 2 SD) obtained in a group of normal dogs (n=18). When the results were analyzed alone, none of the obese dogs presented hyperglycemia. On the other hand, 10 dogs (29.4%) presented absolute hyperinsulinemia (>18.8 uUI/ml), 11 (32.4%) presented relative hyperinsulinemia as demonstrated by the amended insulin to glucose ratio (AIGR>37.7), 10 (32.4%) presented increased HOMA index (>4.82), 7 (20.6%) presented hypercholesterolemia (>307 mg/dL) and 9 (26.5%) presented hypertriacylglyceridemia (>140 mg/dL). The glycemia values in the normal dogs did not differ from those in the obese dogs (T-test, p>0.05). On the other hand, the cholesterol, triacylglycerol, insulin, AIGR and HOMA values in obese dogs were higher than those in normal animals (Mann-Whitney test; p<0.05). Finally, when all the data were considered, 8 of the obese dogs presented, simultaneously, hyperlipidemia and hyperinsulinemia (absolute and relative), in addition to visceral obesity. We conclude that a significant number of animals in the obese client-owned dog group assessed in the present study presented metabolic syndrome, which is an important risk factor for the development of diabetes mellitus, as well as atherosclerosis, a morbid condition that is uncommon in the canine species, but that has been described in the presence of endocrine or metabolic disorders.

**ABSTRACT #280**

A VEGETARIAN DIETARY OIL SOURCE OF γ-LINOLENIC ACID RESULTS IN ARACHIDONIC ACID ACCUMULATION IN FELINE UTERINE TISSUES: EVIDENCE OF DELTA-5 DESATURATION. AJ Chamberlin[1], Y Mitsuhashi[1], KE Bigley[1], JE Bauer[1,2]. 1. Companion Animal Nutrition Lab., 2. Intercollegiate Faculty of Nutrition, Texas A&M University, College Station, TX.

Beyond the fact that cats express low delta-6 desaturase activities, lipid metabolism in this species remains to be fully explored. Essential fatty acid requirements of cats include a preformed source of arachidonic acid (AA) especially during growth and reproduction. Consequently, the ability of reproductive tissues to sustain healthy pregnancies depends on AA availability. The possibility exists that γ-linolenic acid (GLA, 18:3n6) may serve as a precursor of AA in such tissues especially if coupled with chain elongation and a functionally active delta-5 desaturase. Delta-5 desaturase activity has been found to be low in cats but it was undetermined whether it might be sufficient to result in AA synthesis. In the present study, it was hypothesized that uterine and ovarian tissues of cats fed GLA would accumulate AA and other fatty acid metabolites compared to a control diet. In addition, the appearance of dihomogamma-linolenic acid (DGLA, 20:3n6), the delta-5 desaturase substrate, was predicted to accumulate in adipose tissue under this condition as a reservoir for conversion. A group of 19 adult female cats were divided into two groups and fed either a high (n=10) or low (n=9) GLA diet (4.2 vs < 0.1 g/kg diet). Both diets contained similar amounts of LA and minimally adequate AA. The diets were fed for 300 days prior to ovariohysterectomy at which time EDTA plasma and ovarian, uterine, and subcutaneous adipose tissues were collected. Homogenates of each tissue were prepared and frozen in aliquots at -80 C. Total lipids were extracted from the plasma and tissue homogenates followed by phospholipid (PL) fractionation via thin layer chromatography and fatty acid analyses by gas chromatography. The Shapiro-Wilks test was used to determine normal distribution of fatty acid data followed by Student's t test (p<0.05). Plasma PLs were significantly increased in both GLA and DGLA in the high GLA group. Uterine tissue homogenates were found to have statistically significant increased amounts of DGLA and AA. However, ovarian tissue showed an increase of only DGLA. It is concluded that a high GLA diet results in increased AA in uterine, but not ovarian, tissues and thus may supply eicosanoid precursors in support of reproduction. Furthermore, the increase in DGLA may provide an adipose storage reservoir for additional conversion under times of metabolic need. These data support the presence of a functionally active delta-5 desaturase in uterine, but not ovarian, tissues. The findings also suggest that increased dietary GLA may be used to meet the AA requirements for reproduction in cats in the absence of an animal based pre-formed source of AA.

**ABSTRACT #281**

EFFECT OF PARENTERAL L-ALANYL-L-GLUTAMINE SUPPLEMENTATION ON PHAGOCYTIC RESPONSES OF CANINE PERIPHERAL BLOOD POLYMORPHONUCLEAR NEUTROPHILIC LEUKOCYTES EXPOSED TO METHYLPREDNISOLONE SODIUM SUCCINATE. JH Kang, MP Yang. Chungbuk National University College of Veterinary Medicine, Cheongju, Chungbuk, Republic of Korea.

The objective of this study was to examine the effects of parenteral L-alanyl-L-glutamine (Ala-Gln) supplementation on functions of canine peripheral blood polymorphonuclear neutrophilic leukocytes (PMNs) exposed to methylprednisolone sodium succinate (MPSS). The experimental design involved administration of a high dose of MPSS, which is the recommended protocol for dogs with acute spinal cord injury. Fifteen healthy Beagles were randomly assigned into 3 groups and administered an IV infusion with 0.9% NaCl solution (Treatment A), admixture of 0.9% NaCl solution with 8.5% amino acids (2.3 g/kg/day) (Treatment B), or admixture of 0.9% NaCl with Ala-Gln solution (Dipeptiven) (0.5 g/kg/day)-supplemented 8.5% amino acids (1.8 g/kg/day) (Treatment C). The two admixtures were isonitrogenous and isocaloric. The infusion started with the first injection of MPSS and continued until 12 hours after MPSS injections completed. To evaluate PMNs functions, blood samples were collected before IV injections of MPSS (time 0) and 2, 12, and 24 hours after injections ceased. Phagocytic capacity, oxidative burst activity (OBA) and filamentous actin (F-actin) polymerization were measured by use of flow cytometry.

Relative to preinfusion values, the phagocytic capacity, OBA and F-actin polymerization of PMNs in Treatment A and B groups were suppressed by MPSS injections, while the PMNs functions in Treatment C group were not reduced. The PMNs values suppressed by MPSS in Treatment A and B groups were restored 12 hours after the injections concluded. In conclusion, these results suggest that the parenteral Ala-Gln supplementation can modulate the immune functions of canine PMNs affected by MPSS treatment.

**ABSTRACT #282**

THE PHARMACOKINETICS OF MIRTAZAPINE IN HEALTHY CATS. JM Quimby, DL Gustafson, BJ Samber, KF Lunn. Department of Clinical Sciences, Colorado State University, Fort Collins, CO.

This study was designed to determine the pharmacokinetics of mirtazapine, an appetite stimulant, in healthy cats after oral administration of a single 3.75 mg high dose (HD), or 1.88 mg low dose (LD).

Ten cats with normal CBC, chemistry and urinalysis were used. Blood samples were collected prior to and 0.25, 0.5, 1, 4, 8, 24, 48, and 72 hours after oral administration of 3.75 mg (5 cats), or 1.88 mg (5 cats) of mirtazapine. Serum was collected, frozen and analyzed by LC/MS/MS. Mirtazapine, 8-hydroxymirtazapine and glucuronide metabolite concentrations were measured. Non-compartmental pharmacokinetic modeling was performed.

Increased vocalization and affection were the only side effects noted. Mean half-life was 15.4 +/- 4.7 hours (HD) and 10.2 +/- 2.2 hours (LD). Mean peak plasma concentration was 156.5.0 +/- 92.4 ng/ml (HD) and 73.1 +/- 45.5 ng/ml (LD). Mean clearance was 18.0 +/- 3.1 ml/min/kg (HD) and 10.5 +/- 3.6 ml/min/kg (LD). The Mann-Whitney test was used to compare the groups and a statistically significant difference in half-life and clearance was found ($p = 0.03$ for both). The lack of dose proportionality may be due to delay in metabolism at the higher dose, as there was no significant difference between groups in the amount of glucuronidated metabolite. The limited glucuronidation ability of cats may explain this observation.

A single low dose of mirtazapine was well tolerated and resulted in a half-life that is compatible with 24 hour dosing intervals in healthy cats. Higher doses may result in delayed metabolism of the drug.

**ABSTRACT #283**

BIOAVAILABILITY OF A NOVEL, BIOENHANCED PREPARATION OF CURCUMIN IN DOGS. B Antony[1], RK Butchin[2], DW Griffin[2], 1. Arjuna Natural Extracts Ltd, 2. Nutramax Laboratories, Inc., MD.

Curcuminoids, the biologically active components of tumeric, have been shown to have high antioxidant, and anti-inflammatory activity. Both in vitro and in vivo studies on model systems in multiple species have demonstrated a wide range of effects that support the use of curcumin in a wide range of conditions. However, these applications are constrained by the relatively poor bioavailability of curcumin. A comparison was made in dogs of a standard commercial curcumin extract and a unique highly bioavailable enhanced formulation of curcumin extract (NMXCC95™ Nutramax Laboratories, Inc.). Two groups of 3 dogs, weighing 12-15 kg were dosed with 2 gram equivalents of curcumin either as curcumin powdered extract or NMXCC95™ following a 12 hour fast. After a one week washout dogs received the other formulation. Plasma curcumin levels were determined for 0-8 hr samples by HPLC method. The standard curcumin extract had a shorter $T_{max}$ compared to the NXMCC95™ formulation (1.5 hrs vs 3.17 hrs), but the enhanced formulation reached a 3-fold higher $C_{max}$ (296.4 vs 98.6 ng/g) and a 7-fold higher AUC over 8 hrs (1381 vs 199 ng•hr/g). Furthermore the plasma levels with NMXCC95™ remained elevated at 8 hrs (107.1 ± 62.7 ng/g) while levels from standard curcumin extract returned close to zero (6.3 ± 8.2 ng/g). These data show that the enhanced formulation of curcumin NMXCC95™ has a substantially higher bioavailability in dogs than standard powdered curcumin extract and has the potential to serve as an effective control of conditions with an underlying inflammatory basis.

**ABSTRACT #284**

BIOAVAILABILITY OF A NOVEL FORMULATION OF S-ADENOSYLMETHIOINE IN BEAGLE DOGS. DW Griffin, MO Whalen, CR Filburn. Nutramax Laboratories, Inc., Edgewood, MD.

S-adenosylmethionine (SAMe) is used to support liver function both in companion animals and humans, and is also used in humans for mental and joint health. Tablets of SAMe are typically enteric-coated to protect from moisture and enhance stability. In a blinded crossover study a new Denosyl® (Nutramax Laboratories, Inc.) chewable tablet containing a novel, non-hygroscopic, protected formulation of SAMe (NMXSS75™) was administered to beagle dogs to assess its bioavailability relative to an enteric-coated Denosyl SAMe tablet formulation. Six male and 6 female beagles were dosed after an overnight fast with 225 mg SAMe ion (mean 18.75 mg/kg) in one formulation of the supplement, followed by the other formulation 6 days later. Plasma values for SAMe 2 days before dosing and at 0 - 24 hrs were determined using an LC-MS/MS assay. Plasma SAMe increased substantially to $C_{max}$ values of 1,577 ± 920 and 1,964 ± ng/ml for the enteric coated and chewable tablets, respectively. Enteric coated and chewable tablet mean values for $AUC_{0-24}$ (5,741± 2,729 and 7,099 ± 2,313 ng•hr/ml) and $t_{½}$ (2.06 ± 1.17 and 2.01 ± 1.19 hrs) were comparable for both formulations. However, the time course of uptake with the chewable tablet was highly reproducible with an earlier, significantly shorter $T_{max}$ of 2.00 ± 0.0 hrs compared to the more variable uptake of the enteric coated tablet (3.42 ± 2.0 hrs, $p<0.01$). Unlike the enteric coated tablet, in the chewable tablet formulation SAMe was detected in the blood at 30 minutes. Fasting prior to dosing resulted in higher absorption than dosing with a meal, regardless of the type of formulation. These data indicate that this chewable formulation of SAMe can be used in dogs with equivalent bioavailability and substantially improved reproducibility to enteric coated SAMe tablets.

**ABSTRACT #285**

IN VIVO EFFECTS OF FIROCOXIB, MELOXICAM AND TEPOXALIN ADMINISTRATION ON EICOSANOID PRODUCTION IN TARGET TISSUES OF NORMAL CATS. LA Goodman, BT Torres, LR Reynolds, SC Budsberg. University of Georgia College of Veterinary Medicine, Athens, GA.

The purpose of this study was to investigate the in vivo activity of firocoxib, meloxicam and tepoxalin in normal cats by measuring eicosanoid production within target tissues. Eight normal adult neutered male cats were used in this blinded, randomized, crossover study. Cats were treated with firocoxib (1 mg/kg PO q 24 h), meloxicam (0.05 mg/kg PO q 24 h), tepoxalin (5.0 mg/kg PO q 12 h), or placebo for 8 days. Blood samples and gastric and duodenal mucosal biopsies were collected on days 0 (baseline), 3, and 8 of each dosing period. Thromboxane B 2 (TXB2) concentrations were measured in serum and prostaglandin E2 (PGE2) and leukotriene B4 (LTB4) levels were measured in plasma. PGE1 and PGE2 synthesis, and LTB4 levels were determined in the mucosal biopsy specimens. A 21 day washout period was observed between treatments. Repeated measures analyses were performed with significance set at $p<0.05$.

In the blood, firocoxib and meloxicam decreased plasma PGE2 levels compared to baseline on both days. Firocoxib and meloxicam decreased PGE2 compared to placebo on day 3. Tepoxalin did not decrease PGE2 in the plasma compared to baseline or placebo on either day. Firocoxib showed no difference in serum TXB2 compared to baseline or placebo on both days. Tepoxalin decreased TXB2 compared to baseline, placebo and firocoxib on both days. No decrease was noted in plasma LTB4 concentrations in cats administered placebo, firocoxib, meloxicam, or tepoxalin at any time. Firocoxib decreased pyloric PGE1 synthesis on day 8 compared to baseline. Meloxicam showed no significant differences from baseline or placebo on either day. Tepoxalin decreased duodenal PGE1 synthesis on both days compared to baseline. Neither firocoxib or meloxicam decreased pyloric PGE2 synthesis compared to baseline or placebo on days 3 and 8. Tepoxalin decreased pyloric PGE2 synthesis compared to baseline on days 3 and 8 and decreased pyloric PGE2 synthesis compared to placebo on day 3. Furthermore, tepoxalin decreased pyloric PGE2 synthesis compared to firocoxib and meloxicam on days 3 and 8. Tepoxalin decreased duodenal PGE2 synthesis compared to baseline on days 3 and 8. Meloxicam decreased pyloric LTB4 on day 3 compared to baseline and placebo. Tepoxalin decreased pyloric mucosal LTB4 on days 3 and 8 compared to baseline.

Firocoxib and meloxicam spared the cyclooxygenase-1(COX-1) enzyme while tepoxalin exhibited COX-1, COX-2 and lipoxagenase-5 (LOX-5) inhibition. Meloxicam may have LOX-5 inhibition capabilities at the dosages used in this study.

**ABSTRACT #286**

INTESTINAL PARASITES OF DOGS ON THE GALAPAGOS ISLANDS.

EN Gingrich[1], AV Scorza[1], MR Lappin[1], EL Clifford[2]. 1. Colorado State University, College of Veterinary Medicine and Biomedical Sciences, Fort Collins, CO. 2. Animal Balance, San Francisco, CA.

Dogs on the Galapagos Islands are a unique population created by isolation from the mainland and regulations prohibiting further importation. The effect of infectious agents of these domestic dogs on the indigenous fauna is largely unknown. The purpose of this study was to determine the prevalence of intestinal parasites in dogs on the Galapagos Islands.

Fecal samples were collected from 97 dogs presented during neutering campaigns on Santa Cruz (n=51), San Cristobal (n=17), and Isabela (n=29) islands. Feces were evaluated for parasites by microscopic examination after zinc sulfate centrifugation flotation as well as by a commercially available IFA for *Cryptosporidium* spp. and *Giardia* spp. Polymerase chain reaction for *Cryptosporidium* spp. DNA and *Giardia* spp. DNA was performed on all positive samples to provide the infecting genotypes.

*Ancylostoma caninum* (57.7%) and *Toxocara canis* (16.5%) were most commonly detected, followed by *Giardia* spp. (5.2%), *Isospora canis* (4.1%), *Sarcocystis canis* (3.1%), and *Cryptosporidium* spp. (1%). Adequate DNA for sequencing was available for one *Giardia* spp. which was shown to be assemblage D.

Despite being isolated, the dogs on the Galapagos have many of the same enteric parasites detected on the mainland of South America. These dogs are not routinely administered anthelmintics or other drugs, but are often allowed to roam the streets and live in close proximity to humans. Parasite prophylaxis is necessary to decrease the parasite burden within the population and to lessen the risk of spread to humans or other animals also inhabiting the islands.

**ABSTRACT #287**

CANADIAN PREVALENCE AND DIAGNOSIS OF GIARDIA INFECTION IN DOGS AND CATS USING A FECAL ANTIGEN TEST AND FECAL SMEAR.

Merle E Olson Bow Valley Research Inc. Calgary, Alberta and the Department of Biomedical Engineering, University of Alberta, Edmonton Alberta.

Giardiasis is a common gastrointestinal infection in dogs and cats but it remains a diagnostic challenge due to intermittent shedding and the small size of the cysts. A fecal antigen test (SNAP® *Giardia* Test kit, IDEXX Laboratories, Inc.) was utilized in Canadian veterinary clinics to determine the prevalence of *Giardia* in dogs (134 clinics) and cats (94 clinics) with gastrointestinal clinical signs. The fecal antigen test is an in-clinic ELISA which detects a highly specific, soluble cyst wall antigen that is released at the time the trophozoites encyst. Results of the fecal antigen test were compared to those of a fecal smear which was the method used by many veterinary clinics. A total of 1871 dogs and 389 cats were enrolled in the study. The presence of fecal antigens were observed in 299 (16.0%) and 30 (7.7%) symptomatic dogs and cats, respectively. Loose or watery diarrhea with increased frequency was the predominant presenting clinical sign of both dogs and cats. *Giardia* cysts were not observed in the fecal smears of 131 (68.2%) fecal antigen positive dogs and 11 (73.3%) fecal antigen positive cats. The fecal antigen test for *Giardia* appears to be a valuable tool in the diagnosis and ultimate control of giardiasis in dogs and cats.

**ABSTRACT #288**

PREVALENCE AND MOLECULAR ANALYSIS OF *GIARDIA* SPP. INFECTIONS IN DOGS IN CHIANG MAI, THAILAND: PRELIMINARY FINDINGS.

S Tangtrongsup, AV Scorza, LR Ballweber, JS Reif, MD Salman, MR Lappin, Colorado State University, Fort Collins, CO.

*Giardia* spp. is a common cause of waterborne diarrhea in human, pets and wildlife animals worldwide. However, the prevalence of this organism in dogs in Thailand and the potential for dogs to serve as a reservoir host is unknown. This study was conducted to explore the prevalence of *Giardia* infection in dogs in this province. The results will be used to determine prevention strategies in Chiang Mai.

A cross-sectional study was designed and 82 canine fecal samples were obtained from client owned dogs, breeding farms and a shelter during June and July 2008. Demographic and geographic data were recorded. Fecal samples were scored using a standardized system. *Giardia* infections were diagnosed using a PCR targeting the GDH gene. Associations of age category (<1 year, 1-7 years and >7 years), type of housing (client-owned, breeding farm, and shelter), fecal score (1 to 7), and *Giardia* positive results were analyzed using Fisher's exact test and odds ratios were estimated.

The prevalence of *Giardia* infections was 15.9%. Positive results were significantly associated with the age categories (p = 0.04). When compared to dogs >7 years, the odds ratio of dogs <1 year was 11.90 (95%CI: 0.61-230.84) and the odds ratio of dogs 1-7 years was 3.36 (95%CI: 0.18-64.1). Fecal score and type of housing were not significantly associated with *Giardia* positive results. The PCR products were genotyped and all isolates were *Giardia intestinalis* assemblages C or D.

In conclusion, *Giardia* infection in young dogs in Chiang Mai was common with dog specific assemblages.

**ABSTRACT #289**

SEROTYPES AND PATHOGENICITY ASSOCIATED WITH ANTIMICROBIAL-INDUCED RESISTANCE IN FECAL *ESCHERICHIA COLI* OF HEALTHY DOGS. N. Debavalya, and DM Boothe. Auburn University, College of Veterinary Medicine, Auburn University, AL.

The purpose of this study was to examine the relationship between resistance and virulence in canine fecal *E coli* isolates multidrug resistance (MDR) versus those not expressing multidrug resistance (NMDR) in response to antimicrobial therapy. We have previously demonstrated that close to 100% of fecal *Escherichia coli* in healthy dogs develop NMDR versus multidrug MDR resistance when treated with amoxicillin (10 mg/kg bid) or enrofloxacin (5 mg/kg sid), respectively, with no resistance emerging untreated controls (n=8 per treatment group).

Isolates from each treatment group (n=10) were subjected to phenotyping using microbroth dilution susceptibility testing to 17 antimicrobial drugs using a customized plate and following CLSI guidelines. Isolates were genotyped based on PFGE. Serotypes (O and H) and virulence factors (7 toxins and 1 adhesin) were then determined in one isolate from each phenotype and genotype in each treatment group.

Amoxicillin treatment resulted in 12 genotypes and 7 phenotypes among resistant isolates (NMDR) whereas enrofloxacin treatment (MDR isolates) resulted in only 1 genotype and 3 phenotypes. Only one genotype and phenotype was yielded from the controls (no resistance). Seven serotypes were detected in 11 NMDR isolates from amoxicillin treatment, whereas one serotype was detected in 3 MDR isolates from enrofloxacin treatment. The MDR serotype (H30:O9) was not similar to those of NMDR (H31:O83, H30:O8, H43:O138, H12:O-negative, H32:O-negative, H11 or 47:O11, and H-positive:O-negative). Control (susceptible) isolates exhibited serotype H4:O5 which also was different from both NMDR and MDR serotypes. Among the virulence genes tested, only Cytotoxic Necrotizing Factors (*cnf*) 1 and 2 were detected (indicating necrotoxigenic *E. coli*) and only in association with 3 phenotypes: one control isolate (*cnf*1), and two in the amoxicillin (NMDR) isolates that had the same serotypes; *cnf*1 in H31:O83 isolates (resistant to beta-lactams and ceftiofur) and *cnf*2 in H-positive (O-negative) isolates (resistant to beta-lactams, ceftiofur and tetracycline). No other toxins or adhesins were expressed in isolates of this study.

These data demonstrate that antimicrobial resistance induced in fecal *E coli* by either amoxicillin or enrofloxacin was not associated with an increase in pathogenicity.

**ABSTRACT #290**

COMPARISON OF QUANTITATIVE PCR AND CONVENTIONAL ENDPOINT PCR FOR AMPLICATION OF PARVOVIRUS DNA IN BLOOD FROM NATURALLY INFECTED AND RECENTLY VACCINATED DOGS. JK Veir, AL Duffy, SW Dow, MR Lappin. Department of Clinical Sciences, Colorado State University, Fort Collins, CO.

Canine parvovirus (CPV) can be detected in feces by antigen ELISA and in blood and feces by PCR. We previously demonstrated that CPV DNA can be readily amplified from the blood of healthy puppies via conventional endpoint PCR (cPCR) as early as day 2 and through day 14 after modified live CPV vaccination. Because viremia is present one to two days after infection in clinically infected dogs as well, differentiation of clinically affected and vaccinated dogs using endpoint PCR is impossible. We hypothesize that vaccination with modified live CPV should produce lower levels of circulating CPV DNA as compared to natural infection, thereby allowing differentiation between vaccinates and diseased dogs. Real time quantitative PCR (qPCR) allows for accurate and reliable quantification of viral load. The objectives of this study were to compare results of cPCR and qPCR assays using blood from vaccinated dogs and to compare the CPV DNA load between vaccinated dogs and naturally infected dogs.

A modified live CPV vaccine was administered SQ to twelve, six week old puppies on day 0. Presence of CPV DNA in whole blood taken from the puppies at days 0, 2, 5, 7, 10, and 14 was evaluated using both cPCR and qPCR. Additionally, the CPV viral load in peripheral blood collected on the day of presentation from naturally infected, fecal CPV antigen positive, clinically ill dogs was determined via qPCR. Results of the log transformed quantitative data were compared via the Kruskal-Wallis test with Dunn's post analysis and sensitivity and specificity calculated via a receiver operator characteristic curve (ROC). Agreement between the two methods was evaluated using the Kappa coefficient.

No parvoviral DNA was detected via either method in any vaccinated dog samples from day 0. Canine parvoviral DNA was detected via both methods in all samples taken on day 14. Kappa values for days 2, 5, 7, and 10 were 0.667, 0.657, 0.824, 0.833, indicating substantial agreement between the two methods in the samples from vaccinated dogs. There were statistical differences in CPV viral load between the vaccinated dogs and the naturally infected dogs at all time points except days 0 and 14 post vaccination ($p < 0.05$) with the naturally infected dogs having higher values. The area of the ROC curve was 0.9646 with a p value of <0.0001 when all samples except day 0 from the vaccinates were defined as the control group and the clinical animals were defined as the diseased group.

Canine parvoviral DNA is detectable by both cPCR and qPCR after vaccination. Additionally, quantitation of viral load in the peripheral blood was able to differentiate between naturally infected, diseased dogs and dogs that were recently vaccinated with a single modified live CPV vaccine. However, viral load following vaccination using other commercially available vaccines must be investigated prior to widespread clinical use of the assay.

**ABSTRACT #291**

THE EPIDEMIOLOGY OF *STREPTOCOCCUS CANIS* INFECTIONS IN DOGS AND CATS. EF Kruger, BA Byrne, PA Pesavento, KF Hurley, JE Sykes. University of California Davis William R. Pritchard Veterinary Medical Teaching Hospital, Davis, CA.

Little is known regarding the degree of genotypic relatedness between *Streptococcus canis* isolates from dogs and cats. The purpose of this study was to determine whether correlations existed between the genotypes of canine and feline *S. canis* isolates as determined using pulsed field gel electrophoresis (PFGE) and different clinical manifestations of disease. Eighty-nine isolates of *Streptococcus canis* that caused specific manifestations of disease in dogs and cats were collected from the University of California, Davis between 1998-2005. Associated clinical manifestations included sepsis, otitis, pyometra, skin infections, respiratory disease, and urinary tract infections. Ten isolates were also collected from a Northern California shelter which experienced an outbreak of *S. canis* infections manifesting as chronic respiratory infections and septic arthritis in cats. Bacterial isolates were characterized by PFGE analysis using the restriction enzyme *SmaI*. The relationships between banding patterns were analyzed using BioNumerics software combined with visual interpretation. Results demonstrated that the feline shelter isolates of *Streptococcus canis* were 99% similar in bacterial PFGE profile suggesting a clonal origin. The remainder of samples differed significantly, with less than 80% similarity in PFGE banding patterns. Thus in the latter group of isolates, there appeared to be no relationship between clinical manifestation of infection and PFGE profile.

**ABSTRACT #292**

BACTERIAL AND FUNGAL COLONIZATION OF PERIPHERAL INTRAVENOUS CATHETERS IN DOGS AND CATS. JP Pages[1], J Seguela[2] [1]Clinique Vétérinaire Croix du Sud, Saint Orens [2]Dept Clinical Sciences, National Veterinary School, Toulouse, France.

Infections associated with intravenous (IV) catheters have been recently shown to be a potential concern in an intensive care unit (ICU) population of 151 dogs and cats, the prevalence of positive catheter-tip culture rates reaching 25% (Marsh-Ng et al, JAAHA 2007;43:13-20). However, this issue in non ICU small animal patients has not been documented. The aim of this prospective study was to assess, in routine clinical settings, the prevalence of positive catheter-tip cultures associated with peripheral IV catheters in dogs and cats

100 peripherally placed IV catheters (20-gauge 1.1x30 mm and 22-gauge 0.9x25 mm) from 13 cats and 78 dogs were used for infusion therapy for various surgical and medical indications. They were aseptically removed and the distal two-thirds were cut and submitted to bacterial and fungal culture. Antimicrobial susceptibility of each isolate was determined.

17/100 bacterial cultures were positive. *Staphylococcus* spp. was the most common isolate (61%). The other isolates were *Klebsiella*, *Citrobacter*, *Escherichia*, *Pseudomonas* and *Enterobacter* spp.. The most frequent antimicrobical resistance were for amoxicillin (13/17), lincomycine (12/17), doxycycline (11/17), flumequine (8/17), tylosin (8/17), enrofloxacin (8/17). Inversely, resistance to gentamicin (3/17), amoxicillin-clavulanate (3/17) and cefalotin (2/17) were less common. *Candida glabrata* was also isolated in one catheter.

In conclusion, colonization of IV catheters with bacteria is frequent but its potential clinical consequences require further investigations.

**ABSTRACT #293**

DETERMINATION OF FOUR IgG SUBCLASSES IN ASYMPTOMATIC DOGS WITH VISCERAL LEISHMANIASIS AND IN ANIMALS VACCINATED AGAINST THE DISEASE. FA Ikeda-Garcia[1], MJ Day[2], VMF Lima[1], FA Rosa[1], LSV Sobrinho[1], AAD Gomes[1], JP Vides[1], SHV Perri[1], M Marcondes[1]. 1- São Paulo State University, Araçatuba, São Paulo, Brazil. 2- Department of Pathology and Microbiology, University of Bristol, Bristol, United Kingdom.

Visceral leishmaniasis in Brazil has been showing a geographical spread since the 80s, reaching big urban areas. In this context, a vaccinal immunoprophylaxis applied to dogs may represent an important control measure of the infection through the implementation of immune protective mechanisms. However, due to the fact that vaccinated animals have the ability of performing seroconversion, routine serological tests cannot differentiate an infected dog from a vaccinated one. This present study aimed to evaluate the four IgG subclasses in asymptomatic dogs naturally infected with *Leishmania chagasi* and in animals which were vaccinated against the disease. In order to meet the aforementioned purpose, three dog groups were used, as such: control (n=45), asymptomatic with visceral leishmaniasis (n=45) and vaccinated against the disease (n=37). The presence of anti-*Leishmania chagasi* IgG in serum was determined by the ELISA technique. In order to determine the four IgG subclasses, monoclonal antibodies were employed. In vaccinated animals the predominant subclass was IgG1, followed by IgG3, IgG2 and IgG4. In asymptomatic dogs the only subclass that was stimulated was IgG1. The IgG1 and IgG2 subclasses did not allow differentiation between dogs vaccinated against visceral leishmaniasis from naturally infected asymptomatic dogs. On the other hand, IgG3 and

IgG4 allowed differentiation of dogs from the two groups because the vaccinated animals had an increase of the two subfractions, whereas the same immunoglobulins from asymptomatic ones were not stimulated.

**ABSTRACT #294**

PERFORMANCE EVALUATION OF SPECIES-SPECIFIC PEPTIDE-BASED ASSAYS FOR DETECTION OF CANINE ANTIBODIES TO *E. CANIS* AND *E. CHAFFEENSIS*. J. Saucier[1], D. Daniluk[1], R. Krah[1], R. Chandrashekar[1], S. Gaunt[2], T. O'Connor[1]. 1. IDEXX Laboratories, Inc., Westbrook, ME. 2. Louisiana State University, Baton Rouge, LA.

Canine ehrlichiosis is caused by several species of tick-borne intracellular bacteria that are distributed worldwide. Infection is characterized by fever, lethargy, weight loss and thrombocytopenia and can lead to death. Early diagnosis and treatment provides the best opportunity for a favorable outcome and typically leads to complete recovery. Species-specific differentiation of *E. canis* and *E. chaffeensis* infection is important as it can lead to different approaches for disease management and treatment.

The goal of the project was to identify peptides that can be used in place of full-length proteins in diagnostic assays for *E. canis* and *E. chaffeensis*. Generally, we have found that peptides were more amenable for use in rapid-format assays and were typically more specific than full-length proteins. We synthesized and compared the reactivity of peptides from several regions of the p16 and p140 proteins of *E. canis* and from the variable length PCR target (VLPT) and p120 proteins of *E. chaffeensis*.

In a series of *E. canis*-experimentally infected dogs, the p16 peptide-based ELISA detected antibody between 13 and 17 days post infection which typically coincided with the appearance of clinical symptoms. The full length p16 recombinant protein ELISA and the synthetic peptide ELISA had identical results for 69 of the 71 dogs (31 positive, 40 negative) in a population of dogs from an *E. canis*-endemic region of the United States.

In *E. chaffeensis*-experimentally infected dogs the VLPT peptide-based ELISA detected antibody 7 days post infection which coincided with the detection of antibody in the IFA and recombinant protein-based ELISA. Twenty-eight of 29 samples reactive in the VLPT-recombinant protein ELISA were reactive in the synthetic peptide ELISA. Samples from *E. canis*-infected dogs and from *E. chaffeensis*-infected dogs did not react in the VLPT and p16 peptide assays, respectively. Samples from *E. ewingii*-experimentally infected dogs did not react in either assay.

These peptides can be used to develop diagnostic assays capable of simultaneous but specific detection of antibody to *E. canis* and *E. chaffeensis* in canine serum, plasma and whole blood.

**ABSTRACT #295**

*EHRLICHIA/ANAPLASMA* SPP. INFECTION AND EXPOSURE RATES IN DOGS FROM OKLAHOMA AND ARKANSAS. T. O'Connor[1], J. Saucier[1], S.E. Little[2], R. Chandrashekar[1]. 1. IDEXX Laboratories, Inc, Westbrook, ME. 2. Dept of Veterinary Pathobiology, Oklahoma State University, Stillwater, OK.

Several species of *Ehrlichia* and *Anaplasma* are capable of infecting dogs resulting in a condition characterized by fever, lethargy, weight loss and anemia that can lead to death. Species-specific identification of *Ehrlichia* is important because it can lead to different approaches for disease management and treatment.

The purpose of this study was to determine the prevalence of *Ehrlichia/Anaplasma* spp. by microscopy and PCR and the seroprevalence of antibodies reactive to *Ehrlichia* spp. by ELISA in a population of febrile and healthy dogs from the Ozark Plateau region of northeastern Oklahoma and northwestern Arkansas. A total of 143 samples were collected from veterinary practices (V3 and V4) (n=88) and shelters (n=55) in the study area. Among the samples from the veterinary practices, 11 (12.5%) were collected from febrile (T>103.4 F) dogs; ticks were noted on 9 (10.2%) of the dogs.

Results are detailed in the table below. PCR results for *E. canis* (*Eca*), *E. chaffeensis* (*Ech*) and *E. ewingii* (*Ee*) are based on conventional nested PCR; serology results for *E. canis*, *B. burgdorferi* (*Bb*), and *A. phagocytophilum* (*Ap*) antibody were obtained using the SNAP® 4Dx® test. Results for *E. ewingii* antibody were obtained using a peptide-based microtiter plate ELISA.

| | Serology | | | | PCR | | |
|---|---|---|---|---|---|---|---|
| **Group** | *Eca* | *Bb* | *Ap* | *Ee* | *Eca* | *Ech* | *Ee* |
| **V3 (n=39)** | 20.5% | 0 | 2.6% | 59.0% | 0 | 2.6% | 23.1% |
| **V4 (n=49)** | 8.2% | 0 | 0 | 55.1% | 0 | 4.1% | 4.1% |
| **Shelter, (n=55)** | 1.8% | 0 | 7.3% | 25.5% | 0 | 1.8% | 3.6% |
| **Total** | 9.1% | 0 | 3.5% | 44.8% | 0 | 2.8% | 9.1% |

Morulae were not found in any of the blood smears (N=143). Overall, 9.1% of the dogs were reactive in the *E. ewingii* PCR and 44.8% in the *E. ewingii* ELISA, supporting previous studies suggesting that *E. ewingii* may be the primary agent of canine ehrlichiosis in this region. The *E. ewingii* ELISA detected 11 of 13 (84.6%) of the PCR positive samples. Differences in positive rates between groups of dogs are likely due to timing of samples collection and relative exposure to *A. americanum* ticks.

**ABSTRACT #296**

EFFICACY OF DOXYCYCLINE TREATMENT IN DOGS NATURALLY INFECTED WITH *ANAPLASMA PHAGOCYTOPHILUM*. Diniz, PPVP[1]; Correa, MT[1]; Chandrashekar, R[2]; Beall, M[2]; Breitschwerdt, EB[1]. 1. Intracellular Pathogens Research Laboratory, North Carolina State University, College of Veterinary Medicine, Raleigh, USA. 2. IDEXX Laboratories, Westbrook, USA.

*A. phagocytophilum* is a tick-transmitted bacterium capable of infecting dogs, cats, horses, human beings and other species. Clinical observations of selected cases, as well as preliminary data from an experimental study, have suggested that doxycycline may not clear the infection in all treated dogs. Since differences in antibiotic susceptibility could occur between experimental strains and wild strains of this organism, this study evaluated the efficacy of doxycycline treatment for canine granulocytic anaplasmosis (CGA) in 11 naturally infected dogs. Dogs presenting signs compatible with CGA at two private clinics in the Northeastern United States (Baxter, MN and Plover, WI) were evaluated every 30 days over a 2-month period: at the initial consultation (day 0); after doxycycline treatment (day 30) and roughly 2 months after the initial consultation (day 60). Each enrolled dog was treated with doxycycline at a dose of 5 mg/kg twice a day for 28 days. Blood-EDTA and serum samples were collected at each evaluation date, and lymph node aspirates (LNA) were collected at days 0 and 60. Additionally, clients and veterinarians answered a clinical and epidemiological questionnaire at each visit. Finally, complete blood cell counts were performed from all enrolled dogs at each evaluation. DNA from blood-EDTA and LNA was extracted and *A. phagocytophilum* DNA was amplified targeting the msp2 gene. Antibodies against *Anaplasma* spp. were detected by Snap® 4Dx®. All 11 dogs were re-evaluated at the first follow-up (range of 24 – 47 days from day 0) and 10 dogs were evaluated for the second follow-up (range of 57 – 94 days). *A. phagocytophilum* DNA was not detected from blood-EDTA or LNA from any of the 11 dogs at day 30 nor from any of the 10 evaluated dogs at day 60. Six of 11 infected dogs seroconverted between day 0 and day 30, while 5 dogs were seroreactive on day 0. Platelet numbers significantly increased from day 0 to days 30 and 60 of the study (mean±SD: day 0 = 98.3 ± 69.3 $x10^3/mm^3$, range 11 – 250 $x10^3/mm^3$; day 30 = 255.7 ± 89.6 $x10^3/mm^3$, range 147 – 420 $x10^3/mm^3$; day 60 = 294.6 ± 92.3 $x10^3/mm^3$, range 205 – 451 $x10^3/mm^3$; $p<0.001$). Body temperature significantly decreased after treatment (mean±SD: day 0 = 104 ± 1.3°F, range 101.3 – 105.4°F; day 30 = 101.8 ± 0.5°F, range 101 – 102.8°F; day 60 = 101 ± 1.2°F, range 98 – 102°F; $p<0.001$). After doxycycline treatment, all dogs fully recovered according to clients' assessment. Our results suggest that doxycycline is effective in controlling *A. phagocytophilum* infection and in reversing clinical signs in dogs in selected areas of the USA; however, a larger population of naturally-infected dogs from several regions needs to be evaluated to confirm these results.

**ABSTRACT #297**

*BARTONELLA* SPP. ASSOCIATED ENDOCARDITIS IN DOGS IN COLORADO AND WYOMING.

A Fenimore[1], M Lappin[1], M Varanat[2], R Maggi[2], E Breitschwerdt[2]. 1. Department of Clinical Sciences, Colorado State University, Fort Collins, CO. 2. North Carolina State University, School of Veterinary Medicine, Department of Clinical Sciences, Raleigh, NC.

Since the mid 1990's, several species of *Bartonella*, including *B. vinsonii*, *B.quintana*, and *B. clarridgeiae* have been identified in dogs diagnosed with infectious endocarditis (IE). The prevalence of *Bartonella* spp. infection in dogs in the referral area for Colorado State University is unknown. The purpose of this study was to evaluate the heart valves of dogs with suspected IE for the presence of *Bartonella* spp. DNA by polymerase chain reaction (PCR).

The medical records system at the Veterinary Medical Center was searched for dogs with a clinical diagnosis of endocarditis that had been admitted from January 1990 to June 2008. DNA was extracted from the available formalin fixed valvular tissues and assessed for *Bartonella* spp. DNA by three PCR methods. The *Bartonella* spp. was determined by genetic sequencing or fluorogenic PCR.

Of the 119 patients with a presumptive diagnosis of endocarditis, 22 had IE based on histopathology results. The histopathology blocks and medical records were available from 9 dogs. *Bartonella henselae* DNA was amplified from the tissues of seven dogs; *B. vinsonii* DNA was amplified concurrently from three dogs. Of the seven dogs, six were from Colorado and one was from Wyoming (*B. henselae* only) and all were greater than five years of age. The dogs were not known to have left the region and fleas or ticks were reported on some dogs.

This is the first report of *B. henselae* associated IE in dogs. The results suggest that *Bartonella* spp. infection can occur in dogs in the western states and these agents should be on the differential list for dogs with suspected IE in the region.

**ABSTRACT #298**

DETECTION OF *BARTONELLA HENSELAE* IGM IN CATS WITH EXPERIMENTALLY INDUCED BARTONELLOSIS.

J Ficociello, C Bradbury, A Morris, MR Lappin. Colorado State University, College of Veterinary Medicine and Biomedical Sciences, Fort Collins, CO.

The temporal appearance of *Bartonella henselae* DNA in blood and *B. henselae* antibodies in serum has usually been studied in cats after parenteral inoculation. However, little is known about the progression of *B. henselae* test results in cats exposed to *B. henselae* infected *Ctenocephalides felis*.

Cats (n = 6) were inoculated with a field strain of *B. henselae* IV or were exposed to fleas allowed to feed on *B. henselae* PCR positive cats (n = 6). Blood and serum samples were collected on week 0, week 2 or 3, and then weekly until week 13. A *Bartonella* spp. IgM ELISA was validated using samples from experimentally inoculated cats. All samples were then assayed in the optimized IgM ELISA, a previously reported IgG ELISA, and a previously reported conventional PCR assay that detects DNA of *B. henselae*.

For cats inoculated IV, the first positive *Bartonella* IgM, IgG, and PCR assay result was detected on week 4 (four cats), week 3 (three cats), and week 3 (six cats), respectively. For cats exposed to fleas, the first positive *Bartonella* IgM, IgG, and PCR assay result was detected on week 8 (one cat), week 4 (one cat), and week 7 (one cat), respectively. All cats in both groups eventually became IgM positive (week 12 for flea cats; week 4 for IV cats) and PCR positive (week 11 for flea cats; week 3 for IV cats). Five IV inoculated cats were IgM negative week 13 and all cat exposed to fleas were IgM positive week 13. All cats infected IV became IgG positive by week 4 and maintained positive titers throughout the study. Of the cats exposed to fleas, five of six were IgG positive by week 12; the sixth cat never developed detectable IgG and died of *B. henselae* myocarditis during week 13.

Differences may exist in *Bartonella* spp. test results in cats based on the route of inoculation. These findings may relate to the infective dose administered or on effects imparted by the vector.

**ABSTRACT #299**

PREVALENCE OF *BARTONELLA* AND HAEMOPLASMA DNA IN SAMPLES FROM NON-OWNED FREE-ROAMING CATS ADMITTED TO ANIMAL SHELTERS IN COLORADO AND FLORIDA. JN Eucher, M Brewer, MR Lappin. From the Department of Clinical Sciences, Colorado State University, Fort Collins, CO.

*Bartonella* spp. and the haemoplasmas (*Mycoplasma haemofelis, 'Candidatus* M. turicensis', *'Candidatus* M. haemominutum') are common feline pathogens. *Ctenocephalides felis* has been shown to be the vector for *Bartonella* spp.. It is hypothesized that *Bartonella* spp. and haemoplasmas are transmitted between cats by bites or scratches. To further elucidate the mechanisms of transmission of these organisms, the prevalence of *Bartonella* spp. and haemoplasma spp. in blood, salivary glands, and tonsils was compared among free-roaming cats admitted to animal shelters in states where fleas are rare (Colorado) or common (Florida).

Blood, salivary glands, and palatine tonsils were collected from non-owned, free-roaming shelter cats in Colorado and Florida within 24 hours after euthanasia. DNA was extracted by use of commercially available kits and then assessed for *Bartonella* spp. and haemoplasma spp. DNA using previously published polymerase chain reaction assays.

The prevalence rates for *Bartonella* spp. DNA in blood (CO = 0 of 20 [0%]; FL = 10 of 18 [55.6%] $p < 0.0001$), salivary gland (CO = 0 of 24 [0%]; FL = 5 of 18 [27.8%]; $p = 0.017$) and tonsil (CO = 0 of 20 [0%]; FL = 6 of 18 [33.3%]; $p = 0.006$) were statistically different between the two states. Haemoplasma spp. DNA was detected in similar numbers of blood samples (CO = 1 of 20 [5%]; FL = 2 of 18 [11.1%]), salivary glands (CO = 3 of 24 [12.5%]; FL = 2 of 18 [11.1%]), and tonsils (CO = 3 of 24 [12.5%]; FL = 2 of 18 [11.1%]).

The presence of *Bartonella* spp. and haemoplasma DNA in salivary glands or tonsils may merely reflect the presence of infected erythrocytes within the tissues. Alternately, the results may support the hypothesis that the organisms are passed in these tissues as a mode of transmission.

**ABSTRACT #300**

ASSOCIATION OF PATHOGENICITY AND RATE OF COINFECTION OF THREE FELINE HEMOTROPIC MYCOPLASMA STRAINS WITH PRESENCE AND SEVERITY OF ANEMIA. C.M. Leutenegger, Thao M, Estrada M, Robertson J, Cornwell D. IDEXX Laboratories, West Sacramento, CA.

Three species of epierythrocytic mycoplasmas (Feline Hemotropic Mycoplasma, FHM) are known to infect cats: *Mycoplasma haemofelis* (Mhf), *Candidatus Mycoplasma haemominutum* (CMh) and *Candidatus Mycoplasma turicensis* (CMt). Infection with Mhf has been associated with severe hemolytic anemia in immunocompetent cats. In contrast, infection with Mhm, a smaller organism, has only rarely been associated with disease in immunocompetent cats and is dependent on the presence of cofactors such as co-infections with other hemotropic mycoplasma strains or retrovirus infections. With intermediate pathogenicity, CMt, a smaller organism similar to CMh, has been described to induce anemia in immunocompetent cats. The purpose of this study was to evaluate the association between strain presence, presence of coinfections and severity of anemia.

Whole blood samples from 500 cats submitted to the IDEXX Reference Laboratory system throughout the US were analyzed using Real-time TaqMan® PCR assays. These assays were developed with high analytical sensitivity and specificity for each strain and results are quantitative. The prevalence of cats mono-infected with each of the three hemotropic strains and the prevalence of cats coinfected with more than one strain was determined. Mycoplasma load for each infection was also determined.

Cats were divided into 3 groups according to hematocrit values: Group 1: <26% (moderately to severely anemic); group 2: 26-28% (mildly anemic); group 3: > 28% (non-anemic). The overall prevalence of FHM in group 1 was 29%. Monoinfections accounted for: MHf: 6.9%, CMh: 13.2%, CMt: 1.7 %. Coinfections: 26.1%. Group 2: overall prevalence 28.6%. Monoinfections accounted for: MHf: 5.5%, CMh: 16.5%, CMt: 1.1%. Coinfections: 19.2 %. Group 3: overall prevalence was 13.5%. Monoinfections accounted for MHf: 0%, CMh: 10.8%, CMt: 2.7%. Coinfections: 0%.

In this study, the presence and severity of anemia correlated with a) the degree of coinfections and b) the presence of Mhf and CMt and c) the Mycoplasma load present in cats. Monoinfections with the least pathogenic strain CMh in group 2 were associated with anemia if the Mycoplasma load exceeded 20,000 Mycoplasma organisms per ml blood.

**ABSTRACT #301**

ANALYTICAL SENSITIVITY AND SPECIFICITY OF A REAL-TIME PCR ASSAY DETECTING PATHOGENIC LEPTOSPIRA IN DOGS BASED ON THE HAP-1 GENE. C.M. Leutenegger, Palaniappan R, Elsemore D, Estrada M, Thao M, Cornwell D. IDEXX Laboratories, West Sacramento, CA.

Leptospirosis is a zoonotic bacterial disease of worldwide importance in both human and veterinary medicine. Disease is caused by serovars of the pathogenic species Leptospira interrogans sensu latu. The consequences of infection in dogs range from subclinical to peracute disease, with typical presentations including acute renal failure sometimes combined with acute hepatic disease. Currently available microscopic agglutination tests (MAT) cannot provide an early diagnosis since MAT relies on antibodies to leptospiral antigens and cannot detect infection until 10-15 days post-exposure. Immunofluorescence and immunoperoxidase staining in the clinical samples are not sensitive and culturing the organisms is difficult and slow due to their fastidious nature.

Renal carriage and urine shedding remains a key objective in the diagnostic workup to determine implication in active disease in dogs and to determine the zoonotic potential from a public health perspective. Serology is commonly described as a poor predictor of leptospiruria or health status. Molecular tools such as real-time PCR have been described to reliably detect pathogenic Leptospira in the urine, even in healthy carrier animals. This study aimed to define the diagnostic sensitivity and specificity of a real-time TaqMan® PCR assay targeting the gene encoding for the hemolysin adapted protein (hap1). This gene has been mapped to pathogenic Leptospira and is absent in non-pathogenic strains.

A total of 72 paired blood and urine samples in duplicate obtained from experimentally infected dogs were utilized in this study. Dogs were infected with Leptospira sensu lato, serovar pomona strain. Genomic DNA (gDNA) was extracted from whole blood and urine samples using an automated extraction system and analyzed in parallel using controls to monitor the amount and quality of the extracted gDNA as well as extraction efficiency using a spiked internal positive control (IPC). DNA samples were free of PCR inhibiting substances as determined by IPC and dilution experiments. Using MAT as the gold standard, real-time PCR had 92% diagnostic sensitivity and 99% diagnostic specificity in this sample set. Advantages of a molecular diagnostic approach are the detection of all pathogenic strains as opposed to those used for a particular MAT procedure, the acute phase and urine diagnostic application, and a limited risk for vaccine interference.

**ABSTRACT #302**

PREVALENCE OF INFECTIOUS AGENTS AND ANTI-ERYTHROCYTE ANTIBODIES IN CATS WITH ANEMIA.

KL Dowers, AG Miller, CB Webb, RF Keegan, AC Avery, PK Kiser, MR Lappin. Colorado State University, Fort Collins, Colorado.

The most common causes of feline immune-mediated hemolytic anemia (IMHA) are infectious agents and neoplasia. Primary IMHA is considered less common, but the Coombs' test is the only available assay to detect anti-erythrocyte (RBC) antibodies and it has low sensitivity and specificity in cats. The study's purpose was to determine the prevalence of infectious agents and anti-RBC antibodies using the Coombs' test and flow cytometry in anemic cats.

Blood collected from 35 anemic cats was evaluated with the Coombs' test at 37C and 4C, and flow cytometry. Infectious agent assays included FeLV antigen, FIV antibody, *Bartonella* spp. antibody, and PCR assays for *Mycoplasma hemofelis*, 'Candidatus *M. haemominutum*', 'Candidatus *M. turicensis*', *Bartonella* spp, *Ehrlichia* spp and *Anaplasma phagocytophilum.*

Of the cats, 57.1% (20/35) were positive for an infectious agent. Of these, 10 cats were positive for anti-RBC antibodies in at least one of the assays. Anti-RBC antibody positive cats by flow cytometry (5/5), Coombs' test at 37C (2/3), or Coombs' test at 4C (9/13) were usually concurrently positive for an infectious agent. There was 88.6% agreement between flow cytometry and the Coombs' at 37C and 66.7% between flow cytometry and the Coombs' at 4C.

Anti-RBC antibodies were detected in 50% of the cats with evidence of exposure to an infectious agent and 88.6% of anti-erythrocyte antibody positive cats were concurrently positive for an infectious agent. Therefore the presence of anti-RBC antibodies did not distinguish between infectious and primary immune-mediated causes of anemia.

**ABSTRACT #303**

*BARTONELLA HENSELAE* ANTIGEN RECOGNITION PATTERNS IN THE SERUM AND AQUEOUS HUMOR OF CATS WITH AND WITHOUT UVEITIS.

CL McInnis, CC Powell, JR Hawley, EJ Ehrhart, MR Lappin. Colorado State University, Fort Collins, CO.

Uveitis is a common ocular syndrome of cats that has multiple causes, including *Bartonella* spp.. However, the organisms are host-adapted and so *Bartonella* spp. antibodies and bacteremia are common in both normal cats and cats with uveitis, making it difficult to prove a clinical association with uveitis using currently available blood tests. The importance of accurately diagnosing ocular bartonellosis lies in the public health significance of this zoonotic disease, as well as in determining the optimal appropriate medical therapy for the individual cat. The objective of this study is to assess serum and aqueous humor *Bartonella* spp. antigenic recognition patterns as a diagnostic method for ocular bartonellosis.

Paired serum and aqueous humor samples were collected from cats with and without uveitis (healthy feral shelter cats from Florida). All samples were assayed in an IgG Western blot immunoassay, using culture grown *B. henselae* as the antigen preparation. The density and apparent molecular masses of any detectable band was determined with a computerized system by an individual blinded to the sample source. A serum or aqueous humor sample was considered positive if a previously defined immunodominant antigen was recognized with an average intensity greater than a predetermined cutoff value. The primary criterion for documenting intraocular production of *Bartonella* spp. antibodies was recognition of an immunodominant antigen by the aqueous humor sample but not the corresponding serum sample. In addition, if an immunodominant antigen was recognized by both the serum and aqueous humor sample and the average band intensity was greater in the aqueous humor sample, intraocular antibody production was documented.

*Bartonella* spp. antibodies were detected in serum of 61 of 84 cats (72.6%) with uveitis and 19 of 19 (100%) of the normal shelter cats. *Bartonella* spp. antibodies were detected in aqueous humor of 20 of 84 cats (23.8%) with uveitis and 0 of 19 (0%) of the normal shelter cats. Based on the criteria described, 10 of 84 cats (11.9%) with uveitis had evidence of intraocular production of antibodies against *Bartonella* spp.

The results of this study suggest that Western blot analysis on paired serum and aqueous humor samples from cats with uveitis may aid in the diagnosis of ocular bartonellosis. Whether these test results can be used to predict clinical response to appropriate medical therapy remains to be proven.

**ABSTRACT #304**

EFFECT OF ORBIFLOXACIN OR DOXYCYCLINE ADMINISTRATION ON *BARTONELLA* SPP. AND HEMOPLASMA TEST RESULTS FROM NATURALLY-EXPOSED CATS.

W Miller, AK Morris, M Lappin. Department of Clinical Sciences Colorado State University, Fort Collins, CO.

*Bartonella* spp. and hemoplasmas are differential diagnoses for cats with fever. Little is known concerning the effects of antibiotic treatment on clinical responses or diagnostic test results in naturally-exposed cats. The purpose of this study was to monitor clinical signs, *Bartonella* spp. IgG titers, and PCR assay results for *Bartonella* spp. and hemoplasmas after antibiotic treatment.

Veterinary clinics in states considered high risk for *Ctenocephalides felis* infestation (TX, FL) participated. Adult cats with a body temperature of > 102.5F without an obvious cause on physical examination were eligible for the study. All cats were assessed by complete blood cell count, FeLV antigen assay, FIV antibody assay, *Bartonella* spp. IgG ELISA, and PCR assays that amplify DNA of *Bartonella* spp., *Mycoplasma haemofelis*, '*Candidatus* M. haemominutum', and '*Candidatus* M. turicensis.' Each cat was administered doxycycline (50 mg, PO, daily for 28 days) or orbifloxacin (22.7 mg, PO, daily for 28 days). Owners were requested to return for followup testing one week and four weeks after finishing antibiotic treatment.

Of the 18 cats entered to date, 15 were administered the entire treatment course and both followup testing and clinical information were available. All cats were negative for FeLV and FIV but 14 had laboratory evidence of *Bartonella* spp. or hemoplasma infection. On day 0, three cats were *Bartonella* PCR +, *Bartonella* IgG +, five cats were *Bartonella* PCR +, *Bartonella* IgG -, five cats were *Bartonella* PCR -, *Bartonella* IgG +, and two cats were negative in both tests (one of these cats seroconverted by day 35). DNA of *M. haemofelis* (one cat) and '*Candidatus* M. haemominutum' (two cats) were amplified from some cats with concurrent evidence of *Bartonella* spp. infection. Clinical signs resolved within days in seven of nine cats (77.8%) administered orbifloxacin and five of six cats (83.3%) administered doxycycline. Two of eight cats with *Bartonella* PCR positive results became negative (one orbifloxacin and one doxycycline). One of eight cats with *Bartonella* spp. IgG in serum became negative (orbifloxacin). Two of three haemoplasma spp. infected cats became negative (one orbifloxacin and one doxycycline).

The combination of serology and PCR assay results is needed to provide the most information concerning the *Bartonella* infection status of individual cats. If these cats were clinically ill from *Bartonella* or hemoplasma spp. infections, both orbifloxacin and doxycycline had similar clinical effects. Administration of these antibiotic protocols does not consistently eliminate *Bartonella* spp. or hemoplasma spp. DNA from the blood of cats.

**ABSTRACT #305**

Abstract Withdrawn

**ABSTRACT #306**

ALPHA ENOLASE AND ANNEXIN A2 ANTIBODIES IN CATS. J Whittemore[1], JR Hawley[2], MR Lappin[2]. 1. University of Tennessee, Knoxville TN. 2. Colorado State University, Fort Collins, CO.

Feline herpesvirus 1, calicivirus, and panleukopenia (FVRCP) vaccine antigens are grown on cell lines. In previous studies, we had shown that cats inoculated with FVRCP vaccines grown on the Crandell-Reese feline kidney (CRFK) cell line developed anti-CRFK antibodies that also cross-reacted with feline renal cell extracts. In addition, some cats that were hypersensitized with CRFK proteins developed interstitial nephritis. By use of immunoprecipitation and protein sequencing, we showed the immunodominant antigens recognized by vaccinated cats or CRFK hypersensitized cats to be alpha-enolase or annexin A2. Spontaneously occurring antibodies to these two enzymes are associated with a variety of autoimmune disorders in people. The purpose of this study was to validate ELISA for detection of anti-alpha enolase and anti-annexin A2 antibodies in cats.

The protein structures of alpha-enolase and annexin A2 are thought to be conserved among species. Commercially available alpha-enolase and annexin A2 were purchased and used as the antigens in indirect ELISAs. Checkerboard titrations using different microELISA plates, binding buffers, antigen concentrations, and secondary antibody concentrations were performed in preliminary experiments. Using serum from vaccinated cats and CRFK hypersensitized cats assayed in six different ELISAs, the coefficient of variation for the optimized alpha-enolase and annexin A2 ELISAs were shown to be 9.7% and 10.1%, respectively. It was determined that a %ELISA value of > 10% was positive for both alpha-enolase and annexin A2 antibodies. To determine whether alpha-enolase and annexin A2 antibodies develop in FVRCP vaccinated cats, serum was collected from young adult cats housed in a SPF facility without vaccination (n = 20) and young adult cats (n = 20) three weeks after the administration of three doses of a commercially available modified live FVRCP vaccine.

While only one of the 20 SPF cats (5%) without FVRCP vaccination developed an alpha-enolase %ELISA value > 10%, 11 of the 20 cats (55%) of FVRCP vaccinated cats were positive ($p = 0.0012$). While only one of the 20 SPF cats (5%) without FVRCP vaccination developed an annexin A2 %ELISA value > 10%, 13 of the 20 cats (65%) of FVRCP vaccinated cats were positive ($p = 0.0001$).

Results of this study document that cats administered this FVRCP vaccine developed antibodies against alpha-enolase and annexin A2. Whether these antibodies are associated with clinical diseases in cats remains to be proven.

**ABSTRACT #307**

PREVALENCE OF FELINE IMMUNODEFFICIENCY VIRUS ANTIBODY AND FELINE LEUKEMIA VIRUS ANTIGEN IN SAMPLES SUBMITTED TO IDEXX REFERENCE LABORATORIES FOR FELINE HEMOTROPIC MYCOPLASMA TESTING. K. Curtis, C. Leutenegger, R Chandrashekar. IDEXX Laboratories, Westbrook, ME.

Three feline hemotropic mycoplasma (FHM) species are the causative agents for infectious anemia in cats; *Mycoplasma haemofelis* (MHf), *Candidatus Mycoplasma haemominutum* (CMh) and *Candidatus Mycoplasma turicensis* (CMt). Studies have indicated old age, male gender and outdoor activity as risk factors for FHM infection. FHM infection has also been found to be associated with feline immunodeficiency virus (FIV) and feline leukemia virus (FeLV) infection. The aim of this study was to evaluate the prevalence of FIV and FeLV infection in cats with suspected cases of hemoplasmosis.

Samples from anemic cats (n=185) submitted to the IDEXX Reference Laboratories for FHM testing by polymerase chain reaction (PCR) were obtained for this evaluation. FIV and FeLV testing was performed using the SNAP® Feline Triple™. In addition, reference laboratory data was obtained for each sample where available. FHM positive PCR results were obtained for 44 (23.8%) of the 185 samples. CMt infection was limited to cats co-infected with CMh; co-infection with MHf and CMh was found in 6 of the 44 FHM positive samples. FHM positive cats ranged in age from 7 months to 16 years (median of 7 years). FIV and FeLV infections were prevalent in these FHM suspect cases, 14/185 (7.6%) and 41/185 (22.2%) respectively. More than a third of the FHM positive cats were co-infected with FIV and/or FeLV.

| FHM | | FIV Positive | FeLV Positive |
|---|---|---|---|
| Positive | 44 | 8 | 11 |
| Negative | 141 | 6 | 30 |

The findings in this study are consistent with previous reports suggesting an association between FHM and retrovirus infection in cats. Because of this association and the potential for similar symptoms, testing for retrovirus as well as FHM is recommended in cases where infectious anemia is suspected.

**ABSTRACT #308**

AN UPDATED EPIDEMIOLOGICAL SURVEY OF FELINE IMMUNODEFICIENCY VIRUS INFECTION IN JAPAN. Y Nakamura[1], Y Nakamura[1], A Ura[1], M Hirata[1], K Nishigaki[2], H Tsujimoto[3], A Setoguchi-Mukai[1] and Y Endo[1]. 1. Kagoshima University, Kagoshima, Japan. 2. Yamaguchi University, Yamaguchi, Japan. 3. The University of Tokyo, Tokyo, Japan.

Feline immunodeficiency virus (FIV) is distributed in domestic cats worldwide. The prevalence of FIV is thought to range from approximately 10 to 30%. The subtypes A, B, C and D of FIV are present in Japan. An epidemiological and molecular based survey covering all over Japan had not been undertaken since an FIV vaccine was introduced to Japan in 2008. In this study, we evaluated the prevalence and distribution of subtypes of FIV in Japan.

Blood samples were collected from cats admitted at 47 hospitals located in each prefecture from March to October in 2008. Cats venturing outside at least once a week were included, however cats stating completely indoors were excluded. Age, gender and chief complaint were recorded at each hospital. The status of FIV infection was confirmed using two methods; a serological test using a SNAP FeLV/FIV combo kit (IDEXX Laboratories) for detection of anti-FIV antibody, and polymerase chain reaction for detection of *env* (V3-V5) and *gag* regions of FIV proviral DNA. The nucleotide sequence of amplified FIV *env*-derived DNA fragments were determined by direct sequencing and phylogenetic analysis was performed by a Neighbor-Joining method to estimate subtypes of FIV.

1,770 cats were finally included in this study; 411 (23.3%) were sero-positive. All kittens less than 6-month old were sero-negative for FIV; however, juvenile cats (0.5-2 years old) revealed 3.4-14.4% prevalence. Adult cats showed a high infection rate of approximately 28-30%. Male cats constituted 71.1% of all sero-positive animals. The infection rate in 716 cats with a history of fighting wounds was 34.9% while that in cats without any history was 15.0%. Infection rates in cats with or without any clinical signs were 28.9% and 12.6%, respectively. Proviral DNA was detected in 404 cases of the 411 sero-positive cats. In addition, nineteen sero-negative cats were positive by PCR analysis. The concordance between SNAP and PCR tests was 98.5%. DNA fragments of the FIV *env* V3-V5 region isolated from 347 cases proceeded to phylogenetic analysis. This analysis revealed that four subtypes (A, B, C and D) of FIV were still distributed in Japan. Regional characteristics were also observed: Subtype A in Hokkaido, Chugoku, Shikoku and southern Kyushu; Subtype B in the Tohoku and Kanto areas, Subtype C in mid-Japan; and Subtype D in the area Kinki and northern Kyushu. In addition, subtype D viruses were possibly divided into two groups, classic D and varied D. The varied type virus was mainly distributed in the Hokuriku area. Limited information about subtype C virus in Japan has previously been available; however, a number of subtype C virus cases were detected in this study.

We identified that FIV is epidemic in Japanese cats, especially in outdoor accessing cats. Even though an FIV vaccine was introduced in Japan, close attention still has to be paid to epidemic and genotypic trends of FIV.

**ABSTRACT #309**

EVALUATION OF CD4+ AND CD8+ T LYMPHOCYTES IN CATS NATURALLY INFECTED WITH FELINE LEUKEMIA VIRUS. FF Gonsales, A Reche Jr, AGT Daniel, SI Miyashiro, D Passarelli, BM Teixeira, MK Hagiwara. University of Sao Paulo, Faculdade de Medicina Veterinaria e Zootecnia, São Paulo, São Paulo, Brasil.

Feline leukemia virus (FeLV) is known as a cause of immune dysfunction and opportunistic infections in naturally and experimentally infected cats. Humoral and cell-mediated immune response abnormalities have been reported by many investigators, implicating in particular T-cell alterations. Cats exposed to FeLV may either become persistently viremic or recover from infection. Viremic cats showed little evidence of an immune response to the virus, and are at very high risk of developing a fatal FeLV-related disease within 2 to 4 years. The pathogenesis of the immune deficit caused by FeLV, however, remains incompletely understood. Changes in T-cell subpopulations in FIV-infected cats, especially decreased CD4+/CD8+ ratios have been reported to be similar to those occurring in HIV infection.

As FeLV infection is known to induce lymphopenia and there is little information on changes in lymphocyte subpopulations, T-cells subpopulations of FeLV infected cats were phenotyped and compared to FeLV-non infected cats in this particular study.

Eleven adult cats (7 male and 4 female, mean age –2.7 yrs) were identified as persistently infected with FeLV (IFA test for FeLV antigen). All cats with exception of one had presented chronic upper respiratory disease, and two of them, also had an associated severe gingivitis. Sixteen healthy adult cats (7 male and 9 female; mean age - 4.7 yrs) sharing the same household were negative for FeLV test. Neither one was infected with feline immunodeficiency virus (FIV). Blood from those cats was collected for CBC and lymphocytes phenotyping in a tube with EDTA–K3 (Vacutainer®). An additional twenty five young adult cats were included as FeLV- non infected controls. Absolute leukocyte counting was done on an authomatized blood analyser (ABC-Vet, ABX Diagnostics, Fr) and lymphocytes subsets were analysed by flow cytometry (FACScan; Becton Dickinson; Mountain View, CA). Mouse anti feline CD4: FITC mAb (Serotec MCA 1346F) and mouse anti-feline CD8+ mAb (Serotec MCA 1347PE) were used for lymphocytes phenotyping. The non parametric Mann-Whitney test was performed to compare both groups.

Decreased lymphocyte cells (2556.01±1697.76 and 6249.93 ± 3666.70 lymphocytes /µL; p = 0.03), T-CD4+ lymphocytes (585.73 ±543.46 and 1648.78±11. 86 / µL; p < 0.001), T-CD8+ lymphocytes (396.33±690.78 and 690.78 ± 461.93 /µL, p = 0.039) and CD4+/CD8+ ratio (2.31 and 3.03; p = 0.018) were observed among FeLV-infected cats compared to non-infected cats. The resulting immunodeficiency might be responsible for chronic upper respiratory infection found in almost all of FeLV-infected cats. It is concluded that a decreased CD4+/CD8+ ratio occurs in FeLV infection and evaluation of CD4+ T lymphocytes might be an useful tool to assess the prognosis of the infection.

**ABSTRACT #310**

CUTANEOUS MYCOFLORA AND CD4+:CD8+ RATIO OF CATS INFECTED WITH FELINE IMMUNODEFICIENCY VIRUS. A Reche Junior, AGT Daniel, TCPL Strauss, CP Taborda, SAV Marques, K Haipek, LJ Oliveira, JM Monteiro, JR Kfoury Junior. School of Veterinary Medicine – University of São Paulo, Brazil.

Opportunistic mycoses occur mainly when immunocompromising conditions are present. FIV-infected cats show a progressive depletion of CD4+ lymphocytes and other immunologic abdnormalities. Considering that FIV-infected cats may have a higher isolation rate of fungal flora than FIV-uninfected cats.

The purpose of the present study was to compare cutaneous mycoflora isolation and CD4+:CD8+ ratio in FIV-infected cats with that in FIV-uninfected cats.

Sixty client-owned cats were examined. Twenty-five were FIV-infected cats and 35 were FIV-uninfected cats. All the 60 cats were FeLV-negative. They were matched as closely as possible for age, sex, breed and lifestyle. None of the 60 cats had dermatologic problems at the moment of specimens collection.

Animals were tested for the retroviruses with a commercially available ELISA test kit. Cutaneous specimens were obtained from the cat coat by rubbing the whole hair with a sterile piece of carpet. Fungi were isolated from specimens using Sabouraud dextrose agar incubated at 27°C. Immunophenotyping of peripheral CD4+ and CD8+ T lymphocytes was done by flow cytometry.

Statistical analyses were done by one-way ANOVA and Tukey-Kramer multiple comparisons test. A $P$ value < 0.05 was considered significant. Results are presented as means±standard deviation (SD) where appropriate.

Fungal colonies were isolated from at least one specimen from 22 of 25 (88%) FIV-infected cats. Among the FIV-uninfected cats fungal colonies were recovered from 13 of 35 (37%) specimens. Dermatophytes were recovered from 2 of 25(8%) FIV-infected cats, from one coat specimen was isolated *Microsporum canis* and from the other *Microsporum gypseum*. Among the FIV-uninfected cats the only dermatophyte isolated was *Microsporum gypseum* in 3 out of 35 (8.5%) cats. *Malassezia* spp was the most commonly isolated organism overall (85.7%). *Malassezia* spp was more commonly isolated from FIV-infected cats, compared with FIV-uninfected cats (68.8% vs 28.6%).

The CD4+ to CD8+ lymphocyte ratio for FIV-infected cats was significantly less than the CD4+ to CD8+ ratio in the FIV-uninfected cats (0.328±0.222 and 1.250±0.507, respectively; $P$ <0.01). The CD4+ to CD8+ lymphocyte ratio for FIV-infected cats with cutaneous overall fungal isolation was significantly less than the CD4+ to CD8+ lymphocyte ratio in the FIV-infected cats but without cutaneous fungal isolation (0.187±0.044 and 0.683±0.085, respectively; $P$ <0.01). We can conclude that immunologic depletion due to retroviral infection might represent a risk factor to cutaneous fungal infection in cats.

**ABSTRACT #311**

PREVALENCE OF FELINE CALICIVIRUS AND HERPESVIRUS TYPE 1 IN MAINE COON CATS WITH CHRONIC GINGIVITIS. A Reche Júnior, AGT Daniel, CA Geraldo Júnior, MVC Albino, MN Arena, A Pellegrino, BM Teixeira, PE Brandão. School of Veterinary Medicine – University of São Paulo, Brazil.

Inflammatory diseases of feline oral cavity are extremely common in daily clinical practice. Factors that may contribute to the development of gingivitis and stomatitis are: diet, oral conformation, specific breed characteristics (juvenile gingivitis of Maine Coons), immune-mediated diseases and infectious diseases, like FIV, FeLV, feline calicivirus (FCV) and feline herpesvirus type 1 (FHV-1). The Maine Coon breed seems to be predisposed to develop chronic gingivitis. The aim of this study was to investigate the prevalence and correlation of the FCV and FHV-1 in Maine Coons with chronic gingivitis.

Seventy-two client-owned cats were placed in one of four groups: group I (n=22) – Maine Coons with chronic gingivitis; group II (n = 23) – Maine Coons without oral disease; group III (n=10) – Other breeds [domestic shorthair, Bengal, Norwegian forest cat, Ragdoll] with chronic gingivitis and group IV (n=17) other breeds without oral disease. These animals were matched as closely as possible for age, sex and lifestyle. All cats of all groups were negative for FIV and FeLV. Oral samples were collected with a sterile swab and RT-PCR was performed to search for FCV and PCR for FHV-1. Fisher's Exact Test was used to statistic analyze. A $P$ value <0.05 was considered significant. Results are presented as percentage of presence of virus.

Seventeen cats (77.2%) of group I, two cats (8%) of group II, none cat in group III and one cat (6%) of group IV were positive for FCV. Only two cats of group IV were positive for FHV-1. Statistically significant difference was obtained ($P$ < 0.0001) between group I and all other groups, for the presence of FCV. No statistically significant difference was obtained for FHV-1. Maine Coon cats of group I showed significant prevalence of FCV in gingival tissue, suggesting that this virus might have some influence on the chronic gingivitis in this breed. Although other researches reinforce the presence of the FCV in the pathogenesis of gingivitis, in this study the virus was found predominantly in Maine Coon cats with gingivitis. Considering that most of the Maine Coons used in the present study were born in catteries and FCV infection is usually a common problem among these cats, the results of the present study may reflect only differences in the environment rather than a Maine Coon predisposition to FCV infection. Although many of the other cats, including the Maine Coons without gingivitis were also born in catteries and did not show the same prevalence for FCV infection. Further studies are necessary to investigate the role of FCV on pathogenesis of chronic gingivitis in Maine Coon cats.

**ABSTRACT #312**

DISEASE PREVALENCE AND CAUSES OF DEATH IN AMERICAN KENNEL CLUB REGISTERED GREYHOUNDS. S Zaldivar, LM Marin, H Hamilton, CG Couto. The Ohio State University College of Veterinary Medicine, Columbus, OH.

Recently, we reported the prevalence of diseases and causes of death in retired racing Greyhounds (RRGs) using a web-based survey. Some of the diseases we see in RRGs appear to be uncommon in American Kennel Club (AKC) Greyhounds, and viceversa. The objective of the current study was to obtain comparable information for AKC Greyhounds, and compare the most relevant pathologies in both groups.

The data were collected through a survey posted online. The AKC announced the study and the owners/breeders completed the survey voluntarily; 214 responses were obtained.

Only 211 questionnaires were analyzed; 3 dogs were excluded because they were RRGs. Forty dogs were dead (19%) at the time of the study. The main cause of death was cancer (7 dogs, 17.5%). The most commonly reported group of diseases were dental (42, 20.2%), digestive (32, 15.5%), and skin (22, 10.5%). On the other hand, the most frequent disorders in RRGs were skeletal (33%), skin (28%), and digestive (18%).

Comparing these results with the data previously reported, we can conclude that in general the distribution of diseases in AKC Greyhounds is different than in retired racing Greyhounds. Cancer was the most common cause of death in both groups, and osteosarcoma (OSA) was the main tumor type found exclusively in retired racers (no AKC Greyhounds had OSA). Behavioral disorders were less frequent in the AKC group. The different results obtained could be explained by a progression of this breed towards two different genetic lines or changes due to environmental factors.

**ABSTRACT #313**

THROMBOELASTOGRAPHIC CHANGES AFTER GONADECTOMY IN RETIRED RACING GREYHOUNDS. P Vilar, N Westendorf, MC Iazbik, L Marín, MA McLoughlin, and C. G Couto. The Ohio State University, Veterinary Clinical Sciences, Columbus, OH.

Postoperative bleeding occurs in 26% of retired racing Greyhounds (RRG) that are spayed or neutered (Lara A et al; 2008). Identifying the patients at risk for prevention of this complication will be extremely valuable. Thromboelastography (TEG®) allows for a global analysis of the hemostatic system providing information about the primary and secondary hemostasis, and the fibrinolytic system. The objective of this study was to evaluate perioperative hemostatic features in RRGs using TEG®, and determine whether we could predict bleeding in Greyhounds that undergo surgical procedures. We evaluated 21 healthy Greyhounds (11 females and 10 males); 8 were classified as "Bleeders" and 13 as "Non-Bleeders" based on a bleeding scoring system we recently validated. Blood samples obtained via jugular venipuncture were collected into one 2.7 ml Vacutainer 1:9 with 3.2% buffered sodium citrate collection tube, and TEG® test were done using citrated native technique as previously described. There were no statistical differences ($p>0.05$) preoperatively between "Bleeders" and "Non-Bleeders" for TEG® parameters, fibrinogen concentration ($p=0.7465$), platelet count ($p=0.673$), or hemoglobin concentration ($p=0.1853$). In the "Non-Bleeders", the "MA" ($p=0.0017$) and "G" ($p=0.0047$) increased significantly after surgery. In contrast, in the "Bleeders" the "R" ($p=0.0315$) and angle ($p=0.0444$) were significantly prolonged and decreased, respectively after surgery. In addition, there were no significant changes in "MA" ($p=0.0873$) or "G" ($p=0.0959$) from the presurgery values. "Non-Bleeder" TEG® values after surgery suggest the formation a stronger clot and a decrease in fibrinolysis, an expected response to surgery also described in humans (Okamura K et al; 2007). "Bleeder" TEG® values after surgery supports the theory of slower clot formation and reduced clot strength, thus failing to reach a reactive hypercoagulable postsurgical state and contributing to the bleeding observed in the breed. Preoperative TEG® failed to predict bleeding in RRG that undergo surgical procedures.

**ABSTRACT #314**

IMMUNOLOGICAL RESPONSES IN PERSISTENTLY BOVINE VIRAL DIARRHEA VIRUS (BVDV) INFECTED ALPACAS. D Bedenice[a], W Davis[b], MJ Hamilton[b], A Grimm[b], E Dubovi[c], L Costa[a]. [a] Cummings School of Veterinary Medicine at Tufts University, North Grafton, MA; [b] Washington State University, Pullman WA; [c] Cornell Diagnostic Laboratory, Ithaca, NY.

The objective of this study was to characterize the effect of chronic and persistent BVDV infection in alpacas on immunological function.

Baseline flow cytometry and serum protein electrophoresis were performed on 9 persistently infected (PI), 16 age-matched control (BVDV PCR negative) and 2 chronically BVD infected alpacas. Chronic infection (CI) was defined by a positive BVD PCR for >3 weeks in the face of rising serum neutralizing BVDV antibodies. To further assess the functional activity of peripheral blood mononuclear cells [PBMC] in four PI, two CI and four control animals, cultured PBMC were separately stimulated with concanavalin A (ConA), phorbol myristate acetate (PMA), phytohemagglutinin (PHA) and pokeweed mitogen (PWM) for 3 days. Cells were subsequently collected and labeled for flow cytometry. Analyses were repeated at two separate time points for all animals. Overall results were reported as mean

(+/- standard deviation) and analyzed via independent samples T-test (P≤0.05).

Significant anemia and hypoproteinemia (reduced albumin but increased $\alpha 2$ globulin fraction), a reduction in T-lymphocytes (CD 4) and relative monocytosis were evident in PI compared to healthy control animals. Furthermore, monocyte, B lymphocyte and $\gamma\delta$ T cell proliferation were observed in mitogen stimulated cells. Superior stimulation of B cells was achieved using PWM (71-144%). PI alpacas consistently showed greater lymphocyte proliferation than control animals when stimulated with ConA, PHA and PWM.

Despite absolute reduction in peripheral T cells, PBMC from persistently BVDV infected alpacas demonstrated comparable, if not superior proliferation in response to in vitro mitogen stimulation.

**ABSTRACT #315**

HUMORAL RESPONSE TO BLUETONGUE VACCINATION IN ALPACAS. CE Whitehead[1], CA Batten[2], J Brownlie[1], B Jackson[1]. 1. The Royal Veterinary College, London, UK. 2. Institute for Animal Health, Pirbright, UK.

Bluetongue virus (serotype 8) was first documented in the UK in September 2007. This study evaluates a new vaccine (Bovilis® BTV8, Intervet) in alpacas. Three groups of alpacas were enrolled in a field trial: Group 1 (10 alpacas) received one dose of vaccine only; Group 2 (10 alpacas) received two doses, 3 weeks apart; Group 3 (6 alpacas, control group) received no vaccine. All alpacas were healthy, adult male and non-pregnant female alpacas kept at pasture. Blood samples were collected at Days 0, 21 (prior to vaccination), and 42. All samples underwent routine hematology and biochemistry, and ELISA and PCR testing for presence of antibody and virus respectively. Seropositive samples were tested for presence of neutralizing antibodies by serum neutralisation (SNT). Soft tissue reactions to the vaccine were common, and one alpaca exhibited transient lameness. All alpacas were seronegative at the start of the study and were negative for virus throughout the period of study. Control alpacas were negative for antibody and virus throughout. After one dose of vaccine (day 21), only 7/20 alpacas were seropositive. At day 42, Group 2 alpacas were all seropositive, had significantly greater percentage inhibition (PI) on ELISA compared with Group 1 alpacas (p<0.0001), and significantly higher serum neutralization titres (p=0.002). This study suggests that the vaccine appears to be safe in healthy alpacas and that it can stimulate an antibody response in alpacas. Further, this study suggests that two doses given 3 weeks apart is an acceptable vaccination strategy in alpacas.

**ABSTRACT #316**

INCIDENCE OF *CLOSTRIDIUM DIFFICILE* IN FECES OF MALE HOLSTEIN VEAL CALVES. MC Costa, LG Arroyo, DL Pearl, JS Weese, H Staempfli. Ontario Veterinary College, University of Guelph, Guelph, Ontario, Canada.

Neonatal calf diarrhea (NCD) is a multifactorial disease complex causing loss to the bovine industry. Recently, *Clostridium difficile* has been identified as potential additional cause for NCD in dairy calves. *Clostridium difficile* has been found in retail beef samples and was associated with sudden death in a group of veal calves. Very little information is available about *C. difficile* colonization in young calves. The objective of this study was to monitor longitudinally *C. difficile* colonization in a group of Holstein male calves reared on a milk replacer diet for meat on one farm.

For this study, 163 male Holstein-Friesen calves originating from several different farms were housed either individually (n=70) or in group pens (n=93). Calves were medicated with oral oxytetracycline for five days after arrival and were fed a complete milk replacer diet. Fecal samples were obtained by rectal swab within 48 hours post arrival and again six days after the first sampling. Several swabs were taken from the environment at the moment of the calves' arrival. Selective culture for *C. difficile* was performed, and isolates were characterized using standard methods.

*C. difficile* was isolated initially from 22/70 (34%) individually housed and in 42/69 (61%) 6 days later (P<0.001). Similarly, *C. difficile* was isolated from 31/93 (33%) group housed calves initially and in 46/93 (49%) 6 days later (P=0.037). Overall, *C. difficile* was isolated from 53/163 (33%) calves initially and from 88/162 (54%) at the second sampling. There was no difference in the prevalence of colonization between management types at either sampling point (P=0.87 and 0.16 respectively). 14 (45%) of group housed calves and 7 (32%) of individually housed calves that were initially positive were negative at the 2nd sampling. No *C. difficile* was grown from any of the 7 samples taken from the environment at the first sampling.

In conclusion the colonization rate of calves in this study was high, even at admission from different source farms and increased significantly shortly after arrival. Comingling, stress and antimicrobial therapy could have contributed to the high incidence. Virtually all calves had some degree of diarrhea at the time of the second sampling. Recently, toxigenic strains of *C. difficile* have been positively associated with diarrhea in dairy calves. Despite the significant increase in the prevalence of the bacterium concomitant with the increase in diarrhea, molecular analysis would be necessary to establish a clear association. Evaluation of typing data will provide additional insight into the epidemiology of *C. difficile* in calves reared in high-density environments and its' role in NCD.

**ABSTRACT #317**

PRAIRIE RATTLESNAKE ENVENOMATION IN 25 NEW WORLD CAMELIDS. JM Sonis, ES Hackett, RJ Callan, TB Hackett, Colorado State University Department of Clinical Sciences, Fort Collins, CO.

Morbidity and mortality from rattlesnake envenomation is regionally specific due to variability in relative toxicity of the species of snake encountered. A previous report of rattlesnake envenomation in New World Camelids from the western coastal United States documented high mortality rates and guarded prognosis for survival. The purpose of this study was to describe clinical findings, treatments, and outcome of New World Camelids following envenomation in the Rocky Mountain region of the United States by Prairie rattlesnakes (*Crotalus viridis viridis*). We hypothesized that Prairie rattlesnake envenomation of New World Camelids would result in a higher survival rate than that reported for rattlesnakes in other regions, and that the survival rate would more closely approximate that for Prairie rattlesnake envenomation in other mammalian species.

Records of New World Camelids evaluated for rattlesnake envenomation from 1992 to 2007 were reviewed retrospectively. Signalment, area of bite, clinical signs, clinicopathologic data, treatment, and outcome were recorded. Continuous data were reported as mean and 95% confidence interval. Fisher's exact test was used to evaluate for an association between platelet count, and treatments of steroids, NSAIDS, and commercial antivenin on outcome. Logistic regression was used to evaluate for an association between age, days of hospitalization, and outcome. For all comparisons, significance was set at a value of $p<0.05$.

Twenty-five New World Camelids experienced envenomation in this time period, consisting of 24 Llamas and 1 Alpaca. Llamas were overrepresented compared to hospital caseload. Mean age of envenomation was 3.8 years [95%CI: 2.8-4.9 years]. Presenting clinical signs included respiratory distress (7), fever (12), tachycardia (10), and tachypnea (12). The face was the most common site of envenomation (20 of 21). Most camelids were treated with intravenous fluids, antibiotic therapy, and NSAIDS. Eight animals required tracheotomy to maintain airway and 8 were treated with supplemental oxygen. Mean hospitalization was 2.4 days [95%CI: 1.4-3.4 days]. Overall survival rate was 72%. Age did not significantly impact outcome. Of the treatments evaluated, only commercial antivenin administration was protective for survival.

In conclusion, mortality rate for prairie rattlesnake envenomation in New World Camelids was lower than that reported in other regions of the United States and similar to that reported for prairie rattlesnake envenomation in horses. Commercial antivenin treatment should be considered for prairie rattlesnake envenomation in this species and may favorably impact outcome.

**ABSTRACT #318**

CONCENTRATIONS OF CARDIAC TROPONIN I MEASURED WITH AN I-STAT®1 ANALYZER IN NORMAL HORSES UNDERGOING A STANDARD TREADMILL PERFORMANCE EXAMINATION. MS Kraus, TG Divers, DV Nydam, SA Jesty, AR Gelzer, ND Ducharme, Cornell University, Ithaca, NY.

Elevated serum concentration of cardiac troponin I (cTnI) is a biomarker for myocardial damage in horses. Treadmill exercise is a common method for evaluation of horses with poor performance, but detection of occult heart disease is difficult. Treadmill exercise may induce increased cTnI concentrations, thereby unmasking myocardial disease. We hypothesize that in healthy horses, treadmill exercise does not induce cTnI concentrations above the normal range. Our objective was to determine cTnI concentrations in normal horses undergoing a standardized treadmill performance examination.

Eleven healthy horses (8 Thoroughbreds, 2 Standardbreds and 1 Warmblood) were exercised using a stepwise incremental treadmill protocol. Blood samples for cTnI were taken prior to exercise and at 5, 60, 180 minutes, and 24 post exercise and then daily for 5 days post exercise. Heparinized plasma samples were frozen at -20° C and analyzed for cTnI concentration utilizing the (i-STAT®1, Heska Corporation). Within each horse the change in cTnI over time was assessed using linear regression. The regression coefficients were then compared to baseline levels of cTnI using a 1 sample Wilcoxon rank test. The p-value was 0.61, indicating that there was no significant change in cTnI concentration over time in these 11 horses. Clinical significance of this study is that cTnI concentrations are not increased in clinically healthy horses undergoing a standardized treadmill exam. Thus, cTnI concentrations above the reference range measured in a horse at any time point following similar treadmill exercise should be considered abnormal.

**ABSTRACT #319**

THE EFFECTS OF DEHYDRATION ON CENTRAL VENOUS PRESSURE IN ADULT HORSES. RD Nolen-Walston, JL Norton, R Boston, C Underwood, J Slack, BL Dallap. New Bolton Center, University of Pennsylvania School of Veterinary Medicine, Kennett Square, PA.

Central venous pressure (CVP) is used in many species to monitor right-sided intravascular volume status especially in critical care medicine. Our hypothesis was that dehydration in adult horses is associated with a proportional reduction in CVP. Ten healthy adult horses underwent central venous catheter placement via the right jugular vein, and CVP readings were obtained using water manometry. The horses were then deprived of water and administered furosemide (1mg/kg IV QID) for up to 36 hours. Weight, CVP, vital signs, PCV, total solids, and serum lactate were monitored at baseline and every 6 hours until a target of 5% dehydration was achieved. Linear regression analysis was used to assess the association of CVP and other clinical parameters with degree of dehydration over time. A value of $P<0.05$ was used to separate statistical differences from those due to the causes explored in this investigation. A significant association was found between CVP and percent dehydration ($P<0.001$), with a decrease in CVP of 2.2 $cmH_2O$ for every percentage point increase in dehydration. Other significant associations between dehydration and parameters measured included increased total protein ($P=0.007$) and increased serum lactate concentration ($P=0.048$). There was no significant association between CVP and heart rate, respiratory rate, rectal temperature, or PCV. In conclusion, there is a strong association between central venous pressure and the degree of dehydration caused by diuretic use and water deprivation. These findings suggest that CVP monitoring may be a useful addition to the clinical evaluation of hydration status in adult horses.

**ABSTRACT #320**

COMPARISON OF WATER MANOMETRY TO TWO ELECTRONIC PRESSURE MONITORS FOR CENTRAL VENOUS PRESSURE MEASUREMENT IN HORSES. JL Norton, C Underwood, J Slack, BL Dallap, RD Nolen-Walston. New Bolton Center, University of Pennsylvania School of Veterinary Medicine, Kennett Square, PA.

Central venous pressure (CVP) has customarily been measured in veterinary patients using water manometry. However, many institutions are now using bedside/stallside electronic monitors in both anesthesia and intensive care units for many aspects of patient monitoring. The aim of this study was to evaluate the agreement of CVP values obtained using water manometry and two commercially available bedside monitoring devices (Datascope Passport LT™ and Medtronic Lifepak 12™). Central venous catheters were placed routinely in 10 healthy adult horses. Measurement sets were taken every 12 hours over the course of 3 days. At each time point, three measurements were obtained with each of the three methods. Data were analyzed using Bland-Altman limits of agreement, with all devices compared pairwise. Compared to water manometry, agreement (bias) of the Passport was -1.94 $cmH_2O$ (limits of agreement, -8.54 to 4.66 $cmH_2O$) and of the Medtronic was -1.83 $cmH_2O$ (limits of agreement, -8.60 to 4.94 $cmH_2O$). When compared to the Passport, agreement of the data obtained with the Medtronic was 0.27 $cmH_2O$ (limits of agreement, -4.39 to 4.93 $cmH_2O$). These data show that both electronic monitors systematically provide measurements that are approximately 2 $cmH_2O$ lower than water manometry; however, differences between the two electronic devices are small enough (less than 0.5 $cmH_2O$) to be considered clinically insignificant. This discrepancy should be taken into account when interpreting data obtained with these monitoring devices. Although water manometry has typically been considered the gold standard technique, this study does not identify which of these methods is more accurate.

**ABSTRACT #321**

REPEATABILITY AND EFFECT OF HEAD POSITION ON CENTRAL VENOUS PRESSURE MEASUREMENT IN STANDING ADULT HORSES.

JL Norton, R Boston, C Underwood, J Slack, BL Dallap, RD Nolen-Walston. New Bolton Center, University of Pennsylvania School of Veterinary Medicine, Kennett Square, PA.

Central venous pressure (CVP) measurements are typically obtained from equine patients using water manometry, but the reliability of these measurements and the effect of changes in head elevation have not been established. In humans, variation in CVP is seen when the patient moves from supine to standing position. We hypothesized that, in the horse, alteration in head position relative to the heart would significantly alter CVP. Jugular CVP catheters were placed in 10 healthy adult horses and measurements were obtained every twelve hours for three days. Readings were taken in triplicate (n=498) at three different head positions: neutral (muzzle level with point of shoulder), elevated (muzzle at the level of the withers) and lowered (muzzle on the ground). Data were analyzed using mixed effect modeling, with significance set at $P<0.05$. Mean CVP in the neutral head position was 9.4 $\pm$ 3.6$cmH_2O$. In regards to repeatability, variation in the "neutral" measurements obtained at each time point was minimal (2%), as compared to variation as a function of day and subject. No significant difference ($P>0.4$) was demonstrated between repetitions at any time point. Head height had a significant and directional effect on CVP. The elevated head position decreased CVP by a mean of -2.0 $\pm$ 6.5$cmH_2O$ ($P<0.001$) while the lowered head position increased CVP by 3.7 $\pm$ 5.5$cmH_2O$ ($P<0.001$). In conclusion, CVP values obtained using water manometry are highly repeatable in adult horses, provided head position is standardized to prevent inconsistent readings.

**ABSTRACT #322**

CARDIAC TROPONIN I CONCENTRATIONS IN HORSES REFERRED FOR COLIC.

OS Diaz, MM Durando, EK Birks, VB Reef. University of Pennsylvania, School of Veterinary Medicine, Kennett Square, PA.

Cardiac dysrhythmias are observed in horses hospitalized with colic. Endotoxemia, hypovolemia, electrolyte and acid-base disturbances commonly associated with colic may predispose these horses to myocardial damage and rhythm disturbances. This study was performed to determine if a relationship exists between a measure of myocardial damage, plasma cardiac troponin I (cTnI), and outcome (survival/non-survival), treatment (medical/surgical), or the presence of dysrhythmias.

111 horses presented to the University of Pennsylvania Large Animal Hospital with a complaint of colic were included. Blood for cTnI was drawn at admission, and 12 and 24 hours post admission or surgery, for medical or surgical colics, respectively. A 24-hour ambulatory ECG was placed the morning after admission (medical colics) or surgery to record cardiac rhythm. Medical records were reviewed to obtain clinical and clinico-pathologic data and outcome. Data at admission were evaluated by ANOVA, chi-square and linear regression where appropriate. For all tests, $P<0.05$ was considered significant.

38/110 horses had an abnormal cTnI at admission. Higher cTnI concentrations were significantly associated with the occurrence of ventricular arrhythmias (VA) ($p=0.01$) and outcome ($p<0.001$), but not with treatment (medical/surgical). An abnormal cTnI concentration was significantly associated with increased HR ($p=0.03$), PCV ($p=0.012$) and lactate concentration ($p=0.002$), and VA ($p=0.012$), surgery ($p=0.032$) and mortality ($p=0.012$).

These data suggest that horses presenting for colic with elevated cTnI concentrations have a less-favorable prognosis for successful recovery from colic. Also, the association between higher cTnI concentrations and VA in horses presenting for colic suggests concurrent myocardial injury in these horses.

**ABSTRACT #323**

ASSOCIATION OF GROWTH HORMONE AND GHRELIN WITH THE ENERGY METABOLISM AND MORTALITY IN SEPTIC FOALS.

RJIM Barsnick[1], SDA Hurcombe[1], WJ Saville[1], NM Slovis[2], PS Marsh[2], PA Smith[1], RE Toribio[1]. 1. The Ohio State University College of Veterinary Medicine, Columbus, OH. 2. Hagyard Equine Medical Institute, Lexington, KY.

Sepsis is a major cause of mortality in newborn foals and has been shown to cause dysregulation of hormones involved in the hypothalamic-pituitary axis and energy metabolism in affected foals. Even though growth hormone is a central factor in energy regulation, information on growth hormone in neonatal foals (healthy or diseased) is limited. Ghrelin is produced by P/D1cells in the gastric mucosa to stimulate short-term hunger and the secretion of growth hormone from the pituitary gland. Both hormones play important roles in energy homeostasis.

The goal of this study was to determine the association between growth hormone and ghrelin concentrations with sepsis, mortality of sepsis, severity of disease and serum glucose and triglyceride concentrations in critically ill foals.

We measured the plasma concentrations of ghrelin and growth hormone in 44 septic, 62 sick non-septic (hospitalized for other diseases) and 19 healthy foals. Sepsis was defined as a sepsis score >11.

Ghrelin concentrations were higher in septic foals than in sick non-septic and healthy foals ($p<0.001$). Growth hormone concentrations were higher in septic and sick non-septic foals than in healthy foals ($p<0.001$). Septic foals had lower serum glucose and higher triglyceride concentrations than healthy foals ($p<0.01$). Among the septic foals, non-survivors had lower glucose and higher triglyceride values than survivors ($p<0.05$). Ghrelin and growth hormone concentrations were positively correlated with sepsis scores as well as triglyceride concentrations, but negatively correlated with glucose concentrations ($p<0.05$). No difference was found in ghrelin and growth hormone concentrations between septic foals that died or survived.

The results of this study suggest that ghrelin and growth hormone concentrations increase in response to systemic inflammation in septic foals, which probably is an attempt to overcome the negative energy balance by stimulating hunger and lipolysis. As septic foals are often anorexic, the neuronal response to ghrelin is likely to be impaired during sepsis leading to a lack of energy intake and a catabolic state in sick foals.

**ABSTRACT #324**

SEASONAL CHANGES IN PLASMA ALPHA-MELANOCYTE-STIMULATING HORMONE AND ADRENOCORTICOTROPIC HORMONE IN RESPONSE TO THYROTROPIN-RELEASING HORMONE ADMINISTRATION IN NORMAL AGED HORSES. RA Funk, AJ Stewart, AA Wooldridge, RJ Kemppainen, EN Behrend, AK Johnson. Auburn University College of Veterinary Medicine, Auburn, Alabama.

Diagnostic tests for equine pituitary pars intermedia dysfunction (PPID) including endogenous adrenocorticotropic hormone (ACTH) and the overnight dexamethasone suppression test (DST) have been demonstrated to be affected by season in the Northeastern United States. Adrenocorticotropic hormone response to thyrotropin-releasing hormone (TRH) is a new and potentially more sensitive diagnostic test for equine PPID, but has not been evaluated for seasonality. The purpose of this study was to evaluate seasonal changes in plasma ACTH and alpha melanocyte-stimulating hormone (αMSH) responses to TRH administration in the Southeastern United States.

Ten healthy aged horses with normal DST (indicating absence of PPID) were administered synthetic TRH (1 mg) intravenously. Adrenocorticotropic hormone and αMSH concentrations were measured at 0, 5, 10, 15, 20, 25, 30, 45, 60, and 180 minutes. Testing was performed in February, July, August, September, October, and November. Mean ACTH and αMSH concentrations at each time point were compared across months using 2-way analysis of variance. Significance was set at $p<0.05$.

Concentrations of ACTH (post-TRH administration) were significantly higher in August, September and October compared to February. Concentrations from July and November were not significantly different than concentrations in any other month. Concentrations of αMSH (post-TRH administration) were significantly higher in August compared to February. Concentrations in July were not significantly different than concentrations in February or August.

Adrenocorticotropic hormone and αMSH responses to TRH administration experience seasonal variation, with increased TRH-stimulated ACTH and αMSH concentrations in the fall. These results support previous evidence of increased activity of the equine adrenocortical axis in the fall months.

**ABSTRACT #325**

EFFECT OF INSULIN INFUSION INTO THE DIGITAL ARTERY ON GLUCOSE CONCENTRATIONS IN THE DIGITAL VEIN IN HEALTHY HORSES. John C. Haffner (1), Kellie A. Fecteau (2), Hugo Eiler (2). (1) Horse Science Center, Middle Tennessee State University, Murfreesboro, TN; (2) Department of Comparative Medicine, College of Veterinary Medicine, The University of Tennessee, Knoxville, TN.

Laminitis is often associated with insulin resistance and carbohydrate metabolism. The objective of this research was to determine whether insulin infused into the digital artery can affect the arterial-venous (A/V) glucose differential in healthy horses as an indication that glucose uptake by the foot is regulated by insulin.

Seven healthy horses were used. Both digital arteries and the lateral digital vein from one foot were catheterized with 18G X 3.2 cm Teflon catheters. Infusion was through the lateral artery by infusion pump. Blood samples were collected concurrently from the medial artery and lateral vein at 1 or 5 min intervals during untreated-control, saline-infusion, and insulin infusion at 0.1, 0.5, 1.0, 5.0 units/min (0.50 ml/min infusion speed). Each infusion period lasted 20 min. Glucose concentrations were determined by a validated glucometer.

Glucose A/V concentration differential were: Mean ± SD, (lowest-highest value), number of samples, % samples with arterial glucose concentrations greater than venous. No treatment-control period: 9.4 ± 8.6 (1-39), n = 32, 91.4%. Saline-infusion period: 9.8 ± 7.6 (1-33), n = 19, 70.3%. Insulin infusion period (all dosages combined): 9.0 ± 5.2 (1-23), n = 78, 98.7%; compared to controls there was no difference ($P> 0.05$). A/V differential was not affected by saline infusion, insulin or hypoglycemia (34-40 mg/dL level). The glucose A/V differential (9-10 mg/dL) demonstrated utilization of glucose by the digit. However, insulin infusion into the digital artery did not increase glucose A/V differential. This suggests that the foot is not affected by insulin.

## ABSTRACT #326

VALIDATION OF A SPECIES SPECIFIC ENZYME-LINKED IMMUNOSORBENT ASSAY FOR MEASUREMENT OF SERUM INSULIN IN HORSES. J Öberg, I Lilliehöök, O Wattle, Å Karlsson, J Bröjer. Swedish University of Agricultural Sciences, Uppsala, Sweden.

Equine serum insulin has previously been analyzed with methods developed for human serum insulin. The performance and results of these human RIA and ELISA methods vary considerably because of different cross-reactivity of the anti-human antibodies and matrix problems. A species optimized, quantitative method (Mercodia Equine Insulin ELISA) for measurement of equine serum insulin has currently been produced by Mercodia, Uppsala, Sweden. It is a direct sandwich ELISA in which two monoclonal antibodies are directed against two separate antigenic determinants on the equine insulin molecule. The detection limit is 0.01µg/L and the detection range is 0.02-1.50 µg/L.

Our study evaluated the precision and linearity of the Mercodia Equine Insulin ELISA and compared it with two methods for human insulin (Mercodia Human Insulin ELISA and Coat-A-Count Insulin RIA kit from DPC, Siemens Diagnostics, Los Angeles, CA, US). Biological relevance of the assay was evaluated by measuring insulin before and after feeding.

Forty healthy horses were sampled before and 90 minutes after feeding. These 80 serum samples were analyzed with all three methods. Pearson´s correlation values were calculated. Precision for the equine ELISA was evaluated by determining the intra- and inter-assay coefficient of variation (CV). Sera from one horse with a mean concentration of 0.116 µg/L was analyzed in duplicate on fifteen different assay runs. Another sample (mean 0.017µg/L) was analyzed in duplicate four times on one assay the same day. One sample with high insulin value (0.998µg/L) was diluted with physiologic saline in five steps. The results were compared with mathematically predicted values and recovery was determined.

Results from the equine ELISA had high correlation both with the human ELISA and the RIA, $r^2 = 0.97$ and 0.97 respectively. Inter-assay CV was 10.7% and intra-assay CV was 4.6%. Recovery was 92-122%. Insulin concentration was significantly greater after feeding compared with the pre-feeding sample in 40 horses (Sign test $p<0.0001$). The median insulin level using the equine ELISA before feeding was 0.07 µg/L compared with 0.29µg/L after feeding. In comparison, the insulin results from human RIA and ELISA increased from median 4.8 and 7.8mU/L to 21.1 and 26.1mU/L respectively.

The Mercodia Equine Insulin ELISA had good precision and linearity, and showed high correlation with both the human ELISA and the Coat-A-Count RIA. This commercial assay's advantages are that it is developed for equine samples and results are quantitative (µg/L) instead of using activity units based on human insulin.

## ABSTRACT #327

HYPERINSULINEMIA IN HORSES WITH PITUITARY PARS INTERMEDIA DYSFUNCTION AND EFFECTS OF HOUSING ON INSULIN RESPONSE DURING A GLUCOSE TOLERANCE TEST. HC Schott II[1], L Groppi[1], S Wismer[1], S Beyerlein[2], LJ McCutcheon[1], P Schenck[2], RJ Geor[1]. 1. Department of Large Animal Clinical Sciences 2. Diagnostic Center for Population and Animal Health, Michigan State University College of Veterinary Medicine, East Lansing, MI.

Insulin resistance (IR) can be a comorbid condition in horses with pituitary pars intermedia dysfunction (PPID) and presence of hyperinsulinemia, supportive of IR, is considered a negative prognostic indicator for long-term survival. Although the prevalence of PPID in older equids likely approaches 20%, the prevalence of concurrent hyperinsulinemia has not been documented. Thus, basal serum insulin concentration was determined (double antibody RIA, Diagnostic Systems Laboratories, Webster, TX) in nine horses with PPID diagnosed by presence of hirsutism and supported by an overnight dexamethasone suppression test. Blood samples were collected between 9-10 am after hay (but no concentrate) had been fed at 8 am. In addition, the insulin response to glucose administration (100 mg/kg, IV bolus) was determined in the nine PPID horses under three conditions: 1) at winter pasture (15-30°F); 2) after an overnight stay in the hospital environment; or 3) after 5 or 6 days in the hospital environment (60-65°F). Diet for this subgroup was mixed grass-alfalfa hay with no concentrate feed or other supplements throughout the study period (December, 2008). While in the hospital environment, horses remained confined in individual box stalls with no forced exercise.

Basal insulin concentration was 133 ± 32 (SE) pmol/L with a range of 14-911 pmol/L. Only one of the nine horses had a basal insulin concentration >300 pmol/L. The insulin response to glucose administration was not different when the subgroup of horses was tested at pasture or after 1 night in the hospital (Table) but insulin concentration at baseline and 5 and 30 min after glucose administration was greater ($p<0.01$) after 5-6 days of hospitalization. These data demonstrate that ambient conditions and perhaps amount of daily exercise can have a substantial impact on testing for IR.

| Housing | pre-glucose administration | 5 min | 30 min |
|---|---|---|---|
| Pasture | 106 ± 28[a] | 402 ± 79[a] | 189 ± 54[a] |
| Hospital 1 day | 90 ± 24[a] | 495 ± 117[a,b] | 325 ± 119[a,b] |
| Hospital 5-6 days | 204 ± 89[b] | 677 ± 152[b] | 504 ± 194[b] |

[*] different letters indicate significant difference ($p<0.01$) within each column

## ABSTRACT #328

CHARACTERIZATION OF THE CDNA SEQUENCE ENCODING FOR THE PLATELET INTEGRIN ALPHAIIB AND BETA3 IN A HORSE WITH GLANZMANN THROMBASTENIA. S Macieira, J Lussier, C Bédard. Faculté de Médecine Vétérinaire, Université de Montréal, Saint-Hyacinthe, QC, Canada.

Glanzmann thrombastenia (TG) is characterized by a defect of platelet aggregation. It is an autosomal recessive genetic disorder caused by an abnormality of the platelet receptor for fibrinogen. The integrin receptor located on the plasma membrane is formed by a complex of alphaIIb and beta3 subunits, encoded by two distinct genes. Recently, we identified a horse with clinical and pathological features of TG. Flow cytometry studies revealed a deficiency of the alphaIIb subunit, suggesting the presence of one or several mutations in the gene encoding for the protein.

The aim of this study was to describe this case of TG at the molecular level by characterizing the cDNAs encoding for the beta3 and alphaIIb.

Total RNA was extracted from platelets and converted into cDNA by reverse transcription and poly-dT. Genomic DNA was extracted from white blood cells. Specific primers for alphaIIb and beta3 were used to amplify by PCR the corresponding cDNA or genomic regions that were further characterized by sequencing and compared by BLAST analysis (GenBank). A point mutation from G to C was identified in exon 2 of alphaIIb causing the substitution of amino acid Arg72, normally found to Pro72. This amino acid change may result into abnormal structural conformations that yield an inactive alphaIIb subunit. The analysis of genomic DNA showed that this horse was homozygous for the missense mutation Arg to Pro. The genomic DNA sequences encoding exon 2 of the dam and the sire were heterozygous for this nucleic acid change and were clinically normal.

**ABSTRACT #329**

COMPARISON OF HEMATOLOGIC AND BIOCHEMICAL VALUES OBTAINED VIA INTRAVENOUS CATHETER VERSUS VENIPUNCTURE IN HOSPITALIZED HORSES. ML May, ME Utter, RD Nolen-Walston. New Bolton Center, University of Pennsylvania School of Veterinary Medicine, Kennett Square, PA.

During hospitalization, horses typically undergo frequent blood sampling for diagnostic testing and monitoring. The nature of veterinary patients and the need for numerous samples makes acquisition from an intravenous catheter both convenient and less stressful to the patient. We hypothesized that there would be no significant difference in the plasma chemistry and hemoconcentration parameters from venous blood samples obtained from a jugular catheter versus direct jugular venipuncture. Paired samples were obtained simultaneously from 23 adult hospitalized horses, of which 14 were receiving intravenous crystalloids, and 9 were receiving no fluid therapy. Where applicable, fluids were stopped for 2 minutes, and a pre-sample of 15ml withdrawn from the catheter and discarded. Using identical syringes and needles, 10ml samples were obtained from both the catheter and the contralateral jugular vein. Samples were submitted for blinded analysis including complete blood count, standard equine plasma chemistry analysis, serum lactate, PCV, and total protein concentration. Data obtained were analyzed using a mixed-model ANOVA and Bonferroni adjustment for multiple comparisons. There was no significant difference associated with sampling method (venipuncture versus catheter) irrespective of fluid administration status in any of the 24 analytes measured. In addition, the absolute differences between sampling methods were subjectively judged to be to be clinically insignificant. In conclusion, blood samples obtained via intravenous catheter have clinically equivalent values to those taken using venipuncture in commonly performed analyses. Further investigation is warranted to establish if this technique is associated with increased complications such as phlebitis or bacteremia.

**ABSTRACT #330**

EFFECTS OF LIPOPOLYSACCHARIDE, LIPOTEICHOIC ACID, AND PEPTIDOGLYCAN ON TUMOR NECROSIS FACTOR PRODUCTION IN EQUINE WHOLE BLOOD CULTURE. JL Day, PJ Johnson, AR Honaker, J Amorim, AE DeClue. University of Missouri College of Veterinary Medicine, Columbia, MO.

Whole blood culture ($C_{wb}$), an *ex vivo* method for studying inflammatory mediator response to stimuli, maintains interactions between populations of blood cells, and plasma and can be used to evaluate anti-inflammatory therapies. Although $C_{wb}$ has been used in horses, dose response curves and Gram-positive microbial products have not been evaluated. The purpose of this study was to determine the concentration at which TNF is maximally stimulated by several microbial products using $C_{wb}$.

Blood, collected from 4 healthy adult mares, diluted 1:2 with RPMI/penicillin/streptomycin was cultured on 12 well plates. The blood was stimulated with various concentrations of lipopolysaccharide (LPS) (5-1000 ng/ml), lipoteichoic acid (LTA) (100-2500 ng/ml), peptidoglycan (PG) (100-2500 ng/ml), or control solution (PBS) and incubated at 37°C for 24 h. A cell killing bioassay was used to measure TNF in the $C_{wb}$ supernatant. Data were analyzed using ANOVA and *post-hoc* Fisher LSD method with $p<0.05$ considered significant.

Multiple concentrations of LPS, LTA and PG significantly stimulated TNF production from the cultured blood. The lowest concentrations that stimulated maximal mean±SD TNF bioactivity for each microbial product compared to control (1.6±0.7 ng/ml) were 50 ng/ml for LPS (10.7±3.4 ng/ml; $p=0.007$), 500 ng/ml for LTA (17.7±8.8 ng/ml; $p=0.039$) and 500 ng/ml for PG (26.1±4.9 ng/ml; $p=0.006$).

In conclusion, $C_{wb}$ is an efficient method of evaluating TNF production in response to LPS, LTA and PG in horses. These data provide optimum stimulant concentrations for use with $C_{wb}$ in future studies investigating anti-inflammatory interventions for both Gram negative and positive infections in this species.

**ABSTRACT #331**

LAMINAR INFLAMMATORY GENE EXPRESSION IN THE CARBOHYDRATE OVERLOAD MODEL OF EQUINE LAMINITIS. BS Leise[1], PJ Johnson[2], RR Faleiros[1,3], SJ Black[4], JK Belknap[1]. 1. The Ohio State University College of Veterinary Medicine, Columbus, OH. 2. University of Missouri College of Veterinary Medicine, Columbia, MO. 3. Universidade Federal de Minas Gerais, Brazil 4. University of Massachusetts, Amherst, MA.

Similar to sepsis-related organ injury in the human patient, laminar injury in the black walnut extract (BWE) model is associated with marked increases in laminar inflammatory indices, including increases in laminar mRNA concentrations of COX-2, IL-1β and IL-6. However, no reports have detailed the inflammatory response in the carbohydrate overload (CHO) laminitis model. The CHO model more closely mimics other clinical causes of acute laminitis (i.e. grain overload, enterocolitis, strangulating intestinal lesions) due to the presence of endotoxemia, the extended time frame of systemic signs, and the higher frequency of structural failure of the laminae. The purpose of this study was to determine if a similar pattern of laminar inflammation characterized by proinflammatory cytokine expression occurs in the CHO model as has been previously reported for the BWE model. Twelve horses were administered 17.6 g of starch (85% corn starch/ 15% wood flour)/kg body weight via nasogastric tube; these horses were anesthetized either after a two degree increase in rectal temperature (DEV group, n=6) or at the onset of Obel grade 1 lameness (OG1 group, n=6). Control horses (CON group, n=7) were anesthetized 24 hours after NG administration of 6 liters of deionized water. Serial physical examinations including gait analysis and complete blood counts were performed. Laminar tissue was collected from horses while under anesthesia, followed by humane euthanasia. Complementary DNA was made from laminar mRNA isolated from 3 different segments of the dorsal laminae of each horse. Real-time PCR for IL-1β, IL-6, TNFα, IFNγ, COX-1 and COX-2 was performed. Increased mRNA concentrations ($P<0.05$) for IL-1β IL-6 and COX-2 were present in laminae from horses with OG1 lameness when compared to the CON and DEV horses. No differences between the groups were found for TNFα, IFNγ, or COX-1 at either the DEV or OG1 time points. In contrast to the BWE model, no increases in IL-1β, IL-6 or COX-2 mRNA concentrations were present in the DEV group. Furthermore, the CHO model also differs from results reported in the oligofructose model where no increase in IL-1β was reported, but an increase in IFNγ was present at the onset of lameness. In conclusion, although the onset of cytokine expression is delayed in the CHO model compared to the BWE model and is slightly different from the oligofructose model, a similar prominent laminar inflammation occurs at onset of lameness, further supporting the need to address inflammatory signaling in the acute clinical case of laminitis.

**ABSTRACT #332**

IL-10 DOWNREGULATES PRODUCTION OF PEROXYNITRITE IN EQUINE MONOCYTE-DERIVED MACROPHAGES: IMPLICATIONS FOR RHODOCOCCAL INFECTION. BA Sponseller, SK Clark, MM de Macedo, DM Wong, DE Jones. College of Veterinary Medicine, Iowa State University, Ames, IA.

*Rhodococcus equi* is an important cause of pneumonia, the leading cause of morbidity and mortality among foals aged one to six months, and a cause of significant economic loss to the equine industry. Infection of foals by *R. equi* is believed to occur within the first few weeks of life. Why some foals are susceptible to infection while adults are not is incompletely understood.

In previous *in vitro* studies we determined that activated neonatal equine macrophages express higher levels of IL-10 cytokine mRNA than adult controls (Sponseller et al., 2008). Similarly, Giguère et al. found that foals infected at approximately 3 weeks of age with virulent *R. equi* $103^{+}$ had increased levels of IL-10 in lung tissue at 3 and 14 days post-inoculation compared to foals infected with the avirulent, plasmid-cured *R. equi* $103^{-}$ and uninfected controls (Giguère et al., 1999). IL-10 down-regulates many effector functions of macrophages, including the release of reactive nitrogen and oxygen intermediates. In addition, studies in mice have convincingly demonstrated that peroxynitrite, a very reactive product of superoxide and nitric oxide, is required for killing of virulent *R. equi* in murine macrophages (Darrah et al., 2000). In fact, neither superoxide nor nitric oxide alone is effective in killing *R. equi*. Interestingly, vap-A and -G genes, contained on the *R. equi*-virulence plasmid, undergo increased transcription when *R. equi* is exposed to oxidative stress, indicating an adaptive bacterial response to such cellular defenses. Thus, we speculate that equine macrophage oxidative responses, and peroxynitrite in particular, are curtailed in the presence of increased expression of IL-10.

The purpose of this study was to determine whether or not IL-10 modulates production of peroxynitrite in equine monocyte-derived macrophages. Replicates of adult monocyte derived macrophage cultures in a 96 well-format were exposed to 0.2, 0.6, and 1.0 μg of anti-IL-10 antibody alone (R&D Systems), or with recombinant equine IFN-γ (R&D Systems), or with IFN-γ + LPS (Sigma-Aldrich). Intracellular peroxynitrite-mediated oxidation of dihydrorhodamine-123 (Invitrogen, Carlsbad, CA) was measured as rhodamine fluorescence with the FluoStar Omega fluorimeter. In all culture conditions, increasing concentrations of anti-IL-10 antibody resulted in increased levels of peroxynitrite, suggesting that a cytokine milieu rich in IL-10 would result in decreased macrophage production of peroxynitrite. Since activated foal macrophages express high levels of IL-10, our results indicate that neonatal foals infected with *R. equi* may have compromised macrophage oxidative responses and an inherent inability to clear infection.

**ABSTRACT #333**

EFFECT OF INTRAVENOUS LIDOCAINE ADMINISTRATION ON LEUKOCYTE EMIGRATION IN THE BLACK WALNUT EXTRACT MODEL OF LAMINITIS. Williams JM [1], Ravis W[2], Loftus J[3], Peroni J[4], Hubbell J[1], Faleiros R[1], Black SJ[3], Belknap JK[1]. [1]Department of Clinical Sciences, The Ohio State University, College of Veterinary Medicine, Columbus, OH; [2]Dept. of Pharmacal Sciences, Auburn University, Auburn, AL; [3]Department of Veterinary and Animal Sciences, University of Massachusetts, Amherst, MA; [4]Dept. of Large Animal Medicine, University of Georgia, Athens, GA.

Leukocyte emigration, the extravasation and migration of leukocytes into target tissue, is a central event in cellular injury resulting in organ failure in human sepsis. Recent research in the black walnut extract (BWE) laminitis model suggests that these animals exhibit a similar systemic inflammation as seen in sepsis, including leukocyte emigration into the laminar tissue that is likely to play a major role in laminar injury in laminitis. In models of human sepsis, intravenous lidocaine has been reported to decrease leukocyte emigration via the inhibition of leukocyte activation, endothelial activation, and expression of proinflammatory cytokines and chemokines. Anti-inflammatory properties of lidocaine in sepsis have also been proposed to be valuable in the treatment of ileus in the equine patient. The purpose of this study was to assess the efficacy of intravenous lidocaine therapy in the inhibition of leukocyte emigration patterns documented to occur in the BWE model of laminitis. Due to the fact that laminar tissue is an integumentary structure, investigators have used skin biopsies in two studies of the

BWE model to demonstrate similar leukocyte extravasation into the skin as occurs in the laminae. Therefore, in this study, immunohistochemistry for a myeloid leukocyte marker, calprotectin, was used on serial skin biopsies to assess the effect of intravenous lidocaine on leukocyte emigration. Twelve horses were administered BWE, and were immediately treated with a constant rate infusion (CRI) of either lidocaine (0.05mg/kg/min, LD group, n=6) or saline (CON group, n=6) for 10 hours (H). Skin samples (6mm diameter) were obtained from 3 areas of the neck (approx. 10 cm. apart) for the following time points after BWE administration: 0H, 3H, and 10H. Compared to the 0H time point, increased integumentary leukocyte (neutrophil and monocyte/macrophage) numbers were present at the 10 H time point ($P<0.05$), with a trend for increased leukocyte numbers at the 3H time point in both the LD and CON groups. No significant differences were present between LD and CON for any time point. These data are consistent with previously presented data from our laboratory showing no effect of lidocaine CRI on laminar cytokine or endothelial adhesion molecule expression or laminar leukocyte concentrations in the same horses. In conclusion, these data indicate that intravenous lidocaine does not effectively inhibit events leading to leukocyte emigration into tissues in disease states characterized by systemic inflammation.

*Previously presented at IVECCS, September 2008*

**ABSTRACT #334**

THE IMMUNE RESPONSE OF FOALS TO NATURAL INFECTION WITH EQUID HERPESVIRUS-2 AND ITS ASSOCIATION WITH FEBRILE ILLNESS. SA Bell, MT Blanchard, JL Stott, N Pusterla, SM Mapes, W Vernau, KD DeJong, NJ MacLachlan. University of California School of Veterinary Medicine, Davis, CA.

Equid herpesvirus-2 (EHV-2) infection of horses is ubiquitous. EHV-2 infection anecdotally has been associated with a variety of disease syndromes in horses, but its true pathogenic significance remains uncertain. Other gammaherpesviruses such as Epstein Barr virus (EBV) have been shown to cause febrile illness in humans related to viral immunopathologic effects. Thus, the purpose of this study was to describe the ontogeny of the immune response of a cohort of 9 foals to natural infection with EHV-2 by evaluating serial complete blood counts, lymphocyte morphology, cytokine gene expression in peripheral blood mononuclear cells (PBMC), viral load in nasal swabs and peripheral blood, and antigen-specific cellular immune responses of PBMC, in conjunction with clinical evaluation of the foals. The occurrence of fever in foals was not related to lymphocytosis or lymphocyte morphology, cytokine gene expression, or viral load, but was significantly associated with increased levels of antigen-specific PBMC in peripheral blood of the foals. These data suggest that cellular immune responses to EHV-2 may lead to an immunologically mediated disease of foals that is perhaps analogous to infectious mononucleosis caused by EBV infection in humans.

*Previously presented at The Second Havemeyer EHV-1 Workshop, Steamboat Springs, CO, September 2008*

**ABSTRACT #335**

USE OF QUANTITATIVE REAL-TIME PCR FOR THE DETECTION OF *SALMONELLA* SP. IN FECAL SAMPLES OF HORSES ADMITTED TO A VETERINARY MEDICAL TEACHING HOSPITAL. Emir Hodzic[1], Barbara A. Byrne[2], Samantha Mapes[3], Spencer S. Jang[4], K. Gary Magdesian[3], N. Pusterla[3]. [1]The Lucy Whittier Molecular and Diagnostic Core Facility, [2]The Department of Medicine and Epidemiology, [3]The Department of Pathology, Microbiology and Immunology and [4]The William R. Pritchard Veterinary Medical Teaching Hospital, School of Veterinary Medicine, University of California, Davis, CA.

The purpose of this study was to compare quantitative real-time PCR with conventional culture in the detection of *Salmonella* sp. from feces of horses admitted to our veterinary hospital.

A quantitative real-time PCR assay was developed for the detection of *Salmonella* sp. in the feces of 911 equids admitted to a veterinary hospital. Fresh feces and feces following a 24 hour enrichment step in selenite broth were tested by conventional culture and real-time PCR targeting the *invA* gene of *Salmonella* sp. The detection limit for the real-time PCR assay was 3 and 10 *Salmonella* sp. organisms, when spiked samples were purified from selenite broth and feces, respectively. The analytical specificity was 100% based on the detection of a panel of 40 *Salmonella* serotypes from 5 serogroups and the lack of cross-reactivity with non-related microorganisms.

*Salmonella* sp. was not cultured from any of the fresh feces; however, *Salmonella* was cultured from the enrichment broth of 6 out of 911 fecal submissions (0.6%). *Salmonella* was detected by real-time PCR in 3 (0.3%) and 22 (2.4%) fecal and enrichment broth samples, respectively. Six broth samples were dually positive by culture and real-time PCR, while 16 additional broth samples tested positive only by real-time PCR. The *Salmonella* load in enrichment broth samples varied from 3 to 861,037 *Salmonella invA* gene copies/μl of DNA. Broth samples testing positive by both culture and real-time PCR had significantly higher pathogen load than broth samples testing only positive by real-time PCR. The PCR assay used in this study achieved an overall relative accuracy of 98%, a relative sensitivity of 100% and a relative specificity of 98% when compared to conventional culture.

The use of real-time PCR has the potential to reduce turn-around-time, with results being available within 22-28 hours from the time of sample collection. Further, the use of absolute quantitation may aid in the assessment of the infectious nature of hospitalized animals and may be an excellent alternative to conventional culture methods for surveillance and research studies.

**ABSTRACT #336**

COMPARISON OF PCR TESTING FOR MANAGEMENT OF AN OUTBREAK OF EHVI AT A RACETRACK. [1]AF Roy, [2]SC Eades, [2]RS McConnico, [2]FM Andrews, Louisiana Animal Disease Diagnostic Laboratory and [2]School of Veterinary Medicine, Louisiana State University, Baton Rouge, Louisiana.

An outbreak of Equine Herpes Virus I (EHV-I) involving racing Thoroughbred horses at a Louisiana racetrack afforded the opportunity to evaluate diagnostic testing techniques used to manage the outbreak situation. EHVI virus was determined to be the wild-type and not the neuropathogenic strain of the virus even though the index horse showed abnormal neurologic signs. A total of 74 exposed horses and 3 additional febrile and/or neurologic horses were sampled. Nasal swabs were passed at least 6 inches (15 cm) into the horse's nasal passages via the ventral meatus and swirled or held in place for at least 30-60 seconds to absorb respiratory secretions. Jugular venipuncture blood samples were taken for serology and molecular testing of the buffy coat. A single-step real-time PCR was performed on blood and nasal swabs using TaqMan reporter probes for discrimination between neuropathogenic and non-neuropathogenic strains of EHV-l. The realtime PCR was performed using an Applied Biosystems 7500 real-time PCR system. A 96 well format was used and the turnaround time from extraction to results is less than 12 hours. This single-step real-time PCR was compared with a nested PCR performed at another laboratory. The single-step real-time PCR identified more horses with EHV-1 viral DNA in nasal swabs than did the nested PCR and appears to be more sensitive. All PCR positive samples were identified as the non-neuropathogenic strain of EHV-1 Serological testing was performed using paired, acute and convalescent sera with an Indirect Fluorescent Antibody procedure.

**ABSTRACT #337**

SALMONELLA ORANIENBURG IN A LARGE ANIMAL VETERINARY TEACHING HOSPITAL. KJ Mitchell, KJ Cummings, LD Rodriguez Rivera, M Wiedmann, P McDonough, DM Ainsworth, TJ Divers, L Cavaney, GA Perkins, LD Warnick, Cornell University, Ithaca, NY.

To describe an outbreak of multidrug-resistant *Salmonella* Oranienburg within the Cornell University Equine Farm Animal Hospital (CU-EFAH) and discuss control and prevention measures.

Medical records of patients that were fecal positive for *S.* Oranienburg in the CU-EFAH from 2006-2008 were evaluated. Pulsed-field gel electrophoresis (PFGE) was performed on isolates.

*Salmonella* Oranienburg was identified in 28 animals: 3 cows, 5 alpacas, and 20 horses. The index case was admitted 8/15/2006; an additional 26 cases occurred in 2007, with a cluster between 09-10/2007. One case occurred in 01/2008, with no further cases for 12 months. Six animals negative during previous hospitalization were positive on re-admission, with 3/6 returning due to clinical signs of infection. Salmonellosis developed in 18/28 cases, five animals were euthanized. Antimicrobial sensitivity patterns revealed resistance to multiple antimicrobials. The minimum inhibitory concentration for enrofloxacin was $\geq$ 1 (intermediate sensitivity) for 18/28 cases, with one isolate showing *in vitro* resistance. The PFGE patterns from 23/26 animals were identical.

Environmental disinfection and culturing increased during the cluster of Salmonella cases in the fall of 2007. *S.* Oranienburg was isolated from the environment four times (from 611 cultures) during 2007 and once in 02/2008. Additional hospital-wide biosecurity measures were implemented, including twice weekly fecal cultures of all patients.

*Salmonella* Oranienburg is of concern given the multidrug resistance and evidence of high morbidity associated with infection. The PFGE data support hospital-acquired infection in the majority of cases. Improving biosecurity has led to elimination of animal infection and environmental positives for over a year.

**ABSTRACT #338**

PARADOXICAL ACIDURIA IN HORSES: 37 CASES (2000-2008). HC Schott II[1], JR ROSSETTO[2]. 1. Michigan State University College of Veterinary Medicine, East Lansing, MI. 2. The Ohio State University College of Veterinary Medicine, Columbus, OH.

Paradoxical aciduria (PA) is well recognized in ruminants with gastrointestinal disorders accompanied by hypochloremic metabolic alkalosis, hypokalemia, and decreased circulating volume. The proposed renal mechanisms that contribute to these clinicopathological abnormalities include increased bicarbonate reabsorption (consequent to hypochloremia) and maximal sodium reabsorption, partly via exchange with hydrogen ions (into the tubular lumen). In urine samples collected from dehydrated horses within the first few hours of hospitalization, the authors have also observed PA. Thus, we retrospectively examined 37 case records of horses that were >1 year of age, had a urine sample collected within the initial 6 hours of hospitalization that was acidic (pH<7), and had a venous blood pH >7.35 on admission blood gas analysis. Horses with primary renal disease were excluded. Urine pH ranged from 4.0-6.5 (reagent strip analysis), venous blood pH ranged from 7.36-7.49, and specific gravity ranged from 1.004-1.037 (no correlations found). Primary medical disorders included enterocolitis (16), colic (8), liver disease (2), neurological disorders (2), metritis (2), myopathy (1), inflammatory bowel disease (1), cellulitis (1), heart failure (1), pleuropneumonia (1), red maple leaf toxicosis (1), and fever of unknown origin (1). A common feature of all horses was partial to complete anorexia ranging from 1-14 days. Serum biochemical abnormalities included hypoproteinemia (20), azotemia (18), chloride ($Cl^-$) <100 mmol/L (19), total calcium <10 mg/dL (17), sodium ($Na^+$) <130 mmol/L (12), and potassium ($K^+$) <3.0 (9). The only significant correlations detected were between $Na^+$ and $Cl^-$ (0.80, p<0.01), $Cl^-$ and creatinine (-0.43, p<0.01), and $Na^+$ and creatinine (-0.36, p<0.02). Electrolyte and creatinine concentrations were measured in 18 urine samples and $Na^+$, $Cl^-$, and $K^+$ ranged from 10-148, 15-162, and 13-112 mmol/L, respectively, resulting in fractional clearances of 0.05-7.1%, 0.14-8.4%, and 6.3-69% for $Na^+$, $Cl^-$, and $K^+$, respectively. The only urine measure significantly correlated to urine pH was fractional $Na^+$ clearance (0.51, p<0.04), providing support that PA in horses may also be a consequence of increased renal $Na^+$ reabsorption.

**ABSTRACT #339**

DIAGNOSTIC UTILITY OF COMPUTED TOMOGRAPHY OF THE HEAD IN HORSES AFFECTED BY NEUROLOGIC DISORDERS.

C Sogaro-Robinson[1], VA Lacombe[2], SM Reed[3]. 1. Presque Isle, ME. 2. The Ohio State University, Columbus, OH. 3. Rood and Riddle, Lexington, KY.

The use of computed tomography (CT) and contrast-enhanced CT to image the head is common in horses presented for a wide variety of neurologic conditions. However, the validity of CT as a diagnostic indicator of neurological diseases in horses is unknown.

The validity of CT was estimated by comparing clinical, clinicopathologic, and histopathologic findings to CT findings in 11 horses presented for neurologic clinical signs, for which pre- and post-contrast CT images of the head were reviewed.

All horses with abnormal head CT had abnormal neurologic examinations on presentation and 90% (10/11) of horses exhibited abnormal mentation during hospitalization. Ten horses had abnormal postmortem examinations of the head, and CT imaging identified histopathologic lesions in 6 of these cases, which included: skull fractures, temporohyoid osteoarthropathy, large intracranial masses (oligodendroglioma, nasal adenocarcinoma, choleosteatomas) and acute vascular event. In 17% (1/6) of these CT studies, abnormal findings (meningeal congestion) were only detected after intravenous iodinated contrast injection. In 4 cases, pre- and post-contrast CT imaging failed to identify histopathologic lesions, such as encephalitis, meningitis, brainstem hemorrhage and local gliosis. CT sensitivity was 60% (95% confidence interval [CI]: 27-86), which was not greater than the sensitivity of CSF analysis (78%, 95% CI: 40-96). Postmortem examination of the head was unremarkable in 1 horse, for which CT imaging was normal (specificity, 100%).

Although it showed an excellent specificity, CT had a limited sensitivity, in detecting inflammatory disorders and small soft tissue lesions in the equine brain, which was not increased by contrast enhancement.

**ABSTRACT #340**

THE EFFECT OF METFORMIN ON INSULIN SENSITIVITY IN HORSES.

AM Firshman, Hustace, JL, Peterson K, Mata J. College of Veterinary Medicine, Oregon State University, Corvallis, OR.

Metformin is an oral antihyperglycemic drug used to enhance peripheral tissue sensitivity to insulin in humans. Metformin has been insufficiently studied in horses and conflicting results regarding efficacy have been reported. Equine Metabolic Syndrome (EMS) is a common condition that results in abnormal adipose tissue distribution, elevated plasma lipids, laminitis, and infertility. Horses with EMS are commonly glucose intolerant and insulin resistant. Treatment for horses with EMS includes feeding a low glycemic index diet and increasing exercise. Treatment with metformin in horses with EMS is appealing due to its potentially beneficial method of action and low cost. We hypothesized that insulin-stimulated glucose uptake could be increased by administering metformin to horses using dosages obtained from a preliminary pharmacokinetic study. The specific aim was to determine if metformin increases insulin sensitivity compared to placebo.

Eight normal adult horses, fed grass hay, were used in a 2x2 switch back design using 15mg/kg metformin PO q8hrs for 15d or saline placebo. Horses were allowed 6-8hrs of paddock turnout daily with a 2-week washout between treatments. At the end of each treatment period, a 3hr hyperinsulinemic euglycemic clamp was performed.

No effect of treatment was found on the rate of glucose infusion necessary to maintain euglycemia during the clamp compared to placebo ($P$=0.223). The results of this study indicate that 15 days of metformin treatment did not increase insulin sensitivity in normal horses. This lack of response may be related to the poor bioavailability of metformin in horses, however further study of the effect of metformin in insulin resistant individuals is warranted.

**ABSTRACT #341**

THE POSITIVE INTERACTION OF PROPIONATE, GLUCOSE, INSULIN AND CORTISOL IN THE HORSE.

DT Lewis, ML Katz. Department of Veterinary and Animal Science, University of Massachusetts, Amherst MA.

Propionate is the gluconeogenic substance derived through fermentation in the colon of the horse. Propionate has been demonstrated to increase plasma insulin in the pre-ruminal and ruminal sheep, goat and cow and plasma cortisol in the cow. Previously, propionate has been shown to not increase plasma insulin in the horse. During gluconeogenesis, propionate might stimulate hormonal changes similar to glucose. Therefore, this investigation examined the effects of increased plasma propionate on the concentrations of glucose, insulin and cortisol in the horse.

Four miniature horses were fed only low soluble carbohydrate first cutting grass hay and a vitamin mineral supplement for at least one year. Thus, we attempted to maximize gluconeogenesis as the source of glucose in these horses. We increased the plasma level of propionate about two fold in four miniature horses, based on the data of Argenzio & Hintz (1971) who reported a propionate turnover half-time of 8 minutes and an average basal plasma level of 4 mmol/l. Assuming an extracellular fluid volume of 20%, a priming dose of Na propionate in phosphate-buffered saline was administered as an intravenous bolus to increase the plasma propionate concentration to about 8 mmol/l and then a constant rate infusion pump was used to maintain this propionate concentration in the horse for 4 hours. Blood samples were taken every twenty minutes beginning one hour before and during the 4 hour infusion period. The horses had hay available to them during the experiments. Glucose was measured using the glucose-oxidase method and insulin and cortisol were measured by RIA (Coat-A-Count) methods.

Plasma glucose concentration steadily increased after the start of infusion so that plasma glucose was 10-20 mg/dl higher in all horses by 180 minutes after the beginning of propionate infusion. Each horse exhibited a marked rise (four to five fold) in insulin levels immediately upon the initial bolus injection of propionate, which was followed by a return towards baseline by the time of the 20 minute sample. Insulin levels then gradually increased to about two fold in all horses after 180 minutes of propionate infusion. During the propionate infusion plasma cortisol concentrations in all horses rose by 20 – 200% by 20 minutes and increased 200 - 500% by 180 minutes.

This study reveals a transient immediate insulin peak and a steady increase in glucose, insulin and cortisol concentrations in the plasma of the horse throughout propionate infusion, when fed a diet requiring gluconeogenesis to meet glucose needs. This data seems to reveal a direct temporal relationship between propionate and insulin.

**ABSTRACT #342**

ACUTE HYPERINSULINEMIA MODULATES EXPRESSION OF GENES INVOLVED IN GLUCOSE AND FAT METABOLISM IN SUBCUTANEOUS ADIPOSE TISSUE OF HORSES. JK Suagee[1], RJ Geor[2], LJ McCutcheon[2], BA Corl[1], MW Hulver[1]. Virginia Tech, Blacksburg, VA.[1] College of Veterinary Medicine, Michigan State University, East Lansing, MI.[2]

Low insulin sensitivity and/or high insulin concentrations have been associated with laminitic predisposition in ponies and prolonged experimental hyperinsulinemia (> 1000 mU/L) while maintaining euglycemia has induced laminitis in healthy ponies. This study examined the effects of elevated plasma insulin concentrations on expression of genes associated with glucose and fat metabolism in subcutaneous adipose tissue of healthy, non-obese horses.

Ten mares (BCS 6.3 ± 0.5) received a 6 hr infusion of insulin (6 mIU/kg/min) and a saline infusion, 13 d apart, in a balanced cross-over design. Euglycemia was maintained during insulin infusion by variable rate dextrose infusion. Blood samples for measurement of insulin concentrations were collected every 15 min. Following the clamp procedure, biopsy samples of subcutaneous adipose tissue (neck crest) were collected and RNA extracted to determine, by quantitative RT-PCR, the expression of CD36, FATP, GLUT-4, GLUT-1, and the insulin receptor (IR). Statistical analyses were performed using the mixed models procedure (SAS software system), with significance taken at $P < 0.05$.

Mean plasma insulin concentrations during the insulin and saline infusions were 933.5 ± 22.0 and 15.9 ± 21.7 mIU/mL, respectively. Expression of CD36 ($P = 0.002$), FATP ($P = 0.012$), and IR ($P = 0.006$) were decreased in adipose tissue following insulin infusion. The mRNA of GLUT-1 and GLUT-4 were not affected by treatment.

This study demonstrated that a 6-h period of supraphysiological hypersinulinemia decreased expression of genes involved in fatty acid transport (CD36, FATP) and insulin signaling (IR) in the subcutaneous adipose tissue of horses.

**ABSTRACT #343**

INSULIN SENSITIVITY IN PONIES IS INFLUENCED BY SEASON AND PASTURE. KH Treiber[1], RJ Geor[1], PA Harris[2]. 1.Virginia Tech MARE Center, Middleburg, VA. 2.WALTHAM Centre for Pet Nutrition, Melton Mowbry, UK.

Insulin sensitivity is a principle factor in glucose and energy regulation. Reduced insulin sensitivity (i.e. insulin resistance) has been implicated in equine pasture associated laminitis, which develops predominantly in the spring season. Increased risk of laminitis in the spring has been attributed to changes in pasture carbohydrates, however underlying metabolic predisposition may be influenced by other seasonal environmental factors. This study performed the Minimal Model of Glucose and Insulin Dynamics on 12 pasture-kept ponies in northern Virginia in January, March, May, July and September of 2007. Ponies were maintained together on the same pasture for at least 1 week prior to testing each month. On the morning of the test ponies were brought into stalls, jugular catheters were placed and 4 hour insulin-modified (20 mIU/kg) frequently sampled intravenous glucose (200 mg/kg) tolerance tests (FSIGTs) were performed. Ponies were then released into a drylot and fed grass hay. After 1 week on drylot the FSIGT was repeated. Glucose and insulin curves from the FSIGTs were modeled using MinMod Millenium and WinSAAM software to determine insulin sensitivity (SI, $mIU/L^{-1}\cdot min^{-1}\cdot 10^4$).

Table 1. Insulin sensitivity (SI) of ponies reported as median (IQR) and compared by Wilcoxon's matched-pairs signrank test.

| | January | March | May | July | September |
|---|---|---|---|---|---|
| Drylot | $0.17^c$ (0.27) | $0.26^{a,b}$ (0.98) | $0.45^a$ (1.11) | $0.19^{b,c}$ (0.34) | $0.22^{b,c}$ (0.56) |
| Pasture | $0.69^a$ (3.06) | $0.24^a$ (1.38) | $0.16^b$ (0.60) | $0.48^b$ (0.44) | $0.27^b$ (0.55) |
| Enclosure[1] | 0.003 (0/11) | 0.69 (6/6) | 0.019 (11/1) | 0.026 (1/10) | 0.69 (4/8) |

[a,b,c]Values with differing superscripts are different between months in the same row.
[1]Differences between pasture and drylot reported as P-value (negative/positive differences)

Drylot resulted in low SI values, possibly in response to energy or exercise restriction. In May, however, drylot SI was higher than other months, suggesting a spring increase in SI which would promote utilization of carbohydrates available in the pasture, facilitate activity during the breeding season, and potentially protect against laminitis. Conversely, SI on May pasture was numerically lowest, corresponding with clinical cases of laminitis which were successfully treated by restriction to drylot. Reduced SI on May pasture may result from the glycemic or digestive response to exaggerated carbohydrate fractions accumulating in improved pastures in the spring. Pasture SI values in March resembled drylot values, not surprisingly as very little grass was available and ponies were subsisting primarily on supplemented hay. Low SI values on September pasture may reflect a return to a carbohydrate and energy conserving metabolic state in preparation for the winter season. Understanding seasonal adaptations in metabolism and responses to nutritional environment will improve the management of equids, especially ponies, to avoid the development of pasture associated laminitis.

**ABSTRACT #344**

MONONUCLEAR LEUKOCYTES IN THE LAMINAE OF NORMAL HORSES AND THOSE WITH BLACK WALNUT EXTRACT-INDUCED LAMINITIS. RR Faleiros[1,2], GJ Nuovo[1], AD Flechtner[1], JK Belknap[1] 1. Ohio State University, Columbus, Ohio. 2. Universidade Federal de Minas Gerais, Belo Horizonte, Brazil.

The laminar interstitium is reported to have no resident leukocyte populations, but undergo a marked neutrophil infiltration in early stages of laminitis. In addition to neutrophils, both resident lymphocytes and monocyte/macrophages have recently been reported to play important roles in human sepsis-related organ injury. The purpose of this study was to assess monocyte/macrophage and lymphocyte populations in laminae of normal horses and those with black walnut extract (BWE)-induced laminitis. Immunohistochemistry was performed on laminar samples from 20 horses divided equally in 4 groups: control animals (CON), and 3 different time points after BWE administration: 1.5 hours (ETP group), onset of leucopenia (DTP group), and onset of lameness (LAM group). Antibodies against CD3, CD20 and CD163 were used to recognize lymphocytes (T and B), and macrophages (CD163+ indicates resident/"alternatively activated" macrophage), respectively. CON laminae contained $CD3^+$ and $CD20^+$ lymphocytes situated around deep dermal vessels, and $CD163^+$ resident macrophages in both the deep dermis (perivascular orientation) and in dermal laminae (primary and secondary). BWE treatment resulted in no changes in lymphocyte counts, but a marked increase ($P=.0016$) in $CD163^+$ macrophages in the secondary dermal laminae (SDL) in the ETP and DTP groups, returning to CON values in the LAM group. In conclusion, resident leukocytes in the form of lymphocytes and macrophages are present in the laminae of clinically normal horses, and BWE administration induces increases in the numbers of $CD163^+$ macrophages in SDL. Thus, mononuclear cells are likely to play both homeostatic and pathologic roles in laminitis.

**ABSTRACT #345**

Abstract Withdrawn

**ABSTRACT #346**

EVALUATION OF VARIOUS MATRIX METALLOPROTEINASE INHIBITORS (MMPIs) IN THE HORSE. LA Fugler, SC Eades, CE Koch. Louisiana State University School of Veterinary Medicine, Baton Rouge, LA.

The purpose of this study was to establish a non-terminal model of MMP induction in the horse, and to use this model for evaluating various MMPIs that may have potential therapeutic use in the treatment of equine laminitis. Lipopolysaccharide (LPS) was administered intravenously to normal horses, and digital venous plasma samples were collected hourly for 24 hours. Plasma MMP-2 and MMP-9 activities were assessed using zymography. The administration of LPS resulted in significant increases in MMP-2 and MMP-9 concentrations in the digital circulation. This model of MMP induction was then used to evaluate the effects of doxycycline, oxytetracycline, flunixin meglumine, and pentoxifylline on equine MMP inhibition. Horses were treated with the inhibitors every 12 hours beginning 12 hours prior to LPS administration, and digital venous plasma samples were evaluated for MMP-2 and MMP-9 activities using zymography. Pentoxifylline and oxytetracycline appeared to be potent MMP-9 and modest MMP-2 inhibitors in the horse. Flunixin meglumine and doxycycline were potent inhibitors of equine MMP-2, but only weak inhibitors of equine MMP-9. Recent studies suggest that MMP-9 may play an important role in the development of equine laminitis; therefore, these findings warrant the evaluation of pentoxifylline and oxytetracycline as MMPIs in the prevention/treatment of equine laminitis.

**ABSTRACT #347**

ULTRASONOGRAPHIC MEASUREMENT OF SPLEEN VOLUME IN HORSES. C Navas de Solis, JH Foreman, CR Byron, RA Carpenter. University of Illinois, Urbana, Illinois.

Ante-mortem measurement of spleen size might be helpful in clinical determination of splenomegaly. Our hypothesis was that spleen volume can be calculated from percutaneous ultrasonographic measurements in live standing horses.

Eight donated horses free of splenic disease were studied after obtaining IACUC approval. Ultrasonographic spleen volume and Evans blue dye-dilution blood volume (BV) were measured before (baseline) and after intravenous administration of detomidine and epinephrine (separately). Spleen volume was calculated using a standard ellipsoid formula ($VOL_{standard}$) and with the formula modified for the unique shape of the equine spleen ($VOL_{modified}$). Horses were euthanized using pentobarbital sodium, spleens were removed immediately, and spleen measurements were repeated. Descriptive statistics, correlations, and regression analysis were performed. The level of significance was set at $P<0.05$.

Mean $VOL_{standard}$ was 31.4±9.8 L at baseline, 34.0±10.5 L after detomidine, and 5.4±3.1 L after epinephrine. $VOL_{modified}$ was 25.1±9.1 L at baseline, 27.4±9.6 L after detomidine, and 4.2±2.4 L after epinephrine. Post-mortem spleen volume was

14.1±4.6 L and was correlated with $VOL_{standard}$ (r=0.86, P=0.006) and $VOL_{modified}$ (r=0.87, P=0.004). BV was correlated with $VOL_{standard}$ (r=0.81, P=0.014) and $VOL_{modified}$ (r=0.78, P=0.02). PCV after epinephrine was correlated with $VOL_{standard}$ (r=0.95, P=0.0002) and $VOL_{modified}$ (r=0.96, P=0.0001). When compared to baseline, spleen volume was smaller after epinephrine (P=0.009) but not larger after detomidine (P=0.12). BV after detomidine was larger than after epinephrine (P=0.009).

The study demonstrated that splenic size, under various pharmacological influences, can be calculated from ultrasonographic measurements in the live standing horse. PCV after epinephrine was the best indicator of spleen volume.

**ABSTRACT #348**

EFFECTS OF FLUNIXIN MEGLUMINE ON LEUKOCYTE ACTIVATION IN A LOW-DOSE ENDOTOXEMIA MODEL IN HORSES. G Forbes, SR Bailey, S Church, CJ Savage. University of Melbourne, Victoria, Australia.

Endotoxemia remains a major cause of equine morbidity and mortality. Tissue damage and organ dysfunction resulting from endotoxemia are partly brought about by leukocyte activation and their subsequent attachment and migration through the vascular endothelium and into tissues. Cytokines, matrix metalloproteinase enzymes and reactive oxygen species are released. Flunixin meglumine, in addition to its cyclo-oxygenase inhibitory actions as a non-steroidal anti-inflammatory drug, has been postulated to have further anti-endotoxic effects on leukocytes by inhibiting the transcription factor, NFkB. The purpose of this study was to determine the effects of flunixin on clinical signs and leukocyte counts in an established model of equine endotoxemia.

Endotoxin (lipopolysaccharide isolated from *E. Coli* 055:B5 bacteria) was administered intravenously at a dose of 1 ng/kg/min, over 30 min (total dose 30 ng/kg; dissolved in 500 ml sterile saline), into six adult female standardbred horses. Horses were pre-treated with either flunixin meglumine (1.1 mg/kg, intravenously) or saline. Each horse underwent both treatments (in randomised order), with a washout period of at least 14 days between each experiment. Clinical signs (including rectal temperature, heart rate, respiratory rate and mentation) were monitored at frequent intervals (initially every 15 min for 2 hours then every 30 min from 2 hours until 6 hours). Jugular venous blood samples were taken at the same time points. Leukocyte and platelet counts were performed using a Coulter Counter (model Z1; Coulter Electronics Inc.).

The endotoxin caused a small but significant increase in rectal temperature, reaching a peak of 39.18 ±0.15 °C at 210 minutes after the start of the LPS infusion. Other clinical signs were mild, and included transient muscle fasciculations. Blood leukocyte counts decreased from 9.15 ± 1.36 x $10^9$/L at time 0 to a nadir of 3.70 ±0.77 at 105 mins, before recovering back to baseline values by 5 hrs. Flunixin meglumine significantly reduced the peak temperature (maximum 38.48 ±0.14 °C at 120 mins). However, the blood leukocyte counts were not significantly affected by flunixin, reaching a nadir of 3.25 ±0.80 x $10^9$/L and slowly recovering to 5.25 ±1.30 x $10^9$/L at 5 hours.

These data suggest that while flunixin may effectively reduce some of the signs of endotoxemia, including pyrexia, it may not affect leukocyte activation in terms of endothelial adherence and subsequent tissue damage. Even at the recommended anti-inflammatory dose of 1.1 mg/kg, leukocyte margination was unaffected. Further studies are required to examine the cytokines and signalling pathways involved, affected by flunixin in this model; however, this study indicates that other therapeutic agents may be required to effectively inhibit leukocyte activation in endotoxemia.

**ABSTRACT #349**

THE PHARMACOKINETICS OF INTRAVENOUS AND INTRAMUSCULAR BUPRENORPHINE IN THE HORSE. KM Messenger, JL Davis, DH LaFevers, BM Barlow, LP Posner. North Carolina State University, College of Veterinary Medicine, Raleigh, NC.

Pain management options in equine medicine are limited due to a lack of available pharmacokinetic and safety data, expense of treatment, and a high risk of adverse effects with some classes of drugs. Opioid analgesics, particularly butorphanol and morphine are frequently used for moderate to severe pain in horses. However, these drugs have a short duration of action, require frequent dosing, and are associated with adverse effects such as gastrointestinal stasis and behavioral changes. Buprenorphine is a highly potent, longer acting mu opioid agonist analgesic frequently used in human and small animal medicine. In humans, therapeutic effects are achieved at concentrations greater than 0.5-1 ng/mL. Its analgesic effects have been studied in the horse, however its pharmacokinetics in this species are unknown. The purpose of this study was to determine the pharmacokinetics of buprenorphine after intravenous (IV) and intramuscular (IM) administration in healthy horses.

Six healthy adult horses were used in this study. Each horse received either IV or IM buprenorphine at a dose of 0.005 mg/kg in a randomized, crossover design, with a minimum 1 week washout period between experiments. Plasma samples were collected via jugular catheter immediately prior to drug administration, and then at 10, 20, 30, 45, and 60 minutes, and 1.5, 2, 4, 6, 8, 12, and 24 hours after drug administration. Horses were monitored for adverse reactions throughout the study. Plasma buprenorphine concentrations were measured using ultra performance liquid chromatography with electrospray ionization mass spectrometry.

Following IV administration, buprenorphine had a clearance of 7.88 ± 4.47 mL/kg/min, and a plasma half-life ($t_{1/2}$) of 7.14 ± 4.97 hr. The volume of distribution at steady state was 3.39 ± 1.46 L/kg. Following IM administration, the average maximum plasma concentration was 1.74 ± 0.09 ng/mL, which was significantly lower than the highest measured concentration (4.34 ± 1.22 ng/mL) 10 minutes after IV administration ($P < 0.001$). The time to maximum plasma concentration was 0.9 ± 0.69 hr. The plasma $t_{1/2}$ was 7.34 ± 5.06 hr. Bioavailability was highly variable, ranging from 41-93%. After administration of buprenorphine via either route, several of the horses showed signs of mild to moderate excitement. Gut sounds were decreased for 10 ± 2.19 hours in the IV group and 8.67 ± 1.63 hours in the IM group. The time to first defecation was approximately 5 hours after administration in both groups.

Buprenorphine has a long half-life in the horse and was detected in the plasma at concentrations expected to be therapeutic after a dose of 0.005 mg/kg after IV and IM administration. Signs of excitement and gastrointestinal stasis may be noted in horses.

**ABSTRACT #350**

BIOAVAILABILITY OF DETOMIDINE ADMINISTERED TO HORSES AS AN OROMUCOSAL (SUBLINGUAL) GEL AND COMPARISON OF ABSORPTION OF DETOMIDINE BY THE SUBLINGUAL AND INTRAMUSCULAR ROUTES.

H Kaukinen[1], J Aspegrén[1], S Hyyppä[2], JS Salonen[1] PhD, LTamm[1]. [1]Orion Corporation, Turku, Finland. [2] Agrifood Research Finland, Ypäjä, Finland.

Detomidine is a potent and specific alpha-2 adrenoceptor agonist that is widely used as sedative for horses. Injectable detomidine has also been shown to produce effective sedation when given to horses sublingually.

The objective of this open randomised study was to determine the bioavailability of detomidine administered to horses as an oromucosal gel, and to compare the absorption and the sedative effect of detomidine by sublingual and intramuscular routes.

Nine healthy institute-owned horses received approximately 40 μg/kg detomidine either intravenously, intramuscularly (Domosedan® 10 mg/mL solution for injection, Orion Pharma, Espoo, Finland), or as an oromucosal gel (Domosedan Gel® 7.6 mg/mL oromucosal gel, Orion Pharma, Turku, Finland) administered under the tongue with a 7-day withdrawal period between treatments. All horses received three treatments with a three sequences crossover design.

Blood was collected from an indwelling catheter before drug administration and at 10, 20, 30, 40, 60, 120, 180, 240, 300, 360 and 480 minutes post-dosing for the determination of detomidine level in serum. The effect of routes of administration of detomidine on heart rate and rhythm was evaluated, and sedation was assessed by measuring head droop and scoring ataxia at the time points of blood collection. The mean bioavailability of detomidine was 22% (±5.3%) as the oromucosal gel and 38.2% (±7.9%) injected intramuscularly. The sedative effects correlated with drug concentrations in serum with all routes of administration.

Sublingual bioavailability of detomidine is lower than after intramuscular injection, as part of the gel is swallowed and does not reach the systemic circulation. Despite the moderate bioavailability, oromucosal detomidine gel at a dose of about 40 mcg/kg produces safe sedation in horses. The bioavailability is lower than after intramuscular injection, as part of the gel is swallowed and does not reach the systemic circulation. The slower absorption seems to have the advantage of less and milder adverse effects.

**ABSTRACT #351**

PHARMACOKINETICS OF ORALLY ADMINISTERED TERBINAFINE IN HORSES.

Butch KuKanich, Megan Montgomery, and Elizabeth Davis. Kansas State University, Department of Anatomy and Physiology, Manhattan, KS.

The purpose of the study was to assess the pharmacokinetics of the antifungal drug terbinafine administered orally to horses. Terbinafine has a broad spectrum of antifungal effects and with the introduction of generic formulations is no longer cost prohibitive for use in veterinary medicine.

The study was approved by the Animal Care and Use Committee and Kansas State University. Six healthy horses weighing between 446 and 625 kg were administered 20 mg/kg of commercially available terbinafine tablets (250 mg per tablet) per os crushed and mixed with 60 mL light corn syrup. Blood samples were obtained at predetermined intervals for the determination of plasma drug concentrations with liquid chromatography and mass spectrometry. The pharmacokinetic parameters were estimated with noncompartmental analyses.

The plasma profile consisted of an absorption phase followed by a biphasic decline in plasma concentrations. The geometric mean (range) of the half-life (T ½), 8.1 (3.9 – 11.6) hours, clearance per fraction of the drug absorbed (Cl/F), 187 (132-282) mL/min/kg, and volume of distribution per fraction of the drug absorbed (Vd/F) 131 (50-266) L/kg were estimated for terbinafine. The maximum plasma concentration 0.31 (0.21-0.61) mcg/mL occurred at 1.7 (0.75 - 4) hours. The area under the curve extrapolated to infinity was 1.79 (1.18-2.51) hr*mcg/mL. One of the horses during the pilot study (data not presented) pawed at the ground, acted anxious and began circling in her stall 17 minutes after drug administration which resolved spontaneously within 30 minues. One of the horses during the pharmacokinetic study pawed the ground 10 minutes after drug administration, curled its lips and shook its head 10 minutes after drug administration which resolved spontaneously within 30 minute. No other adverse effects were noted.

In conclusion, further studies are indicated to assess the pharmacokinetics and safety of higher and multiple doses of terbinafine in horses. Further studies are also indicated to assess the efficacy of terbinafine for the treatment of fungal diseases in horses.

**ABSTRACT #352**

ASSESSMENT OF CARDIAC FUNCTION, SKELETAL MUSCLE MORPHOLOGY, AND MUSCLE ENZYMES IN HORSES ON LONG-TERM CLENBUTEROL THERAPY.

JA Thompson, SC Eades, AM Chapman, SA Barker, DB Paulsen, RS McConnico, School of Veterinary Medicine, Louisiana State University, Baton Rouge, Louisiana.

Potential for drug abuse exists in the horse industry, due to the perception that clenbuterol, a beta-adrenergic agonist approved for veterinary use as a bronchodilator in horses with reactive airway disease, increases performance and lean muscle mass. Recent studies into the effects of clenbuterol on exercising horses suggest that clenbuterol doses within therapeutic ranges negatively impact aerobic capacity and cardiac function. Studies in murine models show that clenbuterol causes skeletal muscle cell death at high doses, and clinical cases of equine rhabdomyolysis have been documented with clenbuterol overdose. This study examined the effects of oral clenbuterol, given at a dose up to 3.2ug/kg for 14 days, on skeletal and cardiac muscle in clinically healthy horses undergoing treadmill exercise, as compared to a control group.

Twelve clinically healthy Thoroughbred-type horses between the ages of 2 and 10 years old were evaluated. The study was approved by the Louisiana State University Institutional Animal Care and Use Committee. Horses were randomly assigned to either the control group (n=6) or the clenbuterol group (n=6). Animals in the control group received saline by mouth twice daily for 14 days. Horses in the clenbuterol group received clenbuterol as Ventipulmin® syrup (Boehringer Ingelheim Vetmedica, Inc., St. Joseph, MO) by mouth twice daily for 14 days. The dose was incrementally increased during the treatment period to minimize adverse side effects, according to the following schedule: 0.8ug/kg twice daily for 3 days, then 1.6ug/kg twice daily for 3 days, then 2.4ug/kg twice daily for 3 days, then 3.2ug/kg twice daily for 5 days. Horses were subjected to daily submaximal treadmill exercise during the treatment period. Venous blood samples were collected from the jugular vein(s) every 3 days during treatment. Echocardiography was repeated 7 days after beginning treatment. Muscle biopsies were collected before and after treatment for determination of percent myocyte necrosis and apoptosis. Echocardiographic measurements, serum levels of creatinine kinase, aspartate aminotransferase, and cardiac troponin I, and serum clenbuterol levels were measured before, during, and after treatment.

Response variables were compared between treatment groups and across time periods using a Mann-Whitney U test and Friedman's test for repeated nonparametric data, respectively. An adjusted level of significance at $p<0.01$ was used to reduce type I error. Response variables were summarized as median and range.

No significant effect of clenbuterol or exercise on response variables was found between treatment and control groups at any time point, nor within groups over time. This study did not show any adverse effects of clenbuterol treatment on equine cardiac or skeletal muscle at the approved dose level for treatment in horses.

**ABSTRACT #353**

EFFECTS OF CLOPIDOGREL ON PLATELET FUNCTION IN THE HORSE. Dorothy D. Whelchel, Benjamin M. Brainard, Barbara Fortes, James N. Moore, College of Veterinary Medicine, University of Georgia, Athens, Georgia.

Clopidogrel (Plavix®), an antagonist at the platelet P2Y12 ADP receptor, significantly reduces platelet aggregation in many species. This study provided preliminary data regarding the pharmacodynamics of clopidogrel on platelet aggregation in the horse. Specifically, a compounded paste formulation of Clopidogrel was administered (2mg/kg) orally once a day in the morning to 3 healthy horses for 3 days. Blood was collected before and at specified time intervals after the first oral dosing (t=0, 3, 6, 24, 48, 72, and 96 hours). A complete blood count and serum chemistry was performed at the start and conclusion of the dosing interval (t=0 and t=72). Serum serotonin (5-HT) concentrations were measured at each time point using a commercial ELISA. Mean platelet volume increased at 72 hr (9.1 ± 0.8 fl, $p<0.012$) from baseline values (8.0 ± 0.6 fl). ADP-induced platelet aggregation decreased from baseline values (mean ± SD; 72.7 ± 23%) at 72h (23.7 ± 4%; $p<0.002$) and 96h (28.0 ± 3%; $p<0.003$). Collagen-induced platelet aggregation decreased from baseline values (86.3 ± 10%) at 3h (19.7 ± 4%, $p< 0.0001$), 6h (20.7 ± 5% $p<0.0001$), 24h (18.3 ± 5%; $p<0.0001$) and 72h (33.3 ± 13.9%; $p<0.0001$). Serum 5-HT concentrations decreased from baseline (1580.9 ± 164.6 ng/mL) at 48h (126.8 ± 44.3 ng/mL, $p<0.05$). Clopidogrel given at 2 mg/kg PO q 24h appears to decrease both platelet aggregation and secretion of 5-HT after oral administration in healthy horses.

**ABSTRACT #354**

UPPER AND LOWER AIRWAY ENDOSCOPIC FINDINGS AND BRONCHOALVEOLAR LAVAGE-INDUCED LOWER AIRWAY COLLAPSE IN EQUINE RESPIRATORY INFLAMMATORY DISEASES. R Léguillette[1,2], K Koblinger[2], AJ Wasko[1].1. University of Calgary Faculty of Veterinary Medicine, Calgary, AB. 2. Moore&Co Veterinary Services, Calgary, AB.

In this study, we tested the hypotheses -1) that the lung inflammation in equine inflammatory airway disease (IAD) and recurrent airway obstruction (RAO) is correlated to the endoscopic inflammatory and mucus scores of upper and lower airway and -2) that horses with RAO have bronchial collapse during bronchoalveolar lavage (BAL) aspiration.

We used a random population of 138 horses around Calgary of age ranging 1-28 years. Both healthy horses and horses with respiratory symptoms not compatible with a respiratory infection were included in the study. One BAL was performed and recorded on each of the 138 subjects using a video endoscope, followed by a BAL cytological analysis. Each BAL video was viewed and scored on-site by one veterinarian, then *a posteriori* by two veterinarians. Pharyngitis was scored I to IV, pharyngeal mucus accumulation was scored 0 to 5, tracheal and lower airway mucus accumulation were scored 1 to 5, using previously published scales. We made a new scale to grade lower airway inflammation and collapse upon aspiration as 0 (none), 1(moderate) or 2 (severe). RAO was defined as a BAL ≥ 15% neutrophils and ≤ 2% mast cells and ≤ 1% eosinophils. IAD was defined as a BAL > 2% mast cells or >1% eosinophils or 5-15% neutrophils.

Inter-observer correlations for upper airway pharyngitis and mucus was significant ($R^2$=0.54 and $R^2$=0.24 respectively). Inter-observer correlations for tracheal and lower bronchial mucus was significant ($R^2$=0.51 and $R^2$=0.32 respectively). There was no significant inter-observer correlation for neither lower airway inflammation nor airway collapse ($R^2$= 0.05 and $R^2$=0.15 respectively).

Neutophil correlation: When collapse was scored on-site during the BAL procedure it was significantly correlated with neutrophil percentage in the BAL fluid ($R^2$= 0.29). Horses with severe collapse were also 3.8 times more likely to have RAO than to be normal or have IAD.

When scores were averaged between observers, tracheal and bronchial mucus accumulation scores were significantly correlated with neutophils ($R^2$=0.29 and $R^2$=0.37 respectively), but it was not the case for the lower airways inflammation score ($R^2$=0.10).

Pharyngitis and pharynx mucus accumulation are not correlated to lung inflammation. Lower bronchial mucus accumulation is better correlated to the BAL neutrophil percentage than tracheal mucus score, possibly because it is more distal thus closer from the sampling site. Our lower airway inflammation score is too subjective. BAL-induced lower airway collapse is difficult to evaluate *a posteriori* but not on-site. Airways collapse during the BAL procedure is correlated to RAO and neutrophilia possibly due to airway remodeling.

# SPEAKER INDEX

Dr. James Buchanan
University of Pennsylvania
Clinical Studies, SVM
3900 Delancey St.
Philadelphia, PA 19104
84

Dr. Virginia Buechner-Maxwell
Virginia Tech
LA Clinical Sci., VA-MD Regional CVM
Phase II, Duckpond Dr.
Blacksburg, VA 24061
629

Dr. Tony Buffington
The Ohio State University
CVM
601 Vernon L. Tharp St.
Columbus, OH 43210-1089
463, 465

Dr. Anthony Carr
University of Saskatchewan
SA Clinical Sciences, Western CVM
52 Campus Ave.
Saskatoon, SK S7N 5B4
Canada
503

Dr. Wayne Carter
No Address Available
420

Dr. Christopher Cebra
Oregon State University
Clinical Sciences, CVM
Magruder Hall 158
Corvallis, OR 97331
239, 242

Dr. Annie Chen
Washington State University
VTH, Clinical Sciences
100 Grimes Way
Pullman, WA 99164
302

Dr. Dennis Chew
The Ohio State University
CVM
601 Vernon L. Tharp St.
Columbus, OH 43210-1089
610

Dr. Joan Coates
University of Missouri
VMTH, CVM
900 E. Campus Dr.
Columbia, MO 65211
295

Dr. Michael Cockram
Atlantic Veterinary College
University of Prince Edward Island
Health Management
550 University Ave.
Charlottetown, PEI C1A 4PS
Canada
563, 566

Dr. Leah Cohn
University of Missouri
Medicine & Surgery, CVM
900 E. Campus Dr.
Columbia, MO 65211
446, 522, 620

Dr. Anthony Confer
Oklahoma State University
CVM
Rm. 250, McElroy Hall
Stillwater, OK 74078
257, 260

Dr. Peter Constable
Purdue University
Vet. Clinical Sciences School of Vet. Med.
625 Harrison St.
West Lafayette, IN 47907
202, 235

Dr. Audrey Cook
Texas A&M University
CVM, Small Animal Clinical Sciences
4474 TAMU
College Station, TX 77843
429

Dr. Etienne Cote
University of Prince Edward Island
Comp. Animals, Atlantic Vet College
550 University Ave.
Charlottetown, PE C1A 4P3
Canada
105

Dr. Larry Cowgill
University of Califorina
Medicine & Epidemiology, SVM
2108 Tupper Hall
Davis, CA 95616
598

Dr. Ronaldo da Costa
The Ohio State University
CVM, Dept. of Vet. Clinical Sciences
601 Vernon L. Tharp St.
Columbus, OH 43210-1089
284

Dr. Elizabeth Davis
Kansas State University, CVS
1800 Denison Ave.
106A Mosier Hall
Manhattan, KS 66506
131

Dr. Michael Davis
Oklahoma State University
264 McElroy Hall
Stillwater, OK 74078
619, 624

Dr. Lucy Davison
University of Cambridge
Queen's Veterinary School Hospital
Madingley Rd
Cambridge, CB3 0ES
United Kingdom
432

Dr. Eric de Madron
Alta Vista Animal Hospital
2616 Bank St.
Ottawa, ON K1T 1M9
Canada
82

Dr. Teresa DeFrancesco
North Carolina State University
CVM
4700 Hillsborough St.
Raleigh, NC 27606
97

Dr. André Desrochers
Faculté de Médecine Vétérinaire
Clinical Sciences
3200 rue Sicotte, C.P. 5000
St-Hyacinthe, QC J2S 7C6
Canada
228, 265

Dr. A. Ray Dillon
Auburn University
SA Surgery & Medicine, CVM
Auburn, AL 36849
93, 454

Dr. Thomas Divers
Cornell University
LA Medicine, NYS, CVM
Ithaca, NY 14853
129

Dr. Elizabeth Dore
University of Montreal
Centre Hospitalier Universitaire Veterinaire
CP 5000
Saint-Hyacinthe, QC J2S 7C6
Canada
279

Dr. Kristy Dowers
Colorado State University
CVMBS-VTH, Clinical Sciences
300 W. Drake Rd.
Ft. Collins, CO 80523
393

Louise Dunn
Snowgoose Veterinary Management Consulting
12 Snowgoose Cove
Greensboro, NC 27455
569, 570, 571, 573, 574, 576, 577

Dr. Janice Dye
US EPA
NHEERL
ETD/PTB MD:B143-01
Research Triangle Park, NC 27711
614

Dr. Amara Estrada
University of Florida
S.A. Clinical Sciences, CVM
PO Box 100126
Gainesville, FL 32610-0126
100

Dr. Stephen Ettinger
California Animal Hospital
1736 S. Sepulveda Blvd., Ste. B
Los Angeles, CA 90025
77, 78, 91

Dr. Timothy Fan
University of Illinois
SA Clinic, CVM
1008 W. Hazelwood Dr., Rm. 6
Urbana, IL 61802-4714
326, 333, 363

Dr. Amy Fauber
Carolina Veterinary Specialists
2225 Township Rd.
Charlotte, NC 28273
287

Dr. Gilles Fecteau
Universite de Montreal
FacultT de MTdecine VTtTrinaire
3200 Rue Sicotte, CP 5000
St. Hyacinthe, QC J2S 7C6
Canada
265

Dr. Marie-Eve Fecteau
University of Pennsylvania
New Bolton Center, SVM
382 W. Street Rd.
Kennett Square, PA 19348-1692
274

Dr. Deborah Fine
University of Missouri
Veterinary Medicine & Surgery
900 E. Campus Dr.
Columbia, MO 65211
452

Dr. Maria Julia Flaminio
Cornell University
Clinical Sciences, CVM
C3-522 CPC
Ithaca, NY 14853
134

Dr. Thomas Flegel
Universitat Leipzig
Klinik fnr Kleintiere
An den Tierkliniken 23
Leipzig, 4103
Germany
301, 304

Dr. Derek Foster
North Carolina State University
Population Health and Pathobiology
4700 Hillsborough St.
Raleigh, NC 27606
210, 234

Dr. Nicholas Frank
University of Tennessee
LA Clinical Sciences, CVM
2407 River Dr.
Knoxville, TN 37996
185

Dr. Anne French
University of Edinburgh
Faculty of Veterinary Medicine
Royal (Dick) School of Veterinary Stud.
Roslin, Midlothian EH25 9RG
Great Britain
96

Dr. Scott Friedman
No Address Available
578, 583

Dr. Angela Frimberger
Veterinary Oncology Consultants
379 Lake Innes Dr.
Wauchope 2446, NSW
Australia
345, 346, 366, 370, 373, 375

Dr. Martin Furr
Marion duPont Scott Equine Medical Ctr.
Virginia-Maryland Regional CVM
PO Box 1938
Leesburg, VA 20177
127

Dr. Ahmed Galal
University of Toronto
Princess Margaret Hospital
610 University Ave.
Toronto, ON M5G 2M9
Canada

Dr. Oliver Garden
Royal Veterinary College
Reg. T Cell Lab., Inf./Imm. Res. Grp
Camden Campus, Royal College St.
London, NW1 OTU
Great Britain
426

Dr. Anna Gelzer
Cornell University
Cardiology, Clinical Sciences
Ithaca, NY 14853
80

Dr. Christian Gerspach
University of Zurich
Food Animal Department
Winterthurer Strasse 260
Zurich, OH 8057
Switzerland
245, 276

Dr. Steeve Giguere
University of Florida
LA Clinical Sciences, CVM
Box 100136, HSC
Gainesville, FL 32610-0136
152

Dr. Michael Gomez
Saint Francis Hospital
Newborn Intensive Care
Eastern Oklahoma Perinatal Center
6161 South Yale Ave.
Tulsa, OK 74136
221, 223

Dr. Jody Gookin
North Carolina State University
CVM
4700 Hillsborough St.
Raleigh, NC 27606
390

Dr. Sonya Gordon
Texas A&M University
SA Medicine, CVM
4474 TAMU
College Station, TX 77843-4474
102

Dr. Gayle Hallowell
University of Nottingham
School of Vet. Medicine & Science
Sutton Bonington, LE12 5RD
United Kingdom
155, 232

Dr. Mitchell Halperin
University of Toronto
Toronto, ON M5B 1W8
Canada
398, 400

Dr. Katrin Hartmann
Ludwig-Maximilians-University
Munich
Medizinische Kleintierklinik
VeterinSrstrasse 13
Munich, BAVARIA 80539
Germany
384

Dr. Carolyn Henry
University of Missouri
Medicine & Surgery, CVM
900 E. Campus Dr.
Columbia, MO 65211
330, 338

Dr. Thomas Herdt
Michigan State University
LA Clinical Sciences, CVM
E. Lansing, MI 48824
4, 7

Dr. Margarethe Hoenig
University of Georgia
Physiology & Pharmacology, CVM
D.W. Brooks Dr.
Athens, GA 30602
502

Dr. Shannon Holmes
University of Florida
Small Animal Clinical Sciences
2015 S.W. 16th Ave.
Gainesville, FL 32610
662

Dr. Kate Hopper
University of California, Davis
Surgery & Radiologic Sciences, SVM
One Shields Ave.
Davis, CA 95616
630, 633, 636, 639

Dr. David Horohov
Louisiana State University
Dept. Pathobiological Sciences, SVM
Skip Bertman Dr. @ River Rd.
Baton Rouge, LA 70803
181

Dr. Kate Hurley
University of Califonia, Davis
School of Veterinary Medicine
One Shields Ave.
Davis, CA 95616
237, 512

Dr. Gisella Hussey
Colorado State University
300 W. Drake Rd.
Ft. Collins, CO 80523
174

Dr. Nick Jeffery
University of Cambridge
Clinical Vet Medicine
Madingley Rd.
Cambridge, CB3 0ES
Great Britain
318, 320, 322

Dr. Albert Jergens
Iowa State University
Clinical Sciences, CVM
Ames, IA 50010

Dr. Amy Johnson
University of Pennsylvania
New Bolton Center
382 West Street Rd.
Kennett Square, PA 19348
177, 184

Dr. Sanjay Kapil
Oklahoma State University
Virology & Serology
Farm & Ridge Rd.
Stillwater, OK 74078
208, 378, 381

Dr. Heidi Kellihan
University of Wisconsin-Madison
Cardiology
2015 Linden Dr.
Madison, WI 53706
79

Dr. Marie Kerl
University of Missouri
Medicine & Surgery, CVM
379 E. Campus Dr.
Columbia, MO 65211
650, 653

Dr. Chand Khanna
Oncology Service
Friendship Hospital for Animals
4105 Brandywine St. N.W.
Washington, DC 20016
344

Dr. Nathalie Kirschvink
University of Namur
Animal Physiology Veterinary Dept.
rue de Bruxelles 61
Namur, 5000
Belgium
621

Dr. William Kisseberth
The Ohio State University
601 Vernon L. Tharp St.
Columbus, OH 43210
349

Dr. Barbara Kitchell
Michigan State University
SA Clinical Sciences, CVM
D208, VMC
E. Lansing, MI 48824

Dr. Mark Kittleson
University of California, Davis
Medicine & Epidemiology, SVM
1 Shields Ave.,Tupper Hall 2108
Davis, CA 95616
114, 117, 120, 122

Dr. Bruce Kornreich
No Address Available
74

Dr. Charlotte Lacroix
Veterinary Business Advisors, Inc.
24 Coddington Rd.
Whitehouse Station, NJ 08889
47, 50, 53, 57, 62, 64

Dr. Jeffrey Lakritz
The Ohio State University
Clinical Sciences, CVM
601 Vernon L. Tharp St.
Columbus, OH 42310-1089
245, 276

Dr. Michael Lappin
Colorado State University
CVMBS-VTH, Clinical Sciences
300 W. Drake Rd.
Ft. Collins, CO 80523
387, 423

Dr. Robert Larson
University of Misourri-Columbia
A331 Clydesdale Hall
Columbia, MO 65211
196, 199

Dr. Patty Lathan
Mississippi State University
CVM
PO Box 6340
Mississippi State, MS 39762
403

Dr. Pat LeBlanc
Michigan State University
CVM
East Lansing, MI 48824-1314
665

Dr. Mathilde Leclere
Universite de Montreal
Faculty of Vet. Med.
CHUV-Large Animal Hospital
3200 Sicotte, C.P. 5000
Ste-Hyacinthe, QC J2S7C6
Canada
628

Dr. George Lees
Texas A&M University
SA Clinical Sciences, CVM
4474 TAMU
College Station, TX 77843-4474
600

Dr. Jonathan Levine
Texas A&M University
Sm. Animal Clinical Sciences
4474 TAMU
College Station, TX 77843-4474
281, 292, 324

Dr. Michel Lévy
Purdue University
Clinical Sciences
SVM 625 Harrison St.
W. Lafayette, IN 47907-2026
233

Dr. Annette Litster
Purdue University
School of Veterinary Medicine
625 Harrison St.
W. Lafayette, IN 47907
460

Dr. Cheryl London
The Ohio State University, CVM
Vet. Med. Academic Blvd.
1925 Coffey Rd., Rm. 454
Columbus, OH 43210
360

Dr. Clinton Lothrop
University of Alabama at Birmingham
1918 University Blvd.
404 McLm Bldg.
Birmingham, AL 35294-0024
411, 413

Dr. Alan Loynachan
University of Kentucky
Livestock Disease Diagnostic Center
P.O. Box 14125
Lexington, KY 40512-4125
124

Dr. Pamela Lucas
University of Illinois
Veterinary Clinical Medicine
1008 W. Hazelwood Dr.
Urbana, IL 61802
333

Dr. Nicole Luckschander
University of Bern
Clinic for SA/Vet Med.
Langgassstrasse 124
Bern, CH 3012
Switzerland
425

Dr. Virginia Luis Fuentes
University of London
Veterinary Clinical Science, RVC
Hawkshead Ln.
North Mymms, Hatfield, Herts. AL9 7TA
Great Britain
109

Dr. Jody Lulich
University of Minnesota
CVM
1352 Boyd Ave.
St. Paul, MN 55108
606

Dr. Katharine Lunn
Colorado State University
Clinical Sciences
300 W. Drake Rd.
Ft. Collins, CO 80523
173, 174, 549

Dr. Douglass Macintire
Auburn University
Clinical Sciences, CVM
Auburn, AL 36849
528

Dr. André Marette
CHUL Research Center
Lipid Research Unit
2705 Laurier Boul.
Ste-Foy, QC G1V 4G2
Canada
592

Dr. Stanley Marks
University of California, Davis
Medicine & Epidemiology, SVM
Davis, CA 95616
554, 559

Dr. Tomás Martin-Jimenez
University of Tennessee
College of Veterinary Medicine
2407 River Dr.
Knoxville, TN 37996-4550
27

Dr. Nicola Mason
University of Pennsylvania
Rm. 315, Hill Pavilion
380 S. University Ave.
Philadelphia, PA 19104
355

Dr. Robert Mattrey
University of California
Radiology
UCSD Medical Center
200 West Arbor Dr.
San Diego, CA 92103
655, 657

Dr. Fiona Maunsell
University of Florida
Pathobiology
Box 110880
Gainesville, FL 32611
238

Dr. Lara Maxwell
Oklahoma State University
264 McElroy Hall
Stillwater, OK 74078
176, 184

Dr. Melissa Mazan
Tufts University
Clinical Sciences, SVM
200 Westboro Rd.
N. Grafton, MA 01536
625, 626

Dr. Sheila McCullough
VCA Aurora Animal Hospital
215 W. Galena Blvd.
Aurora, IL 60506-4025
507

Dr. Dianne McFarlane
Oklahoma State University
Phys. Sci., Ctr. of Vet. Health Sci.
264 McElroy Hall
Stillwater, OK 74078
176, 184

Dr. Bruce McGorum
Easter Bush Veterinary Centre
Easter Bush, Roslin
Midlothian, EH25 9RG
Great Britain
162

Dr. Catherine McGowan
University of Liverpool
Faculty of Veteriary Science, Equine Division
Leahurst
Neston, Wirral CH64 7TE
United Kingdom
178, 183

Dr. Brendan McKiernan
No Address Available

Dr. Joanne Messick
Cornell University
Clinical Pathology Laboratory
Upper Tower Road
Ithaca, NY 14850-4902
531

Dr. Mireille Meylan
University of Bern
Ruminant Clinic, CVM
Bremgartenstrasse 109a
Bern, CH3012
Switzerland
236

Dr. Jaime Modiano
University of Minnesota
CVM, Cancer Center
455 VMC, MMC6194
1365 Gortner Ave.
St. Paul, MN 55455
353

Dr. Jeffrey Mogil
McGill University
Dept. of Psychology
Montreal, QC H3A 2K6
139, 140

Dr. N. Sydney Moise
Cornell University
NYS, CVM
Ithaca, NY 14853

Dr. Antony Moore
Veterinary Oncology Consultants
379 Lake Innes Dr.
Wauchope, NSW 2446
Australia
345, 346, 366, 370, 373, 375

Dr. Carol Morgan
The University of British Columbia
Interdisciplinary Studies
6201 Cecil Green Park Rd.
Green Commons
Vancouver, BC V6T 1Z1
Canada

Dr. Dawn Morin
University of Illinois
CVM
1008 W. Hazelwood Dr.
Urbana, IL 61802
231, 235

Dr. Paul Morley
Colorado State University
Animal Population Health Institute
Ft. Collins, CO 80523-1681
166

Dr. Anthony Mutsaers
Sunnybrook
Health Sciences Ctr.
S-221 Research Bld
2075 Bayview Ave.
Toronto, ON M4N 3M5
Canada
358

Dr. Stanley Nattel
Montreal Heart Institute
5000, Bélanger St.
Montréal, QC H1T 1C8
Canada

Dr. Sylvain Nichols
No Address Available
225

Dr. Martin Nielsen
No Address Available
137

Dr. Yvette Nout
No Address Available
167

Dr. Dennis O'Brien
University of Missouri
Veterinary Medicine & Surgery
900 East Campus Dr.
Columbia, MO 65211
289, 306

Dr. Natasha Olby
North Carolina State University
CVN
4700 Hillsborough St.
Raleigh, NC 27606
304, 305

Dr. Melissa Paoloni
NIH/National Cancer Institute
COP
37 Convent Dr., Rm. 2144
Bethesda, MD 20892
344, 351

Dr. Vakis Papanastassiou
No Address Available

Dr. Mark Papich
North Carolina State University
CVM
4700 Hillsborough St.
Raleigh, NC 27695
30, 44

Dr. Simon Peek
University of Wisconsin-Madison
Medical Sciences, SVM
2015 Linden Dr.
Madison, WI 53706
206

Dr. Dominique Peeters
University of LiFge
Small Animal Clinic
Bat 44
Sart Tilman, Liege 4000
Belgium
415, 417

Dr. Patricia Pesavento
University of California, Davis
Pathology, Microbiology &
Immunology
1 Shields Ave.
1044 Haring Hall
Davis, CA 95616
512

Dr. Paul Pion
Veterinary Information Network
777 W. Covell Ave.
Davis, CA 95616
111

Dr. Constantin Polychronakos
McGill University Health Centre
Pediatrics and Human Genetics
2300 Tupper, Ste. C244
Montréal, QC H3H1P3
Canada
595

Dr. David Polzin
University of Minnesota
SA Clinical Sciences, CVM
1352 Boyd Ave.
St. Paul, MN 55108
602

Dr. Maria Prado
University of Tennessee
Large Animal Clin. Sci., C-222 VTH
2407 River Dr.
Knoxville, TN 37996-4545
203

Dr. Marc Prentki
CR-CHUM, Technopôle Angus
Montreal Diabetes Research Center
2901, Rachel Est.
Montréal, QC H1W 4A4
Canada

Dr. Jane Quandt
University of Minnesota
VMC, College of Veterinary Medicine
1352 Boyd Ave.
St. Paul, MN 55108-1016
642, 647

Dr. Anant Radhakrishnan
Bluegrass Veterinary Specialists
2568 Richmond Rd.
Lexington, KY 40509
422

Dr. Jacqueline Rand
University of Queensland
Ctr. for Companion Animal Health,
SVS
Brisbane, QLD 4072
Australia
469, 472, 475, 477

Dr. Rose Raskin
Purdue University
Comparative Pathobiology
725 Harrison Street
West Lafayette, IN 47907
540, 543

Dr. Carol Reinero
University of Missouri
Medicine & Surgery, CVM
900 E. Campus Dr., Clydesdale Hall
Columbia, MO 65211
449, 620

Dr. Charlotte Reinhard Bjornvad
University of Copenhagen
Dyrlaegevej 16
Frederiksberg, 1970
Denmark
427

Dr. Keith Richter
Vet. Specialty Hosp. of San Diego
10435 Sorrento Valley Rd.
San Diego, CA 92121
561

Dr. Michael Riggs
University of Arizona
Dept. of Veterinary Science &
Microbiology
Tucson, AZ 85721
215, 218

Dr. Mark Rishniw
Cornell University
NYSCVM Tower Road
VMC 02-015
Ithaca, NY 14853
80, 111

Dr. Carlos Rodriguez, Jr.
San Francisco Veterinary Specialists
600 E. Alabama St.
San Francisco, CA 94110-2008
329

Dr. Mark Rondeau
University of Pennsylvania
Clinical Studies, SVM
3900 Delancey St.
Philadelphia, PA 19104
590

Dr. Kirsten Roomp
MPI for Informatives
Stuhlsatzenhausweg 85
Saarbruecken, 66123
Germany

Dr. Elizabeth Rozanski
Tufts University
SVM
200 Westboro Rd.
N. Grafton, MA 01536
421, 618

Dr. Clare Rusbridge
Stone Lion Veterinary Centre
41 High St.
Wimbledon Village, London, SW19
5AU
Great Britain
303

Dr. Chris Sanchez
University of Florida
LA Clinical Sciences, CVM
Box 100136, HSC
Gainesville, FL 32610-0136
141

Dr. Sherry Sanderson
University of Georgia
Physiology & Pharmacology, CVM
501 DW Brooks Dr.
Athens, GA 30602-7390
498

Dr. Roberto Santilli
Clinica Veterinaria Malpensa
Via Marconi 27
Samarate, Varese 21017
Italy
71

Dr. Scott Schatzberg
University of Georgia
CVM
Athens, GA 30602-7371
292

Dr. Thomas Schermerhorn
Kansas State University
Clinical Sciences, CVM
1800 Denison Ave.
Manhattan, KS 66506
495

Dr. Johan Schoeman
University of Pretoria
Department of Companion Animal Clinical Studies
Faculty of Veterinary Science
Onderstepoort, 110
South Africa
405

Dr. Harold Schott
Michigan State University
LA Clinical Sciences, CVM
D202, VMC
E. Lansing, MI 48824-1314
146

Dr. Colin Schwarzwald
University of Zurich
Vetsuisse Faculty, Equine Hospital
Winterthurerstrasse 260
Zurich, 8057
Switzerland
186, 188, 192

Dr. Gabriela Seiler
University of Pennsylvania
SVM - Section of Radiology
3900 Delancey St.
Philadelphia, PA 19104-6010
334, 659

Dr. Kathy Seino
Washington State University
CVM, Clinical Sciences
PO Box 647010
Pullman, WA 99163
170

Dr. Kim Selting
University of Missouri
Medicine & Surgery, CVM
379 E. Campus Dr., Clydesdale Hall
Columbia, MO 65211
330

Dr. G. Diane Shelton
University of California, San Diego
Pathology 0709
La Jolla, CA 92093-0709
300, 556

Dr. J. Andy Shores
Auburn University, CVM
Clinican Sciences
Hoerlein Hall
Auburn, AL 36849
309

Dr. Tawni Silver
WCVM
University of Saskatchewan
52 Campus Drive
Saskatoon, SK S7N 5B4
Canada
262

Dr. Kenneth Simpson
Cornell University
Clinical Sciences, NYS, CVM
VMC 2001
Ithaca, NY 14853
437

Dr. Nathan Slovis
Hagyard Equine Medical Insititute
McGee Medical Center
4250 Iron Works Rd.
Lexington, KY 40511
124

Dr. Geof Smith
North Carolina State University
Pop. Health & Pathobiology, CVM
4700 Hillsborough St.
Raleigh, NC 27606
234, 271

Dr. Stephanie Smith
University of Illinois
Biochemistry
506 S. Mathews, MC-714
Urbana, IL 61801
234, 546

Dr. John Speciale
No Address Available
312

Dr. Enrico Spugnini
Medico Veterinario
N Iscr Ord Rm 1203
Viale C.T. Odescalchi, 10
Roma, 147
Italy
331

Dr. Subramaniam (Sri) Srikumaran
Washington State University
Microbilogy and Pathology
Pullman, WA 99164
251, 254

Dr. Joerg Steiner
Texas A&M University
GI Lab, Small Animal Clinical Sci.
4474 TAMU
College Station, TX 77843-4474
424, 504

Dr. Rebecca Stepien
University of Wisconsin-Madison
SVM
2015 Linden Dr., W.
Madison, WI 53711
79, 89, 109

Dr. Ashley Stokes
Louisiana State University
SVM
Baton Rouge, LA 70803
145

Dr. Jan Suchodolski
Texas A&M University
Gastrointestinal Lab
4474 TAMU
College Station, TX 77843-4474
408, 424

Dr. Steven Suter
North Carolina State University
CVM Research #308
4700 Hillsborough St.
Raleigh, NC 27606
327, 341

Dr. Kelly Swanson
University of Illinois
Animal Sciences
1207 W. Gregory Dr.
Urbana, IL 61801
15, 18

Dr. Bud Tennant
Cornell University
Clinical Sciences, NYS, CVM
Rm. C2-005
Ithaca, NY 14853
440

Dr. Justin Thomason
Oklahoma State University
Clinical Sciences
1 BVMTH
Stillwater, OK 74078
87

Dr. Craig Thompson
Purdue University
Veterinary Pathobiology
725 Harrison St.
W. Lafayette, IN 47907-2027
537, 539

Dr. Ramiro Toribio
The Ohio State University
CVM
601 Vernon L. Tharp St.
Columbus, OH 43210
149

Dr. Maude Touret
University of Montreal
Faculty of Veterinary Medicine
1500 des Vétérinaires, CP 5000
Saint Hyacinthe, QC J2S 7C6
Canada
343

Dr. Lauren Trepanier
University of Wisconsin-Madison
Medical Sciences, SVM
2015 Linden Dr.
Madison, WI 53706
33

Dr. Carrie Tupper Hume
University of Pennsylvania
Oncology
3900 Delancy St.
Philadelphia, PA 19104
334

Dr. David Twedt
Colorado State University
CVMBS-VTH, Clinical Sciences
300 W. Drake Rd.
Ft. Collins, CO 80523
588

Dr. Jeff Tyler
University of Missouri
Medicine & Surgery, CVM
A314 Clydesale Hall
Columbia, MO 65211
212

Dr. Shelly Vaden
North Carolina State University
Comp. Animal & Spec. Species, CVM
4700 Hillsborough St.
Raleigh, NC 27606
604

Dr. Nicole Van Israel
Animal CardioPuomonary
Consultancy
Masta 11
Masta, BE 4970
Belgium
83

Dr. Robert Van Saun
Penn State University
Dept. of Veterinary Science
115 W.L. Henning Bldg.
University Park, PA 16802-3500
1, 9

Dr. Charles Vite
University of Pennsylvania
VTH, SVM
3850 Spruce St.
Philadelphia, PA 19104
298

Dr. Pamela Walker
Alpaca Jack's Suri Farm
16259 State Route
Findlay, OH 45840
248

Dr. Cynthia Ward
University of Georgia
College of Vet. Med.
501 DW Brooks Dr.
Athens, GA 30602
435

Dr. Amy Warren
No Address Available
585

Dr. Robert Washabau
University of Minnesota
Veterinary Clinical Sciences, CVM
1352 Boyd Ave.
St. Paul, MN 55108
515

Dr. Wegner
University of Pennsylvania
New Bolton Center, SVM
382 W. Street Rd.
Kennett Square, PA 19348
144

Dr. Robert Whitlock
University of Pennsylvania
New Bolton Center, SVM
382 W. Street Rd.
Kennett Square, PA 19348
271

Dr. Ted Whittem
University of Melbourne
Faculty of Veterinary Science
250 Princes Highway
Werribee, VIC 3030
Australia
24

Dr. Melinda Wilkerson
Kansas State University, CVM
Diagnostic Medicine/Pathobiology
1800 Denison Ave.
Manhattan, KS 66506-5606
335

Dr. Michael Willard
Texas A&M University
SA Medicine & Surgery, CVM
College Station, TX 77843-4474
480, 483, 486, 489

Dr. Tamara Wills
No Address Available
552

Dr. Stephen Withrow
Colorado State University
CVMBS-VTH, Animal Cancer Ctr.
300 W. Drake Rd.
Ft. Collins, CO 80523
441

Dr. J. Paul Woods
University of Guelph
Clinical Studies
Ontario Veterinary College
Guelph, ON N1G 2W1
Canada
358

Dr. Amelia Woolums
University of Georgia
LA Medicine, CVM
Athens, GA 30602
257

Dr. Kristine Yee
No Address Available

Dr. Debra Zoran
Texas A&M University
SA Clinical Sciences, CVM
College Station, TX 77843-4474
492

# KEYWORD INDEX

**American College of Veterinary Internal Medicine**
1997 Wadsworth Boulevard Suite A
Lakewood CO 80214-5293

CURRAN ASSOCIATES INC
proceedings
.com